S0-CUN-188

THE CRC HANDBOOK OF THERMAL ENGINEERING

The Mechanical Engineering Handbook Series

Series Editor

Frank Kreith

Consulting Engineer

Published Titles

The CRC Handbook of Mechanical Engineering
Frank Kreith

The Handbook of Fluid Dynamics
Richard W. Johnson

Forthcoming Titles

Microelectromechanical Systems Handbook
Mohamed Gad-el-Hak

Handbook of Heating, Ventalation, Air Conditioning and Refrigeration
Jan F. Kreider

Manufacturing Engineering and Systems Handbook
Jay Lee and Robert Schafrik

Handbook of High Performance Computing
Kim Mish

Handbook of Mixed Waste Management Technology
Chang H. Oh

Handbook of Non-Destructive Testing and Evaluation Engineering
James Tulenko, David Hintenlang

Computational Intelligence in Design and Manufacturing Handbook
Jun Wang, Andrew Kusiak

Handbook of Air Pollution Control Technology
Karl B. Schnelle, Charles A. Brown

THE CRC HANDBOOK OF THERMAL ENGINEERING

Editor-in-Chief

FRANK KREITH

CRC Press
Boca Raton London New York Washington, D.C.

Library of Congress Cataloging-in-Publication Data

The CRC handbook of thermal engineering / edited by Frank Kreith.
p. cm. -- (The mechanical engineering handbook series)
Includes bibliographical references and index.
ISBN 0-8493-9581-X (alk. paper)
1. Heat engineering Handbooks, manuals, etc.
I. Kreith, Frank. II. Series.
TJ260.C69 1999
621.402—dc21 99-38340
CIP

Direct all inquiries to CRC Press LLC, 2000 N.W. Corporate Blvd., Boca Raton, Florida 33431.

International Standard Book Number 0-8493-9581-X
Library of Congress Card Number 99-38340
Printed in the United States of America 2 3 4 5 6 7 8 9 0
Printed on acid-free paper

Acknowledgment

This book is dedicated to professionals in the field of thermal engineering.

I want to express my appreciation for the assistance rendered by members of the Editorial Advisory Board, as well as the lead authors of the various sections. I would also like to acknowledge the assistance of the many reviewers who provided constructive criticism on various parts of this handbook during its development. Their reviews were in the form of written comments as well as telephone calls and e-mails. I cannot remember all the people who assisted as reviewers, and rather than mention a few and leave out others, I am thanking them as a group. There are, of course, some special individuals without whose dedication and assistance this book would not have been possible. They include my editorial assistant, Bev Weiler, and the editors at CRC — Norm Stanton, Bob Stern, and Maggie Mogck. My wife, Marion, helped keep track of the files and assisted with other important facets of this handbook.

But the existence of the handbook and its high quality is clearly the work of the individual authors, and I want to express my deep appreciation to each and every one of them for their contribution.

I hope that the handbook will serve as a useful reference on all topics of interest to thermal engineers in their professional lives. But during the planning stages of the book, certain choices had to be made to limit its scope. I realize, however, that the field of thermal engineering is ever-changing and growing. I would, therefore, like to invite engineers who will use this book to give me their input on topics that should be included in the next edition. I would also like to invite readers and users of the handbook to send me any corrections, errors, or omissions they discover, in order that these can be corrected in the next printing.

Frank Kreith
Boulder, Colorado
fkreith@aol.com

Introduction

Industrial research today is conducted in a changing, hectic, and highly competitive global environment. Until about 25 years ago, the R&D conducted in the U.S. and the technologies based upon it were internationally dominant. But in the last 20 years, strong global competition has emerged and the pace at which high technology products are introduced has increased. Consequently, the lifetime of a new technology has shortened and the economic benefits of being first in the marketplace have forced an emphasis on short-term goals for industrial development. To be successful in the international marketplace, corporations must have access to the latest developments and most recent experimental data as rapidly as possible.

In addition to the increased pace of industrial R&D, many American companies have manufacturing facilities, as well as product development activities in other countries. Furthermore, the restructuring of many companies has led to an excessive burden of debt and to curtailment of in-house industrial research. All of these developments make it imperative for industry to have access to the latest information in a convenient form as rapidly as possible. The goal of this handbook is to provide this type of up-to-date information for engineers involved in the field of thermal engineering.

This handbook is not designed to compete with traditional handbooks of heat transfer that stress fundamental principles, analytical approaches to thermal problems, and elegant solutions of traditional problems in the thermal sciences. The goal of this handbook is to provide information on specific topics of current interest in a convenient form that is accessible to the average engineer in industry. The handbook contains in the first three chapters sufficient background information to refresh the reader's memory of the basic principles necessary to understand specific applications. The bulk of the book, however, is devoted to applications in thermal design and analysis for technologies of current interest, as well as to computer solutions of heat transfer and thermal engineering problems.

The applications treated in the book have been selected on the basis of their current relevance to the development of new products in diverse fields such as food processing, energy conservation, bioengineering, desalination, measurement techniques in fluid flow and heat transfer, and other specific topics. Each application section stands on its own, but reference is made to the basic introductory material as necessary. The introductory material is presented in such a manner that it can be referred to and used by several authors of application sections. For the convenience of the reader, each author has been requested to use the same nomenclature in order to help the reader in the transition from material in some of the basic chapters to the application chapters. But wherever necessary, authors have defined special symbols in their chapters.

A special feature of this handbook is an introduction to the use of the Second Law rather than the First Law of Thermodynamics in analysis, optimization, and economics. This approach has been widely used in Europe and Asia for many years, but has not yet penetrated engineering education and usage in the U.S. The Second Law approach will be found particularly helpful in analyzing and optimizing thermal systems for the generation and/or conservation of energy.

The material for this handbook has been peer reviewed and carefully proofread. However, in a project of this magnitude with authors from varying backgrounds and different countries, it is unavoidable that errors and/or omissions occur. As the editor, I would, therefore, like to invite the professional engineers who use this book to give me their feedback on topics that should be included in the next edition. I would also greatly appreciate it if any readers who find an error would contact me by e-mail in order for the manuscript to be corrected in the next printing. Since CRC Press expects to update the book frequently, both in hard copy and on CD-ROM, errors will be corrected and topics of interest will be added promptly.

Frank Kreith
fkreith@aol.com
Boulder, CO

Nomenclature

Symbol	Quantity	Unit		Dimensions (MLtT)
		SI	English	
a	Velocity of sound	m/s	ft/s	$L\ t^{-1}$
a	Acceleration	m/s^2	ft/s^2	$L\ t^{-2}$
A	Area: A_c, cross-sectional area; A_p, projected area of a body normal to the direction of flow; A_q, area through which rate of heat flow is q; A_g, surface area; A_o, outside surface area; A_i, inside surface area; A_f, fin surface area	m^2	ft^2	L^2
b	Breadth or width	m	ft	L
c	Specific heat; c_p, specific heat at constant pressure; c_v, specific heat at constant volume	J/kg K	Btu/lb_m °R	$L^2\ t^{-2}\ T^{-1}$
C	Constant or Coefficient; C_D, total drag coefficient; C_f, skin friction coefficient; C_{fx}, local value of C_f at distance x, from leading edge; $\overline{C}_f$, average value of C_f	none	none	—
C	Thermal capacity	J/K	Btu/°F	$M\ L^2\ t^{-2}\ T^{-1}$
$\dot{C}$	Hourly heat capacity rate; $\dot{C}_c$, hourly heat capacity rate of colder fluid in a heat exchanger; $\dot{C}_h$, hourly heat capacity of hotter fluid; C^*, ratio of heat capacity rates in heat exchangers	W/K	Btu/hr°F	$M\ L^2\ t^{-1}\ T^{-1}$
D	Diameter, D_H, hydraulic diameter; D_o, outside diameter; D_i, inside diameter	m	ft	L
e	Base of natural or Napierian logarithm	none	none	—
e	Total energy per unit mass	J/kg	Btu/lb_m $L^2\ t^{-2}$	—
E	Total energy	J	Btu	$M\ L^2\ t^{-2}$
E	Emissive power of a radiating body; E_b, emissive power of a blackbody	W/m^2	$Btu/hr{\cdot}ft^2$	$M\ t^{-2}$
E_λ	Monochromatic emissive power per micron at wavelength λ	W/m μm	$Btu/hr{\cdot}ft^2$ micron	$M\ t^{-2}\ L^{-1}$
f	Darcy friction factor for flow through a pipe or duct	none	none	—
f′	Friction coefficient for flow over banks of tubes	none	none	—
F	Force; F_B, buoyant force	N	lb	$M\ L\ t^{-2}$
F_T	Temperature factor	none	none	—
$F_{1\text{-}2}$	Geometric shape factor for radiation from one blackbody to another	none	none	—
g	Acceleration due to gravity	m/s^2	ft/s^2	$L\ t^{-2}$
g_c	Dimensional conversion factor	$1.0\ kg{\cdot}m/N{\cdot}s^2$	$32.2\ ft{\cdot}lb_m/lb{\cdot}s^2$	
G	Mass velocity or flow rate per unit area	$kg/s{\cdot}m^2$	$lb_m/hr{\cdot}ft^2$	$M\ L^{-2}\ t^{-1}$
G	Irradiation incident on unit surface in unit time	W/m^2	$Btu/hr{\cdot}ft^2$	$M\ L^{-2}\ t^{-1}$
h	Enthalpy per unit mass	J/kg	Btu/lb_m	$L^2\ t^{-2}$

Symbol	Quantity	Unit: SI	Unit: English	Dimensions (MLtT)
h	Local heat transfer coefficient; $\bar{h}$, average heat transfer coefficient $\bar{h} = \bar{h}_c + \bar{h}_r$; h_b, heat transfer coefficient of a boiling liquid; h_c, local convection heat transfer coefficient; $\bar{h}_c$, average heat transfer coefficient; $\bar{h}_r$, average heat transfer coefficient for radiation	W/m²·K	Btu/hr·ft²·°F	$M\ t^{-3}\ T^{-1}$
h_{fg}	Latent heat of condensation or evaporation	J/kg	Btu/lb_m	$L^2\ t^{-2}$
H	Head, elevation of hydraulic grade line	m	ft	L
i	Angle between sun direction and surface normal	rad	deg	—
I	Moment of inertia	m^4	ft^4	L^4
I	Intensity of radiation	W/sr	Btu/hr unit solid angle	$M\ L^2\ t^{-3}$
I_λ	Intensity per unit wavelength	W/sr·μm	Btu/hr·sr micron	$M\ L\ t^{-3}$
J	Radiosity	W/m^2	$Btu/hr \cdot ft^2$	$M\ L^{-2}\ t^{-1}$
k	Thermal conductivity; k_s, thermal conductivity of a solid; k_f, thermal conductivity of a fluid; k_g, thermal conductivity of a gas	W/m·K	Btu/hr·ft°F	$M\ L^{-2}\ t^{-1}\ T^{-1}$
K	Thermal conductance; k_k, thermal conductance for conduction heat transfer; k_c, thermal convection conductance; K_r, thermal conduction for radiation heat transfer	W/K	Btu/hr·ft°F	$M\ t^{-1}\ T^{-1}$
K	Bulk modulus of elasticity	Pa	lb/ft^2	$M\ L^{-1}\ t^{-2}$
log	Logarithm to the base 10	none	none	—
ln	Logarithm to the base e	none	none	—
l	Length, general or characteristic length of a body	m	ft	L
L	Lift	N	lb	$M\ L\ t^{-2}$
L_f	Latent heat of solidification	J/kg	Btu/lb_m	$L^2\ t^{-2}$
$\dot{m}$	Mass flow rate	kg/s	lb_m/s	$M\ t^{-1}$
m	Mass	kg	lb_m	M
M	Molecular weight	gm/gm mole	lb_m/lb mole	—
$\dot{M}$	Momentum per unit time	N	lb	MLt^{-2}
n	Manning roughness factor	none	none	—
n	Number of moles	none	none	—
NPSH	Net positive suction head	m	ft	L
N	Number in general; number of tubes, etc.	none	none	—
p	Static pressure; p_c, critical pressure; p_A, partial pressure of component A	N/m^2	psi or lb/ft^2 or atm	$M\ L^{-1}\ t^{-2}$
P	Wetted perimeter or height of weir	m	ft	L
q	Discharge per unit width	m^2/s	ft^2/s	$L^2\ t^{-1}$
q	Rate of heat flow; q_k, rate of heat flow by conduction; q_r, rate of heat flow by radiation; q_c, rate of heat flow by convection; q_b, rate of heat flow by nucleate boiling	W	Btu/hr	$M\ L^2\ t^{-3}$
q'''	Rate of heat generation per unit volume	W/m^3	$Btu/hr \cdot ft^3$	$M\ L^{-1}\ t^{-3}$
q''	Rate of heat generation per unit area (heat flux)	W/m^2	$Btu/hr \cdot ft^2$	$M\ t^{-3}$
Q	Quantity of heat	J	Btu	$M\ L^2\ t^{-3}$
r	Radius; r_H, hydraulic radius; r_i, inner radius; r_o, outer radius	m	ft	L
R	Thermal resistance; R_c, thermal resistance to convection heat transfer; R_k, thermal resistance to conduction heat transfer; R_r, to radiation heat transfer	K/W	hr°F/Btu	$L\ T\ M^{-1}$

Symbol	Quantity	Unit SI	Unit English	Dimensions (MLtT)
R_e	Electrical resistance	ohm	ohm	—
R	Perfect gas constant	8.314 J/K·kg mole	1545 ft·lb_f/lb·mole°F	$L^2 t^{-2} T^{-1}$
s	Entropy per unit mass	J/kg·K	ft·lb/lb_m·°R	$L^2t^{-2}T^{-1}$
S	Entropy	J/K	ft·lb/°R	$ML^2t^{-2}T^{-1}$
S_L	Distance between centerlines of tubes in adjacent longitudinal rows	m	ft	L
S_T	Distance between centerlines of tubes in adjacent transverse rows	m	ft	L
t	Time	s	hr or s	t
T	Temperature; T_b, temperature of bulk of fluid; T_f, mean film temperature; T_s, surface temperature, T_o, temperature of fluid far removed from heat source or sink; T_m, mean bulk temperature of fluid flowing in a duct; T_M, temperature of saturated vapor; T_{sl}, temperature of a saturated liquid; T_{fr}, freezing temperature; T_t, liquid temperature; T_{as}, adiabatic wall temperature	K or °C	°F or R	T
u	Internal energy per unit mass	J/kg	Btu/lb_m L^2 t^{-2}	
u	Velocity in x direction; u′, instantaneous fluctuating x component of velocity; $\bar{u}$, average velocity	m/s	ft/s or ft/hr	$L t^{-1}$
u*	Shear stress velocity	m/s	ft/s	Lt^{-1}
U	Internal energy	J	Btu	ML^2t^{-2}
U	Overall heat transfer coefficient	W/m^2K	Btu/hr·ft^2°F	$M t^{-3} T^{-1}$
U_∞	Free-stream velocity	m/s	ft/s	$L t^{-1}$
v	Specific volume	m^3/kg	ft^3/lb_m	$L^3 M^{-1}$
v	Velocity in y direction; v′, instantaneous fluctuating y component of velocity	m/s	ft/s or ft/hr	$L t^{-1}$
V	Volume	m^3	ft^3	L^3
$\dot{V}$	Volumetric flow rate	m^3/s	ft^3/s	$L^3 t^{-1}$
W_s	Shaft work	m·N	ft·lb	ML^2t^{-2}
$\dot{W}$	Rate of work output or power	W	Btu/hr	$M L^2 t^{-3}$
x	Coordinate or distance from the leading edge; x_c, critical distance from the leading edge where flow becomes turbulent	m	ft	L
x	Quality	percent	percent	none
y	Coordinate or distance from a solid boundary measured in direction normal to surface	m	ft	L
z	Coordinate	m	ft	L
Z	Ratio of hourly heat capacity rates in heat exchangers	none	none	—

Greek Symbols

Symbol	Quantity	Unit SI	Unit English	Dimensions (MLtT)
α	Absorptivity for radiation, α_λ, monochromatic absorptivity at wavelength λ	none	none	—
α	Thermal diffusivity = $k/\rho c$	m^2/s	ft^2/s	$L^2 t^{-1}$
β	Temperature coefficient of volume expansion	1/K	1/R	T^{-1}
β_k	Temperature coefficient of thermal conductivity	1/K	1/R	T^{-1}
γ	Specific heat ratio, c_p/c_v	none	none	—
Γ	Circulation	m^2	ft^2	L^2t^{-1}

Symbol	Quantity	Unit SI	Unit English	Dimensions (MLtT)
Γ	Body force per unit mass	N/kg	lb/lb_m	$L\ t^{-2}$
Γ_c	Mass rate of flow of condensate per unit breadth $= \dot{m}/\pi D$ for a vertical tube	kg/s·m	lb_m/hr·ft	$M\ L^{-2}\ t^{-1}$
δ	Boundary-layer thickness; δ_h, hydrodynamic boundary-layer thickness; δ_{th}, thermal boundary-layer thickness	m	ft	L
Δ	Difference between values	none	none	—
ε	Heat exchanger effectiveness	none	none	—
ϵ	Roughness height	m	ft	L
ϵ	Emissivity for radiation; ϵ_λ, monochromatic emissivity at wavelength λ; ϵ_ϕ, emissivity in direction ϕ			
ϵ_H	Thermal eddy diffusivity	m^2/s	ft^2/s	$L^2\ t^{-1}$
ϵ_M	Momentum eddy diffusivity	m^2/s	ft^2/s	$L^2\ t^{-1}$
ζ	Ratio of thermal to hydrodynamic boundary-layer thickness, δ_h/δ_{th}	—	—	—
η	Efficiency; η_f, fin efficiency	none	none	—
λ	Wavelength; λ_{max}, wavelength at which monochromatic emissive power $E_{b\lambda}$ is a maximum	μm	micron	L
μ	Absolute viscosity	$N \cdot s/m^2$	lb/ft·s	$M\ L^{-1}\ t^{-1}$
ν	Kinematic viscosity, μ/ρ	m^2/s	ft^2/s	$L^2\ t^{-1}$
ν_f	Frequency of radiation	1/s	1/s	t^{-1}
Φ	Velocity potential	m^2/s	ft^2/s	$L^2\ t^{-1}$
ρ	Mass density, 1/v; ρ_l, density of liquid; ρ_v, density of vapor	kg/m^3	$lb_m ft^3$	$M\ L^{-3}$
τ	Shearing stress, τ_s, shearing stress at surface; τ_w, shear at wall of a tube or a duct	N/m^2	lb/ft^2	$M\ L^{-1}\ t^{-2}$
τ	Transmissivity for radiation	none	none	—
σ	Stefan-Boltzmann constant	W/m^2K^4	Btu/hr ft^2R^4	$M\ t^{-3}\ T^{-4}$
σ	Surface tension	N/m	lb/ft	$M\ t^{-2}$
ϕ	Angle	rad	rad	—
ψ	Stokes' stream function	m^3/s	ft^3/s	$L^3\ t^{-1}$
ω	Angular velocity	1/s	1/s	t^{-1}
ω	Solid angle	sr	steradian	—

Dimensionless Numbers	
Bi	Biot number
Ec	Eckert number
Eu	Euler number
Fo	Fourier modulus
Fr	Froude number
Gz	Graetz number
Gr	Grahsof number
Ja	Jakob number
Kn	Knudsen number
M	Mach number
Nu	Average Nusselt number; Nu_D, average diameter Nusselt number; Nu_x, local Nusselt number
Pe	Peclet number

Pr	Prandtl number
Ra	Rayleigh number
Re	Reynolds number; Re_x, local value of Re at a distance x from leading edge; Re_D, diameter Reynolds number; Re_b, bubble Reynolds number
Θ	Boundary Fourier modulus or dimensionless time
St	Stanton number
We	Weber number

Miscellaneous

$a > b$	a great than b
$a < b$	a smaller than b
$\propto$	Proportional sign
$\simeq$	Approximately equal sign
∞	Infinity sign
Σ	Summation sign

Subscripts

c = critical condition
i = inlet
f = fin
u = unit quantities
w = wall or properties at wall temperature
c.s. = control surface
c.v. = control volume
o = stagnation or standard state condition; outlet or outside
1,2 = inlet and outlet, respectively, of control volume

Note: Those symbols and subscripts that are not included in the above list are defined in the text.

Editor-in-Chief

Dr. Frank Kreith is Professor Emeritus of Engineering at the University of Colorado and currently serves as the ASME Legislative Fellow for Energy and Environment at the National Conference of State Legislatures in Denver, CO. In this capacity, he provides technical assistance on engineering and science topics such as energy management, waste disposal, environmental protection, and utility restructuring to legislators and their staff in all 50 state governments.

Previously, he was a research engineer at the Jet Propulsion Laboratory from 1945 to 1949 and a Guggenheim Fellow at Princeton University from 1950 to 1951. Between 1951 and 1977, Dr. Kreith taught mechanical engineering at the University of California at Berkeley, Lehigh University, and the University of Colorado.

From 1978 to 1988, Dr. Kreith was Chief of Thermal Research and Senior Research Fellow at the Solar Energy Research Institute, currently the National Renewable Energy Laboratory. During his tenure at SERI, he participated in the Presidential Domestic Energy Review, the White House Forum on Domestic Energy Policy, and edited the *ASME Journal of Solar Energy Engineering*. In 1995, he participated in the White House Forum on Technology for a Sustainable Future. He has served as a national lecturer for Sigma Xi and is currently a distinguished lecturer for the American Society of Mechanical Engineers.

Dr. Kreith is the recipient of the ASME Heat Transfer Memorial Award (1972), the ASME Worcester R. Warner Medal (1981), the Distinguished Service Award of the Solar Energy Research Institute (1983), the Max Jakob Memorial Award of ASME/AIChE (1986), the Charles Greeley Abbott Award of the American Solar Energy Society (1988), the ASME Energy Resource Technology Award (1989), the Ralph Coates Roe Medal of ASME (1992), and the Professional and Scholarly Excellence Award of the Association of American Publishers (1995). In 1997, he was awarded the Washington Award by a consortium of seven engineering societies for "unselfish and preeminent service in advancing human progress."

He is the author of textbooks on heat transfer, nuclear power, solar energy, and energy management. He has edited handbooks on energy conservation, solid waste management, and energy efficiency. He has also published more than 120 peer-reviewed articles on various mechanical engineering topics.

Dr. Kreith has had wide experience in mechanical engineering as teacher and consultant for academia, industry, and governments all over the world. His assignments have included consultancies for NATO, the U.S. Agency for International Development, the United Nations, the National Academy of Engineering, and the U.S. Department of Energy. Dr. Kreith is a member of Pi Tau Sigma, Sigma Xi, a Life Fellow of ASME, and a Fellow of AAAS.

Advisory Board

Contributors

Randall F. Barron
Louisiana Tech University
Ruston, Louisiana

Kenneth J. Bell
Oklahoma State University
Stillwater, Oklahoma

Stanley A. Berger
University of California
Berkeley, California

Arthur E. Bergles
Rensselear Polytechnic Institute
Troy, New York

Robert F. Boehm
University of Nevada
Las Vegas, Nevada

Massimo Capobianchi
Gonzaga University
Spokane, Washington

Van P. Carey
University of California
Berkeley, California

John C. Chen
Lehigh University
Bethlehem, Pennsylvania

Stuart W. Churchill
University of Pennsylvania
Philadelphia, Pennsylvania

Raymond Cohen
Purdue University
West Lafayette, Indiana

Kenneth R. Diller
University of Texas
Austin, Texas

Ibrahim Dincer
King Fahd University of Petroleum and Minerals
Dhabran, Saudi Arabia

Donald L. Fenton
Kansas State University
Manhattan, Kansas

Kenneth E. Goodson
Stanford University
Stanford, California

Eckhard Groll
Purdue University
West Lafayette, Indiana

Frank Hagin
Colorado School of Mines
Golden, Colorado

William H. Harden
Ingersoll-Rand Company
Clemmon, North Carolina

Kenneth E. Hickman
York International Corporation
York, Pennsylvania

K. G. Terry Hollands
University of Waterloo
Waterloo, Ontario, Canada

Thomas F. Irvine, Jr.
State University of New York
Stony Brook, New York

Harold R. Jacobs
CEEMS
Bothell, Washington

Yogesh Jaluria
Rutgers State University
New Brunswick, New Jersey

Jungho Kim
University of Maryland
College Park, Maryland

Moncef Krarti
University of Colorado
Boulder, Colorado

Ajay Kumar
NASA Langley Research Center
Hampton, Virginia

Pradeep Lall
Motorola
Libertyville, Illinois

Noam Lior
University of Pennsylvania
Philadelphia, Pennsylvania

Alan T. McDonald
Purdue University
West Lafayette, Indiana

Anthony F. Mills
University of California
Los Angeles, California

Dilip K. Mistry
Ingersoll-Rand Company
Clemmon, North Carolina

Michael F. Modest
Pennsylvania State University
University Park, Pennsylvania

Robert J. Moffat
Stanford University
Stanford, California

Michael J. Moran
Ohio State University
Columbus, Ohio

Earl Muir
Copeland Corporation
Sidney, Ohio

Paul Norton
National Renewable Energy Laboratory
Golden, Colorado

Jeff Nowobilski
Praxair, Inc.
Tonawanda, New York

John A. Pearce
University of Texas
Austin, Texas

Donald W. Radford
Colorado State University
Ft. Collins, Colorado

George Raithby
University of Waterloo
Waterloo, Ontario, Canada

Rolf D. Reitz
University of Wisconsin
Madison, Wisconsin

Mihir Sen
University of Notre Dame
South Bend, Indiana

Ramesh K. Shah
Delphi Harrison Thermal Systems
Lockport, New York

Henry Shaw
New Jersey Institute of Technology
Newark, New Jersey

Sherif A. Sherif
University of Florida
Gainesville, Florida

N.V. Suryanarayana
Michigan Technological University
Houghton, Michigan

Larry W. Swanson
Simulation Sciences, Inc.
Laguna, California

Timothy W. Tong
Colorado State University
Ft. Collins, Colorado

Kirtan K. Trivedi
Exxon Research and Engineering Company
Florham Park, New Jersey

George Tsatsaronis
Institut für Energietechnik
Technische Universität
Berlin, Germany

J. Paul Tullis
Utah State University
Logan, Utah

Jonathan W. Valvano
University of Texas
Austin, Texas

Frank M. White
University of Rhode Island
Kingston, Rhode Island

K. T. Yang
University of Notre Dame
South Bend, Indiana

David W. Yarbrough
Tennessee Technical University
Cookeville, Tennessee

Contents

SECTION 1 Engineering Thermodynamics

SECTION 2 Fluid Mechanics

SECTION 3 Heat and Mass Transfer

SECTION 4 Applications

SECTION 5 Numerical Analysis and Computational Tools

APPENDICES

1

Engineering Thermodynamics

Michael J. Moran
The Ohio State University

George Tsatsaronis
Technische Universität Berlin

Although various aspects of what is now known as thermodynamics have been of interest since antiquity, formal study began only in the early 19th century through consideration of the motive power of *heat*: the capacity of hot bodies to produce *work*. Today the scope is larger, dealing generally with *energy and*

0-8493-9581-X/00/$0.00+$.50

entropy, and with relationships among the *properties* of matter. Moreover, in the past 25 years engineering thermodynamics has undergone a revolution, both in terms of the presentation of fundamentals and in the manner that it is applied. In particular, the second law of thermodynamics has emerged as an effective tool for engineering analysis and design.

1.1 Fundamentals

Classical thermodynamics is concerned primarily with the macrostructure of matter. It addresses the gross characteristics of large aggregations of molecules and not the behavior of individual molecules. The microstructure of matter is studied in kinetic theory and statistical mechanics (including quantum thermodynamics). In this chapter, the classical approach to thermodynamics is featured.

Basic Concepts and Definitions

Thermodynamics is both a branch of physics and an engineering science. The scientist is normally interested in gaining a fundamental understanding of the physical and chemical behavior of fixed, quiescent quantities of matter and uses the principles of thermodynamics to relate the *properties* of matter. Engineers are generally interested in studying *systems* and how they interact with their *surroundings*. To facilitate this, engineers have extended the subject of thermodynamics to the study of systems through which matter flows.

System

In a thermodynamic analysis, the *system is* the subject of the investigation. Normally the system is a specified quantity of matter and/or a region that can be separated from everything else by a well-defined surface. The defining surface is known as the *control surface* or *system boundary*. The control surface may be movable or fixed. Everything external to the system is the *surroundings*. A system of fixed mass is referred to as a *control mass* or as a *closed system*. When there is flow of mass through the control surface, the system is called a *control volume*, or *open*, *system*. An *isolated* system is a closed system that does not interact in any way with its surroundings.

State, Property

The condition of a system at any instant of time is called its *state*. The state at a given instant of time is described by the properties of the system. A *property* is any quantity whose numerical value depends on the state but not the history of the system. The value of a property is determined in principle by some type of physical operation or test.

Extensive properties depend on the size or extent of the system. Volume, mass, energy, and entropy are examples of extensive properties. An extensive property is additive in the sense that its value for the whole system equals the sum of the values for its parts. *Intensive* properties are independent of the size or extent of the system. Pressure and temperature are examples of intensive properties.

A *mole* is a quantity of substance having a mass numerically equal to its molecular weight. Designating the molecular weight by $\mathcal{M}$ and the number of moles by n, the mass m of the substance is $m = n\mathcal{M}$. One kilogram mole, designated kmol, of oxygen is 32.0 kg and one pound mole (lbmol) is 32.0 lb. When an extensive property is reported on a unit mass or a unit mole basis, it is called a *specific* property. An overbar is used to distinguish an extensive property written on a per-mole basis from its value expressed per unit mass. For example, the volume per mole is $\bar{v}$, whereas the volume per unit mass is v, and the two specific volumes are related by $\bar{v} = \mathcal{M}v$.

Process, Cycle

Two states are identical if, and only if, the properties of the two states are identical. When any property of a system changes in value there is a change in state, and the system is said to undergo a *process*. When a system in a given initial state goes through a sequence of processes and finally returns to its initial state, it is said to have undergone a *cycle*.

Phase and Pure Substance

The term *phase* refers to a quantity of matter that is homogeneous throughout in both chemical composition and physical structure. Homogeneity in physical structure means that the matter is all *solid,* or all *liquid*, or all *vapor* (or, equivalently, all *gas). A* system can contain one or more phases. For example, a system of liquid water and water vapor (steam) contains *two* phases. A *pure substance* is one that is uniform and invariable in chemical composition. *A* pure substance can exist in more than one phase, but its chemical composition must be the same in each phase. For example, if liquid water and water vapor form a system with two phases, the system can be regarded as a pure substance because each phase has the same composition. The nature of phases that coexist in equilibrium is addressed by the *phase rule* (Section 1.3, Multicomponent Systems).

Equilibrium

Equilibrium means a condition of balance. In thermodynamics the concept includes not only a balance of forces, but also a balance of other influences. Each kind of influence refers to a particular aspect of thermodynamic (complete) equilibrium. *Thermal* equilibrium refers to an equality of temperature, *mechanical* equilibrium to an equality of pressure, and *phase* equilibrium to an equality of chemical potentials (Section 1.3, Multicomponent Systems). *Chemical* equilibrium is also established in terms of chemical potentials (Section 1.4, Reaction Equilibrium). For complete equilibrium, the several types of equilibrium must exist individually.

To determine if a system is in thermodynamic equilibrium, one may think of testing it as follows: isolate the system from its surroundings and watch for changes in its observable properties. If there are no changes, it may be concluded that the system was in equilibrium at the moment it was isolated. The system can be said to be at an *equilibrium state.* When a system is *isolated,* it cannot interact with its surroundings; however, its state can change as a consequence of spontaneous events occurring internally as its intensive properties, such as temperature and pressure, tend toward uniform values. When all such changes cease, the system is in equilibrium. At equilibrium. temperature and pressure are uniform throughout. If gravity is significant, a pressure variation with height can exist, as in a vertical column of liquid.

Temperature

A scale of temperature independent of the *thermometric substance* is called a *thermodynamic* temperature scale. The Kelvin scale, a thermodynamic scale, can be elicited from the second law of thermodynamics (Section 1.1, The Second Law of Thermodynamics, Entropy). The definition of temperature following from the second law is valid over all temperature ranges and provides an essential connection between the several *empirical* measures of temperature. In particular, temperatures evaluated using a *constant-volume gas thermometer* are identical to those of the Kelvin scale over the range of temperatures where gas thermometry can be used.

The empirical *gas scale* is based on the experimental observations that (1) at a given temperature level all gases exhibit the same value of the product $p\bar{v}$ (p is pressure and $\bar{v}$ the specific volume on a molar basis) if the pressure is low enough, and (2) the value of the product $p\bar{v}$ increases with the temperature level. On this basis the gas temperature scale is defined by

$$T = \frac{1}{\bar{R}} \lim_{p \to 0} (p\bar{v})$$

where T is temperature and $\bar{R}$ is the *universal gas constant.* The absolute temperature at the *triple point of water* (Section 1.3, *P-v-T* Relations) is fixed by international agreement to be 273.16 K on the *Kelvin* temperature scale. $\bar{R}$ is then evaluated experimentally as $\bar{R}$ = 8.314 kJ/kmol · K (1545 ft · lbf/lbmol · °R).

The *Celsius termperature scale* (also called the centigrade scale) uses the degree Celsius (°C), which has the same magnitude as the Kelvin. Thus, temperature *differences* are identical on both scales. However, the zero point on the Celsius scale is shifted to 273.15 K, as shown by the following relationship between the Celsius temperature and the Kelvin temperature:

$$T(°C) = T(K) - 273.15 \tag{1.1}$$

On the Celsius scale, the triple point of water is 0.01°C and 0 K corresponds to –273.15°C.

Two other temperature scales are commonly used in engineering in the U.S. By definition, the *Rankine scale,* the unit of which is the degree Rankine (°R), is proportional to the Kelvin temperature according to

$$T(°R) = 1.8T(K) \tag{1.2}$$

The Rankine scale is also an absolute thermodynamic scale with an absolute zero that coincides with the absolute zero of the Kelvin scale. In thermodynamic relationships, temperature is always in terms of the Kelvin or Rankine scale unless specifically stated otherwise.

A degree of the same size as that on the Rankine scale is used in the *Fahrenheit scale,* but the zero point is shifted according to the relation

$$T(°F) = T(°R) - 459.67 \tag{1.3}$$

Substituting Equations 1.1 and 1.2 into Equation 1.3 gives

$$T(°F) = 1.8T(°C) + 32 \tag{1.4}$$

This equation shows that the Fahrenheit temperature of the *ice point* (0°C) is 32°F and of the *steam point* (100°C) is 212°F. The 100 Celsius or Kelvin degrees between the ice point and steam point corresponds to 180 Fahrenheit or Rankine degrees.

To provide a standard for temperature measurement taking into account both theoretical and practical considerations, the International Temperature Scale of 1990 (ITS-90) is defined in such a way that the temperature measured on it conforms with the thermodynamic temperature, the unit of which is the Kelvin, to within the limits of accuracy of measurement obtainable in 1990. Further discussion of ITS-90 is provided by Preston-Thomas (1990).

The First Law of Thermodynamics, Energy

Energy is a fundamental concept of thermodynamics and one of the most significant aspects of engineering analysis. Energy can be *stored* within systems in various macroscopic forms: kinetic energy, gravitational potential energy, and internal energy. Energy can also be *transformed* from one form to another and *transferred* between systems. For closed systems, energy can be transferred by *work* and *heat transfer.* The total amount of energy is *conserved* in all transformations and transfers.

Work

In thermodynamics, the term *work* denotes a means for transferring energy. Work is an effect of one system on another that is identified and measured as follows: work is done by a system on its surroundings if the *sole effect* on everything external to the system *could have been* the raising of a weight. The test of whether a work interaction has taken place is not that the elevation of a weight is actually changed, nor that a force actually acted through a distance, but that the sole effect *could be* the change in elevation of a weight. The magnitude of the work is measured by the number of standard weights that could have been raised. Since the raising of a weight is in effect a force acting through a distance, the work concept of mechanics is preserved. This definition includes work effects such as is associated with rotating shafts, displacement of the boundary, and the flow of electricity.

Work done *by* a system is considered positive: $W > 0$. Work done *on* a system is considered negative: $W < 0$. The time rate of doing work, or *power,* is symbolized by $\dot{W}$ and adheres to the same sign convention.

Energy

A closed system undergoing a process that involves only work interactions with its surroundings experiences an *adiabatic* process. On the basis of experimental evidence, it can be postulated that *when a closed system is altered adiabatically, the amount of work is fixed by the end states of the system and is independent of the details of the process.* This postulate, which is one way the *first law of thermodynamics* can be stated, can be made regardless of the type of work interaction involved, the type of process, or the nature of the system.

As the work in an adiabatic process of a closed system is fixed by the end states, an extensive property called *energy* can be defined for the system such that its change between two states is the work in an adiabatic process that has these as the end states. In engineering thermodynamics the change in the energy of a system is considered to be made up of three macroscopic contributions: the change in *kinetic energy, KE,* associated with the motion of the system *as a whole* relative to an external coordinate frame, the change in gravitational *potential energy, PE,* associated with the position of the system *as a whole* in the Earth's gravitational field, and the change in *internal energy, U,* which accounts for all other energy associated with the system. Like kinetic energy and gravitational potential energy, internal energy is an extensive property.

In summary, the change in energy between two states of a closed system in terms of the work W_{ad} of an adiabatic process between these states is

$$\left(KE_2 - KE_1\right) + \left(PE_2 - PE_1\right) + \left(U_2 - U_1\right) = -W_{ad} \tag{1.5}$$

where 1 and 2 denote the initial and final states, respectively, and the minus sign before the work term is in accordance with the previously stated sign convention for work. Since any arbitrary value can be assigned to the energy of a system at a given state 1, no particular significance can be attached to the value of the energy at state 1 or at *any* other state. Only *changes* in the energy of a system have significance.

The specific energy (energy per unit mass) is the sum of the specific internal energy, u, the specific kinetic energy, $\mathrm{v}^2/2$, and the specific gravitational potential energy, gz, such that

$$\text{specific energy} = u + \frac{\mathrm{v}^2}{2} + \mathrm{g}\mathrm{z} \tag{1.6}$$

where the velocity v and the elevation z are each relative to specified datums (often the Earth's surface) and g is the acceleration of gravity.

A property related to internal energy u, pressure p, and specific volume v is *enthalpy,* defined by

$$h = u + pv \tag{1.7a}$$

or on an extensive basis

$$H = U + pV \tag{1.7b}$$

Heat

Closed systems can also interact with their surroundings in a way that cannot be categorized as work, as, for example, a gas (or liquid) contained in a closed vessel undergoing a process while in contact with a flame. This type of interaction is called a *heat interaction,* and the process is referred to as *nonadiabatic.*

A fundamental aspect of the energy concept is that energy is conserved. Thus, since a closed system experiences precisely the same energy change during a nonadiabatic process as during an adiabatic process between the same end states, it can be concluded that the *net* energy transfer to the system in each of these processes must be the same. It follows that heat interactions also involve energy transfer. Denoting the amount of energy transferred *to* a closed system in heat interactions by Q, these considerations can be summarized by the *closed system energy balance:*

$$\left(U_2 - U_1\right) + \left(KE_2 - KE_1\right) + \left(PE_2 - PE_1\right) = Q - W \tag{1.8}$$

The closed system energy balance expresses the conservation of energy principle for closed systems of all kinds.

The quantity denoted by Q in Equation 1.8 accounts for the amount of energy transferred to a closed system during a process by means other than work. On the basis of an experiment, it is known that such an energy transfer is induced only as a result of a temperature difference between the system and its surroundings and occurs only in the direction of decreasing temperature. This means of energy transfer is called an *energy transfer by heat.* The following sign convention applies:

$$Q > 0: \text{ heat transfer } to \text{ the system}$$

$$Q < 0: \text{ heat transfer } from \text{ the system}$$

The time rate of heat transfer, denoted by $\dot{Q}$, adheres to the same sign convention.

Methods based on experiment are available for evaluating energy transfer by heat. These methods recognize two basic transfer mechanisms: *conduction* and *thermal radiation.* In addition, theoretical and empirical relationships are available for evaluating energy transfer involving *combined* modes such as *convection.* Further discussion of heat transfer fundamentals is provided in Chapter 3.

The quantities symbolized by W and Q account for *transfers* of energy. The terms *work* and *heat* denote different *means* whereby energy is transferred and not *what* is transferred. Work and heat are not properties, and it is improper to speak of work or heat "contained" in a system. However, to achieve economy of expression in subsequent discussions, W and Q are often referred to simply as work and heat transfer, respectively. This less formal approach is commonly used in engineering practice.

Power Cycles

Since energy is a property, over each cycle there is no net change in energy. Thus, Equation 1.8 reads for *any* cycle

$$Q_{cycle} = W_{cycle}$$

That is, for *any* cycle the net amount of energy received through heat interactions is equal to the net energy transferred out in work interactions. A *power cycle,* or *heat* engine, is one for which a net amount of energy is transferred out by work: $W_{cycle} > 0$. This equals the net amount of energy transferred in by heat.

Power cycles are characterized both by addition of energy by heat transfer, Q_A, and inevitable rejections of energy by heat transfer, Q_R:

$$Q_{cycle} = Q_A - Q_R$$

Combining the last two equations,

$$W_{cycle} = Q_A - Q_R$$

The *thermal efficiency* of a heat engine is defined as the ratio of the net work developed to the total energy added by heat transfer:

$$\eta = \frac{W_{cycle}}{Q_A} = 1 - \frac{Q_R}{Q_A} \tag{1.9}$$

The thermal efficiency is strictly less than 100%. That is, some portion of the energy Q_A supplied is invariably rejected $Q_R \neq 0$.

The Second Law of Thermodynamics, Entropy

Many statements of the second law of thermodynamics have been proposed. Each of these can be called a statement of the second law *or* a corollary of the second law since, if one is invalid, all are invalid. In every instance where a consequence of the second law has been tested directly or indirectly by experiment it has been verified. Accordingly, the basis of the second law, like every other physical law, is experimental evidence.

Kelvin-Planck Statement

The Kelvin-Plank statement of the second law of thermodynamics refers to a *thermal reservoir.* A thermal reservoir is a system that remains at a constant temperature even though energy is added or removed by heat transfer. A reservoir is an idealization, of course, but such a system can be approximated in a number of ways — by the Earth's atmosphere, large bodies of water (lakes, oceans), and so on. Extensive properties of thermal reservoirs, such as internal energy, can change in interactions with other systems even though the reservoir temperature remains constant, however.

The Kelvin-Planck statement of the second law can be given as follows: *It is impossible for any system to operate in a thermodynamic cycle and deliver a net amount of energy by work to its surroundings while receiving energy by heat transfer from a single thermal reservoir.* In other words, a *perpetual-motion machine of the second kind* is impossible. Expressed analytically, the Kelvin-Planck statement is

$$W_{cycle} \leq 0 \qquad (\text{single reservoir})$$

where the words *single reservoir* emphasize that the system communicates thermally only with a single reservoir as it executes the cycle. The "less than" sign applies when *internal irreversibilities* are present as the system of interest undergoes a cycle and the "equal to" sign applies only when no irreversibilities are present.

Irreversibilities

A process is said to be *reversible* if it is possible for its effects to be eradicated in the sense that there is some way by which *both* the system and its surroundings can be *exactly restored* to their respective initial states. A process is *irreversible* if there is no way to undo it. That is, there is no means by which the system and its surroundings can be exactly restored to their respective initial states. A system that has undergone an irreversible process is not necessarily precluded from being restored to its initial state. However, were the system restored to its initial state, it would not also be possible to return the surroundings to their initial state.

There are many effects whose presence during a process renders it irreversible. These include, but are not limited to, the following: heat transfer through a finite temperature difference; unrestrained expansion of a gas or liquid to a lower pressure; spontaneous chemical reaction; mixing of matter at different compositions or states; friction (sliding friction as well as friction in the flow of fluids); electric current flow through a resistance; magnetization or polarization with hysteresis; and inelastic deformation. The term *irreversibility* is used to identify effects such as these.

Irreversibilities can be divided into two classes, *internal* and *external.* Internal irreversibilities are those that occur within the system, while external irreversibilities are those that occur within the surroundings, normally the immediate surroundings. As this division depends on the location of the boundary there is some arbitrariness in the classification (by locating the boundary to take in the immediate surroundings, all irreversibilities are internal). Nonetheless, valuable insights can result when this distinction between irreversibilities is made. When internal irreversibilities are absent during a process, the process is said to be *internally reversible.* At every intermediate state of an internally reversible process of a closed system, all intensive properties are uniform throughout each phase present: the temperature, pressure, specific volume, and other intensive properties do not vary with position. The discussions to follow compare the actual and internally reversible process concepts for two cases of special interest.

For a gas as the system, the work of expansion arises from the force exerted by the system to move the boundary against the resistance offered by the surroundings:

$$W = \int_1^2 F dx = \int_1^2 pA dx$$

where the force is the product of the moving area and the pressure exerted by the system there. Noting that Adx is the change in total volume of the system,

$$W = \int_1^2 p dV$$

This expression for work applies to both actual and internally reversible expansion processes. However, for an internally reversible process p is not only the pressure at the moving boundary but also the pressure of the entire system. Furthermore, for an internally reversible process the volume equals mv, where the specific volume v has a single value throughout the system at a given instant. Accordingly, the work of an internally reversible expansion (or compression) process is

$$W = m\int_1^2 p dv \tag{1.10}$$

When such a process of a closed system is represented by a continuous curve on a plot of pressure vs. specific volume, the area *under* the curve is the magnitude of the work per unit of system mass (area a-b-c′-d′ of Figure 1.3, for example).

Although improved thermodynamic performance can accompany the reduction of irreversibilities, steps in this direction are normally constrained by a number of practical factors often related to costs. For example, consider two bodies able to communicate thermally. With a *finite* temperature difference between them, a spontaneous heat transfer would take place and, as noted previously, this would be a source of irreversibility. The importance of the heat transfer irreversibility diminishes as the temperature difference narrows; and as the temperature difference between the bodies vanishes, the heat transfer approaches *ideality.* From the study of heat transfer it is known, however, that the transfer of a finite amount of energy by heat between bodies whose temperatures differ only slightly requires a considerable amount of time, a large heat transfer surface area, or both. To approach *ideality,* therefore, a heat transfer would require an exceptionally long time and/or an exceptionally large area, each of which has cost implications constraining what can be achieved practically.

Carnot Corollaries

The two corollaries of the second law known as *Carnot* corollaries state: (1) the thermal efficiency of an irreversible power cycle is always less than the thermal efficiency of a reversible power cycle when each operates between the same two thermal reservoirs; (2) all reversible power cycles operating between the same two thermal reservoirs have the same thermal efficiency. A cycle is considered *reversible* when there are no irreversibilities within the system as it undergoes the cycle, and heat transfers between the system and reservoirs occur ideally (that is, with a vanishingly small temperature difference).

Kelvin Temperature Scale

Carnot corollary 2 suggests that the thermal efficiency of a reversible power cycle operating between two thermal reservoirs depends only on the temperatures of the reservoirs and not on the nature of the substance making up the system executing the cycle or the series of processes. With Equation 1.9 it can be concluded that the ratio of the heat transfers is also related only to the temperatures, and is independent of the substance and processes:

$$\left(\frac{Q_C}{Q_H}\right)_{\substack{rev\\cycle}} = \psi(T_C, T_H)$$

where Q_H *is* the energy transferred to the system by heat transfer from a *hot* reservoir at temperature T_H, and Q_C is the energy rejected from the system to a *cold* reservoir at temperature T_C. The words *rev cycle* emphasize that this expression applies only to systems undergoing reversible cycles while operating between the two reservoirs. Alternative temperature scales correspond to alternative specifications for the function ψ in this relation.

The *Kelvin temperature scale is* based on $\psi(T_C, T_H) = T_C/T_H$. Then

$$\left(\frac{Q_C}{Q_H}\right)_{\substack{rev\\cycle}} = \frac{T_C}{T_H} \tag{1.11}$$

This equation defines only a ratio of temperatures. The specification of the Kelvin scale is completed by assigning a numerical value to one standard reference state. The state selected is the same used to define the *gas scale:* at the triple point of water the temperature is specified to be 273.16 K. If a reversible cycle is operated between a reservoir at the reference-state temperature and another reservoir at an unknown temperature T, then the latter temperature is related to the value at the reference state by

$$T = 273.16\left(\frac{Q}{Q'}\right)_{\substack{rev\\cycle}}$$

where Q is the energy received by heat transfer from the reservoir at temperature T, and Q' is the energy rejected to the reservoir at the reference temperature. Accordingly, a temperature scale is defined that is valid over all ranges of temperature and that is independent of the thermometric substance.

Carnot Efficiency

For the special case of a reversible power cycle operating between thermal reservoirs at temperatures T_H and T_C on the Kelvin scale, combination of Equations 1.9 and 1.11 results in

$$\eta_{max} = 1 - \frac{T_C}{T_H} \tag{1.12}$$

called the *Carnot efficiency.* This is the efficiency of *all* reversible power cycles operating between thermal reservoirs at T_H and T_C. Moreover, it is the *maximum theoretical* efficiency that any power cycle, real or ideal, could have while operating between the same two reservoirs. As temperatures on the Rankine scale differ from Kelvin temperatures only by the factor 1.8, the above equation may be applied with either scale of temperature.

The Clausius Inequality

The Clausius inequality provides the basis for introducing two ideas instrumental for quantitative evaluations of processes of systems from a second law perspective: *entropy* and *entropy generation.* The Clausius inequality states that

$$\oint \left(\frac{\delta Q}{T} \right)_b \le 0 \tag{1.13a}$$

where δQ represents the heat transfer at a part of the system boundary during a portion of the cycle, and T is the absolute temperature at that part of the boundary. The symbol δ is used to distinguish the differentials of *nonproperties,* such as heat and work, from the differentials of properties, written with the symbol d. The subscript b indicates that the integrand is evaluated at the boundary of the system executing the cycle. The symbol $\oint$ indicates that the integral is to be performed over all parts of the boundary and over the entire cycle. The Clausius inequality can be demonstrated using the Kelvin-Planck statement of the second law, and the significance of the inequality is the same: the equality applies when there are no internal irreversibilities as the system executes the cycle, and the inequality applies when internal irreversibilities are present.

The Clausius inequality can be expressed alternatively as

$$\oint \left(\frac{\delta Q}{T} \right)_b = -S_{gen} \tag{1.13b}$$

where S_{gen} can be viewed as representing the *strength* of the inequality. The value of S_{gen} is positive when internal irreversibilities are present, zero when no internal irreversibilities are present, and can never be negative. Accordingly, S_{gen} is a measure of the irreversibilities present within the system executing the cycle. In the next section, S_{gen} is identified as the *entropy* generated (or *produced*) by internal irreversibilities during the cycle.

Entropy and Entropy Generation

Entropy

Consider two cycles executed by a closed system. One cycle consists of an internally reversible process A from state 1 to state 2, followed by an internally reversible process C from state 2 to state 1. The other cycle consists of an internally reversible process B from state 1 to state 2, followed by the same process C from state 2 to state 1 as in the first cycle. For these cycles, Equation 1.13b takes the form

$$\left(\int_1^2 \frac{\delta Q}{T} \right)_A + \left(\int_2^1 \frac{\delta Q}{T} \right)_C = -S_{gen} = 0$$

$$\left(\int_1^2 \frac{\delta Q}{T} \right)_B + \left(\int_2^1 \frac{\delta Q}{T} \right)_C = -S_{gen} = 0$$

where S_{gen} has been set to zero since the cycles are composed of internally reversible processes. Subtracting these equations leaves

$$\left(\int_1^2 \frac{\delta Q}{T} \right)_A = \left(\int_1^2 \frac{\delta Q}{T} \right)_B$$

Since A and B are arbitrary, it follows that the integral of $\delta Q/T$ has the same value for *any* internally reversible process between the two states: the value of the integral depends on the end states only. It can be concluded, therefore, that the integral defines the change in some property of the system. Selecting the symbol S to denote this property, its change is given by

$$S_2 - S_1 = \left(\int_1^2 \frac{\delta Q}{T} \right)_{\substack{int \\ rev}} \tag{1.14a}$$

where the subscript *int rev* indicates that the integration is carried out for any internally reversible process linking the two states. This extensive property is called *entropy.*

Since entropy is a property, the change in entropy of a system in going from one state to another is the same for *all* processes, both internally reversible and irreversible, between these two states. In other words, once the change in entropy between two states has been evaluated, this is the magnitude of the entropy change for *any* process of the system between these end states.

The definition of entropy change expressed on a differential basis is

$$dS = \left(\frac{\delta Q}{T} \right)_{\substack{int \\ rev}} \tag{1.14b}$$

Equation 1.14b indicates that when a closed system undergoing an internally reversible process *receives* energy by heat transfer, the system experiences an *increase* in entropy. Conversely, when energy is *removed* from the system by heat transfer, the entropy of the system *decreases.* This can be interpreted to mean that an entropy transfer is *associated* with (or accompanies) heat transfer. The direction of the entropy transfer is the same as that of the heat transfer. In an *adiabatic* internally reversible process of a closed system the entropy would remain constant. A constant entropy process is called an *isentropic* process.

On rearrangement, Equation 1.14b becomes

$$(\delta Q)_{\substack{int \\ rev}} = TdS$$

Then, for an internally reversible process of a closed system between state 1 and state 2,

$$Q_{\substack{int \\ rev}} = m \int_1^2 Tds \tag{1.15}$$

When such a process is represented by a continuous curve on a plot of temperature vs. specific entropy, the area *under* the curve is the magnitude of the heat transfer per unit of system mass.

Entropy Balance

For a cycle consisting of an actual process from state 1 to state 2, during which internal irreversibilities are present, followed by an internally reversible process from state 2 to state 1, Equation 1.13b takes the form

$$\int_1^2 \left(\frac{\delta Q}{T} \right)_b + \int_2^1 \left(\frac{\delta Q}{T} \right)_{\substack{int \\ rev}} = -S_{gen}$$

where the first integral is for the actual process and the second integral is for the internally reversible process. Since no irreversibilities are associated with the internally reversible process, the term S_{gen} accounting for the effect of irreversibilities during the cycle can be identified with the actual process only.

Applying the definition of entropy change, the second integral of the foregoing equation can be expressed as

$$S_1 - S_2 = \int_2^1 \left(\frac{\delta Q}{T}\right)_{\substack{int\\rev}}$$

Introducing this and rearranging the equation, the *closed system entropy balance* results:

$$\underset{\text{entropy change}}{\underline{S_2 - S_1}} = \underset{\text{entropy transfer}}{\underline{\int_1^2 \left(\frac{\delta Q}{T}\right)_b}} + \underset{\text{entropy generation}}{\underline{S_{gen}}} \tag{1.16}$$

When the end states are fixed, the entropy change on the left side of Equation 1.16 can be evaluated independently of the details of the process from state 1 to state 2. However, the two terms on the right side depend explicitly on the nature of the process and cannot be determined solely from knowledge of the end states. The first term on the right side is associated with heat transfer to or from the system during the process. This term can be interpreted as the *entropy transfer associated with (or accompanying) heat transfer*. The direction of entropy transfer is the same as the direction of the heat transfer, and the same sign convention applies as for heat transfer: a positive value means that entropy is transferred into the system, and a negative value means that entropy is transferred out.

The entropy change of a system is not accounted for solely by entropy transfer, but is also due to the second term on the right side of Equation 1.16 denoted by S_{gen}. The term S_{gen} is positive when internal irreversibilities are present during the process and vanishes when internal irreversibilities are absent. This can be described by saying that entropy is *generated* (or produced) within the system by the action of irreversibilities. The second law of thermodynamics can be interpreted as specifying that entropy is generated by irreversibilities and conserved only in the limit as irreversibilities are reduced to zero. Since S_{gen} measures the effect of irreversibilities present within a system during a process, its value depends on the nature of the process and not solely on the end states. Entropy generation is *not* a property.

When applying the entropy balance, the objective is often to evaluate the entropy generation term. However, the value of the entropy generation for a given process of a system usually does not have much significance by itself. The significance is normally determined through comparison. For example, the entropy generation within a given component might be compared to the entropy generation values of the other components included in an overall system formed by these components. By comparing entropy generation values, the components where appreciable irreversibilities occur can be identified and rank ordered. This allows attention to be focused on the components that contribute most heavily to inefficient operation of the overall system.

To evaluate the entropy transfer term of the entropy balance requires information regarding both the heat transfer and the temperature on the boundary where the heat transfer occurs. The entropy transfer term is not always subject to direct evaluation, however, because the required information is either unknown or undefined, such as when the system passes through states sufficiently far from equilibrium. In practical applications, it is often convenient, therefore, to enlarge the system to include enough of the immediate surroundings that the temperature on the boundary of the *enlarged system* corresponds to the ambient temperature, T_{amb}. The entropy transfer term is then simply Q/T_{amb}. However, as the irreversibilities present would not be just those for the system of interest but those for the enlarged system, the entropy generation term would account for the effects of internal irreversibilities within the system *and* external irreversibilities present within that portion of the surroundings included within the enlarged system.

A form of the entropy balance convenient for particular analyses is the *rate form:*

$$\frac{dS}{dt} = \sum_j \frac{\dot{Q}_j}{T_j} + \dot{S}_{gen} \tag{1.17}$$

where dS/dt is the time rate of change of entropy of the system. The term $\dot{Q}_j / T_j$ represents the time rate of entropy transfer through the portion of the boundary whose instantaneous temperature is T_j. The term $\dot{S}_{gen}$ accounts for the time rate of entropy generation due to irreversibilities within the system.

For a system *isolated* from its surroundings, the entropy balance is

$$\left(S_2 - S_1\right)_{isol} = S_{gen} \tag{1.18}$$

where S_{gen} is the total amount of entropy generated within the isolated system. Since entropy is generated in all actual processes, the only processes of an isolated system that actually can occur are those for which the entropy of the isolated system increases. This is known as the *increase of entropy principle.*

1.2 Control Volume Applications

Since most applications of engineering thermodynamics are conducted on a control volume basis, the control volume formulations of the mass, energy, and entropy balances presented in this section are especially important. These are given here in the form of *overall* balances. Equations of change for mass, energy, and entropy in the form of differential equations are also available in the literature (see, e.g., Bird et al., 1960).

Conservation of Mass

When applied to a control volume, the principle of mass conservation states: *The time rate of accumulation of mass within the control volume equals the difference between the total rates of mass flow in and out across the boundary.* An important case for engineering practice is one for which inward and outward flows occur, each through one or more ports. For this case the conservation of mass principle takes the form

$$\frac{dm_{cv}}{dt} = \sum_i \dot{m}_i - \sum_e \dot{m}_e \tag{1.19}$$

The left side of this equation represents the time rate of change of mass contained within the control volume, $\dot{m}_i$ denotes the mass flow rate at an inlet, and $\dot{m}_e$ is the mass flow rate at an outlet.

The *volumetric flow rate* through a portion of the control surface with area dA is the product of the velocity component normal to the area, $\mathrm{v_n}$, times the area: $\mathrm{v_n}\,dA$. The *mass* flow rate through dA is $\rho(\mathrm{v_n}\,dA)$. The mass rate of flow through a port of area A is then found by integration over the area

$$\dot{m} = \int_A \rho \mathrm{v_n}\, dA$$

For *one-dimensional* flow the intensive properties are uniform with position over area A, and the last equation becomes

$$\dot{m} = \rho \mathrm{v} A = \frac{\mathrm{v}A}{v} \tag{1.20}$$

where v denotes the specific volume and the subscript n has been dropped from velocity for simplicity.

Control Volume Energy Balance

When applied to a control volume, the principle of energy conservation states: *The time rate of accumulation of energy within the control volume equals the difference between the total incoming rate of energy transfer and the total outgoing rate of energy transfer.* Energy can enter and exit a control volume by work and heat transfer. Energy also enters and exits with flowing streams of matter. Accordingly, for a control volume with one-dimensional flow at a single inlet and a single outlet,

$$\frac{d(U + KE + PE)_{cv}}{dt} = \dot{Q}_{cv} - \dot{W} + \dot{m}\left(u_i + \frac{\mathrm{v}_i^2}{2} + \mathrm{g}z_i\right) - \dot{m}\left(u_e + \frac{\mathrm{v}_e^2}{2} + \mathrm{g}z_e\right) \tag{1.21}$$

where the underlined terms account for the specific energy of the incoming and outgoing streams. The terms $\dot{Q}_{cv}$ and $\dot{W}$ account, respectively, for the net rates of energy transfer by heat and work over the boundary (control surface) of the control volume.

Because work is always done on or by a control volume where matter flows across the boundary, the quantity $\dot{W}$ of Equation 1.21 can be expressed in terms of two contributions: one is the work associated with the force of the fluid pressure as mass is introduced at the inlet and removed at the exit. The other, denoted as $\dot{W}_{cv}$, includes *all other* work effects, such as those associated with rotating shafts, displacement of the boundary, and electrical effects. The work rate concept of mechanics allows the first of these contributions to be evaluated in terms of the product of the pressure force, pA, and velocity at the point of application of the force. To summarize, the work term $\dot{W}$ of Equation 1.21 can be expressed (with Equation 1.20) as

$$\begin{aligned} \dot{W} &= \dot{W}_{cv} + (p_e A_e)\mathrm{V}_e - (p_i A_i)\mathrm{V}_i \\ &= \dot{W}_{cv} + \dot{m}_e(p_e v_e) - \dot{m}_i(p_i v_i) \end{aligned} \tag{1.22}$$

The terms $\dot{m}_i\,(pv_i)$ and $\dot{m}_e\,(p_e v_e)$ account for the work associated with the pressure at the inlet and outlet, respectively, and are commonly referred to as *flow work.*

Substituting Equation 1.22 into Equation 1.21, and introducing the specific enthalpy h, the following form of the control volume energy rate balance results:

$$\frac{d(U + KE + PE)_{cv}}{dt} = \dot{Q}_{cv} - \dot{W}_{cv} + \dot{m}_i\left(h_i + \frac{\mathrm{V}_i^2}{2} + \mathrm{g}z_i\right) - \dot{m}_e\left(h_e + \frac{\mathrm{V}_e^2}{2} + \mathrm{g}z_e\right) \tag{1.23}$$

To allow for applications where there may be several locations on the boundary through which mass enters or exits, the following expression is appropriate:

$$\frac{d(U + KE + PE)_{cv}}{dt} = \dot{Q}_{cv} - \dot{W}_{cv} + \sum_i \dot{m}_i\left(h_i + \frac{\mathrm{V}_i^2}{2} + \mathrm{g}z_i\right) - \sum_e \dot{m}_e\left(h_e + \frac{\mathrm{V}_e^2}{2} + \mathrm{g}z_e\right) \tag{1.24}$$

Equation 1.24 is an *accounting* rate balance for the energy of the control volume. It states that the time rate of accumulation of energy within the control volume equals the difference between the total rates of energy transfer in and out across the boundary. The mechanisms of energy transfer are heat and work, as for closed systems, and the energy accompanying the entering and exiting mass.

Control Volume Entropy Balance

Like mass and energy, entropy is an extensive property. And like mass and energy, entropy can be transferred into or out of a control volume by streams of matter. As this is the principal difference between the closed system and control volume forms, the control volume entropy rate balance is obtained by modifying Equation 1.17 to account for these entropy transfers. The result is

$$\underbrace{\frac{dS_{cv}}{dt}}_{\substack{\text{rate of}\\\text{entropy}\\\text{change}}} = \underbrace{\sum_j \frac{\dot{Q}_j}{T_j} + \sum_i \dot{m}_i s_i - \sum_e \dot{m}_e s_e}_{\substack{\text{rate of}\\\text{entropy}\\\text{transfer}}} + \underbrace{\dot{S}_{gen}}_{\substack{\text{rate of}\\\text{entropy}\\\text{generation}}} \tag{1.25}$$

where dS_{cv}/dt represents the time rate of change of entropy within the control volume. The terms $\dot{m}_i s_i$ and $\dot{m}_e s_e$ account, respectively, for rates of entropy *transfer* into and out of the control volume associated with mass flow. One-dimensional flow is assumed at locations where mass enters and exits. $\dot{Q}_j$ represents the time rate of heat transfer at the location on the boundary where the instantaneous temperature is T_j; and $\dot{Q}_j / T_j$ accounts for the associated rate of entropy *transfer.* $\dot{S}_{gen}$ denotes the time rate of entropy *generation* due to irreversibilities *within* the control volume. When a control volume comprises a number of components, $\dot{S}_{gen}$ is the sum of the rates of entropy generation of the components.

Control Volumes at Steady State

Engineering systems are often idealized as being at *steady state,* meaning that all properties are unchanging in time. For a control volume at steady state, the identity of the matter within the control volume changes continuously, but the total amount of mass remains constant. At steady state, Equation 1.19 reduces to

$$\sum_i \dot{m}_i = \sum_e \dot{m}_e \tag{1.26a}$$

The energy rate balance of Equation 1.24 becomes, at steady state,

$$0 = \dot{Q}_{cv} - \dot{W}_{cv} + \sum_i \dot{m}_i \left(h_i + \frac{V_i^2}{2} + gz_i \right) - \sum_e \dot{m}_e \left(h_e + \frac{V_e^2}{2} + gz_e \right) \tag{1.26b}$$

At steady state, the entropy rate balance of Equation 1.25 reads

$$0 = \sum_j \frac{\dot{Q}_j}{T_j} + \sum_i \dot{m}_i s_i - \sum_e \dot{m}_e s_e + \dot{S}_{gen} \tag{1.26c}$$

Mass and energy are conserved quantities, but entropy is not generally conserved. Equation 1.26a indicates that the total rate of mass flow into the control volume equals the total rate of mass flow out of the control volume. Similarly, Equation 1.26b states that the total rate of energy transfer into the control volume equals the total rate of energy transfer out of the control volume. However, Equation 1.26c shows that the rate at which entropy is transferred out *exceeds* the rate at which entropy enters, the difference being the rate of entropy generation within the control volume owing to irreversibilities.

Applications frequently involve control volumes having a single inlet and a single outlet, as, for example, the control volume of Figure 1.1 where heat transfer (if any) occurs at T_b: the temperature, or a suitable average temperature, on the boundary where heat transfer occurs. For this case the mass rate balance, Equation 1.26a, reduces to $\dot{m}_i = \dot{m}_e$. Denoting the common mass flow rate by $\dot{m}$, Equations 1.26b and 1.26c read, respectively,

$$0 = \dot{Q}_{cv} - \dot{W}_{cv} + \dot{m}\left[\left(h_i - h_e\right) + \left(\frac{V_i^2 - V_e^2}{2} \right) + g\left(z_i - z_e\right) \right] \tag{1.27a}$$

$$0 = \frac{\dot{Q}_{cv}}{T_b} + \dot{m}\left(s_i - s_e\right) + \dot{S}_{gen} \tag{1.28a}$$

When Equations 1.27a and 1.28a are applied to particular cases of interest, additional simplifications are usually made. The heat transfer term $\dot{Q}_{cv}$ is dropped when it is insignificant relative to other energy

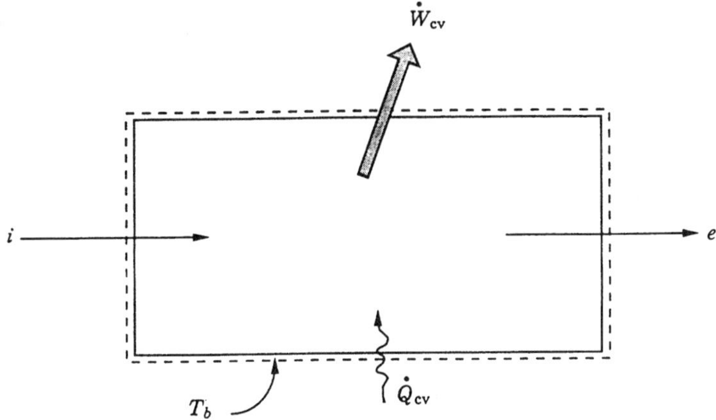

FIGURE 1.1 One-inlet, one-outlet control volume at steady state.

transfers across the boundary. This may be the result of one or more of the following: (1) the outer surface of the control volume is insulated, (2) the outer surface area is too small for there to be effective heat transfer, (3) the temperature difference between the control volume and its surroundings is small enough that the heat transfer can be ignored, (4) the gas or liquid passes through the control volume so quickly that there is not enough time for significant heat transfer to occur. The work term $\dot{W}_{cv}$ drops out of the energy rate balance when there are no rotating shafts, displacements of the boundary, electrical effects, or other work mechanisms associated with the control volume being considered. The changes in kinetic and potential energy of Equation 1.27a are frequently negligible relative to other terms in the equation.

The special forms of Equations 1.27a and 1.28a listed in Table 1.1 are obtained as follows: when there is no heat transfer, Equation 1.28a gives

$$s_e - s_i = \frac{\dot{S}_{gen}}{\dot{m}} \geq 0 \qquad (1.28b)$$

$$\text{(no heat transfer)}$$

Accordingly, when irreversibilities are present within the control volume, the specific entropy increases as mass flows from inlet to outlet. In the ideal case in which no internal irreversibilities are present, mass passes through the control volume with no change in its entropy — that is, *isentropically.*

For no heat transfer, Equation 1.27a gives

$$\dot{W}_{cv} = \dot{m}\left[\left(h_i - h_e\right) + \left(\frac{\mathrm{v}_i^2 - \mathrm{v}_e^2}{2}\right) + g\left(z_i - z_e\right)\right] \qquad (1.27b)$$

A special form that is applicable, at least approximately, to *compressors*, *pumps*, and *turbines* results from dropping the kinetic and potential energy terms of Equation 1.27b, leaving

$$\dot{W}_{cv} = \dot{m}\left(h_i - h_e\right) \qquad (1.27c)$$

$$(\textit{compressors, pumps, } \text{and } \textit{turbines})$$

TABLE 1.1 Energy and Entropy Balances for One-Inlet, One-Outlet Control Volumes at Steady State and No Heat Transfer

Energy balance

$$\dot{W}_{cv} = \dot{m}\left[(h_i - h_e) + \left(\frac{\mathrm{v}_i^2 - \mathrm{v}_e^2}{2}\right) + g(z_i - z_e)\right] \quad (1.27b)$$

Compressors, pumps, and turbines[a]

$$\dot{W}_{cv} = \dot{m}(h_i - h_e) \quad (1.27c)$$

Throttling

$$h_e \cong h_i \quad (1.27d)$$

Nozzles, diffusers[b]

$$\mathrm{v}_e = \sqrt{\mathrm{v}_i^2 + 2(h_i - h_e)} \quad (1.27f)$$

Entropy balance

$$s_e - s_i = \frac{\dot{S}_{gen}}{\dot{m}} \geq 0 \quad (1.28b)$$

[a] For an ideal gas with constant c_p, Equation 1′ of Table 1.7 allows Equation 1.27c to be written as

$$\dot{W}_{cv} = \dot{m}c_p(T_i - T_e) \quad (1.27c')$$

The power developed in an *isentropic process* is obtained with Equation 5′ of Table 1.7 as

$$\dot{W}_{cv} = \dot{m}c_pT_i\left[1 - (p_e/p_i)^{(k-1)/k}\right] \quad (s = c) \quad (1.27c'')$$

where $c_p = kR/(k-1)$.

[b] For an ideal gas with constant c_p, Equation 1′ of Table 1.7 allows Equation 1.27f to be written as

$$\mathrm{v}_e = \sqrt{\mathrm{v}_i^2 + 2c_p(T_i - T_e)} \quad (1.27f')$$

The exit velocity for an *isentropic process* is obtained with Equation 5′ of Table 1.7 as

$$\mathrm{v}_e = \sqrt{\mathrm{v}_i^2 + 2c_pT_i\left[1 - (p_e/p_i)^{(k-1)/k}\right]} \quad (s = c) \quad (1.27f'')$$

where $c_p = kR/(k-1)$.

In *throttling devices* a significant reduction in pressure is achieved simply by introducing a restriction into a line through which a gas or liquid flows. For such devices $\dot{W}_{cv} = 0$ and Equation 1.27c reduces further to read

$$h_e \cong h_i \quad (\textit{throttling process}) \quad (1.27d)$$

That is, upstream and downstream of the throttling device, the specific enthalpies are equal.

A *nozzle* is a flow passage of varying cross-sectional area in which the velocity of a gas or liquid increases in the direction of flow. In a *diffuser,* the gas or liquid decelerates in the direction of flow. For such devices, $\dot{W}_{cv} = 0$. The heat transfer and potential energy change are also generally negligible. Then Equation 1.27b reduces to

$$0 = h_i - h_e + \frac{\mathrm{v}_i^2 - \mathrm{v}_e^2}{2} \quad (1.27e)$$

Solving for the outlet velocity

$$\mathrm{v}_e = \sqrt{\mathrm{v}_i^2 + 2(h_i - h_e)} \qquad (1.27\mathrm{f})$$

$$(nozzle,\ diffuser)$$

Further discussion of the flow-through nozzles and diffusers is provided in Chapter 2.

The mass, energy, and entropy rate balances, Equations 1.26, can be applied to control volumes with multiple inlets and/or outlets, as, for example, cases involving heat-recovery steam generators, feedwater heaters, and counterflow and crossflow heat exchangers. Transient (or unsteady) analyses can be conducted with Equations 1.19, 1.24, and 1.25. Illustrations of all such applications are provided by Moran and Shapiro (2000).

Example 1

A turbine receives steam at 7 MPa, 440°C and exhausts at 0.2 MPa for subsequent process heating duty. If heat transfer and kinetic/potential energy effects are negligible, determine the steam mass flow rate, in kg/hr, for a turbine power output of 30 MW when (a) the steam quality at the turbine outlet is 95%, (b) the turbine expansion is internally reversible.

Solution. With the indicated idealizations, Equation 1.27c is appropriate. Solving, $\dot{m} = \dot{W}_{cv}/(h_i - h_e)$. Steam table data (Table A.5) at the inlet condition are h_i = 3261.7 kJ/kg, s_i = 6.6022 kJ/kg · K.

(a) At 0.2 MPa and x = 0.95, h_e = 2596.5 kJ/kg. Then

$$\dot{m} = \frac{30\ \mathrm{MW}}{(3261.7 - 2596.5)\,\mathrm{kJ/kg}}\left(\frac{10^3\ \mathrm{kJ/sec}}{1\ \mathrm{MW}}\right)\left(\frac{3600\ \mathrm{sec}}{1\ \mathrm{hr}}\right)$$

$$= 162{,}357\ \mathrm{kg/hr}$$

(b) For an internally reversible expansion, Equation 1.28b reduces to give $s_e = s_i$. For this case, h_e = 2499.6 kJ/kg (x = 0.906), and $\dot{m}$ = 141,714 kg/hr.

Example 2

Air at 500°F, 150 lbf/in.^{2}, and 10 ft/sec expands adiabatically through a nozzle and exits at 60°F, 15 lbf/in.^{2}. For a mass flow rate of 5 lb/sec determine the exit area, in in.^{2}. Repeat for an isentropic expansion to 15 lbf/in.^{2}. Model the air as an ideal gas (Section 1.3, Ideal Gas Model) with specific heat c_p = 0.24 Btu/lb · °R (k = 1.4).

Solution. The nozzle exit area can be evaluated using Equation 1.20, together with the ideal gas equation, $v = RT/p$:

$$A_e = \frac{\dot{m}v_e}{\mathrm{v}_e} = \frac{\dot{m}(RT_e/p_e)}{\mathrm{v}_e}$$

The exit velocity required by this expression is obtained using Equation 1.27f′ of Table 1.1,

$$\mathrm{v}_e = \sqrt{\mathrm{v}_i^2 + 2c_p(T_i - T_e)}$$

$$= \sqrt{\left(\frac{10\ \mathrm{ft}}{s}\right)^2 + 2\left(0.24\frac{\mathrm{Btu}}{\mathrm{lb}\cdot\mathrm{R}}\right)\left(\frac{778.17\ \mathrm{ft}\cdot\mathrm{lbf}}{1\ \mathrm{Btu}}\right)(440°\mathrm{R})\left(\frac{32.174\ \mathrm{lb}\cdot\mathrm{ft/sec}^2}{1\ \mathrm{lbf}}\right)}$$

$$= 2299.5\ \mathrm{ft/sec}$$

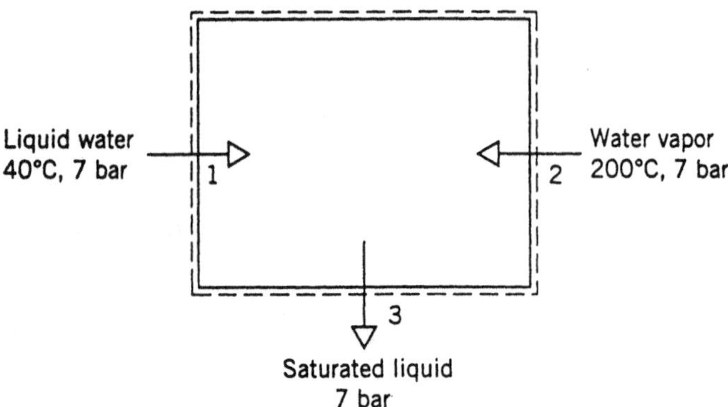

FIGURE 1.2 Open feedwater heater.

Finally, with $R = \overline{R}/\mathcal{M} = 53.33$ ft · lbf/lb · °R,

$$A_e = \frac{\left(5\dfrac{\text{lb}}{\text{sec}}\right)\left(53.3\dfrac{\text{ft}\cdot\text{lbf}}{\text{lb}\cdot°\text{R}}\right)(520°\text{R})}{\left(2299.5\dfrac{\text{ft}}{\text{sec}}\right)\left(15\dfrac{\text{lbf}}{\text{in.}^2}\right)} = 4.02 \text{ in.}^2$$

Using Equation 1.27f″ in Table 1.1 for the isentropic expansion,

$$\text{v}_e = \sqrt{(10)^2 + 2(0.24)(778.17)(960)(32.174)\left[1-\left(\frac{15}{150}\right)^{0.4/1.4}\right]}$$

$$= 2358.3 \text{ ft/sec}$$

Then $A_e = 3.92$ in.^{2}.

Example 3

Figure 1.2 provides steady-state operating data for an open feedwater heater. Ignoring heat transfer and kinetic/potential energy effects, determine the ratio of mass flow rates, $\dot{m}_1/\dot{m}_2$.

Solution. For this case Equations 1.26a and 1.26b reduce to read, respectively,

$$\dot{m}_1 + \dot{m}_2 = \dot{m}_3$$

$$0 = \dot{m}_1 h_1 + \dot{m}_2 h_2 - \dot{m}_3 h_3$$

Combining and solving for the ratio $\dot{m}_1/\dot{m}_2$,

$$\frac{\dot{m}_1}{\dot{m}_2} = \frac{h_2 - h_3}{h_3 - h_1}$$

Inserting steam table data, in kJ/kg, from Table A.5,

$$\frac{\dot{m}_1}{\dot{m}_2} = \frac{2844.8 - 697.2}{697.2 - 167.6} = 4.06$$

Internally Reversible Heat Transfer and Work

For one-inlet, one-outlet control volumes at steady state, the following expressions give the heat transfer rate and power in the absence of internal irreversibilities:

$$\left(\frac{\dot{Q}_{cv}}{\dot{m}}\right)_{\substack{int\\rev}} = \int_1^2 Tds \tag{1.29}$$

$$\left(\frac{\dot{W}_{cv}}{\dot{m}}\right)_{\substack{int\\rev}} = -\int_1^2 \mathrm{v}dp + \frac{\mathrm{V}_1^2 - \mathrm{V}_2^2}{2} + g(z_1 - z_2) \tag{1.30a}$$

(see, e.g., Moran and Shapiro, 2000).

If there is no significant change in kinetic or potential energy from inlet to outlet, Equation 1.30a reads

$$\left(\frac{\dot{W}_{cv}}{\dot{m}}\right)_{\substack{int\\rev}} = -\int_1^2 \mathrm{v}dp \qquad (\Delta ke = \Delta pe = 0) \tag{1.30b}$$

The specific volume remains approximately constant in many applications with liquids. Then Equation 1.30b becomes

$$\left(\frac{\dot{W}_{cv}}{\dot{m}}\right)_{\substack{int\\rev}} = -v(p_2 - p_1) \qquad (v = \text{constant}) \tag{1.30c}$$

When the states visited by a unit of mass flowing without irreversibilities from inlet to outlet are described by a continuous curve on a plot of temperature vs. specific entropy, Equation 1.29 implies that the area under the curve is the magnitude of the heat transfer per unit of mass flowing. When such an ideal process is described by a curve on a plot of pressure vs. specific volume, as shown in Figure 1.3, the magnitude of the integral $\int vdp$ of Equations 1.30a and 1.30b is represented by the area a-b-c-d *behind* the curve. The area a-b-c′-d′ *under* the curve is identified with the magnitude of the integral $\int pdv$ of Equation 1.10.

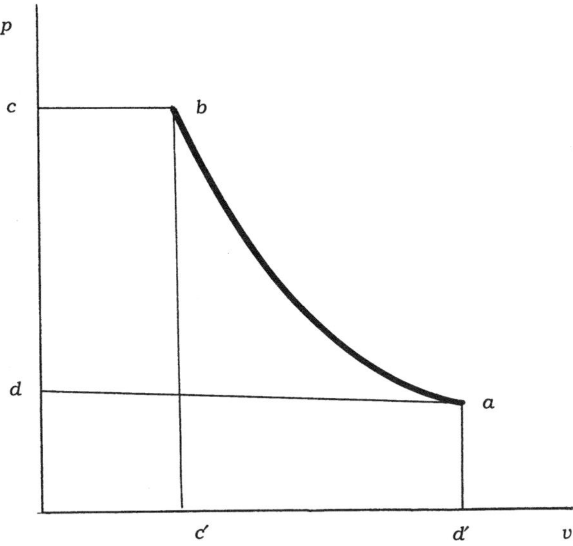

FIGURE 1.3 Internally reversible process on p–v coordinates.

1.3 Property Relations and Data

Pressure, temperature, volume, and mass can be found experimentally. The relationships between the specific heats c_v and c_p and temperature at relatively low pressure are also accessible experimentally, as are certain other property data. Specific internal energy, enthalpy, and entropy are among those properties that are not so readily obtained in the laboratory. Values for such properties are calculated using experimental data of properties that are more amenable to measurement, together with appropriate property relations derived using the principles of thermodynamics. In this section property relations and data sources are considered for *simple compressible systems*, which include a wide range of industrially important substances.

Property data are provided in the publications of the *National Institute of Standards and Technology* (formerly the U.S. Bureau of Standards), of professional groups such as the *American Society of Mechanical Engineering (ASME),* the *American Society of Heating. Refrigerating, and Air Conditioning Engineers (ASHRAE)*, and the *American Chemical Society,* and of corporate entities such as *Dupont* and *Dow Chemical.* Handbooks and property reference volumes such as included in the list of references for this chapter are readily accessed sources of data. Property data are also retrievable from various commercial online data bases. Computer software is increasingly available for this purpose as well.

Basic Relations for Pure Substances

An energy balance in differential form for a closed system undergoing an internally reversible process in the absence of overall system motion and the effect of gravity reads

$$dU = (\delta Q)_{\substack{int\\rev}} - (\delta W)_{\substack{int\\rev}}$$

From Equation 1.14b, $(\delta Q)_{\substack{int\\rev}} = TdS$. When consideration is limited to *simple compressible systems:* systems for which the only significant work in an internally reversible process is associated with volume change, $(\delta W)_{\substack{int\\rev}} = pdV$, the following equation is obtained:

$$dU = TdS - pdV \tag{1.31a}$$

Introducing enthalpy, $H = U + pV$, the Helmholtz function, $\Psi = U - TS$, and the Gibbs function, $G = H - TS$, three additional expressions are obtained:

$$dH = TdS + Vdp \tag{1.31b}$$

$$d\Psi = -pdV - SdT \tag{1.31c}$$

$$dG = Vdp - SdT \tag{1.31d}$$

Equations 1.31 can be expressed on a per-unit-mass basis as

$$du = Tds - pdv \tag{1.32a}$$

$$dh = Tds + vdp \tag{1.32b}$$

$$d\psi = -pdv - sdT \tag{1.32c}$$

$$dg = vdp - sdT \tag{1.32d}$$

Similar expressions can be written on a per-mole basis.

Maxwell Relations

Since only properties are involved, each of the four differential expressions given by Equations 1.32 is an *exact* differential exhibiting the general form $dz = M(x, y)dx + N(x, y)dy$, where the second mixed partial derivatives are equal: $(\partial M/\partial y) = (\partial N/\partial x)$. Underlying these exact differentials are, respectively, functions of the form $u(s, v)$, $h(s, p)$, $\psi(v, T)$, and $g(T, p)$. From such considerations the *Maxwell relations* given in Table 1.2 can be established.

Example 4

Derive the Maxwell relation following from Equation 1.32a.

TABLE 1.2 Relations from Exact Differentials

Function	*Differential*	*Coefficients*	*Maxwell*
General:			
$z = z(x,y)$	$dz = M(x,y)dx + N(x,y)dy$	$\left(\frac{\partial z}{\partial x}\right)_y = M$	$\left(\frac{\partial M}{\partial y}\right)_x = \left(\frac{\partial N}{\partial x}\right)_y$
		$\left(\frac{\partial z}{\partial y}\right)_x = N$	
Internal energy:			
$u(s,v)$	$du = Tds - pdv$	$\left(\frac{\partial u}{\partial s}\right)_v = T$	$\left(\frac{\partial T}{\partial v}\right)_s = -\left(\frac{\partial p}{\partial s}\right)_v$
		$\left(\frac{\partial u}{\partial v}\right)_s = -p$	
Enthalpy:			
$h(s,p)$	$dh = Tds + vdp$	$\left(\frac{\partial h}{\partial s}\right)_p = T$	$\left(\frac{\partial T}{\partial p}\right)_s = \left(\frac{\partial v}{\partial s}\right)_p$
		$\left(\frac{\partial h}{\partial p}\right)_s = v$	
Helmholtz function:			
$\psi(v,T)$	$d\psi = -pdv - sdT$	$\left(\frac{\partial \psi}{\partial v}\right)_T = -p$	$\left(\frac{\partial p}{\partial T}\right)_v = \left(\frac{\partial s}{\partial v}\right)_T$
		$\left(\frac{\partial \psi}{\partial T}\right)_v = -s$	
Gibbs function:			
$g(T,p)$	$dg = vdp - sdT$	$\left(\frac{\partial g}{\partial p}\right)_T = v$	$\left(\frac{\partial v}{\partial T}\right)_p = -\left(\frac{\partial s}{\partial p}\right)_T$
		$\left(\frac{\partial g}{\partial T}\right)_p = -s$	

Solution. The differential of the function $u = u(s, v)$ is

$$du = \left(\frac{\partial u}{\partial s}\right)_v ds + \left(\frac{\partial u}{\partial v}\right)_s dv$$

By comparison with Equation 1.32a,

$$T = \left(\frac{\partial u}{\partial s}\right)_v, \quad -p = \left(\frac{\partial u}{\partial v}\right)_s$$

In Equation 1.32a, T plays the role of M and $-p$ plays the role of N, so the equality of second mixed partial derivatives gives the Maxwell relation,

$$\left(\frac{\partial T}{\partial v}\right)_s = -\left(\frac{\partial p}{\partial s}\right)_v$$

Since each of the properties T, p, v, and s appears on the right side of two of the eight coefficients of Table 1.2, four additional property relations can be obtained by equating such expressions:

$$\left(\frac{\partial u}{\partial s}\right)_v = \left(\frac{\partial h}{\partial s}\right)_p, \quad \left(\frac{\partial u}{\partial v}\right)_s = \left(\frac{\partial \psi}{\partial v}\right)_T$$

$$\left(\frac{\partial h}{\partial p}\right)_s = \left(\frac{\partial g}{\partial p}\right)_T, \quad \left(\frac{\partial \psi}{\partial T}\right)_v = \left(\frac{\partial g}{\partial T}\right)_p$$

These four relations are identified in Table 1.2 by brackets. As any three of Equations 1.32 can be obtained from the fourth simply by manipulation, the 16 property relations of Table 1.2 also can be regarded as following from this single differential expression. Several additional first-derivative property relations can be derived; see, e.g., Zemansky, 1972.

Specific Heats and Other Properties

Engineering thermodynamics uses a wide assortment of thermodynamic properties and relations among these properties. Table 1.3 lists several commonly encountered properties.

Among the entries of Table 1.3 are the specific heats c_v and c_p. These intensive properties are often required for thermodynamic analysis, and are defined as partial derivations of the functions $u(T, v)$ and $h(T, p)$, respectively,

$$c_v = \left(\frac{\partial u}{\partial T}\right)_v \tag{1.33}$$

$$c_p = \left(\frac{\partial h}{\partial T}\right)_p \tag{1.34}$$

Since u and h can be expressed either on a unit mass basis or a per-mole basis, values of the specific heats can be similarly expressed. Table 1.4 summarizes relations involving c_v and c_p. The property k, the specific heat ratio, is

$$k = \frac{c_p}{c_v} \tag{1.35}$$

TABLE 1.3 Symbols and Definitions for Selected Properties

Property	Symbol	Definition	Property	Symbol	Definition
Pressure	p		Specific heat, constant volume	c_v	$(\partial u/\partial T)_v$
Temperature	T		Specific heat, constant pressure	c_p	$(\partial h/\partial T)_p$
Specific volume	v		Volume expansivity	β	$\frac{1}{v}(\partial v/\partial T)_p$
Specific internal energy	u		Isothermal compressivity	κ	$-\frac{1}{v}(\partial v/\partial p)_T$
Specific entropy	s		Isentropic compressibility	α	$-\frac{1}{v}(\partial v/\partial p)_s$
Specific enthalpy	h	$u + pv$	Isothermal bulk modulus	B	$-v(\partial p/\partial v)_T$
Specific Helmholtz function	ψ	$u - Ts$	Isentropic bulk modulus	B_s	$-v(\partial p/\partial v)_s$
Specific Gibbs function	g	$h - Ts$	Joule-Thomson coefficient	μ_J	$(\partial T/\partial p)_h$
Compressibility factor	Z	pv/RT	Joule coefficient	η	$(\partial T/\partial v)_u$
Specific heat ratio	k	c_p/c_v	Velocity of sound	c	$\sqrt{-v^2(\partial p/\partial v)_s}$

Values for c_v and c_p can be obtained via statistical mechanics using *spectroscopic* measurements. They can also be determined macroscopically through exacting property measurements. Specific heat data for common gases, liquids, and solids are provided by the handbooks and property reference volumes listed among the Chapter 1 references. Specific heats are also considered in Section 1.3 as a part of the discussions of the *incompressible model* and the *ideal gas model.* Figure 1.4 shows how c_p for water vapor varies as a function of temperature and pressure. Other gases exhibit similar behavior. The figure also gives the variation of c_p with temperature in the limit as pressure tends to zero (the ideal gas limit). In this limit c_p increases with increasing temperature, which is a characteristic exhibited by other gases as well

The following two equations are often convenient for establishing relations among properties:

$$\left(\frac{\partial x}{\partial y}\right)_z \left(\frac{\partial y}{\partial x}\right)_z = 1 \tag{1.36a}$$

$$\left(\frac{\partial y}{\partial z}\right)_x \left(\frac{\partial z}{\partial x}\right)_y \left(\frac{\partial x}{\partial y}\right)_z = -1 \tag{1.36b}$$

Their use is illustrated in Example 5.

Example 5

Obtain Equations 2 and 11 of Table 1.4 from Equation 1.

Solution. Identifying x, y, z with s, T, and v, respectively, Equation 1.36b reads

$$\left(\frac{\partial T}{\partial v}\right)_s \left(\frac{\partial v}{\partial s}\right)_T \left(\frac{\partial s}{\partial T}\right)_v = -1$$

Applying Equation 1.36a to each of $(\partial T/\partial v)_s$ and $(\partial v/\partial s)_T$,

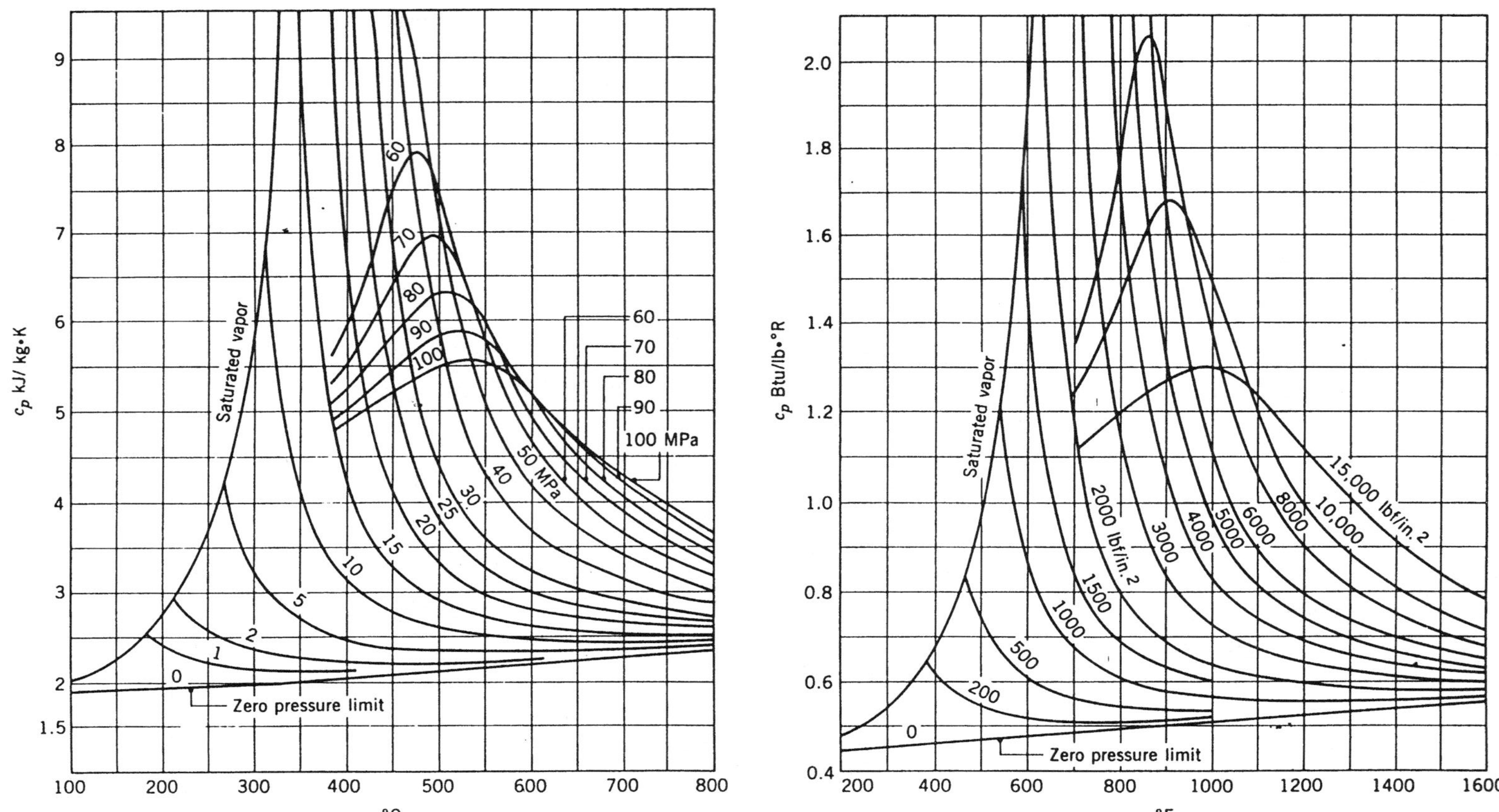

FIGURE 1.4 c_p of water vapor as a function of temperature and pressure. (Adapted from Keenan, J.H., Keyes, F.G., Hill, P.G., and Moore, J.G. 1969 and 1978. *Steam Tables — S.I. Units (English Units)*. John Wiley & Sons, New York.)

TABLE 1.4 Specific Heat Relations[a]

$c_v = \left(\frac{\partial u}{\partial T}\right)_v = T\left(\frac{\partial s}{\partial T}\right)_v$	(1)
$= -T\left(\frac{\partial p}{\partial T}\right)_v \left(\frac{\partial v}{\partial T}\right)_s$	(2)
$c_p = \left(\frac{\partial h}{\partial T}\right)_p = T\left(\frac{\partial s}{\partial T}\right)_p$	(3)
$= T\left(\frac{\partial v}{\partial T}\right)_p \left(\frac{\partial p}{\partial T}\right)_s$	(4)
$c_p - c_v = T\left(\frac{\partial p}{\partial T}\right)_v \left(\frac{\partial v}{\partial T}\right)_p$	(5)
$= -T\left(\frac{\partial v}{\partial T}\right)_p^2 \left(\frac{\partial p}{\partial v}\right)_T$	(6)
$= \frac{Tv\beta^2}{\kappa}$	(7)
$c_p = \frac{1}{\mu_J}\left[T\left(\frac{\partial v}{\partial T}\right)_p - v\right]$	(8)
$c_v = -\frac{1}{\eta}\left[T\left(\frac{\partial p}{\partial T}\right)_v - p\right]$	(9)
$k = \frac{c_p}{c_v} = \left(\frac{\partial v}{\partial p}\right)_T \left(\frac{\partial p}{\partial v}\right)_s$	(10)
$\left(\frac{\partial c_v}{\partial v}\right)_T = T\left(\frac{\partial^2 p}{\partial T^2}\right)_v$	(11)
$\left(\frac{\partial c_p}{\partial p}\right)_T = -T\left(\frac{\partial^2 v}{\partial T^2}\right)_p$	(12)

[a] See, for example, Moran, M.J. and Shapiro, H.N. 2000. *Fundamentals of Engineering Thermodynamics*, 4th ed. Wiley, New York.

$$\left(\frac{\partial s}{\partial T}\right)_v = -\frac{1}{(\partial T/\partial v)_s (\partial v/\partial s)_T} = -\left(\frac{\partial v}{\partial T}\right)_s \left(\frac{\partial s}{\partial v}\right)_T$$

Introducing the Maxwell relation from Table 1.2 corresponding to $\psi(T, v)$,

$$\left(\frac{\partial s}{\partial T}\right)_v = -\left(\frac{\partial v}{\partial T}\right)_s \left(\frac{\partial p}{\partial T}\right)_v$$

With this, Equation 2 of Table 1.4 is obtained from Equation 1, which in turn is obtained in Example 6. Equation 11 of Table 1.4 can be obtained by differentiating Equation 1 with repect to specific volume at fixed temperature, and again using the Maxwell relation corresponding to ψ.

P-v-T Relations

Considerable pressure, specific volume, and temperature data have been accumulated for industrially important gases and liquids. These data can be represented in the form $p = f(v, T)$, called an *equation of state*. Equations of state can be expressed in tabular, graphical, and analytical forms.

P-v-T Surface

The graph of a function $p = f(v, T)$ is a surface in three-dimensional space. Figure 1.5 shows the *p-v-T* relationship for water. Figure 1.5b shows the projection of the surface onto the pressure-temperature plane, called the *phase diagram*. The projection onto the *p–v* plane is shown in Figure 1.5c.

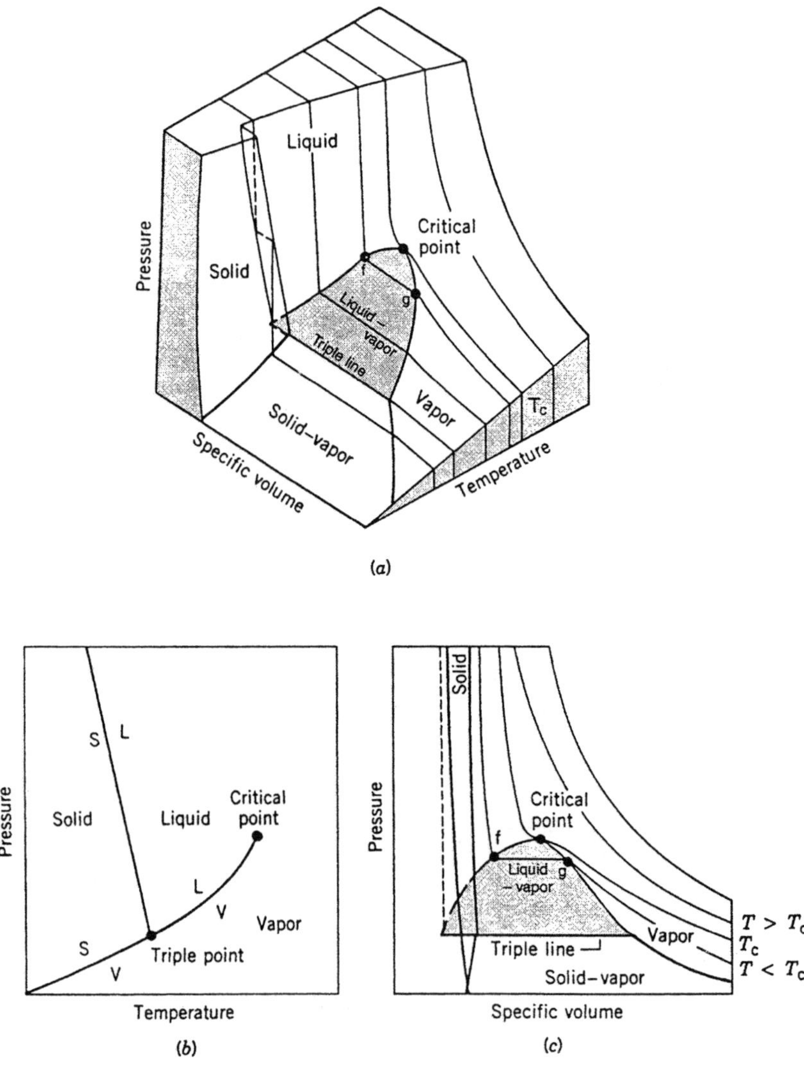

FIGURE 1.5 Pressure-specific volume-temperature surface and projections for water (not to scale).

Figure 1.5 has three regions labeled solid, liquid, and vapor where the substance exists only in a single phase. Between the single phase regions lie *two-phase* regions, where two phases coexist in equilibrium. The lines separating the single-phase regions from the two-phase regions are *saturation lines*. Any state represented by a point on a saturation line is a *saturation state*. The line separating the liquid phase and the two-phase liquid-vapor region is the saturated liquid line. The state denoted by f is a saturated liquid state. The saturated vapor line separates the vapor region and the two-phase liquid-vapor region. The state denoted by g is a saturated vapor state. The saturated liquid line and the saturated vapor line meet at the *critical point.* At the critical point, the pressure is the *critical pressure* p_c, and the temperature is the *critical temperature* T_c. Three phases can coexist in equilibrium along the line labeled *triple line.* The triple line projects onto a point on the phase diagram. The triple point of water is used in defining the Kelvin temperature scale (Section 1.1, Basic Concepts and Definitions; The Second Law of Thermodynamics, Entropy).

When a phase change occurs during constant pressure heating or cooling, the temperature remains constant as long as both phases are present. Accordingly, in the two-phase liquid-vapor region, a line of constant pressure is also a line of constant temperature. For a specified pressure, the corresponding temperature is called the *saturation temperature.* For a specified temperature, the corresponding pressure is called the *saturation pressure.* The region to the right of the saturated vapor line is known as the *superheated vapor region* because the vapor exists at a temperature greater than the saturation temperature for its pressure. The region to the left of the saturated liquid line is known as the *compressed liquid region* because the liquid is at a pressure higher than the saturation pressure for its temperature.

When a mixture of liquid and vapor coexists in equilibrium, the liquid phase is a saturated liquid and the vapor phase is a saturated vapor. The total volume of any such mixture is $V = V_f + V_g$; or, alternatively, $mv = m_f v_f + m_g v_g$, where m and v denote mass and specific volume, respectively. Dividing by the total mass of the mixture m and letting the *mass fraction* of the vapor in the mixture, m_g/m, be symbolized by x, called the *quality*, the apparent specific volume v of the mixture is

$$\begin{aligned} v &= (1-x)v_f + xv_g \\ &= v_f + xv_{fg} \end{aligned} \tag{1.37a}$$

where $v_{fg} = v_g - v_f$. Expressions similar in form can be written for internal energy, enthalpy, and entropy:

$$\begin{aligned} u &= (1-x)u_f + xu_g \\ &= u_f + xu_{fg} \end{aligned} \tag{1.37b}$$

$$\begin{aligned} h &= (1-x)h_f + xh_g \\ &= h_f + xh_{fg} \end{aligned} \tag{1.37c}$$

$$\begin{aligned} s &= (1-x)s_f + xs_g \\ &= s_f + xs_{fg} \end{aligned} \tag{1.37d}$$

For the case of water, Figure 1.6 illustrates the phase change from solid to liquid (melting): *a-b-c*; from solid to vapor (sublimation): *a′-b′-c′*; and from liquid to vapor (vaporization): *a″-b″-c″*. During any such phase change the temperature and pressure remain constant and thus are not independent properties. The *Clapeyron equation* allows the change in enthalpy during a phase change at fixed temperature to be evaluated from *p-v-T* data pertaining to the phase change. For vaporization, the Clapeyron equation reads

$$\left(\frac{dp}{dT}\right)_{sat} = \frac{h_g - h_f}{T(v_g - v_f)} \tag{1.38}$$

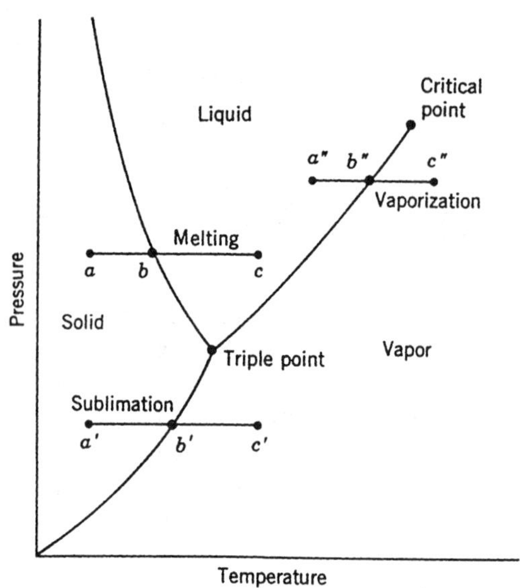

FIGURE 1.6 Phase diagram for water (not to scale).

where $(dp/dT)_{sat}$ is the slope of the saturation pressure-temperature curve at the point determined by the temperature held constant during the phase change. Expressions similar in form to Equation 1.38 can be written for sublimation and melting.

The Clapeyron equation shows that the slope of a saturation line on a phase diagram depends on the signs of the specific volume and enthalpy changes accompanying the phase change. In most cases, when a phase change takes place with an increase in specific enthalpy, the specific volume also increases, and $(dp/dT)_{sat}$ is positive. However, in the case of the melting of ice and a few other substances, the specific volume decreases on melting. The slope of the saturated solid-liquid curve for these few substances is negative, as illustrated for water in Figure 1.6.

Graphical Representations

The intensive states of a pure, simple compressible system can be represented graphically with any two independent intensive properties as the coordinates, excluding properties associated with motion and gravity. While any such pair may be used, there are several selections that are conventionally employed. These include the *p-T* and *p-v* diagrams of Figure 1.5, the *T-s* diagram of Figure 1.7, the *h-s* (Mollier) diagram of Figure 1.8, and the *p-h* diagram of Figure 1.9. The compressibility charts considered next use the compressibility factor as one of the coordinates.

Compressibility Charts

The *p-v-T* relation for a wide range of common gases is illustrated by the generalized compressibility chart of Figure 1.10. In this chart, the compressibility factor, Z, is plotted vs. the *reduced* pressure, p_R, *reduced* temperature, T_R, and *pseudoreduced* specific volume, v'_R, where

$$Z = \frac{p\bar{v}}{\bar{R}T} \tag{1.39}$$

and

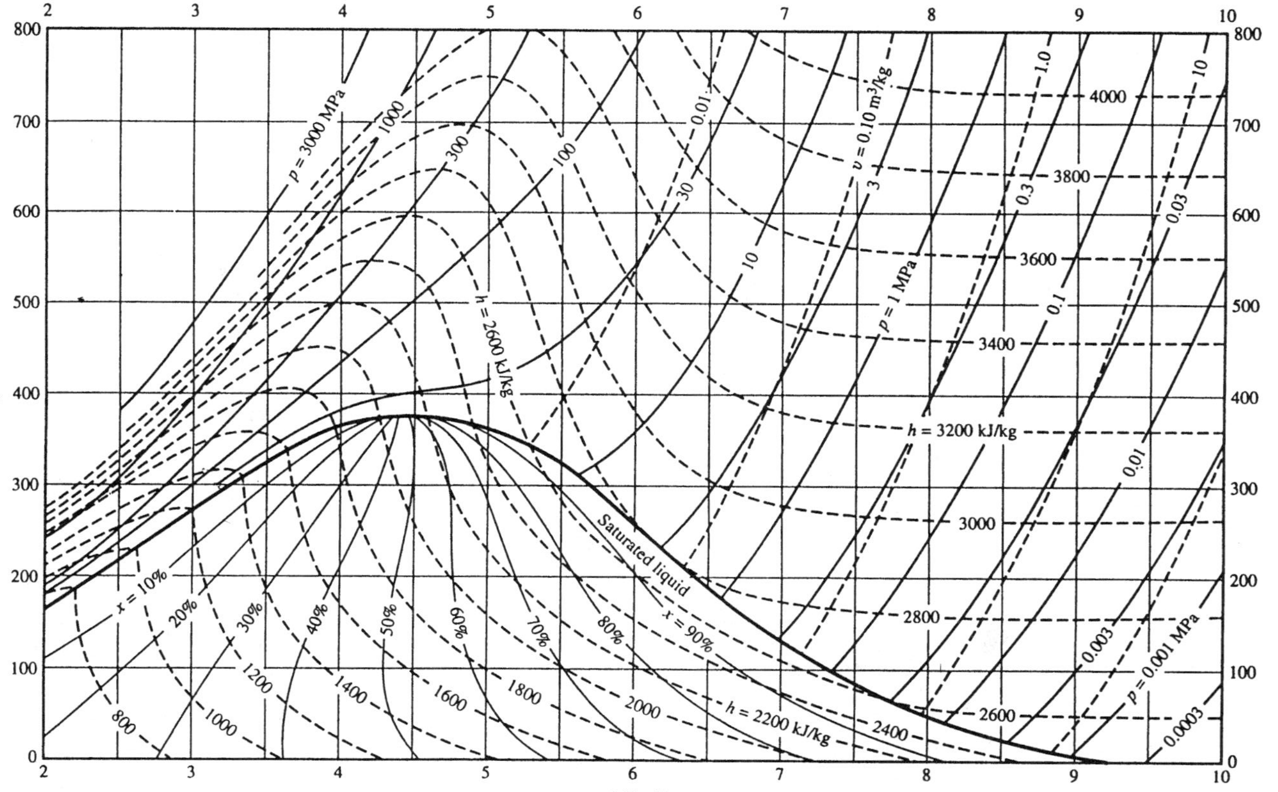

FIGURE 1.7 Temperature-entropy diagram for water. (*Source:* Jones, J.B. and Dugan, R.E. 1996. *Engineering Thermodynamics*, Prentice-Hall, Englewood Cliffs, NJ, based on data and formulations from Haar, L., Gallagher, J.S., and Kell, G.S. 1984. *NBS/NRC Steam Tables.* Hemisphere, Washington, D.C.)

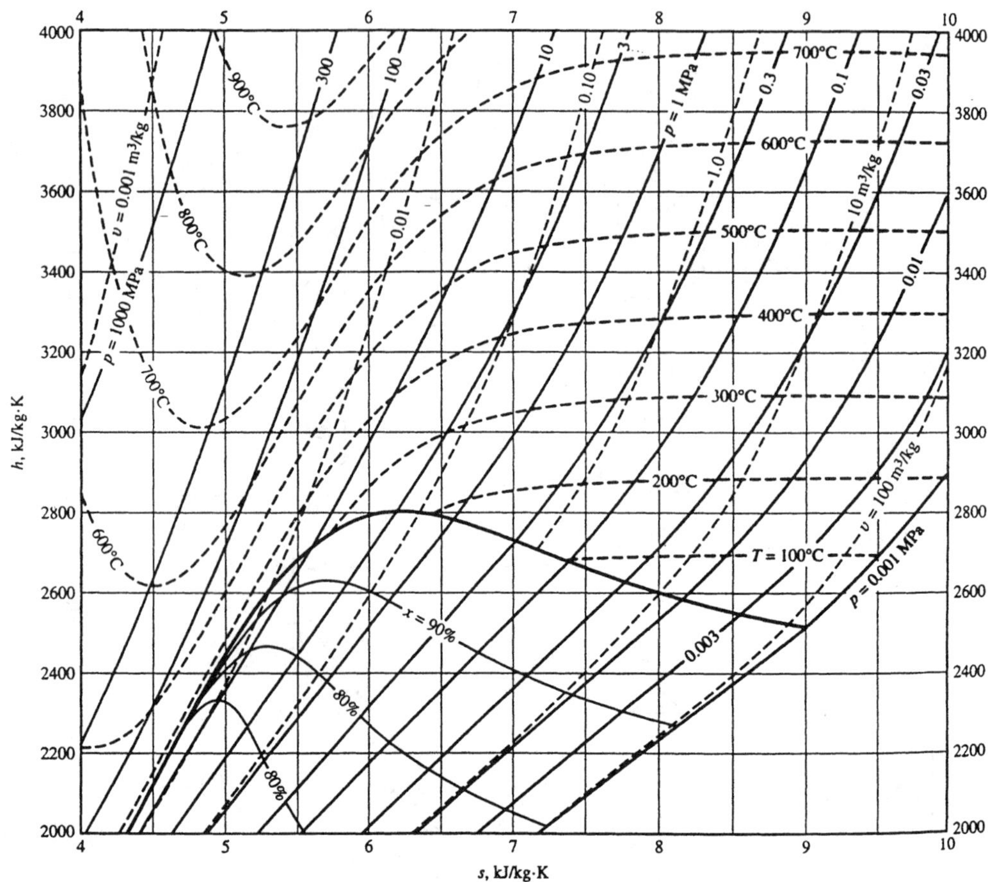

FIGURE 1.8 Enthalpy-entropy (Mollier) diagram for water. (*Source:* Jones, J.B. and Dugan, R.E. 1996. *Engineering Thermodynamics*. Prentice-Hall, Englewood Cliffs, NJ, based on data and formulations from Haar, L., Gallagher, J.S., and Kell, G.S. 1984. *NBS/NRC Steam Tables*. Hemisphere, Washington, D.C.)

$$p_R = \frac{p}{p_c}, \quad T_R = \frac{T}{T_c}, \quad v'_R = \frac{\bar{v}}{\left(\bar{R}T_c/p_c\right)} \tag{1.40}$$

In these expressions, $\bar{R}$ is the universal gas constant and p_c and T_c denote the critical pressure and temperature, respectively. Values of p_c and T_c are given for several substances in Table A.9. The reduced isotherms of Figure 1.10 represent the best curves fitted to the data of several gases. For the 30 gases used in developing the chart, the deviation of observed values from those of the chart is at most on the order of 5% and for most ranges is much less.*

Figure 1.10 gives a common value of about 0.27 for the compressibility factor at the critical point. As the critical compressibility factor for different substances actually varies from 0.23 to 0.33, the chart is inaccurate in the vicinity of the critical point. This source of inaccuracy can be removed by restricting the correlation to substances having essentially the same Z_c values. which is equivalent to including the critical compressibility factor as an independent variable: $Z = f(T_R, p_R, Z_c)$. To achieve greater accuracy

* To determine Z for hydrogen, helium, and neon above a T_R of 5, the reduced temperature and pressure should be calculated using $T_R = T/(T_c + 8)$ and $P_R = p/(p_c + 8)$, where temperatures are in K and pressures are in atm.

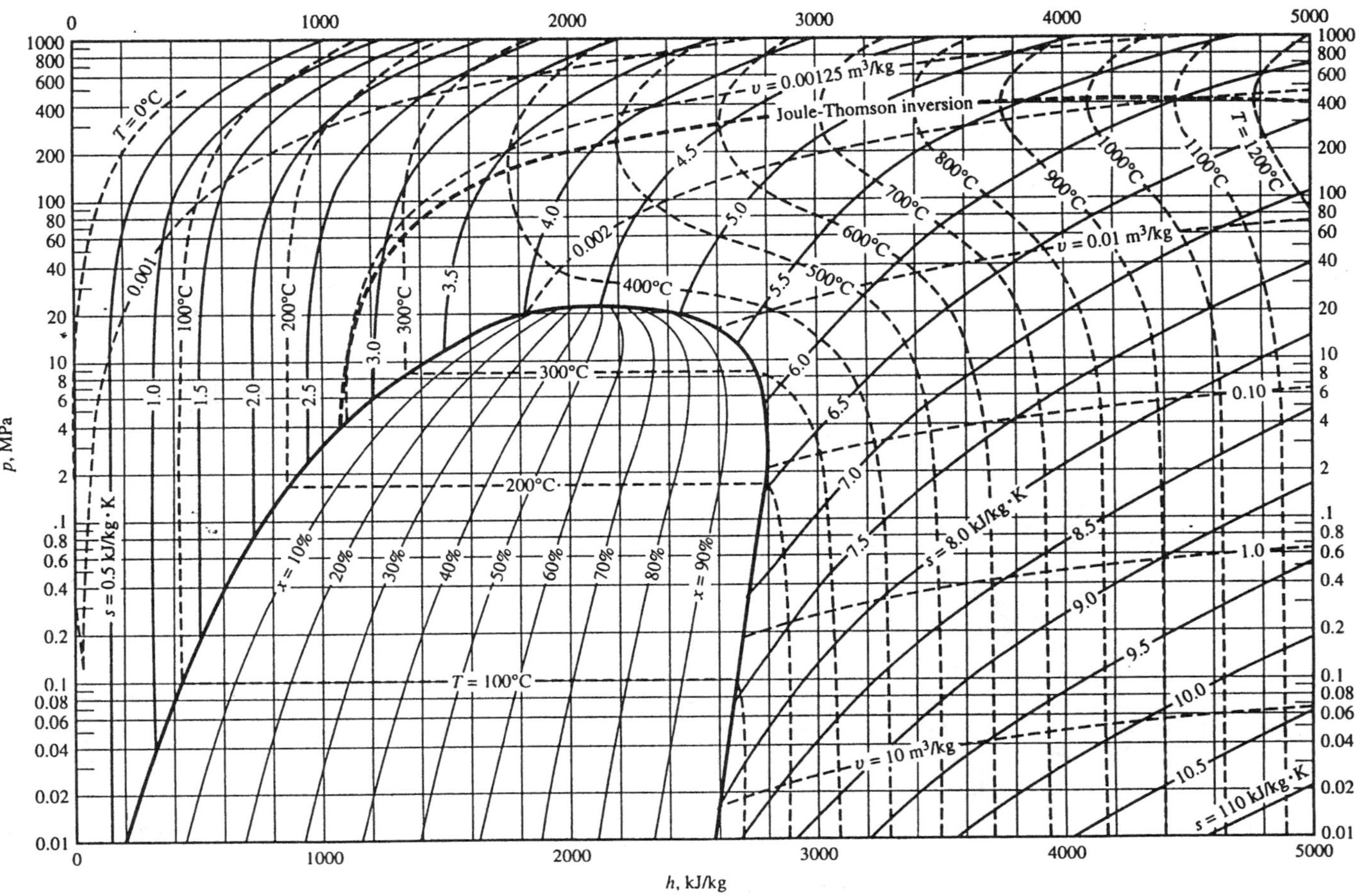

FIGURE 1.9 Pressure-enthalpy diagram for water. (*Source:* Jones, J.B. and Dugan, R.E. 1996. *Engineering Thermodynamics*. Prentice-Hall, Englewood Cliffs, NJ, based on data and formulations from Haar, L., Gallagher, J.S., and Kell, G.S. 1984. *NBS/NRC Steam Tables*. Hemisphere, Washington, D.C.)

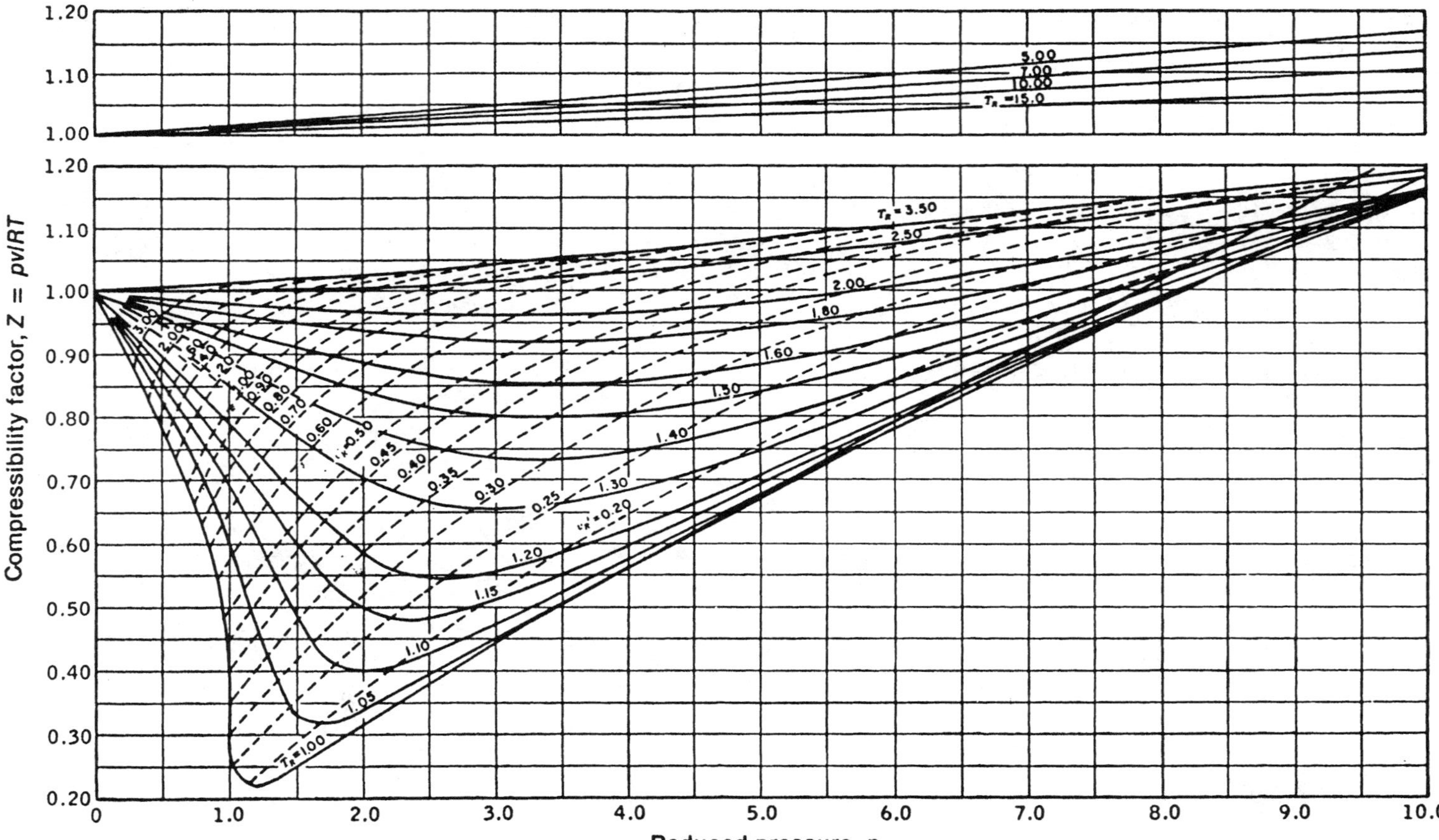

FIGURE 1.10 Generalized compressibility chart ($T_R = T/T_C$, $p_R = p/p_C$, $v'_R = \bar{v}p_C/RT_C$) for $p_R \leq 10$. (*Source:* Obert, E.F. 1960 *Concepts of Thermodynamics*. McGraw-Hill, New York.)

variables other than Z_c have been proposed as a third parameter — for example, the *acentric factor* (see, e.g., Reid and Sherwood, 1966).

Generalized compressibility data are also available in tabular form (see, e.g., Reid and Sherwood, 1966) and in equation form (see, e.g., Reynolds, 1979). The use of generalized data in any form (graphical, tabular, or equation) allows p, v, and T for gases to be evaluated simply and with reasonable accuracy. When accuracy is an essential consideration, generalized compressibility data should not be used as a substitute for p-v-T data for a given substance as provided by computer software, a table, or an equation of state.

Equations of State

Considering the isotherms of Figure 1.10, it is plausible that the variation of the compressibility factor might be expressed as an equation, at least for certain intervals of p and T. Two expressions can be written that enjoy a theoretical basis. One gives the compressibility factor as an infinite series expansion in pressure,

$$Z = 1 + \hat{B}(T)p + \hat{C}(T)p^2 + \hat{D}(T)p^3 + \ldots$$

and the other is a series in $1/\bar{v}$,

$$Z = 1 + \frac{B(T)}{\bar{v}} + \frac{C(T)}{\bar{v}^2} + \frac{D(T)}{\bar{v}^3} + \ldots$$

These expressions are known as *virial expansions*, and the coefficients $\hat{B}$, $\hat{C}$ $\hat{D}$, … and B, C, D … are called *virial coefficients.* In principle, the virial coefficients can be calculated using expressions from statistical mechanics derived from consideration of the force fields around the molecules. Thus far only the first few coefficients have been calculated and only for gases consisting of relatively simple molecules. The coefficients also can be found, in principle, by fitting p-v-T data in particular realms of interest. Only the first few coefficients can be found accurately this way, however, and the result is a *truncated* equation valid only at certain states.

Over 100 equations of state have been developed in an attempt to portray accurately the p-v-T behavior of substances and yet avoid the complexities inherent in a full virial series. In general, these equations exhibit little in the way of fundamental physical significance and are mainly empirical in character. Most are developed for gases, but some describe the p-v-T behavior of the liquid phase, at least qualitatively. Every equation of state is restricted to particular states. The realm of applicability is often indicated by giving an interval of pressure, or density, where the equation can be expected to represent the p-v-T behavior faithfully. When it is not stated, the realm of applicability often may be approximated by expressing the equation in terms of the compressibility factor Z and the reduced properties, and superimposing the result on a generalized compressibility chart or comparing with compressibility data from the literature.

Equations of state can be classified by the number of adjustable constants they involve. The Redlich-Kwong equation is considered by many to be the best of the two-constant equations of state. It gives pressure as a function of temperature and specific volume and thus is *explicit* in pressure:

$$p = \frac{\bar{R}T}{\bar{v} - b} - \frac{a}{\bar{v}(\bar{v} + b)T^{1/2}} \tag{1.41}$$

This equation is primarily empirical in nature, with no rigorous justification in terms of molecular arguments. Values for the Redlich-Kwong constants for several substances are provided in Table A.9. Modified forms of the equation have been proposed with the aim of achieving better accuracy.

Although the two-constant Redlich-Kwong equation performs better than some equations of state having several adjustable constants, two-constant equations tend to be limited in accuracy as pressure (or density) increases. Increased accuracy normally requires a greater number of adjustable constants. For example, the Benedict-Webb-Rubin equation, which involves eight adjustable constants, has been successful in predicting the *p-v-T* behavior of *light hydrocarbons.* The Benedict-Webb-Rubin equation is also explicit in pressure,

$$p = \frac{\bar{R}T}{\bar{v}} + \left(B\bar{R}T - A - \frac{C}{T^2}\right)\frac{1}{\bar{v}^2} + \frac{\left(b\bar{R}T - a\right)}{\bar{v}^3} + \frac{a\alpha}{\bar{v}^6} + \frac{c}{\bar{v}^3 T^2}\left(1 + \frac{\gamma}{\bar{v}^2}\right)\exp\left(-\frac{\gamma}{\bar{v}^2}\right) \tag{1.42}$$

Values of the Benedict-Webb-Rubin constants for various gases are provided in the literature (see, e.g., Cooper and Goldfrank, 1967). A modification of the Benedict-Webb-Rubin equation involving 12 constants is discussed by Lee and Kessler, 1975. Many multiconstant equations can be found in the engineering literature, and with the advent of high speed computers, equations having 50 or more constants have been developed for representing the *p-v-T* behavior of different substances.

Gas Mixtures

Since an unlimited variety of mixtures can be formed from a given set of pure components by varying the relative amounts present, the properties of mixtures are reported only in special cases such as air. Means are available for predicting the properties of mixtures, however. Most techniques for predicting mixture properties are empirical in character and are not derived from fundamental physical principles. The realm of validity of any particular technique can be established by comparing predicted property values with empirical data. In this section, methods for evaluating the *p-v-T* relations for pure components are adapted to obtain plausible estimates for gas mixtures. The case of ideal gas mixtures is discussed in Section 1.3, Ideal Gas Model. In Section 1.3, Multicomponent Systems, some general aspects of property evaluation for multicomponent systems are presented.

The total number of moles of mixture, n, *is* the sum of the number of moles of the components, n_i:

$$n = n_1 + n_2 + \dots n_j = \sum_{i=1}^{j} n_i \tag{1.43}$$

The *relative* amounts of the components present can be described in terms of *mole fractions.* The mole fraction y_i of component i is $y_i = n_i/n$. The sum of the mole fractions of all components present is equal to unity. The apparent molecular weight $\mathcal{M}$ is the mole fraction average of the component molecular weights, such that

$$\mathcal{M} = \sum_{i=1}^{j} y_i \mathcal{M}_i \tag{1.44}$$

The *relative* amounts of the components present also can be described in terms of *mass fractions:* m_i/m, where m_i is the mass of component i and m is the total mass of mixture.

The *p-v-T* relation for a gas mixture can be estimated by applying an equation of state to the overall mixture. The constants appearing in the equation of state are *mixture values* determined with empirical combining rules developed for the equation. For example, mixture values of the constants a and b for use in the Redlich-Kwong equation are obtained using relations of the form

$$a = \left(\sum_{i=1}^{j} y_i a_i^{1/2}\right)^2, \quad b = \sum_{i=1}^{j} y_i b_i \tag{1.45}$$

where a_i and b_i are the values of the constants for component i. Combination rules for obtaining mixture values for the constants in other equations of state are also found in the literature.

Another approach is to regard the mixture as if it were a single pure component having critical properties calculated by one of several mixture rules. *Kay's rule is* perhaps the simplest of these, requiring only the determination of a mole fraction averaged critical temperature T_c and critical pressure p_c:

$$T_c = \sum_{i=1}^{j} y_i T_{c,i}, \quad p_c = \sum_{i=1}^{j} y_i p_{c,i} \tag{1.46}$$

where $T_{c,i}$ and $p_{c,i}$ are the critical temperature and critical pressure of component i, respectively. Using T_c and p_c, the mixture compressibility factor Z is obtained as for a single pure component. The unkown quantity from among the pressure p, volume V, temperature T, and total number of moles n of the gas mixture can then be obtained by solving $Z = pV/n\bar{R}T$.

Additional means for predicting the p-v-T relation of a mixture are provided by empirical mixture rules. Several are found in the engineering literature. According to the *additive pressure rule*, the pressure of a gas mixture is expressible as a sum of pressures exerted by the individual components:

$$p = p_1 + p_2 + p_3 \ldots \Big]_{T,V} \tag{1.47a}$$

where the pressures p_1, p_2, etc. are evaluated by considering the respective components to be at the volume V and temperature T of the mixture. The additive pressure rule can be expressed alternatively as

$$Z = \sum_{i=1}^{j} y_i Z_i \Bigg]_{T,V} \tag{1.47b}$$

where Z is the compressibility factor of the mixture and the compressibility factors Z_i are determined assuming that component i occupies the entire volume of the mixture at the temperature T.

The *additive volume rule* postulates that the volume V of a gas mixture is expressible as the sum of volumes occupied by the individual components:

$$V = V_1 + V_2 + V_3 \ldots \Big]_{p,T} \tag{1.48a}$$

where the volumes V_1, V_2, etc. are evaluated by considering the respective components to be at the pressure p and temperature T of the mixture. The additive volume rule can be expressed alternatively as

$$Z = \sum_{i=1}^{j} y_i Z_i \Bigg]_{p,T} \tag{1.48b}$$

where the compressibility factors Z_i are determined assuming that component i exists at the pressure p and temperature T of the mixture.

Evaluating Δh, Δu, and Δs

Using appropriate specific heat and p-v-T data, the changes in specific enthalpy, internal energy, and entropy can be determined between states of single-phase regions. Table 1.5 provides expressions for such property changes in terms of particular choices of the independent variables: temperature and pressure, and temperature and specific volume.

Taking Equation 1 of Table 1.5 as a representative case, the change in specific enthalpy between states 1 and 2 can be determined using the three steps shown in the accompanying property diagram. This requires knowledge of the variation of c_p with temperature at a fixed pressure p', and the variation of $[v - T(\partial v/\partial T)_p]$ with pressure at temperatures T_1 and T_2:

1-a: Since temperature is constant at T_1, the first integral of Equation 1 in Table 1.5 vanishes, and

$$h_a - h_1 = \int_{p_1}^{p'} \left[v - T(\partial v/\partial T)_p\right] dp$$

a-b: Since pressure is constant at p', the second integral of Equation 1 vanishes, and

$$h_b - h_a = \int_{T_1}^{T_2} c_p(T, p')\, dT$$

b-2: Since temperature is constant at T_2, the first integral of Equation 1 vanishes, and

$$h_2 - h_b = \int_{p'}^{p_2} \left[v - T(\partial v/\partial T)_p\right] dp$$

Adding these expressions, the result is $h_2 - h_1$. The required integrals may be performed numerically or analytically. The analytical approach is expedited when an equation of state explicit in specific volume is known.

Similar considerations apply to Equations 2 to 4 of Table 1.5. To evaluate $u_2 - u_1$ with Equation 3, for example, requires the variation of c_v with temperature at a fixed specific volume v', and the variation of $[T(\partial p/\partial T)_v - p]$ with specific volume at temperatures T_1 and T_2. An analytical approach to performing the integrals is expedited when an equation of state explicit in pressure is known.

As changes in specific enthalpy and internal energy are related through $h = u + pv$ by

$$h_2 - h_1 = (u_2 - u_1) + (p_2 v_2 - p_1 v_1) \tag{1.49}$$

only one of $h_2 - h_1$ and $u_2 - u_1$ need be found by integration. The other can be evaluated from Equation 1.49. The one found by integration depends on the information available: $h_2 - h_1$ would be found when an equation of state explicit in v and c_p as a function of temperature at some fixed pressure is known, $u_2 - u_1$ would be found when an equation of state explicit in p and c_v as a function of temperature at some specific volume is known.

Example 6

Obtain Equation 1 of Table 1.4 and Equations 3 and 4 of Table 1.5.

Solution. With Equation 1.33 and the Maxwell relation corresponding to $\psi(T, v)$ from Table 1.2, Equations 3′ and 4′ of Table 1.5 become, respectively,

$$du = c_v dT + \left(\frac{\partial u}{\partial v}\right)_T dv$$

$$ds = \left(\frac{\partial s}{\partial T}\right)_v dT + \left(\frac{\partial p}{\partial T}\right)_v dv$$

Introducing these expressions for du and ds in Equation 1.32a, and collecting terms,

TABLE 1.5 Δh, Δu, Δs Expressions

Independent properties:	temperature and pressure	temperature and specific volume
Preferred data:	$v(T, p)$, $c_p(T, p)$	$p(T, v)$, $c_v(T, v)$
Property diagram:	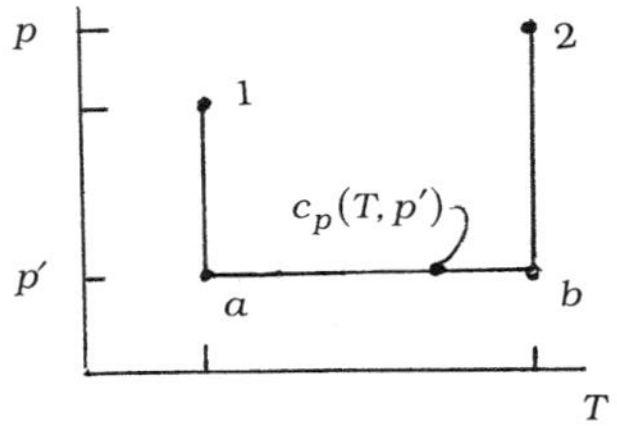	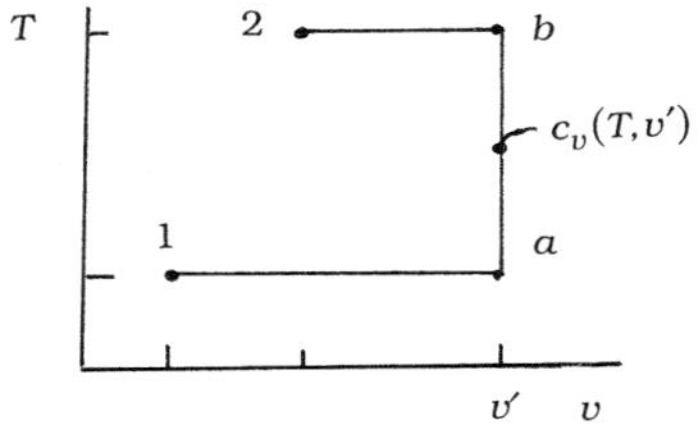

Property expressions:

$h(T, p)$:

$$dh = \underbrace{\left(\frac{\partial h}{\partial T}\right)_p}_{c_p} dT + \underbrace{\left(\frac{\partial h}{\partial p}\right)_T}_{\left[v - T\left(\frac{\partial v}{\partial T}\right)_p\right]} dp \quad (1')$$

$$\Delta h = \int c_p dT + \int \left[v - T\left(\frac{\partial v}{\partial T}\right)_p\right] dp \quad (1)$$

$s(T, p)$:

$$ds = \underbrace{\left(\frac{\partial s}{\partial T}\right)_p}_{\frac{c_p}{T}} dT + \underbrace{\left(\frac{\partial s}{\partial p}\right)_T}_{-\left(\frac{\partial v}{\partial T}\right)_p} dp \quad (2')$$

$$\Delta s = \int \frac{c_p}{T} dT - \int \left(\frac{\partial v}{\partial T}\right)_p dp \quad (2)$$

$u(T, v)$:

$$du = \underbrace{\left(\frac{\partial u}{\partial T}\right)_v}_{c_v} dT + \underbrace{\left(\frac{\partial u}{\partial v}\right)_T}_{\left[T\left(\frac{\partial p}{\partial T}\right)_v - p\right]} dv \quad (3')$$

$$\Delta u = \int c_v dT + \int \left[T\left(\frac{\partial p}{\partial T}\right)_v - p\right] dv \quad (3)$$

$s(T, v)$:

$$ds = \underbrace{\left(\frac{\partial s}{\partial T}\right)_v}_{\frac{c_v}{T}} dT + \underbrace{\left(\frac{\partial s}{\partial v}\right)_T}_{\left(\frac{\partial p}{\partial T}\right)_v} dv \quad (4')$$

$$\Delta s = \int \frac{c_v}{T} dT + \int \left(\frac{\partial p}{\partial T}\right)_v dv \quad (4)$$

$$\left[T\left(\frac{\partial s}{\partial T}\right)_v - c_v\right]dT = \left[\left(\frac{\partial u}{\partial v}\right)_T + p - T\left(\frac{\partial p}{\partial T}\right)_v\right]dv$$

Since T and v are independent, the coefficients of dT and dv must vanish, giving, respectively,

$$\left(\frac{\partial s}{\partial T}\right)_v = \frac{c_v}{T}$$

$$\left(\frac{\partial u}{\partial v}\right)_T = T\left(\frac{\partial p}{\partial T}\right)_v - p$$

The first of these corresponds to Equation 1 of Table 1.4 and Equation 4 of Table 1.5. The second of the above expressions establishes Equation 3 of Table 1.5. With similar considerations, Equation 3 of Table 1.4 and Equations 1 and 2 of Table 1.5 may be obtained.

Fundamental Thermodynamic Functions

A fundamental thermodynamic function is one that provides a complete description of the thermodynamic state. The functions $u(s, v)$, $h(s, p)$, $\psi(T, v)$, and $g(T, p)$ listed in Table 1.2 are fundamental thermodynamic functions.

In principle, all properties of interest can be determined from a fundamental thermodynamic function by differentiation and combination. Taking the function $\psi(T, v)$ as a representative case, the properties v and T, being the independent variables, are specified to fix the state. The pressure p and specific entropy s at this state can be determined by differentiation of $\psi(T, v)$, as shown in Table 1.2. By definition, $\psi = u - Ts$, so specific internal energy is obtained as

$$u = \psi + Ts$$

with u, p, and v known, the specific enthalpy can be found from the definition $h = u + pv$. Similarly, the specific Gibbs function is found from the definition $g = h - Ts$. The specific heat c_v can be determined by further differentiation $c_v = (\partial u/\partial T)_v$.

The development of a fundamental function requires the selection of a functional form in terms of the appropriate pair of independent properties and a set of adjustable coefficients that may number 50 or more. The functional form is specified on the basis of both theoretical and practical considerations. The coefficients of the fundamental function are determined by requiring that a set of selected property values and/or observed conditions be statisfied in a least-squares sense. This generally involves property data requiring the assumed functional form to be differentiated one or more times, for example p-v-T and specific heat data. When all coefficients have been evaluated, the function is tested for accuracy by using it to evaluate properties for which accepted values are known such as *velocity of sound* and *Joule-Thomson* data. Once a suitable fundamental function is established, extreme accuracy in and consistency among the thermodynamic properties are possible. The properties of water tabulated by Keenan et al. (1969) and by Haar et al. (1984) have been calculated from representations of the Helmholtz function.

Thermodynamic Data Retrieval

Tabular presentations of pressure, specific volume, and temperature are available for practically important gases and liquids. The tables normally include other properties useful for thermodynamic analyses, such as internal energy, enthalpy, and entropy. The various *steam tables* included in the references of this chapter provide examples. Computer software for retrieving the properties of a wide range of substances is also available, as, for example, the ASME Steam Tables (1993) and Bornakke and Sonntag (1996).

Increasingly, textbooks come with computer disks providing thermodynamic property data for water, certain refrigerants, and several gases modeled as ideal gases — see, e.g., Moran and Shapiro (1996).

The sample *steam table* data presented in Table 1.6 are representative of data available for substances commonly encountered in mechanical engineering practice. Table A.5 and Figures 1.7 to 1.9 provide *steam table* data for a greater range of states. The form of the tables and figures, and how they are used are assumed to be familiar. In particular, the use of *linear interpolation* with such tables is assumed known.

Specific internal energy, enthalpy, and entropy data are determined relative to arbitrary datums and such datums vary from substance to substance. Referring to Table 1.6a, the datum state for the specific internal energy and specific entropy of water is seen to correspond to saturated liquid water at 0.01°C (32.02°F), the triple point temperature. The value of each of these properties is set to zero at this state. If calculations are performed involving only differences in a particular specific property, the datum cancels. When there are changes in chemical composition during the process, special care should be exercised. The approach followed when composition changes due to chemical reaction is considered in Section 1.4.

Liquid water data (see Table 1.6d) suggests that at fixed temperature the variation of specific volume, internal energy, and entropy with pressure is slight. The variation of specific enthalpy with pressure at fixed temperature is somewhat greater because pressure is explicit in the definition of enthalpy. This behavior for v, u, s, and h is exhibited generally by liquid data and provides the basis for the following set of equations for estimating property data at liquid states from saturated liquid data:

$$v(T,p) \approx v_f(T) \tag{1.50a}$$

$$u(T,p) \approx u_f(T) \tag{1.50b}$$

$$h(T,p) \approx h_f(T) + \underline{v_f\left[p - p_{sat}(T)\right]} \tag{1.50c}$$

$$s(T,p) \approx s_f(T) \tag{1.50d}$$

As before, the subscript f denotes the saturated liquid state at the temperature T, and p_{sat} *is* the corresponding saturation pressure. The underlined term of Equation 1.50c is often negligible, giving $h(T, p) \approx h_f(T)$, which is used in Example 3 to evaluate h_1.

In the absence of saturated liquid data, or as an alternative to such data, the *incompressible model* can be employed:

$$\text{Incompressible model: } \begin{cases} v = \text{constant} \\ u = u(T) \end{cases} \tag{1.51}$$

This model is also applicable to solids. Since internal energy varies only with temperature, the specific heat c_v is also a function of only temperature: $c_v(T) = du/dT$. Although specific volume is constant, enthalpy varies with both temperature and pressure, such that

$$h(T,p) = u(T) + pv \tag{1.52}$$

Differentiation of Equation 1.52 with respect to temperature at fixed pressure gives $c_p = c_v$. The common specific heat is often shown simply as c. Specific heat and density data for several liquids and solids are

TABLE 1.6 Sample Steam Table Data

(a) Properties of Saturated Water (Liquid-Vapor): Temperature Table

		Specific Volume (m^3/kg)		Internal Energy (kJ/kg)		Enthalpy (kJ/kg)			Entropy (kJ/kg · K)	
Temp (°C)	Pressure (bar)	Saturated Liquid ($v_f \times 10^3$)	Saturated Vapor (v_g)	Saturated Liquid (u_f)	Saturated Vapor (u_g)	Saturated Liquid (h_f)	Evap. (h_{fg})	Saturated Vapor (h_g)	Saturated Liquid (s_f)	Saturated Vapor (s_g)
.01	0.00611	1.0002	206.136	0.00	2375.3	0.01	2501.3	2501.4	0.0000	9.1562
4	0.00813	1.0001	157.232	16.77	2380.9	16.78	2491.9	2508.7	0.0610	9.0514
5	0.00872	1.0001	147.120	20.97	2382.3	20.98	2489.6	2510.6	0.0761	9.0257
6	0.00935	1.0001	137.734	25.19	2383.6	25.20	2487.2	2512.4	0.0912	9.0003
8	0.01072	1.0002	120.917	33.59	2386.4	33.60	2482.5	2516.1	0.1212	8.9501

(b) Properties of Saturated Water (Liquid-Vapor): Pressure Table

		Specific Volume (m^3/kg)		Internal Energy (kJ/kg)		Enthalpy (kJ/kg)			Entropy (kJ/kg · K)	
Pressure (bar)	Temp (°C)	Saturated Liquid ($v_f \times 10^3$)	Saturated Vapor (v_g)	Saturated Liquid (u_f)	Saturated Vapor (u_g)	Saturated Liquid (h_f)	Evap. (h_{fg})	Saturated Vapor (h_g)	Saturated Liquid (s_f)	Saturated Vapor (s_g)
0.04	28.96	1.0040	34.800	121.45	2415.2	121.46	2432.9	2554.4	0.4226	8.4746
0.06	36.16	1.0064	23.739	151.53	2425.0	151.53	2415.9	2567.4	0.5210	8.3304
0.08	41.51	1.0084	18.103	173.87	2432.2	173.88	2403.1	2577.0	0.5926	8.2287
0.10	45.81	1.0102	14.674	191.82	2437.9	191.83	2392.8	2584.7	0.6493	8.1502
0.20	60.06	1.0172	7.649	251.38	2456.7	251.40	2358.3	2609.7	0.8320	7.9085

TABLE 1.6 Sample Steam Table Data (continued)

(c) Properties of Superheated Water Vapor

T(°C)	v(m³/kg)	u(kJ/kg)	h(kJ/kg)	s(kJ/kg · K)	v(m³/kg)	u(kJ/kg)	h(kJ/kg)	s(kJ/kg · K)
	p = 0.06 bar = 0.006 MPa (T_{sat} 36.16°C)				p = 0.35 bar = 0.035 MPa (T_{sat} = 72.69°C)			
Sat.	23.739	2425.0	2567.4	8.3304	4.526	2473.0	2631.4	7.7158
80	27.132	2487.3	2650.1	8.5804	4.625	2483.7	2645.6	7.7564
120	30.219	2544.7	2726.0	8.7840	5.163	2542.4	2723.1	7.9644
160	33.302	2602.7	2802.5	8.9693	5.696	2601.2	2800.6	8.1519
200	36.383	2661.4	2879.7	9.1398	6.228	2660.4	2878.4	8.3237

(d) Properties of Compressed Liquid Water

T(°C)	$v \times 10^3$ (m³/kg)	u(kJ/kg)	h(kJ/kg)	s(kJ/kg · K)	$v \times 10^3$ (m³/kg)	u(kJ/kg)	h(kJ/kg)	s(kJ/kg · K)
	p = 25 bar = 2.5 MPa (T_{sat} 223.99°C)				p = 50 bar = 5.0 MPa (T_{sat} = 263.99°C)			
20	1.0006	83.80	86.30	0.2961	0.9995	83.65	88.65	0.2956
80	1.0280	334.29	336.86	1.0737	1.0268	333.72	338.85	1.0720
140	1.0784	587.82	590.52	1.7369	1.0768	586.76	592.15	1.7343
200	1.1555	849.9	852.8	2.3294	1.1530	848.1	853.9	2.3255
Sat.	1.1973	959.1	962.1	2.5546	1.2859	1147.8	1154.2	2.9202

Source: Moran, M.J. and Shapiro, H.N. 2000. *Fundamentals of Engineering Thermodynamics*, 4th ed. Wiley, New York, as extracted from Keenan, J. H., Keyes, F.G., Hill, P.G., and Moore, J.G. 1969. *Steam Tables*. Wiley, New York.

provided in Tables B.2, C.1, and C.2. As the variation of c with temperature is slight, c is frequently taken as constant.

When the incompressible model is applied. Equation 1.49 takes the form

$$\begin{aligned} h_2 - h_1 &= \int_{T_1}^{T_2} c(T)dT + v(p_2 - p_1) \\ &= c_{ave}(T_2 - T_1) + v(p_2 - p_1) \end{aligned} \tag{1.53}$$

Also, as Equation 1.32a reduces to $du = Tds$, and $du = c(T)dT$, the change in specific entropy is

$$\begin{aligned} \Delta s &= \int_{T_1}^{T_2} \frac{c(T)}{T} dT \\ &= c_{ave} \ln \frac{T_2}{T_1} \end{aligned} \tag{1.54}$$

Ideal Gas Model

Inspection of the generalized compressibility chart, Figure 1.10, shows that when p_R is small, and for many states when T_R is large, the value of the compressibility factor Z is close to 1. In other words, for pressures that are low relative to p_c, and for many states with temperatures high relative to T_c, the compressibility factor approaches a value of 1. Within the indicated limits, it may be assumed with reasonable accuracy that $Z = 1$ — that is,

$$p\bar{v} = \bar{R}T \quad \text{or} \quad pv = RT \tag{1.55a}$$

where $R = \bar{R}/\mathcal{M}$ is the *specific* gas constant. Other forms of this expression in common use are

$$pV = n\bar{R}T, \qquad pV = mRT \tag{1.55b}$$

Referring to Equation 3′ of Table 1.5, it can be concluded that $(\partial u/\partial v)_T$ vanishes identically for a gas whose equation of state is *exactly* given by Equation 1.55, and thus the specific internal energy depends only on temperature. This conclusion is supported by experimental observations beginning with the work of Joule, who showed that the internal energy of air at low density depends primarily on temperature.

These considerations allow for an *ideal gas model* of each real gas: (1) the equation of state is given by Equation 1.55 and (2) the internal energy and enthalpy are functions of temperature alone. The real gas approaches the model in the limit of low reduced pressure. At other states the actual behavior may depart substantially from the predictions of the model. Accordingly, caution should be exercised when invoking the ideal gas model lest significant error is introduced.

Specific heat data for gases can be obtained by direct measurement. When extrapolated to zero pressure, ideal gas-specific heats result. Ideal gas-specific heats also can be calculated using molecular models of matter together with data from spectroscopic measurements. Table A.9 provides ideal gas-specific heat data for a number of substances. The following ideal gas-specific heat relations are frequently useful:

$$c_p(T) = c_v(T) + R \tag{1.56a}$$

$$c_p = \frac{kR}{k-1}, \qquad c_v = \frac{R}{k-1} \tag{1.56b}$$

where $k = c_p/c_v$.

With the ideal gas model, Equations 1 to 4 of Table 1.5 give Equations 1 to 4 of Table 1.7, respectively. Equation 2 of Table 1.7 can be expressed alternatively using $s°(T)$ defined by

$$s°(T) \equiv \int_0^T \frac{c_p(T)}{T} dT \tag{1.57}$$

as

$$s(T_2, p_2) - s(T_1, p_1) = s°(T_2) - s°(T_1) - R\ln\frac{p_2}{p_1} \tag{1.58}$$

Expressions similar in form to Equations 1.56 to 1.68 can be written on a molar basis.

TABLE 1.7 Ideal Gas Expressions for Δh, Δu, and Δs

Variable Specific Heats		Constant Specific Heats	
$h(T_2) - h(T_1) = \int_{T_1}^{T_2} c_p(T)dT$	(1)	$h(T_2) - h(T_1) = c_p(T_2 - T_1)$	(1′)
$s(T_2, p_2) - s(T_1, p_1) = \int_{T_1}^{T_2} \frac{c_p(T)}{T} dT - R\ln\frac{p_2}{p_1}$	(2)	$s(T_2, p_2) - s(T_1, p_1) = c_p \ln\frac{T_2}{T_1} - R\ln\frac{p_2}{p_1}$	(2′)
$u(T_2) - u(T_1) = \int_{T_1}^{T_2} c_v(T)dT$	(3)	$u(T_2) - u(T_1) = c_v(T_2 - T_1)$	(3′)
$s(T_2, v_2) - s(T_1, v_1) = \int_{T_1}^{T_2} \frac{c_v(T)}{T} dT + R\ln\frac{v_2}{v_1}$	(4)	$s(T_2, v_2) - s(T_1, v_1) = c_v \ln\frac{T_2}{T_1} + R\ln\frac{v_2}{v_1}$	(4′)
$s_2 = s_1$		$s_2 = s_1$	
$\frac{p_r(T_2)}{p_r(T_1)} = \frac{p_2}{p_1}$	(5)	$\frac{T_2}{T_1} = \left(\frac{p_2}{p_1}\right)^{(k-1)/k}$	(5′)
$\frac{v_r(T_2)}{v_r(T_1)} = \frac{v_2}{v_1}$	(6)	$\frac{T_2}{T_1} = \left(\frac{v_1}{v_2}\right)^{k-1}$	(6′)

For processes of an ideal gas between states having the same specific entropy, $s_2 = s_1$, Equation 1.58 gives

$$\frac{p_2}{p_1} = \frac{\exp[s°(T_2)/R]}{\exp[s°(T_1)/R]}$$

or with $p_r = \exp[s°(T)/R]$

$$\frac{p_2}{p_1} = \frac{p_r(T_2)}{p_r(T_1)} \qquad (s_2 = s_1) \tag{1.59a}$$

A relation between the specific volume and temperatures for two states of an ideal gas having the same specific entropy can also be developed:

$$\frac{v_2}{v_1} = \frac{v_r(T_2)}{v_r(T_1)} \qquad (s_2 = s_1) \tag{1.59b}$$

Equations 1.59 are listed in Table 1.7 as Equations 5 and 6, respectively.

Table A.8 provides a tabular display of h, u, $s°$, p_r, and v_r vs. temperature for air as an ideal gas. Tabulations of $\bar{h}$, $\bar{u}$, and $\bar{s}°$ for several other common gases are provided in Table A.2. Property retrieval software also provides such data; see, e.g., Moran and Shapiro (2000). The use of data from Table A.8 for the nozzle of Example 2 is illustrated in Example 7.

When the ideal gas-specific heats are assumed constant, Equations 1 to 6 of Table 1.7 become Equations 1′ to 6′, respectively. The specific heat c_p is taken as constant in Example 2.

Example 7

Using data from Table A.8, evaluate the exit velocity for the nozzle of Example 2 and compare with the exit velocity for an isentropic expansion to 15 lbf/in.².

Solution. The exit velocity is given by Equation 1.27f

$$\text{V}_e = \sqrt{\text{V}_i^2 + 2\left(h_i - h_e\right)}$$

At 960 and 520°R, Table A.8 gives, respectively, h_i = 231.06 Btu/lb and h_e = 124.27 Btu/lb. Then

$$\begin{aligned} \text{V}_e &= \sqrt{\left(\frac{10\text{ ft}}{s}\right)^2 + 2(231.06 - 124.27)\left(\frac{\text{Btu}}{\text{lb}}\right)\left(\frac{778.17\text{ ft}\cdot\text{lbf}}{1\text{ Btu}}\right)\left(\frac{32.174\text{ lb}\cdot\text{ft/sec}^2}{1\text{ lbf}}\right)} \\ &= 2312.5\text{ ft/sec} \end{aligned}$$

Using Equation 1.59a and p_r data from Table A.8, the specific enthalpy at the exit for an isentropic expansion is found as follows:

$$p_r\left(T_e\right) = p_r\left(T_i\right)\frac{p_e}{p_i} = 10.61\left(\frac{15}{150}\right) = 1.061$$

Interpolating with p_r data, h_e = 119.54 Btu/lb. With this, the exit velocity is 2363.1 ft/sec. The actual exit velocity is about 2% less than the velocity for an isentropic expansion, the maximum theoretical value. In this particular application, there is good agreement in each case between velocities calculated using Table A.8 data and, as in Example 2, assuming c_p constant. Such agreement cannot be expected generally, however. See, for example, the Brayton cycle data of Table 1.15.

Polytropic Processes

An internally reversible process described by the expression pv^n = constant is called a *polytropic process* and *n is* the *polytropic exponent.* Although this expression can be applied with real gas data, it most generally appears in practice together with the use of the ideal gas model. Table 1.8 provides several expressions applicable to polytropic processes and the special forms they take when the ideal gas model is assumed. The expressions for $\int pdv$ and $\int vdp$ have application to work evaluations with Equations 1.10 and 1.30, respectively. In some applications it may be appropriate to determine n by fitting pressure-specific volume data.

Example 8 illustrates both the polytropic process and the reduction in the compressor work achievable by cooling a gas as it is compressed.

Example 8

A compressor operates at steady state with air entering at 1 bar, 20°C and exiting at 5 bar. (a) If the air undergoes a polytropic process with n = 1.3, determine the work and heat transfer, each in kJ/kg of air flowing. Repeat for (b) an isothermal compression and (c) an isentropic compression.

TABLE 1.8 Polytropic Processes: pv^n = Constant[a]

General		Ideal Gas[b]	
$\frac{p_2}{p_1} = \left(\frac{v_1}{v_2}\right)^n$	(1)	$\frac{p_2}{p_1} = \left(\frac{v_1}{v_2}\right)^n = \left(\frac{T_2}{T_1}\right)^{n/(n-1)}$	(1′)
$n = 0$: constant pressure $n = \pm\infty$: constant specific volume		$n = 0$: constant pressure $n = \pm\infty$: constant specific volume $n = 1$: constant temperature $n = k$: constant specific entropy when k is constant	
$n = 1$		$n = 1$	
$\int_1^2 p dv = p_1 v_1 \ln \frac{v_2}{v_1}$	(2)	$\int_1^2 p dv = RT \ln \frac{v_2}{v_1}$	(2′)
$-\int_1^2 v dp = -p_1 v_1 \ln \frac{p_2}{p_1}$	(3)	$-\int_1^2 v dp = -RT \ln \frac{p_2}{p_1}$	(3′)
$n \neq 1$		$n \neq 1$	
$\int_1^2 p dv = \frac{p_2 v_2 - p_1 v_1}{1-n}$ $= \frac{p_1 v_1}{n-1}\left[1 - \left(\frac{p_2}{p_1}\right)^{(n-1)/n}\right]$	(4)	$\int_1^2 p dv = \frac{R(T_2 - T_1)}{1-n}$ $= \frac{RT_1}{n-1}\left[1 - \left(\frac{p_2}{p_1}\right)^{(n-1)/n}\right]$	(4′)
$-\int_1^2 v dp = \frac{n}{1-n}(p_2 v_2 - p_1 v_1)$ $= \frac{n p_1 v_1}{n-1}\left[1 - \left(\frac{p_2}{p_1}\right)^{(n-1)/n}\right]$	(5)	$-\int_1^2 v dp = \frac{nR}{1-n}(T_2 - T_1)$ $= \frac{nRT_1}{n-1}\left[1 - \left(\frac{p_2}{p_1}\right)^{(n-1)/n}\right]$	(5′)

[a] For polytropic processes of closed systems where volume change is the only work mode, Equations 2, 4, and 2′, 4′ are applicable with Equation 1.10 to evaluate the work. When each unit of mass passing through a one-inlet, one-exit control volume at steady state undergoes a polytropic process, Equations 3, 5, and 3′, 5′ are applicable with Equations 1.30a and 1.30b to evaluate the power. Also note that generally, $-\int_1^2 v dp = n\int_1^2 p dv$.

[b]

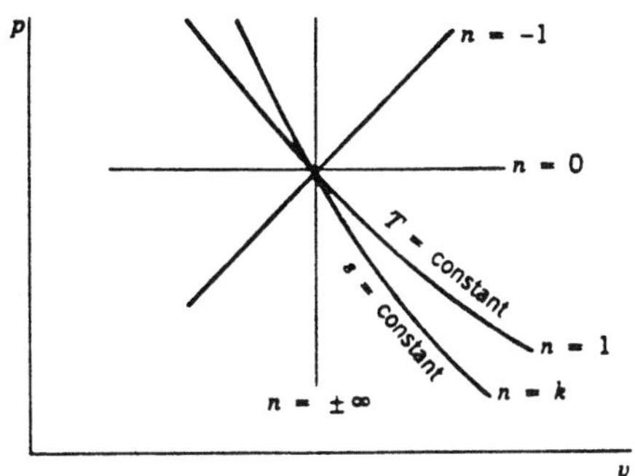

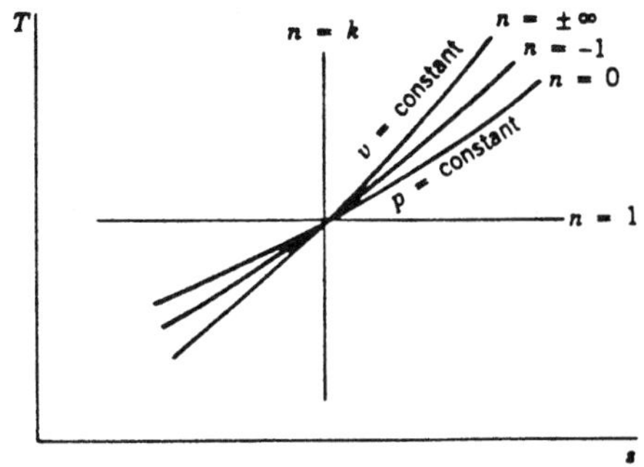

Solution. Using Equation 5′ of Table 1.8 together with Equation 1.30b,

$$\frac{\dot{W}_{cv}}{\dot{m}} = \frac{nRT_1}{n-1}\left[1-\left(\frac{p_2}{p_1}\right)^{(n-1)/n}\right]$$

$$= \left(\frac{1.3}{0.3}\right)\left(\frac{8.314}{28.97}\frac{\text{kJ}}{\text{kg}\cdot\text{K}}\right)(293\text{ K})\left[1-(5)^{0.3/1.3}\right]$$

$$= -163.9\ \frac{\text{kJ}}{\text{kg}}$$

(The area behind process 1-2 of Figure 1.11, area 1-2-a-b, represents the magnitude of the work required, per unit mass of air flowing.) Also, Equation 1′ of Table 1.8 gives T_2 = 425 K.

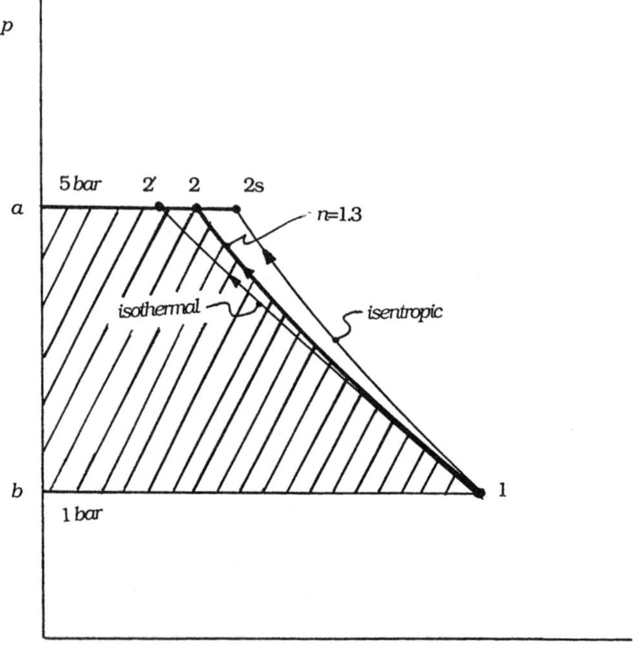

FIGURE 1.11 Internally reversible compression processes.

An energy rate balance at steady state and enthalpy data from Table A.8 gives

$$\frac{\dot{Q}_{cv}}{\dot{m}} = \frac{\dot{W}_{cv}}{\dot{m}} + h_2 - h_1$$

$$= -163.9 + (426.3 - 293.2) = -30.8\frac{\text{kJ}}{\text{kg}}$$

(b) Using Equation 3′ of Table 1.8 together with Equation 1.30b,

$$\frac{\dot{W}_{cv}}{\dot{m}} = -RT\ln\frac{p_2}{p_1}$$

$$= -\left(\frac{8.314}{28.97}\right)(293)\ln 5$$

$$= -135.3\frac{\text{kJ}}{\text{kg}}$$

Area 1-2′-*a*-*b* on Figure 1.11 represents the magnitude of the work required, per unit of mass of air flowing. An energy balance reduces to give $\dot{Q}_{cv}/\dot{m} = \dot{W}_{cv}/\dot{m} = -135.3$ kJ/kg. (c) For an isentropic compression, $\dot{Q}_{cv} = 0$ and an energy rate balance reduces to give $\dot{W}_{cv}/\dot{m} = -(h_{2s} - h_1)$, where 2*s* denotes the exit state. With Equation 1.59a and p_r data, $h_{2s} = 464.8$ kJ/kg ($T_{2s} = 463$K). Then $\dot{W}_{cv}/\dot{m} = -(464.8 - 293.2) = -171.6$ kJ/kg. Area 1-2s-*a*-*b* on Figure 1.11 represents the magnitude of the work required, per unit of mass of air flowing.

Ideal Gas Mixtures

When applied to an ideal gas mixture, the additive pressure rule (Section 1.3, *p*-*v*-*T* Relations) is known as the *Dalton model.* According to this model, each gas in the mixture acts as if it exists separately at the volume and temperature of the mixture. Applying the ideal gas equation of state to the mixture as a whole and to each component *i*, $pV = n\bar{R}T$, $p_iV = n_i\bar{R}T$, where p_i, the *partial pressure* of component *i*, is the pressure that component *i* would exert if n_i moles occupied the full volume *V* at the temperature *T*. Forming a ratio, the partial pressure of component *i* is

$$p_i = \frac{n_i}{n}p = y_i p \tag{1.60}$$

where y_i is the mole fraction of component *i*. The sum of the partial pressures equals the mixture pressure.

The internal energy, enthalpy, and entropy of the mixture can be determined as the sum of the respective properties of the component gases, provided that the contribution from each gas is evaluated at the condition at which the gas exists in the mixture. On a *molar* basis,

$$U = \sum_{i=1}^{j} n_i\bar{u}_i \quad \text{or} \quad \bar{u} = \sum_{i=1}^{j} y_i\bar{u}_i \tag{1.61a}$$

$$H = \sum_{i=1}^{j} n_i\bar{h}_i \quad \text{or} \quad \bar{h} = \sum_{i=1}^{j} y_i\bar{h}_i \tag{1.61b}$$

$$S = \sum_{i=1}^{j} n_i\bar{s}_i \quad \text{or} \quad \bar{s} = \sum_{i=1}^{j} y_i\bar{s}_i \tag{1.61c}$$

The specific heats $\bar{c}_v$ and $\bar{c}_p$ for an ideal gas mixture in terms of the corresponding specific heats of the components are expressed similarly:

$$\bar{c}_v = \sum_{i=1}^{j} y_i\bar{c}_{vi} \tag{1.61d}$$

$$\bar{c}_p = \sum_{i=1}^{j} y_i \bar{c}_{pi} \tag{1.61e}$$

When working on a *mass* basis, expressions similar in form to Equations 1.61 can be written using *mass* and *mass fractions* in place of *moles* and *mole fractions*, respectively, and using u, h, s, c_p, and c_v in place of $\bar{u}, \bar{h}, \bar{s}, \bar{c}_p$, and $\bar{c}_v$, respectively.

The internal energy and enthalpy of an ideal gas depend only on temperature, and thus the $\bar{u}_i$ and $\bar{h}_i$ terms appearing in Equations 1.61 are evaluated at the temperature of the mixture. Since entropy depends on *two* independent properties, the $\bar{s}_i$ terms are evaluated either at the temperature and the partial pressure p_i of component i, or at the temperature and volume of the mixture. In the former case

$$\begin{aligned} S &= \sum_{i=1}^{j} n_i \bar{s}_i(T, p_i) \\ &= \sum_{i=1}^{j} n_i \bar{s}_i(T, x_i p) \end{aligned} \tag{1.62}$$

Inserting the expressions for H and S given by Equations 1.61b and 1.61c into the Gibbs function, $G = H - TS$,

$$\begin{aligned} G &= \sum_{i=1}^{j} n_i \bar{h}_i(T) - T \sum_{i=1}^{j} n_i \bar{s}_i(T, p_i) \\ &= \sum_{i=1}^{j} n_i \bar{g}_i(T, p_i) \end{aligned} \tag{1.63}$$

where the molar-specific Gibbs function of component i is $g_i(T, p_i) = h_i(T) - Ts_i(T, p_i)$. The Gibbs function of i can be expressed alternatively as

$$\begin{aligned} \bar{g}_i(T, p_i) &= \bar{g}_i(T, p') + \bar{R}T \ln(p_i/p') \\ &= \bar{g}_i(T, p') + \bar{R}T \ln(x_i p/p') \end{aligned} \tag{1.64}$$

were p' is some specified pressure. Equation 1.64 is obtained by integrating Equation 1.32d at fixed temperature T from pressure p' to p_i.

Moist Air

An ideal gas mixture of particular interest for many practical applications is *moist air.* Moist air refers to a mixture of dry air and water vapor in which the dry air is treated as if it were a pure component. Ideal gas mixture principles usually apply to moist air. In particular, the *Dalton model is* applicable, and so the mixture pressure p is the sum of the partial pressures p_a and p_v of the dry air and water vapor, respectively.

Saturated air is a mixture of dry air and saturated water vapor. For saturated air, the partial pressure of the water vapor equals $p_{sat}(T)$, which is the saturation pressure of water corresponding to the dry-bulb (mixture) temperature T. The makeup of moist air can be described in terms of the *humidity ratio* (*specific humidity*) and the *relative humidity.* The bulb of a *wet-bulb thermometer* is covered with a wick saturated with liquid water, and the *wet-bulb* temperature of an air-water vapor mixture is the temperature indicated by such a thermometer exposed to the mixture.

When a sample of moist air is cooled at constant pressure, the temperature at which the sample becomes saturated is called the *dew point temperature.* Cooling below the dew point temperature results in the condensation of some of the water vapor initially present. When cooled to a final equilibrium state at a temperature below the dew point temperature, the original sample would consist of a gas phase of dry air and saturated water vapor in equilibrium with a liquid water phase.

Psychrometric charts are plotted with various moist air parameters, including the dry-bulb and wet-bulb temperatures, the humidity ratio, and the relative humidity, usually for a specified value of the mixture pressure such as 1 atm.

Generalized Charts for Enthalpy, Entropy, and Fugacity

The changes in enthalpy and entropy between two states can be determined in principle by correcting the respective property change determined using the ideal gas model. The corrections can be obtained, at least approximately, by inspection of the generalized enthalpy correction and entropy correction charts, Figures 1.12 and 1.13, respectively. Such data are also available in tabular form (see, e.g., Reid and Sherwood, 1966) and calculable using a generalized equation for the compressibility factor (Reynolds, 1979). Using the superscript * to identify ideal gas property values, the changes in specific enthalpy and specific entropy between states 1 and 2 are

$$\bar{h}_2 - \bar{h}_1 = \underline{\bar{h}_2^* - \bar{h}_1^*} - \underline{\bar{R}T_c\left[\left(\frac{\bar{h}^* - \bar{h}}{\bar{R}T_c}\right)_2 - \left(\frac{\bar{h}^* - \bar{h}}{\bar{R}T_c}\right)_1\right]} \tag{1.65a}$$

$$\bar{s}_2 - \bar{s}_1 = \underline{\bar{s}_2^* - \bar{s}_1^*} - \underline{\bar{R}\left[\left(\frac{\bar{s}^* - \bar{s}}{\bar{R}}\right)_2 - \left(\frac{\bar{s}^* - \bar{s}}{\bar{R}}\right)_1\right]} \tag{1.65b}$$

The first underlined term on the right side of each expression represents the respective property change assuming ideal gas behavior. The second underlined term is the correction that must be applied to the ideal gas value to obtain the actual value. The quantities $(\bar{h}^* - \bar{h})/\bar{R}T_c$ *and* $(\bar{s}^* - \bar{s})/\bar{R}$ at state 1 would be read from the respective correction chart or table or calculated, using the reduced temperature T_{R1} and reduced pressure p_{R1} corresponding to the temperature T_1 and pressure p_1 at state 1, respectively. Similarly, $(\bar{h}^* - \bar{h})/\bar{R}T_c$ *and* $(\bar{s}^* - \bar{s})/\bar{R}$ at state 2 would be obtained using T_{R2} and p_{R2}. Mixture values for T_c and p_c determined by applying Kay's rule or some other mixture rule also can be used to enter the generalized enthalpy correction and entropy correction charts.

Figure 1.14 gives the *fugacity* coefficient, *f/p*, as a function of reduced pressure and reduced temperature. The fugacity *f* plays a similar role in determining the specific Gibbs function for a real gas as pressure plays for the ideal gas. To develop this, consider the variation of the specific Gibbs function with pressure at fixed temperature (from Table 1.2)

$$\left.\frac{\partial g}{\partial p}\right)_T = v$$

For an ideal gas, integration at fixed temperature gives

$$g^* = RT \ln p + C(T)$$

where *C(T)* is a function of integration. To evaluate *g* for a real gas, fugacity replaces pressure,

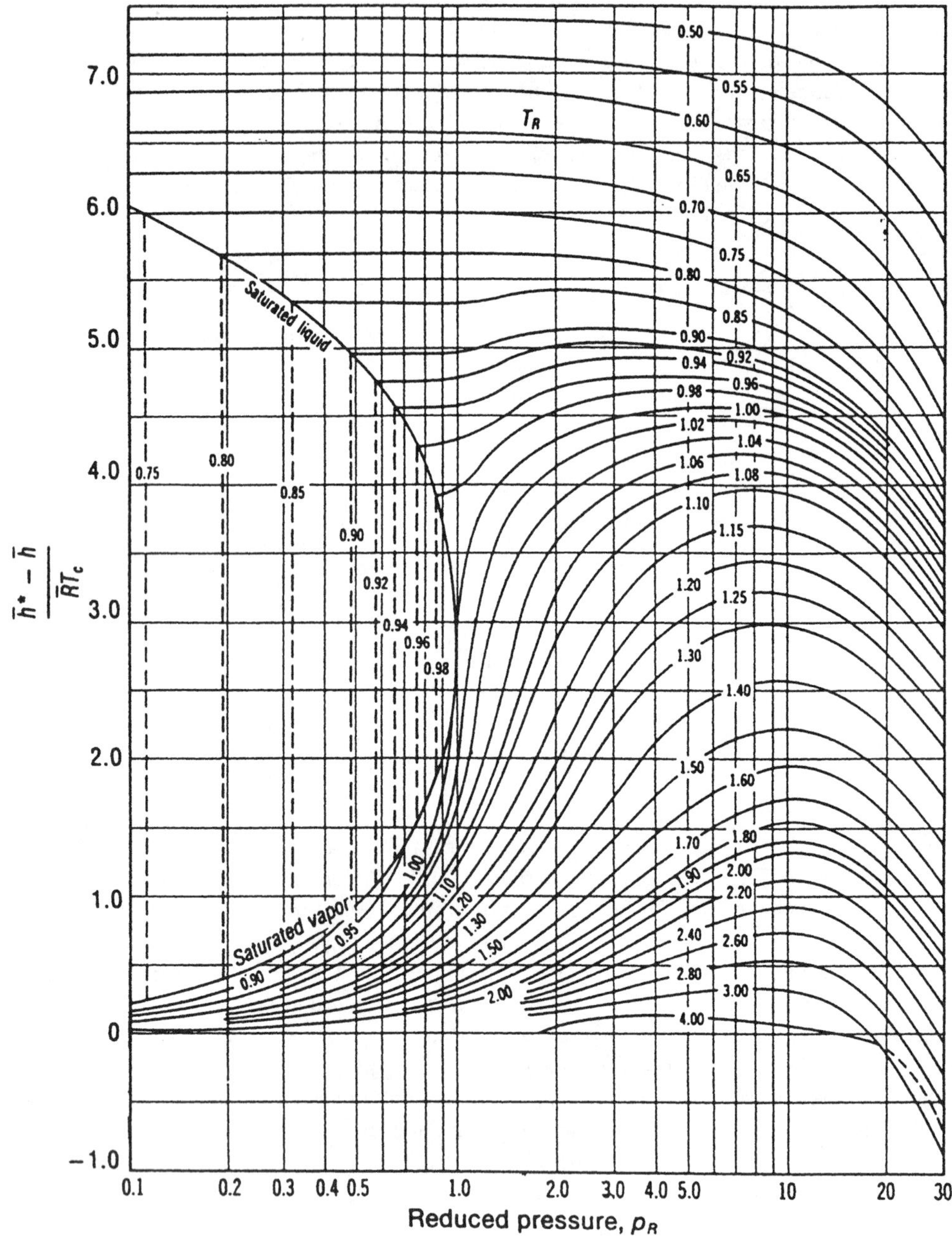

FIGURE 1.12 Generalized enthalpy correction chart. (*Source:* Adapted from Van Wylen, G. J. and Sonntag, R. E. 1986. *Fundamentals of Classical Thermodynamics*, 3rd ed., English/SI. Wiley, New York.)

$$g = RT \ln f + C(T)$$

In terms of the fugacity coefficient the departure of the real gas value from the ideal gas value at fixed temperature is then

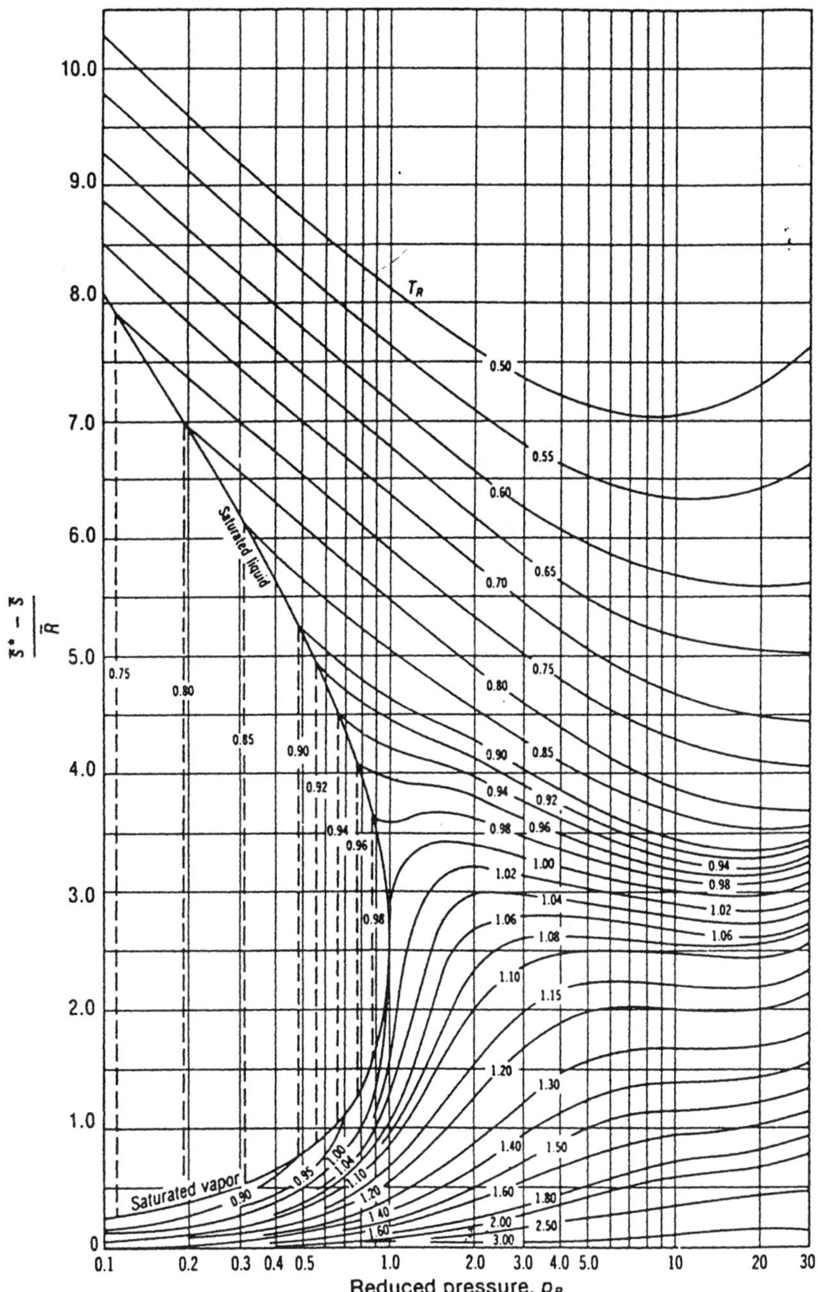

FIGURE 1.13 Generalized entropy correction chart. (*Source:* Adapted from Van Wylen, G. J. and Sonntag, R. E. 1986. *Fundamentals of Classical Thermodynamics*, 3rd ed., English/SI. Wiley, New York.)

$$g - g^* = RT \ln \frac{f}{p} \tag{1.66}$$

As pressure is reduced at fixed temperature, f/p tends to unity, and the specific Gibbs function is given by the ideal gas value.

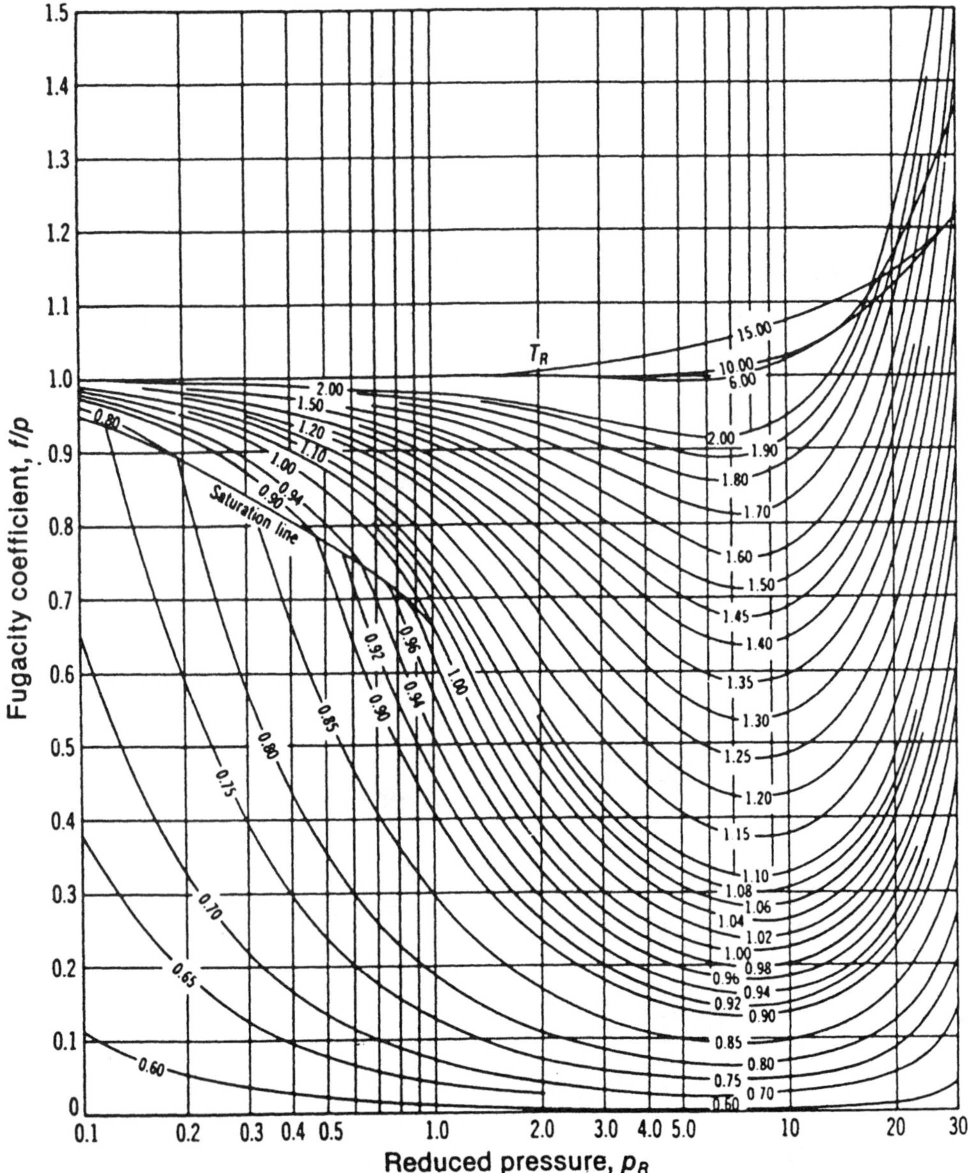

FIGURE 1.14 Generalized fugacity coefficient chart. (*Source:* Van Wylen, G. J. and Sonntag, R. E. 1986. *Fundamentals of Classical Thermodynamics*, 3rd ed., English/SI. Wiley, New York.)

Multicomponent Systems

In this section are presented some general aspects of the properties of multicomponent systems consisting of nonreacting mixtures. For a single phase *multicomponent* system consisting of j components, an extensive property X may be regarded as a function of temperature, pressure, and the number of moles of each component present in the mixture: $X = X(T, p, n_1, n_2, \ldots n_j)$. Since X is mathematically *homogeneous of degree one* in the n's, the function is expressible as

$$X = \sum_{i=1}^{j} n_i \overline{X}_i \tag{1.67}$$

where the *partial molar property* $\overline{X}_i$ is by definition

$$\overline{X}_i = \left.\frac{\partial X}{\partial n_i}\right)_{T,p,n_\ell} \tag{1.68}$$

and the subscript n_ℓ denotes that all n's except n_i are held fixed during differentiation. As $\overline{X}_i$ depends in general on temperature, pressure, and mixture composition: $\overline{X}_i(T, p, n_1, n_2, \ldots n_j)$, the partial molal property $\overline{X}_i$ is an intensive property of the mixture and not simply a property of the *i*th component.

Selecting the extensive property X to be volume, internal energy, enthalpy, entropy, and the Gibbs function, respectively, gives

$$V = \sum_{i=1}^{j} n_i \overline{V}_i, \qquad U = \sum_{i=1}^{j} n_i \overline{U}_i$$

$$H = \sum_{i=1}^{j} n_i \overline{H}_i, \qquad S = \sum_{i=1}^{j} n_i \overline{S}_i \tag{1.69}$$

$$G = \sum_{i=1}^{j} n_i \overline{G}_i$$

where $\overline{V}_i$, $\overline{U}_i$, $\overline{H}_i$, $\overline{S}_i$, and $\overline{G}_i$ denote the respective partial molal properties.

When pure components, each initially at the same temperature and pressure, are mixed, the changes in volume, internal energy, enthalpy, and entropy on mixing are given by

$$\Delta V_{\text{mixing}} = \sum_{i=1}^{j} n_i \left(\overline{V}_i - \bar{v}_i\right) \tag{1.70a}$$

$$\Delta U_{\text{mixing}} = \sum_{i=1}^{j} n_i \left(\overline{U}_i - \bar{u}_i\right) \tag{1.70b}$$

$$\Delta H_{\text{mixing}} = \sum_{i=1}^{j} n_i \left(\overline{H}_i - \bar{h}_i\right) \tag{1.70c}$$

$$\Delta S_{\text{mixing}} = \sum_{i=1}^{j} n_i \left(\overline{S}_i - \bar{s}_i\right) \tag{1.70d}$$

where $\bar{v}_i$, $\bar{u}_i$, $\bar{h}_i$, and $\bar{s}_i$ denote the molar-specific volume, internal energy, enthalpy, and entropy of pure component *i*.

Chemical Potential

The partial molal Gibbs function of the *i*th component of a multicomponent system is the *chemical potential*, μ_i,

$$\mu_i = \bar{G}_i = \left.\frac{\partial G}{\partial n_i}\right)_{T,p,n_\ell} \tag{1.71}$$

Like temperature and pressure, the chemical potential, μ_i is an *intensive* property.

When written in terms of chemical potentials, Equation 1.67 for the Gibbs function reads

$$G = \sum_{i=1}^{j} n_i \mu_i \tag{1.72}$$

For a *single component sysrem*, Equation 1.72 reduces to $G = n\mu$; that is, the chemical potential equals the molar Gibbs function. For an ideal gas mixture, comparison of Equations 1.63 and 1.72 suggests $\mu_i = \bar{g}_i\,(T, p_i)$; that is, the chemical potential of component i in an ideal gas mixture equals its Gibbs function per mole of gas i evaluated at the mixture temperature and the partial pressure of the ith gas of the mixture.

The chemical potential is a measure of the *escaping tendency* of a substance in a multiphase system: a substance tends to move from the phase having the higher chemical potential for that substance to the phase having a lower chemical potential. A necessary condition for *phase equilibrium is* that the chemical potential of each component has the same value in every phase.

The *Gibbs phase rule* gives the number F of independent intensive properties that may be arbitrarily specified to fix the intensive state of a system at equilibrium consisting of N nonreacting components present in P phases: $F = 2 + N - P$. F is called the *degrees of freedom* (or the *variance*). For water as a single component, for example, $N = 1$ and $F = 3 - P$.

- For a single phase, $P = 1$ and $F = 2$: two intensive properties can be varied independently, say temperature *and* pressure, while maintaining a single phase.
- For two phases, $P = 2$ and $F = 1$: only one intensive property can be varied independently if two phases are maintained — for example, temperature *or* pressure.
- For three phases, $P = 3$ and $F = 0$: there are no degrees of freedom; each intensive property of each phase is fixed. For a system consisting of ice, liquid water, and water vapor at equilibrium, there is a unique temperature: 0.01°C (32.02°F) and a unique pressure: 0.6113 kPa (0.006 atm).

The phase rule does not address the relative amounts that may be present in the various phases.

With $G = H - TS$ and $H = U + pV$, Equation 1.72 can be expressed as

$$U = TS - pV + \sum_{i=1}^{j} n_i \mu_i \tag{1.73}$$

from which can be derived

$$dU = TdS - pdV + \sum_{i=1}^{j} \mu_i dn_i \tag{1.74}$$

When the mixture composition is constant, Equation 1.74 reduces to Equation 1.31a.

Ideal Solution

The *Lewis-Randall rule* states that the fugacity $\bar{f}_i$ of each component i in an *ideal solution* is the product of its mole fraction and the fugacity of the pure component, f_i, at the same temperature, pressure, and state of aggregation (gas, liquid, or solid) as the mixture:

$$\bar{f}_i = y_i f_i \qquad \text{(Lewis-Randall rule)} \tag{1.75}$$

The following characteristics are exhibited by an ideal solution: $\bar{V}_i = \bar{v}_i$, $\bar{U}_i = \bar{u}_i$, $\bar{H}_i = \bar{h}_i$. With these, Equations 1.70a, b, and c show that there is no change in volume, internal energy, or enthalpy on mixing pure components to form an ideal solution. The *adiabatic* mixing of different pure components would result in an increase in entropy, however, because such a process is irreversible.

The volume of an ideal solution is

$$V = \sum_{i=1}^{j} n_i \bar{v}_i = \sum_{i=1}^{j} V_i \qquad \text{(ideal solution)} \tag{1.76}$$

where V_i is the volume that pure component *i* would occupy when at the temperature and pressure of the mixture. Comparing Equations 1.48a and 1.76, the *additive volume rule* is seen to be exact for ideal solutions. The internal energy and enthalpy of an ideal solution are

$$U = \sum_{i=1}^{j} n_i \bar{u}_i, \quad H = \sum_{i=1}^{j} n_i \bar{h}_i \qquad \text{(ideal solution)} \tag{1.77}$$

where $\bar{u}_i$ and $\bar{h}_i$ denote, respectively, the molar internal energy and enthalpy of pure component *i* at the temperature and pressure of the mixture. Many gaseous mixtures at low to moderate pressures are adequately modeled by the Lewis Randall rule. The ideal gas mixtures considered in Section 1.3, Ideal Gas Model, is an important special case. Some liquid solutions also can be modeled with the Lewis-Randall rule.

1.4 Combustion

The thermodynamic analysis of reactive systems is primarily an extension of principles presented in Sections 1.1 to 1.3. It is necessary, though, to modify the methods used to evaluate specific enthalpy and entropy.

Reaction Equations

In combustion reactions, rapid oxidation of combustible elements of the fuel results in energy release as combustion products are formed. The three major combustible chemical elements in most common fuels are carbon, hydrogen, and sulfur. Although sulfur is usually a relatively unimportant contributor to the energy released, it can be a significant cause of pollution and corrosion.

The emphasis in this section is on hydrocarbon fuels, which contain hydrogen, carbon, sulfur, and possibly other chemical substances. Hydrocarbon fuels may be liquids, gases, or solids such as coal. Liquid hydrocarbon fuels are commonly derived from crude oil through distillation and cracking processes. Examples are gasoline, diesel fuel, kerosene, and other types of fuel oils. The compositions of liquid fuels are commonly given in terms of mass fractions. For simplicity in combustion calculations, gasoline is often considered to be octane, C_8H_{18}, and diesel fuel is considered to be dodecane, $C_{12}H_{26}$. Gaseous hydrocarbon fuels are obtained from natural gas wells or are produced in certain chemical processes. Natural gas normally consists of several different hydrocarbons, with the major constituent being methane, CH_4. The compositions of gaseous fuels are commonly given in terms of mole fractions. Both gaseous and liquid hydrocarbon fuels can be synthesized from coal, oil shale, and tar sands. The composition of coal varies considerably with the location from which it is mined. For combustion calculations, the makeup of coal is usually expressed as an *ultimate analysis* giving the composition on a mass basis in terms of the relative amounts of chemical elements (carbon, sulfur, hydrogen, nitrogen, oxygen) and ash.

A fuel is said to have burned *completely* if all of the carbon present in the fuel is burned to carbon dioxide, all of the hydrogen is burned to water, and all of the sulfur is burned to sulfur dioxide. In practice, these conditions are usually not fulfilled and combustion is *incomplete*. The presence of carbon monoxide (CO) in the products indicates incomplete combustion. The products of combustion of *actual* combustion reactions and the relative amounts of the products can be determined with certainty only by experimental means. Among several devices for the experimental determination of the composition of products of combustion are the *Orsat analyzer*, *gas chromatograph*, *infrared analyzer*, and *flame ionization detector*. Data from these devices can be used to determine the makeup of the gaseous products of combustion. Analyses are frequently reported on a "dry" basis: mole fractions are determined for all gaseous products as if no water vapor were present. Some experimental procedures give an analysis including the water vapor, however.

Since water is formed when hydrocarbon fuels are burned, the mole fraction of water vapor in the gaseous products of combustion can be significant. If the gaseous products of combustion are cooled at constant mixture pressure, the *dew point temperature* (Section 1.3, Ideal Gas Model) is reached when water vapor begins to condense. Corrosion of duct work, mufflers, and other metal parts can occur when water vapor in the combustion products condenses.

Oxygen is required in every combustion reaction. Pure oxygen is used only in special applications such as cutting and welding. In most combustion applications, air provides the needed oxygen. Idealizations are often used in combustion calculations involving air: (1) all components of air other than oxygen (O_2) *are* lumped with nitrogen (N_2). On a molar basis air is then considered to be 21% oxygen and 79% nitrogen. With this idealization the molar ratio of the nitrogen to the oxygen in combustion air is 3.76; (2) the water vapor present in air may be considered in writing the combustion equation or ignored. In the latter case the combustion air is regarded as *dry*; (3) additional simplicity results by regarding the nitrogen present in the combustion air as inert. However, if high-enough temperatures are attained, nitrogen can form compounds, often termed NO_X, such as nitric oxide and nitrogen dioxide.

Even trace amounts of oxides of nitrogen appearing in the exhaust of internal combustion engines can be a source of air pollution.

The minimum amount of air that supplies sufficient oxygen for the complete combustion of all the combustible chemical elements is the *theoretical*, or *stoichiometic*, amount of air. In practice, the amount of air actually supplied may be greater than or less than the theoretical amount, depending on the application. The amount of air is commonly expressed as the *percent of theoretical air* or the *percent excess* (or *percent deficiency*) of air. The *air-fuel ratio* and its reciprocal *the fuel-air ratio*, each of which can be expressed on a mass or molar basis, are other ways that fuel-air mixtures are described. Another is the *equivalence ratio:* the ratio of the actual fuel-air ratio to the fuel-air ratio for complete combustion with the theoretical amount of air. The reactants form a *lean* mixture when the equivalence ratio is less than unity and a *rich* mixture when the ratio is greater than unity.

Example 9

Methane, CH_4, is burned with dry air. The molar analysis of the products on a dry basis is CO_2, 9.7%; CO, 0.5%; O_2, 2.95%; and N_2, 86.85%. Determine (a) the air-fuel ratio on both a molar and a mass basis, (b) the percent of theoretical air, (c) the equivalence ratio, and (d) the dew point temperature of the products, in °F, if the pressure is 1 atm.

Solution.

(a) The solution is conveniently conducted on the basis of 100 lbmol of dry products. The chemical equation then reads

$$a\mathrm{CH_4} + b(\mathrm{O_2} + 3.76\mathrm{N_2}) \rightarrow 9.7\mathrm{CO_2} + 0.5\mathrm{CO} + 2.95\mathrm{O_2} + 86.85\mathrm{N_2} + c\mathrm{H_2O}$$

where N_2 is regarded as inert. Water is included in the products together with the assumed 100 lbmol of dry products. Balancing the carbon, hydrogen, and oxygen, the reaction equation is

$$10.2\mathrm{CH_4} + 23.1(\mathrm{O_2} + 3.76\mathrm{N_2}) \rightarrow 9.7\mathrm{CO_2} + 0.5\mathrm{CO} + 2.95\mathrm{O_2} + 86.85\mathrm{N_2} + 20.4\mathrm{H_2O}$$

The nitrogen also balances, as can be verified. This checks the accuracy of both the given product analysis and the calculations conducted to determine the unknown coefficients. Exact closure cannot be expected with measured data, however. On a molar basis, the air-fuel ratio is

$$\overline{\mathrm{AF}} = \frac{23.1(4.76)}{10.2} = 10.78 \frac{\mathrm{lbmol(air)}}{\mathrm{lbmol(fuel)}}$$

On a mass basis

$$\mathrm{AF} = (10.78)\left(\frac{28.97}{16.04}\right) = 19.47 \frac{\mathrm{lb(air)}}{\mathrm{lb(fuel)}}$$

(b) The balanced chemical equation for the complete combustion of methane with the *theoretical* amount of air is

$$\mathrm{CH_4} + 2(\mathrm{O_2} + 3.76\mathrm{N_2}) \rightarrow \mathrm{CO_2} + 2\mathrm{H_2O} + 7.52\mathrm{N_2}$$

The theoretical air-fuel ratio on a molar basis is

$$\left(\overline{\mathrm{AF}}\right)_{theo} = \frac{2(4.76)}{1} = 9.52 \frac{\mathrm{lbmol(air)}}{\mathrm{lbmol(fuel)}}$$

The percent theoretical air is then

$$\% \text{ theoretical air} = \frac{(\overline{AF})}{(\overline{AF})_{theo}}$$

$$= \frac{10.78}{9.52} = 1.13(113\%)$$

(c) Equivalence ratio = $(\overline{FA})/(\overline{FA})_{theo}$ = 9.52/10.78 = 0.88. The reactants form a lean mixture.

(d) To determine the dew point temperature requires the partial pressure p_v of the water vapor. The mole fraction of the water vapor is

$$y_v = \frac{20.4}{100 + 20.4} = 0.169$$

Since p = 1 atm, p_v = 0.169 atm = 2.48 lbf/in.^{2}. With p_{sat} = 2.48 lbf/in.^{2}, the corresponding saturation temperature from the steam tables is 134°F. This is the dew point temperature.

Property Data for Reactive Systems

Tables of thermodynamic properties such as the steam tables provide values for the specific enthalpy and entropy relative to some arbitrary datum state where the enthalpy (or alternatively the internal energy) and entropy are set to zero. When a chemical reaction occurs, however, reactants disappear and products are formed, and it is generally no longer possible to evaluate $\Delta\bar{h}$ and $\Delta\bar{s}$ so that these arbitrary datums cancel. Accordingly, special means are required to assign specific enthalpy and entropy for application to reacting systems.

Property data suited for the analysis of reactive systems are available from several sources. The encyclopedic *JANAF Thermochemical Tables* is commonly used. Data for a wide range of substances are retrievable from Knacke et al. (1991), which provides both tabular data and analytical expressions readily programmable for use with personal computers of the specific heat, enthalpy, entropy, and Gibbs function. Textbooks on engineering thermodynamics also provide selected data, as, for example, Moran and Shapiro (2000).

Enthalpy of Formation

An enthalpy datum for reacting systems can be established by assigning arbitrarily a value of zero to the enthalpy of the *stable elements* at a *standard reference state* where the temperature is T_{ref} = 298.15 K (25°C) and the pressure is p_{ref}, which may be 1 bar or 1 atm depending on the data source. The term stable simply means that the particular element is chemically stable. For example, at the standard state the stable forms of hydrogen, oxygen, and nitrogen are H_2, O_2, and N_2 and not the monatomic H, O, and N.

The molar enthalpy of a *compound* at the standard state equals its *enthalpy of formation*, symbolized here by $\bar{h}_f^\circ$. The enthalpy of formation is the energy released or absorbed when the compound is formed from its elements, the compound and elements all being at T_{ref} and p_{ref}. The enthalpy of formation may be determined by application of procedures from statistical thermodynamics using observed spectroscopic data. The enthalpy of formation also can be found in principle by measuring the heat transfer in a reaction in which the compound is formed from the elements. In this chapter, the superscript ° is used to denote p_{ref}. For the case of the enthalpy of formation, the reference temperature T_{ref} is also intended by this symbol. Table 1.9 gives the values of the enthalpy of formation of various substances at 298 K and 1 atm.

The molar enthalpy of a substance at a state other than the standard state is found by adding the molar enthalpy change $\Delta\bar{h}$ between the standard state and the state of interest to the molar enthalpy of formation:

TABLE 1.9 Enthalpy of Formation, Gibbs Function of Formation, and Absolute Entropy of Various Substances at 298 K and 1 atm

$\bar{h}_f^\circ$ and $\bar{g}_f^\circ$ (kJ/kmol), $\bar{s}^\circ$ (kJ/kmol•K)

Substance	Formula	$\bar{h}_f^\circ$	$\bar{g}_f^\circ$	$\bar{s}^\circ$
Carbon	C(s)	0	0	5.74
Hydrogen	H_2(g)	0	0	130.57
Nitrogen	N_2(g)	0	0	191.50
Oxygen	O_2(g)	0	0	205.03
Carbon monoxide	CO(g)	–110,530	–137,150	197.54
Carbon dioxide	CO_2(g)	–393,520	–394,380	213.69
Water	H_2O(g)	–241,820	–228,590	188.72
	H_2O(l)	–285,830	–237,180	69.95
Hydrogen peroxide	H_2O_2(g)	–136,310	–105,600	232.63
Ammonia	NH_3(g)	–46,190	–16,590	192.33
Oxygen	O(g)	249,170	231,770	160.95
Hydrogen	H(g)	218,000	203,290	114.61
Nitrogen	N(g)	472,680	455,510	153.19
Hydroxyl	OH(g)	39,460	34,280	183.75
Methane	CH_4(g)	–74,850	–50,790	186.16
Acetylene	C_2H_2(g)	226,730	209,170	200.85
Ethylene	C_2H_4(g)	52,280	68,120	219.83
Ethane	C_2H_6(g)	–84,680	–32,890	229.49
Propylene	C_3H_6(g)	20,410	62,720	266.94
Propane	C_3H_8(g)	–103,850	–23,490	269.91
Butane	C_4H_{10}(g)	–126,150	–15,710	310.03
Pentane	C_5H_{12}(g)	–146,440	–8,200	348.40
Octane	C_8H_{18}(g)	–208,450	17,320	463.67
	C_8H_{18}(l)	–249,910	6,610	360.79
Benzene	C_6H_6(g)	82,930	129,660	269.20
Methyl alcohol	CH_3OH(g)	–200,890	–162,140	239.70
	CH_3OH(l)	–238,810	–166,290	126.80
Ethyl alcohol	C_2H_5OH(g)	–235,310	–168,570	282.59
	C_2H_5OH(l)	–277,690	174,890	160.70

Source: Adapted from Wark, K. 1983. *Thermodynamics*, 4th ed. McGraw-Hill, New York, as based on JANAF Thermochemical Tables, NSRDS-NBS-37, 1971; *Selected Values of Chemical Thermodynamic Properties*, NBS Tech. Note 270-3, 1968; and *API Research Project 44*, Carnegie Press, 1953.

$$\bar{h}(T,p) = \bar{h}_f^\circ + \left[\bar{h}(T,p) - \bar{h}\left(T_{ref}, p_{ref}\right)\right] = \bar{h}_f^\circ + \Delta\bar{h} \tag{1.78}$$

That is, the enthalpy of a substance is composed of $\bar{h}_f^\circ$, associated with the formation of the substance from its elements, and $\Delta\bar{h}$, associated with a change of state at constant composition. An arbitrarily chosen datum can be used to determine $\Delta\bar{h}$, since it is a *difference* at constant composition. Accordingly, $\Delta\bar{h}$ can be evaluated from sources such as the steam tables and the ideal gas tables.

The *enthalpy of combustion*, $\bar{h}_{RP}$, is the difference between the enthalpy of the products and the enthalpy of the reactants, each on a per-mole-of-fuel basis, when complete combustion occurs and both reactants and products are at the same temperature and pressure. For hydrocarbon fuels the enthalpy of combustion is negative in value since chemical internal energy is liberated in the reaction. The *heating value* of a fuel is a positive number equal to the magnitude of the enthalpy of combustion. Two heating values are recognized: the *higher* heating value and the *lower* heating value. The higher heating value is obtained when all the water formed by combustion is a liquid; the lower heating value is obtained when all the water formed by combustion is a vapor. The higher heating value exceeds the lower heating value by the energy that would be required to vaporize the liquid water formed at the specified temperature. Heating values are typically reported at a temperature of 25°C (77°F) and a pressure of 1 bar (or 1 atm). These values also depend on whether the fuel is a liquid or a gas. A sampling is provided on a unit-mass-of-fuel basis in Table 1.10.

TABLE 1.10 Heating Values in kJ/kg of Selected Hydrocarbons at 25°C

		Higher Value[a]		Lower Value[b]	
Hydrocarbon	**Formula**	**Liquid Fuel**	**Gas. Fuel**	**Liquid Fuel**	**Gas. Fuel**
Methane	CH_4	—	55,496	—	50,010
Ethane	C_2H_6	—	51,875	—	47,484
Propane	C_3H_8	49,973	50,343	45,982	46,352
n-Butane	C_4H_{10}	49,130	49,500	45,344	45,714
n-Octane	C_8H_{18}	47,893	48,256	44,425	44,788
n-Dodecane	$C_{12}H_{26}$	47,470	47,828	44,109	44,467
Methanol	CH_3OH	22,657	23,840	19,910	21,093
Ethanol	C_3H_5OH	29,676	30,596	26,811	27,731

[a] H_2O liquid in the products.
[b] H_2O vapor in the products.

In the absence of work $\dot{W}_{cv}$ and appreciable kinetic and potential energy effects, the energy liberated on combustion is transferred from a reactor at steady state in two ways: the energy accompanying the exiting combustion products and by heat transfer. The temperature that would be achieved by the products in the limit of adiabatic operation is the *adiabatic flame* or *adiabatic combustion* temperature.

For a specified fuel and specified temperature and pressure of the reactants, the *maximum* adiabatic flame temperature is realized for complete combustion with the theoretical amount of air. Example 10 provides an illustration. The measured value of the temperature of the combustion products may be several hundred degrees below the calculated maxunum adiabatic flame temperature, however, for several reasons including the following: (1) heat loss can be reduced but not eliminated; (2) once adequate oxygen has been provided to permit complete combustion, bringing in more air dilutes the combustion products, lowering the temperature; (3) incomplete combustion tends to reduce the temperature of the products, and combustion is seldom complete; (4) as result of the high temperatures achieved, some of the combustion products may dissociate. Endothermic dissociation reactions also lower the product temperature.

Absolute Entropy

A common datum for assigning entropy values to substances involved in chemical reactions is realized through the *third law* of thermodynamics, which is based on experimental observations obtained primarily from studies of chemical reactions at low temperatures and specific heat measurements at temperatures approaching absolute zero. The third law states that the entropy of a pure crystalline substance is zero at the absolute zero of temperature, 0 K or 0°R. Substances not having a pure crystalline structure have a nonzero value of entropy at absolute zero.

The third law provides a datum relative to which the entropy of each substance participating in a reaction can be evaluated. The entropy relative to this datum is called the *absolute* entropy. The change in entropy of a substance between absolute zero and any given state can be determined from measurements of energy transfers and specific heat data or from procedures based on statistical thermodynamics and observed molecular data. Table 1.9 and Tables A.2 and A.8 provide absolute entropy data for various substances. In these tables, $p_{ref} = 1$ atm.

When the absolute entropy is known at pressure p_{ref} and temperature T, the absolute entropy at the same temperature and *any* pressure p can be found from

$$\bar{s}(T,p) = \bar{s}\left(T,p_{ref}\right) + \left[\bar{s}(T,p) - \bar{s}\left(T,p_{ref}\right)\right] \tag{1.79}$$

For an ideal gas, the second term on the right side of Equation 1.79 can be evaluated by using Equation 1.58, giving

$$\bar{s}(T,p) = \bar{s}^{\circ}(T) - \bar{R}\ln\frac{p}{p_{ref}} \qquad \text{(ideal gas)} \tag{1.80}$$

In this expression, $\bar{s}^{\circ}(T)$ denotes the absolute entropy at temperature T and pressure p_{ref}.

The entropy of the ith component of an *ideal gas mixture* is evaluated at the mixture temperature T and the *partial* pressure p_i: $\bar{s}_i(T, p_i)$. For the ith component, Equation 1.80 takes the form

$$\begin{aligned}\bar{s}_i\left(T,p_i\right) &= \bar{s}_i^{\circ}(T) - \bar{R}\ln\frac{p_i}{p_{ref}} \\ &= \bar{s}_i^{\circ}(T) - \bar{R}\ln\frac{y_i p}{p_{ref}} \qquad \text{(ideal gas)}\end{aligned} \tag{1.81}$$

where $\bar{s}_i^{\circ}(T)$ is the absolute entropy of component i at temperature T and p_{ref}.

Example 10

Liquid octane at 25°C, 1 atm enters a well insulated reactor and reacts with dry air entering at the same temperature and pressure. For steady-state operation and negligible effects of kinetic and potential energy, determine the temperature of the combustion products for complete combustion with the theoretical amount of air, and (b) the rates of entropy generation and exergy destruction, each per kmol of fuel.

Solution. For combustion of liquid octane with the theoretical amount of air, the chemical equation is

$$C_8H_{18}(l) + 12.5O_2 + 47N_2 \rightarrow 8CO_2 + 9H_2O(g) + 47N_2$$

(a) At steady state, the control volume energy rate balance reduces to read

$$0 = \cancelto{0}{\frac{\dot{Q}_{cv}}{\dot{n}_F}} - \cancelto{0}{\frac{\dot{W}_{cv}}{\dot{n}_F}} + \sum_R n_i\left(\bar{h}_f^{\circ} + \Delta\bar{h}\right)_i - \sum_P n_e\left(\bar{h}_f^{\circ} + \Delta\bar{h}\right)_e$$

where R denotes reactants, P denotes products, and the symbols for enthalpy have the same significance as in Equation 1.78. Since the reactants enter at 25°C, the corresponding $(\Delta\bar{h})_i$ terms vanish, and the energy rate equation becomes

$$\sum_P n_e\left(\Delta\bar{h}\right)_e = \sum_R n_i\bar{h}_{fi}^{\circ} - \sum_P n_e\bar{h}_{fe}^{\circ}$$

Introducing coefficients from the reaction equation, this takes the form

$$\begin{aligned}8\left(\Delta\bar{h}\right)_{CO_2} + 9\left(\Delta\bar{h}\right)_{H_2O(g)} + 47\left(\Delta\bar{h}\right)_{N_2} &= \left[\left(\bar{h}_f^{\circ}\right)_{C_8H_{18}(l)} + 12.5\left(\bar{h}_f^{\circ}\right)_{O_2} + 47\left(\bar{h}_f^{\circ}\right)_{N_2}\right] \\ &- \left[8\left(\bar{h}_f^{\circ}\right)_{CO_2} + 9\left(\bar{h}_f^{\circ}\right)_{H_2O(g)} + 47\left(\Delta\bar{h}_f^{\circ}\right)_{N_2}\right]\end{aligned}$$

Using data from Table 1.9 to evaluate the right side,

$$8\left(\Delta\bar{h}\right)_{CO_2} + 9\left(\Delta\bar{h}\right)_{H_2O(g)} + 47\left(\Delta\bar{h}\right)_{N_2} = 5{,}074{,}630 \text{ kJ/kmol (fuel)}$$

Each $\Delta\bar{h}$ term on the left side of this equation depends on the temperature of the products, T_p, which can be solved for iteratively as $T_p = 2395$ K.

(b) The entropy rate balance on a per-mole-of-fuel basis takes the form

$$0 = \sum_j \frac{\dot{Q}_j/T_j}{\dot{n}_F}^{0} + \bar{s}_F + \left(12.5\bar{s}_{O_2} + 47\bar{s}_{N_2}\right) - \left(8\bar{s}_{CO_2} + 9\bar{s}_{H_2O(g)} + 47\bar{s}_{N_2}\right) + \frac{\dot{S}_{gen}}{\dot{n}_F}$$

or on rearrangement,

$$\frac{\dot{S}_{gen}}{\dot{n}_F} = \left(8\bar{s}_{CO_2} + 9\bar{s}_{H_2O(g)} + 47\bar{s}_{N_2}\right) - \bar{s}_F - \left(12.5\bar{s}_{O_2} + 47\bar{s}_{N_2}\right)$$

The absolute entropy of liquid octane from Table 1.9 is 360.79 kJ/mol · K. The oxygen and nitrogen in the combustion air enter the reactor as components of an ideal gas mixture at T_{ref}, p_{ref}. With Equation 1.81, where $p = p_{ref}$, and absolute entropy data from Table 1.9,

$$\bar{s}_{O_2} = \bar{s}^{\circ}_{O_2}\left(T_{ref}\right) - \bar{R}\ln y_{O_2}$$

$$= 205.03 - 8.314\ln 0.21 = 218.01 \text{ kJ/kmol} \cdot \text{K}$$

$$\bar{s}_{N_2} = \bar{s}^{\circ}_{N_2}\left(T_{ref}\right) - \bar{R}\ln y_{N_2}$$

$$= 191.5 - 8.314\ln 0.79 = 193.46 \text{ kJ/kmol} \cdot \text{K}$$

The product gas exits as a gas mixture at 1 atm, 2395 K with the following composition: y_{CO_2} = 8/64 = 0.125, $y_{H_2O(g)}$ = 9/64 = 0.1406, y_{N_2} = 47/64 = 0.7344. With Equation 1.81, where $p = p_{ref}$, and absolute entropy data at 2395 K from Table A.2,

$$\bar{s}_{CO_2} = 320.173 - 8.314\ln 0.125 = 337.46 \text{ kJ/kmol} \cdot \text{K}$$

$$\bar{s}_{H_2O} = 273.986 - 8.314\ln 0.1406 = 290.30 \text{ kJ/kmol} \cdot \text{K}$$

$$\bar{s}_{N_2} = 258.503 - 8.314\ln 0.7344 = 261.07 \text{ kJ/kmol} \cdot \text{K}$$

Inserting values, the rate of entropy generation is

$$\frac{\dot{S}_{gen}}{\dot{n}_F} = 8(337.46) + 9(290.30) + 47(261.07) - 360.79 - 12.5(218.01) - 47(193.46)$$

$$= 5404 \text{ kJ/kmol} \cdot \text{K}$$

Using Equation 1.87 and assuming $T_0 = 298$ K, the rate of exergy destruction is $\dot{E}_D/\dot{n}_F = 1.61 \times 10^6$ kJ/kmol.

Gibbs Function of Formation

Paralleling the approach used for enthalpy, a value of zero is assigned to the Gibbs function of each stable element at the standard state. The *Gibbs function of formation* of a compound equals the change in the Gibbs function for the reaction in which the compound is formed from its elements. Table 1.9 provides Gibbs function of formation data of various substances at 298 K and 1 atm.

The Gibbs function at a state other than the standard state is found by adding to the Gibbs function of formation the change in the specific Gibbs function $\Delta\bar{g}$ between the standard state and the state of interest:

$$\bar{g}(T,p) = \bar{g}_f^\circ + \left[\bar{g}(T,p) - \bar{g}\left(T_{ref}, p_{ref}\right)\right] = \bar{g}_f^\circ + \Delta\bar{g} \tag{1.82a}$$

where

$$\Delta\bar{g} = \left[\bar{h}(T,p) - \bar{h}\left(T_{ref}, p_{ref}\right)\right] - \left[T\bar{s}(T,p) - T_{ref}\bar{s}\left(T_{ref}, p_{ref}\right)\right] \tag{1.82b}$$

The Gibbs function of component i in an ideal gas mixture is evaluated at the partial pressure of component i and the mixture temperature.

As an application, the maximum theoretical work that can be developed, per mole of fuel consumed, is evaluated for the control volume of Figure 1.15, where the fuel and oxygen each enter in separate streams and carbon dioxide and water each exit separately. All entering and exiting streams are at the same temperature T and pressure p. The reaction is complete:

$$C_aH_b + \left(a + \frac{b}{4}\right)\mathrm{O}_2 \rightarrow a\mathrm{CO}_2 + \frac{b}{2}\mathrm{H}_2\mathrm{O}$$

This control volume is similar to idealized devices such as a reversible fuel *cell* or a *van't Hoff equilibrium box.*

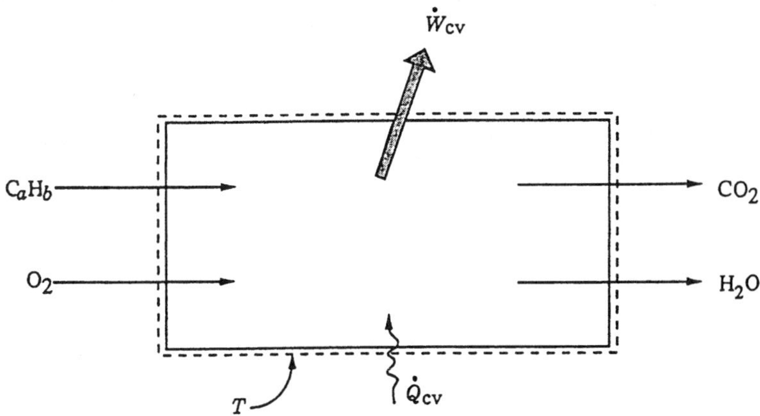

FIGURE 1.15 Device for evaluating maximum work.

For steady-state operation, the energy rate balance reduces to give

$$\frac{\dot{W}_{cv}}{\dot{n}_F} = \frac{\dot{Q}_{cv}}{\dot{n}_F} + \bar{h}_F + \left(a + \frac{b}{4}\right)\bar{h}_{\mathrm{O}_2} - a\bar{h}_{\mathrm{CO}_2} - \frac{b}{2}\bar{h}_{\mathrm{H}_2\mathrm{O}}$$

where $\dot{n}_F$ denotes the molar flow rate of the fuel. Kinetic and potential energy effects are regarded as negligible. If heat transfer occurs only at the temperature T, an entropy balance for the control volume takes the form

$$0 = \frac{\dot{Q}_{cv}/\dot{n}_F}{T} + \bar{s}_F + \left(a + \frac{b}{4}\right)\bar{s}_{O_2} - a\bar{s}_{CO_2} - \frac{b}{2}\bar{s}_{H_2O} + \frac{\dot{S}_{gen}}{\dot{n}_F}$$

Eliminating the heat transfer term from these expressions, an expression for the maximum theoretical value of the work developed per mole of fuel is obtained when the entropy generation term is set to zero:

$$\left(\frac{\dot{W}_{cv}}{\dot{n}_F}\right)_{\substack{int\\rev}} = \left[\bar{h}_F + \left(a + \frac{b}{4}\right)\bar{h}_{O_2} - a\bar{h}_{CO_2} - \frac{b}{2}\bar{h}_{H_2O}\right](T,p) - T\left[\bar{s}_F + \left(a + \frac{b}{4}\right)\bar{s}_{O_2} - a\bar{s}_{CO_2} - \frac{b}{2}\bar{s}_{H_2O}\right](T,p)$$

This can be written alternatively in terms of the enthalpy of combustion as

$$\left(\frac{\dot{W}_{cv}}{\dot{n}_F}\right)_{\substack{int\\rev}} = -\bar{h}_{RP}(T,p) - T\left[\bar{s}_F + \left(a + \frac{b}{4}\right)\bar{s}_{O_2} - a\bar{s}_{CO_2} - \frac{b}{2}\bar{s}_{H_2O}\right](T,p) \tag{1.83a}$$

or in terms of Gibbs functions as

$$\left(\frac{\dot{W}_{cv}}{\dot{n}_F}\right)_{\substack{int\\rev}} = \left[\bar{g}_F + \left(a + \frac{b}{4}\right)\bar{g}_{O_2} - a\bar{g}_{CO_2} - \frac{b}{2}\bar{g}_{H_2O}\right](T,p) \tag{1.83b}$$

Equation 1.83b is used in the solution to Example 11.

Example 11

Hydrogen (H_2) and oxygen (O_2), each at 25°C, 1 atm, enter a fuel cell operating at steady state, and liquid water exits at the same temperature and pressure. The hydrogen flow rate is 2×10^{-4} kmol/sec and the fuel cell operates isothermally at 25°C. Determine the maximum theoretical power the cell can develop, in kW.

Solution. The overall cell reaction is $H_2 + 1/2\ O_2 \rightarrow H_2O(\ell)$, and Equations 1.83 are applicable. Selecting Equation 1.83b, and using Gibbs function data from Table 1.9,

$$\left(\frac{\dot{W}_{cv}}{\dot{n}_F}\right)_{\substack{int\\rev}} = \left(\bar{g}_{H_2} + \frac{1}{2}\bar{g}_{O_2} - \bar{g}_{H_2O(\ell)}\right)(25°\text{C, 1 atm})$$

$$= 0 + \frac{1}{2}(0) - (-237{,}180) = 237{,}180 \text{ kJ/kmol}$$

Then

$$\left(\dot{W}_{cv}\right)_{\substack{int\\rev}} = \left(237{,}180 \frac{\text{kJ}}{\text{kmol}}\right)\left(2 \times 10^{-4} \frac{\text{kmol}}{\text{s}}\right)\left(\frac{\text{kW}}{1\text{kJ/s}}\right) = 47.4 \text{ kW}$$

Reaction Equilibrium

Let the objective be to determine the equilibrium composition of a system consisting of five gases A, B, C, D, and E, at a temperature T and pressure p, subject to a chemical reaction of the form

$$\nu_A A + \nu_B B \leftrightarrow \nu_C C + \nu_D D$$

where the ν's are stoichiometric coefficients. Component E is assumed to be inert and thus does not appear in the reaction equation. The equation suggests that at equilibrium the tendency of A and B to form C and D is just balanced by the tendency of C and D to form A and B.

At equilibrium, the temperature and pressure would be uniform throughout the system. Additionally, the *equation of reaction equilibrium* must be satisfied:

$$\nu_A \mu_A + \nu_B \mu_B = \nu_C \mu_C + \nu_D \mu_D \tag{1.84a}$$

where the μ's are the chemical potentials (Section 1.3, Multicomponent Systems) of A, B, C, and D in the equilibrium mixture. In principle, the composition that would be present at equilibrium for a given temperature and pressure can be determined by solving this equation.

For ideal gas mixtures, the solution procedure is simplified by using the *equilibrium constant $K(T)$* and the following equation:

$$\begin{aligned} K(T) &= \frac{y_C^{\nu_C} y_D^{\nu_D}}{y_A^{\nu_A} y_B^{\nu_B}} \left(\frac{p}{p_{ref}} \right)^{\nu_C + \nu_D - \nu_A - \nu_B} \\ &= \frac{n_C^{\nu_C} n_D^{\nu_D}}{n_A^{\nu_A} n_B^{\nu_B}} \left(\frac{p/p_{ref}}{n} \right)^{\nu_C + \nu_D - \nu_A - \nu_B} \end{aligned} \tag{1.84b}$$

where y_A, y_B, y_C, and y_D denote the mole fractions of A, B, C, and D in the equilibrium mixture and $n = n_A + n_B + n_C + n_D + n_E$, where the n's denote the molar amounts of the gases in the mixture. Tabulations of $K(T)$ for each of several reactions of the form Equation 1.84a are provided in Table 1.11. An application of Equation 1.84b is provided in Example 12.

Example 12

One kmol of CO reacts with the theoretical amount of dry air to form an equilibrium mixture of CO_2, CO, O_2, and N_2 at 2500 K, 1 atm. Determine the amount of CO in the equilibrium mixture, in kmol.

Solution. The reaction of CO with the theoretical amount of dry air to form CO_2, CO, O_2, and N_2 is

$$\mathrm{CO} + \frac{1}{2}\mathrm{O_2} + 1.88\mathrm{N_2} \rightarrow z\mathrm{CO} + \frac{z}{2}\mathrm{O_2} + (1-z)\mathrm{CO_2} + 1.88\mathrm{N_2}$$

where z is the amount of CO, in kmol, present in the equilibrium mixture. The total number of moles n is

$$n = z + \frac{z}{2} + (1-z) + 1.88 = \frac{5.76 + z}{2}$$

At equilibrium $CO_2 \leftrightarrow CO + 1/2\ O_2$; and Equation 1.84b takes the form

$$K = \frac{z(z/2)^{1/2}}{1-z} \left[\frac{p/p_{ref}}{(5.76+z)/2} \right]^{1/2}$$

where $p/p_{ref} = 1$. At 2500 K, Table 1.11 gives $K = 0.0363$. Solving iteratively, z = 0.175.

TABLE 1.11 Logarithms to the Base 10 of the Equilibrium Constant *K*

	$\log_{10} K$								
Temp (K)	$H_2 \Leftrightarrow 2H$	$O_2 \Leftrightarrow 2O$	$N_2 \Leftrightarrow 2N$	$^1/_2O_2 + {}^1/_2N_2 \Leftrightarrow NO$	$H_2O \Leftrightarrow H_2 + {}^1/_2O_2$	$H_2O \Leftrightarrow OH + {}^1/_2H_2$	$CO_2 \Leftrightarrow CO + {}^1/_2 O_2$	$CO_2 + H_2 \Leftrightarrow CO + H_2O$	Temp (°R)
298	−71.224	−81.208	−159.600	−15.171	−40.048	−46.054	−45.066	−5.018	537
500	−40.316	−45.880	−92.672	−8.783	−22.886	−26.130	−25.025	−2.139	900
1000	−17.292	−19.614	−43.056	−4.062	−10.062	−11.280	−10.221	−0.159	1800
1200	−13.414	−15.208	−34.754	−3.275	−7.899	−8.811	−7.764	+0.135	2160
1400	−10.630	−12.054	−28.812	−2.712	−6.347	−7.021	−6.014	+0.333	2520
1600	−8.532	−9.684	−24.350	−2.290	−5.180	−5.677	−4.706	+0.474	2880
1700	−7.666	−8.706	−22.512	−2.116	−4.699	−5.124	−4.169	+0.530	3060
1800	−6.896	−7.836	−20.874	−1.962	−4.270	−4.613	−3.693	+0.577	3240
1900	−6.204	−7.058	−19.410	−1.823	−3.886	−4.190	−3.267	+0.619	3420
2000	−5.580	−6.356	−18.092	−1.699	−3.540	−3.776	−2.884	+0.656	3600
2100	−5.016	−5.720	−16.898	−1.586	−3.227	−3.434	−2.539	+0.688	3780
2200	−4.502	−5.142	−15.810	−1.484	−2.942	−3.091	−2.226	+0.716	3960
2300	−4.032	−4.614	−14.818	−1.391	−2.682	−2.809	−1.940	+0.742	4140
2400	−3.600	−4.130	−13.908	−1.305	−2.443	−2.520	−1.679	+0.764	4320
2500	−3.202	−3.684	−13.070	−1.227	−2.224	−2.270	−1.440	+0.784	4500
2600	−2.836	−3.272	−12.298	−1.154	−2.021	−2.038	−1.219	+0.802	4680
2700	−2.494	−2.892	−11.580	−1.087	−1.833	−1.823	−1.015	+0.818	4860
2800	−2.178	−2.536	−10.914	−1.025	−1.658	−1.624	−0.825	+0.833	5040
2900	−1.882	−2.206	−10.294	−0.967	−1.495	−1.438	−0.649	+0.846	5220
3000	−1.606	−1.898	−9.716	−0.913	−1.343	−1.265	−0.485	+0.858	5400
3100	−1.348	−1.610	−9.174	−0.863	−1.201	−1.103	−0.332	+0.869	5580
3200	−1.106	−1.340	−8.664	−0.815	−1.067	−0.951	−0.189	+0.878	5760
3300	−0.878	−1.086	−8.186	−0.771	−0.942	−0.809	−0.054	+0.888	5940
3400	−0.664	−0.846	−7.736	−0.729	−0.824	−0.674	+0.071	+0.895	6120
3500	−0.462	−0.620	−7.312	−0.690	−0.712	−0.547	+0.190	+0.902	6300

Source: Based on data from the JANAF Thermochemical Tables, NSRDS-NBS-37, 1971.

1.5 Exergy Analysis

The method of *exergy analysis* (*availability analysis*) presented in this section enables the location, cause, and true magnitude of energy resource waste and loss to be determined. Such information can be used in the design of new energy-efficient systems and for improving the performance of existing systems. Exergy analysis also provides insights that elude a purely first-law approach. For example, on the basis of first-law reasoning alone, the condenser of a power plant may be mistakenly identified as the component primarily responsible for the plant's seemingly low overall performance. An exergy analysis correctly reveals not only that the condenser loss is relatively unimportant (see the last two rows of the Rankine cycle values of Table 1.15), but also that the steam generator is the principal site of thermodynamic inefficiency owing to combustion and heat transfer irreversibilities within it.

When exergy concepts are combined with principles of engineering economy, the result is known as *thermoeconomics* or *exergoeconomics*. Thermoeconomics allows the real cost sources at the component level to be identified: capital investment costs, operating and maintenance costs, and the costs associated with the destruction and loss of exergy. Optimization of thermal systems can be achieved by a careful consideration of such cost sources. From this perspective thermoeconomics *is exergy-aided cost minimization.*

Discussions of exergy analysis and thermoeconomics are provided by Bejan et al. (1996), Moran (1989), and Moran and Shapiro (2000). In this section salient aspects are presented. Also see Sections 1.8 to 1.10.

Defining Exergy

An opportunity for doing work exists whenever two systems at different states are placed in communication because, in principle, work can be developed as the two are allowed to come into equilibrium. When one of the two systems is a suitably idealized system called an *environment* and the other is some system of interest, *exergy* is the maximum theoretical useful work (shaft work or electrical work) obtainable as the systems interact to equilibrium, heat transfer occurring with the environment only. (Alternatively, exergy is the minimum theoretical useful work required to form a quantity of matter from substances present in the environment and to bring the matter to a specified state.) Exergy is a measure of the *departure* of the state of the system from that of the environment, and is therefore an attribute of the system and environment together. Once the environment is specified, however, a value can be assigned to exergy in terms of property values for the system only, so exergy can be regarded as an extensive property of the system.

Exergy can be destroyed and generally is not conserved. A limiting case is when exergy would be completely destroyed, as would occur if a system were to come into equilibrium with the environment *spontaneously* with no provision to obtain work. The capability to develop work that existed initially would be completely wasted in the spontaneous process. Moreover, since no work needs to be done to effect such a spontaneous change, the value of exergy can never be negative.

Environment

Models with various levels of specificity are employed for describing the environment used to evaluate exergy. Models of the environment typically refer to some portion of a system's surroundings, the intensive properties of each phase of which are uniform and do not change significantly as a result of any process under consideration. The environment is regarded as composed of common substances existing in abundance within the Earth's atmosphere, oceans, and crust. The substances are in their stable forms as they exist naturally, and there is no possibility of developing work from interactions — physical or chemical — between parts of the environment. Although the intensive properties of the environment are assumed to be unchanging, the extensive properties can change as a result of interactions with other systems. Kinetic and potential energies are evaluated relative to coordinates in the environment, all parts of which are considered to be at rest with respect to one another.

For computational ease, the temperature T_0 and pressure p_0 of the environment are often taken as standard-state values, such as 1 atm and 25°C (77°F). However, these properties may be specified differently depending on the application. T_0 and p_0 might be taken as the average ambient temperature and pressure, respectively, for the location at which the system under consideration operates. Or, if the system uses atmospheric air, T_0 might be specified as the average air temperature. If both air and water from the natural surroundings are used, T_0 would be specified as the lower of the average temperatures for air and water.

Dead States

When a system is in equilibrium with the environment, the state of the system is called the *dead state.* At the dead state, the conditions of mechanical, thermal, and chemical equilibrium between the system and the environment are satisfied: the pressure, temperature, and chemical potentials of the system equal those of the environment, respectively. In addition, the system has no motion or elevation relative to coordinates in the environment. Under these conditions, there is no possibility of a spontaneous change within the system or the environment, nor can there be an interaction between them. The value of exergy is zero.

Another type of equilibrium between the system and environment can be identified. This is a restricted form of equilibrium where only the conditions of mechanical and thermal equilibrium must be satisfied. This state of the system is called the *restricted dead state.* At the restricted dead state, the fixed quantity of matter under consideration is imagined to be sealed in an envelope impervious to mass flow, at zero velocity and elevation relative to coordinates in the environment, and at the temperature T_0 and pressure p_0.

Exergy Balances

Exergy can be transferred by three means: exergy transfer associated with work, exergy transfer associated with heat transfer, and exergy transfer associated with the matter entering and exiting a control volume. All such exergy transfers are evaluated relative to the environment used to define exergy. Exergy is also destroyed by irreversibilities within the system or control volume.

Exergy balances can be written in various forms, depending on whether a closed system or control volume is under consideration and whether steady-state or transient operation is of interest. Owing to its importance for a wide range of applications, an exergy rate balance for control volumes at steady state is presented next.

Control Volume Exergy Rate Balance

At steady state, the control volume exergy rate balance takes the form

$$0 = \underbrace{\sum_j \dot{E}_{q,j} - \dot{W}_{cv} + \sum_i \dot{E}_i - \sum_e \dot{E}_e}_{\text{rates of exergy transfer}} - \underbrace{\dot{E}_D}_{\text{rate of exergy destruction}} \tag{1.85a}$$

or

$$0 = \sum_j \left(1 - \frac{T_0}{T_j}\right)\dot{Q}_j - \dot{W}_{cv} + \sum_i \dot{m}_i e_i - \sum_e \dot{m}_e e_e - \dot{E}_D \tag{1.85b}$$

$\dot{W}_{cv}$ has the same significance as in Equation 1.22: the work rate excluding the flow work. $\dot{Q}_j$ is the time rate of heat transfer at the location on the boundary of the control volume where the instantaneous temperature is T_j. The associated rate of exergy transfer is

$$\dot{E}_{q,j} = \left(1 - \frac{T_0}{T_j}\right)\dot{Q}_j \tag{1.86}$$

As for other control volume rate balances, the subscripts i and e denote inlets and outlets, respectively. The exergy transfer rates at control volume inlets and outlets are denoted, respectively, as $\dot{E}_i = \dot{m}_i e_i$ and $\dot{E}_e = \dot{m}_e e_e$. Finally, $\dot{E}_D$ accounts for the time rate of exergy destruction due to irreversibilities within the control volume. The exergy destruction rate is related to the entropy generation rate by

$$\dot{E}_D = T_0 \dot{S}_{gen} \tag{1.87}$$

The specific exergy transfer terms e_i and e_e are expressible in terms of four components: physical exergy e^{PH}, kinetic exergy e^{KN}, potential exergy e^{PT}, and chemical exergy e^{CH}:

$$e = e^{PH} + e^{KN} + e^{PT} + e^{CH} \tag{1.88}$$

The first three components are evaluated as follows:

$$e^{PH} = \left(h - h_0\right) - T_0\left(s - s_0\right) \tag{1.89a}$$

$$e^{KN} = \frac{1}{2}\mathrm{v}^2 \tag{1.89b}$$

$$e^{PT} = gz \tag{1.89c}$$

In Equation 1.89a, h_0 and s_0 denote, respectively, the specific enthalpy and specific entropy at the restricted dead state. In Equations 1.89b and 1.89c, v and z denote velocity and elevation relative to coordinates in the environment, respectively. The chemical exergy e^{CH} is considered next.

Chemical Exergy

To evaluate the chemical exergy, the exergy component associated with the departure of the chemical composition of a system from that of the environment, the substances comprising the system are referred to the properties of a suitably selected set of environmental substances. For this purpose, alternative models of the environment have been developed. For discussion, see, for example, Moran (1989) and Kotas (1995).

Exergy analysis is facilitated, however, by employing a *standard environment* and a corresponding table of *standard chemical exergies.* Standard chemical exergies are based on standard values of the environmental temperature T_0 and pressure p_0 — for example, 298.15 K (25°C) and 1 atm, respectively. A standard environment is also regarded as consisting of a set of reference substances with standard concentrations reflecting as closely as possible the chemical makeup of the natural environment. The reference substances generally fall into three groups: gaseous components of the atmosphere, solid substances from the lithosphere, and ionic and noninonic substances from the oceans. The chemical exergy data of Table 1.12 correspond to two alternative standard exergy reference environments, called here model I and model II, that have gained acceptance for engineering evaluations.

Although the use of standard chemical exergies greatly facilitates the application of exergy principles, the term *standard* is somewhat misleading since there is no one specification of the environment that suffices for all applications. Still, chemical exergies calculated relative to alternative specifications of the environment are generally in good agreement. For a broad range of engineering applications the simplicity and ease of use of standard chemical exergies generally outweigh any slight lack of accuracy that might result. In particular, the effect of slight variations in the values of T_0 and p_0 about the values used to determine the standard chemical exergies reported in Table 1.12 can be neglected.

The literature of exergy analysis provides several expressions allowing the chemical exergy to be evaluated in particular cases of interest. The molar chemical exergy of a gas mixture, for example, can be evaluated from

$$\bar{e}^{CH} = \sum_{i=1}^{j} y_i \bar{e}_i^{CH} + \bar{R}T_0 \sum_{i=1}^{j} y_i \ln y_i \tag{1.90}$$

where $\bar{e}_i^{CH}$ is the molar chemical exergy of the ith component.

Example 13

Ignoring the kinetic and potential exergies, determine the exergy rate, in kJ/kg, associated with each of the following streams of matter:

(a) Saturated water vapor at 20 bar.
(b) Methane at 5 bar, 25°C.

Let T_0 = 298 K, p_0 = 1.013 bar (1 atm).

Solution. Equation 1.88 reduces to read

$$e = (h - h_0) - T_0(s - s_0) + e^{CH}$$

(a) From Table A.5, h = 2799.5 kJ/kg, s = 6.3409 kJ/kg · K. At T_0 = 298 K (25°C), water would be a liquid; thus with Equations 1.50c and 1.50d, $h_0 \approx$ 104.9 kJ/kg, $s_0 \approx$ 0.3674 kJ/kg · K. Table 1.12 (model I) gives e^{CH} = 45/18.02 = 2.5 kJ/kg. Then

$$\begin{aligned} e &= (2799.5 - 104.9) - 298(6.3409 - 0.3674) + 2.5 \\ &= 914.5 + 2.5 = 917.0 \text{ kJ/kg} \end{aligned}$$

Here the specific exergy is determined predominately by the physical component.

(b) Assuming the ideal gas model for methane, $h - h_0 = 0$. Also, Equation 1.58 reduces to give $s - s_0 = -R\ln p/p_0$. Then, Equation 1.88 reads

$$e = RT_0 \ln p/p_0 + e^{CH}$$

With e^{CH} = 824,350/16.04 = 51,393.4 kJ/kg from Table 1.12 (model I),

$$\begin{aligned} e &= \left(\frac{8.314}{16.04}\frac{\text{kJ}}{\text{kg} \cdot \text{K}}\right)(298 \text{ K})\ln\frac{5}{1.013} + 51{,}393.4\frac{\text{kJ}}{\text{kg}} \\ &= 246.6 + 51{,}393.4 \\ &= 51{,}640 \text{ kJ/kg} \end{aligned}$$

Here the specific exergy is determined predominately by the chemical component.

TABLE 1.12 Standard Molar Chemical Exergy, e^{CH} (kJ/kmol), of Various Substances at 298 K and p_0

Substance	Formula	Model I[a]	Model II[b]
Nitrogen	$N_2(g)$	640	720
Oxygen	$O_2(g)$	3,950	3,970
Carbon dioxide	$CO_2(g)$	14,175	19,870
Water	$H_2O(g)$	8,635	9,500
	$H_2O(l)$	45	900
Carbon (graphite)	C(s)	404,590	410,260
Hydrogen	$H_2(g)$	235,250	236,100
Sulfur	S(s)	598,160	609,600
Carbon monoxide	CO(g)	269,410	275,100
Sulfur dioxide	$SO_2(g)$	301,940	313,400
Nitrogen monoxide	NO(g)	88,850	88,900
Nitrogen dioxide	$NO_2(g)$	55,565	55,600
Hydrogen sulfide	$H_2S(g)$	799,890	812,000
Ammonia	$NH_3(g)$	336,685	337,900
Methane	$CH_4(g)$	824,350	831,650
Ethane	$C_2H_6(g)$	1,482,035	1,495,840
Methanol	$CH_3OH(g)$	715,070	722,300
	$CH_3OH(l)$	710,745	718,000
Ethyl alcohol	$C_2H_5OH(g)$	1,348,330	1,363,900
	$C_2H_5OH(l)$	1,342,085	1,357,700

[a] Ahrendts, J. 1977. Die Exergie Chemisch Reaktionsfähiger Systeme, *VDI-Forschungsheft*. VDI-Verlag, Dusseldorf, 579. Also see Reference States, *Energy — The International Journal*, 5: 667–677, 1980. In Model I, p_0 = 1.019 atm. This model attempts to impose a criterion that the reference environment be in equilibrium. The reference substances are determined assuming restricted chemical equilibrium for nitric acid and nitrates and unrestricted thermodynamic equilibrium for all other chemical components of the atmosphere, the oceans, and a portion of the Earth's crust. The chemical composition of the gas phase of this model approximates the composition of the natural atmosphere.

[b] Szargut, J., Morris, D. R., and Steward, F. R. 1988. *Energy Analysis of Thermal, Chemical, and Metallurgical Processes*. Hemisphere, New York. In Model II, p_0 = 1.0 atm. In developing this model a reference substance is selected for each chemical element from among substances that contain the element being considered and that are abundantly present in the natural environment, even though the substances are not in completely mutual stable equilibrium. An underlying rationale for this approach is that substances found abundantly in nature have little economic value. On an overall basis, the chemical composition of the exergy reference environment of Model II is closer than Model I to the composition of the natural environment, but the equilibrium criterion is not always satisfied.

The small difference between p_0 = 1.013 bar and the value of p_0 for model I has been ignored.

Exergetic Efficiency

The exergetic efficiency (second law efficiency, effectiveness, or rational efficiency) provides a true measure of the performance of a system from the thermodynamic viewpoint. To define the exergetic efficiency both a *product* and a *fuel* for the system being analyzed are identified. The product represents the desired result of the system (power, steam, some combination of power and steam, etc.). Accordingly, the definition of the product must be consistent with the purpose of purchasing and using the system.

The fuel represents the resources expended to generate the product and is not necessarily restricted to being an actual fuel such as a natural gas, oil, or coal. Both the product and the fuel are expressed in terms of exergy.

For a control volume at steady state whose exergy rate balance reads

$$\dot{E}_F = \dot{E}_P + \dot{E}_D + \dot{E}_L$$

the exergetic efficiency is

$$\varepsilon = \frac{\dot{E}_P}{\dot{E}_F} = 1 - \frac{\dot{E}_D + \dot{E}_L}{\dot{E}_F} \tag{1.91}$$

where the rates at which the fuel is supplied and the product is generated are $\dot{E}_F$ and $\dot{E}_P$, respectively. $\dot{E}_D$ and $\dot{E}_L$ denote the rates of exergy destruction and exergy loss, respectively. Exergy is destroyed by irreversibilities within the control volume, and exergy is lost from the control volume via stray heat transfer, material streams vented to the surroundings, and so on. The exergetic efficiency shows the percentage of the fuel exergy provided to a control volume that is found in the product exergy. Moreover, the difference between 100% and the value of the exergetic efficiency, expressed as a percent, is the percentage of the fuel exergy wasted in this control volume as exergy destruction and exergy loss.

To apply Equation 1.91, decisions are required concerning what are considered as the fuel and the product. Table 1.13 provides illustrations for several common components. Similar considerations are used to write exergetic efficiencies for systems consisting of several such components, as, for example, a power plant.

Exergetic efficiencies can be used to assess the thermodynamic performance of a component, plant, or industry relative to the performance of *similar* components, plants, or industries. By this means the performance of a gas turbine, for instance, can be gauged relative to the typical present-day performance level of gas turbines. A comparison of exergetic efficiencies for *dissimilar* devices — gas turbines and heat exchangers, for example — is generally not significant, however.

The exergetic efficiency is generally more meaningful, objective, and useful than other efficiencies based on the first or second law of thermodynamics, including the thermal efficiency of a power plant, the isentropic efficiency of a compressor or turbine, and the effectiveness of a heat exchanger. The thermal efficiency of a cogeneration system, for instance, is misleading because it treats both work and heat transfer as having equal thermodynamic value. The isentropic turbine efficiency (Equation 1.95a) does not consider that the working fluid at the outlet of the turbine has a higher temperature (and consequently a higher exergy that may be used in the next component) in the actual process than in the isentropic process. The heat exchanger effectiveness fails, for example, to identify the exergy destruction associated with the pressure drops of the heat exchanger working fluids.

Example 14

Evaluate the exergetic efficiency of the turbine in part (a) of Example 1 for T_0 = 298 K.

Solution. The exergetic efficiency from Table 1.13 is

$$\varepsilon = \frac{\dot{W}}{\dot{E}_1 - \dot{E}_2} = \frac{\dot{W}}{\dot{m}(e_1 - e_2)}$$

Using Equations 1.88 and 1.89a, and noting that the chemical exergy at 1 and 2 cancels,

TABLE 1.13 The Exergetic Efficiency for Selected Components at Steady State[a]

	Turbine or Expander	Extraction Turbine	Compressor, Pump, or Fan	Heat Exchanger[b]	Mixing Unit	Gasifier or Combustion Chamber	Boiler
Component	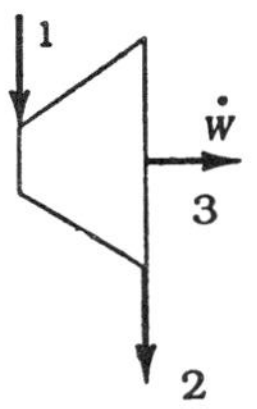 1, 2, 3, $\dot{W}$	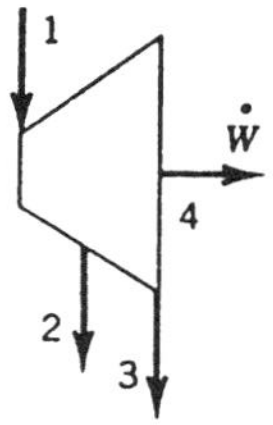 1, 2, 3, 4, $\dot{W}$	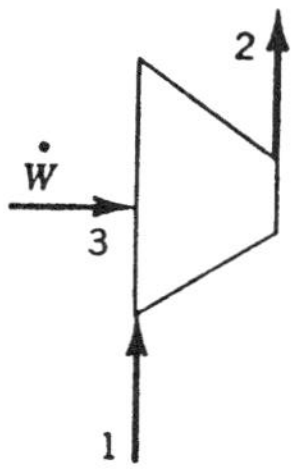 1, 2, 3, $\dot{W}$	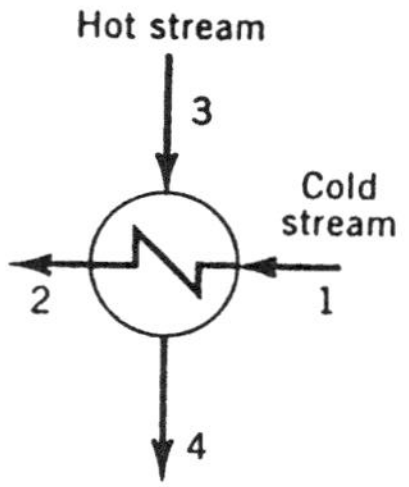 Hot stream 3, 4; Cold stream 1, 2	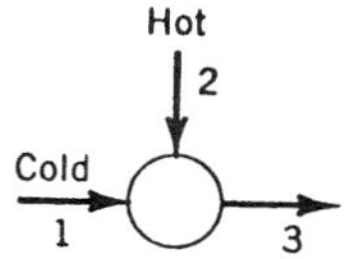 Cold 1, Hot 2, 3	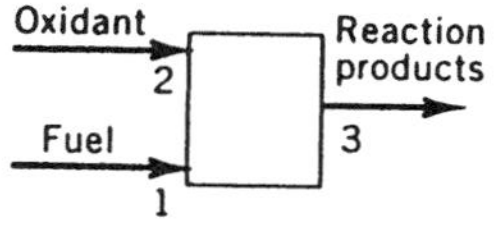 Oxidant 2, Fuel 1, Reaction products 3	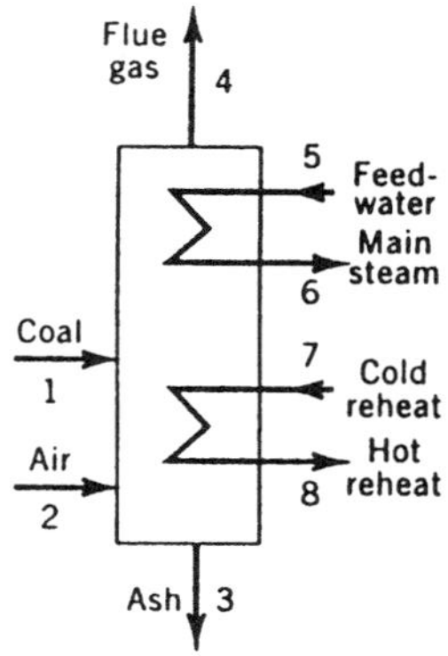 Coal 1, Air 2, Ash 3, Flue gas 4, Feed-water 5, Main steam 6, Cold reheat 7, Hot reheat 8
$\dot{E}_P$	$\dot{W}$	$\dot{W}$	$\dot{E}_2 - \dot{E}_1$	$\dot{E}_2 - \dot{E}_1$	$\dot{E}_3$	$\dot{E}_3 - \dot{E}_2$	$(\dot{E}_6 - \dot{E}_5) + (\dot{E}_8 - \dot{E}_7)$
$\dot{E}_F$	$\dot{E}_1 - \dot{E}_2$	$\dot{E}_1 - \dot{E}_2 - \dot{E}_3$	$\dot{W}$	$\dot{E}_3 - \dot{E}_4$	$\dot{E}_1 + \dot{E}_2$	$\dot{E}_1$	$(\dot{E}_1 - \dot{E}_2) + (\dot{E}_3 - \dot{E}_4)$
ε	$\frac{\dot{W}}{\dot{E}_1 - \dot{E}_2}$	$\frac{\dot{W}}{\dot{E}_1 - \dot{E}_2 - \dot{E}_3}$	$\frac{\dot{E}_2 - \dot{E}_1}{\dot{W}}$	$\frac{\dot{E}_2 - \dot{E}_1}{\dot{E}_3 - \dot{E}_4}$	$\frac{\dot{E}_3}{\dot{E}_1 + \dot{E}_2}$	$\frac{\dot{E}_3 - \dot{E}_2}{\dot{E}_1}$	$\frac{(\dot{E}_6 - \dot{E}_5) + (\dot{E}_8 - \dot{E}_7)}{(\dot{E}_1 - \dot{E}_2) - (\dot{E}_3 - \dot{E}_4)}$

[a] For discussion, see Bejan et al. (1996).

[b] This definition assumes that the purpose of the heat exchanger is to heat the cold stream ($T_1 \geq T_0$). If the purpose of the heat exchanger is to provide cooling ($T_3 \geq T_0$), then the following relations should be used: $\dot{E}_P = \dot{E}_4 - \dot{E}_3$ and $\dot{E}_F = \dot{E}_1 - \dot{E}_2$.

$$\varepsilon = \frac{\dot{W}}{\dot{m}\left[\left(h_1 - h_2\right) - T_0\left(s_1 - s_2\right)\right]}$$

Since $\dot{W} = \dot{m}(h_1 - h_2)$,

$$\varepsilon = \frac{\dot{W}}{\dot{W} + \dot{m}T_0\left(s_2 - s_1\right)}$$

Finally, using data from Example 1 and $s_2 = 6.8473$ kJ/kg · K,

$$\varepsilon = \frac{30 \text{ MW}}{30 \text{ MW} + \left(\dfrac{162{,}357}{3600}\dfrac{\text{kg}}{\text{s}}\right)(298 \text{ K})(6.8473 - 6.6022)\left(\dfrac{\text{kJ}}{\text{kg} \cdot \text{K}}\right)\left(\dfrac{1 \text{ MW}}{10^3 \text{ kJ/sec}}\right)}$$

$$= \frac{30 \text{ MW}}{(30 + 3.29)\text{MW}} = 0.9(90\%)$$

Introduction to Exergy Costing

Since exergy measures the true thermodynamic values of the work, heat, and other interactions between the system and its surroundings as well as the effect of irreversibilities within the system, exergy is a rational basis for assigning costs. This aspect of thermoeconomics is called *exergy costing*. An introduction to exergy costing is given in the present discussion. A detailed development of exergy costing is provided in Sections 1.8 to 1.10 together with allied concepts and case studies.

Referring to Figure 1.16 showing a steam turbine-electric generator at steady state, the total cost to produce the electricity and exiting steam equals the cost of the entering steam plus the cost of owning and operating the device. This is expressed by the *cost rate balance* for the turbine-generator:

$$\dot{C}_e + \dot{C}_2 = \dot{C}_1 + \dot{Z} \tag{1.92a}$$

where $\dot{C}_e$ is the cost rate associated with the electricity, $\dot{C}_1$ and $\dot{C}_2$ are the cost rates associated with the entering steam and exiting steam, respectively, and $\dot{Z}$ accounts for the cost rate associated with owning and operating the system, each *annualized* in $ per year.

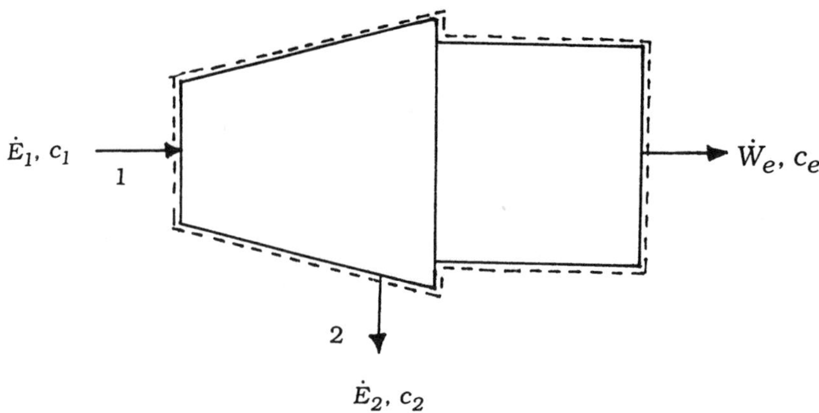

FIGURE 1.16 Steam turbine/electric generator used to discuss exergy costing.

With exergy costing, the cost rates $\dot{C}_1$, $\dot{C}_2$, and $\dot{C}_e$ are evaluated in terms of the associated rate of exergy transfer and a *unit cost.* Equation 1.92a then appears as

$$c_e \dot{W}_e + c_2 \dot{E}_2 = c_1 \dot{E}_1 + \dot{Z} \tag{1.92b}$$

The coefficients c_1, c_2, and c_e in Equation 1.92b denote the *average* cost per unit of exergy for the associated exergy rate. The unit cost c_1 of the entering steam would be obtained from exergy costing applied to the components upstream of the turbine. Assigning the same unit cost to the exiting steam: $c_2 = c_1$ on the basis that the purpose of the turbine-generator is to generate electricity and thus all costs associated with owning and operating the system should be charged to the power, Equation 1.92b becomes

$$c_e \dot{W}_e = c_1 \left(\dot{E}_1 - \dot{E}_2 \right) + \dot{Z} \tag{1.92c}$$

The first term on the right side accounts for the cost of the net exergy used and the second term accounts for cost of the system itself. Introducing the exergetic efficiency from Table 1.13, the unit cost of the electricity is

$$c_e = \frac{c_1}{\varepsilon} + \frac{\dot{Z}}{\dot{W}_e} \tag{1.93}$$

This equation shows, for example, that the unit cost of electricity would increase if the exergetic efficiency were to decrease owing to a deterioration of the turbine with use.

Example 15

A turbine-generator with an exergetic efficiency of 90% develops 7×10^7 kW · hr of electricity annually. The annual cost of owning and operating the system is $\$2.5 \times 10^5$. If the average unit cost of the steam entering the system is \$0.0165 per kW · hr of exergy, evaluate the unit cost of the electricity.

Solution. Substituting values into Equation 1.93,

$$\begin{aligned} c_e &= \frac{\$0.0165/\text{kW} \cdot \text{h}}{0.9} + \frac{\$2.5 \times 10^5/\text{year}}{7 \times 10^7 \text{ kW} \cdot \text{h/year}} \\ &= 0.0183 + 0.0036 = \$0.0219/\text{kW} \cdot \text{h} \end{aligned}$$

1.6 Vapor and Gas Power Cycles

Vapor and gas power systems develop electrical or mechanical power from energy sources of chemical, solar, or nuclear origin. In *vapor* power systems the *working fluid*, normally water, undergoes a phase change from liquid to vapor, and conversely. In *gas* power systems, the working fluid remains a gas throughout, although the composition normally varies owing to the introduction of a fuel and subsequent combustion. The present section introduces vapor and gas power systems.

The processes taking place in power systems are sufficiently complicated that idealizations are typically employed to develop tractable thermodynamic models. The *air standard analysis* of gas power systems considered later in the present section is a noteworthy example. Depending on the degree of idealization, such models may provide only qualitative information about the performance of the corresponding real-world systems. Yet such information is frequently useful in gauging how changes in major operating parameters might affect actual performance. Elementary thermodynamic models can also provide simple settings to assess, at least approximately, the advantages and disadvantages of features proposed to improve thermodynamic performance.

Rankine and Brayton Cycles

In their simplest embodiments vapor power and gas turbine power plants are represented conventionally in terms of four components in series, forming, respectively, the Rankine cycle and the Brayton cycle shown schematically in Table 1.14. The thermodynamically ideal counterparts of these cycles are composed of four internally reversible processes in series: two isentropic processes alternated with two constant pressure processes. Table 1.14 provides property diagrams of the actual and corresponding ideal cycles. Each actual cycle is denoted 1-2-3-4-1; the ideal cycle is 1-2s-3-4s-1. For simplicity, pressure drops through the boiler, condenser, and heat exchangers are not shown. Invoking Equation 1.29 for the ideal cycles, the heat added per unit of mass flowing is represented by the area *under* the isobar from state 2s to state 3: area a-2s-3-b-a. The heat rejected is the area *under* the isobar from state 4s to state 1: area a-1-4s-b-a. Enclosed area 1-2s-3-4s-1 represents the net heat added per unit of mass flowing. For any power cycle, the net heat added equals the net work done.

Expressions for the principal energy transfers shown on the schematics of Table 1.14 are provided by Equations 1 to 4 of the table. They are obtained by reducing Equation 1.27a with the assumptions of negligible heat loss and negligible changes in kinetic and potential energy from the inlet to the outlet of each component. All quantities are positive in the directions of the arrows on the figure. Using these expressions, the thermal efficiency is

$$\eta = \frac{(h_3 - h_4) - (h_2 - h_1)}{h_3 - h_2} \tag{1.94}$$

To obtain the thermal efficiency of the ideal cycle, h_{2s} replaces h_2 and h_{4s} replaces h_4 in Equation 1.94.

Decisions concerning cycle operating conditions normally recognize that the thermal efficiency tends to increase as the average temperature of heat addition increases and/or the temperature of heat rejection decreases. In the Rankine cycle, a high average temperature of heat addition can be achieved by superheating the vapor prior to entering the turbine, and/or by operating at an elevated steam-generator pressure. In the Brayton cycle an increase in the compressor pressure ratio p_2/p_1 tends to increase the average temperature of heat addition. Owing to materials limitations at elevated temperatures and pressures, the state of the working fluid at the turbine inlet must observe practical limits, however. The turbine inlet temperature of the Brayton cycle, for example, is controlled by providing air far in excess of what is required for combustion. In a Rankine cycle using water as the working fluid, a low temperature of heat rejection is typically achieved by operating the condenser at a pressure below 1 atm. To reduce

TABLE 1.14 Rankine and Brayton Cycles

Rankine Cycle	**Brayton Cycle**

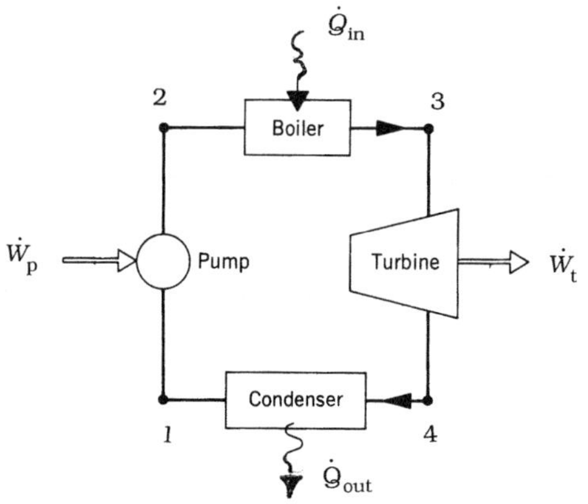

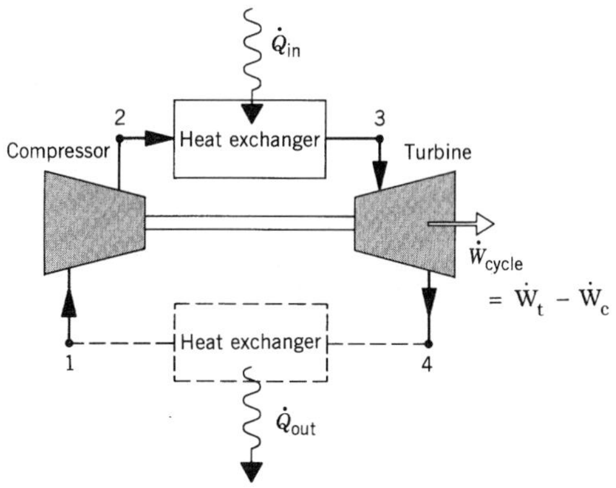

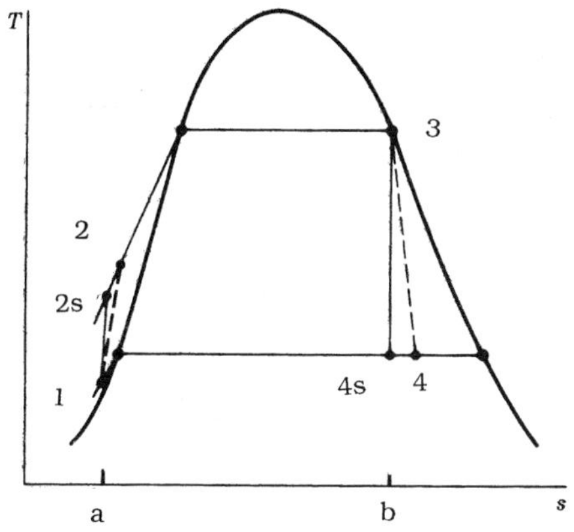

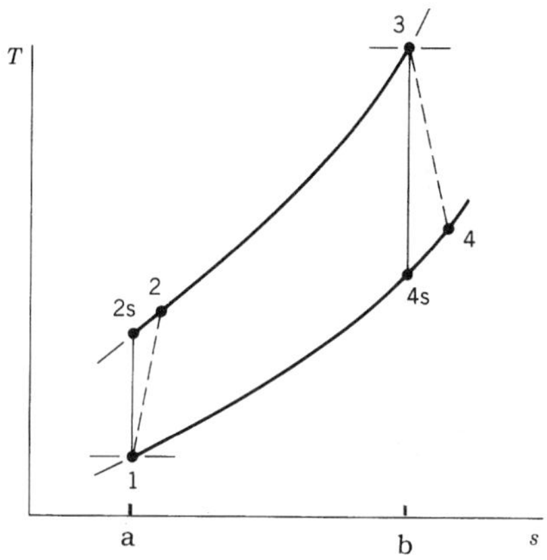

$$\left.\begin{matrix}\dot{W}_p \\ \dot{W}_c\end{matrix}\right\} = \dot{m}\left(h_2 - h_1\right) \quad (>0) \tag{1}$$

$$\dot{Q}_{in} = \dot{m}\left(h_3 - h_2\right) \quad (>0) \tag{2}$$

$$\dot{W}_t = \dot{m}\left(h_3 - h_4\right) \quad (>0) \tag{3}$$

$$\dot{Q}_{out} = \dot{m}\left(h_1 - h_4\right) \quad (>0) \tag{4}$$

erosion and wear by liquid droplets on the blades of the Rankine cycle steam turbine, at least 90% quality should be maintained at the turbine exit: $x_4 > 0.9$.

The *back work ratio*, bwr, is the ratio of the work required by the pump or compressor to the work developed by the turbine:

$$bwr = \frac{h_2 - h_1}{h_3 - h_4} \tag{1.95}$$

As a relatively high specific volume vapor expands through the turbine of the Rankine cycle and a much lower specific volume liquid is pumped, the back work ratio is characteristically quite low in vapor power plants — in many cases on the order of 1 to 2%. In the Brayton cycle, however, both the turbine and compressor handle a relatively high specific volume gas, and the back ratio is much larger, typically 40% or more.

The effect of friction and other irreversibilities for flow-through turbines, compressors, and pumps is commonly accounted for by an appropriate *isentropic efficiency.* The isentropic turbine efficiency is

$$\eta_t = \frac{h_3 - h_4}{h_3 - h_{4s}} \tag{1.95a}$$

The isentropic compressor efficiency is

$$\eta_c = \frac{h_{2s} - h_1}{h_2 - h_1} \tag{1.95b}$$

In the isentropic pump efficiency, η_p, which takes the same form as Equation 1.95b, the numerator is frequently approximated via Equation 1.30c *as* $h_{2s} - h_1 \approx v_1 \Delta p$, where Δp is the pressure rise across the pump.

Simple gas turbine power plants differ from the Brayton cycle model in significant respects. In actual operation, excess air is continuously drawn into the compressor, where it is compressed to a higher pressure; then fuel is introduced and combustion occurs; finally the mixture of combustion products and air expands through the turbine and is subsequently discharged to the surroundings. Accordingly, the low-temperature heat exchanger shown by a dashed line in the Brayton cycle schematic of Table 1.14 is not an actual component, but included only to account formally for the cooling in the surroundings of the hot gas discharged from the turbine.

Another frequently employed idealization used with gas turbine power plants is that of an *air-standard analysis.* An air-standard analysis involves two major assumptions: (1) as shown by the Brayton cycle schematic of Table 1.14, the temperature rise that would be brought about by combustion is effected instead by a heat transfer from an external source; (2) the working fluid throughout the cycle is air, which behaves as an ideal gas. In a *cold* air-standard analysis the specific heat ratio k for air is taken as constant. Equations 1 to 6 of Table 1.7 together with data from Table A.8 apply generally to air-standard analyses. Equations 1′ to 6′ of Table 1.7 apply to cold air-standard analyses, as does the following expression for the turbine power obtained from Table 1.1 (Equation 27c″):

$$\dot{W}_t = \dot{m}\frac{kRT_3}{k-1}\left[1 - \left(p_4/p_3\right)^{(k-1)/k}\right] \tag{1.96}$$

(Equation 1.96 also corresponds to Equation 5′ of Table 1.8 when $n = k$.) An expression similar in form can be written for the power required by the compressor.

For the simple Rankine and Brayton cycles of Table 1.14 the results of sample calculations are provided in Table 1.15. The Brayton cycle calculations are on an air-standard analysis basis.

Otto, Diesel, and Dual Cycles

Although most gas turbines are also internal combustion engines, the name is usually reserved to *reciprocating* internal combustion engines of the type commonly used in automobiles, trucks, and buses. Two principal types of reciprocating internal combustion engines are the *spark-ignition* engine and the *compression-ignition* engine. In a spark-ignition engine a mixture of fuel and air is ignited by a spark

TABLE 1.15 Sample Calculations for the Rankine and Brayton Cycles of Table 1.14

Rankine Cycle

Given data: $p_1 = p_4 = 8$ kPa (saturated liquid at 1)
$T_3 = 480°C$ (superheated vapor at 3)
$p_2 = p_3 = 8$ MPa
$\dot{W}_{net} = 100$ MW
Ideal cycle: $\eta_t = \eta_p = 100\%$
Actual cycle: $\eta_t = 85\%$, $\eta_p = 70\%$

Parameter	Ideal Cycle	Actual Cycle
x_4	0.794	0.873
h_2 (kJ/kg)	181.9[a]	185.4
$\dot{m}$ (kg/h)	2.86×10^5	3.38×10^5
η (%)	39.7	33.6
$\dot{Q}_{out}$ (MW)	151.9	197.6
$\dot{E}_{q,out}$ (MW)[b]	8.2	10.7

[a] $h_{2s} \approx h_1 + v_1\Delta p$
[b] Equation 1.86 with $T_0 = 298$ K, $T_j = T_{sat}$ (8 kPa) = 315 K

Brayton Cycle

Given data: $p_1 = p_4 = 1$ bar
$p_2 = p_3 = 10$ bar
$T_3 = 1400$ K
$\eta_t = \eta_c = 100\%$

Parameter	Air-Standard Analysis	Cold Air-Standard Analysis $k = 1.4$
T_2 (K)	574.1	579.2
T_4 (K)	787.7	725.1
$\dot{W}_{net}/\dot{m}$ (kJ/kg)	427.2	397.5
η (%)	45.7	48.2
bwr	0.396	0.414

plug. In a compression ignition engine air is compressed to a high-enough pressure and temperature that combustion occurs spontaneously when fuel is injected.

In a *four-stroke* internal combustion engine, a piston executes four distinct strokes within a cylinder for every two revolutions of the crankshaft. Figure 1.17 gives *a* pressure-displacement diagram as it might be displayed electronically. With the intake valve open, the piston makes an *intake stroke* to draw a fresh charge into the cylinder. Next, with both valves closed, the piston undergoes a *compression stroke* raising the temperature and pressure of the charge. A combustion process is then initiated, resulting in a high-pressure, high-temperature gas mixture. A *power* stroke follows the compression stroke, during which the gas mixture expands and work is done on the piston. The piston then executes an *exhaust stroke* in which the burned gases are purged from the cylinder through the open exhaust valve. Smaller engines operate on *two-stroke* cycles. In two-stroke engines, the intake, compression, expansion, and

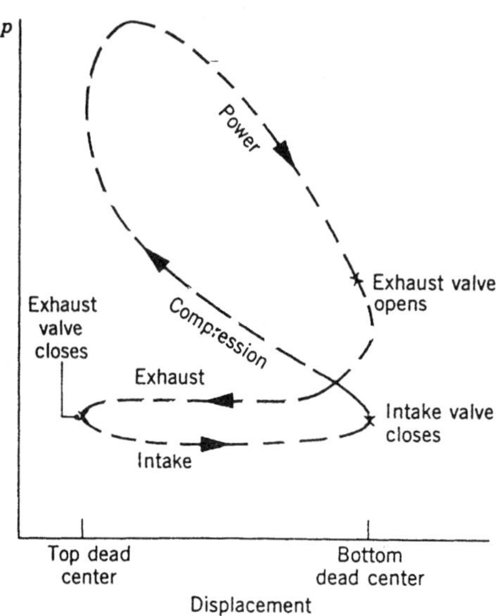

FIGURE 1.17 Pressure-displacement diagram for a reciprocating internal combustion engine.

exhaust operations are accomplished in one revolution of the crankshaft. Although internal combustion engines undergo *mechanical* cycles, the cylinder contents do not execute a *thermodynamic* cycle, since matter is introduced with one composition and is later discharged at a different composition.

A parameter used to describe the performance of reciprocating piston engines is the *mean effective pressure*, or mep. The mean effective pressure is the theoretical constant pressure that, if it acted on the piston during the power stroke, would produce the same *net* work as actually developed in one cycle. That is,

$$\text{mep} = \frac{\text{net work for one cycle}}{\text{displacement volume}}$$

where the displacement volume is the volume swept out by the piston as it moves from the top dead center to the bottom dead center. For two engines of equal displacement volume, the one with a higher mean effective pressure would produce the greater net work and, if the engines run at the same speed, greater power.

Detailed studies of the performance of reciprocating internal combustion engines may take into account many features, including the combustion process occurring within the cylinder and the effects of irreversibilities associated with friction and with pressure and temperature gradients. Heat transfer between the gases in the cylinder and the cylinder walls and the work required to charge the cylinder and exhaust the products of combustion also might be considered. Owing to these complexities, accurate modeling of reciprocating internal combustion engines normally involves computer simulation.

To conduct *elementary* thermodynamic analyses of internal combustion engines, considerable simplification is required. A procedure that allows engines to be studied *qualitatively* is to employ an *air-standard analysis* having the following elements: (1) a fixed amount of air modeled as an ideal gas is the system; (2) the combustion process is replaced by a heat transfer from an external source and generally represented in terms of elementary thermodynamic processes; (3) there are no exhaust and intake processes as in an actual engine: the cycle is completed by a constant-volume heat rejection process; (4) all processes are internally reversible.

The processes employed in air-standard analyses of internal combustion engines are selected to represent the events taking place within the engine simply and mimic the appearance of observed pressure-displacement diagrams. In addition to the constant volume heat rejection noted previously, the compression stroke and at least a portion of the power stroke are conventionally taken as isentropic. The heat addition is normally considered to occur at constant volume, at constant pressure, or at constant volume followed by a constant pressure process, yielding, respectively, the Otto, Diesel, and Dual cycles shown in Table 1.16. Referring to Table 1.16, the ratio v_1/v_2 is the *compression ratio, r*. For the Diesel cycle, the ratio v_3/v_2 is the *cutoff ratio*, r_c.

Reducing the closed system energy balance, Equation 1.8, gives the following expressions for heat and work applicable in each case shown in Table 1.16:

$$\frac{W_{12}}{m} = u_1 - u_2 \qquad (< 0)$$

$$\frac{W_{34}}{m} = u_3 - u_4 \qquad (> 0)$$

$$\frac{Q_{41}}{m} = u_1 - u_4 \qquad (< 0)$$

Table 1.16 provides additional expressions for work, heat transfer, and thermal efficiency identified with each case individually. The thermal efficiency, evaluated from Equation 1.9, takes the form

$$\eta = 1 - \frac{|Q_{41}/m|}{Q_A/m}$$

Equations 1 to 6 of Table 1.7 together with data from Table A.8, apply generally to air-standard analyses. In a *cold* air-standard analysis the specific heat ratio k for air is taken as constant. Equations 1′ to 6′ of Table 1.7 apply to cold air-standard analyses, as does Equation 4′ of Table 1.8, with $n = k$ for the isentropic processes of these cycles.

As all processes are internally reversible, areas on the *p-v* and *T-s* diagrams of Table 1.16 can be interpreted, respectively, as work and heat transfer. Invoking Equation 1.10 and referring to the *p-v* diagrams, the areas under process 3-4 of the Otto cycle, process 2-3-4 of the Diesel cycle, and process x-3-4 of the Dual cycle represent the work done by the gas during the power stroke, per unit of mass. For each cycle, the area under the isentropic process 1-2 represents the work done on the gas during the compression stroke, per unit of mass. The enclosed area of each cycle represents the net work done per unit mass. With Equation 1.15 and referring to the *T-s* diagrams, the areas under process 2-3 of the Otto and Diesel cycles and under process 2-x-3 of the Dual cycle represent the heat added per unit of mass. For each cycle, the area under the process 4-1 represent the heat rejected per unit of mass. The enclosed area of each cycle represents the net heat added, which equals the net work done, each per unit of mass.

Carnot, Ericsson, and Stirling Cycles

Three thermodynamic cycles that exhibit the Carnot efficiency (Equation 1.12) are the Carnot, Ericsson, and Stirling cycles shown in Figure 1.18. Each case represents a reversible power cycle in which heat is added from an external source at a constant temperature T_H (process 2-3) and rejected to the surroundings at a constant temperature T_C (process 4-1). Carnot cycles can be configured both as vapor power cycles and as cycles executed by a gas in a piston-cylinder assembly (see, e.g., Moran and Shapiro, 2000). Carnot cycles also can be executed in systems where a capacitor is charged and discharged, a paramagnetic substance is magnetized and demagnetized, and in other ways. Regardless of the type of device and the working substance used, the Carnot cycle always has the same four internally reversible processes in series: two isentropic processes alternated with two isothermal processes.

TABLE 1.16 Otto, Diesel, and Dual Cycles

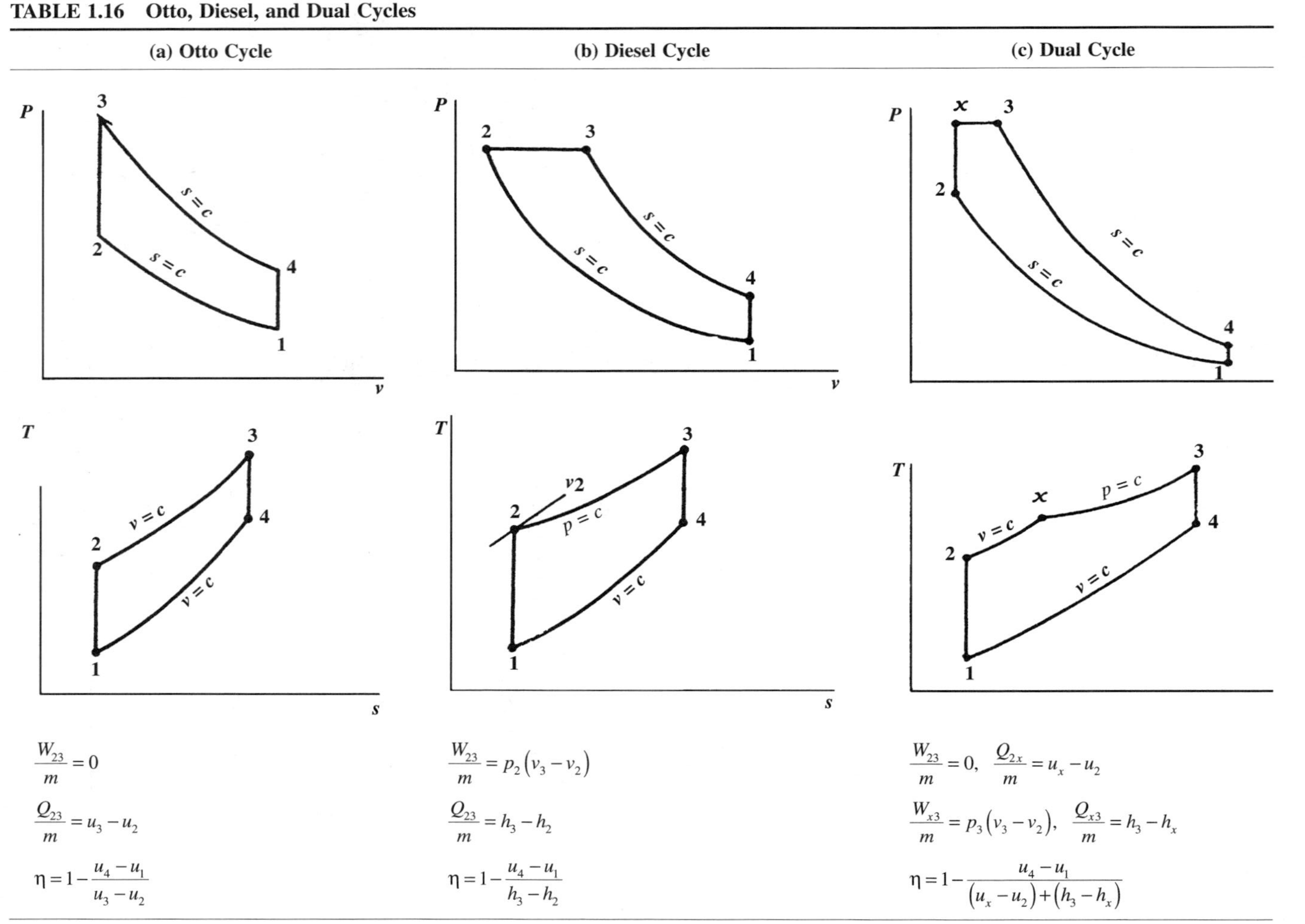

(a) Otto Cycle	(b) Diesel Cycle	(c) Dual Cycle
$\frac{W_{23}}{m} = 0$	$\frac{W_{23}}{m} = p_2(v_3 - v_2)$	$\frac{W_{23}}{m} = 0, \quad \frac{Q_{2x}}{m} = u_x - u_2$
$\frac{Q_{23}}{m} = u_3 - u_2$	$\frac{Q_{23}}{m} = h_3 - h_2$	$\frac{W_{x3}}{m} = p_3(v_3 - v_2), \quad \frac{Q_{x3}}{m} = h_3 - h_x$
$\eta = 1 - \frac{u_4 - u_1}{u_3 - u_2}$	$\eta = 1 - \frac{u_4 - u_1}{h_3 - h_2}$	$\eta = 1 - \frac{u_4 - u_1}{(u_x - u_2) + (h_3 - h_x)}$

(A)

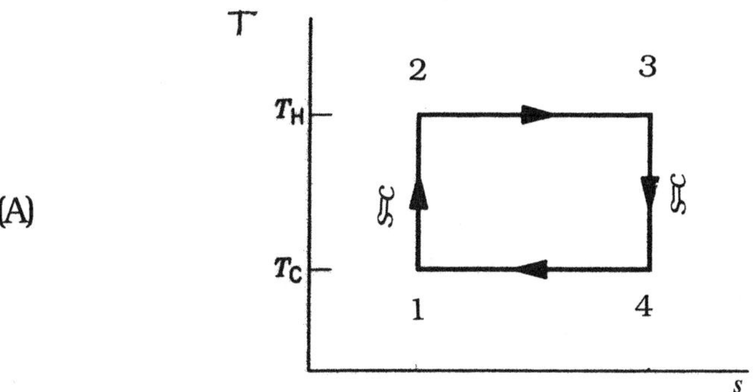

(B)

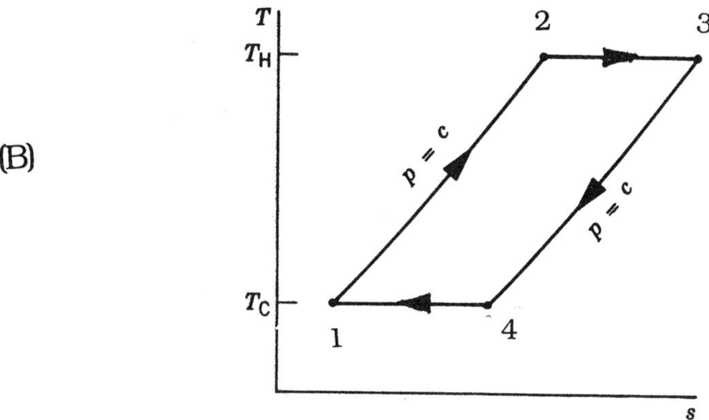

(C)

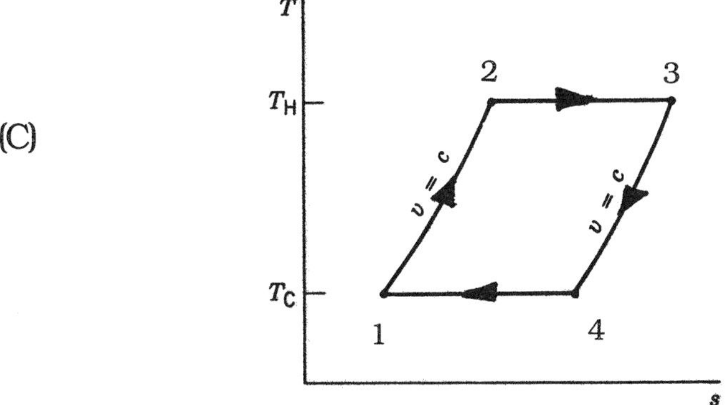

FIGURE 1.18 (A) Carnot, (B) Ericsson, and (C) Stirling cycles.

The Ericsson and Stirling cycles also consist of four internally reversible processes in series: heating from state 1 to state 2 (at constant pressure in the Ericsson cycle and at constant volume in the Stirling cycle), isothermal heating from state 2 to state 3 at temperature T_H, cooling from state 3 to state 4 (at constant pressure in the Ericsson cycle and at constant volume in the Stirling cycle), and isothermal cooling from state 4 to state 1 at temperature T_C. An ideal regenerator allows the heat input required for process 1-2 to be obtained from the heat rejected in process 3-4. Accordingly, as in the Carnot cycle all the heat added externally occurs at T_H and all of the heat rejected to the surroundings occurs at T_C.

The Ericsson and Stirling cycles are principally of theoretical interest as examples of cycles that exhibit the same thermal efficiency as the Carnot cycle: Equation 1.12. However, a practical engine of the piston-cylinder type that operates on a closed regenerative cycle having features in common with the Stirling cycle has been under study in recent years. This engine, known as the *Stirling engine*, offers the opportunity for high efficiency together with reduced emissions from combustion products because the combustion takes place externally and not within the cylinder as in internal combustion engines. In the Stirling engine, energy is transferred to the working fluid from products of combustion, which are kept separate. It is an *external* combustion engine.

1.7 Guidelines for Improving Thermodynamic Effectiveness

Thermal design frequently aims at the most effective system from the cost viewpoint. Still, in the cost optimization process, particularly of complex energy systems, it is often expedient to begin by identifying a design that is nearly optimal thermodynamically; such a design can then be used as a point of departure for cost optimization. Presented in this section are guidelines for improving the use of fuels (natural gas, oil, and coal) by reducing sources of thermodynamic inefficiency in thermal systems. Further discussion is provided by Bejan et al. (1996).

To improve thermodynamic effectiveness it is necessary to deal directly with inefficiencies related to exergy destruction and exergy loss. The primary contributors to exergy destruction are chemical reaction, heat transfer, mixing, and friction, including unrestrained expansions of gases and liquids. To deal with them effectively, the principal sources of inefficiency not only should be understood qualitatively, but also determined quantitatively, at least approximately. Design changes to improve effectiveness must be done judiciously, however, for the cost associated with different sources of inefficiency can be different. For example, the unit cost of the electrical or mechanical power required to provide for the exergy destroyed owing to a pressure drop is generally higher than the unit cost of the fuel required for the exergy destruction caused by combustion or heat transfer.

Since chemical reaction is a significant source of thermodynamic inefficiency, it is generally good practice to minimize the use of combustion. In many applications the use of combustion equipment such as boilers is unavoidable, however. In these cases a significant reduction in the combustion irreversibility by conventional means simply cannot be expected, for the major part of the exergy destruction introduced by combustion is an inevitable consequence of incorporating such equipment. Still, the exergy destruction in practical combustion systems can be reduced by minimizing the use of excess air and by preheating the reactants. In most cases only a small part of the exergy destruction in a combustion chamber can be avoided by these means. Consequently, after considering such options for reducing the exergy destruction related to combustion, efforts to improve thermodynamic performance should focus on components of the overall system that are more amenable to betterment by cost-effective conventional measures. In other words, *some exergy destructions and energy losses can be avoided, others cannot. Efforts should be centered on those that can be avoided.*

Nonidealities associated with heat transfer also typically contribute heavily to inefficiency. Accordingly, unnecessary or cost-ineffective heat transfer must be avoided. Additional guidelines follow:

- The higher the temperature T at which a heat transfer occurs in cases where $T > T_0$, where T_0 denotes the temperature of the environment (Section 1.5), the more valuable the heat transfer and, consequently, the greater the need to avoid heat transfer to the ambient, to cooling water, or to a refrigerated stream. Heat transfer across T_0 should be avoided.
- The lower the temperature T at which a heat transfer occurs in cases where $T < T_0$, the more valuable the heat transfer and, consequently, the greater the need to avoid direct heat transfer with the ambient or a heated stream.
- Since exergy destruction associated with heat transfer between streams varies inversely with the temperature level, the lower the temperature level, the greater the need to minimize the stream-to-stream temperature difference.
- Avoid the use of intermediate heat transfer fluids when exchanging energy by heat transfer between two streams

Although irreversibilities related to friction, unrestrained expansion, and mixing are often secondary in importance to those of combustion and heat transfer, they should not be overlooked, and the following guidelines apply:

- Relatively more attention should be paid to the design of the lower temperature stages of turbines and compressors (the last stages of turbines and the first stages of compressors) than to the remaining stages of these devices.

- For turbines, compressors, and motors, consider the most thermodynamically efficient options.
- Minimize the use of throttling; check whether power recovery expanders are a cost-effective alternative for pressure reduction.
- Avoid processes using excessively large thermodynamic driving forces (differences in temperature, pressure, and chemical composition). In particular, minimize the mixing of streams differing significantly in temperature, pressure, or chemical composition.
- The greater the mass rate of flow, the greater the need to use the exergy of the stream effectively.
- The lower the temperature level, the greater the need to minimize friction.

Flowsheeting or *process simulation* software can assist efforts aimed at improving thermodynamic effectiveness by allowing engineers to readily model the behavior of an overall system, or system components, under specified conditions and do the required thermal analysis, sizing, costing, and optimization. Many of the more widely used flowsheeting programs: ASPEN PLUS, PROCESS, and CHEMCAD are of the *sequential-modular* type. SPEEDUP is a popular program of the *equation-solver* type. Since process simulation is a rapidly evolving field, vendors should be contacted for up-to-date information concerning the features of flowsheeting software, including optimization capabilities (if any). As background for further investigation of suitable software, see Biegler (1989) for a survey of the capabilities of 15 software products.

1.8 Exergoeconomics

Exergoeconomics is an exergy-aided cost-reduction method that combines exergy and cost analyses to provide the designer or operator of an energy conversion plant with information not available through conventional energy, exergy, or cost analyses. Exergoeconomics is also a powerful tool for understanding the interconnections between thermodynamics and economics and, thus, the behavior of an energy conversion plant from the cost viewpoint. For a more detailed presentation of exergoeconomics, readers may refer to the following references: Bejan et al. (1996), Tsatsaronis (1993), and Tsatsaronis et al. (1984, 1985, 1986). Sections 1.2 through 1.5 have outlined the principles for conducting detailed thermodynamic evaluations of thermal systems. In particular, techniques have been developed for evaluating the thermodynamic inefficiencies of these systems: exergy destructions and exergy losses. However, we often need to know the cost of such inefficiencies. Knowledge of these costs is very useful for improving the cost-effectiveness of the system — that is, for reducing the costs of the final products produced by the system.

In addition, if a system has more than one product, as for example the net power and saturated vapor of the cogeneration system shown in Figure 1.19, we would want to know the production cost for each product. This is a common problem in chemical plants where electrical power, chilled water, compressed air, and steam at various pressure levels are generated in one department and sold to another. The plant operator wants to know the true cost at which each of the utilities is generated; these costs are then charged to the appropriate final products according to the type and amount of each utility used to generate a final product. In the design of a thermal system, such cost allocation assists in pinpointing cost-ineffective processes and operations and in identifying technical options that might improve the cost-effectiveness of the system.

Accordingly, the objective of an exergoeconomic analysis might be to (1) calculate separately the costs of each product generated by a system having more than one product, (2) understand the cost formation process and the flow of costs in the system, (3) optimize specific variables in a single component, or (4) minimize the costs associated with the overall system.

An exergoeconomic analysis must be preceded by an exergy analysis (see Section 1.5) and an economic analysis (see Bejan et al. [1996] and Section 1.10). Because of the variation of costs from year to year, when we evaluate the design of a thermal system from the cost viewpoint we must use the cost levelization approach. Therefore, the cost values used throughout the following discussion are levelized costs. For conciseness, the term levelized is omitted, however.

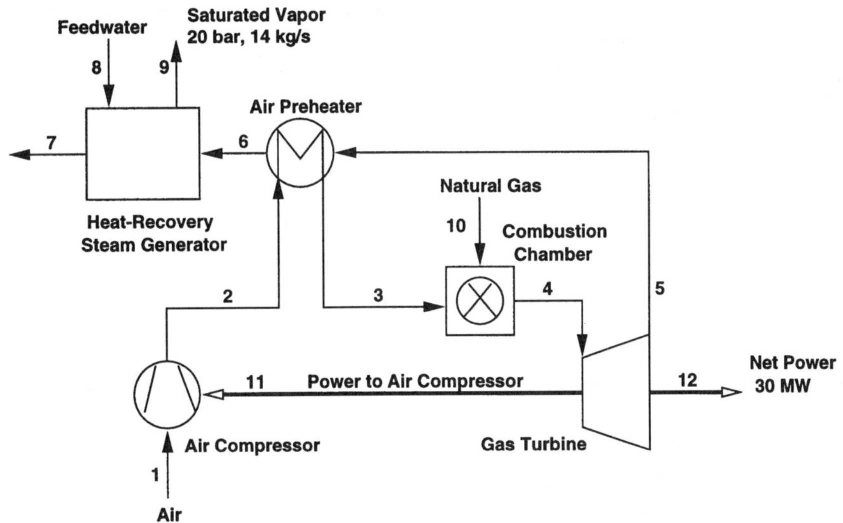

FIGURE 1.19 Cogeneration system.

Cost accounting in exergoeconomics calls for the use of cost balances. Cost balances can be formulated for an overall system (subscript tot) operating at steady state:

$$\dot{C}_{P,tot} = \dot{C}_{F,tot} + \dot{Z}_{tot}^{CI} + \dot{Z}_{tot}^{OM} \tag{1.97}$$

and for each component (see Equations 1.102 and 1.103). Equation 1.97 indicates that the cost rate associated with the product of the system ($\dot{C}_P$) equals the total rate of expenditures made to generate the product, namely, the fuel cost rate ($\dot{C}_F$) and the cost rates associated with capital investment ($\dot{Z}^{CI}$) and operating and maintenance (O&M) ($\dot{Z}^{OM}$). Here, and throughout Sections 1.8, 1.9, and 1.10, the terms *fuel* and *product* are used in the sense introduced in Section 1.5. When referring to a single stream associated with a fuel or product, the expression *fuel stream* or *product stream* is used. The rates $\dot{Z}^{CI}$ and $\dot{Z}^{OM}$ are calculated by dividing the annual contribution of capital investment and the annual operating and maintenance costs, respectively, by the number of time units (usually hours or seconds) of system operation per year. The sum of these two variables is denoted by $\dot{Z}$:

$$\dot{Z} = \dot{Z}^{CI} + \dot{Z}^{OM} \tag{1.98}$$

In this section we discuss the basic elements of exergoeconomics, which include exergy costing, cost balances, and means for costing various exergy transfers. The exergoeconomic variables defined below are used in the evaluation and optimization of the design and operation of thermal systems.

Exergy Costing

For a system operating at steady state there may be a number of entering and exiting streams as well as both heat and work interactions with the surroundings. Associated with these transfers of matter and energy are exergy transfers into and out of the system and exergy destructions caused by the irreversibilities within the system. Since exergy measures the *true* thermodynamic value of such effects, and costs should only be assigned to commodities of value, it is meaningful to use exergy as a basis for assigning costs in thermal systems. Indeed, exergoeconomics rests on the notion that exergy is the *only* rational basis for assigning costs to the interactions a thermal system experiences with its surroundings and to the sources of inefficiencies within it. We refer to this approach as *exergy costing.*

In exergy costing, a cost rate is associated with each exergy transfer. Thus, for entering and exiting streams of matter with exergy transfers $\dot{E}_i$ and $\dot{E}_e$, respectively, power $\dot{W}$, and exergy transfer associated with heat transfer, $\dot{E}_q$, we write, respectively,

$$\dot{C}_i = c_i \dot{E}_i, \quad \dot{C}_e = c_e \dot{E}_e \tag{1.99}$$

$$\dot{C}_w = c_w \dot{W}, \quad \dot{C}_q = c_q \dot{E}_q \tag{1.100}$$

where c_i, c_e, c_w, and c_q denote *average* costs per unit of exergy, for example, in dollars per gigajoule ($/GJ).

Exergy costing does not necessarily imply that costs associated with streams of matter are related *only* to the exergy rate of each respective stream. Nonexergy-related costs also can affect the total cost rate associated with material streams. Examples include the cost rates associated with a treated water stream at the outlet of a water treatment unit, an oxygen or nitrogen stream at the outlet of an air separation unit, a limestone stream supplied to a gasifier or fluidized-bed reactor, iron feedstock supplied to a metallurgical process, and an inorganic chemical fed to a chemical reactor. Accordingly, when significant nonexergy-related costs occur, the total cost rate associated with the material stream *j*, denoted by $\dot{C}_j^{\mathrm{TOT}}$, is given by

$$\dot{C}_j^{\mathrm{TOT}} = \dot{C}_{\mathrm{j}} + \dot{C}_{\mathrm{j}}^{\mathrm{NE}} \tag{1.101}$$

where $\dot{C}_j$ is the cost rate directly related to the exergy of stream *j* (e.g., Equation 1.99) and $\dot{C}_j^{\mathrm{NE}}$ is the cost rate due to nonexergetic effects. The term $\dot{C}_j^{\mathrm{NE}}$ represents a convenient way for charging nonexergy-related costs from one component to other components that should bear such costs. More details about $\dot{C}_j^{\mathrm{NE}}$ are given in Bejan et al. (1996) and Tsatsaronis et al. (1986).

Cost Balance

Exergy costing usually involves cost balances formulated for each component separately. A cost balance applied to the *k*th system component indicates that the sum of cost rates associated with all exiting exergy transfers equals the sum of cost rates of all entering exergy transfers plus the appropriate charges due to capital investment ($\dot{Z}_{\mathrm{k}}^{\mathrm{CI}}$) and operating and maintenance expenses ($\dot{Z}_{\mathrm{k}}^{\mathrm{OM}}$). The sum of the last two terms is denoted by ($\dot{Z}_{\mathrm{k}}$). For example, for a component receiving a heat transfer (subscript q) and generating power (subscript w), we write

$$\sum_e \dot{C}_{e,k} + \dot{C}_{w,k} = \dot{C}_{q,k} + \sum_i \dot{C}_{i,k} + \dot{Z}_k \tag{1.102}$$

This equation simply states that the total cost of the exiting exergy transfers equals the total expenditure to obtain them: the cost of the entering exergy streams plus the capital and other costs. When a component receives power (as in a compressor or a pump) the term $\dot{C}_{w,k}$ would move with its positive sign to the right side of this expression. The term $\dot{C}_{q,k}$ would appear with its positive sign on the left side if there is a heat transfer *from* the component. Cost balances are generally written so that all terms are positive.

Introducing the cost rate expressions of Equations 1.99 and 1.100, Equation 1.102 becomes

$$\sum_e \left(c_e \dot{E}_e\right)_k + c_{w,k}\dot{W}_k = c_{q,k}\dot{E}_{q,k} + \sum_i \left(c_i \dot{E}_i\right)_k + \dot{Z}_k \tag{1.103}$$

The exergy rates exiting and entering the kth component denoted by $\dot{E}_{e,k}$, $\dot{W}_k$, $\dot{E}_{q,k}$, and $\dot{E}_{i,k}$ are calculated in an exergy analysis conducted at a previous stage. The term $\dot{Z}_k$ is obtained by first calculating the capital investment and O&M costs associated with the kth component and then computing the levelized values of these costs per unit of time (year, hour, or second) of system operation (see, for example, Bejan et al. [1996] and Tsatsaronis et al. [1984, 1986]).

The variables in Equation 1.103 are the levelized costs per unit of exergy for the exergy transfers associated with the kth component: $c_{e,k}$, $c_{w,k}$, $c_{q,k}$, and $c_{i,k}$. In analyzing a component, we may assume that the costs per exergy unit are known for all entering streams. These costs are known from the components they exit or, if a stream enters the overall system consisting of all components under consideration, from the purchase cost of this stream. Consequently, the unknown variables to be calculated with the aid of the cost balance for the kth component are the costs per exergy unit of the exiting streams: $c_{e,k}$ and, if power or useful heat are generated in that component, the cost per unit of exergy associated with the transfer of power $c_{w,k}$ or heat $c_{q,k}$. Some auxiliary equations are usually necessary to calculate these costs, as discussed next.

Auxiliary Costing Equations

Various approaches for formulating the auxiliary equations are suggested in the literature. However, the method recommended for obtaining an objective set of auxiliary equations consistent with the definitions of fuel and product is detailed by Lazzaretto and Tsatsaronis (1996, 1997) and summarized in the following three steps:

Step 1: Exergy Streams

All material and energy streams crossing the boundaries of the component being considered should be first identified, and the exergy values associated with these streams should be calculated.

Step 2: Definition of Fuel and Product

In evaluating the performance of a component, it is generally meaningful and appropriate to operate with exergy differences associated with each material stream between the inlet and outlet of the component. For example, in defining the product of a heat exchanger operating above ambient temperature, we consider only the exergy addition to the cold stream and not the sum of the exergies associated with the material streams at the outlet. Similarly, for defining the fuel of the heat exchanger we consider only the exergy removal from the hot stream and not the sum of the exergies associated with the material streams at the inlet. Exergy differences (exergy additions to or removals from a stream) should be applied to all material streams undergoing a change of physical exergy (Equation 1.89a) and to some material streams undergoing a chemical exergy (Equation 1.90) conversion. This approach has been used in developing Table 1.13 for all cases except the gasifier/combustion chamber, which is considered next.

In many cases involving conversion of chemical exergy (e.g., conversion of chemical exergy of a solid fuel in chemical and thermal exergy through a gasification process), the purpose of owning and operating the component dictates that the chemical exergy at the outlet is considered on the product side and the chemical exergy of the fuel stream at the inlet on the fuel side. Thus, in the definition of the exergetic efficiency of a gasifier or combustion chamber in Table 1.13, the exergy of the "fuel" for the component equals the value of the exergy of the entering fuel stream.

Accordingly, when considering the fuel and product of a component, a decision must be made for each exergy stream with respect to whether an exergy difference or just the exergy values at the inlet or outlet should be included in the definitions of fuel and product. Then, the product consists of all the exergy values to be considered at the outlet plus all the exergy increases between inlet and outlet (i.e., the exergy additions to the respective material streams). Similarly, the fuel consists of all the exergy values to be considered at the inlet plus all the exergy decreases between inlet and outlet (i.e., the exergy removals from the respective material streams).

Step 3: Auxiliary Equations

In general, if there are N_e exergy streams exiting the component being considered, we have N_e unknowns and only one equation, the costs balance (Equation 1.103). Therefore, we need to formulate $N_e - 1$ auxiliary equations. This is accomplished with the aid of the F and P rules presented next:

- The F rule refers to the *removal of exergy* from an exergy stream within the component being considered. The F rule states that the total cost associated with this removal of exergy must be equal to the average cost at which the removed exergy was supplied to the same stream in upstream components. The number of auxiliary equations provided by the F rule is always equal to the number ($N_{e,F}$) of exiting exergy streams that are considered in the definition of the fuel for the component.
- The P rule refers to the *supply of exergy* to an exergy stream within the component being considered and to the costing of streams associated with the product. The P rule states that each exergy unit is supplied to any stream associated with the product at the *same average cost*, c_P. This cost is calculated from the cost balance and the equations obtained by applying the F rule. The number of auxiliary equations provided by the P rule is always equal to $N_{e,P} - 1$, where $N_{e,P}$ is the number of exiting exergy streams that are included in the product definition.

Since the total number of exiting streams (N_e) is equal to the sum ($N_{e,F} + N_{e,P}$), the F and P rules together provide the required $N_e - 1$ auxiliary equations.

General Example

The general application of these steps may be demonstrated with the aid of Figure 1.20.

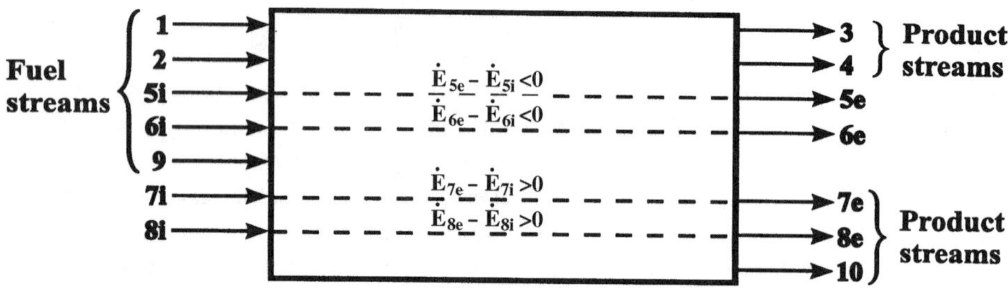

FIGURE 1.20 Schematic of a component in a thermal system to define fuel, product, and auxiliary equations.

Step 1: Referring to Figure 1.20, there are seven exergy streams (1, 2, and 5 through 9) entering the component (subscript i) and seven exergy streams (3 through 8 and 10) exiting the component (subscript e). The streams shown in this figure are selected to cover all situations that might be encountered. In an actual component, however, not all of the streams shown in Figure 1.20 exist.

Step 2: The exergy streams 1 through 4 are associated with the chemical exergy or total exergy of the corresponding material streams. The purpose of owning and operating the component dictates that the entering streams 1 and 2 should be part of the fuel, whereas the exiting streams 3 and 4 should be part of the product. For the exergy streams with the numbers 5, 6, 7, and 8, the purpose of the component dictates the consideration of the respective exergy differences between outlet and inlet. These are positive for streams 7 and 8 and negative for streams 5 and 6. Streams 9 at the inlet and 10 at the outlet represent exergy streams associated with the transport of mechanical, electrical, or thermal energy. We conclude that exergy streams 1, 2, 5, 6, and 9 are associated with the fuel, whereas streams 3, 4, 7, 8, and 10 are associated with the product of the component. Thus, the fuel and product of the component, respectively, are

$$\dot{E}_F = \dot{E}_1 + \dot{E}_2 + \left(\dot{E}_{5i} - \dot{E}_{5e}\right) + \left(\dot{E}_{6i} - \dot{E}_{6e}\right) + \dot{E}_9 \tag{1.104}$$

$$\dot{E}_P = \dot{E}_3 + \dot{E}_4 + \left(\dot{E}_{7e} - \dot{E}_{7i}\right) + \left(\dot{E}_{8e} - \dot{E}_{8i}\right) + \dot{E}_{10} \tag{1.105}$$

Step 3: In accordance with Equation 1.103, the cost balance for the component is

$$\dot{C}_3 + \dot{C}_4 + \dot{C}_{5e} + \dot{C}_{6e} + \dot{C}_{7e} + \dot{C}_{8e} + \dot{C}_{10} = \dot{C}_1 + \dot{C}_2 + \dot{C}_{5i} + \dot{C}_{6i} + \dot{C}_{7i} + \dot{C}_{8i} + \dot{C}_9 + \dot{Z} \tag{1.106}$$

By grouping the terms associated with fuel and product, we obtain

$$\dot{C}_3 + \dot{C}_4 + \left(\dot{C}_{7e} - \dot{C}_{7i}\right) + \left(\dot{C}_{8e} - \dot{C}_{8i}\right) + \dot{C}_{10} = \dot{C}_1 + \dot{C}_2 + \left(\dot{C}_{5e} - \dot{C}_{5i}\right) + \left(\dot{C}_{6e} - \dot{C}_{6i}\right) + \dot{C}_9 + \dot{Z} \tag{1.107}$$

The F rule states that the total cost rate associated with the removal of exergy from stream 5: $(\dot{C}_{5i} - \dot{C}_{5e})$ must be equal to the *average* cost at which the removed exergy: $(\dot{E}_{5i} - \dot{E}_{5e})$ was supplied to that stream in upstream components. Since each exergy unit was supplied to stream 5 at the average cost of c_{5i}, the F rule for stream 5 becomes

$$\dot{C}_{5i} - \dot{C}_{5e} = c_{5i}\left(\dot{E}_{5i} - \dot{E}_{5e}\right) \tag{1.108}$$

From this equation we obtain

$$c_{5e} = c_{5i} \tag{1.109}$$

Similarly, for stream 6

$$c_{6e} = c_{6i} \tag{1.110}$$

The P rule states that each exergy unit is *supplied* to all streams associated with the product at the *same* average cost, c_P. This rule leads to the following equations:

$$c_3 = c_4 = \frac{\dot{C}_{7e} - \dot{C}_{7i}}{\dot{E}_{7e} - \dot{E}_{7i}} = \frac{\dot{C}_{8e} - \dot{C}_{8i}}{\dot{E}_{8e} - \dot{E}_{8i}} = c_{10} = c_P \tag{1.111}$$

Since we assume that the cost rate $\dot{Z}$ and all costs associated with all entering streams are known, we can calculate the unknowns $\dot{C}_3$, $\dot{C}_4$, $\dot{C}_{5e}$, $\dot{C}_{6e}$, $\dot{C}_{7e}$, $\dot{C}_{8e}$, and $\dot{C}_{10}$ by solving the system of Equations 1.107, 1.109, 1.110, and 1.111.

Note that Equation 1.107 may be written as

$$c_P \dot{E}_P = \dot{C}_1 + \dot{C}_2 + \left(\dot{C}_{5i} - \dot{C}_{5e}\right) + \left(\dot{C}_{6i} - \dot{C}_{6e}\right) + \dot{C}_9 + \dot{Z} \tag{1.112}$$

or

$$c_P \dot{E}_P = c_F \dot{E}_F + \dot{Z} \tag{1.113}$$

where $\dot{E}_F$, $\dot{E}_P$, and c_P are given in Equations 1.104, 1.105 and 1.111, respectively. The variable c_F denotes the average cost at which each exergy unit of $\dot{E}_F$ is supplied to the component. Equation 1.112 demonstrates that the average cost (c_P) at which each exergy unit is supplied to all streams associated with the product can be calculated by combining the cost balance (Equation 1.112) with the equations obtained from the F rule (Equations 1.109 and 1.110). After calculating the value of c_P, Equations 1.111 can be used to calculate the cost associated with each exiting stream included in the definition of the product. This suggests that the cost balance (Equation 1.112) should always be used to calculate the value of c_P with the aid of the auxiliary equations obtained from the F rule.

Exergoeconomic Variables and Evaluation

The cost balances together with the auxiliary equations formulated for each plant component form a system of linear equations, the solution of which provides the cost per exergy unit and the cost rates associated with all exergy streams in the system. The remaining exergoeconomic variables are calculated from these cost rates and the known exergy rates using the equations for defining the respective exergoeconomic variables, as discussed next.

The exergoeconomic evaluation is conducted at the system component level using the following variables for the kth component. From the exergy analysis we know the

- Rate of exergy destruction $\dot{E}_{D,k}$:

$$\dot{E}_{D,k} = \dot{E}_{F,k} - \dot{E}_{P,k} - \dot{E}_{L,k} \tag{1.114}$$

- Exergetic efficiency ε_k:

$$\varepsilon_k = \dot{E}_{P,k} \big/ \dot{E}_{F,k} = 1 - \left(\dot{E}_{D,k} + \dot{E}_{L,k}\right) \big/ \dot{E}_{F,k} \tag{1.115}$$

- Exergy destruction ratio y_k:

$$y_{D,k} = \dot{E}_{D,k} \big/ \dot{E}_{F,tot} \tag{1.116}$$

In addition, we calculate the following variables from the exergoeconomic analysis:

- Cost per unit of fuel exergy $c_{F,k}$:

$$c_{F,k} = \dot{C}_{F,k} \big/ \dot{E}_{F,k} \tag{1.117}$$

- Cost per unit of product exergy $c_{P,k}$:

$$c_{P,k} = \dot{C}_{P,k} \big/ \dot{E}_{P,k} \tag{1.118}$$

- Cost rate associated with exergy destruction $\dot{C}_{D,k}$:

$$\dot{C}_{D,k} = c_{F,k}\, \dot{E}_{D,k} \tag{1.119}$$

- Cost rate associated with exergy losses $\dot{C}_{L,k}$:

$$\dot{C}_{L,k} = c_{F,k}\, \dot{E}_{L,k} \tag{1.120}$$

- Cost rate associated with capital investment $\dot{Z}_k^{CI}$
- Cost rate associated with operating and maintenance expenses $\dot{Z}_k^{OM}$

- Sum $\dot{Z}_k$ of the cost rates associated with capital investment and O&M expenses:

$$\dot{Z}_k = \dot{Z}_k^{CI} + \dot{Z}_k^{OM} \tag{1.121}$$

- Relative cost difference r_k:

$$r_k = \frac{c_{P,k} - c_{F,k}}{c_{F,k}} = \frac{1-\varepsilon_k}{\varepsilon_k} + \frac{\dot{Z}_k}{c_{F,k}\,\dot{E}_{P,k}} \tag{1.122}$$

- Exergoeconomic factor f:

$$f_k = \frac{\dot{Z}_k}{\dot{Z}_k + c_{F,k}\left(\dot{E}_{D,k} + \dot{E}_{L,k}\right)} \tag{1.123}$$

The value of $c_{F,k}$ depends on the relative position of the kth component in the system and on the interconnections between the kth component and the remaining components. As a general rule, the closer the kth component to the product (fuel) stream of the overall system, the larger (smaller) the value of $c_{F,k}$.

An exergoeconomic analysis

1. Identifies and compares the real cost sources in a system: Equations 1.119 to 1.121.
2. Illustrates the cost formation process within a system.
3. Calculates separately the cost at which each product stream is generated.
4. Most importantly, assists in the effective cost minimization in a thermal system, particularly its design.

Examples of design optimization are given in Bejan et al. (1996), Lin and Tsatsaronis (1993), Tsatsaronis (1993), and Tsatsaronis et al. (1984, 1985, 1991, 1992, 1994). This point is discussed in Section 1.9.

1.9 Design Optimization

Design optimization of a thermal system means the modification of the structure and the design parameters of a system to minimize the total levelized cost of the system products under boundary conditions associated with available materials, financial resources, protection of the environment, and government regulation, together with the safety, reliability, operability, availability, and maintainability of the system. A truly optimized system is one for which the magnitude of every significant thermodynamic inefficiency (exergy destruction and exergy loss) is justified by considerations related to costs or is imposed by at least one of the above boundary conditions. A *thermodynamic* optimization, which aims only at minimizing the thermodynamic inefficiencies, may be considered as a subcase of design optimization.

An appropriate formulation of the optimization problem is usually the most important and sometimes the most difficult step of a successful optimization study. In optimization problems we separate the independent variables into *decision variables* and *parameters*. The values of the decision variables are amenable to change. The values of the parameters are fixed by the particular application. In optimization studies, only the decision variables may be varied. The parameters are independent variables that are each given one specific and unchanging value in any particular model statement. The variables whose values are calculated from the independent variables using a mathematical model are the dependent variables.

However, the optimization of thermal systems involves complexities that render conventional mathematical optimization approaches ineffective. The reasons include the following:

- Some of the input data and functions required for the thermodynamic and, particularly, the economic model might not be available or might not be in the required form. For example, it is not always possible to express the purchased-equipment costs as a function of the appropriate thermodynamic decision variables.

- A significant decrease in the product costs may be achievable only through changes in the structure of the system, but such changes are seldom elicited from conventional optimization techniques focusing on the optimization of a particular structure. Moreover, it is not always practical to develop a mathematical optimization model for every promising system structure.
- Even if all the required information is available, the complexity of the system might not allow a satisfactory mathematical model to be formulated and solved in a reasonable time.

In such cases the application of exergoeconomic techniques may provide significant benefits for the optimization process. The more complex the thermal system the larger are the expected benefits, particularly when chemical reactions are involved. The interactions of exergoeconomics with several other areas during the optimization procedure are shown schematically in Figure 1.21. Exergoeconomics uses results from the synthesis, cost analysis, and simulation of thermal systems and provides useful information for the evaluation and optimization of these systems as well as for the application of expert systems to improve the design and operation of such systems.

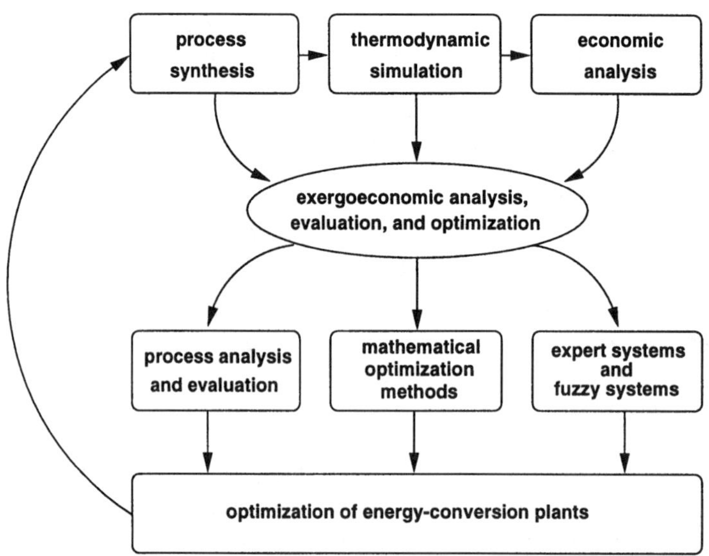

FIGURE 1.21 Interactions of exergoeconomics with other areas of engineering and optimization procedure.

The following section presents the main features of a general methodology that can be used to evaluate and iteratively optimize the design of a thermal system.

An Iterative Exergoeconomic Procedure for Optimizing the Design of a Thermal System

The conventional approach to optimization is to iteratively optimize subsystems and/or ignore the influence of some structural changes and decision variables. An effective alternative approach for the optimization of complex systems is the following iterative exergoeconomic optimization technique that consists of seven steps:

1. In the first step a workable design is developed. The guidelines presented in Section 1.7 and in Bejan et al. (1996), Lin and Tsatsaronis (1993), Linnhoff et al. (1982), Sama (1993), Tsatsaronis (1993), and Tsatsaronis and Pisa (1994) may assist in developing a workable design that is

relatively close to the optimal design. Thus, the use of these guidelines can reduce the total number of required iterations.

2. A detailed exergoeconomic analysis and, if necessary, a pinch analysis are conducted for the design configuration developed in the previous step. The results are used to determine design changes that are expected to improve the design being considered. In this step, and in steps 3 through 7, we consider only changes in the decision variables that affect both the exergetic efficiency and the investment costs. The remaining decision variables are optimized in step 8.
3. A component optimization may be conducted for one or two components in isolation for which the sum of the cost rates ($\dot{Z}_k + \dot{C}_{D,k}$) is *significantly* higher than the same sum for the remaining components. For this, we assume that the costs per exergy unit remain constant for all inlet streams. Step 3 is meaningful only for components in which each of the terms $\dot{Z}_k$ and $\dot{C}_{D,k}$ has a significant contribution to the costs associated with the respective component. If not, step 3 should either be omitted or, preferably, replaced by (a) an efficiency maximization procedure when $\dot{C}_{D,k}$ is the dominating cost rate, or (b) an investment cost minimization procedure when $\dot{Z}_k$ is the dominating cost rate. Another approach is discussed by Tsatsaronis et al. (1991, 1994). The sum ($\dot{Z}_k + \dot{C}_{D,k}$) is a measure of the economic importance of the kth component. Therefore, the components should be considered in order of descending value of this sum. The quality of this information is significantly improved when we consider only *avoidable* costs associated with both the capital investment and the cost of exergy destruction in the kth component (Tsatsaronis and Park, 1999).
4. The exergoeconomic factor f_k is used to identify the major cost source (capital investment or cost of exergy destruction).
   a. When the f_k value is high, investigate whether it is cost effective to reduce the capital investment of the kth component at the expense of the component efficiency.
   b. When the f_k value is low, try to improve the component efficiency by increasing the capital investment.
5. Eliminate any subprocesses that increase the exergy destruction or exergy loss without contributing to the reduction of capital investment or of the fuel costs for other components.
6. Consider improving the exergetic efficiency of a component if it has a relatively low exergetic efficiency or relatively large values for the rate of exergy destruction, the exergy destruction ratio, or the exergy loss ratio.
7. Based on the results from steps 2 through 6, a new design is developed and the value of the objective function for this design is calculated. In comparison with the previous design, if this value has been improved we may decide to proceed with another iteration that involves steps 2 through 7. If, however, the value of the objective function is not better in the new design than in the previous one, we may either revise some design changes and repeat steps 2 through 7 or proceed with step 8.
8. In this step, we use an appropriate mathematical optimization technique to optimize the decision variables that affect the costs but not the exergetic efficiency. At the end of this step, the cost-optimal design is obtained.
9. Finally, a parametric study may be conducted to investigate the effect on the optimization results of some parameters used and/or assumptions made in the optimization procedure.

When applying this methodology, it is important to recognize that the values of all thermoeconomic variables depend on the component types: heat exchanger, compressor, turbine, pump, chemical reactor, and so forth. Accordingly, whether a particular value is judged to be high or low can be determined only with reference to a particular class of components. Application of fuzzy inference systems could be very useful in making such judgments (Cziesla and Tsatsaronis, 1999). It is also important to consider the effects of contemplated design changes in one component on the performance of the remaining components. These effects may be determined either by inspection of the system flowsheets or by using a simulation program.

A cogeneration system studied by Frangopoulos (1994), Tsatsaronis and Pisa (1994), Valero et al. (1994), and von Spakovsky (1994) is used in the following case to demonstrate the application of steps 1, 2, 4, and 7. Examples of exergoeconomic evaluations and improvements of complex thermal systems are discussed by Tsatsaronis et al. (1991, 1992, 1994).

Case Study

Figure 1.19 presents the base-case design of a cogeneration system that develops a net power output of 30 MW and provides 14 kg/s of saturated water vapor at 20 bar. Not all data assumed for this system are realistic. The investment costs have been artificially increased to demonstrate the application of the exergoeconomic methodology to a variety of components. The optimization of an actual cogeneration system would be significantly easier because the components of a gas turbine system would not be optimized individually, as done here.

The first five columns of Table 1.17 show relevant thermodynamic and economic data. The second column of Table 1.18 shows the assumed purchased-equipment costs (PEC) for each component in the base-case design. These costs are obtained from the cost equations given in Appendix B of Bejan et al. (1996). The remaining direct costs, as well as the indirect costs, are estimated. The total capital investment of the cogeneration system in the base case is approximately \$46 million in mid-1994 dollars. Table 7.9 in Bejan et al. (1996) summarizes the parameters and assumptions used in the economic analysis, which is based on the revenue-requirement method (EPRI Technical Assesment Guide, 1991).

TABLE 1.17 Mass Flow Rate, Temperature, Pressure, Exergy Rate, and Cost Data for the Streams of the Cogeneration System

State	Stream	Mass Flow Rate $\dot{m}$ (kg/s)	Temperature T (K)	Pressure p (bar)	Exergy Flow Rate $\dot{E}$ (MW)	Cost Flow Rate $\dot{C}$ (\$/h)	Cost per Exergy Unit c (\$/GJ)
1	Air	91.28	298.1	1.01	0.000	0	0
2	Air	91.28	603.7	10.13	27.538	2756	27.80
3	Air	91.28	850.0	9.62	41.938	3835	25.40
4	Combustion products	92.92	1520.0	9.14	101.454	5301	14.51
5	Combustion products	92.92	1006.2	1.10	38.782	2026	14.51
6	Combustion products	92.92	779.8	1.07	21.752	1137	14.51
7	Combustion products	92.92	426.9	1.01	2.773	145	14.51
8	Water	14.00	298.1	20.00	0.062	0	0
9	Water	14.00	485.6	20.00	12.810	1256	27.23
10	Methane	1.64	298.1	12.00	84.994	1398	4.57
11	Power to air compressor	—	—	—	29.662	2003	18.76
12	Net power	—	—	—	30.000	2026	18.76

TABLE 1.18 Values of the Purchased-Equipment Costs (PEC) and the Thermoeconomic Variables for the Base Design Case (T_3 = 850 K; T_4 = 1520 K; p_2/p_1 = 10; $\eta_{sc} = \eta_{st}$ = 0.86)[a]

Component	PEC (10^6 \$)	ε (%)	$\dot{E}$ (MW)	y_D (%)	c_F (\$/GJ)	c_P (\$/GJ)	$\dot{C}_D$ (\$/h)	$\dot{Z}$ (\$/h)	$\dot{C}_D + \dot{Z}$ (\$/h)	r (%)	f (%)
Combustion Chamber	0.34	80.37	25.48	29.98	11.45	14.51	1050	68	1118	26.7	6.1
Gas Turbine	3.74	95.20	3.01	3.54	14.51	18.76	157	753	910	29.2	82.7
Air Compressor	3.73	92.84	2.12	2.50	18.76	27.80	143	753	896	48.2	84.0
HRSG	1.31	67.17	6.23	7.33	14.51	27.36	326	264	590	88.5	44.8
Air Preheater	0.94	84.55	2.63	3.09	14.51	20.81	137	189	326	43.4	57.9

[a] For the overall plant, we have $\dot{C}_{P,tot}$ = \$3617/h and $\dot{C}_{L,tot} = \dot{C}_7$ = \$145/h.

The year-by-year economic analysis results in the levelized annual costs for fuel ($10.4 × 10^6), operating and maintenance ($6.0 × 10^6), and carrying charges ($10.5 × 10^6) for a levelization time period of 10 years. These values are the corresponding levelized current-dollar costs obtained for the base case. The levelized costs are used as input data for the thermoeconomic analysis and optimization. The cost flow rates in the system are obtained by dividing the levelized annual costs by the number of hours of system operation per year.

The methodology introduced above will now be applied to the case-study cogeneration system. The objectives are to identify the effects of the design variables on the costs and suggest values of the design variables that would make the system more cost-effective. The key design variables — the decision variables — for the cogeneration system are the compressor pressure ratio p_2/p_1, the isentropic compressor efficiency η_{sc}, the isentropic turbine efficiency η_{st}, the temperature of the air entering the combustion chamber T_3, and the temperature of the combustion products entering the gas turbine T_4.

First Design Case

The following nominal values of the decision variables correspond to the first workable design (base-case design) developed for the cogeneration system of Figure 1.19 and Tables 1.17 and 1.18:

$$p_2/p_1 = 10,\ \eta_{sc} = \eta_{st} = 0.86,\ T_3 = 850\ K,\ T_4 = 1520\ K.$$

The last two columns of Table 1.17 and the last 10 columns of Table 1.18 summarize the values of the thermoeconomic variables calculated for each component of the cogeneration system for the base-case design. In accord with the methodology presented, the components are listed in order of descending value of the sum ($\dot{C}_D + \dot{Z}$).

The combustion chamber, the gas turbine, and the air compressor have the highest values of the sum ($\dot{Z} + \dot{C}_D$) and, therefore, are the most important components from the thermoeconomic viewpoint. The low value of the exergoeconomic factor f for the combustion chamber shows that the costs associated with the combustion chamber are almost exclusively due to exergy destruction. A part of the exergy destruction in a combustion chamber can be avoided by preheating the reactants and by reducing the heat loss and the excess air, but this usually leads only to a small reduction. For simplicity, we assume here that the heat loss cannot be further reduced. The excess air is determined by the desired temperature T_4 at the inlet to the gas turbine. The temperature T_4 is a key design variable for it affects both the performance of the entire system (exergy destruction in the combustion chamber, gas turbine, air preheater, and heat-recovery steam generator (HRSG), and exergy loss associated with stream 7) and the investment costs of the components.

An increase in the heat transfer rate in the air preheater, achieved through an increase in temperature T_3, also results in a decrease of the exergy destruction in the combustion chamber. Thus, the temperature T_3 is also a key design variable because, in addition to the combustion chamber, it affects the exergy loss associated with stream 7 as well as the performance and investment costs of the air preheater and the heat-recovery steam generator. Holding all other decision variables constant, the higher the temperature T_3 the smaller the average temperature difference in the air preheater and the heat-recovery steam generator. A decrease in the average temperature difference in these heat exchangers results in an increase in both the exergetic efficiency and the capital investment for each heat exchanger. In summary, by considering measures for reducing the high cost rate associated with the exergy destruction in the combustion chamber of the cogeneration system, two key design variables have been identified: temperatures T_3 and T_4. An increase in these temperatures reduces the $\dot{C}_D$ value for the combustion chamber and other components but increases their capital investment costs.

Turning next to the gas turbine, which has the second highest value of the sum ($\dot{Z} + \dot{C}_D$), the relatively large value of the factor f suggests that the capital investment and O&M costs dominate. The capital investment cost of the gas turbine depends on temperature T_4, pressure ratio p_2/p_1, and isentropic efficiency η_{st}. To reduce the high $\dot{Z}$ value associated with the gas turbine, a reduction in the value of at least one of these variables should be considered.

The air compressor has the highest f value and the second-highest relative cost difference r among all components. Thus, we would expect the cost-effectiveness of the entire system to improve if the $\dot{Z}$ value for the air compressor is reduced. This may be achieved by reducing the pressure ratio p_2/p_1 and/or the isentropic compressor efficiency η_{sc}.

The heat-recovery steam generator has the lowest exergetic efficiency and the highest r value among all the components. As the f value indicates, almost 45% of the relative cost difference is caused by the $\dot{Z}$ value in this component, with the remaining 55% caused by exergy destruction. Thus, we might conclude that a decrease of the exergy destruction in the HRSG could be cost-effective for the entire system, even if this would increase the investment cost associated with this component. The exergy destruction in the HRSG can be reduced by decreasing the values of T_6 and T_7. A decrease in the value of T_7 also results in a decrease in the exergy loss from the total system. In terms of the decision variables, temperatures T_6 and T_7 may be reduced by increasing T_5 and/or decreasing T_4 at fixed values of the remaining decision variables.

The relatively high value of f in the air preheater suggests a reduction in the investment cost of this component. This can be achieved by decreasing T_3. It should be noted, however, that changes suggested by the evaluation of this component should only be considered if they do not contradict changes suggested by components with a larger value of $(\dot{C}_D + \dot{Z})$.

Summarizing the foregoing conclusions, the following changes in the design variables are expected to improve the cost-effectiveness of the system:

- Increase the value of T_3, as suggested by the evaluation of the combustion chamber and HRSG.
- Decrease the pressure ratio p_2/p_1 (and thus p_4/p_5) and the isentropic efficiencies η_{sc} and η_{st}, as suggested by the evaluation of the air compressor and gas turbine.
- Maintain T_4 fixed, since we get contradictory indications from the evaluation of the combustion chamber on one side and the gas turbine and HRSG on the other side.

Second Design Case

Contemplating the effects of changes made in accordance with the above list in the values of the design variables used in the first iteration, the following new values are selected for the second design case:

$$T_3 = 870\ K,\ T_4 = 1520\ K\ (\text{unchanged}),\ p_2/p_1 = 9,\ \eta_{sc} = 85\%,\ \eta_{st} = 85\%.$$

The new values of the thermoeconomic variables for each component are summarized in Table 1.19. In the new design, the combustion chamber, the gas turbine, and the air compressor also have the highest values of the sum $(\dot{C}_D + \dot{Z})$ and, therefore, are still the most important components from the thermoeconomic viewpoint. The high cost rate associated with the combustion chamber can be reduced by increasing the values of T_3 and T_4. In the evaluation of the cogeneration system we should, however, consider that the value of the combustion chamber will always be the highest among the $(\dot{C}_D + \dot{Z})$ values for the components of the cogeneration system.

The gas turbine now has the highest f value. The reduction in this value from 82.7% in the base design to 81.0% in the new design is relatively small compared with a target value of below 75%. This observation suggests (1) a significant decrease in the values of η_{st} and/or p_2/p_1 — that is, a decrease greater than the decrease in these variables in the previous step: from 86 to 85% and from 10 to 9, respectively; and (2) a reduction in the value of T_4. Note that the decrease in the T_4 value contradicts the corresponding suggestion from the combustion chamber.

The high values of the exergoeconomic factor f and the relative cost difference r for the air compressor suggest a decrease in the values of the decision variables p_2/p_1 and η_{sc}.

The anticipated increase in the exergetic efficiency of the HRSG (see first iteration) was not realized because of the interdependence of the components. The reduction in the values of p_2/p_1, η_{sc}, and η_{st} for the compressor and the turbine leads to an increase in the temperature differences (and, therefore, a decrease in the exergetic efficiency) of the HRSG. Thus, the HRSG thermoeconomic evaluation suggests that the T_3 value increases and the T_4 value decreases.

TABLE 1.19 Component Data for the Second Iteration Case: (T_3 = 870 K; T_4 = 1520 K; p_2/p_1 = 9; η_{sc} = 85%; η_{st} = 85%)[a]. The Values Given in Parentheses Refer to the Third Iteration Case Considered Here (T_3 = 910 K; T_4 = 1480 K; p_2/p_1 = 7; η_{sc} = 0.83)[b]

Component	ε (%)	$\dot{E}_D$ (MW)	y_D (%)	c_F ($/GJ)	c_P ($/GJ)	$\dot{C}_D$ ($/h)	$\dot{Z}$ ($/h)	$\dot{C}_D + \dot{Z}$ ($/h)	r (%)	f (%)
Combustion chamber	80.3 (81.3)	25.93 (27.47)	29.77 (29.92)	10.50 (9.42)	13.26 (11.71)	980 (931)	72 (55)	1052 (986)	26.3 (24.4)	6.8 (5.5)
Gas turbine	94.9 (94.3)	3.18 (3.69)	3.66 (4.01)	13.26 (11.71)	16.97 (13.75)	152 (155)	647 (296)	799 (451)	28.0 (17.5)	81.0 (65.6)
Air compressor	92.1 (90.5)	2.34 (2.99)	2.69 (3.25)	16.97 (13.75)	23.96 (18.38)	143 (148)	546 (324)	689 (472)	41.2 (33.6)	79.2 (68.7)
HRSG	66.6 (67.6)	6.40 (6.10)	7.35 (6.65)	13.26 (11.71)	25.60 (23.51)	305 (257)	261 (284)	566 (541)	93.1 (100.7)	46.1 (52.5)
Air preheater	84.7 (85.6)	3.15 (4.97)	3.62 (4.90)	13.26 (11.71)	18.94 (16.53)	150 (190)	206 (275)	356 (464)	42.9 (41.2)	57.8 (59.2)
Overall plant	49.1 (46.6)	41.01 (44.79)	47.09 (48.79)	4.57 (4.57)	21.80 (19.06)	675 (736)	1922 (1424)	2597 (2160)	377.0 (317.2)	74.0 (65.9)

[a] For the overall plant in the new design case, we have $\dot{C}_{P,tot}$ = \$3355/h and $\dot{C}_{L,tot} = \dot{C}_7$ = \$157/h.
[b] For the overall plant in this design case, we have $\dot{C}_{P,tot}$ = \$2934/h and $\dot{C}_{L,tot} = \dot{C}_7$ = \$167/h.

The relatively high value of f in the air preheater suggests a reduction in the T_3 value. As noted in the first iteration, however, changes suggested by the evaluation of this component should only be considered if they do not contradict changes suggested by components with a higher value of the sum $(\dot{C}_D + \dot{Z})$.

Summarizing the foregoing suggestions from the thermoeconomic evaluation of each component, the following changes in the decision variables are expected to improve the cost-effectiveness of the cogeneration system:

- Increase the value of T_3, as suggested by the evaluation of the combustion chamber and HRSG.
- Decrease the pressure ratio p_2/p_1 and the isentropic efficiencies η_{sc} and η_{st}, as suggested by the evaluations of the air compressor and gas turbine.
- Decrease the temperature T_4, as suggested from the evaluations of the gas turbine and the HRSG.

Third Design Case

To illustrate the effect of the suggested changes in the decision variables on the overall costs, we use the following new set of values for the design variables:

$$T_3 = 910\ K,\ T_4 = 1480\ K,\ p_2/p_1 = 7,\ \eta_{sc} = 0.83,\ \text{and}\ \eta_{st} = 0.83$$

The results from the thermoeconomic analysis for the last set of values are summarized in Table 1.19 by the values given in parentheses. A comparison of the corresponding values shown in Table 1.19 demonstrates the improvement in the cost-effectiveness of the last design case. As a result of these changes, the value of the objective function $\dot{C}_{P,tot}$ is reduced from \$3355/h to \$2934/h. The $\dot{C}_7$ value has increased from \$157/h in the new case to \$167/h in the last case. This increase, however, is outweighed by the decreases in the values of $(\dot{C}_{D,k} + \dot{Z}_k)$.

Additional Iterations

Additional iterations conducted in a similar way are necessary to further decrease the value of the objective function and/or establish a nearly optimal design case. The cost-optimal values of the decision variables (Bejan et al., 1996) are

$$T_3 = 910\ K,\ T_4 = 1463.0\ K,\ p_2/p_1 = 5.77,\ \eta_{sc} = 0.811,\ \text{and}\ \eta_{st} = 0.845.$$

With these values we obtain the objective function $\dot{C}_{P,tot}$ = \$2870/h. For the cost-optimal case, the exergetic efficiency of the overall system is 45.0%, the cost rate associated with the exergy loss is \$205/h, and the pinch temperature difference in the heat-recovery steam generator is 49.7 K.

1.10 Economic Analysis of Thermal Systems

Generally, each company has a preferred approach for conducting an economic analysis and calculating the costs associated with the products generated in a thermal system. Here only a few general introductory comments are made. A detailed discussion of an economic analysis is given in Bejan et al. (1996).

A complete economic analysis consists of the following three steps:

Step1. Estimate the required total capital investment.

Step 2. Determine the economic, financial, operating, and market-associated parameters for the detailed cost calculation.

Step 3. Calculate the costs of all products generated by the system.

These steps are now discussed in turn.

Estimation of Total Capital Investment

The capital needed to purchase the land, build all the necessary facilities, and purchase and install the required machinery and equipment for a system is called the *fixed-capital investment.* The fixed-capital investment represents the total system cost, assuming a zero-time design and construction period (overnight construction). The *total capital investment* is the sum of the fixed-capital investment and other outlays (e.g., startup costs, working capital, costs of licensing, research and development, as well as interest during construction).

The costs of all permanent equipment, materials, labor, and other resources involved in the fabrication, erection, and installation of the permanent facilities are the *direct costs.* The *indirect costs* (e.g., costs associated with engineering, supervision, and construction, including contractor's profit and contingencies) do not become a permanent part of the facilities but are required for the orderly completion of the project. The fixed-capital investment is the sum of all direct and indirect costs.

The *purchased equipment costs* are estimated with the aid of vendors' quotations, quotations from experienced professional cost estimators, calculations using extensive cost databases, or estimation charts. If necessary, the cost estimates of purchased equipment must be adjusted for size (with the aid of scaling exponents) and for time (with the aid of cost indices). The remaining direct costs are associated with equipment installation, piping, instrumentation, controls, electrical equipment and materials, land, civil structural and architectural work, and service facilities. These direct costs, the indirects costs, and the other outlays, if they cannot be estimated directly, are calculated as a percentage of the purchased equipment costs or alternatively as a percentage of the fixed-capital investment.

For more detailed information on the estimation of the total capital investment, readers should consult the references by Bejan et al. (1996), Chauvel et al. (1976), Douglas (1988), Garrett (1989), Guthrie (1969, 1970, 1974), Humphreys (1991), Perry and Chilton (1973), Peters and Timmerhaus (1991), Ulrich (1984), and Valle-Riestra (1983).

Principles of Economic Evaluation

Decisions about expenditures generally require consideration of the *time value of money*: if an amount of money P (*present value*) is deposited in an account earning i percent interest per time period and the interest is compounded at the end of each of n time periods, the amount will grow to a *future value* (F):

$$F = P(1+i)^n \tag{1.124}$$

TABLE 1.20 Summary of Basic Formulas and Factors Used in Economic Analysis

Formula[a]	Factor
$\frac{F}{P}=\left(1+i_{eff}\right)^n$	Single-payment compound-amount factor (SPCAF)
$\frac{P}{F}=\frac{1}{\left(1+i_{eff}\right)^n}$	Single-payment present-worth factor or single-payment discount factor (SPDF)
$\frac{F}{A}=\frac{\left(1+i_{eff}\right)^n-1}{i_{eff}}$	Uniform-series compound-amount factor (USCAF)
$\frac{A}{F}=\frac{i_{eff}}{\left(1+i_{eff}\right)^n-1}$	Uniform-series sinking fund factor (USSFF)
$\frac{P}{A}=\frac{\left(1+i_{eff}\right)^n-1}{i_{eff}\left(1+i_{eff}\right)^n}$	Uniform-series present-worth factor (USPWF)
$\frac{A}{P}=\frac{i_{eff}\left(1+i_{eff}\right)^n}{\left(1+i_{eff}\right)^n-1}$	Capital-recovery factor (CRF)
$\frac{C_K}{P_0}=\frac{\left(1+i_{eff}\right)^n}{\left(1+i_{eff}\right)^n-1}$	Capitalized-cost factor (CCF)
$\frac{A}{P_0}=\frac{k\left(1-k^n\right)}{1-k}CRF$	Constant-escalation levelization factor (CELF)

[a] In these expressions, cost-of-money compounding and ordinary annuities are assumed. The exponent n denotes the number of years and i_{eff} is the effective rate of return. The factor k is defined by Equation 1.132.

In engineering economy, the unit of time is usually taken as the year. If compounding occurs p times per year ($p \geq 1$) for a total number of n years ($n \geq 1$) and i is the *annual rate of return* (or *cost-of-money rate*), Equation 1.124 becomes

$$F = P\left(1+\frac{i}{p}\right)^{np} \tag{1.125}$$

Then the annual effective rate of return is

$$i_{eff} = \left(1+\frac{i}{p}\right)^{p} - 1 \tag{1.126}$$

and the general equation for calculating the future value becomes

$$F = P\left(1+i_{eff}\right)^{n} \tag{1.127}$$

When Equation 1.127 is used to calculate the present value (P) of a future amount (F), the term i_{eff} is called effective *discount rate*. Table 1.20 summarizes some basic formulas and factors used in economic analysis.

Annuities

An annuity is a series of *equal-amount* money transactions occurring at equal time intervals or periods (usually years). The most common type of annuity is the *ordinary annuity,* which involves money transactions occurring at the end of each period.

If an amount A is deposited at the end of each period in an account earning i_{eff} percent per period, the future sum F (*amount of the annuity*) accrued at the end of the *n*th period is

$$F = A\frac{\left(1+i_{eff}\right)^n - 1}{i_{eff}} \tag{1.128}$$

The term $[(1+i_{eff})^n - 1]/i_{eff}$ is called the *uniform-series compound amount factor* and the reciprocal, $i_{eff}/[(1 + i_{eff})^n - 1]$, is called the *uniform-series sinking fund factor.* These terms are also listed in Table 1.20.

The *present value* or *present worth of an annuity* (P) is defined as the amount of money that would have to be invested at the beginning of the first time interval at an effective compound rate of return per period i_{eff} to yield a total amount at the end of the last time interval equal to the amount of the annuity:

$$\frac{P}{A} = \frac{\left(1+i_{eff}\right)^n - 1}{i_{eff}\left(1+i_{eff}\right)^n} \tag{1.129}$$

The expression on the right side of this equation is called the *uniform-series present-worth factor.* The reciprocal of this factor is the *capital-recovery factor* (CRF, see Table 1.20), which may be used to convert the expenditures associated with capital investment into an annuity.

Cost Escalation

The *real escalation rate* of an expenditure is the annual rate of expenditure change caused by factors such as resource depletion, increased demand, and technological advances. The *nominal* (or *apparent*) *escalation rate* (r_n) is the total annual rate of change in an expenditure and includes the effects of both the real escallation rate (r_r) and the *inflation rate* (r_i):

$$\left(1+r_n\right) = \left(1+r_r\right)\left(1+r_i\right) \tag{1.130}$$

In the analysis of thermal systems we often assume that all costs except fuel costs and the prices of byproducts change annually with the average inflation rate (i.e., $r_r = 0$). When fossil-fuel costs are expected over a long period of future years to increase on the average faster than the predicted inflation rate, a positive real escalation rate for fuel costs may be appropriate for the economic analysis of the thermal system being studied.

Levelization

Cost escalation applied to an expenditure over an *n*-year period results in a nonuniform cost schedule in which the expenditure at any year is equal to the previous year expenditure multiplied by $(1 + r_n)$. The *constant-escalation levelization factor* (CELF, see Table 1.20) is used to express the relationship between the value of the expenditure at the beginning of the first year (P_0) and an equivalent annuity (A), which is now called a *levelized value.* The levelization factor depends on both the effective annual cost-of-money rate i_{eff} and the nominal escalation rate r_n:

$$\frac{A}{P_0} = CELF = \frac{k\left(1-k^n\right)}{1-k}CRF \tag{1.131}$$

where

$$k = \frac{1 + r_n}{1 + i_{eff}} \tag{1.132}$$

The concept of levelization is general and is defined as the use of time-value-of-money arithmetic to convert a series of varying quantities to a financially equivalent constant quantity (annuity) over a specified time interval. In exergoeconomics, the concept of levelization is applied to calculate the levelized annual fuel costs, the levelized annual operating and maintenance expenses, and the levelized total cost associated with the capital investment. Among all available approaches, the *total revenue requirement method* (Bejan et al., 1996; Tsatsaronis et al., 1984, 1986) is the most detailed and most appropriate method for that purpose.

Parameters and Assumptions

To conduct a detailed economic analysis several parameters must be specified and assumptions must be made for the entire life of the system being analyzed. These include:

- The estimated total capital investment.
- Current fuel cost and operating and maintenace expenditures.
- Average capacity factors, average general inflation rate, and average nominal escalation rate of each expenditure.
- Beginning and length of the design and construction period and of the operation period.
- Allocation of investment expenditures to the individual years of design and construction.
- Plant financing sources and associated required returns on capital.
- Tax rates.
- Depreciation method used for tax purposes.
- Insurance cost.

These parameters and assumptions are discussed in Bejan et al. (1996) and the EPRI Technical Assessment Guide (1991).

Calculation of the Product Costs

In a conventional economic analysis, we must identify among all product streams of a thermal system the *main product* stream, the cost of which is calculated from the cost balance. To proceed in this way, it is necessary to make assumptions with respect to the *selling prices* of all the remaining product streams, which are now *byproduct* streams.

However, when exergoeconomics is applied to a thermal system, the *total* annual levelized costs associated with owning and operating the system are apportioned among the product streams using the exergy costing approach and the auxiliary costing equations as discussed in Section 1.8. Thus, the cost associated with each product stream is calculated directly and there is no need to make assumptions about the prices of the byproducts.

References

Ahrendts, J. Reference states, *Energy Int. J.* 5: 667-677, 1980.

ASHRAE Handbook 1993 Fundamentals. American Society of Heating, Refrigerating, and Air Conditioning Engineers, Atlanta, GA, 1993.

ASME Steam Tables, 6th ed., ASME Press, Fairfield, NJ, 1993.

Baasel, W.D., *Preliminary Chemical Engineering Plant Design*, 2nd ed., Van Nostrand Reinhold, New York, 1990.

Bejan, A., Tsatsaronis, G., and Moran, M., *Thermal Design and Optimization*, John Wiley & Sons, New York, 1996.

Biegler, L.T., Chemical process simulation, *Chem. Eng. Progr.,* October: 50-61, 1989.

Bird, R.B., Stewart, W.E., and Lightfoot, E.N., Transport Phenomena. John Wiley & Sons, New York, 1960.

Bolz, R.E. and Tuve, G.L., Eds., *Handbook of Tables for Applied Engineering Science,* 2nd ed., Chemical Rubber Company, Cleveland, 1973.

Bornakke, C. and Sonntag, R.E., *Tables of Thermodynamic and Transport Properties,* John Wiley & Sons, New York, 1996.

Chauvel, A. et al., *Manual of Economic Analysis of Chemical Processes,* McGraw-Hill, New York, 1976.

Cooper, H.W. and Goldfrank, J.C., B-W-R constants and new correlations, *Hydrocarbon Proc.* 46(12): 141-146, 1967.

Cziesla, F. and Tsatsaronis, G., Iterative exergoeconomic evaluation and improvement of thermal power plants using fuzzy inference systems, *Proc. Int. Conf. Efficiency, Costs, Optimization, Simulation and Environmental Aspects of Energy Systems*, Tokyo, June 8-10, 380-385, 1999.

Douglas, J.M., *Conceptual Design of Chemical Processes*, McGraw-Hill, New York, 1988.

Edgar, T.F. and Himmelblau, D.M., *Optimization of Chemical Processes*, McGraw-Hill, New York, 1988.

EPRI Technical Assessment Guide (TAG™), Electric Power Research Institute, TR-100281, 3, Rev. 6, 1991.

Frangopoulos, C.A., Application of the thermoeconomic functional approach to the CGAM problem, *Energy Int. J.* 19, 323-342, 1994.

Garrett, D.E., *Chemical Engineering Economics*, Van Nostrand Reinhold, New York, 1989.

Gray, D.E., Ed., *American Institute of Physics Handbook,* McGraw-Hill, New York, 1972.

Guthrie, K.M., Data and techniques for preliminary capital cost estimating, *Chem. Eng.,* March 24, 114-142, 1969.

Guthrie, K.M., Capital and operating costs for 54 chemical processes, *Chem. Eng.,* June 15, 140-156, 1970.

Guthrie, K.M., *Process Plant Estimating, Evaluation and Control,* Craftsman, Solana Beach, CA, 1974.

Haar, L., Gallagher, J.S., and Kell, G.S., *NBS/NRC Steam Tables*, Hemisphere, New York, 1984.

Handbook of Chemistry and Physics, annual ed., CRC Press, Boca Raton, FL.

Humphreys, K.K., *Jelen's Cost and Optimization Engineering*, 3rd ed., McGraw-Hill, New York, 1991.

JANAF Thermochemical Tables, 3rd ed., American Chemical Society and the American Institute of Physics for the National Bureau of Standards, Washington, D.C., 1986.

Jones, J.B. and Dugan, R.E., *Engineering Thermodynamics*. Prentice-Hall, Englewood Cliffs, NJ, 1996.

Keenan, J.H., Chao, J., and Kaye, J., *Gas Tables International Version,* 2nd ed., John Wiley & Sons, New York (1980, English Units; 1983, SI Units).

Keenan, J.H., Keyes, F.G., Hill, P.G., and Moore, J.G., *Steam Tables*, John Wiley & Sons, New York, 1978.

Knacke, O., Kubaschewski, O., and Hesselmann, K., *Thermochemical Properties of Inorganic Substances*, 2nd ed., Springer-Verlag, Berlin, 1991.

Kotas, T.J., *The Exergy Method of Thermal Plant Analysis*, Krieger, Melbourne, FL, 1995.

Lazzaretto, A. and Tsatsaronis, G., A general process-based methodology for exergy costing, in Duncan, A. B., Fiszdon, J., O'Neal, D., and Den Braven, K., Eds. *Proc. ASME Advanced Energy Sys. Div.,* AES Vol. 36. ASME, New York, 1996, 413-428.

Lazzaretto, A. and Tsatsaronis, G., On the quest for objective equations in exergy costing, in Ramalingam, M.L., Lage, J.G., Mei, V.C., and Chapman, J.N., Eds. *Proc. ASME Advanced Energy Sys. Div.,* AES Vol. 37, ASME, New York, 1997, 413-428.

Lee, B.I. and Kessler, M.G., A generalized thermodynamic correlation based on three-parameter corresponding states, *AIChE J.,* 21, 510-527, 1975.

Liley, P.E., Reid, R.C., and Buck, E., Physical and chemical data, in *Perrys' Chemical Engineers Handbook.* Perry, R.H. and Green, D.W., Eds., 6th ed., McGraw-Hill, New York, 1984.

Liley, P.E., Thermodynamic properties of substances, in *Marks' Standard Handbook for Mechanical Engineers,* Avallone, E.A. and Baumeister, T., Eds., 9th ed., McGraw-Hill, New York, 1987.

Lin, L. and Tsatsaronis, G., Cost optimization of an advanced IGCC power plant concept design, in *Thermodynamics and the Design, Analysis and Improvement of Energy Systems*, Richter, H.J., Ed., AES Vol. 10, ASME, New York, 1993, 156-166.

Linnhoff, B. et al., *A User Guide on Process Integration for the Efficient Use of Energy*, Institution of Chemical Engineers, Rugby, Warks, U.K., 1982.

Moran, M.J., *Availability Analysis — A Guide to Efficient Energy Use,* ASME Press, New York, 1989.

Moran, M.J. and Shapiro, H.N., *Fundamentals of Engineering Thermodynamics*, 4th ed., John Wiley & Sons, New York, 2000.

Moran, M.J. and Shapiro, H.N., *IT: Interactive Thermodynamics,* Computer software to accompany *Fundamentals of Engineering Thermodynamics,* 4th ed. Developed by Intellipro Inc., John Wiley & Sons, New York, 2000.

Obert, E.F., *Concepts of Thermodynamics,* McGraw-Hill, New York, 1960.

Papalambros, P. Y. and Wilde, D. J., *Principles of Optimal Design-Modeling and Computation*, Cambridge University Press, Cambridge, U.K., 1988.

Perry, J.H. and Chilton, C.H., *Chemical Engineer's Handbook*, 5th ed., McGraw-Hill, New York, 1973.

Peters, M.S. and Timmerhaus, K.D., *Plant Design and Economics for Chemical Engineers*, 4th ed., McGraw-Hill, New York, 1991.

Preston-Thomas, H., The International Temperature Scale of 1990 (ITS 90), *Metrologia,* 27: 3-10, 1990.

Reid, R.C. and Sherwood, T.K., *The Properties of Gases and Liquids,* 2nd ed., McGraw-Hill, New York, 1996.

Reid, R.C., Prausnitz, J.M., and Poling, B.E., *The Properties of Gases and Liquids,* 4th ed., McGraw-Hill, New York, 1987.

Reklaitis, G.V., Ravindran, A., and Ragsdell, K.M., *Engineering Optimization*, John Wiley & Sons, New York, 1983.

Reynolds, W.C., Thermodynamic Properties in SI — Graphs, Tables and Computational Equations for 40 Substances, Department of Mechanical Engineering, Stanford University, Palo Alto, CA, 1979.

Sama, D.A., The use of the second law of thermodynamics, in J. Szargut, Z. Kolenda, G. Tsatsaronis, and A. Ziebik, Eds., *Proc. Int. Conf. ENSEC '93*, ASME, New York, 53-76, 1993.

Stephan, K., Tables, in *Dubbel Handbook of Mechanical Engineering,* Beitz, W. and Kuttner, K.-H., Eds., Springer-Verlag, London, 1994.

Szargut, J., Morris, D.R., and Steward, F.R., *Exergy Analysis of Thermal, Chemical, and Metallurgical Processes,* Hemisphere, New York, 1988.

Tsatsaronis, G., Thermoeconomic analysis and optimization of energy systems, *Prog. Energy Combust. Sci.,* 19, 227-257, 1993.

Tsatsaronis, G., Krause, A., Lin, L., and Tawfik, T., Thermoeconomic Evaluation of the Design of a Pressurized Fluidized-Bed Hydroretorting Plant, Final Report prepared for the Institute of Gas Technology and the Department of Energy, DE-AC21-87MC11089, 1992.

Tsatsaronis, G., Lin, L., Pisa, J., and Tawfik, T., Thermoeconomic Design Optimization of a KRW-Based IGCC Power Plant. Final Report prepared for the U.S. Department of Energy, Morgantown Energy Technology Center, DE-FC21-98MC26019, 1991.

Tsatsaronis, G., Lin, L., Pisa, J., and Tawfik, T., Optimization of an IGCC power plant. I and II, in R.F. Boehm et al., Eds., *Thermodynamics and the Design, Analysis and Improvement of Energy Systems*, AES Vol. 27, ASME, New York, 37-67, 1992.

Tsatsaronis, G., Lin, L., and Tawfik, T.J., Exergoeconomic evaluation of a KRW-based IGCC power plant, *Eng. Gas Turbines Power,* 116, 300-306, 1994.

Tsatsaronis, G. and Park, M.H., On avoidable and unavoidable exergy destruction, *Proc. Int. Conf. Efficiency, Costs, Optimization, Simulation and Environmental Aspects of Energy Systems,* Tokyo, June 8-10, 116-121, 1999.

Tsatsaronis, G. and Pisa, J., Exergoeconomic evaluation and optimization of energy systems. *Energy Int. J.* 19, 287-321, 1994.

Tsatsaronis, G. and Winhold, M., Thermoeconomic Analysis of Power Plants. EPRI AP-3651, RP 2029-8. Final Report. Electric Power Research Institute, Palo Alto, CA, 1984.

Tsatsaronis, G. and Winhold, M., Exergoeconomic analysis and evaluation of energy conversion plants. *Energy Int. J.* 10, 69-94, 1985.

Tsatsaronis, G., Winhold, M., and Stojanoff, C.G., Thermoeconomic Analysis of a Gasification-Combined-Cycle Power Plant. EPRI AP-4734, RP 2029-8. Final Report. Electric Power Research Institute, Palo Alto, CA, 1986.

Ulrich, G.D., *A Guide to Chemical Engineering Process Design and Economics*, John Wiley & Sons, New York, 1984.

Valero, A., Lozano, M.A., Serra, L., and Torres, C., Application of the exergetic cost theory to the CGAM problem, *Energy Int. J.* 19, 365-381, 1994.

Valle-Riestra, F.J., *Project Evaluation in the Chemical Process Industries*, McGraw-Hill, New York, 1983.

Van Wylen, G.J., Sonntag, R.E., and Bornakke, C., *Fundamentals of Classical Thermodynamics,* 4th ed., John Wiley & Sons, New York, 1994.

von Spakovksy, M.R., Application of engineering functional analysis to the analysis and optimization of the CGAM problem, *Energy Int. J.* 19, 343-364, 1994.

Wark, K., *Thermodynamics,* 4th ed., McGraw-Hill, New York, 1983.

Zemansky, M.W., Thermodynamic symbols, definitions, and equations, in *American Institute of Physics Handbook*, D.E. Gray, Ed., McGraw-Hill, New York, 1972.

2

Fluid Mechanics

Frank Kreith, Editor
Engineering Consultant
University of Colorado

Stanley A. Berger
University of California, Berkeley

Stuart W. Churchill
University of Pennsylvania

J. Paul Tullis
Utah State University

Frank M. White
University of Rhode Island

Alan T. McDonald
Purdue University

Ajay Kumar
NASA Langley Research Center

John C. Chen
Lehigh University

Thomas F. Irvine, Jr.
State University of New York, Stony Brook

Massimo Capobianchi
Gonzaga University

0-8493-9581-X/00/$0.00+$.50

2.1 Fluid Statics

Stanley A. Berger

Equilibrium of a Fluid Element

If the sum of the external forces acting on a fluid element is zero, the fluid will be either at rest or moving as a solid body — in either case, we say the fluid element is in equilibrium. In this section we consider fluids in such an equilibrium state. For fluids in equilibrium the only internal stresses acting will be normal forces, since the shear stresses depend on velocity gradients, and all such gradients, by the definition of equilibrium, are zero. If one then carries out a balance between the normal surface stresses and the body forces, assumed proportional to volume or mass, such as gravity, acting on an elementary prismatic fluid volume, the resulting equilibrium equations, after shrinking the volume to zero, show that the normal stresses at a point are the same in all directions, and since they are known to be negative, this common value is denoted by $-p$, p being the pressure.

Hydrostatic Pressure

If we carry out an equilibrium of forces on an elementary volume element *dxdydz*, the forces being pressures acting on the faces of the element and gravity acting in the $-z$ direction, we obtain

$$\frac{\partial p}{\partial x} = \frac{\partial p}{\partial y} = 0, \quad \text{and} \quad \frac{\partial p}{\partial z} = -\rho g = -\gamma \tag{2.1.1}$$

The first two of these imply that the pressure is the same in all directions at the same vertical height in a gravitational field. The third, where γ is the specific weight, shows that the pressure increases with depth in a gravitational field, the variation depending on $\rho(z)$. For homogeneous fluids, for which ρ = constant, this last equation can be integrated immediately, yielding

$$p_2 - p_1 = -\rho g(z_2 - z_1) = -\rho g(h_2 - h_1) \tag{2.1.2}$$

or

$$p_2 + \rho g h_2 = p_1 + \rho g h_1 = \text{constant} \tag{2.1.3}$$

where h denotes the elevation. These are the equations for the hydrostatic pressure distribution.

When applied to problems where a liquid, such as the ocean, lies below the atmosphere, with a constant pressure p_{atm}, h is usually measured (positive) downward from the ocean/atmosphere interface and p at any distance h below this interface differs from p_{atm} by an amount

$$p - p_{\text{atm}} = \rho g h \tag{2.1.4}$$

Pressures may be given either as *absolute pressure,* pressure measured relative to absolute vacuum, or *gauge pressure,* pressure measured relative to atmospheric pressure.

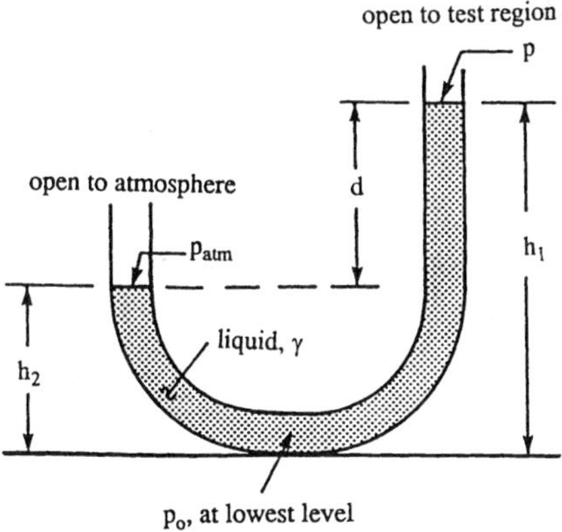

FIGURE 2.1.1 U-tube manometer.

Manometry

The hydrostatic pressure variation may be employed to measure pressure differences in terms of heights of liquid columns — such devices are called manometers and are commonly used in wind tunnels and a host of other applications and devices. Consider, for example the U-tube manometer shown in Figure 2.1.1 filled with liquid of specific weight γ, the left leg open to the atmosphere and the right to the region whose pressure p is to be determined. In terms of the quantities shown in the figure, in the left leg

$$p_0 - \rho g h_2 = p_{\text{atm}} \tag{2.1.5a}$$

and in the right leg

$$p_0 - \rho g h_1 = p \tag{2.1.5b}$$

the difference being

$$p - p_{\text{atm}} = -\rho g (h_1 - h_2) = -\rho g d = -\gamma d \tag{2.1.6}$$

which determines p in terms of the height difference $d = h_1 - h_2$ between the levels of the fluid in the two legs of the manometer.

Hydrostatic Forces on Submerged Objects

The force acting on a submerged object due to the hydrostatic pressure is given by

$$\boldsymbol{F} = \iint p\, \boldsymbol{dA} = \iint p \cdot \boldsymbol{n}\, dA = \iint \rho g h\, \boldsymbol{dA} + p_0 \iint \boldsymbol{dA} \tag{2.1.7}$$

where h is the variable vertical depth of the element $\boldsymbol{dA}$ and p_0 is the pressure at the surface. In turn we consider plane and nonplanar surfaces.

Forces on Plane Surfaces

Consider the planar surface A at an angle θ to a free surface shown in Figure 2.1.2. The force on one side of the planar surface, from Equation (2.1.7), is

$$\boldsymbol{F} = \rho g \boldsymbol{n} \iint_A h\, dA + p_0 A \boldsymbol{n} \tag{2.1.8}$$

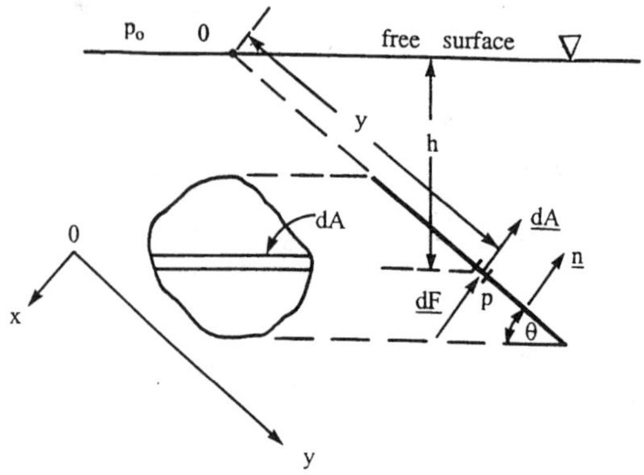

FIGURE 2.1.2 Hydrostatic force on a plane surface.

but $h = y \sin\theta$, so

$$\iint_A h\,dA = \sin\theta \iint_A y\,dA = y_c A \sin\theta = h_c A \tag{2.1.9}$$

where the subscript c indicates the distance measured to the centroid of the area A. The total force (on one side) is then

$$\boldsymbol{F} = \gamma h_c \boldsymbol{A} + p_0 \boldsymbol{A} \tag{2.1.10}$$

Hence, the magnitude of the force is independent of the angle θ, and is equal to the pressure at the centroid, $\gamma h_c + p_0$, times the area. If we use gauge pressure, the term $p_0 A$ in Equation (2.1.10) is dropped.

Since p is not evenly distributed over A, but varies with depth, $\boldsymbol{F}$ does not act through the centroid. The point of action of $\boldsymbol{F}$, called the *center of pressure*, can be determined by considering moments in Figure 2.1.2. The moment of the hydrostatic force acting on the elementary area dA about the axis perpendicular to the page passing through the point 0 on the free surface is

$$y\,dF = y(\gamma y \sin\theta\,dA) = \gamma y^2 \sin\theta\,dA \tag{2.1.11}$$

so if y_{cp} denotes the distance to the center of pressure,

$$y_{cp} F = \gamma \sin\theta \iint y^2\,dA = \gamma \sin\theta\, I_x \tag{2.1.12}$$

where I_x is the moment of inertia of the plane area with respect to the axis formed by the intersection of the plane containing the planar surface and the free surface (say $0x$). Dividing by $F = \gamma h_c A = \gamma y_c \sin\theta A$ gives

$$y_{cp} = \frac{I_x}{y_c A} \tag{2.1.13}$$

By using the parallel axis theorem, $I_x = I_{xc} + Ay_c^2$, where I_{xc} is the moment of inertia with respect to an axis parallel to $0x$ passing through the centroid, Equation (2.1.13) becomes

$$y_{cp} = y_c + \frac{I_{xc}}{y_c A} \tag{2.1.14}$$

which shows that, in general, the center of pressure lies below the centroid.

Similarly, we find x_{cp} by taking moments about the y axis, specifically

$$x_{cp} F = \gamma \sin\theta \iint xy\, dA = \gamma \sin\theta\, I_{xy} \tag{2.1.15}$$

or

$$x_{cp} = \frac{I_{xy}}{y_c A} \tag{2.1.16}$$

where I_{xy} is the product of inertia with respect to the x and y axes. Again, by the parallel axis theorem, $I_{xy} = I_{xyc} + Ax_c y_c$, where the subscript c denotes the value at the centroid, so Equation (2.1.16) can be written

$$x_{cp} = x_c + \frac{I_{xyc}}{y_c A} \tag{2.1.17}$$

This completes the determination of the center of pressure (x_{cp}, y_{cp}). Note that if the submerged area is symmetrical with respect to an axis passing through the centroid and parallel to either the x or y axes that $I_{xyc} = 0$ and $x_{cp} = x_c$; also that as y_c increases, $y_{cp} \rightarrow y_c$.

Centroidal moments of inertia and centroidal coordinates for some common areas are shown in Figure 2.1.3.

Forces on Curved Surfaces

On a curved surface the forces on individual elements of area differ in direction so a simple summation of them is not generally possible, and the most convenient approach to calculating the pressure force on the surface is by separating it into its horizontal and vertical components.

A free-body diagram of the forces acting on the volume of fluid lying above a curved surface together with the conditions of static equilibrium of such a fluid column leads to the results that:

1. The horizontal components of force on a curved submerged surface are equal to the forces exerted on the planar areas formed by the projections of the curved surface onto vertical planes normal to the directions of these components, the lines of action of these forces being calculated as described earlier for planar surfaces; and
2. The vertical component of force on a curved submerged surface is equal in magnitude to the weight of the entire column of fluid lying above the curved surface, and acts through the center of mass of this volume of fluid.

Since the three components of force, two horizontal and one vertical, calculated as above, need not meet at a single point, there is, in general, no single resultant force. They can, however, be combined and considered to be equivalent to a single force at any arbitrary point of application together with a moment about that point.

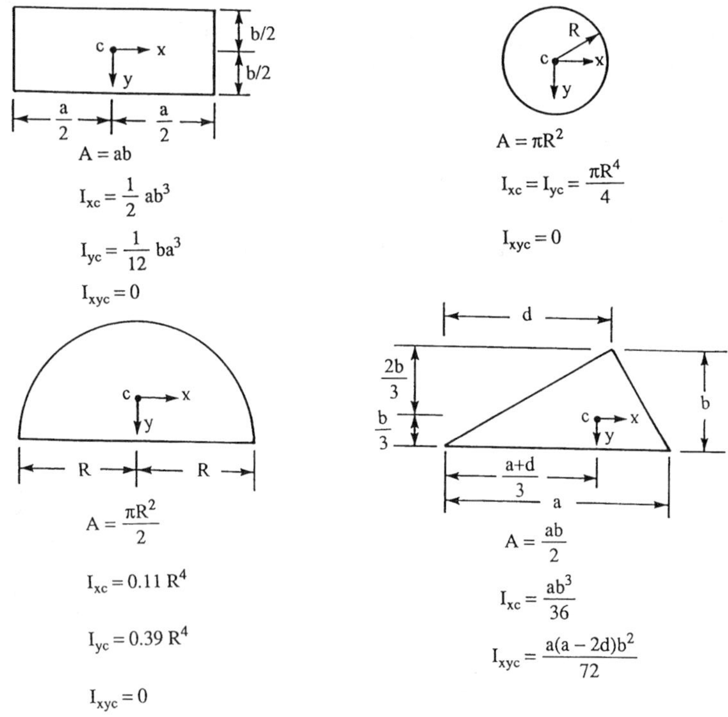

FIGURE 2.1.3 Centroidal moments of inertia and coordinates for some common areas.

Hydrostatic Forces in Layered Fluids

All of the above results which employ the linear hydrostatic variation of pressure are valid only for homogeneous fluids. If the fluid is heterogeneous, consisting of individual layers each of constant density, then the pressure varies linearly with a different slope in each layer and the preceding analyses must be remedied by computing and summing their separate contributions to the forces and moments.

Buoyancy

The same principles used above to compute hydrostatic forces can be used to calculate the net pressure force acting on completely submerged or floating bodies. These laws of buoyancy, the principles of Archimedes, are that:

1. A completely submerged body experiences a vertical upward force equal to the weight of the displaced fluid; and
2. A floating or partially submerged body displaces its own weight in the fluid in which it floats (i.e., the vertical upward force is equal to the body weight).

The line of action of the buoyancy force in both (1) and (2) passes through the centroid of the displaced volume of fluid; this point is called the *center of buoyancy.* (This point need not correspond to the center of mass of the body, which could have nonuniform density. In the above it has been assumed that the displaced fluid has a constant γ. If this is not the case, such as in a layered fluid, the magnitude of the buoyant force is still equal to the weight of the displaced fluid, but the line of action of this force passes through the center of gravity of the displaced volume, not the centroid.)

If a body has a weight exactly equal to that of the volume of fluid it displaces, it is said to be *neutrally buoyant* and will remain at rest at any point where it is immersed in a (homogeneous) fluid.

Stability of Submerged and Floating Bodies

Submerged Body

A body is said to be in stable equilibrium if when given a slight displacement from the equilibrium position the forces thereby created tend to restore it back to its original position. The forces acting on a submerged body are the buoyancy force, F_B, acting through the center of buoyancy, denoted by CB, and the weight of the body, W, acting through the center of gravity denoted by CG (see Figure 2.1.4). We see from Figure 2.1.4 that if the CB lies above the CG a rotation from the equilibrium position creates a restoring couple which will rotate the body back to its original position — thus, this is a *stable* equilibrium situation. The reader will readily verify that when the CB lies below the CG, the couple that results from a rotation from the vertical increases the displacement from the equilibrium position — thus, this is an *unstable* equilibrium situation.

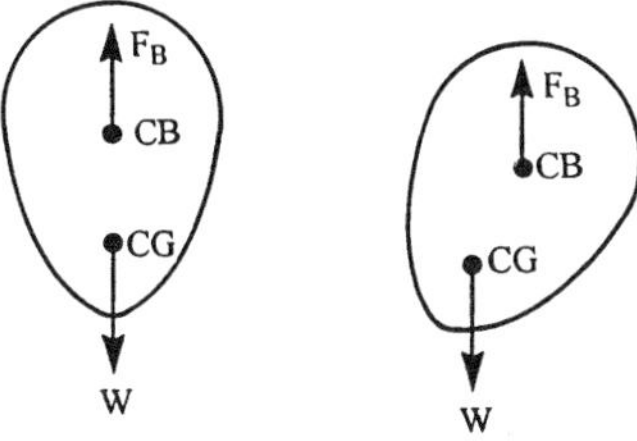

FIGURE 2.1.4 Stability for a submerged body: (left) equilibrium position, (right) slightly rotated.

Partially Submerged Body

The stability problem is more complicated for floating bodies because as the body rotates the location of the center of buoyancy may change. To determine stability in these problems requires that we determine the location of the *metacenter*. This is done for a symmetric body by tilting the body through a small angle $\Delta\theta$ from its equilibrium position and calculating the new location of the center of buoyancy CB′; the point of intersection of a vertical line drawn upward from CB′ with the line of symmetry of the floating body is the metacenter, denoted by M in Figure 2.1.5, and it is independent of $\Delta\theta$ for small

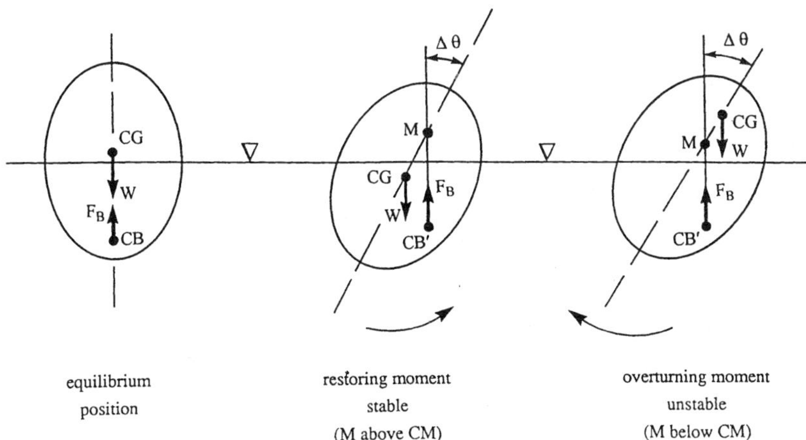

FIGURE 2.1.5 Stability for a partially submerged body.

angles. If M lies above the CG of the body, we see from Figure 2.1.5 that rotation of the body leads to a restoring couple, whereas M lying below the CG leads to a couple which will increase the displacement. Thus, the stability of the equilibrium depends on whether M lies above or below the CG. The directed distance from CG to M is called the *metacentric height*, so equivalently the equilibrium is stable if this vector is positive and unstable if it is negative; stability increases as the metacentric height increases. For geometrically complex bodies, such as ships, the computation of the metacenter can be quite complicated.

Pressure Variation in Rigid-Body Motion of a Fluid

In rigid-body motion of a fluid all the particles translate and rotate as a whole, there is no relative motion between particles, and hence no viscous stresses since these arc proportional to velocity gradients. The equation of motion is then a balance among pressure, gravity, and the fluid acceleration, specifically,

$$\nabla p = \rho(\boldsymbol{g} - \boldsymbol{a}) \tag{2.1.18}$$

where $\boldsymbol{a}$ is the uniform acceleration of the body. Equation (2.1.18) shows that the lines of constant pressure, including a free surface if any, are perpendicular to the direction $\boldsymbol{g} - \boldsymbol{a}$. Two important applications of this are to a fluid in uniform linear translation and rigid-body rotation. While such problems are not, strictly speaking, fluid statics problems, their analysis and the resulting pressure variation results are similar to those for static fluids.

Uniform Linear Acceleration

For a fluid partially filling a large container moving to the right with constant acceleration $\boldsymbol{a} = (a_x, a_y)$ the geometry of Figure 2.1.6 shows that the magnitude of the pressure gradient in the direction $\boldsymbol{n}$ normal to the accelerating free surface, in the direction $\boldsymbol{g} - \boldsymbol{a}$, is

$$\frac{dp}{dn} = \rho\left[a_x^2 + \left(g + a_y\right)^2\right]^{1/2} \tag{2.1.19}$$

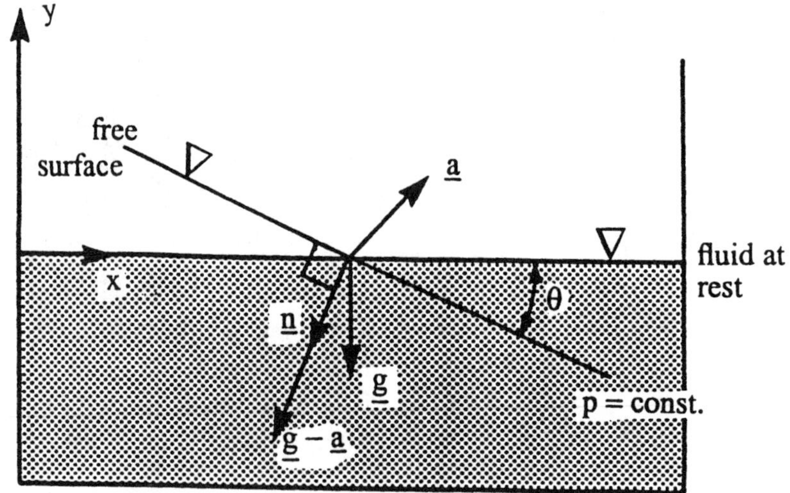

FIGURE 2.1.6 A fluid with a free surface in uniform linear acceleration.

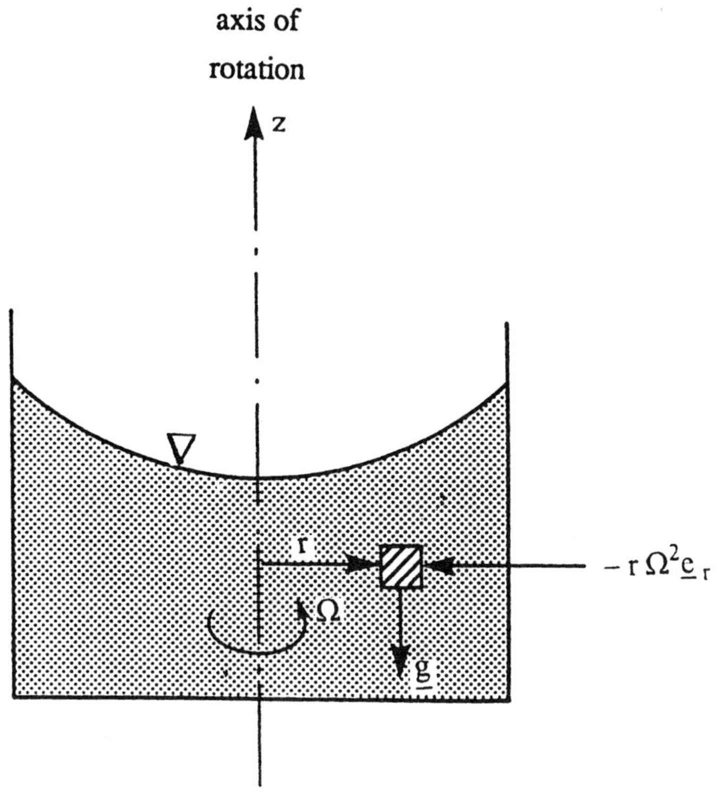

FIGURE 2.1.7 A fluid with a free surface in rigid-body rotation.

and the free surface is oriented at an angle to the horizontal

$$\theta = \tan^{-1}\left(\frac{a_x}{g + a_y}\right) \tag{2.1.20}$$

Rigid-Body Rotation

Consider the fluid-filled circular cylinder rotating uniformly with angular velocity $\Omega = \Omega e_z$ (Figure 2.1.7). The only acceleration is the centripetal acceleration $\Omega \times (\Omega \times r) = -r\Omega^2 e_r$, so Equation 2.1.18 becomes

$$\nabla p = \frac{\partial p}{\partial r} e_r + \frac{\partial p}{\partial z} e_z = \rho(g - a) = \rho\left(r\Omega^2 e_r - g e_z\right) \tag{2.1.21}$$

or

$$\frac{\partial p}{\partial r} = \rho r \Omega^2, \qquad \frac{\partial p}{\partial z} = -\rho g = -\gamma \tag{2.1.22}$$

Integration of these equations leads to

$$p = p_o - \gamma z + \frac{1}{2}\rho r^2 \Omega^2 \tag{2.1.23}$$

where p_o is the reference pressure (at r = z = 0). This result shows that at any fixed r the pressure varies hydrostatically in the vertical direction, while the constant pressure surfaces, including the free surface, are paraboloids of revolution.

Further Information

The reader may find more detail and additional information on the topics in this section in any one of the many excellent introductory texts on fluid mechanics, such as

White, F.M. 1999. *Fluid Mechanics,* 4th ed., McGraw-Hill, New York.

Munson, B.R., Young, D.F., and Okiishi, T.H. 1998. *Fundamentals of Fluid Mechanics,* 3rd ed., John Wiley & Sons, New York.

2.2 Equations of Motion and Potential Flow

Stanley A. Berger

Integral Relations for a Control Volume

Like most physical conservation laws those governing the motion of a fluid apply to material particles or systems of such particles. This so-called Lagrangian viewpoint is generally not as useful in practical fluid flows as an analysis based on the flow through fixed (or deformable) control volumes — the Eulerian viewpoint. The relationship between these two viewpoints can be deduced from the Reynolds transport theorem, from which we also most readily derive the governing integral and differential equations of motion.

Reynolds Transport Theorem

The *extensive* quantity B, a scalar, vector, or tensor, is defined as any property of a given quantity of fluid (e.g., momentum, energy) and b as the corresponding value per unit mass (the *intensive* value). The Reynolds transport theorem for a moving and arbitrarily deforming control volume CV, with boundary CS (see Figure 2.2.1), states that

$$\frac{d}{dt}\left(B_{\text{system}}\right) = \frac{d}{dt}\left(\iiint_{\text{CV}} \rho b \, d\upsilon\right) + \iint_{\text{CS}} \rho b\left(\boldsymbol{V}_r \cdot \boldsymbol{n}\right) dA \tag{2.2.1}$$

where B_{system} is the total quantity of B in the system (any mass of fixed identity), $\boldsymbol{n}$ is the outward normal to the CS, $\boldsymbol{V}_r = \boldsymbol{V}(\boldsymbol{r}, t) - \boldsymbol{V}_{\text{CS}}(\boldsymbol{r}, t)$, the velocity of the fluid particle, $\boldsymbol{V}(\boldsymbol{r}, t)$, relative to that of the CS, $\boldsymbol{V}_{\text{CS}}(\boldsymbol{r}, t)$, and d/dt on the left-hand side is the derivative following the fluid particles, i.e., the fluid mass comprising the system. The theorem states that the time rate of change of the total B in the system is equal to the rate of change within the CV plus the net flux of B through the CS. To distinguish between the d/dt which appears on the two sides of Equation (2.2.1) but which have different interpretations, the derivative on the left-hand side, following the system, is denoted by D/Dt and is called the material derivative. This notation is used in what follows. For any function $f(x, y, z, t)$,

$$\frac{Df}{Dt} = \frac{\partial f}{\partial t} + \boldsymbol{V} \cdot \nabla f$$

For a CV fixed with respect to the reference frame, Equation (2.2.1) reduces to

$$\frac{D}{Dt}\left(B_{\text{system}}\right) = \frac{d}{dt}\iiint_{\substack{\text{CV}\\ \text{(fixed)}}} (\rho b) \, d\upsilon + \iint_{\text{CS}} \rho b(\boldsymbol{V} \cdot \boldsymbol{n}) \, dA \tag{2.2.2}$$

(The time derivative operator in the first term on the right-hand side may be moved inside the integral, in which case it is then to be interpreted as the partial derivative $\partial/\partial t$.)

Conservation of Mass

If we apply Equation (2.2.2) for a fixed control volume, with B_{system} the total mass in the system, then since conservation of mass requires that $DB_{\text{system}}/Dt = 0$ there follows, since $b = B_{\text{system}}/m = 1$,

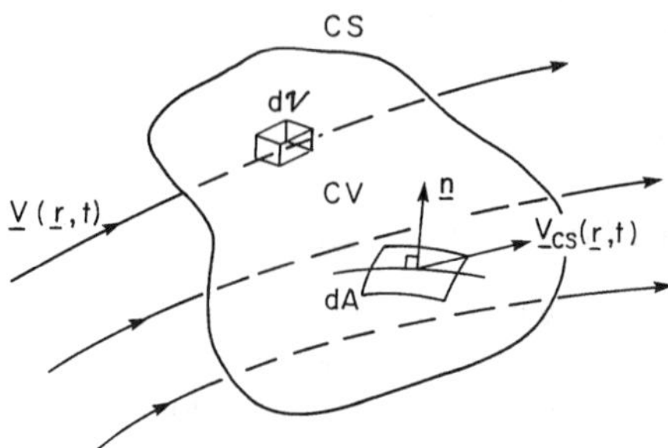

FIGURE 2.2.1 Control volume.

$$\iiint_{\substack{\text{CV}\\(\text{fixed})}} \frac{\partial \rho}{\partial t}\, d\upsilon + \iint_{\text{CS}} \rho(\boldsymbol{V}\cdot\boldsymbol{n})\, dA = 0 \tag{2.2.3}$$

This is the integral form of the conservation of mass law for a fixed control volume. For a steady flow, Equation (2.2.3) reduces to

$$\iint_{\text{CS}} \rho(\boldsymbol{V}\cdot\boldsymbol{n})\, dA = 0 \tag{2.2.4}$$

whether compressible or incompressible. For an incompressible flow, ρ = constant, so

$$\iint_{\text{CS}} (\boldsymbol{V}\cdot\boldsymbol{n})\, dA = 0 \tag{2.2.5}$$

whether the flow is steady or unsteady.

Conservation of Momentum

The conservation of (linear) momentum states that

$$\boldsymbol{F}_{total} \equiv \sum(\text{external forces acting on the fluid system}) = \frac{D\boldsymbol{M}}{Dt} \equiv \frac{D}{Dt}\left(\iiint_{\text{system}} \rho \boldsymbol{V}\, d\upsilon\right) \tag{2.2.6}$$

where $\boldsymbol{M}$ is the total system momentum. For an arbitrarily moving, deformable control volume it then follows from Equation (2.2.1) with b set to $\boldsymbol{V}$,

$$\boldsymbol{F}_{\text{total}} = \frac{d}{dt}\left(\iiint_{\text{CV}} \rho \boldsymbol{V}\, d\upsilon\right) + \iint_{\text{CS}} \rho \boldsymbol{V}(\boldsymbol{V}_r\cdot\boldsymbol{n})\, dA \tag{2.2.7}$$

This expression is only valid in an inertial coordinate frame. To write the equivalent expression for a noninertial frame we must use the relationship between the acceleration $\boldsymbol{a}_I$ in an inertial frame and the acceleration $\boldsymbol{a}_R$ in a noninertial frame,

$$a_I = a_R + \frac{d^2 R}{dt^2} + 2\Omega \times V + \Omega \times (\Omega \times r) + \frac{d\Omega}{dt} \times r \tag{2.2.8}$$

where R is the position vector of the origin of the noninertial frame with respect to that of the inertial frame, Ω is the angular velocity of the noninertial frame, and r and V the position and velocity vectors in the noninertial frame. The third term on the right-hand side of Equation (2.2.8) is the Coriolis acceleration, and the fourth term is the centripetal acceleration. For a noninertial frame Equation (2.2.7) is then

$$\begin{aligned} F_{\text{total}} - \iiint_{\text{system}} \left[\frac{d^2 R}{dt^2} + 2\Omega \times V + \Omega \times (\Omega \times r) + \frac{d\Omega}{dt} \times r \right] \rho \, d\upsilon &= \frac{D}{Dt}\left(\iiint_{\text{system}} \rho V \, d\upsilon \right) \\ &= \frac{d}{dt}\left(\iiint_{\text{CV}} \rho V \, d\upsilon \right) + \iint_{\text{CS}} \rho V \cdot (V_r \cdot n) \, dA \end{aligned} \tag{2.2.9}$$

where the frame acceleration terms of Equation (2.2.8) have been brought to the left-hand side because to an observer in the noninertial frame they act as "apparent" body forces.

For a fixed control volume in an inertial frame for steady flow it follows from the above that

$$F_{\text{total}} = \iint_{\text{CS}} \rho V (V \cdot n) \, dA \tag{2.2.10}$$

This expression is the basis of many control volume analyses for fluid flow problems.

The cross product of r, the position vector with respect to a convenient origin, with the momentum Equation (2.2.6) written for an elementary particle of mass dm, noting that $(dr/dt) \times V = 0$, leads to the integral moment of momentum equation

$$\sum M - M_I = \frac{D}{Dt} \iiint_{\text{system}} \rho (r \times V) \, d\upsilon \tag{2.2.11}$$

where ΣM is the sum of the moments of all the external forces acting on the system about the origin of r, and M_I is the moment of the apparent body forces (see Equation (2.2.9)). The right-hand side can be written for a control volume using the appropriate form of the Reynolds transport theorem.

Conservation of Energy

The conservation of energy law follows from the first law of thermodynamics for a moving system

$$\dot{Q} - \dot{W} = \frac{D}{Dt}\left(\iiint_{\text{system}} \rho e \, d\upsilon \right) \tag{2.2.12}$$

where $\dot{Q}$ is the rate at which heat is added to the system, $\dot{W}$ the rate at which the system works on its surroundings, and e is the total energy per unit mass. For a particle of mass dm the contributions to the specific energy e are the internal energy u, the kinetic energy $V^2/2$, and the potential energy, which in the case of gravity, the only body force we shall consider, is gz, where z is the vertical displacement opposite to the direction of gravity. (We assume no energy transfer owing to chemical reaction as well as no magnetic or electric fields.) For a fixed control volume it then follows from Equation (2.2.2) [with $b = e = u + (V^2/2) + gz$] that

$$\dot{Q} - \dot{W} = \frac{d}{dt}\left(\iiint_{CV} \rho\left(u + \frac{1}{2}V^2 + gz \right) d\upsilon \right) + \iint_{CS} \rho\left(u + \frac{1}{2}V^2 + gz \right)(\boldsymbol{V}\cdot\boldsymbol{n})\, dA \tag{2.2.13}$$

Problem

An incompressible fluid flows through a pump at a volumetric flow rate $\hat{Q}$. The (head) loss between sections 1 and 2 (see Figure 2.2.2) is equal to $\beta\rho V_1^2/2$ (V is the average velocity at the section). Calculate the power that must be delivered by the pump to the fluid to produce a given increase in pressure, $\Delta p = p_2 - p_1$.

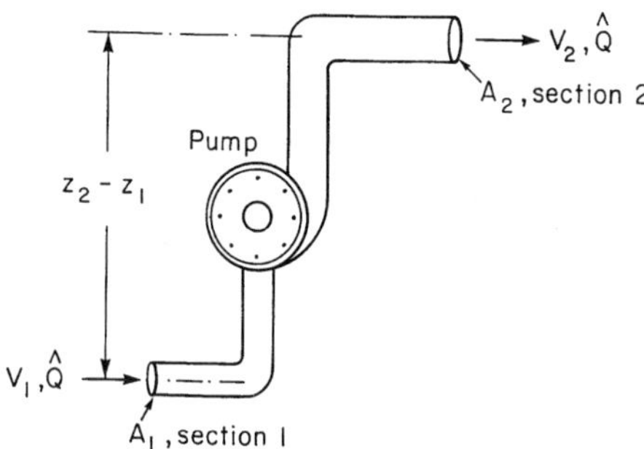

FIGURE 2.2.2 Pump producing pressure increase.

Solution: The principal equation needed is the energy Equation (2.2.13). The term $\dot{W}$, the rate at which the system does work on its surroundings, for such problems has the form

$$\dot{W} = -\dot{W}_{\text{shaft}} + \iint_{CS} p\boldsymbol{V}\cdot\boldsymbol{n}\, dA \tag{P.2.2.1}$$

where $\dot{W}_{\text{shaft}}$ represents the work done on the fluid by a moving shaft, such as by turbines, propellers, fans, etc., and the second term on the right side represents the rate of working by the normal stress, the pressure, at the boundary. For a steady flow in a control volume coincident with the physical system boundaries and bounded at its ends by sections 1 and 2, Equation (2.2.13) reduces to ($u \equiv 0$),

$$\dot{Q} + \dot{W}_{\text{shaft}} - \iint_{CS} p\boldsymbol{V}\cdot\boldsymbol{n}\, dA = \iint_{CS} \left(\frac{1}{2}\rho V^2 + \gamma z \right)(\boldsymbol{V}\cdot\boldsymbol{n})\, dA \tag{P.2.2.2}$$

Using average quantities at sections 1 and 2, and the continuity Equation (2.2.5), which reduces in this case to

$$V_1 A_1 = V_2 A_2 = \hat{Q} \tag{P.2.2.3}$$

we can write Equation (P.2.2.2) as

$$\dot{Q} + \dot{W}_{\text{shaft}} - (p_2 - p_1)\hat{Q} = \left[\frac{1}{2}\rho\left(V_2^2 - V_1^2\right) + \gamma\left(z_2 - z_1\right) \right]\hat{Q} \tag{P.2.2.4}$$

$\dot{Q}$, the rate at which heat is added to the system, is here equal to $-\beta\rho V_1^2/2$, the head loss between sections 1 and 2. Equation (P.2.2.4) then can be rewritten

$$\dot{W}_{\text{shaft}} = \beta\rho\frac{V_1^2}{2} + (\Delta p)\hat{Q} + \frac{1}{2}\rho\left(V_2^2 - V_1^2\right)\hat{Q} + \gamma\left(z_2 - z_1\right)\hat{Q}$$

or, in terms of the given quantities,

$$\dot{W}_{\text{shaft}} = \frac{\beta\rho\hat{Q}^2}{A_1^2} + (\Delta p)\hat{Q} + \frac{1}{2}\rho\frac{\hat{Q}^3}{A_2^2}\left(1 - \frac{A_2^2}{A_1^2}\right) + \gamma\left(z_2 - z_1\right)\hat{Q} \tag{P.2.2.5}$$

Thus, for example, if the fluid is water ($\rho \approx 1000$ kg/m^3, $\gamma = 9.8$ kN/m^3), $\hat{Q} = 0.5$ m^3/sec, the heat loss is $0.2\rho V_1^2/2$, and $\Delta p = p_2 - p_1 = 2 \times 10^5$N/m^2 = 200 kPa, $A_1 = 0.1$ m^2 = $A_2/2$, $(z_2 - z_1) = 2$ m, we find, using Equation (P.2.2.5)

$$\begin{aligned}
\dot{W}_{\text{shaft}} &= \frac{0.2(1000)(0.5)^2}{(0.1)^2} + \left(2\times 10^5\right)(0.5) + \frac{1}{2}(1000)\frac{(0.5)^3}{(0.2)^2}(1-4) + \left(9.8\times 10^3\right)(2)(0.5) \\
&= 5{,}000 + 10{,}000 - 4{,}688 + 9{,}800 = 20{,}112 \text{ Nm/sec} \\
&= 20{,}112 \text{ W} = \frac{20{,}112}{745.7}\text{hp} = 27 \text{ hp}
\end{aligned}$$

Differential Relations for Fluid Motion

In the previous section the conservation laws were derived in integral form. These forms are useful in calculating, generally using a control volume analysis, gross features of a flow. Such analyses usually require some *a priori* knowledge or assumptions about the flow. In any case, an approach based on integral conservation laws cannot be used to determine the point-by-point variation of the dependent variables, such as velocity, pressure, temperature, etc. To do this requires the use of the differential forms of the conservation laws, which are presented below.

Mass Conservation–Continuity Equation

Applying Gauss's theorem (the divergence theorem) to Equation (2.2.3) we obtain

$$\iiint_{\substack{\text{CV}\\ \text{(fixed)}}} \left[\frac{\partial\rho}{\partial t} + \nabla\cdot(\rho\boldsymbol{V})\right] d\upsilon = 0 \tag{2.2.14}$$

which, because the control volume is arbitrary, immediately yields

$$\frac{\partial\rho}{\partial t} + \nabla\cdot(\rho\boldsymbol{V}) = 0 \tag{2.2.15}$$

This can also be written as

$$\frac{D\rho}{Dt} + \rho\nabla\cdot\boldsymbol{V} = 0 \tag{2.2.16}$$

using the fact that

$$\frac{D\rho}{Dt} = \frac{\partial \rho}{\partial t} + \boldsymbol{V} \cdot \nabla \rho \qquad (2.2.17)$$

Special cases:

1. Steady flow $[(\partial/\partial t)\ (\ ) \equiv 0]$

$$\nabla \cdot (\rho \boldsymbol{V}) = 0 \qquad (2.2.18)$$

2. Incompressible flow ($D\rho/Dt \equiv 0$)

$$\nabla \cdot \boldsymbol{V} = 0 \qquad (2.2.19)$$

Momentum Conservation

We note first, as a consequence of mass conservation for a system, that the right-hand side of Equation (2.2.6) can be written as

$$\frac{D}{Dt}\left(\iiint_{\text{system}} \rho \boldsymbol{V}\, d\upsilon \right) \equiv \iiint_{\text{system}} \rho \frac{D\boldsymbol{V}}{Dt}\, d\upsilon \qquad (2.2.20)$$

The total force acting on the system which appears on the left-hand side of Equation (2.2.6) is the sum of body forces $\boldsymbol{F}_b$ and surface forces $\boldsymbol{F}_s$. The body forces are often given as forces per unit mass (e.g., gravity), and so can be written

$$\boldsymbol{F}_b = \iiint_{\text{system}} \rho \boldsymbol{f}\, d\upsilon \qquad (2.2.21)$$

The surface forces are represented in terms of the second-order stress tensor* $\underline{\underline{\sigma}} = \{\sigma_{ij}\}$, where σ_{ij} is defined as the force per unit area in the i direction on a planar element whose normal lies in the j direction.** From elementary angular momentum considerations for an infinitesimal volume it can be shown that σ_{ij} is a symmetric tensor, and therefore has only six independent components. The total surface force exerted on the system by its surroundings is given by

$$\boldsymbol{F}_s = \iint_{\substack{\text{system}\\ \text{surface}}} \underline{\underline{\sigma}} \cdot \boldsymbol{n}\, dA, \text{ with } i-\text{component } F_{s_i} = \iint \sigma_{ij} n_j\, dA \qquad (2.2.22)$$

The integral momentum conservation law Equation (2.2.6) can then be written

$$\iiint_{\text{system}} \rho \frac{D\boldsymbol{V}}{Dt}\, d\upsilon = \iiint_{\text{system}} \rho \boldsymbol{f}\, d\upsilon + \iint_{\substack{\text{system}\\ \text{surface}}} \underline{\underline{\sigma}} \cdot \boldsymbol{n}\, dA \qquad (2.2.23)$$

---

*We shall assume the reader is familiar with elementary Cartesian tensor analysis and the associated subscript notation and conventions. The reader for whom this is not true should skip the details and concentrate on the final principal results and equations given at the ends of the next few subsections.

**This assignment of roles to the first and second subscripts of the stress tensor is a convention and is far from universal. Frequently, their roles are reversed, the first subscript denoting the direction of the normal to the planar element, the second the direction of the force.

The application of the divergence theorem to the last term on the right-side of Equation (2.2.23) leads to

$$\iiint_{\text{system}} \rho \frac{DV}{Dt}\, d\upsilon = \iiint_{\text{system}} \rho f \, d\upsilon + \iiint_{\text{system}} \nabla \cdot \underline{\underline{\sigma}} \, d\upsilon \qquad (2.2.24)$$

where $\nabla \cdot \underline{\underline{\sigma}} \equiv \{\partial\sigma_{ij}/x_j\}$. Since Equation (2.2.24) holds for any material volume, it follows that

$$\rho \frac{DV}{Dt} = \rho f + \nabla \cdot \underline{\underline{\sigma}} \qquad (2.2.25)$$

(With the decomposition of $\boldsymbol{F}_{\text{total}}$ above, Equation (2.2.10) can be written

$$\iiint_{\text{CV}} \rho f \, d\upsilon + \iint_{\text{CS}} \underline{\underline{\sigma}} \cdot \boldsymbol{n} \, dA = \iint_{\text{CS}} \rho \boldsymbol{V}(\boldsymbol{V} \cdot \boldsymbol{n}) \, dA \qquad (2.2.26)$$

If ρ is uniform and $\boldsymbol{f}$ is a conservative body force, i.e., $\boldsymbol{f} = -\nabla\Psi$, where Ψ is the force potential, then Equation (2.2.26), after application of the divergence theorem to the body force term, can be written

$$\iint_{\text{CS}} \left(-\rho\Psi \boldsymbol{n} + \underline{\underline{\sigma}} \cdot \boldsymbol{n}\right) dA = \iint_{\text{CS}} \rho \boldsymbol{V}(\boldsymbol{V} \cdot \boldsymbol{n}) \, dA \qquad (2.2.27)$$

It is in this form, involving only integrals over the surface of the control volume, that the integral form of the momentum equation is used in control volume analyses, particularly in the case when the body force term is absent.

Analysis of Rate of Deformation

The principal aim of the following two subsections is to derive a relationship between the stress and the rate of strain to be used in the momentum Equation (2.2.25). The reader less familiar with tensor notation may skip these sections, apart from noting some of the terms and quantities defined therein, and proceed directly to Equations (2.2.38) or (2.2.39).

The relative motion of two neighboring points P and Q, separated by a distance η, can be written (using $\boldsymbol{u}$ for the local velocity)

$$\boldsymbol{u}(Q) = \boldsymbol{u}(P) + (\nabla \boldsymbol{u})\boldsymbol{\eta}$$

or, equivalently, writing $\nabla \boldsymbol{u}$ as the sum of antisymmetric and symmetric tensors,

$$\boldsymbol{u}(Q) = \boldsymbol{u}(P) + \frac{1}{2}\left((\nabla \boldsymbol{u}) - (\nabla \boldsymbol{u})^*\right)\boldsymbol{\eta} + \frac{1}{2}\left((\nabla \boldsymbol{u}) + (\nabla \boldsymbol{u})^*\right)\boldsymbol{\eta} \qquad (2.2.28)$$

where $\nabla \boldsymbol{u} = \{\partial u_i/\partial x_j\}$, and the superscript * denotes transpose, so $(\nabla \boldsymbol{u})^* = \{\partial u_j/\partial x_i\}$. The second term on the right-hand side of Equation (2.2.28) can be rewritten in terms of the *vorticity*, $\nabla \times \boldsymbol{u}$, so Equation (2.2.28) becomes

$$\boldsymbol{u}(Q) = \boldsymbol{u}(P) + \frac{1}{2}(\nabla \times \boldsymbol{u}) \times \boldsymbol{\eta} + \frac{1}{2}\left((\nabla \boldsymbol{u}) + (\nabla \boldsymbol{u})^*\right)\boldsymbol{\eta} \qquad (2.2.29)$$

which shows that the local velocity field consists of a rigid-body translation, a rigid-body rotation with angular velocity $^1/_2\ (\nabla \times \boldsymbol{u})$, and a velocity or rate of deformation. The coefficient of η in the last term in Equation (2.2.29) is defined as the rate-of-strain tensor and is denoted by $\underline{\underline{e}}$, in subscript form

$$e_{ij} = \frac{1}{2}\left(\frac{\partial u_i}{\partial x_j} + \frac{\partial u_j}{\partial x_i}\right) \tag{2.2.30}$$

From $\underline{\underline{e}}$ we can define a rate-of-strain central quadric, along the principal axes of which the deforming motion consists of a pure straining extension or contraction.

Relationship Between Forces and Rate of Deformation

We are now in a position to determine the required relationship between the stress tensor $\underline{\underline{\sigma}}$ and the rate of deformation. Assuming that in a static fluid the stress reduces to a (negative) hydrostatic or thermodynamic pressure, equal in all directions, we can write

$$\underline{\underline{\sigma}} = -p\underline{\underline{I}} + \underline{\underline{\tau}} \quad \text{or} \quad \sigma_{ij} = -p\delta_{ij} + \tau_{ij} \tag{2.2.31}$$

where $\underline{\underline{\tau}}$ is the viscous part of the total stress and is called the deviatoric stress tensor, $\underline{\underline{I}}$ is the identity tensor, and δ_{ij} is the corresponding Kronecker delta ($\delta_{ij} = 0$ if $i \neq j$; $\delta_{ij} = 1$ if $i = j$). We make further assumptions that (1) the fluid exhibits no preferred directions; (2) the stress is independent of any previous history of distortion; and (3) that the stress depends only on the local thermodynamic state and the kinematic state of the immediate neighborhood. Precisely, we assume that $\underline{\underline{\tau}}$ is linearly proportional to the first spatial derivatives of $\boldsymbol{u}$, the coefficient of proportionality depending only on the local thermodynamic state. These assumptions and the relations below which follow from them are appropriate for a Newtonian fluid. Most common fluids, such as air and water under most conditions, are Newtonian, but there are many other fluids, including many which arise in industrial applications, which exhibit so-called non-Newtonian properties. The study of such non-Newtonian fluids, such as viscoelastic fluids, is the subject of the field of rheology.

With the Newtonian fluid assumptions above, and the symmetry of $\underline{\underline{\tau}}$ which follows from the symmetry of $\underline{\underline{\sigma}}$, one can show that the viscous part $\underline{\underline{\tau}}$ of the total stress can be written as

$$\underline{\underline{\tau}} = \lambda(\nabla \cdot \boldsymbol{u})\underline{\underline{I}} + 2\mu\underline{\underline{e}} \tag{2.2.32}$$

so the total stress for a Newtonian fluid is

$$\underline{\underline{\sigma}} = -p\underline{\underline{I}} + \lambda(\nabla \cdot \boldsymbol{u})\underline{\underline{I}} + 2\mu\underline{\underline{e}} \tag{2.2.33}$$

or, in subscript notation

$$\sigma_{ij} = -p\delta_{ij} + \lambda\left(\frac{\partial u_k}{\partial x_k}\right)\delta_{ij} + \mu\left(\frac{\partial u_i}{\partial x_j} + \frac{\partial u_j}{\partial x_i}\right) \tag{2.2.34}$$

(the Einstein summation convention is assumed here, namely, that a repeated subscript, such as in the second term on the right-hand side above, is summed over; note also that $\nabla \cdot \boldsymbol{u} = \partial u_k/\partial x_k = e_{kk}$.) The coefficient λ is called the "second viscosity" and μ the "absolute viscosity," or more commonly the "dynamic viscosity," or simply the "viscosity." For a Newtonian fluid λ and μ depend only on local thermodynamic state, primarily on the temperature.

We note, from Equation (2.2.34), that whereas in a fluid at rest the stress is an isotropic normal stress (equal to p in all directions), this is not the case for a moving fluid, since in general $\sigma_{11} \neq \sigma_{22} \neq \sigma_{33}$. To have an analogous quantity to p for a moving fluid we define the pressure in a moving fluid as the negative mean normal stress, denoted, say, by $\bar{p}$

$$\bar{p} = -\frac{1}{3}\sigma_{ii} \tag{2.2.35}$$

(σ_{ii} is the trace of $\underline{\underline{\sigma}}$ and an invariant of $\underline{\underline{\sigma}}$, independent of the orientation of the axes). From Equation (2.2.34)

$$\bar{p} = -\frac{1}{3}\sigma_{ii} = p - \left(\lambda + \frac{2}{3}\mu\right)\nabla \cdot \boldsymbol{u} \tag{2.2.36}$$

For an incompressible fluid $\nabla \cdot \boldsymbol{u} = 0$ and hence $\bar{p} \equiv p$. The quantity $(\lambda + {}^2/_3\mu)$ is called the bulk viscosity. If one assumes that the deviatoric stress tensor τ_{ij} makes no contribution to the mean normal stress, it follows that $\lambda + {}^2/_3\mu = 0$, so again $\bar{p} = p$. This condition, $\lambda = -{}^2/_3\mu$, is called the Stokes assumption or hypothesis. If neither the incompressibility nor the Stokes assumptions are made, the difference between $\bar{p}$ and p is usually still negligibly small because $(\lambda + {}^2/_3\mu)\ \nabla \cdot \boldsymbol{u} << p$ in most fluid flow problems. If the Stokes hypothesis is made, as is often the case in fluid mechanics, Equation (2.2.34) becomes

$$\sigma_{ij} = -p\delta_{ij} + 2\mu\left(e_{ij} - \frac{1}{3}e_{kk}\delta_{ij}\right) \tag{2.2.37}$$

The Navier–Stokes Equations

Substitution of Equation (2.2.33) into (2.2.25), since $\nabla \cdot (\phi \underline{\underline{I}}) = \nabla\phi$, for any scalar function ϕ, yields (replacing $\boldsymbol{u}$ in Equation (2.2.33) by $\boldsymbol{V}$)

$$\rho\frac{D\boldsymbol{V}}{Dt} = \rho\boldsymbol{f} - \nabla p + \nabla(\lambda\nabla \cdot \boldsymbol{V}) + \nabla \cdot \left(2\mu\underline{\underline{e}}\right) \tag{2.2.38}$$

These equations are the Navier–Stokes equations (although the name is as often given to the full set of governing conservation equations). With the Stokes assumption ($\lambda = -{}^2/_3\mu$), Equation (2.2.38) becomes

$$\rho\frac{D\boldsymbol{V}}{Dt} = \rho\boldsymbol{f} - \nabla p + \nabla \cdot \left[2\mu\left(\underline{\underline{e}} - \frac{1}{3}e_{kk}\underline{\underline{I}}\right)\right] \tag{2.2.39}$$

If the Eulerian frame is not an inertial frame, then one must use the transformation to an inertial frame either using Equation (2.2.8) or the "apparent" body force formulation, Equation (2.2.9).

Energy Conservation — The Mechanical and Thermal Energy Equations

In deriving the differential form of the energy equation we begin by assuming that heat enters or leaves the material or control volume by heat conduction across the boundaries, the heat flux per unit area being $\boldsymbol{q}$. It then follows that

$$\dot{Q} = -\iint \boldsymbol{q} \cdot \boldsymbol{n}\, dA = -\iiint \nabla \cdot \boldsymbol{q}\, d\upsilon \tag{2.2.40}$$

The work-rate term $\dot{W}$ can be decomposed into the rate of work done against body forces, given by

$$-\iiint \rho \boldsymbol{f} \cdot \boldsymbol{V}\, d\upsilon \tag{2.2.41}$$

and the rate of work done against surface stresses, given by

$$-\iint_{\substack{\text{system}\\\text{surface}}} \boldsymbol{V} \cdot \left(\underline{\underline{\sigma}} \boldsymbol{n}\right) dA \tag{2.2.42}$$

Substitution of these expressions for $\dot{Q}$ and $\dot{W}$ into Equation (2.2.12), use of the divergence theorem, and conservation of mass lead to

$$\rho \frac{D}{Dt}\left(u + \frac{1}{2}V^2\right) = -\nabla \cdot \boldsymbol{q} + \rho \boldsymbol{f} \cdot \boldsymbol{V} + \nabla \cdot \left(\boldsymbol{V}\underline{\underline{\sigma}}\right) \tag{2.2.43}$$

(note that a potential energy term is no longer included in e, the total specific energy, as it is accounted for by the body force rate-of-working term $\rho \boldsymbol{f} \cdot \boldsymbol{V}$).

Equation (2.2.43) is the total energy equation showing how the energy changes as a result of working by the body and surface forces and heat transfer. It is often useful to have a purely thermal energy equation. This is obtained by subtracting from Equation (2.2.43) the dot product of $\boldsymbol{V}$ with the momentum Equation (2.2.25), after expanding the last term in Equation (2.2.43), resulting in

$$\rho \frac{Du}{Dt} = \frac{\partial V_i}{\partial x_j}\sigma_{ij} - \nabla \cdot \boldsymbol{q} \tag{2.2.44}$$

With $\sigma_{ij} = -p\delta_{ij} + \tau_{ij}$, and the use of the continuity equation in the form of Equation (2.2.16), the first term on the right-hand side of Equation (2.2.44) may be written

$$\frac{\partial V_i}{\partial x_j}\sigma_{ij} = -\rho \frac{D\left(\dfrac{p}{\rho}\right)}{Dt} + \frac{Dp}{Dt} + \Phi \tag{2.2.45}$$

where Φ is the rate of dissipation of mechanical energy per unit mass due to viscosity, and is given by

$$\Phi \equiv \frac{\partial V_i}{\partial x_j}\tau_{ij} = 2\mu\left(e_{ij}e_{ij} - \frac{1}{3}e_{kk}^2\right) = 2\mu\left(e_{ij} - \frac{1}{3}e_{kk}\delta_{ij}\right)^2 \tag{2.2.46}$$

With the introduction of Equation (2.2.45), Equation (2.2.44) becomes

$$\rho \frac{De}{Dt} = -p\nabla \cdot \boldsymbol{V} + \Phi - \nabla \cdot \boldsymbol{q} \tag{2.2.47}$$

or

$$\rho \frac{Dh}{Dt} = \frac{Dp}{Dt} + \Phi - \nabla \cdot \boldsymbol{q} \tag{2.2.48}$$

where $h = e + (p/\rho)$ is the specific enthalpy. Unlike the other terms on the right-hand side of Equation (2.2.47), which can be negative or positive, Φ is always nonnegative and represents the increase in internal energy (or enthalpy) owing to irreversible degradation of mechanical energy. Finally, from elementary thermodynamic considerations

$$\frac{Dh}{Dt} = T\frac{DS}{Dt} + \frac{1}{\rho}\frac{Dp}{Dt}$$

where S is the entropy, so Equation (2.2.48) can be written

$$\rho T\frac{DS}{Dt} = \Phi - \nabla \cdot \boldsymbol{q} \tag{2.2.49}$$

If the heat conduction is assumed to obey the Fourier heat conduction law, so $\boldsymbol{q} = -k\nabla T$, where k is the thermal conductivity, then in all of the above equations

$$-\nabla \cdot \boldsymbol{q} = \nabla \cdot (k\nabla T) = k\nabla^2 T \tag{2.2.50}$$

the last of these equalities holding only if k = constant.

In the event the thermodynamic quantities vary little, the coefficients of the constitutive relations for $\underline{\underline{\sigma}}$ and $\boldsymbol{q}$ may be taken to be constant and the above equations simplified accordingly.

We note also that if the flow is incompressible, then the mass conservation, or continuity, equation simplifies to

$$\nabla \cdot \boldsymbol{V} = 0 \tag{2.2.51}$$

and the momentum Equation (2.2.38) to

$$\rho\frac{D\boldsymbol{V}}{Dt} = \rho\boldsymbol{f} - \nabla p + \mu\nabla^2\boldsymbol{V} \tag{2.2.52}$$

where ∇^2 is the Laplacian operator. The small temperature changes, compatible with the incompressibility assumption, are then determined, for a perfect gas with constant k and specific heats, by the energy equation rewritten for the temperature, in the form

$$\rho c_v\frac{DT}{Dt} = k\nabla^2 T + \Phi \tag{2.2.53}$$

Boundary Conditions

The appropriate boundary conditions to be applied at the boundary of a fluid in contact with another medium depends on the nature of this other medium — solid, liquid, or gas. We discuss a few of the more important cases here in turn:

1. *At a solid surface:* $\boldsymbol{V}$ and T are continuous. Contained in this boundary condition is the "no-slip" condition, namely, that the tangential velocity of the fluid in contact with the boundary of the solid is equal to that of the boundary. For an inviscid fluid the no-slip condition does not apply, and only the normal component of velocity is continuous. If the wall is permeable, the tangential velocity is continuous and the normal velocity is arbitrary; the temperature boundary condition for this case depends on the nature of the injection or suction at the wall.

2. *At a liquid/gas interface:* For such cases the appropriate boundary conditions depend on what can be assumed about the gas the liquid is in contact with. In the classical liquid free-surface problem, the gas, generally atmospheric air, can be ignored and the necessary boundary conditions are that (a) the normal velocity in the liquid at the interface is equal to the normal velocity of the interface and (b) the pressure in the liquid at the interface exceeds the atmospheric pressure by an amount equal to

$$\Delta p = p_{\text{liquid}} - p_{\text{atm}} = \sigma\left(\frac{1}{R_1} + \frac{1}{R_2}\right) \tag{2.2.54}$$

where R_1 and R_2 are the radii of curvature of the intercepts of the interface by two orthogonal planes containing the vertical axis. If the gas is a vapor which undergoes nonnegligible interaction and exchanges with the liquid in contact with it, the boundary conditions are more complex. Then, in addition to the above conditions on normal velocity and pressure, the shear stress (momentum flux) and heat flux must be continuous as well.

For interfaces in general the boundary conditions are derived from continuity conditions for each "transportable" quantity, namely continuity of the appropriate intensity across the interface and continuity of the normal component of the flux vector. Fluid momentum and heat are two such transportable quantities, the associated intensities are velocity and temperature, and the associated flux vectors are stress and heat flux. (The reader should be aware of circumstances where these simple criteria do not apply, for example, the velocity slip and temperature jump for a rarefied gas in contact with a solid surface.)

Vorticity in Incompressible Flow

With μ = constant, ρ = constant, and $\boldsymbol{f} = -\boldsymbol{g} = -g\boldsymbol{k}$ the momentum equation reduces to the form (see Equation (2.2.52))

$$\rho\frac{D\boldsymbol{V}}{Dt} = -\nabla p - \rho g\boldsymbol{k} + \mu\nabla^2\boldsymbol{V} \tag{2.2.55}$$

With the vector identities

$$(\boldsymbol{V}\cdot\nabla)\boldsymbol{V} = \nabla\left(\frac{V^2}{2}\right) - \boldsymbol{V}\times(\nabla\times\boldsymbol{V}) \tag{2.2.56}$$

and

$$\nabla^2\boldsymbol{V} = \nabla(\nabla\cdot\boldsymbol{V}) - \nabla\times(\nabla\times\boldsymbol{V}) \tag{2.2.57}$$

and defining the *vorticity*

$$\zeta \equiv \nabla\times\boldsymbol{V} \tag{2.2.58}$$

Equation (2.2.55) can be written, noting that for incompressible flow $\nabla\cdot\boldsymbol{V} = 0$,

$$\rho\frac{\partial\boldsymbol{V}}{\partial t} + \nabla\left(p + \frac{1}{2}\rho V^2 + \rho g z\right) = \rho\boldsymbol{V}\times\zeta - \mu\nabla\times\zeta \tag{2.2.59}$$

The flow is said to be *irrotational* if

$$\zeta \equiv \nabla \times V = 0 \tag{2.2.60}$$

from which it follows that a *velocity potential* Φ can be defined

$$V = \nabla\Phi \tag{2.2.61}$$

Setting $\zeta = 0$ in Equation (2.2.59), using Equation (2.2.61), and then integrating with respect to all the spatial variables, leads to

$$\rho\frac{\partial\Phi}{\partial t} + \left(p + \frac{1}{2}\rho V^2 + \rho gz\right) = F(t) \tag{2.2.62}$$

(the arbitrary function $F(t)$ introduced by the integration can either be absorbed in Φ, or is determined by the boundary conditions). Equation (2.2.62) is the unsteady *Bernoulli equation* for irrotational, incompressible flow. (Irrotational flows are always potential flows, even if the flow is compressible. Because the viscous term in Equation (2.2.59) vanishes identically for $\zeta = 0$, it would appear that the above Bernoulli equation is valid even for viscous flow. Potential solutions of hydrodynamics are in fact exact solutions of the full Navier–Stokes equations. Such solutions, however, are not valid near solid boundaries or bodies because the no-slip condition generates vorticity and causes nonzero ζ; the potential flow solution is invalid in all those parts of the flow field that have been "contaminated" by the spread of the vorticity by convection and diffusion. See below.)

The curl of Equation (2.2.59), noting that the curl of any gradient is zero, leads to

$$\rho\frac{\partial\zeta}{\partial t} = \rho\nabla\times(V\times\zeta) - \mu\nabla\times\nabla\times\zeta \tag{2.2.63}$$

but

$$\begin{aligned}\nabla^2\zeta &= \nabla(\nabla\cdot\zeta) - \nabla\times\nabla\times\zeta \\ &= -\nabla\times\nabla\times\zeta\end{aligned} \tag{2.2.64}$$

since div curl ( ) ≡ 0, and therefore also

$$\nabla\times(V\times\zeta) \equiv \zeta(\nabla V) + V\nabla\cdot\zeta - V\nabla\zeta - \zeta\nabla\cdot V \tag{2.2.65}$$

$$= \zeta(\nabla V) - V\nabla\zeta \tag{2.2.66}$$

Equation (2.2.63) can then be written

$$\frac{D\zeta}{Dt} = (\zeta\cdot\nabla)V + \nu\nabla^2\zeta \tag{2.2.67}$$

where $\nu = \mu/\rho$ is the kinematic viscosity. Equation (2.2.67) is the vorticity equation for incompressible flow. The first term on the right, an inviscid term, increases the vorticity by vortex stretching. In inviscid, two-dimensional flow both terms on the right-hand side of Equation (2.2.67) vanish, and the equation reduces to $D\zeta/Dt = 0$, from which it follows that the vorticity of a fluid particle remains constant as it moves. This is Helmholtz's theorem. As a consequence it also follows that if $\zeta = 0$ initially, $\zeta \equiv 0$ always;

i.e., *initially irrotational flows remain irrotational* (for inviscid flow). Similarly, it can be proved that $D\Gamma/Dt = 0$; i.e., the circulation around a material closed circuit remains constant, which is Kelvin's theorem.

If $\nu \neq 0$, Equation (2.2.67) shows that the vorticity generated, say, at solid boundaries, diffuses and stretches as it is convected.

We also note that for steady flow the Bernoulli equation reduces to

$$p + \frac{1}{2}\rho V^2 + \rho g z = \text{constant} \tag{2.2.68}$$

valid for steady, irrotational, incompressible flow.

Stream Function

For two-dimensional planar, incompressible flows ($\boldsymbol{V} = (u, v)$), the continuity equation

$$\frac{\partial u}{\partial x} + \frac{\partial v}{\partial y} = 0 \tag{2.2.69}$$

can be identically satisfied by introducing a stream function ψ, defined by

$$u = \frac{\partial \psi}{\partial y}, \qquad v = -\frac{\partial \psi}{\partial x} \tag{2.2.70}$$

Physically ψ is a measure of the flow between streamlines. (Stream functions can be similarly defined to satisfy identically the continuity equations for incompressible cylindrical and spherical axisymmetric flows; and for these flows, as well as the above planar flow, also when they are compressible, but only then if they are steady.) Continuing with the planar case, we note that in such flows there is only a single nonzero component of vorticity, given by

$$\boldsymbol{\zeta} = (0, 0, \zeta_z) = \left(0, 0, \frac{\partial v}{\partial x} - \frac{\partial u}{\partial y}\right) \tag{2.2.71}$$

With Equation (2.2.70)

$$\zeta_z = -\frac{\partial^2 \psi}{\partial x^2} - \frac{\partial^2 \psi}{\partial y^2} = -\nabla^2 \psi \tag{2.2.72}$$

For this two-dimensional flow Equation (2.2.67) reduces to

$$\frac{\partial \zeta_z}{\partial t} + u\frac{\partial \zeta_z}{\partial x} + v\frac{\partial \zeta_z}{\partial y} = v\left(\frac{\partial^2 \zeta_z}{\partial x^2} + \frac{\partial^2 \zeta_z}{\partial y^2}\right) = v\nabla^2 \zeta_z \tag{2.2.73}$$

Substitution of Equation (2.2.72) into Equation (2.2.73) yields an equation for the stream function

$$\frac{\partial(\nabla^2 \psi)}{\partial t} + \frac{\partial \psi}{\partial y}\frac{\partial(\nabla^2 \psi)}{\partial x} - \frac{\partial \psi}{\partial x}\frac{\partial}{\partial y}(\nabla^2 \psi) = \nu \nabla^4 \psi \tag{2.2.74}$$

where $\nabla^4 = \nabla^2 (\nabla^2)$. For uniform flow past a solid body, for example, this equation for Ψ would be solved subject to the boundary conditions:

$$\begin{aligned} &\frac{\partial \psi}{\partial x} = 0, \quad \frac{\partial \psi}{\partial y} = V_\infty \quad \text{at infinity} \\ &\frac{\partial \psi}{\partial x} = 0, \quad \frac{\partial \psi}{\partial y} = 0 \quad \text{at the body (no-slip)} \end{aligned} \tag{2.2.75}$$

For the special case of irrotational flow it follows immediately from Equations (2.2.70) and (2.2.71) with $\zeta_z = 0$, that ψ satisfies the Laplace equation

$$\nabla^2 \psi = \frac{\partial^2 \psi}{\partial x^2} + \frac{\partial^2 \psi}{\partial y^2} = 0 \tag{2.2.76}$$

Inviscid Irrotational Flow: Potential Flow

For irrotational flows we have already noted that a velocity potential Φ can be defined such that $\boldsymbol{V} = \nabla\Phi$. If the flow is also incompressible, so $\nabla \cdot \boldsymbol{V} = 0$, it then follows that

$$\nabla \cdot (\nabla \Phi) = \nabla^2 \Phi = 0 \tag{2.2.77}$$

so Φ satisfies Laplace's equation. (Note that unlike the stream function ψ, which can only be defined for two-dimensional flows, the above considerations for Φ apply to flow in two and three dimensions. On the other hand, the existence of ψ does not require the flow to be irrotational, whereas the existence of Φ does.)

Since Equation (2.2.77) with appropriate conditions on $\boldsymbol{V}$ at boundaries of the flow completely determines the velocity field, and the momentum equation has played no role in this determination, we see that inviscid irrotational flow — *potential theory* — is a purely kinematic theory. The momentum equation enters only after Φ is known in order to calculate the pressure field consistent with the velocity field $\boldsymbol{V} = \nabla\Phi$.

For both two- and three-dimensional flows the determination of Φ makes use of the powerful techniques of potential theory, well developed in the mathematical literature. For two-dimensional planar flows the techniques of complex variable theory are available, since Φ may be considered as either the real or imaginary part of an analytic function (the same being true for ψ, since for such two-dimensional flows Φ and ψ are conjugate variables.)

Because the Laplace equation, obeyed by both Φ and ψ, is linear, complex flows may be built up from the superposition of simple flows; this property of inviscid irrotational flows underlies nearly all solution techniques in this area of fluid mechanics.

Problem

A two-dimensional inviscid irrotational flow has the velocity potential

$$\Phi = x^2 - y^2 \tag{P.2.2.6}$$

What two-dimensional potential flow does this represent?

Solution. It follows from Equations (2.2.61) and (2.2.70) that for two-dimensional flows, in general

$$u = \frac{\partial \Phi}{\partial x} = \frac{\partial \psi}{\partial y}, \quad v = \frac{\partial \Phi}{\partial y} = -\frac{\partial \psi}{\partial x} \tag{P.2.2.7}$$

It follows from Equation (P.2.2.6) that

$$u = \frac{\partial \psi}{\partial y} = 2x, \qquad \nu = -\frac{\partial \psi}{\partial x} = -2y \tag{P.2.2.8}$$

Integration of Equation (P.2.2.8) yields

$$\psi = 2xy \tag{P.2.2.9}$$

The streamlines, ψ = constant, and equipotential lines, Φ = constant, both families of hyperbolas and each family the orthogonal trajectory of the other, are shown in Figure 2.2.3. Because the x and y axes are streamlines, Equations (P.2.2.6) and (P.2.2.9) represent the inviscid irrotational flow in a right-angle corner. By symmetry, they also represent the planar flow in the upper half-plane directed toward a stagnation point at $x = y = 0$ (see Figure 2.2.4). In polar coordinates (r, θ), with corresponding velocity components (u_r, u_θ), this flow is represented by

$$\Phi = r^2 \cos 2\theta, \qquad \psi = r^2 \sin 2\theta \tag{P.2.2.10}$$

with

$$\begin{aligned} u_r &= \frac{\partial \Phi}{\partial r} = \frac{1}{r}\frac{\partial \psi}{\partial \theta} = 2r\cos 2\theta \\ u_\theta &= \frac{1}{r}\frac{\partial \Phi}{\partial \theta} = -\frac{\partial \psi}{\partial r} = -2r\sin 2\theta \end{aligned} \tag{P.2.2.11}$$

For two-dimensional planar potential flows we may also use complex variables, writing the complex potential $f(z) = \Phi + i\psi$ as a function of the complex variable $z = x + iy$, where the complex velocity is given by $f'(z) = w(z) = u - i\nu$. For the flow above

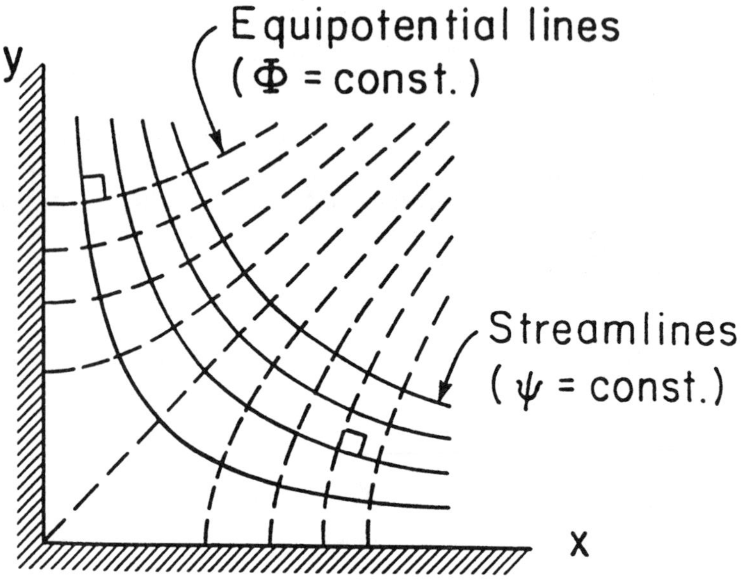

FIGURE 2.2.3 Potential flow in a 90° corner.

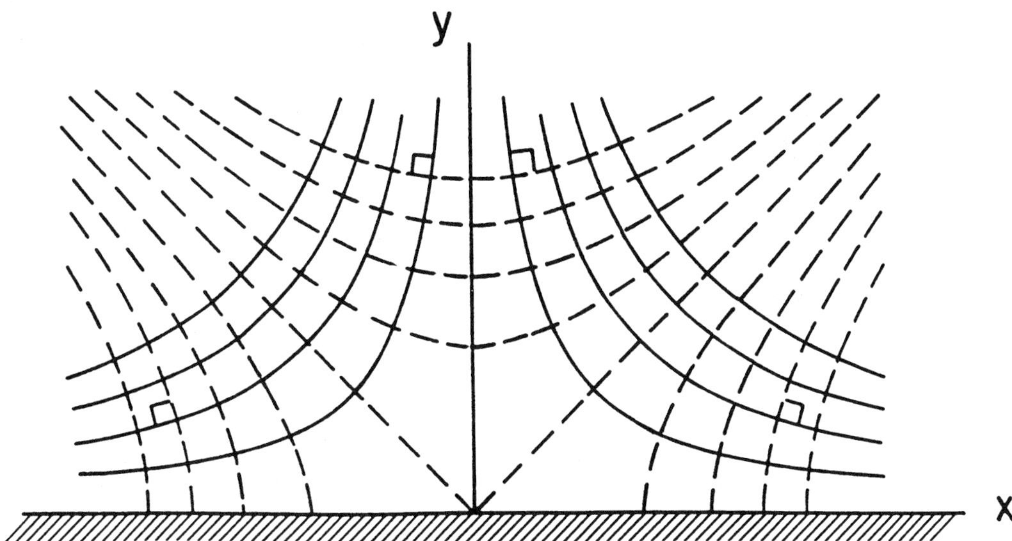

FIGURE 2.2.4 Potential flow impinging against a flat (180°) wall (plane stagnation-point flow).

$$f(z) = z^2 \tag{P.2.2.12}$$

Expressions such as Equation (P.2.2.12), where the right-hand side is an analytic function of z, may also be regarded as a conformal mapping, which makes available as an aid in solving two-dimensional potential problems all the tools of this branch of mathematics.

Further Information

More detail and additional information on the topics in this section may be found in more advanced books on fluid dynamics, such as

Batchelor, G.K. 1967. *An Introduction to Fluid Dynamics,* Cambridge University Press, Cambridge, England.

Warsi, Z.U.A. 1999. *Fluid Dynamics. Theoretical and Computational Approaches,* 2nd ed., CRC Press, Boca Raton, FL.

Sherman, F.S. 1990. *Viscous Flow,* McGraw-Hill, New York.

Panton, R.L. 1996. *Incompressible Flow,* 2nd ed., John Wiley & Sons, New York.

Pozrikidis, C. 1997. *Introduction to Theoritical and Computational Fluid Dynamics,* Oxford University Press, New York.

2.3 Similitude: Dimensional Analysis and Data Correlation

Stuart W. Churchill

Dimensional Analysis

Similitude refers to the formulation of a description for physical behavior that is general and independent of the individual dimensions, physical properties, forces, etc. In this subsection the treatment of similitude is restricted to *dimensional analysis*; for a more general treatment see Zlokarnik (1991). The full power and utility of dimensional analysis is often underestimated and underutilized by engineers. This technique may be applied to a complete mathematical model or to a simple listing of the variables that define the behavior. Only the latter application is described here. For a description of the application of dimensional analysis to a mathematical model see Hellums and Churchill (1964).

General Principles

Dimensional analysis is based on the principle that all additive or equated terms of a complete relationship between the variables must have the same net dimensions. The analysis starts with the preparation of a list of the individual dimensional variables (dependent, independent, and parametric) that are presumed to define the behavior of interest. The performance of dimensional analysis in this context is reasonably simple and straightforward; the principal difficulty and uncertainty arise from the identification of the variables to be included or excluded. If one or more important variables are inadvertently omitted, the reduced description achieved by dimensional analysis will be incomplete and inadequate as a guide for the correlation of a full range of experimental data or theoretical values. The familiar band of plotted values in many graphical correlations is more often a consequence of the omission of one or more variables than of inaccurate measurements. If, on the other hand, one or more irrelevant or unimportant variables are included in the listing, the consequently reduced description achieved by dimensional analysis will result in one or more unessential dimensionless groupings. Such excessive dimensionless groupings are generally less troublesome than missing ones because the redundancy will ordinarily be revealed by the process of correlation. Excessive groups may, however, suggest unnecessary experimental work or computations, or result in misleading correlations. For example, real experimental scatter may inadvertently and incorrectly be correlated in all or in part with the variance of the excessive grouping.

In consideration of the inherent uncertainty in selecting the appropriate variables for dimensional analysis, it is recommended that this process be interpreted as *speculative* and subject to correction on the basis of experimental data or other information. Speculation may also be utilized as a formal technique to identify the effect of eliminating a variable or of combining two or more. The latter aspect of speculation, which may be applied either to the original listing of dimensional variables or to the resulting set of dimensionless groups, is often of great utility in identifying possible limiting behavior or dimensionless groups of marginal significance. The systematic speculative elimination of all but the most certain variables, one at a time, followed by regrouping, is recommended as a general practice. The additional effort as compared with the original dimensional analysis is minimal, but the possible return is very high. A general discussion of this process may be found in Churchill (1981).

The minimum number of independent dimensionless groups i required to describe the fundamental and parametric behavior is (Buckingham, 1914)

$$i = n - m \tag{2.3.1}$$

where n is the number of variables and m is the number of fundamental dimensions such as mass M, length L, time θ, and temperature T that are introduced by the variables. The inclusion of redundant dimensions such as force F and energy E that may be expressed in terms of mass, length, time, and temperature is at the expense of added complexity and is to be avoided. (Of course, mass could be replaced by force or temperature by energy as alternative fundamental dimensions.) In some rare cases i is actually greater than $n - m$. Then

$$i = n - k \tag{2.3.2}$$

where k is the maximum number of the chosen variables that cannot be combined to form a dimensionless group. Determination of the minimum number of dimensionless groups is helpful if the groups are to be chosen by inspection, but is unessential if the algebraic procedure described below is utilized to determine the groups themselves since the number is then obvious from the final result.

The *particular* minimal set of dimensionless groups is arbitrary in the sense that two or more of the groups may be multiplied together to any positive, negative, or fractional power as long as the number of independent groups is unchanged. For example, if the result of a dimensional analysis is

$$\phi\{X,Y,Z\} = 0 \tag{2.3.3}$$

where X, Y, and Z are independent dimensionless groups, an equally valid expression is

$$\phi\left\{XY^{1/2}, Z/Y^2, Z\right\} = 0 \tag{2.3.4}$$

Dimensional analysis itself does not provide any insight as to the best choice of equivalent dimensionless groupings, such as between those of Equations (2.3.3) and (2.3.4). However, isolation of each of the variables that are presumed to be the most important in a separate group may be convenient in terms of interpretation and correlation. Another possible criterion in choosing between alternative groupings may be the relative invariance of a particular one. The functional relationship provided by Equation (2.3.3) may equally well be expressed as

$$X = \phi\{Y,Z\} \tag{2.3.5}$$

where X is implied to be the dependent grouping and Y and Z to be independent or parametric groupings.

Three primary methods of determining a minimal set of dimensionless variables are (1) by inspection; (2) by combination of the residual variables, one at a time, with a set of chosen variables that cannot be combined to obtain a dimensionless group; and (3) by an algebraic procedure. These methods are illustrated in the examples that follow.

Example 2.3.1: Fully Developed Flow of Water Through a Smooth Round Pipe

Choice of Variables. The shear stress τ_w on the wall of the pipe may be postulated to be a function of the density ρ and the dynamic viscosity μ of the water, the inside diameter D of the pipe, and the space-mean of the time-mean velocity u_m. The limitation to fully developed flow is equivalent to a postulate of independence from distance x in the direction of flow, and the specification of a smooth pipe is equivalent to the postulate of independence from the roughness e of the wall. The choice of τ_w rather than the pressure drop per unit length $-dP/dx$ avoids the need to include the acceleration due to gravity g and the elevation z as variables. The choice of u_m rather than the volumetric rate of flow V, the mass rate of flow w, or the mass rate of flow per unit area G is arbitrary but has some important consequences as noted below. The postulated dependence may be expressed functionally as $\phi\{\tau_w, \rho, \mu, D, u_m\} = 0$ or $\tau_w = \phi\{\rho, \mu, D, u_m\}$.

Tabulation. Next prepare a tabular listing of the variables and their dimensions:

	τ_w	ρ	μ	D	u_m
M	1	1	1	0	0
L	–1	–3	–1	1	1
θ	–2	0	–1	0	–1
T	0	0	0	0	0

Minimal Number of Groups. The number of postulated variables is 5. Since the temperature does not occur as a dimension for any of the variables, the number of fundamental dimensions is 3. From Equation (2.3.1), the minimal number of dimensionless groups is 5 – 3 = 2. From inspection of the above tabulation, a dimensionless group cannot be formed from as many as three variables such as D, μ, and ρ. Hence, Equation (2.3.2) also indicates that $i = 5 - 3 = 2$.

Method of Inspection. By inspection of the tabulation or by trial and error it is evident that only two independent dimensionless groups may be formed. One such set is

$$\phi\left\{\frac{\tau_w}{\rho u_m^2}, \frac{D u_m \rho}{\mu}\right\} = 0$$

Method of Combination. The residual variables τ_w and μ may be combined in turn with the noncombining variables ρ, D, and u_m to obtain two groups such as those above.

Algebraic Method. The algebraic method makes formal use of the postulate that the functional relationship between the variables may in general be represented by a power series. In this example such a power series may be expressed as

$$\tau_w = \sum_{i=1}^{N} A_i \rho^{a_i} \mu^{b_i} D^{c_i} u_m^{d_i}$$

where the coefficients A_i are dimensionless. Each additive term on the right-hand side of this expression must have the same net dimensions as τ_w. Hence, for the purposes of dimensional analysis, only the first term need be considered and the indexes may be dropped. The resulting highly restricted expression is $\tau_w = A\rho^a\mu^b D^c u_m^d$. Substituting the dimensions for the variables gives

$$\frac{M}{L\theta^2} = A\left(\frac{M}{L^3}\right)^a\left(\frac{M}{L\theta}\right)^b L^c\left(\frac{L}{\theta}\right)^d$$

Equating the sum of the exponents of M, L, and θ on the right-hand side of the above expression with those of the left-hand side produces the following three simultaneous linear algebraic equations: $1 = a + b$; $-1 = -3a - b + c + d$; and $-2 = -b - d$, which may be solved for a, c, and d in terms of b to obtain $a = 1 - b$, $c = -b$, and $d = 2 - b$. Substitution then gives $\tau_w = A\rho^{1-b}\mu^b D^{-b} u_m^{2-b}$ which may be regrouped as

$$\frac{\tau_w}{\rho u_m^2} = A\left(\frac{\mu}{D u_m \rho}\right)^b$$

Since this expression is only the first term of a power series, it should *not* be interpreted to imply that $\tau_w/\rho u_m^2$ is necessarily proportional to some power at $\mu/Du_m\rho$ but instead only the equivalent of the expression derived by the method of inspection. The inference of a power dependence between the dimensionless groups is the most common and serious error in the use of the algebraic method of dimensional analysis.

Speculative Reductions. Eliminating ρ from the original list of variables on speculative grounds leads to

$$\phi\left\{\frac{\tau_w D}{\mu u_m}\right\} = 0$$

or its exact equivalent:

$$\frac{\tau_w D}{\mu u_m} = A$$

The latter expression with $A = 8$ is actually the exact solution for the laminar regime ($Du_m\rho/\mu < 1800$). A relationship that does not include ρ may alternatively be derived directly from the solution by the method of inspection as follows. First, ρ is eliminated from one group, say $\tau_w/\rho u_m^2$, by multiplying it with $Du_m\rho/\mu$ to obtain

$$\phi\left\{\frac{\tau_w D}{\mu u_m}, \frac{Du_m\rho}{\mu}\right\} = 0$$

The remaining group containing ρ is now simply dropped. Had the original expression been composed of three independent groups each containing ρ, that variable would have to be eliminated from two of them before dropping the third one.

The relationships that are obtained by the speculative elimination of μ, D, and u_m, one at a time, do not appear to have any range of physical validity. Furthermore, if w or G had been chosen as the independent variable rather than u_m, the limiting relationship for the laminar regime would not have been obtained by the elimination of ρ.

Alternative Forms. The solution may also be expressed in an infinity of other forms such as

$$\phi\left\{\frac{\tau_w D^2\rho}{\mu^2}, \frac{Du_m\rho}{\mu}\right\} = 0$$

If τ_w is considered to be the principal dependent variable and u_m the principal independent variable, this latter form is preferable in that these two quantities do not then appear in the same grouping. On the other hand, if D is considered to be the principal independent variable, the original formulation is preferable. The variance of $\tau_w/\rho u_m^2$ is less than that of $\tau_w D/\mu u_m$ and $\tau_w D^2\rho/\mu^2$ in the turbulent regime while that of $\tau_w D/\mu u_m$ is zero in the laminar regime. Such considerations may be important in devising convenient graphical correlations.

Alternative Notations. The several solutions above are more commonly expressed as

$$\phi\left\{\frac{f}{2}, \mathrm{Re}\right\} = 0$$

$$\phi\left\{\frac{f\,\mathrm{Re}}{2}, \mathrm{Re}\right\} = 0$$

or

$$\phi\left\{\frac{f\mathrm{Re}^2}{2},\mathrm{Re}\right\}=0$$

where $f = 2\tau_w/\rho u_m^2$ is the *Fanning friction factor* and $Re = Du_m\rho/\mu$ is the *Reynolds number.*

The more detailed forms, however, are to be preferred for purposes of interpretation or correlation because of the explicit appearance of the individual, physically measurable variables.

Addition of a Variable. The above results may readily be extended to incorporate the roughness e of the pipe as a variable. If two variables have the same dimensions, they will always appear as a dimensionless group in the form of a ratio, in this case e appears most simply as e/D. Thus, the solution becomes

$$\phi\left\{\frac{\tau_w}{\rho u_m^2},\frac{Du_m\rho}{\mu},\frac{e}{D}\right\}=0$$

Surprisingly, as contrasted with the solution for a smooth pipe, the speculative elimination of μ and hence of the group $Du_m\rho/\mu$ now results in a valid asymptote for $Du_m\rho/\mu \to \infty$ and all finite values of e/D, namely,

$$\phi\left\{\frac{\tau_w}{\rho u_m^2},\frac{e}{D}\right\}=0$$

Example 2.3.2: Fully Developed Forced Convection in Fully Developed Flow in a Round Tube

It may be postulated for this process that $h = \phi\{D, u_m, \rho, \mu, k, c_p\}$, where here h is the local heat transfer coefficient, and c_p and k are the specific heat capacity and thermal conductivity, respectively, of the fluid. The corresponding tabulation is

	h	D	u_m	ρ	μ	k	c_p
M	1	0	0	1	1	1	0
L	0	1	1	–3	–1	1	2
θ	–3	0	–1	0	–1	–3	–2
T	–1	0	0	0	0	–1	–1

The number of variables is 7 and the number of independent dimensions is 4, as is the number of variables such as D, u_m, ρ, and k that cannot be combined to obtain a dimensionless group. Hence, the minimal number of dimensionless groups is 7 – 4 = 3. The following acceptable set of dimensionless groups may be derived by any of the procedures illustrated in Example 1:

$$\frac{hD}{k}=\phi\left\{\frac{Du_m\rho}{\mu},\frac{c_p\mu}{k}\right\}$$

Speculative elimination of μ results in

$$\frac{hD}{k}=\phi\left\{\frac{Du_m\rho c_p}{k}\right\}$$

which has often erroneously been inferred to be a valid asymptote for $c_p\mu/k \to 0$. Speculative elimination of D, u_m, ρ, k, and c_p individually also does not appear to result in expressions with any physical validity. However, eliminating c_p and ρ or u_m gives a valid result for the laminar regime, namely,

$$\frac{hD}{k} = A$$

The general solutions for flow and convection in a smooth pipe may be combined to obtain

$$\frac{hD}{k} = \phi\left\{\frac{\tau_w D^2\rho}{\mu^2}, \frac{c_p\mu}{k}\right\}$$

which would have been obtained directly had u_m been replaced by τ_w in the original tabulation. This latter expression proves to be superior in terms of speculative reductions. Eliminating D results in

$$\frac{h\mu}{k(\tau_w\rho)^{1/2}} = \phi\left\{\frac{c_p\mu}{k}\right\}$$

which may be expressed in the more conventional form of

$$\mathrm{Nu} = \mathrm{Re}\left(\frac{f}{2}\right)^{1/2}\phi\{\mathrm{Pr}\}$$

where Nu = hD/k is the *Nusselt number* and Pr = $c_p\mu/k$ is the *Prandtl number*. This result appears to be a valid asymptote for Re $\to \infty$ and a good approximation for even moderate values (>5000) for large values of Pr. Elimination of μ as well as D results in

$$\frac{h}{c_p(\tau_w\rho)^{1/2}} = A$$

or

$$\mathrm{Nu} = A\,\mathrm{Re}\,\mathrm{Pr}\left(\frac{f}{2}\right)^{1/2}$$

which appears to be an approximate asymptote for Re $\to \infty$ and Pr $\to 0$. Elimination of both c_p and ρ again yields the appropriate result for laminar flow, indicating that ρ rather than u_m is the meaningful variable to eliminate in this respect.

The numerical value of the coefficient A in the several expressions above depends on the mode of heating, a true variable, but one from which the purely functional expressions are independent. If j_w, the heat flux density at the wall, and $T_w - T_m$, the temperature difference between the wall and the bulk of the fluid, were introduced as variables in place of $h \equiv j_w/(T_w - T_m)$, another group such as $c_p(T_w - T_m)(D\rho/\mu)^2$ or $\rho c_p(T_w - T_m)/\tau_w$ or $c_p(T_w - T_m)/u_m^2$, which represents the effect of viscous dissipation, would be obtained. This effect is usually but not always negligible.

Example 2.3.3: Free Convection from a Vertical Isothermal Plate

The behavior for this process may be postulated to be represented by

$$h = \phi\left\{g, \beta, T_w - T_\infty, x, \mu, \rho, c_p, k\right\}$$

where g is the acceleration due to gravity, β is the volumetric coefficient of expansion with temperature, T_∞ is the unperturbed temperature of the fluid, and x is the vertical distance along the plate. The corresponding tabulation is

	h	g	β	$T_w - T_\infty$	x	μ	ρ	c_p	k
M	1	0	0	0	0	1	1	0	1
L	0	1	0	0	1	–1	–3	2	1
θ	–3	–2	0	0	0	–1	0	–2	–3
T	–1	0	–1	1	0	0	0	–1	1

The minimal number of dimensionless groups indicated by both methods is 9 – 4 = 5. A satisfactory set of dimensionless groups, as found by any of the methods illustrated in Example 1, is

$$\frac{hx}{k} = \phi\left\{\frac{\rho^2 g x^3}{\mu^2}, \frac{c_p\mu}{k}, \beta(T_w - T_\infty), c_p(T_w - T_\infty)\left(\frac{\rho x}{\mu}\right)^2\right\}$$

It may be reasoned that the buoyant force which generates the convective motion must be proportional to $\rho g\beta(T_w - T_\infty)$, thus, g in the first term on the right-hand side must be multiplied by $\beta(T_w - T_\infty)$, resulting in

$$\frac{hx}{k} = \phi\left\{\frac{\rho^2 g\beta(T_w - T_\infty)x^3}{\mu^2}, \frac{c_p\mu}{k}, \beta(T_w - T_\infty), c_p(T_w - T_\infty)\left(\frac{\rho x}{\mu}\right)^2\right\}$$

The effect of expansion other than on the buoyancy is now represented by $\beta(T_w - T_\infty)$, and the effect of viscous dissipation by $c_p(T_w - T_\infty)(\rho x/\mu)^2$. Both effects are negligible for all practical circumstances. Hence, this expression may be reduced to

$$\frac{hx}{k} = \phi\left\{\frac{\rho^2 g\beta(T_w - T_\infty)x^3}{\mu^2}, \frac{c_p\mu}{k}\right\}$$

or

$$\mathrm{Nu}_x = \phi\{\mathrm{Gr}_x, \mathrm{Pr}\}$$

where $\mathrm{Nu}_x = hx/k$ and $\mathrm{Gr}_x = \rho^2 g\beta(T_w - T_\infty)x^3/\mu^2$ is the *Grashof number.*

Elimination of x speculatively now results in

$$\frac{hx}{k} = \left(\frac{\rho^2 g\beta(T_w - T_\infty)x^3}{\mu^2}\right)^{1/3}\phi\{\mathrm{Pr}\}$$

or

$$\mathrm{Nu}_x = \mathrm{Gr}_x^{1/3}\phi\{\mathrm{Pr}\}$$

This expression appears to be a valid asymptote for $\mathrm{Gr}_x \to \infty$ and a good approximation for the entire turbulent regime. Eliminating μ speculatively rather than x results in

$$\frac{hx}{k} = \phi\left\{\frac{\rho^2 c_p^2 g\beta(T_w - T_\infty)x^3}{k^2}\right\}$$

or

$$\text{Nu}_x = \phi\left\{\text{Gr}_x \text{Pr}^2\right\}$$

The latter expression appears to be a valid asymptote for $\text{Pr} \to 0$ for all Gr_x, that is, for both the laminar and the turbulent regimes. The development of a valid asymptote for large values of Pr requires more subtle reasoning. First $c_p\mu/k$ is rewritten as $\mu/\rho\alpha$ where $\alpha = k/\rho c_p$. Then ρ is eliminated speculatively except as it occurs in $\rho g\beta(T_w - T_\infty)$ and $k/\rho c_p$. The result is

$$\frac{hx}{k} = \phi\left\{\frac{c_p \rho^2 g\beta(T_w - T_\infty)x^3}{\mu k}\right\}$$

or

$$\text{Nu}_x = \phi\left\{\text{Ra}_x\right\}$$

where

$$\text{Ra}_x = \frac{c_p \rho^2 g\beta(T_w - T_\infty)x^3}{\mu k} = \text{Gr}_x \text{Pr}$$

is the *Rayleigh number.* The above expression for Nu_x appears to be a valid asymptote for $\text{Pr} \to \infty$ and a reasonable approximation for even moderate values of Pr for all Gr_x, that is, for both the laminar and the turbulent regimes.

Eliminating x speculatively from the above expressions for small and large values of Pr results in

$$\text{Nu}_x = A\left(\text{Gr}_x \text{Pr}^2\right)^{1/3} = A\left(\text{Ra}_x \text{Pr}\right)^{1/3}$$

and

$$\text{Nu}_x = B\left(\text{Gr}_x \text{Pr}\right)^{1/3} = B\left(\text{Ra}_x\right)^{1/3}$$

The former appears to be a valid asymptote for $\text{Pr} \to 0$ and $\text{Gr}_x \to \infty$ and a reasonable approximation for very small values of Pr in the turbulent regime, while the latter is well confirmed as a valid asymptote for $\text{Pr} \to \infty$ and $\text{Gr}_x \to \infty$ and as a good approximation for moderate and large values of Pr over the entire turbulent regime. The expressions in terms of Gr_x are somewhat more complicated than those in terms of Ra_x, but are to be preferred since Gr_x is known to characterize the transition from laminar to turbulent motion in natural convection just as Re_D does in forced flow in a channel. The power of speculation combined with dimensional analysis is well demonstrated by this example in which valid asymptotes are thereby attained for several regimes.

Correlation of Experimental Data and Theoretical Values

Correlations of experimental data are generally developed in terms of dimensionless groups rather than in terms of the separate dimensional variables in the interests of compactness and in the hope of greater generality. For example, a complete set of graphical correlations for the heat transfer coefficient h of Example 2.3.2 above in terms of each of the six individual independent variables and physical properties might approach book length, whereas the dimensionless groupings both imply that a single plot with one parameter should be sufficient. Furthermore, the reduced expression for the turbulent regime implies that a plot of Nu/Re $f^{1/2}$ vs. Pr should demonstrate only a slight parametric dependence on Re or Re $f^{1/2}$. Of course, the availability of a separate correlation for f as a function of Re is implied.

Theoretical values, that is, ones obtained by numerical solution of a mathematical model in terms of either dimensional variables or dimensionless groups, are presumably free from imprecision. Even so, because of their discrete form, the construction of a correlation or correlations for such values may be essential for the same reasons as for experimental data.

Graphical correlations have the merit of revealing general trends, of providing a basis for evaluation of the choice of coordinates, and most of all of displaying visually the scatter of the individual experimental values about a curve representing a correlation or their behavior on the mean. (As mentioned in the previous subsection, the omission of a variable may give the false impression of experimental error in such a plot.) On the other hand, correlating equations are far more convenient as an input to a computer than is a graphical correlation. These two formats thus have distinct and complementary roles; both should generally be utilized. The merits and demerits of various graphical forms of correlations are discussed in detail by Churchill (1979), while the use of logarithmic and arithmetic coordinates, the effects of the appearance of a variable in both coordinates, and the effects of the distribution of error between the dependent and independent variable are further illustrated by Wilkie (1985).

Churchill and Usagi (1972; 1974) proposed general usage of the following expression for the formulation of correlating equations:

$$y^n\{x\} = y_0^n\{x\} + y_\infty^n\{x\} \tag{2.3.6}$$

where $y_o\{x\}$ and $y_\infty\{x\}$ denote asymptotes for small and large values of x, respectively, and n is an arbitrary exponent. For convenience and simplicity, Equation (2.3.6) may be rearranged in either of the following two forms:

$$\left(Y(x)\right)^n = 1 + Z^n\{x\} \tag{2.3.7}$$

or

$$\left(\frac{Y\{x\}}{Z\{x\}}\right)^n = 1 + \frac{1}{Z^n\{x\}} \tag{2.3.8}$$

where $Y\{x\} \equiv y\{x\}/y_o\{x\}$ and $Z\{x\} \equiv y_\infty\{x\}/y_o\{x\}$. Equations (2.3.6), (2.3.7), and (2.3.9) are hereafter denoted collectively as the CUE (Churchill–Usagi equation). The principle merits of the CUE as a canonical expression for correlation are its simple form, generality, and minimal degree of explicit empiricism, namely, only that of the exponent n, since the asymptotes $y_o\{x\}$ and $y_\infty\{x\}$ are ordinarily known in advance from theoretical considerations or well-established correlations. Furthermore, as will be shown, the CUE is quite insensitive to the numerical value of n. Although the CUE is itself very simple in form, it is remarkably successful in representing closely very complex behavior, even including the dependence on secondary variables and parameters, by virtue of the introduction of such dependencies through $y_o\{x\}$ and $y_\infty\{x\}$. In the rare instances in which such dependencies are not represented in the asymptotes, n may be correlated as a function of the secondary variables and/or parameters. Although the CUE usually produces very close representations, it is empirical and not exact. In a few instances, numerical values of n have been derived or rationalized on theoretical grounds, but even then some degree of approximation is involved. Furthermore, the construction of a correlating expression in terms of the CUE is subject to the following severe limitations:

1. The asymptotes $y_o\{x\}$ and $y_\infty\{x\}$ must intersect once and only once;
2. The asymptotes $y_o\{x\}$ and $y_\infty\{x\}$ must be free of singularities. Even though a singularity occurs beyond the asserted range of the asymptote, it will persist and disrupt the prediction of the CUE, which is intended to encompass all values of the independent variable x; and
3. The asymptotes must both be upper or lower bounds.

In order to avoid or counter these limitations it may be necessary to modify or replace the asymptotes with others. Examples of this process are provided below. A different choice for the dependent variable may be an option in this respect. The suitable asymptotes for use in Equation (2.3.6) may not exist in the literature and therefore may need to be devised or constructed. See, for example, Churchill (1988b) for guidance in this respect. Integrals and derivatives of the CUE are generally awkward and inaccurate, and may include singularities not present or troublesome in the CUE itself. It is almost always preferable to develop a separate correlating equation for such quantities using derivatives or integrals of $y_o\{x\}$ and $y_\infty\{x\}$, simplified or modified as appropriate.

The Evaluation of *n*

Equation (2.3.6) may be rearranged as

$$n = \frac{\ln\left\{1 + \left(\dfrac{y_\infty\{x\}}{y_0\{x\}}\right)^n\right\}}{\ln\left\{\dfrac{y\{x\}}{y_0\{x\}}\right\}} \tag{2.3.9}$$

and solved for n by iteration for any known value of $y\{x\}$, presuming that $y_o\{x\}$ and $y_\infty\{x\}$ are known. If $y\{x^*\}$ is known, where x^* represents the value of x at the point of intersection of the asymptotes, that is, for $y_o\{x\} = y_\infty\{x\}$, Equation (2.3.9) reduces to

$$n = \frac{\ln\{2\}}{\ln\left\{\dfrac{y\{x^*\}}{y_0\{x^*\}}\right\}} \tag{2.3.10}$$

and iterative determination of n is unnecessary.

A graphical and visual method of evaluation of n is illustrated in Figure 2.3.1 in which $Y\{Z\}$ is plotted vs. Z for $0 \le Z \le 1$ and $Y\{Z\}/Z$ vs. $1/Z$ for $0 \le 1/Z \le 1$ in arithmetic coordinates with n as a parameter. Values of $y\{x\}$ may be plotted in this form and the best overall value of n selected visually (as illustrated in Figure 2.3.2). A logarithmic plot of $Y\{Z\}$ vs. Z would have less sensitivity relative to the dependence on n. (See, for example, Figure 1 of Churchill and Usagi, 1972.) Figure 2.3.1 explains in part the success of the CUE. Although y and x may both vary from 0 to ∞, the composite variables plotted in Figure 2.3.1 are highly constrained in that the compound independent variables Z and $1/Z$ vary only between 0 and 1, while for $n \ge 1$, the compound dependent variables $Y\{Z\}$ and $Y\{Z\}/Z$ vary only from 1 to 2.

Because of the relative insensitivity of the CUE to the numerical value of n, an integer or a ratio of two small integers may be chosen in the interest of simplicity and without significant loss of accuracy. For example, the maximum variance in Y (for $0 \le Z \le 1$) occurs at $Z = 1$ and increases only $100(2^{1/20} - 1) = 3.5\%$ if n is decreased from 5 to 4. If $y_o\{x\}$ and $y_\infty\{x\}$ are both lower bounds, n will be positive,

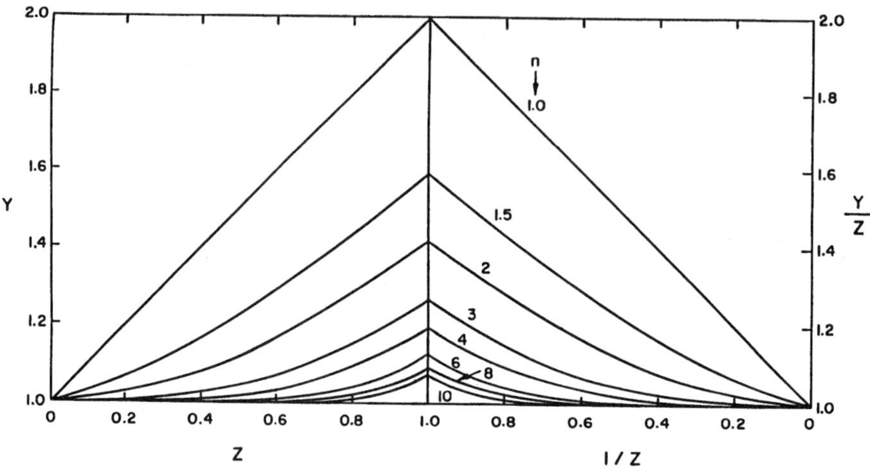

FIGURE 2.3.1 Arithmetic, split-coordinate plot of Equation 2.3.10. (From Churchill, S.W. and Usagi, R. *AIChE J.* 18(6), 1123, 1972. With permission from the American Institute of Chemical Engineers.)

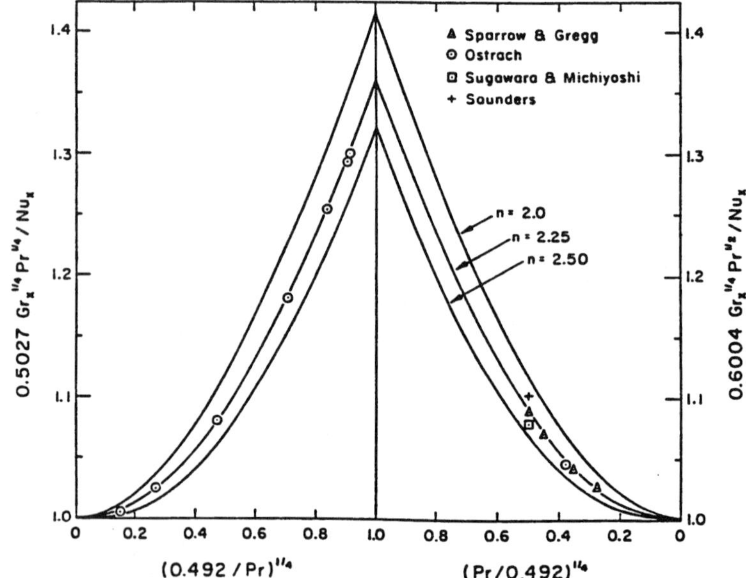

FIGURE 2.3.2 Arithmetic, split-coordinate plot of computed values and experimental data for laminar free convection from an isothermal vertical plate. (From Churchill, S.W. and Usagi, R. *AIChE J.* 18(6), 1124, 1972. With permission from the American Institute of Chemical Engineers.)

and if they are both upper bounds, n will be negative. To avoid extending Figure 2.3.1 for negative values of n, $1/y\{x\}$ may simply be interpreted as the dependent variable.

Intermediate Regimes

Equations (2.3.6), (2.3.7), and (2.3.8) imply a slow, smooth transition between $y_o\{x\}$ and $y_\infty\{x\}$ and, moreover, one that is symmetrical with respect to $x^*(Z = 1)$. Many physical systems demonstrate instead a relatively abrupt transition, as for example from laminar to turbulent flow in a channel or along a flat plate. The CUE may be applied serially as follows to represent such behavior if an expression $y_i\{x\}$ is postulated for the intermediate regime. First, the transition from the initial to the intermediate regime is represented by

$$y_1^n = y_0^n + y_i^n \tag{2.3.11}$$

Then the transition from this combined regime to the final regime by

$$y^m = y_1^m + y_\infty^m = \left(y_0^n + y_i^n\right)^{m/n} + y_\infty^m \tag{2.3.12}$$

Here, and throughout the balance of this subsection, in the interests of simplicity and clarity, the functional dependence of all the terms on x is implied rather than written out explicitly. If y_o is a lower bound and y_i is implied to be one, y_1 and y_∞ must be upper bounds. Hence, n will then be positive and m negative. If y_o and y_i are upper bounds, y_1 and y_∞ must be lower bounds; then n will be negative and m positive. The reverse formulation starting with y_∞ and y_1 leads by the same procedure to

$$y^n = y_0^n + \left(y_i^m + y_\infty^m\right)^{n/m} \tag{2.3.13}$$

If the intersections of y_i with y_o and y_∞ are widely separated with respect to x, essentially the same pair of values for n and m will be determined for Equations (2.3.12) and (2.3.13), and the two representations for y will not differ significantly. On the other hand, if these intersections are close in terms of x, the pair of values of m and n may differ significantly and one representation may be quite superior to the other. In some instances a singularity in y_o or y_∞ may be tolerable in either Equation (2.3.12) or (2.3.13) because it is overwhelmed by the other terms. Equations (2.3.12) and (2.3.13) have one hidden flaw. For $x \to 0$, Equation (2.3.12) reduces to

$$y \to y_0\left[1+\left(\frac{y_\infty}{y_0}\right)^m\right]^{1/m} \tag{2.3.14}$$

If y_o is a lower bound, m is necessarily negative, and values of y less than y_o are predicted. If y_o/y_∞ is sufficiently small or if m is sufficiently large in magnitude, this discrepancy may be tolerable. If not, the following alternative expression may be formulated, again starting from Equation (2.3.11):

$$\left(y^n - y_0^n\right)^m = y_i^{nm} + \left(y_\infty^n - y_0^n\right)^m \tag{2.3.15}$$

Equation (2.3.15) is free from the flaw identified by means of Equation (2.3.14) and invokes no additional empiricism, but a singularity may occur at $y_\infty = y_o$, depending on the juxtapositions of y_o, y_i, and y_∞. Similar anomalies occur for Equation (2.3.13) and the corresponding analog of Equation (2.3.14), as well as for behavior for which $n < 0$ and $m > 0$. The preferable form among these four is best chosen by trying each of them.

One other problem with the application of the CUE for a separate transitional regime is the formulation of an expression for $y_i\{x\}$, which is ordinarily not known from theoretical considerations. Illustrations of the empirical determination of such expressions for particular cases may be found in Churchill and Usagi (1974), Churchill and Churchill (1975), and Churchill (1976, 1977), as well as in Example 2.3.5 below.

Example 2.3.4: The Pressure Gradient in Flow through a Packed Bed of Spheres

The pressure gradient at asymptotically low rates of flow (the creeping regime) can be represented by the Kozeny–Carman equation, $\Phi = 150\ \mathrm{Re}_p$, and at asymptotically high rates of flow (the inertial regime) by the Burke–Plummer equation, $\Phi = 1.75\,(\mathrm{Re}_p)^2$, where $\Phi = \rho\varepsilon^2 d_p(-dP_f/dx)\mu^2(1-\varepsilon)$, $\mathrm{Re}_p = d_p u_o \rho/\mu(1-\varepsilon)$, d_p = diameter of spherical particles, m, ε = void fraction of bed of spheres, dP_f/dx = dynamic pressure gradient (due to friction), Pa/m, and u_o = superficial velocity (in absence of the spheres), m/sec. For the origin of these two asymptotic expressions see Churchill (1988a). They both have a theoretical structure, but the numerical coefficients of 150 and 1.75 are basically empirical. These equations are both lower bounds and have one intersection. Experimental data are plotted in Figure 2.3.3, which has the form of Figure 2.3.1 with $Y = \Phi/150\,\mathrm{Re}_p$, $Y/Z = \Phi/(1.75\,\mathrm{Re}_p)^2$ and $Z = 1.75\,\mathrm{Re}_p^2/150\,\mathrm{Re}_p = \mathrm{Re}_p/85.7$. A value of $n = 1$ is seen to represent these data reasonably well on the mean, resulting in

$$\Phi = 150\mathrm{Re}_p + 1.75\left(\mathrm{Re}_p\right)^2$$

which was originally proposed as a correlating equation by Ergun (1952) on the conjecture that the volumetric fraction of the bed in "turbulent" flow is proportional to Re_p. The success of this expression in conventional coordinates is shown in Figure 2.3.4. The scatter, which is quite evident in the arithmetic split coordinates of Figure 2.3.3, is strongly suppressed in a visual sense in the logarithmic coordinates of Figure 2.3.4.

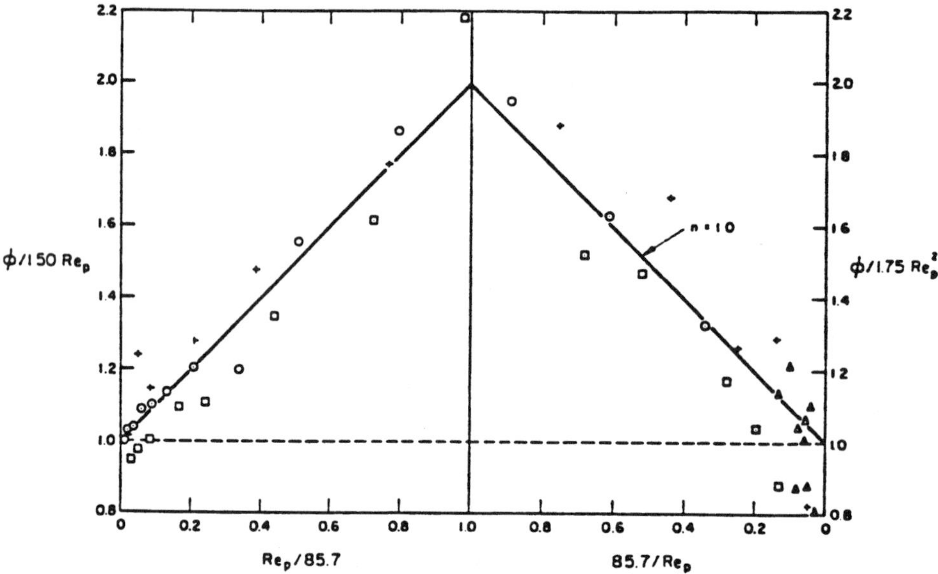

FIGURE 2.3.3 Arithmetic, split-coordinate plot of experimental data for the pressure drop in flow through a packed bed of spheres. (From Churchill, S.W. and Usagi, R. *AIChE J.* 18(6), 1123, 1972. With permission from the American Institute of Chemical Engineers.)

Example 2.3.5: The Friction Factor for Commercial Pipes for All Conditions

The serial application of the CUE is illustrated here by the construction of a correlating equation for both smooth and rough pipes in the turbulent regime followed by combination of that expression with ones for the laminar and transitional regimes.

The Turbulent Regime. The Fanning friction factor, f_F, for turbulent flow in a smooth round pipe for asymptotically large rates of flow (say $\mathrm{Re}_D > 5000$) may be represented closely by the empirical expression:

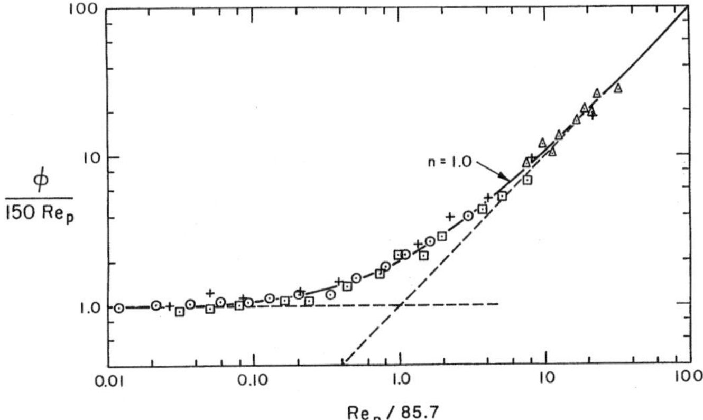

FIGURE 2.3.4 Logarithmic correlation of experimental data for the pressure drop in flow through a packed bed of spheres. (From Churchill, S.W. and Usagi, R. *AIChE J.* 18(6), 1123, 1972. With permission from the American Institute of Chemical Engineers.)

$$\left(\frac{2}{f_F}\right)^{1/2} = 1.70 - \frac{310}{\mathrm{Re}_D\left(\frac{f_F}{2}\right)^{1/2}} + \left(\frac{100}{\mathrm{Re}_D\left(\frac{f_F}{2}\right)^{1/2}}\right)^2 + \frac{1}{0.436}\ln\left\{\mathrm{Re}_D\left(\frac{f}{2}\right)^{1/2}\right\}$$

A corresponding empirical representation for naturally rough pipe is

$$\left(\frac{2}{f_F}\right)^{1/2} = 4.46 + \frac{1}{0.436}\ln\left\{\frac{D}{e}\right\}$$

Direct combination of these two expressions in the form of the CUE does not produce a satisfactory correlating equation, but their combination in the following rearranged forms:

$$e^{0.436(2/f_F)^{1/2}} = 2.10\ \mathrm{Re}_D\left(\frac{f_F}{2}\right)^{1/2} - e^{-135/\mathrm{Re}_D(f_F/2)^{1/2}} + e^{\left(43.6/\mathrm{Re}_D\left(\frac{f_F}{2}\right)^{1/2}\right)^2}$$

and

$$e^{0.430}(2/f_F)^{1/2} = 6.69\left(\frac{D}{e}\right)$$

with $n = -1$ results in, after the reverse rearrangement,

$$\left(\frac{2}{f}\right)^{1/2} = 1.70 + \frac{310}{\mathrm{Re}_D\left(\frac{f_F}{2}\right)^{1/2}} + \left(\frac{100}{\mathrm{Re}_D\left(\frac{f_F}{2}\right)^{1/2}}\right)^2 + \frac{1}{0.436}\ln\left\{\frac{\mathrm{Re}_D\left(\frac{f_F}{2}\right)^{1/2}}{1+0.301\left(\frac{e}{D}\right)\mathrm{Re}_D\left(\frac{f_F}{2}\right)^{1/2}}\right\}$$

The exact equivalent of this expression in structure but with the slightly modified numerical coefficients of 0.300, 2.46, and 0.304 was postulated by Colebrook (1938–1939) to represent his own experimental data. The coefficients of the expression given here are presumed to be more accurate, but the difference in the predictions of f_F with the two sets of coefficients is within the band of uncertainty of the experimental data. The turbulent regime of the "friction-factor" plot in most current textbooks and handbooks is simply a graphical representation of the Colebrook equation. Experimental values are not included in such plots since e, the effective roughness of commercial pipes, is simply a correlating factor that forces agreement with the Colebrook equation. Values of e for various types of pipe in various services are usually provided in an accompanying table, that thereby constitutes an integral part of the correlation.

The Laminar Region. The Fanning friction factor in the laminar regime of a round pipe ($\mathrm{Re}_d < 1800$) is represented exactly by the following theoretical expression known as Poiseuille's law: $f_F = 16/\mathrm{Re}_D$. This equation may be rearranged as follows for convenience in combination with that for turbulent flow:

$$\left(\frac{2}{f_F}\right)^{1/2} = \frac{\mathrm{Re}_D\left(f_F/2\right)^{1/2}}{8}$$

The Transitional Regime. Experimental data as well as semitheoretical computed values for the limiting behavior in the transition may be represented closely by $(f_F/2) = (\mathrm{Re}_D/37500)^2$. This expression may be rewritten, in terms of $(2/f_F)^{1/2}$ and $\mathrm{Re}_D(f_F/2)^{1/2}$, as follows:

$$\left(\frac{f_F}{2}\right)^{1/2} = \left(\frac{37500}{\mathrm{Re}_D\left(f_F/2\right)^{1/2}}\right)^{1/2}$$

Overall Correlation. The following correlating equation for all $\mathrm{Re}_D(f_F/2)^{1/2}$ and e/D may now be constructed by the combination of the expressions for the turbulent and transition regimes in the form of the CUE with $n = 8$, and then that expression and that for the laminar regime with $n = -12$, both components being chosen on the basis of experimental data and predicted values for the full regime of transition:

$$\left(\frac{2}{f_F}\right)^{1/2} = \left[\left(\frac{8}{\mathrm{Re}_D\left(f_F/2\right)^{1/2}}\right)^{12} + \left(\left[\frac{37500}{\mathrm{Re}_D\left(f_F/2\right)^{1/2}}\right]\right.\right.$$

$$\left.\left. + \left|1.706 - \frac{322.4}{\mathrm{Re}_D\left(\frac{f_F}{2}\right)^{1/2}} + \left(\frac{95.2}{\mathrm{Re}_D\left(\frac{f_F}{2}\right)^{1/2}}\right)^2 + \frac{1}{0.436}\ln\left\{\frac{\mathrm{Re}_D\left(\frac{f_F}{2}\right)^{1/2}}{1+0.301\left(\frac{e}{D}\right)\mathrm{Re}_D\left(\frac{f_F}{2}\right)^{1/2}}\right\}\right|^8\right)^{-3/2}\right]^{-1/12}$$

The absolute value signs are only included for aesthetic reasons; the negative values of the logarithmic term for very small values of $\mathrm{Re}_D(f_F/2)^{1/2}$ do not affect the numerical value of $(2/f_F)^{1/2}$ in the regime in which they occur. This overall expression appears to have a complicated structure, but it may readily be recognized to reduce to its component parts when the corresponding term is large with respect to the other two. It is insensitive to the numerical values of the two arbitrary exponents. For example, doubling their values would have almost no effect on the predictions of $(f_F/2)^{1/2}$. The principal uncertainty is associated with the expression for the transition regime, but the overall effect of the corresponding term is very small. The uncertainties associated with this correlating equation are common to most graphical correlations and algebraic expressions for the friction factor, and are presumed to be fairly limited in magnitude and to be associated primarily with the postulated value of e. Although the overall expression is explicit in $\mathrm{Re}_D(f_F/2)^{1/2}$ rather than Re_D, the latter quantity may readily be obtained simply by multiplying the postulated value of $\mathrm{Re}_D(f_F/2)^{1/2}$ by the computed values of $(2/f_F)^{1/2}$.

References

Buckingham, E. 1914. On physically similar systems; illustrations of the use of dimensional equations. *Phys. Rev., Ser. 2,* 4(4):345–375.

Churchill, S.W. 1976. A comprehensive correlating equation for forced convection from plates. *AIChE J.* 22(2):264–268.

Churchill, S.W. 1977. Comprehensive correlating equations for heat, mass and momentum transfer in fully developed flow in smooth tubes. *Ind. Eng. Chem. Fundam.* 16(1):109–116.

Churchill, S.W. 1979. *The Interpretation and Use of Rate Data. The Rate Process Concept,* rev. printing, Hemisphere Publishing Corp., Washington, D.C.

Churchill, S.W. 1981. The use of speculation and analysis in the development of correlations. *Chem. Eng. Commun.* 9:19–38.

Churchill, S.W. 1988a. Flow through porous media, Chapter 19 in *Laminar Flows. The Practical Use of Theory,* pp. 501–538, Butterworths, Boston.

Churchill, S.W. 1988b. Derivation, selection, evaluation and use of asymptotes. *Chem. Eng. Technol.* 11:63–72.

Churchill, S.W. and Churchill, R.U. 1975. A general model for the effective viscosity of pseudoplastic and dilatant fluids. *Rheol. Acta.* 14:404–409.

Churchill, S.W. and Usagi, R. 1972. A general expression for the correlation of rates of transfer and other phenomena. *AIChE J.* 18(6):1121–1128.

Churchill, S.W. and Usagi, R. 1974. A standardized procedure for the production of correlations in the form of a common empirical equation. *Ind. Eng. Chem. Fundam.* 13(1):39–44.

Colebrook, C.R. 1938–1939. Turbulent flow in pipes with particular reference to the transition region between the smooth and rough pipe laws. *J. Inst. Civ. Eng.* 11(5024):133–156.

Ergun, S. 1952. Fluid flow through packed beds. *Chem. Eng. Prog.* 48(2):81–96.

Hellums, J.D. and Churchill, S.W. 1964. Simplifications of the mathematical description of boundary and initial value problems. *AIChE J.* 10(1):110–114.

Wilkie, D. 1985. The correlation of engineering data reconsidered. *Int. J. Heat Fluid Flow.* 8(2):99–103.

Zlokarnik, M. 1991. *Dimensional Analysis and Scale-Up in Chemical Engineering.* Springer-Verlag, Berlin.

2.4 Hydraulics of Pipe Systems

J. Paul Tullis

Basic Computations

Equations

Solving fluid flow problems involves the application of one or more of the three basic equations: continuity, momentum, and energy. These three basic tools are developed from the law of conservation of mass, Newton's second law of motion, and the first law of thermodynamics.

The simplest form of the continuity equation is for one-dimensional incompressible steady flow in a conduit. Applying continuity between any two sections gives

$$A_1V_1 = A_2V_2 = Q \tag{2.4.1}$$

For a variable density the equation can be written

$$\rho_1A_1V_1 = \rho_2A_2V_2 = \dot{m} \tag{2.4.2}$$

in which A is the cross-sectional area of the pipe, V is the mean velocity at that same location, Q is the flow rate, ρ is the fluid density, and $\dot{m}$ is the mass flow rate. The equations are valid for any rigid conduit as long as there is no addition or loss of liquid between the sections.

For steady state pipe flow, the momentum equation relates the net force in a given direction (F_x) acting on a control volume (a section of the fluid inside the pipe), to the net momentum flux through the control volume.

$$F_x = \rho_2A_2V_2V_{2x} - \rho_1A_1V_1V_{1x} \tag{2.4.3}$$

For incompressible flow this equation can be reduced to

$$F_x = \rho Q\left(V_{2x} - V_{1x}\right) \tag{2.4.4.}$$

These equations can easily be applied to a three-dimensional flow problem by adding equations in the y and z directions.

A general form of the energy equation (see Chapter 1) applicable to incompressible pipe or duct flow

$$\frac{P_1}{\gamma} + Z_1 + \frac{V_1^2}{2g} = \frac{P_2}{\gamma} + Z_2 + \frac{V_2^2}{2g} - H_p + H_t + H_f \tag{2.4.5}$$

The units are energy per unit weight of liquid: ft · lb/lb or N · m/N which reduce to ft or m. The first three terms are pressure head (P/γ), elevation head (Z) (above some datum), and velocity head ($V^2/2g$). The last three terms on the right side of the equation are the total dynamic head added by a pump (H_p) or removed by a turbine (H_t) and the friction plus minor head losses (H_f). The sum of the first three terms in Equation 2.4.5 is defined as the total head, and the sum of the pressure and elevation heads is referred to as the piezometric head.

The purpose of this section is to determine the pressure changes resulting from incompressible flow in pipe systems. Since pipes of circular cross sections are most common in engineering application, the analysis in this section will be performed for circular geometry. However, the results can be generalized for a pipe of noncircular geometry by substituting for the diameter D in any of the equations, the hydraulic diameter, D_h, defined as

$$D_h = 4 \times \frac{\text{the cross sectional area}}{\text{the wetted perimeter}}$$

The analysis in this section can also be applied to gases and vapors, provided the Mach number in the duct does not exceed 0.3. For greater values of the Mach number, the compressibility effect becomes significant and the reader is referred to Section 2.7 on compressible flow.

Fluid Friction

The calculation of friction loss in pipes and ducts depends on whether the flow is laminar or turbulent. The Reynolds number is the ratio of inertia forces to viscous forces and is a convenient parameter for predicting if a flow condition will be laminar or turbulent. It is defined as

$$\text{Re}_D = \frac{\rho VD}{\mu} = \frac{VD}{\nu} \tag{2.4.6}$$

in which V is the mean flow velocity, D diameter, ρ fluid density, μ dynamic viscosity, and ν kinematic viscosity.

Friction loss (H_f) depends on pipe diameter (d), length (L), roughness (e), fluid density (ρ) or specific weight (γ), viscosity (ν), and flow velocity (V). Dimensional analysis can be used to provide a functional relationship between the friction loss H_f, pipe dimensions, fluid properties, and flow parameters. The resulting equation is called the Darcy–Weisbach equation:

$$H_f = \frac{fLV^2}{2gd} = \frac{fLQ^2}{1.23gD^5} \tag{2.4.7}$$

The friction factor f is a measure of pipe roughness. It has been evaluated experimentally for numerous pipes. The data were used to create the Moody friction factor chart shown as Figure 2.4.1. For Re < 2000, the flow in a pipe will be laminar and f is only a function of Re_D. It can be calculated by

$$f = \frac{64}{\text{Re}_D} \tag{2.4.8}$$

At Reynolds numbers between about 2000 and 4000 the flow is unstable as a result of the onset of turbulence (critical zone in Figure 2.4.1). In this range, friction loss calculations are difficult because it is impossible to determine a unique value of f. For Re > 4000 the flow becomes turbulent and f is a function of both Re and relative pipe roughness (e/d). At high Re, f eventually depends only on e/d; defining the region referred to as fully turbulent flow. This is the region in Figure 2.4.1 where the lines for different e/d become horizontal (e is the equivalent roughness height and d pipe diameter). The Re_D at which this occurs depends on the pipe roughness. Laminar flow in pipes is unusual. For example, for water flowing in a 0.3-m-diameter pipe, the velocity would have to be below 0.02 m/sec for laminar flow to exist. Therefore, most practical pipe flow problems are in the turbulent region.

Using the Moody chart in Figure 2.4.1 to get f requires that Re and e/d be known. Calculating Re is direct if the water temperature, velocity, and pipe diameter are known. The problem is obtaining a good value for e. Typical values of e are listed in Figure 2.4.1. These values should be considered as guides only and not used if more–exact values can be obtained from the pipe supplier.

Since roughness may vary with time due to buildup of solid deposits or organic growths, f is also time dependent. Manufacturing tolerances also cause variations in the pipe diameter and surface roughness. Because of these factors, the friction factor for any pipe can only be approximated. A designer is required to use good engineering judgment in selecting a design value for f so that proper allowance is made for these uncertainties.

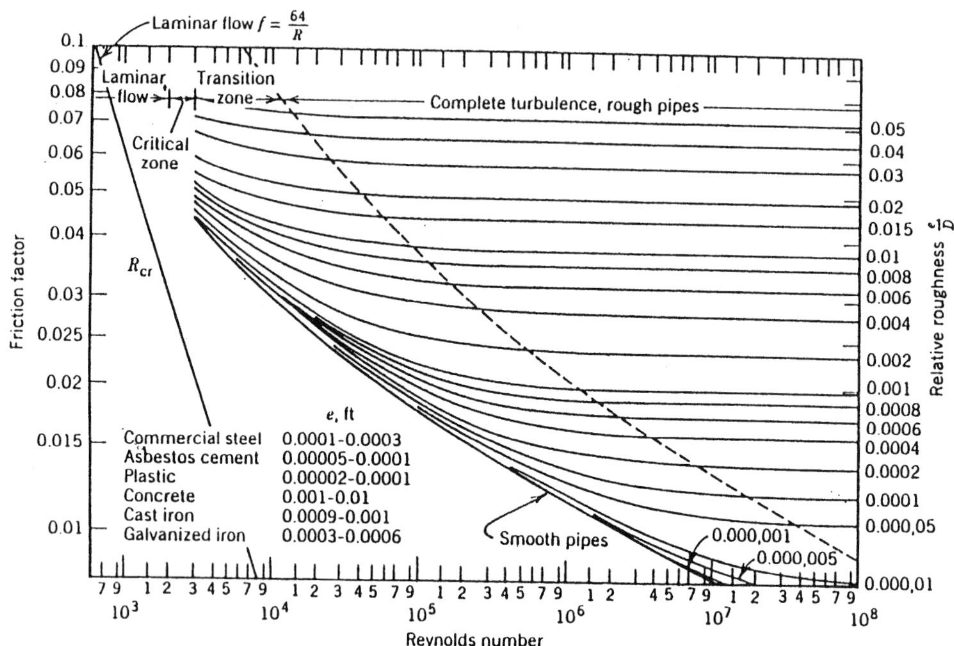

FIGURE 2.4.1 The Moody diagram.

For noncircular pipes, the only change in the friction loss equation is the use of an equivalent diameter — based on the hydraulic radius (R), i.e., $d = 4R$ — in place of the circular pipe diameter d. R is the ratio of the flow area to the wetter perimeter.

Wood (1966) developed equations which can be used in place of the Moody diagram to estimate f for Re > 10^4 and $10^{-5} < k < 0.04$ ($k = e/d$).

$$f = a + b\text{Re}^{-c} \tag{2.4.9}$$

$$a = 0.094k^{0.225} + 0.53k, \quad b = 88k^{0.44}, \quad c = 1.62k^{0.134}$$

The practical problem is still obtaining a reliable value for e. It cannot be directly measured but must be determined from friction loss tests of the pipe.

An exact solution using the Darcy–Weisbach equation can require a trial-and-error solution because of the dependency of f on Re if either the flow or pipe diameter are not known. A typical approach to solving this problem is to estimate a reasonable fluid velocity to calculate Re and obtain f from the Moody chart or Equation (2.4.9). Next, calculate a new velocity and repeat until the solution converges. Converging on a solution is greatly simplified with programmable calculators and a variety of software available for computer.

For long gravity flow pipelines, the starting point in selecting the pipe diameter is to determine the smallest pipe that can pass the required flow without friction loss exceeding the available head. For pumped systems, the selection must be based on an economic analysis that compares the pipe cost with the cost of building and operating the pumping plant.

Local Losses

Flow through valves, orifices, elbows, transitions, etc. causes flow separation which results in the generation and dissipation of turbulent eddies. For short systems containing many bends, valves, tees, etc. local or minor losses can exceed friction losses. The head loss h_l associated with the dissipation caused by a minor loss is proportional to the velocity head and can be accounted for as a minor or local loss using the following equation.

$$h_l = K_l \frac{Q^2}{\left(2gA_m^2\right)} \tag{2.4.10}$$

in which K_l is the minor loss coefficient and A_m is the area of the pipe at the inlet to the local loss. The loss coefficient K_l is analogous to fL/d in Equation 2.4.7.

The summation of all friction and local losses in a pipe system can be expressed as

$$H_f = h_f + h_l \tag{2.4.11}$$

$$H_f = \left[\sum\left(\frac{fL}{2gdA_p^2}\right) + \sum\left(\frac{K_l}{2gA_m^2}\right)\right]Q^2 = CQ^2 \tag{2.4.12}$$

in which

$$C = \sum\left(\frac{fL}{2gdA_p^2}\right) + \sum\left(\frac{K_l}{2gA_m^2}\right) \tag{2.4.13}$$

It is important to use the correct pipe diameter for each pipe section and local loss.

In the past some have expressed the local losses as an equivalent pipe length: $L/d = K_l/f$. It simply represents the length of pipe that produces the same head loss as the local or minor loss. This is a simple, but not a completely accurate method of including local losses. The problem with this approach is that since the friction coefficient varies from pipe to pipe, the equivalent length will not have a unique value. When local losses are truly minor, this problem becomes academic because the error only influences losses which make up a small percentage of the total. For cases where accurate evaluation of all losses is important, it is recommended that the minor loss coefficients K_l be used rather than an equivalent length.

The challenging part of making minor loss calculations is obtaining reliable values of K_l. The final results cannot be any more accurate than the input data. If the pipe is long, the friction losses may be large compared with the minor losses and approximate values of K_l will be sufficient. However, for short systems with many pipe fittings, the local losses can represent a significant portion of the total system losses, and they should be accurately determined. Numerous factors influence K_l. For example, for elbows, K_l is influenced by the shape of the conduit (rectangular vs. circular), by the radius of the bend, the bend angle, the Reynolds number, and the length of the outlet pipe. For dividing or combining tees or Y-branches, the percent division of flow and the change in pipe diameter must also be included when estimating K_l. One factor which is important for systems where local losses are significant is the interaction between components placed close together. Depending on the type, orientation, and spacing of the components, the total loss coefficient may be greater or less than the simple sum of the individual K_l values.

Comparing the magnitude of $\Sigma(fL/2gdA_p^2)$ to $\Sigma(K_l/2gA_m^2)$ will determine how much care should be given to the selection of the K_l values. Typical values of K_l are listed in Table 2.4.1 (Tullis, 1989). When more comprehensive information on loss coefficients is needed, the reader is referred to Miller (1990).

TABLE 2.4.1 Minor Loss Coefficients

Item	K_l Typical Value	Typical Range
Pipe inlets		
Inward projecting pipe	0.78	0.5–0.9
Sharp corner-flush	0.50	—
Slightly rounded	0.20	0.04–0.5
Bell mouth	0.04	0.03–0.1
Expansions[a]	$(1 - A_1/A_2)^2$ (based on V_1)	
Contractions[b]	$(1/C_c - 1)^2$ (based on V_2)	

A_2/A_1	0.1	0.2	0.3	0.4	0.5	0.6	0.7	0.8	0.9
C_c	0.624	0.632	0.643	0.659	0.681	0.712	0.755	0.813	0.892

Item	K_l Typical Value	Typical Range
Bends[c]		
Short radius, $r/d = 1$		
90	—	0.24
45	—	0.1
30	—	0.06
Long radius, $r/d = 1.5$		
90	—	0.19
45	—	0.09
30	—	0.06
Mitered (one miter)		
90	1.1	—
60	0.50	0.40–0.59
45	0.3	0.35–0.44
30	0.15	0.11–0.19
Tees	[c]	—
Diffusers	[c]	—
Valves		
Check valve	0.8	0.5–1.5
Swing check	1.0	0.29–2.2
Tilt disk	1.2	0.27–2.62
Lift	4.6	0.85–9.1
Double door	1.32	1.0–1.8
Full-open gate	0.15	0.1–0.3
Full-open butterfly	0.2	0.2–0.6
Full-open globe	4.0	3–10

[a] See Streeter and Wylie, 1975, p. 304.
[b] See Streeter and Wylie, 1975, p. 305.
[c] See Miller, 1990.
[d] See Kalsi Engineering and Tullis Engineering Consultants, 1993.

Pipe Design

Pipe Materials

Materials commonly used for pressure pipe transporting liquids are ductile iron, concrete, steel, fiberglass, PVC, and polyolefin. Specifications have been developed by national committees for each of these pipe materials. The specifications discuss external loads, internal design pressure, available sizes, quality of materials, installation practices, and information regarding linings. Standards are available from the following organizations:

American Water Works Association (AWWA)
American Society for Testing and Materials (ASTM)
American National Standards Institute (ANSI)
Canadian Standards Association (CSA)

Federal Specifications (FED)
Plastic Pipe Institute (PPI)

In addition, manuals and other standards have been published by various manufacturers and manufacturer's associations. All of these specifications and standards should be used to guide the selection of pipe material. ASCE (1992) contains a description of each of these pipe materials and a list of the specifications for the various organizations which apply to each material. It also discusses the various pipe-lining materials available for corrosion protection.

For air- and low-pressure liquid applications one can use unreinforced concrete, corrugated steel, smooth sheet metal, spiral rib (sheet metal), and HDPE (high-density polyethylene) pipe. The choice of a material for a given application depends on pipe size, pressure requirements, resistance to collapse from internal vacuums and external loads, resistance to internal and external corrosion, ease of handling and installing, useful life, and economics.

Pressure Class Guidelines

Procedures for selecting the pressure class of pipe vary with the type of pipe material. Guidelines for different types of materials are available from AWWA, ASTM, ANSI, CSA, FED, PPI and from the pipe manufacturers. These specifications should be obtained and studied for the pipe materials being considered.

The primary factors governing the selection of a pipe pressure class are (1) the maximum steady state operating pressure, (2) surge and transient pressures, (3) external earth loads and live loads, (4) variation of pipe properties with temperature or long-time loading effects, and (5) damage that could result from handling, shipping, and installing or reduction in strength due to chemical attack or other aging factors. The influence of the first three items can be quantified, but the last two are very subjective and are generally accounted for with a safety factor which is the ratio of the burst pressure to the rated pressure.

There is no standard procedure on how large the safety factor should be or on how the safety factor should be applied. Some may feel that it is large enough to account for all of the uncertainties. Past failures of pipelines designed using this assumption prove that it is not always a reliable approach. The procedure recommended by the author is to select a pipe pressure class based on the internal design pressure (IDP) defined as

$$\text{IDP} = \left(P_{\text{max}} + P_s\right)\text{SF} \tag{2.4.14}$$

in which P_{max} is the maximum steady state operating pressure, P_s is the surge or water hammer pressure, and SF is the safety factor applied to take care of the unknowns (items 3 to 5) just enumerated. A safety factor between 3 and 4 is typical.

The maximum steady state operating pressure (P_{max}) in a gravity flow system is usually the difference between the maximum reservoir elevation and the lowest elevation of the pipe. For a pumped system it is usually the pump shutoff head calculated based on the lowest elevation of the pipe.

Surge and transient pressures depend on the specific pipe system design and operation. Accurately determining P_s requires analyzing the system using modern computer techniques. The most commonly used method is the "Method of Characteristics" (Tullis, 1989; Wylie and Streeter, 1993). Some of the design standards give general guidelines to predict P_s that can be used if a detailed transient analysis is not made. However, transients are complex enough that simple "rules of thumb" are seldom accurate enough. Transients are discussed again in a later subsection.

Selection of wall thickness for larger pipes is often more dependent on collapse pressure and handling loads than it is on burst pressure. A thin-wall, large-diameter pipe may be adequate for resisting relatively high internal pressures but may collapse under negative internal pressure or, if the pipe is buried, the soil and groundwater pressure plus live loads may be sufficient to cause collapse even if the pressure inside the pipe is positive.

External Loads

There are situations where the external load is the controlling factor determining if the pipe will collapse. The magnitude of the external load depends on the diameter of the pipe, the pipe material, the ovality (out of roundness) of the pipe cross section, the trench width, the depth of cover, the specific weight of the soil, the degree of soil saturation, the type of backfill material, the method used to backfill, the degree of compaction, and live loads. The earth load increases with width and depth of the trench, and the live load reduces with depth of cover. The cumulative effect of all these sources of external loading requires considerable study and analysis.

There are no simple guidelines for evaluating external pipe loads. Because of the complexity of this analysis, the default is to assume that the safety factor is adequate to account for external loads as well as the other factors already mentioned. One should not allow the safety factor to replace engineering judgment and calculations. One option to partially compensate for the lack of a detailed analysis is to use a higher-pressure class of pipe in areas where there will be large live loads or where the earth loading is unusually high. One should consider the cost of a pipe failure caused by external loads compared with the cost of using a thicker pipe or the cost of performing a detailed analysis. Those interested in the details of performing calculations of earth loading should be Spranger and Handy, 1973.

Limiting Velocities

There are concerns about upper and lower velocity limits. If the velocity is too low, problems may develop due to settling of suspended solids and air being trapped at high points and along the crown of the pipe. The safe lower velocity limit to avoid collecting air and sediment depends on the amount and type of sediment and on the pipe diameter and pipe profile. Velocities greater than about 1 m/sec (3 ft/sec) are usually sufficient to move trapped air to air release valves and keep the sediment in suspension.

Problems associated with high velocities are (1) erosion of the pipe wall or liner (especially if coarse suspended sediment is present), (2) cavitation at control valves and other restrictions, (3) increased pumping costs, (4) removal of air at air release valves, (5) increased operator size and concern about valve shaft failures due to excessive flow torques, and (6) increased risk of hydraulic transients. Each of these should be considered before making the final pipe diameter selection. A typical upper velocity for many applications if 6 m/sec (20 ft/sec). However, with proper pipe design and analysis (of the preceding six conditions), plus proper valve selection, much higher velocities can be tolerated.

Valve Selection

Valves serve a variety of functions. Some function as isolation or block valves that are either full open or closed. Control valves are used to regulate flow or pressure and must operate over a wide range of valve openings. Check valves prevent reverse flow, and air valves release air during initial filling and air that is collected during operation and admit air when the pipe is drained.

Control Valves

For many flow control applications it is desirable to select a valve that has linear control characteristics. This means that if you close the valve 10%, the flow reduces about 10%. Unfortunately, this is seldom possible since the ability of a valve to control flow depends as much on the system as it does on the design of the valve. The same valve that operates linearly in one system may not in another.

Valve selection — Selecting the proper flow control valve should consider the following criteria:

1. The valve should not produce excessive pressure drop when full open.
2. The valve should control over at least 50% of its movement.
3. The operating torque must not exceed the capacity of the operator or valve shaft and connections at any flow connection.
4. The valve should not be subjected to excessive cavitation.

5. Pressure transients caused by valve operation should not exceed the safe limits of the system.
6. Some valves should not be operated at very small openings. Other valves should be operated near full open.

Controllability. To demonstrate the relationship between a valve and system, consider a butterfly valve that will be used to control the flow between two reservoirs with an elevation difference of ΔZ. System A is a short pipe (0.3 m dia., 100 m long, ΔZ = 10 m) where pipe friction is small $fL/2gdA_p^2 = 46.9$, and System B is a long pipe (0.3 m dia., 10,000 m long, ΔZ = 200 m) with high friction $fL/2gdA_p^2 =$ 4690. Initially, assume that the same butterfly valve will be used in both pipes and it will be the same size as the pipe. The flow can be calculated using the energy equation (Equation 2.4.5) and the system loss equation (Equation (2.4.12):

$$Q = \sqrt{\frac{\Delta Z}{\left[\sum\left(\frac{fL}{2gdA_p^2}\right) + \sum\left(\frac{K_l}{2gA_m^2}\right)\right]}} \tag{2.4.15}$$

For the valve, assume that the K_l full open is 0.2 and at 50% open it is 9.0. Correspondingly, $K_l/2gA_m^2$ = 1.905 and 85.7. For System A, the flow with the valve full open will be 0.453 m^3/sec and at 50% open 0.275 m^3/sec, a reduction of 39%. Repeating these calculations over the full range of valve openings would show that the flow for System A reduces almost linearly as the valve closes.

For System B, the flow with the valve full open will be 0.206 m^3/sec and at 50% open 0.205 m^3/sec, a reduction of less than 1%. For System B the valve does not control until the valve loss, expressed by $K_l/2gA_m^2$ becomes a significant part of the friction term (4690). The same valve in System B will not start to control the flow until it has closed more than 50%. A line-size butterfly valve is obviously not a good choice for a control valve in System B. One solution to this problem is to use a smaller valve. If the butterfly valve installed in System B was half the pipe diameter, it would control the flow over more of the stroke of the valve.

The range of opening over which the valve controls the flow also has a significant effect on the safe closure time for control valves. Transient pressures are created when there is a sudden change in the flow. Most valve operators close the valve at a constant speed. If the valve does not control until it is more than 50% closed, over half of the closing time is wasted and the effective valve closure time is less than half the total closing time. This will increase the magnitude of the transients that will be generated.

Torque. To be sure that the valve shaft, connections, and operator are properly sized, the maximum torque or thrust must be known. If the maximum force exceeds operator capacity, it will not be able to open and close the valve under extreme flow conditions. If the shaft and connectors are underdesigned, the valve may fail and slam shut causing a severe transient.

For quarter-turn valves, the force required to operate a valve consists of seating, bearing, and packing friction, hydrodynamic (flow) forces, and inertial forces. These forces are best determined experimentally. A key step in applying experimental torque information is the determination of the flow condition creating maximum torque. This requires that the system be analyzed for all possible operating conditions and valve openings. For a given size and type of valve, the flow torque depends on the torque coefficient (which is dependent on the specific valve design) and the pressure drop which, in turn, depends on the flow. In short systems where there is little friction loss and high velocities, a quarter-turn valve will see maximum torques at large openings where the flow is high. In long systems with high reservoir heads and smaller velocities, the same valve will see maximum torque at small openings where the pressure drop is high.

One situation where it is easy to overlook the condition causing maximum torque is with parallel pumps. Each pump normally will have a discharge control valve. The maximum system flow occurs with all three pumps operating. However, the flow and the torque for any of the pump discharge valves is maximum for only one pump operating. One specific example (Tullis, 1989) showed that the torque on a butterfly valve was three times higher when one pump was operating compared with three pumps operating in parallel.

Cavitation. Cavitation is frequently an important consideration in selection and operation of control valves. It is necessary to determine if cavitation will exist, evaluate its intensity, and estimate its effect on the system and environment. Cavitation can cause noise, vibration, and erosion damage and can decrease performance. The analysis should consider the full range of operation. Some valves cavitate worst at small openings and others cavitate heavily near full open. It depends on both the system and the valve design. If cavitation is ignored in the design and selection of the valves, repairs and replacement of the valves may be necessary. Information for making a complete cavitation analysis is beyond the scope of this section. Detailed information on the process to design for cavitation is contained in Tullis (1989, 1993).

The first step in a cavitation analysis is selecting the acceptable level of cavitation. Experimental data are available for four limits: incipient (light, intermittent noise), critical (light, steady noise), incipient damage (pitting damage begins), and choking (very heavy damage and performance drops off). Limited cavitation data are available for each of these limits (Tullis, 1989, 1993). Choosing a cavitation limit depends on several factors related to the operating requirements, expected life, location of the device, details of the design, and economics. For long-term operation of a control valve in a system where noise can be tolerated, the valve should never operate beyond incipient damage. In systems where noise is objectionable, critical cavitation would be a better operating limit.

Using a choking cavitation as a design limit is often misused. It is generally appropriate as a design limit for valves that only operate for short periods of time, such as a pressure relief valve. The intensity of cavitation and the corresponding noise vibration and erosion damage at the valve are at their maximum just before a valve chokes. If the valve operates beyond choking (sometimes referred to as supercavitation), the collapse of the vapor cavities occurs remote from the valve. Little damage is likely to occur at the valve, but farther downstream serious vibration and material erosion problems can occur.

If the cavitation analysis indicates that the valve, orifice, or other device will be operating at a cavitation level greater than can be tolerated, various techniques can be used to limit the level of cavitation. One is to select a different type of valve. Recent developments in valve design have produced a new generation of valves that are more resistant to cavitation. Most of them operate on the principle of dropping the pressure in stages. They usually have multiple paths with numerous sharp turns or orifices in series. Two limitations to these valves are that they often have high pressure drops (even when full open), and they are only usable with clean fluids.

A similar approach is to place multiple conventional valves in series or a valve in series with orifice plates. Proper spacing of valves and orifices placed in series is important. The spacing between valves depends upon the type. For butterfly valves it is necessary to have between five and eight diameters of pipe between valves to prevent flutter of the leaf of the downstream valve and to obtain the normal pressure drop at each valve. For globe, cone, and other types of valves, it is possible to bolt them flange to flange and have satisfactory operation.

For some applications another way to suppress cavitation is to use a free-discharge valve that is vented so cavitation cannot occur. There are valves specifically designed for this application. Some conventional valves can also be used for free discharge, if they can be adequately vented.

Cavitation damage can be suppressed by plating critical areas of the pipe and valve with cavitation-resistant materials. Based on tests using a magnetostriction device, data show that there is a wide variation in the resistance of the various types of material. Limited testing has been done on the erosion resistance of different materials and coating to cavitation in flowing systems. The available data show that there is less variation in the damage resistance of materials in actual flowing systems. However, experience has shown the plating parts of the valve with the right material will extend valve life.

Injecting air to suppress cavitation is a technique which has been used for many years with varying degrees of success. The most common mistake is placing the air injection port in the wrong location so the air does not get to the cavitation zone. If an adequate amount of air is injected into the proper region, the noise, vibrations, and erosion damage can be significantly reduced. The air provides a cushioning effect reducing the noise, vibration, and erosion damage. If the system can tolerate some air being injected, aeration is usually the cheapest and best remedy for cavitation.

Transients. Transient pressures can occur during filling and flushing air from the line, while operating valves, and when starting or stopping pumps. If adequate design provisions and operational procedures are not established, the transient pressure can easily exceed the safe operating pressure of the pipe. A system should be analyzed to determine the type and magnitudes of possible hydraulic transients. The basic cause is rapid changes in velocity. The larger the incremental velocity change and the faster that change takes place, the greater will be the transient pressure. If the piping system is not designed to withstand the high transient pressures, or if controls are not included to limit the pressure, rupture of the pipe or damage to equipment can result.

All pipelines experience transients. Whether or not the transient creates operational problems or pipe failure depends upon its magnitude and the ability of the pipes and mechanical equipment to tolerate high pressures without damage. For example, an unreinforced concrete pipeline may have a transient pressure head allowance of only a meter above its operating pressure before damage can occur. For such situations even slow closing of control valves or minor interruptions of flow due to any cause may create sufficient transient pressures to rupture the pipeline. In contrast, steel and plastic pipes can take relatively high transient pressures without failure.

Transients caused by slow velocity changes, such as the rise and fall of the water level in a surge tank, are called surges. Surge analysis, or "rigid column theory" involves mathematical or numerical solution of simple ordinary differential equations. The compressibility of the fluid and the elasticity of the conduit are ignored, and the entire column of fluid is assumed to move as a rigid body.

When changes in velocity occur rapidly, both the compressibility of the liquid and the elasticity of the pipe must be included in the analysis. This procedure is often called "elastic" or "waterhammer" analysis and involves tracking acoustic pressure waves through the pipe. The solution requires solving partial differential equations.

An equation predicting the head rise ΔH caused by a sudden change of velocity $\Delta V = V_2 - V_1$ can be derived by applying the unsteady momentum equation to a control volume of a section of the pipe where the change of flow is occurring. Consider a partial valve closure which instantly reduces the velocity by an amount ΔV. Reduction of the velocity can only be accomplished by an increase in the pressure upstream from the valve. This creates a pressure wave of magnitude ΔH which travels up the pipe at the acoustic velocity **a**. The increased pressure compresses the liquid and expands the pipe. The transient head rise due to an incremental change in velocity is

$$\Delta H = -\mathbf{a}\Delta V/g, \quad \text{for } \mathbf{a} \gg \Delta V \tag{2.4.16}$$

This equation is easy to use for multiple incremental changes of velocity as long as the first wave has not been reflected back to the point of origin.

The derivation of Equation (2.4.16) is based on an assumption of an instant velocity change. For a valve closing at the end of the pipe, instant closure actually refers to a finite time. It is the longest time that a valve can be closed and still cause a pressure rise equal to that of an instant closure. Normally, it is equal to $2L/a$ sec (which is the time required for the first pressure wave to travel to and from the other end of the pipe of length L); the head rise at the valve will be the same as if the valve were closed instantly. The $2L/a$ time is therefore often the instant closure time.

For a valve at the end of a long pipeline, the instant closure time can be considerably greater than $2L/a$. This is because when the friction loss coefficient fL/d is much greater than the loss coefficient for the valve K_l, the valve can be closed a long way before the flow changes. This dead time must be added to the $2L/a$ time to identify the actual instant closure time. To avoid the maximum potential transient pressure rise, the valve must be closed much slower than the instant closure time.

Computational techniques for estimating transient pressures caused by unsteady flow in pipelines are too complex to be done with simple hand calculations. The solution involves solving partial differential equations based on the equations of motion and continuity. These equations are normally solved by the method of characteristics. This technique transforms the equations into total differential equations. After integration, the equations can be solved numerically by finite differences. This analysis provides equations that can be used to predict the flow and head at any interior pipe section at any time (Tullis, 1989; Wiley and Streeter, 1993).

To complete the analysis, equations describing the boundary conditions are required. Typical boundary conditions are the connection of a pipe to a reservoir, a valve, changes in pipe diameter or material, pipe junctions, etc. Friction loss is included in the development of the basic equations and minor losses are handled as boundary conditions. The analysis properly models friction and the propagation and reflections of the pressure wave. It can also be used for surge calculations.

It is recommended that every pipe system should have at least a cursory transient analysis performed to identify the possibility of serious transients and decide whether or not a detailed analysis is necessary. If an analysis indicates that transients are a problem, the types of solutions available to the engineer include

1. Increasing the closing time of control valves.
2. Using a smaller valve to provide better control.
3. Designing special facilities for filling, flushing, and removing air from pipelines.
4. Increasing the pressure class of the pipeline.
5. Limiting the flow velocity.
6. Using pressure relief valves, surge tanks, air chambers, etc.

Check Valves

Selecting the wrong type or size of check valve can result in poor performance, severe transients, and frequent repairs (Kalsi, 1993). Proper check valve selection requires understanding the characteristics of the various types of check valves and analyzing how they will function as a part of the system in which they will be installed. A check valve that operates satisfactorily in one system may be totally inadequate in another. Each valve type has unique characteristics that give it advantages or disadvantages compared with the others. The characteristics of check valves that describe their hydraulic performance and which should be considered in the selection process include

1. Opening characteristics, i.e., velocity vs. disk position data.
2. Velocity required to fully open and firmly backseat the disk.
3. Pressure drop at maximum flow.
4. Stability of the disk at partial openings.
5. Sensitivity of disk flutter to upstream disturbances.
6. Speed of valve closure compared with the rate of flow reversal of the system.

Disk stability varies with flow rate, disk position, and upstream disturbances and is an important factor in determining the useful life of a check valve. For most applications it is preferable to size the check valve so that the disk is fully open and firmly backseated at normal flow rates. One of the worst design errors is to oversize a check valve that is located just downstream from a disturbance such as a pump, elbow, or control valve. If the disk does not fully open, it will be subjected to severe motion that will accelerate wear. To avoid this problem, it may be necessary to select a check valve that is smaller than the pipe size.

The transient pressure rise generated at check valve closure is another important consideration. The pressure rise is a function of how fast the valve disk closes compared with how fast the flow in the system reverses. The speed that the flow in a system reverses depends on the system. In systems where rapid flow reversals occur, the disk can slam shut causing a pressure transient (Thorley, 1989).

The closing speed of a valve is determined by the mass of the disk, the forces closing the disk, and the distance of travel from full open to closed. Fast closing valves have the following properties: the disk (including all moving parts) is lightweight, closure is assisted by springs, and the full stroke of the disk is short. Swing check valves are the slowest-closing valves because they violate all three of these criteria; i.e., they have heavy disks, no springs, and long disk travel. The nozzle check valve is one of the fastest-closing valves because the closing element is light, is spring loaded, and has a short stroke. The silent, duo, double door, and lift check valves with springs are similar to nozzle valves in their closing times, mainly because of the closing force of the spring.

Systems where rapid flow reversals occur include parallel pumps, where one pump is stopped while the others are still operating, and systems that have air chambers or surge tanks close to the check valve. For these systems there is a high-energy source downstream from the check valve to cause the flow to quickly reverse. As the disk nears its seat, it starts to restrict the reverse flow. This builds up the pressure, accelerates the disk, and slams it into the seat. Results of laboratory experiments, field tests, and computer simulations show that dramatic reductions in the transient pressures at disk closure can be achieved by replacing a slow-closing swing check valve with a fast-acting check valve. For example, in a system containing parallel pumps where the transient was generated by stopping one of the pumps, the peak transient pressure was reduced from 745 to 76 kPa when a swing check was replaced with a nozzle check valve. Such a change improved performance and significantly reduced maintenance.

Air Valves

There are three types of automatic air valves: (1) air/vacuum valves, (2) air release valves, and (3) combination valves. The air/vacuum valve is designed for releasing air while the pipe is being filled and for admitting air when the pipe is being drained. The valve must be large enough that it can admit and expel large quantities of air at a low pressure differential. The outlet orifice is generally the same diameter as the inlet pipe.

These valves typically contain a float, which rises and closes the orifice as the valve body fills with water. Once the line is pressurized, this type of valve cannot reopen to remove air that may subsequently accumulate until the pressure becomes negative, allowing the float to drop. If the pressure becomes negative during a transient or while draining, the float drops and admits air into the line. For thin-walled pipes that can collapse under internal vacuums, the air/vacuum valves should be sized for a full pipe break at the lowest pipe elevation. The vacuum valve must supply an air flow equal to the maximum drainage rate of the water from the pipe break and at an internal pipe pressure above the pipe collapse pressure.

The critical factor in sizing air/vacuum valves is usually the air flow rate to protect the pipe from a full pipe break. Since a pipe is filled much slower than it would drain during a full break, the selected valve will be sized so that the air is expelled during filling without pressurizing the pipe. Sizing charts are provided by manufacturers.

Air release valves contain a small orifice and are designed to release small quantities of pressurized air that are trapped during filling and that accumulate after initial filling and pressurization. The small orifice is controlled by a plunger activated by a float at the end of a lever arm. As air accumulates in the valve body, the float drops and opens the orifice. As the air is expelled, the float rises and closes off the orifice. Sizing air release valves requires an estimate of the amount of pressurized air that must be expelled. This is determined by the filling procedure and any source of air that can be admitted into the pipe or be degassed from the liquid during operation.

The combination valve is actually two valves: a large valve that functions as an air/vacuum valve and a small one that functions as an air release valve. The installation can either consist of an air/vacuum valve and an air release valve plumbed in parallel, or the two can be housed in a single valve body. Most air valve installations require combination valves.

One caution is that manual air release valves should be avoided because improper operation of them can be very dangerous. If the system is pressurized with the manual air valves closed, the trapped air will be pressurized to full system pressure. When the air valve is manually opened, releasing the pressurized air can cause rapid acceleration of the liquid and generate serious transients when the water is decelerated as it hits the air valve. If manual air valves are installed, they should be very small so the air release rate is controlled to a safe rate.

Locating air valves in a piping system depends on the pipe profile, pipe length, and flow rates. Preferably, pipes should be laid to grade with valves placed at the high points or at intervals if there are no high points. One should use engineering judgment when defining a high point. If the pipe has numerous high points that are close together, or if the high points are not pronounced, it will not be necessary to have an air valve at each high point. If the liquid flow velocity is above about 1 m/sec (3 ft/sec), the flowing water can move the entrained air past intermediate high points to a downstream air valve. Releasing the air through an air valve prevents any sizable air pockets under high pressure from forming in the pipe. Trapped air under high pressure is extremely dangerous.

Velocity of the flow during filling is important. A safe way to fill a pipe is to limit the initial fill rate to an average flow velocity of about 0.3 m/sec (1 ft/sec) until most of the air is released and the air/vacuum valves close. The next step is to flush the system at about 1 m/sec (3 ft/sec), at a low system pressure, to flush the remaining air to an air release valve. It is important that the system not be pressurized until the air has been removed. Allowing large quantities of air under high pressure to accumulate and move through the pipe can generate severe transients. This is especially true if the compressed air is allowed to pass through a control valve or manual air release valve. When pressurized air flows through a partially open valve, the sudden acceleration and deceleration of the air and liquid can generate high pressure transients.

Pump Selection

Optimizing the life of a water supply system requires proper selection, operation, and maintenance of the pumps. During the selection process, the designer must be concerned about matching the pump performance to the system requirements and must anticipate problems that will be encountered when the pumps are started or stopped and when the pipe is filled and drained. The design should also consider the effect of variations in flow requirements, and also anticipate problems that will be encountered due to increased future demands and details of installation.

Selecting a pump for a particular service requires matching the system requirements to the capabilities of the pump. The process consists of developing a system equation by applying the energy equation to evaluate the pumping head required to overcome the elevation difference, friction, and minor losses. For a pump supplying water between two reservoirs, the pump head required to produce a given discharge can be expressed as

$$Hp = \Delta Z + H_f \tag{2.4.17}$$

or

$$Hp = \Delta Z + CQ^2 \tag{2.4.18}$$

in which the constant C is defined by Equation (2.4.13).

Figure 2.4.2 shows a system curve for a pipe having an 82-m elevation lift and moderate friction losses. If the elevation of either reservoir is a variable, then there is not a single curve but a family of curves corresponding to differential reservoir elevations.

The three pump curves shown in Figure 2.4.2 represent different impeller diameters. The intersections of the system curve with the pump curves identify the flow that each pump will supply if installed in that system. For this example both A and B pumps would be a good choice because they both operate at or near their best efficiency range. Figure 2.4.2 shows the head and flow that the B pump will produce

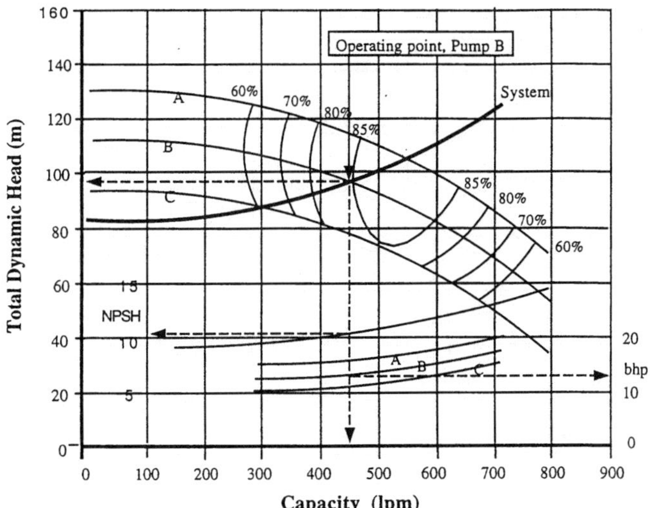

FIGURE 2.4.2 Pump selection for a single pump.

when operating in that system are 97 m and 450 L/m. The net positive suction head (NPSH) and brake horsepower (bhp) are obtained as shown in the figure.

The selection process is more complex when the system demand varies, either due to variations in the water surface elevation or to changing flow requirements. If the system must operate over a range of reservoir elevations, the pump should be selected so that the system curve, based on the mean (or the most frequently encountered) water level, intersects the pump curve near the midpoint of the best efficiency range. If the water level variation is not too great, the pump may not be able to operate efficiently over the complete flow range.

The problem of pump selection also becomes more difficult when planning for future demands or if the pumps are required to supply a varying flow. If the flow range is large, multiple pumps or a variable-speed drive may be needed. Recent developments in variable-frequency drives for pumps make them a viable alternative for systems with varying flows. Selection of multiple pumps and the decision about installing them in parallel or in series depend on the amount of friction in the system. Parallel installations are most effective for low-friction systems. Series pumps work best in high-friction systems.

For parallel pump operation the combined two pump curve is constructed by adding the flow of each pump. Such a curve is shown in Figure 2.4.3 (labeled 2 pumps). The intersection of the two-pump curve with the system curve identifies the combined flow for the two pumps. The pump efficiency for each pump is determined by projecting horizontally to the left to intersect the single-pump curve. For this example, a C pump, when operating by itself, will be have an efficiency of 83%. With two pumps operating, the efficiency of each will be about 72%. For the two pumps to operate in the most efficient way, the selection should be made so the system curve intersects the single-pump curve to the right of its best efficiency point.

Starting a pump with the pipeline empty will result in filling at a very rapid rate because initially there is little friction to build backpressure. As a result, the pump will operate at a flow well above the design flow. This may cause the pump to cavitate, but the more serious problem is the possibility of high pressures generated by the rapid filling of the pipe. Provisions should be made to control the rate of filling to a safe rate. Start-up transients are often controlled by starting the pump against a partially open discharge valve located near the pump and using a bypass line around the pump. This allows the system to be filled slowly and safely. If the pipe remains full and no air is trapped, after the initial filling, subsequent start-up of the pumps generally does not create any serious problem. Adequate air release valves should be installed to release the air under low pressure.

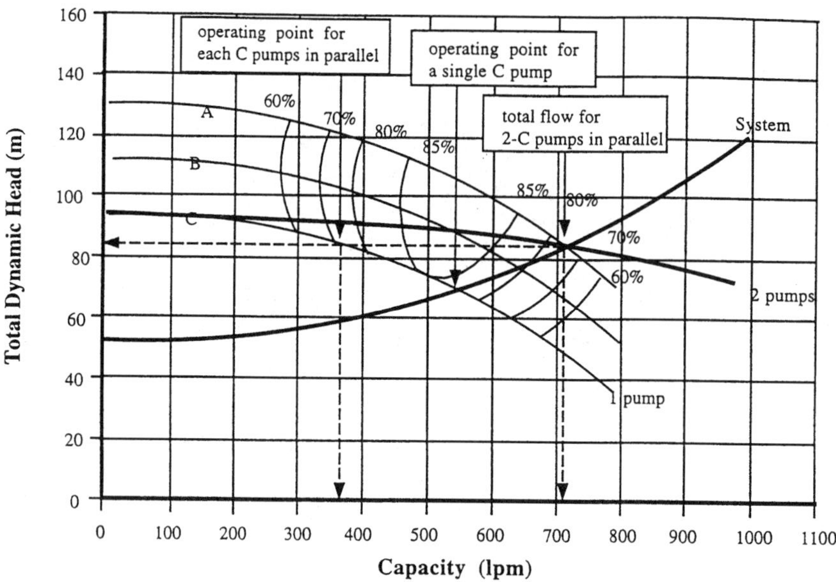

FIGURE 2.4.3 Selection of parallel pumps.

For some systems, stopping the pump, either intentionally or accidentally, can generate high pressures that can damage the pipe and controls. If the design process does not consider these potential problems, the system may not function trouble free. Downtime and maintenance costs may be high. Not all systems will experience start-up and shutdown problems, but the design should at least consider the possibility. The problem is more severe for pipelines that have a large elevation change and multiple high points. The magnitude of the transient is related to the length and profile of the pipeline, the pump characteristics, the magnitude of the elevation change, and the type of check valve used. The downsurge caused by stopping the pump can cause column separation and high pressures due to flow reversals and closure of the check valves. Surge-protection equipment can be added to such systems to prevent damage and excessive maintenance.

Another operational problem occurs with parallel pumps. Each pump must have a check valve to prevent reverse flow. When one of the pumps is turned off, the flow reverses almost immediately in that line because of the high manifold pressure supplied by the operating pumps. This causes the check valve to close. If a slow-closing check valve is installed, the flow can attain a high reverse velocity before the valve closes, generating high pressure transients.

Numerous mechanical devices and techniques have been used to suppress pump shutdown transients. These include increasing the rotational inertia of the pump, use of surge tanks or air chambers near the pump, pressure relief valves, vacuum-breaking valves, and surge-anticipating valves. Selection of the proper transient control device will improve reliability, extend the economic life of the system, and reduce maintenance. Failure to complete a transient analysis and include the required controls will have the opposite effect. A system is only as good as it is designed to be.

Other Considerations

Feasibility Study

Designing pipelines, especially long transmission lines, involves more than just determining the required type and size of pipe. A starting point for major projects is usually a feasibility study which involves social, environmental, political, and legal issues, as well as an economic evaluation of the engineering alternatives developed during the preliminary design. The preliminary design should identify the scope of the project and all major features that influence the cost or viability. Since local laws, social values, and environmental concerns vary significantly between geographic areas, the engineer must be aware of the problems unique to the area.

Choices that affect the economics of the project include alternative pipe routes, amount of storage and its effect on reliability and controllability of flow, choice of pipe material, diameter and pressure class, provision for future demands, etc. In making decisions one must consider both the engineering and economic advantages of the alternatives. Reliability, safety, maintenance, operating, and replacement costs must all be given their proper value. The analysis should consider (1) expected life of the pipe, which is a function of the type of pipe material and the use of linings or protective coatings; (2) economic life, meaning how long the pipe will supply the demand; (3) planning for future demand; (4) pumping cost vs. pipe cost; and (5) provisions for storage.

During the feasibility study only a general design has been completed so a detailed analysis of all hydraulic problems and their solutions is not available. Even so, it is necessary to anticipate the need for special facilities or equipment and problems such as safe filling, provisions for draining, cavitation at control valves, and transient problems caused by valve or pump operation. Provisions should be made for the cost of the detailed analysis, design, and construction costs required to control special operational problems. Attention should also be given to costs associated with winterizing, stream crossings, highways crossing, special geologic or topographic problems, and any other items that would have a significant influence on the cost, reliability, or safety of the project.

Storage

The purposes of storage tanks and intermediate reservoirs include (1) to supply water when there is a temporary interruption of flow from the supply, (2) to provide supplemental water during peak periods, (3) to sectionalize the pipe to reduce mean and transient pressures, (4) to maintain pressure (elevated storage), and (5) to simplify control. Storage also has a significant impact on the control structures, pumping plants, and general operation of the pipeline. If there is adequate storage, large fluctuations in demand can be tolerated. Any mismatch in supply and demand is made up for by an increase or decrease in storage, and valves in the transmission main will require only infrequent adjustments to maintain storage. Pumps can be activated by level controls at the storage tank and not by fluctuations in demand so they can operate for long periods near their design point.

If there is no storage, the system may have to provide continuous fine adjustment of the flow to provide the required flow within safe pressure limits. For gravity systems this may require automatic pressure- or flow-regulating valves. For pumped systems, the variations in flow can cause constant-speed centrifugal pumps to operate both below and above their design point where power consumption is high, efficiency is low, and where there is more chance of operational problems. Selection of a variable-frequency drive can avoid these problems. The selection of multiple pumps vs. a variable-speed pump is primarily a economic decision.

Thrust Blocks

Any time there is a change of pipe alignment, an unbalanced force is developed. The force required to restrain the pipe can be calculated with the two-dimensional, steady state momentum equation. For buried pipelines, this force can be transmitted to the soil with a thrust block. Determining the size of the block and, consequently, the bearing surface area depends on pipe diameter, fluid pressure, deflection angle of the pipe, and bearing capacity of the soil. A convenient monograph for sizing thrust blocks was published in the *Civil Engineering* in 1969 (Morrison, 1969).

References

ASCE. 1992. *Pressure Pipeline Design for Water and Wastewater.* Prepared by the Committee on Pipeline Planning of the Pipeline Division of the American Society of Civil Engineers, New York.

Kalsi Engineering and Tullis Engineering Consultants. 1993. *Application Guide for Check Valves in Nuclear Power Plants*, Revision 1, NP-5479. Prepared for Nuclear Maintenance Applications Center, Charlotte, NC.

Miller, D.S. 1990. *Internal Flow Systems — Design and Performance Prediction*, 2nd ed. Gulf Publishing Company, Houston.

Morrison, E.B. 1969. Monograph for the design of thrust blocks. *Civil Eng.,* 39, June, 55–51.

Spanger, M.G. and Handy, R.L. 1973. *Soil Engineering,* 3rd ed. Intext Eductational Publishers, New York, Chap. 25 and 26.

Streeter, V.L. and Wylie, E.B. 1975. *Fluid Mechanics,* 6th ed. McGraw-Hill, New York, 752 pp.

Thorley, A.R.D. 1989. Check valve behavior under transient flow conditions: a state-of-the-art review. *ASME*, 111, Vol. 2, June. *J. Fluids Engineering: Transactions of the ASME,* pp. 173–183.

Tullis, J.P. 1989. *Hydraulics of Pipelines — Pumps, Valves, Cavitation, Transients*, John Wiley & Sons, New York.

Tullis, J.P. 1993. Cavitation Guide for Control Valves, NUREG/CR-6031, U.S. Nuclear Regulatory Commission, Washington, D.C.

Wood, D.J. 1966. An explicit friction factor relationship. *Civil Eng.,* 36, December, 60–61.

Wylie, E.B. and Streeter, V.L. 1993. *Fluid Transients in Systems,* Prentice-Hall, Englewood Cliffs, N.J.

Further Information

Bean, H.S., ed. 1971. *Fluid Meters, Their Theory and Application,* 6th ed. The American Society of Mechanical Engineers, New York.

Handbook of PVC-Design and Construction. 1979. Uni-Bell Plastic Pipe Association, Dallas.

King, H.W. 1954. *Handbook of Hydraulics — For the Solution of Hydraulic Problems,* 4th ed. Revised by E.F. Brater. McGraw-Hill, New York.

Stephenson, D. 1981. *Pipeline Design for Water Engineers,* 2nd ed. Elsevier Scientific Publishing Company, Distributed in the U.S. by Gulf Publishing Company, Houston.

Stutsman, R.D. 1993. Steel Penstocks, ASCE Manuals and Reports on Engineering Practice No. 79, Energy Division, American Society of Civil Engineers, New York.

Watkins, R.K. and Spangler, M.G., Some Characteristics of the Modulus of Passive Resistance of Soil: A Study in Similitude. Highway Research Board Proceedings Vol. 37, 1958, pp. 576–583.

Tullis, J.P. 1996. Valves, in *The Engineering Handbook,* Dorf, R.C., Ed., CRC Press, Boca Raton, FL.

2.5 Open Channel Flow*

Frank M. White

Definition

The term *open channel flow* denotes the gravity-driven flow of a liquid with a free surface. Technically, we may study any flowing liquid and any gas interface. In practice, the vast majority of open channel flows concern water flowing beneath atmospheric air in artificial or natural channels.

The geometry of an arbitrary channel is shown in Figure 2.5.1. The area A is for the water cross section only, and b is its top width. The wetted perimeter P covers only the bottom and sides, as shown, not the surface (whose air resistance is neglected). The water depth at any location is y, and the channel slope is θ, often denoted as $S_o = \sin\theta$. All of these parameters may vary with distance x along the channel. In unsteady flow (not discussed here) they may also vary with time.

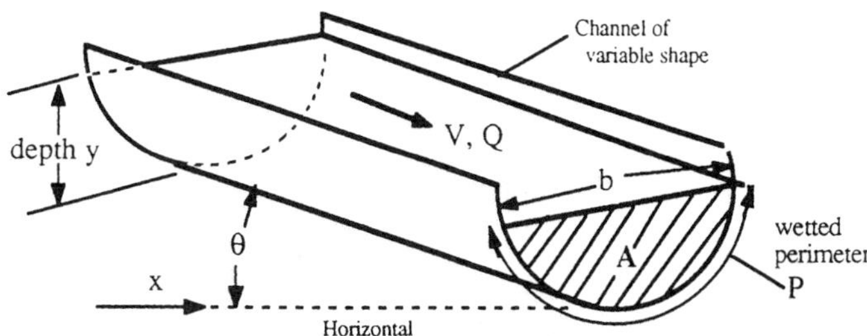

FIGURE 2.5.1 Definition sketch for an open channel.

Uniform Flow

A simple reference condition, called *uniform flow,* occurs in a long straight prismatic channel of constant slope S_o. There is no acceleration, and the water flows at constant depth with fluid weight exactly balancing the wetted wall shear force: $\rho gLA \sin\theta = \tau_w PL$, where L is the channel length. Thus, $\tau_w = \rho g R_h S_o$, where $R_h = A/P$ is called the *hydraulic radius* of the channel. If we relate wall shear stress to the Darcy friction factor f, $\tau_w = (f/8)\rho V^2$, we obtain the basic uniform flow open channel relation:

$$\text{Uniform flow:} \quad V = \sqrt{\frac{8g}{f}}\sqrt{R_h S_o}, \quad \text{where} \quad \sqrt{\frac{8g}{f}} = C = \text{Chézy coefficient} \tag{2.5.1}$$

Antoine Chézy first derived this formula in 1769. It is satisfactory to base f upon the pipe-flow Moody diagram (Figure 2.4.1) using the hydraulic diameter, $D_h = 4R_h$, as a length scale. That is, $f = fcn\,(VD_h/\nu, \varepsilon/D_h)$ from the Moody chart. In ordinary practice, however, engineers assume fully rough, high-Reynolds-number flow and use Robert Manning's century-old correlation:

$$C \approx \frac{\zeta}{n} R_h^{1/6}, \quad \text{or} \quad V_{\text{uniform}} \approx \frac{\zeta}{n} R_h^{2/3} S_o^{1/2} \quad \text{and} \quad Q = VA \tag{2.5.2}$$

where ζ is a conversion factor equal to 1.0 in SI units and 1.486 in English units. The quantity n is Manning's roughness parameter, with typical values, along with the associated roughness heights ε, listed in Table 2.5.1.

* Nomenclature appears at end of this section.

TABLE 2.5.1 Average Roughness Parameters for Various Channel Surfaces

		Average Roughness Height ε	
	n	ft	mm
Artificial lined channels			
Glass	0.010 ± 0.002	0.0011	0.3
Brass	0.011 ± 0.002	0.0019	0.6
Steel; smooth	0.012 ± 0.002	0.0032	1.0
Painted	0.014 ± 0.003	0.0080	2.4
Riveted	0.015 ± 0.002	0.012	3.7
Cast iron	0.013 ± 0.003	0.0051	1.6
Cement; finished	0.012 ± 0.002	0.0032	1.0
Unfinished	0.014 ± 0.002	0.0080	2.4
Planed wood	0.012 ± 0.002	0.0032	1.0
Clay tile	0.014 ± 0.003	0.0080	2.4
Brickwork	0.015 ± 0.002	0.012	3.7
Asphalt	0.016 ± 0.003	0.018	5.4
Corrugated metal	0.022 ± 0.005	0.12	37
Rubble masonry	0.025 ± 0.005	0.26	80
Excavated earth channels			
Clean	0.022 ± 0.004	0.12	37
Gravelly	0.025 ± 0.005	0.26	80
Weedy	0.030 ± 0.005	0.8	240
Stony; cobbles	0.035 ± 0.010	1.5	500
Natural channels			
Clean and straight	0.030 ± 0.005	0.8	240
Sluggish, deep pools	0.040 ± 0.010	3	900
Major rivers	0.035 ± 0.010	1.5	500
Floodplains			
Pasture, farmland	0.035 ± 0.010	1.5	500
Light brush	0.05 ± 0.02	6	2000
Heavy brush	0.075 ± 0.025	15	5000
Trees	0.15 ± 0.05	?	?

Critical Flow

Since the surface is always atmospheric, pressure head is not important in open channel flows. Total energy E relates only to velocity and elevation:

$$\text{Specific energy } E = y + \frac{V^2}{2g} = y + \frac{Q^2}{2gA^2}$$

At a given volume flow rate Q, the energy passes through a minimum at a condition called *critical flow*, where $dE/dy = 0$, or $dA/dy = b = gA^3/Q^2$:

$$A_{\text{crit}} = \left(\frac{bQ^2}{g}\right)^{1/3} \qquad V_{\text{crit}} = \frac{Q}{A_{\text{crit}}} = \left(\frac{gA_{\text{crit}}}{b}\right)^{1/2} \tag{2.5.3}$$

where b is the top-surface width as in Figure 2.5.1. The velocity V_{crit} equals the speed of propagation of a surface wave along the channel. Thus, we may define the Froude number Fr of a channel flow, for any cross section, as $Fr = V/V_{\text{crit}}$. The three regimes of channel flow are

$$\text{Fr} < 1\text{: subcritical flow;} \quad \text{Fr} = 1\text{: critical flow;} \quad \text{Fr} > 1\text{: supercritical flow}$$

There are many similarities between Froude number in channel flow and Mach number in variable-area duct flow (see Section 2.6).

For a rectangular duct, $A = by$, we obtain the simplified formulas

$$V_{\text{crit}} = \sqrt{gy} \qquad \text{Fr} = \frac{V}{\sqrt{gy}} \tag{2.5.4}$$

independent of the width of the channel.

Example 2.5.1

Water (ρ = 998 kg/m^3, μ = 0.001 kg/m · sec) flows uniformly down a half-full brick 1-m-diameter circular channel sloping at 1°. Estimate (a) Q; and (b) the Froude number.

Solution 2.5.1 (a). First compute the geometric properties of a half-full circular channel:

$$A = \frac{\pi}{8}(1\ \text{m})^2 = 0.393\ \text{m}^2; \quad P = \frac{\pi}{2}(1\ \text{m}) = 1.57\ \text{m}; \quad R = \frac{A}{P} = \frac{0.393}{1.57} = 0.25\ \text{m}$$

From Table 2.5.1, for brickwork, $n \approx 0.015$. Then, Manning's formula, Equation (2.5.2) predicts

$$V = \frac{\zeta}{n} R_h^{1/6} S_o^{1/2} = \frac{1.0}{0.015}(0.25)^{1/6}(\sin 1°)^{1/2} \approx 3.49\,\frac{\text{m}}{\text{sec}}; \quad Q = 3.49(0.393) \approx \mathbf{1.37}\,\frac{\mathbf{m^3}}{\mathbf{sec}} \quad \textit{Solution 2.5.1(a)}$$

The uncertainty in this result is about ±10%. The flow rate is quite large (21,800 gal/min) because 1°, although seemingly small, is a substantial slope for a water channel.

One can also use the Moody chart. With $V \approx 3.49$ m/sec, compute Re = $\rho V D_h/\mu \approx 3.49$ E6 and $\varepsilon/D_h \approx 0.0037$, then compute $f \approx 0.0278$ from the Moody chart. Equation (2.5.1) then predicts

$$V = \sqrt{\frac{8g}{f} R_h S_o} = \sqrt{\frac{8(9.81)}{0.0278}(0.25)(\sin 1°)} \approx 3.51\,\frac{\text{m}}{\text{sec}}; \quad Q = VA \approx \mathbf{1.38}\,\frac{\mathbf{m^3}}{\mathbf{sec}}$$

Solution 2.5.1 (b). With Q known from part (a), compute the critical conditions from Equation (2.5.3):

$$A_{\text{crit}} = \left(\frac{bQ^2}{g}\right)^{1/3} = \left[\frac{1.0(1.37)^2}{9.81}\right]^{1/3} = 0.576\ \text{m}^2, \quad V_{\text{crit}} = \frac{Q}{A_{\text{crit}}} = \frac{1.37}{0.576} = 2.38\,\frac{\text{m}}{\text{sec}}$$

Hence

$$\text{Fr} = \frac{V}{V_{\text{crit}}} = \frac{3.49}{2.38} \approx \mathbf{1.47}\ (\text{supercritical}) \quad \textit{Solution 2.5.1(b)}$$

Again the uncertainty is approximately ±10%, primarily because of the need to estimate the brick roughness.

Hydraulic Jump

In gas dynamics (Section 2.6), a supersonic gas flow may pass through a thin normal shock and exit as a subsonic flow at higher pressure and temperature. By analogy, a supercritical open channel flow may pass through a *hydraulic jump* and exit as a subcritical flow at greater depth, as in Figure 2.5.2. Application of continuity and momentum to a jump in a rectangular channel yields

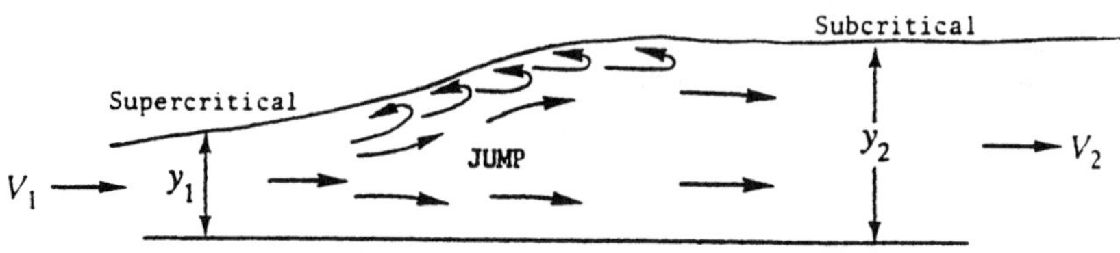

FIGURE 2.5.2 A two-dimensional hydraulic jump.

$$V_2 = V_1 \frac{y_1}{y_2} \quad y_2 = \frac{y_1}{2}\left[-1 + \sqrt{1 + 8\mathrm{Fr}_1^2}\right] \text{ where } \mathrm{Fr}_1 = \frac{V_1}{\sqrt{g y_1}} > 1 \tag{2.5.5}$$

Both the normal shock and the hydraulic jump are dissipative processes: the entropy increases and the effective energy decreases. For a rectangular jump,

$$\Delta E = E_1 - E_2 = \frac{(y_2 - y_1)^3}{4 y_1 y_2} > 0 \tag{2.5.6}$$

For strong jumps, this loss in energy can be up to 85% of E_1. The second law of thermodynamics requires $\Delta E > 0$ and $y_2 > y_1$ or, equivalently, $\mathrm{Fr}_1 > 1$,

Note from Figure 2.5.2 that a hydraulic jump is not thin. Its total length is approximately four times the downstream depth. Jumps also occur in nonrectangular channels, and the theory is much more algebraically laborious.

Weirs

If an open channel flow encounters a significant obstruction, it will undergo rapidly varied changes which are difficult to model analytically but can be correlated with experiment. An example is the *weir* in Figure 2.5.3 (colloquially called a *dam*), which forces the flow to deflect over the top. If $L \ll Y$, the weir is termed *sharp-crested*; if $L = O(Y)$ it is *broad-crested*. Small details, such as the upper front corner radius or the crest roughness, may be significant. The crest is assumed level and of width b into the paper.

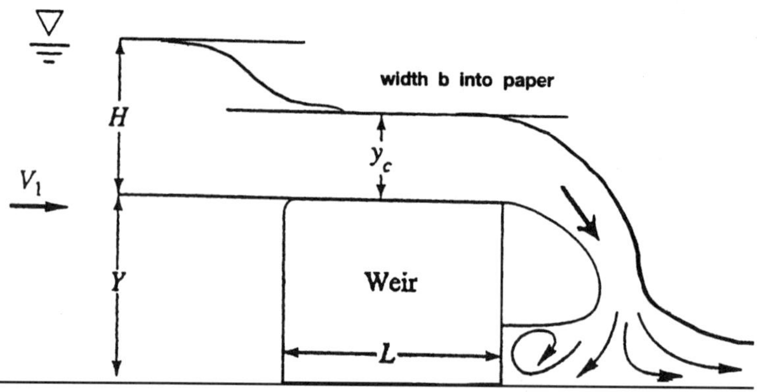

FIGURE 2.5.3 Geometry and notation for flow over a weir.

If there is a free overfall, as in Figure 2.5.3, the flow accelerates from subcritical upstream to critical over the crest to supercritical in the overfall. There is no flow when the excess upstream depth $H = 0$. A simple Bernoulli-type analysis predicts that the flow rate Q over a wide weir is approximately proportional to $bg^{1/2}H^{3/2}$. An appropriate correlation is thus

$$Q_{\text{weir}} = C_d b g^{1/2} H^{3/2}, \quad \text{where } C_d = \text{dimensionless weir coefficient} \tag{2.5.7}$$

If the upstream flow is turbulent, the weir coefficient depends only upon geometry, and Reynolds number effects are negligible. If the weir has sidewalls and is narrow, replace width b by $(b - 0.1H)$.

Two recommended empirical correlations for Equation (2.5.7) are as follows:

$$\begin{aligned} &\text{Sharp-crested:} \quad C_d \approx 0.564 + 0.0846\frac{H}{Y} \quad \text{for} \quad \frac{L}{Y} < 0.07 \\ &\text{Broad-crested:} \quad C_d \approx 0.462 \quad \text{for} \quad 0.08 < \frac{H}{L} < 0.33 \end{aligned} \tag{2.5.8}$$

These data are for wide weirs with a sharp upper corner in front. Many other weir geometries are discussed in the references for this section. Of particular interest is the sharp-edged vee-notch weir, which has no length scale b. If 2θ is the total included angle of the notch, the recommended correlation is

$$\text{Vee-notch, angle } 2\theta: \quad Q \approx 0.44 \tan\theta \, g^{1/2} H^{5/2} \quad \text{for} \quad 10° < \theta \le 50° \tag{2.5.9}$$

The vee-notch is more sensitive at low flow rates (large H for a small Q) and thus is popular in laboratory measurements of channel flow rates.

A weir in the field will tend to spring free and form a natural *nappe*, or air cavity, as in Figure 2.5.3. Narrow weirs, with sidewalls, may need to be aerated artificially to form a nappe and keep the flow from sliding down the face of the weir. The correlations above assume nappe formation.

Gradually Varied Flow

Return to Figure 2.5.1 and suppose that (y, A, b, P, S_o) are all variable functions of horizontal position x. If these parameters are slowly changing, with no hydraulic jumps, the flow is termed *gradually varied* and satisfies a simple one-dimensional first-order differential equation if Q = constant:

$$\frac{dy}{dx} \approx \frac{S_o - S}{1 - \dfrac{V^2 b}{gA}}, \quad \text{where } V = \frac{Q}{A} \quad \text{and} \quad S = \frac{f}{D_h}\frac{V^2}{2g} = \frac{n^2 V^2}{\zeta^2 R_h^{4/3}} \tag{2.5.10}$$

The conversion factor $\zeta^2 = 1.0$ for SI units and 2.208 for English units. If flow rate, bottom slope, channel geometry, and surface roughness are known, we may solve for $y(x)$ for any given initial condition $y = y_o$ at $x = x_o$. The solution is computed by any common numerical method, e.g., Runge–Kutta.

Recall from Equation (2.5.3) that the term $V^2b/(gA) \equiv \text{Fr}^2$, so the sign of the denominator in Equation (2.5.10) depends upon whether the flow is sub- or supercritical. The mathematical behavior of Equation (2.5.10) differs also. If Fr is near unity, the change dy/dx will be very large, which probably violates the basic assumption of "gradual" variation.

For a given flow rate and local bottom slope, two reference depths are useful and may be computed in advance:

(a) The *normal* depth y_n for which Equation (2.5.2) yields the flow rate:
(b) The *critical* depth y_c for which Equation (2.5.3) yields the flow rate.

Comparison of these two, and their relation to the actual local depth y, specifies the type of solution curve being computed. The five bottom-slope regimes (mild M, critical C, steep S, horizontal H, and adverse A) create 12 different solution curves, as illustrated in Figure 2.5.4. All of these may be readily generated by a computer solution of Equation 2.5.10. The following example illustrates a typical solution to a gradually varied flow problem.

Example 2.5.2

Water, flowing at 2.5 m³/sec in a rectangular gravelly earth channel 2 m wide, encounters a broad-crested weir 1.5 m high. Using gradually varied theory, estimate the water depth profile back to 1 km upstream of the weir. The bottom slope is 0.1°.

Solution. We are given Q, Y = 1.5 m, and b = 2 m. We may calculate excess water level H at the weir (see Figure 2.5.3) from Equations (2.5.7) and (2.5.8):

$$Q = 2.5\frac{\text{m}^3}{\text{sec}} = C_d b_{\text{eff}} g^{1/2} H^{3/2} = 0.462(2.0 - 0.1H)(9.81)^{1/2} H^{3/2}, \quad \text{solve for } H \approx \mathbf{0.94}\text{ m}$$

Since the weir is not too wide, we have subtracted 0.1 H from b as recommended. The weir serves as a "control structure" which sets the water depth just upstream. This is our initial condition for gradually varied theory: $y(0) = Y + H = 1.5 + 0.94 \approx 2.44$ m at $x = 0$. Before solving Equation (2.5.10), we find the normal and critical depths to get a feel for the problem:

$$\text{Normal depth:} \quad Q = 2.5\frac{\text{m}^3}{\text{sec}} = \frac{1.0}{0.025}(2.0y_n)\left(\frac{2.0y_n}{2.0 + 2y_n}\right)^{2/3}\sqrt{\sin 0.1°}, \quad \text{solve } y_n \approx \mathbf{1.14}\text{ m}$$

$$\text{Critical depth:} \quad A_c = 2.0y_c = \left(\frac{bQ^2}{g}\right)^{1/3} = \left[\frac{2.0(2.5)^2}{9.81}\right]^{1/3}, \quad \text{solve } y_c \approx \mathbf{0.54}\text{ m}$$

We have taken $n \approx 0.025$ for gravelly earth, from Table 2.5.1. Since $y(0) > y_n > y_c$, we are on a mild slope $M - 1$ "backwater" curve, as in Figure 2.5.4. For our data, Equation (2.5.10) becomes

$$\frac{dy}{dx} \approx \frac{S_o - n^2Q^2/\left(\zeta^2 A^2 R_h^{4/3}\right)}{1 - Q^2 b/\left(gA^3\right)}$$

where $Q = 2.5$, $b = 2$, $\zeta = 1$, $A = 2y$, $S_o = \sin 0.1°$, $R_h = 2y/(2 + 2y)$, $g = 9.81$, $y(0) = 2.44$ at $x = 0$.

Integrate numerically backward, that is, for $\Delta x < 0$, until $x = -1$ km $= -1000$ m. The complete solution curve is shown in Figure 2.5.5. The water depth decreases upstream and is approximately $y \approx 1.31$ m at $x = -1000$ m. If slope and channel width remain constant, the water depth asymptotically approaches the normal depth y_n far upstream.

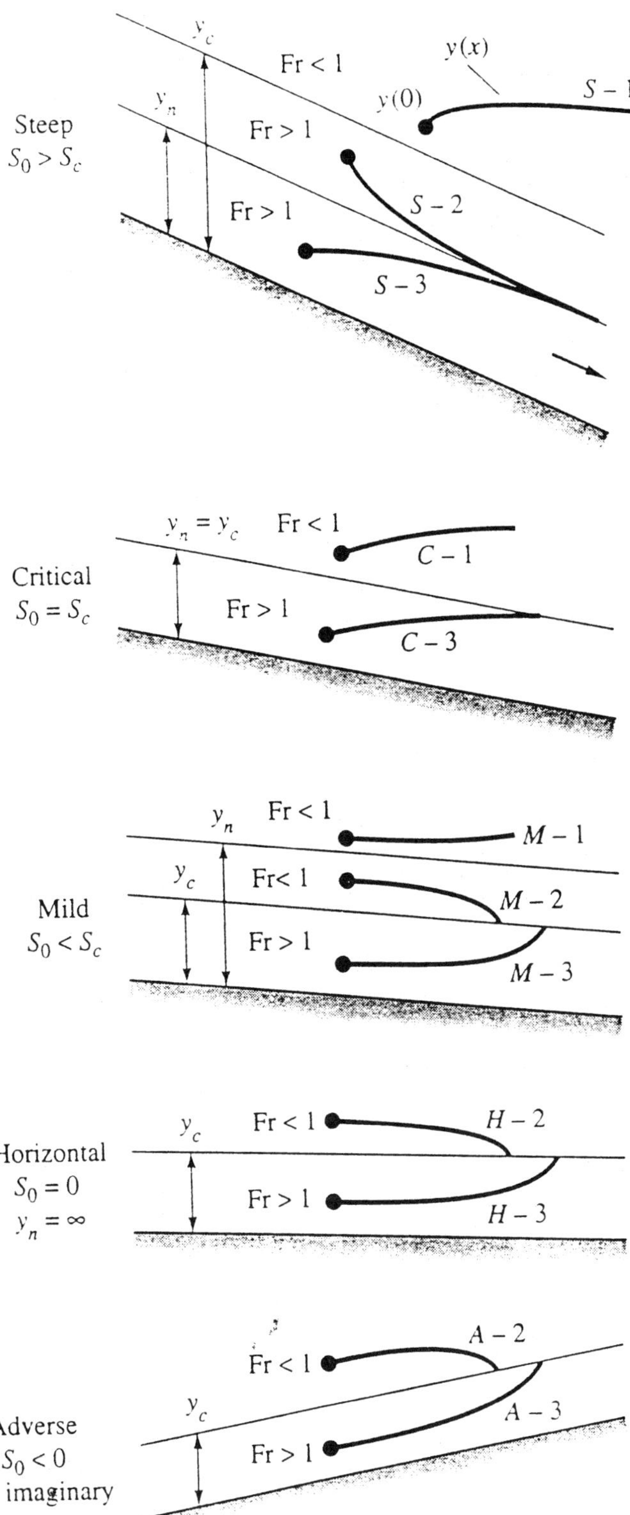

FIGURE 2.5.4 Classification of solution curves for gradually varied flow.

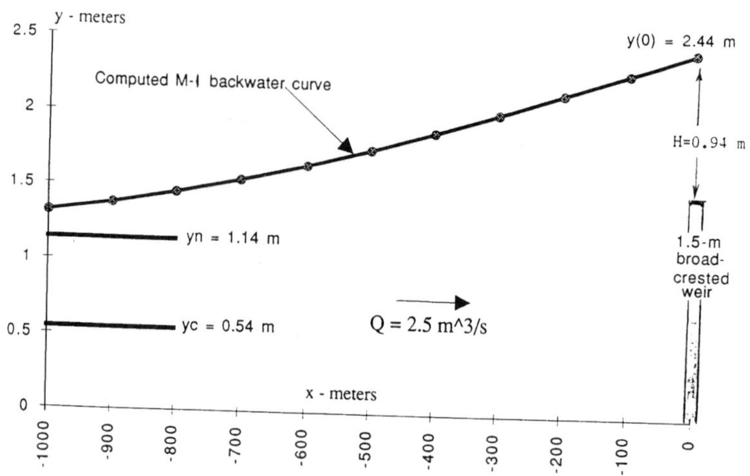

FIGURE 2.5.5 Backwater solution curve for Example 2.5.2.

References

Ackers, P. et al. 1978. *Weirs and Flumes for Flow Measurement*, John Wiley & Sons, New York.

Bos, M.G. 1985. *Long-Throated Flumes and Broad-Crested Weirs,* Martinus Nijhoff (Kluwer), Dordrecht, The Netherlands.

Bos, M.G., Replogle, J.A., and Clemmens, A.J. 1984. *Flow-Measuring Flumes for Open Channel Systems,* John Wiley & Sons, New York.

Brater, E.F. 1976. *Handbook of Hydraulics,* 6th ed., McGraw-Hill, New York.

Chow, V.T. 1959. *Open Channel Hydraulics,* McGraw-Hill, New York.

French, R.H. 1985. *Open Channel Hydraulics,* McGraw-Hill, New York.

Henderson, F.M. 1966. *Open Channel Flow,* Macmillan, New York.

Sellin, R.H.J. 1970. *Flow in Channels*, Gordon & Breach, London.

Spitzer, D.W. (Ed.). 1991. *Flow Measurement: Practical Guides for Measurement and Control*, Instrument Society of America, Research Triangle Park, NC.

2.6 External Incompressible Flows

Alan T. McDonald

Introduction and Scope

Potential flow theory (Section 2.2) treats an incompressible *ideal fluid* with zero viscosity. There are no shear stresses; pressure is the only stress acting on a fluid particle. Potential flow theory predicts no drag force when an object moves through a fluid, which obviously is not correct, because all real fluids are viscous and cause drag forces. The objective of this section is to consider the behavior of viscous, incompressible fluids flowing over objects.

A number of phenomena that occur in external flow at high Reynolds number over an object are shown in Figure 2.6.1. The freestream flow divides at the stagnation point and flows around the object. Fluid at the object surface takes on the velocity of the body as a result of the no-slip condition. Boundary layers form on the upper and lower surfaces of the body; flow in the boundary layers is initially laminar, then **transition** to turbulent flow may occur (points "T").

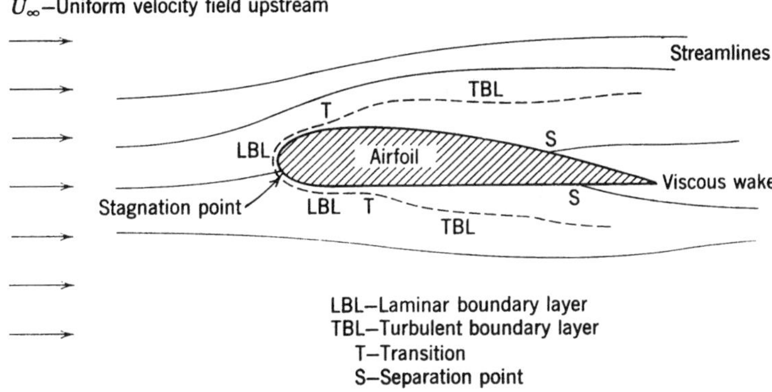

FIGURE 2.6.1 Viscous flow around an airfoil (boundary layer thickness exaggerated for clarity).

Boundary layers thickening on the surfaces cause only a slight displacement of the streamlines of the external flow (their thickness is greatly exaggerated in the figure). **Separation** may occur in the region of increasing pressure on the rear of the body (points "S"); after separation boundary layer fluid no longer remains in contact with the surface. Fluid that was in the boundary layers forms the viscous *wake* behind the object.

The Bernoulli equation is valid for steady, incompressible flow without viscous effects. It may be used to predict pressure variations outside the boundary layer. Stagnation pressure is constant in the uniform inviscid flow far from an object, and the Bernoulli equation reduces to

$$p_\infty + \frac{1}{2}\rho V^2 = \text{constant} \tag{2.6.1}$$

where p_∞ is pressure far upstream, ρ is density, and V is velocity. Therefore, the local pressure can be determined if the local freestream velocity, U, is known.

Boundary Layers

The Boundary Layer Concept

The **boundary layer** is the thin region near the surface of a body in which viscous effects are important. By recognizing that viscous effects are concentrated near the surface of an object, Prandtl showed that only the Euler equations for inviscid flow need be solved in the region outside the boundary layer. Inside the boundary layer, the elliptic Navier-Stokes equations are simplified to boundary layer equations with parabolic form that are easier to solve. The thin boundary layer has negligible pressure variation across it; pressure from the freestream is impressed upon the boundary layer.

Basic characteristics of all laminar and turbulent boundary layers are shown in the developing flow over a flat plate in a semi-infinite fluid. Because the boundary layer is thin, there is negligible disturbance of the inviscid flow outside the boundary layer, and the **pressure gradient** along the surface is close to zero. Transition from laminar to turbulent boundary layer flow on a flat plate occurs when Reynolds number based on x exceeds $Re_x = 500{,}000$. Transition may occur earlier if the surface is rough, pressure increases in the flow direction, or separation occurs. Following transition, the turbulent boundary layer thickens more rapidly than the laminar boundary layer as a result of increased shear stress at the body surface.

Boundary Layer Thickness Definitions

Boundary layer disturbance thickness, δ, is usually defined as the distance, y, from the surface to the point where the velocity within the boundary layer, u, is within 1% of the local freestream velocity, U. As shown in Figure 2.6.2, the boundary layer velocity profile merges smoothly and asymptotically into the freestream, making δ difficult to measure. For this reason and for their physical significance, we define two integral measures of boundary layer thickness. Displacement thickness, δ^*, is defined as

$$\frac{\delta^*}{\delta} = \int_0^\infty \left(1 - \frac{u}{U}\right) d\left(\frac{y}{\delta}\right) \tag{2.6.2}$$

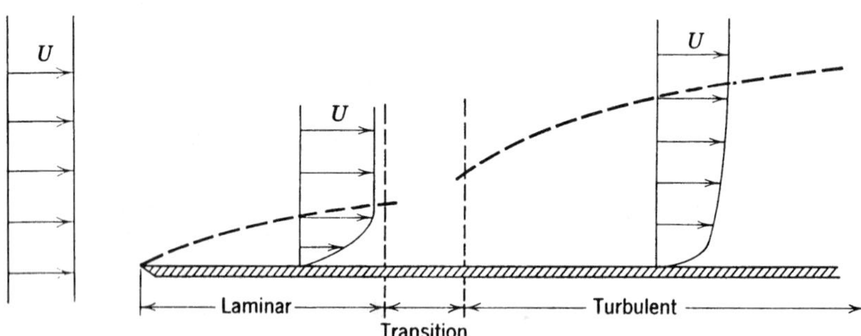

FIGURE 2.6.2 Boundary layer on a flat plate (vertical thickness exaggerated for clarity).

Physically, δ^* is the distance the solid boundary would have to be displaced into the freestream in a frictionless flow to produce the mass flow deficit caused by the viscous boundary layer. Momentum thickness, θ, is defined as

$$\frac{\theta}{\delta} = \int_0^\infty \frac{u}{U}\left(1 - \frac{u}{U}\right) d\left(\frac{y}{\delta}\right) \tag{2.6.3}$$

Physically, θ is the thickness of a fluid layer, having velocity U, for which the momentum flux is the same as the deficit in momentum flux within the boundary layer (momentum flux is momentum per unit time passing a cross section).

Because δ^* and θ are defined in terms of integrals for which the integrand vanishes in the freestream, they are easier to evaluate experimentally than disturbance thickness δ.

Exact Solution of the Laminar Flat-Plate Boundary Layer

Blasius obtained an exact solution for laminar boundary layer flow on a flat plate. He assumed a thin boundary layer to simplify the streamwise momentum equation. He also assumed *similar* velocity profiles in the boundary layer, so that when written as $u/U = f(y/\delta)$, velocity profiles do not vary with x. He used a similarity variable to reduce the partial differential equations of motion and continuity to a single third-order ordinary differential equation.

Blasius used numerical methods to solve the ordinary differential equation. Unfortunately, the velocity profile must be expressed in tabular form. The principal results of the Blasius solution may be expressed as

$$\frac{\delta}{x} = \frac{5}{\sqrt{\mathrm{Re}_x}} \tag{2.6.4}$$

and

$$C_f = \frac{\tau_w}{\frac{1}{2}\rho U^2} = \frac{0.664}{\sqrt{\mathrm{Re}_x}} \tag{2.6.5}$$

These results characterize the laminar boundary layer on a flat plate; they show that laminar boundary layer thickness varies as $x^{1/2}$ and wall shear stress varies as $1/x^{1/2}$.

Approximate Solutions

The Blasius solution cannot be expressed in closed form and is limited to laminar flow. Therefore, approximate methods that give solutions for both laminar and turbulent flow in closed form are desirable. One such method is the *momentum integral equation* (MIE), which may be developed by integrating the boundary layer equation across the boundary layer or by applying the streamwise momentum equation to a differential control volume (Fox and McDonald, 1992). The result is the ordinary differential equation

$$\frac{d\theta}{dx} = \frac{\tau_w}{\rho U^2} - \left(\frac{\delta^*}{\theta} + 2\right)\frac{\theta}{U}\frac{dU}{dx} \tag{2.6.6}$$

The first term on the right side of Equation (2.6.6) contains the influence of wall shear stress. Since τ_w is always positive, it always causes θ to increase. The second term on the right side contains the pressure gradient, which can have either sign. Therefore, the effect of the pressure gradient can be to either increase or decrease the rate of growth of boundary layer thickness.

Equation (2.6.6) is an ordinary differential equation that can be solved for θ as a function of x on a flat plate (zero pressure gradient), provided a reasonable shape is assumed for the boundary layer velocity profile and shear stress is expressed in terms of the other variables. Results for laminar and turbulent flat-plate boundary layer flows are discussed below.

Laminar Boundary Layers. A reasonable approximation to the laminar boundary layer velocity profile is to express u as a polynomial in y. The resulting solutions for δ and τ_w have the same dependence on x as the exact Blasius solution. Numerical results are presented in Table 2.6.1. Comparing the approximate and exact solutions shows remarkable agreement in view of the approximations used in the analysis. The trends are predicted correctly and the approximate values are within 10% of the exact values.

Turbulent Boundary Layers. The turbulent velocity profile may be expressed well using a power law, $u/U = (y/\delta)^{1/n}$, where n is an integer between 6 and 10 (frequently 7 is chosen). For turbulent flow it is

TABLE 2.6.1 Exact and Approximate Solutions for Laminar Boundary Layer Flow over a Flat Plate at Zero Incidence

Velocity Distribution $\frac{u}{U} = f\left(\frac{y}{\delta}\right) = f(\eta)$	$\frac{\theta}{\delta}$	$\frac{\delta^*}{\delta}$	$a = \frac{\delta}{x}\sqrt{\mathbf{Re}_x}$	$b = C_f\sqrt{\mathbf{Re}_x}$
$f(\eta) = 2\eta - \eta^2$	2/15	1/3	5.48	0.730
$f(\eta) = 3/2\ \eta - 1/2\ \eta^3$	39/280	3/8	4.64	0.647
$f(\eta) = \sin(\pi/2\ \eta)$	$(4 - \pi)/2\pi$	$(\pi - 2)/\pi$	4.80	0.654
Exact	0.133	0.344	5.00	0.664

not possible to express shear stress directly in terms of a simple velocity profile; an empirical correlation is required. Using a pipe flow data correlation gives

$$\frac{\delta}{x} = \frac{0.382}{\mathrm{Re}_x^{1/5}} \tag{2.6.7}$$

and

$$C_f = \frac{\tau_w}{\frac{1}{2}\rho U^2} = \frac{0.0594}{\mathrm{Re}_x^{1/5}} \tag{2.6.8}$$

These results characterize the turbulent boundary layer on a flat plate. They show that turbulent boundary layer thickness varies as $x^{4/5}$ and wall shear stress varies as $1/x^{1/5}$.

Approximate results for laminar and turbulent boundary layers are compared in Table 2.6.2. At a Reynolds number of 1 million, wall shear stress for the turbulent boundary layer is nearly six times as large as for the laminar layer. For a turbulent boundary layer, thickness increases five times faster with distance along the surface than for a laminar layer. These approximate results give a physical feel for relative magnitudes in the two cases.

TABLE 2.6.2 Thickness and Skin Friction Coefficient for Laminar and Turbulent Boundary Layers on a Flat Plate

Reynolds	Boundary Layer Thickness/x		Skin Friction Coefficient		Turbulent/Laminar Ratio	
Number	Laminar BL	Turbulent BL	Laminar BL	Turbulent BL	BL Thickness	Skin Friction
2E + 05	0.0112	0.0333	0.00148	0.00517	2.97	3.48
5E + 05	0.00707	0.0277	0.000939	0.00431	3.92	4.58
1E + 06	0.00500	0.0241	0.000664	0.00375	4.82	5.64
2E + 06	0.00354	0.0210	0.000470	0.00326	5.93	6.95
5E + 06	0.00224	0.0175	0.000297	0.00272	7.81	9.15
1E + 07	0.00158	0.0152	0.000210	0.00236	9.62	11.3
2E + 07	0.00112	0.0132	0.000148	0.00206	11.8	13.9
5E + 07	0.000707	0.0110	0.0000939	0.00171	15.6	18.3

Note: BL = boundary layer.

The MIE cannot be solved in closed form for flows with nonzero pressure gradients. However, the role of the pressure gradient can be understood qualitatively by studying the MIE.

Effect of Pressure Gradient

Boundary layer flow with favorable, zero, and adverse pressure gradients is depicted schematically in Figure 2.6.3. (Assume a thin boundary layer, so flow on the lower surface behaves as external flow on

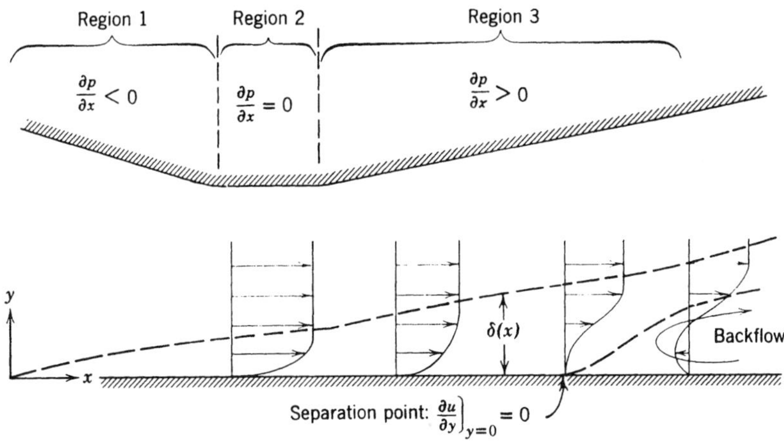

FIGURE 2.6.3 Boundary layer flow with presssure gradient (thickness exaggerated for clarity).

a surface, with the pressure gradient impressed on the boundary layer.) The pressure gradient is favorable when $\partial p/\partial x < 0$, zero when $\partial p/\partial x = 0$, and adverse when $\partial p/\partial x > 0$, as indicated for Regions 1, 2, and 3.

Viscous shear always causes a net retarding force on any fluid particle within the boundary layer. For zero pressure gradient, shear forces alone can never bring the particle to rest. (Recall that for laminar and turbulent boundary layers the shear stress varied as $1/x^{1/2}$ and $1/x^{1/5}$, respectively; shear stress never becomes zero for finite x.) Since shear stress is given by $\tau_w = \mu\ \partial u/\partial y)_{y=0}$, the velocity gradient cannot be zero. Therefore, flow cannot separate in a zero pressure gradient; shear stresses alone can never cause flow separation.

In the favorable pressure gradient of Region 1, pressure forces tend to maintain the motion of the particle, so flow cannot separate. In the adverse pressure gradient of Region 3, pressure forces oppose the motion of a fluid particle. An adverse pressure gradient is a necessary condition for flow separation.

Velocity profiles for laminar and turbulent boundary layers are shown in Figure 2.6.2. It is easy to see that the turbulent velocity profile has much more momentum than the laminar profile. Therefore, the turbulent velocity profile can resist separation in an adverse pressure gradient better than the laminar profile.

The freestream velocity distribution must be known before the MIE can be applied. We obtain a first approximation by applying potential flow theory to calculate the flow field around the object. Much effort has been devoted to calculation of velocity distributions over objects of known shape (the "direct" problem) and to determination of shapes to produce a desired pressure distribution (the "inverse" problem). Detailed discussion of such calculation schemes is beyond the scope of this section; the state of the art continues to progress rapidly.

Drag

Any object immersed in a viscous fluid flow experiences a net force from the shear stresses and pressure differences caused by the fluid motion. *Drag* is the force component parallel to, and *lift* is the force component perpendicular to, the flow direction. *Streamlining* is the art of shaping a body to reduce fluid dynamic drag. Airfoils (hydrofoils) are designed to produce lift in air (water); they are streamlined to reduce drag and thus to attain high lift–drag ratios.

In general, lift and drag cannot be predicted analytically for flows with separation, but progress continues on computational fluid dynamics methods. For many engineering purposes, drag and lift forces are calculated from experimentally derived coefficients, discussed below.

Drag coefficient is defined as

$$C_D = \frac{F_D}{\frac{1}{2}\rho V^2 A} \tag{2.6.9}$$

where $^1/_2\rho V^2$ is dynamic pressure and A is the area upon which the coefficient is based. Common practice is to base drag coefficients on projected *frontal area* (Fox and McDonald, 1992).

Similitude was treated in Section 2.3. In general, the drag coefficient may be expressed as a function of Reynolds number, Mach number, Froude number, relative roughness, submergence divided by length, and so forth. In this section we consider neither high-speed flow nor free-surface effects, so we will consider only Reynolds number and roughness effects on drag coefficient.

Friction Drag

The total friction drag force acting on a plane surface aligned with the flow direction can be found by integrating the shear stress distribution along the surface. The drag coefficient for this case is defined as friction force divided by dynamic pressure and *wetted area* in contact with the fluid. Since shear stress is a function of Reynolds number, so is drag coefficient (see Figure 2.6.4). In Figure 2.6.4, transition occurs at $\mathrm{Re}_x = 500{,}000$; the dashed line represents the drag coefficient at larger Reynolds numbers. A number of empirical correlations may be used to model the variation in C_D shown in Figure 2.6.4 (Schlichting, 1979).

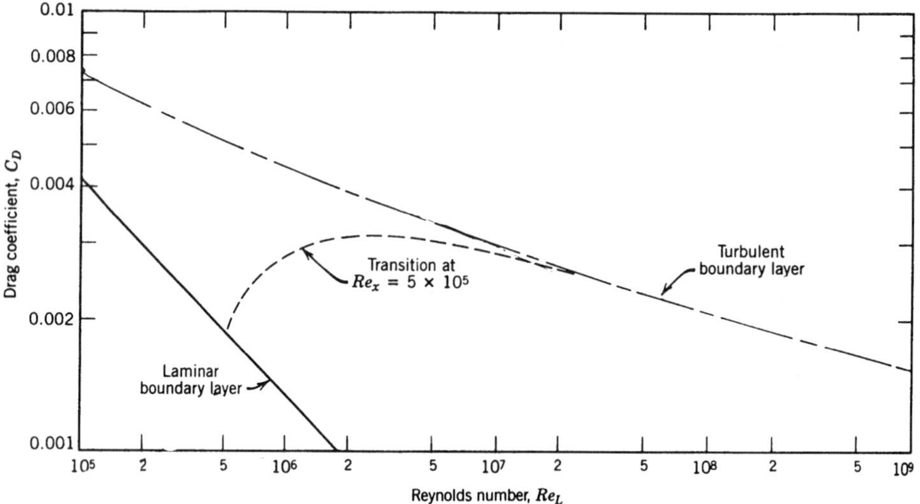

FIGURE 2.6.4 Drag coefficient vs. Reynolds number for a smooth flat plate parallel to the flow.

Extending the laminar boundary layer line to higher Reynolds numbers shows that it is beneficial to delay transition to the highest possible Reynolds number. Some results are presented in Table 2.6.3; drag is reduced more than 50% by extending laminar boundary layer flow to $\mathrm{Re}_L = 10^6$.

Pressure Drag

A thin flat surface normal to the flow has no area parallel to the flow direction. Therefore, there can be no friction force parallel to the flow; all drag is caused by pressure forces. Drag coefficients for objects with sharp edges tend to be independent of Reynolds number (for Re > 1000), because the separation points are fixed by the geometry of the object. Drag coefficients for selected objects are shown in Table 2.6.4.

Rounding the edges that face the flow reduces drag markedly. Compare the drag coefficients for the hemisphere and C-section shapes facing into and away from the flow. Also note that the drag coefficient

TABLE 2.6.3 Drag Coefficients for Laminar, Turbulent, and Transition Boundary Layers on a Flat Plate

Reynolds Number	Drag Coefficient Laminar BL	Turbulent BL	Transition	Laminar/ Transition	% Drag Reduction
2E + 05	0.00297	0.00615	—	—	—
5E + 05	0.00188	0.00511	0.00189	—	—
1E + 06	0.00133	0.00447	0.00286	0.464	53.6
2E + 06	0.000939	0.00394	0.00314	0.300	70.0
5E + 06	0.000594	0.00336	0.00304	0.195	80.5
1E + 07	0.000420	0.00300	0.00284	0.148	85.2
2E + 07	0.000297	0.00269	0.00261	0.114	88.6
5E + 07	0.000188	0.00235	0.00232	0.081	9.19

Note: BL = Boundary layer.

TABLE 2.6.4 Drag Coefficient Data for Selected Objects (Re > 1000)

Object	Diagram		$C_D(\text{RE}^* \gtrsim 10^3)$
Square prism	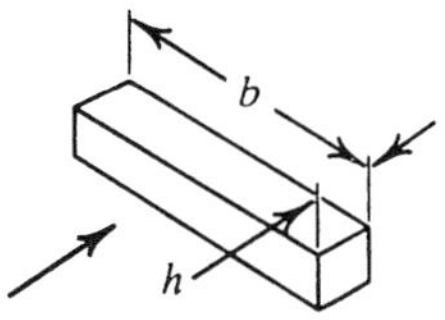	$b/h = \infty$	2.05
		$b/h = 1$	1.05
Disk			1.17
Ring			1.20[b]
Hemisphere (open end facing flow)	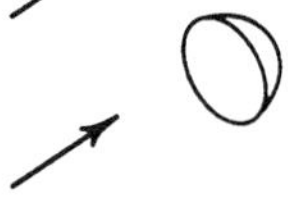		1.42
Hemisphere (open end facing downstream)	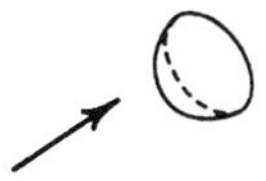		0.38
C-section (open side facing flow)	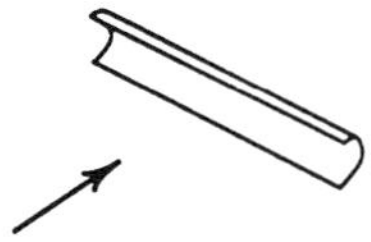		2.30
C-section (open side facing downstream)	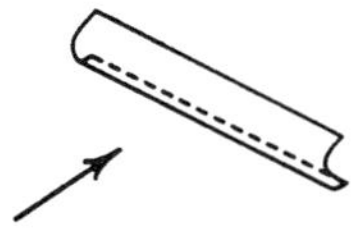		1.20

[a] Data from Hoerner, 1965.
[b] Based on ring area.

for a two-dimensional object (long square cylinder) is about twice that for the corresponding three-dimensional object (square cylinder with $b/h = 1$).

Friction and Pressure Drag: Bluff Bodies

Both friction and pressure forces contribute to the drag of *bluff bodies* (see Shapiro, 1960, for a good discussion of the mechanisms of drag). As an example, consider the drag coefficient for a smooth sphere shown in Figure 2.6.5. Transition from laminar to turbulent flow in the boundary layers on the forward portion of the sphere causes a dramatic dip in drag coefficient at the *critical Reynolds number* ($Re_D \approx 2 \times 10^5$). The turbulent boundary layer is better able to resist the adverse pressure gradient on the rear of the sphere, so separation is delayed and the wake is smaller, causing less pressure drag.

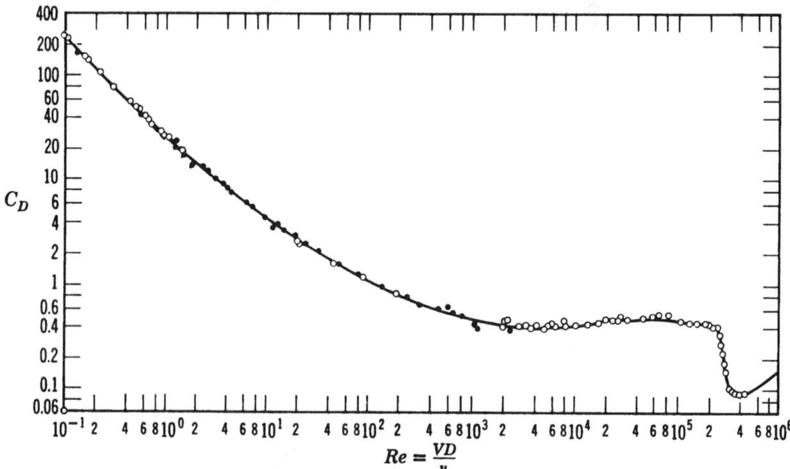

FIGURE 2.6.5 Drag coefficient vs. Reynolds number for a smooth sphere. (From Schlichting, H. 1979. *Boundary Layer Theory,* 7th ed., McGraw-Hill, New York. With permission.)

Surface roughness (or freestream disturbances) can reduce the critical Reynolds number. Dimples on a golf ball cause the boundary layer to become turbulent and, therefore, lower the drag coefficient in the range of speeds encountered in a drive.

Streamlining

Streamlining is adding a faired tail section to reduce the extent of separated flow on the downstream portion of an object (at high Reynolds number where pressure forces dominate drag). The adverse pressure gradient is taken over a longer distance, delaying separation. However, adding a faired tail increases surface area, causing skin friction drag to increase. Thus, streamlining must be optimized for each shape.

Front contours are of principal importance in road vehicle design; the angle of the back glass also is important (in most cases the entire rear end cannot be made long enough to control separation and reduce drag significantly).

Lift

Lift coefficient is defined as

$$C_L = \frac{F_L}{\frac{1}{2}\rho V^2 A} \tag{2.6.10}$$

Note that lift coefficient is based on projected *planform area.*

Airfoils

Airfoils are shaped to produce lift efficiently by accelerating flow over the upper surface to produce a low-pressure region. Because the flow must again decelerate, inevitably there must be a region of adverse pressure gradient near the rear of the upper surface (pressure distributions are shown clearly in Hazen, 1965).

Lift and drag coefficients for airfoil sections depend on Reynolds number and *angle of attack* between the chord line and the undisturbed flow direction. The *chord line* is the straight line joining the leading and trailing edges of the airfoil (Abbott and von Doenhoff, 1959).

As the angle of attack is increased, the minimum pressure point moves forward on the upper surface and the minimum pressure becomes lower. This increases the adverse pressure gradient. At some angle of attack, the adverse pressure gradient is strong enough to cause the boundary layer to separate completely from the upper surface, causing the airfoil to *stall*. The separated flow alters the pressure distribution, reducing lift sharply.

Increasing the angle of attack also causes the the drag coefficient to increase. At some angle of attack below stall the ratio of lift to drag, the *lift–drag* ratio, reaches a maximum value.

Drag Due to Lift

For wings (airfoils of finite span), lift and drag also are functions of aspect ratio. Lift is reduced and drag increased compared with infinite span, because end effects cause the lift vector to rotate rearward. For a given geometric angle of attack, this reduces effective angle of attack, reducing lift. The additional component of lift acting in the flow direction increases drag; the increase in drag due to lift is called *induced drag*.

The effective aspect ratio includes the effect of planform shape. When written in terms of effective aspect ratio, the drag of a finite-span wing is

$$C_D = C_{D,\infty} + \frac{C_L^2}{\pi ar} \tag{2.6.11}$$

where ar is effective aspect ratio and the subscript ∞ refers to the infinite section drag coefficient at C_L. For further details consult the references.

The lift coefficient must increase to support aircraft weight as speed is reduced. Therefore, induced drag can increase rapidly at low flight speeds. For this reason, minimum allowable flight speeds for commercial aircraft are closely controlled by the FAA.

Boundary Layer Control

The major part of the drag on an airfoil or wing is caused by skin friction. Therefore, it is important to maintain laminar flow in the boundary layers as far aft as possible; laminar flow sections are designed to do this. It also is important to prevent flow separation and to achieve high lift to reduce takeoff and landing speeds. These topics fall under the general heading of boundary layer control.

Profile Shaping

Boundary layer transition on a conventional airfoil section occurs almost immediately after the minimum pressure at about 25% chord aft the leading edge. Transition can be delayed by shaping the profile to maintain a favorable pressure gradient over more of its length. The U.S. National Advisory Committee for Aeronautics (NACA) developed several series of profiles that delayed transition to 60 or 65% of chord, reducing drag coefficients (in the design range) 60% compared with conventional sections of the same thickness ratio (Abbott and von Doenhoff, 1959).

Flaps and Slats

Flaps are movable sections near the trailing edge of a wing. They extend and/or deflect to increase wing area and/or increase wing camber (curvature), to provide higher lift than the clean wing. Many aircraft also are fitted with leading edge slats which open to expose a slot from the pressure side of the wing to the upper surface. The open slat increases the effective radius of the leading edge, improving maximum lift coefficient. The slot allows energized air from the pressure surface to flow into the low-pressure region atop the wing, energizing the boundary layers and delaying separation and stall.

Suction and Blowing

Suction removes low-energy fluid from the boundary layer, reducing the tendency for early separation. Blowing via high-speed jets directed along the surface reenergizes low-speed boundary layer fluid. The objective of both approaches is to delay separation, thus increasing the maximum lift coefficient the wing can achieve. Powered systems add weight and complexity; they also require bleed air from the engine compressor, reducing thrust or power output.

Moving Surfaces

Many schemes have been proposed to utilize moving surfaces for boundary layer control. Motion in the direction of flow reduces skin friction, and thus the tendency to separate; motion against the flow has the opposite effect. The aerodynamic behavior of sports balls — baseballs, golf balls, and tennis balls — depends significantly on aerodynamic side force (lift, down force, or side force) produced by spin. These effects are discussed at length in Fox and McDonald (1992) and its references.

Computation vs. Experiment

Experiments cannot yet be replaced completely by analysis. Progress in modeling, numerical techniques, and computer power continues to be made, but the role of the experimentalist likely will remain important for the foreseeable future.

Computational Fluid Dynamics (CFD)

Computation of fluid flow requires accurate mathematical modeling of flow physics and accurate numerical procedures to solve the equations. The basic equations for laminar boundary layer flow are well known. For turbulent boundary layers generally it is not possible to resolve the solution space into sufficiently small cells to allow direct numerical simulation. Instead, empirical models for the turbulent stresses must be used. Advances in computer memory storage capacity and speed (e.g., through use of massively parallel processing) continue to increase the resolution that can be achieved.

A second source of error in CFD work results from the numerical procedures required to solve the equations. Even if the equations are exact, approximations must be made to discretize and solve them using finite-difference or finite-volume methods. Whichever is chosen, the solver must guard against introducing numerical instability, round-off errors, and numerical diffusion (Hoffman, 1992).

Role of the Wind Tunnel

Traditionally, wind tunnel experiments have been conducted to verify the design and performance of components and complete aircraft. Design verification of a modern aircraft may require expensive scale models, several thousand hours of wind tunnel time at many thousands of dollars an hour, and additional full-scale flight testing.

New wind tunnel facilities continue to be built and old ones refurbished. This indicates a need for continued experimental work in developing and optimizing aircraft configurations.

Many experiments are designed to produce baseline data to validate computer codes. Such systematic experimental data can help to identify the strengths and weaknesses of computational methods.

CFD tends to become only indicative of trends when massive zones of flow separation are present. Takeoff and landing configurations of conventional aircraft, with landing gear, high-lift devices, and flaps extended, tend to need final experimental confirmation and optimization. Many studies of vertical takeoff and vectored thrust aircraft require testing in wind tunnels.

Defining Terms

Boundary layer: Thin layer of fluid adjacent to a surface where viscous effects are important; viscous effects are negligible outside the boundary layer.

Drag coefficient: Force in the flow direction exerted on an object by the fluid flowing around it, divided by dynamic pressure and area.

Lift coefficient: Force perpendicular to the flow direction exerted on an object by the fluid flowing around it, divided by dynamic pressure and area.

Pressure gradient: Variation in pressure along the surface of an object. For a *favorable* pressure gradient, pressure *decreases* in the flow direction; for an *adverse* pressure gradient, pressure *increases* in the flow direction.

Separation: Phenomenon that occurs when fluid layers adjacent to a solid surface are brought to rest and boundary layers depart from the surface contour, forming a low-pressure *wake* region. Separation can occur only in an *adverse pressure gradient.*

Transition: Change from laminar to turbulent flow within the boundary layer. The location depends on distance over which the boundary layer has developed, pressure gradient, surface roughness, freestream disturbances, and heat transfer.

References

Abbott, I.H. and von Doenhoff, A.E. 1959. *Theory of Wing Sections, Including a Summary of Airfoil Data.* Dover, New York.

Fox, R.W. and McDonald, A.T. 1992. *Introduction to Fluid Mechanics,* 4th ed. John Wiley & Sons, New York.

Hazen, D.C. 1965. *Boundary Layer Control,* film developed by the National Committee for Fluid Mechanics Films (NCFMF) and available on videotape from Encyclopaedia Britannica Educational Corporation, Chicago.

Hoerner, S.F. 1965. *Fluid-Dynamic Drag,* 2nd ed. Published by the author, Midland Park, NJ.

Hoffman, J.D. 1992. *Numerical Methods for Engineers and Scientists.* McGraw-Hill, New York.

Schlichting, H. 1979. *Boundary-Layer Theory,* 7th ed. McGraw-Hill, New York.

Shapiro, A.H. 1960. *The Fluid Dynamics of Drag,* film developed by the National Committee for Fluid Mechanics Film (NCFMF) and available on videotape from Encyclopaedia Britannica Educational Corporation, Chicago.

Further Information

A comprehensive source of basic information is the *Handbook of Fluid Dynamics,* edited by Victor L. Streeter (McGraw-Hill, New York, 1960).

Timely reviews of important topics are published in the *Annual Review of Fluid Mechanics* series (Annual Reviews, Inc., Palo Alto, CA.). Each volume contains a cumulative index.

ASME (American Society of Mechanical Engineers, New York, NY) publishes the *Journal of Fluids Engineering* quarterly. *JFE* contains fluid machinery and other engineering applications of fluid mechanics.

The monthly *AIAA Journal* and bimonthly *Journal of Aircraft* (American Institute for Aeronautics and Astronautics, New York) treat aerospace applications of fluid mechanics.

2.7 Compressible Flow

Ajay Kumar

Introduction

This section deals with compressible flow. Only one- or two-dimensional steady, inviscid flows under perfect gas assumption are considered. Readers are referred to other sources of information for unsteady effects, viscous effects, and three-dimensional flows.

The term *compressible flow* is routinely used to define variable density flow which is in contrast to incompressible flow, where the density is assumed to be constant throughout. In many cases, these density variations are principally caused by the pressure changes from one point to another. Physically, the *compressibility* can be defined as the fractional change in volume of the gas element per unit change in pressure. It is a property of the gas and, in general, can be defined as

$$\tau = \frac{1}{\rho}\frac{d\rho}{dp}$$

where τ is the compressibility of the gas, ρ is the density, and p is the pressure being exerted on the gas. A more precise definition of compressibility is obtained if we take into account the thermal and frictional losses. If during the compression the temperature of the gas is held constant, it is called the isothermal compressibility and can be written as

$$\tau_T = \frac{1}{\rho}\left(\frac{\partial \rho}{\partial p}\right)_T$$

However, if the compression process is reversible, it is called the isentropic compressibility and can be written as

$$\tau_s = \frac{1}{\rho}\left(\frac{\partial \rho}{\partial p}\right)_s$$

Gases in general have high compressibility (τ_T for air is 10^{-5} m^2/N at 1 atm) as compared with liquids (τ_T for water is 5×10^{-10} m^2/N at 1 atm).

Compressibility is a very important parameter in the analysis of compressible flow and is closely related to the *speed of sound*, a, which is the velocity of propagation of small pressure disturbances and is defined as

$$a^2 = \left(\frac{\partial p}{\partial \rho}\right)_s \quad \text{or} \quad a = \sqrt{\left(\frac{\partial p}{\partial \rho}\right)_s}$$

In an isentropic process of a perfect gas, the pressure and density are related as

$$\frac{p}{\rho^\gamma} = \text{constant}$$

Using this relation along with the perfect gas relation $p = \rho RT$, we can show that for a perfect gas

$$a = \sqrt{\gamma RT} = \sqrt{\frac{\gamma p}{\rho}}$$

where γ is the ratio of specific heats at constant pressure and constant volume, R is the gas constant, and T is the temperature. For air under normal conditions, γ is 1.4 and R is 287 m^2/sec^2 K so that the speed of sound for air becomes $a = 20.045\sqrt{T}$ m/sec where T is in kelvin.

Another important parameter in compressible flows is the *Mach number, M,* which is defined as the ratio of the gas velocity to the speed of sound or

$$M = \frac{V}{a}$$

where V is the velocity of gas. Depending upon the Mach number of the flow, we can define the following flow regimes:

$M \ll 1$ Incompressible flow

$M < 1$ Subsonic flow

$M \approx 1$ Transonic flow

$M > 1$ Supersonic flow

$M \gg 1$ Hypersonic flow

Subsonic through hypersonic flows are compressible in nature. In these flows, the velocity is appreciable compared with the speed of sound, and the fractional changes in pressure, temperature, and density are all of significant magnitude. We will restrict ourselves in this section to subsonic through flows only.

Before we move on to study these flows, let us define one more term. Let us consider a gas with static pressure p and temperature T, traveling at some velocity V and corresponding Mach number M. If this gas is brought isentropically to stagnation or zero velocity, the pressure and temperature which the gas achieves are defined as *stagnation pressure* p_0 and *stagnation temperature* T_0 (also called total pressure and total temperature). The speed of sound at stagnation conditions is called the *stagnation speed of sound* and is denoted as a_0.

One-Dimensional Flow

In one-dimensional flow, the flow properties vary only in one coordinate direction. Figure 2.7.1 shows two streamtubes in a flow. In a *truly one-dimensional flow* illustrated in Figure 2.7.1(a), the flow variables are a function of x only and the area of the stream tube is constant. On the other hand, Figure 2.7.1(b) shows a flow where the area of the stream tube is also a function of x but the flow variables are still a function of x only. This flow is defined as the *quasi-one-dimensional flow.* We will first discuss the truly one-dimensional flow.

In a steady, truly one-dimensional flow, conservation of mass, momentum, and energy leads to the following simple algebraic equations.

$$\rho u = \text{constant}$$

$$p + \rho u^2 = \text{constant} \tag{2.7.1}$$

$$h + \frac{u^2}{2} + q = \text{constant}$$

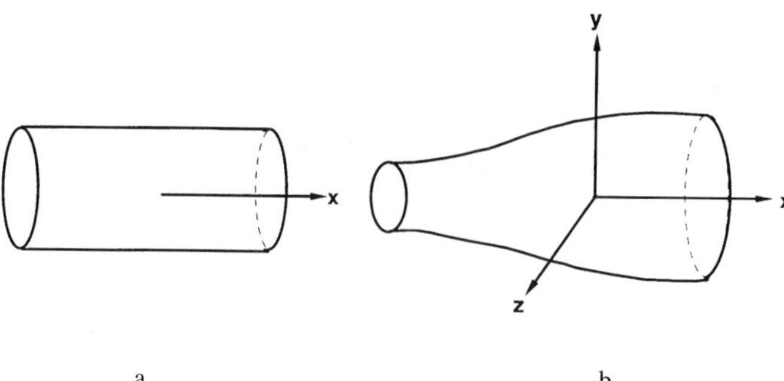

FIGURE 2.7.1 (a) One-dimensional flow; (b) quasi-one-dimensional flow.

where q is the heat added per unit mass of the gas. These equations neglect body forces, viscous stresses, and heat transfer due to thermal conduction and diffusion. These relations given by Equation 2.7.1, when applied at points 1 and 2 in a flow with no heat addition, become

$$\begin{aligned} \rho_1 u_1 &= \rho_2 u_2 \\ p_1 + \rho_1 u_1^2 &= p_2 + \rho_2 u_2^2 \\ h_1 + \frac{u_1^2}{2} &= h_2 + \frac{u_2^2}{2} \end{aligned} \tag{2.7.2}$$

The energy equation for a calorically perfect gas, where $h = c_p T$, becomes

$$c_p T_1 + \frac{u_1^2}{2} = c_p T_2 + \frac{u_2^2}{2}$$

Using $c_p = \gamma R/(\gamma - 1)$ and $a^2 = \gamma RT$, the above equation can be written as

$$\frac{a_1^2}{\gamma - 1} + \frac{u_1^2}{2} = \frac{a_2^2}{\gamma - 1} + \frac{u_2^2}{2} \tag{2.7.3}$$

Since Equation (2.7.3) is written for no heat addition, it holds for an adiabatic flow. If the energy equation is applied to the stagnation conditions, it can be written as

$$c_p T + \frac{u^2}{2} = c_p T_0$$

$$\frac{T_0}{T} = 1 + \frac{\gamma - 1}{2} M^2 \tag{2.7.4}$$

It is worth mentioning that in arriving at Equation (2.7.4), only adiabatic flow condition is used whereas stagnation conditions are defined as those where the gas is brought to rest isentropically. Therefore, the definition of stagnation temperature is less restrictive than the general definition of stagnation conditions. According to the general definition of isentropic flow, it is a reversible adiabatic flow. This definition is needed for the definition of stagnation pressure and density. For an isentropic flow,

$$\frac{p_0}{p} = \left(\frac{\rho_0}{\rho}\right)^{\gamma} = \left(\frac{T_0}{T}\right)^{\gamma/(\gamma-1)} \tag{2.7.5}$$

From Equations 2.7.4 and 2.7.5, we can write

$$\frac{p_0}{p} = \left(1 + \frac{\gamma - 1}{2} M^2\right)^{\gamma/(\gamma-1)} \tag{2.7.6}$$

$$\frac{\rho_0}{\rho} = \left(1 + \frac{\gamma - 1}{2} M^2\right)^{1/(\gamma-1)} \tag{2.7.7}$$

Values of stagnation conditions are tabulated in Anderson (1982) as a function of M for $\gamma = 1.4$.

Normal Shock Wave

A shock wave is a very thin region (of the order of a few molecular mean free paths) across which the static pressure, temperature, and density increase whereas the velocity decreases. If the shock wave is perpendicular to the flow, it is called a *normal shock wave.* The flow is supersonic ahead of the normal shock wave and subsonic behind it. Figure 2.7.2 shows the flow conditions across a normal shock wave which is treated as a discontinuity. Since there is no heat added or removed, the flow across the shock wave is adiabatic. By using Equations 2.7.2 the normal shock equations can be written as

$$\begin{aligned} \rho_1 u_1 &= \rho_2 u_2 \\ p_1 + \rho_1 u_1^2 &= p_2 + \rho_2 u_2^2 \\ h_1 + \frac{u_1^2}{2} &= h_2 + \frac{u_2^2}{2} \end{aligned} \tag{2.7.8}$$

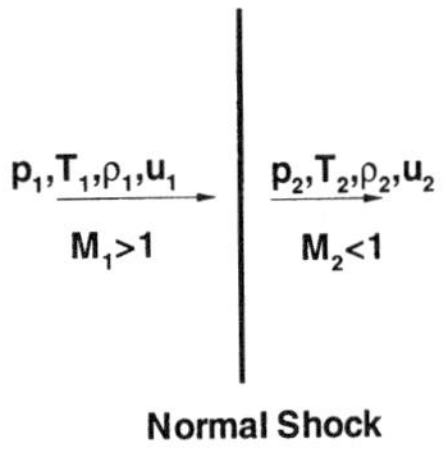

FIGURE 2.7.2 Flow conditions across a normal shock.

Equations (2.7.8) are applicable to a general type of flow; however, for a calorically perfect gas, we can use the relations $p = \rho RT$ and $h = c_p T$ to derive a number of equations relating flow conditions downstream of the normal shock to those at upstream. These equations (also known as Rankine–Hugoniot relations) are

$$\begin{aligned} \frac{p_2}{p_1} &= 1 + \frac{2\gamma}{\gamma+1}\left(M_1^2 - 1\right) \\ \frac{\rho_2}{\rho_1} &= \frac{u_1}{u_2} = \frac{(\gamma+1)M_1^2}{2 + (\gamma-1)M_1^2} \end{aligned} \tag{2.7.9}$$

$$\frac{T_2}{T_1} = \frac{h_2}{h_1} = \left[1 + \frac{2\gamma}{\gamma+1}\left(M_1^2 - 1\right)\right]\left[\frac{2 + (\gamma-1)M_1^2}{(\gamma+1)M_1^2}\right]$$

$$M_2^2 = \frac{1 + \frac{\gamma-1}{2}M_1^2}{\gamma M_1^2 - \frac{\gamma-1}{2}}$$

Again, the values of p_2/p_1, ρ_2/ρ_1, T_2/T_1, etc. are tabulated in Anderson (1982) as a function of M_1 for $\gamma = 1.4$. Let us examine some limiting cases. As $M_1 \to 1$, Equations 2.7.9 yield $M_2 \to 1$, $p_2/p_1 \to 1$, $\rho_2/\rho_1 \to 1$, and $T_2/T_1 \to 1$. This is the case of an extremely weak normal shock across which no finite changes occur. This is the same as the sound wave. On the other hand, as $M_1 \to \infty$, Equations (2.7.9) yield

$$M_2 \to \sqrt{\frac{\gamma-1}{2\gamma}} = 0.378; \quad \frac{\rho_2}{\rho_1} \to \frac{\gamma+1}{\gamma-1} = 6; \quad \frac{p_2}{p_1} \to \infty; \quad \frac{T_2}{T_1} \to \infty$$

However, the calorically perfect gas assumption no longer remains valid as $M_1 \to \infty$.

Let us now examine why the flow ahead of a normal shock wave must be supersonic even though Equations (2.7.8) hold for $M_1 < 1$ as well as $M_1 > 1$. From the second law of thermodynamics, the entropy change across the normal shock can be written as

$$s_2 - s_1 = c_p \ln\frac{T_2}{T_1} - R\ln\frac{p_2}{p_1}$$

By using Equations (2.7.9) it becomes

$$s_2 - s_1 = c_p \ln\left\{\left[1 + \frac{2\gamma}{\gamma+1}\left(M_1^2 - 1\right)\right]\left[\frac{2 + (\gamma-1)M_1^2}{(\gamma+1)M_1^2}\right]\right\} - R\ln\left[1 + \frac{2\gamma}{\gamma+1}\left(M_1^2 - 1\right)\right] \qquad (2.7.10)$$

Equation (2.7.10) shows that the entropy change across the normal shock is also a function of M_1 only. Using Equation (2.7.10) we see that

$$\begin{aligned} s_2 - s_1 &= 0 \quad \text{for} \quad M_1 = 1 \\ &< 0 \quad \text{for} \quad M_1 < 1 \\ &> 0 \quad \text{for} \quad M_1 > 1 \end{aligned}$$

Since it is necessary that $s_2 - s_1 \geqslant 0$ from the second law, $M_1 \geq 1$. This, in turn, requires that $p_2/p_1 \geqslant 1$, $\rho_2/\rho_1 \geqslant 1$, $T_2/T_1 \geqslant 1$, and $M_2 \leqslant 1$.

We now examine how the stagnation conditions change across a normal shock wave. For a calorically perfect gas, the energy equation in Equations (2.7.9) gives

$$c_p T_{01} = c_p T_{02} \quad \text{or} \quad T_{01} = T_{02}$$

In other words, the total temperature remains constant across a stationary normal shock wave.

Let us now apply the entropy change relation across the shock using the stagnation conditions.

$$s_2 - s_1 = c_p \ln \frac{T_{02}}{T_{01}} - R \ln \frac{p_{02}}{p_{01}}$$

Note that entropy at stagnation conditions is the same as at the static conditions since to arrive at stagnation conditions, the gas is brought to rest isentropically. Since $T_{02} = T_{01}$,

$$s_2 - s_1 = -R \ln \frac{p_{02}}{p_{01}}$$

$$\frac{p_{02}}{p_{01}} = e^{-(s_2 - s_1)/R} \tag{2.7.11}$$

Since $s_2 > s_1$ across the normal shockwave, Equation (2.7.11) gives $P_{02} < P_{01}$ or, in other words, the total pressure decreases across a shock wave.

One-Dimensional Flow with Heat Addition

Consider one-dimensional flow through a control volume as shown in Figure 2.7.3. Flow conditions going into this control volume are designated by 1 and coming out by 2. A specified amount of heat per unit mass, q, is added to the control volume. The governing equations relating conditions 1 and 2 can be written as

$$\begin{gathered} \rho_1 u_1 = \rho_2 u_2 \\ p_1 + \rho_1 u_1^2 = p_2 + \rho_2 u_2^2 \\ h_1 + \frac{u_1^2}{2} + q = h_2 + \frac{u_2^2}{2} \end{gathered} \tag{2.7.12}$$

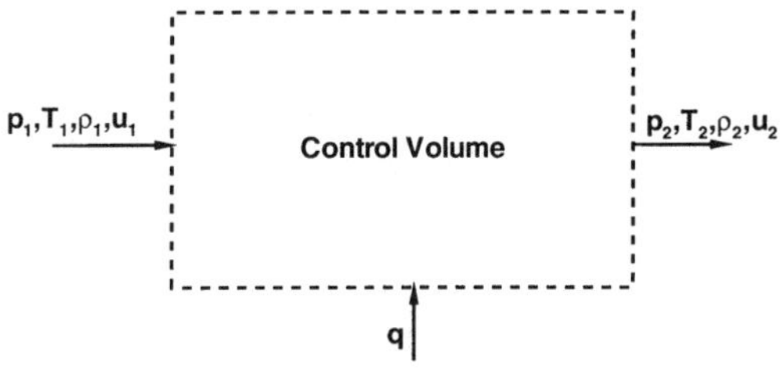

FIGURE 2.7.3 One-dimensional control volume with heat addition.

The following relations can be derived from Equation (2.7.12) for a calorically perfect gas

$$q = c_p \left(T_{02} - T_{01}\right) \tag{2.7.13}$$

$$\frac{p_2}{p_1} = \frac{1 + \gamma M_1^2}{1 + \gamma M_2^2} \tag{2.7.14}$$

$$\frac{T_2}{T_1} = \left(\frac{1+\gamma M_1^2}{1+\gamma M_2^2}\right)^2 \left(\frac{M_2}{M_1}\right)^2 \tag{2.7.15}$$

$$\frac{\rho_2}{\rho_1} = \left(\frac{1+\gamma M_2^2}{1+\gamma M_1^2}\right)^2 \left(\frac{M_1}{M_2}\right)^2 \tag{2.7.16}$$

Equation (2.7.13) indicates that the effect of heat addition is to directly change the stagnation temperature T_0 of the flow. Table 2.7.1 shows some physical trends which can be obtained with heat addition to subsonic and supersonic flow. With heat extraction the trends in Table 2.7.1 are reversed.

TABLE 2.7.1 Effect of Heat Addition on Subsonic and Supersonic Flow

	$M_1 < 1$	$M_1 > 1$
M_2	Increases	Decreases
p_2	Decreases	Increases
T_2	Increases for $M_1 < \gamma^{-1/2}$ and decreases for $M_1 > \gamma^{-1/2}$	Increases
u_2	Increases	Decreases
T_{02}	Increases	Increases
p_{02}	Decreases	Decreases

Figure 2.7.4 shows a plot between enthalpy and entropy, also known as the Mollier diagram, for one-dimensional flow with heat addition. This curve is called the Rayleigh curve and is drawn for a set of given initial conditions. Each point on this curve corresponds to a different amount of heat added or removed. It is seen from this curve that heat addition always drives the Mach numbers toward 1. For a certain amount of heat addition, the flow will become sonic. For this condition, the flow is said to be *choked*. Any further increase in heat addition is not possible without adjustment in initial conditions. For example, if more heat is added in region 1, which is initially supersonic, than allowed for attaining Mach 1 in region 2, then a normal shock will form inside the control volume which will suddenly change the conditions in region 1 to subsonic. Similarly, in case of an initially subsonic flow corresponding to region 1′, any heat addition beyond that is needed to attain Mach 1 in region 2, the conditions in region 1′ will adjust to a lower subsonic Mach number through a series of pressure waves.

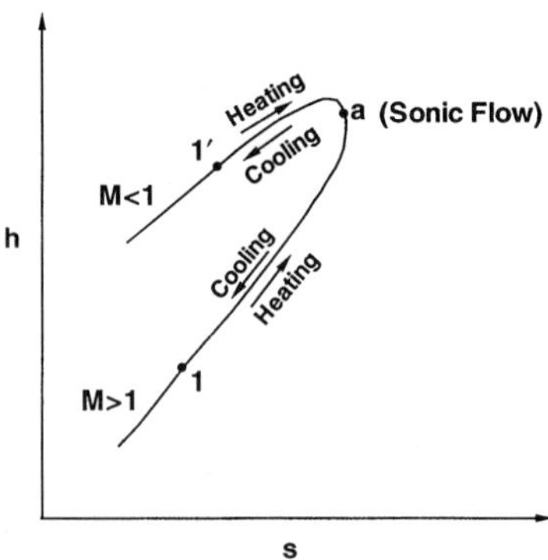

FIGURE 2.7.4 The Rayleigh curve.

Similar to the preceding heat addition or extraction relationships, we can also develop relationships for one-dimensional steady, adiabatic flow but with frictional effects due to viscosity. In this case, the momentum equation gets modified for frictional shear stress. For details, readers are referred to Anderson (1982).

Quasi-One-Dimensional Flow

In quasi-one-dimensional flow, in addition to flow conditions, the area of duct also changes with x. The governing equations for quasi-one-dimensional flow can be written in a differential form as follows using an infinitesimal control volume shown in Figure 2.7.5.

$$d(\rho u A) = 0 \tag{2.7.17}$$

$$dp + \rho u \, du = 0 \tag{2.7.18}$$

$$dh + u \, du = 0 \tag{2.7.19}$$

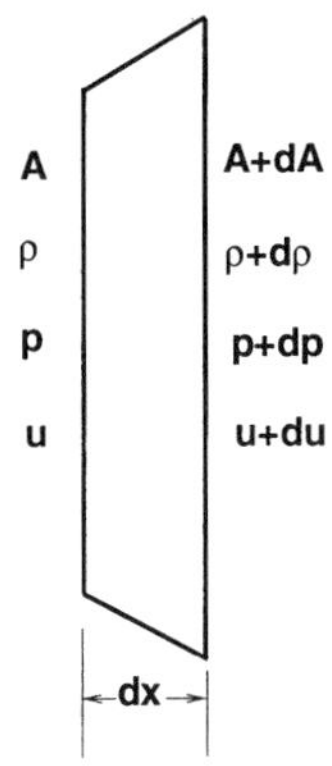

FIGURE 2.7.5 Control volume for quasi-one-dimensional flow.

Equation 2.7.17 can be written as

$$\frac{d\rho}{\rho} + \frac{du}{u} + \frac{dA}{A} = 0 \tag{2.7.20}$$

which can be further written as follows for an isentropic flow:

$$\frac{dA}{A} = \left(M^2 - 1\right)\frac{du}{u} \tag{2.7.21}$$

Some very useful physical insight can be obtained from this area–velocity relation.

- For subsonic flow ($0 \leq M < 1$), an increase in area results in decrease in velocity, and vice versa.
- For supersonic flow ($M > 1$), an increase in area results in increase in velocity, and vice versa.
- For sonic flow ($M = 1$), $dA/A = 0$, which corresponds to a minimum or maximum in the area distribution, but it can be shown that a minimum in area is the only physical solution.

Figure 2.7.6 shows the preceding results in a schematic form.

It is obvious from this discussion that for a gas to go isentropically from subsonic to supersonic, and vice versa, it must flow through a convergent–divergent nozzle, also known as the de Laval nozzle. The minimum area of the nozzle at which the flow becomes sonic is called the throat. This physical observation forms the basis of designing supersonic wind tunnels shown schematically in Figure 2.7.7. In general, in a supersonic wind tunnel, a stagnant gas is first expanded to the desired supersonic Mach number. The supersonic flow enters the test section where it passes over a model being tested. The flow

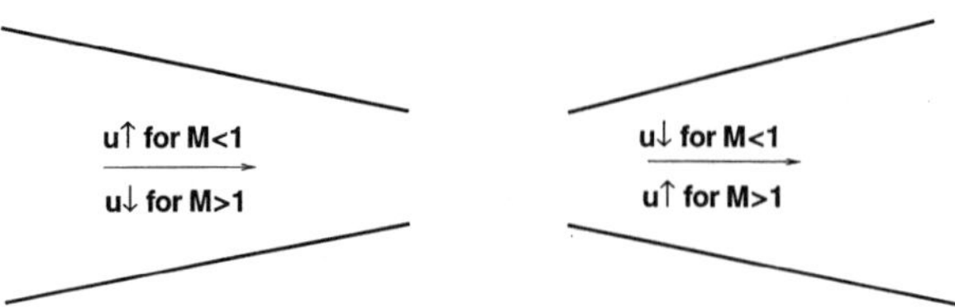

FIGURE 2.7.6 Compressible flow in converging and diverging ducts.

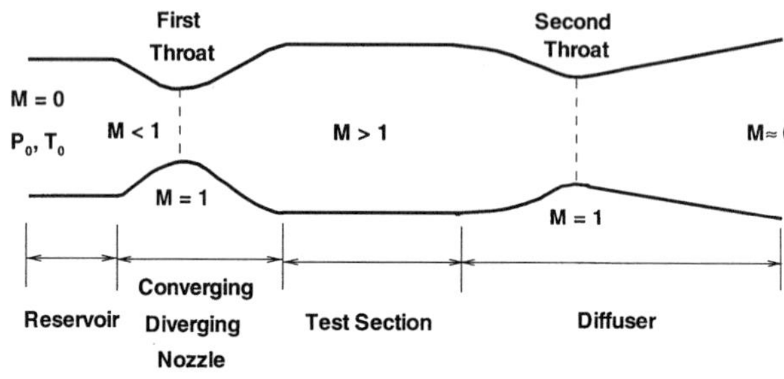

FIGURE 2.7.7 Schematic of a typical supersonic wind tunnel.

then is slowed down by compressing it through a second convergent–divergent nozzle, also known as a diffuser, before it is exhausted to the atmosphere.

Now, using the equations for quasi-one-dimensional flow and the isentropic flow conditions, we can derive a relation for the area ratio that is needed to accelerate or decelerate the gas to sonic conditions. Denoting the sonic conditions by an asterisk, we can write $u^* = a^*$. The area is denoted as A^*, and it is obviously the minimum area for the throat of the nozzle. From Equation (2.7.17) we have

$$\rho u A = \rho^* u^* A^*$$

$$\frac{A}{A^*} = \frac{\rho^* u^*}{\rho u} = \frac{\rho^*}{\rho_0}\frac{\rho_0}{\rho}\frac{u^*}{u} \tag{2.7.22}$$

Under isentropic conditons,

$$\frac{\rho_0}{\rho} = \left(1 + \frac{\gamma - 1}{2} M^2\right)^{1/(\gamma-1)} \tag{2.7.23}$$

$$\frac{\rho_0}{\rho^*} = \left(1 + \frac{\gamma - 1}{2}\right)^{1/(\gamma-1)} = \left(\frac{\gamma + 1}{2}\right)^{1/(\gamma-1)} \tag{2.7.24}$$

Also, $u^*/u = a^*/u$. Let us define a Mach number $M^* = u/a^*$. M^* is known as the *characteristic Mach number* and it is related to the local Mach number by the following relation:

$$M^{*2} = \frac{\frac{\gamma + 1}{2} M^2}{1 + \frac{\gamma - 1}{2} M^2} \tag{2.7.25}$$

Using Equations (2.7.23) through (2.7.25) in Equation (2.7.22) we can write

$$\left(\frac{A}{A^*}\right)^2 = \frac{1}{M^2}\left[\left(\frac{2}{\gamma+1}\right)\left(1+\frac{\gamma-1}{2}M^2\right)\right]^{(\gamma+1)/(\gamma-1)} \tag{2.7.26}$$

Equation (2.7.26) is called the area Mach number relation. Figure 2.7.8 shows a plot of A/A^* against Mach number. A/A^* is always ≥ 1 for physically viable solutions.

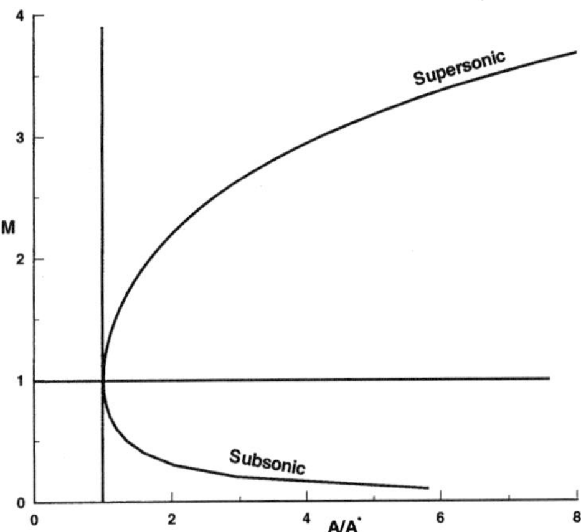

FIGURE 2.7.8 Variation of area ratio A/A* as a function of Mach number for a quasi-one-dimensional flow.

The area Mach number relation says that for a given Mach number, there is only one area ratio A/A^*. This is a very useful relation and is frequently used to design convergent–divergent nozzles to produce a desired Mach number. Values of A/A^* are tabulated as a function of M in Anderson (1982).

Equation (2.7.26) can also be written in terms of pressure as follows:

$$\frac{A}{A^*} = \frac{\left[1-\left(\dfrac{p}{p_0}\right)^{(\gamma-1)/\gamma}\right]^{1/2}\left(\dfrac{p}{p_0}\right)^{1/\gamma}}{\left(\dfrac{\gamma-1}{2}\right)^{1/2}\left(\dfrac{2}{\gamma+1}\right)^{(\gamma+1)/2(\gamma-1)}} \tag{2.7.27}$$

Nozzle Flow

Using the area relations, we can now plot the distributions of Mach number and pressure along a nozzle. Figure 2.7.9 shows pressure and Mach number distributions along a given nozzle and the wave configurations for several exit pressures. For curves a and b, the flow stays subsonic throughout and the exit pressure controls the flow in the entire nozzle. On curve c, the throat has just become sonic, and so the pressure at the throat, and upstream of it, can decrease no further. There is another exit pressure corresponding to curve j ($p_j < p_c$) for which a supersonic isentropic solution exists. But if the pressure lies between p_c and p_j, there is no isentropic solution possible. For example, for an exit pressure p_d, a shock will form in the nozzle at location s which will raise the pressure to $p_{d'}$ and turn the flow subsonic. The pressure will then rise to p_d as the subsonic flow goes through an increasing area nozzle. The location, s, depends on the exit pressure. Various possible situations are shown in Figure 2.7.9. It is clear that if the exit pressure is equal to or below p_f, the flow within the nozzle is fully supersonic. This is

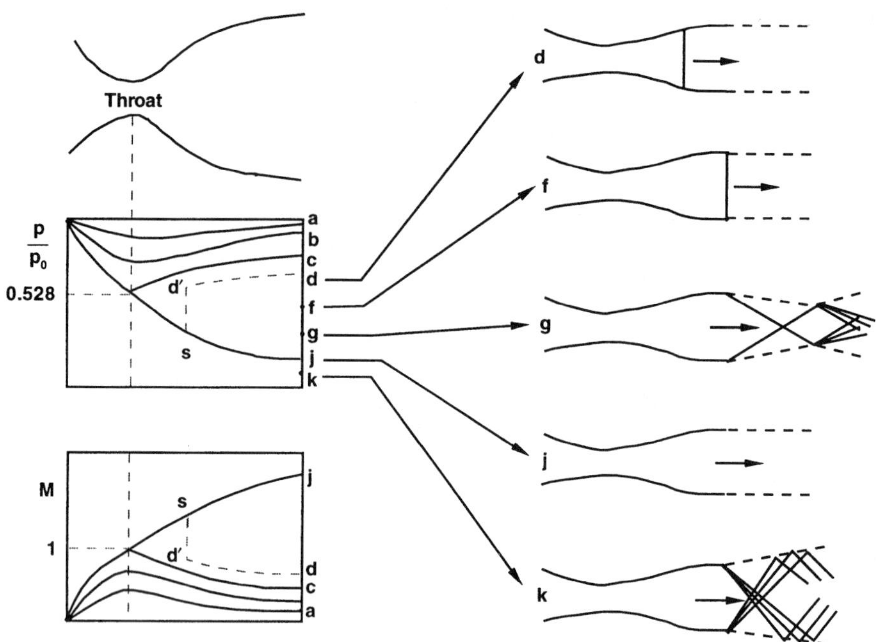

FIGURE 2.7.9 Effect of exit pressure on flow through a nozzle.

the principle used in designing supersonic wind tunnels by operating from a high-pressure reservoir or into a vacuum receiver, or both.

Diffuser

If a nozzle discharges directly into the receiver, the minimum pressure ratio for full supersonic flow in the test section is

$$\left(\frac{p_0}{p_E}\right)_{\min} = \frac{p_0}{p_f}$$

where p_f is the value of p_E at which the normal shock stands right at the nozzle exit. However, by adding an additional diverging section, known as a diffuser, downstream of the test section as shown in Figure 2.7.10 it is possible to operate the tunnel at a lower pressure ratio than p_0/p_f. This happens because the diffuser can now decelerate the subsonic flow downstream of the shock isentropically to a stagnation pressure p_0'. The pressure ratio required then is the ratio of stagnation pressures across a normal shock wave at the test section Mach number. In practice, the diffuser gives lower than expected recovery as a result of viscous losses caused by the interaction of shock wave and the boundary layer which are neglected here.

The operation of supersonic wind tunnels can be made even more efficient; i.e., they can be operated at even lower pressure ratios than p_0/p_0', by using the approach shown in Figure 2.7.7 where the diffuser has a second throat. It can slow down the flow to subsonic Mach numbers isentropically and, ideally, can provide complete recovery, giving $p_0' = p_0$. However, due to other considerations, such as the starting process of the wind tunnel and viscous effects, it is not realized in real life.

Two-Dimensional Supersonic Flow

When supersonic flow goes over a wedge or an expansion corner, it goes through an oblique shock or expansion waves, respectively, to adjust to the change in surface geometry. Figure 2.7.11 shows the two

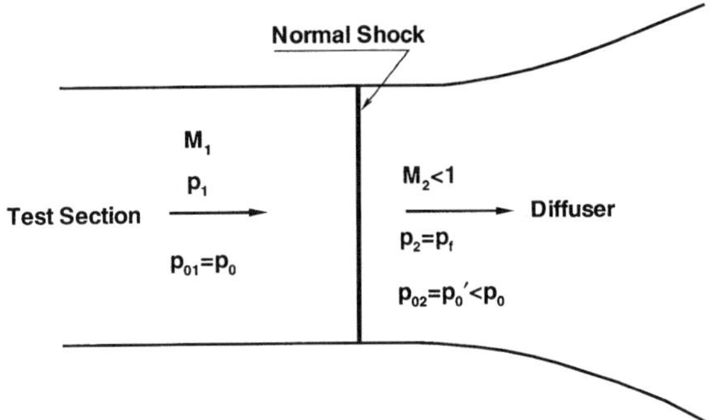

FIGURE 2.7.10 Normal shock diffuser.

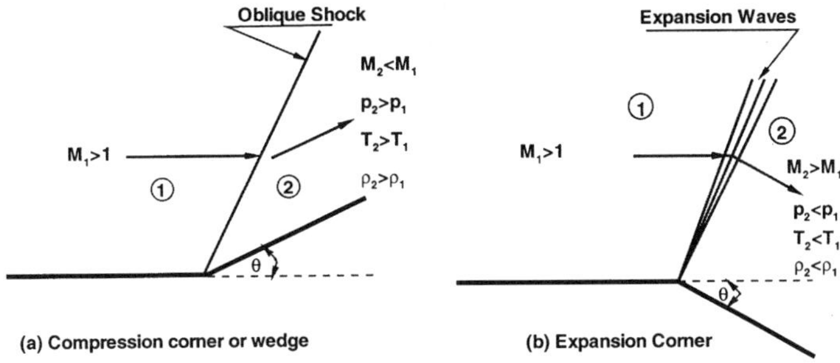

FIGURE 2.7.11 Supersonic flow over a corner.

flow situations. In Figure 2.7.11(a) an oblique shock abruptly turns the flow parallel to the wedge surface. The Mach number behind the shock is less than ahead of it, whereas the pressure, temperature, and density increase. In the case of an expansion corner, oblique expansion waves smoothly turn the flow to become parallel to the surface downstream of the expansion corner. In this case, the Mach number increases, but the pressure, temperature, and density decrease as the flow goes through the expansion corner. Oblique shocks and expansion waves occur in two- and three-dimensional supersonic flows. In this section, we will restrict ourselves to steady, two-dimensional supersonic flows only.

Oblique Shock Waves

The oblique shock can be treated in the same way as the normal shock by accounting for the additional velocity component. If a uniform velocity v is superimposed on the flow field of the normal shock, the resultant velocity ahead of the shock can be adjusted to any flow direction by adjusting the magnitude and direction of v. If v is taken parallel to the shock wave, as shown in Figure 2.7.12, the resultant velocity ahead of the shock is $w_1 = \sqrt{u_1^2 + v_1^2}$ and its direction from the shock is given by $\beta = \tan^{-1}(u_1/v)$. On the downstream side of the shock, since u_2 is less than u_1, the flow always turns toward the shock. The magnitude of u_2 can be determined by the normal shock relations corresponding to velocity u_1 and the magnitude of v is such that the flow downstream of the shock turns parallel to the surface. Since imposition of a uniform velocity does not affect the pressure, temperature, etc., we can use normal shock relations with Mach number replaced in them to correspond to velocity u_1 or u_1/a_1, which is nothing but $M_1 \sin \beta$. Thus, oblique shock relations become

$$\frac{p_2}{p_1} = 1 + \frac{2\gamma}{\gamma + 1}\left(M_1^2 \sin^2 \beta - 1\right) \tag{2.7.28}$$

$$\frac{\rho_2}{\rho_1} = \frac{(\gamma+1)M_1^2 \sin^2\beta}{(\gamma-1)M_1^2 \sin^2\beta + 2} \tag{2.7.29}$$

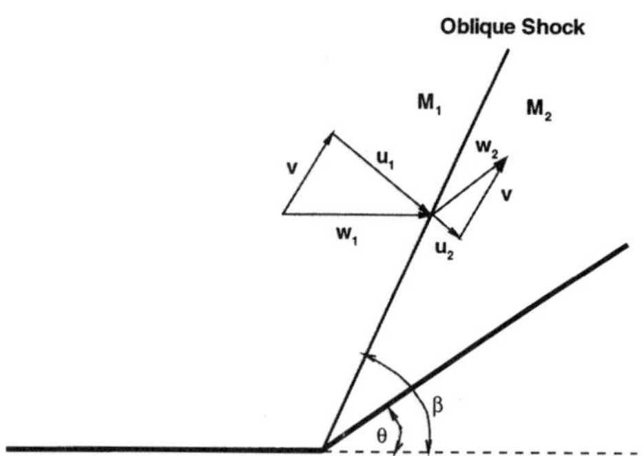

FIGURE 2.7.12 Oblique shock on a wedge.

$$\frac{T_2}{T_1} = \frac{a_2^2}{a_1^2} = \left[1 + \frac{2\gamma}{\gamma+1}\left(M_1^2 \sin^2\beta - 1\right)\right]\left[\frac{2 + (\gamma-1)M_1^2 \sin^2\beta}{(\gamma+1)M_1^2 \sin^2\beta}\right] \tag{2.7.30}$$

The Mach number M_2 (= w_2/a_2) can be obtained by using a Mach number corresponding to velocity u_2 (= $w_2 \sin(\beta - \theta)$) in the normal shock relation for the Mach number. In other words,

$$M_2^2 \sin^2(\beta - \theta) = \frac{1 + \frac{\gamma-1}{2} M_1^2 \sin^2\beta}{\gamma M_1^2 \sin^2\beta - \frac{\gamma-1}{2}} \tag{2.7.31}$$

To derive a relation between the wedge angle θ and the wave angle β, we have from Figure 2.7.12

$$\tan\beta = \frac{u_1}{v} \quad \text{and} \quad \tan(\beta - \theta) = \frac{u_2}{v}$$

so that

$$\frac{\tan(\beta-\theta)}{\tan\beta} = \frac{u_2}{u_1} = \frac{\rho_1}{\rho_2} = \frac{(\gamma-1)M_1^2 \sin^2\beta + 2}{(\gamma+1)M_1^2 \sin^2\beta}$$

This can be simplified to

$$\tan\theta = 2\cot\beta \frac{M_1^2 \sin^2\beta - 1}{M_1^2(\gamma + \cos 2\beta) + 2} \tag{2.7.32}$$

Dennard and Spencer (1964) have tabulated oblique shock properties as a function of M_1. Let us now make some observations from the preceding relations.

From the normal shock relations, $M_1 \sin\beta \geq 1$. This defines a minimum wave angle for a given Mach number. The maximum wave angle, of course, corresponds to the normal shock or $\beta = \pi/2$. Therefore, the wave angle β has the following range

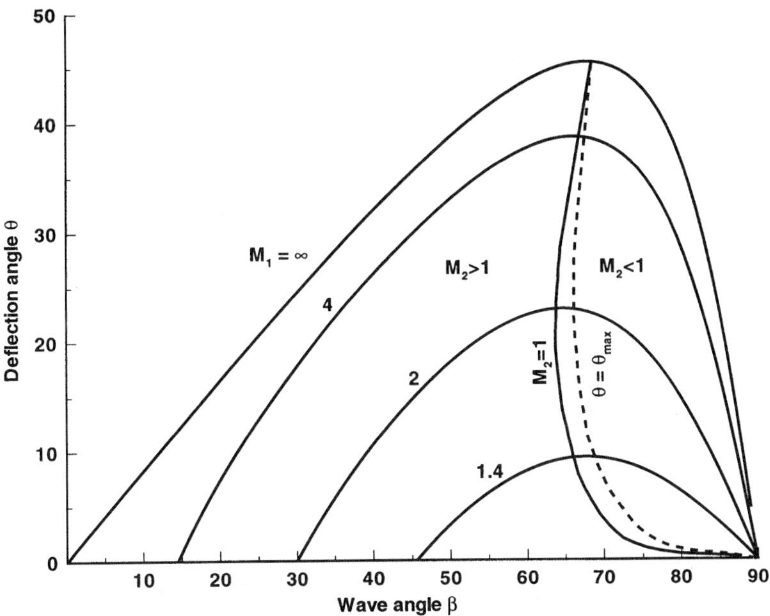

FIGURE 2.7.13 Oblique shock characteristics.

$$\sin^{-1}\frac{1}{M} \leq \beta \leq \frac{\pi}{2} \tag{2.7.33}$$

Equation 2.7.32 becomes zero at the two limits of β. Figure 2.7.13 shows a plot of θ against β for various values of M_1. For each value of M_1, there is a maximum value of θ. For $\theta < \theta_{max}$, there are two possible solutions having different values of β. The larger value of β gives the stronger shock in which the flow becomes subsonic. A locus of solutions for which $M_2 = 1$ is also shown in the figure. It is seen from the figure that with weak shock solution, the flow remains supersonic except for a small range of θ slightly smaller than θ_{max}.

Let us now consider the limiting case of θ going to zero for the weak shock solution. As θ decreases to zero, β decreases to the limiting value μ, given by

$$M_1^2 \sin^2 \mu - 1 = 0$$

$$\mu = \sin^{-1}\frac{1}{M_1} \tag{2.7.34}$$

For this angle, the oblique shock relations show no jump in flow quantities across the wave or, in other words, there is no disturbance generated in the flow. This angle μ is called the *Mach angle* and the lines at inclination μ are called *Mach lines.*

Thin-Airfoil Theory

For a small deflection angle Δθ, it can be shown that the change in pressure in a flow at Mach M_1 is given approximately by

$$\frac{\Delta p}{p_1} \approx \frac{\gamma M_1^2}{\sqrt{M_1^2 - 1}}\Delta\theta \tag{2.7.35}$$

This expression holds for both compression and expansion. If Δp is measured with respect to the freestream pressure, p_1, and all deflections to the freestream direction, we can write Equation (2.7.35) as

$$\frac{p - p_1}{p_1} = \frac{\gamma M_1^2}{\sqrt{M_1^2 - 1}}\theta \tag{2.7.36}$$

where θ is positive for a compression and negative for expansion. Let us define a pressure coefficient C_p, as

$$C_p = \frac{p - p_1}{q_1}$$

where q_1 is the dynamic pressure and is equal to $\gamma p_1 M_1^2/2$. Equation (2.7.36) then gives

$$C_p = \frac{2\theta}{\sqrt{M_1^2 - 1}} \tag{2.7.37}$$

Equation (2.7.37) states that the pressure coefficient is proportional to the local flow deflection. This relation can be used to develop supersonic thin-airfoil theory. As an example, for a flat plate at angle of attack α_0 (shown in Figure 2.7.14), the pressure coefficients on the upper and lower surfaces are

$$C_p = \mp\frac{2\alpha_0}{\sqrt{M_1^2 - 1}}$$

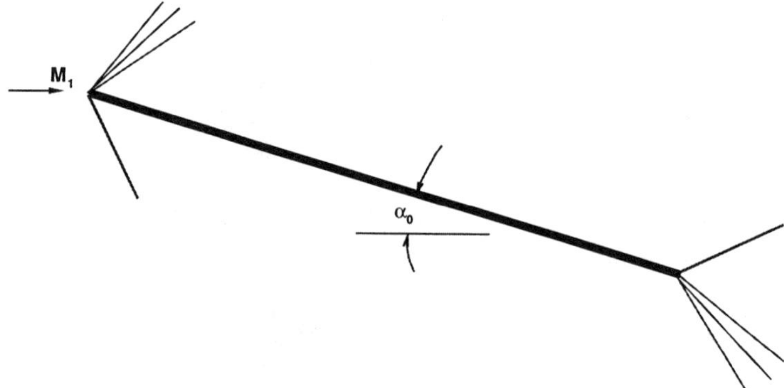

FIGURE 2.7.14 Lifting flat plate.

The lift and drag coefficients can be written as

$$C_L = \frac{(p_L - p_U)c\cos\alpha_0}{q_1 c} = \left(C_{p_L} - C_{p_U}\right)\cos\alpha_0$$

$$C_D = \frac{(p_L - p_U)c\sin\alpha_0}{q_1 c} = \left(C_{p_L} - C_{p_U}\right)\sin\alpha_0$$

where c is the chord length of the plate. Since α_0 is small, we can write

$$C_L = \frac{4\alpha_0}{\sqrt{M_1^2 - 1}}, \qquad C_D = \frac{4\alpha_0^2}{\sqrt{M_1^2 - 1}} \tag{2.7.38}$$

A similar type of expression can be obtained for an arbitrary thin airfoil that has thickness, camber, and angle of attack. Figure 2.7.15 shows such an airfoil. The pressure coefficients on the upper and lower surfaces can be written as

$$C_{P_U} = \frac{2}{\sqrt{M_1^2 - 1}} \frac{dy_U}{dx}, \qquad C_{P_L} = \frac{2}{\sqrt{M_1^2 - 1}} \left(-\frac{dy_L}{dx} \right) \tag{2.7.39}$$

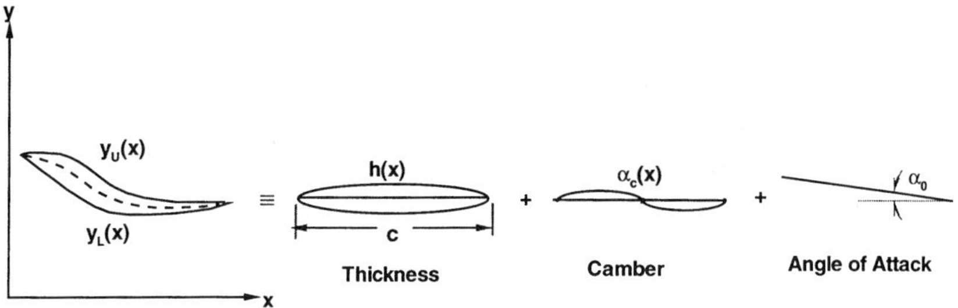

FIGURE 2.7.15 Arbitrary thin airfoil and its components.

For the thin airfoil, the profile may be resolved into three separate components as shown in Figure 2.7.15. The local slope of the airfoil can be obtained by superimposing the local slopes of the three components as

$$\begin{aligned} \frac{dy_U}{dx} &= -\left(\alpha_0 + \alpha_c(x)\right) + \frac{dh}{dx} = -\alpha(x) + \frac{dh}{dx} \\ \frac{dy_L}{dx} &= -\left(\alpha_0 + \alpha_c(x)\right) - \frac{dh}{dx} = -\alpha(x) - \frac{dh}{dx} \end{aligned} \tag{2.7.40}$$

where $\alpha = \alpha_0 + \alpha_c\ (x)$ is the local total angle of attack of the camber line. The lift and drag for the thin airfoil are given by

$$L = q_1 \int_0^c \left(C_{P_L} - C_{P_U} \right) dx$$

$$D = q_1 \int_0^c \left[C_{P_L} \left(-\frac{dy_L}{dx} \right) + C_{P_U} \left(\frac{dy_U}{dx} \right) \right] dx$$

Let us define an average value of $\alpha\ (x)$ as

$$\bar{\alpha} = \frac{1}{c} \int_0^c \alpha(x)\, dx$$

Using Equation (2.7.40) and the fact that $\bar{\alpha}_0 = \alpha$ and $\bar{\alpha}_c = 0$ by definition, the lift and drag coefficients for the thin airfoil can be written as

$$C_L = \frac{4\alpha_0}{\sqrt{M_1^2 - 1}}$$

$$C_D = \frac{4}{\sqrt{M_1^2 - 1}}\left[\overline{\left(\frac{dh}{dx}\right)^2} + \overline{\alpha_C^2(x)} + \alpha_0^2\right] \tag{2.7.41}$$

Equations (2.7.41) show that the lift coefficient depends only on the mean angle of attack whereas the drag coefficient is a linear combination of the drag due to thickness, drag due to camber, and drag due to lift (or mean angle of attack).

References

Anderson, J.D. 1982. *Modern Compressible Flow,* McGraw-Hill, New York.

Dennard, J.S. and Spencer, P.B. 1964. Ideal-Gas Tables for Oblique-Shock Flow Parameters in Air at Mach Numbers from 1.05 to 12.0. NASA TN D-2221.

Liepmann, H.W. and Roshko, A. 1966. *Elements of Gas Dynamics,* John Wiley & Sons, New York.

Further Information

As mentioned in the beginning, this section discussed only one- or two-dimensional steady, inviscid compressible flows under perfect gas assumption. Even this discussion was quite brief because of space limitations. For more details on the subject as well as for compressible unsteady viscous flows, readers are referred to Anderson (1982) and Liepmann and Roshko (1966).

2.8 Multiphase Flow

John C. Chen

Introduction

Classic study of fluid mechanics concentrates on the flow of a single homogeneous phase, e.g., water, air, steam. However, many industrially important processes involve simultaneous flow of multiple phases, e.g., gas bubbles in oil, wet steam, dispersed particles in gas or liquid. Examples include vapor–liquid flow in refrigeration systems, steam–water flows in boilers and condensers, vapor–liquid flows in distillation columns, and pneumatic transport of solid particulates. In spite of their importance, multiphase flows are often neglected in standard textbooks. Fundamental understanding and engineering design procedures for multiphase flows are not nearly so well developed as those for single-phase flows. An added complexity is the need to predict the relative concentrations of the different phases in the multiphase flows, a need that doesn't exist for single-phase flows.

Inadequate understanding not withstanding, a significant amount of data have been collected and combinations of theoretical models and empirical correlations are used in engineering calculations. This knowledge base is briefly summarized in this section and references are provided for additional information. While discussions are provided of solid–gas flows and solid–liquid flows, primary emphasis is placed on multiphase flow of gas–liquids since this is the most often encountered class of multiphase flows in industrial applications.

A multiphase flow occurs whenever two or more of the following phases occur simultaneously: gas/vapor, solids, single-liquid phase, multiple (immiscible) liquid phases. Every possible combination has been encountered in some industrial process, the most common being the simultaneous flow of vapor/gas and liquid (as encountered in boilers and condensers). All multiphase flow problems have features which are characteristically different from those found in single-phase problems. First, the relative concentration of different phases is usually a dependent parameter of great importance in multiphase flows, while it is a parameter of no consequence in single-phase flows. Second, the spatial distribution of the various phases in the flow channel strongly affects the flow behavior, again a parameter that is of no concern in single-phase flows. Finally, since the density of various phases can differ by orders of magnitude, the influence of gravitational body force on multiphase flows is of much greater importance than in the case of single-phase flows. In any given flow situation, the possibility exists for the various phases to assume different velocities, leading to the phenomena of slip between phases and consequent interfacial momentum transfer. Of course, the complexity of laminar/turbulent characteristics occurs in multiphase flows as in single-phase flows, with the added complexity of interactions between phases altering the laminar/turbulent flow structures. These complexities increase exponentially with the number of phases encountered in the multiphase problem. Fortunately, a large number of applications occur with just two phase flows, or can be treated as pseudo-two-phase flows.

Two types of analysis are used to deal with two-phase flows. The simpler approach utilizes homogeneous models which assume that the separate phases flow with the same identical local velocity at all points in the fluid. The second approach recognizes the possibility that the two phases can flow at different velocities throughout the fluid, thereby requiring separate conservation equations for mass and momentum for each phase. Brief descriptions of both classes of models are given below.

Fundamentals

Consider n phases in concurrent flow through a duct with cross-sectional area A_c. Fundamental quantities that characterize this flow are

$\dot{m}_i$ = mass flow rate of ith phase

u_i = velocity of ith phase

α_i = volume fraction of ith phase in channel

Basic relationships between these and related parameters are

$$G_i = \text{mass flux of } i\text{th phase} = \frac{\dot{m}_i}{A_c} \tag{2.8.1}$$

$$v_i = \text{superficial velocity of } i\text{th phase} = \frac{G_i}{\rho_i} \tag{2.8.2}$$

$$u_i = \text{actual velocity of } i\text{th phase} = \frac{v_i}{\alpha_i} \tag{2.8.3}$$

$$x_i = \text{flow quality of } i\text{th phase} = \frac{\dot{m}_i}{\sum_i^n m_i} = \frac{G_i}{\sum_{i=1}^n G_i} \tag{2.8.4}$$

$$\alpha_i = \text{volume fraction of } i\text{th phase} = \frac{\left(\frac{x_i}{\rho_i u_i}\right)}{\sum_{i=1}^n \left(\frac{x_i}{\rho_i u_i}\right)} \tag{2.8.5}$$

In most engineering calculations, the above parameters are defined as average quantities across the entire flow area, A_c. It should be noted, however, that details of the multiphase flow could involve local variations across the flow area. In the latter situation, G_i, v_i, and α_i are often defined on a local basis, varying with transverse position across the flow area.

Pressure drop along the flow channel is associated with gravitational body force, acceleration forces, and frictional shear at the channel wall. The total pressure gradient along the flow axis can be represented as

$$\frac{dP}{dz} = \left(\frac{dP}{dz}\right)_g + \left(\frac{dP}{dz}\right)_a + \left(\frac{dP}{dz}\right)_f \tag{2.8.6}$$

where

$$\left(\frac{dP}{dz}\right)_g = -g\cos\theta \cdot \sum_{i=1}^{n} \alpha_i \rho_i \tag{2.8.7}$$

θ = angle of channel from vertical

and

$$\left(\frac{dP}{dz}\right)_a = -\sum_{i=1}^{n} G_i \frac{du_i}{dz} \tag{2.8.8}$$

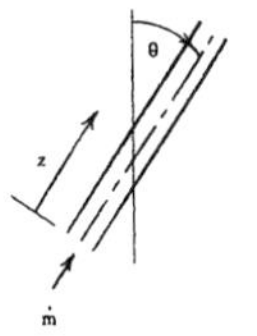

$$\left(\frac{dP}{dz}\right)_f = -\frac{\rho u^2}{2D} f \tag{2.8.9}$$

ρ = density of multiphase mixture

$$= \sum_{i=1}^{n} \rho_i \alpha_i \tag{2.8.10}$$

u = an average mixture velocity

$$= \frac{1}{\rho} \sum_{i=1}^{n} G_i \tag{2.8.11}$$

f = equivalent Darcy friction factor for the multiphase flow

In applications, the usual requirement is to determine pressure gradient (dP/dz) and the volume fractions (α_i). The latter quantities are of particular importance since the volume fraction of individual phases affects all three components of the pressure gradient, as indicated in Equations (2.8.7) to (2.8.11). Correlations of various types have been developed for prediction of the volume fractions, all but the simplest of which utilize empirical parameters and functions.

The simplest flow model is known as the homogeneous equilibrium model (HEM), wherein all phases are assumed to be in neutral equilibrium. One consequence of this assumption is that individual phase velocities are equal for all phases everywhere in the flow system:

$$u_i = u \quad \text{for all } i \tag{2.8.12}$$

This assumption permits direct calculation of the volume fractions from known mass qualities:

$$\alpha_i = \frac{x_i}{\rho_i \sum_{i=1}^{n} \left(\frac{x_i}{\rho_i}\right)} \tag{2.8.13}$$

The uniform velocity for all phases is the same as mixture velocity:

$$u = \frac{1}{\rho}\sum_{i=1}^{n} G_i \tag{2.8.14}$$

where

$$\frac{1}{\rho} = \sum_{i=1}^{n}\left(\frac{x_i}{\rho_i}\right) \tag{2.8.15}$$

This homogeneous model permits direct evaluation of all three components of axial pressure gradient, if flow qualities (x_i) are known:

$$\left(\frac{dP}{dz}\right)_g = -\frac{g\cos\theta}{\sum_{i=1}^{n}\left(\frac{x_i}{\rho_i}\right)} \tag{2.8.16}$$

$$\left(\frac{dP}{dz}\right)_a = -\left(\sum_{i=1}^{n} G_i\right)\cdot\frac{du}{dz} \tag{2.8.17}$$

$$\left(\frac{dP}{dz}\right)_f = -\frac{\rho u^2}{2D_f}\cdot f \tag{2.8.18}$$

where u and ρ are given by Equations (2.8.14) and (2.8.15).

Predicting the coefficient of friction (f to clear) remains a problem, even in the homogeneous model. For cases of fully turbulent flows, experience has shown that a value of 0.02 may be used as a first-order approximation for (f to clear). More-accurate estimates require empirical correlations, specific to particular classes of multiphase flows and subcategories of flow regimes.

The following parts of this section consider the more common situations of two-phase flows and describe improved design methodologies specific to individual situations.

Gas–Liquid Two-Phase Flow

The most common case of multiphase flow is two-phase flow of gas and liquid, as encountered in steam generators and refrigeration systems. A great deal has been learned about such flows, including delineation of flow patterns in different flow regimes, methods for estimating volume fractions (gas void fractions), and two-phase pressure drops.

Flow Regimes

A special feature of multiphase flows is their ability to assume different spatial distributions of the phases. These different flow patterns have been classified in flow regimes, which are themselves altered by the direction of flow relative to gravitational acceleration. Figures 2.8.1 and 2.8.2 (Delhaye, 1981) show the flow patterns commonly observed for co-current flow of gas and liquid in vertical and horizontal channels, respectively. For a constant liquid flow rate, the gas phase tends to be distributed as small bubbles at low gas flow rates. Increasing gas flow rate causes agglomeration of bubbles into larger slugs and plugs. Further increasing gas flow rate causes separation of the phases into annular patterns wherein liquid concentrates at the channel wall and gas flows in the central core for vertical ducts. For horizontal ducts, gravitational force tends to drain the liquid annulus toward the bottom of the channel, resulting

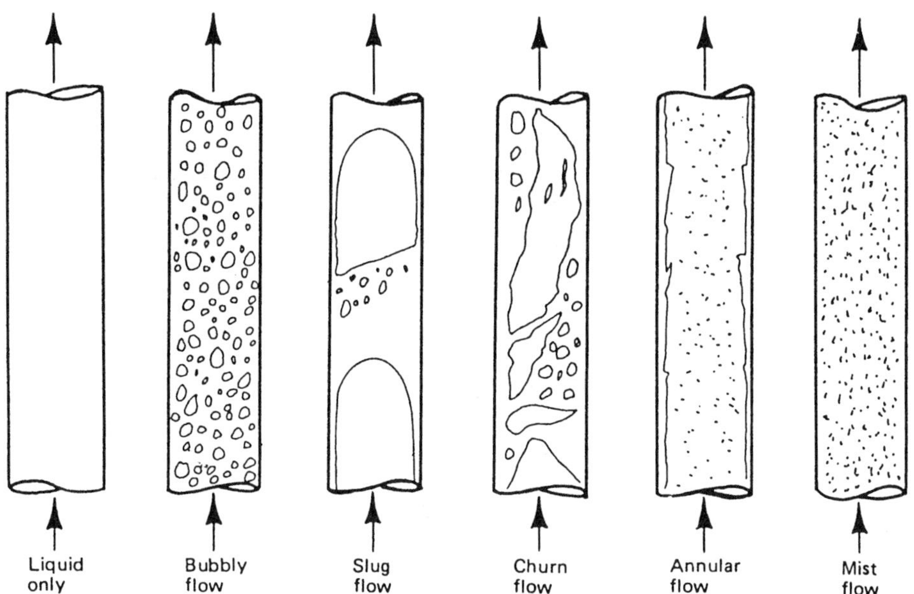

FIGURE 2.8.1 Flow patterns in gas–liquid vertical flow. (From Lahey, R.T., Jr. and Moody, F.I. 1977. *The Thermal Hydraulics of a Boiling Water Nuclear Reactor,* The American Nuclear Society, LaGrange, IL. With permission.)

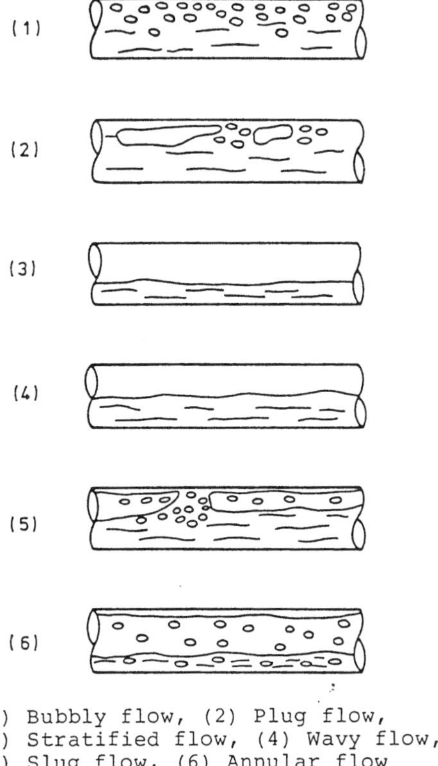

FIGURE 2.8.2 Flow patterns in gas–liquid horizontal flow.

in stratified and stratified wavy flows. This downward segregation of the liquid phase can be overcome by kinetic forces at high flow rates, causing stratified flows to revert to annular flows. At high gas flow rates, more of the liquid tends to be entrained as dispersed drops; in the limit one obtains completely dispersed mist flow.

Flow pattern maps are utilized to predict flow regimes for specific applications. The first generally successful flow map was that of Baker (1954) for horizontal flow, reproduced here in Figure 2.8.3. For vertical flows, the map of Hewitt and Roberts (1969), duplicated in Figure 2.8.4, provides a simple method for determining flow regimes. Parameters used for the axial coordinates of these flow maps are defined as follows:

$$\lambda = \left(\frac{\rho_g \rho_\ell}{\rho_a \rho_w}\right)^{1/2} \tag{2.8.19}$$

$$\psi = \left(\frac{\sigma_w}{\sigma}\right)\left[\left(\frac{\mu_\ell}{\mu_w}\right)\left(\frac{\rho_w}{\rho_\ell}\right)^2\right]^{1/3} \tag{2.8.20}$$

$$j = \text{volumetric flux, } \frac{G}{\rho} \tag{2.8.21}$$

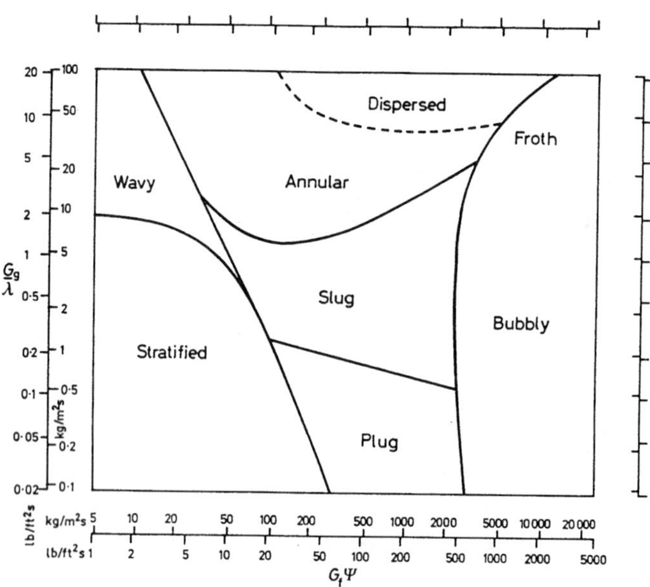

FIGURE 2.8.3 Flow pattern map for horizontal flow (Baker, 1954). (From Collier, J.G. 1972. *Convective Boiling and Condensation,* McGraw-Hill, London. With permission.)

Void Fractions

In applications of gas–liquid flows, the volume fraction of gas (α_g) is commonly called "void fraction" and is of particular interest. The simplest method to estimate void fraction is by the HEM. From Equation (2.8.13), the void fraction can be estimated as

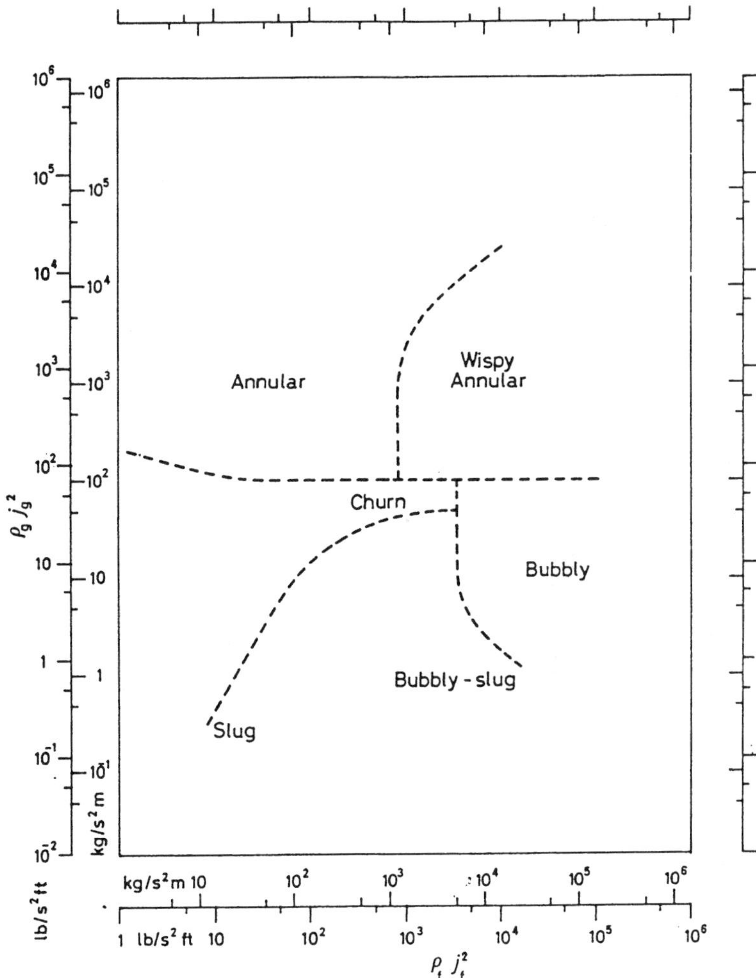

FIGURE 2.8.4 Flow pattern map for vertical flow (Hewitt and Roberts, 1969). (From Collier, J.G. 1972. *Convective Boiling and Condensation,* McGraw-Hill, London. With permission.)

$$\alpha_g = \frac{x_g}{x_g + \left(1 - x_g\right)\frac{\rho_g}{\rho_\ell}} \tag{2.8.22}$$

where α_g, x_g, ρ_g, ρ_ℓ are cross-sectional averaged quantities.

In most instances, the homogenous model tends to overestimate the void fraction. Improved estimates are obtained by using separated-phase models which account for the possibility of slip between gas and liquid velocities. A classic separated-phase model is that of Lockhart and Martinelli (1949). The top portion of Figure 2.8.5 reproduces the Lockhart–Martinelli correlation for void fraction (shown as α) as a function of the parameter X which is defined as

$$X = \left[\left(\frac{dP}{dz}\right)_{f\ell} \div \left(\frac{dP}{dz}\right)_{fg}\right]^{1/2} \tag{2.8.23}$$

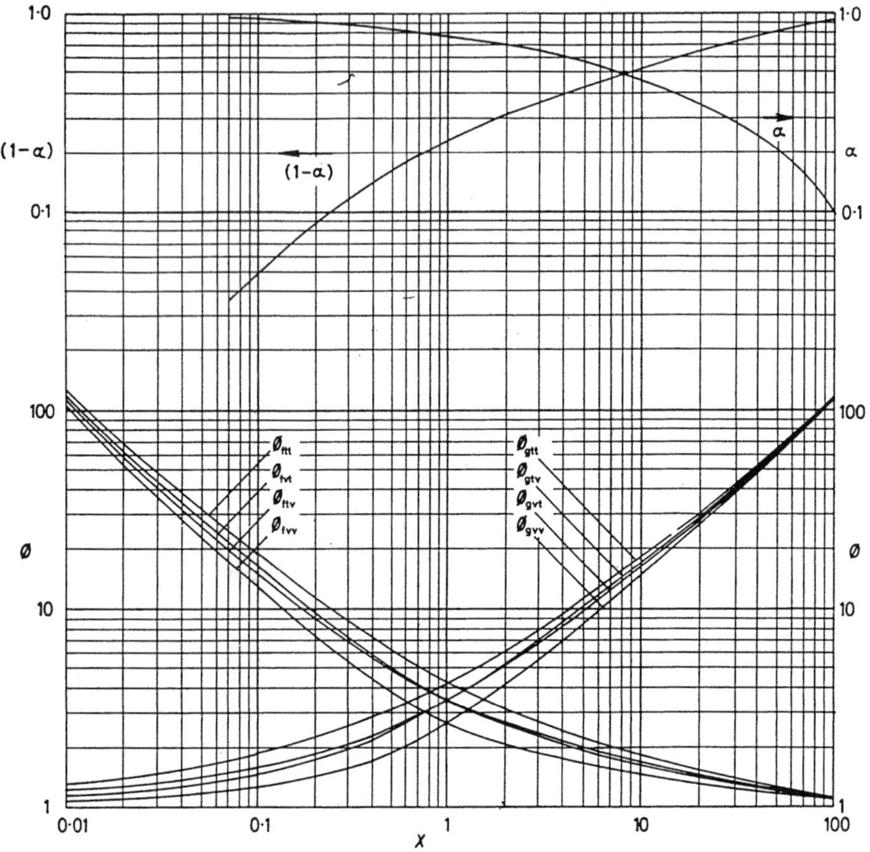

FIGURE 2.8.5 Correlations for void fraction and frictional pressure drop (Lockhart and Martinelli, 1949). (From Collier, J.G. 1972. *Convective Boiling and Condensation,* McGraw-Hill, London. With permission.)

where

$$\left(\frac{dP}{dz}\right)_{f\ell} = \text{frictional pressure gradient of liquid phase flowing alone in channel}$$

$$\left(\frac{dP}{dz}\right)_{fg} = \text{frictional pressure gradient of gas phase flowing alone in channel}$$

Often, flow rates are sufficiently high such that each phase if flowing alone in the channel would be turbulent. In this situation the parameter X can be shown to be

$$X_{tt} = \left(\frac{1-x_g}{x_g}\right)^{0.9} \left(\frac{\rho_g}{\rho_\ell}\right)^{0.5} \left(\frac{\mu_\ell}{\mu_g}\right)^{0.1} \tag{2.8.24}$$

Another type of separated-phase model is the drift-flux formulation of Wallis (1969). This approach focuses attention on relative slip between phases and results in slightly different expressions depending on the flow regime. For co-current upflow in two of the more common regimes, the drift-flux model gives the following relationships between void fraction and flow quality:

Bubbly flow or churn-turbulent flow:

$$\alpha_g = \frac{x_g}{\left(\frac{u_o \rho_g}{G}\right) + C_o \left[x_g + \left(1 - x_g\right)\frac{\rho_g}{\rho_\ell}\right]} \tag{2.8.25}$$

Dispersed drop (mist) flow:

$$x_g = \frac{1 - \left(1 - \alpha_g\right)\left(\frac{u_o \rho_\ell}{G}\alpha_g^2 + 1\right)}{1 - \left(1 - \alpha_g\right)\left(1 - \frac{\rho_\ell}{\rho_g}\right)} \tag{2.8.26}$$

where u_o= terminal rise velocity of bubble, in bubbly flow, or terminal fall velocity of drop in churn-turbulent flow

C_o = an empirical distribution coefficient $\simeq$ 1.2

Pressure Drop

Equations (2.8.16) through (2.8.18) permit calculation of two-phase pressure drop by the homogeneous model, if the friction coefficient (f) is known. One useful method for estimating (f) is to treat the entire two-phase flow as if it were all liquid, except flowing at the two-phase mixture velocity. By this approach the frictional component of the two-phase pressure drop becomes

$$\left(\frac{dP}{dz}\right)_f = \left[1 + x_g\left(\frac{\rho_\ell}{\rho_g} - 1\right)\right] \cdot \left(\frac{dP}{dz}\right)_{f\ell G} \tag{2.8.27}$$

where $(dP/dz)_{f\ell G}$ = frictional pressure gradient if entire flow (of total mass flux G) flowed as liquid in the channel.

The equivalent frictional pressure drop for the entire flow as liquid, $(dP/dz)_{f\ell g}$, can be calculated by standard procedures for single-phase flow. In using Equations (2.8.16) through (2.8.18), the void fraction would be calculated with the equivalent homogeneous expression Equation (2.8.13).

A more accurate method to calculate two-phase pressure drop is by the separated-phases model of Lockhart and Martinelli (1949). The bottom half of Figure 2.8.5 shows empirical curves for the Lockhart–Martinelli frictional multiplier, ϕ:

$$\phi_i = \left[\left(\frac{dP}{dz}\right)_f \div \left(\frac{dP}{dz}\right)_{fi}\right]^{1/2} \tag{2.8.28}$$

where (i) denotes either the fluid liquid phase (f) or gas phase (g). The single-phase frictional gradient is based on the ith phase flowing alone in the channel, in either viscous laminar (v) or turbulent (t) modes. The most common case is where each phase flowing alone would be turbulent, whence one could use Figure 2.8.5 to obtain

$$\left(\frac{dP}{dz}\right)_f = \text{frictional pressure gradient for two-phase flow}$$

$$= \phi_{gtt}^2 \cdot \left(\frac{dP}{dz}\right)_{fg} \tag{2.8.29}$$

where $(dP/dz)_{fg}$ is calculated for gas phase flowing alone and $X = X_{tt}$ as given by Equation (2.8.24).

The correlation of Lockhart–Martinelli has been found to be adequate for two-phase flows at low-to-moderate pressures, i.e., with reduced pressures less than 0.3. For applications at higher pressures, the revised models of Martinelli and Nelson (1948) and Thom (1964) are recommended.

Gas–Solid, Liquid–Solid Two-Phase Flows

Two-phase flows can occur with solid particles in gas or liquid. Such flows are found in handling of granular materials and heterogeneous reaction processing. Concurrent flow of solid particulates with a fluid phase can occur with various flow patterns, as summarized below.

Flow Regimes

Consider vertical upflow of a fluid (gas or liquid) with solid particles. Figure 2.8.6 illustrates the major flow regimes that have been identified for such two-phase flows. At low flow rates, the fluid phase percolates between stationary particles; this is termed flow through a fixed bed. At some higher velocity a point is reached when the particles are all suspended by the upward flowing fluid, the drag force between particles and fluid counterbalancing the gravitational force on the particles. This is the point of minimum fluidization, marking the transition from fixed to fluidized beds. Increase of fluid flow rate beyond minimum fluidization causes instabilities in the two-phase mixture, and macroscopic bubbles or channels of fluid are observed in the case of gaseous fluids. In the case of liquid fluids, the two-phase mixture tends to expand, often without discrete bubbles or channels. Further increase of fluid velocity causes transition to turbulent fluidization wherein discrete regions of separated phases (fluid slugs or channels and disperse suspensions of particles) can coexist. Depending on specific operating conditions (e.g., superficial fluid velocity, particle size, particle density, etc.), net transport of solid particles with the flowing fluid can occur at any velocity equal to or greater than that associated with slug flow and turbulent flow. Further increases in fluid velocity increase the net transport of solid particles. This can occur with large-scale clusters of solid particles (as exemplified by the fast fluidization regime) or with dilute dispersions of solid particles (as often utilized in pneumatic conveying). For engineering application of fluid–solid two-phase flows, the important thresholds between flow regimes are marked by the fluid velocity for minimum fluidization, terminal slip, and saltation threshold.

Minimum Fluidization

The transition from flow through packed beds to the fluidization regime is marked by the minimum fluidization velocity of the fluid. On a plot pressure drop vs. superficial fluid velocity, the point of minimum fluidization is marked by a transition from a linearly increasing pressure drop to a relatively constant pressure drop as shown in Figure 2.8.7 for typical data, for two-phase flow of gas with sand particles of 280 μm mean diameter (Chen, 1996). The threshold fluid velocity at minimum fluidization is traditionally derived from the Carman–Kozeny equation,

$$U_{mf} = \frac{\left(\rho_s - \rho_f\right)(\phi\, dp)^2 g}{150\mu_f} \cdot \frac{\alpha_{mf}^2}{\left(1 - \alpha_{mf}\right)} \tag{2.8.30}$$

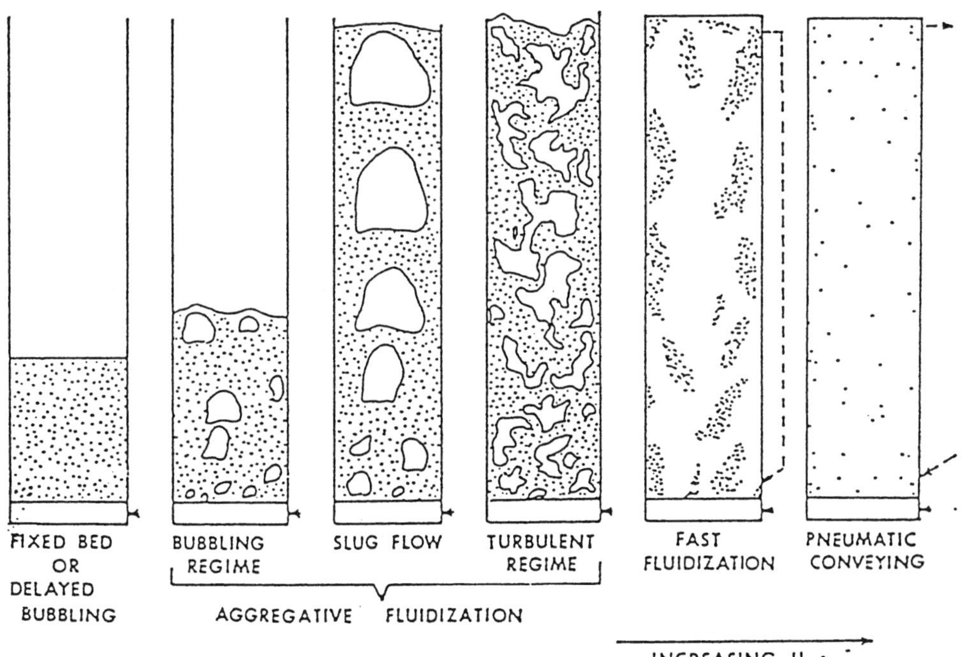

FIGURE 2.8.6 Flow patterns for vertical upflow of solid particles and gas or liquid. (From Chen, J.C. 1994. *Proc. Xth Int. Heat Transfer Conf.,* Brighton, U.K., 1:369–386. With permission.)

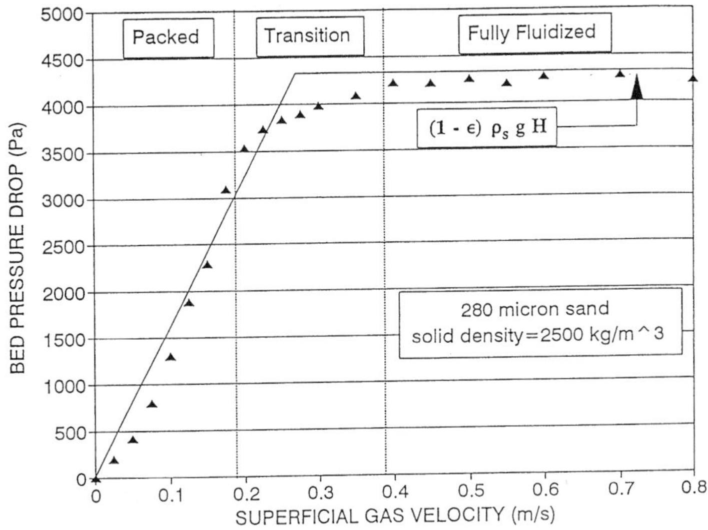

FIGURE 2.8.7 Transition at minimum fluidization. (From Chen, J.C. 1996. In *Annual Review of Heat Transfer,* Vol. VII, Begal House, Washington, D.C. With permission.)

where ϕ= sphericity of particles (unity for spherical particles)

α_{mf} = volumetric fraction of fluid at minimum fluidization

Small, light particles have minimum fluidization voidage (α_{mf}) of the order 0.6, while larger particles such as sand have values closer to 0.4.

An alternative correlation for estimating the point of minimum fluidization is that of Wen and Yu (1966):

$$\frac{U_{mf} d_p \rho_f}{\mu_f} = (33.7 + 0.041 Ga)^{0.5} - 33.7 \tag{2.8.31}$$

where $Ga = \rho_f d_p^3 (\rho_s - \rho_f) g / \mu_f^2$.

When the fluid velocity exceeds U_{mf}, the two-phase mixture exists in the fluidized state in which the pressure gradient is essentially balanced by the gravitational force on the two-phase mixture:

$$\frac{dP}{dz} = g\left[\alpha_s \rho_s + \alpha_f \rho_f\right] \tag{2.8.32}$$

This fluidized state exists until the fluid velocity reaches a significant fraction of the terminal slip velocity, beyond which significant entrainment and transport of the solid particles occur.

Terminal Slip Velocity

For an isolated single particle the maximum velocity relative to an upflowing fluid is the terminal slip velocity. At this condition, the interfacial drag of the fluid on the particle exactly balances the gravitational body force on the particle:

$$U_t = \left(U_f - U_s\right)_t = \left[\frac{4 d_p \left(\rho_s - \rho_f\right)}{3 \rho_f} \cdot \frac{1}{C_D}\right]^{1/2} \tag{2.8.33}$$

where C_D = coefficient of drag on the particle.

The coefficient of drag on the particle (C_D) depends on the particle Reynolds number:

$$\mathrm{Re}_p = \frac{\rho_f d_p \left(U_f - U_s\right)}{\mu_f} \tag{2.8.34}$$

The following expressions may be used to estimate C_D as appropriate:

$$\begin{aligned} C_D &= \frac{32}{\mathrm{Re}_p}, & \mathrm{Re}_p &\le 1 \\ C_D &= \frac{18.5}{\mathrm{Re}_p^{0.67}}, & 1 \le \mathrm{Re}_p &\le 10^3 \end{aligned} \tag{2.8.35}$$

Pneumatic Conveying

A desirable mode of pneumatic conveying is two-phase flow with solid particles dispersed in the concurrent flowing fluid. Such dispersed flows can be obtained if the fluid velocity is sufficiently high. For both horizontal and vertical flows, there are minimum fluid velocities below which saltation of the solid particles due to gravitational force occurs, leading to settling of the solid particles in horizontal channels and choking of the particles in vertical channels. Figures 2.8.8 and 2.8.9 for Zenz and Othmer (1960) show these different regimes of pneumatic conveying for horizontal and vertical transport, respectively. Figure 2.8.8 shows that for a given rate of solids flow (W) there is a minimum superficial fluid velocity below which solid particles tend to settle into a dense layer at the bottom of the horizontal channels. Above this saltation threshold, fully dispersed two-phase flow is obtained. In the case of vertical

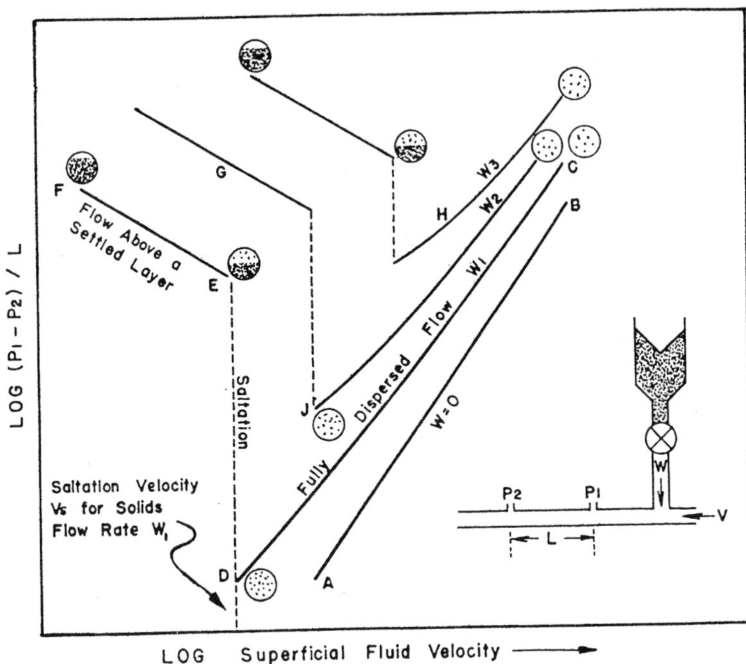

FIGURE 2.8.8 Flow characteristics in horizontal pneumatic conveying. (From Zeng, F.A. and Othmer, D.F. 1960. *Fluidization and Fluid-Particle Systems,* Reinhold, New York. With permission.)

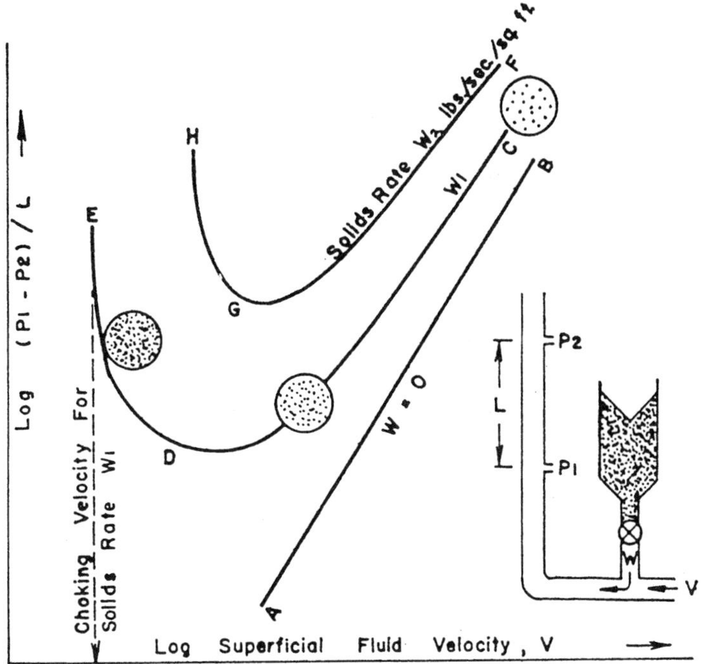

FIGURE 2.8.9 Flow characteristics in vertical pneumatic conveying. (From Zeng, F.A. and Othmer, D.F. 1960. *Fluidization and Fluid-Particle Systems,* Reinhold, New York. With permission.)

transport illustrated in Figure 2.8.9, there is a minimum fluid velocity below which solid particles tend to detrain from the two-phase suspension. This choking limit varies not only with particle properties but also with the actual rate of particle flow. Well-designed transport systems must operate with superficial fluid velocities greater than these limiting saltation and choking velocities.

Zenz and Othmer (1960) recommend the empirical correlations represented in Figure 2.8.10 estimating limiting superficial fluid velocities at incipient saltation or choking, for liquid or gas transport of uniformly sized particles. Note that these correlations are applicable for either horizontal or vertical concurrent flow. Figure 2.8.10 is duplicated from the original source and is based on parameters in engineering units, as noted in the figure. To operate successfully in dispersed pneumatic conveying of solid particles, the superficial fluid velocity must exceed that determined from the empirical correlations of Figure 2.8.10.

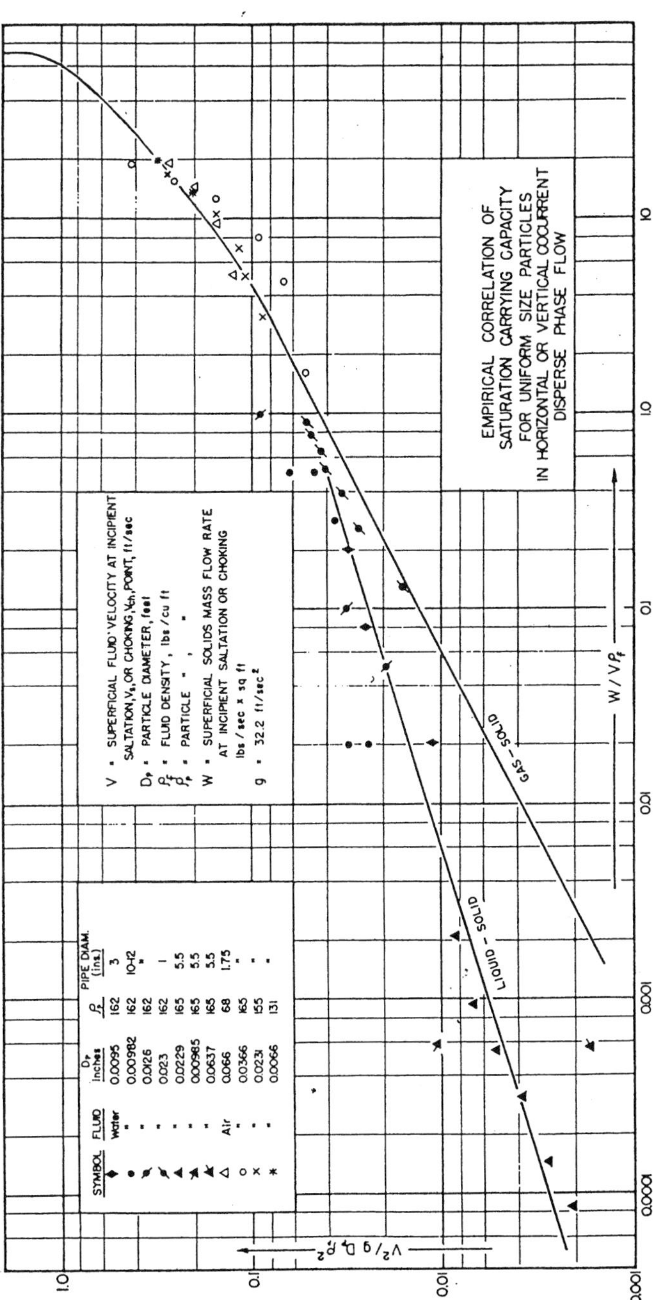

FIGURE 2.8.10 Correlations for limiting velocities in pneumatic conveying. (From Zeng, F.A. and Othmer, D.F. 1960. *Fluidization and Fluid-Particle Systems*, Reinhold, New York. With permission.)

References

Baker, O. 1954. Design of pipelines for simultaneous flow of oil and gas, *Oil Gas J.*

Chen, J.C. 1994. Two-phase flow with and without phase changes: suspension flows. Keynote lecture, *Proc. Xth Int. Heat Transfer Conf.,* Brighton, U.K., 1:369–386.

Chen, J.C. 1996. Heat transfer to immersed surfaces in bubbling fluidized beds, in *Annual Review of Heat Transfer*, Vol. VII, Bengel House, Washington, D.C.

Collier, J.G. 1972. *Convective Boiling and Condensation,* McGraw-Hill, London.

Delhaye, J.M. 1981. Two-phase flow patterns, in *Two-Phase Flow and Heat Transfer,* A.E. Bergles, J.G. Collier, J.M. Delhaye, G.F. Newitt, and F. Mayinger, Eds., Hemisphere Publishing, McGraw-Hill, New York.

Hewitt, G.F. and Roberts, D.N. 1969. Studies of Two-Phase Flow Patterns by Simultaneous X-Ray and Flash Photography, Report AERE-M 2159.

Lahey, R.T., Jr. and Moody, F.I. 1977. *The Thermal Hydraulics of a Boiling Water Nuclear Reactor,* The American Nuclear Society, La Grange, IL.

Lockhart, R.W. and Martinelli, R.C. 1949. Proposed correlation of data for isothermal two-phase two-component flow in pipes, *Chem. Eng. Progr.,* 45:39.

Martinelli, R.C. and Nelson, D.B. 1984. Prediction of pressure drop during forced-circulation boiling of water, *Trans. ASME,* 70:695–702.

Thom, J.R.S. 1964. Prediction of pressure drop during forced circulation boiling of water, *Int. J. Heat Mass Transfer,* 7:709–724.

Wallis, G.B. 1969. *One-Dimensional Two-Phase Flow,* McGraw-Hill, New York.

Wen, C.Y. and Yu, Y.H. 1966. A generalized method of predicting the minimum fluidization velocity, *AIChE J.,* 12:610–612.

Zenz, F.A. and Othmer, D.F. 1960. *Fluidization and Fluid-Particle Systems,* Reinhold, New York.

2.9 New-Newtonian Flows

Thomas F. Irvine, Jr. and Massimo Capobianchi

Introduction

An important class of fluids exists which differ from Newtonian fluids in that the relationship between the shear stress and the flow field is more complicated. Such fluids are called non-Newtonian or rheological fluids. Examples include various suspensions such as coal–water or coal–oil slurries, food products, inks, glues, soaps, polymer solutions, etc.

An interesting characteristic of rheological fluids is their large "apparent viscosities". This results in laminar flow situations in many applications, and consequently the engineering literature is concentrated on laminar rather than turbulent flows. It should also be mentioned that knowledge of non-Newtonian fluid mechanics and heat transfer is still in an early stage and many aspects of the field remain to be clarified.

In the following sections, we will discuss the definition and classification of non-Newtonian fluids, the special problems of thermophysical properties, and the prediction of pressure drops in both laminar and turbulent flow in ducts of various cross-sectional shapes for different classes of non-Newtonian fluids.

Classification of Non-Newtonian Fluids

It is useful to first define a Newtonian fluid since all other fluids are non-Newtonian. Newtonian fluids possess a property called viscosity and follow a law analogous to the Hookian relation between the stress applied to a solid and its strain. For a one-dimensional Newtonian fluid flow, the shear stress at a point is proportional to the rate of strain (called in the literature the *shear rate*) which is the velocity gradient at that point. The constant of proportionality is the dynamic viscosity, i.e.,

$$\tau_{y,x} = \mu \frac{du}{dy} = \mu\dot{\gamma} \tag{2.9.1}$$

where x refers to the direction of the shear stress y the direction of the velocity gradient, and $\dot{\gamma}$ is the shear rate. The important characteristic of a Newtonian fluid is that the dynamic viscosity is independent of the shear rate.

Equation (2.9.1) is called a constitutive equation, and if $\tau_{x,y}$ is plotted against $\dot{\gamma}$, the result is a linear relation whose slope is the dynamic viscosity. Such a graph is called a *flow curve* and is a convenient way to illustrate the viscous properties of various types of fluids.

Fluids which do not obey Equation (2.9.1) are called non-Newtonian. Their classifications are illustrated in Figure 2.9.1 where they are separated into various categories of purely viscous time-independent or time-dependent fluids and viscoelastic fluids. Viscoelastic fluids, which from their name possess both viscous and elastic properties (as well as memory), have received considerable attention because of their ability to reduce both drag and heat transfer in channel flows. They will be discussed in a later subsection.

Purely viscous time-independent fluids are those in which the shear stress in a function only of the shear rate but in a more complicated manner than that described in Equation (2.9.1). Figure 2.9.2 illustrates the characteristics of purely viscous time-independent fluids. In the figure, (a) and (b) are fluids where the shear stress depends only on the shear rate but in a nonlinear way. Fluid (a) is called pseudoplastic (or shear thinning), and fluid (b) is called dilatant (or shear thickening). Curve (c) is one which has an initial yield stress after which it acts as a Newtonian fluid, called Bingham plastic, and curve (d), called Hershel-Buckley, also has a yield stress after which it becomes pseudoplastic. Curve (e) depicts a Newtonian fluid.

Figure 2.9.3 shows flow curves for two common classes of purely viscous time-dependent non-Newtonian fluids. It is seen that such fluids have a hysteresis loop or memory whose shape depends

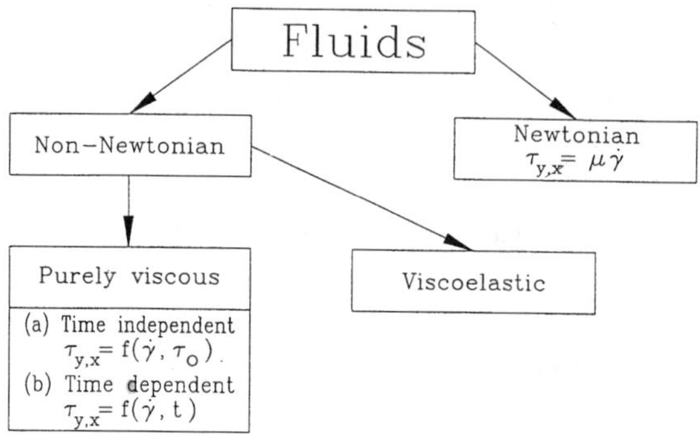

FIGURE 2.9.1 Classification of fluids.

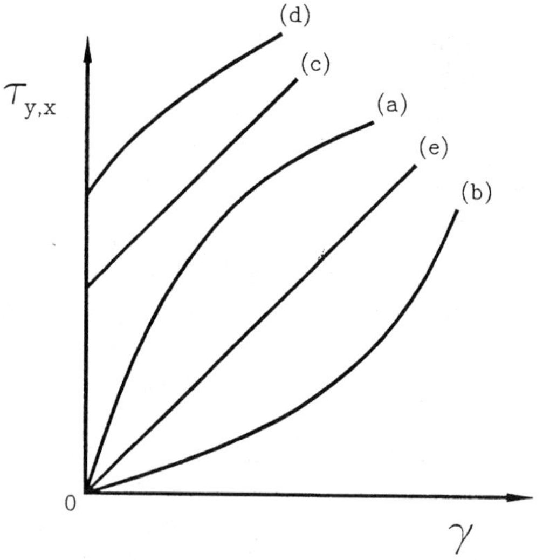

FIGURE 2.9.2 Flow curves of purely viscous, time-independent fluids: (a) pseudoplastic; (b) dilatant; (c) Bingham plastic; (d) Hershel–Buckley; (e) Newtonian.

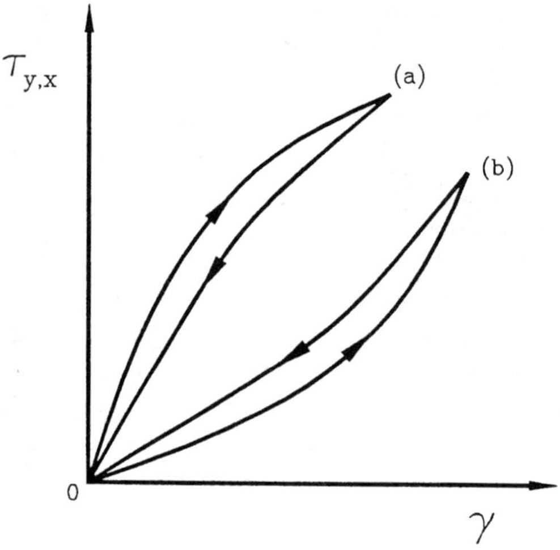

FIGURE 2.9.3 Flow curves for purely viscous, time-dependent fluids: (a) thixotropic; (b) rheopectic.

upon the time-dependent rate at which the shear stress is applied. Curve (a) illustrates a pseudoplastic time-dependent fluid and curve (b) a dilatant time-dependent fluid. They are called, respectively, thixotropic and rheopectic fluids and are complicated by the fact that their flow curves are difficult to characterize for any particular application.

Apparent Viscosity

Although non-Newtonian fluids do not have the property of viscosity, in the Newtonian fluid sense, it is convenient to define an apparent viscosity which is the ratio of the local shear stress to the shear rate at that point.

$$\mu_a = \frac{\tau}{\dot{\gamma}} \tag{2.9.2}$$

The apparent viscosity is not a true property for non-Newtonian fluids because its value depends upon the flow field, or shear rate. Nevertheless, it is a useful quantity and flow curves are often constructed with the apparent viscosity as the ordinate and shear rate as the abscissa. Such a flow curve will be illustrated in a later subsection.

Constitutive Equations

A constitutive equation is one that expresses the relation between the shear stress or apparent viscosity and the shear rate through the rheological properties of the fluid. For example, Equation (2.9.1) is the constitutive equation for a Newtonian fluid.

Many constitutive equations have been developed for non-Newtonian fluids with some of them having as many as five rheological properties. For engineering purposes, simpler equations are normally satisfactory and two of the most popular will be considered here.

Since many of the non-Newtonian fluids in engineering applications are pseudoplastic, such fluids will be used in the following to illustrate typical flow curves and constitutive equations. Figure 2.9.4 is a qualitative flow curve for a typical pseudoplastic fluid plotted with logarithmic coordinates. It is seen in the figure that at low shear rates, region (a), the fluid is Newtonian with a constant apparent viscosity of μ_o (called the *zero shear rate viscosity*). At higher shear rates, region (b), the apparent viscosity begins to decrease until it becomes a straight line, region (c). This region (c) is called the power law region and is an important region in fluid mechanics and heat transfer. At higher shear rates than the power law region, there is another transition region (d) until again the fluid becomes Newtonian in region (e). As discussed below, regions (a), (b), and (c) are where most of the engineering applications occur.

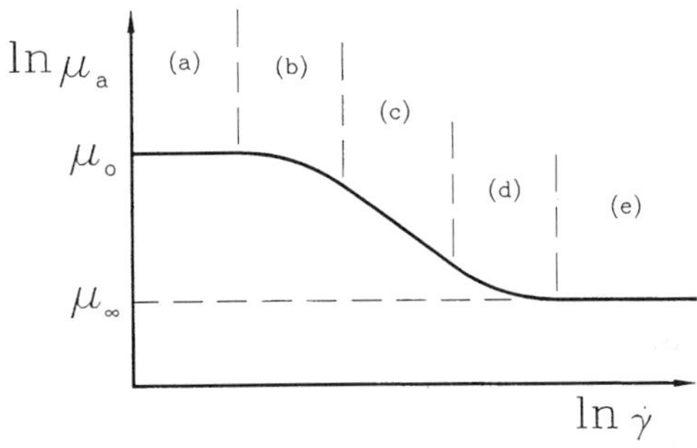

FIGURE 2.9.4 Illustrative flow curve for a pseudoplastic fluid (a) Newtonian region; (b) transition region I: (c) power law region; (d) transition region II; (e) high-shear-rate Newtonian region.

Power Law Constitutive Equation

Region (c) in Figure 2.9.4, which was defined above as the power law region, has a simple constitutive equation:

$$\tau = K\dot{\gamma}^n \tag{2.9.3}$$

or, from Equation (2.9.2):

$$\mu_a = K\dot{\gamma}^{n-1} \tag{2.9.4}$$

Here, K is called the fluid consistency and n the flow index. Note that if $n = 1$, the fluid becomes Newtonian and K becomes the dynamic viscosity. Because of its simplicity, the power law constitutive equation has been most often used in rheological studies, but at times it is inappropriate because it has several inherent flaws and anomalies. For example, if one considers the flow of a pseudoplastic fluid ($n < 1$) through a circular duct, because of symmetry at the center of the duct the shear rate (velocity gradient) becomes zero and thus the apparent viscosity from Equation (2.9.4) becomes infinite. This poses conceptual difficulties especially when performing numerical analyses on such systems. Another difficulty arises when the flow field under consideration is not operating in region (c) of Figure 2.9.4 but may have shear rates in region (a) and (b). In this case, the power law equation is not applicable and a more general constitutive equation is needed.

Modified Power Law Constitutive Equation

A generalization of the power law equation which extends the shear rate range to regions (a) and (b) is given by

$$\mu_a = \frac{\mu_o}{1 + \frac{\mu_o}{K}\dot{\gamma}^{1-n}} \tag{2.9.5}$$

Examination of Equation (2.9.5) reveals that at low shear rates, the second term in the denominator becomes small compared with unity and the apparent viscosity becomes a constant equal to μ_O. This represents the Newtonian region in Figure 2.9.4. On the other hand, as the second term in the denominator becomes large compared with unity, Equation (2.9.5) becomes Equation (2.9.4) and represents region (c), the power law region. When both denominator terms must be considered, Equation (2.9.5) represents region (b) in Figure 2.9.4.

An important advantage of the modified power law equation is that it retains the rheological properties K and n of the power law model plus the additional property μ_o. Thus, as will be shown later, in the flow and heat transfer equations, the same dimensionless groups as in the power law model will appear plus an additional dimensionless parameter which describes in which of the regions (a), (b), or (c) a particular system is operating. Also, solutions using the modified power law model will have Newtonian and power law solutions as asymptotes.

Equation (2.9.5) describes the flow curve for a pseudoplastic fluid ($n < 1$). For a dilatant fluid, ($n > 1$), an appropriate modified power law model is given by

$$\mu_a = \mu_o\left[1 + \frac{K}{\mu_o}\dot{\gamma}^{n-1}\right] \tag{2.9.6}$$

Many other constitutive equations have been proposed in the literature (Skelland, 1967; Cho and Hartnett, 1982; Irvine and Karni, 1987), but the ones discussed above are sufficient for a large number of engineering applications and agree well with the experimental determinations of rheological properties.

Rheological Property Measurements

For non-Newtonian fluids, specifying the appropriate rheological properties for a particular fluid is formidable because such fluids are usually not pure substances but various kinds of mixtures. This means that the properties are not available in handbooks or other reference materials but must be measured for each particular application. A discussion of the various instruments for measuring rheological properties is outside the scope of the present section, but a number of sources are available which describe different rheological property measurement techniques and instruments: Skelland (1967), Whorlow (1980), Irvine and Karni (1987), and Darby (1988). Figure 2.9.5 is an illustration of experimental flow curves measured with a falling needle viscometer and a square duct viscometer for polymer solutions of different concentrations. Also shown in the figure as solid lines is the modified power law equation used to represent the experimental data. It is seen that Equation (2.9.5) fits the experimental data within ±2%. Table 2.9.1 lists the rheological properties used in the modified power law equations in Figure 2.9.5. It must be emphasized that a proper knowledge of these properties is vital to the prediction of fluid mechanics and heat transfer phenomena in rheological fluids.

TABLE 2.9.1 Rheological Properties Used in the Modified Power Law Equations in Figure 2.9.5 for Three Polymer Solutions of CMC-7H4

CMC	K (N · sec^n/m^2)	n	μ_o (N · sec/m^2)n
5000 wppm	2.9040	0.3896	0.21488
2500 wppm	1.0261	0.4791	0.06454
1500 wppm	0.5745	0.5204	0.03673

Source: Park, S. et al., *Proc. Third World Conf. Heat Transfer, Fluid Mechanics, and Thermodynamics,* Vol. 1, Elsevier, New York, 1993, 900–908.

Fully Developed Laminar Pressure Drops for Time-Independent Non-Newtonian Fluids

Modified Power Law Fluids

This important subject will be considered by first discussing modified power law fluids. The reason is that such solutions include both friction factor–Reynolds number relations and a shear rate parameter. The latter allows the designer to determine the shear rate region in which his system is operating and thus the appropriate solution to be used, i.e., regions (a), (b), or (c) in Figure 2.9.4.

For laminar fully developed flow of a modified power law fluid in a circular duct, the product of the friction factor and a certain Reynolds number is a constant depending on the flow index, n, and the shear rate parameter, β.

$$f_D \cdot \mathrm{Re}_m = \mathrm{constant}(n, \beta) \tag{2.9.7}$$

where f_D is the Darcy friction factor and Re_m the modified power law Reynolds number, i.e.,

$$f_D = \frac{2\dfrac{\Delta p}{L} D_H}{\rho \bar{u}^2} \quad \text{(Darcy friction factor)}^*$$

* It should be noted that the Fanning friction factor is also used in the technical literature. The Fanning friction factor is $^1/_4$ of the Darcy friction factor, and will be characterized by the symbol f_F.

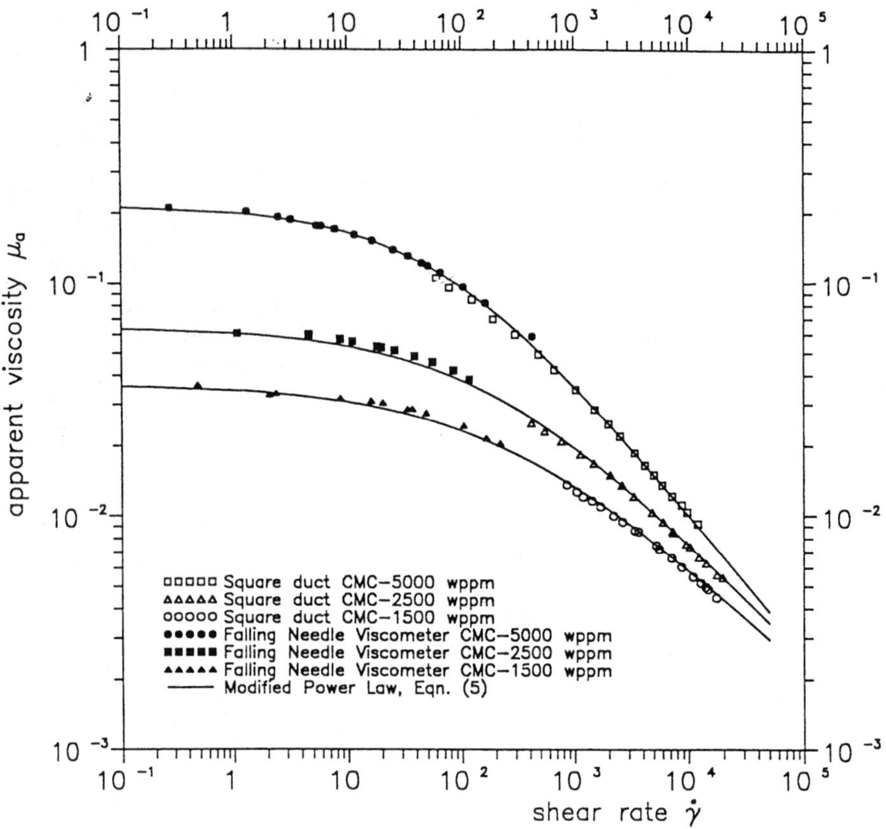

FIGURE 2.9.5 Experimental measurements of apparent viscosity vs. shear rate for polymer solutions (CMC-7H4) at different concentrations. (From Park, S. et al., in *Proc. Third World Conf. Heat Transfer, Fluid Mechanics, and Thermodynamics,* Vol. 1, Elsevier, New York, 1993, 900–908.

$$\mathrm{Re}_m = \frac{\rho \bar{u} D_H}{\mu^*}$$

$$\mu^* = \frac{\mu_o}{1+\beta}$$

$$\beta = \frac{\mu_o}{K}\left(\frac{\bar{u}}{D_H}\right)^{1-n}$$

where β is the shear rate parameter mentioned previously which can be calculated by the designer for a certain operating duct ($\bar{u}$ and d) and a certain pseudoplastic fluid (μ_o, K, n). The solution for a circular tube has been calculated by Brewster and Irvine (1987) and the results are shown in Figure 2.9.6 and in Table 2.9.2. Referring to Figure 2.9.6, we can see that when the $\log_{10} \beta$ is less than approximately –2, the duct is operating in region (a) of Figure 2.9.4 which is the Newtonian region and therefore classical Newtonian solutions can be used. Note that in the Newtonian region, Re_m reverts to the Newtonian Reynolds number given by

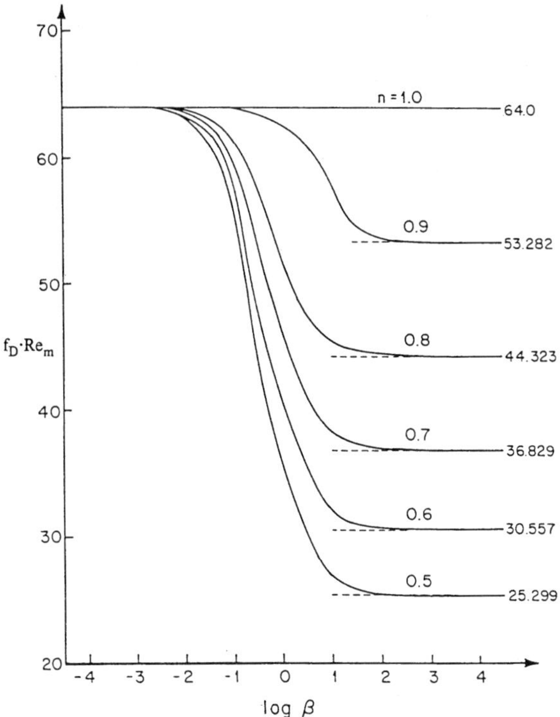

FIGURE 2.9.6 Product of friction factor and modified Reynolds number vs. $\log_{10} \beta$ for a circular duct. (From Brewster, R.A. and Irvine, T.F., Jr., *Wärme und Stoffübertragung,* 21, 83–86, 1987.

TABLE 2.9.2 Summary of Computed Values of $f_D \cdot Re_m$ for Various Values of n and β for a Circular Duct

β	$f_D \cdot Re_m$ for Flow Index: n =					
	1.0	**0.9**	**0.8**	**0.7**	**0.6**	**0.5**
10^{-5}	64.000	64.000	64.000	64.000	63.999	63.999
10^{-4}	64.000	63.999	63.997	63.995	63.993	63.990
10^{-3}	64.000	63.987	63.972	63.953	63.930	63.903
10^{-2}	64.000	63.873	63.720	63.537	63.318	63.055
10^{-1}	64.000	62.851	61.519	59.987	58.237	56.243
10^{0}	64.000	58.152	52.377	46.761	41.384	36.299
10^{1}	64.000	54.106	45.597	38.308	32.082	26.771
10^{2}	64.000	53.371	44.458	36.985	30.716	25.451
10^{3}	64.000	53.291	44.336	36.845	30.573	25.314
10^{4}	64.000	53.283	44.324	36.831	30.559	25.300
10^{5}	64.000	53.282	44.323	36.830	30.557	25.299
Exact solution	64.000	53.282	44.323	36.829	30.557	25.298

Source: Brewster, R.A. and Irvine, T.F., Jr., *Wärme und Stoffübertragung,* 21, 83–86, 1987. With permission.

$$\text{Re}_N = \frac{\rho \bar{u} D_H}{\mu_o} \tag{2.9.8}$$

When the value of $\log_{10} \beta$ is approximately in the range $-2 \le \log_{10} \beta \le 2$, the duct is operating in the transition region (b) of Figure 2.9.4 and the values of $f_D \cdot \text{Re}_m$ must be obtained from Figure 2.9.6 or from Table 2.9.2.

When $\log_{10} \beta$ is greater than approximately 2, the duct is operating in the power law region (c) of Figure 2.9.4 and power law friction factor Reynolds number relations can be used. They are also indicated in Figure 2.9.6 and Table 2.9.2. In this region, Re_m becomes the power law Reynolds number given by

$$\text{Re}_g = \frac{\rho \bar{u}^{2-n} D_H^n}{K} \tag{2.9.9}$$

For convenience, Brewster and Irvine (1987) have presented a correlation equation which agrees within 0.1% with the results tabulated in Table 2.9.2.

$$f_D \cdot \text{Re}_m = \frac{\dfrac{1+\beta}{\beta}}{\dfrac{1}{64} + \dfrac{\beta}{2^{3n+3}\left(\dfrac{3n+1}{4n}\right)^n}} \tag{2.9.10}$$

Thus, Equation (2.9.10) contains all of the information required to calculate the circular tube laminar fully developed pressure drop for a pseudoplastic fluid depending upon the shear rate region(s) under consideration, i.e., regions (a), (b), or (c) of Figure 2.9.4. Note that in scaling such non-Newtonian systems, both Re_m and β must be held constant. Modified power law solutions have been reported for two other duct shapes. Park et al. (1993) have presented the friction factor–Reynolds number relations for rectangular ducts and Capobianchi and Irvine (1992) for concentric annular ducts.

Power Law Fluids

Since the power law region of modified power law fluids ($\log_{10} \beta \ge 2$) is often encountered, the friction factor–Reynolds number relations will be discussed in detail in this subsection.

An analysis of power law fluids which is most useful has been presented by Kozicki et al. (1967). Although the method is approximate, its overall accuracy (±5%) is usually sufficient for many engineering calculations. His expression for the friction factor–Reynolds number product is given by

$$f_D \cdot \text{Re}^* = 2^{6n} \tag{2.9.11}$$

where

$$\text{Re}^* = \text{Kozicki Reynolds number,} \quad \text{Re}^* = \frac{\text{Re}_g}{\left[\dfrac{a+bn}{n}\right]^n 8^{n-1}} \tag{2.9.12}$$

and a and b are geometric constants which depend on the cross-sectional shape of the duct. For example, for a circular duct, $a = 0.25$ and $b = 0.75$. Values of a and b for other duct shapes are tabulated in Table 2.9.3. For additional duct shapes in both closed and open channel flows, Kozicki et al. (1967) may be consulted.

TABLE 2.9.3 Constants *a* and *b* for Various Duct Geometrics Used in the Method Due to Kozicki et al. (1967)

Geometry	α^*	*a*	*b*
Concentric annuli, $\alpha^* = \frac{d_i}{d_o}$	0.1	0.4455	0.9510
	0.2	0.4693	0.9739
	0.3	0.4817	0.9847
	0.4	0.4890	0.9911
	0.5	0.4935	0.9946
	0.6	0.4965	0.9972
	0.7	0.4983	0.9987
	0.8	0.4992	0.9994
	0.9	0.4997	1.0000
	1.0[a]	0.5000	1.0000
Rectangular, $\alpha^* = \frac{c}{h}$	0.0	0.5000	1.0000
	0.25	0.3212	0.8482
	0.50	0.2440	0.7276
	0.75	0.2178	0.6866
	1.00	0.2121	0.8766
Elliptical, $\alpha^* = \frac{c}{h}$	0.00	0.3084	0.9253
	0.10	0.3018	0.9053
	0.20	0.2907	0.8720
	0.30	0.2796	0.8389
	0.40	0.2702	0.8107
	0.50	0.2629	0.7886
	0.60	0.2575	0.7725
	0.70	0.2538	0.7614
	0.80	0.2515	0.7546
	0.90	0.2504	0.7510
	1.00[b]	0.2500	0.7500
	2ϕ (deg)		
Isosceles triangular (2ϕ)	10	0.1547	0.6278
	20	0.1693	0.6332
	40	0.1840	0.6422
	60	0.1875	0.6462
	80	0.1849	0.6438
	90	0.1830	0.6395
	N		
Regular polygon (*N* sides)	4	0.2121	0.6771
	5	0.2245	0.6966
	6	0.2316	0.7092
	8	0.2391	0.7241

[a] Parallel plates.

[b] Circle.

Source: Irvine, T.F., Jr. and Karni, J., in *Handbook of Single Phase Convective Heat Transfer,* John Wiley & Sons, New York, 1987, pp 20.1–20.57.

Fully Developed Turbulent Flow Pressure Drops

In a number of engineering design calculations for turbulent flow, the shear rate range falls in region (c) of Figure 2.9.4. Thus, power law relations are appropriate for such pressure drop calculations.

Hartnett and Kostic (1990) have investigated the various correlations which have appeared in the literature for circular tubes and have concluded that for a circular tube the relation proposed by Dodge and Metzner (1959) is the most reliable for pseudoplastic fluids. It is given by

$$\frac{1}{f_F^{1/2}} = \frac{4.0}{n^{0.75}} \cdot \log_{10}\left[\text{Re}'_g \left(f_F\right)^{1-(1/2n)}\right] - \frac{0.40}{n^{1.2}} \tag{2.9.13}$$

where f_F is the Fanning friction factor and

$$\text{Re}'_g = \text{Re}_g \left[\frac{8^{1-n}}{\left[\frac{3n+1}{4n}\right]^n}\right] \tag{2.9.14}$$

Figure 2.9.7 is a graphical representation of Equation (2.9.13) which indicates the Dodge and Metzner experimental regions by solid lines, and by dashed lines where the data are extrapolated outside of their experiments.

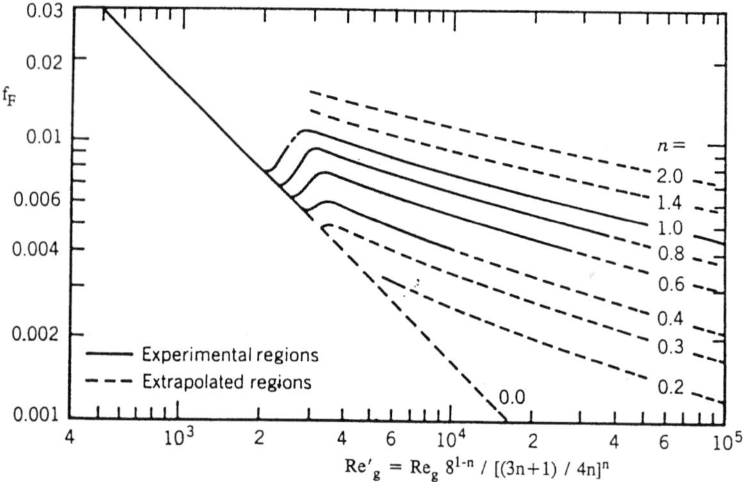

FIGURE 2.9.7 Dodge and Metzner relation between Fanning friction factor and Re'_g. (From Dodge, D.W. and Metzner, A.B., *AIChE J.*, 5, 189–204, 1959.)

For noncircular ducts in turbulent fully developed flow, only a limited amount of experimental data are available. Kostic and Hartnett (1984) suggest the correlation:

$$\frac{1}{f_F^{1/2}} = \frac{4}{n^{0.75}} \cdot \log_{10}\left[\text{Re}^* \left(f_F\right)^{1-(1/2n)}\right] - \frac{0.40}{n^{0.5}} \tag{2.9.15}$$

where f_F is again the Fanning friction factor and Re^* is the Kozicki Reynolds number:

$$\text{Re}^* = \frac{\text{Re}_g}{\left[\frac{a+bn}{n}\right]^n 8^{n-1}} \tag{2.9.16}$$

and a and b are geometric constants given in Table 2.9.3.

Viscoelastic Fluids

Fully Developed Turbulent Flow Pressure Drops

Viscoelastic fluids are of interest in engineering applications because of reductions of pressure drop and heat transfer which occur in turbulent channel flows. Such fluids can be prepared by dissolving small amounts of high-molecular-weight polymers, e.g., polyacrylamide, polyethylene oxide (Polyox), etc., in water. Concentrations as low as 5 parts per million by weight (wppm) result in significant pressure drop reductions. Figure 2.9.8 from Cho and Hartnett (1982) illustrates the reduction in friction factors for Polyox solutions in a small-diameter capillary tube. It is seen that at zero polymer concentration the data agree with the Blasius equation for Newtonian turbulent flow. With the addition of only 7 wppm of Polyox, there is a significant pressure drop reduction and for concentrations of 70 wppm and greater all the data fall on the Virk line which is the maximum drag-reduction asymptote. The correlations for the Blasius and Virk lines as reported by Cho and Hartnett (1982) are

$$f_F = \frac{0.079}{\mathrm{Re}^{1/4}} \quad \text{(Blasius)} \tag{2.9.17}$$

$$f_F = 0.20\,\mathrm{Re}_a^{-0.48} \quad \text{(Virk)} \tag{2.9.18}$$

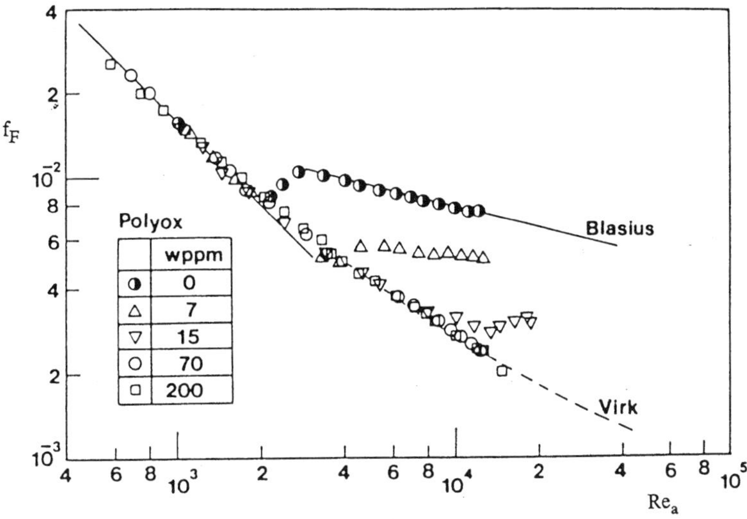

FIGURE 2.9.8 Reduction in friction factors for polyethylene oxide (Polyox) solutions in a small-diameter capillary tube. (From Cho, Y.I. and Harnett, J.P., *Adv. Heat Transfer,* 15, 59–141, 1982. With permission.)

At the present time, no generally accepted method exists to predict the drag reduction between the Blasius and Virk lines. Kwack and Hartnett (1983) have proposed that the amount of drag reduction between those two correlations is a function of the Weissenberg number, defined as

$$w_s = \frac{\lambda \bar{u}}{D_H} \tag{2.9.19}$$

where λ = characteristic time of the viscoelastic fluid. They present correlations which allow the friction factor to be estimated at several Reynolds numbers between the Blasius and Virk lines.

Fully Developed Laminar Flow Pressure Drops

The above discussion on viscoelastic fluids has only considered fully developed turbulent flows. Laminar fully developed flows can be considered as nonviscoelastic but purely viscous non-Newtonian. Therefore, the method of Kozicki et al. (1967) may be applied to such situations once the appropriate rheological properties have been determined.

Nomenclature

a = duct shape geometric constant
b = duct shape geometric constant
c = duct width (see Table 2.9.3) (m)
d_i = concentric annuli inner diameter (see Table 2.9.3) (m)
d_o = concentric annuli outer diameter (see Table 2.9.3) (m)
f_D = Darcy friction factor
f_F = Fanning friction factor
h = duct height (see Table 2.9.3) (m)
K = fluid consistency (Ns^n/m^2)
n = flow index
N = number of sides in polygon (see Table 2.9.3)
Re_g = generalized Reynolds number,

$$\mathrm{Re}_g = \frac{\rho \bar{u}^{2-n} D_H^n}{K}$$

Re_m = modified power law Reynolds number,

$$\mathrm{Re}_m = \frac{\rho \bar{u} D_H}{\mu^*}$$

Re_N = modified power law Reynolds number Newtonian asymptote,

$$\mathrm{Re}_N = \frac{\rho \bar{u} D_H}{\mu_o}$$

Re_a = apparent Reynolds number

$$\mathrm{Re}_a = \frac{\mathrm{Re}_g}{\left(\frac{3n+1}{4n}\right)^{n-1} 8^{n-1}}$$

Re^* = Kozicki Reynolds number

$$\mathrm{Re}^* = \frac{\rho \bar{u}^{2-n} D_H^n}{K\left[\frac{a+bn}{n}\right]^n 8^{n-1}}$$

Re'_g = Metzner Reynolds number

$$Re'_g = Re_g\left[\frac{8^{1-n}}{\left[\frac{3n+1}{4n}\right]^n}\right]$$

$\bar{u}$ = average streamwise velocity (m/sec)
t = time (sec)
w_s = Weissenberg number
x = direction of shear stress (m)
y = direction of velocity gradient (m)

Greek

α^* = duct aspect ratio in Table 2.9.3
β = shear rate parameter

$$\beta = \frac{\mu_o}{K}\left(\frac{\bar{u}}{D_H}\right)^{1-n}$$

$\dot{\gamma}$ = shear rate (L/sec)
ΔP = presure drop (N/m²)
λ = characteristic time of viscoelastic fluid (sec)
μ_a = apparent viscosity (N · sec/m²)
μ_o = zero shear rate viscosity (N · sec/m²)
μ_∞ = high shear rate viscosity (N · sec/m²)
μ^* = reference viscosity

$$\mu^* = \frac{\mu_o}{1+\beta} \quad \left(\mathrm{N}\cdot\mathrm{sec/m^2}\right)$$

τ_o = yield stress (N/m²)
$\tau_{y,x}$ = shear stress (N/m²)
ϕ = half apex angle (see Table 2.9.3) (°)

References

Brewster, A.A. and Irvine, T.F. Jr. 1987. Similtude considerations in laminar flow of power law fluids in circular ducts, *Wärme und Stoffübertagung,* 21:83–86.

Capobianchi, M. and Irvine, T.F. Jr. 1992. Predictions of pressure drop and heat transfer in concentric annular ducts with modified power law fluids, *Wärme und Stoffübertagung,* 27:209–215.

Cho, Y.I. and Hartnett, J.P. 1982. Non-Newtonian fluids in circular pipe flow, in *Adv. Heat Transfer,* 15:59–141.

Darby, R. 1988. Laminar and turbulent pipe flows of non-Newtonian fluids, in *Encyclopedia of Fluid Mechanics,* Vol. 7, Gulf Publishing, Houston, 7:20–53.

Dodge, D.W. and Metzner, A.B. 1959. Turbulent flow of non-Newtonian systems, *AIChE J.,* 5:189–204.

Harnett, J.P. and Kostic, M. 1990. Turbulent Friction Factor Correlations for Power Law Fluids in Circular and Non-Circular Channels, *Int. Comm. Heat and Mass Transfer,* 17:59–65.

Irvine, T.F. Jr. and Karni, J. 1987. Non-Newtonian fluid flow and heat transfer, in *Handbook of Single Phase Convective Heat Transfer,* pp. 20-1–20-57, John Wiley & Sons, New York.

Kostic, M. and Hartnett, J.P. 1984. Predicting turbulent friction factors of non-Newtonian fluids in noncircular ducts, *Int. Comm. Heat and Mass Transfer,* 11:345–352.

Kozicki, W., Chou, C.H., and Tiu, C. 1967. Non-Newtonian flow in ducts of arbitrary cross-sectional shape, *Can. J. Chem. Eng.,* 45:127–134.

Kwack, E.Y. and Hartnett, J.P. 1983. Empirical correlations of turbulent friction factors and heat transfer coefficients for viscoelastic fluids, *Int. Comm. Heat and Mass Transfer,* 10:451–461.

Park, S., Irvine, Jr., T.F., and Capobianchi, M. 1993. Experimental and numerical study of friction factor for a modified power law fluid in a rectangular duct, *Proc. Third World Conf. Heat Transfer, Fluid Mechanics, and Thermodynamics,* Vol. 1, Elsevier, New York, 1:900–908.

Skelland, A.H.P. 1967. *Non-Newtonian Flow and Heat Transfer,* John Wiley & Sons, New York.

Whorlow, R.W. 1980. *Rheological Techniques,* Halsted Press, New York.

Further Information

It is not possible to include all of the interesting non-Newtonian topics in a section of this scope. Other items which may be of interest and importance are listed below along with appropriate references: hydrodynamic and thermal entrance lengths, Cho and Hartnett (1982); non-Newtonian flow over external surfaces, Irvine and Karni (1987); chemical, solute, and degradation effects in viscoelastic fluids, Cho and Harnett (1982); general references, Skelland (1967), Whorlow (1980), and Darby (1988).

3

Heat and Mass Transfer

Frank Kreith, Editor
Engineering Consultant
University of Colorado

Robert F. Boehm
University of Nevada-Las Vegas

George D. Raithby
University of Waterloo

K. G. Terry Hollands
University of Waterloo

N. V. Suryanarayana
Michigan Technological University

Thomas F. Irvine, Jr.
State University of New York, Stony Brook

Massimo Capobianchi
Gonzaga University

Michael F. Modest
Pennsylvania State University

Van P. Carey
University of California at Berkeley

John C. Chen
Lehigh University

Noam Lior
University of Pennsylvania

Anthony F. Mills
University of California at Los Angeles

3.1 Conduction Heat Transfer

Robert F. Boehm

Introduction

Conduction heat transfer phenomena are found virtually throughout the physical world and the industrial domain. The analytical description of this heat transfer mode is one of the best understood processes. Some of the bases of the understanding of conduction date back to early history. It was recognized that by invoking certain relatively minor simplifications, mathematical solutions resulted directly. Some of these were easily formulated. What transpired over the years was a vigorous development of applications to a broad range of processes. Perhaps no single work better summarizes the wealth of these studies than does the book by Carslaw and Jaeger (1959). It provides solutions to a broad range of problems,

0-8493-9581-X/00/$0.00+$.50

ranging from topics related to the cooling of the earth to the current-carrying capacities of wires. The general analyses given there have been applied to a range of modern-day problems from laser heating to temperature control systems.

Today, conduction heat transfer is still an active area of research and application. A great deal of interest has been shown in recent years in topics like contact resistance — where a temperature difference develops between two solids that do not have perfect contact with each other. Additional issues of current interest include non-Fourier conduction, where the processes occur so fast that the equations described below do not apply. Also, the problems related to transport in minaturized systems are garnering a great deal of interest. Increased interest has also been focused on ways of handling composite materials, where the ability to conduct heat is very directional.

Much of the work in conduction analysis is now accomplished by the use of sophisticated computer codes. These tools have given the heat-transfer analyst the capability to solve problems in inhomogenous media with very complicated geometries and with very involved boundary conditions. It is still important to understand analytical methods for determining the performance of conducting systems. At the minimum these can be used as calibrations for numerical codes.

Fourier's Law

The basis of conduction heat transfer is **Fourier's Law**. This law involves the idea that the heat flux is proportional to the temperature gradient in any direction n. **Thermal conductivity**, k, a property of materials that is temperature dependent, is the constant of proportionality.

$$q_k = -kA\frac{\partial T}{\partial n} \tag{3.1.1}$$

In many systems the area A is a function of the distance in the direction n. One important extension is that this can be combined with the First Law of Thermodynamics to yield the **Heat Conduction Equation**. For constant thermal conductivity, this is given as:

$$\nabla^2 T + \frac{\dot{q}_G}{k} = \frac{1}{\alpha}\frac{\partial T}{\partial t} \tag{3.1.2}$$

In this equation, α is the thermal diffusivity, and $\dot{q}_G$ is the internal heat generation per unit volume. Some problems, typically steady-state, one-dimensional formulations where only the heat flux is desired, can be easily solved using Equation (3.1.1). Most conduction analyses are performed with Equation (3.1.2). In the latter — a more general approach — the temperature distribution is found from this equation and the appropriate boundary conditions. Then the heat flux, if desired, is found at any location using Equation (3.1.1). Normally, it is the temperature distribution that is of most importance. For example, it may be desirable to determine through analysis if a material will reach some critical temperature, like its melting point. Less frequently, the heat flux is the desired objective.

While there are times when one needs only to understand the temperature response of a structure, often the engineer is faced with a need to increase or decrease heat transfer to some specific level. Examination of the thermal conductivity of materials gives some insight to the range of possibilities that exist through simple conduction.

Of the more common engineering materials, pure copper exhibits one of the highest abilities to conduct heat, with a thermal conductivity approaching 400 W/m^2 K. Aluminum, also considered to be a good conductor, has a thermal conductivity a little over half that of copper. In order to increase the heat transfer above the values possible through simple conduction, more sophisticated designs are necessary that incorporate a variety of other heat transfer modes like convection and phase change.

Decreasing the rate of heat transfer is accomplished with the use of insulation. A discussion of this follows.

Insulations

Insulating materials are used to decrease heat flow and thus decrease surface temperatures. These materials are found in a variety of forms, typically *loose fill*, *batt*, and *rigid*. Even a gas, like air, can be a good insulator if it can be kept from moving when it is heated or cooled. A vacuum is an excellent insulator. Usually, though, the engineering approach to insulation is the addition of a low-conducting material to the surface. While there are many chemical forms, costs, and maximum operating temperatures of common types of insulators, it seems that when higher operating temperatures are required, the thermal conductivity and cost of the insulation will often also be higher.

Loose-fill insulation includes such materials as milled alumina-silica (maximum operating temperature of 1260°C and thermal conductivities in the range of 0.1–0.2 W/m² K) and perlite (maximum operating temperature of 980°C and thermal conductivities in the range of 0.05–1.5 W/m² K). Batt type insulation includes one of the more common types — glass fiber. This type of insulation comes in a variety of densities which, in turn, have a profound effect on the thermal conductivity. Thermal conductivities for glass fiber insulation can range from about 0.03–0.06 W/m² K. Rigid insulations show a very wide range of types and performance characteristics. For example, a rigid insulation in foam form, polyurethane, is light in weight, shows a very low thermal conductivity (about 0.02 W/m² K), but has a maximum operating temperature only up to about 120°C. Rigid insulations in refractory form show quite different characteristics. For example, alumina brick is quite dense, has a thermal conductivity of about 2 W/m² K, but can remain operational to temperatures around 1760°C. Many insulations are characterized in the book edited by Guyer (1989).

Often, commercial insulation systems designed for high-temperature operation use a layered approach. Temperature tolerance may be critical. Perhaps a refractory material is applied in the highest-temperature region, an intermediate-temperature foam insulation is used in the middle section, and a high-performance, lower-operating-temperature insulation is used on the outer side near ambient conditions.

Analyses can be performed showing the effects of temperature variations of thermal conductivity. However, the most frequent approach is to assume that the thermal conductivity is constant at some temperature between the two extremes experienced by the insulation.

Plane Wall at Steady State

Consider steady-state heat transfer in a plane wall of thickness L, but of very large extent in both other directions. The wall has temperature T_1 on one side and T_2 on the other. If the thermal conductivity is considered to be constant, then Equation (3.1.1) can be integrated directly to give the following result:

$$q_k = \frac{kA}{L}\left(T_1 - T_2\right) \tag{3.1.3}$$

This can be used to determine the steady-state heat transfer through slabs.

An electrical circuit analog is widely used in conduction analyses. This is accomplished by considering the temperature difference to be analogous to a voltage difference, the heat flux to be like current flow, and the remainder of Equation (3.1.3) to be like a thermal resistance. The latter is seen to be

$$R_k = \frac{L}{kA} \tag{3.1.4}$$

Heat transfer through walls made of layers of different types of materials can be easily found by summing the resistances in series or parallel form, as appropriate.

In the design of systems, seldom is a surface temperature specified or known. More often, the surface is in contact with a bulk fluid, whose temperature is known at some distance from the surface. Convection from the surface is then represented by Newton's Law of Cooling:

$$q = \bar{h}_c A\left(T_s - T_\infty\right) \tag{3.1.5}$$

This equation can also be represented as a temperature difference divided by a thermal resistance, which is $1/\bar{h}_c A$. It can be shown that a very low surface resistance, as might be represented by phase-change phenomena, effectively imposes the fluid temperature directly on the surface. Hence, usually a *known* surface temperature results from a fluid temperature being imposed directly on the surface through a very high heat-transfer coefficient. For this reason, in the later results given here, particularly those for transient systems, a convective boundary will be assumed. For steady-state results this is less important because of the ability to add resistances via the circuit analogy.

Long, Cylindrical Systems at Steady State

For long (L), annular systems at steady-state conditions with constant thermal conductivities, the following two equations are the appropriate counterparts to Equations (3.1.3) and (3.1.4). The heat transfer can be expressed as

$$q_k = \frac{2\pi Lk}{\ln\left[r_2/r_1\right]}\left(T_1 - T_2\right) \tag{3.1.6}$$

Here r_1 and r_2 represent the radii of the annular sections. A thermal resistance for this case is as shown below.

$$R_k = \frac{\ln\left[r_2/r_1\right]}{2\pi Lk} \tag{3.1.7}$$

Overall Heat Transfer Coefficient

The **overall heat transfer coefficient** concept is valuable in several aspects of heat transfer. It involves a modified form of Newton's law of cooling, as noted above, and it is written as:

$$Q = \bar{U}A\Delta T \tag{3.1.8}$$

In this formulation, $\bar{U}$ is the overall heat transfer coefficient based upon the area A. Because the area for heat transfer in a problem can vary (as with a cylindrical geometry), it is important to note that the $\bar{U}$ is dependent upon which area is selected. The overall heat transfer coefficient is usually found from a combination of thermal resistances. Hence, for a common-series combination-circuit analog, the $\bar{U}A$ product is taken as the sum of the resistances.

$$\bar{U}A = \frac{1}{\sum_{i=1}^{n} R_i} = \frac{1}{R_{\text{total}}} \tag{3.1.9}$$

For an example of the use of this concept, see Figure 3.1.1.

For steady-state conditions, the product $\bar{U}A$ remains constant for a given heat transfer and overall temperature difference. This can be written as

$$\bar{U}_1 A_1 = \bar{U}_2 A_2 = \bar{U}_3 A_3 = \bar{U}A \tag{3.1.10}$$

If the inside area, A_1, is chosen as the basis, the overall heat transfer coefficient can then be expressed as

$$\bar{U}_1 = \frac{1}{\dfrac{1}{\bar{h}_{c,i}} + \dfrac{r_1 \ln\left(r_2/r_1\right)}{k_{\text{pipe}}} + \dfrac{r_1 \ln\left(r_3/r_2\right)}{k_{\text{ins}}} + \dfrac{r_1}{r_3 \bar{h}_{c,o}}} \tag{3.1.11}$$

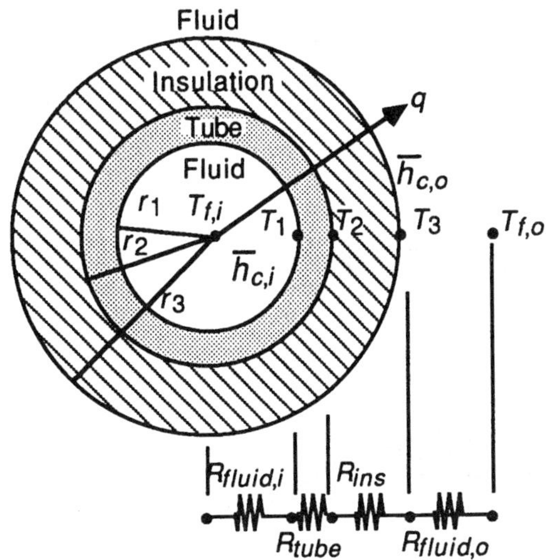

FIGURE 3.1.1 An insulated tube with convective environments on both sides.

Critical Thickness of Insulation

Sometimes, adding insulation can cause an increase in heat transfer. This circumstance should be noted so that one can apply it when desired and design around it when an insulating effect is needed. Consider the circumstance shown in Figure 3.1.1. Assume that the temperature on the outside of the tube (inside of the insulation) is known. This could be determined if the inner heat transfer coefficient is very large and the thermal conductivity of the tube is large. In this case, the inner fluid temperature will be almost the same temperature as the inner surface of the insulation. Alternatively, this could be applied to a coating (say an electrical insulation) on the outside of a wire. By forming the expression for the heat transfer in terms of the variables shown in Equation (3.1.11), and examining the change of heat transfer with variations in r_3 (that is, the thickness of the insulation). While simple results are given in many texts (showing the critical radius as the ratio of the insulation thermal conductivity to the heat transfer coefficient on the outside), Sparrow (1970) considered a heat transfer coefficient that varies as

$$\bar{h}_{c,o} \sim r_3^{-m} \; |T_3 - T_{f,o}|^n.$$

For this case, it is found that the heat transfer is maximized at:

$$r_3 = r_{\text{crit}} = \left[(1-m)/(1+n)\right]\frac{k_{\text{ins}}}{\bar{h}_{c,o}} \qquad (3.1.12)$$

By examining the orders of magnitude of m, n, k_{ins}, and $\bar{h}_{\text{c,o}}$ the critical radius is often found to be on the order of a *few millimeters*. Hence, additional insulation on small-diameter cylinders such as narrow-gauge electrical wires could actually increase the heat dissipation. On the other hand, the addition of insulation to large-diameter pipes and ducts will almost always decrease the heat transfer rate.

Internal Heat Generation

The analysis of temperature distributions and the resulting heat transfer in the presence of volume heat sources is required in some circumstances. These include phenomena such as nuclear fission processes, Joule heating, and microwave energy deposition. Consider first a slab of material 2 l thick but otherwise very large, with internal generation. The outside of the slab is kept at temperature T_1. To find the temperature distribution within the slab, the thermal conductivity is assumed to be constant. Equation (3.1.2) reduces to the following:

$$\frac{d^2T}{dx^2} + \frac{\dot{q}_G}{k} = 0 \tag{3.1.13}$$

Solving this equation by separating the variables, integrating twice, and applying boundary conditions gives

$$T(x) - T_1 = \frac{\dot{q}_G L^2}{2k}\left[1 - \left(\frac{x}{L}\right)^2\right] \tag{3.1.14}$$

A similar type of analysis for a long cylindrical element of radius r_1 gives

$$T(r) - T_1 = \frac{\dot{q}_G r_1^2}{4k}\left[1 - \left(\frac{r}{r_1}\right)^2\right] \tag{3.1.15}$$

Two additional cases will be given. Both involve the situation when the rate of heat generation is dependent upon the local temperature, according to the following relationship:

$$\dot{q}_G = \dot{q}_{G,o}\left[1 + \beta\left(T - T_o\right)\right] \tag{3.1.16}$$

For a plane wall of 2 l thickness, and a temperature of T_1 specified on each surface

$$\frac{T(x) - T_o + 1/\beta}{T_1 - T_o + 1/\beta} = \frac{\cos \mu x}{cox \mu l} \tag{3.1.17}$$

For a similar situation in a long cylinder with a temperature of T_1 specified on the outside radius r_1

$$\frac{T(r) - T_o + 1/\beta}{T_1 - T_o + 1/\beta} = \frac{J_o(\mu r)}{J_o(\mu r_1)} \tag{3.1.18}$$

In Equation (3.1.18), the J_o is the typical notation for the Bessel function. Variations of this function are tabulated in Abramowitz and Stegun (1964) or are available on many computer systems. In both of the cases shown above the following holds

$$\mu \equiv \sqrt{\frac{\beta \dot{q}_{G,o}}{k}}$$

Fins

Fins are widely used to enhance the heat transfer (usually convective, but also radiative) from a surface. This is particularly true when the surface is in contact with a gas. Fins are used on air-cooled engines, electronic cooling forms, as well as in a number of other applications. Since the heat transfer coefficient tends to be low in gas convection, area is added to the surface in the form of fins to decrease the convective thermal resistance.

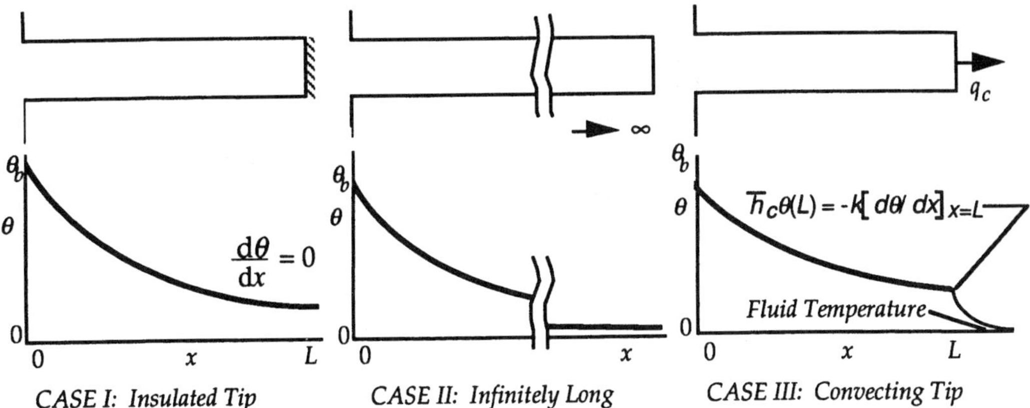

FIGURE 3.1.2 Three typical cases for one-dimensional, constant-cross-section fins are shown.

The simplest fins to analyze are those usually found in practice — one-dimensional and constant in cross section. In simple terms, to be one dimensional the fins have to be long compared to a transverse dimension. Three cases are normally considered for analysis, and these are shown in Figure 3.1.2. They are the insulated-tip fin, the very long fin, and the convecting-tip fin.

For Case I, the solution to the governing equation and the application of the boundary conditions of the known temperatures at the base and the insulated tip yields:

$$\text{Case I:} \qquad \theta = \theta_b = \frac{\cosh m(L-x)}{\cosh mL} \tag{3.1.19}$$

For the infinitely long case, the following simple form results.

$$\text{Case II:} \qquad \theta(x) = \theta_b e^{-mx} \tag{3.1.20}$$

The final case yields the following result:

$$\text{Case III:} \qquad \theta(x) = \theta_b \frac{mL\cosh m(L-x) + \text{Bi}\sinh m(L-x)}{mL\cosh mL + \text{Bi}\sinh mL} \tag{3.1.21}$$

where $\text{Bi} \equiv \bar{h}_c L/k$.

In all three of the cases given, the following definitions apply:

$$\theta \equiv T(x) - T_\infty, \quad \theta_b \equiv T(x=0) - T_\infty, \quad \text{and} \quad m^2 \equiv \frac{\bar{h}_c P}{kA}$$

Here A is the cross section of the fin parallel to the wall. P is the perimeter around that area.

To find the amount of heat removed in any of these cases, the temperature distribution is used in Fourier's law, Equation (3.1.1). For most fins that truly fit the one-dimensional assumption (i.e., long compared to their transverse dimensions), all three equations will yield results that don't differ significantly.

Two performance indicators are found in the fin literature. The **fin efficiency** is defined as the ratio of the actual heat transfer rate from a fin to the heat-transfer rate from an ideal fin.

$$\eta \equiv \frac{q_{\text{actual}}}{q_{\text{ideal}}} \tag{3.1.22}$$

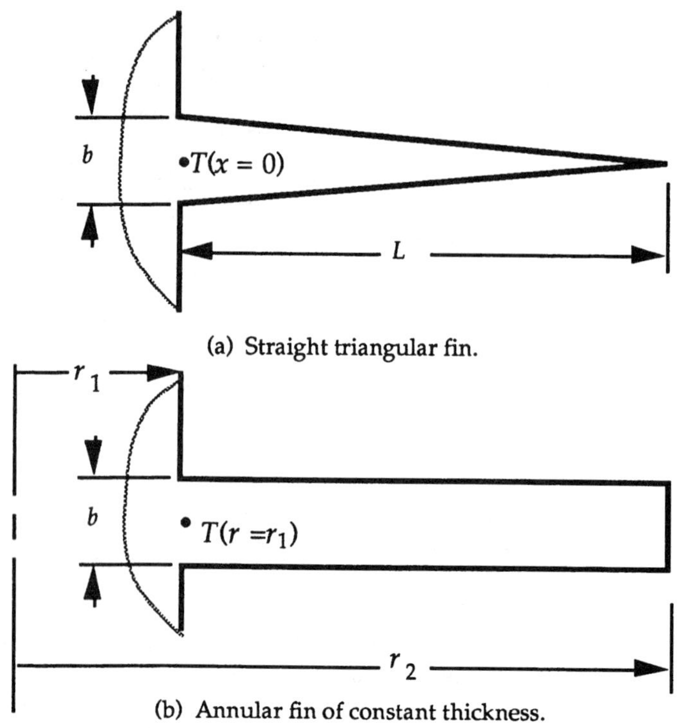

FIGURE 3.1.3 Two examples of fins with a cross-sectional area that varies with distance from the base.

The ideal heat transfer is found from convective gain or loss from an area the same size as the fin surface area, all at a temperature T_b. Fin efficiency is normally used to tabulate heat transfer results for various types of fins, including those with nonconstant area or not meeting the one-dimensional assumption. An example of the former can be developed from a result given by Arpaci (1966). Consider a straight fin of triangular profile, as shown in Figure 3.1.3. The solution is found in terms of modified Bessel functions of the first kind. Tabulations are given in Abramowitz and Stegun (1964).

$$\eta = \frac{I_1\left(2\tilde{m}L^{1/2}\right)}{\tilde{m}L^{1/2}I_o\left(2\tilde{m}L^{1/2}\right)} \tag{3.1.23}$$

Here, $\tilde{m} \equiv \sqrt{2\bar{h}_c L/kb}$.

The **fin effectiveness**, ε, is defined as the heat transfer from the fin compared to the bare-surface transfer through the same base area.

$$\varepsilon = \frac{q_{\text{actual}}}{q_{\text{bare base}}} = \frac{q_f}{\bar{h}_c A\left(T_b - T_\infty\right)} \tag{3.1.24}$$

Carslaw and Jaeger (1959) give an expression for the effectiveness of a fin of constant thickness around a tube (see Figure 3.1.3). This is given as ($\tilde{\mu} \equiv \sqrt{2\bar{h}_c/kb}$):

$$\varepsilon = \frac{2}{\tilde{\mu}b}\frac{I_1\left(\tilde{\mu}r_2\right)K_1\left(\tilde{\mu}r_1\right) - K_1\left(\tilde{\mu}r_2\right)I_1\left(\tilde{\mu}r_1\right)}{I_o\left(\tilde{\mu}r_1\right)K_1\left(\tilde{\mu}r_2\right) + K_o\left(\tilde{\mu}r_1\right)I_1\left(\tilde{\mu}r_2\right)} \tag{3.1.25}$$

Here the notations I and K denote Bessel functions that are given in Abramowitz and Stegun (1964).

Fin effectiveness can be used as one indication whether or not fins should be added. A rule of thumb indicates that if the effectiveness is less than about three, fins should not be added to the surface.

Transient Systems

Negligible Internal Resistance

Consider the transient cooling or heating of a body with surface area A and volume V. This is taking place by convection through a heat transfer coefficient $\bar{h}_c$ to an ambient temperature of T_∞. Assume the thermal resistance to conduction inside the body is significantly less than the thermal resistance to convection (as represented by Newton's law of cooling) on the surface of the body. This ratio is denoted by the **Biot Number**, Bi:

$$\text{Bi} = \frac{R_k}{R_c} = \frac{\bar{h}_c (V/A)}{k} \tag{3.1.26}$$

The temperature (which will be uniform throughout the body at any time for this situation) response with time for this system is given by the following relationship. Note that the shape of the body is not important — only the ratio of its volume to its area matters.

$$\frac{T(t) - T_\infty}{T_o - T_\infty} = e^{-\frac{\bar{h}_c A t}{\rho V c}} \tag{3.1.27}$$

Typically this will hold for the Biot Number being less than (about) 0.1.

Bodies with Significant Internal Resistance

When a body is being heated or cooled transiently in a convective environment, but the internal thermal resistance of the body cannot be neglected, the analysis becomes more complicated. Only simple geometries (a symmetrical plane wall, a long cylinder, a composite of geometrical intersections of these geometries, or a sphere) with an imposed step change in ambient temperature are addressed here.

The first geometry considered is a large slab of minor dimension 2 l. If the temperature is initially uniform at T_o, and at time 0+ it begins convecting through a heat transfer coefficient to a fluid at T_∞, the temperature response is given by:

$$\theta = 2\sum_{n=1}^{\infty} \left(\frac{\sin \lambda_n L}{\lambda_n L + \sin \lambda_n L \cos \lambda_n L} \right) \exp\left(-\lambda_n^2 L^2 \text{Fo}\right) \cos\left(\lambda_n x\right) \tag{3.1.28}$$

and the λ_n are the roots of the transcendental equation: $\lambda_n L \tan \lambda_n L = \text{Bi}$. The following definitions hold:

$$\text{Bi} \equiv \frac{\bar{h}_c L}{k} \qquad \text{Fo} \equiv \frac{\alpha t}{L^2} \qquad \theta \equiv \frac{T - T_\infty}{T_o - T_\infty}$$

The second geometry considered is a very long cylinder of diameter $2R$. The temperature response for this situation is

$$\theta = 2\text{Bi} \sum_{n=1}^{\infty} \frac{\exp\left(-\lambda_n^2 R^2 \text{Fo}\right) J_o\left(\lambda_n r\right)}{\left(\lambda_n^2 R^2 + \text{Bi}^2\right) J_o\left(\lambda_n R\right)} \tag{3.1.29}$$

Now the λ_n are the roots of $\lambda_n R\, J_1(\lambda_n R) - \text{Bi}\, J_o(\lambda_n R) = 0$, and

$$\text{Bi} = \frac{\bar{h}_c R}{k} \qquad \text{Fo} = \frac{\alpha t}{R^2} \qquad \theta = \frac{T - T_\infty}{T_o - T_\infty}$$

The common definition of Bessel's functions applies here.

For the similar situation involving a solid sphere, the following holds:

$$\theta = 2\sum_{n=1}^{\infty} \frac{\sin(\lambda_n R) - \lambda_n R\cos(\lambda_n R)}{\lambda_n R - \sin(\lambda_n R)\cos(\lambda_n R)} \exp\left(-\lambda_n^2 R^2 \text{Fo}\right) \frac{\sin(\lambda_n r)}{\lambda_n r} \tag{3.1.30}$$

and the λ_n are found as the roots of $\lambda_n R \cos\lambda_n R = (1 - \text{Bi}) \sin\lambda_n R$. Otherwise, the same definitions hold as were given for the cylinder.

Solids that can be envisioned as the geometrical intersection of the simple shapes described above can be analyzed with a simple product of the individually shape solutions. For these cases, the solution is found as the product of the dimensionless temperature functions for each of the simple shapes, with appropriate distance variables taken in each solution. This is illustrated in the right-hand diagram in Figure 3.1.4. For example, a very long rod of rectangular cross section can be seen as the intersection of two large plates. A short cylinder represents the intersection of an infinitely long cylinder and a plate. The temperature at any location within the short cylinder is

$$\theta_{2R,2L\ \text{Rod}} = \theta_{\text{Infinite}\ 2R\ \text{Rod}}\theta_{2L\ \text{Plate}} \tag{3.1.31}$$

Details of the formulation and solution of the partial differential equations in heat conduction are found in the text by Arpaci (1966).

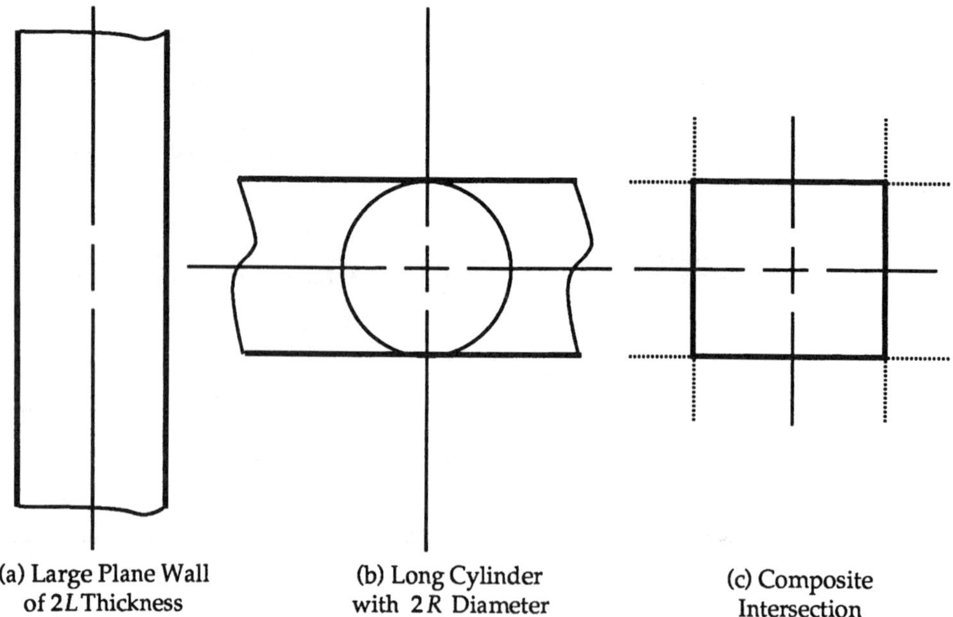

FIGURE 3.1.4 Three types of bodies that can be analyzed with methods given in this section.

Finite Difference Analysis of Conduction

Today, numerical solution of conduction problems is the most-used analysis approach. Two general techniques are applied for this: those based upon finite difference ideas and those based upon finite element concepts. General numerical formulations are introduced in Chapter 5. Here, we briefly introduce some finite difference concepts as they might be applied in heat conduction analysis.

First, consider formulation ideas. Consider the two-dimensional form of Equation (3.1.2), neglecting heat generation. This is given by:

$$\frac{\partial^2 T}{\partial x^2} + \frac{\partial^2 T}{\partial y^2} = \frac{1}{\alpha}\frac{\partial T}{\partial t} \tag{3.1.32}$$

By using finite approximations to the first and second derivatives, the following results:

$$\frac{T_{i+1,j} - 2T_{i,j} - T_{i-1,j}}{\Delta x^2} + \frac{T_{i,j+1} - 2T_{i,j} - T_{i,j-1}}{\Delta y^2} = \frac{T'_{i,j} - T_{i,j}}{\alpha \Delta t} \tag{3.1.33}$$

In this notation, i denotes node center locations in the x direction, j indicates node center locations in the y direction, and $T_{i,j} = T(i, j, t)$, and $T'_{i,j} = T(i, j, t+\Delta t)$. As written, the form is denoted as the *explicit* formulation because the forward difference is used for the time derivative (a new time-step temperature is found in terms of all of the old time-step temperatures). If the time derivative is taken as the backward difference, then the new (and, hence, unknown) temperatures at all surrounding nodes are used with the known (old) temperature at each node to find the temperatures at the new time. Because each equation contains a number of unknowns that must be found simultaneously, the result is termed the *implicit* formulation. This is shown below for a general interior node.

$$\frac{T'_{i+1,j} - 2T'_{i,j} - T'_{i-1,j}}{\Delta x^2} + \frac{T'_{i,j+1} - 2T'_{i,j} - T'_{i,j-1}}{\Delta y^2} = \frac{T'_{i,j} - T_{i,j}}{\alpha \Delta t} \tag{3.1.34}$$

A formal differencing approach, where finite differences are generated directly from partial derivatives, can be carried out for each application. However, the possibility of unusual geometries and mixing of boundary conditions often results in situations that are not easily differenced. For this reason, an energy balance method often yields an approach that can be more easily applied physically.

Attention is drawn to a one-dimensional slab (very large in two directions compared to the thickness). The slab is divided across the thickness into smaller subslabs (denoted as nodes in what follows), and this is shown in Figure 3.1.5. All nodes are of thickness Δx except for the two boundaries where the thickness is $\Delta x/2$. A characteristic temperature for each node is assumed to be represented by the temperature at the node center. Of course this assumption becomes more accurate as the size of the node becomes smaller. On the periphery of the object being analyzed the location representing the temperature of the node is located on the boundary of the object, and this results in fractionally sized nodes at these positions. Hence, with the two boundary node centers located exactly on the boundary, a total of n nodes are used ($n - 2$ full, interior nodes and half-nodes on each of the two boundaries).

In the analysis, a general interior node i (this applies to all nodes 2 through $n - 1$) is considered for an overall energy balance. Conduction in from node $i - 1$ and from node $i + 1$, as well as any heat generation present, is assumed to be energy per unit time flowing into the node. This is then equated to the time rate of change of energy within the node. A backward difference on the time derivative is applied here, and the notation $T_i' \equiv T_i(t+\Delta t)$ is used. The balance gives the following on a per-unit-area basis:

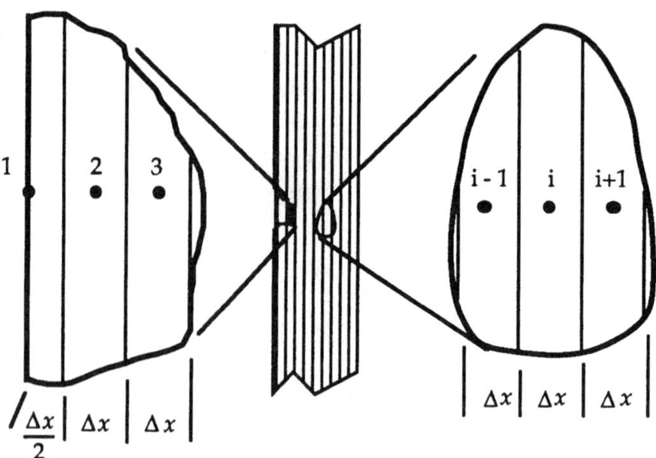

FIGURE 3.1.5 A one-dimensional finite differencing of a slab with a general interior node and one surface node detailed.

$$\frac{T'_{i-1} - T'_i}{\Delta x/k_-} + \frac{T'_{i+1} - T'_i}{\Delta x/k_+} + \dot{q}_{G,i}\,\Delta x = \rho\,\Delta x\,c_p\,\frac{T'_i - T_i}{\Delta t} \tag{3.1.35}$$

In this equation different thermal conductivities have been used to allow for possible variations in properties throughout the solid.

The analysis of the boundary nodes will depend upon the nature of the conditions there. For the purposes of illustration, convection will be assumed to be occurring off of the boundary at node 1. A balance similar to Equation (3.1.35), but now for node 1, gives the following:

$$\frac{T'_\infty - T'_1}{1/h_c} + \frac{T'_2 - T'_1}{\Delta x/k_+} + \dot{q}_{G,1}\,\frac{\Delta x}{2} = \rho\frac{\Delta x}{2}c_p\,\frac{T'_1 - T_1}{\Delta t} \tag{3.1.36}$$

If the heat transfer situation in the slab is symmetrical about the slab center line, consideration of this aspect allows only half of the slab to be analyzed. Consideration of the half-node located on the left-hand side of the centerline yields:

$$\frac{T'_{n-1} - T'_n}{\Delta x/k_+} + 0 + \dot{q}_{G,1}\,\frac{\Delta x}{2} = \rho\frac{\Delta x}{2}c_p\,\frac{T'_n - T_n}{\Delta t} \tag{3.1.37}$$

Here, the zero shown as the second term denotes the fact that no energy flows across the centerline.

After all n equations are written, it can be seen that there are n unknowns represented in these equations: the temperature at all nodes. If one or both of the boundary conditions is known in terms of a specified temperatures, this will decrease the number of equations and unknowns by one or two, respectively. To determine the temperature as a function of time, the time step is arbitrarily set, and all the temperatures are found by simultaneous solution at $t = 0 + \Delta t$. This is denoted by the matrix system shown below. For the situation shown here, the thermal conductivity is taken to be constant throughout the slab, and no internal heat generation is considered. With these simplifications, the system of equations to be solved (using Equations 3.1.35 through 3.1.37) is

$$\begin{bmatrix} (1+2Fo+2\,Fo\,Bi) & -Fo & 0 & 0 & \cdots & 0 \\ -Fo & (1+2Fo) & -Fo & 0 & \cdots & 0 \\ \cdots & -Fo & (1+2\,Fo) & -Fo & \cdots & 0 \\ \cdots & \cdots & \cdots & \cdots & \cdots & \cdots \\ \cdots & \cdots & \cdots & -Fo & (1+2Fo) & -Fo \\ \cdots & \cdots & \cdots & \cdots & -2Fo & (1+2Fo) \end{bmatrix} \begin{bmatrix} T_1' \\ T_2' \\ T_3' \\ \cdots \\ T_{n-1}' \\ T_n' \end{bmatrix} = \begin{bmatrix} 2FoBiT_\infty' + T_1 \\ T_2 \\ T_3 \\ \cdots \\ T_{n-1} \\ T_n \end{bmatrix} \quad (3.1.38)$$

The time is then advanced by Δt and the temperatures are again found by simultaneous solution.

The finite difference approach just outlined using the backward difference for the time derivative is termed the *implicit* technique, and it results in an n x n system of linear simultaneous equations. If the forward difference is used for the time derivative, then only one unknown will exist in each equation. This gives rise to what is called an *explicit* or "marching" solution. While this type of system is more straightforward to solve because it deals with only one equation at a time with one unknown, a *stability criterion* must be considered which limits the time step relative to the distance step.

Two- and three-dimensional problems are handled in conceptually the same manner. One-dimensional heat fluxes between adjoining nodes are again considered. Now there are contributions from each of the dimensions represented. Details are outlined in the book by Jaluria and Torrance (1986).

Defining Terms

Biot Number: Ratio of the internal (conductive) resistance to the external (convective) resistance from a solid exchanging heat with a fluid.

Fin: Additions of material to a surface to increase area and thus decrease the external thermal resistance from convecting and/or radiating solids.

Fin effectiveness: Ratio of the actual heat transfer from a fin to the heat transfer from the same cross-sectional area of the wall without the fin.

Fin efficiency: Ratio of the actual heat transfer from a fin to the heat transfer from a fin with the same geometry but completely at the base temperature.

Fourier's law: The fundamental law of heat conduction. Relates the local temperature gradient to the local heat flux, both in the same direction.

Heat conduction equation: A partial differential equation in temperature, spatial variables, time, and properties that, when solved with appropriate boundary and initial conditions, describes the variation of temperature in a conducting medium.

Overall heat transfer coefficient: The analogous quantity to the heat transfer coefficient found in convection (Newton's law of cooling) that represents the overall combination of several thermal resistances, both conductive and convective.

Thermal conductivity: The property of a material that relates a temperature gradient to a heat flux. Dependent upon temperature.

References

Abramowitz, M. and Stegun, I., Handbook of Mathematical Functions with Formulas, Graphs, and Mathematical Tables. National Bureau of Standards, Applied Mathematics Series 55, U.S. Government Printing Office, Washington, D.C., 1964.

Arpaci, V., *Conduction Heat Transfer*, Addison-Wesley, Reading, MA, 1966.

Carslaw, H. S. and Jaeger, J. C., *Conduction of Heat in Solids*, 2nd ed., Oxford University Press, London, 1959.

Guyer, E. (Ed.), Part 3, Thermal Insulations, in *Handbook of Applied Thermal Design*, McGraw-Hill, New York, 1989.

Jaluria, Y. and Torrance, K., *Computational Heat Transfer*, Hemisphere, New York, 1986.
Sparrow, E., Reexamination and Correction of the Critical Radius for Radial Heat Conduction, *AIChE J.*, 16(1), 149, 1970.

For Further Information

The references listed above will give the reader an excellent introduction to analytical formulation and solution (Arpaci), material properties (Guyer), and numerical formulation and solution (Jaluria and Torrance). Current developments in conduction heat transfer appear in several publications, including the *Journal of Heat Transfer*, *International Journal of Heat and Mass Transfer*, and *Numerical Heat Transfer*.

3.2 Convection Heat Transfer

3.2.1 Natural Convection

George D. Raithby and K.G. Terry Hollands

Introduction

Natural convection heat transfer occurs when the convective fluid motion is induced by density differences that are themselves caused by the heating. An example is shown in Figure 3.2.1(A), where a body at surface temperature T_s transfers heat at a rate q to ambient fluid at temperature $T_\infty < T_s$.

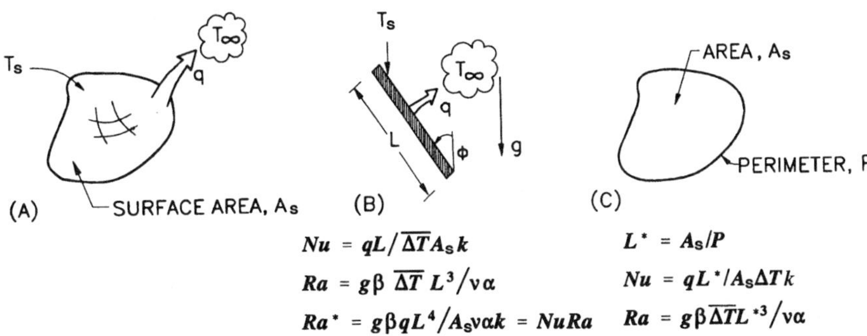

FIGURE 3.2.1 (A) Nomenclature for external heat transfer. (A) General sketch; (B) is for a tilted flat plate, and (C) defines the length scale for horizontal surfaces.

In this section, correlations for the average Nusselt number are provided from which the heat transfer rate q from surface area A_s can be estimated. The Nusselt number is defined as

$$\text{Nu} = \frac{\bar{h}_c L}{k} = \frac{qL}{A_s \Delta T k} \tag{3.2.1}$$

where $\Delta T = T_s - T_\infty$ is the temperature difference driving the heat transfer. A dimensional analysis leads to the following functional relation:

$$\text{Nu} = f(\text{Ra, Pr, geometric shape, boundary conditions}) \tag{3.2.2}$$

For given thermal boundary conditions (e.g., isothermal wall and uniform T_∞), and for a given geometry (e.g., a cube), Equation (3.2.2) states that Nu depends only on the Rayleigh number, Ra, and Prandtl number, Pr. The length scales that appear in Nu and Ra are defined, for each geometry considered, in a separate figure. The fluid properties are generally evaluated at T_f, the average of the wall and ambient temperatures. The exception is that β, the temperature coefficient of volume expansion, is evaluated at T_∞ for external natural convection (Figures 3.2.1 to 3.2.3) in a *gaseous* medium.

The functional dependence on Pr is approximately independent of the geometry, and the following Pr-dependent function will be useful for laminar heat transfer (Churchill and Usagi, 1972):

$$\bar{C}_\ell = 0.671\Big/\left(1 + (0.492/\text{Pr})^{9/16}\right)^{4/9} \tag{3.2.3}$$

C_t^V and C_t^H are functions, defined in Equations 3.2.4 and 3.2.5, will be useful for turbulent heat transfer:

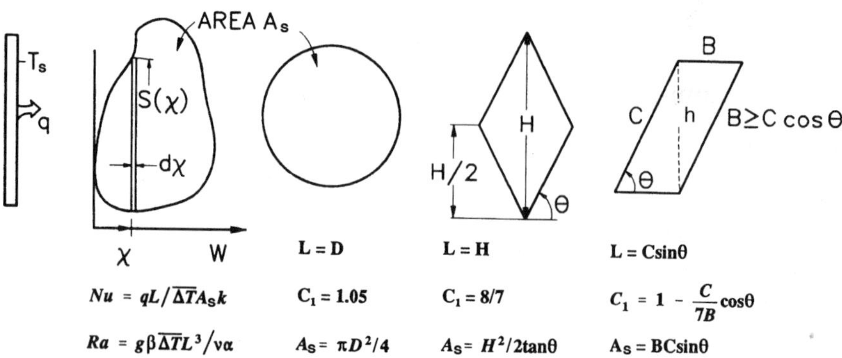

FIGURE 3.2.2 Nomenclature for heat transfer from planar surfaces of different shapes.

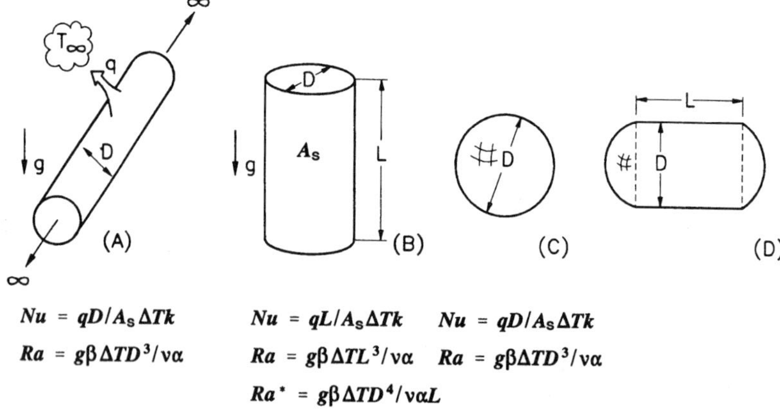

FIGURE 3.2.3 Definitions for computing heat transfer from a long circular cylinder (A), from the lateral surface of a vertical circular cylinder (B), from a sphere (C), and from a compound body (D).

$$C_t^V = 0.13\text{Pr}^{0.22}\Big/\left(1 + 0.61\text{Pr}^{0.81}\right)^{0.42} \tag{3.2.4}$$

$$C_t^H = 0.14\left(\frac{1 + 0.0107\text{Pr}}{1 + 0.01\text{Pr}}\right) \tag{3.2.5}$$

The superscripts *V* and *H* refer to the vertical and horizontal surface orientation.

The Nusselt numbers for fully laminar and fully turbulent heat transfer are denoted by Nu_ℓ and Nu_t, respectively. Once obtained, these are blended (Churchill and Usagi, 1972) as follows to obtain the equation for Nu:

$$\text{Nu} = \left(\left(\text{Nu}_\ell\right)^m + \left(\text{Nu}_t\right)^m\right)^{1/m} \tag{3.2.6}$$

The blending parameter *m* depends on the body shape and orientation.

The equation for Nu_ℓ in this section is usually expressed in terms of Nu^T, the Nusselt number that would be valid if the thermal boundary layer were thin. The difference between Nu_ℓ and Nu^T accounts for the effect of the large boundary layer thicknesses encountered in natural convection.

It is assumed that the wall temperature of a body exceeds the ambient fluid temperature ($T_s > T_\infty$). For $T_s < T_\infty$ the same correlations apply with ($T_\infty - T_s$) replacing ($T_s - T_\infty$) for a geometry that is rotated

180° relative to the gravitational vector; for example, the correlations for a horizontal heated upward-facing flat plate applies to a cooled downward-facing flat plate of the same planform.

Correlations for External Natural Convection

This section deals with problems where the body shapes in Figures 3.2.1 to 3.2.3 are heated while immersed in a quiescent fluid. Different *cases* are enumerated below.

1. *Isothermal Vertical (ϕ = 0) Flat Plate, Figure 3.2.1B.* For heat transfer from a vertical plate (Figure 3.2.1B), for $1 < \text{Ra} < 10^{12}$,

$$\text{Nu}^T = \overline{C}_\ell \text{Ra}^{1/4} \qquad \text{Nu}_\ell = \frac{2.0}{\ln\left(1 + 2.0/\text{Nu}^T\right)} \tag{3.2.7}$$

$$\text{Nu}_t = C_t^V \text{Ra}^{1/3} \Big/ \left(1 + 1.4 \times 10^9 \,\text{Pr}/\text{Ra}\right)$$

$\overline{C}_\ell$ and C_t^V are given by Equations (3.2.3) and (3.2.4). Nu is obtained by substituting Equation (3.2.7) expressions for Nu_ℓ and Nu_t into Equation (3.2.6) with $m = 6$.

2. *Vertical Flat Plate with Uniform Heat Flux, Figure 3.2.1B.* If the plate surface has a constant (known) heat flux, rather than being isothermal, the objective is to calculate the average temperature difference, $\overline{\Delta T}$, between the plate and fluid. For this situation, and for $15 < \text{Ra}^* < 10^5$,

$$\text{Nu}^T = \overline{G}_\ell \left(\text{Ra}^*\right)^{1/5} \qquad \text{Nu}_\ell = \frac{1.83}{\ln\left(1 + 1.83/\text{Nu}^T\right)} \qquad \text{Nu}_t = \left(C_t^V\right)^{3/4} \left(\text{Ra}^*\right)^{1/4} \tag{3.2.8a}$$

$$\overline{G}_\ell = \frac{6}{5}\left(\frac{\text{Pr}}{4 + 9\sqrt{\text{Pr}} + 10\text{Pr}}\right)^{1/5} \tag{3.2.8b}$$

Ra^* is defined in Figure 3.2.1B and C_t^V is given by Equation (3.2.4). Find Nu by inserting these expressions for Nu_ℓ and Nu_t into Equation (3.2.6) with $m = 6$. The $\overline{G}_\ell$ expression is due to Fujii and Fujii (1976).

3. *Horizontal Upward-Facing (ϕ = 90°) Plates, Figure 3.2.1C.* For horizontal isothermal surfaces of various platforms, correlations are given in terms of a lengthscale L^* (Goldstein et al., 1973), defined in Figure 3.2.1C. For $\text{Ra} \geq 1$,

$$\text{Nu}^T = 0.835\overline{C}_\ell \text{Ra}^{1/4} \qquad \text{Nu} = \frac{1.4}{\ln\left(1 + 1{:}4/\text{Nu}^T\right)} \qquad \text{Nu}_t = C_t^H \text{Ra}^{1/3} \tag{3.2.9}$$

Nu is obtained by substituting Nu_ℓ and Nu_t from Equation 3.2.9 into Equation 3.2.6 with m = 10. For non-isothermal surfaces, replace ΔT by $\overline{\Delta T}$.

4. *Horizontal Downward-Facing (ϕ = –90°) Plates, Figure 3.2.1C.* For horizontal downward-facing plates of various planforms, the main buoyancy force is into the plate so that only a very weak force drives the fluid along the plate; for this reason, only laminar flows have been measured. For this case, the following equation applies for $\text{Ra} < 10^{10}$, $\text{Pr} \geq 0.7$:

$$\text{Nu}^T = H_\ell \text{Ra}^{1/5} \qquad H_\ell = \frac{0.527}{\left[1 + (1.9/\text{Pr})^{9/10}\right]^{2/9}} \qquad \text{Nu} = \frac{2.45}{\ln\left(1 + 2.45/\text{Nu}^T\right)} \tag{3.2.10}$$

H_ℓ fits the analysis of Fujii et al. (1973).

5. *Inclined Plates, Downward Facing* ($-90° \le \phi \le 0$), *Figure 3.2.1B*. First calculate q from *Case 1* with g replaced by g cos ϕ; then calculate q from *Case 4* (horizontal plate) with g replaced by g sin ($-\phi$), and use the maximum of these two values of q.
6. *Inclined Plates, Upward Facing* ($0 \le \phi \le 90$), *Figure 3.2.1B*. First calculate q from *Case 1* with g replaced by g cos ϕ; then calculate q from *Case 3* with g replaced by g sin ϕ, and use the maximum of these two values of q.
7. *Vertical and Tilted Isothermal Plates of Various Planform, Figure 3.2.2*. The line of constant χ in Figure 3.2.2 is the line of steepest ascent on the plate. Provided all such lines intersect the plate edges just twice, as shown in the figure, the thin-layer (Nu^T) heat transfer can be found by subdividing the body into strips of width $\Delta\chi$, calculating the heat transfer from each strip, and adding. For laminar flow from an isothermal vertical plate, this results in

$$\mathrm{Nu}^T = C_1 \bar{C}_\ell \mathrm{Ra}^{1/4} \qquad C_1 \equiv \left(\frac{L^{1/4}}{A} \int_0^W S^{3/4}\, d\chi \right) \tag{3.2.11}$$

Symbols are defined in Figure 3.2.2, along with L and calculated C_1 values for some plate shapes. If the plate is vertical, follow the procedure in *Case 1* above (isothermal vertical flat plate) except replace the expression for Nu^T in Equation (3.2.7) by Equation (3.2.11). If the plate is tilted, follow the procedure described in *Case 5* or *6* (as appropriate) but again use Equation (3.2.11) for Nu^T in Equation (3.2.7)
8. *Horizontal Cylinders, Figure 3.2.3A*. For a long, horizontal circular cylinder use the following expressions for Nu_ℓ and Nu_t:

$$\mathrm{Nu}^T = 0.772 \bar{C}_\ell \mathrm{Ra}^{1/4} \qquad \mathrm{Nu}_\ell = \frac{2f}{\left(1 + 2f/\mathrm{Nu}^T\right)} \qquad \mathrm{Nu}_t = \bar{C}_t \mathrm{Ra}^{1/3} \tag{3.2.12}$$

$\bar{C}_t$ is given in the table below. For Ra > 10^{-2}, $f = 0.8$ can be used, but for 10^{-10} < Ra < 10^{-2} use $f = 1 - 0.13/(\mathrm{Nu}^T)^{0.16}$. To find Nu, the values of Nu_ℓ and Nu_t from Equation (3.2.12) are substituted into Equation (3.2.6) with m = 15 (Clemes et al., 1994).

$\bar{C}_t$ for Various Shapes and Prandtl Numbers

Pr→	0.01	0.022	0.10	0.71	2.0	6.0	50	100	2000
Horizontal cylinder	0.077	0.81	0.90	0.103	0.108	0.109	0.100	0.097	0.088
Spheres	0.074	0.078	0.088	0.104	0.110	0.111	0.101	0.97	0.086

9. *Vertical Cylinders* (ϕ = 90°), *Figure 3.2.3B*. For high Ra values and large diameter, the heat transfer from a vertical cylinder approaches that for a vertical flat plate. Let the Nu^T and Nu_ℓ equations for a vertical flat plate of height L, Equation (3.2.7), be rewritten here as Nu_p^T and Nu_p, respectively. At smaller Ra and diameter, transverse curvature plays a role which is accounted for in the following equations:

$$\mathrm{Nu}_\ell = \frac{0.9\xi \mathrm{Nu}_p}{\ln(1 + 0.9\xi)} \qquad \xi = \frac{2L/D}{Nu_p^T} \tag{3.2.13}$$

These equations are valid for purely laminar flow. To obtain Nu, blend Equation (3.2.13) for Nu_ℓ with Equation (3.2.7) for Nu_t using Equation (3.2.6) with m = 10.

10. *Spheres, Figure 3.2.3C.* For spheres use Equation (3.2.6), with $m = 6$, and with

$$\text{Nu}_\ell = 2 + 0.878\bar{C}_\ell \text{Ra}^{1/4} \quad \text{and} \quad \text{Nu}_t = \bar{C}_t \text{Ra}^{1/3} \tag{3.2.14}$$

The table above contains $\bar{C}_t$ values.

11. *Combined Shapes, Figure 3.2.3D.* For combined shapes, such as the cylinder in Figure 3.2.3D with spherical end caps, calculate the heat transfer from the cylinder of length L (*Case 8*), the heat transfer from a sphere of diameter D (*Case 10*) and add to obtain the total transfer. Other shapes can be treated in a similar manner.

Correlations for Open Cavities

Examples of this class of problem are shown in Figure 3.2.4. Walls partially enclose a fluid region (cavity) where boundary openings permit fluid to enter and leave. Upstream from its point of entry, the fluid is at the ambient temperature, T_∞. Since access of the ambient fluid to the heated surfaces is restricted, some of the heated surface is starved of cool ambient to which heat can be transferred. As the sizes of the boundary openings are increased, the previous class of problems is approached; for example, when the plate spacing in Figure 3.2.4A (*Case 12*) becomes very large, the heat transfer from each vertical surface is given by *Case 1*.

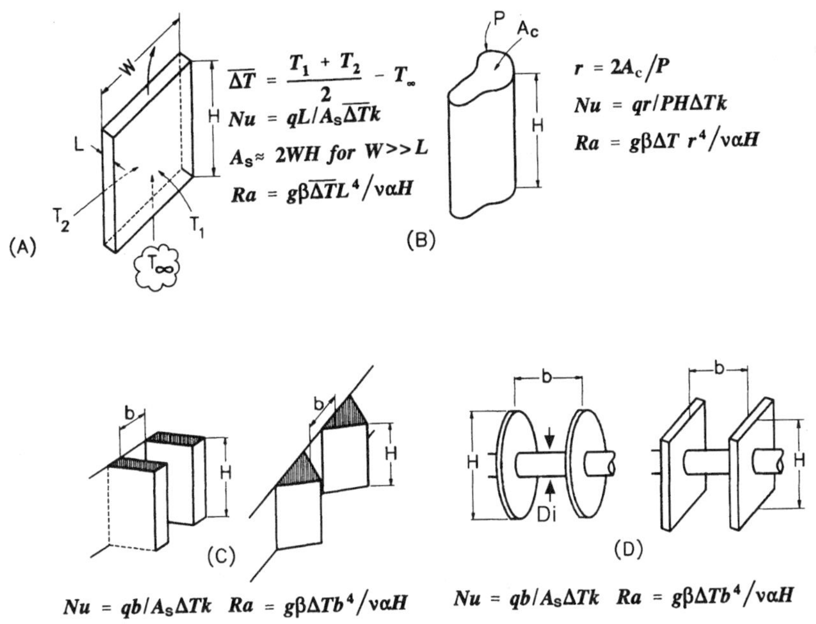

FIGURE 3.2.4 Nomenclature for various open-cavity problems.

12. *Isothermal Vertical Channels, Figure 3.2.4A and B.* Figure 3.2.4A shows an open cavity bounded by vertical walls and open at the top and bottom. The large opposing plates are isothermal, at temperatures T_1 and T_2, respectively, and the spacing between these plates is small. $\overline{\Delta T}$ is the average temperature difference between the plates and T_∞, as shown in Figure 3.2.4A, but T_1 and T_2 must not straddle T_∞. For this case

$$\text{Nu} = \left(\left(\frac{\text{Ra}}{f\,\text{Re}} \right)^m + \left(C_1 \bar{C}_\ell \text{Ra}^{1/4} \right)^m \right)^{1/m} \qquad \text{Ra} \le 10^5 \tag{3.2.15}$$

where fRe is the product of friction factor and Reynolds number for fully developed flow through, and C_1 is a constant that accounts for the augmentation of heat transfer, relative to a vertical flat plate (*Case 1*), due to the chimney effect. The fRe factor accounts for the cross-sectional shape (Elenbaas, 1942a). Symbols are defined in Figure 3.2.4A and B; in the Nu equation, q is the total heat transferred to the ambient fluid from all heated surfaces.

For the parallel plate channel shown in Figure 3.2.4(A), use f Re = 24, $m = -1.9$, and for gases $C_1 \approx 1.2$. It should be noted, however, that C_1 must approach 1.0 as Pr increases or as the plate spacing increases. For channels of circular cross section (Figure 3.2.4B) fRe = 16, $m = -1.03$, and for gases $C_1 \approx 1.17$. For other cross-sectional shapes like the square (fRe = 14.23), hexagonal (fRe = 15.05), or equilateral triangle (fRe = 13.3), use Equation (3.2.15) with the appropriate fRe, and with $m = -1.5$, and $C_1 \approx 1.2$ for gases.

The heat transfer per unit cross-sectional area, q/A_c, for a given channel length H and temperature difference, passes through a maximum at approximately $\mathrm{Ra}_{\max}$, where

$$\mathrm{Ra}_{\max} = \left(\frac{f\mathrm{Re}C_1\bar{C}_\ell}{2^{1/m}} \right)^{4/3} \tag{3.2.16}$$

$\mathrm{Ra}_{\max}$ provides the value of hydraulic radius $r = 2A_c/P$ at this maximum.

13. *Isothermal Triangular Fins, Figure 3.2.4C.* For a large array of triangular fins (Karagiozis et al., 1994) in air, for $0.4 < \mathrm{Ra} < 5 \times 10^5$

$$\mathrm{Nu} = \bar{C}_\ell \mathrm{Ra}^{1/4} \left[1 + \left(\frac{3.26}{\mathrm{Ra}^{0.21}} \right)^3 \right]^{-1/3} \qquad 0.4 < \mathrm{Ra} < 5 \times 10^5 \tag{3.2.17}$$

In this equation, b is the average fin spacing (Figure 3.2.4C), defined such that bL is the cross-sectional flow area between two adjacent fin surfaces up to the plane of the fin tips. For Ra < 0.4, Equation (3.2.17) underestimates the convective heat transfer. When such fins are mounted horizontally (vertical baseplate, but the fin tips are horizontal), there is a substantial reduction of the convective heat transfer (Karagiozis et al., 1994).

14. *U-Channel Fins, Figure 3.2.4C.* For the fins most often used as heat sinks, there is uncertainty about the heat transfer at low Ra. By using a conservative approximation applying for Ra < 100 (that underestimates the real heat transfer), the following equation may be used:

$$\mathrm{Nu} = \left[\left(\frac{\mathrm{Ra}}{24} \right)^{-2} + \left(C_1 \bar{C}_\ell \mathrm{Ra} \right)^{-2} \right]^{-0.5} \tag{3.2.18}$$

For air C_1 depends on aspect ratio of the fin as follows (Karagiozis, 1991):

$$C_1 = \left[1 + \left(\frac{H}{b} \right), 1.16 \right]_{\min} \tag{3.2.19}$$

Equation (3.2.18) agrees well with measurements for Ra > 200, but for smaller Ra it falls well below data because the leading term does not account for heat transfer from the fin edges and for three-dimensional conduction from the entire array.

15. *Circular Fins on a Horizontal Tube, Figure 3.24D.* For heat transfer from an array of circular fins (Edwards and Chaddock, 1963), for $H/D_i = 1.94$, $5 < \mathrm{Ra} < 10^4$, and for air,

$$\text{Nu} = 0.125\text{Ra}^{0.55}\left[1 - \exp\left(-\frac{137}{\text{Ra}}\right)\right]^{0.294} \tag{3.2.20}$$

A more general, but also more complex, relation is reported by Raithby and Hollands (1985).

16. *Square Fins on a Horizontal Tube, Figure 3.2.4D.* Heat transfer (Elenbaas, 1942b) from the square fins (excluding the cylinder that connects them) is correlated for gases by

$$\text{Nu} = \left[\left(\text{Ra}^{0.89}/18\right)^m + \left(0.62\text{Ra}^{1/4}\right)^m\right]^{1/m} \qquad m = -2.7 \tag{3.2.21}$$

Heat Transfer in Enclosures

This section deals with cavities where the bounding walls are entirely closed, so that no mass can enter or leave the cavity. The fluid motion inside the cavity is driven by natural convection, which enhances the heat transfer among the interior surfaces that bound the cavity.

17. *Extensive Horizontal Layers, Figure 3.2.5A.* If the heated plate, in a horizontal parallel-plate cavity, is on the top (0 = 180°), heat transfer is by conduction alone, so that Nu = 1. For heat transfer from below (θ = 0°) (Hollands, 1984):

$$\text{Nu} = 1 + \left[1 - \frac{1708}{\text{Ra}}\right]^{\bullet}\left[k_1 + 2\left(\frac{\text{Ra}^{1/3}}{k_2}\right)^{1-\ln\left(\text{Ra}^{1/3}/k_2\right)}\right] + \left[\left(\frac{\text{Ra}}{5830}\right)^{1/3} - 1\right]^{\bullet} \tag{3.2.22}$$

where

$$[x]^{\bullet} = (x,0)_{\max} \qquad k_1 = \frac{1.44}{1 + 0.018/\text{Pr} + 0.00136/\text{Pr}^2} \qquad k_2 = 75\exp\left(1.5\text{Pr}^{-1/2}\right) \tag{3.2.23}$$

The equation has been validated for Ra < 10^{11} for water, Ra < 10^{8} for air, and over a smaller Ra range for other fluids. Equation (3.2.22) applies to extensive layers: $W/L \geq 5$. Correlations for nonextensive layers are provided by Raithby and Hollands (1985).

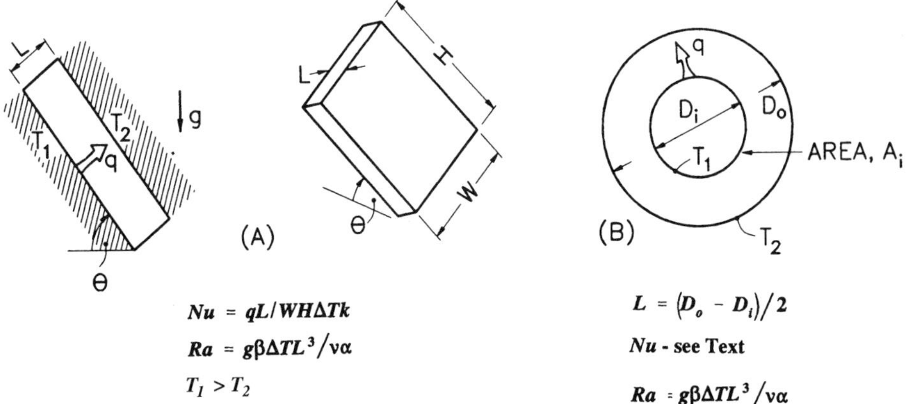

FIGURE 3.2.5 Nomenclature for enclosure problems.

18. *Vertical Layers, Figure 3.2.5(A), with $\theta = 90°$. W/L > 5.* For a vertical, gas-filled (Pr ≈ 0.7) cavity with H/L ≥ 5, the following equation closely fits the data, for example that of Shewen et al. (1996) for $Ra(H/L)^3 \leq 5 \times 10^{10}$ and $H/L \geq 40$.

$$Nu_1 = \left[1 + \left(\frac{0.0665 Ra^{1/3}}{1 + \left(\frac{9000}{Ra}\right)^{1.4}}\right)^2\right]^{1/2} \qquad Nu_2 = 0.242\left(Ra\frac{L}{H}\right)^{0.273} \qquad Nu = \left[Nu_1, Nu_2\right]_{max} \tag{3.2.24}$$

For Pr ≥ 4, the following equation is recommended (Seki et al., 1978) for $Ra(H/L)^3 < 4 \times 10^{12}$

$$Nu = \left[1, 0.36 Pr^{0.051}\left(\frac{L}{H}\right)^{0.36} Ra^{0.25}, 0.084 Pr^{0.051}\left(\frac{L}{H}\right)^{0.1} Ra^{0.3}\right]_{max} \tag{3.2.25a}$$

and for $Ra\,(H/L)^3 > 4 \times 10^{12}$

$$Nu = 0.039 Ra^{1/3} \tag{3.2.25b}$$

19. *Tilted Layers, Figure 3.25A, with $0 \leq \theta \leq 90°$, W/L > 8.* For gases (Pr ≈ 0.7), $0 \leq \theta \leq 60°$ and $Ra \leq 10^5$ (Hollands et al., 1976), use

$$Nu = 1 + 1.44\left[1 - \frac{1708}{Ra\cos\theta}\right]^{\bullet}\left[1 - \frac{1708(\sin 1.8\theta)^{1.6}}{Ra\cos\theta}\right] + \left[\left(\frac{Ra\cos\theta}{5830}\right)^{1/3} - 1\right]^{\bullet} \tag{3.2.26}$$

See equation (3.2.23) for definition of $[x]°$. For $60° \leq \theta \leq 90°$ linear interpolation is recommended using Equations (3.2.24) for $\theta = 90°$ and (3.2.26) for $\theta = 60°$.

20. *Concentric Cylinders, Figure 3.2.5B.* For heat transfer across the gap between horizontal concentric cylinders, the Nusselt number is defined as $Nu = q' \ln(D_o/D_i)/2\pi k\Delta T$ where q' is the heat transfer per unit length of cylinder. For $Ra \leq 8 \times 10^7$, $0.7 \leq Pr \leq 6000$, $1.15 \leq D/D_i \leq 8$ (Raithby and Hollands, 1975)

$$Nu = \left[0.603\bar{C}_\ell \frac{\ln(D_o/D_i) Ra^{1/4}}{\left[(L/D_i)^{3/5} + (L/D_o)^{3/5}\right]^{5/4}}, 1\right]_{max} \tag{3.2.27}$$

For eccentric cylinders, see Raithby and Hollands (1985).

21. *Concentric Spheres, Figure 3.2.5B.* The heat transfer between concentric spheres is given by the following equation (Raithby and Hollands, 1975) for $Ra \leq 6 \times 10^8$, $5 \leq Pr \leq 4000$, $1.25 < D_o/D_i \leq 2.5$,

$$Nu = \frac{qL}{D_i D_o k\Delta T} = \left[1.16\bar{C}_\ell\left(\frac{L}{D_i}\right)^{1/4} \frac{Ra^{1/4}}{\left[(D_i/D_o)^{3/5} + (D_o/D_i)^{4/5}\right]^{5/4}}, 1\right]_{max} \tag{3.2.28}$$

For eccentric spheres, see Raithby and Hollands (1985).

Example Calculations

Problem 1: Heat Transfer from Vertical Plate, Figure 3.2.6A. For the vertical isothermal surface in Figure 3.2.6A with $T_s = 40°C$, $H_1 = 1$ m, $H_2 = 1$ m, $W_1 = 1$ m, $W_2 = 1$ m and for an ambient air temperature of $T_\infty = 20°C$ (at 1 atm), find the heat transfer from one side of the plate.

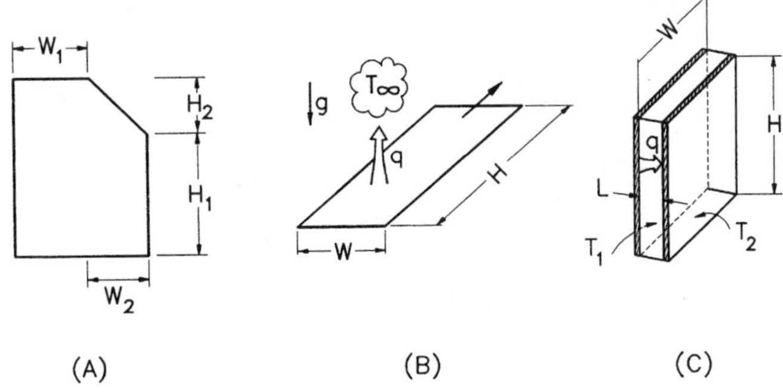

FIGURE 3.2.6 Sketches for example problems.

Properties: At $T_f = (T_w + T_\infty)/2 = 30°C$ and atmospheric pressure for air: $\nu = 1.59 \times 10^{-5}$ m²/sec, $\alpha = 2.25 \times 10^{-5}$ m²/sec, Pr = 0.71, k = 0.0263 W/mK. At T_∞, $\beta \approx 1/T_\infty = 1(273 + 20) = 0.00341$ K^{-1}.

Solution: For the geometry shown in Figure 3.2.6A:

$$A_s = (H_1 + H_2)W_1 + \left(H_1 + \frac{H_2}{2}\right)W_2 = 3.5 \text{ m}^2 \quad \text{(plate surface area)}$$

$$\int_0^{W_1+W_2} S^{3/4}\, d\chi = (H_1 + H_2)^{3/4} W_1 + \frac{4}{7}\frac{W_2}{H_2}\left[(H_1 + H_2)^{7/4} - H_1^{7/4}\right] = 3.03 \text{ m}^{7/4}$$

$$L^{1/4} = (H_1 + H_2)^{1/4} = 1.19 \text{ m}^{1/4}$$

$$C_1 = \frac{L^{1/4}\int_0^{W_1+W_2} S^{3/4}\, d\chi}{A_s} = \frac{1.19 \times 3.03}{3.5} = 1.03$$

$$\text{Ra} = \frac{g\beta_\infty L^3 (T_w - T_\infty)}{\nu\alpha} = \frac{9.81 \times 0.00341 \times 2^3 \times (40-20)}{1.59 \times 10^{-5} \times 2.25 \times 10^{-5}} = 1.50 \times 10^{10}$$

$\overline{C}_\ell = 0.514$ from Equation (3.2.3); $C_t = C_t^V = 0.103$ from Equation (3.2.4). $\text{Nu}^T = C_1 \overline{C}_\ell \text{Ra}^{1/4} = 185$ from Equation (3.2.11).

$$\left.\begin{aligned} \text{Nu}_l &= \frac{2.0}{\ln(1 + 2.0/\text{Nu}^T)} = 186 \\ \text{Nu}_t &= C_t^V \text{Ra}^{1/3} / (1 + 1.4 \times 10^9 \text{Pr}/\text{Ra}) = 238 \end{aligned}\right\} \text{(from Equation (3.2.7))}$$

$$\text{Nu} = \frac{qL}{A\Delta T k} = \left(\text{Nu}_\ell^6 + \text{Nu}_t^6\right)^{1/6} = 246$$

from Equation (3.2.6) with $m = 6$.

$$q = \frac{A_s \Delta T k \mathrm{Nu}}{\mathrm{L}} = \frac{3.5 \times 20 \times 0.0263 \times 246}{2} = 226W$$

Comments on Problem 1: Since $\mathrm{Nu}_\ell < \mathrm{Nu}_t$, the heat transfer is primarily turbulent. Do not neglect radiation. Had the surface been specified to be at constant heat flux, rather than isothermal, the equations in this section can be used to find the approximate average temperature difference between the plate and fluid.

Problem 2: Heat Transfer from Horizontal Strip, Figure 3.2.6B. Find the rate of heat loss per unit length from a very long strip of width $W = 0.1$ m with a surface temperature of $T_s = 70°\text{C}$ in water at $T_\infty = 30°\text{C}$.

Properties: At $T_f = (T_s + T_\infty)1/2 = 50°\text{C}$

$$\nu = 5.35 \times 10^{-7}\ \text{m}^2/\text{sec} \qquad \alpha = 1.56 \times 10^{-7}\ \text{m}^2/\text{sec} \qquad \text{Pr} = 3.42$$

$$k = 0.645\ \text{W/mK} \qquad \beta = 2.76 \times 10^{-4}\ \text{K}^{-1}$$

Solution: This problem corresponds to *Case 3* and Figure 3.2.1C.

$$C_t^H = 0.14$$

from Equation 3.2.5 and $\bar{C}_l = 0.563$ from Equation (3.2.3).

$$L^* = \lim_{H \to \infty}\left(\frac{WH}{2W + 2H}\right) = \frac{W}{2} = 0.05\ \text{m}$$

From Figure 3.2.1C and Equation (3.2.9)

$$\text{Ra} = \frac{g\beta\Delta T L^{*3}}{\nu\alpha} = 1.62 \times 10^8 \qquad \text{Nu}^T = 0.835\bar{C}_\ell \text{Ra}^{1/4} = 55.6$$

$$\text{Nu}_\ell = \frac{1.4}{\ln\left(1 + 1.4/\text{Nu}^T\right)} = 56.3 \qquad \text{Nu}_t = C_t^H \text{Ra}^{1/3} = 76.3$$

$$\text{Nu} = \frac{q}{WH\Delta T}\frac{L^*}{k} = \left(\text{Nu}_\ell^{10} + \text{Nu}_t^{10}\right)^{0.1} = 76.7$$

$$q/H = \frac{W\Delta T k \text{Nu}}{\text{L}^*} = 3960\ \text{W/m-length}$$

Comments: Turbulent heat transfer is dominant. Radiation can be ignored (since it lies in the far infrared region where it is not transmitted by the water).

Problem 3: Heat Loss Across a Window Cavity, Figure 3.2.6C. The interior glazing is at temperature $T_1 = 10°\text{C}$, the exterior glazing at $T_2 = -10°\text{C}$, the window dimensions are $W = 1$ m, $H = 1.7$ m, and the air gap between the glazings is $L = 1$ cm and is at atmospheric pressure. Find the heat flux loss across the window.

Properties: At $\bar{T} = T_1 + T_2/2 = 0°C = 273K$

$$\nu = 1.35\times10^{-5}\ \text{m}^2/\text{sec} \qquad \alpha = 1.89\times10^{-5}\ \text{m}^2/\text{sec} \qquad \text{Pr} = 0.71$$

$$k = 0.024\ \text{W/mK} \qquad \beta = 1/273 = 3.66\times10^{-3}\ \text{K}^{-1}$$

Solution: The appropriate correlations are given in *Case 18* and by Equation (3.2.24).

$$\text{Ra} = \frac{g\beta(T_1 - T_2)L^3}{\nu\alpha} = \frac{9.81\times3.66\times10^{-3}\times20\times(0.01)^3}{1.35\times10^{-5}\times1.89\times10^{-5}} = 2.81\times10^3$$

$$\text{Nu}_1 = \left[1 + \left\{\frac{0.0665\text{Ra}^{1/3}}{1+\left(\frac{9000}{\text{Ra}}\right)^{1.4}}\right\}^2\right]^{1/2} = 1.01$$

$$\text{Nu}_2 = 0.242\left(\text{Ra}\frac{L}{H}\right)^{0.273} = 0.242\left(2.81\times10^3\times\frac{0.01}{1.7}\right)^{0.273} = 0.520$$

$$\text{Nu} = \frac{qL}{WH(T_1 - T_2)k} = \left(\text{Nu}_1,\text{Nu}_2\right)_{\max} = 1.01$$

$$q/WH = \frac{\text{Nu}(T_1 - T_2)k}{L} = \frac{1.01\times20\times0.024}{0.01} = 48.5\ \text{W/m}^2$$

Comments: For pure conduction across the air layer, Nu = 1.0. For the calculated value of Nu = 1.01, convection must play little role. For standard glass, the heat loss by radiation would be roughly double the natural convection value just calculated.

Special Nomenclature

Note that nomenclature for each geometry considered is provided in the figures that are referred to in the text.

$\bar{C}_\ell$ = function of Prandtl number, Equation (3.2.3)
C_t^V = function of Prandtl number, Equation (3.2.4)
C_t^H = function of Prandtl number, Equation (3.2.5)
$\bar{C}_t$ = surface averaged value of C_t, page 3–39
$\overline{\Delta T}$ = surface averaged value of $T_w - T_\infty$

References

Churchill, S.W. 1983. *Heat Exchanger Design Handbook*, Sections 2.5.7 to 2.5.10, E.V. Schlinder, Ed., Hemisphere Publishing, New York.

Churchill S.W. and Usagi, R. 1972. A general expression for the correlation of rates of transfer and other phenomena, *AIChE J.*, 18, 1121–1128.

Clemes, S.B., Hollands, K.G.T., and Brunger, A.P. 1994. Natural convection heat transfer from horizontal isothermal cylinders, *J. Heat Transfer*, 116, 96–104.

Edwards, J.A. and Chaddock, J.B. 1963. An experimental investigation of the radiation and free-convection heat transfer from a cylindrical disk extended surface, *Trans., ASHRAE*, 69, 313–322.

Elenbaas, W. 1942a. The dissipation of heat by free convection: the inner surface of vertical tubes of different shapes of cross-section, *Physica*, 9(8), 865–874.

Elenbaas, W. 1942b. Heat dissipation of parallel plates by free convection, *Physica*, 9(1), 2–28.

Fujii, T. and Fujii, M. 1976. The dependence of local Nusselt number on Prandtl number in the case of free convection along a vertical surface with uniform heat flux, *Int. J. Heat Mass Transfer*, 19, 121–122.

Fujii, T., Honda, H., and Morioka, I. 1973. A theoretical study of natural convection heat transfer from downward-facing horizontal surface with uniform heat flux, *Int. J. Heat Mass Transfer*, 16, 611–627.

Goldstein, R.J., Sparrow, E.M., and Jones, D.C. 1973. Natural convection mass transfer adjacent to horizontal plates, *Int. J. Heat Mass Transfer*, 16, 1025–1035.

Hollands, K.G.T. 1984. Multi-Prandtl number correlations equations for natural convection in layers and enclosures, *Int. J. Heat Mass Transfer*, 27, 466–468.

Hollands, K.G.T., Unny, T.E., Raithby, G.D., and Konicek, K. 1976. Free convection heat transfer across inclined air layers, *J. Heat Transfer*, 98, 189–193.

Incropera, F.P. and DeWitt, D.P. 1990. *Fundamentals of Heat and Mass Transfer*, 3rd ed., John Wiley & Sons, New York.

Karagiozis, A. 1991. An Investigation of Laminar Free Convection Heat Transfer from Isothermal Finned Surfaces, Ph.D. Thesis, Department of Mechanical Engineering, University of Waterloo.

Karagiozis, A., Raithby, G.D., and Hollands, K.G.T. 1994. Natural convection heat transfer from arrays of isothermal triangular fins in air, *J. Heat Transfer*, 116, 105–111.

Kreith, F. and Bohn, M.S. 1993. *Principles of Heat Transfer.* West Publishing, New York.

Raithby, G.D. and Hollands, K.G.T. 1975. A general method of obtaining approximate solutions to laminar and turbulent free convection problems, in *Advances in Heat Transfer*, Irvine, T.F. and Hartnett, J.P., Eds., Vol. 11, Academic Press, New York, 266–315.

Raithby, G.D. and Hollands, K.G.T. 1985. *Handbook Heat Transfer*, Chap. 6: Natural Convection, Rohsenow, W.M., Hartnett, J.P., and Ganic, E.H., Eds., McGraw-Hill, New York.

Seki, N., Fukusako, S., and Inaba, H. 1978. Heat transfer of natural convection in a rectangular cavity with vertical walls of different temperatures, *Bull. JSME.*, 21(152), 246–253.

Shewen, E., Hollands, K.G.T., and Raithby, G.D. 1996. Heat transfer by natural convection across a vertical air cavity of large aspect ratio, *J. Heat Transfer*, 118, 993–995.

Further Information

There are several excellent heat transfer textbooks that provide fundamental information and correlations for natural convection heat transfer (e.g., Kreith and Bohn, 1993; Incropera and DeWitt, 1990). The correlations in this section closely follow the recommendations of Raithby and Hollands (1985), but that reference considers many more problems. Alternative equations are provided by Churchill (1983).

3.2.2 Forced Convection — External Flows

N.V. Suryanarayana

Introduction

In this section we consider heat transfer between a solid surface and an adjacent fluid which is in motion relative to the solid surface. If the surface temperature is different from that of the fluid, heat is transferred as forced convection. If the bulk motion of the fluid results solely from the difference in temperature of the solid surface and the fluid, the mechanism is natural convection. The velocity and temperature of the fluid far away from the solid surface are the free-stream velocity and free-stream temperature. Both are usually known or specified. We are then required to find the heat flux from or to the surface with specified surface temperature or the surface temperature if the heat flux is specified. The specified temperature or heat flux either may be uniform or may vary. The convective heat transfer coefficient h is defined by

$$q'' = h\left(T_s - T_\infty\right) \tag{3.2.29}$$

In Equation (3.2.29) with the local heat flux, we obtain the local heat transfer coefficient, and with the average heat flux with a uniform surface temperature we get the average heat transfer coefficient. For a specified heat flux the local surface temperature is obtained by employing the local convective heat transfer coefficient.

Many correlations for finding the convective heat transfer coefficient are based on experimental data which have some uncertainty, although the experiments are performed under carefully controlled conditions. The causes of the uncertainty are many. Actual situations rarely conform completely to the experimental situations for which the correlations are applicable. Hence, one should not expect the actual value of the heat transfer coefficient to be within better than ±10% of the predicted value.

Many different correlations to determine the convective heat transfer coefficient have been developed. In this section only one or two correlations are given. For other correlations and more details, refer to the books given in the bibliography at the end of this section.

Flat Plate

With a fluid flowing parallel to a flat plate, changes in velocity and temperature of the fluid are confined to a thin region adjacent to the solid boundary — the boundary layer. Several cases arise:

1. Flows without or with pressure gradient
2. Laminar or turbulent boundary layer
3. Negligible or significant viscous dissipation (effect of frictional heating)
4. $\Pr \geq 0.7$ (gases and most liquids) or $\Pr \ll 1$ (liquid metals)

Flows with Zero Pressure Gradient and Negligible Viscous Dissipation

When the free-stream pressure is uniform, the free-stream velocity is also uniform. Whether the boundary layer is laminar or turbulent depends on the Reynolds number Re_x ($\rho U_\infty x/\mu$) and the shape of the solid at entrance. With a sharp edge at the leading edge (Figure 3.2.7) the boundary layer is initially laminar but at some distance downstream there is a transition region where the boundary layer is neither totally laminar nor totally turbulent. Farther downstream of the transition region the boundary layer becomes turbulent. For engineering applications the existence of the transition region is usually neglected and it is assumed that the boundary layer becomes turbulent if the Reynolds number, Re_x, is greater than the critical Reynolds number, Re_{cr}. A typical value of 5×10^5 for the critical Reynolds number is generally accepted, but it can be greater if the free-stream turbulence is low and lower if the free-stream turbulence is high, the surface is rough, or the surface does not have a sharp edge at entrance. If the entrance is blunt, the boundary layer may be turbulent from the leading edge.

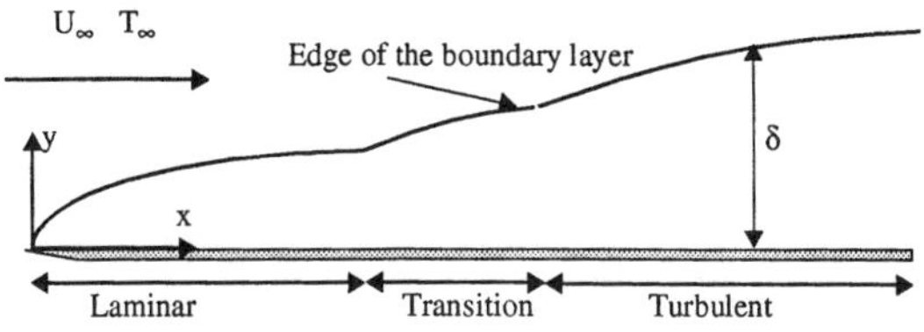

FIGURE 3.2.7 Flow of a fluid over a flat plate with laminar, transition, and turbulent boundary layers.

Temperature Boundary Layer

Analogous to the velocity boundary layer there is a temperature boundary layer adjacent to a heated (or cooled) plate. The temperature of the fluid changes from the surface temperature at the surface to the free-stream temperature at the edge of the temperature boundary layer (Figure 3.2.8).

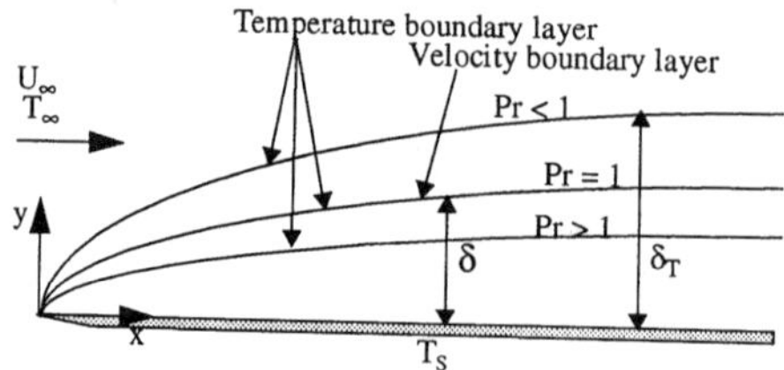

FIGURE 3.2.8 Temperature boundary layer thickness relative to velocity boundary layer thickness.

The velocity boundary layer thickness δ depends on the Reynolds number Re_x. The thermal boundary layer thickness δ_T depends both on Re_x and Pr

$\text{Re}_x < \text{Re}_{cr}$:

$$\frac{\delta}{x} = \frac{5}{\sqrt{\text{Re}_x}} \qquad \begin{array}{ll} \text{Pr} > 0.7 & \dfrac{\delta}{\delta_T} = \text{Pr}^{1/3} \\ \text{Pr} \ll 1 & \dfrac{\delta}{\delta_T} = \text{Pr}^{1/2} \end{array} \tag{3.2.30}$$

$\text{Re}_{cr} < \text{Re}_x$:

$$\frac{\delta}{x} = \frac{0.37}{\text{Re}_x^{0.2}} \qquad \delta \approx \delta_T \tag{3.2.31}$$

Viscous dissipation and high-speed effects can be neglected if $\text{Pr}^{1/2}\,\text{Ec}/2 \ll 1$. For heat transfer with significant viscous dissipation see the section on flow over flat plate with zero pressure gradient: Effect of High Speed and Viscous Dissipation. The Eckert number Ec is defined as $\text{Ec} = U_\infty^2 / C_p(T_s - T_\infty)$.

With a rectangular plate of length L in the direction of the fluid flow the average heat transfer coefficient h_L with uniform surface temperature is given by

$$h_L = \frac{1}{L}\int_0^L h_x \, dx$$

Laminar Boundary Layer (Figure 3.2.9) *($Re_x < Re_{cr}$, $Re_L < Re_{cr}$)*: With heating or cooling starting from the leading edge the following correlations are recommended. Note: in all equations evaluate fluid properties at the film temperature defined as the arithmetic mean of the surface and free-stream temperatures unless otherwise stated.

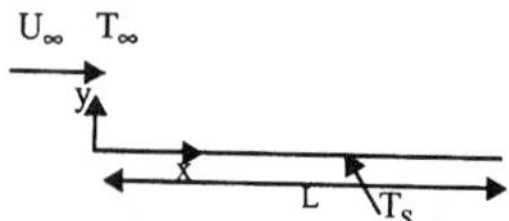

FIGURE 3.2.9 Heated flat plate with heating from the leading edge.

Local Heat Transfer Coefficient (Uniform Surface Temperature)
The Nusselt number based on the local convective heat transfer coefficient is expressed as

$$\mathrm{Nu}_x = f_{\mathrm{Pr}} \mathrm{Re}_x^{1/2} \tag{3.2.32}$$

The classical expression for f_{Pr} is 0.564 $\mathrm{Pr}^{1/2}$ for liquid metals with very low Prandtl numbers, $0.332\mathrm{Pr}^{1/3}$ for $0.7 < \mathrm{Pr} < 50$ and $0.339\mathrm{Pr}^{1/3}$ for very large Prandtl numbers. Correlations valid for all Prandtl numbers developed by Churchill (1976) and Rose (1979) are given below.

$$\mathrm{Nu}_x = \frac{0.3387\mathrm{Re}_x^{1/2}\,\mathrm{Pr}^{1/3}}{\left[1+\left(\frac{0.0468}{\mathrm{Pr}}\right)^{2/3}\right]^{1/4}} \tag{3.2.33}$$

$$\mathrm{Nu}_x = \frac{\mathrm{Re}_x^{1/2}\,\mathrm{Pr}^{1/2}}{\left(27.8+75.9\,\mathrm{Pr}^{0.306}+657\,\mathrm{Pr}\right)^{1/6}} \tag{3.2.34}$$

In the range $0.001 < \mathrm{Pr} < 2000$, Equation (3.2.33) is within 1.4% and Equation (3.2.34) is within 0.4% of the exact numerical solution to the boundary layer energy equation.

Average Heat Transfer Coefficient
The average heat transfer coefficient is given by

$$\mathrm{Nu}_L = 2\mathrm{Nu}_{x=L} \tag{3.2.35}$$

From Equation 3.2.35 it is clear that the average heat transfer coefficient over a length L is twice the local heat transfer coefficient at x = L.

Uniform Heat Flux

Local Heat Transfer Coefficient
Churchill and Ozoe (1973) recommend the following single correlation for all Prandtl numbers.

$$\mathrm{Nu}_x = \frac{0.886\mathrm{Re}_x^{1/2}\,\mathrm{Pr}^{1/2}}{\left[1+\left(\frac{\mathrm{Pr}}{0.0207}\right)^{2/3}\right]^{1/4}} \tag{3.2.36}$$

Note that for surfaces with uniform heat flux the local convective heat transfer coefficient is used to determine the local surface temperature. The total heat transfer rate being known, an average heat transfer coefficient is not needed and not defined.

Turbulent Boundary Layer *($Re_x > Re_{cr}$, $Re_L > Re_{cr}$)*: For turbulent boundary layers with heating or cooling starting from the leading edge use the following correlations:

Local Heat Transfer Coefficient

$\mathrm{Re}_{cr} < \mathrm{Re}_x < 10^7$:

$$\mathrm{Nu}_x = 0.0296\mathrm{Re}_x^{4/5}\,\mathrm{Pr}^{1/3} \tag{3.2.37}$$

$10^7 < \mathrm{Re}_x$:

$$\mathrm{Nu}_x = 1.596\mathrm{Re}_x\left(\ln \mathrm{Re}_x\right)^{-2.584}\mathrm{Pr}^{1/3} \tag{3.2.38}$$

Equation (3.2.38) is obtained by applying Colburn's j factor in conjunction with the friction factor suggested by Schlicting (1979).

In laminar boundary layers, the convective heat transfer coefficient with uniform heat flux is approximately 36% higher than with uniform surface temperature. With turbulent boundary layers, the difference is very small and *the correlations for the local convective heat transfer coefficient can be used for both uniform surface temperature and uniform heat flux.*

Average Heat Transfer Coefficient

If the boundary layer is initially laminar followed by a turbulent boundary layer at $\mathrm{Re}_x = \mathrm{Re}_{cr}$, the following correlations for $0.7 < \mathrm{Pr} < 60$ are suggested:

$\mathrm{Re}_{cr} < \mathrm{Re}_L < 10^7$:

$$\mathrm{Nu}_L = \left[0.664\mathrm{Re}_L^{1/2} + 0.037\left(\mathrm{Re}_L^{4/5} - \mathrm{Re}_{cr}^{4/5}\right)\right]\mathrm{Pr}^{1/3} \tag{3.2.39}$$

If $\mathrm{Re}_{cr} < \mathrm{Re}_L < 10^7$ and $\mathrm{Re}_{cr} = 10^5$, Equation 3.2.39 simplifies to

$$\mathrm{Nu}_L = \left(0.037\mathrm{Re}_L^{4/5} - 871\right)\mathrm{Pr}^{1/3} \tag{3.2.40}$$

$10^7 < \mathrm{Re}_L$ and $\mathrm{Re}_{cr} = 5 \times 10^5$:

$$\mathrm{Nu}_L = \left[1.963\mathrm{Re}_L\left(\ln \mathrm{Re}_L\right)^{-2.584} - 871\right]\mathrm{Pr}^{1/3} \tag{3.2.41}$$

Uniform Surface Temperature — Pr > 0.7: Unheated Starting Length

If heating does not start from the leading edge as shown in Figure 3.2.10, the correlations have to be modified. Correlation for the local convective heat transfer coefficient for laminar and turbulent boundary layers are given by Equations (3.2.42) and (3.2.43) (Kays and Crawford, 1993) — the constants in Equations (3.2.42) and (3.2.43) have been modified to be consistent with the friction factors. These correlations are also useful as building blocks for finding the heat transfer rates when the surface temperature varies in a predefined manner. Equations (3.2.44) and (3.2.45), developed by Thomas (1977), provide the average heat transfer coefficients based on Equations (3.2.42) and (3.2.43).

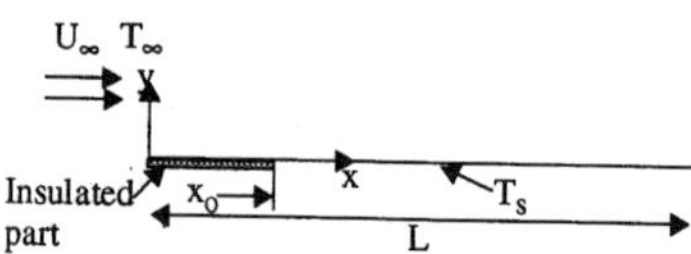

FIGURE 3.2.10 Heated flat plate with unheated starting length.

Local Convective Heat Transfer Coefficient

$\mathrm{Re}_x < \mathrm{Re}_{cr}$:

$$\mathrm{Nu}_x = \frac{0.332\mathrm{Re}_x^{1/2}\mathrm{Pr}^{1/3}}{\left[1-\left(\frac{x_o}{x}\right)^{3/4}\right]^{1/3}} \tag{3.2.42}$$

$\mathrm{Re}_x > \mathrm{Re}_{cr}$:

$$\mathrm{Nu}_x = \frac{0.0296\mathrm{Re}_x^{4/5}\mathrm{Pr}^{3/5}}{\left[1-\left(\frac{x_o}{x}\right)^{9/10}\right]^{1/9}} \tag{3.2.43}$$

Average Heat Transfer Coefficient over the Length (L – x_o)

$Re_L < Re_{cr}$:

$$h_{L-x_o} = \frac{0.664\mathrm{Re}_L^{1/2}\mathrm{Pr}^{1/3}\left[1-\left(\frac{x_o}{L}\right)^{3/4}\right]^{2/3} k}{L-x_o}$$
$$= 2\frac{1-\left(\frac{x_o}{L}\right)^{3/4}}{1-x_o/L} h_{x=L} \tag{3.2.44}$$

In Equation (3.2.44) evaluate $h_{x=L}$ from Equation (3.2.42).

$Re_{cr} = 0$:

$$h_{L-x_o} = \frac{0.037\mathrm{Re}_L^{4/5}\mathrm{Pr}^{3/5}\left[1-\left(\frac{x_o}{L}\right)^{9/10}\right]^{8/9} k}{L-x_o}$$
$$= 1.25\frac{1-\left(x_o/L\right)^{9/10}}{1-x_o/L} h_{x=L} \tag{3.2.45}$$

In Equation (3.2.45) evaluate $h_{x=L}$ from Equation (3.2.43).

Flat Plate with Prescribed Nonuniform Surface Temperature

The linearity of the energy equation permits the use of Equations (3.2.42) through (3.2.45) for uniform surface temperature with unheated starting length to find the local heat flux and the total heat transfer rate by the principle of superposition when the surface temperature is not uniform. Figure 3.2.11 shows the arbitrarily prescribed surface temperature with a uniform free-stream temperature of the fluid. If the surface temperature is a differentiable function of the coordinate *x*, the local heat flux can be determined by an expression that involves integration (refer to Kays and Crawford, 1993). If the surface temperature can be approximated as a series of step changes in the surface temperature, the resulting expression for the local heat flux and the total heat transfer rate is the summation of simple algebraic expressions. Here the method using such an algebraic simplification is presented.

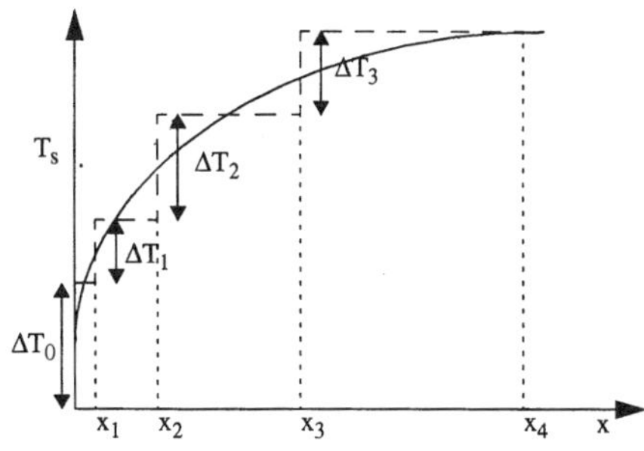

FIGURE 3.2.11 Arbitrary surface temperature approximated as a finite number of step changes.

The local convective heat flux at a distance x from the leading edge is given by

$$q''_x = \sum_1^n h_{xi} \Delta T_{si} \tag{3.2.46}$$

where h_{xi} denotes the local convective heat transfer coefficient at x due to a single step change in the surface temperature ΔT_{si} at location $x_i(x_i < x)$. Referring to Figure 3.2.11, the local convective heat flux at x $(x_3 < x < x_4)$ is given by

$$q''_x = h_x(x,0)\Delta T_o + h_x(x,x_1)\Delta T_1 + h_x(x,x_2)\Delta T_2 + h_x(x,x_3)\Delta T_3$$

where $h_x(x, x_1)$ is the local convective heat transfer coefficient at x with heating starting from x_1; the local convective heat transfer is determined from Equation (3.2.42) if the boundary layer is laminar and Equation (3.2.43) if the boundary layer is turbulent from the leading edge. For example, $h_x(x, x_2)$ in the third term is given by

$$\mathrm{Re}_x < \mathrm{Re}_{cr} \qquad h_x(x,x_2) = \frac{0.332\left(\frac{\rho U_\infty x}{\mu}\right)^{1/2} \mathrm{Pr}^{1/3}}{\left[1-\left(\frac{x_2}{x}\right)^{3/4}\right]^{1/3}} \frac{k}{x}$$

$$\mathrm{Re}_{cr} = 0 \qquad h_x(x,x_2) = \frac{0.0296\left(\frac{\rho U_\infty x}{\mu}\right)^{4/5} \mathrm{Pr}^{3/5}}{\left[1-\left(\frac{x_2}{x}\right)^{9/10}\right]^{1/9}} \frac{k}{x}$$

The procedure for finding the total heat transfer rate from $x = 0$ to $x = L$ is somewhat similar. Denoting the width of the plate by W,

$$\frac{q}{W} = \sum \bar{h}_{L-x_i} \Delta T_i (L - x_i) \tag{3.2.47}$$

where $\bar{h}_{L-x_i}$ is the average heat transfer coefficient over the length $L - x_i$ due to a step change ΔT_i in the surface temperature at x_i. For example, the heat transfer coefficient in the third term in Equation (3.2.47) obtained by replacing x_o by x_2 in Equation (3.2.44) or (3.2.45) depending on whether $\mathrm{Re}_L < \mathrm{Re}_{cr}$ or $\mathrm{Re}_{cr} = 0$.

Flows with Pressure Gradient and Negligible Viscous Dissipation

Although correlations for flat plates are for a semi-infinite fluid medium adjacent to the plate, most applications of practical interest deal with fluid flowing between two plates. If the spacing between the plates is significantly greater than the maximum boundary layer thickness, the medium can be assumed to approach a semi-infinite medium. In such a case if the plates are parallel to each other and if the pressure drop is negligible compared with the absolute pressure, the pressure gradient can be assumed to be negligible. If the plates are nonparallel and if the boundary layer thickness is very much smaller than the spacing between the plates at that location, the medium can still be considered as approaching a semi-infinite medium with a non-negligible pressure gradient. In such flows the free-stream velocity (core velocity outside the boundary layer) is related to the pressure variation by the Bernoulli equation:

$$\frac{p}{\rho} + \frac{U_\infty^2}{2} + zg = \text{constant}$$

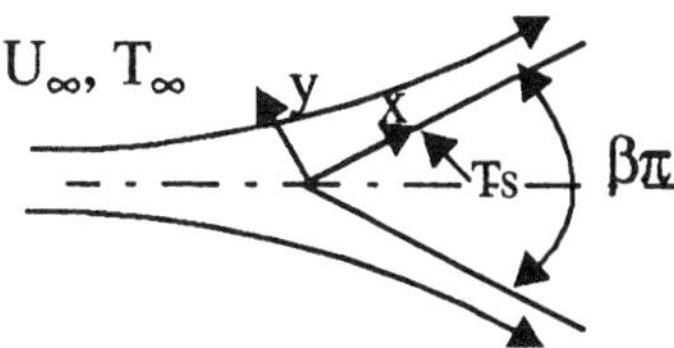

FIGURE 3.2.12 Flow over a wedge. $\beta\pi$ is the wedge angle.

Another situation where the free-stream velocity varies in the direction of flow giving rise to a pressure gradient is flow over a wedge. For the family of flows for which the solutions are applicable, the free-stream velocity at the edge of the boundary layer is related to the x-coordinate by a power law, $U_\infty = cx^m$. Flows over semi-infinite wedges (Figure 3.2.12) satisfy that condition. The exponent m is related to the wedge angle $\beta\pi$

$$\beta = \frac{2m}{1+m} \qquad m = \frac{\beta}{2-\beta}$$

With laminar boundary layers, the boundary layer thickness, friction factor, and Nusselt numbers are defined by

$$\frac{\delta}{x} = \frac{c_1}{\sqrt{\text{Re}_x}} \qquad \frac{C_{fx}}{2} = \frac{\tau_w}{\rho U_\infty^2} = \frac{c_2}{\sqrt{\text{Re}_x}} \qquad \text{Nu}_x = c_3 \text{Re}_x^{1/2}$$

The values of c_1, c_2, and c_3 are available in Burmeister (1993). For example, for $\beta = 0.5$ (wedge angle $= 90°$), $m = 1/3$, $c_1 = 3.4$, $c_2 = 0.7575$, and $c_3 = 0.384$ for Pr = 0.7, and $c_3 = 0.792$ for Pr = 5. Re_x is based on $U_\infty = cx^m$; the free-stream velocity is not uniform.

Uniform Temperature: Flat Plate with Injection or Suction with External Flows of a Fluid Parallel to the Surface

Injection (Figure 3.2.13) or suction has engineering applications. When the free-stream temperature of the fluid is high, as in gas turbines, a cooling fluid is introduced into the mainstream to cool the surface. If the cooling fluid is introduced at discrete locations (either perpendicular to the surface or at an angle), it is known as film cooling. If a fluid is introduced or withdrawn through a porous medium, it is known as transpiration (Figure 3.2.13). An application of suction is to prevent boundary layer separation.

Analytical solutions for a laminar boundary layer with transpiration suction or blowing are available if the velocity perpendicular to the surface varies in the following manner:

$$\text{v}_o = \text{constant } x^{(m-1)/2}$$

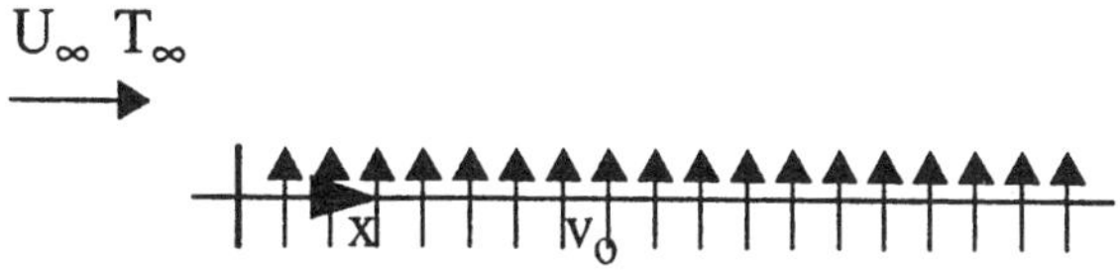

FIGURE 3.2.13 Flat plate with transpiration injection.

Solutions are limited to the cases of the injected fluid being at the same temperature as the surface and the injected fluid being the same as the free-stream fluid. Positive values of v_o indicate blowing and negative values indicate suction. Values of $Nu_x/Re_x^{1/2}$ for different values of Pr and for different values of blowing or suction parameter are given in Kays and Crawford (1993).

For example, for a laminar boundary layer over a flat plate with a fluid (Pr = 0.7) the value of $Nu_x/Re_x^{1/2}$ is 0.722 for $(v_o/U_\infty)\sqrt{\rho U_\infty x/\mu} = -0.75$ (suction) and 0.166 for $(v_o/U_\infty)\sqrt{\rho U_\infty x/\mu} = 0.25$ (blowing). Heat transfer coefficient increases with suction which leads to a thinning of the boundary layer. Blowing increases the boundary layer thickness and decreases the heat transfer coefficient.

For *turbulent boundary layers* Kays and Crawford (1993) suggest the following procedure for finding the friction factor and convective heat transfer coefficient. Define friction blowing parameter B_f and heat transfer blowing parameter B_h as

$$B_f = \frac{v_o/U_\infty}{C_f/2} \tag{3.2.48}$$

$$B_h = \frac{v_o/U_\infty}{\text{St}} = \frac{\dot{m}''/G_\infty}{\text{St}} \tag{3.2.49}$$

where

v_o = velocity normal to the plate
U_∞ = free-stream velocity
$\dot{m}''$ = mass flux of the injected fluid at the surface (ρv_o)
G_∞ = mass flux in the free stream (ρU_∞)
St = Stanton number = $Nu_x/Re_x Pr = h/\rho U_\infty c_p$

The friction factors and Stanton number with and without blowing or suction are related by

$$\frac{C_f}{C_{fo}} = \frac{\ln(1+B_f)}{B_f} \tag{3.2.50}$$

$$\frac{\text{St}}{\text{St}_o} = \frac{\ln(1+B_h)}{B_h} \tag{3.2.51}$$

In Equations (3.2.50) and (3.2.51) C_{fo} and St_o are the friction factor and Stanton number with $v_o = 0$ (no blowing or suction), and C_f and St are the corresponding quantities with blowing or suction at the same $Re_x(\rho U_\infty x/\mu)$.

For the more general case of variable free-stream velocity, temperature difference, and transpiration rate, refer to Kays and Crawford (1993).

Flow over Flat Plate with Zero Pressure Gradient: Effect of High-Speed and Viscous Dissipation

In the boundary layer the velocity of the fluid is reduced from U_∞ to zero at the plate leading to a reduction in the kinetic energy of the fluid. Within the boundary layer there is also the work done by viscous forces; the magnitude of the such viscous work is related to the velocity of the fluid, the velocity gradient, and the viscosity of the fluid. The effect of such a reduction in the kinetic energy and the viscous work is to increase the internal energy of the fluid in the boundary layer. The increase in the internal energy may be expected to lead to an increase in the temperature; but because of the heat transfer to the adjacent fluid the actual increase in the internal energy (and the temperature) will be less than the sum of the decrease in the kinetic energy and viscous work transfer; the actual temperature increase depends on the decrease in the kinetic energy, the viscous work transfer, and the heat transfer from the fluid. The maximum temperature in the fluid with an adiabatic plate is known as the adiabatic wall temperature (which occurs at the wall) and is given by

$$T_{aw} = T_\infty + r\frac{U_\infty^2}{2C_p} \tag{3.2.52}$$

In Equation (3.2.52) r is the recovery factor and is given by Eckert and Drake (1972).

Laminar boundary layer $0.6 < \text{Pr} < 15$ $r = \text{Pr}^{1/2}$

Turbulent boundary layer $r = \text{Pr}^{1/3}$

Equation (3.2.52) can be recast as

$$\frac{T_{aw} - T_\infty}{T_s - T_\infty} = \frac{r}{2}\frac{U_\infty^2}{C_p(T_s - T_\infty)} \tag{3.2.53}$$

From Equation (3.2.53) the maximum increase in the fluid temperature as a fraction of the difference between the plate and free-stream temperatures is given by r Ec/2. With air flowing over a plate at 500 m/sec, the increase in the temperature of the air can be as high as 105°C. With T_s = 40°C and T_∞ = 20°C, the temperature of the air close to the plate can be higher than the plate temperature. It is thus possible that although the plate temperature is higher than the free-stream temperature, the heat transfer is from the air to the plate. At a Mach number greater than 0.1 for gases, viscous dissipation becomes significant.

The temperature profiles for high-speed flows for different values of T_s are shown in Figure 3.2.14. In high-speed flows, as heat transfer can be to the plate even if the plate temperature is greater than the fluid temperature, the definition of the convective heat transfer coefficient given in Equation (3.2.29) is not adequate. On the other hand, as the heat transfer is always from the plate if $T_s > T_{aw}$, the adiabatic wall temperature is more appropriate as the reference temperature. Thus, in high-speed flows the definition of the convective heat transfer coefficient is given by

$$q'' = h(T_s - T_{aw}) \tag{3.2.54}$$

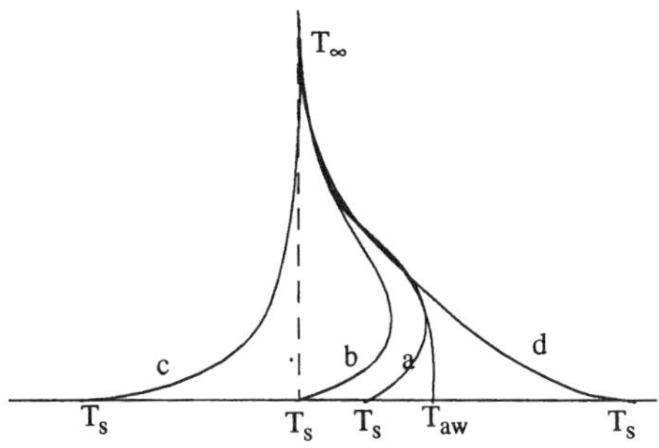

FIGURE 3.2.14 Temperature profiles for high-speed flows: (a) $T_\infty < T_s < T_{aw}$; (b) $T_s = T_\infty$; (c) $T_s \ll T_\infty$; (d) $T_s > T_{aw}$.

Equation (3.2.54) is consistent with Equation (3.2.29) as the adiabatic wall temperature equals the free-stream temperature if the effects of viscous dissipation and reduced kinetic energy in the boundary layer are neglected. With the adiabatic wall temperature as the fluid reference temperature for the definition of the convective heat transfer coefficient, equations for low speeds can also be used for high-speed flows. Because of the greater variation in the fluid temperature in the boundary layer, the variation of properties due to temperature variation becomes important. It is found that the correlations are best approximated if the properties are evaluated at the reference temperature T^* defined by Eckert (1956):

$$T^* = 0.5\left(T_s + T_\infty\right) + 0.22\left(T_{aw} - T_\infty\right) \tag{3.2.55}$$

With properties evaluated at the reference temperature given by Equation (3.2.55), Equation (3.2.56) through (3.2.61) are applicable to high-speed flows with Prandtl numbers less than 15. It should be noted that the adiabatic wall temperatures in the laminar and turbulent regions are different affecting both the temperature at which the properties are evaluated and the temperature difference for determining the local heat flux. Therefore, when the boundary layer is partly laminar and partly turbulent, an average value of the heat transfer coefficient is not defined as the adiabatic wall temperatures in the two regions are different. In such cases the heat transfer rate in each region is determined separately to find the total heat transfer rate.

Evaluate properties at reference temperature given by Equation (3.2.55):

Laminar — Local: $\mathrm{Re}_x < \mathrm{Re}_{cr}$

$$\mathrm{Nu}_x = 0.332\mathrm{Re}_x^{1/2}\mathrm{Pr}^{1/3} \tag{3.2.56}$$

Average: $\mathrm{Re}_L < \mathrm{Re}_{cr}$

$$\mathrm{Nu}_L = 0.664\mathrm{Re}_L^{1/2}\mathrm{Pr}^{1/3} \tag{3.2.57}$$

Turbulent — Local: $10^7 > \mathrm{Re}_x > \mathrm{Re}_{cr}$

$$\mathrm{Nu}_x = 0.0296\mathrm{Re}_x^{4/5}\mathrm{Pr}^{1/3} \tag{3.2.58}$$

Local: $10^7 < \mathrm{Re}_x < 10^9$

$$\mathrm{Nu}_x = 1.596\mathrm{Re}_x\left(\mathrm{lnRe}_x\right)^{-2.584}\mathrm{Pr}^{1/3} \tag{3.2.59}$$

Average: $\mathrm{Re}_{cr} = 0,\ \mathrm{Re}_L < 10^7$

$$\mathrm{Nu}_L = 0.037\mathrm{Re}_L^{4/5}\mathrm{Pr}^{1/3} \tag{3.2.60}$$

Average: $\mathrm{Re}_{cr} = 0,\ 10^7 < \mathrm{Re}_L < 10^9$

$$\mathrm{Nu}_L = 1.967\mathrm{Re}_L\left(\mathrm{lnRe}_L\right)^{-2.584}\mathrm{Pr}^{1/3} \tag{3.2.61}$$

When the temperature variation in the boundary layer is large, such that the assumption of constant specific heat is not justified, Eckert (1956) suggests that the properties be evaluated at a reference temperature corresponding to the specific enthalpy i^* given by

$$i^* = 0.5\left(i_s + i_\infty\right) + 0.22\left(i_{aw} - i_\infty\right) \tag{3.2.62}$$

where i is the specific enthalpy of the fluid evaluated at the temperature corresponding to the subscript. Equation (3.2.62) gives the same values as Equation (3.2.55) if C_p is constant or varies linearly with temperature.

At very high speeds the gas temperature may reach levels of temperatures that are sufficient to cause disassociation and chemical reaction; these and other effects need to be taken into account in those cases.

Flow over Cylinders, Spheres, and Other Geometries

Flows over a flat plate and wedges were classified as laminar or turbulent, depending on the Reynolds number, and correlations for the local and average convective heat transfer coefficients were developed. But flows over cylinders (perpendicular to the axis) and spheres are more complex. In general, the flow over cylinders and spheres may have a laminar boundary layer followed by a turbulent boundary layer

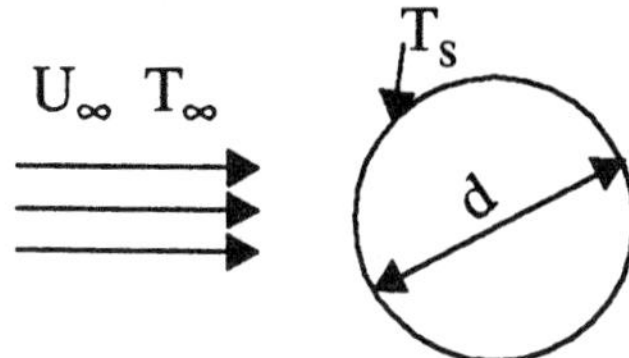

FIGURE 3.2.15 A fluid stream in cross flow over a cylinder.

and a wake region depending on the Reynolds number with the diameter as the characteristic length (Figure 3.2.15). Because of the complexity of the flow patterns, only correlations for the average heat transfer coefficients have been developed.

Cylinders: Use the following correlation proposed by Churchill and Bernstein (1977): $Re_d\, Pr > 0.2$. Evaluate properties at $(T_s + T_\infty)/2$:

$$Re_d > 400{,}000: \qquad Nu_d = 0.3 + \frac{0.62 Re_d^{1/2} Pr^{1/3}}{\left[1 + (0.4/Pr)^{2/3}\right]^{1/4}} \left[1 + \left(\frac{Re_d}{282{,}000}\right)^{5/8}\right]^{4/5} \tag{3.2.63}$$

$$10{,}000 < Re_d < 400{,}000: \quad Nu_d = 0.3 + \frac{0.62 Re_d^{1/2} Pr^{1/3}}{\left[1 + (0.4/Pr)^{2/3}\right]^{1/4}} \left[1 + \left(\frac{Re_d}{282{,}000}\right)^{1/2}\right] \tag{3.2.64}$$

$$Re_d < 10{,}000: \qquad Nu_d = 0.3 + \frac{0.62 Re_d^{1/2} Pr^{1/3}}{\left[1 + (0.4/Pr)^{2/3}\right]^{1/4}} \tag{3.2.65}$$

For flow of liquid metals, use the following correlation suggested by Ishiguro et al. (1979):

$$1 < Re_d Pr < 100 \qquad Nu_d = 1.125\left(Re_d Pr\right)^{0.413} \tag{3.2.66}$$

For more information on heat transfer with flow over cylinders, refer to Morgan (1975) and Zukauskas (1987).

Spheres: For flows over spheres (Figure 3.2.16) use one of the following three correlations.

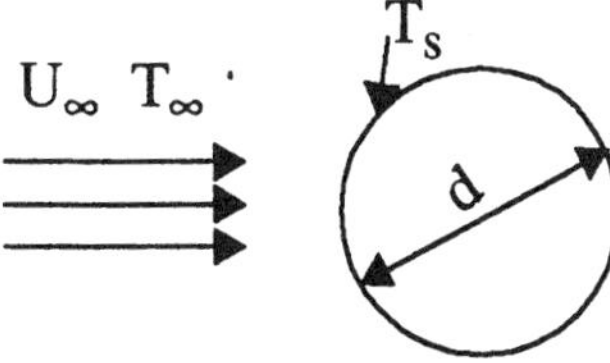

FIGURE 3.2.16 A fluid flowing over a sphere.

1. Whitaker (1972): Evaluate properties at T_∞ except μ_s at T_s.

$$3.5 < Re_d < 76{,}000 \qquad 0.71 < Pr < 380 \qquad 1 < \mu/\mu_s < 3.2$$

$$\mathrm{Nu}_d = 2.0 + \left(0.4\mathrm{Re}_d^{1/2} + 0.06\mathrm{Re}_d^{2/3}\right)\mathrm{Pr}^{2/5}\left(\frac{\mu}{\mu_s}\right)^{1/4} \tag{3.2.67}$$

2. Achenbach (1978): Evaluate properties at $(T_s + T_\infty)/2$:

$100 < \mathrm{Re}_d < 2\times10^5$ $\quad$ $\mathrm{Pr} = 0.71$

$$\mathrm{Nu}_d = 2 + \left(0.25\mathrm{Re}_d + 3\times10^{-4}\mathrm{Re}_d^{1.6}\right)^{1/2} \tag{3.2.68}$$

$4\times10^5 < \mathrm{Re}_d < 5\times10^6$ $\quad$ $\mathrm{Pr} = 0.71$

$$\mathrm{Nu}_d = 430 + 5\times10^{-3}\mathrm{Re}_d + 0.25\times10^{-9}\mathrm{Re}_d^2 - 3.1\times10^{-17}\mathrm{Re}_d^3 \tag{3.2.69}$$

3. Liquid Metals: From experimental results with liquid sodium, Witte (1968) proposed

$$3.6\times10^4 < \mathrm{Re}_d < 1.5\times10^5 \qquad \mathrm{Nu}_d = 2 + 0.386\left(\mathrm{Re}_d\mathrm{Pr}\right)^{1/2} \tag{3.2.70}$$

Other Geometries: For geometries other than cylinders and spheres, use Equation (3.2.71) with the characteristic dimensions and values of the constants given in the Table 3.2.1.

$$\mathrm{Nu}_D = c\mathrm{Re}_D^m \tag{3.2.71}$$

Although Equation (3.2.71) is based on experimental data with gases, its use can be extended to fluids with moderate Prandtl numbers by multiplying Equation (3.2.71) by $(\mathrm{Pr}/0.7)^{1/3}$.

Heat Transfer across Tube Banks

When tube banks are used in heat exchangers, the flow over the tubes in the second and subsequent rows of tubes is different from the flow over a single tube. Even in the first row the flow is modified by the presence of the neighboring tubes. The extent of modification depends on the spacing between the tubes. If the spacing is very much greater than the diameter of the tubes, correlations for single tubes can be used. Correlations for flow over tube banks when the spacing between tubes in a row and a column is not much greater than the diameter of the tubes have been developed for use in heat-exchanger applications. Two arrangements of the tubes are considered — aligned and staggered as shown in Figure 3.2.17. The nomenclature used in this section is shown in the figure.

For the average convective heat transfer coefficient with tubes at uniform surface temperature, from experimental results, Zukauskas (1987) recommends correlations of the form:

$$\mathrm{Nu}_d = c\left(\frac{a}{b}\right)^p \mathrm{Re}_d^m \mathrm{Pr}^n \left(\frac{\mathrm{Pr}}{\mathrm{Pr}_s}\right)^{0.25} \tag{3.2.72}$$

In Equation (3.2.72) all properties are evaluated at the arithmetic mean of the inlet and exit temperatures of the fluid, except Pr_s which is evaluated at the surface temperature T_s. The values of the constants c, p, m, and n are given in Table 3.2.2 for in-line arrangement and in Table 3.2.3 for staggered arrangement.

TABLE 3.2.1 Values of *c* and *m* in Equation (3.2.71)

Geometry	Re_D	c	m
	5000-100 000	0.092	0.675
	2500-8000	0.160	0.699
	5000-100 000	0.222	0.588
	2500-7500	0.261	0.624
	5000-19500	0.144	0.638
	19 500-100 000	0.035	0.782
	5000-100 000	0.138	0.638
	2500-15 000	0.224	0.612
	3000-15 000	0.085	0.804
	4000-15 000	0.205	0.731
Charcteristic dimension is the equivalent circular diameter = Perimeter/π For example, for a square rod with each side a, D = 4a/π			

From Jakob, 1949. With permission.

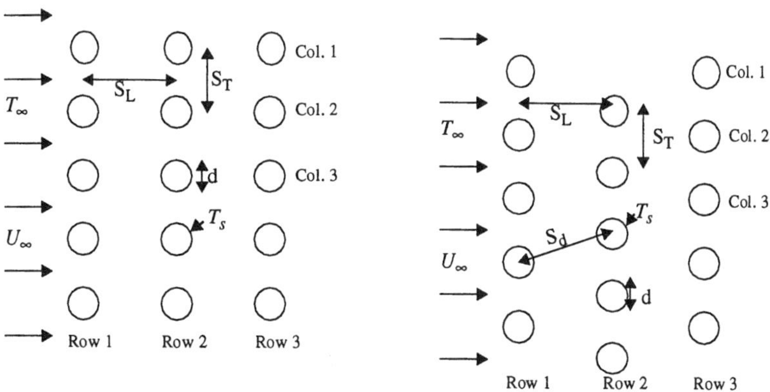

FIGURE 3.2.17 Two arrangements of tube banks. In-line or aligned arrangement on the left and staggered arrangement on the right. ($a = S_T/d$; $b = S_L/d$.)

TABLE 3.2.2 In-Line Arrangement — Values of Constants in Equation (3.2.72) ($p = 0$ in all cases)

Re_d	c	m	n
1–100	0.9	0.4	0.36
100–1000	0.52	0.5	0.36
10^3–2×10^5	0.27	0.63	0.36
$2 \times 10^5 - 2 \times 10^6$	0.033	0.8	0.4

TABLE 3.2.3 Staggered Arrangement — Values of Constants in Equation (3.2.72)

Re_d	c	p	m	n
1–500	1.04	0	0.4	0.36
500–1000	0.71	0	0.5	0.36
10^3–2×10^5	0.35	0.2	0.6	0.36
$2 \times 10^5 - 2 \times 10^6$	0.031	0.2	0.8	0.36

In computing Re_d, the maximum average velocity between tubes is used. The maximum velocities for the in-line and staggered arrangements are given by

$$\text{In-line:} \qquad U_{max} = \frac{U_\infty S_T}{S_T - d} \tag{3.2.73}$$

$$\text{Staggered:} \qquad S_d > \frac{S_T + d}{2} \qquad U_{max} = \frac{U_\infty S_T}{S_T - d} \tag{3.2.74}$$

$$\text{Staggered:} \qquad S_d < \frac{S_T + d}{2} \qquad U_{max} = \frac{U_\infty S_T}{2(S_d - d)} \tag{3.2.75}$$

$$S_d = \left[S_L^2 + \left(\frac{S_T}{2} \right)^2 \right]^{1/2}$$

Equation (3.2.72) is for tube banks with 16 or more rows. When there are fewer than 16 rows, the heat transfer coefficient given by Equation (3.2.72) is multiplied by the correction factor c_1 defined by Equation (3.2.76) and given in Table 3.2.4.

$$\frac{h_N}{h_{16}} = c_1 \tag{3.2.76}$$

where

h_N = heat transfer coefficient with N rows (fewer than 16)
h_{16} = heat transfer coefficient with 16 or more rows

TABLE 3.2.4 Correction Factor c_1 to Be Used with Equation (3.2.76)

	Number of Rows (N)							
Tube Arrangement	**1**	**2**	**3**	**4**	**5**	**7**	**10**	**13**
In-line	0.70	0.80	0.86	0.90	0.93	0.96	0.98	0.99
Staggered	0.64	0.76	0.84	0.89	0.93	0.96	0.98	0.99

Pressure Drop: With tube banks, pressure drop is a significant factor, as it determines the fan power required to maintain the fluid flow. Zukauskas (1987) recommends that the pressure drop be computed from the relation

$$\Delta p = p_i - p_e = N\chi \frac{\rho U_{max}^2}{2} f \tag{3.2.77}$$

where p_i and p_e are the fluid pressures at inlet and exit of the tube banks. The values of χ and f are presented in Figure 3.2.18a. In Figure 3.2.18a the friction factor f for in-line arrangement is presented for different values of b (S_L/d) for $S_L = S_T$. For values of S_L/S_T other than 1, the correction factor χ is given in the inset for different values of $(a - 1)/(b - 1)$. Similarly, the friction factor for staggered arrangement (for equilateral triangle arrangement) and a correction factor for different values of a/b are also given in Figure 3.2.18b. The value of f is for one row of tubes; the total pressure drop is obtained by multiplying the pressure drop for one row by the number of rows, N.

The temperature of the fluid varies in the direction of flow, and, therefore, the value of the convective heat transfer coefficient (which depends on the temperature-dependent properties of the fluid) also varies in the direction of flow. However, it is common practice to compute the total heat transfer rate with the assumption of uniform convective heat transfer coefficient evaluated at the arithmetic mean of the inlet and exit temperatures of the fluid. With such an assumption of uniform convective heat transfer coefficient, uniform surface temperature and constant specific heat (evaluated at the mean fluid temperature), the inlet and exit fluid temperatures are related by

$$\ln\left(\frac{T_s - T_e}{T_s - T_i}\right) = -\frac{hA_s}{\dot{m}c_p} \tag{3.2.78}$$

The heat transfer rate to the fluid is related by the equation

$$q = m \cdot c_p \left(T_e - T_i\right) \tag{3.2.79}$$

Example

A heat exchanger with aligned tubes is used to heat 40 kg/sec of atmospheric air from 10 to 50°C with the tube surfaces maintained at 100°C. Details of the heat exchanger are

Diameter of tubes	25 mm
Number of columns	20
Length of each tube	3 m
$S_L = S_T$	75 mm

Determine the number of rows required.

Solution: Average air temperature = $(T_i + T_e)/2$ = 30°C. Properties of atmospheric air (from Suryanarayana, 1995):

$$\begin{aligned} &\rho = 1.165\ \text{kg/m}^2 && c_p = 1007\ \text{J/kg K} \\ &\mu = 1.865 \times 10^{-5}\ \text{N sec/m}^3 && k = 0.0264\ \text{W/mK} \\ &\text{Pr} = 0.712 && \text{Pr}_s\ (\text{at } 100°\text{C}) = 0.705 \end{aligned}$$

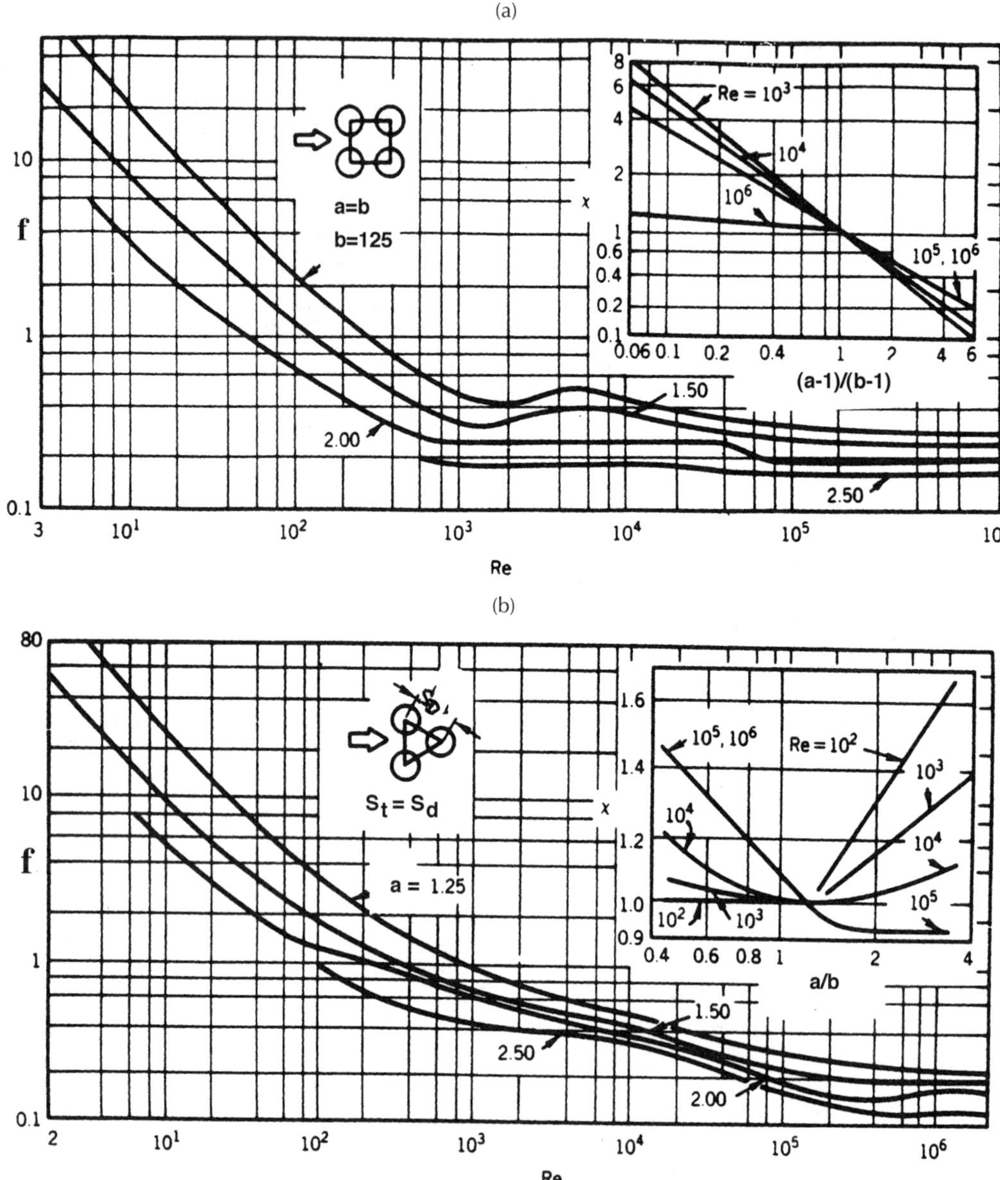

FIGURE 3.2.18 Friction factors for tube banks. (a) In-line arrangement; (b) Staggered arrangement. (From Zukauskas, A., in *Handbook of Single-Phase Convective Heat Transfer*, Kakac, S. et al., Eds., Wiley Interscience, New York, 1987. With permission.)

To find U_{max} we need the minimum area of cross section for fluid flow (Figure 3.2.19).

$$H = 20 \times 0.075 = 1.5 \text{ m}$$

$$A_{min} = 20(0.075 - 0.025) \times 3 = 3 \text{ m}^2$$

$$U_{max} = \frac{\dot{m}}{\rho A_{min}} = \frac{40}{1.165 \times 3} = 11.44 \text{ m/sec}$$

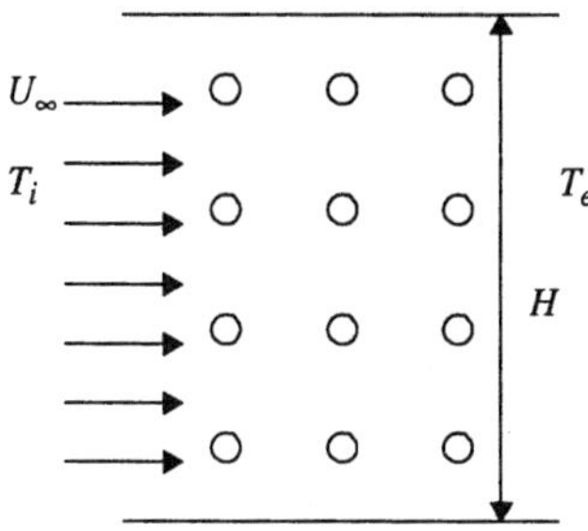

FIGURE 3.2.19 Aligned tube heat exchanger (only a few of the 20 columns and rows are shown).

$$\text{Re}_d = \frac{\rho U_{\max} d}{\mu} = \frac{1.165 \times 11.44 \times 0.025}{1.865 \times 10^{-5}} = 17{,}865$$

With values from Table 3.2.2,

$$\text{Nu}_d = 0.27 \times 17{,}865^{0.63} \times 0.712^{0.36} \left(\frac{0.712}{0.705}\right)^{0.25} = 114.3$$

$$h = \frac{114.3 \times 0.0264}{0.025} = 120.7 \text{ W/m}^2 \text{ K}$$

From Equation 3.2.78,

$$\ln\left(\frac{100 - 50}{100 - 10}\right) = -\frac{120.7 \times A_s}{40 \times 1007} \qquad A_s = \pi \times 0.025 \times 3 \times 20 \times N$$

$$\underline{N = \text{number of rows} = 42}$$

Fan Power: From the first law of thermodynamics (see Chapter 1), the fan power is

$$\dot{W}_F = \dot{m}\left(\frac{p_i}{\rho_i} + \frac{p_e}{\rho_e} + \frac{\mathbf{v}_e^2}{2}\right)$$

p_i and p_e are the pressures at inlet and exit of the heat exchanger and $\mathbf{v}_e$ is the fluid velocity at exit. Assuming constant density evaluated at $(T_i + T_e)/2$ the pressure drop is found from Figure 3.2.18a.

$$\text{Re}_d = 17{,}865: \qquad a = b = S_T/d = 75/25 = 3$$

In Figure 3.2.18, although the friction factor is available for values of b up to 2.5, we will estimate the value of f for $b = 3$. From Figure 3.2.18, $f \approx 0.11$. The correction factor $c = 1$.

$$p_i - p_e = N\chi \frac{\rho U_{\max}^2}{2} f = 42 \times 1 \frac{1.165 \times 11.44^2}{2} \times 0.11 = 352.2 \text{ kPa}$$

$$\mathbf{v}_e = \frac{11.44 \times 50}{75} = 7.63 \text{ m/sec}$$

$$\dot{W}_F = 40\left(352.2 + \frac{7.63^2}{2}\right) = 15{,}250 \text{ W}$$

Heat Transfer with Jet Impingement

Jet impingement (Figure 3.2.20) on a heated (or cooled) surface results in high heat transfer rates, and is used in annealing of metals, tempering of glass, cooling of electronic equipment, internal combustion engines, and in a wide variety of industries — textiles, paper, wood, and so on. Usually, the jets are circular, issuing from a round nozzle of diameter d, or rectangular, issuing from a slot of width w. They may be used singly or in an array. The jets may impinge normally to the heated surface or at an angle. If there is no parallel solid surface close to the heated surface, the jet is said to be free; in the presence of a parallel surface close to the heated surface, the jet is termed confined. In this section only single, free jets (round or rectangular) impinging normally to the heated surface are considered.

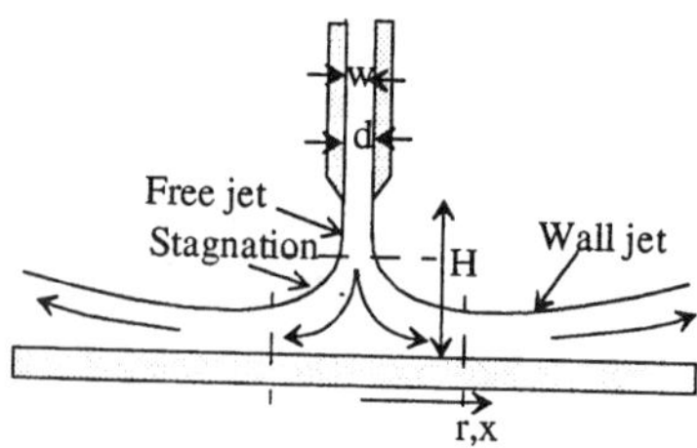

FIGURE 3.2.20 Circular jet of diameter d or a rectangular jet of width w.

Jets may be submerged with the fluid from the nozzle exiting into a body of a fluid (usually the same fluid), for example, air impinging on a surface surrounded by atmospheric air. In submerged jets entrained fluid (the part of the surrounding fluid dragged by the jet) has a significant effect on the flow and heat transfer characteristics of the jet, but the effect of gravity is usually negligible. In free-surface jets — a liquid jet in an atmosphere of air is a good approximation to a free-surface jet — the entrainment effect is usually negligible, but the effect of gravity may be significant.

A jet is usually divided into three regions, a free-jet region, a stagnation region, and a wall-jet region. In the free-jet region the effect of the target surface on the flow is negligible. In the stagnation region the target surface affects the flow field, and the velocity parallel to the surface increases while the velocity component normal to the surface decreases. At the beginning of the stagnation region, the axial velocity of the fluid is very much greater than the radial component (or the x-component) of the velocity. The stagnation region is followed by the wall-jet region where the radial component (or the x-component) of the velocity is much greater than the axial velocity.

The heat transfer coefficient is a function of H/d (or H/w), $\text{Re}_d(\rho v_j d/\mu)$ or $(\rho v_j 2w/\mu)$, and Pr and depends on the region (stagnation or wall jet), whether it is submerged or nonsubmerged and whether the flow adjacent to the plate is laminar or turbulent. Some of the heat transfer correlations suggested by different researchers are given below. All the correlations are for single jets.

Submerged Jets: Single Circular Jets

$$\text{Re}_d = \frac{4\dot{m}}{\pi d\mu} \qquad \text{Nu}_d = \frac{hd}{k} \qquad \dot{m} = \text{mass rate of flow of fluid}$$

Average heat transfer coefficients up to radius r (Martin, 1990):

$$\mathrm{Nu}_d = 2\frac{d}{r}\frac{1-1.1d/r}{1+0.1(H/d-6)d/r}\left[\mathrm{Re}_d\left(1+\frac{\mathrm{Re}_d^{0.55}}{200}\right)\right]^{1/2}\mathrm{Pr}^{0.42} \tag{3.2.80}$$

Range of validity:

$$2000 \le \mathrm{Re}_d \le 400{,}000 \qquad 2.5 \le r/d \le 7.5 \qquad 2 \le H/d \le 12$$

Local convective heat transfer coefficient at radius r (Webb and Ma, 1995):

$$\mathrm{Nu}_d = 1.29\mathrm{Re}_d^{1/2}\mathrm{Pr}^{0.4}\left\{\left[\frac{\tanh(0.88r/d)}{r/d}\right]^{-8.5}+\left[1.69\left(\frac{r}{d}\right)^{-1.07}\right]^{-17}\right\}^{-1/17} \tag{3.2.81}$$

Submerged Jets: Single Rectangular Jet

$$\mathrm{Re}_w = \frac{\rho v_j 2w}{\mu} = \frac{2\dot{m}}{\mu} \qquad \dot{m} = \text{mass rate of flow per unit length of jet}$$

$$\mathrm{Nu}_w = \frac{h2w}{k}$$

Average heat transfer coefficient (Martin, 1990):

$$\mathrm{Nu}_w = \frac{1.53\mathrm{Pr}^{0.42}\mathrm{Re}_w^m}{\dfrac{x}{2w}+\dfrac{H}{2w}+1.39} \tag{3.2.82}$$

$$m = 0.695 - \left[\frac{x}{2w}+\left(\frac{H}{2w}\right)^{1.33}+3.06\right]^{-1}$$

Free-Surface Jets: Single Circular Jet. Correlations are given in Table 3.2.5 (Liu et al., 1991 and Webb and Ma, 1995).

For more information on jet impingement heat transfer, refer to Martin (1977) and Webb and Ma (1995) and the references in the two papers.

Bibliography

ASHRAE *Handbook of Fundamentals*, 1993. American Society of Heating, Ventilating and Air Conditioning Engineers, Atlanta, GA.

Hewitt, G.F., Ed. 1990. *Handbook of Heat Exchanger Design*, Hemisphere Publishing, New York.

Incropera, F.P. and Dewitt, D.P. 1990. *Fundamentals of Heat and Mass Transfer*, 3rd ed., John Wiley & Sons, New York.

Kakaç, S., Shah, R.K., and Win Aung, Eds. 1987. *Handbook of Single Phase Convective Heat Transfer*, Wiley-Interscience, New York.

Kreith, F. and Bohn, M.S. 1993. *Principles of Heat Transfer*, 5th ed., PWS, Boston.

Suryanarayana, N.V. 1995. *Engineering Heat Transfer*, PWS, Boston.

TABLE 3.2.5 Correlations for Free-Surface Jets $r_v/d = 0.1773\ \mathrm{Re}_d^{1/3}$

		Nu_d	
$r/d < 0.787$	$0.15 \le \mathrm{Pr} \le 3$	$0.715\mathrm{Re}_d^{1/2}\mathrm{Pr}^{0.4}$	(3.2.83)
	$\mathrm{Pr} > 3$	$0.797\mathrm{Re}_d^{1/2}\mathrm{Pr}^{1/3}$	(3.2.84)
$0.787 < r/d < r_v/d$		$0.632\mathrm{Re}_d^{1/2}\mathrm{Pr}^{1/3}\left(\dfrac{d}{r}\right)^{1/2}$	(3.2.85)
$r_v/d < r/d < r_t/d$		$\dfrac{0.407\mathrm{Re}_d^{1/3}\mathrm{Pr}^{1/3}(d/r)^{2/3}}{\left[\dfrac{0.1713}{(r/d)^2}+\dfrac{5.147}{\mathrm{Re}_d}\dfrac{r}{d}\right]^{2/3}\left[\dfrac{(r/d)^2}{2}+C\right]^{1/3}}$ where $C = -5.051\times10^{-5}\mathrm{Re}_d^{2/3}$; $\dfrac{r_t}{d} = \left\{-\dfrac{s}{2}+\left[\left(\dfrac{s}{2}\right)^2+\left(\dfrac{p}{3}\right)^3\right]^{1/2}\right\}^{1/3} + \left\{-\dfrac{s}{2}+\left[\left(\dfrac{s}{2}\right)^2-\left(\dfrac{p}{3}\right)^3\right]^{1/2}\right\}^{1/3}$; $p = \dfrac{-2C}{0.2058\mathrm{Pr}-1}$ $\quad s = \dfrac{0.00686\mathrm{Re}_d\mathrm{Pr}}{0.2058\mathrm{Pr}-1}$	(3.2.86)
$r > r_t$	$\mathrm{Pr} < 3.9$	$\dfrac{0.25}{\dfrac{1}{\mathrm{Re}_d\mathrm{Pr}}\left[1-\left(\dfrac{r_t}{r}\right)^2\right]\left(\dfrac{r}{d}\right)^2+0.13\dfrac{h}{d}+0.0371\dfrac{h_t}{d}}$ where $h_t = h$ at r_t and $\dfrac{h}{d} = \dfrac{0.1713}{r/d}+\dfrac{5.147}{\mathrm{Re}_d}\left(\dfrac{r}{d}\right)^2$	(3.2.87)

References

Achenbach, E. 1978. *Heat Transfer from Spheres up to Re = 6 × 10⁶*, in *Proc. 6th Int. Heat Transfer Conf.*, Vol. 5, Hemisphere Publishing, Washington, D.C.

Burmeister, L.C. 1993. *Convective Heat Transfer*, Wiley-Interscience, New York.

Churchill, S.W. 1976. A comprehensive correlation equation for forced convection from a flat plate, *AIChE J.* 22(2), 264.

Churchill, S.W. and Bernstein, M. 1977. A correlating equation for forced convection from gases and liquids to a circular cylinder in cross flow, *J. Heat Transfer*, 99, 300.

Churchill, S.W. and Ozoe, H. 1973. Correlations for laminar forced convection with uniform heating in flow over a plate and in developing and fully developed flow in a tube, *J. Heat Transfer*, 18, 78.

Eckert, E.R.G. 1956. Engineering relations for heat transfer and friction in high-velocity laminar and turbulent boundary-layer flow over surfaces with constant pressure and temperature, *Trans. ASME*, 56, 1273.

Eckert, E.R.G. and Drake, M., Jr. 1972. *Analysis of Heat and Mass Transfer*, McGraw-Hill, New York.

Ishiguro, R., Sugiyama, K., and Kumada, T. 1979. Heat transfer around a circular cylinder in a liquid-sodium cross flow, *Int. J. Heat Mass Transfer*, 22, 1041.

Jakob, H., 1949. *Heat Transfer*, John Wiley & Sons, London.

Kays, W.M. and Crawford, M.E. 1993. *Convective Heat and Mass Transfer*, 3rd ed., McGraw-Hill, New York.

Liu, X., Lienhard, v., J.H., and Lombara, J.S. 1991. Convective heat transfer by impingement of circular liquid jets, *J. Heat Transfer*, 113, 571.

Martin, H. 1977. Heat and mass transfer between impinging gas jets and solid surfaces, in *Advances in Heat Transfer*, Hartnett, J.P. and Irvine, T.F., Eds., 13, 1, Academic Press, New York.

Martin, H. 1990. Impinging jets, in *Handbook of Heat Exchanger Design*, Hewitt, G.F., Ed., Hemisphere, New York.

Morgan, Vincent T., 1975. The overall convective heat transfer from smooth circular cylinders, in *Advances in Heat Transfer*, Irvine, T.F. and Hartnett, J.P., Eds., 11, 199, Academic Press, New York.

Rose, J.W. 1979. Boundary layer flow on a flat plate, *Int. J. Heat Mass Transfer*, 22, 969.

Schlichting, H. 1979. *Boundary Layer Theory*, 7th ed., McGraw-Hill, New York.

Suryanarayana, N.V. 1995. *Engineering Heat Transfer*, West Publishing, Minneapolis.

Thomas, W.C. 1977. Note on the heat transfer equation for forced-convection flow over a flat plate with an unheated starting length, *Mech. Eng. News (ASEE)*, 9(1), 19.

Webb, B.W. and Ma, C.F. 1995. Single-phase liquid jet impingement heat transfer, in *Advances in Heat Transfer*, Hartnett, J.P. and Irvine, T.F., Eds., 26, 105, Academic Press, New York.

Witte, L.C. 1968. An experimental study of forced-convection heat transfer from a sphere to liquid sodium, *J. Heat Transfer*, 90, 9.

Zukauskas, A. 1987. Convective heat transfer in cross flow, in *Handbook of Single-Phase Convective Heat Transfer*, Kakaç, S., Shah, R.K., and Win Aung, Eds., Wiley-Interscience, New York.

3.2.3 Forced Convection — Internal Flows

N.V. Suryanarayana

Introduction

Heat transfer to (or from) a fluid flowing inside a tube or duct is termed *internal forced convection*. The fluid flow may be laminar or turbulent. If the Reynolds number based on the average velocity of the fluid and diameter of the tube ($\rho vd/\mu$) is less than 2100 (Reynolds numbers in the range of 2000 to 2300 are cited in different sources), the flow is laminar. If the Reynolds number is greater than 10,000, the flow is turbulent. The flow with a Reynolds number in the range 2100 to 10,000 is considered to be in the transitional regime. With heating or cooling of the fluid, there may or may not be a change in the phase of the fluid. Here, only heat transfer to or from a single-phase fluid is considered.

Fully Developed Velocity and Temperature Profiles. When a fluid enters a tube from a large reservoir, the velocity profile at the entrance is almost uniform as shown in Figure 3.2.21. The fluid in the immediate vicinity of the tube surface is decelerated and the velocity increases from zero at the surface to u_c at a distance δ from the surface; in the region $r = 0$ to $(R - \delta)$ the velocity is uniform. The value of δ increases in the direction of flow and with constant fluid density the value of the uniform velocity u_c increases. At some location downstream, δ reaches its maximum possible value, equal to the radius of the tube, and from that point onward the velocity profile does not change.

The region where δ increases, i.e., where the velocity profile changes, is known as the entrance region or hydrodynamically developing region. The region downstream from the axial location where δ reaches its maximum value and where the velocity profile does not change is the fully developed velocity profile or hydrodynamically fully developed region. Similarly, downstream of the location where heating or cooling of the fluid starts, the temperature profile changes in the direction of flow. But beyond a certain distance the dimensionless temperature profile does not change in the direction of flow. The region where the dimensionless temperature profile changes is the thermally developing region or the thermal entrance region, and the region where the dimensionless temperature profile does not change is the thermally

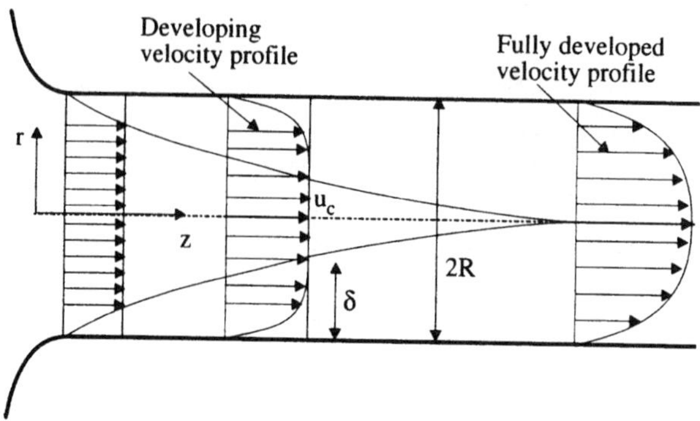

FIGURE 3.2.21 Developing and fully developed velocity profiles.

fully developed region. For simultaneously developing velocity and temperature profiles in laminar flows, the hydrodynamic and thermal entrance lengths are given by

$$\frac{L_e}{d} = 0.0565\text{Re}_d \tag{3.2.88}$$

$$\frac{L_{e,th}}{d} = 0.053\text{Re}_d\text{Pr} \qquad \text{Uniform heat flux} \tag{3.2.89}$$

$$\frac{L_{e,th}}{d} = 0.037\text{RePr} \qquad \text{Uniform surface temperature} \tag{3.2.90}$$

In most engineering applications, with turbulent flows, correlations for fully developed conditions can be used after about 10 diameters from where the heating starts.

Convective Heat Transfer Coefficient and Bulk Temperature. The reference temperature for defining the convective heat transfer coefficient is the bulk temperature T_b and the convective heat flux is given by

$$q'' = h(T_s - T_b) \tag{3.2.91}$$

The bulk temperature T_b is determined from the relation

$$T_b = \frac{\int_{A_c} \rho v C_p T dA_c}{\int_{A_c} \rho v C_p dA_c} \tag{3.2.92}$$

where A_c is the cross-sectional area perpendicular to the axis of the tube.

If the fluid is drained from the tube at a particular axial location and mixed, the temperature of the mixed fluid is the bulk temperature. It is also know as the mixing cup temperature. With heating or cooling of the fluid the bulk temperature varies in the direction of flow. In some cases we use the term *mean fluid temperature*, T_m, to represent the arithmetic mean of the fluid bulk temperatures at inlet and exit of the tube.

Heat Transfer Correlations

Laminar Flows — Entrance Region. For laminar flows in a tube with uniform surface temperature, in the entrance region the correlation of Sieder and Tate (1936) is

$$\overline{\mathrm{Nu}_d} = 1.86\left(\frac{\mathrm{Re}_d \mathrm{Pr}}{L/d}\right)^{1/3}\left(\frac{\mu}{\mu_s}\right)^{0.14} \tag{3.2.93}$$

valid for

$$\frac{L}{d} < \frac{\mathrm{Re}_d \mathrm{Pr}}{8}\left(\frac{\mu}{\mu_s}\right)^{0.42} \qquad 0.48 < \mathrm{Pr} < 16,700 \qquad 0.0044 < \frac{\mu}{\mu_s} < 9.75$$

The overbar in the Nusselt number indicates that it is formed with the average heat transfer coefficient over the entire length of the tube. Properties of the fluid are evaluated at the arithmetic mean of the inlet and exit bulk temperatures. In Equation (3.2.93) the heat transfer coefficient was determined from

$$q = \bar{h}\pi dL\left(T_s - \frac{T_{bi} + T_{be}}{2}\right) \tag{3.2.94}$$

Therefore, to find the total heat transfer rate with $\bar{h}$ from Equation (3.2.93) employ Equation (3.2.94).

Laminar Flows — Fully Developed Velocity and Temperature Profiles. Evaluate properties at the bulk temperature

Uniform Surface Temperature $\quad \mathrm{Nu}_d = 3.66 \quad$ (3.2.95)

Uniform Surface Heat Flux $\quad \mathrm{Nu}_d = 4.36 \quad$ (3.2.96)

Turbulent Flows. If the flow is turbulent, the difference between the correlations with uniform surface temperature and uniform surface heat flux is not significant and the correlations can be used for both cases. For turbulent flows, Gnielinsky (1976, 1990) recommends:
Evaluate properties at the bulk temperature.

$$0.6 < \mathrm{Pr} < 2000 \qquad 2300 < \mathrm{Re}_d < 10^6 \qquad 0 < d/L < 1$$

$$\mathrm{Nu}_d = \frac{(f/2)(\mathrm{Re}_d - 1000)\mathrm{Pr}}{1 + 12.7(f/2)^{1/2}\left(\mathrm{Pr}^{2/3} - 1\right)}\left[1 + \left(\frac{d}{L}\right)^{2/3}\right] \tag{3.2.97}$$

$$f = \left[1.58\ln(\mathrm{Re}_d) - 3.28\right]^{-2} \tag{3.2.98}$$

f = friction factor = $2\tau_w/\rho v^2$.

To reflect the effect of variation of fluid properties with temperature, multiply the Nusselt numbers in Equation (3.2.97) by $(T_b/T_s)^{0.45}$ for gases and $(\mathrm{Pr}/\mathrm{Pr}_s)^{0.11}$ for liquids where the temperatures are absolute, and T and Pr with a subscript s are to be evaluated at the surface temperature. The equations can be used to evaluate the heat transfer coefficient in the developing profile region. To determine the heat transfer coefficient in the fully developed region set $d/L = 0$. A simpler correlation (fully developed region) is the Dittus–Boelter (1930) equation. Evaluate properties at T_b.

$$0.7 \le \text{Pr} \le 160 \qquad \text{Re}_d > 10{,}000 \qquad d/L > 10$$

$$\text{Nu}_d = 0.023\text{Re}_d^{4/5}\text{Pr}^n \tag{3.2.99}$$

where $n = 0.4$ for heating ($T_s > T_b$) and $n = 0.3$ for cooling ($T_s < T_b$).

For liquid metals with $\text{Pr} \ll 1$ the correlations due to Sleicher and Rouse (1976) are Uniform surface temperature:

$$\text{Nu}_{d,b} = 4.8 + 0.0156\text{Re}_{d,f}^{0.85}\text{Pr}_s^{0.93} \tag{3.2.100}$$

Uniform heat flux:

$$\text{Nu}_{d,b} = 6.3 + 0.0167\text{Re}_{d,f}^{0.85}\text{Pr}_s^{0.93} \tag{3.2.101}$$

Subscripts b, f, and s indicate that the variables are to be evaluated at the bulk temperature, film temperature (arithmetic mean of the bulk and surface temperatures), and surface temperature, respectively.

In the computations of the Nusselt number the properties (evaluated at the bulk temperature) vary in the direction of flow and hence give different values of h at different locations. In many cases a representative average value of the convective heat transfer coefficient is needed. Such an average value can be obtained either by taking the arithmetic average of the convective heat transfer coefficients evaluated at the inlet and exit bulk temperatures or the convective heat transfer coefficient evaluated at the arithmetic mean of the inlet and exit bulk temperatures. If the variation of the convective heat transfer coefficient is large, it may be appropriate to divide the tube into shorter lengths with smaller variation in the bulk temperatures and evaluating the average heat transfer coefficient in each section.

Uniform Surface Temperature — Relation between the Convective Heat Transfer Coefficient and the Total Heat Transfer Rate: With a uniform surface temperature, employing an average value of the convective heat transfer coefficient the local convective heat flux varies in the direction of flow. To relate the convective heat transfer coefficient to the temperatures and the surface area, we have, for the elemental length Δz (Figure 3.2.22).

$$\dot{m}C_p \frac{dT_b}{dz} = h\frac{dA_s}{dz}\left(T_s - T_b\right) \tag{3.2.102}$$

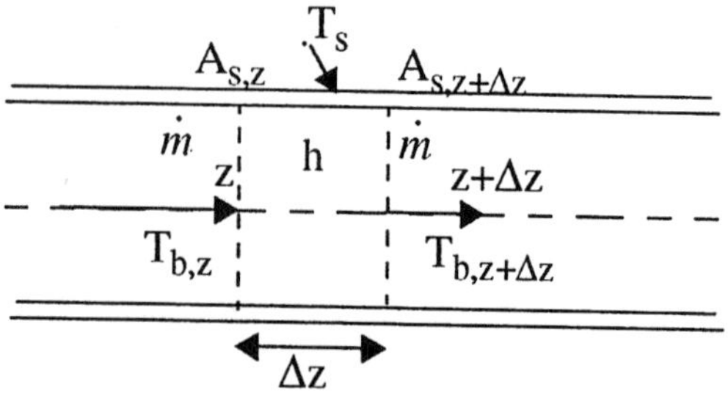

FIGURE 3.2.22 Elemental length of a tube for determining heat transfer rate.

Assuming a suitable average convective heat transfer coefficient over the entire length of the tube, separating the variables, and integrating the equation from $z = 0$ to $z = L$, we obtain

$$\ln \frac{T_s - T_{be}}{T_s - T_{bi}} = -\frac{hA_s}{\dot{m}C_p} \tag{3.2.103}$$

Equation (3.2.103) gives the exit temperature. For a constant-density fluid or an ideal gas, the heat transfer rate is determined from

$$q = \dot{m}C_p\left(T_{be} - T_{bi}\right) \tag{3.2.104}$$

Equation (3.2.103) was derived on the basis of uniform convective heat transfer coefficient. However, if the functional relationship between h and T_b is known, Equation (3.2.102) can be integrated by substituting the relationship. The convective heat transfer coefficient variation with T_b for water in two tubes of different diameters for two different flow rates is shown in Figure 3.2.23. From the figure it is clear that h can be very well approximated as a linear function of T. By substituting such a linear function relationship into Equation (3.2.102), it can be shown that

$$\ln \frac{h_i}{h_e} \frac{T_s - T_{be}}{T_s - T_{bi}} = -\frac{h_s A_s}{\dot{m}C_p} \tag{3.2.105}$$

where h_i, h_e, and h_s are the values of the convective heat transfer coefficient evaluated at bulk temperatures of T_{bi}, T_{be}, and T_s, respectively. Although it has been demonstrated that h varies approximately linearly with the bulk temperature with water as the fluid, the variation of h with air or oil as the fluid is much smaller and is very well approximated by a linear relationship. For other fluids it is suggested that the relationship be verified before employing Equation (3.2.105). [**Note:** It is tempting to determine the heat transfer rate from the relation

$$q = hA_s \frac{\left(T_s - T_{be}\right) + \left(T_s - T_{bi}\right)}{2}$$

Replacing q by Equation (3.2.104) and solving for T_{be} for defined values of the mass flow rate and tube surface area, the second law of thermodynamics will be violated if $hA_s / \dot{m}C_p > 2$. Use of Equation (3.2.103) or (3.2.105) ensures that no violation of the second law occurs however large A_s is.]

Uniform Surface Heat Flux: If the imposed heat flux is known, the total heat transfer rate for a defined length of the tube is also known. From Equation (3.2.104) the exit temperature of the fluid is determined. The fluid temperature at any location in the pipe is known from the heat transfer rate up to that location ($q = q''A_s$) and Equation (3.2.104). The convective heat transfer coefficient is used to find the surface temperature of the tube.

Temperature Variation of the Fluid with Uniform Surface Temperature and Uniform Heat Flux: The fluid temperature variations in the two cases are different. With the assumption of uniform heat transfer coefficient, with a uniform surface temperature the heat flux decreases in the direction of flow leading to a progressively decreasing rate of temperature change in the fluid with axial distance. With uniform heat flux, the surface and fluid temperatures vary linearly except in the entrance region where the higher heat transfer coefficient leads to a smaller difference between the surface and fluid temperatures. The variation of the fluid temperature in the two cases is shown in Figure 3.2.24.

Convective Heat Transfer in Noncircular Tubes

Laminar Flows: The Nusselt numbers for laminar flows have been analytically determined for different noncircular ducts. Some of them can be found in Kakac et al. (1987), Kays and Crawford (1993), and

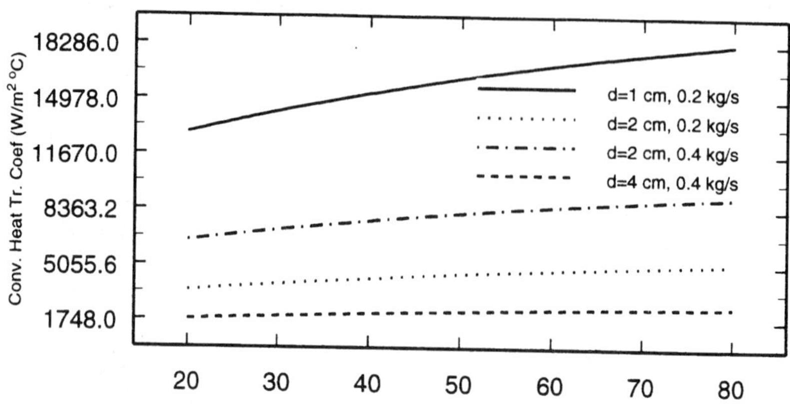

FIGURE 3.2.23 Variation of h with T_b in 1-, 2-, and 4-cm-diameter tubes with water flow rates of 0.2 kg/sec and 0.4 kg/sec with uniform surface temperature.

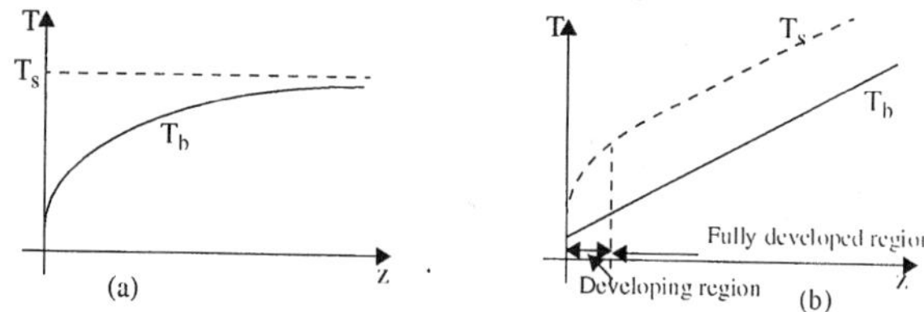

FIGURE 3.2.24 Variation of fluid temperature in a tube with (a) uniform surface temperature and (b) uniform heat flux.

Burmeister (1993). A few of the results are given below. The characteristic length for forming the Reynolds number and Nusselt number is the hydraulic mean diameter defined as

$$d_h = \frac{4 \text{ cross-sectional area}}{\text{wetted perimeter}}$$

Infinite parallel plates: a = spacing between plates, $d_h = 2a$

Both plates maintained at uniform and equal temperatures: Nu = 7.54
Both plates with imposed uniform and equal heat fluxes: Nu = 8.24

Rectangular ducts: a = longer side, b = shorter side, $d_h = 2ab/(a + b)$

b/a	**1**	**0.7**	**0.5**	**0.25**	**0.125**
Uniform surface temperature	2.98	3.08	3.39	4.44	5.6
Uniform heat flux*	3.61	3.73	4.12	5.33	6.49

Equilateral triangle: $d_h = a/3^{1/2}$, a = length of each side

Uniform surface temperature: Nu = 2.35
Uniform surface heat flux:* Nu = 3.0

Coaxial tubes: With coaxial tubes many different cases arise — each tube maintained at uniform but different temperatures, each tube subjected to uniform but different heat fluxes (an insulated

* Uniform axial heat flux but circumferentially uniform surface temperature.

surface is a special case of imposed heat flux being zero), or a combinations of uniform surface temperature of one tube and heat flux on the other. The manner in which the heat transfer coefficient is determined for uniform but different heat fluxes on the two tubes is described below. Define:

$$d_h = 2(r_o - r_i) \qquad r^* = r_i/r_o$$

$$q_i'' = h_i(T_i - T_b) \qquad \mathrm{Nu}_i = \frac{h_i d_h}{k} \qquad q_o'' = h_o(T_o - T_b) \qquad \mathrm{Nu}_o = \frac{h_o d_h}{k}$$

$$q_o'' = 0 \qquad \mathrm{Nu}_{ii} = \frac{h_i d_h}{k} \quad \text{and} \quad q_i'' = 0 \qquad \mathrm{Nu}_{oo} = \frac{h_o d_h}{k}$$

Then

$$\mathrm{Nu}_i = \frac{\mathrm{Nu}_{ii}}{1 - \dfrac{q_o''}{q_i''}\theta_i^*} \quad \text{and} \quad \mathrm{Nu}_o = \frac{\mathrm{Nu}_{oo}}{1 - \dfrac{q_i''}{q_o''}\theta_o^*} \tag{3.2.106}$$

TABLE 3.2.6 Values for Use with Equation (3.2.106)

r^*	Nu_{ii}	Nu_{oo}	θ_i^*	θ_o^*
0.05	17.81	4.792	2.18	0.0294
0.1	11.91	4.834	1.383	0.0562
0.2	8.499	4.883	0.905	0.1041
0.4	6.583	4.979	0.603	0.1823
0.6	5.912	5.099	0.473	0.2455
0.8	5.58	5.24	0.401	0.299
1.0	5.385	5.385	0.346	0.346

Some of the values needed for the computations of Nu_i and Nu_o (taken from Kays and Crawford, 1993) are given in the Table 3.2.6.

For a more detailed information on heat transfer and friction factors for laminar flows in noncircular tubes refer to Kakac et al. (1987).

Turbulent Flows: For noncircular tubes, estimates of the convective heat transfer coefficient can be obtained by employing equations for circular tubes with d_h replacing d in the computations of the Reynolds and Nusselt numbers. To determine the heat transfer coefficients in developing regions and for more-accurate values with turbulent flows in noncircular tubes refer to Kakac et al. (1987) and the references in that book.

Mixed Convection

If the fluid velocity is low, the effect of natural convection becomes significant and the heat transfer rate may be increased or decreased by natural convection. From a review of experimental results, Metais and Eckert (1964) developed maps to delineate the different regimes where one or the other mode is dominant and where both are significant. Figures 3.2.25 and 3.2.26 show the relative significance of natural and forced convection in vertical and horizontal tubes. The maps are applicable for $10^{-2} < \mathrm{Pr}(d/L) < 1$ where d and L are the diameter and the axial length of the tube. The maps show the limits of forced and natural convection regimes. The limits are delineated "in such a way that the actual heat flux under the combined influence of the forces does not deviate by more than 10 percent from the heat flux that

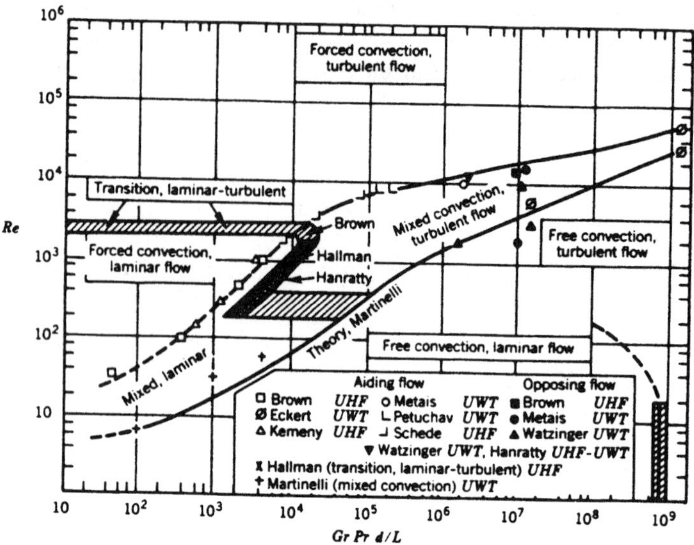

FIGURE 3.2.25 Map delineating forced, mixed, and natural convection — vertical tubes.

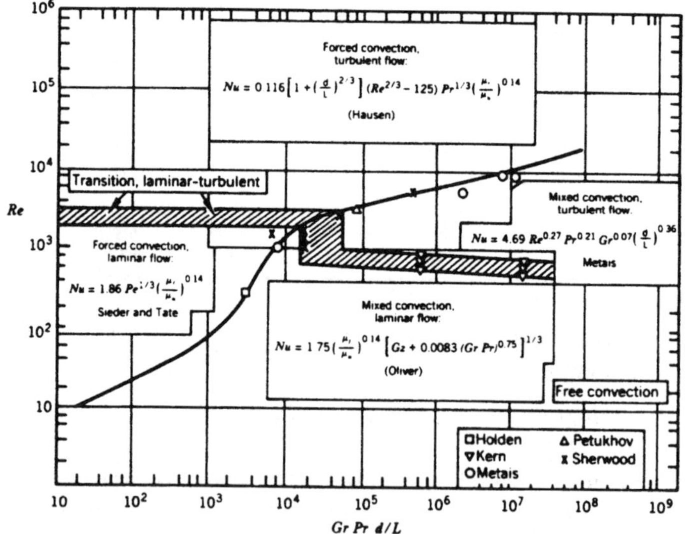

FIGURE 3.2.26 Map delineating forced, mixed, and natural convection — horizontal tubes.

would be caused by the external forces alone or by the body forces alone." The Grashof number is based on the diameter of the tube.

For flows in horizontal tubes, correlations were developed for the mixed convection regime in isothermal tubes by Depew and August (1971) and for uniform heat flux by Morcos and Bergles (1975).

Uniform Surface Temperature. Fully developed velocity profile, developing temperature profile:

$L/d < 28.4$ $\qquad 25 < Gz < 712$ $\qquad 0.7 \times 10^5 < Gr < 9.9 \times 10^5$

μ_s = dynamic viscosity, evaluated at the wall temperature

All other properties at the average bulk temperature of the fluid

$$\text{Gz} = \frac{\dot{m}C_p}{kL} \qquad \text{Gr} = g\beta\Delta T\, d^3/\nu^2$$

$$\text{Nu}_d = 1.75\left[\text{Gz} + 0.12\left(\text{GzGr}^{1/3}\text{Pr}^{0.36}\right)^{0.88}\right]^{1/3}\left(\mu_b/\mu_s\right)^{0.14} \tag{3.2.107}$$

Uniform Heat Flux. Properties at $(T_s + T_b)/2$: $3 \times 10^4 < \text{Ra} < 10^6$, $4 < \text{Pr} < 175$, $2 < hd^2/(k_w t) < 66$, k_w = tube wall thermal conductivity, t = tube wall thickness.

$$\text{Gr}_d^* = g\beta d^4\, q_w''/\left(\nu^2 k\right) \qquad P_w = kd/\left(k_w t\right) \qquad \text{Ra}_d = g\beta\Delta T\, d^3\text{Pr}/\nu^2$$

$$\text{Nu}_d = \left\{4.36^2 + \left[0.145\left(\frac{\text{Gr}_d^*\text{Pr}^{1.35}}{P_w^{0.25}}\right)^{0.265}\right]^2\right\}^{0.5} \tag{3.2.108}$$

In Equation (3.2.107) and (3.2.108) evaluate fluid properties at the arithmetic mean of the bulk and wall temperatures.

References

Burmeister, L.C. 1993. *Convective Heat Transfer*, 2nd ed., Wiley-Interscience, New York.

Depew, C.A. and August, S.E. 1971. Heat transfer due to combined free and forced convection in a horizontal and isothermal tube, *Trans. ASME* 93C, 380.

Dittus, F.W. and Boelter, L.M.K. 1930. Heat transfer in automobile radiators of the tubular type, *Univ. Calif. Pub. Eng.*, 13, 443.

Gnielinsky, V. 1976. New equations for heat and mass transfer in turbulent pipe channel flow, *Int. Chem. Eng.*, 16, 359.

Gnielinsky, V. 1990. Forced convection in ducts, in *Handbook of Heat Exchanger Design*, Hewitt, G.F., Ed., Begell House/Hemisphere, New York.

Kakac, S., Shah, R.K., and Win Aung, Eds. 1987. *Handbook of Single-Phase Convective Heat Transfer*, Wiley-Interscience, New York.

Kays, W.M. and Crawford, M.E. 1993. *Convective Heat and Mass Transfer*, 3rd ed., McGraw-Hill, New York.

Metais, B. and Eckert, E.R.G. 1964. Forced, mixed, and free convection regimes, *Trans. ASME* 86C, 295.

Morcos, S.M. and Bergles, A.E. 1975. Experimental investigation of combined forced and free laminar convection in a horizontal tube, *Trans. ASME* 97C, 212.

Sieder, E.N. and Tate, C.E. 1936. Heat transfer and pressure drop of liquids in tubes, *Ind. Eng. Chem.*, 28, 1429.

Sleicher, C.A. and Rouse, M.W. 1976. A convenient correlation for heat transfer to constant and variable property fluids in turbulent pipe flow, *Int. J. Heat Mass Transfer*, 18, 677.

3.2.4 Convection Heat Transfer in Non-Newtonian Fluids

Thomas F. Irvine, Jr., and Massimo Capobianchi

Introduction

The general characteristics of non-Newtonian fluids are described in Section 2.9 and will not be repeated here. Topics to be included in this section are laminar and turbulent heat transfer in fully developed duct flow, and laminar free convection heat transfer in vertical channels and plates and several other common geometries.

For non-Newtonian flows, except for certain classes of fluids which exhibit a slip phenomenon at solid boundaries, the boundary condition is taken as no-slip or zero velocity at all solid surfaces. For heat transfer analyses, however, the situation is more complicated because there are many different ways to heat a wall, which in turn affects the type of thermal boundary conditions.

In general, the rate of heat transfer from a surface, or the temperature difference between the wall and the fluid, is calculated using the equation $q_c = h_c A_q \Delta T$. Since the heat transfer coefficient can vary considerably for different thermal boundary conditions, it is important that the boundary conditions be specified correctly. Although the number of thermal boundary conditions is in principle infinite, several classical types have been identified and are in common use. They are usually identified in terms of the Nusselt number, $\mathrm{Nu} = h_c L/k$, with a particular subscript. For example, for duct flow, the symbol Nu_T is used to specify the Nusselt number when the wall temperature is constant in both the flow and peripheral directions. Other thermal boundary conditions are described in Table 3.2.7 for duct heat transfer and will be used throughout this section.

TABLE 3.2.7 Thermal Boundary Conditions for Duct Heat Transfer

1.	Constant wall temperature in both the flow and circumferential direction	Nu_T
2.	Constant heat flux in the flow direction and constant temperature in the circumferential direction	Nu_{H1}
3.	Constant heat flux in the flow and circumferential directions	Nu_{H2}
4.	Constant heat flux per unit volume in the wall with circumferential wall heat conduction	Nu_{H4}

It should be noted that because of the symmetry in circular and parallel plate ducts, Nu_{H1} and Nu_{H2} are identical and are referred to simply as Nu_H. Nu_{H4} with wall conduction is a more-complicated problem where the energy equations must be solved simultaneously in both the wall and the fluid. Such problems are called conjugated. In the Nu_{H4} situation, the designer has the flexibility of affecting the heat transfer by varying either or both the characteristics of the duct wall or the convective fluid. In the heat transfer relations to be considered later, care will be taken to identify the proper thermal boundary conditions using the nomenclature in Table 3.2.7.

Laminar Duct Heat Transfer — Purely Viscous, Time-Independent Non-Newtonian Fluids

As discussed in Section 2.9, a convenient and comprehensive constitutive equation for pseudoplastic fluids (flow index, $n < 1$) is the modified power law equation:

$$\mu_a = \frac{\mu_o}{1 + \frac{\mu_o}{K}(\dot{\gamma})^{1-n}} \tag{3.2.109}$$

Equation (3.2.109) has the characteristic that at low shear rates, the equation approaches that for a Newtonian fluid while at large shear rates it describes a power law fluid. In addition, solutions using

Equation (3.2.109) generate a shear rate parameter, β, which describes whether any particular system is in the Newtonian, transitional, or power law region. For duct flow, β is given by

$$\beta = \frac{\mu_o}{K}\left(\frac{\bar{u}}{D_H}\right)^{1-n} \tag{3.2.110}$$

If $\log_{10} \beta > 2$: Power law region
If $\log_{10} \beta < -2$: Newtonian region
If $-2 \le \log_{10} \beta \le 2$: Transition region

For fully developed flow, the characteristic length is the hydraulic diameter, D_H, and the fluid temperature is the "bulk" temperature defined as

$$T_b = \frac{1}{A_c \bar{u}} \int_{A_c} uTdA_c \tag{3.2.111}$$

Figure 3.2.27 illustrates the values of Nu_T vs. β for a circular duct with the flow index, n, as a parameter. It is seen from the figure that the effect of β on Nu_T is only moderate, but for some applications it may be important to know at what value of β the system is operating. The situation is similar for boundary condition Nu_H.

Although Figure 3.2.27 shows the Nusselt number relation graphically, it is convenient to have simple correlation equations to represent the solutions for both boundary conditions. For fully developed Nusselt numbers with values of $0.5 \le n \le 1.0$ and $10^{-4} \le \beta \le 10^4$, Irvine et al. (1988) present the following equation which represents both solutions with a maximum difference of 1.5%:

$$\mathrm{Nu} = \frac{\mathrm{Nu}_N(1+\beta)}{1+\dfrac{\mathrm{Nu}_N \beta}{\mathrm{Nu}_P}} \tag{3.2.112}$$

The Newtonian Nusselt numbers are $Nu_N = 3.6568$ for Nu_T, and $Nu_N = 4.3638$ for Nu_H. In addition, Table 3.2.8 lists the power law Nusselt numbers, Nu_{TP} and Nu_{HP}, for $\log_{10} \beta = 4$.

Graetz solutions for the thermal entrance lengths are also available. They assume that the velocity profile is fully developed at the duct entrance and present the duct lengths required for the Nusselt numbers to reach within 1% of the fully developed values. Figure 3.2.28 shows these thermal entrance lengths for Nu_T thermal boundary condition. The situation is similar for boundary condition Nu_H.

A correlation equation for the thermal entrance lengths for both the Nu_T and Nu_H boundary conditions by Irvine et al. (1988) represents the numerical solutions within 0.5% for $0.5 \le n \le 1.0$ and $-4 \le \log_{10} \beta \le 4$. Table 3.2.9 lists the power law thermal entrance lengths which are needed to evaluate the following correlation equation:

$$x^+_{\mathrm{ent},\beta,n} = \frac{x^+_{\mathrm{ent},N}(1+\beta)}{1+\dfrac{x^+_{\mathrm{ent},N}(\beta)}{x^+_{\mathrm{ent},P}}} \tag{3.2.113}$$

where $x^+_{\mathrm{ent},\beta,n}$ is the modified power law dimensionless entrance length defined as $x^+_{\mathrm{ent},\beta,n} = (x_{\mathrm{ent},\beta,n}/D_H)/Pe$, and $x^+_{\mathrm{ent},N}$ and $x^+_{\mathrm{ent},P}$ are the Newtonian and power law values, respectively. The Newtonian dimensionless entrance lengths are $x^+_{\mathrm{ent},N} = 0.03347$ for Nu_T and $x^+_{\mathrm{ent},N} = 0.04309$ for Nu_H.

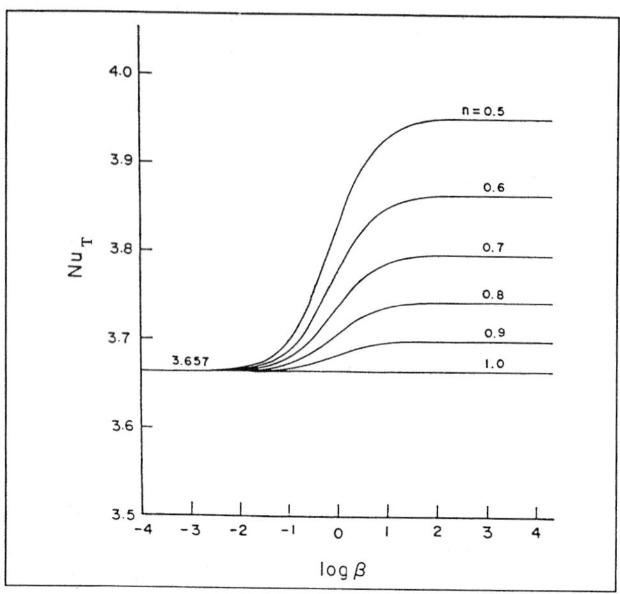

FIGURE 3.2.27 Variation of the fully developed circular duct Nusselt numbers, Nu_T, with the shear rate parameter β and *n*. (From Irvine, T.F., Jr. et al., in *ASME Symposium on Fundamentals of Forced Convection Heat Transfer*, ASME publ. HTD 101, 1988, 123–127. With permission.)

TABLE 3.2.8 Power Law Nu_T and Nu_H Solutions for a Circular Duct ($\log_{10} \beta = 4$)

n	Nu_{TP}	Nu_{HP}
1.0 (Newtonian)	3.6568	4.3638
0.9	3.6934	4.4109
0.8	3.7377	4.4679
0.7	3.7921	4.5385
0.6	3.8605	4.6281
0.5	3.9494	4.7456

Source: Irvine, T.F., Jr. et al., in *ASME Symposium on Fundamentals of Forced Convection Heat Transfer*, ASME publ. HTD 101, 1988, 123–127.

Only one noncircular geometry using the modified power law equation has been published in the archival literature for laminar fully developed heat transfer (Capobianchi and Irvine, 1992). A correlation equation for Nu_{H1} for annuli with constant heat flux at the inner wall and the outer wall insulated is

$$n < 1 \qquad Nu_{H1} = \frac{1+\beta}{\dfrac{1}{Nu_{H1,N}} + \dfrac{\beta}{Nu_{H1,P}}} \tag{3.2.114}$$

Nusselt numbers for square ducts and power law fluids can be found in Chandrupatla and Sastri (1977) and, for isosceles triangular ducts, in Cheng (1984). Thermally developing and thermally developed laminar heat transfer in rectangular channels has been studied by Hartnett and Kostic (1989).

For other cross-sectional shapes, a power law approximate correlation has been proposed by Cheng (1984):

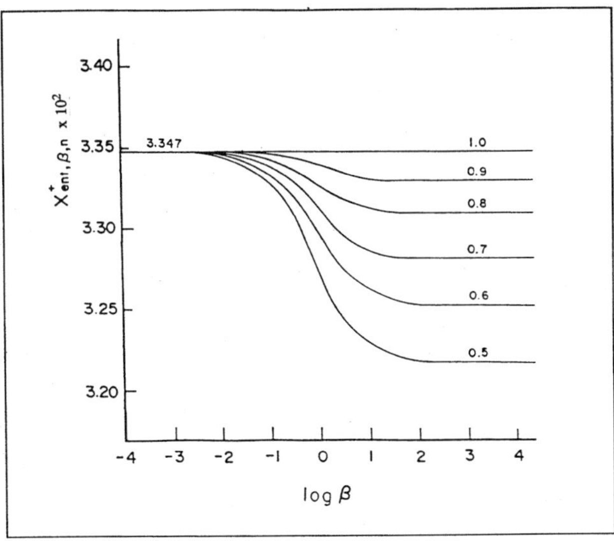

FIGURE 3.2.28 Thermal entrance lengths vs. shear rate parameter β and n for Nu_T in circular ducts. (From Irvine, T.F., Jr. et al., in *ASME Symposium on Fundamentals of Forced Convection Heat Transfer*, ASME publ. HTD 101, 1988, 123–127. With permission.)

TABLE 3.2.9 Values of Circular Duct Thermal Entrance Lengths for Nu_T and Nu_H for Use in Equation 3.2.113

n	Nu_T, $x^+_{ent,P} \times 10^2$	Nu_H, $x^+_{ent,P} \times 10^2$
1.0 (Newtonian)	3.347	4.309
0.9	3.326	4.281
0.8	3.306	4.248
0.7	3.279	4.210
0.6	3.250	4.166
0.5	3.213	4.114

Source: Irvine, T.F., Jr., et al., in *ASME Symposium on Fundamentals of Forced Convection Heat Transfer*, ASME publ. HTD 101, 1988, 123–127.

$$Nu_P = Nu_N \left[\frac{(a+bn)}{(a+b)n} \right]^{1/3} \tag{3.2.115}$$

where a and b are the Kozicki geometric constants listed in Table 3.9.3 in the section on non-Newtonian flows. Equation (3.2.115) applies to any thermal boundary condition. For circular ducts, Equation 3.2.115 predicts the correct solution for both Nu_T and Nu_H.

Turbulent Duct Flow for Purely Viscous Time-Independent Non-Newtonian Fluids

It is known that in turbulent flow, the type of thermal boundary conditions has much less effect than in laminar flow. Therefore, turbulent flow heat transfer investigations are often reported without specifying the thermal boundary conditions. Yoo (1974) has presented an empirical correlation for turbulent heat transfer in circular ducts for purely viscous time-independent power law fluids.

$$\mathrm{St}\,\mathrm{Pr}_a^{2/3} = 0.0152\mathrm{Re}_a^{-0.155} \tag{3.2.116}$$

Equation (3.2.116) describes all of the experimental data available in the literature at the time with a mean deviation of 2.3%. Equation (3.2.116) is recommended in order to predict the turbulent fully developed heat transfer in the ranges $0.2 \le n \le 0.9$ and $3000 \le \mathrm{Re}_a \le 90{,}000$. The Reynolds number and Prandtl numbers in Equation (3.2.116) are based on the apparent viscosity at the wall, μ_a, i.e.,

$$\mathrm{Re}_a = \frac{\rho \bar{u} D_H}{\mu_a} \tag{3.2.117}$$

$$\mathrm{Pr}_a = \frac{\mu_a c_p}{k} \tag{3.2.118}$$

In order to evaluate Equations (3.2.117) and (3.2.118) in terms of the rheological properties and operating parameters, an expression must be obtained for μ_a in terms of these quantities. The value of μ_a is evaluated by considering that μ_a is determined from fully developed laminar circular tube power law fluid flow for which it can be shown that (Irvine and Karni, 1987)

$$\mu_a = K\left(\frac{3n+1}{4n}\right)^{n-1}\left(\frac{8\bar{u}}{D_H}\right)^{n-1} \tag{3.2.119}$$

assuming that the quantities K, n, c_p, and k are constant. It is also of interest that the Prandtl number is no longer a thermophysical property for power law fluids but depends upon the average velocity, $\bar{u}$, and the hydraulic diameter, D_H.

Hartnett and Rao (1987) have investigated fully developed turbulent heat transfer for a rectangular duct with a 2:1 aspect ratio and propose the following equation which generally agreed with their experimental data within ±20%:

$$\mathrm{Nu} = (0.0081 + 0.0149n)\mathrm{Re}_a^{0.8}\mathrm{Pr}_a^{0.4} \tag{3.2.120}$$

Viscoelastic Fluids

An important characteristic of viscoelastic fluids is their large hydrodynamic and thermal entrance lengths. Cho and Hartnett (1982) have reported hydrodynamic entrance lengths of up to 100 diameters and thermal entrance lengths up to 200 to 800 diameters depending upon the Reynolds and Prandtl numbers. These can be compared with Newtonian fluids entrance lengths which are of the order of 10 to 15 diameters. Therefore, care must be used in applying fully developed relations to practical situations.

Cho et al. (1980) reported heat transfer measurements in the thermal entrance region and recommend the following empirical equation for saturated aqueous polymer solutions for $6000 \le \mathrm{Re}_a$ and x/D_H values up to 450:

$$J_H = 0.13\left(x/D_H\right)^{-0.24}\mathrm{Re}_a^{-0.45} \tag{3.2.121}$$

where $J_H = \mathrm{St}\,\mathrm{Pr}_a^{2/3}$ and St - $h_c/\rho c_p \bar{u}$.

All of the reported fully developed turbulent flow heat transfer measurements have been plagued by solute and solvent, thermal entrance, and degradation effects, and thus there is considerable scatter in the results. Degradation effects can be reduced or eliminated by using large amounts of polymer (500 to 10,000 wppm) so that the solution becomes saturated. Cho and Hartnett (1982) attempted to eliminate these effects by using a thermal entrance length of 430 diameters and saturated polymer solutions which should yield maximum heat transfer reductions. Their experimental results for fully developed heat transfer were correlated for a Reynolds number range $3500 \le \mathrm{Re}_a \le 40{,}000$ and concentration solutions of 500 to 5000 wppm of polyacrylamide and polyethylene oxide by

$$J_H = 0.03\mathrm{Re}_a^{-0.45} \tag{3.2.122}$$

For viscoelastic fluids in fully developed (hydrodynamically and thermally) *laminar flow in circular ducts* there is no apparent viscoelastic effect. Thus, the heat transfer relations are the same as those for time-independent fluids such as power law or modified power law fluids. The same situation holds for thermal entrance region heat transfer (Graetz problem). Relations for laminar Nusselt numbers in thermal entrance regions are presented by Cho and Hartnett (1982).

Free Convection Flows and Heat Transfer

Free convection information available in the heat transfer literature up to the present time is concentrated on heat transfer to power law fluids for vertical plates and parallel plate channels. For free convection flows, however, the velocities and thus the shear rates are low and care must be taken that the flow for a particular fluid is in the power law shear rate region before using power law solutions or correlations. Comprehensive review articles on free convection with non-Newtonian fluids have been presented by Shenoy and Mashelkar (1982) and Irvine and Karni (1987).

For a single vertical plate with a modified power law fluid and a thermal boundary condition $\overline{Nu}_T$, in laminar flow, the following relation is recommended by Shenoy and Mashelkar (1982):

$$\overline{Nu}_{TL} = T(n)Gr_{TL}^{1/(2n+2)}\mathrm{Pr}_{TL}^{n/(3n+1)} \tag{3.2.123}$$

where $\overline{Nu}_{TL}$ is the average Nusselt number and

$$\mathrm{Gr}_{TL} = \frac{\rho^2 L^{n+2}}{K^2}\left[g\alpha\left(T_s - T_\infty\right)\right]^{2-n} \tag{3.2.124}$$

$$\mathrm{Pr}_{TL} = \frac{\rho c_p}{k}\left(\frac{K}{\rho}\right)^{2/(n+1)} L^{(n-1)/(2n+2)}\left[g\alpha\left(T_s - T_\infty\right)\right]^{(3n-3)/(2n+2)} \tag{3.2.125}$$

where α is the isobaric thermal expansion coefficient.

In the range $0.5 \le n \le 1$, $T(n)$ can be approximated by

$$T(n) = 0.1636n + 0.5139 \tag{3.2.126}$$

The characteristic dimension in the Nusselt and Grashof numbers is the plate height, L.

For thermal boundary conditions Nu_H, the following relation is also recommended by Shenoy and Mashelkar (1982). Since the heat flux, q_w is specified in this case, the local plate temperature at any x (measured from the bottom of the plate) can be obtained from the local Nusselt number Nu_{Hx}. The heat transfer coefficient is defined in terms of the difference between the wall and free-stream temperatures.

$$\mathrm{Nu}_{Hx} = 0.619\left[\mathrm{Gr}_{Hx}^{(3n+2)/(n+4)}\mathrm{Pr}_{Hx}^{n}\right]^{0.213} \tag{3.2.127}$$

where

$$\mathrm{Gr}_{Hx} = \frac{\rho^2 x^4}{k^2}\left(\frac{g\alpha q_w}{k}\right)^{2-n} \tag{3.2.128}$$

$$\mathrm{Pr}_{Hx} = \frac{\rho c_p}{K}\left(\frac{K}{\rho}\right)^{5/(n+4)} x^{(2n-2)/(n+4)}\left(\frac{g\alpha q_w}{k}\right)^{(3n-3)/(n+4)} \tag{3.2.129}$$

Vertical Parallel Plates

For *power law fluids* and laminar flow, Figure 3.2.29 presents the graphical results of a numerical solution. Of interest are the average Nusselt number $\overline{Nu}_{Tb}$ and the dimensionless average flow velocity between the plates, U_o^+. These are shown on the left and right ordinates respectively in Figure 3.2.29 (Irvine et al., 1982). The characteristic dimension in the Nusselt and Grashof numbers is the plate spacing, *b*. The dimensionless quantities used in Figure 3.2.29 are defined as follows:

$$\overline{Nu}_{Tb} = \frac{\bar{h}_c b}{k} \qquad U_o^+ = \frac{b u_o}{L u^*}$$

$$\mathrm{Pr}_g = \frac{\rho c_p}{k}\left[\frac{v_k^{1/(2-n)}}{\left(\frac{L}{b}\right)^{(1-n)/(2-n)} b^{(2n-2)/(2-n)}}\right] \qquad v_K = \frac{K}{\rho}$$

$$\mathrm{Gr}_g = \frac{g\alpha\left(T_s - T_\infty\right) b^{(n+2)/(2-n)}}{v_K^{2/(2-n)}\left(\frac{L}{b}\right)^{n/(2-n)}} \qquad u^* = \frac{v_K^{1/(2-n)}\, b^{(1-2n)/(2-n)}}{L^{(1-n)/(2-n)}}$$

For vertical parallel plates for the average Nusselt number, $\overline{Nu}_{Hb}$, and the between plate average velocity, Schneider and Irvine (1984) have presented graphical results similar to Figure 3.2.29.

Lee et al. (1998) has presented a numerical solution for laminar flow of a *modified power law fluid* between vertical plates. Lee has also calculated thermal entrance regions and shown that if a parallel plate system is actually operating in the transition region and if the power law solution is used, both the total heat transfer and the velocity between plates can differ by over an order of magnitude. It is important to consider the shear rate parameter in order to determine which free convection solution to use.

Sphere and Horizontal Cylinder — Power Law Fluids

For flow over a sphere, the correlation for power law fluids by Amato and Tien (1976) is

$$\overline{Nu}_{Tr} = CZ^D \tag{3.2.130}$$

where

$$Z = \mathrm{Gr}_{Tr}^{1/(2n+2)}\,\mathrm{Pr}_{Tr}^{n/(3n+1)} \tag{3.2.131}$$

and

$$C = 0.996 \pm 0.120, \quad D = 0.682 \qquad \text{for } Z < 10$$

$$C = 0.489 \pm 0.005, \quad D = 1.10 \qquad \text{for } 10 \le Z \le 40$$

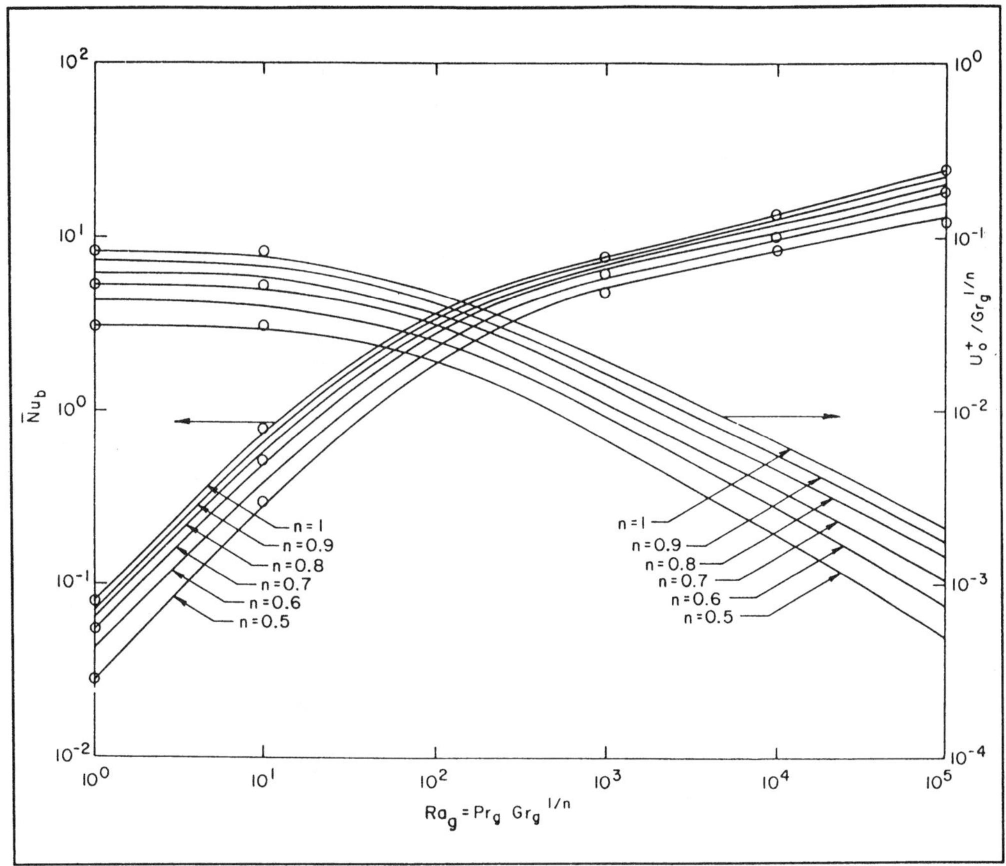

FIGURE 3.2.29 Free convection average Nusselt number, $\bar{Nu}_b$ and dimensionless average velocity U_o^+ between vertical plates for a power law fluid vs. generalized Raleigh number for the Nu_T boundary condition. (From Irvine, T.F., Jr. et al., ASME Paper 82-WA/HT-69, 1982. With permission.)

where the characteristic dimension in all dimensionless variables is the sphere radius, r, and Gr_{Tr} and Pr_{Tr} are defined in Equations (3.2.124) and (3.2.125).

For pseudoplastic fluids flowing over a cylinder, an experimental correlation proposed by Gentry and Worllersheim (1974) for the average Nusselt number, $\bar{Nu}_{TD}$, is

$$\bar{Nu}_{TD} = \frac{\bar{h}_c D}{k} = 1.19\left(\mathrm{Gr}_{TD}\,\mathrm{Pr}_{TD}\right)^{0.2} \tag{3.2.132}$$

where Gr_{TD} and Pr_{TD} are defined as in Equations (3.2.124) and (3.2.125) with the cylinder diameter, D, being used instead of L.

References

Acrivos, A. 1960. A theoretical analysis of laminar natural convection heat transfer to non-Newtonian fluids, *AIChE J.*, 6, 584–590.

Amato, W.S. and Tien, C. 1976. Free convection heat transfer from isothermal spheres in polymer solutions, *Int. J. Heat Mass Transfer*, 19, 1257–1266.

Capobianchi, M. and Irvine, T.F., Jr. 1992. Predictions of pressure drop and heat transfer in concentric annular ducts with modified power law fluids, *Wärme Stoffübertragung*, 27,209–215.

Chandrupatla, A.R. and Sastri, V.M. 1977. Laminar forced convection heat transfer of a non-Newtonian fluid in a square duct, *Int. J. Heat Mass Transfer*, 20, 1315–1324.

Cheng, J.A. 1984. Laminar Forced Convection Heat Transfer of Power Law Fluids in Isosceles Triangular Ducts, Ph.D. Thesis, Mechanical Engineering Department, State University of New York at Stony Brook.

Cho, Y.I. and Hartnett, J.P. 1982. Non-Newtonian fluids in circular pipe flow, *Adv. Heat Transfer*, 15, 59–141.

Cho, Y.I., Ng, K.S., and Hartnett, J.P. 1980. Viscoelastic fluids in turbulent pipe flow — a new heat transfer correlation, *Lett. Heat Mass Transfer*, 7, 347.

Gentry, C.C. and Wollersheim, D.E. 1974. Local free convection to non-Newtonian fluids from a horizontal isothermal cylinder, *ASME J. Heat Transfer*, 96, 3–8.

Hartnett, J.P. and Kostic, M. 1989. Heat transfer to Newtonian and non-Newtonian fluids in rectangular ducts, *Adv. Heat Transfer*, 19, 247–356.

Hartnett, J.P. and Rao, B.K. 1987. Heat transfer and pressure drop for purely viscous non-Newtonian fluids in turbulent flow through rectangular passages, *Wärme Stoffüberetragung*, 21, 261.

Irvine, T.F., Jr. and Karni, J. 1987. Non-Newtonian flow and heat transfer, in *Handbook of Single Phase Convective Heat Transfer*, John Wiley & Sons, New York, 20-1–20-57.

Irvine, T.F., Jr., Wu, K.C., and Schneider, W.J. 1982. Vertical Channel Free Convection to a Power Law Fluid, ASME Paper 82-WA/HT-69.

Irvine, T.F., Jr., Kim, S.C., and Gui, F.L. 1988. Graetz problem solutions for a modified power law fluid, in *ASME Symposium on Fundamentals of Forced Convection Heat Transfer*, ASME publ. HTD 101, pp. 123–127.

Lee, S.R., Irvine, T.F., Jr., and Greene, G.A., 1992, A Computational Analysis of Natural Convection in a Vertical Channel with a Modified Power Law Fluid, Ph.D. Thesis, Mechanical Engineering Department, State University of New York at Stony Brook. Now published in *Proc. 11th Int. Heat Transfer Conf.*, Kyongju, Korea.

Schneider, W.J. and Irvine, T.F., Jr. 1984. Vertical Channel Free Convection for a Power Law Fluid with Constant Heat Flux, ASME Paper 84-HT-16.

Shenoy, A.V. and Mashelkar, R.A. 1982 Thermal convection in non-Newtonian fluids, *Adv. Heat Transfer*, 15, 143–225.

Yoo, S.S. 1974. Heat Transfer and Friction Factors for Non-Newtonian Fluids in Turbulent Pipe Flow, Ph.D. Thesis, University of Illinois at Chicago.

Further Information

Other sources which may be consulted for more detailed information are Cho and Hartnett (1982), Shenoy and Mashelkar (1982), Irvine and Karni (1987), and Hartnett and Kostic (1989).

3.3 Radiation

Michael F. Modest

Nature of Thermal Radiation

All materials continuously emit and absorb radiative energy by lowering or raising their molecular energy levels. This thermal radiative energy may be viewed as consisting of electromagnetic waves or of massless energy parcels, called **photons**. Electromagnetic waves travel through any medium at the speed of light c, which is $c_0 = 2.998 \times 10^8$ m/sec in vacuum and approximately the same in most gases such as air and combustion products. These are characterized by their wavelength λ or frequency ν, which are related by

$$\nu = c/\lambda \tag{3.3.1}$$

The strength and wavelengths of **emission** and **absorption** depend on the temperature and nature of the material.

The ability of photons to travel unimpeded through vacuum and gases makes thermal radiation the dominant mode of heat transfer in vacuum, low-pressure environments, and outer space applications (due to the near absence of conduction and convection). Its temperature dependence [as given by Equation (3.3.3) below] on the other hand, guarantees that radiative heat transfer is of utmost importance in high-temperature applications (including solar radiation: with the sun being a high-temperature heat source at an effective temperature of $T_{sun} = 5762$ K).

When an electromagnetic wave traveling through a gas (or vacuum) strikes the surface of a medium, the wave may be partly or totally reflected, and any nonreflected part will penetrate into the medium. If a wave passes through a medium without any attenuation, the material is called **transparent**. A body with partial attenuation is known as **semitransparent**, and a body through which none of the incoming radiation penetrates is called **opaque**. Most gases are rather transparent to radiation (except for narrow spectral regions, called *absorption bands*), while most solids tend to be strong absorbers for most wavelengths, making them opaque over a distance of a few nanometers (electrical conductors, i.e., metals) to a few micrometers (ceramics, semiconductors), or more (dielectrics).

Blackbody Radiation

The total amount of radiative energy emitted from a surface into all directions above it is termed **emissive power**; we distinguish between **spectral** (at a given wavelength λ, per unit wavelength) and total (encompassing all wavelengths) emissive power. The magnitude of emissive power depends on wavelength λ, temperature T, and a surface property, called **emissivity** ε, which relates the ability of a surface to emit radiative energy to that of an ideal surface, which emits the maximum possible energy (at a given wavelength and temperature). Such an ideal surface is known as a "**blackbody**" or "black surface," since it absorbs all incoming radiation; i.e., it reflects no radiation and is, therefore, invisible ("black") to the human eye. The spectral distribution of the emissive power of a black surface is given by **Planck's law**.

$$E_{b\lambda} = \frac{C_1}{\lambda^5\left[e^{C_2/\lambda T} - 1\right]}, \qquad C_1 = 3.7419 \times 10^{-16}\ \text{Wm}^2, \qquad C_2 = 14{,}388\ \mu\text{mK} \tag{3.3.2}$$

where C_1 and C_2 are sometimes called Planck function constants. The total emissive power of a blackbody is given by

$$E_b = \int_0^{\infty} E_{b\lambda}\, d\lambda = \sigma T^4, \qquad \sigma = 5.670 \times 10^{-8}\ \text{W/m}^2\text{K}^4 \tag{3.3.3}$$

with σ known as the Stefan–Boltzmann constant. Figure 3.3.1 shows the spectral solar irradiation that impinges on Earth, which closely resembles the spectrum of a blackbody at 5762 K. The general behavior of Planck's law is depicted in Figure 3.3.2, together with the fractional emissive power, $f(\lambda T)$, defined as

$$f(\lambda T) = \frac{1}{E_b} \int_0^{\lambda} E_{b\lambda}(\lambda, T)\, d\lambda \tag{3.3.4}$$

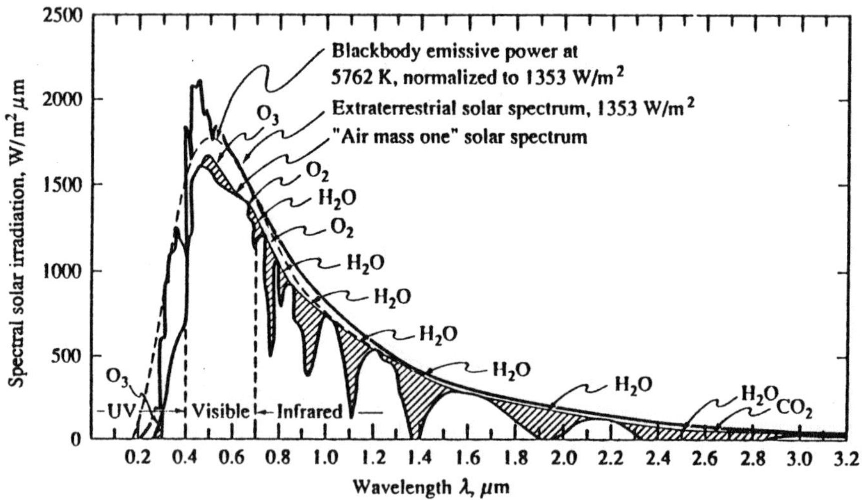

FIGURE 3.3.1 Solar irradiation onto Earth. (From Modest, M., *Radiative Heat Transfer,* McGraw-Hill, New York, 1993. With permission.)

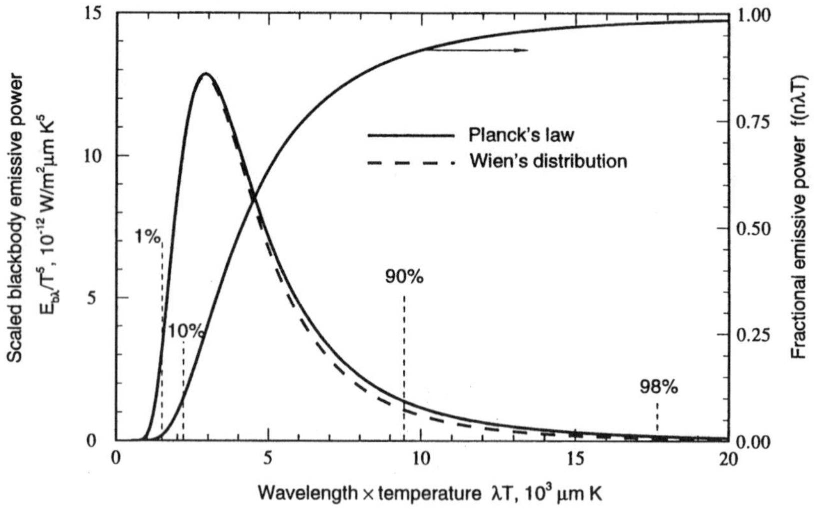

FIGURE 3.3.2 Normalized blackbody emissive power spectrum. (From Modest, M., *Radiative Heat Transfer,* McGraw-Hill, New York, 1993. With permission.)

Note that 90% of all blackbody emission takes place at wavelengths of $\lambda T > 2200$ μmK and at all wavelengths $\lambda T < 9400$ μmK. This implies that — for typical high-temperature heat transfer applications in the range between 1000 and 2000 K — infrared wavelengths in the range 1 μm < λ < 10 μm govern the heat transfer rates. For solar applications shorter wavelengths, down to $\lambda \cong 0.4$ μm are also important. Also shown in Figure 3.3.2 is Wien's law:

$$E_{b\lambda} = \frac{C_1}{\lambda^5} e^{-C_2/\lambda T} \tag{3.3.5}$$

which approximates Planck's law accurately over the part of the spectrum that is important to heat transfer, and that is easier to manipulate mathematically.

Example 3.3.1

What fraction of total solar emission falls into the visible spectrum (0.4 to 0.7 μm)?

Solution: With a solar temperature of 5762 K it follows that for

$$\lambda_1 = 0.4\ \mu\text{m}, \qquad \lambda_1 T_{\text{sun}} = 0.4 \times 5762 = 2304\ \mu\text{mK}$$

and for

$$\lambda_2 = 0.7\ \mu\text{m}, \qquad \lambda_2 T_{\text{sun}} = 0.7 \times 5762 = 4033\ \mu\text{mK}$$

From Figure 3.3.2 we can estimate $f(\lambda_1 T_{\text{sun}}) \cong 12\%$ and $f(\lambda_2 T_{\text{sun}}) \cong 48\%$. Thus, the visible fraction of sunlight is $48 - 12 \cong 36\%$: with a bandwidth of only 0.3 μm the human eye responds to approximately 36% of all emitted sunlight!

Radiative Exchange between Opaque Surfaces

Radiative Properties of Surfaces

Strictly speaking, the surface of an enclosure wall can only reflect radiative energy and allow a part of it to penetrate into the substrate. A surface cannot absorb or emit photons: attenuation takes place inside the solid, as does emission of radiative energy (with some of the emitted energy escaping through the surface into the enclosure). In practical systems, the thickness of the surface layer over which absorption of **irradiation** from inside the enclosure occurs is very small compared with the overall dimension of an enclosure — usually a few nanometers for metals and a few micrometers for most nonmetals. The same may be said about emission from within the walls that escapes into the enclosure. Thus, in the case of opaque walls it is customary to speak of absorption by and emission from a "surface," although a thin surface layer is implied. Four fundamental radiative properties are defined:

Reflectivity, $$\rho \equiv \frac{\text{reflected part of incoming radiation}}{\text{total incoming radiation}} \tag{3.3.6a}$$

Absorptivity, $$\rho \equiv \frac{\text{absorbed part of incoming radiation}}{\text{total incoming radiation}} \tag{3.3.6b}$$

Transmissivity, $$\tau \equiv \frac{\text{transmitted part of incoming radiation}}{\text{total incoming radiation}} \tag{3.3.6c}$$

Emissivity, $$\varepsilon \equiv \frac{\text{energy emitted from a surface}}{\text{energy emitted by a black surface at same temperature}} \tag{3.3.6d}$$

Since all incoming radiation must be reflected, absorbed, or transmitted, it follows that

$$\rho + \alpha + \tau = 1 \tag{3.3.7}$$

In most practical applications surface layers are thick enough to be opaque ($\tau = 0$, leading to $\rho + \alpha = 1$. All four properties may be functions of wavelength, temperature, incoming direction (except emissivity), and outgoing direction (except absorptivity).

Directional Behavior. For heat transfer applications, the dependence on incoming direction for absorptivity (as well as ρ and τ) and outgoing direction for emissivity is generally weak and is commonly neglected; i.e., it is assumed that the surface absorbs and emits diffusely. Then, for an opaque surface, for any given wavelength

$$\varepsilon_\lambda = \alpha_\lambda = 1 - \rho_\lambda \tag{3.3.8}$$

Published values of emissivities are generally either "normal emissivities" (the directional value of ε_λ in the direction perpendicular to the surface) or "hemispherical emissivities" (an average value over all outgoing directions). The difference between these two values is often smaller than experimental accuracy and/or repeatability.

Reflected energy (due to a single, distinct incoming direction) may leave the surface into a single direction ("specular" reflection, similar to reflection from a mirror for visible light), or the reflection may spread out over all possible outgoing directions. In the extreme case of equal amounts going into all directions, we talk about "diffuse" reflection. Smooth surfaces (as compared with the wavelength of radiation) tend to be specular reflectors, while rough surfaces tend to be more or less diffusely reflecting. Analysis is vastly simplified if diffuse reflections are assumed. Research has shown that — except for some extreme geometries and irradiation conditions susceptible to beam channeling (irradiated open cavities, channels with large aspect ratios) — radiative heat transfer rates are only weakly affected by the directional distribution of reflections. Therefore, it is common practice to carry out radiative heat transfer calculations assuming only diffuse reflections.

Spectral Dependence. The emissivity of a surface generally varies strongly and in complex ways with wavelength, depending on the material, surface layer composition, and surface structure (roughness). Therefore, unlike bulk material properties (such as thermal conductivity) the surface emissivity may display significant differences between two ostensibly identical samples, and even for one and the same sample measured at different times (due to surface roughness and contamination). Despite these difficulties, surfaces may be loosely grouped into two categories — metals and nonconductors (dielectrics), and some generalizations can be made.

Polished Metals. Smooth, purely metallic surfaces (i.e., without any nonmetallic surface contamination, such as metal oxides) tend to have very low emissivities in the infrared. For many clean metals $\varepsilon_\lambda < 0.1$ for $\lambda > 2$ μm, and spectral as well as temperature dependence are generally well approximated by the proportionality $\varepsilon_\lambda \propto \sqrt{T/\lambda}$ in the infrared. However, for shorter wavelengths ($\lambda < 1$ μm), emissivity values may become quite substantial, and temperature dependence is usually reversed (decreasing, rather than increasing, with temperature). Typical room temperature behavior of several metals is shown in Figure 3.3.3. Caution needs to be exercised when choosing an emissivity value for a metal surface: unless extraordinary care is taken to keep a polished metal clean (i.e., free from oxidation and/or surface contamination), its emissivity may soon become several times the value of the original, polished specimen (for example, consider the formation of aluminum oxide on top of aluminum, Figure 3.3.3).

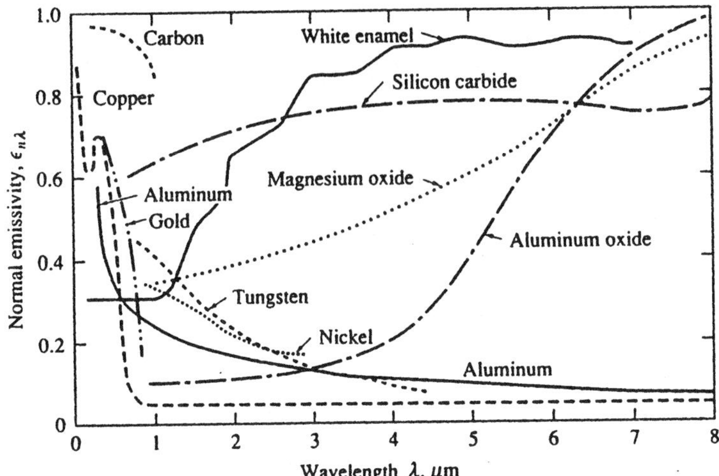

FIGURE 3.3.3 Normal, spectral emissivities for selected materials. (From Modest, M., *Radiative Heat Transfer*, McGraw-Hill, New York, 1993. With permission.)

Ceramics and Refractories. Smooth ceramics tend to have fairly constant and intermediate emissivity over the near- to mid-infrared, followed by a sharp increase somewhere between 4 and 10 μm. At short wavelengths these materials display strong decreases in emissivity, so that a number of them may appear white to the human eye even though they are fairly black in the infrared. The temperature dependence of the emissivity of ceramics is rather weak; generally a slight increase with temperature is observed in the infrared. The spectral emissivity of a few ceramics is also shown in Figure 3.3.3.

Other Nonconductors. The behavior of most electrically nonconducting materials is governed by surface structure, nonhomogeneity, dopants, porosity, flaws, surface films, etc. The emissivity may vary irregularly across the spectrum because of various emission bands, influence of flaws, etc., making any generalization impossible. This irregularity may be exploited to obtain surfaces of desired spectral behavior, so-called selective surfaces. Some selective surfaces (as compared with a common steel) are depicted in Figure 3.3.4. For a solar collector it is desirable to have a high spectral emissivity for short wavelengths $\lambda < 2.5$ μm (strong absorption of solar irradiation), and a low value for $\lambda > 2.5$ μm (to minimize re-emission from the collector). The opposite is true for a spacecraft radiator panel used to reject heat into space.

It is clear that (1) values of spectral surface emissivity are subject to great uncertainty and (2) only a relatively small range of infrared wavelengths are of importance. Therefore, it is often assumed that the surfaces are "**gray**", i.e., the emissivity is constant across (the important fraction of) the spectrum, $\varepsilon_\lambda \neq \varepsilon_\lambda(\lambda)$, since this assumption also vastly simplifies analysis. Table 3.3.1 gives a fairly detailed listing of total emissivities of various materials, defined as

$$\varepsilon(T) = \frac{1}{E_b(T)} \int_0^\infty \varepsilon_\lambda(\lambda, T) E_{b\lambda}(T)\, d\lambda \tag{3.3.9}$$

which may be enlisted for a gray analysis.

View Factors

In many engineering applications the exchange of radiative energy between surfaces is virtually unaffected by the medium that separates them. Such (radiatively) *nonparticipating media* include vacuum as well as monatomic and most diatomic gases (including air) at low to moderate temperature levels (i.e., before ionization and dissociation occurs). Examples include spacecraft heat rejection systems, solar collector systems, radiative space heaters, illumination problems, and so on. It is common practice

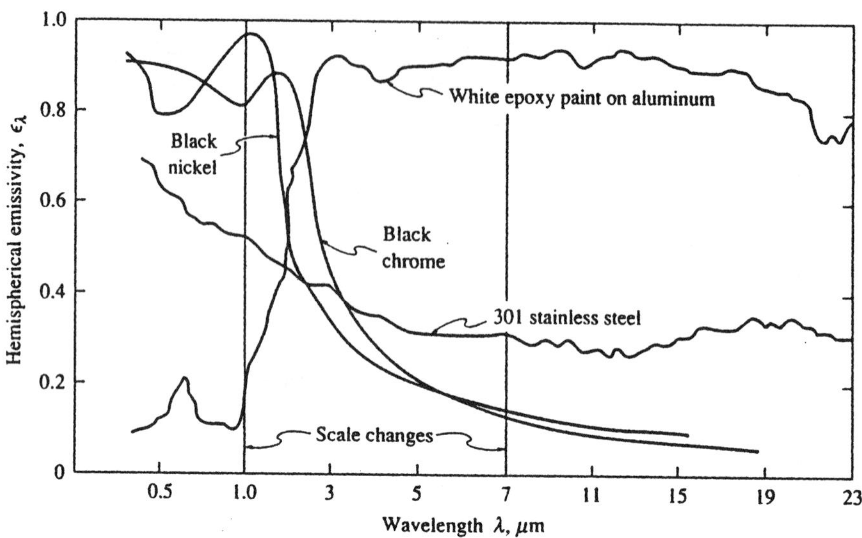

FIGURE 3.3.4 Spectral, hemispherical reflectivities of several spectrally selective surfaces. (From Modest, M., *Radiative Heat Transfer,* McGraw-Hill, New York, 1993. With permission.)

to simplify the analysis by making the assumption of an *idealized enclosure* and/or of *ideal surface properties*. The greatest simplification arises if all surfaces are black: for such a situation no reflected radiation needs to be accounted for, and all emitted radiation is diffuse (i.e., the radiative energy leaving a surface does not depend on direction). The next level of difficulty arises if surfaces are assumed to be gray, diffuse emitters (and, thus, absorbers) as well as gray, diffuse reflectors. The vast majority of engineering calculations are limited to such ideal surfaces, since, particularly, the effects of nondiffuse reflections are usually weak (see discussion in previous section).

Thermal radiation is generally a long-range phenomenon. This is *always* the case in the absence of a participating medium, since photons will travel unimpeded from surface to surface. Therefore, performing a thermal radiation analysis for one surface implies that all surfaces, no matter how far removed, that can exchange radiative energy with one another must be considered simultaneously. How much energy any two surfaces exchange depends in part on their size, separation, distance, and orientation, leading to geometric functions known as **view factors**, defined as

$$F_{i-j} = \frac{\text{diffuse energy leaving } A_i \text{ directly toward and intercepted by } A_j}{\text{total diffuse energy leaving } A_i} \tag{3.3.10}$$

In order to make a radiative energy balance we always need to consider an entire *enclosure* rather than and infinitesimal control volume (as is normally done for other modes of heat transfer, i.e., conduction or convection). The enclosure must be closed so that irradiation from all possible directions can be accounted for, and the enclosure surfaces must be *opaque* so that all irradiation is accounted for, for each direction. In practice, an incomplete enclosure may be closed by introducing artificial surfaces. An enclosure may be idealized in two ways, as indicated in Figure 3.3.5: by replacing a complex geometric shape with a few simple surfaces, and by assuming surfaces to be isothermal with constant (i.e., average) heat flux values across them. Obviously, the idealized enclosure approaches the real enclosure for sufficiently small isothermal subsurfaces.

Mathematically, the view factor needs to be determined from a double integral, i.e.,

$$F_{i-j} = \frac{1}{A_i} \int_{A_i} \int_{A_j} \frac{\cos\theta_i \cos\theta_j}{\pi S_{ij}^2} dA_j \, dA_i \tag{3.3.11}$$

TABLE 3.3.1 Total Emissivity and Solar Absorptivity of Selected Surfaces

	Temperature (°C)	Total Normal Emissivity	Extraterrestrial Solar Absorptivity
Alumina, flame-sprayed	–25	0.80	0.28
Aluminum foil			
As received	20	0.04	
Bright dipped	20	0.025	0.10
Aluminum, vacuum-deposited	20	0.025	0.10
Hard-anodized	–25	0.84	0.92
Highly polished plate, 98.3% pure	225–575	0.039–0.057	
Commercial sheet	100	0.09	
Rough polish	100	0.18	
Rough plate	40	0.055–0.07	
Oxidized at 600°C	200–600	0.11–0.19	
Heavily oxidized	95–500	0.20–0.31	
Antimony, polished	35–260	0.28–0.31	
Asbestos	35–370	0.93–0.94	
Beryllium	150	0.18	0.77
	370	0.21	
	600	0.30	
Beryllium, anodized	150	0.90	
	370	0.88	
	600	0.82	
Bismuth, bright	75	0.34	
Black paint			
Parson's optical black	–25	0.95	0.975
Black silicone	–25–750	0.93	0.94
Black epoxy paint	–25	0.89	0.95
Black enamel paint	95–425	0.81–0.80	
Brass, polished	40–315	0.10	
Rolled plate, natural surface	22	0.06	
Dull plate	50–350	0.22	
Oxidized by heating at 600°C	200–600	0.61–0.59	
Carbon, graphitized	100–320	0.76–0.75	
	320–500	0.75–0.71	
Candle soot	95–270	0.952	
Graphite, pressed, filed surface	250–510	0.98	
Chromium, polished	40–1100	0.08–0.36	
Copper, electroplated	20	0.03	0.47
Carefully polished electrolytic copper	80	0.018	
Polished	115	0.023	
Plate heated at 600°C	200–600	0.57	
Cuprous oxide	800–1100	0.66–0.54	
Molten copper	1075–1275	0.16–0.13	
Glass, Pyrex, lead, and soda	260–540	0.95–0.85	
Gypsum	20	0.903	
Gold, pure, highly polished	225–625	0.018–0.035	
Inconel X, oxidized	–25	0.71	0.90
Lead, pure (99.96%), unoxidized	125–225	0.057–0.075	
Gray oxidized	25	0.28	
Oxidized at 150°C	200	0.63	
Magnesium oxide	275–825	0.55–0.20	
	900–1705	0.20	
Magnesium, polished	35–260	0.07–0.13	
Mercury	0–100	0.09–0.12	
Molybdenum, polished	35–260	0.05–0.08	
	540–1370	0.10–0.18	

TABLE 3.3.1 (continued) Total Emissivity and Solar Absorptivity of Selected Surfaces

	Temperature (°C)	Total Normal Emissivity	Extraterrestrial Solar Absorptivity
	2750	0.29	
Nickel, electroplated	20	0.03	0.22
Polished	100	0.072	
Platinum, pure, polished	225–625	0.054–0.104	
Silica, sintered, powdered, fused silica	35	0.84	0.08
Silicon carbide	150–650	0.83–0.96	
Silver, polished, pure	40–625	0.020–0.032	
Stainless steel			
Type 312, heated 300 hr at 260°C	95–425	0.27–0.32	
Type 301 with Armco black oxide	–25	0.75	0.89
Type 410, heated to 700°C in air	35	0.13	0.76
Type 303, sandblasted	95	0.42	0.68
Titanium, 75A	95–425	0.10–0.19	
75A, oxidized 300 hr at 450°C	35–425	0.21–0.25	0.80
Anodized	–25	0.73	0.51
Tungsten, filament, aged	27–3300	0.032–0.35	
Zinc, pure, polished	225–325	0.045–0.053	
Galvanized sheet	100	0.21	

Source: Modest, M., *Radiative Heat Transfer,* McGraw-Hill, New York, 1993. With permission.

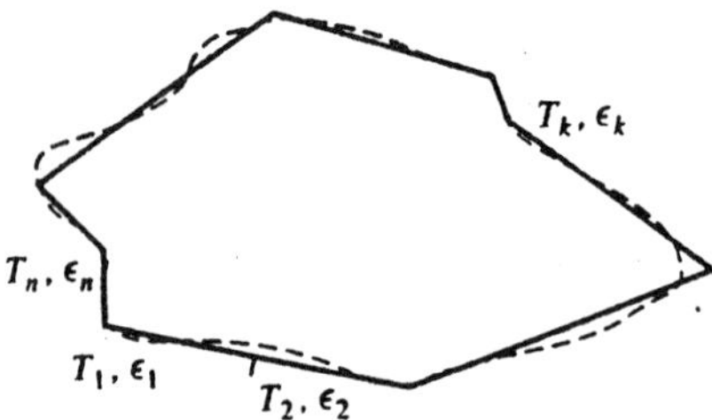

FIGURE 3.3.5 Real and ideal enclosures for radiative transfer calculations. (From Modest, M., *Radiative Heat Transfer,* McGraw-Hill, New York, 1993. With permission.)

where θ_i and θ_j are the angles between the surface normals on A_i and A_j, respectively, and the line (of length S_{ij}) connecting two points on the two surfaces. Analytical solutions to Equation (3.3.11) may be found for relatively simple geometries. A few graphical results for important geometries are shown in Figures 3.3.6 to 3.3.8. More-extensive tabulations as well as analytical expressions may be found in textbooks on the subject area (Modest, 1993; Siegel and Howell, 1992) as well as view factor catalogs (Howell, 1982). For nontrivial geometries view factors must be calculated numerically, either (1) by numerical quadrature of the double integral in Equation (3.3.11), or (2) by converting Equation (3.3.11) into a double-line integral, followed by numerical quadrature, or (3) by a Monte Carlo method (statistical sampling and tracing of selected light rays).

View Factor Algebra. For simple geometries analytical values can often be found by expressing the desired view factor in terms of other, known ones. This method is known as *view factor algebra*, by manipulating the two relations,

Reciprocity rule: $$A_i F_{i-j} = A_j F_{j-i} \qquad (3.3.12)$$

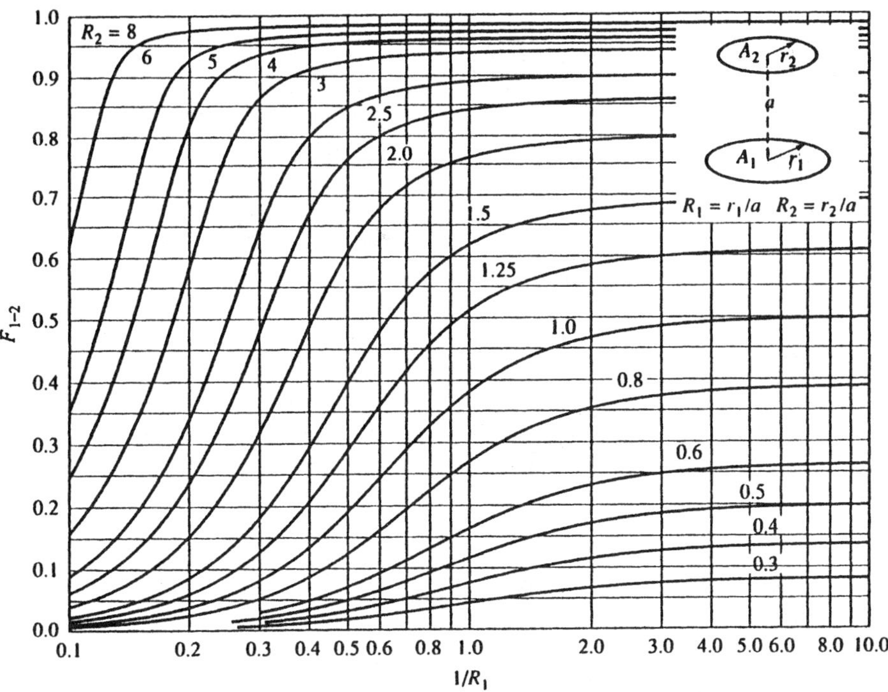

FIGURE 3.3.6 View factor between parallel, coaxial disks of unequal radius. (From Modest, M., *Radiative Heat Transfer,* McGraw-Hill, New York, 1993. With permission.)

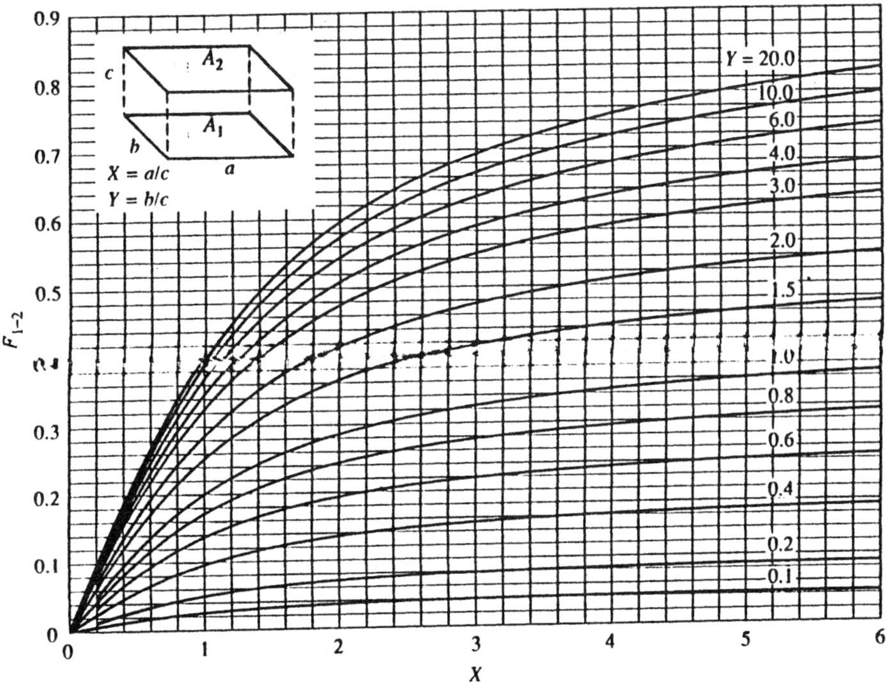

FIGURE 3.3.7 View factor between identical, parallel, directly opposed rectangles. (From Modest, M., *Radiative Heat Transfer,* McGraw-Hill, New York, 1993. With permission.)

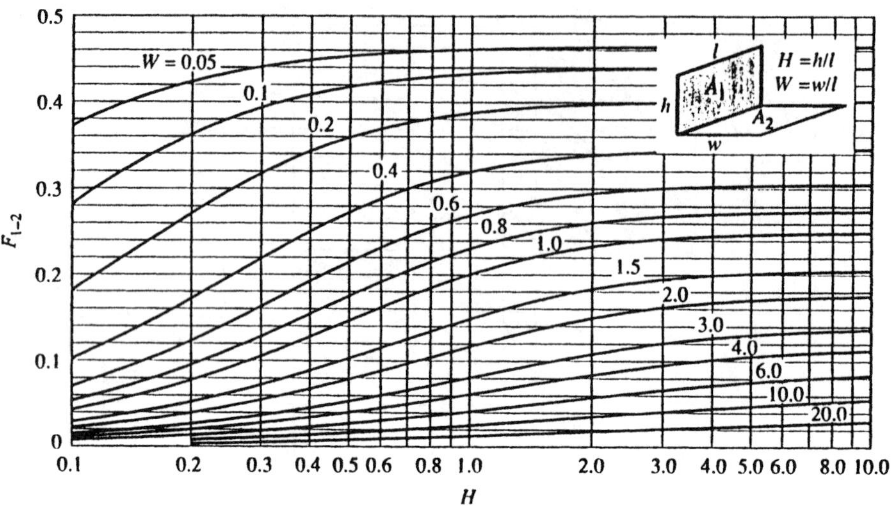

FIGURE 3.3.8 View factor between perpendicular rectangles with common edge. (From Modest, M., *Radiative Heat Transfer*, McGraw-Hill, New York, 1993. With permission.)

Summation rule:

$$\sum_{j=1}^{N} F_{i-j} = 1, \qquad i = 1, N \tag{3.3.13}$$

assuming that the (closed) configuration consists of N surfaces. The reciprocity rule follows immediately from Equation (3.3.11), while the summation rule simply states that the fractional energies leaving surface A_i must add up to a whole.

Example 3.3.2

Assuming the view factor for a finite corner, as shown in Figure 3.3.8 is known, determine the view factor F_{3-4}, between the two perpendicular strips as shown in Figure 3.3.9.

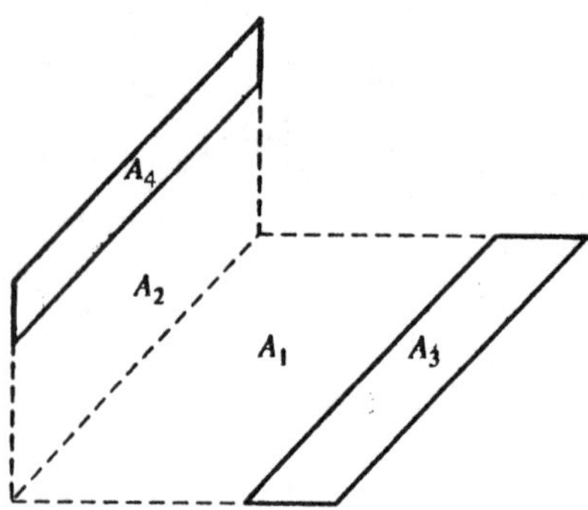

FIGURE 3.3.9 Configuration for Example 3.3.2 (strips on a corner piece). (From Modest, M., *Radiative Heat Transfer*, McGraw-Hill, New York, 1993. With permission.)

Solution. From the definition of the view factor, and since the energy traveling to A_4 is the energy going to A_2 and A_4 minus the one going to A_2, it follows that

$$F_{3-4} = F_{3-(2+4)} - F_{3-2}$$

and, using reciprocity,

$$F_{3-4} = \frac{1}{A_3}\left[\left(A_2 + A_4\right)F_{(2+4)-3} - A_2 F_{2-3}\right]$$

Similarly, we find

$$F_{3-4} = \frac{A_2 + A_4}{A_3}\left(F_{(2+4)-(1+3)} - F_{(2+4)-1}\right) - \frac{A_2}{A_3}\left(F_{2-(1+3)} - F_{2-1}\right)$$

All view factors on the right-hand side are corner pieces and, thus, are known from Figure 3.3.8.

Crossed-Strings Method. A special type of view factor algebra may be used to determine all the view factors in long enclosures with constant cross section. The method is called the crossed-strings method since the view factors can be determined experimentally with four pins, a roll of string, and a yardstick. Consider the configuration in Figure 3.3.10, which shows the cross section of an infinitely long enclosure, continuing into and out of the plane of the figure. Repeatedly applying reciprocity and summation rules allows the evaluation of F_{1-2} as

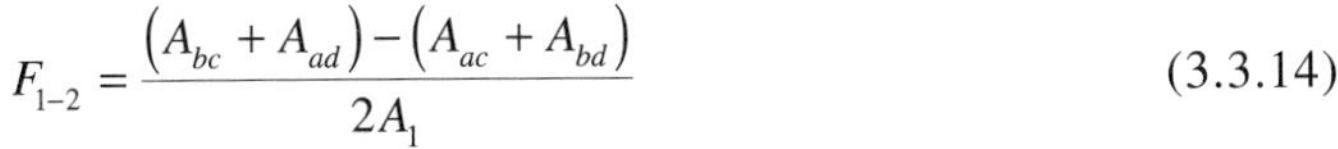

$$F_{1-2} = \frac{\left(A_{bc} + A_{ad}\right) - \left(A_{ac} + A_{bd}\right)}{2A_1} \tag{3.3.14}$$

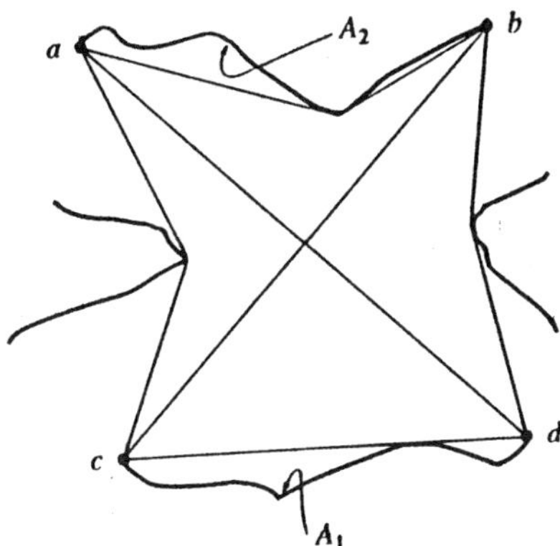

FIGURE 3.3.10 The crossed-strings method for arbitrary two-dimensional configurations. (From Modest, M., *Radiative Heat Transfer,* McGraw-Hill, New York, 1993. With permission.)

where A_{ab} is the area (per unit depth) defined by the length of the string between points a and b, etc. This formula is easily memorized by looking at the configuration between any two surfaces as a generalized "rectangle," consisting of A_1, A_2, and the two sides A_{ac} and A_{bd}. Then

$$F_{1-2} = \frac{\text{diagonals} - \text{sides}}{2 \times \text{originating area}} \tag{3.3.15}$$

Example 3.3.3

Calculate F_{1-2} for the configuration shown in Figure 3.3.11.

Solution. From the figure it is obvious that

$$s_1^2 = (c - d\cos\alpha)^2 + d^2\sin^2\alpha = c^2 + d^2 - 2cd\cos\alpha$$

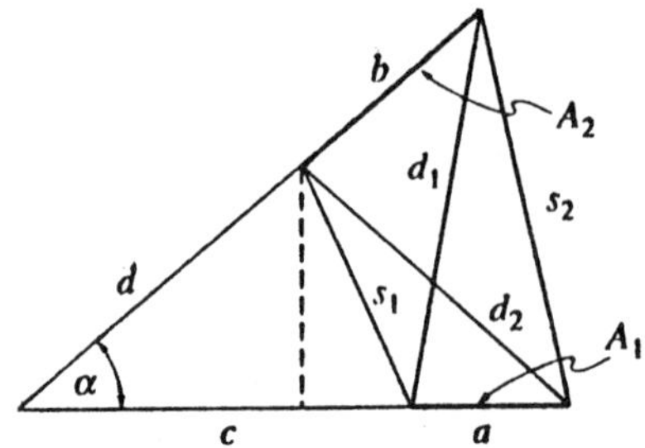

FIGURE 3.3.11 Infinitely long wedge-shaped groove for Example 3.3.3. (From Modest, M., *Radiative Heat Transfer,* McGraw-Hill, New York, 1993. With permission.)

Similarly, we have

$$s_2^2 = (a+c)^2 + (b+d)^2 - 2(a+c)(b+d)\cos\alpha$$

$$d_1^2 = (a+c)^2 + d^2 - 2(a+c)d\cos\alpha$$

$$d_2^2 = c^2 + (b+d)^2 - 2c(b+d)\cos\alpha$$

and

$$F_{1-2} = \frac{d_1 + d_2 - (s_1 + s_2)}{2a}$$

Radiative Exchange between Opaque Surfaces (Net Radiation Method)

Consider an enclosure consisting of N opaque surfaces. The enclosure is closed, or, if not, no surface external to the surface reflects or emits radiation into the enclosure (i.e., the open configuration may be artificially closed by replacing openings with cold, black surfaces); any external radiation entering the enclosure is dealt with individually for each surface [see Equation (3.3.17) below]. All surfaces are assumed to be gray, and emit and reflect diffusely. Traditionally, the **radiosity** J of the surfaces is determined, defined as the total diffuse radiative energy leaving a surface (by emission and reflection),

$$J_i = \varepsilon_1 E_{bi} + \rho_i H_i, \qquad i = 1, N \tag{3.3.16}$$

where H_i is the incoming radiative flux (irradiation) onto surface A_i. This leads to N simultaneous equations for the unknown radiosities, specifically,

$$J_i = \varepsilon_i E_{bi} + (1-\varepsilon_i)\left[\sum_{j=1}^{N} J_i F_{i-j} + H_{oi}\right] \tag{3.3.17a}$$

or

$$J_i = q_i + \sum_{j=1}^{N} J_j F_{i-j} + H_{oi} \tag{3.3.17b}$$

depending on whether surface temperature or surface flux are known on surface A_i. In Equation (3.3.17) H_{oi} is irradiation on surface A_i from outside the enclosure, if any; H_{oi} is always zero for closed configurations, but is useful in the presence of external light sources (such as solar energy, lasers, etc.). The radiosity neither is a useful quantity to determine, nor is there a need to determine it. Eliminating the radiosities from Equations (3.3.17a) and (3.3.17b) leads to N simultaneous equations in temperature (E_{bi}) and heat flux (q_i):

$$\frac{q_i}{\varepsilon_i} - \sum_{j=1}^{N} \left(\frac{1}{\varepsilon_j} - 1 \right) F_{i-j} q_j + H_{oi} = E_{bi} - \sum_{j=1}^{N} F_{i-j} E_{bj} \tag{3.3.18}$$

Note that no artificial closing surfaces ($j > N$) appear in Equation (3.3.18), since for these surfaces $\varepsilon_j = 1$ and $E_{bj} = 0$. Thus, such closing surfaces may simply be ignored in the analysis.

Since Equation (3.3.18) is a set of N equations, this requires that N values of emissive power E_{bi} and/or flux q_i must be given as boundary conditions, in order to solve for the remaining N unknowns. For computer calculations Equation (3.3.18) may be recast in matrix form

$$\mathbf{C} \cdot \mathbf{q} = \mathbf{A} \cdot \mathbf{e_b} - \mathbf{h_o} \tag{3.3.19a}$$

where

$$C_{ij} = \frac{\delta_{ij}}{\varepsilon_j} - \left(\frac{1}{\varepsilon_j} - 1 \right) F_{i-j} \tag{3.3.19b}$$

$$A_{ij} = \delta_{ij} - F_{i-j} \tag{3.3.19c}$$

δ_{ij} is Kronecker's delta, i.e.,

$$\delta_{ij} = \begin{cases} 1 & \text{if } i = j \\ 0 & \text{if } i \neq j \end{cases} \tag{3.3.20}$$

and $\mathbf{q}$, $\mathbf{e_b}$, and $\mathbf{h_o}$ are vectors of the surface heat fluxes q_i, emissive powers E_{bi}, and external irradiations H_{oi} (if any). For example, if the temperatures are given for all the surfaces, and the heat fluxes are to be determined, Equation (3.3.19) is solved by matrix inversion, and

$$\mathbf{q} = \left(\mathbf{C}^{-1} \cdot \mathbf{A} \right) \cdot \mathbf{e_b} - \left(\mathbf{C}^{-1} \cdot \mathbf{h_o} \right) \tag{3.3.21}$$

Example 3.3.4

A right-angled groove, consisting of two long black surfaces of width a, is exposed to solar radiation q_{sol} (Figure 3.3.12). The entire groove surface is kept isothermal at temperature T. Determine the net radiative heat transfer rate from the groove.

Solution. We may employ Equation (3.3.19). However, the enclosure is not closed, and we must close it artificially. We note that any radiation leaving the cavity will not come back (barring any reflection from other surfaces nearby). Thus, our artificial surface should be black. We also assume that, with the exception of the (parallel) solar irradiation, no external radiation enters the cavity. Since the solar irradiation is best treated separately through the external irradiation term H_o, our artificial surface is nonemitting. Both criteria are satisfied by covering the groove with a black surface at 0 K (A_3). Even though we now have three surfaces, the last one does not really appear in Equation (3.3.18) (since $E_{b3} = 0$ and $1/\varepsilon_3 - 1 = 0$):

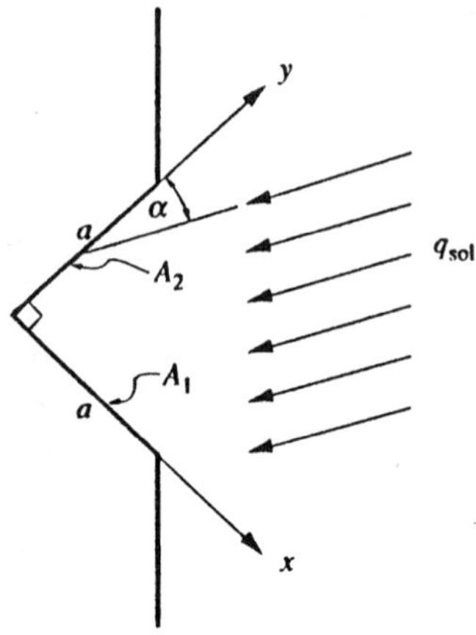

FIGURE 3.3.12 Right-angled groove exposed to solar irradiation, Example 3.3.4. (From Modest, M., *Radiative Heat Transfer,* McGraw-Hill, New York, 1993. With permission.)

$$q_1 = E_{b1} - F_{1-2}E_{b2} - H_{o1} = \sigma T^4\left(1 - F_{1-2}\right) - q_{sol}\cos\alpha$$

$$q_2 = E_{b2} - F_{2-1}E_{b1} - H_{o2} = \sigma T^4\left(1 - F_{2-1}\right) - q_{sol}\sin\alpha$$

From the crossed-strings method, Equation (3.3.15), we find

$$F_{1-2} = \frac{a + a - \left(\sqrt{2}a + 0\right)}{2a} = \frac{1}{2}\left(2 - \sqrt{2}\right) = 0.293 = F_{2-1}$$

and

$$Q' = a\left(q_1 + q_2\right) = a\left[\sqrt{2}\sigma T^4 - q_{sol}\left(\cos\alpha + \sin\alpha\right)\right]$$

Example 3.3.5

Consider a very long duct as shown in Figure 3.3.13. The duct is 30 × 40 cm in cross section, and all surfaces are covered with gray, diffuse surface material. Top and bottom walls are at $T_1 = T_3 = 1000$ K

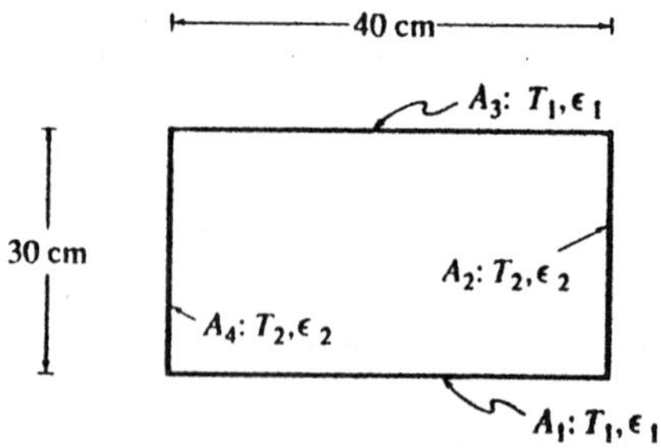

FIGURE 3.3.13 Two-dimensional gray, diffuse duct for Example 3.3.5. (From Modest, M., *Radiative Heat Transfer,* McGraw-Hill, New York, 1993. With permission.)

with $\varepsilon_1 = \varepsilon_3 = 0.3$, while the side walls are at $T_2 = T_4 = 600$ K with $\varepsilon_2 = \varepsilon_4 = 0.8$ as shown. Determine the net radiative heat transfer rates for each surface.

Solution. Using Equation (3.3.18) for $i = 1$ and $i = 2$ and noting that $F_{1-2} = F_{1-4}$ and $F_{2-1} = F_{2-3}$,

$$i = 1: \quad \frac{q_1}{\varepsilon_1} - 2\left(\frac{1}{\varepsilon_2} - 1\right)F_{1-2}q_2 - \left(\frac{1}{\varepsilon_1} - 1\right)F_{1-3}q_1 = 2F_{1-2}\left(E_{b1} - E_{b2}\right)$$

$$i = 2: \quad \frac{q_2}{\varepsilon_2} - 2\left(\frac{1}{\varepsilon_1} - 1\right)F_{2-1}q_1 - \left(\frac{1}{\varepsilon_2} - 1\right)F_{2-4}q_2 = 2F_{2-1}\left(E_{b2} - E_{b1}\right)$$

The view factors are readily evaluated from the crossed-strings method as $F_{1-2} = {}^1/_4$, $F_{1-3} = 1 - 2F_{1-2} = {}^1/_2$, $F_{2-1} = {}^4/_3\, F_{1-2} = {}^1/_3$ and $F_{2-4} = 1 - 2F_{2-1} = {}^1/_3$. Substituting these, as well as emissivity values, into the relations reduces them to the simpler form of

$$\left[\frac{1}{0.3} - \left(\frac{1}{0.3} - 1\right)\frac{1}{2}\right]q_1 - 2\left(\frac{1}{0.8} - 1\right)\frac{1}{4}q_2 = 2 \times \frac{1}{4}\left(E_{b1} - E_{b2}\right)$$

$$-2\left(\frac{1}{0.3} - 1\right)\frac{1}{3}q_1 + \left[\frac{1}{0.8} - \left(\frac{1}{0.8} - 1\right)\right]\frac{1}{3}q_2 = 2 \times \frac{1}{3}\left(E_{b2} - E_{b1}\right)$$

or

$$\frac{13}{6}q_1 - \frac{1}{8}q_2 = \frac{1}{2}\left(E_{b1} - E_{b2}\right)$$

$$-\frac{14}{9}q_1 + \frac{7}{6}q_2 = -\frac{2}{3}\left(E_{b1} - E_{b2}\right)$$

Thus,

$$\left(\frac{13}{6} \times \frac{7}{6} - \frac{14}{9} \times \frac{1}{8}\right)q_1 = \left(\frac{1}{2} \times \frac{7}{6} - \frac{2}{3} \times \frac{1}{8}\right)\left(E_{b1} - E_{b2}\right)$$

$$q_1 = \frac{3}{7} \times \frac{1}{2}\left(E_{b1} - E_{b2}\right) = \frac{3}{14}\sigma\left(T_1^4 - T_2^4\right)$$

and

$$\left(-\frac{1}{8} \times \frac{14}{9} + \frac{7}{6} \times \frac{13}{6}\right)q_2 = \left(\frac{1}{2} \times \frac{14}{9} - \frac{2}{3} \times \frac{13}{6}\right)\left(E_{b1} - E_{b2}\right)$$

$$q_2 = \frac{3}{7} \times \frac{2}{3}\left(E_{b1} - E_{b2}\right) = -\frac{2}{7}\sigma\left(T_1^4 - T_2^4\right)$$

Finally, substituting values for temperatures,

$$Q_1' = 0.4\ \text{m} \times \frac{3}{14} \times 5.670 \times 10^{-8}\ \frac{\text{W}}{\text{m}^2\text{K}^4}\left(1000^4 - 600^4\right)\text{K}^4 = 4230\ \text{W/m}$$

$$Q_2' = -0.3 \text{ m} \times \frac{2}{7} \times 5.670 \times 10^{-8} \frac{\text{W}}{\text{m}^2\text{K}^4} \left(1000^4 - 600^4\right) \text{K}^4 = -4230 \text{ W/m}$$

Note that, for conservation of energy, both heat transfer rates must add up to zero.

Small Body Inside Isothermal Enclosure. An especially simple — but important — case occurs if a small, convex body A_1 (i.e., a surface that cannot "see" itself, or $F_{1-1} = 0$) is totally enclosed by an isothermal enclosure A_2. Then, with $N = 2$ and $F_{1-2} = 1$, Equation (3.3.18) reduces to

$$q_1 = \frac{E_{b1} - E_{b2}}{\dfrac{1}{\varepsilon_1} + \dfrac{A_1}{A_2}\left(\dfrac{1}{\varepsilon_2} - 1\right)} = \frac{\sigma\left(T_1^4 - T_2^4\right)}{\dfrac{1}{\varepsilon_1} + \dfrac{A_1}{A_2}\left(\dfrac{1}{\varepsilon_2} - 1\right)} \tag{3.3.22}$$

If the enclosure is large, i.e., $A_1 \ll A_2$, then Equation (3.3.22) simplifies further to

$$q_1 = \varepsilon_1 \sigma\left(T_1^4 - T_2^4\right) \tag{3.3.23}$$

Radiation Shields. If it is desired to minimize radiative heat transfer between two surfaces, it is common practice to place one or more radiation shields between them (usually thin metallic sheets of low emissivity). If two surfaces A_i and A_j are close together, so that $A_i \cong A_j$ and $F_{i-j} \cong 1$, then the radiative exchange between them is, from Equation (3.3.22),

$$q = \frac{E_{bi} - E_{bj}}{R_{ij}}, \qquad R_{ij} = \frac{1}{\varepsilon_i} + \frac{1}{\varepsilon_j} - 1 \tag{3.3.24}$$

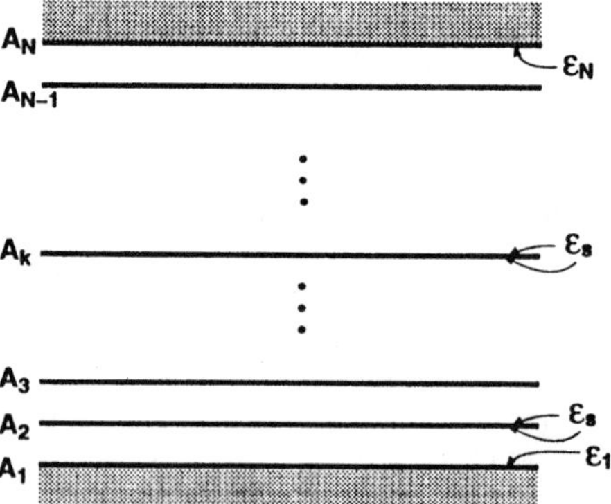

FIGURE 3.3.14 Placement of radiation shields between two large, parallel plates. (From Modest, M., *Radiative Heat Transfer,* McGraw-Hill, New York, 1993. With permission.)

where R_{ij} is termed the *radiative resistance*. Equation (3.3.24) is seen to be analogous to an electrical circuit with "current" q and "voltage potential" $E_{bi} - E_{bj}$. Therefore, expressing radiative fluxes in terms of radiative resistances is commonly known as **network analogy**. The network analogy is a very powerful method of solving one-dimensional problems (i.e., whenever only two isothermal surfaces see each other, such as infinite parallel plates, or when one surface totally encloses another). Consider, for example, two large parallel plates, A_1 and A_N, separated by $N - 2$ radiation shields, as shown in Figure 3.3.14. Let each shield have an emissivity ε_S on both sides. Then, by applying Equation (3.3.24) to any two consecutive surfaces and using the fact that q remains constant throughout the gap,

$$q = \frac{E_{b1} - E_{b2}}{R_{12}} = \cdots = \frac{E_{bk-1} - E_{bk}}{R_{k-1,k}} = \cdots = \frac{E_{bN-1} - E_{bN}}{R_{N-1,N}} = \frac{E_{b1} - E_{bN}}{\sum_{j-2}^{N} R_{j-1,j}} \tag{3.3.25}$$

where

$$R_{j-1,j} = \frac{1}{\varepsilon_{j-1}} + \frac{1}{\varepsilon_j} - 1 \tag{3.3.26}$$

and, if $\varepsilon_2 = \varepsilon_3 \cdots = \varepsilon_{N-1} = \varepsilon_S$,

$$\sum_{j=2}^{N} R_{j-1,j} = \frac{1}{\varepsilon_1} + \frac{1}{\varepsilon_N} - 1 + (N-2)\left(\frac{2}{\varepsilon_S} - 1\right) \tag{3.3.27}$$

Equations (3.3.24) to (3.3.27) are also valid for concentric cylinders, concentric spheres, and similar configurations, as long as $r_N - r_1 \ll r_1$. Also, the relations are readily extended to shields with nonidentical emissivities.

While the network analogy can (and has been) applied to configurations with more than two surfaces seeing each other, this leads to very complicated circuits (since there is one resistance between any two surfaces). For such problems the network analogy is not recommended, and the net radiation method, Equation (3.3.18), should be employed.

Radiative Exchange within Participating Media

In many high-temperature applications, when radiative heat transfer is important, the medium between surfaces is not transparent, but is "participating," i.e., it absorbs, emits, and (possibly) scatters radiation. In a typical combustion process this interaction results in (1) continuum radiation due to tiny, burning soot particles (of dimension <1 μm) and also due to larger suspended particles, such as coal particles, oil droplets, fly ash; (2) banded radiation in the infrared due to emission and absorption by molecular gaseous combustion products, mostly water vapor and carbon dioxide; and (3) chemiluminescence due to the combustion reaction itself. While chemiluminescence may normally be neglected, particulates as well as gas radiation generally must be accounted for.

Radiative Properties of Molecular Gases

When a photon (or an electromagnetic wave) interacts with a gas molecule, it may be absorbed, raising the energy level of the molecule. Conversely, a gas molecule may spontaneously lower its energy level by the emission of an appropriate photon. This leads to large numbers of narrow spectral lines, which partially overlap and together form so-called vibration-rotation bands. As such, gases tend to be transparent over most of the spectrum, but may be almost opaque over the spectral range of a band. The **absorption coefficient** κ_λ is defined as a measure of how strongly radiation is absorbed or emitted along a path of length, L, leading to the spectral absorptivity and emissivity for this path, or

$$\alpha_\lambda = \varepsilon_\lambda = 1 - e^{-\kappa_\lambda L} \tag{3.3.28}$$

Although gases are distinctly nongray, for simple heat transfer calculations we often need to determine the total emissivity for an isothermal path (compare Equation (3.3.9))

$$\varepsilon = \frac{1}{E_b}\int_0^\infty \left(1 - e^{-\kappa_\lambda L}\right) E_{b\lambda}\left(T_g\right) d\lambda \tag{3.3.29}$$

For a mixture of gases the total emissivity is a function of path length L, gas temperature T_g, partial pressure(s) of the absorbing gas(es) p_a, and total pressure p. For the — in combustion applications most important — mixture of nitrogen with water vapor and/or carbon dioxide, the total emissivity may be calculated from Leckner (1972). First, the individual emissivities for water vapor and carbon dioxide, respectively, are calculated separately from

$$\varepsilon\left(p_a L, p, T_g\right) = \varepsilon_0\left(p_a L, T_g\right)\left(\frac{\varepsilon}{\varepsilon_0}\right)\left(p_a L, p, T_g\right) \tag{3.3.30a}$$

$$\left(\frac{\varepsilon}{\varepsilon_0}\right)\left(p_a L, p, T_g\right) = \left[1 - \frac{(a-1)(1-P_E)}{a+b-1+P_E}\exp\left(-c\left[\log_{10}\frac{(p_a L)_m}{p_a L}\right]^2\right)\right] \tag{3.3.30b}$$

$$\varepsilon_0\left(p_a L, T_g\right) = \exp\left[\sum_{i=0}^{N}\sum_{j=0}^{N} c_{ji}\left(\frac{T_g}{T_0}\right)^j\left(\log_{10}\frac{p_a L}{(p_a L)_0}\right)^i\right] \tag{3.3.30c}$$

Here ε_0 is the total emissivity of a reference state, i.e., for the case of $p = 1$ bar and $p_a \to 0$ (but $p_a L > 0$), and the correlation constants a, b, c, c_{ji}, P_E $(p_a L)_0$, $(p_a L)_m$, and T_0 are given in Table 3.3.2 for water vapor and carbon dioxide. (For convenience, plots of ε_0 are given in Figures 3.3.15 for CO_2 and 3.3.16 for H_2O.) The total emissivity of a mixture of nitrogen with both water vapor and carbon dioxide is calculated from

$$\varepsilon_{CO_2+H_2O} = \varepsilon_{CO_2} + \varepsilon_{H_2O} - \Delta\varepsilon \tag{3.3.31}$$

TABLE 3.3.2 Correlation Constants for the Determination of the Total Emissivity for Water Vapor and Carbon Dioxide

Gas	Water Vapor			Carbon Dioxide			
M, N		2,2			2,3		
$c_{00} \cdots c_{N1}$	−2.2118	−1.1987	0.035596	−3.9893	2.7669	−2.1081	0.39163
$\vdots \;\; \ddots \;\; \vdots$	0.85667	0.93048	−0.14391	1.2710	−1.1090	1.0195	−0.21897
$c_{0M} \cdots c_{NM}$	−0.10838	−0.17156	0.045915	−0.23678	0.19731	−0.19544	0.044644
P_E	$(p + 2.56 p_a/\sqrt{t})/p_0$			$(p + 0.28 p_a)/p_0$			
$(p_a L)_m/(p_a L)_0$	$13.2t^2$			$0.054/t^2$, $t < 0.7$; $0.225t^2$, $t > 0.7$			
a	2.144, $t < 0.75$; $1.88 - 2.053 \log_{10} t$, $t > 0.75$			$1 + 0.1/t^{1.45}$			
b	$1.10/t^{1.4}$			0.23			
c	0.5			1.47			

Note: $T_0 = 1000$ K, $p_0 = 1$ bar, $t = T/T_0$, $(p_a L)_0 = 1$ bar cm.

Source: Modest, M., *Radiative Heat Transfer*, McGraw-Hill, New York, 1993. With permission.

$$\Delta\varepsilon = \left(\frac{\zeta}{10.7 + 101\zeta} - 0.0089\zeta^{10.4} \right) \left(\log_{10} \frac{\left(p_{H_2O} + p_{CO_2}\right)L}{\left(p_a L\right)_0} \right)^{2.76} \tag{3.3.32}$$

$$\zeta = \frac{p_{H_2O}}{p_{H_2O} + p_{CO_2}}$$

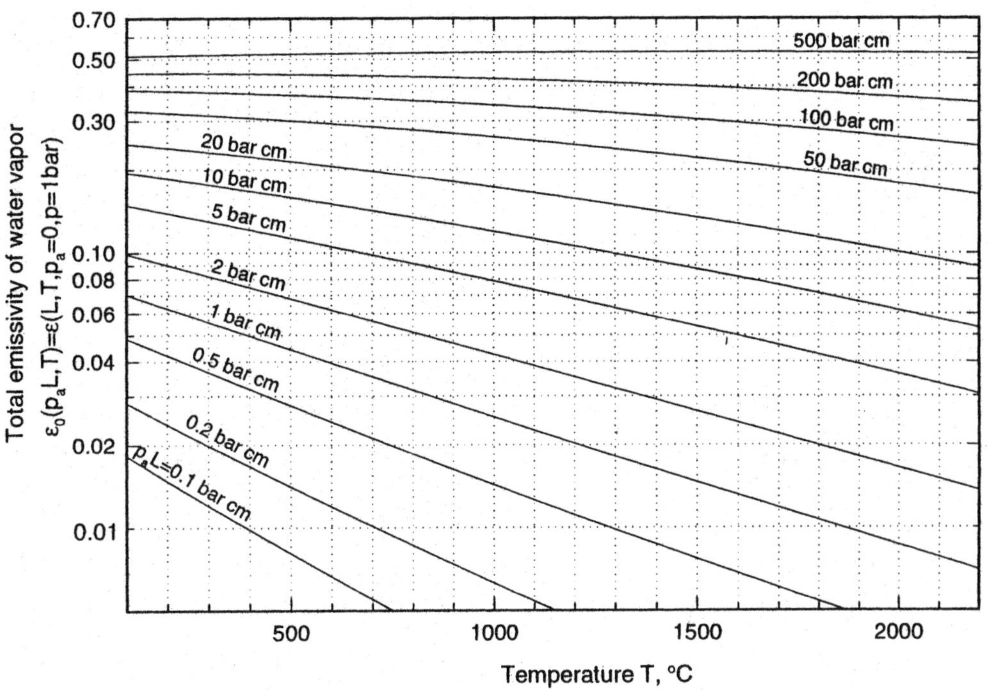

FIGURE 3.3.15 Total emissivity of water vapor at reference state (total gas pressure p = 1 bar, partial pressure of H_2O $p_a \rightarrow 0$). (From Modest, M., *Radiative Heat Transfer,* McGraw-Hill, New York, 1993. With permission.)

where the $\Delta\varepsilon$ compensates for overlap effects between H_2O and CO_2 bands, and the ε_{CO_2} and ε_{H_2O} are calculated from Equation (3.3.30).

If radiation emitted externally to the gas (for example, by emission from an adjacent wall at temperature T_s) travels through the gas, the total amount absorbed by the gas is of interest. This leads to the absorptivity of a gas path at T_g with a source at T_s:

$$\alpha\left(p_a L, p, T_g, T_s\right) = \frac{1}{E_b\left(T_s\right)} \int_0^\infty \left(1 - e^{-\kappa_\lambda\left(T_g\right)L}\right) E_{b\lambda}\left(T_s\right) d\lambda \tag{3.3.33}$$

which for water vapor or carbon dioxide may be estimated from

$$\alpha\left(p_a L, p, T_g, T_s\right) = \left(\frac{T_g}{T_s}\right)^{1/2} \varepsilon\left(p_a L \frac{T_s}{T_g}, p, T_s\right) \tag{3.3.34}$$

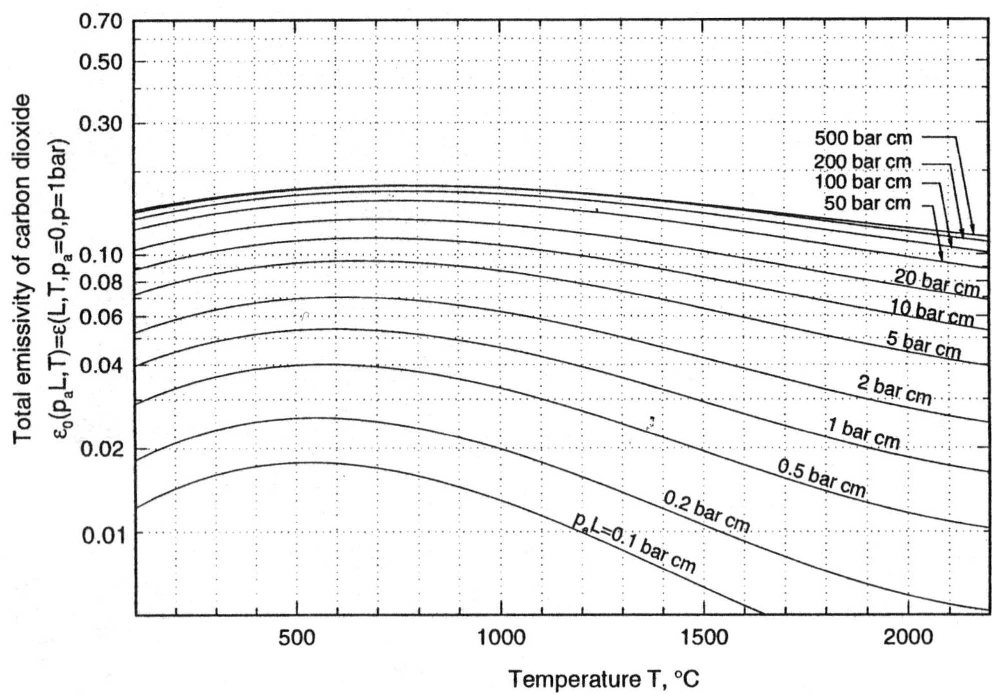

FIGURE 3.3.16 Total emissivity of carbon dioxide at reference state (total gas pressure p = 1 bar, partial pressure of CO_2 $p_a \to 0$). (From Modest, M., *Radiative Heat Transfer,* McGraw-Hill, New York, 1993. With permission.)

where ε is the emissivity calculated from Equation (3.3.30) evaluated at the temperature of the surface T_s, and using an adjusted pressure path length, $p_a L T_s / T_g$. For mixtures of water vapor and carbon dioxide band overlap is again accounted for by taking

$$\alpha_{CO_2+H_2O} = \alpha_{CO_2} + \alpha_{H_2O} - \Delta\varepsilon \tag{3.3.35}$$

with Δε evaluated for a pressure path length of $p_a L T_s / T_g$.

Example 3.3.6

Consider a layer of a gas mixture at 1000 K and 5 bar that consists of 10% carbon dioxide and 70% nitrogen. What is its emissivity for a path length of 1.76 m, and its absorptivity (for the same path) if the layer is irradiated by a source at 1500 K?

Solution. First we calculate the total emissivity of the CO_2 at the reference state (p = 1 bar, $p_a \to 0$). for a length of 1.76 m from Equation (3.3.30c) or Figure 3.3.15. With

$$T_g = 1000 \text{ K} = 727°\text{C} \quad \text{and} \quad p_a L = 0.1 \times 5 \text{ bar} \times 1.76 \text{ m} = 88 \text{ bar cm}$$

one gets, interpolating Figure 3.3.15, $\varepsilon_0 \cong 0.15$. The correction factor in Equation (3.3.30b) is calculated from Table 3.3.2 with $P_E = 5 + 0.28 \times 0.5 = 5.14$, $a = 1.1$, $b = 0.23$, $c = 1.47$, and $(p_a L)_m = 0.225$ bar cm. Thus,

$$\frac{\varepsilon}{\varepsilon_0} = 1 - \frac{0.1 \times (-4.14)}{0.33 + 5.14} \exp\left(-1.47\left(\log_{10}\frac{0.225}{88}\right)^2\right) \cong 1$$

and

$$\varepsilon \cong 0.15$$

To calculate the absorptivity ε_0 must be found for a temperature of

$$T_s = 1500 \text{ K} = 1227°\text{C} \quad \text{and} \quad p_a L \frac{T_s}{T_g} = 88 \times 1500/1000 = 132 \text{ bar cm}$$

From Figure 3.3.15 it follows that $\varepsilon_0 \cong 0.15$ again and, with $\varepsilon/\varepsilon_0$ pretty much unchanged, from Equation (3.3.34),

$$\alpha \cong \left(\frac{1000}{1500}\right)^{1/2} \times 0.15 \times 1.00 = 0.122$$

Radiative Properties of Particle Clouds

Nearly all flames are visible to the human eye and are, therefore, called *luminous* (sending out light). Apparently, there is some radiative emission from within the flame at wavelengths where there are no vibration–rotation bands for any combustion gases. This luminous emission is known today to come from tiny *char* (almost pure carbon) particles, call *soot*, which are generated during the combustion process. The "dirtier" the flame is (i.e., the higher the soot content), the more luminous it is.

Radiative Properties of Soot. Soot particles are produced in fuel-rich flames, or fuel-rich parts of flames, as a result of incomplete combustion of hydrocarbon fuels. As shown by electron microscopy, soot particles are generally small and spherical, ranging in size between approximately 50 and 800 Å (0.005 to 0.08 μm), and up to about 3000 Å in extreme cases. While mostly spherical in shape, soot particles may also appear in agglomerated chunks and even as long agglomerated filaments. It has been determined experimentally in typical diffusion flames of hydrocarbon fuels that the volume percentage of soot generally lies in the range between 10^{-4} to 10^{-6}%.

Since soot particles are very small, they are generally at the same temperature as the flame and, therefore, strongly emit thermal radiation in a continuous spectrum over the infrared region. Experiments have shown that soot emission often is considerably stronger than the emission from the combustion gases.

For a simplified heat transfer analysis it is desirable to use suitably defined mean absorption coefficients and emissivities. If the soot volume fraction f_v is known as well as an appropriate spectral average of the complex index of refraction of the soot, $m = n - ik$, one may approximate the spectral absorption coefficient by (Felske and Tien, 1977).

$$\kappa_\lambda = C_0 \frac{f_v}{\lambda} \qquad C_0 = \frac{36\pi nk}{\left(n^2 - k^2 + 2\right)^2 + 4n^2k^2} \tag{3.3.36}$$

and a total, or spectral-average value may be taken as

$$\kappa_m = 3.72 f_v C_0 T / C_2 \tag{3.3.37}$$

where C_2 = 1.4388 mK is the second Planck function constant. Substituting Equation (3.3.37) into Equation (3.3.29) gives a total soot cloud emissivity of

$$\varepsilon\left(f_v TL\right) = 1 - e^{-\kappa_m L} = 1 - e^{-3.72 C_0 f_v TL/C_2} \tag{3.3.38}$$

Pulverized Coal and Fly Ash Dispersions. To calculate the radiative properties of arbitrary size distributions of coal and ash particles, one must have knowledge of their complex index of refraction as a function of wavelength and temperature. Data for carbon and different types of coal indicate that its real part, n, varies little over the infrared and is relatively insensitive to the type of coal (e.g., anthracite, lignite, bituminous), while the absorptive index, k, may vary strongly over the spectrum and from coal to coal. If the number and sizes of particles are known and if a suitable average value for the complex index of refraction can be found, then the spectral absorption coefficient of the dispersion may be estimated by a correlation given by Buckius and Hwang (1980). Substitution into Equation (3.3.29) can then provide an estimate of the total emissivity. If both soot as well as larger particles are present in the dispersion, the absorption coefficients of all constituents must be added before applying Equation (3.3.29).

Mixtures of Molecular Gases and Particulates. To determine the total emissivity of a mixture it is generally necessary to find the spectral absorption coefficient κ_λ of the mixture (i.e., the sum of the absorption coefficient of all contributors), followed by numerical integration of Equation (3.3.29). However, since the molecular gases tend to absorb only over a small part of the spectrum, to some degree of accuracy

$$\varepsilon_{\text{mix}} \cong \varepsilon_{\text{gas}} + \varepsilon_{\text{particulates}} \tag{3.3.39}$$

Equation (3.3.39) gives an upper estimate since overlap effects result in lower emissivity (compare Equation (3.3.31) for gas mixtures).

Heat Exchange in the Presence of a Participating Medium

The calculation of radiative heat transfer rates through an enclosure filled with a participating medium is a challenging task, to say the least. High-accuracy calculations are rare and a topic of ongoing research. There are, however, several simplistic models available that allow the estimation of radiative heat transfer rates, and relatively accurate calculations for some simple cases.

Diffusion Approximation. A medium through which a photon can only travel a short distance without being absorbed is known as *optically thick*. Mathematically, this implies that $\kappa_\lambda L \gg 1$ for a characteristic dimension L across which the temperature does not vary substantially. For such an optically thick, nonscattering medium the spectral radiative flux may be calculated from

$$\mathbf{q}_\lambda = -\frac{4}{3\kappa_\lambda}\nabla E_{b\lambda} \tag{3.3.40}$$

similar to Fourier's diffusion law for heat conduction. Note that a medium may be optically thick at some wavelengths, but thin ($\kappa_\lambda L \ll 1$) at others (e.g., molecular gases!). For a medium that is optically thick for all wavelengths, Equation (3.3.40) may be integrated over the spectrum, yielding the total radiative flux

$$\mathbf{q} = -\frac{4}{3\kappa_R}\nabla E_b = -\frac{4}{3\kappa_R}\nabla\left(\sigma T^4\right) = -\frac{16\sigma T^3}{3\kappa_R}\nabla T \tag{3.3.41}$$

where κ_R is the suitably averaged absorption coefficient, termed the *Rosseland-mean absorption coefficient*. For a cloud of soot particles, $\kappa_R \cong \kappa_m$ from Equation (3.3.37) is a reasonable approximation. Equation (3.3.41) may be rewritten by defining a "radiative conductivity" k_R,

$$\mathbf{q} = -k_R \nabla T \qquad k_R = \frac{16\sigma T^3}{3\kappa_R} \tag{3.3.42}$$

This form shows that the diffusion approximation is mathematically equivalent to conductive heat transfer with a (strongly) temperature-dependent conductivity.

Note: More accurate calculations show that, in the absence of other modes of heat transfer (conduction, convection), there is generally a temperature discontinuity near the boundaries ($T_{\text{surface}} \neq T_{\text{adjacent medium}}$), and, unless boundary conditions that allow such temperature discontinuities are chosen, the diffusion approximation will do very poorly in the vicinity of bounding surfaces.

Example 3.3.7

A soot cloud is contained between two walls at T_1 = 1000 K and T_2 = 2000 K, spaced 1 m apart. The effective absorption coefficient of the medium is κ_R = 10 m^{-1} and the effective thermal conductivity is k_c = 0.1 W/mK. Estimate the total heat flux between the plates (ignoring convection effects).

Solution. For simplicity we may want to assume a constant total conductivity $k = k_c + k_R$, leading to

$$q = -k\frac{dT}{dx} = k\frac{T_2 - T_1}{L}$$

where k_R must be evaluated at some effective temperature. Choosing, based on its temperature dependence,

$$k_R \cong \frac{8\sigma}{3\kappa_R}\left(T_1^3 + T_2^3\right) = \frac{8 \times 5.670 \times 10^{-8}\ \text{W/m}^2\text{K}^4}{3 \times 10/\text{m}}\left(1000^3 + 2000^3\right)\text{K}^3 = 136\frac{\text{W}}{\text{mK}}$$

gives

$$q = (0.1 + 136)\frac{2000 - 1000}{1}\frac{\text{W}}{\text{m}^2} = 136\frac{\text{kW}}{\text{m}^2\text{K}}$$

Note that (1) conduction is negligible in this example and (2) the surface emissivities do not enter the diffusion approximation. While a more accurate answer can be obtained by taking the temperature dependence of k_R into account, the method itself should be understood as a relatively crude approximation.

Mean Beam Length Method. Relatively accurate yet simple heat transfer calculations can be carried out if an isothermal, absorbing–emitting, but not scattering medium is contained in an isothermal, black-walled enclosure. While these conditions are, of course, very restrictive, they are met to some degree by conditions inside furnaces. For such cases the local heat flux on a point of the surface may be calculated from

$$q = \left[1 - \alpha\left(L_m\right)\right]E_{bw} - \varepsilon\left(L_m\right)E_{bg} \tag{3.3.43}$$

where E_{bw} and E_{bg} are blackbody emissive powers for the walls and medium (gas and/or particulates), respectively, and $\alpha(L_m)$ and $\varepsilon(L_m)$ are the total absorptivity and emissivity of the medium for a path length L_m through the medium. The length L_m, known as the average *mean beam length*, is a directional average of the thickness of the medium as seen from the point on the surface. On a spectral basis Equation (3.3.43) is exact, provided the above conditions are met and provided an accurate value of the (spectral) mean beam length is known. It has been shown that spectral dependence of the mean beam length is weak (generally less than ±5% from the mean). Consequently, total radiative heat flux at the surface may be calculated very accurately from Equation (3.3.43), provided the emissivity and absorptivity of the medium are also known accurately. The mean beam lengths for many important geometries have been calculated and are collected in Table 3.3.3. In this table L_o is known as the geometric mean beam length, which is the mean beam length for the optically thin limit ($\kappa_\lambda \to 0$), and L_m is a spectral average of the mean beam length. For geometries not listed in Table 3.3.3, the mean beam length may be estimated from

$$L_o \cong 4\frac{V}{A} \qquad L_m \cong 0.9L_o \cong 3.6\frac{V}{A} \tag{3.3.44}$$

TABLE 3.3.3 Mean Beam Lengths for Radiation from a Gas Volume to a Surface on Its Boundary

Geometry of Gas Volume	Characterizing Dimension L	Geometric Mean Beam Length L_0/L	Average Mean Beam Length L_m/L	L_m/L_0
Sphere radiating to its surface	Diameter, L = D	0.67	0.65	0.97
Infinite circular cylinder to bounding surface	Diameter, L = D	1.00	0.94	0.94
Semi-infinite circular cylinder to:	Diameter, L = D			
Element at center of base		1.00	0.90	0.90
Entire base		0.81	0.65	0.80
Circular cylinder (height/diameter = 1) to:	Diameter, L = D			
Element at center of base		0.76	0.71	0.92
Entire surface		0.67	0.60	0.90
Circular cylinder (height/diameter = 2) to:	Diameter, L = D			
Plane base		0.73	0.60	0.82
Concave surface		0.82	0.76	0.93
Entire surface		0.80	0.73	0.91
Circular cylinder (height/diameter = 0.5) to:	Diameter, L = D			
Plane base		0.48	0.43	0.90
Concave surface		0.53	0.46	0.88
Entire surface		0.50	0.45	0.90
Infinite semicircular cylinder to center of plane rectangular face	Radius, L = R	—	1.26	—
Infinite slab to its surface	Slab thickness, L	2.00	1.76	0.88
Cube to a face	Edge L	0.67	0.6	0.90
Rectangular 1 × 1 × 4 parallelepipeds:	Shortest edge, L			
To 1 × 4 face		0.90	0.82	0.91
To 1 × 1 face		0.86	0.71	0.83
To all faces		0.89	0.81	0.91

Source: Modest, M., *Radiative Heat Transfer*, McGraw-Hill, New York, 1993. With permission.

where V is the volume of the participating medium and A is its entire bounding surface area.

Example 3.3.8

An isothermal mixture of 10% CO_2 and 90% nitrogen at 1000 K and 5 bar is contained between two large, parallel, black plates, which are both isothermal at 1500 K. Estimate the net radiative heat loss from the surfaces.

Solution. The heat loss may be calculated from Equation (3.3.43), after determining the mean beam length, followed by evaluation of $\varepsilon(L_m)$ and $\alpha(L_m)$. From Table 3.3.3 it is clear that $L_m = 1.76 \times$ thickness of slab = 1.76 m. It turns out that the necessary $\varepsilon(L_m) = 0.15$ and $\alpha(L_m) = 0.122$ have already been calculated in Example 3.3.6. Thus, the heat flux is immediately calculated from Equation (3.3.43) as

$$q = (1 - 0.122)5.670 \times 10^{-8} \times 1500^4 - 0.15 \times 5.670 \times 10^{-8} \times 1000^4$$

$$= 2.44 \times 10^5 \frac{\text{W}}{\text{m}^2} = 244 \ \text{kW/m}^2$$

Defining Terms

Absorptivity: The ability of a medium to absorb (i.e., trap and convert to other forms of energy) incoming radiation; gives the fraction of incoming radiation that is absorbed by the medium.

Absorption coefficient: The ability of a medium to absorb (i.e., trap and convert to other forms of energy) over a unit path length; the reciprocal of the mean distance a photon travels before being absorbed.

Blackbody: Any material or configuration that absorbs all incoming radiation completely. A blackbody also emits the maximum possible amount of radiation as described by Planck's law.

Diffuse surface: A surface that emits and/or reflects equal amounts of radiative energy (photons) into all directions. Or a surface that absorbs and/or reflects equal amounts of radiation independent of incoming direction.

Emissive power: The rate of radiative energy leaving a surface through emission. The maximum amount of emissive power is emitted by a blackbody with a spectral strength described by Planck's law.

Emissivity: The ability of a medium to emit (i.e., convert internal energy into electromagnetic waves or photons) thermal radiation; gives the fraction of emission as compared with a blackbody.

Gray: A medium whose radiative properties (such as absorptivity, emissivity, reflectivity, absorption coefficient) do not vary with wavelength.

Irradiation: Incoming radiative flux onto a surface from outside it.

Network analogy: Expressing radiative heat exchange between surfaces in terms of an electrical network, with heat flux as "current," differences in emissive power as "potentials," and defining radiative resistances.

Opaque medium: A medium of sufficient thickness that absorbs all nonreflected irradiation; no radiation is transmitted through the medium.

Photon: A massless particle carrying energy in the amount of $h\nu$; the quantum mechanical alternative view of an electromagnetic wave carrying radiative energy.

Planck's law: The law describing the spectral distribution of the radiative energy emitted (emissive power) of a blackbody.

Radiosity: Total radiative flux leaving a surface (diffusely), consisting of emitted as well as reflected radiation.

Reflectivity: The ability of an interface, or of a medium or of a composite with a number of interfaces, to reflect incoming radiation back into the irradiating medium.

Semitransparent: See **transparent**.

Spectral value: The value of a quantity that varies with wavelength at a given wavelength; for dimensional quantities the amount per unit wavelength.

Transmissivity: The ability of a medium to let incoming radiation pass through it; gives the fraction of incoming radiation that is transmitted through the medium.

Transparent: The ability of a medium to let incoming radiation pass through it. A medium that lets all radiation pass through it is called transparent, a medium that only allows a part to pass through it is called **semitransparent**.

View factor: The fraction of diffuse radiant energy leaving one surface that is intercepted by another surface.

References

Brewster, M.Q. 1992. *Thermal Radiative Transfer & Properties*, John Wiley & Sons, New York.

Buckius, R.O. and Hwang, D.C. 1980. Radiation properties for polydispersions: application to coal, *J. Heat Transfer*, 102, 99–103.

Felske, J.D. and Tien, C.L. 1977. The use of the Milne-Eddington absorption coefficient for radiative heat transfer in combustion systems, *J. Heat Transfer*, 99(3), 458–465.

Hottel, H.C. and Sarofim, A.F. 1967. *Radiation Transfer*, McGraw-Hill, New York.

Howell, J.R. 1982. *Catalog of Radiation Configuration Factors*, McGraw-Hill, New York.

Leckner, B. 1972. Spectral and total emissivity of water vapor and carbon dioxide, *Combust. Flame*, 19, 33–48.

Modest, M.F. 1993. *Radiative Heat Transfer*, McGraw-Hill, New York.

Ozisik, M.N. 1973. *Radiative Transfer and Interactions with Conduction and Convection*, John Wiley & Sons, New York.

Siegel, R. and Howell, J.R. 1992. *Thermal Radiation Heat Transfer*, 3rd ed., Hemisphere Publishing, New York.

Sparrow, E.M. and Cess, R.D. 1978. *Radiation Heat Transfer*, Hemisphere, New York.

3.4 Phase-Change

3.4.1 Boiling and Condensation

Van P. Carey

Introduction

Liquid-vapor phase-change processes play an important role in many technological applications. The virtually isothermal heat transfer associated with boiling and condensation processes makes their inclusion in power and refrigeration processes highly advantageous from a thermodynamic efficiency standpoint. In addition, the high heat transfer coefficients associated with boiling and condensation have made the use of these processes increasingly attractive in the thermal control of compact devices that have high heat dissipation rates. Applications of this type include the use of boiling heat transfer to cool electronic components in computers and the use of compact evaporators and condensers for thermal control of aircraft avionics and spacecraft environments. Liquid-vapor phase-change processes are also of critical importance to nuclear power plant design, both because they are important in normal operating circumstances and because they dominate many of the accident scenarios that are studied as part of design evaluation.

The heat transfer and fluid flow associated with liquid-vapor phase-change processes are typically among the more complex transport circumstances encountered in engineering applications. These processes have all the complexity of single-phase convective transport, plus additional elements resulting from motion of the interface, nonequilibrium effects, and dynamic interactions between the phases. Due to the highly complex nature of these processes, development of methods to accurately predict the associated heat and mass transfer is often a formidable task.

In this section, commonly used variables not defined in the nomenclature are as follows: q'' = surface heat flux, μ_l = liquid viscosity, μ_v = vapor viscosity, Pr_l = liquid Prandtl number, T_w = wall surface temperature, T_{sat} = saturation temperature, c_{pl} = liquid specific heat, k_v = vapor thermal conductivity, g = gravitational acceleration, and x = mass quality.

Boiling

Three mechanisms that play important roles in boiling processes are (1) surface tension effects, (2) surface wetting characteristics of the liquid, and (3) metastable phase stability.

Anyone who has watched small bubbles rise in a carbonated beverage or a pot of boiling water has undoubtedly noticed that the bubbles are almost perfectly spherical, as if an elastic membrane were present at the interface to pull the vapor into a spherical shape. This apparent interfacial tension or *surface tension* σ is equivalent to an energy stored in the interface region per unit area. The energy excess in this region is due to the slightly larger separation of the liquid phase molecules adjacent to the gas phase.

The magnitude of the surface tension for a substance is directly linked to the strength of intermolecular forces in the material. Nonpolar liquids typically have the lowest surface tension. Water and other **polar molecules** have somewhat higher surface tension, and liquid metals, which exhibit metallic bond attraction, have very high surface tension. The surface tension of water at 20°C is 0.0728 N/m, whereas liquid mercury has a surface tension of 0.484 N/m at the same temperature. The surface tension for any pure liquid varies with temperature. It decreases almost linearly with increasing temperature, vanishing altogether at the critical point where the distinction between the phases disappears.

As a result of the surface tension at the interface, the pressure inside a spherical bubble of radius r must exceed that in the surrounding liquid by $2\sigma/r$:

$$P_v = P_l + \frac{2\sigma}{r} \tag{3.4.1}$$

By using the relation (1) between the pressure in the two phases it can be shown that for the bubble to be in equilibrium with the surrounding liquid, the liquid must actually be superheated above the saturation temperature for the ambient liquid pressure. The amount of required superheating increases as the radius of curvature of the bubble interface decreases.

The wetting characteristics of the liquid are generally quantified in terms of a *contact angle* between the solid surface and the tangent to the interface at the point where it contacts the solid. This angle is measured through the liquid phase, as shown in Figure 3.4.1. In some systems, the wetting angle established at equilibrium may depend on the fluid motion history. For some systems the contact angle established by liquid advancing over a solid surface is larger than that established when a liquid front recedes over the surface. This behavior is referred to as *contact angle hysteresis*. Contact angle hysteresis can have an important effect on boiling and condensation processes, particularly those involving water.

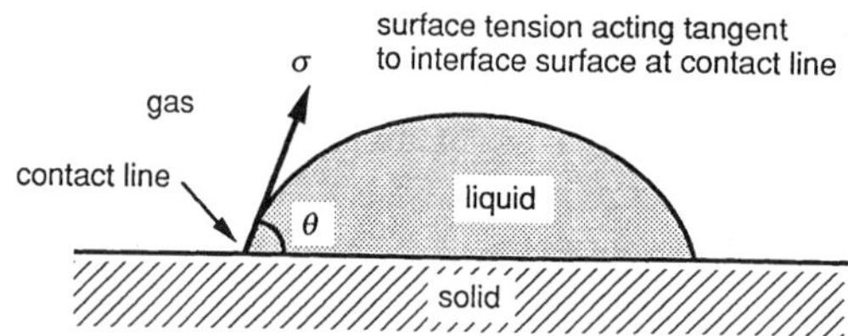

FIGURE 3.4.1 Definition of the contact angle θ.

For a bubble with a specified vapor volume, the contact angle will dictate the radius of curvature of the bubble interface. The wetting behavior in combination with the surface tension effect, thus, determines the level of superheat required for the bubble to be in equilibrium with the surrounding liquid. The liquid must be heated above this superheat level for the bubble to grow. A steady boiling process can be sustained only if the liquid is heated above this threshold superheat level.

It can be shown from basic thermodynamic analysis that a necessary and sufficient condition for phase stability is that

$$\left(\frac{\partial P}{\partial v}\right)_T < 0 \tag{4.3.2}$$

where v is the specific volume. Below the critical temperature, extrapolation of the isotherms for the liquid and vapor phases consistent with an equation of state like the van de Waals equation results in an isotherm shape similar to that shown in Figure 4.3.2.

The locus of points where $(\partial P/\partial v)_T = 0$ are termed *spinodal curves*. Regions of metastable vapor and liquid exist between the saturation curve and the spinodal curves. The effects of surface tension discussed above require that fluid surrounding a vapor bubble be in the metastable superheated liquid region. Predictions of statistical thermodynamics imply that as $(\partial P/\partial v)_T$ approaches zero, the level of fluctuations in a fluid system increases. This, in turn, increases the probability that an embryonic new phase will form as a result of density fluctuations. Initiation of a phase change in this manner is termed *homogeneous nucleation*. Generally, a pure liquid must be heated to nearly 90% of its absolute critical temperature before homogeneous nucleation of vapor bubbles occurs.

In most physical systems of engineering interest, the bulk phase is in contact with solid walls of the containing structures, or solid particulate contaminants. These solid phases may provide nucleation sites where phase change may occur if the system state is driven into the metastable range. Nucleation of vapor bubbles may preferentially occur at low liquid superheat levels in crevices in the solid surface where gas is trapped. This type of nucleation at the solid surface of a containment wall is categorized as *heterogeneous nucleation*. Because solid containment walls usually contain microscopic crevice-type

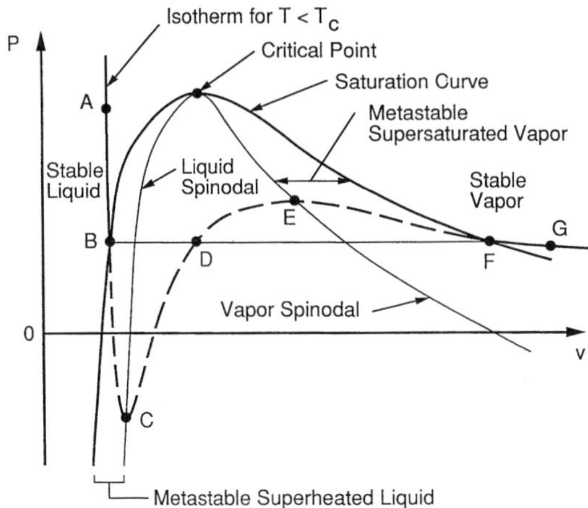

FIGURE 3.4.2 Spinodal lines and metastable regions on a *P*–v diagram.

imperfections, heterogeneous nucleation is more common than homogeneous nucleation in systems where boiling occurs.

Vapor entrapment in crevices of the heated walls of evaporator heat exchangers usually makes it easier to initiate the vaporization process. Vapor bubbles grow from these crevices until buoyancy or drag on the bubbles exceeds the surface tension force holding the droplet to the solid surface. The bubble then releases into the bulk liquid. A small remnant of vapor remains in the crevice after a bubble releases, and this remnant grows in size as further vaporization occurs until the bubble grows out of the crevice again. The result is a cyclic process of bubble growth and release known as the *ebullition cycle*. Crevices at which the ebullition cycle is sustained are said to be active nucleation sites. When the ebullition process occurs at many sites over a heated surface, the overall process is referred to as **nucleate boiling**, which is one possible mode of **pool boiling**.

Pool Boiling

Vaporization of liquid at the surface of a body immersed in an extensive pool of motionless liquid is generally referred to as pool boiling. The nature of the pool boiling process varies considerably depending on the conditions at which boiling occurs. The level of heat flux, the thermophysical properties of the liquid and vapor, the surface material and finish, and the physical size of the heated surface all may have an effect on the boiling process.

The regimes of pool boiling are most easily understood in terms of the so-called boiling curve: a plot of heat flux q'' vs. wall superheat $T_w - T_{sat}$ for the circumstances of interest. Many of the features of the classic pool boiling curve were determined in the early investigations of pool boiling conducted by Nukiyama (1934). Strictly speaking, the classic pool boiling curve defined by the work of this investigator and others applies to well-wetted surfaces for which the characteristic physical dimension L is large compared to the bubble or capillary length scale L_b defined as

$$L_b = \sqrt{\frac{\sigma}{g(\rho_l - \rho_v)}} \tag{3.4.3}$$

The discussion in this section is limited to pool boiling of wetting liquids on surfaces with dimensions large compared with L_b. Additional information on features of the boiling curve when the liquid poorly wets the surface or when L/L_b is small can be found in Carey (1992). To make this discussion concrete, we will assume that the ambient liquid surrounding the immersed body is at the saturation temperature for the ambient pressure. If the surface temperature of the immersed body is controlled and slowly

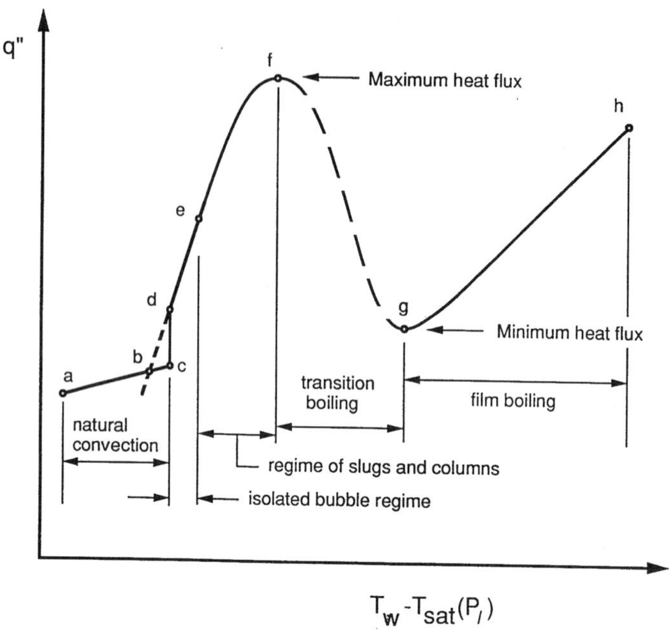

FIGURE 3.4.3 Pool boiling regimes for an independently controlled surface temperature.

increased, the boiling curve will look similar to that shown in Figure 3.4.3. The axes in this plot are logarithmic scales. The regimes of pool boiling encountered for an upward-facing horizontal flat surface as its temperature is increased are also indicated in Figure 3.4.3. The lateral extent of the surface is presumed to be much larger than L_b. At very low wall superheat levels, no nucleation sites may be active and heat may be transferred from the surface to the ambient liquid by natural convection alone and q'' increases slowly with $T_w - T_{sat}$.

Eventually, the superheat becomes large enough to initiate nucleation at some of the cavities on the surface. This *onset of nucleate boiling* (ONB) condition occurs at point *c* in Figure 3.4.3. Once nucleate boiling is initiated, any further increase in wall temperature causes the system operating point to move upward along section *d-f* of the curve in Figure 3.4.3. This portion of the curve corresponds to the nucleate boiling regime. The active sites are few and widely separated at low wall superheat levels. This range of conditions, corresponding to segment *d-e* of the curve, is sometimes referred to as the *isolated bubble regime.*

With increasing surface superheat, more sites become active, and the bubble frequency at each site generally increases. Eventually, the active sites are spaced so closely that bubbles from adjacent sites merge together during the final stages of growth and release. Vapor is being produced so rapidly that bubbles merging together form columns of vapor slugs that rise upward in the liquid pool toward its free surface. This higher range of wall superheat, corresponding to segment *e-f* of the boiling curve in Figure 3.4.3, is referred to as the *regime of slugs and columns.*

Increasing the wall superheat and heat flux within the regime of slugs and columns produces an increase in the flow rate of vapor away from the surface. Eventually, the resulting vapor drag on the liquid moving toward the surface becomes so severe that liquid is unable to reach the surface fast enough to keep the surface completely wetted with liquid. Vapor patches accumulate at some locations and evaporation of the liquid between the surface and some of these patches dries out portions of the surface.

If the surface temperature is held constant and uniform, dry portions of the surface covered with a vapor film will locally transfer a much lower heat flux than wetted portions of the surface where nucleate boiling is occurring. Because of the reduction in heat flux from intermittently dry portions of the surface, the mean overall heat flux from the surface is reduced. Thus, increasing the wall temperature within the slugs and columns region ultimately results in a peaking and rollover of the heat flux. The peak value of heat flux is called the **critical heat flux** (CHF), designated as point f in Figure 3.4.3

If the wall temperature is increased beyond the critical heat flux condition, a regime is encountered in which the mean overall heat flux decreases as the wall superheat increases. This regime, which is usually referred to as the **transition boiling** regime, corresponds to segment f-g on the boiling curve shown in Figure 3.4.3. The transition boiling regime is typically characterized by rapid and severe fluctuations in the local surface heat flux and/or temperature values (depending on the imposed boundary condition). These fluctuations occur because the dry regions are generally unstable, existing momentarily at a given location before collapsing and allowing the surface to be rewetted.

The vapor film generated during transition boiling can be sustained for longer intervals at higher wall temperatures. Because the intermittent insulating effect of the vapor blanketing is maintained longer, the time-averaged contributions of the blanketed locations to the overall mean heat flux are reduced. The mean heat flux from the surface thus decreases as the wall superheat is increased in the transition regime. As this trend continues, eventually a point is reached at which the surface is hot enough to sustain a stable vapor film on the surface for an indefinite period of time. The entire surface then becomes blanketed with a vapor film, thus making the transition to the **film boiling** regime. This transition occurs at point g in Figure 3.4.3.

Within the film boiling regime, the heat flux monotonically increases as the superheat increases. This trend is a consequence of the increased conduction and/or convection transport due to the increased driving temperature difference across the vapor film. Radiative transport across the vapor layer may also become important at higher wall temperatures.

Once a surface is heated to a superheat level in the film boiling regime, if the surface temperature is slowly decreased, in general the system will progress through each of the regimes described above in reverse order. However, the path of the boiling curve may differ significantly from that observed for increasing wall superheat, depending on whether the surface heat flux or temperature is controlled.

Experimental evidence summarized by Witte and Lienhard (1982) implies that the path of the transition boiling curve is determined, to a large degree, by the wetting characteristics of the liquid on the solid surface. For a given wall superheat level in the transition boiling regime, a higher heat flux is generally obtained if the liquid wets the surface than if it poorly wets the surface. For systems that exhibit contact angle hysteresis, the transition boiling curves obtained for decreasing and increasing wall superheat may therefore be somewhat different. The transition boiling curve for decreasing wall superheat may be significantly below that for increasing superheat for such circumstances, as indicated in Figure 3.4.4.

For an electrically heated surface, the rise in temperature associated with the jump from nucleate to film boiling at the critical heat flux is very often large enough to melt component materials and burn out the component. As a result, the critical heat flux is often referred to as the *burnout heat flux* to acknowledge the potentially damaging effects of applying this heat flux level to components cooled by nucleate boiling. Once the jump to film boiling has been made, any further increase in applied heat flux increases the wall superheat, and the system follows basically the same film boiling curve as in the temperature-controlled case.

Correlations of nucleate pool boiling heat transfer data have typically been used as tools to predict nucleate boiling heat transfer in engineering systems and heat exchangers. Many investigators have proposed methods of correlating data of this type; so many, in fact, that a complete discussion of them all could easily fill a major portion of this section. In this section, three of the more commonly used correlation methods will be mentioned. However, before proceeding, two aspects of the interpretation of such correlations are worth noting. First, experimental data indicate that the subcooling of the liquid pool has a negligible effect on the nucleate boiling heat transfer rate. Consequently, the pool boiling correlations are generally regarded as being valid for both subcooled and saturated nucleate boiling. Second, it has also been observed that at moderate to high heat flux levels, a pool boiling heat transfer correlation developed for one heated surface geometry in one specific orientation often works reasonably well for other geometries and/or other orientations. Hence, although a correlation was developed for a

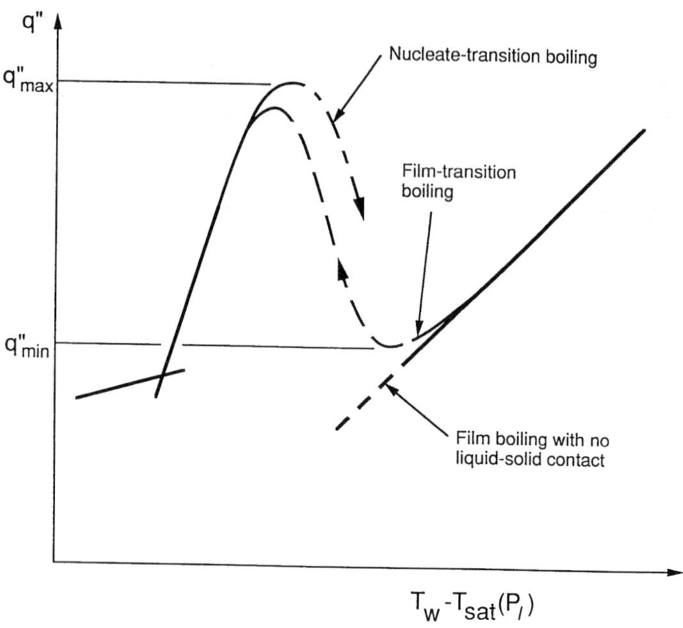

FIGURE 3.4.4 Relative locations of the nucleate transition and film transition portions of the pool boiling curve.

specific geometry and orientation, it may often be used for other geometries, at least at moderate to high heat flux levels.

Having taken note of the above points, a commonly used correlation for nucleate boiling heat transfer developed by Rohsenow (1962) is

$$\frac{q''}{\mu_l h_{fg}}\left[\frac{\sigma}{g(\rho_l-\rho_v)}\right]^{1/2}=\left(\frac{1}{C_{sf}}\right)^{1/r}\mathrm{Pr}_l^{-s/r}\left[\frac{c_{pl}\left[T_w-T_{sat}(P_l)\right]}{h_{fg}}\right]^{1/r} \tag{3.4.4}$$

Values of $r = 0.33$ and $s = 1.7$ are recommended for this correlation, but for water s should be changed to 1.0. The values of C_{sf} in this correlation vary with the type of solid surface and the type of fluid in the system. This empirically accounts for material property and/or wetting angle effects. Recommended values of C_{sf} for specific liquid–solid combinations are given by Rohsenow (1962), but whenever possible, an experiment should be conducted to determine the appropriate value of C_{sf} for the particular solid–liquid combination of interest. If this is not possible, a value of $C_{sf} = 0.013$ is recommended as a first approximation.

As noted previously, the pool boiling curve generally exhibits a maximum heat flux or CHF at the transition between nucleate and transition boiling. This peak value is the maximum level of heat flux from the surface which the system can provide in a nonfilm-boiling mode at a given pressure. The mechanism responsible for the CHF has been the subject of considerable investigation and debate over the past five decades. As the heat flux increases, bubbles generated at the surface coalesce to form vapor columns or jets. Perhaps the most widely cited CHF model postulates that the CHF condition occurs when Helmholtz instability of the large vapor jets leaving the surface distorts the jets, blocking liquid flow to portions of the heated surface. Continued vaporization of liquid at locations on the surface which are starved of replacement liquid than leads to formation of a vapor blanket over part or all of the surface. According to Zuber (1959) for a flat horizontal surface, the predicted maximum heat flux q''_{max} is

$$q''_{\max} = 0.131\rho_v h_{fg}\left[\frac{\sigma_g(\rho_l-\rho_v)}{\rho_v^2}\right]^{1/4} \tag{3.4.5}$$

but Lienhard and Dhir (1973) recommend that the constant 0.131 in the above relation be replaced with 0.141. Other geometries are treated by Lienhard et al. (1973) and Lienhard and Dhir (1973b). An alternative model has been proposed by Haramura and Katto (1983).

Lienhard and Witte (1985) discuss the development and limitations of hydrodynamic CHF theories.

As shown in the Figure 3.4.3, the boundary between the transition boiling regime and the film boiling regime corresponds to a minimum in the heat flux vs. superheat curve. This condition is referred to as the **minimum heat flux** condition, referred to as the *Leidenfront point*. The minimum heat flux corresponds approximately to the lowest heat flux which will sustain stable film boiling.

For an infinite flat (upward-facing) heated surface, vapor generated at the interface during stable film boiling is released as bubbles at the nodes of a standing two-dimensional Taylor wave pattern. The following relation for the minimum heat flux $q''_{\min}$ derived by Zuber (1959) and Berenson (1961).

$$q''_{\min} = 0.09\rho_v h_{fg}\left[\frac{\sigma_g(\rho_l-\rho_v)}{(\rho_l+\rho_v)^2}\right]^{1/4} \tag{3.4.6}$$

$q''_{\min}$ correlations have been developed by Lienhard and Wong (1964) for horizontal cylinders and Gunnerson and Cronenberg (1980) for spheres.

In film boiling, transport of heat across the vapor film from the wall to the interface may occur by convection, conduction, and radiation. The radiation contribution depends on the nature of the solid surface, but when the radiation effect is small, the heat transfer for film boiling is independent of the material properties and finish of the surface. For buoyancy-driven laminar film boiling over a vertical flat isothermal surface in a pool of saturated liquid, the local heat transfer coefficient from the surface can be obtained from the following relation:

$$h = \left[\frac{k_v^3 g\rho_v(\rho_l-\rho_v)h_{fg}}{4\mu_v(T_w-T_{\text{sat}})x}\right]^{1/4} \tag{3.4.7}$$

At low surface temperatures, radiation effects are negligible and consideration of convective transport alone is sufficient to predict the heat transfer. At higher temperatures radiation effects must also be included. If the vapor in the film absorbs and emits radiation at infrared wavelengths, a detailed treatment of the radiation interaction with the vapor may be necessary to accurately predict the film boiling heat transfer.

Additional information mechanisms such as interfacial waves, turbulence, and variable properties is summarized in Carey (1992).

Transition pool boiling has traditionally been interpreted as a combination of nucleate and film boiling alternately occurring over the heated surface, and a model of transition boiling that accounts for contact angle effects has been proposed by Ramilison and Lienhard (1987).

Internal Convective Boiling

Flow boiling in tubes is perhaps the most complex convective process encountered in applications. In most evaporator and boiler applications, the flow is either horizontal or vertically upward. Figure 3.4.5 schematically depicts a typical low-flux vaporization process in a horizontal round tube. In this example liquid enters as subcooled liquid and leaves as superheated vapor. As indicated in Figure 3.4.5, the flow undergoes transitions in the boiling regime and the two-phase flow regime as it proceeds down the tubes. The regimes encountered depend on the entrance conditions and the thermal boundary conditions at the

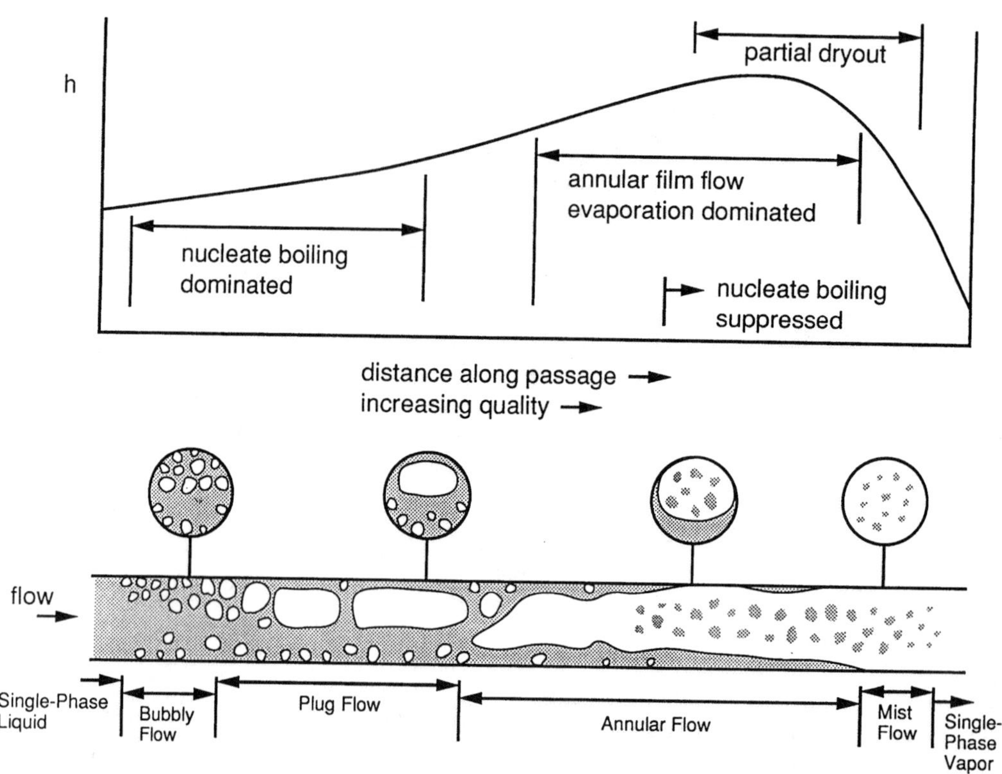

FIGURE 3.4.5 Qualitative variation of the heat transfer coefficient h and flow regime with quality for internal convective boiling in a horizontal tube at moderate wall superheat.

tube wall. At low quality the vaporization process is dominated by nucleate boiling, with convective effects being relatively weak. As the quality increases, the flow quickly enters the annular film flow regime in which convective evaporation of the annular liquid film is the dominant heat transfer mechanism. Often the conditions are such that liquid droplets are often entrained in the core vapor flow during annular flow evaporation. Eventually, the annular film evaporates away, leaving the wall dry. Mist-flow evaporation of entrained liquid droplets continues in the post-dryout regime until only vapor remains. Similar, sequences of flow and boiling regimes occurring in vertical upward flow, as indicated in Figure 3.4.6.

The boiling regime trends shown in Figures 3.4.5 and 3.4.6 are typical for low heat flux vaporization processes. At high wall superheat levels, transition boiling or film boiling can also occur. The transition from nucleate boiling to one of these regimes is termed a *departure from nucleate boiling* (DNB) or the CHF condition. However, the heat transfer performance of an evaporator under transition or film boiling conditions is so poor that equipment is not usually designed to operate under such conditions.

Because low-quality or subcooled flow boiling are nucleate boiling dominated, the heat transfer coefficient in these regimes is often predicted using a nucleate boiling correlation developed as a fit to pool boiling data. The usefulness of such an approach is a consequence of the fact that for most conditions of practical interest, nucleate boiling heat transfer is only weakly affected by liquid subcooling or liquid bulk convection.

For saturated convective boiling prior to dryout, relations to predict the heat transfer coefficient have typically been formulated to impose a gradual suppression of nucleate boiling and gradual increase in liquid film evaporation heat transfer as the quality increases. A number of correlations based on such an approach have been developed. An early correlation of this type developed by Chen (1966) has been widely used. One of the better methods of this type is the recently developed correlation of Kandlikar

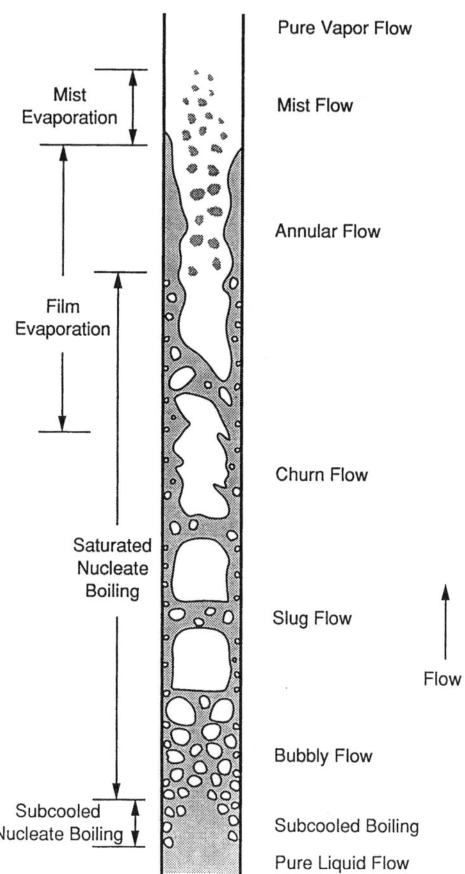

FIGURE 3.4.6 Flow regimes and boiling mechanisms for upflow convective boiling in a vertical tube at moderate wall superheat.

(1989) which has been fit to a broad spectrum of data for both horizontal and vertical tubes. For this method the heat transfer coefficient for a tube of diameter D is given by

$$h = h_l \left[C_1 Co^{C_2} \left(25\mathrm{Fr}_{le} \right)^{C_5} + C_3 Bo^{C_4} F_K \right] \tag{3.4.8}$$

where

$$Co = \left(\frac{1-x}{x} \right)^{0.8} \left(\frac{\rho_v}{\rho_l} \right)^{0.5} \tag{3.4.9}$$

$$Bo = q'' / G h_{fg} \tag{3.4.10}$$

$$Fr_{le} = G^2 / \rho_l^2 g D \tag{3.4.11}$$

and h_l is the single-phase heat transfer coefficient for the liquid phase flowing alone in the tube computed as

$$h_l = 0.023\left(\frac{k_l}{D}\right)\left(\frac{G(1-x)D}{\mu_l}\right)^{0.8} \mathrm{Pr}_l^{0.4} \tag{3.4.12}$$

The constants $C_1 - C_5$ are given in Table 3.4.1. The factor F_K is a fluid-dependent parameter. For water, $F_K = 1$. For R-12 and nitrogen, recommended values of F_K are 1.50 and 4.70, respectively. Values of F_K for a variety of other fluids can be obtained from Kandlikar (1989).

TABLE 3.4.1 Constants for the Correlation of Kandlikar (1987)

	***Co* < 0.65 (Convective Region)**	***Co* ≥ 0.65 (Nucleate Boiling Region)**
C_1	1.1360	0.6683
C_2	–0.9	–0.2
C_3	667.2	1058.0
C_4	0.7	0.7
C_5[a]	0.3	0.3

[a] $C_5 = 0$ for vertical tubes and horizontal tubes with $\mathrm{Fr}_{le} > 0.04$.

Methods for predicting the conditions at which dryout or a DNB transition occurs have typically been empirical in nature. Based on fits to extensive data, Levitan and Lantsman (1975) recommended the following relations for the DNB heat flux and the quality at which dryout occurs during flow boiling of water in a tube with an 8-mm diameter.

$$q''_{\text{crit}} = \left[10.3 - 7.8\left(\frac{P}{98}\right) + 1.6\left(\frac{P}{98}\right)^2\right]\left(\frac{G}{1000}\right)^{1.2\{[0.25(P-98)/98]-x\}} e^{-1.5x} \tag{3.4.13}$$

$$x_{\text{crit}} = \left[0.39 + 1.57\left(\frac{P}{98}\right) - 2.04\left(\frac{P}{98}\right)^2 + 0.68\left(\frac{P}{98}\right)^3\right]\left(\frac{G}{1000}\right)^{-0.5} \tag{3.4.14}$$

In the above relations, q''_{crit} is in MW/m^2, P is the pressure in bar, and G is in kg/m^{2}sec. To obtain values of q''_{crit} and x_{crit} for diameters other than 8 mm, Levitan and Lantsman (1975) recommended that the 8-mm values from the above relations be corrected as follows:

$$q''_{\text{crit}} = \left(q''_{\text{crit}}\right)_{8\text{mm}}\left(\frac{8}{D}\right)^{1/2} \tag{3.4.15}$$

$$x_{\text{crit}} = \left(x_{\text{crit}}\right)_{8\text{mm}}\left(\frac{8}{D}\right)^{0.15} \tag{3.4.16}$$

where D is the diameter in millimeters. A good generalized empirical correlation for predicting dryout or CHF conditions in vertical uniformly heated tubes is that recently proposed by Katto and Ohno (1984).

In many cases, post-dryout mist flow evaporation is driven primarily by convective transport from the tube wall to the gas and then to the entrained droplets. In some circumstances, impingement of droplets onto the heat surface and radiation interactions may also be important. In cases where convection is dominant, predictions of the heat transfer coefficient have been developed by modifying a single-phase correlation for the entire flow as vapor with a correction factor which accounts for the presence of the entrained droplets. Often this correction factor has been presumed to be a function of property ratios. An example of such an approach is the correlation of Dougall and Rohsenow (1963) for which the heat transfer coefficient h is given by

$$\frac{hD}{k_v} = 0.023\left[\left(\frac{GD}{\mu_v}\right)\left(x + \frac{\rho_v}{\rho_l}(1-x)\right)\right]^{0.8} \mathrm{Pr}_{v,\mathrm{sat}}^{0.4} \tag{3.4.17}$$

For further information on mechanisms of convective boiling, see the texts of Collier (1981), Stephan (1992), and Carey (1992).

Condensation

As in the case of boiling, surface tension effects, surface wetting characteristics, and metastable phase stability also can play important roles in condensation processes. As a result of interfacial tension, the pressure inside a spherical liquid droplet of radius r must exceed that in the surrounding liquid by $2\sigma/r$. A consequence of this and basic thermodynamics is that at equilibrium the surrounding vapor must actually be slightly supersaturated. The amount of supersaturation required at equilibrium increases as the radius of curvature of the bubble interface decreases.

For a liquid droplet on a solid surface with a specified volume, the wetting contact angle dictates the radius of curvature of the droplet interface. Because of the linkage between the interface curvature and the required equilibrium supersaturation, the wetting behavior thus determines the level above which the vapor supersaturation must be raised for the droplet to grow. Steady condensation on the droplet interface can be sustained only if the vapor is driven beyond this supersaturation level by cooling or depressurization. For such conditions, the vapor is in the metastable supersaturated range indicated in Figure 3.4.2.

Condensation on external surfaces of a body immersed in a gas phase generally falls into one or two categories: **dropwise condensation** or **film condensation**. In dropwise condensation, the liquid-phase condensate collects as individual droplets which grow in size with time on the cold surface. This mode of condensation is most likely when the liquid poorly wets the solid surface. When the condensation rate is high or the liquid readily wets the surface, a film of liquid condensate covers the solid surface, and the process is referred to as film condensation.

Dropwise Condensation. Dropwise condensation may occur on a solid surface cooled below the saturation temperature of a surrounding vapor when the surface is poorly wetted except at locations where well-wetted contaminant nuclei exist. The poorly wetted surface condition can result from contamination or coating of the surface with a substance which is poorly wetted by the liquid phase of the surrounding vapor. In practice, this can be achieved for steam condensation by (1) injecting a nonwetting chemical into the vapor which subsequently deposits on the surface, (2) introducing a substance such as a fatty (i.e., oleic) acid or wax onto the solid surface, or (3) by permanently coating the surface with a low-surface-energy polymer or a noble metal. The effects of the first two methods are generally temporary, since the resulting surface films eventually are dissolved or eroded away.

During dropwise condensation, the condensate is usually observed to appear in the form of droplets which grow on the surface and coalesce with adjacent droplets. When droplets become large enough, they are generally removed from the surface by the action of gravity or drag forces resulting from the motion of the surrounding gas. As the drops roll or fall from the surface they merge with droplets in their path, effectively sweeping the surface clean of droplets. Droplets then begin to grow anew on the freshly-exposed solid surface. This sweeping and renewal of the droplet growth process is responsible for the high heat transfer coefficients associated with dropwise condensation. Theoretical aspects of dropwise condensation are described in two publications by Tanaka (1975, 1979). A discussion of correlations for the heat transfer coefficient associated with dropwise condensation is provided in the review article by Merte (1973).

External Film Condensation. If the liquid phase fully wets a cold surface in contact with a vapor near saturation conditions, the conversion of vapor to liquid will take the form of film condensation. As the name implies, the condensation takes place at the interface of a liquid film covering the solid surface. Because the latent heat of vaporization must be removed at the interface to sustain the process, the rate of condensation is directly linked to the rate at which heat is transported across the liquid film from the interface to the surface.

The classic integral analysis of Nusselt (1916) for laminar falling-film condensation on a vertical surface considers the physical circumstances shown in Figure 3.4.7. The surface exposed to a motionless ambient of saturated vapor is taken to be isothermal with a temperature below the saturation temperature. Note that although a vertical surface is considered here, the analysis is identical for an inclined surface, except that the gravitational acceleration g is replaced by $g \sin \Omega$, with Ω being the angle between the surface and the horizontal. Because the liquid film flows down the surface because of gravity, this situation is sometimes referred to as *falling-film condensation.*

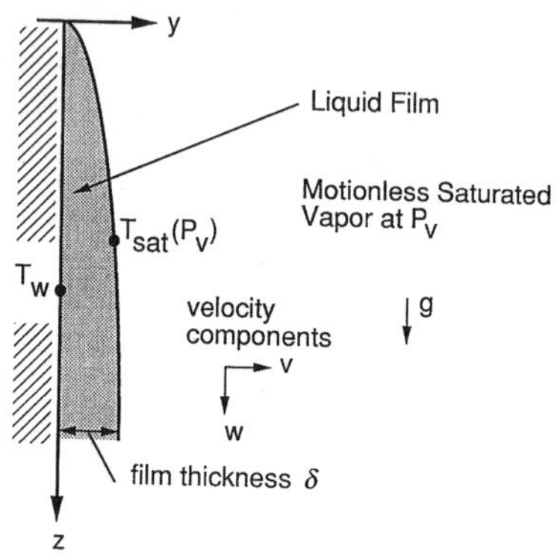

FIGURE 3.4.7 System model for the Nusselt analysis of falling-film condensation.

In its simplest form, the classic Nusselt analysis incorporates the following idealizations: (1) laminar flow, (2) constant properties, (3) that subcooling of liquid is negligible in the energy balance, (4) that inertia effects are negligible in the momentum balance, (5) that the vapor is stationary and exerts no drag, (6) that the liquid-vapor interface is smooth, and (7) that heat transfer across film is only by conduction (convection is neglected). With these idealizations, the following relation for the local heat transfer coefficient h can be obtained

$$\frac{hz}{k_l} = \left[\frac{\rho_l(\rho_l - \rho_v) g h_{fg} z^3}{4 k_l \mu_\ell (T_{\text{sat}} - T_w)} \right]^{1/4} \tag{3.4.18}$$

Modified versions of this analysis have been subsequently developed which relax many of these assumptions. Laminar film condensation on a vertical surface can also be analyzed with a full boundary layer formulation. An example of this type of approach is the analysis presented by Sparrow and Gregg (1959).

The analyses described above do not include two physical mechanisms which can significantly affect the transport: (1) the effects of waves on the liquid-vapor interface and (2) interfacial vapor shear drag on the interface. The effects of interfacial shear have been studied analytically by numerous investigators. The effects of surface waves on laminar film condensation are more difficult to incorporate into theoretical analyses. In general, interfacial waves are expected to enhance convective heat transport in the film since it intermittently thins the film, increases the interfacial area, and induces mixing. Because of these effects, laminar film condensation heat transfer data are often significantly higher than the values predicted by simple boundary layer models.

As for any boundary layer flow, when the film Reynolds number becomes large enough, it is expected that a transition to turbulent flow will occur. Eddy diffusivity models of the resulting turbulent transport have been developed by Seban (1954), Dukler (1960), and others. This methodology was later extended to evaporation of a falling liquid film (see, for example, Mills and Chung, 1973).

Subsequent studies (see, for example, Mills and Chung, 1973) have suggested that the presence of the interface tends to damp larger turbulent eddies near the interface in the liquid film. This implies that a viscous sublayer exists at the interface as well as at the wall. Recent efforts to model turbulent falling-film evaporation and condensation processes have therefore included a variation of the eddy viscosity in which it goes to zero at both the wall and the interface. The analysis tools and correlations described above work reasonably well for values of liquid Prandtl number above 1. However, deviation of the predictions using these methods from heat transfer data for liquid metals can be quite significant.

Because of its importance to the design of tube-and-shell condensers, condensation on the outside of horizontal tubes has been the subject of numerous studies. The length of the tube perimeter over which the condensate flows is usually small for commonly used tubes. Consequently, the film Reynolds number is usually low and the flow in the liquid film is laminar.

With slight modification, the Nusselt (1916) analysis of laminar falling-film condensation over a flat plate can be adapted to film condensation on an isothermal horizontal cylinder. Doing so yields the following relation for the mean heat transfer coefficient:

$$\frac{\bar{h}D}{k_l} = 0.728\left[\frac{(\rho_l - \rho_v)gh_{fg}D^3\mathrm{Pr}_l}{\rho_l \nu_l^2 c_{pl}(T_{\mathrm{sat}} - T_w)}\right]^{1/4} \tag{3.4.19}$$

Selin (1961) found that better agreement with film condensation data for horizontal tubes was obtained by replacing the constant factor in Equation (3.4.19) by 0.61. Correlations similar to the single-tube relation above have also been developed for the average condensation heat transfer coefficient for banks of round tubes.

Analytical treatment of laminar film condensation on a sphere is virtually the same as that for a horizontal cylinder. The only differences result from the angular variation of the body perimeter because of the spherical geometry. A general analytical prediction of the local heat transfer coefficient for laminar film condensation on arbitrary axisymmetric bodies has been developed by Dhir and Lienhard (1971).

Condensation in the Presence of a Noncondensable Gas. In nature and in a number of technological applications, condensation of one component vapor in a mixture may occur in the presence of other noncondensable components. The most common example is the condensation of water vapor in the air on a cold solid surface. If the component gases are considered to be a mixture of independent substances, condensation of one component vapor will occur if the temperature of the surface is below the saturation temperature of the pure vapor at its partial pressure in the mixture. This temperature threshold is referred to as the *dew point* of the mixture.

Because only the vapor is condensed, the concentration of the noncondensable gas at the interface is higher than its value in the far ambient. This, in turn, decreases the partial pressure of the vapor at the interface below its ambient value. The corresponding saturation temperature at the interface is therefore lower than the bulk temperature. The resulting depression of the interface temperature generally reduces the condensation heat transfer rate below that which would result for pure vapor alone under the same conditions. Space limitations here preclude a detailed discussion of the effects of noncondensable gases. The interested reader may find more-extensive discussions of this topic in the references by Collier (1981) and Carey (1992).

Internal Convective Condensation. In most power and refrigeration systems, the flow in the condenser is either horizontal or vertically downward. Figure 3.4.8 schematically depicts a typical condensation process in a horizontal round tube. Superheated vapor enters the tube and at the exit end the liquid is subcooled. At a point some distance downstream of the entrance, vapor begins to condense on the walls of the tube. The location at which this occurs is at or slightly before the bulk flow reaches the equilibrium saturation condition. In most condensers, the liquid readily wets the interior of the tube and at high vapor volume fractions the liquid forms a thin liquid film on the interior wall of the tube.

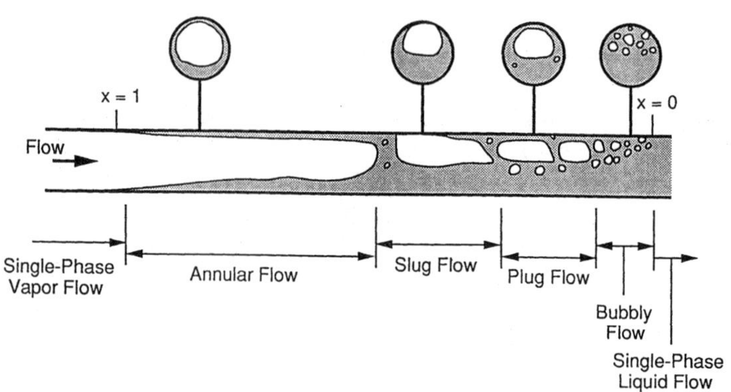

FIGURE 3.4.8 Flow regimes during horizontal cocurrent flow with condensation.

The vapor velocity is generally high at the inlet end of the condenser tube, and the liquid film is driven along the tube by strong vapor shear on the film. At low vapor flow rates, some stratification may occur and the film may be thicker on the bottom of the horizontal tube. At high vapor flow rates, turbulent stresses acting on the liquid may tend to keep the thickness of the liquid film nominally uniform over the perimeter of the tube.

In most condenser applications, shear-dominated annular flow persists to very low qualities and the overwhelming majority of the heat transfer occurs in this regime. The very last stage of the condensation process, corresponding to qualities less than a few percent, may occur in slug, plug, or bubbly two-phase flow. Generally these regimes represent such a small portion of the overall heat transfer in the condenser that some inaccuracy in estimating the heat transfer coefficient for them is tolerated. As a first estimate, the heat transfer coefficient may be predicted using a correlation for pure single-phase liquid flow in the tube at the same total flow rate, or a correlation for annular flow condensation may simply be extrapolated to zero quality.

Because most of the heat duty occurs in the annular flow regime, accurate prediction of the overall heat transfer performance of the condenser requires a predictive methodology that accurately treats the transport in this regime. For this reason, the form of most correlation methods for predicting local convective condensation heat transfer coefficients are optimized to match data in the annular flow regime. One example of such a correlation is the following relation for the local heat transfer coefficient for annular flow condensation proposed by Traviss et al. (1973):

$$\frac{hD}{k_l} = \frac{0.15\text{Pr}_l\text{Re}_l^{0.9}}{F_T}\left[\frac{1}{X_{tt}} + \frac{2.85}{X_{tt}^{0.476}}\right] \tag{3.4.20}$$

where

$$\text{Re}_l = \frac{G(1-x)D}{\mu_l}, \qquad X_{tt} = \left(\frac{1-x}{x}\right)^{0.9}\left(\frac{\rho_v}{\rho_l}\right)^{0.5}\left(\frac{\mu_l}{\mu_v}\right)^{0.1}$$

and F_T is given by

$$F_T = 5\mathrm{Pr}_l + 5\ln\left\{1 + 5\mathrm{Pr}_l\right\} + 2.5\ln\left\{0.0031\mathrm{Re}_l^{0.812}\right\} \quad \text{for} \quad \mathrm{Re}_l > 1125$$

$$= 5\mathrm{Pr}_l + 5\ln\left\{1 + \mathrm{Pr}_l\left(0.0964\mathrm{Re}_l^{0.585} - 1\right)\right\} \quad \text{for} \quad 50 < \mathrm{Re}_l < 1125$$

$$= 0.707\mathrm{Pr}_l\mathrm{Re}_l^{0.5} \quad \text{for} \quad \mathrm{Re}_l < 50$$

Carey (1992) has shown that the generic form of this correlation can be derived from a theoretical model of annular flow condensation in a round tube. Several correlations of this general type have been developed as fits to experimental data; see Carey (1992) for a summary. The predictions of these correlations may vary significantly for a given set of conditions. When possible, a correlation should be selected which has been tested against data for conditions close to those for the application of interest.

A correlation methodology that can be used to predict internal convective condensation heat transfer for slug, plug, or wavy stratified flow has also been proposed by Rossen and Meyers (1965). To predict the overall heat transfer performance of a condenser, methods to predict the local heat transfer coefficient must be combined with a scheme to numerically integrate finite-difference forms of the energy, mass, and momentum balances in the tube. For further information on such schemes see the references by Collier (1981) and Carey (1992) (Figure 3.4.9).

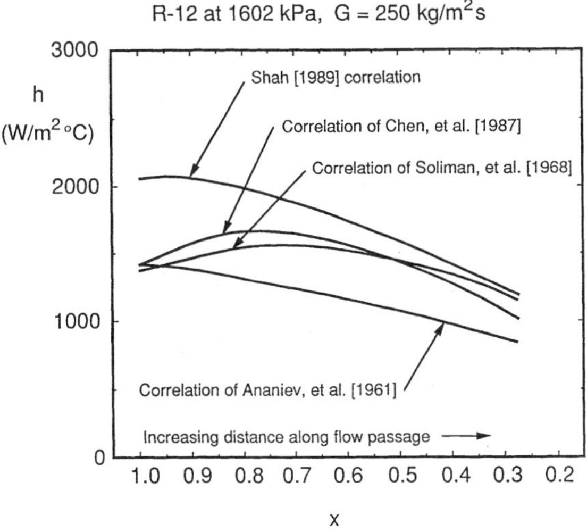

FIGURE 3.4.9 Comparison of the variation of h with x predicted by four correlation methods for internal convective condensation. References cited in this figure are listed in chapter 11 of Carey (1992).

Defining Terms

Critical heat flux (CHF): A maximum heat flux condition that characterizes the transition between nucleate boiling and transition boiling or film boiling.

Dropwise condensation: Condensation of vapor into liquid in discrete droplets, usually attained when a cold surface is poorly wetted by the liquid phase.

Film boiling: Generation of vapor at the interface of a vapor film which entirely covers the hot surface.

Film condensation: Condensation of vapor onto the interface of a liquid film that completely covers a cold surface.

Minimum heat flux: A minimum heat flux condition on the classic boiling curve that characterizes the transition between film boiling and transition boiling. Also, sometimes referred to as the Leidenfrost point, it is a lower bound for heat flux values at which stable film boiling may occur.

Nucleate boiling: Generation of vapor at a hot surface by formation of bubbles at discrete nucleation sites with full liquid wetting of the surface.

Polar molecules: Molecules which have a permanent electric dipole moment. Examples include water and ammonia.

Pool boiling: Generation of vapor at the surface of a hot body immersed in an extensive liquid pool.

Transition boiling: Generation of vapor at a hot surface with intermittent or partial liquid wetting of the surface.

References

Berenson, P.J. 1961. Film boiling heat transfer from a horizontal surface. *J. Heat Transfer*, 83, 351–356.

Carey, V.P. 1992. *Liquid-Vapor Phase Change Phenomena.* Taylor and Francis, Washington, D.C.

Chen, J.C. 1966. Correlation for boiling heat transfer to saturated fluids in convective flow. *Ind. Eng. Chem. Proc. Design and Dev.* 5(3), 322–339.

Collier, J.G. 1981. *Convective Boiling and Condensation*, 2nd ed. McGraw-Hill, New York.

Dhir, V.K. and Lienhard, J. 1971. Laminar film condensation on plane and axisymmetric bodies in nonuniform gravity. *J. Heat Transfer* 93, 97–100.

Dougall, R.S. and Rohsenow, W.M. 1963. Film boiling on the inside of vertical tubes with upward flow of the fluid at low qualities. MIT Report No. 9079-26. MIT, Cambridge, MA.

Dukler, A.E. 1960. Fluid mechanics and heat transfer in vertical falling film systems. *Chem. Eng. Prog. Symp. Ser.* 56(30), 1–10.

Gunnerson, F.S. and Cronenberg, A.W. 1980. On the minimum film boiling conditions for spherical geometries. *J. Heat Transfer* 102,335–341.

Haramura, Y. and Katto, Y. 1983. A new hydrodynamic model of the critical heat flux, applicable widely to both pool and forced convective boiling on submerged bodies in saturated liquids. *Int. J. Heat Mass Transfer* 26, 389–399.

Kandlikar, S.G. 1989. A general correlation for saturated two-phase flow boiling heat transfer inside horizontal and vertical tubes. *J. Heat Transfer* 112, 219–228.

Katto, Y. and Ohno, H. 1984. An improved version of the generalized correlation of critical heat flux for the forced convective boiling in uniformly heated vertical tubes. *Int. Heat Mass Transfer* 21, 1527–1542.

Levitan, L.L. and Lantsman, F.P. 1975. Investigating burnout with flow of a steam-water mixture in a round tube, *Therm. Eng.* (USSR). English trans., 22, 102–105.

Lienhard, J.H. and Dhir, V.K. 1973. Extended hydrodynamic theory of the peak and minimum pool boiling heat fluxes. NASA CR-2270.

Lienhard, J.H. and Witte, L.C. 1985. A historical review of the hydrodynamic theory of boiling. *Rev. Chem. Eng.* 3, 187–277.

Lienhard, J.H. and Wong, P.T.Y. 1964. The dominant unstable wavelength and minimum heat flux during film boiling on a horizontal cylinder. *J. Heat Transfer* 86, 220–226.

Merte, H. 1973. Condensation heat transfer. *Adv. Heat Transfer* 9, 181–272.

Mills, A.F. and Chung, D.K. 1973. Heat transfer across turbulent falling films. *Int. J. Heat Mass Transfer* 16, 694–696.

Nukiyama, S. 1934. The maximum and minimum values of Q transmitted from metal to boiling water under atmospheric pressure. *J. Jpn. Soc. Mech. Eng.* 37, 367–374.

Nusselt, W. 1916. Die Oberflachenkondensation des Wasser dampfes. *Z. Ver. Dtsch. Ininuere* 60, 541–575.

Ramilison, J.M. and Lienhard, J.H. 1987. Transition boiling heat transfer and the film transition regime. *J. Heat Transfer* 109, 746–752.

Rohsenow, W.M. 1962. A method of correlating heat transfer data for surface boiling of liquids. *Trans. ASME* 84, 969–975.

Rossen, H.F. and Meyers, J.A. 1965. Point values of condensing film coefficients inside a horizontal tube. *Chem. Eng. Prog. Symp. Ser.* 61(59), 190–199.

Seban, R. 1954. Remarks on film condensation with turbulent flow. *Trans. ASME* 76, 299–303.
Selin, G. 1961. Heat transfer by condensing pure vapors outside inclined tubes, in *Proc. First Int. Heat Transfer Conf.*, University of Colorado, Boulder, Part II, 279–289.
Sparrow, E.M. and Gregg, J.L. 1959. A boundary-layer treatment of laminar film condensation. *J. Heat Transfer* 81, 13–23.
Stephen, K. 1992. *Heat Transfer in Condensation and Boiling*. Springer-Verlag, New York.
Tanaka, H. 1975. A theoretical study of dropwise condensation. *J. Heat Transfer* 97, 72–78.
Tanaka, H. 1979. Further developments of dropwise condensation theory. *J. Heat Transfer* 101, 603–611.
Traviss, D.P., Rohsenow, W.M., and Baron, A.B. 1973. Forced convection condensation in tubes: a heat transfer correlation for condenser design. *ASHRAE Trans.* 79(I), 157–165.
Witte, L.C. and Lienhard, J.H. 1982. On the existence of two "transition" boiling curves. *Int. J. Heat Mass Transfer* 25, 771–779.
Zuber, N. 1959. Hydrodynamic aspects of boiling heat transfer. AEC Rep. AECU-4439.

Further Information

The texts *Heat Transfer in Condensation and Boiling* by K. Stephan (Springer-Verlag, New York, 1992) and *Liquid-Vapor Phase Change Phenomena* by V.P. Carey (Taylor and Francis, Washington, D.C., 1992) provide an introduction to the physics of boiling and condensation processes. The text by J.G. Collier, *Convective Boiling and Condensation* (2nd ed., McGraw-Hill, New York, 1981), summarizes more-advanced elements of convective boiling and condensation processes. The *ASHRAE Handbook of Fundamentals* (American Society of Heating, Refrigerating, and Air-Conditioning Engineers, Atlanta, GA, 1993) provides some information on boiling and condensation heat transfer and is a good source of thermophysical property data needed to analyze boiling and condensation processes.

3.4.2 Particle Gas Convection

John C. Chen

Introduction

Heat transfer in two-phase systems involving gas and solid particles are encountered in several types of operations important in chemical, power, and environmental technologies. Chief among these are gas fluidized beds which are widely used to achieve either physical processing or chemical reactions that require interfacial contact between gas and solid particles. Currently, fluidized beds operate in either the *bubbling regime* or the *fast-circulating regime*. In the first case, particles are retained in the fluidized bed while the gas passes upward past the particles, partially as rising bubbles. In the second case, gas velocities exceed terminal velocity for the individual particles and the two phases flow through the fluidized bed in cocurrent upward flow. For those applications which require thermal control, convective heat transfer between the fluidized medium and heat transfer surfaces (either immersed tubes or the vessel walls) is an essential element of the process design.

Bubbling Fluidized Beds

Bubbling fluidization occurs when the superficial gas velocity exceeds a critical value wherein the gravitational body force on the solid particles is balanced by the shear force between particles and flowing gas. The superficial gas velocity at this condition, commonly called the minimum fluidization velocity (U_{mf}), marks the boundary between gas flow through packed beds and gas flow in fluidized beds. Wen and Yu (1966) derived the following general equation to estimate U_{mf} for spherical particles:

$$\mathrm{Re}_{mf} = \left[(33.7)^2 + 0.0408\,\mathrm{Ar}\right]^{1/2} - 33.7 \qquad (3.4.21)$$

where

$$\mathrm{Re}_{mf} = \text{particle Reynolds number at } U_{\mathrm{mf}} = \frac{U_{\mathrm{mf}} d_p \rho_g}{\mu_g}$$

$$\mathrm{Ar} = \text{Archimedes number} = \frac{d_p^3 \rho_g (\rho_s - \rho_g) g}{\mu_g^2}$$

Increasing gas velocity beyond minimum fluidization causes the excess gas to collect into discrete bubbles that grow and rise through the fluidized matrix of solid particles. In this bubbling fluidization regime, the total pressure drop over the height of the fluidized bed, H, is equal to the hydrostatic pressure of the solid mass,

$$\Delta P = g\rho_s (1-\varepsilon) H \tag{3.4.22}$$

where ε = volume fraction of gas (void fraction).

Tubes carrying cooling or heating fluids are often immersed in bubbling fluidized beds to extract or add thermal energy. The effective heat transfer coefficient at the surface of such tubes has been the objective of numerous experimental and analytical investigations. Data for the circumferentially averaged heat transfer coefficient for horizontal tubes are shown in Figure 3.4.10 for various types of solid particles. Characteristics representative of such systems are

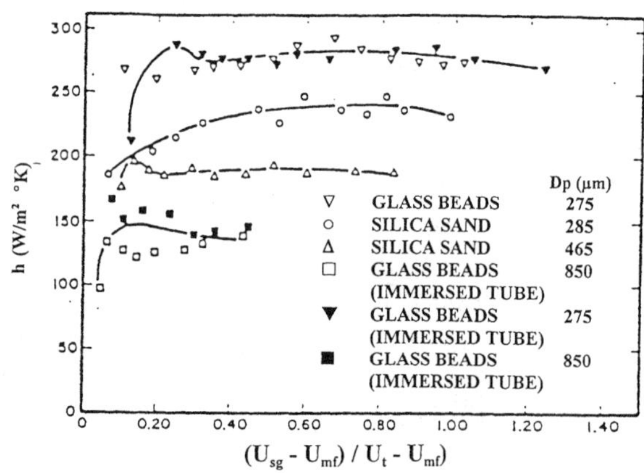

FIGURE 3.4.10 Average heat transfer coefficients for horizontal tubes immersed in bubbling fluidized beds. (From Biyikli, K. et al., 1983. *AIChE J.*, 29(5), 712-716. With permission.)

- The heat transfer coefficient increases sharply as the gas velocity exceeds minimum fluidization velocity.
- After the initial increase, the heat transfer coefficient remains fairly constant over a significant range of the gas velocity beyond minimum fluidization velocity.
- The absolute magnitude of the heat transfer coefficient is severalfold greater than single-phase gas convection at the same superficial velocity.
- The heat transfer coefficient increases as particle size decreases.

Kunii and Levenspiel (1991) have shown that increasing gas pressure and density significantly increases the magnitude of the heat transfer coefficient as well as promoting the occurrence of minimum fluidization at a lower value of superficial gas velocity. The effect of bundle spacing is insignificant at 1-atm pressure but becomes increasingly more important as gas pressure and density increase. The data of Jacob and Osberg (1957) indicate that the convective heat transfer coefficient in fluidized beds increases with increasing thermal conductivity of the gas phase, for any given particle size.

Several different types of correlations have been suggested for predicting convective heat transfer coefficients at submerged surfaces in bubbling fluidized beds. The first type attributes the enhancement of heat transfer to the scouring action of solid particles on the gas boundary layer, thus decreasing the effective film thickness. These models generally correlate a heat transfer Nusselt number in terms of the fluid Prandtl number and a modified Reynolds number with either the particle diameter or the tube diameter as the characteristic length scale. Examples are

Leva's correlation for vertical surfaces and larger particles (Leva and Gummer, 1952);

$$\mathrm{Nu}_{dp} = \frac{h_c d_p}{k_g} = 0.525\left(\mathrm{Re}_p\right)^{0.75} \tag{3.4.23}$$

where

$$\mathrm{Re}_p = \frac{d_p \rho_g U}{\mu_g}$$

Vreedenberg's (1958) correlation for horizontal tubes refers to the particle of diameter D_t.

$$\mathrm{Nu}_{Dt} = \frac{h_c D_t}{k_g} = 420\left(\frac{\rho_s}{\rho_g}\mathrm{Re}_t\right)^{0.3}\left(\frac{\mu_g^2}{g\rho_s^2 d_p^3}\right)^{0.3}\left(\mathrm{Pr}_g\right)^{0.3} \tag{3.4.24}$$

for

$$\left(\frac{\rho_s}{\rho_g}\mathrm{Re}_t\right) > 2250$$

where

$$\mathrm{Re}_t = \frac{D_t \rho_g U}{\mu_g}$$

Molerus and Schweinzer (1989) developed an alternative type of correlation based on the supposition that the heat transfer is dominated by gas convection through the matrix of particles in the vicinity of the heat transfer surface. Their correlation takes the form:

$$\mathrm{Nu} = \frac{h_c d_p}{k_g} = 0.0247(\mathrm{Ar})^{0.4304}(\mathrm{Pr})^{0.33} \tag{3.4.25}$$

Figure 3.4.11 shows comparison of this model with experimental data obtained at three different pressures. The solid curve represents the relationship for fixed beds, while the dashed lines represent the behavior for fluidized beds (i.e., Equation 3.4.25) upon exceeding minimum fluidization.

A third type of model considers the heat transfer surface to be contacted alternately by gas bubbles and packets of packed particles, leading to a surface renewal process for heat transfer. Mickley and Fairbanks (1955) provided the first analysis of this renewal mechanism. Ozkaynak and Chen (1980)

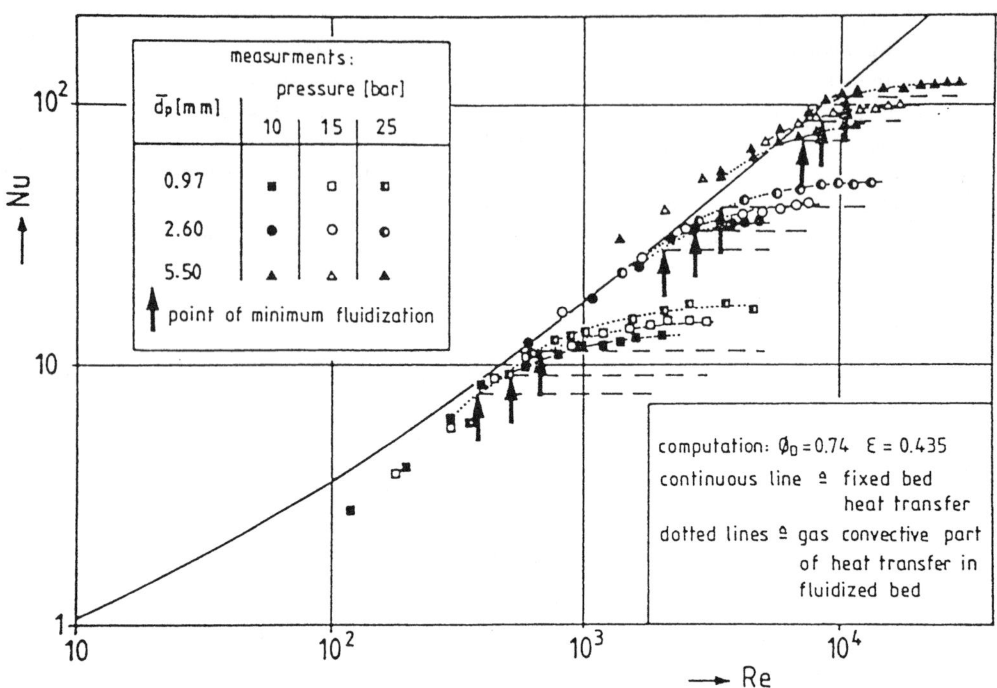

FIGURE 3.4.11 Correlation of Molerus and Schweinzer compared with experimental data (1989).

showed that if experimentally measured values of the packet contact time and residence times are used in the packet model analysis, excellent agreement is obtained.

Fast-Circulating Fluidized Beds

Fast fluidization occurs when the superficial gas velocity exceeds the terminal velocity of the solid particles, causing the particles to be suspended in cocurrent upward flow with the gas. This upward flow occurs in "rise reactors" wherein desired physical or chemical reactions occur. In most applications, the two-phase flow exits the top of the riser into a cyclone where the gas phase is separated and exhausted while the solid particles are captured and returned for reinjection at the bottom of the riser. The volumetric concentration of solid particles in these fast fluidized beds (FFBs) tend to be fairly dilute, often with average concentrations of less than 2%. Heat exchange with the particle/gas suspension is usually accomplished through the vertical wall surfaces or through vertical tubes immersed in the duct.

The heat transfer coefficient at vertical surfaces FFBs has been found to increase with increasing solid concentration, aside from other second-order parametric effects. Figure 3.4.12 shows heat transfer coefficients experimentally measured by Dou et al. (1994) for an FFB operating with sand particles of 124 μm mean diameter. Figure 3.4.12b shows that the heat transfer coefficient increased with solid mass flux, for a constant superficial gas velocity. Figure 3.4.12a shows that the heat transfer coefficient decreased parametrically with superficial gas velocity for a constant solid mass flux. Both figures indicate that heat transfer coefficients decrease with increasing elevation in the riser duct. These three parametric trends are all consistent with the hypothesis that heat transfer in FFBs increases with increasing concentration of the solid phase.

It is generally accepted that the effective heat transfer coefficient for surfaces in FFBs have contributions for gas-phase convection, particle-induced convection, and radiation:

$$h = h_g + h_p + h_r \tag{3.4.26}$$

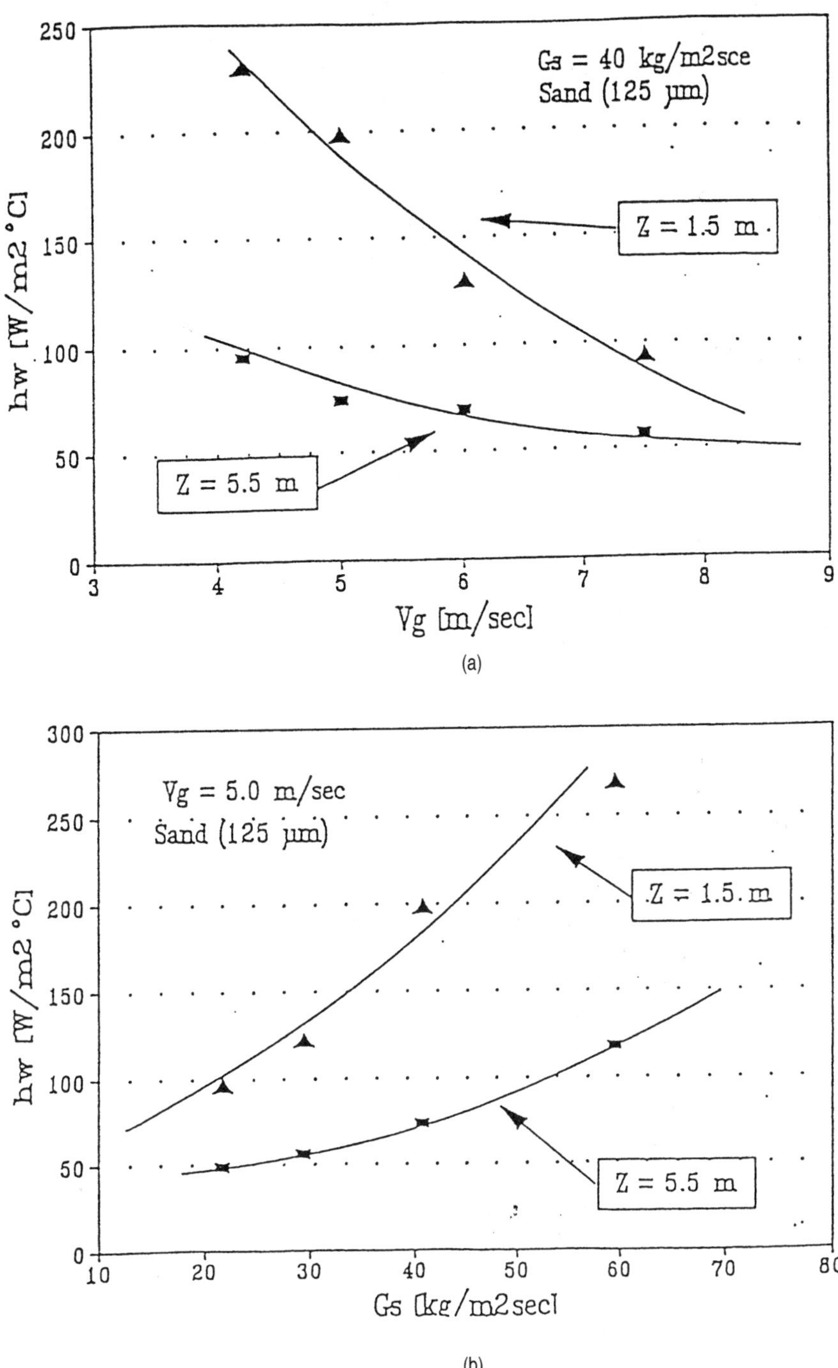

FIGURE 3.4.12 Heat transfer coefficients in fast fluidized beds; Vg is superficial gas velocity, Gs is mass flux of particles, and Z is elevation in FFB. (From Dou, S. et al., 1992. *Fluidization VII,* Engineering Foundation, 793-802. With permission.)

In contrast to the situation in dense-bubbling fluidized beds, the relatively dilute concentration of solid particles in FFBs often results in significant contributions from all three heat transfer mechanisms. The radiation coefficient can be obtained by a gray body model suggested by Grace (1985). The contribution of the gas phase convection (h_g) is commonly estimated based on correlations for gas flow alone at the same superficial gas velocity. Although the presence of particles may alter the turbulence characteristic of this gas flow, any errors caused by this procedure are usually small since h_g is generally smaller than the particle-phase convective coefficient h_p.

For most FFBs, the particle convective contribution to heat transfer is most important and the prediction of h_p is the major concern in thermal design. Unfortunately, mechanistically based models are still lacking and most design methods rely on empirical correlations which often combine the contributions of gas and particle phases into a single convective heat transfer coefficient (h_c). One such correlation proposed by Wen and Miller (1961) is

$$\mathrm{Nu}_{dp} = \frac{h_c d_p}{k_g} = \left(\frac{C_{pp}}{C_{pg}}\right)\left(\frac{\rho_{\mathrm{susp}}}{\rho_p}\right)^{0.3}\left(\frac{V_t}{g d_p}\right)^{0.21} \mathrm{Pr}_g \tag{3.4.27}$$

where V_t = terminal velocity of particle.

Other correlations have been proposed by Fraley (1992) and Martin (1984). These correlations are useful as a starting point but have not yet been verified over wide parametric ranges. Large deviations can occur when compared with measurements obtained outside of the experimental parametric ranges.

References

Biyikli, K., Tuzla, K., and Chen, J.C. 1983. Heat transfer around a horizontal tube in freeboard region of fluidized beds, *AIChE J.,* 29(5), 712–716.

Dou, S., Herb, B., Tuzla, K., and Chen, J.C. 1992. Dynamic variation of solid concentration and heat transfer coefficient at wall of circulating fluidized bed, in *Fluidization VII*, Eds. Potter and Nicklin, Engineering Foundation, 793–802.

Fraley, L.D., Lin, Y.Y., Hsiao, K.H., and Solbakken, A. 1983. ASME Paper 83-HT-92, National Heat Transfer Conference, Seattle.

Grace, J.R. 1985. Heat transfer in circulating fluidized beds, *Circulating Fluidized Bed Technology I*, Peramon Press, New York, 63–81.

Jacob, A. and Osberg, G.L. 1957. Effect of gas thermal conductivity on local heat transfer in a fluidized bed, *Can. J. Chem. Eng.*, 35(6), 5–9.

Kunii, D. and Levenspiel, O. 1991. *Fluidization Engineering*, 2nd ed., Butterworth-Heinemann, Boston.

Leva, M. and Grummer, M. 1952. A correlation of solids turnovers in fluidized systems, *Chem. Eng. Prog.*, 48(6), 307–313.

Martin, H. 1984. *Chem. Eng. Process*, 18, 157–223.

Mickley, H.S. and Fairbanks, D.F. 1955. Mechanism of heat transfer to fluidized beds, *AIChE J.*, 1(3), 374–384.

Molerus, O. and Scheinzer, J. 1989. Prediction of gas convective part of the heat transfer to fluidized beds, in *Fluidization IV*, Engineering Foundation, New York, 685–693.

Ozkaynak, T.F. and Chen, J.C. 1980. Emulsion phase residence time and its use in heat transfer models in fluidized bed, *AIChE J.*, 26(4), 544–550.

Vreedenberg, H.A. 1958. Heat transfer between a fluidized bed and a horizontal tube, *Chem. Eng. Sci.*, 9(1), 52–60.

Wen, C.Y. and Yu, Y.H. 1966. A generalized method for predicting the minimum fluidization velocity, *AIChE J.*, 12(2), 610–612

Wen, C.Y. and MIller, E.N. 1961. *Ind. Eng. Chem.*, 53, 51–53.

3.4.3 Melting and Freezing

Noam Lior

Introduction and Overview

Melting and freezing occur naturally (Lunardini, 1981) as with environmental ice in the atmosphere (hail, icing on aircraft), on water bodies and ground regions at the Earth surface, and in the molten Earth core (Figure 3.4.13). They are also a part of many technological processes, such as preservation of foodstuffs (ASHRAE, 1990, 1993), refrigeration and air-conditioning (ASHRAE, 1990, 1993), snow and ice making for skiing and skating (ASHRAE, 1990), organ preservation and cryosurgery (Rubinsky and Eto, 1990), manufacturing (such as casting, molding of plastics, coating, welding, high-energy beam cutting and forming, crystal growth, electrodischarge machining, electrodeposition) (Flemings, 1974; Cheng and Seki, 1991; Tanasawa and Lior, 1992), and thermal energy storage using solid–liquid phase-changing materials (deWinter, 1990).

(a)

(b)

FIGURE 3.4.13 Melting and freezing in nature. (a) A melting icicle. (b) Frozen lava in Hawaii.

In simple thermodynamic systems (i.e., without external fields, surface tension, etc.) of a pure material, melting or freezing occurs at certain combinations of temperature and pressure. Since pressure typically has a relatively smaller influence, only the fusion (freezing or melting) temperature is often used to identify this phase transition. Fusion becomes strongly dependent on the concentration when the material contains more than a single species. Furthermore, melting and freezing are also sensitive to external effects, such as electric and magnetic fields, in more-complex thermodynamic systems.

The equilibrium thermodynamic system parameters during phase transition can be calculated from the knowledge that the partial molar Gibbs free energies or chemical potentials of each component in the two phases must be equal. One important result of using this principle for simple single-component systems is the Clapeyron equation relating the temperature (T) and pressure (P) during the phase transition, such that

$$\frac{dP}{dT} = \frac{h_{s\ell}}{T\Delta v} \tag{3.4.28}$$

where $h_{s\ell}$ is the enthalpy change from phase A to phase B ($=h_B - h_A$, the latent heat of fusion with appropriate sign) and Δv is the specific volume difference between phases A and B ($= v_B - v_A$). Considering for example that phase A is a solid and B a liquid ($h_{s\ell}$ is then positive), examination of Equation (3.4.28) shows that increasing the pressure will result in an increase in the melting temperature if $\Delta v > 0$ (i.e., when the specific volume of the liquid is higher than that of the solid, which is a property of tin, for example), but will result in a decrease of the melting temperature when $\Delta v < 0$ (for water, for example).

In some materials, called glassy, the phase change between the liquid and solid occurs with a gradual transition of the physical properties, from those of one phase to those of the other. When the liquid phase flows during the process, the flow is strongly affected because the viscosity increases greatly as the liquid changes to solid. Other materials, such as pure metals and ice, and eutectic alloys, have a definite line of demarcation between the liquid and the solid, the transition being abrupt. This situation is easier to analyze and is therefore more thoroughly addressed in the literature.

Gradual transition is most distinctly observed in mixtures. Consider the equilibrium phase diagram for a binary mixture (or alloy) composed of species a and b, shown in Figure 3.4.14. χ is the concentration of species b in the mixture, ℓ denotes the liquid, s the solid, s_a a solid with a lattice structure of species a in its solid phase but containing some molecules of species b in that lattice, and s_b a solid with a lattice structure of species b in its solid phase but containing some molecules of species a in that lattice. "Liquidus" denotes the boundary above which the mixture is just liquid, and "solidus" is the boundary separating the final solid mixture of species a and b from the solid–liquid mixture zones and from the other zones of solid s_a and solid s_b.

For illustration, assume that a liquid mixture is at point 1, characterized by concentration χ_1 and temperature T_1 (Figure 3.4.14), and is cooled (descending along the dashed line) while maintaining the concentration constant. When the temperature drops below the liquidus line, solidification starts, creating a mixture of liquid and of solid s_a. Such a two-phase mixture is called the **mushy zone**. At point 2 in that zone, the solid phase (s_a) portion contains a concentration χ_{2,s_a} of component b, and the liquid phase portion contains a concentration $\chi_{2,\ell}$ of component b. The ratio of the mass of the solid s_a to that of the liquid is determined by the lever rule, and is $(\chi_{2,\ell} - \chi_2)/(\chi_2 - \chi_{2,s_a})$ at point 2. Further cooling to below the solidus line, say to point 3, results in a solid mixture (or alloy) of s_a and s_b, containing concentrations χ_{3,s_A} and χ_{3,s_B} of species b, respectively. The ratio of the mass of the solid s_a to that of s_b is again determined by the lever rule, and is $(\chi_{3,s_b} - \chi_3)/(\chi_3 - \chi_{3,s_a})$ at point 3.

A unique situation occurs if the initial concentration of the liquid is χ_e: upon constant-concentration cooling, the liquid forms the solid mixture $s_a + s_b$ having the same concentration and without the formation of a two-phase zone. χ_e is called the **eutectic concentration**, and the resulting solid mixture (or alloy) is called a *eutectic*.

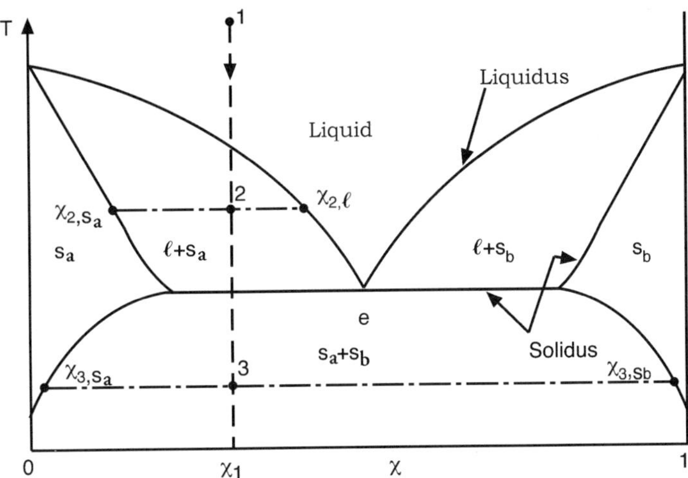

FIGURE 3.4.14 A liquid–solid phase diagram of a binary mixture.

The presence of a two-phase mixture zone with temperature-dependent concentration and phase proportion obviously complicates heat transfer analysis, and requires the simultaneous solution of both the heat and mass transfer equations. Furthermore, the liquid usually does not solidify on a simple planar surface. Crystals of the solid phase are formed at some preferred locations in the liquid, or on colder solid surfaces immersed in the liquid, and as freezing progesses the crystals grow in the form of intricately shaped fingers, called dendrites. This complicates the geometry significantly and makes mathematical modeling of the process very difficult. An introduction to such problems and further references are available in Hayashi and Kunimine (1992) and Poulikakos (1994).

Flow of the liquid phase often has an important role in the inception of, and during, melting and freezing (see Incropera and Viskanta, 1992). The flow may be forced, such as in the freezing of a liquid flowing through or across a cooled pipe, and/or may be due to natural convection that arises whenever there are density gradients in the liquid, here generated by temperature and possibly concentration gradients. It is noteworthy that the change in phase usually affects the original flow, such as when the liquid flowing in a cooled pipe gradually freezes and the frozen solid thus reduces the flow passage, or when the evolving dendritic structure gradually changes the geometry of the solid surfaces that are in contact with the liquid. Under such circumstances, strong coupling may exist between the heat transfer and fluid mechanics, and also with mass transfer when more than a single species is present. The process must then be modeled by an appropriate set of continuity, momentum, energy, mass conservation, and state equations, which need to be solved simultaneously.

More-detailed information about melting and freezing can be found in the monograph by Alexiades and Solomon (1993) and in the comprehensive reviews by Fukusako and Seki (1987) and Yao and Prusa (1989).

Melting and Freezing of Pure Materials

Thorough mathematical treatment of melting and freezing is beyond the scope of this section, but examination of the simplified one-dimensional case for a pure material and without flow effects provides important insights into the phenomena, identifies the key parameters, and allows analytical solutions and thus qualitative predictive capability for at least this class of problems.

In the freezing model, described in Figure 3.4.15, a liquid of infinite extent is to the right ($x > 0$) of the infinite surface at $x = 0$, initially at a temperature T_i higher than the fusion temperature T_f. At time $t = 0$ the liquid surface temperature at $x = 0$ is suddenly lowered to a temperature $T_0 < T_f$, and maintained at that temperature for $t > 0$. Consequently, the liquid starts to freeze at $x = 0$, and the freezing interface (separating in Figure 3.4.15 the solid to its left from the liquid on its right) located

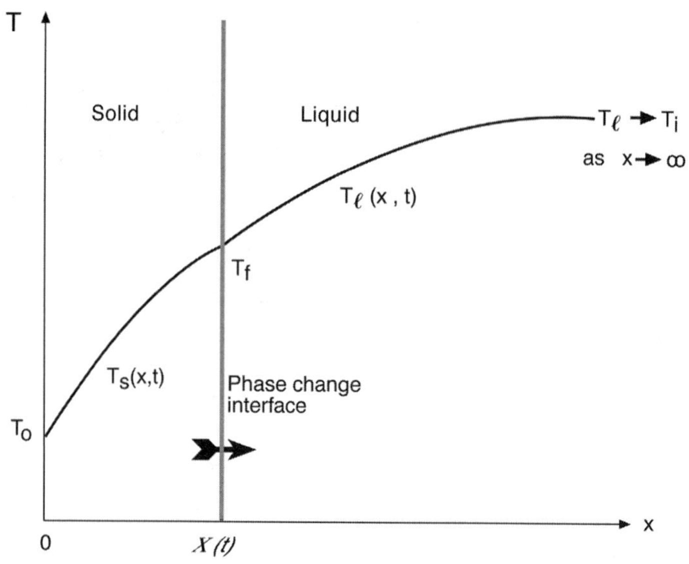

FIGURE 3.4.15 Freezing of semi-inifinite liquid with heat conduction in both phases.

at the position $x = X(t)$ moves gradually to the right (in the positive x direction). We note that in this problem heat is conducted in both phases.

Assuming for simplification that heat transfer is by conduction only — although at least natural convection (Incropera and Viskanta, 1992) and sometimes forced convection and radiation also take place — the governing equations are

In the liquid: The transient heat conduction equation is

$$\frac{\partial T_\ell(x,t)}{\partial t} = \alpha_\ell \frac{\partial^2 T_\ell(x,t)}{\partial x^2} \quad \text{in} \quad X(t) < x < \infty \quad \text{for} \quad t > 0 \tag{3.4.29}$$

$$T_\ell(x,t) = T_i \quad \text{in} \quad x > 0, \quad \text{at} \quad t = 0 \tag{3.4.30}$$

where α_ℓ is the thermal diffusivity of the liquid, with the initial condition and the boundary condition

$$T_\ell(x \to \infty, t) \to T_i \quad \text{for} \quad t > 0 \tag{3.4.31}$$

In the solid: The transient heat conduction equation is

$$\frac{\partial T_s(x,t)}{\partial t} = \alpha_s \frac{\partial^2 T_s(x,t)}{\partial x^2} \quad \text{in} \quad 0 < x < X(t) \quad \text{for} \quad t > 0 \tag{3.4.32}$$

where α_s is the thermal diffusivity of the solid, with the boundary condition

$$T_s(0,t) = T_0 \quad \text{for} \quad t > 0 \tag{3.4.33}$$

The remaining boundary conditions are those of temperature continuity and heat balance at the solid–liquid phase-change interface $X(t)$,

$$T_\ell[X(t)] = T_s[X(t)] = T_f \quad \text{for} \quad t > 0 \tag{3.4.34}$$

$$k_s\left(\frac{\partial T_s}{\partial x}\right)_{[X(t)]} - k_\ell\left(\frac{\partial T_\ell}{\partial x}\right)_{[X(t)]} = \rho h_{s\ell}\frac{dX(t)}{dt} \quad \text{for} \quad t > 0 \tag{3.4.35}$$

where k_s and k_ℓ are the thermal conductivities of the solid and liquid, respectively, ρ is the density (here it is assumed for simplicity to be the same for the liquid and solid), and $h_{s\ell}$ is the latent heat of fusion. The two terms on the left-hand side of Equation (3.4.35) thus represent the conductive heat flux away from the phase-change interface, into the solid at left and the liquid at right, respectively. Energy conservation at the interface requires that the sum of these fluxes leaving the interface be equal to the amount of heat generated due to the latent heat released there, represented by the term on the right-hand side of the equation.

The analytical solution of Equations (3.4.29) to (3.4.35) yields the temperature distributions in the liquid and solid phases,

$$T_\ell(x,t) = T_i - \left(T_i - T_f\right)\frac{\operatorname{erfc}\left(\dfrac{x}{2\sqrt{\alpha_\ell t}}\right)}{\operatorname{erfc}\left(\lambda\sqrt{\alpha_s/\alpha_\ell}\right)} \tag{3.4.36}$$

$$T_s(x,t) = T_0 + \left(T_f - T_0\right)\frac{\operatorname{erfc}\left(\dfrac{x}{2\sqrt{\alpha_s t}}\right)}{\operatorname{erfc}\lambda} \tag{3.4.37}$$

where erf and erfc are the *error function* and the *complementary error function*, respectively, and λ is a constant, obtained from the solution of the equation

$$\frac{e^{\lambda^2}}{\operatorname{erf}\lambda} - \frac{k_\ell}{k_s}\sqrt{\frac{\alpha_s}{\alpha_\ell}}\,\frac{T_i - T_f}{T_f - T_0}\,\frac{e^{(\alpha_s/\alpha_\ell)\lambda^2}}{\operatorname{erfc}\left(\lambda\sqrt{\alpha_s/\alpha_\ell}\right)} = \frac{\lambda\sqrt{\pi}}{\mathrm{Ste}_s} \tag{3.4.38}$$

where Ste is the Stefan number (dimensionless), here defined for the solid as

$$\mathrm{Ste}_s \equiv \frac{c_s\left(T_f - T_0\right)}{h_{s\ell}} \tag{3.4.39}$$

and c_s is the specific heat of the solid. Solutions of Equation (3.4.38) are available for some specific cases in several of the references, and can be obtained relatively easily by a variety of commonly used software packages.

The solution of Equations (3.4.29) to (3.4.35) also gives an expression for the transient position of the freezing interface,

$$X(t) = 2\lambda\left(\alpha_s t\right)^{1/2} \tag{3.4.40}$$

where λ is the solution of Equation 3.4.38, and thus the expression for the rate of freezing, i.e., the velocity of the motion of the solid liquid interface, is

$$\frac{dX(t)}{dt} = \lambda \alpha_s^{1/2} t^{1/2} \tag{3.4.41}$$

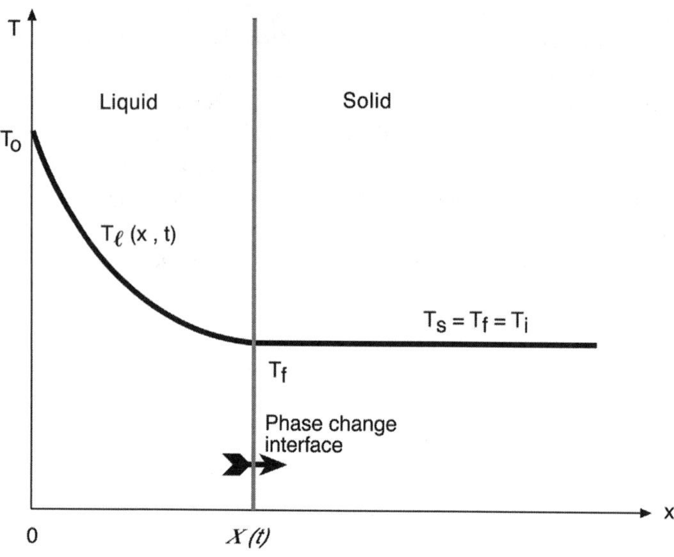

FIGURE 3.4.16 Melting of semi-infinite solid with conduction in the liquid phase only.

For a simple one-dimensional melting example of an analytical solution for melting, consider the semi-infinite solid described in Figure 3.4.16, initially at the fusion temperature T_f. For time $t > 0$ the temperature of the surface (at $x = 0$) is raised to $T_0 > T_f$, and the solid consequently starts to melt there. In this case the temperature in the solid remains constant, $T_s = T_f$, so the temperature distribution needs to be calculated only in the liquid phase. It is assumed that the liquid formed by melting remains motionless and in place. Very similarly to the above-described freezing case, the equations describing this problem are the heat conduction equation

$$\frac{\partial T_\ell(x,t)}{\partial t} = \alpha_\ell \frac{\partial^2 T_\ell(x,t)}{\partial x^2} \quad \text{in} \quad 0 < x < X(t) \quad \text{for} \quad t > 0 \tag{3.4.42}$$

with the initial condition

$$T_\ell(x,t) = T_f \quad \text{in} \quad x > 0, \quad \text{at} \quad t = 0 \tag{3.4.43}$$

the boundary condition

$$T_\ell(0,t) = T_0 \quad \text{for} \quad t > 0 \tag{3.4.44}$$

and the liquid–solid interfacial temperature and heat flux continuity conditions

$$T_\ell[X(t)] = T_f \quad \text{for} \quad t > 0 \tag{3.4.45}$$

$$-k_\ell \left(\frac{\partial T_\ell}{\partial x}\right)_{[X(t)]} = \rho h_{\ell s} \frac{dX(t)}{dt} \quad \text{for} \quad t > 0 \tag{3.4.46}$$

The analytical solution of this problem yields the temperature distribution in the liquid,

$$T_\ell(x,t) = T_0 - \left(T_0 - T_f\right)\frac{\operatorname{erf}\left(\dfrac{x}{2\sqrt{\alpha_\ell t}}\right)}{\operatorname{erf}\lambda'} \quad \text{for} \quad t > 0 \tag{3.4.47}$$

where λ' is the solution of the equation

$$\lambda' e^{\lambda'^2} \operatorname{erf}(\lambda') = \frac{\text{Ste}_\ell}{\sqrt{\pi}} \tag{3.4.48}$$

with Ste_ℓ here defined for the liquid as

$$\text{Ste}_\ell \equiv \frac{c_\ell\left(T_0 - T_f\right)}{h_{s\ell}} \tag{3.4.49}$$

λ' as a function of Ste, for $0 \le \text{Ste} \le 5$, is given in Figure 3.4.17. The interface position is

$$X(t) = 2\lambda'\left(\alpha_\ell t\right)^{1/2} \tag{3.4.50}$$

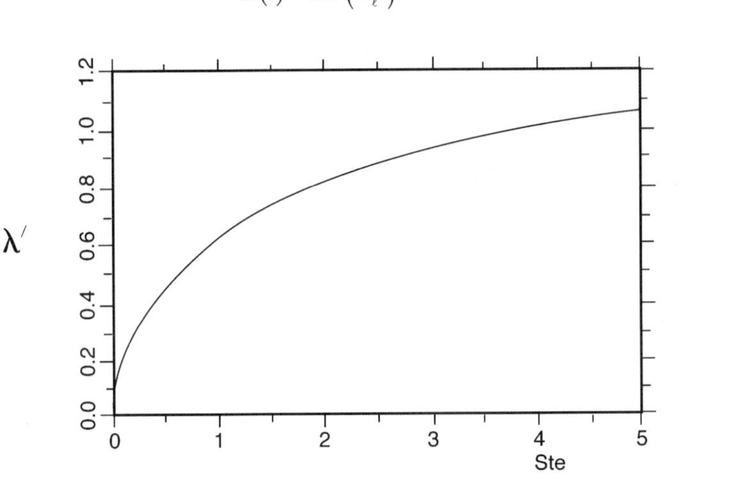

FIGURE 3.4.17 The root λ' of Equation 3.4.48.

The solution of the *freezing* problem under similar conditions, i.e., of a semi-infinite liquid initially at temperature T_f where $T(0,t)$ is abruptly reduced to $T_0 < T_f$ for $t > 0$, is identical to the above if every subscript ℓ is replaced by s and the latent heat $h_{s\ell}$ is replaced by $-h_{s\ell}$.

Example: The temperature of the vertical surface of a large volume of solid paraffin wax used for heat storage, initially at the fusion temperature, $T_i = T_f = 28°\text{C}$, is suddenly raised to 58°C. Any motion in the melt may be neglected. How long would it take for the paraffin to solidify to a depth of 0.1 m? Given properties: $\alpha_\ell = (1.09)\ 10^{-7}$ m²/sec, $\rho_s = \rho_\ell = 814$ kg/m³, $h_{s\ell} = 241$ kJ/kg, $c_\ell = 2.14$ kJ/kg°C. To find the required time we use Equation (3.4.50), in which the value of λ' needs to be determined. λ' is calculated from Equation (3.4.48), which requires the knowledge of Ste_ℓ . From Equation (3.4.49)

$$\text{Ste}_\ell = \frac{(2.14 \text{ kJ/kg°C})(58\text{°C} - 28\text{°C})}{241.2 \text{ kJ/kg}} = 0.266$$

The solution of Equation (3.4.48) as a function of Ste_ℓ is given in Figure 3.4.17, yielding $\lambda \approx 0.4$. By using Equation (3.4.50), the time of interest is calculated by

$$t = \frac{[X(t)]^2}{4\lambda^2\alpha_\ell} = \frac{(0.1 \text{ m})^2}{4(0.4)^2\left[(1.09)10^7 \text{ m}^2/\text{sec}\right]} = (1.43)10^5 \text{ sec} = 39.8 \text{ hr}$$

The axisymmetric energy equation in *cylindrical coordinates*, applicable to both the solid phase and immobile liquid phase (with appropriate assignment of the properties) is

$$\frac{\partial T(r,t)}{\partial t} = \frac{1}{\rho c}\frac{\partial}{\partial r}\left(\frac{k}{r}\frac{\partial T(r,t)}{\partial r}\right) \quad \text{for} \quad t > 0 \tag{3.4.51}$$

and the temperature and heat balance conditions at the solid–liquid phase-change interface $r = R(t)$ are

$$T_\ell[R(t)] = T_s[R(t)] \quad \text{for} \quad t > 0 \tag{3.4.52}$$

$$k_s\left(\frac{\partial T_s}{\partial r}\right)_{R(t)} - k_\ell\left(\frac{\partial T_\ell}{\partial r}\right)_{R(t)} = h_{s\ell}\frac{dR(t)}{dt} \tag{3.4.53}$$

Because of the nature of the differential equations describing nonplanar and multidimensional geometries, analytical solutions are available for only a few cases, such as line heat sources in cylindrical coordinate systems or point heat sources in spherical ones, which have very limited practical application. Other phase-change problems in nonplanar geometries, and in cases when the melt flows during phase change, are solved by approximate and numerical methods (Yao and Prusa, 1989; Alexiades and Solomon, 1993).

Some Approximate Solutions

Two prominent approximate methods used for the solution of melting and freezing problems are the integral method and the *quasi-static* approximation. The integral method is described in Goodman (1964), and only the quasi-static approximation is described here.

To obtain rough estimates of melting and freezing processes quickly, in cases where heat transfer takes place in only one phase, it is assumed in this approximation that effects of sensible heat are negligible relative to those of latent heat (Ste $\rightarrow$ 0), thus eliminating the sensible-heat left-hand side of the energy equations (such as (3.4.29), (3.4.32), and (3.4.51)). This is a significant simplification, since the energy equation then becomes independent of time, and solutions to the steady-state heat conduction problem are much easier to obtain. At the same time, the transient phase-change interface condition (such as Equations (3.4.35) and (3.4.53)) 5 is retained, allowing the estimation of the transient interface position and velocity. This is hence a quasi-static approximation, and its use is shown below.

We emphasize that these are just approximations, without full information on the effect of specific problem conditions on the magnitude of the error incurred when using them. In fact, in some cases, especially with a convective boundary condition, they may produce incorrect results. It is thus necessary to examine the physical viability of the results, such as overall energy balances, when using these approximations.

All of the examples are for melting, but freezing problems have the same solutions when the properties are taken to be those of the solid and $h_{s\ell}$ is replaced everywhere by $-h_{s\ell}$. It is assumed here that the problems are one-dimensional, and that the material is initially at the fusion temperature T_f.

Examples of the Quasi-Static Approximation for Cartesian Coordinate Geometries. Given a semi-infinite solid (Figure 3.4.16), on which a *time-dependent temperature* $T_0(t) > T_f$ *is imposed at* $x = 0$, the above-described quasi-static approximation of Equations (3.4.42) to (3.4.46) easily yields the solution

$$X(t) = \left[2 \frac{k_\ell}{\rho h_{s\ell}} \int_0^t \left[T_0(t) - T_f \right] dt \right]^{1/2} \quad \text{for} \quad t \geq 0 \tag{3.4.54}$$

$$T_\ell(x,t) = T_0(t) - \left[T_0(t) - T_f \right] \frac{x}{X(t)} \quad \text{in} \quad 0 \leq x \leq X(t) \quad \text{for} \quad t \geq 0 \tag{3.4.55}$$

The heat flux needed for melting, $q(x,t)$, can easily be determined from the temperature distribution in the liquid (Equation 3.4.55), which is linear because of the steady-state form of the heat conduction equation in this quasi-static approximation, so that

$$q(x,t) = -k_\ell \frac{dT_\ell(x,t)}{dx} = k_\ell \frac{T_0(t) - T_f}{X(t)} \tag{3.4.56}$$

For comparison of this approximate solution to the exact one (Equations (3.4.47) and (3.4.50)), consider the case where $T_0(t) = T_0 =$ constant. Rearranging to use the Stefan number, Equations (3.4.54) and (3.4.55) become

$$X(t) = 2\left(\text{Ste}_\ell/2\right)^{1/2} \left(\alpha_\ell t\right)^{1/2} \quad \text{for} \quad t > 0 \tag{3.4.57}$$

$$T(x,t) = T_0 - \left[T_0 - T_f \right] \frac{x \Big/ \left[2\left(\alpha_\ell t\right)^{1/2} \right]}{\left(\text{Ste}_\ell/2\right)^{1/2}} \quad \text{in} \quad 0 \leq x \leq X(t) \quad \text{for} \quad t \geq 0 \tag{3.4.58}$$

It is easy to show that λ' in the exact solution (Equation 3.4.48) approaches the value $(\text{Ste}_\ell/2)^{1/2}$ when $\text{Ste}_\ell \to 0$, and that otherwise $\lambda' < (\text{Ste}_\ell/2)^{1/2}$. The approximate solution is therefore indeed equal to the exact one when $\text{Ste}_\ell \to 0$, and it otherwise overestimates the values of both $X(t)$ and $T(x,t)$. While the errors depend on the specific problem, they are confined to about 10% in the above-described case (Alexiades and Solomon, 1993).

For the same melting problem but with the *boundary condition of an imposed time-dependent heat flux* $q_0(t)$,

$$-k_\ell \left(\frac{dT_\ell}{dx} \right)_{0,t} = q_0(t) \quad \text{for} \quad t > 0 \tag{3.4.59}$$

the quasi-static approximate solution is

$$X(t) \equiv \frac{1}{\rho h_{s\ell}} \int_0^t q_0(t)\, dt \quad \text{for} \quad t > 0 \tag{3.4.60}$$

$$T_\ell(x,t) = T_f + \frac{q_0}{k_\ell}\left[\frac{q_0}{\rho h_{s\ell}} t - x\right] \quad \text{in} \quad 0 \le x \le X(t) \quad \text{for} \quad t > 0 \tag{3.4.61}$$

For the same case if the *boundary condition is a convective heat flux* from an ambient fluid at the transient temperature $T_a(t)$, characterized by a heat transfer coefficient $\bar{h}$,

$$-k_\ell\left(\frac{dT_\ell}{dx}\right)_{0,t} = \bar{h}\left[T_a(t) - T_\ell(0,t)\right] \quad \text{for} \quad t \ge 0 \tag{3.4.62}$$

the quasi-static approximate solution is

$$X(t) = -\frac{k_\ell}{\bar{h}} + \left\{\left(\frac{k_\ell}{\bar{h}}\right)^2 + 2\frac{k_\ell}{\rho h_{s\ell}}\int_0^t \left[T_a(t) - T_f\right] dt\right\}^{1/2} \quad \text{for} \quad t \ge 0 \tag{3.4.63}$$

$$T_\ell(x,t) = T_f(t)\left[T_a(t) - T_f\right]\frac{\bar{h}\left[X(t) - x\right]}{\bar{h}X(t) + k_\ell} \quad \text{in} \quad 0 \le x \le X(t) \quad \text{for} \quad t > 0 \tag{3.4.64}$$

Examples of the Quasi-Static Approximation for Cylindrical Coordinate Geometries. It is assumed in these examples that the cylinders are very long and that the problems are axisymmetric. Just as in the Cartesian coordinate case, the energy Equation (3.4.51) is reduced by the approximation to its steady-state form. Here

$$T_\ell(r_i,t) = T_0(t) > T_f \quad \text{for} \quad t > 0 \tag{3.4.65}$$

Consider the *outward-directed melting* of a hollow cylinder due to a temperature imposed at the internal radius r_i. The solution is

$$T_\ell(r,t) = T_f + \left[T_0(t) - T_f\right]\frac{\ln\left[r/R(t)\right]}{\ln\left[r_i/R(t)\right]} \quad \text{in} \quad r_i \le r \le R(t) \quad \text{for} \quad t > 0 \tag{3.4.66}$$

and the transient position of the phase front, $R(t)$, can be calculated from the transcendental equation

$$2R(t)^2 \ln\frac{R(t)}{r_i} = R(t)^2 - r_i^2 + \frac{4k_\ell}{\rho h_{s\ell}}\int_0^t \left[T_0(t) - T_f\right] dt \tag{3.4.67}$$

If the melting for the same case occurs due to the imposition of a *heat flux q_0 at r_p*

$$-k_\ell\left(\frac{dT_\ell}{dx}\right)_{r_i,t} = q_0(t) > 0 \quad \text{for} \quad t > 0 \tag{3.4.68}$$

the solution is

$$T_\ell(r,t) = T_f - \frac{q_0(t) r_i}{k_\ell}\ln\frac{r}{R(t)} \quad \text{in} \quad r_i \le r \le R(t) \quad \text{for} \quad t > 0 \tag{3.4.69}$$

$$R(t) = \left(r_i^2 + 2\frac{r_i}{\rho h_{s\ell}} \int_0^t q_0(t)\, dt \right)^{1/2} \quad \text{for} \quad t > 0 \tag{3.4.70}$$

If the melting for the same case occurs due to the imposition of a *convective heat flux from a fluid at the transient temperature* $T_a(t)$, *with a heat transfer coefficient* $\bar{h}$, *at* r_i

$$-k_\ell \left(\frac{dT_\ell}{dr} \right)_{r_i,t} = \bar{h}\left[T_a(t) - T_f(r_i,t)\right] > 0 \quad \text{for} \quad t > 0 \tag{3.4.71}$$

The solution is

$$T_\ell(r,t) = T_f + \left[T_a(t) - T_f\right] \frac{\ln\left[r/R(t)\right]}{\ln\left[r_i/R(t)\right] - k_\ell/\bar{h} r_i} \quad \text{in} \quad r_i \le r \le R(t) \quad \text{at} \quad t > 0 \tag{3.4.72}$$

with $R(t)$ calculated from the transcendental equation

$$2R(t)^2 \ln \frac{R(t)}{r_i} = \left(1 - \frac{2k_\ell}{\bar{h} r_i} \right)\left[R(t)^2 - r_i^2\right] + \frac{4k_\ell}{\rho h_{s\ell}} \int_0^t \left[T_a(t) - T_f\right] dt \tag{3.4.73}$$

The solutions for *inward melting* of a cylinder, where heating is applied at the outer radius r_o, are the same as the above-described ones for the outward-melting cylinder, if the replacements $r_i \to r_o$, $q_0 \to -q_0$, and $\bar{h} \to -\bar{h}$ are made. If such a cylinder is not hollow, then $r_i = 0$ is used.

Estimation of Freezing and Melting Time

There are a number of approximate formulas for estimating the freezing and melting times of different materials having a variety of shapes. The American Society of Heating, Refrigerating, and Air-Conditioning Engineers (ASHRAE) provides a number of such approximations for estimating the freezing and thawing times of foods (ASHRAE, 1993). For example, if it can be assumed that the freezing or thawing occurs at a single temperature, the time to freeze or thaw, t_f, for a body that has shape parameters P and R (described below) and thermal conductivity k, initially at the fusion temperature T_f, and which is exchanging heat via heat transfer coefficient $\bar{h}$ with an ambient at the constant T_a, can be approximated by Plank's equation

$$t_f = \frac{h_{s\ell}\rho}{\left|T_f - T_a\right|} \left(\frac{Pd}{\bar{h}} + \frac{Rd^2}{k} \right) \tag{3.4.74}$$

where d is the diameter of the body if it is a cylinder or a sphere, or the thickness when it is an infinite slab, and where the shape coefficients P and R for a number of body forms are given in Table 3.4.2. Shape coefficients for other body forms are also available. To use Equation (3.4.74) for freezing, k and ρ should be the values for the food in its frozen state. In thawing, they should be for the unfrozen food. Other simple approximations for melting and thawing times can be found in Cleland et al. (1987).

Example of Using Plank's Equation (3.4.74) for Estimating Freezing Time. Estimate the time needed to freeze a fish, the shape of which can be approximated by a cylinder 0.5 m long having a diameter of 0.1 m. The fish is initially at its freezing temperature, and during the freezing process it is surrounded by air at $T_a = -25°C$, with the cooling performed with a convective heat transfer coefficient $\bar{h} = 68$ W/m² K. For the fish, $T_f = -1°C$, $h_{s\ell} = 200$ kJ/kg, $\rho_s = 992$ kg/m³, and ${}_sk_s = 1.35$ W/m K.

TABLE 3.4.2 Shape Factors for Equation (3.4.74)

Forms	P	R
Slab	1/2	1/8
Cylinder	1/4	1/16
Sphere	1/6	1/24

From ASHRAE, in *Fundamentals*, ASHRAE, Atlanta, 1993, chap. 29. With permission.

By using Table 3.4.2, the geometric coefficients for the cylindrical shape of the fish are $P = {}^1/_2$ and $R = {}^1/_{16}$, while d is the cylinder diameter, = 0.1 m. Substituting these values into Equation (3.4.74) gives

$$t_f = \frac{200{,}000 \cdot 992}{-1-(-25)}\left(\frac{1/4(0.1)}{68} + \frac{1/16(0.1)^2}{1.35}\right) = 6866 \text{ sec} = 1.9 \text{ hr}$$

In fact, freezing or melting of food typically takes place over a range of temperatures, and approximate Plank-type formulas have been developed for various specific foodstuffs and shapes to represent reality more closely than Equation (3.4.74) (ASHRAE, 1993).

Alexiades and Solomon (1993) provide several easily computable approximate equations for estimating the time needed to melt a simple solid body initially at the fusion temperature T_f. It is assumed that conduction occurs in one phase (the liquid) only, that the problems are axi- and spherically symmetric for cylindrical and spherical bodies, respectively, and that the melting process for differently shaped bodies can be characterized by a single geometric parameter, r, in the body domain $0 \le r \le L$, using a shape factor, ω, defined by

$$\omega = \frac{LA}{V} - 1 \tag{3.4.75}$$

where A is the surface area across which the heat is transferred into the body and V is the body volume, to account for the specific body shape:

$$\omega = \begin{cases} 0 & \text{for a slab insulated at one end} \\ 1 & \text{for a cylinder} \\ 2 & \text{for a sphere} \end{cases} \tag{3.4.76}$$

$0 \le \omega \le 2$ always, and ω may be assigned appropriate values for shapes intermediate between the slab, cylinder, and sphere. For example, a football-shaped body, somewhere between a cylinder and sphere, may be assigned $\omega = 1.5$, and a short cylinder with a large diameter-to-height ratio may have $\omega = 0.5$.

For the case where the temperature $T_0 > T_f$ is imposed on the boundary at $t = 0$, the melt time, t_m can be estimated by

$$t_m = \frac{L^2}{2\alpha_\ell (1+\omega)\mathrm{Ste}_\ell}\left[1 + \left(0.25 + 0.17\omega^{0.7}\right)\mathrm{Ste}_\ell\right] \tag{3.4.77}$$

valid for $0 \le \mathrm{Ste}_\ell \le 4$.

If the *heat input is convective*, with a heat transfer coefficient $\bar{h}$ from a fluid at temperature T_a, the approximate melt time is

$$t_m = \frac{L^2}{2\alpha_\ell(1+\omega)\text{Ste}_\ell}\left[1+\frac{2k_\ell}{\bar{h}L}+\left(0.25+0.17\omega^{0.7}\right)\text{Ste}_\ell\right] \tag{3.4.78}$$

valid for $0 \le \text{Ste}_\ell \le 4$ and $\bar{h}L/k_\ell \ge 0.1$, and the temperature, $T(0,t)$, of the surface across which the heat is supplied can be estimated from the implicit time-temperature relationship:

$$t = \frac{\rho c_\ell k_\ell}{2\bar{h}^2\text{Ste}_\ell}\left[1.18\text{Ste}_\ell\left(\frac{T(0,t)-T_f}{T_a-T(0,t)}\right)^{1.83}+\left(\frac{T_a-T_f}{T_a-T(0,t)}\right)^2-1\right] \tag{3.4.79}$$

Both equations (3.4.78) and (3.4.79) are claimed to be accurate within 10%.

The suitability of using several simplified analytical solutions for the estimation of freezing and melting times for more-realistic problems was assessed by Dilley and Lior (1986).

Defining Terms

Eutectic concentration: A concentration of a component of a multicomponent liquid at which the liquid would upon freezing form a solid containing the same concentration, and at which the freezing process is completed at a single temperature.

Mushy zone: The zone composed of both liquid and solid, bounded by the liquidus and solidus curves, in a freezing or melting process.

References

Alexiades, V. and Solomon, A.D. 1993. *Mathematical Modeling of Melting and Freezing Processes,* Hemisphere Publishing, Washington, D.C.

ASHRAE (American Society of Heating, Refrigerating, and Air-Conditioning Engineers). 1993. Cooling and freezing times of foods, in *Fundamentals*, ASHRAE, Atlanta, GA, chap. 29.

ASHRAE (American Society of Heating, Refrigerating, and Air-Conditioning Engineers). 1990. *Refrigeration*, ASHRAE, Atlanta, GA.

Cheng, K.C. and Seki, N., Eds. 1991. *Freezing and Melting Heat Transfer in Engineering,* Hemisphere Publishing, Washington, D.C.

Cleland, D.J., Cleland, A.C., and Earle, R.L. 1987. Prediction of freezing and thawing times for multi-dimensional shapes by simple formulae: Part 1, regular shapes; Part 2, irregular shapes. *Int. J. Refrig.,* 10, 156–166; 234–240.

DeWinter, F. 1990. Energy storage of solar systems; in *Solar Collectors, Energy Storage, and Materials,* MIT Press, Cambridge, MA, Section II.

Dilley, J.F. and Lior, N. 1986. The evaluation of simple analytical solutions for the prediction of freeze-up time, freezing, and melting. *Proc. 8th International Heat Transfer Conf.*, 4, 1727–1732, San Francisco.

Flemings, M.C. 1974. *Solidification Processes*, McGraw-Hill, New York.

Fukusako, S. and Seki, N. 1987. Fundamental aspects of analytical and numerical methods on freezing and melting heat-transfer problems, in *Annual Review of Numerical Fluid Mechanics and Heat Transfer*, Vol. 1, T.C. Chawla, Ed., Hemisphere Publishing, Washington, D.C., chap. 7, 351–402.

Goodman, T.R. 1964. Application of integral methods to transient nonlinear heat transfer, in *Advances in Heat Transfer*, Vol. 1, T.F. Irvine and J.P. Hartnett, Eds., Academic Press, San Diego, 51–122.

Hayashi, Y. and Kunimine, K. 1992. Solidification of mixtures with supercooling, in *Heat and Mass Transfer in Materials Processing*, I. Tanasawa and N. Lior, Eds., Hemisphere Publishing, New York, 265–277.

Incropera, F.P. and Viskanta, R. 1992. Effects of convection on the solidification of binary mixtures, in *Heat and Mass Transfer in Materials Processing*, I. Tanasawa and N Lior, Eds., Hemisphere Publishing, New York, 295–312.

Lunardini, V.J. 1981. *Heat Transfer in Cold Climate*, Van Nostrand-Reinhold, Princeton, NJ.

Poulikakos, D. 1994. *Conduction Heat Transfer,* Prentice-Hall, Englewood Cliffs, NJ.

Rubinsky, B. and Eto, T.K. 1990. Heat transfer during freezing of biological materials, in *Annual Review of Heat Transfer*, Vol. 3, C.L. Tien, Eds., Hemisphere Publishing, Washington, D.C., chap. 1, 1–38.

Tanasawa, I. and Lior, N., Ed. 1992. *Heat and Mass Transfer in Materials Processing*, Hemisphere Publishing, New York.

Yao, L.S. and Prusa, J. 1989. Melting and freezing, in *Advances in Heat Transfer*, Vol. 19, J.P. Hartnett and T.F. Irvine, Eds., Academic Press, San Diego, 1–95.

Further Information

Many textbooks on heat transfer (some listed in the References section above) contain material about melting and freezing, and many technical journals contain articles about this subject. Some of the major journals, classified by orientation, are

General: *ASME Journal of Heat Transfer, International Journal of Heat & Mass Transfer, Numerical Heat Transfer, Canadian Journal of Chemical Engineering, AIChE Journal*

Refrigeration: *Transactions of the ASHRAE, International Journal of Refrigeration, Referigeration, Journal of Food Science, Bulletin of the International Institute of Refrigeration*

Manufacturing: *ASME Journal of Engineering for Industry, Journal of Crystal Growth, Materials Science and Engineering A*

Geophysical, climate, cold regions engineering: *Limnology and Oceanography, Journal of Geophysical Research, ASCE Journal of Cold Regions Engineering, Cold Regions Science and Technology*

Medical: *Cryobiology, ASME Journal of Biomechanical Engineering, Journal of General Physiology*

3.5 Mass Transfer

Anthony F. Mills

Introduction

Mass transfer may occur in a gas mixture, a liquid solution, or a solid solution. There are several physical mechanisms that can transport a chemical species through a phase and transfer it across phase boundaries. The two most important mechanisms are ordinary diffusion and convection. Mass diffusion is analogous to heat conduction and occurs whenever there is a gradient in the concentration of a species. Mass convection is essentially identical to heat convection: a fluid flow that transports heat may also transport a chemical species. The similarity of mechanisms of heat transfer and mass transfer results in the mathematics often being identical, a fact that can be exploited to advantage. But there are some significant differences between the subjects of heat and mass transfer. One difference is the much greater variety of physical and chemical processes that require mass transfer analysis. Another difference is the extent to which the essential details of a given process may depend on the particular chemical system involved, and on temperature and pressure.

In the next subsection, concentrations, velocities, and fluxes are defined, and special attention is paid to phase interfaces where the concentration of a chemical species is almost always discontinuous. Fick's law of ordinary diffusion is introduced in the third section, where other diffusion phenomena are also discussed. The fourth section presents various forms of the species conservation equation. Results for diffusion in a stationary medium are given in the fifth section, and include steady diffusion across a plane wall, transient diffusion in a semi-infinite solid, and diffusion in a porous catalyst. Results for diffusion in a moving medium are given in the sixth section, and the Stefan flow is introduced for diffusion with one component stationary. Also considered are particle combustion, droplet evaporation, and combustion of a volatile liquid hydrocarbon fuel droplet. The last section deals with mass convection. Low mass transfer rate theory is presented and how to exploit the analogy between convective heat and mass transfer is shown. Particular attention is given to situations involving simultaneous heat and mass transfer associated with evaporation or condensation. The section closes by presenting high mass transfer rate theory for convection, and gives engineering calculation methods for boundary layer flows that account for variable property effects.

Concentrations, Velocities, and Fluxes

Definitions of Concentrations

In a gas mixture, or liquid or solid solution, the local *concentration* of a mass species can be expressed in a number of ways. The *number density* of species i in a mixture or solution of n species is defined as

$$\begin{aligned}\text{Number density of species } i &\equiv \text{Number of molecules of } i \text{ per unit volume}\\ &\equiv \mathcal{N}_i \text{ molecules/m}^3\end{aligned} \tag{3.5.1}$$

Alternatively, if the total number of molecules of all species per unit volume is denoted as $\mathcal{N}$, then we define the *number fraction* of species i as

$$n_i \equiv \frac{\mathcal{N}_i}{\mathcal{N}}; \qquad \mathcal{N} = \sum \mathcal{N}_i \tag{3.5.2}$$

where the summation is over all species present, $i = 1,2,\ldots,n$. Equations (3.5.1) and (3.5.2) describe *microscopic* concepts and are used, for example, when the kinetic theory of gases is used to describe transfer processes.

Whenever possible, it is more convenient to treat matter as a continuum. Then the smallest volume considered is sufficiently large for macroscopic properties such as pressure and temperature to have their usual meanings. For this purpose we also require *macroscopic* definitions of concentration. First, on a mass basis,

$$\begin{aligned} \textit{Mass concentration} \text{ of species } i &\equiv \text{partial density of species } i \\ &\equiv \rho_i \ \text{kg/m}^3 \end{aligned} \tag{3.5.3}$$

The total mass concentration is the total mass per unit volume, that is, the density $\rho = \sum \rho_i$. The *mass fraction* of species i is defined as

$$m_i = \frac{\rho_i}{\rho} \tag{3.5.4}$$

Second, on a molar basis,

$$\begin{aligned} \textit{Molar concentration} \text{ of species } i &\equiv \text{number of moles of } i \text{ per unit volume} \\ &\equiv c_i \ \text{kmol/m}^3 \end{aligned} \tag{3.5.5}$$

If M_i (kg/kmol) is the molecular weight of species i, then

$$c_i = \frac{\rho_i}{M_i} \tag{3.5.6}$$

The total molar concentration is the molar density $c = \sum c_i$. The *mole fraction* of species i is defined as

$$x_i \equiv \frac{c_i}{c} \tag{3.5.7}$$

A number of important relations follow directly from these definitions. The mean molecular weight of the mixture of solution is denoted M and may be expressed as

$$M = \frac{\rho}{c} = \sum x_i M_i \tag{3.5.8a}$$

or

$$\frac{1}{M} = \sum \frac{m_i}{M_i} \tag{3.5.8b}$$

There are summation rules

$$\sum m_i = 1 \tag{3.5.9a}$$

$$\sum x_i = 1 \tag{3.5.9b}$$

It is often necessary to have the mass fraction of species *i* expressed explicitly in terms of mole fractions and molecular weights; this relation is

$$m_i = \frac{x_i M_i}{\sum x_j M_j} = x_i \frac{M_i}{M} \tag{3.5.10a}$$

and the corresponding relation for the mole fraction is

$$x_i = \frac{m_i / M_i}{\sum m_j / M_j} = m_i \frac{M}{M_i} \tag{3.5.10b}$$

Dalton's law of partial pressures for an ideal gas mixture states that

$$P = \sum P_i, \quad \text{where} \quad P_i = \rho_i R_i T \tag{3.5.11}$$

Dividing partial pressure by total pressure and substituting $R_i = \mathcal{R}/M_i$ gives

$$\frac{P_i}{P} = \frac{\rho_i}{M_i}\frac{\mathcal{R}T}{P} = c_i \frac{\mathcal{R}T}{P} = x_i \frac{c\mathcal{R}T}{P} = x_i \tag{3.5.12}$$

Thus, for an ideal gas mixture, the mole fraction and partial pressure are equivalent measures of concentration (as also is the number fraction).

A commonly used specification of the composition of dry air is 78.1% N_2, 20.9% O_2, and 0.9% Ar, by volume. (The next largest component is CO_2, at 0.3%.) Since equal volumes of gases contain the same number of moles, specifying composition on a volume basis is equivalent to specifying mole fractions, namely,

$$x_{N_2} = 0.781; \qquad x_{O_2} = 0.209; \qquad x_{Ar} = 0.009$$

The corresponding mass fractions are calculated to be

$$m_{N_2} = 0.755; \qquad m_{O_2} = 0.231; \qquad m_{Ar} = 0.014$$

Concentrations at Interfaces

Although temperature is continuous across a phase interface, concentrations are usually discontinuous. In order to define clearly concentrations at interfaces, we introduce imaginary surfaces, denoted *u* and *s*, on both sides of the real interface, each indefinitely close to the interface, as shown in Figure 3.5.1 for water evaporating into an airstream. Thus, the liquid-phase quantities at the interface are subscripted *u*, and gas-phase quantities are subscripted *s*. If we ignore the small amount of air dissolved in the water, $x_{H_2O,u} = 1$. Notice that the subscript preceding the comma denotes the chemical species, and the subscript following the comma denotes location. To determine $x_{H_2O,s}$ we make use of the fact that, except in extreme circumstances, the water vapor and air mixture at the *s*-surface must be in thermodynamic equilibrium with water at the *u*-surface. Equilibrium data for this system are found in conventional steam tables: the saturation vapor pressure of steam at the water temperature, T_s, ($T_s = T_u$), is the required partial pressure $P_{H_2O,s}$. With the total pressure *P* known, $x_{H_2O,s}$ is calculated as $P_{H_2O,s}/P$. If $m_{H_2O,s}$ is required, Equation (3.5.10a) is used.

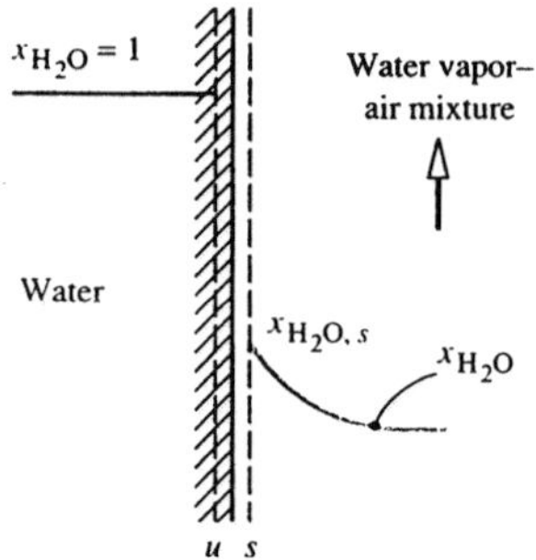

FIGURE 3.5.1 Concentrations at a water-air interface.

For example, at T_s = 320 K, the saturation vapor pressure is obtained from steam tables as 0.10535 × 10^5 Pa. If the total pressure is 1 atm = 1.0133 × 10^5,

$$x_{H_2O,s} = \frac{0.10535 \times 10^5}{1.0133 \times 10^5} = 0.1040$$

$$m_{H_2O,s} = \frac{(0.1040)(18)}{(0.1040)(18) + (1 - 0.1040)(29)} = 0.06720$$

For a gas or solid dissolving in a liquid, equilibrium data are often referred to simply as solubility data, found in chemistry handbooks. Many gases are only sparingly soluble, and for such dilute solutions solubility data are conveniently represented by *Henry's law*, which states that the mole fraction of the gas at the *s*-surface is proportional to its mole fraction in solution at the *u*-surface, the constant of proportionality being the *Henry number*, He. For species *i*,

$$x_{i,s} = \text{He}_i x_{i,u} \tag{3.5.13}$$

The Henry number is inversely proportional to total pressure and is also a function of temperature. The product of Henry number and total pressure is the *Henry constant*, C_{He}, and for a given species is a function of temperature only:

$$\text{He}_i P = C_{\text{He}_i}(T) \tag{3.5.14}$$

Solubility data are given in Table 3.5.1.

TABLE 3.5.1 Henry Constants C_{He} for Dilute Aqueous Solutions at Moderate Pressures ($P_{i,s}/x_{i,u}$ in atm, or in bar = 10^5 Pa, within the accuracy of the data).

Solute	290 K	300 K	310 K	320 K	330 K	340 K
H_2S	440	560	700	830	980	1,140
CO_2	1,280	1,710	2,170	2,720	3,220	—
O_2	38,000	45,000	52,000	57,000	61,000	65,000
H_2	67,000	72,000	75,000	76,000	77,000	76,000
CO	51,000	60,000	67,000	74,000	80,000	84,000
Air	62,000	74,000	84,000	92,000	99,000	104,000
N_2	16,000	89,000	101,000	110,000	118,000	124,000

For example, consider absorption of carbon dioxide from a stream of pure CO_2 at 2 bar pressure into water at 310 K. From Table 3.5.1, C_{He} = 2170 bar; thus

$$\text{He}_{CO_2} = \frac{2170}{2} = 1085; \qquad x_{CO_{2,u}} = \frac{1}{1085} = 9.22 \times 10^{-4}$$

Dissolution of gases into metals is characterized by varied and rather complex interface conditions. Provided temperatures are sufficiently high, hydrogen dissolution is reversible (similar to CO_2 absorption into water); hence, for example, titanium-hydrogen solutions can exist only in contact with a gaseous hydrogen atmosphere. As a result of hydrogen going into solution in atomic form, there is a characteristic square root relation

$$m_{H_2,u} \alpha P_{H_2,s}^{1/2}$$

The constant of proportionality is strongly dependent on temperature, as well as on the particular titanium alloy: for Ti-6Al-4V alloy it is twice that for pure titanium. In contrast to hydrogen, oxygen dissolution in titanium is irreversible and is complicated by the simultaneous formation of a rutile (TiO_2) scale on the surface. Provided some oxygen is present in the gas phase, the titanium-oxygen *phase diagram* (found in a metallurgy handbook) shows that $m_{O_2,u}$ in alpha-titanium is 0.143, a value essentially independent of temperature and O_2 partial pressure. Dissolution of oxygen in zirconium alloys has similar characteristics to those discussed above for titanium.

All the preceding examples of interface concentrations are situations where thermodynamic equilibrium can be assumed to exist at the interface. Sometimes thermodynamic equilibrium does not exist at an interface: a very common example is when a chemical reaction occurs at the interface, and temperatures are not high enough for equilibrium to be attained. Then the concentrations of the reactants and products at the *s*-surface are dependent both on the rate at which the reaction proceeds — that is, the *chemical kinetics* — as well as on mass transfer considerations.

Definitions of Fluxes and Velocities

The mass (or molar) flux of species i is a vector quantity giving the mass (or moles) of species i that pass per unit time through a unit area perpendicular to the vector (Figure 3.5.2). We denote the absolute mass and molar fluxes of species i, that is, relative to stationary coordinate axes, as $\boldsymbol{n}_i$ (kg/m²sec) and $\boldsymbol{N}_i$ (kmol/m²sec), respectively. The absolute mass flux of the mixture (mass velocity) is

$$\boldsymbol{n} = \sum \boldsymbol{n}_i \tag{3.5.15}$$

and the local mass-average velocity is

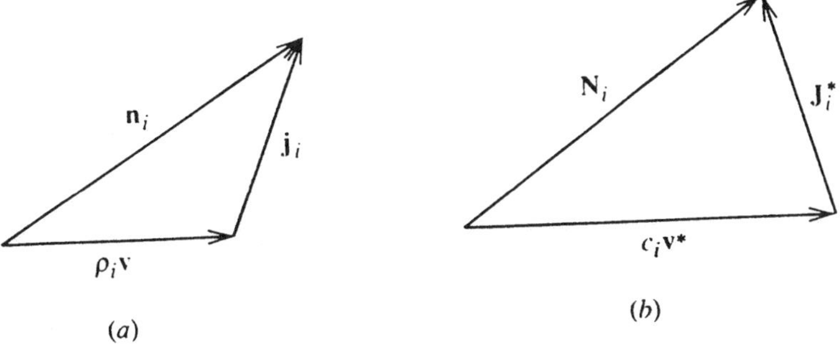

FIGURE 3.5.2 Flux vectors: (a) mass basis, (b) molar basis.

$$\boldsymbol{v} = \frac{\boldsymbol{n}}{\rho} \text{ m/sec} \tag{3.5.16}$$

The velocity v is the velocity that would be measure by a Pitot tube and corresponds to the velocity used in considering pure fluids. On a molar basis, the absolute molar flux of the mixture is

$$\boldsymbol{N} = \sum \boldsymbol{N}_i \tag{3.5.17}$$

and the local molar-average velocity is

$$\boldsymbol{v}^* = \frac{\boldsymbol{N}}{c} \text{ m/sec} \tag{3.5.18}$$

The absolute fluxes of species i have two components. On a mass basis we write

$$\boldsymbol{n}_i = \rho_i \boldsymbol{v} + \boldsymbol{j}_i \tag{3.5.19}$$

where $\rho_i \boldsymbol{n}$ is transport of species i by bulk motion of the fluid at velocity v and is the *convective* component. Thus, $\boldsymbol{j}_i$ is transport of species i relative to the mass average velocity; it is called the *diffusive* component because most commonly it is due to ordinary (concentration) diffusion of the species. On a molar basis the corresponding relation is

$$\boldsymbol{N}_i = c_i \boldsymbol{v}^* + \boldsymbol{J}_i^* \tag{3.5.20}$$

Some important relations follow from these definitions:

$$\sum \boldsymbol{j}_i = \sum \boldsymbol{J}_i^* = 0 \tag{3.5.21}$$

$$\boldsymbol{N}_i = \frac{\boldsymbol{n}_i}{M_i} \tag{3.5.22}$$

$$\boldsymbol{n}_i = \rho_i \boldsymbol{v} + \boldsymbol{j}_i = m_i \sum \boldsymbol{n}_i + \boldsymbol{j}_i \tag{3.5.23a}$$

$$\boldsymbol{N}_i = c_i \boldsymbol{v}^* + \boldsymbol{J}_i^* = x_i \sum \boldsymbol{N}_i + \boldsymbol{J}_i^* \tag{3.5.23b}$$

Mechanisms of Diffusion

Ordinary Diffusion

Fick's law of ordinary diffusion is a linear relation between the rate of diffusion of a chemical species and the local concentration gradient of that species. It is exact for a binary gas mixture, for which the kinetic theory of gases gives

$$\boldsymbol{j}_1 = -\rho \mathcal{D}_{12} \nabla m_1 \text{ kg/m}^2 \text{ sec} \tag{3.5.24a}$$

on a mass basis, and

$$J_1^* = -c\mathcal{D}_{12}\nabla x_1 \text{ kg/m}^2 \text{ sec} \tag{3.5.24b}$$

on a molar basis; $\mathcal{D}_{12}$ (m^2/sec) is the binary diffusion coefficient (or mass diffusivity), and $\mathcal{D}_{21}$ = $\mathcal{D}_{21}$. Equations (3.5.24a) and (3.5.24b) are mathematically equivalent; however, notice that it is incorrect to write

$$j_i = -\mathcal{D}_{12}\nabla\rho_1 \tag{3.5.25}$$

since $\nabla\rho_1 \neq \rho \nabla m_1$ in general. Fick's law in the form of Equations (3.5.24a) and (3.5.24b) is also valid for dilute liquid and solid solutions, for which it is often possible to assume ρ (or c) constant, and then Equation (3.5.25) or its molar equivalent are good approximations.

Ordinary diffusion in multicomponent systems is described by the Stefan–Maxwell equations (Hirschfelder et al., 1954). These equations are difficult to use for engineering analysis. In gas mixtures containing species that do not have widely varying molecular weights, it is possible to model approximately the diffusion process by using an effective binary diffusion coefficient in Fick's law. This coefficient is a suitable average over the species in the mixture, and may be calculated from

$$\mathcal{D}_{1m} = \frac{(1-x_1)}{\sum_{i=2}^{n}(x_i/\mathcal{D}_{1i})}; \qquad x_1 \ll 1 \tag{3.5.26}$$

This equation works well for most mixtures of combustion gases (except those containing appreciable concentrations of H or H_2).

Binary diffusion coefficients at 300 K are of the order of 10^{-5} m^2/sec in gases at 1 atm, 10^{-9} m^2/sec in aqueous solutions, and 10^{-10} to 10^{-13} m^2/sec in solids. However, the product $\rho\mathcal{D}$ or $(c\mathcal{D})$ is, at most, one order of magnitude different for gases and liquids. Data for diffusion coefficients may be found in Tables 3.5.2 through 3.5.5.

Molecules in a gas mixture, and in a liquid or solid solution, can diffuse by mechanisms other than ordinary diffusion governed by Fick's law. *Thermal diffusion* is diffusion due to a temperature gradient and is often called the *Soret effect*. Thermal diffusion is usually negligible compared with ordinary diffusion, unless the temperature gradient is very large. However, there are some important processes that depend on thermal diffusion, the most well known being the large-scale separation of uranium isotopes. *Pressure diffusion* is diffusion due to a pressure gradient and is also usually negligible unless the pressure gradient is very large. Pressure diffusion is the principle underlying the operation of a centrifuge. Centrifuges are used to separate liquid solutions and are increasingly being used to separate gaseous isotopes as well. *Forced diffusion* results from an external force field acting on a molecule. Gravitational force fields do not cause separation since the force per unit mass of a molecule is constant. Forced diffusion occurs when an electrical field is imposed on an electrolyte (for example, in charging an automobile battery), on a semiconductor, or on an ionized gas (for example, in a neon tube or metal-ion laser). Depending on the strength of the electric field, rates of forced diffusion can be very large.

Some interesting diffusion phenomena occur in porous solids. When a gas mixture is in a porous solid, such as a catalyst pellet or silica–gel particle, the pores can be smaller than the mean free path of the molecules. Then, the molecules collide with the wall more often than with other molecules. In the limit of negligible molecule collisions we have *Knudsen diffusion*, also called *free molecule flow* in the fluid mechanics literature. If the pore size approaches the size of a molecule, then Knudsen diffusion becomes negligible and *surface diffusion*, in which adsorbed molecules move along the pore walls, becomes the dominant diffusion mechanism.

TABLE 3.5.2 Diffusion Coefficients in Air at 1 atm (1.013×10^5 Pa)[a]

	Binary Diffusion Coefficient (m^2/sec × 10^4)							
T(K)	O_2	CO_2	CO	C_7H_6	H_2	NO	SO_2	He
200	0.095	0.074	0.098	0.036	0.375	0.088	0.058	0.363
300	0.188	0.157	0.202	0.075	0.777	0.180	0.126	0.713
400	0.325	0.263	0.332	0.128	1.25	0.303	0.214	1.14
500	0.475	0.385	0.485	0.194	1.71	0.443	0.326	1.66
600	0.646	0.537	0.659	0.270	2.44	0.603	0.440	2.26
700	0.838	0.684	0.854	0.364	3.17	0.782	0.576	2.91
800	1.05	0.857	1.06	0.442	3.93	0.978	0.724	3.64
900	1.26	1.05	1.28	0.538	4.77	1.18	0.887	4.42
1000	1.52	1.24	1.54	0.641	5.69	1.41	1.060	5.26
1200	2.06	1.69	2.09	0.881	7.77	1.92	1.440	7.12
1400	2.66	2.17	2.70	1.13	9.90	2.45	1.870	9.20
1600	3.32	2.75	3.37	1.41	12.5	3.04	2.340	11.5
1800	4.03	3.28	4.10	1.72	15.2	3.70	2.850	13.9
2000	4.80	3.94	4.87	2.06	18.0	4.48	3.360	16.6

[a] Owing to the practical importance of water vapor-air mixtures, engineers have used convenient empirical formulas for $\mathcal{D}_{H_2O\,air}$. A formula that has been widely used for many years is

$$\mathcal{D}_{H_2O\,air} = 1.97 \times 10^{-5} \left(\frac{P_0}{P}\right)\left(\frac{T}{T_0}\right)^{1.685} \text{m}^2/\text{sec}; \quad 273\text{ K} < T < 373\text{ K}$$

where $P_0 = 1$ atm; $T_0 = 256$ K. More recently, the following formula has found increasing use. (Marrero, T.R. and Mason, E.A. 1992. Gaseous diffusion coefficients, *J. Phys. Chem. Ref. Data*, 1, 3–118):

$$\mathcal{D}_{H_2O\,air} = 1.87 \times 10^{-10} \frac{T^{2.072}}{P}; \quad 280\text{ K} < T < 450\text{ K}$$

$$= 2.75 \times 10^{-9} \frac{T^{1.632}}{P}; \quad 450\text{ K} < T < 1070\text{ K}$$

for *P* in atmospheres and *T* in kelvins. Over the temperature range 290 to 330 K, the discrepancy between the two formulas is less than 2.5%. For small concentrations of water vapor in air, the older formula gives a constant value of $\text{Sc}_{H_2O\,air} = 0.61$ over the temperature range 273 to 373 K. On the other hand, the Marrero and Mason formula gives values of $\text{Sc}_{H_2O\,air}$ that vary from 0.63 at 280 K to 0.57 at 373 K.

Very small particles of 10^{-3} to 10^{-1} μm size — for example, smoke, soot, and mist — behave much like large molecules. Ordinary diffusion of such particles is called *Brownian motion* and is described in most elementary physics texts. Diffusion of particles due to a temperature gradient is called *thermophoresis* and plays an important role for larger particles, typically in the size range 10^{-1} to 1 μm. Diffusion of particles in a gas mixture due to concentration gradients of molecular species is called *diffusiophoresis. Forced diffusion* of a charged particle in an electrical field is similar to that for an ionized molecular species. Thermal and electrostatic precipitators are used to remove particles from power plant and incinerator stack gases, and depend on thermophoresis and forced diffusion, respectively, for their operation. Diffusion phenomena are unimportant for particles of size greater than about 1 μm in air at 1 atm; the motion of such particles is governed by the laws of Newtonian mechanics. Transport of particles is dealt with in the *aerosol science* literature.

Species Conservation Equation

The principle of conservation of a chemical species is used to derive the *species conservation equation.* On a mass basis this equation is

TABLE 3.5.3 Schmidt Number for Vapors in Dilute Mixture in Air at Normal Temperature, Enthalpy of Vaporization, and Boiling Point at 1 atm[a]

Vapor	Chemical Formula	Sc[b]	h_{fg}, J/kg × 10^{-6}	T_{BP}, K
Acetone	CH_3COCH_3	1.42	0.527	329
Ammonia	NH_3	0.61	1.370	240
Benzene	C_6H_6	1.79	0.395	354
Carbon dioxide	CO_2	1.00	0.398	194
Carbon monoxide	CO	0.77	0.217	81
Chlorine	Cl_2	1.42	0.288	238
Ethanol	CH_3CH_2OH	1.32	0.854	352
Helium	He	0.22	—	4.3
Heptane	C_7H_{16}	2.0	0.340	372
Hydrogen	H_2	0.20	0.454	20.3
Hydrogen sulfide	H_2S	0.94	0.548	213
Methanol	CH_3OH	0.98	1.110	338
Naphthalene	$C_{10}H_8$	2.35[c]	—	491
Nitric oxide	NO	0.87	0.465	121
Octane	C_8H_{18}	2.66	0.303	399
Oxygen	O_2	0.83	0.214	90.6
Pentane	C_5H_{12}	1.49	0.357	309
Sulfur dioxide	SO_2	1.24	0.398	263
Water vapor	H_2O	0.61	2.257	373

[a] With the Clausius-Clapeyron relation, one may estimate vapor pressure as

$$P_{\text{sat}} \simeq \exp\left\{-\frac{Mh_{fg}}{\mathcal{R}}\left(\frac{1}{T}-\frac{1}{T_{BP}}\right)\right\} \text{ atm} \quad \text{for } T \sim T_{BP}$$

[b] The Schmidt number is defined as Sc = $\mu/\rho\mathcal{D} = \nu/\mathcal{D}$. Since the vapors are in small concentrations, values for μ, ρ, and ν can be taken as pure air values.

[c] From a recent study by Cho, C., Irvine, T.F., Jr., and Karni, J. 1992. Measurement of the diffusion coefficient of naphthalene into air, *Int. J. Heat Mass Transfer*, 35, 957–966. Also, h_{sg} = 0.567 × 10^6 J/kg at 300 K.

$$\frac{\partial \rho_i}{\partial t} + \nabla \cdot \boldsymbol{n}_i = \dot{r}_i''' \tag{3.5.27}$$

and on a molar basis

$$\frac{\partial c_i}{\partial t} + \nabla \cdot \boldsymbol{N}_i = \dot{R}_i''' \tag{3.5.28}$$

where $\dot{r}_i'''$ and $\dot{R}_i'''$ are the mass and molar rates of production of species i due to chemical reactions. Summing Equation 3.5.27 over all species gives the mass conservation or continuity equation,

$$\frac{\partial \rho}{\partial t} + \nabla \cdot \rho \boldsymbol{v} = 0 \tag{3.5.29}$$

The molar form is

$$\frac{\partial c}{\partial t} + \nabla \cdot c\boldsymbol{v}^* = \sum_i \dot{R}_i''' \tag{3.5.30}$$

since, in general, moles are not conserved in chemical reactions. A useful alternative form to Equation (3.5.27) can be obtained using Equations (3.5.23a) and (3.5.29) and is

TABLE 3.5.4 Schmidt Numbers for Dilute Solution in Water at 300 K[a]

Solute	Sc	M
Helium	120	4.003
Hydrogen	190	2.016
Nitrogen	280	28.02
Water	340	18.016
Nitric oxide	350	30.01
Carbon monoxide	360	28.01
Oxygen	400	32.00
Ammonia	410	17.03
Carbon dioxide	420	44.01
Hydrogen sulfide	430	34.08
Ethylene	450	28.05
Methane	490	16.04
Nitrous oxide	490	44.02
Sulfur dioxide	520	64.06
Sodium chloride	540	58.45
Sodium hydroxide	490	40.00
Acetic acid	620	60.05
Acetone	630	58.08
Methanol	640	32.04
Ethanol	640	46.07
Chlorine	670	70.90
Benzene	720	78.11
Ethylene glycol	720	62.07
n-Propanol	730	60.09
i-Propanol	730	60.09
Propane	750	44.09
Aniline	800	93.13
Benzoic acid	830	122.12
Glycerol	1040	92.09
Sucrose	1670	342.3

[a] Schmidt number $\mathrm{Sc} = \mu/\rho\mathcal{D}$; since the solutions are dilute, μ and ρ can be taken as pure water values. For other temperatures use $\mathrm{Sc}/\mathrm{Sc}_{300\,\mathrm{K}} \simeq (\mu^2/\rho T)/(\mu^2\rho T)_{300\,\mathrm{K}}$, where μ and ρ are for water, and T is absolute temperature. For chemically similar solutes of different molecular weights use $\mathrm{Sc}_2/\mathrm{Sc}_1 \simeq (M_2/M_1)^{0.4}$. A table of $(\mu^2/\rho T)/(\mu^2\rho T)_{300\,\mathrm{K}}$ for water follows.

T(K)	$(\mu^2/\rho T)/(\mu^2/\rho T)_{300\,\mathrm{K}}$
290	1.66
300	1.00
310	0.623
320	0.429
330	0.296
340	0.221
350	0.167
360	0.123
370	0.097

From Spalding, D.B. 1963. *Convective Mass Transfer*, McGraw-Hill, New York. With permission.

$$\rho\frac{Dm_i}{Dt} = \nabla \cdot j_i + \dot{r}_i''' \tag{3.5.31}$$

where D/Dt is the substantial derivative operator.

If we consider a binary system of species 1 and 2 and introduce Fick's law, Equation (3.5.24a) into Equation (3.5.31), then

TABLE 3.5.5 Diffusion Coefficients in Solids, $\mathcal{D} = \mathcal{D}_o \exp(-E_a/\mathcal{R}T)$

System	$\mathcal{D}_o$, m²/sec	E_a, kJ/kmol
Oxygen-Pyrex glass	6.19×10^{-8}	4.69×10^{4}
Oxygen-fused silica glass	2.61×10^{-9}	3.77×10^{4}
Oxygen-titanium	5.0×10^{-3}	2.13×10^{5}
Oxygen-titanium alloy (Ti-6Al-4V)	5.82×10^{-2}	2.59×10^{5}
Oxygen-zirconium	4.68×10^{-5}	7.06×10^{5}
Hydrogen-iron	7.60×10^{-8}	5.60×10^{3}
Hydrogen-α-titanium	1.80×10^{-6}	5.18×10^{4}
Hydrogen-β-titanium	1.95×10^{-7}	2.78×10^{4}
Hydrogen-zirconium	1.09×10^{-7}	4.81×10^{4}
Hydrogen-Zircaloy-4	1.27×10^{-5}	6.05×10^{5}
Deuterium-Pyrex glass	6.19×10^{-8}	4.69×10^{4}
Deuterium-fused silica glass	2.61×10^{-9}	3.77×10^{4}
Helium-Pyrex glass	4.76×10^{-8}	2.72×10^{4}
Helium-fused silica glass	5.29×10^{-8}	2.55×10^{4}
Helium-borosilicate glass	1.94×10^{-9}	2.34×10^{4}
Neon-borosilicate glass	1.02×10^{-10}	3.77×10^{4}
Carbon-FCC iron	2.3×10^{-5}	1.378×10^{5}
Carbon-BCC iron	1.1×10^{-6}	8.75×10^{4}

Various sources.

$$\rho \frac{Dm_i}{Dt} = \nabla \cdot \left(\rho \mathcal{D}_{12} \nabla m_1\right) + \dot{r}_1''' \tag{3.5.32}$$

When working on a mass basis we define a stationary medium as one in which the mass average velocity ν is zero everywhere. Substituting in Equation (3.5.32) with no chemical reactions and assuming constant properties,

$$\frac{\partial m_1}{\partial t} = \mathcal{D}_{12} \nabla^2 m_1 \tag{3.5.33}$$

which is the *diffusion* equation, and is the mass transfer analog to Fourier's equation for heat conduction. For steady diffusion, Equation (3.5.33) reduces to Laplace's equation

$$\nabla^2 m_1 = 0 \tag{3.5.34}$$

Notice that since properties have been assumed constant, any measure of concentration can be used in Equations (3.5.33) and (3.5.34), for example ρ_1, c_1, and x_1.

Diffusion in a Stationary Medium

Many problems involving diffusion in a stationary medium are governed by the diffusion equation (Equation 3.5.33). Often solutions may be obtained from their heat conduction analogs. Some important cases follow.

Steady Diffusion through a Plane Wall

The mass flow of species 1 across a plane wall of thickness L and cross-sectional area A is

$$\dot{m}_1 = \frac{\rho \mathcal{D}_{12} A}{L}\left(m_{1,u} - m_{1,u'}\right) \text{ kg/m}^2 \text{ sec} \tag{3.5.35}$$

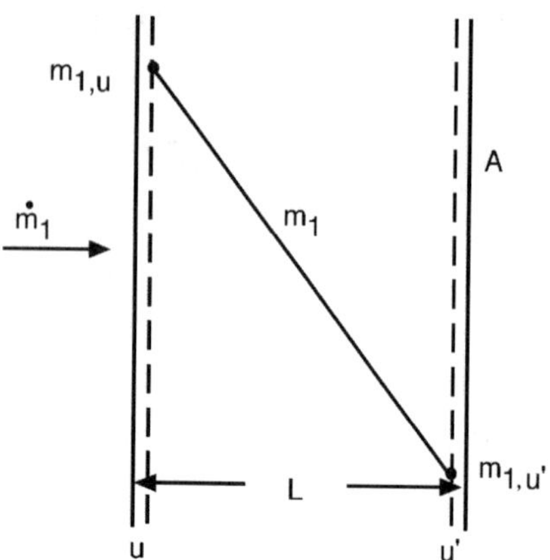

FIGURE 3.5.3 Steady diffusion across a plane wall.

where the u- and u'-surfaces are shown in Figure 3.5.3. Solubility data are required to relate the u- and u'-surface concentrations to s- and s'-surface concentrations. Alternatively for systems that obey Henry's law, a solubility $\mathcal{S}$ can be defined as the volume of solute gas (at STP of 0°C and 1 atm) dissolved in unit volume when the gas is at a partial pressure of 1 atm. Then, defining permeability $\mathcal{P}_{12}$ as the product $\mathcal{D}_{12}\mathcal{S}$, the volume flow of species 1 is

$$\dot{V}_1 = \frac{\mathcal{P}_{12} A}{L}\left(P_{1,s} - P_{1,s'}\right) \text{ m}^3 \text{ (STP)/sec} \tag{3.5.36}$$

where the partial pressures P_1 are in atmospheres. The SI units for permeability are m^3 (STP)/m^{2}sec(atm/m). Permeability and solubility data are given in Table 3.5.6. For example, consider helium at 10^5 Pa contained in a 7056-glass vessel with a 1-mm-thick wall at 680 K. For a surface area of 0.01 m^2, the leakage rate into ambient air is

$$\dot{V} = \frac{\left(1.0 \times 10^{-12}\right)(0.01)}{(0.001)}\left(10^5 - 0\right) = 1.0 \times 10^{-6} \text{ m}^3 \text{ (STP)/sec}$$

where the value $\mathcal{P}_{12}$ was obtained from Table 3.5.6.

In general, mass fractions are discontinuous across phase interfaces. Hence, Equation (3.5.35) cannot be generalized to a number of walls in series by simply adding diffusion resistances. However, equilibrium partial pressures P_1 are continuous, and for two walls A and B, Equation (3.5.36) becomes

$$\dot{V}_1 = \frac{P_{1,s} - P_{1,s'}}{\dfrac{L_A}{\mathcal{P}_{1A} A} + \dfrac{L_B}{\mathcal{P}_{1B} A}} \text{ m}^3 \text{ (STP)/sec} \tag{3.5.37}$$

Transient Diffusion in a Semi-Infinite Solid

The typically low diffusion coefficients characterizing solids result in many situations where concentration changes are limited to a thin region near the surface (of thickness $\delta_c \sim (\mathcal{D}_{12}t)^{1/2}$). Examples include case-hardening of mild steel and coloring of clear sapphires. Details of the geometry are then unimportant

TABLE 3.5.6 Solubility and Permeability of Gases in Solids

Gas	Solid	Temperature, K	$\mathcal{S}$ (m³(STP)/m³ atm) or $\mathcal{S}'$[a]	Permeability[b] m³(STP)/m² sec (atm/m)
H_2	Vulcanized rubber	300	$\mathcal{S} = 0.040$	0.34×10^{-10}
	Vulcanized neoprene	290	$\mathcal{S} = 0.051$	0.053×10^{-10}
	Silicone rubber	300		4.2×10^{-10}
	Natural rubber	300		0.37×10^{-10}
	Polyethylene	300		0.065×10^{-10}
	Polycarbonate	300		0.091×10^{-10}
	Fused silica	400	$\mathcal{S}' \simeq 0\ 035$	
		800	$\mathcal{S}' \simeq 0.030$	
	Nickel	360	$\mathcal{S}' = 0.202$	
		440	$\mathcal{S}' = 0.192$	
He	Silicone rubber	300		2.3×10^{-10}
	Natural rubber	300		0.24×10^{-10}
	Polycarbonate	300		0.11×10^{-10}
	Nylon 66	300		0.0076×10^{-10}
	Teflon	300		0.047×10^{-10}
	Fused silica	300	$\mathcal{S}' \simeq 0.018$	
		800	$\mathcal{S}' \simeq 0.026$	
	Pyrex glass	300	$\mathcal{S}' \simeq 0.006$	
		800	$\mathcal{S}' \simeq 0.024$	
	7740 glass	470	$\mathcal{S} = 0.0084$	4.6×10^{-13}
	(94% SiO_2 + B_2O_3 + P_2O_5,	580	$\mathcal{S} = 0.0038$	1.6×10^{-12}
	5% Na_2O + Li_2 + K_2O,	720	$\mathcal{S} = 0.0046$	6.4×10^{-12}
	1% other oxides)			
	7056 glass	390	$\mathcal{S}' = 0.0039$	1.2×10^{-14}
	(90% SiO_2 + B_2O_3 + P_2O_5,	680	$\mathcal{S}' = 0.0059$	1.0×10^{-12}
	8% Na_2O + Li_2 + K_2O,			
	1% PbO, 5% other oxides)			
O_2	Vulcanized rubber	300	$\mathcal{S} = 0.070$	0.15×10^{-10}
	Silicone rubber	300		3.8×10^{-10}
	Natural rubber	300		0.18×10^{-10}
	Polyethylene	300		4.2×10^{-12}
	Polycarbonate	300		0.011×10^{-10}
	Silicone-polycarbonate copolymer (57% silicone)	300		1.2×10^{-10}
	Ethyl cellulose	300		0.09×10^{-10}
N_2	Vulcanized rubber	300	$\mathcal{S} = 0.035$	0.054×10^{-10}
	Silicone rubber	300		1.9×10^{-12}
	Natural rubber	300		0.062×10^{-10}
	Silicone-polycarbonate copolymer (57% silicone)	300		0.53×10^{-10}
	Teflon	300		0.019×10^{-10}
CO_2	Vulcanized rubber	300	$\mathcal{S} = 0.090$	1.0×10^{-10}
	Silicone rubber	290		21×10^{-10}
	Natural rubber	300		1.0×10^{-10}
	Silicone-polycarbonate copolymer (57% silicone)	300		7.4×10^{-10}
	Nylon 66	300		0.0013×10^{-10}

TABLE 3.5.6 Solubility and Permeability of Gases in Solids

Gas	Solid	Temperature, K	$\mathcal{S}$ (m³(STP)/m³ atm) or $\mathcal{S}'$[a]	Permeability[b] m³(STP)/m² sec (atm/m)
H_2O	Silicone rubber	310		$0.91–1.8 \times 10^{-10}$
Ne	Fused silica	300–1200	$\mathcal{S} \simeq 0.002$	
Ar	Fused silica	900–1200	$\mathcal{S} \simeq 0.01$	

[a] Solubility $\mathcal{S}$ = volume of solute gas (0°C, 1 atm) dissolved in unit volume of solid when the gas is at 1 atm partial pressure. Solubility coefficient $\mathcal{S}' = c_{1,u}/c_{1,s}$.

[b] Permeability $\mathcal{P}_{12} = \mathcal{D}_{12}\mathcal{S}$.

From various sources, including Geankoplis, C.J. 1993. *Transport Processes and Unit Operations*, 3rd ed., Prentice-Hall; Englewood Cliffs, N.J.; Doremus, R.H. 1973. *Glass Science,* Wiley, New York; Altemose, V.O. 1961. Helium diffusion through glass, *J. Appl. Phys.*, 32, 1309–1316. With permission.

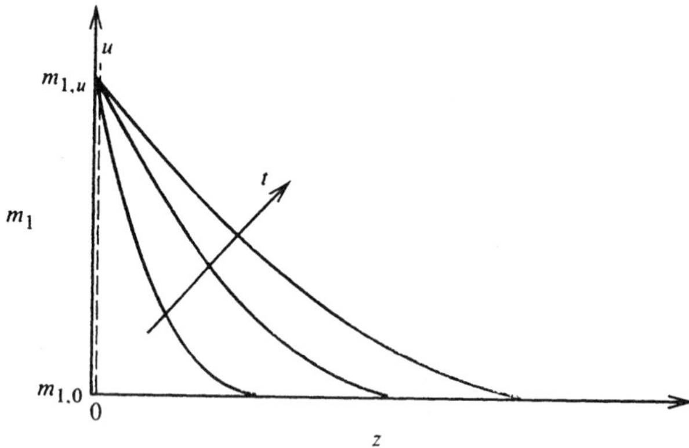

FIGURE 3.5.4 Transient diffusion in a plane slab.

and semi-infinite solid model can be used (Figure 3.5.4). For an initial concentration $m_{1,0}$ and a u-surface concentration suddenly changed to $m_{1,u}$ at time $t = 0$, the concentration distribution $m_1(z,t)$ is

$$\frac{m_1 - m_{1,0}}{m_{1,u} - m_{1,0}} = erfc\frac{z}{\left(4\mathcal{D}_{12}t\right)^{1/2}} \tag{3.5.38}$$

and the dissolution rate is

$$\dot{m}_1 = j_{1,u}A = \rho A\left(\frac{\mathcal{D}_{12}}{\pi t}\right)^{1/2}\left(m_{1,u} - m_{1,0}\right) \text{ kg/sec} \tag{3.5.39}$$

For example, consider a Pyrex glass slab at 800 K suddenly exposed to helium at 10^4 Pa. The molar equivalent to Equation (3.5.39) for an assumed constant solid phase molar concentration c is

$$\frac{\dot{M}_1}{A} = \left(\frac{\mathcal{D}_{12}}{\pi t}\right)^{1/2}\left(c_{1,u} - c_{1,0}\right)$$

From Table 3.5.6, $\mathcal{S}' = c_{1,u}/c_{1,s} \cong 0.024$; hence, $c_{1,u} = (0.024)(10^4)/(8314)(800) = 3.61 \times 10^{-5}$ kmol/m³. From Table 3.5.4, $\mathcal{D}_{12} = 4.76 \times 10^{-8} \exp[-(2.72 \times 10^4)(10^3)/(8314)(800)] = 7.97 \times 10^{-10}$ m²/sec. Hence,

$$\frac{\dot{M}_1}{A} = \left(\frac{7.97\times10^{-10}}{\pi t}\right)^{1/2}\left(3.61\times10^{-5} - 0\right) = 5.75\times10^{-10}/t \text{ kmol/sec}$$

Transient Diffusion in Slabs, Cylinders, and Spheres

Transient heat conduction in slates, cylinders, and spheres with surface convection is dealt with in Section 3.1. The analogous mass diffusion problem for the slab $-L < z < L$ is now considered. On a molar basis the governing differential equation is

$$\frac{\partial x_1}{\partial t} = \mathcal{D}_{12} \frac{\partial^2 x_1}{\partial z^2} \tag{3.5.40}$$

with initial condition $x_1 = x_{1,0}$ at $t = 0$. Boundary conditions are $\partial x_1/\partial z = 0$ at $z = 0$, and at the surface $z = L$,

$$-c\mathcal{D}_{12} \left.\frac{\partial x_1}{\partial z}\right|_{z=L} = \mathcal{G}_{m1}\left(y_{1,s} - y_{1,e}\right) \tag{3.5.41}$$

The convective boundary condition is of the same form as Newton's law of cooling, and defines the mole transfer conductance $\mathcal{G}_{m1}$ (kmol/m^{2}sec) (see also the section on mass and mole transfer conductances). Also, we have followed chemical engineering practice and denoted mole fraction x in the solid (or liquid) phase and y in the liquid (or gas) phase, to emphasize that generally mole fraction is not continuous across a phase interface. For example, consider absorption of a sparingly soluble gas into a liquid for which Henry's law, Equation (3.5.13), applies: then $y_{1,s} = \text{He}x_{1,u}$.

In using heat conduction charts for mass diffusion problems, particular care must be taken with the evaluation of the Biot number. For heat conduction Bi = h_cL/k, where k is the solid conductivity. For mass diffusion the Biot number accounts for the discontinuity in concentration across the phase interface. Using gas absorption into a plane layer of liquid, for example, when Equation (3.5.41) is put into an appropriate dimensionless form, the mass transfer Biot number is seen to be

$$\text{Bi}_m = \frac{\mathcal{G}_{m1}\text{He}L}{c\mathcal{D}_{12}} \tag{3.5.42}$$

For sparingly soluble gases, e.g., O_2 or CO_2 in water, He, and hence Bi$_m$, are very large, and the absorption process is liquid-side controlled; that is, a uniform gas-phase composition can be assumed. Often interface equilibrium data are in graphical or tabular form; then an effective Biot number at the concentration of concern must be used.

For example, consider a 2-mm-diameter droplet of water at 300 K entrained in an air flow at 1 atm pressure containing 1% by volume CO_2. From Table 3.5.5, He = C_{He}= 1710. The liquid phase molar density can be approximated by the pure water value of $c = \rho/M = 996/18 = 55.3$ kmol/m^3. The liquid phase diffusion coefficient is obtained from Table 3.5.4 as $\mathcal{D}_{12} = \nu_{H_2O}/Sc_{12} = 0.87 \times 10^{-6}/420 = 2.07 \times 10^{-9}$ m^2/sec. For negligible relative motion between the droplet and gas, the Sherwood number (see the section on dimensionless groups) is approximately 2.0, and hence the gas phase mole transfer conductance is $\mathcal{G}_{m1} = 2c\mathcal{D}_{12}/\mathcal{D}$. For the gas phase, the molar density $c = \mathcal{P}/\mathcal{R}T = (1.0133 \times 10^5)/(8314)(300) = 0.0406$ kmol/m^3 and $\mathcal{D}_{12} = 0.157 \times 10^{-4}$ m^2/sec from Table 3.5.2. Thus,

$$\mathcal{G}_{m1} = \frac{(2)(0.0406)\left(0.157\times 10^{-4}\right)}{(0.002)} = 6.37\times 10^{-4}\ \text{kmol/m}^2\,\text{sec}$$

From Equation 3.5.42 with $L = R$ the droplet radius, the mass transfer Biot number is

$$\text{Bi}_m = \frac{\left(6.37\times 10^{-4}\right)(1710)(0.001)}{(55.3)\left(2.07\times 10^{-9}\right)} = 9520$$

Thus, even for a small droplet with a relatively large gas-side mole transfer conductance, the absorption process is liquid-side controlled.

Diffusion in a Porous Catalyst

Porous catalysts are used to give a large surface area per unit volume of catalyst surface. Current practice for automobile catalytic convertors is to use a ceramic matrix as a support for a thin porous alumina layer that is impregnated with the catalyst (called a *washcoat*). A typical matrix has passages of hydraulic diameter 1 mm, and the washcoat may be about 20 μm thick. Pore sizes are of the order of 1 μm for which ordinary and Knudsen diffusion resistances are important. A simple model for diffusion in a porous catalyst is

$$J_1 = -c\mathcal{D}_{1,\text{eff}}\nabla x_1 \ \text{kmol/m}^2\,\text{sec} \tag{3.5.43}$$

where the subscript eff denotes an effective diffusivity that accounts for the presence of the solid material. Assuming additive resistances,

$$\frac{1}{\mathcal{D}_{1,\text{eff}}} = \frac{1}{\mathcal{D}_{12,\text{eff}}} + \frac{1}{\mathcal{D}_{K1,\text{eff}}} \tag{3.5.44}$$

and

$$\mathcal{D}_{12,\text{eff}} = \frac{\varepsilon_v}{\tau}\mathcal{D}_{12}; \qquad \mathcal{D}_{K1,\text{eff}} = \frac{\varepsilon_v}{\tau}\mathcal{D}_{K1,\text{eff}} \tag{3.5.45}$$

where ε_v is the volume void fraction and τ is the tortuosity factor (usually between 4 and 8). From the kinetic theory of gases the Knudsen diffusion coefficient is

$$\mathcal{D}_{K1} = 97r_e\left(T/M_1\right)^{1/2} \ \text{m}^2/\text{sec} \tag{3.5.46}$$

for effective pore radius r_e in meters and T in kelvins.

When a chemical reaction takes place within a porous layer, a concentration gradient is set up, and surfaces on pores deep within the pellet are exposed to lower reactant concentrations than surfaces near the pore openings. For a first-order reaction it is straightforward to obtain the concentration distribution. The results of such an analysis are conveniently given in the form of an effectiveness η_p, which is defined as the actual consumption rate of the reactant divided by that for an infinite diffusion coefficient. For a layer of thickness L exposed to reactants on one side, as shown in Figure 3.5.5.

$$\eta_p = \frac{\tanh bL}{bL}; \qquad b = \left(\frac{k'' a_P}{\mathcal{D}_{1,\text{eff}}}\right)^{1/2} \tag{3.5.47}$$

where k'' (m/sec) is the rate constant for a first-order reaction and a_p (m^{-1}) is the catalyst area per unit volume. Notice that this effectiveness is analogous to the efficiency of a heat transfer fin.

For example, consider a 30-μm-thick porous alumina washcoat with a volume void fraction $\varepsilon_v = 0.8$, a tortuosity factor $\tau = 4.0$, average pore radius $r_e = 1$ μm, and catalytic surface area per unit volume $a_p = 7.1 \times 10^5$ cm²/cm³. For carbon monoxide oxidation by copper oxide at 800 K, 1 atm, the rate constant is approximately 4.2×10^{-4} m²/sec. To calculate the effectiveness of the washcoat, we first need to calculate the effective diffusion coefficient $\mathcal{D}_{1,\text{eff}}$:

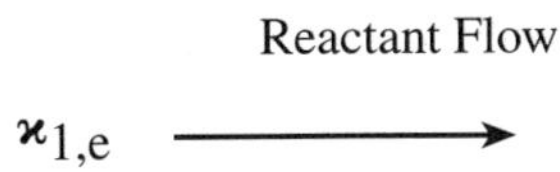

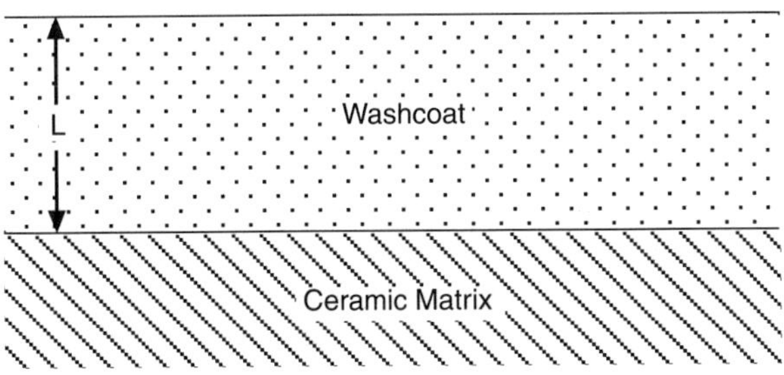

FIGURE 3.5.5 A catalyst layer.

$$\mathcal{D}_{12,\text{eff}} = \frac{\varepsilon_v}{\tau}\mathcal{D}_{12} = \frac{0.8}{4.0}\left(1.06\times10^{-4}\right) = 2.12\times10^{-5}\ \text{m}^2/\text{sec}$$

where $\mathcal{D}_{12}$ is approximated as the CO-air value from Table 3.5.2.

$$\mathcal{D}_{K1,\text{eff}} = \frac{\varepsilon_v}{\tau}\mathcal{D}_{12} = \frac{0.8}{4.0}(97)\left(1\times10^{-6}\right)(800/28)^{1/2} = 1.04\times10^{-4}\ \text{m}^2/\text{sec}$$

$$\frac{1}{\mathcal{D}_{1,\text{eff}}} = \frac{1}{2.12\times10^{-5}} + \frac{1}{1.04\times10^{-4}}; \qquad \mathcal{D}_{1,\text{eff}} = 1.76\times10^{-5}\ \text{m}^2/\text{sec}$$

$$b = \left[\frac{\left(4.2\times10^{-4}\right)\left(7.1\times10^{5}\right)\left(10^{2}\right)}{1.76\times10^{-5}}\right]^{1/2} = 4.2\times10^{4}\ \text{m}^{-1}; \qquad bL = \left(4.2\times10^{4}\right)\left(30\times10^{-6}\right) = 1.236$$

$$\eta_P = \frac{\tanh 1.236}{1.236} = 68.3\%$$

In an automobile catalytic convertor, Equation 3.5.47 applies to the catalyst washcoat. However, the mass transfer problem also involves a convective process for transport of reactants from the bulk flow. Referring to Figure 3.5.6 there are two mass transfer resistances in series, and the consumption rate of species 1 per unit surface area of the washcoat is

$$J_{1,s} = \frac{-x_{1,e}}{\dfrac{1}{L\eta_p k''c} + \dfrac{1}{\mathcal{G}_{m1}}}\ \text{kmol/m}^2\ \text{sec} \tag{3.5.48}$$

where $\mathcal{G}_{m1}$ is the mole transfer conductance describing convective transport to the washcoat surface (see the section on mass and mole transfer conductances). Notice that when $\mathcal{G}_{m1} \ll L\eta_p k''c$ the reaction rate is controlled by mass transfer from the gas stream to the washcoat surface; when $L\eta_p k''c \ll \mathcal{G}_{m1}$, the reaction rate is controlled by diffusion within the washcoat and the kinetics of the reaction.

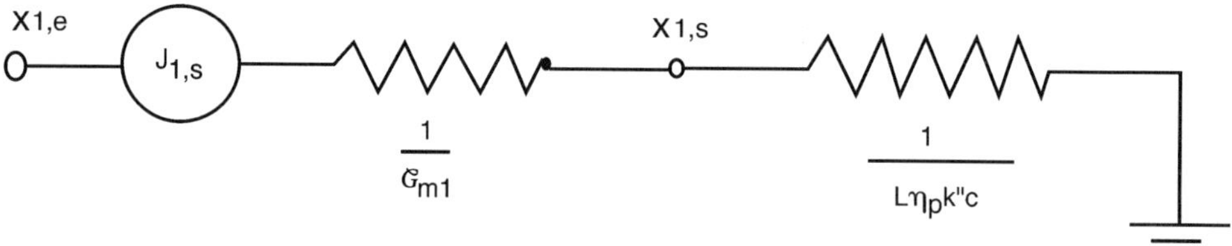

FIGURE 3.5.6 Equivalent circuit for mass transfer in an automobile catalytic convertor.

Diffusion in a Moving Medium

Net mass transfer across a surface results in a velocity component normal to the surface, and an associated convective flux in the direction of mass transfer. This convective flow is called a *Stefan flow*. The solutions of a number of mass transfer problems, involving a Stefan flow induced by the mass transfer process itself, follow. When necessary to obtain an analytical result, properties are assumed constant. Thus, use of these results requires evaluation of properties at a suitable reference state.

Diffusion with One Component Stationary

As an example, consider the simple heat pipe shown in Figure 3.5.7 with the evaporator and condenser located at the ends only (a bad design!). Then, if the working fluid is species 1, and a noncondensable gas is species 2, the concentration distribution is

$$\left(\frac{1-x_1}{1-x_{1,s}}\right)=\left(\frac{1-x_{1,e}}{1-x_{1,s}}\right)^{z/L} \tag{3.5.49}$$

and the vapor flux along the heat pipe is

$$N_1=\frac{c\mathcal{D}_{12}}{L}\ln\frac{1-x_{1,e}}{1-x_{1,s}}\ \text{kmol/m}^2\ \text{sec} \tag{3.5.50}$$

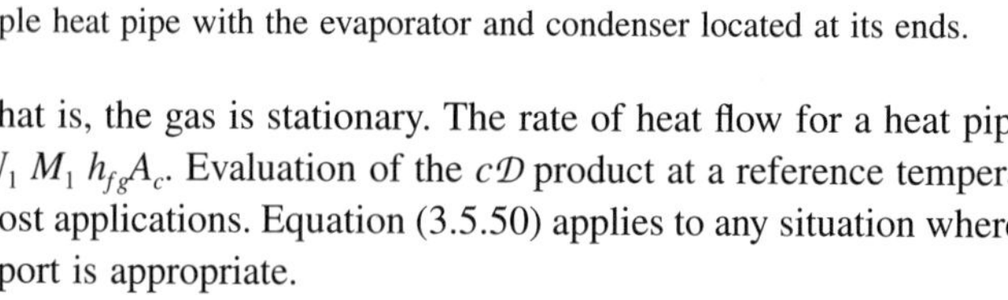

FIGURE 3.5.7 A simple heat pipe with the evaporator and condenser located at its ends.

Notice that $N_2 = 0$; that is, the gas is stationary. The rate of heat flow for a heat pipe of cross-sectional area of A_c is $\dot{Q} = N_1 M_1 h_{fg} A_c$. Evaluation of the $c\mathcal{D}$ product at a reference temperature $T_r = (1/2)(T_s + T_e)$ is adequate for most applications. Equation (3.5.50) applies to any situation where a one-dimensional model of mass transport is appropriate.

Heterogeneous Combustion

As an example, consider a small carbon particle entrained in a high-temperature airstream, as shown in Figure 3.5.8. The surface reaction is $2C + O_2 \rightarrow 2CO$ and there are no reactions in the gas phase. The stoichiometric ratio for the reaction is $r = 4/3$ kg oxygen/kg carbon. The reaction is diffusion controlled at the temperatures under consideration, that is, $m_{O_2,s} \simeq 0$. The mass transfer rate is n_s, which we give the distinctive symbol $\dot{m}''$ since it is usually the desired result of an analysis; in this situation $\dot{m}'' = n_{C,u}$ is the combustion rate of carbon, and for a spherical particle of radius R is given by

$$\dot{m}'' = \frac{\rho \mathcal{D}_{O_2,m}}{R} \ln\left[1 + \frac{m_{O_2,e} - m_{O_2,s}}{m_{O_2,s} + 4/3}\right] = 0.160 \frac{\rho \mathcal{D}_{O_2,m}}{R} \text{ kg/m}^2 \text{ sec} \tag{3.5.51}$$

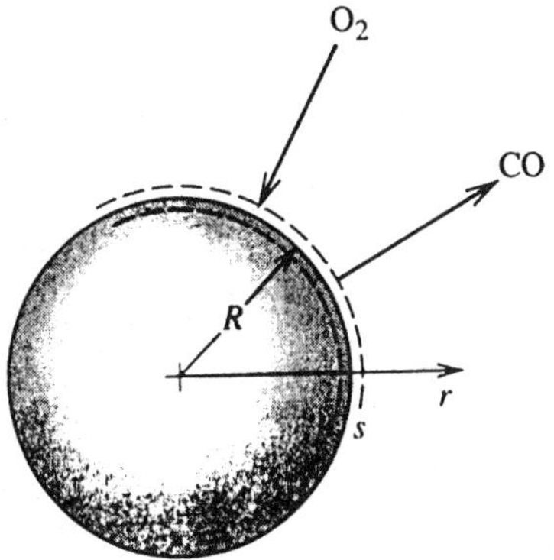

FIGURE 3.5.8 Combustion of a carbon particle in high-temperature air. The surface reaction is 2C + $O_2 \rightarrow$ 2CO.

The carbon particle temperature depends on its radius, and it is required to evaluate the property product $\rho\mathcal{D}$ at an appropriate reference temperature: an energy balance on the particle should be performed by this purpose. The resulting particle lifetime τ is

$$\tau = \frac{\rho_{\text{solid}} D_O^2}{1.28\left(\rho \mathcal{D}_{O_2,m}\right)_r} \text{ sec} \tag{3.5.52}$$

for an initial particle diameter of D_0. Air properties at an average mean film temperature can be used to evaluate $\rho\mathcal{D}_{O_2,m}$.

Consider a 10-μm-diameter carbon particle ignited in an airstream at 1500 K and 1 atm. An energy balance on the particle (including radiation to surroundings at 1500 K) shows that the average temperature of the particle is approximately 2550 K, and, thus, T_r = (1/2)(1500 + 2550) = 2025 K or $\rho \simeq \rho_{\text{air}}$ = 0.175 kg/m³ and $\mathcal{D}_{O_2,m} \simeq \mathcal{D}_{O_2,\text{air}}$ = 4.89 × m²/sec (from Table 3.5.1). Then

$$\tau = \frac{(1810)\left(10 \times 10^{-6}\right)^2}{(1.28)(0.175)\left(4.89 \times 10^{-4}\right)} = 1.65 \times 10^{-3} \text{ sec}$$

Droplet Evaporation

Consider a small droplet of species 1 entrained in a gas stream, species 2 (Figure 3.5.9). This is a simultaneous heat and mass transfer problem, and the mass transfer rate can be obtained by solving simultaneously

$$\dot{m}'' = \frac{\rho \mathcal{D}_{12}}{R} \ln\left(1 + \frac{m_{1,e} - m_{1,s}}{m_{1,s} - 1}\right) = \frac{k/c_{p1}}{R} \ln\left(1 + \frac{c_{p1}\left(T_e - T_s\right)}{h_{fg}}\right) \text{kg/m}^2 \text{ sec} \tag{3.5.53a}$$

$$m_{1,s} = m_{1,s}(T,P) \quad \text{(from vapor−pressure data)} \tag{3.5.53b}$$

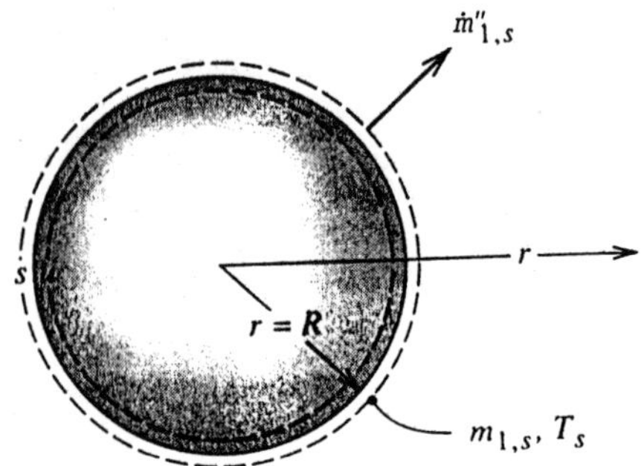

FIGURE 3.5.9 Evaporation of a droplet.

Temperature T_s is the adiabatic vaporization temperature and is essentially the psychrometric wet-bulb temperature. Properties can be evaluated at mean film temperature and composition; alternatively, c_{p1} can be set equal to the reference specific heat and all properties evaluated using Hubbard's $^1/_3$ rule, namely,

$$m_{1,r} = m_{1,s} + (1/3)(m_{1,e} - m_{1,s}) \tag{3.5.54a}$$

$$T_r = T_s + (1/3)(T_e - T_s) \tag{3.5.54b}$$

Droplet Combustion

Figure 3.5.10 shows a schematic of a volatile liquid hydrocarbon fuel droplet burning in air at zero gravity. The flame diameter is typically four to six times the droplet diameter. Heat is transferred from the flame to the droplet and serves to vaporize the fuel. In the flame the vapor reacts with oxygen to form gaseous products, primarily CO_2 and H_2O. When a fuel droplet ignites, there is a short initial transient during which the droplet heats up, until further conduction into the droplet is negligible and the droplet attains a steady temperature (approximately the wet-bulb temperature, which is very close to the boiling point for a typical hydrocarbon fuel). The reaction in the flame can be modeled as a single-step reaction with a constant stoichiometric ratio, r, and heat of combustion Δh_c J/kg of fuel.

The burning (mass transfer) rate of the droplet is given by the Godsave–Spalding formula,

$$\dot{m}'' = \frac{k/c_p}{R} \ln[1 + \mathcal{B}] \text{ kg/m}^2 \text{ sec} \tag{3.5.55}$$

where

$$\mathcal{B} = \frac{m_{ox,e} \Delta h_c / r + c_p (T_e - T_s)}{h_{fg}}$$

is the *mass transfer driving force* (or *transfer number*). The droplet lifetime is then

$$\tau = \frac{\rho_l D_O^2}{8(k/c_p)\ln(1 + \mathcal{B})} \text{ sec} \tag{3.5.56}$$

Based on experimental data for alkane droplets burning in air, Law and Williams (1972) recommend that properties be evaluated at a reference temperature $T_r = (1/2)(T_{BP} + T_{\text{flame}})$ where T_{flame} is the adiabatic flame temperature. The reference specific heat is $c_{pr} = c_{pfu}$, and the reference thermal conductivity is k_r

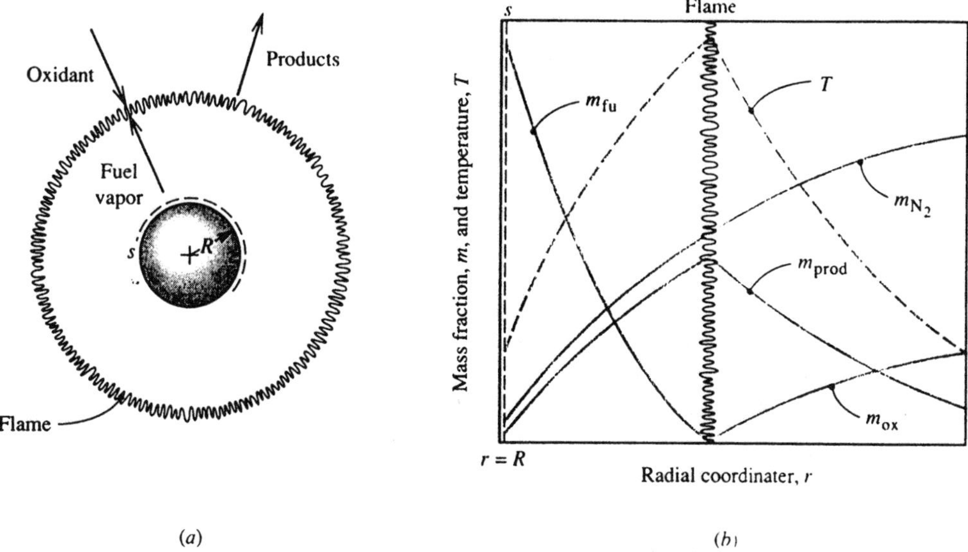

FIGURE 3.5.10 Combustion of a volatile fuel droplet burning in air: (a) schematic showing the flame, (b) concentration and temperature profiles.

$= 0.4k_{fu} + 0.6k_{air}$. Radiation has been ignored in the analysis leading to Equation (3.5.55) but is accounted for in using the Law and Williams reference-property scheme.

For example, consider a 1-mm-diameter *n*-octane droplet burning in air at 1 atm and 300 K, at near zero gravity. For *n*-octane (n-C_8H_{18}), $\rho_l = 611$ kg/m^3, $h_{fg} = 3.03 \times 10^5$ J/kg, $\Delta h_c = 4.44 \times 10^7$ J/kg, and $T_{BP} = 399$ K. The flame temperature is $T_{\text{flame}} = 2320$ K. At a reference temperature of (1/2) ($T_{\text{flame}} + T_{BP}$) = 1360 K, property values of *n*-octane vapor include $k = 0.113$ W/m K, $c_p = 4280$ J/kg K. The reaction is

$$C_8H_{18} + 12.5O_2 \rightarrow 8CO_2 + 9H_2O$$

Hence, the stoichiometric ratio $r = 400/114.2 = 3.50$. Also $m_{ox,e} = 0.231$ and $T_s \cong T_{BP} = 399$ K. Thus, the transfer number is

$$\mathcal{B} = \frac{(0.231)\left(4.44 \times 10^7\right)/(3.50) + 4280(300 - 399)}{3.03 \times 10^5} = 8.27$$

At $T_r = 1360$ K, $k_{\text{air}} = 0.085$ W/m K. Hence,

$$k_r = 0.4k_{fu} + 0.6k_{\text{air}} = (0.4)(0.113) + (0.6)(0.085) = 0.096 \text{ W/m K}$$

and the droplet lifetime is

$$\tau = \frac{(611)\left(1 \times 10^{-3}\right)^2}{(8)(0.096/4280)\ln(1 + 8.27)} = 1.53 \text{ sec}$$

Mass Convection

The terms *mass convection* or *convective mass transfer* are generally used to describe the process of mass transfer between a surface and a moving fluid, as shown in Figure 3.5.11. The surface may be that

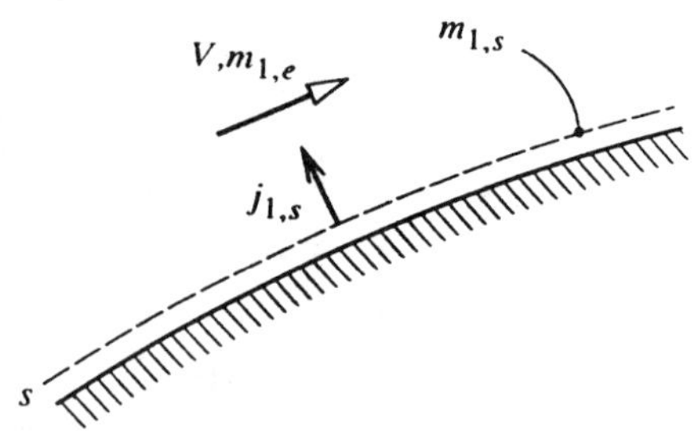

FIGURE 3.5.11 Notation for convective mass transfer into an external flow.

of a falling water film in an air humidifier, of a coke particle in a gasifier, or of a silica-phenolic heat shield protecting a reentry vehicle. As is the case for heat convection, the flow can be *forced* or *natural, internal* or *external,* and *laminar* or *turbulent*. In addition, the concept of whether the mass transfer rate is *low* or *high* plays an important role: when mass transfer rates are low, there is a simple analogy between heat transfer and mass transfer that can be efficiently exploited in the solution of engineering problems.

Mass and Mole Transfer Conductances

Analogous to convective heat transfer, the rate of mass transfer by convection is usually a complicated function of surface geometry and *s*-surface composition, the fluid composition and velocity, and fluid physical properties. For simplicity, we will restrict our attention to fluids that are either binary mixtures or solutions, or situations in which, although more than two species are present, diffusion can be adequately described using effective binary diffusion coefficients, as was discussed in the section on ordinary diffusion. Referring to Figure 3.5.11, we define the *mass transfer conductance* of species 1, g_{m1}, by the relation

$$j_{1,s} = g_{m1}\,\Delta m_1; \qquad \Delta m_1 = m_{1,s} - m_{1,e} \tag{3.5.57}$$

and the units of g_{m1} are seen to be the same as for mass flux (kg/m^{2}sec). Equation (3.5.57) is of a similar form to Newton's law of cooling, which defines the heat transfer coefficient h_c. Why we should not use a similar name and notation (e.g., mass transfer coefficient and h_m) will become clear later. On a molar basis, we define the *mole transfer conductance* of species 1, $\mathcal{G}_{m1}$, by a corresponding relation,

$$J_{1,s} = \mathcal{G}_{m1}\,\Delta x_1; \qquad \Delta x_1 = x_{1,s} - x_{1,e} \tag{3.5.58}$$

where $\mathcal{G}_{m1}$ has units (kmol/m^{2}sec).

Low Mass Transfer Rate Theory

Consider, as an example, the evaporation of water into air, as shown in Figure 3.5.12. The water–air interface might be the surface of a water reservoir, or the surface of a falling water film in a cooling tower or humidifier. In such situations the mass fraction of water vapor in the air is relatively small; the highest value is at the *s*-surface, but even if the water temperature is as high as 50°C, the corresponding value of $m_{H_2O,s}$ at 1 atm total pressure is only 0.077. From Equation (3.5.54) the driving potential for diffusion of water vapor away from the interface is $\Delta m_1 = m_{1,s} - m_{1,e}$, and is small compared to unity, even if the free-stream air is very dry such that $m_{1,e} \simeq 0$. We then say that the mass transfer rate is *low* and the rate of evaporation of the water can be approximated as $j_{1,s}$; for a surface area A,

$$\dot{m}_1 = \left(m_{1,s} n_s + j_{1,s}\right) A \simeq j_{1,s} A \text{ kg/sec} \tag{3.5.59}$$

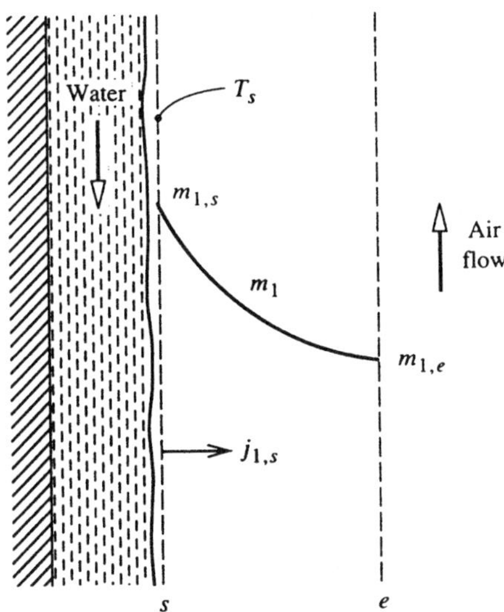

FIGURE 3.5.12 Evaporation of water into an air flow.

In contrast, if the water temperature approaches its boiling point, $m_{1,s}$ is no longer small, and of course, in the limit of $T_s = T_{BP}$, $m_{1,s} = 1$. The resulting driving potential for diffusion Δm_1 is then large, and we say that the mass transfer rate is *high*. Then, the evaporation rate cannot be calculated from Equation (3.5.59), as will be explained in the section on high mass transfer rate theory. For water evaporation into air, the error incurred in using low mass transfer rate theory is approximately $(^1/_2)\ \Delta m_1$, and a suitable criterion for application of the theory to engineering problems is $\Delta m_1 < 0.1$ or 0.2.

A large range of engineering problems can be adequately analyzed assuming low mass transfer rates. These problems include cooling towers and humidifiers as mentioned above, gas absorbers for sparingly soluble gases, and catalysis. In the case of catalysis, the *net* mass transfer rate is actually zero. Reactants diffuse toward the catalyst surface and the products diffuse away, but the catalyst only promotes the reaction and is not consumed. On the other hand, problems that are characterized by high mass transfer rates include condensation of steam containing a small amount of noncondensable gas, as occurs in most power plant condensers; combustion of volatile liquid hydrocarbon fuel droplets in diesel engines and oil-fired power plants, and ablation of phenolic-based heat shields on reentry vehicles.

Dimensionless Groups

Dimensional analysis of convective mass transfer yields a number of pertinent dimensionless groups that are, in general, analogous to dimensionless groups for convective heat transfer. The most important groups are as follows.

1. The Schmidt number, $\mathrm{Sc}_{12} = \mu/\rho\mathcal{D}_{12}$, which is a properties group analogous to the Prandtl number. For gas mixtures, $\mathrm{Sc}_{12} = O(1)$, and for liquid solutions, $\mathrm{Sc}_{12} = O(100)$ to $O(1000)$. There are not fluids for which $\mathrm{Sc}_{12} \ll 1$, as is the case of Prandtl number for liquid metals.
2. The Sherwood number (or mass transfer Nusselt number). $\mathrm{Sh} = \mathcal{g}_{m1}L/\rho\mathcal{D}_{12}$ $(= \mathcal{G}_{m1}L/c\mathcal{D}_{12})$ is a dimensionless conductance.
3. The mass transfer Stanton number $\mathrm{St}_m = \mathcal{g}_{m1}/\rho V$ $(= \mathcal{G}_{m1}/cV)$ is an alternative dimensionless conductance.

As for convective heat transfer, forced convection flows are characterized by a Reynolds number, and natural convection flows are characterized by a Grashof or Rayleigh number. In the case of Gr or Ra it is not possible to replace $\Delta\rho$ by $\beta\Delta T$ since density differences can result from concentration differences (and both concentration and temperature differences for simultaneous heat and mass transfer problems).

Analogy between Convective Heat and Mass Transfer

A close analogy exists between convective heat and convective mass transfer owing to the fact that conduction and diffusion in a fluid are governed by physical laws of identical form, that is, Fourier's and Fick's laws, respectively. As a result, in many circumstances the Sherwood or mass transfer Stanton number can be obtained in a simple manner from the Nusselt number or heat transfer Stanton number for the same flow conditions. Indeed, in most gas mixtures Sh and St_m are nearly equal to their heat transfer counterparts. For dilute mixtures and solutions and low mass transfer rates, the rule for exploiting the analogy is simple: The *Sherwood or Stanton number is obtained by replacing the Prandtl number by the Schmidt number in the appropriate heat transfer correlation.* For example, in the case of fully developed turbulent flow in a smooth pipe

$$\mathrm{Nu}_D = 0.023\mathrm{Re}_D^{0.8}\mathrm{Pr}^{0.4}; \qquad \mathrm{Pr} > 0.5 \tag{3.5.60a}$$

which for mass transfer becomes

$$\mathrm{Sh}_D = 0.023\mathrm{Re}_D^{0.8}\mathrm{Sc}^{0.4}; \qquad \mathrm{Sc} > 0.5 \tag{3.5.60b}$$

Also, for natural convection from a heated horizontal surface facing upward,

$$\overline{\mathrm{Nu}} = 0.54\left(\mathrm{Gr}_L\mathrm{Pr}\right)^{1/4}; \qquad 10^5 < \mathrm{Gr}_L\mathrm{Pr} < 2\times10^7 \ \text{(laminar)} \tag{3.5.61a}$$

$$\overline{\mathrm{Nu}} = 0.14\left(\mathrm{Gr}_L\mathrm{Pr}\right)^{1/3}; \qquad 2\times10^7 < \mathrm{Gr}_L\mathrm{Pr} < 3\times10^{10} \ \text{(turbulent)} \tag{3.5.61b}$$

which for isothermal mass transfer with $\rho_s < \rho_e$ become

$$\overline{\mathrm{Sh}} = 0.54\left(\mathrm{Gr}_L\mathrm{Sc}\right)^{1/4}; \qquad 10^5 < \mathrm{Gr}_L\mathrm{Sc} < 2\times10^7 \ \text{(laminar)} \tag{3.5.62a}$$

$$\overline{\mathrm{Sh}} = 0.14\left(\mathrm{Gr}_L\mathrm{Sc}\right)^{1/3}; \qquad 2\times10^7 < \mathrm{Gr}_L\mathrm{Sc} < 3\times10^{10} \ \text{(turbulent)} \tag{3.5.62b}$$

With evaporation, the condition, $\rho_s < \rho_e$ will be met when the evaporating species has a smaller molecular weight than the ambient species, for example, water evaporating into air. Mass transfer correlations can be written down in a similar manner for almost all the heat transfer correlations given in Section 3.2. There are some exceptions: for example, there are no fluids with a Schmidt number much less than unity, and thus there are no mass transfer correlations corresponding to those given for heat transfer to liquid metals with $\mathrm{Pr} \ll 1$. In most cases it is important for the wall boundary conditions to be of analogous form, for example, laminar flow in ducts. A uniform wall temperature corresponds to a uniform concentration $m_{1,s}$ along the *s*-surface, whereas a uniform heat flux corresponds to a uniform diffusive flux $j_{1,s}$. In chemical engineering practice, the analogy between convective heat and mass transfer is widely used in a form recommended by Chilton and Colburn in 1934, namely, $\mathrm{St}_m/\mathrm{St} = (\mathrm{Sc}/\mathrm{Pr})^{-2/3}$. The Chilton-Colburn form is of adequate accuracy for most external forced flows but is inappropriate for fully developed laminar duct flows.

For example, air at 1 atm and 300 K flows inside a 3-cm-inside-diameter tube at 10 m/sec. Using pure-air properties the Reynolds number is $VD/\nu = (10)(0.03)/15.7 \times 10^{-6} = 1.911 \times 10^4$. The flow is turbulent. Using Equation (3.5.60b) with $\mathrm{Sc}_{12} = 0.61$ for small concentrations of H_2O in air,

$$\mathrm{Sh}_D = (0.023)\left(1.911\times10^4\right)^{0.8}(0.61)^{0.4} = 50.2$$

$$\mathcal{g}_{m1} = \rho\mathcal{D}_{12}\text{Sh}/D = \rho\nu\text{Sh}/\text{Sc}_{12}D = \frac{(1.177)(15.7\times10^{-6})(50.2)}{(0.61)(0.03)} = 5.07\times10^{-2}\ \text{kg/m}^2\ \text{sec}$$

Further insight into this analogy between convective heat and mass transfer can be seen by writing out Equations (3.5.60a) and (3.5.60b) as, respectively,

$$\frac{(h_c/c_p)D}{k/c_p} = 0.023\text{Re}_D^{0.8}\left(\frac{\mu}{k/c_p}\right)^{0.4} \tag{3.5.63a}$$

$$\frac{\mathcal{g}_m D}{\rho\mathcal{D}_{12}} = 0.023\text{Re}_D^{0.8}\left(\frac{\mu}{\rho\mathcal{D}_{12}}\right)^{0.4} \tag{3.5.63b}$$

When cast in this form, the correlations show that the property combinations k/c_p and $\rho\mathcal{D}_{12}$ play analogous roles; these are *exchange coefficients* for heat and mass, respectively, both having units kg/m sec, which are the same as those for dynamic viscosity μ. Also, it is seen that the ratio of heat transfer coefficient to specific heat plays an analogous role to the mass transfer conductance, and has the same units (kg/m^2 sec). Thus, it is appropriate to refer to the ratio h_c/c_p as the *heat transfer conductance*, $\mathcal{g}_h$, and for this reason we should not refer to $\mathcal{g}_m$ as the mass transfer *coefficient.*

Simultaneous Heat and Mass Transfer

Often problems involve simultaneous convective heat and mass transfer, for which the surface energy balance must be carefully formulated. Consider, for example, evaporation of water into air, as shown in Figure 3.5.13. With H_2O denoted as species 1, the steady-flow energy equation applied to a control volume located between the u- and s-surfaces requires that

$$\dot{m}\left(h_{1,s} - h_{1,u}\right) = A\left(q''_{\text{cond}} - q''_{\text{conv}} - q''_{\text{rad}}\right)\ \text{W} \tag{3.5.64}$$

where it has been recognized that only species 1 crosses the u- and s-surfaces. Also, the water has been assumed to be perfectly opaque so that all radiation is emitted or absorbed between the u-surface and the interface.

If we restrict our attention to conditions for which low mass transfer rate theory is valid, we can write $\dot{m}/A \simeq j_{1,s} = \mathcal{g}_{m1}\,(m_{1,s} - m_{1,e})$. Also, we can then calculate the convective heat transfer as if there were no mass transfer, and write $q_{\text{conv}} = h_c(T_s - T_e)$. Substituting in Equation (3.5.64) with $q_{\text{conv}} = -k\partial T/\partial y|_u$, $h_{1,s} - h_{1,u} = h_{fg}$, and rearranging, gives

$$-k\left.\frac{\partial T}{\partial y}\right|_u = h_c\left(T_s - T_e\right) + \mathcal{g}_{m1}\left(m_{1,s} - m_{1,e}\right)h_{fg} + q''_{\text{rad}}\ \text{W/m}^2 \tag{3.5.65}$$

It is common practice to refer to the convective heat flux $h_c(T_s - T_e)$ as the *sensible* heat flux, whereas the term $\mathcal{g}_{m1}\,(m_{1,s} - m_{1,e})\,h_{fg}$ is called the *evaporative* or *latent* heat flux. Each of the terms in Equation (3.5.65) can be positive or negative, depending on the particular situation. Also, the evaluation of the conduction heat flux at the u-surface, $-k\partial T/\partial y|_u$, depends on the particular situation. Four examples are shown in Figure 3.5.13. For a water film flowing down a packing in a cooling tower (Figure 3.5.13b), this heat flux can be expressed in terms of convective heat transfer from the bulk water at temperature T_L to the surface of the film, $-k\partial T/\partial y|_u = h_{cL}\,(T_L - T_s)$. If the liquid-side heat transfer coefficient h_{cL} is large enough, we can simply set $T_s \simeq T_L$, which eliminates the need to estimate h_{cL}. The evaporation process is then *gas-side controlled.* Figure 3.5.13c shows film condensation from a steam-air mixture on the outside of a vertical tube. In this case we can write $k\partial T/\partial y|_u = U(T_s - T_c)$, where T_c is the coolant

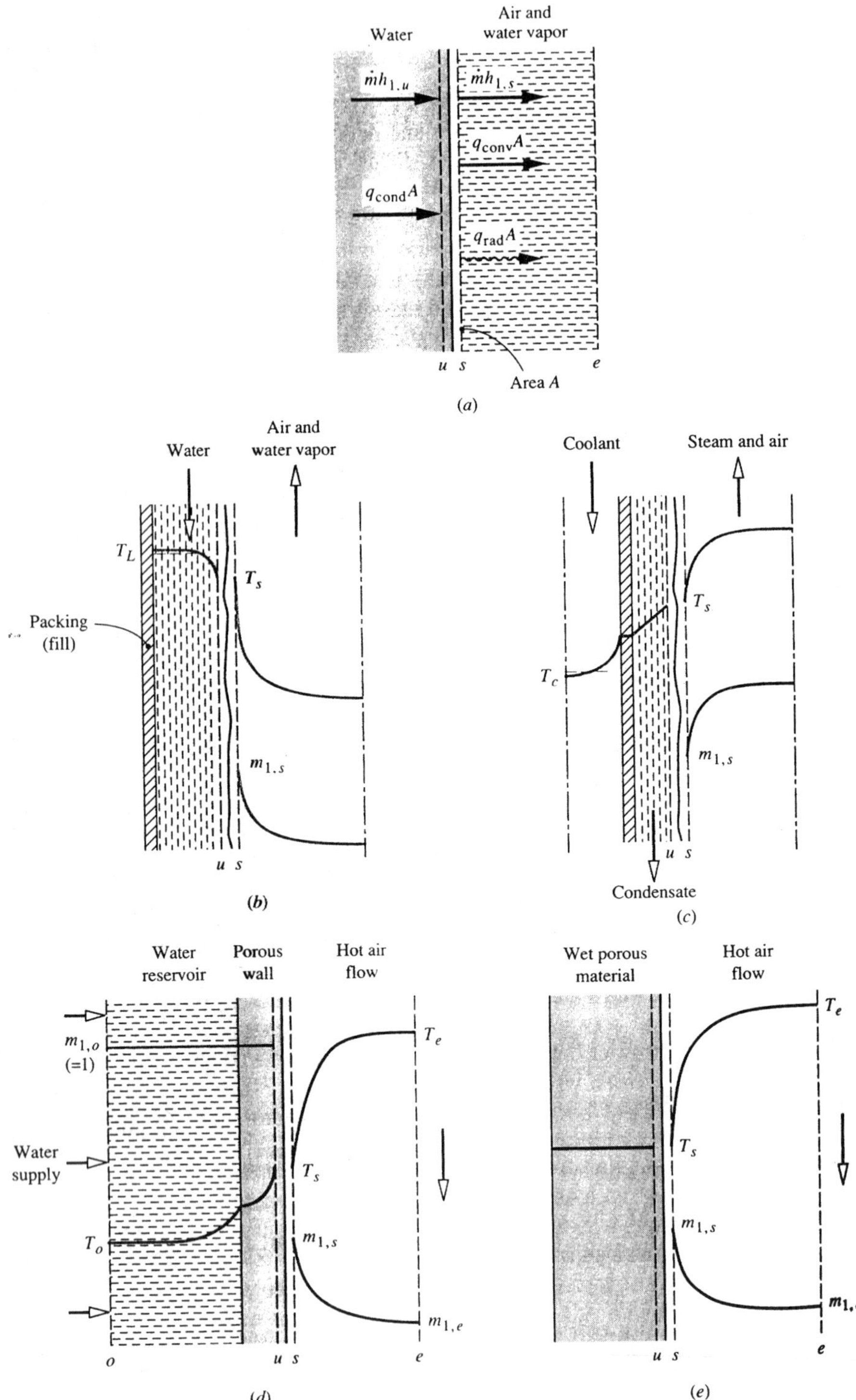

FIGURE 3.5.13 The surface energy balance for evaporation of water into an air stream.

bulk temperature. The overall heat transfer coefficient U includes the resistances of the condensate film, the tube wall, and the coolant. Sweat cooling is shown in Figure 3.5.13d, with water from a reservoir (or *plenum chamber*) injected through a porous wall at a rate just sufficient to keep the wall surface wet. In this case, the conduction across the u-surface can be related to the reservoir conditions by application of the steady-flow energy equation to a control volume located between the o- and u-surfaces. Finally, Figure 3.5.13e shows drying of a wet porous material (e.g., a textile or wood). During the constant-rate period of the process, evaporation takes place from the surface with negligible heat conduction into the solid; then $-k\partial T/\partial y|_u \simeq 0$. The term *adiabatic vaporization* is used to describe evaporation when $q_{\text{cond}} = 0$; constant-rate drying is one example, and the wet-bulb psychrometer is another.

Consider a 1-m-square wet towel on a washline on a day when there is a low overcast and no wind. The ambient air is at 21°C, 1 atm, and 50.5% RH. In the constant-rate drying period the towel temperature is constant, and $q_{\text{cond}} = 0$. An iterative calculation is required to obtain the towel temperature using correlations for natural convection on a vertical surface to obtain h_c and g_{m1}; q_{rad} is obtained as $q_{\text{rad}} = \sigma\varepsilon(T_s^4 - T_e^4)$ with $\varepsilon = 0.90$. The results are $T_s = 17.8°\text{C}$, $h_c = 1.69\ \text{W/m}^2\text{K}$, $g_{m1} = 1.82 \times 10^{-3}\ \text{kg/m}^2\text{sec}$, and the energy balance is

$$q_{\text{cond}} = h_c\left(T_s - T_e\right) + g_{m1}\left(m_{1,s} - m_{1,e}\right)h_{fg} + q_{\text{rad}}$$

$$0 = -5.4 + 21.7 - 16.3\ \text{W/m}^2$$

Evaluation of composition-dependent properties, in particular the mixture specific heat and Prandtl number, poses a problem. In general, low mass transfer rates imply small composition variations across a boundary layer, and properties can be evaluated for a mixture of the free-stream composition at the mean film temperature. In fact, when dealing with evaporation of water into air, use of the properties of dry air at the mean film temperature gives results of adequate engineering accuracy. If there are large composition variations across the boundary layer, as can occur in some catalysis problems, properties should be evaluated at the mean film composition and temperature.

The Wet- and Dry-Bulb Psychrometer

The wet- and dry-bulb psychrometer is used to measure the moisture content of air. In its simplest form, the air is made to flow over a pair of thermometers, one of which has its bulb covered by a wick whose other end is immersed in a small water reservoir. Evaporation of water from the wick causes the wet bulb to cool and its steady-state temperature is a function of the air temperature measured by the dry bulb and the air humidity. The wet bulb is shown in Figure 3.5.14. In order to determine the water vapor mass fraction $m_{1,e}$, the surface energy balance Equation (3.5.66) is used with conduction into the wick and q''_{rad} set equal to zero. The result is

$$m_{1,e} = m_{1,s} - \frac{c_p}{h_{fg}}\left(\frac{\text{Pr}}{\text{Sc}_{12}}\right)^{-2/3}\left(T_e - T_s\right) \tag{3.5.66}$$

Usually $m_{1,e}$ is small and we can approximate $c_p = c_{p\ \text{air}}$ and $(\text{Pr}/\text{Sc}_{12})^{-2/3} = 1/1.08$. Temperatures T_s and T_e are the known measured wet- and dry-bulb temperatures. With T_s known, $m_{1,s}$ can be obtained using steam tables in the usual way. For example, consider an air flow at 1000 mbar with measured wet- and dry-bulb temperatures of 305.0 and 310.0 K, respectively. Then $P_{1,s} = P_{\text{sat}}(T_s) = P_{\text{sat}}(305.0\ \text{K}) = 4714$ Pa from steam tables. Hence, $x_{1,s} = P_{1,s}/P = 4714/10^5 = 0.04714$, and

$$m_{1,s} = \frac{0.04714}{0.04714 + (29/18)(1 - 0.04714)} = 0.02979$$

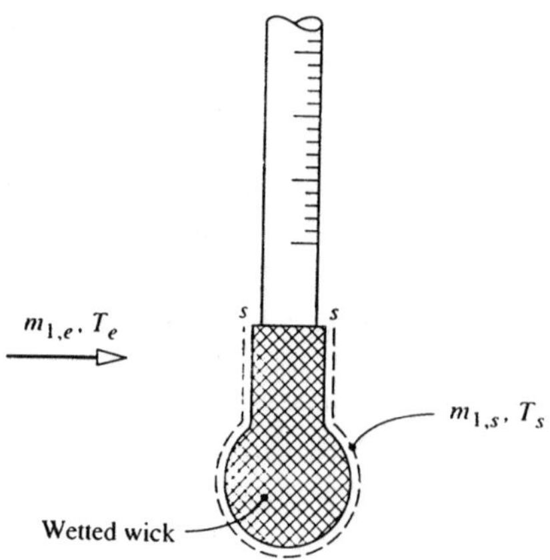

FIGURE 3.5.14 Wet bulb of a wet- and dry-bulb psychrometer.

Also, h_{fg} (305 K) = 2.425 × 10^6 J/kg, and $c_{p\ \mathrm{air}}$ = 1005 J/kg K; thus

$$m_{1,e} = 0.02979 - \frac{1005}{(1.08)(2.425\times10^6)}(310-305) = 0.02787$$

$$x_{1,e} = \frac{0.02787}{0.02787+(18/29)(1-0.02787)} = 0.04415$$

$$P_{1,e} = x_{1,e}P = (0.04415)(10^5) = 4412 \text{ Pa}$$

By definition, the relative humidity is RH = $P_{1,e}/P_{\mathrm{sat}}(T_e)$; RH = 4415/6224 = 70.9%.

In the case of other adiabatic vaporization processes, such as constant-rate drying or evaporation of a water droplet, $m_{1,e}$ and T_e are usually known and Equation (3.5.66) must be solved for T_s. However, the thermodynamic wet-bulb temperature obtained from psychrometric charts or software is accurate enough for engineering purposes.

High Mass Transfer Rate Theory

When there is net mass transfer across a phase interface, there is a convective component of the absolute flux of a species across the *s*-surface. From Equation (3.5.23a) for species 1,

$$n_{1,s} = \rho_{1,s}v_s + j_{1,s} \text{ kg/m}^2 \text{ sec} \tag{3.5.67}$$

During evaporation the convection is directed in the gas phase, with a velocity normal to the surface v_s. When the convective component cannot be neglected, we say that the mass transfer rate is *high*. There are two issues to consider when mass transfer rates are high. First, the rate at which species 1 is transferred across the *s*-surface is not simply the diffusive component $j_{1,s}$ as assumed in low mass transfer rate theory, but is the sum of the convective and diffusive components shown in Equation (3.5.67). Second, the normal velocity component v_s has a *blowing* effect on the concentration profiles, and hence on the Sherwood number. The Sherwood number is no longer analogous to the Nusselt number of conventional heat transfer correlations, because those Nusselt numbers are for situations involving impermeable surfaces, e.g., a metal wall, for which $v_s = 0$.

Substituting for $j_{1,s}$ from Equation (3.5.57) into Equation (3.5.67) gives

$$\dot{m}'' = g_{m1}\frac{m_{1,e} - m_{1,s}}{m_{1,s} - n_{1,s}/\dot{m}''} = g_{m1}\mathcal{B}_{m1} \tag{3.5.68}$$

where $\dot{m}'' = n_s$ is the mass transfer rate introduced in the section on heterogeneous combustion and $\mathcal{B}_{m1}$ is the *mass transfer driving force*. In the special case where only species 1 is transferred, $n_{1,s}/\dot{m}'' = 1$, for example, when water evaporates into air, and dissolution of air in the water is neglected. It is convenient to rewrite Equation (3.5.68) as

$$\dot{m}'' = g^*_{m1}\left(g_{m1}/g^*_{m1}\right)\mathcal{B}_{m1} \text{ kg/m}^2 \text{ sec} \tag{3.5.69a}$$

where

$$g^*_{m1} = \lim_{\mathcal{B}_m \to 0} g_{m1} \tag{3.5.69b}$$

Now g^*_{m1} is the limit value of g_{m1} for zero mass transfer (i.e., $v_s = 0$), and Sh^* can be obtained from conventional heat transfer Nusselt number correlations for impermeable surfaces. The ratio (g_{m1}/g^*_{m1}) is termed a *blowing factor* and accounts for the effect of v_s on the concentration profiles. Use of Equation (3.5.69) requires appropriate data for the blowing factor. For the constant-property laminar boundary layer on a flat plate, Figure 3.5.15 shows the effect of the Schmidt number on the blowing factor. The abscissa is a *blowing parameter* $B_m = \dot{m}''/g^*_m$.

The blowing velocity also affects the velocity and temperature profiles, and hence the wall shear stress and heat transfer. The curve for Sc = 1 in Figure 3.5.15 also gives the effect of blowing on shear stress as τ_s/τ^*_s, and the curve for Sc = 0.7 gives the effect of blowing on heat transfer for air injection into air as h_c/h^*_c (since Pr = 0.7 for air).

Variable Property Effects of High Mass Transfer Rates

High mass transfer rate situations are usually characterized by large property variations across the flow, and hence property evaluation for calculating g_m and h_c is not straightforward. An often-encountered situation is transfer of a single species into an inert laminar or turbulent boundary layer flow. The effect of variable properties can be very large as shown in Figure 3.5.16 for laminar boundary layers, and Figure 3.5.17 for turbulent boundary layers.

A simple procedure for correlating the effects of flow type and variable properties is to use weighting factors in the exponential functions suggested by a constant-property Couette-flow model (Mills, 1995). Denoting the injected species as species i, we have

$$\frac{g_{m1}}{g^*_{m1}} = \frac{a_{mi}B_{mi}}{\exp\left(a_{mi}B_{mi}\right) - 1}; \qquad B_{mi} = \frac{\dot{m}''}{g^*_{mi}} \tag{3.5.70a}$$

or

$$\frac{g_{m1}}{g^*_{mi}} = \frac{\ln\left(1 + a_{mi}\mathcal{B}_{mi}\right)}{a_{mi}\mathcal{B}_{mi}}; \qquad \mathcal{B}_{mi} = \frac{\dot{m}''}{g_{mi}} = \frac{m_{i,e} - m_{i,e}}{m_{i,s} - 1}$$

$$\frac{\tau_s}{\tau^*_s} = \frac{a_{fi}B_f}{\exp\left(a_{fi}B_f\right) - 1}; \qquad B_f = \frac{\dot{m}''u_e}{\tau^*_s} \tag{3.5.70b}$$

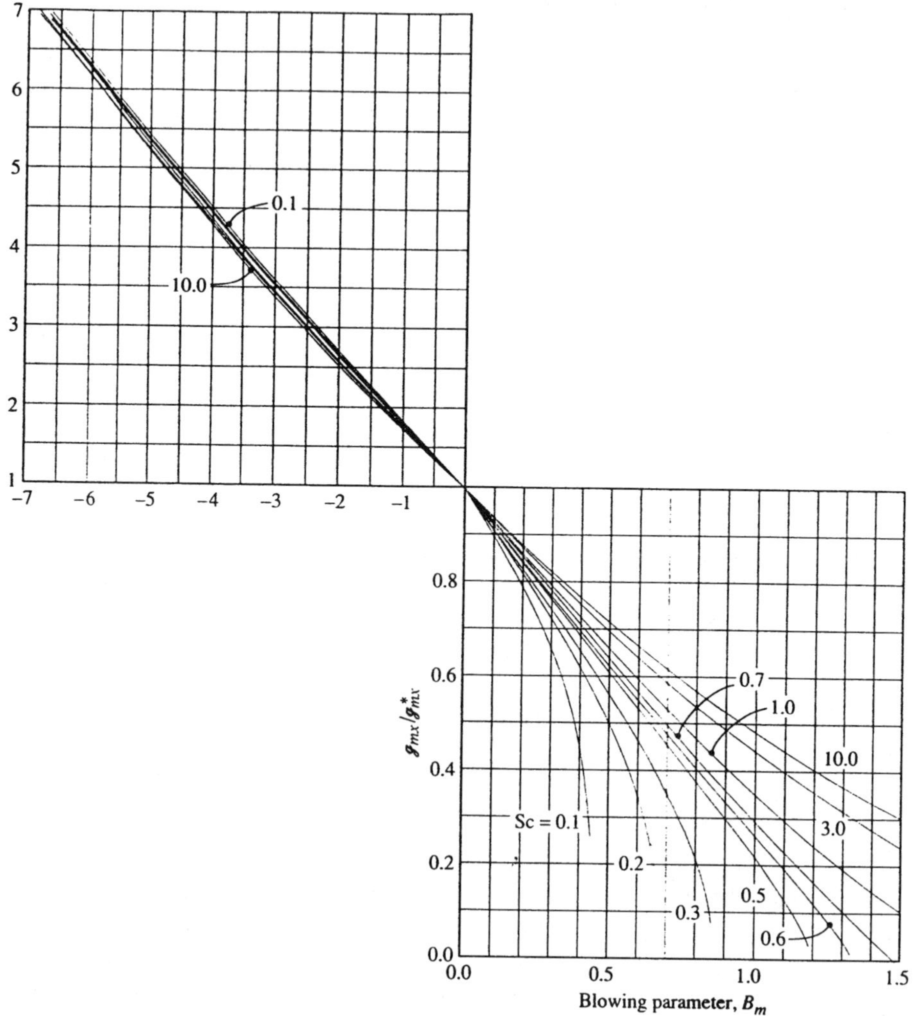

FIGURE 3.5.15 Effect of mass transfer on the mass transfer conductance for a laminar boundary layer on a flat plate: g_m / g^*_m vs. blowing parameter $B_m = \dot{m}'' / g^*_m$.

$$\frac{h_c}{h_c^*} = \frac{a_{hi} B_h}{\exp(a_{hi} B_h) - 1}; \qquad B_h = \frac{\dot{m}'' c_{pe}}{h_c^*} \tag{3.5.70c}$$

Notice that g^*_{mi}, τ^*_s, h^*_c, and c_{pe} are evaluated using properties of the free-stream gas at the mean film temperature. The weighting factor a may be found from exact numerical solutions of boundary layer equations or from experimental data. Some results for laminar and turbulent boundary layers follow.

1. *Laminar Boundary Layers.* We will restrict our attention to low-speed air flows, for which viscous dissipation and compressibility effects are negligible, and use exact numerical solutions of the self-similar laminar boundary layer equations (Wortman, 1969). Least-squares curve fits of the numerical data were obtained using Equations (3.5.70a) to (3.5.70c). Then, the weighting factors for axisymmetric stagnation-point flow with a cold wall ($T_s/T_e = 0.1$) were correlated as

$$a_{mi} = 1.65 (M_{\text{air}} / M_i)^{10/12} \tag{3.5.71a}$$

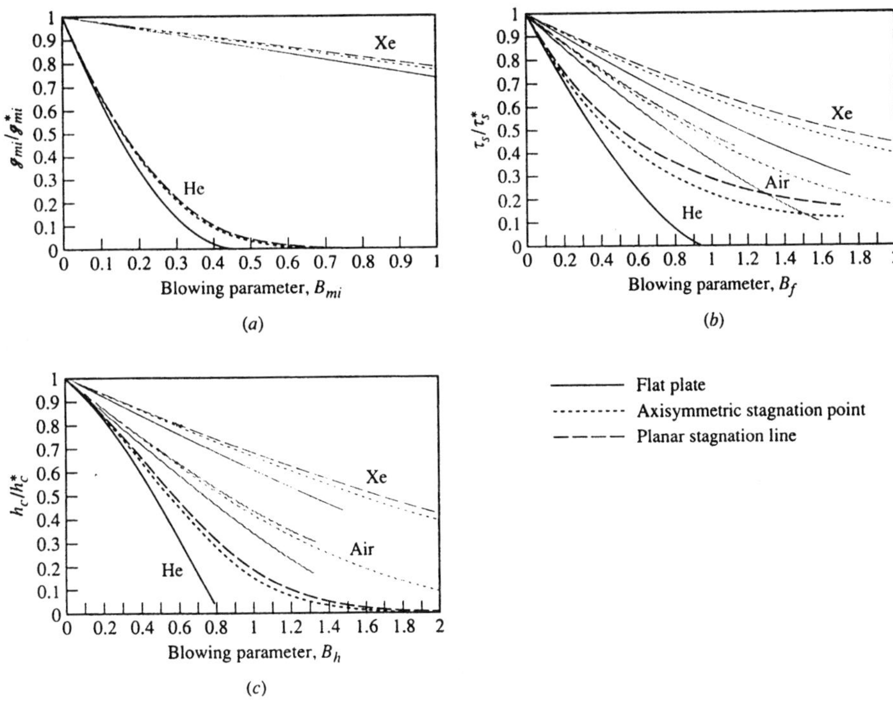

FIGURE 3.5.16 Numerical results for the effect of pressure gradient and variable properties on blowing factors for laminar boundary layers: low-speed air flow over a cold wall ($T_s/T_e = 0.1$) with foreign gas injection: (a) mass transfer conductance, (b) wall shear stress, (c) heat transfer coefficient. (From Wortman, A., Ph.D. dissertation, University of California, Los Angeles, 1969. With permission.)

$$a_{fi} = 1.38\left(M_{\text{air}}/M_i\right)^{5/12} \tag{3.5.71b}$$

$$a_{hi} = 1.30\left(M_{\text{air}}/M_i\right)^{3/12}\left[c_{pi}/\left(2.5\mathcal{R}/M_i\right)\right] \tag{3.5.71c}$$

Notice that $c_{pi}/2.5\mathcal{R}/M_i)$ is unity for a monatomic species. For the planar stagnation line and the flat plate, and other values of the temperature ratio T_s/T_e, the values of the species weighting factors are divided by the values given by Equations (3.5.71a,b,c) to give correction factors G_{mi}, G_{fi}, and G_{hi}, respectively. The correction factors are listed in Table 3.5.7.

The exponential relation blowing factors cannot accurately represent some of the more anomalous effects of blowing. For example, when a light gas such as H_2 is injected, Equation (3.5.70c) indicates that the effect of blowing is always to reduce heat transfer, due to both the low density and high specific heat of hydrogen. However, at very low injection rates, the heat transfer is actually increased, as a result of the high thermal conductivity of H_2. For a mixture, $k \approx \sum x_i k_i$ whereas $c_p = \sum m_i c_{pi}$. At low rates of injection, the mole fraction of H_2 near the wall is much larger than its mass fraction; thus, there is a substantial increase in the mixture conductivity near the wall, but only a small change in the mixture specific heat. An increase in heat transfer results. At higher injection rates, the mass fraction of H_2 is also large, and the effect of high mixture specific heat dominates to cause a decrease in heat transfer.

2. *Turbulent Boundary Layers.* Here we restrict our attention to air flow along a flat plate for Mach numbers up to 6, and use numerical solutions of boundary layer equations with a mixing length turbulence model (Landis, 1971). Appropriate species weighting factors for $0.2 < T_s/T_e < 2$ are

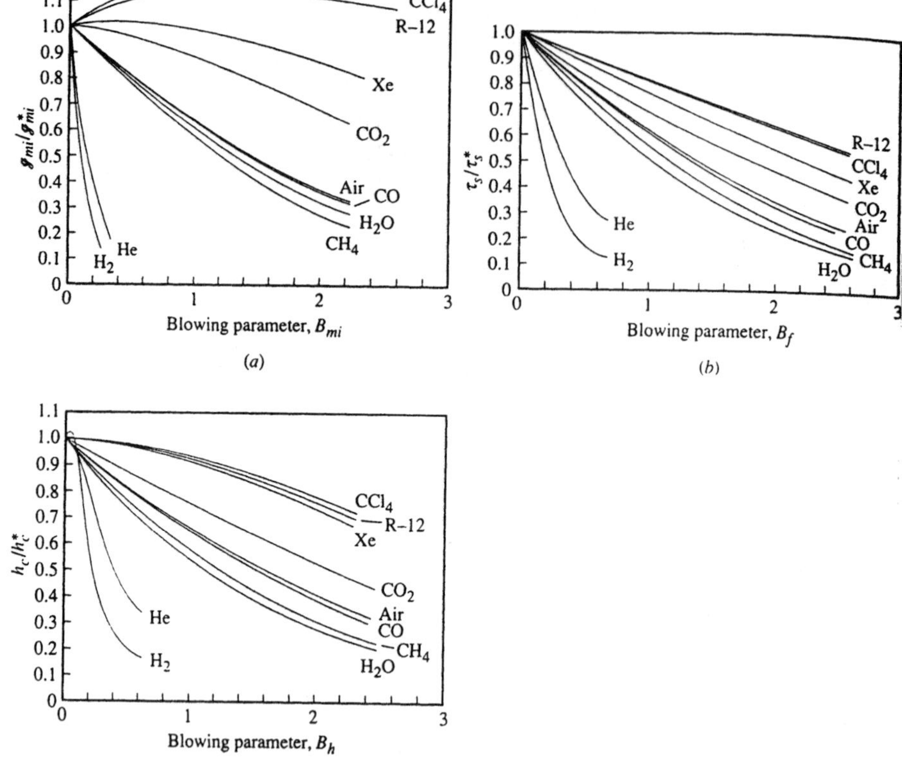

FIGURE 3.5.17 Numerical results for the effect of variable properties on blowing factors for a low-speed turbulent air boundary layer on a cold flat plate ($T_s/T_e = 0.2$) with foreign gas injection: (a) mass transfer conductance, (b) wall shear stress, (c) heat transfer coefficient. (From Landis, R.B., Ph.D. dissertation, University of California, Los Angeles, 1971. With permission.)

$$a_{mi} = 0.79\left(M_{air}/M_i\right)^{1.33} \tag{3.5.72a}$$

$$a_{fi} = 0.91\left(M_{air}/M_i\right)^{0.76} \tag{3.5.72b}$$

$$a_{hi} = 0.86\left(M_{air}/M_i\right)^{0.73} \tag{3.5.72c}$$

In using Equation (3.5.70), the limit values for $\dot{m}'' = 0$ are elevated at the same location along the plate. Whether the injection rate is constant along the plate or varies as $x^{-0.2}$ to give a self-similar boundary layer has little effect on the blowing factors. Thus, Equation (3.5.72) has quite general applicability. Notice that the effects of injectant molecular weight are greater for turbulent boundary layers than for laminar ones, which is due to the effect of fluid density on turbulent transport. Also, the injectant specific heat does not appear in a_{hi} as it did for laminar flows. In general, c_{pi} decreases with increasing M_i and is adequately accounted for in the molecular weight ratio.

Reference State Schemes. The reference state approach, in which constant-property data are used with properties evaluated at some reference state, is an alternative method for handling variable-property effects. In principle, the reference state is independent of the precise property data used and of the

TABLE 3.5.7 Correction Factors for Foreign Gas Injection into Laminar Air Boundary Layers

		$G_{mi}\ T_s/T_e$			$G_{fi}\ T_s/T_e$			$G_{hi}\ T_s/T_e$		
Geometry	**Species**	**0.1**	**0.5**	**0.9**	**0.1**	**0.5**	**0.9**	**0.1**	**0.5**	**0.9**
	H	1.14	1.36	1.47	1.30	1.64	1.79	1.15	1.32	—
	H_2	1.03	1.25	1.36	1.19	1.44	1.49	1.56	1.17	1.32
	He	1.05	1.18	1.25	1.34	1.49	1.56	1.18	1.32	—
	Air	—	—	—	1.21	1.27	1.27	1.17	1.21	—
	Xe	1.21	1.13	1.15	1.38	1.34	1.34	1.19	1.18	—
	CCl_4	1.03	0.95	1.00	1.00	1.03	1.03	1.04	1.04	—
Axisymmetric	H	1.00	1.04	1.09	1.00	0.62	0.45	1.00	0.94	0.54
stagnation point	H_2	1.00	1.06	1.06	1.00	0.70	0.62	1.00	1.00	1.01
	He	1.00	1.04	1.03	1.00	0.66	0.56	1.00	1.00	0.95
	C	1.00	1.01	1.00	1.00	0.79	0.69	1.00	0.99	0.87
	CH_4	1.00	1.01	1.00	1.00	0.88	0.84	1.00	1.00	1.00
	O	1.00	0.98	0.97	1.00	0.79	0.70	1.00	0.98	0.95
	H_2O	1.00	1.01	1.00	1.00	0.82	0.73	1.00	1.00	0.99
	Ne	1.00	1.00	0.98	1.00	0.83	0.75	1.00	0.97	0.95
	Air	—	—	—	1.00	0.87	0.82	1.00	0.99	0.97
	A	1.00	0.97	0.94	1.00	0.93	0.91	1.00	0.96	0.95
	CO_2	1.00	0.97	0.95	1.00	0.96	0.94	1.00	0.99	0.97
	Xe	1.00	0.98	0.96	1.00	0.96	1.05	1.00	1.06	0.99
	CCl_4	1.00	0.90	0.83	1.00	1.03	1.07	1.00	0.96	0.93
	I_2	1.00	0.91	0.85	1.00	1.02	1.05	1.00	0.97	0.94
Planar stagnation line	He	0.96	0.98	0.98	0.85	0.53	0.47	0.93	0.91	0.92
	Air	—	—	—	0.94	0.84	0.81	0.94	0.94	—
	Xe	0.92	0.87	0.83	0.90	0.93	0.95	0.93	0.93	—

Based on numerical data of Wortman (1969). Correlations developed by Dr. D.W. Hatfield.

combination of injectant and free-stream species. A reference state for a boundary layer on a flat plate that can be used in conjunction with Figure 3.5.14 is (Knuth, 1963)

$$m_{1,r} = 1 - \frac{M_2}{M_2 - M_1} \frac{\ln\left(M_e/M_s\right)}{\ln\left(m_{2,e}M_e/m_{2,s}M_s\right)} \tag{3.5.73}$$

$$T_r = 0.5\left(T_e + T_s\right) + 0.2r^*\left(u_e^2/2c_{pr}\right) + 0.1\left[B_{hr} + \left(B_{hr} + B_{mr}\right)\frac{c_{p1} - c_{pr}}{c_{pr}}\right]\left(T_s - T_e\right) \tag{3.5.74}$$

where species 1 is injected into species 2 and r^* is the recovery factor for an impermeable wall. Use of the reference state method is impractical for hand calculations: a computer program should be used to evaluate the required mixture properties.

References

Hirschfelder, J.O., Curtiss, C.F., and Bird, R.B. 1954. *Molecular Theory of Gases and Liquids*, John Wiley & Sons, New York.

Knuth, E.L. 1963. Use of reference states and constant property solutions in predicting mass-, momentum-, and energy-transfer rates in high speed laminar flows, *Int. J. Heat Mass Transfer*, 6, 1–22.

Landis, R.B. 1972. Numerical solution of variable property turbulent boundary layers with foreign gas injection, Ph.D. dissertation, School of Engineering and Applied Science, University of California, Los Angeles.

Law, C.K. and Williams, F.A. 1972. Kinetics and convection in the combustion of alkane droplets, *Combustion and Flame*, 19, 393–405.

Mills, A.F. 1995. *Heat and Mass Transfer*, Richard D. Irwin, Chicago.
Wortman, A. 1969. Mass transfer in self-similar boundary-layer flows, Ph.D. dissertation, School of Engineering and Applied Science, University of California, Los Angeles.

Further Information

Geankoplis, C.J. 1993. *Transport Processes and Unit Operations*, 3rd ed., Prentice-Hall, Englewood Cliffs, NJ. This text gives a chemical engineering perspective on mass transfer.
Mills, A.F. 1995. *Heat and Mass Transfer*, Richard D. Irwin, Chicago. Chapter 11 treats mass transfer equipment relevant to mechanical engineering.
Strumillo, C. and Kudra, T. 1986. *Drying: Principles, Applications and Design*, Gordon and Breach, New York.
Mujamdar, A.S.. Ed. 1987. *Handbook of Industrial Drying*, Marcel Dekker, New York.

4

Applications

Frank Kreith, Editor
Engineering Consultant
University of Colorado

Klaus Timmerhaus, Editor
University of Colorado

Noam Lior
University of Pennsylvania

Henry Shaw
New Jersey Institute of Technology

Ramesh K Shah
Delphi Harrison Thermal Systems

Kenneth J. Bell
Oklahoma State University

Kenneth R. Diller
University of Texas/Austin

Jonathan W. Valvano
University of Texas/Austin

John A. Pearce
University of Texas/Austin

David W. Yarbrough
Tennessee Technical University

Jeff Nowobilski
Praxair, Inc.

Moncef Krarti
University of Colorado

Raymond Cohen
Purdue University

Eckhard Groll
Purdue University

William H. Harden
Ingersoll-Rand Company

Kenneth E. Hickman
York International Corporation

Dilip K. Mistry
Ingersoll-Rand Company

Earl Muir
Copeland Corporation

0-8493-9581-X/00/$0.00+$.50

Robert F. Boehm
University of Nevada

Anthony F. Mills
University of California

Donald W. Radford
Colorado State University

Timothy W. Tong
Colorado State University

Kirtan K. Trivedi
Exxon Research and Engineering Company

Randall F. Barron
Louisiana Tech University

Donald L. Fenton
Kansas State University

Yogesh Jaluria
Rutgers State University

Arthur E. Bergles
Rensselaer Polytechnic Institute

Larry W. Swanson
Simulation Sciences, Inc.

Rolf D. Reitz
University of Wisconsin

Ibrahim Dincer
King Fahd University of Petroleum and Minerals

Kenneth E. Goodson
Stanford University

Pradeep Lall
Motorola

Harold R. Jacobs
CEEMS

Robert J. Moffat
Stanford University

Jungho Kim
University of Maryland

Sherif A. Sherif
University of Florida

Alan T. McDonald
Purdue University

Mihir Sen
University of Notre Dame

K. T. Yang
University of Notre Dame

4.1 Water Desalination

Noam Lior

Introduction and Overview

Water desalination is a process that separates water from a saline water solution. The natural water cycle is the best and most prevalent example of water desalination. Ocean waters evaporate due to solar heating and atmospheric influences; the vapor consisting mostly of fresh water (because of the negligible volatility of the salts at these temperatures) rises buoyantly and condenses into clouds in the cooler atmospheric regions, is transported across the sky by cloud motion, and is eventually deposited back on the earth surface as fresh water rain, snow, and hail. The global freshwater supply from this natural cycle is ample, but many regions on Earth do not receive an adequate share. Population growth, rapidly increasing demand for fresh water, and increasing contamination of the available natural fresh water resources render water desalination increasingly attractive. Water desalinaiton has grown over the last four decades to an output of about 20 million m^3 of fresh water per day, by about 10,000 sizeable land-based water desalination plants.

The salt concentration in the waters being desalted ranges from below 100 ppm wt. (essentially fresh water, when ultrapure water is needed), through several thousand parts per million (brackish waters unsuitable for drinking or agricultural use) and seawater with concentrations between 35,000 and 50,000 ppm. Official salt concentration limits for drinkable water are about 1000 ppm, and characteristic water supplies are restricted to well below 500 ppm, with city water in the U.S. being typically below 100 ppm. Salinity limits for agricultural irrigation waters depend on the type of plant, cultivation, and soil, but are typically below 2000 ppm.

Many ways are availiable for separating water from a saline water solution. The oldest and still prevalent desalination process is distillation. The evaporation of the solution is effected by the addition of heat or by lowering of its vapor pressure, and condensation of these vapors on a cold surface produces fresh water. The three dominant distillation processes are multistage flash (MSF), multieffect (ME), and vapor compression (VC). Until the early 1980s the MSF process was prevalent for desalination. Now membrane processes, especially reverse osmosis (RO), are economical enough to have taken about one third of the market. In all membrane processes separation occurs due to the selective nature of the permeability of a membrane, which permits, under the influence of an external driving force, the passage of either water or salt ions but not of both. The driving force may be pressure (as in RO), electric potential (as in electrodialysis, ED), or heat (as in membrane distillation, MD). A process used for low-salinity solutions is the well-known ion exchange (IE), in which salt ions are preferentially adsorbed onto a material that has the required selective adsorption property and thus reduce the salinity of the water in the solution.

The cost of desalted water is comprised of the capital cost of the plant, the cost of the energy needed for the process, and the cost of operation and maintenance staff and supplies. In large seawater desalination plants the cost of water is about \$1.4 to \$2/m^3, dropping to less than \$1/m^3 for desalting brackish water. A methodology for assessing the economic viability of desalination in comparison with other water supply methods is described by Kasper and Lior (1979). Desalination plants are relatively simple to operate, and progress toward advanced controls and automation is gradually reducing operation expenses. The relative effect of the cost of the energy on the cost of the fresh water produced depends on local conditions, and is up to one half of the total.

The boiling point of a salt solution is elevated as the concentration is increased, and the **boiling point elevation** is a measure of the energy needed for separation. Thermodynamically reversible separation defines the minimal energy requirement for that process. The minimal energy of separation W_{min} in such a process is the change in the Gibbs free energy between the beginning and end of the process, ΔG. The minimal work when the number of moles of the solution changes from n_1 to n_2 is thus

$$W_{min} = \int_{n_1}^{n_2} (\Delta G)\, dn_W \tag{4.1.1}$$

The minimal energy of separation of water from seawater containing 3.45 wt.% salt, at 25°C, is 2.55 kJ/(kg fresh water) for the case of zero fresh water recovery (infinitesimal concentration change) and 2.91 kJ/(kg fresh water) for the case of 25% freshwater recovery. W_{min} is, however, severalfold smaller than the energy necessary for water desalination in practice. Improved energy economy can be obtained when desalination plants are integrated with power generation plants (Aschner, 1980). Such dual-purpose plants save energy but also increase the capital cost and complexity of operation.

Two aspects of the basically simple desalination process require special attention. One is the high-corrosivity of seawater, especially pronounced in the higher-temperature distillation processes, which requires the use of corrosion-resistant expensive materials. Typical materials in use are copper–nickel alloys, stainless steel, titanium, and, at lower temperatures, fiber-reinforced polymers (George et al., 1975). Another aspect is scale formation (Glater et al., 1980; Heitman, 1990). Salts in saline water, particularly calcium sulfate, magnesium hydroxide, and calcium carbonate, tend to precipitate when a certain temperature and concentration are exceeded. The precipitate, often mixed with dirt entering with the seawater and with corrosion products, will gradually plug up pipes, and when depositing on heat transfer surfaces reduces heat transfer rates and thus impairs plant performance. While the ambient-temperature operation of membrane processes reduces scaling, membranes are much more susceptible not only to minute amounts of scaling or even dirt, but also to the presence of certain salts and other compounds that reduce their ability to separate salt from water. To reduce corrosion, scaling, and other problems, the water to be desalted is pretreated. The pretreatment consists of filtration, and may inlude removal of air (deaeration), removal of CO_2 (decarbonation), and selective removal of scale-forming salts (softening). It also includes the addition of chemicals that allow operation at higher temperatures without scale deposition, or which retard scale deposition and/or cause the precipitation of scale which does not adhere to solid surfaces, and that prevent foam formation during the desalination process.

Saline waters, including seawater, contain, besides a variety of inorganic salts, also organic materials and various particles. They differ in composition from site to site, and also change with time due to both natural and person-made causes. Design and operation of desalination plants requires good knowledge of the saline water composition and properties (Fabuss, 1980; Heitman, 1991).

The major water desalination processes that are currently in use or in advanced research stages are concisely described below. Information on detailed modeling can be found in the references.

Distillation Processes

Multistage Flash Evaporation (MSF)

Almost all of the large desalination plants use the MSF process shown schematically in Figure 4.1.1. A photo of an operating plant is shown in Figure 4.1.2. The seawater feed is preheated by internal heat recovery from condensing water vapor during passage through a series of stages, and then heated to its

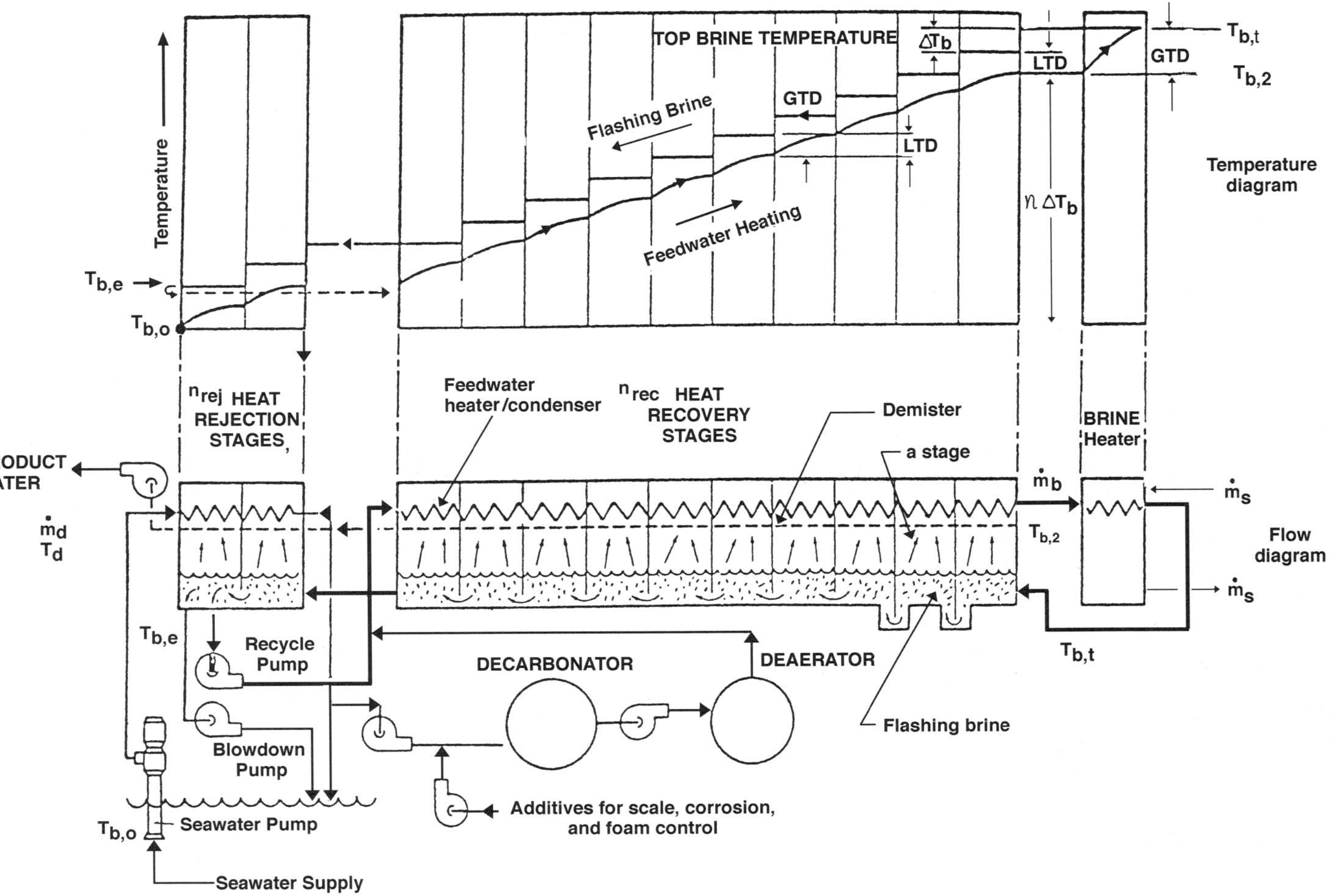

FIGURE 4.1.1 Schematic flow and temperature diagram of the MSF process, for a recirculation type plant.

FIGURE 4.1.2 One of the six units of the 346,000 m^3/day MSF desalination plant Al Taweelah B in Abu Dhabi, United Arab Emirates. (Courtesy of Italimpianti S. p. A.) It is a dual-purpose plant, composed of six identical power and desalination units. Five of the six boilers are seen in the background. The desalination units were in 1996 the largest in the world. They have 17 recovery and 3 reject stages and a performance ratio (PR) of 8.1. The plant also produces 732 MWe of power.

top temperature by steam generated by an external heat source. The hot seawater then flows as a horizontal free-surface stream through a series of "stages," created by vertical walls which separate the vapor space of each stage from the others. These walls allow the vapor space of each stage to be maintained at a different pressure, which is gradually decreased along the flow path due to the gradually decreasing temperature in the condenser/seawater-preheater installed above the free stream. The seawater is superheated by a few degrees celsius relative to the vapor pressure in each stage it enters, and consequently evaporates in each stage along its flow path. The latent heat of the evaporation is supplied by equivalent reduction of the sensible heat of the evaporating water, thus resulting in a gradual lowering of the stream temperature. The evaporation is vigorous, resulting in intensive bubble generation and growth with accompanying stream turbulence, a process known as **flash evaporation** (Lior and Greif, 1980; Miyatake et al., 1992, 1993). One of the primary advantages of the MSF process is the fact that evaporation occurs from the saline water stream and not on heated surfaces (as in other distillation processes such as submerged tube and ME evaporation) where evaporation typically causes scale depostition and thus gradual impairment of heat transfer rates. Also, the fact that the sensible heat of water is much smaller than its latent heat of evaporation, where the specific heat $c_p = 4.182$ kJ/kg/°C change of water temperature and the latent heat is $h_{fg} = 2378$ kJ/kg, and the fact that the top temperature is limited by considerations of scaling and corrosion, dictate the requirement for a very large flow rate of the evaporating stream. For example (in the following, the subscripts b, d, and s refer to brine distillate, and steam, respectively), operating between a typical top temperature $T_{b,t}$ of 90°C at the inlet to the evaporator and an exit temperature $T_{b,e}$ of 40°C corresponding to the ambient conditions, the overall temperature drop of the evaporating stream if 50°C. By using these values, the heat balance between the sensible heat of the water stream, flowing at a mass flow rate $\dot{m}_b$, and the latent heat needed for generating water vapor (distillate) at a mass flow rate $\dot{m}_d$ is

$$\left(\dot{m}_b - \dot{m}_d\right) c_p \left(T_{b,t} - T_{b,e}\right) \approx \dot{m}_d h_{fg} \tag{4.1.2}$$

which yields the brine-to-product mass flow ratio as

$$\frac{\dot{m}_b}{\dot{m}_d} = \frac{h_{fg}}{c_p\left(T_{b,t} - T_{b,e}\right)} + 1 = \frac{2378}{(4.182)(50)} + 1 = 12.37 \tag{4.1.3}$$

Therefore, 12.37 kg of saline water are needed to produce 1 kg of distillate. This high flow rate incurs corresponding pumping equipment and energy expenses, sluggish system dynamics, and, since the stream level depth is limited to about 0.3 to 0.5 m for best evaporation rates, also requires large evaporator vessels with their associated expense.

The generated water vapor rises through a screen ("demister") placed to remove entrained saline water droplets. Rising further, it then condenses on the condenser tube bank, and internal heat recovery is achieved by transferring its heat of condensation to the seawater feed that is thus being preheated. This internal heat recovery is another of the primary advantages of the MSF process. The energy performance of distillation plants is often evaluated by the *performance ratio*, PR, typically defined as

$$\mathrm{PR} \equiv \frac{\dot{m}_d}{\dot{m}_s} \tag{4.1.4}$$

where $\dot{m}_s$ is the mass flow rate of heating steam. Since the latent heat of evaporation is almost the same for the distillate and the heating steam, PR is also the ratio of the heat energy needed for producing one unit mass of product (distillate) to the external heat actually used for that purpose. Most of the heating of the brine stream to the top temperature $T_{b,t}$ is by internal heat recovery, and as seen in Figure 4.1.1, the external heat input is only the amount of heat needed to elevate the temperature of the preheated brine from its exit from the hottest stage at $T_{b,2}$ to $T_{b,t}$. Following the notation in Figure 4.1.1, and using heat balances similar to that in Equation (4.1.3) for the brine heater and flash evaporator, the PR can thus also be defined as

$$\mathrm{PR} = \frac{\dot{m}_b\left(\overline{c_{p,b}}\right)_{e\to t}\left(T_{b,t} - T_{b,e}\right)\Big/h_{fg,b}}{\dot{m}_b\left(\overline{c_{p,b}}\right)_{2\to t}\left(T_{b,t} - T_{b,2}\right)\Big/h_{fg,s}} \approx \frac{T_{b,t} - T_{b,e}}{T_{b,t} - T_{b,2}} \tag{4.1.5}$$

where $(\overline{c_{p,b}})_{e\to t}$ and $(\overline{c_{p,b}})_{2\to t}$ are the specific heats of brine, the first averaged over the temperature range $T_{b,e} \to T_{b,t}$ and the second over $T_{b,2} \to T_{b,t}$. The rightmost expression in Equation (4.1.5) is nearly correct because the specific heat of the brine does not change much with temperature, and the latent heat of evaporation of the brine is nearly equal to the latent heat of condensation of the heating steam. It is obvious from Equation (4.1.5) that PR increases as the top heat recovery temperature $T_{b,2}$ (at the exit from the condenser/brine-preheater) increases. It is also obvious (even from just examining Figure 4.1.1) that increasing the number of stages (matched with a commensurate increase in condenser heat transfer area and assuming no significant change in the overall heat transfer coefficient) for a given $T_{b,t}$, will raise the flash evaporator inlet temperature $T_{b,3}$, which will lead to a rise in $T_{b,2}$ and thus also in the PR.

Assuming that the temperature drop of the flashing brine, ΔT_b, is the same in each stage, the relationship between the number of stages (n) and the performance ratio is

$$\mathrm{PR} = \frac{1}{\dfrac{\mathrm{LTD}}{T_{b,t} - T_{b,e}} + \dfrac{1}{n}} \tag{4.1.6}$$

where LTD is the lowest temperature difference between the flashed vapor and the heated feedwater, in each stage (Figure 4.1.1). Equation (4.1.6) shows that increasing the number of stages increases the PR. This implies that more heat is then recovered internally, which would thus require a larger condenser/brine-preheater heat transfer area. The required heat transfer area, A, *per unit mass of distillate produced* for the entire heat recovery section (composed of n_{rec} stages), and taking average values of the overall vapor-to-feedwater heat transfer coefficient U and LMTD, is thus

$$A = n_{rec} A_n = n_{rec} \frac{h_{b,fg}}{U(\text{LMTD})} \tag{4.1.7}$$

LMTD, the log-mean temperature difference between the vapor condensing on the tubes and the heated brine flowing inside the tubes, for an average stage is

$$\text{LMTD} = \frac{\text{GTD} - \text{LTD}}{\ln \dfrac{\text{GTD}}{\text{LTD}}} = \frac{\left(T_{b,t} - T_{b,2}\right) - \text{LTD}}{\ln\left(\dfrac{T_{b,t} - T_{b,2}}{\text{LTD}}\right)} \tag{4.1.8}$$

where GTD is the greatest temperature difference between the flashing brine and the brine heated in the condenser. The size of the heat transfer area per unit mass of distillate is

$$A = \frac{h_{fg,b}}{U} \frac{n_{rec}}{\left(T_{b,t} - T_{b,e}\right)} \ln\left(\frac{n_{rec}}{n_{rec} - PR}\right) \tag{4.1.9}$$

Examination of this equation will show that the required heat transfer area for the heat recovery section per unit mass of distillate produced, A, increases significantly when PR is increased, and decreases slightly as the number of heat recovery stages, n_{rec}, is increased.

The MSF plant shown in Figure 4.1.1 is of the *recirculation* type, where not all of the brine stream emerging from the last evaporation stage is discharged from the plant (as it would have been in a *once-through* type of plant). A fraction of the emerging brine is mixed with pretreated seawater and recirculated into the condenser of the heat recovery section of the plant. Since only a fraction of the entire stream in this configuration is new seawater, which needs to be pretreated (removal of air and CO_2, i.e., deaeration and decarbonation, and the addition of chemicals that reduce scale deposition, corrosion, and foaming), the overall process cost is reduced. The recirculation plant is also easier to control than the once-through type.

While most of the energy exchange in the plant is internal, steady-state operation requires that energy in an amount equal to all external energy input be also discharged from the plant. Consequently, the heat supplied in the brine heater (plus any pumping energy) is discharged in the heat rejection stages section of the plant (Figure 4.1.1). Assuming an equal temperature drop in each stage, and that the pumping energy can be neglected relative to the heat input in the brine heater, indicates that the ratio of the number of the heat-recovery to heat-rejection stages is approximately equal to the performance ratio PR.

Further detail about MSF desalination can be found in Steinbruchel and Rhinesmith, (1980) and Khan (1986). A detailed design of an MSF plant producing 2.5 million gals. of freshwater per day was published by the U.S. government (Burns and Roe, 1969).

Multi-Effect Distillation (ME)

The principle of the ME distillation process is that the latent heat of condensation of the vapor generated in one effect is used to generate vapor in the next effect, thus obtaining internal heat recovery and good energy efficiency. Several ME plant configurations, most prominently the horizontal tube ME (HTME, shown in Figure 4.1.3) and the vertical tube evaporator (VTE, shown schematically in Figure 4.1.4) are in use. In the HTME, vapor is circulated through a horizontal tube bundle, which is subjected to an external spray of somewhat colder saline water. The vapor flowing in these spray-cooled tubes condenses, and the latent heat of condensation is transferred through the tube wall to the saline water spray striking

FIGURE 4.1.3 Two HTME desalination units, each producing 5000 m^3/day, in St. Croix, U.S. Virgin Islands. (Courtesy of I.D.E. Technologies Ltd.)

the exterior of the tube, causing it to evaporate. The vapor generated thereby flows into the tubes in the next effect, and the process is repeated from effect to effect.

In the VTE the saline water typically flows downward inside vertical tubes and evaporates as a result of condensation of vapor coming from a higher temperature effect on the tube exterior. While internal heat recovery is a feature common to both MSF and ME processes, there are at least three important differences between them. One is that evaporation in the ME process occurs on the heat transfer surfaces (tubes), while in the MSF process it takes place in the free stream. This makes the ME process much more susceptible to scale formation. At the same time, the heat transfer coefficient between the vapor and the preheated brine is lower in the MSF process because the heated brine does not boil. In the ME process it does boil, and it is well known that boiling heat transfer coefficients are significantly higher than those where the heating does not result in boiling. In using direct transfer of latent heat of condensation to latent heat of evaporation, instead of sensible heat reduction to latent heat of evaporation as in MSF, the ME process requires a much smaller brine flow than the MSF. Limiting brine concentration in the last effect to about three times that of the entering seawater, for example, requires a brine flow of only about 1.5 times that of the distillate produced. At the same time, a pump (although much smaller than the two pumps needed in MSF) is needed for each effect.

The PR of ME plants is just sightly lower than the number of effects, which is determined as an optimized compromise between energy efficiency and capital cost. Six effects are typical, although plants with as many as 18 effects have been built.

Further detail about ME desalination can be found in Steinbruchel and Rhinesmith (1980) and Standiford, (1986a).

Vapor Compression Distillation (VC)

As stated earlier, the vapor pressure of saline water is lower than that of pure water at the same temperature, with the pressure difference proportional to the boiling point elevation of the saline water. Desalination is attained here by evaporating the saline water and condensing the vapor on the pure water. Therefore, the pressure of the saline water vapor must be raised by the magnitude of that pressure difference, plus some additional amount to compensate for various losses. This is the principle of the vapor compression desalination method. Furthermore, as shown in Figure 4.1.5, the heat of condensation

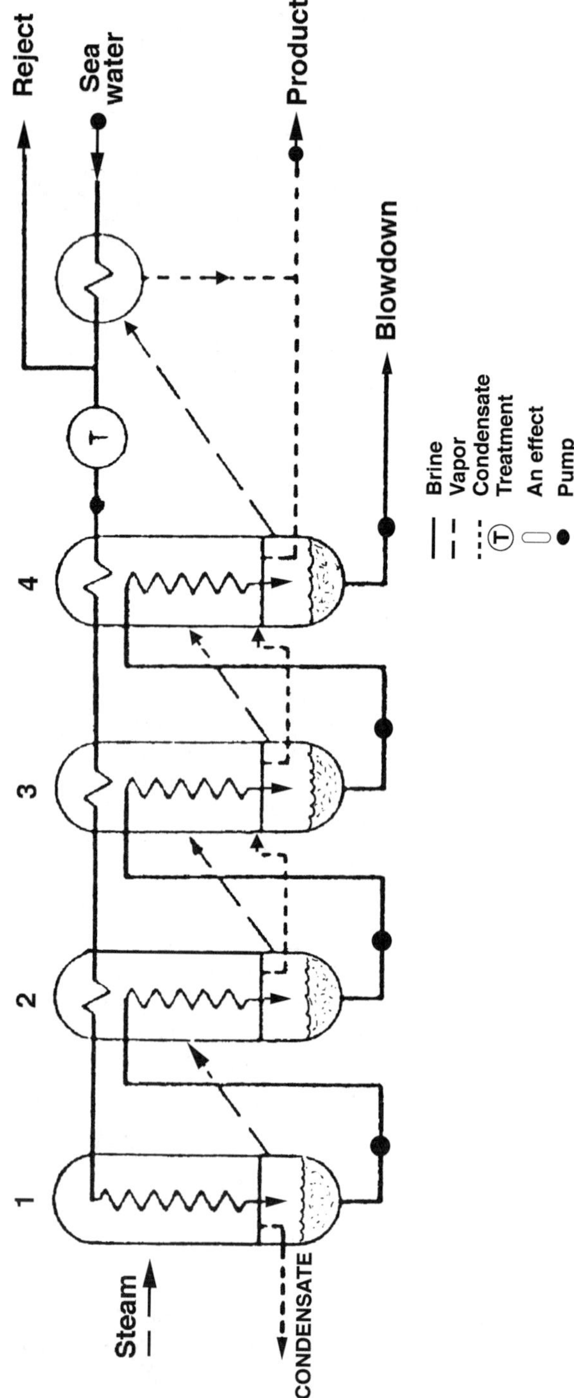

FIGURE 4.1.4 Simplified schematic flow diagram of a typical four-effect VTE desalination plant.

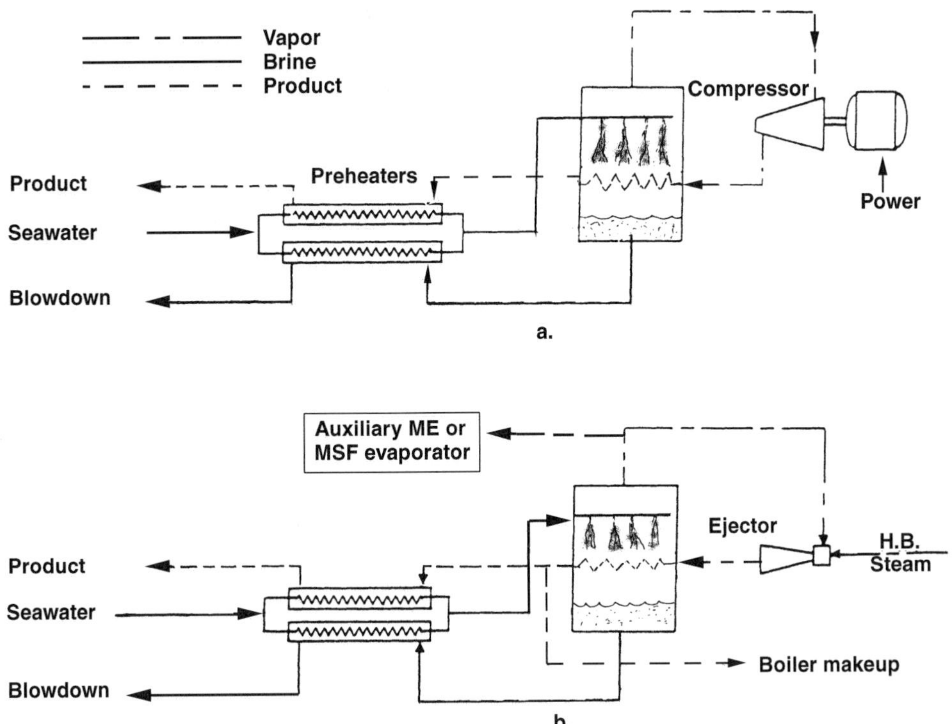

FIGURE 4.1.5 Schematic flow diagram of a basic horizontal-tube VC desalination plant (a) with mechanical, motor-driven compressor; (b) with a thermo-compressor, using an ejector.

of the compressed vapor is recovered internally by using it to evaporate the saline water. Additional heat recovery is obtained by transferring heat from the concentrated brine effluent and the produced freshwater (which need to be cooled down to as close to ambient conditions as possible anyway) to the feed saline water which is thus preheated. The schematic flow diagram in Figure 4.1.5 shows a design in which the preheated seawater is sprayed onto a bank of horizontal tubes carrying condensing compressed vapor at a temperature higher than that of the seawater. The spray thus evaporates on contact with the exterior of the tube and provides the cooling needed for the internal condensation. Considering the fact that the energy required for vapor compression over a typical overall temperature difference of 4°C and a vapor compressor efficiency of 0.8 is 34 kJ/kg (easily calculated from an enthalpy balance), and that the latent heat of condensation is about 2400 kJ/kg, one can see that a small amount of compression energy enables a large amount of heat to be used internally for desalination. One can thus envisage the VC plant as a large flywheel, wheeling a large amount of energy around at the expense of a small amount needed for sustaining its motion.

The compressor can be driven by electric motors, gas or steam turbines, or internal combustion (usually diesel) engines. The compressor can also be a steam-driven ejector (Figure 4.1.5b), which improves plant reliability because of its simplicity and absence of moving parts, but also reduces its efficiency because an ejector is less efficient than a mechanical compressor. In all of the mentioned thermally driven devices, turbines, engines, and the ejector, the exhaust heat can be used for process efficiency improvement, or for desalination by an additional distillation plant.

Figure 4.1.6 shows a multi-effect VC plant. Using more than a single effect reduces the vapor volume that needs to be compressed. Furthermore, the overall required heat transfer area is also decreased because much of the single-phase heat transfer process in the preheater of the single-effect plant is

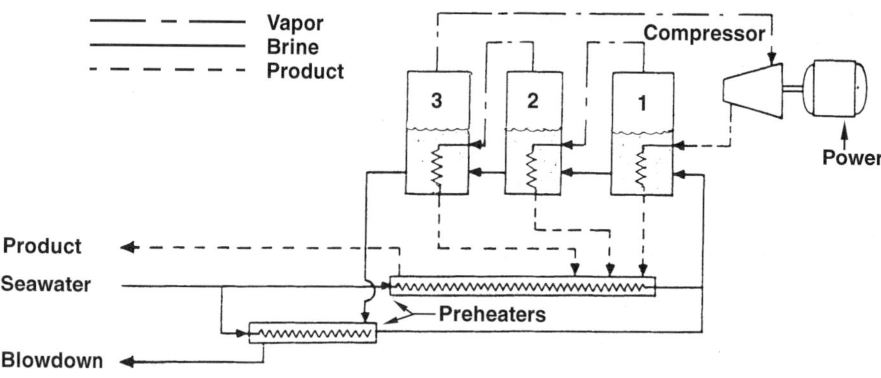

FIGURE 4.1.6 Schematic flow diagram of a ME vapor compression submerged-tube desalinaton plant with three effects.

replaced by the high-heat-transfer condensation–evaporation processes in the effects. Although the ME feature also increases the required compression ratio, the cost of produced water is reduced overall.

Further detail about VC desalination can be found in Steinbruchel and Rhinesmith (1980), Khan (1986), and Standiford, (1986b).

Solar Distillation

The benefits of using the nonpolluting and practically inexhaustible energy of the sun for water desalination are obvious. Furthermore, many water-poor regions also have a relatively high solar flux over a large fraction of the time. The major impediment in the use of solar energy is economical: the diffuse nature of solar energy dictates the need for constructing a large solar energy collection area. For example, assuming a single-effect solar still efficiency of 50% (which is the upper practical limit for conventional designs), the still would produce at most about 3.5 to 4.8 kg fresh water per m^2 per day, or a 208 to 286 m^2 solar still would be required to produce 1 m^3 of fresh water per day. More realistic still efficiencies increase the area requirement about twofold.

Shown in Figure 4.1.7, a typical solar still consists of a saline water container in which the water is exposed to the sun and heated by it. The temperature rise to above ambient causes net evaporation of

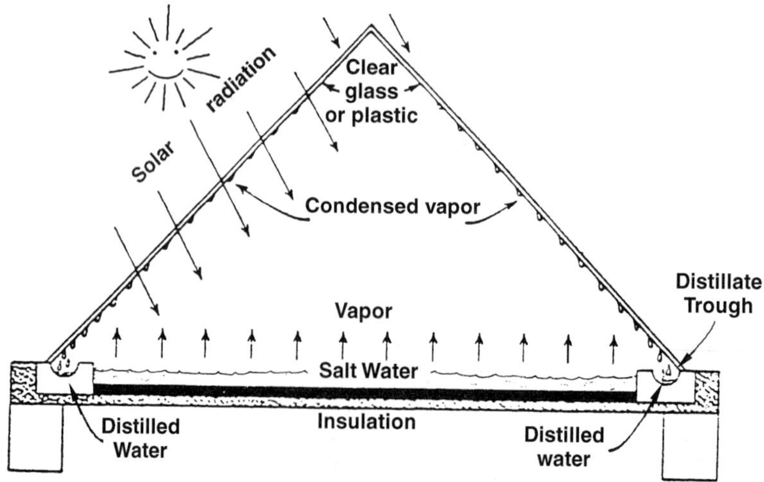

FIGURE 4.1.7 A typical basin-type solar still.

the saline water, thus separating pure water vapor from the solution. The vapor condenses on the colder cover, and this destilled water flows to collection troughs.

Solar stills of the type depicted in Figure 4.1.7, in many sizes and constructional variants, have been built and used successfully in many countries in the world. They are simple, easy to construct, reliable, and require very little maintenance although in some regions the covers must be cleaned frequently from accumulated dust or sand.

Since the heat of condensation in single-effect stills of the type shown in Figure 4.1.7 is lost to the ambient, more energy-efficient operation can obviously be achieved in a multi-effect design, where the heat of condensation is used to evaporate additional saline water. A number of such stills were built and tested successfully, but are not commercially competitive yet.

Solar stills integrate the desalination and solar energy collection processes. Another approach to solar desalination is to use separately a conventional desalination process and a suitable solar energy supply system for it. Any compatible desalination and solar energy collection processes could be used. Distillation, such as MSF or ME, can be used with heat input from solar collectors, concentrators, or solar ponds (Hoffman, 1992; Glueckstern, 1995). Net average solar energy conversion efficiencies of solar collectors (Rabl, 1985; Lior, 1991) are about 25% and of solar ponds (Lior, 1993) about 18%, similar to the efficiencies of solar stills, but the MSF or ME plants can operate at preformance ratios of 10 or more, thus basically increasing the freshwater production rate by at least tenfold, or reducing the required solar collection area by at least tenfold for the same production rate.

Solar or wind energy can also be used for desalination processes that are driven by mechanical or electrical power, such as VC, RO, and ED. The solar energy can be used to generate the required power by a variety of means, or photovoltaic cells can be used to convert solar energy to electricity directly.

Freeze Desalination

It is rather well known that freezing of saline water solutions is an effetive separation process in that it generates ice crystals that are essentially salt-free water, surrounded by saline water of higher concentration. This process requires much less energy than distillation, and the problems of corrosion and scaling are markedly reduced due to the much lower operating temperatures. Several pilot plants were constructed and have proven concept viability. Nevertheless, the process has not yet reached commercial introduction for several reasons, such as the difficulty in developing efficient and economical compressors for vapor with the extremely high specific volume at the low process pressure, and difficulties in maintaining the vacuum system leak free and in effecting reliable washing of the ice crystals. A review of freeze desalination processes is given by Tleimat (1980).

Membrane Separation Processes

Reverse Osmosis (RO)

Separation of particulate matter from a liquid by applying pressure to the liquid and passing it through a porous membrane, whereby particles larger than the pore size remain on the upstream side of the membrane and the liquid flows to its downstream side, is well known as *filtration*. Semipermeable very dense membranes that actually separate salt molecules (ions) from the water, by similarly keeping the salt on the upstream side and allowing the pressurized pure water to flow through the membrane, were developed in the 1950s. The reverse of this process, **osmosis**, is well known: for example, if a membrane is placed to separate water from an aqueous salt solution, and the membrane is semipermeable (here meaning that it permits transfer of water only, not the salt components in the aqueous solution), the water will tend naturally to migrate through this membrane into the salt solution. Osmosis is, for example, the major mass transport phenomenon across living cells. The driving force for this water flux is proportional to the concentration difference between the two sides of the membrane, and is exhibited as the so-called **osmotic pressure**, which is higher by 2.51 MPa on the water side of the membrane for typical seawater at 25°C. If a pressure higher than the osmotic pressure is applied on the saline solution side of the membrane, the water flux can be reversed to move pure water across the membrane from the saline solution side to the pure water one. This process is called *reverse osmosis* (and sometimes *hyperfiltration*), and is the basic principle of RO desalination

Unlike filtration of particulates, the selective "filtration" of the water in RO is not due to the relationship of the membrane pore size to the relative sizes of the salt and water molecules. Rather, one way to explain the process is that the very thin active surface layer of the membrane forms hydrogen bonds with water molecules and thus makes them unavailable for dissolving salt. Salt thus cannnot penetrate through that layer. Water molecules approaching that layer are, however, transported through it by forming such hydrogen bonds with it and in that process displacing water molecules that were previously hydrogen bonded at these sites. The displaced water molecules then move by capillary action through the pores of the remainder of the membrane, emerging at its other side.

The most prevalent membrane configurations used in RO plants are of the spiral-wound or hollow-fiber types. The basic spiral-wound-type module (Figure 4.1.8) is made of two sheets placed upon each other and rolled together in an increasing diameter spiral around a cylindrical perforated tube. One of the sheets is in the form of a sandwich typically composed of five layers bonded together along three edges. The two outer layers are the semipermeable membranes. Each of them is backed by a porous material layer for mechanical strength, and the very central layer is a thicker porous material layer that takes up the produced fresh water. The second sheet is a porous mesh through which the high-pressure saline water feed is passed in an axial direction. Product water separates from the saline solution and permeates through the two adjacent semipermeable membranes into the central product water–carrying layer, which conducts it spirally to the unbonded edge of the "sandwich" and to the inner perforated tube. The semipermeable membranes are typically made from cellulose acetate, and more recently from composites of several polymers.

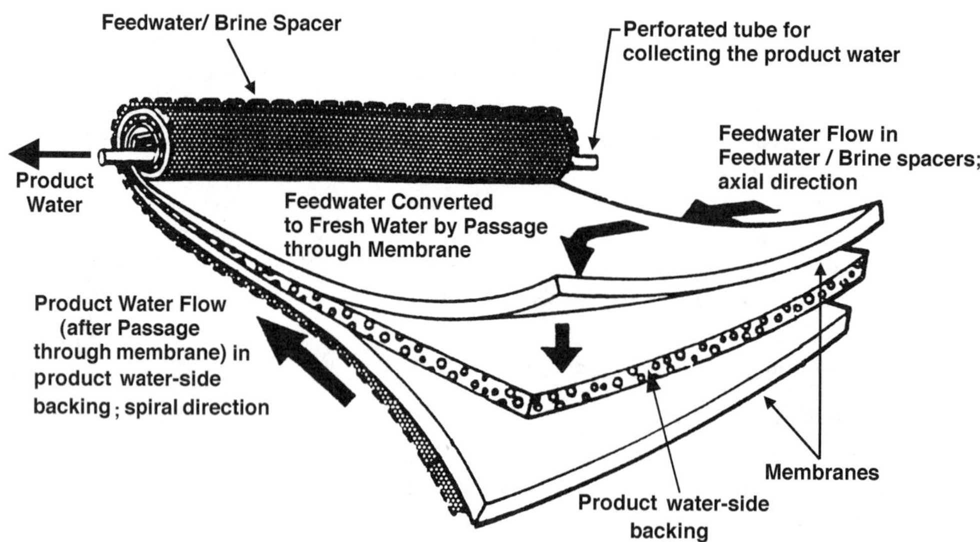

FIGURE 4.1.8 A spiral-wound RO membrane element.

Hollow fiber modules have a configuration similar to a shell-and-tube heat exchanger, with the fibers taking the place of the tubes. A very large number of typicallly 25 to 250 μm outside-diameter semi-permeable hollow fibers (wall thickness typically 5 to 50 μm) are bundled together and placed in a saline water pressure vessel. The hollow core of each fiber is sealed on one end. The pressurized saline water is brought into the module (through a central porous feed tube, Figure 4.1.9) to circulate on the exterior surface of the fibers, and water permeates through the fiber wall into its hollow core, through which it flows to a permeate collection manifold at the open end of the fiber bundle. The increasingly concentrated

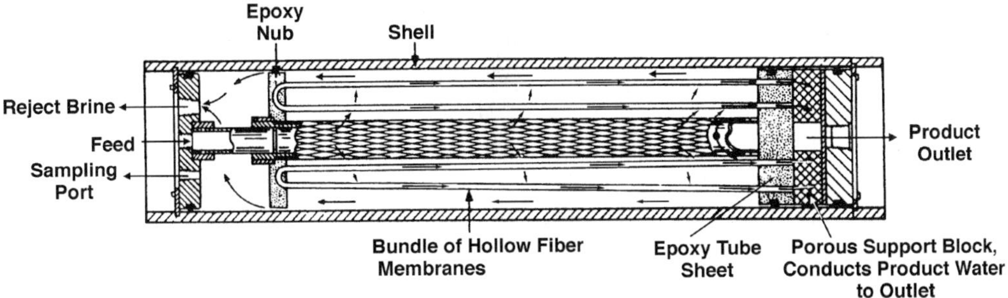

FIGURE 4.1.9 A hollow-fiber RO membrane module. (Du Pont Permasep™.)

saline water flows radially and is discharged at the exterior shell of the bundle. The hollow fibers are typically made of polyamide or cellulose triacetate, and offer about 20 fold more surface (separation) area per unit volume than the spiral-wound configuration.

The basic approximate equation for the separation process gives the water flux $\dot{m}_w''$ (kg/m²sec) across an RO membrane, in the absence of fouling, as

$$\dot{m}_w'' = K_{pe} K_{cf} \left[\left(P_f - P_p \right) - \left(\pi_f - \pi_p \right) \right] \tag{4.1.10}$$

where

K_{pe} *water permeability constant* of the membrane (in kg/m²sec Pa), typically increasing strongly as the temperature rises: a plant designed to operate at 20°C may produce up to 24% more water if the water temperature is 28°C,

K_{cf} *compaction correction factor* (dimensionless) which corrects for the fact that the flux is reduced due to densification of the barrier layer (a phenomenon similar to creep) of the membrane, and which increases with the operating pressure and temperature. It is often calculated by the relationship

$$K_{cf} = BC(T)C(P)C(t) \tag{4.1.11}$$

where B is a constant,

$C(T)$ represents the temperature dependence of the Compaction Correction Factor for the particular membrane of interest,

$C(P)$ represents its pressure dependence: while a higher pressure difference across the membrane is shown in Equation (4.1.10) to increase the water flux, higher feed pressure (P_f) also tends to compact the membrane and thus reduce its water flux, typically according to

$$C(P) = P_f^n \tag{4.1.12}$$

where n is a negative number,

and where the time dependence $C(t)$ is represented by

$$C(t) = t^m \tag{4.1.13}$$

where t is the operating time (say, in days) and m is a negative number depending on the membrane.

P water or saline solution pressure (Pa),

π osmotic pressure (Pa),

and the subscripts f and p pertain to the saline feed water and to the desalted product water, respectively.

The required membrane area A can be estimated by

$$A = \frac{\dot{m}_p}{\dot{m}''_p f} \tag{4.1.14}$$

where $\dot{m}_p$ is the freshwater mass production rate of the plant (kg/sec), and f ($0 < f \leq 1.0$) is the *area utilization factor* that corrects for the fact that the membrane surface is incompletely in contact with the saline water feed stream due to the porous mesh and other devices, such as turbulence promoters, placed in the feed stream path; in a good design $f > 0.9$.

Examination of Equation (4.1.10) shows that water separation rate increases with the water permeability constant K_{pe}. Unfortunately, so does the salt flux across the membrane, resulting in a saltier product. An approximation for this salt flow is

$$\dot{m}_s = KK_s\left(C_{fm} - C_p\right) \tag{4.1.15}$$

where

$\dot{m}_s$ salt mass transfer rate across the membrane, kg/sec,
K a proportionality constant, dimensionless,
K_s *salt permeation constant,* kg/sec, which increases with pressure and temperature.

The salinity of the product water (C_p) can be estimated by the formula

$$C_p = K_{cp}(1 - \eta)\overline{C} \tag{4.1.16}$$

where

K_{cp} *concentration polarization coefficient,* $\equiv C_{fm}/\overline{C}$ is a measure of the increase of the feedwater salinity at the membrane wall beyond that of the bulk solution,
C_{fm} salt concentration at the membrane wall,
$\overline{C}$ bulk salinity of the saline water feed, $\approx (C_f + C_r)/2$,
C_r salt concentration of the reject brine,
η *salt rejection factor,* $\equiv$ (amount of salts rejected by the membrane)/(amount of salts in the brine feed).

The pressure to be used for RO depends on the salinity of the feed water, the type of membrane, and the desired product purity. It ranges form about 1.5 MPa for low feed concentrations or high-flux membranes, through 2.5 to 4 MPa for brackish waters, and to 6 to 8.4 MPa for seawater desalination. In desalination of brackish water, typical product water fluxes through spiral-wound membranes are about 600 to 800 kg/(m^{2}day) at a recovery ratio (RR) of 15% and an average salt rejection of 99.5%, where

$$\mathrm{RR} = \frac{\dot{m}_p}{\dot{m}_f} \cong 1 - \frac{C_f}{C_r} \tag{4.1.17}$$

The fluxes in hollow-fiber membranes used in seawater desalination are 20- to 30-fold smaller, but the overall RO system size does not increase, because the hollow-fiber membranes have a much larger surface area per unit volume. The RR and salt rejection ratio are similar to those of spiral-wound membranes.

Since the concentrated reject brine is still at high pressure, it is possible to recover energy by passing this brine through hydraulic turbines, and thus reduce the overall energy consumption by up to 20%. The energy requirements of seawater RO desalination plants with energy recovery are about 5 to 9 kWh, or 18 to 33 MJ, of mechanical or electric power per m^3 fresh water produced. In comparison, the MSF desalination process requires about 120 to 280 MJ of heat and about 15 MJ of mechanical/electric power (for pumping and auxiliaries) per m^3. The energy requirement of the RO process is thus smaller than that of the MSF process even if the RO energy requirement is multiplied by the thermal-to-mechanical (or electrical) power conversion factor of 3 to 4. The specific *exergy* consumption of the MSF process using 120°C steam is about 2- to 3-fold higher than that of the RO process, but becomes comparable in magnitude if the steam temperature is lowered to 80°C.

The life of membranes is affected by gradual chemical decomposition or change. For example, cellulose acetate membranes **hydrolyze** with time. The rate of hydrolysis has a steep minimum at a solution pH of 4.5 to 5.0, and increases drastically with temperature.

Membranes are susceptible to plugging by dirt and to deterioration in their selectivity caused by various species present in the saline water. Careful pretreatment of the feed water is therefore necessary. It typically consists of clarification, filtration, chlorination for destroying organic matter and microorganisms, removal of excess chlorine to prevent membrane oxidation, and dosing with additives to prevent calcium sulfate scaling and foam formation. Periodical chemical or mechanical cleaning is also necessary. Pretreatment and cleaning are significant and increasing fractions of the RO plant capital and operating costs.

Further detail about RO desalination can be found in Sourirajan and Matsuura (1985) and Amjad (1993).

Electrodialysis (ED)

In ED, the saline solution is placed between two membranes, one permeable to cations only and the other to anions only. A direct electrical current is passed across this system by means of two electrodes, cathode and anode, exposed to the solution (Figure 4.1.10). It causes the cations in the saline solution to move toward the cathode, and the anions to the anode. As shown in Figure 4.1.10, the anions can leave the compartment in their travel to the anode because the membrane separating them from the anode is permeable to them. Cations would similarly leave the compartment toward the cathode. The exit of these ions from the compartment reduces the salt concentration in it, and increases the salt concentration in the adjacent compartments. Tens to hundreds of such compartments are stacked together in practical ED plants, leading to the creation of alternating compartments of fresh and salt-concentrated water. ED is a continuous-flow process, where saline feed is continuously fed into all compartments and the product water and concentrated brine flow out of alternate compartments. The flow along the membranes also improves the mass transport there, and the separators between the membranes are constructed to provide good flow distribution and mixing on the membrane surfaces. Membrane sizes are roughly 0.5 × 1 m, spaced about 1 mm apart. Many typed of polymers are used to manufacture these ion-exchange selective membranes, which are often reinforced by strong fabrics made from other polymers or glass fibers.

Careful and thorough feed water pretreatment similar to that described in the section on RO is required. Pretreatment needs and operational problems of scaling are diminished in the electrodialysis reversal (EDR) process, in which the electric current flow direction is periodically reversed (say, three to four times per hour), with simultaneous switching of the water flow connections. This also reverses the salt concentration buildup at the membrane and electrode surfaces, and prevents concentrations that cause the precipitation of salts and scale deposition.

The voltage used for ED is about 1 V per membrane pair, and the current flux is of the order of 100 A/m^2 of membrane surface. The total power requirement increases with the feed water salt concentration, amounting to about 10 MW/m^3 product water per 1000 ppm reduction in salinity. About half this power is required for separation and half for pumping. Many plant flow arrangements exist, and their description can be found, along with other details about the process, in Shaffer and Mintz (1980) and Heitman (1991).

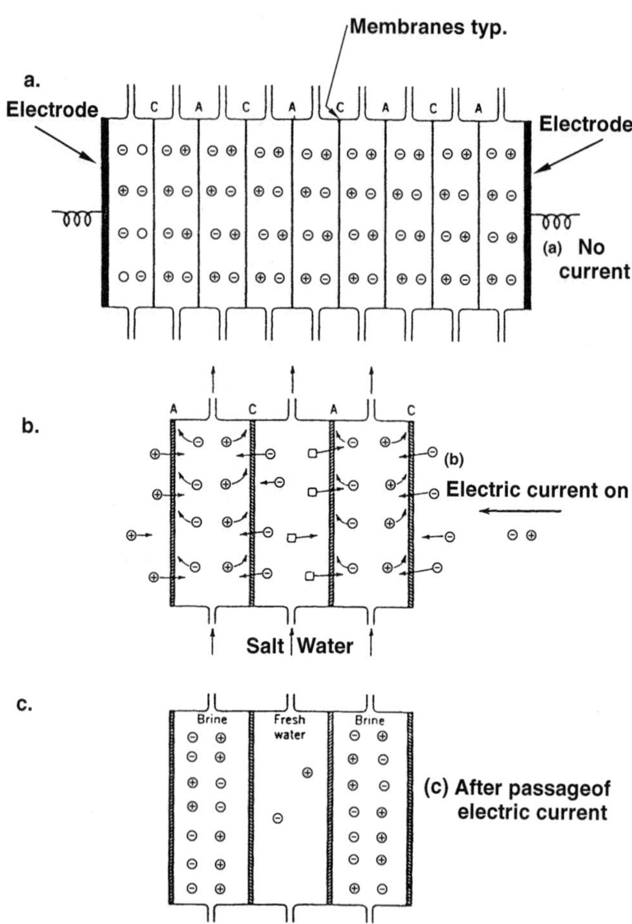

FIGURE 4.1.10 The ED process. *C* and *A* are cation- and anion-permeable membranes, respectively. Application of electric current causes ion transport in a way that salt is depleted in alternate compartments, and enriched in the remaining ones.

Defining Terms

Boiling point elevation: The number of degrees by which the boiling point temperature of a solution is higher than that of the pure solute at the same pressure.

Flash evaporation: An evaporation process that occurs when a liquid with a free surface is exposed to its vapor, where the vapor is below the saturation pressure corresponding to the temperature of the liquid. The process is typically vigorous, accompanied by rapid growth of bubbles and associated turbulence in the liquid.

Hydrolysis: Decompostion in which a compound is split into other compounds by taking up the elements of water.

Osmosis: The diffusion process of a component of a solution (or mixture) across a semipermeable membrane, driven by the concentration difference (or gradient) of that component across the membrane.

Osmotic pressure: The minimal pressure that has to be applied to the solution (mixture) on the lower concentration side of a membrane permeable to one solution component, for stopping the osmosis of that component through the membrane.

References

Amjad, Z., Ed. 1993. *Reverse Osmosis: Membrane Technology, Water Chemistry and Industrial Applications*. Van Nostrand Reinhold, New York.

Aschner, F.S. 1980. Dual purpose plants, in *Principles of Desalination*, 2nd ed., Part A, K.S. Spiegler and A.D.K. Laird, Eds., Academic Press, New York, chap. 5, 193–256.

Burns and Roe, Inc, 1969. *Universal Design—Report and User's Manual on Design of 2.5 Million Gallon per Day Universal Desalting Plant*, Vols. I–V, U.S. Department of the Interior, O.S.W. Contract No. 14-01-0001-955, Washington, D.C.

Fabuss, B.M. 1980. Properties of seawater, in *Principles of Desalination*, 2nd ed., Part B, K. S. Spiegler and A.D.K. Laird, Eds., Academic Press, New York, Appendix 2, 765–799.

George P.F., Manning, J.A., and Schrieber, C.F. 1975. *Desalinatin Materials Manual*. U.S. Department of the Interior, Office of Saline Water, Washington, D. C.

Glater, J., York, J.L., and Campbell, K.S. 1980. Scale formation and prevention, in *Principles of Desalination*, 2nd ed., Part B, K.S. Spiegler and A.D.K. Laird, Eds., Academic Press, New York, chap. 10, 627–678.

Glueckstern, P. 1995, Potential uses of solar energy for seawater desalination, *Desalination*, 101, 11–20.

Heitman, H.-G. 1990. *Saline Water Processing*, VCH Publications, New York.

Hoffman, D. 1992. The application of solar energy for large scale sea water desalination, *Desalination*, 89, 115–184.

Kasper, S.P. and Lior, N. 1979. A methodology for comparing water desalination to competitive freshwater transportation and treatment, *Desalination*, 30, 541–552.

Khan, A.S. 1986. *Desalination Processes and Multistage Flash Distillation Practice*, Elsevier, Amsterdam.

Lior, N., Ed. 1986. *Measurements and Control in Water Desalinaion*, Elsevier, Amsterdam.

Lior, N. 1991. Thermal theory and modeling of solar collectors, in *Solar Collectors, Energy Storage, and Materials*, F. de Winter, Ed., MIT Press, Cambridge, MA, chap. 4, 99–182.

Lior, N. 1993. Research and new concepts, in *Active Solar Systems*, G.O.G. Löf, Ed., MIT Press, Cambridge, MA, chap. 17, 615–674.

Lior, N. and Greif, R, 1980. Some basic observations on heat transfer and evaporation in the horizontal flash evaporator, *Desalination*, 33, 269–286.

Miyatake, O., Hashimoto, T., and Lior, N. 1992. The liquid flow in multi-stage flash evaporators, *Int. J. Heat Mass Transfer*, 35, 3245–3257.

Miyatake, O., Hashimoto, T., and Lior, N. 1993. The relationship between flow pattern and thermal non-equilibrium in the multi-stage flash evaporation process, *Desalination*, 91, 51–64.

M.W. Kellogg Co. 1975. *Saline Water Data Conversion Engineering Data Book*, 3rd ed., U.S. Department of the Interior, Office of Saline Water Contract No. 14-30-2639, Washington, D.C.

Rabl, A. 1985. *Active Solar Collectors and Their Applications*, Oxford University Press, New York.

Shaffer, L.H. and Mintz, M.S. 1980. Electrodialysis, in *Principles of Desalination*, 2nd ed., Part A, K.S. Spiegler and A.D.K. Laird, Eds., Academic Press, New York, chap. 6, 257–357.

Sourirajan, S. and Matsuura, T., Eds. 1985. *Reverse Osmosis and Ultrafiltration*, ACS Symposium Series 281, American Chemical Society, Washington, D.C.

Spiegler, K.S. and El-Sayed, Y.M. 1994. *A Desalination Primer*. Balaban Desalination Publications, Mario Negri Sud Research Institute, 66030 Santa Maria Imbaro (Ch), Italy.

Spiegler, K.S. and Laird, A.D.K., Eds. 1980. *Principles of Desalination*, 2nd ed., Academic Press, New York.

Standiford, F.C. 1986a. Control in multiple effect desalination plants, in *Measurements and Control in Water Desalination*, N. Lior, Ed., Elsevier, Amsterdam, chap. 2.2, 263–292.

Standiford, F.C. 1986b. Control in vapor compression evaporators, in *Measurements and Control in Water Desalination*, N. Lior, Ed., Elsevier, Amsterdam, chap. 2.3, 293–306.

Steinbruchel, A.B. and Rhinesmith, R.D. 1980. Design of distilling plants, in *Principles of Desalination*, 2nd ed., Part A, K.S. Spiegler and A.D.K. Laird, Eds., Academic Press, New York, chap. 3, 111–165.

Tleimat, B.W. 1980. Freezing methods, in *Principles of Desalination*, 2nd ed., Part B, K.S. Spiegler and A.D.K. Laird, Eds., Academic Press, New York, chap. 7, 359–400.

Further Information

The major texts on water desalination written since the 1980s are Spiegler and Laird (1980), Khan, (1986) (contains many practical design aspects), Lior (1986) (on the measurements and control aspects), Heitman (1990) (on pretreatment and chemistry aspects), and Spiegler and El-Sayed (1994) (an overiew primer). Extensive data sources are provided in George et al. (1975) and M. W. Kellog (1975).

The two major professional journals in this field are *Desalination, The International Journal on the Science and Technology of Desalting and Water Purification* and *Membrane Science*, which often addresses membrane-based desalination processes, both published by Elsevier, Amsterdam.

The major professional society in the field is the International Desalination Association (IDA) headquartered at P.O. Box 387, Topsfield, MA 01983. IDA regularly organizes international conferences, promotes water desalinaton and reuse technology, and is now publishing a trade magazine *The International Desalination & Water Reuse Quarterly.*

The *Desalination Directory* by M. Balaban Desalination Publications, Mario Negri Sud Research Institute, 66030 Santa Maria Imbaro (Ch), Italy, lists more than 5000 individuals and 2000 companies and institutions in the world of desalinaiton and water reuse.

Two useful (though by now somewhat dated) books on desalination are by Howe, E. D. 1974. *Fundamentals of Water Desalination*, Marcel Dekker, New York, and by Porteous, A. 1975. *Saline Water Distillation Processes*, Longman, London.

Much information on oceans and seawater properties is avaiable in the book by Riley, J. P. and Skinner, Eds. 1975. *Chemical Oceanography*, Academic Press, New York.

4.2 Environmental Heat Transfer

Henry Shaw

Introduction

Environmental heat transfer studies the impact of human activity on the delicate balance between the quantity of solar energy affecting our planet Earth and the heat radiating back into space. To understand the implications of global heat transfer, we must investigate the inner workings of Earth's climate. Earth supports life because of its distance from the sun and the composition of the atmosphere. However, our climate may be changing, threatening to send this delicate balance out of control. To understand how climate can change, we must learn how it depends on factors that are under our influence. This chapter will address the principal factors that control Earth's climate and view the implications of human activity in modifying climate. The development of planetary heat exchange follows the excellent description provided by R.P. Turco (1997). A simple model is presented that can be used to estimate equilibrium climate change as a function of the buildup of infrared-absorbing gases in the atmosphere. This model uses as input the atmospheric buildup of carbon dioxide (CO_2) due to fossil fuel use and population growth to estimate the magnitude of the "greenhouse effect."

Global Climate

Global climate is the weather as reported by meteorologists averaged over extended periods of time and over the surface of Earth. The global average temperature is used as the figure of merit in projecting future climate because it is the most stable climate parameter. The period for averaging generally involves decades or longer. Consequently, the average global temperature does not vary significantly on a year-to-year basis. Typically, the surface air temperature can change by as much as 20°C from day to night. Averaging over the entire globe eliminates the effect of the day-night cycle, since the same total area of the globe is always under illumination. The global average temperature is a good indicator of the total energy that Earth has absorbed from the sun. The atmosphere, land, and oceans have enormous capacities to store heat. Accordingly, these reservoirs of heat maintain the average temperature over long time intervals despite fluctuations in the global heat input or loss. The temperature changes are caused by variations in the sun's energy reaching Earth and movements of air masses with different meteorological characteristics that include temperature, moisture, clouds, and precipitation. If the temperature over a much larger region were averaged, then much less variability would be observed. In a similar vein, periodic disruptions of oceanic flow and temperature in the southern Pacific, known as El Nino/Southern Oscillation (ENSO) or La Nina, cause major weather disruptions due to buildup of hot or cold water masses, respectively, during these relatively local occurrences. The atmosphere and tropical Pacific Ocean interact in such a way that wind and water currents periodically create large pools of warm surface water over the eastern Pacific Ocean. This is a major factor contributing to the important El Nino Southern Oscillation (ENSO) phenomenon, which periodically triggers anomalous and destructive weather around the Pacific Ocean basin.

Climate studies focus on the global balance of energy. The source of energy is the sun. The absorption of solar energy by Earth, flow of energy through Earth reservoirs, and eventual loss of energy back to space are the processes that are of concern in environmental heat transfer. There are a number of other concerns that can be attributed to environmental heat transfer on a more localized level, such as the effect of:

- Fluorochlorocarbons (CFCs), ideal heat transfer fluids used in refrigeration and air conditioning, on the stratospheric ozone layer.
- Deposition of ammonium nitrate and ammonium sulfate on boiler heat transfer surfaces due to the reaction of ammonia (added for NO_x control) with SO_2 and NO_2.
- SO_2 control in reducing atmospheric sulfur concentration needed for cloud seeding, etc.

These potential areas of environmental heat transfer will not be discussed at this time.

The sun generates energy by the process of nuclear fusion. The atoms are fused together to form heavier nuclei. When nuclei fuse, huge amounts of energy are released. The nuclear fusion of hydrogen atoms (H) to form helium atoms (He), for example, is the basic process that powers the sun. An empirical way of explaining this process is through the concept of binding energy or mass defect per nuclear particle. Figure 4.2.1 relates the binding energies of all nuclear particles to their atomic mass. Iron ($_{26}Fe^{56}$) is considered the most stable element and the fusion of two atoms lighter than iron, e.g., $_1H^2$ + $_1H^3$ (the two heavy isotopes of hydrogen, deuterium, and tritium) give up energies on the order of 17 MeV. Nuclear fission, which is the process operating in all nuclear reactors, is similarly explained as uranium ($_{92}U^{235}$) splitting due to the capture of a thermal neutron. The uranium fragments or fission products have higher binding energies because they have lower mass, producing 200 MeV per fission. Nuclear processes will be discussed later as alternatives to fossil fuel combustion for producing power without emitting infrared absorbing gases.

Average Temperature of Earth

The balance of energy held in various heat reservoirs determines the average temperature of Earth at any moment. Two overall processes control the total heat content of these reservoirs, viz., the absorption of energy from incoming sunlight and the emission of thermal radiation back into space.

Figure 4.2.2a compares the spectra of sunlight and terrestrial heat radiation. The sun is a blackbody with an emission temperature of about 6000 K. The emission peaks at a wavelength of about 0.55 μm (as predicted by Wien's law, $\lambda_{max} = 2987/T$ in μm·K). The absolute intensity of the radiation reaching Earth is controlled by the distance from the sun and the size of the sun itself. The size of the sun is

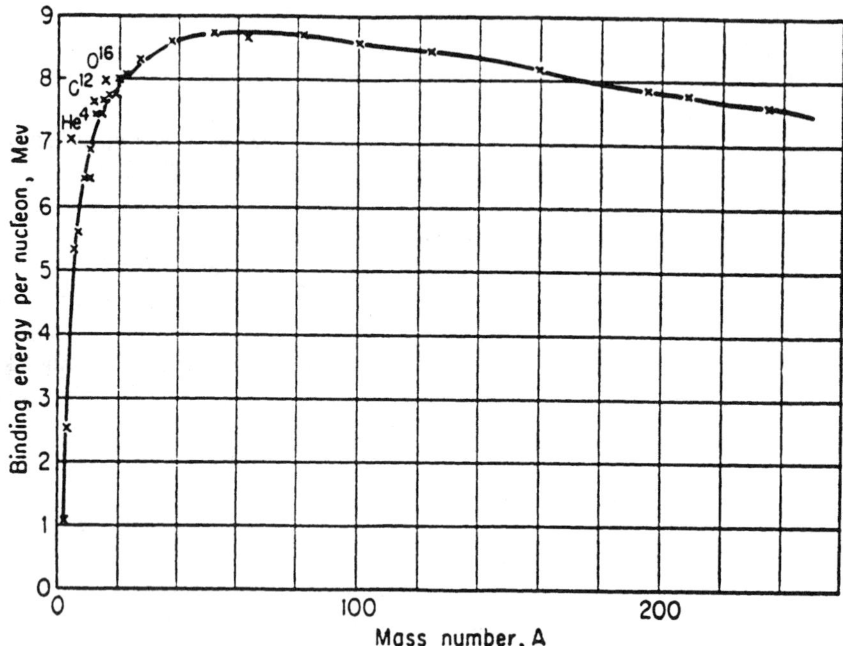

FIGURE 4.2.1 Binding energy per stable isotope (Adapted from Stephenson, 1954).

constant and need not be considered a factor in contemporary climate change. However, the distance of Earth from the sun does vary. Earth's orbit around the sun is not circular, but elliptical. Although the eccentricity of the orbit is very small, amounting to variations in the mean distance from the sun of less than 2%, the variation in distance can be important over a seasonal cycle.

In considering the overall energy balance of Earth, it is helpful to consider the total radiative energy contained in the absorption or emission spectrum. The total energy emitted by the sun is equivalent to the integral of the solar spectrum in Figure 4.2.3, i.e., the total emission is proportional to the area under the radiation emissive curve for sunlight. Similarly, the total energy emitted by Earth as thermal radiation is given by the corresponding area in Figure 4.2.2. These spectrally integrated total emissions are usually expressed as a radiant energy flux in units of energy per unit time per unit area perpendicular to the direction of the source of the radiation. The total energy flux reaching Earth from the sun, denoted f, depends on the distance from Earth to the sun, x_{es}, on the order of 1.5×10^8 km, in the following way:

$$f = \bar{f}\left(\bar{x}_{es}/x_{es}\right)^2 \tag{4.2.1}$$

where: f = the solar constant, or the total energy flux reaching Earth, W/m²
x_{es} = distance between Earth and the sun, m
$\bar{f}, \bar{x}_{es}$ are the average values for the solar constant and the distance from the sun

In order to solve the climate-related heat balance equations, a set of consistent physical characteristic values for Earth is needed. Such a table, assembled by Clark (1982), is included as Table 4.2.1 with data sources.

The solar constant has a value of about 1375 W/m². The energy flux falls off in proportion to x^{-2} as the distance from the sun increases. This is the general law of the way intensity varies with distance for spherical radiators. The relative distance of Earth from the sun varies by about 3.4% during the course of a year. This is a consequence of the eccentricity of Earth's orbit around the sun.

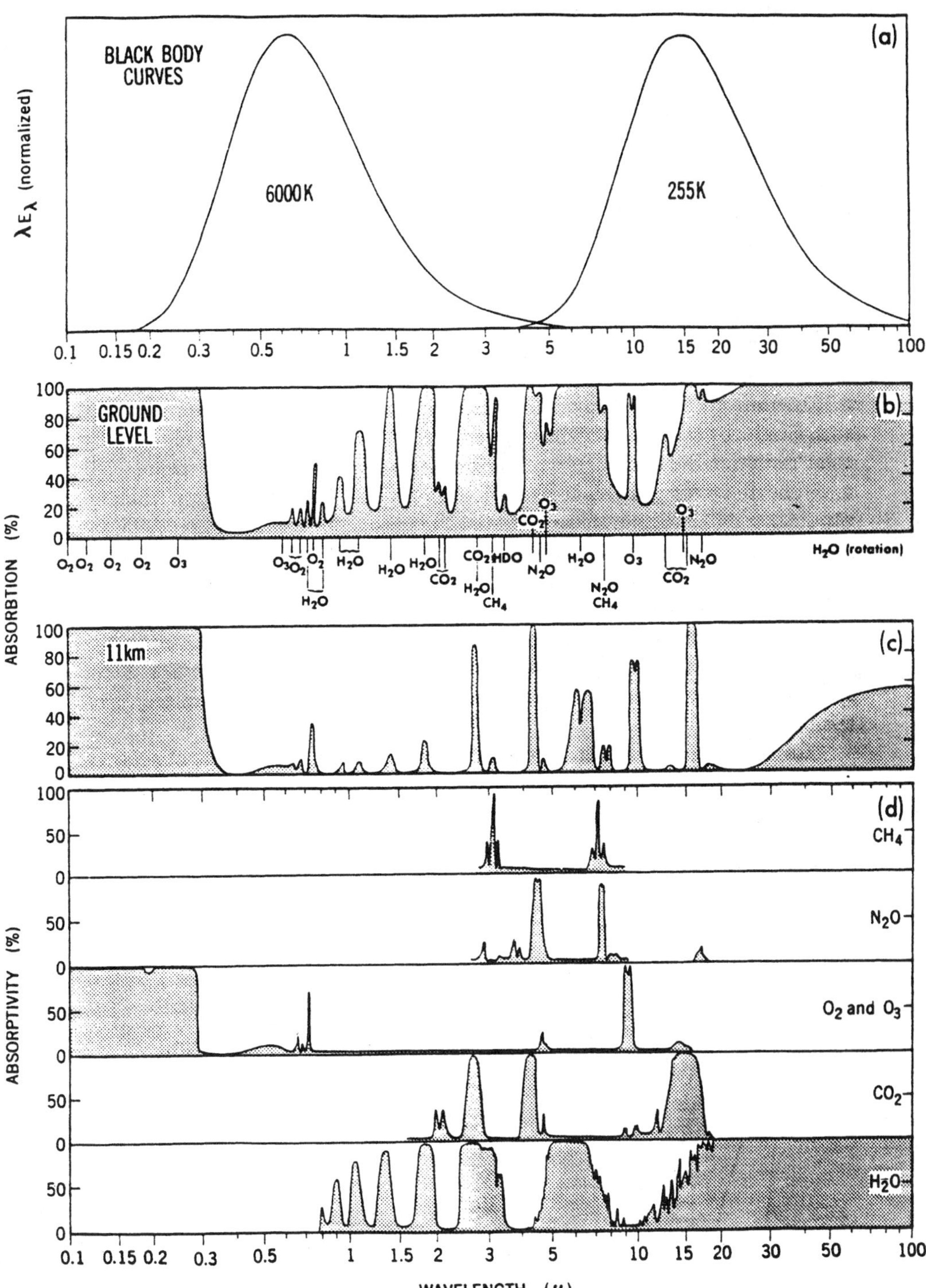

FIGURE 4.2.2 Absorption and emission spectra important in the study of climate (Peixoto and Oort, 1992). (a) Black body curves for the solar radiation (assumed to have a temperature of 255 K): (b) absorption spectra for the entire vertical extent of the atmosphere and (c) for the portion of the atmosphere above 11 km; and (d) absorption spectra for the various atmospheric gases between the top of the atmosphere and the Earth's surface. Updated with data from Fels and Schwarzkopf (1988, personal communication) between 10 and 100 μm.

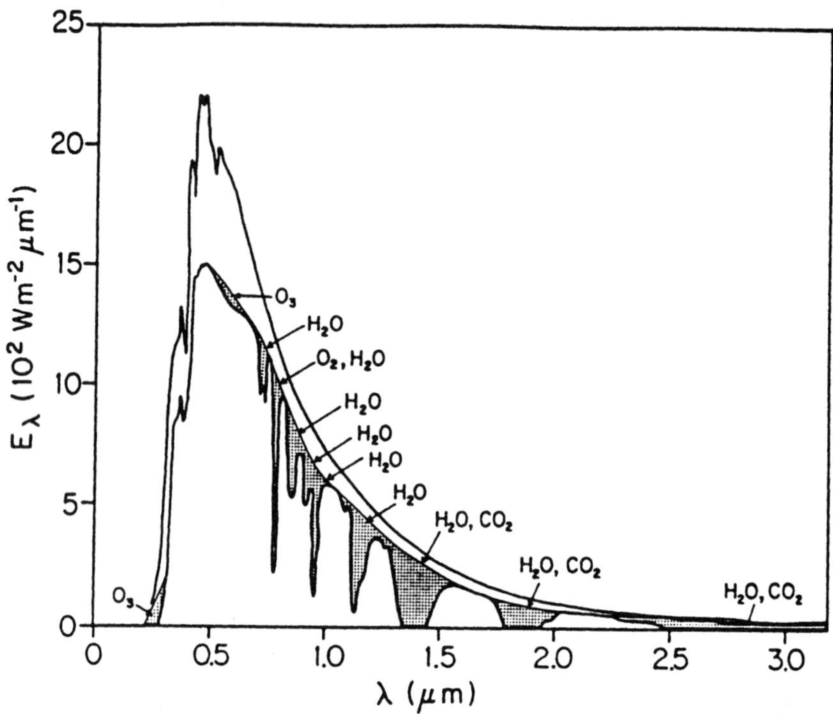

FIGURE 4.2.3 Spectral distribution of solar irradiation at the top of the atmosphere (upper curve) and at sea level (lower curve) for average atmospheric conditions of the sun at its zenith. The shaded areas represent absorption by various atmospheric gases. The unshaded area between the two curves represents the portion of the solar energy backscattered by the air, water vapor, dust, and aerosols and reflected by clouds. 1360 W m^{-2} represents the solar constant. (Adapted from Gast, Handbook of Geophyscis and Space Environments, Air Force Cambridge Research Laboratory, 1965.)

Albedo and Insolation

The fraction of total incident solar energy that is reflected is the albedo of the planet (Budyko, 1974). The albedo represents the reflectance averaged over wavelength as well as over the different angles of incidence from the sun. Earth's planetary albedo is calculated to be about 0.33. The albedo depends on a large number of factors such as the distribution of clouds, quantity of pollution and dust particles in the atmosphere, amount of snow and ice on the surface, wetness of the ground, growth of crops, ocean foam, etc. Due to the complexity of calculating the albedo, satellites are used to measure the planetary averaged albedo directly and have confirmed the value of 0.33. The variation of albedo for various surfaces is given in Table 4.2.2.

Solar insolation is the quantity of solar energy that reaches Earth. This calculation is slightly more complex than taking the difference between the solar constant and the albedo flux. This is a consequence of multiple scattering of incident radiation by the atmosphere, clouds, aerosols, and the surface. The radiation is also absorbed in the atmosphere and never reaches the surface. The fraction of solar energy absorbed by Earth's surface is less than half of insolation. The atmosphere absorbs one quarter and the rest is scattered into space.

For the simple climate analysis presented here, we follow the derivation of Chamberlain (1978). The solar constant is first reduced by the fraction of the incident radiation that is reflected as albedo and then is reduced by a factor of four, corresponding to the ratio of the surface area to the maximum cross-sectional area of Earth.

TABLE 4.2.1 Some Useful Quantities in CO_2 Research

Quantity	Symbol[a]	Value	Source
Solar constant	f	1.375 kW/m^2	1
Earth mass	M	5.976×10^{24} kg	2
Equatorial radius	a	6.378×10^{6} m	2
Polar radius	c	6.357×10^{6} m	2
Mean radius	R	6.371×10^{6} m	3
Surface area	A_e	5.101×10^{14} m^2	4
Land area	A_l	1.481×10^{14} m^2	5
Ocean area	A_s	3.620×10^{14} m^2	6
Mean land elevation	h_l	840 m	5
Mean ocean depth	$\mathrm{h_s}$	3730 m	7
Mean ocean volume	V_s	1.350×10^{18} m^3	6
Ocean mass	M_s	1.384×10^{21} kg	8
Mass of atmosphere	M_a	5.137×10^{18} kg	9
Equatorial surface gravity	g	9.780 m/s^2	2

[a] Symbols generally following reference standards used in Source 10 below.

Sources and Notes:

1. D. V. Hoyt, 1979, The Smithsonian Astrophysical Observatory Solar Constant Program, *Rev. Geophys. Space Physics,* 17:427-458.
2. F. Press and R. Siever, 1974, *Earth,* W. H. Freeman, San Francisco.
3. For sphere of Earth's volume.
4. Calculated form land and ocean data cited here.
5. B. K. Ridley, 1979, *The Physical Environment,* Ellis Horwood, West Sussex, U.K.
6. H. W. Menard and S. M. Smith, 1966, Hypsometry of Ocean Basin Provinces, *J. Geophys. Res.,* 71:4305-4325, adopted as reference standard by Bolin (10).
7. Calculated from volume and area data cited here.
8. Calculated from volume data cited here plus density of 1025 kg/m^3, adopted as reference standard in Bolin (10).
9. K. E. Trenberth, 1981, Seasonal Variations in Global Sea-Level Pressure and the Total Mass of the Atmosphere, *J. Geophys. Res.,* 86:5238-5246; this supersedes value adopted as reference standard by Bolin (10).
10. B. Bolin, Ed., 1981, *Carbon Cycle Modelling,* SCOPE 16, John Wiley & Sons, New York.

Table source: Adapted from Clark, W.C., *Carbon Dioxide Review 1982,* Oxford University Press, New York, 1982.

TABLE 4.2.2 Albedo for Various Surfaces in the Visible Parts of the Spectrum

Surface	Albedo
Sand	0.18–0.28
Grassland	0.16–0.20
Green crops	0.15–0.25
Forests	0.14–0.20
Dense forests	0.50–0.10
Fresh snow	0.75–0.95
Old snow	0.40–0.60
Cities	0.14–0.18

Source: Houghton, H.G., *Physical Meteorology,* MIT Press, Cambridge, MA, 1985.

As can be seen in Figure 4.2.2, the terrestrial emission spectrum is well separated in wavelength from the solar spectrum and lies at much longer wavelengths. This difference is essential in supporting Earth's climate system.

Terrestrial Radiation

Earth emits radiation that can be approximated as blackbody radiation. If we consider the Earth-atmosphere system to be an ideal blackbody, it would exhibit a mean emission temperature of about 255 K. Using Wien's law, the peak emission intensity at this temperature corresponds to a wavelength of roughly 10 μm. Figure 4.2.2a illustrates the ideal spectrum of Earth emissions. The area under this spectrum represents the total energy flux emitted by Earth per unit area of surface, averaged over the entire planet.

The two principal components of Earth's climate system must be balanced. Considering the planet as a whole, the solar energy absorbed must equal the terrestrial energy emitted. Otherwise, there would be a net gain or loss of energy over time, and the temperature of Earth would change accordingly. Because the average planetary temperature is known to be very stable over long periods of time, the heat source and sink must be in a steady state.

The total solar energy absorbed by the planet is determined by a number of factors, including the incident solar energy (solar constant), f, the size of the planet, and the amount of energy reflected (albedo). In terms of these parameters, summarized in Table 4.2.1, the energy source for the climate system is given in Equation (4.2.2).

$$q_{in} = fA_c(1 - \overline{\alpha}_e) \tag{4.2.2}$$

where: q_{in} = incoming solar energy, W
f = the solar constant, 1375 W/m^2
A_c = cross-sectional area of Earth = πR^2, m^2
R = average diameter of Earth, m
$\overline{\alpha}_e$ = average albedo of Earth

The albedo of Earth is the fraction of impinging solar energy reflected to space. Thus, the fraction absorbed that contributes to the climate system is $1 - \overline{\alpha}_e$. The cross-sectional area of Earth is used in Equation (4.2.2) because this is the area that intercepts the solar energy. The energy absorbed from sunlight must be balanced by the emission of heat into space. The total radiative energy flux (at all wavelengths) emitted by a perfect blackbody radiator, per unit surface area of the emitter, is given by the Stefan-Boltzmann law. In this case, the emission of thermal radiation occurs from the entire surface of the planet, i.e., $4\pi R^2$.

$$F_b = A_s \sigma \overline{T}^4 \tag{4.2.3}$$

where: F_b = heat radiation, kW
A_s = surface area of Earth = $4\pi R^2$, m^2
σ = Stefan-Boltzmann constant, 5.672×10^{-11} kW/m$^2 \cdot$ K^4
T = average temperature, K

To estimate the average temperature of Earth at the top of the atmosphere, we equate Equations (4.2.2) and (4.2.3).

$$f(\pi R^2)(1 - \overline{\alpha}_e) = (4\pi R^2)\sigma T^4 \tag{4.2.4}$$

Solving for the temperature, we get Equation (4.2.5):

$$\overline{T} = \left((f/4\sigma)(1 - \overline{\alpha}_e)\right)^{1/4}$$
$$\overline{T} = \left((1.375/4 \times 5.672 \times 10^{-11}\ K^4)(1 - 0.33)\right)^{1/4} = 255\ K \tag{4.2.5}$$

Inherent in Equation (4.2.5) is the requirement that the energy reservoirs of Earth, i.e., atmosphere, oceans, and land, be in equilibrium. The reservoirs respond rapidly to imbalances in energy sources and sinks. The atmosphere can adjust its temperature by several degrees in a matter of days; the oceans, in months. These time scales are short compared with the variations in the climate system of interest, which span decades to centuries. Consequently, we are assured that the climate system is in equilibrium for the conditions assumed in the primitive climate model.

The heat balance represented by Equation (4.2.5) is not arbitrary. If insolation increases, then the heat content of the reservoirs will increase. According to Equation (4.2.6), the temperature of the reservoirs must rise in proportion to the added energy:

$$\Delta Q = \bar{c}_p M \Delta T \qquad (4.2.6)$$

where: ΔQ = added energy, J
$\bar{c}_p$ = average heat capacity, J/kg · K
M = mass of reservoir, kg
ΔT = change in temperature, K

As the temperature of Earth increases, its thermal radiation also increases in proportion to the fourth power of the temperature, according to Equation (4.2.5). This nonlinear response is important because it allows small variations in climate, forcing (solar constant or albedo) to be compensated by much smaller variations in the effective radiation temperature. The difference between the effective temperature of Earth on top of the atmosphere of 255 K, and the average temperature of land and ocean surfaces of 290 K is remarkable because of the implications for supporting life. The difference in temperature is attributed to the presence of an atmosphere that causes two critical alterations of the simple energy balance model. First, the gas molecules absorb and scatter radiation, and second, clouds and small particles scatter, absorb, and emit radiation.

Heat Reservoirs

The overall heat balance that equates the energy input from the sun to the heat emitted by Earth provides the average global temperature on top of the atmosphere. However, the behavior of Earth's climate system is determined by the heat transfer between all the reservoirs on Earth. The thermal properties of major reservoirs are provided in Table 4.2.3. Since we cannot account for every reservoir, we infer climatic behavior in Table 4.2.4 by estimating the average thermal behavior of the principal reservoirs. The information required for Table 4.2.4 is obtained from Tables 4.2.1 and 4.2.3.

The surface reservoirs of Earth are land area = 1.48×10^8 km^2, ocean area = 3.36×10^8 km^2, and cryosphere = 0.26×10^8 km^2, for a total area of 5.101×10^8 km^2. Table 4.2.1 gives the area of the oceans as 3.62×10^8 km^2, which corresponds to the sum of the hydrosphere and cryosphere as used in Table 4.2.4.

Energy storage and flow within the climate system are depicted in Figure 4.2.4. The sizes, masses, and other characteristics of these reservoirs are summarized in Table 4.2.4. Equation (4.2.6) is used to obtain the heat content of each reservoir. The principal reservoirs of heat that affect climate are the atmosphere, surface oceans, and land surfaces. These reservoirs interact on relatively short time scales. The processes that transfer energy between the reservoirs, and the rates of energy exchange, are indicated in Table 4.2.5. The most massive heat reservoir is that of the deep oceans. The deep ocean waters are cold and not readily accessible. A significant change in the temperature of a reservoir is only a small percentage of its absolute temperature. For a reservoir at 300 K, a temperature change on the order of 10 K is very important. One can estimate the time required for such a change by assuming that all factors, other than total energy loss or gain, can be ignored. Equation (4.2.7) provides this simple relationship:

$$\Delta t = c_p M \Delta T / q_L \qquad (4.2.7)$$

where: Δt = total time for a change, s
c_p = heat capacity, J/kg · K
M = mass of reservoir, kg
ΔT = change in temperature, K
q_L = total energy loss rate from the reservoir, W

TABLE 4.2.3 Thermal Properties of Major Reservoirs

Substance	Condition	Density ρ (10^3 kg m^{-3})	Specific Heat c (10^3 J kg^{-1} K^{-1})	Heat Capacity ρc (10^6 J m^{-3} K^{-1})	Thermal Conductivity k (W m^{-1} K^{-1})	Thermal Diffusivity k* (10^{-6} m^2 s^{-1})	Conductive Capacity c[a] (10^3 J m^{-2} K^{-1} $s^{-1/2}$)	Penetration Depth 3d (*m*)	
								Diurnal	Annual
Air	20°C, still	0.0012	1.00	0.0012	0.026	21.5	0.006	2.3	44
	stirred					4×10^{6}[b]	2.4	1×10^3	19×10^3
Water	20°C, still	1.00	4.19	4.19	0.58	0.14	1.57	0.2	3.6
	stirred					130[b,c]	48	5.7	108
Ice	0°C, pure	0.92	2.10	1.93	2.24	1.16	2.08	0.5	10.2
Snow	Fresh	0.10	2.09	0.21	0.08	0.38	0.13	0.3	6.0
Sandy soil	Dry	1.60	0.80	1.28	0.30	0.24	0.63	0.2	4.8
(40% pore space)	Saturated	2.00	1.48	2.98	2.20	0.74	2.56	0.4	8.1
Clay soil	Dry	1.60	0.89	1.42	0.25	0.18	0.60	0.2	3.9
(40% pore space)	Saturated	2.00	1.55	3.10	1.58	0.51	2.21	0.4	6.9
Peat soil	Dry	0.30	1.92	0.58	0.06	0.10	0.18	0.2	3.0
(80% pore space)	Saturated	1.10	3.65	4.02	0.50	0.12	1.39	0.2	3.3

[a] $k^* = k/\rho c$, $c^* = \rho c\sqrt{k^*}$, $d = (Pk^*/\pi)^{1/2}$. [These values are from Oke, T.R., *Boundary Layer Climates,* Halstead, New York (1987)].

[b] The values of k^* for stirred wate and air are, of course, much greater than those for still conditions because turbulent eddy mixing is a more efficient process to transport heat vertically than molecular conduction. In spite of the large uncertainties in specifying k^* for stirred conditions, the implied annual penetration depths of about 100 m for the oceans and 19 km for the atmosphere give reasonable order of magnitude estimates if we compare them with the observed profiles in Figure 4.2.20.

[c] From Munk (1966) based on geochemical data for lower thermocline ($k^* = 1.3$ cm^2 s^{-1}).

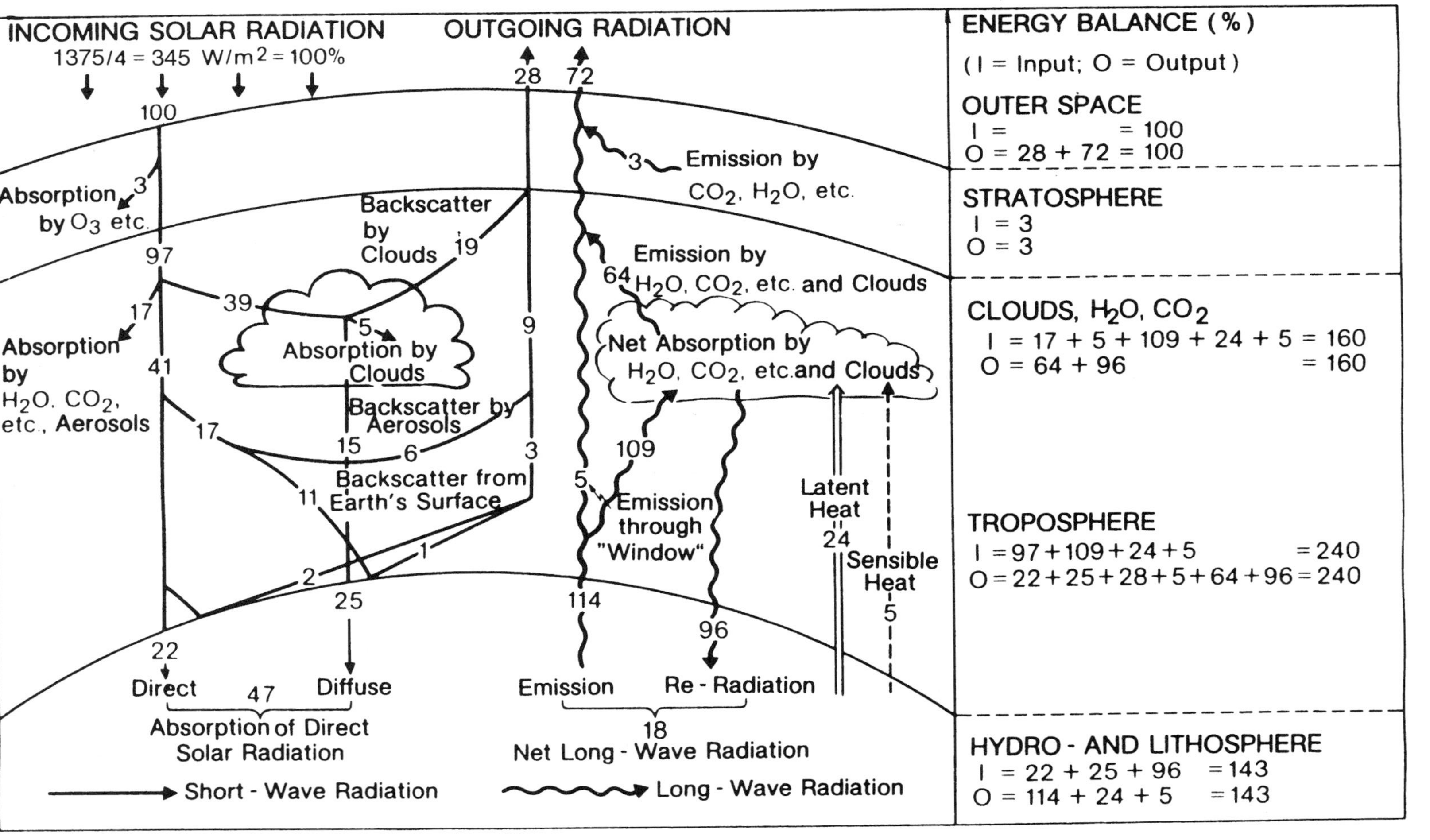

FIGURE 4.2.4 The average radiation and heat balance of the Earth – atmospheric system, relative to 100 units of incoming solar radiation. (Adapted from Rotty, 1975 and Gates, 1979.)

TABLE 4.2.4 Earth's Energy Reservoirs

Reservoir	Volume (km^3)[a]	Mass (Gt)	Temp. (K)	Heat Capacity (J/g · K)	Energy Content (EJ)
Atmosphere[b]	5.0×10^9	5.2×10^6	250	1.0	1.3×10^6
Land					
Surface[c]	3.0×10^4	3.3×10^4	290	3.7	3.5×10^4
Subsurface[c]	1.5×10^5	1.6×10^5	280	3.7	1.7×10^5
Oceans					
Surface[d]	3.3×10^7	3.3×10^7	280	4.2	4.0×10^7
Thermocline[d]	1.7×10^8	1.7×10^8	275	4.2	1.9×10^8
Deep[d]	1.2×10^9	1.2×10^9	270	4.2	1.4×10^9
Cryosphere[e]	5.1×10^7	4.7×10^7	265	2.1	2.6×10^7

[a] Volume is specified in order to estimate the reservoir's heat contents.

[b] The atmosphere is assumed to be 10 km thick (approximately the Troposphere) at density of 1.05 kg/m^3.

[c] The land surface depth is taken to be 0.2 m thick for diurnal response, with a density of 1100 kg/m^3, based on saturated sandy and clay soil with 80% saturated pore space. The deep soil layer is taken as 1.0 m thick, for seasonal variations.

[d] The oceans have a density of 1025 kg/m^3 and average depths of about 100 m for the surface, 0.5 km for the thermocline, and 3.7 km for the deep oceans.

[e] The ice and snow reservoir has a density of 920 kg/m^3 and an average depth of 2 km.

Source: Modified from Turco, R.P., *Earth Under Siege,* Oxford University Press, New York, 1987.

TABLE 4.2.5 Time Required for Reservoirs to Cool by 10 K

Net Heat Change, EW	Surface Ocean	Surface Land	Atmosphere
IR emissions to space	50	20	860
Net IR transfer to atmosphere	160	60	
Sensible heat transfer to atmosphere	50	50	
Latent heat transfer to atmosphere	200	50	
Total	**460**	**180**	**860**
Time, seconds	3×10^7 1 year	70000 1 day	6000 2 h

Data taken from Table 4.2.4 and Platt (1976) can be used with Equation (4.2.7) to estimate the time required to heat or cool the surface of the ocean, land, and atmosphere by 10 K.

The estimated time of cooling the ocean surface reservoir is about 1 year. With a small percentage reduction in insolation, the ocean surface is therefore likely to cool several degrees per decade. Land surface is much more susceptible to rapid temperature change. In the absence of sunlight, the time for land to cool by 10 K is on the order of a day. The atmosphere can cool even faster, i.e., on the order of hours. We know empirically that the ground and surface air can indeed cool overnight.

The Greenhouse Effect

As indicated above, molecules in the atmosphere can absorb radiation at some wavelengths and scatter radiation at all wavelengths by means of Rayleigh scattering (intensity of scattered light is proportional to the inverse of wavelength to the fourth power). Also, molecules experience vibrational and rotational motions that lead to absorption and emission of radiation in the infrared portion of the electromagnetic spectrum. Absorption spectroscopy resolves their spectra into a series of sharp lines, concentrated in broader spectral bands. Molecules composed of different atoms, like carbon dioxide (CO_2) and water vapor (H_2O), tend to have very strong absorption bands. Most of the other molecules, such as oxygen (O_2), nitrogen (N_2), and hydrogen (H_2), have very weak bands.

Figures 4.2.2 and 4.2.3 show the ability of the atmosphere to absorb incoming solar radiation as a function of wavelength throughout the solar and infrared spectrum. Sunlight at a visible wavelength of 0.5 μm is nearly unaffected by atmospheric absorption, whereas at an ultraviolet wavelength of 0.2 μm it is completely absorbed. Surface infrared radiation at a wavelength of about 8 to 9 μm passes unaffected through the atmosphere to space, but at 6 μm escape is not possible.

The longer the path of a photon through the atmosphere, the more likely it will be absorbed. The probability that a photon will be absorbed along the path is expressed as the negative exponential of the optical depth by the Beer–Bouger–Lambert law. Optical depth, at a specific wavelength, is the product of the absorption coefficient, concentration of absorbing molecules, and path length. Consequently, radiation traversing the atmosphere at an angle is more likely to be absorbed than if it were travelling normal to Earth.

Figure 4.2.2 shows that atmospheric absorption in the thermal wavelength regions is dominated by the presence of H_2O and CO_2 absorption bands. Water vapor absorbs both solar near-infrared and thermal longwave radiation. The important absorption bands are in the near-infrared bands, the 6.3-μm vibration-rotation band, and the pure rotation band at wavelengths longer than 15 μm. CO_2 is active mainly in the 15-μm vibration-rotation band. The ozone (O_3) 9.6-μm vibration-rotation band appears in a region without other strong absorptions.

The atmospheric absorption spectrum has two regions through which radiation can travel easily. Outside these regions, the air is quite opaque. The first region spans the ultraviolet, visible, and near-infrared spectrum, from about 0.3 to 0.7 μm in wavelength. The second region is in the infrared region, from about 8 to 13 μm. In climate studies, this thermal region is the one referred to by Luther and Ellingson (1985) as the atmospheric window. The window only has a strong absorption of O_3 (9.6 μm band) and a weaker background absorption due to the water vapor continuum. The greenhouse effect is caused by the relative ease with which solar radiation can reach Earth's surface through the UV visible region of the spectrum, and the difficulty that thermal radiation has in escaping from the surface. Both H_2O and CO_2 behave as ideal greenhouse gases because they are transparent at visible and near-infrared wavelengths, and are opaque at longer infrared wavelengths. As will be discussed later, many other gases have similar radiative properties and can act as effective greenhouse gases.

The emission spectrum of Earth is compared with blackbody radiation at various temperatures in Figure 4.2.5. The actual emission spectrum does not resemble a perfect blackbody spectrum. There is no reason to believe that all the thermal radiation leaving Earth is emitted from the same atmospheric level. Atmospheric temperature varies substantially with height. It follows that the radiation emitted by gases near the surface has a different emission temperature than the radiation emitted by gases at the tropopause, or the upper thermosphere. Figure 4.2.6 shows the structure of the atmosphere up to 100 km. It should be noted that 90% of the mass of the atmosphere is found in the lower 10 km.

The blackbody emission spectra in Figure 4.2.5 has the same intensity as the CO_2 band at 15 μm emission at close to 210 K. Comparing this temperature with the temperature profile in Figure 4.2.6, we see that temperatures in this range can be found in the lower stratosphere and middle mesosphere. Although either is possible, it has been established to be from the mesosphere. Similarly, the region from about 10 to 13 μm falls close to a temperature of 290 K, based on comparisons with the blackbody curves. Such a temperature is found only near the surface. This is consistent with the atmospheric window.

The mechanism of the greenhouse effect can now be explained. At constant solar energy input, the thermal emissions must also remain constant. Thus, the area under the emission curve in Figure 4.2.5 must remain constant. Greenhouse gases partially close the atmospheric window and trap heat at the surface and in the lower atmosphere. To balance this effect, the surface and lower atmosphere must warm and emit thermal radiation at a greater intensity.

The greenhouse effect occurs when the atmosphere absorbs thermal emissions. In accordance with the principle of conservation of energy, the amount of heat absorbed equals the amount reemitted.

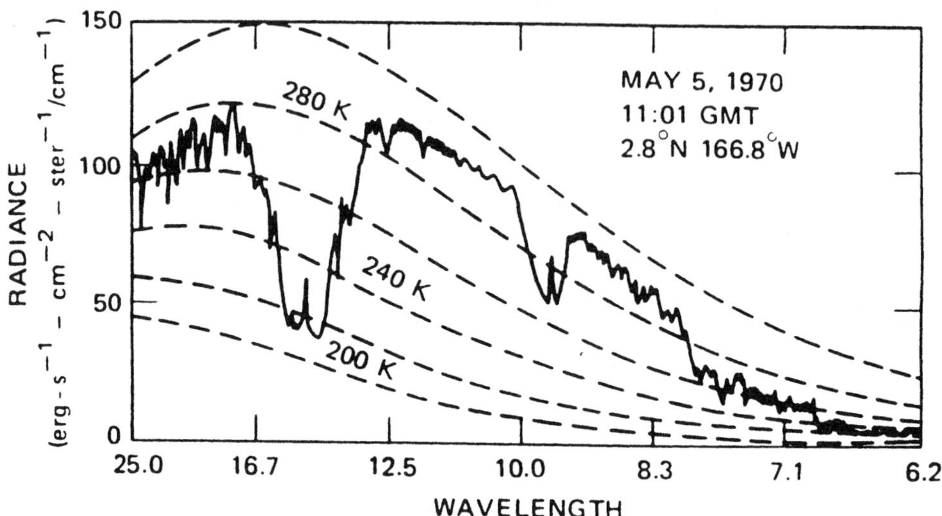

FIGURE 4.2.5 Comparison of satellite-measured infrared emissions of Earth's with blackbody temperature (Clark, 1982). Sample spectra from the Iris satellite, indicating the effective radiation temperature for different wavelengths of thermal infrared radiation. (Adapted from Hanel et al., *J. Geophys. Res.,* 77, 2629-2641, 1972.)

Roughly half of the reemitted heat travels in the same direction as the original thermal radiation, but the other half travels in the opposite direction. Thus, multiple occurrences of absorption and reradiation inhibit the escape of thermal energy to space. The atmosphere absorbs and emits radiation with nearly the same efficiency as determined by Kirchhoff's law, i.e., absorptivity and emissivity of air are equal. Consequently, the primitive climate model is modified by representing the entire atmosphere as an absorber and emitter with the same average emissivity at all wavelengths. This result was derived by Chamberlain (1978).

$$T_s = T_e\left(2/(2-\varepsilon)\right)^{1/4} \tag{4.2.8}$$

where: T_s = the actual average temperature of the surface, K
T_e = the effective blackbody radiation temperature of Earth, K
ε = emissivity ≈ 0.75

The emissivity is about three-quarters of the heat radiated by Earth's surface which is effectively absorbed by the atmosphere. Using this value in Equation (4.2.8) yields an average planetary surface temperature of 287 K, in accord with the measurements. The greenhouse effect of the atmosphere, therefore, explains the difference between the effective blackbody emission temperature of Earth and its surface temperature. Furthermore, most of the radiation leaving Earth originates in the upper atmosphere and not at the surface.

The absorption of trace components in the atmospheric window is explained in great detail in Figure 4.2.7. It should be noted that ozone and the CFCs (referred to as Freons, or F11, F12, F13, and F22) have strong absorptions in the window region, while methane and nitrous oxide absorb near the short-wavelength edge of the window. CO_2 tends to narrow the window from the long-wavelength side. As these greenhouse gases become more concentrated, their absorption bands overlap, narrowing the window. Earth's radiation where absorptions are strong is colder and therefore less intense. The surface and lower atmosphere respond by warming up, intensifying the emission in the more transparent regions of the spectrum to compensate for the loss of intensity in the window.

The greenhouse power of the trace atmospheric components is extremely large. This can be seen even for CFC concentrations of about 1 ppb(v) which produce the same global warming as roughly 300 ppm(v)

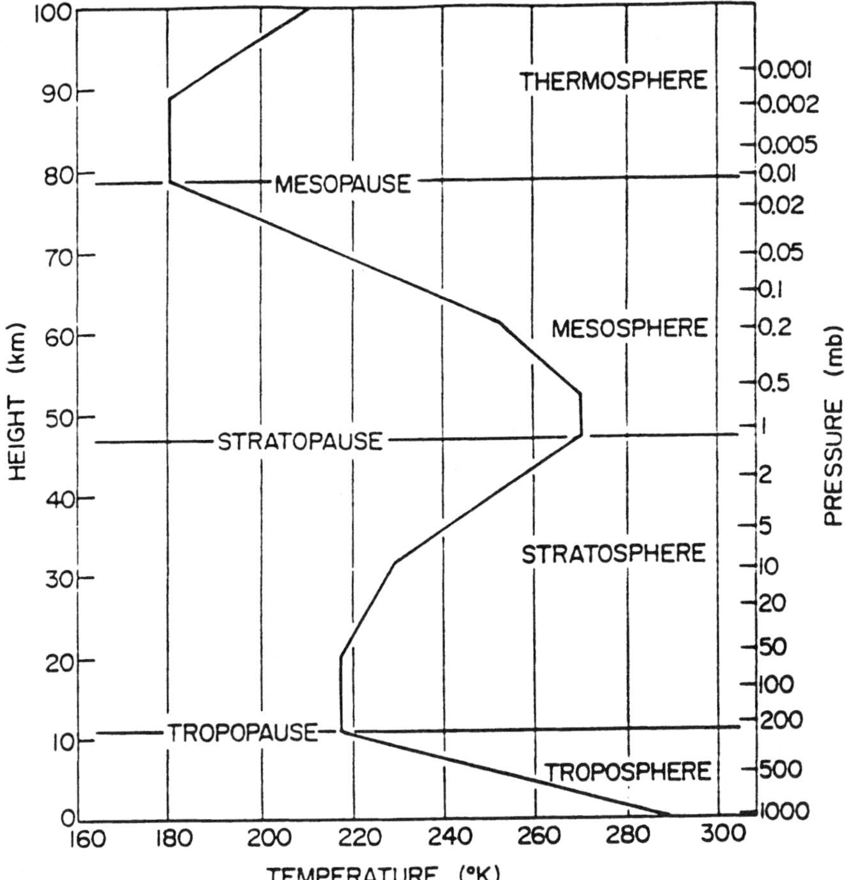

FIGURE 4.2.6 Structure of the atmosphere as a function of height above the surface of Earth and temperature. (From Peixoto J.P., and Dort, A.H., *Physics of Climates,* American Institute of Physics, College Park, MD, 1992. With permission.) Idealized vertical temperature profile according to the U.S. Standard Atmosphere (1976). Also shown are the names commonly used for the various layers and pauses in the atmosphere (Wallace and Hobbs, 1977).

CO_2. In other words, CFCs are about 300,000 times more effective than CO_2 in causing a greenhouse effect.

Clouds have an important effect on both solar and longwave radiation. At infrared wavelengths, clouds may be treated as blackbodies because they contain condensed water in the form of cloud droplets or ice crystals that are excellent infrared absorbers. Clouds absorb all the infrared radiation striking them and emit blackbody radiation at the temperature of the cloud. These rules may be applied to all forms of clouds described in Table 4.2.6.

There is some spectral variation in cloud absorbtivity (or emissivity), but this is significant mainly in the case of very thin cirrus clouds. Clouds scatter solar visible radiation and control Earth's albedo. About 60% of Earth is covered by clouds at any time, and the average albedo of these clouds is roughly 0.50. Thus, most of Earth's albedo can be attributed to clouds. Clouds also absorb solar near-infrared radiation (Figure 4.2.3), which accounts for about half the solar constant. About one-fourth the total insolation is due to near-infrared absorption by clouds. Finally, through multiple scattering, clouds diffuse the solar radiation emerging from their bottom surfaces.

Clouds contribute to the greenhouse effect by absorbing heat emitted by Earth's surface and lower atmosphere. Because they are blackbodies, clouds can absorb heat radiated by Earth. This is referred to as closing the atmospheric window at around 10 μm. The thermal flux emitted by the surface is absorbed

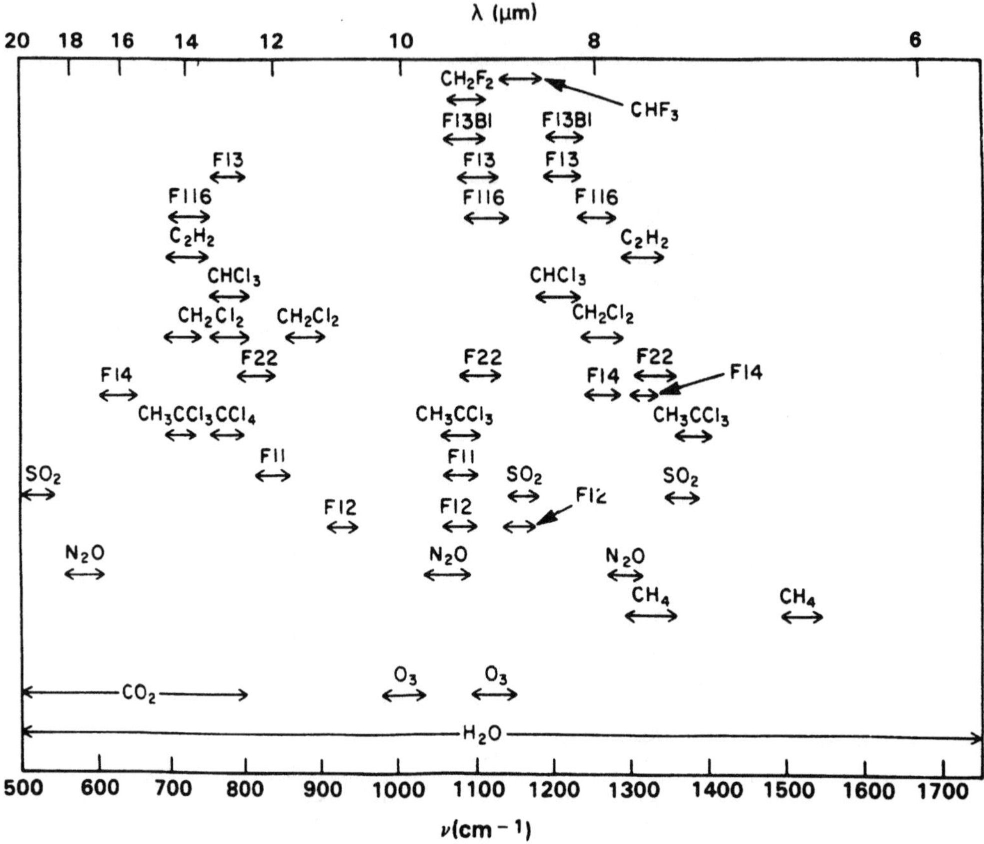

FIGURE 4.2.7 The absorption of trace constituents in the atmosphere window (Watson et al., 1986). Spectral locations of the absorption features of various trace gases. The spectrum between 7 and 13 μm is referred to as the atmospheric "window."

TABLE 4.2.6 Effects of Clouds

Type	Description	Area Covered	Effects on Shortwave	Longwave
Stratus	Low-lying, dense, surface to 2 km	Large	Efficient reflector, albedo = 0.5	Absorb/reradiate, blackbody radiation
Cumulus	Convectively active, separate puffs, forms storm clouds, 2 to 7 km	Localized	Efficient reflector, albedo = 0.5	Absorb/reradiate blackbody radiation
Cirrus	High, thin ice, 5 to 13 km	Large	Inefficient reflector, albedo = 0	Absorb/reradiate blackbody radiation

by the cloud bottom. The cloud bottom emits a flux of the same intensity back to the surface, preventing heat loss from the system. However, the top of the cloud also emits heat out to space. From space, the cloud tops look like the surface. The net effect is a cooling of local climate. In the case of cirrus clouds, the solar albedo is small. The large ice crystals in these clouds are not efficient in reflecting sunlight back to space. However, these clouds remain opaque to longwave radiation and behave as blackbodies. The thermal radiation emitted by the surface and lower atmosphere is absorbed at the cloud bottom. The cloud bottom reradiates the heat back toward the surface. The top of the cloud also radiates heat towards space. Since the cloud top is much colder than the surface, the emission to space is less intense than the original absorbed radiation. These clouds enhance the local greenhouse effect. Consequently, low-level clouds tend to produce net cooling, and cirrus clouds produce a net warming. Overall, the global effect of clouds on Earth's climate is to produce a small net cooling of surface temperatures,

The extent of and variations in global cloud cover are not well understood. What is the distribution of high and low clouds? How do clouds respond to global warming or cooling? Although clouds play a central role in the climate system, we do not know how they respond to a greenhouse effect. The surface of Earth is not very reflective, as seen from the values in Table 4.2.2. The oceans have an albedo of only about 0.1. Because the fraction of land and oceans covered with ice have albedos of 0.7, they are important contributors to the global average albedo. In fact, the increased albedo reinforces the cooling in a positive feedback process. In colder climates, snow and ice migrate toward lower latitudes and down mountain slopes. This enhances the cooling and promotes further migration. However, the system is believed stable because during glacial periods the ice margins eventually retreated. It should be noted that interglacial periods have lasted 10,000 to 20,000 years, as compared with glacial periods that lasted 100,000 years. This suggests that icy climates are more resistant to warming.

The Greenhouse Energy Balance

All the basic elements of the global energy balance, including the solar and thermal components and the greenhouse effect, are depicted in Figure 4.2.4. Of the total solar energy incident on Earth (given as 100% of incoming energy, equivalent to one-quarter of the solar constant, or about 345 W/m^2 when averaged over the entire surface of Earth), scattering from the atmosphere, clouds, and the surface reflects about 28% (the albedo). About 22% (17% + 5%) of the incident solar energy is absorbed in the atmosphere, mainly by water vapor and clouds. The rest of the solar energy, about 47%, is absorbed on the surface and converted to heat. The surface is also heated by thermal radiation emitted downward from the atmosphere and clouds. About 29% of the heat is removed from the surface as latent and sensible heat. The energy used in water vapor evaporation is released when it condenses to water and ice. Thus, when a cloud forms, the latent heat of the condensing water vapor is released and warms the air trapped in the cloud.

Solar heat absorbed on the surface is emitted as longwave radiation. Only 5% of this emission escapes into space through the atmospheric window. Most of the thermal emission from the surface is absorbed by clouds and greenhouse gases. The absorbed longwave radiation, together with the solar radiation directly absorbed into the atmosphere and the latent and sensible heat transferred to the atmosphere from the surface, are converted to heat. The heat is dissipated by reradiating it as longwave radiation. The radiation from cloud tops and from the greenhouse gases above the surface (64%) account for most of the thermal energy escaping Earth. The sum of the direct longwave emissions from the troposphere, with 3% from the stratosphere and the 5% that penetrate the atmospheric window, exactly balance the net solar energy input of 72%. The clouds and greenhouse gases also emit on the order of 96% longwave radiation back to the surface, which contributes to the surface energy balance in the form of the greenhouse effect.

Energy Reservoirs

The aggregate of glaciers and snow is referred to as the cryosphere. The principal source of glacial ice on Earth is the Antarctic continent. The Antarctic ice sheets cover an area the size of Brazil to an average depth of more than 2 km. The volume of ice is enormous, amounting to more than 2×10^7 km^3. Additional volumes of permanent ice are located on Greenland and in a number of mountain glaciers. The Antarctic ice sheets are equivalent in total volume to about 80 m of depth of the world's oceans. If the ice sheets were to melt, the sea level would rise by roughly that amount and coastal regions would be flooded. The most permanent glaciers reside on Antarctica, where the ice may be up to 200,000 years old. Air bubbles trapped in the ice have revealed the history of CO_2 concentrations over the past 160,000 years. The ice also exhibits heavy deposits of sulfate and ash following major volcanic eruptions. The temperature of the planet is recorded as well in the oxygen isotopes long ago frozen into the ice. Ice and snow participate in Earth's climate system in several ways. As we noted in Table 4.2.4, the water tied up as ice on land represents a substantial fraction of the total surface ocean water. Ice- and snow-covered surfaces have a high albedo and so increase the reflectivity of Earth. If a substantial fraction of the surface area of Earth was to become covered with ice, then the planetary albedo would increase significantly and further cooling would result. The frozen Earth might never thaw, and no life as we know it could exist. Ice covering the seas, as in the Arctic Ocean, has a strong influence on the energy balance of these regions. Fortunately, floating ice is an excellent insulator against heat loss. The layer of cold ice isolates the ocean water from the atmosphere, preventing efficient latent heat exchange.

The total energy contained in the ice reservoir is about 2.6×10^7 EJ, comparable to the total energy held in the surface oceans. However, the rate of transfer of energy between the ice sheets and the rest of the climate system is much slower than the rate of transfer between the surface oceans and the atmosphere. Thus, the effects of the cryosphere must be factored into climate changes occurring over periods of 1000 years or longer.

Processes of Climate

The flow of energy among different elements of the climate system can become quite complex. Details concerning these processes extend beyond the requirements for understanding environmental heat transfer. As mentioned above, the radiative processes (scattering, absorption, and emission) dominate the climate system. Physical processes (and material properties) also are important. The conduction of heat in soil and the capacity of water to store heat are significant. The latent heat properties of water are critical. The behavior of cloud particles and the optical properties of clouds and aerosols are also important. Clearly, the dynamics of the atmosphere is fundamental to climate.

Astronomical processes contribute to Earth's climate and climate change. The sun and the processes that control its output of energy are critical. The brightness of the sun and sunspots, which oscillates over periods of about 11 years, are frequently discussed in relation to climate. The orbit of Earth and small perturbations in it over time have been connected to ice ages. According to Milankovitch (1941), Earth's orbital parameters vary over periods of about 20,000, 40,000, and 100,000 years. The paths of asteroids and comets through the solar system and the mechanics of impact of these bodies with Earth may explain periodic climatic disasters in Earth's history, according to Alvarez (1983). All these factors external to Earth itself cause variations in climate.

Chemical processes also play a role in the climate system. The concentrations of greenhouse gases are controlled, to varying degrees, by chemical processes. O_3 is the most chemically active of all the greenhouse compounds and water vapor is the least active. A variety of mechanisms may affect the amounts of climate-active compounds in the atmosphere, including photochemical reactions, biochemical processes, interactions with clouds, geochemical transformations, and industrial processes. In the latter case, relatively small concentrations of compounds such as chlorofluorocarbons were released into the environment without realizing their efficiency for absorbing infrared radiation. Similarly, massive quantities of otherwise natural compounds such as SO_2, CO_2, and N_2O have been emitted without realizing their climatic impacts.

Biological processes are key drivers of climate system. Several greenhouse gases, such as CO_2, CH_4, and N_2O, are partly controlled by biological processes. Photosynthesis and respiration are the fundamental life processes on the planet, and they control surface environmental conditions that affect the absorption and transfer of energy in terrestrial and marine components of the climate system. Biological processes have also been invoked as a potential means of modifying or controlling certain components of the climate system, e.g., the use of iron to stimulate phytoplankton to absorb excess carbon dioxide from the atmosphere.

Climate Variability

Climate variability is the statistical noise in the climate system. It is the change in climate from the average or mean over a time interval of interest. The interval may be a year, decade, century, millennium, or longer. The climate is conveniently measured by the average surface temperature of Earth. A change of less than 0.5°C is considered normal in the current climate that has lasted over 5 centuries.

Volcanic Eruptions, Smoke, Dust, and Haze

Major eruptions produce widespread layers of aerosols, or fine particles, in the stratosphere, which efficiently scatter sunlight and enhance Earth's albedo. Climate cools as a result of the increased albedo. Volcanic aerosols are composed primarily of mineral ash particles ejected by volcanos. Large amounts of ash are emitted and settle over vast regions. The ash falls out within a few months, while volcanic aerosols produced from the stratospheric oxidation of SO_2 and H_2S remain in the stratosphere for years. Volcanic aerosols spread over Earth and create spectacular sunsets and climate change. During the time the aerosols are dispersed over the entire globe, they affect the global radiation balance.

The effect of changes in smoke, dust, and haze concentrations in the upper troposphere or stratosphere is to increase the planetary albedo and reduce the surface temperature. However, the introduction of highly absorbing aerosols in the lower atmosphere (boundary layer) can lower the albedo and warm the surface.

A Simple Model on the Effect of Energy Consumption on Climate Modification

Summary of Model

Per capita energy growth patterns were used to project the growth of atmospheric CO_2. The population of less-developed countries will grow 2.7 times — from 50 to 67% of world population — from 1991 to 2100. Over the same period, world population will grow from 5.0 to 10.3 billion people, and energy use from 15 to 50.7 TW. LDC energy use will grow disproportionately faster, from 20 to 46%, and the energy use in industrialized countries will slow from 30 to 22% of world energy. It is anticipated that the global standard of living will improve substantially while relative energy consumption will decrease on the order of 1.1% per acre due to conservation and efficiency improvements. Nonfossil energy sources consisting mostly of nuclear energy will overtake fossil energy consisting mostly of coal in the year 2075. The growth of CO_2 emissions from 6 to 18.2 GtC per acre will result in an average global temperature increase of 3°C due to this source only. However, CO_2 is only about half the problem. When all infrared-absorbing gases are considered, an average increase of 5.6°C is projected for 2100. This scenario is similar to the IPCC scenario IS 92A.

Introduction to Model

The atmospheric monitoring program by Keeling et al. (1997) shows the level of carbon dioxide in the atmosphere has increased about 16% over the last 39 years and now stands at about 365 ppm. This observed increase is believed to be the continuation of a trend, which began in the middle of the last century with the start of the Industrial Revolution. Fossil fuel combustion, cement manufacturing, and the clearing of virgin forests (deforestation) are considered to be the primary anthropogenic contributors, although the relative contribution of each is uncertain because deforestation appears to have been a net source during some periods of time and a sink during other periods.

Carbon dioxide (CO_2) is the major component of trace atmospheric gases that affect global climate. The other trace gases contained in the atmosphere are water vapor (H_2O), ozone (O_3), methane (CH_4), carbon monoxide (CO), nitrous oxide (N_2O), sulfur hexafluoride (SF_6), and chlorofluorocarbons (CFC). Except for H_2O and O_3, the other trace gases have long lifetimes in the atmosphere. They absorb part of the infrared rays reradiated by Earth and induce atmospheric warming.

Predictions of the climatological impact of a CO_2-induced greenhouse effect draw upon various mathematical models to gauge the global average temperature increase. The scientific community generally discusses the impact in terms of doubling the 1975 atmospheric CO_2 content in order to get

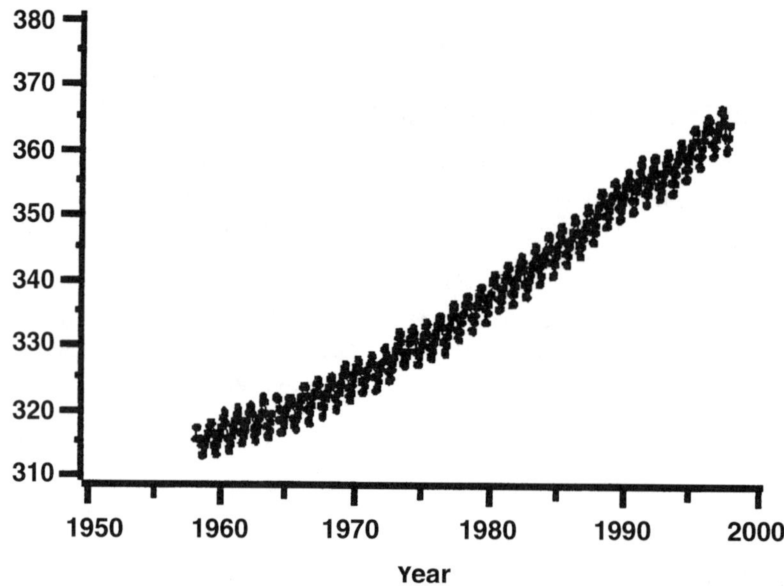

FIGURE 4.2.8 Growth of CO_2 as a function of time: Mauna Loa, Hawaii. (Data from Keeling, D. and Whorf, T. Scripps Institution of Oceanography.)

beyond the normal fluctuations (noise level) of climate data. A scenario was built into the present study for projecting the atmospheric concentration of CO_2 and the other infrared-absorbing trace gases in order to estimate future global average temperatures. The scenario is based on data presented by:

- Rogner (1986) on regional energy consumption
- Global energy requirements by Häfele (1981) with modifications from various industrial projections
- Growth projections of CH_4, N_2O, CFC 11, CFC 12, and other CFCs by Ramanathan (1985) and Wigley (1987), as summarized by Krause et al. (IPSEP, 1989)
- Energy efficiency improvement and conservation measures from various DOE sources

It should be noted that the scenario discussed here is similar to the business-as-usual scenario IS 92A in the IPCC study (Pepper et al., 1992).

Methodology

The impact of per capita energy growth patterns on future energy demand and growth of atmospheric carbon dioxide was evaluated by subdividing the globe into 6 regions of similar sociopolitical background and population growth rate. The 6 key global regions specifically considered are

1. North America (U.S. and Canada) [NA].
2. Middle East (North Africa and Persian Gulf States) [ME].
3. Commonwealth of Independent States and Eastern Europe [CISEE].
4. China and other centrally planned Asiatic economies [CPAE].
5. Industrialized countries (including Western Europe, Australia, New Zealand, Israel, Japan, and South Africa) [IC].
6. Less-developed countries (including all of South and Central America, Central Africa, and the rest of Asia) [LDC].

TABLE 4.2.7 Current Greenhouse Gas Concentrations

Gas	Preindustrial Concentration (1860)	Present Tropospheric Concentration[1]	GWP[2] (100 yr. Time Horizon)	Atmospheric Lifetime (Years)[3]
Carbon dioxide (CO_2) (ppm)	288[4]	364.3[5]	1	120
Methane (CH_4) (ppb)	848[6]	1850[7]/1683[8]	21	12
Nitrous oxide (N_2O) (ppb)	285[9]	311[7]/310[8]	310	120
CFC-11 (trichlorofluoromethane) (CCl_3F) (ppt)	zero	265[7]/261[8]	3800	50
CFC-12 (dichlorotrifluoroethane) ($C_2F_3Cl_3$) (ppt)	zero	535[7]/524[8]	8100	102
CFC-113 (trichlorotrifluoroethane) ($C_2F_3Cl_3$) (ppt)	zero	84[7]/82[8]	4800	85
Carbon tetrachloride (CCl_4) (ppt)	zero	100[7]/98[8]	1400	42
Methyl chloroform (CH_3CCl_3) (ppt)	zero	94[7]/85[8]	360[10]	5
HCFC-22 (chlorodifluoromethane) ($CHClF_2$) (ppt)	zero	126/110[11]	1500	12
Sulfur hexafluoride (SF_6) (ppt)	zero	3.6 NH/3.4 SH[12]	23,900	3200
Perfluoroethane (C_2F_6) (ppt)	zero	4[13]	9200	10,000
Surface ozone (ppb)	25[14]	26[15]/25[16]	17	hours

[1] Present tropospheric concentration estimates are calculated as annual arithmetic averages; ppm = parts per million (10^6), ppb = parts per billion (10^9), ppt = parts per trillion (10^{12}).

[2] The Global Warming Potential (GWP) is typically used to contrast different greenhouse gases relative to CO_2. The GWP provides a simple measure of the relative radiative effects of the emissions of various greenhouse gases and is calculated using the formula:

$$\mathrm{GWP} = \frac{\int_0^n a_i c_i \, dt}{\int_0^n a_{CO_2} c_{CO_2} \, dt}$$

Formula taken from: IPCC (Intergovernmental Panel on Climate Change) 1990: *Climate Change: The IPCC Scientific Assessment,* J.T. Houghton, G.J. Jenkins, and J.J. Ephraums (Eds.). Cambridge University Press, Cambridge, U.K.

GWPs taken from: IPCC, 1996: *Climate Change 1995: The Science of Climate Change,* J.T. Houghton, L.G. Meira Filho, B.A. Callander, N. Harris, A. Kattenberg, and K. Maskell, Cambridge, University Press, Cambridge, U.K. (see page 22 and page 119).

[3] Atmospheric lifetimes (general term used for the "adjustment time", the time scale characterizing the decay of an instantaneous pulse input into the reservoir) taken from: IPCC, 1996: *Climate Change 1995: The Science of Climate Change* (see pp. 92-93).

[4] Historical CO_2 record from the Siple Station, Antarctica ice core.

[5] 1997 Annual average derived from *in situ* sampling at Mt. Cimone, Italy.

[6] Historical CH_4 concentration from the Law Dome "DE08", Antarctica ice core.

[7] Annual arithmetic average based on 1996 monthly, nonpollution values from the Advanced Global Atmospheric Gases Experiment (AGAGE), Mace Head, Ireland monitoring site.

[8] Annual arithmetic average based on 1996 monthly, nonpollution values from the AGAGE, Cape Grim, Tasmania, Australia monitoring site.

[9] Historical N_2O concentration from the Law Dome "BHD", Antarctica ice core.

[10] IPCC, 1994: *Climate Change 1994: Radiative Forcing of Climate Change and an Evaluation of the IPCC IS92 Emission Scenarios,* J.T. Houghton, L.G. Meira Filho, J. Bruce, Hoesung Lee, B.A. Callander, E. Haites, N. Harris, and K. Maskell (Eds.), Cambridge University Press, Cambridge, U.K. (see page 33).

[11] Atmospheric HCFC-22 measurements from flask samples taken at Pt. Barrow, Alaska, U.S. and Cape Grim, Tasmania, Australia, respectively. These measurements were conducted by the Nitrous Oxide and Halocompounds Group (NOAH) of the National Oceanic and Atmospheric Administration (NOAA)/Climate Monitoring and Diagnostics Laboratory (CMDL).

[12] NOAH hemispheric SF_6 estimates derived from 1996 flask and *in situ* air samples at NOAA/CMDL land sites.

[13] IPCC, 1996: *Climate Change 1995* (see page 93).

[14] IPCC, 1996: *Climate Change 1995* (see page 90).

Population and Energy

Although economic considerations are critical in determining per capita energy use, this study did not independently evaluate economics, but used the projections of Edmonds, et al. (1984). It was found that 50% of the world's population currently residing in LDC will grow to 67% by 2100, while world population will grow from 5.0 to 10.3 billion people in the year 2100 (see Figure 4.2.9). Consequently, the world's energy needs will grow from 15 to 50.7 TW (see Figure 4.2.10) while the LDC needs will grow from 20 to 46%, and the IC needs will decrease from 41 to 22% (see Figure 4.2.11). The apparent decrease in energy demand by the IC will not affect the standard of living, because substantial increases in the efficiency of energy utilization and conservation will be implemented (see Figure 4.2.12 and Table 4.2.8). The per capita energy use will decrease in NA as a consequence of energy demand reductions of about 30% by the year 2100 (see Figure 4.2.13). Similar efficiency improvements are projected worldwide. Specific details relating to power plants and automotive efficiency improvements are explicitly presented in Figure 4.2.12 and Table 4.2.8. Other effects such as decreasing the consumption of electricity for lighting, heating, refrigeration, and air conditioning (with fluids other than banned chlorofluorocarbons) were assumed as part of the model based on Geller's (1986) projections and reasonable market penetration and replacement rates.

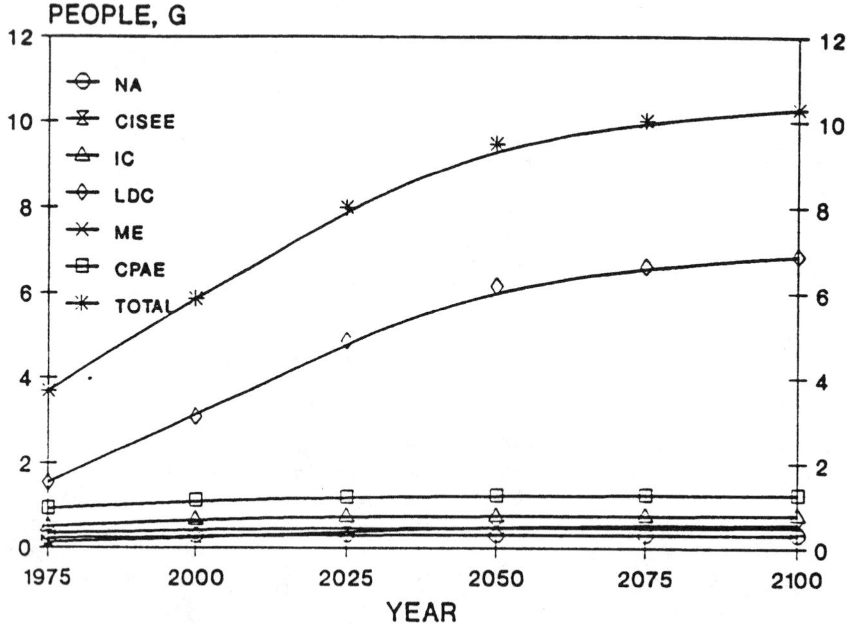

FIGURE 4.2.9 World population projection.

Nonfossil energy sources are projected to grow faster than fossil fuels and cross about the year 2075 (see Figure 4.2.10). After 2025, the predominant fossil fuel will be coal (see Figure 4.2.14) and nonfossil energy source will be nuclear, i.e., either fission or fusion or both (see Figure 4.2.15). The use of renewable energy will grow exponentially, but will not predominate as a nonfossil fuel source because it is very area intensive, thus conflicting with population growth. It should also be noted that most of the petroleum-based energy will be used in transportation and not for stationary power or other industrial energy uses.

Carbon Emissions

The carbon cycle for 1975 (depicted in Figure 4.2.16) is used as the starting point to show the redistribution of carbon among the relevant reservoirs. Carbon emissions are estimated to grow from 6 to

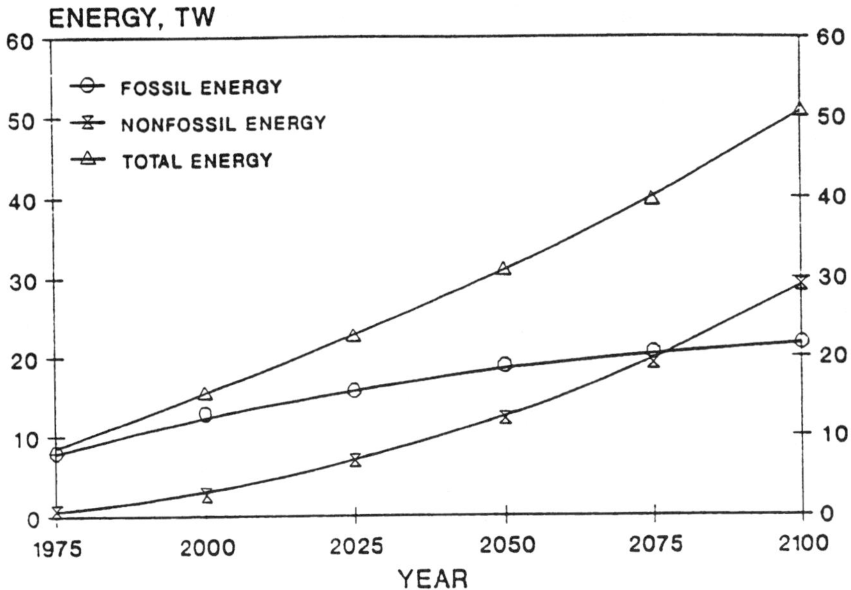

FIGURE 4.2.10 Comparison of total primary fossil and nonfossil fuel energy sources.

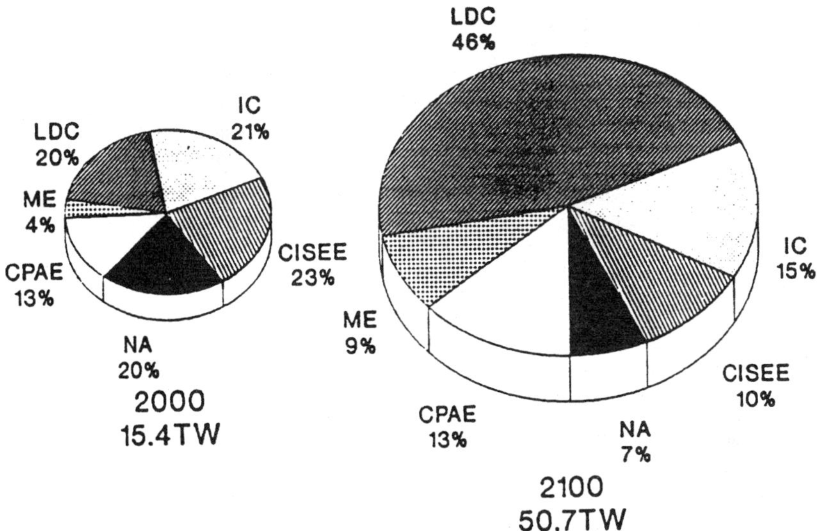

FIGURE 4.2.11 Total primary energy comparison by global region.

18.2 GtC/a in 2100 (see Table 4.2.9 and Figure 4.2.17). The per capita emissions of carbon will decrease for NA and CISEE, and increase for the rest of the world, but the relative order for the six global regions will remain the same between 1975 and 2100 (see Figure 4.2.18). It should be noted, however, that the percentage contribution from the LDC and ME will increase from 23 to 55% (see Figure 4.2.19).

It is projected that the 1975 atmospheric CO_2 concentration will double by 2100, resulting in a global average temperature increase of 3°C due to this constituent only. These results agree well with the Edmonds et al. (1984) median case (B) projections, albeit under different synthetic oil projections. Another global average temperature projection due to CO_2 can only be made on the assumption that the

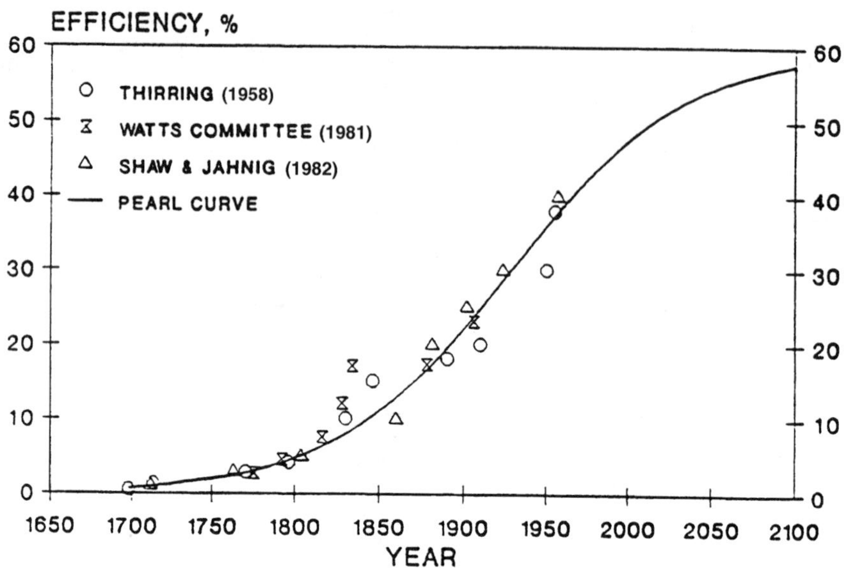

FIGURE 4.2.12 Steam plant thermal efficiency improvement with time.

TABLE 4.2.8 Automotive Carbon Emissions

	Automobile Fuel		Carbon Emissions		Coal[a]
Year	**Electric (miles/kWh)**	**Gasoline (miles/gal)**	**Electric (kg/mile)**	**Gasoline (kg/mile)**	**Power (kg/kWh)**
2000	2.0	30	0.10	0.09	0.202
2050	4.0	50	0.04	0.05	0.171
2100	5.0	60	0.03	0.04	0.162

[a] Power plant efficiences are 45% in 2000, 53% in 2050, and 56% in 2100.

equilibrium global average temperature exceeded the 0.5°C normal temperature fluctuation level in 1980, and the preindustrial concentration was 280 ppm CO_2. The latter projection predicts a temperature increase of 2.4°C in 2100. When the other radiatively and chemically active trace gases are included in a model using the radiative forcing constants for CO_2, CH_4, N_2O, CFC 11, CFC 12, and all other CFCs, then one predicts a temperature increase of 5.6°C in 2100 (see Figure 4.2.20). These estimates do not consider negative feedback mechanisms such as high cloud formation. The question of which predictions and which models best simulate a carbon dioxide and other trace-gases-induced climate change is still being debated by the scientific community. The incremental temperature increase would not be uniform over Earth's surface. The polar caps are likely to see temperature increases on the order of 10°C. There will be little, if any, temperature increase at the equator.

Climate

There is currently no definitive scientific evidence that Earth is warming. If Earth is on a warming trend, we are not likely to detect it before a decade passes. This is about the earliest projection of when the temperature might rise above the 0.5°C needed to get beyond the range of normal temperature fluctuations. On the other hand, if climate modeling uncertainties have exaggerated the temperature rise, it is possible that a greenhouse effect induced by infrared-absorbing gases may not be detected until 2020

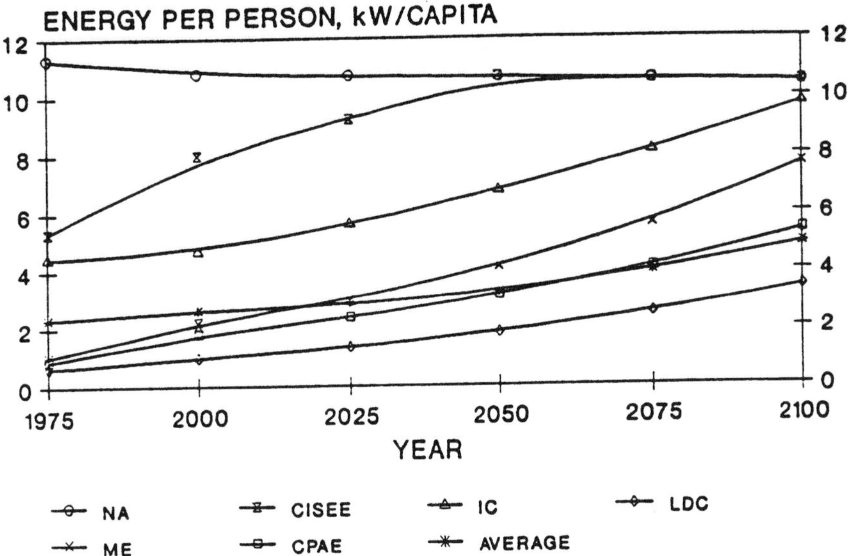

FIGURE 4.2.13 Regional per capita primary energy consumption.

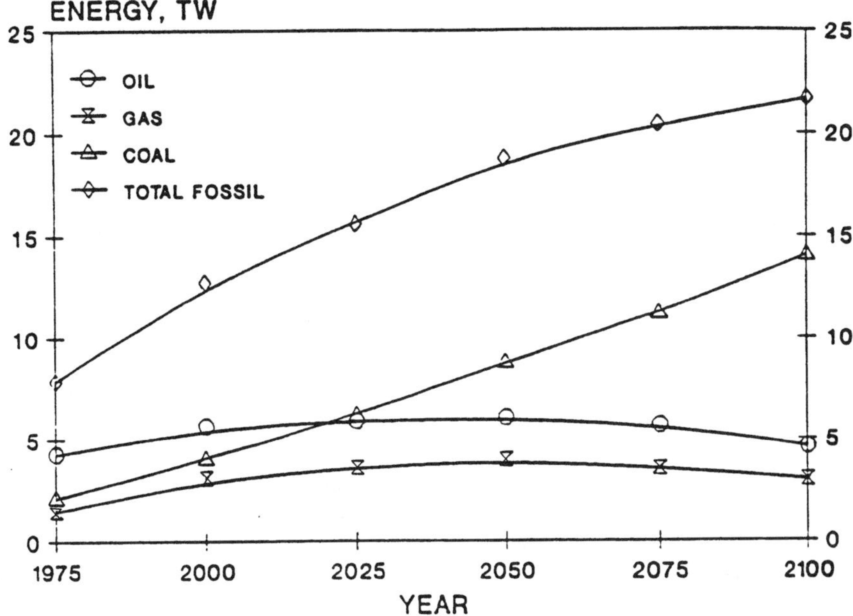

FIGURE 4.2.14 Primary fossil fuel energy sources.

at the earliest. It should be noted that the equilibrium response of climate, as estimated by global average temperature, lags the measured temperature by about two decades due to the thermal inertia of the oceans.

The greenhouse effect is not likely to cause substantial climatic changes until the average global temperature rises at least 1°C above today's level. This could occur in the first to second quarter of the next century. However, there is concern among some scientific groups that once the effects are measurable, they might not be reversible, and little could be done to correct the situation in the short term.

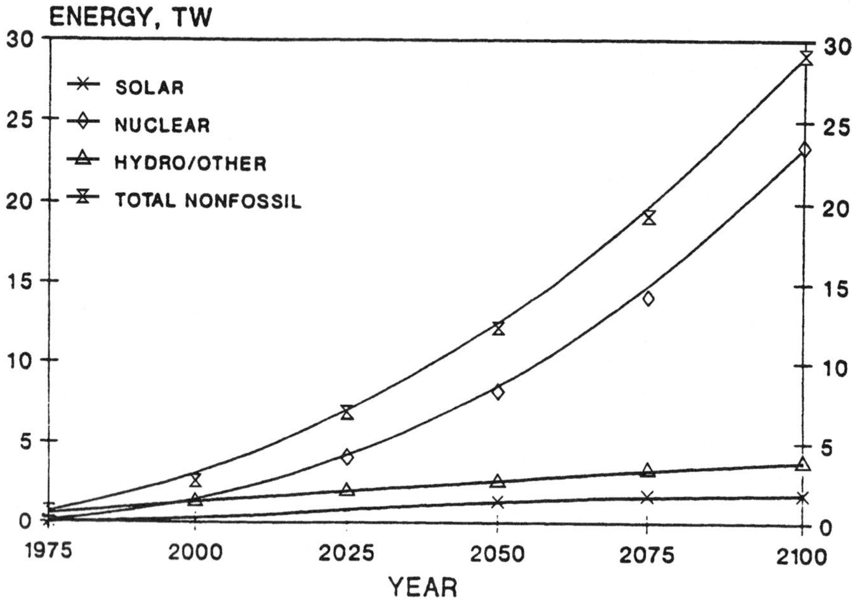

FIGURE 4.2.15 Primary nonfossil fuel energy sources.

Therefore, a number of environmental groups are calling for action now to prevent a potentially undesirable situation from developing in the future.

Mitigation of the greenhouse effect would require major reductions in fossil fuel combustion. Shifting between fossil fuels is not a feasible alternative because of limited long-term supply availability of certain fuels, although oil does produce about 18% less carbon dioxide per kW·h of heat released than coal, and gas about 32% less than oil (see Table 4.2.9). The energy outlook suggests synthetic fuels will have a negligible impact on the growth of CO_2 in the atmosphere, contributing less than 10% of the total carbon dioxide released from fossil fuel combustion by the year 2050. This low level includes the expected contribution from carbonate decomposition, which occurs during shale oil recovery and assumes essentially no efficiency improvement in synthetic fuel processes above those currently achievable. After 2050, however, the contribution of carbonate decomposition may account for as much as half the total carbon emitted from oil utilization, and CO_2-contaminated natural gas from remote deposits may account for half the natural-gas CO_2 emissions.

In addition to improvements in energy efficiency, use of nonfossil fuels, a large international research effort is currently underway to evaluate the feasibility of capturing and sequestering CO_2. Herzog et al. (1997) discuss the current international status of these technologies. The main challenge regarding CO_2 capture technology is to reduce the overall cost by lowering both the energy and the capital cost requirements. One strategy that looks extremely promising is to combine CO_2 removal with advanced coal energy conversion processes that have features which will enable low energy-intensive capture. The major options for CO_2 storage are underground or in the ocean. The capacity of the ocean to accept CO_2 is almost unlimited, but there are questions that still need to be addressed about its effectiveness, including sequestration lifetimes, and the environmental impact associated with increased seawater acidity. While there are diverse niche opportunities for industrial utilization of power plant CO_2, these uses are all small compared to the total quantities of CO_2 emitted by the power sector.

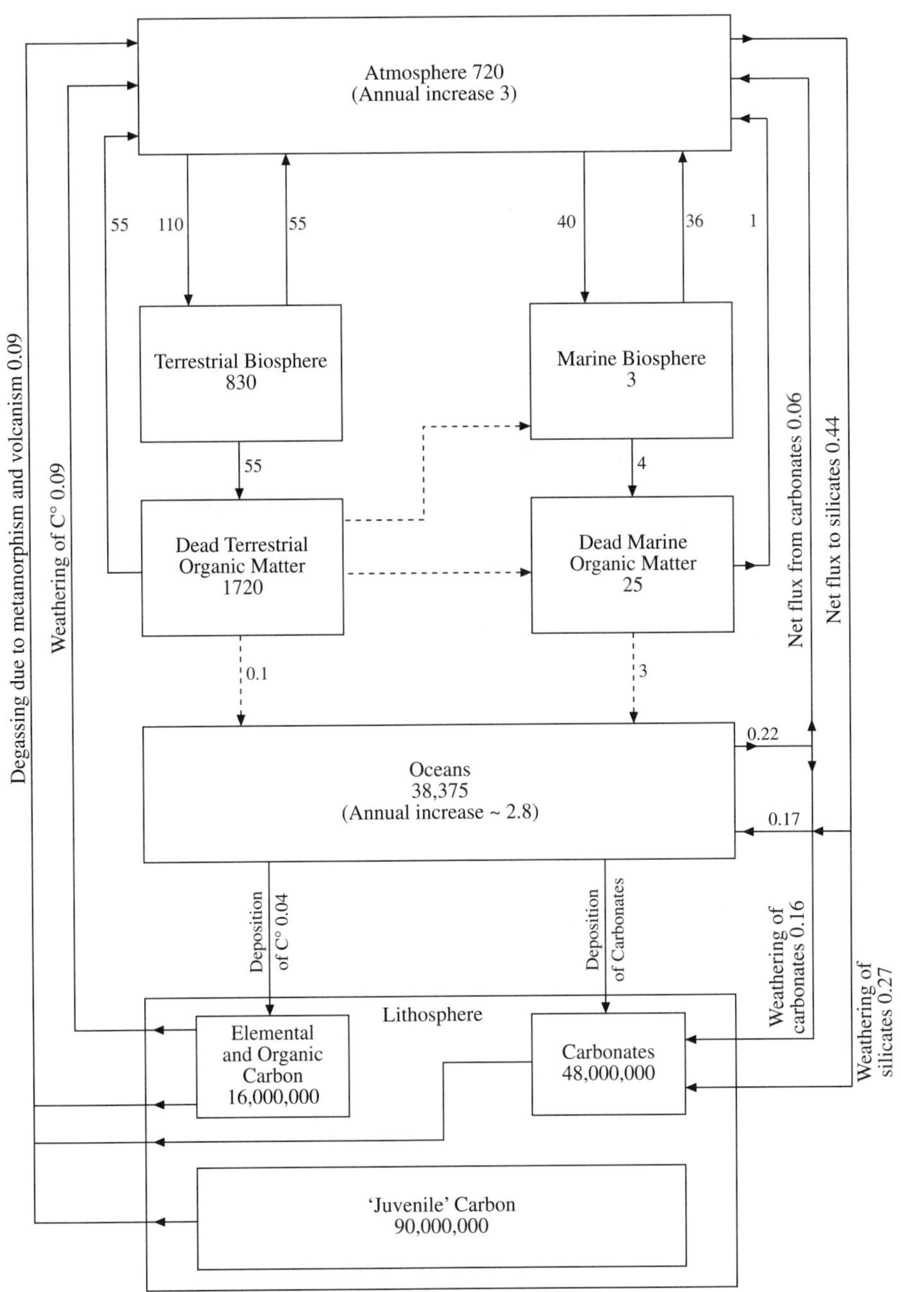

FIGURE 4.2.16 The global carbon cycle. Arrows show the transfer of elemental carbon (C) between the various carbon reservoirs (boxes) and indicate the closed loops that make up the cycle. Transfer rates are given in units of 10^{12} kg of C per year, and reservoir contents in 10^{12} kg of C. Burning of fossil fuels releases an additional 5 x 10^{12} kg yr^{-1} to yield the net increase in atmospheric burden. (Adapted from Holland, H.D., *The Chemistry of the Atmospheres and Oceans*, John Wiley & Sons, Chichester, 1978, and updated with information reviewed by Clark, W.C. (Ed.), *Carbon Dioxide Review: 1982*, Oxford University Press, Oxford, 1982; and by Bolin, B. and Cook, R.B, (Eds.), *The Major Biogeochemical Cycles and Their Interactions*, SCOPE 21, John Wiley & Sons, Chichester, 1983.)

TABLE 4.2.9 Production of Carbon Dioxide from Fossil Fuels per Unit Energy

Fossil Fuel	GtC/TW	gC/MJ
Coal	0.722	22.9
Oil	0.653	20.7
Gas	0.391	12.4

Note: TW = 31.54 EJ/a
= 29.89 Quads/a

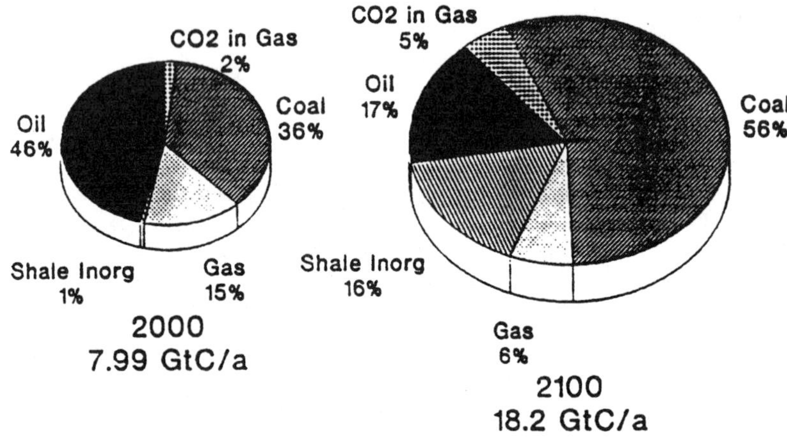

FIGURE 4.2.17 Comparison of carbon emissions by source.

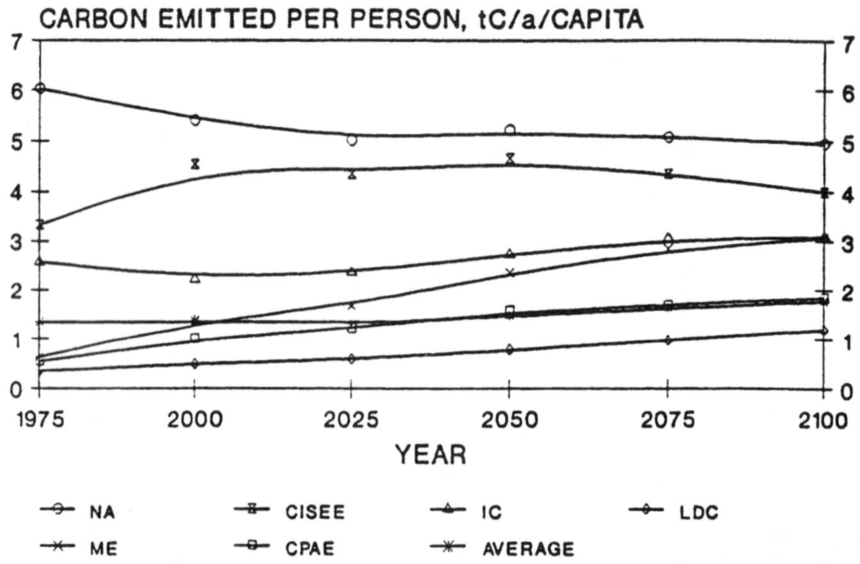

FIGURE 4.2.18 Regional per capita carbon emissions.

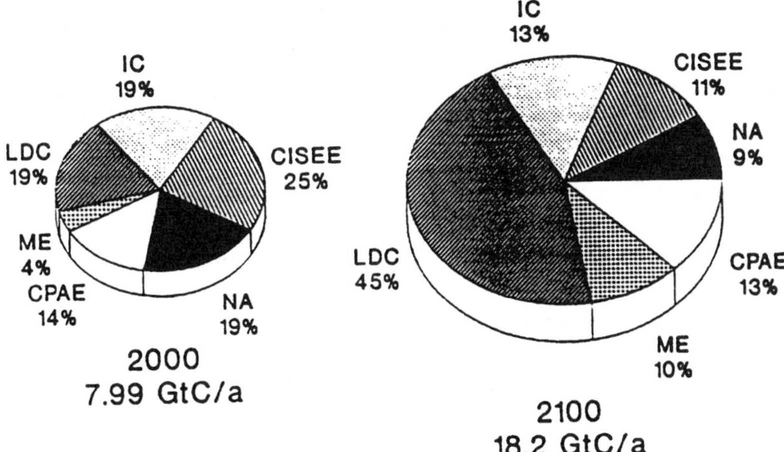

FIGURE 4.2.19 Comparison of carbon emissions by global region.

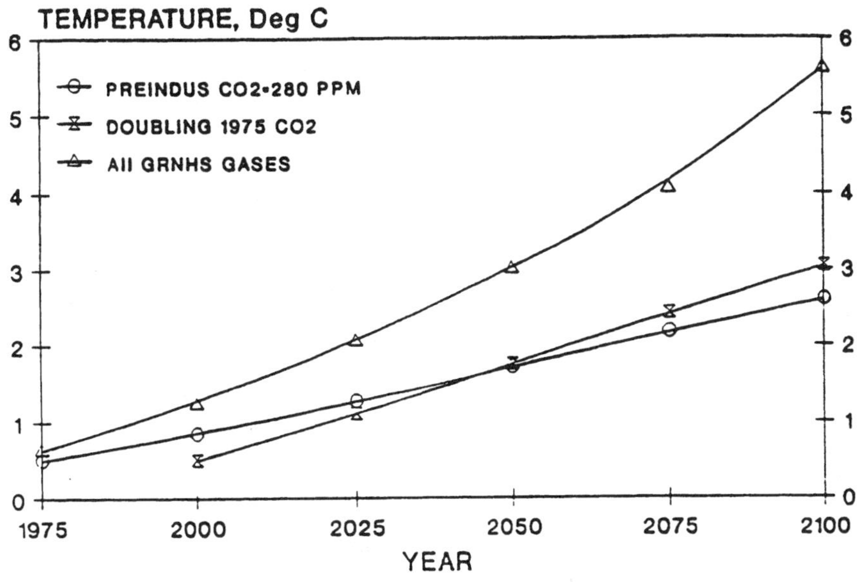

FIGURE 4.2.20 Greenhouse global average temperature increase.

References

Alvarez, L. W., Experimental Evidence that an Asteroid Impact Led to the Extinction of Many Species 65 Million Years Ago, *Proc. Natl. Acad. Sci. U.S.A.*, 80, 627, 1983.

Budyko, M., *Climate and Life*, Academic Press, London, 1974.

Chamberlain, J. W., *Theory of Planetary Atmospheres. An Introduction to their Physics and Chemistry,* Int. Geophys. Ser. 22, Academic Press, New York, 1978.

Clark, W.C., *Carbon Dioxide Review: 1982,* Oxford University Press, New York, 1982.

Edmonds, J. A., Reilly, J., Trabalka, J. R., and Reichle, D. E., An Analysis of Possible Future Atmospheric Retention of Fossil Fuel CO_2, DOE/OR/21400-11 (TRO 13), Department of Energy, Washington, D.C., September, 1984.

Gast, P.R., Solar Electromagnetic Radiation, in Handbook of Geophysics and Space Environments, Air Force Cambridge Research Laboratory, 16.1-16.9, 1965.

Geller, H. S., Energy-efficient appliances: performance issues and policy options, *IEEE Technol. Soc. J.,* 5, 4-10, 1986.

Häfele, W. et al., *Energy in a Finite World: A Global Systems Analysis,* Rep. by the Energy Systems Program Group of IIASA, Ballinger, Cambridge, MA, 1981.

Hanel, R. A., Conrath, B. J., Kunde, V. G., Prabhakara, C., Revah, I., Salomonson V. V., and Wolford, G., The nimbus 4 infrared spectroscopy experiment. I. calibrated thermal emission spectra, *J. Geophys. Res.,* 77, 2629-2641, 1972.

Herzog, H., Drake, E., and Adams, E. C., *CO_2 Capture, Reuse, and Storage Technologies for Mitigating Global Climate Change: A White Paper,* DOE Order No. DE-AF22-96PC01257, MIT Press, Cambridge, MA, 1997

Houghton, H.G., *Physical Meteorology,* MIT Press, Cambridge, MA, 1985.

Hoyt, D.V., The Smithsonian Astrophysical Observatory solar constant program, *Rev. Geophys. Space Physics,* 17, 427-458, 1979.

Krause, F., Bach, W., and Koomey, J., Energy Policy in the Greenhouse, From Warming Fate to Warming Limit: Benchmarks for a Global Climate Convention, Vol. 1, International Project for Sustainable Energy Paths (IPSEP), El Cerrito, CA 94530, 1989.

Luther, F.M. and Ellingson, R.G., Carbon Dioxide and the Radiation Budget, Projecting the Climatic Effect of Increasing Carbon Dioxide, DOE Rep. DOE/ER-0237, McCracken, M.C. and Luther, F.M., Eds., Department of Energy, Washington, D.C., 1985.

Milankovitch, M., Canon of Insolation and the Ice Age Problem, Program for Scientific Translations, National Science Foundation, Arlington, VA, 1941.

Oke, T.R., *Boundary Layer Climates,* 2nd ed., Halstead, New York, 1987.

Peixoto, J. P. and Oort, A. H., *Physics of Climate,* American Institute of Physics, College Park, MD, 1992.

Platt, M.R., *Radiative Processes in Meteorology and Climatology,* Elsevier, New York, 1976.

Ramanathan, V. et al., Trace gas trends and their potential role in climate change, *J. Geophys. Res.,* 90, 5547-5566, 1985.

Rogner, Hans-Holger, Long-term projections and novel energy systems, in *The Changing Carbon Cycle — A Global Analysis,* Trabalka, J.R. and Reichle, D.E., Eds., Springer-Verlag, New York, 1986.

Schneider, S. H. and R. Londer, *The Coevolution of Climate and Life,* Sierra Club Books, San Francisco, 1984.

Shaw, H. and C. E. Jahnig, Environmental Assesment of Advanced Energy Conversion Technologies — Final Report, EXXON/GRU.9DA.82, 1982.

Stephenson, R., *Introduction to Nuclear Engineering,* McGraw-Hill, New York, 1954.

Thring, H., *Energy for Man,* Indiana University Press, Bloomington, Indiana, 1958.

Turco, R.P., *Earth Under Siege,* Oxford University Press, New York, 1997.

UN, Population Bulletin of the United Nations, No. 14, Department of International Economic and Social Affairs, United Nations Secretariat, New York, 1982.

UN, Population Bulletin of the United Nations, Department of International Economic and Social Affairs, United Nations Secretariat, New York, 1984.

Watson, R. T., Geller, M. A., Stolarski, R. S., and Hampson, R. F., Present State of Knowledge in the Upper Atmosphere: Processes that Control Ozone and Other Climatically Important Trace Gases, NASA Ref. Publ. 1162, National Aeronautic and Space Administrator, Washington, D.C., 1986.

Watt Committee on Energy, Factors Determining Energy Costs and an Introduction to the Influence of Electronics, Rep. No. 10, London Science Center, London, September 1981.

Wigley, T. M. L., Relative contributions of different trace gases to the greenhouse effect, *Climate Monitor,* 6, 14-28, 1987.

4.3 Heat Exchangers

Ramesh K. Shah and Kenneth J. Bell

The two major categories of heat exchangers are shell-and-tube exchangers and compact exchangers. Basic constructions of gas-to-gas compact heat exchangers are plate-fin, tube-fin and all prime surface recuperators (including polymer film and laminar flow exchangers), and compact regenerators. Basic constructions of liquid-to-liquid and liquid-to-phase-change compact heat exchangers are gasketed and welded plate-and-frame, welded stacked plate (without frames), spiral plate, printed circuit, and dimple plate heat exchangers.

Shell-and-tube exchangers are custom designed for virtually any capacity and operating condition, from high vacuums to ultrahigh pressures, from cryogenics to high temperatures, and for any temperature and pressure differences between the fluids, limited only by the materials of construction. They can be designed for special operating conditions: vibration, heavy fouling, highly viscous fluids, erosion, corrosion, toxicity, radioactivity, multicomponent mixtures, etc. They are made from a variety of metal and nonmetal materials, and in surface areas from less than 0.1 to 100,000 m^2 (1 to over 1,000,000 ft^2). They have generally an order of magnitude less surface area per unit volume than the compact exchangers, and require considerable space, weight, support structure, and footprint.

Compact heat exchangers have a large heat transfer surface area per unit volume of the exchanger, resulting in reduced space, weight, support structure and footprint, energy requirement and cost, as well as improved process design, plant layout and processing conditions, together with low fluid inventory compared with shell-and-tube exchangers. From the operating condition and maintenance point of view, compact heat exchangers of different constructions are used for specific applications, such as for high-temperature applications (up to about 850°C or 1550°F), high pressure applications (over 200 bars), and moderate fouling applications. However, applications do not involve both high temperature and pressure simultaneously. Plate-fin exchangers are generally brazed, and the largest size currently manufactured is 1.2 × 1.2 × 6 m (4 × 4 × 20 ft). Fouling is one of the major potential problems in many compact exchangers except for the plate heat exchangers. With a large frontal area exchanger, flow maldistribution could be another problem. Because of short transient times, a careful design of controls is required for startup of some compact heat exchangers compared with shell-and-tube exchangers. No industry standards or recognized practice for compact heat exchangers is yet available.

This section is divided into two parts: Compact Heat Exchangers and Shell-and-Tube Exchangers, written by R. K. Shah and K. J. Bell, respectively. In the compact heat exchangers section, the following topics are covered: definition and description of exchangers, heat transfer and pressure drop analyses, heat transfer and flow friction correlations, exchanger design (rating and sizing) methodology, flow maldistribution, and fouling. In the shell-and-tube heat exchangers section, the following topics are covered: construction features, principles of design, and an approximate design method with an example.

Compact Heat Exchangers

Ramesh K. Shah

Introduction

A heat exchanger is a device to provide for transfer of internal thermal energy (enthalpy) between two or more fluids, between a solid surface and a fluid, or between solid particulates and a fluid, in thermal contact without external heat and work interactions. The fluids may be single compounds or mixtures. Typical applications involve heating or cooling of a fluid stream of concern, evaporation or condensation of single or multicomponent fluid stream, and heat recovery or heat rejection from a system. In other applications, the objective may be to sterilize, pasteurize, fractionate, distill, concentrate, crystallize, or control process fluid. In some heat exchangers, the fluids transferring heat are in direct contact. In other heat exchangers, heat transfer between fluids takes place through a separating wall or into and out of a wall in a transient manner. In most heat exchangers, the fluids are separated by a heat transfer surface and do not mix. Such exchangers are referred to as *direct transfer type*, or simply *recuperators*. Exchangers in which there is an intermittent flow of heat from the hot to cold fluid (via heat storage and heat rejection through the exchanger surface or matrix) are referred to as *indirect transfer type* or simply *regenerators*.

The heat transfer surface is a surface of the exchanger core which is in direct contact with fluids and through which heat is transferred by conduction in a recuperator. The portion of the surface which also separates the fluids is referred to as a *primary* or *direct surface*. To increase heat transfer area, appendages known as fins may be intimately connected to the primary surface to provide an *extended*, *secondary*, or *indirect surface*. Thus, the addition of fins reduces the thermal resistance on that side and thereby increases the net heat transfer from the surface for the same temperature difference.

Heat exchangers may be classified according to transfer process, construction, flow arrangement, surface compactness, number of fluids, and heat transfer mechanisms as shown in Figure 4.3.1.

A gas-to-fluid heat exchanger is referred to as a compact heat exchanger if it incorporates a heat transfer surface having a surface area density above about 700 m^2/m^3 (213 ft^2/ft^3) on at least one of the

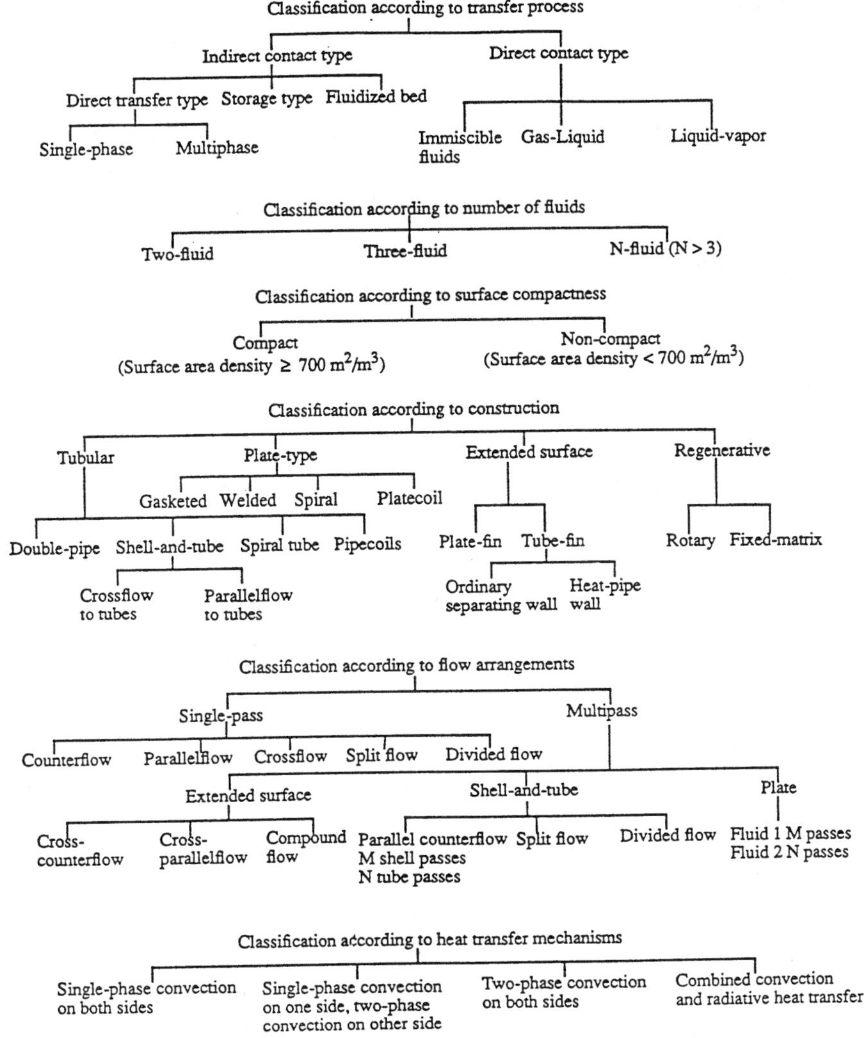

FIGURE 4.3.1 Classification of heat exchangers.

fluid sides which usually has gas flow. It is referred to as a laminar flow heat exchanger if the surface area density is above about 3000 m^2/m^3 (914 ft^2/ft^3), and as a micro heat exchanger if the surface area density is above about 10,000 m^2/m^3 (3050 ft^2/ft^3). A liquid/two-phase heat exchanger is referred to as a compact heat exchanger is the surface area density on any one fluid side is above about 400 m^2/m^3 (122 ft^2/ft^3). A typical process industry shell-and-tube exchanger has a surface area density of less than 100 m^2/m^3 on one fluid side with plain tubes, and two to three times that with the high-fin-density low-finned tubing. Plate-fin, tube-fin, and rotary regenerators are examples of compact heat exchangers for gas flows on one or both fluid sides, and gasketed and welded plate heat exchangers are examples of compact heat exchangers for liquid flows.

Types and Description

Gas-to-Fluid Exchangers.

The important design and operating considerations for compact extended surface exchangers are (1) usually at least one of the fluids is a gas or specific liquid that has low h; (2) fluids must be clean and relatively noncorrosive because of small hydraulic diameter (D_h) flow passages and no easy techniques for mechanically cleaning them; (3) the fluid pumping power (i.e., pressure drop) design constraint is often equally as important as the heat transfer rate; (4) operating pressures and temperatures are somewhat limited compared with shell-and-tube exchangers as a result of the joining of the fins to plates or tubes such as brazing, mechanical expansion, etc.; (5) with the use of highly compact surfaces, the resultant shape of a gas-to-fluid exchanger is one having a large frontal area and a short flow length (the header design of a compact heat exchanger is thus important for a uniform flow distribution among the very large number of small flow passages); (6) the market potential must be large enough to warrant the sizable manufacturing research and tooling costs for new forms to be developed.

Some advantages of plate-fin exchangers over conventional shell-and-tube exchangers are as follows. Compact heat exchangers, generally fabricated from thin metallic plates, yield large heat transfer surface area per unit volume (β), typically up to ten times greater than the 50 to 100 m^2/m^3 provided by a shell-and-tube exchanger for general process application and from 1000 to 6000 m^2/m^3 for highly compact gas side surfaces. Compact liquid or two-phase side surfaces have a β ratio ranging from 500 to 600 m^2/m^3. A compact exchanger provides a tighter temperature control; thus it is useful for heat-sensitive materials, improves the product (e.g., refining fats from edible oil) and its quality (such as a catalyst bed). Also, a compact exchanger could provide rapid heating or cooling of a process stream, thus improving the product quality. The plate-fin exchangers can accommodate multiple (up to 12 or more) fluid streams in one exchanger unit with proper manifolding, thus allowing process integration and cost-effective compact solutions.

Fouling is one of the potential major problems in compact heat exchangers (except for plate-and-frame heat exchangers), particularly those having a variety of fin geometries or very fine circular or noncircular flow passages that cannot be cleaned mechanically. Chemical cleaning may be possible; thermal baking and subsequent rinsing is possible for small-size units. Hence, extended surface compact heat exchangers may not be used in heavy fouling applications.

Liquid-to-Liquid Exchangers.

Liquid-to-liquid and phase-change exchangers are plate-and-frame and welded plate heat exchangers (PHE), spiral plate, and printed circuit exchangers; some of them are described next in some detail along with other compact heat exchangers and their applications.

Plate-Fin Heat Exchangers.

This type of exchanger has "corrugated" fins or spacers sandwiched between parallel plates (referred to as plates or parting sheets) as shown in Figure 4.3.2. Sometimes fins are incorporated in a flat tube with rounded corners (referred to as a formed tube), thus eliminating a need for the side bars. If liquid or phase-change fluid flows on the other side, the parting sheet is usually replaced by a flat tube with or without inserts/webs. Other plate-fin constructions include drawn-cup (see Figure 4.3.3) or tube-and-center configurations. Fins are die- or roll-formed and are attached to the plates by brazing, soldering,

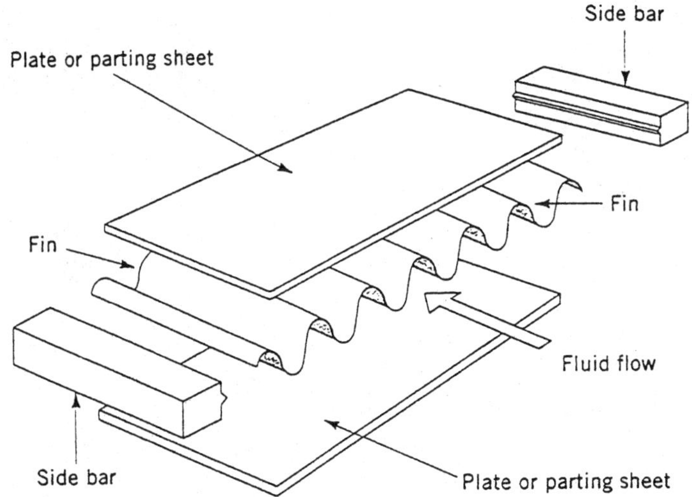

FIGURE 4.3.2 Typical components of a plate-fin exchanger.

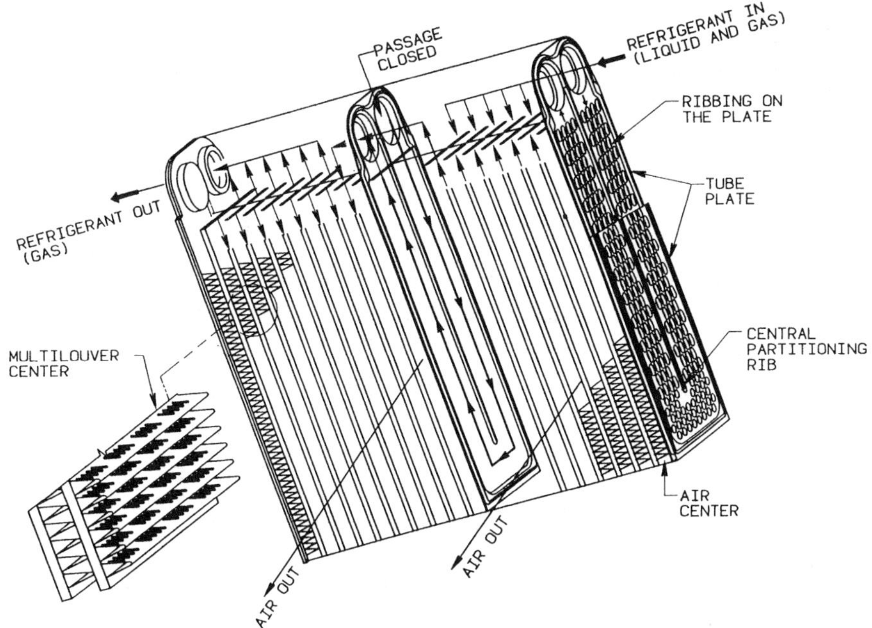

FIGURE 4.3.3 U-channel ribbed plates and multilouver fin automotive evaporator. (Courtesy of Delphi Harrison Thermal Systems, Lockport, NY.)

adhesive bonding, welding, mechanical fit, or extrusion. Fins may be used on both sides in gas-to-gas heat exchangers. In gas-to-liquid applications, fins are usually used only on the gas side; if employed on the liquid side, they are used primarily for structural strength and flow-mixing purposes. Fins are also sometimes used for pressure containment and rigidity.

Plate fins are categorized as (1) plain (i.e., uncut) and straight fins, such as plain triangular and rectangular fins; (2) plain but wavy fins (wavy in the main fluid flow direction); and (3) interrupted fins such as offset strip, louver, and perforated. Examples of commonly used fins are shown in Figure 4.3.4.

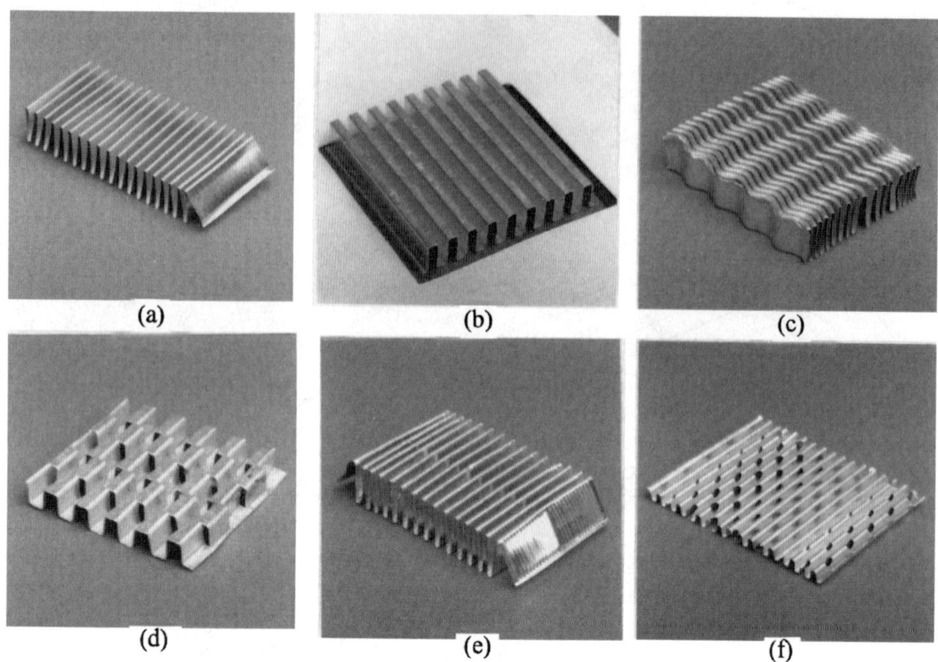

FIGURE 4.3.4 Fin geometries for plate-fin heat exchangers: (a) plain triangular fin, (b) plain rectangular fin, (c) wavy fin, (d) offset strip fin, (e) multilouver fin, and (f) perforated fin.

Plate-fin exchangers have been built with a surface area density of up to about 5900 m^2/m^3 (1800 ft^2/ft^3). There is a total freedom of selecting fin surface area on each fluid side, as required by the design, by varying fin height and fin density. Although typical fin densities are 120 to 700 fins/m (3 to 18 fins/in.), applications exist for as many as 2100 fins/m (53 fins/in.). Common fin thicknesses range from 0.05 to 0.25 mm (0.002 to 0.010 in.). Fin heights range from 2 to 25 mm (0.08 to 1.0 in.). A plate-fin exchanger with 600 fins/m (15.2 fins/in.) provides about 1300 m^2 (400 ft^2/ft^3) of heat transfer surface area per cubic meter volume occupied by the fins. Plate-fin exchangers are manufactured in virtually all shapes and sizes, and made from a variety of materials.

Tube-Fin Heat Exchangers.

In this type of exchanger, round and rectangular tubes are the most common, although elliptical tubes are also used. Fins are generally used on the outside, but they may be used on the inside of the tubes in some applications. They are attached to the tubes by a tight mechanical fit, tension winding, adhesive bonding, soldering, brazing, welding, or extrusion. Fins on the outside of the tubes may be categorized as follows: (1) normal fins on individual tubes, referred to as individually finned tubes or simply as *finned tubes*, as shown in Figures 4.3.5a and 4.3.6; (2) flat or continuous (plain, wavy, or interrupted) external fins on an array of tubes, as shown in Figures 4.3.5b and 4.3.7; (3) longitudinal fins on individual tubes. The exchanger having flat (continuous) fins on tubes has also been referred to as a *plate-fin and tube* exchanger in the literature. In order to avoid confusion with plate-fin surfaces, we will refer to it as a tube-fin exchanger having flat (plain, wavy, or interrupted) fins. Individually finned tubes are probably more rugged and practical in large tube-fin exchangers. Shell-and-tube exchangers sometimes employ low-finned tubes to increase the surface area on the shell side when the shell-side heat transfer coefficient is low compared with the tube side coefficient. The exchanger with flat fins is usually less expensive on a unit heat transfer surface area basis because of its simple and mass-production-type construction features. Longitudinal fins are generally used in condensing applications and for viscous fluids in double-pipe heat exchangers.

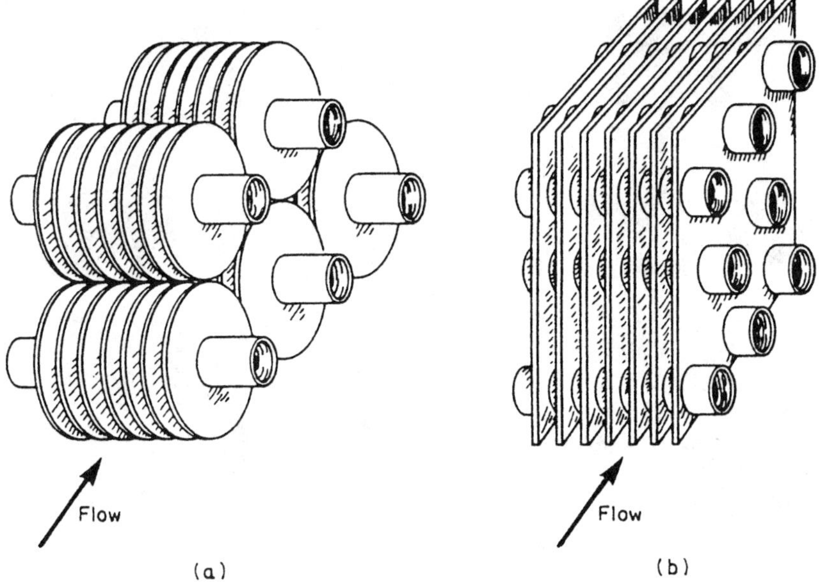

FIGURE 4.3.5 (a) Individually finned tubes, (b) flat or continuous fins on an array of tubes.

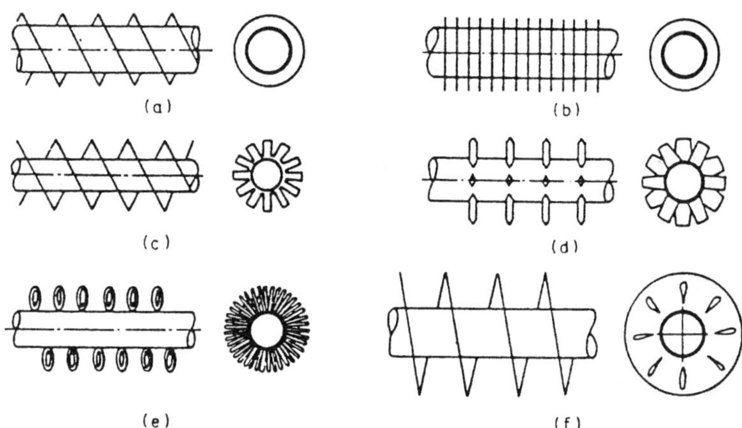

FIGURE 4.3.6 Individually finned tubes: (a) helical, (b) annular disk, (c) segmented, (d) studded, (e) wire loop, and (f) slotted helical.

Tube-fin exchangers can withstand high pressures on the tube side. The highest temperature is again limited by the type of bonding, the materials employed, and the material thickness. Tube-fin exchangers with an area density of about 3300 m^2/m^3 (1000 ft^2/ft^2) are commercially available. On the fin side, the desired surface area can be employed by using the proper fin density and fin geometry. The typical fin densities for flat fins vary from 250 to 800 fins/m (6 to 20 fins/in.), fin thicknesses vary from 0.08 to 0.25 mm (0.003 to 0.010 in.), and fin flow lengths from 25 to 250 mm (1 to 10 in.). A tube-fin exchanger having flat fins with 400 fins/m (10 fins/in.) has a surface area density of about 720 m^2/m^3(220 ft^2/ft^3). These exchangers are extensively used as condensers and evaporators in air-conditioning and refrigeration applications, as condensers in electric power plants, as oil coolers in propulsive power plants, and as air-cooled exchangers (also referred to as a fin-fan exchanger) in process and power industries.

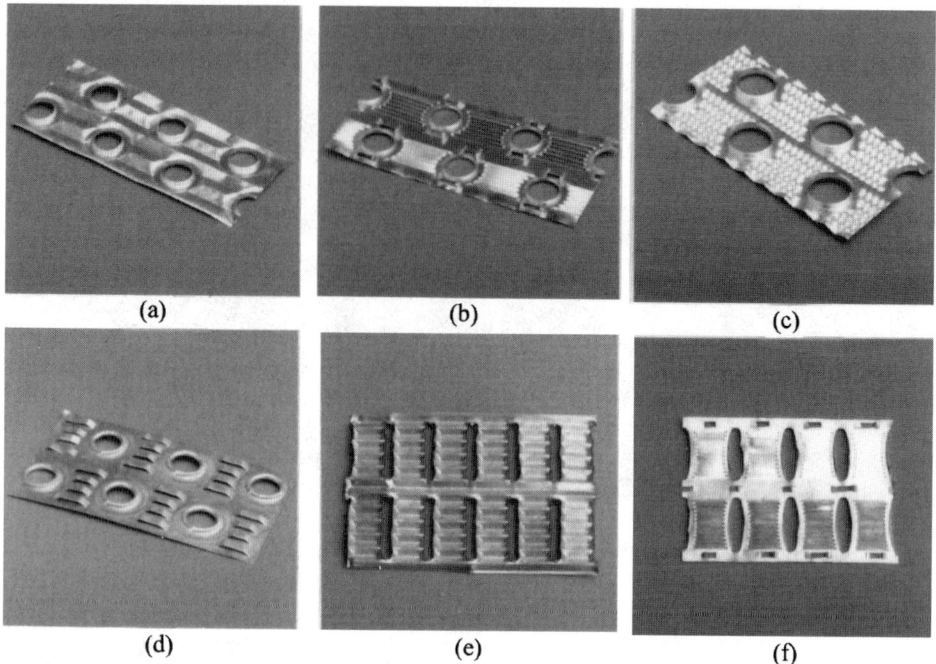

FIGURE 4.3.7 Flat or continuous fins on an array of tubes: On round tubes: (a) wavy fin, (b) multilouver fin, (c) fin with structured surface roughness (dimples), (d) parallel louver fin; (e) louver fin on flat tubes; (f) multilouver fin on elliptical tubes.

Regenerators.

The regenerator is a storage-type exchanger. The heat transfer surface or elements are usually referred to as a matrix in the regenerator. In order to have continuous operation, either the matrix must be moved periodically into and out of the fixed streams of gases, as in a *rotary* regenerator (Figure 4.3.8a), or the gas flows must be diverted through valves to and from the fixed matrices as in a *fixed-matrix* regenerator (Figure 4.3.8b). The latter is also sometimes referred to as a *periodic-flow regenerator* or a *reversible heat accumulator.* A third type of regenerator has a fixed matrix (in the disk form) and the fixed stream of gases, but the gases are ducted through rotating hoods (headers) to the matrix as shown in Figure 4.3.8c. This Rothemuhle regenerator is used as an air preheater in some power-generating plants. The thermodynamically superior counterflow arrangement is usually employed in regenerators.

The **rotary regenerator** is usually a disk type in which the matrix (heat transfer surface) is in a disk form and fluids flow axially. It is rotated by a hub shaft or a peripheral ring gear drive. For a rotary regenerator, the design of seals to prevent leakage of hot to cold fluids and vice versa becomes a difficult task, especially if the two fluids are at significantly differing pressures. Rotating drives also pose a challenging mechanical design problem.

Major advantages of rotary regenerators are the following. For a highly compact regenerator, the cost of the regenerator surface per unit of heat transfer area is usually substantially lower than that for the equivalent recuperator. A major disadvantage of a regenerator is an unavoidable carryover of a small fraction of the fluid trapped in the passage to the other fluid stream just after the periodic flow switching. Since fluid contamination (small mixing) is prohibited with liquids, the regenerators are used exclusively for gas-to-gas heat or energy recovery applications. Cross contamination can be minimized significantly by providing a purge section in the disk and using double-labyrinth seals.

Rotary regenerators have been designed for a surface area density of up to about 6600 m^2/m^3 (2000 ft^2/ft^3), and exchanger effectivenesses exceeding 85% for a number of applications. They can employ thinner stock material, resulting in the lowest amount of material for a given effectiveness and pressure

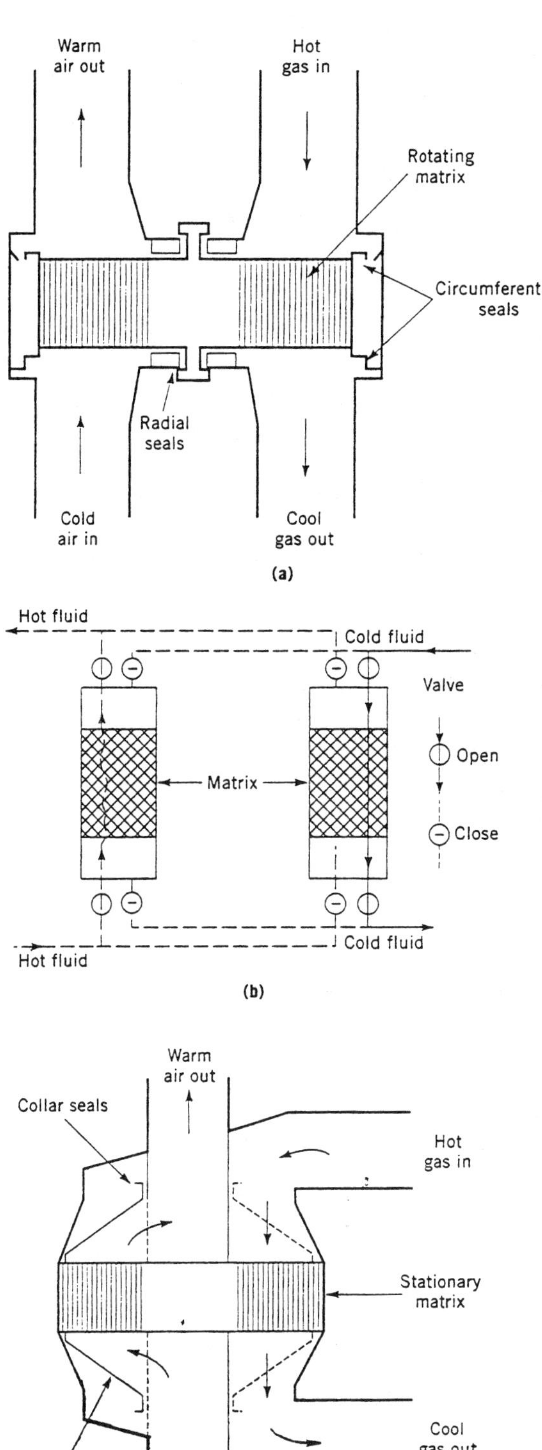

FIGURE 4.3.8 Regenerators: (a) rotary, (b) fixed-matrix, and (c) Rothemuhle.

drop of any heat exchanger known today. The metal rotary regenerators have been designed for continuous inlet temperatures up to about 790°C (1450°F) and ceramic matrices for higher-temperature applications; these regenerators are designed up to 400 kPa or 60 psi pressure differences between hot and cold gases. Plastic, paper, and wool are used for regenerators operating below 65°C (150°F) inlet temperature of the hot gas and 1 atm pressure. Typical regenerator rotor diameters and rotational speeds are as follows: up to 10 m (33 ft) and 0.5 to 3 rpm for power plant regenerators, 0.25 to 3 m (0.8 to 9.8 ft) and up to 10 rpm for air-ventilating regenerators, and up to 0.6 m (24 in.) and up to 18 rpm for vehicular regenerators. Refer to Shah (1994) for the description of **fixed-matrix regenerator**, also referred to as a *periodic-flow*, *fixed bed*, *valved*, or *stationary* regenerator.

Plate-Type Heat Exchangers.

These exchangers are usually built of thin plates (all prime surface). The plates are either smooth or have some form of corrugations, and they are either flat or wound in an exchanger. Generally, these exchangers cannot accommodate very high pressures, temperatures, and pressure and temperature differentials. These exchangers may be further classified as plate, spiral plate, lamella, and plate-coil exchangers as classified in Figure 4.3.1. The plate heat exchanger, being the most important of these, is described next.

The **plate-and-frame** or **gasketed PHE** consists of a number of thin rectangular corrugated or embossed metal plates sealed around the edges by gaskets and held together in a frame as shown in Figure 4.3.9. The plate pack with fixed and movable end covers is clamped together by long bolts, thus compressing the gaskets and forming a seal. Sealing between the two fluids is accomplished by elastomeric molded gaskets (typically 5 mm or 0.2 in. thick) that are fitted in peripheral grooves mentioned earlier. The most conventional flow arrangement is one pass to one pass counterflow with all inlet and outlet connections on the fixed end cover. By blocking flow through some ports with proper gasketing, either one or both fluids could have more than one pass. Also more than one exchanger can be accommodated in a single frame with the use of intermediate connector plates such as up to five "exchangers" or sections to heat, cool, and regenerate heat between raw milk and pasteurized milk in a milk pasteurization application.

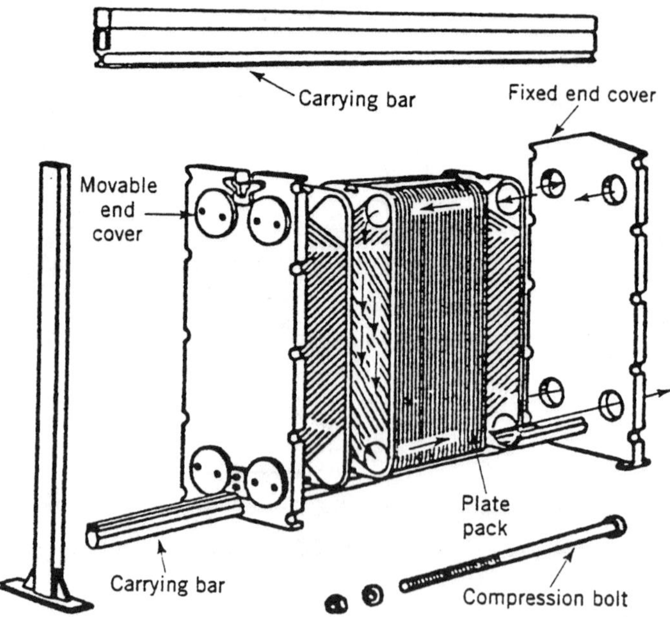

FIGURE 4.3.9 A plate-and-frame or gasketed PHE.

TABLE 4.3.1 Some Geometric and Operating Condition Characteristics of Plate-and-Frame Heat Exchangers

Unit		**Operation**	
Maximum surface area	2500 m^2	Pressure	0.1 to 2.5 MPa
Number of plates	3–700	Temperature	–40 to 260°C
Port size	Up to 400 mm	Maximum port velocity	6 m/sec
Plates		Channel flow rates	0.05 to 12.5 m^3/hr
Thickness	0.5–1.2 mm	Max unit flow rate	2500 m^3/hr
Size	0.03–3.6 m^2	**Performance**	
Spacing	1.5–5 mm	Temperature approach	As low as 1°C
Width	70–1200 mm	Heat exchanger efficiency	Up to 93%
Length	0.6–5 m	Heat transfer coefficients for water-water duties	3000 to 7000 W/m^2 K

Source: Shah, R.K., in *Encyclopedia of Energy Technology and the Environment*, A. Bision and S.G. Boots, Eds., John Wiley & Sons, New York, 1994, 1651–1670. With permission.

Typical PHE dimensions and performance parameters are given in Table 4.3.1 (Shah, 1994). Any metal which can be cold-worked is suitable for PHE applications. The most common plate materials are stainless steel (AISI 304 or 316) and titanium. Plates made from Incoloy 825, Inconel 625, Hastelloy C-276 are also available. Nickel, cupronickel, and monel are rarely used. Carbon steel is not used because of low corrosion resistance for thin plates. The heat transfer surface area per unit volume for plate exchangers ranges from 120 to 660 m^2/m^3 (37 to 200 ft^2/ft^3).

In PHEs, the high turbulence due to plates reduces fouling from about 10 to 25% of that of a shell-and-tube exchanger. High thermal performance can be achieved in plate exchangers because the high degree of counterflow in PHEs makes temperature approaches of up to 1°C (2°F) possible. The high thermal effectiveness (up to about 93%) makes low-grade heat recovery economical. PHEs are most suitable for liquid-liquid heat transfer duties.

Welded PHEs. One of the limitations of gasketed PHE is the presence of the gaskets which restricts the use to compatible fluids and which limits operating temperatures and pressures. In order to overcome this limitation, a number of welded PHE designs have surfaced with a welded pair of plates for one or both fluid sides. However, the disadvantage of such design is the loss of disassembling flexibility on the fluid side where the welding is done. Essentially, welding is done around the complete circumference where the gasket is normally placed. A *stacked plate heat exchanger* is another welded PHE design from Pacinox in which rectangular plates are stacked and welded at the edges. The physical size limitations of PHEs (1.2 m wide × 4 m long max, 4 × 13 ft) are considerably extended to 1.5 m wide × 20 m long (5 × 66 ft) in this exchanger. A maximum surface area of (10,000 m^2 or over 100,000 ft^2) can be accommodated in one unit. The potential maximum operating temperature is 815°C (1500°F) with an operating pressure of up to 20 MPa (3000 psig) when the stacked plate assembly is placed in a cylindrical pressure vessel. For operating pressures below 2 MPa (300 psig) and operating temperatures below 200°C (400°F), the plate bundle is not contained in a pressure vessel, but is bolted between two heavy plates. Some of the applications of this exchanger are catalytic reforming, hydrosulfurization, crude distillation, synthesis converter feed effluent exchanger for methanol, propane condenser, etc.

A number of other PHE constructions have been developed to address some of the limitations of the conventional PHEs. A double-wall PHE is used to avoid mixing of the two fluids. A wide-gap PHE is used for fluids having high fiber content or coarse particles. A graphite PHE is used for highly corrosive fluids. A flow-flex exchanger has plain fins on one side between plates and the other side has conventional plate channels, and is used to handle asymmetric duties (flow rate ratio of 2 to 1 and higher).

A vacuum **brazed PHE** is a compact PHE for high-temperature and high-pressure duties, and it does not have gaskets, tightening bolts, frame, or carrying and guide bars. It simply consists of stainless steel plates and two end plates. The brazed unit can be mounted directly on piping without brackets and foundations.

FIGURE 4.3.10 A section of a printed circuit heat exchanger. (Courtesy of Heatric Ltd., Dorset, U.K.)

Printed Circuit Heat Exchangers. This exchanger, as shown in Figure 4.3.10, has only primary heat transfer surfaces as PHEs. Fine grooves are made in the plate by using the same techniques as those employed for making printed electrical circuits. High surface area densities (650 to 1350 m^2/m^3 or 200 to 400 ft^2/ft^3 for operating pressures of 500 to 100 bar respectively) are achievable. A variety of materials including stainless steel, nickel, and titanium alloys can be used. It has been successfully used with relatively clean gases, liquids and phase-change fluids in chemical processing, fuel processing, waste heat recovery, and refrigeration industries. Again, this exchanger is a new construction with limited special applications currently.

Exchanger Heat Transfer and Pressure Drop Analysis

In this subsection, starting with the thermal circuit associated with a two-fluid exchanger, ε-NTU, P-NTU, and mean temperature difference (MTD) methods used for an exchanger analysis are presented, followed by the fin efficiency concept and various expressions. Finally, pressure drop expressions are outlined for various single-phase exchangers.

Two energy conservation differential equations for a two-fluid exchanger with any flow arrangement are (see Figure 4.3.11 for counterflow)

$$dq = q'' \, dA = -C_h \, dT_h = \pm C_c \, dT_c \tag{4.3.1}$$

where the ± sign depends upon whether dT_c is increasing or decreasing with increasing dA or dx. The local overall rate equation is

$$dq = q'' \, dA = U\left(T_h - T_c\right)_{\text{local}} \, dA = U \, \Delta T \, dA \tag{4.3.2}$$

Integration of Equations (4.3.1) and (4.3.2) across the exchanger surface area results in

$$q = C_h\left(T_{h,i} - T_{h,o}\right) = C_c\left(T_{c,o} - T_{c,i}\right) \tag{4.3.3}$$

and

$$q = UA \, \Delta T_m = \Delta T_m / R_o \tag{4.3.4}$$

where ΔT_m is the true mean temperature difference (or MTD) that depends upon the exchanger flow arrangement and degree of fluid mixing within each fluid stream. The inverse of the overall thermal conductance UA is referred to as the overall thermal resistance R_o as follows (see Figure 4.3.12).

$$R_o = R_h + R_{s,h} + R_w + R_{s,c} + R_c \tag{4.3.5}$$

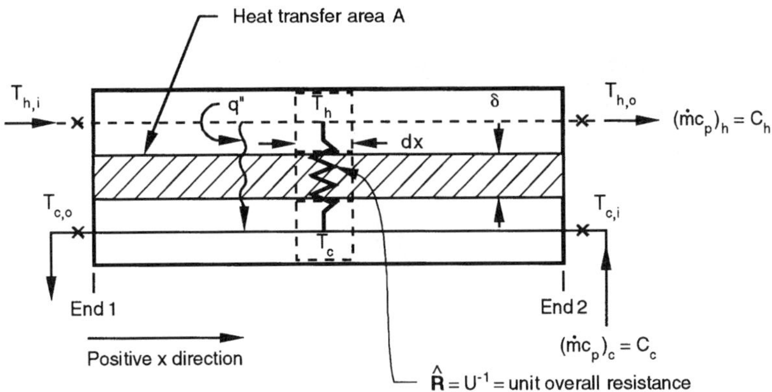

FIGURE 4.3.11 Nomenclature for heat exchanger variables.

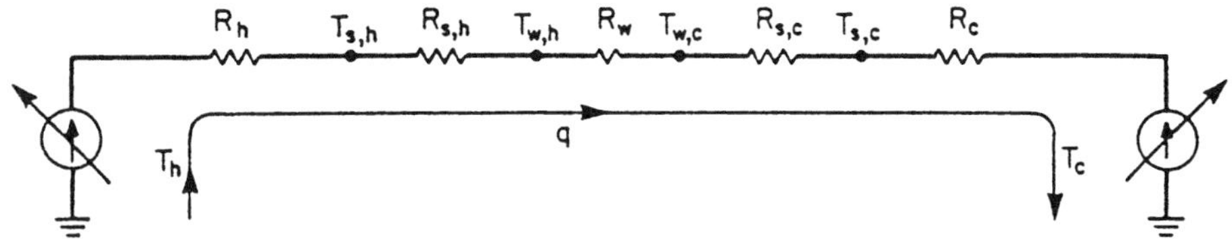

FIGURE 4.3.12 Thermal circuit for heat transfer in an exchanger.

where the subscripts *h*, *c*, *s*, and *w* denote hot, cold, fouling (or scale), and wall, respectively. In terms of the overall and individual heat transfer coefficients, Equation (4.3.5) is represented as

$$\frac{1}{UA} = \frac{1}{\left(\eta_o hA\right)_h} + \frac{1}{\left(\eta_o h_s A\right)_h} + R_w + \frac{1}{\left(\eta_o h_s A\right)_c} + \frac{1}{\left(\eta_o hA\right)_c} \tag{4.3.6}$$

where η_o = the overall surface efficiency of an extended (fin) surface and is related to the fin efficiency η_f, fin surface area A_f, and the total surface area A as follows:

$$\eta_o = 1 - \frac{A_f}{A}\left(1 - \eta_f\right) \tag{4.3.7}$$

The wall thermal resistance R_w of Equation (4.3.5) is given by

$$R_w = \begin{cases} \delta / A_w k_w & \text{for a flat wall} \\ \dfrac{\ln\left(d_o/d_i\right)}{2\pi k_w L N_t} & \text{for a circular tube with a single-layer wall} \\ \dfrac{1}{2\pi L N_t}\left[\sum_j \dfrac{\ln\left(d_{j+1}/d_j\right)}{k_{w,j}}\right] & \text{for a circular tube with a multiple-layer wall} \end{cases} \tag{4.3.8}$$

If one of the resistances on the right-hand side of Equation (4.3.5) or (4.3.6) is significantly higher than the other resistances, it is referred to as the *controlling thermal resistance*. A reduction in the controlling thermal resistance will have much more impact in reducing the exchanger surface area (A) requirement compared with the reduction in A as a result of the reduction in other thermal resistances.

UA of Equation (4.3.6) may be defined in terms of hot or cold fluid side surface area or wall conduction area as

$$UA = U_h A_h = U_c A_c = U_w A_w \tag{4.3.9}$$

When R_w is negligible, $T_{w,h} = T_{w,c} = T_w$ of Figure 4.3.12 is computed from

$$T_w = \frac{T_h + \left[\left(R_h + R_{s,h}\right)/\left(R_c + R_{s,c}\right)\right] T_c}{1 + \left[\left(R_h + R_{s,h}\right)/\left(R_c + R_{s,c}\right)\right]} \tag{4.3.10}$$

When $R_{s,h} = R_{s,c} = 0$, Equation (4.3.10) reduces to

$$T_w = \frac{T_h/R_h + T_c/R_c}{1/R_h + 1/R_c} = \frac{\left(\eta_o hA\right)_h T_h + \left(\eta_o hA\right)_c T_c}{\left(\eta_o hA\right)_h + \left(\eta_o hA\right)_c} \tag{4.3.11}$$

ε-NTU, P-NTU, and MTD Methods. If we consider the fluid outlet temperatures or heat transfer rate as dependent variables, they are related to independent variable/parameters of Figure 4.3.11 as follows.

$$T_{h,o},\ T_{c,o},\ \text{or}\ \ q = \phi\left\{T_{h,i},\ T_{c,i},\ C_c,\ C_h,\ U,\ A,\ \text{flow arrangement}\right\} \tag{4.3.12}$$

Six independent and three dependent variables of Equation (4.3.12) for a given flow arrangement can be transferred into two independent and one dependent dimensionless groups; three different methods are presented in Table 4.3.2 based on the choice of three dimensionless groups. The relationship among three dimensionless groups is derived by integrating Equations (4.3.1) and (4.3.2) across the surface area for a specified exchanger flow arrangement. Such expressions are presented later in Table 4.3.4 for the industrially most important flow arrangements. Now we briefly describe the three methods.

In *the ε-NTU method*, the heat transfer rate from the hot fluid to the cold fluid in the exchanger is expressed as

$$q = \varepsilon C_{\min}\left(T_{h,i} - T_{c,i}\right) \tag{4.3.13}$$

Here the exchanger effectiveness ε is an efficiency factor. It is a ratio of the actual heat transfer rate from the hot fluid to the cold fluid in a given heat exchanger of any flow arrangement to the maximum possible heat transfer rate $q_{\max}$ thermodynamically permitted. The $q_{\max}$ is obtained in a *counterflow* heat exchanger (recuperator) of *infinite surface area* operating with the fluid flow rates (heat capacity rates) and fluid inlet temperatures equal to those of an actual exchanger (constant fluid properties are idealized). As noted in Table 4.3.1, the exchanger effectiveness ε is a function of NTU and C^* in this method. The number of transfer units NTU is a ratio of the overall conductance UA to the smaller heat capacity rate $C_{\min}$. NTU designates the dimensionless "heat transfer size" or "thermal size" of the exchanger. Other interpretations of NTU are given by Shah (1983). The heat capacity rate ratio C^* is simply a ratio of the smaller to the larger heat capacity rate for the two fluid streams. Note that $0 \le \varepsilon \le 1$, $0 \le \text{NTU} \le \infty$ and $0 \le C^* \le 1$.

TABLE 4.3.2 General Functional Relationships and Dimensionless Groups for ε-NTU, P-NTU, and MTD Methods

ε-NTU Method	P-NTU Method[a]	MTD Method[a]
$q = \varepsilon C_{\min}(T_{h,i} - T_{c,i})$	$q = P_1 C_1 \lvert T_{1,i} - T_{2,i} \rvert$	$q = UAF\Delta T_{lm}$
$\varepsilon = \phi(\text{NTU}, C^*, \text{flow arrangement})$	$P_1 = \phi(\text{NTU}_1, R_1, \text{flow arrangement})$	$F = \phi(P, R, \text{flow arrangement})$[b]
$\varepsilon = \dfrac{C_h(T_{h,i} - T_{h,o})}{C_{\min}(T_{h,i} - T_{c,i})} = \dfrac{C_c(T_{c,o} - T_{c,i})}{C_{\min}(T_{h,i} - T_{c,i})}$	$P_1 = \dfrac{T_{1,o} - T_{1,i}}{T_{2,i} - T_{1,i}}$	$F = \dfrac{\Delta T_m}{\Delta T_{\text{lm}}}$
$\text{NTU} = \dfrac{UA}{C_{\min}} = \dfrac{1}{C_{\min}} \displaystyle\int_A U\,dA$	$\text{NTU}_1 = \dfrac{UA}{C_1} = \dfrac{\lvert T_{1,o} - T_{1,i} \rvert}{\Delta T_m}$	$\text{LMTD} = \Delta T_{\text{lm}} = \dfrac{\Delta T_1 - \Delta T_2}{\ln(\Delta T_1/\Delta T_2)}$
$C^* = \dfrac{C_{\min}}{C_{\max}} = \dfrac{(\dot{m}c_p)_{\min}}{(\dot{m}c_p)_{\max}}$	$R_1 = \dfrac{C_1}{C_2} = \dfrac{T_{2,i} - T_{2,o}}{T_{1,o} - T_{1,i}}$	$\Delta T_1 = T_{h,i} - T_{c,o} \quad \Delta T_2 = T_{h,o} - T_{c,i}$

[a] Although P, R, and NTU_1 are defined on fluid side 1, it must be emphasized that all the results of the P-NTU and MTD methods are valid if the definitions of P, NTU, and R are consistently based on C_c, C_s, C_h, or C.

[b] P and R are defined in the P-NTU method.

The P-NTU method represents a variant of the ε-NTU method. The ε-NTU relationship is different depending upon whether the shell fluid is the $C_{\min}$ or $C_{\max}$ fluid in the (stream unsymmetric) flow arrangements commonly used for shell-and-tube exchangers. In order to avoid possible errors and to avoid keeping track of the $C_{\min}$ fluid side, an alternative is to present the temperature effectiveness P as a function of NTU and R, where P, NTU, and R are defined consistently either for Fluid 1 side or Fluid 2 side; in Table 4.3.2, they are defined for Fluid 1 side (regardless of whether that side is the hot or cold fluid side), and Fluid 1 side is clearly identified for each flow arrangement in Table 4.3.4; it is the shell side in a shell-and-tube exchanger. Note that

$$q = P_1 C_1 \lvert T_{1,i} - T_{2,i} \rvert = P_2 C_2 \lvert T_{2,i} - T_{1,i} \rvert \tag{4.3.14}$$

$$P_1 = P_2 R_2 \qquad P_2 = P_1 R_1 \tag{4.3.15}$$

$$\text{NTU}_1 = \text{NTU}_2 R_2 \qquad \text{NTU}_2 = \text{NTU}_1 R_1 \tag{4.3.16}$$

and

$$R_1 = 1/R_2 \tag{4.3.17}$$

In *the MTD method*, the heat transfer rate from the hot fluid to the cold fluid in the exchanger is given by

$$q = UA\,\Delta T_m = UAF\,\Delta T_{\text{lm}} \tag{4.3.18}$$

where ΔT_m the log-mean temperature difference (LMTD), and F the LMTD correction factor, a ratio of true (actual) MTD to the LMTD, where

$$\text{LMTD} = \Delta T_{\text{lm}} = \frac{\Delta T_1 - \Delta T_2}{\ln(\Delta T_1/\Delta T_2)} \tag{4.3.19}$$

Here ΔT_1 and ΔT_2 are defined as

$$\Delta T_1 = T_{h,i} - T_{c,o} \qquad \Delta T_2 = T_{h,o} - T_{c,i} \qquad \text{for all flow arrangements except for parallel flow} \tag{4.3.20}$$

$$\Delta T_1 = T_{h,i} - T_{c,i} \qquad \Delta T_2 = T_{h,o} - T_{c,o} \qquad \text{for parallel flow} \tag{4.3.21}$$

The LMTD represents a true MTD for a counterflow arrangement under the idealizations listed below. Thus, the LMTD correction factor F represents a degree of departure for the MTD from the counterflow LMTD; it does not represent the effectiveness of a heat exchanger. It depends on two dimensionless group P_1 and R_l or P_2 and R_2 for a given flow arrangement.

The relationship among the dimensionless groups of the ε-NTU, P-NTU, and MTD methods are presented in Table 4.3.3. The closed-form formulas for industrially important exchangers are presented in terms of P_1, NTU_1, and R_1 in Table 4.3.4. These formulas are valid under idealizations which include: (1) steady-state conditions; (2) negligible heat losses to the surrounding; (3) no phase changes in the fluid streams flowing through the exchanger, or phase changes (condensation or boiling) occurring at constant temperature and constant effective specific heat; (4) uniform velocity and temperature at the entrance of the heat exchanger on each fluid side; (5) the overall extended surface efficiency η_o is uniform and constant; (6) constant individual and overall heat transfer coefficients; (7) uniformly distributed heat transfer area on each fluid side; (7) the number of baffles as large in shell-and-tube exchangers; (8) no flow maldistribution; and (9) negligible longitudinal heat conduction in the fluid and exchanger wall.

TABLE 4.3.3 Relationships between Dimensionless Groups of the P-NTU and LMTD Methods and Those of the ε-NTU Method

$$P_1 = \frac{C_{\min}}{C_1}\varepsilon = \begin{cases} \varepsilon & \text{for } C_1 = C_{\min} \\ \varepsilon C^* & \text{for } C_1 = C_{\max} \end{cases}$$

$$R_1 = \frac{C_1}{C_2} = \begin{cases} C^* & \text{for } C_1 = C_{\min} \\ 1/C^* & \text{for } C_1 = C_{\max} \end{cases}$$

$$\text{NTU}_1 = \text{NTU}\frac{C_{\min}}{C_1} = \begin{cases} \text{NTU} & \text{for } C_1 = C_{\min} \\ \text{NTU}\,C^* & \text{for } C_1 = C_{\max} \end{cases}$$

$$F = \frac{\text{NTU}_{cf}}{\text{NTU}} = \frac{1}{\text{NTU}(1-C^*)}\ln\left[\frac{1-C^*\varepsilon}{1-\varepsilon}\right] \xrightarrow[C^*=1]{} \frac{\varepsilon}{\text{NTU}(1-\varepsilon)}$$

$$F = \frac{1}{\text{NTU}_1(1-R_1)}\ln\left[\frac{1-RP_1}{1-P_1}\right] \xrightarrow[R_1=1]{} \frac{P_1}{\text{NTU}_1(1-P_1)}$$

The overall heat transfer coefficient can vary as a result of variations in local heat transfer coefficients due to two effects: (1) change in heat transfer coefficients in the exchanger as a result of changes in the fluid properties or radiation due to rise or drop of fluid temperatures and (2) change in heat transfer coefficients in the exchanger due to developing thermal boundary layers; it is referred to as the *length effect.* The first effect due to fluid property variations (or radiation) consists of two components: (1) distortion of velocity and temperature profiles at a given flow cross section due to fluid property variations — this effect is usually taken into account by the so-called property ratio method, with the correction scheme of Equations (4.3.55) and (4.3.56) — and (2) variations in the fluid temperature along the axial and transverse directions in the exchanger depending upon the exchanger flow arrangement — this effect is referred to as the *temperature effect.* The resultant axial changes in the overall mean heat transfer coefficient can be significant; the variations in U_{local} could be nonlinear, dependent upon the type of the fluid. The effect of varying U_{local} can be taken into account by evaluating U_{local} at a few points in the

TABLE 4.5.4 P_1 - NTU_1 Formulas and Limiting Values P_1 and $R_1 = 1$ and $NTU_1 \to \infty$ for Various Exchanger Flow Arrangements[a]

Flow Arrangement	Eq. no.	General formula	Value for $R_1 = 1$	Value for $NTU_1 \to \infty$
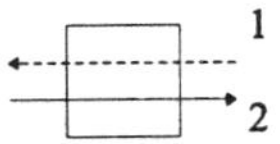	I.1.1	$P_1 = \frac{1-\exp[-NTU_1(1-R_1)]}{1-R_1\exp[-NTU_1(1-R_1)]}$	$P_1 = \frac{NTU_1}{1+NTU_1}$	$P_1 \to 1$ for $R_1 \le 1$ $P_1 \to 1/R_1$ for $R_1 \ge 1$
	I.1.2	$NTU_1 = \frac{1}{(1-R_1)}\ln\left[\frac{1-R_1P_1}{1-P_1}\right]$	$NTU_1 = \frac{P_1}{1-P_1}$	$NTU_1 \to \infty$
Counterflow exchanger, stream symmetric.	I.1.3	$F = 1$	$F = 1$	$F = 1$
	I.2.1	$P_1 = \frac{1-\exp[-NTU_1(1+R_1)]}{1+R_1}$	$P_1 = \frac{1}{2}[1-\exp(-2NTU_1)]$	$P_1 \to \frac{1}{1+R_1}$
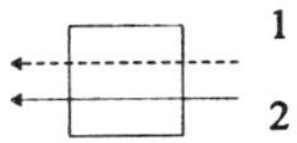	I.2.2	$NTU_1 = \frac{1}{1+R_1}\ln\left[\frac{1}{1-P_1(1+R_1)}\right]$	$NTU_1 = \frac{1}{2}\ln\left[\frac{1}{1-2P_1}\right]$	$NTU_1 \to \infty$
Parallel flow exchanger, stream symmetric.	I.2.3	$F = \frac{(R_1+1)\ln\left[\frac{1-R_1P_1}{1-P_1}\right]}{(R_1-1)\ln[1-P_1(1+R_1)]}$	$F = \frac{2P_1}{(P_1-1)\ln(1-2P_1)}$	$F \to 0$
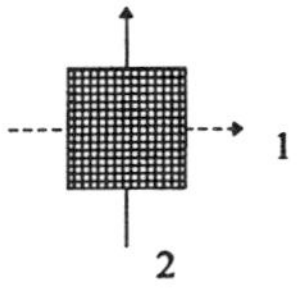 Single-pass crossflow exchanger, both fluids unmixed, stream symmetric	II.1	$P_1 = 1-\exp(NTU_1)$ $-\exp[-(1+R_1)NTU_1]$ $\cdot\sum_{n=1}^{\infty} R_1^n P_n(NTU_1)$ $P_n(y) = \frac{1}{(n+1)!}\sum_{j=1}^{n}\frac{(n+1-j)}{j!}y^{n+j}$	same as Eq. (II.1) with $R_1 = 1$	$P_1 \to 1$ for $R_1 \le 1$ $P_1 \to \frac{1}{R_1}$ for $R_1 \ge 1$

TABLE 4.5.4 (continued) P_1 - NTU_1 Formulas and Limiting Values P_1 and $R_1 = 1$ and $NTU_1 \rightarrow \infty$ for Various Exchanger Flow Arrangements[a]

Flow Arrangement	Eq. no.	General formula	Value for $R_1 = 1$	Value for $NTU_1 \rightarrow \infty$
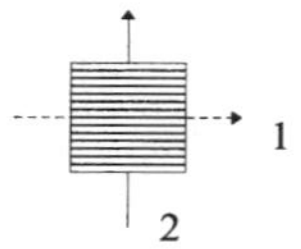	II.2.1	$P_1 = [1 - \exp(-KR_1)] / R_1$ $K = 1 - \exp(-NTU_1)$	$P_1 = 1 - \exp(-K)$	$P_1 \rightarrow \frac{1 - \exp(-R_1)}{R_1}$
	II.2.2	$NTU = \ln\left[\frac{1}{1 + \frac{1}{R_1}\ln(1 - R_1P_1)}\right]$	$NTU_1 = \ln\left[\frac{1}{1 + \ln(1 - P_1)}\right]$	$NTU_1 \rightarrow \infty$
Single-pass crossflow exchanger, fluid 1 unmixed, fluid 2 mixed.	II.2.3	$F = \frac{\ln[(1 - R_1P_1)/(1 - P_1)]}{(R_1 - 1)\ln\left[1 + \frac{1}{R_1}\ln(1 - R_1P_1)\right]}$	$F = \frac{P_1}{(P_1 - 1)\ln[1 + \ln(1 - P_1)]}$	$F \rightarrow 0$
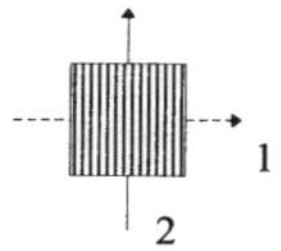	II.3.1	$P = 1 - \exp(-K / R_1)$ $K = 1 - \exp(-R_1NTU_1)$	$P = 1 - \exp(-K)$ $K = 1 - \exp(-NTU_1)$	$P_1 \rightarrow 1 - \exp(-1 / R_1)$
	II.3.2	$NTU_1 = \frac{1}{R_1}\ln\left[\frac{1}{1 + R_1}\ln(1 - P_1)\right]$	$NTU_1 = \ln\left[\frac{1}{1 + \ln(1 - P_1)}\right]$	$NTU_1 \rightarrow \infty$
Single-pass crossflow exchanger, fluid 1 mixed, fluid 2 unmixed.	II.3.3	$F = \frac{\ln(1 - R_1P_1)/(1 - P_1)]}{(1 - 1/R_1)\ln[1 + R_1\ln(1 - P_1)]}$	$F = \frac{P_1}{(P_1 - 1)\ln[1 + \ln(1 - P_1)]}$	$P_1 \rightarrow \frac{1}{1 + R_1}$
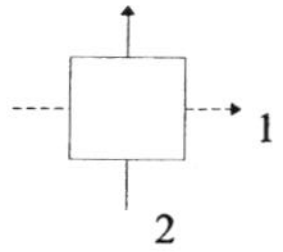 Single-pass crossflow exchanger, both fluids mixed, stream symmetric.	II.4	$P_1 = \left[\frac{1}{K_1} + \frac{R_1}{K_2} - \frac{1}{NTU_1}\right]^{-1}$ $K_1 = 1 - \exp(-NTU_1)$ $K_2 = 1 - \exp(-R_1NTU_1)$	$P_1 = \left[\frac{2}{K_1} - \frac{1}{NTU_1}\right]^{-1}$	$P_1 \rightarrow \frac{1}{1 + R_1}$

TABLE 4.5.4 (continued) P_1 - NTU_1 Formulas and Limiting Values P_1 and $R_1 = 1$ and $NTU_1 \to \infty$ for Various Exchanger Flow Arrangements[a]

Flow Arrangement	Eq. no.	General formula	Value for $R_1 = 1$	Value for $NTU_1 \to \infty$
	III.1.1	$P_1 = \dfrac{2}{1 + R_1 + E \coth(E\, NTU_1 / 2)}$ $E = [1 + R_1^2]^{1/2}$	$P_1 = \dfrac{1}{1 + \coth(NTU_1/\sqrt{2})/\sqrt{2}}$	$P_1 \to \dfrac{2}{1 + R_1 + E}$
	III.1.2	$NTU_1 = \dfrac{1}{E} \ln\left[\dfrac{2 - P_1(1 + R_1 - E)}{2 - P_1(1 + R_1 + E)}\right]$	$NTU_1 = \ln\left[\dfrac{2 - P_1}{2 - 3P_1}\right]$	$NTU_1 \to \infty$
1-2 TEMA E shell-and-tube exchanger, shell fluid mixed, stream symmetric	III.1.3	$F = \dfrac{E \ln[(1 - R_1 P_1)/(1 - P_1)]}{(1 - R_1) \ln\left[\dfrac{2 - P_1(1 + R_1 - E)}{2 - P_1(1 + R_1 + E)}\right]}$	$F = \dfrac{P_1/(1 - P_1)}{\ln[(2 - P_1)/(2 - 3P_1)]}$	$F \to 0$
	IV.1.1	$P_1 = \dfrac{\prod_{i=1}^{n} (1 - R_1 P_{1,A_i}) - \prod_{i=1}^{n} (1 - P_{1,A_i})}{\prod_{i=1}^{n} (1 - R_1 P_{1,A_i}) - R_1 \prod_{i=1}^{n} (1 - P_{1,A_i})}$	$P_1 = \dfrac{\sum_{i=1}^{n} \dfrac{P_{1,A_i}}{1 - P_{1,A_i}}}{1 + \sum_{i=1}^{n} \dfrac{P_{1,A_i}}{1 - P_{1,A_i}}}$	same as Eq. (I.1.1) counterflow
	IV.1.2	$R_1 = R_{1,A_i}, \quad i = 1, \ldots, n$	$1 = R_{1,A_i}, \quad i = 1, \ldots, n$	same as Eq. (IV.1.2)
	IV.1.3	$NTU_1 = \sum_{i=1}^{n} NTU_{1,A_i}$	same as for Eq. (IV.1.3)	same as Eq. (IV.1.3)
Series coupling of n exchangers, overall counterflow arrangement. Stream symmetric if all A_1 are stream symmetric.	IV.1.4	$F = \dfrac{1}{NTU_1} \sum_{i=1}^{n} NTU_{1,A_i} F_{A_i}$	same as Eq. (IV.1.4)	same as Eq. (IV.1.4)

[a] In this table, all variables, except P_1, R_1, NTU_1 and F, are local or dummy variables not necessarily related to similar ones defined in the nomenclature and the text. Source: Shah, R.K. and Mueller, A.C., 1988. With permission.

exchanger and subsequently integrating U_{local} values by the Simpson or Gauss method (Shah, 1993). The temperature effect can increase or decrease mean U slightly or significantly, depending upon the fluids and applications. The length effect is important for developing laminar flows for which high heat transfer coefficients are obtained in the thermal entrance region. However, in general it will have less impact on the overall heat transfer coefficient because the other thermal resistances in series in an exchanger may be controlling. The length effect reduces the overall heat transfer coefficient compared with the mean value calculated conventionally (assuming uniform mean heat transfer coefficient on each fluid side). It is shown that this reduction is up to about 11% for the worst case (Shah, 1993).

Shah and Pignotti (1997) have shown that the following are the specific number of baffles beyond which the influence of the finite number of baffles on the exchanger effectiveness is not significantly larger than 2%: $N_b \geq 10$ for 1-1 TEMA E counterflow exchanger; $N_b \geq 6$ for 1-2 TEMA E exchanger for $NTU_s \leq 2$, $R_s \leq 5$; $N_b \geq 9$ for 1-2 TEMA J exchanger for $NTU_s \leq 2$, $R_s \leq 5$; $N_b \geq 5$ for 1-2 TEMA G exchanger for $NTU_s \leq 3$, all R_s; $N_b \geq 11$ for 1-2 TEMA H exchanger for $NTU_s \leq 3$, all R_s. Various shell-and-tube heat exchangers (such as TEMA E, G, H, J, etc.) are classified by the Tubular Exchanger Manufacturers' Association (TEMA, 1988).

If any of the basic idealizations are not valid for a particular exchanger application, the best solution is to work directly with either Equations (4.3.1) and (4.3.2) or their modified form by including a particular effect, and to integrate them over a small exchanger segment numerically in which all of the idealizations are valid.

Fin Efficiency and Extended Surface Efficiency.

Extended surfaces have fins attached to the primary surface on one or both sides of a two-fluid or a multifluid heat exchanger. Fins can be of a variety of geometries — plain, wavy, or interrupted — and can be attached to the inside, outside, or both sides of circular, flat, or oval tubes, or parting sheets. Fins are primarily used to increase the surface area (when the heat transfer coefficient on that fluid side is relatively low) and consequently to increase the total rate of heat transfer. In addition, enhanced fin geometries also increase the heat transfer coefficient compared to that for a plain fin. Fins may also be used on the high heat transfer coefficient fluid side in a heat exchanger primarily for structural strength purposes (for example, for high-pressure water flow through a flat tube) or to provide a thorough mixing of a highly viscous liquid (such as for laminar oil flow in a flat or a round tube). Fins are attached to the primary surface by brazing, soldering, welding, adhesive bonding, or mechanical expansion, or they are extruded or integrally connected to the tubes. Major categories of extended surface heat exchangers are plate-fin (Figures 4.3.2 to 4.3.4) and tube-fin (Figures 4.3.5 to 4.3.7) exchangers. Note that shell-and-tube exchangers sometimes employ individually finned tubes — low finned tubes (similar to Figure 4.3.5a but with low-height fins) (Shah, 1985).

The concept of fin efficiency accounts for the reduction in temperature potential between the fin and the ambient fluid due to conduction along the fin and convection from or to the fin surface depending upon the fin cooling or heating situation. The fin efficiency is defined as the ratio of the actual heat transfer rate through the fin base divided by the maximum possible heat transfer rate through the fin base which would be obtained if the entire fin were at the base temperature (i.e., its material thermal conductivity were infinite). Since most of the real fins are "thin", they are treated as one-dimensional (1-D) with standard idealizations used for the analysis (Huang and Shah, 1992). This 1-D fin efficiency is a function of the fin geometry, fin material thermal conductivity, heat transfer coefficient at the fin surface, and the fin tip boundary condition; it is not a function of the fin base or fin tip temperature, ambient temperature, and heat flux at the fin base or fin tip in general. Fin efficiency formulas for some common fins are presented in Table 4.3.5 (Shah, 1985). Huang and Shah (1992) also discuss the influence on η_f if any of the basic idealizations used in the fin analysis are violated.

The fin efficiency for flat fins (Figure 4.3.5b) is obtained by a sector method (Shah, 1985). In this method, the rectangular or hexagonal fin around the tube (Figures 4.3.7a and b) or its smallest symmetrical section is divided into n sectors (Figure 4.3.13). Each sector is then considered as a circular fin with the radius $r_{e,i}$ equal to the length of the centerline of the sector. The fin efficiency of each sector is

TABLE 4.3.5 Fin Efficiency Expressions for Plate-Fin and Tube-Fin Geometries of Uniform Fin Thickness

Geometry	Fin Efficiency Formula $m_i = \left[\frac{2h}{k_f\delta_i}\left(1+\frac{\delta_i}{l_f}\right)\right]^{1/2} \quad E_1 = \frac{\tanh(m_i l_1)}{m_i l_1} \quad i = 1, 2$
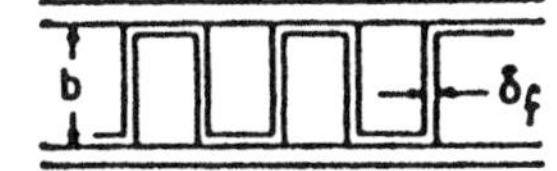 Plain, wavy, or offset strip fin of rectangular cross section	$\eta_f = E_1$ $l_1 = \frac{b}{2} - \delta_1 \quad \delta_1 = \delta_f$
 Triangular fin heated from one side	$\eta_f = \dfrac{hA_1(T_0 - T_a)\frac{\sinh(m_1 l_1)}{m_1 l_1} + q_e}{\cosh(m_1 l_1)\left[hA_1(T_0 - T_a) + q_e \frac{T_0 - T_a}{T_1 - T_a}\right]}$
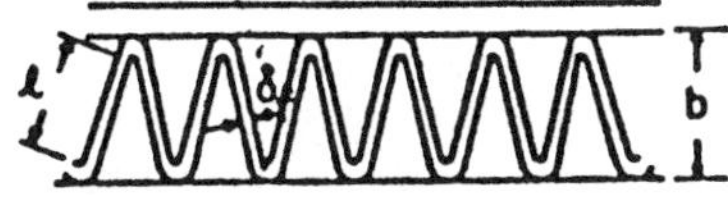 Plain, wavy, or louver fin of triangular cross section	$\eta_f = E_1$ $l_1 = \frac{l}{2} \quad \delta_1 = \delta_f$
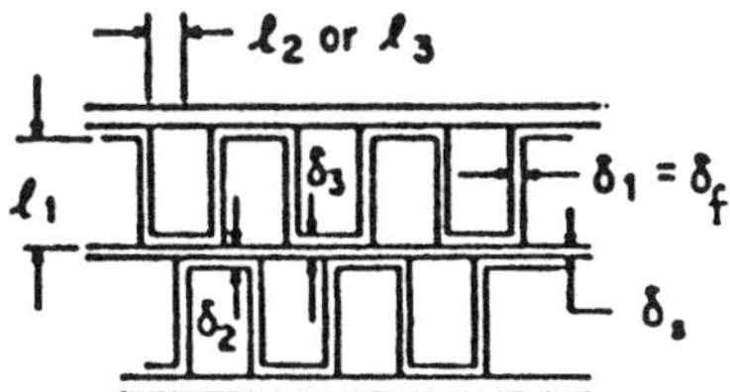 Double sandwich fin	$\eta_f = \frac{E_1 l_1 + E_2 l_2}{l_1 + l_2} \frac{1}{1 + m_1^2 E_1 E_2 l_1 l_2}$ $\delta_1 = \delta_f \quad \delta_2 = \delta_3 = \delta_f + \delta_s$ $l_1 = b - \delta_f + \frac{\delta_s}{2} \quad l_2 = l_3 = \frac{p_f}{2}$
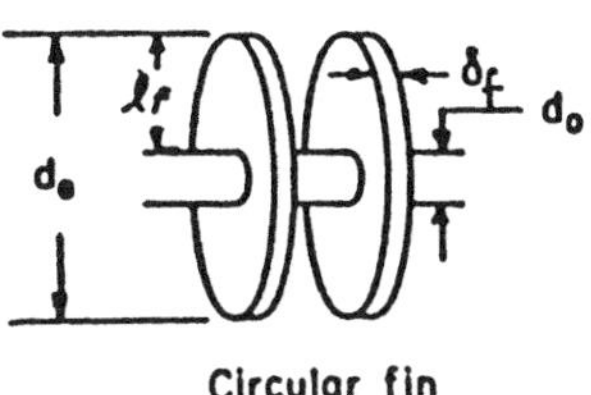 Circular fin	$\eta_j = \begin{cases} a(ml_e)^{-b} & \text{for } \Phi > 0.6 + 2.257(r^*)^{-0.445} \\ \frac{\tanh\Phi}{\Phi} & \text{for } \Phi \le 0.6 + 2.257(r^*)^{-0.445} \end{cases}$ $a = (r^*)^{-0.246} \quad \Phi = ml_e(r^*)^{\exp(0.13ml_e - 1.3863)}$ $b = \begin{cases} 0.9107 + 0.0893r^* & \text{for } r^* \le 2 \\ 0.9706 + 0.17125\ln r^* & \text{for } r^* > 2 \end{cases}$ $m = \left(\frac{2h}{k_f\delta_f}\right)^{1/2} \quad l_e = l_j + \frac{\delta_f}{2} \quad r^* = \frac{d_e}{d_o}$
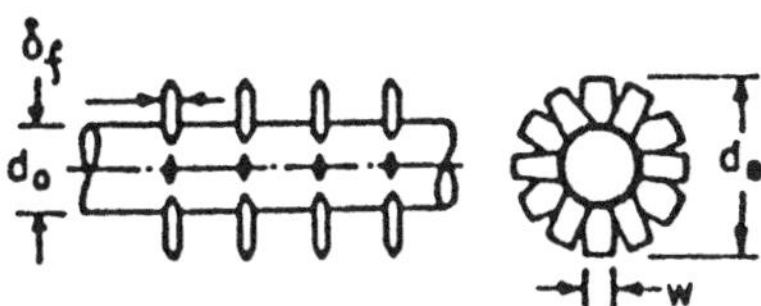 Studded fin	$\eta_j = \frac{\tanh(ml_e)}{ml_e}$ $m = \left[\frac{2h}{k_f\delta_f}\left(1+\frac{\delta_f}{w}\right)\right]^{1/2} \quad l_e = l_j + \frac{\delta_f}{2} \quad l_f = \frac{(d_e - d_o)}{2}$

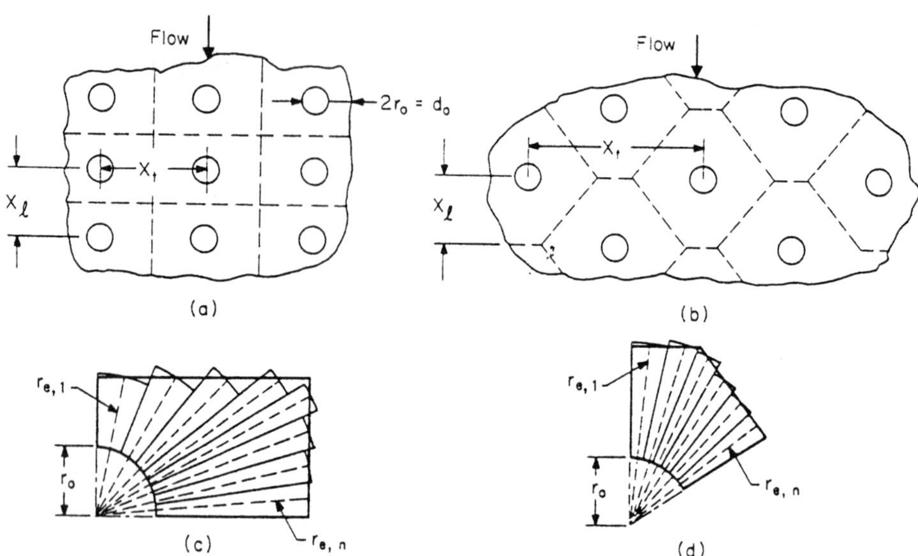

FIGURE 4.3.13 A flat fin over (a) an in-line and (b) staggered tube arrangement; the smallest representative segment of the fin for (c) an in-line and (d) a staggered tube arrangement.

subsequently computed using the circular fin formula of Table 4.3.5. The fin efficiency η_f for the whole fin is then the surface area weighted average of $\eta_{f,i}$ of each sector.

$$\eta_f = \frac{\sum_{i=1}^{n} \eta_{f,i} A_{f,i}}{\sum_{i=1}^{n} A_{f,i}} \tag{4.3.22}$$

Since the heat flow seeks the path of least thermal resistance, actual η_f will be equal to or higher than that calculated by Equation (4.3.22); hence, Equation (4.3.22) yields a somewhat conservative value of η_f.

The η_f values of Table 4.3.5 or Equation (4.3.22) are not valid in general when the fin is thick, when it is subject to variable heat transfer coefficients or variable ambient fluid temperature, or when it has a temperature depression at the fin base. See Huang and Shah (1992) for details. For a thin rectangular fin of constant cross section, the fin efficiency as presented in Table 4.3.5 is given by

$$\eta_f = \frac{\tanh(m\ell)}{m\ell} \tag{4.3.23}$$

where $m = [2h(1 + \delta_f/\ell_f)/k_f\delta_f]^{1/2}$. For a thick rectangular fin of constant cross section, the fin efficiency (a counterpart of Equation (4.3.23) is given by (Huang and Shah, 1992)

$$\eta_f = \frac{\left(\text{Bi}^+\right)^{1/2}}{K\text{Bi}} \tanh\left[K\left(\text{Bi}^+\right)^{1/2}\right] \tag{4.3.24}$$

where $\text{Bi}^+ = \text{Bi}/(1 + \text{Bi}/4)$, $\text{Bi} = (h\delta_f/2k_f)^{1/2}$, $K = 2\ell/\delta_f$. Equation (4.3.23) is accurate (within 0.3%) for a “thick” rectangular fin having $\eta_f > 80\%$; otherwise, use Equation (4.3.24) for a thick fin.

In an extended-surface heat exchanger, heat transfer takes place from both the fins ($\eta_f < 100\%$) and the primary surface ($\eta_f = 100\%$). In that case, the total heat transfer rate is evaluated through a concept of extended surface efficiency η_o defined as

$$\eta_o = \frac{A_p}{A} + \eta_f \frac{A_f}{A} = 1 - \frac{A_f}{A}\left(1 - \eta_f\right) \tag{4.3.25}$$

where A_f is the fin surface area, A_p is the primary surface area, and $A = A_f + A_p$. In Equation (4.3.25), heat transfer coefficients over the finned and unfinned surfaces are idealized to be equal. Note that $\eta_o \geq \eta_f$ and η_o is always required for the determination of thermal resistances of Equation (4.3.5) in heat exchanger analysis.

Pressure Drop Analysis.

Usually a fan, blower, or pump is used to flow fluid through individual sides of a heat exchanger. Due to potential initial and operating high cost, low fluid pumping power requirement is highly desired for gases and viscous liquids. The fluid pumping power $\wp$ is approximately related to the core pressure drop in the exchanger as (Shah, 1985).

$$\wp = \frac{\dot{m}\Delta p}{\rho} \approx \begin{cases} \dfrac{1}{2g_c}\dfrac{\mu}{\rho^2}\dfrac{4L}{D_h} f\,\mathrm{Re} & \text{for laminar flow} \quad (4.3.26) \\[2ex] \dfrac{0.046}{2g_c}\dfrac{\mu^{0.2}}{\rho^2}\dfrac{4L}{D_h}\dfrac{\dot{m}^{2.8}}{A_0^{1.8} D_h^{0.2}} & \text{for turbulent flow} \quad (4.3.27) \end{cases}$$

It is clear from Equations (4.2.26) and (4.2.27) that the fluid pumping power is strongly dependent upon the fluid density ($\wp \propto 1/\rho^2$) particularly for low-density fluids in laminar and turbulent flows, and upon the viscosity in laminar flow. In addition, the pressure drop itself can be an important consideration when blowers and pumps are used for the fluid flow since they are head limited. Also for condensing and evaporating fluids, the pressure drop affects the heat transfer rate. Hence, the pressure drop determination in the exchanger is important.

The pressure drop associated with a heat exchanger consists of (1) core pressure drop, and (2) the pressure drop associated with the fluid distribution devices such as inlet and outlet manifolds, headers, tanks, nozzles, ducting, and so on, which may include bends, valves, and fittings. This second Δp component is determined from Idelchik (1994) and Miller (1990). The core pressure drop may consist of one or more of the following components depending upon the exchanger construction: (1) friction losses associated with fluid flow over heat transfer surface (this usually consists of skin friction, form (profile) drag, and internal contractions and expansions, if any); (2) the momentum effect (pressure drop or rise due to fluid density changes) in the core; (3) pressure drop associated with sudden contraction and expansion at the core inlet and outlet, and (4) the gravity effect due to the change in elevation between the inlet and outlet of the exchanger. The gravity effect is generally negligible for gases. For vertical flow through the exchanger, the pressure drop or rise ("static head") due to the elevation change is given by

$$\Delta p = \pm \frac{\rho_m g L}{g_c} \tag{4.3.28}$$

Here the "+" sign denotes vertical upflow (i.e., pressure drop), the "–" sign denotes vertical downflow (i.e., pressure rise). The first three components of the core pressure drop are now presented for plate-fin, tube-fin, plate, and regenerative heat exchangers.

Plate-fin heat exchangers. For the plate-fin exchanger (Figure 4.3.2), all three components are considered in the core pressure drop evaluation as follows:

$$\frac{\Delta p}{p_i} = \frac{G^2}{2g_c}\frac{1}{p_i\rho_i}\left[\left(1-\sigma^2+K_c\right)+f\frac{L}{r_h}\rho_i\left(\frac{1}{\rho}\right)_m+2\left(\frac{\rho_i}{\rho_o}-1\right)-\left(1-\sigma^2-K_e\right)\frac{\rho_i}{\rho_o}\right] \tag{4.3.29}$$

where f is the Fanning friction factor, K_c and K_e are flow contraction (entrance) and expansion (exit) pressure loss coefficients (see Figure 4.3.14), and σ is a ratio of minimum free flow area to frontal area. K_c and K_e for four different entrance flow passage geometries are presented by Kays and London (1984). The entrance and exit losses are important at low values of σ and L (short cores), high values of Re, and for gases; they are negligible for liquids. The values of K_c and K_e apply to long tubes for which flow is fully developed at the exit. For partially developed flows, K_c and K_e is higher than that for fully developed flows. For interrupted surfaces, flow is never a fully developed boundary layer type. For highly interrupted fin geometries, the entrance and exit losses are generally small compared to the core pressure drop and the flow is well mixed; hence, K_c and K_e for Re $\to \infty$ should represent a good approximation. The mean specific volume v_m or $(1/\rho)_m$ in Equation (4.3.29) is given as follows. For liquids with any flow arrangement, or for a perfect gas with $C^* = 1$ and any flow arrangement (except for parallel flow),

$$\left(\frac{1}{\rho}\right)_m = v_m = \frac{v_i + v_o}{2} = \frac{1}{2}\left(\frac{1}{\rho_i}+\frac{1}{\rho_o}\right) \tag{4.3.30}$$

where v is the specific volume in m^3/kg. For a perfect gas with $C^* = 0$ and any flow arrangement,

$$\left(\frac{1}{\rho}\right)_m = \frac{\tilde{R}}{p_{\text{ave}}}T_{\text{lm}} \tag{4.3.31}$$

Here $\tilde{R}$ is the gas constant in J/(kg K), $p_{\text{ave}} = (p_i + p_o)/2$, and $T_{\text{lm}} = T_{\text{const}} + \Delta T_{\text{lm}}$ where T_{const} is the mean average temperature of the fluid on the other side of the exchanger; the LMTD ΔT_{lm} is defined in Table 4.3.2. The core frictional pressure drop in Equation (4.3.29) may be approximated as

$$\Delta p \approx \frac{4fLG^2}{2g_c D_h}\left(\frac{1}{\rho}\right)_m \tag{4.3.32}$$

Tube-fin heat exchangers. The pressure drop inside a circular tube is computed using Equation (4.3.29) with proper values of f factors, and K_c and K_e from Figure (4.3.18) for circular tubes.

For flat fins on an array of tubes (see Figure 4.3.5b), the components of the core pressure drop (such as those in Equation (4.3.29) are the same with the following exception: the core friction and momentum effect take place within the core with $G = \dot{m}/A_0$, where A_o is the minimum free-flow area within the core, and the entrance and exit losses occur at the leading and trailing edges of the core with the associated flow area A_0' such that

$$\dot{m} = GA_o = G'A_o' \quad \text{or} \quad G'\sigma' = G\sigma \tag{4.3.33}$$

where σ' is the ratio of free-flow area to frontal area at the fin leading edges. The pressure drop for flow normal to a tube bank with flat fins is then given by

$$\frac{\Delta p}{p_i} = \frac{G^2}{2g_c}\frac{1}{p_i\rho_i}\left[f\frac{L}{r_h}\rho_i\left(\frac{1}{\rho}\right)_m+2\left(\frac{\rho_i}{\rho_o}-1\right)\right]+\frac{G'^2}{2g_c}\frac{1}{p_i\rho_i}\left[\left(1-\sigma'^2-K_c\right)-\left(1-\sigma'^2-K_e\right)\frac{\rho_i}{\rho_o}\right] \tag{4.3.34}$$

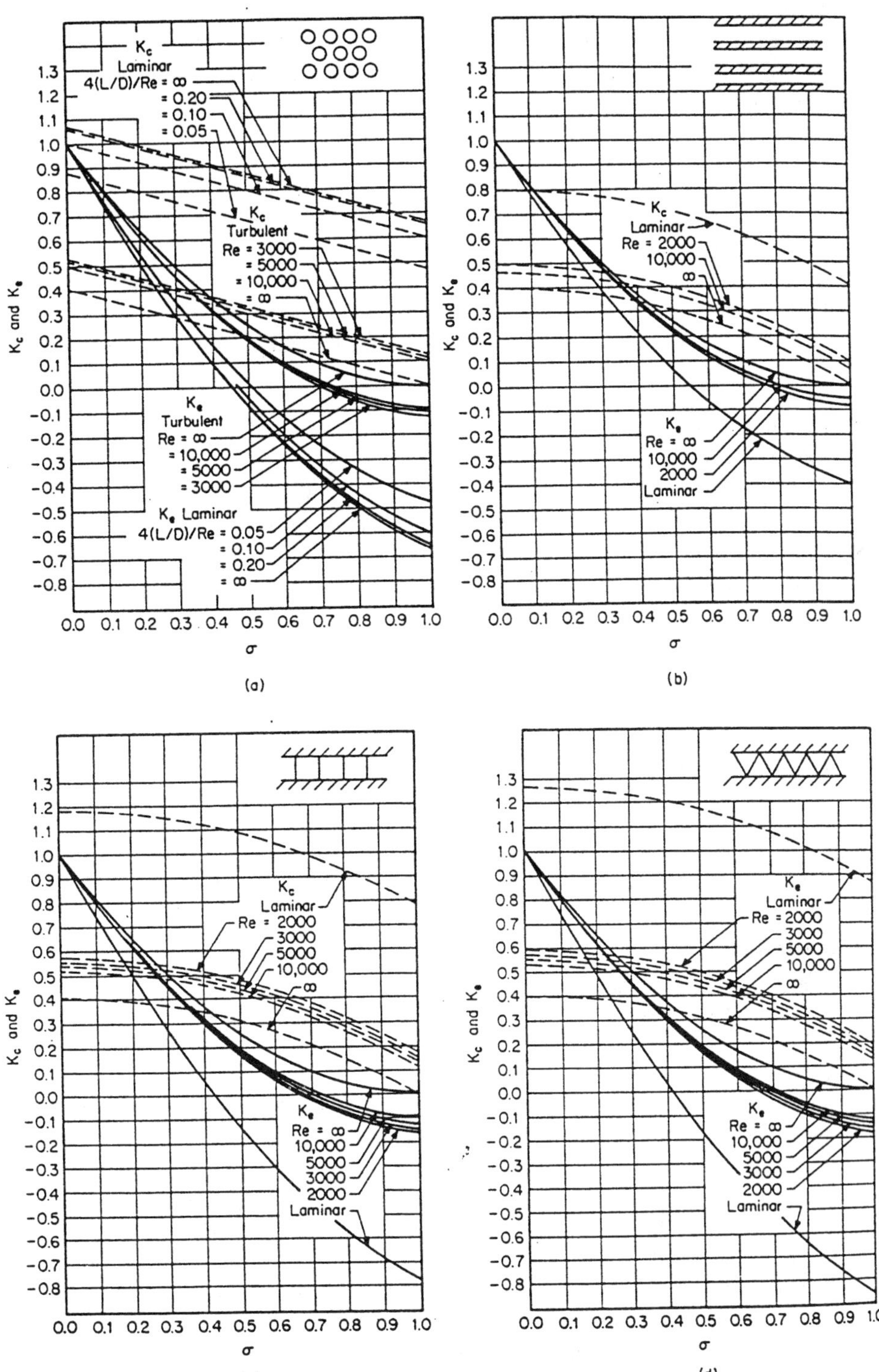

FIGURE 4.3.14 Entrance and exit pressure loss coefficients: (a) circular tubes, (b) parallel plates, (C) square passages, and (d) triangular passages. (From Kays, W.M. and London, A.L., *Compact Heat Exchangers*, 3rd ed., McGraw-Hill, New York, 1984. With permission.) For each of these flow passages, shown in the inset, the fluid flows perpendicular to the plane of the paper into the flow passages.

For individually finned tubes as shown in Figure 4.3.5a, flow expansion and contraction take place along each tube row, and the magnitude is of the same order as that at the entrance and exit. Hence, the entrance and exit losses are generally lumped into the core friction factor. Equation (4.3.29) then reduces for individually finned tubes to

$$\frac{\Delta p}{p_i} = \frac{G^2}{2g_c} \frac{1}{p_i \rho_i} \left[f \frac{L}{r_h} \rho_i \left(\frac{1}{\rho} \right)_m + 2 \left(\frac{\rho_i}{\rho_o} - 1 \right) \right] \tag{4.3.35}$$

Regenerators. For regenerator matrices having cylindrical passages, the pressure drop is computed using Equation (4.3.29) with appropriate values of f, K_c, and K_e. For regenerator matrices made up of any porous material (such as checkerwork, wire, mesh, spheres, copper wools, etc.), the pressure drop is calculated using Equation (4.3.35) in which the entrance and exit losses are included in the friction factor f.

Plate heat exchangers. Pressure drop in a PHE consists of three components: (1) pressure drop associated with the inlet and outlet manifolds and ports, (2) pressure drop within the core (plate passages), and (3) pressure drop due to the elevation change. The pressure drop in the manifolds and ports should be kept as low as possible (generally < 10%, but it is found as high as 25 to 30% of higher in some designs). Empirically, it is calculated as approximately 1.5 times the inlet velocity head per pass. Since the entrance and exit losses in the core (plate passages) cannot be determined experimentally, they are included in the friction factor for the given plate geometry. The pressure drop (rise) caused by the elevation change for liquids is given by Equation (4.3.28). Hence, the pressure drop on one fluid side in a PHE is given by

$$\Delta p = \frac{1.5 G^2 N_p}{2 g_c \rho_i} + \frac{4 f L G^2}{2 g_c D_e} \left(\frac{1}{\rho} \right)_m + \left(\frac{1}{\rho_o} - \frac{1}{\rho_i} \right) \frac{G^2}{g_c} \pm \frac{\rho_m g L}{g_c} \tag{4.3.36}$$

where N_p is the number of passes on the given fluid side and D_e is the equivalent diameter of flow passages (usually twice the plate spacing). Note that the third term on the right-hand side of the equality sign of Equation (4.3.36) is for the momentum effect which is generally negligible in liquids.

Heat Transfer and Flow Friction Correlations

Accurate and reliable surface heat transfer and flow friction characteristics are a key input to the exchanger heat transfer and pressure drop analyses or to the rating and sizing problems (Shah, 1985). Some important analytical solutions and empirical correlations are presented next for selected exchanger geometries.

The heat transfer rate in laminar duct flow is very sensitive to the thermal boundary condition. Hence, it is essential to identify carefully the thermal boundary condition in laminar flow. The heat transfer rate in turbulent duct flow is insensitive to the thermal boundary condition for most common fluids (Pr > 0.7); the exception is liquid metals (Pr < 0.03). Hence, there is generally no need to identify the thermal boundary condition in turbulent flow for all fluids except liquid metals.

Fully developed laminar flow analytical solutions for some duct shapes of interest in compact heat exchangers are presented in Table 4.3.6 for three important thermal boundary conditions denoted by the subscripts $H1$, $H2$, and T (Shah and London, 1978; Shah and Bhatti, 1987). Here, $H1$ denotes constant axial wall heat flux with constant peripheral wall temperature, $H2$ denotes constant axial and peripheral wall heat flux, and T denotes constant wall temperature. The entrance effects, flow maldistribution, free convection, property variation, fouling, and surface roughness all affect fully developed analytical solutions. In order to account for these effects in real plate-fin plain fin geometries having fully developed flows, it is best to reduce the magnitude of the analytical Nu by at least 10% and to increase the value of the analytical fRe by 10% for design purposes.

TABLE 4.3.6 Solutions for Heat Transfer and Friction for Fully Developed Flow-Through Specified Ducts

Geometry ($L/D_h > 100$)	Nu_{H1}	Nu_{H2}	Nu_{τ}	fRe	j_{H1}/f [a]	Nu_{H1}/Nu_{τ}
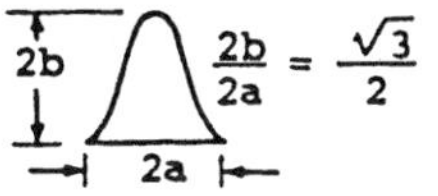 Triangle, $\frac{2b}{2a} = \frac{\sqrt{3}}{2}$	3.014	1.474	2.39[b]	12.630	0.269	1.26
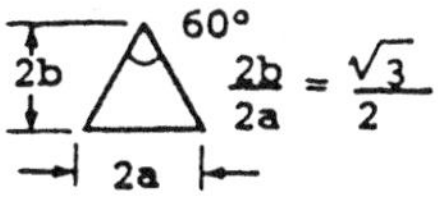 Equilateral triangle (60°), $\frac{2b}{2a} = \frac{\sqrt{3}}{2}$	3.111	1.892	2.47	13.333	0.263	1.26
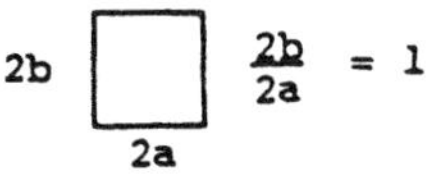 Square, $\frac{2b}{2a} = 1$	3.608	3.091	2.976	14.227	0.286	1.21
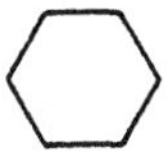 Hexagon	4.002	3.862	3.34[b]	15.054	0.299	1.20
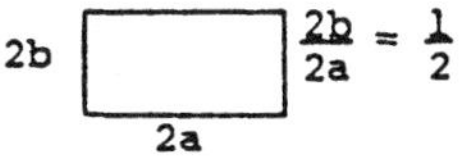 Rectangle, $\frac{2b}{2a} = \frac{1}{2}$	4.123	3.017	3.391	15.548	0.299	1.22
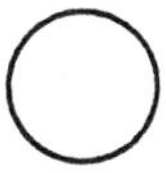 Circle	4.364	4.364	3.657	16.000	0.307	1.19
 Rectangle, $\frac{2b}{2a} = \frac{1}{4}$	5.331	2.94	4.439	18.233	0.329	1.20
 Rectangle, $\frac{2b}{2a} = \frac{1}{8}$	6.490	2.94	5.597	20.585	0.355	1.16
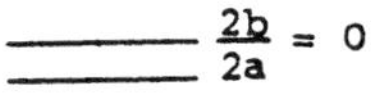 Parallel plates, $\frac{2b}{2a} = 0$	8.235	8.235	7.541	24.000	0.386	1.09
 Parallel plates, one side insulated, $\frac{b}{a} = 0$	5.385	—	4.861	24.000	0.253	1.11

[a] This heading is the same as $Nu_{H1}\,Pr^{-1/3}/f$ Re with Pr = 0.7.

[b] Interpolated values.

The initiation of transition flow, the lower limit of the critical Reynolds number (Re_{crit}), depends upon the type of entrance (e.g., smooth vs. abrupt configuration at the exchanger flow passage entrance). For a sharp square inlet configuration, Re_{crit} is about 10 to 15% lower than that for a rounded inlet configuration. For most exchangers, the entrance configuration would be sharp. Some information on Re_{crit} is provided by Ghajar and Tam (1994).

Transition flow and fully developed turbulent flow Fanning friction factors (within ±2% accuracy) are given by Bhatti and Shah (1987) as

$$f = A + B\mathrm{Re}^{-1/m} \tag{4.3.37}$$

where

$$A = 0.0054, \quad B = 2.3\times10^{-8}, \quad m = -2/3, \quad \text{for} \quad 2100 \le \mathrm{Re} \le 4000$$

$$A = 0.00128, \quad B = 0.1143, \quad m = 3.2154, \quad \text{for} \quad 4000 \le \mathrm{Re} \le 10^7$$

The transition flow and fully developed turbulent flow Nusselt number correlation for a circular tube is given by Gnielinski as reported in Bhatti and Shah (1987) as

$$\text{Nu} = \frac{(f/2)(\text{Re} - 1000)\text{Pr}}{1 + 12.7(f/2)^{1/2}\left(\text{Pr}^{2/3} - 1\right)} \tag{4.3.38}$$

which is accurate within about ±10% with experimental data for $2300 \le \text{Re} \le 5 \times 10^6$ and $0.5 \le \text{Pr} \le 2000$.

A careful observation of accurate experimental friction factors for all noncircular smooth ducts reveals that ducts with laminar $f\text{Re} < 16$ have turbulent f factors lower than those for the circular tube; whereas ducts with laminar $f\text{Re} > 16$ have turbulent f factors higher than those for the circular tube (Shah and Bhatti, 1988). Similar trends are observed for the Nusselt numbers. Within ±15% accuracy, Equations (4.3.37) and (4.3.38) for f and Nu can be used for noncircular passages with the hydraulic diameter as the characteristic length in f, Nu, and Re; otherwise, refer to Bhatti and Shah (1987) for more accurate results for turbulent flow.

For hydrodynamically and thermally developing flows, the analytical solutions are boundary condition dependent (for laminar flow heat transfer only) and geometry dependent. The hydrodynamic entrance lengths for developing laminar and turbulent flows are given by Shah and Bhatti (1987) and Bhatti and Shah (1987) as

$$\frac{L_{hy}}{D_h} = \begin{cases} 0.0565\text{Re} & \text{for laminar flow } (\text{Re} \le 2100) \quad (4.3.39) \\ 1.359\text{Re}^{1/4} & \text{for turbulent flow } \left(\text{Re} \ge 10^4\right) \quad (4.3.40) \end{cases}$$

Analytical results are useful for well-defined constant-cross-sectional surfaces with essentially unidirectional flows. The flows encountered in heat exchangers are generally very complex having flow separation, reattachment, recirculation, and vortices. Such flows significantly affect Nu and f for the specific exchanger surfaces. Since no analytical or accurate numerical solutions are available, the information is derived experimentally. Kays and London (1984) and Webb (1994) present most of the experimental results reported in the open literature. In the following, empirical correlations for only some important surfaces are summarized due to space limitations.

Plate-Fin Extended Surfaces.

Offset strip fins. This is one of the most widely used enhanced fin geometries (Figure 4.3.15) in aircraft, cryogenics, and many other industries that do not require mass production. This surface has one of the highest heat transfer performances relative to the friction factor. The most comprehensive correlations for j and f factors for the offset strip fin geometry is provided by Manglik and Bergles (1995) as follows.

$$j = 0.6522\text{Re}^{-0.5403}\left(\frac{s}{h'}\right)^{-0.1541}\left(\frac{\delta_f}{l_f}\right)^{0.1499}\left(\frac{\delta_f}{s}\right)^{-0.0678} \times\left[1 + 5.269 \times 10^{-5}\text{Re}^{1.340}\left(\frac{s}{h'}\right)^{0.504}\left(\frac{\delta_f}{l_f}\right)^{0.456}\left(\frac{\delta_f}{s}\right)^{-1.055}\right]^{0.1} \tag{4.3.41}$$

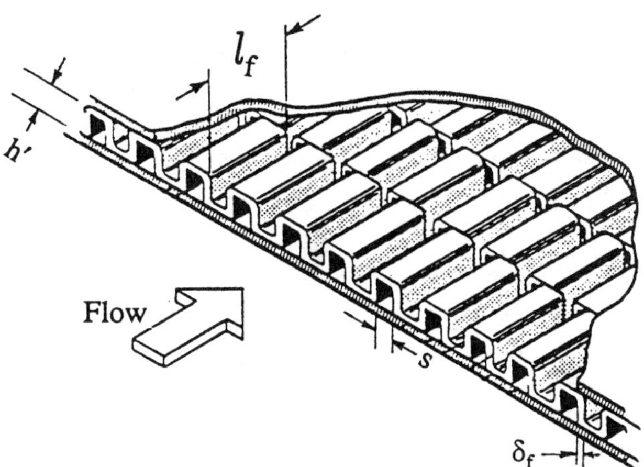

FIGURE 4.3.15 An offset strip fin geometry.

$$f = 9.6243\text{Re}^{-0.7422}\left(\frac{s}{h'}\right)^{-0.1856}\left(\frac{\delta_f}{l_f}\right)^{0.3053}\left(\frac{\delta_f}{s}\right)^{-0.2659} \times\left[1+7.669\times10^{-8}\text{Re}^{4.429}\left(\frac{s}{h'}\right)^{0.920}\left(\frac{\delta_f}{l_f}\right)^{3.767}\left(\frac{\delta_f}{s}\right)^{0.236}\right]^{0.1} \tag{4.3.42}$$

where

$$D_h = 4A_o\Big/\left(A/l_f\right) = 4sh'l_f\Big/\left[2\left(sl_f + h'l_f + \delta_f h'\right) + \delta_f s\right] \tag{4.3.43}$$

Geometric symbols in Equation (4.3.43) are shown in Figure 4.3.15.

These correlations predict the experimental data of 18 test cores within ±20% for $120 \leq \text{Re} \leq 10^4$. Although all the experimental data for these correlations are obtained for air, the j factor takes into consideration minor variations in the Prandtl number, and the above correlations should be valid for $0.5 < \text{Pr} < 15$.

Louver fins. Louver or multilouver fins are extensively used in the auto industry because of their mass production manufacturability and hence lower cost. The louver fin has generally higher j and f factors than those for the offset strip fin geometry, and also the increase in the friction factors is in general higher than the increase in the j factors. However, the exchanger can be designed for higher heat transfer and the same pressure drop compared to that with the offset strip fins by a proper selection of exchanger frontal area, core depth, and fin density. Published literature and correlations on the louver fins are summarized by Webb (1994) and Cowell et al. (1995), and the understanding of flow and heat transfer phenomena is summarized by Cowell et al. (1995). Because of the lack of systematic studies reported in the open literature on modern louver fin geometries, no correlation can be recommended for the design purpose.

Tube-Fin Extended Surfaces.

Two major types of tube-fin extended surfaces as shown in Figure 4.3.5 are (1) individually finned tubes and (2) flat fins (also sometimes referred to as plate fins) with or without enhancements/interruptions on an array of tubes. An extensive coverage of the published literature and correlations for these extended surfaces are provided by Webb (1994), Kays and London (1984), and Rozenman (1976). Empirical correlations for some important geometries are summarized below.

Individually finned tubes. This fin geometry, helically wrapped (or extruded) circular fins on a circular tube as shown in Figure 4.3.5a, is commonly used in process and waste heat recovery industries. The following correlation for *j* factors is recommended by Briggs and Young (see Webb, 1994) for individually finned tubes on staggered tube banks.

$$j = 0.134\text{Re}_d^{-0.319}\left(s/l_f\right)^{0.2}\left(s/\delta_f\right)^{0.11} \tag{4.3.44}$$

where l_f is the radial height of the fin, δ_f the fin thickness, $s = p_f - \delta_f$ is the distance between adjacent fins and p_f is the fin pitch. Equation (4.3.44) is valid for the following ranges: $1100 \le \text{Re}_d \le 18{,}000$, $0.13 \le s/l_f \le 0.63$, $1.01 \le s/\delta_f \le 6.62$, $0.09 \le l_f/d_o \le 0.69$, $0.011 \le \delta_f/d_o \le 0.15$, $1.54 \le X_t/d_o \le 8.23$, fin root diameter d_o between 11.1 and 40.9 mm, and fin density N_f ($= 1/p_f$) between 246 and 768 fin/m. The standard deviation of Equation (4.3.44) with experimental results was 5.1%.

For friction factors, Robinson and Briggs (see Webb, 1994) recommended the following correlation:

$$f_{tb} = 9.465\text{Re}_d^{-0.316}\left(X_t/d_o\right)^{-0.927}\left(X_t/X_d\right)^{0.515} \tag{4.3.45}$$

Here $X_d = (X_t^2 + X_l^2)^{1/2}$ is the diagonal pitch, and X_t and X_l are the transverse and longitudinal tube pitches, respectively. The correlation is valid for the following ranges: $2000 \le \text{Re}_d \le 50{,}000$, $0.15 \le s/l_f \le 0.19$, $3.75 \le s/\delta_f \le 6.03$, $0.35 \le l_f/d_o \le 0.56$, $0.011 \le \delta_f/d_o \le 0.025$, $1.86 \le X_t/d_o \le 4.60$, $18.6 \le d_o \le 40.9$ mm, and $311 \le N_f \le 431$ fin/m. The standard deviation of Equation (4.3.45) with correlated data was 7.8%

For crossflow over low-height finned tubes, a simple but accurate correlation for heat transfer is given by Ganguli and Yilmaz (1987) as

$$j = 0.255\text{Re}_d^{-0.3}\left(d_e/s\right)^{-0.3} \tag{4.3.46}$$

A more accurate correlation for heat transfer is given by Rabas and Taborek (1987). Chai (1988) provides the best correlation for friction factors:

$$f_{tb} = 1.748\text{Re}_d^{-0.233}\left(\frac{l_f}{s}\right)^{0.552}\left(\frac{d_o}{X_t}\right)^{0.599}\left(\frac{d_o}{X_1}\right)^{0.1738} \tag{4.3.47}$$

This correlation is valid for $895 < \text{Re}_d < 713{,}000$, $20 < \theta < 40°$, $X_t/d_o < 4$, $N \ge 4$, and θ is the tube layout angle. It predicts 89 literature data points within a mean absolute error of 6%; the range of actual error is from –16.7 to 19.9%.

Flat plain fins on a staggered tubebank. This geometry, as shown in Figure 4.3.5b, is used in air-conditioning/refrigeration industry as well as where the pressure drop on the fin side prohibits the use of enhanced/interrupted flat fins. An inline tubebank is generally not used unless very low fin side pressure drop is the essential requirement. Heat transfer correlation for Figure 4.3.5b flat plain fins on staggered tubebanks is provided by Gray and Webb (see Webb, 1994) as follows for four or more tube rows.

$$j_4 = 0.14\text{Re}_d^{-0.328}\left(X_t/X_l\right)^{-0.502}\left(s/d_o\right)^{0.031} \tag{4.3.48}$$

For the number of tube rows N from 1 to 3, the j factor is lower and is given by

$$\frac{j_N}{j_4} = 0.991\left[2.24\text{Re}_d^{-0.092}(N/4)^{-0.031}\right]^{0.607(4-N)} \tag{4.3.49}$$

Gray and Webb (see Webb, 1994) hypothesized the friction factor consisting of two components: one associated with the fins and the other associated with the tubes as follows.

$$f = f_f\frac{A_f}{A} + f_t\left(1 - \frac{A_f}{A}\right)\left(1 - \frac{\delta_f}{p_f}\right) \tag{4.3.50}$$

where

$$f_f = 0.508\text{Re}_d^{-0.521}\left(X_t/d_o\right)^{1.318} \tag{4.3.51}$$

and f_t (defined the same way as f) is the Fanning friction factor associated with the tube and can be determined from Eu of Figure 19 of Zukaukas (1987) as $f_t = \text{EuN}(X_t - d_o)/\pi d_o$. Equation (4.3.50) correlated 90% of the data for 19 heat exchangers within ±20%. The range of dimensionless variables of Equations (4.3.50) and (4.3.51) are $500 \le \text{Re} \le 24{,}700$, $1.97 \le X_t/d_o \le 2.55$, $1.7 \le X_l/d_o \le 2.58$, and $0.08 \le s/d_o \le 0.64$.

Exchanger Design Methodology

The problem of heat exchanger design is complex and multidisciplinary (Shah, 1991). The major design considerations for a new heat exchanger include process/design specifications, thermal and hydraulic design, mechanical design, manufacturing and cost considerations, and trade-offs and system-based optimization, as shown in Figure 4.3.16 with possible strong interactions among these considerations as indicated by double-sided arrows. The thermal and hydraulic design methods are mainly analytical, and the structural design is analytical to some extent. Most of the other major design considerations involve qualitative and experience-based judgments, trade-offs, and compromises. Therefore, there is no unique solution to designing a heat exchanger for given process specifications. Further details on this design methodology is given by Shah (1991).

Two important heat exchanger design problems are the rating and sizing problems. Determination of heat transfer and pressure drop performance of either an existing exchanger or an already sized exchanger is referred to as the rating problem. The objective here is to verify vendor's specifications or to determine the performance at off-design conditions. The rating problem is also sometimes referred to as the performance problem. In contrast, the design of a new or existing type of exchanger is referred to as the sizing problem. In a broad sense, it means the determination of the exchanger construction type, flow arrangement, heat transfer surface geometries and materials, and the physical size of an exchanger to meet the specified heat transfer and pressure drops. However, from the viewpoint of quantitative thermal-hydraulic analysis, we will consider that the selection of the exchanger construction type, flow arrangement, and materials has already been made. Thus, in the sizing problem, we will determine the physical size (length, width, height) and surface areas on each side of the exchanger. The sizing problem is also sometimes referred to as the design problem.

The step-by-step solution procedures for the rating and sizing problems for counterflow and crossflow single-pass plate-fin heat exchangers have been presented with a detailed illustrative example by Shah (1981). Shah (1988a) presented further refinements in these procedures as well as step-by-step procedures for two-pass cross-counterflow plate-fin exchangers, and single-pass crossflow and two-pass cross-counterflow tube-fin exchangers. Also, step-by-step solution procedures for the rating and sizing problems for rotary regenerators (Shah, 1988b), heat pipe heat exchangers (Shah and Giovannelli, 1988) and PHEs (Shah and Wanniarachchi, 1991) are available. As an illustration, the step-by-step solution procedures will be covered here for a single-pass crossflow exchanger.

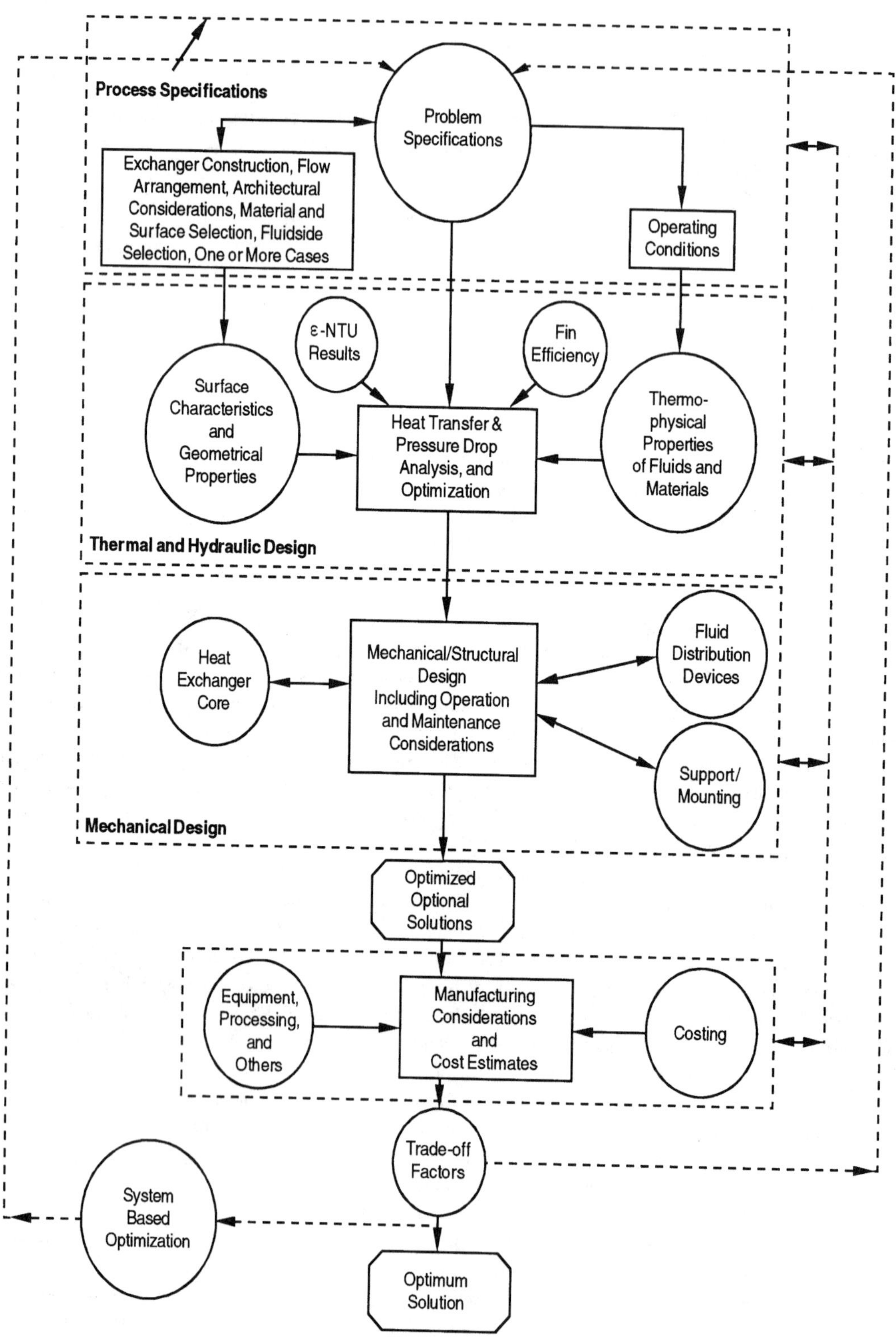

FIGURE 4.3.16 Heat exchanger design methodology.

Rating Problem for a Crossflow Plate-Fin Exchanger.

Following is a step-by-step procedure for rating a crossflow plate-fin exchanger. Inputs to the rating problem for a two-fluid exchanger are the exchanger construction, flow arrangement and overall dimensions, complete details on the materials and surface geometries on both sides including their nondimensional heat transfer and pressure drop characteristics (j and f vs. Re), fluid flow rates, inlet temperatures, and fouling factors. The fluid outlet temperatures, total heat transfer rate, and pressure drops on each side of the exchanger are then determined as the rating problem solution.

1. Determine the surface geometric properties on each fluid side. This includes the minimum free-flow area A_o, heat transfer surface area A (both primary and secondary), flow lengths L, hydraulic diameter D_h, heat transfer surface area density β, the ratio of minimum free-flow area to frontal area σ, fin length l_f, and fin thickness δ for fin efficiency determination, and any specialized dimensions used for heat transfer and pressure drop correlations.
2. Compute the fluid bulk mean temperature and fluid thermophysical properties on each fluid side. Since the outlet temperatures are not known for the rating problem, they are estimated initially. Unless it is known from past experience, assume an exchanger effectiveness as 60 to 75% for most single-pass crossflow exchangers, or 80 to 85% for single-pass counterflow exchangers. For the assumed effectiveness, calculate the fluid outlet temperatures.

$$T_{h,o} = T_{h,i} - \varepsilon\left(C_{\min}/C_h\right)\left(T_{h,i} - T_{c,i}\right) \tag{4.3.52}$$

$$T_{c,o} = T_{c,i} - \varepsilon\left(C_{\min}/C_c\right)\left(T_{h,i} - T_{c,i}\right) \tag{4.3.53}$$

Initially, assume $C_c/C_h = \dot{m}_c/\dot{m}_h$ for a gas-to-gas exchanger, or $C_c/C_h = \dot{m}_c c_{p,c}/\dot{m}_h c_{p,h}$ for a gas-to-liquid exchanger with very approximate values of c_p for the fluids in question.

For exchangers with $C^* > 0.5$ (usually gas-to-gas exchangers), the bulk mean temperatures on each fluid side will be the arithmetic mean of the inlet and outlet temperatures on each fluid side (Shah, 1981). For exchangers with $C^* > 0.5$ (usually gas-to-gas exchangers), the bulk mean temperature on the $C_{\max}$ side will be the arithmetic mean of inlet and outlet temperatures; the bulk mean temperature on the $C_{\min}$ side will be the log-mean average temperature obtained as follows:

$$T_{m,C_{\min}} = T_{m,C_{\max}} \pm \Delta T_{\mathrm{lm}} \tag{4.3.54}$$

where ΔT_{lm} is the LMTD based on the terminal temperatures (see Equation 4.3.19). Use the plus sign if the $C_{\min}$ side is hot; otherwise, use the negative sign.

Once the bulk mean temperature is obtained on each fluid side, obtain the fluid properties from thermophysical property books or from handbooks. The properties needed for the rating problem are μ, c_p, k, Pr, and ρ. With this c_p, one more iteration may be carried out to determine $T_{h,o}$ or $T_{c,o}$ from Equation (4.3.52) or (4.3.53) on the $C_{\max}$ side, and subsequently T_m on the $C_{\max}$ side, and refine fluid properties accordingly.

3. Calculate the Reynolds number Re = GD_h/μ and/or any other pertinent dimensionless groups (from the basic definitions) needed to determine the nondimensional heat transfer and flow friction characteristics (e.g., j or Nu and f) of heat transfer surfaces on each side of the exchanger. Subsequently, compute j or Nu and f factors. Correct Nu (or j) for variable fluid property effects (Shah, 1981) in the second and subsequent iterations from the following equations.

For gases:
$$\frac{\mathrm{Nu}}{\mathrm{Nu}_{\mathrm{cp}}} = \left[\frac{T_w}{T_m}\right]^{n'} \qquad \frac{f}{f_{\mathrm{cp}}} = \left[\frac{T_w}{T_m}\right]^{m'} \tag{4.3.55}$$

For liquids: $$\frac{\text{Nu}}{\text{Nu}_{\text{cp}}} = \left[\frac{\mu_w}{\mu_m}\right]^{n'} \qquad \frac{f}{f_{\text{cp}}} = \left[\frac{\mu_w}{\mu_m}\right]^{m'} \tag{4.3.56}$$

where the subscript cp denotes constant properties, and m' and n' are empirical constants provided in Table 4.3.7. Note that T_w and T_m in Equations (4.3.55) and (4.3.56) and in Tables 4.3.7a and b and are absolute temperatures.

TABLE 4.3.7a Property Ratio Method Exponents of Equations (4.3.55) and (4.3.56) for Laminar Flow

Fluid	Heating	Cooling
Gases	$n' = 0.00$, $m' = 1.00$ for $1 < T_w/T_m < 3$	$n' = 0.0$, $m' = 0.81$ for $0.5 < T_w/T_m < 1$
Liquids	$n' = -0.14$, $m' = 0.58$ for $\mu_w/\mu_m < 1$	$n' = -0.14$, $m' = 0.54$ for $\mu_w/\mu_m > 1$

Source: Shah, R.K., in *Heat Exchangers: Thermal-Hydraulic Fundamentals and Design*, S. Kakaç et al., Eds., Hemisphere Publishing, Washington, D.C., 1981. With permission.

TABLE 4.3.7b Property Ratio Method Correlations of Exponents of Equations (4.3.55) and (4.3.56) for Turbulent Flow

Fluid	Heating	Cooling
Gases	$\text{Nu} = 5 + 0.012\,\text{Re}^{0.83}\,(\text{Pr} + 0.29)\,(T_w/T_m)^n$ $n = -[\log_{10}(T_w/T_m)]^{1/4} + 0.3$ for $1 < T_w/T_m < 5$, $0.6 < \text{Pr} < 0.9$, $10^4 < \text{Re} < 10^6$, and $L/D_h > 40$	$n' = 0$
	$m' = -0.1$ for $1 < T_w/T_m < 2.4$	$m' = -0.1$ (tentative)
Liquids	$n' = -0.11$[a] for $0.08 < \mu_w/\mu_m < 1$	$n' = -0.25$[a] for $1 < \mu_w/\mu_m < 40$
	$f/f_{\text{cp}} = (7 - \mu_m/\mu_w)/6$[b] or $m' = 0.25$ for $0.35 < \mu_w/\mu_m < 1$	$m' = 0.24$[b] for $1 < \mu_w/\mu_m < 2$

[a] Valid for $2 \le \text{Pr} \le 140$, $10^4 \le \text{Re} \le 1.25 \times 10^5$.
[b] Valid for $1.3 \le \text{Pr} \le 10$, $10^4 \le \text{Re} \le 2.3 \times 10^5$.
Source: Shah, R.K., in *Heat Exchangers: Thermal-Hydraulic Fundamentals and Design*, S. Kakaç et al., Eds., Hemisphere Publishing, Washington, D.C., 1981. With permission.

4. From Nu or j, compute the heat transfer coefficients for both fluid streams.

$$h = \text{Nu}\,k/D_h = jGc_p\text{Pr}^{-2/3} \tag{4.3.57}$$

Subsequently, determine the fin efficiency η_f and the extended surface efficiency η_o

$$\eta_f = \frac{\tanh m\ell}{m\ell} \quad \text{where} \quad m^2 = \frac{h\tilde{P}}{k_f A_k} \tag{4.3.58}$$

where $\tilde{P}$ is the wetted perimeter of the fin surface.

$$\eta_o = 1 - \frac{A_f}{A}\left(1 - \eta_f\right) \tag{4.3.59}$$

Also calculate the wall thermal resistance $R_w = \delta/A_w k_w$. Finally, compute the overall thermal conductance UA from Equation (4.3.6) knowing the individual convective film resistances, wall thermal resistances, and fouling resistances, if any.

5. From the known heat capacity rates on each fluid side, compute $C^* = C_{min}/C_{max}$. From the known UA, determine NTU = UA/C_{min}. Also calculate the longitudinal conduction parameter λ. With the known NTU, C^*, λ, and the flow arrangement, determine the exchanger effectiveness ε from either closed-form equations of Table 4.3.4 or tabular/graphical results from Kays and London (1984).
6. With this ε, finally compute the outlet temperatures from Equations (4.3.52) and (4.3.53). If these outlet temperatures are significantly different from those assumed in Step 2, use these outlet temperatures in Step 2 and continue iterating Steps 2 to 6, until the assumed and computed outlet temperatures converge within the desired degree of accuracy. For a gas-to-gas exchanger, most probably one or two iterations will be sufficient.
7. Finally, compute the heat duty from

$$q = \varepsilon C_{min}\left(T_{h,i} - T_{c,i}\right) \tag{4.3.60}$$

8. For the pressure drop calculations, first we need to determine the fluid densities at the exchanger inlet and outlet (ρ_i and ρ_o) for each fluid. The mean specific volume on each fluid side is then computed from Equation (4.3.30).

   Next, the entrance and exit loss coefficients, K_c and K_e, are obtained from Figure 4.3.14 for known σ, Re, and the flow passage entrance geometry.

   The friction factor on each fluid side is corrected for variable fluid properties using Equation (4.3.55) or (4.3.56). Here, the wall temperature T_w is computed from

$$T_{w,h} = T_{m,h} - \left(R_h + R_{s,h}\right)q \tag{4.3.61}$$

$$T_{w,c} = T_{m,c} + \left(R_c + R_{s,c}\right)q \tag{4.3.62}$$

   where the various resistance terms are defined by Equation (4.3.6).

The core pressure drops on each fluid side are then calculated from Equation (4.3.29). This then completes the procedure for solving the rating problem.

Sizing Problem for a Crossflow Plate-Fin Exchanger.
As defined earlier, we will concentrate here to determine the physical size (length, width, and height) of a single-pass crossflow exchanger for specified heat duty and pressure drops. More specifically, inputs to the sizing problem are surface geometries (including their nondimensional heat transfer and pressure drop characteristics), fluid flow rates, inlet and outlet fluid temperatures, fouling factors, and pressure drops on each side.

For the solution to this problem, there are four unknowns — two flow rates or Reynolds numbers (to determine correct heat transfer coefficients and friction factors) and two surface areas — for the two-fluid crossflow exchanger. The following four equations — Equations (4.3.63), (4.3.65), and (4.3.67 — are used to solve iteratively the surface areas on each fluid side: UA in Equation (4.3.63) is determined from NTU computed from the known heat duty or ε and C^*; G in Equation (4.3.65) represents two equations, for Fluids 1 and 2 (Shah, 1988a); and the volume of the exchanger in Equation (4.3.67) is the same based on the surface area density of Fluid 1 or Fluid 2.

$$\frac{1}{UA} \approx \frac{1}{\left(\eta_o hA\right)_h} + \frac{1}{\left(\eta_o hA\right)_c} \tag{4.3.63}$$

Here we have neglected the wall and fouling thermal resistances. This equation in nondimensional form is given by

$$\frac{1}{\text{NTU}} = \frac{1}{\text{ntu}_h\left(C_h/C_{\min}\right)} + \frac{1}{\text{ntu}_c\left(C_c/C_{\min}\right)} \tag{4.3.64}$$

$$G_i = \left[\frac{2g_c\Delta p}{\text{Deno}}\right]_i^{1/2} \qquad i = 1,2 \tag{4.3.65}$$

where

$$\text{Deno}_i = \left[\frac{f}{j}\frac{\text{ntu}}{\eta_o}\text{Pr}^{2/3}\left(\frac{1}{\rho}\right)_m + 2\left(\frac{1}{\rho_o} - \frac{1}{\rho_i}\right) + \left(1 - \sigma^2 + K_c\right)\frac{1}{\rho_i} - \left(1 - \sigma^2 - K_e\right)\frac{1}{\rho_o}\right]_i \tag{4.3.66}$$

$$V = \frac{A_1}{\alpha_1} = \frac{A_2}{\alpha_2} \tag{4.3.67}$$

In the iterative solutions, the first time one needs ntu_h and ntu_c to start the iterations. These can be either determined from the past experience or by estimations. If both fluids are gases or both fluids are liquid, one could consider that the design is "balanced," i.e., that the thermal resistances are distributed approximately equally on the hot and cold sides. In that case, $C_h = C_c$, and

$$\text{ntu}_h \approx \text{ntu}_c \approx 2\text{NTU} \tag{4.3.68}$$

Alternatively, if we have liquid on one side and gas on the other side, consider 10% thermal resistance on the liquid side, i.e.,

$$0.10\left(\frac{1}{UA}\right) = \frac{1}{\left(\eta_o hA\right)_{\text{liq}}} \tag{4.3.69}$$

Then, from Equations (4.3.63) and (4.3.64) with $C_{\text{gas}} = C_{\min}$, we can determine the ntu values on each side as follows:

$$\text{ntu}_{\text{gas}} = 1.11\text{NTU}, \qquad \text{ntu}_{\text{liq}} = 10C^*\text{NTU} \tag{4.3.70}$$

Also note that initial guesses of η_o and j/f are needed for the first iteration to solve Equation (4.3.66). For a good design, consider $\eta_o = 0.80$ and determine an approximate value of j/f from the plot of j/f vs. Re curve for the known j and f vs. Re characteristics of each fluid side surface. The specific step-by-step design procedure is as follows:

1. In order to compute the fluid bulk mean temperature and the fluid thermophysical properties on each fluid side, determine the fluid outlet temperatures from the specified heat duty

$$q = \left(\dot{m}c_p\right)_h\left(T_{h,i} - T_{h,o}\right) = \left(\dot{m}c_p\right)_c\left(T_{c,o} - T_{c,i}\right) \tag{4.3.71}$$

   or from the specified exchanger effectiveness using Equation (4.3.52) and (4.3.53). For the first time, estimate the values of c_p.

For exchangers with $C^* \geq 0.5$, the bulk mean temperature on each fluid side will be the arithmetic mean of inlet and outlet temperatures on each side. For exchangers with $C^* < 0.5$, the bulk mean temperature on the C_{max} side will be the arithmetic mean of the inlet and outlet temperatures on that side and the bulk mean temperature on the C_{min} side will be the log-mean average as given by Equation (4.3.54). With these bulk mean temperatures, determine c_p and iterate one more time for the outlet temperatures if warranted. Subsequently, determine μ, c_p, k, Pr, and ρ on each fluid side.

2. Calculate C^* and ε (if q is given), and determine NTU from the ε-NTU expression, tables, or graphical results for the selected flow arrangement (in this case, it is unmixed–unmixed crossflow, Table 4.3.4). The influence of longitudinal heat conduction, if any, is ignored in the first iteration since we don't know the exchanger size yet.
3. Determine ntu on each side by the approximations discussed with Equations (4.3.68) and (4.3.70) unless it can be estimated from past experience.
4. For the selected surfaces on each fluid side, plot *j/f* vs. Re curve from the given surface characteristics and obtain an approximate value of *j/f*. If fins are employed, assume $\eta_o = 0.80$ unless a better value can be estimated.
5. Evaluate G from Equation (4.3.65) on each fluid side using the information from Steps 1 to 4 and the input value of Δp.
6. Calculate Reynolds number Re, and determine j and f on each fluid side from the given design data for each surface.
7. Compute h, η_f, and η_o using Equations (4.3.57) to (4.3.59). For the first iteration, determine U_1 on Fluid 1 side from the following equation derived from Equations (4.3.6) and (4.3.67).

$$\frac{1}{U_1} = \frac{1}{(\eta_o h)_1} + \frac{1}{(\eta_o h_s)_1} + \frac{\alpha_1/\alpha_2}{(\eta_o h_s)_2} + \frac{\alpha_1/\alpha_2}{(\eta_o h)_2} \tag{4.3.72}$$

where $\alpha_1/\alpha_2 = A_1/A_2$, $\alpha = A/V$, V is the exchanger total volume, and subscripts 1 and 2 denote Fluid 1 and 2 sides. For a plate-fin exchanger, α terms are given by Shah (1981) and Kays and London (1984):

$$\alpha_1 = \frac{b_1\beta_1}{b_1 + b_2 + 2a} \qquad \alpha_2 = \frac{b_2\beta_2}{b_1 + b_2 + 2a} \tag{4.3.73}$$

Note that the wall thermal resistance in Equation (4.3.72) is ignored in the first iteration. In second and subsequent iterations, compute U_1 from

$$\frac{1}{U_1} = \frac{1}{(\eta_o h)_1} + \frac{1}{(\eta_o h_s)_1} + \frac{\delta A_1}{k_w A_w} + \frac{A_1/A_2}{(\eta_o h_s)_2} + \frac{A_1/A_2}{(\eta_o h)_2} \tag{4.3.74}$$

where the necessary geometry information A_1/A_2 and A_1/A_w is determined from the geometry calculated in the previous iteration.

8. Now calculate the core dimensions. In the first iteration, use NTU computed in Step 2. For subsequent iterations, calculate longitudinal conduction parameter λ (and other dimensionless groups for a crossflow exchanger). With known ε, C^*, and λ, determine the correct value of NTU using either a closed-form equation or tabular/graphical results (Kays and London, 1984). Determine A_1 from NTU using U_1 from the previous step and known C_{min}.

$$A_1 = \text{NTU}\, C_{min}/U_1 \tag{4.3.75}$$

and hence

$$A_2 = (A_2/A_1)A_1 = (\alpha_2/\alpha_1)A_1 \tag{4.3.76}$$

A_o from known $\dot{m}$ and G is given by

$$A_{o,1} = (\dot{m}/G)_1 \qquad A_{o,2} = (\dot{m}/G)_2 \tag{4.3.77}$$

so that

$$A_{fr,1} = A_{o,1}/\sigma_1 \qquad A_{fr,2} = A_{o,2}/\sigma_2 \tag{4.3.78}$$

where σ_1 and σ_2 are generally specified for the surface or can be computed for plate-fin surfaces from Shah (1981) and Kays and London (1984):

$$\sigma_1 = \frac{b_1\beta_1 D_{h,1}/4}{b_1 + b_2 + 2\delta} \qquad \sigma_2 = \frac{b_2\beta_2 D_{h,2}/4}{b_1 + b_2 + 2\delta} \tag{4.3.79}$$

Now compute the fluid flow lengths on each side (see Figure 4.3.17) from the definition of the hydraulic diameter of the surface employed on each side.

$$L_1 = \left(\frac{D_h A}{4A_o}\right)_1 \qquad L_2 = \left(\frac{D_h A}{4A_o}\right)_2 \tag{4.3.80}$$

Since $A_{fr,1} = L_2L_3$ and $A_{fr,2} = L_1L_3$, we can obtain

$$L_3 = \frac{A_{fr,1}}{L_2} \quad \text{or} \quad L_3 = \frac{A_{fr,2}}{L_1} \tag{4.3.81}$$

Theoretically, L_3 calculated from both expressions of Equation (4.3.81) should be identical. In reality, they may differ slightly because of the round-off error. In that case, consider an average value for L_3.

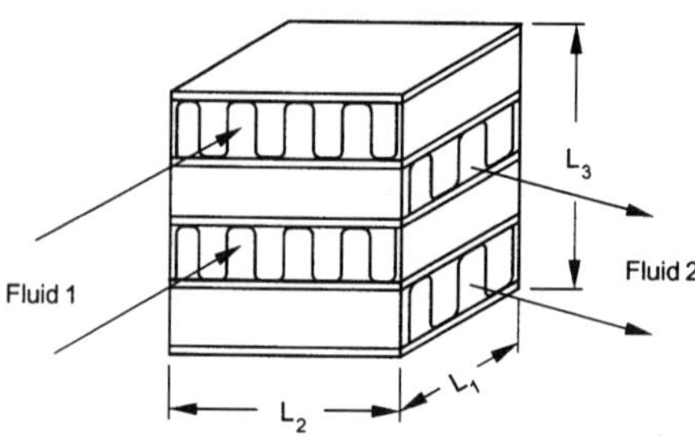

FIGURE 4.3.17 A single-pass crossflow exchanger.

9. Finally, compute the pressure drop on each fluid side, after correcting *f* factors for variable property effects, in a manner similar to Step 8 of the rating problem for a Crossflow Plate Fin Exchanger.
10. If the calculated values of Δp are within and close to input specifications, the solution to the sizing problem is completed. Finer refinements in the core dimensions, such as integer numbers

of flow passages, etc., may be carried out at this time. Otherwise, compute the new value of G on each fluid side using Equation (4.3.29) in which Δp is the input-specified value, and f, K_c, K_e, and geometric dimensions are from the previous iteration.

11. Repeat (iterate) Steps 6 to 10 until both transfer and pressure drops are met as specified. It should be emphasized that since we have imposed no constraints on the exchanger dimensions, the above procedure will yield L_1, L_2, and L_3 for the selected surfaces such that the design will meet exactly the heat duty and pressure drops on both fluid sides.

Flow Maldistribution

In the previously presented heat transfer (ε-NTU, MTD, etc. methods) and pressure drop analyses, it is presumed that the fluid is uniformly distributed through the core. In practice, flow maldistribution does occur to some extent and often severely, and may result in a significant reduction in exchanger heat transfer performance and an increase in the pressure drop. Hence, it may be necessary for the designer to take into account the effect of flow maldistribution causing undesirable performance deterioration up front while designing a heat exchanger.

Some maldistributions are geometry-induced (i.e., the result of exchanger fabrication conditions, such as header design or manufacturing tolerances, or the duct geometry/structure upstream of the exchanger), and other maldistributions are the result of exchanger operating conditions. Gross, passage-to-passage and manifold-induced flow maldistributions are examples of the former category, while viscosity, natural convection, and density-difference-induced flow maldistributions are of the latter category. Flow maldistributions associated with two-phase and multiphase flow are too complex, with only limited information available in the literature. The analysis methods and results for some of the above flow maldistributions for single-phase flows are given by Shah (1985), Mueller and Chiou (1987), and Putnam and Rohsenow (1985).

Fouling in Heat Exchangers

Fouling, Its Effect, and Mechanisms.

Fouling refers to undesired accumulation of solid material (by-products of the heat transfer processes) on heat exchanger surfaces which results in additional thermal resistance to heat transfer, thus reducing exchanger performance. The fouling layer also blocks the flow passage/area and increases surface roughness, thus either reducing the flow rate in the exchanger or increasing the pressure drop or both. The foulant deposits may be loose such as magnetite particles or hard and tenacious such as calcium carbonate scale; other deposits may be sediment, polymers, coking or corrosion products, inorganic salts, biological growth, etc. Depending upon the fluids, operating conditions, and heat exchanger construction, the maximum fouling layer thickness on the heat transfer surface may result in a few hours to a number of years.

Fouling could be very costly depending upon the nature of fouling and the applications. It increases capital costs: (1) oversurfacing heat exchanger, (2) provisions for cleaning, and (3) use of special materials and constructions/surface features. It increases maintenance costs: (1) cleaning techniques, (2) chemical additives, and (3) troubleshooting. It may cause a loss of production: (1) reduced capacity and (2) shutdown. It increases energy losses: (1) reduced heat transfer, (2) increased pressure drop, and (3) dumping dirty streams. Fouling promotes corrosion, severe plugging, and eventual failure of uncleaned heat exchangers. In a fossil-fired exhaust environment, gas-side fouling produces a potential fire hazard in heat exchangers.

The following are the major fouling mechanisms:

- *Crystallization or precipitation fouling* results from the deposition/formation of crystals of dissolved substances from the liquid onto heat transfer surface due to solubility changes with temperature beyond the saturation point. If the deposited layer is hard and tenacious, it is often referred to as scaling. If it is porous and mushy, it is called sludge.

- *Particulate fouling* results from the accumulation of finely divided substances suspended in the fluid stream onto heat transfer surface. If the settling occurs as a result of gravity, it is referred to as sedimentation fouling.
- *Chemical reaction fouling* is defined as the deposition of material produced by chemical reaction (between reactants contained in the fluid stream) in which the heat transfer surface material does not participate.
- *Corrosion fouling* results from corrosion of the heat transfer surface that produces products fouling the surface and/or roughens the surface, promoting attachment of other foulants.
- *Biological fouling* results from the deposition, attachment, and growth of biological organisms from liquid onto a heat transfer surface. Fouling due to microorganisms refers to microbial fouling and fouling due to macroorganisms refers to macrobial fouling.
- *Freezing fouling* results from the freezing of a single-component liquid or higher-melting-point constituents of a multicomponent liquid onto a subcooled heat transfer surface.

Biological fouling occurs only with liquids since there are no nutrients in gases. Also crystallization fouling is not too common with gases since most gases contain few dissolved salts (mainly in mists) and even fewer inverse-solubility salts. All other types of fouling occur in both liquid and gas. More than one mechanism is usually present in many fouling situations, often with synergetic results. Liquid-side fouling generally occurs on the exchanger side where the liquid is being heated, and gas-side fouling occurs where the gas is being cooled; however, reverse examples can be found.

Importance of Fouling.
Fouling in liquids and two-phase flows has a significant detrimental effect on heat transfer with some increase in pressure drop. In contrast, fouling in gases reduces heat transfer somewhat (5 to 10% in general) in compact heat exchangers, but increases pressure drop significantly up to several hundred percent. For example, consider U = 1400 W/m²K as in a process plant liquid-to-liquid heat exchanger. Hence, $R = 1/U$ = 0.00072 m²K/W. If the fouling factors $(r_{s,h} + r_{s,c})$ together amount to 0.00036 (considering a typical TEMA value of the fouling factor as 0.00018), 50% of the heat transfer area requirement A for given q is chargeable to fouling. However, for gas flows on both sides of an exchanger, $U \approx$ 280 W/m²K, and the same fouling factor of 0.00036 would represent only about 10% of the total surface area. Thus, one can see a significant impact on the heat transfer surface area requirement due to fouling in heat exchangers having high U values (such as having liquids or phase-change flows).

Considering the core frictional pressure drop (Equation (4.3.32)) as the main pressure drop component, the ratio of pressure drops of fouled and cleaned exchangers is given by

$$\frac{\Delta p_F}{\Delta p_C} = \frac{f_F}{f_C}\left(\frac{D_{h,C}}{D_{h,F}}\right)\left(\frac{u_{m,F}}{u_{m,C}}\right)^2 = \frac{f_F}{f_C}\left(\frac{D_{h,C}}{D_{h,F}}\right)^5 \tag{4.3.82}$$

where the term after the second equality sign is for a circular tube and the mass flow rates under fouled and clean conditions remain the same. Generally, $f_F > f_C$ due to the fouled surface being rough. Thus, although the effect of fouling on the pressure drop is usually neglected, it can be significant, particularly for compact heat exchangers with gas flows. If we consider $f_F = f_C$, and the reduction in the tube inside diameter due to fouling by only 10 and 20%, the resultant pressure drop increase will be 69 and 205%, respectively, according to Equation (4.3.82) regardless of whether the fluid is liquid or gas!

Accounting of Fouling in Heat Exchangers.
Fouling is an extremely complex phenomenon characterized by a combined heat, mass, and momentum transfer under transient condition. Fouling is affected by a large number of variables related to heat exchanger surfaces, operating conditions, and fluids. Fouling is time dependent, zero at $\tau = 0$; after the induction or delay period τ_d, the fouling resistance is either pseudolinear, falling rate, or asymptotic.

Fouling is characterized by all or some of the following sequential events: initiation, transport, attachment, removal, and aging (Epstein, 1983). Research efforts are concentrated on quantifying these events by semitheoretical models (Epstein, 1978) with very limited success on specific fouling situations. Hence, the current heat exchanger design approach is to use a constant (supposedly an asymptotic) value of the fouling factor $r_s = 1/h_s$. Equation (4.3.6) presented earlier includes the fouling resistances on the hot and cold sides for a nontubular extended-surface exchanger. Here $1/h_s = r_s$ is generally referred to as the *fouling factor*. Fouling factors for some common fluids are presented in Tables 4.3.8 and 4.3.9.

The specification of fouling effects in a process heat exchanger is usually represented in the following form, wherein the combined fouling factor $r_{s,t}$ is the sum of the fouling factors on the hot and cold sides:

Combined fouling factor $$r_{s,t} = \frac{1}{U_C} - \frac{1}{U_F} \tag{4.3.83}$$

Cleanliness factor $$\mathrm{CF} = U_F/U_C \tag{4.3.84}$$

Percentage oversurface $$\%\mathrm{OS} = \left(\frac{A_F}{A_C} - 1\right)100 \tag{4.3.85}$$

Here the subscripts F and C denote fouled and clean exchanger values. From Equation (4.3.6) with $A_h = A_c = A$, $\eta_o = 1$, $\Delta T_{m,F} = \Delta T_{m,C}$, it can be shown that

$$\frac{A_F}{A_C} = \frac{U_C}{U_F} = 1 + U_C r_{s,t} \tag{4.3.86}$$

where $r_{s,t} = r_{s,h} + r_{s,c}$. In heat exchanger design, constant (supposedly an asymptotoic) values of $r_{s,h}$ and $r_{s,c}$ are used. Accordingly, extra heat transfer surface area is provided to take into account the deleterious effect of fouling. Thus, the heat exchanger will be "oversized" for the initial clean condition, "correctly sized" for asymptotic fouling (if it occurs in practice), and "undersized" just before the cleaning operation for nonasymptotic fouling.

Influence of Operating and Design Variables.

Based on operational experience and research over the last several decades, many variables have been identified that have a significant influence on fouling. The most important variables are summarized next.

Flow velocity. Flow velocity is one of the most important variables affecting fouling. Higher velocities increase fluid shear stress at the fouling deposit–fluid interface and increase the heat transfer coefficient; but, at the same time, increased pressure drop and fluid pumping power may erode the surface and may accelerate the corrosion of the surface by removing the protective oxide layer. The fouling buildup in general is inversely proportional to $u_m^{1.5}$. For water, the velocity should be kept above 2 m/sec to suppress fouling, and the absolute minimum should be above 1 m/sec to minimize fouling.

Surface temperature. Higher surface temperatures promote chemical reaction, corrosion, crystal formation (with inverse solubility salts), and polymerization, but reduce biofouling for temperatures above the optimum growth, avoid potential freezing fouling, and avoid precipitation of normal-solubility salts. It is highly recommended that the surface temperature be maintained below the reaction temperature; it should be kept below 60°C for cooling tower water.

Tube material. The selection of the tube material is important from the corrosion point of view which in turn could increase crystallization and biological fouling. Copper alloys can reduce certain biofouling, but their use is limited by environmental concerns with river, ocean, and lake waters.

There are many other variables that affect fouling. It is beyond the scope here, but the reader may refer to TEMA (1988).

TABLE 4.3.8 Fouling Factors for Various Fluid Streams Used in Heat Exchangers

Water Type	Fouling Factors (m² · K)/W
Seawater (43°C maximum outlet)	0.000275 to 0.00035
Brackish water (43°C maximum outlet)	0.00035 to 0.00053
Treated cooling tower water (49°C maximum outlet)	0.000175 to 0.00035
Artificial spray pond (49°C maximum outlet)	0.000175 to 0.00035
Closed-loop treated water	0.000175
River water	0.00035 to 0.00053
Engine jacket water	0.000175
Distilled water or closed-cycle condensate	0.00009 to 0.000175
Treated boiler feedwater	0.00009
Boiler blowdown water	0.00035 to 0.00053
Liquids	
No. 2 fuel oil	0.00035
No. 6 fuel oil	0.0009
Transformer oil	0.000175
Engine lube oil	0.000175
Refrigerants	0.000175
Hydraulic fluid	0.000175
Industrial organic HT fluids	0.000175 to 0.00035
Ammonia	0.000175
Ammonia (oil bearing)	0.00053
Methanol solutions	0.00035
Ethanol solutions	0.00035
Ethylene glycol solutions	0.00035
MEA and DEA solutions	0.00035
DEG and TEG solutions	0.00035
Stable side draw and bottom products	0.000175 to 0.00035
Caustic solutions	0.00035
Gas or Vapor	
Steam (non-oil-bearint)	0.0009
Exhaust steam (oil-bearing)	0.00026 to 0.00035
Refrigerant (oil-bearing)	0.00035
Compressed air	0.000175
Ammonia	0.000175
Carbon dioxide	0.00035
Coal flue gas	0.00175
Natural gas flue gas	0.00090
Acid gas	0.00035 to 0.00053
Solvent vapor	0.000175
Stable overhead products	0.000175
Natural Gas and Petroleum Streams	
Natural gas	0.000175 to 0.00035
Overhead products	0.000175 to 0.00035
Lean oil	0.00035
Rich oil	0.000175 to 0.00035
Natural gasoline and liquefied petroleum gases	0.000175 to 0.00035
Oil Refinery Streams	
Crude and vacuum unit gases and vapors	
Atmospheric tower overhead vapors	0.00017
Light naphthas	0.00017
Vacuum overhead vapors	0.00035

TABLE 4.3.8 (continued) Fouling Factors for Various Fluid Streams Used in Heat Exchangers

Oil Refinery Streams	
Crude and vacuum liquids	
Gasoline	0.00035
Naphtha and light distillates	0.00035 to 0.00053
Kerosene	0.00035 to 0.00053
Light gas oil	0.00035 to 0.00053
Heavy gas oil	0.00053 to 0.0009
Heavy fuel oil	0.00053 to 0.00123
Vacuum tower bottoms	0.00176
Atmospheric tower bottoms	0.00123
Cracking and coking unit streams	
Overhead vapors	0.00035
Light cycle oil	0.00035 to 0.00053
Heavy cycle oil	0.00053 to 0.0007
Light coker gas oil	0.00053 to 0.0007
Heavy coker gas oil	0.00070 to 0.0009
Bottoms slurry oil (1.5 m/sec minimum)	0.00053
Light liquid products	0.00035
Catalytic reforming, hydrocracking, and hydrodesulfurization streams	
Reformer charge	0.00026
Reformer effluent	0.00026
Hydrocharger charge and effluent[a]	0.00035
Recycle gas	0.000175
Liquid product over 50°C (API)[b]	0.000175
Liquid product 30 to 50°C (API)[b]	0.00035
Light ends processing streams	
Overhead vapors and gases	0.000175
Liquid products	0.000175
Absorption oils	0.00035 to 0.00053
Alkylation trace acid streams	0.00035
Reboiler streams	0.00035 to 0.00053

[a] Depending on charge characteristics and storage history, charge fouling resistance may be many times this value.

[b] American Petroleum Institute.

Source: Chenoweth, J., Final Report, HTRI/TEMA Joint Committee to Review the Fouling Section of TEMA Standards, HTRI, Alhambra, CA, 1988. With permission.

Fouling Control and Cleaning Techniques.

Control of fouling should be attempted first before any cleaning method is attempted. For gas-side fouling, one should verify that fouling exists, identify the sequential event that dominates the foulant accumulation, and characterize the deposit. For liquid-side fouling, fouling inhibitors/additives should be employed while the exchanger is in operation; for example, use antidispersant polymers to prevent sedimentation fouling, "stabilizing" compounds to prevent polymerization and chemical reaction fouling, corrosion inhibitors to prevent corrosion fouling, biocide/germicide to prevent biofouling, softeners, acids, and polyphosphates to prevent crystallization fouling.

If the foulant control is not effective, the exchanger must be cleaned either on-line or off-line. On-line cleaning includes flow-driven brushes/sponge balls inside tubes, power-driven rotating brushes inside tubes, acoustic horns/mechanical vibrations for tube banks with gases, soot blowers, and shutting off of the cold gas supply, flowing hot gas, or reversing of the fluids. Off-line cleaning methods, without dismantling the exchanger include chemical cleaning (circulate acid/detergent solutions), circulating of particulate slurry (such as sand and water), and thermal melting of frost layers. Off-line cleaning with a heat exchanger opened includes high-pressure steam or water cleaning, and thermal baking of an exchanger and then rinsing for small heat exchanger modules removed from the container of the modular exchangers.

TABLE 4.3.9 Fouling Factors and Design Parameters for Finned Tubes in Fossil Fuel Exhaust Gases

Type of Flue Gas	Fouling Factor, $m^2 K/W$	Minimum Spacing between Fins, m	Maximum Gas Velocity to Avoid Erosion, m/sec
Clean Gas (Cleaning Devices Not Required)			
Natural Gas	0.0000881–0.000528	0.00127–0.003	30.5–36.6
Propane	0.000176–0.000528	0.00178	—
Butane	0.000176–0.000528	0.00178	—
Gas turbine	0.000176	—	—
Average Gas (Provisions for Future Installation of Cleaning Devices)			
No. 2 oil	0.000352–0.000704	0.00305–0.00384	25.9–30.5
Gas turbine	0.000264	—	—
Diesel engine	0.000528	—	—
Dirty Gas (Cleaning Devices Required)			
No. 6 oil	0.000528–0.00123	0.00457–0.00579	18.3–24.4
Crude oil	0.000704–0.00264	0.00508	—
Residual oil	0.000881–0.00352	0.00508	—
Coal	0.000881–0.00881	0.00587–0.00864	15.2–21.3

Source: Weierman, R.C., 1982. Design of Heat Transfer Equipment for Gas-Side Fouling Service, Workshop on an Assessment of Gas-Side Fouling in Fossil Fuel Exhaust Environments, W.J. Marner and R.L. Webb, Eds., JPL Publ. 82-67, Jet Propulsion Laboratory, California Institute of Technology, Pasadena. With permission.

Nomenclature

- A — total heat transfer area (primary + fin) on one fluid side of a heat exchanger, A_p: primary surface area, A_f: fin surface area, m^2
- A_{fr} — frontal area on one side of an exchanger, m^2
- A_k — total wall cross-sectional area for heat conduction in fin or for longitudinal conduction in the exchanger, m^2
- A_o — minimum free-flow area on one fluid side of a heat exchanger, m^2
- b — plate spacing, $h' + \delta_f$, m
- C — flow stream heat capacity rate with a subscript c or h, $\dot{m}c_p$, W/°C
- C^* — heat capacity rate ratio, C_{min}/C_{max}, dimensionless
- c_p — specific heat of fluid at constant pressure, J/kg K
- D_h — hydraulic diameter of flow passages, $4A_oL/A$, m
- d_e — fin tip diameter of an individually finned tube, m
- d_i, d_o — tube inside and outside diameters, respectively, m
- Eu — N-row average Euler number, $\Delta p/(\rho\, u_m^2\, N/2g_c)$, $\rho \Delta p g_c/(NG^2/2)$, dimensionless
- F — log-mean temperature difference correction factor, dimensionless
- f — Fanning friction factor, $\rho \Delta p g_c D_h/(2LG^2)$, dimensionless
- f_{tb} — Fanning friction factor per tube row for crossflow over a tube bank outside, $\rho \Delta p g_c/(2NG^2)$
- G — mass velocity based on the minimum free flow area, $\dot{m}/A_o$, kg/m^2sec
- g — gravitational acceleration, m^2/sec
- g_c — proportionality constant in Newton's second law of motion, $g_c = 1$ and dimensionless in SI units
- H — fin length for heat conduction from primary surface to either fin tip or midpoint between plates for symmetric heating, m
- h — heat transfer coefficient, W/m^2K

h' height of the offset strip fin (see Figure 4.3.15), m
j Colburn factor, $\mathrm{NuPr}^{-1/3}/\mathrm{Re}$, $\mathrm{StPr}^{2/3}$, dimensionless
k fluid thermal conductivity, W/m K
k_f thermal conductivity of the fin material, W/m K
k_w thermal conductivity of the matrix (wall) material, W/m K
L fluid flow (core or tube) length on one side of an exchanger, m
l fin length for heat conduction from primary surface to the midpoint between plates for symmetric heating, see Table 4.3.5 for other definitions of l, m
l_f offset trip fin length or fin height for individually finned tubes, l_f represents the fin length in the fluid flow direction for an uninterrupted fin with $l_f = L$ in most cases, m
m fin parameter, 1/m
N number of tube rows
N_f number of fins per meter, 1/m
N_t total number of tubes in an exchanger
NTU number of heat transfer units, UA/C_{min}, it represents the total number of transfer units in a multipass unit, $\mathrm{NTU}_s = UA/C_{\mathrm{shell}}$, dimensionless
Nu Nusselt number, hD_h/k, dimensionless
ntu_c number of heat transfer units based on the cold side, $(\eta_o hA)_c/C_c$, dimensionless
ntu_h number of heat transfer units based on the hot side, $(\eta_o hA)_h/C_h$, dimensionless
$\dot{m}$ mass flow rate, kg/sec
P temperature effectiveness of one fluid, dimensionless
$\wp$ fluid pumping power, W
Pr fluid Prandtl number, $\mu c_p/k$, dimensionless
p fluid static pressure, Pa
Δp fluid static pressure drop on one side of heat exchanger core, Pa
p_f fin pitch, m
q heat duty, W
q_e heat transfer rate (leakage) at the fin tip, W
q'' heat flux, $\mathrm{W/m^2}$
R heat capacity rate ratio used in the P-NTU method, $R_1 = C_1/C$, $R_2 = C_2/C_1$, dimensionless
R thermal resistance based on the surface area A, compare Equations (4.3.5) and (4.3.6) for definitions of specific thermal resistances, K/W
Re Reynolds number, GD_h/μ, dimensionless
Re_d Reynolds number, $\rho u_m d_o/\mu$, dimensionless
r_h hydraulic radius, $D_h/4$, $A_o L/A$, m
r_s fouling factor, $1/h_s$, $\mathrm{m^2K/W}$
St Stanton number, h/Gc_p, dimensionless
s distance between adjacent fins, $p_f - \delta_f$ m
T fluid static temperature to a specified arbitrary datum, °C
T_a ambient temperature, °C
T_o fin base temperature, °C
T_l fin tip temperature, °C
U overall heat transfer coefficient, $\mathrm{W/m^2\ K}$
u_m mean axial velocity in the minimum free flow area, m/sec
V heat exchanger total volume, $\mathrm{m^3}$
X_d diagonal tube pitch, m
X_l longitudinal tube pitch, m
X_t transverse tube pitch, m
α ratio of total heat transfer area on one side of an exchanger to the total volume of an exchanger, A/V, $\mathrm{m^2/m^3}$
β heat transfer surface area density, a ratio of total transfer area on one side of a plate-fin heat exchanger to the volume between the plates on that side, $\mathrm{m^2/m^3}$

ε heat exchanger effectiveness, it represents an overall exchanger effectiveness for a multiplass unit, dimensionless

δ wall thickness, m

δ_f fin thickness, m

η_f fin efficiency, dimensionless

η_o extended surface efficiency, dimensionless

λ longitudinal wall heat conduction parameter based on the total conduction area, $\lambda = k_w A_{k,t}/C_{\min}L$, $\lambda_c = k_w A_{k,c}/C_c L$, $\lambda_h = k_w A_{k,h}/C_h L$, dimensionless

μ fluid dynamic viscosity, Pa·s

ρ fluid density, kg/m^3

σ ratio of free flow area to frontal area, A_o/A_{fr}, dimensionless

Subscripts

C clean surface value

c cold fluid side

F fouled surface value

f fin

h hot fluid side

i inlet to the exchanger

o outlet to the exchanger

s scale or fouling

w wall or properties at the wall temperature

1 one section (inlet or outlet) of the exchanger

2 other section (outlet or inlet) of the exchanger

References

Bhatti, M.S. and Shah, R.K. 1987. Turbulent and transition flow convective heat transfer in ducts, in *Handbook of Single-Phase Convective Heat Transfer,* S. Kakaç, R. K. Shah, and W. Aung, Eds., John Wiley & Sons, New York, chap. 4, 166 pp.

Chai, H.C. 1988. A simple pressure drop correlation equation for low finned tube crossflow heat exchangers, *Int. Commun. Heat Mass Transfer*, 15, 95–101.

Chenoweth, J. 1988. Final Report, HTRI/TEMA Joint Committee to Review the Fouling Section of TEMA Standards, HTRI, Alhambra, CA.

Cowell, T.A., Heikal, M.R., and Achaichia, A. 1995. Flow and heat transfer in compact louvered fin surfaces, *Exp. Thermal Fluid Sci.*,10, 192–199.

Epstein, N. 1978. Fouling in heat exchangers, in *Heat Transfer 1978*, Vol. 6, Hemisphere Publishing, New York, 235–254.

Epstein, N. 1983. Thinking about heat transfer fouling: a 5×5 matrix, *Heat Transfer Eng.*, 4(1), 43–56.

Foumeny, E.A. and Heggs, P.J. 1991. *Heat Exchange Engineering*, Vol. 2, *Compact Heat Exchangers: Techniques for Size Reduction*, Ellis Horwood Ltd., London.

Ganguli, A. and Yilmaz, S.B. 1987. New heat transfer and pressure drop correlations for crossflow over low-finned tube banks, *AIChE Symp. Ser. 257*, 83, 9–14.

Ghajar, A.J. and Tam, L.M. 1994. Heat transfer measurements and correlations in the transition region for a circular tube with three different inlet configurations, *Exp. Thermal Fluid Sci.*, 8, 79–90.

Huang, L.J. and Shah, R.K. 1992. Assessment of calculation methods for efficiency of straight fins of rectangular profiles, *Int. J. Heat Fluid Flow*, 13, 282–293.

Idelchik, I.E. 1994. *Handbook of Hydraulics Resistance*, 3rd ed., CRC Press, Boca Raton, FL.

Kakaç, S., Ed. 1991. *Boilers, Evaporators, and Condensers*, John Wiley & Sons, New York.

Kakaç, S., Bergles, A.E., and Mayinger, F. 1981. *Heat Exchangers: Thermal-Hydraulic Fundamentals and Design*, Hemisphere Publishing, Washington, D.C.

Kakaç, S., Shah, R.K., and Bergles, A.E. 1983. *Low Reynolds Number Flow Heat Exchangers*, Hemisphere Publishing, Washington, D.C.

Kakaç, S., Bergles, A.E., and Fernandes, E.O. 1988. *Two-Phase Flow Heat Exchangers: Thermal Hydraulic Fundamentals and Design*, Kluwer Academic Publishers, Dordrecht, Netherlands.

Kays, W.M. and London, A.L. 1984. *Compact Heat Exchangers*, 3rd ed., McGraw-Hill, New York.

Manglik, R.M. and Bergles, A.E. 1995. Heat transfer and pressure drop correlations for the rectangular offset-strip-fin compact heat exchanger, *Exp. Thermal Fluid Sci.*, 10, 171–180.

Miller, D.S. 1990. *Internal Flow Systems, 2nd ed., BHRA (Information Services),* Cranfield, Bedford, U.K.

Mueller, A.C. and Chiou, J.P. 1987. *Review of Various Types of Flow Maldistribution in Heat Exchangers*, Book No. H00394, HTD-Vol. 75, ASME, New York, 3–16.

Putnam, G.R. and Rohsenow, W.M. 1985. Viscosity induced nonuniform flow in laminar flow heat exchangers, *Int. J. Heat Mass Transfer*, 28, 1031–1038.

Rabas, T.J. and Taborek, J. 1987. Survey of turbulent forced-convection heat transfer and pressure drop characteristics of low-finned tube banks in cross flow, *Heat Transfer Eng.*, 8(2), 49–62.

Roetzel, W., Heggs, P.J., and Butterworth, D., Eds. 1991. *Design and Operation of Heat Exchangers*, Springer-Verlag, Berlin.

Rozenman, T. 1976. Heat transfer and pressure drop characteristics of dry cooling tower extended surfaces, Part I: Heat transfer and pressure drop data, Report BNWL-PFR 7-100; Part II: Data analysis and correlation, Report BNWL-PFR 7-102, Battelle Pacific Northwest Laboratories, Richland, WA.

Shah, R.K. 1981. Compact heat exchangers, in *Heat Exchangers: Thermal-Hydraulic Fundamentals and Design*, S. Kakaç, A.E. Bergles, and F. Mayinger, Eds., Hemisphere Publishing, Washington, D.C., 111–151.

Shah, R.K. 1983. Heat Exchanger Basic Design Methods, in *Low Reynolds Number Flow Heat Exchanger*, S. Kakaç, R.K. Shah and A.E. Bergles, Eds., pp. 21–72, Hemisphere, Washington, D.C.

Shah, R.K. 1985. Compact heat exchangers, in *Handbook of Heat Transfer Applications*, 2nd ed., W.M. Rohsenow, J.P. Hartnett, and E.N. Ganic, Eds., McGraw-Hill, New York, Chap. 4, Part 3.

Shah, R.K. 1988a. Plate-fin and tube-fin heat exchanger design procedures, in *Heat Transfer Equipment Design*, R.K. Shah, E.C. Subbarao, and R.A. Mashelkar, Eds., Hemisphere Publishing, Washington, D.C., 255–266.

Shah, R.K. 1988b. Counterflow rotary regenerator thermal design procedures, in *Heat Transfer Equipment Design*, R.K. Shah, E.C. Subbarao, and R.A. Mashelkar, Eds., Hemisphere Publishing, Washington, D.C., 267–296.

Shah, R.K. 1991. Multidisciplinary approach to heat exchanger design, in *Industrial Heat Exchangers*, J.-M. Buchlin, Ed., Lecture Series No. 1991-04, von Kármán Institute for Fluid Dynamics, Rhode Saint Genèse, Belgium.

Shah, R.K. 1993. Nonuniform heat transfer coefficients for heat exchanger thermal design, in *Aerospace Heat Exchanger Technology 1993*, R.K. Shah and A. Hashemi, Eds., Elsevier Science, Amsterdam, Netherlands, 417–445.

Shah, R.K. 1994. Heat exchangers, in *Encyclopedia of Energy Technology and The Environment*, A. Bision and S.G. Boots, Eds., John Wiley & Sons, New York, 1651–1670.

Shah, R.K., Bell, K.J., Mochizuki, S., and Wadekar, V. V., Eds., 1997. *Compact Heat Exchangers for the Process Industries*, Begell House, New York.

Shah, R.K. and Bhatti, M.S. 1987. Laminar convective heat transfer in ducts, in *Handbook of Single-Phase Convective Heat Transfer*, S. Kacaç, R.K. Shah, and W. Aung, Eds., John Wiley & Sons, New York, Chap. 3, 137 pp.

Shah, R.K. and Bhatti, M.S. 1988. Assessment of correlations for single-phase heat exchangers, in *Two-Phase Flow Heat Exchangers: Thermal-Hydraulic Fundamentals and Design*, S. Kakaç, A.E. Bergles, and E.O. Fernandes, Eds., Kluwer Academic Publishers, Dordrecht, The Netherlands, 81–122.

Shah, R.K. and Giovannelli, A.D. 1988. Heat pipe heat exchanger design theory, in *Heat Transfer Equipment Design*, R.K. Shah, E.C. Subbarao, and R.A. Mashelkar, Eds., Hemisphere Publishing, Washington, D.C., 609–653.

Shah, R.K. and Hashemi, A., Eds. 1993. *Aerospace Heat Exchanger Technology*, Elsevier Science, Amsterdam.

Shah, R.K., Kraus, A.D., and Metzger, D.E., Eds., 1990. *Compact Heat Exchangers — A Festschrift for Professor A.L. London*, Hemisphere, Washington, D.C.

Shah, R.K. and London, A.L. 1978. Laminar flow forced convection in ducts, Suppl. 1 to *Advances in Heat Transfer*, Academic Press, New York.

Shah, R.K. and Mueller, A.C. 1988. Heat Exchange, in *Ullmann's Encyclopedia of Industrial Chemistry*, Unit Operations II, vol. B3, chap. 2, 108 pages, VCH, Weinheim, Germany.

Shah, R.K. and Pignotti, A. 1997. The influence of a finite number of baffles on the shell-and-tube heat exchanger performance, *Heat Transfer Eng.*, 18.

Shah, R.K., Subbarao, E.C., and Mashelkar, R.A., Eds. 1988. *Heat Transfer Equipment Design*, Hemisphere Publishing, Washington, D.C.

Shah, R.K. and Wanniarachchi, A.S. 1991. Plate heat exchanger design theory, in *Industrial Heat Exchangers*, J.-M. Buchlin, Ed., Lecture Series No. 1991-04, von Kármán Institute for Fluid Dynamics, Rhode Saint Genèse, Belgium.

Taylor, M.A. 1987. *Plate-Fin Heat Exchangers: Guide to Their Specificaitons and Use*, 1st ed., HTFS, Harwell Laboratory, Oxon, U.K., rev. 1990.

TEMA, 1988. *Standards of the Tubular Exchanger Manufacturers Association*, 7th ed., Tubular Exchanger Manufacturers Association, New York.

Webb, R.L. 1994. *Principles of Enhanced Heat Transfer*, John Wiley & Sons, New York.

Weierman, R.C. 1982. Design of Heat Transfer Equipment for Gas-Side Fouling Service, Workshop on an Assessment of Gas-Side Fouling in Fossil Fuel Exhaust Environments, W.J. Marner and R.L. Webb, Eds., JPL Publ. 82-67, Jet Propulsion Laboratory, California Institute of Technology, Pasadena.

Zukauskas, A. 1987. Convective heat transfer in cross flow, in *Handbook of Single-Phase Convective Heat Transfer*, S. Kacaç, R.K. Shah, and W. Aung, John Wiley & Sons, New York, Chap. 6.

Further Information

Heat exchangers play a crucial and dominant role in many developments related to energy conservation, recovery, utilization, economic development of new energy sources, and environmental issues such as air and water pollution control, thermal pollution, waste disposal, etc. Many new and innovative heat exchangers have been developed for these and many other applications worldwide. A broad overview is provided for various heat exchangers and basic design theory for single-phase heat exchangers. For further details and study, the reader may refer to the following references: Kakaç et al. (1981; 1983; 1988), Taylor (1987), Shah et al. (1990), Foumeny and Heggs (1991), Kakaç (1991), Roetzel et al. (1991), Shah and Hashemi (1993), and Shah et al. (1997).

Shell-and-Tube Heat Exchangers

Kenneth J. Bell

Introduction

A shell-and-tube heat exchanger is essentially a bundle of tubes enclosed in a shell and so arranged that one fluid flows through the tubes and another fluid flows across the outside of the tubes, heat being transferred from one fluid to the other through the tube wall. A number of other mechanical components are required to guide the fluids into, through, and out of the exchanger, to prevent the fluids from mixing, and to ensure the mechanical integrity of the heat exchanger. A typical shell-and-tube heat exchanger is shown in Figure 4.3.18 (TEMA, 1988), but the basic design allows many modifications and special features, some of which are described below.

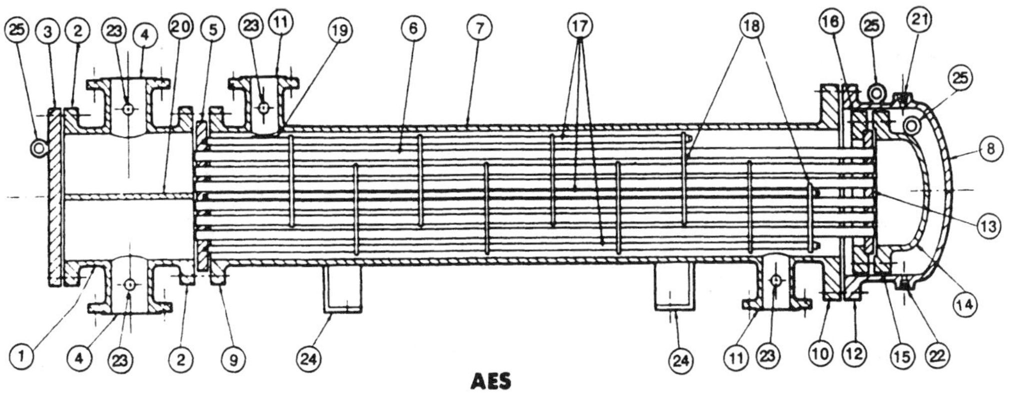

FIGURE 4.3.18 Longitudinal section of a typical shell-and-tube heat exchanger (TEMA AES) with nomenclature. (Modified from TEMA, *Standards* 7th ed., Tubular Exchanger Manufacturers Association, Tarrytown, NY, 1988.)

Nomenclature of Heat Exchanger Components

For the purpose of establishing standard terminology, Figure 4.3.18 illustrates a typical shell and tube of heat exchanger. Typical parts and connections, for illustrative purposes only, are numbered for identification:

1. Stationary Head-Channel
2. Stationary Head Flange-Channel or Bonnet
3. Channel Cover
4. Stationary Head Nozzle
5. Stationary Tubesheet
6. Tubes
7. Shell
8. Shell Cover
9. Shell Flange-Stationary Head End
10. Shell Flange-Rear Head End
11. Shell Nozzle
12. Shell Cover Flange
13. Floating Tubesheet
14. Floating Head Cover
15. Floating Head Cover Flange
16. Floating Head Backing Device
17. Tierods and Spacers
18. Transverse Baffles or Support Plates
19. Impingement Plates
20. Pass Partition
21. Vent Connection
22. Drain Connection
23. Instrument Connection
24. Support Saddle
25. Lifting Lug

Shell-and-tube heat exchangers have been constructed with heat transfer areas from less than 0.1 m^2 (1 ft^2) to over 100,000 m^2 (1,000,000 ft^2), for pressures from deep vacuum to over 1000 bar (15,000 psi), for temperatures from near 0 to over 1400 K (2000°F), and for all fluid services including single-phase heating and cooling and multiphase vaporization and condensation. The key to such flexibility is the wide range of materials of construction, forming and joining methods, and design features that can be built into these exchangers (see Hewitt, Vol. 4, 1998; Saunders, 1988; and Yokell, 1990). Most shell-and-tube heat exchangers are manufactured in conformity with TEMA *Standards* (1988) and the *ASME Boiler and Pressure Vessel Code* (latest edition), but other codes and standards may apply.

Construction Features

In the design process, it is important to consider the mechanical integrity under varying operational conditions and the maintainability (especially cleaning) of the exchanger as equally important with the thermal-hydraulic design.

Tubes. Tubes used in shell-and-tube exchangers range from 6.35 mm ($^1/_4$ in.) to 50.8 mm (2 in.) and above in outside diameter, with the wall thickness usually being specified by the Birmingham wire gauge (BWG). Tubes are generally available in any desired length up to 30 m (100 ft) or more for plain tubes. While plain tubes are widely used, a variety of internally and/or externally enhanced tubes is available to provide special heat transfer characteristics when economically justified (see subsection on enhancement in Section 4.15). Low fin tubes having circumferential fins typically 0.8 to 1.6 mm (0.032 to 0.062 in.) high, spaced 630 to 1260 fins/m (16 to 32 fins/in.) are often employed, especially when the shell-side heat transfer coefficient is substantially smaller than the tube-side coefficient. The outside heat transfer area of a low fin tube is three to six times the inside area, resulting in a smaller heat exchanger shell for the same service, which may offset the higher cost of the tube per unit length.

The tubes are inserted into slightly oversized holes drilled (or, occasionally, punched) through the tubesheets (items 5 and 13, Figure 4.3.18). The tubes are secured by several means, depending upon the mechanical severity of the application and the need to avoid leakage between the streams. In some low-severity applications, the tubes are roller-expanded into smooth holes in the tubesheet. For a stronger joint, two shallow circumferential grooves are cut into the wall of the hole in the tubesheet and the tube roller-expanded into the grooves; to eliminate the possibility of leakage, a seal weld can be run between the outer end of the tube and the tubesheet. Alternatively, the tubes may be strength-welded into the tubesheet.

Tube Supports. It is essential to provide periodic support along the length of the tubes to prevent sagging and destructive vibration caused by the fluid flowing across the tube bank. A secondary role played by the tube supports is to guide the flow back and forth across the tube bank, increasing the velocity and improving the heat transfer on the shell side (but also increasing the pressure drop). The tube support is usually in the form of single segmental baffles (item 18 in Figure 4.3.18) — circular plates with holes drilled to accommodate the tubes and with a segment sheared off to form a "window" or "turnaround" to allow the shell-side fluid to pass from one cross-flow section to the next. The baffles must overlap at least one full row of tubes to give the bundle the necessary rigidity against vibration. When minimizing shell-side pressure drop is not a priority, a baffle cut of 15 to 25% of the shell inside diameter is customary. Baffle spacing is determined first by the necessity to avoid vibration and secondarily to approximately match the free cross-flow area between adjacent baffles to the flow area in the window; i.e., small baffle cuts correspond to closer baffle spacing.

In situations such as low-pressure gas flows on the shell side where pressure drop is severely limited, double segmental and strip baffle arrays can be used. More recently, a helical baffle arrangement has been introduced (Kral et al., 1996) which causes the shell-side fluid to spiral through the exchanger giving improved heat transfer vs. pressure drop characteristics. Where vibration prevention and/or minimum pressure drop are the main concerns, grids of rods or strips can be used (Gentry et al., 1982).

Shells. The shell is the cylinder which confines the shell-side fluid (item 7 in Figure 4.3.18), fitted with nozzles for fluid inlet and exit. Diameters range from less than 50 mm (2 in.) to 3.05 m (10 ft) commonly, and at least twice that value for special applications. In diameters up to 610 mm (24 in.), shells are usually made from standard pipe or tubular goods by cutting to the desired length; in larger sizes, metal plates are rolled to the desired diameter and welded.

A variety of nozzle arrangements are used for special purposes, and TEMA has a standard code to identify the major types, as well as the various front and rear head configurations on the tube side. Figure 4.3.19 shows these configurations with the corresponding code letters.

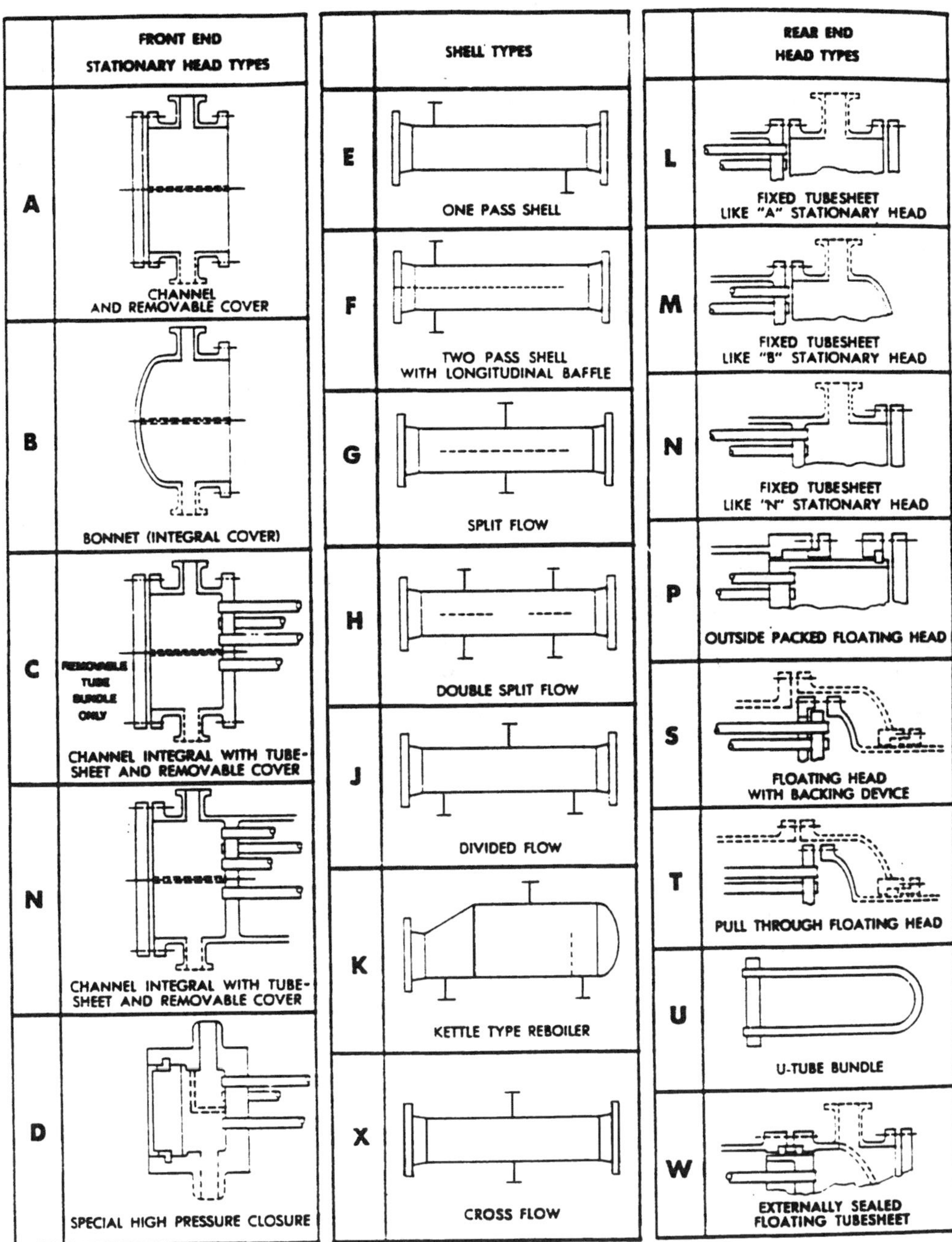

FIGURE 4.3.19 TEMA nomenclature for shell and tube configurations. (From TEMA, *Standards,* 7th ed., Tubular Exchanger Manufacturers Association, Tarrytown, NY, 1988. With permission.)

The E shell (center column, top) has the nozzles on opposite ends of the shell and is the most common configuration. It is used for any of the thermal services (single-phase heating or cooling, vaporization, and condensation). The nozzles may be on opposite sides of the shell as shown, or on the same side; the choice is largely determined by plumbing convenience. The E shell allows countercurrent flow (see below) of the two streams if there is one tube-side pass (i.e., the tube-side fluid flows through all of the tubes in parallel).

The F shell has both nozzles at one end of the shell and uses a longitudinal baffle on the shell side (shown dashed in the drawing) to force the shell-side fluid to flow to the far end of the heat exchanger and then back to the exit nozzle on the other side of the longitudinal baffle. Ideally, this allows countercurrent flow of the two streams if there are two tube-side passes (i.e., the tube-side fluid flows through half of the tubes in one direction, is turned around in the rear head, and returns through the other half of the tubes — see discussion of head types below). However, the longitudinal baffle must be carefully sealed to the shell to prevent leakage of the shell-side fluid across it; this is done by welding the longitudinal baffle to the shell and front tubesheet (which limits some design options) or by using mechanical seals. The F shell is mainly used for sensible heat transfer services.

The G shell has both nozzles at the center of the shell, with a centrally located longitudinal baffle to force the fluid to the ends of the shell before returning. While the G shell is used for all services, its main application is as a shellside vaporizer with either forced or natural (thermosiphon) convection of the boiling fluid; in the latter service, limited leakage across the baffle generally does not greatly degrade the thermal performance and the longitudinal baffle does not need to be perfectly sealed against the shell.

The H shell is effectively a double G shell and is employed in the same services. It is considered when the calculated shell-side pressure drop for a G arrangement is too high and threatens to limit the circulation rate.

The J shell, with one nozzle on top of the shell and two on the bottom, or vice versa, is commonly used in vacuum-condensing applications because of its low pressure drop. Two J shells (one inverted) may be mated in series for long-condensing-range mixtures. The nozzles are usually different diameters, with the large diameter accommodating the inlet vapor. The baffles are vertically cut.

The K shell (or kettle reboiler or flooded chiller) is exclusively intended for vaporization of liquid on the shell side, with a condensing vapor (usually steam) or a hot liquid on the tube side as the heating medium. The tubesheet diameter is large enough to accommodate the tube bundle, but the shell transitions to a larger diameter to allow the vapor to disengage from the liquid pool and exit from the top nozzle. A weir or other level control is used to maintain the liquid level, usually just above the top tubes in the bundle.

The X shell is intended to provide a well-distributed cross flow of the shell-side fluid, the fluid usually entering at the top and exiting at the bottom but occasionally used for upflow or horizontal cross flow. To obtain good distribution, multiple nozzles from a properly designed manifold may be required. Alternatively, the upper tubes in the bundle may be omitted to allow internal redistribution, or a large plenum chamber may be welded to the top of the shell ("vapor dome" or "bathtub nozzle"), or a diverging transition section may be placed between the inlet piping and the top of the shell. The tube supports may be complete circles since there is little or no longitudinal shell-side flow. The X shell gives the lowest shell-side pressure drop of any configuration and is often used for low-pressure vapor condensers.

Front Head. TEMA recognizes several front head designs as shown in the first column of Figure 4.3.19. Any of these designs will get the tube-side fluid into the tubes, but each has special features which recommend it to meet special needs. In Figure 4.3.19 the dashed lines indicate optional features depending upon need.

The A head, a channel with removable cover, bolts directly to the shell flange as shown in Figure 4.3.18, the tubesheet in that case being held between them and sealed with gaskets. Alternatively, the tubesheet may be integral with the shell (see the L rear head in Figure 4.3.19). A removable channel cover permits inspection, cleaning, removal, and replacement of tubes without disturbing the piping. The dashed lines at the center and the lower nozzle indicate that a pass partition plate may be welded in the channel (and gasketed against the tubesheet and channel cover) to provide for two tube-side passes (as shown in Figure 4.3.18 and required by the F shell design). Additional pass partitions may be provided to allow four, six, or any even number of tube-side passes. This permits the designer to utilize the available tube-side pressure drop to increase velocity, improve the heat transfer coefficient, and possibly reduce fouling. A second nozzle is required on the channel for multipass designs.

The B, or bonnet, front head reduces the number of gasketed joints and thus the opportunity for leakage, but it does not permit inspection of the tubes without breaking the piping connection. It is generally less expensive than the A head.

C and N heads retain the removable cover feature of the A head but, respectively, replace the channel-to-tubesheet and the tubesheet-to-shell gasketed connections with welds to minimize leakage possibilities. The N head is particularly used in nuclear services.

The D head is mainly used in feed-water heater applications where tube-side pressures are in the 100 to 400 bar range. The internal partition (optional) need only withstand the 1 to 2 bar pressure drop through the tubes so it can be of lightweight construction. The high-pressure closure against the atmosphere uses a shear key ring to lock the main closure in place.

Rear Head. A variety of rear head designs are used in shell-and-tube exchangers, primarily because of the need to accommodate thermally induced stresses. During operation, the tubes and the shell have different temperatures and therefore will expand (or try to) different amounts, even if there were no residual stresses in the exchanger before start-up and even if the entire exchanger is made out of the same material. The thermal stress problem is exacerbated if there are residual stresses, or if the exchanger is made of different materials, or during transient operation (including start-up and shutdown). If the temperature differences are small, the structure may be able to accommodate the thermal stresses safely; usually, however, it is necessary to make specific provision to allow the shell and the tubes to expand or contract independently. Failure to do so can result in buckling, bending, or even rupture of the shell or the tubes, or destruction of the tube-to-tubesheet joint.

A simple solution is to incorporate an expansion joint or a bellows into the shell (or in certain special applications, into the tube-side piping internal to the shell cover). However, this solution cannot cover the entire range of pressures and temperature differences encountered in practice. Further, it is usually possible to incorporate other desirable features, such as removable bundles, with thermal stress relief in the variety of rear head designs available. These are shown in the last column of Figure 4.3.19.

The L and M rear heads correspond to the A and B front heads previously described. As shown, they require a fixed tubesheet design; that is, the tubesheets are rigidly fastened to the shell, and thermal stress relief, if necessary, must be provided by a shell-side expansion joint or bellows. The tube bundle cannot be removed for inspection or mechanical cleaning on the shell side. However, the outer tube limit (OTL) — the diameter of the tube field circumscribing the outermost tubes in the bundle — can be as little as 0.4 in. (10 mm) less than the inside diameter of a pipe shell and 0.5 in. (12.7 mm) for a rolled shell. Therefore, the tube field can be very full, giving more tubes and minimizing bypass flow. Similar comments apply to the N rear head, except that more clearance must be left between the outermost tubes and the shell.

The type P head uses packing between the skirt on the rear tubesheet and the shell extension to seal the shell-side fluid against leakage. The compression on the packing has to be adjusted to prevent excessive leakage on the one hand and to allow limited movement of the tube-side head on the other, so the shell-side fluid must be benign and cheap (not surprisingly, it is often cooling water). On the other hand, leakage between the two fluids can occur only through tube hole leaks. Because of the tubesheet skirt, clearance between the outermost tubes and the shell must increase compared with types L or M; accordingly, fewer tubes are possible in a given shell, and sealing strips to partially block the bundle-to-shell bypass stream are recommended. When the floating head cover and packing gland are removed, the tube bundle can be pulled out of the shell for inspection and cleaning.

The TEMA S split-ring floating head design uses a split backing ring to hold the floating head cover and its gasket to the tubesheet. The split backing ring is bolted to the cover with a bolt circle outside the diameter of the tubesheet. Therefore, when the split ring is removed, the entire tube bundle may be pulled out of the shell. Tube count is similar to type P design and sealing strips are recommended. Usually, the split-ring floating head is used with an even number of tube passes so that a plain bonnet-type shell cover can be used. However, as shown by the dashed lines in Figure 4.3.19, single tube-side pass design (and countercurrent flow) can be achieved by use of a packing gland on the exit piping through the bonnet; alternatively, a deep bonnet can be used together with an expansion joint or bellows on the tube-side exit piping.

The pull-through floating head, type T, uses a floating head cover that flanges directly to the tubesheet, reducing the possibility of internal leakage compared with type S, but also eliminating more tubes around the periphery. Sealing strips are a virtual necessity. Single tube-side pass design is similar to type S, but is rarely used.

TEMA type U uses a bundle of U tubes and hence requires no rear head at all. The U-tube bundle effectively eliminates the thermal stress problem between shell and tubes, because each tube is free to expand or contract independently. The U bundle is also the cheapest construction because the cost of a second tubesheet is avoided. However, there are a number of drawbacks: designs must have an even number of tube-side passes, mechanical cleaning of the smaller bend radius tubes in the U bend is impossible, individual tubes cannot be replaced except in the outer row, some tube count is lost because of minimum bend limits, and the U bend must be carefully supported against vibration or kept out of the cross-flow stream by placing the shell nozzle upstream of the bend. The tube side in the U bend is susceptible to erosion, especially with two-phase or particulate-containing fluids.

Type W uses two sets of packing, often with a lantern ring in between. This construction is generally limited to benign fluids and low to very moderate pressures and temperatures.

Other Features. Numerous other components are necessary or optional to construction of shell-and-tube exchangers. Probably the most complete discussion is given by Yokell (1990).

Principles of Design

Design Logic. The design of a shell-and-tube exchanger involves the following steps:

1. Selection of a set of design features which are required for mechanical integrity and ease of maintenance, and which will likely lead to satisfying the thermal requirements within the allowable pressure drops, and at lowest cost.
2. Selection of a set of dimensions for the actual exchanger.
3. For the dimensions selected in (2), calculation of the thermal performance of the heat exchanger and both tube-side and shell-side pressure drops, using available rating procedures.
4. Comparison of the thermal performance calculated in (3) with that required and examination of the pressure drops calculated in (3) to ensure that the allowed pressure drops are reasonably used but not exceeded.
5. Adjustment of the dimensions selected in (2) and repetition of steps (3) and (4) until the criteria are satisfied.
6. Completion of the mechanical design to satisfy code requirements.
7. Cost estimation.

Basic Design Equations. The basic design equation for a shell-and-tube exchanger in steady-state service is

$$A^* = \int_0^{q_T} \frac{dq}{U^*\left(T_h - T_c\right)} \tag{4.3.87}$$

where A^* is the heat transfer area required in the heat exchanger, m^2 (ft^2); q_T is the heat transfer rate of the heat exchanger, W (Btu/hr); U^* is the local overall heat transfer coefficient referenced to area A^*, W/m^2 K (Btu/hr ft^2 °F); and T_h and T_c are the local hot and cold stream temperatures, K (°F). The * superscript on A^* and U^* only means that a consistent reference area must be used in defining these terms. For example, for an exchanger with plain tubes, it is customary to use the total outside heat transfer area of all of the tubes in the exchanger, A_o, as the reference area, and then U_o is the overall heat transfer coefficient referenced to A_o. If the exchanger has low-finned tubes, A^* may refer either to the total outside area including fins or to the inside tube heat transfer area; the choice is optional, but must be spelled out. Since T_h and T_c generally vary with the amount of heat transferred (following the first law of thermodynamics, and excepting isobaric phase transition of a pure component) and U^* may vary with local heat transfer conditions, in principle Equation (4.3.87) must be numerically integrated with T_h, T_c, and U^* calculated along the path of integration, and this process is performed by the most-advanced computer-based design methods.

For many applications, certain reasonable assumptions can be made allowing the analytical integration of Equation (4.3.87) to give (Hewitt, Vol. 1, 1998; Hewitt et al., 1994)

$$A^* = \frac{q_T U^*}{(\text{MTD})} \tag{4.3.88}$$

where MTD is the mean temperature difference for the particular flow conditions and configuration. The key assumptions are that there is no significant bypassing of fluid around the heat transfer surface, that the overall heat transfer coefficient is constant, and that the specific heats of the two streams are constant over their temperature ranges in the exchanger; isothermal phase transitions, such as vaporizing or condensing a pure component at constant pressure, are also allowed.

If the two streams are in countercurrent flow, i.e., if they flow always in the opposite direction to one another,

$$\text{MTD} = (\text{LMTD})_{\text{countercurrent}} = \frac{\left(T_{h,i} - T_{c,o}\right) - \left(T_{h,o} - T_{c,i}\right)}{\ln\left(\dfrac{T_{h,i} - T_{c,o}}{T_{h,o} - T_{c,i}}\right)} \tag{4.3.89}$$

where $(\text{LMTD})_{\text{countercurrent}}$ is the "logarithmic mean temperature difference for countercurrent flow" and the subscripts i and o indicate "inlet" and "outlet," respectively. E shells with a single tube-side pass and F shells with two tube-side passes are almost always designed for countercurrent flow. (While the flow between adjacent baffles is basically cross flow, it can be shown that the total shell-side flow pattern is equivalent to countercurrent flow if there are more than three or four baffles).

Very occasionally, usually when close control of tube wall temperatures is required, cocurrent flow is specified, with the two streams flowing in the same direction through the exchanger. For this case,

$$\text{MTD} = (\text{LMTD})_{\text{cocurrent}} = \frac{\left(T_{h,i} - T_{c,i}\right) - \left(T_{h,o} - T_{c,o}\right)}{\ln\left(\dfrac{T_{h,i} - T_{c,i}}{T_{h,o} - T_{c,o}}\right)} \tag{4.3.90}$$

where the symbols have the same meaning as before. $(\text{LMTD})_{\text{countercurrent}}$ is always equal to or greater than $(\text{LMTD})_{\text{cocurrent}}$, so wherever possible, countercurrent design and operation is preferred.

However, most shell-and-tube exchangers have nozzle and tube pass configurations which lead to mixed countercurrent and cocurrent flow regions (as well as cross flow in the X shell). For these cases,

$$\text{MTD} = F(\text{LMTD})_{\text{countercurrent}} \tag{4.3.91}$$

where $(\text{LMTD})_{\text{countercurrent}}$ is calculated from Equation (4.3.89) and F is the "configuration correction factor" for the flow configuration involved. F has been found as a function of dimensionless temperature ratios for most flow configurations of interest and is given in analytical and/or graphical form in the earlier part of this section by Shah and in many heat transfer references (e.g., Hewitt, Vol. 1, 1998).

F is equal to unity for pure countercurrent flow and is less than unity for all other cases; practical considerations limit the range of interest to values above 0.7 at the lowest and more comfortably to values above 0.8. Values of zero or below indicate conditions that violate the second law of thermodynamics.

The Overall Heat Transfer Coefficient. The overall heat transfer coefficient U^*, referenced to the heat transfer area A^*, is related to the individual (film) heat transfer coefficients and the fouling resistances by

$$U^* = \frac{1}{\frac{A^*}{h_i A_i} + R_{fi}\frac{A^*}{A_i} + \frac{A^* \ln(d_o/d_i)}{2\pi N_t L k_w} + R_{fo}\frac{A^*}{A_o} + \frac{A^*}{h_o A_o}} \tag{4.3.92}$$

where h_i and h_o are, respectively, the tube-side and shell-side film heat transfer coefficients, W/m^{2}K (Btu/hr ft^2 °F), each referenced to its corresponding heat transfer area; R_{fi} and R_{fo} the corresponding fouling resistances (see below), m^{2}K/W (hr ft^2 °F/Btu); N_t the total number of tubes in the heat exchanger; L the effective tube length between the inside surfaces of the tubesheets, m (ft); d_o and d_i the outside and inside tube diameters, m (ft); and k_w the thermal conductivity of the tube wall material, W/m K (Btu/hr ft°F). For the special but important case of plain tubes

$$A^* = A_o = N_t(\pi d_o L) \tag{4.3.93}$$

and Equation (4.3.92) reduces to

$$U_o = \frac{1}{\frac{d_o}{h_i d_i} + R_{fi}\frac{d_o}{d_i} + \frac{d_o \ln(d_o/d_i)}{2k_w} + R_{fo} + \frac{1}{h_o}} \tag{4.3.94}$$

If finned tubes are used, the root diameter d_r of the fins replaces d_o in Equation (4.3.92) and A_o includes the surface area of the fins as well as the bare tube surface between the fins; it is also necessary to include a fin efficiency (typically about 0.8 to 0.95) multiplier in the numerators of the last two terms on the right side of Equation (4.3.92) to account for resistance to conduction in the fins. The treatment of fin efficiency is fully developed in Kern and Kraus (1972). Efficiencies of some of the important geometries are given in the earlier half of this section.

Film Heat Transfer Coefficients. Calculation of single-phase tube-side heat transfer coefficients for plain tubes is discussed in Section 3.1; special correlations are required for internally enhanced tubes, see discussion of enhancement in Section 4.15. Intube condensation and vaporization are covered in the subsection on boiling and condensation in Section 3.4.

Shell-side heat transfer calculations are more complex owing to the large number and range of design variables and process conditions that can occur. The most accurate methods are proprietary and computer based. The best known of these methods are those of Heat Transfer Research, Inc. (HTRI), College Station, TX; Heat Transfer and Fluid Flow Services (HTFS), Harwell, U.K.; and B-JAC, Midlothian, VA. For single-phase flow, the Delaware method appears to be the best in the open literature, and it is feasible for both had and computer use; various presentations of the method appear in many references, including Hewitt, Vol. 3 (1998) and Hewitt et al. (1994). These references also give methods for shell-side vaporizing and condensing design. An approximate design procedure is given in the next subsection.

Fouling. Fouling is the formation of any undesired deposit on the heat transfer surface, and it presents an additional resistance to the flow of heat. Several different types of fouling are recognized:

Sedimentation: deposition of suspended material on the surface.
Crystallization: precipitation of solute from supersaturated solutions.
Corrosion: formation of corrosion products on the surface.
Thermal degradation/polymerization: formation of insoluble products by oxidation, charring, and/or polymerization of a process stream.
Biofouling: growth of large organisms (e.g., barnacles) that interfere with flow to or past a heat transfer surface ("macrobiofouling") or small organisms (e.g., algae) that form a fouling layer on the surface ("microbiofouling").

The effect of fouling on design is twofold: Extra surface must be added to the heat exchanger to overcome the additional thermal resistance, and provision must be made to allow cleaning either by chemical or mechanical means. The fouling resistances included in Equation (4.3.92) result in requiring extra surface by reducing U^* (though they do not properly account for the time-dependent nature of fouling) and should be chosen with care. Table 4.3.8, based on the TEMA *Standards* provides some guidance, but prior experience with a given service is the best source of values. Ranges of typical values for major classes of service are included in Table 4.3.10.

Other things being equal, a fouling stream that requires mechanical cleaning should be put in the tubes because it is easier to clean the tube side. If this is not possible or desirable, then a removable bundle with a rotated square tube layout should be chosen to facilitate cleaning.

Pressure Drop. Tube-side pressure drop in plain tubes is discussed in Section 2.4. These calculations are straightforward and quite accurate as long as the tubes are smooth and clean; however, even a small amount of roughening due to corrosion or fouling (sometimes with a significant reduction of flow area) can double or triple tube-side pressure drop. Special correlations are required for internally enhanced tubes.

Calculation of shell-side pressure drop is implicit in the design methods mentioned above for heat transfer. Roughness has less effect on shell-side pressure drop than on tube side, but fouling still may have a very substantial effect if the deposits fill up the clearances between the baffles and the shell and between the tubes and the baffles, or if the deposits are thick enough to narrow the clearances between adjacent tubes. Existing design methods can predict these effects if the thickness of the fouling layer can be estimated.

Limitations of Design. It should be recognized that even under the best of conditions — new, clean exchangers with conventional construction features — heat exchanger design is not highly accurate. The best methods, when compared with carefully taken test data, show deviations of ±20% on overall heat transfer and ±40% on shell-side pressure drop (Palen and Taborek, 1969). These ranges are considerably worsened in fouling services. In these cases, the thermal *system* should be designed for operational flexibility, including carefully chosen redundancy of key components, and easy maintenance.

Approximate Design Method

Because of the complexity of rigorous design methods, it is useful to have an estimation procedure that can quickly give approximate dimensions of a heat exchanger for a specified service. Such a method is given here for purposes of preliminary cost estimation, plant layout, or checking the results of computer output. This method is based upon Equation (4.3.88) with $A^* = A_o$ and $U^* = U_o$ and depends upon rapidly estimating values for q_T, MTD, and U_o. The procedure is as follows:

Estimation of q_T For sensible heat transfer,

$$q_T = \dot{m}_h c_{p,h}\left(T_{h,i} - T_{h,o}\right) = \dot{m}_c c_{p,c}\left(T_{c,o} - T_{c,i}\right) \tag{4.3.95}$$

where $\dot{m}$ is the mass flow rate, c_p the specific heat, and T the stream temperature, with subscripts h and c denoting the hot and cold streams, respectively, and i and o inlet and outlet, respectively.

TABLE 4.3.10 Typical Film Heat Transfer Coefficients for Shell-and-Tube Heat Exchangers

Fluid Conditions		h, W/m²K[a,b]	Fouling resistance, m² K/W[a]
Sensible heat transfer			
Water[c]	Liquid	5000–7500	$1–2.5 \times 10^{-4}$
Ammonia	Liquid	6000–8000	$0–1 \times 10^{-4}$
Light organics[d]	Liquid	1500–2000	$0–2 \times 10^{-4}$
Medium organics[e]	Liquid	750–1500	$1–4 \times 10^{-4}$
Heavy organics[f]	Liquid		
	Heating	250–750	$2–10 \times 10^{-4}$
	Cooling	150–400	$2–10 \times 10^{-4}$
Very heavy organics[g]	Liquid		
	Heating	100–300	$4–30 \times 10^{-3}$
	Cooling	60–150	$4–30 \times 10^{-3}$
Gas[h]	Pressure 100–200 kN/m² abs	80–125	$0–1 \times 10^{-4}$
Gas[h]	Pressure 1 MN/m² abs	250–400	$0–1 \times 10^{-4}$
Gas[h]	Pressure 10 MN/m² abs	500–800	$0–1 \times 10^{-4}$
Condensing heat transfer			
Steam, ammonia	Pressure 10 kN/m² abs, no noncondensables[i,j]	8000–12000	$0–1 \times 10^{-4}$
Steam, ammonia	Pressure 10 kN/m² abs, 1% noncondensables[k]	4000–6000	$0–1 \times 10^{-4}$
Steam, ammonia	Pressure 10 kN/m² abs, 4% noncondensables[k]	2000–3000	$0–1 \times 10^{-4}$
Steam, ammonia	Pressure 100 kN/m² abs, no noncondensables[i,j,k,l]	10000–15000	$0–1 \times 10^{-4}$
Steam, ammonia	Pressure 1 MN/m² abs, no noncondensables[i,j,k,l]	15000–25,000	$0–1 \times 10^{-4}$
Light organics[d]	Pure component, pressure 10 kN/m² abs, no noncondensables[i]	1500–2000	$0–1 \times 10^{-4}$
Light organics[d]	Pressure 10 kN/m² abs, 4% noncondensables[k]	750–1000	$0–1 \times 10^{-4}$
Light organics[d]	Pure component, pressure 100 kN/m² abs, no noncondensables	2000–4000	$0–1 \times 10^{-4}$
Light organics[d]	Pure component, pressure 1 MN/m² abs	3000–4000	$0–1 \times 10^{-4}$
Medium organics[e]	Pure component or narrow condensing range, pressure 100 kN/m² abs[m,n]	1500–4000	$1–3 \times 10^{-4}$
Heavy organics	Narrow condensing range pressure 100 kN/m² abs[m,n]	600–2000	$2–5 \times 10^{-4}$
Light multicomponent mixtures, all condensable[d]	Medium condensing range, pressure 100 kN/m² abs[k,m,o]	1000–2500	$0–2 \times 10^{-4}$
Medium multicomponent mixtures, all condensable[e]	Medium condensing range, pressure 100 kN/m² abs[k,m,o]	600–1500	$1–4 \times 10^{-4}$
Heavy multicomponent mixtures, all condensable[f]	Medium condensing range, pressure 100 kN/m² abs[k,m,o]	300–600	$2–8 \times 10^{-4}$
Vaporizing heat transfer[p,q]			
Water[r]	Pressure < 0.5 MN/m² abs, $\Delta T_{SH,max} = 25$ K	3000–10000	$1–2 \times 10^{-4}$
Water[r]	Pressure > 0.5 MN/m² abs, pressure < 10 MN/m² abs, $\Delta T_{SH,max} = 20$ K	4000–15000	$1–2 \times 10^{-4}$

TABLE 4.3.10 (continued) Typical Film Heat Transfer Coefficients for Shell-and-Tube Heat Exchangers

Fluid Conditions		h, W/m²K[a,b]	Fouling resistance, m² K/W[a]
Ammonia	Pressure < 3 MN/m² abs, $\Delta T_{SH,max}$ = 20 K	3000–5000	$0–2 \times 10^{-4}$
Light organics[d]	Pure component, pressure < 2 MN/m² abs, $\Delta T_{SH,max}$ = 20 K	1000–4000	$1–2 \times 10^{-4}$
Light organics[d]	Narrow boiling range,[s] pressure < 2 MN/m² abs, $\Delta T_{SH,max}$ = 15 K	750–3000	$0–2 \times 10^{-4}$
Medium organics[e]	Pure component, pressure < 2 MN/m² abs, $\Delta T_{SH,max}$ = 20 K	1000–3500	$1–3 \times 10^{-4}$
Medium organics[e]	Narrow boiling range,[s] pressure < 2 MN/m² abs, $\Delta T_{SH,max}$ = 15 K	600–2500	$1–3 \times 10^{-4}$
Heavy organics[f]	Pure component, pressure < 2 MN/m² abs, $\Delta T_{SH,max}$ = 20 K	750–2500	$2–5 \times 10^{-4}$
Heavy organics[g]	Narrow boiling range,[s] pressure < 2 MN/m² abs, $\Delta T_{SH,max}$ = 15 K	400–1500	$2–8 \times 10^{-4}$
Very heavy organics[h]	Narrow boiling range,[s] pressure < 2 MN/m² abs, $\Delta T_{SH,max}$ = 15 K	300–1000	$2–10 \times 10^{-4}$

Source: Hewitt, G.F., Ed., *Heat Exchanger Design Handbook 1998*, Begell House, New York, 1998. With permision.

[a] Heat transfer coefficients and fouling resistances are based on area in contact with fluid. Ranges shown are typical, not all encompassing. Temperatures are assumed to be in normal processing range; allowances should be made for very high or low temperatures.

[b] Allowable pressure drops on each side are assumed to be about 50–100 kN/m² except for (1) low-pressure gas and two-phase flows, where the pressure drop is assumed to be about 5% of the absolute pressure; and (2) very viscous organics, where the allowable pressure drop is assumed to be about 150–250 kN/m².

[c] Aqueous solutions give approximately the same coefficients as water.

[d] Light organics include fluids with liquid viscosities less than about 0.5×10^{-3} Nsec/m², such as hydrocarbons through C_8, gasoline, light alcohols and ketones, etc.

[e] Medium organics include fluids with liquid viscosities between about 0.5×10^{-3} and 2.5×10^{-3} Nsec/m², such as kerosene, straw oil, hot gas oil, and light crudes.

[f] Heavy organics include fluids with liquid viscosities greater than 2.5×10^{-3} Nsec/m², but not more than 50×10^{-3} Nsec/m², such as cold gas oil, lube oils, fuel oils, and heavy and reduced crudes.

[g] Very heavy organics include tars, asphalts, polymer melts, greases, etc., having liquid viscosities greater than about 50×10^{-3} Nsec/m². Estimation of coefficients for these materials is very uncertain and depends strongly on the temperature difference, because natural convection is often a significant contribution to heat transfer in heating, whereas congelation on the surface and particularly between fins can occur in cooling. Since many of these materials are thermally unstable, high surface temperatures can lead to extremely severe fouling.

[h] Values given for gases apply to such substances as air, nitrogen, carbon dioxide, light hydrocarbon mixtures (no condensation), etc. Because of the very high thermal conductivities and specific heats of hydrogen and helium, gas mixtures containing appreciable fractions of these components will generally have substantially higher heat transfer coefficients.

[i] Superheat of a pure vapor is removed at the same coefficient as for condensation of the saturated vapor if the exit coolant temperature is less than the saturation temperature (at the pressure existing in the vapor phase) and if the (constant) saturation temperature is used in calculating the MTD. But see note k for vapor mixtures with or without noncondensable gas.

[j] Steam is not usually condensed on conventional low-finned tubes; its high surface tension causes bridging and retention of the condensate and a severe reduction of the coefficient below that of the plain tube.

TABLE 4.3.10 (continued) Typical Film Heat Transfer Coefficients for Shell-and-Tube Heat Exchangers

Fluid Conditions	h, W/m^{2}K[a,b]	Fouling resistance, m^2 K/W[a]

[k] The coefficients cited for condensation in the presence of noncondensable gases or for multicomponent mixtures are only for very rough estimation purposes because of the presence of mass transfer resistances in the vapor (and to some extent, in the liquid) phase. Also, for these cases, the vapor-phase temperature is not constant, and the coefficient given is to be used with the MTD estimated using vapor-phase inlet and exit temperatures, together with the coolant temperatures.

[l] As a rough approximation, the same relative reduction in low-pressure condensing coefficients due to noncondensable gases can also be applied to higher pressures.

[m] Absolute pressure and noncondensables have about the same effect on condensing coefficients for medium and heavy organics as for light organics. For large fractions of noncondensable gas, interpolate between pure component condensation and gas cooling coefficients.

[n] Narrow condensing range implies that the temperature difference between dew point and bubble point is less than the smallest temperature difference between vapor and coolant at any place in the condenser.

[o] Medium condensing range implies that the temperature difference between dew point and bubble point is greater than the smallest temperature difference between vapor and coolant, but less than the temperature difference between inlet vapor and outlet coolant.

[p] Boiling and vaporizing heat transfer coefficients depend very strongly on the nature of the surface and the structure of the two-phase flow past the surface in addition to all of the other variables that are significant for convective heat transfer in other modes. The flow velocity and structure are very much governed by the geometry of the equipment and its connecting piping. Also, there is a maximum heat flux from the surface that can be achieved with reasonable temperature differences between surface and saturation temperatures of the boiling fluid; any attempt to exceed this maximum heat flux by increasing the surface temperature leads to partial or total coverage of the surface by a film of vapor and a sharp decrease in the heat flux.

Therefore, the vaporizing heat transfer coefficients given in this table are only for very rough estimating purposes and assume the use of plain or low-finned tubes without special nucleation enhancement.

$\Delta T_{SH,max}$ is the maximum allowable temperature difference between surface and saturation temperature of the boiling liquid. No attempt is made in this table to distinguish among the various types of vapor-generation equipment, since the major heat transfer distinction to be made is the propensity of the process stream to foul. Severely fouling streams will usually call for a vertical thermosiphon or a forced-convection (tube-side) reboiler for ease of cleaning.

[q] Subcooling heat load is transferred at the same coefficient as latent heat load in kettle reboilers, using the saturation temperature in the MTD. For horizontal and vertical thermosiphons and forced-circulation reboilers, a separate calculation is required for the sensible heat transfer area, using appropriate sensible heat transfer coefficients and the liquid temperature profile for the MTD.

[r] Aqueous solutions vaporize with nearly the same coefficient as pure water if attention is given to boiling-point elevation, if the solution does not become saturated, and if care is taken to avoid dry wall conditions.

[s] For boiling of mixtures, the saturation temperature (bubble point) of the final liquid phase (after the desired vaporization has taken place) is to be used to calculate the MTD. A narrow-boiling-range mixture is defined as one for which the difference between the bubble point of the incoming liquid and the bubble point of the exit liquid is less than the temperature difference between the exit hot stream and the bubble point of the exit boiling liquid. Wide-boiling-range mixtures require a case-by-case analysis and cannot be reliably estimated by these simple procedures.

For isothermal phase change,

$$q_T = \dot{m} h_{fg} \tag{4.3.96}$$

where $\dot{m}$ is the mass rate of condensation or vaporization and h_{fg} is the latent heat of phase transformation.

For more complex cases, such as partial or multicomponent condensation, more elaborate analyses are required, although this method can still be used with care to give rough estimates.

Estimation of MTD. The first step is to calculate or estimate LMTD$_{\text{countercurrent}}$ from Equation (4.3.89) and then estimate F as follows:

1. If the two streams are in countercurrent flow, $F = 1$.
2. If the two streams are in a combination of countercurrent and cocurrent flows (i.e., multiple tube passes) and the outlet temperatures of the two streams are equal, $F = 0.8$.
3. If the exchanger has multiple passes and $T_{h,o} > T_{c,o}$, then $0.8 < F < 1.0$, with the actual value depending upon the temperature ranges of the two streams and $(T_{h,o} - T_{c,o})$. It is usually sufficiently accurate to take $F = 0.9$ in this case, but a more accurate value can be obtained from the earlier half of this section by Shah.
4. Design of a multiple tube pass exchanger with $T_{h,o} < T_{c,o}$ (i.e., a temperature cross) leads to $F <$ 0.8, which is inefficient, of uncertain inaccuracy, and perhaps even thermodynamically impossible. The problem can be handled with multiple shells in series. Consult Shah's discussion.
5. Then, MTD = $F(\text{LMTD})_{\text{countercurrent}}$, (Equation 4.3.91).

Estimation of U_o. The best way to estimate U_o is to use Equation (4.3.94), together with values of h_o, h_i, $R_{f,o}$, and $R_{f,i}$, chosen from Table 4.3.10. This table includes ranges of values that are typical of the fluids and services indicated assuming normally allowable pressure drops, exchanger construction, and fouling. However, care should be taken in selecting values to consider possible unusual conditions, e.g., especially high or low velocities (implying correspondingly high or low allowable pressure drops), and especially fouling. In selecting values from the table, the user should carefully read the footnotes for each entry.

Calculation of A_o. The total outside tube heat transfer area required in the heat exchanger is now found from Equation (4.3.88).

Estimation of Exchanger Dimensions. Figure 4.3.20 shows the relationship among A_o, effective tube length L, and inside shell diameter for a fully tubed, fixed tubesheet heat exchanger with one tube-side pass, with $^3/_4$ in. (19.05 mm) plain tubes on a $^{15}/_{16}$ in. (23.8 mm) pitch equilateral triangular tube layout. These curves are constructed using tube count tables (e.g., Saunders, 1988). The dashed lines marked 3:1, 6:1, 8:1, 10:1, and 15:1 indicate ratios of tube length to shell inside diameter for guidance in selection. Exchangers of less than 3:1 ratio are expensive because of the large-diameter shell and tubesheet, with more holes to be drilled and tubes rolled and/or welded, and shell-side flow distribution is likely to be poor and lead to excessive fouling. Exchangers greater than 15:1 ratio are probably beyond the point of saving money by reducing shell diameter and number of tubes and may require excessive clear way for pulling the bundle; the bundles may be springy and difficult to handle during maintenance. Most heat exchangers fall into the 6:1 to 10:1 range.

Figure 4.3.20 is a very specific case which is used as a reference. In order to extend its usefulness to other tube diameters, layouts, bundle constructions, etc., Equation (4.3.97) is used:

$$A_o' = A_o F_1 F_2 F_3 \tag{4.3.97}$$

where A_o' is the value to be used with Figure 4.3.20, A_o is the required area calculated from Equation (4.3.88), and

F_1 is the correction factor for the tube layout. $F_1 = 1.00$ for $^3/_4$ in. (19.05 mm) outside diameter tubes on a $^{15}/_{16}$ in. (23.8 mm) triangular pitch. Values of F_1 for other tube diameters and pitches are given in Table 4.3.11.

F_2 is the correction factor for the number of tube-side passes. $F_2 = 1.00$ for one tube-side pass, and Table 4.3.12 gives values of F_2 for more passes.

F_3 is the correction factor for shell construction/tube bundle configuration. $F_3 = 1.00$ for fixed tubesheet, fully tubed shells, and Table 4.3.13 gives values of F_3 for the standard TEMA types.

Once a value of A_o' has been calculated from Equation (4.3.97), enter the ordinate of Figure 4.3.20 at that value and move horizontally, picking off the combinations of shell inside diameter and tube length that meet that requirement. The final choice can then be made from among those possibilities.

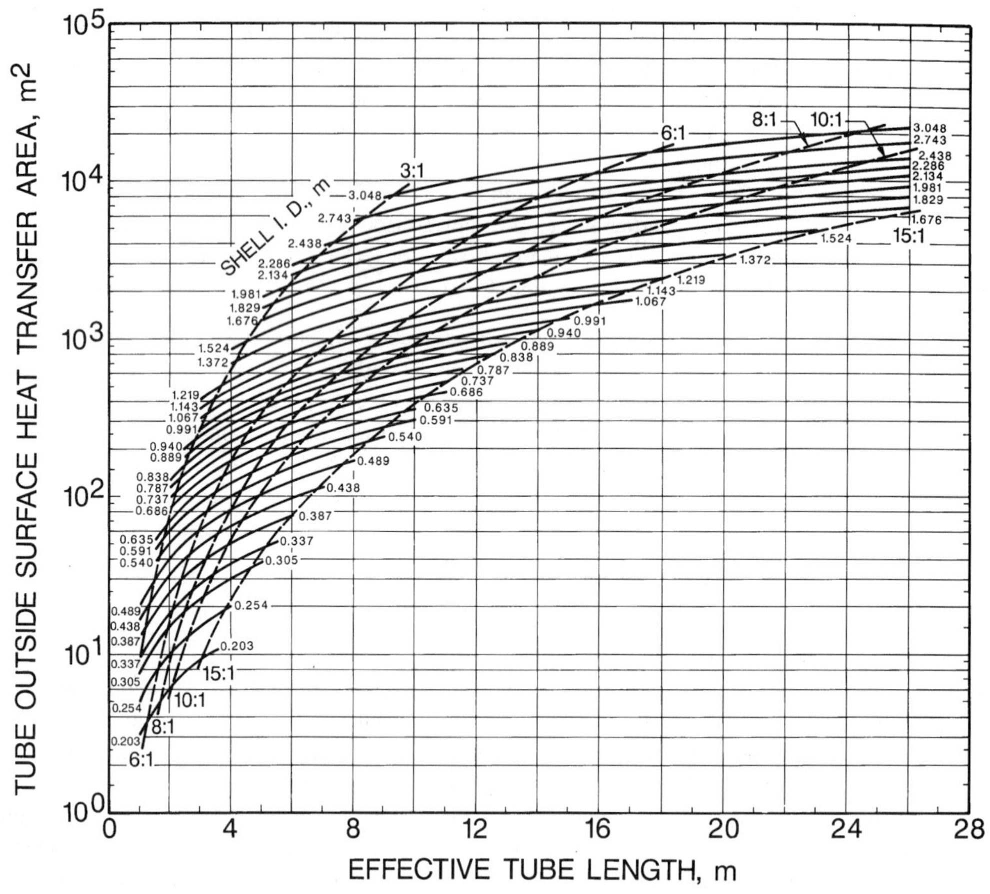

FIGURE 4.3.20 Heat transfer area as a function of shell inside diameter and effective tube length for 19.05 mm ($^3/_4$ in.) tubes on a 23.8 mm ($^{15}/_{16}$ in.) equilateral triangular tube layout, fixed tubesheet, one tube-side pass, fully tubed shell. (From Hewitt, G.F.., Ed. *Heat Exchanger Design Handbook 1998*, Begell House, New York, 1998. With permission.)

Example of the Approximate Design Method

Problem Statement. Estimate the dimensions of a shell-and-tube heat exchanger to cool 100,000 lb_m/hr (12.6 kg/sec) of liquid toluene from 250 to 110°F (121.1 to 43.3°C) using cooling tower water available at 80°F (26.7°C). Use split-ring floating head construction (TEMA S) with $^3/_4$ in. (19.05 mm) outside diameter × 14 BWG (0.083 in. = 2.11 mm wall) low-carbon steel tubes on $^{15}/_{16}$ in. (23.8 mm) equilateral triangular pitch. This construction implies one shell-side pass and an even number of tube-side passes — assume two for the present. Choose cooling water exit temperature of 100°F (37.8°C). Specific heat of toluene is 0.52 Btu/lb_m°F (2177 J/kgK) and viscosity at 180°F (82.2°C) is 0.82 lb_m/ft hr (0.34×10^{-3} Nsec/m^2 or 0.34 cP).

Solution.

$$q_T = \left(100{,}000\ \text{lb}_\text{m}/\text{hr}\right)\left(0.52\ \text{Btu}/\text{lb}_\text{m}\,°\text{F}\right)\left(250-110\right)°\text{F}$$

$$= 7.28\times10^6\ \text{Btu/hr} = 2.14\times10^6\ \text{W}$$

TABLE 4.3.11 Values of F_1 for Various Tube Diameters and Layouts

Tube Outside Diameter, in. (mm)	Tube Pitch, in. (mm)	Layout	F_1
5/8 (15.88)	13/16 (20.6)	→ ◁	0.90
5/8 (15.88)	13/16 (20.6)	→ ◇ □	1.04
3/4 (19.05)	15/16 (23.8)	→ ◁	1.00
3/4 (19.05)	15/16 (23.8)	→ ◇ □	1.16
3/4 (19.05)	1 (25.4)	→ ◁	1.14
3/4 (19.05)	1 (25.4)	→ ◇ □	1.31
1 (25.4)	1 1/4 (31.8)	→ ◁	1.34
1 (25.4)	1 1/4 (31.8)	→ ◇ □	1.54

$$F_1 = \frac{(\text{Heat transfer area / cross-sectional area of unit cell})_{\text{Reference}}}{(\text{Heat transfer area / cross-sectional area of unit cell})_{\text{New Case}}}$$

This table may also be used for low-finned tubing in the following way. The value estimated for h_o from Table 4.3.10 should be multiplied by the fin efficiency (usually between 0.75 and 1 for a good application; 0.85 is a good estimate) and used in Equation 4.3.92 with $A^* = A_o$, the total outside heat transfer area including fins. Then this value of A_o is divided by the ratio of the finned tube heat transfer area to the plain tube area (per unit length). The result of this calculation is used as A_o in Equation 4.3.96 to find A_o' to enter Figure 4.3.20.

Source: Hewitt, G.F., Ed., *Heat Exchanger Design Handbook 1998*, Begell House, New York, 1998. With permission.

TABLE 4.3.12 Values of F_2 for Various Numbers of Tube Side Passes[a]

	F_2 Number of Tube-Side Passes			
Inside Shell Diameter, in. (mm)	**2**	**4**	**6**	**8**
Up to 12 (305)	1.20	1.40	1.80	—
$13^1/_4$ to $17^1/_4$ (337 to 438)	1.06	1.18	1.25	1.50
$19^1/_4$ to $23^1/_4$ (489 to 591)	1.04	1.14	1.19	1.35
25 to 33 (635 to 838)	1.03	1.12	1.16	1.20
35 to 45 (889 to 1143)	1.02	1.08	1.12	1.16
48 to 60 (1219 to 1524)	1.02	1.05	1.08	1.12
Above 60 (above 1524)	1.01	1.03	1.04	1.06

[a] Since U-tube bundles must always have at least two passes, use of this table is essential for U-tube bundle estimation. Most floating head bundles also require an even number of passes.

Source: Hewitt, G.F., Ed., *Heat Exchanger Design Handbook 1998*, Begell House, New York, 1998. With permission.

$$\text{LMTD}_{\text{countercurrent}} = \frac{(250-100)-(110-80)}{\ln\dfrac{250-100}{110-80}} = 74.6°\text{F} = 41.4°\text{C}$$

Since there are at least two tube-side passes, flow is not countercurrent, and $T_{h_o} > T_{c_o}$, estimate $F \approx 0.9$. Therefore, MTD = 0.9 (74.6°F) = 67.1°F = 37.3°C.

TABLE 4.3.13 F_3 for Various Tube Bundle Constructions

Type of Tube Bundle Construction	F_3 Inside Shell Diameter, in. (mm)				
	Up to 12 (305)	13–22 (330–559)	23–36 (584–914)	37–48 (940–1219)	Above 48 (1219)
Split backing ring (TEMA S)	1.30	1.15	1.09	1.06	1.04
Outside packed floating heat (TEMA P)	1.30	1.15	1.09	1.06	1.04
U-Tube* (TEMA U)	1.12	1.08	1.03	1.01	1.01
Pull-through floating head (TEMA T)	—	1.40	1.25	1.18	1.15

[a] Since U-tube bundles must always have at least two tube-side passes, it is essential to use Table 4.3.12 also for this configuration.

Source: Hewitt, G.F., Ed., *Heat Exchanger Design Handbook 1998*, Begell House, New York, 1998. With permission.

Estimation of U_o. Light organic liquid cooled by liquid water. (Note that 1 Btu/hr ft^2 °F = 5.678 W/m^{2}K).

Water (in tubes) h_i	1000 Btu/hr ft^2 °F	5700 W/m^{2}K
Toluene (in shell) h_o	300 Btu/hr ft^2 °F	1700 W/m^{2}K
Tube-side fouling R_{f_i}	0.001 hr ft^2 °F/Btu	1.8 × 10^{-4} m^{2}K/W
Shell-side fouling R_{f_o}	0.0005 hr ft^2 °F/Btu	8.8 × 10^{-5} m^{2}K/W

Tube wall resistance (for estimation purposes, this term can be approximated by x_w/k_w, where x_w is the wall thickness):

$$\frac{x_w}{k_w} = \frac{0.083 \text{ in.}}{(12 \text{ in./ft})(26 \text{ Btu/hr ft °F})} = 2.7\times10^{-4}\,\frac{\text{hr ft}^2\text{ °F}}{\text{Btu}} = 4.6\times10^{-5}\,\frac{\text{m}^2\text{K}}{\text{W}}$$

Then,

$$U_o = \frac{1}{\dfrac{0.750}{1000(0.584)} + \dfrac{0.001(0.750)}{0.584} + 2.7\times10^{-4} + 0.0005 + \dfrac{1}{300}}$$

$$= 150 \text{ Btu/hr ft °F} = 848 \text{ W/m}^2\text{K}$$

$$A_o = \frac{7.28\times10^6 \text{ Btu/hr}}{(150 \text{ Btu/hr ft}^2\text{ °F})(67.1\text{°F})} = 723 \text{ ft}^2 = 67.7 \text{ m}^2$$

Correct for changes in construction features (preliminary examination of Figure 4.3.20 indicates shell inside diameter will be in the range of 500 mm, or 20 in.):

F_1: F_1 = 1.00 since the same tube size and layout is used;
F_2: F_2 = 1.04, assuming two passes;
F_3: F_3 = 1.15, TEMA S construction;
A_o' = (723 ft^2) (1.00)(1.04)(1.15) = 865 ft^2 = 81 m^2.

From Figure 4.3.20, entering at A_o', pick off the following combinations of shell inside diameter and tube length:

Shell Inside Diameter		Effective Tube Length		L/D_s
in.	mm	ft	m	
27	686	6.6	2.0	2.9
25	635	7.5	2.3	3.6
$23^1/_4$	591	9.2	2.8	4.7
$21^1/_4$	540	10.8	3.3	6.1
$19^1/_4$	489	13.1	4.0	8.2
$17^1/_4$	438	16.7	5.1	11.6

Any of these combinations would supply the desired area; the $21^1/_4$ in. (540 mm) and $19^1/_4$ in. (489 mm) would appear to be likely choices.

References

American Society of Mechanical Engineers. 1998. *ASME Boiler and Pressure Vessel Code*, Section VIII. New editions published every 3 years. ASME, New York.

Gentry, C.C., Young, R.K., and Small, W.M. 1982. RODbaffle heat exchanger thermal-hydraulic predictive methods, in *Proceedings of the Seventh International Heat Transfer Conference*, Munich, Germany, 6, 197–202.

Hewitt, G.F.., Ed., 1998. *Heat Exchanger Design Handbook 1998*, Begell House, New York.

Hewitt, G.F., Shires, G.L., and Bott, T.R. 1994. *Process Heat Transfer*, CRC/Begell House, Boca Raton, FL.

Kern, D.Q. and Kraus, A.D. 1972. *Extended Surface Heat Transfer*, McGraw-Hill, New York.

Kral, D., Stehlik, P., Van der Ploeg, H.J., and Master, B.I., 1996. Helical baffles in shell and tube heat exchangers. Part I: Experimental verification, *Heat Transfer Eng.*, 17(1), 93–101.

Palen, J.W. and Taborek, J. 1969. Solution of shell side flow pressure drop and heat transfer by stream analysis method, *Chem. Eng. Prog. Symp. Ser. No. 92, Heat Transfer-Philadelphia,* 65, 53–63.

Saunders, E.A.D. 1988. *Heat Exchangers: Selection, Design, and Construction,* Longman Scientific & Technical/John Wiley & Sons, New York.

Singh, K.P. and Soler, A.I. 1984. *Mechanical Design of Heat Exchangers and Pressure Vessel Components*, Arcturus, Cherry Hill, NJ.

TEMA. 1988. *Standards*, 7th ed., Tubular Exchanger Manufacturers Association, Tarrytown, NY.

Yokell, S. 1990. *A Working Guide to Shell and Tube Heat Exchangers*, McGraw-Hill, New York.

4.4 Bioheat Transfer

Kenneth R. Diller, Jonathan W. Valvano, and John A. Pearce

Abstract

Heat transfer plays a major role in many processes in living systems. This chapter presents a brief review of bioheat transfer from the perspective of those aspects which distinguish it from processes in inanimate systems. Of particular interest are the effects of blood perfusion on temperature distribution, coupling with biorate processes, measurement of thermal transport properties in tissues, therapeutic processes, human thermoregulation, and effects on living tissues of elevated temperatures and subzero temperatures. An extensive compilation of properties for biomaterials is also presented.

Introduction

Analysis and control of heat transfer processes in living systems involve phenomena that are not encountered in systems that are not alive. For example, there is a continuous flow of blood through tissues and organs via a complex network of branching vessels. Heat exchange processes are influenced by vascular geometry, which consists of arteries and veins with countercurrent flow over large portions of the network of pairs. For some physiological conditions, arterial and venous blood temperatures are different, neither is equal to the local tissue temperature, and these temperature differences may vary as a function of many transient physiological and physical parameters. Mechanisms of regulation for the thermal state in tissue are quite nonlinear and have presented a major challenge to understand and model. Nonetheless, thermoregulatory processes are critical to the maintenance of life and must be accounted for in the design of many types of systems that interface with humans and animals.

Although considerable progress has been made in defining and modeling thermal processes in living systems, and especially so over the past 15 years, many important problems remain. This section provides a brief discussion of the present state of knowledge and understanding of some of the more active areas of bioheat transfer. Greater detail and additional areas of application of bioheat transfer may be found among the references cited.

Coupling of Temperature History to Rate Processes

Although the processes of life are observed at the system (macroscopic) and microscopic levels of resolution, the true essence of life lies in the multitudinous biochemical phenomena which occur continuously throughout all organisms. Over the past 50 years the science of biochemistry has elucidated many of the governing life processes at the molecular level. Today the major frontier of knowledge in the life sciences is associated with molecular biology. Increasingly, bioengineers are collaborating with molecular biologists to understand and manipulate the molecules and biochemical processes that constitute the basis of life. Temperature is a primary controlling parameter in the regulation of these rate processes. The study of thermal biology has identified that the rates of nearly all physiological functions are altered by 6 to 10% per degree Celsius over wide range of thermal states produced in response to environment conditions.[1] Likewise, temperature is often altered during therapeutic or diagnostic procedures to produce or measure a targeted effect, based on the fact that a change in local temperature will have a large effect on biochemical process rates. Thus, a knowledge of how temperature can be monitored and/or controlled in living tissues is of great value in both the assessment of normal physiological function and the treatment of pathological states.

In assessing the effects of temperature alterations on biochemical rate processes, two broad categories of state changes can be considered: temperatures both above and below the normal physiological thermal state. An extensive review of both of these thermal domains has been published recently,[2] to which the reader is referred for further details and bibliography.

A summary of some of the most important and recent advances in bioheat transfer will be discussed in this section.

Tissue Thermal Transport Properties

The transport of thermal energy in living tissue is a complex process involving multiple phenomenological mechanisms including conduction, convection, radiation, metabolism, evaporation, and phase change. The equilibrium thermal properties presented here were measured after temperature stability had been achieved.

Thermal probe techniques are used frequently to determine the thermal conductivity and the thermal diffusivity of biomaterials.[3-6] Common to these techniques is the use of a thermistor bead either as a heat source or a temperature sensor. Various thermal diffusion probe techniques[7] have been developed from Chato's first practical use of the thermal probe.[3] Physically, for all of these techniques, heat is introduced to the tissue at a specific location and is dissipated by conduction through the tissue and by convection with blood perfusion.

Thermal probes are constructed by placing a miniature thermistor at the tip of a plastic catheter. The volume of tissue over which the measurement occurs depends on the surface area of the thermistor. Electrical power is delivered simultaneously to a spherical thermistor positioned invasively within the tissue of interest. The tissue is assumed to be homogeneous within the milliliter surrounding the probe. The electrical power and the resulting temperature rise are measured by a microcomputer-based instrument. When the initial tissue temperature is just below the freezing point, the thermistor heat is removed both by conduction and by latent heat. In this situation, the instrument measures effective thermal properties that are the combination of conduction and latent heat. By taking measurements over a range of temperatures, the processes of conduction and latent heat can be separated. When the tissue is perfused by blood, the thermistor heat is removed both by conduction and by heat transfer due to blood flow near the probe. *In vivo*, the instrument measures effective thermal properties that are the combination of conductive and convective heat transfer. Thermal properties are derived from temperature and power measurements using equations that describe heat transfer in the integrated probe/tissue system.

The following five complexities make the determination of thermal properties a technically challenging task. First, tissue heat transfer includes conduction, convection, radiation, metabolism, evaporation, and phase change. It is difficult but necessary to decouple these different heat transfer mechanisms. Second, the mechanical and thermal interactions between the probe and tissue are complex, and must be properly modeled to achieve accurate measurements. When the probe is inserted into living tissue a fluid pool may form around the probe because of the mechanical trauma. Because the probe is most sensitive to the tissue closest to it, the presence of a pool of blood and other fluids will significantly alter the results. Tissue damage due to probe insertion may also occur *in vitro*. Third, the tissue structure is quite heterogeneous within each sample. Thus, the probe (which returns a single measurement value) measures a spatial average of the tissue properties surrounding the active elements. Unfortunately, the spatial average is very nonuniform.[8] The probe is most sensitive to the tissue immediately adjacent to it. It is important to control this effective measurement volume. If the effective volume is too small, then the measurement is highly sensitive to the mechanical/thermal contact between the probe and tissue. If the effective volume is too large, then the measurement is sensitive to the boundary conditions at the surface of the tissue sample. Fourth, there are significant sample to sample and species to species variabilities. One must be careful when extrapolating results obtained in one situation to different situations. Fifth, tissue handling is critical. Thermal properties are dependent on temperature and water content.[9-12] Blood flow, extracellular water, and local metabolism are factors that strongly affect heat transfer in living tissue, but are difficult to determine or control experimentally. Once a tissue dies, if handled improperly there will be significant water fluxes which will affect tissue thermal properties. Tissues should be stored in a slightly hypertonic saline buffer or on a saline-moistened surgical gauze pad and wrapped in aluminum foil and kept at 4°C to minimize tissue mass transfer.

Currently, there is no method to quantify simultaneously the major three parameters: the intrinsic tissue thermal conductivity, k_m, the tissue thermal diffusivity, α_m, and perfusion, w. Either the knowledge of k_m is required prior to the perfusion measurement, or even when k_m is measured in the presence of perfusion, the thermal diffusivity cannot be measured.[6,13]

Background

There are many good reviews of techniques to measure thermal properties.[10,14-16] *Thermophysical Properties of Matter* is a ten-volume set that catalogs thermal properties. Volumes 1 and 3 contain thermal conductivity data, and Volume 10 contains thermal diffusivity data. Extensive reviews of measurement techniques exist as prefaces to each volume of the set. Additional thermal property data can be found in Kreith and Bohn.[17] John Chato has written an excellent chapter in *Heat Transfer in Medicine and Biology*, edited by Shitzer and Eberhart, which reviews techniques to measure thermal properties of biologic materials.[10] Valvano has documented the temperature dependence of tissue thermal properties.[11,12] Duck has written an excellent review chapter on this subject.[18]

The thermal diffusion probe was conceived by Chato.[3,10] Significant developments were obtained by Balasubramaniam, Bowman, Chen, Holmes, and Valvano.[19-27] Patel and Walsh have applied the self-heated thermistor technique to nondestructive surface measurements.[8,24-26] Unfortunately, surface probes are unreliable due to poor probe/tissue contact and uncertain boundary conditions at the tissue surface.[8,24]

Self-heated thermistors have been used to measure perfusion.[4,9,19,20,22-24] Effective thermal conductivity, k_{eff}, is the total ability of perfused tissue to transfer heat in the steady state; k_{eff} is the combination of conduction (due to intrinsic thermal conductivity, k_m) and convection (due to perfusion). Measurements of k_{eff} are very sensitive to perfusion. The limitation of most techniques is that the intrinsic tissue thermal conductivity must be known in order to accurately measure perfusion. Holmes and Chen use a combination of steady-state and transient heating modes to determine perfusion without requiring a no-flow calibration measurement.[13,19,20] The uncertainty of k_m significantly limits the perfusion accuracy.[9]

Measurement of Thermal Conductivity and Diffusivity

Methods

In the constant temperature heating technique, the instrument first measures the baseline tissue temperature, T_0. Then, an electronic feedback circuit applies a variable voltage, V(t), in order to maintain the average thermistor temperature at a predefined constant, T_h. The applied thermistor power includes a steady state and a transient term:

$$P(t) = A + Bt^{-1/2} \tag{4.4.1}$$

where A and B are system constants that depend on the characteristics of the probe and the thermal interaction with the tissue in which it is embedded.

In order to measure thermal conductivity, thermal diffusivity, and tissue perfusion, the relationship between applied thermistor power, P, and resulting thermistor temperature rise, $\Delta T(t) = T_h - T_0$, must be known. In the constant temperature method, ΔT is constant. The thermistor bead is treated as a sphere of radius "a" embedded in a homogeneous medium. Since all media are considered to have constant parameters with respect to time and space, the initial temperature will be uniform when no power is supplied to the probe.

$$T_b = T_m = T_0 = T_a + \frac{Q_{met}}{wc_{bl}} \quad \text{at } t = 0 \tag{4.4.2}$$

where w is the rate of perfusion of blood through the local tissue (ml blood/ml tissue – s) and c_{bl} is the specific heat of blood (kJ/kg-K).

Let $\overline{T}$ be the temperature rise above baseline, $\overline{T} = T - T_0$. Both the thermistor bead temperature rise ($\overline{T}_b$) and the tissue temperature rise ($\overline{T}_m$) are initially zero. The subscripts b and m refer, respectively, to the bead and the tissue (i.e., the surrounding medium).

$$\overline{T}_b = \overline{T}_m = 0 \quad \text{at } t = 0 \tag{4.4.3}$$

Assuming the venous blood temperature equilibrates with the tissue temperature and that the metabolic heat is uniform in time and space, the Pennes' bioheat transfer equation[28] in spherical coordinates is given by:

$$\rho_b c_b \frac{\partial \overline{T}_b}{\partial t} = k_b \frac{1}{r^2}\frac{\partial}{\partial r}\left(r^2 \frac{\partial \overline{T}_b}{\partial r}\right) + \frac{A + Bt^{-1/2}}{\frac{4}{3}\pi a^3} \qquad r < a \tag{4.4.4}$$

$$\rho_m c_m \frac{\partial \overline{T}_m}{\partial t} = k_m \frac{1}{r^2}\frac{\partial}{\partial r}\left(r^2 \frac{\partial \overline{T}_m}{\partial r}\right) - w c_{bl} V_m \qquad r > a \tag{4.4.5}$$

Perfect thermal contact is assumed between the finite-sized spherical thermistor and the infinite homogeneous perfused tissue. At the interface between the bead and the tissue, continuity of thermal flux and temperature leads to the following boundary conditions:

$$\overline{T}_b = \overline{T}_m \quad \text{at } r = a \tag{4.4.6}$$

$$k_b \frac{\partial \overline{T}_b}{\partial r} = k_m \frac{\partial \overline{T}_m}{\partial r} \quad \text{at } r = a \tag{4.4.7}$$

The other necessary boundary conditions are at positions $r \to 0$ and $r \to \infty$. Since no heat is gained or lost at the center of the thermistor:

$$\overline{T}_b = \text{finite}\left(\text{or } k_b \frac{\partial \overline{T}_b}{\partial r} = 0\right) \quad \text{as } r \to 0 \tag{4.4.8}$$

Because the thermistor power is finite and the tissue is infinite, the tissue temperature rise at infinity goes to zero:

$$\overline{T}_m \to 0 \quad \text{as } r \to \infty \tag{4.4.9}$$

It is this last initial condition that allows the Laplace transform to be used to solve the coupled partial differential equations. The Laplace transform converts the partial differential equations into ordinary differential equations that are independent of time t. The steady-state solution allows for the determination of thermal conductivity and perfusion.[20]

$$\overline{T}_b(r) = \frac{A}{4\pi a k_b}\left\{\frac{k_b}{k_m\left(1+\sqrt{z}\right)} + \frac{1}{2}\left[1 - \left(\frac{r}{a}\right)^2\right]\right\} \tag{4.4.10}$$

$$\overline{T}_m(r) = \frac{A}{4\pi r k_m}\left(\frac{e^{(1-r/a)\sqrt{z}}}{1+\sqrt{z}}\right) \tag{4.4.11}$$

where z is a dimensionless Pennes' model perfusion term ($wc_{bl}a^2/k_m$). The measured thermistor response, ΔT, is assumed be the simple volume average of the thermistor temperature:

$$\Delta T = \frac{\int_0^a \bar{T}_b(r) 4\pi r^2 \, dr}{\frac{4}{3}\pi a^3} \tag{4.4.12}$$

Inserting Equation 4.4.10 into Equation 4.4.12 yields the relationship used to measure thermal conductivity, assuming no perfusion.[4]

$$k_m = \frac{1}{\frac{4\pi a \Delta T}{A} - \frac{0.2}{k_b}} \tag{4.4.13}$$

A similar equation allows the measurement of thermal diffusivity, α, from the transient response, again assuming no perfusion.[22]

$$\alpha_m = \left[\frac{a}{\sqrt{\pi}\, B/A \left(1 + 0.2 \frac{k_m}{k_b}\right)} \right]^2 \tag{4.4.14}$$

Rather than using the actual probe radius (a) and probe thermal conductivity (k_b), the following empirical equations are used to calculate thermal properties.

$$k_m = \frac{1}{\frac{c_1 \Delta T}{A} + c_2} \tag{4.4.15}$$

$$\alpha_m = \left[\frac{c_3}{\sqrt{\pi}\, B/A \left(1 + 0.2 \frac{k_m}{c_4}\right)} \right]^2 \tag{4.4.16}$$

The coefficients c_1, c_2, c_3, and c_4 are determined by operating the probe in two materials of known thermal properties. Typically, agar-gelled water and glycerol are used as thermal standards. This empirical calibration is performed at the same temperatures at which the thermal property measurements will be performed.

It is assumed that the baseline tissue temperature, T_0, is constant during the 30-s transient. Patel has shown that if the temperature drift, dT_0/dt, is larger than 0.1°C/min, then significant errors will occur.[8] The electronic feedback circuit forces T_h to a constant. Thus, if T_0 is constant, then ΔT does not vary during the 30-s transient perod.

The applied power, P(t), varies during the 30-s transient. Linear regression is used to calculate the steady-state and transient terms in Equation 4.4.1. Figure 4.4.1 shows some typical responses. The steady-state response (time equals infinity) is a measure of the thermal conductivity. The transient response (slope) indicates the thermal diffusivity.

The time of heating can vary from 10 to 60 s. Shorter heating times are better for small tissue samples and for situations where there is baseline tissue temperature drift. Another advantage of shorter heating times is the reduction in the total time required to make one measurement. Longer heating times increase

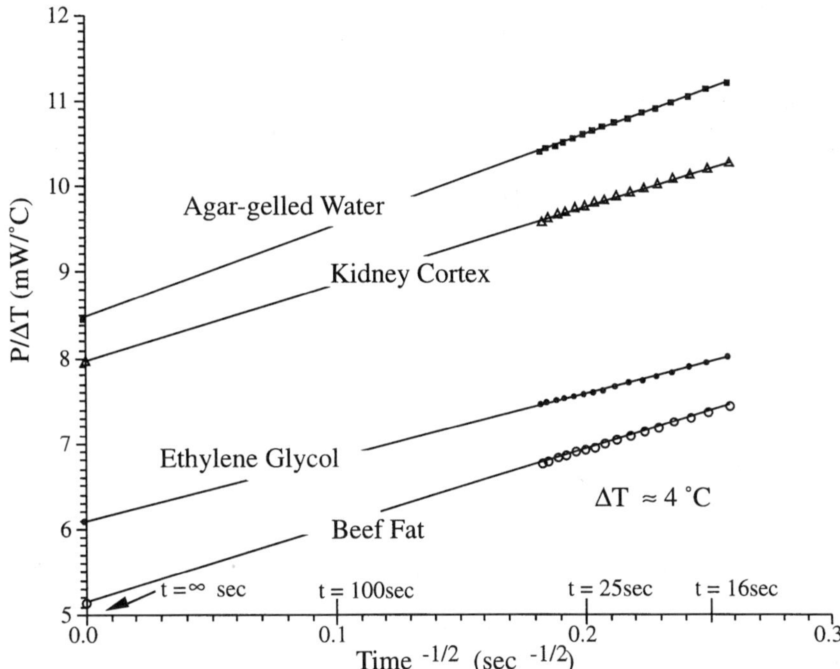

FIGURE 4.4.1 Typical P/ΔT vs. time$^{-1/2}$ data for the constant temperature heating technique. (From Pearce, J.S. and Thomsen, S., in *Optical-Thermal Response of Laser-Irradiated Tissue,* Welch, A.D. and van Germert, M.M.C., Eds., Plenum Press, New York, 1995. With permission.)

the measurement volume and reduce the effect of imperfect thermistor/tissue coupling. Typically, shorter heating times are used *in vivo* because it allows more measurements to be taken over the same time period. On the other hand, longer heating times are used *in vitro* because accuracy is more important than measurement speed.

Probe Design

Thermal probes must be constructed in order to measure thermal properties. The two important factors for the thermal probe are thermal contact and transducer sensitivity. The shape of the probe should be chosen to minimize trauma during insertion. Any boundary layer between the thermistor and the tissue of interest will cause a significant measurement error. The second factor is transducer sensitivity; that is, the slope of the thermistor voltage vs. tissue thermal conductivity. Equation 4.4.13 shows that for a fixed ΔT, k_m, and k_b, the thermistor power (A) increases linearly with probe size (a). Therefore, larger probes are more sensitive to thermal conductivity. Thermometrics P60DA102M and Fenwal 121-102EAJ-Q01 are glass probe thermistors that make excellent transducers. The glass-coated spherical probes provide a large bead size and a rugged, stable transducer. The Thermometrics BR55KA102M and Fenwal 112-102EAJ-B01 bead thermistors also provide excellent results. For large tissue samples multiple thermistors can be wired in parallel, so they act electrically and thermally as one large device. There are two advantages to using multiple thermistors. The effective radius, $a = c_1/4\pi$, is increased from about 0.1 cm for a typical single P60DA102M probe to about 0.5 cm for a configuration of three P60DA102M thermistors. The second advantage is that the three thermistors are close enough to each other that the tissue between the probes will be heated by all three thermistors. This cooperative heating tends to increase the effective measurement volume and reduce the probe/tissue contact error. Good mechanical/thermal contact is critical. The probes are calibrated after they are constructed, so that the thermistor geometry is incorporated into the coefficients c_1, c_2, c_3, and c_4. The same water bath and probe configuration should be used during the calibration and tissue measurements.

Calibration

Calibration is a critical factor when using an empirical technique. For temperatures below 0°C, ice and ethylene glycol are used as thermal standards. For temperatures between 0 and 15°C, agar-gelled water and ethylene glycol can be used as thermal standards. For temperatures between 15 and 75°C, agar-gelled water and glycerol can be used; 1 g of agar per 100 ml of water should be added to prevent convection. The instrument has been used to measure k_m and α_m of various concentrations of agar-gelled water at 37°C. A mixture of water and glycerol can be used to estimate the accuracy of the technique. The mass fraction, m, can be used to determine the true thermal properties of the mixture where k_w and k_g are in mW/(m – k):[29]

$$k_m = m\ k_g + (1-m)k_w + 1.4\ m\ (m-1)\left(k_w - k_g - 2\right) - 0.014\ m\ (m-1)(T - 20°C) \quad (4.4.17)$$

$$\alpha_m = m\ \alpha_g + (1-m)\ \alpha_w \quad (4.4.18)$$

Estimation of Heat Transfer Coefficient

Introduction

The heart is an organ in which it is possible to make experimental measurements of convective transport between flowing blood and tissue. The topologies of the inner surfaces of the heart and blood vessels are very complex. Consequently, it is very difficult to model the heat transfer at the wall analytically. Any attempt to estimate the heat transfer coefficient analytically will be inaccurate because of the complex contours of structures like the endocardium. The experimental method presented in this section to estimate the heat transfer coefficient is very appropriate.

The following method can be used to estimate the convective coefficient due to blood flow inside the chambers of the heart. This method can be applied to blood vessels as well. The method to measure the convective coefficient due to blood flow inside the heart is derived from basic heat transfer analysis. Two miniature thermistor (BR11, Thermometrics Inc.) probes are inserted at fixed locations near the inner surface of the model. T_1 is the measured temperature nearest the endocardial wall, and T_2 is measured 2 mm inside the myocardial wall. The epicardial surface of the heart is maintained at a steady temperature, T_o. Blood at a different temperature T_f is pumped through the heart. The thermistor probes are used to continuously monitor the spatial temperature gradient in the heart muscle. The basic principle used to estimate the heat transfer coefficient, h, is as follows:

$$h \cdot \left(T_{wall} - T_f\right) = k_s \cdot \left.\frac{dT}{d\hat{n}}\right|_{wall}, \quad (4.4.19)$$

where $\hat{n}$ is the direction perpendicular to the surface and k_s is the thermal conductivity of the heart wall. The gradient at the wall of the surface is calculated from the data obtained from the array of thermistors.

Calibration

Ideally, the temperature measured in any experiment should be the temperature of the zero-volume point of interest. But when a measurement is made the temperature sensor has a finite volume and the actual measurement is the volume-averaged measurement of some region around the sensor. Similarly, the gradient measured using two sensors is an estimate of the gradient between the two sensors. These two errors are responsible for the estimate of the heat transfer coefficient measured using Equation 4.4.19 to be vastly different from the correct value. Hence, the probe assembly is first calibrated by operating the probe in situations where the true convective heat transfer coefficient is known. Two calibration factors, c_1 c_2, are added to Equation 4.4.19 to adjust for the finite size of the temperature sensors, and for the fact that the measured temperature gradient is not at the wall surface:

$$h = c_1 \exp\left(c_2 \frac{k_s (T_1 - T_2)}{(T_f - T_1)} \right) \tag{4.4.20}$$

where T_1 and T_2 are measured temperatures in the wall. A cylindrical tube made with a material (e.g., silicone) that has thermal properties close to tissue can be used to calibrate the sensor combination. The key to accurate measurements lies in how close the calibration configuration matches the actual experimental conditions.

Verification

The effect of measurement errors on the estimate of the heat transfer coefficient was studied for a tube of inner radius R_i (0.73 cm) and outer radius of R_o (1.33 cm). The temperature distribution inside the tube wall was analytically determined for water flowing inside and the outer wall temperature held at a constant value of 23°C. The heat transfer coefficient was then calculated from the temperature distribution using the relation in Equation 4.4.19. The effect of the measurement errors on the heat transfer coefficient estimate was analyzed by using values for the temperature and the gradient away from the wall of the tube.

The temperature at any point of distance R from the axis of the tube is given by

$$T = T_o - Bi(T_f - T_w)\ln\left(\frac{R}{R_o}\right), \tag{4.4.21}$$

where the T_w is the inner wall temperature given by

$$T_w = \frac{T_o - Bi \ln\left(\frac{R_i}{R_o}\right) T_f}{1 - Bi \ln\left(\frac{R_i}{R_o}\right)} \tag{4.4.22}$$

where T_o is the outer wall temperature, Bi is the Biot number, calculated as (hR_i/k), h is the heat transfer coefficient, and k is the thermal conductivity of the material of the tube.

In Vitro Studies

The first *in vitro* study used two cylindrical tubes. The first tube was used to calibrate the probe and the second to evaluate measurement accuracy. The exact position of the thermistor array for sensing the temperature must be in a similar location relative to the inner wall for both calibration and measurement. Water at 37°C was made to flow through the tube at a rate of L (liters per minute). The flow was measured using a rotometer type flowmeter (accuracy 2%, Omega Engineering Inc.). The temperature gradients were recorded continuously.

The tube was initially maintained at room temperature. For a smooth tube of cylindrical cross section the relation between Nu, Re, and Pr for turbulent flow conditions is given by:

$$Nu_D = \frac{(f/8)(\mathrm{Re}_D - 1000)\mathrm{Pr}}{1 + 12.7(f/8)^{1/2}(\mathrm{Pr}^{2/3} - 1)}, \tag{4.4.23}$$

where the friction factor, f is

$$f = (0.79 \ln \mathrm{Re}_D - 1.64)^{-2} \tag{4.4.24}$$

This correlation is valid for $0.5 < Pr < 2000$ and $2300 < Re_D < 5 \times 10^6$. The setup was calibrated with a tube of 1.46 cm diameter and then tested with a tube of 1.27 cm diameter. The flow rates were selected to give a range of h values from 800 to 4000 W/m^2-K. The average accuracy is about 10%.

Temperature-Dependent Thermal Properties

Temperature Dependence of Organ Tissue

When modeling heat transfer in situations where the temperature change exceeds 10°C it is important to consider the temperature dependence of the tissue thermal properties. Valvano measured tissue thermal properties as a function of temperature using the constant ΔT thermistor heating technique.[11,12] The results, shown in Table 4.4.1, were derived from *in vitro* measurements taken at 3, 10, 17, 23, 30, 37, and 45°C.

TABLE 4.4.1 Thermal Properties as a Function of Temperature

Tissue	Species	k_0 mW/cm°C	k_1 mW/cm°C^2	α_0 cm^2/s	α_1 cm^2/s°C
Adenocarinoma of the breast	Human	4.194	0.03911	0.001617	–0.000049
Cerebral cortex	Human	5.043	0.00296	0.001283	0.000050
Colon cancer	Human	5.450	(at 19°C)	0.001349	(at 19°C)
Fat of spleen	Human	3.431	–0.00254	0.001321	–0.000002
Liver	Human	4.692	0.01161	0.001279	0.000036
Liver	Pig	4.981	0.00800	0.001240	0.000053
Liver	Rabbit	4.668	0.02601	0.001370	0.000178
Lung	Human	3.080	0.02395	0.001071	0.000082
Lung	Human	4.071	0.01176	0.001192	0.000031
Lung	Pig	2.339	0.02216	0.000695	0.000080
Myocardium	Dog	4.869	0.01332	0.001296	0.000058
Myocardium	Human	4.925	0.01195	0.001289	0.000050
Myocardium	Pig	4.841	0.01333	0.001270	0.000051
Pancreas	Dog	4.790	0.00849	0.001287	0.000062
Pancreas	Human	4.365	0.02844	0.001391	0.000084
Pancreas	Pig	4.700	0.00194	0.001530	0.000130
Renal cortex	Dog	4.905	0.01280	0.001333	0.000039
Renal cortex	Human	4.989	0.01288	0.001266	0.000055
Renal cortex	Pig	4.967	0.01176	0.001284	0.000039
Renal cortex	Rabbit	4.945	0.01345	0.001311	0.000027
Renal medulla	Dog	5.065	0.01298	0.001305	0.000063
Renal medulla	Human	4.994	0.01102	0.001278	0.000055
Renal pelvis	Dog	4.930	0.01055	0.001334	0.000052
Renal pelvis	Human	4.795	0.01923	0.001329	0.000011
Spleen	Human	4.913	0.01300	0.001270	0.000047
Spleen	Rabbit	4.863	0.01267	0.001257	0.000042

Source: Pearce, J.S. and Thomsen, S., in *Optical-Thermal Response of Laser-Irradiated Tissue,* Welch, A.D. and van Germert, M.M.C., Eds., Plenum Press, New York, 1995. With permission.

The animal tissues were measured from freshly sacrificed dogs, rabbits, and pigs. The normal human tissues were obtained by autopsy. The human cancers were freshly excised. The k_0, k_1, α_0, and α_1 values are the linear fit of the thermal properties as a function of temperature.

$$k = k_0 + k_1 T \tag{4.4.25}$$

$$\alpha = \alpha_0 + \alpha_1 T \tag{4.4.26}$$

The average thermal properties of these data are

$$k = 4.574 + 0.01403 T \tag{4.4.27}$$

$$\alpha = 0.001284 + 0.000053 \text{ T} \tag{4.4.28}$$

where conductivity is in mW/cm°C, diffusivity is in cm²/s, and temperature is in °C.

Temperature Dependence of Human Arterial Tissue

Aortic tissue was obtained from a local pathology lab. The thermal probes were placed on the endothelial surface of the aortic wall, and the tissue/probe combination was wrapped in plastic. The tissue surface was kept wet to improve the thermal contact and to prevent drying. The samples were placed in a beaker of saline and the beaker was put into a temperature-controlled water bath. Thermal conductivity and thermal diffusivity were measured 10 times at each temperature 35, 55, 75, and 90°C. The measurement sequence was varied between 35, 55, 75, 90, 95, 75, 55, 35 75, 90, 55, 35 and 55, 35, 90, 75. Measurements were obtained from both normal and diseased tissue. The plaques were categorized by gross visual observation. The calcified plaques were hard and bony. The fibrous plaques were firm but pliable. The fatty plaques were loose and buttery. The results from 54 tissues are presented in Tables 4.4.2 and 4.4.3. The column n refers to the number of tissue samples. The standard deviation is given in the parentheses.

TABLE 4.4.2 Thermal Conductivity (mW/cm-°C) of Human Aorta and Atherosclerotic Plaque

Tissue	n	at 35°C	at 55°C	at 75°C	at 90°C
Normal aorta	12	4.76 (0.41)	5.03 (0.60)	5.59 (0.37)	6.12 (0.12)
Fatty plaque	13	4.84 (0.44)	4.97 (0.49)	5.46 (0.54)	5.88 (0.81)
Fibrous plaque	12	4.85 (0.22)	5.07 (0.30)	5.38 (0.38)	5.77 (0.56)
Calcified plaque	17	5.02 (0.59)	5.26 (0.73)	5.81 (0.82)	6.19 (0.85)

Source: Pearce, J.S. and Thomsen, S., in *Optical-Thermal Response of Laser-Irradiated Tissue,* Welch, A.D. and van Germert, M.M.C., Eds., Plenum Press, New York, 1995. With permission.

TABLE 4.4.3 Thermal Diffusivity (*1000 cm²/s) of Human Aorta and Atherosclerotic Plaque

Tissue	n	at 35°C	at 55°C	at 75°C	at 90°C
Normal aorta	12	1.27 (0.07)	1.33 (0.11)	1.44 (0.10)	1.56 (0.05)
Fatty plaque	13	1.28 (0.05)	1.32 (0.06)	1.41 (0.11)	1.46 (0.15)
Fibrous plaque	12	1.29 (0.03)	1.36 (0.07)	1.41 (0.10)	1.52 (0.20)
Calcified plaque	17	1.32 (0.07)	1.37 (0.12)	1.53 (0.17)	1.66 (0.20)

Source: Pearce, J.S. and Thomsen, S., in *Optical-Thermal Response of Laser-Irradiated Tissue,* Welch, A.D. and van Germert, M.M.C., Eds., Plenum Press, New York, 1995. With permission.

The two-sample *t*-test with $p = .05$ was used to determine significant differences. The tissue thermal properties increased with temperature and were significantly less than those of water. The measurement order did not affect the measured thermal properties. There was no significant difference between the thermal conductivity of normal aorta, fatty plaque, and fibrous plaque. The thermal conductivity and thermal diffusivity of calcified plaque were slightly higher than normal aorta, fatty plaque, and fibrous plaque.

Temperature Dependence of Canine Arterial Tissue

Carotid and femoral arteries were harvested immediately post-mortem. The thermal probes were placed on the endothelial surface of the arterial wall. Thermal conductivity and thermal diffusivity were measured 10 times at each temperature 25, 35, 45, 55, 65, 75, 85, and 95°C. Measurements were obtained only from normal tissue. The results from 18 tissues are summarized below.

Canine femoral artery

$$k(\text{mW/cm-°C}) = 3.688 + 0.0062014 \text{ T (°C)} \tag{4.4.29}$$

$$\alpha\left(cm^2/s\right) = 0.001003 + 0.000001381\ T\ (°C) \tag{4.4.30}$$

Canine carotid artery

$$k\ \left(mW/cm\text{-}°C\right) = 4.480 + 0.0000164\ T\ (°C) \tag{4.4.31}$$

$$\alpha\left(cm^2/s\right) = 0.001159 + 0.000003896\ T\ (°C) \tag{4.4.32}$$

The two-sample t-test with $p = 0.01$ shows that both thermal conductivity and thermal diffusivity are higher in carotid vs. femoral artery. These results could be explained from the fact that the carotid artery contains more collagen than femoral artery. A tissue with a higher percentage of collagen would have lower thermal properties because collagen is a thermal insulator.

Temperature Dependence of Swine Left Ventricle

Swine myocardial samples were harvested immediately post-mortem. The thermal probes were placed on the left ventricular muscle. Thermal conductivity and thermal diffusivity were measured 10 times at each temperature 25, 37, 50, 62, and 76°C. Measurements were obtained only from normal tissue. The results are summarized in Tables 4.4.4 and 4.4.5.

TABLE 4.4.4 Thermal Conductivity ($mW \cdot cm^{-1} \cdot K^{-1}$) of Myocardial Tissue

Temperature	25°C	37°C	50°C	62°C	76°C
	5.23	5.14	5.17	4.39	5.24
	5.07	5.12	4.75	3.30	4.29
	5.30	5.21	5.61	5.67	4.83
	5.43	5.54	4.22	4.16	5.89
	4.68	5.35	4.93	5.33	5.23
	5.25	5.08	4.84	5.70	5.39
	5.27	5.48	4.42	5.11	4.75
	5.28	4.57	4.93	4.99	3.25
	5.86	5.76	5.52	5.03	2.69
	4.78	5.10	5.88	5.30	5.28
	4.75	5.35	5.35	4.67	5.60
	4.92	6.02	5.60	5.49	4.68
Mean	5.15	5.31	5.1	4.93	4.76
Std. Dev.	0.33	0.37	0.51	0.70	0.95

Thermal Properties of Frozen Tissue

The thermal properties of frozen tissue are significantly higher than normal tissue. Valvano measured frozen tissue thermal properties using the constant ΔT thermistor heating technique.[30] The results, shown in Table 4.4.6, were derived from *in vitro* measurements taken at –18, –5, and +0.1°C.

Thermal Properties as a Function of Water and Fat Content

In a global sense, the thermal properties of tissue are determined by the relative concentrations of its constituent parts. Spells found a linear relationship between tissue thermal conductivity and water content:[31]

$$k\ \left(mW/cm\text{-}°C\right) = 0.54 + 5.73\ m_{water}\ \text{for}\ m_{water} > 0.2 \tag{4.4.33}$$

where m_{water} is the mass fraction of water in the tissue. Cooper and Trezek found an empirical relationship between thermal conductivity and mass fractions of water, protein, and fat[32]:

TABLE 4.4.5 Thermal Diffusivity ($cm^2 \cdot s^{-1}$) of Myocardial Tissue

Temperature	25°C	37°C	50°C	62°C	76°C
	0.00151	0.00170	0.00165	0.00159	0.00167
	0.00154	0.00147	0.00203	0.00235	0.00249
	0.00143	0.00165	0.00151	0.00169	0.00166
	0.00146	0.00143	0.00116	0.00191	0.00229
	0.00159	0.00160	0.00176	0.00167	0.00173
	0.00141	0.00178	0.00179	0.00163	0.00185
	0.00165	0.00149	0.00235	0.00143	0.00185
	0.00132	0.00206	0.00179	0.00170	0.00199
	0.00141	0.00144	0.00147	0.00143	0.00062
	0.00168	0.00179	0.00160	0.00180	0.00167
	0.00154	0.00156	0.00173	0.00161	0.00173
	0.00164	0.00138	0.00171	0.00169	0.00192
Mean	0.00152	0.00161	0.00171	0.00171	0.00179
Std. Dev.	0.00012	0.00020	0.00031	0.00025	0.00047

TABLE 4.4.6 Average Thermal Properties

T (°C)	N	M	Species Tissue	k_m(mW/cm°C)	α_m(1000*cm²/s)
+0.1	45	6	Bovine kidney cortex	4.54 (±0.16)	1.18 (±0.09)
−5	15	4	Bovine kidney cortex	15.35 (±1.09)	4.71 (±0.99)
−18	18	3	Bovine kidney cortex	13.72 (±0.73)	6.84 (±0.83)
+0.1	66	9	Bovine liver	4.17 (±0.13)	1.05 (±0.09)
−5	66	9	Bovine liver	13.96 (±2.49)	4.77 (±0.58)
−18	56	8	Bovine liver	9.89 (±0.44)	5.71 (±0.74)
+0.1	48	6	Bovine muscle	4.25 (±0.37)	1.05 (±0.11)
−5	42	7	Bovine muscle	13.93 (±1.23)	5.37 (±0.97)
−18	60	8	Bovine muscle	10.76 (±1.14)	6.84 (±1.10)
+0.1	21	3	Bovine fat	1.93 (±0.12)	0.59 (±0.13)
−5	32	4	Bovine fat	2.66 (±0.38)	0.98 (±0.19)
−18	24	4	Bovine fat	2.80 (±0.53)	1.54 (±0.57)

Note: N is the number of measurements; M is the number of tissues; (±0.xx) is the standard deviation of the average.

Source: Pearce, J.S. and Thomsen, S., in *Optical-Thermal Response of Laser-Irradiated Tissue,* Welch, A.D. and van Germert, M.M.C., Eds., Plenum Press, New York, 1995. With permission.

$$k\left(\text{mW/cm-°C}\right) = \rho \sum_n \frac{k_n m_n}{\rho_n} = \rho\left(6.28\ m_{water} + 1.17\ m_{protein} + 2.31\ m_{fat}\right) \tag{4.4.34}$$

Cooper and Trezek[32] found similar relationships for specific heat and density.

$$c\left(\text{J/g-°C}\right) = \sum_n c_n\ m_n = 4.2\ m_{water} + 1.09\ m_{protein} + 2.3\ m_{fat} \tag{4.4.35}$$

$$\rho\left(\text{g/cm}^3\right) = \frac{1}{\sum_n \frac{m_n}{\rho_n}} = \frac{1}{m_{water} + 0.649\ m_{protein} + 1.227\ m_{fat}} \tag{4.4.36}$$

Effect of Blood Flow on Temperature

Bioheat transfer processes in living tissues are often influenced by blood perfusion through the vascular network. When there is a significant difference between the temperature of blood and the tissue through which it flows, convective heat transport will occur, altering the temperatures of both the blood and the tissue. Perfusion-based heat transfer interaction is critical to a number of physiological processes such as thermoregulation and inflammation.

The blood/tissue thermal interaction is a function of several parameters including the rate of perfusion and the vascular anatomy, which vary widely among the different tissues, organs of the body, and pathology. Appendix B contains an extensive compilation of perfusion rate data for many tissues and organs and for many species. The literature on mathematical modeling of the influence of blood perfusion on bioheat transfer phenomena has been reviewed recently by Charney,[33] and this reference is highly recommended for readers desiring an in-depth presentation of the topic.

The rate of perfusion of blood through different tissues and organs varies over the time course of a normal day's activities, depending on factors such as physical activity, physiological stimulus, and environmental conditions. Further, many disease processes are characterized by alterations in blood perfusion, and some therapeutic interventions result in either an increase or decrease in blood flow in a target tissue. For these reasons, it is very useful in a clinical context to know what the absolute level of blood perfusion is within a given tissue. There are numerous techniques that have been developed for this purpose over the past several decades. In some of these techniques, the coupling between vascular perfusion and local tissue temperature is applied to advantage to assess the flow through local vessels via inverse solution of equations which model the thermal interaction between perfused blood and the surrounding tissue.

Pennes[28] published the seminal work on developing a quantitative basis for describing the thermal interaction between tissue and perfused blood. His work consisted of a series of experiments to measure temperature distribution as a function of radial position in the forearms of nine human subjects. A butt-junction thermocouple was passed completely through the arm via a needle inserted as a temporary guideway, with the two leads exiting on opposite sides of the arm. The subjects were unanesthetized so as to avoid the effects of anesthesia on blood perfusion. Following a period of normalization, the thermocouple was scanned transversely across the mediolateral axis to measure the temperature as a function of radial position within the interior of the arm. The environment in the experimental suite was kept thermally neutral during the experiments. Pennes' data showed a temperature differential of three to four degrees between the skin and the interior of the arm, which he attributed to the effects of metabolic heat generation and heat transfer with arterial blood perfused through the microvasculature.

Pennes proposed a model to describe the effects of metabolism and blood perfusion on the energy balance within tissue. These two effects were incorporated into the standard thermal diffusion equation, which is written in its simplified form as:

$$\rho c \frac{\partial T}{\partial t} = \nabla \cdot k \nabla T + (\rho c)_b \, \omega_b \left(T_a - T\right) + q_{met} \tag{4.4.37}$$

Metabolic heat generation, q_{met}, is assumed to be homogeneously distributed throughout the tissue of interest as the rate of energy deposition per unit volume. It is assumed that the blood perfusion effect is homogeneous and isotropic and that thermal equilibration occurs in the microcirculatory capillary bed. In this scenario, blood enters capillaries at the temperature of arterial blood, T_a, where heat exchange occurs to bring the temperature to that of the surrounding tissue, T. There is assumed to be no energy transfer either before or after the blood passes through the capillaries, so that the temperature at which it enters the venous circulation is that of the local tissue. The total energy exchange between blood and tissue is directly proportional to the density, ρ_b, specific heat, c_b, and perfusion rate, ω_b, of blood through the tissue, and is described in terms of the change in sensible energy of the blood. This thermal transport model is analogous to the process of mass transport between blood and tissue, which is confined primarily to the capillary bed.

A major advantage of the Pennes model is that the added term to account for perfusion heat transfer is linear in temperature, which facilitates the solution of Equation 4.4.37. Since the publication of this work, the Pennes model has been adapted by many researchers for the analysis of a variety of bioheat transfer phenomena. These applications vary in physiological complexity from a simple homogeneous volume of tissue to thermal regulation of the entire human body.[34,35] As more scientists have evaluated the Pennes model for application in specific physiological systems, it has become increasingly clear that many of the assumptions foundational to the model are not valid. For example, Chato,[36] Chen and Holmes,[37] and Weinbaum et al.[38-50] have demonstrated very convincingly that thermal equilibration between perfused blood and local tissue occurs in the precapillary arterioles, and that by the time blood flows into vessels 60 μm in diameter and smaller, the equilibration process is complete. Therefore, no significant heat transfer occurs in the capillary bed; the exchange of heat occurs in the larger components of the vascular tree. The vascular morphology varies considerably among the various organs of the body, which contributes to the need for specific models for the thermal effects of blood flow (as compared to the Pennes model that incorporates no information concerning vascular geometry). It would appear as a consequence of these physiological realities that the validity of the Pennes model is questionable.

Many investigators have developed alternative models for the exchange of heat between blood and tissue. These models have accounted for the effects of vessel size (Chato,[36] Chen and Holmes,[37] Mooibroek and Lagendijk[51]), countercurrent heat exchange (Baish,[52] Huang et al.,[53] Keller and Seiler,[54] Mitchell and Meyers,[55] Mooibroek and Lagendijk[51]), as well as a combination of partial countercurrent exchange and bleed-off perfusion (Weinbaum and Jiji[38-50]). All of these models provided a larger degree of rigor in the analysis, but at the compromise of greater complexity and reduced generality. Some of these models have been the subject of considerable debate concerning their validity and range of appropriate application (Baish et al.,[56] Weinbaum and Jiji,[57] Wissler[58,59]). These studies also led to an increased appreciation of the necessity for a more explicit understanding of the local vascular morphology as it governs bioheat transfer, which has given rise to experimental studies to measure and characterize the three-dimensional architecture of the vasculature in tissues and organs of interest.

It is quite interesting that, in the context of the above studies to improve on the widely applied but questionable Pennes model, the 50th anniversary of the publication of Pennes' research was recognized recently.[60] For this occasion, Wissler[59] returned to Pennes' original data and analysis and reevaluated his work. Given the hindsight of 5 decades of advances in bioheat transfer plus greatly improved computational tools and better constitutive property data, Wissler's analysis pointed out further flaws in Pennes' work which had not been appreciated previously. However, he also showed that much of the criticism that has been directed toward the Pennes model is not justified, in that his improved computations with the model demonstrated a good standard of agreement with the experimental data. Thus, Wissler's conclusion is that "those who base their theoretical calculations on the Pennes model can be somewhat more confident that their starting equations are valid." The quantitative analysis of the effects of blood perfusion on the internal temperature distribution in living tissue remains a topic of active research after a half-century of study.

Limitations of Pennes' Model

Pennes' model was the first major effort in quantifying the heat transfer contribution of perfusion. It was developed for describing the transverse temperature profile in the human forearm. The model is unique in that the perfusion is very simple. The "bioheat equation" was previously shown as Equation 4.4.37.

The limitations of this model arise from the erroneous view of the heat transfer process and its anatomical location. Chen and Holmes' analysis of blood vessel thermal equilibration lengths showed that Pennes' concept is incorrect.[37] The thermal equilibration length is defined as the length at which the difference between the blood and tissue temperature decreases to 1/e of the initial value. They indicated that thermal equilibration occurs predominantly within the terminal arterioles and venules, and that blood is essentially equilibrated prior to entering the capillaries. In considering the contribution of perfusion as a nondirectional term, the directional convective mechanism is neglected. Nor does the model account for specific vascular architecture such as countercurrent arteries and veins. The limitations of Pennes' model have motivated subsequent investigators to develop their own models.

Despite its erroneous concept, the perfusion term of Pennes' model has been widely used, and found to be valid for situations other than the forearm. Its wide usage has been mainly due to its simplicity of implementation, especially in analyses where a closed form analytical solution is sought.[22,61] Investigators have obtained good temperature predictions for the following circumstances:

1. The porcine kidney cortex in the absence of large vessels (diameter >300 μm).[62]
2. The rat liver.[5,22]
3. The capillary bleed-off from large vessels.[42]

In the last case, Charny compared Weinbaum and Jiji's countercurrent model and Pennes' model against the experimental results of Pennes. The simulations found that Pennes' model is valid in the initial branchings of the largest microvessels from the countercurrent vessels (diameter >500 μm) in deep tissue. In this case, the microvessel blood temperature is close to arterial temperature. Arkin et al.[63] provide an explanation of the inconsistencies between the two anatomical sites. They suggest that since blood typically travels down successive generations of the vascular branches before equilibrating with the tissue temperature, Charny's claim actually refers to the collective contribution of numerous smaller thermally significant vessels in a region dominated by the large microvessels. The distinction of being thermally significant is based upon the ratio of thermal equilibration length to actual vessel length (ε):

$$\varepsilon = \frac{L_e}{L} \tag{4.4.38}$$

Along with Xu's observations in porcine kidney,[62] Pennes' model appears to be applicable to regions where the vasculature comprises numerous, small, thermally significant vessels ($\varepsilon \approx 1$).

Continuum Models

Among the continuum formulations of bioheat transfer, the Chen-Holmes model is the most developed. Prior to Chen-Holmes, continuum formulations by Wulff[64] and Klinger[65] addressed the isotropicity of Pennes' perfusion term. However, they did not challenge the Pennes' concept of the site of heat exchange. Chen and Holmes[37] formulated their model after analyzing the significance of blood vessel thermal equilibration length. Through this analysis, they quantitatively showed that the major heat transfer processes occur in the 50- to 500-μm diameter vessels and refuted Pennes' paradigm. In their model, they proposed that larger vessels be modeled separately from smaller vessels and tissue. Larger vessels were distinguished using the ratio of equilibration length to actual vessel length of about one ($\varepsilon \approx 1$) as the criterion. The smaller vessels and tissue were then modeled as a continuum. In a differential control volume of this continuum, they further separated solid tissue from the blood within the vascular space (Figure 4.4.2).

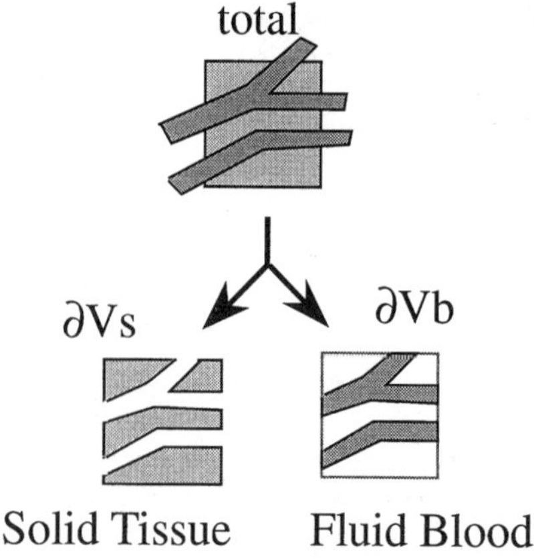

FIGURE 4.4.2 Schematic representation of tissue control volume as used by Chen-Holmes.[8]

Subsequently, the heat transfer mechanisms can be divided into the contributions from (1) nonequilibrated blood in the thermally significant vessels, (2) blood that has equilibrated with the surrounding tissue, and (3) nearly equilibrated blood. The perfusion term of Pennes is then replaced with three terms:

$$\rho c \frac{\partial T}{\partial t} = \nabla \cdot k \nabla T + (\rho c)_b \, \omega^* \left(T_a^* - T\right) - (\rho c)_b \, \vec{u} \cdot \nabla T + \nabla \cdot k_p \nabla T + q_m \qquad (4.4.39)$$

The second through the fourth term on the right hand side arise from each of the three categories described above. The second term, $(\rho c)_b \omega^* (T_a^* - T)$, is similar to Pennes except the perfusion and the arterial temperature are specific to the volume being considered. The $(\rho c)_b \, \vec{u} \cdot \nabla T$ term is a directional convective term due to the net flux of equilibrated blood. Finally, the $\nabla \cdot k_p \nabla T$ term is to account for the contribution of the nearly equilibrated blood in a tissue temperature gradient. The nearly equilibrated blood contributes to small temperature fluctuations within the tissue and the effect is modeled as a tensor "perfusion conductivity"

$$k_p = n(\rho c)_b \, \pi r_b^2 \, \bar{V} \cos^2 \gamma \sum_{i=1}^{\infty} \frac{L_e}{L_e^2 \beta_i^2 + 1} \qquad (4.4.40)$$

which is a function of local average blood flow velocity vector within the vessel ($\bar{V}$) relative angle (γ) between blood vessel direction and the tissue temperature gradient, the number of vessels (n), and vessel radius (r_b). The Fourier integral spectral wave number (β) can be approximated as the inverse of vessel length. The contribution of this conductivity is minimal except when vessels with large equilibration lengths are considered. However, for this situation, Chen and Holmes recommend that these vessels be treated separately. The assumptions which were made for their model include: (1) neglecting the mass transfer between vessel and tissue space, and (2) treating the thermal conductivity and temperature within the tissue-blood continuum as that of the solid tissue, since the vascular volume is much smaller than that of the solid tissue.

The limitation of this model is that given the detail required, the model is not easy to implement. Also, the perfusion conductivity term is difficult to evaluate, and distinction within the continuum model is not well defined. Furthermore, the model does not explicitly address the effect of closely spaced countercurrent artery-vein pairs. This model has been applied to the porcine kidney and was found to predict temperatures similar to Pennes' model, and thus, given the simplicity of the latter, Xu et al.[62] recommended that Pennes be used. Arkin et al.[63] claim that the Chen-Holmes model can be essentially applied to the same tissue region as that for Pennes.

Vasculature-Based Models

Weinbaum-Jiji-Lemons[38-50]

The modeling of countercurrent vasculature, which was not explicitly addressed by the Chen-Holmes model, developed separately from that of the continuum models. Bazett et al.[66] initially presented the countercurrent structure from observations of large arteries and veins in human limbs. The first major quantitative analysis was presented by Mitchell and Myers.[55] It was then followed by the work of Keller and Seiler,[54] which became the predecessor to the Weinbaum-Jiji models. In 1979, Weinbaum and Jiji[43] proposed the initial model of the artery–vein pair as two parallel cylinders of equal diameters with collateral bleed-off in the plane normal to the cylinders. The anatomical configuration is a schematic of an artery and vein pair with branches to the peripheral skin layer (Figure 4.4.3). The contribution of perfusion to heat transfer in tissue was treated as heat transfer in a porous medium. It was considered as a unidirectional convective term normal to the artery–vein pair. Knowledge of vessel density, diameter, and blood velocity was required at the different blood vessel generations.

In 1984, they presented a more thorough model based upon anatomical observations with Lemons.[38,39] This model analyzed three tissue layers of a limb: (1) deep, (2) intermediate, and (3) superficial or

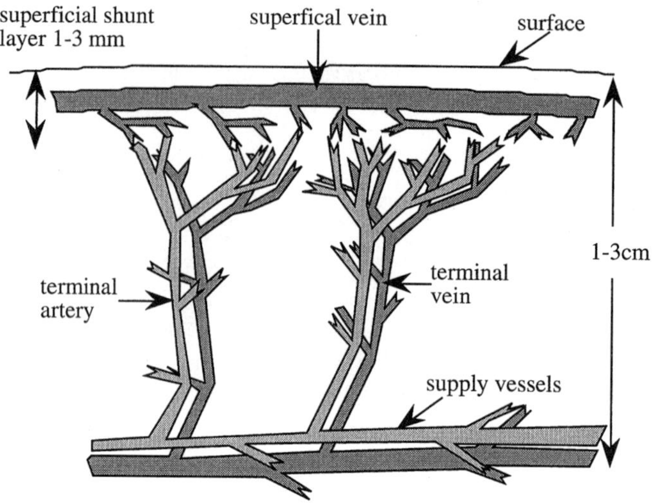

FIGURE 4.4.3 Schematic of artery and vein pair in peripheral skin layer.[43]

cutaneous. For the countercurrent structure of the deep tissue layer, they proposed a system of three coupled equations:

$$(\rho c)_b \pi r_b^2 \overline{V} \cdot \frac{dT_a}{ds} = -q_a \tag{4.4.41}$$

$$(\rho c)_b \pi r_b^2 \overline{V} \cdot \frac{dT_v}{ds} = -q_v \tag{4.4.42}$$

$$\rho c \frac{\partial T}{\partial t} = \nabla \cdot k \nabla T + n\, g (\rho c)_b \cdot (T_a - T_v) - n \pi r_b^2 (\rho c)_b \overline{V} \cdot \frac{d(T_a - T_v)}{ds} + q_m \tag{4.4.43}$$

The first two equations describe the heat transfer of the thermally significant artery and vein, respectively. The third equation refers to the tissue surrounding the artery–vein pair. For this equation, the middle two right-hand-side terms represent the capillary bleed-off energy exchange, and the net heat exchange between the tissue and artery–vein pair, respectively. The capillary bleed-off term is similar to Pennes' perfusion term except the bleed-off mass flow (g) is used. Their analysis showed that the major heat transfer is due to the imperfect countercurrent heat exchange between artery–vein pairs. They quantified the effect of perfusion bleed-off associated with this vascular structure, and showed that Pennes' perfusion formulation is negligible due to the temperature differential.

Assumptions include the following:

1. Neglecting the lymphatic fluid loss, so that the mass flow rate in the artery is equal to that of the vein.
2. Spatially uniform bleed-off perfusion.
3. Heat transfer in the plane normal to the artery–vein pair is greater than that along the vessels (in order to apply the approximation of superposition of a line sink and source in a pure conduction field).
4. A linear relationship for the temperature along the radial direction in the plane normal to the artery and vein.
5. The artery–vein border temperature equals the mean of the artery and vein temperature.
6. The blood exiting the bleed-off capillaries and entering the veins is at the venous blood temperature.

The last assumption has drawn criticism based on studies that indicate the temperature to be closer to tissue.[58,63] Limitations of this model include the difficulty of implementation, and that the artery and vein diameters must be identical. Both of these issues have led to the development of the models described in subsequent sections. Studies using this model have been applied to the peripheral muscle tissue of a limb,[44-46] and the model is accepted as valid for vasculature with diameters <300 μm and $\varepsilon < 0.3$.[63]

Simplified Weinbaum-Jiji (W-J)[40]

In response to the criticism that their previous model is too difficult and complex to apply, Weinbaum and Jiji simplified the three-equation model to a single equation:

$$\rho c \frac{\partial T}{\partial t} = \frac{\partial}{\partial x}\left(k_{eff} \frac{\partial T}{\partial x} \right) + q_{met} \tag{4.4.44}$$

In their simplification, they derived an equation based on the temperature of tissue only. The imperfect countercurrent heat exchange is embodied in an effective conductivity tensor term.

$$k_{eff} = k\left\{ 1 + \frac{n\left[(\rho c)_b \pi r_b^2 \overline{V} \cos\gamma \right]^2}{\sigma_\Delta \cdot k^2} \right\} + q_{met} \tag{4.4.45}$$

The k_{eff} term has similar parameters to the tissue and artery–vein pair heat exchange term in Equation 4.4.43, and a shape factor term (σ_Δ). In order to eliminate the blood temperature from their previous formulation, two major assumptions (the closure conditions) were used:

1. The mean tissue temperature = $(T_a + T_v)/2$.
2. Heat from a paired artery is mostly conducted to the corresponding vein:

$$q_a \approx q_v \approx \sigma_\Delta k \left(T_a - T_v \right). \tag{4.4.46}$$

Both of these assumptions were based upon studies in rabbit thigh muscle from their previous formulation. However, to respond to criticism of these assumptions, they performed further mathematical analysis on ε and provided insights into the limits for applying these assumptions.[47] An obvious limitation of this model is that the local temperatures along the countercurrent artery and vein cannot be calculated. Another limitation is that the model is applicable only in situations where $\frac{L_e}{L} \ll 1$. In the example of analyzing the peripheral tissue in the arm, L is equal to the characteristic radius of the arm.[48] Weinbaum and Lemons[48] admit that this assumption breaks down under the following conditions: (1) if blood flow rates significantly increased in the larger vessel pairs of the peripheral tissue layer, and (2) if deeper muscle tissue, where the diameter of the countercurrent pair vessels are less than 300 μm, are included. This model has been tested in the porcine[62] and canine kidney,[67] and continues to be verified by the Weinbaum group[41,50] and other investigators.[68]

Small Artery Model[69,70]

The small artery model was developed by Anderson in studies of the canine kidney cortex. The model considers the energy balance in a control volume (i, j, k) which contains either an arterial (Q_a) or venous (Q_v) vessel. For a volume with an artery parallel to the "z" coordinate axis, the equation is

$$Q_a = N(VA)_a (\rho c)_b \left(1 + \lambda - 2\lambda\xi \right)\left(\frac{T_z - T_{z-\Delta z}}{\Delta z} \right) \tag{4.4.47}.$$

For a volume with a vein:

$$Q_v = M(VA)_v(\rho c)_b(1+\lambda-2\lambda\xi)\left(\frac{T_{z+\Delta z}-T_z}{\Delta z}\right) \quad (4.4.48).$$

where N and M are the density of the interlobular arteries and veins in the kidney cortex, respectively; ξ refers to the fraction of the total interlobular artery flow within the control volume; in the kidney cortex, $\xi = 1$ at the corticomedullary junction and decreases to $\xi = 0$ at the outer capsule. The total flow within this region accounts for bleed-off from the interlobular arteries through the λ term, where $\lambda = 1$ represents complete bleed-off.

The discrete representation provides for straightforward numerical implementation when the vessel density within the tissue region of interest is known. Model assumptions include:

1. Thermal equilibration length within the volume is much less than vessel length.
2. A linear relation between the effect of bleed-off on arterial flow and location along the length of the vessel.
3. Bleed-off is modeled as a change in the arterial flow.
4. Bleed-off heat transfer is negligible.
5. No major thermally significant vessels (i.e., $\varepsilon << 1$) in the region of interest.

In its initial formulation, arbitrarily oriented vessels which would cause more than one vessel to occupy a control volume were not considered. Branching vessels are also not addressed. Even though this is not an inherent limitation of the model, implementation would be more difficult. Due to its discrete representation, the model cannot solve the inverse problem. The model has been shown to be valid in the canine kidney cortex where there is uniformly oriented countercurrent artery–vein architecture of 70-μm diameter.

Hybrid Models

The lack of an encompassing model which can account for the various tissue structures has led researchers to propose and apply hybrid models. The substantial amount of effort related to the application and investigation of the major models discussed has shown that no one model applies to all the different vascular structures in tissue.[58,63] Wissler[58,59] points to the unlikelihood of a single equation being able to provide a complete description of the heat transfer process in tissue, and thus suggests the use of a combination of equations. The realization of this suggestion is found in Charny's work,[42] in which W-J's and Pennes' models are used to describe peripheral and deep muscle tissue, respectively. When the applicability of each of the major models has been conclusively defined, an algorithm which would enable users to arrive at the appropriate choice of equations given the tissue vasculature of interest would be beneficial.

The applicability of the models discussed requires an understanding of the validity and development of each model. A comparison of the models discussed has been summarized by Charney.[33]

Thermal Measurements of Perfusion

Introduction

Perfusion, the transmission of blood in the microcirculation, is an important factor in surgery, tissue transplants, heart disease, and cancer therapy. Despite its importance, no clinical method of measuring perfusion is currently available for a majority of applications. One technique that shows considerable promise involves the use of self-heated thermistors.[6-13] In this method, a miniature thermistor (0.5 to 2.5 mm in diameter) is placed invasively in the tissue of interest and heated with a predetermined applied power. Since both tissue conduction and perfusion act to carry heat away from the thermistor, the resulting volumetric-average temperature rise in the thermistor bead, ΔT, is related to both the tissue thermal conductivity and perfusion. By knowing the intrinsic tissue conductivity and the apparent conductivity of the tissue (due to both blood flow and conduction), the perfusion rate can be calculated.

At least two difficulties exist with this technique. The first is that the intrinsic tissue conductivity of perfused tissue must be known in order to calculate the perfusion rate. Although one could stop the blood flow to a tissue and measure its conductivity, this is clearly not desirable, nor is it always practical. In order to overcome this problem, Holmes and Chen have proposed techniques that measure perfusion without interrupting blood flow.[13,20]

Perfusion Resolution

The perfusion resolution, δw, is defined as the smallest change in perfusion that can be detected by the instrument. It can be determined theoretically:

$$\delta w = \delta k \cdot \frac{\partial w}{\partial k} \tag{4.4.49}$$

For the constant temperature heating technique using a P60 thermistor, δk is about 0.02 mW/cm-°C, and the sensitivity is about 100 (ml/100g-min)/(mW/cm-°C). Using Equation 4.4.49 gives a perfusion resolution of about 2 ml/100g-min. Due to fluctuations in the baseline tissue temperature, the practical resolution is about 10 ml/100g-min. There are so many experimental and tissue variables that it is extremely important to test these techniques in preparations where the perfusion is known.

Measurement Volume

The measurement volume of a thermistor is a complex function of many factors, including the perfusion rate and vascular anatomy of the tissue of interest. One problem with small thermistors is their small measurement volume. A boundary layer (decoupler) between the thermistor and the tissue causes a significant measurement error. This unwanted boundary layer is often caused by the probe itself during insertion. The larger probes exhibit a smaller error, but are likely to cause a larger decoupler because of the increased trauma during insertion.

Temperature Dependence of Perfusion

Perfusion depends on a wide variety of factors, some local to the tissue (pH, temperature, O_2), some which are external but directly control local flow (parasympathetic, hormones), and some which indirectly affect local flow (heart rate, blood pressure, skin temperature, needs of other organs). A simple experiment studied the effect of local tissue temperature on muscle perfusion. The constant ΔT method was used to measure perfusion in an anesthetized rat. The muscle temperature was manipulated by placing the hind limb into a water bath. Insulation was carefully placed so as to minimize changes to the body temperature. The rectal and neck temperatures were constant while the muscle in the hind limb was heated. Figure 4.4.4 shows the perfusion as a function of tissue temperature for a typical experiment. The dip in perfusion as a function of temperature, as shown in Figure 4.4.4, consistently occurred, but did not always occur at the same temperature. This dip may be due to an anastomotic shunt attempting to regulate the core body temperature.

Figure 4.4.5 presents the averaged results for 10 rats. Because the dip occurred at different temperatures for the various rats, it does not appear in the average. The large standard deviations are due to perfusion differences from one rat to the next. A linear fit to this averaged data gives the following approximation:

$$w = 1.9720\left(1 + 0.059\ T\right) \tag{4.4.50}$$

where w has units of ml/100g-min and T has units of °C.

Yuan et al.[71] measured perfusion and temperatures at various locations within each of the four canine prostates subjected to a transurethral microwave thermal source. The total number of the perfusion sampling points coupled with temperature is 15. Colored microspheres were used to measure perfusion

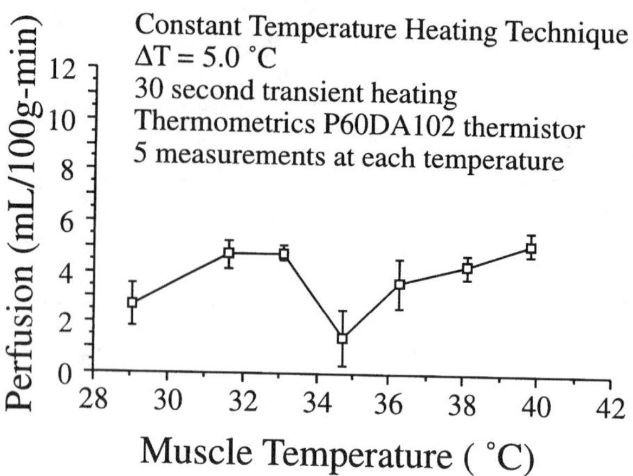

FIGURE 4.4.4 Perfusion vs. muscle temperature during a typical experiment. (From Pearce, J.S. and Thomsen, S., in *Optical-Thermal Response of Laser-Irradiated Tissue,* Welch, A.D. and van Germert, M.M.C., Eds., Plenum Press, New York, 1995. With permission.)

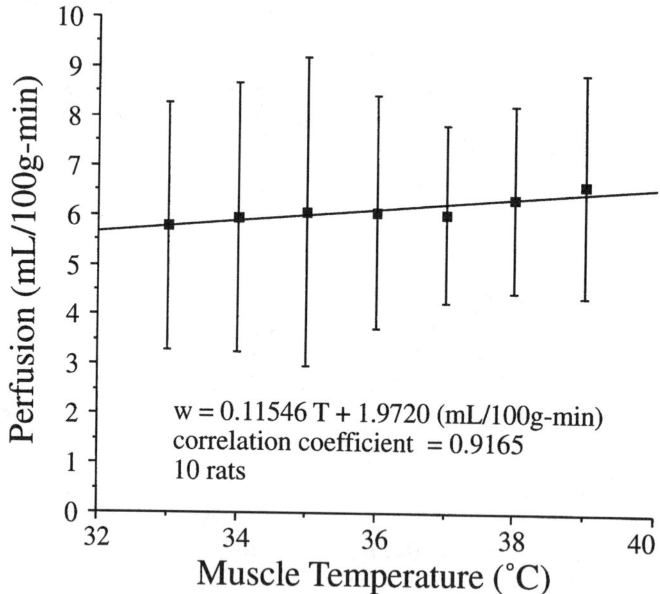

FIGURE 4.4.5 Perfusion vs. muscle temperature averaged over 10 experiments. (From Pearce, J.S. and Thomsen, S., in *Optical-Thermal Response of Laser-Irradiated Tissue,* Welch, A.D. and van Germert, M.M.C., Eds., Plenum Press, New York, 1995. With permission.)

due to their simplicity compared with radioactively labeled microspheres, and because the microsphere trapping method is regarded as a standard. Temperatures were measured using miniature thermistors. The prostate temperatures were raised to 40°C ~ 45°C by 5-W-step increments of the microwave power at hourly intervals to 15 W. Temperatures and perfusion were measured at baseline, and at the beginning and end of each heating interval. Thus, the periods between perfusion samples were approximately either 5 or 60 min. Under baseline conditions, the temperature fluctuations within the prostate were approximately ±0.3°C. A relative dispersion estimate of 15% was derived from one dog for the fluctuations in baseline perfusion. Thus, changes in absolute perfusion and temperature greater than 15% and 0.3°C, respectively, were considered to be substantial changes.

As heating progressed, a variety of substantial changes were observed, but no uniform pattern emerged. However, the measurements included changes typically expected for hyperthermia:

1. An initial perfusion increase associated with elevating the baseline temperature.
2. A perfusion return towards baseline after this initial increase.
3. A dramatic increase in perfusion at elevated temperatures.

The initial perfusion increases were observed in three dogs when the temperatures exceeded 38 ± 3°C (mean ± 1 SD, N = 8). The perfusion increased 34% from a baseline value of 0.59 ± 0.26 ml/g-min over a temperature rise of 1.7 ± 1.3°C. Half of the measurements in the three dogs subsequently showed a decrease in perfusion ranging from 16 to 25%. In two dogs, dramatic perfusion increases as high as 364% were observed with a corresponding decrease in tissue temperature.

The mean perfusion and temperature measured from all dogs at similar instances during the experimental protocol were calculated. Figure 4.4.6 shows the changes as the experiment progressed from baseline conditions.

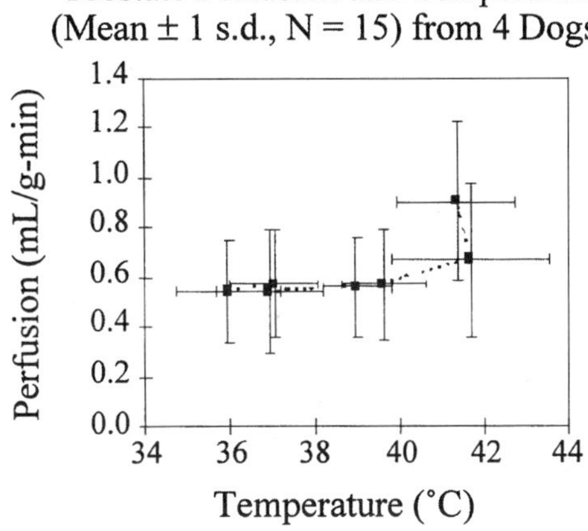

FIGURE 4.4.6 Perfusion vs. prostate temperature.[71]

The mean behavior indicates no substantial change in perfusion until the tissue temperature exceeded 39.6°C, after which the perfusion increased 17%. This occurred over a 5-min period as the nominal microwave power was stepped from 10 to 15 W. The perfusion increased another 35% when the tissue temperature exceeded 41.7°C, and appeared to affect a slight lowering of tissue temperature.

Xu et al.[72] measured perfusion using the pulse-decay self-heated thermistor technique[13,20] in these dog prostates during the same transurethral microwave hyperthermia treatments. Interestingly, in the exact same dogs at roughly the same locations, the perfusion response to temperature measured with the thermal technique was roughly linear with temperature as shown in Figure 4.4.7.

The interdependence between perfusion and temperature was observed in these studies: most notably, a decrease in tissue temperature associated with a dramatic increase of perfusion. Such changes have been modeled previously with Pennes' bioheat equation, using assumed perfusion values and changes for the prostate. The data from this study will provide more realistic estimates of perfusion values and thermoregulation models in hyperthermic canine prostates. Two hypotheses exist that explain the differences between the perfusion responses to hyperthermia as measured by microspheres and the thermal decay probe. The first possibility is that one or both methods have significant measurement errors. A second possibility is that perfusion as measured by spheres trapped in the capillaries is a different parameter than perfusion as measured by enhanced heat transfer within the 70- to 200-μm-diameter vessels.

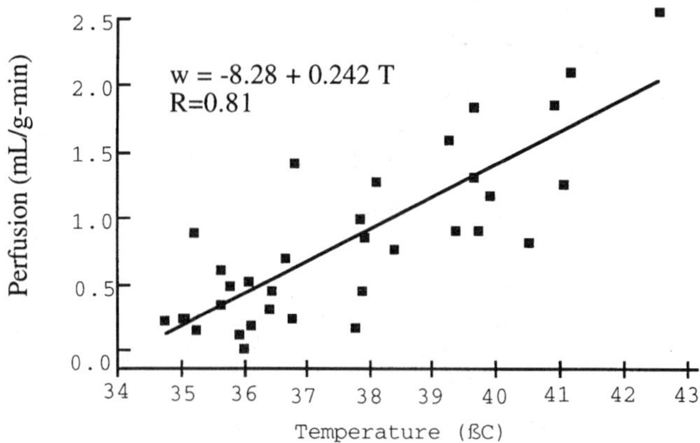

FIGURE 4.4.7 Perfusion vs. prostate temperature.[72]

Human Thermoregulation

Humans possess an elaborate thermal control (thermoregulation) system that ensures maintenance of internal body temperatures near a physiological set point under a large spectrum of environmental conditions and metabolic rate activities. After many years of research, much has been learned of the operation of the human thermoregulatory system and it remains a topic of active investigation. A number of reviews on specific aspects of the physiological basis of thermoregulation are recommended for further background detail.[73-76]

Internal and boundary heat transfer processes are key to thermoregulation. As the operational mechanisms have become better understood, it has been possible to model these functions in an attempt to simulate and predict thermoregulatory behavior and to be able to design systems to interact thermally with the human body (such as an environmental protective garment or a space suit) without compromising the health and safety of the subject.

Physiological Processes of Thermoregulation

The prevailing thermal physiology theory is that the human thermoregulation system operates to maintain the body core temperature at a constant value consistent with that required for regular body function, regardless of the environmental temperature and energy loading. Alternatively, it has been suggested that thermoregulation serves the purpose of maintaining the body's energy balance, with body temperatures being a *result*, not a cause, of the regulation process.[73,77,78]

Many general observations of the thermal characteristics of human thermoregulation are consistent, even if alternative explanations of the underlying operative mechanisms are proffered. For example, temperature and internal energy storage of the human body vary with time of day, metabolic activity, and individuality of the person.[79] The body incorporates a number of energy production and dissipation mechanisms to preserve thermal homeostasis, many of which are controlled by feedback signals based on specific physiological states. Examples of thermoregulatory processes that are governed by feedback are sweating, shivering, and variable localized blood flow.

According to the theory of thermoregulation by control of the body core temperature, the main control center is located in the hypothalamus of the brain, from which multiple reflex responses operate to maintain the body temperature within a narrow range.[80] The signals that activate the hypothalamic temperature-regulating centers come largely from two sources: the temperature-sensitive cells in the anterior hypothalamus and cutaneous temperature receptors. The cells in the anterior hypothalamus sense the temperature of the body core or, specifically, the temperature of arterial blood which passes through the head.

Webb has presented a theory of thermoregulation to maintain an energy balance within the body, as demonstrated by the existence of temperature sensors at several levels in the skin enabling the sensing of heat flow within and from the body.[78] Evidence also supports neurological sensing of thermal gradients which direct thermoregulation behavior. The hypothesis in support of the theory of energy content regulation based on Webb's experimental observations is

> Heat (energy) regulation achieves heat (energy) balance over a wide range of heat (energy) loads. Heat flow to or from the body is sensed, and physiological responses defend the body heat (energy) content. Heat (energy) content varies over a range that is related to heat (energy) load. Changes in body heat (energy) content drive deep body temperatures.[78]

The proposed mechanism of energy-driven thermoregulation balances the constantly changing metabolic energy production and the adjustment of heat losses to maintain the body as a system at steady-state. In contrast, the proposed mechanism of temperature-driven thermoregulation affects a coordination of physiological processes to maintain the body core temperature at a set point.

Thermoregulatory Processes

Conservation of energy for the human body must account for internal metabolic energy production plus multiple mechanisms of environmental heat and work exchange.

$$\Delta E = M - \left(W + Q_{conv} + Q_{cond} + Q_{rad} + Q_{evap} + Q_{resp}\right) \tag{4.4.51}$$

where ΔE = Rate of energy storage in the body (W)
M = Metabolic energy production (W)
W = External work (W)
Q_{conv} = Surface heat loss by convection (W)
Q_{cond} = Surface heat loss by conduction (W)
Q_{rad} = Surface heat loss by radiation (W)
Q_{evap} = Surface heat loss by evaporation (W)
Q_{resp} = Respiratory heat loss (W)

The human body produces energy, exchanges heat with the environment, and loses heat by evaporation of body fluids. Energy is produced in the body by basal (resting), metabolism defined as the minimal metabolism measured at a temperature of thermal neutrality in a resting homeotherm with normal body temperature several hours after a meal and not immediately after hypothermia, and also at an increased rate due to muscle activity, including physical exercise and shivering, and by food intake. Therefore, the total energy production in the body is determined by the energy needed for basic body processes plus any external work. Since the body operates with less than 100% efficiency, only a fraction of the metabolic rate is applied to work, with the remainder dissipated as heat.[79] The mechanical efficiency, η, associated with metabolic energy utilization, is zero for most activities except when the person is performing external mechanical work such as in walking upstairs, lifting something to a higher level, or cycling on an ergometer.[81] When work transferred from the environment is dissipated as heat in the human body, η is negative. An example of this case is walking downstairs.

Convection, radiation, conduction, and evaporation of sweat at the skin surface all facilitate heat transfer from the body. Heat transfer also occurs via the respiratory tract and lungs, although to a lesser extent in human beings as compared with many mammals and reptiles. Storage of energy takes place whenever there is an imbalance of production and dissipation mechanisms. In many instances, such as astronauts in space suits or military personnel in chemical defense garments, energy storage is forced due to the lack of appropriate heat exchange with the environment.[82]

The human thermoregulatory system is quite complex and behaves mathematically in a highly nonlinear manner. It contains multiple sensors, multiple feedback loops, and multiple outputs.[83] The primary mechanisms by which the body responds to control the storage of energy (positive or negative) include the evaporation of sweat, shivering of the muscles, and vasoconstriction and vasodilation of the blood vessels.

Heat transfer internal to the body is due to the conductance which governs the flow of energy between the core, through the tissue, and to the surface. This transport process is governed significantly by peripheral blood flow, the core-skin temperature gradient, and the conductivity of the various body tissues. Transient blood perfusion provides an effective variable path of peripheral energy transport via convection between blood and tissue and countercurrent heat exchange between the arteries and the veins. Blood flow distribution is controlled according to metabolic needs of the body, local tissue temperature, and the need to maintain the appropriate core temperature. When the core becomes too hot, the blood vessels in the skin dilate to allow increased perfusion to the body surface. The blood is cooled by the environment, and the cooler blood is returned to the core. Increased blood flow to the skin surface also enables greater sweat production, adding to the cooling process. In contrast, when the core becomes too cold, blood flow to the skin is constricted to conserve the body's internal energy. Sweating occurs when the arterial blood is elevated in temperature, causing the hypothalamus to increase nerve impulses to the sweat glands. Shivering, on the other hand, is an involuntary response of the skeletal muscles when passive body cooling exceeds metabolic energy production.[80]

Human thermoregulation mechanisms can be divided into two main categories: autonomic and behavioral.[38] Autonomic thermoregulation is controlled primarily by the hypothalamus, whereas behavioral thermoregulation is controlled consciously by humans and includes active movement and adjustment of clothing. Behavioral thermoregulation is associated with conscious temperature sensation as well as with thermal comfort or discomfort. Some engineered systems are designed to interact with the body to enhance its ability to maintain thermoregulation in harsh or hazardous environments. Examples range from the common HVAC systems encountered in buildings and vehicles to sophisticated garments worn during certain military and space missions.

Wissler Model of Human Thermoregulation

Models that accurately incorporate the transient whole-body behavior during thermoregulation for a wide range of states and environmental challenges may be quite useful in describing and predicting this important human physiological function. Moreover, such a model can be used as a design tool in the development of systems with which humans must interact for a variety of work, pathological, and recreational circumstances. The development of models of human thermoregulation has proved to be a daunting task that has been addressed by many researchers. The complexity of the coupling among physiological processes involved in thermoregulation and of the control algorithms has dictated that models that incorporate these multiple effects be solved numerically. Thus, the first realistic thermoregulation models appeared in the 1960s and 1970s from the studies of Wissler,[84,85] Stolwijk,[86,87] Mitchell et al.,[87,88] Hayward et al.,[89,90] and Kuznetz.[91] An important early application of modeling human thermoregulation was the design and development of active thermal control garments to be worn under the space suit during extravehicular activity.[79,82,92,93] Over the years the Wissler model has been improved and updated on a continuous basis, and it is now applied to a very broad spectrum of human thermal control scenarios.[94-97]

The Wissler human thermoregulation model is a transient-state mathematical formulation used to describe human physiological responses to various levels and combinations of exercise and thermal stress. The model computes thermal, cardiovascular, ventilatory, and metabolic changes that occur during a specified period of time. The values for the independent physiological and environmental variables, such as metabolic rate, environmental conditions, and garment cooling liquid flow rate, are specified as inputs to the model. A finite-difference scheme is applied to numerically compute tissue and blood temperatures throughout the body as a function of time.

The model represents the human body as 15 cylindrical elements for the head, thorax, abdomen, and proximal, medial, and distal portions of each arm and leg. A schematic representation of the multielement model of the human body is shown in Figure 4.4.8. Each body element is composed of cylindrical shells that are defined by physical and physiological properties for composite layers of viscera, bone, muscle, fat, and skin. In each of the elements metabolic energy may be generated as appropriate due to basal metabolism, exercise, and shivering. The generated energy is either conducted to adjacent body regions, convected to circulating blood, or stored in the element. Each body element contains an arterial and venous pool and a capillary bed. The pools are interconnected serially between adjacent elements. As venous blood is returned to the heart it passes through the lungs for exchange of gases before entering the arterial system.

FIGURE 4.4.8 Multielement model representing the human body in the Wissler model. (From Wissler, E.H., *Heat Transfer in Medicine and Biology: Analysis and Applications,* Vol. 1, Plenum Press, New York, 325, 1985. With permission.)

A number of signal feedback loops are used to account for thermoregulation within the body. Feedback data are the differences between skin and central head temperatures and their respective specified set-point values. Error signals specified by these differences drive sweat secretion, shivering, and vasomotion responses. The algorithm equations have the general format of a proportional controller which is activated when a defined threshold condition is satisfied. The control equations do not all have the same level of response to error signals. For example, sweating and vasodilation are more responsive to an increase in central head temperature than cutaneous temperature, while vasoconstriction is more responsive to cutaneous temperature.

Thermoregulation is also affected via heat exchange during the venous return flow of blood. Venous flow occurs through either deep or superficial veins, thus influencing the rate of countercurrent heat exchange. Modeling of the venous return provides a realistic simulation of the countercurrent heat exchange that takes place in the extremities during exercise in hot environments and immersion in cold water.

Blood flow rates are simulated to satisfy the level of local metabolic need for oxygen in the tissues. Therefore, a lower limit is set for extremity blood perfusion during cold exposure. The model also incorporates material balances for oxygen, carbon dioxide, and lactate as defined for cardiovascular response. In addition to physiological factors, conditions of the environment with which the subject interacts are accounted for. Physical properties of the environmental fluid and characteristics of the flow field in the vicinity of the subject are specified as they determine conductive and convective heat transfer.

The model allows the user to specify layers of clothing described by physical properties of annular shells outside of but interacting with the skin layer. The physical properties of the garments may change with the accumulation of unevaporated sweat, which may saturate the material.

Transient tissue and blood temperatures are calculated during simulated thermal scenarios by Crank-Nicolson-type finite-difference equations. Updated temperatures and other physiological values are computed every 10 sections. The model allows a simulation to be broken down into any number of scenario subintervals, with the following factors being specified in each interval:

- Basal metabolic rate
- Kind and level of exercise being performed
- Environmental conditions (pressure, temperature, dewpoint, etc.)
- Garment properties
- Use of a fluid-conditioned garment (FCG) (flow rates, inlet temperatures, etc.)

The equations needed to describe a liquid- or air-cooled garment are derived from material and energy balances for the circulating fluid. Sensible heat transfer is treated the same for liquid- and air-cooling situations. However, the evaporation and condensation associated with latent heat transfer must be treated differently for the two fluids. In the case of the liquid-cooled garment, water from sweat and environmental moisture condenses on the cool surface of the garment. On the other hand, air cooling generally facilitates evaporation of sweat by removing water vapor from the garment.

In its early formulation the Wissler model required a large mainframe computer to perform simulations of thermal regulation. However, owing to advances in computational capabilities in recent years, it is now possible to run extensive simulations on a desktop personal computer in an order of magnitude less than real time. The thermoregulation model provides reliable predictions of changes in the human thermal state for many different types of environmental and metabolic stress, and it has been used for the design of new automated personal garments for application in hazardous environments.[97]

Therapeutic Heating

Normal physiologic temperatures range from approximately 30 to 42°C. The average so-called "core" (central abdominal) temperature in mammals varies slightly among species between about 36°C and 38°C (37°C for humans, 38°C for most canines). It was noted above that thermal gradients are inherent in metabolically driven open systems at steady state, and measured "core" temperatures will vary slightly depending on the relative metabolic rate, blood flow, and location of the various organs, and from point to point within an organ. Though core temperature is maintained close to a steady-state value, skin surface temperature at rest in a controlled room environment varies between about 30°C and 34°C, depending on climate and season.

Elevation of body, appendage, or organ temperature to between 40°C and 42°C can be used for analgesia (relief of pain) and to provide the most advantageous environment for injury response (wound healing) processes. Note that there is no external intervention which can be claimed to "promote" wound healing — one may only provide the best conditions under which the natural processes will heal wounds. Thermal intervention has been used for many years for these two purposes.

Heat Generation Modalities

Therapeutic heating can be obtained by surface heat transfer means (hot or cold packs), but the effectiveness is limited to surface injury. Nevertheless, contact heating is the method of choice for such injuries as sprains, muscle strain, and postoperative swelling. The standard response to elevated temperatures is to increase the blood flow, and thus perfusion. One applies cold heat sinks to reduce perfusion in the injured area during the first 12 to 24 h and hot packs or circulating water (up to 45°C) after that to promote perfusion and thus increase the supply of neutrophils and other blood components related to the injury response and wound healing.

Deeper injuries require volumetric heating to attain adequate temperatures. The most effective volumetric heating methods to date have been (1) electromagnetic fields at radio frequencies (RF) or microwave frequencies (MW), and (2) ultrasound (US). Assigned radio frequencies for industrial, medical, and scientific use (ISM) are 6.78, 13.56, 27.12, and 40.68 MHz. Practical ISM frequencies for medical use in the microwave range are 915 MHz and 2.45 GHz. Typical ultrasound heating devices operate between about 500 kHz and 10 MHz. In all three cases there is an engineering trade-off among local volumetric heat generation, q''', thermal heat generation field dimensions, and depth of penetration. Higher frequencies have shorter wavelengths and, in general, are absorbed more strongly and thus penetrate less deeply. Tissues are inhomogeneous and, with only a few exceptions, anisotropic. The electrical and acoustic properties of tissues can vary over several orders of magnitude. Consequently, electromagnetic and acoustic boundary conditions often determine the volume generation term in spite of efforts to shape the field by clever applicator design.

For electromagnetic heating in tissues the volume generation term is essentially governed by the electrical conductivity, σ, the imaginary part of the electric permittivity, ε'', and the magnitude of the local electric field, $|\mathbf{E}|$:

$$q''' = (\sigma + \omega\varepsilon'')|\mathbf{E}|^2 \tag{4.4.52}$$

where ω is the angular frequency (r/s). Heating due to direct absorption from a magnetic field, $\omega\mu'' |\mathbf{H}|^2$ where μ is the magnetic permeability (H/m) is negligible except in the most unusual of circumstances. Yet, RF and MW magnetic fields heat tissue very effectively, too. This is because, by Faraday's law of induction, a time-varying magnetic field will induce an electric field in the tissue, and it is the electric field which actually generates the heat.

Integral form:

$$\oint \mathbf{E} \bullet d\mathbf{L} = -\frac{\partial}{\partial t}\left[\iint \mu\mathbf{H} \bullet d\mathbf{S}\right] \tag{4.4.53}$$

Point form:

$$\nabla \times \mathbf{E} = -\mu \frac{\partial \mathbf{H}}{\partial t} \tag{4.4.54}$$

A comprehensive discussion of these properties and effects in materials may be found in Reference 98. Values of relevant electrical and acoustical properties of tissues are given in Tables 4.4.7 and 4.4.8, respectively.

TABLE 4.4.7 Representative Electric Properties of Tissues

Tissue	Conductivity (S/m)[a]	Real Part, ε_r'	Imaginary Part, ε_r''
Fat	0.05 to 0.09	3.9 to 7.2	0.67 to 1.4
Muscle	0.1 to 1.0	45 to 48	13 to 13.9
Bone	0.006 to 0.05	4.2 to 5.8 (marrow)	0.7 to 1.3 (marrow)

[a] Conductivities are from Reference 99.

Note: Muscle is highly anisotropic, higher conductivity parallel to fibers. Electric permittivity is as measured at 3 GHz; ε (F/m) = $\varepsilon_0[\varepsilon_r' - j\varepsilon_r'']$ where $j = \sqrt{(-1)}$ and ε_0 = free space permittivity, 8.85×10^{-12} F/m.[100]

TABLE 4.4.8 Representative Acoustic Properties of Tissues[100]

Tissue	Velocity, a (m/s)	Absorption at 1 MHz (cm^{-1})
Fat	1450	0.06
Muscle	1585	0.2 to 0.7
Bone	4080	30

Acoustic waves are highly scattered in many tissues, so simple wave-propagation models are inadequate to predict the local acoustic field strength. Nevertheless, a uniform plane-wave description of the heating field often suffices to explain the distribution of the heat generation with acceptable accuracy. The volume heat generation term is determined by the acoustic absorption coefficient, α and the wave power density, E, approximately according to Beer's law:

$$q'''(z) = \alpha E_0 e^{-\alpha z} \tag{4.4.55}$$

where E_0 is the surface power density at z = 0 W/m^2 and α varies according to frequency, f $\alpha \approx f^n$, where $1 < n < 2$. Table 4.4.8 units for α are cm^{-1}. In other references, α may be reported as dB/cm (dB = $10 \log_{10}\{E_{out}/E_{in}\}$).

Physiologic Effects of Local Heating

The general physiological effects of local heating comprise a cascade of responses.[101] As noted in the previous section, systemic response is controlled from the hypothalamus by both neuronal and hormonal signals. Appendage- and organ-level response to local heat may be spinal-cord-mediated; but local heating can trigger local release of bradykinins that induce vascular dilation or constriction, thus affecting temperature rise.

The increase in tissue temperature is accompanied by an increase in cellular metabolism. The arterioles (microscopic arteries, typically about of 40 to 200 μm inner diameter[101]) dilate in response to heating under smooth muscle control. The downstream capillary pressure increases above the homeostatic level of about 25 torr. Two consequences are an increased capillary flow and capillary pressure. First, at higher capillary pressures the gaps between the endothelial cells (thin epithelial cells comprising the vessel wall) tend to widen and, at higher pressures, edema may form (an increase in the fluid in the extracellular compartment).[102,103] Second, higher capillary flow results in rapid clearing of cellular metabolites and increases convection heat transfer, delivery of tissue oxygen, cellular nutrients, antibodies, and leukocytes (white blood cells) including monocytes which are necessary in the healing response. Neutrophils are white blood cells that initially release proteolytic enzymes to digest dead tissue and, later, phagocytose (eat) the debris.[104] Monocytes are other leukocytes that remove damaged and dead cells by phagocytosis.

Tissue Effects: Elevated Temperatures

Exposure to temperatures above normal physiologic ranges (>42°C) can result in measurable irreversible changes in tissue structure or function. Cell death or tissue alterations may be detrimental — for example, skin burns — or beneficial, as in vessel sealing or tumor destruction. Tissues of the central nervous system are the most thermally sensitive, exhibiting irreversible changes for long-term exposures above about 42°C. The specific pathologic end point may be evaluated histologically, histochemically, and/or physiologically. Most assays of thermal alterations are qualitative in nature; however, several end points which are inherently quantitative lend themselves well to rate process descriptions of their thermal kinetics. Even for qualitative processes, thermal kinetic models often provide useful descriptions and so provide helpful insights into the underlying principles of tissue thermal damage.

Here we use "damage" in a generic sense to denote all kinds of irreversible alterations, therapeutic or not. In the Arrhenius models, irreversible thermal damage is exponentially dependent on temperature and linearly dependent on time of exposure. Many damage processes can be modeled as first-order rate processes for which two experimentally derived coefficients are sufficient. Second-order processes require four parameters, again derived from experiment. The first-order rate process models often apply well to the prediction of damage thresholds and less well as the damage becomes complete or severe since several of the fundamental assumptions are violated. In order to be useful in evaluating thermal insult, the kinetic model must be coupled to quantitative pathological analysis. This subsection describes several quantitative markers of thermal damage and experimental methods for estimating relevant kinetic coefficients both in constant temperature and transient thermal history experiments. As expected, transient *in vivo* thermal history data yield a noisy kinetic plot; however, estimates of the appropriate rate coefficients can be made.

Theory of Rate Process Descriptions

The original work on the application of rate process models to thermal damage was reported by Moritz and Henriques in a series of seminal papers entitled "Studies of Thermal Injury" in 1947.[105-108] They applied flowing water at elevated temperatures to pig skin and measured exposure times required to create first-, second-, and third-degree burns. In their work, the damage was quantified using a single parameter, Ω, which ranges on the positive real axis and is calculated from an Arrhenius integral:

$$\Omega(\tau) = \int_0^{\tau} A\, e^{\left[\frac{-E}{RT}\right]} dt \tag{4.4.56}$$

where A is a frequency factor (s^{-1}), τ the total heating time (s), E an activation energy barrier (J/mole), R the universal gas constant (8.314 $J\text{-}mole^{-1}\text{-}K^{-1}$), and T the absolute temperature (K).

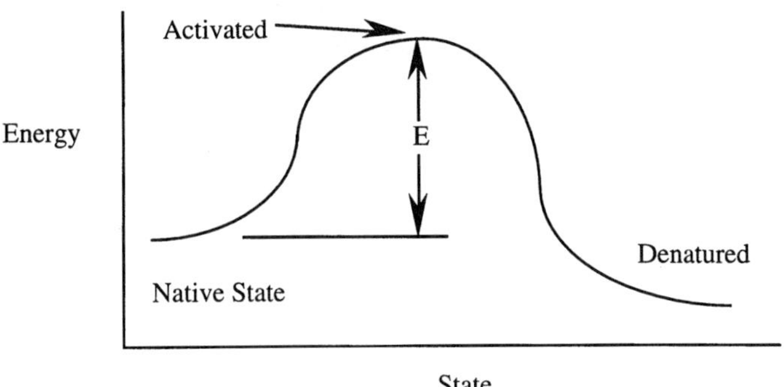

FIGURE 4.4.9 Energy-state diagram of a bimolecular process.

Over the ensuing 50 years many subsequent investigators have modeled this process[109-112] and made experimental measurements of threshold burn conditions in human,[113-115] animal,[116-117] and cellular and molecular[118-120] systems. Although a considerable body of literature has accrued, there is by no means a consensus on how to accurately predict the occurrence of thermal injury over the wide range of conditions that cause burns.

One difficulty with this model is that a single damage parameter inherently lumps all damage processes into one global measure. In multiple process cases, such as in the case of a skin burn, the lowest temperature process saturates the damage measure, Ω, early during an exposure. In this discussion, we review the underlying assumptions and origin of the terms in Equation 4.4.56 and recast the traditional thermal damage parameter, Ω, into a form suitable for evaluation of multiple-process thermal damage effects.

Kinetic Models of Reaction Product Formation Rates

The basis for rate process models of thermal damage may be obtained from chemical reaction kinetics (see, for example, Reference 121). In a typical reaction process, thermally active reactants jump an activation barrier to form products, as illustrated in Figure 4.4.9. In the figure, E is the energy barrier in Equation 4.4.56. The collision theory description of ordinary first-order bimolecular reaction kinetics holds that the reactants are activated by collisions; n* are activated out of n total molecules, and the probability of activation is

$$\frac{n^*}{n} = e^{-\left[\frac{E}{RT}\right]} \tag{4.4.57}$$

In such a process, activated reactants are considered to form an activated "complex" which may either relax to inactivated single reactants or irreversibly progress to form product molecules. The complex has some of the properties of an ordinary molecule and is at least temporarily stable. For reactant molecules A and B, the sequence of formation is

$$[A]+[B] \underset{k_b}{\overset{k_a}{\rightleftarrows}} [AB]^* \xrightarrow{k} \text{Products} \tag{4.4.58}$$

The overall reaction velocity, k (s^{-1}), determines the rate of formation of product and is related to the equilibrium constant for formation of activated complex, K*, by:

$$k = \frac{RT}{Nh} K^* = \frac{RT}{Nh} e^{-\frac{\Delta G^*}{RT}} \tag{4.4.59}$$

where N is Avogadro's number (6.023×10^{23}), h is Planck's constant (6.627×10^{-34} J-s), and ΔG^* is the Gibb's free energy of formation of activated complex. In turn, the free energy of formation is given by:

$$\Delta G^* = \Delta H^* - T\Delta S^* \tag{4.4.60}$$

where ΔH^* is the enthalpy of activation (J/mole) and ΔS^* is the entropy of activation (J/mol-K).

The activation entropy is not calculable except for the simplest possible reactions in the gas phase, and therefore is usually determined from experimental measurements of the reaction velocity and activation enthalpy. The activation enthalpy, ΔH^*, is determined from the observed activation energy, E by:

$$\Delta H^* = E - iRT \tag{4.4.61}$$

where i is 1 for first-order reactions in solution and gases, 2 for second-order, and 3 for third-order reactions.

Unimolecular Process Descriptions

Thermal damage in tissue is a unimolecular process — tissue constituents transition from the native state to the damaged state (Figure 4.4.10). Absolute reaction rate theory can also be used to explain the rate of formation for this process if we assume that a time lag exists between molecular activation and denaturization.[121] During this time lag, the molecules may either denature or relax back to the native state, as illustrated in Figure 4.4.10. Here, ΔH is the enthalpy (internal thermal energy) difference between native state and denatured molecules. The relative barriers are such that, in the thermal damage of tissue, ΔH^* is almost always smaller than ΔH. So, the activation process may be regarded as reasonably likely, and the probability of denatured tissue relaxing back to native state tissue is near enough to zero that it may be regarded as the impossible event in the absence of an energy-consuming healing process. The rate of damage formation is then proportional to only those molecules which remain activated. For a unimolecular process in the native state C, having an activated state, C*, with velocity constants k_a, and k_b:

$$[C] \underset{k_b}{\overset{k_a}{\rightleftarrows}} [C] + [C^*] \tag{4.4.62}$$

the activated complex, [C*] progresses to the damaged state at k_c:

$$[C^*] \xrightarrow{k_c} \text{Damaged state} \tag{4.4.63}$$

so that the rate of disappearance of native state molecules, [C], is given by:

$$-\frac{d[C]}{dt} = k_c [C^*] \tag{4.4.64}$$

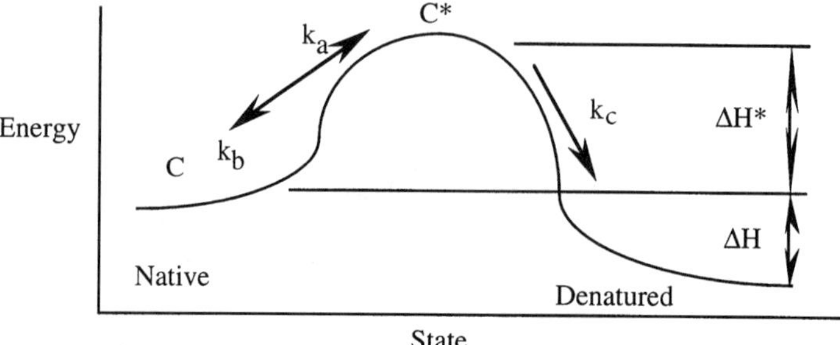

FIGURE 4.4.10 Unimolecular process activation and denaturalization.

Generally, [C*] is neither known nor calculable; however, at sufficiently low concentrations of C* the steady-state principle asserts that for short-lived activated states the rate of formation can be considered equal to the rate of disappearance. The activated state, [C*], forms at a rate k_a $[C]^2$, relaxes back to inactivated at rate k_b [C][C*], and denatures at the rate k_c [C*]. Consequently:

$$k_a[C]^2 = k_b[C][C^*] + k_c[C^*] \tag{4.4.65}$$

and so:

$$[C^*] = \frac{k_a[C]^2}{k_c + k_b[C]} \tag{4.4.66}$$

We actually need an overall reaction velocity, k, which relates [C] to its rate of disappearance:

$$-\frac{d[C]}{dt} = k[C] \tag{4.4.67}$$

There are two limiting cases for Equation 4.4.66: first, the concentration of remaining undamaged material, [C], may be large enough that deactivation at k_b dominates the k_c pathway, so $[C^*] \cong [C]\, k_a/k_b$ for which the overall formation rate, $k = k_c\, k_a/k_b$ and a first-order process results; and second, if the remaining undamaged material concentration, [C], is small, $k_c >> k_b$ [C] and the process is second order since from Equation 4.4.66 $k = k_a$ [C]. In liquid-phase systems with appreciable concentrations of native state molecules, the first condition should apply so the first-order approximation applies. After a long time of exposure at damaging temperatures such that [C] is very small, $k_c >> k_b$ C and a second-order process results:

$$-\frac{d[C]}{dt} = k[C]^2 \tag{4.4.68}$$

where for simplicity the [C] dependence has been removed from k:

First-Order Solution

Equation 4.4.67, then, is a Bernoulli differential equation with the solution:

$$C(\tau) = C(0)e^{\{\int k\, dt\}} \tag{4.4.69}$$

Equations 4.4.59 and 4.4.60 may be used to relate k to ΔH* and ΔS*. It should be noted at this point that the energy barrier, E, (Figure 4.4.9) is in fact ΔH* + RT; however, in practice {ΔH* _ 5 _ 105} >> {RT ≅ 3 × 10^3}, so we may assume that E ≅ ΔH*.

The pre-exponential term in Equation 4.4.59 suggests that it is temperature dependent; however, the linear dependence of A on 1/T is extremely weak compared to the exponential dependence in the final term.

Second-Order Solution

The second-order process of Equation 4.4.67 can be solved by dividing both sides by C^2 and using the substitution that y = 1/C. With that substitution:

$$\frac{dy}{dt} = \frac{dy}{dC}\frac{dC}{dt} = -\frac{1}{C^2}\frac{dC}{dt} = k \tag{4.4.70}$$

and the solution is straightforward:

$$y(\tau) = \int_0^\tau k\,dt + y(0) \tag{4.4.71}$$

or

$$\frac{1}{C(\tau)} - \frac{1}{C(0)} + \int_0^\tau k\,dt \tag{4.4.72}$$

Application of Kinetic Formulations in Thermal Damage Studies

When a quantitative thermal damage phenomenon has been identified, it can be studied as either a first-order or combined first- and second-order process. This subsection inspects the more common first-order analysis in some detail and introduces one form of combined process analysis to extend the usefulness of the method. The specific example of Henriques and Moritz original data is also studied. When the damage end-point is a qualitative measure, application of these kinetic models is closer to a curve-fitting exercise than to a fundamental study of the underlying phenomena. Therefore, precise identification and strict definition of quantifiable damage end-points are required for meaningful analysis.

First-Order Process Analysis

A more useful form of Equation 4.4.56 may obtained by recasting the result into a volume fraction model. In this formulation, C signifies the remaining concentration of native state (undamaged) tissue constituent. Therefore, the physical significance of the traditional damage measure, Ω, is the logarithm of the ratio of the original concentration of native tissue to the remaining native state tissue:

$$\Omega(\tau) = ln\left\{\frac{C(0)}{C(\tau)}\right\} = \int_0^\tau A\,e^{-\left[\frac{E}{RT}\right]}dt \tag{4.4.73}$$

This form of the damage integral has the advantage that it is easily compared to quantitative pathologic end points such as collagen and muscle birefringence loss, collagen hyaline damage, leakage of fluorescent dyes, or cell survival in culture. Using this description, direct comparisons can be made between computer models of $T(x, y, z, \tau)$ and measured histologic damage. A set of coefficients, A and E, is required for each damage process considered in the computer model. For tissue damage processes studied to date, A varies from about 10^{40} to 10^{105} s^{-1}, while E usually ranges from about 10^5 to 10^6 J/mol. Each damage process is then allowed to progress in parallel, driven by the calculated thermal field. This formulation assumes that the individual processes are thermodynamically independent, a reasonable description for identifiable thermal damage processes. The concentration of each of the damage markers, $C(\tau)_i$, is accumulated, and a distributed field description of the predicted histologic end point may be generated.

Functional behavior of the first-order model — The characteristic behavior of the kinetic damage model is that below a threshold temperature the rate of damage accumulation is negligible, but it increases precipitously when this value is exceeded. For purposes of discussion, it is useful to define the critical temperature as the temperature at which the damage accumulation rate, $d\Omega/dt$, is 1.0:

$$T_{crit} = \frac{E}{R\,ln\{A\}} \tag{4.4.74}$$

A damage process with representative coefficients of $A = 1.0 \times 10^{75}$ and $\Delta E = 5 \times 10^5$ has a critical temperature of 74.8°C. Figure 4.4.11 illustrates the damage accumulation rate dependence on temperature for the example process.

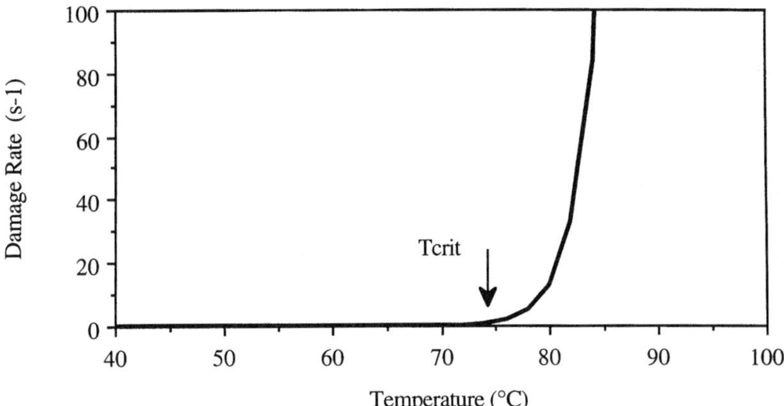

FIGURE 4.4.11 Hypothetical example damage accumulation process has A = $1.0 \times 10_{75}$ and E = 5×10^5 which gives the damage rate, $d\Omega/dt$, and a critical temperature of 74.8°C. (From Pearce, J.S. and Thomsen, S., in *Optical-Thermal Response of Laser-Irradiated Tissue,* Welch, A.D. and van Germert, M.M.C., Eds., Plenum Press, New York, 1995. With permission.)

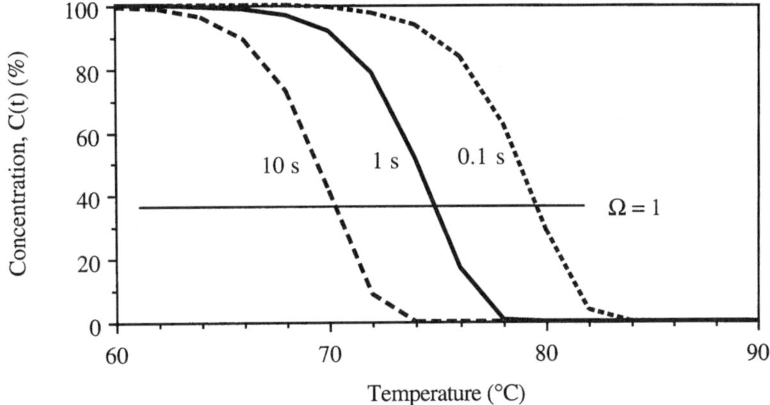

FIGURE 4.4.12 Concentration vs. temperature at various durations of exposure for the hypothetical example process. (From Pearce, J.S. and Thomsen, S., in *Optical-Thermal Response of Laser-Irradiated Tissue,* Welch, A.D. and van Germert, M.M.C., Eds., Plenum Press, New York, 1995. With permission.)

Constant temperature exposures of the example process will result in a decrease in concentration of native state material depending on the time of exposure. Figure 4.4.12 shows the remaining concentration for this process for constant temperature exposures of τ = 0.1, 1.0, and 10 s; the concentration is seen to gradually decrease with increasing temperature for fixed exposure times, as expected. The strong exponential nature of the process is evident as well. Applying this model framework relies heavily on identifying independent damage processes which can be quantitatively measured. From plots of the form of Figure 4.4.12, estimates of A and E may be made. Each curve will give one point on an Arrhenius plot (see above), where $\Omega = 1$. Note the necessity of using exposure times which span several orders of magnitude in order to separate the curves sufficiently to give acceptable accuracy in the determination of the kinetic coefficients.

Though the damage parameter, Ω, cannot be measured directly in histologic section, often a clearer picture of the functional behavior can be obtained from it. The exponential dependence of damage on the inverse of absolute temperature means that the temperature required to obtain comparable damage levels is quite sensitive to time of exposure. For example, we may define a threshold temperature, T_{TH},

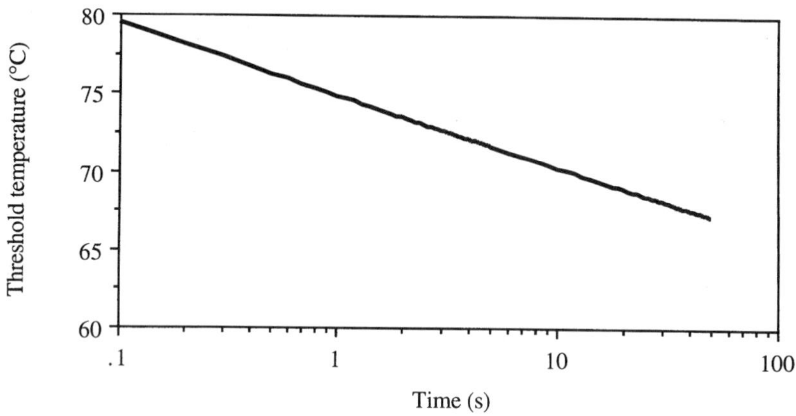

FIGURE 4.4.13 Threshold temperature, T_{TH}, (where $\Omega = 1$) as a function of duration, τ, for constant temperature exposures for the hypothetical example damage process. (From Pearce, J.S. and Thomsen, S., in *Optical-Thermal Response of Laser-Irradiated Tissue,* Welch, A.D. and van Germert, M.M.C., Eds., Plenum Press, New York, 1995. With permission.)

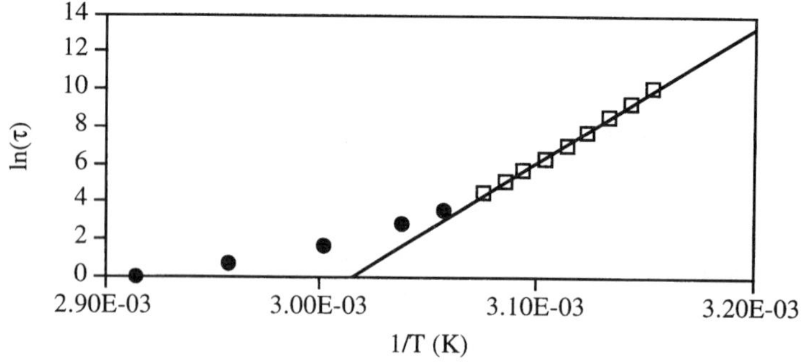

$\ln(\tau) = 72504\ (1/T) - 218.610; \quad r^2 = 0.997$

FIGURE 4.4.14 Plot of Henriques and Moritz[107] data on Arrhenius axes with curve fit line for data at 52°C and lower (open diamonds). For the fit E/R = 72504 and ln {A} = 218.61.

as the temperature at which the damage parameter, Ω, is 1 for a given duration, τ. Assuming a constant temperature exposure, the integral of Equation 4.4.72 reduces to a simple multiplication and the threshold temperature is

$$T_{TH} = \frac{E}{R\left[ln\{\tau\} - ln\left\{\frac{1}{A}\right\}\right]} \tag{4.4.75}$$

Figure 4.4.13 is a plot of the threshold temperature as a function of duration for the example damage process of Figures 4.4.11 and 4.4.12. Note that threshold temperature is approximately exponentially dependent on duration, as expected, with (in this case) a slope of about –4.65°C per decade. So, while 79.5°C is sufficient to result in $\Omega = 1$ at 0.1 s, 102.3°C would be required for a 1 μs exposure for the example process.

Application to the data of Henriques and Moritz[107] — A study of the original skin burn data reported by Henriques and Moritz.[107] Table 4.4.9 by Diller and Klutke[122] shows that the fit to a first order process is weak for the higher-temperature/shorter-exposure time experiments,[122] and that the original values of $A = 3.1 \times 10^{98}$ and $E = 6.27 \times 10^5$ (J/mol)[107] do not fit the data as well as their values of $A = 1.3 \times 10^{95}$ and $E = 6.04 \times 10^5$ (for temperatures less than 52°C). In Figure 4.4.15 the skin burn data are well fit with $A = 87.3 \times 10^{93}$ and $E = 6.03 \times 10^5$ when the last reasonable datum is included (at 52°C).

TABLE 4.4.9 Henriques and Moritz Skin Burn Threshold Data ($\Omega = 1$ Means a Second-Degree Burn Was Observed)[108]

Exposure Time (s)	Temperature (°C)
25,000	44
11,000	45
5,000	46
2,400	47
1,100	48
570	49
300	50
160	51
90	52
35	54
16	56
5	60
2	65
1	70

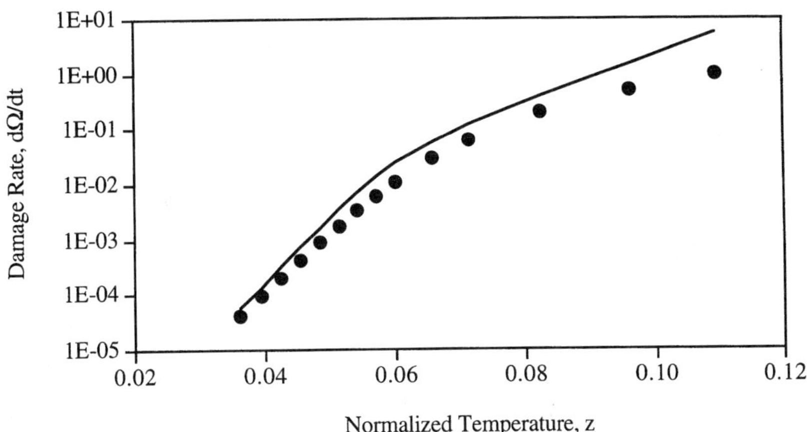

FIGURE 4.4.15 Plot of damage rate for the original Henriques and Moritz data[107] (solid circles) and the enzyme deactivation model of Xu and Qian[123] vs. normalized temperature.

Enzyme Deactivation Model

Xu and Qian[123] presented an enzyme deactivation model for skin burn data which combines some of the first- and second-order aspects in a single calculation. The method is based on a general simple enzyme-catalyzed reaction sequence from substrate, S, to products, P:

$$E_a + S \underset{k^-}{\overset{k^+}{\rightleftharpoons}} E_aS \xrightarrow{k} E_a + P \tag{4.4.76}$$

and

$$E_a \xrightarrow{k_d} E_i \tag{4.4.77}$$

where E_a is activated enzyme, E_i is inactivated denatured enzyme (at velocity k_d), and E_aS is the enzyme-substrate complex. The analysis is completed by assuming that the enzyme inactivation process is slow compared to the main reaction sequence. Nondimensional temperature, z, is used in the analysis, where:

$$z = 1 - \frac{T_0}{T} \tag{4.4.78}$$

The reference temperature, T_0, as used by Xu and Qian was representative of the average resting skin temperature: T_0 = 32.49°C = 305.65 K.

$$\Omega(\tau) = ln\left\{\frac{C(0)}{C(\tau)}\right\} = ln\left\{\frac{E_a(0)}{E_a(\tau)}\right\} = \int_0^{\tau} \frac{A\,e^{-\alpha z}}{1 + Be^{-\beta z}}\,dt \tag{4.4.79}$$

This combined formulation fits the data of Henriques and Moritz (Table 4.4.9) quite well over its entire range if $A = 1.0 \times 10^{-4}$, $\alpha = 100$, $B = 8.0 \times 10^{4}$, and $\beta = 195$, (see Figure 4.4.15).[123]

Histologic Markers of Thermal Damage*

Tissue and cell heating produce several pathologic alterations that can be directly attributed to thermal damage mechanisms. Many of these are measurable using quantitative pathologic techniques, and thus are amenable to description by kinetic rate process models. As discussed above, it is important to recognize that several independent damage processes can be occurring at the same time. Therefore, it is imperative to describe precisely which particular end point will be measured for kinetic analysis. Depending on the tissue, some end points are more easily identified and measured than others. From the pragmatic biological standpoint, cellular and tissue thermal damage end points are divided into two general categories: lethal and nonlethal thermal injury.

Nonlethal Low-Temperature Tissue Effects

Most living cells and tissues can tolerate and survive modest temperature elevations for limited time periods depending on the species and the metabolic status of the individual. Recovery after heating is marked by restoration of normal functions. Most nonlethal thermal injuries of individual cells are secondary either to: (1) heat-induced acceleration of metabolism, (2) thermal inactivation of particular enzymes, and/or (3) rupture of cellular membranes. These alterations lead to disruption of several physiologic reactions of which the most critical are respiration and energy-generating metabolic functions.[123,124] The best indicators of reversible, nonlethal thermal injury in the living cell or organism are physiological tests that monitor heat-sensitive biochemical and metabolic changes rather than morphologic alterations.

Immediate effects — Routine light microscopic techniques applied immediately after heating are not useful for the detection of histologic markers of nonlethal thermal injury. However, enzyme histochemistry, a laborious, somewhat capricious methodology, can be used to reveal inactivation of a few heat-labile enzymes,[125] and transmission electron microscopy (TEM) provides sufficient magnification and resolution to show ruptured cellular membranes and distorted organelles. However, if the ruptures and distortions are frequently seen, then the damage is not likely lethal. Over time, recovery occurs with synthesis of replacement enzyme proteins and new membranes to patch the holes and reconstitute the normal functions of the organelle and cell.

On the other hand, tissues and organs are composed of a diverse population of cells that have (1) varying levels of heat sensitivity, (2) different types of metabolic functions and repair capabilities, and (3) various contributions to the function of the tissue or organ. Unlike single cells, reversible thermal damage of tissues can and usually does involve some cell death, but it occurs at a level that does not impair the function of the organ or the tissues. Like the situation in single cells, histologic assessment of reversible tissue thermal injury immediately after heating requires enzyme histochemistry and TEM to determine the degree and distribution of thermal damage, because this damage is not detectable with routine light microscopic techniques.

---

*The authors acknowledge the contributions to this subsection by Sharon Thomsen, M.D.; portions excerpted from Pearce, J.A. and Thomsen, S., in *Optical-Thermal Response of Laser-Irradiated Tissue,* Welch, A.J. and van Germert, M.M.C., Eds., Plenum Press, New York, 1995. With permission.

Delayed effects — Depending on the tissue, intracellular edema, tissue edema, and hyperemia occur within seconds to hours. These are easily identified delayed organismal responses to nonlethal thermal damage in both single cells and living tissues. Intracellular edema is due to abnormal accumulation of fluids secondary to thermally induced metabolic dysfunction, and resolution of the edema upon repair is a marker of healing. Heat-induced tissue edema and transient hyperemia (increased blood flow and blood vessel dilatation) are mediated by release of vasoactive polypeptides from local inflammatory cells within several seconds of injury. The action of the polypeptides causes blood fluids (primarily plasma) to escape through gaps between the endothelial cells lining the vessel. On the other hand, prolonged hyperemia, like that associated with sunburn, is a delayed response associated with direct endothelial damage. In this case, the histologic indicators of the more severe vascular damage are dilated blood vessels stuffed with red blood cells and, not infrequently, microscopic leakage of the red blood cells through the damaged vessel wall into the adjacent tissue spaces. Recovery, which is secondary to endothelial regeneration and repair, is seen in a few days.[126]

Lethal Low-Temperature Thermal Effects

Cell and tissue death and necrosis — Cell death and subsequent necrosis result when the damage is so severe that the usual repair mechanisms cannot cope and/or the mediators of the repair mechanisms (DNA and RNA transcription enzymes) are thermally destroyed. Again, routine light microscopic techniques cannot be used to identify dead cells immediately after heating, except in the case of severe damage. However, within minutes of lethal thermal damage, TEM reveals the accumulation of chromatin (the DNA-containing chromosomes) at the margins of the nuclei, a recognized TEM marker of cell death, plus more extensive disruption of cellular organelles and limiting cell membranes that are incompatible with survival.[127] Enzyme histochemistry demonstrates enzyme inactivation but, as discussed above, enzyme inactivation by itself can not be used as a hallmark of cell death. After death, cells and tissues undergo necrosis, which is a natural process of disintegration that follows a predictable course over time. Tissue necrosis, the "gold standard" marker of death, is easily recognized in light microscopic sections of lesions 24 to 72 h after the lethal event and, for most tissues, histologic evaluation of the maximal extent of tissue necrosis is seen at about 3 to 5 days.[128-130] After 5 days, the boundaries of necrosis and other markers of thermal coagulation are obscured by the healing processes and precise measurement of the extent of lethal heating can no longer be done.

***In vivo* red zone thermal damage** — When localized electromagnetic, acoustic, or optical heating results in deep zones of necrosis, as when laser light is delivered through an optical fiber or ultrasound or electromagnetic waves are focused, concentric zones of thermal damage form around the source during heating and immediately thereafter. The boundaries between these zones are often distinct and measurable. Depending on the power density and duration of exposure, the three major zones in the lesion which can be seen by the naked eye are (1) a central ablation hole (at high-power densities), (2) a whitish zone of coagulation, and (3) when the blood supply is intact, a surrounding peripheral red zone. The red zone is formed as a result of hemostasis, hemorrhage, and hyperemia (increased blood flow). The vascular responses are intermixed; however, hemostasis — due to direct damage to blood cells and vessels — tends to occur at the inner boundary between the surrounding red zone and the more central white coagulation zone. Studies performed in rat liver and goat breast show that the outer boundary of the red zone corresponds to the maximum extent of lethal thermal damage as determined by tissue necrosis evaluated in surviving animals 3 days after treatment.[128,129]

Microscopically, additional distinct zones of thermal damage can be detected in the white coagulum. Advancing radially inward from the boundary between the red and white zones, one observes the following quantitative histopathological markers of thermal damage:

1. Lethal cellular injury manifested by cell shrinkage, spindling, and hyperchromasia.
2. Collagen hyalinization.
3. Collagen birefringence changes.
4. Muscle birefringence changes.
5. Water-dominated effects.

The changes intensify as the hotter center of the lesion is approached. Additionally, a mass defect created by ablation of the tissue is observed in cases of intense heating, but is not considered here.

Thermal Coagulation

As temperatures rise and/or heating times are prolonged, cellular and tissue structural proteins, which are thermally more stable than the vital, energy-producing enzymes, undergo denaturization and conformational changes — a process defined as thermal coagulation. Coagulation is immediately apparent and always indicates lethal thermal effect. For most tissues, coagulation can be seen with the naked eye as whitening of the tissue associated with increased turgor and opacity. The changes of the egg white while the egg is being fried is an obvious example of thermal coagulation. On the other hand, coagulation of collagen-rich tissues, such as tendon and skin, can be signaled by increasing transparency due, apparently, to decreased optical scattering by the denatured collagen fibrils.

Microscopically, thermal coagulation of cellular and extracellular structural proteins includes an array of morphological alterations that mark a large range of temperatures and exposure times. As tissue temperatures and/or exposure times increase, the coagulative changes become more obvious and light microscopic analysis becomes the investigative tool of choice.

The most useful histologic markers of coagulative thermal damage in tissues are structural alterations of cells and collagens. Thermally coagulated cells and intracellular organelles shrink and undergo characteristic conformational changes. Collagens are a widely dispersed biochemical class of extracellular fibrous proteins that form the supporting scaffolding of nearly all soft tissues and the organic strut system of bone, cartilage, and teeth. These fibrillar proteins swell and form amorphous masses as a unique response to heat.

Cellular shrinkage and conformational changes — Thermally coagulated cells shrink due to denaturation of the elongate cytoplasmic and nuclear proteins that form the three-dimensional scaffolding system of these structures.[130,131] Coagulated surface epithelial cells — such as those that line the gut, bladder, and numerous glands of the body — tend to become elongate (spindle-shaped). The boundary between this change and the normal epithelium can be quite distinct, and thus potentially useful for kinetic analysis.

In contrast, the epithelial cells of the solid organs — such as the liver, pancreas, and kidney — just shrink and do not undergo noticeable conformational change at the light microscopic level. The cellular shrinkage is also due to loss of intracellular water. The coagulated proteins of the desiccated cells become more densely packed, as reflected by dark cytoplasmic and nuclear staining (hyperchromasia) in routinely prepared histologic sections. Unfortunately, these histologic changes are very subtle and, frequently, cannot be distinguished from technical artifacts due to poor fixation. Therefore, they are not reliable markers of thermal damage, especially in the hands of an inexperienced observer.

Collagen coagulation — Thermal denaturization of extracellular fibrous structural proteins, the collagens, is marked histologically by swelling and an amorphous, glassy transformation of the fibers (hyalinization).[128,129,132,133] This transformation is easily seen in those tissues composed of loose networks of collagen fibers, such as the cornea, skin, submucosa of the stomach, intestines, and urinary bladder. However, some collagen fibers, such as those found in arterial media and liver sinusoids, are very thin ("reticulin fibers"); thus detection of hyalinization is beyond the resolution of the light microscope. And, at the other end of the fiber size spectrum, mild hyalinization and fiber swelling are equally difficult to see in tissues composed of thick, densely packed collagen fibers, such as those found in blood vessel adventitia and dermis. Changes of birefringence of thermally coagulated collagen (described below) are sometimes easier to detect than hyalinization in these circumstances.

Thermally Induced Loss of Birefringence

Birefringence is an optical property of some tissues that rotates incident polarized light. Thermal coagulation is associated with partial and total loss of the native birefringence of muscle tissues and certain fibrillar collagens as revealed by transmission polarizing microscopy (TPM).[129,133] The birefringence of muscle is due to the very regular arrangement of the fibrillar contractile protein macromolecules, actin and myosin, that form the contractile unit of the muscle. Collagen birefringence is the result of the molecular structure and regimentation of the tropocollagen molecules from which collagen fibrils are built. Tissue birefringence has two components, intrinsic and form birefringence. Intrinsic birefringence is secondary to the α-helical conformation of the polypeptides that form the individual contractile proteins of muscle and tropocollagen molecules and accounts for 12 to 30% of the total birefringence of these tissues. The remaining birefringence, form birefringence, is related to the crystalline-like array of the actin and myosin molecules in the sarcomere and the regimented longitudinal array of tropocollagens within the collagen fibrils.[131,133-146]

Thermally induced changes of muscle birefringence are associated with dissociation and disruption of the molecules, as revealed by TEM. Partial loss of birefringence in skeletal and cardiac muscle has been identified with the breakdown of the relationships of actin and myosin in the sarcomere (loss of form birefringence). The source of the birefringence of striated muscle is the central portion of the sarcomere, the A or anisotropic band, in which the actin and myosin are arranged in parallel crystalline-like array. The I or isotropic band of the sarcomere includes portions of adjacent sarcomeres that contain actin molecules which are attached to the Z band, an attachment zone for the actin molecules. Total loss of birefringence is associated with increased temperature-time histories and is probably related to the reduction of the individual fibrillar contractile proteins to thermally denatured granular profiles (loss of intrinsic birefringence) as shown in TEM of heated myocardium.[147,148]

Thermally induced birefringence changes in collagen involve decreasing intensity of the birefringent image until it is totally lost at higher temperatures. TEM of heated collagens shows a gradual unraveling of the fibers with disappearance of the characteristic periodicity of collagen, reflecting the dissolution of the regimented organization of native collagen as it is heated.[129,146]

The onset of the birefringence changes of collagens varies depending on the experimental conditions, anatomic site, and the age of the organism. Native tissue collagens reside in complex environments of different proteins, glycoproteins, and cells that are greatly influenced by the local and systemic physiology of the living organism. The intimate relationships of the collagens with these tissue constituents are regulated by water and salt concentrations, ionic and covalent bonds, and hormonal and metabolic effects. Collagens extracted from tissues are not in their natural environment, therefore *in vitro* experiments testing molecular relationships including measurements of birefringence will not predict the native behavior of collagen *in situ* or *in vivo*.[142-144] In addition, as the organism ages, the collagens, especially the Type I collagen found in most tissues, form more molecular cross-links that alter the temperature-time thresholds of fiber dissociation.[143]

Polarizing microscopy can be performed on paraffin sections prepared for routine light microscopy; thus the birefringent images can be correlated directly to morphologic structure. The intensity of the birefringent images is governed by the thickness of the section, the type of dyes used, and in the case of collagens, the collagen fiber thickness.[128,149]

Water-Dominated Effects

The role of water vaporization becomes dominant as the tissue temperatures approach 100°C. The tissue effects depend on: (1) the rate of water vapor formation, (2) the temperature of the water vapor, (3) the rate of diffusion (escape) of the vapor from the tissue, and (4) the tissue mechanical properties.[150-159] All tissues exposed to air will lose water vapor to the atmosphere because of the differences between the partial pressure of water vapor across the tissue/air boundary. Below 100°C water vaporization is a surface phenomenon. As tissues are heated water vapor will be generated and the vapor will diffuse toward the surface to escape. The water vapor is in thermodynamic equilibrium with liquid water and, in the tissues deep to the surface, the relatively low partial pressure gradients favor the liquid over the vapor phase. Meanwhile, at the surface, the tissues dry out as the water vapor escapes leaving the hard, not infrequently brittle, solid tissue components behind.

At or just above 100°C water vapor is generated volumetrically, equilibrium is pushed toward the vapor phase, and, for sufficiently high volume generation rates, q''', more vapor is produced than can escape by simple diffusion. The excess vapor is trapped within the cell and in the extracellular tissue layers forming vacuoles or dissections along planes of mechanical weakness.[159-161] The surface desiccates rapidly because diffusion lengths are short. The temperatures of the tissues that still contain liquid water will either remain close to 100°C or the pressure will rise.

The vapor in the vacuoles will expand quickly compressing the surrounding, rapidly drying tissues that form the vacuolar walls. As the vacuoles expand the walls separating the vapor pockets from each other or the tissue surface become thin. The walls rupture as the force of the increasing pressures of expanding vapor overcomes the mechanical strength of the tissue. The vacuoles coalesce to become larger holes within the tissues or form irregular defects on the surface. When this process takes place over relatively long heating times, the tissue defects are formed as the result of tissue compression, desiccation, and shrinkage, but *not* to loss of dry mass. On the other hand, rapid heating times result in explosive rupture which causes tissue fragments to be ejected from the surface creating irregular craters formed secondary to dry tissue mass loss. These water-dominated events are the same as those when popcorn is made, hence the tissue changes have been called the "popcorn" effect.[151-154,160-162]

The kinetics of water vaporization will not be treated here; see the discussion of phase change in Chapter 2. Histologically, the zone of water vapor formation manifests as a distinct region populated by vacuoles adjacent to ablation craters formed from volume heat sources. In some experiments, for example, electrosurgical cutting and laser ablation at water-absorbed wavelengths, the vacuolized zone usually has a distinct and measurable border.

Damage Rate Process Coefficients

Thermal damage kinetic coefficients are usually determined from constant temperature exposures of relatively long duration. Threshold damage results are selected out of a set of damaged tissue samples for analysis from which estimates of A and E are obtained.

Constant Temperature Exposures

Because of the sensitivity of the damage integral to small changes in temperature, the typical approach in obtaining A and E is to expose the tissue to a constant temperature, identify experiments in which the damage is threshold — i.e., for which $\Omega = 1$, or C(t) = 36.8% of C(0) — and obtain A from the intercept and E from the slope of an Arrhenius plot of ln(t) vs. 1/T for the threshold experiments. If the temperature is held constant the integral reduces to a simple multiplication, and C(t) = 0.368 C(0), then Equation 4.4.69 becomes:

$$ln\{\tau\} = \frac{E}{R}\frac{1}{T} - ln\{A\} \qquad (4.4.80)$$

The usual experimental method is to expose thin slices of tissue to constant temperature in a water bath, by surface application of heated water,[105-108] or on a heated metallic plate for desired time intervals. Table 4.4.10 lists rate coefficients obtained in various experiments and the relative damage rates ($d\Omega/dt$) are compared for several of them in Figure 4.4.16.

When using the kinetic models and measured coefficients, it is imperative that an adequate description of the particular end point be given. This is because different end points in the same tissue will have widely varying critical temperatures and damage accumulation rates. For example, in Table 4.4.10 in addition to the variation in tissue, the end points in the various studies differ markedly.

Retina — The measurements of Welch and Polhamus[163] used the diameter of the edge of the visible lesion formed in bovine retina under argon laser irradiation as the end point. The temperatures were not measured directly, but were determined in separate experiments on retinas in which a correlation between temperature and radius was established using microthermocouples (about 5 μm in diameter) advanced from the posterior surface of the eye to a point just below the retina. The correlates were used to estimate the retinal temperature given laser beam power and duration. The critical temperature for these coefficients is 56.0°C. Takata et al.[164] used a similar decision criterion for retinal damage for shorter exposure times. They fit the data with three sets of coefficients because a single first-order model was not sufficient. It would appear that several parallel damage processes were at work in their study, thus a set of coefficients was required. The critical temperature for the high-temperature set of coefficients is 59.9°C. Birngruber

TABLE 4.4.10 Experimentally Determined Rate Coefficients

	A (s^{-1})	E (J/mol)	Conditions
Retina:			
Takata et al., 1974	0		T < 316 K
	4.3×10^{64}	4.2×10^5	316 < T < 323
	9.3×10^{104}	6.7×10^5	T > 323
Welch and Polhamus, 1984	3.1×10^{99}	6.28×10^5	
Birngruber, 1980, 1985	10^{44}	2.93×10^5	
Skin:			
Henriques, 1947	3.1×10^{98}	6.27×10^5	
Weaver and Stoll, 1967	2.2×10^{124}	7.83×10^5	317 < T < 323 K
	1.8×10^{51}	3.27×10^5	T > 323
Egg:			
Yang et al., 1991	White 3.8×10^{57}	3.85×10^5	
	Yolk 3.05×10^{56}	3.89×10^5	
Aorta (collagen):			
Agah, 1988	5.6×10^{63}	4.3×10^5	
Calcein Leakage, AT-1 cells:			
Bhowmick and Bischof, 1998	1.14×10^5	4.78×10^4	300 < T < 328 K
	2.7×10^{11}	8.78×10^4	T > 328 K
50% Cell Survival, AT-1 cells:			
Bhowmick and Bischof, 1998	1.16×10^{48}	6.58×10^6	300 < T < 328 K
	6.11×10^{12}	9.22×10^4	T > 328 K

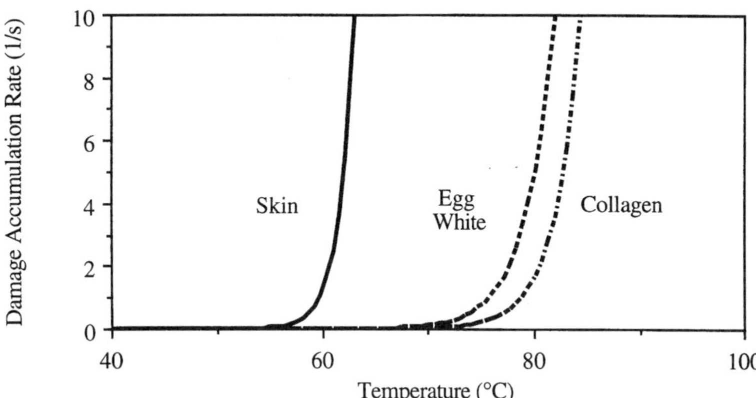

FIGURE 4.4.16 Comparison of selected damage processes from Table 4.4.10: damage rate, dΩ/dt, vs. temperature. (From Pearce, J.S. and Thomsen, S., in *Optical-Thermal Response of Laser-Irradiated Tissue,* Welch, A.D. and van Germert, M.M.C., Eds., Plenum Press, New York, 1995. With permission.)

et al.[165-166] estimated a frequency factor and activation energy from consideration of the thermodynamics of protein and enzyme denaturization processes. Their estimates have a critical temperature of 74.5°C.

Skin — The end point for $\Omega = 1$ in the original studies of Henriques and Moritz[105-108] corresponded to a continuum of cascaded effects. In their study the skin of pigs was exposed, *in vivo*, to flowing water at a controlled temperature for exposure times varying over several orders of magnitude. They calibrated their coefficients so that $\Omega = 0.53$ corresponded to the onset of erythema (characterized as "first degree" in their paper). Then $\Omega = 1.0$ corresponded to a "second degree", or partial thickness burn, and $\Omega = 10^4$ to a full thickness, or "third degree" burn. Their coefficients, $A = 3.1 \times 10^{98}$ and $E = 6.27 \times 10^5$, have a critical temperature of 59.7°C. Weaver and Stoll[111] used similar criteria to those of Henriques and Moritz and applied two sets of coefficients (the upper values applicable above 50°C, as in the Takata et al. study) to match the experimental data. The critical temperature for their highest-temperature coefficient set is 59.4°C.

Egg white and egg yolk — Egg white and yolk were exposed to constant temperature (±0.2°C) in a water bath for varying exposure times.[167] Coagulation was defined as the onset of whiteness (coagulum formation due, apparently, to an increase in scattering in the clear liquid egg white) observed by the naked eye. Water bath temperatures ranged from 60 to 90°C in 4°C increments. At each temperature the time to threshold was measured and plotted in accordance with Equation 4.4.22 and rate coefficients determined from linear regression, as described. Approximately 3 to 5 s were required to obtain whitening in the egg white at 70°C; while in the egg yolk 82°C was required for the same exposure time. The table values have a critical temperatures of 76.0°C for egg white and 86.6°C for egg yolk.

Aorta — We have measured coefficients for collagen damage in excised human aorta wherein the specific damage end point was a measurable and repeatable weakening of the connection between collagen and elastin in the medial layers.[168] The experiments consisted of variable exposure time (15 to 1500 s) contact of the intimal surface with a heated copper block controlled to ±1°C — the adventitial surface was insulated. The weakening was reflected as a tearing between these fibers in histologic section. Of course, tearing in a histologic section can be due to a number of processing variables. In order for an experiment to be acceptable as a damage measure, the tearing region had to be specifically associated with the region of heated block application in multiple adjacent sections. There were 10 thresholds obtained in 90 experiments, from which the damage coefficients were estimated to be: $A = 5.6 \times 10^{63}$ (s^{-1}) and $E = 4.3 \times 10^5$ (J/mol) — the critical temperature for these coefficients is 78.9°C.

The kinetic nature of thermal damage means that the time of exposure is of critical importance when discussing threshold temperatures. For example, it is commonly stated that thermal damage occurs above 45°C — and, in fact, 45°C was identified by Moritz and Henriques as the long-term exposure asymptote for threshold damage. However, at that temperature it would require 9.7 h of exposure to get a burn ($\Omega = 1$). To get a skin burn in 1 s, a temperature of 59°C would be required, and over the 0.1 s typical of the laser activations, 63°C is required. Similarly, using the coefficients for collagen, at 45°C it would require about 73 days of exposure to get $\Omega = 1$; for damage in 1 s, 79°C, and in 0.1 s a constant temperature of 84.5°C is required. So, when comparing the higher temperatures required to obtain collagen denaturization in laser experiments (above about 80°C) to, for example, published values of collagen temperature thresholds (which range from 50 to 60°C), it is important to note that: (1) the times of exposure used in published collagen damage reports are usually in terms of hours, and (2) it is entirely likely that collagen in dilute solution has a different set of kinetic coefficients than collagen *in situ* owing to the complex structural interrelationships inherent in the tissue.

Calcein leakage — Bhowmick and Bischof[120] measured calcein leakage rates from type AT-1 prostate tumor cells exposed to elevated temperatures. Calcein is a fluorescent dye which leaks out of cell membranes at varying rates depending on membrane integrity. The coefficients in Table 4.4.10 correspond to a 50% decrease in intracellular calcein fluorescence intensity. They also reported cell survival rates (again, for 50% survival). These data are reasonably close to $\Omega = 1$ points (36.8% survival or remaining fluorescence intensity). Many other cell survival studies have been reported in literature (see, for example, References 169 and 170).

Transient Thermal History Experiments

Problems with the standard constant temperature approach to determining A and E are that:

1. With only a very few exceptions, the tissue must be excised to perform the experiments disrupting its blood perfusion and activating autolytic (post-mortem degeneration and necrosis) processes.
2. No exposure is truly constant temperature and rise time segments must necessarily be short compared to total exposure time in order to be negligible (restricting one to the study of very slow processes).
3. It is difficult to obtain trustworthy estimates of A since the temperature axis is hyperbolic (1/T) and, even if the exposure times vary over several orders of magnitude, small uncertainties in the slope (E/R) create very large uncertainties in A.
4. The methods are difficult to use in laser damage studies since constant temperature exposures are not possible to obtain.

Working with transient thermal data adds a high level of uncertainty to the damage coefficients since the time of exposure is problematic. Nevertheless, estimates of rate coefficients can be made upon which treatment protocols can be evaluated. We have applied an approximate method with some limited success,[171,172] and describe it here.

As can be seen from Table 4.4.10, most of the damage processes have activation enthalpies on the order of 10^5 J/mol. For the hypothetical example damage process, with $E = 5.0 \times 10^5$ J/mol, at 60°C the rate of damage formation would double if the temperature increased to 61.3°C (about 0.4% temperature increase on the absolute scale). The rate of rise of laser spot heating experiments is usually many degrees per second; therefore, we assign the exposure time as the time during which the tissue temperature is within 2°C of the maximum value. Of course, noise in the determination of temperature creates significant uncertainty in the determination of exposure time, τ, for a calculation; however, technically acceptable experiments can be extracted by inspection of the temperature history. Note that the temperature history at the location of threshold damage is required; so, we determine the surface location of threshold damage in the histologic section, using the methods previously outlined, and then analyze the thermographic record of temperature at that location. We have used this method to make preliminary estimates of damage coefficients for birefringence loss in rabbit myocardium and purpura formation in pig skin irradiated with an argon laser. By referring directly to the recorded thermographic image data, we avoid the uncertainty associated with optical models and properties of the tissue.

Purpura formation — When using lasers for the clinical treatment of dermatologic anomalies such as port wine stain, a variable of major importance in dosimetry planning is the purpura threshold. The purpura threshold is determined by optical and thermal properties of the skin since purpura generation is of photothermal rather than photochemical origin. In controlling the dose, thermal feedback can be used to limit the laser irradiance if the damage process is known. Direct use of thermal feedback has the additional advantage that variations in the tissue optical properties are not important to the control algorithm. Therefore, useful thermodynamic models of purpura formation could provide additional information to be used in treatment planning, and will likely facilitate thermal feedback approaches to dosimetry control.

Twenty experiments were conducted on porcine skin *in vivo* using an argon laser.[171] The spot size was nominally 2 mm and activation times and beam powers ranged from 0.2 to 20 s and 1 to 11 W, respectively. Skin surface temperature was recorded thermographically and transient thermal histories for the radius of purpura formation were determined *a posteriori* at locations determined from gross and histologic observation. The 2°C criterion was used to estimate the exposure time at the threshold radius. The time and temperature were plotted on an Arrhenius plot and A and E determined by linear regression. It should be noted that the purpura coefficients were derived by measuring surface temperature and the active thermal damage process is occurring subsurface at the capillary level. Also, the argon laser wavelength is significantly scattered in dermis and epidermis, so one would expect that the deep temperatures were higher than the skin surface temperatures. Consequently, the coefficients apply to measured surface temperature, not the capillary damage process per se. Of course, when feedback control based on surface temperature is desired, this is the variable of choice. For these data, $A = 4.11 \times 10^{53}$ s^{-1} and $E = 3.39 \times 10^5$ J/mol and the critical temperature is 56.9°C. The data are quite noisy, as can be seen in Figure 4.4.17, not an unexpected result.

Birefringence loss in myocardium — A similar series of experiments was conducted on freshly excised rabbit myocardium less than 10 min after sacrifice.[172] The epicardial surface of the excised left ventricle was exposed to argon laser light. Again, the spot size was 2 mm with durations and beam powers from 0.2 to 20 s and 1 to 12 W, respectively. After exposure the visible lesion was measured and marked across the diameter to guide the histologic sections. Stained sections (Mallory's trichrome and H&E) were analyzed to determine the depth and surface diameter of the zones of partial and total birefringence loss. The temperature histories of the radii of the threshold of birefringence loss were calculated from the recorded thermal imagery. Figure 4.4.18 is an Arrhenius plot of the resulting data. Damage coefficients for this process were estimated to be $A = 3.12 \times 10^{20}$ (s^{-1}) and $E = 1.28 \times 10^5$ (J/mol) with a critical temperature of 52.9°C. Interestingly, in this data set the diameters of the visible lesions equaled those of the birefringence loss zones within the accuracy of the measurements.

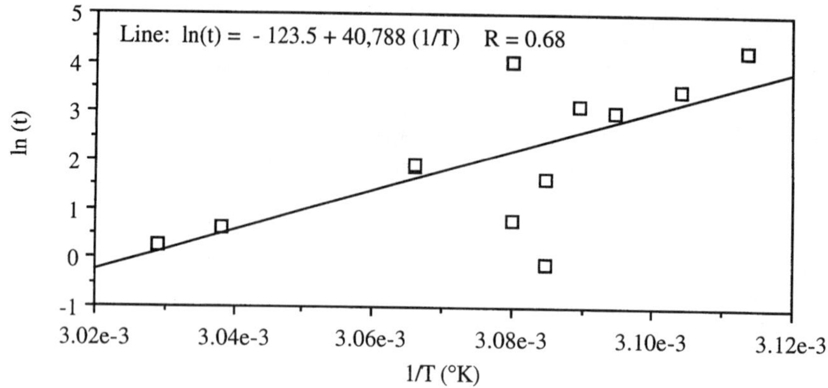

FIGURE 4.4.17 Arrhenius plot for purpura formation *in vivo* in porcine skin. Data derived from surface measurement of temperature resulting from coagulation of deeper blood vessels.

Summary

First-order kinetic models for tissue damage are useful for predicting trends in damage experiments. Unfortunately, there are only a few damage processes for which the frequency factor and energy have been determined *in situ* — critical temperatures for known processes range from about 50 to 90°C. Nevertheless, these models can be used to make direct comparisons between numerical predictions of damage and histologic results; something which cannot, as yet, be achieved any other way. A very careful definition of the particular histologic end point is necessary. To avoid the arbitrariness which characterizes much of the early damage studies, the end point ought to be (1) clearly recognizable and defined, (2) reproducible under varying experimental conditions, (3) easily measured, and (4) involve relatively homogeneous and readily available test tissues. For example, thermally induced birefringence image intensity decrease in skeletal and cardiac muscle is easily seen and measured using polarizing microscopy and optical detectors. Beef and pork loin cuts, chicken breasts, and canine or beef hearts can provide the numerous tissue slices required to accumulate adequate data.

In part because of the effects of biologic inhomogeneities and random fluctuations in tissue characteristics, and in part due to the difficulty of resolving small temperature differences, thermal damage data are inherently noisy. Consequently, the prediction of observed damage boundaries in very long-exposure experiments (on the order of minutes to hours) in the presence of substantial spatial thermal gradients may be frustrating. However, in small spot-size exposures of relatively short duration (up to fractions of seconds) the predicted boundaries compare favorably to those observed. This is probably due to the very steep thermal gradients typical of small spot experiments — that is, a rather large error in actual critical temperature may be swamped out by the very steep thermal gradient so that the location of the predicted damage contour may, in fact, agree fairly well with experimental histologic observation. Certainly, even though the results of a particular experiment may eventually prove impossible to duplicate in numerical models, a rather careful analysis of the trends which one would obtain from changing power, contact temperature, spot size, and duration can be studied in detail in the numerical model and on a spatial scale similar to microscopic observation. Also, the numerical model allows dissection of the transient development of thermal damage — something which cannot be achieved in any other way. So, while there are many uncertainties associated with kinetic models of thermal damage, they can be extremely illuminating and helpful in dosimetry planning.

Tissue Effects: Subzero Temperatures

An important area of bioheat transfer application is in the domain of subzero temperatures. As was noted previously, biochemical rate processes are governed by the local system temperature. Lowering the temperature has the effect of reducing reaction rates, and at sufficiently low temperatures a state of suspended animation can be achieved. Owing to the major aqueous component of physiological fluids, temperatures low enough to effect suspended animation normally result in freezing. The freezing of native biomaterials is nearly always lethal to the affected tissue upon thawing. The affected injury can be used beneficially in cryosurgery for the purpose of destroying a target tissue such as cancer. Alternatively, the tissue can be modified prior to freezing by the introduction of a chemical cryoprotective agent (CPA) to afford protection from freeze-thaw injury. Techniques have been developed for the successful long-term cryopreservation of many biomaterials for subsequent applications, including human transplantation. The response of living biomaterials to freezing and thawing is intimately tied to the thermal history during processing, especially at subzero temperatures. Thus, bioheat transfer analysis has played a key role in the design and development of low-temperature bioprocessing protocols.

Cryopreservation

Living tissues may be frozen to deep subzero temperatures to create a state of suspended animation for indefinite periods and recovered with very minimal loss of viability and function. It is necessary to store tissues at below approximately –120°C so the kinetics of chemical reactions and ice nucleation become infinitesimally small. Successful cryopreservation protocols require that subject tissue be modified prior to cooling to subzero (C) temperatures by addition of a cryoprotective additive (CPA), either to protect against the injurious effects of ice formation or to block the formation of ice so that a glassy state results (vitrification).

Widespread interest has developed in exploiting cryopreservation as a means for reversibly banking a broad spectrum of tissues for transplantation. The seminal paper which first reported this work described the use of glycerol to freeze fowl sperm 50 years ago.[173] Successes were reported in succession for other types of tissues having rather simple cell structures, such as erythrocytes, gametes, and various cells obtained from primary cultures.[174-176] Most of these cryopreservation techniques were derived via largely empirical methods, and starting in the 1970s it came to be realized that the cryopreservation of more complex systems such as multicellular tissues and whole organs require a more rigorous scientific understanding of the mechanisms of the governing biophysical processes and cellular response to freezing and thawing. Since that time engineers have made significant contributions to the developing science of cryobiology, not the least of which has been to identify some of the key biophysical problems to be solved.[177]

Biophysical Basis of Freezing Processes in Cells

Analysis of cryopreservation by freeze/thaw processes is based on addressing phenomena associated with the solidification of aqueous solutions and their resulting effects on embedded living cells surrounded by semipermeable membranes. When an aqueous solution freezes, water is sequestered into the solid ice phase, resulting in concentration of solutes in the residual liquid phase solution. If equilibrium is maintained between the liquid and solid phases, the coupling between temperature and solute concentration is described by the phase diagram for the solution. As shown in Figure 4.4.19, as the temperature is depressed below 0°C the amount of solute in the liquid increases according to a function defined by the liquidus curve until the eutectic state is reached. Thus, cooling a biological tissue until ice is nucleated imposes an osmotic as well as thermal stress.

A consequence of the osmotic stress is that the individual cells in a tissue will lose intracellular water to the solute-enriched environment. (It is assumed that at subzero temperatures the time scales for transport of other molecular species will be negligible in comparison to that for water.) The rate at which this mass transport occurs across the cell membrane is governed by the membrane permeability to water. This transport process has been modeled as a simple diffusion phenomenon.[178]

$$\frac{\partial \Psi}{\partial T} = -\frac{L_p SB\Re T}{\nu}\ln\left(\frac{p_{in}}{p_{ex}}\right) \tag{4.4.81}$$

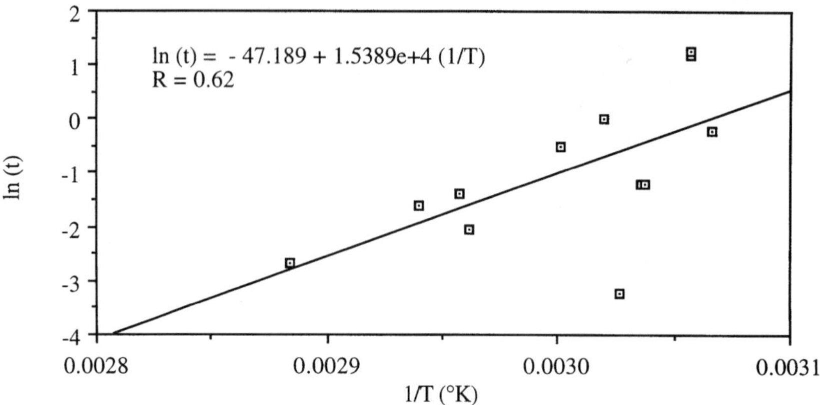

FIGURE 4.4.18 Arrhenius plot for birefringence loss in rabbit myocardium *in vito.*

The Mazur equation predicts the change in the volume, Ψ, of a cell with temperature during freezing at a cooling rate of B in conjunction with the transport of water from within the cell in response to the osmotic differential developed across the membrane as solutes are concentrated during the solidification process. The rate of transport is dependent on the surface area, S, of the cell, the membrane permeability to water, L_p, the molar volume of water, ν, and the ratio of intracellular to extracellular vapor pressures of water. Although the latter term is not a rigorous description of the driving potential for the transport process, it is easily replaced by the transmembrane differential in chemical potential or solution concentration.

Dehydration is not the only process by which equilibrium can be achieved between intracellular and extracellular water during freezing. Ice crystals may nucleate within the cell, forming intracellular ice (IIF). The occurrence of IIF is nearly always lethal to cells and is therefore to be avoided during cryopreservation. The balance between equilibration of water across the cell membrane by osmotic dehydration and by IIF is governed primarily by the magnitude of the cooling rate. As depicted in Figure 4.4.20, at rapid cooling rates there is little opportunity for water to escape from the interior of the cell as the extracellular solute concentration increases with progressive freezing. Eventually, the conditions are satisfied for intracellular nucleation of ice crystals. At slow cooling rates, the cells are able to dehydrate as the extracellular solute concentration increases, and the intracellular water becomes frozen in the extracellular space. The resulting shrinkage of the cell can give rise to injurious chemical and mechanical stresses. In general, optimal survival from cryopreservation is achieved at intermediate cooling rates, the magnitude of which are dictated by the membrane permeability at subzero temperatures.[179]

Cryoprotective Additives (CPA)

The sensitivity of cells and tissues to both IIF and osmotic stress and dehydration can be modified substantially by the prefreezing additions of a CPA. As a consequence, virtually all practical cryopreservation procedures are based on the use of a CPA to which the cell membrane is permeable. During both the CPA addition and removal procedures and the freezing and thawing processes, the applied osmotic stress results in coupled transport between the CPA and water. This coupled process is typically described in terms of irreversible thermodynamics using a format defined by Kedem and Katchalsky in which the phenomenological coefficients are related directly to physically measurable properties of a cell membrane.[180,181] These equations are written as

$$J_V = L_p \Delta p - \sum_{i=1}^{n} \sigma_i \Delta \pi_i \tag{4.4.82}$$

$$J_S = C_S (1 - \sigma) J_V + \omega_S \Delta \pi \tag{4.4.83}$$

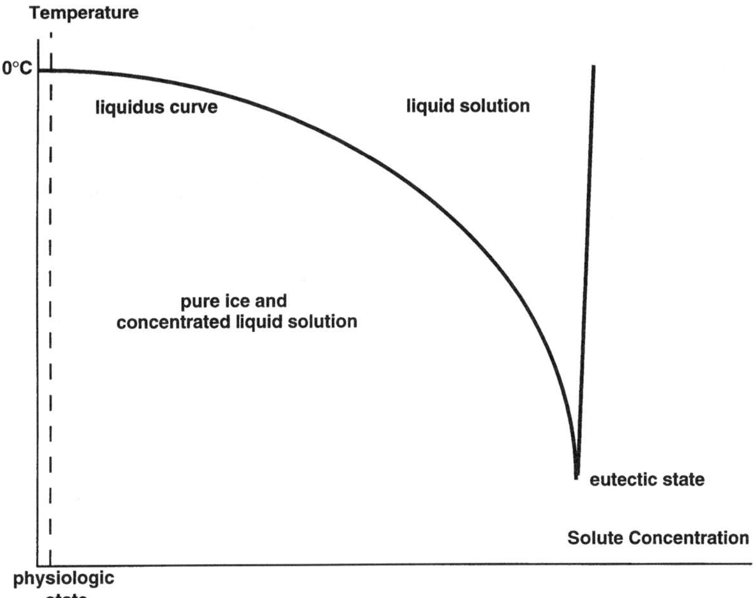

FIGURE 4.4.19 Simplified representation of an aqueous equilibrium phase diagram. Physiological solutions have solute concentrations that are quite small (about 300 mOsm) in comparison with the eutectic state. As freezing progresses, the liquid phase concentration may increase by more than 20-fold, subjecting cells to intense osmotic stress.

L_p is the membrane permeability to water; ω, the permeability to solute (CPA for applications in cryopreservation); and σ, the coupling between the two flows (reflection coefficient). The average concentration of permeable solute between the extracellular and intracellular solutions is given by C_S, and the total volume flux and solute fluxes by J_V and J_S, respectively. Equation 4.4.82 shows a provision for volume flow occurring under the action of multiple solutes.

An alternative, more generalized formulation of this type of phenomenon was also posed by Katchalsky in terms of network thermodynamics.[182] The network thermodynamic model has been applied successfully to the analysis of the cellular response to freezing processes.[183-185] It can readily be applied for inverse solution to quantify the transport coefficients from either freezing data or CPA addition data.[184,186]

Tissues and Organs

The preceding models hold for the analysis of individual cells. However, many important biological systems of interest for cryopreservation consist of organized tissues which have a three-dimensional geometry that exerts a significant effect on the osmotic behavior. For these systems transport occurs not only across the individual cell membranes, but the interstitial volume is also involved in both the transport and storage of water and CPA. In this case the model must be expanded to include both parallel and serial transport and storage for all mobile chemical species. Network thermodynamics provides an effective modeling format for describing these processes.

The cryopreservation of whole organs, which is still in the research stage, will demand even more complex coupled thermal and chemical transport analyses. CPAs are added and removed via perfusion through the vascular network, and this hydrodynamic flow is coupled to the osmotic, diffusional, and viscoelastic energy domains of the organ. The network thermodynamic approach to modeling is very well suited to analysis of these coupled multidomain processes and has been applied successfully to the design of CPA perfusion protocols.[187]

Influence of Cell Membrane Permeability

It is well documented that the permeability of cell membranes to water and to CPA is a strong function of the cell species, and that for all species it is a function of temperature.[181] For example, the water permeability among various species may vary by a factor of 10^3 or greater. In general, membranes are significantly more permeable to water than to common CPAs, although the opposite case may also occur under unusual conditions.[188] As the temperature is reduced, the permeability is depressed dramatically. For most practical purposes, at temperatures below the range of about –20 to –30°C the impedance to membrane transport becomes so high that there can be very little molecular exchange between the intracellular and extracellular compartments.

One of the major challenges in applying models for cell membrane transport during freezing has been developing instrumentation to measure the permeability to water and CPAs at subzero temperatures, both in the presence and absence of ice. The approach that has proven to be most successful is to follow visually on a light microscope the transient size of individual cells or tissues when they are subjected to controlled osmotic stress created either by freezing the specimen[189-191] or by changing the chemical environment.[192,193] An effective step change in the concentration of the extracellular solution is produced when ice is nucleated in the extracellular medium and the temperature is subsequently held constant for the duration of the experiment, or when a new bathing medium is perfused rapidly through the specimen chamber on the stage while the cells or tissues are physically immobilized. By both procedures the concentration of the environmental solution is altered from an initial concentration to a final well-defined state, and the change in cell volume with time can be measured by direct microscopic observation.

The two-dimensional cross-sectional areas of the cells in the micrographs are quantified via digital image analysis and extrapolated to three-dimensional volumes based on an assumption of a consistent geometric morphology throughout the process. These experiments are repeated serially at different temperatures to generate a series of transient volume curves as shown in Figure 4.4.21 for pancreas islets,[194] a multicellular mammalian tissue. A 2 *M* solution of the CPA dimethyl sulfoxide (Me_2SO) was introduced stepwise into the perfusion cryostage to control the osmotic environment of the islets. An initial shrinkage is measured as water leaves the cells more rapidly than the Me_2SO can enter. As time progresses the islet reaches a minimum volume and then begins to swell as Me_2SO and water enter the

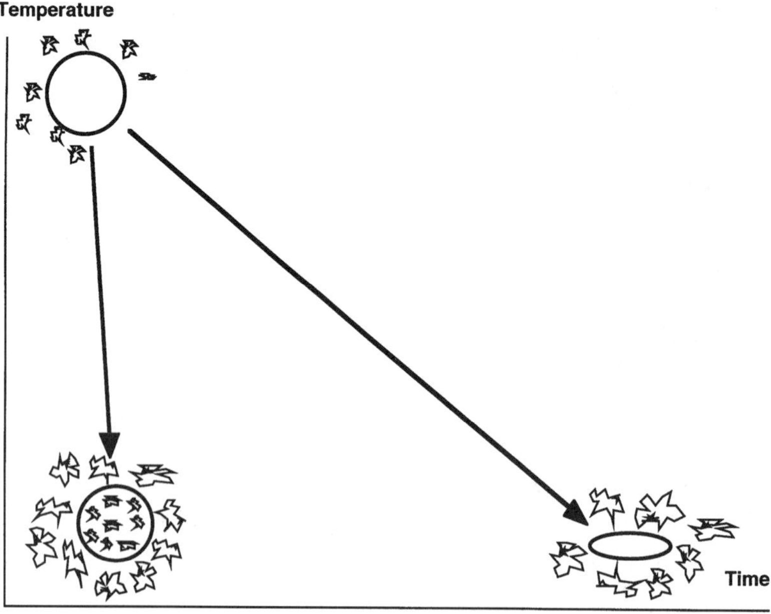

FIGURE 4.4.20 Schematic of the reaction of individual cells to extreme variations in the cooling rate during freezing. At rapid cooling rates, dehydration of the cell is minimal and intracellular ice forms. At slow cooling rates, dehydration is extensive, resulting in loss of intracellular water and morphological distortion.

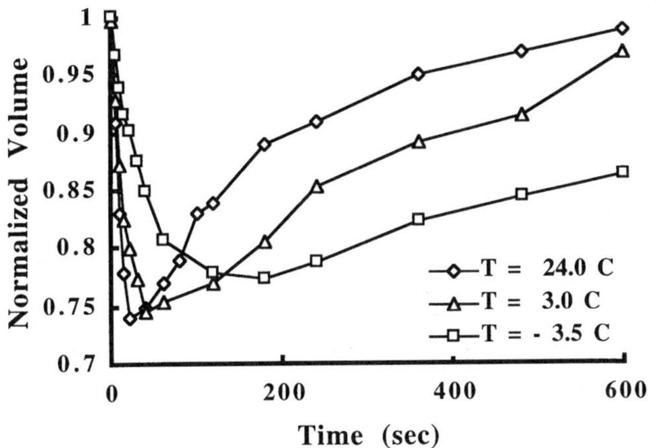

FIGURE 4.4.21 Osmotic response of human pancreas islets to exposure to a 2M solution of dimethylsulfoxide at the indicated temperatures. (From deFreitas, R.C. et al., *Cryobiology,* 35, 230, 1997. With permission.)

cell. The design of protocols for addition and removal of CPAs to tissues is an important and challenging step in developing effective cryopreservation methods. A rational design process is dependent on being able to measure the constitutive transport properties of the tissue of interest and having a model which can be applied to optimize the osmotic process, minimizing stresses (which are potentially lethal) on the cells.

Thus, the utility of the transient volume data is to provide a data base from which Equations 4.4.81 to 4.4.83 can be applied to determine the constitutive transport properties for a cell or tissue of interest by inverse solution techniques. Property values are identified at each temperature for which experimental data exist, and the data for each property are used to fit an Arrhenius-type function to characterize the temperature coefficient.[181] Equation 4.4.84 shows a typical formulation used for this purpose.

$$L_p(T) = L_{p,g}\left(T_g\right)\exp\left[\frac{\Delta E}{\Re}\left(\frac{1}{T} - \frac{1}{T_g}\right)\right] \tag{4.4.84}$$

Experiments for many different cell types have shown that the Arrhenius expression provides a model that describes the temperature dependence of the tissue transport properties well.[181]

In order to fully interpret the transient osmotic data described above, it is necessary to determine the fraction of the cell contents which is unable to cross the cell membrane under the types of osmotic stress encountered during cryopreservation. This component of the cytoplasm is often referred to as the osmotically inactive volume. It can be measured by conducting a series of experiments in which cells are exposed to increasing concentrations of impermeable solute and measuring the equilibrium volume at each state. These volumes are then plotted as a function of reciprocal solute concentration and extrapolated to the intercept of the ordinate, which indicates the volume which the cell would assume without injury when exposed to an infinitely large solute concentration. The complementary normalized fraction is a measure of the initial amount of intracellular water which can participate in the response to extracellular osmotic stress. The data are displayed on a Boyle-van't Hoff plot.[195]

A fundamental issue in the analysis of cryopreservation processes that has only recently begun to be addressed via rigorous quantitative modeling is the osmotic behavior of frozen cells during the thawing process. It has now been demonstrated by application of a model based on Equation 4.4.81 that there are combinations of freeze/thaw protocols for which the warming rate may be the governing component of the thermal history in determining the mechanism and extent of injury.[196] The design of optimal cryoprocessing protocols, therefore, may require an analysis of the effect on a tissue of interest of the complete thermal history, including both cooling and warming.

Cryopreservation by Vitrification

Ice nucleation can be avoided by doping the biological specimen with a high concentration of chemicals to induce a vitreous state.[197] The chemicals act to raise the glass transition temperature to a high value which can be reached at cooling rates consistent with standard refrigeration systems. Two major problems involving transport processes remain to be solved before vitrification can be applied to the cryopreservation of whole organs. One is the process by which a 4 to 5 M solution equilibrates with the entire volume of an organ, including all the constituent cells. Since this high concentration of CPA can be toxic to cells, the addition process is preferably effected at a low temperature, such as 4°C, to reduce the rates of injurious chemical reactions. In addition, the duration of exposure to the CPA prior to further cooling should be minimized to limit the accrued reaction time. However, the trade-off to lowering the temperature is caused by the thermal coefficient of viscosity, which makes it much more difficult to effect the perfusion of CPA solution through the vascular systems of an organ, and the diffusion from the vascular lumen through the interstitial tissue and into the individual cells. Optimal design is dependent upon balancing the coupled hydrodynamic, osmotic, and viscoelastic reaction of the organ to the perfusion process with the biochemical toxic reactions.[197]

Optimization of Heat Transfer during Cryopreservation

The foregoing discussion indicates the importance of the combined osmotic and thermal history at the local cellular level in determining the response to cryopreservation, including the type and extent of injury which may occur. For a given combination of cell type and CPA composition and concentration, the post-thaw viability achieved is governed by the thermal history of the process. The cooling rate determines the frozen state for storage, and the warming rate determines the manifestation of the frozen state on the final level of viability. In general, there are two extremes of behavior during freezing. For relatively slow cooling rates, the rate at which the extracellular solute concentration increases will have a time scale comparable with that for the osmotic dehydration of the cell at high subzero temperatures. Therefore, the cell will lose a considerable fraction of its water and experience extensive dehydration as indicated in Figure 4.4.20. This large loss of volume is known to damage cells by one or more molecular-level mechanisms which have not been fully identified.[179] Increasing the concentration of CPA will provide an osmotic buffer against this type of dehydration-induced injury, as well as possibly providing other means of protection. Alternatively, when the cooling rate is rapid, the temperature is reduced much more rapidly than the cell can respond osmotically (because the membrane permeability becomes greatly reduced at low temperatures, as will be explained subsequently). Therefore, there is no opportunity for intracellular water to osmotically equilibrate with the extracellular water which is being progressively diminished in concentration. When the transmembrane water disequilibrium (as manifested in terms of liquid phase supercooling) reaches a critical value, the conditions will be satisfied for nucleation of ice in the intracellular volume. It is important to note that the cell membrane acts as a barrier to the growth of a continuous ice phase between the intracellular and extracellular compartments. The crystalline structure of intracellular ice will be considerably smaller than that of extracellular ice owing to the greater extent of supercooling prior to nucleation. Consequently, the propensity for recrystallization will be relatively large in the intracellular volume, and this process has been demonstrated to be nearly universally lethal to cells.

The combination of different injury mechanisms associated with rapid and slow cooling rates during freezing gives rise to a two-factor theory of cryoinjury.[179,198] Between the extremes of rapid and slow cooling there will be a range of intermediate rates at which survival will be maximized. The magnitude and breadth of this range is defined uniquely for each cell type as a function of its membrane permeability and sensitivity to injury. Cryopreservation protocols are targeted to produce a cooling process which lies within the optimal range of this survival signature that will produce the highest possible biological function following thawing. The design of a cryopreservation protocol for a given cell type must take into account the combined effects of the CPA composition and concentration plus the cooling and warming rates.

The above analysis of the freezing process is presented from the perspective of the local thermal history for an individual cell. However, practical cryopreservation procedures are conducted either on tissues and organs of finite macroscopic dimensions or on containers of solutions of cells or microscopic tissue suspensions. The consequence is that although the governing mass transport processes occur across microscopic scale dimensions (a single membrane and the associated boundary layers may be less than 0.1 μm), the governing heat transport occurs on a scale measured in millimeters or centimeters. It is this macroscopic heat transfer that drives all components of the cryopreservation process. Because of the macroscopic dimensions and the restriction that the refrigeration source be applied only at the external boundary of the systems during cooling, there can be significant spatial gradients in cooling rate manifested within the system. Also, in general the cooling rate will not be constant over the entire cooling process. The cooling rate should be defined over the range of temperatures most important to determining the frozen state of a cell. The critical temperature range is defined as that for which transport occurs across the cell membrane and for which intracellular ice formation (IIF) is most probable. Analysis of this process involves solution of the classic Stefan moving boundary problem.

For the freezing of biological systems, it is most important to accurately determine the cooling at high subzero temperatures within the mushy zone[199] during which the primary membrane transport processes occur. It is precisely this range of temperatures for which simplifying assumptions concerning the pattern of latent heat release frequently are made in order to render the mathematical solution more tractable. Therefore, it is important to match as well as possible the coupling between temperature and changing solute concentration as defined by the phase diagram. Only when the actual process physics are matched can the cooling rate for the critical temperature range be described accurately. The combination of a nonlinear pattern of latent heat release with temperature and the often complex system geometry dictates that the governing heat transport equations be solved via numerical methods. The thermal histories calculated may then be correlated with the cooling rates on the survival signature and with the membrane mass transport equations, such as Equation 4.4.81, to predict the biological response of a given tissue to a defined boundary cooling protocol.[200]

Thermal Stress Effects

Another analysis problem to be addressed in cryopreservation is that of thermal stresses in an organ that has been vitrified. Although the constitutive properties are yet to be measured, it is apparent that a biological tissue with dimensions on the order of several centimeters will be subject to fracture at even moderate cooling rates below –120°C after a glass phase has formed.[201] A mechanical fracture of a whole organ will render it useless for subsequent transplantation. It is important to measure the viscoelastic properties of biological materials loaded with high concentrations of CPA over the temperature range between the glass transition and storage temperatures (usually at or near liquid nitrogen temperature at –196°C). Complementary to the property measurements is a need for a thermal stress analysis that can be applied with a versatility of geometric details.

Rubinsky et al.[202] have developed an initial model for how mechanical stresses may develop in the interior of an organ during solidification and the subsequent cooling process. Different organs have widely divergent shapes and sizes, and constitutive property data for frozen and vitrified tissues are practically nonexistent. In addition, an organ is likely to be preserved in a container of fluid with which it will be coupled thermally and mechanically. Factors which can be considered in a process design are manipulation of the boundary thermal protocol, shape of the overall container into which an organ is placed, and coordination of the pattern of distributed internal deposition with cooling at the boundary.

Recent Developments and Innovations

Recent innovative experiments by Fowler and Toner[203] have demonstrated that it is possible at the micro (cellular) scale to melt and recool cells containing intracellular ice at a rate that can produce a vitrified state without the introduction of the CPA to block the initial ice nucleation event. The frozen specimen is irradiated with a laser wavelength that is absorbed preferentially by the cell cytoplasm over the extracellular solution. For a solution of individual cells having characteristic diameters of about 6 μm, warming rates on the order of 10^{11}°C/s can be achieved within the cells, while the extracellular matrix remains frozen. When the heating process is terminated in an approximate step-wise manner after 7 ns, heat transfer from the extracellular matrix to the cells produces an intracellular cooling rate of about 10^{6}°C/s, which is adequate to produce a glass state in an aqueous solution. The vitrified cells may then be warmed rapidly to suprafreezing temperatures (>0°C) without nucleation and with no measured injury. If this process is to realize broad clinical or commercial application, there is considerable engineering work to be accomplished in scale-up from these microscopic level processes to applications in larger tissue and organ systems.

Since the initial achievements in the cryopreservation of cells, many applications have been developed in both the biological and medical fields. There are now numerous for-profit concerns that have developed successful techniques for preserving human tissues for subsequent transplantation. Many of these corporations use proprietary processing protocols to preserve specific tissues, and the number of commercial participants in this field has been growing steadily over the past 10 years. In like manner, there are many for-profit and not-for-profit groups which are marketing the cryopreservation of a very broad spectrum of nonhuman tissues. Some of the more prominent applications involve mammalian and amphibian gametes, tissue culture collections, and plant germplasm. A potentially important new area of application will be the cryopreservation of living materials manufactured by the tissue engineering techniques which are being developed.[204] Bioartificial systems such as skin and organs that contain living cells will require a technology for maintaining them in a viable but latent state between the time of manufacture and implementation in a clinical setting. Cryopreservation provides a tool of suitable potential for meeting this unique need, but the processing protocols will have to be designed to ensure successful storage and recovery for each type of manufactured tissue.

Cryosurgery

All of the foregoing discussion has been devoted to cryopreservation, for which the objective is to maximize the survival of cells from the frozen state to ensure their living function after thawing. The antithesis of this work is cryosurgery, for which the objective is to maximize destruction of a target tissue *in situ*.[205] Cryosurgery has long been practiced as an effective means for killing surface lesions, and recently it has been adopted more widely as a tool for treating internal tumors that are difficult to resect mechanically.[206]

Many of the recent gains in the successful application of cryosurgery are due to the development of new imaging methods that enable the surgeon to follow the growth of the solid-liquid interface, to which the zone of cellular destruction is coupled, in real time during the freezing process. The initial imaging methodology adapted for this purpose was ultrasonography.[207] Subsequently, other imaging modalities that can be coupled to the transient temperature field are being developed, including NMR[208] and spectroscopy.[209] Advances in cryosurgical probes that provide more effective removal of heat from tissues via enhanced thermal transport mechanisms have also contributed to the recent growth in cryosurgery.[210] Finally, over the past decade there has been a steady advance in understanding the mechanisms by which freezing and thawing act to cause tissue death as a function of the thermal parameters of the solidification process and tissue properties. Many of these fundamental advances have resulted from the application of bioengineering analysis of the tissue freezing process,[211-213] which portends the potential for engineering to further contribute to the field of cryosurgery.

A new approach to improving the efficacy of cryosurgery is derived from techniques long applied to enhance cryopreservation processes. Namely, the tissue is modified by addition of chemical agent prior to the initiation of freezing. However, for applications in cryosurgery the desired result is an increased level of cell killing. Antifreeze proteins (AFP) are proving to be effective for this purpose.[214-215] AFPs are chemical compounds synthesized by many differing types of plants and animals to provide protection against freezing injury at high subzero temperatures.[216-217] It has also been demonstrated that AFPs modify ice crystals to needle-like shapes that can destroy cells during freezing to deeper subzero temperatures.[218]

In summary, clearly there continue to be many opportunities in low temperature biological heat transfer for contributions with the potential for substantial positive impact on health care delivery.

Acknowledgments

This paper was prepared with the support of National Science Foundation grants CTS-9632378 to K.R. Diller and DBI-9512746 to K.R. Diller, NIH grant #1 R01 HL56143-01 to JWV, and the Joe J. King Professorship in Engineering at the University of Texas at Austin. The perfusion measurements were supported in part by a grant to J.W. Valvano from Urologix Inc. The subsections on convective heat transfer and thermal properties of swine myocardium were derived in part from Naresh C. Bhavaraju's soon to be completed Ph.D. dissertation, and the subsections on thermal models and microsphere perfusion measurements in the canine prostate from David Yuan's soon to be completed Ph.D. dissertation, both at the University of Texas at Austin. The section on human thermoregulation was derived in part from Karen Nyberg's Ph.D dissertation at the University of Texas at Austin. Collaborations with colleagues in writing prior reviews in bioheat transfer are acknowledged. In particular, we note the contribution of Dr. Sharon Thomsen to the subsection on histological markers and Dr. Thomas Ryan for a broad review.[219]

References

1. Johnston, K.A. and Bennett, A.F., Eds., *Animals and Temperature: Phenotypic and Evolutionary Adaptation*, Cambridge University Press, Cambridge, 1996.
2. Diller, K.R., Modeling of bioheat transfer processes at high and low temperatures, *Adv. Heat Trans.*, 22, 157-357, 1992.
3. Chato, J.C., A method for the measurement of thermal properties of biologic materials, Symposium on Thermal Problems in Biotechnology, American Society of Mechanical Engineers, New York, 1968, 16-25.
4. Balasubramaniam, T.A. and Bowman, H.F., Thermal conductivity and thermal diffusivity of biomaterials: a simultaneous measurement technique, *J. Biomech. Eng.*, 99, 148-154, 1977.
5. Valvano, J.W. et al., An isolated rat liver model for the evaluation of thermal techniques to measure perfusion, *J. Biomech. Eng.*, 106, 187-191, 1984.
6. Arkin, H. et al., Thermal pulse decay method for simultaneous measurement of local thermal conductivity and blood perfusion: a theoretical analysis, *J. Biomech. Eng.*, 108, 208-214, 1986.
7. Kress, R., A comparative analysis of thermal blood perfusion measurement techniques, *J. Biomech. Eng.*, 109, 218-225, 1987.
8. Patel, P.A. et al., A finite element analysis of a surface thermal probe, *Thermodynamics, Heat, and Mass Transfer in Biotech.*, American Society of Mechanical Engineers, New York, 1987, 95-102.
9. Bowman, H.F., Estimation of tissue blood flow, in *Heat Transfer in Medicine and Biology: Analysis and Applications,* Vol. 1, Shitzer, A. and Eberhart, R.C., Eds., Plenum Press, New York, 1985, chap. 9, 193-230.
10. Chato, J.C., Measurement of thermal properties of biological materials, in *Heat Transfer in Medicine and Biology: Analysis and Applications,* Vol. 1, Shitzer, A. and Eberhart, R.C., Eds., Plenum Press, New York, 1985, chap. 8, 167-192.
11. Valvano, J.W. et al., Thermal conductivity and diffusivity of biomaterials measured with self-heated thermistors, *Int. J. Thermophys.*, 6, 301-311, 1985.
12. Valvano, J.W. and Chitsabesan, B., Thermal conductivity and diffusivity of arterial wall and atherosclerotic plaque, *Lasers Life Sci.*, 1, 219-229, 1987.
13. Holmes, K.R. and Chen, M.M., *In vivo* tissue thermal conductivity and local blood perfusion measured with heat pulse-decay method, *Adv. Bioeng.*, 113-115, 1980.

14. Touloukian, Y.S. et al., *Thermophysical Properties of Matter: Thermal Conductivity*, Vol. 3, IFI/Plenum, New York, 1970, 120, 209.
15. Touloukian, Y.S. et al., *Thermophysical Properties of Matter: Thermal Conductivity*, Vol. 3, IFI/Plenum, New York, 1970, preface.
16. Touloukian, Y.S. et al., *Thermophysical Properties of Matter: Thermal Diffusivity*, Vo1. 10, IFI/Plenum, New York, 1973, 15a-42a.
17. Kreith, F. and Bohn, M.S. *Principles of Mass Transfer*, 5th ed., West Publishing, St. Paul, MN, 1993.
18. Duck, F.A., *Physical Properties of Tissue: A Comprehensive Reference Book*, Academic Press, London, 1991.
19. Chen, M.M. et al., Pulse-decay method for measuring the thermal conductivity of living tissue, *J. Biomech. Eng.* 103, 253-260, 1981.
20. Holmes, K.R. and Chen, M.M., Local tissue heating, microbead pulse decay technique for heat transfer parameter evaluation, *Measurement of Blood Flow and Local Tissue Energy Production by Thermal Methods*, Muller-Schenburg, Ed., Thieme-Stratton, New York, 1983, 50-56.
21. Valvano, J.W. et al., A finite element analysis of self-heated noninvasive thermistors, *Adv. Bioeng.*, American Society of Mechanical Engineers, New York, 1983.
22. Valvano, J.W. et al., The simultaneous measurement of thermal conductivity, thermal diffusivity and perfusion in small volume of tissue, *J. Biomech. Eng.,* 106, 192-197, 1984.
23. Valvano, J.W., Badeau, A.F., and Pearce, J.A., Simultaneous measurement of intrinsic and effective thermal conductivity, *Heat Transfer in Bioengineering and Medicine*, American Society of Mechanical Engineers, New York, 1987, 31-36.
24. Patel, P.A. et al., A self-heated thermistor technique to measure effective thermal properties from the tissue surface, *J. Biomech. Eng.,* 109, 330-335, 1987.
25. Patel, P.A., Valvano, J.W., and Hayes, L.J., Perfusion measurement by a surface thermal probe, IEEE Engineering in Medicine and Biology, Boston, 1987.
26. Walsh, J.T., A Noninvasive Thermal Method for the Quantification of Tissue Perfusion, M.S. Thesis, M.I.T., Cambridge, 1984.
27. Patera, P.A. et al., Prediction of tissue perfusion from measurement of the phase shift between heat flux and temperature, Paper #79-WA/HT-71, American Society of Mechanical Engineers, New York, 1979.
28. Pennes, H.H., Analysis of tissue and arterial blood temperatures in the resting forearm, *J. Appl. Physiol.,* Vol. 1, pp. 93-122, 1948 (republished for 50th anniversary issue of *J. Appl. Physiol.,* 85, 5-34, 1998).
29. Rastorguev, Y.L. and Ganiev, Y.A., Thermal conductivity of aqueous solutions or organic materials, *Russ. J. Phys. Chem.*, 40, 869-871, 1966.
30. Valvano, J.W., Low temperature tissue thermal properties, in *Low Temperature Biotechnology: Emerging Applications and Engineering Contributions*, McGrath, J.J. and Diller, K.R., Eds., American Society of Mechanical Engineers, New York, 331-346, 1988.
31. Spells, K. E., The thermal conductivities of some biological fluids, *Phys. Med. Biol.*, 5, 139-153, 1960.
32. Cooper, T. E., and Trezck, G. J., Correlation of thermal properties of some human tissues with water content, *Aerosp. Med.*, 42, 24-27, 1971.
33. Charney, C.K., Mathematical models of bioheat transfer, *Adv. Heat Trans.*, 22, 19-155, 1992.
34. Wissler, E.H., Steady-state temperature distribution in man, *J. Appl. Physiol.,* 16, 734-740, 1961.
35. Wissler, E.H., Mathematical simulation of human thermal behavior using whole body models, in *Heat Transfer in Medicine and Biology: Analysis and Applications,* Vol. 1, Shitzer, A. and Eberhart, R.C., Eds., Plenum Press, New York, 1985, chap. 13, 325-373.
36. Chato, J.C., Heat transfer to blood vessels, *J. Biomech. Eng.,* 102, 110-118, 1980.
37. Chen, M.M. and Holmes, K.R., Microvascular contributions in tissue heat transfer, *Ann. N.Y. Acad. Sci.*, 335, 137-150, 1980.
38. Weinbaum, S., Jiji, L., and Lemons, D.E., Theory and experiment for the effect of vascular temperature on surface tissue heat transfer. I. Anatomical foundation and model conceptualization, *J. Biomech. Eng.*, 106, 246-251, 1984.

39. Weinbaum, S., Jiji, L., and Lemons, D.E., Theory and experiment for the effect of vascular temperature on surface tissue heat transfer. II. Model formulation and solution, *J. Biomech. Eng.*, 106, 331-341, 1984.
40. Weinbaum, S. and Jiji, L.M., A new simplified bioheat equation for the effect of blood flow on average tissue temperature, *J. Biomech. Eng.*, 107 131-139, 1985.
41. Zhu, L., Lemons, D.E., and Weinbaum, S., A new approach for predicting the enhancement in the effective conductivity of perfused tissue due to hyperthermia, American Society of Mechanical Engineers, HTD 288: 37-43, 1994.
42. Charny, C.K., Weinbaum, S., and Levin, R.L., An evaluation of the Weinbaum-Jiji bioheat equation for normal and hyperthermic conditions, *J. Biomech. Eng.*, 112, 80-87, 1990.
43. Weinbaum, S. and Jiji, L.M., A two phase theory for the influence of circulation on the heat transfer in surface tissue, *Adv. Bioeng.*, ASME WA/HT-72: 179-182, 1979.
44. Dagan, Z., Weinbaum, S., and Jiji, L.M., Parametric study of the three layer microcirculatory nodel for surface tissue energy exchange, *J. Biomech. Eng.*, 108, 89-96, 1986.
45. Song, W.J., Weinbaum, S., and Jiji, L.M., A combined macro and microvascular model for whole limb heat transfer, *J. Biomech. Eng.*, 110, 259-267, 1988.
46. Song, W.J., Weinbaum, S., and Jiji, L.M., A theoretical model for peripheral heat transfer using the bioheat equation of Weinbaum and Jiji, *J. Biomech. Eng.*, 109, 72-78, 1987.
47. Weinbaum, S. and Jiji, L.M., The matching of thermal fields surrounding countercurrent microvessels and the closure approximation in the Weinbaum-Jiji equation, *J. Biomech. Eng.*, 111, 271-275, 1989.
48. Weinbaum, S. and Lemons, D.E., Heat transfer in living tissue: the search for a blood-tissue energy equation and the local thermal microvascular control mechanism, *BMES Bull.*, 16, 38-43, 1992.
49. Zhu, M., Weinbaum, S., and Lemons, D.E., On the generalization of the Weinbaum-Jiji equation to microvessels of unequal size: the relation between the near field and local average tissue temperatures, *J. Biomech. Eng.*, 110, 74-81, 1988.
50. Lemons, D.E., Weinbaum, S., and Jiji, L.M., Experimental studies on the role of the micro and macro vascular system in tissue heat transfer, *Am. J. Physiol.*, 253, R128, 1987.
51. Mooibroek, J. and Lagendijk, J.J.W., A fast and simple algorithm for the calculation of convective heat transfer by large vessels in 3-dimensional inhomogeneous tissue, *IEEE Trans. Biomed. Eng.*, 38, 490-501, 1991.
52. Baish, J.W., Heat transport by countercurrent blood vessels in the presence of an arbitrary temperature gradient, *J. Biomech. Eng.*, 112, 207-211, 1990.
53. Huang, H.W., Chen, Z.P., and Roemer, R.B., A countercurrent vascular network model of heat transfer in tissues, *J. Biomech. Eng.*, 118, 120-129, 1996.
54. Keller, K.H. and Seiler, L., An analysis of peripheral heat transfer in man, *J. Appl. Physiol.*, 30, 779, 1971.
55. Mitchell, J.W. and Myers, G.E., An analytical model of the countercurrent heat exchange phenomena, *Biophys. J.*, 8, 897-911, 1968.
56. Baish, J.W., Ayyaswamy, P.S., and Foster, K.R., Heat transport mechanisms in vascular tissues: a model comparison, *J. Biomech. Eng.*, 108, 324-331, 1986.
57. Weinbaum, S. and Jiji, L.M., Discussion of papers by Wissler and Baish et al. concerning the Weinbaum-Jiji bioheat equation, *J. Biomech. Eng.*, 109, 234-237, 1987.
58. Wissler, E.H., Comments on Weinbaum and Jiji's discussion of their proposed bioheat equation, *J. Biomech. Eng.*, 109, 355-356, 1987.
59. Wissler, E.H., Pennes' 1948 paper revisited, *J. Appl. Physiol.*, 85, 35-41, 1998.
60. Pennes, H.H., Analysis of tissue and arterial blood temperatures in the resting forearm, *J. Appl. Physiol.*, 1, pp. 93-122, 1948, (republished for 50th anniversary issue of *J. Appl. Physiol.*, 85, 5-34, 1998).
61. Huang, H.W., Chan, C.L., and Roemer, R.B., Analytical solutions of Pennes bioheat transfer equation with a blood vessel, *J. Biomech. Eng.*, 116, 208-212, 1994.

62. Xu, L.X., Chen, M.M., Holmes, K.R., and Arkin, H., The evaluation of the Pennes, the Chen-Holmes, the Weinbaum-Jiji bioheat transfer models in the pig kidney vortex, ASME HTD 189, American Society of Mechanical Engineers, New York, 1991, 15-21.
63. Arkin, H., Xu, L.X., and Holmes, K.R., Recent developments in modeling heat transfer in blood perfused tissues, *IEEE Trans. Biomed. Eng.*, BME-41, 97-107, 1994.
64. Wulff, W., The energy conservation equation for living tissue, *IEEE Trans. Biomed. Eng.*, BME-21, 494-495, 1974.
65. Klinger, H.G., Heat transfer in perfused biological tissue. I. General theory, *Bull. Math. Biol.*, 36, 403-415, 1974.
66. Bazett, H.C. and McGlone, B., Temperature gradients in the tissue in man, *Am. J. Physiol.*, 82, 415, 1927.
67. Valvano, J.W., Nho, S., and Anderson, G.T., Analysis of the Weinbaum-Jiji model of blood flow in the canine kidney cortex for self-heated thermistors, *J. Biomech. Eng.,* 116, 201-207, 1994.
68. Brinck, H. and Werner, J., Estimation of the thermal effect of blood flow in a branching countercurrent network using a three-dimensional vascular model, *J. Biomech. Eng.,* 116, 324-330, 1994.
69. Anderson, G.T. and Valvano, J.W., An interlobular artery and vein based model for self-heated thermistor measurements of perfusion in the canine kidney cortex, Bioheat Transfer – Applications in Hyperthermia, Emerging Horizons in Instrumentation and Modeling, Vol. 12, American Society of Mechanical Engineers, New York, 1989, 29-35.
70. Anderson, G.T. and Valvano, J.W., A small artery heat transfer model for self-heated thermistor measurements of perfusion in the canine kidney cortex, *J. Biomech. Eng.,* 116, 71-78, 1994.
71. Yuan, D.Y., Zhu, L., Holmes, K.R., and Valvano, J.W., Perfusion and temperature measurements in hyperthermic canine prostates, 17th S. Biomed. Eng. Conf., San Antonio, TX, 85, 1998.
72. Xu, L.X., Zhu, L., and Holmes, K.R., Thermoregulation in canine prostate during transurethral microwave hyperthermia. II. Blood flow response, *Int. J. Hyperthermia,* 14, 65-73, 1998.
73. Bligh, J., Regulation of body temperature in man and other mammals, in *Heat Transfer in Medicine and Biology: Analysis and Applications,* Vol. 1, Shitzer, A. and Eberhart, R.C., Eds., Plenum Press, New York, 1985, chap. 2, 15-52.
74. Rowell, L.B. and Wyss, C.R., Temperature regulation in exercising and heat-stressed man, in *Heat Transfer in Medicine and Biology: Analysis and Applications,* Vol. 1, Shitzer, A. and Eberhart, R.C., Eds., Plenum Press, New York, 1985, chap. 3, 53-78.
75. Lipton, J.M, Thermoregulation in pathological states, in *Heat Transfer in Medicine and Biology: Analysis and Applications,* Vol. 1, Shitzer, A. and Eberhart, R.C., Eds., Plenum Press, New York, 1985, chap. 4, 79-106.
76. Heller, H. C. and Glotzbach, S.F., Thermoregulation and sleep, in *Heat Transfer in Medicine and Biology: Analysis and Applications,* Vol. 1, Shitzer, A. and Eberhart, R.C., Eds., Plenum Press, New York, 1985, chap. 5, 107-134.
77. Chappuis, P., Pittet, P., and Jequier, E., Heat storage regulation in exercise during thermal transients, *J. Appl. Physiol.,* 40, 384-392, 1976.
78. Webb, P., The physiology of heat regulation, *Am. J. Physiol.*, 268, R838-R850, 1995.
79. Nunneley, S. A., Water cooled garments: a review, *Space Life Sci.*, 2, 335-360, 1970.
80. Ganong, W. F., *Review of Medical Physiology*, 16th ed., Appleton & Lange, Norwalk, CT, 1993.
81. Fanger, P. O., *Thermal Comfort: Analysis and Applications in Environmental Engineering*, McGraw-Hill, New York, 1970.
82. Fulcher, C.W.G., Control of a Liquid Cooling Garment for Extravehicular Astronauts by Cutaneous and External Auditory Meatus Temperatures, Ph.D. Dissertation, University of Houston, 1970.
83. Hensen, J.L.M., Literature review on thermal comfort in transient conditions, *Build. Environ.*, 25, 309-316, 1990.
84. Wissler, E.H., Steady-state temperature distribution in man, *J. Appl. Physiol.,* 16, 734-740, 1961.
85. Wissler, E.H., Comparison of results obtained from two mathematical models. A simple 14-node model and a complex 250-node model, *J. Physiol. (Paris)*, 63, 455-458, 1970.

86. Stolwijk, J.A.J. and Hardy, J.D., Temperature regulation in man. A theoretical study, *Pflugers Arch.*, 291, 129-162, 1966.
87. Nadel, E.R., Mitchell, J.W., Saltin, B., and Stolwijk, J.A.J., Peripheral modifications to the central drive for sweating, *J. Appl. Physiol.,* 31, 828-833, 1971.
88. Mitchell, J.W., Galvez, T.L., Hengle, J., Myers, G.E., and Siebecker, K.L., Thermal response of human legs during cooling, *J. Appl. Physiol.,* 29, 859-865, 1970.
89. Hayward, J.S., Thermal balance and survival time prediction of man in cold water, *Can. J. Physiol. Pharmacol.*, 53, 21-32, 1975.
90. Hayward, J.S., Eckerson, J.D., and Collis, M.L., Thermoregulatory heat production in man: prediction equation based on skin and core temperatures, *J. Appl. Physiol.,* 42, 377-384, 1977.
91. Kuznetz, L. H., A two-dimensional transient mathematical model of human thermoregulation, *Am. J. Physiol.*, 6, R266-R277, 1979.
92. Shitzer, A., Chato, J.C., and Hertig, B.A., Thermal protective garment using independent regional control of coolant temperature, *Aerosp. Med.*, 44, 49-59, 1973.
93. Shitzer, A. and Chato, J.C., Thermal interaction with garments, in *Heat Transfer in Medicine and Biology: Analysis and Applications,* Vol. 1, Shitzer, A. and Eberhart, R.C., Eds., Plenum Press, New York, 1985, chap. 14, 375-394.
94. Wissler, E.H., Mathematical simulation of human thermal behavior using whole body models, in *Heat Transfer in Medicine and Biology: Analysis and Applications,* Vol. 1, Shitzer, A. and Eberhart, R.C., Eds., Plenum Press, New York, 1985, chap. 13, 325-373.
95. Wissler, E., Simulation of fluid-cooled or heated garments that allow man to function in hostile environments, *Chem. Eng. Sci.*, 41, 1689-1698, 1986.
96. Wissler, E. H., A review of human thermal models, in *Environmental Ergonomics*, Mekjavic, I.B., Banister, E.W., and Morrison., J.B., Eds., Taylor & Francis, New York, 1988, 267-285.
97. Nyberg. K.L., Diller, K.R., and Wissler, E.H., Modeling of human thermal regulation for liquid cooling garment applications, *Adv. Heat Mass Trans. Biotech.*, 37, 119-126, 1997.
98. Roussy, G. and Pearce, J., *Foundations and Industrial Applications of Microwaves and Radio Frequency Fields,* Wiley-Interscience, Chichester, U.K., 1985.
99. Geddes, L.A. and Baker, L.E., The specific resistance of biological material. A compendium of data for the biomedical engineer and physiologist, *Med. Biol. Eng.,* 5, 271-293, 1967.
100. Schwan, H.P., Biophysics of Diathermy, *Therapeutic Heat and Cold*, E. Licht, Ed., Waverly Press, Baltimore, 1972, chap. 3.
101. Fawcett, D.W. *Textbook of Histology,* 12th ed., Chapman & Hall, New York, 1994, 386-409.
102. Stillwell, G.K. General principles of thermal therapy, *Therapeutic Heat and Cold,* E. Licht, Ed., Waverly Press, Baltimore, 1972, chap. 7.
103. Green, D.M. and Diller, K.R. Measurement of burn-induced leakage of macromolecules in living tissue, *Trans. ASME J. Biomech. Eng.,* 100, 153-158, 1978.
104. Cotran, R.S., Kumar, V., and Collins, T., *Pathologic Basis of Disease,* 6th ed., W.B. Saunders, Philadelphia, 1999, 1-112.
105. Moritz, A.R. and Henriques, F.C., Studies of thermal injury. II. The relative importance of time and surface temperature in the causation of cutaneous burns, *Am. J. Pathol.,* 23, 695-720, 1947.
106. Moritz, A.R., Studies of thermal injury. III. The pathology and pathogenesis of cutaneous burns: an experimental study, *Am. J. Pathol.*, 23, 915-934, 1947.
107. Henriques, F.C. and Moritz, A.R., Studies of thermal injury in the conduction of heat to and through skin and the temperatures attained therein: a theoretical and experimental investigation, *Am. J. Pathol.,* 23, 531-549, 1947.
108. Henriques, F.C., Studies of thermal injury. V. The predictability and the significance of thermally induced rate processes leading to irreversible epidermal injury, *Arch. Pathol.,* 23, 489-502, 1947.
109. Büttner, K., Effects of extreme heat and cold on human skin. I. Analysis of temperature changes caused by different kinds of heat application, *J. Appl. Physiol.,* 3, 691-702, 1951.

110. Stoll, A.M., A computer solution for determination of thermal tissue damage integrals from experimental data, *IRE Trans. Med. Electron.*, 7, 355-358, 1960.
111. Weaver, J.A. and Stoll, A.M., Mathematical model of skin exposed to thermal radiation, *Aerosp. Med.,* 40, 24-30, 1967.
112. Takata, A.N., Development of criterion for skin burns, *Aerosp. Med.,* 45, 634-637, 1974.
113. Stoll, A.M. and Green, L.C., Relationship between pain and tissue damage due to thermal radiation, *J. Appl. Physiol.,* 14, 373-382, 1959.
114. Stoll, A.M. and Chianta, M.A., Burn production and prevention in convective and radiant heat transfer, *Aerosp. Med.,* 39, 1232-1238, 1968.
115. Lawrence, J.C. and Bull, J.P., Thermal conditions which cause skin burns, *Eng. Med.*, 5, 61-63, 1976.
116. Büttner, K., Effects of extreme heat and cold on human skin. II. Surface temperature, pain and heat conductivity in experiments with radiant heat, *J. Appl. Physiol.,* 3, 691-702, 1951.
117. Ross, D.C. and Diller, K.R., An experimental investigation of burn injury in living tissue, *J. Heat Trans.*, 98, 292-296, 1976.
118. Moussa, N.A., McGrath, J.J., Cravalho, E.G., and Asimacopoulos, P.J., Kinetics of thermal injury in cells, *J. Biomech. Eng.,* 99, 155-159, 1977.
119. Moussa, N.A., Tell, E.N., and Cravalho, E.G., Time progression of hemolysis of erythrocyte populations exposed to supraphysiological temperatures, *J. Biomech. Eng.,* 101, 213-217, 1979.
120. Bhowmick, S. and Bischof, J.C. Supraphysiological thermal injury in Dunning AT-1 prostate tumor cells, *Advances in Heat and Mass Transfer in Biotechnology,* American Society of Mechanical Engineers, New York, 1998, 77-78.
121. Maron, S.H. and Lando, J.B., *Fundamentals of Physical Chemistry,* Macmillan, New York, 1974.
122. Diller, K.R. and Klutke, G.A., Accuracy analysis of the Henriques model for predicting burn injury, *Advances in Bioheat and Mass Transfer,* American Society of Mechanical Engineers, New York, 1993, 268.
123. Xu, Y. and Qian, R., Analysis of thermal injury process based on enzyme deactivation mechanisms, *Trans. ASME J. Biomech. Eng.,* 117, 462-465, 1995.
124. Anghileri, L.J. and Robert, J., *Hyperthermia in Cancer Treatment,* CRC Press, Boca Raton, FL, 1986.
125. Pearse, A.G.E., *Histochemistry: Theoretical and Applied,* 4th ed., Churchill Livingstone, New York, Vols. 1 and 2, 1980; Vol. 3, 1992.
126. Cotran, S., Kumar, V., and Collins, T., *Pathologic Basis of Disease.* 4th ed., W.B. Saunders, Philadelphia, 1999.
127. Ghadially, F. N., *Ultrastructural Pathology of the Cell and Matrix.* 3rd ed., Butterworth Publishers, Boston, 1988.
128. Thomsen, S., Mapping thermal injury in biologic tissues using quantitative pathologic techniques, *Proc. SPIE,* 3594-09, 82-97, 1999.
129. Thomsen, S., Identification of lethal thermal injury at the time of photothermal treatment, In *Laser-Induced Interstitial Thermotherapy,* Muller, G. and Roggan, A., Eds., SPIE Publishing, Bellingham, WA, 1995, 459-467.
130. Ingber, D.E., The architecture of life, *Sci. Am.*, 278, 48-57, 1998.
131. Fawcett, D., *A Textbook of Histology,* 12th ed., W.B. Saunders, Philadelphia, 1994.
132. Schober, R., Ullrich, F., Sander, T., Dunselen, H., and Hessel, S., Laser-induced alteration of collagen substructures allows microsurgical tissue welding, *Science*, 232, 1421-1422, 1986.
133. Thomsen, S., Pearce, J.A., and Cheong W.-F., Changes in birefringence as markers of thermal damage in tissues, *IEEE Trans. Biomed. Eng.,* BME-36, 1174-1179, 1989.
134. Schmidt, S.J., Die doppelbrechung von karyoplasma, zytoplasma und metaplasma, in *Protoplasma-Monographien,* Vol. II, Verlag von Gebruder Borntraeger, Berlin, 1937, 154-267.
135. Fisher, E., The birefringence of striated and smooth muscles, *J. Cell. Comp. Physiol.,* 23, 110-130, 1944.

136. Ramachandran, G. N. and Ramakrishnan, C. Molecular structure, in *Biochemistry of Collagen.* Ramachandran G.N. and Reddi, A.H., Eds., Plenum Press, New York, 1976, 45-84.
137. Miller, E. J. and Gay, S., Collagens In Structural and Contractile Proteins: Part D, Extracellular Matrix, *Methods Enzymol.,* 144, 3-171, 1987.
138. Cohen, K.I., Diegelmann, R.F., and Linblad, W.J., Eds. Structural and regulatory components of wound healing, in *Wound Healing: Biochemical and Clinical Aspects*, W.B. Saunders, Philadelphia, 1992, 130-237.
139. Peckham, M. and Irving, M., Myosin crossbridge orientation in demembranated muscle fibres studied by birefringence and x-ray diffraction measurements, *J. Mol. Biol.,* 210, 113-126, 1989.
140. Harris, P. and Heath, D., Structure and function of vascular smooth muscle, in *The Human Pulmonary Circulation: Its Form and Function in Health and Disease,* Churchill Livingstone, New York, 1986, 161-82.
141. Canham, P. B., Finlay, H. M., Whittaker, P., and Starkey, J., The tunica musclularis of human brain arteries: three-dimensional measurements of alignment of the smooth muscle mechanical axis by polarized light and the universal stage, *Neurol. Res.,* 8, 66-74, 1986.
142. Hulmes, D.J.S., Miller, A., Parry, D.A.D., Piez, K.A., and Woodhead-Galloway, J., Crystalline regions in collagen fibrils, *J. Mol. Biol.,* 184, 473-477, 1985.
143. Flory, P. and Garrett, R.R., Phase transition in collagen and gelatin systems, *J. Am. Chem. Soc.,* 80, 4836-4845, 1958.
144. Wood, G.C., Spectral changes accompanying the thermal denaturation of collagen, *Biochem. Biophys. Res. Commun.* 13, 95-99, 1963.
145. Deak, Gy and Romhanyi, Gy, The thermal shrinkage process of collagen fibers as revealed by polarization optical analysis of topo-optical staining reactions, *Acta Morphol. Acad. Sci. Hung.,* 15, 195-200, 1967.
146. Lim, J.J., Transition temperature and enthalpy change dependence on stabilizing and destabilizing ions in the helix-coil transition in native tendon collagen, *Biopolymers,* 15, 2371-2381, 1976.
147. Bosman, S., Pickering, J.W., vanMarle, J., and vanGemert, M.J.C., Ultrastructural alterations in heated canine myocardium, *Lasers Surg. Med.,* 17, 39-48, 1995.
148. Thomsen, S., Jacques, S.L., and Flock, S., Microscopic correlates of macroscopic optical property changes during thermal coagulation of myocardium, *Laser-Tissue Interact.,* 1202, 2-11, 1990.
149. Junqueira, L.C.U., Bignolas, G., and Brentani, R.R., Picrosirius staining plus polarization microscopy, a specific method for collagen detection in tissue sections, *Histochem. J.,* 11, 447-455, 1979.
150. McKenzie, A.L., A three-zone model of soft-tissue damage by a CO_2 laser, *Phys. Med. Biol.,* 31, 967-983, 1986.
151. Partovi, F., Izatt, J.A., Cothren, R.M., Kittrell, C., Thomas, J.E., Strikwerda, S., Kramer, J.R., and Feld, M.S., A model for thermal ablation of biological tissue using laser radiation, *Lasers Surg. Med.,* 7, 141-154, 1987.
152. Rastegar, S., Motamedi, M., Welch, A.J., and Hayes, L.J., A theoretical study of the effect of optical properties in laser ablation of tissue, *IEEE Trans. Biomed. Eng.,* BME-36, 1180-1187, 1989.
153. LeCarpentier, G.L., Motamedi, M., Rastegar, S., and Welch, A.J., Simultaneous analysis of thermal and mechanical events during cw laser ablation of biological media, *Proc. SPIE,* 1064, 107-113, 1989.
154. Verdaasdonk, R.M., Borst, C., and vanGemert, M.J.C., Explosive onset of continuous wave laser tissue ablation, *Phys. Med. Biol.,* 35, 1129-44, 1990.
155. Van Leeuwen, T.G., van der Veen, J.J., Verdaasdonk, R.M., and Borst, C., Non-contact tissue ablation by holmium:YAG laser pulsed in blood, *Lasers Surg. Med.,* 11, 26-34, 1991.
156. Gijsbers, G.H.M., Selten, F.M., and vanGemert, M.J.C., CW laser ablation velocities as a function of absorption in and experimental one-dimensional tissue model, *Lasers Surg. Med.,* 11, 287-296, 1991.
157. Zweig, A.D., Infrared tissue ablation: consequences of liquefaction, *Laser-Tissue Interact.,* 1427, 2-8, 1991.

158. Frenz, M.C., Greber, M., Romano, Forrer, V.M., and Weber, H.P., Damage induced by pulsed IR laser radiation at transitions between different tissues, *Laser-Tissue Interact.,* 1427, 9-15, 1991.
159. Pearce, J.A. and Thomsen, S., Kinetic models of tissue fusion processes, *Laser-Tissue Interact.,* 3, 1643, 1992.
160. Zweig, A.D., Meierhofer, B., Muller, O.M., Mischler, C., Romano, V., Frenz, M., and Weber, H.P., Lateral damage along pulsed laser incisions, *Lasers Surg. Med.,* 10**,** 262-274, 1990.
161. Van Leeuwen, T.G., Van Erven, L., Meertens, J. H., Motamedi, M., Post, M. J., and Borst, C., Origin of wall dissections induced by pulsed excimer and mid-infrared laser ablation in the pig, *J. Am. Coll. Cardiol.,* 19:7, 1610-1618, 1992.
162. Dabby, F.W. and U. Paek. High-intensity laser-induced vaporization and explosion of solid material, *IEEE J. Quantum. Electron.,* QE8, 106-111, 1972.
163. Welch, A.J. and Polhamus, G.D., Measurement and prediction of thermal injury in the retina of Rhesus monkey, *IEEE Trans. Biomed. Eng.,* BME-31, 633-644, 1984.
164. Takata, A.N. et al., Thermal model of laser-induced eye damage, Final Rept. USAF School of Aerospace Medicine, Brooks AFB TX, Contract F41609-74-C-0005, IIT Research Institute, Chicago, 1974.
165. Birngruber, R., Hillenkamp, F., and Gabel, V.-P., Theoretical investigations of laser thermal retinal injury, *Health Phys.,* 48(6), 781-796, 1985.
166. Birngruber, R., Thermal modeling in biological tissue, in *Lasers in Biology and Medicine,* Hillenkamp, F., Pratesi, R., and Sacchi, C.A., Eds., Plenum Press, New York, 1980 77-97.
167. Yang, Y., Welch, A.J., and Rylander, H.G., III, Rate process parameters of albumen, *Lasers Surg. Med.,* 11, 188-190, 1991.
168. Agah, R., Quantitative Characterization of Arterial Tissue Thermal Damage, MSE thesis, University of Texas at Austin, 1988.
169. Mixter, G., Delhry, G.P., Derksen, W.L., and Monahan, T., The Influence of time on the death of HeLa cells at elevated temperature, in *Temperature: Its Measurement and Control in Science and Industry,* Vol. 3, Hardy, J.D., Ed., Reinhold, New York, 1963.
170. Padanilam, J., Bischof, J.C., Lee, R.C., Cravalho, E.G., Tompkins, R.G., Yarmush, M.L., and Toner, M., Effectiveness of Poloxamer 188 in arresting calcein leakage from thermally damaged isolated skeletal muscle cells, *Ann. N.Y. Acad. Sci.,* 720, 111-123, 1994.
171. Pearce, J.A., Cheong W.F., Pandit, K., McMurray, T., and Thomsen, S., Kinetic models for coagulation processes: determination of rate coefficients *in vivo, Proc. Lasers Dermatol. Tissue Welding,* 1422, 27-32, 1991.
172. Pearce, J.A., Han, A., Gutierrez, T., and Thomsen, S., Argon laser coagulation of myocardium: the effect of pulse width on threshold temperature, Lasers in Surgery and Medicine Suppl. 2 Proc. ASLMS Meeting, Arlington, VA, 1990, 16.
173. Polge, C., Smith, A.U., and Parkes, A.S., Revival of spermatozoa after vitrification and dehydration at low temperatures, *Nature*, 164, 666, 1949.
174. Lovelock, J.E., The mechanism of the protective action of glycerol against haemolysis by freezing and thawing, *Biochim. Biophys. Acta*, 11, 28-36, 1953.
175. Strumia, M.M., Clawell, L.S., and Strumia, P.V., The preservation of blood for transfusion, *J. Lab. Clin. Med.*, 56, 576-593, 1960.
176. Whittingham, D.G., Leibo, S.P., and Mazur, P., Survival of mouse embryos frozen to –196°C and –296°C, *Science*, 178, 411-414, 1972.
177. McGrath, J.J. and Diller, K.R., Eds., *Low Temperature Biotechnology: Emerging Applications and Engineering Contributions*, American Society of Mechanical Engineers, New York, 1-380, 1988.
178. Mazur, P., Kinetics of water loss from cells at subzero temperatures and the likelihood of intracellular freezing, *J. Gen. Physiol.*, 47, 347-369, 1963.
179. Mazur, P., Equilibrium, quasi-equilibrium and nonequilibrium freezing of mammalian embryos, *Cell Biophys.*, 17, 53-92, 1990.
180. Kedem, O. and Katchalsky, A., Thermodynamic analysis of permeability of biological membranes to nonelectrolytes, *Biochim. Biophys. Acta*, 27, 229-246, 1958.

181. McGrath, J.J., Membrane transport properties, in *Low Temperature Biotechnology: Emerging Applications and Engineering Contributions*, McGrath, J.J. and Diller, K.R., Eds., American Society of Mechanical Engineers, New York, 1988, 273-330.
182. Oster, G.F., Perelson, A.S., and Katchalsky, A., Network thermodynamic modelling of biophysical systems, *Q. Rev. Biophys.,* 6, 1-134, 1973.
183. Diller, K.R., Energy based network modeling for cryobiology, in *Low Temperature Biotechnology: Emerging Applications and Engineering Contributions*, McGrath, J.J. and Diller, K.R., Eds., American Society of Mechanical Engineers, New York, 1988, 189-202.
184. Diller, K.R., Beaman, J.J., Montoya, J.P., and Breedfeld, P.C., Network thermodynamic modeling with bond graphs for membrane transport during cell freezing, *J. Heat Trans.,* 110, 938-945, 1988.
185. deFreitas, R.C., Diller, K.R., Lachenbruch, C.A., and Merchant, F.A., Network thermodynamic model of coupled transport in a multicellular tissue — the islet of Langerhans, *Ann. N.Y. Acad. Sci.*, 858, 191-204, 1998.
186. Walsh, J.R., Diller, K.R., and Brand, J.J., Osmomechanical behavior of plant cells across regions of elastic and osmotic stress, Advances in Heat and Mass Transfer in Biotechnology, ASME HTD 355, BED 37, 15-23, 1997.
187. Lachenbruch, C.A., Pegg, D.E., and Diller, K.R., Sensitivity of kidney perfusion protocol design to physical and physiological parameters, *Ann. N.Y. Acad. Sci.*, 858, 298-309, 1998.
188. Walsh, J.R., Diller, K.R., and Brand, J.J., Measurement and simulation of water transport in algal cells, *J. Biomech. Eng.,* 1999, submitted.
189. Diller, K.R. and Cravalho, E.G., A cryomicroscope for the study of freezing and thawing processes in biological cells, *Cryobiology*, 7, 191-199, 1970.
190. Diller, K.R., Quantitative low temperature optical microscopy of biological systems, *J. Microsc.,* 126, 9-28, 1982.
191. Diller, K.R., Cryomicroscopy, in *Low Temperature Biotechnology: Emerging Applications and Engineering Contributions*, McGrath, J.J. and Diller, K.R., Eds., American Society of Mechanical Engineers, New York, 1988, 347-362.
192. McGrath, J.J., A microscope diffusion chamber for the determination of the equilibrium and non-equilibrium response of individual cells, *J. Microsc.,* 139, 249-263, 1985.
193. Walcerz, D.B. and Diller, K.R., Quantitative light microscopy of combined perfusion and freezing processes, *J. Microsc.,* 161, 297-311, 1991.
194. deFreitas, R.C., Diller, K.R., Lakey, J.R.T., and Rajotte, R.V., Osmotic behavior and transport properties of human islets in a dimethyl sulfoxide solution, *Cryobiology***,** 35, 230-239, 1997.
195. Levin, R.L. Osmotic behavior of cells during freezing and thawing, in *Low Temperature Biotechnology: Emerging Applications and Engineering Contributions*, McGrath, J.J. and Diller, D.R., Eds., American Society of Mechanical Engineers, New York, 1988, 177-188.
196. Kasharin, A.V. and Karlsson, J.O.M., Analysis of mass transport during warming of cryopreserved cells, *Ann. N.Y. Acad. Sci.*, 858, 163-174, 1998.
197. Fahy, G.M., Vitrification, in *Low Temperature Biotechnology: Emerging Applications and Engineering Contributions*, McGrath, J.J. and Diller, D.R., Eds., American Society of Mechanical Engineers, New York, 1988, 113-146.
198. Steponkus, P.L., The role of the plasma membrane in freezing injury and cold acclimation, *Annu. Rev. Plant Physiol.,* 35, 543-584, 1984.
199. Kurz, W. and Fisher, D.J., *Fundamentals of Solidification*, 3rd ed., Trans Tech Publ., Aedermannsdorf, Switzerland, 1992.
200. Hayes, L.J., Diller, K.R., Chang, H.J., and Lee, H.S., Prediction of local cooling rates and cell survival during the freezing of cylindrical specimens, *Cryobiology*, 25, 67-82, 1988.
201. Fahy, G.M., Physical problems with the vitrification of large biological systems, *Cryobiology*, 27, 492-510, 1990.
202. Rubinsky, B., Cravalho, E.G., and Mikic, B., Thermal stresses in frozen organs, *Cryobiology*, 17, 66-73, 1980.

203. Fowler, A.J. and Toner, M., Prevention of hemolysis in rapidly frozen erythrocytes by using a laser pulse, *Ann. N.Y. Acad. Sci.,* 858, 245-252, 1998.
204. Nerem, R.M. and Sambanis, A., Tissue engineering: from biology to biological substitutes, *Tissue Eng.*, 1, 3-13, 1995.
205. Rubinsky, B., Biophysical and bioengineering aspects of cryosurgery, *Cryobiol. Cryotechnol.,* 41, 67-81, 1995.
206. Onik, G. and Rubinsky, B., Cryosurgery: new developments in understanding and technique, in *Low Temperature Biotechnology: Emerging Applications and Engineering Contributions*, McGrath, J.J. and Diller, K.R., Eds., American Society of Mechanical Engineers, New York, 1988, 57-80.
207. Onik, G., Gilbert, J., Hoddick, W., Filly, R., Callen, P., Rubinsky, B., and Christianson, M., Ultrasonic characteristics of frozen liver, *Cryobiology*, 21, 331-335, 1984.
208. Rubinsky, B., Gilbert, J.C., Onik, G.M., Roos, H.S., Wong, S.T.S., and Brennan, K.M., Monitoring cryosurgery in the brain and in the prostate with proton NMR, *Cryobiology*, 30, 191-199, 1993.
209. Otten, D.M., Rubinsky, B., Cheong, W.-F., and Benaron, D.A., Ice front propagation monitoring in tissue by the use of visible-light spectroscopy, *Appl. Opt.*, 37, 6006-6010, 1998.
210. Baust, J.G. and Chang, Z., Underlying mechanisms of damage and new concept in cryosurgical instrumentation, *Cryosurgery: Mechanism and Applications*, IIR Press, Paris, 1995, 21-36.
211. Rubinsky, B. and Pegg, D.E., A mathematical model for the freezing process in biological tissue, *Proc. R. Soc. London*, B234, 343-358, 1989.
212. Bischof, J.C. and Rubinsky, B., Microscale heat and mass transfer of vascular and intracellular freezing in the liver, *J. Heat Trans.*, 115, 1029-1035, 1993.
213. Pazhayannur, P.V. and Bischof, J.C., Measurement and simulation of water transport during freezing in mammalian liver tissue, *J. Biomech. Eng.,* 119, 269-277, 1997.
214. Koushafar, H. and Rubinsky, B., Effect of antifreeze proteins on frozen primary prostatic adenocarcinoma cells, *Urology*, 49, 421-425, 1997.
215. Koushafar, H., Pham, L., Lee, C., and Rubinsky, B., Chemical adjuvant cryosurgery with antifreeze proteins, *J. Surg. Oncol.*, 66, 114-121, 1997.
216. DeVries, A.L., The role of antifreeze glycopeptides and peptides in the freezing avoidance of antarctic fishes, *Comp. Biochem. Physiol.*, 90, 611-621, 1988.
217. Ananthanarayanan, V.S., Antifreeze proteins: structural diversity and mechanism of action, *Life Chem. Rep.*, 7, 1-32, 1989.
218. Ishiguro, H. and Rubinsky, B., Influence of fish antifreeze proteins on the freezing of cell suspensions with cryoprotectant penetrating cells, *Int. J. Heat Mass Trans.*, 41, 1907-1915, 1998.
219. Diller, K.R. and Ryan, T.P., Heat transfer in living systems: current opportunities, *J. Heat Trans.*, 120, 810-829, 1998.
220. Holmes, K.R., Thermal conductivity of selected tissues, in *Biotransport: Heat and Mass Transfer in Living Systems*, Diller, K.R., Ed., New York Academy of Science, New York, 1998.

Appendix A

The following physiological properties were compiled by Professor Kenneth R. Holmes ⟨krholmes@ux1.cso.uiuc.edu⟩ and were published in part previously.[220] The tabulation includes values for both the native thermal conductivity of biomaterials (Appendix A) and the blood perfusion rates for specific tissues and organs (Appendix B). Original sources are documented in the dedicated list of references at the end of each appendix. Knowledge of the perfusion behavior of tissues is important in that the flow of blood can have a direct quantitative effect on the temperature distribution within living tissue.

Thermal Conductivity Data for Specific Tissues and Organs for Humans and Other Mammalian Species

Tissue	k (W/mK)	%H_2O	Appendix Ref.
Kidney			
whole (rabbit)	0.502	81	6
whole (rabbit)	0.495		16
whole (human)	0.543	84	4
cortex (rabbit)	0.465–0.490 (n = 7)	76.6–79.8	1
cortex (dog)	0.491		16
cortex (human)	0.499		16
medulla (rabbit)	0.502–0.544 (n = 7)	82.0–86.0	1
medulla (dog)	0.507		16
medulla (human)	0.499		16
Aorta			
human	0.476 ± 0.041 (SD) (n = 12)		2
Arterial plaque			
fatty	0.484 ± 0.044 (SD) (n = 13)		2
fibrous	0.485 ± 0.022 (SD) (n = 12)		2
calcified	0.502 ± 0.059 (SD) (n = 17)		2
Artery			
femoral (dog)	0.391		2
carotid (dog)	0.448		2
Blood			
whole	0.505		10
whole (human)	0.492 ± 0.009 (SD) (n = 10)		5
plasma	0.582		10
plasma (human)	0.570 ± 0.010 (SD) (n = 10)		5
Spleen			
(human)	0.543	80	4
Liver			
(rabbit)	0.493	72	6
(rat)	0.498–0.528 (n = 4)		3
(sheep)	0.495 ± 0.017 (SD) (n = 9)		5
(dog)	0.550 ± 0.010 (SD) (n = 7)		5
(dog)	0.574		11
(pig)	0.498		16
(human)	0.469		16
(human)	0.564	77	4
Heart			
(rat)	0.521–0.531 (n = 2)		3
(dog)	0.527		11
(dog)	0.487		16
(pig)	0.484		16
(human)	0.493		16
(human)	0.585	81	4
Muscle			
(rat)	0.505–0.530 (n = 7)		3
(cow)	0.410–0.426		7
(pig)	0.518–0.618		7

Tissue	k (W/mK)	%H_2O	Appendix Ref.
(pig)	0.460		17
(sheep)	0.478 ± 0.019 (SD) (n = 10)		5
Fat			
(cow)	0.185–0.233		7
(cow)	0.222		17
(pig)	0.317–0.402		7
(pig)	0.159		17
blubber (Minke whale)	0.200–0.280	3–30	8
blubber (Harp seal)	0.190 ± 0.010		9
(human)	0.201–0.217		17
Brain			
whole	0.527	78	4
cerebral cortex	0.564	83	4
cerebral cortex	0.504		16
white matter	0.502	71	4
Tumor			
periphery	0.511		11
core	0.561		11
colon cancer (human)	0.545		16
Bone	0.410–0.630		15
Skin	0.210–0.410		15
crocodile, middle of back	0.432		12
crocodile, tail, ventral	0.334		12
epidermis			
foot pad (cat)	0.116–0.290		13
(hydrated)	0.295–0.580		13
elephant (freeze/thaw)	0.475	62.5	14
rhinoceros (freeze/thaw)	0.452	60.1	14
giraffe (formalin fixed)	0.442	60.0	14
epidermis (human)	0.209		17
epidermis (pig)	0.209		17
dermis (human)	0.293–0.322		17
Pure water	0.627		17

Note: $c = [\%H_2O + 0.4 * (100 - \%H_2O)] * 41.9$ J/kg K (from Footnote 4).
$\rho = 1.05(10^3)$ kg/m^3 (from Footnote 4).

1. Holmes, K.R., Ryan, W., and Chen, M.M., Thermal conductivity and H_2O content in rabbit kidney cortex and medulla, *J. Therm. Biol.*, 8, 311-313, 1983.
2. Valvano, J.W. and Chitsabesan, B., Thermal conductivity and diffusivity of arterial wall and atherosclerotic plaque. *Lasers Life Sci.*, 1, 219-229, 1987.
3. Valvano, J.W., Allen, J.T., and Bowman, H.F., The simultaneous measurement of thermal conductivity, thermal diffusivity, and perfusion in small volumes of tissue, *ASME 81-WA/HT-21,* 1981.
4. Cooper, T.E. and Trezek, G.J., Correlation of thermal properties of some human tissue with water content, *Aerosp. Med.,* 42, 24-27, 1971.
5. Balasubramaniam, T.A. and Bowman, H.F., Thermal conductivity and thermal diffusivity of biomaterials. A simultaneous measurement technique, *J. Biomech. Eng.*, 99, 148-154, 1977.
6. Grayson, J., Internal calorimetry in the determination of thermal conductivity and blood flow. *J. Physiol.*, 118, 54-72, 1952.
7. Chato, J.C., A method for the measurement of the thermal properties of biological materials. In *Thermal Problems in Biotechnology,* Chato, J.C., Ed., American Society of Mechanical Engineers, New York, 1968.
8. Kvadsheim, P.H., Folkow, L.P., and Blix, A.S., Thermal conductivity of Minke whale blubber, *J. Therm. Biol.*, 21, 123-128, 1996.
9. Kvadsheim, P.H., Folkow, L.P., and Blix, A.S., A new device for measurement of the thermal conductivity of fur and blubber, *J. Therm. Biol.*, 19, 431-435, 1994.
10. Dumas, A. and Barozzi, G.S., Laminar heat transfer to blood flowing in a circular duct, *Int. J Heat Mass Trans.*, 27, 391-398, 1984.

11. Holmes, K.R. and Chen, M.M., Local thermal conductivity of Para-7 fibrosarcoma in hamster. *1979 Advances in Bioengineering*, American Society of Mechanical Engineers, New York, 1979, 147-149.
12. Drane, C.R., The thermal conductivity of the skin of crocodilians, *Comp. Biochem. Physiol.,* 68A, 107-110, 1981.
13. Holmes, K.R. and Adams, T., Epidermal thermal conductivity and stratum corneum hydration in cat footpad, *Am. J. Physiol.*, 228, 1903-1908, 1975.
14. Holmes, K.R., Unpublished data.
15. Chato, J.C., Fundamentals of bioheat transfer, in *Thermal Dosimetry and Treatment Planning*, Gautherie, M., Ed., Springer-Verlag, New York, 1990, 51.
16. Valvano, J.W., Cochran, J.R., and Diller, K.R., Thermal conductivity and diffusivity of biomaterials measured with self-heated thermistors, *Int. J. Thermophys.*, 6, 301-311, 1985.
17. Adapted from Cohen, M.L., Measurement of the thermal properties of human skin. A review, *J. Invest. Dermatol.*, 69, 333-338, 1977, and its references.

Appendix B

Blood Perfusion Data for Specific Tissues and Organs for Humans and Other Mammalian Species

Organ	Blood Flow (ml/min/g)	Species (* = Unanesthetized)	Appendix Ref.	Technique
Brain	0.36 ± 0.04 (SE)	dog (n = 11)	36	μs (pentobarbital)
Brain	0.78 ± 0.06 (SE)	rat (n = 7)	30	μs
Brain	1.52 ± 0.23 (SE)	fetal sheep* (n = 12)	32	μs
Brain	0.82 ± 0.06 (SE)	neonatal sheep* (n = 13)	37	μs
Brain	0.64 ± 0.03 (SE)	sheep* (n = 7)	37	μs
Brain	1.14 ± 0.14 (SD)	rat* (12 mo old)(n = 5)	33	μs
Brain	0.65 ± 0.07 (SD)	rat (n = 10)	33	μs (pentobarbital)
Brain	1.31 ± 0.22 (SD)	rat* (24 mo old)(n = 5)	33	μs
Brain	0.43 ± 0.14 (SD)	rat (n = 7)	33	μs (decerebrate)
Mean Cerebral	0.373 ± 0.062	dog	2	
Mean Cerebral	0.491 ± 0.078	baboon	2	
Cerebrum	0.48 ± 0.03	cat	18	
Cerebrum	0.53 ± 0.06	cat	18	
Cerebrum	0.32 ± 0.06	monkey	18	
Cerebrum	0.39 ± 0.03	monkey	18	
Cerebrum	0.79 ± 0.12	pig*	19	
Cerebrum	1.17 ± 0.11 (SE)	rat* (n = 8)	31	μs
Cerebrum	0.61 ± 0.06 (SE)	rat (n = 7)	31	μs (pentobarbital)
Cerebrum	0.72 ± 0.05 (SE)	rat (n = 7)	31	μs (chlor.-urethan)
Cerebrum	0.50 ± 0.04 (SE)	rat (n = 7)	31	μs (decerebrate)
Cerebrum	0.60 ± 0.06 (SE)	cat (n = 8)	34	μs (α-chloralose)
Cerebrum	1.29 ± 0.03 (SE)	cat (n = 8) (hypercapnic)	34	μs (α-chloralose)
cerebral cortex	0.446 ± 0.061	dog	2	
cerebral cortex	0.447 ± 0.036	dog	3	
cerebral cortex	0.64 ± 0.05	cat	18	
cerebral cortex	0.75 ± 0.11	cat	18	
cerebral cortex	0.40 ± 0.11	monkey	18	
cerebral cortex	0.50 ± 0.11	monkey	18	
frontal cortex	1.06 ± 0.05	rat	4	
parietal cortex	1.07 ± 0.05	rat	4	
occipital cortex	0.98 ± 0.08	rat	4	
cerebral white	0.240 ± 0.037	dog	2	
cerebral white	0.305 ± 0.022	dog	3	
cerebral white	0.20 ± 0.05	monkey	18	
cerebral white	0.21 ± 0.06	monkey	18	
cerebral white	0.29 ± 0.03	cat	18	
cerebral white	0.20 ± 0.03 (SE)	cat (n = 8)	34	μs (α-chloralose)
cerebral white	0.35 ± 0.09 (SE)	cat (n = 8) (hypercapnic)	34	μs (α-chloralose)
White matter	0.164 ± 0.022	baboon	2	
corpus callosum	0.249 ± 0.035	dog	3	
corpus callosum	0.62 ± 0.03	rat	4	
Gray matter	0.552 ± 0.077	baboon	2	
cortical gray	0.87 ± 0.11 (SE)	cat (n = 8)	34	μs (α-chloralose)
cortical gray	1.94 ± 0.37 (SE)	cat (n = 8) (hypercapnic)	34	μs (α-chloralose)
caudate nucleus	0.88 ± 0.03	rat	4	
caudate nucleus	0.65 ± 0.06	cat	18	
caudate nucleus	0.87 ± 0.11	cat	18	
caudate nucleus	0.39 ± 0.05	monkey	18	
caudate nucleus	0.47 ± 0.05	monkey	18	
caudate nucleus	0.85 ± 0.10 (SE)	cat (n = 8)	34	μs (α-chloralose)
caudate nucleus	1.31 ± 0.19 (SE)	cat (n = 8) (hypercapnic)	34	μs (α-chloralose)
Choroid plexus	4.18 ± 0.60 (SE)	neonatal sheep* (n = 13)	37	μs
Choroid plexus	7.53 ± 1.11 (SE)	sheep* (n = 7)	37	μs
Hypothalamus	1.03 ± 0.06	rat	4	

Organ	Blood Flow (ml/min/g)	Species (* = Unanesthetized)	Appendix Ref.	Technique
Pituitary	1.08 ± 0.24 (SE)	neonatal sheep* (n = 13)	37	μs
Pituitary	1.13 ± 0.16 (SE)	sheep* (n = 7)	37	μs
Thalamus	0.92 ± 0.07	rat	4	
thalamus-midbrain	0.62 ± 0.07 (SE)	cat (n = 8)	34	μs (α-chloralose)
thalamus-midbrain	1.64 ± 0.32 (SE)	cat (n = 8) (hypercapnic)	34	μs (α-chloralose)
Hippocampus	0.68 ± 0.06	rat	4	
Cerebellum	0.63 ± 0.04	cat	18	
Cerebellum	0.70 ± 0.07	cat	18	
Cerebellum	0.60 ± 0.08 (SE)	cat (n = 8)	34	μs (α-chloralose)
Cerebellum	1.63 ± 0.31 (SE)	cat (n = 8) (hypercapnic)	34	μs (α-chloralose)
Cerebellum	0.40 ± 0.08	monkey	18	
Cerebellum	0.50 ± 0.02	monkey	18	
Cerebellum	0.78 ± 0.06	pig*	19	
Cerebellum	1.31 ± 0.19 (SE)	rat* (n = 8)	31	μs
Cerebellum	1.02 ± 0.12 (SE)	rat (n = 7)	31	μs (pentobarbital)
Cerebellum	1.04 ± 0.08 (SE)	rat (n = 7)	31	μs (chlor.-urethan)
Cerebellum	0.78 ± 0.07 (SE)	rat (n = 7)	31	μs (decerebrate)
Pons	0.50 ± 0.03 (SE)	cat (n = 8)	34	μs (α-chloralose)
Pons	1.18 ± 0.25 (SE)	cat (n = 8) (hypercapnic)	34	μs (α-chloralose)
Medulla	0.37 ± 0.06 (SE)	cat (n = 8)	34	μs (α-chloralose)
Medulla	1.07 ± 0.20 (SE)	cat (n = 8) (hypercapnic)	34	μs (α-chloralose)
Medulla	0.84 ± 0.04	rat	4	
Brain stem	0.52 ± 0.03	cat	18	
Brain stem	0.58 ± 0.03	cat	18	
Brain stem	0.33 ± 0.06	monkey	18	
Brain stem	0.36 ± 0.03	monkey	18	
Brain stem	0.46 ± 0.06	pig*	19	
Brain stem	0.55 ± 0.05 (SE)	cat (n = 8)	34	μs (α-chloralose)
Brain stem	1.44 ± 0.28 (SE)	cat (n = 8) (hypercapnic)	34	μs (α-chloralose)
Spinal cord	0.22 ± 0.02	cat	18	
Spinal cord	0.31 ± 0.03	cat	18	
Spinal cord	0.20 ± 0.03	monkey	18	
Spinal cord	0.21 ± 0.03	monkey	18	
Nerve, sciatic	0.27 ± 0.03 (SE)	rat (n-14)	29	[^{14}C]iodoantipyrine
Kidney	3.018 ± 0.412 (SE)	dog (n = 11)	36	μs (pentobarbital)
Kidney	3.99 ± 0.88 (SD)	rat* (12 mo old)(n = 5)	33	μs
Kidney	3.64 ± 0.92 (SD)	rat (n = 10)	33	μs (pentobarbital)
Kidney	4.82 ± 1.39 (SD)	rat* (24 mo old)(n = 5)	33	μs
Kidney	2.31 ± 0.82 (SD)	rat (n = 7)	33	μs (decerebrate)
Kidney	1.54 ± 0.12 (SE)	fetal sheep* (n = 12)	32	μs
Kidney	3.00 ± 0.14 (SE)	neonatal sheep* (n = 13)	37	μs
Kidney	4.19 ± 0.32 (SE)	sheep* (n = 7)	37	μs
Kidney	4.32 ± 0.30	dog	3	
Kidney	3.2 ± 0.3	rabbit	17	
Kidney	2.80 ± 0.05	pig*	19	
Kidney	4.7 ± 0.3	rat	13	
Kidney	5.55 (4.87-8.76)	rat (n = 14)	26	
Kidney	6.23 ± 0.49	rat (n = 9)	27	
Kidney	4.38 ± 0.51 (SE)	rat (n = 7)	30	μs
Kidney	4.38 ± 0.44 (SE)	rat (n = 7)	30	μs
Kidney	7.63 ± 0.46 (SE)	rat* (n = 8)	31	μs
Kidney	6.16 ± 0.39 (SE)	rat (n = 7)	31	μs (pentobarbital)
Kidney	5.96 ± 0.42 (SE)	rat (n = 7)	31	μs (chlor.-urethan)
Kidney	5.46 ± 0.53 (SE)	rat (n = 7)	31	μs (decerebrate)
whole cortex	4.0-5.0	human*	6	
whole cortex	3.63 ± 0.54 (range = 3.0-4.4)	pig	8	
whole cortex	3.71 ± 0.284	pig (n = 4)	25	

Organ	Blood Flow (ml/min/g)	Species (* = Unanesthetized)	Appendix Ref.	Technique
whole cortex	4.69 ± 0.347	pig (n = 6)	25	
whole cortex	5.06 ± 0.251	pig (n = 6)	25	
whole cortex	3.25 ± 0.20 (SE)	sheep* (n = 6)	40	μs
outer cortex	3.81 ± 0.23 (SE)	sheep* (n = 6)	40	μs
outer cortex	6.32 ± 0.33	dog	3	
outer cortex	7.56 ± 0.70	dog	11	
outer cortex	4.62	dog	7	
outer cortex	8.84 ± 0.014	rabbit	5	
outer cortex	8.7 ± 0.6	rat	13	
middle cortex	9.53 ± 0.013	rabbit	5	
inner cortex	2.57 ± 0.22 (SE)	sheep* (n = 6)	40	μs
inner cortex	4.38 ± 1.19	dog	11	
inner cortex	3.35 ± 0.42	dog	3	
inner cortex	3.96	dog	7	
inner cortex	3.24 ± 0.004	rabbit	5	
inner cortex	4.7 ± 0.4	rat	13	
outer 1/4 cortex	4.24 ± 0.40	dog	12	
next 1/4	6.24 ± 0.61	dog	12	
next 1/4	3.55 ± 0.45	dog	12	
inner 1/4	1.73 ± 0.13	dog	12	
medulla	0.35 ± 0.09 (SE)	sheep* (n = 6)	40	μs
medulla	0.77 ± 0.064	dog	3	
outer medulla	1.2	human*	6	
outer medulla	1.30	dog	7	
outer medulla	2.10 ± 0.28	dog	11	
inner medulla	0.25	human*	6	
Adrenal gland	7.62 ± 2.55 (SE)	fetal sheep* (n = 12)	32	μs
Adrenal gland	1.83 ± 0.12 (SE)	neonatal sheep (n = 13)	37	μs
Adrenal gland	1.99 ± 0.13 (SE)	sheep* (n = 7)	37	μs
Adrenal gland	1.341 ± 0.297 (SE)	dog (n = 11)	36	μs (pentobarbital)
Muscle				
skeletal	0.027 (average)	human*	16	
skeletal	0.12 ± 0.02 (SE)	neonatal sheep* (n = 13)	37	μs
skeletal	0.04 ± 0.01 (SE)	sheep* (n = 7)	37	μs
skeletal	0.18 ± 0.07 (SD)	rat* (12 mo old)(n = 5)	33	μs
skeletal	0.03 ± 0.01 (SD)	rat (n = 10)	33	μs (pentobarbital)
skeletal	0.22 ± 0.12 (SD)	rat* (24 mo old)(n = 5)	33	μs
skeletal	0.05 ± 0.02 (SD)	rat (n = 7)	33	μs (decerebrate)
skeletal	0.39 ± 0.20 (SD)	rat (n = 12)	41	^{133}Xe (ether)
resting thigh				
1.5 cm depth	0.018 ± 0.011	human*	14	
3.0 cm depth	0.026 ± 0.013	human*	14	
resting thigh	0.020-0.022	human*	15	
vastus medialis & triceps	0.14 ± 0.03	pig*	19	
gracilis	0.049 ± 0.064	dog	2	
masseter	0.09 ± 0.01 (SE)	rat (n = 7)	30	μs
cremaster	0.16 ± 0.03 (SE)	rat (n = 7)	30	μs
cremaster	0.07 ± 0.018 (SE)	rat* (n = 8)	31	μs
cremaster	0.06 ± 0.008 (SE)	rat (n = 7)	31	μs (pentobarbital)
cremaster	0.05 ± 0.013 (SE)	rat (n = 7)	31	μs (chlor.-urethan)
cremaster	0.04 ± 0.012 (SE)	rat (n = 7)	31	μs (decerebrate)
diaphragm	0.93 ± 0.19 (SE)	rat* (n = 8)	31	μs
diaphragm	0.22 ± 0.02 (SE)	rat (n = 7)	31	μs (pentobarbital)
diaphragm	0.24 ± 0.04 (SE)	rat (n = 7)	31	μs (chlor.-urethan)
diaphragm	0.35 ± 0.04 (SE)	rat (n = 7)	31	μs (decerebrate)
diaphragm	0.144 ± 0.017 (SE)	dog (n = 11)	36	μs (pentobarbital)
rectus abdominus	0.18 ± 0.029 (SE)	rat* (n = 8)	31	μs

Organ	Blood Flow (ml/min/g)	Species (* = Unanesthetized)	Appendix Ref.	Technique
rectus abdominus	0.04 ± 0.008 (SE)	rat (n = 7)	31	μs (pentobarbital)
rectus abdominus	0.09 ± 0.025 (SE)	rat (n = 7)	31	μs (chlor.-urethan)
rectus abdominus	0.09 ± 0.027 (SE)	rat (n = 7)	31	μs (decerebrate)
gastrocnemius	0.11 ± 0.022 (SE)	rat* (n = 8)	31	μs (decerebrate)
gastrocnemius	0.04 ± 0.003 (SE)	rat (n = 7)	31	μs (decerebrate)
gastrocnemius	0.15 ± 0.084 (SE)	rat (n = 7)	31	μs (decerebrate)
gastrocnemius	0.06 ± 0.019 (SE)	rat (n = 7)	31	μs (decerebrate)
tibialis anterior	0.36 ± 0.068 (SE)	rat* (n = 8)	31	μs (decerebrate)
tibialis anterior	0.03 ± 0.004 (SE)	rat (n = 7)	31	μs (decerebrate)
tibialis anterior	0.17 ± 0.11 (SE)	rat (n = 7)	31	μs (decerebrate)
tibialis anterior	0.08 ± 0.03 (SE)	rat (n = 7)	31	μs (decerebrate)
psoas	0.26 ± 0.034 (SE)	rat* (n = 8)	31	μs (decerebrate)
psoas	0.04 ± 0.004 (SE)	rat (n = 7)	31	μs (decerebrate)
psoas	0.09 ± 0.037 (SE)	rat (n = 7)	31	μs (decerebrate)
psoas	0.07 ± 0.022 (SE)	rat (n = 7)	31	μs (decerebrate)
heart	3.72 ± 0.45 (SE)	fetal sheep* (n = 12)	32	μs
heart	2.16 ± 0.23 (SE)	neonatal sheep* (n = 13)	37	μs
heart	1.13 ± 0.09 (SE)	sheep* (n = 7)	37	μs
heart	5.45 ± 0.59 (SE)	rat (n = 7)	30	μs
heart	5.31 ± 0.59 (SE)	rat* (n = 8)	31	μs
heart	2.83 ± 0.60 (SE)	rat (n = 7)	31	μs (pentobarbital)
heart	2.23 ± 0.37 (SE)	rat (n = 7)	31	μs (chlor.-urethan)
heart	3.19 ± 0.55 (SE)	rat (n = 7)	31	μs (decerebrate)
heart	6.61 ± 2.20 (SD)	rat* (12 mo old)(n = 5)	33	μs
heart	5.45 ± 1.62 (SD)	rat (n = 10)	33	μs (pentobarbital)
heart	8.66 ± 2.74 (SD)	rat* (24 mo old)(n = 5)	33	μs
heart	3.67 ± 1.03 (SD)	rat (n = 7)	33	μs (decerebrate)
right ventricle	0.92 ± 0.12	pig*	19	
left ventricle	1.45 ± 0.09	pig*	19	
left ventricle	1.58 ± 0.13	dog	20	
left ventricle	1.09 ± 0.35	dog	21	
left ventricle	0.61 ± 0.41	dog	22	
left ventricle	0.88 ± 0.14	dog	23	
epicardium	0.89 ± 0.14	dog	24	
epicardium	0.86 ± 0.14	dog	24	
endocardium	1.14 ± 0.18	dog	24	
endocardium	1.08 ± 0.18	dog	24	
GI				
intestine	0.47 ± 0.05 (SE)	fetal sheep* (n = 12)	32	μs
intestine	0.389 ± 0.04	dog	3	
tract	2.05 ± 0.25	rat* (n = 8)	31	μs
tract	1.31 ± 0.11	rat (n = 7)	31	μs (pentobarbital)
tract	1.26 ± 0.09	rat (n = 7)	31	μs (chlor.-urethan)
tract	1.19 ± 0.14	rat (n = 7)	31	μs (decerebrate)
stomach	1.27 ± 0.21 (SE)	rat* (n = 8)	31	μs
stomach	0.46 ± 0.04 (SE)	rat (n = 7)	31	μs (pentobarbital)
stomach	0.53 ± 0.05 (SE)	rat (n = 7)	31	μs (chlor.-urethan)
stomach	0.52 ± 0.07 (SE)	rat (n = 7)	31	μs (decerebrate)
stomach	1.31 ± 0.30 (SD)	rat* (12 mo old)(n = 5)	33	μs
stomach	0.55 ± 0.21 (SD)	rat (n = 10)	33	μs (pentobarbital)
stomach	1.50 ± 0.45 (SD)	rat* (24 mo old)(n = 5)	33	μs
stomach	0.42 ± 0.21 (SD)	rat (n = 7)	33	μs (decerebrate)
stomach	0.53 ± 0.26 (SD)	dog (n = 12)	39	Rb^{86}(pentobarbital)
duodenum	1.11 ± 0.46 (SD)	dog (n = 12)	39	Rb^{86}(pentobarbital)
duodenum	0.901 ± 0.132 (SE)	dog (n = 11)	36	μs (pentobarbital)
duodenum	3.63 ± 0.70 (SD)	rat* (12 mo old)(n = 5)	33	μs
duodenum	2.33 ± 0.89 (SD)	rat (n = 10)	33	μs (pentobarbital)

Organ	Blood Flow (ml/min/g)	Species (* = Unanesthetized)	Appendix Ref.	Technique
duodenum	3.31 ± 0.64 (SD)	rat* (24 mo old)(n = 5)	33	µs
duodenum	2.00 ± 0.76 (SD)	rat (n = 7)	33	µs (decerebrate)
jejunum	0.642 ± 0.086 (SE)	dog (n = 11)	36	µs (pentobarbital)
jejunum	0.263 ± 0.034 (SE)	cat (n = 12	35	µs (pentobarbital)
ileum	0.498 ± 0.083 (SE)	dog (n = 11)	36	µs (pentobarbital)
small bowel	1.00 ± 0.45 (SD)	dog (n = 12)	39	Rb^{86}(pentobarbital)
small bowel	2.09 ± 0.17 (SE)	neonatal sheep* (n = 13)	37	µs
small bowel	0.81 ± 0.06 (SE)	sheep* (n = 7)	37	µs
small bowel	2.81 ± 0.46 (SE)	rat* (n = 8)	31	µs
small bowel	1.73 ± 0.18 (SE)	rat (n = 7)	31	µs (pentobarbital)
small bowel	1.67 ± 0.13 (SE)	rat (n = 7)	31	µs (chlor.-urethan)
small bowel	1.52 ± 0.20 (SE)	rat (n = 7)	31	µs (decerebrate)
cecum	1.89 ± 0.19 (SE)	rat* (n = 8)	31	µs
cecum	1.52 ± 0.08 (SE)	rat (n = 7)	31	µs (pentobarbital)
cecum	1.51 ± 0.15 (SE)	rat (n = 7)	31	µs (chlor.-urethan)
cecum	1.55 ± 0.22 (SE)	rat (n = 7)	31	µs (decerebrate)
colon	1.24 ± 0.57 (SD)	dog (n = 12)	39	Rb^{86}(pentobarbital)
colon	0.711 ± 0.065 (SE)	dog (n = 11)	36	µs (pentobarbital)
large bowel	0.92 ± 0.06 (SE)	neonatal sheep* (n = 13)	37	µs
large bowel	0.55 ± 0.09 (SE)	sheep* (n = 7)	37	µs
large bowel	1.33 ± 0.29 (SE)	rat* (n = 8)	31	µs
large bowel	0.66 ± 0.09 (SE)	rat (n = 7)	31	µs (pentobarbital)
large bowel	0.59 ± 0.07 (SE)	rat (n = 7)	31	µs (chlor.-urethan)
large bowel	0.70 ± 0.13 (SE)	rat (n = 7)	31	µs (decerebrate)
Pancreas	0.008-0.016	dog	1	
Pancreas	0.545 ± 0.070 (SE)	dog (n = 11)	36	µs (pentobarbital)
Pancreas	0.652 ± 0.26 (SD)	dog (n = 12)	39	Rb^{86}(pentobarbital)
Liver				
hep art. + portal	2.65 ± 0.530	pig (n = 4)	25	
hep art. + portal	2.53 ± 0.184	pig (n = 6)	25	
hep art. + portal	3.17 ± 0.228	pig (n = 6)	25	
hep art. + portal	1.40 ± 0.15 (SE)	rat* (n = 8)	31	µs
hep art. + portal	1.19 ± 0.07 (SE)	rat (n = 7)	31	µs (pentobarbital)
hep art. + portal	1.22 ± 0.09 (SE)	rat (n = 7)	31	µs (chlor-urethan)
hep art. + portal	1.25 ± 0.16 (SE)	rat (n = 7)	31	µs (decerebrate)
via portal vein	1.25	rat	25	
via portal vein	1.19 ± 0.12 (SE)	rat* (n = 8)	31	µs
via portal vein	0.98 ± 0.05 (SE)	rat (n = 7)	31	µs (pentobarbital)
via portal vein	0.98 ± 0.10 (SE)	rat (n = 7)	31	µs (chlor.-urethan)
via portal vein	0.93 ± 0.12 (SE)	rat (n = 7)	31	µs (decerebrate)
via hepatic art.	0.24 ± 0.04	pig	3	
via hepatic art.	0.22 ± 0.04	pig*	19	
via hepatic art.	0.15 ± 0.03	rat (n = 9)	27	
via hepatic art.	0.22 ± 0.05 (SE)	rat* (n = 8)	31	µs
via hepatic art.	0.21 ± 0.05 (SE)	rat (n = 7)	31	µs (pentobarbital)
via hepatic art.	0.23 ± 0.02 (SE)	rat (n = 7)	31	µs (chlor.-urethan)
via hepatic art.	0.31 ± 0.07 (SE)	rat (n = 7)	31	µs (decerebrate)
via hepatic art.	0.10 ± 0.05 (SE)	fetal sheep* (n = 12)	32	µs
via hepatic art.	0.10 ± 0.03 (SE)	neonatal sheep* (n = 13)	37	µs
via hepatic art.	0.14 ± 0.05 (SE)	sheep* (n = 7)	37	µs
via hepatic art.	0.02 ± 0.02 (SD)	rat* (12 mo old)(n = 5)	33	µs
via hepatic art.	0.07 ± 0.04 (SD)	rat (n = 10)	33	µs (pentobarbital)
via hepatic art.	0.03 ± 0.01 (SD)	rat* (24 mo old)(n = 5)	33	µs
via hepatic art.	0.11 ± 0.07 (SD)	rat (n = 7)	33	µs (decerebrate)
via hepatic art.	0.178 ± 0.050 (SE)	dog (n = 11)	36	µs (pentobarbital)
via hepatic art.	0.31 ± 0.30 (SD)	rabbit (n = 5)	42	µs

Organ	Blood Flow (ml/min/g)	Species (* = Unanesthetized)	Appendix Ref.	Technique
Spleen	1.46 ± 0.14	dog	3	
Spleen	1.381 ± 0.199 (SE)	dog (n = 11)	36	μs (pentobarbital)
Spleen	3.60 ± 0.16	pig*	19	
Spleen	1.64 ± 0.14 (SE)	rat* (n = 8)	31	μs
Spleen	1.52 ± 0.22 (SE)	rat (n = 7)	31	μs (pentobarbital)
Spleen	2.35 ± 0.49 (SE)	rat (n = 7)	31	μs (chlor.-urethan)
Spleen	2.01 ± 0.22 (SE)	rat (n = 7)	31	μs (decerebrate)
Spleen	5.43 ± 1.13 (SE)	fetal sheep* (n = 12)	32	μs
Spleen	4.03 ± 0.73 (SE)	neonatal sheep* (n = 13)	37	μs
Spleen	1.88 ± 0.27 (SE)	sheep* (n = 7)	37	μs
Spleen	1.51 ± 0.54 (SD)	rat* (12 mo old)(n = 5)	33	μs
Spleen	0.75 ± 0.37 (SD)	rat (n = 10)	33	μs (pentobarbital)
Spleen	1.13 ± 0.38 (SD)	rat* (24 mo old)(n = 5)	33	μs
Spleen	0.40 ± 0.15 (SD)	rat (n = 7)	33	μs (decerebrate)
Prostate	0.17 (.11-.21)	rat (n = 14)	26	
Prostate	0.18 ± 0.02	rat (n = 9)	27	
Prostate	0.31-0.79	dog (n = 11)	38	(pentobarb, halothane, or Nembutal™)
Bone				
femur	0.096 ± 0.039	rat	9	
tibia	0.083 ± 0.032	rat	9	
femur	0.14 ± 0.01	pig*	19	
Skin	0.20 ± 0.02 (SE)	neonatal sheep* (n = 13)	37	μs
Skin	0.09 ± 0.01 (SE)	sheep* (n = 7)	37	μs
Skin	0.18 ± 0.03 (SE)	rat* (n = 8)	31	μs
Skin	0.12 ± 0.03 (SE)	rat (n = 7)	31	μs (pentobarbital)
Skin	0.15 ± 0.02 (SE)	rat (n = 7)	31	μs (chlor.-urethan)
Skin	0.17 ± 0.02 (SE)	rat (n = 7)	31	μs (decerebrate)
Skin				
in the cold	approx. 0.02	human forearm*	10	
thermoneutral	0.04-0.05	human forearm*	10	
hyperthermic	near or >0.20	human forearm*	10	
thermoneutral	0.9 ± 0.09	pig*	19	
Subcutaneous tissue	0.48 ± 0.15	rat (n = 29)	41	^{133}Xe (ether)
Tumor	0.25 ± 0.11	rat (n = 12)	41	^{133}Xe (ether)
	1.79 ± 0.70	rabbit (n = 5)	42	μs
Fat	0.21 ± 0.06	pig*	19	
Testes	0.45 ± 0.06	rat (n = 9)	27	
Lungs	1.14 ± 0.14 (SE)	fetal sheep* (n = 12)	32	μs

1. Rappaport, A.M., Kawamura, T., Davidson, J.K., Lin, B.J., Ohira, S., Zeigler, M., Coddling, J.A., Henderson, M.J., and Haist, R.E., Effects of hormones and blood flow on insulin output of isolated pancreas in situ, *Am. J. Physiol.*, 221, 343, 1971.
2. Marcus, M.L., Bischof, C.J., and Heistad, D.D., Comparison of microsphere and xenon-133 clearance method in measuring skeletal muscle and cerebral blood flow, *Circ. Res.*, 48, 748-761, 1981.
3. Fan, F-C., Schuessler, G.B., Chen, R.Y.Z., and Chien, S., Determinations of blood flow and shunting of 9- and 15-μm spheres in regional beds. *Am. J. Physiol.*, 237, H25-H33, 1979.
4. Mraovitch, S., Iadecola, C., and Reis, D. J., Vasoconstriction unassociated with metabolism in cerebral cortex elicited by electrical stimulation of the parabrachial nucleus in rat, *J. Cereb. Blood Flow Metab.*, 3 (Suppl. 1), S196-S197, 1983.
5. Bhattacharya, J. and Beilin, L.J., Left ventricular cannulation for microsphere estimation of rabbit renal blood flow, *Am. J. Physiol.*, 238, H736-H739, 1980.
6. Thurau, K. and Levine, D.Z., The renal circulation, in *The Kidney: Morphology, Biochemistry, Physiology,* Rouiller, C. and Muller, A.F., Eds., Academic Press, New York, 1971, 1-70.
7. Passmore, J.C., Neiberger, R.E., and Eden, S.W., Measurement of intrarenal anatomic distribution of krypton-85 in endotoxic shock in dogs, *Am. J. Physiol.*, 232, H54-58, 1977.
8. Gyrd-Hansen, N., Renal clearances in pigs, *Acta Vet. Scand.*, 9, 183-198, 1968.

9. Schoutens, A., Bergmann, P., and Verhas, M., Bone blood flow measured by ^{85}Sr microspheres and bone seeker clearances in the rat, *Am. J. Physiol.*, 236, H1-H6, 1979.
10. Johnson, J.M., Brengelmann, G.L., Hales, J.R.S., Vanhoutte, P.M., and Wenger, C.B., Regulation of the cutaneous circulation, *Fed. Proc.*, 45, 2841-2850, 1986.
11. Neiberger, R.E. and Passmore, J.C., Effects of dopamine on canine intrarenal blood flow distribution during hemorrhage, *Kidney Int.*, 15, 219-226, 1979.
12. Youichi, A., Okahara, T., Kishimoto, T., Yamamoto, K., and Ueda, J., Relationship between intrarenal distribution of blood flow and renin secretion, *Am. J. Physiol.*, 225, 319-323, 1973.
13. Mimran, A. and Casellas, D., Microsphere size and determination of intrarenal blood flow distribution in the rat, *Pflugers Arch.*, 382, 233-240, 1979.
14. Sekins, K.M., Dundore, D., Emery, A.F., Lehmann, J.F., McGrath, P.W., and Nelp, W.B., Muscle blood flow changes in response to 915 MHz diathermy with surface cooling as measured by Xe^{133} clearance, *Arch Phys. Med. Rehabil.*, 61, 105-113, 1980.
15. Lassen, N.A., Lindbjerg, J., and Munck, O., Measurement of bloodflow through skeletal muscle by intramuscular injection of xenon-133, *Lancet*, 1, 686-689, 1964.
16. Keele, C.A. and Neil, E., Eds., *Samson Wright's Applied Physiology*, 12th ed., Oxford Press, London, 1971, 62.
17. Ott, C.E. and Vari, R.C., Renal autoregulation of blood flow and filtration rate in the rabbit, *Am. J. Physiol.*, 237, F479-F482, 1979.
18. Haws, C.W. and Heistad, D.D., Effects of nimodipine on cerebral vasoconstrictor responses, *Am. J. Physiol.*, 247, H170-H176, 1984.
19. Tranquilli, W.J., Parks, C.M., Thurmon, J.C., Benson, G.J., Koritz, G.D., Manohar, M., and Theodorakis, M.C., Organ blood flow and distribution of cardiac output in nonanesthetized swine, *Am. J. Vet. Res.*, 43, 895-897, 1982.
20. Hernandez, E.J., Hoffman, J.K., Fabian, M., Siegel, J.H., and Eberhart, R.C., Thermal quantification of regional myocardial perfusion and heat generation, *Am. J. Physiol.*, 236, H345-H355, 1979.
21. Drake, A.J. and Noble, M.I.M., Myocardial blood flow measured by carbonized microspheres before and after cardiac denervation, *Bibl. Anat.*, 15, 53-56, 1976.
22. Baer, R.W., Payne, B.D., Verrier, E.D., Vlahakes, G.J., Molodowitch, D., Uhlig, P.N., and Hoffman, J.I.E., Increased number of myocardial blood flow measurements with radionuclide-labeled microspheres, *Am. J. Physiol.*, 246, H418-434, 1984.
23. Crystal, G.J., Boatwright, R.B., Downey, H.F., and Bashour, F.A., Shunting of microspheres across the canine coronary circulation, *Am. J. Physiol.*, 236, H7-H12, 1979.
24. Capurro, N.L., Goldstein, R.E., Aamodt, R., Smith, H.J., and Epstein, S.E., Loss of microspheres from ischemic canine cardiac tissue, *Circ. Res.*, 44, 223-227, 1979.
25. Lovell, R.A., The Toxicity of Microcystin-LR in Swine and Mice, Ph.D. thesis, University of Illinois, 1989.
26. Damber, J.-E., Bergh, A., Daehlin, L., Petrow, V., and Landstrom, M., Effects of 6-methylene progesterone on growth, morphology, and blood flow of the Dunning R3327 prostatic adenocarcinoma, *The Prostate*, 20, 187-197, 1992.
27. Jonsson, O., Widmark, A., Grankvist, K., Damber, J.-E., and Henriksson, R., Effects of clonidine-induced hypertension on blood flows in prostatic adenocarcinoma (Dunning R3327) and normal tissue, *The Prostate*, 20, 225-232, 1992.
28. Gores, G.J., Kost, L.J., and LaRusso, N.F., The isolated perfused rat liver: conceptual and practical considerations, *Hepatology*, 6, 511-517, 1986.
29. Rundquist, I., Smith, Q.R., Michel, M.E., Ask, P., Öberg, P.A., and Rapoport, S.I., Sciatic nerve blood flow measured by laser Doppler flowmetry and [^{14}C] iodoantipyrine, *Am. J. Physiol.*, 248, H311-H317, 1985.
30. Proctor, K.G. and Busija, D.W., Relationships among arteriolar, regional, and whole organ blood flow in cremaster muscle, *Am. J. Physiol.*, 249, H34-H41, 1985.
31. Seyde, W.C., McGowan, L., Lund, N., Duling, B., and Longnecker, D.E., Effects of anesthetics on regional hemodynamics in normovolemic and hemorrhaged rats, *Am. J. Physiol.*, 249, H164-H173, 1985.
32. Crandell, S.S., Fisher, D.J., and Morriss, F.H., Jr., Effects of ovine maternal hyperglycemia on fetal regional blood flows and metabolism, *Am. J. Physiol.*, 249, E454-E460, 1985.
33. Tuma, R.F., Irion, G.L., Vasthare, U.S., and Heinel, L.A., Age-related changes in regional blood flow in the rat, *Am. J. Physiol.*, 249, H485-H491, 1985.
34. Baumbach, G.I. and Heistad, D.D., Heterogeneity of brain blood flow and permeability during acute hypertension, *Am. J. Physiol.*, 249, H629-H637, 1985.
35. Kvietys, P.R., Shepherd, A.P., and Granger, D.N., Laser-Doppler, H_2 clearance, and microsphere estimates of mucosal blood flow, *Am. J. Physiol.*, 249, G221-G227, 1985.
36. Kapin, M.A. and Ferguson, J.L., Hemodynamic and regional circulatory alterations in dog during anaphylactic challenge, *Am. J. Physiol.*, 249, H430-H437, 1985.
37. Koehler, R.C., Traystman, R.J., and Jones, M.D., Jr., Regional blood flow and O_2 transport during hypoxic and CO hypoxia in neonatal and adult sheep, *Am. J. Physiol.*, 248, H118-H124, 1985.

38. Andersson, L., Dahn, I., Nelson, K.E., and Norgren, A., Method for measuring prostate blood flow with Xenon^{133} in the dog, *Invest. Urol.*, 5, 140-148, 1967.
39. Goodhead, B., Acute pancreatitis and pancreatic blood flow, *Surg. Gynecol. Obstet.*, 129, 331-340, 1969.
40. Reddy, V.K., Zamora, C.S., Frandle, K.A., and Samson, M.D., Regional renal blood flow in ewes, *Am. J. Vet. Res.*, 42, 1802-1803, 1981.
41. Mattsson, J., Alpsten, M., Appelgren, L., and Peterson, H.-I., Influence of Noradrenalin on local tumor blood flow, *Eur. J. Cancer,* 16, 99-102, 1980.
42. Burton, M.A., Kelleher, D.K., Gray, B.N., and Morgan, C.K., Effect of temperature on liver tumour blood flow, *Eur. J. Cancer,* 26, 999, 1990.

4.5 Thermal Insulation

David W. Yarbrough and Jeff Nowobilski

Introduction

In Fourier's law of steady-state heat conduction in one dimension, Q = –k A (dT/dx), k is an assumed constant material property which relates the flow of heat to the temperature gradient in the material. This constant is known as the material thermal conductivity. In many cases the thermal conductivity can be treated as a constant, but for more general cases it can be a function of the temperature and the direction of the heat flow. Thermal insulation design is an area of engineering that deals with the minimization or control of the temperature or heat flux from a body through proper use and design of the material thermal conductivity or insulation surrounding the body.

Thermal insulation is used either to minimize the energy loss from a body or area such as a building, or to control the surface temperature of a body such as the outer shell of a furnace, for safety reasons. It can also be used to control an object's rate of temperature change such as a biological material, while being frozen or thawed.

Thermal insulation is important for all temperature ranges, from near absolute zero in cryogenic applications up to the highest melting point materials which can be used as insulation, ~3000 K. Thermal insulation beyond the highest melting point is no longer a material consideration, since solid materials can no longer exist in that condition and still function as insulators, but rather become purely a heat transfer problem in radiation and convection. This is the case in gas plasmas where temperatures exceed 10,000 K.

The thermal conductivity of materials that could conceivably be used as insulation is given as a function of temperature in Figure 4.5.1. At very low temperatures the conductivity will approach zero for all materials. At a few degrees above absolute zero the conductivity of some pure materials such as copper and aluminum has a large peak in thermal conductivity of several thousand W/m-K and then quickly decreases. As the temperature is increased, the conductivity of metals increases up about 50 W/m-K. The thermal conductivity of materials can range from 1E- W/m-K to 2000 W/m-K. This

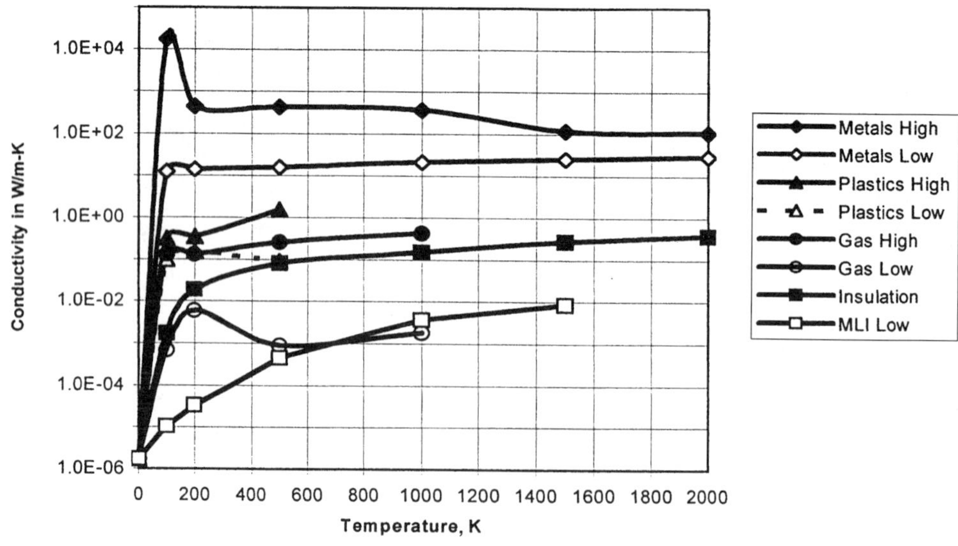

FIGURE 4.5.1 Material thermal conductivity ranges.

wide range of thermal conductivity provides the means to control the temperature or heat flow of objects or areas.

Heat Transfer in Thermal Insulation

Efficient use of thermal insulation requires an understanding of the heat transfer that occurs in insulation material. The heat transfer can be considered to occur by three assumed independent mechanisms: conduction through the solid material or gas in the insulation, gas convection, and radiation. Each will be discussed separately and then combined to give an overall apparent thermal conductivity correlation. Additional discussion can be found in References 1 through 3.

Solid Conduction

Solid conduction occurs as heat is transferred through the solid portions of the insulation material, from the hot to the cold boundaries. It is characterized by the thermal conductivity of the continuous solid material of the insulation. In a fibrous insulation this is the conductivity of the material which forms the fibers, the type of contact between the fibers, and the number of fibers per unit volume, i.e., the insulation density. The solid conductivity of the insulation can be modeled by the following formula:

$$k_s = A\ k_m \tag{4.5.1}$$

where k_s = the solid conductivity of the insulation material
k_m = the solid conductivity of the solid phase material
A = empirical constant which can be related to the load on the insulation and the nature of the contacts or structure of the solid phase and temperature.

The reader is referred to References 4 through 6 for models that describe this interaction.

Gas Conduction and Convection

Gas convection is normally suppressed in most practical industrial insulation materials because of the insulation density in a fibrous or open-cell insulation, the closed-cell properties of some solid insulation, or the gas being at a low density in evacuated insulation. In some insulation systems, such as double- or triple-pane windows, or low-density open-cell or fibrous insulation, convection can be of importance and will be included in the manufacturers' thermal conductivity data sheets. Gas conduction is of importance in most insulations and will be briefly discussed.

Gas conduction at higher-pressure ranges will follow a similar equation as solid conduction.

$$k_g = B\ k'_g \tag{4.5.2}$$

where k_g = the gas conduction component of the insulation
k'_g = the gas thermal conductivity at the mean temperature
B = an empirical constant

This shows the insulation conductivity can be reduced by replacing the air in the insulation with a lower-conductivity heavy gas such as CO_2, argon, or a fluorocarbon. The conductivity of the insulation can be increased if the air in the insulation is replaced by a higher-conductivity gas such as helium or hydrogen. The conductivity ratio of krypton to air is 0.21 while helium to air is 5.7 at STP (273 K and 1 atm pressure).

Trapping a fuel such as hydrogen or an oxidizer such as oxygen in the insulation can be very dangerous and precautions must be taken to purge or vent the insulation space as a safety measure. In a cryogenic system there is also the possibility of condensing liquid air in the insulation. The liquid air is enriched in oxygen and will be hazardous if the insulation is combustible.

The previous correlation (Equation 4.5.2) is valid as long as the mean free path of the gas is much shorter than the mean spacing between solid particles in fibrous or powder insulation or the pore size in a cellular insulation. In such cases the gas molecules collide with each other more often than with the insulation surfaces and the gas behaves as if it were at the mean temperature of the insulation. As the gas pressure is reduced, the mean free path of the gas increases and the effective gas conductivity decreases, and at low pressure it is proportional to the gas pressure, namely;

$$k_g = C\, k'_g\, L\, P \tag{4.5.3}$$

where C = a constant which is dependent on the gas at a reference pressure and temperature
L = average distance a gas molecule moves before striking a surface in the porous insulation; this is smaller than the mean free path of the gas.
P = the absolute gas pressure
k'_g = the gas conductivity at the mean insulation temperature and the gas reference pressure, usually taken as 1 atm

As the gas pressure is lowered, the gas conductivity at some point becomes negligible compared to the radiation and solid conduction component of the insulation. If the insulation interparticle spacing is made smaller than the mean free path of the gas at atmospheric pressure, then the effective gas conductivity in the insulation will be less than the conductivity of the gas at the mean temperature and pressure of the insulation. This is the mechanism by which low-density aerogel insulation can attain conductivity less than air at ambient pressure and the mean insulation temperature. More detailed discussions can be found in References 7 to 9.

Radiation Heat Transfer

The basics of thermal radiation are discussed in Chapter 3 along with the radiation mechanism through insulation containing multiple layers of low emittance radiation barriers. In powder, cellular or fibrous insulation systems the radiation resistance is provided by reflection, scattering, and absorption.[10,11] The details depend on the optical properties of the material, the size of the particle, and the wavelength of the radiation, which is related to the temperature of the material. In some circumstances the radiation conductivity for a particulate insulation can be expressed as

$$k_r = \sigma\, T^3/D \tag{4.5.4}$$

where T = absolute temperature in K
σ = Stefan-Boltzmann constant $5.67{*}10^{-8}$ W/m^2-K^4
D = insulation extinction coefficient per unit thickness

Multilayer insulations consist of multiple low-emissivity metal foils separated by a low-solid conductivity spacer. The radiation conductivity for multilayer insulation can be expressed as

$$k_r = \sigma\, \varepsilon\, T^3/N \tag{4.5.5}$$

where ε = the emittance of the foil radiation barrier, and N = the number of radiation barriers per unit thickness.

This shows that in multifoil systems, the radiation heat transfer can be reduced by increasing the number of the radiation layers. The emissivity for metals at ambient temperature that are used as radiation barriers range from aluminum 0.02, gold 0.01, and molybdenum 0.1.

Combined Thermal Conductivity

The combined point thermal conductivity of an insulation can be expressed by:

$$k = k_s + k_g + k_r \tag{4.5.6}$$

or

$$k = A\ k_m + B\ k'_g + \sigma\ T^3/D \tag{4.5.7}$$

The average apparent conductivity of the insulation between two temperatures becomes

$$k_a = A\ k_m + B\ k'_g + \sigma\left(T_2^4 - T_1^4\right)/\left(4\ D*\left(T_2 - T_1\right)\right) \tag{4.5.8}$$

Application of this equation to specific types of insulation systems is discussed in the following sections.

Most manufactures report thermal conductivity as the average between two temperatures — typically ambient and a higher or lower temperature. The relation between the point thermal conductivity and the average is

$$k_a = \frac{1}{T_2 - T_1}\int_{T_1}^{T_2} k(t)dt \tag{4.5.9}$$

Insulation Systems (Nonvacuum and Vacuum)

Powder Systems

Powder insulation consists of particulate materials such as perlite, vermiculite, beaded Styrofoam, microcell, and some types of aerogels. The advantage of this type of insulation is that the particulate material can generally be poured or blown into the space which is to be insulated and will flow to some extent around equipment and piping. The material can also be easily removed by gravity flow, fluidized transfer, or evacuation. This is an advantage in complex process equipment that has to be accessible for repair or maintenance. The disadvantage of powder insulation is that a container has to be placed around the item that is to be insulated in order to keep the powder insulation in place and in a subambient temperature system to eliminate water vapor migration into the powder and condensing or freezing. If the latter occurs, the thermal conductivity of the insulation will be degraded.

The thermal performance of the insulation is limited by the conduction of the gas filling the pores of the powder. The solid conductivity component is reduced by the multiple point contacts between particles and by selecting a material with low intrinsic solid conductivity. In some powders the particles are not solid but are porous, such as perlite or vermiculite, or are agglomerates of smaller particles, such as in fumed silica. Radiation in powder insulation can be reduced through the selection of the proper particle size or the addition of opacifiers such as carbon or reflective metal flakes of aluminum or copper.[12,13]

Some of the most common powder insulation material properties and thermal conductivies between various temperature ranges are given in Table 4.5.1. The thermal conductivity curves of perlite, vermiculite, and aerogel at various densities are shown in Figure 4.5.2. As shown in Figure 4.5.2, the conductivities of gas-filled perlite and vermiculite powders at the same temperature do not vary greatly from one another since both are dominated by the gas filling the insulation pores. The aerogel powder insulation has a pore size approaching the mean free path of the gas, resulting in greatly reduced thermal conductivity.[14,15]

To reduce the conductivity, the gas can be removed from the insulation. This will reduce the apparent conductivity of the insulation by an order of magnitude or more. The level of vacuum which has to be achieved to reduce the gas conductivity to a negligible value is of practical importance. Therefore, the thermal conductivity of various powder insulations are presented as a function of the gas pressure. The evacuated thermal conductivity curves for perlite, aerogel, and microcel with 10% aluminum flakes are shown in Figure 4.5.3. The differences in the powder conductivity and the effect of particle size can be seen in these figures.

TABLE 4.5.1 Powder Insulation

Materials	Maximum Operating Temperature (K)	Density (kg/m³)	Hot Side Temp (K)	Cold Side Temp (K)	Apparent Thermal Conductivity (W/m-K)
Perlite	1255	112	327	294	0.0418
Vermiculite	1033	144	327	294	0.0677
Silica aerogel	977	80	327	294	0.0303
Silica aerogel with 15 wt% carbon	977	116	327	294	0.0135
Cork	366	208	327	294	0.049
Diatomaceous silica	1144	240	327	294	0.0577

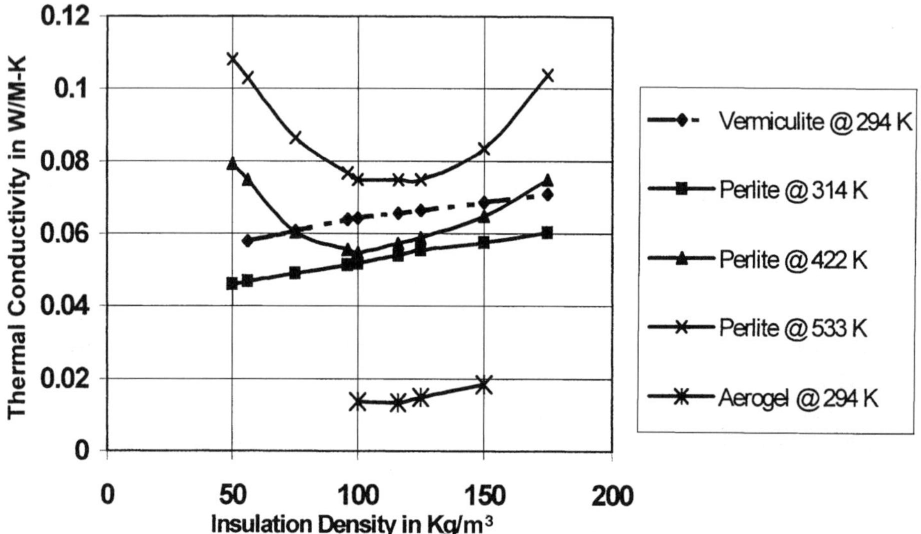

FIGURE 4.5.2 Powder insulation thermal conductivity.

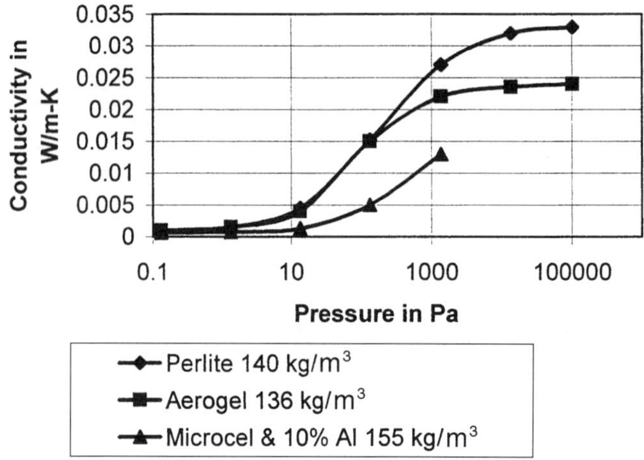

FIGURE 4.5.3 Evacuated powder thermal conductivity from 300 to 77 K.

Handling considerations of powder insulation deserve special mention due to the small particle size of many of the insulations, which can result in airborne dust hazards while loading the insulation powder. If the insulation powder is combustible, the airborne dust can pose an explosion hazard.

Nonvacuum perlite powder insulation is used in cryogenic process cold boxes. Evacuated perlite powder insulation is used in mobile carriers for cryogenic liquid oxygen and nitrogen. Evacuated carbon powder insulation is used in some metal thermos bottles. Powders opacified with metal flakes to reduce radiation heat transfer were used in early small portable liquid oxygen and nitrogen containers, but were later replaced with higher-performing evacuated multifoil insulation. The effect of the addition of opacifiers to perlite, Cab-o-sil, and aerogel, is shown in Figure 4.5.4.

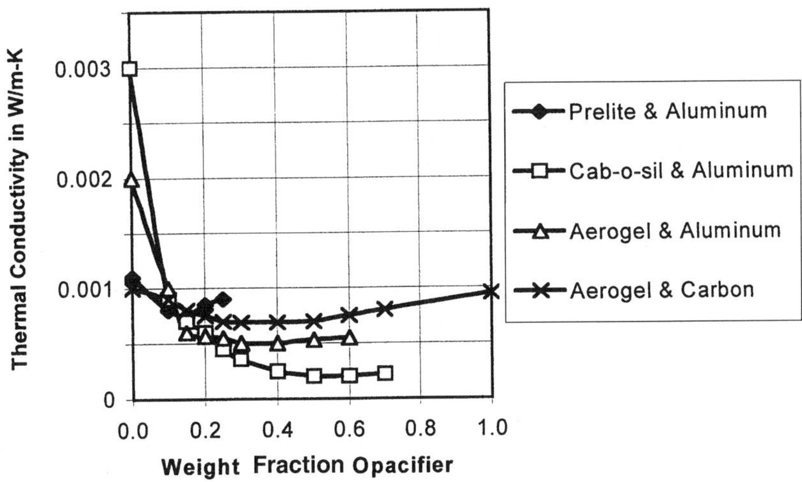

FIGURE 4.5.4 Evacuated opacified powder insulation thermal conductivity.

Evacuated Multifoil Insulation

Evacuated multifoil insulation has the lowest thermal conductivity of any insulation. This is achieved by removing the gas from the insulation space, using very low solid-conductivity lightweight fiberglass or plastic spacers to separate the radiation shields, and reducing the radiation heat transfer with many low emittance metal foils or metalized Mylar™ films. This type of insulation has been used successfully from liquid helium temperatures to 1400 K with the correct selection of materials.[16,19]

The selection of material is critical for the insulation to achieve a low thermal conductivity and to maintain its performance over the life of the system. Problems that can occur are out-gassing or off-gassing of the materials, which can degrade the vacuum or condense on the cooler radiation barrier surfaces and increase the foil emittance. Reactions between the radiation barriers and the spacer material can also increase the solid conduction and reduce the foil emittance. In high-temperature systems the insulation material can be altered as the temperature decreases through the insulation to achieve a lower overall thermal conductivity. Multifoil insulation systems, which have been used successfully in the past, are listed in Table 4.5.2 along with the operating temperature range.

Commercial Fibrous and Solid Insulation Systems for Moderate and High Temperatures

Introduction

This subsection will deal with systems operating at temperatures above 300 K. Processes operating at moderate or high temperatures are insulated to conserve energy, reduce risk of injury, and control temperature. A variety of products are available for a given temperature range and the thermal properties of a given type of insulation at a given temperature can vary significantly among vendors. It is important,

TABLE 4.5.2 Vacuum Multifoil Insulation

Materials		Maximum Operating Density (K)	Layer Density (layers/cm)	Density (kg/m³)	Hot Side Temp (K)	Cold Side Temp (K)	Apparent Thermal Conductivity (W/m-K)
Radiation Barrier	**Spacer**						
Aluminum foil	Glass paper	750	24	87	294	77	0.000036
Aluminum foil	Glass mat	750	4	19	294	77	0.00017
Aluminized Mylar™	None	370	55	50	294	77	0.000042
Copper foil	Quartz paper	1000	31	339	1000	294	0.00095
Nickel foil	Quartz cloth	1200	42	384	1200	294	0.0021
Molybdenum foil	Zirconia cloth	1600	50	459	1500	294	0.0026

Note: Vacuum level = 0.1 μm.

therefore, to verify the temperature limits and the measured thermal conductivity for the specific product under consideration.

Health and Safety

OSHA guidelines limit the operating temperatures of accessible surfaces. In some regions state or local codes may also apply. Safety considerations for systems operating above ambient temperature are often taken to be more crucial than energy savings. The skin damage resulting from hot surface contact depends on a combination of factors such as surface temperature, contact time, and properties of the surface material such as thermal conductivity, specific heat, and density.[20,21] A suggested design limit for personnel protection is to limit accessible surface temperatures to about 52°C for highly conductive (metal) surfaces and 65°C for nonconductive surfaces.[22]

In some cases, insulating materials contain binders or additives which may volatilize near the upper temperature limit of their range of operation. This upper limit temperature may be approached for the insulation immediately adjacent to a hot surface. The out-gassing is usually limited to the first few hours of heating but must be taken into consideration, especially for insulations used in enclosed areas. The manufacturer's literature should always be examined for warnings about initial off-gassing of products.

Heat Loss Calculations

Process piping and flat furnace insulations are the most common types of insulation to be considered. Virtually all exposed surfaces operating at temperatures above about 60°C are candidates for thermal insulation. Economic and safety considerations become more important as the temperature increases.

The calculated heat loss from high-temperature surfaces requires consideration of the entire system, inside material to outside air. In many process flow situations, steam lines, for example, the inside heat transfer coefficient is large so the thermal resistance attributable to the inside film coefficient can be neglected. Similarly, the thermal resistance of metal walls and pipes are generally neglected in comparison with the thermal resistance of even a minimal thickness of insulation. The thermal conductivity or thermal resistance of insulation materials to be applied to an outside surface, the external heat transfer coefficient for conduction-convection and the loss to the surroundings due to radiation are of primary importance.

Discussions of the calculation of heat loss from pipes or flat surfaces at temperatures above ambient are discussed elsewhere in this handbook and in texts on heat transfer. A "standard practice" for calculating heat gain or heat loss complete with supporting computer code has been developed by the American Society for Testing and Materials (ASTM).[23] *The Standard Practice* published by the ASTM forms the basis for a program entitled "3E Plus, Insulation Thickness Computer Program" and is obtainable from the North American Insulation Manufacturers Association.[24]

The temperature change across moderate- or high-temperature insulation can be very large. Since the thermal conductivity of insulations used in such systems increases with temperature, it is necessary to use either mean values for the insulation properties over the range of temperatures to be encountered or use temperature-dependent expressions in Fourier's law for calculating steady-state heat loss from hot systems.

Heat loss calculations require reliable thermal properties for the insulations that are used, exterior surface emittances, and knowledge of the exterior conditions such as air temperature and wind speed. The following discussion contains selected thermal data for a number of materials. Products that are selected for use should be validated by thermal test data obtained in accordance with ASTM C 177,[25] ASTM C 335,[26] ASTM C 518,[27] or C 1033.[28]

The heat transfer through thermal insulations used to reduce heat loss are often described by expressions like Equation 4.5.10:

$$k_a(T) = a + bT + cT^3, \tag{4.5.10}$$

where the T^3 term is associated with radiative transport, and the linear part of the equation is related to conduction. The radiative term should be written in terms of absolute temperature and the coefficients a, b, and c are positive. Equations 4.5.11 to 4.5.13 are also used to correlate thermal conductivity data with temperature.

$$k_a(T) = a + bT + cT^2 \tag{4.5.11}$$

$$k_a(T) = \exp(a + bT) \tag{4.5.12}$$

$$k_a(T) = a_i + b_iT \qquad i = 1, 2, 3 \tag{4.5.13}$$

Care must be exercised in interpreting laboratory k_a data, especially when large temperature differences are anticipated.[29] Measured k_a are commonly reported as the average of the cold and hot side temperatures. This value is then taken to be a point value for the apparent thermal conductivity. The measurement, however, was actually for the mean value of k_a on the temperature interval of the test. In the case of large measurement intervals, data are best described by an integral method.[30]

Thermal Insulations

A variety of thermal insulating materials are available for use on surfaces operating above 300 K. The maximum allowable temperature for an insulation type is the first consideration in selecting an insulation for a particular application. Table 4.5.3 contains a list of commonly used insulations, an indication of the maximum application temperature, and a reference to a standard specification. The table contains materials used to insulate pipes, furnaces, and process vessels.

The standard specifications for insulations 1 through 12 in Table 4.5.3 include sufficient data to determine coefficients for Equation 4.5.10. The coefficients are listed in Table 4.5.4 along with apparent thermal conductivities at selected temperatures.

Table 4.5.5 contains k_a values for calcium silicate and perlite pipe insulation. These data are representative of commercially available insulations. Table 4.5.6 contains k_a values for several fibrous-type insulations that are representative of commercially available products. Since the data are from various manufacturers, it is not possible to interpolate density.

Insulation Application

Thermal Insulation Products and Installation

The thermal data contained in the previous paragraphs are representative of the various types of insulation that are available for pipes and flat surfaces. The technical data for a specific product should be consulted before design calculations are completed. The performance of thermal insulations of all types requires installation in accordance with manufacturer's instructions and regular maintenance to remove and replace damaged sections.

TABLE 4.5.3 Commonly Used Industrial Insulations

Number and Type of Insulation		Maximum Temperature (K)	ASTM Standard Specification[a]
1	Mineral fiber pipe (Type 1)	727	C 547
2	Mineral fiber pipe (Type 2)	923	C 547
3	Mineral fiber pipe (Type 3)	923	C 547
4	Mineral fiber board (Type 1A,1B)	505	C 612
5	Mineral fiber board (Type 2)	727	C 612
6	Mineral fiber board (Type 3)	811	C 612
7	Mineral fiber board (Type 4)	922	C 612
8	Mineral fiber board (Type 5)	1255	C 612
9	Calcium silicate (Type 1)	922	C 533
10	Calcium silicate (Type 2)	1200	C 533
11	Perlite block and pipe	922	C 610
12	Cellular glass	700	C 552
13	Polystyrene	348	C 578
14	Polyisocyanurate board	366	C 1289
15	Loose-fill perlite	1033	C 549
16	Loose-fill vermiculite	1033	C 516

[a] These standard specifications can be found in Vol. 04.06 of the *1998 Annual Book of ASTM.* Standards published by the American Society for Testing and Materials, West Conshohocken, PA.

TABLE 4.5.4 Thermal Conductivity Data Taken from ASTM Standard Specifications[a]

	Coefficients for Equation 4.5.10			k_a (W/m-K)			
Insulation[b]	a[c]	b	c	300 K	500 K	700 K	900 K
1	–0.1342 E-1	0.1405 E-3	0.1666 E-9	0.033	0.078	0.142	—
2	0.1763 E-1	0.3736 E-4	0.2579 E-9	0.036	0.069	0.132	0.239
3	0.1763 E-1	0.3736 E-4	0.2579 E-9	0.036	0.069	0.132	0.239
4 (I-A)	0.1131 E-1	0.4574 E-4	0.4704 E-9	0.038	0.093	—	—
4 (I-B)	–0.2721 E-2	0.1182 E-3	0.1711 E-9	0.038	0.078	—	—
5	0.1224 E-1	0.4380 E-4	0.4296 E-9	0.037	0.088	0.190	—
6	0.2223 E-1	0.4020 E-5	0.5123 E-9	0.037	0.088	0.201	—
7	0.9717 E-2	0.6254 E-4	0.2300 E-9	0.035	0.070	0.132	0.234
8	–0.2303 E-1	0.1888 E-3	0.7477 E-10	0.036	0.081	0.135	0.201
9	0.1910 E-1	0.1230 E-3	0.1488 E-10	0.056	0.082	0.110	0.141
10	0.4241 E-1	0.1015 E-3	–0.2550 E-10	0.072	0.090	0.105	0.115
11	0.4082 E-1	0.9171 E-4	0.9363 E-10	0.072	0.098	0.137	0.192
12	0.1793 E-1	0.7351 E-4	0.3796 E-9	0.050	0.102	0.200	—
13	18 kg/m^3			0.038	—	—	—
	26 kg/m^3			0.029	—	—	—
	48 kg/m^3			0.029	—	—	—
14	Type 1 Class 1	Foil faced		0.022	—	—	—
15	Expanded	105 kg/m^3		0.053[31]	—	—	—
16	Expanded	122 kg/m^3		0.068[31]	—	—	—

[a] Maximum k_a values
[b] Refer to Table 4.5.1
[c] E-1 means $\times 10^{-1}$

In general, the thermal resistance of porous or cellular insulation decreases if water is sorbed into the insulation. Thermal insulation is covered with mastics or metal sheathing to prevent water uptake.

Vacuum Maintenance

In order to achieve the design thermal conductivity of a vacuum-insulated system over the life of the system, the vacuum level of the insulation must be maintained at the design value. This is achieved by

TABLE 4.5.5 Thermal Conductivity Data for Calcium Silicate and Perlite Pipe Insulations

Insulation Type (density kg/m³)	Coeffecients for Equation 4.5.9			k_a (W/m-K)		
	a	b	c	300 K	500 K	700 K
Calcium Silicate						
(below 130)	–0.5770 E-2[a]	0.1356 E-3	0.6982 E-10	0.0368	0.0708	0.113
(130–190)	0.4841 E-1	–0.3478 E-4	0.2031 E-9	0.0435	0.0564	0.0937
(190–260)	0.3998 E-1	0.3480 E-4	0.1165 E-9	0.0536	0.0719	0.104
(260–400)	0.6243 E-1	–0.1467 E-4	0.1838 E-9	0.0630	0.0781	0.115
(400–560)	0.5573 E-1	0.7523 E-4	0.2692 E-10	0.0790	0.0967	0.118
Perlite						
(below 160)	0.9902 E-2	0.1543 E-3	0.1233 E-10	0.0565	0.0886	0.122
(above 160)	0.8481 E-1	–0.6834 E-4	0.2992 E-9	0.0724	0.0880	0.140

[a] E-2 means $\times 10^{-2}$

TABLE 4.5.6 Thermal Conductivity Data for Fibrous Pipe and Board Insulation

Insulation Type (density)	Coeffecients for Equation 4.5.9			k_a (W/m-K)		
	a	b	c	300 K	400 K	500 K
Mineral fiber board						
(64 kg/m³)	–0.1074 E-1[a]	0.1279 E-3	0.2621 E-9	0.0347	0.0572	0.0860
(96 kg/m³)	0.2173 E-1	0.1827 E-4	0.2796 E-9	0.0348	0.0469	0.0658
(160 kg/m³)	0.1185 E-1	0.6506 E-4	0.1375 E-9	0.0351	0.0467	0.0616
Mineral fiber pipe						
(128 kg/m³)	0.5427 E-2	0.7864 E-4	0.1472 E-9	0.0330	0.0463	0.0631
Fiberglass pipe						
(64 kg/m³)	0.7111 E-2	0.5306 E-4	0.2805 E-9	0.0306	0.0463	0.0687
(224 kg/m³)	0.4276 E-1	0.3082 E-4	0.1446 E-9	0.0559	0.0643	0.0762
Ceramic fiber pipe						
(218 kg/m³)	0.8725 E-1	-0.1180 E-3	0.3667 E-9	0.0618	0.0635	0.0741

[a] E-1 means $\times 10^{-1}$

checking the joints in the enclosure to eliminate air leaks into the vacuum space, calculating the gas permeation through the wall of the enclosure, off-gassing the insulation for a sufficient period of time before sealing the system, and through the addition of active chemical getters and adsorbents.

In cryogenic liquid oxygen, nitrogen, and argon systems, oxygen, nitrogen, and water can be adsorbed using molecular sieves placed on the cold wall of the system. Hydrogen, which will out-gas from the metal walls of the enclosure and from the metal radiation foils, will not be adsorbed in significant quantities until the cold side attains a temperature that is below liquid hydrogen. Therefore, an active metal getter has to be placed in the vacuum space. These materials combine chemically with the reactive gas to remove it from the vacuum space. SAES Getters Inc. produce commercial getters for this purpose.

In ambient and elevated temperature systems adsorption will not work. The only methods available for vacuum maintenance are chemical getters or very long evacuations combined with high-temperature out-gassing of the insulation, or the use of insulation with very small pore sizes that can tolerate high vacuum levels.

Vacuum maintenance materials and temperature levels are given in Table 4.5.7. References 32 and 33 are sources of additional information.

Acknowledgment

Insulation product data were generously provided by the Pabco division of the Fibreboard Corporation, the Rock Wool Manufacturing Co., and the North American Insulation Manufacturers Association.

TABLE 4.5.7 Vacuum Maintenance Materials

Material	Temperature Range Low	Temperature Range High	Gettered Gas	Capacity (torr-cm³/g)	Activation Temperature
Palladium oxide	273 K	600 K	H_2	158,000	NA
SAES St 707	273	973	H_2 at 298 K	1100	720 to 620 K
Zr 70%, V 24.6%, Fe 5.4%			CO at 298 K	40	
			N_2 at 298 K	9	
			at 750 K activation		
SAES St 172	273	973	H_2 at 298 K	420	673 to 1173 K
St 707 + zirconium powder			CO at 298 K	14	
			at 750 K activation		
SAES flash barium getter			H_2 at 298 K	110,000	
$BaAl_4$ + Nickel			CO at 298 K	14,000	
			N_2 at 298 K	26,000	
			O_2 at 298 K	50,000	
			at 750 K activation		

References

1. Kagan, M.G., *Thermal Insulation In Cryogenic Engineering,* IPST Press, Jerusalem, 1969.
2. Flynn, T.M., *Cryogenic Engineering,* Marcel Dekker, New York, 1997.
3. Barron, R., *Cryogenic Systems,* McGraw-Hill, New York, 1966
4. Strong, H.M., Bundy, F.P., and Bovenkerk, H.P., Flat Panel Vacuum Thermal Insulation, *J. Appl. Phys.,* 31(1), 1960.
5. Nayak, A.l. and Tien, C.l., Lattice-Vacancy Analysis for Packed-Sphere Conductance, *Prog. Astronaut. Aeronaut.,* 56, 113-125, 1977.
6. Hammond, M.B., An Analytical Model For Determining the Thermal Conductivity of Closed-Cell Foar Insulation, *Adv. Cryog. Eng.,* 15, 332-341, 1969.
7. Dushman, *Scientific Foundations of Vacuum Technique,* John Wiley & Sons, New York, 1966.
8. Kistler, S.S., The Relation Between Heat Conductivity and Structure in Silica Aerogel, *J. Phys. Chem.,* 39, 79-85, 1935.
9. Yarbrough, D.W., Literature Review: Heat Transfer Through Two-Phase Insulation Systems Consisting of Powders in a Continuous Gas Phase, Oak Ridge National Laboratory, Rep. ORNL/M-2426, December 1992, Oak Ridge, TN.
10. Larkin, and Churchill, Heat Transfer by Radiation Through Porous Insulation, *AIChE J.,* 5(4), 467-474, 1959.
11. Verschoor, J.D. and Greebler, P., Heat Transfer By Gas Conduction and Radiation In Fiberous Insulation, *Trans. ASME,* 74(6), 961-967, 1952.
12. Kropshot, R.H. and Burgess, R.W., Perlite for Cryogenic Insulation, *Adv. Cryog. Eng.,* 8, 425-436, 1963.
13. Hunter, B.J. and Kropshot, R.H., Metal Powder Additives in Evacuated Powder Insulation, *Adv. Cryog. Eng.,* 5, 146-156, 1960.
14. Smith, D.M., Aerogel-based Thermal Insulation, *J. Non-Crystl. Solids,* 225, 254-259, 1998.
15. Deissler, R.G. and Boegly, Y.S., An Investigation of Effective Thermal Conductivities of Powders in Various Gases, *Trans. ASME,* 80(7), 1417, 1958.
16. Matsh, L.C., Advances in Multilayer Insulations, *Adv. Cryog. Eng.,* 7, 413-418, 1962.
17. Kropshot, R.H., Multiple-Layer Insulations for Cryogenic Application, *Cryogenics,* 1(3), 171-177, 1961.
18. NCR-2 Sales Literature, King-Seely Thermos Co., Winchester, MA.

19. Grunert, W.E. et al., Research and Development in a Thermal Insulation Study, Final Report #ALO 3632-43 Sept. 1967 to Sept. 1970, U.S. Atomic Energy Commission, AT(29-2)-2832, Washington, D.C., 1970.
20. ASTM C 1055, Standard Guide for Heated System Surface Conditions that Produce Contact Burn Injuries, *1998 Annual Book of ASTM Standards,* Vol. 04.06, American Society for Testing and Materials, West Conshohocken, PA, 1998, 557-563.
21. ASTM C 1057, Standard Practice for Determination of Skin Contact Temperature from Heated Surfaces Using a Mathematical Model and Thermesthesiometer, *1998 Annual Book of ASTM Standards,* Vol. 04.06, American Society for Testing and Materials, West Conshohocken, PA, 1998, 564-569.
22. Anon., Industrial Insulation for Systems Operating Above Ambient Temperature, ORNL/M-4678, Oak Ridge National Laboratory, Oak Ridge, TN, 1995.
23. ASTM C 680, Standard Practice for Determination of Heat Gain or Loss and the Surface Temperatures of Insulated Pipe and Equipment Systems by the Use of a Computer Program, *1998 Annual Book of ASTM Standards,* Vol. 04.06, American Society for Testing and Materials, West Conshohocken, PA, 1998, 277-328.
24. 3E Plus, Insulation Thickness Computer Program, North American Insulation Manufacturers Association, 44 Central Center Plaza, Alexandria, VA 22314, Jan. 1994.
25. ASTM C 177, Standard Test Method for Steady-State Heat Flux Measurements and Thermal Transmission Properties by Means of the Guarded-Hot-Plate Apparatus, *1998 Annual Book of ASTM Standards,* Vol. 04.06, American Society for Testing and Materials, West Conshohocken, PA, 1998, 20-41.
26. ASTM C 335, Standard Test Method for Steady-State Heat Transfer Properties of Horizontal Pipe Insulation, *1998 Annual Book of ASTM Standards,* Vol. 04.06, American Society for Testing and Materials, West Conshohocken, PA, 1998, 82-92.
27. ASTM C 518, Standard Test Method for Steady-State Heat Flux Measurements and Thermal Transmission Properties by Means of the Heat Flow Meter Apparatus, *1998 Annual Book of ASTM Standards,* Vol. 04.06, American Society for Testing and Materials, West Conshohocken, PA, 1998, 163-174.
28. ASTM C 1033, Standard Test Method for Steady-State Heat Transfer Properties of Pipe Isulation Installed Vertically, *1998 Annual Book of ASTM Standards,* Vol. 04.06, American Society for Testing and Materials, West Conshohocken, PA, 1998, 505-512.
29. Langlais, C., Thermal Gradients Effect on Thermal Property Measurements, *J. Thermal Insulation,* 11, 189-195, 1998.
30. Hust, J. G. and Langford, A. B., Comments of the Measurements of Thermal Conductivity and Presentation of a Thermal Conductivity Integral Method, *Int. J. Thermophysics,* 3(1), 1982.
31. Rosenow, W. M., Harnett, J. P., and Cho, Y. I., *Handbook of Heat Transfer,* 3rd ed., McGraw-Hill, New York, 1998.
32. SAES Getters, St 707 Non-evaporable Getters Activateable at Low Temperatures, SAES Getters U.S., 1122 East Cheyenne Mt. Blvd., Colorado Springs, CO.
33. Boffito, C. et al., Gettering in Cryogenic Applications, *J. Vac. Sci. Technology,* 5(6), 1987.

4.6 Energy Audit for Buildings

Moncef Krarti

Abstract

This section provides an overview of an energy audit procedure that is suitable for commercial and industrial buildings. Today, an energy audit is performed by various energy service companies to improve the energy efficiency of various building systems. In particular, energy auditing has a vital role in evaluating the success of performance-contracting projects. There are several types of energy audits that are commonly performed by energy service engineers with various degrees of complexity. Here, we describe briefly the key aspects of a detailed energy audit and provides a comprehensive and systematic approach to identify and recommend cost-effective energy conservation measures for buildings.

Introduction

Since the oil embargo of 1973, significant improvements have been made in the efficiency of new buildings. However, the vast majority of the existing stock of buildings are more than a decade old and do not meet current energy-efficient construction standards (EIA, 1991). Therefore, energy retrofits of existing buildings will be required for decades to come if the overall energy efficiency of the building stock is to be meet the standards.

Investing to improve the energy efficiency of buildings provides an immediate and relatively predictable positive cash flow resulting from lower energy bills. In addition to the conventional financing options available to owners and building operators (such as loans and leases), other methods are available to finance energy retrofits for buildings. One of these methods that is becoming increasingly common is performance contracting, in which payment for a retrofit project is contingent upon its successful outcome. Typically, an energy service company (ESCO) assumes all the risks for a retrofit project by performing the engineering analysis and obtaining the initial capital to purchase and install equipment needed for energy efficiency improvements. Energy auditing is an important step used by energy service companies to ensure the success of their performance contracting projects.

Currently, several large industrial and commercial buildings have established internal energy management programs, based on energy audits, to reduce waste in energy use or to comply with the specifications of some regulations and standards. Other building owners and operators take advantage of available financial incentives typically offered by utilities or state agencies to perform energy audits and implement energy conservation measures.

In the 1970s, building energy retrofits consisted of simple measures such as shutting off lights, turning down heating temperatures, turning up air conditioning temperatures, and reducing the hot water temperatures. Today, building energy management infers complete control of the energy systems and consumption within a facility. Therefore, the energy auditor should be aware of key energy issues such as the subtleties of electrical utility rate structures and of the latest building energy efficiency technologies and their applications.

This section describes a general but systematic procedure for energy auditing that is suitable for both commercial buildings and industrial facilities. Some of the commonly recommended energy conservation measures are briefly discussed. A case study for an office building is presented to illustrate the various tasks involved in an energy audit. Finally, an overview is provided to outline the existing methods for measurement and verification of energy savings incurred by the implementation of energy conservation measures.

Types of Energy Audits

The term "energy audit" is widely used and may have different meanings depending on the energy service companies. Energy auditing of buildings can range from a short walk-through of the facility to a detailed analysis with hourly computer simulation. Generally, four types of energy audits can be distinguished as briefly described below.

Walk-Through Audit

This audit consists of a short on-site visit of the facility to identify areas where simple and inexpensive actions can provide immediate energy and/or operating cost savings. Some engineers refer to these types of actions as operating and maintenance (O&M) measures. Examples of O&M measures include lowering heating set-point temperatures, replacing broken windows, insulating exposed hot water or steam pipes, and adjusting boiler fuel-air ratio.

Utility Cost Analysis

The main purpose of this type of audit is to carefully analyze the operating costs of the facility. Typically, the utility data over several years are evaluated to identify the patterns of energy use, peak demand, weather effects, and the potential for energy savings. To perform this analysis, it is recommended that the energy auditor conducts a walk-through survey to get acquainted with the facility and its energy systems.

It is important that the energy auditor understands clearly the utility rate structure that applies to the facility for several reasons, including:

- Checking the utility charges and ensure that no mistakes were made in calculating the monthly bills. Indeed, the utility rate structures for commercial and industrial facilities can be quite complex, with ratchet charges and power factor penalties.
- Determining the most dominant charges in the utility bills. For instance, peak demand charges can be a significant portion of the utility bill, especially when ratchet rates are applied. Peak usage shaving measures can then be recommended to reduce these demand charges.
- Identifying if the facility can benefit from using other utility rate structures or purchase cheaper fuel and reduce its operating costs. This analysis can provide a significant reduction in the utility bills, especially with the electrical industry deregulation and the advent of real time pricing (RTP) rate structures.

Moreover, the energy auditor can determine whether or not the facility is prime for energy retrofit projects by analyzing the utility data. Indeed, the energy use of the facility can be normalized and compared to indices (for instance, the energy use per unit of floor area for commercial buildings — or per unit of a product for industrial facilities).

Standard Energy Audit

The standard audit provides a comprehensive analysis of the energy systems of the facility. In addition to the activities described for the walk-through audit and the utility cost analysis described above, the standard energy audit includes the development of a baseline for the energy use of the facility and the evaluation of the energy savings and the cost effectiveness of appropriately selected energy conservation measures. The step by step approach of the standard energy audit is similar to that of the detailed energy audit, which is described in the following subsection.

Typically, simplified tools are used in the standard energy audit to develop baseline energy models and to predict the energy savings of energy conservation measures. Among these tools are the degree-day methods, and linear regression models (Fels, 1986). In addition, a simple payback analysis is generally performed to determine the cost-effectiveness of energy conservation measures.

Detailed Energy Audit

This is the most comprehensive but also time-consuming energy audit type. Specifically, the detailed energy audit include the use of instruments to measure energy use for the whole building and/or for some energy systems within the building (for instance, by end uses: lighting, office equipment, fans, chiller, etc.). In addition, sophisticated computer simulation programs are typically employed for detailed energy audits to evaluate and recommend energy retrofits for the facility.

The techniques available to perform measurements for an energy audit are diverse. During an on-site visit, hand-held and clamp-on instruments can be used to determine the variance of some building parameters such as the indoor air temperature, the luminance level, and the electrical energy use. When long-term measurements are needed, sensors are typically used and are connected to a data acquisition system so measured data can be stored and be remotely accessible. Recently, nonintrusive load monitoring (NILM) techniques have been proposed (Shaw et al., 1998). The NILM technique can determine the real-time energy use of the significant electrical loads in a facility by using only a single set of sensors at the facility service entrance. The minimal effort associated with using the NILM technique compared to the traditional multimetering approach (which requires a separate set of sensors to monitor energy consumption for each end use) makes the NILM a very attractive and inexpensive load-monitoring technique for energy service companies and facility owners.

The computer simulation programs used in the detailed energy audit typically provide the energy use distribution by load type (i.e., energy use for lighting, fans, chillers, boilers, etc.). They are often based on dynamic thermal performance of the building energy systems and usually require a high level of engineering expertise and training. These simulation programs range from those based on the bin method (Knebel, 1983) to those that provide hourly building thermal and electrical loads such as DOE-2 (LBL, 1980).

In the detailed energy audit, more rigorous economical evaluation of the energy conservation measures are generally performed. Specifically, the cost-effectiveness of energy retrofits may be determined based on the life-cycle cost (LCC) analysis rather than the simple payback period analysis. LCC analysis takes into account a number of economic parameters such interest, inflation, and tax rates.

General Procedure for a Detailed Energy Audit

To perform an energy audit, several tasks are typically carried out depending on the type of the audit and the size and function of the building. Some of the tasks may have to be repeated, reduced in scope, or even eliminated based on the findings of other tasks. Therefore, the execution of an energy audit is often not a linear process and is rather iterative. However, a general procedure can be outlined for most buildings.

Step 1: Building and Utility Data Analysis

The main purpose of this step is to evaluate the characteristics of the energy systems and the patterns of energy use for the building. The building characteristics can be collected from the architectural/mechanical/electrical drawings and/or from discussions with building operators. The energy use patterns can be obtained from a compilation of utility bills over several years. Analysis of the historical variation of the utility bills allows the energy auditor to determine any seasonal and weather effects on the building energy usage. Some of the tasks that can be performed in this step are presented below, with the key goals expected from each task noted in italics:

- Collect at least 3 years of records of utility data [*to identify a historical energy use pattern*]
- Identify the fuel types used (electricity, natural gas, oil, etc.) [*to determine the fuel type that accounts for the largest energy use*]
- Determine the patterns of fuel use by fuel type [*to identify the peak demand for energy use by fuel type*]
- Understand utility rate structure (energy and demand rates) [*to evaluate if the building is penalized for peak demand and if cheaper fuel can be purchased*]
- Analyze the effect of weather on fuel consumption
- Perform utility energy use analysis by building type and size (building signature can be determined including energy use per unit area [*to compare against typical indices*]

Step 2: Walk-Through Survey

This step should identify potential energy savings measures. The results of this step are important since they determine if the building warrants any further energy auditing work. Some of the tasks involved in this step are

- Identify the customer's concerns and needs
- Check the current operating and maintenance procedures
- Determine the existing operating conditions of major energy use equipment (lighting, HVAC systems, motors, etc.)
- Estimate the occupancy, equipment, and lighting (energy use density and hours of operation)

Step 3: Baseline for Building Energy Use

The main purpose of this step is to develop a base-case model that represents the existing energy use and operating conditions for the building. This model will be used as a reference to estimate the energy savings due to appropriately selected energy conservation measures. The major tasks to be performed during this step are

- Obtain and review architectural, mechanical, electrical, and control drawings
- Inspect, test, and evaluate building equipment for efficiency, performance, and reliability
- Obtain all occupancy and operating schedules for equipment (including lighting and HVAC systems)
- Develop a baseline model for building energy use
- Calibrate the baseline model using the utility data and/or metered data

Step 4: Evaluation of Energy-Saving Measures

In this step, a list of cost-effective energy conservation measures is determined using both energy savings and economic analysis. To achieve this goal, the following tasks are recommended:

- Prepare a comprehensive list of energy conservation measures (using the information collected in the walk-through survey)
- Determine the energy savings due to the various energy conservation measures pertinent to the building by using the baseline energy use simulation model developed in Step 3
- Estimate the initial costs required to implement the energy conservation measures
- Evaluate the cost-effectiveness of each energy conservation measure using an economical analysis method (simple payback or life-cycle cost analysis)

Tables 4.6.1 and 4.6.2 provide summaries of the energy audit procedure recommended, respectively, for commercial buildings and for industrial facilities. Energy audits for thermal and electrical systems are separated since they are typically subject to different utility rates.

Common Energy Conservation Measures

In this subsection some energy conservation measures (ECMs) commonly recommended for commercial and industrial facilities are briefly discussed. It should be noted that the list of ECMs presented below does not pretend to be exhaustive nor comprehensive. It is provided merely to indicate some of the options that the energy auditor can consider when performing an energy analysis of a commercial or an industrial facility. However, it is strongly advised that the energy auditor keeps abreast of any new technologies that can improve the facility energy efficiency. Moreover, the energy auditor should recommend the ECMs only after he performs an economical analysis for each ECM.

Building Envelope

For some buildings, the envelope (i.e., walls, roofs, floors, windows, and doors) can have an important impact on the energy used to condition the facility. The energy auditor should determine the actual characteristics of the building envelope. During the survey, a sheet for the building envelope should be established to include information such as materials of construction (for instance, the level of insulation in walls, floors, and roofs) and the area and number of various assemblies of the envelope (for instance, the type and the number of panes for the windows should be noted). In addition, comments on the repair needs and recent replacement should be noted during the survey.

TABLE 4.6.1 Energy Audit Summary for Residential and Commercial Buildings

Thermal Systems	Electric Systems
Utility Analysis	
Thermal energy use profile (building signature) Thermal energy use per unit area (or per student for schools or per bed for hospitals) Thermal energy use distribution (heating, DHW, process, etc.) Fuel types used Weather effect on thermal energy use Utility rate structure	Electrical energy use profile (building signature) Electrical energy use per unit area (or per student for schools or per bed for hospitals) Electrical energy use distribution (cooling, lighting, equipment, fans, etc.) Weather effect on electrical energy use Utility rate structure (energy charges, demand charges, power factor penalty, etc.)
On-Site Survey	
Construction materials (thermal resistance type and thickness) HVAC system type DHW system Hot water /steam use for heating Hot water/steam for cooling Hot water/steam for DHW Hot water/steam for specific applications (hospitals, swimming pools, etc.)	HVAC system type Lighting type and density Equipment type and density Energy use for heating Energy use for cooling Energy use for lighting Energy use for equipment Energy use for air handling Energy use for water distribution
Energy Use Baseline	
Review architectural, mechanical, and control drawings Develop a base-case model (using any baselining method ranging from very simple to more detailed tools) Calibrate the base-case model (using utility data or metered data)	Review architectural, mechanical, electrical, and control drawings Develop a base-case model (using any baselining method ranging from very simple to more detailed tools) Calibrate the base-case model (using utility data or metered data)
Energy Conservation Measures	
Heat recovery system (heat exchangers) Efficient heating system (boilers) Temperature setback EMCS HVAC system retrofit DHW use reduction Cogeneration	Energy efficient lighting Energy efficient equipment (computers) Energy efficient motors HVAC system retrofit EMCS Temperature setup Energy efficient cooling system (chiller) Peak demand shaving Thermal energy storage system Cogeneration Power factor improvement Reduction of harmonics

Some of the commonly recommended energy conservation measures to improve the thermal performance of building envelope are

1. *Addition of Thermal Insulation.* For building surfaces without any thermal insulation, this measure can be cost effective.
2. *Replacement of Windows.* When windows represent a significant portion of the exposed building surfaces, using more energy-efficient windows (high R-value, low-emissivity glazing, airtight, etc.) can be beneficial in both reducing the energy use and improving the indoor comfort level.
3. *Reduction of Air Leakage.* When the infiltration load is significant, leakage areas of the building envelope can be reduced by simple and inexpensive weatherstripping techniques.

TABLE 4.6.2 Energy Audit Summary for Industrial Facilities

Thermal Systems	Electric Systems
Utility Analysis	
Thermal energy use profile (building signature) Thermal energy use per unit of a product Thermal energy use distribution (heating, process, etc.) Fuel types used Analysis of the thermal energy input for specific processes used in the production line (such as drying) Utility rate structure	Electrical energy use profile (building signature) Electrical energy use per unit of a product Electrical energy use distribution (cooling, lighting, equipment, process, etc.) Analysis of the electrical energy input for specific processes used in the production line (such as drying) Utility rate structure (energy charges, demand charges, power factor penalty, etc.)
On-Site Survey	
List of equipment that use thermal energy Perform heat balance of the thermal energy Monitor thermal energy use of all or part of the equipment Determine the by-products of thermal energy use (such as emissions and solid waste)	List of equipment that use electrical energy Perform heat balance of the electrical energy Monitor electrical energy use of all or part of the equipment Determine the by-products of electrical energy use (such pollutants)
Energy Use Baseline	
Review mechanical drawings and production flow charts Develop a base-case model using (any baselining method) Calibrate the base-case model (using utility data or metered data)	Review electrical drawings and production flow charts Develop a base-case model (using any baselining method) Calibrate the base-case model (using utility data or metered data)
Energy Conservation Measures	
Heat recovery system Efficient heating and drying system EMCS HVAC system retrofit Hot water and steam use reduction Cogeneration (possibly with solid waste from the production line)	Energy efficient motors Variable speed drives Air compressors Energy efficient lighting HVAC system retrofit EMCS Cogeneration (possibly with solid waste from the production line) Peak demand shaving Power factor improvement Reduction of harmonics

The energy audit of the envelope is especially important for residential buildings. Indeed, the energy use from residential buildings is dominated by weather since heat gain and/or loss from direct conduction of heat or from air infiltration/exfiltration through building surfaces accounts for a major portion (50 to 80%) of the energy consumption. For commercial buildings, improvements to the building envelope are often not cost-effective due to the fact that modifications to the building envelope (replacing windows, adding thermal insulation in walls) typically are very expensive. However, it is recommended to systematically audit the envelope components not only to determine the potential for energy savings but also to ensure the integrity of its overall condition. For instance, thermal bridges, if present, can lead to heat transfer increase and to moisture condensation. The moisture condensation is often more damaging and costly than the increase in heat transfer since it can affect the structural integrity of the building envelope.

Electrical Systems

For most commercial buildings and a large number of industrial facilities, electrical energy cost constitutes the dominant part of the utility bill. Lighting, office equipment, and motors are the electrical systems that consume the major part of energy usage in commercial and industrial buildings.

1. *Lighting.* Lighting for a typical office building represents, on average, 40% of the total electrical energy use. There are a variety of simple and inexpensive measures to improve the efficiency of lighting systems. These measures include the use of energy-efficient lighting lamps and ballasts, the addition of reflective devices, delamping (when the luminance levels are above the recommended levels by the standards), and the use of daylighting controls. Most lighting measures are especially cost-effective for office buildings for which payback periods are less than 1 year.
2. *Office Equipment.* Office equipment constitutes the fastest growing part of the electrical loads, especially in commercial buildings. Office equipment includes computers, fax machines, printers, and copiers. Today, there are several manufacturers that provide energy efficient office equipment (such as those that comply with U.S. EPA Energy Star specifications). For instance, energy-efficient computers automatically switch to a low-power "sleep" mode or off mode when not in use.
3. *Motors.* The energy cost to operate electric motors is a significant part of the operating budget of any commercial and industrial building. Measures to reduce the energy cost of using motors include reducing operating time (turning off unnecessary equipment), optimizing motor systems, using controls to match motor output with demand, using variable speed drives for air and water distribution, and installing energy-efficient motors. Table 4.6.3 provides typical efficiencies for several motor sizes.

TABLE 4.6.3 Typical Efficiencies of Motors

Motor Size (HP)	Standard Efficiency (%)	High Efficiency (%)
1	72	81
2	76	84
3	77	89
5	80	89
7.5	82	89
10	84	89
15	86	90
20	87	90
30	88	91
40	89	92
50	90	93

In addition to the reduction in the total facility electrical energy use, retrofits of the electrical systems decrease the cooling loads and, therefore, further reduce the electrical energy use in the building. These cooling energy reductions, as well as possible increases in thermal energy use (for space heating), should be accounted for when evaluating the cost-effectiveness of improvements in lighting and office equipment.

HVAC Systems

The energy use due to HVAC systems can represent 40% of the total energy consumed by a typical commercial building. The energy auditor should obtain the characteristics of major HVAC equipment to determine the condition of the equipment, its operating schedule, its quality of maintenance, and its control procedures. A large number of measures can be considered to improve the energy performance of both primary and secondary HVAC systems. Some of these measures are listed below:

1. *Setting up/back thermostat temperatures.* When appropriate, set-back of heating temperatures can be recommended during unoccupied periods. Similarly, set-up of cooling temperatures can be considered.

2. *Retrofit of constant air volume systems.* For commercial buildings, variable air volume (VAV) systems should be considered when the existing HVAC systems rely on constant-volume fans to condition part or the entire building.
3. *Installation of heat recovery systems.* Heat can be recovered from some HVAC equipment. For instance, heat exchangers can be installed to recover heat from air handling unit (AHU) exhaust air streams and from boiler stacks.
4. *Retrofit of central heating plants.* The efficiency of a boiler can be drastically improved by adjusting the fuel-air ratio for proper combustion. In addition, installation of new energy-efficient boilers can be economically justified when old boilers are to be replaced.
5. *Retrofit of central cooling plants*: Currently, there are several chillers that are energy efficient and easy to control and operate and are suitable for retrofit projects.

It should be noted that there is a strong interaction between various components of a heating and cooling system. Therefore, a whole-system analysis approach should be followed when retrofitting a building HVAC system. Optimizing the energy use of a central cooling plant (which may include chillers, pumps, and cooling towers) is one example of using a whole-system approach to reduce the energy use for heating and cooling buildings.

Compressed Air Systems

Compressed air has become an indispensable tool for most manufacturing facilities. Its uses range from air-powered hand tools and actuators to sophisticated pneumatic robotics. Unfortunately, staggering amounts of compressed air are wasted in a large number of facilities. It is estimated that only about 20 to 25% of input electrical energy is delivered as useful compressed air energy. Leaks are reported to account for 10 to 50% of the waste while misapplication accounts for 5 to 40% of the loss of compressed air (Howe and Scales, 1998).

To improve the efficiency of compressed air systems, the auditor can consider several issues including whether compressed air is the right tool for the job (for instance, electric motors are more energy efficient than air-driven rotary devices), how the compressed air is applied (for instance, lower pressures can be used to supply pneumatic tools), how it is delivered and controlled (for instance, the compressed air needs to be turned off when the process is not running), and how the compressed air system is managed (for each machine or process, the cost of compressed air needs to be known to identify energy and cost savings opportunities).

Energy Management Controls

Because of the steady decrease in the cost of computer technology, automated control of a wide range of energy systems within commercial and industrial buildings is becoming increasingly popular and cost effective. An energy management and control system (EMCS) can be designed to control and reduce the building energy consumption within a facility by continuously monitoring the energy use of various equipment and making appropriate adjustments. For instance, an EMCS can automatically monitor and adjust indoor ambient temperatures, set fan speeds, open and close air handling unit dampers, and control lighting systems.

If an EMCS is already installed in the building, it is important to recommend a system tune-up to ensure that the controls are operating properly. For instance, the sensors should be calibrated regularly in accordance with manufacturers' specifications. Poorly calibrated sensors may cause an increase in heating and cooling loads and may reduce occupant comfort.

Indoor Water Management

Water and energy savings can be achieved in buildings by using water-saving equipment instead of the conventional fixtures for toilets, faucets, showerheads, dishwashers, and clothes washers. Savings can also be achieved by eliminating leaks in pipes and fixtures.

Table 4.6.4 provides the typical water usage of conventional and water-efficient fixtures. In addition, Table 4.6.4 indicates the hot water consumption by each fixture as a fraction of the total water usage. With water-efficient fixtures, a savings of 50% of water use can be achieved.

TABLE 4.6.4 Characteristics of Water-Using Fixtures

End Use	Conventional Fixtures	Water-Efficient Fixtures	Usage Pattern	Hot Water Usage (%)
Toilets	3.5 gal/flush	1.6 gal/flush	4 flushes/person/day	0
Showers	5.0 gal/min	2.5 gal/min	5 min/shower	60
Faucets	4.0 gal/min	2.0 gal/min	2.5 min/person/day	50
Dishwashers	14.0 gal/load	8.5 gal/load	0.17 loads/person/day	100
Clothes washers	55.0 gal/load	42.0 gal/load	0.3 loads/person/day	25
Leaks	10% total use	2% total use	N/A	50

New Technologies

The energy auditor may consider the potential of implementing and integrating new technologies within the facility. It is, therefore, important that the energy auditor understands these new technologies and knows how to apply them. Among the new technologies that can be considered for commercial and industrial buildings include:

Building Envelope Technologies

Recently, several materials and systems have been proposed to improve the energy efficiency of the building envelope, especially windows, including:

- Spectrally selective glasses which can optimize solar gains and shading effects
- Chromogenic glazings which change their properties automatically depending on temperature and/or light-level conditions (similar to sunglasses that become dark in sunlight)
- Building integrated photovoltaic panels that can generate electricity while absorbing solar radiation and reducing heat gain through the building envelope (typically roofs)

Light Pipe Technologies

While the use of daylighting is straightforward for perimeter zones that are near windows, it is not usually feasible for interior spaces, particularly those without skylights. Recent but still emerging technologies allow one to "pipe" light from roof or wall-mounted collectors to interior spaces that are not close to windows or skylights.

HVAC Systems and Controls

Several strategies can be considered for energy retrofits, including:

- Heat recovery technologies such as rotary heat wheels and heat pipes can recover 50 to 80% of the energy used to heat or cool ventilation air supplied to the building
- Desiccant-based cooling systems are now available and can be used in buildings with large dehumidification loads during long periods (such as hospitals, swimming pools, and supermarket fresh produce areas)
- Geothermal heat pumps can provide an opportunity to take advantage of the heat stored underground to condition building spaces
- Thermal energy storage (TES) systems offer a mean of using less-expensive off-peak power to produce cooling or heating to condition the building during on-peak periods; several optimal control strategies have been developed in recent years to maximize the cost savings of using TES systems

Cogeneration

This is not really a new technology. However, recent improvements in its combined thermal and electrical efficiency have made cogeneration cost effective in several applications including institutional buildings such as hospitals and universities.

Case Study

To illustrate the energy audit process described above, a case study is presented below. The activities performed for each step of the energy audit are briefly described. For more details, refer to Kim et al. (1998). The building analyzed in this case study is a medium size office building located in Seoul, Korea.

Step 1: Building and Utility Data Analysis

The first step in the building energy audit process is to collect all available information about the energy systems and the energy use pattern of the building. This information was collected before the field survey, in particular, from the architectural, mechanical, and electrical drawings and utility bills.

Building Characteristics

The building is a 26-story office building with a 2-story penthouse and 4-story basement. It is located in downtown Seoul, Korea. The structure of the building consists of modular concrete and steel frame. The building area is 3920 m^2 and the site area is 6555 m^2. Single-glazed windows are installed throughout the building. Figure 4.6.1 shows a typical floor plan of the building. Table 4.6.5 describes the various construction materials used throughout the building.

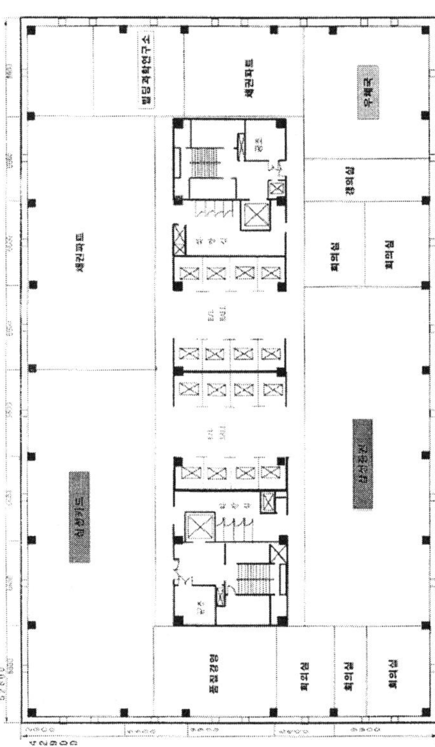

FIGURE 4.6.1 Typical floor plan of the audited office building.

Energy Use

Figure 4.6.2 summarizes the monthly electrical energy use of the building for 1993. The monthly average dry-bulb outdoor air temperature recorded for 1993 is also presented in Figure 4.6.2. It is clear that the electrical energy use increases during the summer months (June through October) when the outdoor temperatures are high. During the other months, the electrical energy use is almost constant and can be attributed mostly to lighting and office equipment loads. Preliminary analysis of the metered building energy indicates that natural gas consumption is inconsistent from month to month. For example, gas consumption during January is six times higher than during December, even though the weather conditions are similar for both months. Therefore, the recorded data for natural gas were considered unreliable and only metered electrical energy data were analyzed.

Step 2: On-Site Survey

A one-day field survey was conducted with the assistance of the building operator in the summer of 1996. During the survey, several useful and revealing information and engineering data were collected. For instance:

TABLE 4.6.5 Building Construction Materials

Component	Materials
Exterior wall	5 cm tile 16 cm concrete 2.5 cm foam insulation 0.6 cm finishing material
Roof	5 cm lightweight concrete 15 cm concrete 2.5 cm foam insulation
Interior wall	2 cm finishing cement mortar 19 cm concrete block
Glazing	1.2 cm thick single-pane glazing
Underground wall	25 cm concrete asphalt shingle air-space 10 cm brick 2 cm finishing cement mortar
Underground floor	15 cm concrete asphalt shingle 12 cm concrete 2 cm finishing cement mortar

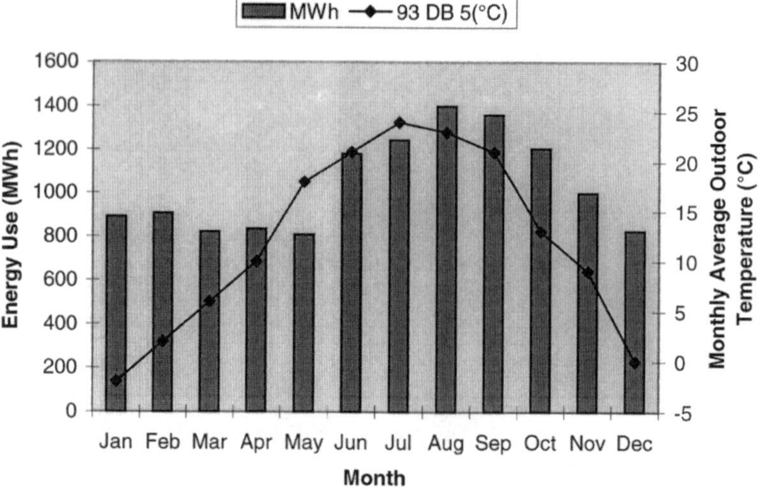

FIGURE 4.6.2 Monthly actual electrical energy consumption.

- It was found that the building had been retrofitted with energy-efficient lighting systems. The measurement of luminance levels throughout the working areas indicated adequate lighting. To determine an estimate of the energy use for lighting, the number and type of lighting fixtures were recorded.
- It was observed that the cooling and heating temperature set-points were 25.5°C and 24.5°C, respectively. However, indoor air temperature and relative humidity measurements during the field survey revealed that during the afternoon the thermal conditions were uncomfortable in several office spaces, with average air dry-bulb temperature of 28°C and relative humidity of 65%. A discussion with the building operator indicated that the chillers were no longer able to meet the cooling loads following the addition of several computers in the building during the last few years. As a solution, an ice storage tank was then added to reduce the peak cooling load.

- It was discovered during the survey that the building is heated and cooled simultaneously by two systems: constant air volume (CAV) and fan coil unit (FCU) systems. The CAV system is complemented by the FCU system as necessary. Two air-handling units serve the entire building, and about 58 FCUs are on each floor.
- The heating and cooling plant consists of three boilers, six chillers, three cooling towers, and one ice storage tank. The capacities of the boilers and the chillers are

  Boilers: 13 MBtu/h (2 units) and 3.5 MBtu/h (1 unit)

  Chillers: 215 tons (5 units) and 240 tons (1 unit)

  The thermal energy storage (TES) system consists of a brine ice-on-coil tank. The hours of charging and discharging are 10 and 13, respectively. The TES system is currently controlled using simple and nonpredictive storage-priority controls.

The internal heat gain sources are shown in Table 4.6.6. Operating schedules were based on the discussion with the building operators and on observations during the field survey.

TABLE 4.6.6 Internal Heat Gain Level for the Office Building

Internal Heat Gain	Design Load
Occupancy	17 m²/person Latent heat gain: 45 W Sensible heat gain: 70 W
Lighting	14 W/m²
Equipment	16 W/m²
Ventilation	14.7 CFM/person

Step 3: Energy Use Baseline Model

To model the building using DOE-2, each floor was divided into two perimeter and two core zones. Figure 4.6.3 shows the zone configuration used to model the building floors. The main reason for this zoning configuration is the lack of flexibility in the DOE-2 SYSTEMS (Energy Information Agency, 1980) program. Although the actual building is conditioned by the combination of constant air volume (CAV) and fan coil unit (FCU) systems, the SYSTEM module of DOE-2.1E cannot model two different types of HVAC systems serving one zone. Therefore, a simplification has been made to simulate the actual HVAC system of the building. This simplification consists of the following: the perimeter zone is conditioned by the FCUs, while the core zone is conditioned by CAV. Since all the FCUs are located at the perimeter, this simplification is consistent with the actual HVAC system operation.

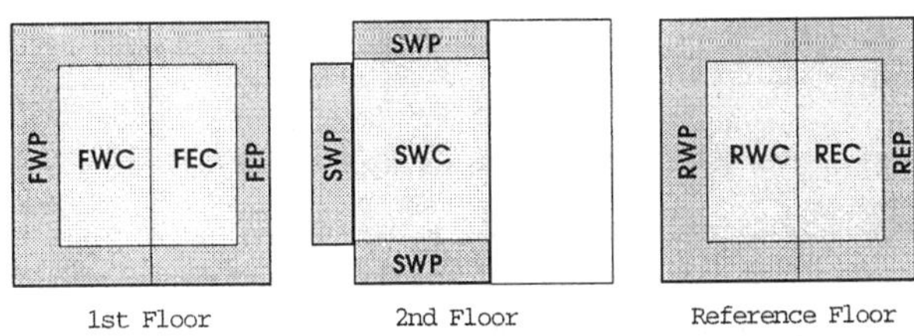

FIGURE 4.6.3 Building zoning configuration for DOE-2 computer simulation.

Figure 4.6.4 shows the monthly electrical energy consumption predicted by the DOE-2 base model and the actual energy use recorded in 1993 for the building. It shows that DOE-2 predicts the actual energy use pattern of the building fairly well, except for the months of September and October. The difference between the annual metered energy use in the building and the annual predicted electricity use by the DOE-2 base-case model is about 762 MWh. DOE-2 predicts that the building consumes 6% more electricity than the actual metered annual energy use. To perform the DOE-2 base-case model, a TRY-type weather file of Seoul was created using the raw weather data collected for 1993. Using the DOE-2 base-case model, a number of ECOs can now be evaluated.

Figure 4.6.5 shows the distribution by end uses of the building energy consumption. The electrical energy consumption of the building is dominated by lighting and equipment. The electricity utilized for lighting and office equipment represents about 75% of the total building electric consumption. As

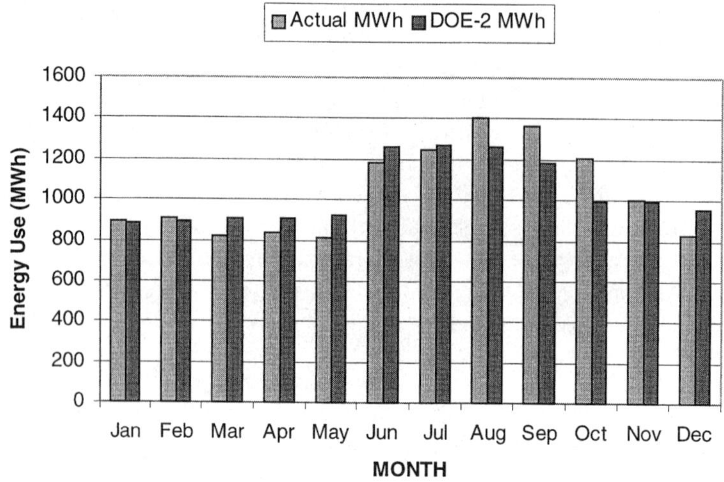

FIGURE 4.6.4 Comparison of the DOE-2 prediction and actual building electrical energy use.

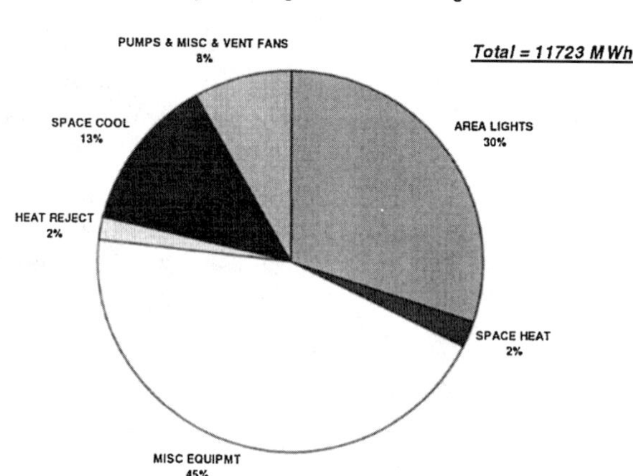

FIGURE 4.6.5 Electricity use distribution.

mentioned earlier, a recent lighting retrofit was installed in the building, comprising electronic ballasts and energy-efficient fluorescent fixtures. Therefore, it was decided not to consider a lighting retrofit as an ECO for this study. The electricity consumption for cooling is about 13.1%. The ECOs selected for this building aim mostly at reducing the cooling loads.

Step 4: Evaluation of Energy Conservation Opportunities (ECOS)

Based on the evaluation of the energy use pattern of the building, several energy conservation opportunities (ECOs) for the building were analyzed. Among the ECOs considered in the study, six of them successfully reduced energy consumption:

1. *ECO #1: CAV to VAV conversion.* The present AHU fans are all constant-speed fans. They supply conditioned air through a constant-volume air supply system to the selected zones. The system is designed to supply enough air to heat or cool the building under design conditions. Under non-design conditions, more air than needed is supplied. Changing the system to a variable air volume (VAV) system would reduce the amount of air supplied by the AHUs and result in less energy to condition the various zones. For this ECO, a constant-volume reheat fan system assigned to the core zones in the building was changed to a VAV system. In particular, VAV boxes, controlled by

the space thermostat, vary the amount of conditioned air supply to the building zones to control the indoor temperature. Both labor and equipment costs were included in the estimation of the payback period for this measure.

2. *ECO #2: Optimal ice storage control.* The current TES system is operated using a nonpredictive storage-priority control. To improve the benefits of the TES system, a near-optimal controller is suggested. This ECO is analyzed to determine if the cost of electrical energy consumption for the building is reduced when a near-optimal control strategy is used. To determine the savings of this option, the simulation environment developed by Henze et al. (1997) is used. This simulation environment is based on a dynamic programming technique and determines the best operating controls for the TES system given the cooling and noncooling load profiles and electrical rate structure. No DOE-2 simulation is performed for this ECO. The results of the dynamic programming simulation indicate that an energy reduction of 5% can be achieved using a near-optimal control in lieu of the storage-priority control. To implement this near-optimal control, a predictor is required to determine future building cooling/noncooling loads. An example of such a predictor could be based on neural networks (Kreider et al., 1995). The labor cost and the initial cost of adding some sensors and a computer were considered in determining the payback period for this measure.
3. *ECO #3: Glazing retrofit.* For this building, low-e glazing systems were considered to reduce the internal heat gain due to solar radiation. Thus, the cooling load is reduced. In addition, the increased U-value of the glazing reduces the heating load. For this ECO, the existing single-pane windows with the glass conductance of 6.17 W/m^2-K and the shading coefficient of 0.69 were changed to double-pane windows of 1.33 W/m^2-K and 0.15 shading coefficient.
4. *ECO #4: Indoor temperature set-back/set-up.* In this ECO, the impact of the indoor temperature setting on energy use was analyzed using the DOE-2 simulation program; the heating temperature was changed from 24.5°C to 22.5°C and the cooling temperature was changed from 25.5°C to 27.5°C. There is only a labor cost associated with this measure.
5. *ECO #5: Motor replacement.* Increasing the efficiency of the motors for fans and pumps can reduce the total electric energy consumption in the building. In this ECO, the existing efficiencies for the motors was assumed to range from 0.85 (10 HP) to 0.90 (50 HP). Energy efficient motors have efficiencies that range from 0.91 (10 HP) to 0.95 (50 HP). Only the differential cost was considered in the economic analysis.
6. *ECO #6: Daylighting control plan.* A continuous dimming control would regulate the light level so that the luminance level inside the zones remains constant. The electricity consumption of the building can be significantly reduced, while the gas consumption can be slightly increased because of the reduced space heat gain from the lighting system. For this ECO, a daylighting system with continuous dimming control was considered for perimeter office zones. The cost of adding the sensors was considered.

The impact of the selected ECOs on the electricity use in the building as predicted by DOE-2 simulations is shown in Table 4.6.7. Based on these results, converting CAV to VAV and implementing a daylighting control system with dimming control reduced the total electricity consumption of the building 5.2% and 7.3%, respectively. These savings are significant considering that the electricity consumption for the cooling plant alone is about 13.1% of the total electricity consumption of the building.

The economic analysis was performed using the utility rate of Seoul, Korea. Table 4.6.7 presents the energy cost savings of ECOs in Korean currency (1000 Won = $1). In addition to the electricity cost, the natural gas cost was also included in the economic analysis. The natural gas is used only for heating the building. The economic analysis shows that the VAV conversion reduces the total energy consumption by more than 10%, and the daylighting control saves about 6% of the total energy costs.

Recommendations

The results for economic analysis, the VAV conversion (ECO #1), adjustment of temperature set-point (ECO #4), and the daylighting control (ECO #6) are the recommended energy saving opportunities to

TABLE 4.6.7 Economic Analysis of the ECOs

	Electricity Cost (MWon)	LNG Cost (MWon)	Total Cost (MWon)	Capital Cost (MWon)	Saving (%)	Savings (MWon)	Payback Period (years)
Base Case	984.4	139.1	1123.5	—	—	—	—
ECO #1	940.8	49.8	990.5	465.5	11.8	133.0	3.5
ECO #2	979.1	139.1	1118.2	42.4	0.6	5.3	8.0
ECO #3	977.9	126.9	1104.8	280.5	1.7	18.7	15.0
ECO #4	983.7	106.4	1090.1	16.7	3.0	33.4	0.5
ECO #5	972.6	138.7	1111.4	60.5	1.1	12.1	5.0
ECO #6	911.6	144.8	1056.4	268.4	6.0	67.1	4.0

be implemented for the audited office building. In particular, the VAV conversion reduces the total energy consumption more than 10%, and the daylighting control saves about 6% of the total energy consumption.

Verification Methods of Energy Savings

Energy conservation retrofits are deemed cost-effective based on predictions of the amount of energy and money a retrofit will save. However, several studies have found that large discrepancies exist between actual and predicted energy savings. Due to the significant increase in the activities of energy service companies (ESCOs), the need became evident for standardized methods for measurement and verification of energy savings. This interest has led to the development of the *North American Energy Measurement and Verification Protocol* published in 1996 and later expanded and revised under the *International Performance Measurement and Verification Protocol.*

In principle, the measurement of the retrofit energy savings can be obtained by simply comparing the energy use during pre- and post-retrofit periods. Unfortunately, the change in energy use between the pre- and post-retrofit periods is not only due to the retrofit itself but also to other factors such as changes in weather conditions, levels of occupancy, and HVAC operating procedures. It is important to account for all these changes to accurately determine the retrofit energy savings.

Several methods have been proposed to measure and verify savings of implemented energy conservation measures in commercial and industrial buildings. Some of these techniques are briefly described below.

Regression Models

The early regression models used to measure savings adapted the Variable-Base Degree Day (VBDD) method. Among these early regression models, the PRInceton Scorekeeping Method (PRISM) which uses measured monthly energy consumption data and daily average temperatures to calibrate a linear regression model and determine the best values for nonweather-dependent consumption, the temperature at which the energy consumption began to increase due to heating or cooling (the change-point or base temperature), and the rate at which the energy consumption increased. Several studies have indicated that the simple linear regression model is suitable for estimating energy savings for residential buildings. However, subsequent work has shown that the PRISM model does not provide accurate estimates for energy savings for most commercial buildings (Ruch and Claridge, 1992).

Single-variable (temperature) regression models require the use of at least four-parameter segmented linear or change-point regressions to be suitable for commercial buildings. Katipamula et al. (1994) proposed multiple linear regression models to include as independent variables internal gain, solar radiation, wind, and humidity ratio, in addition to the outdoor temperature. For the buildings considered in their analysis, Katipamula et al. found that wind and solar radiation have small effects on the energy consumption. They also found that internal gains have generally modest impact on energy consumption. Katipamula et al. (1998) discuss in more detail the advantages and the limitations of multivariate regression modeling.

Time Variant Models

Several techniques have been proposed to include the effect of time variation of several independent variables on estimating the energy savings due to retrofits of building energy systems. Among these techniques are the artificial neural networks (Krarti et al., 1998), Fourier series (Dhar et al., 1998), and nonintrusive load monitoring (Shaw et al., 1998). These techniques are very involved and require a high level of expertise and training.

Summary

An energy audit of commercial and industrial buildings encompasses a wide variety of tasks and requires expertise in a number of areas to determine the best energy conservation measures suitable for an existing facility. This section provided a description of a general but systematic approach to perform energy audits. If followed carefully, the approach helps facilitate the process of analyzing a seemingly endless array of alternatives and complex interrelationships between the building and its energy system components.

References

Dhar, A., Reddy, T.A., and Claridge, D.E., Modeling hourly energy use in commercial buildings with Fourier series functional forms, *ASME J. Solar Energy Eng.,* 120, 217-223, 1998.

Energy Information Agency, Energy Facts 1991, DOE/EIA-0469(91), Department of Energy, Washington, D.C., 1991.

Energy Information Agency, DOE-2 Users Guides, Ver. 2.1, Report No. LBL-8689, Lawrence Livermore National Laboratory, Berkeley, CA, 1980.

Fels, J., Special issue devoted to measuring energy savings: the scorekeeping approach, *Energy Buildings,* 12(2), 113-127, 1988.

Henze, G.P., Krarti, M., and Brandemuehl, M.J., A simulation environment for the analysis of ice storage control, *Int. J. HVAC Res.,* 3, 128-148, 1997.

Howe, B. and Scales, B., Beyond leaks: demand-side strategies for improving compressed air efficiency, *Energy Eng.,* 95, 31-39, 1998.

Katipamula, S., Reddy, T.A., and Claridge, D.E., Development and Application of Regression Models to Predict Cooling Energy Consumption in Large Commercial Buildings, 1994 ASME-JSEME-JSES Int. Solar Energy Conf., San Francisco, 1994, 307-322.

Katipamula, S., Reddy, T.A., and Claridge, D.E., Multivariate regression modeling, *ASME J. Solar Energy Eng.,* 120, 177-184, 1998.

Kim, K., Yoon, H., Lee, E., Choi, S., and Krarti, M., Building Energy Performance Simulations to Evaluate Energy Conservation Measures, ASME Solar Energy Eng. Conf. Proc., 1998, 45-55.

Knebel, D.E., *Simplified Energy Analysis Using the Modified Bin Method,* American Society of Heating, Refrigeration, and Air-Conditioning Engineers, Atlanta, GA, 1983.

Krarti, M., Kreider, J.F., Cohen, D., and Curtiss, P., Estimation of energy savings for building retrofits using neural networks, *ASME J. Solar Energy Eng.,* 120, 211-216, 1998.

Kreider, J.F., Claridge, D.Z., Curtis, P., Dodier, R., Haberl, J.S., and Krarti, M., Building energy use prediction and system identification using recurrent neural networks, *J. Solar Energy Eng.,* 117, 161-166, 1995.

Kreider, J.F. and Rabl, A., *Heating and Cooling of Buildings,* McGraw-Hill, New York, 1994.

LBL, *DOE-2 User Guide, Version 2.1,* Report No. LBL-8689, Rev. 2, Lawrence Livermore National Laboratory, Berkeley, CA, 1980.

Ruch, D. and Claridge, D.E., A four-parameter change-point model for predicting energy consumption in commercial buildings, *ASME J. Solar Energy Eng.,* 104, 177-184, 1992.

Shaw, S.R., Abler, C.B., Lepard, R.F., Luo, D., Leeb, S.B., and Norford, L.K., Instrumentation for high performance non-intrusive electrical load monitoring, *ASME J. Solar Energy Eng.,* 120, 177-184, 1998.

4.7 Compressors

Raymond Cohen, Eckhard Groll, William H. Harden, Kenneth E. Hickman, Dilip K. Mistry, and Earl Muir

Introduction

Compressors for air and other gases are used for many purposes in industrial, commercial, and residential applications. For example, gas compressors in industrial applications are used for refrigeration, air conditioning, heating, pipeline conveying, natural gas gathering, catalytic cracking, polymerization, and other chemical processes. While the gas compressor may be similar in appearance to an air compressor commonly seen at a construction site, they are specially designed to meet a chemical process requirement of pressure, temperature, or flow, and are capable of handling the specific gases or gas mixtures to be used. Rollins (1989) provides a detailed compilation of the available knowledge about air and gas compressors, compressed air systems, pneumatic tools, and the uses of compressed air.[1]

This section contains a review, including descriptions of certain design features and characteristics, of the most widely used and commercially available compressors. Two major applications are considered: (1) compressors that are commonly used in vapor compression cycles for refrigeration and air-conditioning applications, and (2) compressors that are commonly used to compress air. The most complete set of technical papers about compressors can be found in the Proceedings of the Purdue Compressor Conferences. References 2 through 15 contain 10,558 pages (1377 papers).

In refrigeration and air-conditioning systems, the compressor is one of the four major components that are used to implement the widely used vapor compression cycle. The other three components are the condenser, evaporator, and expansion device. The compressor compresses the refrigerant and circulates it in a continuous cycle through the closed system. ASHRAE provides a detailed compilation of the available knowledge about refrigerant compressors.[16] The compressors for these applications have been chosen for this section because they are representative of most of the compressors made in large quantities.

Compressed air is essential to almost all manufacturing and construction enterprises. The compressor manufacturing industry is so large that air compressors have been selected for special consideration in this section.

The capacity of gas and air compressors is specified according to their delivered flow rate, whereas the capacity of refrigerating compressors is specified according to their cooling capacity.

There are two basic types of compressors: positive displacement compressors and dynamic compressors. Positive displacement compressors increase the pressure of the refrigerant vapor, air, or other gas by reducing the volume of the compression chamber by means of work usually applied to the compressor's shaft mechanism. Dynamic compressors increase the pressure by a continuous transfer of angular momentum from the rotating parts of the compressor to the fluid, followed by the conversion of this momentum into a pressure rise. Both types are described in this section.

Positive Displacement Compressors and Application to Refrigeration and Air Conditioning

Positive displacement compressors depend on a volume change for the compression process, in contrast to dynamic compressors, which depend on acceleration, deceleration, and conversion of velocity into pressure. They are designed for a wide range of flow rates and suction and discharge pressures, and can be used in many applications. Typically, they are used for refrigeration and air-conditioning applications, and for compressing air and other gases for virtually all uses except aircraft propulsion.

When used for cooling and heating, compressors, whether classified as positive displacement compressors or dynamic compressors, are used in air conditioning, refrigeration, and heat pump systems. As the key of these vapor compression system applications, the compressor raises the pressure of the working fluid so that it can be condensed to liquid at the heat sink temperature, expanded or throttled to reduce the pressure, and evaporated into vapor to absorb heat from the refrigerated source. In these systems, refrigeration compressors are a component in a closed loop system which is tightly sealed (often hermetically) to prevent refrigerant loss. These systems are often sealed for life without service or lubricant changes, operating anywhere from 300 to 8000 hours per year with a life expectancy of 5 to 15 years; i.e., operating from 10,000 to 120,000 hours with no maintenance.

When used in air-conditioning, refrigeration, and heat pumps, they must be designed for a variety of refrigerants and compatible lubricants. Commonly used refrigerants include hydrochlorofluorocarbons such as R-22, or hydrofluorocarbons such as R-410A, R-407C, R-404A, and R-134a, or ammonia and various hydrocarbons.

Positive displacement compressors are classified according to their construction; e.g., whether or not they are enclosed in a hermetic housing. They are also classified according to the kinematic arrangement of their parts to accomplish the change in volume.

The particular types of refrigerant compressors to be described in this section are identified according to how they fit into the range on the application chart. This chart is shown in Figure 4.7.1, which shows the application across the top line, from the smallest for household uses to the largest for possible use in freezing plants and for air-conditioning tall buildings.[17] Typically, compressors are specified by the horsepower or the kilowatts needed to drive them, or their refrigerating capability measured in tons of refrigeration, Btu/hr, or kW.

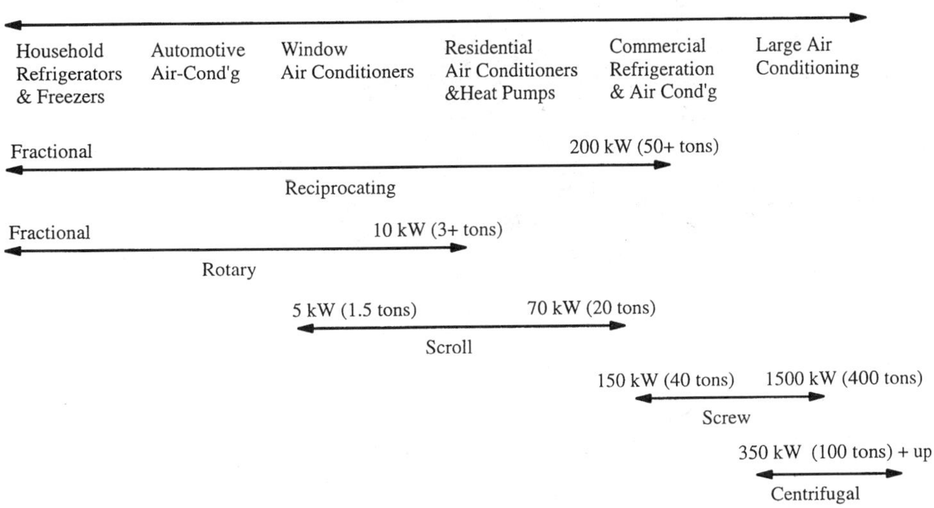

FIGURE 4.7.1 Range of applications of refrigerating compressors. (Data from International Institute of Refrigeration, Paris.)

Compressor Type by Construction

Hermetic Compressors

Hermetic compressors, used for many refrigeration applications, have both the electric motor and the compressor inside a welded steel housing, providing a true hermetic seal. The motor shaft is normally connected directly to the compressor shaft. The housing is a steel shell that is welded at the joints to ensure the hermetic seal. If the gas in the housing is suction gas, then the compressor is called a low-side compressor because of the low suction pressure. If the gas in the housing is discharge gas, then the compressor is called a high-side compressor. A high-side design can often be very advantageous by serving as a source of high-pressure oil in the sump to lubricate the compressor components. When oil is injected into or present during the compression process for cooling or sealing, the high-side design provides a cavity where the oil can be separated from the discharge gas before it is delivered from the compressor to the condenser. Advantages of the low-side compressor are improved motor and oil cooling and lighter housing designs because of lower pressure. Hermetic shells are limited to small sizes because of the high cost of producing the larger ones made by a deep-drawing process. Therefore, hermetic compressors are used for smaller capacities. Figure 4.7.2 shows a hermetic reciprocating compressor with part of the shell cut away.

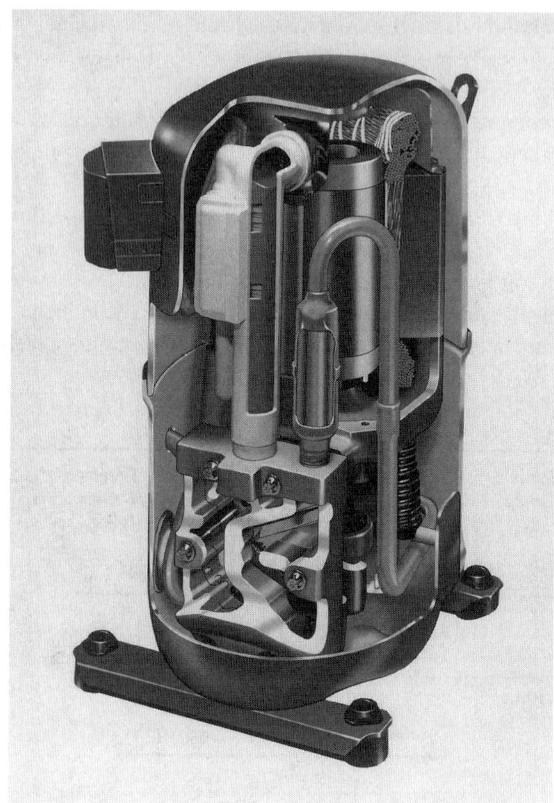

FIGURE 4.7.2 Cutaway drawing of a hermetic reciprocating compressor. (Courtesy of the Copeland Corporation, Sydney, OH.)

Semi-Hermetic Compressors

Semi-hermetic compressors are typically used for medium cooling capacities. They are similar to hermetic compressors except that the outer housing is normally bolted together so that the compressor is accessible for service. An advantage of this design is the ease of construction of multicylinder, large compressors without the need for the expensive deep-drawing manufacturing process. The externally bolted joints are sealed with gaskets that are a source of refrigerant leakage over a long period of time. Figure 4.7.3 shows a cross-sectional drawing of a typical semi-hermetic reciprocating compressor with part of the housing cut away. The electric motor drive is cooled by suction gas as well as by conduction through the motor housing. In many applications fans are used to force air over the motor housing to aid in cooling. Figure 4.7.4 shows that the semi-hermetic reciprocating compressor provides excellent efficiency as compared to other positive displacement compressors in the medium capacity range. Semi-hermetic reciprocating compressors are normally rugged and durable machines; however, they tend to have relatively high vibration and noise levels.

FIGURE 4.7.3 Cutaway drawing of a semi-hermetic reciprocating compressor. (Courtesy of the Copeland Corporation, Sydney, OH.)

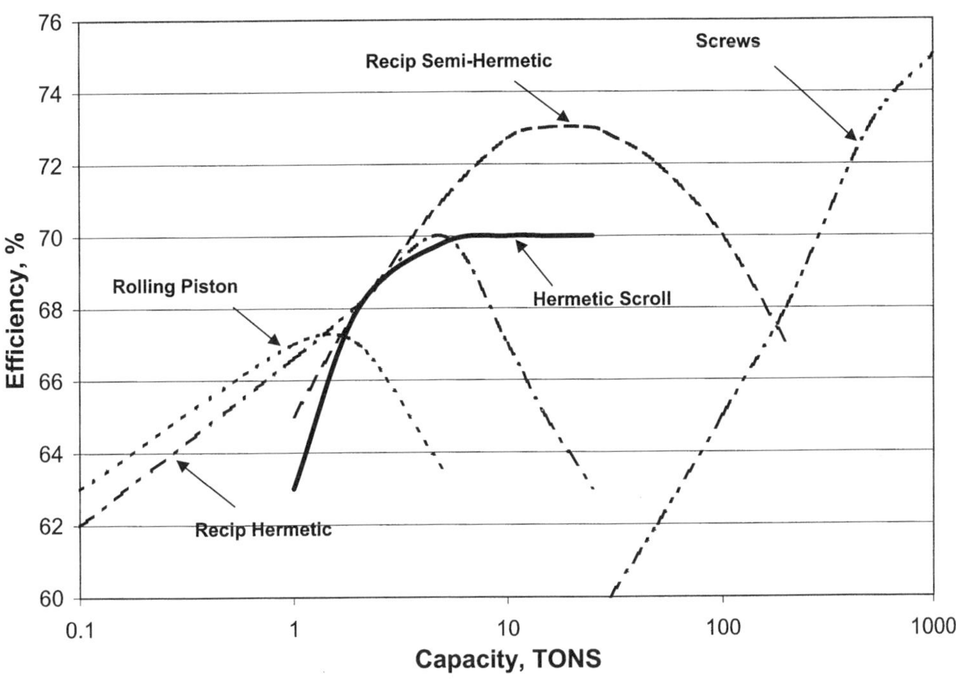

FIGURE 4.7.4 Performance and capacity range for positive displacement refrigeration compressors. (Courtesy of the Copeland Corporation, Sydney, OH.)

Open Drive Compressors

Open drive compressors have their motors external to their compressor housings. The shaft connecting the motor and compressor passes through the compressor housing, requiring a shaft seal. The open drive compressor is more sensitive to leakage than the semi-hermetic compressor due to the use of the shaft seal. Open drive compressors are usually used for larger capacities, and often for automotive applications where the power to drive them is taken from the engine.

The semi-hermetic and open drive types make it easier to have multiple cylinders, which is a popular option for larger capacities and higher pressure ratios. The open drive compressors offer more flexibility, especially when repairs are needed. Installing the larger sizes in tandem is a low-cost way to obtain multiple capacities for improved part load efficiencies.

Compressor Type by Kinematic Linkages

Reciprocating Compressors

The reciprocating compressor is the most widely used positive displacement compressor, although the other types described here are being produced in ever-increasing numbers.

Most reciprocating compressors are single-acting machines that use a piston inside a cylinder. The piston is usually driven by a crankshaft through a connecting rod and wrist pin to compress the gas. However, other driving mechanisms are used sometimes; e.g., scotch-yoke mechanisms or electromagnetic drives. Figure 4.7.5 shows a cross-sectional drawing of a typical reciprocating compressor. The motor is often cooled by suction gas, although conduction through the shell also plays a part in the cooling. In many larger applications, fans are used to force air over the motor housing to aid cooling.

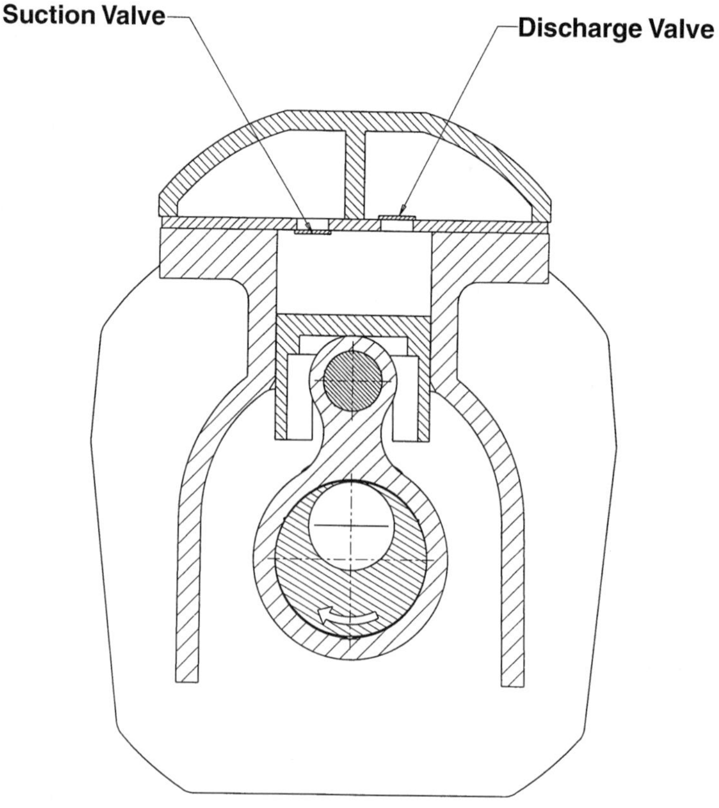

FIGURE 4.7.5 Cross-sectional drawing of a typical reciprocating compressor. (Courtesy of the Copeland Corporation, Sydney, OH.)

Discharge and suction mufflers are used to reduce gas pulsations and sound. Sometimes mufflers are an internal component in the compressor. Reciprocating compressors are normally mounted on a spring suspension system external to the housing and/or sometimes inside an hermetic shell to help reduce vibration and noise.

The reciprocating compressor operates by drawing refrigerant through a pressure-operated suction valve into the cylinder during the downward stroke of the piston. At bottom dead center, when the cylinder has reached its maximum volume and the piston starts its upward travel, the suction valve closes. The gas is then compressed until it reaches the desired discharge pressure, at which time a pressure-operated discharge valve opens, allowing the compressed gas to flow into the compressor's discharge plenum and then into the system.

The volume inside the cylinder at top dead center is called the clearance volume. The clearance volume can never be reduced to zero, and is typically 1 to 5% of the displaced volume due to piston movement. As the piston begins its downward travel, this gas must first be expanded back to suction conditions before the suction valve opens and new suction gas can enter the cylinder; thereby reducing the volumetric efficiency and capacity of the compressor. Single-stage reciprocating compressors are used in applications with pressure ratios ranging from 1.5:1 to 15:1, and two-stage units for applications with pressure ratios of 10:1 and above. At pressure ratios about 15:1, the capacity of high clearance volume compressors is reduced and can approach zero.

Reciprocating compressors with piston ring seals can wear significantly but still provide outstanding performance. Therefore, even as wear increases, they often continue to operate adequately after long service. They are usually considered easy to repair compared to other types, and there is a large cadre of service personnel available.

Capacity modulation can be accomplished several ways in reciprocating compressors. On multicylinder compressors, some suction valves are often blocked. In some cases the suction process is throttled by passing discharge gas back to the suction side of the system. Sometimes the speed of the compressor is varied by mechanical or electrical means, although designers must be careful with valve designs to ensure that they work adequately over the range of speeds used.

Hermetic reciprocating compressors for refrigeration and air-conditioning applications are produced in sizes ranging from a fraction, to up through 25 tons of refrigeration (88 kW). They are typically used in residential, commercial, and industrial air conditioning, heat pump and refrigeration applications with capacities ranging from ¼ to over 200 tons of refrigeration (from 0.88 to over 703.4 kW). Reciprocating compressors may have up to 10 cylinders in radial, in-line, or multiple bank arrangements.

Computer-aided design has greatly improved the performance and efficiency of reciprocating compressors, primarily due to new and innovative valve, port, and flow passage designs. Computer-aided design studies are also used to improve internal heat transfer, so that further improvements in performance and efficiency can be expected. However, it appears that other designs have potentially inherent superior characteristics for better reliability, volumetric efficiency, pressure oscillations, etc.

Rotary Compressors

Rotary compressors are produced in two general types: the rolling piston compressor and the rotating vane compressor. The most common, shown in Figure 4.7.6, is known as the rolling piston or fixed vane rotary compressor. It is commonly used in small refrigeration and air conditioning applications. The second type, the rotating vane compressor, is shown in Figure 4.7.7. It is most commonly used in automotive applications.

The ***rolling piston compressor*** is sometimes called a fixed (or stationary) vane rotary compressor because the vane does not rotate with the rotor, but rather slides in a fixed slot in the stationary part of the compressor. The rolling piston name is given to the cylindrical part of the compressor that is mounted on an eccentric shaft since it appears to roll on the cylindrical surface of the cylinder wall. Further, its position is the controlling element for the compression volume that is formed by the rolling piston, the fixed vane, and the cylinder wall, thereby having a similar function as the piston in a reciprocating compressor. During the suction part of the cycle, suction gas is drawn through a suction port into the

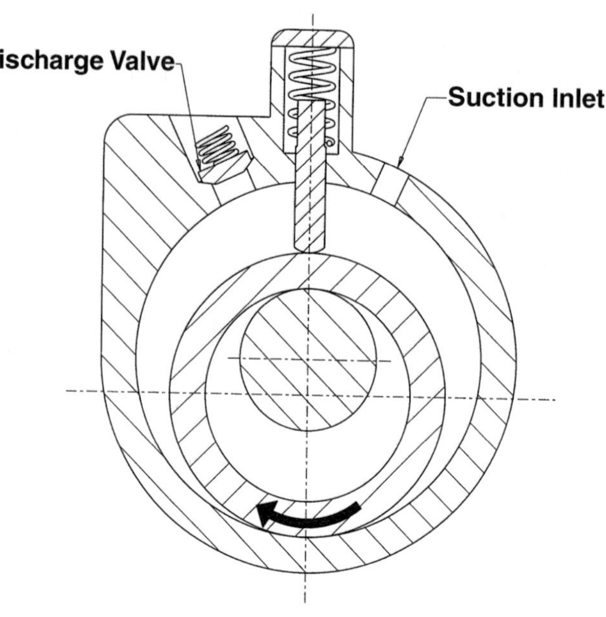

FIGURE 4.7.6 Cross-sectional drawing of a typical rolling piston (fixed vane) rotary compressor. (Courtesy of the Copeland Corporation, Sydney, OH.)

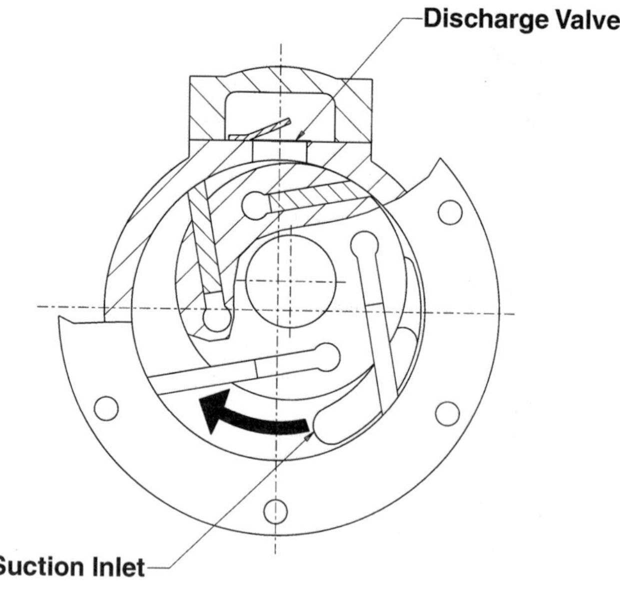

FIGURE 4.7.7 Cross-sectional drawing of a typical rotating vane rotary compressor. (Courtesy of the Copeland Corporation, Sydney, OH.)

rotary chamber, increasing volume just described. During the suction process, the compression and discharge processes take the place in the decreasing volume on the opposite side of the piston and vane. Discharge flow is controlled by use of a pressure-operated discharge valve. An advantage of rolling piston compressors is their small size for a given capacity. However, since rotary compressors are intolerant to contamination, they are usually limited to uses for which the systems are not expected to be opened for servicing. Furthermore, any wear in the compressor, whether it be the bearings, the shaft, the roller, or the vane, increases the clearances, and therefore has a significant effect on the compressor performance. The sensitivity to wear offsets the advantage on overall reliability of not having a suction valve when compared to the reciprocating compressors.

To simplify the construction and lubrication system for rolling piston compressors, high-side designs are common. However, that limits their pressure ratio without extra cooling because of high motor and oil temperatures.

In air conditioning and refrigeration uses, rolling piston compressors are produced with capacities varying from 600 to 60,000 Btu/hr (0.18 to 17.6 kW). Although there are mechanical means to vary the capacity of rolling piston compressors, variable-speed motors are normally used to achieve capacity modulation.

Pressure ratios range from 2:1 to 10:1. Performance data are shown in Figure 4.7.4. The rolling piston compressor has comparatively high efficiencies, especially when fabricated for units with less than 24,000 Btu/hr (7 kW) cooling capacity. Since it does not have a reciprocating piston, there is relatively little linear vibration and noise compared to reciprocating compressors. The compressor and motor are normally rigidly mounted in the compressor shell and, therefore, it has relatively high torsional vibration because of the single-cylinder torque pulse. To reduce the torque pulsations, especially at low speeds, compressors are produced sometimes with two out-of-phase cylinder assemblies.

The ***rotating vane rotary compressor***, shown in Figure 4.7.7, has a rotating rotor with slots that contain several vanes. The vanes are pressed against the cylinder wall as they slide in and out of the rotor as the assembly rotates. The shaft center is eccentric from the center of the cylinder; therefore, the volume of the gas in the pockets between the rotor, vanes, and cylinder varies as the unit rotates. Suction gas enters through a suction port. Discharge gas flow is commonly controlled with a discharge valve. The rotating vane compressor is subject to wear at several sealing surfaces, but is somewhat self-compensating for the wear. It can be more tolerant to contamination than the rolling piston compressor. This type of compressor has cooling capacities ranging from 12,000 to 120,000 Btu/hr (3.5 to 35 kW). It is used primarily in automotive air-conditioning applications.

Scroll Compressors

The scroll compressor is the newest of the various types of compressors to be used for the mass market. However, it is rapidly gaining market share. Figure 4.7.8 is a cross section of a typical hermetic scroll compressor. The steel shell is an integral part of the compressor-motor assembly. An internal spring suspension is not required because of the low vibration levels produced by the scroll mechanism. Suction gas is normally used to cool the electric motor. Cooling is also accomplished with direct conduction from the motor to the shell and then to the atmosphere.

The compression mechanism consists of two matching spiral scroll members opposite each other as shown in Figure 4.7.9. One of the scrolls is typically fixed and the other, driven by a crankshaft, moves in an orbit within the fixed scroll without rotation. Figure 4.7.10 illustrates the process of compression in gas pockets formed by contact of the two scrolls at several radial points. The refrigerant gas enters the lateral openings at the outer diameter between the two scrolls. The lateral openings are then sealed as the orbiting scroll moves, forming two trapped gas pockets indicated by the shaded area at the end of the suction stroke. The gas is then compressed due to the decreasing volume of the pockets as the scroll continues its orbiting motion, moving the compressed gas toward the center of the scroll elements. As the gas reaches the center, a discharge port is opened which allows the compressed gas to move to the discharge chamber of the compressor and into the system.

The development of advanced machining techniques made high-volume manufacturing of scroll compressors possible. Although the addition of compliant mechanisms make the scroll compressor somewhat complicated, there are many inherent technical advantages.

The discharge and suction processes in the scroll compressor are completely separated; therefore, there is no clearance volume effect on the volumetric efficiency and capacity of the compressor. As a result, the volumetric efficiencies of scroll compressors are almost independent of the pressure ratio for a given suction condition (i.e., suction density), due to the lack of a clearance volume effect.

FIGURE 4.7.8 Cutaway drawing of a hermetic scroll compressor. (Courtesy of the Copeland Corporation, Sydney, OH.)

FIGURE 4.7.9 A pair of matching scroll members. (Courtesy of the Copeland Corporation, Sydney, OH.)

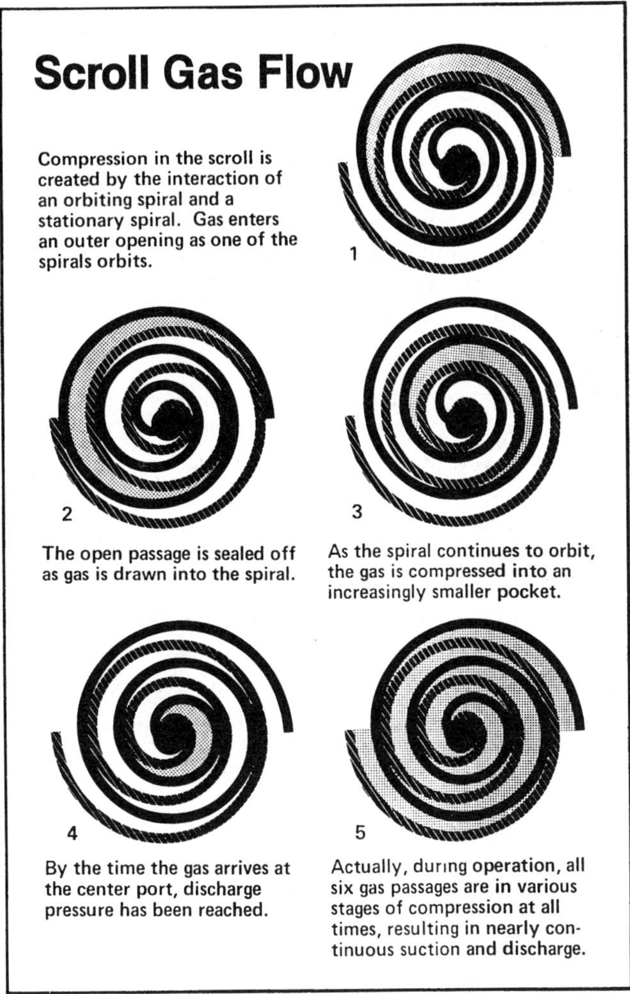

FIGURE 4.7.10 Gas flow in a scroll compressor. (Courtesy of the Copeland Corporation, Sydney, OH.)

Heat transfer between the suction and discharge processes is also greatly reduced, therefore reducing the heat transferred to the suction gas and improving overall efficiency. The scroll compressor has a constant suction volume flow rate regardless of its operating pressure ratio and the volumetric efficiency remains high, between 98% and 100%. By placing a port in the scroll member, access can be obtained to a point in the compression process. This port can be used to inject vapor or liquid in the middle of compression, making it possible to compound the refrigeration cycle to take advantage of an economizer cycle to achieve high system efficiency and capacity.

There are no valves or valve losses, and the absence of valves improves reliability. An almost constant steady flow and relatively low heat transfer gives the scroll compressor the capability of high efficiency. Being dynamically balanced and without valves allows the scroll to operate efficiently at speeds much higher (up to 10,000 rpm) than the rotary or reciprocating compressors, and over a broader range. These features, plus the nearly steady torque and smaller pressure oscillations, result in very low vibration and noise for scroll machines.

Capacity modulation is achieved by varying the speed of the compressor or by delaying the beginning of the compression process. This is done by opening a port that allows gas to flow back to the suction side during the initial compression process.

By the use of a compliant mechanism, scroll compressors are very tolerant to ingesting liquid and dirt without failure or significant loss in performance. The wear of bearings and other parts does not affect the sealing and has almost no effect on the performance of the compressor. Compliance is accomplished by holding the two scrolls together by means of gas pressure and centrifugal forces, not the mechanical structure.

Scroll compressors are typically used in air conditioners and commercial refrigeration equipment. Current production sizes range in cooling capacities from 1 to 25 tons of refrigeration (3.5 to 88 kW). Like the rotary compressors, scroll compressors are smaller in size than reciprocating compressors for the same capacity. They can be designed for the high pressure ratios needed for heat pump applications.

The increasing market share of the scroll compressors is due to their high reliability, low noise, and three performance advantages with respect to reciprocating compressors. These are all due to the almost constant volumetric efficiency with respect to pressure ratio.

1. In refrigeration applications, a smaller-displacement compressor can be used to provide adequate refrigerant capacity at very low evaporation temperatures or very high pressure ratios. This allows the compressor to operate efficiently over a much broader operating range with the same motor.
2. In heat pump applications, the heating capacity at low ambient heating conditions and high pressure ratios is higher, resulting in a higher heating seasonal performance factor (HSPF), a measure of the performance of heat pumps.
3. In air conditioning applications, the system is normally sized to ensure adequate cooling at relatively high ambient temperatures. When operating at normal or lower ambient conditions, the capacity of the scroll compressor does not increase rapidly, resulting in less excess capacity. Therefore, there is a reduced load on the coils, which increases overall system efficiency; i.e., which increases the seasonal energy efficiency ratio (SEER). The SEER is a measure of the performance of an air conditioner.

Screw Compressors

Screw compressors are available in two basic configurations, namely: the Lysholm type with twin screws, or a single screw type with gate rotors.

The ***twin screw compressor*** type is the most common. It utilizes two intermeshing helical rotors, one a male and one a female, as shown in Figure 4.7.11, to achieve the compression process. The male rotor, which is generally the driving rotor, commonly has four lobes, while the female rotor has six lobes. The suction and discharge processes are controlled by ports that are opened and closed by the rotor without

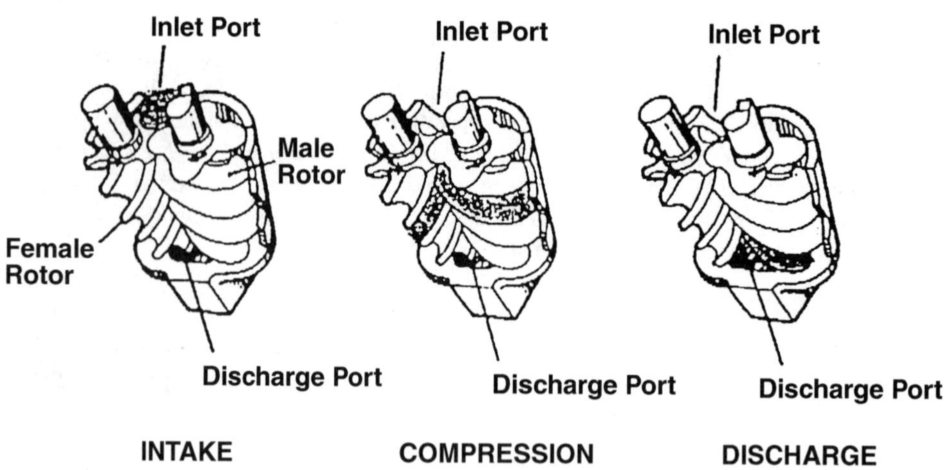

FIGURE 4.7.11 Drawings of twin-screw compressor elements. (Courtesy of the Copeland Corporation, Sydney, OH.)

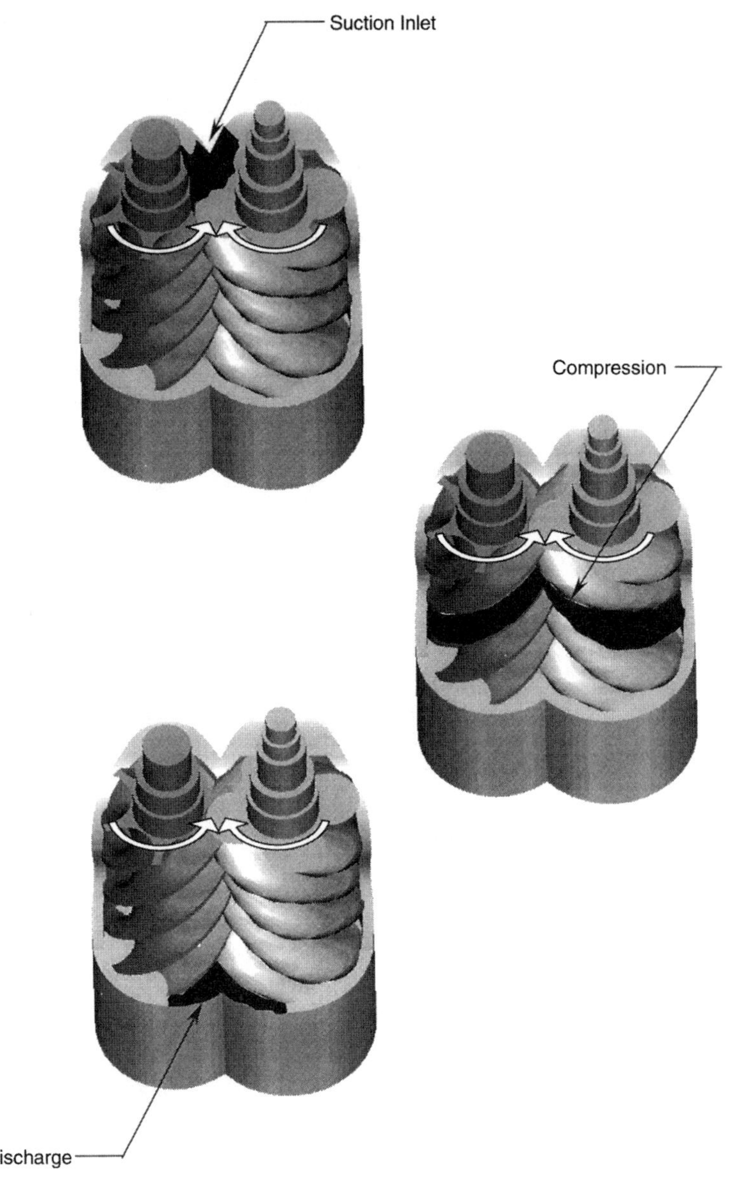

FIGURE 4.7.11 (continued).

the use of valves, as indicated in Figure 4.7.11. The port design helps reduce flow losses and provides a built-in volume ratio that gives peak performance at the corresponding pressure ratio.

The bearing loads of the twin screw compressor are relatively high and the small space between the shafts limits the diameter of the bearing that can be utilized. Normally, antifriction bearings are used on the twin screws, which results in a finite life under the high loads and often requires replacement during the life of the compressor. The overall noise level of the twin screw compressor tends to be high and is at a higher frequency than the other positive displacement compressors. With only two major moving parts, the twin screw has high reliability.

The difficulty encountered in providing a good seal between the rotors and housing is usually met with the use of a large amount of oil (typically called "oil flooding") during the compression process. The oil also serves as a coolant during the compression process. Depending on the rotor design and application, the quantity of oil can be substantial and must be separated from the discharge gas flow in most applications, usually by the use of oil separators. Thus, oil circulation can have a large effect on efficiency and the superheat of the compressor in air conditioning and refrigeration applications. At lower capacities it is difficult to maintain good sealing. The result is poorer performance at lower capacities and that has traditionally limited applying screw compressors in air conditioning and refrigeration applications below 50 tons of refrigeration (175 kW).

During operation, the capacity of the twin screw compressor is varied to match load requirements by a slide valve that extends the suction port, delaying the beginning of the compression process and thus reducing the amount of inlet gas available for compression. Some screw compressors are also equipped with valves capable of varying the size of the discharge port and, thus, have the ability to vary the volume ratio.

The ***single screw compressor*** utilizes one helical rotor and two gate rotors, as shown in Figure 4.7.12. The gate rotor teeth sweep the screw groves to achieve compression. This unique interface differentiates the single screw compressor from the twin screw compressor. The metallic screw is in contact with plastic gate rotors in such a manner that the compression process, although requiring cooling, does not necessarily require oil flooding for sealing. Thus, the single screw compressor does not require significant amounts of oil and cooling can be accomplished by the injection of a liquid refrigerant.

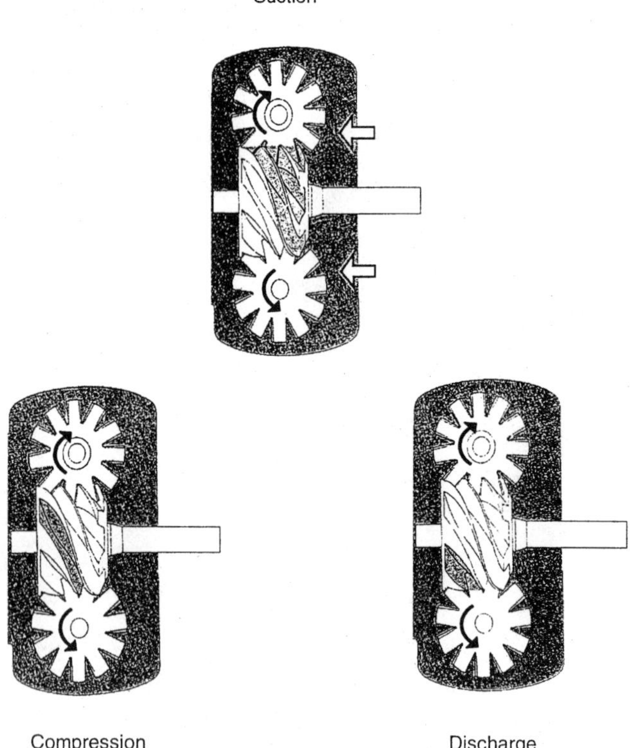

FIGURE 4.7.12 Drawings of a single-screw compressor. (Courtesy of the Copeland Corporation, Sydney, OH.)

The single screw helical rotor has balanced loading for the two compression pockets, and, therefore, does not have the large bearing load limitations of the twin screw compressor. The lower bearing loads contributes to lower noise levels than the twin screw compressors. Varying the capacity of the single screw requires two slide valves, one for each compression pocket.

Screw compressors are made as semi-hermetic compressors, and as open drive compressors. Both twin and single screw compressors can be utilized to compress a variety of gases. Both designs are amenable to capacity control by varying shaft speed as well as by use of slide valves to delay the start of compression, thereby reducing the amount of inlet gas available for compression.

When used for air conditioning and refrigeration, screw compressors, similar to scroll compressors, can be designed so that vapor or liquid can be injected in the middle of compression, making it possible to compound the refrigeration cycle with the economizer concept. This is accomplished with a port in the housing that provides access to a point in the compression process.

Screw compressors vary in capacity from 15 to more than 400 tons of refrigeration (53 to more than 1400 kW). Most screw compressors are made in the range of 50 to 300 tons of refrigeration (175 to 1050 kW). An advantage of screw compressors with respect to reciprocating compressors above 15 tons of refrigeration (53 kW) is the significantly fewer number of parts and their relatively small size. Screw compressors have also become competitive with centrifugal compressors up to 400 tons of refrigeration (1400 kW) because of their superior part load performance. The part load performance of screw compressors is generally 8 to 10% more efficient than for centrifugal technology. Also, screw compressors do not have the surge problems inherent with centrifugal compressors.

Dynamic Compressors

Dynamic compressors belong to the family of turbomachines that include fans, propellers, and turbines. Unlike positive displacement compressors which depend on a volume change for the compression process, centrifugal and axial compressors use the dynamic effect of velocity change, acceleration followed by deceleration and static pressure recovery, for the compression process. Thus, the technology used is different for centrifugal and axial compressors than for positive displacement compressors.

There are many basic references about the technologies of dynamic compressors. Five references are suggested. The basic theory is covered in References 18 and 19, and design is covered in Reference 20. A reference is suggested for centrifugal compressors (Reference 21), and one suggested for axial flow compressors (Reference 22).

A key equation used in the analysis of dynamic compressor performance is Euler's pump and turbine equation:

$$W_s = u_2 C_{\theta 2} - u_1 C_{\theta 1} \tag{4.7.1}$$

where W_s is the shaft work input per unit mass of fluid; u, the blade tip velocity; and C_θ, the tangential component of the fluid velocity leaving the blade. This equation, derived from the law of conservation of angular momentum, can be used with the aid of velocity diagrams (see Figure 4.7.15) to show the energy transfer mechanisms in centrifugal and axial flow compressors.

Centrifugal Compressors

Basic Description

In a centrifugal compressor, the suction gas enters the rotating element, or impeller, in the axial direction and is discharged in a radial direction at a higher velocity. The change in diameter through the impeller increases the velocity of the gas flow. This dynamic pressure is then converted to static pressure through a diffusion process, which generally begins within the impeller and ends in a radial diffuser outboard of the impeller. Figure 4.7.13 is a cross-sectional drawing of a typical single-stage centrifugal compressor.

A centrifugal compressor can be a single-stage compressor, having only one impeller, or it can be a multistage compressor, having two or more impellers mounted in the same casing. The suction gas generally passes through a set of adjustable inlet guide vanes (prerotation vanes) or an external suction damper before it enters the rotating impeller. The guide vanes (or the suction damper) are used to control the flow rate through the compressor. The high-velocity gas discharging from the impeller enters a radial diffuser, which can have vanes or be without vanes. Diffusers with vanes are typically used in compressors designed to provide high pressure ratios. These vanes are generally fixed, but they can also be adjustable. Adjustable diffuser vanes can be used for capacity modulation either in lieu of or in conjunction with the inlet guide vanes.

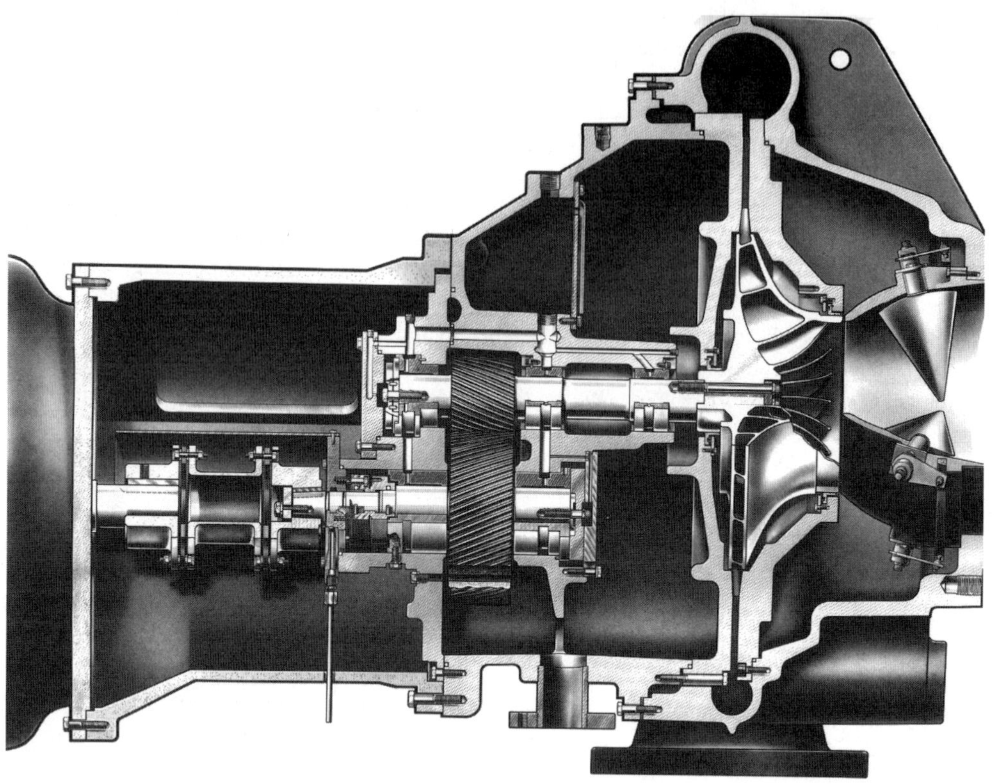

FIGURE 4.7.13 Drawing of a typical single-stage centrifugal compressor. (Courtesy of York International Corporation, York, PA).

Centrifugal compressors are comparatively simple machines with few moving parts and moderate manufacturing tolerances. Yet they have high efficiency and are very robust. If the compressor's driver allows, the compressor speed can be varied to control the pressure ratio. Centrifugal compressors have a broad flow-handling range. At a given speed, the upper limit on flow is set by stall in the diffuser or impeller passages. The lower limit on flow is marked by surge — an unsteady flow condition marked by increased noise and flow reversal through the machine.

For multistage compressors, the gas discharged from the first stage is cooled and directed to the inlet of the second stage through a return channel. The return channel contains a set of fixed flow-straightening vanes or an additional set of adjustable inlet guide vanes. Once the gas reaches the last stage, it is discharged in a volute or collector chamber. From there, the high-pressure gas flows to the compressor discharge connection.

Design Characteristics

Centrifugal compressors range in flow rate from 150 cfm (0.071 m^3/s) in automobile turbochargers to 25,000 cfm (12 m^3/s) in large industrial process compressors. Pressure ratios per stage range up to 6:1 for air compressors, with higher values attained in research equipment. Figure 4.7.14 shows a typical performance map (head rise vs. flow rate) for such a machine.[25] The horizontal axis is proportional to the suction volume flow rate and the vertical axis is proportional to head (head is the isenthalpic enthalpy increase of the gas flowing through the impeller). The term "head" carries over from pump design nomenclature and is expressed in "feet" or "meters" even for compressors. Head is an index of the pressure ratio across the compressor.

The plot also shows contours of constant efficiency, η. The head, flow, efficiency, and tip speed, u, all are nondimensionalized in the plot by using reference values. The reference value in each case is the value of the parameter at the design/rating point of the machine.

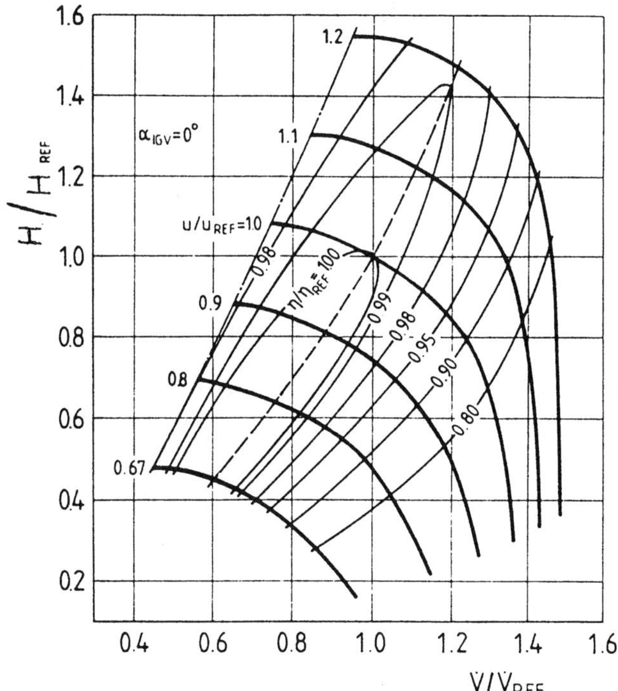

FIGURE 4.7.14 Centrifugal compressor performance map. (From the American Society of Mechanical Engineers, ASME Paper No. 86-GT-127. With permission.)

The right portions of the head-vs.-flow characteristics curves drop because of choking (explained below). The left end of each head-vs.-flow curve is limited by surge, also explained below. The surge limit, shown as a dot-dash line, connects the head/flow curves for different tip speeds. A dashed line in the center of the efficiency contours represents the loci of the "best efficiency points" at all operating speeds.

Performance maps such as the one in Figure 4.7.14 are essential tools for engineers applying compressors in particular applications. An application dictates the flow and head requirements the compressor must satisfy. The map shows the efficiency obtainable from the compressor for this flow and head, and the speed at which it must run.

Most compressors operate over a range of conditions in their applications. The performance map allows the application engineer to determine whether any requirements of the application will cause the compressor to operate too closely to surge or stall limits. If so, speed may have to be changed, prerotation vanes may have to be turned, or a different compressor design may have to be selected.

Operational limits for centrifugal compressors may be set either by structural limits or by flow parameters. For low molecular weight gases such as air, the compressor speed must remain within the stress limits of the impeller materials. Impeller tip speeds commonly fall below 1500 ft/s (460 m/s). For gases with higher molecular weights and lower acoustic velocities, the compressor speed is limited by the compressible flow phenomenon at the impeller and diffuser inlets. Impeller tip speeds commonly fall below a machine's Mach number of 1.5. Mach number here is defined as the impeller tip speed divided by the compressor inlet gas speed of sound. The maximum pressure level for centrifugal compressors is set by the casing design, and can reach several thousand psi (pounds per square inch) for multistage gas compressors used in the process industry.

The bearings in centrifugal compressors can be antifriction bearings in small machines. Plain journal bearings are commonly used in larger machines. Multistage compressors have more complicated bearing systems when operation above the first critical speed is required. The bearings are lubricated by a pressurized oil supply. Suitable seals are provided to minimize or eliminate lubricating oil carryover into the gas being compressed.

Centrifugal compressors are most commonly driven by electric motors with gearboxes to step up the speed to the level required for the compressor's duty. Steam or gas turbines drive some centrifugal compressors. Reciprocating engines are used in a few applications, generally with speed-increasing gears.

The efficiency of centrifugal compressors can be quite high under appropriate application conditions. Isentropic efficiencies of 80 to 83% can be attained in single-stage machines. Sources of losses that reduce efficiency are flow friction (due to high gas velocities), separation and mixing losses, windage in clearance spaces, bearing losses, and effects of high Mach number flows (compressibility and local shocks). Losses arising from internal heat transfer generally are negligible in these machines.

Centrifugal compressors continue to be the dominant choice for very large refrigeration sizes, e.g., applications above 1500 kW (427 tons of refrigeration) cooling capacity. This is driven by unchallenged overall system efficiencies in these sizes. Centrifugal compressors are also very simple with few moving parts. They have manufacturing tolerances two orders of magnitude greater than equivalent screw compressors do. Therefore, centrifugal compressors have relatively low manufacturing costs with high reliability. In addition, centrifugal compressors for large-size applications operate within a very narrow band of operating head pressure, thereby minimizing problems associated with surge.

Development Directions

In the future, IPLV (integrated part load value) will become more important, requiring centrifugal compressors to operate efficiently over a broader range of applications. The main challenge for the next generation of centrifugal compressors will be to maintain the full load efficiency at current levels while meeting the part load challenges of the future.

With the advent of ceramic bearings and hydrostatic bearings coupled with liquid refrigerant used as the lubricant, the elimination of oil management systems and hydrodynamic bearings can be expected. Therefore, oilless compression remains a viable technology for the centrifugal compressor. Indeed, magnetic bearings are offered in some process industry compressors.

Since 1993, centrifugal compressors are being redesigned for CFC refrigerant replacements. Many opportunities exist for increased efficiency and improved thermal aerodynamics. Utilizing the technology of the high-speed, small centrifugal turbines, even smaller sizes of centrifugal compressors are possible. Other opportunities exist for improving reliability and serviceability, for reducing noise and vibration, for increasing the use of electronics for sensing and control, the use of inverters for speed control, and for improving the manufacturing processes to reduce costs.

Flow and Performance Characteristics

Figure 4.7.15 shows the velocity vector diagram for a centrifugal compressor. The velocity diagram indicates that work input to the impeller accelerates the flow to a high velocity at its tip. A static pressure rise may occur across the impeller to an extent dictated by the impeller geometry and gas density change. The diffuser passages downstream of the impeller may be without vanes or may have vanes to decelerate the flow and increase its static pressure. A scroll or volute collects the flow at the radial diffuser exit. A conical diffuser often is located at the collector exit to achieve additional static pressure rise.

Prerotation vanes, fixed or adjustable, may be located upstream of the impeller inlet. These vanes can be used to adjust the impeller work input and inlet flow angles. Also, when turned to large angles, the prerotation vanes act as a throttling device, which may allow the compressor to operate at low mass-flow rates where otherwise surging would occur.

The impeller discharge velocity triangle includes the slip factor, S. This factor accounts for the mass-averaged deviation of the flow through the impeller. The flow does not leave at the geometric blade angle because of irrotational flow in the absolute reference frame, creating the "slip".

Equation 4.7.1 shows that the work input to a centrifugal compressor calls for maximizing the product $u_2C_{\theta 2}$. The upper limit on the value of u_2 is set by the strength of the impeller material or by the acoustic velocity of the gas being compressed. Air compressors are limited by impeller strength while refrigerant compressors are limited by flow velocities approaching the sonic limit at the impeller eye (w_1) or at the

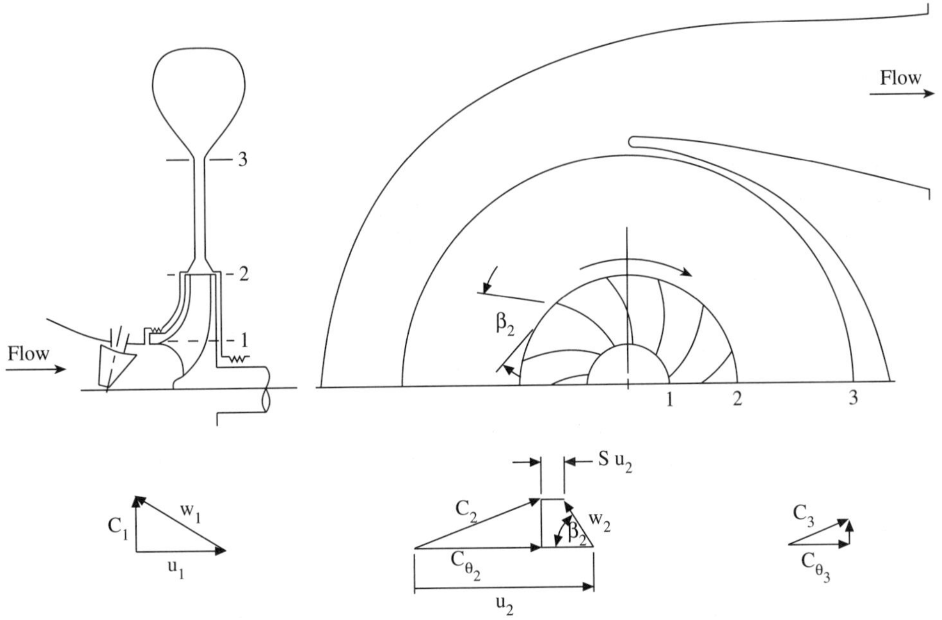

FIGURE 4.7.15 Velocity vectors for a centrifugal compressor.

inlet of vaned diffusers (related to C_2). Note that w is the relative velocity between C and u (see Figure 4.7.15).

The limit for the value of $C_{\theta 2}$ is set by the limit on the value of w_2 and the choice of the impeller discharge blade angle, β_2. Impeller passages act as diffusers, with a maximum limit of 1.4 to 1.6 set for the value of the diffusion ratio, w_1/w_2, to avoid significant stall losses. The implication of the blade angle is discussed below.

The performance map of a centrifugal compressor for work, head, and losses is illustrated qualitatively in Figure 4.7.16. The compressor losses are at a minimum when flow angles throughout the compressor match the design conditions. At lower flow rates, losses increase as impeller, diffuser, and scroll boundary layer flows stall. Stall with high loss also occurs in the impeller and scroll at high flow rates. Ultimately,

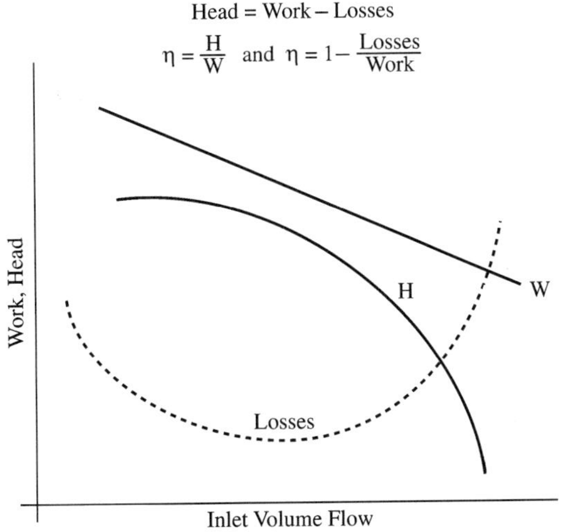

FIGURE 4.7.16 Centrifugal compressor performance at constant speed.

surge (low flow) and choke (high flow) limit the compressor's operating range. Surge is a major flow instability in the compressor, accompanied by complete flow reversal, which occurs when stall losses become so great that the compressor no longer can sustain the discharge pressure imposed upon it by the external system. Choking occurs at high flow rates when stall losses within the compressor become larger than the work input can sustain. In high-performance compressors, choking may be affected by added losses from transonic flow disturbances.

Figure 4.7.16 applies to an impeller with "backward leaning" vanes, i.e., $\beta_2 < 90°$. Figure 4.7.15 shows that if $\beta_2 = 90°$, the radial component of C_2 will have no impact on the level of $C_{\theta 2}$. This is because w_2 is radial in direction, if β_2 is equal to 90°. The slip factor, S, essentially remains constant for varying flow rates in a particular impeller. Thus, $C_{\theta 2}$ becomes a maximum that affects the surge limit of a centrifugal compressor and reduces the stable flow range. In Figure 4.7.16 the line representing work input, W, will be horizontal, not sloping downward with increasing flow. With the work input, W, as a horizontal line combined with the loss characteristics shown, the curve representing the head, H, will drop at low flow rates. Thus, surge will occur at higher flow rates for a radial bladed impeller than for a backward leaning one. A radial impeller generates maximum head but exhibits reduced range of flow between surge and choke limits.

Axial Flow Compressors

Basic Description

In an axial flow compressor, the suction gas enters and the discharge gas leaves the compressor in the axial direction. As shown by the cross section in Figure 4.7.17, axial compressors have a small frontal (inlet) area but are comparatively long in the axial dimension because multiple stages are required for most applications.[24]

Axial flow compressors are used primarily as components of gas turbines for power generation or aircraft propulsion. A few are used in process gas compression service where large volume flow rates are required. Typical flow rates for axial flow compressors range from 25,000 cfm (12 m^3/s) in small aircraft engines to 1 million cfm (470 m^3/s) for large stationary plants. The number of stages range from 7 to about 17, with the upper limit set by the dynamic considerations of the rotor.

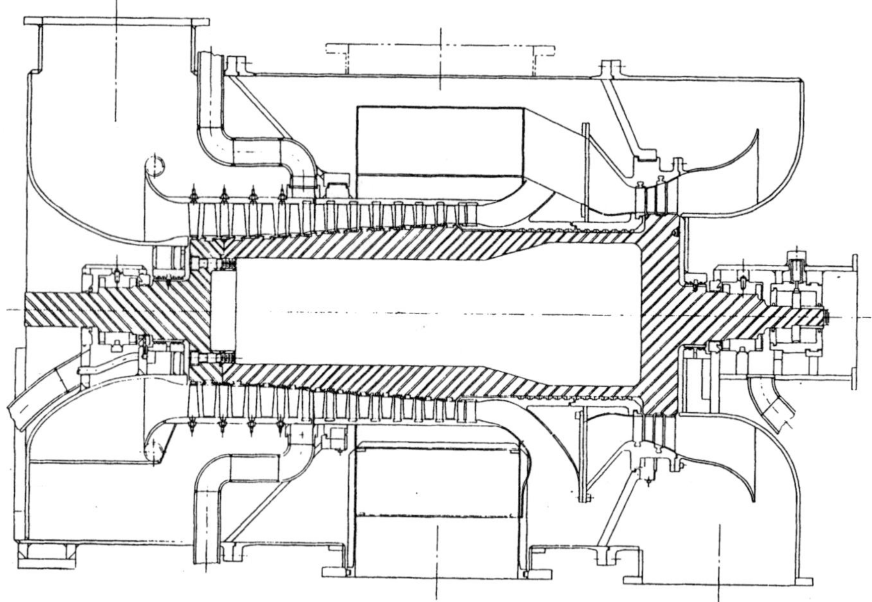

FIGURE 4.7.17 Cross-sectional drawing of an axial-flow compressor. (From the American Society of Mechanical Engineers, ASME Paper No. 58-SA-25. With permission.)

Design Characteristics

Axial compressors are complex machines with high precision required in the manufacturing of the blades. Each blade row is made up of thin, carefully profiled airfoil surfaces, which must have adequate strength to withstand large aerodynamic forces. Consequently, the flow range is narrow for axial compressors as indicated by the typical performance map shown in Figure 4.7.18.[24] The narrow range of the nearly vertical pressure ratio (or head) vs. flow curves in Figure 4.7.18 contrasts with the broader flow range exhibited by the head vs. flow curves of the centrifugal compressor shown in Figure 4.7.14.[25]

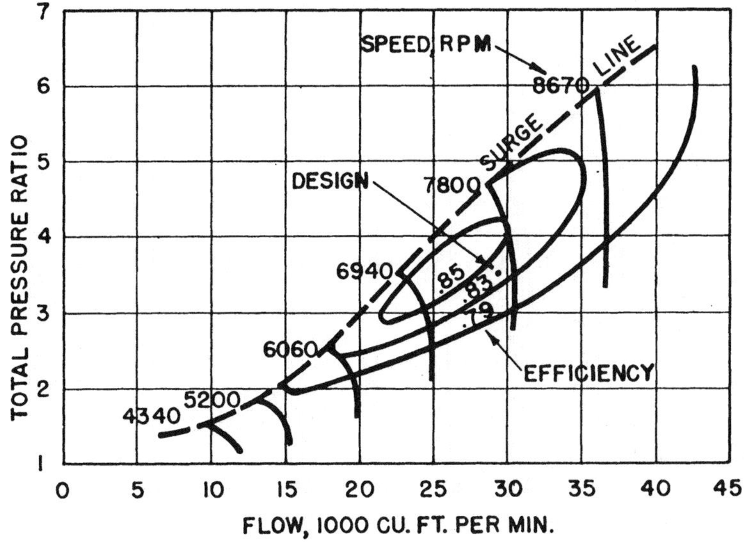

FIGURE 4.7.18 Typical performance map of an axial flow compressor. (From the American Society of Mechanical Engineers, ASME Paper No. 58-SA-25. With permission.)

Flow control in axial compressors is achieved by using variable stator vanes and, in engine applications, speed control. The flow range is limited by blade stall on both the high- and low-flow sides of the performance map. Axial compressors can exhibit surge and also rotating stall patterns under low flow conditions — both are unsteady flow conditions. Sometimes, interstage flow bleeds are used to broaden the flow range of these machines.

Compressible flow (Mach number) effects due to the high flow velocities in axial flow compressors commonly limit the pressure ratio that is achieved per stage.

Tight clearances between rotating blade tips and casing surfaces are required to minimize tip losses. Thus, most axial compressors employ antifriction bearings because of their small clearances. Pressurized lubrication systems are provided, with seals that assure little or no oil carryover into the compressed gas stream.

Gas or steam turbines are the most common drivers for axial compressors. Electric motors with speed-increasing gears sometimes are used in stationary applications.

Axial compressors are employed where high efficiency is essential. Isentropic efficiencies exceed 80% in most applications. Sources of losses which reduce efficiency include flow friction (resulting from high velocities), losses associated with transonic flow, tip clearance losses, flow separation and mixing losses, flow deviation losses in downstream stages, windage losses in disc clearance spaces, and bearing losses. Losses due to internal heat transfer are negligible in axial compressors.

Flow and Performance Characteristics

Figure 4.7.19 shows the velocity vector diagram for an axial flow compressor. The work input in axial compressors occurs as a result of a series of small changes in flow direction taking place at high-velocity

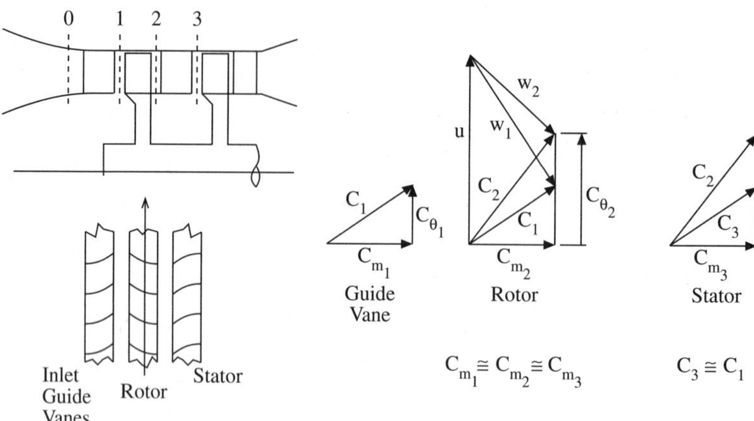

FIGURE 4.7.19 Velocity vectors for an axial flow compressor.

levels with no significant change in radius. Small static pressure increases commonly occur across each blade row, downstream of the inlet guide vanes.

Axial compressor blade design practices at first were closely related to aircraft airfoil design practices. In the U.S. an NACA airfoil series, and in the U.K., a circular arc airfoil series, were used by compressor designers as jet engines advanced in the 1950s. Wind tunnel tests of cascades of blades provided correlation of lift (angle change) and drag (losses) as a function of blade geometry, pitch/chord ratio, and stagger (blade angle relative to the axis).

Stationary blade rows in axial compressors may be fitted with mechanisms to change their angular position. Inlet guide vanes and the first several stator blade rows may be so equipped to allow the compressor to operate over a wider flow range or range of pressure ratios. Downstream stages sometimes have bleed-off ports to accommodate changes in operating conditions that otherwise might cause subsequent stages to stall.

The basic performance map of a single stage (stator and rotor blade row) of an axial compressor is similar to the performance map of the centrifugal compressor shown in Figure 4.7.16. The axial compressor's flow range and work input per stage are much lower than for a centrifugal compressor because of the narrow range of blade inlet flow angles tolerated by the airfoils before stall occurs at positive or negative angles of attack. Axial compressors often run at speeds such that flow velocities approach the velocity of sound in the gas being compressed. Transonic flow can occur at points along the blade rows, increasing the likelihood of stall as flow angles depart from design values.

For axial compressors, Equation 4.7.1 shows that the work input is a result of the product $u(C_{\theta 2} - C_{\theta 1})$ because u, the blade tip velocity, essentially is constant through each blade row. The velocity triangles in Figure 4.7.19 show that both the rotor and stator blade rows act as diffusers. The limits on the value of $(C_{\theta 2} - C_{\theta 1})$ are set by the limits on the amount of diffusion, C_3/C_2 or w_2/w_1 that can be accomplished without high losses caused by diffuser stall.

The velocity, u, has a maximum limit set either by rotor material strength or by the acoustic velocity in the gas being compressed. As values of flow velocities, C_2 and w_1, approach the acoustic velocity, losses increase due to oblique shocks and flow redirection in local regions of transonic flow.

The operating range of a complete multistage axial flow compressor is limited by rotating stall (low flow) and choking (high flow). Rotating stall is a phenomenon that occurs when the stall of one blade (or several blades) on a given blade row causes adjacent blades on the same rotor row to stall. The resulting stall "patch" rotates around the circumference of the blade row. Downstream blade rows are affected by the upstream stall zone and may undergo their own rotating stall, leading to stall for the compressor as a whole. At high flow rates, stall of one or more blade rows creates high losses that prevent the rows from handling increased flow rates against the pressure rise imposed on the compressor by the external system.

Air Compressors

Air compressors can be classified according to whether they are portable or stationary, displacement or dynamic, oil-free or lubricated. An air compressor unit is usually sold including an "airend" (compression module), driver, controls, a cooling system, and a safety system. An enclosure to reduce noise is also available in most cases. A special foundation is normally not required. "Packaged compressors" are self-contained, unlike refrigeration compressors.

Compressor Types for Various Applications

Factors affecting the selection of an air compressor are initial price, operating cost, reliability, warranty, and the availability of service and parts. Figure 4.7.20 shows the ranges of pressure and capacity typical of different types of compressors used in compressed air systems. The range of different types sometimes overlaps. The reciprocating compressors, such as the one shown in Figure 4.7.21, are used for low cost and low capacity applications. For industrial applications below about 2500 cfm (1.2 m^3/s) airflow rate, the dominant type is the oil-flooded screw compressor. These are usually purchased as a packaged compressor, such as shown in Figure 4.7.22. For oil-free applications, dry screw or reciprocating compressors are the most popular choices, with dry screw compressors taking an increasingly larger share of the market. Reciprocating compressors are also used where higher pressures are needed. Above 2500 cfm (1.2 m^3/s) airflow rate, centrifugal compressors, such as shown in Figure 4.7.23, dominate, up to the sizes where axial flow compressors become practical.

The smallest portable compressors, i.e., less than 50 cfm (0.0235 m^3/s) airflow rate, are usually built with reciprocating compressors driven by small gasoline engines. Above this range, most portable air compressors utilize an oil-flooded rotary twin screw design driven by a diesel engine. Low-pressure air compressors with air discharge pressures below 200 psig (1480 kPa) are usually single stage designs, whereas higher-pressure compressors have multiple stages. Oil-free portable compressors are available with two-stage rotary screw or centrifugal compressors. Most portable compressors have air flow rates from 100 to 1800 cfm (0.047 to 0.85 m^3/s), producing air delivery pressures of 100 to 150 psig (790 to 1140 kPa).

Nearly oil-free air can be obtained from oil-flooded compressors by after-cooling the air and then passing it through a series of very fine coalescing filters to remove the remnants of the oil aerosols.

Performance Considerations

Ratings

Air compressors are rated on delivered airflow, discharge pressure, and power consumption. The airflow is stated in terms of inlet volumetric flow, sometimes called "free air delivery." It is determined in a test by measuring the mass flow rate at the discharge and dividing by the inlet density. For a bare compressor, the inlet is clearly defined, but for a packaged compressor the inlet can be specified to be at either the package or compression module inlet. The stated flow will usually be lower if the package inlet is used, because the temperature is lower with a higher density. Test codes for measuring compressor performance have been developed by the European Committee of Manufacturers of Compressors, Vacuum Pumps and Pneumatic Tools (PNEUROP), and in the U.S. by the Compressed Air and Gas Institute (CAGI). These test codes have recently been included in ISO 1217, which is gradually becoming the industry standard in Europe and the U.S. This ISO standard specifies a standard package inlet condition of 20°C (68°F), 100 kPa (14.5 psia), and 0% relative humidity. The unit of flow in the ISO system is m^3/s, whereas in the U.S. ft^3/min (cfm) is generally used. Power is normally expressed in kilowatts or horsepower. For portable compressors, fuel consumption is stated instead of power.

Capacity and Pressure Control

Capacity control is an important performance characteristic, since it is a major factor contributing to how efficiently the compressor operates at part load. In addition to on-off control, compressors are usually fitted with an inlet valve. This inlet valve controls the airflow to the compressor by either opening, throttling, or closing completely in response to pressure-sensing controls. Some compressors are

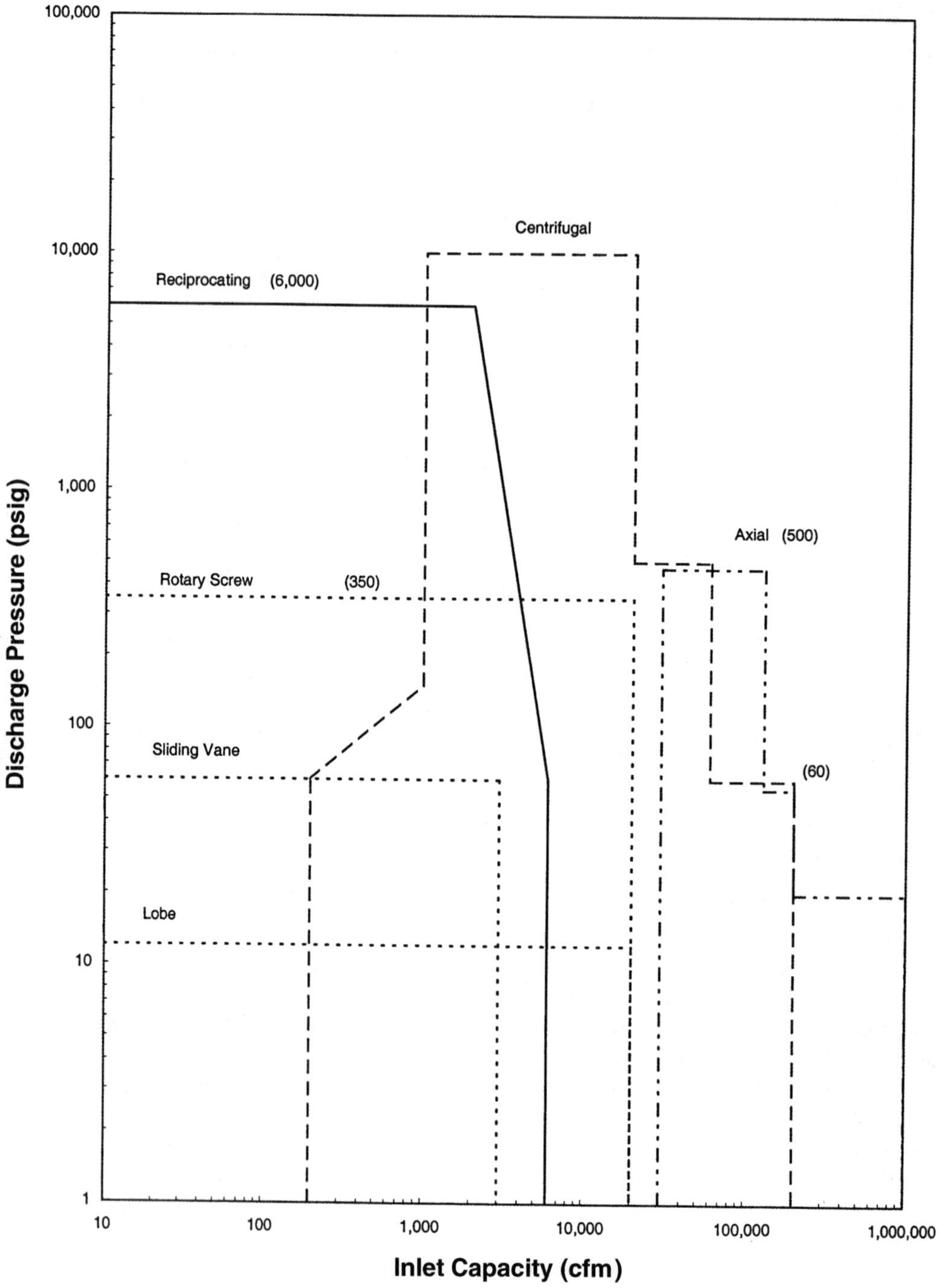

FIGURE 4.7.20 Pressure-capacity chart showing the effective ranges of most air compressors. (Courtesy of Ingersoll-Rand Corp., Woodcliff, NJ.)

equipped with a check valve that allows the receiver tank to be depressurized during periods of low demand. This reduces unloaded power by reducing the discharge pressure. Speed control is used on engine-driven compressors, and, increasingly, on those driven by electric motors. Dry screw compressors are not throttled at the inlet, since this increases the pressure ratio and, thus, the temperature rise across

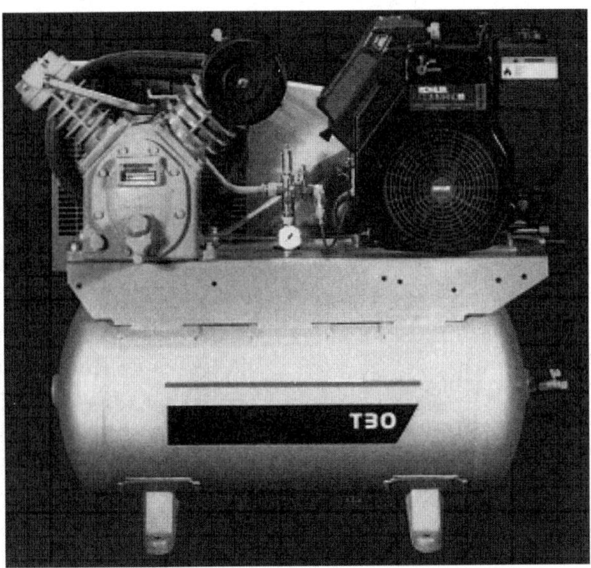

FIGURE 4.7.21 Two-stage reciprocating air compressor package with horizontal receiver tank. (Courtesy of Ingersoll-Rand Corp., Woodcliff, NJ.)

FIGURE 4.7.22 Interior of an air-cooled, two-stage, 200 hp oil-flooded screw compressor package. (Courtesy of Ingersoll-Rand Corp., Woodcliff, NJ.)

a stage. These compressors are usually equipped with a check valve so the discharge can be depressurized. The inlet valve is either full open (when loaded) or full closed with the discharge at ambient pressure.

Ambient Temperature

Since the heat of compression must be carried away by a cooling system, all compressors have a limiting ambient temperature (LAT) above which the internal temperatures exceed their limits. Higher LAT values are possible with better cooling systems. The engine or motor cooling system also affects the LAT.

FIGURE 4.7.23 Three 5700 cfm, 1250 hp centrifugal air compressors supplying oil-free air to a textile facility. (Courtesy of Ingersoll-Rand Corp., Woodcliff, NJ.)

Portable compressors usually have a temperature switch at the compressor discharge and another for the engine coolant. Exceeding the set point of either switch will cause the compressor to shut down.

Compressors also have low-temperature limits. Typical minimum operating temperatures are 35°F (1.7°C) for stationary compressors, and –10°F (–23.3°C) for portables. A low ambient compressor option is usually available for stationary compressors operating outside in cold climates. The low-temperature limits are due to the potential of the condensate freezing in the pneumatic control system components and in the aftercooler. For portable compressors, engine operation at cold temperatures can be the factor that determines the low-temperature limit. Startup at low temperatures must be carefully controlled to ensure that the compressor receives adequate lubrication. Preheating the oil will sometimes allow the compressor to be started at lower temperatures. In general, dry screw compressors have a more narrow ambient temperature range than oil-flooded screw compressors.

Ambient Pressure and Altitude

Except for dry screw and dynamic compressors, the ambient pressure is usually not a limiting factor from the compression module standpoint. As the altitude increases the ambient pressure and temperature decrease. As a result, the power requirement of the compressor also decreases. This more than offsets the lower cooling efficiency when operating with less dense ambient air. For portable compressors, the maximum altitude is usually limited by the engine, which loses power capability with altitude. Dry screw and dynamic compressors must operate at lower discharge pressures at higher altitudes, to keep the pressure ratio from exceeding the limit for a particular unit.

Start-Stop Frequency

While on-off operation is an efficient mode of capacity control, it is limited by the maximum start-stop frequency. This is typically six per hour for an electric motor, to prevent overheating the motor starter components. Diesel engines must operate long enough between starts to recharge the battery. In addition, the starter motor and ring gear will wear out prematurely if the start-stop frequency is too high.

Compressor Speeds and Drives

Oil-flooded screw compressors operate most efficiently at a tip speed of about 35 m/s (115 ft/s). At higher tip speeds the efficiency decreases and the noise increases. These "airends" are usually not operated with tip speeds above 50 m/s (164 ft/s). Dry screw compressors must operate at tip speeds near 100 m/s (328 ft/s), to minimize back-flow leakage. As a result they are much noisier than oil-flooded screw compressors and require more acoustical treatment.

In general, a large, slow-running compressor is more efficient and quieter than a small, high-speed compressor. However, the initial cost is higher. As compressor design advances, the trend is to smaller, faster-running compressors, with better silencing techniques and more sophisticated controls to keep operating costs down.

Most portable air compressors are driven by diesel engines, except those at the lower end of the airflow rates (below 100 cfm [0.047 m^3/s]), where gasoline engines are often used. Electric motor-driven portables are used in some areas. Stationary compressors are driven by electric motors. The majority of them operate at either 1800 rpm with a 60-Hz line frequency or at 1500 rpm where the line frequency is 50 Hz. Very large compressors in process plants are often driven by steam turbines.

Most portable compressors use a drive coupling to transmit the torque from the engine to the compression module, although small compressors are often belt driven. The drive coupling can be a troublesome component and must be carefully selected to minimize operational problems. Ideally, the compressor should be rugged enough to withstand the torque pulses from the engine, so that a torsionally stiff drive coupling can be used. This is often not possible, however, especially with the larger compressors, and a torsionally soft coupling must be used to isolate the "airend" from the engine. In these cases, the engine speed must be maintained well above the torsional critical speed to prevent drive train failures. The torque pulses from an electric motor are mild compared to those from a diesel engine. As a result, the coupling is often eliminated by directly coupling the compressor and motor. This is accomplished by mounting the compressor drive gear on the motor shaft. If a coupling is used, it can be torsionally stiff.

Maintenance Requirements

One of the primary reasons for the popularity of oil-flooded screw compressors is their ability to operate continuously for many years with only routine maintenance. Filter maintenance is most important, since the incoming air is mixed with the oil and any particles in the incoming air end up in the oil. This same oil is used to lubricate the bearings, and particles in the oil will decrease bearing life. Typically, a 25-μm (0.64 μm) air filter is specified, with cleaning or changing required when the restriction exceeds 25 in. of water (6.2 kPa). The oil system also has a 25-μm (0.64 μm) filter, as well as an oil separation element, which captures particles as small as 1 μm (0.025 μm). Any contamination entering the system ends up in the separator element, thus shortening its life. The use of finer air and oil filters will increase the time between separator element changes. This is usually desirable, since changing the air and oil filters is much easier than changing the separator element.

The coolers must also be kept clean. Dirty or fouled coolers reduce the LAT, sometimes so much that the compressor shuts down because of overheating. Another undesirable consequence of dirty coolers is a shorter bearing life due to higher oil temperatures.

Regular filter maintenance is just as important on dry screw and dynamic compressors. A restricted air inlet reduces the compressor inlet pressure and raises the pressure ratio across it. This can cause overheating of a dry screw compressor and surge in a dynamic compressor.

Thermal Issues

Performance

Assuming adiabatic compression and applying the first law of thermodynamics, the rate of work of compression for a single stage compressor is given by

$$\dot{W} = \dot{m} c_p (T_2 - T_1) \tag{4.7.2}$$

where $\dot{m}$ is the mass flow rate of the compressed air; c_p, the specific heat; T_2, the outlet temperature; and T_1, the inlet temperature. Treating the air as a perfect gas, this becomes

$$\dot{W} = P_1 \dot{V}_1 \frac{k}{k-1} \left[\left(\frac{P_2}{P_1} \right)^{\frac{k-1}{k}} - 1 \right] \frac{1}{\eta} \tag{4.7.3}$$

where k is the specific heat ratio; P_1, the absolute inlet pressure, $\dot{V}_1$, the inlet volumetric flow rate; P_2, the absolute discharge pressure; and η, the adiabatic efficiency. For air, k = 1.4 and c_p = 0.24 Btu/(lb_m °F) (c_p = 1.004 kJ/[kg K]). In practice, the input power and flow rate for a bare compressor (i.e., without inlet filters or discharge restriction) are measured on a dynamometer, and the adiabatic efficiency is calculated from Equation 4.7.3. This efficiency includes the power lost in the gears and bearings. Positive displacement compressors, including oil-flooded screw compressors, are almost always analyzed using the adiabatic model. Comparing adiabatic efficiencies at different pressure ratios is not a valid procedure. The efficiency should be measured at each pressure ratio of interest.

Examination of Equation 4.7.3 shows that the required power input depends only on the inlet flow rate, the inlet pressure, the pressure ratio, and the k-value. It does not depend on the inlet temperature of the gas or its molecular weight. In two-stage compressors, the power to the second stage can be reduced by intercooling; i.e., cooling the air between the first and second stages. The lower the temperature at the entrance of stage 2, the lower will be V_1 for that stage. By adding even more stages and intercooling before each additional stage, isothermal compression is approached, which requires less power than adiabatic compression. The better the intercooling, the more efficient is the overall compression process. The power increases approximately 1% for every 10°F (5.6°C) rise in inlet temperature to each stage. It is also important to keep the interstage pressure drop as small as possible, as pressure drop increases the pressure ratio for the next stage.

The inlet flow rate depends mainly on the speed, although it decreases slightly as the pressure ratio increases. In oil-flooded screw compressors, the air is heated by hot injection oil, which in turn decreases the inlet density of the air, thus decreasing the mass flow rate. A reasonable approximation for this effect is a 1% decrease in flow per 10°F (5.6°C) rise in the difference between the oil injection and air inlet temperatures. Since the injection oil temperature is held constant by the thermostat in cold weather, this means the flow decreases about 1% for every 10°F (5.6°C) drop in ambient temperature.

Water Condensation

All air contains moisture, the amount being influenced by pressure, temperature, and proximity to oceans, lakes and rivers. The water content of saturated air (100% relative humidity) is given by

$$W_{sat} = 0.622 \frac{P_{sat}}{P - P_{sat}} \tag{4.7.4}$$

where W_{sat} is the humidity ratio, in lb_m water vapor/lb_m dry air; P_{sat}, the saturation pressure of water vapor at the temperature of the air; and P, the absolute pressure of the air-water vapor mixture. P_{sat} can be obtained from steam tables. As Equation 4.7.4 indicates, the water-holding capacity of air decreases if the air is compressed, unless P_{sat} is raised by increasing the temperature. The dew point temperature is the temperature at which W_{sat} is equal to the specific humidity of the ambient air. For example, ambient air at 14.7 psia (101 kPa), 80°F (26.7°C), and 60% relative humidity has a saturation pressure of 0.51 psia (3.52 kPa) and a vapor pressure of 0.60 × 0.51 = 0.31 psia (2.14 kPa). The humidity ratio is W = 0.622 × 0.31/(14.7 – 0.31) = 0.013 lb_m water/lb_m dry air. If the air is compressed to 200 psig (1480 kPa) and then cooled to 100°F (37.8°C), P_{sat} = 0.95 psia (6.55 kPa) and the water-holding capacity is W_{sat} = 0.622 × 0.95/(214.7 – 0.95) = 0.0028 lb_m water/lb_m dry air. This is much lower than the entering humidity, so the amount of condensation is 0.013 – 0.0028 = 0.010 lb_m water/lb_m dry air (0.01 kg/kg). At 1000 cfm (472 m^3/s) or 75 $lb_{m,air}$/min (34 kg_{air}/min), the condensation rate is 0.010 × 75 = 0.75 $lb_{m,water}$/min (0.34 kg_{water}/min) or 5 gal_{water}/hr (18.9 $liter_{water}$/hr). If the compressed air temperature is 180°F (82.2°C), instead of 100°F (37.8°C), the saturation pressure is 7.51 psia (51.8 kPa) and W_{sat} = 0.022 lb_m water/lb_m dry air. At this temperature, no condensation takes place. Repeating this calculation for different values of the temperature, one finds that W_{sat} is equal to 0.013 lb_m water/lb_m dry air at 160°F (71.1°C). Thus, the dew point temperature is 160°F (71.1°C).

Condensation in oil-flooded screw compressors is particularly objectionable because the water will mix with the oil and reduce its lubricating qualities. Using a thermostat in the oil system to keep the "airend" discharge temperature above a minimum level prevents condensation. A minimum discharge temperature of 180°F (82°C) is usually adequate for pressures up to 350 psig (2500 kPa). Once the oil has been separated from the air, the air can be cooled in an aftercooler and the condensation removed in a water separator. Since the air is saturated when it leaves the aftercooler, and still above ambient temperature, further condensation will take place in the air line. For this reason, most plants employ an air dryer downstream of the compressor.

Reduction of Energy Use

Compressed air can account for 10 to 20% of a manufacturing facility's electricity bill.[4] Energy efficiency for compressed air systems has reached a new height recently with the announcement in 1998 by the U.S. Department of Energy of the Compressed Air Challenge. This initiative is designed to build awareness among users of compressed air about the benefits and approaches for improving and maintaining compressed air system efficiency. Some basic methods for improving efficiency are given by Edwards[23] as follows:

1. Reduce unnecessary demand by eliminating leaks, by using condensate drain valves that open only when a predetermined amount of water has accumulated, by regulating all point-of-use operations at the lowest possible pressure, and by modifying and/or eliminating "blowoffs." Often the output of a low-pressure blower or fan can replace compressed air for "blowoff" operations.
2. Shut off the air supply to "off-line" production equipment. If one point-of-use requires air pressure at a much higher pressure than the rest of the system, consider using a pressure booster or using a separate dedicated system at that point.
3. Use storage tanks to supply air for intermittent events that require large amounts of air for a short time. Storage can eliminate the need for additional compressors to handle peak demands.
4. Reduce pressure drop by having adequate line sizes.
5. Select the compressor sizes and control their operation so that a compressor is either off, or running at rated load. No more than one compressor at a time should operate at part load.
6. Utilize the waste heat.

References

1. Rollins, J. P., Ed., *Compressed Air and Gas Handbook*, 5th ed., Compressed Air and Gas Institute, Cleveland, OH, 1989.
2. Soedel, W., Ed., Proc. 1972 Purdue Compressor Technol. Conf., 90 papers, 568 pages, Purdue University, Ray W. Herrick Laboratories, West Lafayette, IN, July 25-27, 1972.
3. Soedel, W., Ed., Proc. 1974 Purdue Compressor Technol. Conf., 73 papers, 485 pages, Purdue University, Ray W. Herrick Laboratories, West Lafayette, IN, July 10-12, 1974.
4. Hamilton, J. F., Ed., Proc. 1976 Purdue Compressor Technol. Conf., 81 papers, 551 pages, Purdue University, Ray W. Herrick Laboratories, West Lafayette, IN, July 6-9, 1976.
5. Hamilton, J. F., Ed., Proc. 1978 Purdue Compressor Technol. Conf., 55 papers, 400 pages, Purdue University, Ray W. Herrick Laboratories, West Lafayette, IN, July 19-21, 1978.
6. Soedel, W., Ed., Proc. 1980 Purdue Compressor Technol. Conf., 69 papers, 420 pages, Purdue University, Ray W. Herrick Laboratories, West Lafayette, IN, July 23-25, 1980.
7. Cohen, R., Ed., Proc. 1982 Purdue Compressor Technol. Conf., 63 papers, 446 pages, Purdue University, Ray W. Herrick Laboratories, West Lafayette, IN, July 21-23, 1982.
8. Cohen, R., Ed., Proc. 1984 Int. Compressor Eng. Conf. Purdue, 91 papers, 651 pages, Purdue University, Ray W. Herrick Laboratories, West Lafayette, IN, July 11-13, 1984.
9. Hamilton, J. F. and Cohen, R., Eds., Proc. 1986 Int. Compressor Eng. Conf. Purdue, Vol. I, II, and III, 84 papers, 1201 pages, Purdue University, Ray W. Herrick Laboratories, West Lafayette, IN, August 4-7, 1986.
10. Soedel, W., Ed., Proc. 1988 Int. Compressor Eng. Conf. Purdue, Vol. I and II, 89 papers, 666 pages, Purdue University, Ray W. Herrick Laboratories, West Lafayette, IN, July 18-21, 1988.

11. Soedel, W., Ed., Proc. 1990 Int. Compressor Eng. Conf. Purdue, Vol. I and II, 109 papers, 982 pages, Purdue University, Ray W. Herrick Laboratories, West Lafayette, IN, July 17-20, 1990.
12. Hamilton, J. F., Ed., Proc. 1992 Int. Compressor Eng. Conf. Purdue, Vol. I, II, III and IV, 160 papers, 1558 pages, Purdue University, Ray W. Herrick Laboratories, West Lafayette, IN, July 14-17, 1992.
13. Soedel, W., Ed., Proc. 1994 Int. Compressor Eng. Conf. Purdue, Vol. I and II, 127 papers, 824 pages, Purdue University, Ray W. Herrick Laboratories, West Lafayette, IN, July 19-22, 1994.
14. Soedel, W., Ed., Proc. 1996 Int. Compressor Eng. Conf. Purdue, Vol. I and II, 133 papers, 856 pages, Purdue University, Ray W. Herrick Laboratories, West Lafayette, IN, July 23-26 1996.
15. Soedel, W., Ed., Proc. 1998 Int. Compressor Eng. Conf. Purdue, Vol. I and II, 153 papers, 950 pages, Purdue University, Ray W. Herrick Laboratories, July 14-17, 1998.
16. *ASHRAE Handbook: HVAC Systems and Equipment,* American Society of Heating Refrigerating, and Air Conditioning Engineers, Inc., Atlanta, GA, 1996, chap. 34.
17. Cohen, R. and Groll, E. A., Update on Refrigerant Compressors in Light of CFC Substitutes, *Bull. Int. Inst. Refrig.,* 96/5, LXXVI, 1996.
18. Cumpsty, N. A., *Compressor Aerodynamics*, Longman, Harlow, U.K., 1989.
19. Shephard, D. G., *Principles of Turbomachinery,* Macmillan, New York, 1956.
20. Wilson, D. G., *The Design of High-Efficiency Turbomachinery and Gas Turbines*, MIT Press, Cambridge, MA, 1984.
21. Ferguson, T. B., *The Centrifugal Compressor Stage*, Butterworth & Co. Ltd., London, 1963.
22. Horlock, J. H., *Axial Flow Compressors*, Butterworth & Co. Ltd., London, 1958.
23. Edwards, P., Basic efficiency steps help total plant electrical costs, *Facilities Eng.,* 25, 9, 1998.
24. Reisweber, R. C., Glessner, J. W., and Shields, J. R., Design and Development of a Supercharger for a Pressure-Fired Boiler, ASME Paper No. 58-SA-25, American Society of Mechanical Engineers, New York, 1958.
25. Simon, H., Wallmann, T., and Moenk, T., Improvements in Performance Characteristics of Single-Stage and Multistage Centrifugal Compressors by Simultaneous Adjustment of Inlet Guide Vanes and Diffuser Vanes, ASME Paper No. 86-GT-127, American Society of Mechanical Engineers, New York, 1986.

4.8 Pumps and Fans

Robert F. Boehm

Introduction

Pumps are devices that impart a pressure increase to a liquid. Fans are used to increase the velocity of a gas, but this is also accomplished through an increase in pressure. The pressure rise found in pumps can vary tremendously, and this is a very important design parameter along with the liquid flow rate. This pressure rise can range from simply increasing the elevation of the liquid to increasing the pressure hundreds of atmospheres. Fan applications, on the other hand, generally deal with small pressure increases. In spite of this seemingly significant distinction between pumps and fans, there are many similarities in the fundamentals of certain types of these machines as well as with their application and theory of operation.

The appropriate use of pumps and fans depends upon the proper choice of device and the proper design and installation for the application. A check of sources of commercial equipment shows that many varieties of pumps and fans exist. Each of these had special characteristics that must be appreciated for achieving proper function. Preliminary design criteria for choosing between different types is given by Boehm (1987).

As is to be expected, the wise applications of pumps and fans requires knowledge of fluid flow fundamentals. Unless the fluid mechanics of a particular application are understood, the design could be less than desirable.

In this section, pump and fan types are briefly defined. In addition, typical application information is given. Also, some ideas from fluid mechanics that are especially relevant to pump and fan operation are reviewed.

Pumps

Raising of water from wells and cisterns is the earliest form of pumping (a very detailed history of early applications is given by Ewbank, 1842). Modern applications are much broader, and these find a wide variety of machines in use. Modern pumps function on one of two principles. By far the majority of pump installations are of the *velocity head* type. In these devices, the pressure rise is achieved by giving the fluid a movement. At the exit of the machine, this movement is translated into a pressure increase. The other major type of pump is called *positive displacement*. These devices are designed to increase the pressure of the liquid while essentially trying to compress the volume. A categorization of pump types has been given by Krutzsch (1986), and an adaptation of this is shown below.

I. Velocity head
   A. Centrifugal
      1. Axial flow (single or multistage)
      2. Radial flow (single or double suction)
      3. Mixed flow (single or double suction)
      4. Peripheral (single or multistage)
   B. Special Effect
      1. Gas lift
      2. Jet
      3. Hydraulic ram
      4. Electromagnetic
II. Positive displacement
   A. Reciprocating
      1. Piston, plunger
         a. Direct acting (simplex or duplex)

b. Power (single or double acting, simplex, duplex, triplex, multiplex)
2. Diaphragm (mechanically or fluid driven, simplex or multiplex)
B. Rotary
1. Single rotor (vane, piston, screw, flexible member, peristaltic)
2. Multiple rotor (gear, lobe, screw, circumferential piston)

In the next subsection, some of the more common pumps are described.

Centrifugal and Other Velocity Head Pumps

Centrifugal pumps are used in more industrial applications than any other kind of pump. This is primarily because these pumps offer low initial and upkeep costs. Traditionally, pumps of this type have been limited to low-pressure-head applications, but modern pump designs have overcome this problem unless very high pressures are required. Some of the other good characteristics of these types of devices include smooth (nonpulsating) flow and the ability to tolerate nonflow conditions.

The most important parts of the centrifugal pump are the *impeller* and *volute*. An impeller can take on many forms, ranging from essentially a spinning disk to designs with elaborate vanes. The latter is usual. Impeller design tends to be somewhat unique to each manufacturer, as well as finding a variety of designs for a variety of applications. An example of an impeller is shown in Figure 4.8.1. This device imparts a radial velocity to the fluid that has entered the pump perpendicular to the impeller. The volute (there may be one or more) performs the function of slowing the fluid and increasing the pressure. A good discussion of centrifugal pumps is given by Lobanoff and Ross (1992).

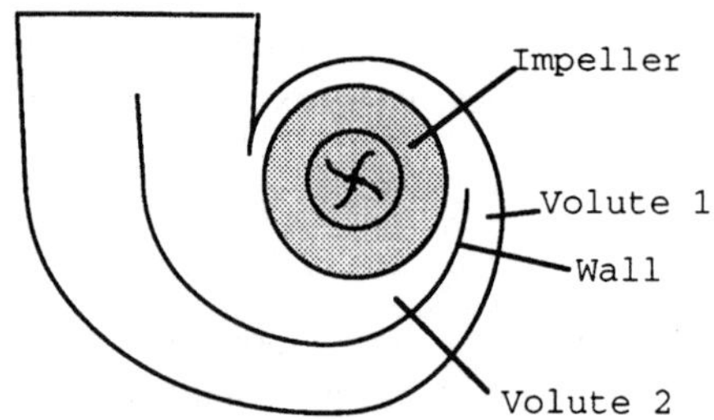

FIGURE 4.8.1. A schematic of a centrifugal pump is shown. The liquid enters perpendicular to the figure, and a radial velocity is imparted by clockwise spin of the impeller.

A very important factor in the specification of a centrifugal pump is the *casing orientation* and *type*. For example, the pump can be oriented vertically or horizontally. Horizontal mounting is most common. Vertical pumps usually offer benefits related to ease of priming and reduction in required net positive suction head (see discussion below). This type also requires less floor space. Submersible and immersible pumps are always of the vertical type. Another factor in the design is the way the casing is split, and this has implications about ease of manufacture and repair. Casings that are split perpendicular to the shaft are called *radially split*, while those split parallel to the shaft axis are denoted as *axially split*. The latter can be *horizontally split* or *vertically split*. The number of *stages* in the pump greatly affects the pump-output characteristics. Several stages can be incorporated into the same casing, with an associated increase in pump output. Multistage pumps are often used for applications with total developed head over 50 atm.

Whether or not a pump is self-priming can be important. If a centrifugal pump is filled with air when it is turned on, the initiation of pumping action may not be sufficient to bring the fluid into the pump. Pumps can be specified with features that can minimize priming problems.

There are other types of velocity head pumps. *Jet pumps* increase pressure by imparting momentum from a high-velocity liquid stream to a low-velocity or stagnant body of liquid. The resulting flow then goes through a diffuser to achieve an overall pressure increase. *Gas lifts* accomplish a pumping action by a drag on gas bubbles that rise through a liquid.

Positive-Displacement Pumps

Positive-displacement pumps demonstrate high discharge pressures and low flow rates. Usually, this is accomplished by some type of pulsating device. A piston pump is a classic example of positive-displacement machines. Rotary pumps are one type of positive-displacement device that do not impart pulsations to the existing flow (a full description of these types of pumps is given by Turton, 1994). Several techniques are available for dealing with pulsating flows, including use of double-acting pumps (usually of the reciprocating type) and installation of pulsation dampeners.

Positive-displacement pumps usually require special seals to contain the fluid. Costs are higher both initially and for maintenance compared with most pumps that operate on the velocity head basis. Positive-displacement pumps demonstrate an efficiency that is nearly independent of flow rate, in contrast to the velocity head type (see Figure 4.8.2 and the discussion related to it below).

Reciprocating pumps offer very high efficiencies, reaching 90% in larger sizes. These types of pumps are more appropriate for pumping abrasive liquids (e.g., slurries) than are centrifugal pumps.

A characteristic of positive displacement pumps which may be valuable is that the output flow is proportional to pump speed. This allows this type of pump to be used for metering applications. Also a positive aspect of these pumps is that they are self-priming, except at initial start-up.

Very high head pressures (often damaging to the pump) can be developed in positive-displacement pumps if the downstream flow is blocked. For this reason, a pressure-relief-valve bypass must always be used with positive-displacement pumps.

Pump/Flow Considerations

Performance characteristics of the pump must be considered in system design. Simple diagrams of pump applications are shown in Figure 4.8.2. First, consider the left-hand figure. This represents a flow circuit, and the pressure drops related to the piping, fittings, valves, and any other flow devices found in the circuit must be estimated using the laws of fluid mechanics. Usually, these resistances (pressure drops) are found to vary approximately with the square of the liquid flow rate. Typical characteristics are shown in Figure 4.8.3. Most pumps demonstrate a flow vs. pressure rise variation that is a positive value at zero flow and decreases to zero at some larger flow. Positive-displacement pumps, as shown on the right-hand side of Figure 4.8.3, are an exception to this in that these devices usually cannot tolerate a zero flow. An important aspect to note is that a closed system can presumably be pressurized. A contrasting situation and its implications are discussed below.

The piping diagram show on the right-hand side of Figure 4.8.2 is a once-through system, another frequently encountered installation. However, the leg of piping through "pressure drop 1" shown there can have some very important implications related to *net positive suction head*, often denoted as **NPSH**. In simple terms, NPSH indicates the difference between the local pressure and the thermodynamic saturation pressure at the fluid temperature. If NPSH = 0, the liquid can vaporize, and this can result in a variety of outcomes from noisy pump operation to outright failure of components. This condition is called **cavitation**. Cavitation, if it occurs, will first take place at the lowest pressure point within the piping arrangement. Often this point is located at, or inside, the inlet to the pump. Most manufacturers specify how much NPSH is required for satisfactory operation of their pumps. Hence, the actual NPSH (denoted as **NPSHA**) experienced by the pump must be larger than the manufacturer's required NPSH (called **NPSHR**). If a design indicates insufficient NPSH, changes should be made in the system, possibly including alternative piping layout, including elevation and/or size, or use of a pump with smaller NPSH requirements.

The manufacturer should be consulted for a map of operational information for a given pump. A typical form is shown in Figure 4.8.4. This information will allow the designer to select a pump that satisfied the circuit operational requirements while meeting the necessary NPSH and most-efficient-operation criteria.

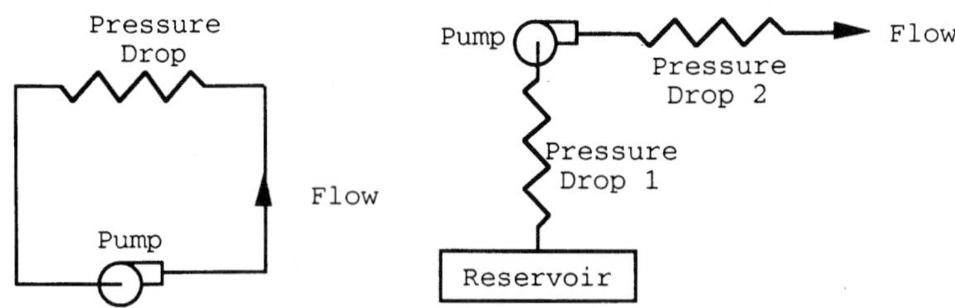

FIGURE 4.8.2. Typical pump applications, either in circuits or once-through arrangements, can be represented as combined fluid resistances as shown. The resistances are determined from fluid mechanics analyses.

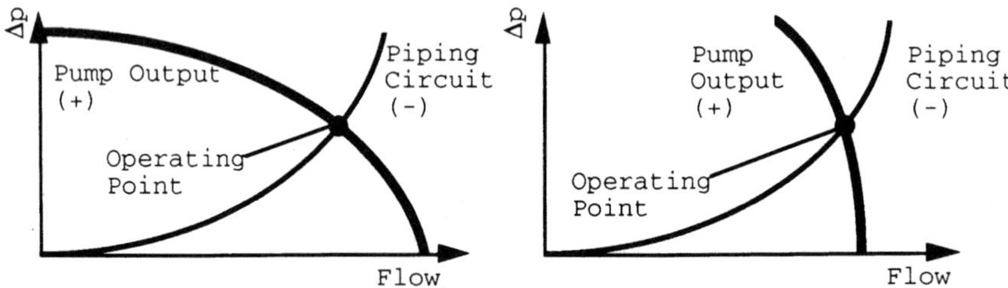

FIGURE 4.8.3. An overlay of the pump flow vs. head curve with the circuit piping characteristics gives the operating state of the circuit. A typical velocity head pump characteristic is shown on the left, while a positive-displacement pump curve is shown on the right.

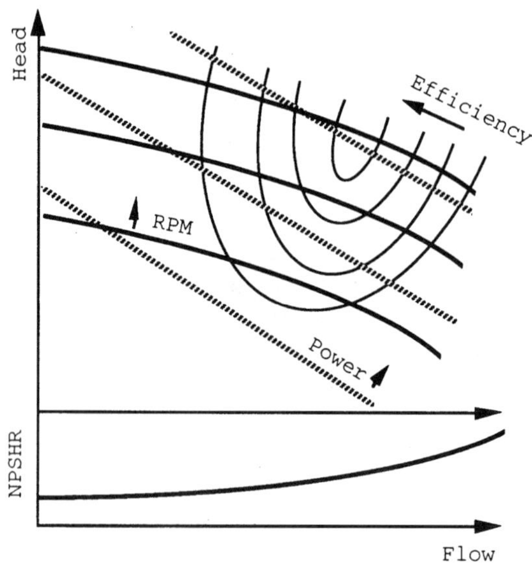

FIGURE 4.8.4. A full range of performance information should be available from the pump manufacturer, and this may include the parameters shown.

Several options are available to the designer for combining pumps in systems. Consider a comparison of the net effect between operating pumps in series or operating the same two pumps in parallel. Examples of this for pumps with characteristics such as centrifugal units are shown in Figure 4.8.5. It is clear that

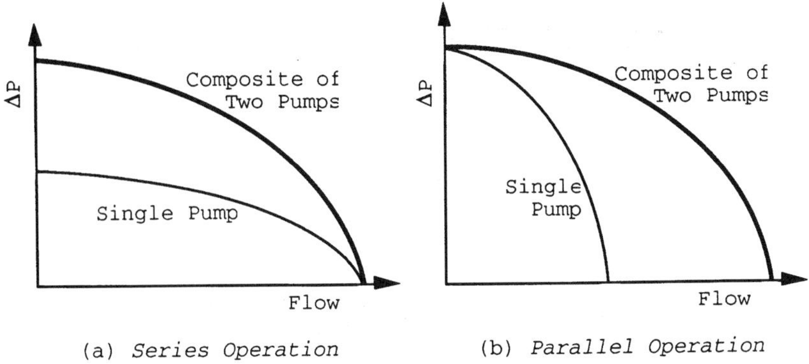

FIGURE 4.8.5. Series (a) and parallel (b) operation of centrifugal pumps are possible. The resultant characteristics for two identical pumps are shown.

one way to achieve high pumping pressures with centrifugal pumps is to place a number of units in series. This is a related effect to what is found in *multistage* designs.

Fans

As noted earlier, fans are devices that cause air to move. This definition is broad and can include a flapping palm branch, but the discussion here deals only with devices that impart air movement due to *rotation of an impeller inside a fixed casing.* In spite of this limiting definition, a large variety of commercial designs are included.

Fans find application in many engineering systems. Along with the chillers and boilers, they are the heart of heating, ventilating, and air conditioning (HVAC) systems. When large physical dimensions of a unit are not a design concern (usually the case), centrifugal fans are favored over axial flow units for HVAC applications. Many types of fans are found in *power plants.* Very large fans are used to furnish air to the boiler, as well as to draw or force air through cooling towers and pollution-control equipment. *Electronic cooling* finds applications for small units. Even automobiles have several fans in them. Because of the great engineering importance of fans, several organizations publish rating and testing criteria (see, for example, ASME, 1990).

Generally fans are classified according to how the air flows through the impeller. These flows may be *axial* (essentially a propeller in a duct), *radial* (conceptually much like the centrifugal pumps discussed earlier), *mixed,* and *cross.* While there are many other fan designations, all industrial units fit one of these classifications. Mixed-flow fans are so named because both axial and radial flow occur on the vanes. Casings for these devices are essentially like those for axial-flow machines, but the inlet has a radial-flow component. On cross-flow impellers, the gas traverses the blading twice.

Characteristics of fans are shown in Figure 4.8.6. Since velocities can be high in fans, often both the total and the static pressure increases are considered. While both are not shown on this figure, the curves have similar variations. Of course the total ΔP will be greater than will the static value, the difference being the velocity head. This difference increases as the volume flow increases. At zero flow (the shutoff point), the static and total pressure difference values are the same. Efficiency variation shows a sharp optimum value at the design point. For this reason, it is critical that fan designs be carefully tuned to the required conditions.

A variety of vane type are found on fans, and the type of these is also used for fan classification. Axial fans usually have vanes of airfoil shape or vanes of uniform thickness. Some vane types that might be found on a centrifugal (radial-flow) fan are shown in Figure 4.8.7.

One aspect that is an issue in choosing fans for a particular application is fan efficiency. Typical efficiency comparisons of the effect of blade type on a centrifugal fan are shown in Figure 4.8.8. Since velocities can be high, the value of aerodynamic design is clear. Weighing against this are cost and other factors.

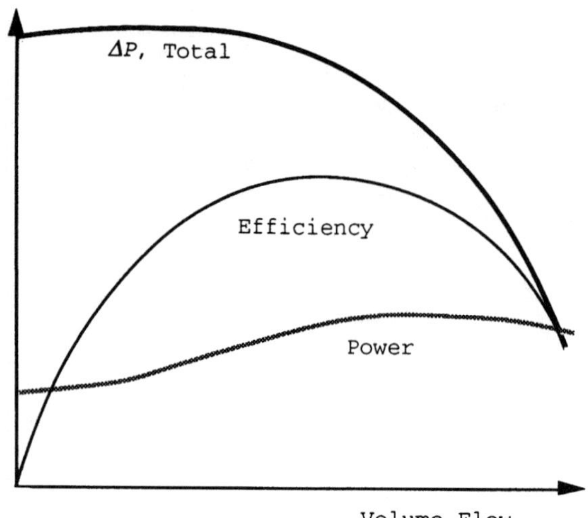

FIGURE 4.8.6. Shown are characteristics of a centrifugal fan. The drawbacks to operating away from optimal conditions are obvious from the efficiency variation.

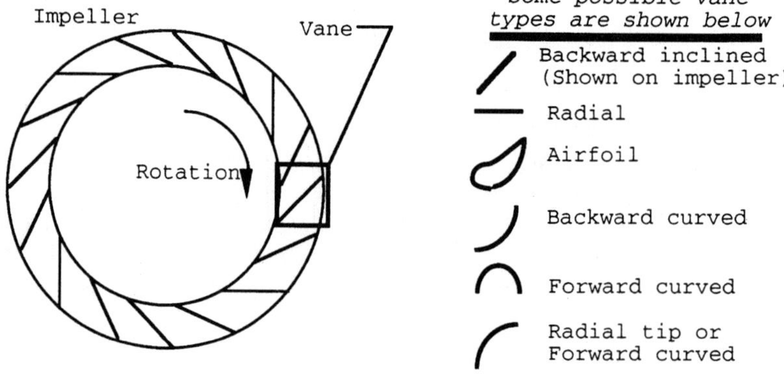

FIGURE 4.8.7. A variety of vane types that might be used on a centrifugal fan are shown.

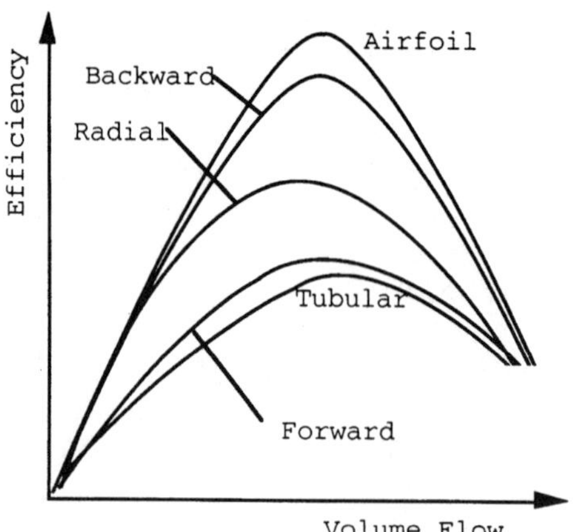

FIGURE 4.8.8. Efficiency variation with volume flow of centrifugal fans for a variety of vane types is shown.

An additional aspect that may be important in the choice of fans is noise generation. This may be most critical in HVAC applications. It is difficult to describe noise characteristics in brief terms because of the frequency-dependent nature of these phenomena. However, a comparison of specific sound power level (usually denoted by K_w) shows backward-curved centrifugal fans with aerodynamic blades perform best among the designs. Details of noise characteristics are given elsewhere (ASHRAE, 1991).

While each type of fan has some specific qualities for certain applications, most installations use centrifugal (radial-flow) fans. A primary exception is for very-high-flow, low-pressure-rise situations where axial (propeller) fans are used.

Similarities exist between fans and pumps because the fluid density essentially does not vary through either type of machine. Of course, in pumps this is because a liquid can be assumed to be incompressible. In fans, a gas (typically air) is moved with little pressure change. As a result, the gas density can be taken to be constant. Since most fans operate near atmospheric pressure, the ideal gas assumptions can be used in determining gas properties.

Flow control in fan applications, where needed, is a very important design concern. Methods for accomplishing this involve use of dampers (either on the inlet or on the outlet of the fan), variable pitch vanes, or variable speed control. Dampers are the least expensive to install, but also the most inefficient in terms of energy use. Modern solid state controls for providing a variable frequency power to the drive motor is becoming the preferred control method, when a combination of initial and operating costs is considered.

Defining Terms

Cavitation: Local liquid conditions allow vapor voids to form (boiling).
NPSH: Net positive suction head is the difference between the local absolute pressure of a liquid and the thermodynamic saturation pressure of the liquid based upon the temperature of the liquid. Applies to the inlet of a pump.
NPSHA: Actual net positive suction head is the NPSH at the given state of operation of a pump.
NPSHR: Required net positive suction head is the amount of NPSH required by a specific pump for a given application.

References

ASHRAE, 1991. *ASHRAE Handbook 1991, HVAC Applications,* American Society of Heating, Refrigerating, and Air Conditioning Engineers, Atlanta, Chapter 42.

ASME, 1990. *ASME Performance Test Codes, Code on Fans*, ASME PTC 11-1984 (reaffirmed 1990), American Society of Mechanical Engineers, New York.

Boehm, R.F. 1987. *Design Analysis of Thermal Systems*, John Wiley and Sons, New York, 17–26.

Ewbank, T. 1842. *A Description and Historical Account of Hydraulic and Other Machines for Raising Water*, 2nd ed., Greeley and McElrath, New York.

Krutzsch, W.C. 1986. Introduction: classification and selection of pumps, in *Pump Handbook*, 2nd ed., I. Karassik et al., Eds., McGraw-Hill, New York, Chapter 1.

Lobanoff, V. and Ross, R. 1992. *Centrifugal Pumps: Design & Application*, 2nd ed., Gulf Publishing Company, Houston.

Turton, R.K. 1994. *Rotodynamic Pump Design*, Cambridge University Press, Cambridge, England.

Further Information

Dickson, C. 1988. *Pumping Manual*, 8th ed., Trade & Technical Press, Morden, England.

Dufour, J. and Nelson, W. 1993. *Centrifugal Pump Sourcebook*, McGraw-Hill, New York.

Fans. 1992. In *1992 ASHRAE Handbook, HVAC Systems and Equipment*, American Society of Heating, Refrigerating, and Air Conditioning Engineers, Atlanta, GA, Chapter 18.

Garay, P.N. 1990. *Pump Application Book*, Fairmont Press, Liburn, GA.

Krivchencko, G.I. 1994. *Hydraulic Machines, Turbines and Pumps*, 2nd ed., Lewis Publishers, Boca Raton, FL.

Stepanoff, A.J. 1993. *Centrifugal and Axial Flow Pumps: Theory, Design, and Application* (Reprint Edition), Krieger Publishing Company, Malabar, FL.

4.9 Cooling Towers

Anthony F. Mills

Introduction

In a wet cooling tower, water is evaporated into air with the objective of cooling the water stream. Both natural- and mechanical-draft towers are popular, and examples are shown in Figure 4.9.1. Large natural-draft cooling towers are used in power plants for cooling the water supply to the condenser. Smaller mechanical-draft towers are preferred for oil refineries and other process industries, as well as for central air-conditioning systems and refrigeration plant. Figure 4.9.1a shows a natural draft *counterflow* unit in which the water flows as thin films down over a suitable packing, and air flows upward. In a natural-draft tower the air flows upward due to the buoyancy of the warm, moist air leaving the top of the packing. In a mechanical-draft tower, the flow is forced or induced by a fan. Since the air inlet temperature is usually lower than the water inlet temperature, the water is cooled both by evaporation and by sensible heat loss. For usual operating conditions the evaporative heat loss is considerably larger than the sensible heat loss. Figure 4.9.1b shows a mechanical draft cross-flow unit. Figure 4.9.2 shows a natural-draft cross-flow tower for a power plant.

Packing Thermal Performance

Counterflow units. Merkel's method (Merkel, 1925) for calculating the number of transfer units required to cool the water stream, for specified inlet and outlet water temperatures and inlet air condition is (Mills, 1995)

$$N_{tu} = \frac{g_m S}{\dot{m}_L} = \int_{h_{L,\text{in}}}^{h_{L,\text{out}}} \frac{dh_L}{h_s - h_G} \tag{4.9.1}$$

$$h_G = h_{G,\text{in}} + \left(\dot{m}_L / \dot{m}_G\right)\left(h_L - h_{L,\text{out}}\right) \tag{4.9.2}$$

$$h_s\left(P,T_s\right) = h_s\left(P,T_L\right) \tag{4.9.3}$$

It is imperative that the usual enthalpy datum states be used, namely, zero enthalpy for dry air and liquid water at 0°C. Table 4.9.1 gives enthalpy data for 1 atm pressure. The important assumptions required to obtain this result include

1. A Lewis number of unity;
2. Low mass transfer rate theory is valid;
3. The liquid-side heat transfer resistance is negligible, that is, $T_s \simeq T_L$;
4. The amount of water evaporated is small compared with the water and air flow rates.

The method is accurate up to temperatures of about 60°C; comparisons with more exact results are usually within 3 to 5%, and seldom show errors greater than 10%. Notice that the method does not give the outlet state of the air; however, in situations encountered in practice, the outlet air can be assumed to be saturated for the purposes of calculating its density. It is possible to extend Merkel's method to include a finite liquid-side heat transfer resistance, but such refinement is seldom warranted. For typical operating conditions the bulk liquid temperature is seldom more than 0.3 K above the interface temperature.

Cross-flow units. Figure 4.9.3 shows a schematic of a cross-flow packing. If we assume that both the liquid and gas streams are unidirectional, and that there is no mixing in either stream, then use of Merkel's assumptions leads to the following pair of differential equations (Mills, 1995):

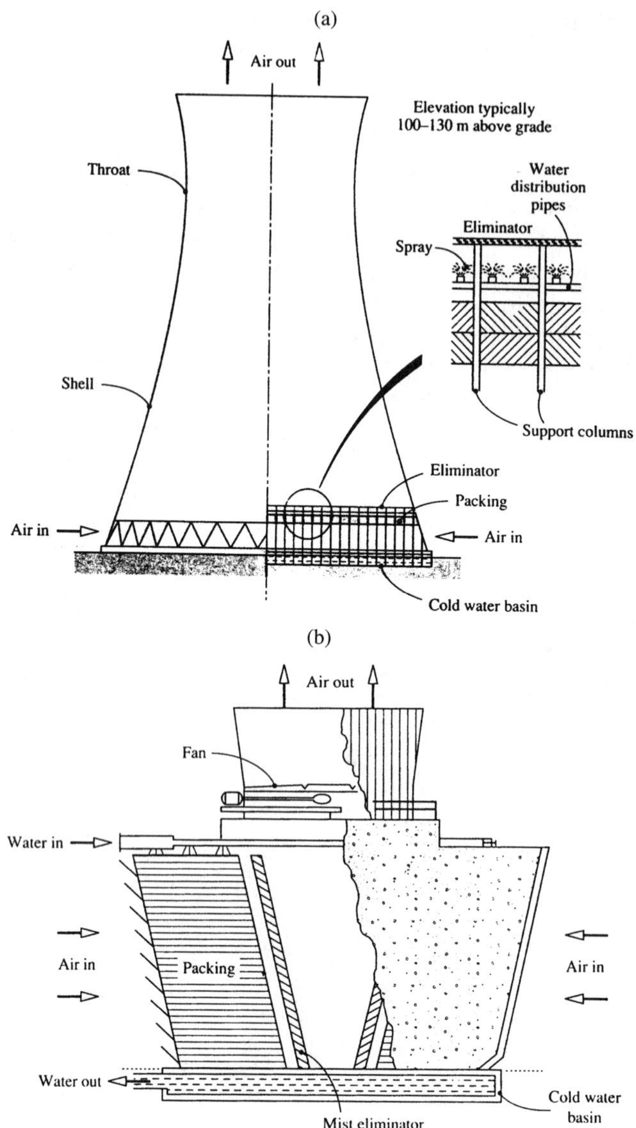

FIGURE 4.9.1 (a) A natural-draft counterflow cooling tower for a power plant. (b) A cross-flow cooling tower for an air-conditioning system.

$$\frac{\partial h_G}{\partial x} = \frac{g_m a}{G}\left(h_s - h_G\right) \tag{4.9.4}$$

$$\frac{\partial h_L}{\partial y} = -\frac{g_m a}{L}\left(h_s - h_G\right) \tag{4.9.5}$$

Also $h_s = h_s(h_L)$ for a negligible liquid-side heat transfer resistance and the required boundary conditions are the inlet enthalpies of both streams. Equations (4.9.4) and (4.9.5) are solved numerically and the solution used to evaluate the average enthalpy of the outlet liquid,

FIGURE 4.9.2 A natural-draft cross-flow cooling tower for a power plant.

$$\bar{h}_{L,\text{out}} = \frac{1}{X}\int_0^X h_{L,\text{out}}\, dx \tag{4.9.6}$$

Substituting in an exchanger energy balance on the liquid stream gives the heat transfer as

$$q = \dot{m}_L\left(h_{L,\text{in}} - h_{L,\text{out}}\right) \tag{4.9.7}$$

Sample calculation. Consider a counterflow unit that is required to cool water from 40 to 26°C when the inlet air is at 10°C, 1 atm, and saturated. We will calculate the number of transfer units required for balanced flow, that is, $\dot{m}_G/\dot{m}_L = 1$. Equation (4.9.1) is to be integrated numerically, with h_G obtained from Equation 4.9.2. The required thermodynamic properties can be obtained from Table 4.9.1. Using Table 4.9.1, $h_{G,\text{in}} = h_{\text{sat}}$ (10°C) = 29.15 kJ/kg, $h_{L,\text{out}} = h_L$(26°C) = 109.07 kJ/kg. Substituting in Equation (4.9.2),

$$h_G = 29.15 + \left(h_L - 109.07\right)$$

Choosing 2°C intervals for convenient numerical integration, the table on page 4-256 is constructed, with h_L and $h_s = h_s(T_L)$ also obtained from Table 4.9.1.

TABLE 4.9.1 Thermodynamic Properties of Water Vapor-Air Mixtures at 1 atm

		Specific Volume, m³/kg		Enthalpy[a,b] kJ/kg		
Temp., °C	Saturation Mass Fraction	Dry Air	Saturated Air	Liquid Water	Dry Air	Saturated Air
10	0.007608	0.8018	0.8054	42.13	10.059	29.145
11	0.008136	0.8046	0.8086	46.32	11.065	31.481
12	0.008696	0.8075	0.8117	50.52	12.071	33.898
13	0.009289	0.8103	0.8148	54.71	13.077	36.401
14	0.009918	0.8131	0.8180	58.90	14.083	38.995
15	0.01058	0.8160	0.8212	63.08	15.089	41.684
16	0.01129	0.8188	0.8244	67.27	16.095	44.473
17	0 01204	0.8217	0.8276	71.45	17.101	47.367
18	0 01283	0.8245	0.8309	75.64	18.107	50.372
19	0.01366	0.8273	0.8341	79.82	19.113	53.493
20	0.01455	0.8302	0.8374	83.99	20.120	56.736
21	0.01548	0.8330	0.8408	88.17	21.128	60.107
22	0.01647	0.8359	0.8441	92.35	22.134	63.612
23	0.01751	0.8387	0.8475	96.53	23.140	67.259
24	0.01861	0.8415	0.8510	100.71	24.147	71.054
25	0.01978	0.8444	0.8544	104.89	25.153	75.004
26	0.02100	0.8472	0.8579	109.07	26.159	79.116
27	0.02229	0.8500	0.8615	113.25	27.166	83.400
28	0.02366	0.8529	0.8650	117.43	28.172	87.862
29	0.02509	0.8557	0.8686	121.61	29.178	92.511
30	0.02660	0.8586	0.8723	125.79	30.185	97.357
31	0.02820	0.8614	0.8760	129.97	31.191	102.408
32	0.02987	0.8642	0.8798	134.15	32.198	107.674
33	0.03164	0.8671	0.8836	138.32	33.204	113.166
34	0.03350	0.8699	0.8874	142.50	34.211	118.893
35	0.03545	0.8728	0.8914	146.68	35.218	124.868
36	0.03751	0.8756	0.8953	150.86	36.224	131.100
37	0.03967	0.8784	0.8994	155.04	37.231	137.604
38	0.04194	0.8813	0.9035	159.22	38.238	144.389
39	0.04432	0.8841	0.9077	163.40	39.245	151.471
40	0.04683	0.8870	0.9119	167.58	40.252	158.862
41	0.04946	0.8898	0.9162	171.76	41.259	166.577
42	0.05222	0.8926	0.9206	175.94	42.266	174.630
43	0.05512	0.8955	0.9251	180.12	43.273	183.037
44	0.05817	0.8983	0.9297	184.29	44.280	191.815
45	0.06137	0.9012	0.9343	188.47	45.287	200.980
46	0.06472	0.9040	0.9391	192.65	46.294	210.550
47	0.06842	0.9068	0.9439	196.83	47.301	220.543
48	0.07193	0.9097	0.9489	201.01	48.308	230.980
49	0.07580	0.9125	0.9539	205.19	49.316	241.881

[a] The enthalpies of dry air and liquid water are set equal to zero at a datum temperature of 0°C.
[b] The enthalpy of an unsaturated water vapor-air mixture can be calculated as $h = h_{\text{dry air}} + (m_1/m_{1,\text{sat}})(h_{\text{sat}} - h_{\text{dry air}})$.

Using the trapezoidal rule,

$$\int_{h_{L,\text{out}}}^{h_{L,\text{in}}} \frac{dh_L}{h_s - h_G} = \frac{8.36}{2}\Big[0.02001 + 2(0.01986 + 0.01942 + 0.01871 + 0.01776$$

$$+ 0.01662 + 0.01536) + 0.01404\Big]$$

$$= 1.043$$

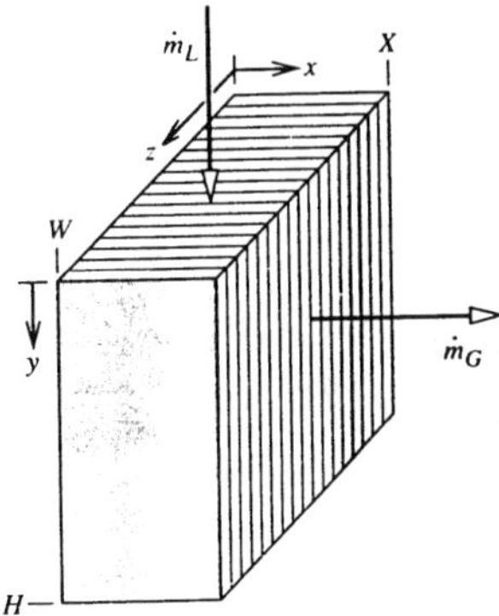

FIGURE 4.9.3 Schematic of a cross-flow cooling tower packing showing the coordinate system.

T_L, °C	h_L, kJ/kg	h_G, kJ/kg	h_s, kJ/kg	$h_s - h_g$, kJ/kg	$\frac{1}{h_s - h_g}$
26	109.07	29.15	79.12	49.97	0.02001
28	117.43	37.51	87.86	50.35	0.01986
30	125.79	45.87	97.36	51.49	0.01942
32	134.15	54.23	107.67	53.44	0.01871
34	142.50	62.58	118.89	56.31	0.01776
36	150.86	70.94	131.10	60.16	0.01662
38	159.22	79.30	144.39	65.09	0.01536
40	167.58	87.66	158.86	71.20	0.01404

From Equation (4.9.1), N_{tu} = 1.043. Also, by using Table 4.9.1, $T_{G,\text{out}}$ = 27.9° for saturated outlet air.

Thermal-Hydraulic Design of Cooling Towers

The thermal-hydraulic design of a mechanical-draft cooling tower is relatively straightforward. The flow rate ratio $\dot{m}_L/\dot{m}_G$ can be specified and varied parametrically to obtain an optimal design, for which the size and cost of the packing is balanced against fan power requirements and operating cost. Data are required for mass transfer conductances and friction for candidate packings. Tables 4.9.2a and b give correlations for a selection of packings. In Table 4.9.2b, the mass transfer conductance is correlated as $g_m a/L$, where a is the transfer area per unit volume and $L = \dot{m}_L/A_{fr}$ is the superficial mass velocity of the water flow (also called the *water loading* on the packing). Similarly, we define $G = \dot{m}_G/A_{fr}$. Typical water loadings are 1.8 to 2.7 kg/m² sec, and superficial air velocities fall in the range 1.5 to 4 m/sec. No attempt is made to correlate g_m and a separately. The number of transfer units of a packing of height H is then

$$N_{tu} = \frac{g_m S}{\dot{m}_L} = \frac{g_m a H}{L} \tag{4.9.8}$$

The correlations are in terms of dimensionless mass velocities L^+ and G^+, and a *hot water correction* T^+_{HW}. The hot water correction accounts for a number of factors, such as errors associated with Merkel's method, deviations from low mass transfer rate theory at higher values of T_s, and fluid property dependence on temperature. Frictional resistance to air flow through the packings is correlated as a *loss coefficient* $N = \Delta P/(\rho V^2/2)$ per unit height or depth of packing, as a function of L^+ and G^+. The velocity V is superficial gas velocity. No hot water correction is required.

In a natural-draft tower, the thermal and hydraulic performance of the tower are coupled, and the flow rate ratio $\dot{m}_L/\dot{m}_G$ cannot be specified *a priori*. The buoyancy force producing the air flow depends on

TABLE 4.9.2a Packings for Counterflow and Cross-Flow Cooling Towers: Designations and Descriptions

Counterflow Packings

1. Flat asbestos sheets, pitch 4.45 cm
2. Flat asbestos sheets, pitch 3.81 cm
3. Flat asbestos sheets, pitch 3.18 cm
4. Flat asbestos sheets, pitch 2.54 cm
5. 60° angle corrugated plastic, Munters M12060, pitch 1.17 in.
6. 60° angle corrugated plastic, Munters M19060, pitch 1.8 in.
7. Vertical corrugated plastic, American Tower Plastics Coolfilm, pitch 1.63 in.
8. Horizontal plastic screen, American Tower Plastics Cooldrop, pitch 8 in. 2 in. grid
9. Horizontal plastic grid, Ecodyne shape 10, pitch 12 in.
10. Angled corrugated plastic, Marley MC67, pitch 1.88 in.
11. Dimpled sheets, Toschi Asbestos-Free Cement, pitch 0.72 in.
12. Vertical plastic honeycomb, Brentwood Industries Accu-Pack, pitch 1.75 in.

Cross-Flow Packings

1. Doron V-bar, 4 × 8 in. spacing
2. Doron V-bar, 8 × 8 in. spacing
3. Ecodyne T-bar, 4 × 8 in. spacing
4. Ecodyne T-bar, 8 × 8 in. spacing
5. Wood lath, parallel to air flow, 4 × 4 in. spacing
6. Wood lath, perpendicular to air flow, 4 × 4 in. spacing
7. Marley α-bar, parallel to air flow, 16 × 4 in. spacing
8. Marley ladder, parallel to air flow, 8 × 2 in. spacing

the state of the air leaving the packing which in turn depends on $\dot{m}_L/\dot{m}_G$ and the inlet air and water states. An iterative solution is required to find the operating point of the tower. The buoyancy force available to overcome the shell and packing pressure drops is

$$\Delta P^B = g\left(\rho_a - \rho_{G,\text{out}}\right)H \tag{4.9.9}$$

where ρ_a is the ambient air density and H is usually taken as the distance from the bottom of the packing to the top of the shell. The various pressure drops are conveniently expressed as

$$\Delta P_i = N_i \frac{\rho_{Gi} V_i^2}{2} \tag{4.9.10}$$

Where N_i is the loss coefficient and V_i is the air velocity at the corresponding location. The pressure drops are associated with the shell, the packing, the mist eliminators, supports and pipes, and the water spray below the packing. Some sample correlations are given in Table 4.9.3.

Water loadings in counterflow natural-draft towers typically range from 0.8 to 2.4 kg/m^2 sec, and superficial air velocities range from 1 to 2 m/sec. The ratio of base diameter to height may be 0.75 to 0.85, and the ratio of throat to base diameter 0.55 to 0.65. The height of the air inlet is usually 0.10 to 0.12 times the base diameter to facilitate air flow into the tower. In practice the air flow distribution in natural-draft towers is not very uniform. However, the assumption of uniform air and water flows in our model of counterflow packing is adequate for most design purposes.

Cost-optimal design of cooling towers requires consideration of the complete power or refrigeration system. For refrigeration, the economics are determined mainly by the operating cost of the chiller (Kintner-Meyer and Emery, 1955).

TABLE 4.9.2b Mass Transfer and Pressure Drop Correlations for Cooling Towers

Packing Number	C_1, m^{-1}	n_1	n_2	n_3	C_2, m^{-1}	n_4	n_5
Counterflow Packings: $L_0 = G_0 =$ 3.391 kg/m² sec							
1	0.289	–0.70	0.70	0.00	2.72	0.35	–0.35
2	0.361	–0.72	0.72	0.00	3.13	0.42	–0.42
3	0.394	–0.76	0.76	0.00	3.38	0.36	–0.36
4	0.459	–0.73	0.73	0.00	3.87	0.52	–0.36
5	2.723	–0.61	0.50	–0.34	19.22	0.34	0.19
6	1.575	–0.50	0.58	–0.40	9.55	0.31	0.05
7	1.378	–0.49	0.56	–0.35	10.10	0.23	–0.04
8	0.558	–0.38	0.48	–0.54	4.33	0.85	–0.60
9	0.525	–0.26	0.58	–0.45	2.36	1.10	–0.64
10	1.312	–0.60	0.62	–0.60	8.33	0.27	–0.14
11	0.755	–0.51	0.93	–0.52	1.51	0.99	0.04
12	1.476	–0.56	0.60	–0.38	6.27	0.31	0.10
Cross-Flow Packings: $L_0 =$ 8.135 kg/m² sec, $G_0 =$ 2.715 kg/m² sec							
1	0.161	–0.58	0.52	–0.44	1.44	0.66	–0.73
2	0.171	–0.34	0.32	–0.43	1.97	0.72	–0.82
3	0.184	–0.51	0.28	–0.31	1.38	1.30	0.22
4	0.167	–0.48	0.20	–0.29	1.25	0.89	0.07
5	0.171	–0.58	0.28	–0.29	3.18	0.76	–0.80
6	0.217	–0.51	0.47	–0.34	4.49	0.71	–0.59
7	0.213	–0.41	0.50	–0.42	3.44	0.71	–0.85
8	0.233	–0.45	0.45	–0.48	4.89	0.59	0.16

Correlations (SI units)

Mass transfer: $\dfrac{g_m a}{L\left[\text{kg/m}^2\,\text{sec}\right]} = C_1\left(L^+\right)^{n_1}\left(G^+\right)^{n_2}\left(T^+_{\text{HW}}\right)^{n_3}$; Pressure drop: $\dfrac{N}{H \text{ or } X} = C_2\left(L^+\right)^{n}+\left(G^+\right)^{n_5}$

where $L^+ = \dfrac{L}{L_0}$, $G^+ = \dfrac{G}{G_0}$, $T^+_{\text{HW}} = \dfrac{1.8T_{L,\text{in}}[°\text{C}]+32}{110}$

Sources: Lowe, H.J. and Christie, D.G. 1961. "Heat transfer and pressure drop data on cooling tower packings, and model studies of the resistance of natural draft towers to airflow" Paper 113, *International Developments in Heat Transfer, Proc. of the International Heat Transfer Conference*, Boulder, CO, ASME, New York; Johnson, B.M., Ed. 1990. *Cooling Tower Performance Prediction and Improvement*, Vols. 1 and 2, EPRI GS-6370, Electric Power Research Institute, Palo Alto, CA. With permission.

Cooling Tower Behavior

There are a number of computer programs available that use variations of Merkel's method to calculate the cooling tower performance, for example, TEFRI (Bourillot, 1983), VERA2D-84 (Mujamdar et al., 1985), CTOWER (Mills, 1995). These programs can be used to perform parametric studies to obtain the response of cooling towers to environmental, duty, and design changes. However, before using such programs, some thought should be given to the important characteristics of cooling tower behavior. For this purpose, it is useful to consider a graphical representation of Merkel's theory for a counterflow tower. Figure 4.9.4 shows a chart with moist air enthalpy plotted vs. water enthalpy (or, equivalently, water temperature) at 1 atm pressure. The *saturation curve* $h_s(T_s)$ is the enthalpy of saturated air. The *operating lines* $h_G(h_L)$ are given by Equation (4.9.2) and relate the air enthalpy to the water enthalpy at each location in the packing. The slope of an operating line is L/G. Since the assumption $T_s = T_L$ is made in Merkel's method, vertical lines on the chart connect h_s and h_G at each location in the packing. The driving force for enthalpy transfer, $(h_s - h_G)$, is the vertical distance between the saturation curve and

TABLE 4.9.3 Pressure Drop Correlations for Cooling Tower Shells, Sprays, Supports, and Mist Eliminators

1. Shell (natural draft counterflow):

$$N = 0.167\left(\frac{D_B}{b}\right)^2$$

where D_B *is* the diameter of the shell base and b is the height of the air inlet.

2. Spray (natural-draft counterflow):

$$N = 0.526(Z_p[\text{m}] + 1.22)\ (\dot{m}_L/\dot{m}_G)^{1.32}$$

3. Mist eliminators:
$N = 2–4$

4. Support columns, pipes, etc. (natural-draft counterflow):
$N = 2–6$

5. Fan exit losses for mechanical-draft towers (velocity based on fan exit area):

$$N = 1.0,\ \text{forced draft}$$

$$\simeq 0.5,\ \text{induced draft, depending on diffuser design}$$

6. Miscellaneous losses for mechanical-draft towers (velocity based on packing crosssectional area):
$N \simeq 3$

Note: N is the loss coefficient defined by Equation 4.9.10, with velocity based on cross-sectional area for air flow underneath the packing in items 1 through 4.

Sources: Lowe, H.J. and Christie, D.G. 1961. Heat transfer and pressure drop data on cooling tower packings, and model studies of the resistance of natural draft towers to airflow. Paper 113, *International Developments in Heat Transfer Proc. of the International Heat Transfer Conference*, Boulder, CO, ASME, New York; Singham, J.R. 1990. Natural draft towers, in *Hemisphere Handbook of Heat Exchanger Design,* Sec. 3.12.3, Hewitt, G.E., Coord. Ed., Hemisphere, New York. With permission.

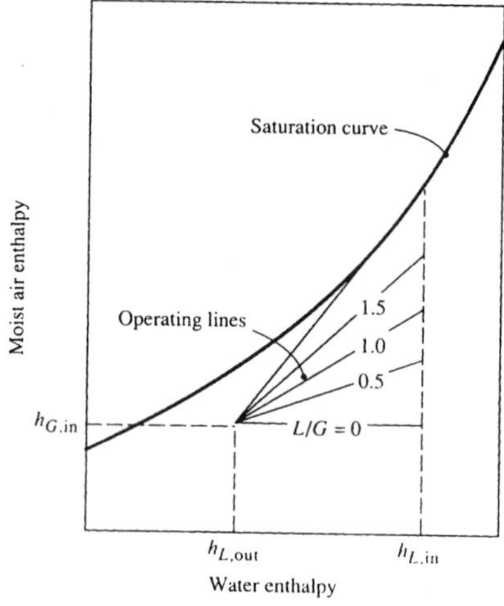

FIGURE 4.9.4 Counterflow cooling tower operating lines for various water-to-air flow-rate ratios shown on an enthalpy chart.

the operating line. The integral in Equation (4.9.1) averages the reciprocal of this distance. By using this chart, a number of observations about cooling tower behavior can be made.

1. Figure 4.9.4 shows the effect of L/G for fixed water inlet and outlet temperatures, and fixed inlet air temperature and humidity. If we imagine L to be fixed as well, we see that as G decreases, the driving forces decrease, and so a larger NTU is required.
2. The minimum NTU required corresponds to $L/G = 0$, that is, an infinite air flow rate, for which the operating line is horizontal.

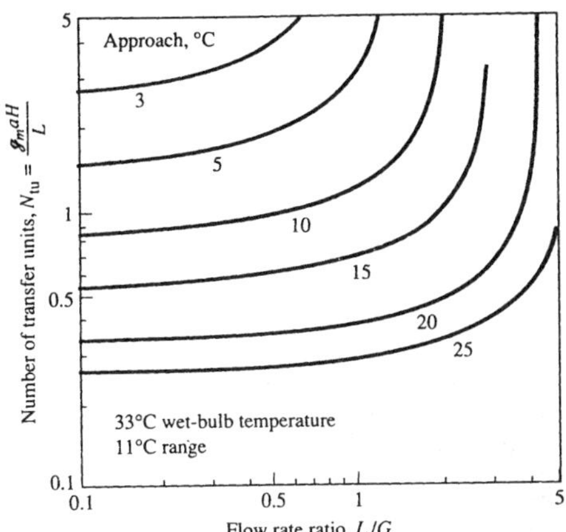

FIGURE 4.9.5 Example of cooling demand curves for a specified wet-bulb temperature and range: NTU vs. flow rate ratio for a fixed approach.

3. Due to the curvature of the operating line, it is possible for the operating line to be tangent to the saturation curve. The indicated NTU is then infinite, which tells us that the air flow rate must be increased in order to achieve the desired water cooling range.
4. For a mechanical-draft tower, the optimal value of *L/G* lies between the two limits described in items 2 and 3 above. If *L/G* is large, the required height of packing is large, and the capital cost will be excessive. If *L/G* is small, the fan power will be excessive (since fan power is proportional to air volume flow rate times pressure drop).

Range and Approach

Cooling tower designers and utility engineers have traditionally used two temperature differences to characterize cooling tower operation. The *range* is the difference between the water inlet and outlet temperatures (also called simply the hot and cold water temperatures). The *approach* is the difference between the outlet water temperature and the wet-bulb temperature of the entering (ambient) air. The approach characterizes cooling tower performance; for a given inlet condition, a larger packing will produce a smaller approach to the ambient wet-bulb temperature, and hence a lower water outlet temperature. (The water cannot be cooled below the ambient wet-bulb temperature.) The approach concept is useful because the ambient dry-bulb temperature has little effect on performance at usual operating conditions (for a specified wet-bulb temperature).

Cooling Demand Curves

Electrical utility engineers have found it convenient to use charts of *cooling demand curves* to evaluate packing specifications. Figure 4.9.5 is an example of such a chart, on which the required NTU, for a given inlet air wet-bulb temperature and range, is plotted vs. *L/G* with the approach as a parameter. Such a plot is possible since the inlet air dry-bulb temperature has only a small effect under usual operating conditions. Now, if it is possible to correlate the mass transfer conductance as

$$\frac{g_m a}{L} = C\left(\frac{L}{G}\right)^{-n} \tag{4.9.11}$$

the NTU of a packing of height H is

$$\frac{g_m S}{\dot{m}_L} = \frac{g_m aH}{L} = C\left(\frac{L}{G}\right)^{-n} H \tag{4.9.12}$$

Equation (4.9.12) can also be plotted on the chart to give the *packing capability line*. For a required approach, the *operating point* of the tower is the intersection of the cooling demand curve and packing capability line. Charts of cooling demand curves are available (Cooling Tower Institute, 1967; Kelly, 1976). Correlations of the form of Equation (4.9.11) do not necessarily fit experimental data well. A dependence $g_m a \propto L^{1-n} G^n$ is implied and, in the past, experimental data were often forced to fit such a relation. If the $g_m a$ correlation does not have the form of Equation (4.9.11), the NTU cannot be plotted as a line on a cooling demand chart.

With the almost universal use of computers and the availability of suitable computer programs, one can expect less use of cooling demand charts in the future. The major sources of error in the predictions made by these programs are related to nonuniform air and water flow, and the correlations of packing mass transfer and pressure drop experimental data. The experimental data are obtained in small-scale test rigs, in which it is impossible to simulate many features of full-size towers — for example, nonuniform flow due to entrance configuration, nonuniform wetting of the packing, and, in the case of counterflow towers, the effect of spray above the packing and rain below the packing. Furthermore, since testing of packings in small-scale test rigs is itself not easy, considerable scatter is seen in such test data. Correlations of the data typically have root mean square errors of 10 to 20%.

Legionnaires' Disease

Legionnaires' disease is a form of pneumonia caused by a strain of legionnella bacteria (sero group I). Smokers and sick people are particularly vulnerable to the disease. Major outbreaks have occurred at conventions and in hospitals, for which the source of the bacteria has been traced to cooling towers of air-conditioning systems. The bacteria require nutrients such as algae or dead bacteria in sludge, and thrive if iron oxides are present. However, properly designed, installed, and maintained cooling towers have never been implicated in an outbreak of the disease. Key requirements to be met include the following:

1. Mist (drift) eliminators should be effective.
2. The tower should be located so as to minimize the possibility of mist entering a ventilation system.
3. Corrosion in the tower and water lines should be minimized by use of glass fiber, stainless steel, and coated steel.
4. The design should facilitate inspection and cleaning, to allow early detection and remedy of sludge buildup.
5. Water treatment and filtration procedures should meet recommended standards.

References

Bourillot, C. 1983. *TEFRI: Numerical Model for Calculating the Performance of an Evaporative Cooling Tower*, EPRI CS-3212-SR, Electric Power Research Institute, Palo Alto, CA.

Cooling Tower Institute, 1967. *Cooling Tower Performance Curves*, the Institute, Houston.

Kelly, N.W. 1976. *Kelly's Handbook of Cross-Flow Cooling Tower Performance*, Neil W. Kelly and Associates, Kansas City, MO.

Kintner-Meyer, M. and Emery, A.F. 1995. Cost-optimal design of cooling towers, *ASHRAE J.*, April, 46–55.

Merkel, F. 1925. Verdunstungskühlung, *Forschungsarb. Ing. Wes.*, no. 275.

Mills, A.F. 1995. *Heat and Mass Transfer*, Richard D. Irwin, Chicago.

Majumdar, A.K., Singhal, A.K., and Spalding, D.B. 1985. *VERA2D-84: A Computer Program for 2-D Analysis of Flow, Heat and Mass Transfer in Evaporative Cooling Towers*, EPRI CS-4073, Electric Power Research Institute, Palo Alto, CA.

Further Information

Baker, D. 1984. *Cooling Tower Performance*, Chemical Publishing Company, New York.

Johnson, B.M. Ed. 1990. *Cooling Tower Performance Prediction and Improvement*, Vols. 1 and 2, EPRI GS-6370, Electric Power Research Institute, Palo Alto, CA.

Singham, J.R. 1990. Natural draft towers, in *Hemisphere Handbook of Heat Exchanger Design*, Section 3.12.3, Hewitt, G.E., Coord Ed., Hemisphere Publishing, New York.

Stoeker, W.F. and Jones, J.W. 1982. *Refrigeration and Air Conditioning*, 2nd ed., McGraw-Hill, New York.

Webb, R.L. 1988. A critical review of cooling tower design methods, in *Heat Transfer Equipment Design*, Shah, R.K., Subba Rao, E.C., and Mashelkar, R.A., Eds., Hemisphere Publishing, Washington, D.C.

4.10 Heat Transfer in Manufacturing

Donald W. Radford and Timothy W. Tong

Introduction

In many manufacturing processes, thermal energy is either used to promote material transformations or generated as a byproduct of the process. In either case, the effects of heat transfer need to be understood in order to achieve the desired product results. There is a rich body of literature related to heat transfer in manufacturing. Presented in this section are only some highlights dealing with several commonly used manufacturing methods. The aim is to provide sufficient information for practitioners to establish guidelines for designing manufacturing processes for their applications.

Casting

Effect of Heat Transfer in the Castings

Casting and solidification of metals are critically dependent upon temperature, rate of cooling, and thermal mass. Thus, heat transfer between the solidifying metal, the mold, and the surroundings is a key factor in the properties of the final solidified metal. Rates of heat extraction determine the form and shape of the solidification interface. Further, cooling rates also determine the subsequent grain size of the solidified metal. The direction of heat flow out of the solidifying metal determines the direction of the solidification interface and the resulting orientation of the grains in the final metal casting. Direction of heat flow out of a mold also impacts where the last metal freezes and where solidification shrinkage takes place. Particularly in shape casting, it is critical to control the form and location of solidification shrinkage to ensure a completely filled mold and therefore a completely formed part.

When we discuss the effects of heat transfer during solidification of a pure metal the primary concerns often center around grain size, grain orientation, and shrinkage. As depicted in Figure 4.10.1, a roughly planar solidification interface progresses from the mold wall along the thermal gradients to the center

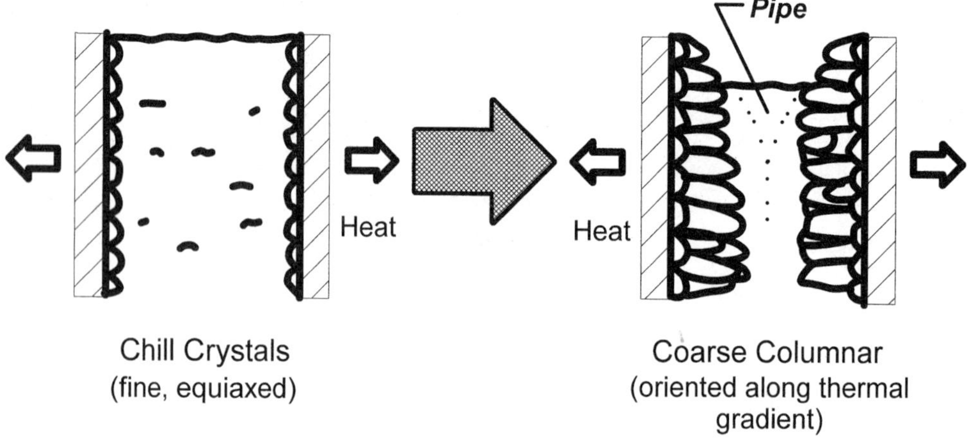

FIGURE 4.10.1 Planar solidification interface.

of the casting. By adjusting the thermal conductivity of localized areas of the mold wall, preferential heat transfer can be generated which allows the development of directionally solidified grain structure. Such areas of enhanced heat transfer in the mold wall are termed chills (see Figure 4.10.2). These chills can be used, in addition to generating directional growth, to enhance mold filling and ensure that solidification shrinkage is concentrated away from shape-critical regions. However, the effects of heat transfer become much more complicated during the freezing of an alloy.

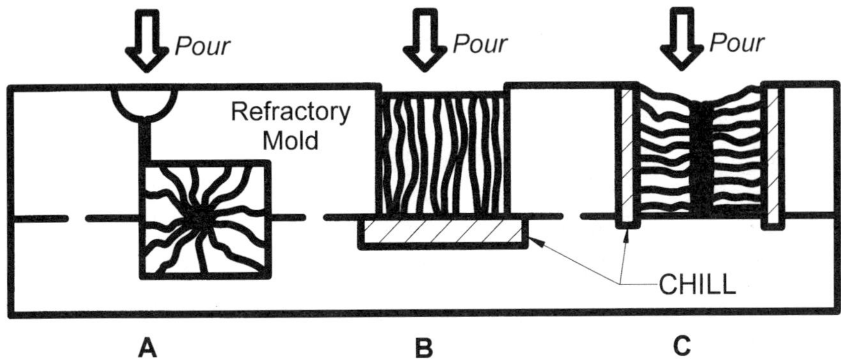

FIGURE 4.10.2 Different mold configurations produce different grain growth.

Freezing of an Alloy

The freezing of an alloy is complicated by the segregation of the alloy constituents. Since in real castings, solidification of molten metal takes place under nonequilibrium conditions, diffusion in the solid state is not fast enough to equilibrate the composition. Cooling at rates faster than the equilibrium cooling rate leads to composition gradients in the casting and influences the way that solid crystals grow from the liquid. These concentration gradients can exist over short or long distances within the casting. Concentration gradients over short distance result in microsegregation, while gradients over large distances are termed macrosegregation. One approach for describing this phenomenon is based on the concepts of constitutional undercooling (ΔT_c), which is a result of liquid temperature gradients that lead to liquid temperatures that are lower than the equilibrium freezing temperature.

Figure 4.10.3 illustrates that ΔT_c occurs ahead of the solidification interface. Therefore, it can be used to predict the form of the solidification interface. Changes to the heat transfer characteristics of the process can then be made to adjust the form of solidification and develop a grain structure that is optimized for a specific application. The concentration gradients in the liquid can extend over distances of a few millimeters. Since diffusion is negligible in the solid and incomplete mixing occurs in the liquid, the liquid adjacent to the solid-liquid interface has a different composition than the bulk melt. The first solid to form is solvent rich, with the subsequent solid being less solvent rich until the steady-state composition is reached. This steady-state composition is the same composition as the original metal melt. The liquid adjacent to the solidification interface, therefore, is solute rich and the last material to

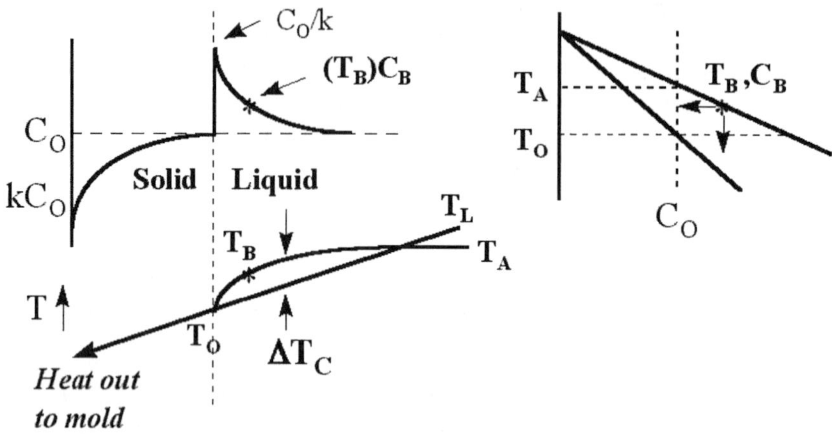

FIGURE 4.10.3 Constitutional undercooling, ΔT_c.

solidify may be highly enriched in solute. This solute-rich material has a lower melting point and poorer mechanical properties, which can lead to hot shortness as well as intergranular melting in the cast part.

Changes in the temperature gradient within the liquid can have a significant effect on the form and degree of segregation in the cast part. By adjusting the heat transfer from the mold, the degree of constitutional undercooling (ΔT_c) can also be changed, with a direct effect on the form of the solidification interface and on the microstructure of the final cast part. Figure 4.10.4 shows how segregation could be affected by the cooling rate. A higher cooling rate results in less segregation due to a shorter mixing time. The converse is true for a lower cooling rate.

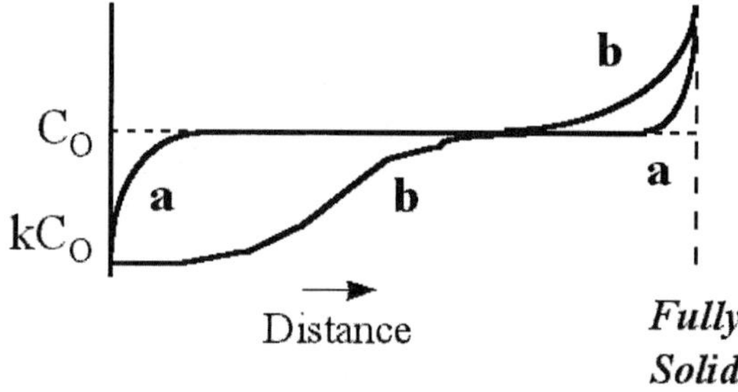

FIGURE 4.10.4 Effect of growth rate on segregation: (a) shorter mixing time, less solute segregation, (b) longer mixing time, more solute segregation.

Small ΔT_c, which can be related to rapid cooling from enhanced heat transfer, results in cellular growth with small ridges being formed on the leading edge of the planar solidification interface (Figure 4.10.5). A moderate ΔT_c results in dendritic growth and, subsequently, in a columnar grain structure (Figure 4.10.6), while large ΔT_c can lead to finer dendrites that may result in a transition from columnar to equiaxed grains (Figure 4.10.7). At very large levels of ΔT_c, independent nucleation results and equiaxed grains grow, imparting more uniform properties on the casting (Figure 4.10.8). Alloys with wide freezing temperature ranges and slow freezing favor large ΔT_c. Thus, by adjusting the heat transfer of the process, the growing grain structure can be modified. Unfortunately, as the melt solidifies against the mold wall, the heat transfer characteristics of the mold are affected. In a real metal alloy casting, grains are often of mixed form due to changing ΔT_c. Consequently, the properties of the casting change from the mold wall toward the center of the casting.

In addition to the variations seen in the as-cast grain structure, changes in the amount of ΔT_c related to variations in heat transfer also result in varying degrees of segregation within the casting. Segregation

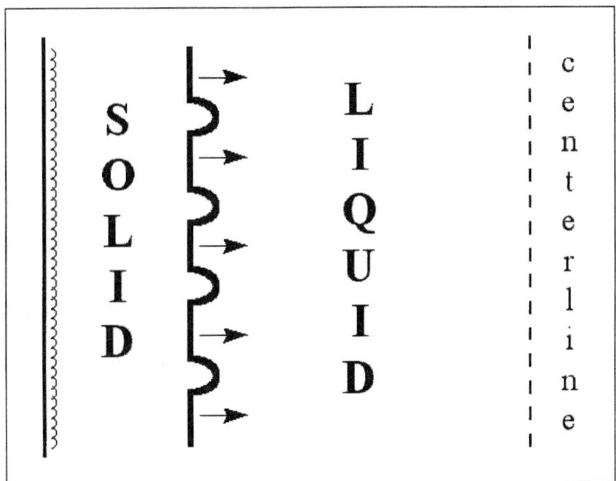

FIGURE 4.10.5 Small ΔT_c results in cellular growth.

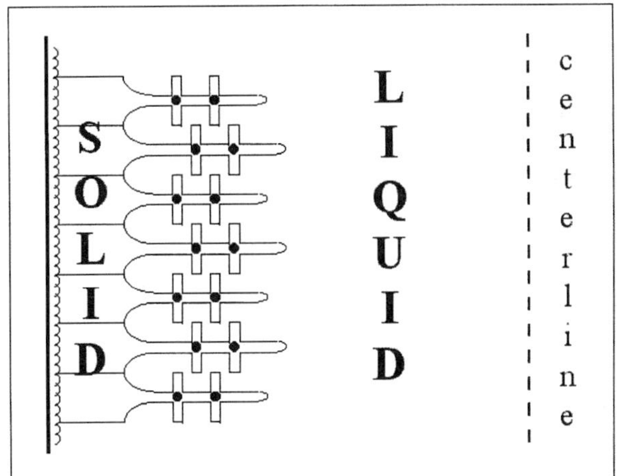

FIGURE 4.10.6 Moderate ΔT_c leads to dendritic growth.

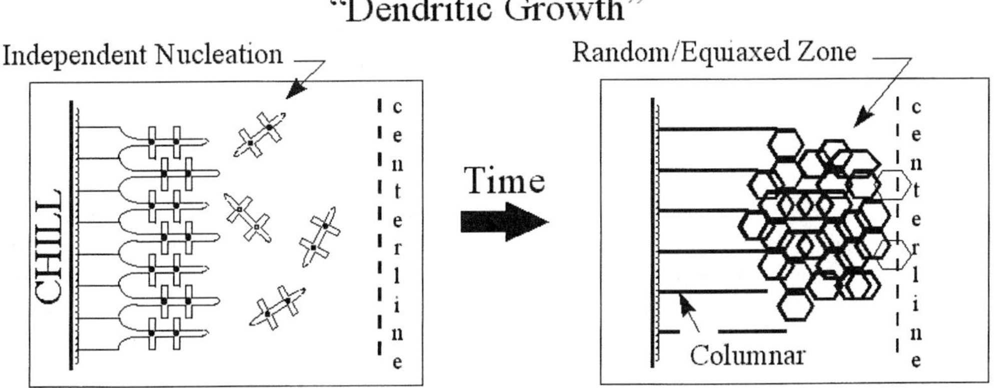

FIGURE 4.10.7 Large ΔT_c results in transition to independent nucleation.

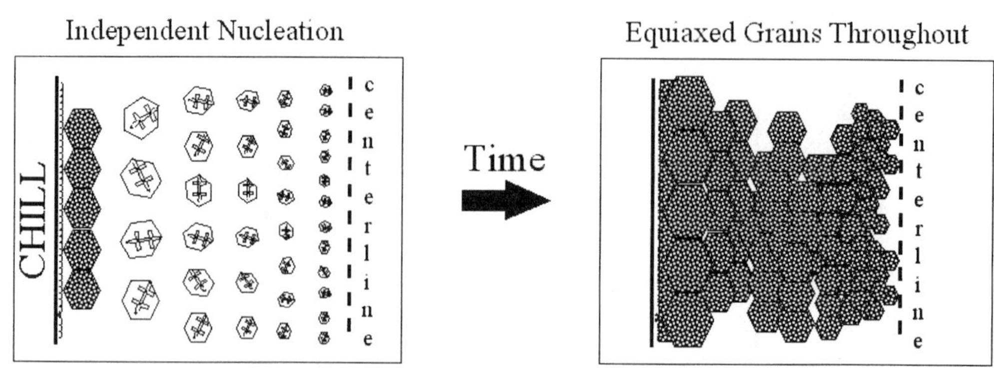

FIGURE 4.10.8 Very large ΔT_c leads to independent nucleation.

across the dimension of a grain, or microsegregation, can usually be eliminated using a post-casting homogenization heat treatment. However, if a homogenizing heat treatment is not performed, problems such as in-service galvanic corrosion, embrittlement, or hot shortness may result. A composition gradient over large distances is termed macrosegregation and leads to nonuniform properties throughout the cast section. These nonuniform properties can lead to variations in the response to heat treatment and can carry over manufacturing problems to the wrought products. If macrosegregation exists, then changes to the casting alloy, or a reduction in cast section size is required.

Residual stresses can also result from thermal gradients within the casting. Materials of high thermal conductivity will show lower thermal gradients, resulting in a lower susceptibility to the generation of harmful residual stresses. Unaccounted-for residual stresses can lead to cold cracking, or if less severe, to the buildup of internal stresses which may later lead to component distortion. Typical solutions to relieve the effects of residual stresses involve slow cooling rates or potentially a stress relief anneal. In either case, low heating and cooling rates are necessary.

Primary Processes

Primary casting processes are used to produce simple shapes and cast billets and ingots, which can then be reformed into other shapes for final use. The shapes may be remelted or may be formed to final shape by other means, such as deformation processing. Primary casting processes have evolved during the past several decades from a batch mode to continuous casting operations. Due to the high volume of cast material produced via continuous casting, quality issues have become very important. To improve the quality of continuous-cast materials, there has been significant modeling during the past decade. These numerical models generally account for both the heat transfer and fluid flow within the solidifying material.

During solidification thermal gradients play a critical role in the generation of the cast microstructure. Thus, by adjusting the heat transfer of the molds, and the process, the microstructure can be readily affected. While this is true in all castings, it becomes very important in continuous casting processes because of the short periods of time available for solidification. Researchers have developed relatively complex numerical models which predict the temperatures and temperature gradients throughout the solidifying material. Heat transfer models, which include the mold material, the air gap, flux, the solidifying shell, and the molten metal, are used to improve the quality and reproducibility of continuous-cast sections. Quality is important at this stage of processing because defects that are introduced during solidification may cause major problems throughout the post-processing operations and can lead to significant lost production.

Secondary Processes (Shape Casting)

There are many forms of shape casting practiced today. However, for the purposes of this section, shape casting may be separated into gravity-fed vs. pressure-fed molding. Pressure-fed processes are often used in the molding of parts with relatively thin sections which would be difficult to fill adequately by gravity. Even though cooling rates of small pressure-fed, die cast components can be very high, the solidification front often takes on a dendritic nature. Yet, rather than acting as impediments to the inflow of molten metal, these dendrites are broken free by the pressure and act as independent nuclei, resulting in fine-grain, as-cast structures. Also, because fresh liquid metal is continually forced into the die in these pressure-fed castings, shrinkage and shrinkage porosity become less of an issue.

In gravity-fed shape casting, mold filling and shrinkage porosity become greater potential problems. Shrinkage porosity generally results when ΔT_c is high and the mold design forces shrinkage to occur in a location which is not adjacent to fresh liquid metal. To overcome these problems, changes in the mold material, additions of chills or insulated regions, or design changes to the size and shape of feeders which supply fresh liquid metal may be necessary.

Corrective Measures for Shrinkage Porosity

One approach to overcoming shrinkage porosity is to increase the freezing rate of the metal. This is done by adjusting the heat transfer of the mold, or of localized regions of the mold, to reduce the degree of ΔT_c. This results in a localized shrinkage, which can then be dealt with by other means.

Another solution for solidification shrinkage is the use of risers, or feeders, such as that shown in Figure 4.10.9(a). The riser provides added hydrostatic head during the freezing of a casting and is the last location to freeze. In this way, the final material to freeze remains completely in the riser and the associated shrinkage cavity is also contained in the riser. The riser is subsequently cut from the casting and remelted. The design of a riser is based on the simple principle that the freezing time of a casting in a homogeneous mold is proportional to the square of the ratio of the volume of metal to the total mold/casting interface area $[(V/A)^2]$. This relationship is readily derived from heat transfer modeling and is verified experimentally.

Modifying the mold using a special insulated sleeve around the riser region to delay its solidification can reduce riser size. For complex shapes, or large shapes, multiple risers as shown in Figure 4.10.9(b) may be required to overcome "center line" shrinkage porosity if ΔT_c is large. The larger ΔT_c, the greater the number of risers that will be necessary. For castings with large section changes, and thick sections separated by thinner sections, each of the heavy sections will require a riser. In general, casting design is based on the ability to solidify in a specified manner, and since solidification is highly dependent upon heat extraction, large section changes can pose severe difficulties. In certain cases of mold and part design, a chill can effectively replace a riser. A chill such as that illustrated in Figure 4.10.9(c) can be used in localized hot spots. It can also be applied to generate directional grain structures during solidification.

Welding

Welding processes can be subdivided into categories dependent upon whether the base metal melts and becomes part of the joint material, or is joined without melting by the addition of a filler material. If the base metal does melt, the joint is considered a fusion weld and the parent materials, plus any filler material, mix and solidify in the "fusion zone." In this case many of the same concerns related to heat transfer which were considered in the discussion of casting, also apply. However, in addition to the effects in the fusion zone, it is also important to realize that a broader region of the parent material is exposed to high temperatures and to significant thermal gradients. This region outside the fusion zone, but still affected by the thermal processing, is called the heat-affected zone (HAZ). To better understand the role of heat transfer in the welding of metals it is necessary to discuss the effects in both the fusion zone and the HAZ.

Heat Input

Unlike the solidification of castings which start out with a relatively uniform material temperature, welding processes transfer the heat into the material in a localized fashion. Therefore, heat is transferred, in a welding process, from the weld region into the remainder of the part, or parts, being welded. This

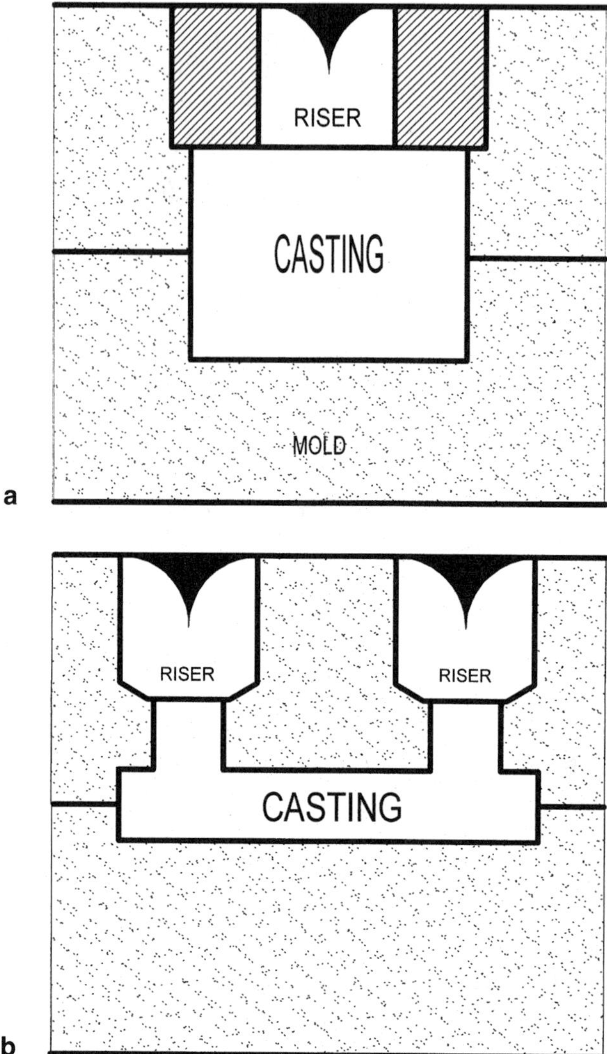

FIGURE 4.10.9 Different mold geometries for controlling porosity and grain growth: (a) incorporation of a riser, (b) use of multiple risers, and (c) combination of a riser and a chill.

means that relatively large thermal gradients can occur over a single component. The resulting thermal gradients can be extremely steep depending upon the rate of heat energy input and the thermal mass of the complete component.

Fusion Zone

Many welding processes are considered fusion processes. Fusion welding processes include stick welding, torch welding, TIG, MIG, electron beam, laser beam, and others. For each of these processes the heat source is sufficient to melt a region of parent material, which must then resolidify in a way very similar to that of a casting. During the solidification of the fusion zone an as-cast microstructure is developed which follows the same trends as discussed in the previous subsection on casting. Grains once again grow along the direction of heat flow and, in general, are relatively large and columnar. The fusion zone microstructure can be adjusted through changes in the cooling rate; however, this is a more difficult variable to control in welding processes than in casting. Approaches to controlling the thermal gradients include pre- and post-heating of the parent materials.

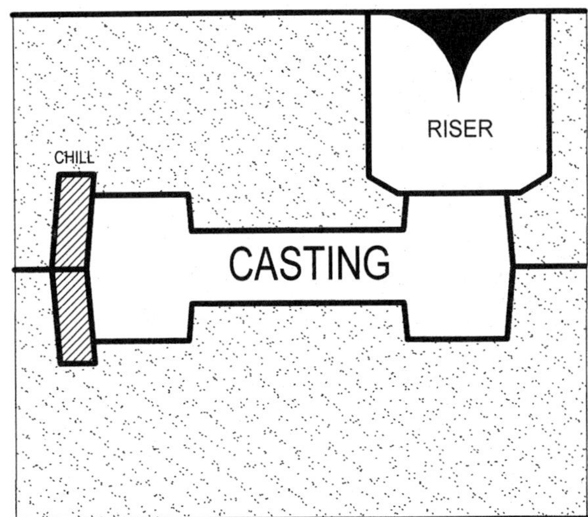

FIGURE 4.10.9 (continued).

Details affecting the size and shape of the fusion zone are strongly dependent upon the type of welding process and its inherent energy transfer capability. Very high-energy processes such as electron beam welding enable relatively narrow and deep fusion zones. In such high-energy processes, the thermal energy is highly focused. Both the fusion zone and the HAZ are relatively small. However, the thermal gradients introduced can be quite large. For these high-energy processes the microstructure in the fusion zone represents a more rapidly cooled as-solidified structure, with the potential of finer grain size. Open-flame welding, or torch welding, employs a much lower energy source. As a result, temperature gradients are smaller but the ultimate temperatures at a distance from the fusion zone are higher. It follows that the extent of the HAZ and the region of affected parent microstructure become large.

Heat-Affected Zone

In all fusion welds there is a region around the weld zone where the parent metal has been heated, in the solid state, to temperatures sufficient to cause structural and property changes. This region is called the HAZ. Heat transfer in the HAZ of a welded component determines the temperature profile and cooling rates. Based on the knowledge, or prediction, of the temperature profiles and cooling rates, microstructural variation can be estimated. Prediction of the form of the changing microstructure can then be used to determine if metallurgical problems are to be expected, and whether welding process modifications must be made. The extent of the HAZ is dependent upon many factors, including the heat source energy, the thermal mass of parent material, the thermal conductivity of the parent, and the workpiece geometry.

The impact that the varying temperatures and cooling rates of the HAZ have upon the material microstructure is dependent upon the parent material and its pre-weld condition. For instance, two pure aluminum plates might be joined by MIG welding. The resulting microstructural variation across the weld zone and the HAZ is shown schematically in Figure 4.10.10.

As indicated in the Figure 4.10.11, the microstructural variation can be quite severe, with yield strengths dropping to levels below half that of the unaffected parent. While this microstructure is highly affected by the welding process, the performance change is relatively uniform from the center of the fusion zone outward, not showing any localized regions of severely degraded properties. However, the resulting microstructure is directly related to the type of strengthening mechanisms available in the material and, therefore, to the material itself. Thus, it is useful to discuss some specific cases in which the effects in the HAZ can become quite deleterious, with the potential for localized failure. In addition to cold working, other common strengthening mechanisms should be discussed. Many aluminum alloys

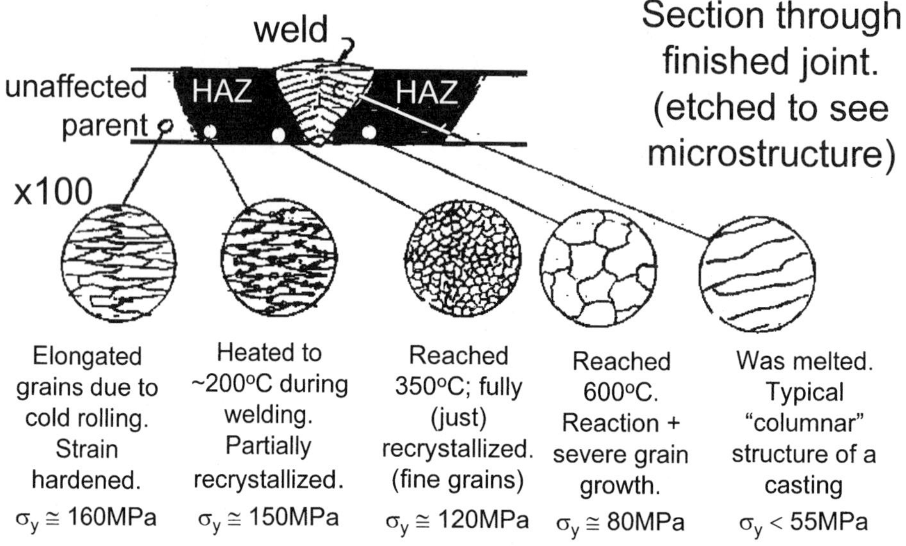

FIGURE 4.10.10 Variation of microstructure in heat-affected zone (cold-worked aluminum).

are strengthened via precipitation hardening. The majority of steels are strengthened by phase transformation effects related to quenching and the resulting variations on a Martensite structure.

For a material such as aluminum, which is precipitation hardened, mechanical property variation becomes more complicated. Property degradation in precipitation-hardened alloys generally results from excessive exposure to elevated temperatures, leading to over-aging and the loss of strength. Since both time and temperature play an important role in the resulting mechanical properties, knowledge of the thermal history is critical. The minimum strength is seen in the HAZ, due to an optimal combination of both peak temperature and the time at that temperature. In this case the over-aging is most severe and it is not uncommon to see weld failures in the HAZ. Figure 4.10.12 indicates the variation of strength and ductility moving from the center the fusion zone toward the affected precipitation-hardened parent. It should be noted that the low strength in the fusion zone could often be combated through the selection of suitable filler metal.

Low-alloy carbon steel responds somewhat differently to the temperature profile within the weld zone (Figure 4.10.13). Again, choosing a higher-performance alloy as a filler metal can often enhance the fusion zone performance; however, effects in the HAZ are more difficult to control. The most basic effect noted is similar to that of the cold-worked case where the grain size is quite large in the HAZ near the fusion zone resulting in relatively low strength. However, since no cold work was included, the grain size variation is relatively uniform, ranging from large grains at the fusion interface to relatively fine grain as the unaffected parent is approached. The secondary effect for this low-alloy carbon steel is related to the onset of the Martensite transformation, due to high cooling rates in HAZ. Since the cooling rate is highly dependent upon the relative ease of heat transfer, the extent of the transition to Martensite will be determined by many factors including parent thickness, thermal conductivity, and geometry. If cooling rates within the HAZ are sufficiently high and cause Martensite to form, this region will be highly embrittled and very susceptible to weld failure. Thus, it becomes obvious that knowledge

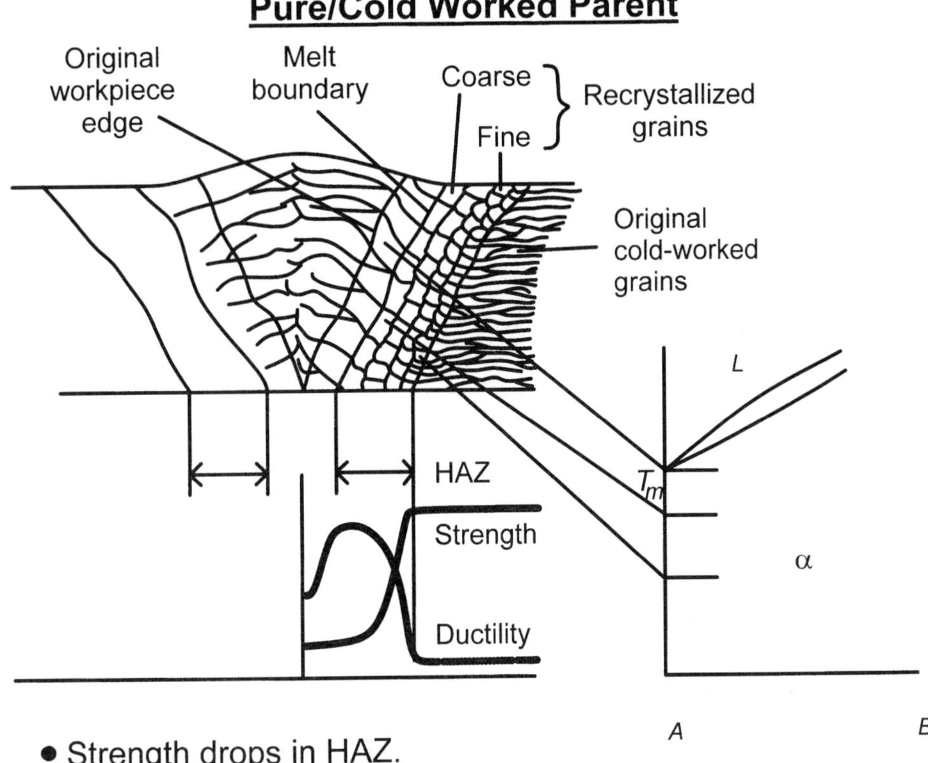

FIGURE 4.10.11 Mechanical property variation in the heat-affected zone (cold-worked parent).

of the peak temperatures, thermal response time, cooling rate, and material composition throughout the HAZ is necessary.

Thermal History

To predict the deleterious effects of welding heat on the parent material microstructure, and, therefore, performance, it is necessary to investigate the thermal history. Specifically, the important components of the thermal history are maximum temperature attained, thermal response time, and the heating and cooling rates. Since heat input is highly localized, heat transfer into the parent material is affected by the rate and direction of the welding operation. Conceptually, Figure 4.10.14 indicates an instantaneous temperature distribution in a thin metal plate during fusion welding. The "F" refers to the fusion zone. It is important to realize that the isotherms are not centered on the fusion zone due to the traveling velocity along the welding direction. This also means that the temperature gradients are steepest at the leading edge of the weld. On the trailing edge of the weld the solidified weld bead is still a source of heat, flowing outward into the parent material.

Heat transfer theory permits a prediction of thermal history through use of appropriate models of the welding process, starting from the equation for 3-D heat conduction in a homogeneous, isotropic solid:

$$\frac{\partial^2 T}{\partial x^2} + \frac{\partial^2 T}{\partial y^2} + \frac{\partial^2 T}{\partial z^2} = \frac{1}{\alpha}\frac{\partial T}{\partial t} \qquad (4.10.1)$$

where α = thermal diffusivity = $\frac{\mathrm{k}(\text{thermal conductivity})}{\mathrm{c}(\text{specific heat})\,\rho(\text{density})}$.

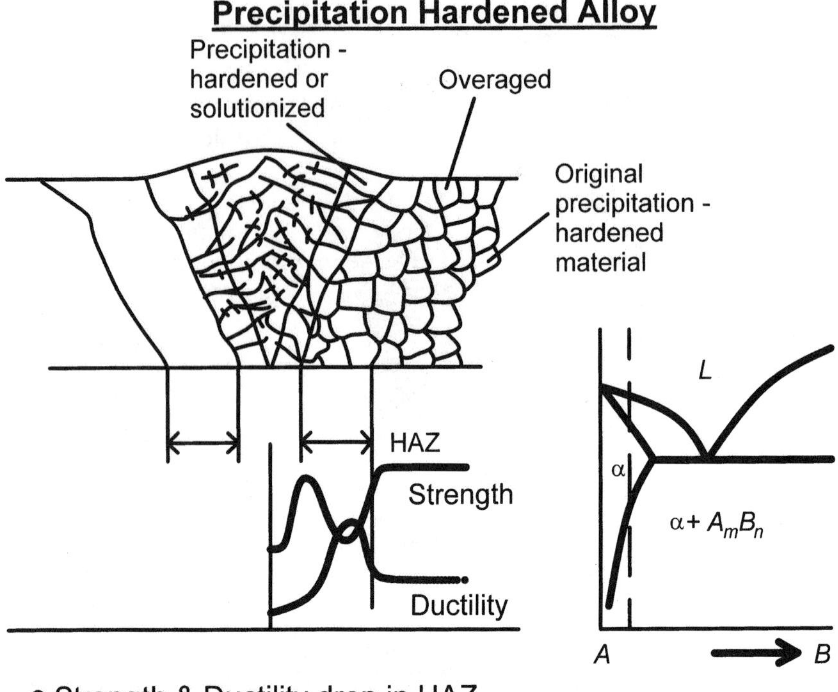

FIGURE 4.10.12 Mechanical property variation in the heat-affected zone (precipitation-hardened parent).

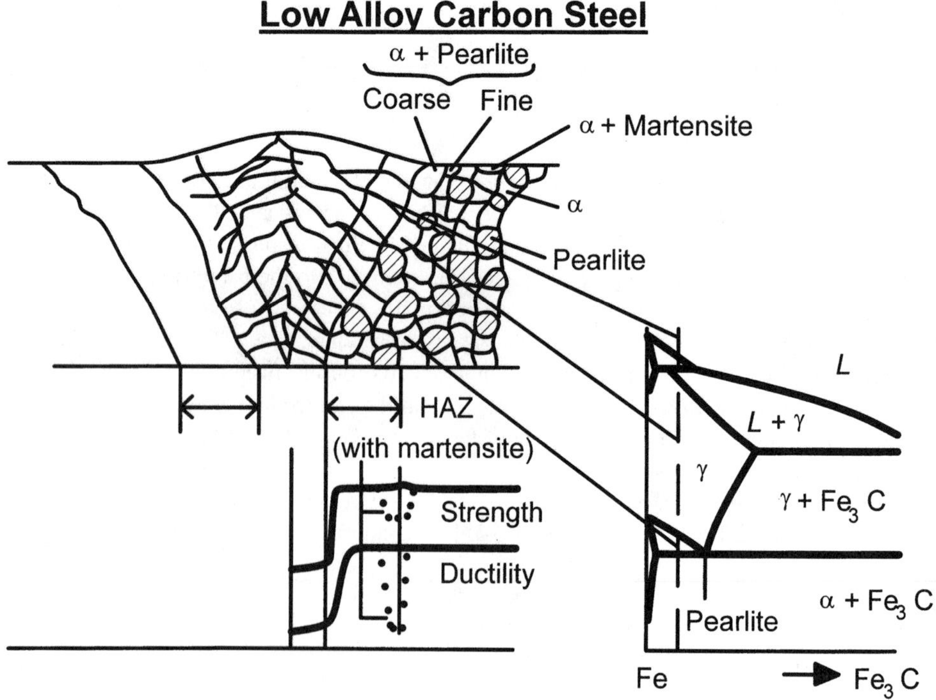

FIGURE 4.10.13 Mechanical property variation in the heat-affected zone (low-alloy carbon steel).

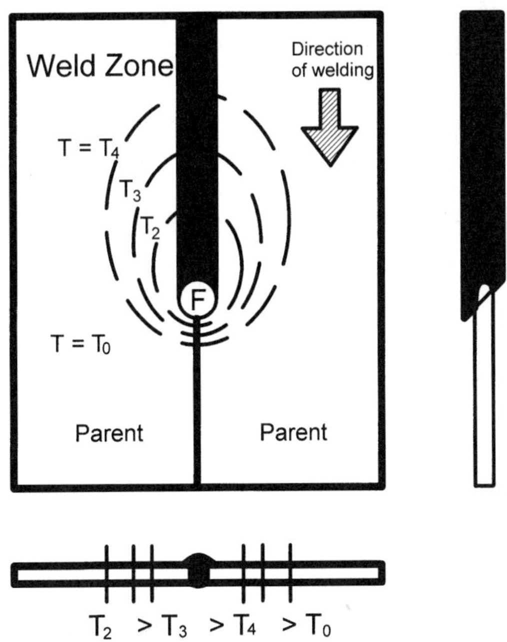

FIGURE 4.10.14 Temperature distribution in a metal plate during fusion welding.

Solutions of this equation are available for a variety of models assuming infinite or semi-infinite plates. Two cases of interest are the moving point source of heat and the moving line source of heat discussed next.

Moving Point Source of Heat

The moving point source is a reasonable description of welding thick sections, or for depositing a weld bead on a plate (see Figure 4.10.15). Relative motion between the source and the plate, in the x-direction, at a velocity, v, is assumed. By applying symmetry to the steady-state solution for a constant rate of

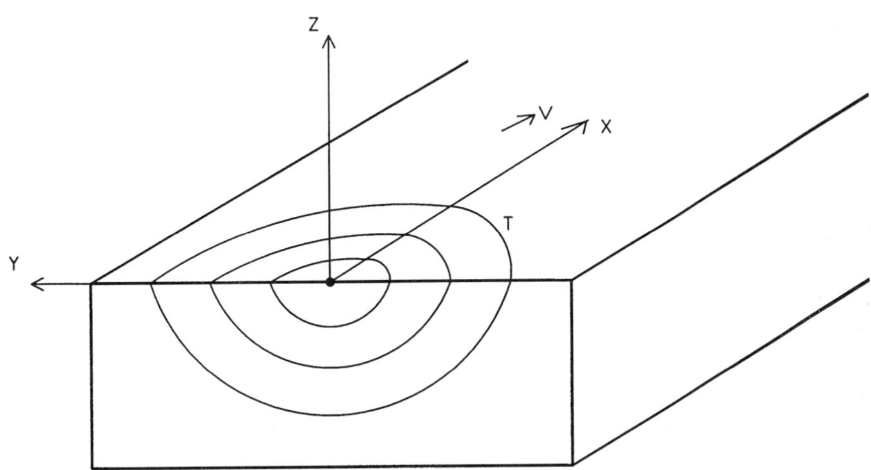

FIGURE 4.10.15 Moving point heat source.

heat input, q, and an infinite medium (Carslaw and Jaeger, 1959), one can obtain the following steady-state solution for a semi-infinite medium bounded by the plane $z = 0$

$$(T - T_0) = \frac{q}{2\pi k r} \exp\left(\frac{-v(r-x)}{2\alpha}\right) \tag{4.10.2}$$

where $r = \sqrt{x^2 + y^2 + z^2}$ and T_o is the initial temperature. In arc welding processes,

$$q \propto \frac{EI}{v}$$

where E = the welding voltage and I = the welding current. Values of $\frac{EI}{\text{v}}$ vary based on the process and the heat source. For stick welding, $\frac{EI}{\text{v}}$ ranges from approximately 1 to 3 kJ/mm, while for a welding process such as electroslag welding, the value can exceed 100 kJ/mm.

Moving Line Source of Heat

For modeling thin sections, a line source of heat is much more appropriate because the heat input is uniform throughout the thickness. This approach would model a single-pass butt weld in thin sheet. In this case, the problem is 2-D heat flow with substantial heat transfer only in the x and y directions (Figure 4.10.16). Again, the steady-state solution can be generated (Carslaw and Jaeger, 1959) as

$$(T - T_0) = \frac{q}{2\pi k} K_0\left(\frac{vr}{2\alpha}\right) \exp\frac{vx}{2\alpha} \tag{4.10.3}$$

where $K_0(z)$ is the modified Bessel function of the second kind of order zero.

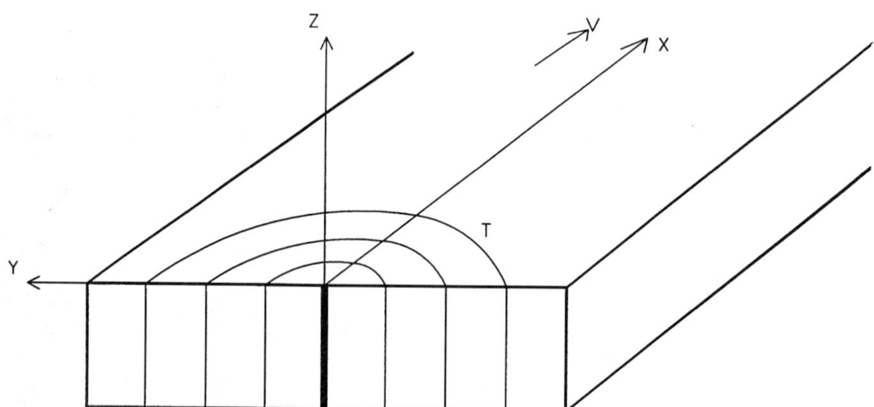

FIGURE 4.10.16 Moving line heat source.

Solutions are also available for intermediate geometries, and for stationary heat sources that might be used in modeling spot-welding. The solutions are limited by the fact that the heat source is not truly a point or a line, requiring an offset of the origin. They are also limited by the thermal properties of the material (k, ρ, c) that vary with temperature, and that the workpieces are not semi-infinite. Yet, the expressions still result in reasonably good agreement with experimental data. Computer solutions can overcome most of these limitations and generate quite complex models, allowing improved accuracy, particularly in modeling complex geometries.

Summary of Thermal Effects

Heating and Cooling Rates

As indicated previously by the closer spacing of the temperature isotherms, heating and cooling rates are at their maximum at the downstream edge of the fusion zone. Heating and cooling rates increase with welding speed, v, and with higher heat input and a correspondingly smaller fusion zone. For instance, in arc welding, the cooling rate is inversely proportional to heat input, EI/v. In addition, materials with high thermal conductivity also show increased rates of heat dissipation in the parent material. This results in the need for higher-energy heat sources. Furthermore, higher thermal conductivity results in a larger HAZ. Heat-affected zones in conductive materials such as copper can extend four to five times beyond that of typical HAZ in steel.

Other material properties such as the coefficient of thermal expansion can also have a significant effect on the success of welding operations. Higher thermal expansion coefficients result in the generation of high stresses that are conducive to distortion and to cracking in crack-sensitive alloys.

Distance from Weld Axis

The peak temperature, as well as heating and cooling rates, decrease with increasing distance from the weld axis. For any given location, the peak temperature increases with the rate of heat input, q, but the cooling rate decreases with increasing heat input.

Section Size and Weld Geometry

As the parent material section increases in size the effective heat sink is also increased. Therefore, higher temperature gradients result, but the time at peak temperature is reduced. For more complex geometries, such as fillet welds (T or L joints), there are additional paths for heat flow, and the part is a more effective heat sink due to the geometry. Finally, it should be noted that the geometry changes from a 2-D to a 3-D heat flow with increasing thickness.

To account for the changing section and geometry the concept of a thermal severity number (TSN) is often introduced, where *TSN = combined thickness of heat flow paths (inches)* $\times$ *4*. To visualize the effect of section size and weld geometry using the TSN, consider that a butt weld joining two 0.25-in.-thick plates results in a TSN of 2, while a fillet weld joining a 1-in.-thick plate to a 2-in.-thick plate results in a TSN value of 20. Obviously, the temperature gradients for the fillet-welded joint of the thicker plates would make the joining operation more challenging. It should also be noted that joining a thin plate to a thick plate is a more difficult problem due to the greater "heat sink" effect of the thick plate. Consequently, increased heat input is required and it has the negative effect of rapidly melting through the thinner material. This can be a difficult welding operation, but preheating the thick plate can alleviate the problem.

Pre- and Post-Heat Treatment of Welds

As has been discussed, thermal gradients introduced by the welding operation can have significant effects on the parent material, and ultimately on the product success. To ensure acceptable performance of the completed joints additional operations can be performed, either in preparation for welding or after the welding process is complete. Post-welding heat treatments can be used for practically sized fabrications to readjust the microstructure of the component. This can be quite a good approach, particularly for steels, where the performance can be modified through phase transformation.

Another often-used technique to reduce thermal gradients is to preheat the parent metal prior to welding. By increasing the base temperature, thermal gradients are reduced and the thermal severity is lessened. This approach can be very effective for materials that are sensitive to rapid cooling, but are not significantly affected by extended times at temperature. Again, for steels this can be an effective solution to welding problems. Multipass welds can often result in microstructures resembling those of preheated materials.

Heat Treatment

The term *heat treatment* is used to describe a variety of processes that are applied to modify the properties of metals with the goal of generating improvements in hardness, strength, ductility, or toughness. One thing that all heat treatments have in common is that they all rely on the transfer of heat to or from the workpiece material. Heat treatments can be subdivided into two categories, nonhardening thermal processes and hardening processes (Mangonon, 1999). Depending upon the specific heat treatment, different aspects of thermal processing must be controlled. Nonhardening processes generally rely on relatively long times at temperature and are often considered isothermal processes, while it is more common that the control of thermal gradients is critical in the application of the principles of heat transfer for hardening processes.

Nonhardening Thermal Processes

Nonhardening thermal processes include processes such as thermal stress relief, annealing, and normalizing. Residual stresses can be generated by many manufacturing processes, ranging from casting and welding to mechanical deformation. Such stresses can lead to component distortion, cracking, stress corrosion, and in some cases delayed failure. By holding the workpiece at an elevated temperature the internal stresses diminish. Thermal stress relief consists of uniformly heating the workpiece to a temperature below the recrystallization temperature, in the recovery range, holding it for a predetermined period of time related to the severity of the stress state, and then uniformly cooling it. As in all nonhardening thermal processes, temperature gradients are to be minimized and the processes are typically considered isothermal.

An annealing heat treatment is performed to soften a material after significant amounts of cold work have been performed. This softening may be desired to enable further mechanical working, or to adjust the metallurgical properties for a specific application. Annealing is performed at temperatures above the recrystallization temperature and results in a softer, more equiaxed grain structure. The process, like thermal stress relief, involves relatively slow heating and cooling, with the time at optimum temperature used to control the final grain structure and size. Long periods of time at high temperature lead to the fully annealed condition, while shorter times at the same temperature can be said to yield subcritical annealing. The temperatures required for such a heat treatment are often given based on equilibrium heating and cooling rates. Since real processes do require finite heating and cooling rates, it must be noted that the critical temperatures will be shifted up with increasing heating rates and down with increased cooling rates.

Normalizing is a term generally reserved for discussions regarding steels. In this nonhardening heat treatment, the cooling rate is specified as natural or air cooled. A normalized steel generally exhibits a uniform fine-grained microstructure which enhances its homogeneity and the machinability; the normalization process is often applied after a shape has been generated by casting or forging. The strength and hardness of a normalized steel may either increase or decrease, depending upon the prior state of the workpiece material.

Hardening Heat-Treating Processes

Hardening heat treatment processes are only applicable to materials which show multiple phases at low temperature, can be locked into a nonequilibrium microstructure by rapid cooling, and can be driven toward a desirable microstructural state during subsequent heating. The results of such a sequence are generally considered the principal strengthening approaches for metals. For steels, the Martensitic transformation is critical and is driven by the ability to rapidly cool the workpiece, generating the nonequilibrium structure, Martensite. Other materials, such as aluminum, are strengthened through a mechanism described as precipitation hardening, and still others grow stronger by solid solution strengthening. However, the ability to lock in the nonequilibrium structure and then subsequently adjust the material microstructure via thermally induced diffusion is shared among these hardening heat-treatment processes.

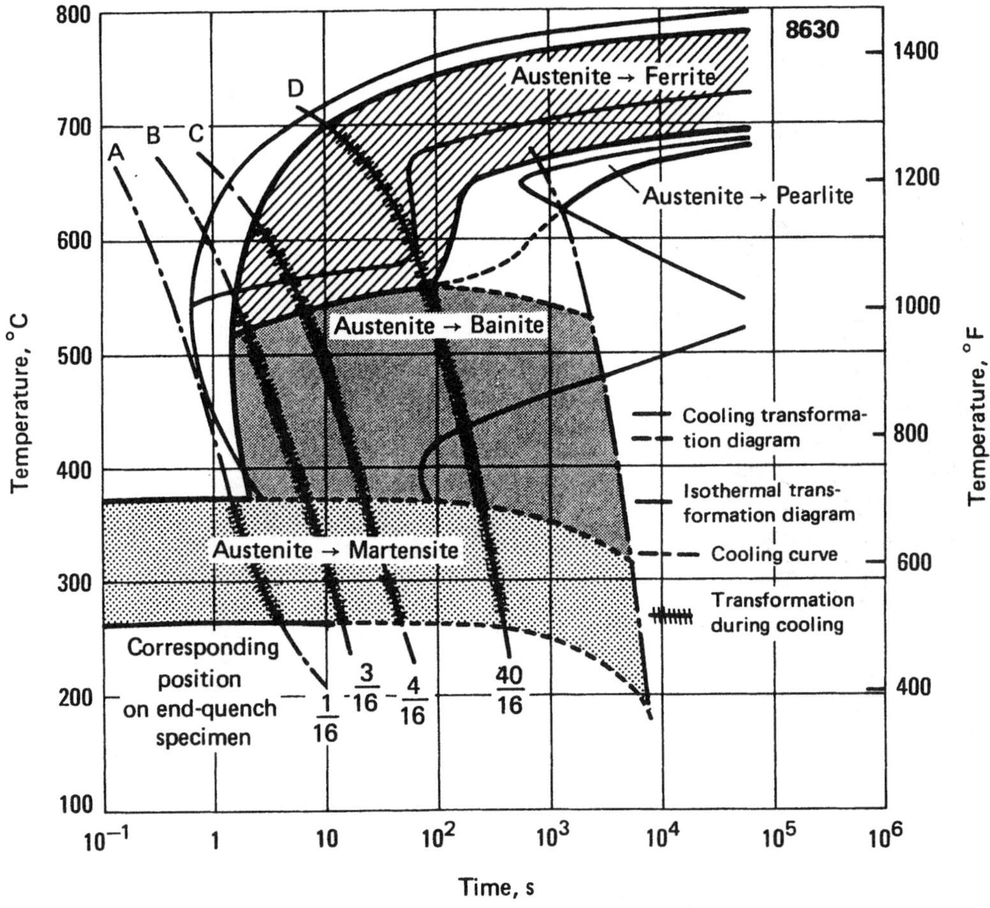

FIGURE 4.10.17 Continuous cooling transformation for 8630 steel. (From *ASTM Handbook,* Vol. 4, American Society for Testing and Materials, West Conshohocken, PA, 1991. With permission.)

For carbon steels it is critical to be able to generate high cooling rates to inhibit equilibrium phase transformation. Rapid cooling results in the generation of Martensite, an exceedingly hard and brittle nonequilibrium structure. A subsequent tempering heat treatment yields performance and microstructures which are otherwise unobtainable. This approach enables a combination of high strength and toughness which are not possible through equilibrium processes and is the basis for the application of steel as the principal structural material in modern engineering design.

Since high cooling rates are critical in quench and temper processes, the transfer of heat from the workpiece to the surroundings is important. Continuous cooling transformation (CCT) diagrams, such as that shown in Figure 4.10.17, are used to determine necessary cooling rates. While attaining critical cooling rates in thin sections is reasonably easy, the actual cooling rate in thicker sections varies with position within the workpiece material. Changes in the cooling medium, and, therefore, the surface cooling rate, can only be used to accomplish a certain degree of improvement in the depth of hardening. To ultimately improve the hardenability, it is generally necessary to choose an alloy developed to have a high hardenability, or in other words, a material that can be fully hardened at relatively slow cooling rates. Steels with increased carbon content, such as 1080 vs. 1034, and for even thicker sections of high-alloy contents such as 5140 and 9261, result in CCT curves that are shifted to the right, thus allowing slower cooling rates to still attain Martensite (see Figure 4.10.18). Tempering of this Martensite, under thermal conditions that are again often modeled as isothermal, results in a fine distribution of carbides

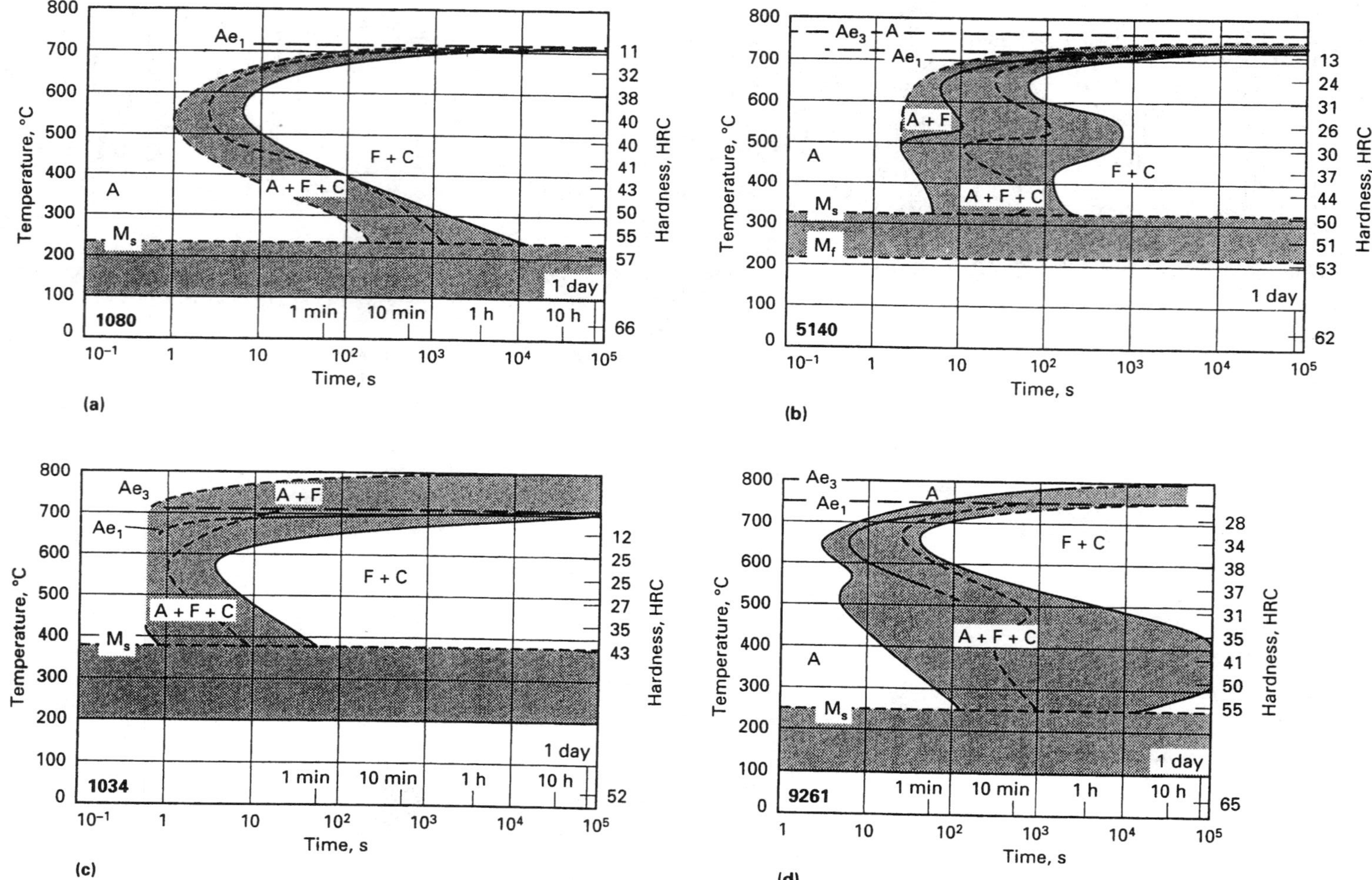

FIGURE 4.10.18 Comparison of continuous cooling transformation showing the effect of increasing carbon content steel. (From *ASTM Handbook*, Vol. 4, American Society for Testing and Materials, West Conshohocken, PA, 1991. With permission.)

which, by interfering with dislocation motion, increase the toughness and serviceability of the material without sacrificing the strength.

Aluminum alloys are often heat-treated in a fashion similar to that of steel; however, the specifics of the microstructural changes are quite different. In precipitation-hardened aluminum alloys, the workpiece is quenched from the single-phase region in the same way as during the heat treatment of steel to lock in a nonequilibrium microstructure. However, the resulting nonequilibrium aluminum structure is a soft and supersaturated solid solution, unlike the hard and brittle Martensite formed during the thermal processing of steel. Yet, it is still critical that this nonequilibrium structure be obtained throughout the complete workpiece section. In this case, the enhanced thermal conductivity of aluminum aids in allowing larger sections to achieve satisfactory cooling rates. The supersaturated solid solution of Al-Cu and Al-Si alloys is then heated to an elevated temperature within the two phase region and held isothermally for a period of time to allow the development of a fine, coherent precipitate. These hard precipitates interfere with dislocation motion and result in substantial strengthening of the aluminum. This is the basis of modern high-performance aerospace aluminum alloys, first developed in the early 1900s and which contributed to the success of the lighter-than-air rigid airships, the dirigibles.

In both the steel and aluminum cases described, increased times of heat treatment for the quenched material results in undue microstructural change and diminished mechanical properties. This is considered over-aging in the precipitation-hardened aluminum alloys and leads to reduced strength and ductility. Since aging temperatures for aluminum alloys are relatively low, usually in the range of 200°C, the useful upper temperature of these alloys is also limited. Extended use at temperatures even just above 135°C can lead to substantial performance drops in a matter of weeks for an aluminum alloy such as 2014 (see Figure 4.10.19).

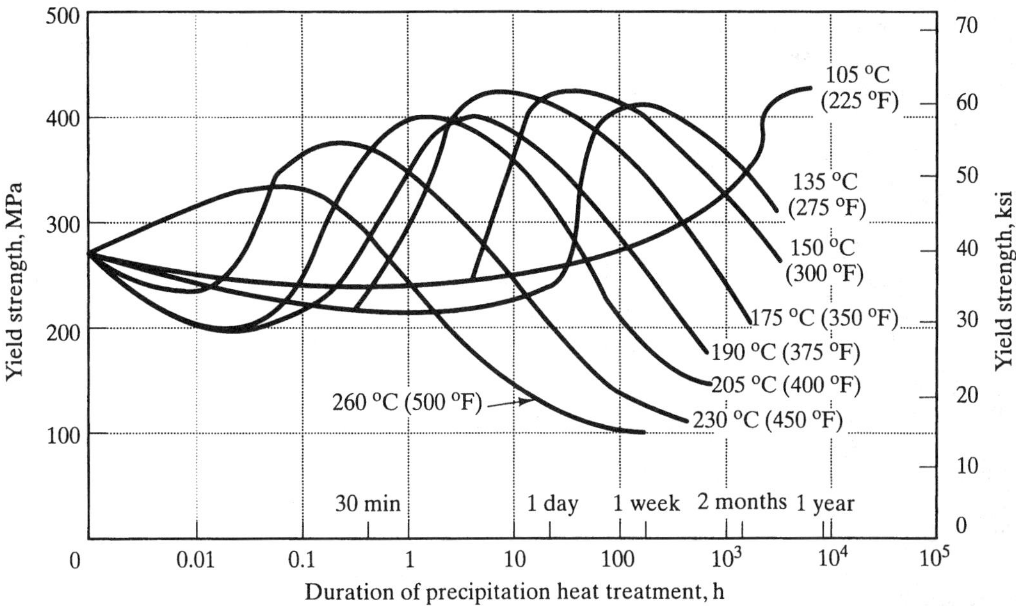

FIGURE 4.10.19 Thermal history of a precipitation-hardened aluminum alloy. (From Mangonon, P.L., *The Principles of Materials Selection for Engineering Design,* Prentice-Hall, Englewood Cliffs, NJ, 1999. With permission.)

Considerable information about heat treatment is available through many sources, including the American Society for Testing and Materials (ASTM, 1991), that can be consulted to determine the specific parameters required to generate final material properties and to understand the hardenability of individual materials. These sources typically include CCT diagrams for common materials and discuss heat treatment trends for specialized materials not typical of larger material families.

Machining

Machining to shape and cutting can both be considered material removal processes. Traditional machining processes remove material through mechanical processes. In these conventional material removal procedures, heating and heat transfer is introduced at the tool/workpiece interface and is the result of the chip generation process. Nontraditional machining processes such as electrical discharge machining (EDM) or laser beam machining (LBM) also result in workpiece heating. For these noncontact processes, material is either eroded away by electric discharge or vaporized using LBM. Since very high heat fluxes can be present, high heating and cooling rates often result, and with them come metallurgical problems such as quench cracking (Kalpakjian, 1995). Thus, process parameters, including rates and power, must be determined based on the thermal properties of the workpiece to minimize the impact of such rapid heating and cooling.

In traditional metal removal operations localized heating of the workpiece and subsequent rapid cooling due to the thermal mass of the material can also result in surface character modification. In general, the surface is hardened by rapid cooling which can lead to surface cracking and spalling, or at least to increased hardness and difficulty in further material removal. In steels, this surface hardening is often referred to as a "white layer," indicating the generation of Martensite due to the rapid cool-down after the cutting tool has passed by. However, the effects of temperature on traditional machining processes are probably often associated with decreased cutting tool life and associated cost increase.

Since the cutting tool must be harder than the workpiece under all conditions if workpiece material removal is to be successful, any thermal conditioning of the cutting tool which reduces its hardness will compromise its useful life. The energy involved in the cutting process is highly localized at the cutting tool tip and is primarily converted to heat. The heat generation rises rapidly as the cutting speed is increased. Because of this, excess speed is always considered the biggest factor in decreased tool life. The heat generated is concentrated in the forming chip, especially at high cutting speeds. Since this chip is in direct contact with the rake face of the cutting tool as the material is removed, the adjacent tool material temperature increases. The temperature increase of the rake face of the cutting tool can be approximated by

$$T_T \cong E\left(\frac{vh}{k\rho c}\right)^{1/2} \tag{4.10.4}$$

where E = cutting energy and h is undeformed chip thickness.

While cutting fluids can be used as a coolant to help increase tool life, the high pressures at the rake face make the introduction of fluids as lubricants difficult. However, the introduction of cutting fluids to cool the forming chip can be effective in maintaining lower tool temperatures. In many cases the only solutions to tool life maintenance are decreased speeds and enhanced cutting tool materials. Enhanced cutting tool materials are generally defined as materials that retain hardness at higher use temperatures. These can be steels with alloying ingredients which form more stable carbides, or may in fact be nonmetallic cutting tool materials such as cermets and ceramics.

Tool life prediction is based on the speed, feed, and depth of cut, but changes in tool life are often given as a power law relationship based strictly on speed due to much greater heating from speed increases.

$$ut^n = C \tag{4.10.5}$$

where

u = cutting speed (m/min)
t = tool life (min)
C = cutting speed for 1 min of life
n is a characteristic of the tool material

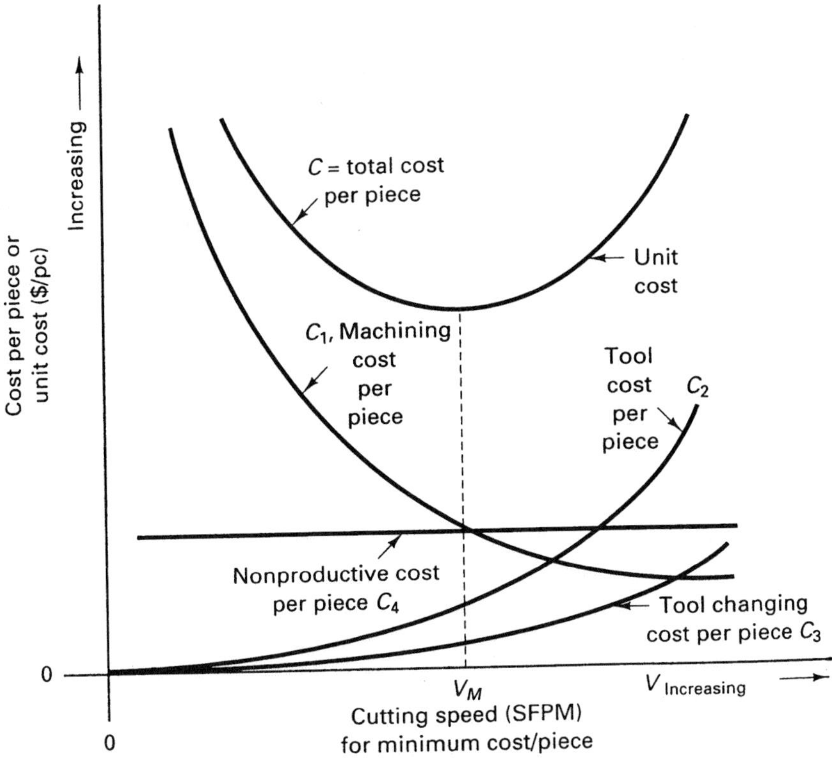

FIGURE 4.10.20 Dependency of unit cost on cutting speed. (From DeGarmo, E.P., Black, J.T., and Kosher, R.A., *Materials and Processes in Manufacturing,* 8th ed., Prentice-Hall, Englewood Cliffs, NJ, 1997. With permission.)

In general terms the tool life relationship can include the effects of speed, feed, and depth of cut, as follows

$$t = \frac{K}{u^{1/n_1} f^{1/n_2} w^{1/n_3}} \tag{4.10.6}$$

where

$n_1 < n_2 < n_3$
u = cutting speed (m/min)
f = feed rate (m/min)
w = depth of cut (m/rev)
K = material-dependent constant

However, since n_1 is generally significantly smaller than n_2 and n_3, speed is the controlling factor. Therefore, changes to cutting processes that reduce tool heating will increase cutting tool life. Figure 4.10.20 illustrates the relationship between the unit cost and cutting speed. There is a base cost associated with handling. Machining cost decreases with an increase in cutting speed because cutting time is reduced. Tool costs and tool changing costs are greater for higher cutting speed due to increased wear. When all these costs are accounted for, there exist an optimum speed for minimizing the unit cost, as shown in the diagram.

Deformation Processing

Bulk deformation processes such as forging, extrusion, and rolling can be carried out at either cold-work or hot-work temperatures. However, in the majority of cases the processes are considered isothermal and heat transfer is not a primary issue. In hot-work processing, the heat is used to lower the flow stress of the workpiece and to reduce the amount of force required to generate the necessary deformation. Under hot-work conditions attention does have to be paid to the temperature increase in the equipment performing the mechanical deformation. If the tooling temperature is the same as the workpiece temperature, the process is considered isothermal.

One concern related to heat transfer in deformation is the efficiency of working and the heat rise that correspond to low process efficiencies. To reshape a material, plastic deformation must be accomplished. Therefore, the work going to plastic deformation is desirable. However, elastic work, redundant work, and work to overcome friction also exist. Thus, efficiency of working can be described in terms of the desired work vs. the total work performed. Efficiency is then

$$\eta = \frac{U_{PL}}{U_{EL} + U_{PL} + U_f + U_\phi} = \frac{U_{PL}}{U_T} \tag{4.10.7}$$

where

U_{PL} = plastic work
U_{EL} = elastic work
U_f = work to overcome friction
U_ϕ = redundant work
U_T = total work

Processes range in efficiency from roughly 20% for extrusion, which exhibits large amounts of redundant work and frictional work, to 90% for cold rolling. Of these work contributions, the sum of the plastic work, the frictional work, and the redundant work primarily result in heat generation. It is usual practice to consider that the work expended to overcome friction leads to heating of the tooling, while adiabatic heating of the workpiece is related to the plastic and redundant work contributions. Based on these assumptions, the maximum theoretical temperature rise of the workpiece can be computed as

$$\Delta T = \frac{U_{PL} + U_\phi}{\rho c J} \tag{4.10.8}$$

where J = 4.18 Nm/cal. From this equation, a maximum theoretical temperature rise of as much as 280°C can be generated in an aluminum workpiece and as much as 570°C in titanium. Such temperature rises are of greatest concern during hot work operations because they can lead to localized melting of the workpiece. Finally, it should be noted that as the working velocities and strain rates increase, less time will be available for heat transfer away from the working zone and localized melting will become more of a concern.

Plastics Molding

Processing of plastics generally involves thermal energy. Unlike metallic materials, plastics have poor thermal conductivities. This can result in a large heat buildup in the material and can lead to problems of excessive localized shrinkage and possible material degradation. The specific effects of heat buildup depend on whether the plastic is a thermoset or a thermoplastic.

Thermoplastics Processing

Thermoplastics are polymers that do not cross-link and, therefore, can be reheated and remelted many times. Polymers, including polyethylenes and polyetherimides, are thermoplastics. Heating reduces the viscosity of these materials and allows them to be molded at elevated temperatures and then cooled to lock in the new shape. The primary heat-induced problems that can be encountered in processing thermoplastics are related to regions within the parts that have significant differential volumes. The regions of greater volume cool more slowly and thus, as in casting metals, results in localized shrinkage and even fracture. Typical solutions revolve around mold redesign to reduce the section variation.

In general, cooling rates are not of great consequence in the processing of thermoplastics as long as the molds and part shape are correctly designed. However, some thermoplastics are semicrystalline in nature. For these thermoplastics, cooling rates define the degree of crystallinity, which in turn determines many of the mechanical and physical properties of the resulting polymer component. Enhanced crystallinity often leads to improved chemical resistance, enhanced mechanical property retention at elevated temperatures, and even higher absolute mechanical performance. Unfortunately, generating controllable amounts of crystallinity is often difficult in practice and as a result many semicrystalline polymers are modified to change the kinetics of the reaction and decrease the likelihood of crystalline transformation.

Thermoset Processing

Thermosets are polymers that are cross-linked during processing. The cross-linking results in a material that cannot be remelted and reprocessed and, as a result, must be molded to final shape. Polymers such as phenolics, epoxies, and vinylesters are examples of thermosets. Thermosets may be processed at room temperature or at elevated temperatures, depending on the specific material and the form of the reaction that leads to cross-linking. During cross-linking, heat is evolved. Since these polymers have inherently poor thermal conductivity, the temperature rise due to cross-linking, particularly in a large volume of the material, can lead to thermal degradation. To slow such rapid exothermic reactions, cures are in some cases carried out at reduced temperature. However, as the materials are spread over larger areas, such as thinner films or as matrix materials in composites, heat transfer out of the material is more rapid and the peak temperatures drop to manageable levels.

Again for thermosets, increased temperatures generally lead to added shrinkage. For a specific polymer, the degree of shrinkage is generally enhanced with increased cure (cross-link) rates. Thus, cure rates are generally controlled to minimize shrinkage and property degradation, while still keeping process times to a reasonable duration. Much research has been performed on thermoset curing as these materials form the basis for the vast majority of polymer matrix composites in service today.

Thermal Spray Deposition

Thermal spray is a method to create protective surface coatings or alter surface properties. Near net-shape manufacturing can also be accomplished using this technique. Typically, liquid droplets are sprayed onto a substrate. Solidification of the droplets could take place partially during flight or entirely after they have arrived at the substrate. Due to the momentum of the droplets, splats are formed upon impact of the droplets. The shape of the splat is affected by the properties of the droplet, surface roughness, and substrate temperature. There is experimental evidence that low substrate temperatures result in star-shaped splats, whereas high substrate temperatures (400°C) produce disc-shape splats (Dykhuizen, 1994). The size of the splat directly impacts the coating thickness as well as the coating properties. Dykhuizen (1994) reviewed splat size models developed before 1994. Madejski (1976, 1983) presented models that accounted for surface tension and solidification. Recently, Zhang (1999) developed a model that accounts for the effects of thermal contact resistance as well. Zhang's model can be summarized as follows

$$\frac{1}{\mathrm{Re}}\left(\frac{\xi_m}{1.18}\right)^5+\frac{3(1-m)\left(\xi_m^2-1\right)}{We}+\omega S\left(\frac{\xi_m}{1.15}\right)^{2.5}=1.0 \tag{4.10.9}$$

where ξ_m is the flattening ratio defined as the splat radius compared to the droplet initial radius (R_o). The Reynolds number is defined as $Re = \frac{2\rho v_o R_o}{\mu}$ with v_o and μ being the initial droplet velocity and viscosity, respectively. The Weber number is defined as $We = \frac{2\rho v_o R_o}{\sigma_1}$ where σ_l is the surface tension of the melt. In Equation 4.10.9, ω is a constant that assumes values between 0.6 and 1.0 and S is given as

$$S = \sqrt{\left(\frac{2Ja}{RePr}\right)\left(\frac{k_s}{k_l}\right)}$$

where k_s and k_l are the thermal conductivities of the deposited layer and the melt, respectively. The Jacob number, Ja, is defined as

$$Ja = \frac{c\left(T_f - T_{se}\right)}{h_{fg}}$$

where T_f and T_{si} are the fusion temperature and the initial substrate temperature, respectively, and h_{fg} is the latent heat of vaporization. The Prandtl number, Pr, is defined as $\Pr = \frac{\nu_l}{\alpha_l}$ with ν_l and α_l being the kinematic viscosity and thermal diffusivity, respectively, for the melt. The coating thickness is given by

$$s = -\left(\frac{k_s R''_{t,c}}{R_o}\right) + \sqrt{\left(\frac{k_s R''_{t,c}}{R_o}\right)^2 + S^2 t} \tag{4.10.10}$$

where $R''_{t,c}$ is the thermal contact resistance, and t is time. Equations 4.10.9 and 4.10.10 are the two equations that can be used to estimate the splat size and the coating thickness.

References

Heat Treating, in *American Society of Testing and Materials Handbook,* Vol. 4, West Conshohocken, PA, 1991.

Carslaw, H.S. and Jaeger, J.C., Conduction of Heat in Solids, 2nd ed., Oxford University Press, Oxford, 1959.

DeGarmo, E.P., Black, J.T., and Kosher, R.A., *Materials and Processes in Manufacturing,* 8th ed., Prentice-Hall, Englewood Cliffs, NJ, 1997.

Dykhuizen, R.C., Review of impact and solidification of molten thermal spray droplets, *J. Thermal Spray Technol.,* 3, 351-361, 1994.

Kalpakjian, S., *Manufacturing Engineering and Technology,* 3rd ed., Addison-Wesley, Reading, MA, 1995.

Mangonon, P.L., *The Principles of Materials Selection for Engineering Design,* Prentice-Hall, Englewood Cliffs, NJ, 1999.

Madejski, J., Solidification of droplets on a cold surface, *Int. J. Heat Mass Transfer,* 19, 1009-1013, 1976.

Madejski, J., Droplets on impact with a solid surface, *Int. J. Heat Mass Transfer,* 26, 1095-1098, 1983.

Zhang, H., Theoretical analysis of spreading and solidification of molten droplet during thermal spray deposition, *Int. J. Heat Mass Transfer,* 42, 2499-2508, 1999.

4.11 Pinch Point Analysis

Kirtan K. Trivedi

Introduction

Many industrial processes require addition and removal of heat to streams within the process. This is notably the case in the chemical process industry and power generation. In the chemical process industry, raw materials are heated or cooled to the appropriate reaction temperature. Heat may be added or removed to carry out the reaction at the specified condition of temperature and pressure. Heat addition and removal achieve product separation. The products may be heated or cooled to the correct temperature for storage and transportation. Normally, for a large petrochemical complex cogeneration scheme with a condensing cycle or a combined cycle is used to provide external heat and power required for the site. In the utility plant, air, demineralized water, and boiler feed water streams need to be heated and the flue gas and condensate streams need to be cooled. Thus, at all times heat is either added or removed in a variety of process streams by utilities or by heat exchange between process streams. In an integrated plant this is normally achieved with a heat exchanger network (HEN). A large fraction of the capital cost of many process plants is attributed to heat recovery networks.

The aim of a designer is to synthesize a near-optimal configuration for a process. This indicates that a proper trade-off between the capital invested and the operating cost of the plant should be achieved. The capital cost depends on the type, number, and size of units utilized to satisfy the design objectives. A substantial part of the operating cost usually depends upon the utilities consumed. To reduce these costs, the process designer should aim for an economic combination having a nearly theoretical minimum number of heat exchanger units and aim to recover the maximum possible heat with them. An obvious way to recover heat is by exchanging it between hot process streams that need to be cooled, and cold process streams that need to be heated, in addition to the heating and cooling utilities. Furthermore, the designer should also investigate the operability of the final design.

The above objectives can be achieved by synthesizing a good heat exchanger network. However, a very large number of alternatives exist. Because of this, it will be highly rewarding to synthesize quickly and systematically the best possible alternatives. It is now possible to do this with the aid of *Pinch Technology.*

Pinch technology refers to a large and growing set of methods for analyzing process energy requirements in order to find economically optimal and controllable designs. Considerable development has taken place in pinch technology during the past 2 decades, mainly due to the efforts of Linnhoff and coworkers.[1] Pinch technology has proved to be effective and is successfully applied to process integration that encompasses overall plant integration including heat exchanger networks and heat and power integration or cogeneration. To date, there are 65 concepts used by this technology. However, due to the limited scope of this section, only the fundamental concepts are discussed here.

Industrial applications of this technology include capital cost reduction, energy cost reduction, emissions reduction, operability improvement, and yield improvement for both new process design and revamped process design. Imperial Chemical Industries (ICI) where this technology was first developed, reported an averaged energy saving of about $11,000,000 per year (about 30%) in processes previously thought optimized. The payback time was typically in the order of 12 months. Union Carbide showed even better results. Studies conducted by Union Carbide on nine projects showed average savings of 50% with an average payback period of 6 months. Savings in energy cost of about $8,000,000 per year was achieved by Union Carbide on these nine projects. BASF reports energy savings of 25% obtained by application of pinch technology to their Ludwigshafen site in Germany. Over a period of 3 years about 150 projects were undertaken by BASF. Energy saving of 10% with a payback period of 2 years is reported by applying pinch technology to the Caltex Refinery situated in Milnerton, Cape, South Africa. The energy consumption before the study was 100 MW for the whole refinery. A newsfront article in *Chemical Engineering*[2] gives more details about the experience of various companies in using pinch technology and the benefits obtained.

Fundamental Principles and Basic Concepts

Pinch technology is based on thermodynamic principles. Hence, in this section we will review the important thermodynamic principles and some basic concepts.

Temperature-Enthalpy Diagram

Whenever there is a temperature change occurring in a system, the enthalpy of the system will change. If a stream is heated or cooled, then the amount of heat absorbed or liberated can be measured by the amount of the enthalpy change. Thus, $Q = \Delta H$. For sensible heating or cooling at constant pressure where $CP = mc_p$,

$$\Delta H = CP\ \Delta T \tag{4.11.1}$$

For latent heating or cooling, $\Delta H = m\lambda$ (specific latent heat of condensation or evaporation). If we assume that the temperature change for latent heating or cooling is 1°C, then $CP = m\lambda$.

Equation (4.11.1) enables us to represent a heating or cooling process on a temperature-enthalpy diagram. The abscissa is the enthalpy and the ordinate is the temperature. The slope of the line is (1/CP). Figure 4.11.1a shows a cold stream being heated from a temperature of 20°C to 80°C with CP = 2.0 kW/°C.

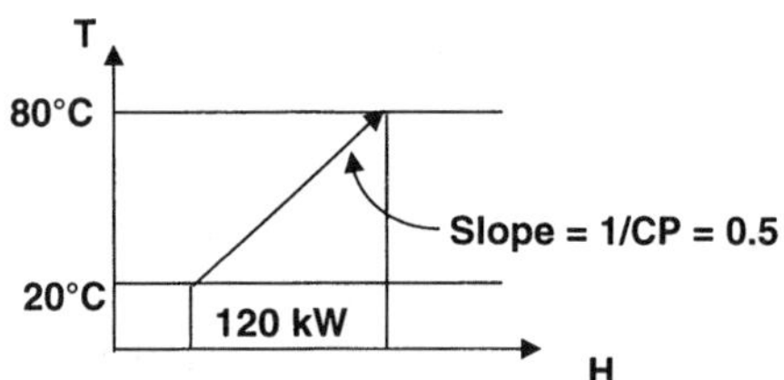

(a) Representation of a cold stream on the T-H diagram.

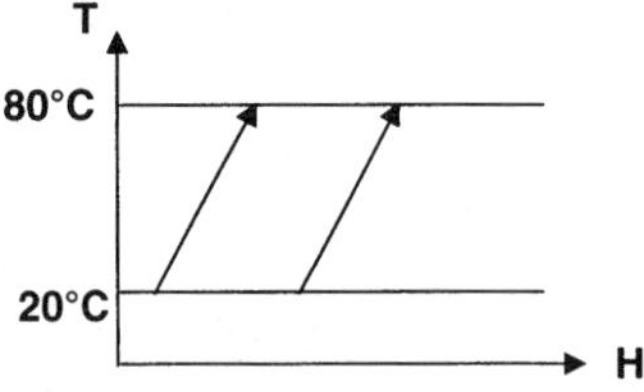

(b) A stream can be moved horizontally on the T-H diagram in a given temperature interval.

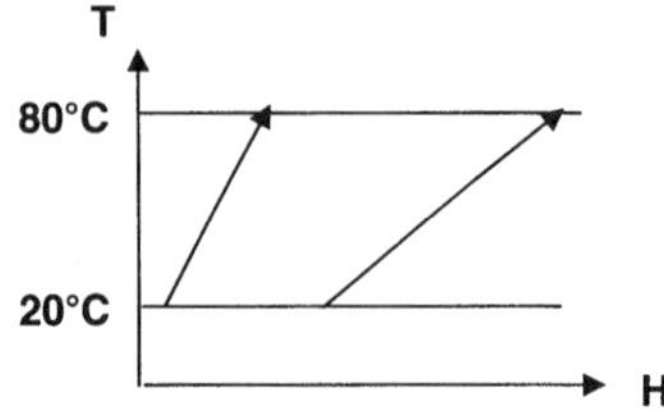

(c) Two different streams in the same temperature interval.

FIGURE 4.11.1 The temperature enthalpy diagram.

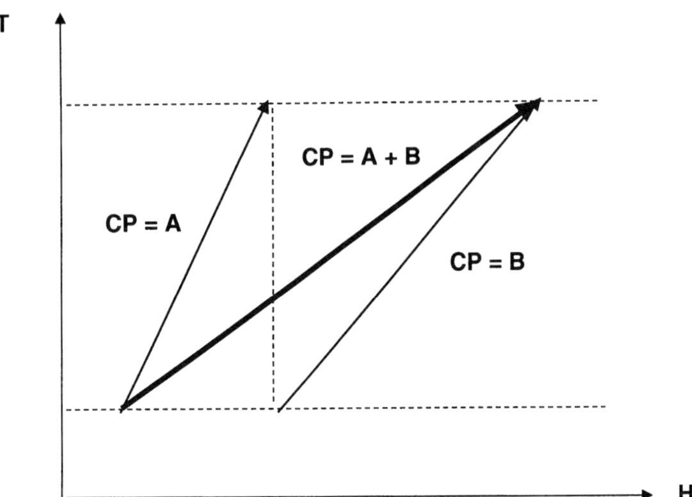

FIGURE 4.11.2 Composite stream obtained from two different streams in the same temperature interval.

As enthalpy is a relative function, the stream can be drawn anywhere on the enthalpy scale as long as it is between its starting and target temperatures and has the same enthalpy change. Figure 4.11.1b shows such a case. Thus, one of the important advantages of representing a stream on the temperature enthalpy plot is that the stream can be moved horizontally between the same temperature interval. Figure 4.11.1c shows two different streams in the same temperature interval.

Two streams can be easily added on the temperature-enthalpy plot to represent a composite of the two streams. Figure 4.11.2 shows how to obtain the composite of two streams on this plot. This feature will be used in later sections to predict the minimum utility requirement for multistream problems.

A heat exchanger is represented by two heat curves on the temperature-enthalpy diagram, as shown in Figure 4.11.3a. This figure also shows how we will represent heat exchangers in a grid diagram form Figure 4.11.3b and the conventional form Figure 4.11.3c. In the grid diagram form, the hot stream goes from left to right and the cold stream goes from right to left. If the flow is countercurrent in the exchanger then the temperature will decrease from left to right. Two circles connected by a vertical line represent the exchanger. The advantages of the grid diagram will become apparent when we discuss the design of heat exchanger networks. The exchanger has a temperature difference at the hot end and another at the cold end. The smaller of these two temperature differences is called Δt_{min}.

Some Definitions

A *match* between a hot and a cold stream indicates that heat transfer is taking place between the two streams. A match between two streams is physically achieved via a heat exchanger *unit.* The number of heat exchanger units impacts the plot plan and determines the piping and the foundation cost.

For reasons of fouling, mechanical expansion, size limitation, cleaning, improved heat transfer coefficients, etc., many process heat exchangers are the shell-and-tube type with 1 shell pass and 2 tube passes. Often what appears as a single match between two streams in a heat exchanger network representation is actually installed as several 1-2 exchangers in series or parallel. The term *shell* will be used to represent a single 1-2 shell-and-tube heat exchanger. Please refer to Section 4.3 for more discussion on 1-2 shell-and-tube heat exchangers.

Software

Pinch technology is a state-of-the-art technology for process integration and design. Different researchers at various universities have written programs, but only a handful of commercial programs are available. Table 4.11.1 shows the commercially available programs for process integration.

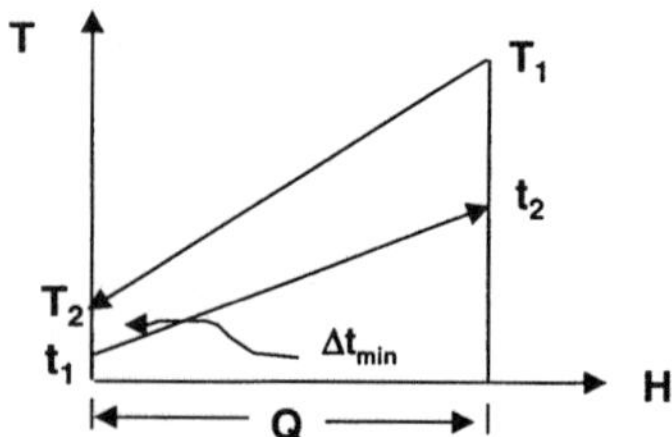

(a) Representation of an exchanger on the T-H diagram.

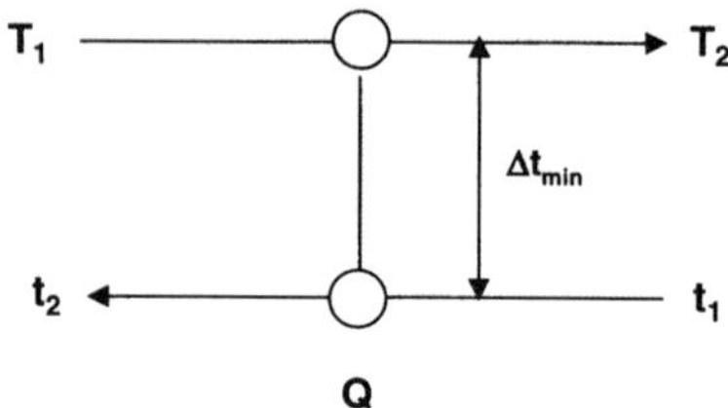

(b) Grid diagram representation of an exchanger.

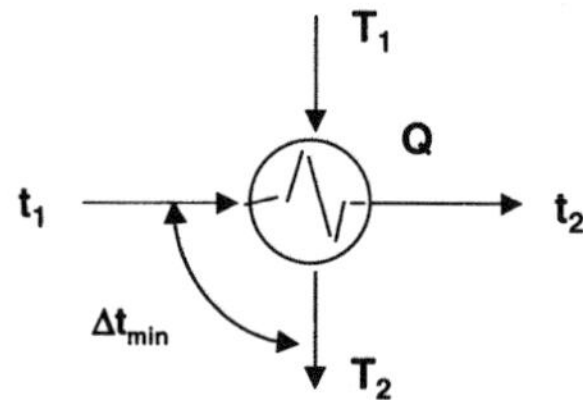

(c) Conventional representation of an exchanger.

FIGURE 4.11.3 Heat exchanger representation.

At the Department of Process Integration, University of Manchester Institute of Science and Technology (UMIST), a consortium funded by various companies conducts research and develops the majority of the concepts of pinch technology. The software developed by the consortium is available to its members only. The consortium is called the **P**rocess **I**ntegration **R**esearch **C**onsortium (PIRC).

Optimization Variables and Heat Exchanger Network Design Philosophy

The objective of the heat exchanger network synthesis problem is to design a network that meets an economic criterion such as minimum total annualized cost. The total annualized cost is the sum of the annual operating cost (which consists mainly of energy costs) and annualized capital cost.

The capital cost of a network primarily depends on the total surface area, the number of shells, and the number of units that will be installed. The capital cost also depends on the individual type of heat exchangers and their design temperature, pressure, and material of construction.

If we include the pressure drop incurred in a heat exchanger, then the capital and operating cost of the pumps will also have to be taken into account. We will limit the scope of this section by not including stream pressure drop constraints.

Figure 4.11.4 shows some of the variables that affect the optimization of a heat exchanger network. To make the network operable or flexible, extra cost may be incurred. This cost may be in the form of added equipment or use of extra utilities or use of additional area for some of the exchangers in the network.

From the above discussion, it is clear that the synthesis of heat exchanger networks is a multivariable optimization problem.

TABLE 4.11.1 Commercially Available Computer Programs for Pinch Technology

Software name	Marketed by
Supertarget	Linnhoff-March Limited, Targeting House, Gadbrook Park Northwich Cheshire, CW 9742, U.K. Phone: (44) 1606 815100 Fax: (44) 1606 815151
Aspen Pinch	Aspen Technology Inc., Ten Canal Park, Cambridge, MA 02141 Phone: (617) 577 0100 Fax: (617) 577 0303
Heatnet	Britain's National Engineering Laboratory, East Kilbride, Scotland
HEXTRAN	Simulation Sciences Inc., 601 S. Valencia Ave., Brea, CA 92621 Phone: (714) 579 0354 Fax: (714) 579 0412
DPI Software	Department of Process Integration, UMIST, Manchester, U.K. Phone: (44) 161 200 4382 Fax: (44)161 236 7439
HX-Net	Hyprotech Ltd., 300 Hyprotech Center, 1110 Center Street North, Calgary T2E 2R2, Alberta, Canada Phone: (403) 520 6122, Fax: (403) 520 6060

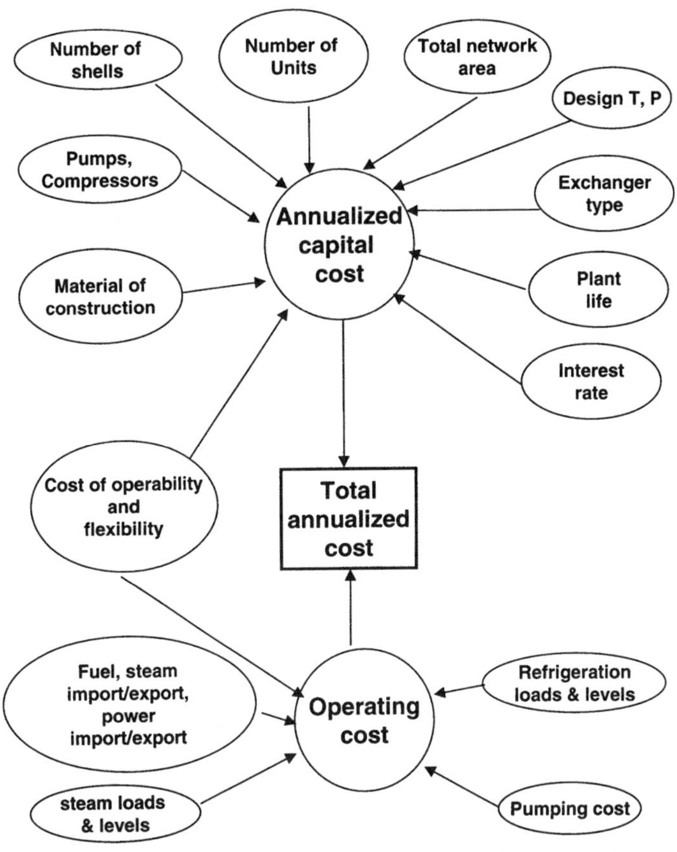

FIGURE 4.11.4 Optimization variables in the design of heat exchanger networks.

To better understand the design process, consider a simple problem consisting of two streams. The first stream needs to be cooled from 160 to 40°C with a mass heat flow capacity, CP, of 1 kW/°C. The second stream needs to be heated from 50 to 180°C with a CP of 2 kW/°C. The economic basis for this design is given in Table 4.11.2. The objective is to design an optimal heat exchanger network. The direct heat recovered between the two streams and the external utilities required to achieve the target temperature will determine the optimal solution. It is assumed that all process streams as well as utility streams have a film heat transfer coefficient of 1 kW/°C m^2.

TABLE 4.11.2 Economic Data for the Design Problem

Heat Exchangers		
Installed cost per shell ($)	=	10,000 $A^{0.6}$ (m^2)
Utility Data		
Cost of using hot oil	=	68 ($/kW.yr), ($2.36/MMBtu)
Cooling water	=	2.5 ($/kW.yr), ($0.09/MMBtu)
Temperature range of hot oil	=	320–310°C
Temperature range of cooling water	=	10–20°C
Plant Data		
Interest rate	=	10%/yr
Lifetime	=	5 yr
Operation time	=	8000 h/yr
Calculated capital recovery factor (CRF)	=	0.2638/yr

An intuitive approach for designing the heat exchanger network is to assume a value for the minimum approach temperature, Δt_{min}, for all the process-process exchangers within the network, i.e., no exchanger in the network will have an approach temperature less than the assumed value of Δt_{min}. For the first iteration, let $\Delta t_{min} = 10$. The network for this Δt_{min} is shown in Figure 4.11.5. For the network structure shown in Figure 4.11.5, Table 4.11.3 shows the details of the network design at different values of Δt_{min}. Figure 4.11.6 shows the optimization curve for this problem.

The design procedure used is summarized in Figure 4.11.7. It is an iterative procedure and can take a long time to find the optimal solution. For multiple stream problems, at a given value of Δt_{min}, a number of different designs can be feasible. Hence, a better approach is needed.

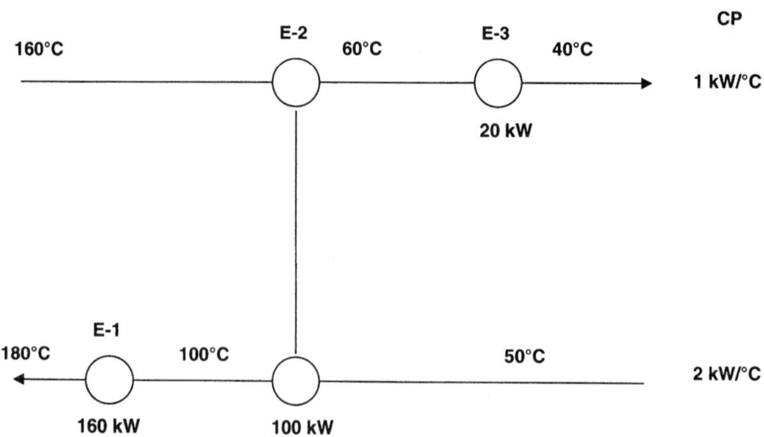

FIGURE 4.11.5 Network for the two stream problem, Δt_{min} = 10°C.

TABLE 4.11.3 Network Design Details for the Two Stream Problem

Δt_{min}	Total Annualized Cost ($/yr)	Design Parameter	E-1	E-2	E-3
10	32,388	Duty, kW	160	100	20
		Number of shells	1	3	1
		Area, m²	1.98	8.29	1.18
		Capital cost $/yr	3974	14,566	2919
		Operating cost $/yr	10,879		50
20	28,808	Duty, kW	170	90	30
		Number of shells	1	2	1
		Area, m²	2.08	5.43	1.59
		Capital cost $/yr	4089	9606	3479
		Operating cost $/yr	11,559		75
30	27,971	Duty, kW	180	80	40
		Number of shells	1	2	1
		Area, m²	2.17	3.62	1.92
		Capital cost $/yr	4199	7531	3902
		Operating cost $/yr	12,239		100
40	26,659	Duty, kW	190	70	50
		Number of shells	1	1	1
		Area, m²	2.26	2.97	2.21
		Capital cost $/yr	4305	5069	4241
		Operating cost $/yr	12,919		125
50	26,728	Duty, kW	200	60	60
		Number of shells	1	1	1
		Area, m²	2.35	2.04	2.46
		Capital cost $/yr	4407	4049	4523
		Operating cost $/yr	13,599		150
60	27,026	Duty, kW	210	50	70
		Number of shells	1	1	1
		Area, m²	2.44	1.45	2.68
		Capital cost $/yr	4504	3303	4764
		Operating cost $/yr	14,279		175

The design can never violate the laws of thermodynamics. The philosophy adopted in pinch technology is to establish targets for the various optimization variables based on thermodynamic principles. These targets set the boundaries and constraints for the design problem. Further, targets help us identify the various trade-offs between the optimization parameters. They help us in obtaining a "birds-eye view" of the solution space and identify the optimal values of the optimization parameters. Once the optimal values are identified, design the network at these values. This approach will always lead to an optimal design. This approach is summarized in Figure 4.11.8. Setting targets ahead of design eliminates the iterations required. Based on the stream data and using thermodynamic principles, targets can be easily set as discussed below.

To understand the interaction between the different optimization variables, consider the heat curves with $\Delta t_{min} = 10$ as shown in Figure 4.11.9a, for the simple two stream system discussed above. Three different sections can be easily identified on this diagram. Section 1 represents the cold utility requirement, Section 2 represents the process-process heat exchange, and Section 3 represents the hot utility requirement. So from this set of heat curves we can calculate the utility requirements for the system. Again for each section we can calculate the area required, as we know the duty for each section and the terminal temperatures. Further, we can deduce that each section will require one unit. We can use Bell's method[3] to estimate number of shells required for each unit that represents the different sections. In Section 2, three shells will be required. The heat curves for the hot and cold streams establish targets

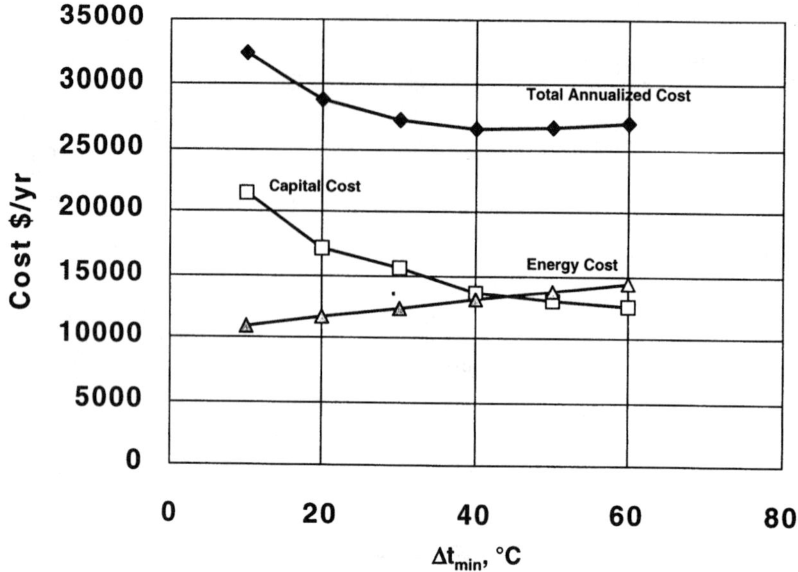

FIGURE 4.11.6 Optimization curve for the two stream problem.

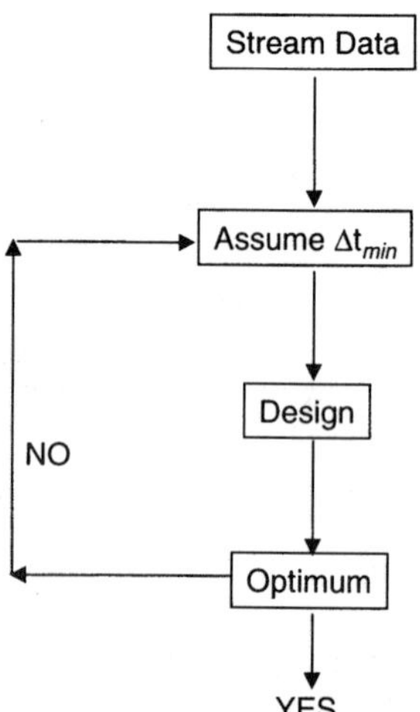

FIGURE 4.11.7 Intuitive procedure for heat exchanger network design.

for energy, area, units, and shells. These are the major components that contribute towards the annualized cost of the network.

Now for the same system, let us increase the Δt_{min} to 40°C. The new set of heat curves is shown in Figure 4.11.9b. The above exercise can now be repeated to obtain targets for energy, area, units, and shells. When we increase the value of Δt_{min} the utility requirement will increase and the total network area required will be decreased. The number of units for this system remains the same, but the number of shells required will decrease. Section 2 now only requires one shell. Thus, as the value of Δt_{min} increases, the utility cost increases and the capital cost decreases. This indicates that Δt_{min} is the single variable that fixes the major optimization variables shown in Figure 4.11.4. Hence, we can reduce the multivariable optimization problem to a single-variable optimization problem. This single variable is Δt_{min}. At the optimum value of Δt_{min}, the other optimization variables, viz., number of shells, number of units, total network area requirement, and the hot and cold utility requirements will also be optimal. For

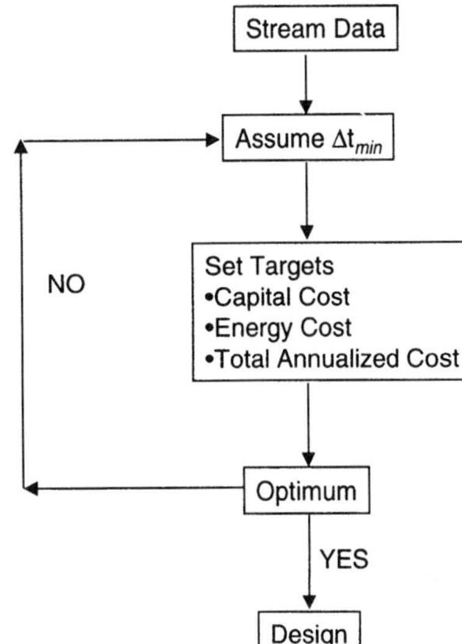

FIGURE 4.11.8 Pinch technology approach for heat exchanger network design.

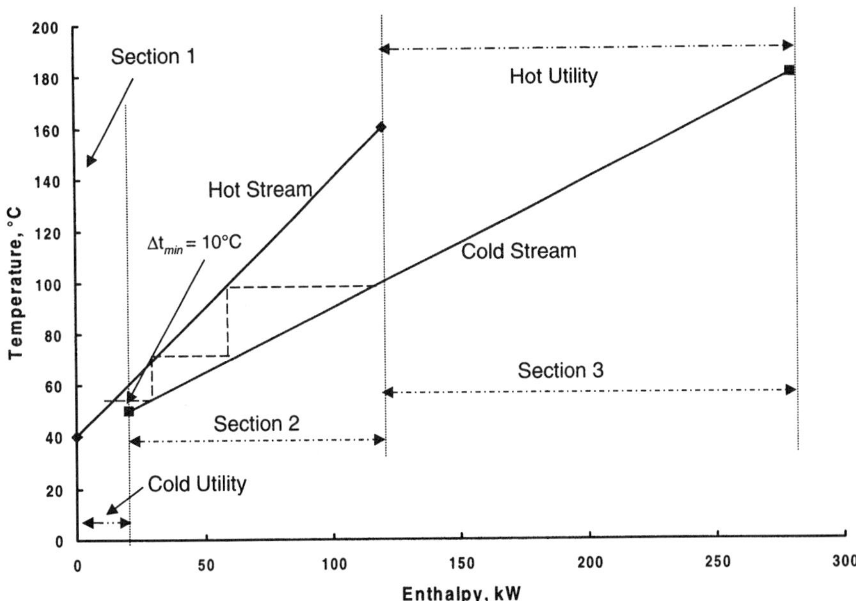

FIGURE 4.11.9a Effect of Δt_{min} on the energy, area, units, and shells required for a heat exchange system.

a multistream problem the same conclusion can be easily derived. Hence, for the multistream problem optimization discussed in the next subsections we will develop methods to optimize the value of Δt_{min}.

The targets obtained by the above procedure are based on thermodynamics. Further, they are established ahead of design using stream data only and, hence, help us identify where to initialize our design. This is a very powerful concept as it eliminates the need to develop a number of different designs to find the optimal design. The targeting procedure identifies the optimal region ahead of design. A design can be easily developed based on the design principles discussed later. Some local optimization can be conducted to account for other factors such as operability, flexibility and constructability.

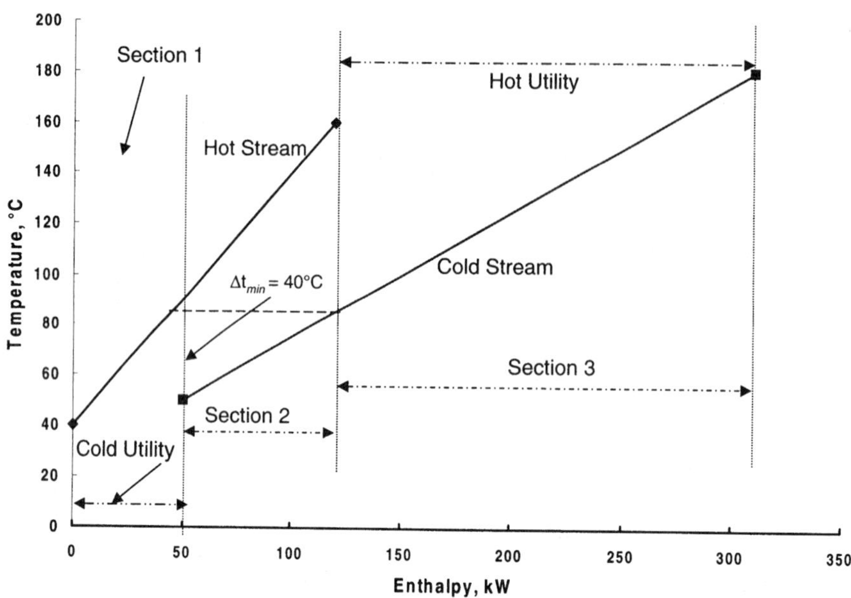

FIGURE 4.11.9b Effect of Δt_{min} on the energy, area, units, and shells required for a heat exchange system.

Multistream Design Problem

Consider a four-stream problem. Table 4.11.4 lists the starting and target temperatures of all the streams involved in the flowsheet. It also shows the CP values. While pressure drop constraints on the individual streams determine the film heat transfer coefficients, we shall not take into account this constraint. Instead, we use (unrealistically) the same heat transfer coefficient value of 1 kW/m² K for all streams. Our objective is to design an optimum heat exchanger network for this process using the economics outlined in Table 4.11.2. Further, let Δt_{min} denote the minimum temperature difference between any hot process stream and any cold process stream in any exchanger in the network. We shall use this problem to illustrate the concepts of pinch technology.

TABLE 4.11.4 Stream Data for a Design Problem

Stream Number	Stream Name	T_s (°C)	T_t (°C)	CP (kW/°C)
1	C1	20	160	40
2	C2	120	260	60
3	H1	180	20	45
4	H2	280	60	30

Targets for Optimization Parameters

Energy Targets

Composite Curves

Let us plot the heat curves for all the hot streams on the T-H diagram (see Figure 4.11.10). We can divide the diagram into a number of temperature intervals, defined by the starting and target temperatures for all the streams. Between two adjacent temperatures we can calculate the total heat content of all the streams that are present in this temperature interval. For example, between 180°C and 60°C the sum total of the heat available is calculated as:

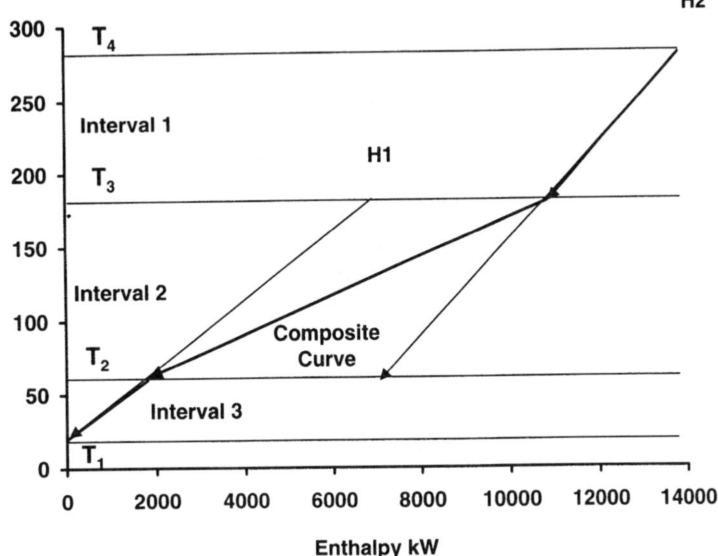

FIGURE 4.11.10 Construction of the hot composite curve.

$$\Delta H_2 = (T_3 - T_2)\sum_j CP_j$$

$$\Delta H_2 = (180 - 60)(45 + 30) = 9000kW$$

A composite curve that represents the heat content of all the hot streams is obtained by summing the heat available in each of these temperature intervals, as shown in Figure 4.11.10. Similarly, a composite curve for all the cold streams can be obtained. These two composite curves can be used as the heat curves for the whole process.

To obtain the energy target, fix the hot composite curve and move the cold composite curve horizontally till the shortest vertical distance between the two curves is equal to the value of Δt_{min}. The "overshoot" of the cold composite curve is the minimum hot utility requirement and the "overshoot" of the hot composite curve is the minimum cold utility requirement.[4] For Δt_{min} = 30°C, the minimum hot utility requirement is 4750 kW and the cold utility requirement is 4550 kW (see Figure 4.11.11).

The Problem Table

The above procedure for obtaining the energy targets using composite curves is time-consuming and clumsy. An alternative method based on thermodynamic principles was developed. Hohmann[4] called it the feasibility table. Linnhoff and Flower[5] independently developed the problem table. The problem table algorithm is easy and involves no trial and error.

The algorithm consist of the following steps:

1. Select a value of Δt_{min}. Since we have already established the targets using composite curves for Δt_{min} = 30°C, we shall use that value here.
2. Convert the actual terminal temperatures into interval temperatures as follows:

   for the hot streams: $T_{int} = T_{act} - \Delta t_{min}/2$
   for the cold streams: $T_{int} = T_{act} + \Delta t_{min}/2$

   where T_{int} = interval temperature and T_{act} = actual stream temperature.

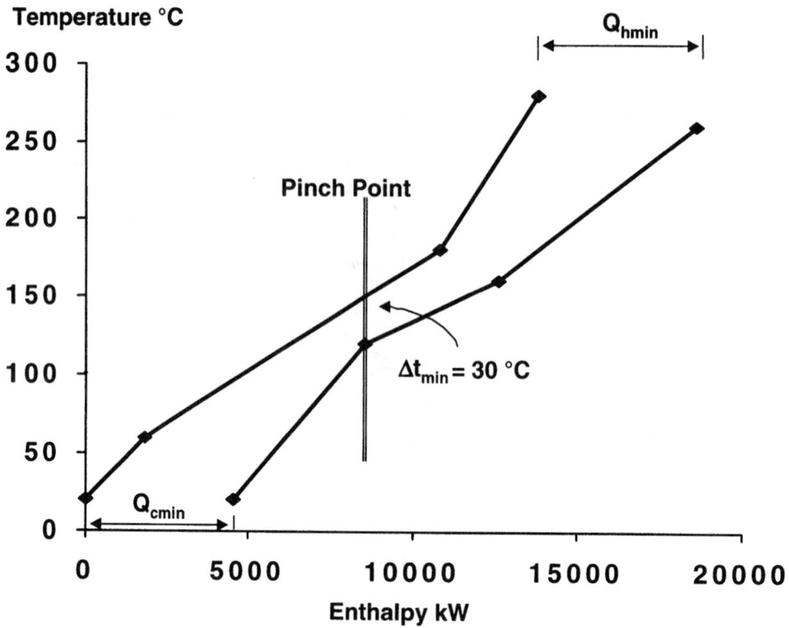

FIGURE 4.11.11 Composite curves for the design problem.

The interval temperatures now have the allowance for Δt_{min}. This modification guarantees that for a given T_{int}, the actual temperature difference between the hot and cold stream will always be greater than or equal to Δt_{min}.

3. All the interval temperatures for both the hot and cold streams are sorted in descending order and the duplicate intervals are removed (see Table 4.11.5).
4. For each interval, an enthalpy balance is made. The enthalpy balance for interval *i* is calculated using the following equation:

$$\Delta H_i = \left(T_{\text{int},i} - T_{\text{int},i+1}\right)\left(\sum_j^{N_{streams}} CP_{c_j} - \sum_j^{N_{streams}} CP_{h_j}\right)$$

where ΔH_i = net heat surplus or deficit in interval *i*
CP_c = mass specific heat of a cold stream
CP_h = mass specific heat of a hot stream

For the example problem, this calculation is shown in Table 4.11.6. Within each interval the enthalpy balance will indicate that there is either a heat deficit or surplus or the interval is in heat balance. A heat surplus is negative and a heat deficit is positive.

The second law of thermodynamics only allows us to cascade heat from a higher temperature to a lower temperature. Thus, a heat deficit from any interval can be satisfied in two possible ways — by using an external utility or by cascading the surplus heat from a higher temperature interval. If no external utility is used, the heat cascade column of Table 4.11.6 can be constructed. All intervals in this heat cascade have negative heat flows. This is thermodynamically impossible in any interval (it would mean heat is flowing from a lower to a higher temperature). To correct this situation, we take the largest negative flow in the cascade and supply that amount of external hot utility at the highest temperature interval. This modification will make the heat cascade feasible, i.e., none of the intervals will have

TABLE 4.11.5 Generation of Temperature Intervals for Δt_{min} = 30°C

Stream No.	Actual Temperature (°C) T_s	Actual Temperature (°C) T_t	Interval Temperature (°C) T_s	Interval Temperature (°C) T_t	Interval Number	Ordered Interval Temperatures (°C)
1	20	160	35	175	1	275
2	120	260	135	275	2	265
3	180	20	165	5	3	175
4	280	60	265	45	4	165
					5	135
					6	45
					7	35
					8	5

negative heat flows. The amount of external heat supplied to the first interval is the minimum amount of hot utility required. The surplus heat in the last interval is the minimum amount of cold utility required.

For the example under consideration, the minimum hot utility required is 4750 kW and minimum cold utility required is 4550 kW. These are the same targets obtained from the composite curves. It is clear, however, that the problem table algorithm is an easier, quicker, and more exact method for setting the energy targets. One can easily set up a spreadsheet to implement the problem table algorithm. Further, once the energy targets are obtained, the composite curves can be easily drawn to visualize the heat flows in the system. It should be noted that the absolute minimum utility targets for a fixed flowsheet are determined by the case of $\Delta t_{min} = 0$.

TABLE 4.11.6 The Problem Table

Stream No. / CP		1 / 40	2 / 60	3 / 45	4 / 30					
Temperature Interval	**Interval Number**					ΔT_{int}	$\Sigma CP_{cj} - \Sigma CP_{hj}$	ΔH_{int}	**Heat Cascade**	**Corrected Heat Cascade**
Hot Utility–275	0								0	4750
275–265	1		↑		↓	10	60	600	–600	4150
265–175	2	↑	↑		↓	90	30	2700	–3300	1450
175–165	3	↑	↑		↓	10	70	700	–4000	750
165–135	4	↑	↑	↓	↓	30	25	750	–4750	0
135–45	5	↑		↓	↓	90	–35	–3150	–1600	3150
45–35	6	↑		↓		10	–5	–50	–1550	3200
35–5	7			↓		30	–45	–1350	–200	4550
5–Cold Utility	8									4550

Capital Cost Targets

To find the optimal value of Δt_{min} ahead of design, we need to set the targets for the capital and energy costs. As seen before, the capital cost target for a given Δt_{min} depends on the total area of the network, number of shells required in the network, and the number of units required in the network. We will now establish these targets.

Target for Minimum Total Area for the Network

The composite curves can be divided into different enthalpy intervals at the discontinuities in the hot and cold composite curves. Between any two adjacent enthalpy intervals, if the hot streams in that interval transfer heat only to the cold streams in that interval and vice versa, then we say that vertical heat transfer takes place along the composite curves.[6] This mode of heat transfer models the pure countercurrent heat exchange.

The equation for establishing the minimum area target is based on a complex network called the "spaghetti" network. This network models the vertical heat transfer on the composite curves. In this

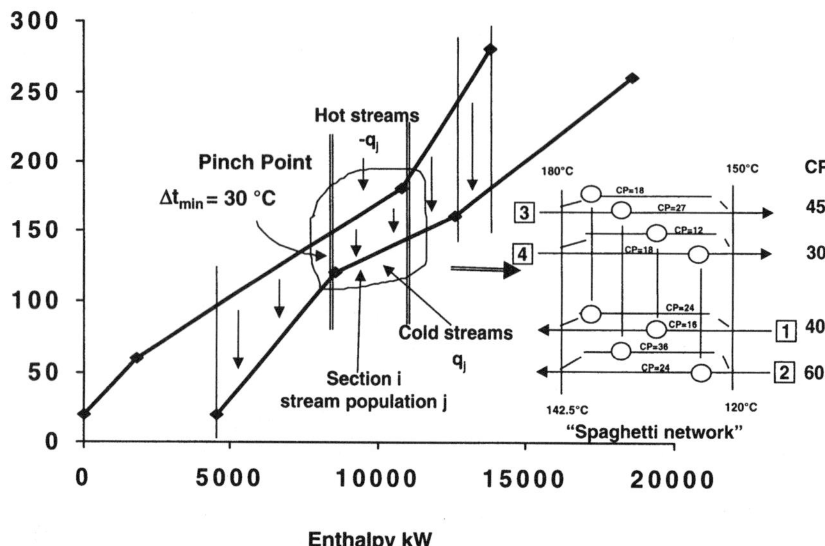

FIGURE 4.11.12 Enthalpy intervals for area targeting and the spaghetti network that mimics vertical heat transfer.

network, for any enthalpy interval defined by the discontinuities in the composite curves, each hot stream is split into the number of cold streams in that enthalpy interval. Each cold stream in the enthalpy interval is also split into the number of hot streams in that interval. A match is made between each hot stream and each cold stream. Figure 4.11.12 shows this network. The area target is obtained by the following equation:

$$A_{\min} = \sum_{i}^{N_{\text{intervals}}} \frac{1}{\Delta T_{LMTDi}} \left(\sum_{j}^{N_{streams}} \frac{q_j}{h_j} \right)_i$$

where A_{min} = total minimum area for the network
ΔT_{LMTDi} = ΔT_{LMTD} for enthalpy interval i
q_j = heat content of stream j in enthalpy interval i
h_j = film plus fouling heat transfer coefficient of stream j in enthalpy interval i

The above equation gives a minimum total surface area for any system where the process streams have uniform heat transfer coefficients. Townsend and Linnhoff[6] claim that for nonuniform heat transfer coefficients the above equation gives an useful approximation of the minimum area, with errors being typically within 10%.

Minimum Number of Units Target

The minimum number of units is given by:[7]

$$u = N + L - s$$

where u = number of units including heaters and coolers
N = number of streams including utilities
L = number of loops
s = number of separate networks

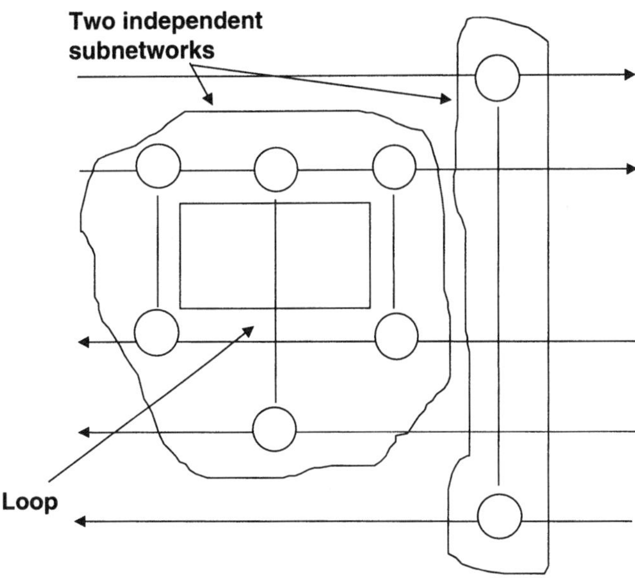

FIGURE 4.11.13 Loops and independent subnetworks in a heat exchanger network.

A loop is a closed path through the network and its effect is to increase the number of units. For a network to have a minimum number of units, the number of loops should be zero. Figure 4.11.13 shows a loop on the gird diagram. Normally for a given stream system, only a single network exists and if the number of loops is assumed to be zero, then the above equation can be reduced[4] to $u = N - 1$.

Figure 4.11.13 illustrates the occurrence of independent networks. The total network consists of two independent subnetworks. Thus, for the system shown in this figure, N = 5, L = 1, $s = 2$, and $u = 4$.

Target for Minimum Number of Shells

Trivedi et al.[8] have developed a method for estimating the total number of shells required for the heat exchanger network. The method starts by setting the energy targets for the selected value of Δt_{min}. The hot and cold utilities are added to the process stream system and the composite curves for the process and the utility streams are constructed. No external utility requirement will be needed for these sets of composite curves. Hence, these are called the balanced composite curves. The method estimates the number of shells based on the balanced composite curves. The method involves two steps:

Step 1: Estimate the total number of shells required by the cold process streams and the cold utility streams (see Figure 4.11.14a).

- Commencing with a cold stream target temperature, a horizontal line is drawn until it intercepts the hot composite curve. From that point a vertical line is dropped to the cold composite curve. This section, defined by the horizontal line, represents a single exchanger shell in which the cold stream under consideration gets heated without the possibility of a temperature cross. In this section, the cold stream will have at least one match with a hot stream. Thus, this section implies that the cold stream will require at least one shell. Further, it ensures that log mean temperature (LMTD) correction factor, $F_t \geq 0.8$.
- Repeat the procedure until a vertical line intercepts the cold composite curve at or below the starting temperature of that particular stream.
- The number of horizontal lines will be the number of shells the cold stream is likely to require to reach its target temperature.
- Repeat the procedure for all the cold streams including the cold utility streams.
- The sum of the number of shells for all the cold streams is the total number of shells required by the cold streams to reach their respective target temperatures.

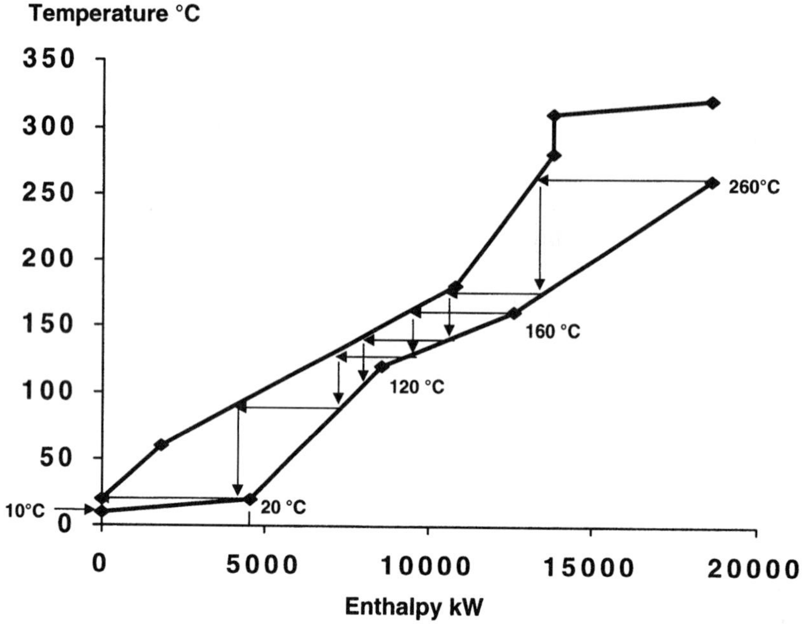

FIGURE 4.11.14a Shell targeting for cold streams on the balanced composite curves. Cold streams will require eight shells.

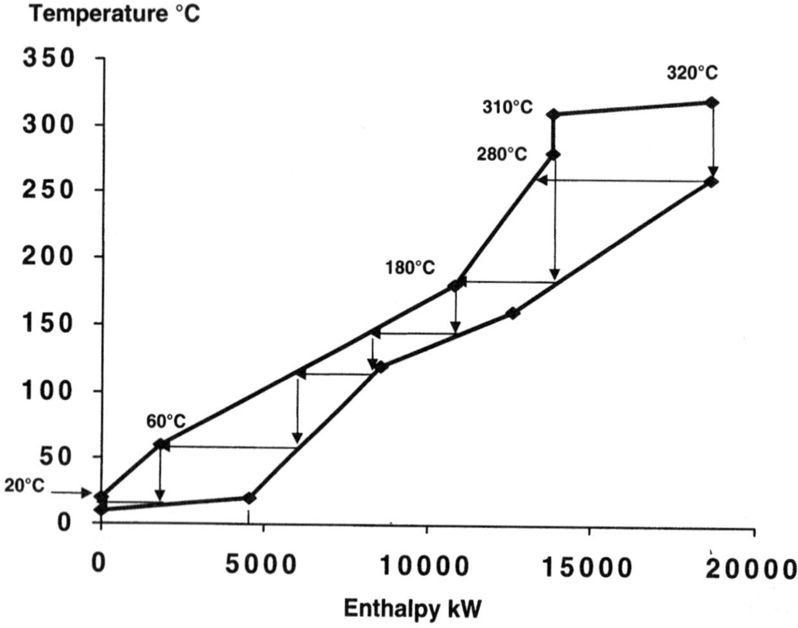

FIGURE 4.11.14b Shell targeting for hot streams on the balanced composite curves. Hot streams will require nine shells.

Step 2: Estimate the total number of shells required by the hot process streams and the hot utility streams (see Figure 4.11.14b).

- Starting from the hot stream initial temperature, drop a vertical line on the balanced composite curve until it intercepts the cold composite curve. From this point, construct the horizontal and vertical lines until a horizontal line intercepts the hot composite curve at or below the hot stream target temperature.

- The number of horizontal lines will be the number of shells required by the hot streams for heat exchange in the network.
- Repeat the procedure for all the hot streams including the hot utility streams.
- The sum of the number of shells required by the hot streams would be the total number of shells required by the hot streams to reach their respective target temperatures.

The quasi-minimum number of shells required in the network would be the larger of either the total number of shells required by the hot streams or the total number of shells required by the cold streams.

Optimum ΔT_{min} Value

The above procedures establish, for a selected Δt_{min} value, targets for minimum energy, minimum area, minimum number of units, and minimum number of shells. These targets, along with the cost data, can be translated into capital and energy cost for the network. The targets can be evaluated at different values of Δt_{min} to obtain a birds-eye view of the solution space and the optimal value of Δt_{min} ahead of design. This philosophy was first proposed by Hohmann[4] and later developed by Linnhoff and Ahmad[9] as Supertargeting.

For the design problem data discussed above, the different targeting curves are shown in Figure 4.11.15a-f . It is seen from the total annualized cost curve, Figure 4.11.15f, that the optimal region is very flat. While selecting a value of Δt_{min} from the optimal region, a couple of points should be kept in mind. Different values of Δt_{min} lead to different topologies. Hence, we should take into account different factors that can affect the final cost. These factors include the nature of the composite curves and the problem constraints that have significance in the network synthesis and refinement. The optimal value of Δt_{min} selected for this problem is 40°C and the target values for energy, area, units, shells, and annualized total cost are given in Table 4.11.7.

The Pinch Point

On the composite curves, there is one or more enthalpy value for which the two composite curves are Δt_{min} apart. For the example under consideration (see Figure 4.11.11), this occurs at a hot stream temperature of 150°C and cold stream temperature of 120°C. This is also the fourth temperature interval in the problem table. In this interval, the heat cascade has zero heat flow, i.e., no heat is transferred across this interval when minimum hot utility is used. This interval, identified by a zero in the corrected heat cascade column, is referred to as the pinch point.[7,10] The significance of the pinch point is now clear — for the minimum external utility requirements do not transfer heat across the pinch point. Any extra amount of external heat that is put into the system above the minimum will be transferred across the pinch point and will be removed by the cold utility.[11]

Cross Pinch Principle[11]

For a given value of Δt_{min}, if the network is using Q_h units of hot utility and if Q_{hmin} is the minimum energy target then,

$$Q_h = Q_{h\min} + \alpha$$

If the network uses Q_c units of cold utility and if Q_{cmin} is the minimum energy target then,

$$Q_c = Q_{c\min} + \alpha$$

where α = the amount of cross pinch heat transfer

Significance of the Pinch Point

The pinch point divides the stream system into two independent subsystems. The subsystem above the pinch point is a net heat sink and the subsystem below the pinch point is a net heat source. Thus, for a

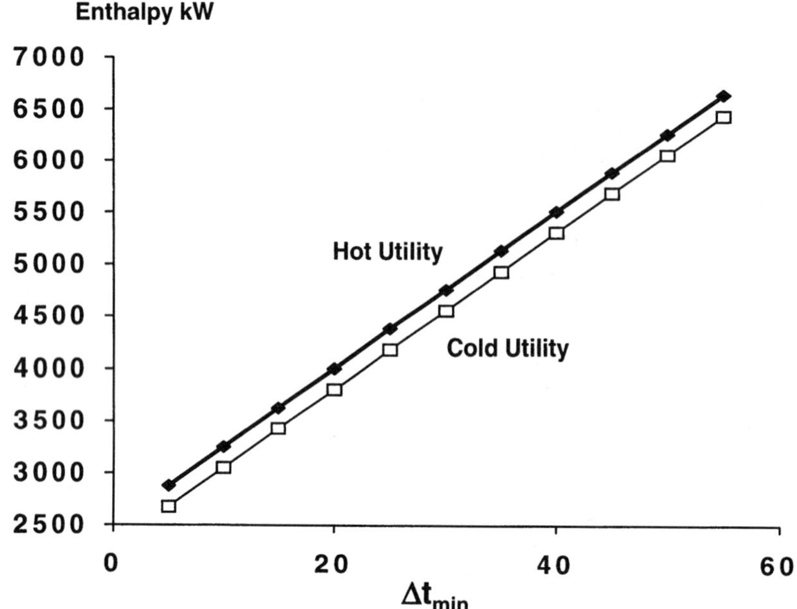

FIGURE 4.11.15a Minimum energy targeting plot.

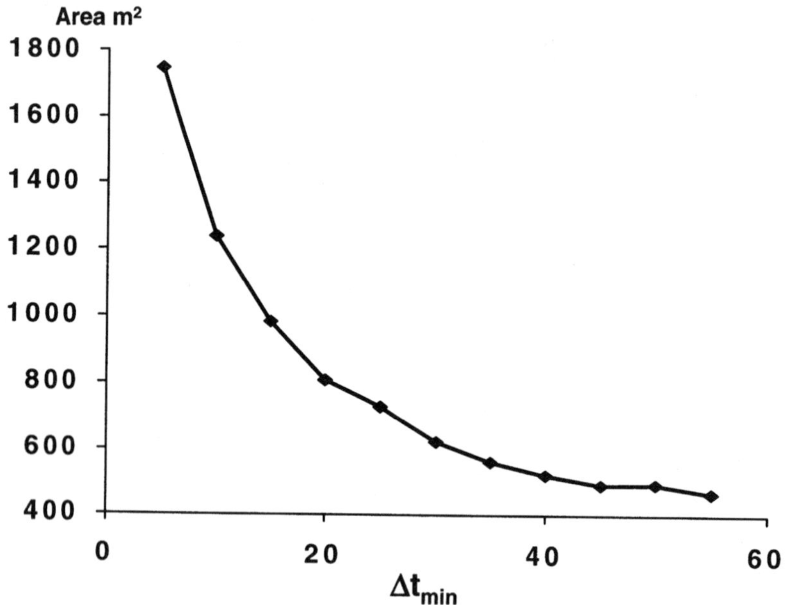

FIGURE 4.11.15b Minimum network area targeting plot.

system of hot and cold streams, to design a network that meets the minimum utility targets the following rules set by the pinch principle should be followed:[11]

- Do not transfer heat across the pinch point
- Do not use hot utility below the pinch point
- Do not use cold utility above the pinch point

We shall use these principles to design the network for the optimal value of Δt_{min}.

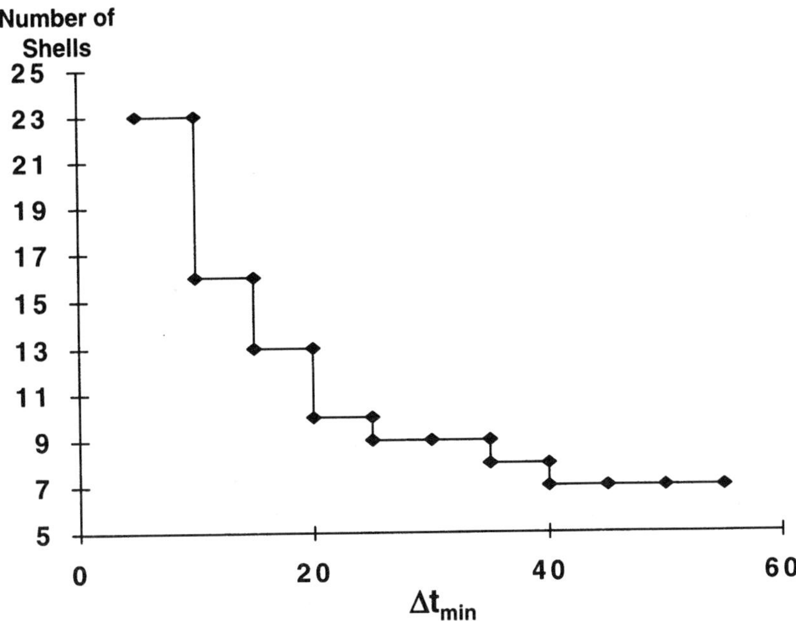

FIGURE 4.11.15c Targeting plot for minimum number of shells.

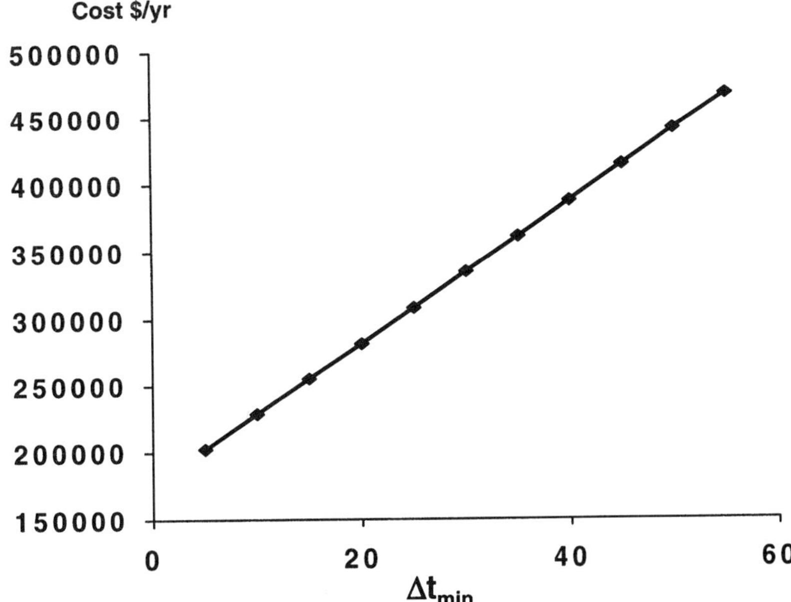

FIGURE 4.11.15d Minimum energy cost targeting plot.

Network Design

Network Representation on the Grid Diagram

If we attempt to design the network using the conventional flowsheet format, any changes in the design may lead to redrawing the flowsheet. Hence, we shall use the grid diagram to represent and design the network. Figure 4.11.16 shows the grid diagram for the design problem assuming all hot streams are

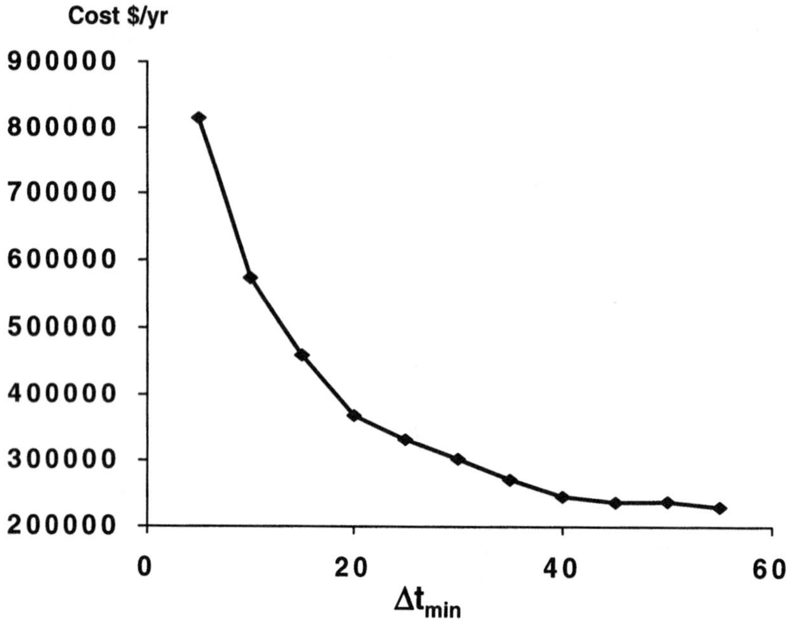

FIGURE 4.11.15e Minimum annualized capital cost targeting plot. (From *The Chemical Engineer,* the Institute of Chemical Engineers, 30, 1987. With permission.)

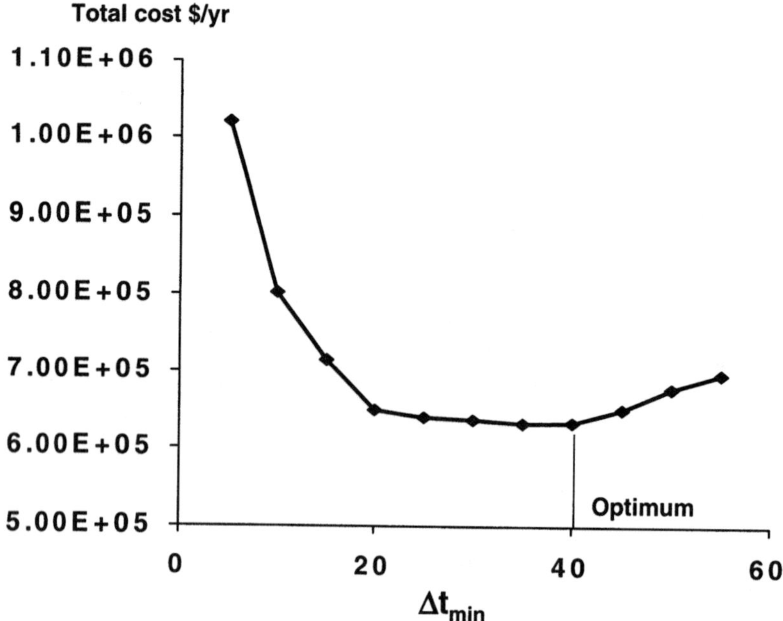

FIGURE 4.11.15f Targeting plot for total annualized cost. (From Linnhoff, B. With permission.)

cooled with cooling utility and all cold streams are heated with hot utility. On this grid diagram we can place a heat exchanger match between two streams without redrawing the whole stream system. The grid representation reflects the countercurrent nature of heat transfer that makes it easier to check temperature feasibility of the match that is placed. Furthermore, we can easily represent the pinch point on the grid diagram, as shown in Figure 4.11.16.

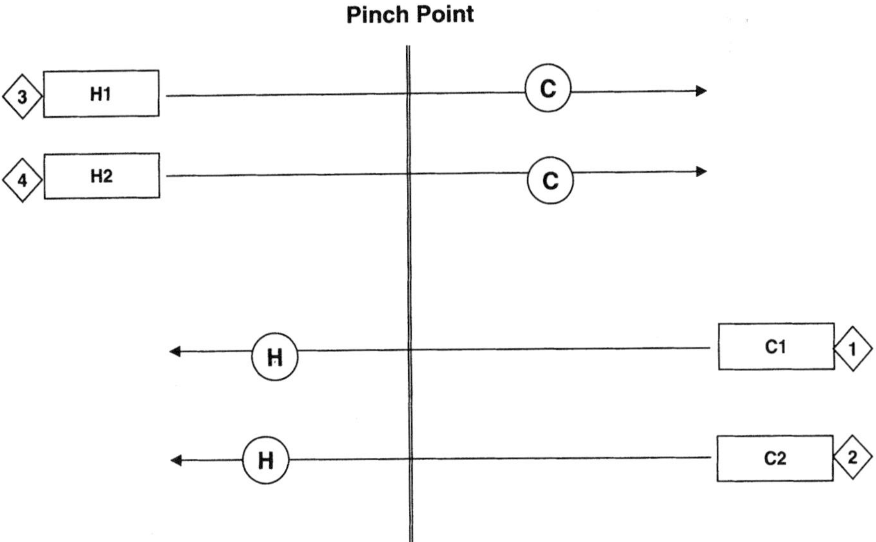

FIGURE 4.11.16 Grid diagram representation of the stream in the design problem. (From Linnhoff, B. With permission.)

Pinch Point and Network Design

We will now continue with the design of the example problem. Figure 4.11.11 shows the composite curves with $\Delta t_{min} = 30°C$ for the process flowsheet. It also shows the pinch point. The pinch point is the most constrained region of the composite curves. From optimization principles, we know that for a constrained problem, the optimal solution is located at the point formed by the intersection of multiple constraints. Hence, if we can satisfy the constraints at that point then we are guaranteed an optimal design. Thus, we should start the design where the problem is most constrained viz. the pinch point.[11] The pinch point divides the problem into two independent subnetworks — above the pinch subnetwork that requires only hot utility and below the pinch subnetwork that requires only cold utility.

Pinch Design Rules and Maximum Energy Recovery

Let us make some observations at the pinch point. Immediately above the pinch point[11]

$$\sum CP_h \leq \sum CP_c$$

and immediately below the pinch point,

$$\sum CP_h \geq \sum CP_c$$

This condition defines the pinch point that is the constraint that the designer has to satisfy. Thus, each and every match in the sink subnetwork, at the pinch point, should be placed such that $CP_h \leq CP_c$. Similarly, each and every match in the source subnetwork, at the pinch point should be placed such that $CP_c \leq CP_h$. Figure 4.11.17 shows these conditions on the T-H diagram.

Sometimes the above condition may not be satisfied. Stream splitting should be undertaken in such a situation.

Figure 4.11.18[12] shows the algorithms that are developed for placing matches immediately at the pinch point for both the above and below pinch subnetworks. N_h and N_c denote the number of hot and cold streams at the pinch point.

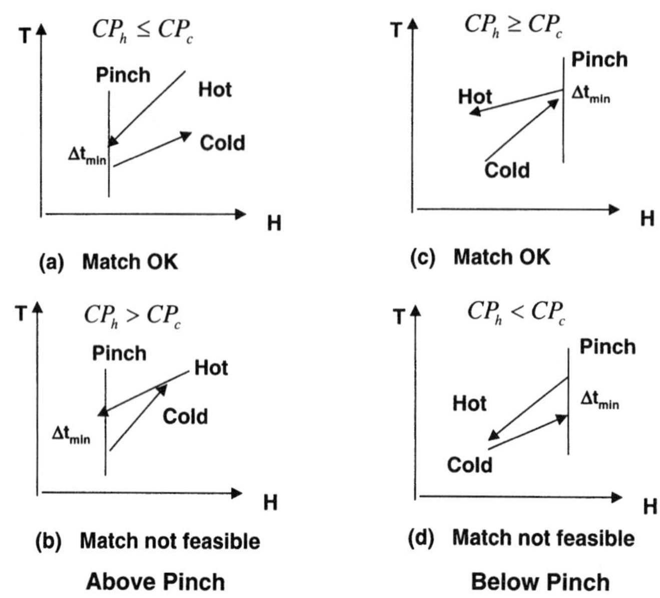

FIGURE 4.11.17 CP matching rules at the pinch point.

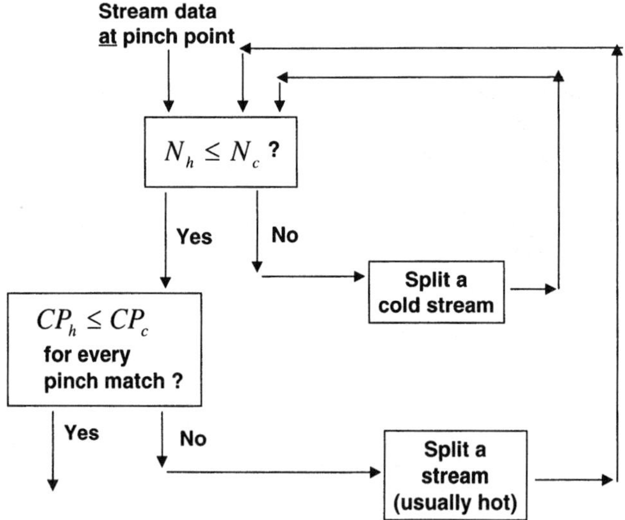

FIGURE 4.11.18a Algorithm for sink subnetwork design at the pinch point.

The matches and stream splitting can be easily identified with the "CP table."[11] Consider the stream system shown in Figure 4.11.19. The CP table is shown in the same figure. The two rows in this table contains the condition that needs to be satisfied for the subnetwork under consideration. The CP values of the hot and cold streams are arranged in a descending order in the columns. The stream numbers are shown in brackets adjacent to the CP values. For the sink subnetwork under consideration, the CP value for the hot streams is listed in the left column and the CP value for the cold stream is listed in the right column. There is no feasible match for hot stream 1. Hence, it will have to be split. The hot stream 2 can match with either stream 3 or 4. Once we split hot stream 1, we violate the stream population constraint. To satisfy this constraint we will have to split a cold stream. Either stream 3 or 4 can be split. The designer can use his judgment and decide which stream should be split depending on controllability

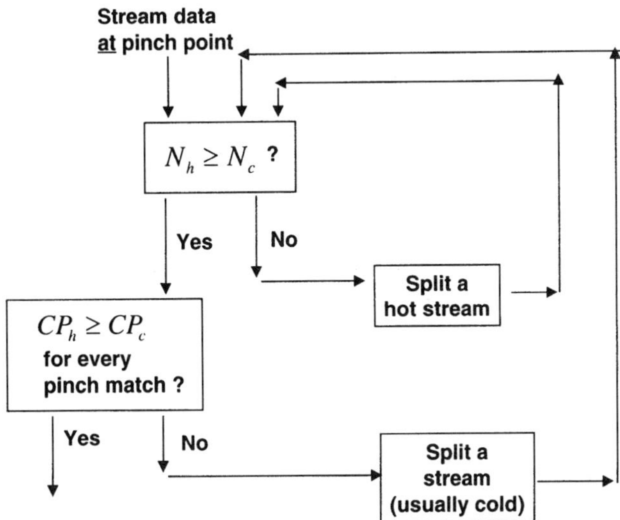

FIGURE 4.11.18b Algorithm for source subnetwork design at the pinch point.

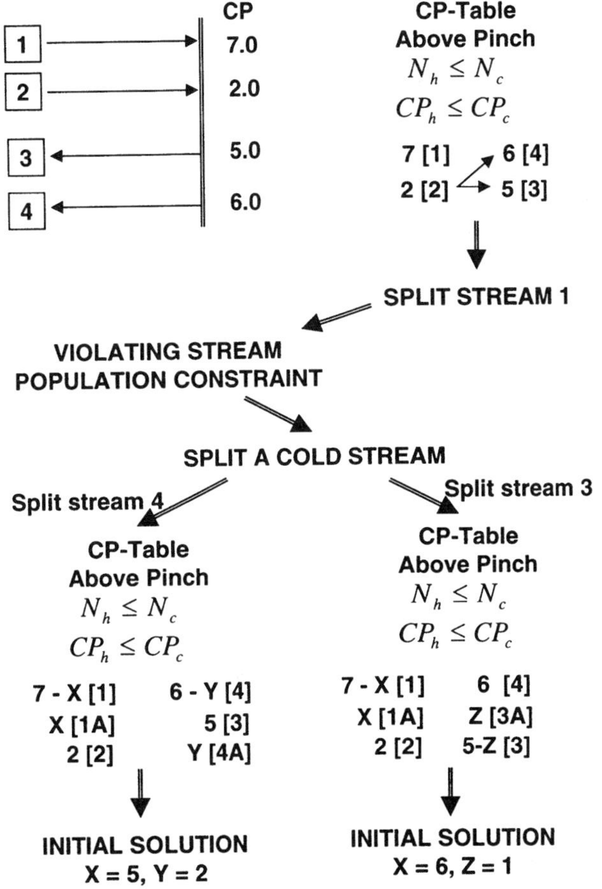

FIGURE 4.11.19 Identifying matches and stream splitting using the CP table.

and other physical constraints. For example, it is not advisable to split a stream that may have a two-phase flow.

Two different designs will be obtained depending on which stream is split. The hot stream is split into two streams having CP values of X and 7-X. In the first alternative, we split stream 4 into streams having CP values of Y and 6-Y. In the second option we split stream 3 into streams having CP values of Z and 5-Z. To find the initial values of X, Y, and Z it is recommended that all matches except for one are set for CP equality. Thus, for the first option set, X = 5 and Y = 2. For the second option set, X = 6 and Z = 1. This is an initial solution. The values of X, Y, and Z can be adjusted to obtain an optimal design.

Once the matches are identified at the pinch, the heat loads for these matches are fixed to maximize the heat exchange to the limit of heat load of either the hot stream or the cold stream. This will eliminate (or "tick-off") that stream from the analysis. Fixing the heat loads for match with this heuristic will also help us to minimize the number of units and thus the installation cost.[11]

Matches away from the pinch point do not have to satisfy the conditions outlined above as the system is not constrained in this region. Matches are easily identified away from the pinch point.

The network obtained will satisfy the energy targets set. Such a network is called maximum energy recovery (MER) network.[11]

Local Optimization — Energy Relaxation

The optimal region is generally very flat and the optimal value of Δt_{min} is not a single point but a region. The targeting exercise helps identify this region. Once a value of Δt_{min} is selected and the network is designed using the methodology outlined above, there is still some scope for further optimization. This is achieved by a process called energy relaxation.[11]

The number of units in the design obtained using the above procedure will generally be greater than the target value. This is due to the presence of loops. If we can eliminate the loops in a network then the heat transfer area is concentrated on fewer matches and, hence, will decrease the piping and foundation requirements. This will tend to decrease the capital cost of the network. However, the some energy penalty may be incurred.

Identifying loops in a large network cannot be done visually. Trivedi et al.[13] have proposed LAPIT, **L**oop **A**nd **P**ath **I**dentification **T**ree, for identifying loops and paths present in the network. A path is the connection between a heater and a cooler via process-process matches.

Loops can be broken by removal of a unit in a loop and redistributing the load of the unit among the remaining units of the loop. Some exchangers may result having a very small Δt_{min} when a loop is broken. The Δt_{min} across such exchangers can be increased by increasing the utility consumption along a path that consists of a heater, the unit having a small value of Δt_{min}, and a cooler.

Loop-breaking is a very complex optimization process. Trivedi et al.[13,14] have proposed a detail method that systematically breaks loop and identifies options available at each step of the process. The method is based on **LO**op **N**etwork **I**nteraction and load **T**ransfer **A**nalysis (LONITA)[14] and on a best-first-search procedure.[13]

Summary of the Design Procedure

The pinch design procedure can now be summarized. Establish a value of the optimum Δt_{min} using the targeting techniques. This identifies the region in which the design should be initialized. Using the pinch design procedure, design the network. Reduce the number of units using the loop-breaking and energy-relaxation techniques outlined above.

Example

The above principles and procedures are used for the design of the network for the multisteam design problem. The value of optimum Δt_{min} is 40°C. The final network MER design is shown in Figure 4.11.20. Note that the annualized capital cost of this network is only about 4% higher than the target value of $692,700. This design is in the neighborhood of the predicted optimum.

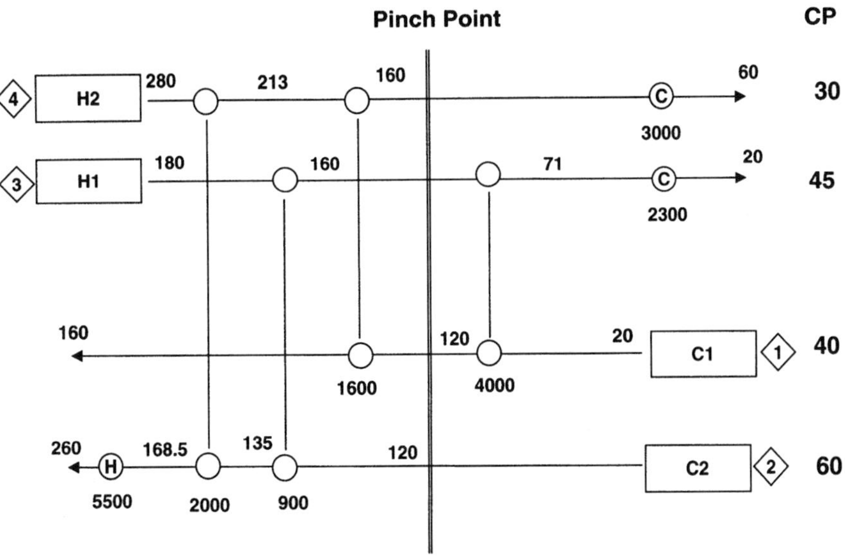

FIGURE 4.11.20 Pinch design for the design problem. Total annualized cost: $720,800.

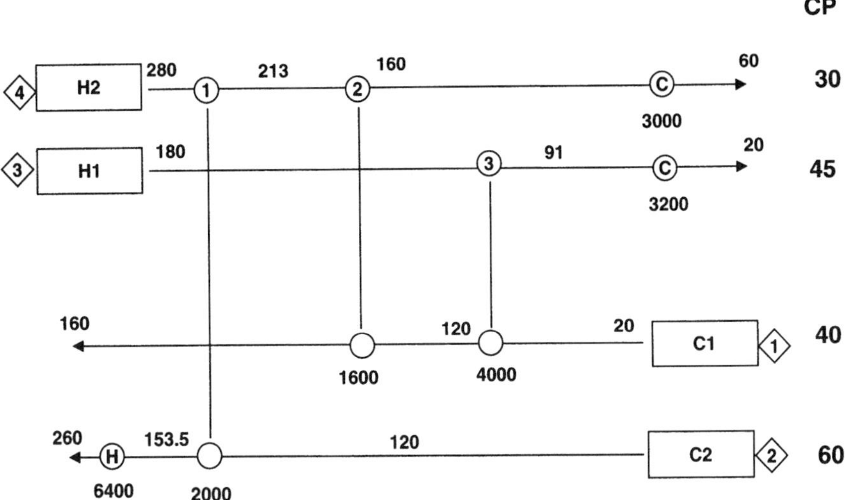

FIGURE 4.11.21 Energy relaxation of the pinch design in Figure 4.11.20. Total annualized cost: $733,700.

The design after energy relaxation is shown in Figure 4.11.21. This design has one less unit than the MER design in Figure 4.11.20. Hence, this will decrease the piping, installation, and maintenance cost. The cost of the new design is about 2% higher than the MER design. Since we are using simplified models for costing, the cost of the design in Figure 4.11.21 is within the errors of cost estimating. However, this design will be cheaper when the detail cost is evaluated as it has fewer number of units.

Table 4.11.7 compares the target values of the various optimization parameters with actual design values for Figures 4.11.20 and 4.11.21.

Selection of Utility Loads and Levels

The annual operating cost depends on the amount and type of utilities used. In a complex process such as an ethylene plant, there would be about four to five steam levels and about seven to eight refrigeration

TABLE 4.11.7 Comparison of Target and Design Values for Different Network Optimization Variables at Δt_{min} = 40°C

		Figure 4.11.20 MER Design		Figure 4.11.21 Design	
	Target Value	Design Value	% Deviation from Target	Design Value	% Deviation from Target
Total network area, m^2	687	779	13	708	3
Total hot utility consumption, kW	5500	5500		6400	16
Total cold utility consumption, kW	5300	5300		6200	17
Number of units	7	7		6	
Number of shells	8	9		7	
Energy cost, $/yr	387,250	387,250		450,700	16
Annualized capital cost $/yr	305,407	333,576	9	282,996	-7
Total annualized cost, $/yr	692,657	720,826	4	733,696	6

levels. High-pressure steam is generated within the process and is let down to other steam levels via steam turbines that generate power. Steam from these levels is used for heating the process. The power generated from the turbines is used by the compressors and pumps and to generate various levels of refrigeration needed. The question faced is what pressure levels of steam to use and what is the load on each level. Also, what are the best temperature levels of the refrigeration and what will be their respective loads.

Pinch technology helps us answer these questions in a very simple manner using the grand composite curves (GRCC).[15] The GRCC is the curve that shows the heat demand and supply within each temperature interval. This curve is derived from the problem table (refer to Table 4.11.6). In the problem table, we had modified the stream starting and target temperatures depending on the value of Δt_{min}. From now on we shall refer these modified temperatures as shifted temperature. The heat flows between two adjacent shifted temperature intervals can be plotted on the shifted-temperature-enthalpy plot. Figure 4.11.22 shows the heat cascade and how the GRCC is developed for the design problem. The GRCC gives us a graphical representation of heat flows taking place in the system. At the pinch point the heat flow is zero.

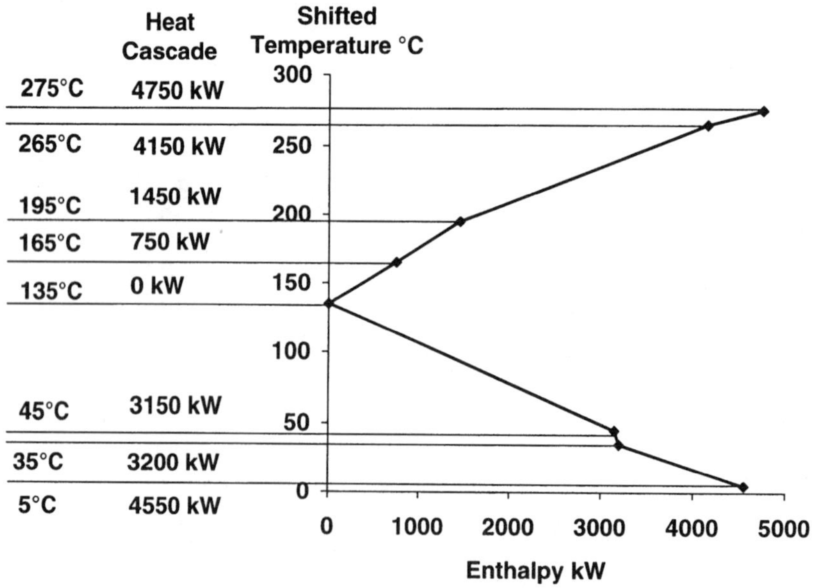

FIGURE 4.11.22 Grand composite curve for the design problem.

The GRCC is piecewise linear. The slope of this curve also changes from interval to interval. A line with a positive slope indicates that the system in that region needs external heat. A line with a negative slope indicates that there is surplus heat available within that temperature interval that can be cascaded down within the system and used at a lower temperature interval. It is very clear from the GRCC that above the pinch region is a heat sink and below the pinch region is a heat source.

To further explain the importance of the grand composite curve and the kind of information that can be extracted from it, consider the curve shown in Figure 4.11.23. Consider the section AB that is between the shifted temperature interval of 395°C and 380°C. This section demands external heat of 2000 kW. Hence, we should place a hot utility such as high-pressure steam or hot oil to supply it.

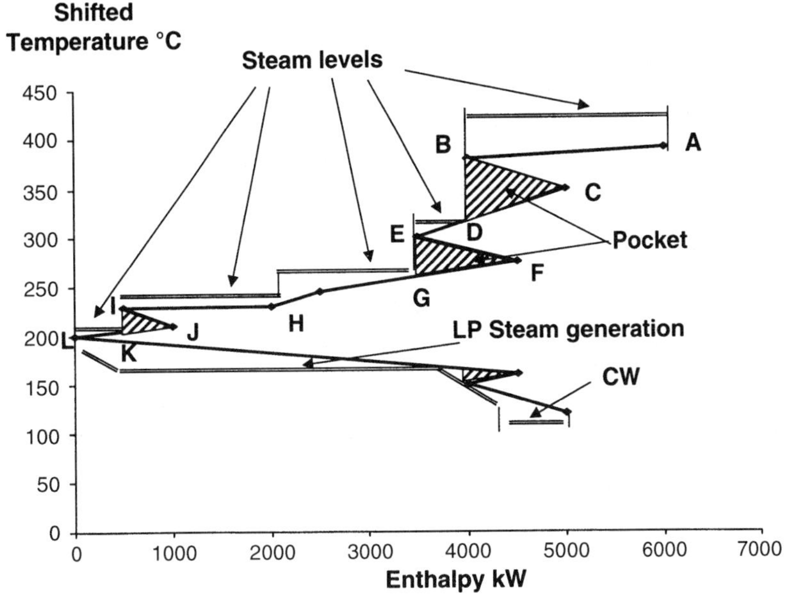

FIGURE 4.11.23 Selection of utility loads and levels.

Next, consider the section BC of the GRCC between the shifted temperature intervals of 380°C and 350°C. This section has a heat surplus and so we can use it elsewhere in the system. We can drop a vertical line from B to meet the GRCC at point D. The section CD of the GRCC that is a heat deficit region of the system can now be satisfied by the heat surplus from the BC section. This section of the GRCC is called a "pocket" and the process is self-sufficient with respect to energy in this region. The section DE now needs external heat. This heat can be supplied at any temperature ranging from the highest available temperature to a minimum temperature corresponding to point D.

Following the same logic, EFG will be a "pocket." In section GHI one hot utility level can be used at a shifted temperature level of point G with a total duty of 3000 kW, or two levels can be used — one at 260°C with a duty of 1500 kW and the other at the shifted temperature level of point H, i.e., 240°C with a duty of 1500 kW. The choice is dictated by the trade-off between the power requirement, capital investment, and complexity of the design. Using only one level will make the design of the utility system simpler and the capital cost of the heat exchanger smaller due to higher temperature approaches. On the other hand, if there is demand for power then using two levels will produce more power.

A similar economic trade-off will be required for supplying the external heating requirement in section KL of the composite curve. Point L is the pinch point. Below point L, heat needs to be rejected into a cooling utility such as an air cooler or cooling water. Also, in the below pinch section of the process we can address the question, "Is it possible to raise steam at some temperature? If so, how much?"

For example, for the process GRCC shown in Figure 4.11.23, we want to find out how much low-pressure superheated steam can be generated. The saturation temperature of the low-pressure steam is 170°C and boiler feed water is available at 130°C. The superheat is 10°C. Using a simple trial and error procedure, we can find out how much steam will be generated.

- Assume the amount of steam that is generated.
- Develop a heat curve for the low-pressure steam generation on the shifted temperature scale.
- As generation of low-pressure steam will be a cold stream, the temperature of the stream will be increased by $\Delta t_{min}/2$.
- Keep on increasing the amount of steam generated till the steam generation heat curve touches the process GRCC at any point.

Once the utility levels are decided, introduce them into the stream data and obtain the balanced composite curves. The number of pinch points will increase. In addition to the original process pinch point, each utility level will introduce at least one pinch point. A balanced grid diagram that includes all the utility streams and all the pinch points identified on the balanced composite curve can now be used along with the network design algorithms to develop a network that achieves the target set.

Data Extraction

Process integration studies start from a base case flowsheet. This flowsheet may be existing or may be developed from the designer's experience. To conduct pinch analysis properly it is important to extract the flowrate, temperature, and heat duty data correctly.

The stream target and starting temperatures should be chosen so that we do not generate the original flowsheet.[12] To illustrate this, consider the flowsheet shown in Figure 4.11.24. If we extract the data as two streams then we might end up with the original flowsheet. If the drum temperature is not important then we can consider it to be one stream and we stand a chance for finding new matches. The drum and the pump can then be kept at a natural break point in the system if the process allows.

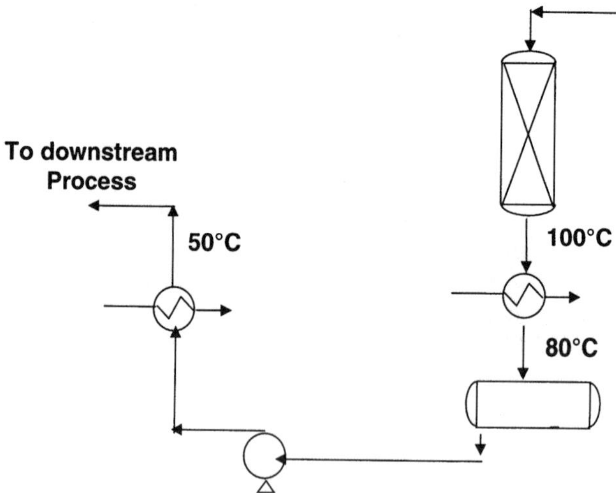

FIGURE 4.11.24 Flowsheet for data extraction example.

While extracting data extra care should be taken when streams are mixing nonisothermally.[12] Consider the system shown in Figure 4.11.25. Stream A is being cooled to the mixed temperature and stream B is being heated to the mixed temperature. This happens due to mixing the streams. The mixed stream is then heated to a higher temperature. If we extract the data as shown in Figure 4.11.25a and if the

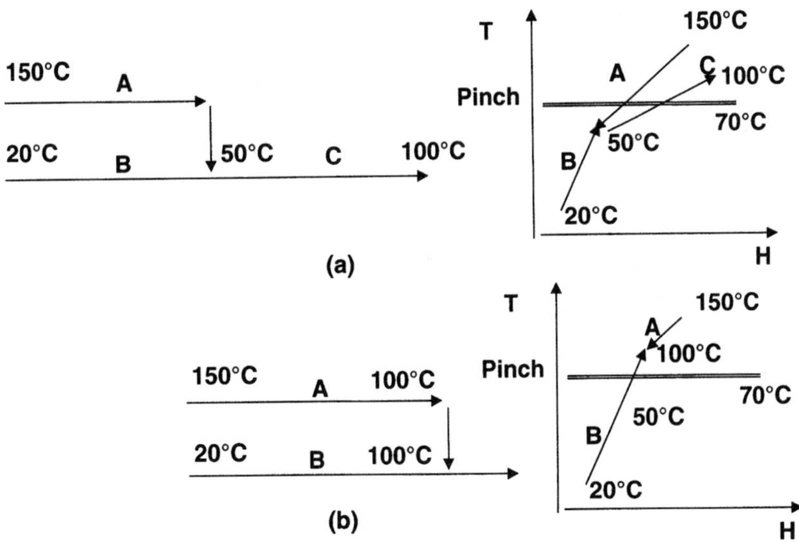

FIGURE 4.11.25 Data extraction for streams that are mixing.

pinch temperature is 70°C, then we are inherently transferring heat across the pinch point by mixing the two streams. In the process of mixing we are cooling stream A and then subsequently heating it up again. The correct way to extract the data is shown in Figure 4.11.25b.

When a stream is split and the split streams have two different target temperatures then each stream is considered as two separate streams.

Process Integration and Recent Developments

Until now, we have discussed the design and optimization of heat exchanger networks. But heat exchanger networks are a part of a whole process. For the process to be optimally designed, all the unit operations should be properly integrated. Generally, each unit operation is individually optimized. It is a misconception that if each individual unit is individually optimized then the resulting process is also optimized. Each unit operation interacts with others in the process. Hence, for the process to be optimized, each unit operation should be properly integrated.

Pinch technology has developed systematic ways to integrate complex processes. Methods are available for appropriate placement and integration of heat engines, heat pumps, and various other unit operations within the process. This technology has also established guidelines for process modifications, developing shaftwork targets, total site-wide integration, revamping existing processes, and much more. Trivedi[16] discusses the details of these methods and procedures.

References

1. Linnhoff B., Use Pinch Analysis to Knock Down Capital Costs and Emissions, *Chem. Eng. Prog.*, 33, August, 1994.
2. Samdaniust, G. and Moore, S., Pinch Technology: Doing More With Less, *Chem. Eng.*, 43, July 1993.
3. Bell, K.J., Estimate S & T Exchanger Design Fast, *Oil Gas J.*, 59, Dec. 4, 1978.
4. Hohmann, E.C., Optimum Networks for Heat Exchange, Ph. D. dissertation, University of Southern California, Los Angeles, 1971.
5. Linnhoff, B. and Flower, J. R., Synthesis of Heat Exchanger Networks, *AIChE J.*, 633, July 1978.
6. Townsend, D.W. and Linnhoff, B., Surface Area Targets for Heat Exchanger Networks, IChemE 11th Annu. Res. Meet., Bath University, U.K., April 1984.

7. Linnhoff, B., Mason, D.R., and Wardle, I., Understanding Heat Exchanger Networks, *Comp. Chem. Eng.*, 3, 295, 1979.
8. Trivedi, K.K., Roach, J.R., and O'Neill, B.K., Shell Targeting in Heat Exchanger Networks, *AIChE J.,* 33(12), 2087, 1987.
9. Linnhoff, B. and Ahmad, S., Supertargeting: Optimum Synthesis of Energy Management Systems, *J. Energy Resour. Technol.*, 111, 121, 1989.
10. Umeda, T., Itoh, J., and Shiroko, K., Heat Exchange System Synthesis, *Chem. Eng. Prog.*, 70, July 1978.
11. Linnhoff, B. and Hindmarsh, E., The Pinch Design Method for Heat Exchanger Networks, *Chem. Eng. Sci.*, 38(5), 745, 1983.
12. Linnhoff, B., Townsend, D.W., Boland, D., Hewitt, G.F., Thomas, B.E.A., Guy, A.R., and Marshland, R.H., *Users Guide on Process Integration for the Efficient Use of Energy,* Institution of Chemical Engineers, Rugby, Warwickshire, U.K., 1982.
13. Trivedi, K.K., O'Neill, B.K., Roach, J.R., and Wood, R.M., A Best-First Search Method for Energy Relaxation, *Eng. Optimization,* 16, 291, 1990.
14. Trivedi, K.K., O'Neill, B.K., Roach, J.R., and Wood, R.M., Systematic Energy Relaxation in MER Heat Exchanger Networks, *Comp. Chem. Eng.*, 14(6), 601, 1990.
15. Townsend, D.W. and Linnhoff, B., Heat and Power Networks in Process Design. I. Criteria for Placement of Heat Engines and Heat Pumps in Process Networks. II. Design Procedure for Equipment Selection and Process Matching, *AIChE J.,* 29(5), 742, 1983.
16. Trivedi, K.K., Process Energy Efficiency: Pinch Technology, *CRC Handbook of Energy Efficiency,* Kreith, F. and West, R., Eds., CRC Press LLC, Boca Raton, FL, 1996.

4.12 Cryogenic Systems

Randall F. Barron

Introduction

Cryogenic engineering is the technical field that is concerned with developing and utilizing low-temperature techniques, processes, and phenomena. The point on the temperature scale at which conventional refrigeration ends and cryogenics begins is somewhat arbitrary. In the 1950s, engineers and scientists at the National Bureau of Standards (now called NIST) in Boulder, CO, suggested that the field of cryogenics be defined as that temperature region below –150°C (123 K or –240°F) (Scott, 1959). This point was selected because the refrigerants used in air conditioning systems and domestic refrigerators boil at temperatures above –150°C; whereas, the gases utilized for cryogenic applications, such as oxygen, nitrogen, hydrogen, and helium, boil at temperatures below –150°C.

Heat transfer in cryogenic systems is an important factor in the design of all low-temperature systems. The cost of removing energy from a low-temperature region is significant. A power input of approximately 14 kW (19 hp) would be required to drive a refrigerator (cryocooler) removing 1 kW (3412 Btu/hr) from a space at 90 K (–298°F) and rejecting the energy to the ambient surroundings. The average cost of the cryocooler is about \$38,000 (in 1974 dollars) or about \$415,000 in 1999 dollars (Strobridge, 1974). Because of the high cost of removing energy from a cryogenic system, heat transfer is examined closely in the design of these systems.

Air Liquefaction

Commercial systems for the production of liquid nitrogen, liquid oxygen, liquid argon, and liquid neon use atmospheric air as the raw stock. The first commercial air liquefaction system was developed in Germany by Karl von Linde (1897) for the production of oxygen-enriched air for use in the steel-making industries (Ruhemann, 1949). Hampson (1895) developed a similar system with a more efficient heat exchanger. A schematic of the Linde-Hampson system is shown in Figure 4.12.1.

In the Linde-Hampson liquefaction system, air is compressed to 200 atm (20.3 MPa or 2940 psia), and the high-pressure stream is cooled in a counterflow heat exchanger. The cold gas is finally expanded through an expansion valve or *Joule-Thomson* valve to a pressure on the order of atmospheric pressure. At the exit of the expansion valve, the stream is in the two-phase (liquid-vapor) condition. The liquid is collected in the liquid receiver, and the cold vapor is returned through the heat exchanger to provide cooling for the incoming warm gas stream.

The fraction of the gas from the compressor that is liquefied (the *liquid yield*) for the Linde-Hampson system is given by:

$$y = \frac{m_f}{m} = \frac{(h_1 - h_2) - (1-\varepsilon)(h_1 - h_g)}{(h_1 - h_f) - (1-\varepsilon)(h_1 - h_g)} \tag{4.12.1}$$

where h_1 = enthalpy of the low-pressure stream at pressure p_1 and temperature $T_1 = T_2$; h_2 = enthalpy of the high-pressure stream at the inlet at the warm end of the exchanger; h_g is the enthalpy of the vapor returning from the liquid receiver; and ε is the heat exchanger effectiveness. The power requirement per unit mass flow rate through the compressor is given by:

$$-\frac{W}{m} = \frac{T_2(s_1 - s_2) - (h_1 - h_2)}{\eta_{co}} \tag{4.12.2}$$

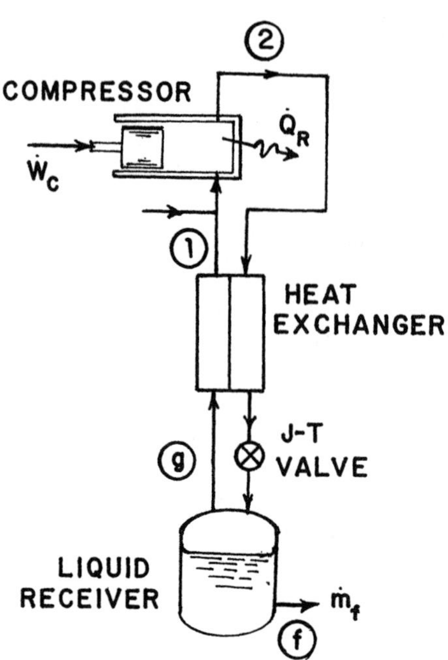

FIGURE 4.12.1 Linde-Hampson liquefaction system.

where s_1 is the entropy of the low-pressure stream at pressure p_1 and temperature $T_1 = T_2$; s_2 is the entropy of the high-pressure stream at the inlet of the warm end of the exchanger; and η_{CO} is the compressor overall efficiency.

The thermodynamic performance of the air liquefier may be improved by using an expansion engine or turbine to remove energy from the warm stream (Claude, 1913). In the Claude liquefaction system, shown in Figure 4.12.2, the gas is compressed to a pressure on the order of 40 atm (4.05 MPa or 590 psia) and cooled in the first counterflow heat exchanger. After the first heat exchanger, a fraction (60 to 80%) of the high-pressure stream is bypassed through the expander and expanded to a pressure on the order of atmospheric pressure. The cold stream is mixed with the exiting gas stream to provide cooling in the second heat exchanger. The portion of the stream not bypassed is cooled in the third heat exchanger and expanded through the J-T valve into the two-phase region. The liquid is passed to the liquid receiver, and the vapor is returned through the heat exchangers to provide cooling for the incoming warm stream of gas (Davies, 1949).

The liquid yield for the Claude liquefaction system is given by:

$$y = \frac{m_f}{m} = \frac{(h_1 - h_2) + \eta_{ad}(m_e/m)(h_3 - h_e)}{(h_1 - h_f)} \tag{4.12.3}$$

where m_e = mass flow rate through the expander, η_{ad} = adiabatic efficiency of the expander, h_3 = enthalpy of the gas at the expander inlet, h_e = enthalpy of the gas at the expander outlet pressure for an isentropic expansion from the inlet conditions. If the expander power is utilized to drive the last stage of the compressor, the net power requirement per unit mass compressed is given by:

$$-\frac{W}{m} = \frac{T_2(s_1 - s_2) - (h_1 - h_2)}{\eta_{co}} - x\ \eta_e (h_3 - h_e) \tag{4.12.4}$$

where η_{co} = compressor overall efficiency and η_e = expander overall efficiency.

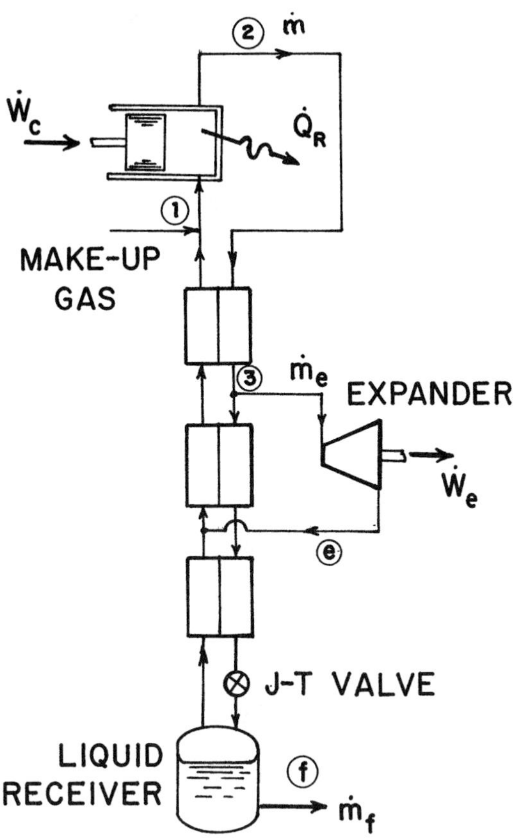

FIGURE 4.12.2 Claude liquefaction system.

A further improvement in the thermodynamic performance of both the Linde-Hampson and Claude liquefaction systems may be achieved by precooling the warm gas stream after exiting the compressor (Barron, 1985). The precooling may be accomplished by using a refrigerant, such as ammonia, that can be condensed under pressure at ambient temperature.

Hydrogen Liquefaction

If hydrogen gas at a temperature above 205 K (–88°F) is expanded through an expansion valve, the gas temperature at the valve exit will be higher than the inlet temperature of the hydrogen gas. This phenomenon occurs for all gases at temperatures above the maximum inversion temperature (205 K for hydrogen), the temperature above which the Joule-Thomson coefficient is always negative. The Joule-Thomson coefficient is defined by:

$$\mu_{JT} = \left(\frac{\partial T}{\partial p}\right)_h \tag{4.12.5}$$

The basic Linde-Hampson system cannot be used for hydrogen gas liquefaction because expansion through the J-T valve in the system would not result in a temperature decrease. A liquid nitrogen-precooled Linde-Hampson system can be used to avoid this problem. Because the Claude system depends on the expander to provide a significant portion of the cooling effect, the Claude system is widely used for hydrogen liquefaction. Expansion of any gas through an expander will result in a temperature decrease, because energy is removed from the gas as mechanical work in the expander.

Hydrogen can exist in two molecular forms: *ortho-hydrogen*, in which the nuclear spins of the two protons in the molecule are in the same direction, and *para-hydrogen*, in which the nuclear spins of the two protons in the molecule are in opposite directions. At room temperature and higher, the equilibrium composition for hydrogen gas is 75% o-H_2 and 25% p-H_2. This mixture is called *normal hydrogen.* At the normal boiling point of hydrogen, the equilibrium composition is 0.2% o-H_2 and 99.8% p-H_2. Since the conversion requires a few hours in the absence of a catalyst, the liquid initially produced in a liquefier would have the n-H_2 composition, if no provision was made to promote the conversion. The conversion from o-H_2 to p-H_2 is an exothermal process; therefore, significant evaporation would occur as the liquid n-H_2 achieved the equilibrium composition (Scott et al., 1964). Commercial hydrogen liquefaction systems use catalysts, such as chromium oxide on alumina pellets, hydroxide gel, and charcoal/silica gel, to complete the conversion process before the liquid is sent to the storage containers.

Helium Liquefaction

Helium is one of the most difficult (i.e., expensive) gases to liquefy because the maximum inversion temperature for helium is only 45 K (-379°F). Prior to 1946, low-temperature research laboratories generally designed and constructed their own liquefiers (Croft, 1961), some of which used liquid hydrogen as a precoolant. Samuel Collins (1947) developed a helium liquefier, shown in Figure 4.12.3, that was a modification of the basic Claude liquefier. The commercial introduction of the Collins liquefier had a significant positive impact on the ease and economy of helium liquefaction.

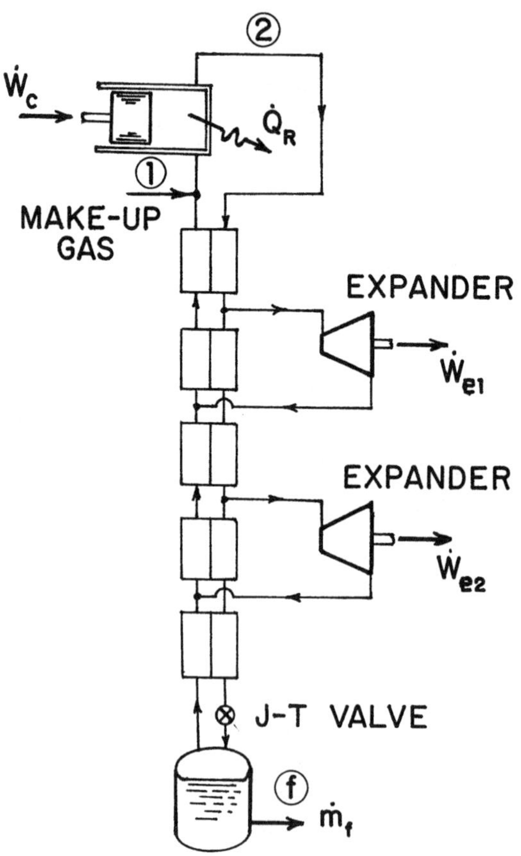

FIGURE 4.12.3 Collins helium liquefaction system.

In the Collins liquefier, helium gas is compressed to a pressure of 1275 kPa (185 psia), and passed through the first heat exchanger. After leaving the first heat exchanger, a portion (about 16%) of the stream is bypassed through the first expander. The gas temperature at the expander inlet is between 40 and 45 K (–388 to –379°F). The remainder of the helium gas flows through the second heat exchanger, and leaves the exchanger at about 15 K (–433°F). A fraction (about 56%) of this stream is bypassed through a second expander after the helium has passed through a third exchanger. The remaining flow passes through two more exchangers and expands through the J-T valve, in which a portion of the stream is liquefied. The vapor formed during the expansion process is returned through the heat exchangers to provide cooling for the incoming gas stream. Although it is not necessary, liquid nitrogen precooling is usually used to improve the liquid yield and offset some of the heat exchanger's inefficiencies.

Collins liquefiers have been constructed with as many as five expanders, depending on the design inlet pressure of the helium gas.

Cryocoolers

Cryogenic refrigeration systems, or *cryocoolers*, are used to maintain cryogenic temperatures in a wide variety of applications, including superconducting magnets (solenoids) and motors, advanced electronic systems, and space environmental simulation chambers. The Joule-Thomson (J-T) cryocooler provides refrigeration by liquefying a portion of the working fluid and allowing the cold liquid to absorb energy through evaporation. Other cryocoolers, such as the Stirling cryocooler, the Gifford-McMahon (GM) cryocooler, and the Vuilleumier (VM) cryocooler, produce a cold gas stream to provide cooling. The magnetic refrigerator uses the magnetocaloric effect in connection with a paramagnetic salt to provide cooling at temperatures below the 1 K level. The ^{3}He-^{4}He dilution refrigerator operates on a physical principle similar to the conventional absorption refrigerator; however, the two isotopes of helium are used to maintain temperatures in the range between 0.005 K and 0.100 K.

Joule-Thomson Cryocooler

Any liquefaction system that uses the expansion valve or J-T valve to produce low temperatures may be classified as a J-T cryocooler. A schematic of the basic J-T cryocooler is shown in Figure 4.12.4. The refrigeration effect, Q_a/m, for the J-T cryocooler is give by:

$$Q_a/m = (h_1 - h_2) - (1-\varepsilon)(h_1 - h_g) \qquad (4.12.6)$$

The quantities are the same as those used in Equation 4.12.1.

The liquid in the evaporator boils at a constant temperature, so the temperature level achieved by the cryocooler is dependent on the liquid used as the working fluid. Liquid nitrogen may be used as the coolant for the temperature range from about 65 to 115 K. The low-temperature limit is set by the triple point for the working fluid. If the evaporator pressure is lowered below the triple point pressure, nitrogen "snow" forms in the evaporator. Unless special heat transfer surfaces are provided, the heat transfer between solid nitrogen and a surface, such as the evaporator tube wall, is much less effective than boiling heat transfer that would occur for the liquid phase. The high-temperature limit is set by the critical temperature for the working fluid. If the evaporator is operated at pressures above the critical pressure, there is no liquid to evaporate.

Because the J-T cryocooler relies on the positive Joule-Thomson effect for production of low temperatures, it cannot be used with hydrogen, helium, or neon gases unless the system is precooled below the maximum inversion temperature for the respective gas.

Microminiature J-T cryocoolers have been developed for cooling infrared sensors and for thermal imaging systems (Little, 1990). These systems use nitrogen-hydrocarbon gas mixtures as the working fluid to produce high refrigeration capacities and more rapid cool-down. The refrigeration capacity of the J-T cryocooler using the gas mixture has been found to be as much as five times that for the system using nitrogen alone.

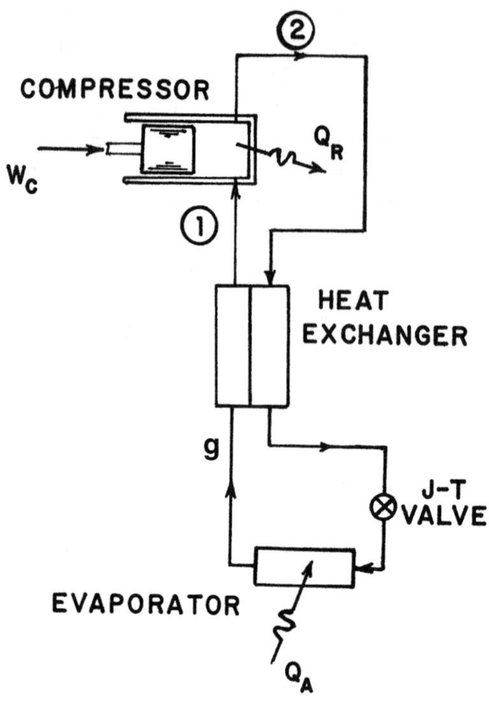

FIGURE 4.12.4 Joule-Thomson cryocooler (refrigerator).

Stirling Cryocooler

The Stirling cycle was originally developed by Robert Stirling in 1816 for use in a hot-air engine (Walker, 1980). Today, Stirling cryocoolers are used in many cryogenic thermal applications, including cooling spacecraft instrument systems, infrared thermal imaging systems, and superconducting elements.

A schematic of the Stirling cryocooler is shown in Figure 4.12.5. The unit consists of a cylinder enclosing the power piston and the displacer piston. The two chambers at either end of the cryocooler are connected through a regenerator, which is one of the critical components of the system.

The sequence of processes for the Stirling cycle is as follows. First, the working fluid (typically, helium gas) is compressed by the power piston while energy is rejected to cooling coils to achieve a near-isothermal compression. Next, the displacer piston moves the high-pressure gas through the regenerator (which was cooled during a prior process), and the gas is cooled to a lower operating temperature.

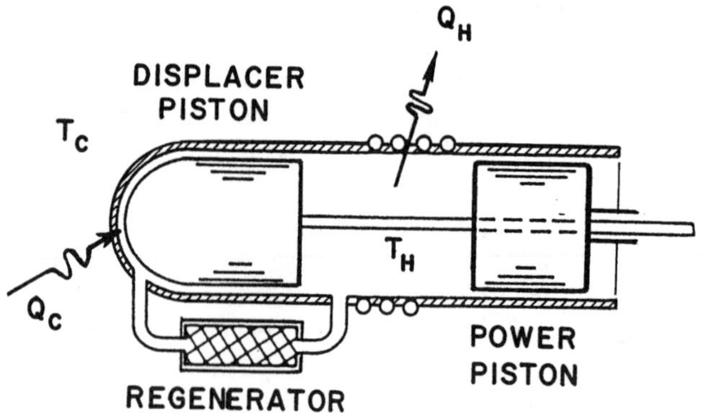

FIGURE 4.12.5 Stirling cycle refrigerator.

During this process, the power piston is practically stationary, so the cooling process approximates a constant-volume process. The cold high-pressure gas is expanded by motion of both the power piston and displacer piston, while energy is added to the gas from the low-temperature region at the cold end of the cylinder to achieve near-isothermal expansion. Finally, the cold low-pressure gas is forced back through the regenerator by motion of the displacer piston (with the power piston practically stationary), and the gas is warmed to its initial condition.

If all processes were ideal, the Stirling refrigerator would have the same coefficient of performance as that of the Carnot cycle,

$$COP_{id} = \frac{Q_a}{-W_{net}} = \frac{T_c}{T_h - T_c} \tag{4.12.7}$$

T_h is the temperature of the gas at the hot end, and T_c is the temperature of the gas at the cold end of the cylinder. Frictional energy dissipation, pressure drop through the regenerator, and the inefficiency of the regenerator interact to reduce the actual COP to approximately 35% of the Carnot value for a typical Stirling cryocooler, depending on the temperature at the cold end. An analysis of the operation of the Stirling cryocooler is available in the literature (Walker, 1983a).

Vuilleumier (VM) Cryocooler

The VM cryocooler was developed by Rudolph Vuilleumier in 1918 for use in low-temperature refrigeration systems (Vuilleumier, 1918). A schematic of the VM system is shown in Figure 4.12.6. The VM cryocooler is similar to the Stirling cryocooler, except the VM system is driven by a thermal compression process instead of a mechanical drive. The VM refrigerator consists of a Stirling engine driving a Stirling refrigerator, with a common ambient-temperature expansion space for the two units.

The COP for the ideal VM cryocooler is a function of the gas temperature in the cold chamber (T_c), the gas temperature in the hot chamber (T_h), and the gas temperature in the intermediate or ambient-temperature chamber (T_o).

$$COP_{id} = \frac{Q_c}{Q_h} = \frac{T_c(T_h - T_o)}{T_h(T_o - T_c)} \tag{4.12.8}$$

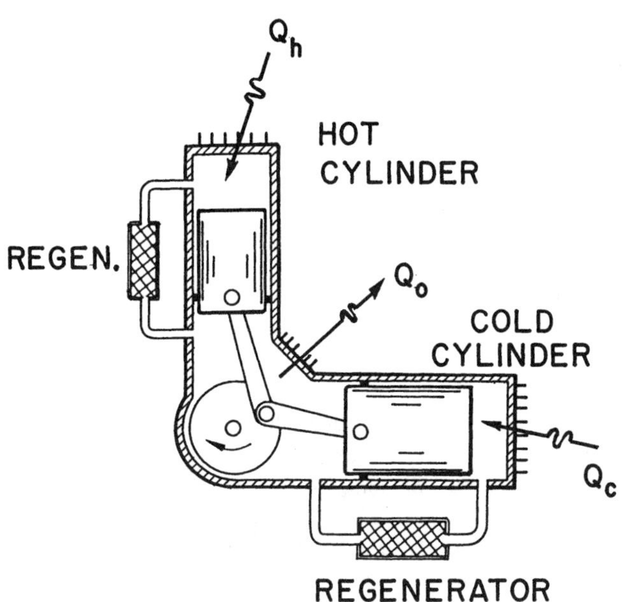

FIGURE 4.12.6 Vuilleumier (VM) cryocooler.

Q_h is the energy added to the gas in the hot chamber from a source at T_h, and Q_c is the heat added to the gas in the cold chamber from a source at T_c.

An advantage of the VM cryocooler is that the thermal input may be provided by a wide variety of sources, including solar energy and isotope thermal energy. This feature makes the VM cryocooler attractive for cryogenic cooling in long-term space exploration and in applications where mechanical vibrations of a drive engine must be avoided (Walker, 1983a). Because the VM refrigerator operates at a relatively low speed (approximately 600 rpm), the reliability of the system is particularly good. For example, one VM cryocooler was operated continuously and unattended for 20,000 hours (833 days) (Pitcher, 1973).

Gifford-McMahon (GM) Cryocooler

The GM cryocooler was developed by McMahon and Gifford (1959) for use as a miniature cryogenic refrigerator. A schematic of the GM cryocooler is shown in Figure 4.12.7. The sequence of processes for the GM cryocooler is as follows.

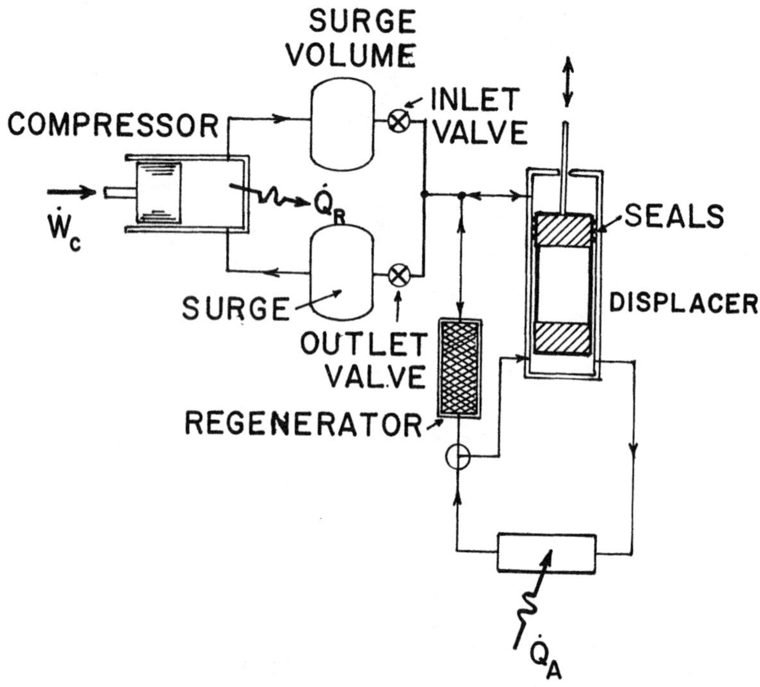

FIGURE 4.12.7 Gifford-McMahon (GM) cryocooler.

Pressure build-up process — With the displacer piston at the bottom of the cylinder and the outlet valve closed, the inlet valve is opened to allow the working fluid (typically helium gas) to enter the upper expansion space.

Intake stroke — The displacer piston is moved from the bottom of the cylinder to the top, while the inlet valve remains open and the outlet valve remains closed. The warm gas originally in the upper space is moved through the regenerator to the lower space. The gas is cooled as it flows through the regenerator, and additional gas is drawn in through the inlet valve to maintain the gas pressure constant within the system.

Expansion process — With the displacer at the top of the cylinder, the inlet valve is closed and the outlet valve is slowly opened. The gas within the lower space expands adiabatically, and the gas temperature decreases.

Exhaust stroke — The low-temperature gas is forced from the lower space by moving the displacer to the bottom of the cylinder while the outlet valve is open. The gas absorbs energy from the space being cooled.

Regeneration process — The gas flows back through the regenerator, where it is warmed, and the cycle is repeated.

There are several advantages of the GM cryocooler which have promoted its widespread use in a variety of cryogenic applications, including cooling of cold traps in vacuum systems and cooling of thermal radiation shields in space cryostats. All valves and seals in the GM cryocooler operate at ambient temperature; therefore, problems with low-temperature seals and valves are eliminated. The GM cryocooler uses a storage-type heat exchanger or regenerator instead of a conventional heat exchanger. Because of the back and forth motion of the gas through the regenerator, impurities deposited during the intake stroke are swept out of the system during the exhaust stroke, thus problems with impurity accumulation are minimized in the GM system.

The GM system is well suited for multistaging (Gifford and Hoffman, 1961). In some cryogenic systems, refrigeration is required at multiple temperature levels. For example, thermal radiation shields may be used to thermally surround a region to be maintained at a low temperature. In this case, the GM cryocooler may be operated with two stages, one cooling the shields at 77 K, and the other providing the refrigeration at 20 K. The multistage GM refrigerator generally has a better thermodynamic performance than a single-stage unit operating between the same upper and lower temperature limits.

Magnetic Refrigerator

If a liquid or gas is used as the refrigerator working material, the operating temperature is limited to about 1 K. All materials other than helium (^{4}He) and the isotope ^{3}He are in the solid phase at temperatures below 1 K. For helium, the vapor pressure is low in this temperature range. At 0.5 K, for example, the vapor pressure of ^{4}He is 1.9 mPa, and the vapor pressure of ^{3}He is 22.2 Pa. It is apparent that a refrigerator operating with these low pressures in the evaporator is impractical for large refrigeration capacities, because the flow passages would be quite large for even moderate mass flow rates.

To break the 1 K temperature "barrier", the magnetocaloric effect (a decrease in magnetic field applied to a material results in a decrease in temperature) may be used with a paramagnetic salt to achieve temperatures in the 0.1 K range. Iron ammonium alum has been used as the working material in magnetic refrigerators. The cycle for a magnetic refrigerator is shown in Figure 4.12.8.

The external magnetic field is increased from 0.14 to 0.72 T while the working salt is in thermal communication through a thermal "valve" with a liquid helium bath at 1.15 K. Heat transfer occurs between the working salt and the helium bath to maintain constant temperature. Next, the thermal valve is closed, thermally isolating the working salt, and the magnetic field is decreased from 0.72 to 0.10 T. The temperature of the working salt will decrease from 1.15 K to about 0.16 K as a result of the magnetocaloric effect. The thermal valve between the working salt and the region to be cooled is opened, allowing heat transfer from the low-temperature region to the working salt while the magnetic field is decreased from 0.10 to 0 T. The working salt temperature remains constant during this process. Finally, the thermal valves are closed, thermally isolating the working salt, while the magnetic field is increased to the initial value of 0.14 T, and the cycle is repeated.

The refrigeration effect for the magnetic refrigerator is

$$Q_a/m = T_3(s_4 - s_3) \tag{4.12.9}$$

where the state-points correspond to those shown in Figure 4.12.8. The magnetic work per unit mass for the working salt is

$$W_{net}/m = (T_1 - T_3)(s_4 - s_3) \tag{4.12.10}$$

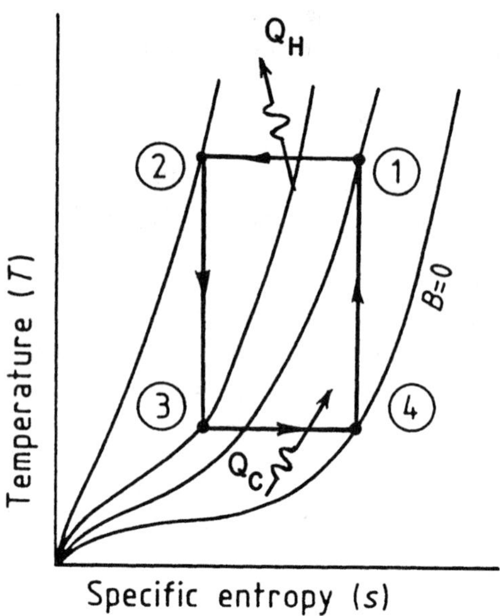

FIGURE 4.12.8 Thermodynamic diagram for the magnetic refrigerator cycle.

The COP for the ideal magnetic refrigerator is the same as that for a Carnot refrigerator operating between the same two temperature limits.

Irreversibilities reduce the thermodynamic performance of the actual system below that of the ideal system. For example, the Daunt-Heer-Barnes refrigerator (Heer et al., 1954) had a refrigeration capacity of 27.5 μW at a low temperature of 0.4 K; whereas, the ideal system operating under the same conditions would have a refrigeration capacity of 34.4 μW.

One type of thermal valve used in the magnetic refrigerator is a thin lead strip conductor. At 1 K, lead is superconducting in zero applied field, and the thermal conductivity in the superconducting state is about 0.025 times the thermal conductivity in the normal state. The valve may be "closed" (little heat transfer through the lead strip) by lowering the magnetic field around the lead strip until the material becomes superconducting. The valve may be "opened" by increasing the magnetic field around the switch to drive the lead into the normal state, during which the thermal conductivity is large. The energy dissipated during the opening and closing of the magnetic thermal valves is zero; whereas, mechanical switches result in excessive energy dissipation when the contact is opened or closed.

The magnetic refrigerator has also been constructed in a wheel form for use in the temperature range between 2 and 10 K (Barclay et al., 1985). In the magnetic wheel refrigerator, the magnetic fields are fixed and the paramagnetic material is moved into and out of the field. Helium gas is usually used as the heat transfer medium.

Dilution Refrigerator

The ^{3}He–^{4}He dilution refrigerator is widely used for the temperature range between 5 mK and 0.1 K. It was discovered in 1956 that a phase separation occurred for ^{3}He–^{4}He liquid mixtures below a temperature of about 0.86 K. The phase separation is accompanied by a cooling effect or heat of "unmixing", similar to that which occurs for ammonia/water and lithium bromide/water mixtures used in conventional absorption refrigeration systems. A schematic of the dilution refrigerator is shown in Figure 4.12.9.

The gas (which is practically pure ^{3}He) is compressed in a vacuum pump from about 4 Pa to 4 kPa. The gas is cooled in a heat exchanger and a liquid helium bath at 4.2 K, and then is condensed in a liquid helium bath at 1.2 K. The liquid helium ^{3}He is expanded through a capillary tube constriction and cooled further in the still, which operates at 0.6 K. The liquid is further cooled in a heat exchanger

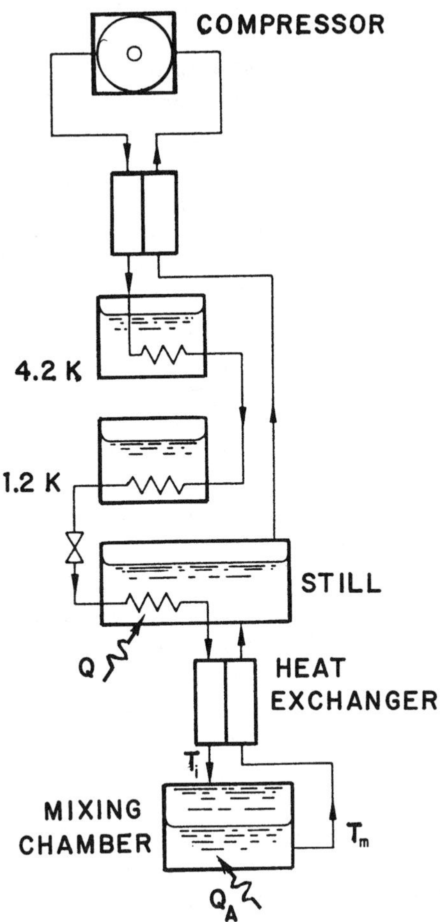

FIGURE 4.12.9 ^{3}He–^{4}He dilution refrigerator.

before it enters the mixing chamber, where ^{4}He is mixed with the ^{3}He at temperatures between 50 mK and 5 mK.

The liquid in the mixing chamber separates into two liquid phases: a less-dense concentrated ^{3}He mixture, and a more-dense diluted ^{3}He mixture. Heat is absorbed from the low-temperature region during this process. The dilute mixture returns through the heat exchanger to the still, where heat is added to evaporate the ^{3}He from the mixture. The mole fraction of ^{3}He in the liquid in the still is about 1%, and the composition of the vapor is between 95 and 98% ^{3}He.

One of the critical components of the dilution refrigerator is the heat exchanger between the still and the mixing chamber. This heat exchanger operates with a temperature difference at the cold end of about 70 mK for a mixing chamber temperature of 40 mK, yet the heat exchanger must have a high effectiveness for net refrigeration to be produced. Heat exchangers with large surface areas per unit volume have been developed using sintered metal elements within the flow passages (Frossati, 1978).

Cryogenic Heat Exchanger Types

One of the critical components of many cryogenic systems, such as liquefiers and cryocoolers, is the heat exchanger. In some conventional systems, such as regenerative gas turbine power plants, the system will operate even if the heat exchanger is not highly effective, say, less than 50%. In contrast, a cryogenic liquefier will produce no liquid if the heat exchanger efficiency is less than approximately 85% (Barron, 1985).

There are many different configurations for nonstorage cryogenic heat exchangers or *recuperators*, but they may generally be classified as tubular exchangers, plate-fin exchangers, or perforated plate exchangers. The storage heat exchanger or regenerator is also widely used in cryogenic systems because of its special characteristics.

The most simple tubular heat exchanger is the concentric-tube or double-pipe heat exchanger. This is the type of heat exchanger used by Linde in his original air liquefier. The heat exchanger consists of a small inner tube, in which the high-pressure stream usually flows, coaxial with a larger tube, with the low-pressure stream flowing in the annular space. The unit is usually wound in a helical spiral to conserve space, and the coil is often placed within an evacuated container to reduce heat transfer from ambient conditions.

A wire or plastic spacer may be spirally wound on the inner tube of the concentric-tube heat exchanger. The use of the spacer allows the inner pipe to be fixed within the outer pipe, and causes the fluid in the annular space to follow a longer helical path. This configuration results in an increase in the fluid velocity and an increase in the heat transfer coefficient, in comparison with the basic concentric tube arrangement. The use of the spiral spacer also results in an increase in the fluid pressure drop. This unit is usually wound in a helical spiral to conserve space.

A unique extended-surface heat exchanger design involving flow in concentric tubes was developed by Samuel Collins (1947) for the Collins helium liquefier. The exchanger consisted of several concentric copper tubes with an edge-wound copper ribbon helix wrapped in the annular spaces between the tubes. The helix was soft-soldered to both sides of the annular space in which it was wound. The ribbon helix acted as a fin to increase the heat transfer surface area of the exchanger. These heat exchangers have been manufactured with as many as 10 separate flow passages.

The classic heat exchanger used in large-scale air and LNG liquefaction systems is the Giauque-Hampson heat exchanger. The heat exchanger consists of carefully spaced helices of small-diameter tubing wound in several layers around a core cylinder or mandrel. The mandrel provides mechanical stability and support during manufacture and operation of the heat exchanger. The minimum diameter of the core cylinder is determined by the diameter at which flattening of the smaller tubes would occur during winding around the mandrel. Successive layers of tubes are wound in opposite directions and separated by spacing strips (Abadzic and Scholz, 1973). The tubes are attached to headers at both ends of the heat exchanger to allow the stream to be brought into or out of the exchanger in a single pipe. A close-fitting outer sheath is placed around the outside of the coil, and the entire unit is thermally insulated.

The high-pressure stream flows inside the small tubes in the Giauque-Hampson exchanger, and the low-pressure stream travels in a crossflow over the small tubes in the annular space between the mandrel and the outer casing. Several layers of tubes may be used to provide multiple parallel paths for the high-pressure stream, thereby reducing the frictional pressure drop. The Giauque-Hampson heat exchanger may be designed for three or more streams, where the high-pressure and intermediate-pressure streams flow inside the small tubes, for example. To maintain a uniform flow distribution for the streams flowing inside the small tubes, the length of each flow passage is usually made the same from layer to layer by varying the helical pitch of each successive layer.

It is quite important to assure that the radial spacing of the small tubes is uniform; otherwise, the low-pressure stream will tend to follow the path through the widest spacing of least frictional resistance, and the flow will not be uniformly distributed over the cross section of the heat exchanger. Giauque, at the University of California, solved the spacing problem by using punched brass spacer strips to maintain uniform tube spacing. A thin strip of cellulose acetate was placed between the spacer and the tube as the tube was being wound. The acetate was dissolved by acetone after the tubes had been wound, and a controlled, reproducible tube spacing was achieved.

Plate-fin heat exchangers consist of stacks of alternate layers of corrugated die-formed metal sheets (the fins) separated by flat metal separation sheets (the plates). The flow is introduced to and collected from the finned flow passages through headers welded or brazed at each end of the core stack. Several fin configurations have been used, including straight fins, wavy fins, herringbone fins, and serrated fins. Various stacking arrangements may be used to produce flow configurations such as counterflow, cross-flow, and multipass heat exchangers. The counterflow configuration is most commonly used in cryogenic systems because of the better thermal performance of a counterflow exchanger for a given heat transfer surface area.

A schematic of the perforated plate heat exchanger is shown in Figure 4.12.10. The perforated plate exchanger consists of a series of parallel perforated plates separated by low-thermal-conductivity spacers or gaskets (Fleming, 1969). The plates are constructed of a high-thermal-conductivity metal, such as aluminum or copper. The spacers are bonded (sealed) to the plates to prevent leakage between the two streams. Older models of the perforated plate exchangers used a plastic material for the spacers to reduce the problem of longitudinal conduction. More recent models have used stainless steel spacers to achieve better mechanical reliability (Hendricks, 1996). The configuration for the perforated plate exchanger is usually counterflow.

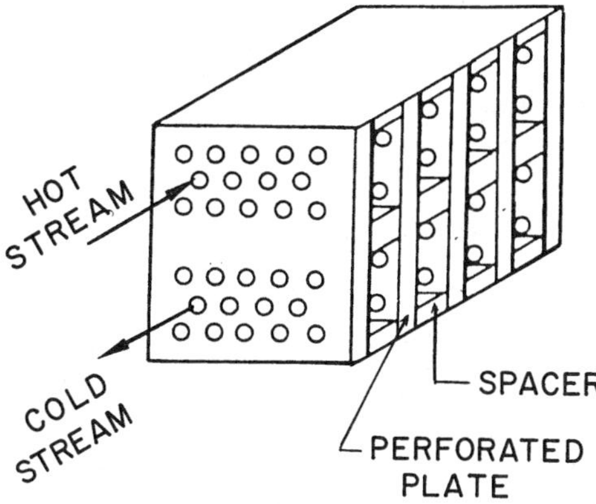

FIGURE 4.12.10 Perforated plate heat exchanger.

Small perforations (diameters ranging from 1.5 mm or 1/16 in. to less than 0.4 mm or 1/64 in.) are made into the plates so that a large heat transfer coefficient is achieved. The ratio of the plate thickness (length of the hole in the plate) to the diameter of the hole is on the order of 0.75, so the thermal and hydrodynamic boundary layers do not become fully developed within the perforations, which results in high heat transfer coefficients and correspondingly high friction factors.

Wire screens have been used to replace the perforated plates in some compact heat exchangers used in small cryocoolers (Lins and Elkan, 1975; Vonk, 1968). The fluid streams flow through the wire layers, and heat is exchanged between the two streams through conduction along the wires. The wire screens are separated by plastic or resin-impregnated paper spacers, or stainless steel metallic spacers that are diffusion-bonded or brazed to the perforated plates.

Cryogenic Heat Exchanger Design Problems

Three problems may become significant in the design of heat exchangers for cryogenic service:

1. The effect of variation of fluid properties with temperature — variation of the fluid specific heat is particularly important (Chowdhury and Sarangi, 1984; Soyars, 1991).
2. The effect of longitudinal conduction along the separating surface between the two streams — this effect is especially critical for short heat exchangers that must have a high efficiency (Kroeger, 1967).
3. The effect of flow maldistribution — the performance deterioration due to flow distribution is most pronounced for high-efficiency heat exchangers (Fleming, 1967).

In many conventional heat exchanger designs, the specific heats of the fluids do not change significantly with temperature; however, for gases such as hydrogen in the cryogenic temperature range, the specific heat may vary as much as 40% between the normal boiling point and ambient temperatures. The specific heat variation can result in a thermal pinch effect within the heat exchanger, in which the temperature difference between the two fluids is smaller than that for constant specific heats.

The problem of thermal pinch may be alleviated by "unbalancing" the heat exchanger by increasing the flow of the hydrogen stream (if possible). An alternate approach would be to bleed off a portion of the flow at some intermediate point within the heat exchanger to change the capacity rate ratio.

Longitudinal conduction in conventional heat exchangers is usually not a serious problem because of the long length of the heat exchanger. Miniature cryogenic heat exchangers, on the other hand, may have lengths as short as 50 mm. One solution to the problem of performance degradation due to longitudinal conduction is to use thin-walled tubes of a relatively low-thermal-conductivity material (stainless steel, for example). Longitudinal conduction in regenerators may be practically eliminated by using several short sections or discrete elements, such as spheres and layers of screen, as the matrix material.

One technique for solving the problem of uneven flow distribution involves the use of flow constrictions on the warm side (Cowans, 1974). If the temperature of the warm stream in one of the parallel-flow passages increases because of a higher flow rate in the passage, the constriction presents a higher flow resistance as a result of the lower fluid density, thereby reducing the warm fluid stream flow through that passage and balancing the flow distribution.

Regenerators

In the conventional heat exchanger or recuperator, the two fluids exchanging energy flow in separate flow channels and are separated by a solid surface such as the tube wall (Giauque-Hampson heat exchanger) or a plate (plate-fin heat exchanger). In a regenerator, the same space is occupied alternately by the hot fluid and the cold fluid. Energy is first transferred from the hot stream into a solid material, called the regenerator packing material or *matrix*. When the cold fluid flows through the matrix during the next part of the cycle, the energy is transferred from the solid material to the cold stream.

The temperature of the fluid within the regenerator and the temperature of the regenerator matrix are functions of both position within the regenerator and time of operation. As a result, the performance of a regenerator is a function of the mass of the matrix, the specific heat of the matrix material, and the frequency of switching of the flow through the regenerator, in addition to the other parameters (capacity rate ratio and NTU) for an ordinary heat exchanger (Schmidt and Willmott, 1981).

Although it would be possible to design a regenerator for crossflow or parallel flow, the counterflow arrangement is most often used because counterflow is more thermally efficient than any other configuration. There are two general types of counterflow regenerators: (1) fixed-bed or valved-type regenerators and (2) rotary regenerators. The flow through the rotary regenerator may be either radial or axial (Kays and London, 1984). Another type of regenerator used in some miniature cryocoolers is the gap regenerator (Daniels and du Pre, 1971). The fluid flows through the small clearance space or gap between the displacer and cylinder wall, and energy is alternately stored in and released from the solid walls. A modification of the gap regenerator involves filling the small gap with thin foil or small tubes (straws) (Walker, 1983b).

The primary advantages of regenerators over ordinary heat exchangers (recuperators) include the following:

1. A much larger heat transfer surface area per unit volume can be achieved for a regenerator. For example, as much as 4000 m^2 of surface area per cubic meter of volume (1200 ft^2/ft^3) can be obtained in a regenerator, compared with 1475 m^2/m^3 (450 ft^2/ft^3) for a compact plate-fin heat exchanger.
2. Regenerators are generally more simple to fabricate than recuperators. In its most simple form, a regenerator could consist of a tube or pipe filled with lead shot. As a result of the simplicity of fabrication, regenerators tend to cost less than a recuperator for the same heat transfer rate.

3. Because of the alternate direction of flow of the hot and cold streams through the same flow passage, the regenerator tends to be self-cleaning. Impurity buildup problems are not nearly as severe in a regenerator as those encountered in ordinary heat exchangers.

The primary disadvantages of regenerators, compared with regular heat exchangers, include:

1. There is always some mixing of the warm and cold streams during the switching process. This "carry-over" may not be important if the two streams are the same material; however, excessive carry-over does degrade the regenerator thermal performance.
2. For the rotary regenerators, leakage at the dynamic seals may become appreciable. This problem is particularly acute when the two streams are at significantly different pressures.

The regenerator acts as a "thermal sponge" or "thermal flywheel" to store energy during the period of flow of the hot stream and deliver this energy back to the cold stream when it flows through the regenerator. The effectiveness of the storage process is related to the thermal capacity (mass × specific heat per unit mass) of the matrix material. If the specific heat is excessively small, the thermal capacity of the matrix material is insufficient to produce efficient energy storage, and the resulting regenerator thermal performance is poor.

The specific heat of metals generally decreases rapidly as the temperature is lowered into the cryogenic range. For example, the specific heat of copper drops to about one-half of the room temperature value when the temperature is lowered to 77 K (139°R). For cryogenic regenerators, lead is a relatively good matrix material because the specific heat of lead does not change appreciably until the temperature is lowered below 90 K (162°R). At a temperature of 21.5 K (38.7°R), the specific heat of lead becomes one-half of the room-temperature value. However, even lead has a small thermal capacity at liquid helium temperatures. For example, the specific heat for lead is 1.11 J/kg-K (0.00027 Btu/lbm-°F) at 4.2 K; whereas, the specific heat for helium gas at 4.2 K (7.6°R) and 4 atm (405 kPa or 58.8 psia) is 3500 J/kg-K (0.84 Btu/lbm-°F), or about 3200 times that of lead on a mass basis. It is apparent that lead becomes a poor regenerator matrix material for regenerator temperatures below about 10 K or 18°R, because its thermal capacity becomes quite small in this temperature range.

Several rare-earth materials exhibit high maximums in their specific heat curves in the temperature region below 20 K (36°R) (Daniels and du Pre, 1971; Kuriyama et al., 1990; Seshake et al., 1991; Tsukagoshi et al., 1996). Some rare-earth materials that have been used as very-low-temperature regenerator matrices include europium sulfide (EuS), erbium-nickel (Er_3Ni), and erbium cobalt (ErCo).

A variety of structures has been used as the matrix material for cryogenic regenerators. Some materials include packed spheres, dense-mesh wire screen (either stacked, punched, or sintered), spiral-wound corrugated metal ribbons (Fraenkl matrix), and etched foil (Yaron et al., 1996).

Cryogenic Insulation

The performance of the insulation in cryogenic fluid storage and transfer systems is important in determining the effectiveness of the system. Additional information on insulation systems for both high-temperature and cryogenic applications may be found in Section 4.5, *Thermal Insulation.* Some of the insulation types used for cryogenic service include expanded closed-cell foam, gas-filled powders and fibrous materials, vacuum alone, evacuated powders and fibrous materials, microsphere insulation, and multilayer insulation (MLI).

Closed-cell foam and gas-filled powder insulations have been used extensively in storage systems for LNG (Forbes and Martel, 1963; Tatro and Bennett, 1975). The thermal conductivity of these insulations is of the same order of magnitude as the thermal conductivity of air (35 mW/m-K or 0.02 Btu/h-ft-°F). Foams and gas-filled powder insulations are generally less expensive than other cryogenic insulations. One disadvantage of the foam insulation for cryogenic service is the large difference between the thermal expansion coefficient for the foam and the metal of the storage vessel on which the insulation is applied. A vapor barrier cover is required for both the foam insulation and the powder insulation to prevent migration of water vapor into the insulation.

By evacuating a powder insulation, such as perlite or silica aerogel, to pressures on the order of 10^{-3} torr (133 mPa), the thermal conductivity may be reduced to approximately 1.2 mW/m-K (0.7×10^{-3} Btu/h-ft-°F). A rigid vacuum jacket is required when this insulation is used. Evacuated powder insulations are used extensively for liquid nitrogen and liquid oxygen storage vessels (dewars) (Kropschot and Burgess, 1963).

One technique for reducing the radiant heat transfer through evacuated powder insulations is to add highly reflective opaque metal flakes (Riede and Wang, 1960). For a silica aerogel evacuated powder insulation with 40 to 50% by weight of copper flakes, the thermal conductivity is 0.35 mW/m-K (0.2×10^{-3} Btu/h-ft-°F).

Because opacified powders can show degradation in the performance when the metallic flakes become concentrated in one region of the insulation, attention was directed toward the use of small, coated glass spheres or *microspheres* as a cryogenic insulation (Cunnington and Tien, 1973). Microsphere insulation consists of packed hollow glass spheres, ranging in size from 0.015 to 0.150 mm in diameter, coated on the outer surface with a reflective metallic film. The insulation is evacuated to a pressure on the order of 10^{-4} torr (13.3 mPa). The thermal conductivity of microsphere insulation in cryogenic service ranges between 0.2 and 0.4 mW/m-K (1.2×10^{-4} to 2.3×10^{-4} Btu/h-ft-°F), which is the same order of magnitude as the thermal conductivity for the opacified powder insulations.

The original (*circa* 1960) MLI consisted of alternating layers of highly reflective foil (aluminum or copper) and a low-thermal-conductivity spacer sheet, such as fiberglass paper or Dacron fabric (Black et al., 1960; Riede and Wang, 1960). Because the foil was fragile, it has been largely replaced by aluminized Mylar. In some MLI, the reflecting layers are separated by a crinkled or embossed aluminized Mylar sheet to replace the fragile fiberglass paper separator.

Multilayer insulations must be evacuated to pressures on the order of 10^{-5} torr (1.33 mPa) to be effective. The thermal conductivity of MLI is approximately 0.043 mW/m-K (2.5×10^{-5} Btu/h-ft-°F), which represents an order of magnitude improvement over opacified powders and microsphere insulation.

The extremely low thermal conductivity of the MLI may be explained by the fact that all modes of heat transfer (solid and gaseous conduction, convection, and thermal radiation) are reduced to small values. Solid conduction is minimized through the use of the fibrous spacer with no binding agent or the crinkled plastic that allows thermal contact at relatively few points. Gaseous conduction is practically eliminated by evacuating the insulation, and there is no gaseous convection at these pressure levels. Thermal radiation is reduced by using several layers (about 28 layers per centimeter of thickness or 70 layers per inch of thickness) of highly reflective foil or metal deposited on a plastic film. The thermal conductivity for a well-evacuated MLI may be determined from:

$$k = \left[h_c + \sigma e\left(T_h^2 + T_c^2\right)\left(T_h + T_c\right)/(2-e)\right]\Big/\left(N/\Delta x\right) \tag{4.12.11}$$

where h_c is the solid conductance for the spacer material, σ is the Stefan-Boltzmann constant (56.69 nW/m²-K⁴), *e* is the emissivity of the radiation shields, $(N/\Delta x)$ is the number of layers per unit thickness, and T_h and T_c are the boundary temperatures of the insulation (absolute temperatures). Expressions for the solid conductance for various types of MLI are available in the literature (McIntosh, 1994; Spradley et al., 1990).

One of the problems associated with MLI is that effective evacuation of the residual gas from within the insulation blanket is difficult. It has been shown (Skurlock and Saull, 1976) that release of gas from the shields (out-gassing) can also introduce significant quantities of gas within the insulation. MLI using carbon-filled glass-fiber paper for the space material have been developed to reduce this problem. The carbon acts as an adsorbent (getter) to absorb the residual gas when the insulation is cooled on one side to cryogenic temperatures. The thermal conductivity of this MLI is 0.014 mW/m-K (0.85×10^{-5} Btu/h-ft-°F).

References

Abadzic, E.E. and Scholz, H.W. Coiled Tubular Heat Exchangers, in *Advances in Cryogenic Engineering,* Vol. 18, Plenum Press, New York, 1973, pp. 42-51.

Barclay, J.A., Stewart, W.F., Overton, W.C., Chamdler, R.J., and Harkleroad, O.D., Experimental Results on a Low-Temperature Magnetic Refrigerator, in *Advances in Cryogenic Engineering,* Vol. 31, Plenum Press, New York, 1985, pp. 743-752.

Barron, R.F. *Cryogenic Systems,* 2nd ed., Oxford University Press, New York, 1985, pp. 127-131.

Black, I.A., Fowle, A.A. and Glaser, P.E., Development of High-Efficiency Insulation, in *Advances in Cryogenic Engineering,* Vol. 5, Plenum Press, New York, 1960, pp. 181-188.

Chowdhury, K. and Sarangi, S., The Effect of Variable Specific Heat of the Working Fluid on the Performance of Counterflow Heat Exchangers, *Cryogenics,* 12, 679, 1972.

Claude, G., *Liquid Air, Oxygen and Nitrogen,* P. Blakiston Son and Co., Philadelphia, 1913.

Collins, S.C., *Rev. Sci. Instrum.,* 18, 1957, 1947.

Cowans, K.W., A Countercurrent Heat Exchanger That Compensates Automatically for Maldistribution of Flow in Parallel Channels, in *Advances in Cryogenic Engineering,* 19, Plenum Press, New York, 1974, pp. 437-444.

Croft, A.J., Helium Liquefiers, in *Progress in Cryogenics,* Vol. 3, Mendelssohn, K., Ed., Academic Press, New York, 1961, pp. 1-21.

Cunnington, G.R. and Tien, C.L., Heat Transfer in Microsphere Insulation, in *Advances in Cryogenic Engineering,* Vol. 18, Plenum Press, New York, 1973, pp. 103-111.

Davies, M., *Physical Principles of Gas Liquefaction and Low Temperature Rectification,* Longman, London, 1949, pg. 37.

Daniels, A. and du Pre, F.K., Triple-Expansion Stirling Cycle Refrigerator, in *Advances in Cryogenic Engineering,* Vol. 16, Plenum Press, New York, 1971, pp. 178-184.

Fleming, R.B., The Effect of Flow Distribution in Parallel Channels of Counterflow Heat Exchangers, in *Advances in Cryogenic Engineering,* Vol. 12, Plenum Press, New York, 1967, pp. 352-362.

Fleming, R.B., A Compact Perforated-Plate Heat Exchanger, in *Advances in Cryogenic Engineering,* Vol. 14, Plenum Press, New York, 1969, pp. 197-204.

Forbes, F.W. and Martel, C.R., Expandable Cryogenic Pipe, in *Advances in Cryogenic Engineering,* Vol. 8, Plenum Press, New York, 1963, pp. 411-416.

Frossati, G., Proc. LT1, *J. Phys. (Paris) Collq. C65,* 39(8), 1578, 1978.

Gifford, W.E. and Hoffman, T.E., A New Refrigeration System for 4.2 K, in *Advances in Cryogenic Engineering,* Vol. 6, Plenum Press, New York, 1961, pp. 82-94.

Hampson, W., British Patent 10165, 1895.

Heer, C.V., Barnes, C.B., and Daunt, J.G., *Rev. Sci. Instrum.,* 25, 1088, 1954.

Hendricks, J.B., A New Method for Producing Perforated Plate Recuperators, in *Advances in Cryogenic Engineering,* Vol. 41, Plenum Press, New York, 1996, pp. 1329-1337.

Kays, W.M. and London, A.L., *Compact Heat Exchangers,* 3rd ed., McGraw-Hill, New York, 1984, pp. 186-279; 29-31.

Kroeger, P.G., Performance Deterioration in High Effectiveness Heat Exchangers Due To Axial Heat Conduction Effects, in *Advances in Cryogenic Engineering,* Vol. 12, Plenum Press, New York, 1967, pp. 363-372.

Kropschot, R.H. and Burgess, R.W., Perlite for Cryogenic Insulation, in *Advances in Cryogenic Engineering,* Vol. 8, Plenum Press, New York, 1963, pp. 425-436.

Kuriyama, T., Hakamada, R., Nakagome, H., Tokai, Y., Sashi, M., Li, R., Yoshida, O., Matsumoto, K., and Hashimoto, T., High Efficient Two-Stage GM Refrigerator With Magnetic Material in the Liquid Helium Temperature Region, in *Advances in Cryogenic Engineering,* Vol. 35, Plenum Press, New York, 1990, pp. 1261-1269.

Linde, K.V., German Patent 88824. See also *Z. Kalteind.*, 4, 23, 1897.

Lins, R.C. and Elkan, M.A., Design and Fabrication of Compact, High-Effectiveness Cryogenic Heat Exchangers Using Wire Mesh Surfaces, in *Advances in Cryogenic Engineering,* Vol. 20, Plenum Press, New York, 1975, pp. 283-299.

Little, W.A., Advances in Joule-Thomson Cooling, in *Advances in Cryogenic Engineering,* Vol. 35, Plenum Press, New York, 1990, pp. 1305-1314.

McIntosh, G.E., Layer by Layer MLI Calculation Using a Separated Mode Equation, in *Advances in Cryogenic Engineering,* Vol. 39, Plenum Press, New York, 1994, pp. 1683-1690.

McMahon, H.O. and Gifford, W.E., A New Low-Temperature Gas Expansion Cycle, Part 1, in *Advances in Cryogenic Engineering,* Vol. 5, Plenum Press, New York, 1960, pp. 354-367.

Pitcher, G.K., Mechanical Life of Space Cryocoolers, Closed Cycle Cryogenic Cooler Technology and Applications, Vol. 1, AFFDL-TR-73-149, Wright-Patterson AFB, Dayton, Ohio, 1973, pp. 211-224.

Riede, P.M. and Wang, D.I.-J., Characteristics and Applications of Some Superinsulations, in *Advances in Cryogenic Engineering,* Vol. 5, Plenum Press, New York, 1960, pp. 209-215.

Ruhemann, M., *The Separation of Gases,* 2nd ed., Oxford University Press, London, 1949, pg. 141.

Schmidt, F.W. and Willmott, A.J., *Thermal Energy Storage and Regeneration,* McGraw-Hill, New York, 1981, pp. 105-127.

Scott, R.B., *Cryogenic Engineering,* D. Van Nostrand, Princeton, NJ, 1959, pg. 1.

Scott, R.B., Denton, W.H., and Nichols, C.M., *Technology and Uses of Liquid Hydrogen,* Macmillan, New York, 1964, pp. 90-92.

Scurlock, R.G. and Saull, B., Development of Multilayer Insulations With Thermal Conductivities Below 0.1 μW/cm-K, *Cryogenics,* 16, 303, 1976.

Seshake, H., Eda, T., Matsumoto, K., Hashimoto, T., Kuriyama, T., and Nakagome, H., Analysis of Rare Earth Compound Refrigerators Operating at 4 K, in *Advances in Cryogenic Engineering,* Vol. 37B, Plenum Press, New York, 1991, pp. 995-1001.

Soyars, W.M., The Applicability of Constant Property Analysis in Cryogenic Helium Heat Exchangers, in *Advances in Cryogenic Engineering,* Vol. 37A, Plenum Press, New York, 1991, pp. 217-223.

Spradley, I.E., Nast, T.C., and Frank, D.J., Experimental Studies of MLI Systems at Very Low Boundary Temperatures, in *Advances in Cryogenic Engineering,* Vol. 35, Plenum Press, New York, 1990, pp. 477-486.

Strobridge, T.R., Cryogenic Refrigerators — An Updated Survey, National Bureau of Standards Technical Note 655, U.S. Government Printing Office, Washington, D.C., 1974, pg. 1-12.

Tatro, R.E. and Bennett, F.O., Internal Insulation for LNG, in *Advances in Cryogenic Engineering,* Vol. 20, Plenum Press, New York, 1975, pp. 315-326.

Tsukagoshi, T., Nitta, H., Yoshida, A., Matsumoto, K., Hashimoto, T., Kuriyama, T., Takahashi, M., Ohtani, Y., and Nakagome, H., Refrigeration Capacity of a GM Refrigerator With Magnetic Regenerator Materials, in *Advances in Cryogenic Engineering,* Vol. 41B, Plenum Press, New York, 1996, pp. 1623-1630.

Vonk, G., A New Type of Compact Heat Exchanger with a High Thermal Efficiency, in *Advances in Cryogenic Engineering,* Vol. 13, Plenum Press, New York, 1968, pp. 582-589.

Vuilleumier, R., U.S. Patent No. 1,275,507, 1918.

Walker, G., *Stirling Engines,* Oxford University Press, London, 1980.

Walker, G., *Cryocoolers, Part I: Fundamentals,* Plenum Press, New York, 1983a, pp. 186-187.

Walker, G., *Cryocoolers, Part 2: Applications,* Plenum Press, New York, 1983b, pp. 52-54.

Walters, K.G. and Fairbank, W.M., *Physics Reviews,* 103, 262, 1956.

Yaron, R., Shokralla, S., Yuan, J., Bradley, P.E., and Radebaugh, R., Etched Foil Regenerator, in *Advances in Cryogenic Engineering,* Vol. 41B, Plenum Press, New York, 1996, pp. 1339-1346.

Further Information

Other sources which may be consulted for additional information about cryogenic systems and heat transfer at cryogenic temperatures include the following:

Ackermann, R.A., *Cryogenic Regenerative Heat Exchangers,* Plenum Press, New York, 1997.
Barron, R.F., *Cryogenic Systems,* 2nd ed., Oxford University Press, New York, 1985.
Barron, R.F., *Cryogenic Heat Transfer,* Taylor and Francis, Washington, D.C., 1999.
Timmerhaus, K.D. and Flynn, T.M., *Cryogenic Process Engineering,* Plenum Press, New York, 1989.
Walker, G., *Cryocoolers,* Part 1 and Part 2, Plenum Press, New York, 1983.

4.13 Air-Conditioning Systems

Donald L. Fenton

Introduction

The term *air conditioning* refers to all equipment and processes used in maintaining comfortable and healthy indoor air suitable for human occupancy. This includes the adjustment or conditioning of the air, the distribution of air, air motion in the occupied space, and cleanliness of the air. This section of the handbook treats the thermal processes important in maintaining the temperature and humidity of the spaces occupied by humans. The acronym HVAC&R signifies heating, ventilating, air conditioning, and refrigerating where all these activities fall under the umbrella of air conditioning. Consequently, air conditioning is not restricted to only the cooling and dehumidifying of air.

Maintaining comfortable indoor air conditions requires understanding the behavior of the air and water vapor mixture — a subject called psychrometry. A space occupied by humans is enclosed by an insulated structure, which limits the energy gains in the summer and losses in the winter. These energy gains and losses are referred to as cooling loads and heating loads, respectively. They are the energy transfer rates that the heating and refrigerating equipment must meet to ensure indoor comfort. These loads are based on the determination of the severe, but likely, outdoor weather conditions typical for the specific geographical location. The introduction of conditioned air to the occupied space must be sufficient to offset the summer and winter loads. The heating and cooling equipment is generally located a distance away from the space, necessitating an air duct system for the conveyance of the conditioned air.

The inch-pound unit system is used, reflecting its prevalence in the HVAC&R industry in the U.S. Conversion to SI units may be made using suitable factors.

Properties of Moist Air

The thermodynamic properties of moist air and how these properties may change is the subject generally referred to as psychrometrics. Moist air is a binary mixture of dry air and water vapor. Harrison (1965) reports that the composition of dry air is nitrogen, 78.084%(v/v); oxygen, 20.9476%(v/v); argon, 0.934%(v/v); carbon dioxide, 0.0314%(v/v); neon, 0.001818%(v/v); helium, 0.000524%(v/v); methane, 0.00015% (v/v); sulfur dioxide, 0 to 0.0001%(v/v); hydrogen, 0.00005%(v/v); and other negligible gases such as krypton, xenon, and ozone. The apparent molecular weight of the dry air gas mixture is 28.9645. The dry air gas constant is computed by

$$R_a = \frac{R}{M} = \frac{1545.32}{28.9645} = 53.352 \frac{ft\ lbf}{lbm\ °R} \tag{4.13.1}$$

where the molecular weight value is based on carbon having a molecular weight of exactly 12.

Moist air is a mixture of dry air and water vapor. All other contaminants that may be present in atmospheric air are excluded from moist air. These contaminants may include smoke, pollen, dust, and gaseous pollutants generally present in ambient or outside air. The quantity of water vapor in moist air varies from none to a maximum amount that depends on the temperature and pressure of the air. This condition of maximum water vapor content is called saturation, which is an equilibrium condition between the moist air and condensed water. The molecular weight of water is 18.01528 and its gas constant is

$$R_{H_2O} = \frac{1545.32}{18.01528} = 85.778 \frac{ft\ lbf}{lbm\ °R} \tag{4.13.2}$$

The barometric pressure and temperature of atmospheric air changes with increasing altitude above sea level. Standard atmospheric conditions are defined at sea level where the pressure is 29.921 in. of mercury (in.Hg) and the temperature is 59°F. The U.S. Standard Atmosphere defines the variation of pressure and temperature with altitude and is generally used to estimate these properties at different altitudes. The troposphere (lower atmosphere) is taken to consist of dry air behaving as an ideal gas. The acceleration due to gravity is also taken to be a constant, 32.1740 ft/s^2. The pressure and temperature are given by

$$T = 59 - 0.00356616Z \tag{4.13.3}$$

$$p = 29.92\left(1 - 6.8753 \times 10^{-6} Z\right)^{5.2559} \tag{4.13.4}$$

where Z is the altitude (ft), p is the barometric pressure (in.Hg), and T is the temperature (°F). Equations (4.13.3) and (4.13.4) are accurate from sea level to an altitude of 36,000 ft. More extensive information including higher altitudes, other physical data, and SI units may be obtained from NASA (1976).

Thermodynamic Properties of Saturated Water

The thermodynamic saturated state is discussed in Chapter 3 of this handbook. Relationships giving the thermodynamic properties of saturated water have been developed by Hyland and Wexler (1983a) over the temperature range from –80 to 300°F. The determination of these properties is based on an ideal temperature scale almost identical to the generally used practical temperature scale used for measurements. As an example, the standard boiling temperature of water (at 14.696 psia or 29.921 in.Hg) is 211.95°F rather than the usual value of 212°F on the practical scale. The enthalpy and entropy properties of saturated liquid water have values defined as zero at the triple point, 32.018°F. Water at a temperature between its triple point and critical point may coexist in two states, liquid and vapor. When these phases are in equilibrium, all the physical features of the system remain constant with time and the vapor and liquid are saturated.

Over liquid water, the saturation pressure for the temperature range from 32 to 392°F is

$$\ln\left(p_{ws}\right) = \frac{C_1}{T} + C_2 + C_3 T + C_4 T^2 + C_6 \ln T \tag{4.13.5}$$

where $C_1 = -1.044039 \times 10^4$, $C_2 = -1.1294650 \times 10^1$, $C_3 = -2.7022355 \times 10^{-2}$, $C_4 = 1.2890360 \times 10^{-5}$, $C_5 = -2.4780681 \times 10^{-9}$, and $C_6 = 6.5459673$, and where T is the absolute temperature (°R) and the subscript "ws" indicates saturated water. The saturation pressure over ice in terms of temperature for a range from –148 to 32°F is

$$\ln\left(p_{ws}\right) = \frac{C_7}{T} + C_8 + C_9 T + C_{10} T^2 + C_{11} T^3 + C_{12} T^4 + C_{13} \ln T \tag{4.13.6}$$

where $C_7 = -1.0214165 \times 10^4$, $C_8 = -4.8932428$, $C_9 = -5.3765794 \times 10^{-3}$, $C_{10} = 1.9202377 \times 10^{-7}$, $C_{11} = 3.5575832 \times 10^{-10}$, $C_{12} = -9.0344688 \times 10^{-14}$, and $C_{13} = 4.1635019$. The ASHRAE *Handbook of Fundamentals* (1997) gives the thermodynamic properties of water at saturation.

Thermodynamic Properties of Moist Air

Hyland and Wexler (1983a; 1983b) developed formulas for the thermodynamic properties of moist air based on the "thermodynamic temperature scale," an ideal temperature scale almost identical to the generally used practical scale used for physical measurements. The ASHRAE *Handbook of Fundamentals* (1997) gives the thermodynamic properties of moist air based on the Hyland and Wexler equations. These thermodynamic properties are

Temperature (°F): T, related to the absolute temperature (°R) by T (°R) = T (°F) + 459.67.

Humidity Ratio: W_s, the condition when the water vapor phase is in thermodynamic equilibrium with the condensed water at a certain temperature and pressure. The humidity ratio for a given moist air condition may have a humidity ratio, W, such that it has a value of zero or greater but less than or equal to W_s.

Specific Volume (ft^3/lb): v_a, dry air specific volume.

Saturated Moist Air Specific Volume (ft^3/lb): v_s, volume of saturated moist air per pound of dry air.

Specific Enthalpy of Dry Air (Btu/lb dry air): h_a, reference value is zero at 0°F and 14.696 psia.

Saturated Moist Air Specific Enthalpy Air (Btu/lb dry air): h_s.

Condensed Water Specific Enthalpy (Btu/lb water): h_w, specific enthalpy of water in equilibrium with saturated air at a certain temperature and pressure. Reference value is zero at triple point of water and saturation pressure.

Water Vapor Pressure (psia): p_w, the actual partial pressure in moist air exerted by the water vapor.

Saturated Water Vapor Pressure (psia): p_s water vapor pressure in saturated air. Very nearly the same as the saturation vapor pressure of pure water, p_{ws}. Therefore, p_s and p_{ws} may be interchanged for one another.

Barometric Pressure (psia): p, atmospheric pressure.

Specific Entropy of Dry Air (Btu/lb°R): s_a, reference value is zero at 0°F and 14.696 psia.

Saturated Moist Air Specific Entropy (Btu/lb°R): s_s.

Condensed Water Specific Entropy (Btu/lb°R): s_w, specific entropy of water in equilibrium with saturated air. Reference value is zero at triple point of water and saturation pressure.

Numerical values of the properties of moist air are calculated on the basis of several assumptions. These are

1. The gaseous phase behaves as a mixture of ideal gases.
2. The liquid and solid phases contain no dissolved gases.
3. Equilibrium between the condensed phase and the gas mixture is not affected by the presence of the other component when the mixture and the condensed phase are at the same pressure and temperature.

Several additional properties and related mathematical relationships are useful when considering calculations relating problems regarding moist air. These are

Dry-Bulb Temperature (°F): T, temperature of air as indicated by a thermometer.

Thermodynamic Wet-Bulb Temperature (°F): T*, the temperature of water (may be liquid or solid) evaporating into moist air at a certain dry-bulb temperature T and humidity ratio W, that brings the air to saturation adiabatically at the same temperature T* maintaining the pressure at a constant value (Howell et al., 1998).

Wet-Bulb Temperature (°F): T_{wb}, the equilibrium temperature of water in a wet wick surrounding the bulb of a thermometer placed in an air stream. A psychrometer is used to measure simultaneously the dry-bulb and the wet-bulb temperatures using two thermometers. While this process involves the simultaneous transfer of heat and mass from the bulb, it is different than the adiabatic saturation process defining the thermodynamic wet-bulb temperature. However, the correction needed to convert wet-bulb temperatures to thermodynamic wet-bulb temperatures is small.

$$W = \frac{m_w}{m_a} \tag{4.13.7}$$

Humidity Ratio: the humidity ratio is calculated as the mass of water divided by the dry air in a sample of moist air.

$$W = \frac{0.62198\, p_w}{(p - p_w)} \tag{4.13.8}$$

Manipulation of the psychrometric and thermodynamic properties yields

$$W = \frac{(1093 - 0.556t^*)W_S^* - 0.240(T - T^*)}{1093 + 1.805T - 4.186T^*} \tag{4.13.9}$$

where W_s^* is the saturated humidity ratio if saturated at the wet-bulb temperature and where T and T* are in °F.

Degree of Saturation: μ, the ratio of the actual humidity ratio divided by the saturated air humidity ratio W_s at the same pressure and temperature.

$$\mu = \frac{W}{W_S} \tag{4.13.10}$$

Relative Humidity: φ, the ratio of the mole fraction water vapor in a given sample of moist air divided by the water vapor mole fraction of a sample at the same pressure and temperature. Using the assumption that the gas mixture may be considered as a mixture of ideal gases gives

$$\phi = \frac{p_w}{p_{ws}} \tag{4.13.11}$$

Dew-Point Temperature (°F): T_d, the temperature of a sample of moist air after having become saturated at the same pressure and humidity ratio as the original sample of moist air. Also, the dew-point temperature is the saturation temperature corresponding to the saturation pressure as listed in the thermodynamic properties of saturated water. For the temperature range from 32 to 150°F, the dew-point temperature is given by,

$$T_d = 100.45 + 33.193 \ln p_w + 2.319(\ln p_w)^2 + 0.17074(\ln p_w)^3 + 1.2063(p_w)0.1984 \tag{4.13.12}$$

and for temperatures less than 32°F,

$$T_d = 71.98 + 24.873 \ln\, p_w + 0.8927(\ln p_w)^2 \tag{4.13.13}$$

by ASHRAE (1997). Other properties include specific volume and enthalpy. The specific volume of moist air is given in units of dry air where the atmospheric pressure is the sum of the dry air and water vapor partial pressures. Thus,

$$\nu = \frac{R_a T}{(p - p_w)} \tag{4.13.14}$$

where p is the atmospheric pressure and p_w is the water vapor partial pressure. Combining the expression for W above gives an alternate relationship for ν sometimes useful in calculations,

$$\nu = 0.7543(T + 459.67)\frac{(1 + 1.6078W)}{p}. \tag{4.13.15}$$

The enthalpy is equal to the sum of the enthalpies of the dry air and the water vapor, or

$$h = h_a + Wh_g \tag{4.13.16}$$

where h_a is the enthalpy of the dry air and h_g is the specific enthalpy for saturated water vapor at the temperature of the moist air mixture. Approximate values for h_a (Btu/lb) and h_g (Btu/lb) are

$$h_a = 0.240T \tag{4.13.17}$$

$$h_g = 1061 + 0.444T \tag{4.13.18}$$

where T is the dry-bulb temperature. The resulting expression for the enthalpy h (Btu/lb) is

$$h = 0.240T + W(1061 + 0.444T) \tag{4.13.19}$$

The above relationships for the moist air mixture are generally adequate for air-conditioning design calculations.

Psychrometric Chart

The psychrometric chart displays simultaneously a number of the thermodynamic properties of moist air. An example psychrometric chart is shown in Figure 4.13.1, where several variables are plotted against temperature and the humidity ratio. Processes involving heat transfer and the addition or removal of moisture are readily shown on the psychrometric chart and so the chart is an aid in visualizing the process. Furthermore, the changes that occur in the moist air properties may be observed from the chart,

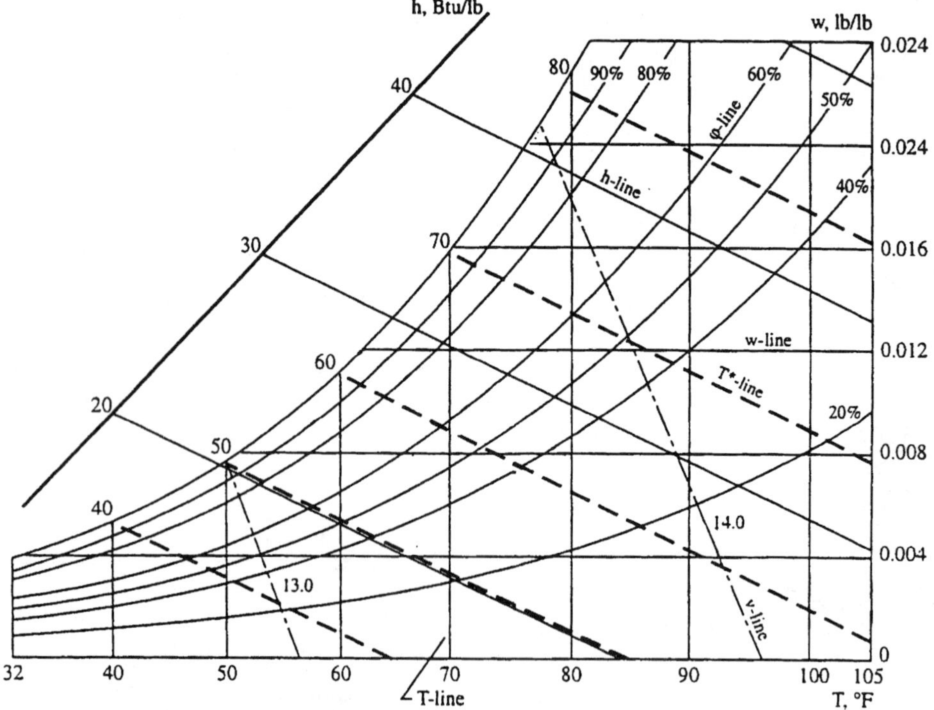

FIGURE 4.13.1 Psychrometric chart. (From Kreith, F., Ed., *The CRC Handbook of Mechanical Engineering,* CRC Press LLC, Boca Raton, FL, 1998. With permission.)

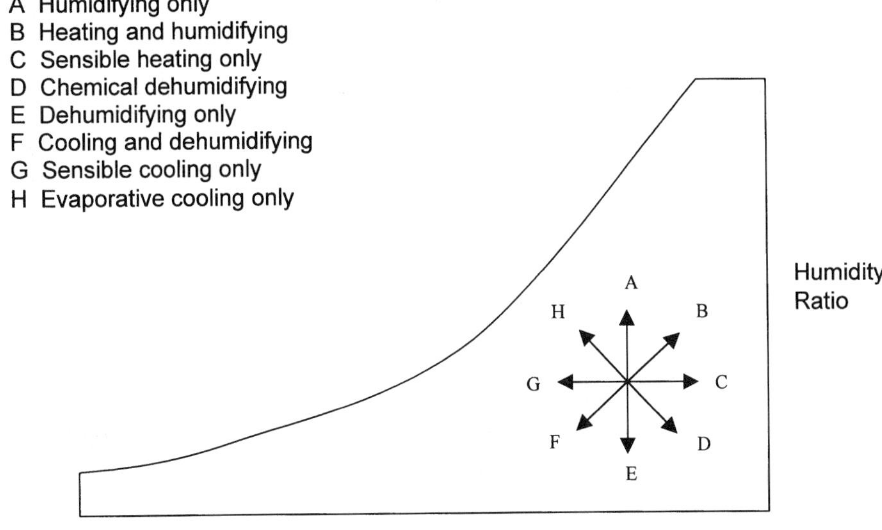

FIGURE 4.13.2 Air conditioning process directions on a psychrometric chart.

particularly changes in dry-bulb temperature, humidity ratio, and enthalpy. Observe that the horizontal movement across the chart involves only a change in dry-bulb temperature or a sensible change. On the other hand, a vertical movement on the chart represents a pure latent change where the dry-bulb temperature remains constant but the humidity ratio changes. Figure 4.13.2 shows a schematic view of the psychrometric chart where the process directions common to air conditioning are shown.

We now consider several examples where the use of the psychrometric chart is demonstrated.

Cooling and Dehumidifying Air

Moist air, when cooled to a temperature below its dew point, will condense a portion of the water vapor initially in the mixture. Equipment that could be used to carry out the cooling and dehumidifying process is shown in Figure 4.13.3, where a refrigerant enters and leaves an evaporator coil. The air inlet and exit conditions are defined as well as the condensed water drained from the pan below the coil. If the system operates on a steady-state flow basis (all variables remain constant with time) then the process may be described by the following water mass and energy equations:

$$\dot{m}_a W_1 = \dot{m}_a W_2 + m_w \tag{4.13.20}$$

$$\dot{m}_a h_1 = \dot{m}_a h_2 + \dot{m}_w h_w + q \tag{4.13.21}$$

where q is the heat transfer from the moist air to the coil. These two equations may be rearranged to give

$$\dot{m}_w = \dot{m}_a \left(W_1 - W_2\right) \tag{4.13.22}$$

$$q = \dot{m}_a \left[\left(h_1 - h_2\right) - \left(W_1 - W_2\right) h_w\right] \tag{4.13.23}$$

where the equation for q indicates that the cooling and dehumidifying process includes sensible and latent heat transfer. The sensible and latent heat transfer quantities are

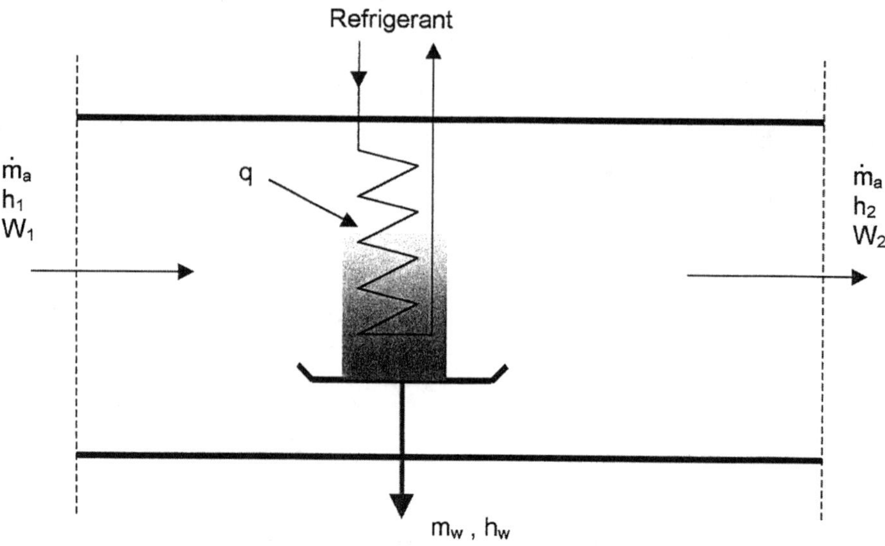

FIGURE 4.13.3 Schematic diagram of cooling and dehumidifying equipment.

$$q(sensible) = \dot{m}_a(h_1 - h_2) = \dot{m}_a c_p(T_1 - T_2) \tag{4.13.24}$$

$$q(latent) = \dot{m}_a(W_1 - W_2)h_{fg} \tag{4.13.25}$$

Figure 4.13.2 shows the direction of this process on a psychrometric chart where the process is noted to be a line sloping downward to the left. The sensible and latent heat transfers are noted by the horizontal and vertical changes in the condition of the air when passing from the inlet to the outlet, respectively.

Heating and Humidifying Air

The equipment shown in Figure 4.13.4 may be used to heat and humidify moist air during the winter heating season. Water mass and energy balances on the system yield

$$\dot{m}_a W_1 + \dot{m}_w = \dot{m}_a W_2 \tag{4.13.26}$$

$$\dot{m}_a h_1 + q + \dot{m}_w h_w = \dot{m}_a h_2 \tag{4.13.27}$$

which may be solved to give the humidity ratio and enthalpy of the exit air.

Adiabatic Mixing of Two Streams

Two streams of moist air converging into one stream are shown in Figure 4.13.5. If there is no heat transfer from the ducts to the surroundings, the mixing process is adiabatic, which is closely approximated by insulated ducts. The equations describing this process include an air mass balance, a water mass balance, and an energy balance, respectively.

$$\dot{m}_{a1} + \dot{m}_{a2} = \dot{m}_{a3} \tag{4.13.28}$$

$$\dot{m}_{a1} W_1 + \dot{m}_{a2} W_2 = \dot{m}_{a3} W_3 \tag{4.13.29}$$

$$\dot{m}_{a1} h_1 + \dot{m}_{a2} h_2 = \dot{m}_{a3} h_3 \tag{4.13.30}$$

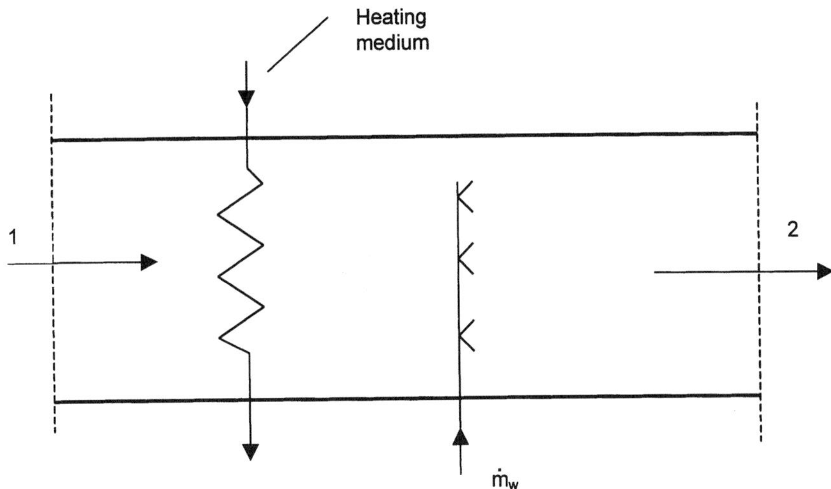

FIGURE 4.13.4 Schematic diagram of typical heating and humidifying equipment.

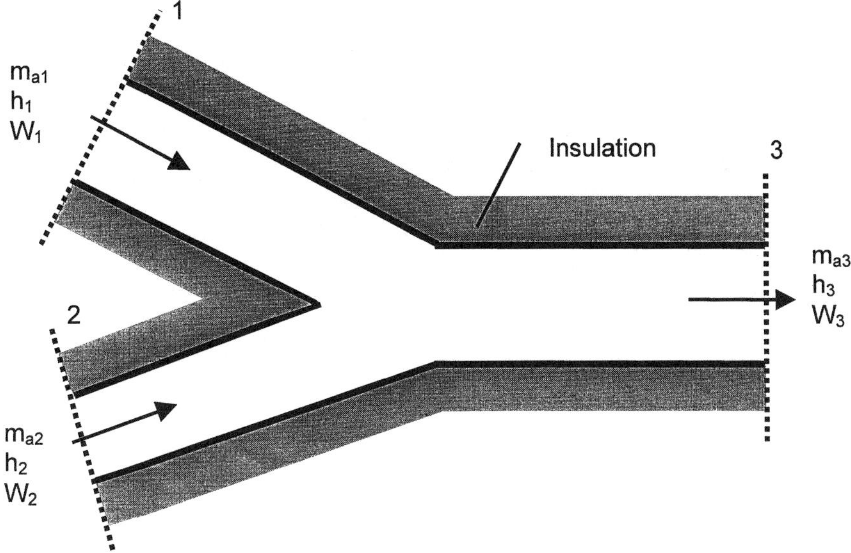

FIGURE 4.13.5 Adiabatic mixing of two streams of moist air.

The above three equations indicate that the leaving humidity ratio and enthalpy are weighted averages of the entering corresponding values. In practice, the approximation is made that the exit temperature is also a weighted average of the entering temperatures where the error, due to the variation of the specific heats for the moist air, is typically less than 1%. Consequently, the exit condition of the moist air lies on a straight line connecting the two entering conditions, or that

$$\frac{\dot{m}_{a2}}{\dot{m}_{a1}} \cong \frac{distance(1)to(3)}{distance(2)to(3)} \tag{4.13.31}$$

Figure 4.13.6 shows the adiabatic mixing process path on a schematic of the psychrometric chart. The above three equations may also be combined to give

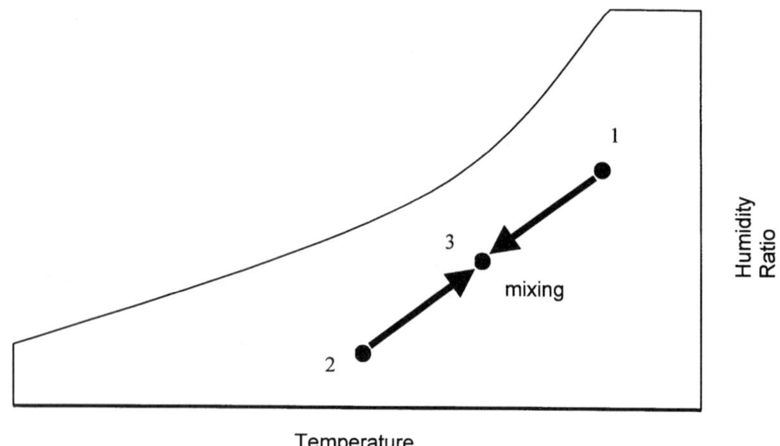

FIGURE 4.13.6 Adiabatic mixing of two streams.

$$\frac{\dot{m}_{a2}}{\dot{m}_{a1}} = \frac{h_3 - h_1}{h_2 - h_3} = \frac{W_3 - W_1}{W_2 - W_3} \cong \frac{T_3 - T_1}{T_2 - T_3} \tag{4.13.32}$$

Example — Consider the mixing of two streams of moist air where one entering stream has a volumetric flow rate of 5000 cfm with a 40°F dry-bulb temperature and a 35°F wet-bulb temperature. The other inlet stream has a volumetric air flow rate of 15,000 cfm with 75°F dry-bulb temperature and 50% relative humidity. Determine the exit conditions for the leaving air stream.

Solution — Label the 5000 cfm air stream as "1" and the 15,000 cfm stream as "2." The exit stream is labeled as "3." From the psychrometric chart, Figure 4.13.1, the following information is observed:

$$v_1 \cong 12.65\, ft^3/lb,\ h_1 \cong 14.0\, Btu/lb$$

$$v_2 \cong 13.68\, ft^3/lb,\ h_2 \cong 28.0\, Btu/lb$$

Thus,

$$\frac{\dot{m}_{a1}}{\dot{m}_{a2}} = \frac{\dfrac{\dot{V}_1}{v_1}}{\dfrac{\dot{V}_2}{v_2}} = \frac{\dfrac{5000}{12.65}}{\dfrac{15000}{13.68}} = 0.3605$$

$$0.3605 = \frac{h_3 - h_2}{h_1 - h_3} = \frac{h_3 - 28.0}{14.0 - h_3}$$

$$h_3 = 24.3\ Btu/lb$$

and from the psychrometric, observe that T_3 is approximately 66°F.

Example — Consider the air-conditioned space shown in Figure 4.13.7, where moist air leaves at 80°F dry-bulb temperature and 65°F thermodynamic wet-bulb temperature. Due to human occupancy, the moisture gain to the space occurs at the rate of 10.5 lb/h as saturated water vapor at 90°F. The space

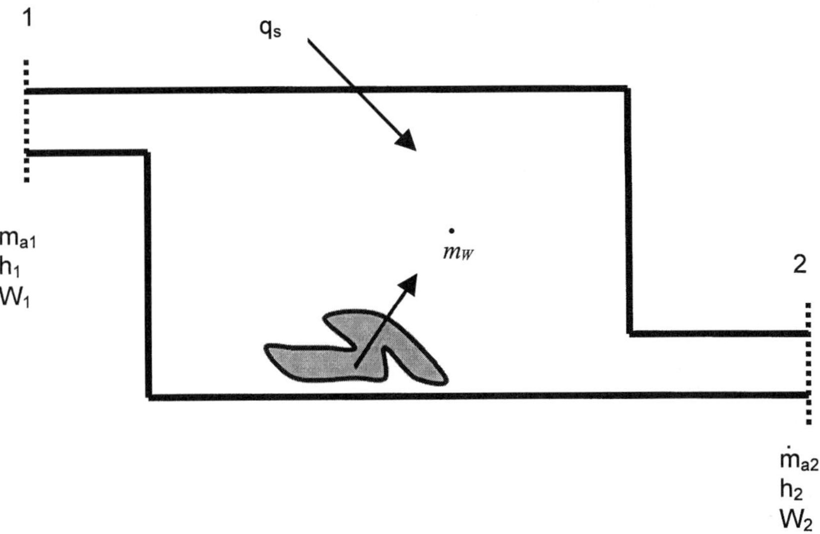

FIGURE 4.13.7 Schematic diagram of an air-conditioned space.

also has a sensible gain of 20,000 Btu/h. Moist air is introduced into the space at 55°F. Estimate the needed thermodynamic wet-bulb temperature and the volumetric flow rate of the supply air.

Solution — The sensible heat gain does not involve any heat transfer due to changes in moisture content of the air. The water vapor added to the space adds energy equal to its specific enthalpy. The governing equations involve air, water, and energy balances

$$\dot{m}_{a1} = \dot{m}_{a2} = \dot{m}_a$$

$$\dot{m}_{a1} W_1 + \dot{m}_w = \dot{m}_{a2} W_2$$

$$\dot{m}_{a1} h_1 + \dot{m}_w h_w + q_s = \dot{m}_a h_2$$

where q_s is the added sensible energy, $m_w h_w$ is the latent load due to the moisture content change, and $\dot{m}_a$ is the dry air flow rate entering and leaving the space. Rewriting the energy balance gives

$$q_s + \dot{m}_w h_w = \dot{m}_a (h_2 - h_1)$$

and rearranging the water balance yields

$$\dot{m}_w = \dot{m}_a (W_2 - W_1).$$

Compute the ratio:

$$\frac{\textit{Total energy added to space}}{\textit{Total water added to space}} = \frac{q_s + \dot{m}_w h_w}{\dot{m}_w}$$

$$= \frac{\dot{m}_a (h_2 - h_1)}{\dot{m}_a (W_2 - W_1)} = \frac{\Delta h}{\Delta W}$$

where the ratio is observed to also equal the enthalpy change divided by the humidity ratio change. A protractor on the psychrometric chart (Figure 4.14.8) shows the slopes associated with the direction that this process takes on the chart. All the possible supply air conditions lie on this line. From the saturated water thermodynamic property tables, h_w = 1100.4 Btu/lb. Thus,

$$\frac{\Delta h}{\Delta W} = \frac{20,000 + (10.5)(1100.4)}{10.5} = 3005 \cong 3000$$

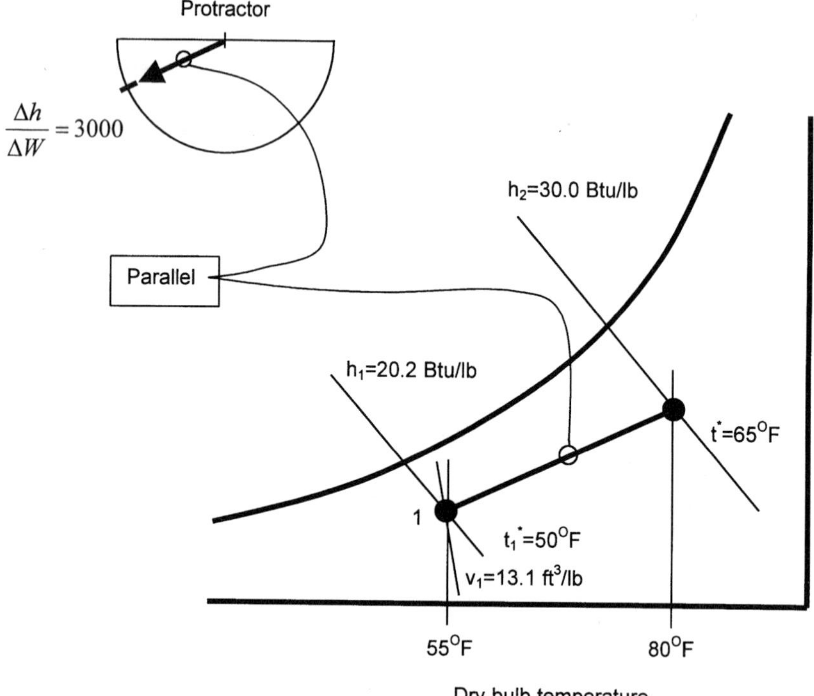

FIGURE 4.14.8 Schematic of the psychrometric chart for an air-conditioned space.

Draw a line through state "2" at a slope equal to Δh/ΔW = 3000. Where this line intersects the supply dry-bulb temperature identifies the wet-bulb temperature, which is approximately 50°F. To obtain the volumetric flow rate at state "1," first calculate the mass flow rate.

$$\dot{m}_a = \frac{q_s + \dot{m}_w h_w}{(h_2 - h_1)}$$

Estimate from the psychrometric chart that h_1 = 20.2 Btu/lb (dry air) and h_2 = 30.0 Btu/lb (dry air). Thus,

$$\dot{m}_a = \frac{20,000 + (10.5)(1100.4)}{(30.0 - 20.2)} = 3220 \frac{lb(dry\ air)}{h} \ or\ 53.7 \frac{lb(dry\ air)}{\min}.$$

The volumetric flow rate is

$$\dot{V}_1 = \dot{m}_a \mathrm{v}_1$$

and from the psychrometric chart, v_1 = 13.1 ft³/lb (dry air). Thus,

$$\dot{V}_1 = (53.7)(13.1) = 703 \frac{ft^3}{\text{min}}.$$

Thermal Comfort Conditions

The now widely accepted definition for human comfort is "that condition of mind that expresses satisfaction with the thermal environment" (ASHRAE 1992). Thus, the quantification of human comfort is nebulous in that a person's feelings are important. A number of inside environmental conditions influence human thermal comfort and include dry-bulb temperature, wet-bulb temperature, dew-point temperature, humidity ratio, air velocity, and the mean radiant temperature. The mean radiant temperature is a representative temperature for all the individual surfaces that exist in the occupied space. The amount and type of clothing affects human comfort. Furthermore, several environmental temperature indices are used to assist with the quantification of thermal comfort. These indices are equivalent temperature, effective temperature, humid operative temperature, the globe temperature, and the resultant temperature, which is applied sometimes in the U.K. As an example of one of these temperature indices, the human operative temperature T_{oh} is the temperature when 100% relative humidity yields the same total heat loss from the skin as would occur for the actual environment (ASHRAE *Handbook of Fundamentals*, 1997). Specifically,

$$T_{oh} = T_o + w i_m LR\left(p_v - p_{oh,s}\right) \tag{4.13.33}$$

where T_o is the operative temperature, w is the wet skin fraction, i_m is the moisture permeability index, LR is the Lewis relation, p_v is the partial pressure of water vapor, and $p_{oh,s}$ is the saturated vapor pressure at T_{oh}. The operative temperature is given by

$$T_o = \frac{h_r T_{mr} + h_c T_a}{h_r - h_c} \tag{4.13.34}$$

and the Lewis relation is given by

$$LR = \frac{h_e}{h_c} \tag{4.13.35}$$

where h_c is the convective heat transfer coefficient, h_r is the linearized radiative heat transfer coefficient, h_e is the evaporative heat transfer coefficient, T_{mr} is the mean radiant temperature of surfaces in the space, and T_a is the space air temperature. Note that the operative temperature is the average of the mean radiant and air temperatures weighted by their respective heat transfer coefficients. While these equations and definitions reveal the complexities associated with quantifying human comfort parameters, additional details are needed to make calculations. Refer to the ASHRAE *Handbook of Fundamentals* (1997) for more information.

ASHRAE Standard 55 (1994) defines comfort zones which specify the conditions for reasonable comfort for human occupants either sedentary or slightly active. These comfort zones carefully incorporate the results obtained through research and are different for winter and summer due to the clothing typically worn during these periods. The boundaries of the comfort zones are not sharp, with the upper and lower humidity levels being less precise than the left and right boundaries associated with temperature. Indoor conditions of temperature and humidity for the conditioned space are usually selected to comply with the ASHRAE Standard 55 (1994) limits. The specific influence of humidity on human comfort is explored in greater detail by Berglund (1998).

Several models have been reported in the literature that predict the level of thermal comfort that is likely to occur in a given situation and space. Fanger (1982) developed a predictive method based on a steady-state energy balance and Gagge (1971, 1986) devised a two-node model suitable for predicting steady-state human comfort responses. These two predictive models are summarized in the ASHRAE *Handbook of Fundamentals* (1997).

Load Calculations

In the context of HVAC (heating, ventilating, and air conditioning) the term load is applied to a building to designate the thermal energy removal or addition rate needed to maintain the desired inside air conditions. For summer months, the removal of thermal energy from the space is required, typically performed by a refrigeration system. Spaces during winter months generally require the addition of thermal energy to maintain the desired inside conditions. Both summer cooling and winter heating involve systems accomplishing the conditioning of the inside air. While the phrase "thermal energy" is technically correct for explaining the air warming and cooling processes, it is common in the industry to use the term "heat" to refer to this energy. The discussion here follows the industry practice of using the term heat.

For cooling loads originating from human comfort requirements in buildings, the absorption of solar radiation into the space is significant. Special procedures have been developed by ASHRAE to accurately predict cooling loads. The current procedures (Pederson et al. 1998) are improvements of earlier methods (ASHRAE 1975; McQuiston and Spitler, 1992). The cooling load is a transient phenomenon — the heat transfer rate through the space envelope varies with time. An enclosed space where the interior temperature is controlled by a single thermostat is usually referred to as a zone. A large nonresidential building may be composed of many zones sharing walls between them.

The heat balance concept is essentially the application of a control volume around a zone of a building. To be specific, a heat balance zone is an air volume, uniform in temperature, which also includes all the heat transfer and thermal storage surfaces bounding or inside the air volume. While the concept of a zone originates with the thermal behavior of the space, several rooms may comprise a zone. Family residential buildings are usually treated as being one zone. Observe that the wall treatments on the interior surfaces and the furnishings characterize the heat balance zone. The typical zone is a 12-surface volume that is used in determining loads where complex geometry of an actual building is reduced to a simple shape accurately duplicating the zone's thermal features. The important information includes:

- Orientation of each surface and configuration area
- Construction features of each surface
- Outside and inside surface environmental conditions

The heat balance approach to load calculations necessitates the use of a digital computer. While it is possible to perform approximate load calculations (more so with the heat load), it is now considered impractical due to the prevalent accessibility of computers.

The migration of moisture through the materials comprising the building's envelope is of concern as well. The accumulation of moisture within the walls of a structure must be avoided to maintain structural integrity, insulating capacity, and indoor air quality. The growth of molds and other biological items are accelerated by the presence of trapped moisture in walls, which in turn degrade the interior air quality. Hosni et al. (1998) have investigated this subject concerning the special concerns that result from high moisture climates such as southern Texas in the U.S. Additionally, Burch and Chi (1997) have developed a moisture migration model, used and partially experimentally verified by Hosni et al. (1998), quantifying the water migration rates in wall structures.

Cooling Load

The cooling load associated with a zone is difficult to accurately predict. For this reason, ASHRAE has continued to improve the available methods (ASHRAE *Handbook of Fundamentals*, 1997) for the prediction of cooling load. The use of the heat balance concept, recently developed by Pederson et al.

(1998), has simplified the calculation procedures while maintaining adequate accuracy. This is accomplished by performing four heat balances as described below. The cooling load calculations are sufficiently complex that digital computing machines are needed to make the calculations.

Heat Balance on Outside Surface

A heat balance applied to the outer surface of an exterior wall includes three sources of heat: direct and diffuse radiation from the sun (shortwave radiation), radiation from the environment (longwave radiation), and convection heat transfer. The heat balance on the outside surface is

$$q''_{asol} + q''_{lwr} + q''_{co} - q''_{ko} = 0 \tag{4.13.36}$$

where q''_{asol} is the absorbed direct and diffuse solar radiant heat flux (Btu/h ft^2), q''_{lwr} is the net longwave radiant flux from the air and objects surrounding the outside surface, q''_{co} is the convective heat flux from the air to the surface, and q''_{ko} is the conductive heat flux leaving the surface entering the wall material. All the heat transfer terms are positive with the exception of the conduction term. Inside the wall, the temperature gradient is negative causing the conduction heat flux term to be positive. The solar radiation, longwave radiation, and convection heat fluxes are sometimes combined using a procedure resulting in the sol-air temperature (ASHRAE *Handbook of Fundamentals*, 1997). However, ASHRAE's implementation of the heat balance method involves straightforward models for solar, environmental, and outside convection to determine the outside surface heat balance (ASHRAE *Handbook of Fundamentals*, 1997).

- q''_{asol} is determined by procedures identified by ASHRAE (*Handbook of Fundamentals*, 1997).
- q''_{lwr} is estimated using the traditional radiant energy exchange models between surfaces. ASHRAE provides guidance giving values for the sky temperature, ground temperature, surface temperature, surface absorptance, and ground view factors.
- q''_{co} is calculated using Newton's law of cooling where the temperature difference between the outside air and outside surface temperature is given by

$$q''_{co} = h_{co}\left(T_{air} - T_{os}\right) \tag{4.13.37}$$

  where h_{co} is the outside convection heat transfer coefficient dependent on the velocity of the air next to the surface, and the temperatures T_{air} and T_{os} are temperatures of the outside air and outside surface.

The required outdoor conditions are the dry-bulb and the wet-bulb temperatures. The ASHRAE *Handbook of Fundamentals* (1997) shows these temperatures for a number of locations throughout the U.S. and the world. Cooling design day dry- and wet-bulb temperatures for numerous geographical locations are compiled from weather data. Summer climatic data are based on hourly weather observations over at least a 12-year period. The summer month conditions are reported on the basis of 0.04, 1.0, and 2.0 annual percentiles. These annual percentiles represent, respectively, 35, 88, and 175 h from a year of 8760 h. Typically, July 21 in the Northern Hemisphere is taken as the summer design day.

Interior Wall Conduction

Heat conduction through the material comprising the wall may be determined in several ways. Prevalent methods now include numerical finite difference, numerical finite element, transform methods, and time series methods. All of these methods introduce a time-dependence feature into the cooling load calculation where the inside surface and outside surface temperatures and heat transfer rates are time dependent. Direct analysis of the heat transfer has inputs consisting of two temperature functions and outputs consisting of the two heat fluxes. While any of the four mentioned methods yield usable results, Pederson et al. (1998) employ a conduction transfer function (CTF) procedure for reasons of adequate generality and good computation speed. The CTF factors in this formulation may be considered as response factors associated with the wall.

Heat Balance on Inside Surface

A heat balance on the inside surfaces involves four heat transfer terms: conduction through the wall material, convection to the inside air, shortwave radiant absorption and reflection, and longwave radiant exchange. The heat balance on the inside surface is

$$q''_{ki} + q''_{lwx} + q''_{sw} + q''_{lws} + q''_{sol} + q''_{ci} = 0 \tag{4.13.38}$$

where

q''_{ki} = the conduction heat flux to the inside surface from the interior of the wall material
q''_{lwx} = the net longwave radiant flux between the zone surfaces which involves emittance and absorption from all surfaces in the zone space
q''_{sw} = the net shortwave radiant flux to the inside surface from lights
q''_{lws} = the longwave radiant heat flux from the equipment in the zone
q''_{sol} = the transmitted solar radiation flux absorbed at the inside surface
q''_{ci} = the convective heat flux from the surface to the interior air in the zone

A common approximation used in cooling load calculations, also used by Pederson et al. (1998), is to treat the air as completely transparent to longwave radiation.

Furnishings that may be present inside the zone influences the time response characteristics of the space. This influence is complicated to quantify and is under investigation. The furniture in a space adds surface area which participates in the radiant and convective heat exchange processes tending to reduce the time response. On the other hand, the furnishings also add thermal mass to the space, which increases the time response of the space. The inside surface heat balance permits these two features to be accommodated in a reasonable fashion in that both the surface area and the thermal mass are included in the heat transfer processes.

The longwave radiation from internal sources is traditionally estimated by establishing a division between the radiation and the convection heat transfer for the heat that originates from the equipment. The portion assigned to radiation is then distributed among the surfaces in some manner (ASHRAE *Handbook of Fundamentals*, 1997). While this approach is awkward and deviates from the heat balance approach, it is practical in that it avoids the specification of location and temperature of all equipment in the space. The shortwave radiation from the lights and the transmitted solar energy is also distributed to the surfaces in the space in some manner (ASHRAE *Handbook of Fundamentals*, 1997).

The convective heat flux from the inside surface is calculated using Newton's law of cooling

$$q''_{ci} = h_{ci}\left(T_a - T_{is}\right) \tag{4.13.39}$$

where h_{ci} is the inside surface convective heat transfer coefficient, T_a is the zone air temperature, and T_{is} is the inside surface temperature. Numerical values for h_{ci} are available (ASHRAE *Handbook of Fundamentals*, 1997), but are based on carefully controlled natural convection experiments. Present-day conditioned spaces are mechanically ventilated, suggesting that the actual inside surface convective heat transfer coefficients may be somewhat different.

Inside Air Heat Balance

The thermal capacitance is generally neglected for the heat balance applied to the air in the space. With this simplification, the air heat balance is performed on a quasi-steady basis for each calculation time increment. The four heat transfers associated with the heat balance are

$$q_{ci} + q_{ce} + q_{iv} + q_{sys} = 0 \tag{4.13.40}$$

where q_{ci} is the heat transfer rate from the surfaces in the zone, q_{ce} is the convection heat transfer rate originating from the internal loads, q_{iv} is sensible load from the infiltration of outside air, and q_{sys} is the heat transfer to the HVAC system. The convection from the internal surfaces is obtained by summing the contributions from each surface:

$$q_{ci} = \sum_{i=1}^{n} h_{ci} A_i \left(T_a - T_{is}\right) \tag{4.13.41}$$

where n is the number of surfaces in the space, h_{ci} is the convection heat transfer for surface "i," A_i is the surface area "i" from which the heat transfer q_{ci} occurs, T_{is} is the temperature of surface "i," and T_a is the inside air temperature. The convection heat transfer from the internal loads q_{ce} is simply added to the heat balance on the air in the space.

This is not strictly correct because the temperatures of these surfaces that produce the internal loads actually add the heat to the air by convection. Unfortunately, this is very difficult to incorporate in the heat balance calculation because the detailed information necessary is usually not obtainable. With the air heat balance, any air that enters the space from the outside is assumed to mix instantaneously with the inside air. The ASHRAE *Handbook of Fundamentals* (1997) presents the latest information on the methods available for estimating the air infiltration rate. Sometimes, the infiltration rate is converted to a parameter equal to the number of air changes per hour (ACH) which would be included in the air heat balance applying the outside air temperature for that time increment.

Heating Load

Air conditioning of a space usually includes the design of a heating system. The heating system provides thermal energy that is then delivered to the space, maintaining the desired space temperature. Additionally, moisture may be added to maintain the relative humidity within acceptable limits.

The actual design of the heating system is dependent on the largest expected heat loss from the space to be heated. If the space is comprised of multiple spaces, each separately controlled by a thermostat, the heat loss is the sum of the largest individual heat losses for each space. Contributions to the heat loss include:

- Heat transfer through walls, floor, ceiling, glass, and all other surfaces
- Heat needed to warm the air from the outside entering the heated space
- Heat required to warm or thaw materials brought into the space

The outdoor temperature, wind velocity, and sunlight vary with time which causes the heat loss rate from the space to also vary with time. However, when determining the heat loss rate, or heat load, the outside temperature is selected as a design temperature originating from climatic data for that location. During winter, the coldest days are typically those that involve cloudy and stormy weather where the outdoor temperature has been noted to remain fairly constant. As a consequence, the heat loss rate from the space is also nearly constant. This is especially true when the internal heat gains are small. For this reason, the design heat load is determined on the basis of steady-state heat transfer and one-design outdoor temperature.

The heating system, as regulated by the thermostat, provides only the quantity of heat required to replace the heat that is lost. As the weather conditions change, the amount of heat supplied will vary. The heating system is not designed to withstand the most severe weather conditions ever recorded. To do so would result in a heating system that would rarely operate at full capacity. Rather, it is considered acceptable to not meet the inside temperature requirement for short periods of time. Consequently, the design outdoor temperature and the thermal characteristics of the zone govern the heat load for the space. A related activity is to estimate the energy consumed by the heating system over a year. To do this, simulation studies are made using hourly weather conditions where the heat loss is then estimated on an hourly basis. The sum of all the hourly heat losses over the heating season is the annual energy consumption needed to heat the space.

Many thermostats incorporate a temperature setback feature to reduce energy consumption of the heating system. The temperature is lowered several degrees at night and raised during the day when the space is occupied. It turns out that the early morning hours is the time when the heat load is the largest. Additionally, the morning time is typically when the thermostat calls for increasing the temperature. As a consequence, the heating system may need considerable time to meet the indoor temperature demand, during which time the occupants may experience discomfort. The time interval needed to warm the space after thermostat setback may be estimated using the heat balance method (Pederson et al., 1998).

The determination of the heat load assumes that no solar energy enters the space, internal heat gains are negligible, and that the outdoor and indoor temperatures are constant. The calculation of the design heat load from a space is as follows:

1. Choose outdoor design conditions: temperature, humidity, wind direction, and wind speed.
2. Choose indoor design conditions.
3. Determine temperatures in any adjacent unheated spaces, if any.
4. Choose the heat transmission coefficients and calculate the heat loss rates for the walls, floors, ceilings, glass, doors, and all other surfaces where heat is transferred to the outside.
5. Calculate the heat load resulting from the infiltration of air and all other outdoor air entering the space.
6. Sum all the contributions to the heat load from transmission and infiltration.

Outdoor Design Conditions

Heating design-day dry and wet bulb temperatures along with wind speed for numerous geographical locations are compiled from recent data by ASHRAE (*Handbook of Fundamentals*, 1997). Winter climatic data are based on hourly weather observations over at least a 12-year period. Winter month conditions are reported on the basis of 99.6 and 99.0 annual percentiles. These annual percentiles represent, respectively, 35 and 88 h from a year of 8760 h. For example, Asheville, NC has a heating dry-bulb temperature equal to 11°F at the 99.6 percentile, which means that for 35 h out of the year the temperature is likely to be below 11°F. Other climatic data, if available for a particular condition, suggesting lower dry-bulb temperatures and/or higher wind velocities should be considered in selecting the design conditions.

Indoor Design Conditions

Generally, the design dry-bulb temperature is chosen from the lower portion of the acceptable range, therefore preventing oversizing of the heating system. A heating system operating at conditions less than the design load does so at less efficiency. For this reason, a heating design dry-bulb temperature equal to 70°F is usually chosen. The relative humidity level is typically set at maximum value of 30% to provide, on the one hand, a healthy environment, while on the other hand assure moisture integrity of the building's envelope.

Heat Transmission Loss

Heat transfer through all the building envelope materials comprising the walls, ceiling, roof, floors, windows, and doors is sensible and is calculated using

$$q = UA\left(T_i - T_o\right) \tag{4.13.42}$$

where U is the overall heat transfer coefficient or U-factor (Btu/h ft^2°F), A is the normal surface area through which the heat transfer passes (ft^2), T_i is the inside design temperature (°F), and T_o is the outside design temperature (°F). The U and A values are generally different for walls, ceilings, roofs, and windows and so separate calculations are needed for each. Heat transmission loss through floor slabs and below-grade walls is determined using

$$q = U_s P(T_i - T_o) \tag{4.13.43}$$

where U_s is the overall heat transfer coefficient based on the slab perimeter (Btu/h ft^2°F) and P is the slab perimeter (ft). Approximate numerical values for U_s are given by ASHRAE (*Handbook of Fundamentals*, 1997).

Infiltration

Every building envelope leaks inside air to the outside. This results in a heat loss because the air that is lost is replaced by cold outside air. The load on the heating system is dependent on the amount of heat needed to warm the infiltrated outside air to the inside design temperature and on the moisture added to raise the humidity to the indoor design condition. The sensible heating load is determined by

$$q_s = c_p \dot{V} \rho (T_i - T_o) \tag{4.13.44}$$

where c_p is the specific heat of the air (Btu/lb °F), $\dot{V}$ is the volumetric flow rate of air entering the building (ft^3/h), and ρ is the air density at temperature T_o (lb/ft^3). The volumetric air flow rate entering the building depends on the wind speed and direction and on the size, among other factors, of the openings through which the infiltration air passes. The ASHRAE *Handbook of Fundamentals* (1997) carefully explains the determination of the volumetric flow due to infiltration. The latent heat loss due to infiltration is given by

$$q_l = \dot{V} \rho (W_i - W_o) h_{fg} \tag{4.13.45}$$

where ρ is the air density at temperature T_i (lb/ft^3), W_i is the indoor air humidity ratio, W_o is the outdoor air humidity ratio, and h_{fg} is the latent heat of vapor at temperature T_i (Btu/lb).

Air Distribution System Heat Losses

When the ducts conveying warm air pass through spaces outside the conditioned space, heat losses will occur. The heat loss from supply air ducts contributes to the load for the space. While these losses may be reduced by insulating the duct, the heat loss cannot be completely eliminated. The heat loss through a duct may be estimated by

$$q = UA_s \, \Delta(T_a - T_u)_m \tag{4.13.46}$$

where U is the overall heat transfer coefficient (Btu/h ft^2°F), A_s is the outside surface area of the duct (ft^2), $\Delta(T_a - T_u)_m$ is the mean temperature difference between the air in the duct, T_a and the temperature of the unconditioned space outside the duct, T_u. For a relatively short duct that is overlaid with 1 or 2 in. of insulation and covered with a reflective material, the mean temperature difference may be approximated by the simple difference between the supply air temperature and the unconditioned space temperature. In the HVAC industry, designers typically assume a duct heat loss equal to 2 to 5% of the sensible load of the conditioned space.

Auxiliary Heat Gains

Within a conditioned space, numerous sources of heat may exist. Examples of common heat sources are lights, motors, machinery, and people. Usage patterns of equipment and people occupancy levels influence the magnitude of the auxiliary heat gain rate. However, as a practical matter, auxiliary heat gains are ignored in those situations where the contribution to the total load is small. An exception is industrial space where equipment, lighting, and occupancy may result in a significant heat gain. ASHRAE (*Handbook of Fundamentals*, 1997) provides guidance in assessing the contributions from a number of auxiliary sources.

The ASHRAE *Handbook of Fundamentals* (1997) gives specific procedures for space heat load predictions for both residential and nonresidential buildings. The heat losses that occur through residential building slabs, crawl spaces, attics and the like are well characterized. With nonresidential buildings, ASHRAE provides the latest information on auxiliary heat gains that may be helpful for an accurate prediction of the space heat load.

Refrigeration

The refrigeration system, sometimes called a heat pump, is the equipment in the HVAC system that provides the cooling for the indoor space. Refrigeration is applied to many fields all sharing the need for a source of low temperature. The application of refrigeration to air conditioning involves the removal of heat from the space, thereby maintaining the desired inside temperature. The refrigeration system accomplishes the absorption of heat from the conditioned space by maintaining a surface in the space at a temperature lower than that of the space. Two refrigeration system types are prevalent in air conditioning: the vapor compression system and the absorption system. The vapor compression system is available and popular in sizes applicable for single rooms to those appropriate for large commercial buildings composed of many individual zone spaces.

The capacity of a refrigeration system may be given in terms of kW or Btu/h. However, by tradition dating back to the early 1800s when ice was harvested for use during the summer in the U.S., the term "ton of refrigeration" was used. It is still used today. A one-ton refrigeration plant has the capacity to produce one ton of ice per day (24 h). The latent heat of ice is 144 Btu/lb and so the capacity of a one-ton plant is equivalent to 12,000 Btu/h or 200 Btu/min.

A measure of a refrigeration system's performance is called the coefficient of performance or COP, defined by

$$COP = \frac{useful\ refrigerating\ rate}{net\ energy\ input\ rate}. \tag{4.13.47}$$

Another measure is the refrigerating efficiency, η_R which is the ratio of the system's actual COP to that system's ideal COP,

$$\eta_R = \frac{COP}{(COP)_{ideal}} \tag{4.13.48}$$

The power input required to operate the refrigeration system relative to the refrigerating rate or "effect" is useful when comparing refrigeration equipment of different capacities. A common parameter is the horsepower per ton, given by

$$hp/ton = \frac{4.715}{COP}. \tag{4.13.49}$$

Carnot Refrigeration Cycle

The Carnot cycle is a theoretical cycle that is useful because no cycle can be constructed operating under the same temperature conditions with a greater operating efficiency. All the processes in the Carnot cycle are thermodynamically reversible (i.e., all the heat transfers occur at zero temperature difference, no friction occurs in any of the components). This cycle consists of two isothermal processes and two adiabatic processes. The Carnot refrigeration cycle appears as a rectangle on a temperature-entropy

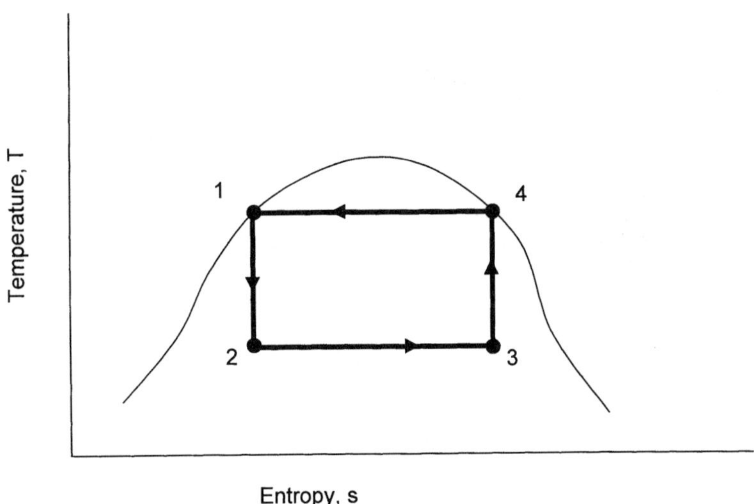

FIGURE 4.13.9 Temperature entropy diagram for the Carnot refrigeration cycle.

diagram shown in Figure 4.13.9. Observe that the cycle moves through the processes in a counterclockwise direction. Heat absorption occurs during the process occurring between points "2" and "3" and that heat rejection occurs between points "4" and "1." Heat absorption takes place in or adjacent to the space kept cool and heat rejection is made at a location outside the conditioned space. The expansion process identified as process "1" to "2" extracts work from the expanding fluid which in turn partially drives the compressor. The COP for the Carnot cycle is

$$COP = \frac{T_c}{T_h - T_c} \tag{4.13.50}$$

where T_c is the cold temperature and T_h is the hot temperature. The hot temperature is the temperature of the surrounding environment to which the heat is rejected.

Figure 4.13.9 also shows the saturated liquid and vapor lines for the working fluid used in the Carnot refrigeration cycle. The heat absorption and rejection processes represented by horizontal lines in the diagram may be achieved by working fluids undergoing a phase change. This cycle, however, is not practical for two important reasons. The compression process, from "3" to "4," occurs with liquid entering the suction side (inlet) of the compressor. The liquid, if allowed to enter the compressor, will cause very severe damage. The other reason involves the expansion process from "1" to "2." Here, the Carnot cycle requires that work be extracted from the working fluid as it expands, flashing into a mixture of vapor and liquid. With one exception at the present time (Brasz, 1995), expanders capable of operating in a wet expansion are not practical. The Carnot cycle remains important as a conceptual cycle in that it serves as a standard by which other cycles may be compared. No refrigeration cycle may have an efficiency greater than that of the Carnot cycle operating between the same cold and hot temperatures.

Vapor Compression Cycle

A practical refrigeration cycle is the vapor compression cycle which consists of the four components shown in Figure 4.13.10. The expander in the Carnot cycle, process "1" to "2," is replaced by an expansion device. Generally, an expansion device is a valve providing a restriction to the flow of liquid refrigerant. The compressor applied to the process from "3" to "4," operates at a rate sufficient to pump the low-pressure vapor from the evaporator to a high-pressure vapor capable of condensing in the condenser. Thus, the regulation of refrigerant flow to the evaporator by the expansion valve and the pumping action of the compressor involves a cycle with a "low-pressure side" and a "high-pressure side".

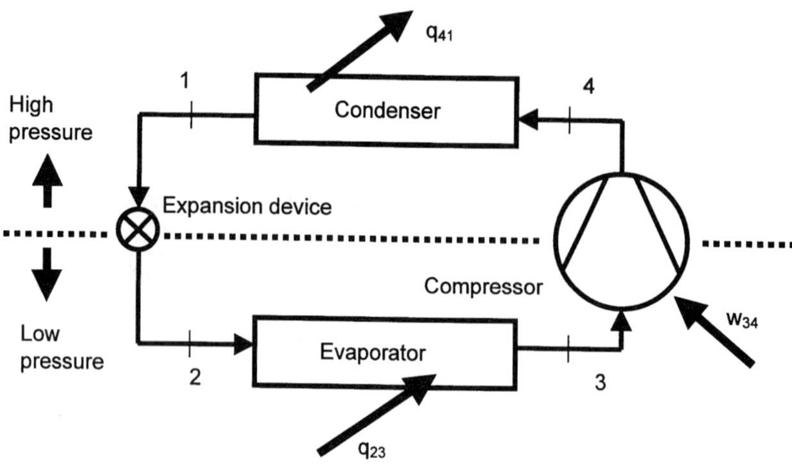

FIGURE 4.13.10 Components for the basic vapor compression refrigeration cycle.

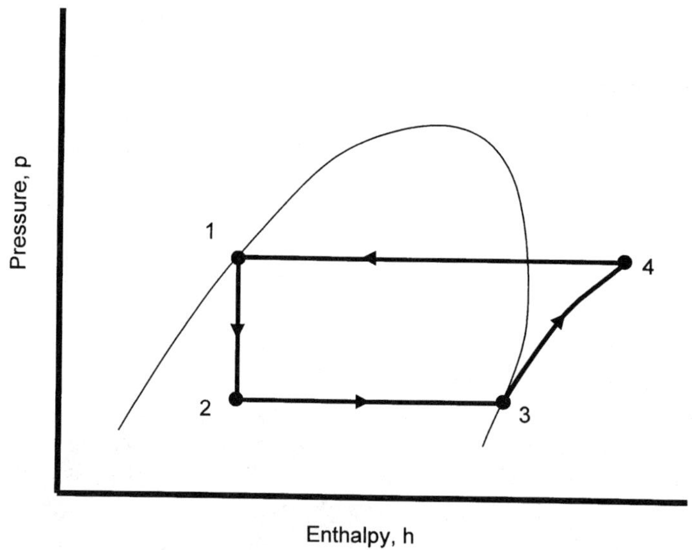

FIGURE 4.13.11 Pressure-enthalpy diagram for the vapor compression refrigeration cycle.

An ideal vapor compression cycle operates at the conditions revealed by the pressure-enthalpy diagram shown in Figure 4.13.11. The refrigerant leaving the condenser at state "1" is a saturated liquid. With the sudden pressure drop in the expansion valve, the refrigerant is a saturated mixture of liquid and vapor entering the evaporator. As the liquid refrigerant boils inside the evaporator, it absorbs heat. Only saturated vapor leaves the evaporator and enters the compressor. In this way, no liquid enters the compressor, thereby assuring long life. Ideal or isentropic compression of the refrigerant vapor takes place in the compressor, where the vapor is superheated at the discharge of the compressor. The saturation temperature corresponding to the discharge pressure must be greater than the temperature of the air or water to which the heat from the condenser is rejected. In the condenser, desuperheating occurs first; next, the vapor is condensed to a liquid. In an ideal condenser, no subcooling occurs and so the liquid leaves the condenser as saturated liquid at the condensing pressure. Furthermore, no pressure drop occurs in any of the piping transporting the refrigerant between the components nor in the evaporator and condenser. Only two pressures exist in an ideal vapor compression cycle: low and high pressure. The COP for the vapor compression cycle may be computed using the following relationships:

$$q_{23} = h_3 - h_2 \tag{4.13.51}$$

$$w_{34} = h_4 - h_3 \tag{4.13.52}$$

$$q_{41} = h_1 - h_4 \tag{4.13.53}$$

$$q_{23} + w_{34} = q_{41} \tag{4.13.54}$$

$$COP = \frac{h_3 - h_1}{h_4 - h_3} \tag{4.13.55}$$

where q represents heat transfer, w the input of work to the compressor, and h the thermodynamic property enthalpy.

In reality, there is some pressure drop in the piping connecting the components and in the evaporator and condenser as well. Pressure drop in the piping represents a loss and a departure from the ideal cycle. Pressure drop in the evaporator causes a "glide" or change in the temperature of the boiling liquid as it progresses through the evaporator. Typically, depending on the particular refrigerant in use, the glide may be several degrees fahrenheit. A similar situation would occur in the condenser. The design of vapor compression refrigeration cycles requires the estimation of the pressure losses in the piping. Information on these pressure drops is given in the ASHRAE *Handbook of Refrigeration* (1998). For the most part, single-component refrigerants (i.e., R-134a) do not exhibit a significant glide in conventional systems. However, refrigerants consisting of blends of chemicals may have larger glides in temperature and may require accommodation of this feature to accurately predict performance of the cycle.

Estimation of an actual cycle's performance may be based upon consideration of an ideal cycle operating between the same temperatures but having an isentropic efficiency representing actual compressor performance (Stoecker, 1998). Actual compressor isentropic efficiencies typically fall in the range from 70 to 80%. However, the subcooling that may occur in the condenser, the superheat at the exit of the evaporator, and the pressure drops in the piping may be readily accounted for in cycle performance calculations when thermodynamic tables of the refrigerant are available. The ASHRAE *Handbook of Fundamentals* (1997) is a source of thermodynamic property data for a number of refrigerants.

Compressors

The compressor is one of the four basic components of the vapor compression cycle. The compressor has two important functions in the cycle: to increase the pressure of the refrigerant vapor from the evaporator so that it will condense, and to maintain the low-side pressure. The evaporator and condenser are heat exchangers facilitating the heat transfer in conjunction with a phase change of the refrigerant. The performance and design of heat exchangers are covered in detail on the heat transfer sections of this handbook.

Compressors fall into two main categories based on their operating principles. These are the positive displacement compressor and the aerodynamic, or more simply, the dynamic compressor. With the positive displacement compressor, the vapor enters the suction port and is subsequently geometrically trapped. The decreasing volume of the trapped vapor due to the expenditure of work causes the compression process to occur. In the dynamic compressor, the vapor enters the suction port whereupon the pressure of the vapor increases as a result of the high rotating velocity which produces a large centrifugal force.

Reciprocating Compressors

The reciprocating compressor is a common positive displacement compressor in the refrigeration industry manufactured in power input capacities from one-tenth of a horsepower to over several hundred horsepower. Compression is achieved by a reciprocating piston inside a cylinder along with the alternate opening and closing of suction and discharge valves. The smaller-capacity compressors are generally single-cylinder units and the large machines use multiple cylinder arrangements where the in-line, V, W, and radial configurations are available. Figure 4.13.12 is a diagram of a typical single-cylinder reciprocating compressor.

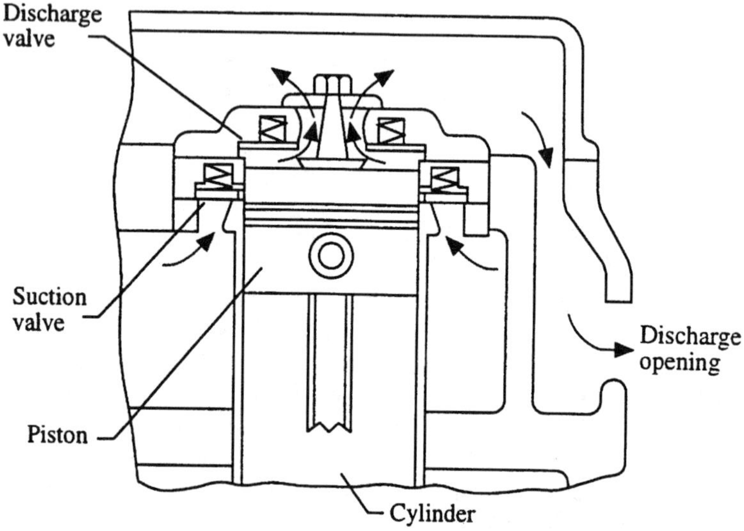

FIGURE 4.13.12 Typical reciprocating compressor cross section. (From Kreith, F., Ed., *The CRC Handbook of Mechanical Engineering,* CRC Press LLC, Boca Raton, FL, 1998. With permission.)

When the compressor is driven by an electric motor, several configurations are available. One is the "open-type compressor" where the crankshaft protrudes out from the compressor's housing. The motor is externally coupled to the compressor shaft. Another configuration is the "hermetically sealed compressor" where the compressor is inside a sealed container. The hermetic compressor is prevalent in small electrically driven units applied to residential air conditioners and smaller commercial units. The hermetic compressor has only the suction line, the discharge line, and the electrical power cable passing through the container's shell or "can." The absence of moving parts provides for a long-term and reliable seal. Another configuration is the "semihermetic compressor" which is identical to the hermetic compressor except that the heads are accessible for maintenance.

Performance characteristics of the reciprocating compressor include:

- *Pressure ratio*: discharge pressure divided by the suction pressure (both absolute pressures)
- *Displacement*: volume swept out by the piston as it moves over the extent of its travel in the cylinder, V_d
- *Displacement rate*: the rate of volume swept by the piston as the compressor operates, $\dot{V}_d$
- *Actual volumetric efficiency*: the actual refrigerant vapor volume flow rate at the suction port divided by the compressor's displacement rate, $\dot{V}_{actual}/\dot{V}_d$
- *Isentropic efficiency*: ratio of the isentropic compression work to the actual compression work

$$\eta_c = \frac{h_{4s} - h_3}{h_4 - h_3} \tag{4.13.56}$$

where h is the enthalpy and subscripts 3 and 4 correspond to the suction and discharge states shown in Figure 4.13.10.

The isentropic efficiency includes the influence of pressure losses across the suction and discharge valves and the heat transfer to the refrigerant vapor from the compressor's internal surfaces. Friction originating from the rubbing of the mechanical parts (e.g., piston rings against the cylinder wall) is also included in the isentropic efficiency. However, leakage is not accounted for in the isentropic efficiency.

Accessories and subcomponents for the reciprocating compressor include a form of capacity control to meet varying refrigerating loads. Several methods are used to achieve capacity control in reciprocating compressors:

- Cycling the compressor on and off to satisfy the required capacity, usually applied to small systems
- Throttling the suction line, good control but inefficient
- Returning a portion of the discharge vapor to the suction line, good control, but inefficient and the compressor may operate hot
- Multicylinder compressors may deactivate cylinders, opening the suction valve and closing the discharge valve; called "cylinder unloading"
- Changing compressor operating speed

Small compressors are lubricated and cooled with a simple oil splash system. Lubrication oil that is miscible in the refrigerant must be conveyed by the refrigerant throughout the entire cycle. The halocarbon refrigerants use miscible oils and attention must be directed at the sizing of all components and piping to ensure that the velocities are sufficiently high everywhere to prevent the oil from accumulating anywhere in the system.

Scroll Compressor

The newest arrival among positive displacement compressors, it has widespread application in residential, commercial, and automotive applications. The range of input power that the scroll compressor can accommodate is from about 1 to 20 hp. Refer to Section 4.7 on compressors for details of the scroll compressor.

Capacity control is available using variable-speed motor drives and porting holes in the fixed scroll member to adjust the displacement of the compressor. A control device sequentially opening the ports and venting the refrigerant vapor to the suction port progressively reduces capacity. Capacity control may also be achieved using suction throttling. The isentropic efficiency of the scroll compressor is comparable to the efficiencies of the other compressor types, including the reciprocating compressor. A limiting refrigeration capacity for the scroll compressor is about 40 tons due to increasing size of the scroll.

Centrifugal Compressor

The centrifugal compressor accomplishes compression by imparting a high velocity to the refrigerant vapor and converting this velocity, actually kinetic energy, to a pressure. The compressor consists of an impeller driven to a high rotating speed. The vapor enters the suction port of the compressor through a set of guide vanes that direct the flow to the center of the impeller. The blades on the impeller are curved in such a way to direct the vapor to flow outward, by centrifugal force, at ever-increasing speeds. Leaving the impeller, the vapor flow enters the diffuser, where a portion of the kinetic energy gained is converted to pressure. Figure 4.13.13 shows a cross-sectional view of a two-stage centrifugal compressor. A two-stage centrifugal compressor utilizes two impellers to perform the compression.

The performance of the centrifugal compressor is obtained by equating the power applied to the impeller to the isentropic work of compression. After rearrangement and simplification (Stoecker, 1982),

$$V_t = \sqrt{g_c J(\Delta h_s)} \tag{4.13.57}$$

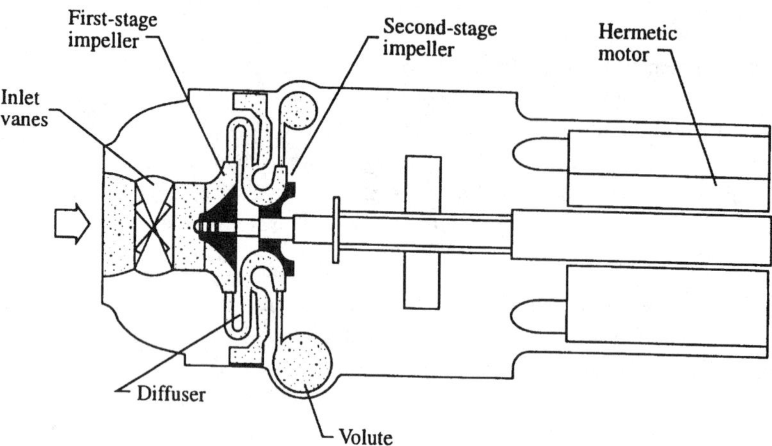

FIGURE 4.13.13 Cross-section view of a centrifugal compressor. (From Kreith, F., Ed., *The CRC Handbook of Mechanical Engineering,* CRC Press LLC, Boca Raton, FL, 1998. With permission.)

where V_t is the tangential velocity (ft/s), g_c is a constant equal to 32.17 lb-ft/lbf-s^2, J is a constant equal to 778 ft lbf/Btu, and Δh_s is the enthalpy change for the isentropic work. This expression, while not accurate, does give reasonable approximations. The above expression shows that the refrigerant selected for a centrifugal compressor affects the diameter of the impeller. It turns out that the tip speed of the impeller is limited to approximately 800 ft/s, the actual value depending on the material used for the impeller. Refrigerants with thermodynamic properties similar to those of R-123 are desirable because the diameter of the impeller is smaller. Typically, centrifugal compressors using R-123 require only one compression stage and still have an acceptable tip speed.

A common application of the centrifugal compressor is the chilling of water for a central air-conditioning system. The refrigerating capacity where this compressor is applied ranges from about 50 to 2500 tons. Isentropic efficiencies for centrifugal compressors are usually in the range from 70 to 80%. Capacity control is accomplished by partially closing the inlet guide vanes upstream from the first impeller or varying the rotating speed of the impeller. Other methods of capacity control, such as bypassing the discharge gas, are not used, because they are not efficient.

Rotary Screw Compressor

This compressor is a positive displacement compressor where the compression process occurs in conjunction with the rotation of one or two screws. Details concerning the construction and operation of screw compressors are given in the compressor section of this handbook.

Performance of the screw compressor may be given in terms of the compression volume ratio, V_i, which is defined by

$$V_i = \frac{\textit{internal volume when suction port covered}}{\textit{internal volume when discharge port uncovered}} \tag{4.13.58}$$

which gives,

$$\text{Pressure ratio} = \frac{p_{dis}}{p_{suc}} = V_i^{\,k} \tag{4.13.59}$$

where k is the ratio of specific heats c_p/c_v. The above equation is only valid when the compression is isentropic. When the compression process is not isentropic, the exponent k is replaced with exponent n which is not equal to k. When the temperature of the vapor remains constant during the compression, the value of n is 1. For actual rotary screw compressors, the value of n lies between k and 1 (Stoecker, 1998).

The discharge pressure for a rotary screw compressor may not match the pressure in the discharge piping. When these pressures are not equal, the adjustment that the vapor undergoes, over- or under-compression, results in losses. Variation of the compressor's volume ratio is possible through the addition of a slide valve that changes the extent of compression that occurs by changing the size of the discharge port. With a variable volume ratio, the over- and undercompression losses are eliminated. Capacity control is achieved by the addition of another slide valve that allows a portion of the suction vapor to bypass back to suction. Oil injection is used to seal the small spaces (approximately 0.0005 in. tolerance) between the lobes and gullies on the screws. Because the oil is heated during compression, it must be cooled to maintain its lubricating properties.

The isentropic efficiency for the screw compressor is about the same as that for the reciprocating compressor. Screw compressors of the same capacity as reciprocating compressors are smaller in size due to their higher rotating speeds. Refrigerating capacities are available from 20 tons to over 1000 tons and are prevalent in the industrial refrigeration field. Recently, screw compressors have been applied to large water-chilling units for air-conditioning applications in commercial buildings.

Absorption Cycle

The absorption refrigeration cycle differs from the vapor compression cycle in that the mechanical work used to compress the vapor is not needed. Rather, the vapor compression process is replaced by a series of processes whereby the refrigerant vapor is first absorbed into a liquid solution, the strong liquid solution is pumped to a higher pressure, and then driven from the generator as a vapor. To accomplish the absorption of the refrigerant and its subsequent release, heat is transferred out of the solution and transferred into the solution. The liquid pump, while requiring the input of work, requires only a small fraction of the work necessary to compress the vapor leaving the evaporator. The diagram in Figure 4.13.14 shows an ideal absorption refrigeration cycle. Observe that three of the vapor compression

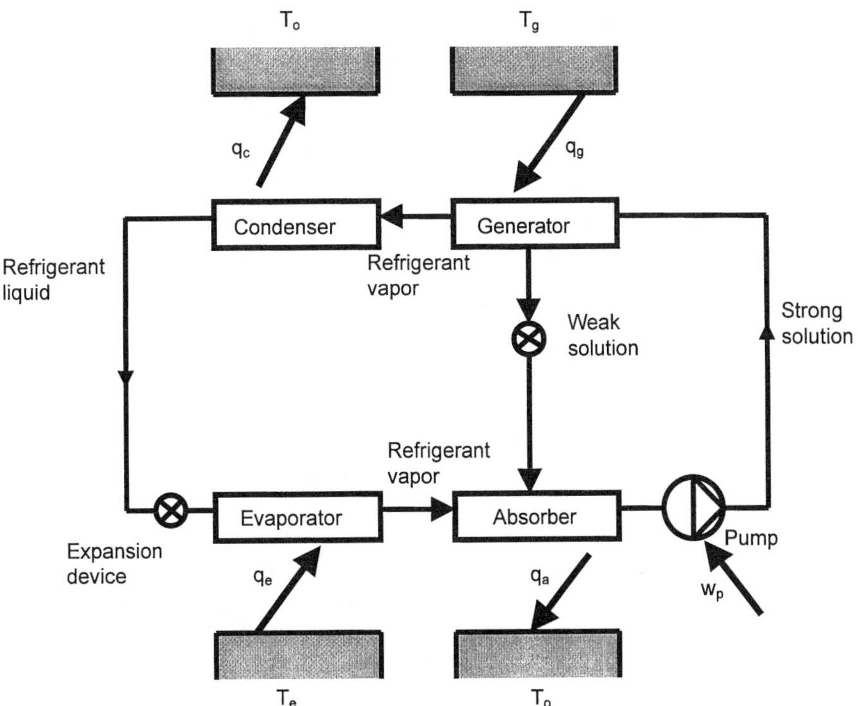

FIGURE 4.13.14 Ideal absorption refrigeration cycle.

cycle's components are also present in the absorption cycle. However, the compressor is replaced by the absorber, liquid pump, and generator which constitutes the path that the liquid solution takes as it circulates through in the cycle.

The working fluid in the absorption cycle is a binary fluid which is a homogeneous mixture of two chemicals. The performance of the absorption cycle depends on the affinity that the two components of the binary mixture have for each other. The pair commonly used for chilling water for air-conditioning applications is lithium bromide and water (Kuehn et al., 1998). In this binary mixture, the refrigerant is water which limits the cycle to temperatures above 32°F. Ammonia–water is another binary pair used in absorption cycles where ammonia is the refrigerant. The ammonia–water pair is typically used in industrial applications where temperatures lower than 32°F must be achieved (Bogart, 1981).

Absorption cycles provide an advantage when waste heat is available. Sources of waste heat may include heat rejected by an electrical power-generating plant, a large internal combustion engine, or an array of solar collectors. Even though the overall performance of the absorption cycle may be significantly less than that of the vapor compression cycle, when waste heat is available it may be competitive on a cost basis. Bosnjakovic (1965) developed the following expression for the maximum COP possible from the ideal absorption cycle shown in Figure 4.13.14:

$$(COP)_{\max} = \frac{T_e\left(T_g - T_o\right)}{T_g\left(T_o - T_e\right)} \tag{4.13.60}$$

Refrigerant Selection

Vapor-compression refrigeration cycles used for the air conditioning of commercial buildings, private homes, and automobiles for human comfort have become common in many parts of the world. One factor promoting the widespread application of the vapor-compression cycle was the development of refrigerants composed of halocarbons, all of which are nontoxic and nonflammable. Halocarbons are chemical compounds containing one or more atoms from the halogen family, namely, chlorine, fluorine, and bromine, in conjunction with at least one carbon atom. The technical term for these compounds is chlorofluorocarbons, generally abbreviated to CFCs. This group of refrigerants were called Freons, soon after their development in the 1930s. Early members of this new group were R-12 and R-11. They have gained prevalent use serving the refrigeration and air-conditioning industry for over 50 years.

During the 1960s, measurements of very small quantities of R-11 were obtained in a number of remote locations on the Earth, including at sea. Questions were raised by many as to whether or not CFC refrigerants present in the atmosphere have an influence on it. The answer became clear as a result of investigations conducted in the 1970s and 1980s. The survival time of a CFC in the atmosphere is very long as a result of its high chemical stability. This permits the CFC molecules to rise to the upper atmosphere and move through the ozone (O_3) layer. In this region of the atmosphere ultraviolet radiation from the sun is present, which causes the chlorine atom to break free and chemically react with the ozone-producing diatomic oxygen (O_2) and chlorine. The chlorine is then free to react again with another ozone molecule, thereby destroying it. This reaction continues for the life of the chlorine atom in the upper atmosphere.

The measurement of serious ozone depletion over Antarctica observed during the spring season of the year and the mounting evidence indicating significant ozone depletion over North America, Europe, and Japan caused a number of countries to make an agreement in Montreal, Canada through the United Nations. This agreement, created in 1987, is called the "Montreal Protocol on Substances that Deplete the Ozone Layer," and has been revised several times since. As a result of this agreement, CFC production in developed countries ceased in the year 1996. The Montreal Protocol also includes agreements that will result in the future phase-out of another group of refrigerants — the hydrogenated halocarbons referred to as HCFCs. A prominent refrigerant in this group is R-22, a refrigerant used widely in air-conditioning systems. At the present time, production curtailment is scheduled for the year 2020 by the Montreal

Protocol (Van der Ree, 1998). Another halocarbon class, the hydrofluorocarbons or HFCs, does not contain chlorine and, therefore, is not ozone depleting. HFCs are not regulated by the Montreal Protocol.

As a consequence of the Montreal Protocol, R-11 and R-12, which are both CFCs, cannot be used in new refrigeration equipment. In response, the refrigeration industry has developed alternative refrigerants and is conducting research that will lead to additional alternative refrigerants. The selection of suitable refrigerants and the attendant concerns with safety and the environment are decisions that the designer must make.

Selection of Refrigerants

When considering vapor compression refrigeration cycles, two overall system requirements must be considered. These are the temperature of the air-conditioned space and the temperature of the environment to which the heat must be rejected. Generally, the temperature of the boiling refrigerant in the evaporator must be about 20°F below that of the conditioned space. The pressure in the low-side of the system should be greater than atmospheric pressure, preventing air from entering the system due to leaks. Generally, the evaporator pressure is very near the saturation pressure which establishes the temperature of the refrigerant. Refrigerants R-11 and R-123 are exceptions in that evaporator pressures are typically less than 10 psia. The condenser presents an additional system constraint because at the condensing pressure which is very near the saturation pressure, the corresponding saturation temperature of the refrigerant must be about 20°F higher than the environment where the heat is rejected.

Table 4.13.1 lists several physical properties of refrigerants used in air conditioning. The critical pressure is given for the refrigerants. The condensing pressure in a conventional vapor compression refrigeration system must be somewhat less so that the refrigerant vapor may condense at nearly constant pressure in the condenser. Table 4.13.2 is extracted from the ASHRAE *Handbook of Fundamentals* (1997) and shows the relative performance of some refrigerants when operating under the same conditions in ideal vapor compression cycles. From the COP values reported in the Table 4.13.2, observe that most are about the same. R-170 (ethane) is the lowest at 2.72 and R-11 is the highest at 5.02. Compressor discharge temperature is highest for ammonia and represents a disadvantage of ammonia particularly when evaporator temperatures are low and the compression ratio is high. However, R-11 and R-123 should be excluded from the comparison as these refrigerants are uniquely suited to the centrifugal compressor due to their large specific volume resulting in reasonable wheel tip speeds. Otherwise, ammonia has the highest COP at 4.77, which is due to its high latent heat.

The environmental impact of some refrigerants is given in Table 4.13.3 where two characteristics have been selected: ozone depletion potential (ODP) and global warming potential (GWP). ODP is the relative capability of the refrigerant to destroy ozone in the atmosphere of the earth. The ODP for R-11 is defined as 1.00. GWP is the capability of the refrigerant to warm the atmosphere of the Earth relative to that caused by carbon dioxide over a period of 100 years. Chemicals in the atmosphere have a unique average residence time which in turn influences the GWP.

Table 4.13.3 indicates that CFCs have high ODP, HCFCs have significantly lower ODP, and HFCs have zero ODP. Halocarbon refrigerants should not be vented to the atmosphere in most countries. GWP values indicate that refrigerants, including the HFCs, represent an important segment of the chemicals that contribute to global warming due to the high values. In the future, restrictions on the use of refrigerants high in GWP may occur.

ASHRAE Classification of Refrigerants

ASHRAE, through Standard 34-1992 “Number Designation and Safety Classification of Refrigerants,” publishes refrigerant classifications as determined by the standard and is periodically updated to include new developments. The ASHRAE classification is based on two refrigerant characteristics: toxicity and flammability. Two levels, “A” (nontoxic) and “B” (toxic), comprise the toxicity classifications. A refrigerant is classified as toxic when exposed to humans at a concentration equal to 400 ppm (v/v) in air causes toxicity (Gopalnarayanan, 1998). Flammability is divided into three classes: highly flammable, moderately flammable, and nonflammable which are designated “3”, “2”, and “1”, respectively. The

TABLE 4.13.1 Physical Properties of Refrigerants Used for Air-Conditioning

ASHRAE Designation	Name	Chemical Formula	Molecular Weight	Boiling Pt. (°F) at 14.7 psia	Freezing Pt. (°F)	Critical Temp. (°F)	Critical Press. (psia)
R-11 (CFC)	Trichloromonofluoromethane	CCl_3F	137.37	74.87	–168.0	388.4	639.5
R-12 (CFC)	Dichlorodifluoromethane	CCl_2F_2	120.93	–21.62	–252	233.6	596.9
R-22 (HCFC)	Chlorodifluoromethane	$CHClF_2$	86.48	–41.36	–256	204.8	721.9
R-123 (HCFC)	2,2-dichloro-1,1,1-trifluoroethane	$CHCl_2CF_3$	152.93	82.17	–160.87	362.82	532.87
R-134a (HFC)	Tetrafluoroethane	CH_2FCF_3	102.3	–15.08	–141.9	214.0	589.8
401A (Blend)	R-22/152a/124 (53/13/34)		94.44	–27.6[a]	—	226.4	667.2
404A (Blend)	R-125/143a/134a (44/52/4)		97.60	–51.7[a]	—	161.8	541.0
407C (Blend)	R-32/125/134a (23/25/52)		86.2	–46.5[a]	—	187.0	670.1
410A (Blend)	R-32/125 (50/50)		72.56	–58.9[a]	—	162.5	719.4
R-32	Difluoromethane	CH_2F_2	52.02	–61.1	–213	173.14	845.6
R-124	Chlorotetrafluoroethane	$CHClFCF_3$	8.26	–326.47	252.5		
R-125	Pentafluoroethane	CHF_2CF_3	120.03	–55.43	–153.67	151.34	526.57
R-143a	1,1,1-trifluoroethane	CF_3CH_3	84.04	–53.3	—	164.5	555.5
R-152a	Difluoroethane	CHF_2CH_3	66.05	–13.0	–178.6	236.3	652
R-717 (Inorganic)	Ammonia	NH_3	17.03	–28.0	–107.9	271.4	1657
R-744 (Inorganic)	Carbon dioxide	CO_2	44.01	–109.2	–69.9	87.9	1070.0
R-170 (Hydrocarbon)	Ethane	C_2H_6	30.07	–127.85	–297	90.0	709.8
R-290 (Hydrocarbon)	Propane	C_3H_8	44.10	–43.73	–305.8	206.3	617.4

[a] For refrigerant blends, the boiling (bubble point temperature) is tabulated.

Source: ASHRAE *Handbook of Fundamentals,* 1997; United Nations, 1994.

TABLE 4.13.2 Performance Characteristics of Several High Temperature Refrigerants Per Ton of Refrigeration

Refrigerant	Evap. Press. (psia)	Cond. Press. (psia)	Net Refrig. Effect (Btu/lb)	Mass Flow Rate (lb/min per ton)	Suction Vapor Specific Volume (ft³/lb)	Volume Rate Leaving Evap. (ft³/min)	Compr. Discharge Temp. (°F)	COP
R-11	2.9	18.3	67.21	2.98	12.24	36.43	110	5.02
R-123	2.3	15.9	61.19	3.27	14.08	46.02	94	4.84
R-12	26.5	108.0	50.25	3.98	1.46	5.83	100	4.75
R-134a	23.8	111.6	64.77	3.09	1.95	6.02	108	4.41
R-22	43.0	172.9	69.90	2.86	1.34	3.55	128	4.67
R-125	58.9	228.1	37.69	5.31	0.628	3.33	108	3.67
R-717	34.2	168.8	474.20	0.422	8.18	3.45	210	4.77
R-744	332.4	1045.4	57.75	3.46	0.264	0.914	156	2.81
R-170	236.4	674.7	69.27	2.89	0.534	1.54	123	2.72
R-290	42.4	156.8	169.60	1.66	2.46	4.09	98	4.41

Source: ASHRAE *Handbook of Fundamentals,* 1997.

TABLE 4.13.3 Flammability and Environmental Characteristics of Selected Refrigerants

Refrigerant	Chemical Formula	ODP	GWP (CO_2 at 100 yr)	Atmos. Life (yr)	TLV (ppm)	LFL (%)	Safety Group
R-11 (CFC)	CCl_3F	1.00	4000	50	1000	None	A1
R-12 (CFC)	CCl_2F_2	1.00	8500	102	1000	None	A1
R-22 (HCFC)	$CHClF_2$	0.055	1700	13.3	1000	None	A1
R-123 (HCFC)	$CHCl_2CF_3$	0.02	93	1.4	10[a]	None	A1
R-134a (HFC)	CH_2FCF_3	0	1300	14	1000	None	A1
R-401A	R-22/152a/124	0.037	—	—	—	None	A1[b]
R-404a	R-125/143a/134a	0.00	3260	—	1000	None	—[c]
R-407C	R-32/125/134a	0.00	—	—	1000	None	—[c]
R-410A	R-32/125	0.00	—	—	1000	None	—[c]
R-717 (Inorganic)	NH_3	0	0	<1	25	14.8	B2
R-744 (Inorganic)	CO_2	0	1	50-200	5000	None	A1
R-170 (Hydrocarbon)	C_2H_6	0	3	—	s.a.[b]	3.3	A3
R-290 (Hydrocarbon)	C_3H_8	0	3	—	s.a.[b]	2.1	A3

[a] Established by ASHRAE Standard 34-1992.
[b] Provisional classification, additional test data pending.
[c] ASHRAE safety classification not established by Standard 34-1992.
Source: United Nations, 1994.

term flammability refers to capability of a mixture of the refrigerant vapor and air to propagate a flame. Table 4.13.3 includes the ASHRAE safety classification for several refrigerants.

The prefix "R" is used to designate a refrigerant. As mentioned earlier, letter designations (i.e., CFC) are also applied to halocarbon refrigerants. The numbering of halocarbon refrigerants and some hydrocarbons is dependent on the atoms that make up the molecule. The first digit on the right is the number of fluorine atoms. The second digit from the right is one more than the number of hydrogen atoms, and the third digit from the right is one less than the number of carbon atoms. When the third digit is zero, it is not written. With inorganic compounds serving as refrigerants, the first number on the left is "7." The next digits on the right is equal to the molecular weight of the compound. An example is R-744 (carbon dioxide). When the first digit is "4", the refrigerant is a blend (or zeotrope) of two or more compounds. The two digits to the right provide the constituents that comprise the mixture, and the letter the order in which the designation was made. When a "5" is designated, the refrigerant is a mixture of two compounds which form an azeotrope, which is a special mixture in that neither constituent can be recovered by distillation alone. The digits to the right of "5" indicate the historical order of that azeotrope's commercial introduction.

Table 4.13.3 also lists values for the refrigerant's vapor toxicity via the threshold limit value or "TLV." It is defined as the time weighted average exposure that a worker can have to a substance for a normal 40-h work week without any adverse health effects.

Exergy Analysis

The ideal vapor compression refrigeration cycle has undergone exergy analyses (Li, 1996; Bejan, 1995; Kotas, 1995). These analyses are based on the second law of thermodynamics where the property exergy is used to track the losses that occur in the cycle. The second law efficiency for the cycle, defined in the thermodynamics section of this handbook, is relatively small. Calculations applied to idealized cycles show that the greatest exergy loss occurs in the condenser, the next largest loss being in the expansion device. The compressor, considered isentropic, has by far the smallest exergy loss. However, in an actual cycle, the exergy losses in the compressor, condenser, and evaporator are roughly the same at typical air-conditioning temperatures. The least exergy loss occurs in the expansion device in actual cycles (Kotas, 1995). Further considerations regarding the importance of exergy analysis are given in the discussions of Kotas (1995) and Bejan (1995).

Energy Distribution Systems

After the cooling and heating loads are determined for the building or application, the air distribution system must be designed resulting in the selection of equipment. The psychrometric processes discussed earlier play an important role in determining the magnitude of the sensible and latent loads for the space. For example, if winter heating is required, the air temperature and moisture content must be high enough to absorb the entire heating load for the space. The air passing through the space is cooled and dehumidified, which are basic psychrometric processes. For commercial buildings, fresh outdoor air is generally required by national and local codes. To provide the fresh air, the return air is usually mixed with outdoor air before it enters the heating and humidifying equipment. For summer cooling, similar psychrometric processes occur: cooling and dehumidification. In both situations the outdoor air, the return air, and the space load are all important in establishing the needed psychrometric processes.

All-Air Systems

Air-conditioning systems that psychrometrically prepare the air and transport that air through air ducts are called all-air systems. These systems may be classified into two types: single-duct systems and dual-duct systems. Single-duct systems have one air flow path in which all the heating and cooling coils are inserted in series. These systems have a common air duct distribution system resulting in one temperature supplied to all the terminal outlets. In contrast, the dual-duct system has the main heating and cooling coils positioned in parallel. Two variations of the dual-duct system are used. One is to maintain the separation of the cool and warm air using two air duct distribution systems — one for the cool air and the other for the warm air. This is a traditional dual-duct system. The other system is to have a dedicated supply air duct for each zone with the mixing of the supply air to the needed conditions occurring at a main mixing damper array. This setup is called a multizone air duct system.

Single-Duct Systems

With single-duct systems, two basic categories characterize most systems: constant air volume and variable air volume.

Constant Volume — The supply air temperature changes in response to the space load while the air flow rate is maintained constant. The simplest version would be used for a single-zone system where the supply air equipment provides air directly or through air ducts to the space. For more complex systems, multiple-zone reheat is one version that accommodates the requirements of several spaces, each controlled by a separate thermostat. Additionally, multiple-zone systems can simultaneously provide heating and cooling to the zones as needed. Reheat of the air, either preconditioned primary air or recirculated air, occurs with these systems.

Small multiple-zone systems incorporate only low air pressures in the ducts and locate the reheat coils in the ducts near each zone. When the system is more complex (i.e., many zones), the primary distribution air ducts have higher pressures to reduce size and cost. Pressure reduction equipment is installed to maintain constant air flow volume for each zone. These systems use conditioned air from centrally located equipment, usually at a constant cold temperature. However, to reduce the quantity of reheat required and conserve energy, the cold temperature may be adjusted by the controls.

A modification to the constant volume reheat system incorporates a bypass box replacing the reheat coil. The change becomes a constant volume primary system with a variable air volume secondary system. To meet the space load, the amount of zone supply air is adjusted by dumping the excess air to the return and simply bypassing the zone altogether. Bypass systems are limited in application to small multiple-zone systems, typically when energy consumption is less important.

Variable Air Volume — The supply air temperature remains constant while the amount of air supplied adjusts to the changing requirements of the space by means of a VAV (variable air volume) terminal device. In operation, the air supply temperature is fairly constant, causing the VAV system to be well suited to perimeter and interior zones. An opportunity for energy savings exists when using VAV systems for perimeter zones where solar heating results in a reduction of the supply air flow rate to make temperature changes. Variable air volume systems may use individual or common fans, individual or common zone temperature control, and may or may not include auxiliary heating equipment.

Humidity control with VAV systems is more difficult than with constant volume systems. Approaches sometimes taken are to set a lower limit on the air flow rate, provide reheat to prevent overcooling with reasonable relative humidity, and provide ventilation using outdoor air. A typical lower limit on the air flow rate is 50% of the original flow rate.

The VAV system typically uses reheat when the heating occurs at the terminal unit a small distance upstream from the space. This system establishes the following control hierarchy: first, reduce the air flow volume, and second, add heat at the reheat coil. Relative to constant volume reheat, VAV systems consume less energy by reducing the primary air that must be cooled and by reducing the secondary air that needs heating. The terminal unit is adjusted to operate at a predetermined throttling ratio which is arrived at by establishing the lowest air flow rate sufficient to:

- Meet the heating load
- Limit the relative humidity (superseded by independent ventilation system)
- Provide acceptable air movement
- Provide required ventilation air (superseded by independent ventilation system)

Variations in the terminal units include provision for the induction of air from the space and have the advantage of recovering a portion of the heat from the lights. The terminal unit may also be fan-powered in either a series or parallel flow arrangement. The advantage with fan-powered units is the larger circulation rate that is possible through the space at lower loads.

Dual-Duct Systems

All the supply air is conditioned in centrally located equipment which in turn is conveyed to the individual spaces by two main parallel ducts. Cold air is conveyed in one duct and warm air in the other. In a dual-duct box near the space, the warm and cold air are mixed by a valve in proportion to satisfy the space load. An advantage of the dual-duct system is that no pipes are required to feed the reheat coils, avoiding the possibility of leaks into the conditioned space. However, energy consumption for the dual-duct system is greater than that of the VAV single-duct system. Dual-duct systems may be either constant volume or VAV.

Constant Volume — Single-fan, no reheat resembles a single-duct system except for the face-and-bypass damper at the cooling coil. This damper bypasses a mixture of recirculated and outdoor air in response to changes in the heat load. A disadvantage of this configuration is the imprecise control temperaure and humidity in the occupied space. An alternative is the single-fan reheat configuration where the single reheat coil is placed in the central unit.

Variable Air Volume — VAV systems utilizing dual-ducts mix the cold and warm air yielding different volume combinations. One configuration employs a single fan capable of handling both the cold and hot ducts. Typically, the cold deck is controlled to hold a constant temperature while the temperature of the hot deck is allowed to vary depending on the outside conditions. The hot deck temperature is increased when the outside temperature is low and the humidity high so that the flow over the cold deck increases to improve dehumidification. Another version is the dual fan VAV dual-duct system where each supply fan is independently controlled. The cold deck temperature is maintained constant by a refrigeration unit in conjunction with providing minimum fresh air, or by an economizer when the outside temperature is less than the setpoint temperature for the cold deck. The hot deck may still obtain heat from the recirculated air or from the heating coil when the recirculated air is too cold.

Air-and-Water Systems

Air-and-water systems use both water and air to convey the energy to and from the conditioned space. The supplied air originates from a central room where the heating and cooling equipment is placed and is called "primary air." The supplied water is called "secondary water." This system, as with the systems described earlier, require a terminal unit for each zone. The terminal unit may be a fan coil device or an induction unit. Sometimes, an ordinary air outlet in combination with a radiant panel may serve as a terminal unit.

The purpose of the primary air for heating is to provide fresh air to comply with ventilation requirements. When needed, water may be added to the warm air to improve comfort. For cooling, the primary air is dehumidified in order to achieve comfortable temperature and humidity in the space. The amount of primary air supplied to each space or zone is dependent on the fresh air ventilation requirement and the sensible cooling load at design conditions. Thus, the space humidity is controlled by the primary air. The secondary water is transported to the terminal units to supply a coil with either warm or cold water. When properly operating during the cooling season, the coil is dry, which greatly reduces odors and bacterial growth, while at the same time extending the life of the coil. However, condensate drains are recommended for all air-and water systems.

Secondary water is supplied using a pump and piping arrangement. At the central plant, the water is typically cooled by heat exchange with the central chilled water plant. Air-and-water systems are classified as two-pipe, three-pipe, or four-pipe systems. All these piping arrangements accomplish the goal of supplying chilled or warmed water to the coil.

- A two-pipe system consists of one supply pipe and one return pipe for either warm or cold water supply temperatures. Seasonal changeover of these systems is relatively difficult.
- A three-pipe system consists of one cold water supply, one warm water supply, and a common return pipe. The mixing of the warm and cold water at the terminal unit generates significant energy losses.
- A four-pipe system consists of a cold water supply and return and a warm water supply and return. This arrangement is relatively efficient and changeover between seasons is convenient.

All-Water Systems

Another system is the all-water system, which may also be employed to heat and cool a space depending on the outdoor conditions. The air in the space is heated and cooled by any combination of the three basic heat transfer mechanisms: conduction, convection, and radiation. Applications include fan-coil units, panels on the floor, walls, or ceiling, and baseboard radiation. For further information on this subject, consult the ASHRAE *Handbook of Systems and Equipment* (1996).

References

Anon., 1995 Assessment of the UNEP Montreal Protocol on Substances that Deplete the Ozone Layer. Refrigeration, Air Conditioning and Heat Pumps Technical Options Committee, Parties to the Montreal Protocol, United Nations, New York.

ASHRAE, *Procedure for Determining Heating and Cooling Loads for Computerizing Energy Calculations, Algorithms for Building Heat Transfer Subroutines,* American Society of Heating, Refrigerating, and Air-Conditioning Engineers, Atlanta, GA, 1975.

ASHRAE, *Thermal Environmental Conditions for Human Occupancy,* ANSI/ASHRAE Standard 55-1992, American Society of Heating, Refrigerating, and Air-Conditioning Engineers, Atlanta, GA, 1992.

ASHRAE, Number Designation and Safety Classification of Refrigerants, ANSI/ASHRAE Standard 34-1992, American Society of Heating, Refrigerating, and Air-Conditioning Engineers, Atlanta, GA, 1992.

ASHRAE, Addendum 55a. ANSI/ASHRAE Standard 55-1992, American Society of Heating, Refrigerating, and Air-Conditioning Engineers, Atlanta, GA, 1994.

ASHRAE *Handbook of Systems and Equipment,* American Society of Heating, Refrigerating, and Air-Conditioning Engineers, Atlanta, GA, 1996.

ASHRAE *Handbook of Fundamentals,* American Society of Heating, Refrigerating, and Air-Conditioning Engineers, Atlanta, GA, 1997.

ASHRAE *Handbook of Refrigeration,* American Society of Heating, Refrigeration, and Air-Conditioning Engineers, Atlanta, GA, 1998.

Berglund, L.G., Comfort and Humidity, *ASHRAE J.,* August, 1988.

Bejan, A., *Entropy Generation Minimization,* CRC Press, Boca Raton, FL, 1995.

Bogart, M., *Ammonia Absorption Refrigeration in Industrial Processes.* Gulf Publishing, Houston, TX, 1981.

Bosnjakovic, F., *Technical Thermodynamics*, Holt, Reinehart & Winston, New York, 1965.

Brasz, J.J., Improving the Refrigeration Cycle With Turbo-Expanders, IIR Cong., The Hague, The Netherlands, 1995.

Burch, D.M. and Chi, J., Moist: A PC Program for Predicting Heat and Moisture Transfer in Building Envelopes, Release 3.0, NIST Spec. Publ. 917, U.S. Government Printing Office, Washington, D.C., 1997.

Fanger, P.O., *Thermal Comfort,* Kreiger Publishing, Malabar, FL, 1982.

Gage, C., Supermarkets, Int. Inst. Ammonia Refrigeration (IIAR) Annu. Meet., IIAR, Washington, D.C., 1996.

Gagge, A.P., Fobelets, A.P., and Berglund, L.G., A Standard Predictive Index for Human Response to the Thermal Environment, *ASHRAE Trans.,* 92(1), 709-731, 1986.

Gagge, A.P., Stolwijk, J., and Nishi, Y., An Effective Temperature Scale Based on a Simple Model of Human Physiological Regulatory Response, *ASHRAE Trans.,* 77(1), 247-62, 1971.

Gopalnarayanan, S., Choosing the Right Refrigerant, in *Mechanical Engineering,* American Society of Mechanical Engineers, New York, 1998.

Harrison, L.P., Fundamental Concepts and Definitions Relating to Humidity, in *Science and Industry,* Wexler, A. and Wildhack, W.H., Eds., Reinhold, New York, 1965.

Howell, H.H., Sauer, H.J. Jr., and Coad, W.J. *Principles of Heating, Ventilating and Air Conditioning.* ASHRAE, Atlanta, GA, 1998.

Hyland, R.W. and Wexler, A., Formulations for the Thermodynamic Properties of the Saturated Phases of H_2O from 173.14 K to 473.15 K, *ASHRAE Trans.*, 89(2A), 500-519, 1983a.

Hyland, R.W. and Wexler, A., Formulations for the Thermodynamic Properties of Dry Air from 173.15 K to 372.15 K, and of Saturated Moist Air from 173.15 K to 372.15 K, at Pressures to 5 MPa, *ASHRAE Trans.*, 89(2A), 520-35, 1983b.

Hosni, M.H., Sipes, J.M., and Wallis, M.H., Controlling Moisture in Walls Exposed to Hot and Humid Climates, Final Rep., ASHRAE Res. Project 864-RP, American Society of Heating, Refrigerating, and Air-Conditioning Engineers, Atlanta, GA, 1998.

Hwang, Y., Ohadi, M., and Radermacher, R., Natural Refrigerants, in *Mechanical Engineering,* American Society of Mechanical Engineers, New York, 1998.

Kotas, T.J., *The Exergy Method of Thermal Plant Analysis,* Krieger Publishing, Malabar, FL, 1995.

Kuehn, T.H., Ramsey, J.W., and Threlkeld, J.L., *Thermal Environmental Engineering,* 3rd ed., Prentice-Hall, Englewood Cliffs, NJ, 1998.

Li, K.W., *Applied Thermodynamics: Availability Method and Energy Conversion,* Taylor & Francis, Washington, D.C., 1996.

McQuiston, F.C. and Spitler, J.D., *Cooling and Heating Load Calculation Manual*, American Society of Heating, Refrigerating, and Air-Conditioning Engineers, Atlanta, GA, 1992.

NASA, U.S. Standard Atmosphere. National Oceanic and Atmospheric Administration, National Aeronautics and Space Administration, and the U.S. Air Force, Superintendent of Documents. U.S. Government Printing Office, Washington, D.C., 1976.

Pederson, C.O., Fisher, D.E., Spiltler, J.D., and Liesen, R.J., *Load Calculation Principles,* American Society of Heating, Refrigerating, and Air-Conditioning Engineers, Atlanta, GA, 1998.

Stoecker, W.F., *Refrigeration and Air Conditioning,* McGraw-Hill, New York, 1982.

Stoecker, W.F., *Industrial Refrigeration Handbook*, McGraw-Hill, New York, 1998.

United Nations, 1994 Report of the Refrigeration, Air Conditioning and Heat Pumps Technical Options Committee, Montreal Protocol on Substances that Deplete the Ozone Layer, United Nations Environmental Program, New York, 1994.

Van der Ree, H., Replacement of R-22, *Bull. Int. Inst. Refrig.,* 75(1), 5-17, 1998.

4.14 Optimization of Thermal Systems

Yogesh Jaluria

Introduction

Optimization is by no means a new concept. In our daily lives, we attempt to optimize by seeking to obtain the largest amount of goods or output per unit expenditure, this being the main idea behind clearance sales and competition. As we well know, most students continually try to achieve the best grades with the least amount of work, hopefully without violating the constraints imposed by ethics and regulations. The worth of various items, including consumer products like televisions, automobiles and cameras, vacation trips, advertisements, and even education per dollar spent, is often quoted to indicate the cost-effectiveness of these items. Different measures of quality, such as durability, finish, dependability, corrosion resistance, strength, and speed are included in these considerations, often on the basis of actual consumer inputs, as is the case with publications such as *Consumer Reports*. Thus, a buyer, who may be a student (or a parent) seeking an appropriate college for higher education, a couple looking for a cruise, or a young professional searching for his first dream car, may use information available on the best buy for their money to make their choice.

The need to optimize is similarly very important in the design of systems, particularly thermal systems, and has become crucial in recent times due to growing global competition. It no longer is enough to obtain a workable or feasible thermal system that performs the desired tasks and meets the given constraints. At the very least, several workable designs should be generated and the final design, one that minimizes or maximizes an appropriately chosen quantity, be selected from these. In general, there are many competing factors that affect the performance and cost of a system. Therefore, if the governing parameters that characterize a system are varied, an optimum can often be obtained in quantities such as power per unit fuel input, cost, efficiency, energy consumption per unit output, and other features of the system. Different product characteristics may be of particular interest in different applications and the most important and relevant ones may be employed for optimization. For instance, weight is particularly important in aerospace and aeronautical applications, acceleration in automobiles, energy consumption in refrigerators, and the flow rate in a water-pumping system. Thus, these characteristics may be chosen for minimization or maximization.

Workable designs are obtained over the allowable ranges of the design variables in order to satisfy the given requirements and constraints. A unique solution is generally not obtained and different system designs may be generated for a given application. We may call this region over which acceptable designs are obtained as the domain of workable designs, given in terms of the physical variables in the problem. Figure 4.14.1 shows, qualitatively, a sketch of such a domain in terms of design variables x_1 and x_2, where these may be physical quantities such as the diameter and length of the shell in a shell-and-tube heat exchanger. Then, any design in this domain is an acceptable or workable design and may be selected for the problem at hand. Optimization, on the other hand, tries to find the best solution, one which minimizes or maximizes a feature or quantity of particular interest in the application under consideration. Local extrema may be present at different points in the domain of acceptable designs. However, only one global optimal point, which yields the minimum or maximum in the entire domain, is found to arise in most applications, as sketched in the figure. It is this optimal design that is sought.

The optimization process is expected to yield an optimal design or a subdomain in which the optimum lies and the final system design is obtained on the basis of this solution. The design variables are generally not taken as exactly equal to those obtained from the optimal solution, but are changed somewhat to use more convenient sizes, dimensions, and standard items available from the manufacturer. For instance, an optimal dimension of 4.65 m may be taken as 5.0 m, a 8.34 kW motor as a 10 kW motor, or a 1.8 kW heater as a 2.0 kW heater, because items with these specifications may be readily available, rather than having these specially made. An important concept which is used at this stage to finalize the design variables is *sensitivity*, which indicates the effect of changing a given variable on the output or performance

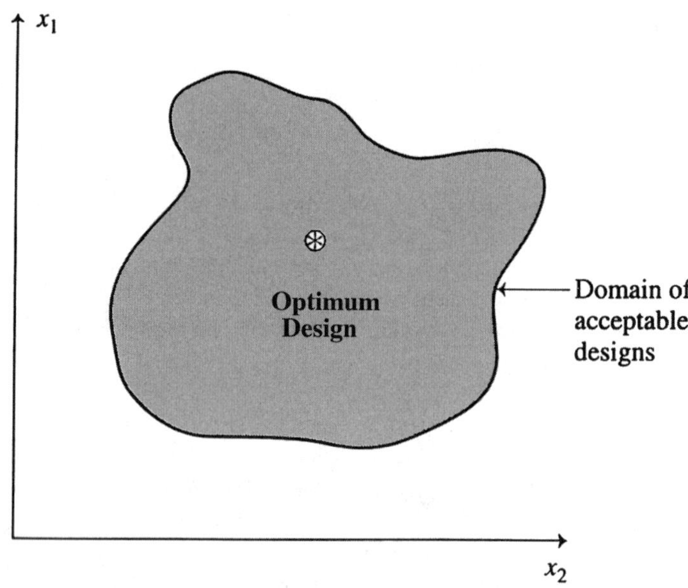

FIGURE 4.14.1 The optimum design in a domain of acceptable designs.

of the system. In addition, safety factors are employed to account for inaccuracies and uncertainties in the modeling, simulation, and design, as well as for fluctuations in operating conditions and other unforeseen circumstances. Some changes may also be made due to fabrication or material limitations. On the basis of all these considerations, the final system design is obtained and communicated to various interested parties, particularly those involved in fabrication and prototype development.

Generally, optimization of a system is directed at its hardware, i.e., at the geometry, dimensions, materials, and components. The hardware refers to the fixed parts of the system; components that can not be easily varied and items which determine the overall specifications of the system. However, the system performance is also dependent on operating conditions, such as temperature, pressure, flow rate, heat input, etc. These conditions can generally be varied quite easily, though over ranges that are determined by the hardware. Therefore, the output of the system as well as the costs incurred may also be optimized with respect to the operating conditions. Such an optimum may be given in terms of the conditions for obtaining the highest efficiency or output. For instance, the temperature setting for optimal output from an air conditioner or a refrigerator may be given as a function of the ambient conditions.

This section presents the important considerations that govern the optimization of a thermal system. Special considerations that arise for thermal systems, such as the thermal efficiency, energy losses, and heat input are discussed. Important questions regarding the implementation of the optimal solution, such as sensitivity analysis, dependence on the model, effect of quantity chosen for optimization, and selection of design variables for the final design, are considered. Many specialized books have been written on optimization in design; for instance, those by Fox (1971), Siddal (1982), Vanderplaats (1984), Arora (1989), and Papalambros and Wilde (1989). A few books are also directed at the design and optimization of thermal systems, such as those by Stoecker (1989), Bejan et al. (1996), and Jaluria (1998). Books are also available on the basic aspects of optimization, such as those by Beveridge and Schechter (1970) and Beightler et al. (1979). These books may be consulted for further details on optimization techniques and their application to design.

Basic Concepts

We can now proceed to formulate the basic problem for the optimization of a thermal system. Since the optimal design must satisfy the given requirements and constraints, the designs considered as possible candidates must be acceptable or workable ones. This implies that the search for an optimal design is carried out in the domain of acceptable designs. For example, if several acceptable designs for a refrigeration system are obtained by applying the design process, the best design from these may be selected on the basis of a chosen criterion such as energy consumption per unit heat removed. The conceptual design is kept fixed so that optimization is carried out within a given concept. Generally, different concepts are considered at the early stages of the design process and a particular conceptual design is selected on the basis of prior experience, environmental impact, material availability, etc. However, if a satisfactory design is not obtained with a particular conceptual design, the design and optimization process may be repeated, starting with a different conceptual design.

Objective Function

Any optimization process requires specification of a quantity or function which is to be minimized or maximized. This function is known as the *objective function* and it represents the aspect or feature which is of particular interest in a given circumstance. Though different costs, including initial and maintenance costs and profit are the most commonly used quantities to be optimized, many other aspects are employed for optimization, depending on the system and the application. The objective functions that are optimized in thermal systems are frequently based on the following characteristics:

- Weight
- Size, volume
- Rate of energy consumption
- Heat transfer rate
- Efficiency
- Overall profit
- Costs incurred
- Environmental effects
- Durability and dependability
- Safety
- System performance, output delivered

The weight is of particular interest in transportation systems, such as airplanes and automobiles. Therefore, an electronic system designed for an airplane may be optimized in order to have the smallest weight while it meets the requirements for the task. Similarly, the size of the air-conditioning system for environmental control of a house may be minimized in order to require the least amount of space. Energy consumption per unit output is particularly important for thermal systems and is usually indicative of the efficiency of the system. Frequently, this is given in terms of the energy rating of the system, thus specifying the power consumed for operation under given conditions. Refrigeration, heating, drying, air conditioning, and many such consumer-oriented systems are generally optimized to achieve the minimum rate of energy consumption for specified operating conditions. Costs and profits are always important considerations and efforts are made to minimize the former and maximize the latter. The output is also of particular interest in many thermal systems, such as manufacturing processes and automobiles. However, even if one wishes to maximize the thrust, torque, or power delivered by a motor vehicle, cost is still a very important consideration. Therefore, in many cases the objective function is based on the output per unit cost. Similarly, other relevant measures of performance are considered in terms of the costs involved. Environmental effects, safety, product quality, and several other such aspects are important in various applications and may also be considered for optimization.

Let us denote the objective function which is to be optimized by U, where U is a function of the n independent variables in the problem $x_1, x_2, x_3, \ldots, x_n$. Then the objective function and the optimization process may be expressed as

$$U = U(x_1, x_2, x_3, \ldots, x_n) \rightarrow U_{opt} \tag{4.14.1}$$

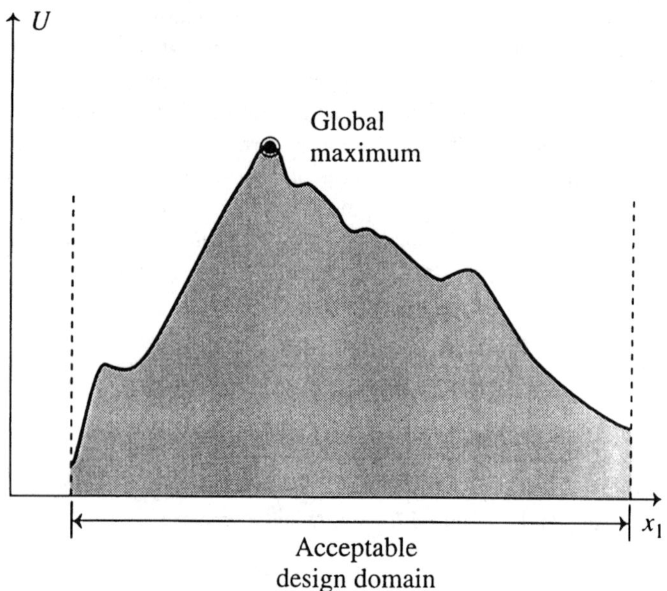

FIGURE 4.14.2 Global maximum of the objective function U in an acceptable design domain.

where U_{opt} denotes the optimal value of U. The x's represent the design variables as well as the operating conditions, which may be changed to obtain a workable or optimal design. Physical variables such as height, thickness, material properties, heat flux, temperature, pressure, and flow rate may be varied over allowable ranges to obtain an optimum design, if such an optimum exists. A minimum or a maximum in U may be sought, depending on the nature of the objective function. Therefore, we would seek to maximize the output, minimize the cost, maximize the output per unit cost, and so on.

The process of optimization involves finding the values of the different design variables for which the objective function is minimized or maximized, without violating the constraints. Figure 4.14.2 shows a sketch of a typical variation of the objective function U with a design variable x_1, over its acceptable range, where U may be output per unit cost of the system and x_1 the size. It is seen that though there is an overall, or global, maximum in U (x_1), there are several local maxima or minima. Our interest lies in obtaining this global optimum. However, the local optima can often confuse the true optimum, making its determination difficult. It is necessary to distinguish between local and global optima so that the best design is obtained over the entire domain.

Constraints

The constraints in a given design problem arise due to limitations on the ranges of the physical variables, and due to the basic conservation principles that must be satisfied. The restrictions on the variables may arise due to the space, equipment, and materials being employed. These may restrict the dimensions of the system, the highest temperature that the components can safely attain, allowable pressure, material flow rate, force generated, and so on. Minimum values of the temperature may be indicated for thermoforming of a plastic and for ignition to occur in an engine. Thus, both minimum and maximum values of the design variables may be indicated.

The constraints limit the domain in which the workable or optimal design lies. Figure 4.14.3 shows a few examples in which the boundaries of the design domain are determined by constraints arising from material or space limitations. For instance, in heat treatment of steel, the minimum temperature needed for the process T_{min} is given, along with the maximum allowable temperature T_{max} at which the material will be damaged. Similarly, the maximum pressure p_{max} in a metal extrusion process is fixed

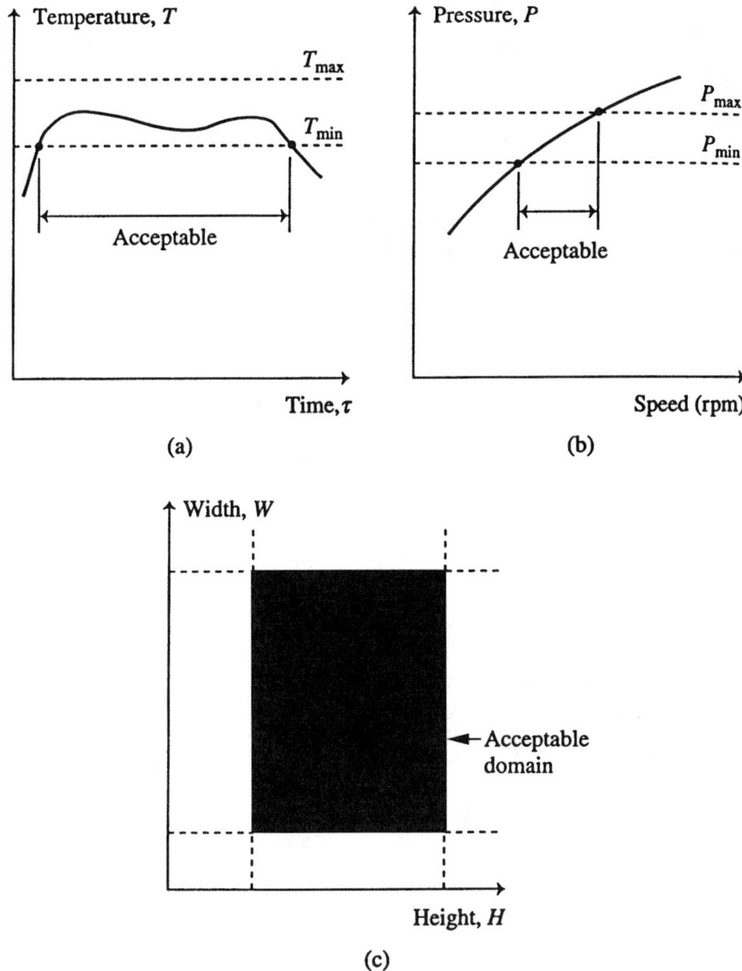

FIGURE 4.14.3 Boundaries of the acceptable design domain specified by limitations on the variables for (a) heat treatment, (b) metal extrusion, and (c) cooling of electronic equipment.

by strength considerations of the extruder and the minimum by the flow stress needed for the process to occur. The limitations on the dimensions W and H define the domain in an electronic system.

Many constraints arise because of the conservation laws, particularly those related to mass, momentum, and energy in thermal systems. Thus, under steady-state conditions, the mass inflow into the system must equal the mass outflow. This condition gives rise to an equation which must be satisfied by the relevant design variables, thus restricting the values that may be employed in the search for an optimum. Similarly, energy balance considerations are very important in thermal systems and may limit the range of temperatures, heat fluxes, dimensions, etc. that may be used. Several such constraints are often satisfied during modeling and simulation since the governing equations are based on the conservation principles. Then the objective function being optimized has already taken these constraints into account. In such cases, only the additional limitations that define the boundaries of the design domain are left to be considered.

There are two types of constraints, *equality* constraints and *inequality* constraints. As the name suggests, equality constraints are equations which may be written as

$$G_1 (x_1, x_2, x_3, \dots, x_n) = 0$$
$$G_2 (x_1, x_2, x_3, \dots, x_n) = 0$$
$$\vdots \qquad (4.14.2)$$
$$G_m (x_1, x_2, x_3, \dots, x_n) = 0$$

Similarly, inequality constraints indicate the maximum or minimum value of a function and may be written as

$$H_1 (x_1, x_2, x_3, \dots, x_n) \le C_1$$
$$H_2 (x_1, x_2, x_3, \dots, x_n) \le C_2$$
$$H_3 (x_1, x_2, x_3, \dots, x_n) \ge C_3$$
$$\vdots \qquad (4.14.3)$$
$$H_\ell (x_1, x_2, x_3, \dots, x_n) \ge C_\ell$$

Therefore, either the upper or the lower limit may be given for an inequality constraint. Here, the C's are constants or known functions. The m equality and ℓ inequality constraints are given for a general optimization problem in terms of the functions G and H, which are dependent on the n design variables $x_1, x_2, \dots, x_n$. Thus, the constraints in Figure 4.14.3 may be given as $T_{min} \le T \le T_{max}$, $p_{min} \le p \le p_{max}$, and so on.

Equality constraints are most commonly obtained from conservation laws, e.g., for a steady flow circumstance in a control volume, we may write

$$\Sigma \text{ (mass flow rate)}_{in} - \Sigma \text{ (mass flow rate)}_{out} = 0, \qquad (4.14.4a)$$

or,

$$\Sigma (\rho V A)_{in} - \Sigma (\rho V A)_{out} = 0 \qquad (4.14.4b)$$

where ρ is the mean density of the material, V the average velocity, A the cross-sectional area, and Σ denotes the sum of flows in and out of several channels, as sketched in Figure 4.14.4. Similarly, equations for energy balance and momentum-force balance may be written. The conservation equations may be employed in their differential or integral forms, depending on the detail needed in the problem.

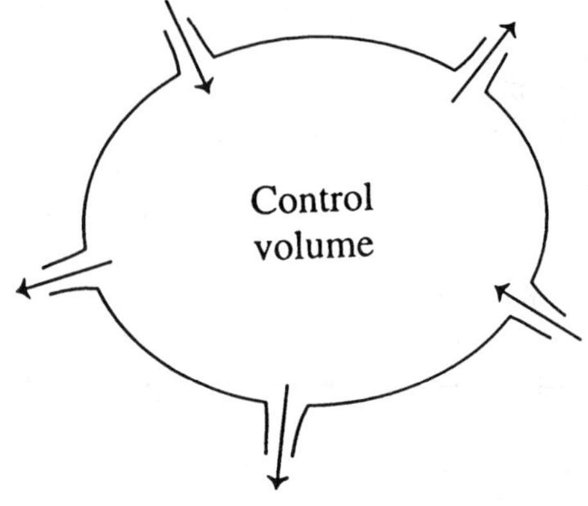

FIGURE 4.14.4 Inflow and outflow of material and energy in a fixed control volume.

It is generally easier to deal with equalities than with inequalities, since many methods are available to solve different types of equations and systems of equations, whereas no such schemes are available for inequalities. Therefore, inequalities are often converted into equations before applying optimization methods. A common approach employed to convert an inequality into an equation is to use a value larger than the constraint if a minimum is specified and a value smaller than the constraint if a maximum is given. For instance, the constraints may be changed as given below

$$H_1 (x_1, x_2, x_3, \ldots, x_n) \leq C_1 \text{ becomes } H_1 (x_1, x_2, x_3, \ldots, x_n) = C_1 - \Delta C_1 \quad (4.14.5a)$$

$$H_3 (x_1, x_2, x_3, \ldots, x_n) \geq C_3 \text{ becomes } H_3 (x_1, x_2, x_3, \ldots, x_n) = C_3 + \Delta C_3 \quad (4.14.5b)$$

where ΔC_1 and ΔC_3 are chosen quantities, often known as *slack* variables, that indicate the difference from the specified limits. Though any finite values of these quantities will satisfy the given constraints, generally the values are chosen on the basis of the characteristics of the given problem and the critical nature of the constraint. Frequently, a fraction of the actual limiting value is used as the slack to obtain the corresponding equation. For instance, if 200°C is given as the limiting temperature for a plastic, a deviation of, say, 10°C may be taken as acceptable to convert the inequality into an equation.

Operating Conditions Vs. Hardware

It has been mentioned earlier that the process of optimization may be applied to a system so that the design, given in terms of the hardware, is optimized. Most of our discussion on optimization will focus on the system so that the corresponding hardware, which includes dimensions, materials, components, etc. is varied to obtain the best design with respect to the chosen objective function. However, it is worth reiterating that once a system has been designed, its performance and characteristics are also functions of the operating conditions. Therefore, it may be possible to obtain conditions under which the system performance is optimum. For instance, if we are interested in minimum fuel consumption of a motor vehicle, we may be able to determine a speed, such as 88 km/h (55 mi/h), at which this condition is met. Similarly, the optimum setting for an air conditioner, at which the efficiency is maximum, may be determined as, say, 22.2°C (72°F) or the revolutions per minute (rpm) of a motor as 125 for optimal performance.

The operating conditions vary from one application to another and from one system to the next. The range of variation of these conditions is generally fixed by the hardware. Therefore, if a heater is chosen for the design of a furnace, the heat input and temperature ranges are fixed by the specifications of the heater. Similarly, a pump or a motor may be used to deliver an output over the range for which it can be satisfactorily operated. The operating conditions in thermal systems are commonly specified in terms of the following variables:

- Heat input rate
- Temperature
- Pressure
- Mass or volume flow rate
- Speed, revolutions per minute (rpm)
- Chemical composition

Thus, imposed temperature and pressure, as well as the rate of heat input, may be varied over the allowable ranges for a system such as a furnace or a boiler. The volume or mass flow rate is chosen, along with the speed (revolutions per minute) for a system like a diesel engine or gas turbine. The chemical composition is important in specifying the chosen inlet conditions for a chemical reactor, such as a food extruder where the moisture content is an important variable.

All such variables that characterize the operation of a given thermal system may be set at different values, over the ranges determined by the system design, and thus affect the system output. It is useful to determine the optimum operating conditions and the corresponding system performance. The approach to optimize the output or performance in terms of the operating conditions is similar to that employed for the hardware design and optimization. The model is employed to study the dependence of the system performance on the operating conditions and an optimum is chosen using the methods discussed here.

Therefore, the general mathematical formulation for the optimization of a system may be written as $U (x_1, x_2, x_3, \ldots, x_n) \rightarrow U_{opt}$, with

$$G_i (x_1, x_2, x_3, \ldots, x_n) = 0, \text{ for } i = 1, 2, 3, \ldots, m$$

and (4.14.6)

$$H_i (x_1, x_2, x_3, \ldots, x_{n)} \leq \text{ or } \geq C_i, \text{ for } i = 1, 2, 3, \ldots, l$$

If the number of equality constraints m is equal to the number of independent variables n, the constraint equations may simply be solved to obtain the variables and there is no optimization problem. If $m > n$, the problem is over-constrained and a unique solution is not possible since some constraints have to be discarded to make $m \leq n$. If $m < n$, an optimization problem is obtained. This is the case considered here.

Optimization Methods

There are several methods that may be employed for solving the mathematical problem given by Equation (4.14.6) for optimizing a thermal system. Each approach has its limitations and advantages over the others. The choice of method largely depends on the nature of the equations representing the objective function and the constraints. It also depends on whether the mathematical formulation is expressed in terms of explicit functions or if numerical solutions are to be obtained to determine the variation of the objective function and the constraints with the design variables. Because of the complicated nature of typical thermal systems, numerical solutions of the governing equations are often needed. However, in several cases detailed numerical results are generated from a model of the system and these are curve-fitted to obtain algebraic equations to represent the characteristics of the system. Optimization of the system may then be undertaken on the basis of these relatively simple algebraic expressions and equations. Let us now consider the commonly used methods for optimization and discuss the nature and type of equations to which these may be applied.

Calculus Methods

The use of calculus for determining the optimum is based on derivatives of the objective function and of the constraints. The derivatives are used to indicate the location of a minimum or a maximum. At a local optimum, the slope is zero, as sketched in Figure 4.14.5 for U varying with a single design variable x_1 or x_2. The equations and expressions that formulate the optimization problem must be continuous and

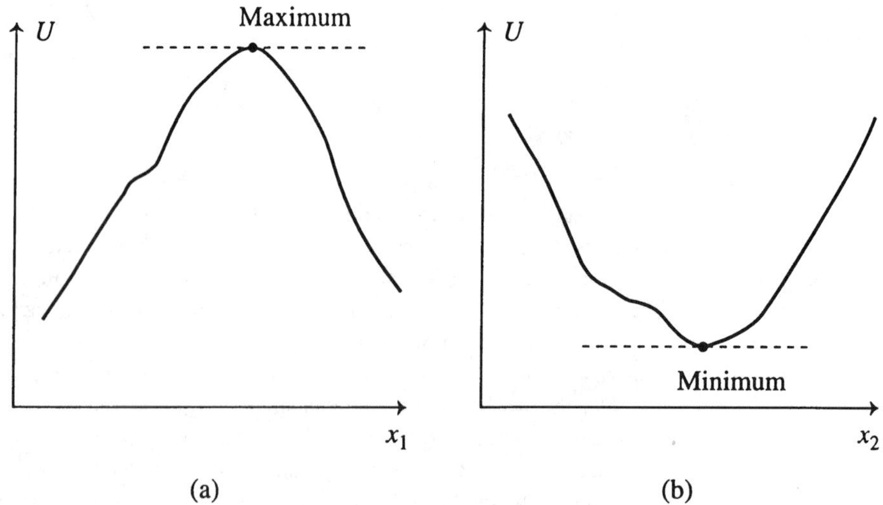

FIGURE 4.14.5 Maximum or minimum in the objective function U, varying with a single independent variable x_1 or x_2.

well-behaved, so that these are differentiable over the design domain. An important method which employs calculus for optimization is the method of *Lagrange multipliers*. This method basically converts the preceding problem of finding the minimum or maximum into the solution of a system of algebraic equations, thus providing a convenient scheme to determine the optimum. The objective function and the constraints are combined into a new function Y, known as the Lagrange expression and defined as

$$Y(x_1, x_2, \ldots, x_n) = U(x_1, x_2, \ldots, x_n) + \lambda_1 G_1(x_1, x_2, \ldots, x_n) + \lambda_2 G_2(x_1, x_2, \ldots, x_n) + \ldots + \lambda_m G_m(x_1, x_2, \ldots, x_n) \tag{4.14.7}$$

where the λ's are unknown parameters, known as *Lagrange multipliers*. Then, according to this method, the optimum occurs at the solution of the system of equations formed by the following equations

$$\frac{\partial Y}{\partial x_1} = 0 \quad \frac{\partial Y}{\partial x_2} = 0 \quad \ldots \quad \frac{\partial Y}{\partial x_n} = 0 \tag{4.14.8a}$$

$$\frac{\partial Y}{\partial \lambda_1} = 0 \quad \frac{\partial Y}{\partial \lambda_2} = 0 \quad \ldots \quad \frac{\partial Y}{\partial \lambda_m} = 0 \tag{4.14.8b}$$

This implies that the optimum is obtained by solving the following system of equations

$$\begin{aligned}
&\frac{\partial U}{\partial x_1} + \lambda_1 \frac{\partial G_1}{\partial x_1} + \lambda_2 \frac{\partial G_2}{\partial x_1} + \cdots + \lambda_m \frac{\partial G_m}{\partial x_1} = 0 \\
&\frac{\partial U}{\partial x_2} + \lambda_1 \frac{\partial G_1}{\partial x_2} + \lambda_2 \frac{\partial G_2}{\partial x_2} + \cdots + \lambda_m \frac{\partial G_m}{\partial x_2} = 0 \\
&\cdots\cdots\cdots \\
&\frac{\partial U}{\partial x_n} + \lambda_1 \frac{\partial G_1}{\partial x_n} + \lambda_2 \frac{\partial G_2}{\partial x_n} + \cdots + \lambda_m \frac{\partial G_m}{\partial x_n} = 0 \\
&G_1(x_1, x_2, x_3, \ldots, x_n) = 0 \\
&G_2(x_1, x_2, x_3, \ldots, x_n) = 0 \\
&\ldots \\
&G_m(x_1, x_2, x_3, \ldots, x_n) = 0
\end{aligned} \tag{4.14.9}$$

If the objective function U and the constraints G_i are continuous and differentiable, a system of algebraic equations is obtained. Since there are m equations for the constraints and n equations are derived from the Lagrange expression, a total of $m+n$ simultaneous equations are obtained. The unknowns are the m multipliers, corresponding to the m constraints, and the n independent variables. Therefore, this system may be solved to obtain the values of the independent variables, that define the optimum as well as the multipliers. Analytical methods for solving a system of algebraic equations may be employed if linear equations are obtained and/or when the number of equations is small, typically up to around five. For nonlinear equations and for large sets, numerical methods are generally more appropriate. The optimum value of the objective function is then determined by substituting the values obtained for the independent variables into the expression for U. The optimum is often represented by asterisks, i.e., x_1^*, x_2^*, …, x_n^* and U^*.

As an example, consider an objective function U of the form

$$U = A\,(x)^a + B\,(y)^b \tag{4.14.10}$$

with a constraint of the form $(x)^c\,(y)^d = E$, which may be written as

$$G = (x)^c\,(y)^d - E = 0 \tag{4.14.11}$$

Here E, the coefficients A and B, and the exponents a, b, c, and d are assumed to be known constants. Such expressions are frequently encountered in thermal systems. For example, U may be the overall cost and x and y the pump needed and head pipe diameter, respectively, in a water flow system. The pressure decreases as the diameter increases, resulting in lower cost for the pump and the cost for the pipe increases. This gives rise to a relationship such as Equation (4.14.11). For this single-constraint problem, Equation (4.14.9) may be written as $\nabla U + \lambda \nabla G = 0$ and $G = 0$, where ∇ is the gradient vector (Jaluria, 1998). Contours of constant U may be drawn along with the constraint curve on a x-y plane, as sketched in Figure 4.14.6a. Then the optimum is indicated by the location where the constant U contour becomes tangential to the constraint curve, thus aligning the ∇U and ∇G vectors and satisfying the preceding vector equation. For the simple case when all the constants in these expressions are unity, i.e., $U = x + y$ and $G = x\,y - 1 = 0$, the constant U contours are straight lines and the constraint curve is given by $x = 1/y$, as sketched in Figure 4.14.6b. The optimum is at $x^* = 1.0$ and $y^* = 1.0$ and the optimum value U^* is 2.0 for this case.

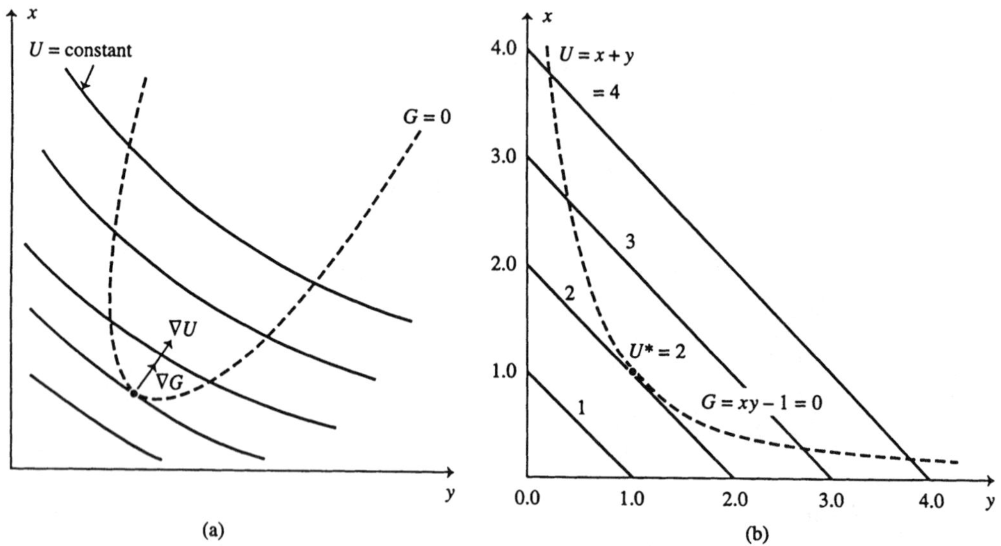

FIGURE 4.14.6 (a) Optimization of the simple constrained problem given by Equations (4.14.10) and (4.14.11), and (b) the results when all the constants in the given expressions are unity.

It can be shown that the Lagrange multiplier λ is the negative of the sensitivity coefficient S_c, which indicates the rate of change in the optimum with change in the constraint. This analysis can easily be extended to multiple constraints and larger number of independent variables. It can then be shown that $\lambda_1 = -(S_c)_1$, $\lambda_2 = -(S_c)_2$, …, $\lambda_m = -(S_c)_m$, where the subscript numbers refer to the different constraint equations. Thus, the method of Lagrange multipliers not only yields the optimum but also the sensitivity coefficients with respect to the various constraints in the problem. The Lagrange multiplier λ may also be related to the change in the objective function U, at the optimum, with the constraint parameter E as $\lambda = -\partial U/\partial E$. Therefore, the values of the Lagrange multipliers obtained in a given problem may be used to guide slight variations in the constraints to choose appropriate sizes, dimensions, etc., that are easily available or more convenient to fabricate.

Example

As a practical example, consider an electronic circuitry in which the power source may be considered as a square with side dimension L in meters. It is desired to minimize the heat transfer from the power supply. The heat transfer coefficient h in W/m^2K is given by the expression

$$h = (2 + 10\ L^{1/2})\ \Delta T^{1/4}\ L^{-1}$$

where ΔT is the temperature difference in K from the ambient. A constraint arises due to the strength of the bond that attaches the power supply to the electronic circuit board as $L\ \Delta T = 5.6$. We need to calculate the side dimension L of the square that would minimize the total heat loss, solving the problem as a constrained one.

The rate of heat loss $\dot{Q}$ from the power supply is the objective function which is to be minimized and is given by the expression

$$\dot{Q} = h\ A\ \Delta T = (2\ L + 10\ L^{3/2})\ \Delta T^{5/4}$$

since area $A = L^2$. The constraint is $L\ \Delta T = 5.6$.

Therefore, the optimum is given by the equations

$$\Delta T^{5/4}\left(2 + 15\ L^{1/2}\right) + \lambda\ \Delta T = 0$$

$$\frac{5}{4}\left(2\ L + 10\ L^{3/2}\right)\Delta T^{1/4} + \lambda\ L = 0$$

$$L\ \Delta T = 5.6$$

These equations can be solved to yield the optimum as

$$L^* = 0.04\ \text{m};\ \Delta T^* = 140;\ \dot{Q}^* = 77.05\ \text{W};\ \lambda = -17.2$$

It can be shown that if the constraint parameter is increased from 5.6 to 5.7, the heat transfer rate $\dot{Q}$ becomes 78.77, i.e., an increase of 1.72. This is expected since the sensitivity coefficient $S_c = -\lambda = 17.2$. Therefore, the effect of relaxing the constraint on the optimum value can be determined. This problem can also be solved as an unconstrained one by substituting ΔT in terms of L from the constraint into the expression for $\dot{Q}$. The solution becomes simpler, but λ is not obtained.

The range of application of calculus methods to the optimization of thermal systems is somewhat limited because of complexities that commonly arise in these systems. Numerical solutions are often needed to characterize the behavior of the system and implicit, nonlinear equations that involve variable material properties are frequently encountered. However, curve fitting may be employed in some cases to yield algebraic expressions that closely approximate the system and material characteristics. If these expressions are continuous and easily differentiable, calculus methods may be conveniently applied to yield the optimum. However, only equality constraints can be considered. Therefore, inequalities must be converted to equations, as discussed earlier, to apply this approach. The method of Lagrange multipliers provides information, through the multipliers, on the sensitivity of the optimum with respect to changes in the constraints. In view of these features, it is worthwhile to apply the calculus methods whenever possible. However, curve-fitting often requires extensive data that may involve detailed experimental measurements or numerical simulations of the system. Since this may demand a considerable amount of effort and time, particularly for thermal systems, it is often better to use other methods of optimization that require a relatively smaller number of simulations.

Search Methods

As the name suggests, these methods involve selection of the best solution from a number of workable designs. If the design variables can only take on certain fixed values, different combinations of these variables may be considered to obtain possible acceptable designs. If these variables can be varied continuously over their allowable ranges, a finite number of acceptable designs may be generated by changing the variables. In either case, a number of workable designs are obtained and the optimal design selected from these. In the simplest approach, the objective function is calculated at uniformly spaced locations in the domain, selecting the design with the optimum value. This approach, known as *exhaustive search*, is not very imaginative and is clearly an inefficient method to optimize a system. As such, it is generally not used for practical systems. However, the basic concept of selecting the best design from a set of acceptable designs is an important one and is used even if a detailed optimization of the system is not undertaken. Sometimes, an unsystematic search, based on prior knowledge of the system, and known as an heuristic search, is carried out instead.

Importance

In many thermal systems of practical interest, the design variables are not continuous functions but assume finite values over their acceptable ranges. This is largely due to the limited number of materials and components available for design. Finite numbers of components such as pumps, blowers, fans, compressors, heat exchangers, heaters, and valves are generally available from the manufacturers at given specifications. Even though additional, intermediate, specifications can be obtained if these are custom made, it is much cheaper and more convenient to consider what is readily available and base the system design on these. Similarly, a finite number of different materials may be considered for the system parts, leading to a finite number of discrete design choices.

In order to obtain an acceptable design, the design process which involves modeling, simulation, and evaluation of the design is followed. Results from the simulation are also needed to determine the effect of the different design variables on the objective function. Because of the effort needed to simulate typical thermal systems, a systematic search strategy is necessary so that the number of simulation runs is kept at a minimum. Each run or set of runs must be used to move closer to the optimum. Random or unsystematic searches where many simulation runs are carried out over the design domain are very inefficient and impractical.

Search methods can be used for a wide variety of problems, ranging from very simple problems with unconstrained single-variable optimization to extremely complicated systems with many constraints and variables. Because of their versatility and easy application, these methods are the ones most commonly used for optimizing thermal systems. In addition, these methods can be used to improve the design even if a complete optimization process is not undertaken. For instance, if an acceptable design has been obtained, the design variables may be varied from the values obtained, in the vicinity of the acceptable design. This allows one to search for a better solution, as given by improvement in the objective function. Similarly, several acceptable designs may be generated during the design process. Again, the best among these is selected as the optimum in the given domain.

It is obvious that search methods provide very important and useful approaches to extract the optimum design and to improve existing designs. We will focus on systematic search schemes which may be used to determine the optimum design in a region whose boundaries are defined by limitations on the design variables. In order to illustrate the different methods, relatively simple expressions are employed here for which search methods are not necessary and simpler schemes such as the calculus methods can easily be employed. However, this is only for illustration purposes and in actual practice each test run or simulation would generally involve considerable time and effort.

There are several approaches that may be employed in search methods, depending on whether a constrained or an unconstrained problem is being considered and on whether the problem involves a single variable or multiple variables. The classification of these methods and the major techniques are discussed here.

Elimination Methods

In these methods, the domain in which the optimum lies is gradually reduced by eliminating regions that are determined not to contain the optimum. We start with the design domain defined by the acceptable ranges of the variables. This region is known as the initial *interval of uncertainty*. Therefore, the region of uncertainty in which the optimum lies is reduced until a desired interval is achieved. Appropriate values of the design variables are chosen from this interval to obtain the optimal design. For single-variable problems, the main search methods based on elimination are

- Exhaustive search
- Dichotomous search
- Fibonacci search
- Golden section search

All these approaches have their own characteristics, advantages, and applicability. These methods can also be used for multivariable problems by applying the approach to one variable at a time.

Uniform exhaustive search — As the name suggests, this method employs uniformly distributed locations over the entire domain to determine the objective function. The number of runs n is chosen and the initial range L_0 of variable x is subdivided by placing n points uniformly over the domain. Therefore, n+1 subdivisions, each of width $L_0/(n+1)$, are obtained. At each of these n points, the objective function U (x) is evaluated through simulation of the system. The interval containing the optimum is obtained by eliminating regions where inspection indicates that it does not lie. Thus, if a maximum in the objective function is desired, the region between the location where the smaller value of U (x) is obtained in two runs and the nearest boundary is eliminated, as shown in Figure 4.14.7 in terms of the results from three runs. In Figure 4.14.7a, the region beyond C and that before A are eliminated, thus reducing the domain in which the maximum lies to the region between A and C. Similarly, in Figure 4.14.7b, the region between the lower domain boundary and point B is eliminated.

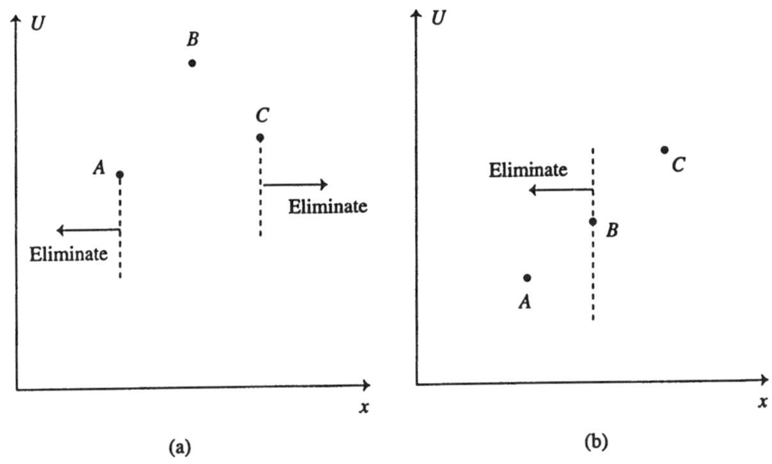

FIGURE 4.14.7 Elimination of regions in the search for a maximum.

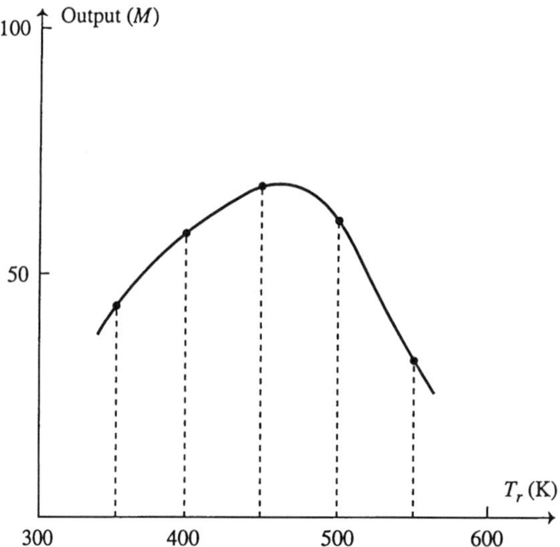

FIGURE 4.14.8 Uniform exhaustive search for the maximum in the output M in a chemical reactor, with the temperature T_r as the independent variable.

Consider a chemical manufacturing plant in which the temperature T_r in the reactor determines the output M by shifting the equilibrium of the reaction. If the temperature can be varied over the range 300 to 600 K, the initial region of uncertainty is 300 K. The maximum output in this range is to be determined. If five trial points or runs are chosen, i.e., n = 5, the range is subdivided into six intervals, each of width 50 K, as shown in Figure 4.14.8. The output is computed from a simulation of the system at the chosen points and the results obtained are shown. From inspection, the maximum output must lie in the interval $400 < T_r < 500$. Therefore, the interval of uncertainty has been reduced from 300 to 100 as a result of five runs. In general, the final region of uncertainty L_f is

$$L_f = \frac{L_o}{\left(\frac{n+1}{2}\right)} = \frac{2\,L_o}{n+1} \tag{4.14.12}$$

since two subintervals, out of a total of n+1, contain the optimum.

The reduction of the interval of uncertainty is generally expressed in terms of the reduction ratio R, defined as

$$R = \frac{\text{Initial energy of uncertainty}}{\text{Final interval of uncertainty}} = \frac{L_o}{L_f} \tag{4.14.13}$$

For the uniform exhaustive search method, the reduction ratio is

$$R = \frac{n+1}{2} \tag{4.14.14}$$

Therefore, the number of trial runs n needed for obtaining a desired interval of uncertainty may be determined from this equation. For instance, if in the above example, the region containing the optimum is to be reduced to 30 K, the reduction ratio is 10 and the number n of trial runs needed to accomplish this is 19.

The exhaustive search method is not a very efficient strategy to determine the optimum, since it covers the entire domain uniformly. However, it does reveal the general characteristics of the objective function being optimized, particularly whether it is unimodal or not, whether there is indeed an optimum, and whether it is a maximum or a minimum. Therefore, though inefficient, this approach is useful for circumstances where the basic trends of the objective function are not known because of the complexity of the problem or because it is a new problem with little prior information. It is not unusual to encounter thermal systems with unfamiliar characteristics of the chosen objective function. The exhaustive search helps in defining the optimization problem more sharply than the original formulation. Only a small number of runs may be made initially to determine the behavior of the function. Using the information thus obtained, one of the more efficient approaches, presented below, may then be selected for optimization.

Dichotomous search — In a dichotomous search, trial runs are carried out in pairs, separated by a relatively small amount ε, in order to determine whether the objective function is increasing or decreasing. Therefore, the total number of runs must be even. Again, the function is assumed to be unimodal in the design domain and regions are eliminated using the values obtained in order to reduce the region of uncertainty which contains the maximum or the minimum. The dichotomous search method may be implemented in the following two ways.

Uniform dichotomous search. In this case, the pairs of runs are spread evenly over the entire domain. Therefore, the approach is similar to the exhaustive search method, except that pairs of runs are used in each case. Each pair is separated by a small amount ε in the independent variable. Considering the earlier example and sketched in Figure 4.14.8, the total design domain stretches from 300 to 600 K. We may decide to use four runs, placing one pair at 400 K and the other at 500 K, with a separation ε of 10 K in each case. As seen in Figure 4.14.9, the left pair allows us to eliminate the region from the left boundary to point A and the right pair the region beyond point b. Here, the pairs A, a and B, b are located at equal distance on either side of the chosen values of 400 and 500 K, with a separation of 10 K. The separation ε must be larger than the error in fixing the value of the variable in order to obtain accurate and repeatable results.

For n runs or simulations, the initial range L_0 is divided into (n/2)+1 subintervals, neglecting the region between a single pair. Since the final interval of uncertainty L has the width of a single subdivision, the reduction ratio R is obtained, neglecting the separation ε, as

$$R = \frac{n}{2} + 1 \tag{4.14.15}$$

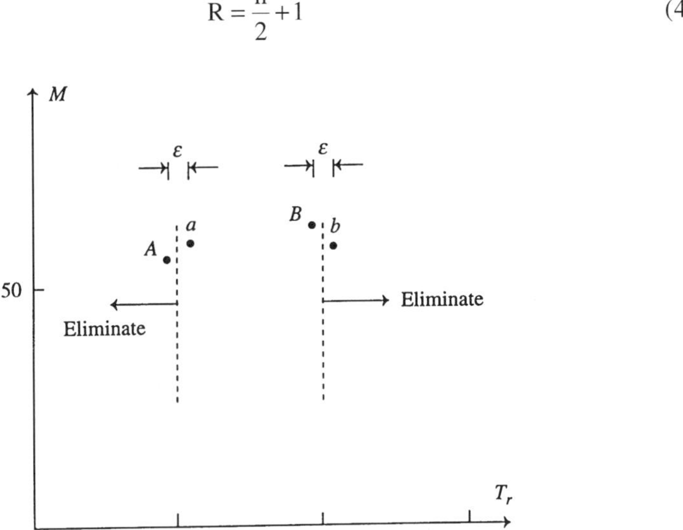

FIGURE 4.14.9 Uniform dichotomous search for maximum in M.

Therefore, the initial interval of uncertainty is reduced to one-third, or $400 < T_r < 500$ K, after four runs. With exhaustive search, 40% of the domain is left after four runs, as seen from Equation (4.14.13). Therefore, the uniform dichotomous search is slightly faster in convergence than the uniform exhaustive search. However, the sequential dichotomous search, given below, is a considerable improvement over both of these.

Sequential dichotomous search. As before, this method uses pairs of experiments or simulations to ascertain whether the function is increasing or decreasing and thus reduce the interval containing the optimum. However, it also uses the information gained from one pair of runs to choose the next pair. The first pair is located near the middle of the given range and about half the domain is eliminated. The next pair is then located near the middle of the remaining domain and the process repeated. This process is continued till the desired interval of uncertainty is obtained. Since pairs of runs are used, the total number of runs is even.

Considering, again, the example used earlier, let us locate the first pair of points, A and a, on either side of $T_r = 450$ K with a separation of 10 K. Since we are seeking a maximum in the output and since $M_a > M_A$, where M_a and M_A are the values of the objective function at these two points, the region to the left is eliminated and the new interval of uncertainty is $450 < T_r < 600$ K. The next pair, B and b, is then placed at the middle of this domain, i.e., at $T_r = 525$ K, as shown in Figure 4.14.10. Again, by inspection since $M_B > M_b$, the region to the right of the pair is eliminated, leaving the interval $450 < T_r < 525$. Therefore, the interval of uncertainty is reduced to 25%, or one-fourth, of its initial value. With each pair, the region of uncertainty is halved. Therefore, neglecting the separation ε, the interval is halved n/2 times, where n is the total number of runs and is an even number. Therefore, the reduction ratio is obtained as $R = 2^{n/2}$. This implies that an even number of runs may be chosen *a priori* to reduce the given initial range to obtain the desired accuracy in the selection of the variable for optimal design.

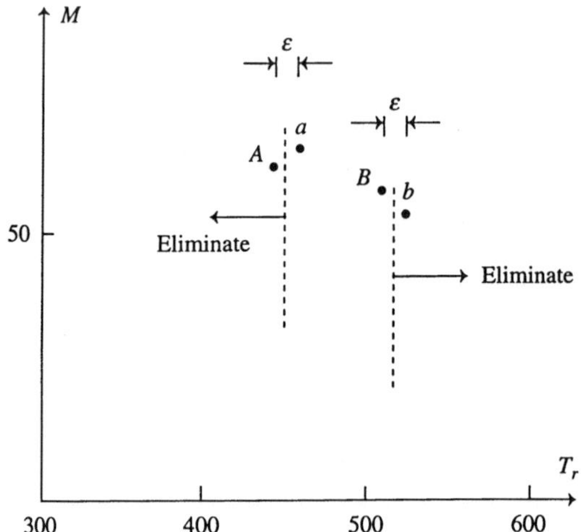

FIGURE 4.14.10 Sequential dichotomous search for maximum in M.

Fibonacci search — The Fibonacci search is a very efficient technique to narrow the domain in which the optimum value of the design variable lies. It uses a sequential approach based on the Fibonacci series, which is a series of numbers derived by Fibonacci, a mathematician in the thirteenth century. The series is given by the expression

$$F_n = F_{n-2} + F_{n-1}, \qquad (4.14.16)$$

where $F_0 = F_1 = 1$.

Therefore, the first two numbers in the series are unity and the nth number is the sum of the preceding two numbers. Thus, the Fibonacci series may be written as

n:	0	1	2	3	4	5	6	7	8	9	10	…
$\mathbf{F_n}$:	1	1	2	3	5	8	13	21	34	55	89	…

It can be seen from this series that the numbers increase rapidly as n increases. The fact that, for $n \geq 2$, each number is a sum of the last two numbers is used advantageously to distribute the trial runs or experiments.

The method starts by choosing the total number of runs n. This choice is based on the reduction ratio, as given earlier. The initial range of values L_o is assumed as given. Then the Fibonacci search places the first two runs at a distance $d_1 = (F_{n-2}/F_n)\, L_o$ from either end of the initial interval. For n = 5, this implies placing the runs at $d_1 = F_3/F_5 = (3/8)\, L_o$ from the two ends of the range. The simulation of the system is carried out at these two values of the design variable and the corresponding objective function determined. The values obtained are used to eliminate regions from further consideration, as discussed earlier and shown in Figure 4.14.7. The remaining interval of width L is now considered and runs are carried out at a distance of d_2 from each end of this interval, where $d_2 = (F_{n-3}/F_{n-1})\, L$. The location of one of the runs coincides with those from the previous runs due to the nature of the series, and only one additional simulation is needed for the second set of points. Again, regions are eliminated from further consideration and points for the next iteration are placed at distance d_3 from the two ends of the new interval, where $d_3 = (F_{n-4}/F_{n-2})\, L$, L being the width of this interval. Thus, the region of uncertainty is reduced. This process is continued till the nth run is reached. This run is placed just to the right of the simulation near the middle of the interval left and thus the region is further halved to yield the final interval of uncertainty L_f.

The following simple example illustrates this procedure. Figure 4.14.11 shows the results for a heating system in which the objective function U(x) is the heat delivered per unit energy consumed. The independent variable x represents the temperature setting and has an initial range of 0 to 8. A maximum in U is desired to operate the system most efficiently. The objective function is given as $U(x) = 6 + 17x - 2x^2$.

Golden section search — The golden section search method is derived from the Fibonacci search method and, though not as efficient, is often more convenient to use. It is based on the fact that the ratio of two

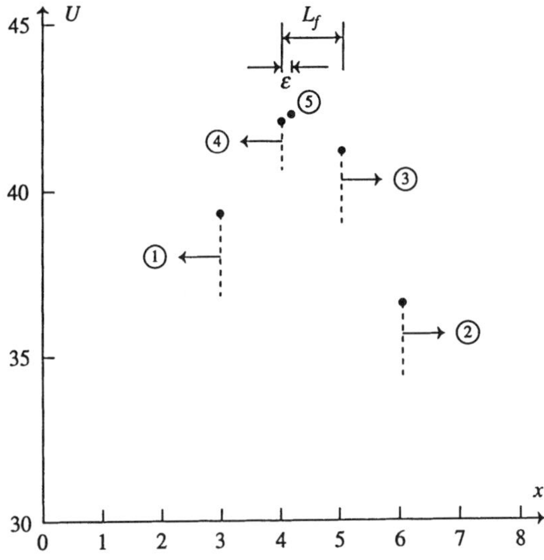

FIGURE 4.14.11 Use of Fibonacci method to reduce the interval of uncertainty with $U = 6 + 17x - 2x^2$.

successive Fibonacci numbers is approximately 0.618 for n > 8, i.e., $F_{n-1}/F_n = 0.618$. This ratio has been known for a long time and was of interest to the ancient Greeks as an aesthetic and desirable ratio of lengths in their constructions. The ratio of the height to the base of the Great Pyramid is also 0.618. The reciprocal of this ratio is 1.618, which has also been used as a number with magical properties. The term for the method itself comes from Euclid, who called the ratio the golden mean and pointed out that a length divided in this ratio results in the same ratio between the larger and smaller segments (Vanderplaats, 1984; Dieter, 1991). The trial runs are placed at 0.618 L from the ends of the interval at each step.

Comparison of different elimination methods — The reduction ratio R, defined in Equation (4.14.13), gives the ratio of the initial interval of uncertainty to the interval obtained after n runs. Therefore, it is a measure of the efficiency of the method. It can also be used to select the number of runs needed to obtain a desired uncertainty in locating the optimum. The reduction ratios for the various methods presented here for the optimization of a single-variable problem are given in Table 4.14.1.

TABLE 4.14.1 Reduction Ratios for Single Variable Search Methods

	General Formula	n = 5	n = 12
Uniform exhaustive	(n + 1)/2	3	6.5
Uniform dichotomous	(n + 2)/2	3.5	7.0
Sequential dichotomous	$2^{n/2}$	5.66	64
Fibonacci	F_n	8	233
Golden section		6.86	199

Here, the effect of the separation ε between pairs of runs on the reduction ratio is neglected. If ε is retained, the final interval can be shown to be

$$L_f = \frac{L_o}{F_n} + \varepsilon, \text{ for Fibonacci search} \tag{4.14.17}$$

$$L_f = \frac{L_o}{2^{n/2}} + \varepsilon\left(1 - \frac{1}{2^{n/2}}\right), \text{ for sequential dichotomous search} \tag{4.14.18}$$

when the second point of the pair is always located to the right of the first point at a separation of ε (Stoecker, 1989). Thus, the reduction ratios given in Table 4.14.1 are obtained when ε is neglected. The corresponding results are also shown graphically in Figure 4.14.12.

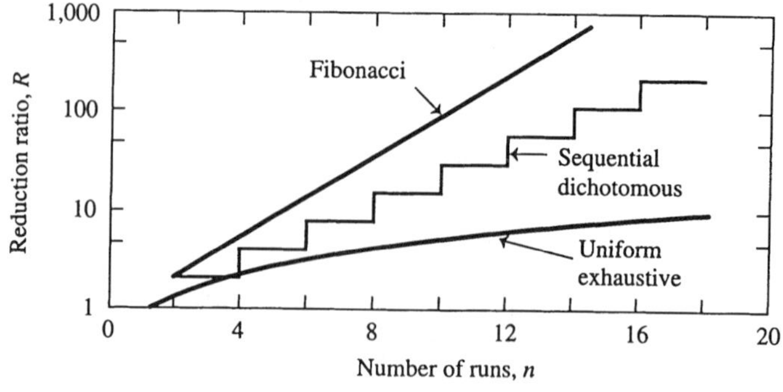

FIGURE 4.14.12 Reduction ratio R as a function of the number of runs n for different elimination search methods.

It is clearly seen that the Fibonacci search is an extremely efficient method and, therefore, is widely used. It is particularly valuable in multivariable optimization problems which are based on single-variable searches and in the optimization of large and complicated systems that require substantial computing time and effort for each simulation run. For small and relatively simple systems, exhaustive search provides a convenient, though not very efficient, approach to optimization. In addition, exhaustive search is useful in determining the nature of the function, its behavior over the given range, and whether it is unimodal or not. Consequently, the exhaustive search coupled with one of the other more efficient methods is often employed for new and complicated problems where little information is available on the nature and behavior of the chosen objective function.

Hill-Climbing Techniques

These methods are based on finding the shortest way to the peak of a hill, which represents the maximum of the objective function. A modification of the approach may be used to locate a valley, which represents the minimum. The calculation always moves so that the objective function improves with each step. Though more involved than the elimination methods, hill-climbing techniques are generally more efficient, requiring a smaller number of iterations to achieve the optimal design. These methods are applied to multivariable problems, for which the most important hill-climbing techniques are

- Lattice search
- Univariate search
- Steepest ascent/descent method

Though these methods are discussed in detail for relatively simple two-variable problems, they can easily be extended to a larger number of independent variables. Derivatives are needed for the steepest ascent/descent method, thus limiting its applicability to continuous and differentiable functions. The other methods mentioned above, though generally less efficient than steepest ascent, are applicable to essentially any system including those that involve discrete and discontinuous values.

Lattice search — This method is based on calculating the objective function U in the neighborhood of a central point and then moving the central point to the location which has the largest value of U, if the search is for a maximum. Thus, the calculation moves in the direction of increasing value of the objective function for locating a maximum. The maximum is reached when the value at the central point is higher than the values at its neighboring points. Though the search for a maximum in U is considered here, a similar procedure may be followed for a minimum, moving the calculation in the direction of decreasing value of the objective function.

Univariate search — This approach involves optimizing the objective function with respect to one variable at a time. Therefore, the multivariable problem is reduced to a series of single-variable optimization problems, with the process converging to the optimum as the variables are alternated. This procedure is shown graphically in Figure 4.14.13. A starting point is chosen on the basis of available information on the system or as a point close to the middle of the region. First, one of the variables, say x, is held constant and the function is optimized with respect to the other variable y. Point A represents the optimum thus obtained. Then y is held constant at the value at point A and the function is optimized with respect to x to obtain the optimum given by point B. Again, x is held constant at the value at point B and y is varied to obtain the optimum, given by point C. This process is continued, alternating the variable which is held constant, till the optimum is attained. This is indicated by the change in the objective function, from one step to the next, becoming less than a chosen convergence criterion or tolerance. Therefore, the two-variable problem is reduced to two single-variable problems applied alternately. The basic procedure can easily be extended to three or more independent variables. In solving the single variable problem, the search methods presented earlier, such as Fibonacci and golden section search, may be used. This provides a very useful method for optimizing thermal systems, particularly those that have discrete values for the design variables and those that have to be simulated for each trial run. Efficient search methods, rather than exhaustive search, are of interest in such cases. Calculus

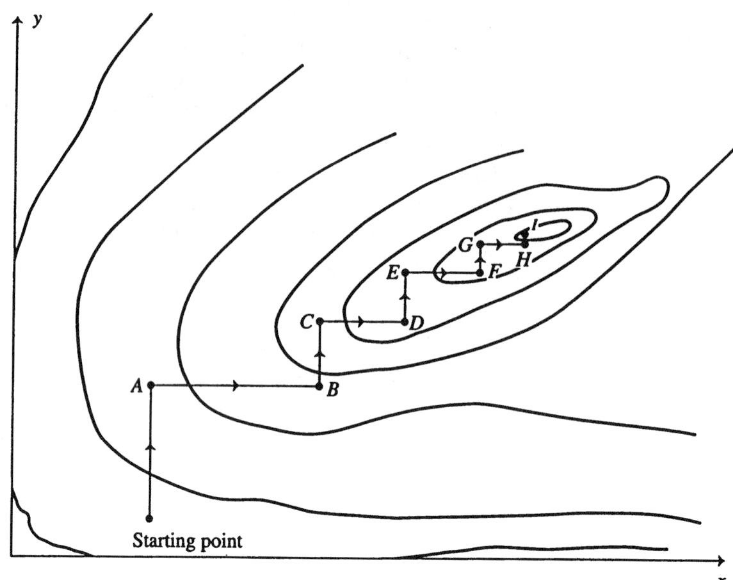

FIGURE 4.14.13 Sketch showing the various steps in the univariate search method.

methods may also be used if continuous, differentiable functions are involved. There are certain circumstances where univariate search may fail, such as those where ridges and very sharp changes occur in the objective function (Stoecker, 1989). However, by varying the starting point, interval of search, and method for single-variable search, such difficulties can often be overcome.

Steepest ascent/descent method — The steepest ascent/descent method is a very efficient search method for multivariable optimization and is widely used for a variety of applications including thermal systems. It is a hill-climbing technique in that it attempts to move towards the peak for maximizing the objective function, or towards the valley for minimizing the objective function over the shortest possible path. The method is termed as steepest ascent in the former case and as steepest descent in the latter. At each step, starting with the initial trial point, the direction in which the objective function changes at the greatest rate is chosen for moving the location of the point which represents the design on the multivariable space. Figure 4.14.14 shows this movement schematically on a hill as well as on a two-variable contour plot. Since the search always moves in the direction of greatest rate of change of U, the number of trial runs needed to reach the optimum is expected to be small and the method to be very efficient. However, it does require the evaluation of gradients in order to determine the appropriate direction of motion, limiting the application of the method to problems where the gradients can be obtained accurately and easily. Numerical differentiation may be used if an algebraic expression is not available for the objective function, which is often the case for thermal systems.

It is known from mathematics that the gradient vector ∇U is normal to the constant U contour line in a two-variable space, to the constant U surface in a three-variable space, and so on. Since the normal direction represents the shortest distance between two contour lines, the direction of the gradient vector ∇U is the direction in which U changes at the greatest rate. For a multivariable problem, the gradient vector may be written as

$$\nabla U = \frac{\partial U}{\partial x_1}\mathbf{i}_1 + \frac{\partial U}{\partial x_2}\mathbf{i}_2 + \frac{\partial U}{\partial x_3}\mathbf{i}_3 + \cdots + \frac{\partial U}{\partial x_n}\mathbf{i}_n \tag{4.14.19}$$

where $\mathbf{i}_1, \mathbf{i}_2, \ldots, \mathbf{i}_n$ are the unit vectors in the $x_1, x_2, \ldots, x_n$ directions, respectively. At each trial point, the gradient vector is determined and the search is moved along this vector, the direction being chosen so that U increases if a maximum is sought, or U decreases if a minimum is of interest.

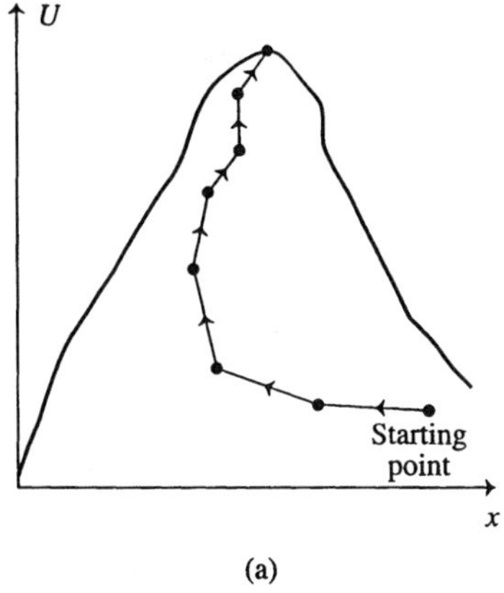

(a)

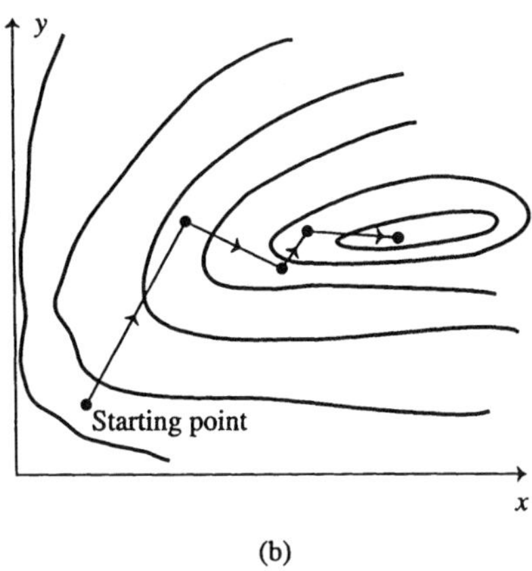

(b)

FIGURE 4.14.14 Steepest ascent method, shown in terms of the climb towards the peak of a hill and also in terms of constant U contours.

The direction represented by the gradient vector is given by the relationship between the changes in the independent variables. Denoting these by Δx_1, Δx_2, ..., Δx_n, we have from vector analysis

$$\frac{\Delta x_1}{\partial U/\partial x_1} = \frac{\Delta x_2}{\partial U/\partial x_2} = \frac{\Delta x_3}{\partial U/\partial x_3} = \cdots \frac{\Delta x_n}{\partial U/\partial x_n} \tag{4.14.20}$$

Therefore, if Δx_1 is chosen, the changes in the other variables must be calculated from these equations. Also Δx_1 is taken as positive or negative, depending on whether U increases or decreases with x_1 and whether a maximum or a minimum is sought. For a maximum in U, Δx_1 is chosen so that U increases, i.e., Δx_1 is positive if $\partial U/\partial x_1$ is positive and negative if $\partial u/\partial x_1$ is negative. The partial derivatives, such as $\partial U/\partial x_1$, are generally obtained numerically by using the expression

$$\frac{\partial U}{\partial x_1} = \frac{U(x_1 + h, x_2, \ldots, x_n) - U(x_1, x_2, \ldots, x_n)}{h} \tag{4.14.21}$$

where h is a small change in x_1. Similarly, the other partial derivatives may be evaluated. If an algebraic expression is available for the objective function, for instance from curve-fitting of numerical simulation results, calculus can be used more advantageously to evaluate these derivatives.

There are two ways of moving the trial point. In the first case, we could choose the magnitude of the step size in terms of one of the variables, say Δx_1, calculate the changes in the remaining variables, and determine the new values of these variables. At the new point the gradient vector is again determined and the point is again moved in the direction of ∇U. This procedure is continued till the optimum is reached, as indicated by a negligible change in the objective function. Again, for a maximum in U, Δx_1 is taken as positive if $\partial U/\partial x_1$ is positive and negative if the latter is negative, these conditions being reversed for a minimum in U.

The second approach is to move the trial point along the direction of the gradient vector till an optimum is reached. This becomes the new trial point and the gradient vector is evaluated, the new direction of movement determined, and the trial point is moved in this direction till, again, an optimum is reached. This procedure is continued till the overall optimum is attained. This approach is the one shown in Figure 4.14.14b. Since the calculation of the gradients may be time consuming, the second approach is often preferred because fewer calculations of the gradient are needed. In addition, the first approach could run into a problem if the objective function U varies very slowly or rapidly with the variable, say x_1, whose step size is chosen.

Example —As an example, consider the objective function U, which represents the cost of a fan and duct system, is given in terms of the design variables x and y, where x represents the fan capacity and y the duct length, as

$$U = \frac{x^2}{6} + \frac{4}{xy} + 3y$$

both x and y being positive. Let us apply the first approach of steepest ascent, given in the preceding discussion, to obtain the optimum.

The partial derivatives in terms of the independent variables x and y are

$$\frac{\partial U}{\partial x} = \frac{2x}{6} - \frac{4}{x^2 y}$$

$$\frac{\partial U}{\partial x} = -\frac{4}{xy^2} + 3$$

To move the trial point in the direction of ∇U, the following relationship applies

$$\frac{\Delta x}{\partial U/\partial x} = \frac{\Delta y}{\partial U/\partial y}$$

Therefore, Δx may be chosen and Δy calculated from this equation. If $\partial U/\partial x$ is positive, Δx is taken as positive for search for a maximum in U. In the present case, we want a minimum in U. Therefore, Δx is taken as positive if $\partial U/\partial x$ is negative.

For the first approach, the derivatives are calculated at each point obtained by changing x by Δx and y by Δy, where Δy is obtained from the above relationship between ΔX and Δy. The starting point is taken as x = y = 0.5. The results obtained for different values of Δx are

Δx	No. of Iterations	x	y	U
0.5	3	2.0	0.699	5.625
0.1	20	2.5	0.731	5.423
0.05	40	2.5	0.731	5.423
0.01	205	2.55	0.723	5.422
0.005	410	2.55	0.722	5.422

Clearly, only a few iterations are needed to reach close to the optimum, but many more are needed to obtain it with a high level of accuracy. Since the final design is not the exact optimum but near it, so that standard available items may be used for the system, there is no reason to insist on a very high accuracy for the optimum. It can easily be shown, by varying x or y from the optimum, that a minimum in U is attained. Univariate search or calculus methods can also be employed for this simple problem.

Constrained Optimization

The techniques mentioned earlier are particularly useful for unconstrained optimization problems. However, many of these can also be used, with some modifications, for constrained problems, which are generally much more difficult to solve. The constraints must be satisfied while searching for the optimum. This restricts the movement toward the optimum. The constraints may also define the acceptable design domain. Two important schemes for optimizing constrained problems are (1) the penalty function method and (2) searching along a constraint.

Penalty function method — The basic approach of this method is to convert the constrained problem into an unconstrained one by constructing a composite function using the objective function and the constraints. Different methods are obtained depending on the way the composite function is constructed. The penalty function method uses certain parameters, known as penalty parameters, that penalize the composite function for violation of the constraints. The penalty is larger if the violation is greater. The composite function is then optimized using any of the techniques applicable for unconstrained problems. The penalty parameters are varied and the new composite function optimized. The process is continued until there is no significant change in the optimum when the penalty parameters are varied.

Let us consider the optimization problem given by the equations

$$U\,(x_1, x_2, x_3, \ldots, x_n) \rightarrow \text{minimum/maximum} \tag{4.14.22}$$

$$G_i\,(x_1, x_2, x_3, \ldots, x_n) = 0, \text{ where } i = 1, 2, 3, \ldots, m \tag{4.14.23}$$

where only equality constraints are considered and, therefore, $n > m$. The composite function, also known as the penalty function, may be formulated in many ways. A commonly used formulation is given here. If a maximum in U is being sought, a new objective function V is defined as

$$V = U - \left[r_1 (G_1)^2 + r_2 (G_2)^2 + r_3 (G_3)^2 + \cdots + r_m (G_m)^2\right] \tag{4.14.24}$$

and if a minimum in U is desired, the new objective function is defined as

$$V = U + \left[r_1 (G_1)^2 + r_2 (G_2)^2 + r_3 (G_3)^2 + \cdots + r_m (G_m)^2\right] \tag{4.14.25}$$

Therefore, the squares of the constraints are included in the new objective function, V, which is known as the penalty function. The use of the square ensures that the magnitude of the violation of a constraint is considered, and not its positive or negative value that may cancel out with the violation in other constraints. Here the r's are scalar quantities that vary the importance given to the various constraints and are known as penalty parameters. They may all be taken as equal or different. Higher values may be taken for the constraints which are critical and smaller for those that are not as important.

If the penalty parameters are all taken as zero, the constraints have no effect on the solution and, therefore, the constraints are not satisfied. On the other hand, if these parameters are taken as large, the constraints are satisfied but the convergence to the optimum is slow. Therefore, by varying the penalty parameters we can vary the rate of convergence and the effect of the different constraints on the solution. The general approach is to start with small values of the penalty parameters and gradually increase these as the G's, which represent the constraints, become small. This implies going gradually and systematically from an unconstrained problem to a constrained one. The values of the G's at a point in the iteration

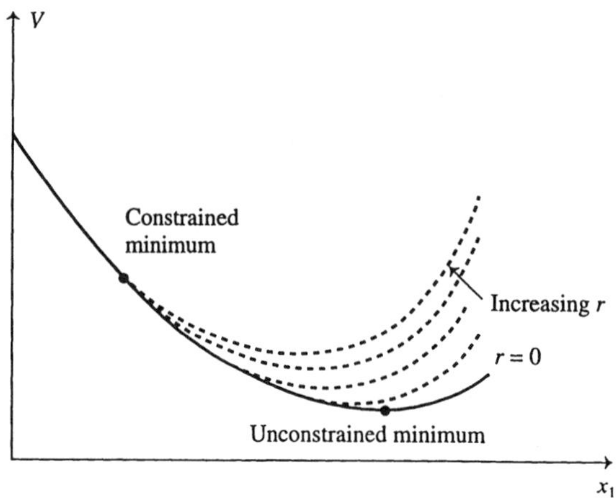

FIGURE 4.14.15 Sketch showing the penalty function method for the combined objective function V and different values of the penalty parameter r.

may also be used to choose the penalty parameters, using larger values for larger G's so that these are driven more rapidly towards zero. Figure 4.14.15 shows schematically the effect of the penalty parameter on the penalty function and on the minimum obtained. Clearly, the unconstrained minimum is obtained at r = 0 and at small values of r. The constrained minimum is attained at larger values of r.

Search along a constraint — Several methods for optimization of constrained problems are based on reaching the constraint and then moving along the constraint in order to search for the optimum. These include the gradient projection method, the generalized reduced gradient method, and the hemstitching method (Arora, 1989; Stoecker, 1989). All of these methods are quite similar, in that they search for the optimum while staying on or close to the constraint, though there are differences in their implementation. Inequality constraints generally determine the feasible region in which the optimum is sought and the search is carried out along the equality constraints so that optimum satisfies all the constraints. Also, inequality constraints can be converted to equalities by the use of slack variables, that ensure that the given limits are not violated, as mentioned earlier. However, these methods are best suited to problems for which the gradients of the objective function and the constraints are defined and easy to determine, analytically or numerically.

Let us first consider the *hemstitching* method. The main steps involved in this method are

- Start with a trial point
- Move toward and reach the constraint(s)
- Move tangentially along the constraint
- Bring point back to the constraint(s)

The direction of the tangential move is chosen so that the objective function increases if a maximum is being sought and decreases if a minimum is of interest. The application of this method depends on the number of variables and the number of equality constraints. It is useful in a variety of thermal problems that can be represented by continuous functions (Stoecker, 1989).

For the simplest case of a single constraint in a two-variable space, the basic approach involves choosing an initial guess or starting point in the feasible domain. We then move to the constraint and obtain a point on the constraint. From this point we move tangentially to the constraint. This takes the trial point off the constraint in nonlinear optimization problems and the next step is used to bring the point back to the constraint. This process is repeated, moving along the constraint, till the optimum value of the objective function is obtained. If two constraints are involved in a three-variable space, the movement of the trial point is along the tangent to both the constraints. This approach is applicable for all cases in which the number of variables n is greater than the number of constraints m by one. If this

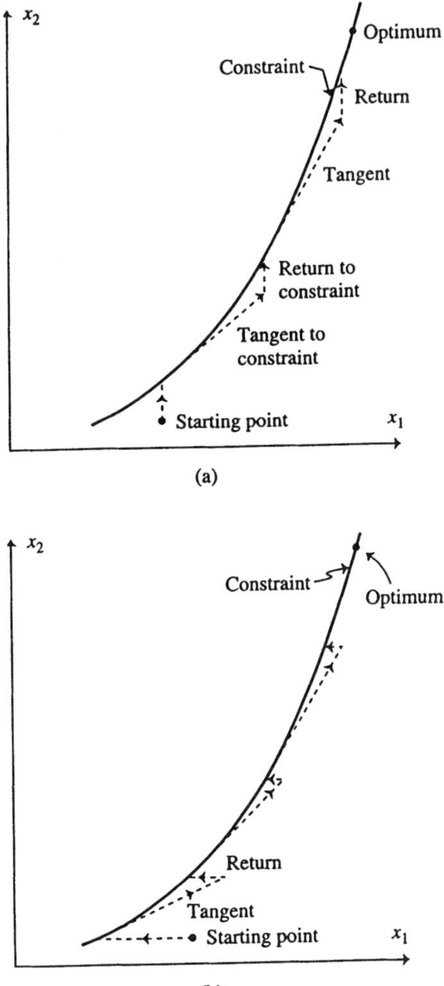

FIGURE 4.14.16 The hemstitching method with return to the constraint obtained by keeping (a) x_1 fixed, and (b) x_2 fixed.

difference is greater than one, the move may be made in a direction that yields the greatest change in the objective function.

Figure 4.14.16 shows the hemstitching method for a two-variable, single-constraint problem. The first step involves reaching the constraint by keeping one of the two variables, x_1 or x_2, fixed, and varying the other till the constraint is satisfied. For example, if the constraint is

$$x_1^2 \, x_2^3 = 35 \tag{4.14.26}$$

we can keep either x_1 or x_2 fixed to obtain the value of the other variable at the constraint as

$$x_1 = \left(\frac{35}{x_2^3}\right)^{1/2} \quad \text{or} \quad x_2 = \left(\frac{35}{x_1^2}\right)^{1/3} \tag{4.14.27}$$

Therefore, a point on the constraint can be located. These equations can also be used to return to the constraint if a move tangential to the constraint takes the point away from the constraint, as shown in Figure 4.14.16 for the two schemes of keeping x_1 or x_2 fixed.

The exact optimum is generally not obtained even for continuous functions, since only a finite number of iterations are used. However, in actual engineering practice, components, materials, and even dimensions are not available as continuous quantities but as discrete steps. For instance, a heat exchanger would typically be available for discrete heat transfer rates such as 50, 100, 200 kW, etc. The cost may be assumed to be a discrete distribution rather than a continuous variation. Similarly, the costs of items like pumps and compressors are discrete functions of the size. Different materials involve distinct sets of properties and not continuous variations of thermal conductivity, specific heat, or other thermal properties. Search methods can easily be applied to such circumstances, whereas calculus methods demand continuous functions. Consequently, search methods are extensively used for the optimization of thermal systems.

Linear and Dynamic Programming

Programming as applied here simply refers to optimization. Linear programming is an important optimization method and is extensively used in industrial engineering, operations research, and many other disciplines. However, the approach can be applied only if the objective function and the constraints are all linear. Large systems of variables can be handled by this method, such as those encountered in air traffic control, transportation networks, supply and utilization of raw materials, etc. However, as we well know, thermal systems are typically represented by nonlinear equations. Consequently, linear programming is not very important in the optimization of thermal systems.

As an example, consider an optimization problem given by the objective function $U = 4x + 3y$, with the constraints given as $x + 3y \leq 200$ and $3x + 2y \leq 300$. This problem can be solved graphically, as shown in Figure 4.14.17. The feasible domain is given by the constraints and contours of constant U can be drawn to obtain the optimum. It is found that the optimum arises at the intersection of the two constraints and thus is on the boundary of the feasible region. In linear programming, the optimum is always obtained on the boundary of the feasible domain, particularly at the corners or vertices of the feasible domain. This is a particular feature of linear programming, and most efficient algorithms such as the *simplex algorithm* seek to move rapidly along the boundary, from vertex to vertex, to obtain the

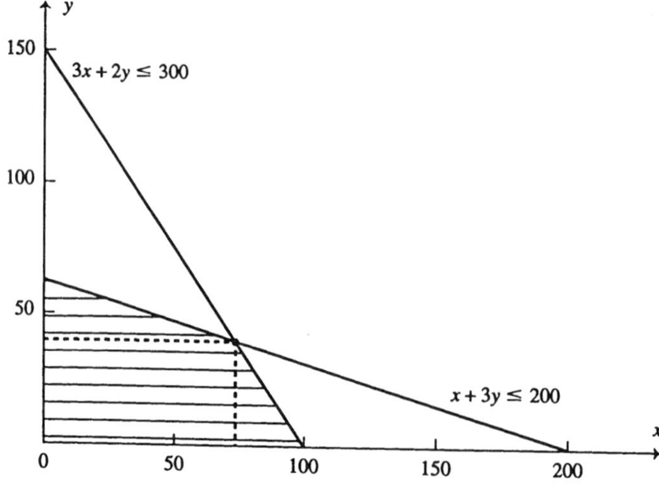

FIGURE 4.14.17 Graphical solution to the linear programming problem posed above.

optimum. Books on linear programming by Hadley (1962) and Gass (1975), and on optimization methods by Haug and Arora (1979) and Reklaitis et al. (1983), among others, may be consulted for further details.

Dynamic programming is used to obtain the best path through a series of stages or steps to achieve a given task, for instance, the optimum configuration of a manufacturing line, the best path for the flow of hot water in a building, and the best layout for transport of coal in a power plant. Therefore, the result obtained from dynamic programming is not a point where the objective function is optimum but is a curve or path over which the function is optimized. Figure 4.14.18 illustrates the basic concept by means of a sketch. Several paths connect the points A and B. The optimum path is the one over which the given objective function, say, total transportation cost, is minimized by obtaining the optimal paths to different stages and using these for total path optimization. Though unique optimal solutions are generally obtained in practical systems, multiple solutions are possible and additional considerations, such as safety, convenience, availability of items, etc. are used to choose the best design. Clearly, there are a few circumstances of interest in thermal systems where dynamic programming may be used to obtain the best layout to minimize losses and reduce costs. The book by Denardo (1982) on dynamic programming and others on optimization methods may be consulted for further information on this method.

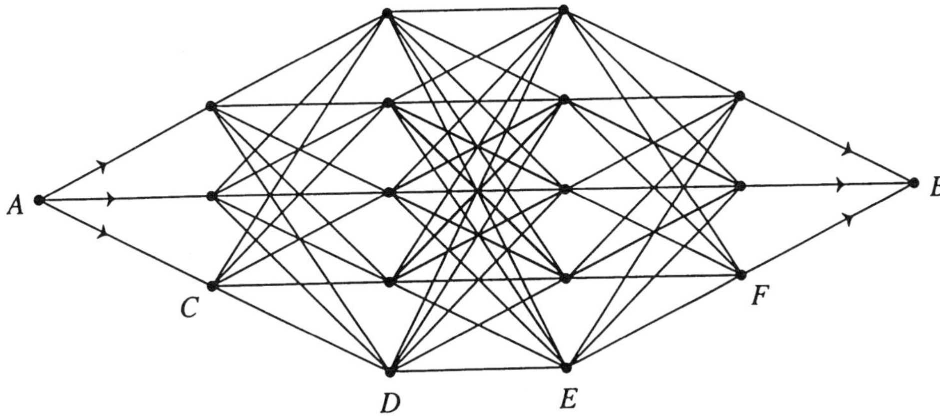

FIGURE 4.14.18 Dynamic programming for choosing the optimum path from the many different paths to go from point A to point B.

Geometric Programming and Other Methods

Geometric programming is an optimization method which can be applied if the objective function and the constraints can be written as sums of polynomials. The independent variables in these polynomials may be raised to positive or negative, integer or noninteger, exponents, e.g.,

$$U = a\, x_1^2 + b\, x_2^{1.2} + c\, x_1\, x_2^{-0.5} + d \tag{4.14.28}$$

where a, b, c, and d are constants, which may also be positive or negative, and x_1 and x_2 are the independent variables. Curve-fitting of experimental data and numerical results for thermal systems often leads to polynomials and power-law variations. Therefore, geometric programming is particularly useful for the optimization of thermal systems if the function to be optimized and the system characteristics can be represented as sums of polynomials. If the method is applicable in a particular case, the optimal solution and even the sensitivity of the solution to changes in the constraints are often obtained directly and with very little computational effort.

Applicability

As mentioned above, geometric programming is applicable to problems in which both the objective function and the constraints can be expressed as sums of polynomials of the independent variables. A few examples of the objective function, $U(x_1, x_2, \ldots, x_n)$, in unconstrained problems, that can be treated by geometric programming are

$$U = 2\,x_2^{3/2} + x_1^2 + 7\,x_1\,x_2^{1/3} \tag{4.14.29}$$

$$U = 550 + 105\,x_1 + \frac{120000}{x_1^3} - 10\,x_1^{1.4} \tag{4.14.30}$$

$$U = x_1\,x_2^2 + 3\frac{x_1}{x_3^{1.3}} - 3.7\,x_2\,x_3^2 + 8\,x_2^{2.2} \tag{4.14.31}$$

Similarly, for constrained problems, the following form is suitable for geometric programming

$$U = 4\,x_1^2\,x_2^3 + 8\,x_1^{-1/2}\,x_2^{-1/3} - 6\,x_1^{0.6} + 3\,x_2 \tag{4.14.32}$$

with the constraint

$$x_1\,x_2^{1.2} = 20 \tag{4.14.33}$$

Therefore, fractional or integral, positive or negative, exponents and coefficients may be considered. Since such polynomial or power-law representations are very common in thermal systems, geometric programming is a useful technique for optimizing these systems.

Degree of Difficulty — An important parameter that determines how the method is to be applied and whether it can be used to yield the optimum directly and without detailed analysis is the degree of difficulty D, which is defined as

$$D = N - (n + 1) \tag{4.14.34}$$

where N is the total number of terms in the objective function and in the constraints and n is the number of independent variables. Since the addition of a constant to the objective function does not affect the location of the optimum, only the terms containing polynomials are counted. For instance, the degree of difficulty for the problem given by Equation (4.14.29) is $D = 3 - (2 + 1) = 0$, since there are three polynomial-containing terms and two independent variables x_1 and x_2. Similarly, the degree of difficulty D for Equations (4.14.30) and (4.14.31) are obtained as, respectively, 1 and 0. For the problem given by Equations (4.14.32) and (4.14.33), there are five polynomial terms in the objective function and the constraint and only two variables, resulting in a degree of difficulty D of 2.

Geometric programming is particularly useful when the degree of difficulty is zero since the optimum value of the objective function can be written straightaway, without resorting to any analysis, and the independent variables can be derived from this result. It is then a very simple method to use. In some cases, if D is not zero, terms may be combined to reduce the degree of difficulty to zero. We shall consider only the circumstance of $D = 0$ here because if D is not zero and cannot be reduced to zero, the application of geometric programming involves the solution of nonlinear equations and becomes quite complicated. Other techniques such as search and calculus methods may be easier to use in this case. For further details, see Duffin et al. (1967), Zener (1971), Wilde (1978), and Stoecker (1989).

Unconstrained Optimization

Let us first consider the application of geometric programming to unconstrained problems. Since we are interested in problems with degree of difficulty zero, the number of terms must be greater than the number of variables by one.

Single independent variable — The objective function U may be written in terms of the independent variable x as

$$U = A\ x^a + B\ x^b \tag{4.14.35}$$

The two terms may be denoted as u_1 and u_2, where $u_1 = Ax^a$ and $u_2 = Bx^b$. Therefore, these terms represent the individual contributions to the overall objective function. For instance, the cost of producing an item using a manufacturing system may be taken as the sum of the contributions due to initial and operating costs, both of which are functions of the capacity or size of the system. As the system size increases, the initial costs increase, but the operating and maintenance costs per item decrease. According to geometric programming, the optimum value of the objective function U^* in the above equation is given by the expression

$$U^* = \left(\frac{A\ x^a}{w_1}\right)^{w_1} \left(\frac{B\ x^b}{w_w}\right)^{w_2} \tag{4.14.36}$$

with

$$w_1 + w_2 = 1 \tag{4.14.37}$$

and

$$a\ w_1 + b\ w_2 = 0 \tag{4.14.38}$$

where w_1 and w_2 are parameters to be determined from Equations (4.14.37) and (4.14.38). The latter equation results in the elimination of the independent variable x in Equation (4.14.36), yielding

$$U^* = \left(\frac{A}{w_1}\right)^{w_1} \left(\frac{B}{w_2}\right)^{w_2} \tag{4.14.39}$$

Also, it can be shown that

$$w_1 = \frac{u_1^*}{u_1^* + u_2^*} = \frac{u_1^*}{U^*} \tag{4.14.40}$$

and

$$w_2 = \frac{u_2^*}{u_1^* + u_2^*} = \frac{u_2^*}{U^*} \tag{4.14.41}$$

Therefore, w_1 and w_2 represent the weights or relative contributions of the two terms to the total objective function at the optimum. This is an important piece of information obtained in geometric programming and may be used for further improvement in the system design. From the above equations for w_1 and w_2, the independent variable x at the optimum may be determined, since $u_1 = Ax^a$ and $u_2 = Bx^b$.

As an example of this procedure, consider a manufacturing system in which rectangular boxes of length x, height y, and width z (in m) and open at the top are used for storing and conveying material. The cost of material and fabrication is \$150 per unit surface area in m^2 and the cost of storage varies inversely as the volume, being 10^3 per unit volume in m^3. Let us formulate the optimization problem for minimizing the cost and obtain the optimum by using geometric programming.

The objective function U is the total cost of material/fabrication and storage/conveying. Therefore, it may be written as

$$U(x, y, z) = 150(xz + 2xy + 2yz) + \frac{1000}{xyz}$$

since the top of area xz is open. All the terms are polynomials and the degree of difficulty D = 4 – (3 + 1) = 0. Therefore, the optimum value of the cost is given by

$$U^* = \left(\frac{150}{w_1}\right)^{w_1} \left(\frac{300}{w_2}\right)^{w_2} \left(\frac{300}{w_3}\right)^{w_3} \left(\frac{1000}{w_4}\right)^{w_4}$$

with

$$w_1 + w_2 + w_3 + w_4 = 1$$

$$w_1 + w_2 - w_4 = 0$$

$$w_2 + w_3 - w_4 = 0$$

$$w_1 + w_3 - w_4 = 0$$

The last three equations ensure that the independent variables x, y, and z are eliminated from the objective function at the optimum, resulting in the above expression for U^*, and the first equation ensures that the sum of the w's is unity. This system of linear equations may be solved easily to yield $w_1 = w_2 = w_3 = 1/5$, and $w_4 = 2/5$. This implies that the last term is twice as important as each of the other terms. Therefore, the optimum value of the objective function is

$$U^* = \left(\frac{150}{1/5}\right)^{1/5} \left(\frac{300}{1/5}\right)^{1/5} \left(\frac{300}{1/5}\right)^{1/5} \left(\frac{1000}{2/5}\right)^{2/5} = \$1601.86$$

The independent variables x, y and z may be obtained from the equations $150\,xz = w_1 U^*$, $300\,xy = w_2 U^*$, and $300\,yz = w_3 U^*$, which gives x = 1.46 m, y = 0.73 m, and z = 1.46 m at the optimum. It can easily be confirmed that the cost obtained is a minimum by changing the variables slightly, away from the optimum.

This problem can also be solved by the use of calculus methods for optimization. However, as mentioned earlier, geometric programming involves the solution of a set of linear equations, whereas calculus methods require solving nonlinear equations. This is a substantial advantage in most practical problems. Also, the weighting factors obtained as part of the solution indicate the relative importance of the various terms and can be used to improve the design by focusing on the dominant terms.

Constrained Optimization

Geometric programming can also be used for optimizing systems with equality constraints. The degree of difficulty is again taken as zero, so that the total number of polynomial terms in the objective function and the constraints is greater than the number of independent variables by one. Let us consider the constrained optimization problem given by the objective function

$$U = u_1 + u_2 + u_3 \tag{4.14.42}$$

subject to the constraint

$$u_4 + u_5 = 1 \tag{4.14.43}$$

with x_1, x_2, x_3 and x_4 as the four independent variables, resulting in zero degree of difficulty. The unity on the right-hand side of Equation (4.14.43) can be obtained by normalizing the equation if a numerical term other than unity appears in the equation, which is often the case. Following the treatment given in the preceding section, the objective function and the constraint may be written as

$$U = \left(\frac{u_1}{w_1}\right)^{w_1} \left(\frac{u_2}{w_2}\right)^{w_2} \left(\frac{u_3}{w_3}\right)^{w_3} \tag{4.14.44}$$

with $w_1 + w_2 + w_3 = 1$ and $w_i = \frac{u_i}{U}$

$$1 = \left(\frac{u_4}{w_4}\right)^{w_4} \left(\frac{u_5}{w_5}\right)^{w_5} \tag{4.14.45}$$

with $w_4 + w_5 = 1$ and $w_4 = \frac{u_4}{1}$, $w_5 = \frac{u_5}{1}$.

Equation (4.14.45) may be raised to the power of an arbitrary constant p and the objective function may be written as

$$U = \left(\frac{u_1}{w_1}\right)^{w_1} \left(\frac{u_2}{w_2}\right)^{w_2} \left(\frac{u_3}{w_3}\right)^{w_3} \left(\frac{u_4}{w_4}\right)^{pw_4} \left(\frac{u_5}{w_5}\right)^{pw_5} \tag{4.14.46}$$

In order to eliminate the independent variables from U*, we get four linear equations in w_1, w_2, w_3, and p. Also, $w_1 + w_2 + w_3 = 1$ and $p(w_4 + w_5) = p$.

These linear equations may be solved for w_1, w_2, w_3, w_4, w_5, and p. Therefore, the optimum value of the objective function is given by Equation (4.14.46) and the independent variables are obtained from the expressions for the weighting factors, as done before. The sensitivity coefficient $S_c = -\lambda = -p\, U^*$ and has the same physical interpretation as discussed earlier for the Lagrange multiplier method, i.e., it is the negative of the rate of change in the optimum with respect to a change in the adjustable parameter E in the constraint $G = g - E = 0$. The above approach may easily be extended to more than one constraint as long as the degree of difficulty is zero. For examples on the use of the method for constrained optimization, see Stoecker (1989) and Jaluria (1998).

Several other optimization methods have been developed in recent years because of the strong need to optimize systems and processes. Many of these are particularly suited to specific applications and may not be easily applied to thermal systems. Among these are shape, trajectory, and structural optimization methods, which involve specialized techniques for finding the desired optimum. Frequently, finite element solution procedures are linked with the relevant optimization strategy. Iterative shapes, trajectories, or structures are generated, starting with an initial design. For monotonically increasing or decreasing objective functions and constraints, a method known as the monotonicity analysis has been developed for optimization. This approach focuses on the constraints and the effect these have on the optimum.

Optimization of Thermal Systems

Important Considerations

Thermal systems are particularly concerned with energy and fluid flow. Therefore, the objective function is frequently based on energy consumption, which involves considerations of energy losses, efficiency of the system and its components, energy exchange with the environment, fuel consumed, etc. A useful objective function is the rate of energy consumption per unit output, where the output may be power delivered, heat removed, products manufactured, and so on. The design which requires the least amount of energy per unit output is then the optimum. Similarly, the system that delivers the largest output per unit energy consumption is optimum. Since energy consumption can be expressed in terms of cost, this objective function can also be considered as the output per unit cost.

Similar considerations often apply to fluid flow, where again we would try to minimize the energy consumed. This implies minimizing the rate of flow and fluid losses, particularly if a closed system is needed for preserving the purity and if the fluid is expensive. A lower pressure head generally translates into lower cost of the pumping system and is desirable. Therefore, some of the physical quantities that are often *maximized* in thermal systems may be listed as

- Efficiency
- Output per unit energy (or fuel) consumption
- Output per unit cost
- Heat removal rate in electronic systems

whereas the quantities that are *minimized* may be listed as

- Energy losses
- Energy input for cooling systems
- Pressure head for fluid flow
- Flow rate of fluid
- Fluid losses
- Rate of energy (or fuel) consumed per unit output

In thermal systems, the constraints arise largely from conservation laws for mass, momentum, and energy, and from limitations of the material, space, and equipment being used, as discussed earlier. However, these lead to nonlinear, multiple, coupled, partial differential equations, with complicated geometries and boundary conditions in typical systems of practical interest. Other complexities may also arise due to the material characteristics, combined thermal transport mechanisms, etc. as discussed elsewhere in this handbook. The main problem that arises due to these complexities is that the simulation of the system for each set of conditions requires a considerable amount of time and effort. Therefore, it is usually necessary to minimize the number of simulation runs needed for optimization. For relatively simple thermal systems, numerical or experimental simulation results may be used, with curve-fitting employed to obtain algebraic expressions and equations to characterize the behavior of the system. Then the optimization problem becomes quite straightforward and many of the available methods can be used to extract the optimum.

Unfortunately, this approach is possible in only a few simple, and often impractical, circumstances. For common practical systems, numerical modeling is employed to obtain the simulation results, as needed, to approach the optimum.

Different Types of Thermal Systems

Thermal systems cover a very wide range of applications. Different concerns, constraints, and requirements arise in different types of systems. Therefore, the objective function and the nature of the constraints would generally vary with the application. Though costs and overall profit or return are frequently optimized, other quantities are also of interest and are used. Let us consider some of the common types of thermal systems and discuss the corresponding optimization problems.

Manufacturing systems — The objective function is typically the number of items produced per unit cost. It could also be the amount of material processed in heat treatment, casting, crystal-growing, extrusion, or forming. The number of solder or welding joints made, length of material cut in gas or laser cutting, or the length of fiber drawn may also be used, depending on the application. Again, the output per unit cost or the cost for a given output may also be used as the objective function. The constraints are often given on the temperature and pressure due to material limitations. Conservation principles and equipment limitations restrict the flow rates, cutting speed, draw speed, etc.

Energy systems — The amount of power produced per unit cost is the most important measure of success in energy systems and, therefore, is an appropriate quantity to be optimized. The overall thermal efficiency is another important variable that may be optimized. Most of the constraints arise from conservation laws. However, environmental and safety considerations also lead to important limitations on items, such as the water outlet temperature and flow rate from the condensers of a power plant. Material and space limitations will also provide some constraints on the design variables.

Electronic systems — The rate of thermal energy removed from the system as well as this rate per unit cost are important design requirements and, thus, may be used as objective functions. The cost of the system may also be minimized while ensuring that the temperature requirements of the components are satisfied. The weight and volume are important considerations in portable systems and in systems used in planes and rockets. These may also be chosen for optimization. Besides the constraints due to conservation principles, space and material limitations generally restrict the temperatures, fluid flow rates, and dimensions in the system.

Transportation systems — The torque, thrust, or power delivered is an important consideration in these systems. Therefore, these quantities as such or per unit cost may be maximized. This may also be taken as the output per unit fuel consumed. The costs for a given output in thrust, acceleration, etc. may also be chosen for minimization. The thermal efficiency of the system is another important aspect which may be maximized. The constraints are largely due to material, weight and size limitations, besides those due to conservation laws. Thus, the temperature, pressure, dimensions, and fuel consumption rate may be restricted within specified limits.

Heating and cooling systems — The amount of heat removed or provided per unit cost is a good measure of the effectiveness of these systems and may be chosen for maximization. The system cost as well as the operating cost, which largely includes the energy costs, may be minimized while satisfying the requirements. The thermal efficiency of the system may be maximized for optimum performance. Besides those due to conservation laws, most of the constraints arise due to space limitations. Weight constraints are important in mobile systems. Fluid constraints lead to constraints on the temperature and pressure in the system.

Heat transfer and fluid flow equipment — The rate of heat transfer and the total flow rate are important considerations in these systems. These quantities thus may be used for optimization. The heat transfer or flow rate per unit equipment or operating costs may also be considered. The resulting temperature of a fluid being heated or cooled, the efficiency of the equipment, energy losses, etc. may also be chosen as objective functions. Space limitations often provide the main constraints on dimensions. Constraints due to weight are also important in many cases, particularly in automobiles. Conservation laws provide constraints on temperatures and flow rates.

The above discussion serves to illustrate the diversity of the objective function and the constraints in the wide range of applications that involve thermal systems. Even though costs and profit are important concerns in engineering systems, other quantities such as output, efficiency, environmental effect, etc. also provide important considerations that may be used effectively in the optimization process. Clearly, the above list is not exhaustive. Many other objective functions, constraints, and applications can be considered, depending on the nature and type of thermal system being optimized.

Examples

Most of the thermal systems mentioned in the preceding discussion can be formulated for optimization. Different objective functions can be chosen for different problems, though costs and output can be used to represent most circumstances. A few examples of optimization problems for thermal systems are considered below.

Example 1 — An important problem in power generation is heat rejection. Bodies of water such as lakes and ponds are frequently used for cooling the condensers. The distance x between the inflow into the cooling pond and its outflow is a very important variable that determines the performance and cost of the system (see Figure 4.14.19). If x increases, the cost increases because of increased distance for

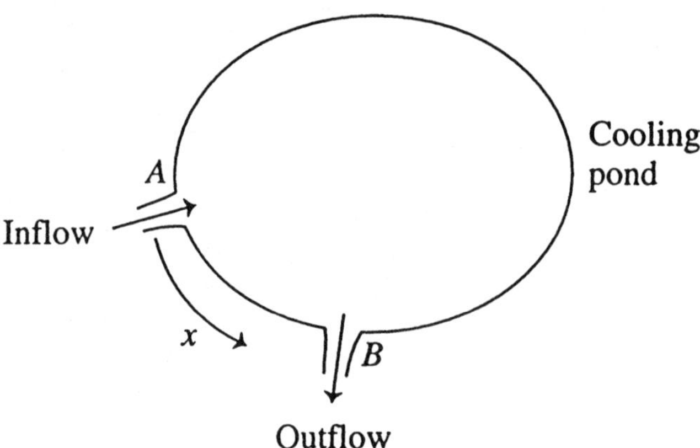

FIGURE 4.14.19 Heat rejection from a power plant to a cooling pond, with x as the distance between the inflow and outflow.

pumping the cooling water. As x decreases, the hot water discharged into the lake can recirculate to the outflow, raising the temperature there. This increases the temperature of the cooling water entering the condensers of the power plant. This, in turn, raises the temperature at which heat rejection occurs and thus lowers the thermal efficiency of the plant, as is well known from thermodynamics. Therefore, an increase in x increases the cost of the piping and pumps, while a decrease in x increases the cost of power generation by lowering the thermal efficiency. If the objective function U is taken as cost per unit of generated power, we may write

$$U(x) = F_1(x) - F_2(x) \tag{4.14.47}$$

where $F_1(x)$ and $F_2(x)$ are costs related to piping and efficiency of the system. This implies that an optimum distance x may be obtained for minimum costs.

This is a very complicated problem since the model involves turbulent multidimensional flow, complex geometries, varying ambient conditions, and several combined modes of heat transfer. The problem has to be solved numerically, with many simplifications, to obtain the desired inputs for design and optimization. Constraints due to conservation principles are already taken into account in the numerical simulation. However, limitations on x due to the shape and size of the pond define an acceptable design domain. If the numerical simulation results are curve-fitted to yield expressions of the form

$$U(x) = A\,x^a + B\,x^b - C\,x^c \tag{4.14.48}$$

where A, B, C, a, b, and c are constants obtained from curve-fitting, calculus methods can easily be obtained to determine the optimum. However, this is a very time-consuming process and a more appropriate approach would be search methods where x is varied over the given domain and selective simulation runs are carried out at chosen locations to determine the optimum. This has been an important problem for the power industry for many years and has resulted in many different designs to obtain the highest efficiency-to-cost ratio.

Example 2 — In an automobile, the drag force on the vehicle due to its motion in air increases with its speed V. The engine efficiency η also varies with the speed due to the higher revolutions per minute of the engine and increased fuel flow rate. The efficiency initially increases and then decreases at large V due to the effect on the combustion process. If the cost per mile of travel is taken as the objective function U, then we may write

$$U(V) = A\,F_1(V) + \frac{B}{F_2(V)} \tag{4.14.49}$$

where F_1 (V) represents the drag force and F_2 (V) the engine efficiency. An increase in drag force increases the cost and an increase in efficiency reduces the cost. The constants A and B represent the effect of these quantities on the cost.

Again, this is a very complicated numerical simulation problem because of the transient three-dimensional problem involving turbulent flow and combustion. The constraints due to the conservation principles are already accounted for in the simulation. The physical limitations on the speed V, say, from 0 to 200 km/hr for common vehicles, may be used to define the domain. If the simulation results are curve-fitted with algebraic expressions, we may use calculus methods, as dU/dV = 0, to obtain the optimum. Search methods are more appropriate since only a limited number of simulations are needed at chosen values of V to extract the optimum.

Example 3 — In many processes, such as optical fiber drawing, hot rolling, continuous casting, and extrusion, the material is cooled by the flow of a cooling fluid, such as inert gases in optical fiber drawing, at velocity V_1, while the material moves at velocity V_2, as shown in Figure 4.14.20. Numerical simulation may be used to obtain the temperature decay with distance for different values of these variables. The temperature decay increases with increasing V_1 because of accelerated cooling, but decreases with increasing V_2 since the time available for heat removal in the cooling section of length L decreases at higher speed.

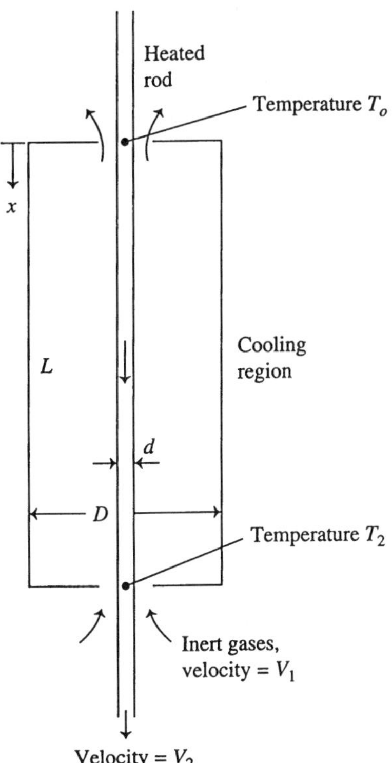

FIGURE 4.14.20 Dependence of temperature decay with distance x on the velocity V_1 of inert gases and velocity V_2 of the heated moving rod.

Numerical simulation may be used to solve this combined conduction–convection problem and obtain the inputs needed for design and optimization of the cooling system. If the cost per unit length of processed material is taken as the objective function U, we may write

$$U(V_1, V_2) = \frac{F_1(V_2)}{F_2(V_2)} + F_3(V_1) \tag{4.14.50}$$

where the function F_1 represents the costs for feeding and pick-up of the material, F_2 represents the productivity, and F_3 represents the cost of the inert gas and of the flow arrangement. Limitations on V_1 and V_2 due to physical considerations define the domain. Constraints due to mass and energy balances are part of the model. Search methods can be used for obtaining the optimum values of V_1 and V_2. Calculus methods and geometric programming may be applicable if the simulation results are curve-fitted to obtain closed-form expressions for the above functions.

The preceding discussion presents a few important considerations in formulating the optimization problem. Clearly, the physical nature of the system and any available information on its characteristics are important in developing a useful and practical approach to optimization. Similar treatment may be extended to the many different systems considered earlier. The resulting optimization problem is then solved by the various methods presented here.

Additional Considerations

There are several other important aspects associated with the optimization process and with the implementation of the optimal design obtained. These considerations are common to all the different approaches and address the practical issues involved in optimization. Since our interest lies in the optimum design which is both feasible and practical, it is necessary to include several additional aspects in the overall design and optimization of thermal systems. These include the choice of variables for optimization and sensitivity analysis. Generally, effort is made to focus on the dominant variables, often based on sensitivity analysis, which is also used to select the final values of the design variables for a realistic optimal design. It is also necessary to examine the dependence of the optimum on the chosen objective function. This involves trade-offs in the design so that the most important aspects are chosen for optimization. Since optimization is carried out for a given conceptual design, it is sometimes necessary to change the conceptual design in order to achieve a more desirable optimum. A consideration of the second law of thermodynamics can also be used as the basis for optimizing a thermal system. This treatment considers entropy generation and attempts to minimize it, as considered in detail by Bejan (1982, 1995). Many of these considerations are important and need to be included in the optimization process for certain thermal systems, as outlined by Jaluria (1998).

Conclusions

We have considered the basic formulation for optimization and different methods that are available for solving these problems. Several physical problems have been mentioned as examples to illustrate the general approach. We can thus summarize these aspects as related to the optimization of thermal systems.

Several different optimization methods have been discussed here. Some of these have only a limited applicability with respect to thermal systems. Calculus methods require continuous functions that can be differentiated and geometric programming requires sums of polynomials to characterize the system. Therefore, both of these methods can be used only if the system behavior can be represented by explicit, closed-form, expressions. This is possible only for very simple systems with few components and idealized behavior, or for cases where curve-fitting is employed to obtain representative equations. The latter approach is used for many systems, particularly for thermodynamic systems such as heating and cooling systems, where curve-fitting can be effectively employed to represent material and process characteristics.

Linear programming is of little interest in the optimization of thermal systems since it is rare to obtain linear equations to represent the constraints and the objective function. Because of the intrinsic nonlinear behavior of most thermal processes, it is generally not feasible to linearize the governing equations while maintaining the accuracy and validity of the representation. Dynamic programming is of greater interest in thermal systems, particularly in networking of channels for material and fluid flow. However, it is still of limited use in the optimization of common thermal systems.

Search methods constitute the most important optimization strategy for thermal systems. Many different approaches have been developed and are particularly appropriate for different problems. However, the underlying idea is to generate a number of designs, which are also called trials or iterations, and to select the best among these. Effort is made to keep the number of trials small, often going to the next iteration only if necessary. This is a very desirable feature with respect to thermal systems since each trial may take a considerable amount of computational effort.

Search methods may also be combined with other methods in order to accelerate convergence or approach to the optimum. For instance, calculus methods may be used at certain stages to narrow the domain in which the optimum lies. Trials for the search method are then used to provide information for extracting the derivatives and other relevant quantities. Prior knowledge on the optimum for similar systems may also be used to develop *heuristic* rules to accelerate the search.

References

Arora, J.S., *Introduction to Optimum Design*, McGraw-Hill, New York, 1989.

Beightler, C.S., Phillips, D.T., and Wilde, D.J., *Foundations of Optimization*, 2nd ed., Prentice-Hall, Englewood Cliffs, NJ, 1979.

Bejan, A., *Entropy Generation Through Heat and Fluid Flow*, John Wiley & Sons, New York, 1982.

Bejan, A., *Entropy Generation Minimization*, CRC Press, Boca Raton, FL, 1995.

Bejan, A., Tsatsaronis, G., and Moran, M., *Thermal Design and Optimization*, John Wiley & Sons, New York, 1996.

Beveridge, G.S.G. and Schechter, R.S., *Optimization: Theory and Practice*, McGraw-Hill, New York, 1970.

Denardo, E.V., *Dynamic Programming Models and Applications*, Prentice-Hall, Englewood Cliffs, NJ, 1982.

Dieter, G.E., *Engineering Design: A Materials and Processing Approach*, 2nd ed., McGraw-Hill, New York, 1991.

Duffin, R.J., Peterson, E.L., and Zener, C.M., *Geometric Programming*, John Wiley & Sons, New York, 1967.

Fox, R.L., *Optimization Methods for Engineering Design*, Addison-Wesley, Reading, MA, 1971.

Gass, S.I., *Linear Programming*, 4th ed., McGraw-Hill, New York, 1975.

Hadley, G.H., *Linear Programming*, Addison-Wesley, Reading, MA, 1962.

Haug, E.J. and Arora, J.S., *Applied Optimal Design*, John Wiley & Sons, New York, 1979.

Jaluria, Y., *Design and Optmization of Thermal System,* McGraw-Hill, New York, 1998.

Papalambros, P.Y. and Wilde, D.J., *Principles of Optimal Design*, Cambridge University Press, New York, 1989.

Reklaitis, G.V., Ravindran, A., and Ragsdell, K.M., *Engineering Optimization Methods and Applications*, John Wiley & Sons, New York, 1983.

Siddall, J.N., *Optimal Engineering Design*, Marcel Dekker, New York, 1982.

Stoecker, W.F., *Design of Thermal Systems*, 3rd ed., McGraw-Hill, New York, 1989.

Vanderplaats, G.N., *Numerical Optimization Techniques for Engineering Design*, McGraw-Hill, New York, 1984.

Wilde, D.J., *Globally Optimum Design,* Wiley-Interscience, New York, 1978.

Zener, C.M., *Engineering Design by Geometric Programming*, Wiley-Interscience, New York, 1971.

4.15 Heat Transfer Enhancement

Arthur E. Bergles

Introduction

Energy- and materials-saving considerations, as well as economic incentives, have led to efforts to produce more efficient heat exchange equipment. Common thermal-hydraulic goals are to reduce the size of a heat exchanger required for a specified heat duty, to upgrade the capacity of an existing heat exchanger, to reduce the approach temperature difference for the process streams, or to reduce the pumping power.

The study of improved heat transfer performance is referred to as heat transfer *enhancement*, *augmentation*, or *intensification*. In general, this means an increase in heat transfer coefficient. Attempts to increase "normal" heat transfer coefficients have been recorded for more than a century, and there is a large store of information. A survey (Bergles et al., 1995) cites 5676 technical publications, excluding patents and manufacturers' literature. The literature has expanded rapidly since 1955.

Enhancement techniques can be classified either as passive methods, which require no direct application of external power (Figure 4.15.1), or as active methods, which require external power. The effectiveness of both types of techniques is strongly dependent on the mode of heat transfer, which may range from single-phase free convection to dispersed-flow film boiling. Brief descriptions of these methods follow.

Treated surfaces involve fine-scale alternation of the surface finish or coating (continuous or discontinuous). They are used for boiling and condensing; the roughness height is below that which affects single-phase heat transfer.

Rough surfaces are produced in many configurations ranging from random sand-grain-type roughness to discrete protuberances. See Figure 4.15.1a. The configuration is generally chosen to disturb the viscous sublayer rather than to increase the heat transfer surface area. Application of rough surfaces is directed primarily toward single-phase flow.

Extended surfaces are routinely employed in many heat exchangers. See Figure 4.15.1a to d. Work of special interest to enhancement is directed toward improvement of heat transfer coefficients on extended surfaces by shaping or perforating the surfaces.

Displaced enhancement devices are inserted into the flow channel so as indirectly to improve energy transport at the heated surface. They are used with forced flow. See Figure 4.15.1e and f.

Swirl-flow devices include a number of geometric arrangements or tube inserts for forced flow that create rotating and/or secondary flow: coiled tubes, inlet vortex generators, twisted-tape inserts, and axial-core inserts with a screw-type winding.

Surface-tension devices consist of wicking or grooved surfaces to direct the flow of liquid in boiling and condensing.

Additives for liquids include solid particles and gas bubbles in single-phase flows and liquid trace additives for boiling systems.

Additives for gases are liquid droplets or solid particles, either dilute-phase (gas-solid suspensions) or dense-phase (fluidized beds).

Mechanical aids involve stirring the fluid by mechanical means or by rotating the surface. Surface "scraping," widely used for batch processing of viscous liquids in the chemical process industry, is applied to the flow of such diverse fluids as high-viscosity plastics and air. Equipment with rotating heat exchanger ducts is found in commercial practice.

Surface vibration at either low or high frequency has been used primarily to improve single-phase heat transfer.

Fluid vibration is the practical type of vibration enhancement because of the mass of most heat exchangers. The vibrations range from pulsations of about 1 Hz to ultrasound. Single-phase fluids are of primary concern.

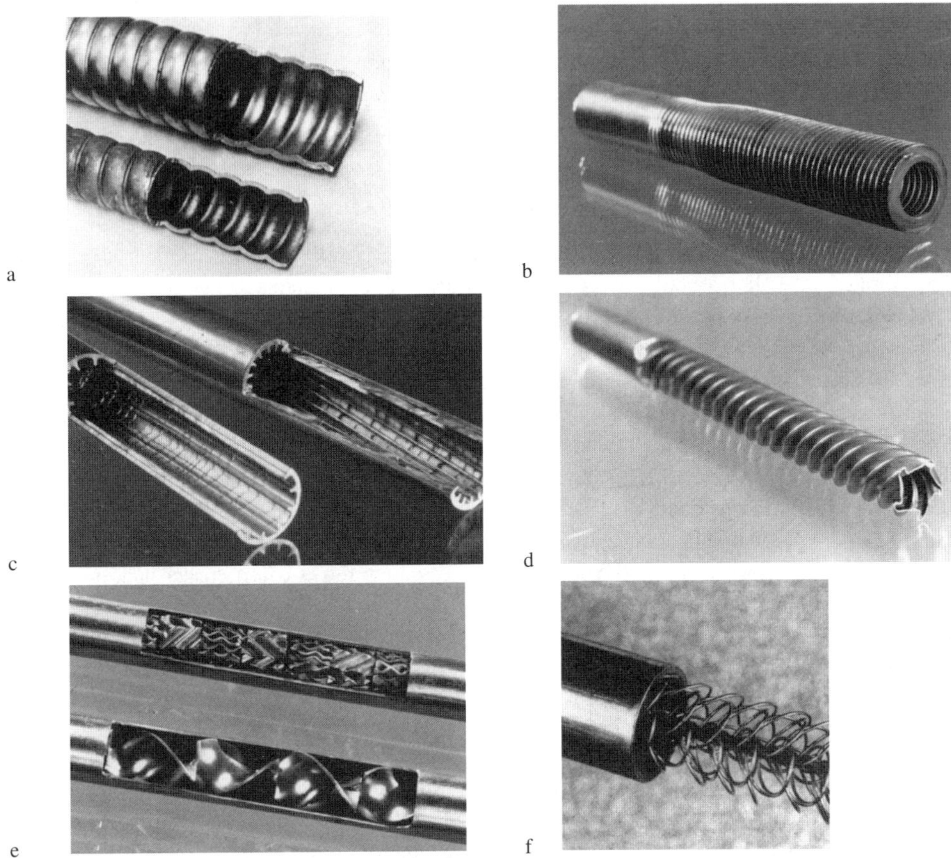

FIGURE 4.15.1 Enhanced tubes for augmentation of single-phase heat transfer. (a) Corrugated or spirally indented tubes with internal protuberances. (b) Integral external fins. (c) Integral internal fins. (d) Deep spirally fluted tube. (e) Static mixer inserts. (f) Wire-wound insert.

Electrostatic fields (DC or AC) are applied in many different ways to dielectric fluids. Generally speaking, electrostatic fields can be directed to cause greater bulk mixing or fluid or disruption of fluid flow in the vicinity of the heat transfer surface, which enhances heat transfer.

Injection is utilized by supplying gas to a stagnant or flowing liquid through a porous heat transfer surface or by injecting similar fluid upstream of the heat transfer section. Surface degassing of liquids can produce enhancement similar to gas injection. Only single-phase flow is of interest.

Suction involves vapor removal, in nucleate or film boiling, or fluid withdrawal, in single-phase flow, through a porous heated surface.

Two or more of the above techniques may be utilized simultaneously to produce an enhancement that is larger than either of the techniques operating separately. This is termed *compound enhancement.*

It should be emphasized that one of the motivations for studying enhanced heat transfer is to assess the effect of an inherent condition on heat transfer. Some practical examples include roughness produced by standard manufacturing, degassing of liquids with high gas content, surface vibration resulting from rotating machinery or flow oscillations, fluid vibration resulting from pumping pulsation, and electrical fields present in electrical equipment.

The surfaces in Figure 4.15.1 have been used for both single-phase and two-phase heat transfer enhancement. The emphasis is on effective and cost-competitive (proved or potential) techniques that have made the transition from the laboratory to commercial heat exchangers.

Single-Phase Free Convection

With the exception of the familiar technique of providing extended surfaces, the passive techniques have little to offer in the way of enhanced heat transfer for free convection. This is because the velocities are usually too low to cause flow separation or secondary flow.

The restarting of thermal boundary layers in interrupted extended surfaces increases heat transfer so as to more than compensate for the lost area.

Mechanically aided heat transfer is a standard technique in the chemical and food industries when viscous liquids are involved. The predominant geometry for surface vibration has been the horizontal cylinder, vibrated either horizontally or vertically. Heat transfer coefficients can be increased tenfold for both low-frequency/high-amplitude and high-frequency/low-amplitude situations. It is, of course, equally effective and more practical to provide steady forced flow. Furthermore, the mechanical designer is concerned that such intense vibrations could result in equipment failures.

Since it is usually difficult to apply surface vibrations to practical equipment, an alternative technique is utilized whereby vibrations are applied to the fluid and focused toward the heated surface. With proper transducer design, it is also possible to improve heat transfer to simple heaters immersed in gases or liquids by several hundred percent.

Electric fields are particularly effective in increasing heat transfer coefficients in free convection. Dielectrophoretic or electrophoretic (especially with ionization of gases) forces cause greater bulk mixing in the vicinity of the heat transfer surface. Heat transfer coefficients may be improved by as much as a factor of 40 with electrostatic fields up to 100,000 V. Again, the equivalent effect could be produced at lower capital cost and without the voltage hazard by simply providing forced convection with a blower or fan.

Single-Phase Forced Convection

The present discussion emphasizes enhancement of heat transfer *inside* ducts that are primarily of circular cross section. Typical data for turbulence promoters inserted inside tubes are shown in Figure 4.15.2. As shown in Figure 4.15.2a, the promoters produce a sizable elevation in the Nusselt number, or heat transfer coefficient, at constant Reynolds number, or velocity. However, in Figure 4.15.2b, there is an accompanying large increase in the friction factor.

Surface roughness has been used extensively to enhance forced convection heat transfer. Integral roughness may be produced by the traditional manufacturing processes of machining, forming, casting,

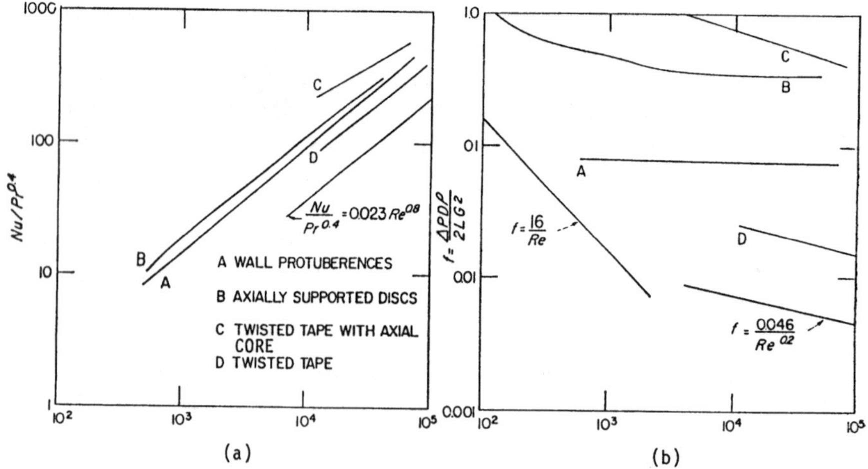

FIGURE 4.15.2 Typical data for turbulence promoters inserted inside tubes: (a) heat transfer data, (b) friction data. (From Bergles, A.E., *Progress in Heat and Mass Transfer,* Pergamon, Oxford, U.K., 1969. With permission.)

or welding. Various inserts can also provide surface protuberances. In view of the infinite number of possible geometric variations, it is not surprising that, even after more than 300 studies, no completely satisfactory unified treatment is available.

In general, the maximum enhancement of laminar flow with many of the techniques is the same order of magnitude, and seems to be independent of the wall boundary condition. The enhancement with some rough tubes, corrugated tubes, inner-fin tubes, various static mixers, and twisted-type inserts is about 200%. The improvements in heat transfer coefficient with turbulent flow in rough tubes (based on nominal surface area) are as much as 250%. Analogy solutions for sand-grain-type roughness and for square-repeated-rib roughness have been proposed. A statistical correlation is also available for heat transfer coefficient and friction factor.

The following correlations are recommended for tubes with transverse or helical repeated ribs (Figure 4.15.1a) with turbulent flow (Ravigururajan and Bergles, 1985):

$$\mathrm{Nu}_{D_i,a}/\mathrm{Nu}_{D_i,s} = \left\{1+\left[2.64\mathrm{Re}^{0.036}\left(e/D_i\right)^{0.212}\left(\left(p/D_i\right)^{-0.21}\right)\left(\alpha/90\right)^{0.29}\left(\mathrm{Pr}\right)^{-0.024}\right]^7\right\}^{1/7} \tag{4.15.1}$$

$$f_a/f_s = \left\{\left[\left[29.1\mathrm{Re}_{D_i}^{\left(0.67-0.06p/D_i-0.49\alpha/90\right)} \times \left(e/D_i\right)^{\left(0.37-0.157p/D_i\right)} \times \left(p/D_i\right)^{\left(-1.66\times10^{-6}\mathrm{Re}_{D_i}-0.33\alpha/90\right)}\right.\right.\right.$$
$$\left.\left.\left.\times\left(\alpha/90\right)^{\left(4.59+4.11\times10^{-6}\mathrm{Re}_{D_i}-0.15p/D_i\right)} \times \left(1+\frac{2.94}{n}\right)\sin\beta\right]^{15/16}\right]^{16/15}\right\} \tag{4.15.2}$$

where the subscript a refers to the enhanced tube and the subscript s refers to the smooth tube. The special symbols are given as follows: e = protuberance height; p = repeated-rib pitch; α = spiral angle for helical ribs, °; n = number of sharp corners facing the flow; and β = contact angle of rib profile, °.

Also,

$$\mathrm{Nu}_s = 0.5f\,\mathrm{Re}_{D_i}\,\mathrm{Pr}\Big/\left(1+12.7(0.5f)^{0.5}\,\mathrm{Pr}^{0.667}-1\right)$$

and

$$f_s = \left(1.58\ln\mathrm{Re}_{D_i} - 3.28\right)^{-2}\,*$$

Much work has been done to obtain the enhanced heat transfer of parallel angled ribs in short rectangular channels, simulating the interior of gas turbine blades. Jets are frequently used for heating, cooling, and drying in a variety of industrial applications. A number of studies have reported that roughness elements of the transverse-repeated-rib type mitigate the deterioration in heat transfer downstream of stagnation.

Extended surfaces can be considered "old technology" as far as most applications are concerned. The real interest now is in increasing heat transfer coefficients on the extended surface. Compact heat exchangers of the plate-fin or tube-and-center variety use several enhancement techniques: offset strip fins, louvered fins, perforated fins, or corrugated fins. Coefficients are several hundred percent above the smooth-tube values; however, the pressure drop is also substantially increased, and there may be vibration and noise problems.

* The Fanning friction factor is used throughout this section.

For the case of offset strip fins the following correlations are recommended for calculating the j and f characteristics (Manglik and Bergles, 1990)

$$j_h = 0.6522\mathrm{Re}_h^{-0.5403}\,\alpha^{-0.1541}\,\delta^{0.1499}\,\gamma^{-0.0678} \times \left[1 + 5.269\times10^{-5}\mathrm{Re}_h^{1.340}\,\alpha^{0.504}\,\delta^{0.456}\,\gamma^{-1.055}\right]^{0.1} \quad (4.15.3)$$

$$f_h = 9.6243\mathrm{Re}_h^{-0.7422}\,\alpha^{-0.1856}\,\delta^{0.3053}\,\gamma^{-0.2659} \times \left[1 + 7.669\times10^{-8}\mathrm{Re}_h^{4.429}\,\alpha^{0.920}\,\delta^{3.767}\,\gamma^{0.236}\right]^{0.1} \quad (4.15.4)$$

where j_H (the heat transfer j-factor $\mathrm{Nu}_H/\mathrm{Re}_H\mathrm{Pr}^{1/3}$), and f_h, and Re_h are based on the hydraulic diameter given by

$$D_h = 4shl/\left[2(sl + hl + th) + ts\right] \quad (4.15.5)$$

Special symbols are α = aspect ratio s/h, δ = ratio t/l, γ = ratio t/s, s = lateral spacing of strip fin, h = strip fin height, l = length of one offset module of strip fins, and t = fin thickness.

These equations are based on experimental data for 18 different offset strip-fin geometries, and they represent the data continuously in the laminar, transition, and turbulent flow regions.

Internally finned circular tubes are available in aluminum and copper (or copper alloys). Correlations (for heat transfer coefficient and friction factor) are available for laminar flow, for both straight and spiral continuous fins.

Turbulent flow in tubes with straight or helical fins (Figure 4.15.1c) was correlated by (Carnavos, 1979)

$$\mathrm{Nu}_h = 0.023\mathrm{Pr}^{0.4}\mathrm{Re}_h^{0.8}\left[\frac{A_c}{A_{ci}}\right]^{0.1}\left[\frac{A_{s,i}}{A_s}\right]^{0.5}(\sec\alpha)^3 \quad (4.15.6)$$

$$f_h = 0.046\mathrm{Re}_h^{-0.2}\left[\frac{A_c}{A_{c,i}}\right]^{0.5}(\sec\alpha)^{0.75} \quad (4.15.7)$$

where $A_{c,i}$ is based on the maximum inside (envelope) flow area, $A_{s,i}$ is based on the maximum inside (envelope) surface area, and α the spiral angle for helical fins, °.

A numerical analysis of turbulent flow in tubes with idealized straight fins was reported. The necessary constant for the turbulence model was obtained from experimental data for air. Further improvements in numerical techniques are expected, so that a wider range of geometries and fluids can be handled without resort to extensive experimental programs.

Many proprietary surface configurations have been produced by deforming the basic tube. The "convoluted," "corrugated," "spiral," or "spirally fluted" tubes (Figure 4.15.1a) have multiple-start spiral corrugations, which add area, along the tube length. A systematic survey of the single-tube performance of condenser tubes indicates up to 400% increase in the nominal inside heat transfer coefficient (based on diameter of a smooth tube of the same maximum inside diameter); however, pressure drops on the water side are about 20 times higher.

Displaced enhancement devices are typically in the form of inserts, within elements arranged to promote transverse mixing (static mixers, Figure 4.15.1e). They are used primarily for viscous liquids, to promote either heat transfer or mass transfer. Displaced promoters are also used to enhance the radiant heat transfer in high-temperature applications. In the flue-tube of a hot-gas-fired hot water heater, there is a trade-off between radiation and convection. Another type of displaced insert generates vortices, which enhance the downstream flow. Delta-wing and rectangular wing promoters, both co-rotating and counterrotating, have been studied. Wire-loop inserts (Figure 4.15.1f) have also been used for enhancement of laminar and turbulent flow.

Twisted-tape inserts have been widely used to improve heat transfer in both laminar and turbulent flow. Correlations are available for laminar flow, for both uniform heat flux and uniform wall temperature conditions. Turbulent flow in tubes with twisted-tape inserts has also been correlated. Several studies have considered the heat transfer enhancement of a decaying swirl flow, generated, say, by a short twisted-tape insert.

Performance Evaluation Criteria for Single-Phase Forced Convection in Tubes

Numerous, and sometimes conflicting, factors enter into the ultimate decision to use an enhancement technique: heat duty increase or area reduction that can be obtained, initial cost, pumping power or operating cost, maintenance cost (especially cleaning), safety, and reliability, among others. These factors are difficult to quantitize, and a generally acceptable selection criterion may not exist. It is possible, however, to suggest some performance criteria for preliminary design guidance. As an example, consider the basic geometry and the pumping power fixed, with the objective of increasing the heat transfer. The following ratio is then of interest

$$R_3 = \left(\frac{h_a}{h_s}\right)_{D_i,L,N,P,T_{\text{in}},\Delta T} = \frac{\left(\text{Nu}/\text{Pr}^{0.4}\right)_a}{\left(\text{Nu}/\text{Pr}^{0.4}\right)_s} = \frac{q_a}{q_s} \tag{4.15.8}$$

where P = pumping power, T_{in} = inlet bulk temperature of fluid, and ΔT = average wall-fluid temperature difference.

With the pumping power (neglecting entrance and exit losses) given as

$$P = NVA_c 4f(L/D)\rho V^2/2 \tag{4.15.9}$$

and

$$f_s = 0.046/\text{Re}_s^{0.2} \tag{4.15.10}$$

$$A_{c,a} f_a \text{Re}_a^3 = 0.046 A_{c,s} \text{Re}_s^{2.8} \tag{4.15.11}$$

The calculation best proceeds by picking $\text{Re}_{D_i,a}$, and reading $\text{Nu}_{D_i,a}/\text{Pr}^{0.4}$ and f_a. $\text{Re}_{D_i,s}$ is then obtained from Equation (4.15.11) and $\text{Nu}_{D_i,s}/\text{Pr}^{0.4}$ obtained from a conventional, empty-tube correlation. The desired ratio of Equation (4.15.8) is then obtained. Typical results are presented in Figure 4.15.3 for a repeated-rib roughness (Bergles et al., 1974).

Active and Compound Techniques for Single-Phase Forced Convection

Under active techniques, mechanically aided heat transfer in the form of surface scraping can increase forced convection heat transfer. Surface vibration has been demonstrated to improve heat transfer to both laminar and turbulent duct flow of liquids. Fluid vibration has been extensively studied for both air (loudspeakers and sirens) and liquids (flow interrupters, pulsators, and ultrasonic transducers). Pulsations are relatively simple to apply to low-velocity liquid flows, and improvements of several hundred percent can be realized.

Some very impressive enhancements have been recorded with electrical fields, particularly in the laminar-flow region. Improvements of at least 100% were obtained when voltages in the 10-kV range were applied to transformer oil. It is found that even with intense electrostatic fields, the heat transfer enhancement disappears as turbulent flow is approached in a circular tube with a concentric inner electrode.

Compound techniques are a slowly emerging area of enhancement that holds promise for practical applications, since heat transfer coefficients can usually be increased above each of the several techniques

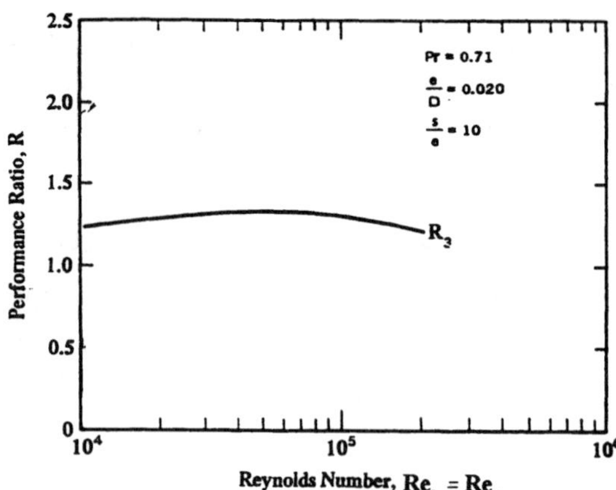

FIGURE 4.15.3 Constant pumping power performance criterion applied to repeated rib roughness.

acting alone. Some examples that have been studied are as follows: rough tube wall with twisted-tape inserts, rough cylinder with acoustic vibrations, internally finned tube with twisted-tape inserts, finned tubes in fluidized beds, externally finned tubes subjected to vibrations, rib-roughened passage being rotated, gas-solid suspension with an electrical field, fluidized bed with pulsations of air, and a rib-roughened channel with longitudinal vortex generation.

Pool Boiling

Selected passive and active enhancement techniques have been shown to be effective for pool boiling and flow boiling/evaporation. Most techniques apply to nucleate boiling; however, some techniques are applicable to transition and film boiling.

It should be noted that phase-change heat transfer coefficients are relatively high. The main thermal resistance in a two-fluid heat exchanger often lies on the non-phase-change side. (Fouling of either side can, of course, represent the dominant thermal resistance.) For this reason, the emphasis is often on enhancement of single-phase flow. On the other hand, the overall thermal resistance may then be reduced to the point where significant improvement in the overall performance can be achieved by enhancing the two-phase flow. Two-phase enhancement would also be important in double-phase-change (boiling/condensing) heat exchangers.

As discussed elsewhere, surface material and finish have a strong effect on nucleate and transition pool boiling. However, reliable control of nucleation on plain surfaces is not easily accomplished. Accordingly, since the earliest days of boiling research, there have been attempts to relocate the boiling curve through use of relatively gross modification of the surface. For many years, this was accomplished simply by area increase in the form of low helical fins. The subsequent tendency was to structure surfaces to improve the nucleate boiling characteristics by a fundamental change in the boiling process. Many of these advanced surfaces are being used in commercial shell-and-tube boilers.

Several manufacturing processes have been employed: machining, forming, layering, and coating. In Figure 4.15.4a standard low-fin tubing is shown. Figure 4.15.4c depicts a tunnel-and-pore arrangement produced by rolling, upsetting, and brushing. An alternative modification of the low fins is shown in Figure 4.15.4d, where the rolled fins have been split and rolled to a T shape. Further modification of the internal, Figure 4.15.4e, or external, Figure 4.15.4f, surface is possible. Knurling and rolling are involved in producing the surface shown in Figure 4.15.4g. The earliest example of a commercial structured surface, shown in Figure 4.15.4b is the porous metallic matrix produced by sintering or brazing small particles. Wall superheat reductions of up to a factor of ten are common with these surfaces. The advantage is not only a high nucleate boiling heat transfer coefficient, but the fact that boiling can take place at very low temperature differences.

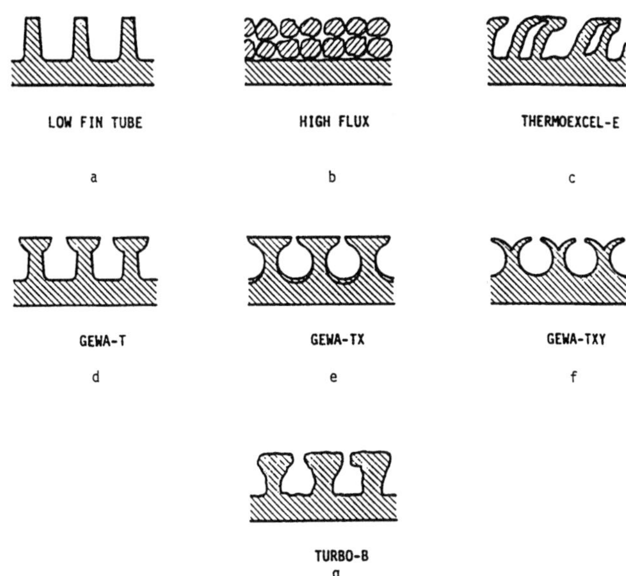

FIGURE 4.15.4 Examples of commercial structured boiling surfaces. (From Pate, M.B. et al., in *Compact Heat Exchangers*, Hemisphere Publishing, New York, 1990. With permission.)

These structured boiling surfaces, developed for refrigeration and process applications, have been used as "heat sinks" for immersion-cooled microelectronic chips.

The behavior of tube bundles is often different with structured-surface tubes. The enhanced nucleate boiling dominates, and the convective boiling enhancement, found in plain tube bundles, does not occur.

Active enhancement techniques include heated surface rotation, surface wiping, surface vibration, fluid vibration, electrostatic fields, and suction at the heated surface. Although active techniques are effective in reducing the wall superheat and/or increasing the critical heat flux, the practical applications are very limited, largely because of the difficulty of reliably providing the mechanical or electrical effect.

Compound enhancement, which involves two or more techniques applied simultaneously, has also been studied. Electrohydrodynamic enhancement was applied to a finned tube bundle, resulting in nearly a 200% increase in the average boiling heat transfer coefficient of the bundle, with a small power consumption for the field.

Convective Boiling/Evaporation

The structured surfaces described in the previous section are generally not used for in-tube vaporization, because of the difficulty of manufacture. One notable exception is the high-flux surface in a vertical thermosiphon reboiler. The considerable increase in the low-quality, nucleate boiling coefficient is desirable, but it is also important that more vapor is generated to promote circulation.

Helical repeated ribs and helically coiled wire inserts have been used to increase vaporization coefficients and the dry-out heat flux in once-through boilers.

Numerous tubes with internal fins, either integral or attached, are available for refrigerant evaporators. Original configurations were tightly packed, copper, offset strip fin inserts soldered to the copper tube or aluminum, star-shaped inserts secured by drawing the tube over the insert. Examples are shown in Figure 4.15.5. Average heat transfer coefficients (based on surface area of smooth tube of the same diameter) for typical evaporator conditions are increased by as much as 200%. A cross-sectional view of a typical "microfin" tube is included in Figure 4.15.5. The average evaporation boiling coefficient is increased 30 to 80%. The pressure drop penalties are less; that is, lower percentage increases in pressure drop are frequently observed.

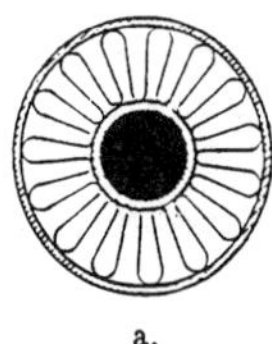

FIGURE 4.15.5 Inner-fin tubes for refrigerant evaporators: (a) Strip-fin inserts, (b) Star-shaped inserts, (c) Microfin.

Twisted-tape inserts are generally used to increase the burnout heat flux for subcooled boiling at high imposed heat fluxes $10^7 - 10^8$ W/m^2, as might be encountered in the cooling of fusion reactor components. Increases in burnout heat flux of up to 200% were obtained at near atmospheric pressure.

Vapor-Space Condensation

As discussed elsewhere, condensation can be either filmwise or dropwise. In a sense, dropwise condensation is enhancement of the normally occurring film condensation by surface treatment. The only real application is for steam condensers, because nonwetting coatings are not available for most other working fluids. Even after much study, little progress has been made in developing permanently hydrophobic coatings for practical steam condensers. The enhancement of dropwise condensation is pointless, because the heat transfer coefficients are already so high.

Surface extensions are widely employed for enhancement of condensation. The integral low fin tubing (Figure 4.15.4a), used for kettle boilers, is also used for horizontal tube condensers. With proper spacing of the fins to provide adequate condensate drainage, the average coefficients can be several times those of a plain tube with the same base diameter. These fins are normally used with refrigerants and other organic fluids that have low condensing coefficients, but which drain effectively, because of low surface tension.

The fin profile can be altered according to mathematical analysis to take full advantage of the Gregorig effect, whereby condensation occurs mainly at the tops of convex ridges. Surface tension forces then pull the condensate into concave grooves, where it runs off. The average heat transfer coefficient is greater than that for an axially uniform film thickness. The initial application was for condensation of steam on vertical tubes used for reboilers and in desalination. According to numerical solutions, the optimum geometry is characterized by a sharp fin tip, gradually changing curvature of the fin surface from tip to root, wide grooves between fins to collect condensate, and periodic condensate strippers. Figure 4.15.6 schematically presents the configuration.

Recent interest has centered on three-dimensional surfaces for horizontal-tube condensers. The considerable improvement relative to low fins or other two-dimensional profiles is apparently due to multidimensional drainage at the fin tips. Other three-dimensional shapes include circular pin fins, square pins, and small metal particles that are bonded randomly to the surface.

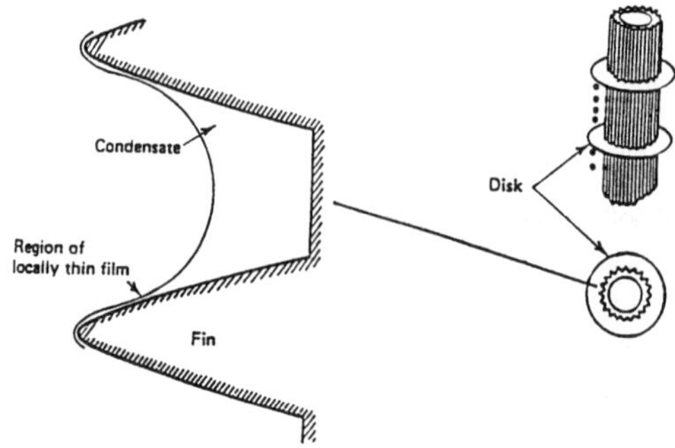

FIGURE 4.15.6 Recommended flute profile and schematic of condensate strippers.

Convective Condensation

This final section on enhancement of the various modes of heat transfer focuses on in-tube condensation. The applications include horizontal kettle-type reboilers, moisture separator reheaters for nuclear power plants, and air-conditioner condensers.

Internally grooved or knurled tubes, deep spirally fluted tubes, random roughness, conventional inner-fin tubes have been shown to be effective for condensation of steam and other fluids.

The microfin tubes mentioned earlier have also been applied successfully to in-tube condensing. As in the case of evaporation, the substantial heat transfer improvement is achieved at the expense of a lesser percentage increase in pressure drop. By testing a wide variety of tubes, it has been possible to suggest some guidelines for the geometry, e.g., more fins, longer fins, and sharper tips; however, general correlations are not yet available. Fortunately for heat-pump operation, the tube that performs best for evaporation also performs best for condensation.

Twisted-tape inserts result in rather modest increases in heat transfer coefficient for complete condensation of either steam or refrigerant. The pressure drop increases are large because of the large wetted surface. Coiled tubular condensers provide a modest improvement in average heat transfer coefficient.

References

Bergles, A.E. 1969. Survey and evaluation of techniques to augment convective heat and mass transfer, in *Progress in Heat and Mass Transfer*, Vol. 1, Pergamon, Oxford, England.

Bergles, A.E. 1988. Some perspectives on enhanced heat transfer — second generation heat transfer technology, *J. Heat Transfer*, 110, 1082–1096.

Bergles, A.E. 1997. Heat transfer enhancement — the encouragement and accommodation of high heat fluxes. *J. Heat Transfer*, 119, 8–19.

Bergles, A.E. 1998. Techniques to enhance heat transfer, in *Handbook of Heat Transfer Applications*, W.M. Rohsenow, J.P. Hartnett, and Y.I. Cho, Eds., McGraw-Hill, New York, 11.1-11.76.

Bergles, A.E., Blumenkrantz, A.R., and Taborek, J. 1974. Performance evaluation criteria for enhanced heat transfer surfaces, in *Heat Transfer 1974*, The Japan Society of Mechanical Engineers, Tokyo, Vol. II, 234–238.

Bergles, A.E., Jensen, M.K., and Shome, B. 1995. Bibliography on enhancement of convective heat and mass transfer, RPI Heat Transfer Laboratory Report, HTL-23.

Carnavos, T.C. 1979. Heat transfer performance of internally finned tubes in turbulent flow, in *Advances in Advanced Heat Transfer*, ASME, New York, 61–67.

Manglik, R.M. and Bergles, A.E. 1990. The thermal-hydraulic design of the rectangular offset-strip-fin compact heat exchanger, in *Compact Heat Exchangers*, Hemisphere Publishing, New York, 123–149.

Pate, M.B., Ayub, Z.H., and Kohler, J. 1990. Heat exchangers for the air-conditioning and refrigeration industry: state-of-the-art design and technology, in *Compact Heat Exchangers*, Hemisphere Publishing, New York, 567–590.

Ravigururajan, S. and Bergles, A.E. 1985. General Correlations for Pressure Drop and Heat Transfer for Single-Phase Turbulent Flow in Internally Ribbed Tubes, in *Augmentation of Heat Transfer in Energy Systems*, HTD-Vol. 52, ASME, New York, 9–20.

Thome, J.R. 1990. *Enhanced Boiling Heat Transfer*, Hemisphere Publishing, New York.

Webb, R.L. 1994. *Principles of Enhanced Heat Transfer*, John Wiley & Sons, New York.

Further Information

This section gives some indication as to why heat transfer enhancement is one of the fastest growing areas of heat transfer. Many techniques are available for improvement of the various modes of heat transfer. Fundamental understanding of the transport mechanism is growing, but, more importantly, design correlations are being established. Many effective and cost-competitive enhancement techniques have made the transition from the laboratory to commercial heat exchangers.

Broad reviews of developments in enhanced heat transfer are available (Bergles, 1988; Thome, 1990; Webb, 1994; Bergles, 1997). Also, several journals, especially *Heat Transfer Engineering, Enhanced Heat Transfer,* and *International Journal of Heating, Ventilating, Air-Conditioning and Refrigerating Research*, feature this technology.

4.16 Heat Pipes

Larry W. Swanson

Introduction

The heat pipe is a vapor-liquid phase-change device that transfers heat from a hot reservoir to a cold reservoir using **capillary forces** generated by a **wick** or porous material and a working fluid. Originally conceived by Gaugler in 1944, the operational characteristics of heat pipes were not widely publicized until 1963 when Grover and his colleagues at Los Alamos Scientific Laboratory independently reinvented the concept. Since then many types of heat pipes have been developed and used by a wide variety of industries.

Figure 4.16.1 shows a schematic of a heat pipe aligned at angle ψ relative to the vertical axis (gravity vector). The heat pipe is composed of a container lined with a wick that is filled with liquid near its saturation temperature. The vapor-liquid interface, usually found near the inner edge of the wick, separates the liquid in the wick from an open vapor core. Heat flowing into the evaporator is transferred through the container to the liquid-filled wicking material, causing the liquid to evaporate and vapor to flow into the open core portion of the evaporator. The capillary forces generated by the evaporating interface increase the pressure difference between the vapor and liquid. The vapor in the open core flows out of the evaporator through the adiabatic region (insulated region) and into the condenser. The vapor then condenses, generating capillary forces similar, although much less in magnitude, to those in the evaporator. The heat released in the condenser passes through the wet wicking material and container

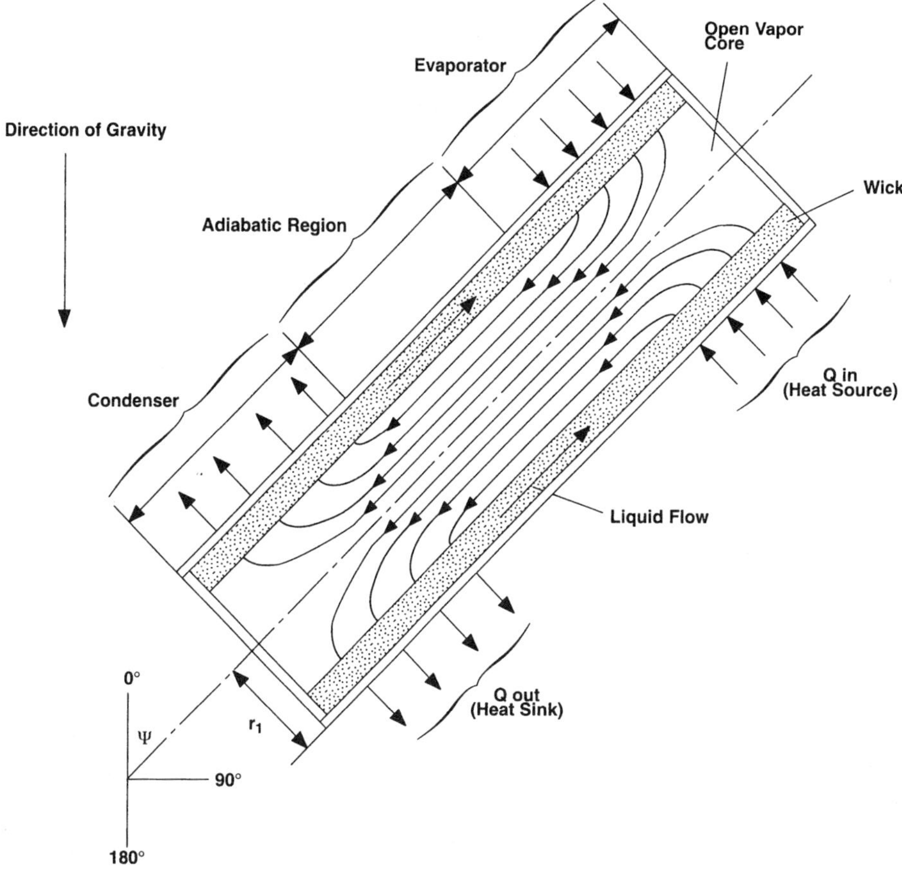

FIGURE 4.16.1 Schematic of a typical heat pipe.

out into the cold reservoir. The condensed liquid is then pumped, by the liquid pressure difference due to the net capillary force between the evaporator and condenser, out of the condenser back into the evaporator. Proper selection and design of the pipe container, working fluid, and wick structure are essential to the successful operation of a heat pipe. The **heat transfer limitations, effective thermal conductivity**, and axial temperature difference define the operational characteristics of the heat pipe.

Heat Pipe Container, Working Fluid, and Wick Structures

The container, working fluid, and wick structure of a heat pipe determine its operational characteristics. One of the most important considerations in choosing the material for the heat pipe container and wick is its compatibility with the working fluid. Degradation of the container or wick and contamination of the working fluid due to chemical reaction can seriously impair heat pipe performance. For example, noncondensable gas created during a chemical reaction eventually can accumulate near the end of the condenser, decreasing the condensation surface area. This reduces the ability of the heat pipe to transfer heat to the external heat sink. The material and geometry of the heat pipe container also must have a high burst strength, low weight, high thermal conductivity, and low porosity.

Using the proper working fluid for a given application is another critical element of proper heat pipe operation. The working fluid must have good thermal stability properties at the specified operational temperature and pressure. The operational temperature range of the working fluid has to lie between its triple point and its critical point for liquid to exist in the wicking material. The **wettability** of the working fluid contributes to its capillary pumping and priming capability. High-surface-tension fluids are commonly used in heat pipes because they provide the capillary pumping and wetting characteristics necessary for proper operation. Other desirable thermophysical properties include a high liquid thermal conductivity, high latent heat of vaporization, low liquid viscosity, and a low vapor viscosity. Table 4.16.1 gives the thermophysical properties for three typical heat pipe working fluids that span a fairly wide operating temperature range. The thermophysical properties for other heat pipe working fluids can be obtained from Dunn and Reay (1982) and Peterson (1994).

The wick structure and working fluid generate the capillary forces required to (1) pump liquid from the condenser to the evaporator and (2) keep liquid evenly distributed in the wicking material. Heat pipe wicks can be classified as either homogeneous wicks or composite wicks. Homogeneous wicks are composed of a single material and configuration. The most common types of homogeneous wicks include wrapped screen, sintered metal, axial groove, annular, crescent, and arterial. Composite wicks are composed of two or more materials and configurations. The most common types of composite wicks include variable screen mesh, screen-covered groove, screen slab with grooves, and screen tunnel with grooves. Regardless of the wick configuration, the desired material properties and structural characteristics of heat pipe wick structures are a high thermal conductivity, high wick porosity, small capillary radius, and high wick permeability. Table 4.16.2 gives the geometric properties of some commonly used homogeneous wicks. The properties of other wick structures, including nonhomogenous types, can be obtained from Peterson (1994). The container, wick structure, and working fluid are used to determine the heat transfer limitations of heat pipes.

Heat Transfer Limitations

Heat pipes undergo various heat transfer limitations depending on the working fluid, the wick structure, the dimensions of the heat pipe, and the heat pipe operational temperature. Figure 4.16.2 gives a qualitative description of the various heat transfer limitations, which include vapor-pressure, sonic, entrainment, capillary, and boiling limitations. The composite curve enclosing the shaded region in Figure 4.16.2 gives the maximum heat transfer rate of the heat pipe as a function of the operational temperature. The figure shows that as the operational temperature increases, the maximum heat transfer rate of the heat pipe is limited by different physical phenomena. As long as the operational heat transfer rate falls within the shaded region, the heat pipe will function properly.

TABLE 4.16.1 Thermophysical Properties of Some Heat-Pipe Fluids

Temperature (°C)	Latent Heat (kJ /kg)	Liquid Density (kg/m³)	Vapor Density (kg/m³)	Liquid Thermal Conductivity (W/m°C)	Liquid Viscosity (cP)	Vapor Viscosity (cP, × 10²)	Vapor Pressure (bars)	Vapor Specific Heat (kJ/kg°C)	Liquid Surface Tension (N/m × 10²)
				Methanol					
–50	1194	843.5	0.01	0.210	1.700	0.72	0.01	1.20	3.26
–30	1187	833.5	0.01	0.208	1.300	0.78	0.02	1.27	2.95
–10	1182	818.7	0.04	0.206	0.945	0.85	0.04	1.34	2.63
10	1175	800.5	0.12	0.204	0.701	0.91	0.10	1.40	2.36
30	1155	782.0	0.31	0.203	0.521	0.98	0.25	1.47	2.18
50	1125	764.1	0.77	0.202	0.399	1.04	0.55	1.54	2.01
70	1085	746.2	1.47	0.201	0.314	1.11	1.31	1.61	1.85
90	1035	724.4	3.01	0.199	0.259	1.19	2.69	1.79	1.66
110	980	703.6	5.64	0.197	0.211	1.26	4.98	1.92	1.46
130	920	685.2	9.81	0.195	0.166	1.31	7.86	1.92	1.25
150	850	653.2	15.90	0.193	0.138	1.38	8.94	1.92	1.04
				Water					
20	2448	998.0	0.02	0.603	1.00	0.96	0.02	1.81	7.28
40	2402	992.1	0.05	0.630	0.65	1.04	0.07	1.89	7.00
60	2359	983.3	0.13	0.649	0.47	1.12	0.20	1.91	6.66
80	2309	972.0	0.29	0.668	0.36	1.19	0.47	1.95	6.26
100	2258	958.0	0.60	0.680	0.28	1.27	1.01	2.01	5.89
120	2200	945.0	1.12	0.682	0.23	1.34	2.02	2.09	5.50
140	2139	928.0	1.99	0.683	0.20	1.41	3.90	2.21	5.06
160	2074	909.0	3.27	0.679	0.17	1.49	6.44	2.38	4.66
180	2003	888.0	5.16	0.669	0.15	1.57	10.04	2.62	4.29
200	1967	865.0	7.87	0.659	0.14	1.65	16.19	2.91	3.89
				Potassium					
350	2093	763.1	0.002	51.08	0.21	0.15	0.01	5.32	9.50
400	2078	748.1	0.006	49.08	0.19	0.16	0.01	5.32	9.04
450	2060	735.4	0.015	47.08	0.18	0.16	0.02	5.32	8.69
500	2040	725.4	0.031	45.08	0.17	0.17	0.05	5.32	8.44
550	2020	715.4	0.062	43.31	0.15	0.17	0.10	5.32	8.16
600	2000	705.4	0.111	41.81	0.14	0.18	0.19	5.32	7.86
650	1980	695.4	0.193	40.08	0.13	0.19	0.35	5.32	7.51
700	1960	685.4	0.314	38.08	0.12	0.19	0.61	5.32	7.12
750	1938	675.4	0.486	36.31	0.12	0.20	0.99	5.32	6.72
800	1913	665.4	0.716	34.81	0.11	0.20	1.55	5.32	6.32
850	1883	653.1	1.054	33.31	0.10	0.21	2.34	5.32	5.92

The vapor-pressure limitation (or viscous limitation) in heat pipes develops when the pressure drop in the vapor core reaches the same order of magnitude as the vapor pressure in the evaporator. Under these conditions, the pressure drop due to flow through the vapor core creates an extremely low vapor pressure in the condenser preventing vapor from flowing in the condenser. A general expression for the vapor-pressure limitation is (Dunn and Reay, 1982)

$$Q_{\mathrm{vp,max}} = \frac{\pi r_{\mathrm{v}}^4 h_{fg} \rho_{v,e} P_{\mathrm{v},e}}{12 \mu_{v,e} l_{\mathrm{eff}}} \tag{4.16.1}$$

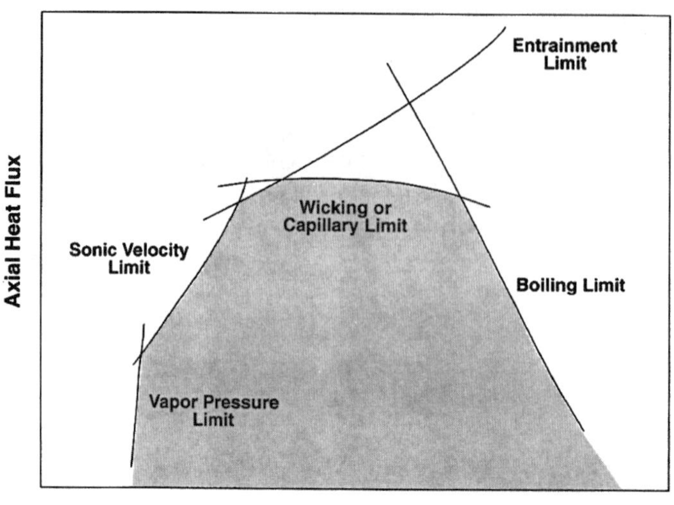

FIGURE 4.16.2 Heat transfer limitations in heat pipes.

where r_v is the cross-sectional radius of the vapor core (m), h_{fg} is the latent heat of vaporization (J/kg), $\rho_{v,e}$ is the vapor density in the evaporator (kg/m^3), P_{ve} is the vapor pressure in the evaporator (Pa), and $\mu_{v,e}$ is the vapor viscosity in the evaporator (N sec/m^2). l_{eff} is the effective length of the heat pipe (m) equal to $l_{\text{eff}} = 0.5(l_e + 2l_a + l_c)$. The vapor-pressure limitation can occur during the start-up of heat pipes at the lower end of the working-fluid-temperature range.

The sonic limitation also can occur in heat pipes during start-up at low temperatures. The low temperature produces a low vapor density, thereby reducing the speed of sound in the vapor core. Thus, a sufficiently high mass flow rate in the vapor core can cause sonic flow conditions and generate a shock wave that chokes the flow and restricts the pipes ability to transfer heat to the condenser. Dunn and Reay (1982) give an expression for the sonic limitation that agrees very well with experimental data,

$$Q_{s,\max} = 0.474 A_v h_{fg} \left(\rho_v P_v\right)^{1/2} \tag{4.16.2}$$

where A_v is the cross-sectional area of the vapor core (m^2). The sonic limitation should be avoided because large temperature gradients occur in heat pipes under choked-flow conditions.

The entrainment limitation in heat pipes develops when the vapor mass flow rate is large enough to shear droplets of liquid off the wick surface causing dry-out in the evaporator. A conservative estimate of the maximum heat transfer rate due to entrainment of liquid droplets has been given by Dunn and Reay (1982) as

$$Q_{e,\max} = A_v h_{fg} \left[\frac{\rho_v \sigma_l}{2 r_{c,\text{ave}}} \right]^{1/2} \tag{4.16.3}$$

where σ_l is the surface tension (N/m) and $r_{c,\text{ave}}$ is the average capillary radius of the wick. Note that for many applications $r_{c,\text{ave}}$ is often approximated by $r_{c,e}$.

The capillary limitation in heat pipes occurs when the net capillary forces generated by the vapor-liquid interfaces in the evaporator and condenser are not large enough to overcome the frictional pressure losses due to fluid motion. This causes the heat pipe evaporator to dry out and shuts down the transfer of heat from the evaporator to the condenser. For most heat pipes, the maximum heat transfer rate due to the capillary limitation can be expressed as (Chi, 1976).

$$Q_{c,\max} = \left[\frac{\rho_l \sigma_l h_{fg}}{\mu_l}\right]\left[\frac{A_w K}{l_{\text{eff}}}\right]\left(\frac{2}{r_{c,e}} - \left[\frac{\rho_l}{\sigma_l}\right] g L_t \cos\Psi\right) \tag{4.16.4}$$

where K is the wick permeability (m^2), A_w is the wick cross-sectional area (m^2), ρ_l is the liquid density (m^3), μ_l is the liquid viscosity (N sec/m^2), $r_{c,e}$ is the wick capillary radius in the evaporator (m), g is the acceleration due to gravity (9.8 m/sec^2), and L_t is the total length of the pipe (m). For most practical operating conditions, this limitation can be used to determine maximum heat transfer rate in heat pipes.

The boiling limitation in heat pipes occurs when the degree of liquid superheat in the evaporator is large enough to cause the nucleation of vapor bubbles on the surface of the wick or the container. Boiling is usually undesirable in heat pipes because local hot spots can develop in the wick, obstructing the flow of liquid in the evaporator. An expression for the boiling limitation is (Chi, 1976)

$$Q_{b,\max} = \frac{4\pi l_{\text{eff}} k_{\text{eff}} T_v \sigma_v}{h_{fg} \rho_l \ln(r_i/r_v)}\left(\frac{1}{r_n} - \frac{1}{r_{c,e}}\right) \tag{4.16.5}$$

where k_{eff} is the effective thermal conductivity of the composite wick and working fluid (W/m K), T_v is the vapor saturation temperature (K), r_i is the inner container radius (m), r_n is the nucleation radius (equal to 2.00×10^{-6} m in the absence of noncondensable gas).

Effective Thermal Conductivity and Heat Pipe Temperature Difference

One key attribute of the heat pipe is that it can transfer a large amount of heat while maintaining nearly isothermal conditions. The temperature difference between the external surfaces of the evaporator and the condenser can be determined from the following expression

$$\Delta T = R_t Q \tag{4.16.6}$$

where R_t is the total thermal resistance (K/W) and Q is the heat transfer rate (W). Figure 4.16.3 shows the thermal resistance network for a typical heat pipe and the associated thermal resistances. In most cases, the total thermal resistance can be approximated by

$$R_t = R_1 + R_2 + R_3 + R_5 + R_7 + R_8 + R_9 \tag{4.16.7}$$

The reader is referred to Peterson (1994) for the specific mathematical relationships used to calculate each thermal resistance. The effective thermal conductivity of the heat pipe is defined as the heat transfer rate divided by the temperature difference between the heat source and heat sink,

$$k_{\text{eff}} = \frac{L_t}{R_t A_t} \tag{4.16.8}$$

where A_t is the overall cross-sectional area of the pipe (m^2). Under normal operating conditions, the total thermal resistance is relatively small, making the external surface temperature in the evaporator approximately equal to that in the condenser. Thus, the effective thermal conductivity in a heat pipe can be very large (at least an order of magnitude larger than that of aluminum).

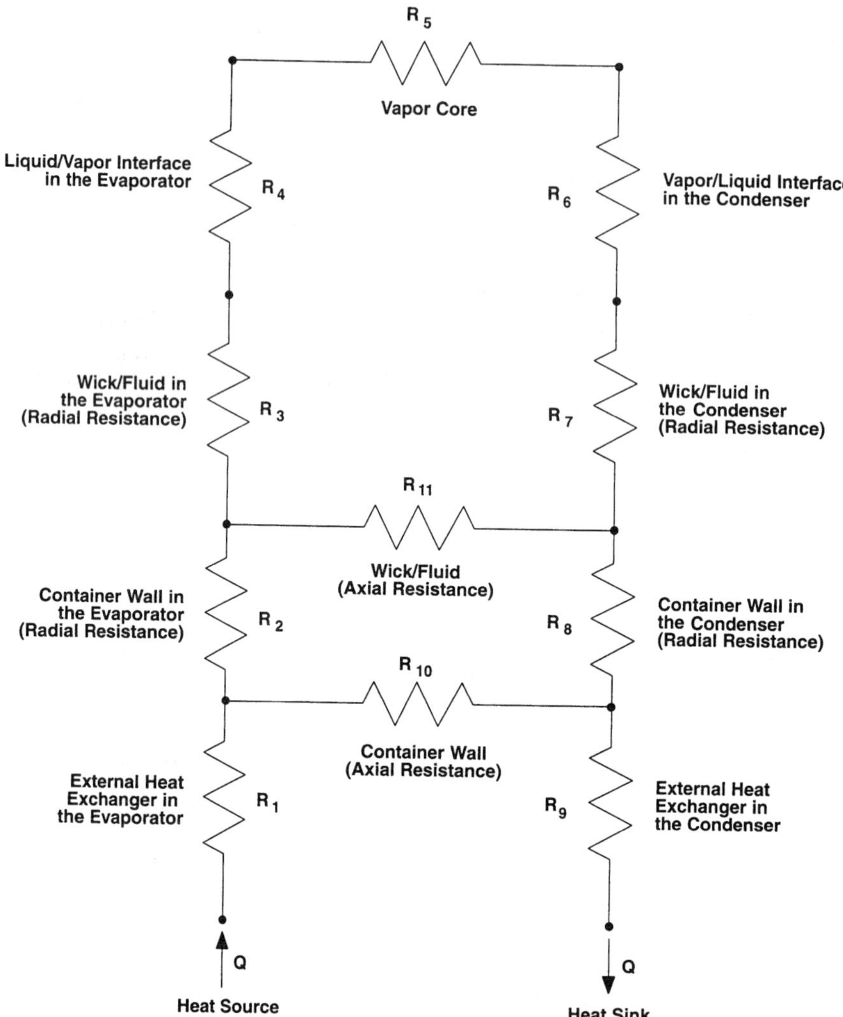

FIGURE 4.16.3 Thermal resistance network in a heat pipe.

Design Example

Design a water heat pipe to transport 80 W of waste heat from an electronics package to cooling water. The heat pipe specifications are

1. Axial orientation — complete gravity-assisted operation (condenser above the evaporator; ψ = 180°).
2. Maximum heat transfer rate — 80 W.
3. Nominal operating temperature — 40°C.
4. Inner pipe diameter — 3 cm.
5. Pipe length — 25 cm evaporator length, 50 cm adiabatic section, and 25 cm condenser length.

The simplest type of wick structure to use is the single-layer wire mesh screen wick shown in Table 4.16.2. The geometric and thermophysical properties of the wick have been selected as (this takes some forethought)

$d = 2.0 \times 10^{-5}$ m

$w = 6.0 \times 10^{-5}$ m

$\frac{1}{2N} = r_c = {}^1/_2(2.0 \times 10^{-5} + 6 \times 10^{-5}) = 4.0 \times 10^{-5}$ m

$\varepsilon = 1$

$$k_{\text{eff}} = k_1 = 0.630 \ \frac{\text{W}}{\text{mK}}$$

$t_w = 1.0 \times 10^{-3}$ m

$$K = \frac{t_w^2}{12} = \frac{\left(1 \times 10^{-3}\right)^2}{12} = 8.33 \times 10^{-8} \ \text{m}^2$$

The other heat pipe geometric properties are

$r_v = r_i - t_w = 0.015 - 0.001 = 0.014$ m

$$l_{\text{eff}} = \frac{0.25 + 0.25}{2} + 0.5 = 0.75 \ \text{m}$$

$L_t = 0.25 + 0.50 + 0.25 + 1.0$ m

$A_w = \pi(r_i^2 - r_v^2) = \pi[(0.015)^2 - (0.014)^2] = 9.11 \times 10^{-5}$ m^2

$A_v = \pi r_v^2 = \pi(0.014)^2 = 6.16 \times 10^{-4}$ m^2

The thermophysical properties of water at 40°C are (see Table 4.16.1):

$\rho_l = 992.1$ kg/m^3

$\rho_v = 0.05$ kg/m^3

$\sigma_l = 2.402 \times 10^6$ J/kg

$\mu_l = 6.5 \times 10^{-3}$ kg/m sec

$\mu_v = 1.04 \times 10^{-4}$ kg/m sec

$P_v = 7000$ Pa

The various heat transfer limitations can now be determined to ensure the heat pipe meets the 80 W heat transfer rate specification. The vapor-pressure limitation is

$$Q_{vp,\max} = \frac{\pi(0.014)^4\left(2.402 \times 10^6\right)(0.05)(7000)}{12\left(1.04 \times 10^{-4}\right)(0.75)} = 1.08 \times 10^5 \ \text{W} \tag{4.16.9}$$

The sonic limitation is

$$\begin{aligned} Q_{s,\max} &= 0.474\left(6.16 \times 10^{-4}\right)\left(2.402 \times 10^6\right)\left[(0.05)(7000)\right]^{1/2} \\ &= 1.31 \times 10^4 \ \text{W} \end{aligned} \tag{4.16.10}$$

The entrainment limitation is

$$\begin{aligned} Q_{e,\max} &= \left(6.16 \times 10^{-4}\right)\left(2.402 \times 10^6\right)\left[\frac{(0.05)(0.07)}{2\left(4.0 \times 10^{-5}\right)}\right]^{1/2} \\ &= 9.79 \times 10^3 \ \text{W} \end{aligned} \tag{4.16.11}$$

TABLE 4.16.2 Physical Properties of Wick Structures (continued)

Wick Type[a]	Thermal Conductivity	Porosity	Minimum Capillary Radius	Permeability
Single-layer wire mesh screens (heat-pipe axis in the plane of the paper in this sketch) 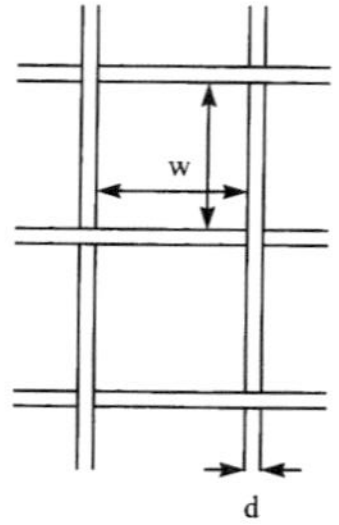 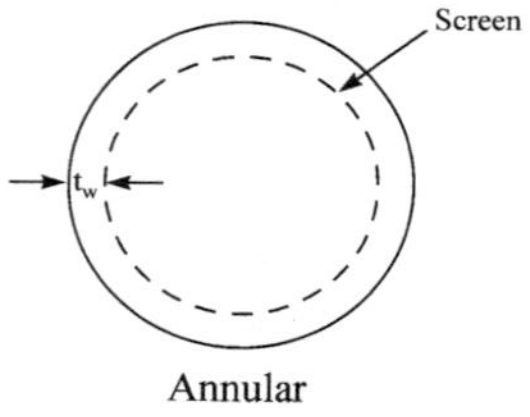 w, d Screen, t_w Annular $1/N = d + w$ N = number of aperatures per unit length	$k_{eff} = k_e$	$\varepsilon = 1$	$r_c = {}^1/_{(2N)}$	$K = t_w^2/12$
 t_w Multiple wire mesh screens,[b] plain or sintered (screen dimensions as for single layers illustrated above)	$k_{eff} = \frac{k_e\left[k_e + k_s - (1-\varepsilon)(k_e - k_s)\right]}{k_e + k_s + (1-\varepsilon)(k_e - k_s)}$	Estimated from $\varepsilon = 1 - (\pi N d)/4$	$r_c = {}^1/_{(2N)}$	$k = \frac{d^2\varepsilon^2}{122(1-\varepsilon)^2}$

t_w	Unconsolidated packed spherical particles (d = average particle diameter)	Plain	$k_{\text{eff}} = \frac{k_e\left[2k_e + k_s - 2(1-\varepsilon)(k_e - k_s)\right]}{2k_e + k_s + (1-\varepsilon)(k_e - k_s)}$	Estimated from (assuming cubic packing) ε = 0.48	$r_c = 0.21d$	$k = \frac{d^2\varepsilon^2}{150(1-\varepsilon)^2}$
		Sintered	$k_{\text{eff}} = \frac{k_e\left[2k_s + k_e - 2\varepsilon(k_s - k_e)\right]}{2k_s + k_e + \varepsilon(k_s - k_e)}$			
t_w	Sintered metal fibers (d = fiber diameter)		$k_{\text{eff}} = \varepsilon^2 k_e (1-\varepsilon)^2 k_s + \frac{4\varepsilon(1-\varepsilon)k_e k_s}{k_e + k_s}$	Use manufacturers data	$r_c = \frac{d}{2(1-\varepsilon)}$	$k = C_1 \frac{y^2 - 1}{y^2 - 1}$ where $y = 1 + \frac{C_2 d^2 \varepsilon^3}{(1-\varepsilon)^2}$; $C_1 = 6.0 \times 10^{-10}\ \text{m}^2$; $C_2 = 3.3 \times 10^7\ 1/\text{m}^2$

[a] The axis of the pipe and direction of fluid flow are normal to the paper.

[b] These wicks are positioned so that the layers follow the contours of the iner surface of the pipe wall.

Revised from Peterson, G.P., *An Introduction to Heat Pipes Modeling, Testing, and Applications*, John Wiley & Sons, New York, 1994.

Noting that cos ψ = –1, the capillary limitation is

$$Q_{c,\max} = \left[\frac{(992.1)(0.07)(2.402\times10^6)}{6.5\times10^{-3}}\right]\left[\frac{(9.11\times10^{-5})(8.33\times10^{-8})}{0.75}\right]\left[\frac{2}{4.0\times10^{-5}}+\frac{992.1}{0.07}9.8(1.0)\right] \tag{4.16.12}$$

$$= 4.90\times10^4 \text{ W}$$

Finally, the boiling limitation is

$$Q_{b,\max} = \frac{4\pi(0.75)(0.63)(313)(0.07)}{(2.402\times10^6)(992.1)\ln\left(\frac{0.015}{0.014}\right)}\left[\frac{1}{2.0\times10^{-6}}-\frac{1}{4.0\times10^{-5}}\right] \tag{4.16.13}$$

$$= 0.376 \text{ W}$$

All of the heat transfer limitations, with the exception of the boiling limitation, exceed the specified heat transfer rate of 80 W. The low value of 0.376 W for the boiling limitation strongly suggests that the liquid will boil in the evaporator and possibly cause local dry spots to develop. The reason the liquid boils is because the effective thermal conductivity of the wick is equal to the conductivity of the liquid, which is very low in this case. Because the liquid is saturated at the vapor-liquid interface, a low effective thermal conductivity requires a large amount of wall superheat which, in turn, causes the liquid to boil. This problem can be circumvented by using a high conductivity wire mesh or sintered metal wick, which greatly increases the effective conductivity. It should be noted, however, that because porous wicks have lower permeabilities, the capillary limitation should be lower as well. Let's try a sintered particle wick made of copper with the following properties (see Table 4.16.2):

$d = 1.91 \times 10^{-4}$ m

$r_{c,3} = 0.21d = 4.0 \times 10^{-5}$ m (same as before)

$\varepsilon = 0.48$

$$K = \frac{(1.91\times10^{-4})^2(0.48)}{150(1-0.48)^2} = 2.07\times10^{-10} \text{ m}^2$$

$$k_s = 400\frac{\text{W}}{\text{mK}} \text{ (copper)}$$

$$k_1 = 0.630\frac{\text{W}}{\text{mK}} \text{ (water)}$$

$$k_{\text{eff}} = \frac{400[2(400)+0.63-2(0.48)(400-0.63)]}{2(400)+0.63+0.48(400-0.63)} = 168 \text{ W/mK}$$

All other geometric and thermophysical properties are the same. The heat transfer limitations affected by the change in wick structure are the capillary and boiling limitations. The sintered metal wick produces a capillary limitation of

$$Q_{c,\max} = \left[\frac{(992.1)(0.07)(2.402\times10^6)}{6.5\times10^{-3}}\right]\left[\frac{(9.11\times10^{-5})(2.07\times10^{-10})}{0.75}\right]\left[\frac{2}{4.0\times10^{-5}}+\frac{992.1}{0.07}9.8(1.0)\right] \tag{4.16.14}$$

$$= 122 \text{ W}$$

The boiling limitation for the sintered wick is

$$Q_{b,\max} = \frac{4\pi(0.75)(168)(313)(0.07)}{\left(2.402\times10^{6}\right)(992.1)\ln\left(\dfrac{0.015}{0.014}\right)}\left[\frac{1}{2.0\times10^{-6}}-\frac{1}{4.0\times10^{-5}}\right] \tag{4.16.15}$$

$$= 100 \text{ W}$$

This design now meets all the specifications defined in the problem statement.

Application of Heat Pipes

Heat pipes have been applied to a wide variety of thermal processes and technologies. It would be an impossible task to list all the applications of heat pipes; therefore, only a few important industrial applications are given in this section. In the aerospace industry, heat pipes have been used successfully in controlling the temperature of vehicles, instruments, and space suits. Cryogenic heat pipes have been applied in (1) the electronics industry for cooling various devices (e.g., infrared sensors, parametric amplifiers) and (2) the medical field for cryogenic eye and tumor surgery. Heat pipes have been employed to keep the Alaskan tundra frozen below the Alaskan pipeline. Other cooling applications include (1) turbine blades, generators, and motors; (2) nuclear and isotope reactors; and (3) heat collection from exhaust gases, solar and geothermal energy.

In general, heat pipes have advantages over many traditional heat-exchange devices when (1) heat has to be transferred isothermally over relatively short distances, (2) low weight is essential (the heat pipe is a passive pumping device and therefore does not require a pump), (3) fast thermal-response times are required, and (4) low maintenance is mandatory.

Defining Terms

Capillary force: The force caused by a curved vapor-liquid interface. The interfacial curvature is dependent on the surface tension of the liquid, the contact angle between the liquid wick structure, the vapor pressure, and the liquid pressure.

Effective thermal conductivity: The heat transfer rate divided by the temperature difference between the evaporator and condenser outer surfaces.

Heat transfer limitations: Limitations on the axial heat transfer capacity imposed by different physical phenomena (i.e., vapor pressure, sonic, entrainment, capillary, and boiling limitations).

Wettability: The ability of a liquid to spread itself over a surface. A wetting liquid spreads over a surface whereas a nonwetting liquid forms droplets on a surface.

Wick: A porous material used to generate the capillary forces that circulate fluid in a heat pipe.

References

Chi, S.W. 1976. *Heat Pipe Theory and Practice*, Hemisphere Publishing, Washington, D.C.

Dunn, P.D. and Reay, D.A. 1982. *Heat Pipes*, 3rd ed., Pergamon Press, Oxford, U.K.

Gaugler, R.S. 1944. Heat Transfer Device. U.S. Patent No. 2350348.

Grover, G.M. 1963. Evaporation-Condensation Heat Transfer Device. U.S. Patent No. 3229759.

Peterson, G.P. 1994. *An Introduction to Heat Pipes Modeling, Testing, and Applications*, John Wiley & Sons, New York.

Further Information

Recent developments in heat pipe research and technology can be found in the proceedings from a number of technical conferences: (1) The International Heat Pipe Conference (2) The National Heat Transfer Conference, (3) The ASME Winter Annual Meeting, (4) The AIAA Thermophysics Conference.

Books particularly useful for the design of heat pipes include (1) *Heat Pipe Design Handbook* by Brennan and Kroliczek available from B&K Engineering in Baltimore, M.D. (2) *The Heat Pipe* by Chisholm available from Mills and Boon Limited in London, England, and (3) *Heat Pipes: Construction and Application* by Terpstra and Van Veen available from Elsevier Applied Science in New York, N.Y.

An additional book particularly strong in heat pipe theory is *The Principles of Heat Pipes* by Ivanovskii, Sorokin, and Yagodkin available from Clarendon Press in Oxford, England.

4.17 Liquid Atomization and Spraying

Rolf D. Reitz

Sprays are involved in many practical applications, including in the process industries (e.g., spray drying, spray cooling, powdered metals); in treatment applications (e.g., humidification, gas scrubbing); in coating applications (e.g., surface treatment, spray painting, and crop spraying); in spray combustion (e.g., burners, furnaces, rockets, gas turbines, diesel and port fuel injected engines); and in medicinal and printing applications. To be able to describe sprays it is necessary to obtain a detailed understanding of spray processes.

In the simplest case, the liquid to be sprayed is injected at a high velocity through a small orifice. Atomization is the process whereby the injected liquid is broken up into droplets. Atomization has a strong influence on spray vaporization rates because it increases the total surface area of the injected liquid greatly. Fast vaporization may be desirable in certain applications, but undesirable in others, where the liquid is required to impinge on a target. The trajectories of the spray drops are governed by the injected momentum of the drop, drag forces, and interactions between the drops and the surrounding gas. Control of these and other spray processes can lead to significant improvements in performance and in quality of product, and to reduction of emission of pollutants.

Spray Characterization

Practical atomizers generate sprays with a distribution of drop sizes, with average sizes in the diameter range from a few microns (1 μm = 10^{-6} m) to as large as 0.5 mm. It is important to quantify the details of the distribution depending on the application. For example, the smaller drops in a spray vaporize fast, and this is helpful to control ignition processes in some combustion systems. On the other hand, the large drops carry most of the mass and momentum of the injected liquid and these drops are able to penetrate into the high-pressure gases in engine combustion chambers. Typical average drop sizes for broad classes of sprays are shown schematically in Figure 4.17.1. It should be noted that the terminology used to describe sprays in Figure 4.17.1 is qualitative and is not universally agreed upon.

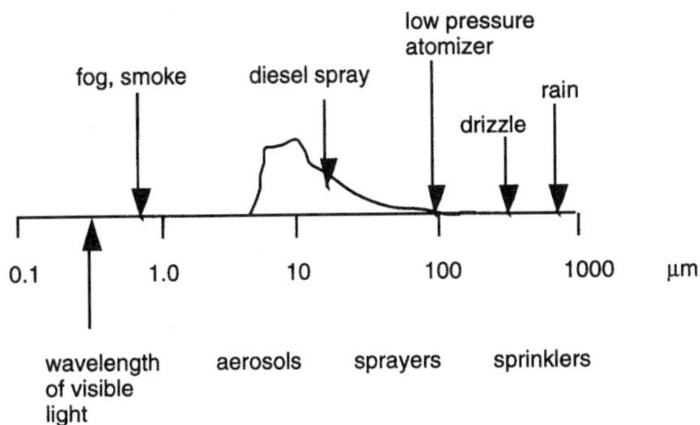

FIGURE 4.17.1 Typical average spray drop sizes for various classes of sprays. A representative size distribution is depicted for the diesel spray.

Methods for characterizing the size distribution of spray drops are discussed in References 1 and 2. A probability distribution function, $F(D)$, is introduced that represents the fraction of drops per unit diameter range about the diameter, D, as shown in Figure 4.17.2. The spray drop sizes span a range from a minimum diameter, D_{min}, to a maximum diameter, D_{max}. It is also convenient to introduce a mean or average drop diameter instead of having to specify the complete drop size distribution. The number median drop diameter (NMD) represents that drop whose diameter is such that 50% of the drops in the spray have sizes less than this size. Spray drop size distribution data can also be represented as a volume

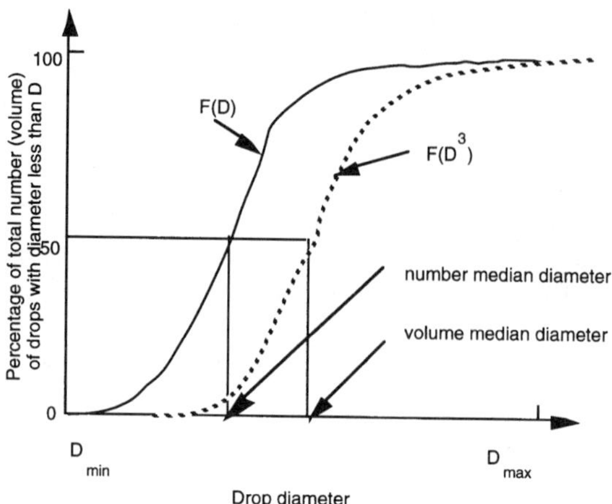

FIGURE 4.17.2 Cumulative spray drop number and volume distributions.

(or mass) distribution function, $F(D^3)$; this gives more weight to the large drops in the distribution. In this case, a volume median diameter (VMD) or a mass median diameter (MMD) can also be defined, as indicated in Figure 4.17.2.

Various other mean diameters are also in common use. These are summarized using the standard notation of Mugele and Evans[2] as

$$\left(D_{jk}\right)^{j-k} = \frac{\int_{D_{\min}}^{D_{\max}} D^j f(D)\, dD}{\int_{D_{\min}}^{D_{\max}} D^k f(D)\, dD} \tag{4.17.1}$$

where $f(D) = dF(D)/dD$ is the drop size probability density function (usually normalized such that $\int_{D_{\min}}^{D_{\max}} f(D)dD = 1$). Commonly used mean diameters are D_{10} (i.e., $j = 1$, $k = 0$, sometimes called the length mean diameter[3] and D_{32} (i.e., $j = 3$, $k = 2$, called the Sauter mean diameter or SMD). The Sauter mean diameter has a useful physical interpretation in combustion applications since drop vaporization rates are proportional to the surface area of the drop. It represents the size of that drop that has the same volume-to-surface area ratio as that of the entire spray.

Several distribution functions have been found to fit experimental data reasonably well. Among these are the Nukiyama–Tanasawa and the Rosin–Rammler distributions which have the general form[3] $f(D) = aD^p \exp(\{-bD\}^q)$, where the constants a, p, b, and q characterize the size distribution. The higher the parameter, q, the more uniform the distribution, and typically $1.5 < q < 4$. Other distributions have been proposed which consist of logarithmic transformations of the normal distribution, such as $f(D) = a \exp(-y^2/2)$, where $y = \delta \ln(\eta D/(D_{\max} - D))$, and a, δ, and η are constants. In this case, the smaller δ, the more uniform the size distribution. It should be noted that there is no theoretical justification for any of these size distributions. Spray drop size distributions can be measured nonintrusively by using optical laser diffraction and phase/Doppler instruments. A discussion of these techniques and their accuracy is reviewed by Chigier.[4]

Atomizer Design Considerations

Atomization is generally achieved by forcing a liquid or a liquid–gas mixture through a small hole or slit under pressure to create thin liquid sheets or jets moving at a high relative velocity with respect to the surrounding ambient gas. Desirable characteristics of atomizers include the ability to atomize the liquid over a wide range of flow rates, low power requirements, and low susceptibility to blockage or fouling. In addition, atomizers should produce consistent sprays with uniform flow patterns in operation.

Atomizers can be broadly characterized as those producing hollow cone or solid cone sprays, as depicted in Figure 4.17.3. In solid cone (or full cone) sprays the spray liquid is concentrated along the spray axis, Figure 4.17.3a. These sprays are useful in applications requiring high spray penetration, such as in diesel engines. In hollow cone sprays the axis region is relatively free of drops, giving wide spray dispersal, Figure 4.17.3b. These sprays are often used in furnaces, gas turbines, and spray-coating applications.

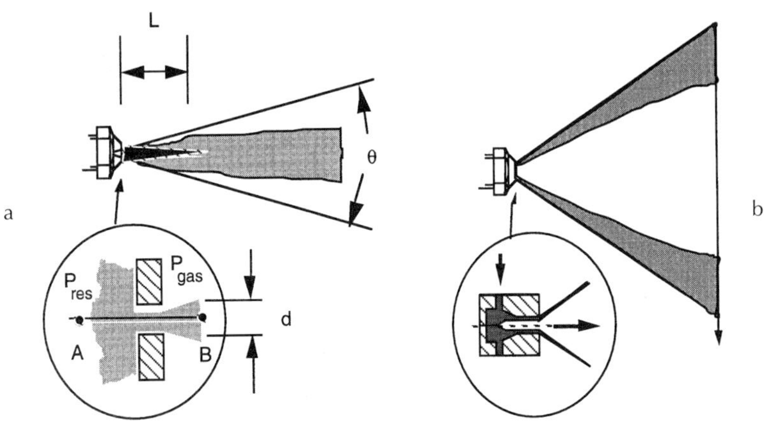

FIGURE 4.17.3 Schematic diagram of (a) solid cone and (b) hollow cone pressure atomizer sprays.

Many different atomizer designs are found in applications. Common atomizer types include pressure, rotary, twin-fluid (air-assist, air-blast, effervescent), flashing, electrostatic, vibratory, and ultrasonic atomizers, as discussed next.

Atomizer Types

In *pressure atomizers* atomization is achieved by means of a pressure difference, $\Delta P = P_{res} - P_{gas}$, between the liquid in the supply reservoir pressure, P_{res}, and the ambient medium pressure, P_{gas}, across a nozzle. The simplest design is the plain orifice nozzle with exit hole diameter, d, depicted in Figure 4.17.3a. The liquid emerges at the theoretical velocity $U = \sqrt{2\Delta P/\rho_{liquid}}$, the (Bernoulli) velocity along the streamline A–B in Figure 4.17.3a, where ρ_{liquid} is the density of the liquid. The actual injection velocity is less than the ideal velocity by a factor called the discharge coefficient, C_D, which is between 0.6 and 0.9 for plain hole nozzles. C_D accounts for flow losses in the nozzle.

Four main jet breakup regimes have been identified, corresponding to different combinations of liquid inertia, surface tension, and aerodynamic forces acting on the jet, as shown in Figure 4.17.4. At low injection pressures the low-velocity liquid jet breaks up due to the unstable growth of long-wavelength waves driven by surface tension forces (Rayleigh regime). As the jet velocity is increased, the growth of disturbances on the liquid surface is enhanced because of the interaction between the liquid and the ambient gas (the first and second wind-induced breakup regimes). At high injection pressures the high-velocity jet disintegrates into drops immediately after leaving the nozzle exit (atomization regime). Criteria for the boundaries between the regimes are available.[5] Aerodynamic effects are found to become very important relative to inertial effects when the jet Weber number, $We_j > 40$, where $We_j = \rho_{gas}U^2d/\sigma$, ρ_{gas} is the gas density, and σ is the liquid surface tension.

Experiments show that the unstable growth of surface waves is aided by high relative velocities between the liquid and the gas, and also by high turbulence and other disturbances in the liquid and gas flows, and by the use of spray liquids with low viscosity and low surface tension.

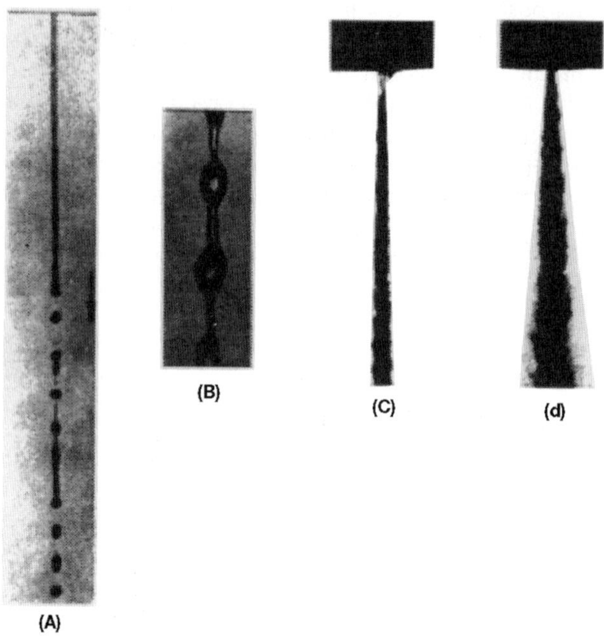

FIGURE 4.17.4 (a) Rayleigh breakup. Drop diameters are larger than the jet diameter. Breakup occurs many nozzle diameters downstream of nozzle. (b) First wind-induced regime. Drops with diameters of the order of jet diameter. Breakup occurs many nozzle diameters downstream of nozzle. (c) Second wind-induced regime. Drop sizes smaller than the jet diameter. Breakup starts some distance downstream of nozzle. (d) Atomization regime. Drop sizes much smaller than the jet diameter. Breakup starts at nozzle exit.

Liquid breakup characteristics such as the spray drop size, the jet breakup length, and the spray angle have been related to the unstable wave growth mechanism. The wavelengths and growth rates of the waves can be predicted using results from a linear stability analysis with[6]

$$\frac{\Lambda}{2} = 9.02\frac{\left(1+0.45Z^{0.5}\right)\left(1+0.4T^{0.7}\right)}{\left(1+0.87\,\mathrm{We}_2^{1.67}\right)^{0.6}} \tag{4.17.2a}$$

$$\Omega\left(\frac{\rho_1 a^3}{\sigma}\right)^{0.5} = \frac{0.34+0.38\,\mathrm{We}_2^{1.5}}{(1+Z)\left(1+1.4T^{0.6}\right)} \tag{4.17.2b}$$

where Λ is the wavelength, Ω is the growth rate of the most unstable surface wave, and a is the liquid jet radius. The maximum wave growth rate increases, and the corresponding wavelength decreases with increasing Weber number, $\mathrm{We}_2 = \rho_{gas}U^2a/\sigma$, where U is the relative velocity between the liquid and the gas. The liquid viscosity appears in the Ohnesorge number, $Z = \mathrm{We}_1^{1/2}/\mathrm{Re}_1$. Here, the Weber number We_1 is based on the liquid density, the Reynolds number is $\mathrm{Re}_1 = Ua/\nu_1$, ν_1 is the liquid viscosity, and the parameter $T = Z\,\mathrm{We}_2^{1/2}$. The wave growth rate is reduced and the wavelength is increased as the liquid viscosity increases.

The size of the drops formed from the breakup process is often assumed to be proportional to the wavelength of the unstable surface waves in modeling studies.[6] However, the drop sizes in the primary breakup region near the nozzle exist have also been found to be influenced by the length scale of the energy-containing eddies in the turbulent liquid flow.[7] There is uncertainty about atomization mechanisms since spray measurements are complicated by the high optical density of the spray in the breakup region (e.g., see Figure 4.17.4(d)). As the drops penetrate into the ambient gas, they interact with each other through collisions and coalescence, and the spray drop size changes dynamically within the spray as a result of secondary breakup and vaporization effects. The drop trajectories are determined by complex drop drag, breakup, and vaporization phenomena, and by interactions with the turbulent gas flow.[6]

High-pressure diesel sprays are intermittent and are required to start and stop quickly without dribble between injections. This is accomplished by means of a plunger arrangement that is actuated by a cam and spring system in mechanical "jerk" pump systems (Figure 4.17.5). Modern electronic injectors include electromagnetic solenoids that permit the duration and injection pressure to be varied independently of each other and of engine speed. Experiments on diesel-type injector nozzles show that the penetration distance, S, of the tip of the spray at time, t, after the beginning of the injection is given by[8]

$$\begin{aligned} S &= 0.39Ut\left(\rho_{\text{liquid}}/\rho_{\text{gas}}\right)^{1/2} \qquad &\text{for } t < t_b \\ S &= 2.46\sqrt{U\,dt}\,\left(\rho_{\text{liquid}}/\rho_{\text{gas}}\right)^{1/4} \qquad &\text{for } t > t_b \end{aligned} \tag{4.17.3}$$

where the "breakup time" is $t_b = 40.5d(\rho_{\text{liquid}}/\rho_{\text{gas}})^{1/2}/U$. The jet breakup length (see Figure 4.17.3(a)), $L = Ut_b$ is independent of the injection velocity. On the other hand, for low-speed jets, or for jets injected into a low-gas-density environment, $t_b = 1.04C\ (\rho_{\text{liquid}}\ d^3/\sigma)^{1/2}$, where C is a constant typically between 12 and 16 and σ is the surface tension. In this case L increases with the injection velocity.[9] The functional form of the above jet breakup time and length correlations can be derived for an inviscid liquid in the limits of large and small Weber number, We_2 from the unstable wave growth rate in Equation (4.17.2) with $t_b \sim \Omega^{-1}$.

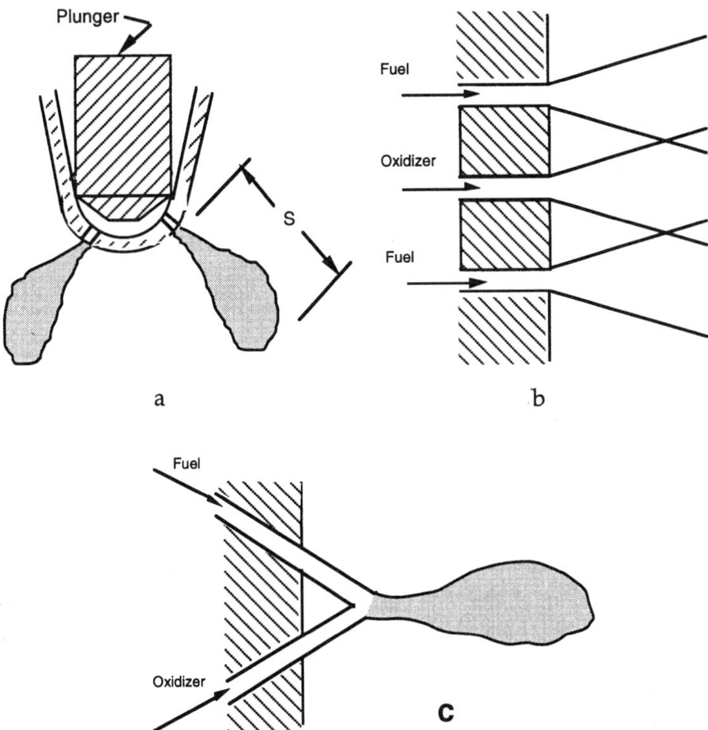

FIGURE 4.17.5 (a) Diesel injector multihole spray nozzle, (b) showerhead, and (c) doublet impingement nozzles.

For high-speed diesel-type jets in the atomization regime the resulting spray diverges in the form of a cone with cone angle, θ, that is usually in the range from 5 to 20°. θ increases with gas density following $\tan\theta = A(\rho_{gas}/\rho_{liquid})^{1/2}$, where A is a constant that depends on the nozzle passage length and (weakly) on the injection velocity.[9] Very high injection pressures are required to produce small drops. In diesel engines ΔP is typically as high as 200 Mpa, and drops are produced with mean diameters of the order of 10 μm (see Figure 4.17.1). Drop size correlations have been proposed for plain-orifice sprays, such as that presented in Table 4.17.1.[3] Note, however, that these correlations do not account for the fact that the spray drop size varies with time, and from place to place in the spray. Moreover, experimental correlations often do not include some parameters that are known to influence spray drop sizes, such as the nozzle passage length and its entrance geometry. Therefore, overall drop size correlations should only be used with caution.

TABLE 4.17.1 Representative Drop Size Correlations for Various Spray Devices (Dimensional quantities are in SI units, kg, m, s)

Device	Correlation	Notes
Plain orifice	$\text{SMD} = 3.08\nu_l^{0.385}\left(\rho_{liquid}\sigma\right)^{0.737}\rho_{gas}^{0.06}\,\Delta P^{-0.54}$	Use SI units
Fan spray	$\text{SMD} = 2.83d_h\left(\sigma\mu_{liquid}^2/\rho_{gas}d_h^3\,\Delta P^2\right)^{0.25}$ $+0.26d_h\left(\sigma\rho_{liquid}/\rho_{gas}d_h\,\Delta P\right)^{0.25}$	d_h = nozzle hydraulic diameter
Rotary atomizer	$\text{SMD} = 0.119Q^{0.1}\sigma^{0.5}/Nd^{0.5}\rho_{liquid}^{0.4}\mu_{liquid}^{0.1}$	N = rotational speed (rev/sec), Q = volumetric flow rate, $A_{inj}\,U$
Pressure swirl	$\text{SMD} = 4.52\left(\sigma\mu_{liquid}^2/\rho_{gas}\,\Delta P^2\right)^{0.25}(t\cos\theta)^{0.25}$ $+0.39\left(\sigma\rho_{liquid}/\rho_{gas}\,\Delta P\right)^{0.25}(t\cos\theta)^{0.75}$ $t = 0.0114A_{inj}\rho_{liquid}^{1/2}d\cos\theta$	t = film thickness; θ = cone angle, d = discharge orifice diameter
Twin fluid/air blast	$\text{SMD} = 0.48d\left(\sigma/\rho_{gas}U^2d\right)^{0.4}(1+1/\text{ALR})^{0.4}$ $+0.15d\left(\mu_{liquid}^2/\sigma\rho_{liquid}d\right)^{0.5}(1+1/\text{ALR})$	ALR = air-to-liquid mass ratio
Prefilming air blast	$\text{SMD} = (1+1/\text{ALR})\left[0.33d_h\left(\sigma/\rho_{gas}U^2d_p\right)^{0.6} + 0.068d_h\left(\mu_{liquid}^2/\sigma\rho_{liquid}d_p\right)^{0.5}\right]$	d_h = hydraulic diameter, d_p = prefilmer diameter, Figure 4.17.9
Ultrasonic	$\text{SMD} = \left(4\pi^3\sigma/\rho_{liquid}\omega^2\right)^{1/3}$	ω = vibration frequency

Source: Lefebvre, A.H., *Atomization and Sprays*, Hemisphere Publishing, New York, 1989. With permission.

The plain orifice design is also used in twin-fluid-type liquid rocket engines in showerhead and doublet designs (Figures 4.17.5b and 4.17.5c). In the case of doublet nozzles, shown in Figure 4.17.6c, the impinging jets create unstable liquid sheets which break up to produce the sprays. Drop size correlations are available for liquid sheets such as those formed by discharging the liquid through a rectangular slit (see *fan spray*, Table 4.17.1). Thin liquid sheets or slits lead to the production of small drops. The breakup mechanism of liquid sheets is also thought to involve the unstable growth of surface waves due to surface tension and aerodynamic forces.[5]

In *rotary atomizers* centrifugal forces are used to further enhance the breakup process. In this case the liquid is supplied to the center of a spinning disk and liquid sheets or ligaments are thrown off the edges of the disk. The drop size depends on the rotational speed of the disk, as indicated in Table 4.17.1.

A spinning wheel or cup (turbobell) is used in some spray-painting applications. The spray shape is controlled by supplying a coflowing stream of "shaping-air."

Centrifugal forces also play a role in the breakup mechanism of *pressure swirl* atomizers (*simplex* nozzles). These atomizers give wider spray cone angle than plain orifice nozzles, and are available in hollow cone and solid cone designs. As depicted in Figure 4.17.3(b) the spray liquid enters a swirl chamber tangentially to create a swirling liquid sheet. The air core vortex within the swirl chamber plays an important role in determining the thickness of the liquid sheet or film at the nozzle exit. This type of nozzle produces relatively coarse sprays. A representative SMD correction is listed in Table 4.17.1. The spray cone angle depends on the ratio of the axial and tangential liquid velocity components at the exit of the nozzle. This type of atomizer is not well suited for use in transient applications because it tends to dribble at start-up and to shut down when the air core is not fully formed.

The basic drawback of all pressure atomizers is that the flow rate depends on the square root of ΔP. The volumetric flow rate is $Q = A_{inj}\ U$, where A_{inj} is the liquid flow area at the nozzle exit, so that a factor of 20 increase in flow rate (a typical turndown ratio from idle to full load operation of a gas turbine engine) requires a factor of 400 increase in injection pressure.

This difficulty has led to so-called wide-range atomizer designs such as those shown in Figure 4.17.6. The *duplex* nozzle features two sets of tangential swirl ports; the primary (or pilot) supplies fuel at low flow rates, while the secondary ports become operational at high flow rates. Another variation is the *dual-orifice* nozzle which is conceptually two simplex nozzles arranged concentrically, one supplying the primary flow and the other supplying the secondary flow. The *spill-return* nozzle is a simplex nozzle with a rear passage that returns fuel to the injection pump. In this design the flow rate is controlled by the relative spill amount, and there are no small passages to become plugged. However, the fuel is always supplied at the maximum pressure which increases the demands on the injection pump. But high swirl is always maintained in the swirl chamber and good atomization is achieved even at low flow rates.

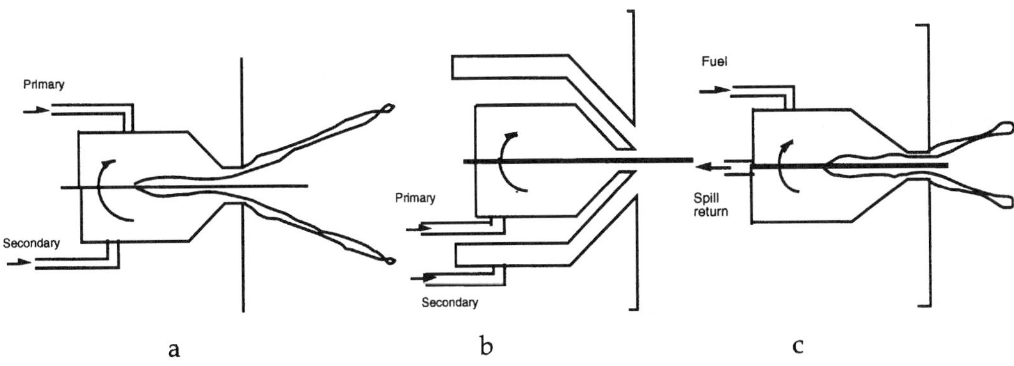

FIGURE 4.17.6 (a) Duplex, (b) dual orifice, and (c) spill-return-type nozzle designs.

In *twin-fluid injectors* atomization is aided by a flow of high-velocity gas through the injector passages. The high-velocity gas stream impinges on a relatively low-velocity liquid either internally (in *internal-mixing* nozzles, Figure 4.17.7) or externally (in *external-mixing* designs, Figure 4.17.8). The liquid and gas flows are typically swirled in opposite directions by means of swirl vanes to improve atomization. *Air-assist* refers to designs that use a relatively small amount of air at high (possibly sonic) velocities. *Air-blast* refers to designs that use large quantities of relatively low-velocity air which often supplies some of the air to help decrease soot formation in combustion systems[3] (Figure 4.17.9).

In *flashing* and *effervescent* atomizers a two-phase flow is passed through the injector nozzle exit. In the former the bubbles are generated by means of a phase change which occurs as the liquid, containing a dissolved propellant gas or vapor, undergoes the pressure drop through the nozzle. This process is

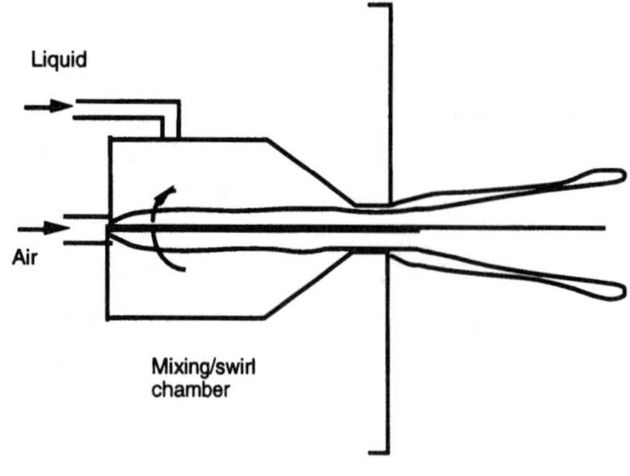

FIGURE 4.17.7 Internal-mixing twin-fluid injector design.

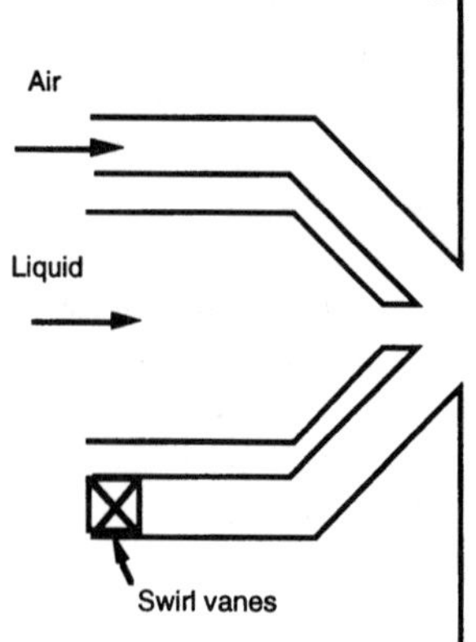

FIGURE 4.17.8 External-mixing twin-fluid injector design.

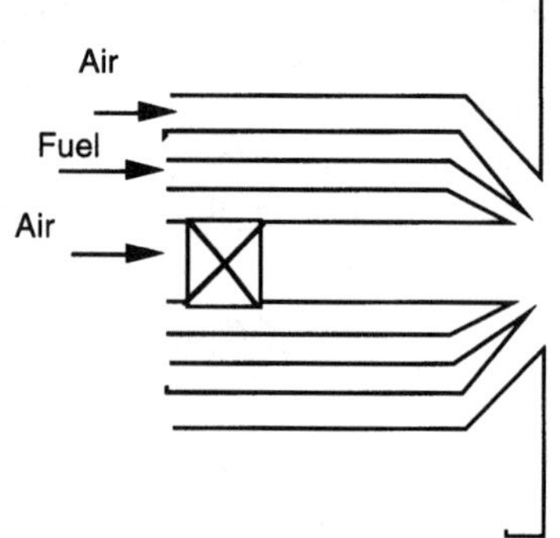

FIGURE 4.17.9 Prefiliming air blast atomizer.

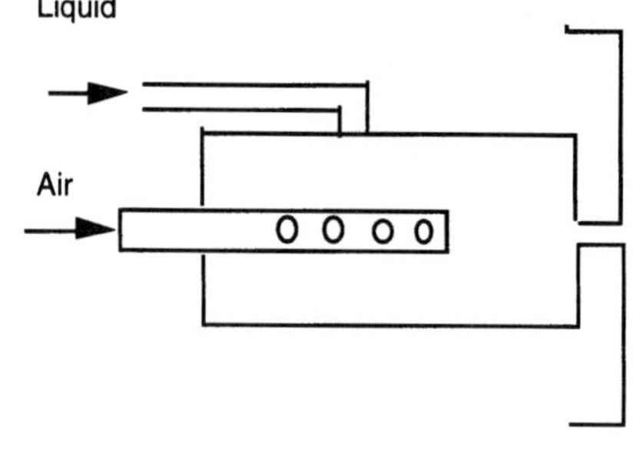

FIGURE 4.17.10 Internal-mixing, effervescent atomizer.

exploited in many household spray cans, but has the disadvantage of releasing the propellant gas required for atomization into the atmosphere. In the so-called effervescent atomizer, air bubbles are introduced into the liquid upstream of the exit orifice, as depicted in Figure 4.17.10. The spray quality is found to depend weakly on the air bubble size and is independent of the nozzle exit diameter. This makes internal-mixing, air-assist atomizers very attractive for use with high-viscosity fluids and slurries where nozzle plugging would otherwise be a problem.[3]

In *electrostatic* atomizers the spray liquid is charged by applying a high-voltage drop across the nozzle. The dispersion of the spray drops is increased by exploiting electrical repulsive forces between the droplets. An electrostatic charge on the drops is also helpful in spray-coating applications, such as in automotive spray painting using electrostatic turbobell sprayers, since the charged drops are attracted to an oppositely charged target surface.

Other atomizer types include *vibratory* and *ultrasonic* atomizers (or *nebulizers*), where the drops are formed by vibrating the injector nozzle at high frequencies and at large amplitudes to produce short-wavelength disturbances to the liquid flow. Ultrasonic atomizers are used in inhalation therapy where very fine sprays (submicron sizes) are required, and an available representative drop size correlation is also listed in Table 4.17.1.

References

1. American Society for Testing and Materials (ASTM) Standard E799. 1988. Data Criteria and Processing for Liquid Drop Size Analysis.
2. Mugele, R. and Evans, H.D. 1951. Droplet size distributions in sprays, *Ind. Eng. Chem.*, 43, 1317–1324.
3. Lefebvre, A.H. 1989. *Atomization and Sprays*, Hemisphere Publishing, New York.
4. Chigier, N.A. 1983. Drop size and velocity instrumentation, *Prog. Energ. Combust. Sci.*, 9, 155–177.
5. Chigier, N. and Reitz, R.D. 1996. Regimes of jet breakup, in *Progress in Astronautics and Aeronautics Series*, K. Kuo, Ed., AIAA, New York, Chapter 4, pp. 109–135.
6. Reitz, R.D. 1988. Modeling atomization processes in high-pressure vaporizing sprays, *Atomiz. Spray Technol.*, 3, 309–337.
7. Wu, P-K., Miranda, R.F., and Faeth, G.M. 1995. Effects of initial flow conditions on primary breakup of nonturbulent and turbulent round liquid jets, *Atomiz. Sprays*, 5, 175–196.
8. Hiroyasu, H. and Arai, M. 1978. Fuel spray penetration and spray angle in diesel engines, *Trans. JSAE*, 34, 3208.
9. Reitz, R.D. and Bracco, F.V. 1986. Mechanisms of breakup of round liquid jets, in *The Encyclopedia of Fluid Mechanics,* Vol. 3, N. Cheremisnoff, Ed., Gulf Publishing, Houston, TX, Chapter 10, 233–249.

Further Information

Information about recent work in the field of atomization and sprays can be obtained through participation in the Institutes for Liquid Atomization and Spraying Systems (ILASS-Americas, -Europe, -Japan, -Korea). These regional ILASS sections hold annual meetings. An international congress (ICLASS) is also held biennially. More information is available on the ILASS-Americas homepage at http://ucicl.eng.uci.edu/ilass. Affiliated with the ILASS organizations is the Institute's Journal publication *Atomization and Sprays* published by Begell House, Inc., New York.

4.18 Thermal Processing in Food Preservation Technologies

Ibrahim Dincer

Introduction

Food preservation is a very important in food processing applications for prolonging the shelf life of food products and maintaining their quality in terms of color, texture, and flavor. The term *food preservation* covers a wide range of techniques from short-term techniques (e.g., cooking, cold storage) to long-term techniques (e.g., canning, freezing, drying), leading to the thermal processing in food industry.

The principles governing the engineering aspects of food processing are the same as those applied in any engineering field. However, it is the application of these principles that distinguishes the various engineering applications. The structure and properties of food products as well as their biological and microbiological characteristics create the need for unique consideration of the thermal engineering principles applied.

Research on food preservation technologies usually focuses on one of two aspects: the selection of process conditions to minimize product quality loss or the development of engineering design methodologies to ensure that the required process conditions can be achieved at low cost. The latter aspect heat transfer is usually the major concern. Heat must be removed from the product by chilling or freezing, and the product subsequently has to be maintained at a low constant temperature in storage. There are also important mass transfer effects, particularly evaporation from the product surface (water loss).

During storage and transportation food products are exposed to changes which may affect various quality parameters such as wholesomeness, nutritional value, taste, texture, and appearance. These changes are caused by spoilage microorganisms; biochemical, chemical, and physical reactions; and complicated interactions between them. In order to maintain the quality at the desired level it, therefore, is of great importance to implement food preservation techniques. For this reason, much research effort has been focused on food preservation methods and their applications. In practice, several food preservation methods are available and these methods can be used individually or together in order to achieve the following main objectives:

- Prevention or delay of microbial decomposition
- Prevention or delay of self-decomposition of the food product
- Prevention of damage caused by insects, animals, mechanical processing, etc.

Heating Process and Methods

The methods of heat application for the thermal processing of food products may be classified as sterilization, pasteurization, blanching, cooking, drying, and irradiation (Dincer, 1997).

Sterilization — Heat sterilization is a widely used high-temperature (above 100°C) thermal process in food preservation and involves the controlled use of heat to increase or decrease the rates of biochemical and/or microorganism reactions in food products in sealed containers. Its success depends on the application of heat for a definite duration of time and at a definite temperature under the specific conditions required to obtain a commercially sterile product.

Pasteurization — This is a thermal process in which high-temperature heat (i.e., usually 60 to 100°C) is applied to kill part but not all of the microorganisms present. After the heat treatment, it is essential that the products (e.g., milk) are cooled down promptly. The heating may be provided by several sources such as steam, hot water, dry heat, electricity, etc.

Blanching — Many food products are blanched or scalded for a short period using either hot water or steam (above 96°C) prior to canning, packaging, freezing, or dehydration in order to provide the enhancement or retention of the green color of certain vegetables, inactivation of enzymes, reduction of the number of microorganisms, and softening of the tissues to aid packaging. Application of the blanching method is dependent on the product, packing method, packing size, etc.

Cooking — It is a thermal process including several forms of heat such as baking, boiling, frying, roasting, broiling, and stewing (above 100°C) and its main aim is to produce a more palatable food. Cooking the food results in destruction or reduction of microorganisms and inactivation of undesirable enzymes.

Drying — Drying is a thermal process (generally, above 50°C) based on removal of water from the food products, so that microorganisms and enzymes cannot find enough water to be active. In this method, the main issue is that the water content of the food must be lowered to a value at which the activities of food spoilage and food poisoning microorganisms are inhibited. Generally, this value is less than 25%. The level necessary to preserve foods is established by the term water activity or equilibrium relative humidity. For example, in a fresh fruit, the water activity is very close to 1; thus rapid growth of microbes that cause deterioration occurs. On the other hand, for a dried food it ranges between 0.2 and 0.6.

Irradiation — The irradiation treatment of food products has been studied for decades. More recently, infrared and microwave radiation techniques have been introduced for heating food products and microwave ovens are now commonplace in commercial and domestic kitchens. Today, an intense radiative energy known as ionizing radiation is the basis of potentially significant technology for processing food products. Ionizing energy includes gamma rays generated by radioisotopes, and x-rays and electron beams generated by particle accelerators. Food irradiation (mostly at 3 to 4°C) is applied to reduce the number of pathogens, leading to increased food safety, and the number of spoilage organisms, leading to an extension of shelf life. It is an effective method for reducing postharvest food losses, ensuring hygienic quality of the food, and facilitating wider trade of certain food items.

Cooling Process and Methods

Preservation by drying, salting, smoking, cooking, etc. is often nutritionally less desirable and alters the taste of the product; but is generally not expensive, and is a useful process. However, applying cooling processes results in foods of higher nutritional value. In practice, all low- and high-temperature cooling applications are sometimes represented with the term *refrigeration*, including cooling, freezing, cool and cold storage, controlled atmosphere storage, and transport cooling. This section will focus on the cooling process and methods and will provide some brief information on freezing, cool and cold storage, controlled atmosphere storage, and transport cooling.

The cooling of fruits and vegetables implies removal of the field heat before processing, transporting, or storing. Cooling inhibits growth of decay-producing microorganisms and restricts enzymatic and respiratory activity during the postharvest holding period, inhibits water loss, reduces ethylene production, and reduces the sensitivity of produce to ethylene. The holding period may be the relatively short time required to transport, sell, or process the product, or it may include a long-term storage period as well. It is important to mention that slowing down metabolism can give rise to physiological disorders which are called *cold storage injuries*. For this reason, the cooling temperature and exposure period must be suitable for the produce held. The main objective of this treatment is to reduce the rates of biochemical and microbiological changes in order to prevent spoilage of produce, maintain its quality (all possible preharvest freshness and flavor), and extend its storage life.

Cooling of food products as quickly as possible after harvest has become a widely used method for maintaining the quality, preventing spoilage, and maximizing the postharvest life. This thermal processing is very common in the food industry, and the design of equipment for such processing, for example, cooling, depends heavily upon knowledge of the thermal properties of the foods to be processed. This knowledge is also important for economic reasons, when energy balances are being considered in the design of process equipment.

The widely used cooling methods are hydrocooling, air cooling, hydraircooling, and vacuum cooling, and their selection is dependent upon some criteria, such as cost, convenience, effectiveness, applicability, efficient use, operation conditions, and personal preference as well as product requirements. Product physiology in terms of the harvest maturity and ambient temperature during harvesting has a major effect on the cooling requirements and cooling methods. Further information on specific fruits and vegetables and their cooling technologies can be found in Holdsworth (1983), ASHRAE (1990), and Dincer (1997).

Hydrocooling

It is a commonly used cooling method for quickly removing field heat of fruits and vegetables. It has several advantages, e.g., simplicity, effectiveness, rapidity, and cost, and some disadvantages, e.g., packing and/or handling difficulties, heavy contamination with soil, plant, sap, plant debris, and disease organisms, and some effluent problems.

Hydrocooling of products is accomplished by flooding, spraying, or by immersing the product or product load in a chilled-water pool. Flooding refers to the showering of products with a liberal supply of cold water flowing under a gravity head from overhead flood pans. Spraying is accomplished by overhead nozzles. The products to be cooled are immersed in, flooded with, or sprayed with cool water (especially near 0°C). There are two basic immersing hydrocooling techniques: the flow-through system and the batch system. In the flow-through system, the product is conveyed in a continuous flow through the cooling tunnel, either in bulk lots or in packages, and cooling may be accomplished by flooding, spraying, immersion, or parts of each depending on the particular type of hydrocooler. In the batch system, the products or batches are simply quenched in a cooling water tank. Figure 4.18.1 shows such a system consisting of a conventional mechanical refrigerating unit and a cooling tank.

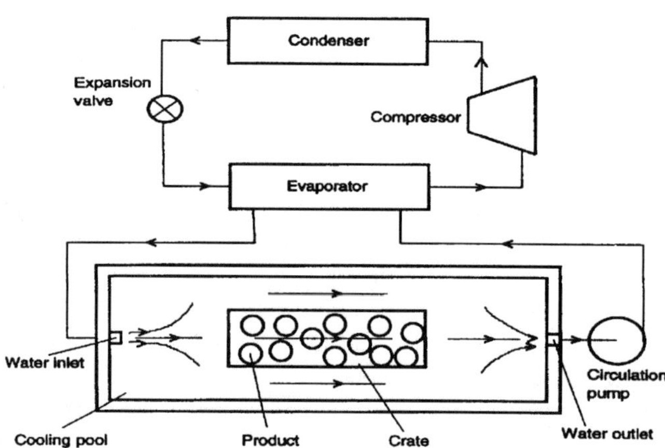

FIGURE 4.18.1 An immersion-type hydrocooling system and its components. (From Dincer, I., *Heat Transfer in Food Cooling Applications,* Taylor & Francis, Washington, D.C., 1997. With permission.)

The type of hydrocooling system selected for fruits and vegetables depends upon (1) the product, whether it is in bulk or shipping containers, (2) the type of container, whether it is handled individually or in unit loads, and (3) personal preference. Products such as asparagus, peas, sweet corn, radishes, carrots, cantaloupes, peaches, and tart cherries are successfully cooled by this method (ASHRAE, 1989).

Forced-Air Cooling

It is the most widely used system because it is simple, economical, sanitary, and relatively noncorrosive to equipment. Its major disadvantages are the dangers of excessive dehydration and possibility of freezing the product if air temperatures below 0°C are used. Some products cooled with air are meat, citrus fruits, grapes, cantaloupes, green beans, plums, nectarines, sweet cherries, and apricots. Commercially, forced-air cooling of food products is performed

- With air circulated in cold rooms adapted for this purpose
- In rail cars or highway vans using special portable cooling equipment that cools the product load while it is transported

- With air forced through the voids of bulk products moving through a cooling tunnel on continuous conveyors or air-cooled refrigerated rail cars or highway vans
- On continuous conveyors in wind tunnels
- By the forced-air method of passing air through the containers by pressure differential

Each of these methods is suitable for certain commodities when properly applied. Figure 4.18.2 shows two schematic configurations for air cooling of products in containers.

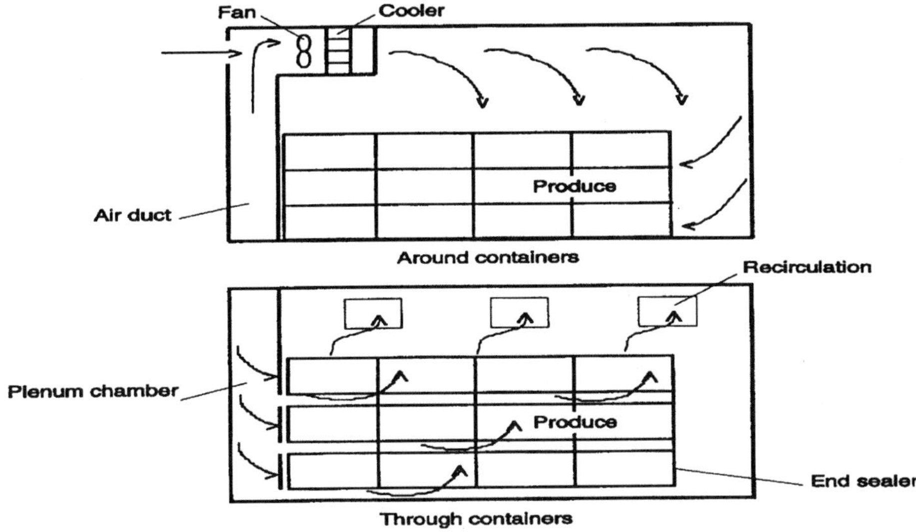

FIGURE 4.18.2 Two forced-air cooling configurations. (From Dincer, I., *Heat Transfer in Food Cooling Applications,* Taylor & Francis, Washington, D.C., 1997. With permission.)

To design an efficient and effective forced-air cooling system for products, refrigeration engineers must have knowledge of the many variables which affect cooling rate and cooling cost. Factors affecting cooling rate for a particular product include product size, shape, and thermal properties, and configuration of the product during cooling (i.e., products may be cooled in well-ventilated bulk containers prior to packing or they may be cooled after packing in shipping containers such as fiberboard cartons). Other important design factors include depth of product load during cooling; initial product temperature; desired final product temperature; and the relative humidity, temperature, and flow rate of the cooling air. Since these factors affect cooling rate, they also have a direct influence on cooling cost. Other factors influencing cost are unit costs for cooling space, refrigeration, and air-handling equipment; electric power; labor; interest rate; and time of operation per year. In practice, many cooling systems do not achieve the proper amount of cooling and/or they do not operate as efficiently and economically as they could. This results in wasted energy. Baird et al. (1988) discussed all components and assumptions associated with the mathematical model and its applications.

Hydraircooling

This system is combination of hydro- and air-cooling systems and utilizes a mixture of air and water as the cooling fluid (Figure 4.18.3). The reason this system was introduced is that most vegetables before shipment are hydrocooled, using large amounts of water in flow-through or batch systems. Hydrocooling systems are reasonably effective but leave "hot spots" throughout the product load, e.g., in loads of sweet corn and celery that are packed in wire-bound crates. Also, existing hydrocooling systems are not desirable for the cooling of vegetables stacked in fiberboard. Vacuum cooling can be utilized, but it is limited particularly to leafy vegetables and is more expensive owing to its high energy requirement. Because of these disadvantages, the hydraircooling technique was developed. It is conducted by applying

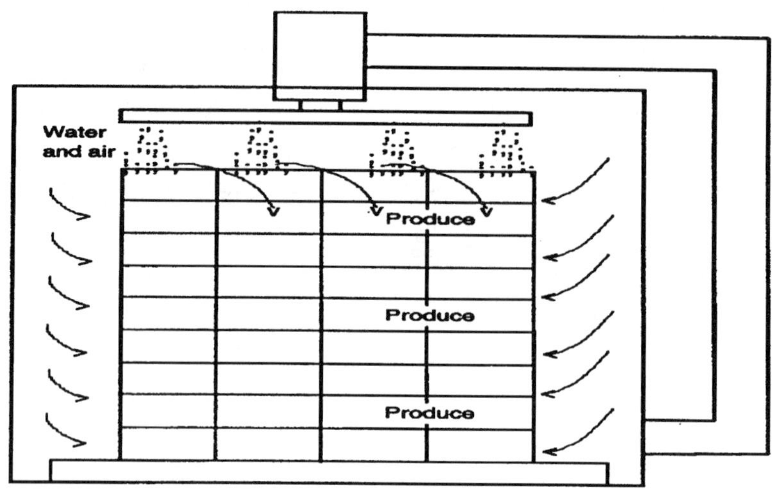

FIGURE 4.18.3 Schematic of a hydraircooling system. (From Dincer, I., *Heat Transfer in Food Cooling Applications,* Taylor & Francis, Washington, D.C., 1997. With permission.)

a mixture of chilled water and cold air to the products. It is a particularly effective method of cooling vegetables in pallet loads and is suitable for other vegetables such as cabbage, bell peppers, and cucumbers that are not normally hydrocooled.

Vacuum Cooling

Vacuum cooling is an alternative method for the rapid removal of field heat from produce to bring it to the storage temperature. It is extremely effective for products possessing a large surface to mass ratio (e.g., leafy vegetables), and an ability to release internal water readily. For suitable products, vacuum cooling is advantageous because of its rapidity and economy (especially reductions in the cost of labor and packaging and in product damage). The main disadvantage is the necessity for a high capital investment. Equipment needs include a large and strong chamber, a pressure-reduction device, and a water condensation device.

The four types of vacuum cooling systems that use water as refrigerant are *steam ejector, centrifugal, rotary,* and *reciprocating*. Among these systems, the steam ejector type of vacuum cooler (Figure 4.18.4) is best suited for displacing the extremely high volumes of water vapor encountered at the low pressures needed, has few moving parts, and doesn't require a compressor to condense the water vapor. However, its usage is limited due to inherent mechanical difficulties at the high rotative speeds required to generate the low pressures. Rotary and reciprocal vacuum coolers make it possible to generate the low pressures required for vacuum cooling and have the advantage of portability. However, they have low volumetric

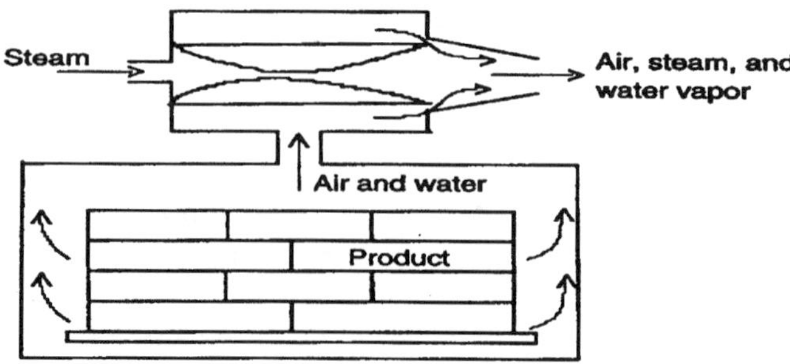

FIGURE 4.18.4 Schematic of a steam ejector-type vacuum cooling system. (From Dincer, I., *Heat Transfer in Food Cooling Applications,* Taylor & Francis, Washington, D.C., 1997. With permission.)

capacity and separate refrigeration systems to condense much of the water vapor that evaporates off the product. Most fruits and vegetables have a large content of water. When the produce is subjected to a suitable vacuum (about 4 to 5 mmHg), some of this water is evaporated, taking its heat of vaporization from the produce and thereby cooling it. In fact, to cool produce by 10°C requires on average the evaporation of water equivalent to 1.8% of the weight of the produce (Dincer, 1997). The extent of cooling is proportional to the water evaporated. Regardless of the product, the temperature is lowered about 5°C for each 1% reduction in water content. Prewetting is useful for many products, especially those that have high initial temperatures and those that retain substantial amounts of the added water on their surfaces until the vacuum is applied. Vacuum cooling is most successfully applied to flowers, mushrooms, and thin leafy produce, e.g., lettuce. It is not suitable for fruits, except some of the berries, particularly strawberries.

Food Freezing

Freezing is one of the most important food-cooling processes for long-term preservation. It reduces the temperature of all parts of the food product below the freezing point. For most food products, the final quality of the frozen products is better if the freezing is done rapidly, especially if the 0°C to –5°C zone is passed through rapidly and if the temperature is reduced to and maintained at a sufficiently low level. Frozen fruits and vegetables have no further metabolic activity. Freezing permits most perishable foods to be kept for several months or longer, depending on the kind of product and the level of the temperature. Several steps are involved before freezing the food or are combined with the freezing process. For example, in a vegetable freezing process, the steps to be taken are selecting, sorting, washing, blanching, and packaging prior to freezing. Freezing equipment is mainly divided into three types, based on the heat transfer situation: air-blast freezers, contact freezers, and cryogenic freezers.

The rate of freezing of foods is dependent primarily on several factors such as the freezing method, the temperature, the circulation of air or refrigerant (or cryogenics), the size and shape of package, and the kind of product. There are two basic ways to achieve food freezing: quick or slow freezing. Quick freezing is the process by which the temperature of foods is lowered to –20°C within 30 min (Jay, 1996), and it may be achieved by direct immersion or indirect contact of foods with the refrigerants (or cryogenics) and the use of air-blast freezing. Slow freezing is the process whereby the desired temperature is achieved within 3 to 72 h and is commonly used in domestic freezers. Quick freezing has some advantages over slow freezing from the standpoint of overall product quality owing the fact that smaller ice crystals are formed and internal metabolic imbalances are avoided.

Cool and Cold Storage

In the past, the differentiation between cold and cool storages was in terms of the storage of food products with ice or without ice. In broad terms, cool stores operate at air cooling temperatures above –2°C (typically, –2 to 10°C) and the cold stores operate below –2°C (typically –15 to –30°C) (Cleland, 1990). These storage technologies are applied for food products in order to maximize shelf life and thereby minimize the rate of senescence.

Most perishable food products, including eggs, dairy products, meats, seafood, fruits, and vegetables, may be held in cold storage for a limited time with little change from their original condition. Enzymatic and microbial changes in the food products are not prevented but are greatly slowed down. Maintenance of the ideal product temperature and minimization of water loss are both of importance. These are achieved by maintaining uniform conditions (i.e., temperature, relative humidity, and air velocity) in the storage environment. Design and operational factors such as layout of the store, insulation levels, door protection devices, frequency of door use, air-cooling coil and fan designs, associated control system design, air flow patterns, and product stacking arrangements can all influence the uniformity of environmental conditions and, therefore, the rate of change of product quality.

In the storage operation, the following points should be kept in mind: (1) minimizing the exposure of products to ambient temperatures, (2) laying out methods of handling and routes, and (3) never leaving the doors open after personnel or products have passed through them. The proper management of cool- and cold-storage rooms principally involves avoiding wide temperature fluctuations, keeping the cooling coils properly defrosted, and preventing heat from entering the room.

In the cool and cold storage of perishable foods, the following optimum conditions are of considerable significance: optimum temperature, optimum relative humidity, condensation of water vapor on the product, optimum air movement, optimum stacking, and sanitation.

Controlled Atmosphere (CA) Storage

The controlled atmosphere is derived from the fact that the composition of the atmospheric gases in contact with the products is controlled at precise levels during storage or transportation in order to prolong and extend the storage and market life of fresh fruits and vegetables. The major and obvious benefit of controlled atmosphere systems, as a supplement to refrigeration, is the extended storage life of food products. The immediate implication of this is that food products can be shipped longer distances and sea transportation becomes a viable alternative to air freight.

In CA storage, the oxygen (O_2) level is reduced and carbon dioxide (CO_2) increased. Lower respiration occurs, thus extending the life of some products in storage and providing additional distribution and marketing possibilities for fresh products. It is important to maintain the proper relationship of O_2 and CO_2 depending on the product and temperatures of storage. In most instances, O_2 and CO_2 are maintained between 2 and 5% volume, with nitrogen (N_2) comprising the remainder of the mixture. The optimum gas composition varies for each commodity and cultivar. The composition of an ordinary cold storage room atmosphere varies within the range 19 to 21% O_2 and 0 to 2% CO_2 depending on the temperature, kind, and quantity of food products stored, and the air-tightness of the structure (Dewey, 1983). Possibly 10% of all commercial apples grown for fresh market purposes are stored in a controlled atmosphere, and optimum operating conditions are 0 to 8% CO_2, 1 to 7% O_2, and from –1 to 4.5°C. Recently, CA applications to different fruits and vegetables have received considerable attention.

Refrigerated Transport

Large amounts of fruits and vegetables are lost every year, not only because of the lack of cooling and refrigerated transport facilities, but also because of the incorrect choice of temperature and humidity, and sometimes due to the mixing of different commodities in the same container.

The world's annual production of perishable food products is estimated by the United Nations Food and Agriculture Organization to be over 450 million tons. Of this, only about 10% is traded internationally; the rest is consumed locally or wasted. About 61% (28.8 million tons) of the total international traded amount (47.3 million tons) is transported by sea. Of the remainder, about 34% (16.3 million tons) is transported by land and about 5% (2.2 million tons) by air. It is estimated that of 28.8 million tons of perishable cargo, about 20.1 million tons (i.e., 70%) is shipped in reefer vessels and 8.7 million tons (i.e., 30%) in containers (Dincer, 1997). Therefore, for perishable products the transport from producer to consumer is one of the main steps in the cold chain and is essential for the quality of products. The packing is designed to protect the products during transport and distribution. Refrigerated facilities play a significant role when transporting perishable food products to marketplaces; the products may be on-board for up to a month. Therefore, good temperature control and air distribution is vital for product quality. So, extensive research is required to make refrigerated transportation capable of performing at its maximum potential level.

The requirements for the product environment during transportation and distribution are generally the same as for cold storage. But the time of exposure is normally much shorter and technical difficulties in maintaining low temperatures are greater. Also, there is a need for better temperature control in transportation. In transport by sea, some problems (e.g., mainly temperature rise) in maintaining satisfactory conditions are relatively slight and the situation has been improved considerably over the past few years. The types of insulation and the refrigeration unit used are much the same as in any stationary cold store. Suitable ships are now available to satisfy every requirement for any type of refrigerated food. In transport by land, more difficult problems from a technical and economic standpoint arise. The small, individual carriers using the road or rail trucks set narrow limits on the design complexity and cost of the cooling equipment. A high degree of reliability is required. Air transport of valuable products is common today, especially for products with a short storage life.

Heat Generation

Living fruits and vegetables produce heat due to respiration and, therefore, the rate of heat generation is one of the most important parameters affecting the postharvest cooling of fresh fruits and vegetables. This tends to increase the product's surface temperature and thereby is a driving force for moisture transfer. The effect is generally small for moderate vapor pressure differences. On the other hand, it may become a dominant factor at humidities close to saturation. In addition to heat generation, respiration produces additional weight loss due to CO_2 evolution. The net loss is that of carbon and is different from transpiration moisture loss. Most researchers have determined the heat generation rate by measuring CO_2 gas evolution from the product.

The cooling process for freshly harvested fruits and vegetables is complicated by the heat generation from their respirational activity. The relation between heat generation and temperature is assumed to involve breakdown of glucose as follows:

$$C_6H_{12}O_6 + 6\ O_2 \rightarrow 6\ CO_2 + 6\ H_2O + 2817\ \text{kJ} \tag{4.18.1}$$

As with most chemical reactions, the heat generation of fruits and vegetables (Q) is an exponential function of absolute temperature (T) as given below (Gogus et al., 1972):

$$Q = A\ \exp(BT) \quad \text{W/kg} \tag{4.18.2}$$

where A and B take different values for each fruit and vegetable. The values of A and B for several fruits and vegetables are given in tabular form in Gogus et al.(1972). The coefficients of the exponent B are in the range of 0.07 to 0.14 1/K, except for potatoes, for which 0.034 1/K was found.

Moisture Loss (Transpiration)

Transpiration of fresh fruits and vegetables is a mass transfer operation that involves the transport of water vapor from the surface of the product to the air medium. The main elements of transpiration are the transpiration coefficient and transpiration rate as follows:

- *The transpiration coefficient* is expressed as the mass of moisture transpired per unit area of product, per unit environmental water vapor pressure deficit per unit area, kg/m^{2}skPa (also expressed per unit mass of product)
- *The transpiration rate* is the mass of moisture transpired per unit area of product per unit time (also expressed per unit mass of product)

A better understanding of the mechanism of transpiration from fruits and vegetables to the surrounding air should help in developing new systems for handling, transport, and storage and improving existing systems so that moisture loss from fresh produce can be reduced and the initial quality at the time of harvest can be kept for a longer period of time. Knowledge of the transpiration rate of the stored fruits and vegetables is essential for a rational analysis of desired medium conditions during their cold storage.

The transpiration rate is affected by a number of factors that particularly influence the driving force and resistance terms such as water vapor pressure deficit, condition of water, air flow velocity, respiratory heat generation, product shape and size, product surface area and structure, product maturity, product seeds, product skin and tissue permeability, dissolved substances in water, evaporative cooling, package microenvironments, physical and physiological conditions, and packaging. Discussions of all these factors and phenomena and their interrelationships with transpiration are given by Sastry (1985).

Cooling Process Parameters

In food processing, there are many situations in which the temperature at any point in the product is a function of time and its coordinate, leading to the *transient heat transfer.* The most notable food processing examples of transient heat transfer are heating, cooling, freezing, drying, blanching, etc. For practical food cooling applications, the factors that influence the temperature change and cooling rate are temperature and flow rate of cooling medium (coolant), thermal properties of food product, and physical dimensions and shape of food product.

The primary objective of a cooling study is to produce usable data and technical information that will help improve the existing cooling systems and provide optimum operation conditions. People who work in the food cooling industry need the following: (1) a procedure for analyzing cooling process parameters, (2) a procedure for using cooling data to design cooling systems for efficient food cooling applications, and (3) a basic data documentation protocol.

Regardless of the type of cooling technique, knowledge and determination of the cooling process parameters are essential to provide efficient and effective food cooling at the micro- and macroscales. Some major design process factors for a food cooling process are

- Cooling process conditions in terms of temperature, flow rate, and relative humidity
- Arrangement of the individual products and/or product batches
- Depth of the product load in the cooling medium
- Initial and final product temperatures

The parameters in terms of the cooling coefficient, lag factor, half-cooling time, and seven-eighths cooling time (Figure 4.18.5) are the most important and meaningful variables for a food cooling process, and these are used to evaluate and present the cooling rate data and cooling behavior of the food products. A number of experimental and modeling studies to determine these parameters for various food products, particularly for fruits and vegetables, have been undertaken (see Dincer, 1995a; Wade, 1984).

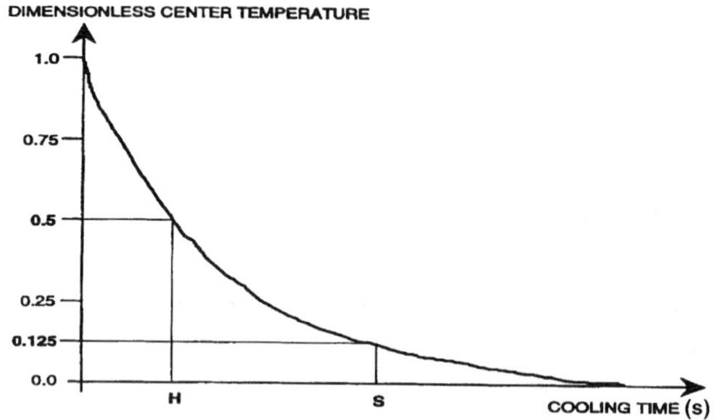

FIGURE 4.18.5 Representation of cooling times on a cooling profile. (From Dincer, I., *Heat Transfer in Food Cooling Applications,* Taylor & Francis, Washington, D.C., 1997. With permission.)

Cooling coefficient — This is an indication of the cooling capability of a food product subject to cooling and denotes the change in the product temperature per unit time of cooling for each degree temperature difference between the product and its surroundings.

Lag factor — It is a function of the size and shape and the thermal properties of the product, such as the effective heat transfer coefficient, thermal conductivity, and thermal diffusivity, and quantifies the resistance to heat transfer from the product to its surroundings in relation to the Biot number.

Half-cooling time — The half-cooling time is the time required to reduce the product temperature by one-half of the difference in temperature between the product and the cooling medium. The food cooling rate data are formed in the cooling times.

Seven-eighths cooling time — The seven-eighths cooling time, which is one of most meaningful parameters, describes the cooling rate in terms of the time required to reduce the product temperature by seven-eighths of the difference in temperature between the product and the cooling medium.

Methodology for Cooling Process Parameters

Here we introduce the semi-experimental method for the cooling process parameters. In this method, it is considered that transient heat transfer occurs with the constant thermal and physical properties of the product and cooling medium (Dincer, 1995a; Wade, 1984).

The dimensionless temperature (θ) is generally expressed in the form of an exponential equation, including the cooling parameters in terms of the cooling coefficient (C, s^{-1}) and lag factor (G) as

$$\theta = G\exp(-Ct) \tag{4.18.3}$$

where the dimensionless center temperature in terms of the initial temperature of product (T_i, °C) and the medium temperature (T_a, °C) is

$$\theta = (T - T_a)/(T_i - T_a) \tag{4.18.4}$$

where T is the center temperature of product at any time, °C.

From the definition of the half-cooling time, by substituting θ = 0.5 into Equation 4.18.3, the half-cooling time (H, s) becomes

$$H = \ln(2G)/C \tag{4.18.5}$$

Also, by substituting θ = 0.125 into Equation 4.18.3, the seven-eighths cooling time (S, s) becomes

$$S = \ln(8G)/C \tag{4.18.6}$$

Fourier-Reynolds Correlations

Using the general expression of the Fourier number, $Fo = (at/Y^2)$, we introduce the Fourier number for the half-cooling time and the Fourier number for the seven-eighths cooling time for food cooling applications as follows:

$$Fo_H = ah/Y^2 \Rightarrow H = Fo_H Y^2/a \tag{4.18.7}$$

$$Fo_S = aS/Y^2 \Rightarrow S = Fo_S Y^2/a \tag{4.18.8}$$

where a stands for thermal diffusivity, m^2/s and Y stands for characteristic dimensions (i.e., L for slab and R for cylinder and sphere), m.

The Fourier-Reynolds correlations to estimate half-cooling times and seven-eighths cooling times of food products (with the characteristic dimensions varying 0.01 to 0.15 m, respectively) being cooled with air, for the range 100 < Re < 100000, are

$$Fo_H = 42.465Re^{-054426} \tag{4.18.9}$$

$$Fo_S = 125.21Re^{-0.53913} \tag{4.18.10}$$

where $Re = U2Y/\nu$. Here, U is the flow velocity of coolant over the product, m/s, and ν is the kinematic viscosity, m/s^2.

The half-cooling times and seven-eighths cooling times of food products cooled with water are

$$Fo_H = 0.3693Re^{-0.11871} \quad (4.18.11)$$

$$Fo_S = 1.2951Re^{-0.16003} \quad (4.18.12)$$

For thermal diffusivity of the product, the following Riedel correlation is of common use, depending on the product's moisture content (*W*, % by weight).

$$a = 0.088 \times 10^{-6} + (a_w - 0.088 \times 10^{-6})W \quad (4.18.13)$$

where a_w is thermal diffusivity of water at the product temperature (i.e., 0.148×10^{-6} m²/s at 25°C).

After inserting the estimated thermal diffusivity into Equations 4.18.7 and 4.18.8, we can estimate the half-cooling time and seven-eighths cooling time, without making any experimental temperature measurement.

Example

For a better understanding and verification of the above correlations we present this example. A watermelon (T_i = 25.5°C, R = 0.1 m, m = 5.17 kg) is cooled in a cold-water flowing tank at T_a = 0.5°C and U = 0.05 m/s. The center temperatures during the cooling process were measured with a data acquisition unit until the center temperatures reached 5°C as the storage temperature (Figure 4.18.6). Let's calculate the cooling times with the following experimental and table data: $a = 1.4098 \times 10^{-7}$ m²/s and $\nu = 0.8937 \times 10^{-6}$ m²/s (for details, see Dincer and Dost, 1996).

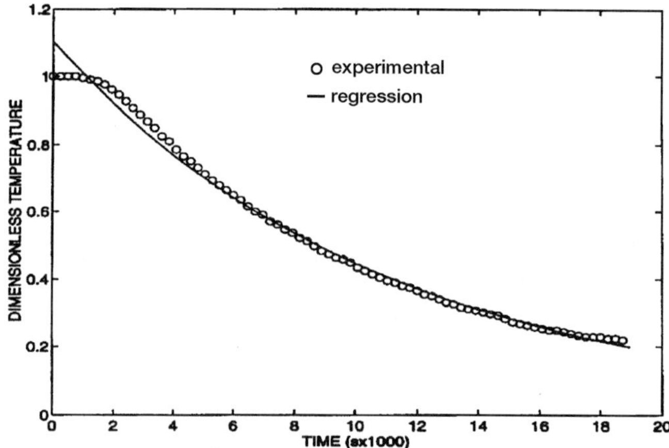

FIGURE 4.18.6 Measured and calculated center temperature distributions of a watermelon.

Solution Procedure:

- $\theta = 1.102065\exp(-0.0000903367t)$, with a correlation coefficient of 0.996
- $Re = UD/\nu = 0.05 \times 0.2/0.8937 \times 10^{-6} = \underline{11198.43}$
- $Fo_H = 0.3693Re^{-0.11871} = 0.3693 \times 11198.43^{-0.11871} = \underline{0.1221}$
- $Fo_S = 1.2951Re^{-0.16003} = 1.2951 \times 11198.43^{-0.16003} = \underline{0.2943}$
- $H = Fo_H Y^2/a = 0.1221 \times 0.1^2/1.4098 \times 10^{-7} = \underline{8660.8\ s}$
- $S = Fo_S Y^2/a = 0.2943 \times 0.1^2/1.4098 \times 10^{-7} = \underline{20875.3\ s}$

Cooling Heat Transfer Parameters

Significant factors in the design of a refrigerated food chain are the cooling heat transfer parameters (e.g., specific heat, thermal conductivity, thermal diffusivity, and heat transfer coefficient) for food products. It is essential that the design engineer knows the quantity of heat to be released and the time taken to remove it; consequently, cooling heat loads are calculated based on the effects of refrigeration on the product batches or the quality changes during storage. During the past three decades there has been continuing interest in analyzing the transient heat transfer that takes place during food cooling applications in order to provide the optimum processing conditions.

Specific Heat

The specific heat is usually modeled in the form $c_p = X + YW$ where X and Y are constants and W is water content, % by weight, depending on the product and the technique used. In the past, several correlations were developed by various researchers and these can be found in Sweat (1986). Among these, Siebel's model is the oldest and the best one for temperatures above freezing, in terms of the water content of the food product:

$$c_p = (0.837 + 3.349\ W) \times 1000 \tag{4.18.14}$$

Variability in composition is a major characteristic of food materials. This is true for those constituents, especially water content, that affect specific heat. All the models tend to converge at 100% water content. The specific heat of water is 4187 J/kg°C. There is a greater variation as the water content decreases, approaching zero. The basis for this variation is the composition of the specific foods modeled by the specific expression. Also, the specific heats of proteins, fats, and carbohydrates are different. In the literature there are some other general correlations including different components of food products, e.g., the Heldman and Singh model:

$$c_p = (1.424\ C + 1.549\ P + 1.675\ F + 0.837\ A + 4.187\ W) \times 1000 \tag{4.18.15}$$

and the Choi and Okos model:

$$c_p = (1.547\ C + 1.711\ P + 1.928\ F + 0.908\ A + 4.180\ W) \times 1000 \tag{4.18.16}$$

where F, S, C, P, and A represent fat, solid, carbohydrate, protein, and ash contents.

It is clear that Equations 4.18.15 and 4.18.16 should give accurate results, owing to inclusion of the main components of the food products. As Sweat (1986) pointed out, however, there is still some error in prescribing the specific heat of a given component. In addition to the equations given above, there are some that include a temperature term in specific heat models. The most common of these models is given below in the Fernandez-Martin and Montes model:

$$c_p = \left[(1.370 + 0.0113\ T) + (1 - W)\right] \times 1000 + 4190\ W \tag{4.18.17}$$

The most comprehensive data sets for specific heat values for a wide range of food products appeared in ASHRAE (1989) and Dincer (1997).

Several experimental methods are available for measurement of specific heats of food products, namely, the differential scanning calorimeter technique, vacuum bottle calorimeter technique, radiation calorimeter technique, guarded-plate technique, and adiabatic agricultural calorimeter technique. Detailed information on experimental instruments and procedures for measuring and calculating specific heat can be found in Mohsenin (1980) and Sweat (1986). Among these methods, the differential scanning calorimeter technique and vacuum bottle calorimeter technique have received much attention in the food industry.

Thermal Conductivity

It is well known that most of cooling heat transfer parameter models are empirical or semiempirical rather than theoretical. In this respect, there are a number of correlations in the form $k = X + YW$ for obtaining the thermal conductivities for food products as developed for specific heats. The following sweat model is a correlation particularly for fruits and vegetables:

$$k = (0.148 + 0.493\ W) \tag{4.18.18}$$

Sweat (1986) also gives the following model for meats and fish:

$$k = (0.080 + 0.52\ W) \tag{4.18.19}$$

For most liquid foods and many solid food products, in addition to water content it is necessary to take into account the protein, carbohydrate, fat (or oil), and ash contents as given in the following models by Domiguez et al. and Sweat, respectively:

$$k = (0.6\ W + 0.2\ P - 0.245\ C + 0.18\ F) \tag{4.18.20}$$

$$k = (0.58\ W + 0.155\ P - 0.25\ C + 0.16\ F + 0.135\ A) \tag{4.18.21}$$

The equations presented above are valid for a narrow temperature range above freezing because the temperature terms are not considered. The main issue on the accuracy of these correlations is having accurate values for the thermal conductivity of pure components. This is easy for the water and oil fractions, but very difficult for the other fractions. In fact, the thermal conductivity values of protein and carbohydrates vary depending on their chemical and physical forms. To obtain more accurate results, it is important to know the thermal conductivity of the specific protein and carbohydrate.

Thermal conductivity measurements are divided into two main categories: those using steady-state heat transfer and those using unsteady-state heat transfer. Each category includes several techniques. A classification of the experimental thermal conductivity determination methods is presented in detail in Mohsenin (1980).

Thermal Diffusivity

The thermal diffusivity is defined as the ratio of the thermal conductivity to the product of the specific heat and density, as follows:

$$a = k/\rho c_p \tag{4.18.22}$$

In the literature, numerous studies on the measurement/estimation of thermal diffusivities of food products have been undertaken by various researchers (see Uno and Hayakawa, 1980; Dincer, 1998). A common approach for measuring thermal diffusivity of food products involves temperature vs. time measurements of a regular-shaped test sample, initially at a uniform temperature, placed in surroundings maintained at a constant temperature different from the initial temperature of the sample. Riedel (1969) developed an expression (see Equation 4.18.13) to estimate thermal diffusivity values of a wide range of foods depending on the moisture content of the food. It should be noted that Equation 4.18.13 may be available to estimate thermal diffusivity of foods, but its usage must be limited to water contents above 40% by mass (ASHRAE, 1989). Martens (1980) studied the effects of water, fat, protein, carbohydrate, and temperature on the thermal diffusivity of foods and found that temperature and water content are the main factors affecting thermal diffusivity, as follows:

$$a = [0.057363W + 0.000288(T + 273)] \times 10^{-6} \tag{4.18.23}$$

There are some experimental techniques for determining thermal diffusivities of foods. For example, Dickerson (1965) developed a transient heating technique for this purpose. It was very creative but requires too much time and too large a sample to be useful today. Choi and Okos (1983) measured thermal diffusivities of foods by means of a line-heat-source thermal conductivity probe with an auxiliary thermocouple. This technique is good for liquid foods, where the distance between the line-heat-source probe and the auxiliary thermocouple (on a second probe) is fixed; however, for solid foods, the distance between the two probes varies because the probes may move laterally during insertion. Because the results of this technique are quite sensitive to the distance between the probes, this technique is limited to applications in which the distance between the probes does not vary or the distance can be precisely measured. Bhowmik and Hayakawa (1979) used a thermal diffusivity cell, similar to the one used by Dickerson. It imposes a step functional change on the temperature of the surrounding medium and needs a simple instrumental set-up for experimentally measuring the temperature data that are used for the calculation of thermal diffusivity. During the past three decades several computational and semiexperimental methods and models have appeared.

Dincer (1998) has undertaken some experimental and theoretical studies dealing with the development of the thermal diffusivity models for geometrically shaped products being cooled. Applications of this model were made using measured experimental center temperature data carried out for slab, spherical, and cylindrical products. It can be concluded that the developed models are capable of determining the thermal diffusivities of such products during cooling in a simple and effective manner. In these studies, three cases of the Biot number (i.e., $Bi = hY/k$) such as $Bi < 0.1$, $0.1 \leq Bi \leq 100$, and $Bi > 100$ were taken into consideration and the following models were developed:

$$a = kEC/h \quad \text{for } Bi < 0.1 \tag{4.18.24}$$

where E is the ratio of volume to area, m (E = 2L for slab, R/2 for cylinder, and R/3 for sphere),

$$a = CY^2/\mu_1^2 \quad \text{for } 0.1 \leq Bi \leq 100 \text{ and } Bi > 10 \tag{4.18.25}$$

Here, the values of μ_1, roots of the characteristic equations, can be determined from the following newly developed correlations for $0.1 \leq Bi \leq 100$:

For slab: $\mu_1 = \text{atan}\,(0.64\ \text{Bi} + 0.38)$ for $0.1 \leq Bi \leq 100$

For cylinder: $\mu_1 = [(0.72)\ \ln(6.8\ \text{Bi} + 1)]^{1/1.4}$ for $0.1 \leq Bi \leq 10$

For cylinder: $\mu_1 = [\ln(1.74\ \text{Bi} + 147.3)]^{1/1.2}$ for $10 \leq Bi \leq 100$

For sphere: $\mu_1 = [(1.12)\ \ln(4.9\ \text{Bi} + 1)]^{1/1.4}$ for $0.1 \leq Bi \leq 10$

For sphere: $\mu_1 = [(1.66)\ \ln(2.2\ \text{Bi} + 152.4)]^{1/1.2}$ for $10 \leq Bi \leq 100$

where the Biot numbers can be calculated using the lag factors via the following models:

For slab: $\text{Bi} = 1.3\ \text{lnG}/(0.2533 - \text{lnG})$

For cylinder: $\text{Bi} = 1.7\ \text{lnG}/(0.5066 - \text{lnG})$

For sphere: $\text{Bi} = 2.1\ \text{lnG}/(0.7599 - \text{lnG})$

For the case $Bi > 100$, the following characteristic equations can be used directly:

For slab: $\mu_1 = \pi/2$

For cylinder: $J_0(\mu_1) = 1 - (\mu_1^2/2^2) + (\mu_1^4/2^2 \times 4^2) \cdots$

For sphere: $\mu_1 = \pi$

Heat Transfer Coefficient

According to the Newton's law of cooling, the effective heat transfer coefficient is a proportionality constant relating the heat flux from a surface to the temperature difference between the surface and the fluid stream moving past the surface. Actually, the heat transfer coefficient is not a direct property of the food product being cooled; however, it defines the rate of heat convection from the surface of a food product. It is mainly dependent upon the flow velocity of coolant, but also depends on several variables such as the fluid properties of the coolant and the shape, size, and surface texture of the product, as well as the temperature difference between the surface and the coolant. Very limited studies on the determination of heat transfer coefficients for food products exposed to cooling have been obtained. Most of them are based on the use of the existing Nusselt-Reynolds correlations and the rest covers mathematical models or correlations or approximations for heat transfer coefficients.

Practically, there are three common methods of determining the heat transfer coefficients, as follows:

- The steady-state measurement of surface temperature for a given heat dissipation of the product
- The measurement of the transient temperature of the product during cooling
- The measurement of heat flux at the surface of the product

Dincer (1994a, 1998) has undertaken some experimental and theoretical studies dealing with the development of the heat transfer coefficient models for products with regular shapes subject to cooling. This methodology utilizes the experimental center temperature data carried out for slab, spherical, and cylindrical products as done for the thermal diffusivity as presented above. Three cases of the Biot number (i.e., $Bi = hY/k$) such as $Bi < 0.1$, $0.1 \leq Bi \leq 100$, and $Bi > 100$ were taken into consideration and the following models were developed.

For $Bi < 0.1$:

$$h = kEC/a \quad \text{for slab, cylinder, and sphere.} \tag{4.18.26}$$

For $0.1 \leq Bi \leq 100$:

$$h = (k/L)[(1.3 \ln G)/(0.2533 - \ln G)] \quad \text{for slab} \tag{4.18.27}$$

$$h = (k/R)[(1.7 \ln G)/(0.5066 - \ln G)] \quad \text{for cylinder} \tag{4.18.28}$$

$$h = (k/R)[(2.1 \ln G)/(0.7599 - \ln G)] \quad \text{for sphere} \tag{4.18.29}$$

For $Bi > 100$:

$$h = \rho c_p a\, Bi/Y \quad \text{for slab, cylinder, and sphere} \tag{4.18.30}$$

The technique presented above may be extended to include products of irregular shapes. In this regard, some relevant geometric indexes must be determined. For instance, the geometry index for a sphere is 1; if this value decreases, it becomes eccentric.

In the present models, the most important point is that the cooling process parameters and the lag factor are incorporated into the thermal diffusivity and the heat transfer coefficient models. These cooling process parameters are of importance for evaluating and representing a cooling process. It is apparent that these parameters have direct effects on the thermal diffusivity and the heat transfer coefficient.

Example

In this example we present the application of the thermal diffusivity and heat transfer coefficient models using the experimental center temperatures of single dry figure as slab, fresh cucumber as cylinder, and fresh pear as sphere. The details of the experimental apparatus and procedure are given in Dincer (1998). The experimental conditions are

For slab: $T_i = 21.6°C$, $T_f = 18°C$, $T_a = -22.0°C$, $L = 0.007$ m, air cooling.
For cylinder: $T_i = 21.5°C$, $T_f = 4°C$, $T_a = 4°C$, $U = 1.5$ m/s, $R = 0.014$ m, air cooling.
For sphere: $T_i = 22.0°C$, $T_f = 4°C$, $T_a = 4°C$, $U = 1.25$ m/s, $R = 0.012$ m, air cooling.

The solution procedure used is as follows:

- The cooling coefficients and lag factors of the products are determined using the methodology presented earlier
- The Biot numbers are calculated by inserting the lag factors into the equation given above and the roots of the characteristic equations by inserting the calculated Biot numbers
- The thermal diffusivity values of the products are determined using Equation 4.18.26
- The heat transfer coefficients of the products are determined using Equations 4.18.27 and 4.18.29

Experimental measurements and model results are tabulated in Table 4.18.1.

As can be seen from table, the thermal diffusivity and heat transfer coefficient results determined from measurements agree well with the model results (within ±13%). Differences less than ±20% are considered reasonable for practical heat transfer problems. All the uncertainties stayed within ±5.0%. The Biot numbers took values between 0.1 and 100 based on the lag factors. The results indicate that the models presented here can be used reliably for a wide range of practical cooling applications. As a result, it can be concluded that the models presented here are simple and powerful tools for the determination of the thermal diffusivities and heat transfer coefficients for food products subject to cooling.

TABLE 4.18.1 Cooling Heat Transfer Parameters Obtained for the Products

	Slab	Cylinder	Sphere
C (1/s)	0.00059	0.00031	0.00045
G (–)	1.04317	1.04651	1.03147
Bi (–)	0.26	0.17	0.12
μ_1 (–)	0.500	0.625	0.655
k (W/m°C)	0.28	0.68	0.59
a_e (m²/s)	9.88×10^8	14.56×10^8	13.78×10^8
a_m (m²/s)	11.57×10^8	15.55×10^8	15.12×10^8
h_e (W/m²°C)	9.1	9.49	5.31
h_m (W/m²°C)	10.45	8.25	4.33

Source: Dincer, I., *Heat Mass Transfer,* 33, 383, 1998. With permission.

Heat Transfer Coefficient and Cooling Coefficient Correlations

Recently, Dincer and Dost (1996) introduced the development of simpler effective heat transfer coefficient correlations for food products, especially fruits and vegetables, cooled in both water and air, in terms of the cooling coefficient. In this study they used the same modeling technique, but went one step further for correlating the experimental heat transfer coefficient values obtained earlier (Dincer, 1995b) by using a least-squares curve-fitting method. These correlations were found with high correlation coefficients and are as follows.

For products cooled with water:

$$h = 27.356 \exp(1381.836C) \tag{4.18.31}$$

For products cooled with air:

$$h = 13.5497\exp(256.9278C) \tag{4.18.32}$$

These models provide less accurate results, but at the same time are practical and extremely useful to get some rough idea about the process and the magnitudes of the heat transfer coefficients.

Effective Nusselt-Reynolds Correlations

In this subsection we are jumping one step ahead to provide newer and simpler correlations, so-called *effective Nusselt-Reynolds correlations*. There are a large number of Nu-Re correlations available in the heat transfer literature leading to steady-state heat transfer. However, during cooling of food products unsteady-state heat transfer occurs. Using these existing Nu-Re correlations for this purpose may cause discrepancies of up to ± 50%. For this reason, the development of effective Nu-Re correlations is required to estimate more accurate effective heat transfer coefficients for food products, which will be helpful in system design and process optimization.

Two recent publications (Dincer, 1994b, 1995b) deal with effective Nusselt-Reynolds correlations developed for spherical and cylindrical fruits and vegetables, which are the most common shapes. These cover both experimental and theoretical studies. In the experimental work, the center temperature distributions of several fruits and vegetables being cooled with air flow were measured and used as process data for the modeling. In the theoretical study, the modeling technique (as given above) was used. Also, some experimental cooling data were taken from the literature. Then the heat transfer coefficients were employed to obtain Nusselt numbers and the Reynolds numbers were calculated. Thus, $Nu/Pr^{1/3}$ data were correlated against the Reynolds number. Detailed information on the effective heat transfer coefficients, data reduction methods, and experimental conditions, as well as the Nu-Re graphs can be found in Dincer (1994b, 1995b, 1997). The effective Nu-Re correlations developed for $100 < Re < 100{,}000$ are

For fruits and vegetables cooled with water:

$$Nu/Pr^{1/3} = 0.267188 \times Re^{0.432436} \tag{4.18.33}$$

For fruits and vegetables cooled with air:

$$Nu/Pr^{1/3} = 1.263870 \times Re^{0.449284} \tag{4.18.34}$$

where $Nu = h2Y/k_a$.

The results presented here show that new Nu-Re correlations are very useful sources for practical cooling applications and provide reliable results. The utilization of these correlations saves money and time for refrigeration engineers and technicians and provides effective and efficient operations. However, further research needs to be carried out to develop other Nusselt-Reynolds correlations for different food commodities.

Beside these Nu-Re correlations to determine effective heat transfer coefficients, Dincer (1996) developed a new Nusselt-Dincer correlation for the range $10^4 < Di < 10^6$:

$$Nu = 2.2893 \times 10^4 \; Di^{1.0047} \tag{4.18.35}$$

where Di is the Dincer number that shows the effect of the flow velocity of the cooling fluid on the cooling coefficient for food products with regular or irregular shapes, as follows:

$$Di = U/CY \tag{4.18.36}$$

Here, this ratio of these quantities defines the relative magnitude of the cooling capability of the food product being cooled in a fluid flow and is the connecting link between the flow velocity and the cooling rate.

Consequently, the Nu-Di correlations are extremely useful from the standpoint of suggesting how fluid properties and cooling process parameters affect each other. It is well known that the Nu-Re correlations are used particularly for determining the heat transfer coefficients for the corresponding products. In this case, if we go one step further, we can say that the Nu-Di correlations also can be used for determining the heat transfer coefficients for products.

References

ASHRAE, *Handbook of Fundamentals*, American Society of Heating, Refrigerating, and Air-Conditioning Engineers, Atlanta, GA, 1989.

ASHRAE, *Handbook of Refrigeration*, American Society of Heating, Refrigerating, and Air-Conditioning Engineers, Atlanta, GA, 1990.

Baird, C.D., Gaffney, J.J., and Talbot, M.T., Design criteria for efficient and cost effective forced air cooling systems for fruits and vegetables, *ASHRAE Trans.,* 94(1), 1434-1454, 1988.

Bhowmik, S.R. and Hayakawa, K., A new method for determining the apparent thermal diffusivity of thermally conductive food, *J. Food Sci.* 44(2), 469-474, 1979.

Choi, Y. and Okos, M.R., The thermal properties of tomato juice concentrates, *Trans. ASAE,* 26(1), 305-311, 1983.

Cleland, A.C., *Food Refrigeration Processes: Analysis, Design and Simulation*, Elsevier, London, 1990.

Dewey, D.H., Controlled atmosphere storage of fruits and vegetables, *Developments in Food Preservation-2,* Thorne, S., Ed., Applied Science Publishers, London, 1983, chap. 1, 1-24.

Dickerson, R.W., Jr., An apparatus for the measurement of thermal diffusivity of foods, *Food Technol.,* 19(5), 198-204, 1965.

Dincer, I. Surface heat transfer coefficients of cylindrical food products cooled with water, *J. Heat Transfer,* 116(3), 764-767, 1994a.

Dincer, I., Development of new effective Nusselt-Reynolds correlations air-cooling of spherical and cylindrical products, *Int. J. Heat Mass Transfer,* 37(17), 2781-2787, 1994b.

Dincer, I., An effective method for analyzing precooling process parameters, *Int. J. Energy Res.,* 19, 95-102, 1995a.

Dincer, I. Development of new correlations for forced convection heat transfer during cooling of products, *Int. J. Energy Res.,* 19(9), 791801, 1995b.

Dincer, I., Development of a new number (the Dincer number) for forced-convection heat transfer in heating and cooling applications, *Int. J. Energy Res.,* 20(5), 419-422, 1996.

Dincer, I., *Heat Transfer in Food Cooling Applications*, Taylor & Francis, Washington, D.C., 1997.

Dincer, I., Thermal properties of solid objects subject to refrigeration: a simple approach, *Heat Mass Transfer,* 33, 383-387, 1998.

Dincer, I. and Dost, S., New correlations for heat transfer coefficients during direct cooling of products, *Int. J. Energy Res.,* 20, 587-594, 1996.

Gogus, A.Y., Akyurt, M., and Yavuzkurt, S., Unsteady cooling of unit loads with exponential heat generation, *Int. Inst. Refrigeration-Annexe* 1972(1), 227-239, 1972.

Holdsworth, S.D., *The Preservation of Fruit and Vegetable Food Products*, Macmillan, London, 1983.

Jay, J.M., *Modern Food Microbiology,* 5th ed., Chapman & Hall, New York, 1996.

Martens, T., Mathematical Model of Heat Processing in Flat Containers, Ph.D. thesis, Katholeike University, Leuven, Belgium, 1980.

Mohsenin, N.N., *Thermal Properties of Foods and Agricultural Materials*, Gordon and Breach, New York, 1980.

Riedel, L., Measurements of thermal diffusivity of foodstuffs rich in water, *Kaltetechnik-Klimatisierung,* 21(11), 315-316, 1969.

Sastry, S.K., Moisture losses from perishable commodities: recent research and developments, *Int. J. Refrig.,* 8, 343-346, 1985.

Sweat, V.E., Thermal properties of foods, in *Engineering Properties of Foods*, Rao, M.A. and Rizvi, S.S.H., Eds., Marcel Dekker, New York, 1986, 49-87.

Uno, J. and Hayakawa, K., A method for estimating thermal diffusivity of heat conduction food in a cylindrical can, *J. Food Sci.,* 45, 692-695, 1980.

Wade, N.L., Estimation of the refrigeration capacity required to cool horticultural produce, *Int. J. Refrig.,* 7, 358-366, 1984.

Further Information

Other sources on thermal processing in food preservation technologies which may be consulted for more detailed information are Mohsenin (1980), Cleland (1990), and Dincer (1997).

4.19 Thermal Conduction in Electronic Microstructures

Kenneth E. Goodson

Introduction

Submicrometer dimensions are the hallmark of micromachined transistors, semiconductor lasers, sensors, and actuators. For many of these devices, heat transfer has a large impact on performance and reliability. The small spatial and temporal scales of the heat transfer processes can render inappropriate much of the theory and experimental technology common for larger systems, such as continuum heat-diffusion theory and infrared imaging. This problem has motivated much research in the last decade on thermal engineering at small scales, much of which has focused on heat conduction theory, thin-film thermal conductivity measurements, and the thermometry of microstructures. This section summarizes research progress on these topics with a focus on the simplest and most effective theoretical relationships and experimental techniques.

There is a wealth of microdevices for which heat conduction is important. Figure 4.19.1 shows a scanning electron micrograph (SEM) of a circuit interconnect structure, in which the temperature rise is governed by the distribution of Joule heating in the metal bridges and thermal conduction in the surrounding dielectric layers. The thermal engineering of interconnect structures must account for the increasing number of metal layers and the continuous introduction of novel dielectric materials, whose thermal properties are generally not known. Figure 4.19.2 is a transmission electron micrograph of a metal-silicon contact region in an integrated circuit, which has failed because of the temperature rise

FIGURE 4.19.1 Scanning electron microscopy (SEM) image of the six-layer copper interconnect structure for integrated electronic circuits. (Courtesy of the IBM Corporation.)

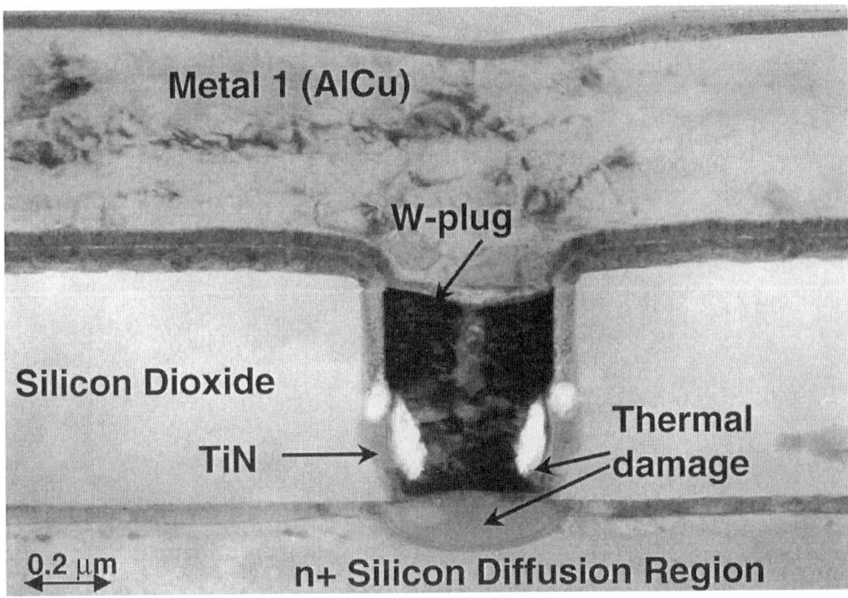

FIGURE 4.19.2 Transmission electron microscopy (TEM) image of a metal-silicon contact, which has failed during a brief electrical pulse. The heat generated in the tungsten plug raises the temperature and induces severe atomic diffusion. (From Banerjee, K. et al., *Proc. Int. Reliab. Phys. Symp.*, 1996.)

during a brief pulse of electrical current. Such current pulses are caused by electrostatic-discharge (ESD), which can occur during chip manufacture and packaging. Figure 4.19.3 shows a cross-sectional micrograph of a high-voltage transistor for a smart-power circuit. High-voltage and high-power transistors are found in *smart-power circuits* for vehicles and are particularly susceptible to thermal failure because of the large rates of heat generation and the electrical noise they must withstand. In this particular device, heat diffuses laterally in a single-crystal silicon layer of thickness as low as 200 nm. Figure 4.19.4 shows an electron micrograph of a silicon cantilever, in which the heat is conducted by a suspended, single-crystal silicon layer. This device is used for high-density thermomechanical data storage, during which Joule heating in the cantilever softens the surface of an organic substrate and atomic-scale forces form a data bit of diameter below 50 nm. The bit dimensions and writing rate are governed by conduction along the cantilever and the tip.

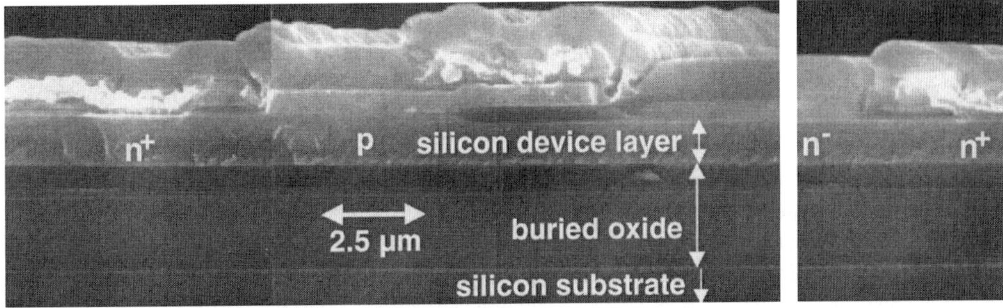

FIGURE 4.19.3 Cross-sectional SEM image of a high-voltage silicon-on-insulator transistor. Electrons flow from the source to the drain and generate heat within the silicon layer, which can be as thin as 200 nm. This device, which is called a lateral diffusion metal-oxide-semiconductor (LDMOS) transistor, is used to block hundreds of volts for smart-power electronic circuits. (From Leung, Y.K. et al., *IEEE Electron Device Lett.*, 18, 414-416, 1997.)

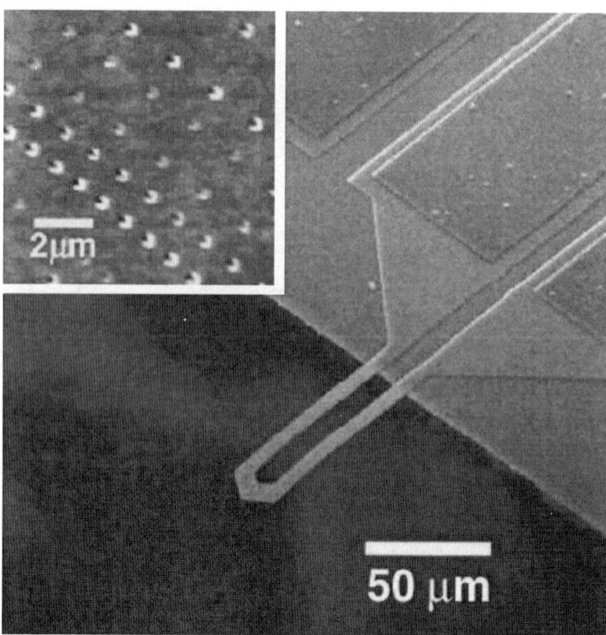

FIGURE 4.19.4 SEM images of the silicon cantilever used for high-density thermomechanical data storage by Chui et al. (1998). An electrical bias current along the cantilever legs induces Joule heating, which causes the tip to locally indent an organic substrate. The inset is a scanning electron micrograph of a substrate with several of these indentations. This data storage technology promises bit diameters below 50 nm.

The small scales of micromachined electronic devices complicate the simulation of heat conduction. Heat conduction is particularly important in these devices and is influenced by the close proximity of interfaces and associated changes in material stoichiometry and structural quality. Continuum heat-diffusion simulations fail when the dimensions of the device are comparable with or smaller than the mean free path of the heat carriers. This criterion is based on the *electron* mean free path in metal films and devices, which is approximately 5 to 50 nm at room temperature. For dielectric and semiconducting materials, heat is carried predominantly by the coupled vibrations of atoms, whose energy quanta are *phonons*. The phonon mean free path in crystalline materials can approach 300 nm at room temperature. Electron and phonon scattering at material interfaces impedes heat conduction and causes anisotropy and a size dependence of the observed effective thermal conductivities within a device. Even in devices with dimensions much larger than the relevant mean free path, the material stoichiometry and structural quality can differ vastly from those found in bulk materials. This situation necessitates independent measurements of material properties in micromachined structures closely resembling the targeted devices. The measurement of temperature fields in devices must offer exceptionally high spatial and temporal resolution, which is inaccessible using conventional infrared imaging and thermocouple thermometry.

The relevance of the thermal engineering of microstructures can best be appreciated from a perspective that includes all of the heat transfer processes in electronic systems. These include convection from and conduction within the packaging for the device and microscale conduction within or very near the micromachined transistor, sensor, or actuator. These heat transfer processes occur with timescales and within regions of dimensions varying by orders of magnitude, as indicated by the thermal circuit in Figure 4.19.5. The peak temperatures typically occur in electronic microstructures, such as semiconductor devices and interconnects, where heat is generated. The thermal resistance $R_{micro/nano}$ is governed by thermal conduction over a distance between 0.5 and 10 μm, and C_{device} or C_{metal} is the total effective heat capacity of the region heated significantly above the substrate temperature. The timescale for device or interconnect cooling, which can be estimated as the product of the relevant resistance and capacitance, varies from 10 nanosecond (ns) to 5 microseconds (μs).

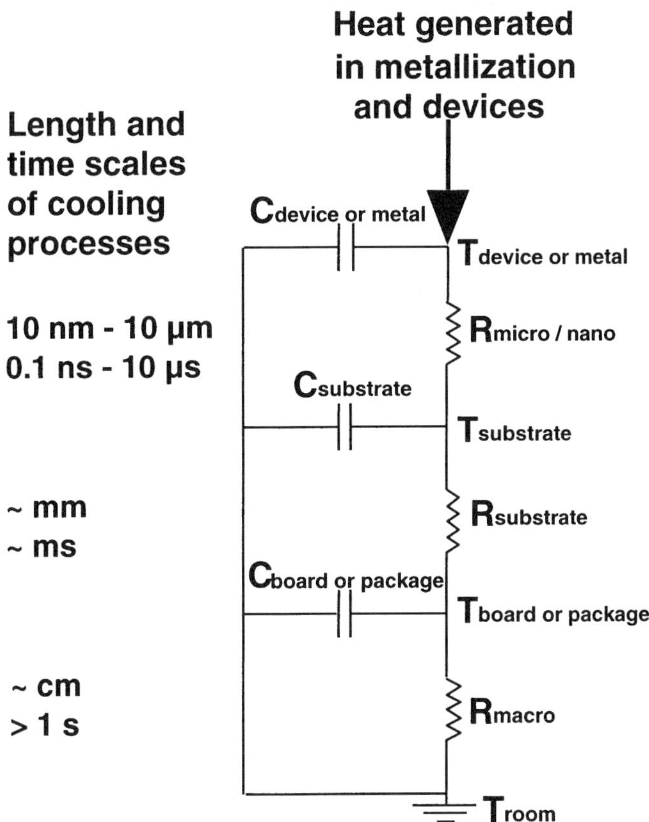

FIGURE 4.19.5 Approximate thermal circuit showing the hierarchy of thermal resistances and capacitances in an electronic system. The temperature rise is governed by the flow of heat power (W) through resistances ($K\ W^{-1}$) and into capacitances ($J\ K^{-1}$).

Heat conduction within electronic microstructures is important if $R_{micro/nano}$ is comparable with or larger than the other resistances in the diagram. There are important examples where the micro/nano resistance is especially large, including silicon-on-insulator (SOI) circuits (e.g., Peters, 1993), in which a buried silicon dioxide layer strongly impedes conduction cooling of transistors (Goodson et al., 1995a). The buried silicon dioxide offers important benefits for both VLSI and power circuits because it diminishes the contribution of the substrate to the electrical capacitance of transistors and interconnects. SOI technology is also attractive for smart-power circuits, such as those containing the high-voltage transistor in Figure 4.19.3, because it allows near-perfect dielectric isolation of power and logic devices. The microscopic resistance is also important in cantilever sensors and actuators, such as the thermomechanical indentor shown in Figure 4.19.4. The cantilever structure provides a small area for conduction cooling, which augments the resistance. Another example is multilevel interconnect structures such as that shown in Figure 4.19.1, in which $R_{micro/nano}$ is augmented by the low-thermal-conductivity passive material (Hunter, 1997a, 1997b). The microscale resistance can be quite large in multilayer semiconducting structures, including those based on GaAs/AlGaAs or Si/Ge, due to the close proximity and large number of interfaces in these materials. This is particularly important for photonic devices (Chen, 1996) and high-frequency amplifiers, whose figures of merit include the maximum power-handling capability.

Figure 4.19.5 indicates that microscale conduction can be very important when interconnects and devices are subjected to brief, intense heating pulses. If a large quantity of heat is generated during a time less than the time constant of device or interconnect cooling to the substrate, then heat travels only micrometers or less from the surface of the chip during the heating pulse. The energy, therefore, must be absorbed by a very small volume, which dramatically increases the temperature rise. Such brief heating phenomena are often the result of electrical overstress (EOS), which is a major reliability concern for both high-power circuits and compact low-power circuits in computers. Electrostatic discharge (ESD) can induce a current pulse into the circuit terminals of a duration between about 1 and 150 ns and of a magnitude as high as 5 A (e.g., Amerasekera et al., 1992).

This section summarizes the theory and experiments that are used to study heat conduction in electronic microstructures. The following subsection describes simulation approaches for the temperature fields in microstructures, including the basics of solid-state theory. After that, theory and measurement approaches for the effective thermal conductivities of thin electronic films are presented. Finally, this section provides an overview of the available metrology for determining thermal conducitivities of thin films and temperature distributions in microstructures.

Simulation Hierarchy for Solid-Phase Heat Conduction

Heat conduction governs the temperature distributions in many micromachined electronic structures. For many macroscopic heat conduction problems, it is possible to obtain accurate predictions without knowing the mechanism responsible for heat transport. For isotropic materials, the heat flux is related to the temperature gradient by Fourier's law,

$$\mathbf{q}'' = -k\nabla T \tag{4.19.1}$$

where k is the thermal conductivity. Equation 4.19.1 can be used to accurately predict the temperature distribution in macroscopic structures without knowledge of what is carrying the heat, as long as the appropriate values of the thermal conductivity are taken from experimental data.

The value of the conductivity can be interpreted by considering the microscopic mechanism of transport, which depends on the medium. Molecular motion carries heat in gases, for example, for which the kinetic theory yields (e.g., Vincenti and Kruger, 1986; Rohsenow and Choi, 1961)

$$k = \frac{1}{3} C \mathrm{v} \Lambda \tag{4.19.2}$$

The conductivity is proportional to the heat capacity per unit volume C [J m^{-3} K^{-1}], the mean molecular speed v, and the mean free path Λ of molecules between consecutive collisions. In solid metals and many metallic alloys, heat conduction is dominated by the motion of electrons. In dielectric and semiconducting materials, the coupled vibrations of atoms are responsible for heat conduction. Quantum theory accounts for atomic vibrational energy in discrete units called phonons, which can be visualized as packets of energy traveling at speeds near that of sound in the material. Conduction in solids can be simulated using the kinetic theory of gases with expressions analogous to Equation 4.19.2 for electrons and phonons.

Accurate simulations of microstructures must consider the heat conduction mechanism. Figure 4.19.6 summarizes the hierarchy of simulation approaches for the semiconducting and dielectric regions in electronic microstructures. The regimes of applicability are distinguished using the mean free path Λ and the wavelength λ of phonons, which are the dominant heat carriers in these materials. In semiconducting crystals such as silicon, the phonon mean free path and wavelength near room temperature are approximately 100 and 1 nm, respectively. Many simulations can effectively use the continuum heat diffusion theory based on Fourier's law, Equation 4.19.1, if care is taken to use accurate values of the thermal conductivities and interface resistances of the constituent films. Phonon transport theory becomes important when the dimensions of a structure are comparable with the phonon mean free path. For this regime, solutions to the Boltzmann transport equation for phonons account for scattering on interfaces.

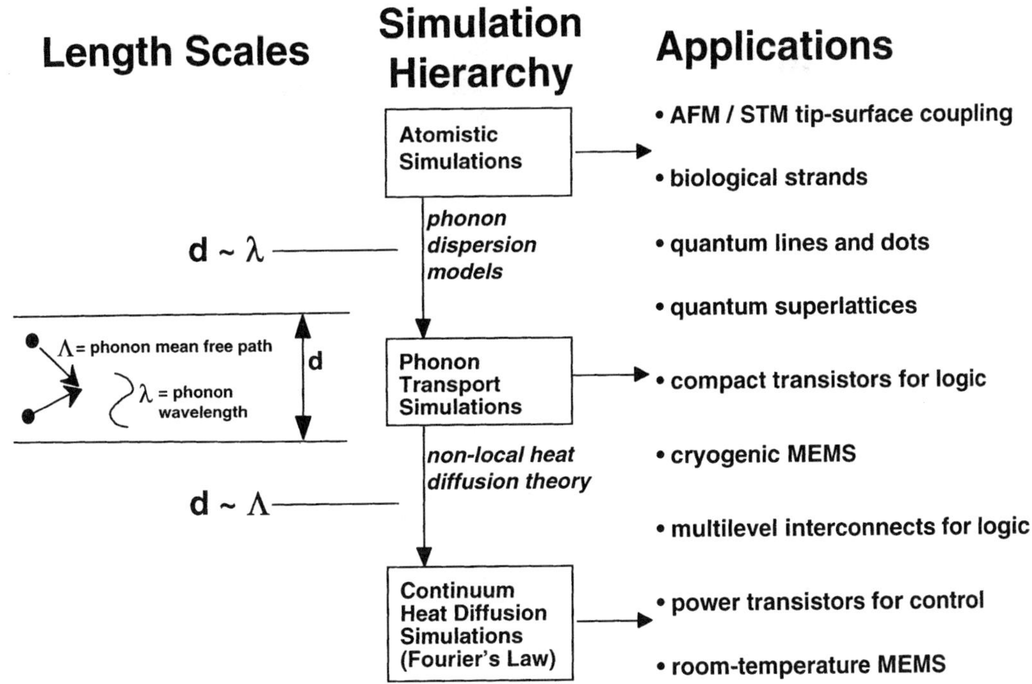

FIGURE 4.19.6 Hierarchy of thermal simulation methods for dielectric and semiconducting regions in electronic microstructures. Most of the heat generated by electrical currents in microstructures is released in semiconducting materials. Disordered dielectric materials, which serve for electrical isolation, can in many cases dominate the thermal resistance.

Nonlocal diffusion theory incorporates solutions to the Boltzmann equation for phonons into the classical heat diffusion equation through effective thermal conductivities. When device dimensions are comparable with the phonon wavelength, it becomes necessary to account for the interference and diffraction of atomic vibrational waves. Because the phonon wavelength is not much larger than the interatomic separation, rigorous treatment of conduction at these scales is inherently atomistic in nature. Atomistic calculations determine the impact of small dimensions on the dispersion relationships of atomic vibrational waves.

The following subsections present the basic behavior and thermal properties of phonons and electrons. Next, we present formal transport theory including the Boltzmann equation, which is used to derive constitutive laws for heat conduction in solids. This is followed by a more detailed discussion of the application of these modeling approaches to simulations of electronic microstructures. This section is not intended to provide a comprehensive overview of the thermal physics of solids, for which the reader should refer to one of several texts (e.g., Kittel, 1986; Ashcroft and Mermin, 1976; Ziman, 1960). Rather, this section summarizes the concepts and equations used most commonly in microscale thermal conduction simulations.

Basic Properties of Phonons

Atomistic simulations yield information about the behavior of lattice vibrational waves and the resulting thermal properties of crystalline solids. The primary goal is to determine the dispersion relationships for these waves, which relate the vibrational frequency to the wavelength and direction of propagation. In their simplest form, atomistic calculations consider a one-dimensional linear chain of atoms with mass M separated by the distance a, as depicted in Figure 4.19.7a. The ionic bonds between neighboring atoms are modeled by a linear spring with constant K. This is a large simplification of the situation in a real crystal, in which atoms are bonded with several neighbors in a complex three-dimensional lattice.

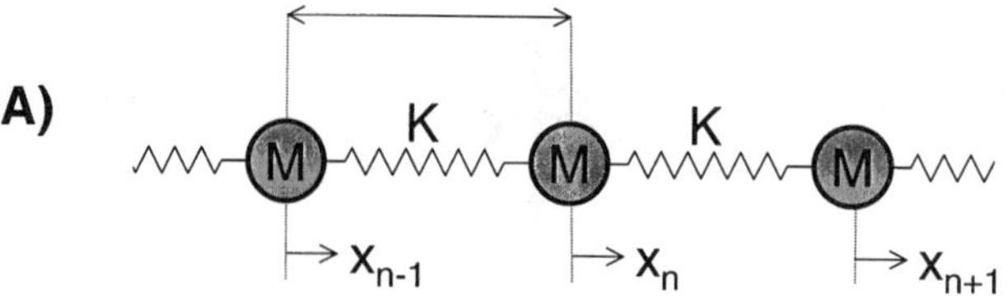

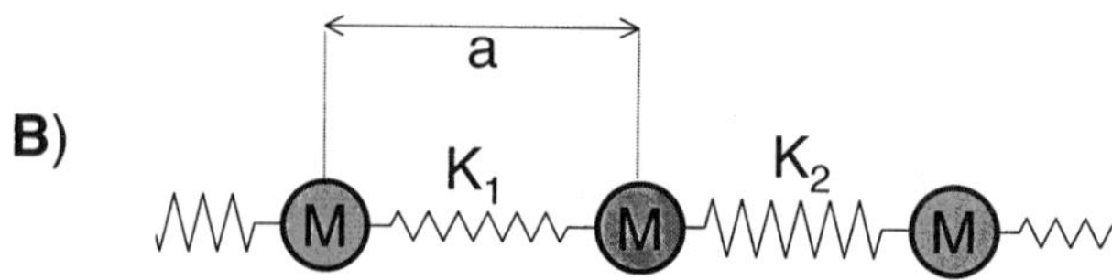

FIGURE 4.19.7 The simplest atomistic simulations involve a linear chain of atoms with mass M. The atoms are linked by springs with constant K as shown in (A). A slightly more detailed simulation uses two atoms within a unit cell, linked alternatively by spring constants K_1 and K_2 as shown in (B).

However, much can be learned from a linear chain of atoms. The acceleration of atom n is related to its position x_n through the forces exerted by the nearest-neighbor atoms,

$$M\frac{d^2x_n}{dt^2} = K(x_{n+1} - x_n) + K(x_{n-1} - x_n) \tag{4.19.3}$$

Equation 4.19.3 is solved by

$$x_n = A\exp(ink_w a - i\omega t) \tag{4.19.4}$$

where A is the amplitude of vibrations and a is the interatomic separation. Equation 4.19.4 is a propagating wave for the displacement of atoms from their equilibrium positions. The wave is described by the wavenumber $k_w = 2\pi/\lambda$, where λ is the wavelength, and angular frequency ω. The wave *dispersion relationship* satisfying Equation 4.19.3 is

$$\omega = \sqrt{\frac{2K}{M}}\left[1 - \cos(k_w a)\right]^{1/2} \tag{4.19.5}$$

Figure 4.19.8 plots this dispersion relationship. The *phase velocity* of any lattice wave, which describes the rate of propagation of a given length of atomic displacement, is ω/k_w. For the purposes of energy transport by a packet of such waves, clustered about a given wavelength, we are more interested in the *group velocity*

$$v = \frac{\partial\omega}{\partial k_w} \tag{4.19.6}$$

For wavelengths long compared to the interatomic separation a, the group velocity becomes independent of the wavenumber

$$v = \sqrt{\frac{K}{M}}\,a; \quad a << \lambda \tag{4.19.7}$$

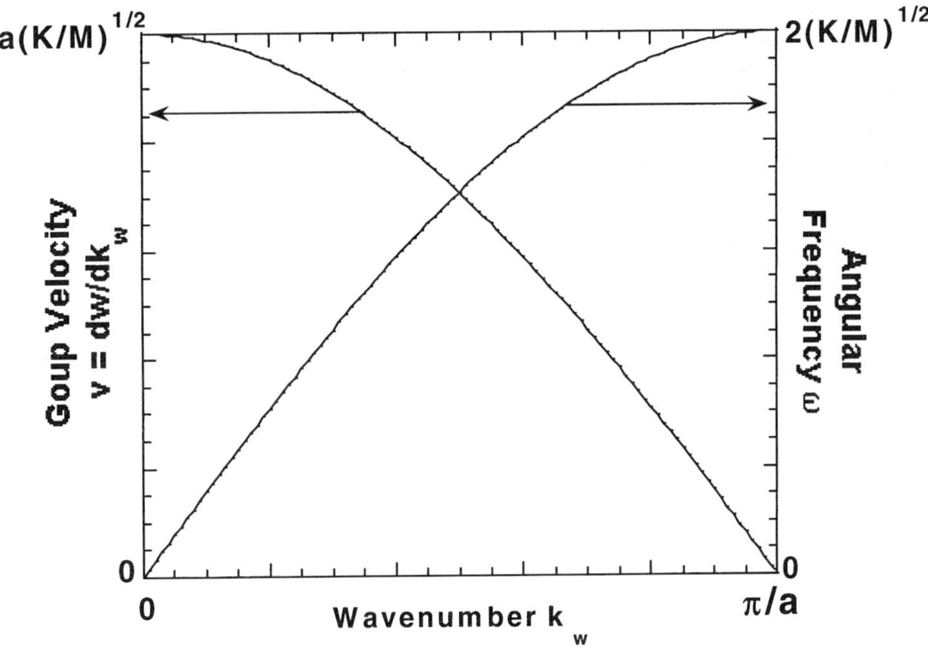

FIGURE 4.19.8 Dispersion relationship and group velocity for the linear chain of atoms in Figure 4.19.7a.

In a three-dimensional crystal, the group velocity for long wavelengths governs the propagation of sound. For wavelengths comparable with the interatomic spacing a, Equations 4.19.5 and 4.19.6 show that the group velocity approaches zero because the vibrations of neighboring atoms oppose each other and are out of phase by 180°. For wavelengths shorter than $2a$, it can be shown that the atomic displacements are redundant with those for a solution with wavelength longer than $2a$ (e.g., Kittel, 1986). The wavenumbers $k_w = \pm\pi/a$ therefore bound the useful dispersion relationship, which lies within the first *Brillouin zone*.

The next level of complexity for atomistic calculations connects atoms with springs of two different constants, K_1 and K_2. The atoms lie along a linear chain as in Figure 4.19.7b. This is a more realistic model of the situation within a three-dimensional crystal, in which the basic repeating unit, called the *unit cell*, often includes two or more atoms. In this case, the first Brillouin zone is bounded by $k_w = \pm\pi/2a$ and contains two solutions. The two frequencies and group velocities for the long wavelength limit ($k_w a << 1$) are

$$\omega = \sqrt{\frac{2(K_1 + K_2)}{M}}; \quad \nu = 0 \qquad \textit{optical branch} \tag{4.19.8}$$

$$\omega = \sqrt{\frac{(K_1 K_2)}{2M(K_1 + K_2)}}\, k_w a; \quad \nu = \sqrt{\frac{(K_1 K_2)}{2M(K_1 + K_2)}}\, a \qquad \textit{acoustic branch} \tag{4.19.9}$$

The vibrations with finite frequency in the long-wavelength limit have zero group velocity. This branch of solutions accounts for the motion of two atoms within a given unit cell with respect to each other, approximately 180° out of phase. Because these vibrations can be caused by oscillating electric fields they belong to the *optical branch* of solutions. Equation 4.19.9 presents a limit not dissimilar to that in Equation 4.19.5 and yields a finite group velocity. Because these vibrational waves carry sound in solids, they belong to the *acoustic* branch.

The situation in a three-dimensional crystal is far more complicated because of the varying atomic positions along different directions within the lattice. The situation is also complicated by the long-range interactions of atoms and the fact that the spring constants can be different for compression (longitudinal) and shear (transverse) waves. Figure 4.19.9 plots the dispersion relationship for silicon along a single crystallographic direction, which can be measured using neutron scattering. One conclusion that may be drawn from this plot is that optical phonons have relatively small group velocities, and therefore can be expected to contribute little to the conduction of heat according to Equation 4.19.2. Another conclusion is that a large fraction of the transverse acoustic branch in silicon also yields relatively small group velocities. The dispersion relationships can be interpreted using atomistic calculations like those presented here, which estimate spring constants through comparison with the data.

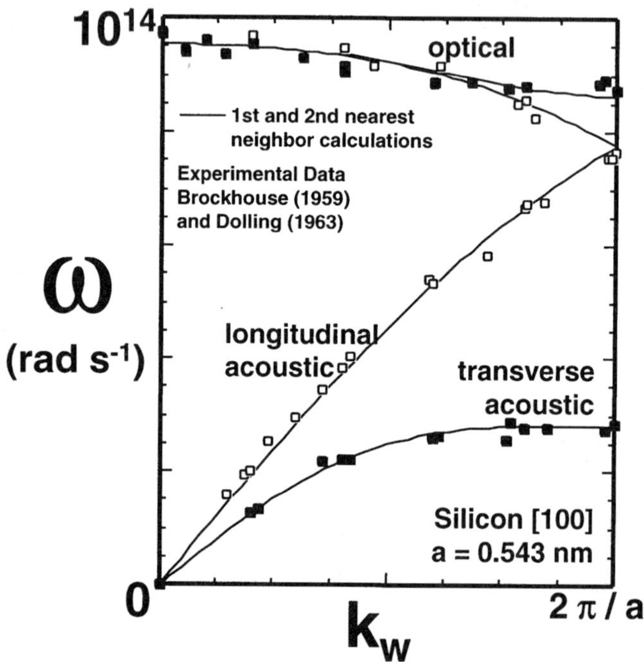

FIGURE 4.19.9 Dispersion relationship along the (100) direction in silicon. The atomistic calculations are compared with neutron-scattering data silicon. (Adapted from M. Asheghi, Stanford University.)

The atomistic simulation approach in Figure 4.19.5 is yielding helpful information about dispersion relationships for vibrational waves, which are needed for heat transport theory as will be described later. In addition, atomistic simulations can be used to directly predict rates of heat transfer in solids. The atomistic approach is influencing our understanding of phonon-interface interactions, for example, which have a large impact on thermal conduction. Atomistic simulations examined the role of inelastic phonon scattering on heat transport across diamond-metal interfaces (Stoner and Maris, 1993). Atomistic simulations were used to study phonon dispersion and modified anharmonic coupling in superlattices (e.g., Srivastava, 1994), which are important when the superlattice period is comparable with the wavelengths of a significant fraction of thermal phonons. Related work is approaching the dispersion relationship from the perspective of wave theory by predicting the impact of phonon interference and confinement in compact electronic structures. For most micromachined structures, phonon interference and modified dispersion phenomena are far less important than phonon scattering on imperfections and thin-film interfaces. These can be considered using the transport theory introduced later.

Everything discussed before this point about lattice vibrational waves can be interpreted using the classical laws of physics. The quantum mechanical behavior of the lattice is observed through the finite number of vibrational states available in a crystal, which are distinguished by their wavevector $\mathbf{k}_w$. The wavevector defines the wave direction as well as the wavelength, because its magnitude is the wavenumber $k_w = 2\pi/\lambda$. Another important implication of quantum theory is that each vibrational state contains energy in discrete increments according to

$$u = \left(n + \frac{1}{2}\right)\hbar\omega \tag{4.19.10}$$

where u is the total energy in the state, $\hbar = 1.05459 \times 10^{-34}$ Js is Planck's constant divided by 2π, and n is the occupation number. The occupation numbers of the vibrational states, therefore, dictate the total energy density in the solid. When the solid is in equilibrium at a given temperature, the average occupation number obeys the Planck distribution function

$$\langle n \rangle = \frac{1}{\exp\left(\dfrac{\hbar\omega}{k_B T}\right) - 1} \tag{4.19.11}$$

where $k_B = 1.308 \times 10^{-23}$ J K^{-1} is the Boltzmann constant. Equation 4.19.11 also governs the occupation of states available to electromagnetic radiation by its photon energy quanta and is the basis of the more common Planck distribution function used in thermal radiation theory. The total energy density per unit volume in the atomic lattice is the sum over all available states of the energy per state, Equation 4.19.10, with values of the occupation number evaluated for the temperature using Equation 4.19.11. This yields

$$U = \frac{1}{V} \sum_{polarizations} \sum_{\mathbf{k}_w} \left(\langle n \rangle_{\omega(k_w)} + \frac{1}{2}\right)\hbar\omega(\mathbf{k}_w) \cong \sum_{polarizations} \int_o^\infty \left(\langle n \rangle_\omega + \frac{1}{2}\right)\hbar\omega\, D(\omega) d\omega \tag{4.19.12}$$

where the first summation accounts for the three phonon polarizations (two transverse, one longitudinal) that can possess a given value of the wavevector $\mathbf{k}_w$. The second expression approximates the wavevector sum using an integral over angular frequency and the density of phonon states $D(\omega)$ per unit volume and frequency. The product $D(\omega)\, d\omega$ has units (m^{-3}) and is the number of states per unit volume with frequency between ω and $\omega + d\omega$. The total volume of the crystal is V.

For heat conduction calculations we are interested in the heat capacity per unit volume of the crystal, which plays a central role in the kinetic theory according to Equation 4.19.2. The specific heat per unit volume is

$$C = \frac{\partial U}{\partial T} = \sum_{polarizations} \int_o^\infty \frac{\partial \langle n \rangle}{\partial T} \hbar\omega\, D(\omega) d\omega \tag{4.19.13}$$

Equation 4.19.13 assumes that the dispersion relationship does not vary with temperature, which enforces a constant volume condition on this calculation of the heat capacity per unit volume. The density of states can be calculated from the detailed dispersion relationships in the crystal, such as that shown for silicon in Figure 4.19.9 (e.g., Kittel, 1986). However, many useful calculations can be performed using simple closed-form models.

The *Einstein model* assumes that all vibrations have the same angular frequency ω_E. Given the three polarization states (two transverse and one longitudinal) associated with each frequency, the density of states can then be written

$$D(\omega) = 3N\delta(\omega - \omega_E) \tag{4.19.14}$$

where N is the density per unit volume of unit cells in the atomic lattice. In the Einstein model, Equation 4.19.13 collapses to

$$C = 3Nk_B\left(\frac{\theta_E}{T}\right)^2 \frac{\exp(\theta_E/T)}{\left[\exp(\theta_E/T)-1\right]^2} \tag{4.19.15}$$

where the Einstein temperature is $\theta_E = \hbar\,\omega_E/k_B$. The high-temperature limit is

$$C = 3Nk_B;\ T >> \theta_E \tag{4.19.16}$$

which is consistent with experimental data for the volumetric heat capacity and the law of Dulong and Petit (e.g., Kittel, 1986). However, at low temperatures Equation 4.19.15 yields a specific heat that varies too rapidly with decreasing temperature and strongly underpredicts the experimental data. The problem is that many vibrational states exist with frequencies below ω_E and are more easily accessed and occupied at low temperatures.

The *Debye model* uses a more complicated expression for the density of states, which solves this problem. The model assumes an isotropic linear dispersion relationship, $\omega = \text{v}\, k_w$, where v is equal to both the group and phase velocities of the vibrational waves. The density of states with a given polarization is

$$D(\omega) = \frac{\omega^2}{2\pi^2 \text{v}^3};\ \omega \le \omega_D \tag{4.19.17}$$

$$D(\omega) = 0;\ \omega > \omega_D \tag{4.19.18}$$

The maximum frequency ω_D is dictated by the total number of states to be assigned within a unit volume. If the number density of lattice unit cells is N, then the cutoff frequency

$$\omega_D = \text{v}\sqrt[3]{6\pi^2 N} \tag{4.19.19}$$

sets the total number of states equal to the number of unit cells. Because there can be multiple atoms per unit cell, the cutoff frequency is often calculated using the total number of atoms per unit volume, N_a rather than N. This approach yields the correct total number of states in the crystal and is reasonably effective at predicting the measured heat capacity. There is a characteristic *Debye temperature* associated with this cutoff frequency,

$$\theta_D = \frac{\hbar\omega_D}{k_B} = \frac{\hbar \text{v}}{k_B}\sqrt[3]{6\pi^2 N_a} \tag{4.19.20}$$

Table 4.19.1 provides Debye temperatures, group velocities, unit-cell number densities, and atomic number densities for several solids. The heat capacity is calculated by combining Equations 4.19.13 and 4.19.17 to 4.19.20, which yield

$$C = 9N_a k_B\left(\frac{T}{\theta_D}\right)^3 \int_0^{\theta_D/T} \frac{x^4 \exp(x)}{\left[\exp(x)-1\right]^2}\,dx \tag{4.19.21}$$

where x serves as an integration variable here. Equation 4.19.21 is compared with experimental data for diamond in Figure 4.19.10. The high and low temperature limits of the heat capacity are

TABLE 4.19.1 Lattice Properties of Solids

Solid	Atomic Density N_a (10^{28} m^{-3})	Mass Density (10^3 kg m^{-3})	Longitudinal Phonon Velocity v (10^3 m s^{-1})	Transverse Phonon Velocity v (10^3 m s^{-1})	Debye Temperature θ_Δ (K)
Si	5.00	2.33	8.97	5.33	645
C (diamond)	17.6	3.51	17.5	12.8	2230
SiO_2	6.62	2.66	6.09	4.10	492
GaAs	—	5.32	5.24	2.48	345
Al	6.02	2.70	6.24	3.04	428
Cu	8.45	8.93	4.91	2.50	343
Au	5.90	19.3	3.39	1.29	165
Ag	5.85	10.5	3.78	1.74	225

Source: Data adapted from Swartz and Pohl, 1989 and Kittel, 1986.

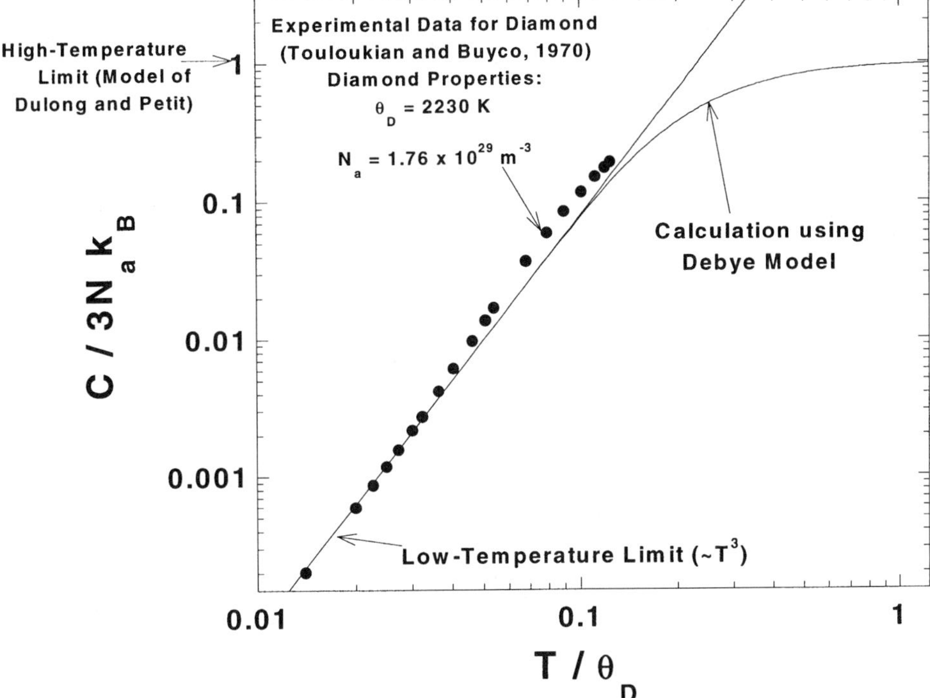

FIGURE 4.19.10 Volumetric heat capacity data for diamond compared with the predictions of the Debye model.

$$C \cong \frac{12\pi^4}{5} N_a k_B \left(\frac{T}{\theta_D}\right)^3; \quad T << \theta_D \tag{4.19.22}$$

$$C \cong 3N_a k_B; \quad T >> \theta_D \tag{4.19.23}$$

The high-temperature limit is consistent with the Einstein model and the law of Dulong and Petit.

The Debye model is somewhat less successful at predicting thermal conduction, because it does not properly distinguish between the group velocities of the acoustic and optical branches. One remedy for this problem is to assign the Debye cutoff based on the number density of unit cells, N, and to account for the optical phonons using the Einstein model with an appropriately averaged characteristic frequency.

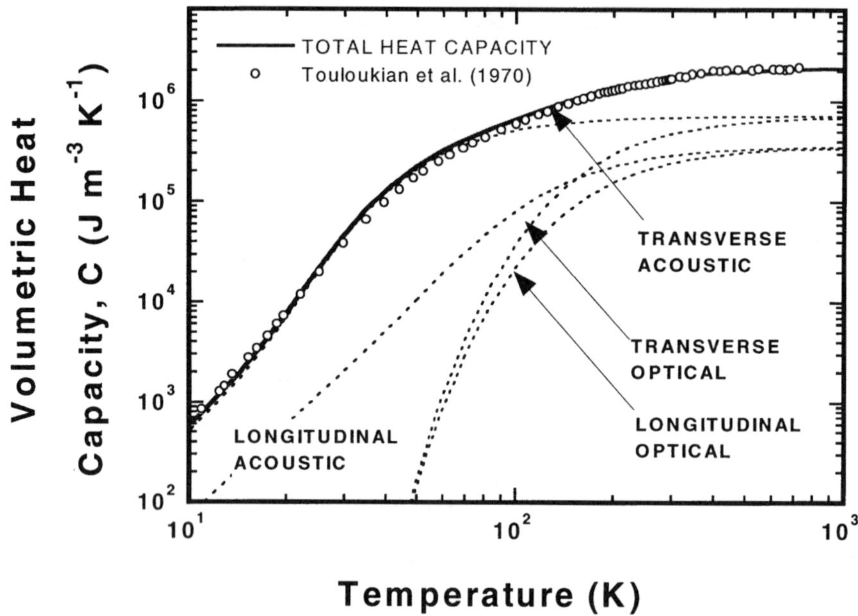

FIGURE 4.19.11 Contributions of the phonon branches to the volumetric heat capacity of silicon. (Adapted from M. Asheghi, Stanford University.)

Another approach is to calculate the density of states explicitly from the dispersion relationships, such that $D(\omega)$ varies continuously with the wavevector and is different for each branch. This approach yields the greatest precision in calculating the heat capacity. Figure 4.19.11 plots the result of this approach for silicon, showing the relative contributions of the branches to the specific heat.

Basic Properties of Electrons in Metals

Electron transport in solids is modeled by integrating the Schroedinger wave equation subjected to a periodic potential function representing the atomic lattice (e.g., Ashcroft and Mermin, 1976). The solutions yield the electronic band structure in the material, which describes the variation of the electron energy with the electron wavevector. The wavevector description of electrons recognizes that they behave as waves while traveling through the atomic lattice. The electronic band structure is analogous to the dispersion relationship for phonons, because it relates the energy and group velocity of electron waves to the direction of propagation and wavelength. A detailed discussion of band structure and electron properties is beyond the scope of this treatment and can be found in basic texts on the solid state.

Here, we are most concerned with electron conduction in metals, in which electrons behave much as free charged particles in a vacuum. In contrast to phonons, which can multiply occupy a given state, the Pauli exclusion principle restricts occupation of electron states to two, with opposite spin directions. At zero temperature the free electrons fill the states with the lowest energy in the system up to the *Fermi energy*

$$E_F = \frac{\hbar^2}{2m}\left(3\pi^2 N_e\right)^{2/3} \tag{4.19.24}$$

where N_e is the number density of free electrons in the metal and $m = 9.10956 \times 10^{-31}$ kg is the free electron mass. Because of the Pauli exclusion principle, only those electron states with energy close to the Fermi level can contribute significantly to heat conduction. Electrons occupying states with lower energies travel within the lattice but yield no net charge or heat current because the states with opposite direction are occupied as well. This makes the velocity of electrons near the Fermi level very important,

TABLE 4.19.2 Electron Properties in Common Metals

Metal	Electron Concentration N_e (10^{28} m^{-3})	Fermi Energy E_F (10^{-19} J)	Fermi Velocity v (10^6 m s^{-1})	Sommerfeld Parameter γ (J m^{-3} K^{-2})	Electrical Conductivity, 300K σ (10^5 Ω^{-1} m^{-1})	Thermal Conductivity, 300 K κ (W m^{-1} K^{-1})	Electron Mean Free Path, 300 K Λ (nm)
Rb	1.15	2.96	0.81	46.0	0.80	58	16
Ag	5.85	8.78	1.39	62.8	6.21	429	49
Au	5.90	8.83	1.39	71.4	4.55	317	32
Cu	8.45	11.2	1.57	97.5	5.88	401	26
In	11.49	13.8	1.74	322	1.14	82	1.5
Pb	13.20	15.0	1.82	653	0.48	35	0.3
Al	18.06	18.6	2.02	405	3.65	237	2.9

Note: The Fermi energy and velocity are calculated using equations in the text. The Sommerfeld parameter and the electrical and thermal conductivities are from experimental data. The mean free path is calculated using $k = CvL/3 = \gamma T v \Lambda/3$.

Source: Data derived from Kittel, 1986.

$$v_F = \left(\frac{2E_F}{m}\right)^{1/2} = \left(\frac{\hbar}{m}\right)\left(3\pi^2 N_e\right)^{1/3} \tag{4.19.25}$$

Table 4.19.2 provides the properties of electrons in a variety of metals. The volumetric heat capacity of electrons is calculated using

$$C = \int_0^\infty \left(E - E_F\right)\frac{df}{dT} D(E)dE \tag{4.19.26}$$

where the density of electron states is

$$D(E) = \frac{1}{2\pi^2}\left(\frac{2m}{\hbar^2}\right)^{3/2} E^{1/2} \tag{4.19.27}$$

Electrons follow the Fermi-Dirac distribution function

$$f = \frac{1}{\exp\left(\frac{E-\mu}{k_B T}\right)+1} \tag{4.19.28}$$

where the chemical potential μ is very nearly equal to the Fermi energy E_F at practical temperatures. The chemical potential is calculated in a manner that conserves the number density of electrons in the metal. This number conservation relates the chemical potential to the temperature according to

$$\frac{\mu}{E_F} = \left[1 - \frac{1}{3}\left(\frac{\pi k_B T}{2E_F}\right)^2\right] \tag{4.19.29}$$

At room temperature in most metals, the ratio $k_B T/E_F$ is much smaller than unity. The distribution function for electrons in aluminum near the Fermi energy is plotted using Equations 4.19.28 and 4.19.29 in Figure 4.19.12. As the temperature increases, electrons are excited to higher energy states within a band above and below the Fermi energy that is comparable with $k_B T$. Because the Fermi energy is much larger

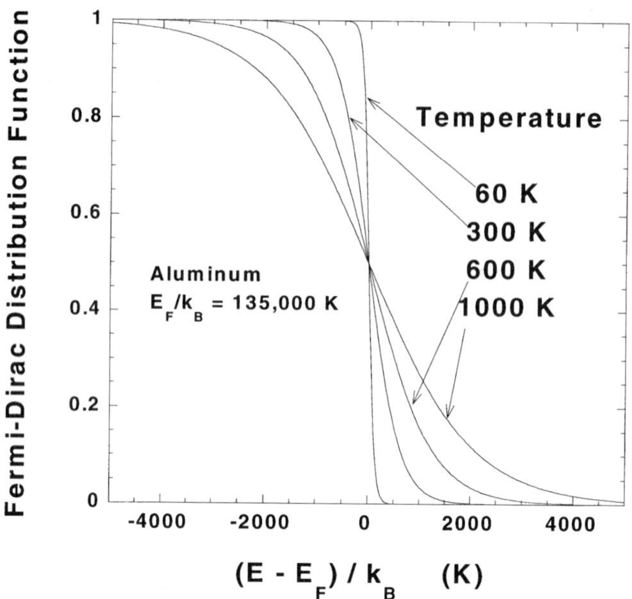

FIGURE 4.19.12 Fermi-Dirac distribution function for the occupation of electron states in aluminum, plotted using Equations 4.19.28 and 4.19.29.

than k_BT, only a small fraction of the electrons participate in thermal excitation and, therefore, contribute to the specific heat.

The volumetric heat capacity of electrons increases linearly with temperature in a metal. In a very qualitative interpretation, the number of electrons available for excitation is proportional to the width of the energy band k_BT. The energy difference they experience due to the excitation is comparable to k_BT. As a result, the total electron energy increases according to a term proportional to the square of the temperature, and the heat capacity, therefore, is linearly proportional to temperature. The exact expression follows from Equations 4.19.25 through 4.19.29:

$$C = \frac{\pi^2 k_B^2 NT}{2E_F} \equiv \gamma T \tag{4.19.30}$$

which defines the Sommerfeld parameter γ. Table 4.19.2 includes experimental values of the Sommerfeld parameter for a variety of metals, which agree qualitatively with the predictions of Equation 4.19.30 (Kittel, 1986).

Transport Theory for Phonons and Electrons

The preceding two subsections describe the basic thermal physics of the equilibrium states of lattice vibrational waves and electrons. Heat conduction is analyzed by considering the relative motion of these particles in conditions departing from equilibrium. Because a detailed simulation of the motion of these particles often is computationally infeasible, many practical calculations use distribution functions f. In this approach, f denotes the number of electrons or phonons in a given state denoted by wavevector $\mathbf{k}_w$ and position $\mathbf{r}$. The evolution of the distribution function with time and space is described using the Boltzmann equation

$$\frac{\partial f}{\partial t} + \mathbf{v} \cdot \nabla_r f + \frac{\partial \mathbf{k}_w}{\partial t} \cdot \nabla_k f = \left(\frac{\partial f}{\partial t}\right)_{Scat} \tag{4.19.31}$$

The gradient operators yield derivatives in physical space, described using the subscript **r**, and wavevector space, described using the subscript $\mathbf{k}_w$. The group velocity **v** of particles usually depends on the wavevector $\mathbf{k}_w$ and has magnitude v. The time rate of change of the distribution function (first term on left) is caused by spatial gradients in the distribution function (second term on left), external forces that change the wavevector (third term on left), and by scattering (first term on right). The third term on the left is only relevant for electrons, for which an electric field vector **E** induces changes in the electron state according to

$$\frac{\partial \mathbf{k}_w}{\partial t} = \frac{e}{\hbar}\mathbf{E} \tag{4.19.32}$$

where $e = 1.60219 \times 10^{-19}$ C is the proton charge. The scattering term on the right of Equation 4.19.31 can be expressed as an integral over initial and final states for the electrons or phonons involving the probabilities for the transitions and the distribution functions (e.g., Zimon, 1960). This integral is often modeled using the equilibrium distribution function and the relaxation time approximation,

$$\left(\frac{\partial f}{\partial t}\right)_{Scat} = \frac{f_{EQ} - f}{\tau} \equiv \frac{-f_{DE}}{\tau} \tag{4.19.33}$$

The equilibrium distribution function f_{EQ} at the local temperature is calculated using either Equation 4.19.11 for phonons or Equation 4.19.28 for electrons. It must be noted that for electrons, the value of f must lie between zero and unity. For phonons, the occupation number can be arbitrarily large and f is positive and unbounded. The distribution function $f_{DE} = f - f_{EQ}$ is the departure of the particle system from equilibrium. The relaxation time τ describes the rate at which the number of particles in the state would return to equilibrium in the absence of spatial gradients. Because there are a variety of scattering mechanisms, it is helpful to break the scattering rate τ^{-1} into components using *Matthiessen's rule*

$$\tau^{-1} = \sum_j \tau_j^{-1} \tag{4.19.34}$$

The sum in Equation 4.19.34 is over all of the significant scattering mechanisms. For phonons, the summation accounts for phonon-phonon scattering, phonon scattering on imperfections or impurities, and phonon-electron scattering (Ziman, 1960). For electrons, the summation is dominated by scattering on phonons and imperfections. Equation 4.19.34 can be augmented to account for scattering on thin-film and grain boundaries using $\tau \sim d/\mathrm{v}$, where d is the characteristic film or grain size and v is the velocity magnitude. Subsequent sections of this review will show that more detailed models are available for handling scattering due to film and superlattice interfaces and grain boundaries. For superconducting materials, Equation 4.19.34 can be augmented to account for phonon scattering with the fraction of the electron gas remaining in the normal state at a given temperature (Uher, 1990).

The Boltzmann equation can be integrated most simply for the steady state and when spatial derivatives of the departure from equilibrium, f_{DE}, are negligible. The second condition is satisfied when the boundaries do not significantly perturb the distribution function, which essentially requires that the medium is large compared to the particle free path. Under these circumstances, we may write

$$\mathbf{v} \cdot \nabla_r f \cong \mathbf{v} \cdot \nabla_r T \frac{\partial f_{EQ}}{\partial T} \tag{4.19.35}$$

and

$$\frac{\partial \mathbf{k}_{\mathrm{w}}}{\partial t} \cdot \nabla_{\mathrm{k}} f \cong -e\mathbf{v} \cdot \mathbf{E} \frac{\partial f_{EQ}}{\partial E} \tag{4.19.36}$$

where Equation 4.19.36 is relevant only for electrons. Under these approximations, the Boltzmann equation reduces to

$$f_{DE} = \mathbf{v} \cdot \nabla_{\mathrm{r}} T \frac{\partial f_{EQ}}{\partial T} \tau - e\mathbf{v} \cdot \mathbf{E} \frac{\partial f_{EQ}}{\partial E} \tau \tag{4.19.37}$$

For electrons, the electrical current density is calculated using

$$\mathbf{j} = \frac{1}{V} \sum_{\text{all states in } V} -e\mathbf{v}Ef_{DE} \tag{4.19.38}$$

For electrons and phonons, the heat flux vector is

$$\mathbf{q}'' = \frac{1}{V} \sum_{\text{all states in } V} \mathbf{v}Ef_{DE} \tag{4.19.39}$$

where E is the particle energy at a given state. The summations over all available states are usually performed using integrals in wavevector space and an appropriate density of particle states, as discussed in the two preceding subsections. For the electron system, Equations 4.19.38 and 4.19.39 yield the Onsager relations governing thermoelectricity in metals (Ziman, 1960; Rohsenow and Choi, 1961), with

$$\mathbf{j} = L_{EE}\mathbf{E} + L_{ET}\nabla T \tag{4.19.40}$$

$$\mathbf{q}'' = L_{TE}\mathbf{E} + L_{TT}\nabla T \tag{4.19.41}$$

A helpful feature of electron transport in metals, as was discussed above, is that transport is dominated by electrons with energy very nearly equal to the Fermi energy. As a result, the coefficients in Equations 4.19.40 and 4.19.41 are dominated by the properties of electrons near the Fermi energy. The electrical conductivity is calculated in the absence of temperature gradients,

$$\sigma = \mathbf{j}/\mathbf{E} = L_{EE} = N_e e^2 \tau / m^* \tag{4.19.42}$$

The thermoelectric power is $P = -L_{ET}/L_{EE}$ and the Peltier coefficient is L_{TE}/L_{EE}. Although a detailed discussion of thermoelectricity is beyond the scope of this treatment, the reader is referred to a helpful text on the subject (Pollock, 1991).

The thermal conductivity of a metal is calculated for the situation when no electrical currents are allowed by setting Equation 4.19.40 to zero. This yields

$$k = -\left(L_{TT} - \frac{L_{TE}L_{ET}}{L_{EE}} \right) \cong \frac{1}{3} C v_F^2 \tau = \frac{1}{3} C v_F \Lambda \tag{4.19.43}$$

where v_F is the Fermi velocity, C is the heat capacity per unit volume of electrons (excluding the lattice contribution), and $\Lambda = v_F\tau$ is the electron mean free path. A direct consequence of the Onsager relations for metals, when interpreted with the help of kinetic theory and Equation 4.19.37, is that the ratio of the thermal and electrical conductivities can be simply related if the relaxation times for both processes are assumed to be identical. The resulting relationship is independent of material and called the Wiedemann-Franz-Lorenz (WFL) law (e.g., Kittel, 1986)

$$\frac{k}{\sigma} = L_0 T \tag{4.19.44}$$

where the Lorenz number is

$$L_0 = \frac{\pi^2 k_B^2}{3e^2} = 2.45 \times 10^{-8}\ \mathrm{W\,\Omega\,K^{-2}} \tag{4.19.45}$$

At room temperature in nearly pure aluminum, copper, gold, and tungsten, the Lorenz number has been observed experimentally to be 2.20, 2.23, 2.35, and 3.04 × 10^{-8} W Ω K^{-2}, respectively. The WFL law is theoretically sound at temperatures above about 50% of the Debye temperature in metals. At lower temperatures, scattering on phonons more strongly reduces the electron relaxation time for charge transport than it reduces the electron relaxation time for heat transport, such that the Lorenz number decreases compared to its room-temperature value. This problem disappears for films with a large defect density, films that are very thin and, in general, below about 10 K, where phonon scattering is overwhelmed by scattering on static imperfections and interfaces. These scattering mechanisms preserve the equality of the relaxation times for charge and heat transport. Because the electrical conductivity is usually more easily measured than the thermal conductivity, the WFL law is very powerful for studying thermal conduction by metal regions in microstructures.

The thermal conductivity by phonons, which dominates in dielectric and semiconducting materials, is calculated using Equations 4.19.37 and 4.19.39. The term accounting for the electric field is not used here. Phonons over a broad range of frequencies can contribute to heat conduction, such that the final expression for the thermal conductivity involves an integral over the dimensionless phonon frequency (e.g., Berman, 1976)

$$k = \sum_{polarizations} \frac{1}{3} \int_0^{\theta_D/T} C_x v^2 \tau dx \tag{4.19.46}$$

where $x = \hbar\omega/k_B T$ is the dimensionless angular phonon frequency and C_x is a frequency-dependent weighting function for the heat capacity per unit volume. When Debye theory is used and the temperature is substantially below the Debye temperature in the solid, the weight contributed by the specific heat function is strongest for phonons with energy near 4 k_BT. More detailed calculations can account for phonon dispersion through more detailed models for the frequency dependence of C_x and v. The summation is over the two transverse and the single longitudinal phonon polarizations, for which the frequency-dependent specific heat function and Debye temperatures θ_D can be calculated from the properties of shear and longitudinal acoustic waves, respectively. Equation 4.19.46 neglects the directional dependence of the acoustic properties in a solid.

For highly disordered materials, a large number of the phonons carrying heat near room temperature are not well defined because of their short relaxation times. For this reason, the phonon model is not well suited for linking diffusion theory with atomistic calculations, as depicted in Figure 4.19.6. At temperatures above the thermal-conductivity plateau region, several heat conduction mechanisms have been proposed to be active in addition to that associated with the propagating modes accommodated by Equation 4.19.46. Some studies considered the hopping of localized vibrational excitations, which occurs via anharmonic interactions with phonons. Other research focused on heat transport by diffusive as opposed to localized vibrational excitations. A recent model considered the random walk of thermal energy between localized oscillators, a mechanism that may be considered as equivalent to diffusive transport. This model has been particularly useful for predicting the minimum thermal conductivity of highly disordered crystals, and yields a reasonable estimate for the thermal conductivity (Cahill et al., 1992)

$$k_{\min} = \left(\frac{\pi}{6}\right)^{1/3} k_B N_a^{2/3} \sum_{\text{polarizations}} v \left(\frac{T}{\theta_D}\right)^2 \int_0^{\theta_D/T} \frac{x^3 \exp(x)}{\left[\exp(x)-1\right]^2} dx \tag{4.19.47}$$

Equation 4.19.47 yields predictions of 1.04 and 0.99 W m^{-1} K^{-1} for amorphous silicon dioxide and silicon, which are both to be compared with the experimental data of 1.4 and 1.05 W m^{-1} K^{-1} for those materials in bulk form.

Moments of the Boltzmann Equations

The moments of the Boltzmann equation are especially relevant for simulating semiconductor devices. Figure 4.19.13 shows the interaction of energy carriers in semiconductor devices. Energy is generally introduced into the material by an electric field, which accelerates electrons. In semiconductors, charge transport must be interpreted by considering two distinct electron energy bands. At zero temperature, electrons fully occupy the *valence* band and are excited across the band energy gap and into the *conduction* band at finite temperature. The maximum energy in the valence band and the minimum energy in the conduction band are separated by the energy gap, E_g, which at room temperature is about 1.1 and 1.4 eV in silicon and gallium arsenide, respectively. The presence of electrons in the conduction band contributes free electrons to the system, and their absence from the valence band yield *holes* in the occupation of the available states. The net motion of electrons in the valence band can be modeled using positively charged holes (e.g., Ashcroft and Mermin, 1976). The electrons and holes gain energy from the electric field, scatter with the atomic lattice, and generate phonons. The recombination and generation of electron and hole pairs releases and absorbs energy in an amount comparable to the gap energy of the semiconductor. The energy can be released as electromagnetic radiation, whose energy quanta are photons, which is common in semiconductor lasers.

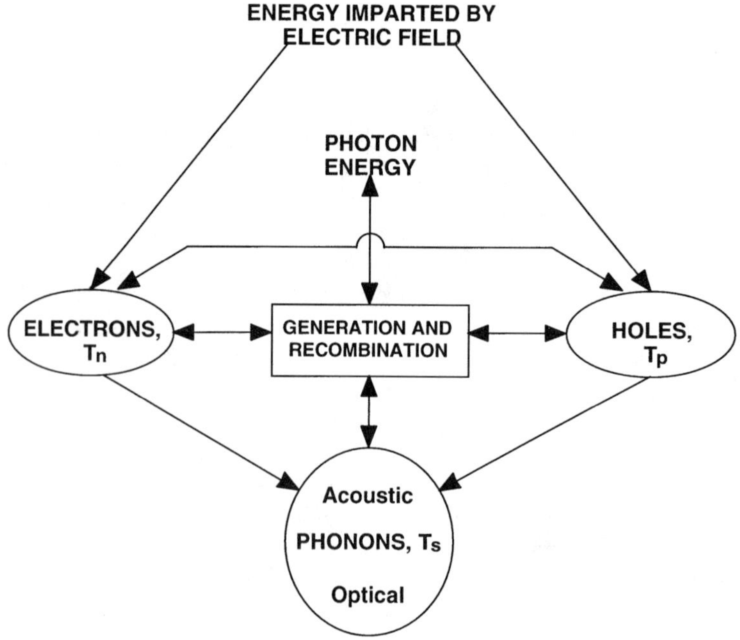

FIGURE 4.19.13 Energy carriers and exchange mechanisms in semiconductor devices. (From Goodson, K.E. et al., *Microscale Energy Transport,* Taylor & Francis, New York, 1998. With permission.)

The distinction between the acoustic and optical phonon branches lies mainly in their group velocities. As discussed previously, the group velocity of acoustic phonons is comparable with the speed of sound and much larger than that of optical phonons. Although both types of phonons scatter with electrons and holes, only acoustic phonons contribute strongly to thermal conduction. Acoustic phonons, therefore, are essential for removing the heat generated during the operation of semiconductor devices.

A detailed simulation of a semiconductor device would solve Boltzmann equations for the electron, hole, and phonon distribution functions. This approach maintains a complete description of the distribution functions of these carriers and their evolution in time and space. A more practical approach is to solve conservation equations based on the Boltzmann equation. These are moments of the transport equations that enforce conservation of charge, momentum, and energy without requiring detailed knowledge of the distribution functions. It is often assumed that electrons and holes each satisfy equilibrium distribution functions, but with temperatures that are not necessarily equal to that of the phonons. This accounts for the severe departure from equilibrium that can be induced among the various energy carriers in compact semiconductor devices. For the case of electrons, these moments are often called the hydrodynamic equations (e.g., Bloetekjaer, 1970; Lai and Majumdar, 1995)

$$\frac{\partial n}{\partial t} + \nabla \cdot (n\mathbf{v}_{\mathrm{n}}) = r_g - r_r \tag{4.19.48}$$

$$\frac{\partial \mathbf{p}_{\mathrm{n}}}{\partial t} + \nabla \cdot (\mathbf{v}_{\mathrm{n}}\mathbf{p}_{\mathrm{n}}) = -en\mathbf{E} - \nabla(nk_B T_n) - \frac{nm_n^*\mathbf{v}_{\mathrm{n}}}{\tau_{n,M}} \tag{4.19.49}$$

$$\frac{\partial W_n}{\partial t} + \nabla \cdot (\mathbf{v}_{\mathrm{n}}W_n) = -en\mathbf{v}_{\mathrm{n}} \cdot \mathbf{E} - \nabla \cdot (\mathbf{v}_{\mathrm{n}}nk_B T_n) + \nabla \cdot (k_n \nabla T_n) - \frac{\left(W_n - \frac{3}{2}nk_B T_s\right)}{\tau_{n-s,E}} - r_r\frac{W_n}{n} - r_g E_{imp} \tag{4.19.50}$$

Equation 4.19.48 is the number conservation equation for electrons in the semiconducting material, whose number density per unit volume is n. The electron temperature is T_n and the phonon temperature is T_s. An analogous set of equations can be derived for holes. The average electron velocity vector $\mathbf{v}_n$ is related to the electron current density through $\mathbf{j}_n = -en\,\mathbf{v}_n$. Equations 4.19.49 and 4.19.50 are momentum and energy balances in which the electron momentum and energy densities are $\mathbf{p}_n = n\,m_n^*\,\mathbf{v}_{\mathrm{n}}$ and W_n, respectively. The first two terms on the right of the momentum balance in Equation 4.19.49 account for acceleration due to an applied field and a net accumulation of electrons of higher momentum per unit volume due to a gradient in the electron temperature. The final term on the right accounts for momentum loss due to scattering processes and is governed by the electron momentum relaxation time, $\tau_{n,M}$. Terms with similar explanations on the right of the energy balance in Equation 4.19.50 are augmented by an energy diffusion term proportional to the electron conductivity, k_n, which can be approximately calculated for the semiconductor from its electrical conductivity using the Wiedemann-Franz-Lorenz law. The energy balance also accounts for the net loss of energy due to electron-hole recombination at the rate r_r and impact ionization at the rate r_g. Impact ionization results from electron-ion collisions, which lift an electron to a higher energy state. Each ionization event absorbs energy from the electron gas in the amount E_{imp}. Further expansion of the recombination terms would distinguish between Auger processes, which transmit energy to the electron gas, and processes involving phonon and photon emission (e.g., Sze, 1981).

A large electric field accelerates electrons into states of much higher energy than are occupied in equilibrium at the lattice temperature. Although much of this excess energy is carried by electrons traveling in the direction of the electric field and is described by the average electron velocity vector $\mathbf{v}_n$, electron-electron scattering causes a significant fraction to be distributed without order among the electron wavevectors. This is reflected by the definition for the electron energy density,

$$W_n = n\left(\frac{3}{2}k_B T + \frac{1}{2}m_n^*\left|\mathbf{v}_\mathbf{n}\right|^2\right) \tag{4.19.51}$$

The first term on the right accounts for electron kinetic energy that is distributed among electron states according to a Maxwellian distribution at the temperature T_n. The second term on the right accounts for excess energy due to net motion of the carriers in a given direction. The electron temperature in the first term is related to the quasi Fermi level for electrons and the electron concentration by means of Equation 4.19.6. The hole temperature, T_p, similarly quantifies a directionally averaged departure of the holes from equilibrium at the lattice temperature. The electron and hole temperatures strongly influence the rate of energy transfer to phonons and the phonon thermal conductivity.

Since phonons can be created and destroyed, their number is not conserved. Because they can exchange momentum with the crystal lattice through Umklapp scattering processes (e.g., Kittel, 1986), a momentum conservation equation is of little use. Only the energy moment of the phonon Boltzmann equation is relevant. For example, Sverdrup et al. (1999) have used the method of discrete ordinates to integrate Equation 4.19.31 for semiconductor devices to determine the impact of interface scattering and small dimensions of the domain of lattice heating compared to the mean free path. However, for practical simulations it is more feasible to integrate the classical heat diffusion equation,

$$C_S \frac{\partial T_s}{\partial t} = \nabla \cdot \left(k_s \nabla T_s\right) + q''' \tag{4.19.52}$$

where k_s is the phonon thermal conductivity, C_s is the phonon heat capacity per unit volume, and the heat generation term is given by

$$q''' = \frac{\left(W_n - \frac{3}{2}nk_B T_s\right)}{\tau_{n-s,E}} + \frac{\left(W_p - \frac{3}{2}pk_B T_s\right)}{\tau_{p-s,E}} + r_r\left(\frac{W_n}{n} + \frac{W_p}{p} + E_g\right) \tag{4.19.53}$$

The first and second terms in Equation 4.19.53 account for heat generation due to scattering with electrons and holes. The second and third terms couple Equation 4.19.52 with Equation 4.19.50 and an analogous equation, not provided here, for the hole energy density W_p. The final term on the right accounts for phonon generation due to the recombination of carriers. This term needs to be reduced if Auger or radiative processes are considered explicitly, because these processes transfer energy to electrons and photons, respectively, rather than to phonons.

The solution of multiple moments of the Boltzmann equations has found broad application for bipolar and field-effect transistors. The solution of the charge and energy moments of the electron Boltzmann equation together with the energy moment of the phonon Boltzmann equation is available through commercial software, such as PISCES-2ET (e.g., Yu et al., 1994; Shur, 1990). This simulation approach has been applied to study ESD failure in silicon field-effect transistors (Beebe et al., 1994) and the importance of self-heating effects in bipolar and SOI field-effect transistors (e.g., Apanovich et al., 1994, 1995). The analysis of Apanovich et al. (1995) is distinguished by its detailed treatment of heat generation due to electron-hole recombination processes, which are most important for bipolar transistors. These authors compared their predictions for the breakdown collector-base voltage in a submicrometer bipolar transistor with predictions using drift-diffusion theory (e.g., Sze, 1981). While the current-voltage predictions using the two different techniques agree well for low values of the collector-emitter voltage, which corresponds to low electric fields, the breakdown collector-emitter voltage is underpredicted by about 40% using drift-diffusion theory. This is attributed to the nonlocal nature of impact ionization, which is better simulated by independently balancing the electron and phonon energies. Apanovich et al. (1995) found that although self-heating of the transistor was important, temperature gradients *within* the transistor were not very important. The temperature rise was dominated by the thermal resistances that were assumed to prevail between the transistor and the environment, which are analogous to the macroscopic resistances in Figure 4.19.5. This indicates that relatively simple simulations using the heat diffusion equation and an isothermal lattice in the channel may be appropriate when coupled with more rigorous simulations of energy generation and transport by the free charge carriers.

Diffusion Theory

Most simulations of electronic microstructures use heat-diffusion theory, which is based on the Fourier heat-conduction law, together with measured or predicted effective thermal conductivity values for constituent films. The electron system is simulated using drift diffusion theory, in which the local electron and hole current densities are dictated by gradients in the electrostatic potential, the number density, and the temperature. The electrostatic potential is calculated using the Poisson equation. Under these circumstances the heat conduction problem collapses to

$$C\frac{\partial T}{\partial t} = \nabla \cdot (k \nabla T) + q''' \tag{4.19.54}$$

where there is now a single temperature and conductivity for the solid and the heat generation rate is (Wachutka, 1994)

$$q''' = \left[\frac{|\mathbf{j}_n|^2}{en\mu_n} + \frac{|\mathbf{j}_p|^2}{ep\mu_p}\right] + (r_r - r_g)\left[E_{Fn} - E_{Fp} + eT(P_p - P_n)\right] - T(\mathbf{j}_n \cdot \nabla P_n + \mathbf{j}_p \cdot \nabla P_p) - S_R \tag{4.19.55}$$

Equation 4.19.55 is strictly valid only in the steady state, but gives perspective on all of the heat generation mechanisms that can occur in a semiconducting material. The first term on the right accounts for resistive heating due to electron and hole scattering processes and dominates, for example, in field-effect transistors. This is the Joule heating component with separate contributions from electrons and hole current densities, $\mathbf{j}_n$ and $\mathbf{j}_p$, respectively. The electrical mobility of each carrier type is denoted here using μ with an appropriate subscript. The second term on the right accounts for heat released due to net recombination of electron-hole pairs and is most important for diodes and bipolar transistors. The largest component of heat release is given by the difference in the Fermi energies of the two carrier types, $E_{Fn} - E_{Fp}$. These Fermi energies are analogous to that previously described for electrons in metals. The third term on the right is the rate of Thomson heating, which results from current flow along a gradient in the thermoelectric power P of electrons or holes. Thomson heating is most important near gradients in the impurity concentration, which influence P_n and P_p. A large fraction of practical device simulations neglect the Thomson component and some can also avoid calculation of the impact of carrier recombination.

Drift-diffusion simulations have been applied for detailed studies of transistor breakdown (e.g., Amerasekera et al., 1993) and is also available commercially (e.g., the TRENDY software package, see Wolbert et al., 1994). The results of Apanovich et al. (1995) place the drift diffusion theory in question for highly nonequilibrium studies. However, since the Boltzmann equation moments and the drift diffusion models are both phenomenological, it can be appropriate to tailor critical parameters, such as the high field mobility and the impact ionization rate, to account for the nonequilibrium conditions. The weakness of existing transistor simulation capabilities is the large number of unknown microscopic parameters, such as the distinct relaxation times for electron momentum and energy transfer to the lattice. Although a solution using more governing equations may be more rigorous in principle, the practical relevance is often strongly degraded by the uncertainty of the many phenomenological parameters.

Lumped heater simulations treat some element of an electronic microstructure as isothermal in solutions to Equation 4.19.54 (e.g., Goodson et al., 1995a). This approach can benefit from the wealth of analytical techniques available for solution of the macroscopic heat diffusion equation, Equation 4.19.55, which are documented with sufficient detail elsewhere in this handbook. Figure 4.19.14 provides an example application of this calculation approach for a silicon-on-insulator transistor, for which conduction cooling can be modeled as occurring along extended one-dimensional regions. In this case, the heat diffusion equation collapses to a set of simultaneous equations for extended surfaces with linked end conditions,

$$\frac{d^2T}{dx^2} - \frac{T - T_{sub}}{L_H^2} = 0 \tag{4.19.56}$$

where the healing length along the silicon, aluminum, and polysilicon extended surfaces, or *fins*, is

$$L_H \cong \sqrt{\frac{k_{fin}\, d_{fin}\, d_{oxide}}{k_{oxide}}} \tag{4.19.57}$$

An exact form of Equation 4.19.57 uses a shape factor for the spreading of heat in the oxide underlying each extended surface. Figure 4.19.14 depicts a model geometry showing the extended surfaces representing the source and drain regions, the gate, and the metallization. The temperature rise in the channel was calculated using the simultaneous solution of Equations 4.19.56 written for the three types of extended surfaces. The resulting temperature rise can be used to modify parameters in current-voltage models for the transistors (e.g., Su et al., 1994). Lumped device models are very useful for circuit-level simulations. Their accuracy depends very sensitively, however, on the accuracy with which thin-film thermal conductivities can be determined. The determination of thin-film thermal conductivities is the subject of the next subsection.

Thermal Conduction Properties of Electronic Films

Most electronic microstructures consist of lithographically patterned films of thickness much smaller than their in-plane dimensions. The temperature distributions in these structures can be predicted with reasonable accuracy using the classical heat diffusion equation if care is taken to determine the thermal conductivities of the films and the thermal resistances at their mutual interfaces. For this reason, the problem of simulating temperature fields in electronic microstructures can, in many cases, be reduced to that of determining the thermal conductivities and interface resistances of their constituent semiconducting, electrically insulating, and metal films.

Although many electronic materials can be produced and thermally characterized in bulk form, in thin-film form their thermal properties can be dramatically different. This is illustrated in Figure 4.19.15 for nearly pure, single-crystal silicon films, in which phonon-interface scattering reduces the in-plane conductivity. While the data in Figure 4.19.15 show the most significant reduction at low temperatures, more recent data illustrated that phonon-interface scattering causes a conductivity reduction by approximately a factor of two for layers of thickness near 100 nm at room temperature and above (Ju and Goodson, 1999a). The thermal conductivity is not a size-independent parameter for films of thickness comparable with the phonon mean free path, in which phonon-interface scattering is important. This problem is discussed in greater detail with the definition of effective thin-film conductivities in the following subsection.

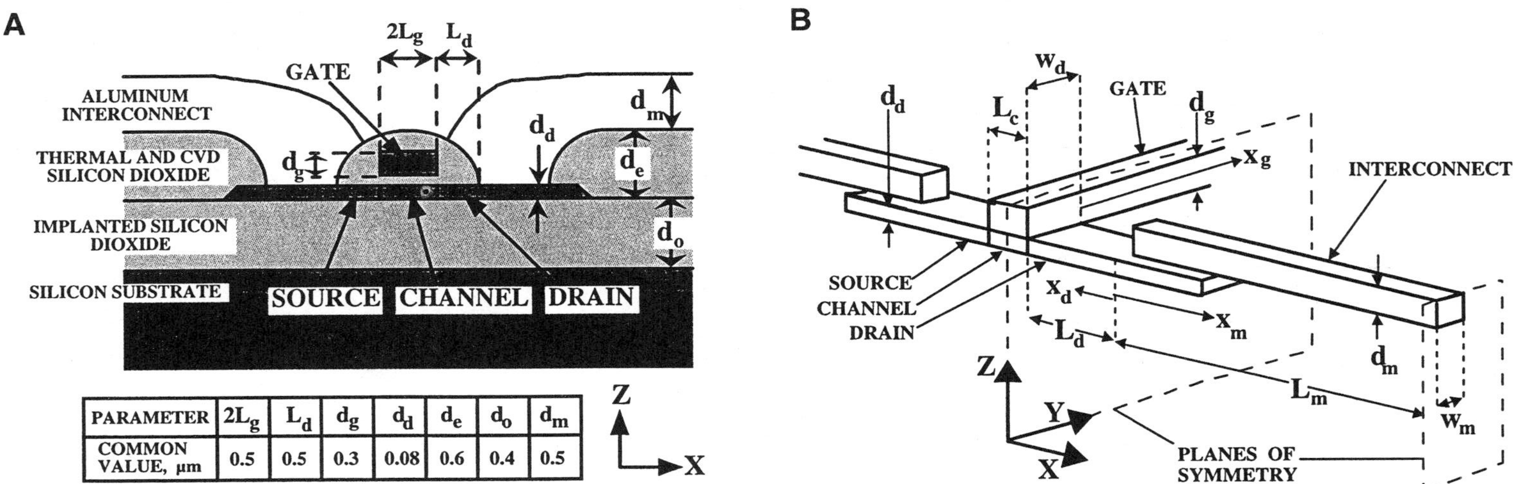

PARAMETER	$2L_g$	L_d	d_g	d_d	d_e	d_o	d_m
COMMON VALUE, μm	0.5	0.5	0.3	0.08	0.6	0.4	0.5

FIGURE 4.19.14 Cross-sectional schematic (a) and model geometry (b) for a silicon-on-insulator transistor. The temperature distribution in calculated through the simultaneous solution of several one-dimensional differential equations for extended surfaces. (From Goodson, K.E. et al., *ASME J. Heat Transfer,* 117, 574, 1995a. With permission.)

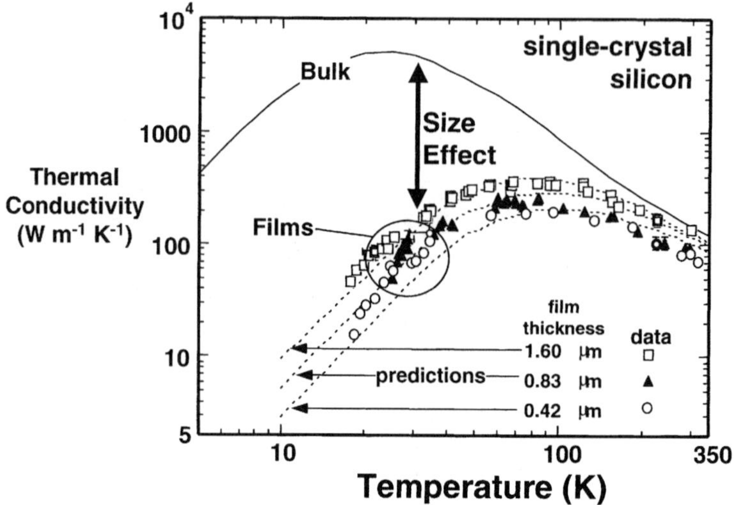

FIGURE 4.19.15 Temperature-dependent thermal conductivity data along silicon films. (Adapted from Asheghi, M. et al., *ASME J. Heat Transfer,* 120, 30, 1998.) (The predictions are based on the thermal conductivity integral for phonons, Equation 4.19.46, together with a solution to the Boltzmann equation accounting for phonon-interface scattering, Equation 4.19.60.)

The discrepancy between bulk and film properties can be large at room temperature for films whose fabrication process introduces imperfections, impurities, or structural modifications. The local thermal conductivity within the film can be strongly nonhomogeneous and anisotropic, even for materials that are isotropic in bulk form. This is illustrated for fully amorphous polyimide and polycrystalline diamond films in Figure 4.19.16. The orientation of grain boundaries in the polycrystalline film favors out-of-plane conduction, and the variation of grain size within the layer yields nonhomogeneity. For the organic

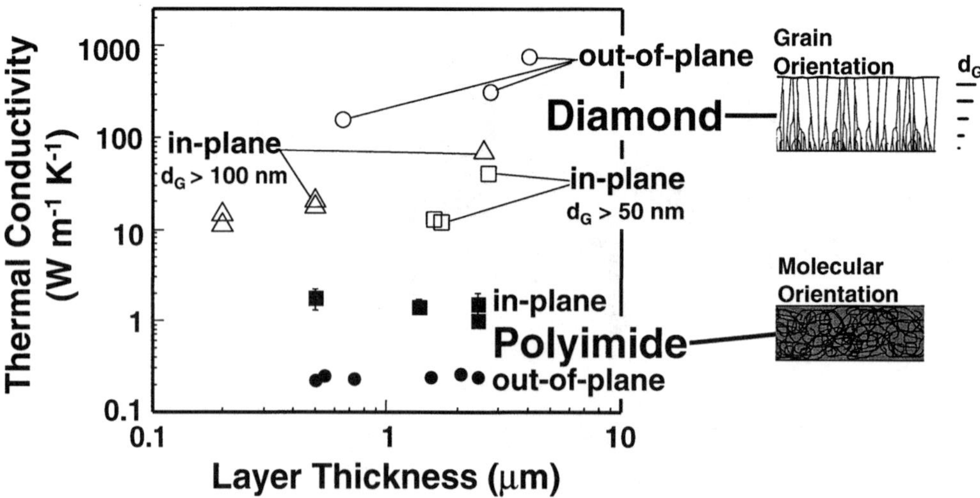

FIGURE 4.19.16 Thickness-dependent thermal conductivity data for diamond and polyimide films. (Data from Touzelbaev and Goodson, 1998; Verhoeven et al., 1997; Kurabayashi et al., 1999.)

film, molecular strand orientation favors in-plane conduction and yields substantial anisotropy that varies little with film thickness.

This section summarizes experimental data and analytical tools available for determining the thermal conductivities and interface resistances of electronic films of thickness between 50 nm and 5 μm. For additional detail on the thermal transport properties of electronic films, the readers are directed to several recent overview articles (Goodson and Ju, 1998; Cahill, 1997), as well as articles focusing on thin-film interface resistances (Swartz and Pohl, 1989), experimental methods (Goodson and Flik, 1994), diamond films (Graebner, 1993; Touzelbaev and Goodson, 1998), superlattices (Hyldgaard and Mahan, 1997; Chen, 1998), and organic films (Kurabayashi et al., 1999; Bauer and Dereggi, 1996; Rogers et al., 1994).

Definitions of the Effective Thermal Conductivities in Films

The thermal conductivity measured in films is not necessarily a property of the film material. Because of electron and phonon scattering at interfaces and interface thermal resistances, the apparent film conductivities often depend on the direction of heat flow (even for isotropic materials), on the film thickness, and on the properties of the film boundaries. For this reason, measurements yield *effective* thermal conductivities for films, which are applicable only for heat propagation in a given direction for films of a given thickness fabricated using a given set of processing conditions. The measurement of effective thermal conductivities is therefore a very important and frequent task.

The effective conductivities are defined here using the geometry in Figure 4.19.17. The effective in-plane thermal conductivity, for conduction along the film, is

$$k_a = -\frac{q_x}{d\,L}\left(\frac{dT}{dx}\right)^{-1} \tag{4.19.58}$$

where x is the coordinate along the film, q_x is the heat power conducted along the film in the x direction, d is the film thickness, and L (not shown in Figure 4.19.17) is the depth of the film normal to the plane illustrated in Figure 4.19.17. The effective out-of-plane thermal conductivity, for conduction normal to the film, is defined to include the volume resistance of the film as well as the thermal resistance between the layer and the bounding media,

$$k_n = \frac{q_y\,d}{w\,L\left(T_0 - T_1\right)} \tag{4.19.59}$$

where q_y is the heat power conducted through the film width w (not shown in Figure 4.19.17) and the temperatures T_0 and T_1 are those of the media just outside of the film interfaces.

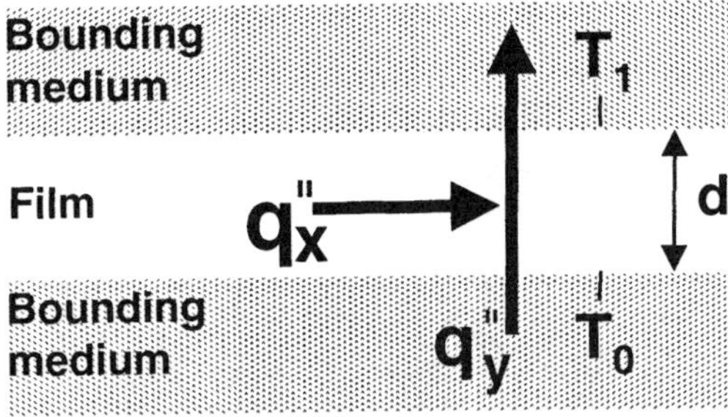

FIGURE 4.19.17 Geometry for defining the effective thermal conductivities of thin films.

Equations 4.19.58 and 4.19.59 define properties of practical relevance for device simulations. Equation 4.19.58 acknowledges that it is exceedingly difficult to extract film volume and interface resistances independently. In films of thickness comparable with the heat carrier mean free path, the distinction between volume and interface resistances is no longer meaningful in the macroscopic sense. This is usually not a problem because the temperature distribution in microstructures is far more sensitive to the total out-of-plane thermal resistance than to the distribution of resistance within the film. For high-conductivity layers, the in-plane conductivity is most important because these layers govern lateral conduction in multilayer structures. This is the case for the silicon layer in the SOI transistor in Figure 4.19.3, for example. For films with low thermal conductivities, such as amorphous oxides and organic films, the out-of-plane conductivity is often more important. These films dominate the thermal resistance in the y direction. Membrane thermal isolation structures, such as those used for infrared bolometers (Richards, 1994), are an important exception to this rule. For these devices, the in-plane conductivity of a suspended amorphous oxide film governs the sensitivity and time constant.

Single-Crystal Semiconducting Films and Superlattices

The in-plane conductivities of semiconducting monolayers are of practical relevance, because they tend to be much larger than the conductivities of surrounding passive regions in micromachined structures. In-plane conduction was most extensively investigated for silicon films of thickness between 70 nm and about 1 μm in silicon-on-insulator (SOI) wafers. SOI technology offers complete dielectric isolation and promises enhanced performance for semiconductor devices in both high-power and low-power circuits. Phonon-interface scattering has a dramatic impact on heat conduction in SOI thin films below room temperature.

Figure 4.19.15 showed the predictions and data of Asheghi et al. (1998), who measured the thermal conductivities of SOI layers at temperatures between 20 and 300 K. The predictions are based on the model of Holland (1963) for conduction in silicon, which is very similar to Equation 4.19.46, and account for phonon-interface scattering. The relaxation time for the film is reduced by a solution to the Boltzmann transport equation for the in-plane transport direction (Sondheimer, 1952):

$$\frac{\tau_{film}}{\tau} = 1 - \frac{3(1-p)}{2\delta} \int_1^{\infty} \left(\frac{1}{\xi^3} - \frac{1}{\xi^5} \right) \frac{1-\exp(-\delta\xi)}{1-p\exp(-\delta\xi)} d\xi \tag{4.19.60}$$

where $\delta = d/\Lambda = d/v\tau$, d is the layer thickness, and p is the fraction of phonons that are specularly reflected by the film boundaries. Equation 4.19.60 with $p = 1$ yields the ideal limit of completely specular reflection, for which no reduction in the conductivity is expected. The use of $p = 0$ yields completely diffuse scattering, which minimizes the conductivity for a given film thickness. While there is reasonable agreement between the predictions and data for the thickest layer, with $d = 1.5$ μm, the disagreement is substantial for the thinner layers. One possible explanation for the discrepancy is larger concentrations of imperfections in the film than in the bulk material. The general trend of the conductivity with temperature is predicted with a reasonable degree of accuracy, which indicates that Equations 4.19.46 and 4.19.60 describe the dominant physics of the interface scattering problem in a single film.

It can be helpful to examine the impact of interface scattering directly on the phonon relaxation time, independent of the thermal conductivity integral in Equation 4.19.46. For computational purposes it can be useful to note that for completely diffuse scattering at the interfaces, $p = 0$, Equation 4.19.60 collapses to

$$\frac{\tau_{film}}{\tau} = 1 - \frac{3}{2\delta}\left[\frac{1}{4} - \varepsilon_3(\delta) + \varepsilon_5(\delta)\right] \tag{4.19.61}$$

where the exponential integral is defined as

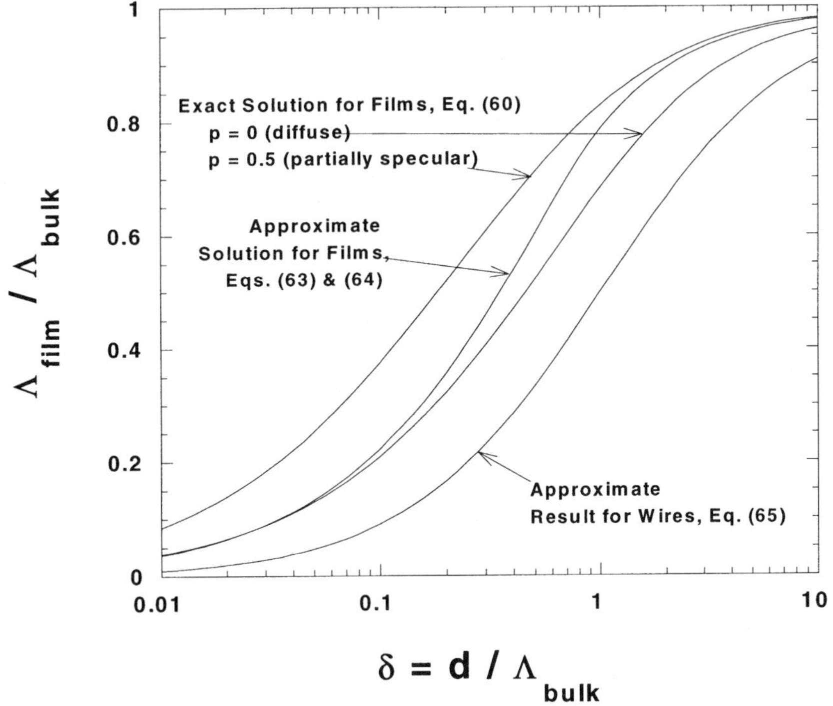

FIGURE 4.19.18 Influence of interface scattering on the relaxation time in planar thin films of thickness d and a circular wire with diameter d. The interfaces reduce the relaxation time τ and the mean free path $\Lambda = v\tau$.

$$\varepsilon_n \equiv \int_1^{\infty} \zeta^{-n} \exp(-\zeta\delta) d\zeta \tag{4.19.62}$$

Figure 4.19.19 plots Equation 4.19.60 together with a closed-form approximation provided by Flik and Tien (1990) also for completely diffuse scattering. For $p = 0$, Equation 4.19.60 agrees within 20% with

$$\frac{\tau_{film}}{\tau} = 1 - \frac{2}{3\pi\delta}; \quad \delta > 1 \tag{4.19.63}$$

$$\frac{\tau_{film}}{\tau} = 1 - \frac{2(1-S^3)}{3\pi\delta} + \frac{2\delta}{\pi} \ln\left(\frac{1+\delta+S}{1+\delta-S}\right) - \frac{2}{\pi} \arccos(\delta); \quad \delta < 1 \tag{4.19.64}$$

where $S = (1 - \delta^2)^{0.5}$. Figure 4.19.18 shows that the reduction in the mean free path of a film decreases dramatically within decreasing film thickness. The predictions for a thin film Equations 4.19.60, 4.19.63, and 4.19.64, yield a weaker reduction in the relaxation time than can be expected for conduction along bridges with comparable thickness and width. Scattering on the side boundaries substantially reduces the relaxation time for this situation. This can be appreciated through an approximate result for a circular wire (e.g., Ziman, 1960)

$$\frac{\tau_{film}}{\tau} = \frac{1}{(1+\delta^{-1})} \tag{4.19.65}$$

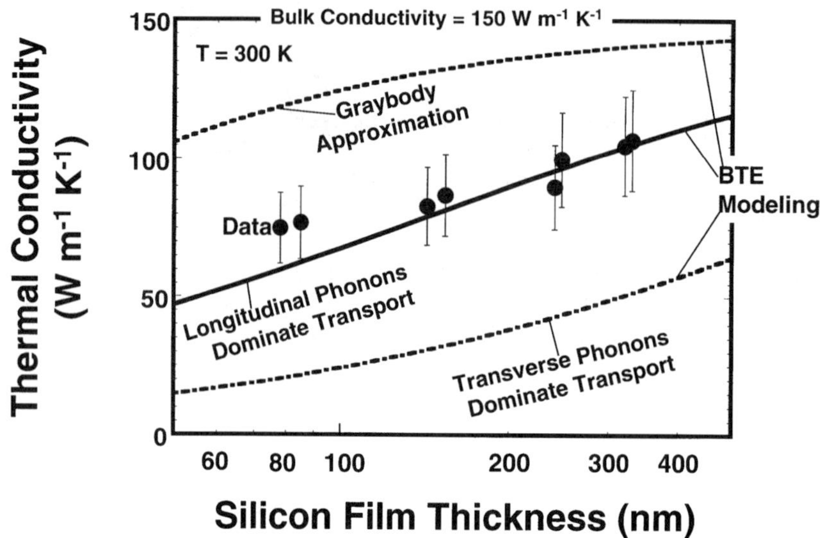

FIGURE 4.19.19 Room-temperature effective thermal conductivities along thin silicon films. (Adapted from Ju, Y.S. and Goodson, K.E., *Appl. Phys. Lett.,* 1999a.)

where δ in this case is the ratio of the wire diameter and the mean free path. Equation 4.19.65 approximates within 5% the exact solution to the Boltzmann equation for conduction along a wire with diffuse scattering. It should be noted that Equation 4.19.65 can be directly derived from Matthiessen's rule, Equation 4.19.34, when a relaxation time of v/d is substituted, where f in this case is the wire diameter.

Conduction modeling is most poorly understood at elevated temperatures, where the complexity and anisotropy of the phonon dispersion relationship more strongly influence transport. The complications become substantial near about 100 K for GaAs, Ge, and AlAs, and near about 200 K for Si due to its higher Debye temperature. Figure 4.19.19 plots room-temperature in-plane thermal conductivity data for silicon layers in SOI substrates as thin as 74 nm, after the work of Ju and Goodson (1999a). The thermal conductivity is smaller by as much as 50% than the bulk value and decreases slowly with decreasing film thickness. Figure 4.19.19 also includes two simple model calculations based on Equation 4.19.60 that assume totally diffuse scattering, $p = 0$. A graybody approximation overpredicts the data because it neglects the spectral dependence of phonon scattering. The second model assumes that either longitudinal or transverse phonons dominate conduction in the bulk material. While either assumption can be used to interpret the thermal conductivity data of bulk samples, the data for thin films lend some support to the hypothesis that longitudinal phonons dominate. This hypothesis also receives support from the highly dispersive nature of high-frequency transverse acoustic phonons in silicon, which reduces their group velocities.

The thin-film boundary conditions for phonon transport analysis warrant careful consideration. Equation 4.19.60 neglects participation by the bounding materials, which are amorphous silicon dioxide for the case of the SOI film studies described here. Using the theory of radiation reflection with interfaces, an expression for the specularity factor can be derived which decreases exponentially with the square of the ratio of the characteristic interface roughness, η, to the phonon wavelength. It can be argued that because phonons are scattered so strongly in the silicon dioxide, this material acts as a diffuse black phonon emitter and absorber with $p = 0$. When the mean free paths in neighboring films are comparable, which is the case in many superlattices, this approximation and Equation 4.19.4 are no longer valid.

Epitaxially grown high-T_c superconducting films, such as $YBa_2Cu_3O_7$, $EuBa_2Cu_3O_7$, and $BiSr_2Ca_1Cu_2O_8$, exhibit highly anisotropic thermal conduction properties due to their orthorhombic unit cell and the large density of oriented imperfections (Uher, 1990). These films are promising for low-loss interconnects and Josephson junctions in hybrid superconductor/semiconductor circuits, as well as for thermal radiation detectors. While there are few if any data available for the thermal conductivity along thin superconducting films, there have been several measurements of the thermal interface resistance between films and their underlying substrate materials (Prasher and Phelan, 1997). The bulk properties and anisotropy have been well documented, and models have been developed for the conductivity reduction in films with the *c* axis oriented normal to the substrate (Richardson et al., 1992; Goodson and Flik, 1993). These calculations needed to consider the simultaneous contribution of electrons and phonons, because both carriers are significant at temperatures above a few tens of Kelvin. The electron contribution in $YBa_2Cu_3O_7$ was found to be approximately 50% of the total conductivity at the transition temperature, 90 K. The reductions in electron and phonon transport were estimated independently using Equations 4.19.46, 4.19.63, and 4.19.64. Figure 4.19.20 plots the temperature dependence of the thermal conductivity for films of varying thickness. This calculation neglected the anisotropy of the acoustic properties of the sample, which could be expected to diminish the size effect, and the spectral dependence of the phonon mean free path, which will underestimate the size effect. However, this model illustrates the anticipated physics. The reduction is substantial for films thinner than about 100 nm and is dominated by the impact of interfaces on the phonon contribution. The electron thermal conductivity and associated thin-film size effect is strongly reduced at low temperatures due to the increasing concentration of Cooper pairs, which do not contribute to heat conduction.

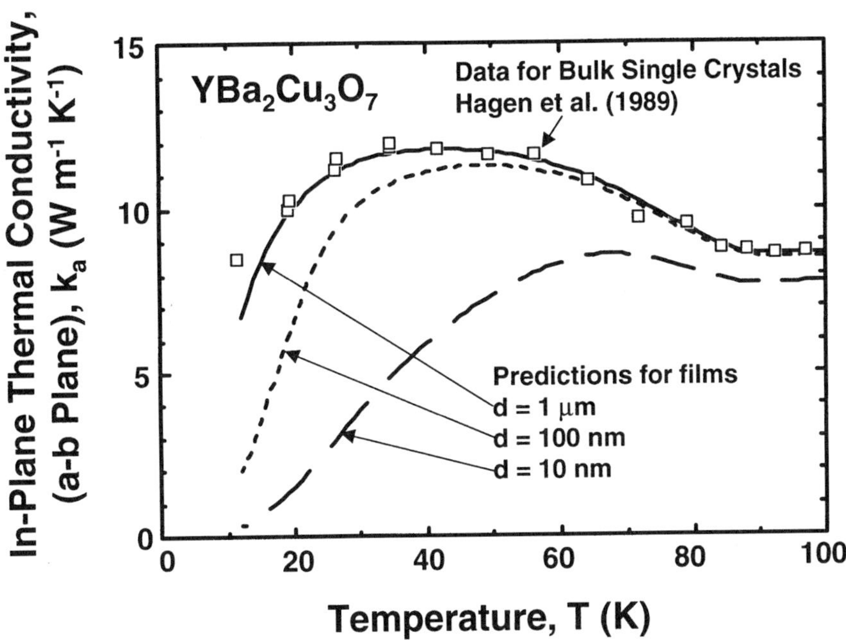

FIGURE 4.19.20 Data for the thermal conductivity in a bulk high-temperature superconducting sample and predictions for the conductivity in thin films. (Adapted from Goodson, K.E. and Flik, M.I., *ASME J. Heat Transfer,* 115, 17, 1993.)

Expressions accounting for lateral transport along crystalline superlattices based on the Boltzmann transport equation consider the degree of specularity in phonon exchange at the periodic interfaces (e.g., Chen, 1998). For completely specular reflection at the interfaces of materials with similar elastic constants, the thermal conductivity approaches that of parallel films of the two materials without interface scattering, which is simply an average of the two bulk conductivities. This is analogous to the predictions of Equation 4.19.60 for a single film with $p = 1$. Completely specular reflection is highly unlikely at room temperature given the relatively short wavelengths of phonons, even for the best superlattices, such that predictions for partially or fully diffuse reflection are expected to be more appropriate. Another complicating phenomenon at superlattice interfaces is the possibility of the confinement in one layer of high-frequency phonons, whose energies lie beyond the frequency maximum of the acoustic branches in the bounding solid.

Although heat conduction normal to single-crystal films is being investigated most extensively for superlattices, there remains a very large gap between phonon transport predictions and data. This problem owes much to large concentrations of imperfections, which occur during the growth process due to the mismatch of lattice constants in the constitutent films. Superlattices also highlight any problems with the assumed phonon-interface scattering behavior. The key distinctions are between specular and diffuse reflection and transmission at an interface, which has already been discussed, and between elastic and inelastic scattering. Elastic scattering assumes that phonons striking an interface must depart with the same frequency, such that there is no energy transfer among different modes. Fully elastic scattering underpredicts the rate of heat conduction by neglecting the anharmonic interfacial coupling among different modes, which can be simulated using atomistic calculations. Fully inelastic scattering overpredicts the rate of heat conduction because it neglects the confinement of phonons in layers with a higher Debye temperature. Chen (1998) has illustrated the impact of these boundary conditions by solving the Boltzmann transport equation. The simplest limiting result neglects scattering within the layers and assumes fully diffuse, fully inelastic scattering, yielding

$$k_n = \left(\frac{C_1 v_1 C_2 v_2}{C_1 v_1 + C_2 v_2} \right) \left(\frac{d_1 + d_2}{2} \right) \tag{4.19.66}$$

where the subscripts 1 and 2 denote parameters of the two repeating films in the superlattice. Equation 4.19.66 is very approximate because it neglects the frequency-dependence of scattering and does not explicitly distinguish between the transport characteristics of optical and acoustic phonons. Equations 4.19.66 and more rigorous calculations are compared with the data of Lee et al. (1997) for pure Si/Ge superlattices in Figure 4.19.21. The model is consistent with the data for the smaller superlattice period, although the results for thicker films disagree substantially. This may result from larger concentrations of imperfections within the layer than are present in the bulk materials, which may occur when

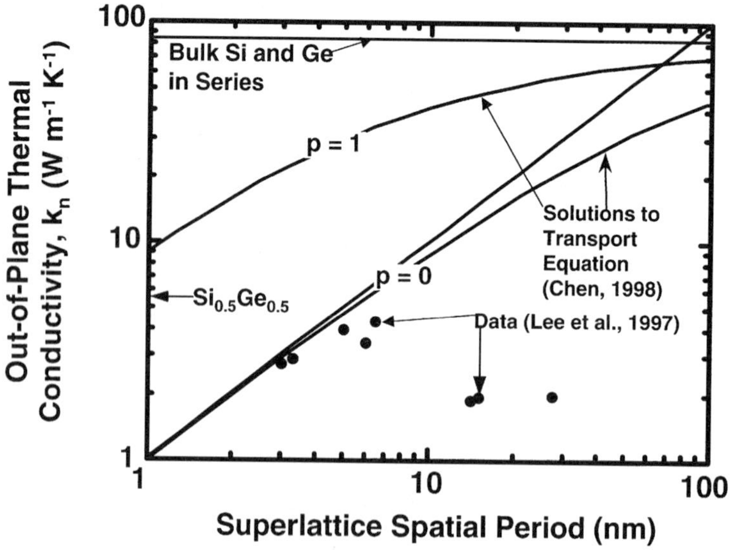

FIGURE 4.19.21 Predictions and data for the out-of-plane thermal conductivity in superlattices. (Adapted from Chen, G. and Neagu, M., *Appl. Phys. Lett.*, 71, 2761, 1997.)

the superlattice period is too large to allow the material to sustain lattice-mismatch strain. Figure 4.19.21 supports the conclusion that interface scattering is predominantly diffuse. The data also suggest that the highly strained superlattices based on pure films contains a much higher concentration of imperfections than superlattices containing alloys. More modeling and measurements must accompany material fabrication work to resolve the role of the lattice imperfections in reducing the thermal conductivity.

Polycrystalline Dielectric and Semiconducting Films

In polycrystalline films, internal scattering can overwhelm scattering on material boundaries. Increased rates of internal scattering result from the grain boundaries and, in some cases, from lattice imperfections and impurities that occur with high densities near grain boundaries. The polycrystalline thin-film material whose thermal properties have received the greatest attention is chemical vapor-deposited diamond (Graebner, 1993; Touzelbaev and Goodson, 1998). The orientation and minimum size of diamond grains are governed by the details of the deposition process, in particular the nucleation method, the substrate temperature, and the composition of the process gases. This subsection uses the data for these films for a representative study of the impact of enhanced phonon scattering in polycrystalline dielectric films.

The simplest approach to modeling phonon scattering on grain boundaries is to augment the scattering rate in Equation 4.19.46 by

$$\tau_G^{-1} = B\frac{v}{d_G} \tag{4.19.67}$$

where B is a dimensionless parameter that increases with the phonon reflection coefficient at grain boundaries and varies with the grain shape. The mean free path calculated using Equation 4.19.67 is $\Lambda = d_G/B$, where d_G is the characteristic grain dimension. For phonons with wavelengths that are long compared to the strained region around the grain boundary, the reflection coefficient is independent of the phonon wavelength and comparable with $(\Delta v/v)^2$, where Δv is the difference in acoustic velocities in the crystal directions for the incident and refracted paths (Ziman, 1960). The minimum grain size d_{G0}, which varies from a few tens of nanometers to about a micrometer, is approximately given by the inverse square root of the nucleation density. For films deposited from the gas phase, this density is controlled by the substrate material and the nucleation method.

Equation 4.19.67 is overly simplistic for relatively defective polycrystalline films. For example, Equation 4.19.67 is questionable for diamond films because there is no corresponding signature in the measured temperature dependence of the low-temperature thermal conductivity. At temperatures above a few degrees Kelvin, the rate of phonon scattering on distinct grain boundaries, which is independent of the phonon frequency, was found to be much smaller than the rate of scattering on other types of defects, e.g., impurities, which depends strongly on the phonon frequency. The densities of these imperfections are closely coupled to the size and orientation of grain boundaries in the films. If these imperfections occur predominantly near grain boundaries, the phonon scattering rate can still be coupled to the grain size d_G. This concept was framed mathematically by Goodson (1996), who defined the dimensionless *grain-boundary scattering strength*,

$$\eta_G = \sum_j \sigma_j n_j \tag{4.19.68}$$

The frequency-dependent cross section and number density per unit grain-boundary area of various imperfections j are denoted as σ_j and n_j, respectively. The scattering rate for randomly-oriented grains is given by Equation 4.19.67 with $B = 2\eta_G$, which assumes that defects associated with a given grain boundary are distributed uniformly within grains. This simple form overpredicts the scattering rate for transport normal to films with columnar grains, for which imperfections near grain boundaries can have a relatively small influence on conduction. An average of phonon free paths over all directions in a columnar grain is well approximated by

$$\tau_G^{-1} = \frac{2v}{\pi d_G}\left[1 - \exp\left(-\frac{\pi^2}{4}\eta_G\right)\right] \tag{4.19.69}$$

which yields a scattering rate with an upper bound dictated by the grain size.

Figure 4.19.22 compares the data for the vertical thermal resistance of thin layers with calculations using Equations 4.19.46 and 4.19.69. The scattering strength is governed by point and extended defects with cross sections taken from research on thicker diamond films (Graebner, 1993). The calculations are reasonably successful for the data included in the plot, which were measured for films deposited at high temperatures. However, agreement for films deposited at lower temperatures requires the use of an internal scattering term, which is independent of the grain size (Touzelbaev and Goodson, 1998). There is only one study, to our knowledge, reporting both the in-plane and out-of-plane thermal conductivities of a given set of diamond films. Verhoeven et al. (1997) observed a particularly large degree of anisotropy for films with predominantly heteroepitaxial grains, with the in-plane conductivity smaller by about one order of magnitude than the out-of-plane conductivity.

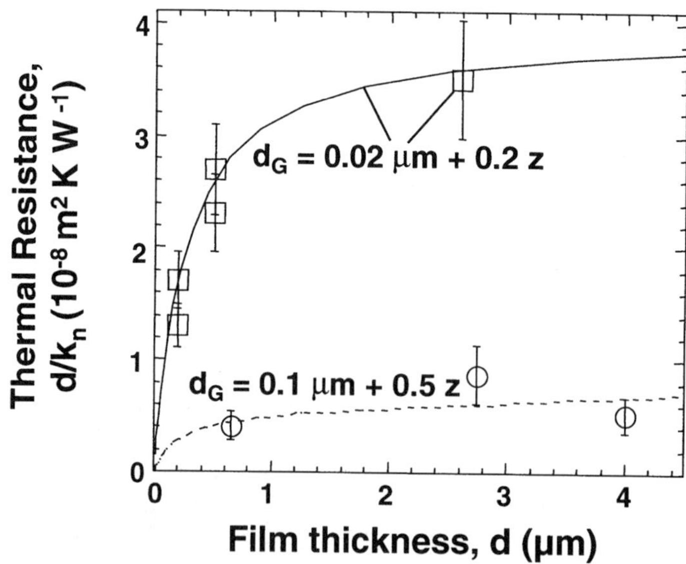

FIGURE 4.19.22 Predictions and data for the out-of-plane thermal conductivity of polycrystalline diamond films. (Adapted from Touzelbaev, M.N. and Goodson, K.E., *Diamond Relat. Mater.*, 7, 1, 1998, and Verhoeven, H. et al., *Diamond Relat. Mater.*, 6, 298, 1997.)

Amorphous Oxide and Organic Films

Process-dependent material structure and stoichiometry influence the thermal conductivities and volumetric heat capacities of highly disordered films, including those of amorphous glasses and organic materials. The best representative data available at present are for silicon-dioxide and polyimide films. Figure 4.19.16 shows that polyimide films exhibit anisotropic conductivities due to the partial alignment of molecular strands in the film plane, which is sensitive to the spin-coating parameters. The conduction of atomic vibrational energy is more effective by means of the electronic bonds coupling atoms along a molecule than by the forces acting between neighboring molecules. The volumetric heat capacity in silicon dioxide, which influences transient heat conduction, can depend strongly on the processing conditions through their impact on the final concentrations of impurities such as water and silanol. The modeling of heat conduction in disordered films is far more approximate than that of crystalline films, and is limited to estimates of the minimum conductivity (Cahill et al., 1992) and simple models for the anisotropy organic films (Kurabayashi and Goodson, 1999).

Dielectric films are often deposited at low temperatures to avoid diffusion or other failure mechanisms in micromachined structures. The low deposition temperature can lead to microstructures or stoichiometry very different from those found in bulk samples. The influence of these differences on thermal transport properties is important not only for process optimization but also for study of the minimum thermal conductivity. There have been numerous studies of the out-of-plane thermal conductivities of silicon-dioxide films prepared using varying methods. These studies reported reductions in the conductivity by as much as 50% from the value in bulk fused silica. Figure 4.19.23 summarizes representative data on silicon dioxide films at room temperature, for which density data are also available. Shown as a dotted line is the prediction of the effective medium theory (e.g., Ju and Goodson, 1999b), that models the density deficit and thermal conductivity reduction using spherical microvoids uniformly distributed within a fully dense matrix. The thermal conductivities of the silicon dioxide films cannot be explained by the porosity alone. One explanation is that the density deficit occurs at an atomic scale and is not well modeled by a theory for macroscopic pores. Recent measurements indicated that, when corrected for impurity and density deficit, the volumetric heat capacity of LPCVD silicon dioxide films does not depend strongly on processing history (Ju and Goodson, 1999b). Interesting observations have been made about the impact of high-temperature annealing on the local atomic structure and thermal conductivity of CVD silicon dioxide films (Goodson et al., 1993). Following a high-temperature anneal, the root-mean-square deviations of bond lengths, which can serve as a measure of the degree of disorder, diminish considerably and approach those of thermally grown silicon dioxide films. This structural change is usually accompanied by a corresponding increase in the thermal conductivity.

The thermal properties of organic films are expected to be highly sensitive to their chemical composition and structural configuration. These materials can be semicrystalline, such that the conductivity should depend on the degree of orientation of crystalline regions and of the molecular strands in the

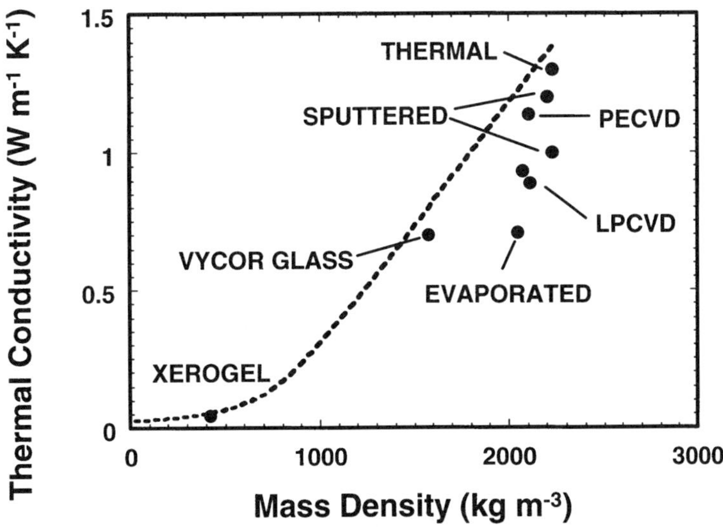

FIGURE 4.19.23 Data for the out-of-plane thermal conductivity in silicon dioxide films for which density data are also available (Cahill et al., 1994; Cahill and Allen, 1994; Lee and Cahill, 1997). The data are compared with predictions and bulk data for porous samples. (Data from Cahill et al., 1990 and Einarsrud et al., 1993.) (Graph adapted from Ju, Y.S. and Goodson, K.E., *J. Appl. Phys.*, in press.)

amorphous regions. The nature of disorder in this material renders most of the modeling concepts discussed in this review inappropriate at room temperature. The relevant length-scales for observing size effects, such as the anisotropy shown for polyimide films in Figure 4.19.16, are not known. While the molecular weight and the molecular radius of gyration might be assumed to play a role, there has been no systematic study of thermal transport in organic films as a function of these parameters. Past studies of the conductivity focused primarily on demonstrating anisotropy and were not accompanied by detailed spectroscopy. However some progress can be made using the extensive literature on heat conduction in bulk organic materials, which has been reviewed (e.g., Ward, 1975).

The series model can be used to interpret thermal conductivity anisotropy data for organic films. The molecular strands are modeled locally within the material as elongated segments with anisotropic intrinsic conductivities. The longitudinal conductivity $k_{a,f}$ describes the transport of atomic vibrational energy along a strand, and the lower out-of-plane conductivity $k_{a,f}$ is used to model the weaker coupling between neighboring strands. While these two conductivities would be extremely difficult to observe experimentally for a single strand, they can be considered equal to the strand-parallel and strand-normal conductivities in a polymer consisting of perfectly oriented molecules. The anisotropy in the material is then given by (Henning, 1967)

$$\frac{1}{k_n} = \frac{1}{2}\left[\left(\frac{1}{k_{a,f}} + \frac{1}{k_{n,f}}\right) - \left(\frac{1}{k_{a,f}} - \frac{1}{k_{n,f}}\right)\left\langle \cos^2 \varphi \right\rangle\right] \tag{4.19.70}$$

$$\frac{1}{k_a} = \left(\frac{1}{k_{a,f}} - \frac{1}{k_{n,f}}\right)\left\langle \cos^2 \varphi \right\rangle + \frac{1}{k_{n,f}} \tag{4.19.71}$$

The angle φ is between the film normal and a given strand direction. The operator < > yields the average cosine considering all strands and is calculated from an assumed orientation distribution function for the polymer strands and reduces to one-third for perfectly random orientation. Figure 4.19.24 shows

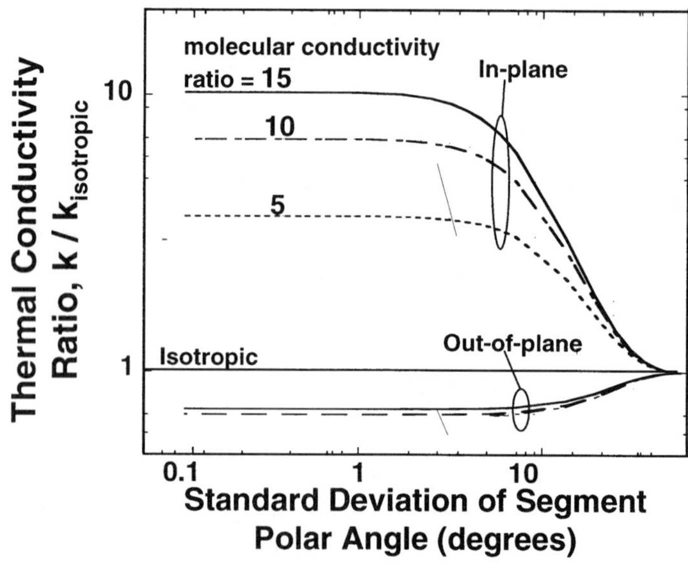

FIGURE 4.19.24 Predicted impact of the molecular orientation on the thermal conductivity anisotropy in organic films. (Adapted from Kurabayashi, K. and Goodson, K.E., *Proc. 5th ASME/JSME Thermal Eng. Joint Conf.*, San Diego, CA, March 1999.)

that by assuming a Gaussian distribution function together with Equations 4.19.70 and 4.19.71, the conductivity anisotropy can be predicted from the standard deviation in the angle of molecular segments with respect to the in-plane direction (Kurabayashi and Goodson, 1999). These results will be much more helpful when more basic modeling principles can be substituted for the intrinsic strand conductivities. Modeling is even more complicated for semicrystalline polymers, in which the properties and volume fractions of the crystalline and amorphous regions add additional degrees of freedom.

Interface Resistance

In macroscopic systems, incomplete contact can be a very important contributor to the thermal resistance at interfaces. For microstructures consisting of deposited films, complete contact between adjacent layers is frequently achieved. For this reason, interface resistances, in general, are smaller and dominated by different mechanisms: (1) phonon-interface scattering (Swartz and Pohl, 1989), and (2) near-interfacial disorder (e.g., Touzelbaev and Goodson, 1998). Because of the very large interface-to-volume ratios in micromachined structures, interface resistances can play a large or even dominant role in the total thermal resistance of the system.

The resistance due to phonon-interface scattering occurs at the mutual boundaries of dielectric and semiconducting films, since phonons are the dominant heat carriers in these materials, as well as for interfaces of metallic films with dielectric or semiconducting materials. Although electrons carry heat in the metal, the transport across the interface is brought about by phonons. One can anticipate an analogous resistance at the interface of two metallic films due to electron scattering, but this resistance is relatively small has received little attention from the research community.

Lattice vibrational waves are reflected or refracted at the interface, much as sound waves at a wall or a boundary between two materials. The resulting resistance is often attributed to *acoustic mismatch* and calculated using

$$\frac{1}{R_C} = \frac{1}{4}\sum_j \int_0^\infty \alpha_1 \hbar\omega D_{1,j}(\omega) \frac{\partial f_{EQ}}{\partial T} d\omega \tag{4.19.72}$$

where $D_{1,j}(\omega)$ is the density of states for phonons in material 1 of polarization j. The transmission coefficient for phonons striking the interface from material 1 is α_1.

For an atomically perfect interface, the transmission coefficients depend on the angle of incidence and can be calculated using the equations of strain continuity at the interface and conservation of momentum (e.g., Swartz and Pohl, 1989). This calculation is rather involved and is questionable for interfaces with significant roughness, for which phonon reflection or transmission can be diffuse. The diffuse mismatch model was derived for this situation. This model assumes that the transmission coefficient is governed by the relative abilities of the two interfaces to transport phonons of a given frequency away from the interface. The quantitative differences between the predictions of the acoustic mismatch and diffuse mismatch models are relatively small, such that the diffuse assumption is used most frequently for engineering calculations. The transmission coefficient for the diffuse mismatch model is

$$\alpha_1 = \frac{\sum_j v_{2,j} D_{2,j}(\omega)}{\sum_j v_{1,j} D_{1,j}(\omega) + \sum_j v_{2,j} D_{2,j}(\omega)} \tag{4.19.73}$$

The expression for the resistance can be simplified using the Debye model for the density of states in both materials, Equations 4.19.17 and 4.19.18, which simplifies Equation 4.19.73 to

$$\alpha_1 = \frac{\frac{2}{v^2_{2,transverse}} + \frac{1}{v^2_{2,longitudinal}}}{\left[\frac{2}{v^2_{1,transverse}} + \frac{1}{v^2_{1,longitudinal}}\right] + \left[\frac{2}{v^2_{2,transverse}} + \frac{1}{v^2_{2,longitudinal}}\right]} \tag{4.19.74}$$

The transmission in this model is assumed to be negligible for frequencies above the minimum Debye frequency of the two materials.

The diffuse mismatch model yields reasonably good agreement with experimental data for a wide variety of interfaces below 77 K. At higher temperatures, two complicating effects become important. First, many phonons experience inelastic collisions at the interface, which can augment energy transmission (e.g., Stoner and Maris, 1993). Another complication is the growing importance of near-interfacial disorder with increasing temperature. Thin films rarely have homogeneous microstructural quality and purity and, in general, the poorer material is concentrated near interfaces. While a detailed simulation might attempt to resolve the varying thermal resistance in volume elements near an interface, a practical compromise is to model the near-interfacial disorder as an interface resistance. Note that this approach assumes that the interface resistance can be modeled independently from the resistance of the volume of that material. If transport normal to a film is important, and the mean free path of phonons in the film is comparable to the film thickness, it is no longer possible to isolate resistance contributions of the interface and the volume. For this situation, it is recommended that calculations employ effective film conductivities for a given thickness and direction of transport.

Measurement Techniques

The small dimensions of electronic microstructures make experimental data particularly important, because the governing equations and many of the appropriate assumptions for simulations are still the subject of research. Small dimensions often make experiments more challenging by requiring new regimes of spatial and temporal resolution. This section provides an overview of the experimental techniques that determine the effective thermal conductivities and interface resistances of thin electronic films as well as the temperature distributions in electronic microstructures. A comprehensive overview of the film thermal conductivity measurement techniques can also be obtained by reading the articles of Graebner (1993), Goodson and Flik (1994), Cahill (1997), and Goodson and Ju (1999). More information about microscale thermometry techniques can be obtained from the reviews of Majumdar et al. (1996) and Goodson et al. (1998).

Thin-Film Thermal Properties

When measuring film thermal conductivities, care must be taken to obtain data that are appropriate for simulating practical devices. Many techniques use a film structure that does not resemble the functional device and, in many cases, requires processing that alters the purity or structural quality of the film. For example, several techniques measure the in-plane thermal conductivity using free-standing films, which experience a different stress history and are exposed to more contaminants than those within a device. We focus here on techniques that are deemed to be most appropriate for use in conjunction with electronic device design.

The techniques use either lasers or electrical currents for heating, which offer contrasting advantages. Laser heating avoids contact with the sample film, which helps for small samples and when adhesion is a problem. It is nearly impossible to consistently know the precise magnitude of incident laser power absorbed by the film, which renders the absolute magnitude of the temperature response of little use. However, the precise temporal control of intensity, which is made possible through picosecond and nanosecond pulsed lasers and optical modulators, allows thermal properties to be extracted from the temporal response without knowledge of the magnitude. Electrical heating is applied using currents in conducting bridges, which are patterned on the film. The Joule heating power can be accurately measured and controlled, in some cases with temporal resolution below 1 μs. The electrical techniques require careful design and patterning of bridges on the film surface.

Steady-state electrical heating and thermometry in patterned bridges have been used to measure k_n for films with lower thermal conductivities than the underlying substrate. This measurement approach was originally developed by Swartz and Pohl (1987) for determining the thermal resistance at the film-substrate interface using the structure shown in Figure 4.19.25a. The Joule heating in one metal bridge raises its temperature (T_{bridge}) and that of the underlying substrate ($T_{substrate}$). The interface or contact resistance is given by

$$\frac{1}{R_C} = \frac{q_y}{wL\left(T_{bridge} - T_{substrate}\right)} \tag{4.19.75}$$

T_{bridge} is the temperature of the heater bridge and $T_{substrate}$ is the temperature of the substrate immediately below this bridge. The quantity q_y is the total Joule heating power in the bridge of width w and length L. The substrate temperature $T_{substrate}$ is determined using electrical-resistance thermometry in the neighboring bridge, which carries a very low current and generates relatively little power. Because the heating power in the neighboring bridge is negligible, there is no significant temperature difference across the interface of the neighboring film and the substrate. However, there is a temperature variation in the substrate between the locations directly beneath the two bridges. This is considered using a solution to the two-dimensional heat-diffusion equation based on the thermal conductivity of the substrate. This technique for measuring the interface resistance can also be used to measure the out-of-plane conductivities of films (Goodson et al., 1993; 1994; Asheghi et al., 1998). The interface is replaced by an electrically insulating film of thickness d, the measurement procedure remains essentially the same, and the measured thermal resistance is that of the film and its interfaces. When the film thickness is not much smaller than the width of the bridge, lateral spreading of heat in the film increases the observed effective conductivity. This can be considered by using an approximate solution to the two-dimensional heat equation in the film.

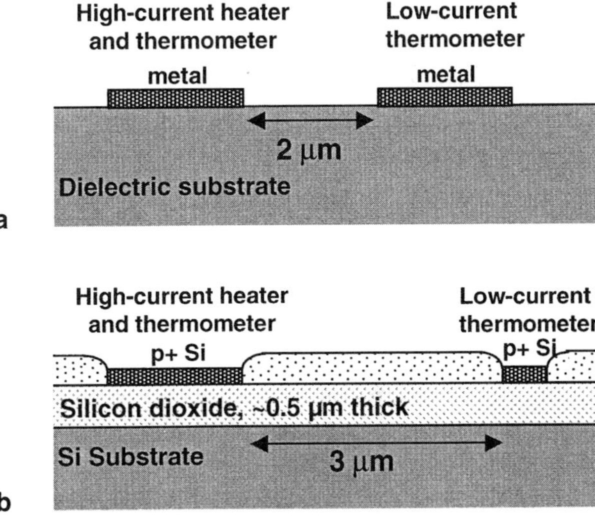

FIGURE 4.19.25 Representative structures for steady-state electrical measurements of (a) the interface thermal resistance (Swartz and Pohl, 1987) and (b) the out-of-plane conductivity for thin films. (Adapted from Goodson et al., *J. Heat Transer,* 116, 317, 1994.)

The steady-state parallel bridge approach illustrated in Figure 4.19.25 is most accurate when used with high thermal conductivity substrates and when the bridge dimensions and separation can be made very narrow. These conditions render the correction for substrate heating less important compared to the temperature difference across the film or interface. That the measured resistance should be large compared with that of the substrate between the two bridges,

$$R_C \ or \ \frac{d}{k_n} >> \frac{w}{k_{substrate}} \tag{4.19.76}$$

While the steady-state dual bridge method is relatively easy to apply, electrical techniques achieve lower experimental uncertainty and more versatility using transient heating and thermometry. Many of these extend the 3-omega method (Cahill, 1990; Lee and Cahill, 1997), for which a representative structure and electrical signal flowchart are provided in Figure 4.19.26. In this method a signal generator is used to sustain an electrical current in the bridge at angular frequency ω, which induces Joule heating at the frequency 2ω and amplitude *P*. The bridge temperature, therefore, fluctuates at the frequency 2ω with a phase shift and amplitude governed by the thermal properties of the structure. The voltage along the line has a dominant component at the frequency ω, governed by the current, and a much smaller component at the frequency 3ω, which is governed by the temperature-induced fluctuations in the electrical resistance at 2ω. The 3-omega technique uses a bridge circuit and a lock-in amplifier to probe the 3ω component, which is used to extract the amplitude and phase of temperature oscillations.

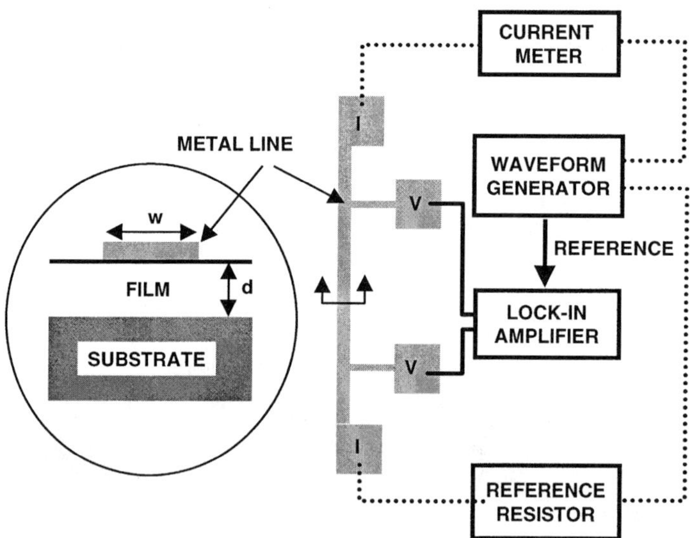

FIGURE 4.19.26 Representative structure and signal chart for transient electrical measurements of film thermal conductivities using the 3-omega method of Cahill, 1997. (Drawing prepared by Y. S. Ju, Stanford University.)

The 3-omega method is simplest to apply for angular frequencies large enough such that heat penetrates fully through the film during each heating period, yet small enough such that heat does not penetrate to the bottom boundary of the substrate. This yields a requirement for the heating frequency 2ω in terms of estimated conductivities *k*, volumetric heat capacities *C*, and thicknesses *d* of the substrate and film,

$$\left(\frac{k}{C\,d^2}\right)_{substrate} << 2\omega << \left(\frac{k}{C\,d^2}\right)_{film} \tag{4.19.77}$$

where the quantities in parentheses are characteristic heat diffusion frequencies for the substrate and film. The solution to the heat-diffusion equation for the temperature amplitude ΔT resulting from heat generation at the rate $P\exp(2\,i\,\omega t)$ is

$$\frac{\Delta T}{P/L} = \frac{1}{\pi k_{substrate}} \int_0^{\infty} \frac{\sin^2(xw/2)}{(xw/2)^2 \left(x^2 + \dfrac{2i\omega C_{substrate}}{k_{substrate}} \right)} dx + \frac{d}{wk_n} \tag{4.19.78}$$

where L is the length of the measurement section between the voltage probes. Note that the film normal conductivity k_n influences only the frequency-independent component to the signal, the second term on the right. The accuracy of the 3-omega technique improves if the film width is made as narrow as possible, since this augments the relative influence of the film resistance on the temperature rise. When the frequency is sufficiently large such that the diffusion length in the substrate is substantially larger than the bridge width, specifically,

$$2\omega << \frac{4k_{substrate}}{C_{substrate}\, w^2} \tag{4.19.79}$$

then Equation 4.19.78 may be simplified to

$$\frac{\Delta T}{P/L} = \frac{1}{\pi k_{substrate}} \left[\frac{1}{2} \ln\left(\frac{4\, k_{substrate}}{C_{substrate}\, w^2} \right) + 0.92 - \frac{1}{2}\ln(2\omega) - \frac{i\pi}{4} \right] + \frac{d}{wk_n} \tag{4.19.80}$$

This shows that the thermal resistance of the substrate contributes a frequency-dependent component to the temperature amplitude, which can be plotted as a line on a log-linear plot. If the heat capacity of the substrate is known with reasonable accuracy, then the substrate conductivity can be extracted using the temperature amplitude measured at two different frequencies. Even when the condition in Equation 4.19.79 is not strictly satisfied, Equation 4.19.80 may be applied with reasonable accuracy by adjusting the numerical constant within the brackets. Cahill (1997) recommends 1.05.

While the 3-omega method has been used primarily to measure the out-of-plane conductivity, careful choice of layer dimensions and operating frequencies can yield other properties in samples that are sufficiently thick. Recent progress includes measurements of the in-plane conductivity of polyimide (Ju et al., 1999; Kurabayashi et al., 1999) and thin silicon films (Ju and Goodson, 1999a). The measurement on thin silicon films used a buried silicon dioxide layer beneath the silicon to promote lateral heat conduction in the structure. When heating period is comparable with the heat diffusion time normal to the film, it is possible to extract the volumetric heat capacity of the film (Ju and Goodson, 1999b). It must be emphasized that these specialized applications of the 3-omega method involve substantially more care with experimental uncertainty analysis and in general require involved numerical solutions to the heat diffusion equation in the frequency domain. These measurements are powerful but very difficult to implement correctly.

Pulsed laser heating and laser-reflectance thermometry have been used for noncontact measurements of the out-of-plane conductivity, as in Figure 4.19.27. The instrumentation and analysis differs greatly depending on the duration of the heating pulse. Measurements with picosecond-scale heating, e.g., from a Ti:Sapphire laser, have been applied to superlattices (Capinski and Maris, 1996) and diamond-like carbon (Morath et al., 1994). These measurements use pump-probe laser diagnostics for the temperature and analysis must decouple the responses from the near-surface temperature rise and subsurface propagation. The analysis must also consider the nondiffusive nature of heat transport in the sample films and the disequilibrium between electrons, which can absorb much of the radiation, and the lattice.

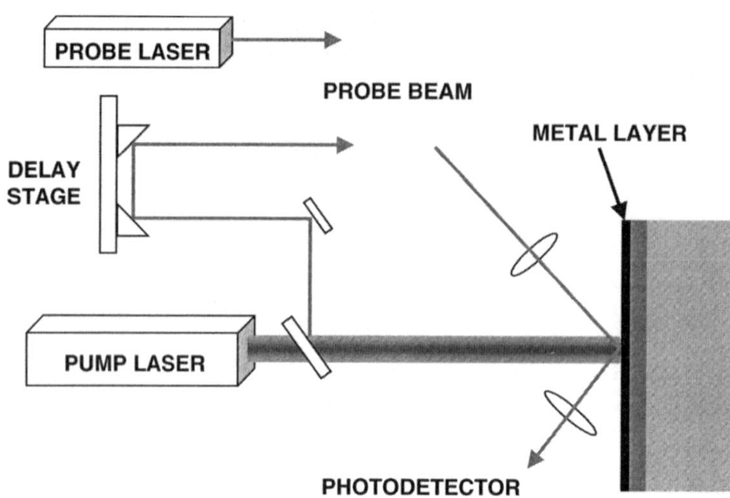

FIGURE 4.19.27 Pump-probe measurements of thermal properties in thin films. (Drawing prepared by Y. S. Ju, Stanford University.)

The instrumentation and extraction are simpler for measurements using nanosecond-scale heating, e.g., from a Nd:YAG laser, during which acoustic waves are fully attenuated. Nanosecond heating and continuous time-domain laser-reflectance thermometry yielded the out-of-plane conductivities of silicon dioxide (Kaeding et al., 1993) and polycrystalline diamond (Goodson et al., 1995b) films of thicknesses down to a few hundred nanometers. The drawbacks of this approach include the time-domain noise from the detector and associated electronics. Depending on the relative thicknesses of the absorbing layer and the underlying film, the measurements can be insensitive to the volumetric heat capacity of the sample layer.

One benefit of laser heating is that the brief pulse duration helps to decouple the properties of the substrate from the film. This facilitates measurements on films with much higher thermal conductivities than the underlying substrate, such as the diamond film with $k_n = 600$ W m^{-1} K^{-1} investigated by Verhoeven et al. (1997) on a silicon substrate. Another advantage is that compared with the electrical methods described in the previous paragraph, these optical methods require relatively little sample preparation. However, the data interpretation can be more involved and the experimental uncertainties are often larger.

The lateral thermal conductivity of films on substrates has been measured using the transient thermal grating technique. Pulsed laser radiation interferes on the sample surface, yielding a harmonic spatial variation of energy absorption. This yields spatially varying temperature and thermomechanical displacement fields, whose temporal decay is detected by the deflection of an incident probe beam. The temporal decay is governed mainly by the lateral thermal conductivity of the film and substrate and the volumetric heat capacity within a depth near the spatial period of energy deposition, which can be varied by altering the difference in incident angles between the two probes.

Temperature Fields in Devices

There has been much progress in recent years on transistor and interconnect thermometry (e.g., Goodson et al., 1998). Much of this work has used either high-resolution optical diagnostics or the scanning probe microscope, which offer improved temporal and spatial resolution, respectively. Optical methods achieve higher temporal resolution by probing the device without contributing to the thermal mass or the electrical capacitance of the system. The electrical techniques are in some cases more precisely calibrated and can offer higher spatial resolution, but in some cases require extra micromachining steps. Techniques based on scanning probe microscopy offer unprecedented spatial resolution. However, all of the techniques make compromises among spatial and temporal resolution and accuracy. These techniques also differ vastly in the amount of sample preparation, such as special micromachining steps, that is required before making the measurements. This situation motivates careful attention to the choice of the best technique for a give problem.

A number of studies have used electrical-resistance thermometry in patterned bridges made of metals or doped semiconducting material. This approach yields a well-calibrated temperature measurement within a precisely defined volume. Maloney and Khurana (1985) and Banerjee et al. (1996) performed transient electrical-resistance thermometry of interconnects subjected to current pulses of duration of less than 1 μs. An approach for field-effect transistors is to modify the gate such that it serves as a temperature-sensing bridge (e.g., Goodson et al., 1995a), yielding the structure shown in Figure 4.19.28. This approach offers the benefit of accurate calibration and a precise understanding of the domain where temperature is being measured. However, these techniques usually yield only a single temperature in the device rather than a distribution. A related strategy is to use the electrical behavior of a device, such as the electrical conductance of a transistor, for sensing temperature (e.g., Cain et al., 1992; Zweidlinger et al., 1996). This approach requires minimal changes to the device and can be calibrated with reasonable accuracy. However, the temperature obtained for the device is a weighted average over a domain that is often difficult to define precisely.

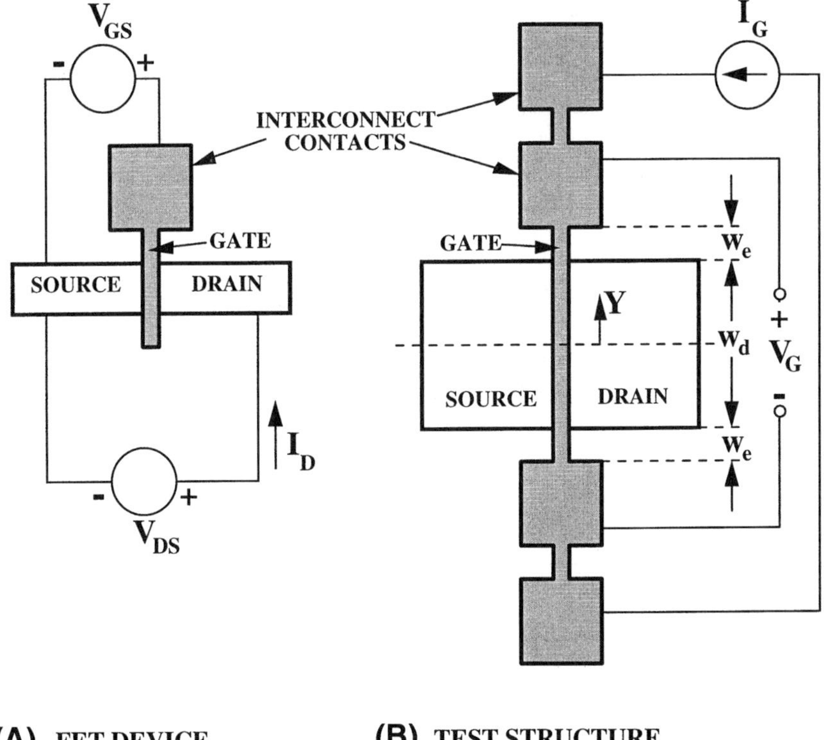

FIGURE 4.19.28 Comparison of a field-effect transistor and an experimental structure for measuring the maximum device temperature. (Adapted from Goodson, K.E. et al., *ASME J. Heat Transfer,* 117, 574, 1995a.)

Scanning probe microscopy has led to a variety of innovations in surface thermometry. These methods scan a micromachined tip with radius of curvature as low as 5 nm across the surface of the sample and use atomic-scale forces or an electrical tunneling current as feedback for the tip-surface separation. The scanning thermal profiler of Williams and Wickramsinghe (1986) qualitatively mapped variations in surface thermal properties. Because this method used the tip-surface conductance as a method for maintaining constant separation, it has difficulty decoupling surface topographical features from variations in the temperature. This approach was refined using a thermocouple junction at the tip of an atomic force microscope probe by Majumdar et al. (1993; 1996). This technique maintains the tip-surface separation using the force-feedback mechanism of the atomic force microscope, while using the thermocouple signal for measurements of temperature. The data obtained using this approach for semiconductor laser and transistor structures promise unrivaled spatial resolution, well below 100 nm, and have been documented extensively in review articles (e.g., Majumdar et al., 1996). These methods are ideally suited for obtaining images with high spatial resolution. However, thermometry methods based on scanning probe microscopy have proven very difficult to calibrate and, therefore, are most useful for imaging qualitative distributions of temperature.

Optical methods use radiation interaction with either the surface of a device or interconnect, in some cases probed through transparent passivation, or with a specially deposited thin layer on the surface. Far-field laser reflectance thermometry calibrates the temperature dependence of the surface reflectivity of an interconnect or device structure. Because of their small radiation penetration depths in metals are especially well suited for laser-reflectance imaging. The metal yields a precisely defined location for the measurement and minimizes the interaction of the probe beam with the electrical currents flowing in the circuit. This motivated Ju et al. (1997) to include an aluminum layer within the overlying passivation of a SOI power transistor. Figure 4.19.29 shows the temperature distribution across the corner of an aluminum interconnect on an organic passivation layer. The temperature image was captured 100 ns after the initiation of an electrical current pulse in the interconnect, and is compared with a coupled analysis of the temperature and potential distributions. Additional optical methods include subwavelength near-field laser-reflectance imaging (Goodson and Asheghi, 1997), thermometry using the Raman effect (Ostermeier et al., 1992), photothermal displacement imaging (Martin and Wickramsinghe, 1987), and fluorescence thermal imaging (Barton and Tangyunyong, 1996), whose differing advantages are discussed in the review of Goodson et al. (1998).

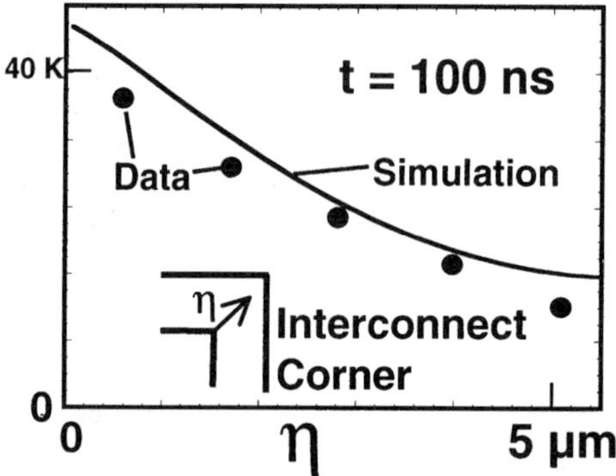

FIGURE 4.19.29 Temperature distribution across the corner of a patterned aluminum alloy interconnect subjected to a brief electrical current pulse. (Adapted from Ju, Y.S. and Goodson, K.E., *IEEE Electron Device Lett.*, 18, 512, 1997b.)

Summary

The decreasing dimensions of micromachined electronic structures motivate research on new regimes of simulation and measurement tools for heat conduction. This section has provided an overview of the simulation methods together with key elements of the solid-state physics needed to understand them. The goal has been to emphasize the most important equations and governing relationships which are used most frequently in practice. This chapter section has also summarized the experimental techniques that are most effective at measuring temperature distributions, thermal conductivities, and interface resistances in electronic microstructures. A particularly large challenge for the near future includes the further integration of many of these techniques into the research and development infrastructure of the semiconductor industry. It is hoped that this overview will facilitate this process.

Acknowledgments

This chapter section has benefitted strongly from the interactions of the author with many of his graduate and postgraduate students at Stanford University, including Mehdi Asheghi, Y. Sungtaek Ju, Dr. Katsuo Kurabayashi, and Maxat N. Touzelbaev. Financial support from the Semiconductor Research Corporation through Contracts SJ-461, PJ-357, and MJ-653 is appreciated. The author also acknowledges support from the ONR Young Investigator Program and the NSF CAREER Program.

References

Amerasekera, A., Chang. M.C., Seitchik, J.A., Chatterjee, A., Mayaram, K., and Chern, J.H., Self-Heating Effects in Basic Semiconductor Structures, *IEEE Trans. Electron Devices*, 30, 1836-1843, 1993.

Apanovich, Y., Blakey, P., Cottle, R., Lyumkis, E., Polsky, B., and Shur, A., Numerical Simulations of Ultra-Thin SOI Transistor using Non-Isothermal Energy-Balance Model, *Proc. IEEE Int. SOI Conf.,* Nantucket, MA, October 3-6, 1994, pp. 33-34.

Apanovich, Y., Blakey, P., Cottle, R., Lyumkis, E., Polsky, B., Shur, A., and Tcherniaev, A., Numerical Simulation of Submicrometer Devices Including Coupled Nonlocal Transport and Nonisothermal Effects, *IEEE Trans. Electron Devices*, 42, 890-898, 1995.

Ashcroft, N.W. and Mermin, N.D., *Solid State Physics*, W.B. Saunders, Orlando, FL, 1976.

Asheghi, M., Touzelbaev, M.N., Goodson, K.E., Leung, Y.K., and Wong, S.S., Temperature-Dependent Thermal Conductivity of Single-Crystal Silicon Layers in SOI Substrates, *ASME J. Heat Transfer*, 120, 30-36, 1998.

Banerjee, K., Amerasekera, A., and Hu, C., Characterization of VLSI Circuit Interconnect Heating and Failure under ESD Conditions, *Proc. Int. Reliab. Phys. Symp.,* 1996, pp. 237-245.

Barton, D.L. and Tangyunyong, Fluorescent Microthermal Imaging-Theory and Methodology for Achieving High Thermal Resolution Images, *Microelectron. Eng.*, 31, 271-279, 1996.

Bauer, S. and Dereggi, A.S., Pulsed Electrothermal Technique for Measuring the Thermal-Diffusivity of Dielectric Films on Conducting Substrates, *J. Appl. Phys.*, 80, 6124-6128, 1996.

Berman, R., *Thermal Conduction in Solids*, Oxford University Press, Oxford, U.K., 1976.

Beebe, S., Rotella, F., Sahul, Z., Yergeau, D., McKenna, G., So, L., Yu, Z., Wu, K.C., Kan, E., Mcvittie, J., and Dutton, R.W., Next Generation Stanford TCAD-PISCES 2ET and SUPREM 007, *Proc. IEEE Int. Electron Devices Meet.,* San Francisco, CA, December 11-14, 1994.

Bloetekjaer, K., Transport Equations for Electrons in Two-Valley Semiconductors, *IEEE Trans. Electron Devices*, ED-17, 38-47, 1970.

Cahill, D.G., Thermal Conductivity Measurement from 30-K to 750-K: The 3-Omega Method, *Rev. Sci. Instrum.*, 61, 802-808, 1990.

Cahill, D.G., Tait, R.H., Stephens, R.B., Watson, S.K., and Pohl. R.O., *Thermal Conductivity*, Plenum Press, New York, 1990, pp. 3-16.

Cahill, D.G., Watson, S.K., and Pohl, R.O., Lower Limit to the Thermal Conductivity of Disordered Crystals, *Phys. Rev. B*, 36, 6131-6140, 1992.

Cahill, D.G., Kativar, M., and Abelson, J.R., Thermal Conductivity of Alpha-SiH Thin Films, *Phys. Rev. B*, 50, 6077-6081, 1994.

Cahill, D.G. and Allen, T.H., Thermal Conductivity of Sputtered and Evaporated SiO_2 and TiO_2 Optical Coatings, *Appl. Phys. Lett.*, 65, 309-311, 1994.

Cahill, D.G., Heat Transport in Dielectric Thin-Films and at Solid-Solid Interfaces, *Microscale Thermophys. Eng.*, 1, 85-109, 1997.

Cain, B.M., Goud, P.A., and Englefield, C.G., Electrical Measurement of the Junction Temperature of an RF Power Transistor, *IEEE Trans. Instrum. Meas.*, 41, 663-665, 1992.

Capinski, W.S. and Maris, H.J., Thermal Conductivity of GaAs/AlAs Superlattices, *Physica B*, 220, 699-701, 1996.

Chen, G., Heat Transfer in Micro- and Nanoscale Photonic Devices, *Annu. Rev. Heat Transfer*, 7, 1-38, 1996.

Chen, G. and Neagu, M., Thermal Conductivity and Heat Transfer in Superlattices, *Appl. Phys. Lett.*, 71, 2761-2763, 1997.

Chen, G., Thermal-Conductivity and Ballistic-Phonon Transport in the Cross-Plane Direction of Superlattices, *Phys. Rev. B*, 57, 14958-14973, 1998.

Chui, B., Stowe, T.D., Ju, Y.S., Goodson, K.E., Mamin, H.J., Terris, B.D., Ried, R.P., and Rugar, D., Low-Stiffness Silicon Cantilevers with Integrated Heaters and Piezoresistive Sensors for High-Density AFM Thermomechanical Data-Storage, *J. MicroElectroMech. Syst.*, 7, 69-78, 1998.

Einarsrud, M.A., Haereid, S., and Wittwer, V., Some Thermal and Optical Properties of a New Transparent Silica Xerogel Material with Low Density, *Sol. Energy Mater. Sol. Cells*, 31, 341-347, 1993.

Flik, M.I. and Tien, C.L., Size Effect on the Thermal Conductivity of High-T_c Thin-Film Superconductors, *ASME J. Heat Transfer*, 112, 872-881, 1990.

Goodson, K.E. and Flik, M.I., Electron and Phonon Thermal Conduction in Epitaxial High-T_c Superconducting Films, *ASME J. Heat Transfer*, 115, 17-25, 1993.

Goodson, K.E., Flik, M.I., Su, L.T., and Antoniadis, D.A., Annealing-Temperature Dependence of the Thermal Conductivity of LPCVD Silicon-Dioxide Layers, *IEEE Electron Device Lett.*, 14, 490-492, 1993.

Goodson, K.E. and Flik, M.I., Solid-Layer Thermal Conductivity Measurement Techniques, *Appl. Mech. Rev.*, 47, 101-112, 1994.

Goodson, K.E, Flik, M.I., Su, L.T., and Antoniadis, D.A., Prediction and Measurement of the Thermal Conductivity of Amorphous Dielectric Layers, *J. Heat Transfer*, 116, 317-324, 1994.

Goodson, K.E., Flik, M.I., Su, L.T., and Antoniadis, D.A., Prediction and Measurement of Temperature Fields in Silicon-on-Insulator Circuits, *ASME J. Heat Transfer*, 117, 574-581, 1995a.

Goodson, K.E., Kaeding, O.W., Roesler, M., and Zachai, Experimental Investigation of Thermal Conduction Normal to Diamond-Silicon Boundaries, *J. Appl. Phys.*, 77, 1385-1392, 1995b.

Goodson, K.E., Thermal Conduction in Nonhomogeneous CVD Diamond Layers in Electronic Microstructures, *ASME J. Heat Transfer*, 118, 279-286, 1996.

Goodson, K.E., and Asheghi, M., Near-Field Optical Thermometry, *Microscale Thermophys. Eng.*, 1, 225-235, 1997.

Goodson, K.E., Ju, Y.S., and Asheghi, M., Thermal Phenomena in Semiconductor Devices and Interconnects, in *Microscale Energy Transport*, Tien, C.L. et al., Eds., Taylor & Francis, New York, 1998, pp. 229-293.

Goodson, K.E. and Ju, Y.S., Heat Conduction in Novel Electronic Films, *Annual Review of Materials Science*, Kaufmann, E.N. et al., Eds., Annual Reviews, Palo Alto, CA, 261-293, 1999.

Graebner, J.E., Thermal Conductivity of CVD Diamond: Techniques and Results, *Diamond Films Technol.*, 3, 77-130, 1993.

Hagen, S.J., Wang, Z.Z., and Ong, N.P., Anisotropy of the Thermal Conductivity of $YBa_2Cu_3O_{7-y}$, *Phys. Rev. B*, 40, 9389-9392, 1989.

Henning, J., Anisotropy and Structure in Uniaxially Stretched Amorphous High Polymers, *J. Polym. Sci. C*, 16, 2751-2761, 1967.

Holland, M.G., Analysis of Lattice Thermal Conductivity, *Phys. Rev.*, 132, 2461-2471, 1963.

Hunter, W.R., Self-Consistent Solutions for Allowed Interconnect Current-Density. I. Implications for Technology Evolution, *IEEE Trans. Electron Devices*, 44, 304-309, 1997a.

Hunter, W.R., Self-Consistent Solutions for Allowed Interconnect Current-Density. II. Application to Design Guidelines, *IEEE Trans. Electron Devices*, 44, 304-309, 1997b.

Hyldgaard P. and Mahan, G.D., Phonon Superlattice Transport, *Phys. Rev. B*, 57, 14958-14973, 1997.

Ju, Y.S., Kaeding, O.W., Leung, Y.K., Wong, S.S., and Goodson, K.E., Short-Timescale Thermal Mapping of Semiconductor Devices, *IEEE Electron Device Lett.*, 18, 169-171, 1997a.

Ju, Y.S. and Goodson, K.E., Thermal Mapping of Interconnects Subjected to Brief Electrical Stresses, *IEEE Electron Device Lett.*, 18, 512-514, 1997b.

Ju, Y.S. and Goodson, K.E., Short-Time-Scale Thermal Mapping of Microdevices Using a Scanning Thermoreflectance Technique, *ASME J. Heat Transfer*, 120, 306-313, 1998.

Ju, Y.S., Kurabayashi, K., and Goodson, K.E., Thermal Characterization of Anisotropic Thin Fielectric Films Using Harmonic Joule Heating, *Thin Solid Films*, 339, 160-164, 1999.

Ju, Y.S. and Goodson, K.E., Phonon Scattering in Silicon Films of Thickness Below 100 nm, *Appl. Phys. Lett.*, 74, 3005-3007, 1999.

Ju, Y.S. and Goodson, K.E., Process-Dependent Thermal Transport Properties of Silicon Dioxide Films Deposited Using Low-Pressure Chemical-Vapor-Deposited, *J. Appl. Phys.*, 85, 7130-7134, 1999.

Kaeding, O.W., Skurk, H., and Goodson, K.E., Thermal Conduction in Metallized Silicon-Dioxide Layers on Silicon *Appl. Phys. Lett.*, 65, 1629-1631, 1993.

Kittel, C., *Introduction to Solid-State Physics*, John Wiley & Sons, New York, chap. 6, 1986.

Kurabayashi, K., Asheghi, M., Touzelbaev, M.N., and Goodson, K.E., Measurement of the Thermal Conductivity Anisotropy in Polyimide Films, *IEEE/ASME J. MicroElectroMech. Syst.*, in press.

Kurabayashi, K. and Goodson, K.E., Prediction of the Thermal Conductivity Anisotropy in Linear-Chain Polyimide Films, *Proc. 5th ASME/JSME Thermal Eng. Joint Conf.,* San Diego, CA, March 14-19, 1999. Submitted to *J. Appl. Phys.*

Lai, J.M. and Majumdar, A., Concurrent Thermal and Electrical Modeling of Sub-Micrometer Silicon Devices, *J. Appl. Phys.*, 79, 7353-7361, 1995.

Lee, S.M. and Cahill, D.G., Heat-Transport in Thin Dielectric Films, *J. Appl. Phys.*, 81, 2590-2595, 1997.

Lee, S.M., Cahill, D.G., and Venkatasubramanian, R., Thermal Conductivity of Si-Ge Superlattices, *Appl. Phys. Lett.*, 70, 2957-2959, 1997.

Leung, Y.K., Paul, A.K., Goodson, K.E., Plummer, J.D., and Wong, S.S., Heating Mechanisms of LDMOS and LIGBT in Ultrathin SOI, *IEEE Electron Device Lett.*, 18, 414-416, 1997.

Majumdar, A., Carrejo, J.P., and Lai, J., Thermal Imaging Using the Atomic Force Microscope, *Appl. Phys. Lett.*, 62, 2501-2503, 1993.

Majumdar, A., Luo, K., Shi, Z., and Varesi, J., Scanning Thermal Microscopy at Nanometer Scales: A New Frontier in Experimental Heat Transfer, *Exp. Heat Transfer*, 9, 83-103, 1996.

Maloney, T.J. and Khurana, N., Transmission line pulsing techniques for Circuit Modeling of ESD Phenomena, *Proc. EOS/ESD Symp.,* pp. 49-54, 1985.

Martin, Y. and Wickramsinghe, H.K., Study of Dynamic Current Distribution in Logic Circuits by Joule Displacement Microscopy, *Appl. Phys. Lett.*, 50, 167-168, 1987.

Morath, C.J., Maris, H.J., Cuomo, J.J., Pappas, D.L., Grill, A., Patel, V.V., Doyle, J.P., and Saenger, K.L., Picosecond Optical Studies of Amorphous Diamond and Diamond-Like Carbon: Thermal Conductivity and Longitudinal Sound Velocity, *J. Appl. Phys.*, 76, 2636-2640, 1994.

Ostermeier, R., Brunner, K., Abstreiter, G., and Weber, W., Temperature Distribution in Si-MOSFET's Studied by Micro-Ramn Spectroscopy, *IEEE Trans. Electron Devices*, 39, 858-863, 1992.

Peters, L., SOI Takes Over Where Silicon Leaves Off, *Semicond. Int.*, 16, 48-51, 1993.

Pollock, D.D., *Thermocouples: Theory and Properties*, CRC Press, Boca Raton, FL, 1991.

Prasher, R.S. and Phelan, P.E., Review of Thermal-Boundary Resistance of High-Temperature Superconductors, *J. Supercond.*, 10, 473-484, 1997.

Richards, P.L., Bolometers for Infrared and Millimeter Waves, *J. Appl. Phys.*, 76, 1-24, 1994.

Richardson, R.A., Peacor, S.D., Uher, C., and Nori, F., $YBa_2Cu_3O_{7-\delta}$ Films: Calculation of the Thermal Conductivity and Phonon Mean Free Path, *J. Appl. Phys.*, 72, 4788-4791, 1992.

Rogers, J.A., Yang, Y., and Nelson, K.A., Elastic-Modulus and In-Plane Thermal Diffusivity Measurements in Thin Polyimide Films Using Symmetry-Selective Real-Time Impulsive Stimulated Thermal Scattering, *Appl. Phys. A*, 58, 523-534, 1994.

Rohsenow, W.M. and Choi, H.Y., *Heat, Mass, and Momentum Transfer*, Prentice-Hall, Englewood Cliffs, NJ, 1961.

Shur, M., *Physics of Semiconductor Devices*, Prentice Hall, Englewood Cliffs, NJ, 1990.

Sondheimer, E.H., The Mean Free Path of Electrons in Metals, *Adv. Phys.*, 1, 1-42, 1952.

Srivastava, G.P., *The Physics of Phonons*, Hilger, New York, 1994.

Stoner, R.J. and Maris, H.J., Kapitza Conductance and Heat-Flow Between Solids at Temperatures from 50 to 300 K, *Phys. Rev. B*, 48, 16373-16387, 1993.

Su, L.T., Antoniadis, D.A., Arora, N.D., Doyle, B.S., and Krakauer, D.B., SPICE Model and Parameters for Fully-Depleted SOI MOSFET's Including Self-Heating, *IEEE Electron Device Lett.*, 15, 374-376, 1994.

Sverdrup, P.G., Ju, Y.S., and Goodson, K.E., Impact of Heat Source Localization on Conduction Cooling of Silicon-on-Insulator Devices, paper presented at the Int. Conf. Modeling Simulation Microsystems, Puerto Rico, April 19-22, 1999.

Swartz, E.T. and Pohl, R.O., Thermal Resistance at Interfaces, *Appl. Phys. Lett.*, 51, 2200-2202, 1987.

Swartz, E.T. and Pohl, R.O., Thermal Boundary Resistance, *Rev. Mod. Phys.*, 61, 605-668, 1989.

Sze, S.M., *Physics of Semiconductor Devices*, John Wiley & Sons, New York, 1981.

Touzelbaev, M.N. and Goodson, K.E., Applications of Micron-Scale Diamond Layers for the IC and MEMS Industries, *Diamond Relat. Mater.*, 7, 1-14, 1998.

Uher, C., Thermal Conductivity of High-T_c Superconductors, *J. Supercond.*, 3, 337-389, 1990.

Verhoeven, H., Boettger, E., Floter, A., Reiss, H., and Zachai, R., Thermal Resistance and Electrical Insulation of Thin Low-Temperature-Deposited Diamond Films, *Diamond Relat. Mater.*, 6, 298-302, 1997.

Vincenti, W.G. and Kruger, C.H., *Introduction to Physical Gas Dynamics,* Krieger, Malabar, FL, 1986.

Yu, Z., Chen, D., So, L., and Dutton, R.W., *PISCES-2ET — Two-Dimensional Device Simulation for Silicon and Heterostructures*, Stanford University Press, Stanford, CA, 1994.

Wachutka, G.K., Rigorous Thermodynamic Treatment of Heat Generation and Conduction in Semiconductor Device Modeling, *IEEE Trans. Computer-Aided Design*, 9, 1141-1149, 1990.

Ward, I.M., *Structure and Properties of Oriented Polymers*, John Wiley & Sons, New York, 1975.

Williams, C.C. and Wickramsinghe, H.K., Scanning Thermal Profiler, *Appl. Phys. Lett.*, 49, 1587-1589, 1986.

Wolbert, P.B., Wachutka, G.K., Krabbenborg, B.H., and Mouthaan, T.J., Nonisothermal Device Simulation using the 2-D Numerical Process/Device Simulator TRENDY and Application to SOI-Devices, *IEEE Trans. Computer-Aided Design Integrated Circuits*, 13, 293-302, 1994.

Zweidlinger, D.T., Fox, R.M., Brodsky, J.S., Jung, T., and Lee, S., Thermal Impedance Extraction for Bipolar Transistors, *IEEE Trans. Electron Devices*, 43, 342-346, 1996.

Ziman, J.M., *Electrons and Phonons*, Oxford University Press, Oxford, U.K., 1960.

4.20 Cooling in Electronic Applications

Pradeep Lall

Introduction

The focus of this section is on cooling in electronic applications. The need and applicability of cooling to electronics has been assessed in detail with discussions on material limits, device performance parameters, and failure mechanisms. The fallacies with the current methodology for characterization of the functional relationship between temperature and electronic reliability have been examined. It has been shown that material limits alone seldom justify the need for cooling. Temperature is more closely linked to degradation in performance. Lowering of the temperature alone may or may not improve reliability — since the majority of the failure mechanisms are based on temperature cycling, the spatial temperature gradient, and time-dependent temperature change. A methodology for evaluating the effects of temperature on reliability, determining the maximum operating temperature, and characterizing the functional relationship between temperature and electronic reliability has been demonstrated.

The cooling needs and constraints vary with the specific product application requirement. The application areas have been broadly categorized into device-level, including PCB and IC packages and system-level including space and size constraints and environment constraints. The space and size constrained devices have been categorized into handheld communication devices and portable computing applications. The environmentally constrained devices include outdoor telecommunication enclosures and high-altitude electronics. For each electronic application the temperature constraints and their bases have been discussed in detail. Typical material sets and heat transfer schemes and their relative effectiveness has also been examined.

Application aspects of cooling in electronics have also been reported in detail. It is not the intent of this section to teach the reader basics of heat transfer — the reader is referred to previous chapters and sections in the book or other basic texts on heat transfer for an in-depth treatment of the subject.

Understanding the Role of Temperature in Design

There has been a common belief that reliable electronics can be achieved by lowering the temperature. Typically, statements such as "lower the temperature by 10°C and double the reliability" are used to characterize the effect of temperature on reliability. This belief in the harmful effects of temperature has woven itself into today's screening and thermal management processes. It is important to emphasize that cooling should be resorted to only when necessary. In this section, we will discuss some of the motivations for incorporating cooling in design and their validity.

The motivation for cooling electronics is based on three major premises

1. Material degrades or causes a fire hazard.
2. Higher temperatures activate failure mechanisms and decreases reliability.
3. Higher temperatures degrade performance.

Determination of requirements for the cooling and design operation temperature is based on the reliability benefits, performance benefits, material limits, and auditability.

Material Limits

Materials typically used in electronics will withstand sustained operation at 125°C. Most electronics today uses 63Sn/36Pb solder which melts at 183°C. Typical reflow temperature is in the neighborhood of 210 to 230°C. Based on the historical precedence of products built using this profile over the decades, it is fairly well known that typical materials including molded-package electronics, silicon circuits, discretes, printed-circuit boards, etc. do not undergo permanent degradation when exposed to temperatures of up to 220°C. Table 4.20.1 shows the material limits for most of the common materials used in electronics. A brief review of the temperature limits indicates that material limits alone would seldom justify the need for cooling.

TABLE 4.20.1 Material Degradation Limits of Electronic Materials Based on Temperature Stress Alone

	Typical Material Sets	Material Degradation Limit Due to Temperature Stress ONLY (°C)
	Integrated Circuits	
Mold compound	Epoxy	T_g = 155–175; T_{max} = 215–245
	Silicone RTV, lead encapsulation	T_{max} = 260; T_{min} = –54
Wirebond	Gold	T_m = 1064.2
	Copper	T_m = 1084.7
	Aluminum	T_m = 660.31
Flip-chip	95Pb/5Sn	
	Electro-deposited nickel	T_m = 1455
TAB	Polymide	T_g = 332–344; T_{max} = 344–360
	Copper metallization	T_m = 1084.7
Interposer substrate	BT-Epoxy	T_g= 220–230; T_{max} = 150–170
	Elastomer	
	Polyimide	T_g = 332–344; T_{max} = 344–360
Package I/O	Alloy42 (lead frame)	
	Copper (lead frame)	T_m = 1084.7
	63Sn/37Pb solder balls	T_m = 183
Metallization	Copper	T_m = 1084.7
	Aluminum	T_m = 660.31
Chip	Silicon	T_{max} = 300–350; T_m = 1414
	Gallium arsenide	
	Discretes	
Spring contacts	Beryllium copper	
	Phosphor bronze	
Capacitors, resistors, inductors	Ceramic (BeO)	T_m = 2527
	Ceramic (Al_2O_3)	T_{max} = 1000
	Copper	T_m = 1084.7
	Polyimide	T_g = 332–344; T_{max} = 344–360
	Assembly-Level	
Interconnections	Solder joints (63Sn/37Pb)	T_m = 183
	Copper metallization	T_m = 1084.7
Printed circuit board	FR-4	T_g = 130–140
	High T_g laminate materials	T_g = 175–180
Plastic casings	ABS	T_{max} = 85
	PC/ABS	T_{max} = 90

Note: T_g = glass-transition temperature; T_{max} = maximum continuous service temperature; T_m = melting temperature.

Impact on Performance

The parameters which are sensitive to temperature depend on device technology and are referred to as the temperature sensitive electrical parameters (TSEP). Table 4.20.2 shows the TSEP vs. device technology.

TABLE 4.20.2 Temperature Sensitive Electrical Parameter vs. Device Technology

Device Technology	Temperature-Sensitive Electrical Parameter (TSEP)
MOSFET	V_{th}, $V_{DS(on)}$, $R_{DS(on)}$
IGBT	V_{th}, $V_{CE(s)}$
Bipolar	V_{BE}, $V_{CE(s)}$
Diode	V_F

The allowable junction temperature, T_J, is determined based on allowable values of the TSEP. The allowable variation in the values of the TSEP allow the device to function without significant degradation in performance. The reference temperature, T_R, is ordinarily measured with a thermocouple at the point of interest. The junction and reference measured temperature values are used to calculate the thermal resistance and the allowable power dissipation. The interested reader is referred to Lall et al. (1997) for a more detailed treatment of the electrical parameter variations vs. temperature.

Impact on Reliability

The impact of temperature on reliability is typically the most misunderstood factor. Accelerated tests including burn-in are often run at high temperatures in the hope that the high temperature will target the dominant failure mechanisms and reveal the suitability of a technology or design for a stress environment. It is often the case that the failure mechanisms precipitated by accelerated testing do not correspond to field life. Because of the nonlinearity of the stress domain of the failure mechanisms, high-temperature accelerated stress may not exist in the equipment operating range of –55 to 125°C. A stress domain is the set of values of stress magnitude over which the failure mechanism can be extrapolated in accordance with a single functional relationship. Linear stress domain signifies that the tendency of occurrence of any one failure mechanism does vary uniformly with change in stress. Conversely, a nonlinear domain signifies that, the accelerated test may not accurately predict how the device will fail during normal use. While high magnitudes of a stress may cause the device to fail instantly, low values of the same stress may not result in failure in any reasonable duration of time.

The naïve practitioner may often ignore the nonlinearity of the stress domain and extrapolate the test results to the operating temperature range of the equipment. Further, overall device reliability is often incorrectly characterized by functional relationships such as the Arrhenius or Eyring models, assigned an activation energy, and then used to extrapolate the test results to operating conditions. One fairly obvious reason for the continuous dependence on this methodology is the imposing simplicity of extrapolation.

Further, several of the failure mechanisms may not be dependent on a steady-state temperature stress, but are instead driven by temperature cycling, spatial temperature gradient, and time-dependent temperature change. Such mechanisms have a minor or no dependence on the absolute value of temperature. Lowering the temperature does not reduce the probability of failure by such mechanisms.

Due to the limited scope of the section it is impossible to go into the details of temperature effects on reliability. Table 4.20.3 shows a summary of the variation in the temperature dependence of typical failure mechanisms in the temperature range of –55 to 400°C. The interested reader is referred to Lall et al. (1997) for an in-depth treatment on the subject.

Thermal Characteristics of Printed Circuit Boards (PCB)

A conventional printed circuit board (PCB) is typically a layered composite consisting of copper foil and a glass-reinforced polymer (FR-4). Figures 4.20.1 and 4.20.2 show the x-section of a four-layer printed circuit board with metal layers and plated-through holes and blind vias.

Figure 4.20.3 shows the mathematical equivalent of the printed-circuit board cross section. Azar and Graebner (1996) and Graebner (1995) have shown that for many thermal calculations, it is convenient to treat the PCB as an homogeneous material, with two different effective thermal conductivities — one describing heat flow within the plane (in-plane) and another for heat flow through the thickness of the plane (through). The equations for calculation each of these conductivities, given the values for the thickness and thermal conductivity of each layer, are presented below:

$$K_{in-plane} = \frac{\sum_{i=1}^{N} K_i t_i}{\sum_{i=1}^{N} t_i}$$

$$K_{through} = \frac{\sum_{i=1}^{N} t_i}{\sum_{i=1}^{N} \frac{t_i}{K_i}}$$

TABLE 4.20.3 Dominant Temperature Dependency Over Steady-State Temperatures from –55 to 500°C

Package Element	Mechanism	Nature of Temperature Dependence		
		–55°C < T < 150°C	150°C < T < 400°C	T > 400°C
		Assembly		
Solder joints	Thermal fatigue	ΔT	ΔT 150°C < T < 183°C None, for T > 183°C 62Sn/36Pb/2Ag solder melts	None 62Sn/36Pb/2Ag solder melts for T > 183°C
	Creep-rupture	σ, T Dominant dependence on stress, σ	σ, T 150°C < T < 183°C, Dominant dependence on stress, σ None, for T > 183°C 62Sn/36Pb/2Ag solder melts	None, 62Sn/36Pb/2Ag solder melts for T > 183°C.
	Intermetallic growth	T, t	T, t	T, t
Underfill	Delamination	ΔT	ΔT	ΔT
PCB	Plated-through holes	ΔT	ΔT ΔT dependence accelerated for $T > T_g$	ΔT ΔT dependence accelerated for $T > T_g$
	High-density interconnect vias	ΔT	ΔT ΔT dependence accelerated for $T > T_g$	ΔT ΔT dependence accelerated for $T > T_g$
	Conductive filament formation	Moisture, voltage, T	Moisture, voltage, T	Moisture, voltage, T
		IC Package		
Wire	Flexure fatigue	Δ T	Δ T	Δ T
Wire bond	Shear fatigue	Δ T	Δ T	Δ T
	Kirkendall voiding	Temperature independent	T dependent for T > 150°C	T dependent
Beam Lead (Elastomer-on-flex packages)	Flexure fatigue	ΔT	ΔT	ΔT
Underfill	Delamination at chip-underfill interface	ΔT	ΔT	ΔT
Die	Die fracture	Δ T	Δ T	Δ T
	Electrical overstress	Temperature independent	T dependent for T > 160°C	T dependent
Die adhesive	Die adhesive fatigue	Δ T	Δ T	Δ T
Encapsulant	Encapsulant reversion (plastic package only)	Temperature independent	T dependent for T > 300°C	T dependent
	Encapsulant cracking (plastic package only)	Δ T, ∂ T/∂ t (≥25°C/s)	T dependent (for T > 215°C), Δ T, ∂ T/∂ t, (≥25°C/s)	T dependent Δ T ∂ T/∂ t
Package	Stress corrosion	∂ T/∂ t	∂ T/∂ t	∂ T/∂ t
Die metallization	Corrosion	Mildly T dependent, ∂ T/∂ t	∂ T/∂ t	∂ T/∂ t
	Electromigration	Structural nonuniformity dependent, ∇T	T dependent (for T > 150°C), ∇T dependent	T dependent (for T > 150°C), ∇T dependent
	Hillock formation	Δ T	Δ T	T dependent (for T > 400°C)
	Metallization migration	Temperature independent	Temperature independent	T dependent (for T > 500°C)

Package Element	Mechanism	Nature of Temperature Dependence: –55°C < T < 150°C	150°C < T < 400°C	T > 400°C
	Contact spiking	Temperature independent	Temperature independent	T dependent (for T > 400°C)
	Constraint Cavitation	T dependent	T^{-1} dependent	T^{-1} dependent
	Slow trapping	Temperature independent	T dependent (for T > 175°C)	T dependent
	Electrostatic discharge	Temperature independent (in presence of protection circuits)	Temperature independent (in presence of protection circuits)	Temperature independent (in presence of protection circuits)
	Time-dependent dielectric breakdown	Voltage dependent, Weak T dependence	Voltage dependent, Weak T dependence	Voltage dependent, Weak T dependence
Device	Ionic contamination	T^{-1} dependence (device not operational)	T^{-1} dependence (device not operational)	T^{-1} dependence (device not operational)
	Forward second breakdown	Temperature independent	T dependent (for T > 160°C)	T dependent
	Reverse second breakdown	Mild T^{-1} dependence	Mild T^{-1} dependence	Mild T^{-1} dependence
	Surface charge spreading	Temperature independent	T dependent (for T > 150°C)	T dependent (for T > 150°C)
Device/oxide interface	Hot electrons	T^{-1} dependent	Temperature independent	Temperature independent

Note: Temperature dependency includes T (steady-state temperature), Δ T (temperature cycle), ∇T (temperature gradient), ∂ T/∂ t (time-dependent temperature change).

Source: Lall, P., Pecht, M., and Hakim, E., *Influence of Temperature on Microelectronic and System Reliability,* CRC Press, Boca Raton, FL, 1997.

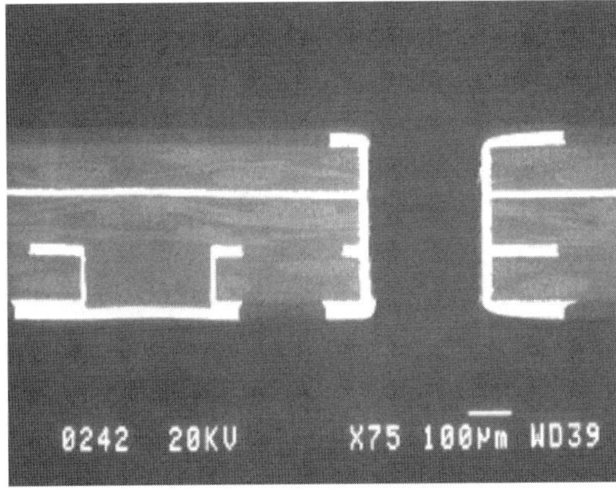

FIGURE 4.20.1 X-section of a four-layer printed circuit board. (Courtesy of Motorola.)

where K is the conductivity in (W/mK) and t is the thickness in (m). The subscripts in-plane, through, and I indicate the in-plane, through, and the individual-layer components, respectively. The above equations assume complete copper planes — which is typically not the case in most designs (except for layers with ground planes). The nonground-plane layers are accounted for by calculating the fractional coverage of copper by

$$K_i = f_i K_{Cu}$$

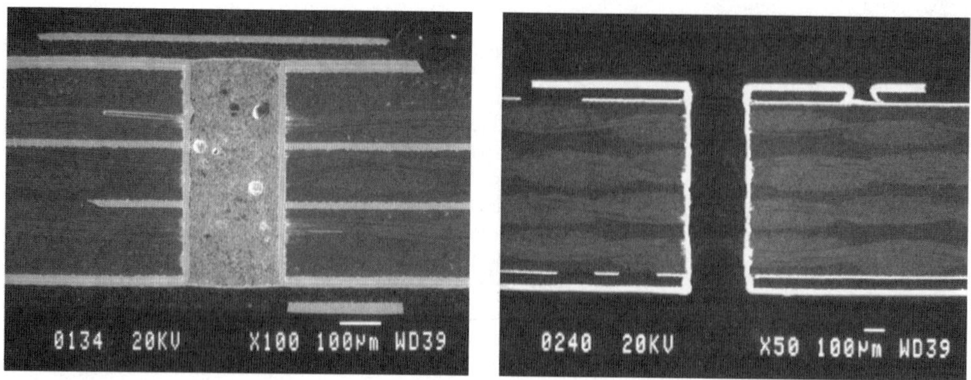

FIGURE 4.20.2 X-section of high-density interconnect PCBs with and without filled vias. (Courtesy of Motorola.)

FIGURE 4.20.3 Mathematical equivalent of the printed circuit board x-section.

where f_i is the fractional coverage of copper for layer i. Figure 4.20.4 shows the in-plane and through conductivity for a 55-mil-thick 2-layer, 4-layer, 6-layer, 8-layer, and 10-layer PCB with ½ oz/ft² copper on the internal layers (assuming that the layers consist only of copper and FR-4, i.e., no vias), where the thermal conductivity for copper is 390 W/mK and thermal conductivity for FR4 is 0.25 W/mK. Figure 4.20.4 indicates that the $K_{in\text{-}plane}$ is two orders of magnitude higher than $K_{through}$. Heat flow through the copper layers dominates the heat transfer because of the large differential in the thermal conductivities

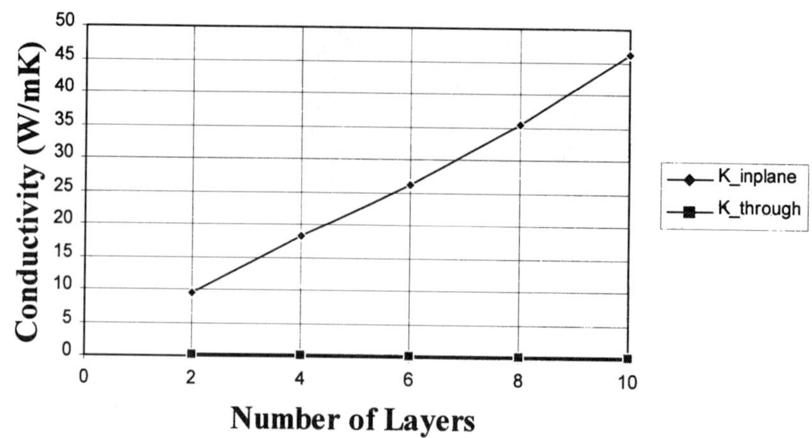

FIGURE 4.20.4 Conductivity vs. number of layers for a 55-mil-thick FR-4 PCB with ½ oz/ft² copper on the internal layers.

of the copper and FR-4 layers. For the same reason, a typical PCB is not an efficient conductor of heat, leading to significant local variations in the PCB temperature in the vicinity of heat-generating components. Thermal conduction is not very efficient for heat flow in a direction lacking a continuous copper path.

Thermal Characteristics of Electronic Packages

Package Architecture

Most electronic packages consist of a silicon chip bonded to an interposer or a lead frame through wirebonds or controlled-collapse solder joints. The package I/O are either solder balls on the substrate or leads on the leadframe. Table 4.20.4 shows a schematic of package cross section and package characteristics for common chip-scale package families including (1) Flex-Substrate BGA, (2) Elastomer-on-Flex package, (3) Partial-Array Chip-Scale Package, and (4) Rigid or Flex Interposer Chip-Scale Package. Other conventional designs include TSOP, QFP, SOIC, etc. The interested reader is referred to the IPC-SM-782 for a more detailed description of the standard package types.

TABLE 4.20.4 Schematic of Package X-Section and Characteristics for Some Chip-Sale Packages

Package Schematic	Characteristics
Chip Mold Attach Polymide Solder PC Board Flex-Substrate BGA	***Flex-Substrate BGA*** is a ball-grid array format chip-scale package. The chip is wirebonded to the single-sided flex-polymide substrate (75 μm thick). Package I/O are 63Sn/37Pb solder balls, 0.45 mm diameter, 0.8 mm pitch which go through holes in the substrate and connect to the solder pads on the chip side of the polyimide substrate.
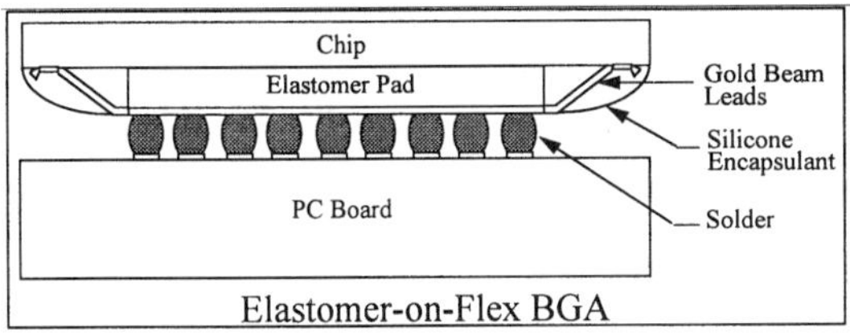 Elastomer-on-Flex BGA	***Elastomer-on-Flex Package*** involves a chip flipped on to an elastomeric substrate. All the chip I/O are routed to a single-sided flexible circuit which is bonded to the other side of the elastomeric substrate by beam leads (gold or copper 18 μm thick). The beam leads are extensions of the printed circuit on the flexible circuit, which are connected to 63Sn/37Pb solder balls. Elastomer-on-Flex BGA typically has 0.3 mm (11.81 mil) diameter solder balls at 0.65 mm pitch.
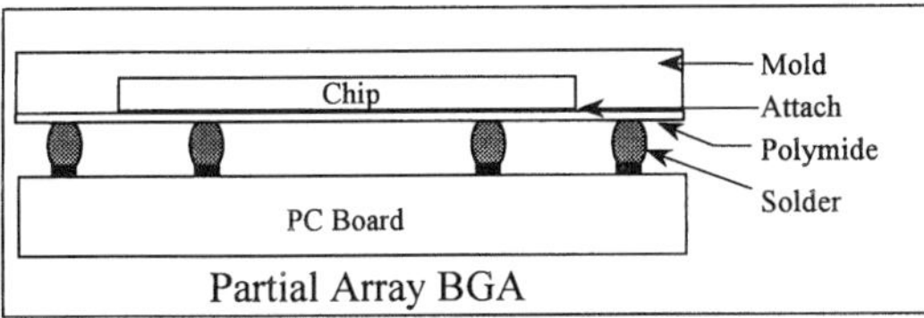 Partial Array BGA	***Partial-Array Chip-Scale Package*** is a ball-grid array format chip-scale package. The chip wirebonded to a multilayer organic substrate. Package I/O are 63Sn/37Pb solder balls, 0.5 mm diameter, 0.8 mm pitch, which go through holes in the substrate and connect to the solder pads on the chip side of the polyimide substrate.
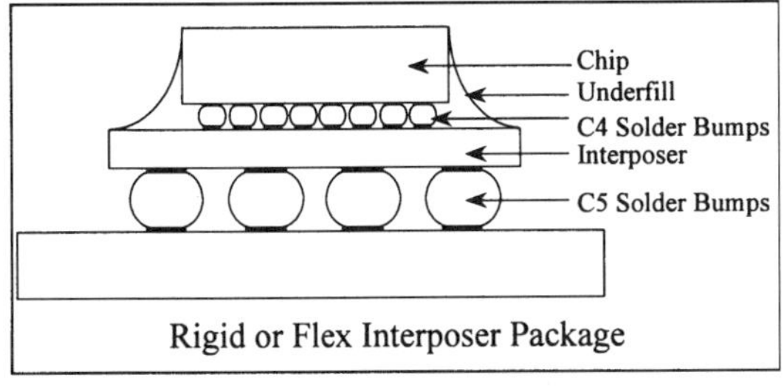 Rigid or Flex Interposer Package	***Rigid or Flex-Interposer Chip-Scale Package*** The rigid or flex interposer packages consists of a solder-bumped integrated circuit (IC) which is flip-chip bonded to an (rigid or flex) interposer substrate approximately 8 mils thick and then underfilled with an encapsulant. Chip I/Os are routed to package I/Os through the interposer substrate to solder bumps (approx. 22.2 mils in diameter on a 32-mil pitch) attached to the bottom side of the interposer substrate.

Materials and Properties

Table 4.20.5 shows the typical material sets used in integrated circuit packages and their thermal conductivities. Die attach bonding is used to mechanically attach the die to the lead frame, and also to provide a heat flow path to the lead frame. A typical adhesive is a thermally conductive epoxy composite with a thermal conductivity of 0.8 W/mK. The epoxy adhesive layer is typically 25 μm thick. Wire bonding is the most commonly used technique in the plastic package interconnection technology. Gold wire is the most preferred material, because of its high oxidation resistance and low work-hardening characteristics. The thermal conductivity of the gold wire is 319 W/m°C. A majority of plastic packages use Novolac-based epoxies. The molding compound is modified by additives that improve its mechanical strength, adhesion to package components, chemical resistance, electrical resistance, coefficient of expansion match, and thermal and moisture resistance in the operational temperature range. The thermal conductivity of a typical Novolac-based epoxy is 0.67 W/mK with a glass transition temperature of 150°C. The predominant semiconductor material in the electronics industry today is silicon. The thermal conductivity of silicon is temperature dependent. Figure 4.20.5 shows the thermal conductivity of silicon vs. temperature in the range of 75 to 150°C.

TABLE 4.20.5 Thermal Conductivity of Typical Packaging Materials

Package Element	Thermal Conductivity (W/mK)
Molding compound	
Novolac epoxy	0.36–2.09
Die attach	
Epoxy	0.8
Solder joints	
63Sn/37Pb	50
Lead frame	
Alloy42	14.7
Cu-Zr	380
Cu-Fe	200–260
Cu-Mg	344
Wirebonds	
Gold	319
Copper	390

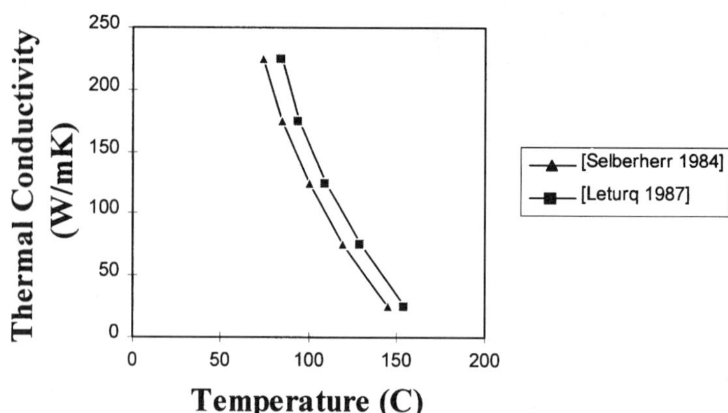

FIGURE 4.20.5 Thermal conductivity of silicon.

Understanding Basic Semiconductor Thermal Parameters

Key parameters used to describe the thermal capability of a semiconductor device include

T_J, junction temperature
T_C, case temperature
T_A, ambient temperature
TSP, temperature-sensitive parameter
T_R, reference temperature (i.e., case or ambient)
$R_{\theta JR}$, junction-to-reference thermal resistance
$R_{\theta JC}$, junction-to-case thermal resistance
$R_{\theta JA}$, junction-to-ambient thermal resistance
$R_{\theta JR(t)}$, junction-to-reference transient thermal resistance
P_D, power dissipation.

The parameters describe the steady-state thermal capability of a power semiconductor device. The thermal behavior of a device is often represented by an electrical equivalent circuit. This circuit consists of a resistor–capacitor network as shown in Figure 4.20.6. Resistors R_1, R_2, and R_3 are all analogous to individual thermal resistances.

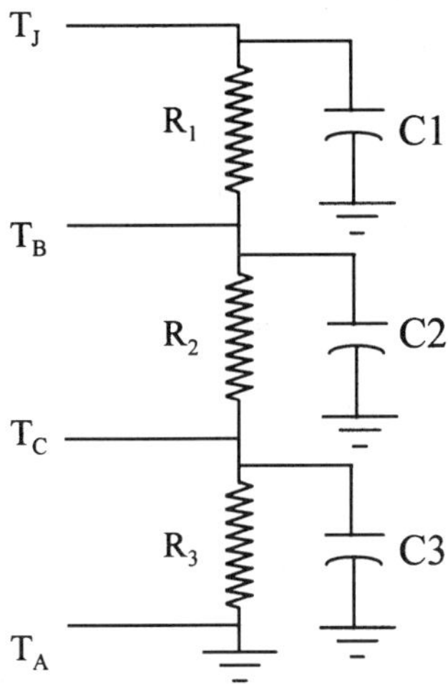

FIGURE 4.20.6 Thermal electrical equivalent circuit heat generated in a device's junction flows from a higher temperature region through each resistor–capacitor pair to a lower temperature region. (Courtesy of Motorola.)

Resistor R_1 is the thermal resistance from the device's junction to its die-bond. Resistor R_2 is the thermal resistance from the die-bond to the device's case. Resistor R_3 is the thermal resistance from the device's case to ambient. The thermal resistance from the junction to some reference point is equal to the sum of the individual resistors between the two points. For instance, the thermal resistance $R_{\theta JC}$ from junction-to-case is equal to the sum of resistors R_1 and R_2. The thermal resistance $R_{\theta JA}$ from junction-to-ambient, therefore, is equal to the sum of resistors R_1, R_2, and R_3. The capacitors model the transient thermal response of the circuit. When heat is instantaneously applied and or generated, there is a charging effect that takes place. This response follows an RC time constant determined by the resistor-capacitor thermal network. Thermal resistance, at a given time, is called transient thermal resistance, $R_{\theta JR(t)}$.

$$R_{\theta JR} = \frac{T_J - T_R}{P_D}$$

$$P_D = \frac{T_{max} - T_R}{R_{\theta JR}}$$

$$T_J = P_D R_{\theta JR} + T_R$$

Generally, T_J is measured utilizing an electrical characteristic of the device that is repeatable and an accurate function of temperature. A key factor in understanding thermal resistance measurements is that if the application configuration is different from that of the measured device, the thermal results will be different. The thermal data on manufacturers' data sheets is designed for comparison with other manufacturers' devices and to give the user a place to begin their thermal management solution.

The thermal performance of power-plastic ball grid array (PBGA) packages cannot be adequately quantified by the traditional $R_{\theta JA}$ measurements. Power-PBGA packages are typically designed thermal vias and "thermal balls," which couples the thermal performance of the package to that of the board. Mulgaonkar et al. (1994) have shown that the thermal performance of the package is dominated by the thermal performance of the printed circuit board.

Figure 4.20.7 shows the comparison of thermal performance between a 100-lead 14 × 14 mm plastic QFP package and a 119-lead 14 × 22 mm PBGA package mounted on a single-layer and a four-layer board. There are 32 vias from the die paddle of PBGA to the array of 21 thermal balls. The thermal balls are soldered to an array of board pads. The board pads are connected to the ground plane in the printed circuit board with 32 vias. The planes in the circuit board are solid 1 oz copper. While the two packages demonstrate very similar thermal performance as measured on the standard single-layer printed circuit board, there is a substantial difference in the performance on the four-layer boards. PBGA packages are more closely coupled to the printed circuit board than QFPs in terms of their sensitivity to temperature and power dissipation of neighboring components. The higher thermal conductivity of the four-layer board with two solid 1 oz planes causes more of the board to act as a heat sink. The effect is enhanced for the PBGA packages because there is a metal conduction path from the die pad to the ground plane of the printed circuit board. This path has much lower thermal resistance than the equivalent path for the QFP.

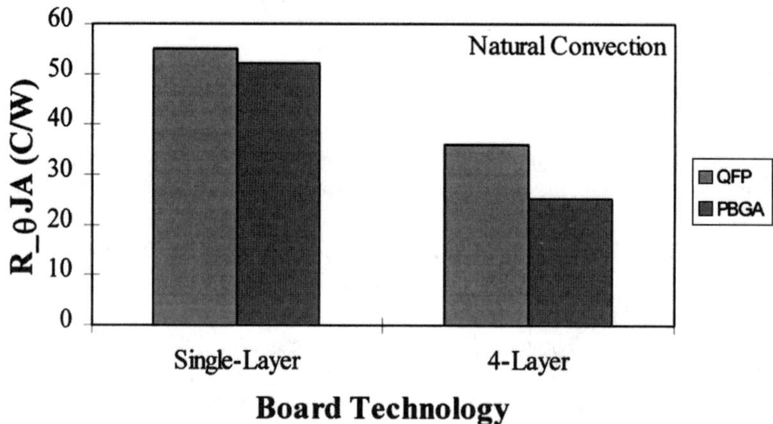

FIGURE 4.20.7 Comparison of thermal performance between a 100-lead 14 × 14 mm plastic QFP package and a 119-lead 14 × 22 mm PBGA package mounted on single-layer and a four-layer boards. (Courtesy of Motorola.)

Figure 4.20.8 shows the $R_{\theta JA}$ vs. $(T_J - T_R)/P_D$ for a PBGA on a populated board subjected to natural convection and various conditions of forced convection, 1 m/s and 2 m/s. The thermal performance of PBGAs is linearly dependent on the board temperature rise above ambient (divided by package power dissipation). Mulgaonkar et al. (1994) represented the thermal performance of electronic packages with

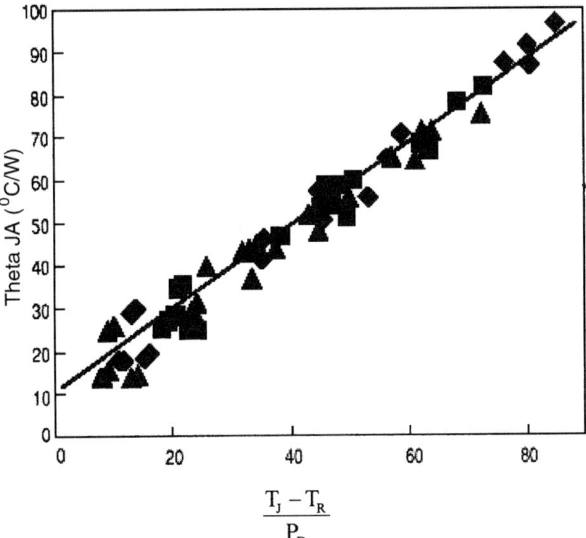

FIGURE 4.20.8 $R_{\theta JA}$ vs. $(T_J\text{-}T_R)/P_D$ for a PBGA subjected to natural convection and various conditions of forced convection, 1 m/s, and 2 m/s. (Courtesy of Motorola.)

equations of straight lines of the form "y = mx + c", where "m" is the slope and "c" is the intercept (Table 4.20.6).

TABLE 4.20.6 Parameters for Thermal Performance of Electronic Packages

	Intercept	Slope
Natural convection	9.8	0.997
Forced 1 m/s	10.7	0.971
Forced 2 m/s	10.7	0.954

Source: Courtesy of Motorola.

TABLE 4.20.7 Effect of Board Configuration on the $R_{\theta JA}$

Board Type	θ_{JA} (C/W) Natural Convection	1 m/s Forced Convection
Single-layer board	52	41
4-Layer board	24	19
8-Components at 1 W	56–62	46–49
8-Components at 2 W		45–49
16-Components at 1 W	104	84

Source: Courtesy of Motorola.

A more traditional way to examine the data is to use a table of θ_{JA} determined by a variety of techniques (Table 4.20.7). The junction temperature depends on the environment, including the conductivity of the board and the power dissipation of surrounding components. The single component on a multilayer board represents the lower extreme, with the higher extreme represented by packages mounted closely together on both sides of the board. The values obtained from the single component on a multilayer board would predict a lower value of the junction temperatures for most applications than would be observed in the typical case with substantial power dissipation in other devices on the board. The effect of the board's power density on the junction temperature is shown by the doubling of the observed θ_{JA}

when the packages are mounted on both sides of the board compared to the single-sided board (Table 4.20.7). Mounting the packages on both sides of the board effectively halves the area available for power dissipation for each package (Mulgaonkar et al., 1994).

The traditional θ_{JA} is useful for comparing package performance and as a preliminary estimate to determine whether further analysis is needed. It gives no information to account for the range of thermal performance given as examples in Table 4.20.7 above. The board temperature is determined by a full board level thermal simulation. There are a number of commercial software codes 6-8 that perform a board level thermal solution with varying degrees of sophistication. These range in sophistication from the 2½ dimensional finite difference or finite element codes to the computational fluid dynamics codes that simultaneously solve the conduction and the fluid flow convection. For all these simulation codes, a simplified thermal model for the package is required.

Thermal Interface Materials

Integrated circuits are kept within their operating temperature limits by transferring the generated heat to the ambient environment. This is often done by attaching a heat sink to the semiconductor package surface. When the electronic component and heat-sink surfaces are brought together, typically, less than 1% of the surfaces make physical contact. Most of the surface area is separated by a layer of interstitial air. The reason is that all surfaces have a certain roughness due to microscopic hills and valleys. Superimposed on this surface, roughness is a macroscopic nonplanarity in the form of a concave, convex, or twisted shape. When the two such surfaces are brought together, only the hills of the surfaces come into physical contact. Some heat is conducted through the physical contact points, but much more has to transfer through the air gaps. Air is a poor conductor of heat, and is replaced by a conductive thermal interface. Interface materials fill the voids and grooves created by imperfect surface finish of two mating surfaces, thus forcing air out and improving surface contact and the conduction of heat across the interface.

Typical thermal interface materials include greases, reactive compounds, elastomers, and pressure-sensitive adhesive films. All are designed to conform to surface irregularities. Some have secondary properties and functions, as well. Thermal greases and epoxies provide the lowest interface resistance, but they are pastes and require care in handling. Elastomers eliminate handling problems but they sometimes require high compressive loads even with well-prepared surfaces. Thermal tapes offer great convenience but their gap filling properties are limited.

Interpretation of Material Data

The thermal properties of interface materials are be measured by either ASTM test standard D5470–93 or military specification MIL–1–49456A. Both methods are recognized throughout the industry as standards. High contact pressures (300 psi for ASTM and 500 psi for MIL SPEC) are used to reduce the effects of interface resistance generated by trapped air. The data gathered at high pressures are then used as the published thermal performance characteristics.

In typical microelectronics packaging applications, the contact pressure varies between 10 to 50 psi and data provided at 300 psi will be an underestimation of material thermal resistance. Any pressure exceeding 50 psi might damage the packaging itself. Typically, the manufacturers do not publish low-contact pressure data because of the case dependency of interface resistance. The thermal resistance of the interface material is a function of the contact pressure.

Types of Thermal Interface Materials

Thermal Greases

Thermal greases are made by dispersing thermally conductive ceramic fillers in silicone or hydrocarbon oils to form a paste. Table 4.20.8 shows some typical grease materials and their thermal conductivity. Sufficient grease is applied to one of the mating surfaces such that when pressed against the other surface, the grease flows into all voids to eliminate the interstitial air. Excess grease flows out past the edges and the thinnest possible thermal joint is formed as the two surfaces come into contact at their high points.

TABLE 4.20.8 Typical Thermal Greases and their Thermal Conductivities

Thermal Grease	Thermal Conductivity (W/mK) @ 300 psi
Zinc oxide/silicone based	0.56
Al_2O_3/nonsilicone	2.78
AlN/nonsilicone	3.47

Joint integrity has to be maintained with spring clips or mounting hardware. Thermal greases are notoriously "user unfriendly," but provide very low thermal resistance between reasonably flat surfaces. Grease does not provide electrical insulation between the two surfaces, and excess grease that flows from the joint should be cleaned up to prevent contamination problems. Greased joints can also dry out with time, resulting in increased thermal resistance.

Thermal Epoxies

Epoxies are useful in cases where a permanent thermal bond is needed, but soldering is not an option. There are three general types (1) silver-filled epoxy can be used from –55 to +125°C, (2) Al_2O_3, and (3) AlN (Table 4.20.9). Both Al_2O_3 and AlN types can operate up to 130°C for a limited amount of time. Thermally conductive compounds are an improvement on thermal grease as these compounds are converted to a cured rubber film after application at the thermal interface. Initially, these compounds flow as freely as grease to eliminate the air voids and reduce the thermal resistance of the interface. After the interface is formed, the compounds cure with heat to a rubbery state and also develop secondary properties such as adhesion. Formulations with adhesive properties do not require mechanical fasteners to maintain the integrity of the joint. Since the binder cures to a rubber, these compounds do not have the migration or the dry joint problems associated with thermal greases. Compounds can be used to fill large gaps where greases would bleed from the joint on account of their migratory nature. Cleanup is also simple as excess material is easily removed after it has been cured to a rubber. Cure times range from 24 h at 22°C, to 2 h at 65°C for silver-filled epoxies, to 1 h at 85°C for Al_2O_3 and AlN-filled epoxies.

TABLE 4.20.9 Typical Thermal Epoxies and their Thermal Conductivities

Thermal Epoxy	Cure Time	Thermal Conductivity (W/mK)
Silver filled	24 h at 22°C	
	2 h at 65°C	1.39
Al_2O_3 filled	1 h at 85°C	0.93
AlN filled	1 h at 85°C	2.78

Elastomer Pads

Thermally conductive elastomers are silicone elastomer pads filled with thermally conductive ceramic particles, often reinforced with woven glass fiber or dielectric film for added strength. These elastomers are available in thicknesses from about 0.1 to 5 mm and hardnesses from 5 to 85 Shore A. Unlike compounds and greases, elastomer pads provide electrical insulation and can be used between surfaces that are at different electrical potential. They are typically used under discrete power devices where electrical isolation is required.

Elastomers do not flow freely like the greases or similar compounds, but will deform if sufficient compressive load is applied to conform to surface irregularities. At low pressures, the elastomer cannot fill the voids between the surfaces and the thermal interface resistance is high. As pressure is increased, more of the microscopic voids are filled by the elastomer and the thermal resistance decreases. For most high-durometer materials, mounting pressures around 300 to 500 psi eliminate the interstitial voids and reduce interface resistance to a minimum. Mounting pressure must be permanently maintained by using fasteners or springs to hold the two surfaces together.

Table 4.20.10 shows the thermal performance of some common conductive elastomers. Despite the 3 to 6 W/m-K thermal conductivity, the thermal resistance at low pressures is in the neighborhood of 1.8°C/W, twice that of grease. Typical contact area at low pressures is in the neighborhood of 30%.

TABLE 4.20.10 Typical Elastomer Pad Materials and their Thermal Conductivities

Elastomer Pads	Thermal Conductivity (W/mK)
Paraffin/Al	3.46 to 6.54
97% Graphite	3.38 to 5.62
Al_2O_3	1.63 to 4.33

Adhesive Tapes

Thermally conductive adhesive tapes are double-sided pressure-sensitive adhesive films filled with sufficient ceramic powder to balance their thermal and adhesive properties. The adhesive tape is usually supported by either an aluminum foil or a polyimide film for strength and ease of handling. Polyimide support also provides electrical insulation. Adhesive tapes perform much like the elastomeric films, in that they also require some initial mating pressure to conform to irregularities in the mating surfaces. They are also unable to fill large gaps between nonflat surfaces. However, once the joint is formed, the adhesive tapes require no mechanical support to maintain the mechanical or thermal integrity of the interface.

Adhesive tapes provide convenience in attaching a heat sink to a semiconductor package because, unlike liquid adhesives, no cure time is required. The film is applied to one of the surfaces, usually to the heat sink, and it is then forced into contact with the semiconductor package to complete the thermal joint. The application pressure is typically 10 to 50 psi for a few seconds. The bond thus formed can be considered permanent and the heat sink is reliably attached to the semiconductor.

However, this convenience comes at a price. The thermal resistance for adhesive tapes is only slightly better than a joint without any thermal interface material. This is because the thermal tapes do not fill gaps as well as liquids, and thermal joints made with tapes will normally include considerable interstitial air gaps. For the most part, the quality of the two joining surfaces will determine the amount of contact that can be achieved and the thermal performance that can be expected. The high shear strength of these thermal tapes means that reliable joints between heat sinks and semiconductors can be achieved, even with poor surfaces and no mechanical fasteners.

Computers

Portable computing applications bring with them challenges of size, weight, and power consumption. The microprocessor (CPU) is the major source of heat dissipation in a portable computer. Cooling systems, in general, increase the size and weight of the portable computer.

Understanding Temperature Limits

Heat from a portable computer is passively dissipated by warming the outside surface (skin) of the plastic or the housing material over the ambient air and actively dissipated by using a fan and heat exchanger. Most fan designs can remove about 4 to 8 W of heat and the rest is passively dissipated by natural convection and radiation from the outside surface. The passive dissipation of the heat requires elevation of the notebook skin temperature over the air temperature. Typically, as a guideline the portable computer skin temperature is kept lower than 50°C. Typical ambient operation temperatures are in the neighborhood of 25 to 35°C, an environment that applies to most air-conditioned places. Most personal computer chassis result in an internal 10°C temperature increase over ambient temperature. In such an environment, the portable computer skin temperature will be 40 to 50°C (assuming 15°C above ambient). Since the human body temperature is 37.1°C, temperatures in the neighborhood of 40°C are perceived as normal and temperatures above 50°C are perceived as hot to the touch. The temperature specifications for many keyboards are generally in 55 to 60°C range and, thus, the keyboard operation will be at the specification limit. Similar logic can apply to other temperature-sensitive components (e.g., HDD, CD-ROM) inside the notebook (Intel, 1998b; AMD, 1998).

Figure 4.20.9 shows the maximum temperature vs. component for a portable computer. Notebook computers are generally put together with standard components of varying maximum ambient temperatures. The failure mechanisms are device dependent. Floppy disk drives and compact disk drives start to degrade in the neighborhood of 50 to 55°C. The media in floppy disk drives warps and sticks to the casing at high temperatures. CD-ROM drives have a laser diode aligned to the compact disc — large dimensional changes produces misalignment and malfunction.

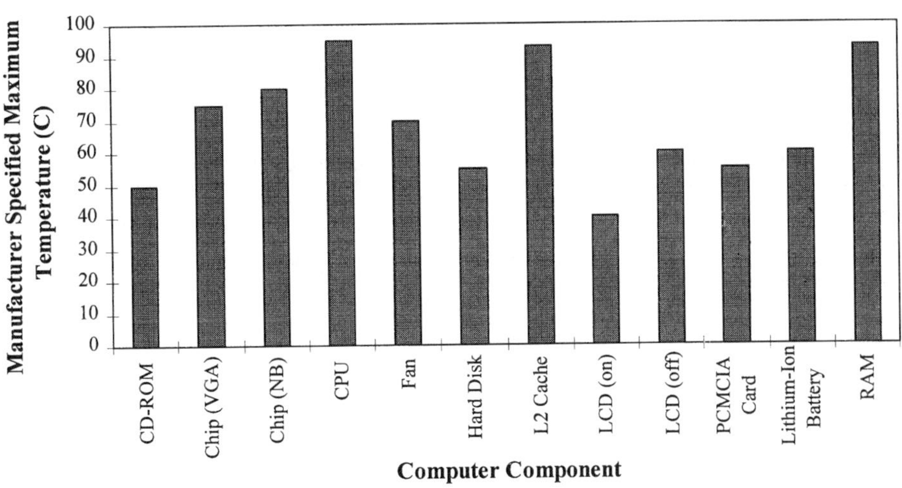

FIGURE 4.20.9 Manufacturer-specified maximum temperature for personal computer components. (Adapted from Sterner, J., *Proc. IMAPS Int. Sys. Packaging Symp.*, 107-112, 1997.)

The temperature dependence of hard drive performance is often alluded to by manufacturers of HDD cooling products. However, temperature (below a certain temperature limit, T_{Limit}) does not directly affect the performance of the drive in any significant way. Seek times and data transfer rates generally stay the same with variation in temperature below T_{Limit}. A rise in temperature, however, does require thermal recalibration, producing a very small performance degradation. Many SCSI drives operating at 10,000 rpm, have a thermal recalibration loop. A rise in the ambient temperature causes the drive to warm up, the platters to expand, and the geometry to change, thus making recalibration necessary. During recalibration, the hard drive makes a short pause to re-calibrate itself. Generally, hard drives operating in cooler environments re-calibrate less frequently. However, new-generation drives with embedded servos do not have thermal recalibration loops and thus their performance is totally unaffected by the operating temperature (up to a certain temperature limit). Hard disk drives are high-precision machines with very tight tolerances. Large temperature variations ($T_{ambient} > T_{limit}$) cause dimensional changes which cannot be compensated for, causing malfunctions. Typically, the upper limit (T_{limit}) of hard drive temperature is in the neighborhood of 60°C. Cooling is generally recommended for all hard drives with rotation speeds higher than 7200 rpm (Steinbrecher, 1998)

Figure 4.20.10 shows the power dissipation typical actively and passively cooled portable computers. The computers compared in Figure 4.20.10 have different equipment. Figure 4.20.11 shows the computer surface temperature vs. power dissipation for typical actively and typical passively cooled portable

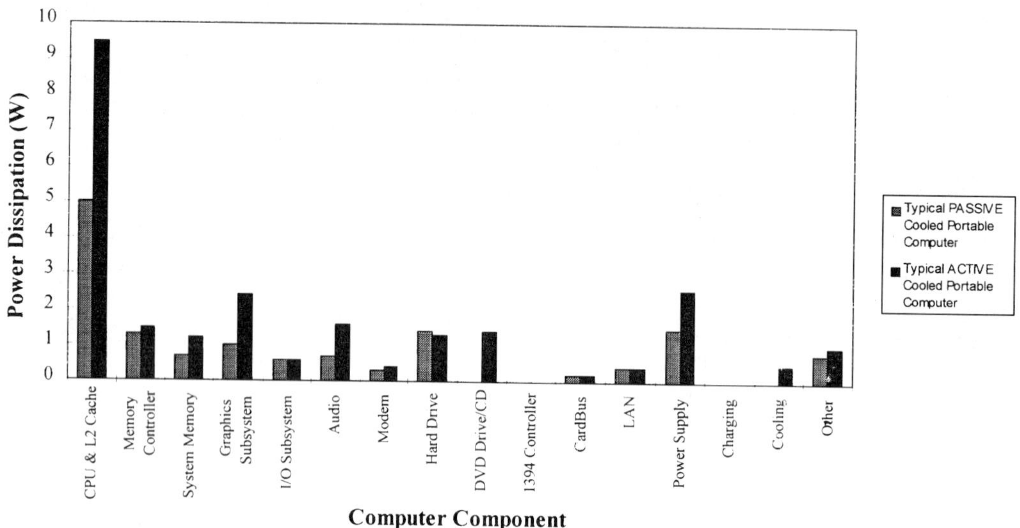

FIGURE 4.20.10 Power dissipation in typical actively and passively cooled portable computers. The computers compared in the figure have different equipment. Actively cooled portable computers can dissipate more power for any temperature specification of the computer casing temperature. (Based on data from Intel, 1998b.)

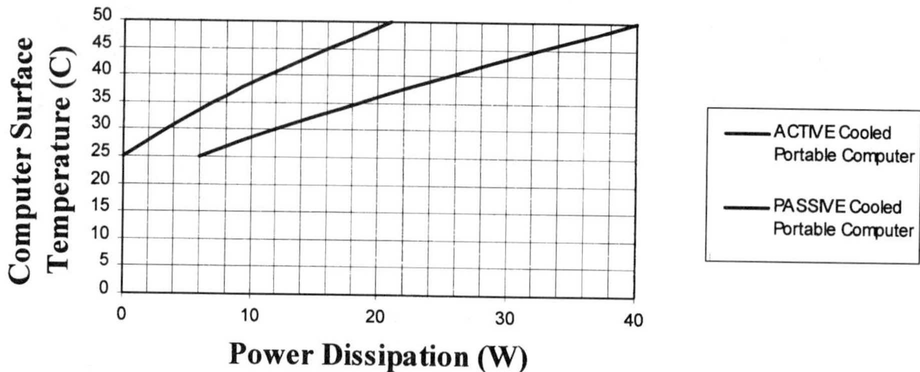

FIGURE 4.20.11 Computer surface temperature vs. power dissipation for typical actively cooled and passively cooled portable computer for an ambient temperature of 25°C. The computers compared in the figure have different equipment. (Based on data from Intel, 1998b.)

computers for an ambient temperature of 25°C. The bottom surface of the computer is insulated. The top (keyboard) and side walls are used as a radiator. The notebook casing is made of ABS/PC plastic or a painted metal chassis (emissivity = 0.95) and has an isothermal casing temperature. The data presented in Figures 4.20.10 and 4.20.11 are for an actively cooled portable computer $10 \times 12.3 \times 1.0$ in. base unit (excluding lid thickness) and a passively cooled portable computer with a base size of $8.2 \times 10.2 \times 0.65$ in.

In general, actively cooled portable computers can dissipate more power than passively cooled portable computers and still meet the temperature specification for the computer casing temperature (Figure 4.20.10). While the CPU is not the only component dissipating heat, it is certainly one of the major contributors to an increase in computer surface temperature. Other major contributors are the graphics card, memory, hard drives, and power supplies. For operation at 15°C above ambient temperature (25°C), about 15 to 17 W of total platform power (entire notebook power minus the display power) can be managed passively in a typical notebook. The allowable power dissipation for an actively cooled portable computer is in the neighborhood of 24 to 25 W (Figure 4.20.11).

Casing Materials

External skin temperature of portable computers is designed to be safe to human touch. For most purposes temperatures above 50°C are deemed unsafe to human touch. Computer casings are generally made of ABS (acrylnitrite-butadiene-styrene) resin, PC/ABS blend (polycarbonate/ABS), or magnesium alloy. The maximum temperature for continuous operation for both ABS and PC/ABS casings is in the range of 85 to 90°C. Table 4.20.11 shows the properties for some typical ABS and PC/ABS blends used for computer housings.

TABLE 4.20.11 Properties for Some Typical ABS and PC/ABS Blends Used for Computer Housings

Material Properties	ABS acrylonitrile-butadiene-styrene	PC/ABS Polycarbonate/acrylonitrile-butadiene-styrene
Tensile strength — yield (psi)	5300–6500	8200–9100
Flexural strength — yield (psi)	9400–11000	12500–14800
Modulus of elasticity (psi)	3.40e+05 to 3.77e+05	3.20e+05 to 3.90+05
CTE (ppm/°C)	100–140	80 Normal-to-flow; 76 Parallel-to-flow
Deflection temperature at 0.46 MPA (°C)	90	100
Deflection temperature at 1.8 MPA (°C)	85	90
Max. continuous service temperature (°C)	85	90
Thermal conductivity (W/m-K)	0.14–0.30	0.2–0.25
Specific heat (J/g-°C)	1.2–1.4	1.2–1.4

The ABS resin has low thermal conductivity and the molding process constraints limit the wall thickness of ABS enclosure to a minimum of 1.2 to 1.4 mm. Acrylonitrile-butadiene-styrene (ABS) thermoplastic resin is widely recognized as an engineering material offering a good balance of properties centering around toughness, hardness, and rigidity. Resin grades in this product family consist of a blend of an elastomeric component and an amorphous thermoplastic component. The elastomeric component is usually polybutadiene or a butadiene copolymer. The amorphous thermoplastic component is SAN: a copolymer of styrene and acrylonitrile. The flexibility offered by the use of the three-monomer system (A, B, and S) allows tailoring of the property profile.

A variation of the ABS material includes blends of amorphous polycarbonate (PC) and acrylonitrile-butadiene-styrene terpolymer (ABS), which combines the most desirable properties of both materials, resulting in an optimum balance of performance, processability, and cost. By varying the ratio of PC and ABS, resins can be tailored to meet the specific property requirements of a wide range of high-performance applications from automotive body panels and computer housings to instrument panels and cellular phones. Cycoloy resins offer the processability of ABS together with the mechanical properties and impact and heat resistance of polycarbonate. These thermoplastic alloys are formulated for exceptional flow for filling thin wall sections and complex parts.

In contrast, magnesium alloy can be injection-molded to a wall thickness of 0.7 mm. Magnesium's best attributes are its high strength-to-weight ratio and light in weight. As a structural material it can be alloyed with several other elements including aluminum, manganese, rare-earth elements, silver, zinc, and zirconium. The various alloys have unusually high strength and are light weight. Magnesium is most commonly found alloyed with aluminum where its addition to aluminum increases hardness and improves corrosion resistance. Magnesium also possesses superior heat radiation properties, heat stability, electromagnetic interference shielding properties, and damping capabilities. Recent developments in thin-wall casting combined with magnesium's other characteristics are leading to increased applications in the electronics field for cellular telephones, video cameras, and personal computers (casings, battery shields, etc.). Table 4.20.12 shows the properties for typical magnesium alloys used for computer housings.

TABLE 4.20.12 Properties for Typical Magnesium Alloys Used for Computer Housings

	Magnesium Alloy	
Material Properties	AM60A-F	AZ91A-F
Tensile strength — yield (ksi)	16.68	21.76
Poisson's ratio	0.35	0.35
Modulus of elasticity (psi)	6.52e+06	6.52e+06
CTE (ppm/°C)	25.6	26
Thermal conductivity (W/m-K)	62	72
Specific heat (J/g-°C)	1	1.05

Typical Heat Transfer Schemes

Figure 4.20.12 shows some of the typical active and passive heat transfer schemes used for portable computing applications. Figure 4.20.12 (left) shows a passive heat transfer scheme using a heat pipe along with a heat spreader to conduct the heat from the microprocessor. Figure 4.20.12 (right) shows an active heat transfer scheme using a finned heat sink and fan.

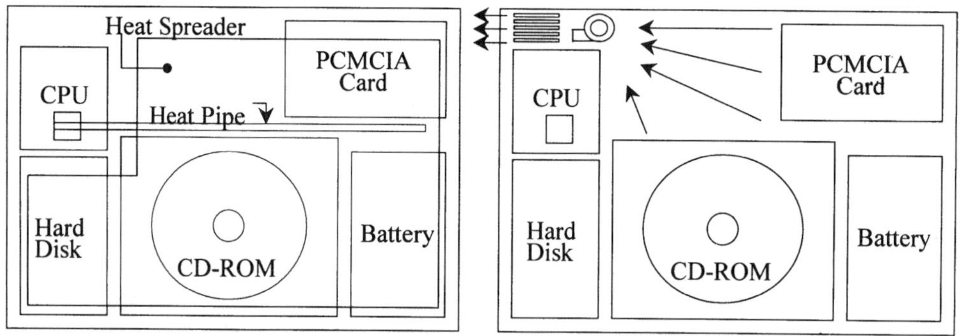

FIGURE 4.20.12 Some typical heat transfer schemes for portable computers. (Adapted from Sterner, J., *Proc. IMAPS Int. Sys. Packaging Symp.*, 107-112, 1997.)

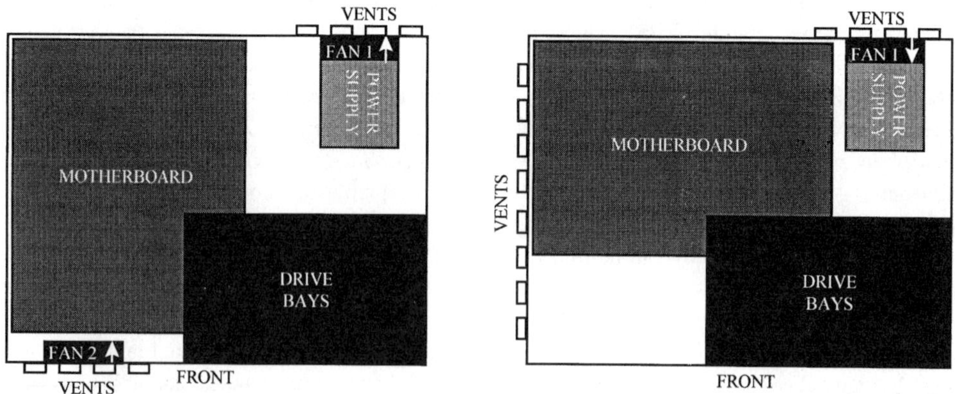

FIGURE 4.20.13 Some common fan configurations for AT and ATX form-factor desktop computer cases (Data from Steinbrecher, 1998; AMD, 1998).

In desktop computers, fans are used to evacuate the hot air from the case. Figure 4.20.13 shows some of the common fan configurations in AT and ATX form-factor cases. Case fans come in many flavors, but usual sizes are 80 × 80 mm and 92 × 92 mm. The size and location of the fan depends on the computer case. Most desktop computers come in AT or ATX form-factor cases. In general most designs use some combination of fans blowing air into and out of the computer case. The idea being that the CPU located directly in the airflow will get cooled. However, as the power supply gets hot, the fan blows warm air towards the CPU. There are two possibilities for installing the fan: either above the power supply on the back (especially common with ATX cases), or on the lower part of the front side (common with AT cases). Care is exercised in determining the direction of air flow to make sure that (1) the fans do not create an air short-circuit, which prevents circulation of air through the computer case, and (2) the fans do not try to work against each other and create a vacuum inside the case.

The increase in system performance with each new processor generation increases the system level thermal demands. In order to meet these thermal needs, the current systems require either multiple fans or heat sinks to maintain the necessary junction temperatures for the baseboard components. Need for multiple fans can be avoided by using fan ducts. The fan duct distributes the air from a single fan to all parts of the system that require cooling, thus avoiding the need for multiple fans. A recommended path for a low-cost solution in these systems is to provide a fan duct that blows external air onto the processor core components (processor, chipset, graphics, graphics controller, and memory). (See Intel, 1998a.)

Handheld Communication Devices

Mobile communications products such as cellular phones, pagers, two-way radios, and personal digital assistants (PDAs) are being driven to increased functionality and smaller form factors by technological enhancements at device, package, and system levels.

Understanding Application Requirements

Typically, handheld products involve plastic cases which have been snap-fitted or screwed together — space is at premium. The electronic packages are assembled in small, confined enclosures; the stand-alone thermal performance of the packages does not guarantee similar levels of performance in the phone. The problem is further complicated by the fact that the plastic housings for handheld products have to remain cool enough to be safe to human touch.

Pagers, typically, are receivers, and their active components dissipate low power (a few milliwatts) over a short operational time (several milliseconds). Cooling is generally not a concern. However, newer pagers, including two-way paging, involve transmitting signals with a power dissipation of approximately 1 W over an extended period of operation. Phones generate most of their heat in the power amplifier module (PA). The power amplifier can dissipate power in the range of 1 to 3 W. Typically, cellular phones are two-board designs including the main RF board and a keypad board. Typically, the keypad board has the front keypad popples for the phone and very few if any components. The power amplifier module and other ICs are mounted on the main RF board. Two-way radios are similar to cellular phones except the power dissipation requirements, in general, are significantly larger. Handheld two-way radio PA may generate 1 to 10 W of power depending on the mode of operation. Two-way radios, like the cellular phones, are two-board designs including a main RF board and a keypad board. The PA, in general, is located on the main RF board.

Thermal Management: Some Test Cases

The choice of thermal management schemes is severely limited because of the space, size, weight, and cooling system power requirement constraints. The added weight of carrying a heavy cooling system in a handheld communication device is not an attractive option — let alone put up with the fan noise while talking on the phone or two-way radio. Cooling system power requirements further increase battery size and reduce standby and talk time. Passive cooling schemes, in general, are preferred — most of the heat generated in handheld portable products by the integrated circuits is transmitted to the externals, predominantly by conduction and to a lesser extent by natural convection and radiation.

Table 4.20.13 shows some of the common passive heat transfer schemes implemented by providing a solid conduction path between the power amplifier module (PA) and the product casing. The solid conduction path between the PA and the system housing can be either a metal block, or a thermally

TABLE 4.20.13 Thermal Enhancement Configurations

Base Case	Thermal Pad on Back Side of PA	Cu Foil	Thermal Pad on Front Side of PA	Cu Foil with Heat Spreader	Cu Spring	Thermal Pad with Heat Spreader	Cu Spring with Heat Spreader

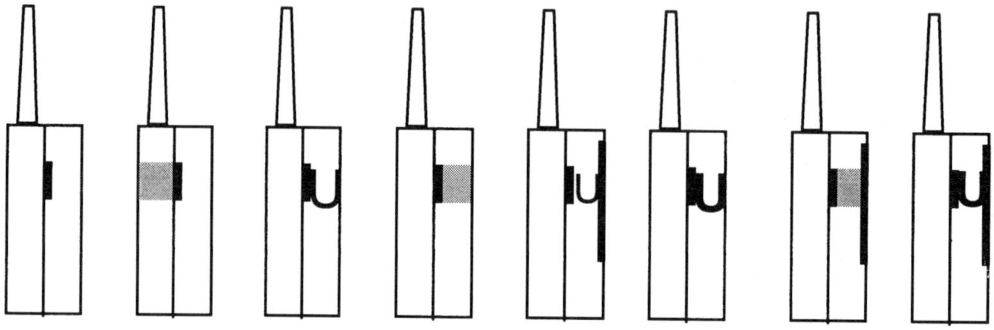

Source: Courtesy of Motorola.

conductive pad (epoxy or silicone with a thermal conductivity of ~2 W/mK), a metal spring (~0.8 mm thick), or a thin metal foil (~0.04 mm thick). The material used for the metal block, spring, or foil could be either copper or aluminum. The metal block, conductive pad, or spring is simply inserted into the gap and the contact is maintained by the press fit. An adhesive tape may be required to attach the foil to the PA and housing. The metal foil can also be used as a heat spreader on the inside surface of the system case. If the power dissipated by the PA can be effectively transferred to the case wall (such as by a solid conduction path), then the heat spreader can quickly distribute the local heat flux over a larger area, reducing component temperatures.

Figure 4.20.14 shows the PA temperature vs. power dissipation at the PA for various thermal enhancement techniques. Lee et al. (1998) showed that the least effective enhancement was made by positioning the thermal pad between the board (on the back side of the PA) and the case. The configuration with the thermal pad directly contacting the PA reduced PA temperature by 34%. The copper spring, due to its higher thermal conductivity and larger heat transfer cross-sectional area, provided the best solid conduction path. The copper foil provided an intermediate enhancement. In addition to the solid conduction path, adding the heat spreader to the case further reduced the PA temperature by an average of 26%. A combination of a thin metallic spring and a heat spreader on the enclosure provided the highest enhancement, with a 68% reduction in PA temperature.

Casing Materials

Typically, handheld portable product casings are made from ABS or PC/ABS blends. Materials are chosen based on various attributes including thin-wall molding, mold release, elastic modulus, and thermal conductivity. Table 4.20.14 shows the mechanical and thermal properties for typical handheld product casing materials.

Impact of Temperature on Handheld Product Reliability

In general, high temperature does not drive the reliability of handheld portable products. A high-temperature bake is often used to cure moisture-related problems in printed circuit boards and packages. A 24-h bake at 150°C prevents popcorn or blistering of the packages or the printed circuit board during reflow. However, a high temperature during operation does impact the handheld product casing temperature.

Product reliability has a dominant dependence on thermal cycling instead of high steady-state temperature. Thermal fatigue is a wear-out mechanism as a result of product usage (e.g., making phone

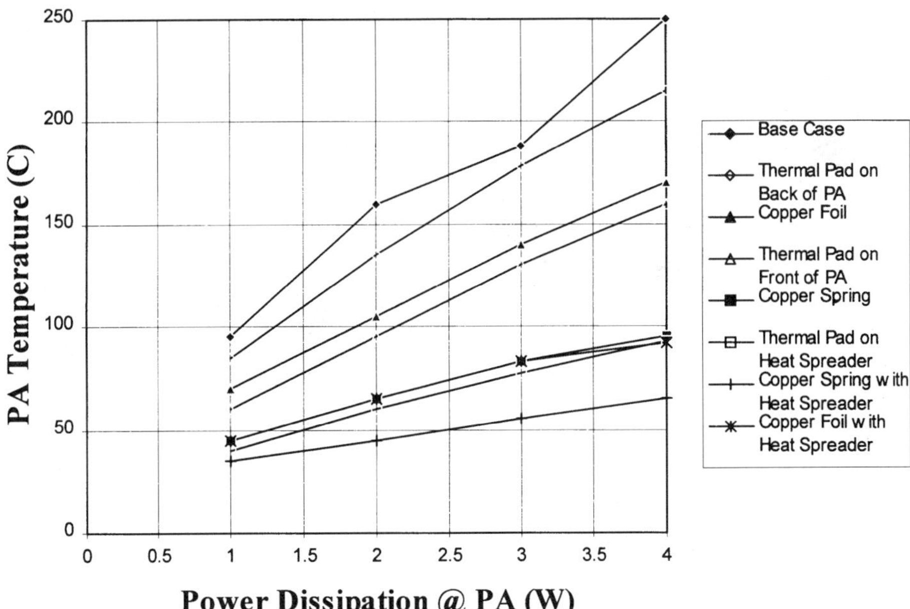

FIGURE 4.20.14 Comparison of various thermal enhancement techniques (the average experimental uncertainty is ±4%). (Courtesy of Motorola.)

TABLE 4.20.14 Material Properties of the Housing for a Typical Handheld Portable Product

Material Properties	
Tensile strength — yield (ksi)	8.0–9.0
Flexural strength — yield (ksi)	11.5–14.0
Modulus of elasticity (psi)	3.40E+05
Thermal conductivity (W/m-°C)	0.27
CTE (ppm/°F)	38
Specific heat (BTU/lb-°F)	0.3
Density (lb/in^3)	0.043

calls; operating the device in home, office, and car; leaving phone in the car, etc.) in different environments and product power cycling (turning on and off). Figures 4.20.15, 4.20.16, and 4.20.17 show the dominant chip-scale package failures caused by thermal fatigue in package form factors typically used in handheld communication products.

Figure 4.20.15 shows a x-section of a flex-substrate BGA after thermal fatigue failure. The 48 I/O, 0.8 mm pitch Flex-Substrate BGA packages (SRAM) were soldered onto a 0.8-mm-thick, immersion Ni/Au finish PC board. The board assemblies were subjected to LLTS, –55 to 125°C, 5.8 c/h. The primary crack propagation site is at the package-to-solder joint interface.

Partial-array BGAs fail primarily through solder joint failure. Figure 4.20.16 shows an x-section of a partial-array BGA package (48 I/O, 0.5 mm pitch) after thermal fatigue failure. The 48 I/O, 0.5 mm pitch partial-array BGA was soldered to a 1.2-mm-thick OPC finish PC board and subjected to LLTS, –55 to 125°C, 5.8 c/h. The primary crack propagation site is at the package substrate-to-solder joint interface.

Elastomer-on-Flex packages fail in thermal fatigue, primarily due to the fracture of the beam leads internal to the package. Figure 4.20.17 shows x-sections for Elastomer-on-Flex Packages (48 I/O, 0.65 mm and 0.75 mm pitch) flash-soldered onto 0.8-mm-thick Ni/Au finish PC boards. Beam-lead failure typically occurs in the neighborhood of 400 cycles (vendor A) to 650 cycles (vendor B). The board assemblies were subjected to LLTS, –55 to 125°C, 5.8 c/h. The interested reader is referred to

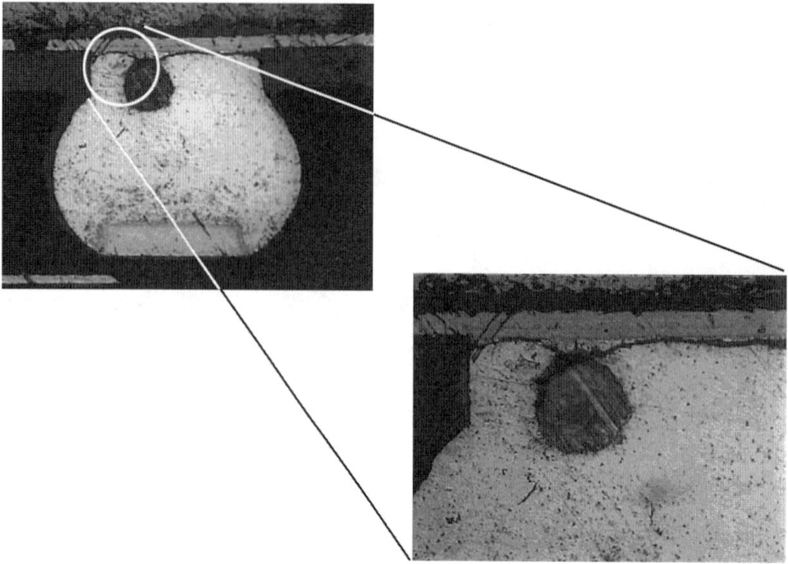

FIGURE 4.20.15 X-section of Flex-Substrate BGA thermal fatigue failure — the primary crack propagation site is at the solder joint to package interface.

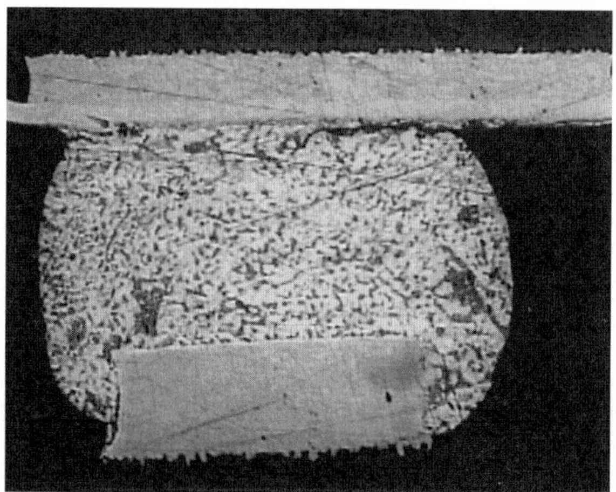

FIGURE 4.20.16 X-section of 0.5 mm pitch package after thermal fatigue failure. Notice the crack propagation at the package-to-solder interface.

Lall et al. (1998) for a more detailed discussion of nonlinear reliability modeling and failure mechanisms for the Elastomer-on-Flex packages.

Figure 4.20.18 shows the thermal fatigue reliability for the partial-array BGA (0.5 mm pitch, 48 I/O), Elastomer-on-Flex package (0.65 mm pitch, 48 I/O), Flex-substrate BGA (0.8 mm pitch, 48 I/O). Weibull distribution distinguishes between failure mechanisms by a difference in the slopes of the distributions. The similarity in the slopes of the Flex-substrate BGA and the partial-array BGA packages is because of similarity in the mechanisms of failure — solder joint fatigue. The difference in the slope of the Elastomer-on-Flex package, is because of a different failure mechanism — beam lead failure. Figure 4.20.19 shows the normalized accelerated test time-to-first-failure of the packages (LLTS -55 to 125°C, 5.8 c/h) vs. package type.

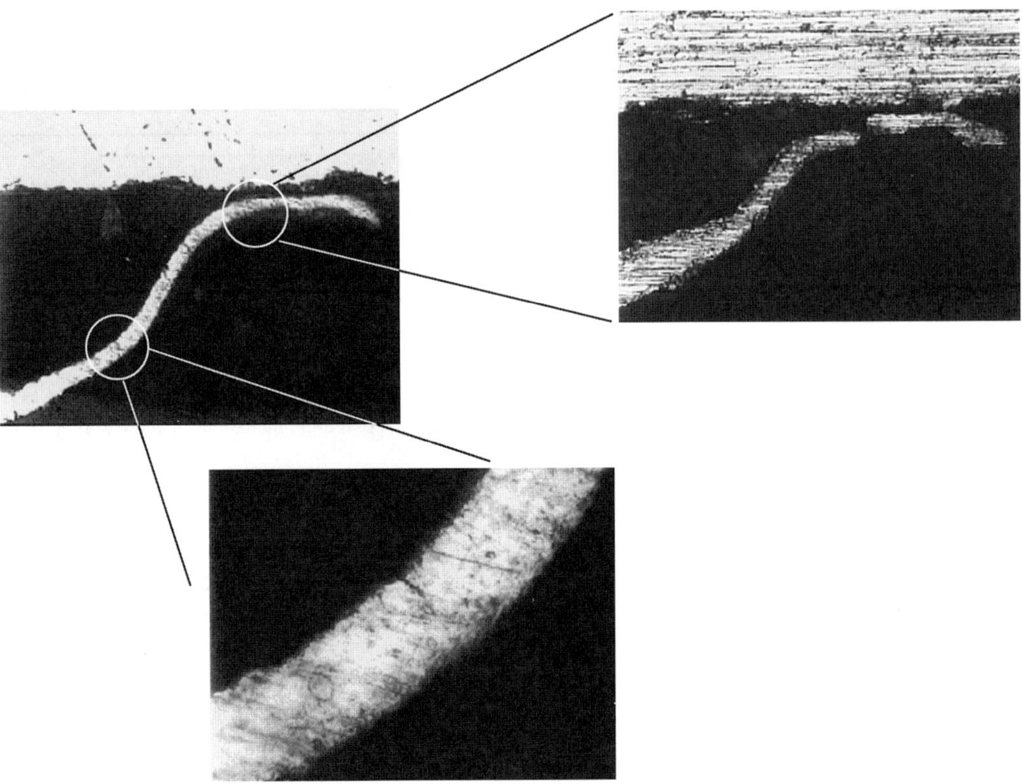

FIGURE 4.20.17 X-section of the Elastomer-on-Flex package after thermal fatigue failure. The solder joints do not fail — the dominant failure mechanism is the thermal fatigue failure of the beam leads internal to the package.

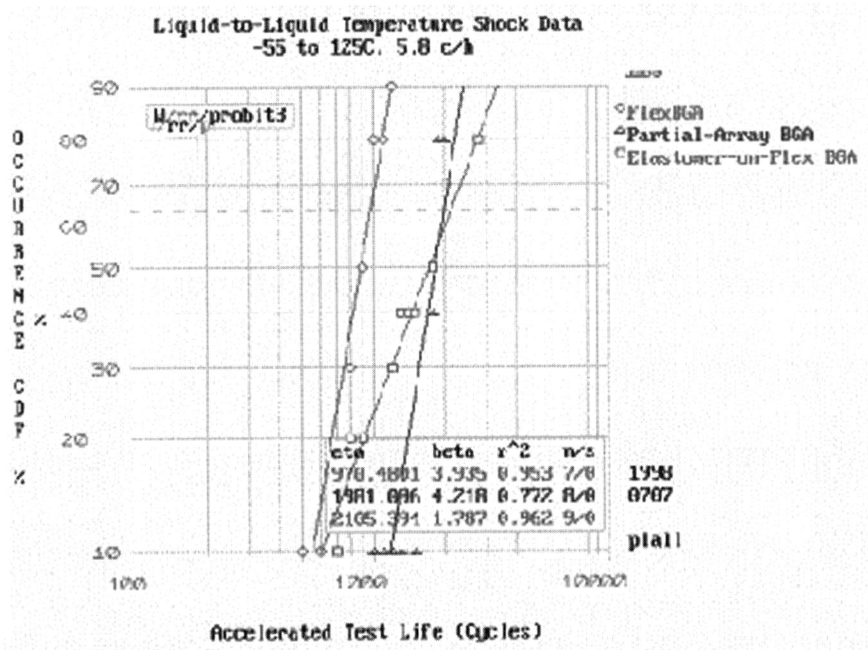

FIGURE 4.20.18 Weibull distributions for the Flex-substrate BGA package, Elastomer-on-Flex package, and the partial-array BGA package subjected to LLTS, –55 to 125°C, 5.8 c/h.

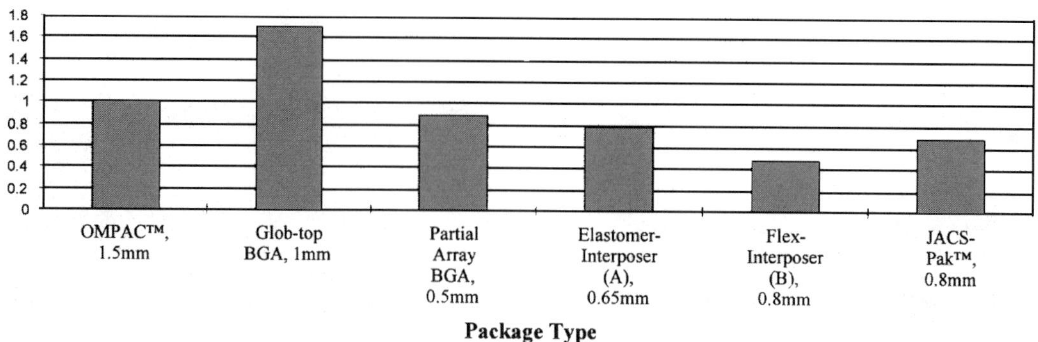

FIGURE 4.20.19 Comparison of the thermal fatigue reliability for the partial-array BGA (0.5 mm pitch, 48 I/O), Elastomer-on-Flex package (0.65 mm pitch, 48 I/O), Flex-substrate BGA (0.8 mm pitch, 48 I/O), and Glob-top BGA (1 mm pitch, 196 I/O) vs. OMPAC™ (1.5 mm pitch, 68 I/O).

Acceleration Transforms

Environmental usage profile has been developed based on the phone usage profile, typical environmental extremes, and the measured thermal behavior of the radio. The profiles (Figure 4.20.20) have been developed for two distinct environmental conditions — extreme (heat or cold). Nonlinear finite element simulations have been run to evaluate the field life and correlate the LLTS cycles to years of field life.

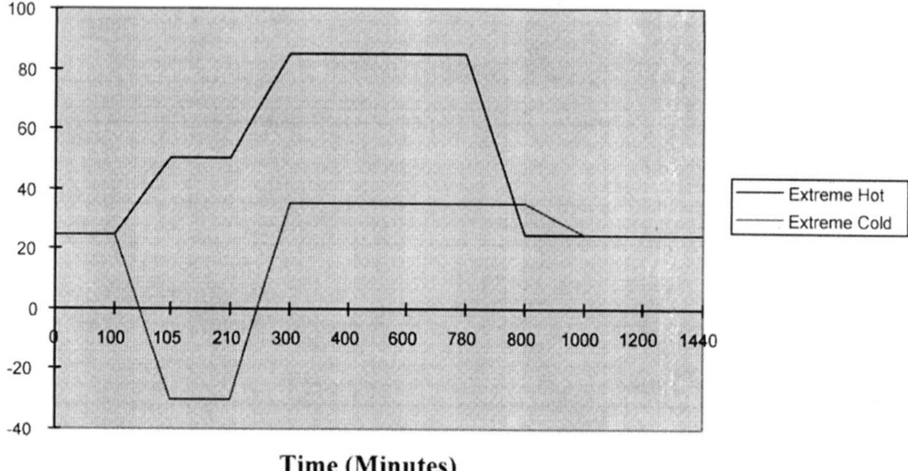

FIGURE 4.20.20 Schematic of temperature profile for extreme hot and extreme cold environments (Data from Lall et al., 1998).

Acceleration transforms have been derived based on nonlinear finite element model predictions. An acceleration transform quantifies the acceleration factor for an accelerated test environment compared to actual field use environments. The acceleration transform is in turn determined by several variables including the component and board geometry, materials, architecture, and most importantly the field profile. Ideally, one would like to have one acceleration transform which can be used to evaluate the field life once the life in accelerated test has been evaluated.

Unfortunately there is no universal acceleration transform — further, there can be no such universal relationship. The reason being that the variables which influence this acceleration transform vary from component to component and, thus, can be derived only for a class of components, e.g., plastic BGAs, ceramic BGAs, TSOPs, pendulum, etc. in specific product architectures. Care has to be exercised in using such a simplification of the damage mechanics as the "acceleration transform."

The acceleration transform lumps into itself several geometry and architecture variables integral to the component, solder joint geometry, and the assumed field profile. If any of the variables change, the acceleration transform has no option but to change. For this reason, acceleration transforms make sense for a particular set of conditions and for particular components. The field profile is the temperature-time plot of the field environment for the radio. The acceleration transform represents the relationship between the accelerated test life and field life for a particular accelerated test and particular field profile. Solder is viscoplastic in nature — which means that it is sensitive to absolute values of temperature, temperature ramp rate, and time spent at temperature. Changes in either the field profile, or accelerated test conditions changes the damage sustained in solder, and thus the joint reliability. **For this reason, the acceleration transforms presented here are valid for the accelerated test conditions and environmental profiles for which they have been derived.**

Figure 4.20.21 shows the acceleration transforms for the 0.5 mm pitch, 48 I/O partial-array package and the Flex-substrate BGA 0.8 mm pitch, 48 I/O package. The vertical bars indicate the time-to-first-failure for the package in each environment. The acceleration transforms can be used to calculate the accelerated test requirement for each package or inversely predict field life in any field use environment or any combination thereof. It is known from Miner's superposition rule that

$$\sum_{i=1}^{m} \frac{n_i}{N_i} = 1$$

where n_i is the number of cycles in a particular environment, and N_i is the number of cycles to failure in the same environment. Miner's superposition rule states that the cumulative damage at end of life will be equal to 1. For example, the accelerated test requirement for a 10-year field life can be calculated, based on Figure 4.20.21 and Miner's superposition, as follows: n_i = 10 years, N_i = 15 years for extreme hot environment (m = 1), and N_i = 1025 for accelerated test environment. We want to find n_i = ?:

$$n_i = \frac{10}{15} \times (1025) = 684 \text{ Cycles}$$

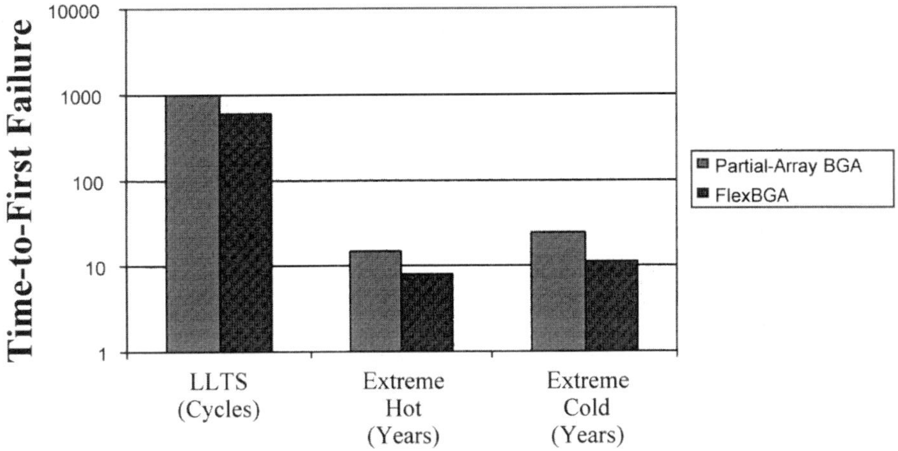

FIGURE 4.20.21 Acceleration transforms for partial-array BGA (0.5 mm pitch, 48 I/O package) and Flex-substrate BGA (0.8 mm pitch, 48 I/O package) subjected to thermal fatigue. (Data from Lall et al., 1998)

Thus, 684 cycles of LLTS are required for the 0.5 mm pitch, 48 I/O package to guarantee a 5-year life in extreme hot. The above method can also be used to calculate the accelerated test requirement for any combination of the above environments. For example, we want to calculate the accelerated test requirement for a product which will be in extreme hot for 6 years and in extreme cold for 4 years: $n_1 = 6$ years; $N_1 = 15$ years for extreme hot environment; $n_2 = 4$ years; $N_2 = 25$ years for extreme cold environment; and $N_i = 1025$ for accelerated test environment. We want to find $n_i = ?$:

$$n_i = \left(\frac{6}{15} + \frac{4}{25}\right) \times (1025) = 574 \text{ Cycles}$$

Thus, 574 cycles of LLTS are required for the 0.5 mm pitch, 48 I/O package to guarantee a 10-year life (6 years in extreme hot and 4 years in extreme cold.) (Lall et al., 1998)

Outdoor Telecommunication Electronics

Telecommunications electronics is extremely decentralized — located away from the controlled and protected environment of the central office — outside the plant (OSP).

Understanding the Environment

The OSP environment contains rain and humidity, dust and pollutants, significant daily and annual temperature swings, wide solar heat-load variations, and physical abuse. To protect electronics from this hostile environment, they are housed in a variety of OSP enclosures, which can range from small buildings (30 m^3) to small boxes (3×10^{-3} m^3) that are mounted on the homes of customers. Heat densities within OSP enclosures can be large, often rivaling those of electronics designed for controlled indoor environments. Insolation is the amount of solar energy that is absorbed and transmitted into the enclosure.

A second source of heat in an outdoor enclosure is the solar energy that is absorbed and transmitted into the enclosure. The total insolation includes direct shortwave radiation from the sun, diffuse shortwave radiation that has been scattered by the atmosphere, and any shortwave radiation that is reflected from the ground or nearby surfaces. It is a complicated function of the orientation of the enclosure with respect to the earth-sun line, the amount of moisture, dust and contaminants in the air, and the solar reflectivity and location of nearby surfaces (e.g., landscaping, fences and buildings). Procedures for calculating the total shortwave radiation for an enclosure with a known location and orientation are well documented (see ASHRAE, 1977; HVAC, 1987).

Typically, outdoor enclosures are designed for maximum solar load at the maximum ambient air temperature and the maximum internal heat load (Bellcore, 1993). Maximum solar load is a function of the enclosure location and orientation. Figure 4.20.22 shows the total insolation incident on a 2 ft × 4 ft × 6 ft enclosure located in a low-reflecting open field environment (i.e., no surrounding reflecting surfaces). The shortwave absorptance of the enclosure surface affects the solar load absorbed into the enclosure. Table 4.20.15 shows typical values of the shortwave absorptance and longwave emittance for common enclosure surface treatments including polished aluminum and oil-based paints (white, light green, and light gray). Shortwave absorptance controls the fraction of incident solar heat absorbed into the enclosure, and the longwave emittance controls the heat that is dissipated by radiation to the surroundings.

Hot surfaces radiate heat continuously in proportion to the fourth power of their absolute temperature (Stefan-Boltzman law); the net heat radiated depends on the longwave emissivity of the surface and on the emissivities and temperatures of the surroundings.

The major heat sink for the heat dissipated from the enclosure is the local ambient air. The temperature of this sink varies widely with location, time of year, and time of day. Typical specification on the ambient air temperature range is –40°C to + 46°C (–40°F to + 115°F) (Bellcore, 1993). Sky temperature is always lower than ambient air temperature, especially when the air contains little water vapor; this is the reason that frost occurs on terrestrial surfaces even when the air temperature is well above freezing.

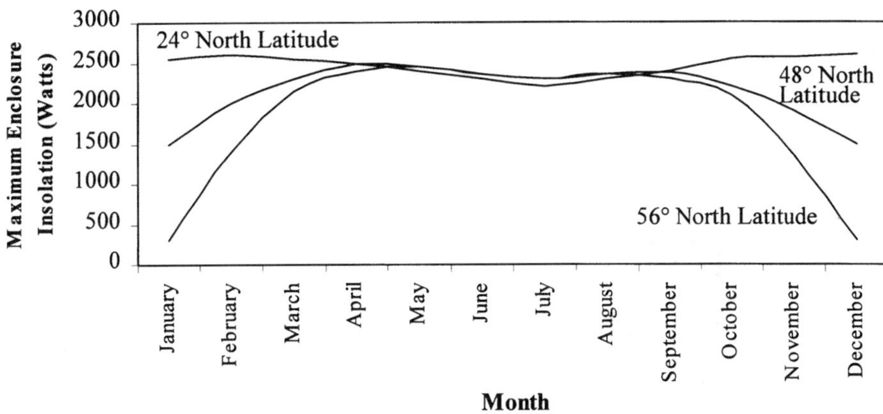

FIGURE 4.20.22 Maximum enclosure insolation vs. calendar month and latitude.

TABLE 4.20.15 Typical Values of Shortwave Absorptance for Common Enclosure Surface Treatments

Surface	Shortwave Absorptance	Longwave Emittance
Polished aluminum	0.03	0.05
White(oil-based paint)	0.2	0.9
Light green (oil-based paint)	0.5	0.9
Light gray (oil-based paint)	0.75	0.9

The atmosphere/remote sky is most effective as a radiant heat sink during the spring and fall, when dew point and dry-bulb temperatures are relatively low, and in dry dessert climates. It is not effective under overcast conditions or summer conditions of high dew point/dry-bulb temperatures.

Typical Heat Transfer Methods

Passive cooling methods are predominantly used in outdoor enclosures because of their simplicity and lack of maintenance. The most common cooling method used today in the OSP is relatively simple. The enclosure is unventilated to protect the internal equipment from rain, dust, and contaminants in the outside air. The internal heat is transferred primarily by convection to the inside surfaces of the enclosure, by conduction through the walls of the enclosure, and then by convection and radiation to the external heat sinks. Other passive methods include natural (free) convection in conjunction with phase-change materials (PCM) and solar reflectors. PCMs are substances that change phase, most often from solid to liquid, as they absorb heat. PCMs are selected for the temperature at which they change phase and for the latent heat associated with phase change. PCMs are sometimes used in conjunction with thermosiphons. Typical PCMs for high-temperature applications include waxes, salts, and paraffins. Water (ice) is used for low-temperature applications. The PCM is kept inside or attached to the enclosure in appropriately designed and sealed reservoirs (Figure 4.20.23). PCMs takes advantage of thermal inertia and phase change effects. For example, an enclosure with PCMs during daylight hours will absorb heat through the cabinet walls and protect the electronics within the enclosure from overheating. The heat absorbed during the day will then be released to the outside world at night when it is cooler.

PCMs can also be incorporated into assisted systems for the cooling of enclosures. In order to enhance cooling, PCMs can be incorporated into a heat-exchanger structure in which two fluids that are at different temperatures are separated by a PCM (possibly in encapsulated form). Hotter, internal air is first circulated through the heat exchanger and cooled by transferring its energy to the PCM, which slowly changes phase. This will occur during the hottest part of the day when external air temperatures prevent the use of outside air to cool the enclosure. Later in the day, when outside air temperatures drop, outside air is

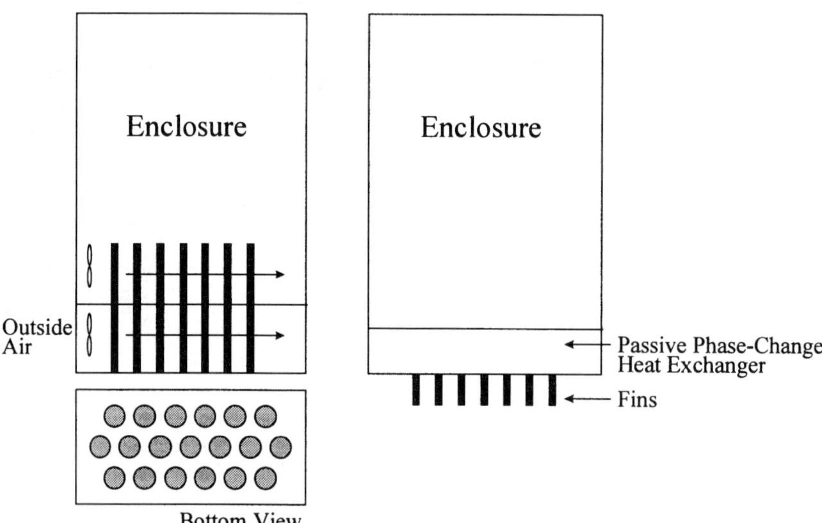

FIGURE 4.20.23 Heat transfer scheme for an outdoor enclosure using phase-change materials. (From Marongiu, M.J. and Clarksean, R., *Electron. Cooling,* 4(1), January, 1998. With permission.)

brought in to remove the heat stored in the PCM. Figure 4.20.23 shows a sketch for this simplified heat exchanger (Maringiu and Clarksean, 1998).

High-Altitude Airborne Electronics

The heat transfer scheme is chosen based on the electronic module power dissipation surface density. Free convection is favored for power dissipation densities less than 500 W/m^2 (0.05 W/cm^2) on the surface of a fully populated PCB. Forced convection is used for power dissipation densities up to 4000 W/m^2. Heat pipes, liquid cooling, and boiling heat transfer is favored for power dissipation densities greater than 4000 W/m^2 (0.4 W/cm^2). This upper limit for forced air cooling is dictated by the fact that heat sinks cannot be used on the circuits due to the lack of space in avionics racks. The boards are designed for a minimum pitch.

Understanding the Environment

Figure 4.20.24 shows the decrease in density (kg/m^3) with increase in altitude (km). Air pressure at altitudes greater than 21 km (70,000 ft) is so low that very little air is available for cooling. The volumetric flow (m^3/s) needed to cool the unit increases with an increase in altitude. The static air temperature decreases with an increase in altitude. In most hot-day cases, the air temperature decreases with altitude. Typically, the reduction in temperature only occurs up to approximately 12 km (40,000 ft). On most cold days the temperature actually increases with altitude in the first 3 km then the temperature begins to decrease. Cold days are not of great concern because the hot day defines the design point for worst-case cooling conditions.

Impact of Equipment Location

Location of the equipment on aircraft affects electronic cooling designs, based on the different qualities of air. Equipment located in the pressurized bay may be cooled by forced air cooling using a fan. The flow rate (Q) can be calculated from the overall power dissipation (P_d) and air temperature increase through the equipment by

$$Q = \frac{P_d}{\left(T_{out} - T_{in}\right) m_p c_p}$$

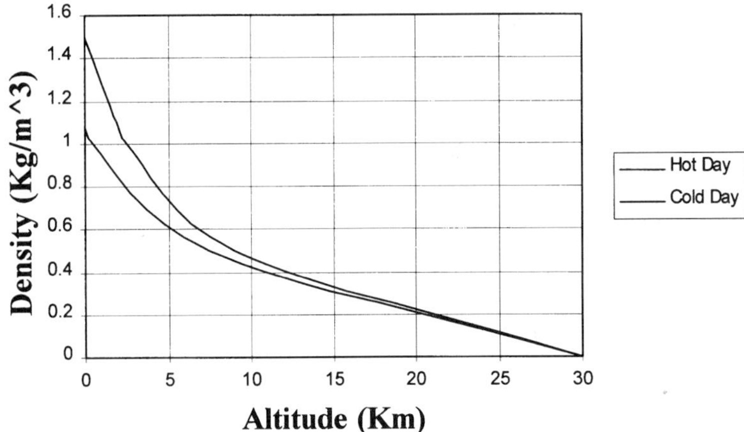

FIGURE 4.20.24 Variation in air density with altitude (Marthinuss J.E., Jr. and Hall, G.T., *Electron. Cooling,* 4(3), Sept., 1998. With permission.)

where T is the air temperature increase through the equipment, and the subscripts *out* and *in* indicate the air temperature at the outlet and the inlet, m_p is the air specific mass, and c_p is the specific heat. The pressure drop can be calculated at the flow rate Q [$\Delta P = f(Q)$]. A fan can be selected for the application based from a manufacturer's catalogue using these two characteristics (Q, ΔP). Air velocity distribution obtained on each circuit should provide (1) a convective heat transfer coefficient sufficient to limit the temperature gradient between the components and the air, and (2) provide uniformity of air flow distribution on each board. Local heat sinks can also be used on high-power components if space is available. Alternatively, a more powerful fan can be selected.

Electronics located in depressurized bays experience a decrease in air density with increase in altitude. The air density decreases with the decrease in atmospheric pressure. The air mass flow rate decreases with the decrease in air density. This means that a fan with a higher volumetric flow is needed. Lower mass flow rate increases the outlet air temperature going through the equipment. This may or may not be compensated for by the lower static air temperature at the inlet at higher altitudes.

Fighter aircraft typically have an environment control system which can provide cool air to the electronic equipment. Flow rates and pressure drops obtained in such an environment is sufficient to cool the electronics efficiently. The main problem of forced convection cooling in such situation is due to air pollution. This air contains particles such as dust, sand, metals, and fluid droplets such as oil, kerosene and liquid water. Unfiltered air used for electronics cooling increases the risk of micro short-circuits, abrasion, or chemical degradation of the components. Cooling with unfiltered air is obtained by avoiding contact between the components and the pollutants — air is passed through cold plates installed at the board level or at the rack level.

Figure 4.20.25 shows a compact, finned cold plate embedded in a double-sided circuit. This is an air-flow-thru module. The heat generated by the electronic components is transferred first by conduction through the board and then by forced air convection. The forced convection heat transfer coefficient is enhanced by the compact fins of the cold plate. Air is usually directed along the width of the circuit boards in order to limit the pressure drop.

Figure 4.20.26 shows a rack which has two cold plates cooled by the environmental control system air. The rack includes circuit boards which are cooled by conduction enhanced by a metal plate (generally aluminum or copper) in contact with the wall cold plates through thermal clamps located on each side of the circuit.

Direct air forced convection on components may authorize power dissipation densities on a fully populated board up to 3000 W/m^2, with an air velocity in the range of 5 m/s. This type of design requires nonpolluted air and a powerful fan on account of the high velocity of 5 m/s which will generate high

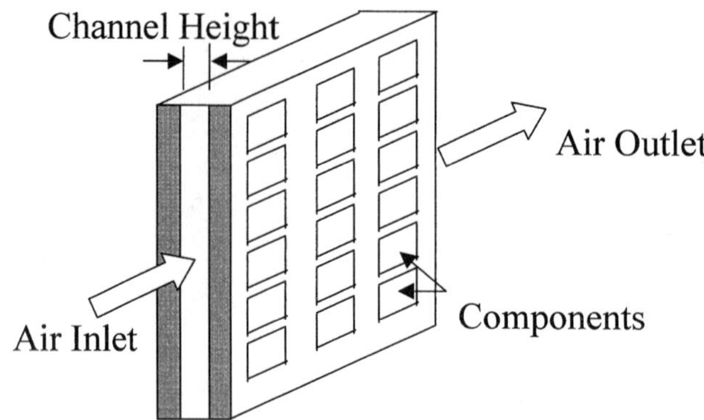

FIGURE 4.20.25 Schematic of a typical air flow-through module. (From Assouad, Y. et al., *Electron Cooling,* 3(2), May 1997. With permission.)

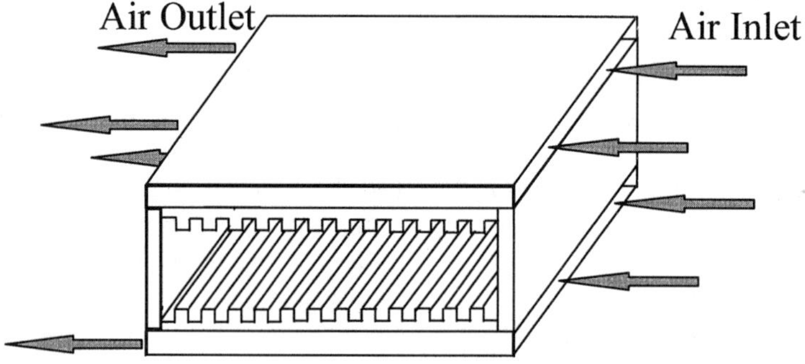

FIGURE 4.20.26 Schematic of a typical air-cooled rack for conduction cooled electronic modules. (From Assouad, Y. et al., *Electron Cooling,* 3(2), May 1997. With permission.)

pressure drops. An AFT may allow power densities up to 3000 W/m^2 or more, but with a very large pressure drop penalty. For 3000 W/m^2, the pressure drop will be in the order of 5 mbar for a 100-mm-wide module, which yields for a 200-mm-wide classical module a total pressure drop of 10 mbar, not accounting for the distribution and collecting air chambers (Assouad et al., 1997). Conduction-cooled modules in an air-cooled rack are limited to power densities up to 1500 W/m^2, due to the parabolic temperature gradient from conduction in the metal plate embedded in the PCB.

Fan Selection

Fan selection is based on attainable flow rate and pressure drop at the higher operating altitudes. The attainable flow rate is determined by the fan supply curve and the pressure drop is determined by the system pressure drop curve. The intersection of the two curves determines the operating point.

Conventional fans run at a constant speed. The operating point of a conventional fan varies with density since the pressure drop and flow rate curves vary with air density. The volumetric flow rate for a conventional fan is constant, while the pressure output is drastically different at different altitudes. The decrease in density produces a decrease in pressure drop and the volume flow rate,

$$\frac{Q_{h1}}{Q_{h2}} = \left(\frac{D_{h1}}{D_{h2}}\right)^3 \left(\frac{\omega_{h1}}{\omega_{h2}}\right)$$

where Q is the volume flow rate, D is the fan diameter, ω is the fan's rotational velocity (rpm), and the subscripts h1 and h2 indicate fan operation at heights h1 and h2.

$$\frac{\Delta P_{h1}}{\Delta P_{h2}} = \left(\frac{D_{h1}}{D_{h2}}\right)^3 \left(\frac{\omega_{h1}}{\omega_{h2}}\right)^3 \left(\frac{\rho_{h1}}{\rho_{h2}}\right)$$

where ΔP is the pressure drop, D is the fan diameter, ω is the fan's rotational velocity (rpm), ρ is the air density, and the subscripts h1 and h2 indicate fan operation at heights h1 and h2. Since the fan diameter and fan rotational velocity is constant, the equations can be reduced to

$$Q_{h1} = Q_{h2}$$

$$\Delta P_{h1} = \Delta P_{h2}\left(\frac{\rho_{h1}}{\rho_{h2}}\right)$$

The first equation shows that the volumetric flow rate is constant with height for a conventional fan. The following equation shows that the pressure output of the fan is directly proportional to the density of the air. Thus, the mass flow rate will drop even though the volumetric flow rate is constant since the two are related as follows:

$$Q_h = \frac{m_h}{\rho_h}$$

where Q is the volume flow rate, is the mass flow rate, ρ is the air density, and the subscript h denotes the altitude.

A slip fan is very different from a conventional fan. A slip fan allows the blades to slip or run at different speeds from the motor driving the fan, which allows the fan to work under many different altitudes and changing density conditions. The pressure flow rate curves for a slip fan at different altitudes are more consistent than a conventional or non-slip fan.

Summary

The need and applicability of cooling techniques in electronics has been discussed in terms of material limits, performance, and reliability. The use of acceleration transforms for determining the required operational temperature for temperature-dependent mechanisms has been discussed. The thermal management of electronic systems has been discussed based on application temperature constraints. Typical heat transfer schemes for each application environment have been discussed.

References

AMD (Advanced Micro Devices), AMD-K6 Processor Thermal Solution Design, Appl. Note, Doc. Number 21085, Rev. H, Amendment 0, November, 1998.

ASHRAE, *Handbook and Product Directory Fundamentals,* American Society of Heating, Refrigerating, and Air-Conditioning Engineers, Atlanta, GA, 1977, chap. 26.

Assouad, Y., Caplot, M., and Gautier, T., Advanced Cooling Techniques Comparison For Airborne Electronic Circuits, Semitherm, Austin, TX, January, 1997.

Assouad, Y., Forced Convection Cooling Of Airborne Electronics, *Electron. Cooling,* 3(2), May, 1997.

Azar, K. and Graebner, J.E., Experimental Determination of Thermal Conductivity of Printed Wiring Boards, Proc. SEMI-THERM XII Conf., Austin, TX, March, 1996, pp. 169-182.

Banerji, K. and Darveaux, R., Effect of Aging on the Strength of and Ductility of Controlled Collapse Solder Joints, First Int. Conf. Micros. Mech. Prop. Aging Mater., TMS, Warrendale, PA, 1993.

Bellcore, TA-NWT-000487, Generic Requirements for Electronic Equipment Cabinets, Issue 2, June 1993.

Bradley, E., Lall, P., and Banerji, K., Effect of Thermal Aging on the Microstructure and Reliability of Ball Grid Array (BGA) Solder Joints, in Proc. Surf. Mount. Int. 1996, San Jose, CA, September 8-12, 1996, pp. 95-106.

Cowan, J.P., *Handbook of Environmental Acoustics,* Van Nostrand Reinhold, New York, 1994.

Coyne, J.C., An Approximate Thermal Model for Outdoor Electronics Cabinets, *Bell Syst. Tech. J.,* 61, February, 1982.

Darveaux, R., Banerji, K., Mawer, A., and Dody, G., Reliability of Plastic Ball Grid Array Assembly, Ball Grid Array Technology, Lau, J., Ed., McGraw Hill, New York, 1994, chap. 13.

Graebner, J.E., Thermal Conductivity of Printing Wiring Boards, Technical Brief, *Electron. Cooling,* 1(2), 27, 1995.

HVAC Systems and Applications, *ASHRAE Handbook,* American Society of Heating, Refrigerating, and Air-Conditioning Engineers, Atlanta, GA, 1987, chap. 47.

Intel Corporation, Low-Profile Fan Duct Board and Chassis Specification and Low Profile Fan Duct System Ingredients Specification, available from http://developer.intel.com/ial/sdt/fanduct.htm, Rev. 1.0, September, 1998a.

Intel Corporation, Mobile Power Guidelines 2000, available from http://developer.intel.com/design/mobile/intelpower/, Rev. 0.8, September 16, 1998b.

INTELEC, Proceedings are available from the IEEE, New York.

Kraus, A.D. and Bar-Cohen, A., *Thermal Analysis and Control of Electronic Equipment,* Hemisphere, New York, 1983.

Lall, P. and Banerji, K., Assembly-Level Reliability Characterization of Chip-Scale Packages, Electron. Components Tech. conf. (ECTC), Seattle, Washington, D.C., May 25-28, Sess. 13, Pap. 2, 1998.

Lall, P. and Banerji, K., Assembly-Level Reliability of Flex-Substrate BGA, Elastomer-on-Flex Packages, and 0.5 mm Pitch Partial Array Packages, Surf. Mount. Int. Conf. (SMI), San Jose, CA, August 23-27, 1998.

Lall, P. and Banerji, K., Solder Joint Design for Optimum Process and Reliability, Surf. Mount. Int. (SMI) Conf., San Jose, CA, Sept 9-11, 1997, pp. 97-108.

Lall, P., Gold, G., Miles, B., Banerji, K., Thompson, P., Koehler, C., and Adhihetty, I., Reliability Characterization of the SLICC Package, Electron. Components Technol. Conf. (ECTC), Orlando, FL, May 28-31, 1996, pp.1202-1210.

Lall, P., Pecht, M., and Hakim, E., *Influence of Temperature on Microelectronic and System Reliability,* CRC Press, Boca Raton, FL, 1997.

Lee, T.T., Chambers, B., and Ramakrishna, K., Thermal Management of Handheld Telecommunication Products, *Electron. Cooling,* 4(2), May, 1998.

Leturq P. et al., A New Approach to Thermal Analysis of Power Devices, *IEEE Trans. Electron Devices,* ED24, 1147-1156, 1987.

Mahidhara, R., Solberg, V., DiStefano, T., and Greathouse, S., Solder Joint Integrity in Tessera's μBGA Package, in *Design and Reliability of Solders and Solder Interconnections,* TMS, Warrendale, PA, 1997, 403-412.

Marongiu, M.J. and Clarksean, R., Thermal Management of Outdoor Enclosures using Phase Change Materials, *Electron. Cooling,* 4(1), January, 1998.

Marthinuss, J.E. Jr. and Hall, G.T., Cooling Electronics At High Altitudes Made Easy, *Electron. Cooling,* 4(3), September, 1998.

Mulgaonkar, S. and Joiner, B., Thermal Performance of PBGA Packages for Next Generation FSRAM Devices, Motorola Semiconductor Tech. Data AN1232/D, 1994.

Selberherr, S., *Analysis and Simulation of Semiconductor Devices,* Springer-Verlag, New York, 1984, 119.

Steinbrecher, T., *Heat-Sink Guide,* http://www.heatsink-guide.com/, 1998.

Sterner, J., Thermal Design Challenge in a Pentium Notebook Computer, *Proc. IMAPS Int. Sys. Packaging Symp.,* 107-112, 1997.

Selected U.S. Electronic Thermal Management Patents

Housing with recirculation control for use with banded axial-flow fans. Yapp, M.G., Needham, MA; Houten, R.V., Winchester, MA; Hickey, R.I., Concord, NH; Patent 5,489,186, Feb. 6, 1996.

Solid-state imaging device having temperature sensor; Miyaguchi, Kazuhisa; Hamamatsu, Jpn, Muraki, Tetsuhiko, Hamamatsu, Jpn; Patent 5,508,740, April 16, 1996.

Active vaned passage casing treatment, Hobbs, D.E., South Windsor, CT; Patent 5,431,533, July 11, 1995.

Appliance for attaching heat sink to pin grid array and socket, Perugini, M.N., Monroe, CT, Romatzick, Milford, D.H. Jr., CT; Patent 5,396,402, March 7, 1995.

Cast casing treatment for compressor blades, Privett, J.D., North Palm Beach, FL, Byrne, W.P., Jupiter, FL, and Nolcheff, N.A., Palm Beach Gardens, FL, Patent 5,474,417, Dec. 12, 1995.

Centrifugal compressor with a flow-stabilizing casting, Heinrich, K., Nussbaumen, Switzerland, Patent 5,466,118, Nov. 14, 1995.

Clip for clamping heat sink module to electronic module, Harmon, R.A., Hudson, MA; and Urrata, G., Wakesfield, MA; Patent 5,381,305, Jan. 10, 1995.

Colored liquid crystal display having cooling, Hyatt, G.P., La Palma, CA; Patent 5,398,041, Mar. 14, 1995.

Epoxy molding composition for surface mount applications, Gallagher, M.K., Lansdale, PA; Petti, M.A., North Wale, PA; Patent 5,434,199, July 18, 1995.

Flexible thermal transfer apparatus for cooling electronic components, Tousignant, L.A., Shoreview, MN; Patent 5,411,077, May 2, 1995.

Heat sink, Voorhes, D.W., Winchester, MA; Goldman, R.D., Stoughton, MA; and Lopez, R.R., Boxford, MA; Patent 5,390,734, Feb. 21, 1995.

Heat sink attachment assembly, Jordan, W.D., Dallas, TX; and Smithers, M.C., Lewisville, TX; Patent 5,396,338, Jan. 31, 1995.

Heat sink assembly for solid state devices, Tata, P., Johnston, D., and Rife, W.B., Greenville, RI; Patent 5,397,919, Mar. 14, 1995.

High performance centrifugal pump having an open-faced impeller, Lehe, J.-R., Vernon, France; Beaurain, A., Chambly, France; Bosson, R., Vernon, France; Tiret, E., La Chapelle, Reanville, France; Patent 5,427,498, June 27, 1995.

Liquid crystal display having conductive cooling, Hyatt, G.P., La Palma, CA; Patent 5,432,526, July 11, 1995.

Organic coatings with ion reactive pigments especially for active metals, Mosser, M.F., Sellersville, PA; Harvey, W.A., III, MontClare, PA; Patent 5,409,970, April 25, 1995.

Powder epoxy resin coating composition, Matsuzaki, K., Yono, Japan; Ono, K., Tokyo, Japan; Iwamoto, S., Soka, Japan; Osa, M., Saitama, Japan; and Watanabe, T., Soka, Japan; Patent 5,418,265, May 23, 1995.

Ultrathin film thermocouples and method of manufacture, Rimai, D.S., Webster, NY, Anne, R.E., Caledonia, NY; and Bowen, R.C., Rochester, NY; Patent 5,411,600, May 2, 1995.

Custom conformal heat sinking device for electronic circuit cards and methods of making the same, Santilli, M.A., Lake Shastina, CA; and Morgan, W.Z., Dallas, TX; Patent 5,403,973, April 4, 1995.

Active thermal isolation for temperature responsive sensors, Martinson, S.D, Hampton, V., Gray, D.L., Newport News, VA; Carraway, D.L., Virginia Beach, VA; and Reda, D.C., San Jose, CA; Patent 5,311,772, May 17, 1994.

Air distribution system and manifold for cooling electronic components, Bartilson, B.W., Chippewa Falls, WI and Jirak, J.J., Jim Falls, WI; Patent 5,321,581, June 14, 1994.

Apparatus for an air impingement heat sink using secondary flow generators, Lindquist, S.E., Boylston, MA and Bailey, D.A., Concord, MA; Patent 5,304,845, April 19, 1994.

Circuit with built-in heat sink, Mehta, A.M., Plantation, FL and Desai, V.D., Plantation, FL; Patent 5,307,519, April 26, 1994.

Clip-on heat sink, Villani, A., Shrewsbury, MA; Patent 5,329,426, July 12, 1994.

Clip-on heat sink and method of cooling a computer chip package, Beijer, G., Moorpark, CA; Patent 5,295,043, March 15, 1994.

Cooling device for a magnetron, Lim, J.H., Buk-ku Daeku-si, Republic of Korea; Patent 5,325,266, June 28, 1994.

Cooling system for integrated circuits, Hastings, R.J., Houston, TX and Davis, C.E., Houston, TX; Patent 5,285,108, Feb. 8, 1994.

Disk augmented heat transfer system, Maling, G.C., Jr., Schmidt, R.R., Poughkeepsie, NY; Patent 5,335,143, August 2, 1994.

Dual active surface, miniature, plug-type heat flux gauge, Liebert, C.H., Middleburg Heights, OH and Koch, J., Jr., Medina, OH; Patent 5,314,247, May 24, 1994.

Fin assembly for an integrated circuit, Chen, C.-C., and Yi Lan Hsien, China (Taiwan), Patent 5,287,249, Feb. 15, 1994.

Flow control method and means, Freeman, C., Farnsfield, U.K.; Day, I.J., Toft, U.K.; and Wright, W.B., Kegworth, U.K.; Patent 5,340,271, Aug. 23, 1994.

Flow control method and means, Freeman, C., Farnsfield, U.K. and Day, I.J., Cambridge, U.K.; Patent 5,275,528, Jan. 4, 1994.

Heat sinking apparatus for surface mountable power devices, Kent, K.L., Woodstock, IL and Glomski, J.D., Arlington Heights, IL; Patent 5,365,399, Nov. 15, 1994.

Heat sink, method of manufacturing the same, and device of manufacturing the same, Nishiguchi, Masanori, Yokohama, Japan; Patent 5,339,215, Aug. 16, 1994.

Heat sink plate for multiple semi-conductors, Villaume, H.F., Intervale, NH; Patent 5,285,350, Feb. 8, 1994.

Heart sink mounting system for semiconductor devices, Goeschel, F.G., Mt. Clemens, MI; Lanting, M.L., Portage, MI; and McConnell, A.M., Dearborn, MI; Patent 5,283,467, Feb. 1, 1994.

Heat sink and cover for tab integrated circuits, Selna, E., Mountain View, CA; Ettehadish, E., Albany, CA; and LaGassa, J., Cupertino, CA; Patent 5,280,409, Jan. 18, 1994.

Heat sink mounting apparatus, Smithers, M.C., Lewisville, TX; Patent 5,276,585, Jan. 4, 1994.

Hybrid cooling system for electronic components, Fox, L., Boxboro, MA and Wade, P.C., Shirley, MA; Patent 5,285,347, Feb. 8, 1994.

Integrated circuit package having a cooling mechanism, Omeozawa, K., Tokyo, Japan; Patent 5,329,426, July 12, 1994.

Integral heat sink-terminal member structure of hybrid integrated circuit assembly and method of fabricating hybrid integrated circuit assembly using such structure, Matsumoto, H., Itami, Japan; Patent 5,291,372, Mar. 1, 1994.

Integral heat pipe, heat exchanger and clamping plate, Ettehadieh, E., Albany, CA; Patent 5,329,993, July 19, 1994.

Low profile integrated heat sink and fan assembly, Bailey, N.W., Sacramento, CA; Patent 5,309,983, May 10, 1994.

Metal matrix composite heat transfer device and method, Monesano, M.J., Fairfax, VA; Patent 5,287,248, Feb. 15, 1994.

Method for repeatable temperature measurement using surface reflectivity, Thakur, R.P.S., Boise, ID; Sandhu, G.S., Boise, ID; and Martin, A.L., Boise, ID; Patent 5,350,236, Sept. 27, 1994.

Method of making microchanneled heat exchangers utilizing sacrificial cores, Hoopman, T.L., River Falls, WI and Krinke, H.L., Marine, WA; Patent 5,317,805, June 7, 1994.

Mist supercooling of a heated surface, Lee, R.S.L., St. James, NY; Patent 5,311,931, May 17, 1994.

Monolith heating element containing electrically conductive tin oxide-containing coatings, Clough, T.J., Santa Monica, CA; Grosvenor, V.L., Topanga, CA; and Pinsky, M., Thousand Oaks, CA; Patent 5,304,783, April 19, 1994.

Package clip on heat sink, Blomquist, M.L., Newbury Park, CA; Patent 5,367,433, Nov. 22, 1994.

Performance characteristics stabilization in a radial compressor, Foerster, A., Worms, Germany; Engels, B., Weisenheim am Sand, Germany; Hauck, P., Ludwigshafen, Germany, Patent 5,333,990, Aug. 2, 1994.

Pin-fin heat sink, Brady, K.J. Millington, NJ and Cohn, C., Wayne, NJ; Patent 5,299,090, March 29, 1994.

Printed circuit board with cooling monitoring system, Bartilson, B.W., Chippewa Falls, WI and Schlimme, E.F., Chippewa Falls, WI; Patent 5,281,026, Jan. 25, 1994.

Remote temperature and/or temperature difference measuring device, Barral, J.-P., Seyssins, France; Million, B., Saint-MartinD'Hres, France; Colomb, F., Grenoble, France; Patent 5,352,039, Oct. 4, 1994.

Rotor case treatment, Koff, S.G., Cambridge, MA; Nikkanen, J.P., West Hartford, CT; and Mazzawy, R.S., South Glastonbury, CT; Patent 5,308,225, May 3, 1994.

Self clamping heat sink assembly, Brauer, E.A., Indianapolis, IN; Patent 5,309,979, May 10, 1994.

Surface-to-air heat exchanged for electronic devices, Pitasi, M.J., Newbury, MA; Patent 5,293,930, March 1, 1994.

System for repeatable temperature measurement using surface reflectivity, Thakur, R.P.S., Sandhu, G.S., and Martin, A.L., Boise, ID; Patent 5,364,187, Nov. 15, 1994.

Thermal transfer plate and integrated circuit chip or other electrical component assemblies including such plate, Hopfer, A.N., Park Ridge, IL; Patent 5,282,111, Jan. 25, 1994.

Thermopile differential thermal analysis sensor, Schaefer, J.W., Wilmington, DE and Danley, R.L., Collingswood, NJ; Patent 5,288,147, Feb. 22, 1994.

Thermal joint, Gruber, P.A., Mohegan Lake, NY and Zingher, A.R., White Plains, NY; Patent 5,291,371, March 1, 1994.

Vaned shroud for centrifugal compressor, Palmer, D.L., Cave Creek, AZ; Patent 15, 5,277,541, Jan. 11, 1994.

High heat flux evaporative spray cooling, Tilton, D..E., and Tilton, C.L., Kent, WA; Patent 5,220,804, June 22, 1993.

Integral heat pipe, heat exchanger, and clamping plate, Davidson, H.L., San Carlos, CA and Ettenhadieh, E., Albany, CA; Patent 5,253,702, Oct. 19, 1993.

Metal heat sink baseplate for a semiconductor device, Della Bosca, P., Brugherio, Italy and Massironi, A., Concorezzo, Italy; Patent 5,229,918, July 20, 1993.

Microprocessor heat dissipation apparatus for a printed circuit board, Koenen, D.J., Houston, TX; Patent 5,272,599, Dec. 21, 1993.

Plasma momentum meter for momentum flux measurements, Zonca, F., Rome, Italy; Cohen, S.A., Hopewell, NJ; Bennett, T., Princeton, NJ; and Timberlake, J.R., Allentown, NJ; Patent 5,239,563, Aug. 24, 1993.

Radiation detector suitable for tympanic temperature measurement, Pompei, F., Wellesley Hills, MA and Looney, J.M., Jr., Clearwater, FL; Patent 5,271,407, Dec. 21, 1993.

Rapid method for determining critical vapor pressure, Wood, R.W., Elkhorn, WI; Hansen, L.D., Orem, UT; and Crawford, J.W., Salem, UT; Patent 5,266,492, Nov. 30, 1993.

Remote high temperature insulator-less heat-flux gauge, Noel, B.W., Espanola, NM; Patent 5,273,359, Dec. 28, 1993.

Single flow turbopump with integrated boosting, Girault, J.-P., Tourny, France; Patent 5,232,333, Aug. 3, 1993.

Stacking heatpipe for three-dimensional electronic packaging, Davidson, H.L., San Carlos, CA and Nishtala, S., Santa Clara, CA; Patent 5,181,167, Jan. 19, 1993.

Toughened fiber-reinforced composites, Qureshi, S.P., Alpharetta, GA; Hoffman, R.E., Cumming, GA; and Newman-Evans, R.H., Alpharetta, GA; Patent 5,268,223, Dec. 7, 1993.

Translating wedge heat sink, Sewell, M.W., Dahlgren, VA; Patent 5,218,517, June 8, 1993.

Exposure control method for adjusting the temperature of a workpiece holding chuck attracting surface based on memorized data, Fujioka, H., Atsugi, Japan; Patent 5,134,436, July 28, 1992.

Pivotal heat sink assembly, Frankeny, J.A., Taylor, TX; Frankeny, R.F., Hermann, K., and Wustrau, R., Austin, TX; Patent 5,161,087, Nov. 3, 1992.

Electrical component assembly with heat sink, Cocconi, A.G., Glendora, CA; Patent 5,060,112, Oct. 22, 1991.

Flow meter for measuring fluid flow with multiple temperature sensors, Ishikawa, K., Kyoto, Japan, Mihira, H., Kyoto, Japan; Kimura, N., Kyoto, Japan; and Yamaguchi, M., Kyoto, Japan; Patent 5,048,332, Sept. 17, 1991.

Heat sink for an electric circuit board, Andersson, B., Jonkoping, Sweden; Danielsson, H., Linkoping, Sweden; Johansson, H., Angstromm Anstroml, Sweden; Andersen, H., Vargon, Sweden; Patent 5,077,638, Dec. 31, 1991.

Heat sink device for electronics modules packaged in cylindrical castings, Sewell, M.W., Fredericksburg, VA; Patent 5,060,115, Oct. 22, 1991.

High conduction cooling module having internal fins and complaint interfaces for VLSI chip technology, Horvath, J.L., Poughkeepsie, NY; Biskeborn, R.G., Pawling, NY; and Harvilchuck, J.M., Billings, NY; Patent 5,052,481, 1991.

Localized circuit card cooling device, Yiu, J.T., Chandler, AZ; Patent 4,931,904, June 5, 1990.

Temperature controller for semiconductor device, Watanabe, H., Kanagawa, Japan; Patent 4,689,659, Aug. 25, 1987.

Mechanical heat transfer device, Altoz, F.E., Catonsville, MD and Winn, W.H., Linthicum, MD; Patent 4,273,183, June 16, 1981.

Bibliography

Ajiki, T., Sugumoto, M., Higuchi, H., and Kumada, S., A New Cyclic Biased T.H.B. Test for Power Dissipating IC's. Int. Reliab. Phys. Symp., 1979, 118-126.

Alexander, D.R., An Electrical Overstress Failure Model for Bipolar Semiconductor Components. *IEEE Trans Components, Hybrids, Manuf. Technol.,* 1, 345-353, 1978.

Arrhenius, S., *Z. Physik. Chem.,* 4, 226, 1889.

Beasley, K., New Standards for Old, *Qual. Reliab. Eng. Int.,* 6, 289-294, 1990.

Berry, R.S., Rice, S.A., and Ross, J., *Physical Chemistry,* John Wiley & Sons, New York, 1980.

Blanks, H.J., Arrhenius and the Temperature Dependence of Non-Constant Failure Rate, *Qual. Reliab. Eng. Int.,* 6, 259-265, 1990.

Blech, I.A. and Meieran, E.S. Direct Transmission Electron Microscope Observations of Electrotransport in Aluminum Films, *Appl. Phys. Lett.*, 11, 263, 1967.

Bowles, J., A Survey of Reliability Prediction Procedures for Microelectronic Devices, *IEEE Trans. Reliab.,* 41(1), March, 1992, 2-12.

Boyko, K.C. and Gelarch, D.L. Time Dependent Dielectric Breakdown of 210A Oxides, 27th Annu. Reliab. Phys. Symp., IEEE, 1989, 1-8.

Canali, C, Fatini, F., Gaviraghi, S., and Senin, A., Reliability Problems in TTL-LS Devices, *Microelectron. Reliab.,* 21, 637-651, 1981.

Clark, I.D., Harrison, L.G., Kondratiev, V.N., Szabo, Z.G., and Wayne, R.P., *The Theory of Kinetics,* Bamford, C.H. and Tipper, C.F.H., Eds., Elsevier, Amsterdam, 1979.

CNET, Recueil de Donnees de Fiabilitie du CNET (Collection of Reliability DATA from CNET), Centre National d'Etudes des Telecommunications (National Center for Telecommunication Studies), 1983.

Crook, D.A., *Techniques for Evaluating Long Term Oxide Reliability At Wafer Level,* IEDM Technical Digest, 1978, 444-448.

Crook, D.A., Method of Determining Reliability Screens for Time Dependent Dielectric Breakdown, 17th Annu. Reliab. Phys. Symp., 1979, 1-7.

DiStefano, T.H. and Shatzkes, M., Dielectric Stability and Breakdown in Wide Bandgap Insulators, *J. Vac. Sci. Technol.,* 12, 37, 1975.

Evans, M.G. and Polanyi, M., *Trans. Faraday Soc. (London),* 34, 11, 1938.

Eyring, H., Lin, S.H., and Lin, S.M., *Basic Chemical Kinetics,* John Wiley & Sons, NY, 1980.

Fukuzawa, I., Ishiguro, S., and Nanbu, S. Moisture Resistance Degradation of Plastic LSI's By Reflow Soldering, Proc. 23rd Annu. Int. Reliab. Phys. Symp., 1985, 192-197.

Gaffeny, J. Internal Lead Fatigue Through Thermal Expansion in Semiconductor Devices. *IEEE Trans. Electronic Devices,* ED-15, 617, 1968.

Hakim, E.B., Reliability Prediction: Is Arrhenius Erroneous?, *Solid State Technol.,* August, 1990, 57.

Handbook of Reliability Data for Components Used in Telecommunication Systems, Issue 4 (HRD4), British Telecom, January, 1987.

Harari, E., Dielectric Breakdown in Electrically Stressed Thin Films of Thermal SiO_2, *J. Appl. Phys.*, 49, 2478, 1978.

Harman, G.G., Metallurgical Failure Modes of Wire Bonds, 12th Int. Reliab. Phys. Symp., 1974, 131-141.

Hart, A., Teng, T.T., and McKenna, A. Reliability Influences from Electrical Overstress on LSI Devices, 18th Annu. Proc. Int. Reliab. Phys. Symp., 1980, 190-196.

Holland, S., Chen, I.C., Ma, T.P., and Hu, C., Physical Models for Gate Oxide Breakdown, *IEEE Electron Device Lett.*, 5, 302, 1984.

Honda, K., Ohsawa, A., and Toyokura, N., Breakdown in Silicon Oxides: Correlation with Cu Precipitates, *Appl. Phys. Lett.*, 45, 270, 1984.

Hsu, F. C. and Chiu, K. Y. Temperature Dependence of Hot Electron Induced Degradation in MOSFET's, *IEEE Electron Devices Lett.,* EDL-5, 1984.

Hu, J., Pecht, M., and Dasgupta, A., A Probabilistic Approach for Predicting Thermal Fatigue Life of Wirebonding in Microelectronics, *ASME J. Electron. Packag.,* 113, 275, 1991.

Hu, C., Hot Electron Effects in MOSFET'S, *IEDM Technical Digest,* 1983, 176.

IITRI/Honeywell, VHSIC/VHSIC-Like Reliability Prediction Modeling, Contract No F30602-86-C-0261, 15 January, 1988.

Klinger, D.J., On the Notion of Activation Energy in Reliability: Arrhenius, Eyring, and Thermodynamics, 1991 Proc. Annu. Reliab. Maintain. Symp., 1991, pp. 295-300.

Ko, P.K., Hot Electron Effects in MOSFET'S, Ph.D dissertation, EECS Dept., University of California, Berkeley, 1982.

Kuo, W. and Kuo, Y., Facing the Headache of Early Failures: State of the Art Review of Burn-in Decisions, *Proc. IEEE,* 71, 1257-1266, 1983.

LaCombe, D.J., Dening, D.C., and Christou, A., A New Failure Mechanism in Thin Gold Films at Elevated Temperature, 20th Proc. Int. Reliab. Phys. Symp., 1982, 81-87.

Lall, P., Pecht, M., and Hakim, E., *Estimating the Role of Temperature on Microelectronic Device Reliability,* John Wiley & Sons, New York, 1994.

Lee, J., Chen, I.C., and Hu, C., Modeling and Characterization of Gate Oxide Reliability. *IEEE Trans. Electron Devices*, 35, 2268, 1988.

Lee, J., Chen, I.C., and Hu, C., Statistical Modeling of Silicon Dioxide Reliability, 26th Annu. Int. Reliab. Phys. Symp., 1988, 131-138.

Liehr, M., Bronner, G.B., and Lewis, J.E., Stacking-Fault-Induced Defect Creation in SiO_2 on Si(100), *Appl. Phys. Lett.*, 52, 1892, 1988.

Lin, P.S.D., Marcus, R.B., and Sheng, T.T., Leakage and Breakdown in Thin Oxide Capacitors: Correlation with Decorated Stacking Faults, *J. Electrochem. Soc.,* 130, 1878, 1983.

Lloyd, J.R., Shatzkes, M., and Challaner, D.C., Kinetic Study of Electromigration Failure in Cr/Al-Cu Thin Film Conductors Covered with Polyimide and the Problem of the Stress Dependent Activation Energy. IEEE Int. Reliab. Phys. Symp., 216-225, 1988.

LSI Logic, Reliability Manual and Data Summary, Milpitas, CA, 1990, 12-18.

Machiels, F., Lijbers, G., Allaire, R., and Poiblaud, G., Derating of Results According to Various Models and Prediction of Lifetimes of Plastic SMD's under Humidity Stress Conditions, 5th Int. Conf. Qual. Electron. Components, Failure, Prevention, Detection, and Analysis, Bordeaux, France, 869-879, 1991.

Matsumoto, H., Sawada, K., Asai, S., Hirayama, M., and Nagasawa, K., Effect of Long Term Stress on IGFET Degradations Due to Hot Electron Trapping, *IEEE Trans. Electron Devices*, ED-28, 923-928, 1981.

McAteer, O.J., *Electrostatic Discharge Control,* McGraw-Hill, New York, 1989.

McLinn, J.A., Constant Failure Rate: A Paradigm Transition. *Qual. Reliab. Eng. Int.,* 6, 237-241, 1990.

McPherson, J.W. and Baglee, D.A., Acceleration Factors for Thin Gate Oxide Stressing, 23rd Annu. Reliab. Phys. Symp., IEEE, 1985, 1-5.

MIL-HDBK 217F, Reliability Prediction of Electronic Equipment, MIL-HDBK 217F, U.S. Department of Defense, Washington D.C., 1991.

Moss, R.Y., Caution: Electrostatic Discharge at Work, *IEEE Trans. Component Hybrids Manuf. Technol.,* 5, 512-515, 1982.

Ning, T.H., Cook, P.W., Dennard, R.H., Osburn, C.M., Schuster, S.E., and Yu, H.W., 1 mm MOSFET VLSI Technology. IV. Hot Electrons Design Constraints, *IEEE Trans. Electron Devices,* ED-26, 346-353, 1979.

NTT, Nippon Telegraph and Telephone Corporation, Standard Reliability Table for Semiconductor Devices, March, 1985.

O'Connor, P.D.T., Reliability Prediction: Help or Hoax?, *Solid State Technol.,* August, 1990, 59-61.

Pancholy, R.K. and Jhnoki, T., CMOS/SOS Gate Potection Networks, *IEEE Trans. Electronic Devices,* ED-25, 917-925, 1978.

Pecht, M., Lall, P., and Hakim, E.B., The Influence of Temperature on Integrated Circuit Failure Mechanisms, *Qual. Reliab. Eng. Int. J.,* 8, 167-175, 1992.

Pecht, M. and Ramappan, V., Are Components Still the Major Problem: A Review of Electronic System and Device Field Failure Returns, *IEEE Trans. Component Hybrids Manuf. Technol.,* 15(6), 1992, pp.1-5.

Pecht, M. and Ko, W., A Corrosion Rate Equation for Microelectronic Die Metallization, *Int. J. Hybrid Microelectron.,* 13, 41-52, 1990.

Pecht, M., U.S.-Japanese Technology Evaluation Center Visit, Private Communications, 1993.

Pecht, M., Lall, P., and Dasgupta, A., A Failure Prediction Model for Wire Bonds, Proc. 1989 Int. Symp. Hybrids Microelectron., 1989, 607-613.

Phillips, W.E., Microelectronic Ultrasonic Bonding, Harman, G.G., Ed., National Bureau of Standards, Spec. Publ. 400-2, 80-86, U.S. Government Printing Office, Washington, D.C., 1974.

Ravi, K.V. and Philosky, E.M., Reliability Improvement of Wire Bonds Subjected to Fatigue Stresses, 10th Annu. Proc. IEEE Reliab Phys. Symp., 1972, 143-149.

Reif, F., *Fundamentals of Statistical and Thermal Physics,* McGraw Hill, New York, 1965.

Ricco, B., Azbel, M. Ya., and Brodsky, M.H., Novel Mechanism for Tunneling and Breakdown of Thin SiO_2 Films, *Phys. Rev. Lett.,* 51, 1795, 1983.

Rome Air Development Center, RAC (Reliability Analysis Center), IC Quality Grades: Impact on System Reliability and Life Cycle Cost, SOAR-3, 1985.

Runayan, W.R., *Silicon Semiconductor Technology,* McGraw-Hill, New York, 1965.

Schnable, G.L., Failure Mechanisms in Microelectronic Devices. *Microelectron. Reliab.,* Artech House Inc., 1, 25-87, 1988.

Schwarzenberger, A.P., Ross, C.A., Evetts, J.E., and Greer, A.L., Electromigration in Presence of a Temperature Gradient: Experimental Study and Modeling, *J. Electron. Mater.,* 17, 473-478, 1988.

Setliff, J.E., A Review of Commercial Microcircuit Qualification and Reliability Methodology, Proc. 1991 Adv. Microelectron. Technol., Qual., Reliab. Logistics Workshop, August 13-15, Seattle, WA, 1991, 325-335.

Shirley, G.C. and Hong, C.E., Optimal Acceleration of Cyclic THB Tests for Plastic Packages Devices, 29th Int. Reliab. Phys. Symp., 1991, 12-21.

Siemens, Standard SN29500, Reliabilty and Quality Specification Failure Rates of Components, 1986.

Smith, J.S., El Overstress Failure Analysis in Microcircuits, 16th Annu. Proc. Int. Reliab. Phys. Symp., 1978, 41-46.

Stojadinovic, N.D., Failure Physics of Integrated Circuits: A Review. *Microelectron. Reliab.,* 23, 609-707, 1983.

Swartz, G.A., Gate Oxide Integrity of NMOS Transistor Arrays, *IEEE Trans. Electron Devices*, ED-33, 1826-1829, 1986.

Sze, S.M., *Physics of Semiconductor Devices,* 2nd ed., John Wiley & Sons, New York, 1981, 30-47.

Takeda, E., Nakagome, Y., Kume, H., Suzuki, N., and Asai, S., Comparison of Characteristic of N-Channel and P-Channel MOSFET's for VLSI, *IEEE Trans. Electron. Devices,* ED-30, 675-680, 1983.

Thomas, R.W. and Calabrese, D.W., Phenomenological Observations on Electromigration, 21st Proc. Int. Reliab. Phys. Symp., 1983, 1-9.

Villela, F. and Nowakowski, M.F., Investigation of Fatigue Problems in 1-mil Diameter Thermocompression and Ultrasonic Bonding of Aluminum Wire, NASA Technical Memorandum, NASA TM-X-64566, National Aeronatucis and Space Administration, Washington, D.C., 1970.

Wendt, H., Cerva, H., Lehmann, V., and Pamler, W., Impact of Copper Contamination on the Quality of Silicon Oxides, *J. Appl. Phys.*, 65, 2402, 1989.

Wigner, E.P., *Trans. Faraday Soc. (London),* 34, 29, 1938.

Witzmann, S. and Giroux, Y., Mechanical Integrity of the IC Device Package: A Key Factor in Achieving Failure Free Product Performance, Trans. First Int. High Temperature Electron. Conf., Albuquerque, June, 1991, 137.

Wong, K.L., The Bathtub Does Not Hold Water Any More, *Qual. Reliab. Int.,* 6, 279-282, 1990.

Wong, K.L., What Is Wrong With the Existing Reliability Prediction Methods?, *Qual. Reliab. Eng. Int.,* 6, 251-257, 1990.

Woods, M.H. and Rossenburg, S., EPROM Reliability; Parts I and II. *Electron. Microelectron. Reliab.,* 53, 133-141, 1980.

Wunsch, D.C. and Bell, R.R., Determination of Threshold Failure Levels of Semiconductor Diodes and Transistors Due to Pulse Voltages, *IEEE Trans. Electron Devices*, NS-15, 244-259, 1968.

4.21 Direct Contact Heat Transfer

Harold R. Jacobs

Introduction

Direct contact heat transfer refers to the transport of thermal energy between two or more mass streams without separation by intervening walls. Generally, the interaction is between only two streams and there is generally a degree of mass transfer as well. In the above definition the words mass streams are used; however, the term direct contact can also refer to batch processing such as in a stirred reactor. Nonetheless, the study of continuous processing is more likely. In process heat transfer, direct contactors are used in absorption refrigeration, distillation, electric power generation (as open feedwater heaters and in conjunction with dry cooling towers in the so-called Heller cycle), food processing, drying, emergency cooling of nuclear reactors, quenching of hot gases, and in wet cooling towers. Other process equipment includes condensers such as barometric condensers, and spray condensers. The process of liquid spray combustion in a variety of applications also is a direct contact heat transfer process through the fact that the fuel is injected as a spray of drops which must be preheated to the point of evaporation. In general, the vaporization process, normally required as the combustion takes place, fits the definition of direct contact heat transfer. Many environmental processes such as the formation of ice, snow, or rain utilize the same physics as in other direct contact processes. Thus, direct contact heat transfer is ubiquitous.

In process-type direct contact heat and mass transfer, we normally deal with only two mass streams, although three or more can occur. For example, in some pipeline problems and in the extraction of heavy oils in the petroleum industry, water, oil, and air are sometimes encountered. This is often the case in off-shore production.

In process heat transfer equipment the mass streams can be cocurrent, countercurrent, or even cross flow. The streams may be composed of, for example, systems that are liquid-liquid, liquid-vapor, liquid-gas, liquid-solid, gas-solid, or even solid-solid. The streams can be composed of the same chemicals, as in the case of steam-water, or can be partially miscible or completely so. One or both streams can be disperse (made up of defined masses, particles, drops, bubbles) or continuous.

As noted previously, heat transfer may be accompanied by mass transfer such as in the case of dryers, distillation columns, wet cooling towers, condensers, and in combustion systems. The possibilities are legion. Thus, in this section we must limit ourselves to a few fundamental processes. The reader is referred to the literature or specific review articles such as Sideman (1966), Sideman and Moalem-Maron (1982), Jacobs (1988), articles by Jacobs (1985, 1995) and proceedings of the ICHMT conferences (1990 and 1997).

The fundamental processes to be included here are sensible heat transfer between parallel fluid streams in co- or counterflow; sensible heat transfer to a disperse flow of particles, drops, or bubbles; flow with condensation, and flow with evaporation with similar geometries to that for sensible heat transfer.

Heat Transfer between Continuous Parallel Streams

Falling Films

In cooling towers, wetted wall towers, packed beds, etc. a liquid is injected into the device in a number of ways, resulting in a thin film of liquid running over solid vertical or inclined surfaces in the form of a coherent thin film or in rivulets under the influence of gravity. Such flows are generally classified under the name of falling films. (In this handbook some previous discussion is included in earlier sections.) The solid substructure can take any variety of forms including a bed of spheres, raschid rings, burl saddles, or a structured array of metal chevrons, for example. The packing can be in an organized array or random and can include particles of different sizes and be of a variety of materials, generally wettable by one of the streams.

The initial work on falling films can be traced to Nusselt (1916). Since that time, many studies have been performed. Hasson (1964) investigated a vertically falling film while neglecting the effects of surface tension, interfacial shear, and pressure gradients. He obtained a universal nondimensional curve for the developing film thickness. Murty and Sastri (1974) studied the problem of a film exiting a slot flowing down an inclined plane of defined height. The surrounding gas was assumed to be quiescent, and pressure variation and surface tension were assumed to be negligible. Assuming a smooth interface and continuity of velocity and shear at the interface, they determined the thickness of the film as a function of the length along the plate. Murty and Sarma (1976) presented an expression for film thickness for either laminar or turbulent flow down an inclined wall with a cocurrent gas stream and incorporated interfacial shear and gravity effects in the nondimensional film thickness. Experimental nondimensional velocity profiles were used to present a universal, nondimensional film thickness. Tekic et al. (1984) repeated the model of Murty and Sastri (1974) but incorporated normal and tangential stresses on the film interface and neglected the interfacial shear of the still air. Their results presented a family of curves for liquids with different physical properties, but underpredicted the film entrance length.

While cocurrent flows are interesting, they result in minimum heat transfer when compared to countercurrent flows. Countercurrent flows, however, can strip the film from the surface, depending on the properties of the fluid external to the film. Since the external fluids are typically a gas or a vapor, the velocities when stripping occurs are normally quite high. Nevertheless, this is a common problem in cooling towers and other devices, and in the case of cooling towers can lead to the need for excessive make-up water and a highly visible plume. Usually the falling film develops a wavy interface, followed by the formation of larger waves on its surface prior to entrainment of liquid in the gas stream. The waves can be initiated by either Tollmein-Schlichting or Kelvin-Helmholtz instabilities. Ostrach and Koestel (1965) discussed these and other instabilities associated with two-phase flows. The Tollmein-Schlichting waves are associated with transition from laminar to turbulent flows while the Kelvin-Helmholtz instabilities are interfacial phenomena resulting from the shear at the interface due to relative velocity differences . The latter are applicable to otherwise laminar flows.

Ishii and Grolmes (1975) discuss entrainment of liquid into the gas stream. For the case of vertical falling films in countercurrent gas flow, possible causes of entrainment are shearing droplets from the top of roll waves and the formation of large-amplitude bulges in the liquid. Unstable films are particularly troublesome in wetted wall columns (which are simply vertical pipes with a gas flowing up the center and an annular film flowing down the surface). Such devices have been reviewed by Bharathan (1988). Tien and Liu (1979) outline an overview of the theory and experiment. More recently, McQuillan et al. (1985) and Bankoff and Lee (1986) reviewed these apparatuses and established a data bank of experimental flooding points. Stefan and Mayinger (1990) studied such systems at high gas pressure.

Heat and mass transport to falling films were reviewed by Seban (1978). He discussed heating at the wall, evaporation, absorption, and the effects of film thickness for laminar and turbulent flows, but omitted interfacial shear effects. Experimental work since Seban's review include that of Segev et al. (1981) and Lee and Bankoff (1983). Theoretical studies were conducted by Jacobs and Fletcher (1988) and Jacobs and Dolan (1994) of the counterflow falling film problem for laminar flows with heat transfer from the gas to a film flowing over an adiabatic wall. Thus, their models apply to packed beds and other devices where the supporting surfaces are only to provide for the formation of the film. (In fact, the packing continuously reforms the film in lengths sufficiently small so as to maximize the local heat transfer coefficient on an individual packing element.) The first paper used integral methods and was applicable for only very small and large values of the gas velocity. The asymptotic solutions indicated that for a given countercurrent velocity a minimum would occur in the heat transfer coefficient. The second paper utilized the boundary layer equations for the film and the Navier-Stokes equations for the external gas stream, and was able to define the extent of the phenomenon. The nature of the flow is shown in Figures 4.21.1a/b. Comparison of the results are shown in Figure 4.21.2. It is clear that both analyses predict a region of minimum heat transfer. Figure 4.21.3 shows a comparison of the mean

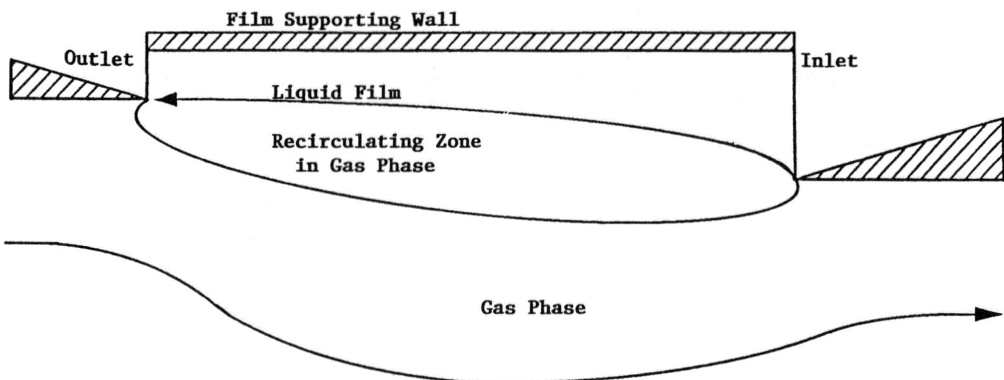

FIGURE 4.21.1a Flow of gas over falling liquid film with decreasing thickness. (From Jacobs, H.R. and Dolan, S.M., *J. Energy, Heat Mass Transfer,* 16, 77, 1994. With permission.)

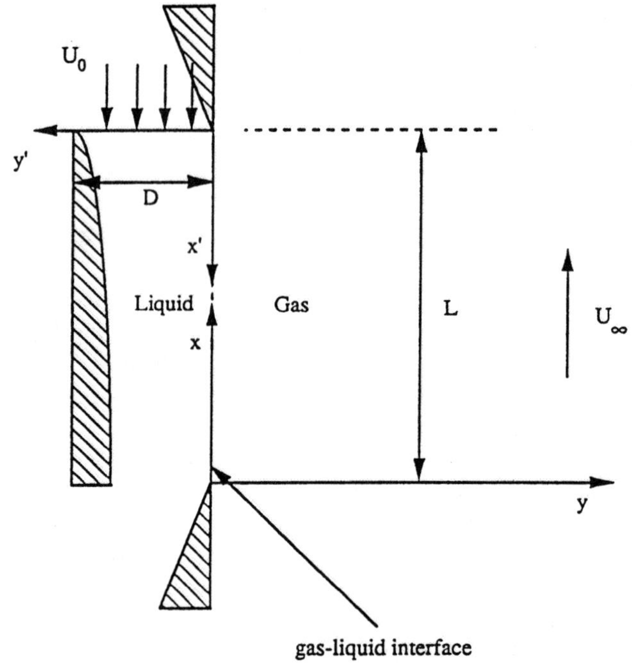

FIGURE 4.21.1b Geometry used in mathematical model for falling film and gas regions. (From Jacobs, H.R. and Dolan, S.M., *J. Energy, Heat Mass Transfer,* 16, 77, 1994. With permission.)

interfacial friction coefficient, and with Figure 4.21.2, indicates an overprediction of drag and an underprediction of heat transfer by the integral model. The results indicate that a gas velocity of at least 10 times that of the mean film should be used, but that the velocity should be less than that necessary to cause entrainment. The heat transfer in the gas liquid film system is controlled by the gas-side heat transfer for the case of sensible heat transfer.

No solution currently exists for the wavy film countercurrent flow situation. Subject to the lack of experimental data, it is recommended that one assume that the surface waves should only influence the heat transfer internal to the film. For situations where the thermal penetration is less than the film thickness, the average Nusselt number for the internal resistance is

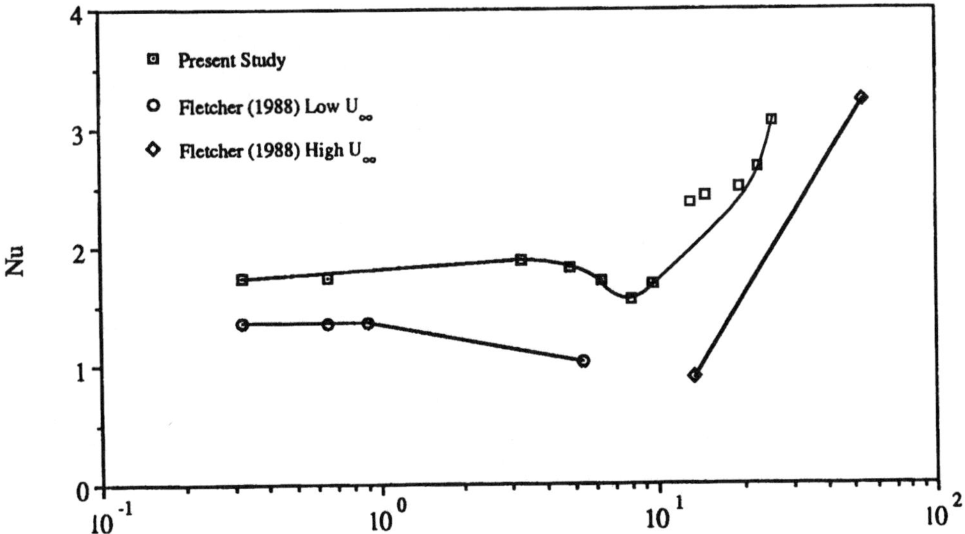

FIGURE 4.21.2 Relationship between nondimensional gas velocity and average Nusselt number, U_∞ = gas velocity, U_o = liquid velocity at inlet (From Jacobs, H.R. and Dolan, S.M., *J. Energy, Heat Mass Transfer,* 16, 77, 1994. With permission.)

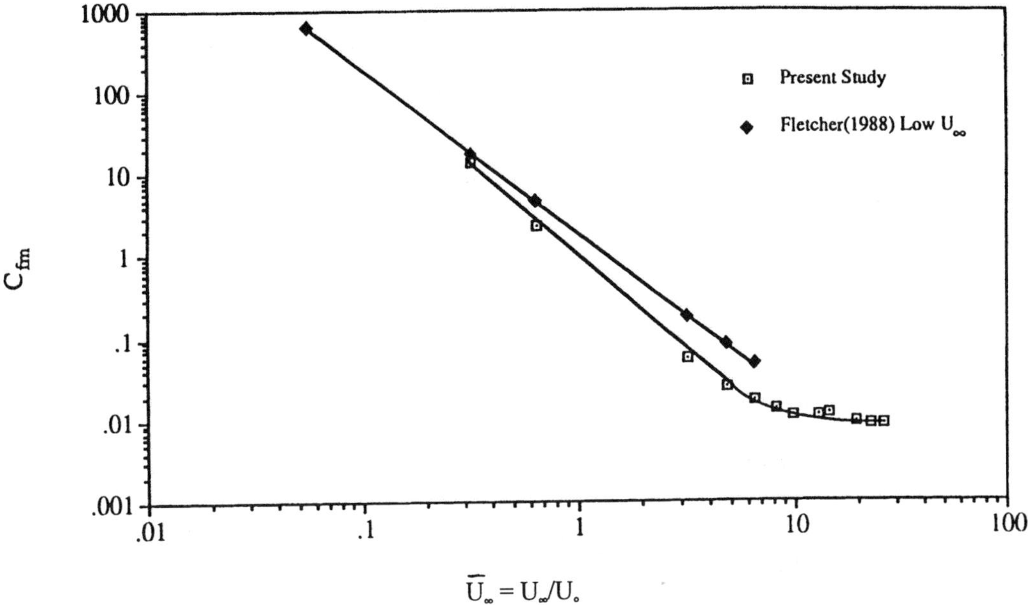

FIGURE 4.21.3 Effect of U_∞ changes on frictional drag.

$$\mathrm{Nu} = \sqrt{\frac{6}{\pi}}\,\mathrm{Re}^{1/2}\,\mathrm{Pr}^{1/2} \tag{4.21.1}$$

where $\mathrm{Re} = \frac{uD}{\nu}$ is the film thickness and u is the mean film velocity. For a wavy film the heat transfer coefficient will be increased by 15% (Levich, 1962) based on an assumed similarity to gas absorption. Similarly for turbulent films, mass transfer analogs should offer some assistance although there exist few controlled experiments (Levich, 1962).

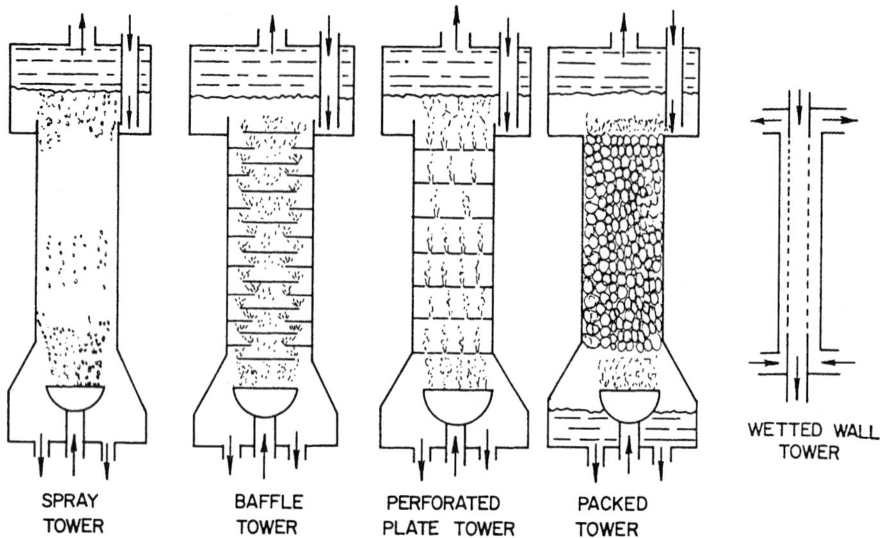

FIGURE 4.21.4 Schematic of various types of direct contact counterflow devices.

Heat Transfer to Jets and Sheets

In many types of process equipment (Figure 4.21.4), the apparatus is simply a vessel containing a few partial trays which are solid or perforated. The trays have a lip so as to hold a quantity of liquid sufficient to provide the needed pressure head to form a series of jets or a sheet of fluid falling under gravity. Such devices are used in humidifiers and direct contact condensers or as mass transfer devices in dealing with gases or vapors.

In such equipment the heat transfer process is described as heat transfer to jets and sheets rather than films, as the fluid is not supported where the primary heat transfer takes place (on the jets and sheets). The series of trays act as mixers. The liquid is caught on a tray and leaves it at its mean temperature. The reforming of jets or sheets provides for a higher heat transfer coefficient than would occur for a stream that fell the entire length of the column and, thus, requires a shorter column to accomplish the required energy transfer. The major disadvantage is a greater pressure loss in the device than for an empty vessel (which is called a spray column).

Typically, the hydrodynamics of a jet or sheet can be assumed to be either a slug flow or fully developed. Thus, the analysis of the heat transfer via the energy equation follows that of a Graetz problem where only the axial component of the advection term and the conduction term across the jet or sheet remain. External to the jet or sheet the heat transfer coefficient must be determined. In most cases values for flow over a solid surface can be used. The flow over a jet or sheet in most column designs is crossflow and, thus, the external heat transfer coefficient is independent of axial location. This allows one to adapt the Graetz-type solutions to solving for the mean axial temperature of the jet or sheet, as was done by Jacobs and Eden (1986). For liquid jets or sheets exposed to a gas or vapor, especially for sensible heat transfer, the external resistance to heat flow may be controlling, and it may be possible to assume negligible gradient across the jet or sheet. Otherwise, the fully developed internal thermal resistance can be used for axial lengths greater than the thermal developing region; however, to minimize the internal resistance requires the use of shorter jets or sheets and requires the distance between trays to be less than the thermal entrance length, as predicted by the Graetz solution. For turbulent flows in the liquid, the maximum heat transfer would neglect the internal resistance altogether. Where the external thermal resistance is not controlling and the stream is turbulent, it is advisable to use a Graetz-type solution with an effective thermal conductivity given by

$$k_{\epsilon H} = k + \frac{\rho c \epsilon_t}{Pr_t} \tag{4.21.2}$$

The above discussion applies to smooth, laminar surfaces. As with thin films, waves will eventually occur on the surface of these streams, leading to a better mixing in the jets or sheets. The wavy regime typically decreases the internal resistance by as much as 15%. Eventually, the jets and or sheets will break down into drops or a spray in a manner similar to that discussed in Section 4.17. While it is possible to have a jet break down into a fine spray, if the jet Reynolds number is sufficiently high, a freely falling jet will typically break down into drops that are twice the diameter of the jet or twice the thickness of a sheet. In many instances this does not provide a sufficient increase in surface area to result in higher heat transfer rates than would occur in a continuous sheet, although internal mixing may increase. Thus, an analysis for a continuous jet or sheet may be applicable even after breakup.

Sensible Heat Transfer to Dispersed Media: Drops, Particles, Bubbles

Sensible heat transfer to disperse media from a continuous phase is dependent upon the proximity of other particles, drops, or bubbles and to their own geometry, as well as their velocity through the continuous surrounding media and its thermal properties. For solid particles, the heat transfer is generally described in terms of an equivalent sphere. An equivalent diameter is obtained from

$$D_{equiv} = \frac{6V}{A_s} \tag{4.21.3}$$

For drops and bubbles, the problem is more difficult than for solid particles, as they are easily distorted, subject to both breakup and coalescence, and may have significant internal circulation depending on how they are formed. Close proximity to other drops or bubbles, as associated with volume fraction or holdup equaling 5% or greater, will make drops that would otherwise distort tend to remain spherical. Further, the presence of impurities tends to stabilize the interfaces of bubbles with the surrounding media no matter whether the impurities are in the dispersed or continuous phase. Thus, models of drops within swarms, based on transport in single bubbles typically overpredict heat and mass transfer since they overpredict drop or bubble distortion and internal circulation. Drop dynamics based on drag for single spheres will be in error, as the drag will vary for drops in a swarm due to differences in the surrounding flow. In the midst of the array the spheres have a greater drag, but at the front the drag can be less than for a single sphere and the drag at the rear can be higher than that within the array for staggered or random arrays. For inline arrays, the drag can be greatest at the front of a column, leading to a bunching up of particles.

For holdups less than 5%, assuming rigid spheres, the expression of Ahmad and Yovanovich (1994) for external to the sphere heat transfer is recommended:

$$Nu_D = 2.0 + \frac{0.775\, Re_D^{1/2}\, Pr^{1/2} \big/ \sqrt{2\gamma + 1}}{\left[1 + \left(\frac{1}{(2\gamma+1)^3 P_r}\right)^{1/6}\right]} \tag{4.21.4}$$

where:

$$\gamma = \frac{1.0}{Re_D^{1/4}} \quad \text{for } \gamma < 1 \tag{4.21.5}$$

and $\gamma = 1.0$ for $\gamma > 1$.

For holdups greater than 5%, Wilson and Jacobs (1993) recommended for isothermal spheres in the midst of a swarm:

$$Nu_D = \left(1 - 0.186(\phi - 0.42)\right)\left(0.877\ Re_D^{1/2} + 0.152\ Re_D^{2/3}\right)Pr^{1/3} \tag{4.21.6}$$

where the velocity is the superficial velocity through the bed. This expression compares well with the correlation of Whitaker (1972) for a densely packed bed as well as for single spheres.

More recently, Evans (1994) studied the heat transfer to particles at the leading and trailing edge of a hexagonal array of spheres. Using a partial spherical shell with slip at the surface to define the effect of the surrounding particles, he showed the effect mentioned earlier for random arrays. Sirignano (1993) has studied inline arrays for up to three particles deep and showed the effects mentioned earlier for such particles. While inline effects may apply to a single column of bubbles, the hydrodynamic instability of such a flow makes the staggered array of Evans (1994) more practical in attempting to model the flows.

An experimental model for the behavior of drops or particles in swarms was recently presented by Wagner and Stichlmair (1997). They provide the following expressions for swarm velocity relative to the continuous phase as a function of the Archimedes number, Ar, void fraction, and so-called fluid number, K:

$$\frac{Re_{ss}}{K_f^{0.086}} = \frac{4.55\ Ar\left(1 - \epsilon_d\right)^{4.65} / K_f^{0.1}}{\left[15.6 + Ar\left(1 - \epsilon_d\right)^{4.65} / K_f^{0.1}\right]^{1.5}} \tag{4.21.7}$$

where $Re_{ss} = \dfrac{W_{ss} D_p \rho_c}{\mu_c}$, $K_f = \sigma^3 \rho_c^3 / \left(\mu_c^4 g\, \Delta p\right)$, $Ar = d_p^3 \rho_c\, \Delta\rho g / \mu_c^2$

and w_{ss} is the superficial velocity of the multiphase system.

Figure 4.21.5 presents a comparison of experimental data with the correlation of Equation 4.21.7, and Figure 4.21.6 shows the flooding limit obtained.

The internal to the drop or particle heat transfer is based on the one-dimensional conduction problem in a sphere. This, of course, must be modified for internal circulation and geometrical distortion. Grace (1983) provides a mapping of drop distortion for single drops of a pure fluid in a second pure immiscible fluid, as shown in Figure 4.21.7. Impurities, however, tend to immobilize the interface and the proximity of other drops will reduce any recirculation behind the drops and return them to sphericity. Jacobs and Eden (1986) provide for a correction to the molecular diffusivity which can account for both internal circulation and distortion in drops in a swarm or dense array as shown in Figure 4.21.8. For a single column of bubbles, Calderbank and Korchinski (1956) also provide an effective thermal diffusivity based on studies of bromobenzene in glycerol. Jacobs and Eden's studies were for isobutane in a geothermal brine.

For rapidly internally circulating drops, a number of investigators have used the Hadamard (1911) model for flow inside a sphere. Of these models, Sideman (1966) recommends

$$Nu_{di} = 2 + 1.13\left(Pe\,K_v\right)^{0.5} \tag{4.21.8}$$

where

$$K_v = 1 - \frac{1.45\left(2 + 3\mu_d/\mu_c\right)}{1 + \dfrac{\rho_d\, \mu_d^{1/2}}{\rho_c\, \mu_c^{1/2}}} Re_{cD}^{1/2} \tag{4.21.9}$$

and subscripts c and d refer to the continuous and drop fluid properties, respectively.

The drop temperature is predicted from

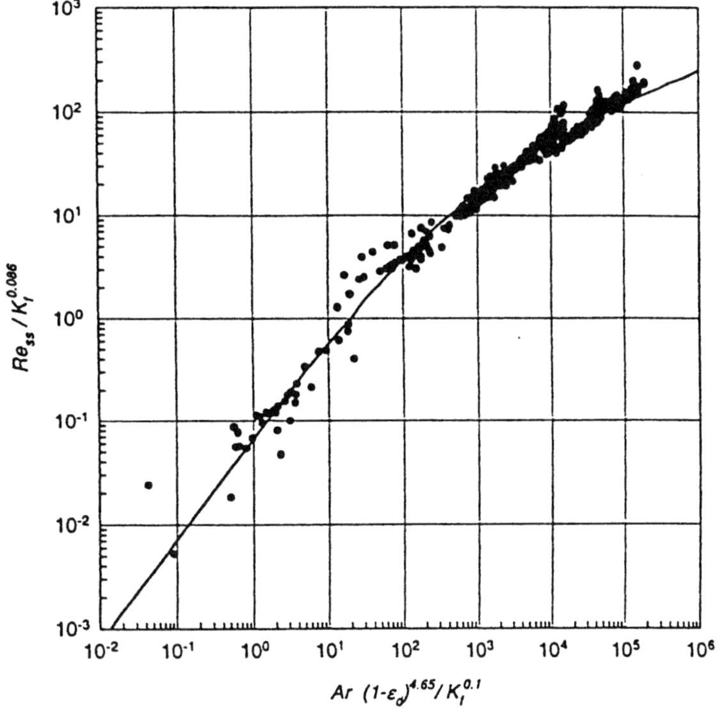

FIGURE 4.21.5 Comparison of Equation 4.21.7 with experimental data.

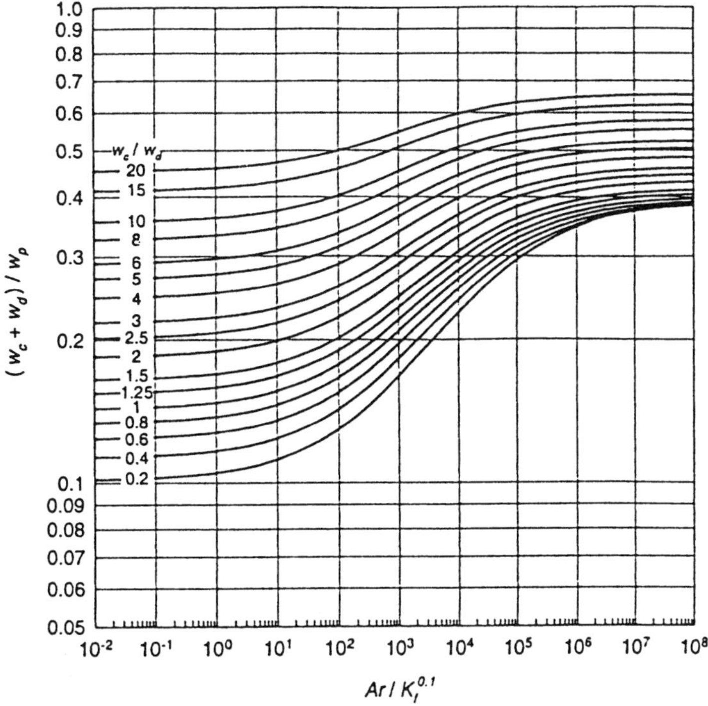

FIGURE 4.21.6 Flooding rates.

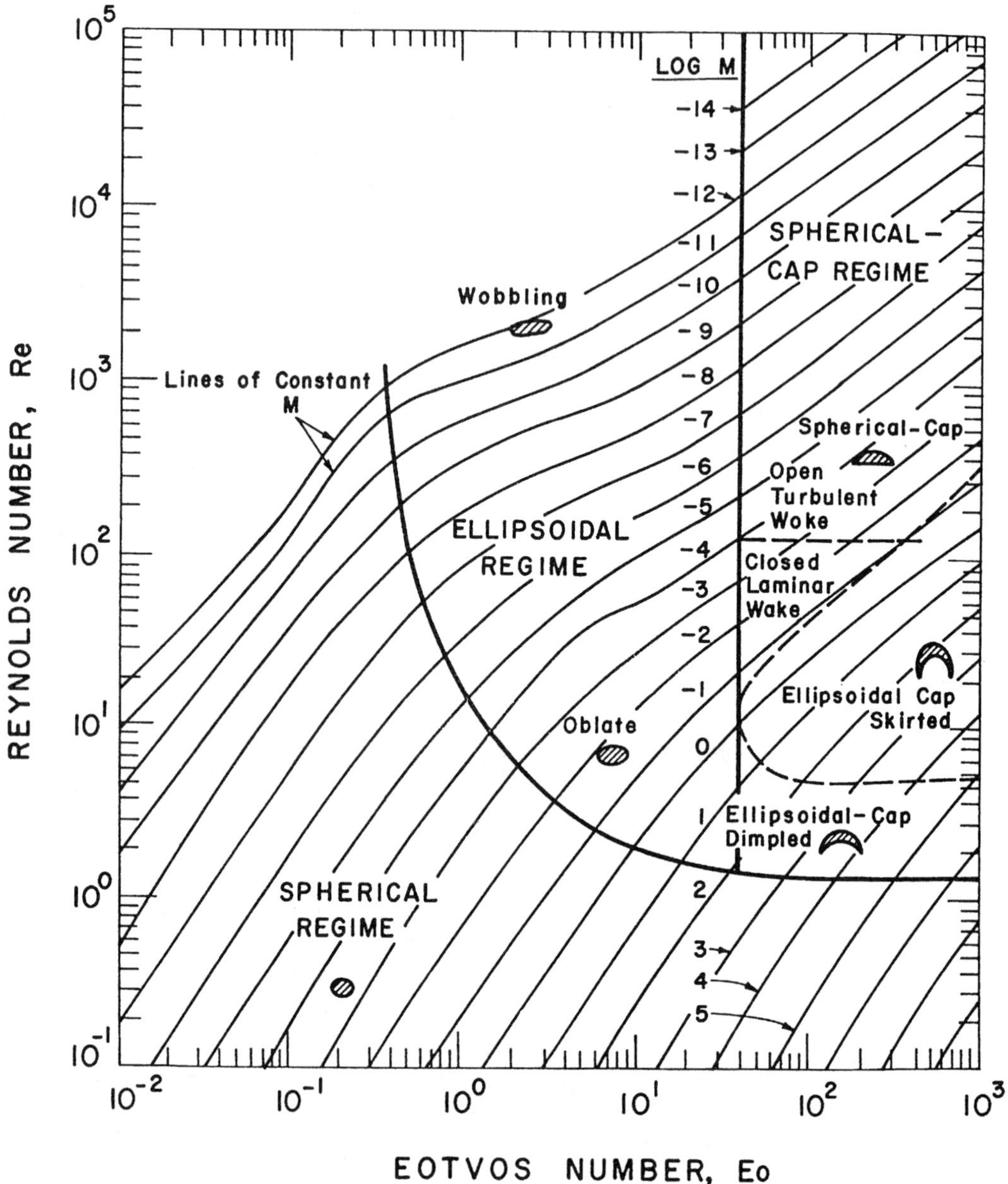

FIGURE 4.21.7 Distortion map as a function of Reynolds and Eötvos numbers. (From Grace, J.R., *Handbook of Fluids in Motion,* Butterfield Group, Ann Arbor, MI, 1983. With permission.)

$$\frac{\rho_d C_{Pd} D}{6} \frac{dT_d}{dt} = U_d \left(T_d - T_c\right) \tag{4.21.10}$$

where U_o is the overall surface-based heat transfer coefficient. When the drop is not strongly circulating the spherical conduction problem must be solved. The temperature distribution in a sphere subject to a constant heat transfer coefficient and fixed continuous phase temperature is given by

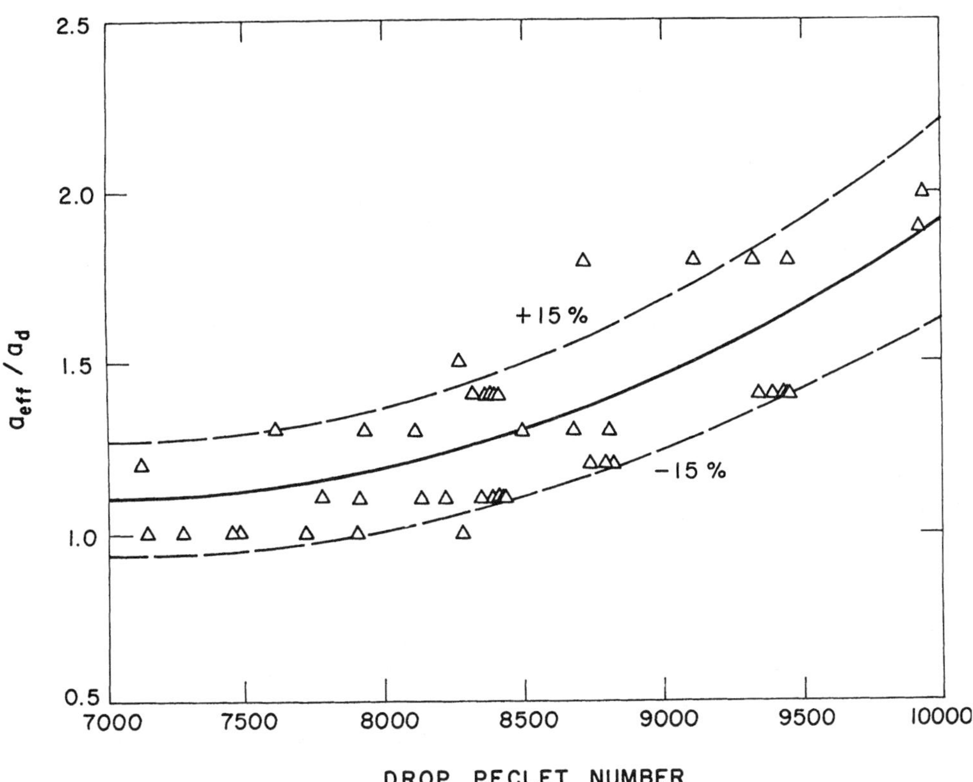

FIGURE 4.21.8 Ratio of effective thermal diffusivity as a function of drop Peclet number. (From Jacobs, H.R. and Eden, T.J., *Proc. 8th Int. Heat Conf.,* Vol. 6, 3013, 1986. With permission.)

$$\frac{T - T_\infty}{T_o - T_\infty} = 2\sum_{n=1}^{\infty} \frac{Bi \sin\mu_n e^{-\mu_n F_o}}{\left(\mu_n - \sin\mu_n \cos\mu_n\right)} \frac{\sin\left(\mu_n r/R\right)}{\mu_n r/R} \tag{4.21.11}$$

where T_o is the initial sphere temperature, Bi is the Biot modulus, Fo is the Fourier modulus, and $\mu_n = \lambda_n R$. Here μ_n are the zeros of μ_n cos μ_n = (1-Bi) sin μ_n. Solution of Equation 4.21.11 can be obtained from the so-called Heisler charts, or calculated.

In the time interval 0 to t, the ratio between the total heat transfer q from or to the sphere, and its initial internal energy qi = 4/3R c(T – T) is

$$\frac{q}{q_i} = 6\sum_{n=1}^{\infty} \frac{\sin\mu_n - \mu_n \cos\mu_n}{\mu_n^3\left(\mu_n - \mu_n \cos\mu_n\right)} \left(1 - e^{-\mu_n^2 F_o}\right) \tag{4.21.12}$$

Use of Equation 4.21.11 or 4.21.12 with values of α_{eff} and an appropriate heat transfer coefficient allows for the prediction of sensible heat transfer to individual drops or particles for low holdup. Equation 4.21.11 can be used as the Kernal in a Duhamel's superposition integral to solve for the temperature of a drop moving through a continuous phase, the temperature of which is variable with the trajectory of the drop or particle.

Direct Contact Heat Transfer with Change of Phase

Evaporation and Boiling

In Section 3 the topics of phase change from a vapor to a liquid and the reverse are introduced. These topics apply to direct contact heat transfer as well; however, the effects of solid surfaces do not come into play. The interface where the heat transfer, and thus the mass transfer to a different phase, takes place is between two independent mass streams.

At the beginning of Section 3.4.1, Van Carey introduces the concept of metastable liquids and the role of surface tension. Consider a gas bubble in a liquid. Van Carey shows that if the gas is a saturated vapor and the liquid is the same fluid, that the liquid must be superheated or in a metastable state if the pressure balance, including surface tension effects, is to be in equilibrium.

For pure liquids, where there exists a continuous phase with temperature sufficiently greater than the saturation temperature of a second fluid immiscible with the first, it is possible to demonstrate the existence of metastable states. If several subcooled drops of a fluid are injected into a second, the drops will have a pressure greater than the surroundings. The drops will gradually heat and will rise to a temperature equal to or nearly equal to the continuous phase. If the temperature is below the Spinodal limit (maximim superheat limit for a liquid) of the drops, nothing may happen. For example if the drops are pentane and the fluid is pure water, the drops may rise to the surface after a period of time and then evaporate. If the water temperature is near the Spinodal lmit, the drops will rise, but before reaching the surface may instantaneously vaporize in an explosive way. The presence of impurities in either phase will cause the drops to vaporize and form two-phase bubbles below the surface. The impurities act as nucleation sites for the phase change just as impurities in a metal or surface flaws serve as nucleation sites for boiling from solid surfaces. Some early experiments to illustrate boiling nucleation were with mercury films on a surface with an oil as the vaporizing medium. They were used to illustrate the need for nucleation sites. Thus, although the thermodynamics of direct contact boiling is related to conventional boiling, the effect of the interfaces between liquids makes them different and metastable states perhaps play a more important role. In most industrial situations, impurites are present and alter the situation for direct contact boiling.

Four basic problems are encountered in direct contact boiling/evaporation. These are (1) boiling at the interface between two parallel streams, (2) evaporation of drops of a spray from an immiscible liquid, (3) evaporation of drops within a liquid or gas/vapor mixture, and (4) the evaporation of the continuous phase around the dispersed phase. The first problem is related directly to film boiling and follows its analyses. The second problem is of interest due to a number of things. First, if the interfacial properties are such that the liquid drops wet the liquid below them, there will exist a metastable state with the liquid in the drops having their phase change only from their upper surfaces. Such a situation was addressed by Kodres, Jacobs, and Boehm (1980, 1981) who showed photographically the difference between this situation and that of Liedenfrost boiling, where the drop is held above the surface by the vaporizing film. (See *Numerical Heat Transfer,* 3:21-34, 1998.) Development of a model for the evaporation rate must be accomplished numerically and depends on the fluid properties.

The evaporation of a drop in a gas stream is dependent upon the nature of the gas phase. For the evaporation of a drop into a gas made up primarily of different substances; for example, a fuel drop in air combustion products requires the solution of the mass diffusion problem coupled with the heat transfer. The same is true of water drops in an air stream as related to the design of cooling towers discussed in Section 4.9. However, the evaporation of drops into a superheated stream of its own vapor is considerably easier to analyze, depending on the drop size. Carey and Hawks (1995) report for the evaporation of large drops in their own vapor

$$\mathrm{Nu}_D = 2\,\frac{\ln\left(\mathrm{Ja}_v - 1\right)}{\mathrm{Ja}_v} \tag{4.21.13}$$

where $J_{av} = e_{pv}(T_\infty - T_R)/h_{fg}$ and for microdrops

$$Nu_D = 2\,\frac{\ln(Ja_v - 1)}{Ja_v}\left[1 + \frac{0.4\,Ja_v}{1 + 10\,Ja_v^2}\left(\frac{K_{np}}{0.001 + K_{nD}}\right)\cdot\sigma_T(1+\gamma)(9\gamma+5)K_{nD} + \sigma_T(1+\gamma)\right] \quad (4.21.14)$$

where Kn is the mean free path of a molecule divided by the drop diameter, γ is the ratio of specific heat, and ϕ_r is the accommodation coefficient. The reader is referred to Sanders and Dwyer (1987) for further review of individual droplet processes. They present a comprehensive numerical model of single-droplet dynamics with and without combustion. Their studies include mass transfer with accelerating and decelerating drops. They solve the complete Navier-Stokes equations, energy equations, mass conservation equations, and equations of state.

The last evaporation/boiling model deals with droplets vaporizing in an immiscible liquid. Such a system was investigated by Coban and Boehm (1986) and by numerous others for application to desalination, geothermal power production, and production of electricity using solar ponds, etc. A volatile fluid is injected into a hot immiscible continuous stream. Unlike evaporation within a gas stream, here the vapor stays with the drop. The problem has been studied by Klipstein (1963), Simpson et al. (1974), Sideman (1966), and more recently, Raina and co-workers (1982, 1984a, 1984b, 1985), and Battya et al. (1983). The models are all single-drop/bubble studies and are assumed to strongly depend on the external heat transfer and surface tension effects. In most of the systems studied, a drop of hydrocarbon was placed in water, the temperature of which was well above the boiling point of the hydrocarbon. For such systems the surface of the two-phase bubble is wetted by the hydrocarbon liquid. This suggests that the hydrocarbon must reach a degree of superheat before evaporation occurs, but not the explosive condition close to the Spinodal line. Various models assume that the liquid settles to the bottom of the drop, as in the early models of Sideman, or to totally wet the walls with a uniform film. The former models do not reproduce well the experimental results while the latter model, first proposed by Simpson, does. It results in

$$h_o = \frac{2.57\left(\frac{D(t)}{Do}\right)^{1/6}}{1 + 0.206\left(\frac{D_t}{Do}\right)^{5/12}}\;\frac{Kw}{m^2oK} \quad (4.21.15)$$

This equation predicts an increasing heat transfer in the early stages of evaporation and a decrease in the latter stages.

The Sideman (1966) model depended only on the external heat transfer coefficient and fraction of the bubble surface wetted by the remaining liquid. It predicts the heat transfer decreases with time as it ignores heat transfer through the vapor. Coban and Boehm (1986) account for heat transfer through the vapor as well as directly to the liquid, resulting in

$$Nu_{ID} = 2\,Pe_c^{1/3}\left\{0.466\left(\pi - B + \sin\frac{dB}{dt}\right)^{2/3} + \frac{5\,K_{dv}}{R}(1 - 0.466)\left(\pi - B + \sin\frac{dB}{dt}\right)^{2/3}\right\} \quad (4.21.16)$$

where c indicates the continuous phase, d the drop, and l and v are properties of the liquid and vapor, respectively. The above model assumes no sloshing of the liquid and B is the segment angle defining the surface of the liquid in the bottom of the drop. The expression is for the overall heat transfer.

The last evaporation phenomenum addressed is for a hot liquid drop or particle in a subcooled liquid with subsequent vaporization of the coolant. Kazimi et al. (1974) dealt with this as a transient film boiling problem but neglected the effects of both gravity and continuous phase motion. Dhir and Purohit

(1977) solved the problem of subcooled film boiling from spheres including buoyancy effects. They predict

$$Nu_{Dv} = Nu_D + Nu_{nc}\frac{Pr\,Sc}{Pr_l\,Sh\,\mu} + \frac{Pr_v\,\sigma}{Sh}\frac{\left(T_w^4 - T_{SAT}^4\right)D}{h_{fg}\,\mu_v} \tag{4.21.17}$$

where

$$Nu_o = 0.8\left(\frac{g\rho_v\left(\rho_2 - \rho_v\right)h_{fg}\,D^3}{M_v\,k_v\left(T_w - T_\infty\right)}\right)^{1/4} \tag{4.21.18}$$

and

$$Nu_{nc} = 0.9\left(\frac{g\rho_L\,c_{pL}\,B_o\left(T_{SAT} - T_\infty\right)D^3}{\mu_L\,K_L}\right)^{1/2} \tag{4.21.19}$$

for the average Nusselt number for a sphere in a quiescent pool.

For forced flow over a sphere or a falling sphere

$$Nu = Nu_o = 0.8\,Re_{cD}^{1/2}\left[1 + \frac{Sc\,Pr\,v}{Sh\,Pr_L\,\mu_c}\right] \tag{4.21.20}$$

where $S_c = \dfrac{C_{PL}\left(T_{SAT} - T_\infty\right)}{h_{fg}}$ and $Sh = \dfrac{C_{pr}\left(T_w = T_\infty\right)}{h_{fg}}$.

For boiling in cryogenic fluids, Hendricks and Baumeister (1969) investigated boiling of nitrogen and helium from a sphere. They show a dependence on the Bond number, $Bo = (P_L - P_r)g\,R^2/(\sigma gc)$, where σ is the surface tension. The results are shown in Figure 4.21.9, where Nu′ and Nu – 2 and $Rd' = \dfrac{h_{fg} + 0.5C_p\,\Delta T}{C_p\,\Delta T}$.

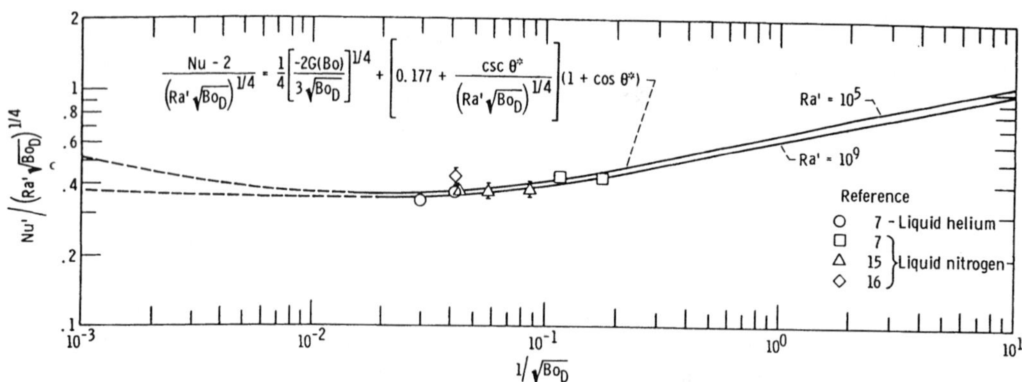

FIGURE 4.21.9 Experimental and predicted results for film boiling off submerged spheres. (From Hendricks, R.C. and Baumeister, K.J., Film Boiling from Submerged Spheres, NASA, Washington, D.C., 1969.)

Direct Contact Condensation

Direct contact condensation occurs in many industrial processes including reflux columns and in separators as well as in open feedwater heaters, etc. Jacobs (1988) provides a significant discussion in his review article on *Direct Contact Heat Transfer for Process Technologies*. The first industrial use of direct contact condensers was in steam engines and dates back prior to the turn of the century. The first design article was written by Hausbrand and was published in English in 1933. In a review of the literature, Jacobs and Fannar (1977) categorized direct contact condensers as being "drop type," " jet and sheet type," and "film type."

Drop type refers to condensation on a spray of coolant that is injected into a chamber filled with vapor or a gas-vapor mixer with the intent to condense the vapor. The coolant can be the same fluid as the vapor or could be immiscible with the condensate. Kutateladze (1952) was the first to recommend the drops be assumed spherical and that the heat transfer is dominated by the transient conduction within the drops when dealing with a pure vapor. This assumption is reasonable in most situations as injectors used in direct contact condensers produce drops in the size range of 0.25 to 2.5 mm in diameter. Such drops fall at sufficiently slow velocity that they remain spherical with little or no internal circulation. The Kutateladze model neglects the added mass. While such an analysis is appropriate for a Jakob number tending to infinity or time to zero, it overpredicts the heat transfer. Jacobs and Cook (1978) developed a theoretical model that accounted for the added mass and predicts the experiments of Ford and Lekic (1973) exactly.

The model development shows that the final to initial drop size is given by

$$R_f/R_I = \left(1 + \frac{1}{Ja}\right)^{1/3} \tag{4.21.21}$$

where

$$Ja = \frac{\rho_L h_{fg}}{\rho_d C_{\rho d}\left(T_{SAT} - T_o\right)} \tag{4.21.22}$$

in which the subscripts L and d refer to the condensate liquid and drop, respectively. Thus, one can deal with a drop being of a different fluid than the condensate. Their solution solves the problem of conduction in a sphere subject to the boundary condition

$$q'' = \frac{k_c\left(T_{SAT} - T_{RI}\right)}{R(t) - Ri} \tag{4.21.23}$$

The nondimensional radius at time t is given as

$$\frac{T_{mean}(t)}{T_{SAT} - T_i} = 1 + \frac{6}{\pi}\sum_{n=1}^{\infty}\frac{1}{n^2}\exp\left(-n^2\pi^2\frac{4at}{D_o^2}\right) \tag{4.21.24a}$$

This is a solution of a spherical Stefan-type problem assuming that the condensate film is thin. It is appropriate for large values of Ja. In the limit as Ja tends to infinity

$$\frac{R(t)}{R_I} = 1 + \frac{3}{Ja}\int_0^{R_i}\left(\frac{r}{R_I}\right)^2\frac{\left(T_{(r,t)} - T_i\right)}{\left(T_{SAT} - T_i\right)}\frac{dr}{Ri} \tag{4.21.24b}$$

For conditions where there are noncondensibles present, it is necessary to solve the flow field external to the drops. Such studies have been conducted by Ayyaswamy, Chung, and Huang (1981 to 1987) who also included the effects of internal circulation. They found the latter to have negligible effects, except at extremely high velocities or low interfacial tension. Celata et al. (1990) attempted to form uniform drops by application of a high-frequency acoustic field. At Re as low as 100, but for Pe of 1000 to 5000, they found heat transfer coefficients higher than predicted for infinite Pe. The best fit to their data was obtained using Equation 4.21.24; however, even it underpredicted the early temperature rise, indicating even stronger mixing than predicted by the Hadamard model for flow inside a sphere.

The influence of noncondensibles on condensation rates is not unlike that predicted for boundary layer type problems, with as little as 5% noncondensibles reducing the heat transfer by as much as 50 to 100%. More studies of this influence are needed.

In addition to accurately predicting the performance of a spray type (drop type condenser), it is required to have data on drop size distribution for a given spray nozzle. Brown (1951) and Isachenko and Kushnyrev (1974) present some work, but more is needed. For example, in order to design a barometric condenser the vapor velocity must not exceed the terminal velocity of the smallest drop; yet to fully utilize the cooling capacity of the spray, the size of the largest drops must be known. From a practical view, this means that the diameter of the condenser for a simple spray column is set by the smallest drops and its height by the largest.

Collapsing Bubble Condensation

Direct contact condensation can be accomplished by injecting a jet of bubbles into a continuous liquid coolant. Depending on the jet velocity, the jet can break down quickly or remain coherent for some time. If the jet is injected at close to sonic velocities, the latter is true. Of course, the vapor can bubble through in a series of bubble trains as well. Thus, single vapor bubbles as well as jets have been studied. Jet flow is impacted by its orientation with gravity, with vertical jets foreshortened or lengthened depending on whether gravity is opposing or favorable. Horizontal jets are typically curved due to buoyancy effects. The first definitve study was by Weimer et al. (1973). Chen and Faeth (1982, 1983) and Sun (1985) used sophisticated two-fluid numerical models to analyze jet collapse, yet they do not adequately predict the heat transfer between the phases (Eden, 1995). Thus, dependence on experimental studies is still required.

The study of bubble collapse is not only of interest as it applies to condensers, but also for other applications. Sideman (1982) reported a wide range of experimental data. Floreschuetz and Chao (1965) modeled a drop collapsing in its own vapor. Jacobs et al. (1978) were the first to account for the internal resistance of the condensate in the collapse of a saturated vapor bubble in an immiscible liquid. They show that the collapse rate for a pure vapor rising at a constant velocity is

$$\frac{D(t)}{D_o} = \left[1 - 0.901\left(\mathrm{Re}_L\, P_r\right)^{1/2} \phi J_a^* \frac{\rho_f}{\rho_v} \mathrm{Fo}_L\right] \tag{4.21.25}$$

where

$$\phi = \sqrt{\left(C_1^2 + C_1\right)} - C_1 \tag{4.21.26}$$

and

$$C_1 = 0.75 \frac{\left(\rho k C_p\right)_f}{\left(\rho k C_p\right)_1} \tag{4.21.27}$$

Jacobs and Major (1982) dealt with bubbles containing a mixture of vapor and a noncondensible gas. They showed that for small bubbles less than or equal to 3.5 mm in radius, the heat transfer was controlled by diffusion, but larger bubbles have good internal circulation due to distortion and oscillation and their analysis can be accomplished by assuming perfect mixing during the collapse.

The effect of adjacent bubbles in a collapsing train was dealt with by Moalem-Maron et al. (1972). Golafshani (1983) improved on the model, using the more sophisticated model of Jacobs and Major (1982). For pure vapor bubbles they found little influence based on the models, indicating that the external thermal resistance dominated the heat transfer.

Despite the relative acceptance of the theories, Johnson et al. (1983) and Sudhoff (1982) found that experiments with bubble trains are plagued by a so-called persistent bubble. Johnson et al. carried out experiments in a very deep heat exchanger. The visual observations indicated that the persistent bubble finally disappears and they conclude that this is due to absorption of the noncondensible gases present by the surrounding liquid. So far, no other investigators have offered hypotheses to explain this phenomenum; however, if the arguments of Johnson et al. are valid, it may be possible that a heat exchanger could be designed with a depth such that condensation may be accomplished and noncondensible gases readily separated.

Condensation on Jets and Sheets

Barometric condensers are designed in the form of spray columns, baffle tray columns, or sieve tray columns. When designed as spray columns, a nozzle forms a coarse spray as discussed in the preceding section. When designed as a baffle tray column, the coolant flows into a tray, and then over its lip in a sheet to the tray below, and subsequently to the next tray and so forth until it and the condensate form a pool at the bottom of the condenser.

In a sieve tray column the coolant leaves the tray through perforations as a series of jets. Both sheets and jets can break down into lesser streams and/or drops as discussed by Bogy (1979). The jets are laminar but instabilities can make them wavy or even turbulent prior to breaking into drops. If the jets or sheets remain laminar or just wavy, they can be analyzed as a Graetz-type problem similar to the analysis for sensible heat transfer. Hasson et al. (1964a,b) applied such analyses to condensing steam on sheets annd fan jets with reasonable success. Jacobs and Nadig (1987) solved the problems for an axisymmetric jet or a sheet including both the resistance of the added condensate mass and the effects of noncondensibles for the vapor and coolant being the same or immiscible fluids. They presented curves of coolant capacity utilization rather than heat transfer coefficients for design purposes. Results for steam-water are presented in Figures 4.21.10a/b. Knowing the noncondensible concentration, and the degree of utilization, the Graetz number and, thus, the length of the sheet can be determined. Extension of this work to wavy or turbulent flow is desirable; however, it is likely that waves will not increase the heat transfer by more than 15% based on the sensible heating information reported earlier for jets and falling films.

When as little as 2% noncondensibles are present, the heat transfer is controlled by the external diffusion of vapor to the interface, thus turbulence in the liquid will also have small effects.

Liu et al. (1989) carried out experiments with round jets, which ultimately broke down and formed drops whose diameter were twice that of the jet. Their results, shown in Figures 4.21.11a/b indicate that the heat transfer for condensation was little different than if the jet had not fragmented. This was because although the surface area was greater, the rate of diffusion was smaller, thus negating the former. Kim and Mills (1988) studied condensation on turbulent jets and noted that the external to the jet heat transfer resistance governed, as noted above for sheets.

Direct Contact Condensation Falling Films

In addition to the barometric condensers mentioned in the preceding subsections, some condensers have been designed as packed beds. The packing can be varied from structured to nonstructured types and be made of a variety of materials. The materials are chosen so that the coolant wets the packing and flows across it as a thin film which is reformed as it drips from one level of packing to the next in a manner analogous to condensate dripping from one tube to the next in a surface condenser.

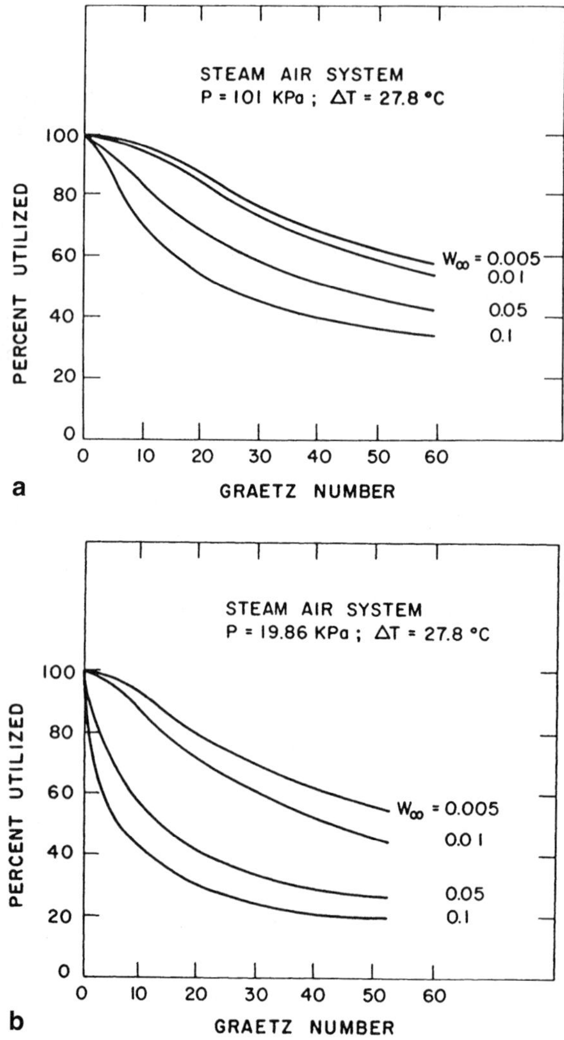

FIGURE 4.21.10 (a) Effect of W_∞ on percent coolant utilized for condensation on a sheet, P = 10/KPa; (b) effect of W_∞ on percent coolant utilized for condensation on a sheet, P = 19.86 KPa. (From Jacobs, H.R. and Nadig, R., *ASME J. Heat Transfer,* 109, 1013, 1987. With permission.)

The modeling of falling films of coolant with subsequent condensation are similar to those for sensible heat transfer mentioned previously, but considerably more complicated, especially with noncondensibles present. Seven equations must be solved simultaneously. Even for a pure vapor the problem is difficult, as shown by Jacobs and Bogart (1980). Jacobs and Nadig (1983) deal with noncondensibles being present as well as the coolant and condensate being immiscible. Jacobs et al. (1982) studied film condensation on a thin film flowing over a sphere. Using their results, they present a correlation for the mean fluid temperature leaving the sphere as

$$\frac{T_m - T_{SAT}}{T_o - T_{SAT}} = 0.03\ Ja^{-0.4}\overline{R}^{-1/2}\ Re^{1/3}\ Pr^{0.36} \tag{4.21.28}$$

for the coolant and condensate being the same. In the above, $\overline{R}$ is $R\Big/\left(\frac{\nu^2}{g}\right)^{1/3}$ and R is the radius of the sphere.

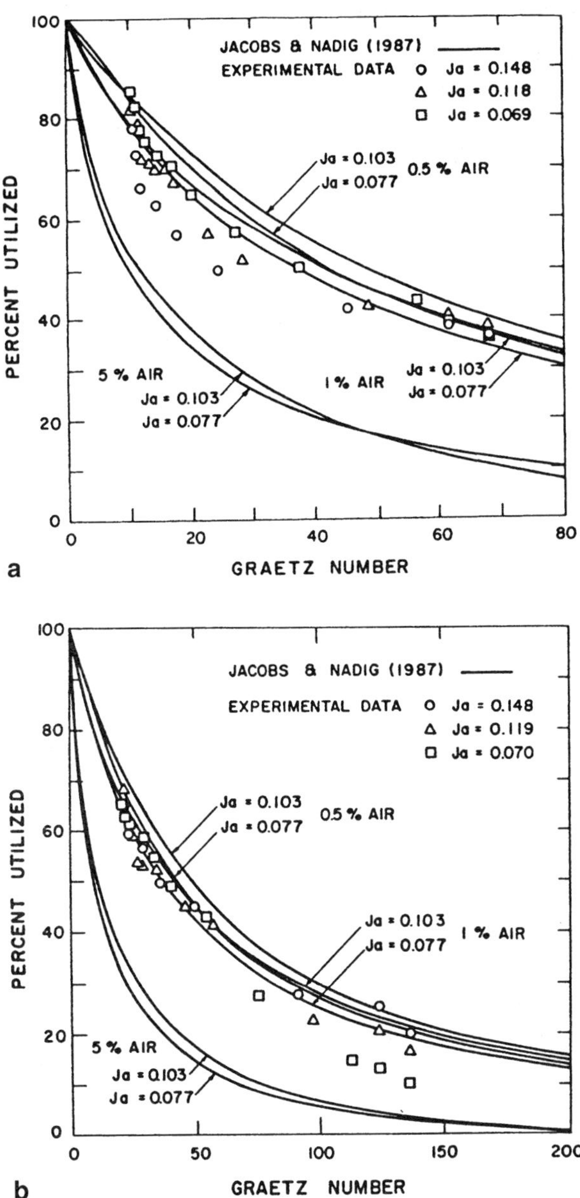

FIGURE 4.21.11 (a) Percent coolant utilized for 12.7 mm head of water; (b) percent coolant utilized for 50.8 mm head of water. (From Lui, T.L., Jacobs, H.R., and Chen, K., *ASME J. Heat Transfer,* 111, 585, 1989. With permission.)

The results showed very good agreement when compared to the experimental data of Finklestein and Tamir (1976).

Repeated layers of spheres with reforming films which flow onto the succeeding layer with a new To equal to Tm at the base of the previous sphere provide a model for a packed bed of spheres.

Jacobs et al. (1979) and Bharathan and Althof (1984) carried out experiments with a variety of packings and relative flow rates of vapor and coolants that account for a nonstagnant vapor. Celata et al. (1989) studied steam condensation on a turbulent film of water. They indicate that laminar models can be used when modifying the liquid thermal conductivity as

$$k_{eff} = k_f \left(0.21\ Re^{0.34} / Pr^{0.24}\right) \tag{4.21.29}$$

Direct Contact Heat Transfer with Solidification

Direct contact with solidification is found in several applications including the manufacture of small beads where a molten spray is injected into a coolant. (Typical of this is the formation of lead and steel shot for shotguns.) It is also used in freezing of food, freeze-desalting of brine, and freeze-coating of thin films on wires. Johnson and Pallone (1973) investigated freeze-desalting by injecting a cold refrigerant into brine. Ice formed at the interface and was recovered with the refrigerant. Cheung et al. (1987, 1990) studied freeze-coating of a moving wire with the intent of forming a uniform coating. These problems are all similar to condensation problems assuming, for example, condensation on a drop or condensation on a jet or sheet with the velocity in the sream being slug flow. The difference in the analysis is that the Jakob number is replaced by the Stefan number. Freezing of liquid drops follows the Stefan problem in conduction, depending on the circulation within the drops and external shear. For metal drops, the high surface tension eliminates the circulation and the solution is in fact a spherical Stefan problem. Of course, no analytical solution exists and numerical or integral methods must be used to solve the nonlinear ptoblem.

Summary

The length of this section on direct contact heat transfer does not provide for a complete development of the subject. However, the references cited in the text will provide further insight to the multitude of complex problems that are direct contact in nature. The reader has been referred to other sections within the text as well. Of particular importance are the development of models for multiphase flow. So-called two-fluid models are often used to determine the heat transfer in direct contact systems. These models are typically one-dimensional Eulerian models. However, when dealing with one or more dispersed phases in a continuum, a so-called Lagrangian-Eulerian model also is used. The author and others have developed computer codes to deal with a range of such problems; nonetheless, continued work is needed, especially for problems where the holdup of the dispersed phase is large.

References

Ahmad, G.R. and Yovannovich, M.M., Approximate Analytical Solution of Forced Convection Heat Transfer from Isothermal Spheres for All Prandtl Numbers, *ASME J. Heat Transfer,* 116, 838-843, 1994.

Arpaci, V. *Conduction Heat Transfer,* Addison-Wesley, Reading, MA, 1966.

Bankoff, S.G. and Lee, S.C., A Critical Review of the Flooding Literature, *Multiphase Sci. Technol.,* 2, 95-180, 1986.

Battya, P., Raghaven, V.R., and Seetharamu, K.N., A Theoretical Analysis of Direct Contact Evaporation of a Moving Drop in an Immiscible Liquid, *Proc. 3rd Int. Conf. Numerical Methods in Thermal Problems,* 1983.

Bharathan, D., Direct Contact Evaporation, *Direct Contact Heat Transfer,* Kreith, F. and Boehm, R.F., Eds., Hemisphere, New York, 1988, chap. 11.

Bharathan, D. and Althof, J., An Experimental Study of Steam Condensation on Water in Countercurrent Flow in the Presence of Inert Gases, ASME Paper No. 84-WA-Sol-25, American Society of Mechanical Engineers, New York, 1984.

Bogy, D.B., Drop Formation in a Circular Jet, *Annu. Rev. Fluid Mech.,* 11, 207-228, 1979.

Brown, G., Heat Transmission During Condensation of Steam on a Spray of Water Drops, Institute of Mechanical Engineers, General Discussion on Heat Transfer, 1951, pp. 49-51.

Calderbank, P.H. and Korchinski, I.J.O., *Chem. Eng. Sci.,* 6, 65-78, 1956.

Carslaw, and Jaeger, *Conduction of Heat in Solids,* Oxford University Press, Oxford, U.K., 1959.

Carey, V.P. and Hawks, N.E., Stochastic Modeling of Molecular Transport to an Evaporating Microdroplet in a Superheated Gas, *ASME J. Heat Transfer,* 117, 432-439, 1995.

Celata, G.P., Cumo, M., Farello, G.E., and Focandi, G., A Theoretical and Experimental Study of Direct Contact Condenstation on Water in Turbulent Flow, *Exp. Heat Transfer,* 2(2), 153-172, 1989.

Celata, G.P., Cumo, M., D'Annibale, F., and Farello, G.E., Direct Contact Condensatino of Steam on Droplets, *Int. J. Multiphase Flow,* 1990.

Chen, L.D. and Faeth, G.M., Condensation of Submerged Vapor Jets in Subcooled Liquids, *ASME J. Heat Transfer,* 104, 774-780, 1982.

Chen, L.D. and Faeth, G.M., Structure of Turbulent Reacting Gas Jets Submerged in Liquid Metals, *Combust. Sci. Technol.,* 31, 277-296, 1983.

Cheung, F.B. and Cha, S.W., Flow-Freezing Interaction During Freeze Coating on a Non-Isothermal Moving Cylinder, ASME Paper No. 87-Ht-4, American Society of Mechanical Engineers, New York, 1987.

Cheung, F.B., Solidification on a Chilled Continuous Surface Moving in a Parallel Free Stream, *ASME J. Heat Transfer,* 112, 521-523, 1990.

Chung, J.N. and Ayyaswamy, P.S., Material Removal Associated with Condensation on a Droplet in Motion, *Int. J. Multiphase Flow,* 7, 329-342, 1981a.

Chung, J.N. and Ayyaswamy, P.S., Laminar Condensation Heat and Mass Transfer of a Moving Drop, *AlChE J.,* 27, 327-347, 1981b.

Chung, J.N., Ayyaswamy, P.S., and Sabhal, S.S., Laminar Condensation on a Moving Drop, I. Singular Perturbation Technique, *J. Fluid Mech.,* 139, 105-130, 1984a.

Chung, J.N., Ayyaswamy, P.S., and Sadhal, S.S., Laminar Condensation on a Moving Drop. II. Numerical Solutions, *J. Fluid Mech.,* 139, 131-144, 1984b.

Coban, R. and Boehm, R.F., Numerical and Experimental Modelings of Three Phase Direct Contact Heat Exchangers, Proc. 8th Int. Heat Transfer Conf., Toronto, 1986, 3019-3024.

Dhir, V.K. and Purohit, G.P., Subcooled Film Boiling Heat Transfer from Spheres, ASME Paper No. 77-HT-78, American Society of Mechanical Engineers, New York, 1977.

Dukler, A.E., Fluid Mechanics and Heat Transfer in Vertical Falling Film Systems, AlChE Chemical Engineering Progress Series — Heat Transfer, STORRS, 56, No. 30, pp. 1-10, 1960.

Eden, T.J., Bouyant Condensing Vapor Jets, Ph.D. dissertation, Penn State University, University Park, PA, 1995.

Evans, M.S., Numerical Solution of Axisymmetric Flow and Heat Transfer from Spherical Particles at the Leading and Trailing Edges of an Array, MSME thesis, Pennsylvania State University, University Park, PA, 1994.

Faghri, A., Prediction of Heat and Mass Transfer for Absorption of a Gas to a Liquid Film, *Numerical Methods Therm. Prob.,* III, 981-990, 1983.

Faghri, A. and Saban, R.A., Heat Transfer in Wavy Liquid Films, *Int. J. Heat Mass Transfer,* 28(3), 506-509, 1985.

Faghri, A. and Seban, R.A., Heat and Mass Transfer to a Turbulent Liquid Film, *Int. J. Heat Mass Transfer,* 31(4), 891-894, 1998.

Finkelstein, Y. and Tamir, A., Interfacial Heat Transfer Coefficients of Various Vapors in Direct Contact Condensation, *Chem. Eng. J.,* 12, 199-209, 1976.

Floreschuetz, L.W. and Chao, B.T., On the Mechanics of Vapor Bubble Collapse, *ASME J. Heat Transfer,* 87, 209-220, 1965.

Ford, J.D. and Lekic, A., Rate of Growth of Drops During Condensation, *Int. J. Heat Mass Transfer,* 16, 61-66, 1973.

Golafshani, M., Bubble Type Direct Contact Condensers, MS thesis, Department of Mechanical and Industrial Engineering, University of Utah, Salt Lake City, Utah, 1983.

Grace, J.R., Hydrodynamics of Liquid Drops in Immiscible Liquids, *Handbook of Fluids in Motion,* Butterfield Group, Ann Arbor, MI, 1983, 1003-1025.

Gunn, D.J., Transfer of Heat and Mass to Particles in Fixed and Fluidized Beds, *Int. J. Heat Mass Transfer,* 21, 467-476, 1978.

Hadamard, J., Mouvement Permanent Lent d'Une Sphere Liquide et Visquesuse dans Une Liquide Visqueux, *C. R. Acad. Sci. Paris,* 152, 1735-1738, 1911.

Hasson, D., Luss, D., and Peck, R., Theoretical Analysis of Vapor Condensation on Liquid Jets, *Int. J. Heat Mass Transfer,* 7, 969-981, 1964a.

Hasson, D., Luss, D., and Naron, V., An Experimental Study of Steam Condensing on a Laminar Water Sheet, *Int. J. Heat Mass Transfer,* 7, 983-1001, 1964b.

Hausbrand, E., *Condensing and Cooling Apparatus,* 5th ed., Van Nostrand Reinhold, New York, 1933.

Hendricks, R.C. and Baumeister, K.J., Film Boiling from Submerged Spheres, NASA TN D-5124, National Aeronautics and Space Administration, Washington, D.C., 1969.

Hikita, H. and Ishimi, K., Frictional Pressure Drop for Laminar Gas Streams in Wetted Wall Columns with Cocurrent and Countercurrent Gas-Liquid Flow, *J. Chem. Eng. Jpn.,* 9(5), 357-361, 1976a.

Hikita, H. and Ishimi, K., Mass Transfer into Turbulent Gas Streams in Wetted Wall Columns with Cocurrent and Countercurrent Gas Liquid Flow, *J. Chem. Eng. Jpn.,* 9(5), 362-367, 1976b.

Huang, L.J. and Ayyaswamy, P.S., Drag Coefficients Associated with a Moving Drop Experiencing Condensation, *ASME J. Heat Transfer,* 109, 1003-1006, 1987a.

Huang, L.J. and Ayyaswamy, P.S., Heat and Mass Transfer Associated with a Spray Drop Experiencing Condensation: A Fully Transient Analysis, *Int. J. Heat Mass Transfer,* 30, 881-891, 1987.

Incropera, F. and DeWitt, D.P., Fundamentals of Heat and Mass Transfer, John Wiley & Sons, New York, 1985.

Isachenko, V.P. and Kushnyrev, V.J., Condensation Heat Transfer in Dispersed Liquid Spray, Proc. 5th Int. Heat Transfer Conf., Vol. III, pp. 217-220, 1974.

Ishii, M., *Thermo-fluid Dynamics of Two-Phase Flow,* Eyrolles, Paris, Scientific and Medical Publication of France, 1975.

Jacobs, H.R., Fannir, H., and Beggs, G.C., Collapse of a Bubble Vapor in an Immiscible Liquid, Proc. 6th Int. Heat Transfer Conf., Vol. 2, 383-388, 1978.

Jacobs, H.R., Direct Contact Condensation, *Direct Contact Heat Transfer,* Kreith, F. and Boehm, R.F., Eds., Hemisphere, New York, 1988a, 223-234.

Jacobs, H.R., Thermal and Hydraulic Design of Direct Contact Spray Columns for Use in Extracting Heat from Geothermal Brines, *Direct Contact Heat Transfer,* Kreith, F. and Boehm, R.F., Eds., Hemisphere, New York, 1988b, 343-370.

Jacobs, H.R. and Bogart, J.A., Condensation on an Immiscible Thin Film, ASME Paper No. 80-HT-110, American Society of Mechanical Engineers, New York, 1980.

Jacobs, H.R., Bogart, J.A., and Pensel, R.W., Condensation on a Thin Film Flowing Over an Adiabatic Sphere, Proc. 7th Int. Heat Transfer Conf., Vol. 5, pp. 89-94, 1982.

Jacbos, H.R. and Cook, D.S., Direct Contact Condensation on a Non-Circulating Drop, Proc. 6th Int. Heat Transfer Conf., Vol. 3, pp. 389-393, 1978.

Jacobs, H.R. and Dolan, S.M., The Transfer of Heat and Momentum Between a Laminar Falling Film and a Countercurrent Gas Stream, *J. Energy, Heat Mass Transfer,* 16, 77-85, 1994.

Jacobs, H.R. and Eden, T.J., Direct Contact Heat Transfer in a Sieve Tray Column, Proc. 8th Int. Heat Transfer Conf., Vol. 6, pp. 3013-3018, 1986.

Jacobs, H.R. and Fannir, H., Direct Contact Condensers: A Literative Survey, USDOE Report, DOE/1523-3, Department of Energy, Washington, D.C., 1977.

Jacobs, H.R. and Fletcher, M.J., Heat Transfer from a Falling Film to a Countercurrent Gas Stream, Proc. 1988 ASME National Heat Transfer Conf., HTD Vol. 96, No. 3, pp. 281-289, 1988.

Jacobs, H.R. and Major, B.H., The Effects of Noncondensible Gases on Bubble Condensation in a Immiscible Liquid, *ASME J. Heat Transfer,* 104, 487-492, 1982.

Jacobs, H.R. and Nadig, R., Condensation on an Immiscible Falling Film in the Presence of a Non-condesible Gas, *Heat Exchangers for Two Phase Applications,* ASME HTD-Vol. 27, American Society of Mechanical Engineers, New York, 1983, 99-106.

Jacobs, H.R. and Nadig, R., Condensation on Coolant Jets and Sheets Including the Effects of Noncondensible Gases, *ASME J. Heat Transfer,* 109, 1013-1020, 1987.

Jacobs, H.R., Thomas, K.D., and Boehm, R.F., Direct Contact Condensation of Immiscible Fluids in Tacked Beds, *Condensation Heat Transfer,* American Society of Mechanical Engineers, New York, 1979, 103-110.

Johnson, K.M., Jacobs, H.R., and Boehm, R.F., Collapse Height for Condensing Vapor Bubbles in an Immiscible Liquid, Proc. Joint ASME/JSME Heat Transfer Conf., Vol. 2, pp. 155-163, 1983.

Joung, R.J., Yang, K.T., and Novotny, J.L., Vapor Liquid Interactions in High Velocity Vapor Jet Condensing in a Coaxial Water Flow, Proc. 5th Int. Heat Transfer Conf., Vol. 3, pp. 226-230, 1974.

Kazami, M.S., Todreas, N.E., Lanning, D.D., and Rohsenow, W.M., A Theoretical Study of the Dynamics Growth of a Vapor Film Around a Hot Sphere in a Coolant, Proc. 1974 Int. Heat Transfer Conf., Vol. IV, pp. 91-95, 1974.

Kim, S. and Mills, A.F., Condensation on Coherent Liquid Jets. I. Experimental Studies, *ASME J. Heat Transfer,* III, 1068-1074, 1989a.

Kim, S. and Mills, A.F., Condensation on Coherent Liquid Jets. II. A Theoretical Study, *ASME J. Heat Transfer,* III, 1075-1082, 1989b.

Klipstein, D.H., Heat Transfer to a Vaporizing Drop, D. Sci. thesis, Massachusetts Institute of Technology, Cambridge, MA, 1963.

Kodres, C.A., Jacobs, H.R., and Boehm, R.F., A numerical method for determining direct contact heat transfer takes to a superheated, evaporating, floating droplet, *J. Numer. Heat Transfer,* 3, 21-34, 1980.

Kronig, R. and Brink, J.C., On the Theory of Extraction from Falling Droplets, *Appl. Sci. Res.,* AZ, 142, 1950.

Kutateladze, S.S., Heat Transfer in Boiling and Condensation, Moscow, English Transl. AEC TR-3770, U.S. Atomic Energy Commission, Washington, D.C., 1952.

Law, C.K., Heat and Mass Transfer in Combustion: Fundamental Concepts and Analytical Techniques, *Prog. Energy Combust. Sci.,* 10, 295-318, 1984.

Lee, S.C. and Bankoff, S.G., Stability of Steam-Water Countercurrent Flow in an Inclined Channel: Flooding, *ASME J. Heat Transfer,* 105, 713-718, 1983.

Levich, V.G., *Physiochemical Hydrodynamics,* Prentice Hall, Englewood Cliffs, NJ, 1962.

List, E.J., Turbulent Jets and Plumes, *Annual Review of Fluid Mechanics,* Academic Press, New York, 1982, 189-212.

Liu, T.L., Jacobs, H.R., and Chen, K., An Experimental Study of Direct Contact Condensation on a Fragmenting Circular Jet, *ASME J. Heat Transfer,* 111, 585-588, 1989.

McQuillan, K.W., Whalley, P.B., and Hewitt, G.F., Flooding in Vertical Two Phase Flow, *Int. J. Multiphase Flow,* II(6), 741-760, 1985.

Moalem-Maron, D., Sideman, S., Hetsroni, G., and Orell, A., Condensation of Bubble Trains: An Approximate Solution, *Prog. Heat Mass Transfer,* 6, 155-177, 1972.

Murty, N.S. and Sarma, V.M.K., Direct Contact Heating of Laminar Fall Liquid Jet, *Int. J. Heat Mass Transfer,* 19, 115-117, 1976.

Murty, N.S. and Sastri, V.K.M., Condensation on a Falling Laminar Liquid Film, Proc. 5th Int. Heat Transfer Conf., Vol. III, 231-235, 1974.

Niyogi, K.K. and Pavlenco, G.F., Heat and Mass Transfer in Wetted Surface Cooling Towers, Proc. 6th Int. Heat Transfer Conf., Vol. 2, 97-103, 1978.

Nusselt, W., Die Oberflachenkondensation des Wasserdamfes, Z. Ver Dt Ing., Vol. 60, pp. 541-569, 1916.

Ostrach, S. and Koestal, A., Film Instabilities in Two-Phase Flow, *AlChE J.,* II, 294-303, 1965.

Prakash, S. and Sirignano, W.A., Theory of Convection Droplet Vaporization with Unsteady Heat Transfer in the Circulating Phase, *Int. J. Heat Mass Transfer,* 23, 253-268, 1980.

Raina, G. and Grover, P., Direct Contact Heat Transfer with Change of Phase: Theoretical Model, *AlChE J.,* 28, 515-517, 1982.

Raina, G. and Grover, P., Direct Contact Heat Transfer with Change of Phase: Theoretical Model Incorporating Sloshing Effects, *AlChE J.,* 31, 507-510, 1985.

Raina, G. and Wanchoo, R., Direct Contact Heat Transfer with Phase Change: Theoretical Expression for Instantaneous Velocity of a Two-Phase Bubble, *Int. Commun. Heat Mass Transfer,* 11, 227-237, 1984a.

Raina, G., Wanchoo, R., and Grover, P., Direct Heat Transfer with Phase Change: Motion of Evaporating Droplets, *AlChE J.,* 30, 835-837, 1984b.

Rangel, R.H. and Sirignano, W.A., Vaporization Ignition and Combustion of Two Parallel Fuel Droplet Streams, *Proc. 1987 ASME/JSME Therm. Eng. Joint Conf.,* Vol. 6, pp. 27-34, 1987.

Ranz, W.E. and Marshall, W.R., *Chem. Eng. Prog.,* 48, 141-173, 1952.

Rao, V.D. and Sarma, P.K., Condensation Heat Transfer on Laminar Liquid Film, *ASME J. Heat Transfer,* 106, 518-523, 1984.

Renksizbulut, M. and Yuen, M.C., Numerical Study of Droplet Evaporation in a High Temperature Stream, *ASME J. Heat Transfer,* 105, 389-403, 1983.

Renz, V. and Odental, H.P., Numerical Predication of Heat and Mass Transfer During Condensation From a Turbulent Gas Vapor Stream Onto a Vertical Film, *Condensation Heat Transfer,* American Society of Mechanical Engineers, New York, 1979, 27-34.

Roberson, M.E., Mass Transfer in Packed Beds of Spherical Particles, MS thesis, Department of Mechanical and Industrial Engineering, University of Utah, Salt Lake City, UT, 1985.

Roberson, M.E. and Jacobs, H.R., An Experimental Study of Mass Transfer in Packed Beds as an Analog to Convective Heat Transfer, Proc. 9th Int. Heat Transfer Conf., Vol. 5, pp. 189-194, 1990.

Rose, P.M. and Kintner, R.C., Mass Transfer from Large Oscillating Drops, *AlChE J.,* 12(3), 530-534, 1966.

Sanders, B.R. and Dwyer, H.H., Modeling Unsteady Droplet Combustion Processes, Proc. ASME/JSME Therm. Eng. Joint Conf., Vol. 1, pp. 3-10, 1987.

Schroder, J.J. and Pohl, U., Minimum Flowrates and Rewetting Rates of Falling Films, Proc. 7th Int. Heat Transfer Conf., Vol. 5, pp. 83-88, 1982.

Seban, R.A., Transport to Falling Films, Proc. 6th Int. Heat Transfer Conf., Vol. 6, pp. 417-428, 1978.

Shulman, H.l. and Margolis, J.E., Performance of Packed Columns, *AlChE J.,* 3, 157-161, 1957.

Segev, A., Flanigan, L.J., Kurth, R.E., and Collier, R.P., Experimental Study of Countercurrent Steam Condensation, *ASME J. Heat Transfer,* 103, 307-311, 1981.

Sideman, S., Direct Contact Heat Transfer Between Immisicible Liquids, *Advances in Chemical Engineering,* Academic Press, New York, 1966, 207-286.

Sideman, S. and Isenberg, J., Direct Contact Heat Transfer with Change of Phase: Bubble Growth in Three-Phase Systems, *Desalination,* 2, 207-214, 1967.

Sideman, S. and Moalem-Maron, D., Direct Contact Condensation, *Advances in Heat Transfer,* Academic Press, New York, 1982, 228-276.

Sideman, S. and Taitel, Y., Direct Contact Heat Transfer with Change of Phase: Evaporation of Drops in an Immiscible Liquid Medium, *Int. J. Heat Mass Transfer,* 7, 1273-1289, 1964.

Simpson, H.C., Beggs, G.C., and Nazir, M., Evaporation of Butane Drops in Brine, *Desalination,* 15, 11-23, 1974.

Sirignano, W.A., Fuel Droplet Vaporization and Spray Combustion Theory, *Prog. Energy Combus. Sci.,* 9, 291-322, 1983.

Stefan, M. and Mayinger, F., Countercurrent Flow Limitation in Vertical Ducts at High System Pressure, Proc. 1990 Int. Heat Transfer Conf., Vol. 6, pp. 47-52, 1990.

Sudhoff, B., Director Warmeubergang bei der Kondensation in Blap Finsaulen, Ph.D. dissertation, Universitat Dortmund, Federal Republic of Germany, 1982.

Sun, T.Y., A Theoretical and Experimental Study of Noncondensible Turbulent Bubbly Jets, Ph.D. dissertation, Department of Mechanical Engineering. Pennsylvania State University, University Park, PA, 1985.

Sundararajan, T. and Ayyaswamy, P.S., Hydrodynamics and Heat Transfer Associated with Condensation on a Moving Drop: Solutions for Intermediate Reynolds Numbers, *J. Fluid Mech.,* 149, 33-58, 1984.

Sundararajan, T. and Ayyaswamy, P.S., Heat and Mass Transfer Associated with Condensation on a Moving Drop: Solution for Intermediate Reynolds Numbers by a Boundary Layer Formulation, *ASME J. Heat Transfer,* 107, 409-416, 1985a.

Sundararajan, T. and Ayyaswamy, P.S., Numerical Evaluation of Heat and Mass Transfer to a Moving Drop Experiencing Condensation, *Numerical Heat Transfer,* 8, 689-706, 1985b.

Tamir, A. and Rachmileu, I., Direct Contact Condensation on a Thin Film of Water, *Int. J. Heat Mass Transfer,* 17, 1241-1251, 1976.

Taitel, Y. and Tamir, A., Condensation in the Presence of a Noncondensible Gas in Direct Contact, *Int. J. Heat and Mass Transfer,* 12, 1157-1169, 1969.

Tekic, M.N., Prosarac, D., and Petrovic, D., Entrance Region Lengths of Laminar Falling Films, *Chem. Eng. Sci.,* 39, 165-167, 1984.

Thompson, R.J. and Jacobs, H.R., Mass and Simultaneous Convective Heat Transfer for Flow in a Packed Bed of Spherical Particles, *AlChE Symp. Ser.,* 80(236), 240-248, 1984.

Tien, C.L. and Liu, C.P. Survey on Vertical Two-Phase Countercurrent Flooding, EPRI NP-984, Electrical Power Research Institute, Palo Alto, CA, 1979.

Wagner, I. and Stichlmair, J., Motion of Swarms of Droplets, Proc. Int. Symp. Liquid-Liquid Two Phase Flow Transport Phenomena, Int. Center Heat and Mass Transfer, Antalya, Turkey, 1997, 87-96.

Wassel, A.T. and Mills, A.F., Turbulent Falling Film Evaporators and Condensers for Open Cycle Ocean Thermal Energy Conversion, *Advancement in Heat Exchangers,* Hemisphere, New York, 1982.

Weimer, J.C., Faeth, G.M., and Olson, D.R., Penetration of Vapor Jets in Subcooled Liquids, *AlChE J.,* 19(3), 552-558, 1973.

Weinberg, S., Heat Transfer to Low Pressure Sprays of Water in a Steam Atmosphere, *Proc. Inst. Mech. Eng.,* 1B, 240-253, 1952.

Whitaker, S., Forced Convection Heat Transfer Correlations for Flow in Pipes, Past Flat Plates, Single Cylinders, Single Spheres and for Flow in Packed Beds and Tube Bundles, *AlChE J.,* 18, 361-371, 1972.

Wilson, M.R. and Jacobs, H.R., A Numerical Solution of the Heat Transfer from a Dense Array of Spherical Particles, 1993 ANS Proc. National Heat Transfer Conf., HTC Vol. 7, pp. 225-233, 1993.

4.22 Temperature and Heat Transfer Measurements

Robert J. Moffat

There are two different kinds of material to consider with respect to experimental methods: the unit operations of measurement (transducers and their environmental errors) and the strategy of experimentation. This section deals only with the unit operations: transducers, their calibrations, and corrections for environmental errors.

Temperature Measurement

An International Practical Temperature Scale (IPTS) has been defined in terms of a set of fixed points (melting points of pure substances) along with a method for interpolating between the fixed points. The IPTS agrees with the thermodynamic temperature scale within a few degrees Kelvin over most of its range. The IPTS is the basis for all commerce and science, and all calibrations are made with respect to the IPTS temperature. The scale is revised periodically.

Accurate calibrations are not enough to ensure accurate data, however. If a sensor has been installed to measure a gas temperature or a surface temperature, any difference between the sensor temperature and the measurement objective due to heat transfer with the environment of the sensor is an "error." In most temperature-measuring applications, the environmental errors are far larger than the calibration tolerance on the sensor and must be dealt with just as carefully as the calibration.

Thermocouples

Any pair of thermoelectrically dissimilar materials can be used as a thermocouple. The pair need only be joined together at one end and connected to a voltage-measuring instrument at the other to form a usable system. A thermocouple develops its signal in response to the temperature difference from one end of the pair to the other. The temperature at one end, known as the *reference junction* end, must be known accurately before the temperature at the other end can be deduced from the voltage.

Thermocouples are the most commonly used electrical output sensors for temperature measurement. With different materials for different ranges, thermocouples have been used from cryogenic temperatures (a few Kelvin) to over 3000 K. In the moderate temperature range, ambient to 1200°C, manufacturer's quoted calibration accuracy can be as good as ±3/8% of reading (referred to 0°C) for precision-grade base metal thermocouples. Broader tolerances apply at very high temperature and very low temperatures. Thermocouple signals are DC voltages in the range from a few microvolts to a few tens of microvolts per degree C. Because of their low signal levels, thermocouple circuits must be protected from ground loops, galvanic effects, and from pickup due to electrostatic or electromagnetic interactions with their surroundings. Thermocouples are low-impedance devices. Multiple channels of thermocouples can be fed to a single voltage reader using low-noise-level scanners or preamplifiers and electronic multiplexers.

The alloys most frequently used for temperature measurement are listed in Table 4.22.1. These alloys have been developed, over the years, for the linearity, stability, and reproducibility of their EMF vs. temperature characteristics and for their high-temperature capability.

Calibration data for thermocouples are periodically reviewed by the National Institutes of Science and Technology based on the then-current IPTS. Values in Table 4.22.1 illustrate the approximate levels which can be expected, and are from the National Bureau of Standards Monograph 125. Maximum temperatures listed in this table are estimates consistent with a reasonable service lifetime. Allowable atmosphere refers to the composition in contact with the thermoelements themselves. Accuracy estimates are provided for two levels of precision: standard grade and precision grade where these data are available.

Noble metal and refractory metal thermocouples are often used with substitute lead wires, as a cost-saving measure. These lead wires, described in Table 4.22.2 are cheaper and easier to handle than the high temperature thermocouples. They have the same temperature–EMF characteristics as their primary thermoelements, but only over the range of temperatures the lead wires will usually encounter (up to a few hundred degrees C). Except for the substitute alloys, thermocouple extension wires have the same

TABLE 4.22.1 Application Characteristics of Some Common Thermocouple Alloys

Max T °F	Max T °C	Allowable Atmos. (Hot)	Material Names	ANSI Type[a]	Color Code	Output mV/100°F	Accuracy, %	
							Standard[a]	Precision[a]
5072	2800	Inert, H_2, vacuum	Tungsten/tungsten 26% rhenium	—	—	0.86	—	—
5000	2760	Inert, H_2, vacuum	Tungsten 5% rhenium/tungsten 26% rhenium	—	—	0.76	—	—
4000	2210	Inert, H_2	Tungsten 3% rhenium/tungsten 35% rhenium	—	—	0.74	—	—
3720	1800	Oxidizing[b]	Platinum 30% rhodium/platinum 6% rhodium	B	—	0.43	1/2	1/4
2900	1600	Oxidizing[b]	Platinum 13% rhodium/platinum	R	—	0.64	1/4	1/4
2800	1540	Oxidizing[b]	Platinum 10% rhodium/platinum	S	—	0.57	1/4	1/4
2372	1300	Oxidizing[b,c]	Platinel II (5355)/Platinel II (7674)	—	—	2.20	5/8	—
2300	1260	Oxidizing	Chromel/Alumel,[d] Tophel/Nial,[e] Advance T1/T2,[f] Thermo-Kanathal P/N[g]	K	Yellow red	2.20	4°F, or 3/4%	2°F, or 3/8%
1800	980	Reducing[a]	Chromel/constantan	E	Purple red	4.20	1/2	3/8
1600	875	Reducing	Iron/constantan	J	White red	3.00	4°F, or 3/4%	2°F, or 3/8%
750	400	Reducing	Copper/constantan	T	Blue red	2.50	3/4	3/8

[a] Per ANSI C96.1 Standard.
[b] Avoid contact with carbon, hydrogen, metallic vapors, silica, reducing atmosphere.
[c] @ Engelhard Corp.
[d] @ Hoskins Mfg. Co.
[e] Wilber B. Driver Co.
[f] Driver-Harris Co.
[g] The Kanthal Corp.

TABLE 4.22.2 Substitute Material Extension Wires for Thermocouples

Thermocouple Material	Thermocouple Type[a]	Extension Wire, Type[a]	Color for (+) Wire	Color for (–) Wire	Overall Color
Tungsten/tungsten 26% rhenium	—	Alloys 200/226[b]	—	—	—
Tungsten 5% rhenium/tungsten 26% rhenium	—	Alloys (405/426)[b]	White	Red	Red[b]
Tungsten 3% rhenium/tungsten 25% rhenium	—	Alloys (203/225)[b]	White/yellow	White/red	Yellow/red[b]
Platinum/platinum rhodium	S, R	SX, SR	Black	Red	Green
Platinel II-5355/Platinel II-7674	—	P2X[d]	Yellow	Red	Black[d]
Chromel/Alumel, Tophel/Nial, Advance, Thermokanthal[c]	K	KX	Yellow	Red	Yellow
Chromel/constantan	E	EX	Purple	Red	Purple
Iron/constantan	J	JX	White	Red	Black
Copper/constantan	T	TX	Blue	Red	Blue

[a] ANSI, except where noted otherwise.
[b] Designations affixed by Hoskins Mfg. Co.
[c] Registered trade mark names.
[d] Engelhard Mfg. Co.

composition as thermocouple wires, differing only in the type of insulation and the accuracy of calibration, which is not held as closely for extension wire as for thermocouple-grade wire.

Any instrument capable of reading low DC voltages (on the order of millivolts) with 5 to 10 μV resolution will suffice for temperature measurements. *Galvanometric measuring instruments* can be used, but, since they draw current, the voltage available at the terminals of the instrument depends not only on the voltage output of the thermocouple loop but also on the resistance of the instrument and the loop together. Such instruments are normally marked to indicate the external resistance for which they have been calibrated. *Potentiometric instruments*, either manually balanced or automatically balanced, draw no current when in balance, hence can be used with thermocouple loops of any arbitrary resistance without error. High-input impedance *voltmeters* draw very low currents and, except for very high resistance circuits, are not affected by the loop resistance.

Thermocouple Theory

Equation (4.22.1) is the general form describing the EMF generated in a two-wire thermocouple (Moffat, 1962). The same form can be derived from either the free-electron theory of metals or from thermodynamic arguments alone: the output of a thermocouple can be described as the sum of a set of terms, one arising in each wire in the circuit.

The junctions do not generate the EMF: they are merely electrical connections between the wires. For a two-wire circuit,

$$\text{EMF} = \int_0^L \varepsilon_1 \frac{dT}{dx} dx + \int_L^0 \varepsilon_2 \frac{dT}{dx} dx \tag{4.22.1}$$

where

ε_1 and ε_2 = the total thermoelectric power of materials 1 and 2, respectively, mV/C. The value of ε is equal to the sum of the Thomson coefficient and the temperature derivative of the Peltier coefficient for the material.

T = temperature, C

x = distance along the wire, m

L = length of the wire, m

This form for expressing the output of a two-wire circuit applies regardless of whether the wires are uniform in composition or not. If a circuit contained four wires (two thermocouple wires and two extension wires), then Equation (4.22.1) would be written with four terms, one for each length of wire.

When the wire is uniform in composition and both wires begin at (T_o) and both end at (T_L) the two terms can be collected into one integral:

$$\text{EMF} = \int_{T_0}^{T_L} (\varepsilon_1 - \varepsilon_2)\, dT \tag{4.22.2}$$

The EMF–temperature (E–T) tables produced by NIST and others are "solutions" to Equation (4.22.2) and can be used only when the following three conditions are met:

1. The thermoelectric power, ε, is not a function of position; i.e., the wires are homogeneous;
2. There are only two wires in the circuit;
3. Each wire begins at T_o and ends at T_L

When the circuit consists entirely of pairs of materials, Equation 4.22.2 can be used directly as the basis for understanding the source of the EMF. As an example, consider the three-pair system shown in Figure 4.22.1. For that circuit, Equation (4.22.2) would have three terms: one for each pair. The total EMF generated by the circuit would be the sum of the EMFs generated in the thermocouple pair and in the extension wire pair. The pair of copper wires would not contribute to the net EMF, assuming the two copper wires were perfectly matched. The EMF contributed by each pair would be proportional to the temperature difference from end to end of that pair, as shown in Equation (4.22.3) and (4.22.4).

$$\text{EMF} = \int_{T_1}^{T_2} (\varepsilon_{cu} - \varepsilon_{cu})\, dT + \int_{T_2}^{T_3} (\varepsilon_+ - \varepsilon_-)_{\text{LEADS}}\, dT + \int_{T_3}^{T_4} (\varepsilon_+ - \varepsilon_-)_{\text{TC}}\, dT \tag{4.22.3}$$

$$\text{EMF} = 0 + (T3 - T2)(\varepsilon_+ - \varepsilon_-)_{\text{LEADS}} + (T4 - T3)(\varepsilon_+ - \varepsilon_-)_{\text{TC}} \tag{4.22.4}$$

Most thermocouple circuits consist only of pairs of wires and can be understood in terms of these two equations, but some require a more detailed treatment. A graphical method of analysis is available, based on Equation (4.22.1).

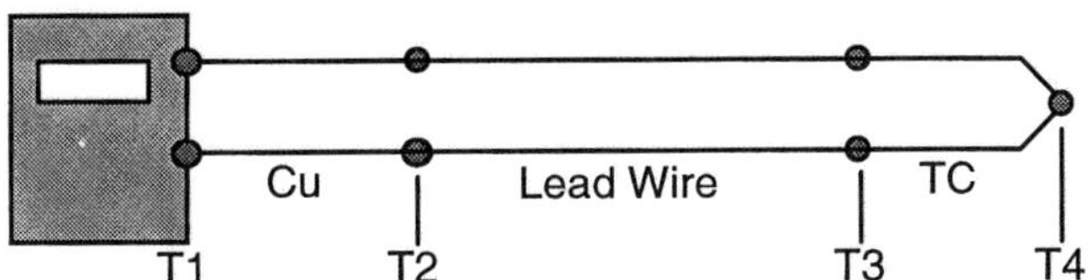

FIGURE 4.22.1 A three-pair circuit.

The temperature–EMF calibrations of the more common materials are shown in Figure 4.22.2 derived from NBS Monograph 125 and other sources. This figure provides the input data for a simple graphical technique for describing the EMF generation in a circuit. Each curve in Figure 4.22.2 represents the output which would be derived from a thermocouple made of material X used with platinum when the cold end is held at 0°C and the hot end is held at *T*.

Those elements commonly used as "first names" for thermocouple pairs, i.e., Chromel (Chromel-Alumel), iron (-constantan), copper (-constantan), have positive slopes in Figure 4.22.2.

The simplest thermocouple circuit for temperature measurement consists of two wires joined together at one end (forming the "measuring junction") with their other ends connected directly to a measuring instrument, as shown in the upper portion of Figure 4.22.3. The EMF generation in this circuit is graphically represented in the lower portion, an E–T diagram, using the data in Figure 4.22.2. The E–T

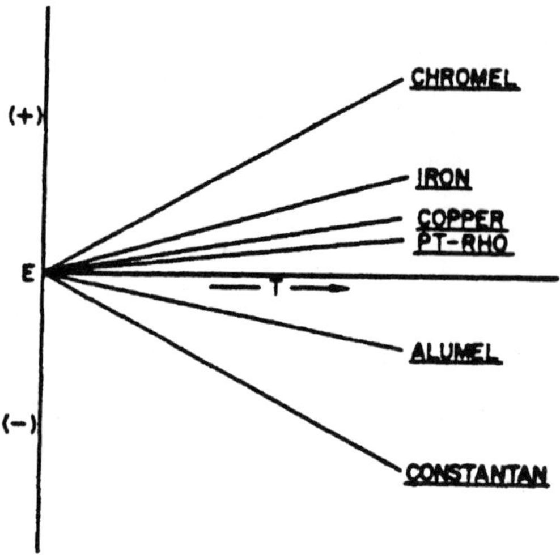

FIGURE 4.22.2 E–T calibrations for several common thermocouple materials.

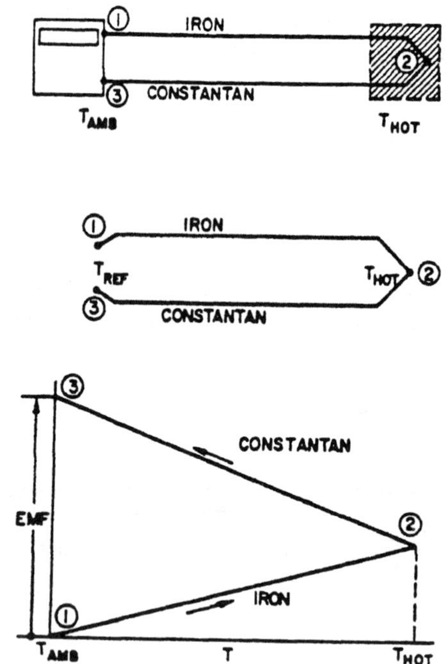

FIGURE 4.22.3 E–T diagram of a thermocouple using an ambient reference.

diagram is used for examining the EMF generated in the circuit, to be certain that it arises only from the desired thermocouple materials, and that all segments of the circuit are properly connected. The E–T diagram is not used for evaluating the output — that is done using the tables after the circuit has been shown to be correctly wired.

To construct an E–T diagram, first sketch the physical system considered and assign a number to each "point of interest" on that sketch and a temperature. On E–T coordinates, locate point 1 at 0/mV and at its assigned temperature. Then start a line from point 1, moving toward the temperature of point 2, and copying the shape of the calibration curve of the iron wire (see Figure 4.22.2). From 2 to 3, use the constantan calibration curve. The difference in elevation of points 1 and 3 describes the net EMF generated in the circuit between points 1 and 3, and describes the polarity. When point 3 lies physically above point 1 in the E–T diagram, it is, by convention, electrically negative with respect to point 1.

The simple triangular shape shown in Figure 4.22.3 identifies a proper circuit. Any thermocouple circuit whose E–T diagram is equivalent to that is appropriate for temperature measurement and its EMF may be interpreted using the conventional tables. Any circuit whose E–T diagram is not equivalent to the pattern circuit should be rewired.

Thermocouples generate their signal in response to the temperature difference between the two ends of the loop. For accurate measurements, the temperature of the "reference junction" must be known. Laboratory users often use an ice bath made from a good-quality Dewar flask or vacuum-insulated bottle of at least 1 pt capacity, as shown in Figure 4.22.4. The flask should be filled with finely crushed ice and then flooded with water to fill the interstices between the ice particles. The reference thermocouple is inserted into a thin-walled glass tube containing a small amount of silicone oil and submerged six or eight diameters into the ice pack. The oil assures good thermal contact between the thermocouple junction and the ice/water mixture. The tube should be sealed at the top to prevent atmospheric moisture from condensing inside it, which would cause corrosion when using iron-constantan thermocouples. Figure 4.22.5 shows in iron-constantan thermocouple circuit with an ice bath. The individual thermocouple wires are connected to copper wires in the ice bath, and the two copper wires taken to the voltmeter. The lower portion of this figure shows the E–T diagram for this circuit, and proves that the output of this circuit is entirely due to the temperature difference from end to end of the iron-constantan loop: the two copper wires do not contribute to the output.

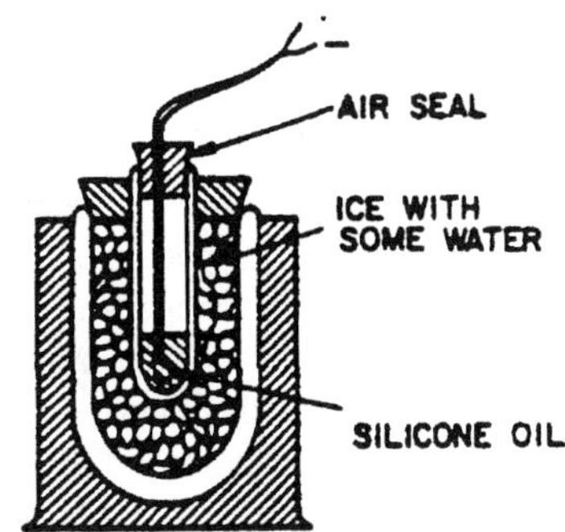

FIGURE 4.22.4 Characteristics of a good ice bath.

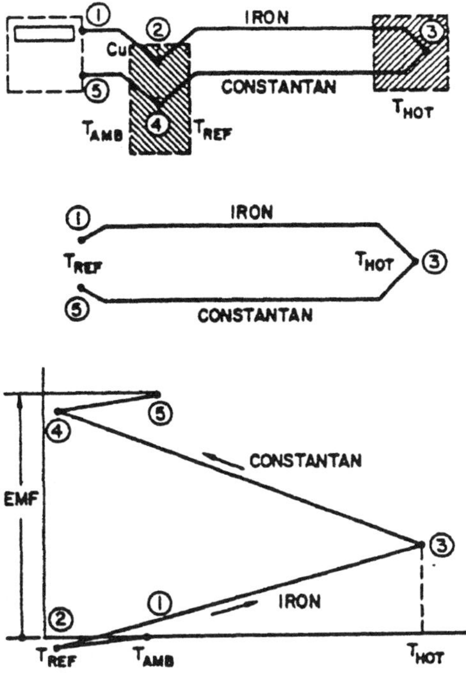

FIGURE 4.22.5 A thermocouple circuit using an ice bath reference, and its E–T diagram.

Calibration

Thermocouple calibrations are provided by the wire manufacturers to tolerances agreed upon industry-wide, as summarized in Table 4.22.1. These tolerances include two components: the uncertainty in the average slope of the calibration curve of the wire, and the effects of local inhomogeneities in the wire. It is difficult to improve the accuracy of a thermocouple by calibrating it. For a truly significant calibration, the thermocouple would have to be exposed to the same temperature during calibration, at every point along it, that it would encounter in service. In an oven calibration, most of the signal is generated in the material at the mouth of the oven, as could be recognized by considering the temperature gradient distribution along the wire. The material inside the oven contributes little or nothing to the signal.

Thermistors

Thermistors are electrical resistance temperature transducers whose resistance varies inversely, and exponentially, with temperature. The resistance of a 5000 Ω thermistor may go down by 200 Ω for each degree C increase in temperature in the vicinity of the initial temperature. Interrogated by a 1.0 mA current source, this yields a signal of 200 mV/°C. As a consequence of this large signal, thermistors are frequently used in systems where high sensitivity is required. It is not uncommon to find thermistor data logged to the nearest 0.001°C. This does not mean that the data are accurate to 0.001°C, simply that the data are readable to that precision. A thermistor probe is sensitive to the same environmental errors which afflict any immersion sensor: its accuracy depends on its environment.

Thermistor probes can be used between –183°C (the oxygen point) and +327°C (the lead point) but most applications are between –80 and +150°C. The sensitivity of a thermistor (i.e., the percent change in resistance per degree C change in thermistor temperature) varies markedly with temperature, being highest at cryogenic temperatures.

Thermistor probes range in size from 0.25-mm spherical beads (glass covered) to 6-mm-diameter steel-jacketed cylinders. Lead wires are proportionately sized. Disks and pad-mounted sensors are available in a wide range of shapes, usually representing a custom design "gone commercial." Aside from the unmounted spherical probes and the cylindrical probes, there is nothing standard about the probe shapes.

Calibration

Thermistor probes vary in resistance from a few hundred ohms to megohms. Probe resistance is frequently quoted at 25°C, with no power dissipation in the thermistor. The commercial range is from about 2000 to 30,000 Ω. Representative values of the sensitivity coefficient (% change in resistance per degree C) is given in Table 4.22.3 and resistance values themselves, in Table 4.22.4.

TABLE 4.22.3 Thermistor Temperature Coefficient Variations with Temperature

Temp. °C	Condition	ΔR/R, %
–183	Liquid oxygen	–61.8
–80	Dry ice	–13.4
–40	Frozen mercury	–9.2
0	Ice point	–6.7
25	Room temperature	–5.2
100	Boiling water	–3.6
327	Melting lead	–1.4

Proprietary probes are available which "linearize" thermistors by placing them in combination with other resistors to form a circuit whose overall resistance varies linearly with temperature over some range. These compound probes can be summed, differenced, and averaged as can any linear sensor. Modern manufacturing practices allow matched sets to be made, interchangeable within +/–0.1°C.

TABLE 4.22.4 Thermistor Resistance Variation with Temperature

Temp., °C	Res., Ω	Temp., °C	Res., Ω
–80	1.66 M	0	7355
–40	75.79 K	25	2252
–30	39.86 K	100	152.8
–20	21.87 K	120	87.7
–10	12.46 K	150	41.9

Thermal Characteristics

Thermistor probes are generally interrogated using a low current, either AC or DC. A level of about 10 μA would be typical. With a probe resistance of 10 K Ω, 0.01 W must be dissipated into its surrounding material. This current results in the probe running slightly above the temperature of the medium into which it is installed: the "self-heating" effect. Since thermistors are often used where very small changes in temperature are important, even small amounts of self-heating may be important.

The self-heating response is discussed in terms of the "dissipation constant" of the probe, in milliwatts per degree C. The dissipation constant depends on the thermal resistance between the thermistor and its surroundings. For fluid-sensing probes, the self-heating varies with velocity and thermal conductivity, while for solid immersion probes, it varies with the method of attachment and type of substrate.

Dissipation constants for representative probes are given in Table 4.22.5. The self-heating effect must be considered in calibration as well as in use.

The transient response of a thermistor is more complex than that of a thermocouple and, size for size, they are not as well suited to transient measurements.

TABLE 4.22.5 Representative Thermal Dissipation Constants for Two Thermistor Probe Designs

Environment	1.0-cm Disk	5.0-cm Cylinder
Still air	8 mW/C	1 mW/C
Still oil	55	—
Still water	—	3.5
Oil at 1 m/sec	250	—

Thermistor probes are sold with calibration tables of resistance vs. temperature at some specified accuracy, on the order of +/–0.1 or 0.2 K, depending on the grade of probe purchased. These tables are typically in increments of 1 K. For computer interpretation, they should be fit to the Steinhart-Hart form[2] and the coefficients determined for least error.

$$\frac{1}{T} = A_0 + A_1 \ln(R) + A_3 \ln\left(R^3\right) \tag{4.22.5}$$

Resistance Temperature Detectors

The terms *resistance temperature detector* (RTD) and *resistance thermometer* are used interchangeably to describe temperature sensors containing either a fine wire or a thin film metallic element whose resistance increases with temperature. In use, a small current (AC or DC) is passed through the element, and its resistance measured. The temperature of the element is then deduced from the measured resistance using a calibration equation or table lookup.

RTDs are used both for standards and calibration laboratories and for field service. Field-service probes are generally encased in stainless steel protective tubes with either wire or film elements bonded to sturdy support structures. They are made to take considerable physical abuse. Laboratory standard-grade probes are often enclosed in quartz tubes, with the resistance wire mounted in a strain-free manner on a delicate mandrel.

High-quality resistance thermometers have been used as defining instruments over part of the range of the IPTS. Because of this association with high-precision thermometry, resistance thermometers in general have acquired a reputation for high precision and stability. Commercial probes, however, are far different in design from the standards-grade probes, and their stability and precision depend on their design and manufacture.

RTDs are often recommended for single-point measurements in steady-state service at temperatures below 1000°C where longtime stability and traceable accuracy are required and where reasonably good heat transfer conditions exist between the probe and its environment.

They are not recommended for use in still air, or in low-conductivity environments. RTDs self-heat, which causes an error when the probes are used in a situation with poor heat transfer. They are not recommended for transient service or dynamic temperature measurements unless specifically designed for such service. The probes tend to have complex transient characteristics and are not amenable to simple time-constant compensation.

Physical Characteristics

The physical characteristics of any given resistance thermometer represent a compromise between two opposing sets of requirements. For accuracy, repeatability, and speed of response, a delicate, low-mass sensing element is desired, supported in a strain-free manner in good thermal contact with its surroundings. For durability, a rugged sensor is indicated, mounted firmly to a sturdy structure inside a robust, sealed protection tube.

Both the short-term calibration (resistance vs. specimen temperature) and the long-term stability (drift) are directly affected by the mechanical configuration of the probe. The electrical resistance of the sensing element is a function of its temperature and state of mechanical strain (Figure 4.22.6).

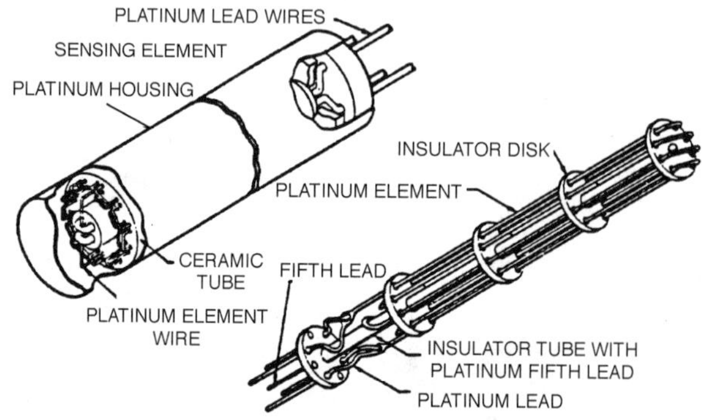

FIGURE 4.22.6 Slack-wire platinum resistance thermometer.

The sensing elements used in field-service RTD probes range from thin metallic films deposited on rectangular ceramic wafers (0.5 × 1.0 × 2.0 mm) with pigtail leads (0.25 mm diameter and 2.5 cm long) to glass-encapsulated, wire-wound mandrels (4 mm in diameter and 2.0 cm long), again with pigtail leads. Bonding the sensor to its support provides good mechanical protection to the element, but subjects the element to strain due to thermal expansion. As long as this process is repeatable, the calibration is stable.

Electrical Characteristics

RTDs are available commercially with resistances from 20 to 20,000 Ω with 100 Ω being common. Bifilar windings are frequently used in wire-wound elements, to reduce the electrical noise pickup. This is more important in the quartz-jacketed probes than in those with stainless steel protection tubes. Twisted pair lead wires are recommended.

Thermal Characteristics

Figure 4.22.7 shows a simplified cross section of a typical resistance thermometer and a thermal circuit which can be used to discuss its behavior. In forming such a thermal circuit model, each element of the probe is described by its resistive and capacitive attributes following conventional heat transfer practice. The principal components are

- The external thermal resistance per unit length;
- The thermal capacitance of the protective tube per unit length, C_T;
- The radial internal thermal resistance between the sensor and the protective tube, R_{int};
- The capacitance of the sensor element and its support, C_{sensor};
- The axial internal thermal resistance of the stem, per unit length, R_T.

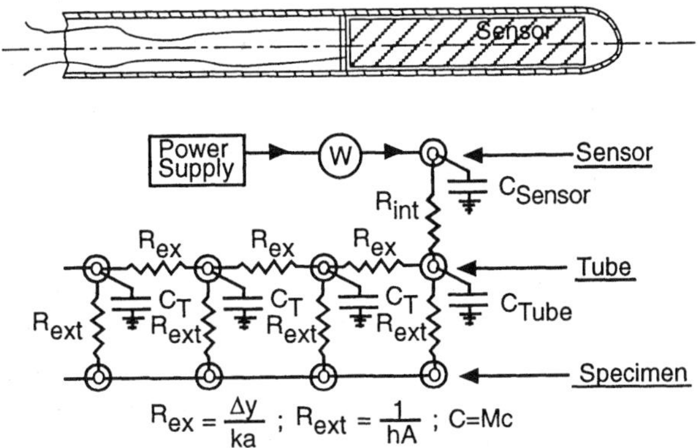

FIGURE 4.22.7 Thermal circuit representation of a typical resistance thermometer.

This circuit can be used to predict the temperature distribution within the probe both at steady state and during transients and can be refined, if needed, by subdividing the resistance and capacitance entities.

Steady-State Self-Heating

Interrogating an RTD by passing a current through it dissipates power in the element, shown in Figure 4.22.7 as W, which goes off as heat transfer through the internal and external resistances. This self-heating causes the sensing element to stabilize at a temperature higher than its surroundings and constitutes an "error" if the intent is to measure the surrounding temperature. The amount of the self-heating error depends on three factors:

- The amount of power dissipated in the element
- The internal thermal resistance of the probe, as a consequence of its design
- The external thermal resistance between the surface of the probe and the surrounding material

The self-heating temperature rise is given by Equation (4.22.6):

$$T_{sens} - T_{surr} = W\left(R_{int} + R_{ext}\right) \tag{4.22.6}$$

The internal thermal resistance of a probe, R_{int}, measured in degree C per watt, describes the temperature rise of the sensing element above the surface temperature of the probe, per unit of power dissipated. The internal thermal resistance can be deduced from measurements of the sensor temperature at several different current levels when the probe is maintained in a well-stirred ice bath, where the external thermal resistance is very low. The slope of the apparent temperature vs. power dissipated line, °C/W, is the internal thermal resistance. When an RTD is used in a gas or liquid, the external resistance between the probe and its surroundings must be estimated from standard heat transfer data. The external resistance is 1/(hA), °C/W.

A typical cylindrical probe exposed to still air will display self-heating errors on the order of 0.1 to 1.0°C per mW (commercial probes of 1.5 to 5 mm in diameter). At 1 m/sec air velocity, the self-heating error is reduced to between 0.03 and 0.3°C. In water at 1 m/sec velocity, the self-heating effect would be reduced by a factor of four or five compared to the values in moving air, depending on the relative importance of the internal and the external thermal resistances.

Calibration and Drift

The relationship between resistance and temperature must be determined for each probe or acquired from the manufacturer. Generally speaking, the reported values will require interpolation.

The resistance–temperature characteristic of a probe may drift (i.e., change with time) while the probe is in service. Manufacturers of laboratory-grade probes will specify the expected drift rate, usually in terms of the expected error in temperature over an interval of time. Two sample specifications are "0.01 C per 100 hours" for a low-resistance, high-temperature probe (0.22 Ω at 0°C, 1100°C maximum service temperature) and "0.01 C per year" for a moderate-resistance, moderate-temperature probe (25.5 Ω at 0°C, 250°C maximum service temperature). Drift of the resistance-temperature relationship takes place more rapidly at high temperatures.

Radiation Devices

Surface temperatures and gas temperatures can be deduced from radiation measurements. Surface-temperature measurements are based on the emitted infrared energy, while gas-temperature measurements use specific emission lines from the gas itself or from a tracer inserted into the gas.

Commercial surface-temperature measurement systems (single-point) are available, at low cost, which can measure temperature to +/–1% of reading, above 38°C, if the emissivity of the surface is known. The device referenced requires a spot size of 1.25 cm diameter, viewed from 75 cm. Spectroscopic gas-temperature measurements can be accurate to +/–3 or 4% of reading, but require a significant investment in effort as well as equipment (on the order of 1 to 2 years and $100,000 to $200,000). Several techniques based on Raman scattering have been used in combustion systems. Planar-lasar-induced fluorescence has shown considerable promise as one of the newer methods.

Infrared emission from a surface is described by two laws: the Stefan Boltzmann law describing the total emitted radiation as a function of temperature, and Planck's law describing its distribution as a function of temperature. These laws form the basis for all radiation-based surface-temperature detectors.

Early radiometers focused the total infrared energy on a thermopile bolometer and used the temperature rise across its calibrated heat loss path to measure the incident energy flux. Solid-state photon detectors have replaced thermopile bolometers as the detector of choice. Such a detector will respond to any photon having energy above a certain level (specific to the detector). Since the energy of a photon is inversely proportional to its wavelength, detectors respond to all wavelengths below some value. Modern detectors use band-pass filters to limit the wavelength band of photons admitted to the detector and rely on Planck's law to infer the temperature from the energy flux:

$$E_{b,\lambda} = \frac{C_1 \lambda^{-5}}{e^{C_2/\lambda T} - 1} \tag{4.22.7}$$

where $E_{b,\lambda}$ = radiated power at the wavelength λ, W/m²
T = temperature, K
C_1 = 3.743 × 10⁸, Wμm⁴/m²
C_2 = 1.4387 × 10⁴, μmK

Commercial radiation temperature detectors use different wave bands for different temperature ranges, with different detectors for each band. The emissivity of the surface must be known, as a function of temperature, in the wavelength band used by the detector.

Radiation detectors are vulnerable to interference from four sources: low signal-to-noise ratio at low temperatures (below a few hundred degrees C); radiation from the surroundings reflecting into the detector (also usually more important at low temperatures); low spatial resolution (also more evident at low temperatures); uncertainty in the emissivity of the surface (at all temperatures); and absorption of radiation into water vapor and CO_2 in the line of sight (at any temperature).

A fiber-optic blackbody temperature detector system is offered by the Luxtron Corporation for standards room and field service above 300°C. The unit consists of a blackbody capsule fiber-optically coupled to a filtered, band-limited photon detector. Accuracy of 0.01 to 0.1°C is claimed, depending on temperature level.

A fluoroptic temperature-measuring system is also offered by the same company, for use only at lower temperatures (–200 to +450°C). This system uses an ultraviolet-stimulated phosphor on the end of an optical fiber as its sensor. The fluorescent signal from the phosphor decays with time, and its "time constant" is a function of temperature. Accuracy of +/–0.5°C is claimed for measurements within +/–50°C of a calibration point, or +/–1°C within 100°C.

Temperature-Sensitive Paints, Crayons, and Badges

Temperature-sensitive paints, crayons, and badges are available from several suppliers (Omega Engineering, Inc., Stamford, CT, and others in Germany and Japan). Each undergoes an irreversible change (e.g., a change in color or a change from solid to liquid) at one specified temperature. With a range of paints, temperatures from ambient to about 1500°C can be covered. The accuracy generally quoted is about +/–1% of level, although melting standards are available to +/–0.5°C.

The phase-change materials melt at well-defined temperatures, yielding easily discernible evidence that their event temperature has been exceeded. When more than one phase-change paint is applied to the same specimen, there can be interference if the melt from the low-melting paint touches the high-melting material. Color change materials do not interfere, but are more difficult to interpret. The calibration of high-temperature paints (both phase change and color change) may shift when they are used on heavily oxidized materials, due to alloying of the oxide with the paint. Recommended practice is to calibrate the paints on specimens of the application material. The event temperature which will cause transformation depends on the time at temperature: short exposure to a high temperature often has the same effect as long exposure to a lower temperature.

The paints and crayons are nonmetallic and, therefore, tend to have higher emissivities for thermal radiation than metals. They should be used only over small areas of metallic surfaces, compared with the metal thickness, or else their different emissivities may lead to a shift in the operating temperature of the parts.

The principal disadvantages of the paints and crayons are that they require visual interpretation, which can be highly subjective, and they are one-shot, irreversible indicators which respond to the highest temperature encountered during the test cycle. They cannot record whether the peak was reached during normal operation or during soak-back.

Liquid crystals can be divided into three groups, depending on their molecular arrangements: (1) smectic, (2) nematic, and (3) cholesteric. Most of the temperature-sensitive liquid crystals now in use are cholesteric: made from esters of cholesterol. Their molecules are arranged in planar layers of molecules with their long axes parallel and in the plane of the layer. The molecules in each layer are rotated with respect to those in its neighboring layers by about 15 min of arc in a continuous, helical pattern along an axis normal to the layers.

The colors reflected from cholesteric liquid crystals are thought to be due to Bragg diffraction from the aligned layers. The "wrap angle" between adjacent layers increases with temperature; hence, the color of the liquid crystal shifts toward short wavelengths (toward blue) as the temperature is raised. The color can also be affected by electric fields, magnetic fields, pressure, shear stress, and some chemical vapors.

Warm cholesterics are colorless liquids and they pass through a series of bright colors as they are heated through their "color-play" temperature band. The first color to appear is a deep red, followed by yellow, green, blue, and violet. Further heating yields a colorless liquid again. This cycle is reversible and repeatable, and the color–temperature relationship can be calibrated.

Liquid crystals selectively reflect only a small fraction of the incident light; hence, to enhance the brightness of the color image, they must be backed up with black paint or a nonreflecting surface.

A typical calibration is shown in Figure 4.22.8 for liquid crystals painted over black paint on an aluminum calibration strip. The upper part of Figure 4.22.8 describes the color variation, while the lower part shows the imposed linear temperature distribution. The hot end is blue, the cold end is red. Color-play intervals range from 0.5 to 10.0°C. Liquid crystals whose color-play interval is on the order of 0.5 to 2.0°C are often referred to as *narrow-band* materials, while those whose interval extends to 5.0 to 10°C are called *wide band.* Narrow-band images are easy to interpret by eye. Wide-band images show only subtle variations of color for small changes in temperature, and accurate work requires digital image handling or multiple images taken with different filters.

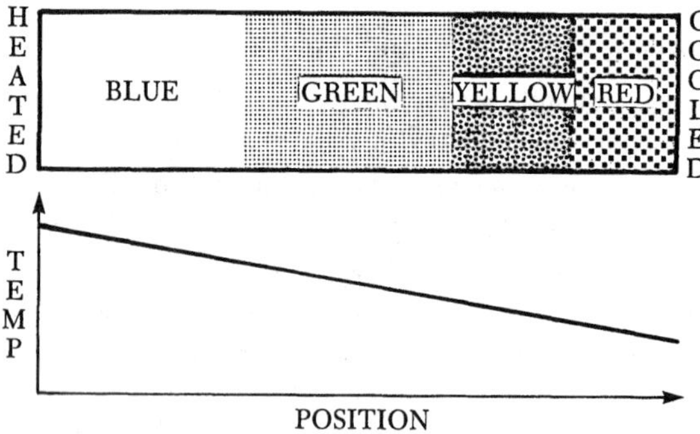

FIGURE 4.22.8 Schematic representation of a calibration strip.

Several different narrow-band liquid crystals can be mixed together to make a single, multi-event paint covering a wide range of temperatures, provided their color-play intervals do not overlap. Such a mixture yields a set of color-play bands, one for each component.

Calibration

Liquid crystals are sold by event temperature and color-play bandwidth, with a nominal accuracy of +/–1°C on the event temperature. In many applications, especially if the image is to be visually interpreted, no further calibration is needed.

The accuracy attainable with a liquid crystal is related to the width of the color-play interval. With narrow-band material (a color-play interval of about 1.0°C), visual interpretation can be done with an uncertainty of 0.25 to 0.5°C. With digital image interpretation, spectrally controlled lighting and appropriate corrections for reflected light interference, the uncertainty can be held below 0.25°C.

Early users reported that the perceived color of a liquid crystal depended on both the lighting angle and the viewing angle. This dependence can be eliminated by using a light source along the line of sight (coaxial viewing and illumination).

Multiple-Event Paints

Several narrow-band paints can be mixed together to make a single paint with all the characteristics of each component, if their color-play intervals do not overlap. Each component retains its original calibration and acts independently of the other components.

Figure 4.22.9 shows the image from a five-event paint used to map the adiabatic wall temperature isotherms around a heated block in mixed convection. The outermost isotherm is 30°C, and the events are spaced apart at 5°C intervals. Determination of the temperatures from a multiple-event image requires that the temperature be known at one point in the image.

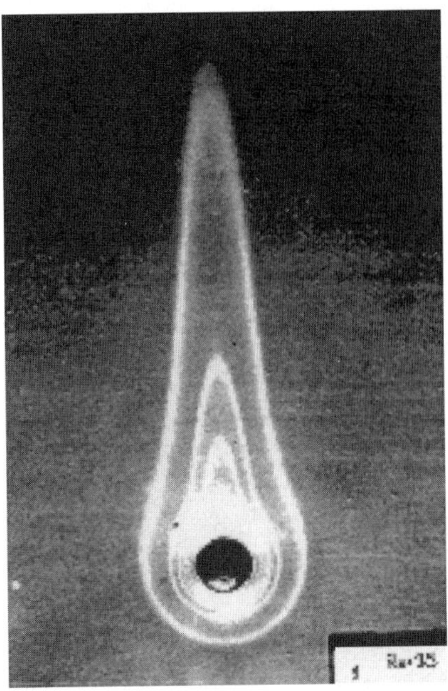

FIGURE 4.22.9 Multi-event liquid crystal used to visualize the isotherm pattern above a heated spot in mixed convection.

Liquid Crystals in Water

Liquid crystals can be used to mark the temperature distribution in water and some other liquids by adding a small quantity of encapsulated liquid crystal material to the liquid and photographing the color distribution using planar lighting. Velocity and temperature distributions can be determined by photographing the liquid crystal particles using a known exposure time. The temperature is deduced from the particle color, and the velocity by the length of the streak its image forms. Figure 4.22.10 shows the velocity and temperature distributions in a shear-driven, water-filled cavity 30 sec after the impulsive start of belt motion. In this view, the belt is at the top of the image, and moved from left to right. The water was stably stratified initially, with the top being 4°C hotter than the bottom. This technique was demonstrated by Rhee et al. (1984) and has been used by several workers.

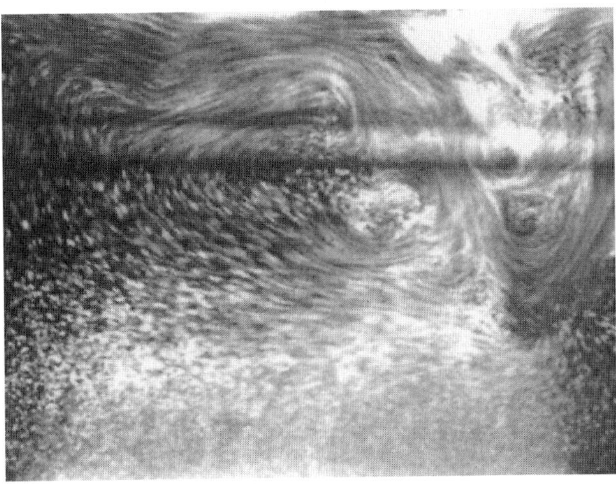

FIGURE 4.22.10 Liquid crystal visualization of the velocity and temperature distribution in a water-filled tank.

Image Processing

Several schemes have been proposed to remove the subjectivity from interpretation of liquid crystal images. Akino et al. (1989), and others, have processed RGB video images of narrow-band images using multiple filters to extract images of specified isochromes, related to temperatures through a calibration. Hollingsworth et al. (1989) processed RGB images of wide-band images using chromaticity coordinates (hue, saturation, and intensity) and extracted temperature at each pixel, rather than along isochromes.

Heat Flux

Heat flux to or from a surface can be measured directly, using heat flux meters, or inferred from an overall energy balance, or inferred from temperature–time measurements at the surface or within the body. There are no primary standards for heat flux measurement.

Three general classes of heat flux meters are in common use: slug calorimeters, planar heat flux gauges (sometimes called Schmidt–Boelter gauges), and circular foil gauges (sometimes called Gardon gauges). Sensitivities range from microvolts per kW/m^2 to millivolts per W/m^2. Planar gauges can be used for radiant or convective heat loads. Circular foil gauges should be used only for radiant loads.

Slug Calorimeter

The slug calorimeter is an energy balance transducer consisting of a known mass of material instrumented so that its temperature can be measured. A simple version is shown in Figure 4.22.11. If losses are negligibly small and the mass and the specific heat are constant, the instantaneous heat flux is deduced from

$$\dot{q}''_{\text{in}} A = Mc \frac{\partial T}{\partial t} \tag{4.22.8}$$

where T = Average temperature of the slug, C
M = Mass of the slug, kg
c = Specific heat, J/kg·C
A = Face area, m^2
τ = Time

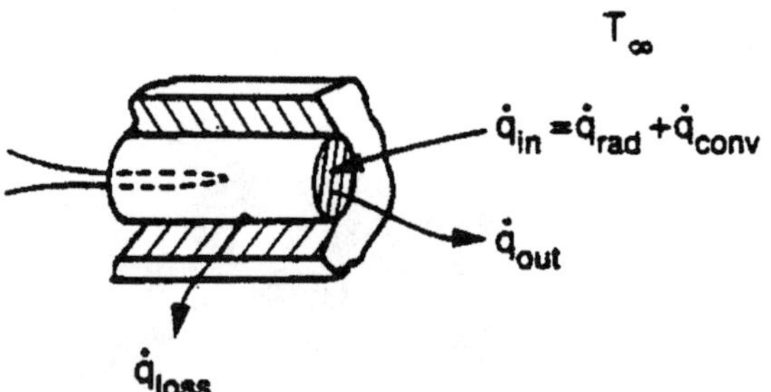

FIGURE 4.22.11 A simple slug calorimeter.

The variation of slug temperature with time is used to infer net heat transfer rate to the gauge. Slug calorimeters are used mainly when the heat flux, or the heat transfer coefficient, is expected to be relatively constant. They are of less value when the input flux changes arbitrarily because of the inaccuracies inherent in differentiating the signals.

Planar Heat Flux Gauge

Planar heat flux gauges use Fourier's law to deduce the heat flux from a steady-state measurement of the temperature difference across a thin sheet of thermally insulating material. The planar gauge geometry is shown in Figure 4.22.12. The working equation for a planar gauge is

FIGURE 4.22.12 A typical planar heat flux gauge.

$$\text{EMF} = N\varepsilon\,\Delta T = \frac{N\varepsilon t}{k}\dot{q}'' \tag{4.22.9}$$

where N = number of junction pairs,
e = thermoelectric power of the thermoelement, mV/C
t = thickness of the insulator, m
k = conductivity of the insulator, W/m·C
q'' = heat flux through the gauge, W/m^2

The figure shows one thermocouple junction on the top and one on the bottom surface of the insulator. Most gauges use multiple junctions. The thermoelements may be wire (down to 0.025 mm diameter) or thin films deposited on the insulator (10 to 20 Å). The assembly is usually sandwiched between two sheets of protective material to form an integral unit. Up to 150°C application temperature, these units are often made of Kapton, and provided with a contact adhesive. They may be as thin as 0.15 mm overall.

Gauges should not be removed and reinstalled without recalibration, as the act of removing them from the surface may delaminate the gauge, changing its thermal resistance, and therefore its calibration.

Circular Foil Gauges

A circular foil gauge consists of a thin circular disk of metal supported by its edge from a structure of constant and uniform temperature. The circular foil gauge is often called a Gardon gauge. A constantan foil is often used, with a copper support structure. Two copper wires complete the circuit: one attached to the center of the foil disk and one to the support structure. The copper–constantan thermocouple thus formed produces an EMF determined by the temperature difference from the center of the foil disk to its rim. That temperature difference is directly proportional to the average heat flux on the disk. A cross-sectional view of a circular foil gauge is shown in Figure 4.22.13.

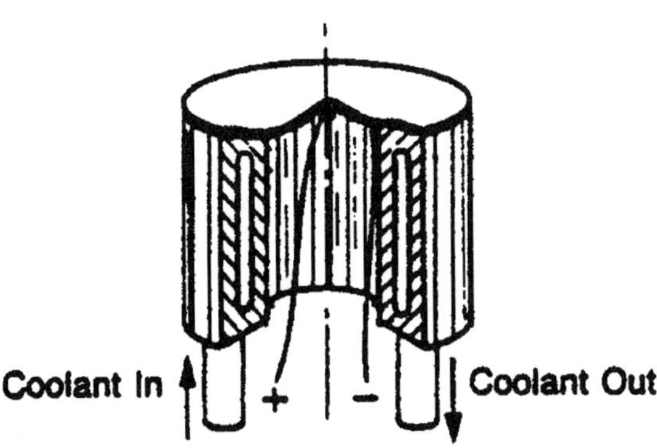

FIGURE 4.22.13 A water-cooled circular foil gauge (Gardon gauge).

The working equation for a circular foil gauge is

$$\text{EMF} = \varepsilon\frac{R^2}{4kt}\dot{q}'' \tag{4.22.10}$$

where ε = thermoelectric power, mV/C

R = radius of the disk, m
k = thermal conductivity of the disk, W/m·C
t = thickness of the disk, m
q'' = heat flux absorbed by the disk, W/m^2 (must be uniform)

The output signal is thus directly proportional to the heat flux on the disk. Cooling passages are frequently built into the support structure to maintain the edge of the disk (the heat sink for the foil disk) at constant temperature.

Calibration

Calibration of the Gardon-type heat flux meters is most easily done by comparison, using a radiation calibrator.

Planar gauges can be calibrated either by conduction or radiation, but the results will depend on the calibration method for some guages.

Sensor Environmental Errors

Temperature sensors generate signals in response to their own temperatures, but are usually installed to measure the temperature of some fluid or solid. There is heat transfer between the sensor and all of its surroundings, with the result that the sensor usually equilibrates at some temperature different from the fluid or solid it is installed in. This difference is considered an error in the measurement.

Similarly, heat flux gauges are generally installed so one can infer the heat flux which would have been there had the gauge not altered the system behavior. But heat flux gauges do disturb the system, and the heat flux at the gauge location, when the gauge is there, may be significantly different from that which would have been there without the gauge. This system disturbance effect must also be considered an error.

Steady-State Errors in Gas-Temperature Measurement

All immersion-type temperature sensors (thermocouples, resistance detectors, and thermistors) are subject to the same environmental errors, which are frequently larger than the calibration errors of the sensors. Large probes are usually affected more than small ones; hence, RTDs and thermistors (selected by investigators who wish to claim high accuracy for their data) are more vulnerable to environmental errors (due to their larger size and their self-heating errors). This aspect of accuracy is sometimes overlooked.

Sensor installations for gas-temperature measurements should be checked for all three of the usual steady-state environmental errors: velocity error, radiation error, and conduction error. The same equations apply to all sensors, with appropriate dimensions and constants.

$$\text{velocity error:} \quad E_v = (1-\alpha)\frac{V^2}{2g_c J c_p} \tag{4.22.11}$$

$$\text{radiation error:} \quad E_r = \frac{\sigma\varepsilon}{h}\left(T_{\text{sens}}^4 - T_{\text{surr}}^4\right) \tag{4.22.12}$$

$$\text{conduction error:} \quad E_c = \frac{T_{\text{gas}} - T_{\text{mount}}}{\cosh\left[L\sqrt{\dfrac{hA_c}{kA_k}}\right]} \tag{4.22.13}$$

where E_v = velocity error, °
α = recovery factor, —
V = velocity, ft/sec

g_c = universal gravitational constant
J = Joules constant, ff/bf/Btu
c_p = specific heat, Btu/lbm, °F

and E_r = radiation error, °R
σ = Stefan-Boltzmann constant
ε = emissivity
h = heat transfer coefficient, Btu/secft^2, °F
T_{sens} = indicated temperature, °R
T_{surr} = surrounding temperature, °R
E_c = conduction error, °R
T_{gas} = gas temperature, °R
T_{mount} = mount temperature, °R
L = length of exposed junction, ft
h = heat transfer coefficient, Btu/secft^2, °F
A_c = heat transfer area, ft^2
k = thermal conductivity, Btu/secft
A_k = conduction area, ft^2

Velocity error depends upon the recovery factor, which varies with the Prandtl number of the fluid. The Prandtl numbers of most liquids are greater than 1; hence, the recovery factor α is greater than 1 and probes tend to read higher than the stagnation temperature in high-speed liquid flows. With thermistors and RTDs in liquids, the self-heating effect and the velocity error both tend to cause high readings. In gases, where the Prandtl number is less than 1, the two effects are of opposite sign and may partly cancel each other.

Radiation and conduction errors vary inversely with the heat transfer coefficient. Both tend to be larger for larger-diameter probes since, all other factors remaining the same, the heat transfer coefficient will be lower for a large-diameter probe. This results in larger radiation and conduction errors. In liquids, radiation error is not a problem, but velocity error and conduction error may both be significant. Conduction error becomes a problem in liquid-temperature measurements when thermowells are used. The depth of immersion of the well is frequently too short to eliminate conduction error.

Steady-State Errors in Solid and Surface-Temperature Measurements

When probes are used to measure solid temperature by inserting them into a hole in the specimen, they are subject to conduction errors proportional to their size and conductivity. A general rule of thumb is to keep the insertion depth at least 20 times the diameter (or wall thickness) of the probe. This assumes a close-fitting hole, backfilled with a material with higher thermal conductivity than air. For more-exact advice regarding a specific installation, a careful thermal circuit analysis of the installation should be developed, and its results used to guide the selection of diametrical clearance, backfill materials, and penetration depth.

A thermocouple attached to a hot surface surrounded by cooler fluid will exchange heat with the fluid by convection and with the surrounding solids by radiation. Heat lost from the thermocouple must be made up from the surface by conduction, which will result in a cold spot at the point of attachment.

Figure 4.22.14 shows the system disturbance error caused by a surface-attached thermocouple, as a fraction of the maximum possible error for the installation. If the surface is irradiated (e.g., by heating lamps), the irradiation will raise the surface temperature, but will also affect the system disturbance error. The effect on the system disturbance error caused by turning on the irradiation is similar to that of raising the temperature of the surrounding fluid to a new value, T_∞°,

where T_{ind} = indicated temperature from an otherwise error-free thermocouple, °C
T_s = undisturbed substrate temperature, °C
T_∞° = effective fluid temperature, °C

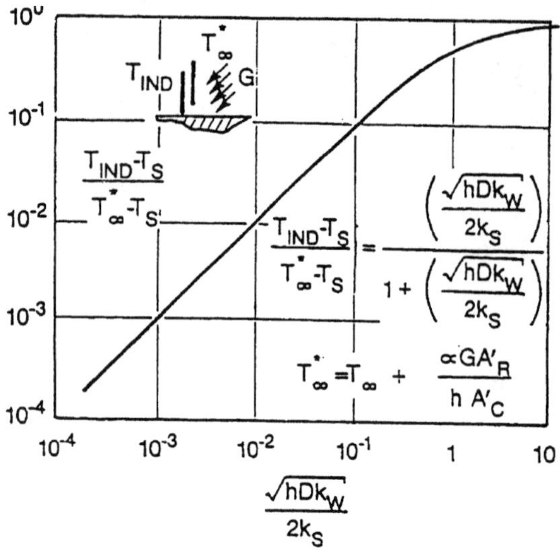

FIGURE 4.22.14 System disturbance errors caused by an attached thermocouple (worst case).

h = heat transfer coefficient, TC to fluid, W/m, °C
D = outside diameter of TC, m
k_w = effective thermal conductivity of TC, W/m, °C
k_s = thermal conductivity of substrate, W/m, °C

The effective gas temperature is defined in terms of the incident irradiation and the heat transfer coefficient as

$$T_\infty^* = T_\infty + \frac{\alpha G A_R}{h A_C} T_\infty^* \tag{4.22.14}$$

where T_∞ = actual gas temperature, °C
α = absorptivity of the TC for thermal radiation
G = incident thermal radiation flux, W/m²
A_R/A_c = ratio of irradiated surface to convective surface
h = heat transfer coefficient between the TC and the gas, W/m°C

Steady-State Errors in Heat Flux Gauges for Convective Heat Transfer

If the gauge is not flush with the surface, it may disturb the flow, and if it is not at the same temperature as the surface, it will disturb the heat transfer. Thus, the gauge may properly report the heat flux which is present when the gauge is present, but that may be significantly different from the heat flux which would have been there if the gauge had not been there.

For planar gauges, both effects are usually small. The thermal resistance of such a gauge is generally small, and they are thin enough to avoid disturbing most flows. Circular foil gauges pose a more serious problem, since they are often cooled significantly below the temperature of the surrounding surface. Dropping the wall temperature at the gauge location can significantly increase the local heat load in two ways: one due to the fact that, for a given value of h, a cold spot receives a higher heat load from the gas stream. The second effect arises because the value of the heat transfer coefficient itself depends on the local wall temperature distribution: a local cold spot under a hot gas flow will experience a higher heat transfer coefficient than would have existed had the surface been of uniform temperature.

Evaluating the Heat Transfer Coefficient

The heat transfer coefficient is a defined quantity, given by

$$h = \frac{\dot{q}''_{\text{conv}}}{\left(T_o - T_{\text{ref}}\right)} \tag{4.22.15}$$

where h = heat transfer coefficient, W/m², °C
$\dot{q}''_{\text{conv}}$ = convective heat flux, W/m²
T_o = temperature of the considered surface, °C
T_{ref} = temperature used as reference for this definition, °C

Different reference temperatures are conventionally used for different situations:

- T_∞: The free-stream temperature. Used for isolated objects of uniform temperature in a uniform free stream, where an average value of h is desired which describes the overall heat transfer between the object and the flow. Also used in boundary layer heat transfer analyses where local values of h are needed to deal with locally varying conditions.
- T_m: The mixed mean fluid temperature. Used for internal flows where the intent of the calculation is to describe the changes in mixed mean fluid temperature (e.g., heat exchangers).
- $T_{\text{adiabatic}}$: The adiabatic surface temperature. Used for isolated objects or small regions of uniform temperature in either internal or external flows, where the overall thermal boundary conditions are neither uniform heat flux nor uniform temperature.

For a given data set, the value of the heat transfer coefficient will depend on the reference temperature chosen, and h should be subscripted to inform later users which reference was used: e.g., h_∞, h_m, or $h_{\text{adiabatic}}$.

Direct Methods

The two most commonly used methods for measuring the heat transfer coefficient are both derived from the same energy balance equation:

$$hA\left(T_o - T_{\text{ref}}\right) = \dot{e}_{\text{in}} + \dot{q}_{\text{cond,in}} + \dot{q}_{\text{rad,in}} - Mc\frac{dT}{d\tau} \tag{4.22.16}$$

where h = the heat transfer coefficient, W/m, °C
A = the area available for convective transport, m²
T_{ref} = the reference temperature used in defining h, °C
T_o = the average surface temperature over the area A, °C
$\dot{e}_{\text{in}}$ = externally provided input, W
$\dot{q}_{\text{cond,in}}$ = net energy conducted in, W
$\dot{q}_{\text{rad,in}}$ = net energy radiated in, W
$Mc\ dT/d\tau$ = rate of increase of thermal energy stored within the system, W

Steady State

In the steady-state method, the transient term is zero (or nearly so), and h is determined by measuring the input power and the operating temperature, and correcting for losses. Equation (4.22.16) can be applied to differentially small elements or to whole specimens. The considered region must be reasonably uniform in temperature, so the energy storage term and the convective heat transfer term use the same value.

For tests of isolated objects, or embedded calorimeter sections, steady-state tests usually use high-conductivity specimens (e.g., copper or aluminum) with embedded electric heaters. The resulting value of h is the average over the area of the specimen. The Biot number, hL/k, for the specimen should be low (on the order of 0.01 or less) if only one temperature sensor is used in the specimen, so the surface temperature can be determined from the embedded sensor.

If a single heated element is used within an array of unheated elements, the resulting heat transfer coefficient is implicitly defined as $h_{adiabatic}$ and should be identified as such. Heat transfer coefficients measured with single-active-element tests cannot be used with the mixed mean fluid temperature.

When the variation of h over a surface is required, one common steady-state technique is to stretch a thin foil (stainless steel, or carbon impregnated paper, or gold deposited on polycarbonate) over an insulating substrate, and electrically heat the foil surface. Liquid crystals or infrared techniques can be used to map the surface temperature, from which the heat transfer coefficient distribution can be determined. The "heated foil with liquid crystal" approach was used by Cooper et al. in 1975 to measure heat transfer coefficients, and has since been used by many others. Hippensteele et al. (1985) have made extensive use of the foil technique in studies of gas turbine heat transfer. An example of their work on the end wall of a turbine cascade is shown in Figure 4.22.15.

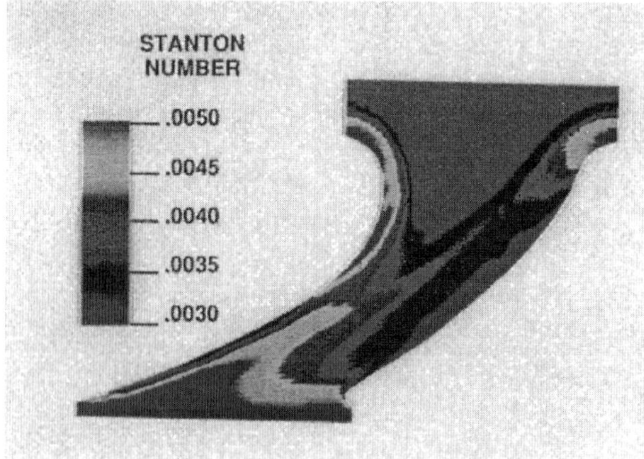

FIGURE 4.22.15 Heat transfer coefficient distribution on the end wall of a turbine cascade. (From Hippensteele, S.A. et al., NASA Technical Memorandum 86900, March, 1985. With permission.)

Hollingsworth et al. (1989) used a stainless steel foil heater for a study in air for an electronics cooling application, illustrated in Figure 4.22.16.

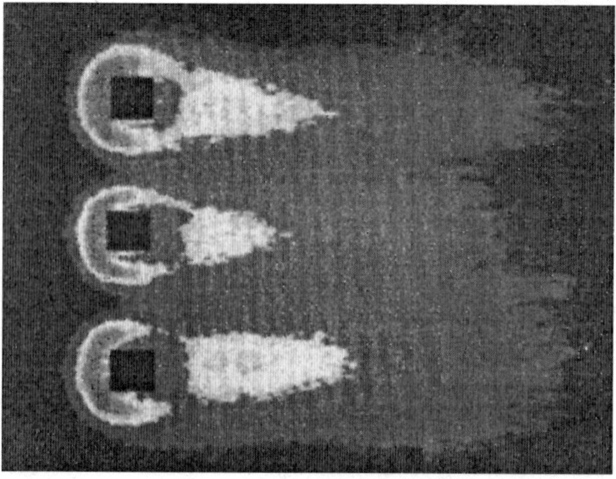

FIGURE 4.22.16 Visualization of the heat transfer coefficient distribution on a heated plate around three unheated cubes.

Another steady-state technique which reveals the distribution of h on the surface was introduced by den Ouden and Hoogendoorn (1974) and is currently in use by Meinders (1996). It uses a uniform and constant-temperature substrate (originally, a tank of warm water, now a copper block) covered with a layer of known thermal resistance (originally, a plate of glass, now a thin layer of epoxy). The surface was painted with liquid crystals (now visualized using infrared imaging) and the surface-temperature distribution determined. The inner (uniform) and outer (measured) temperature distributions are then used as boundary conditions to a three-dimensional conduction solver which calculates the total heat flux at each point on the surface. The total heat flux is corrected for radiation to yield the net convective transport at each point, from which h can be determined.

This method appears to have an advantage in accuracy over the heated foil technique because of the more accurate handling of substrate conduction.

Transient Lumped Parameter Systems

In the lumped parameter transient method, the specimen is assumed to be uniform in temperature at every instant through the transient. The power, $\dot{e}_{in}$, in Equation (4.22.16) is usually zero, although that is not necessary (one could simply change the power level at time zero to initiate the transient). At time zero, a transient is initiated, and the response curve recorded.

The data can be interpreted, and the validity of the first-order assumption tested at the same time by plotting $(T - T_{\text{final}})/(T_{\text{initial}} - T_{\text{final}})$ on the log scale of semilog coordinates, with time on the algebraic scale. If the line is straight, then the system is first order and the characteristic time can be determined from any two points on the line by

$$\tau = \frac{(t_2 - t_1)}{\ln\left\{\dfrac{T_{\text{fin}} - T_1}{T_{\text{fin}} - T_2}\right\}} \qquad (4.22.17)$$

where τ = characteristic time, Mc/hA, sec
t_1 = time at the first instant
t_2 = time at the second instant
T_1 = specimen temperature at the first instant, °C
T_2 = specimen temperature at the second instant, °C
T_{fin} = specimen temperature after a long time (fluid temperature), °C

The heat transfer coefficient is extracted from the time-constant definition.

Indirect Methods

An increasingly popular method is the extraction of h from surface-temperature variations after a step in flow temperature using an inverse calculation method (see the section on inferential methods of heat flux measurement). The simplest inverse method assumes one-dimensional conduction into an infinitely thick plate of constant material properties. Even highly irregular geometries can be studied with this technique, if the streamwise extent of the specimen is small and the testing time is short. A short time interval is necessary so the penetration of the thermal wave is limited to a thin layer near the surface. The short streamwise extent is necessary so the temperature response of the surface upstream does not alter the thermal step applied to the downstream surface. This technique has been used to determine the heat transfer coefficient distribution on the inside walls of passages of irregular shape, by making the passage in a transparent material.

Naphthalene Sublimination

The equations for mass diffusion are similar to those for heat transfer, except for replacing the Prandtl number in the heat transfer equation by the Schmidt number in the diffusion equation. Thus, one could expect that the distribution of the mass transfer coefficients on a surface would mimic the distribution of the heat transfer coefficients.

The most commonly used analog technique is naphthalene sublimination. As early as 1940, the mass transfer/heat transfer similarity was used to estimate the heat transfer coefficient distribution. Naphthalene is a solid material which sublimes at a reasonable rate in air at ambient temperature. Specimens can be cast in naphthalene with good precision, and the recession of the surface mapped as a function of position and time using automated or semiautomated measuring equipment. The surface recession over a known interval of time is a measure of the mass transfer rate, from which the mass transfer coefficient can be deduced.

Naphthalene experiments are generally done at uniform temperature; hence, a uniform vapor pressure exists at the surface. This corresponds to the heat transfer situation of heat transfer from a uniform temperature surface. No counterpart of the uniform heat flux situation has been produced using naphthalene, nor have there been experiments corresponding to variable wall temperature.

Naphthalene sublimation experiments do not suffer from any counterpart of the conduction heat transfer in the substrate. Conduction makes it difficult to work near discontinuities in wall temperature in a heat transfer experiment. Details of the fine structure of mass transfer near obstructions and discontinuities can be resolved in naphthalene experiments, but those details might not exist in a heat transfer process. The Prandtl number of air is much lower than the Schmidt number of naphthalene diffusing in air; hence, thermal conduction would tend to blur out sharp gradients in the temperature field more than diffusion would blur out gradients in naphthalene concentration.

The Schmidt number of naphthalene in air is about 2.5, far different than the Prandtl number of air (0.71); hence, the mass transfer coefficient deduced from a naphthalene experiment is not numerically equal to the heat transfer coefficient which would have existed at those conditions. The usual recommendation is to adjust for the Prandtl number of Schmidt number using a relation of the form:

$$\mathrm{St}\,\mathrm{Pr}^{2/3} = f\{\mathrm{Re}\} = \mathrm{Sh}_j \mathrm{Sc}_j^{2/3} \tag{4.22.18}$$

based on laminar results. That recommendation has not been seriously tested by experiments in turbulent and separated flows. By using nominal values of the Schmidt number and Prandtl number, the heat transfer Stanton number would be 2.3 times higher than the measured Sherwood number and an uncertainty of 10% in that ratio would alter the inferred heat transfer coefficient by 23%.

System Performance Matching. Sometimes the "effective average heat transfer coefficient" for a system is inferred from the overall behavior of the system, e.g., estimating h from the effectiveness of a heat exchanger. Values deduced by this means cannot be expected to agree well with direct measurements unless a very sophisticated system description model is used.

References

Moffat, R.J., The gradient approach to thermocouple circuitry, *Temperature, Its Measurement and Control in Science and Industry*, Rienhold, New York, 1962.

Steinhart, J.S. and Hart, S.R., Calibration curves for thermistors, *Deep Sea Res.*, 15, 497, 1968.

Rhee, H.S., Koseff, J.R., and Street, R.L., Flow visualization of a recirculating flow by rheoscopic liquid and liquid crystal techniques, *Exp. Fluids*, 2, 57–64, 1984.

Hollingsworth, K., Boehman, A.L., Smith, E.G., and Moffat, R.J., Measurement of temperature and heat transfer coefficient distributions in a complex flow using liquid crystal thermography and true-color image processing, in *Coll. Pap. Heat Transfer, ASME HTD*, 123, 35–42, Winter Annual Meeting, 1989.

Cooper, T.E., Field, R.J., and Meyer, J.F., Liquid crystal thermography and its application to the study of convective heat transfer, *J. Heat Transfer*, 97, 442–450, 1975.

Hippensteele, S.A., Russell, L.M., and Torres, F.J., Local Heat Transfer Measurements on a Large Scale Model Turbine Blade Airfoil Using a Composite of a Heater Element and Liquid Crystals, NASA Technical Memorandum 86900, March 1985.

den Ouden, C. and Hoogendoorn, C.J., Local convective heat transfer coefficients for jets impinging on a plate: experiments using a liquid crystal technique, in *Proc. of the 5th Int. Heat Transfer Conf.*, Vol. 5, AIChE, New York, 1974, 293–297.

Personal Communication from Erwin Meinders, March 1996. Work in progress at the Technical University of Delft under Prof. Hanjalic.

Akino, N. and Kunugi, T., *ASME HTD,* Vol. 112, 1989.

4.23 Flow Measurement

Jungho Kim, Sherif A. Sherif, and Alan T. McDonald

Introduction

Flow measurements are perhaps the most common type of measurements made. Many types of flow measuring devices are available. The choice of meter type depends on the required accuracy, range, cost, ease of reading or data reduction, and service life. The simplest and cheapest device that gives the required accuracy should be chosen.

Flow measurement can be classified into four broad types: (1) direct methods, (2) flow restriction methods, (3) linear flow meters, and (4) traversing methods. Each of these are discussed in the subsections below.

Direct Methods

Tanks can be used to determine the flow rate for steady liquid flows by measuring the volume or mass of liquid collected during a known time interval. If the time interval is long enough, flow rates may be determined precisely by using tanks. Compressibility must be considered in gas volume measurements. It is not practical to measure the mass of gas, but a volume sample can be collected by placing an inverted "bell" over water and holding the pressure constant by counterweights. No calibration is required when volume measurements are set up carefully; this is a great advantage of direct methods.

Restriction Flow Meters for Flow in Ducts

Most restriction flow meters for internal flow (except the laminar flow element) are based on acceleration of a fluid stream through some form of nozzle, shown schematically in Figure 4.23.1. Flow separating from the sharp edge of the nozzle throat forms a recirculation zone shown by the dashed lines downstream from the nozzle. The main flow stream continues to accelerate from the nozzle throat to form a vena contracta at Section 2 and then decelerates again to fill the duct. At the vena contracta, the flow area is a minimum, the flow streamlines are essentially straight, and the pressure is uniform across the channel section. The theoretical flow rate is

$$\dot{m}_{\text{theoretical}} = \frac{A_2}{\sqrt{1-\left(A_2/A_2\right)^2}} \sqrt{2\rho\left(p_1 - p_2\right)} \tag{4.23.1}$$

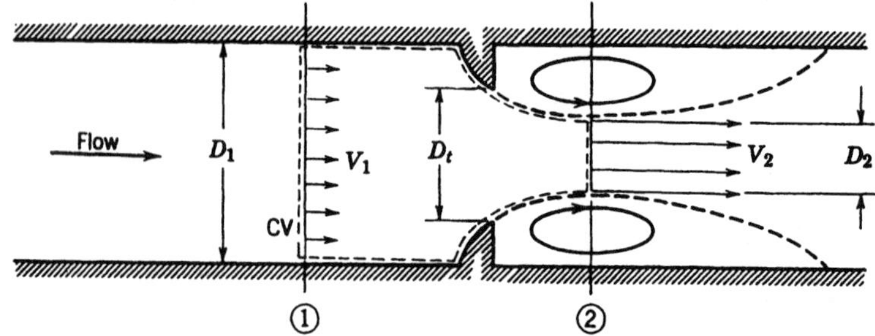

FIGURE 4.23.1 Internal flow through a generalized nozzle, showing control volume used for analysis.

Equation 4.23.1 shows the general relationship for a restriction flow meter: Mass flow rate is proportional to the square root of the pressure differential across the meter taps. This relationship limits the flow rates that can be measured accurately to approximately a 4:1 range.

Several factors limit the utility of Equation 4.23.1 for calculating the actual mass flow rate through a meter. The actual flow area at Section 2 is unknown when the vena contracta is pronounced (e.g., for orifice plates when D_t is a small fraction of D_1). The velocity profiles approach uniform flow only at large Reynolds numbers. Frictional effects can become important (especially downstream from the meter) when the meter contours are abrupt. Finally, the location of the pressure taps influences the differential pressure reading, $p_1 - p_2$.

The actual mass flow rate is given by

$$\dot{m}_{\text{actual}} = \frac{CA_t}{\sqrt{1-\left(A_2/A_2\right)^2}}\sqrt{2\rho\left(p_1 - p_2\right)} \tag{4.23.2}$$

where C is an empirical discharge coefficient.

If $\beta = D_t/D_1$, then $(A_t/A_1)^2 = (D_t/D_1)^4 = \beta^4$, and

$$\dot{m}_{\text{actual}} = \frac{CA_t}{\sqrt{1-\beta^4}}\sqrt{2\rho\left(p_1 - p_2\right)} \tag{4.23.3}$$

where $1/(1 - \beta^4)^{1/2}$ is the velocity correction factor. Combining the discharge coefficient and velocity correction factor into a single *flow coefficient*,

$$K \equiv \frac{C}{\sqrt{1-\beta^4}} \tag{4.23.4}$$

yields the mass flow rate in the form:

$$\dot{m}_{\text{actual}} = KA_t\sqrt{2\rho\left(p_1 - p_2\right)} \tag{4.23.5}$$

Test data can be used to develop empirical equations to predict flow coefficients vs. pipe diameter and Reynolds numbers for standard metering systems. The accuracy of the equations (within specified ranges) is often adequate enough to use the meter without calibration. Otherwise, the coefficients must be measured experimentally.

For the turbulent flow regime ($Re_D > 4000$), the flow coefficient may be expressed by an equation of the form:

$$K = K_\infty + \frac{1}{\sqrt{1-\beta^4}}\frac{b}{\text{Re}_{D_1}^n} \tag{4.23.6}$$

where subscript ∞ denotes the flow coefficient at infinite Reynolds numbers and constants b and n allow for scaling to finite Reynolds numbers. Correlating equations and curves of flow coefficients vs. Reynolds number are given for specific metering elements in the next three subsections following the general comparison of the characteristics of the orifice plate, flow nozzle, and venturi meters in Table 4.23.1 (see Fox and McDonald, 1992).

Flow meter coefficients reported in the literature have been measured with fully developed turbulent velocity distributions at the meter inlet. When a flow meter is installed downstream from a valve, elbow, or other disturbance, a straight section of pipe must be placed in front of the meter. Approximately 10 diameters of straight pipe upstream are required for venturi meters, and up to 40 diameters for orifice plate or flow nozzle meters. Some design data for incompressible flow are given below. The same basic methods can be extended to compressible flows.

TABLE 4.23.1 Characteristics of Orifice, Flow Nozzle, and Venturi Flow Meters

Flow Meter Type	Diagram	Head Loss	Cost
Orifice		High	Low
Flow nozzle		Intermediate	Intermediate
Venturi		Low	High

Orifice Plates

The orifice plate (Figure 4.23.2) may be clamped between pipe flanges. Since its geometry is simple, it is low in cost and easy to install or replace. The sharp edge of the orifice will not foul with scale or suspended matter. However, suspended matter can build up at the inlet side of a concentric orifice in a horizontal pipe; an eccentric orifice may be placed flush with the bottom of the pipe to avoid this difficulty. The primary disadvantages of the orifice are its limited capacity and the high permanent head loss caused by uncontrolled expansion downstream from the metering element.

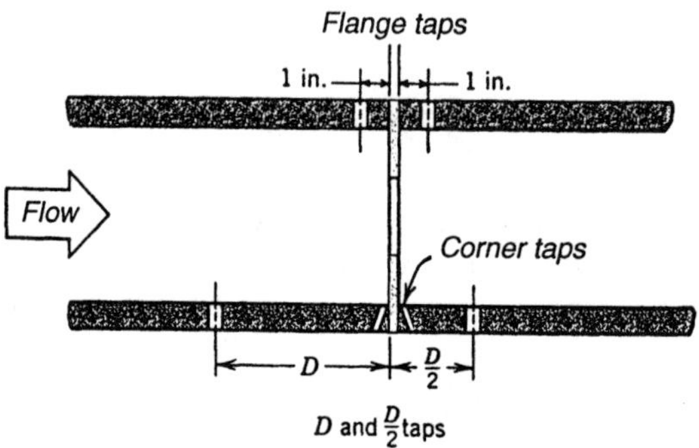

FIGURE 4.23.2 Orifice geometry and pressure tap locations. (From Miller, R.W., *Flow Measurement Engineering Handbook,* 3rd ed., McGraw-Hill, New York, 1996. With permission.)

Pressure taps for orifices may be placed in several locations as shown in Figure 4.23.2 (see Miller, 1996 for additional details). Since the location of the pressure taps influences the empirically determined flow coefficient, one must select handbook values of K consistent with the pressure tap locations.

The correlating equation recommended for a concentric orifice with corner taps is

$$C = 0.5959 + 0.0312\,\beta^{2.1} - 0.184\,\beta^{8} + \frac{91.71\,\beta^{2.5}}{\mathrm{Re}_{D_1}^{0.75}} \tag{4.23.7}$$

Equation 4.23.7 predicts orifice discharge coefficients within ±0.6% for $0.2 < \beta < 0.75$ and for $10^4 < \mathrm{Re}_{D1} < 10^7$. Some flow coefficients calculated from Equation 4.23.7 are presented in Figure 4.23.3. Flow coefficients are relatively insensitive to Reynolds number for $\mathrm{Re}_{D1} > 10^5$ when $\beta > 0.5$.

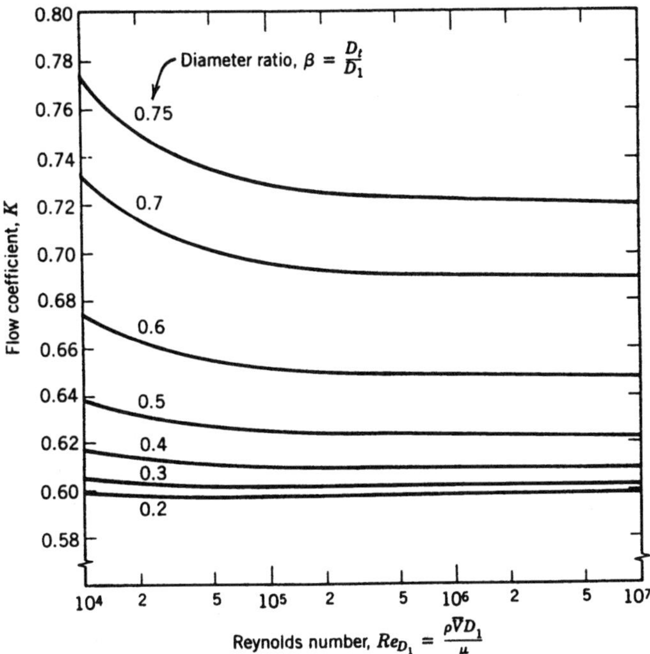

FIGURE 4.23.3 Flow coefficients for concentric orifices with corner taps.

Flow Nozzles

Flow nozzles may be used as metering elements in either plenums or ducts, as shown in Figure 4.23.4; the nozzle section is approximately a quarter ellipse. Design details and recommended locations for pressure taps are given in *ASHRAE Handbook Fundamentals* (1981).

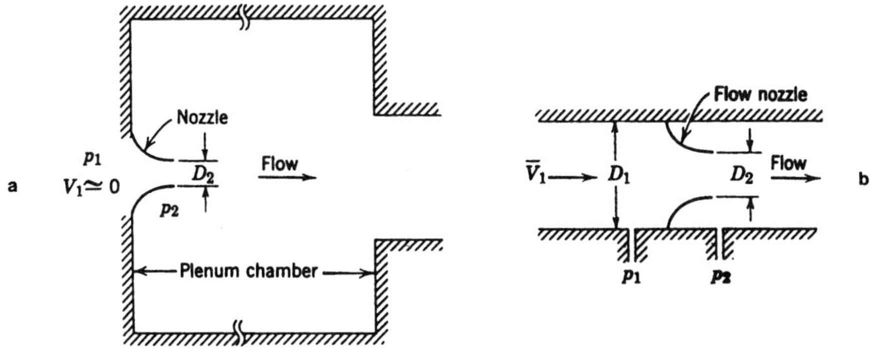

FIGURE 4.23.4 Typical installations of nozzle flow meters: (a) plenum, (b) in duct.

A similar correlating equation is available for orifice plates with D and D/2 taps. Flange taps require a different correlation for every line size. Pipe taps, located at 2 – 1/2 D and 8D, no longer are recommended. The correlating equation recommended for ASME long-radius flow nozzles is

$$C = 0.9975 - \frac{6.53\,\beta^{0.5}}{\mathrm{Re}_{D_1}^{0.5}} \tag{4.23.8}$$

Equation 4.23.8 predicts discharge coefficients for flow nozzles within ± 2.0% for $0.25 < \beta < 0.75$ for $10^4 < Re_{D1} < 10^7$. Some flow coefficients calculated from Equation 4.23.8 are presented in Figure 4.23.5. (K can be greater than 1 when the velocity correction factor exceeds 1.) For plenum installation, nozzles may be fabricated from spun aluminum, molded fiberglass, or other inexpensive materials. Typical flow coefficients are in the range $0.95 < K < 0.99$; the larger values apply at high Reynolds numbers. Thus, the mass flow rate can be computed within approximately ±2% using K = 0.97.

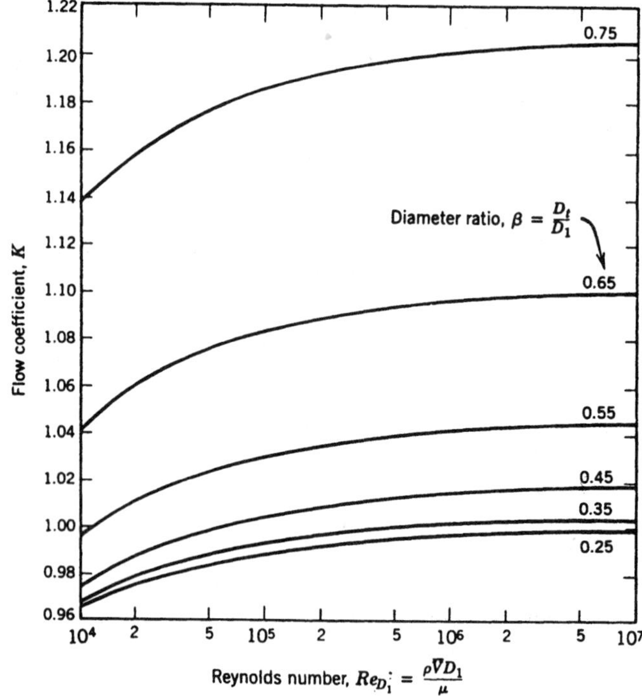

FIGURE 4.23.5 Flow coefficients for ASME long-radius flow nozzles.

Venturi Meters

Venturi meters are generally made from castings machined to close tolerances to duplicate the performance of the standard design, so they are heavy, bulky, and expensive. The conical diffuser section downstream from the throat gives excellent pressure recovery; overall head loss is low. Venturi meters are self-cleaning because of their smooth internal contours.

Experimentally measured discharge coefficients for venturi meters range from 0.980 to 0.995 at high Reynolds numbers ($Re_{D1} > 2 \times 10^5$). Thus, C = 0.99 can be used to calculate mass flow rate within about ±1% at high Reynolds numbers. Consult manufacturers' literature for specific information at Reynolds numbers below 10^5.

Orifice plates, flow nozzles, and venturis all produce pressure drops proportional to the flow rate squared, according to Equation 4.23.4. In practice, a meter must be sized to accommodate the largest flow rate expected. Because the pressure drop vs. flow rate relationship is nonlinear, a limited range of flow rates can be measured accurately. Flow meters with single throats usually are considered for flow rates over a 4:1 range.

Unrecoverable head loss across a metering element may be expressed as a fraction of the differential pressure across the element. Unrecoverable head losses are shown in Figure 4.23.6.

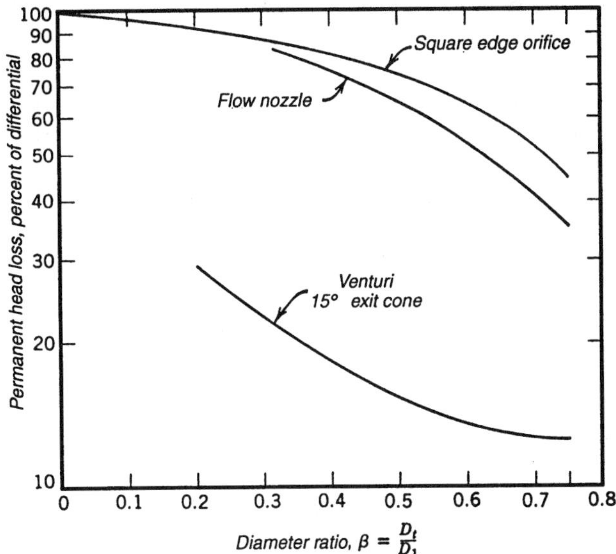

FIGURE 4.23.6 Permanent head loss produced by various flow metering elements. (From Miller, R.W., *Flow Measurement Engineering Handbook,* 3rd ed., McGraw-Hill, New York, 1996. With permission.)

Linear Flow Meters

The output of flow meters is typically calibrated to read a volume flow rate of a specified liquid or gas under certain conditions. For example, the volume flow rate of a liquid could be expressed in terms of liters per minute (LPM) of water, while that of a gas could be expressed in terms of standard cubic feet per minute (SCFM) of air. When using flow meters to determine flow rate at conditions different from the calibration conditions, corrections to the output must be made to obtain the true flow rate.

Rotameters

A typical rotameter consists of a float contained within a diverging tube (Figure 4.23.7). The float is raised by a combination of drag and buoyancy. The height to which the float rises in the tube can be made to vary by varying the shape of the diverging tube — tubes with linear or logarithmic variations can easily be fabricated.

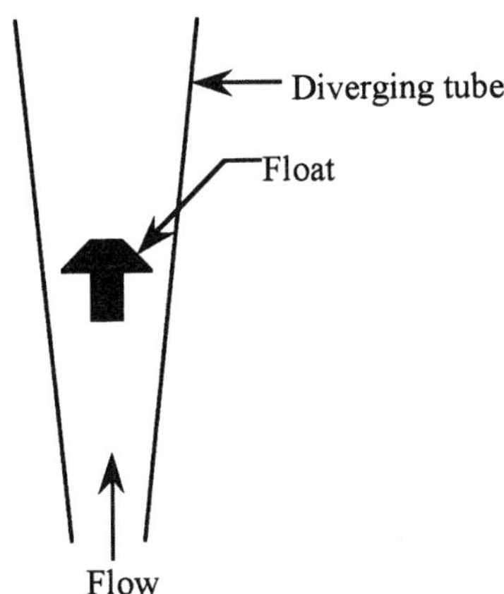

FIGURE 4.23.7 Schematic of rotameter operation.

Glass tube rotameters can be used to temperatures as high as 200°C and pressures up to 20 atm, depending on the diameter of the tube. Plastic tube rotameters are also available for use at lower temperatures and pressures. Flow rates at very high temperatures and pressures can be measured using metal tubes — detectors mounted on the outside of the tube are used to determine the float position.

Rotameters usually come with calibrations based on water or air at STP. The output is generally in SCFM of air or LPM of water. For measuring flow at other conditions or for other fluids, corrections must be made to determine the flow rate. The response of a rotameter can be determined by performing a force balance on the float. The height of the float is determined by a balance between the drag force on the float and the effective weight of the float:

$$F_D = \left(\rho_{float} - \rho_{fluid}\right) g V_{float} \tag{4.23.9}$$

where the drag force is given by

$$F_D = C_D A_{float} \frac{\rho_{fluid} u_{fluid}^2}{2} \tag{4.23.10}$$

Combining Equations 4.23.9 and 4.23.10 and solving for the fluid velocity u_{fluid} yields

$$u_{fluid} = \sqrt{\frac{2g\left(\rho_{float} - \rho_{fluid}\right) V_{float}}{C_d A_p \rho_{fluid}}} \tag{4.23.11}$$

The mass flow rate is given by

$$\dot{m} = \rho_{fluid} A_{meter} u_{fluid} \tag{4.23.12}$$

where A_{meter} is the area between the float and the tube. Combining Equations 4.23.11 and 4.23.12 yields

$$\dot{m} = C\rho_{fluid} A_{meter} \sqrt{\frac{\rho_{float}}{\rho_{fluid}} - 1} \tag{4.23.13}$$

where it has been assumed that the drag coefficient, C_D, is constant. Many floats are designed for constant C_D over a wide flow range.

Consider now the case where a given flow meter indicates the same flow rate for two different fluid flows, i.e., A_{meter} is constant for both cases. Then,

$$\frac{\dot{m}_1}{\dot{m}_2} = \frac{\rho_{fluid,1}}{\rho_{fluid,2}} \frac{\sqrt{\rho_{float}/\rho_{fluid,1} - 1}}{\sqrt{\rho_{float}/\rho_{fluid,2} - 1}} \tag{4.23.14}$$

Since $\dot{m} = \rho_{fluid} \dot{V}$,

$$\frac{\dot{V}_1}{\dot{V}_2} = \frac{\sqrt{\rho_{float}/\rho_{fluid,1} - 1}}{\sqrt{\rho_{float}/\rho_{fluid,2} - 1}} \tag{4.23.15}$$

For cases where $\rho_{float} >> \rho_{fluid}$ (gas flows), the above equation reduces to

$$\frac{\dot{V}_1}{\dot{V}_2} = \frac{\sqrt{\rho_{fluid,2}}}{\sqrt{\rho_{fluid,1}}} \tag{4.23.16}$$

Example: Assume we have oil with a specific gravity of 0.8 flowing through a rotameter with a float made of steel. The steel has a specific gravity of 8.0. If the meter was calibrated for water, what is the correction that must be made to the indicated flow rate in order to obtain the true flow rate of oil?

Solution: Because the fluid density is not negligible compared to the float density, Equation 4.23.15 must be used.

$$\frac{\dot{V}_{oil}}{\dot{V}_{water}} = \frac{\sqrt{\rho_{float}/\rho_{oil} - 1}}{\sqrt{\rho_{float}/\rho_{water} - 1}} = \frac{\sqrt{8.0/0.8 - 1}}{\sqrt{8.0/1.0 - 1}} = 1.13$$

The true flow rate of oil, therefore, would be 1.13 times that indicated.

Target Flow Meters

These meters measure the drag force on a disk or target placed in the flow. The meter may consist simply of a hinged disk in the flow with a dial indicator showing the amount of deflection. More sophisticated meters use a strain gauge or a bellows to measure the force. Target meters can be used in dirty flows, and can also be used to sense the flow direction. Target meters are subject to the same corrections as for rotameters.

Turbine Flow Meters

Turbine flow meters consist of a free-running, vaned impeller mounted in a cylindrical section of tube. A cutaway of some turbine flow meters is shown in Figure 4.23.8. The rate of angular rotation is determined by counting the number of blade passages per unit time magnetically, mechanically, optically, or by some other method. No penetrations or seals in the duct are usually required, enabling turbine flow meters to measure flow rates in corrosive or toxic liquids. They are generally used for liquid flows, and are characterized by high accuracy and wide dynamic range. Reading accuracies of 0.5% over a 30:1 range are common. If units with ball bearings are used, they should be inspected occasionally to check for bearing wear. The life of the unit can be dramatically increased by operating the unit well below the maximum rated speed, however. The response of turbine flow meters can be affected by

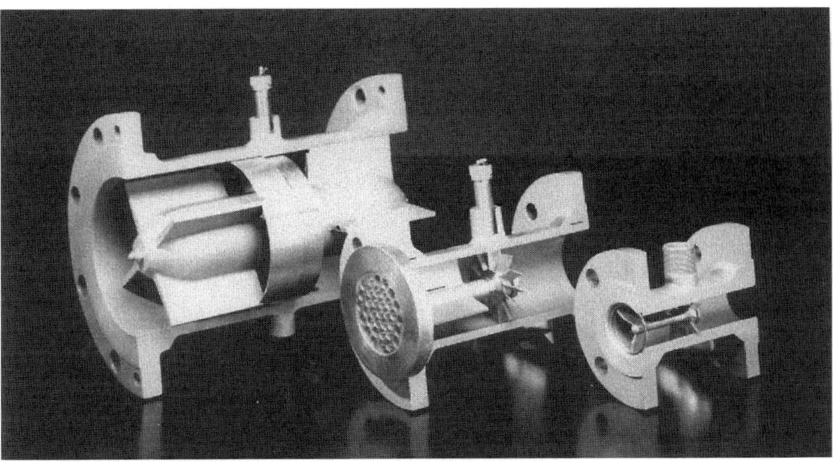

FIGURE 4.23.8 Some turbine flow meters. (Courtesy of Smith Meter, Erie, PA.)

viscosity. A field calibration should be performed when the working fluid has a viscosity significantly higher than that for water.

Paddlewheel flow meters are an inexpensive alternative to turbine flow meters when extreme accuracies are not required. The rate of rotation of a paddlewheel partially immersed in the flow is used as a measure of the flow rate. Unlike turbine flow meters, paddlewheels can be used in dirty flows.

Positive Displacement Meters

Positive displacement meters measure flow rate by separating the flow into discrete increments that are moved from one side of the flow meter to the other. An example of a positive displacement meter is shown in Figure 4.23.9. Positive displacement meters can be used to measure the flow rate of high-viscosity fluids.

FIGURE 4.23.9 A cutaway of a positive displacement meter. (Courtesy of Smith Meter, Erie, PA.)

Vortex Shedding Meters

When a bluff body is placed in a flow, vortices can be shed downstream of the body (Figure 4.23.10). Vortex shedding meters measure the frequency of the asymmetric pressure variations that occur on the body as the vortices are shed and relate this to the flow rate. The vortex shedding frequency is governed by the Strouhal number (St)

FIGURE 4.23.10 Vortices shedding off a body placed in a moving stream.

$$St = \frac{f_s d}{V} \tag{4.23.17}$$

where f_s = vortex shedding frequency
d = characteristic dimension of the body
V = velocity of the flow

Experiments have shown that St is constant within 1% for Reynolds numbers between 10^4 and 10^6. The frequency is directly proportional to the velocity (and thus the flow rate) around the body in this range. These flow meters are insensitive to dirty flows and can be used under a wide variety of conditions because they have no moving parts.

Thermal Mass Flow Meters

These determine flow rate of clean gases by passing the fluid through a precision tube that is heated in one section, and the temperature of the tube wall upstream and downstream of the heated section is measured (Figure 4.23.11). The difference in temperature between the two sensors is a measure of the mass flow rate. Calibration is usually performed using nitrogen and a correction factor is applied for other gases.

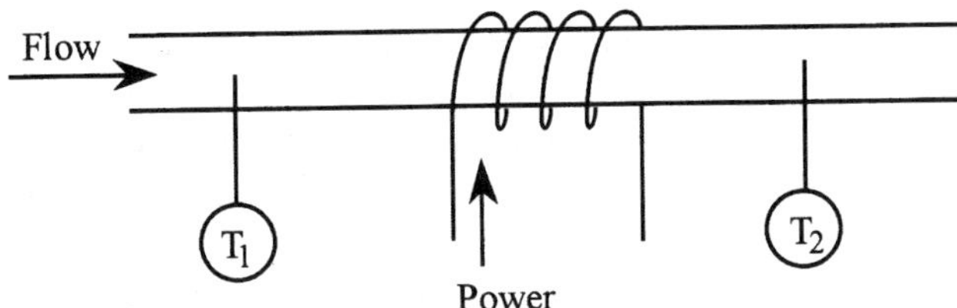

FIGURE 4.23.11 Schematic of thermal mass flow meter operation.

An alternate scheme measures the temperature of the gas, and power is supplied to the downstream sensor such that its temperature is kept a fixed amount above the upstream sensor. The feedback circuit needed to do this is similar to that used in hot-wire anemometry. The amount of power supplied to the downstream sensor is a measure of the gas flow rate. The advantages of these meters are that they give a direct indication of the flow rate without requiring corrections for temperature or pressure variations, they have no moving parts, and are noninvasive.

Ultrasonic Flow Meters

These meters determine flow rate by measuring the Doppler shift of an ultrasonic signal due the movement of particles in the flow (Figure 4.23.12). Particles or bubbles in the fluid are usually required. Recently, meters have been developed that can measure the flow of clean fluids by measuring the velocity of

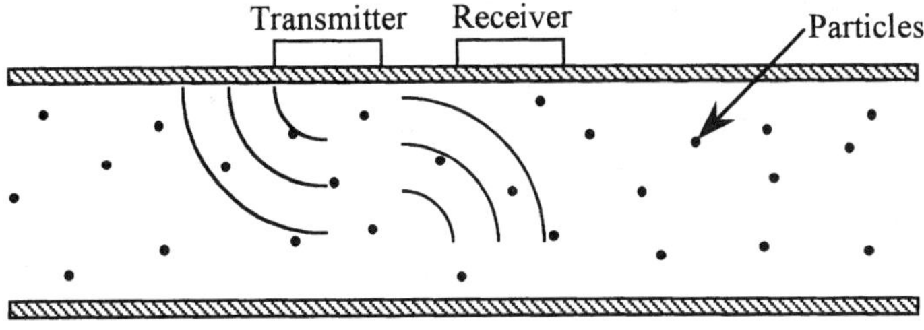

FIGURE 4.23.12 Schematic of ultrasonic flow meter.

turbulent swirls in the liquid. The meter is noninvasive, and can measure flow rate of dirty fluids or slurries.

The transit time method relies on measuring the difference in the time it takes for a signal to traverse the flow in one direction compared to a signal traveling in the opposite direction. Because the difference in transit times are so small, the time difference is usually measured as a phase difference between the two sound signals. When the particles move at flow speed, the frequency shift is proportional to flow speed. For a suitably chosen path, output is proportional to volume flow rate. Ultrasonic meters may require calibration in place. One or two transducers may be used.

Laminar Flow Elements (LFE)

Laminar flow meters operate by passing a gas through a bundle of tubes or a honeycomb (Figure 4.23.13). The small diameter of these flow passages in the matrix laminarize the flow, and the pressure drop between the inlet and exit of the tubes is measured. If the flow passages were circular, the pressure drop across the element would be

$$\Delta P = \frac{128\mu l \dot{V}}{\pi D^4} \tag{4.23.18}$$

and is linear with the volumetric flow rate. Because the flow passages are not necessarily circular, calibration of the flow meter is usually required. Corrections are also needed for changes in fluid viscosity and temperature. The gas must be clean. The LFE may be used with reasonable accuracy over a 10:1 flow rate range. A LFE costs approximately as much as a venturi, but is much lighter and smaller. Thus, the LFE is widely used in applications where compactness and extended range are important.

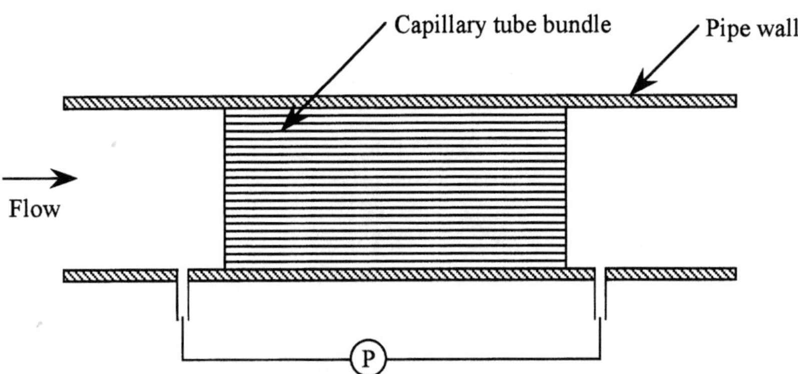

FIGURE 4.23.13 Schematic of laminar flow element.

Coriolis Meters

These meters are true mass flow meters. The concept behind coriolis meters is shown in Figure 4.23.14. Consider the case where the flow moves through a U-tube that is vibrated at its resonant frequency using an electromagnetic coil. If fluid flows into the tube when the tube is moving upward, the fluid needs to be accelerated and will cause the tube to bend down. The fluid that has traveled around the bend at this time already has an upward momentum and needs to be decelerated, causing this part of the tube to bend upward. The tube, therefore, will twist in the opposite direction during the downward motion. The difference in velocity of the two legs of the vibrating tubes is a measure of the mass flow rate. The density of the fluid flowing in the tube can be measured by measuring the change in resonant frequency of the tube. The instrument measures mass flow rate directly and, thus, is ideal for two-phase or liquid–solid flow measurements. Pressure drop across the coriolis meter may be high, but its useful flow rate range is 100:1.

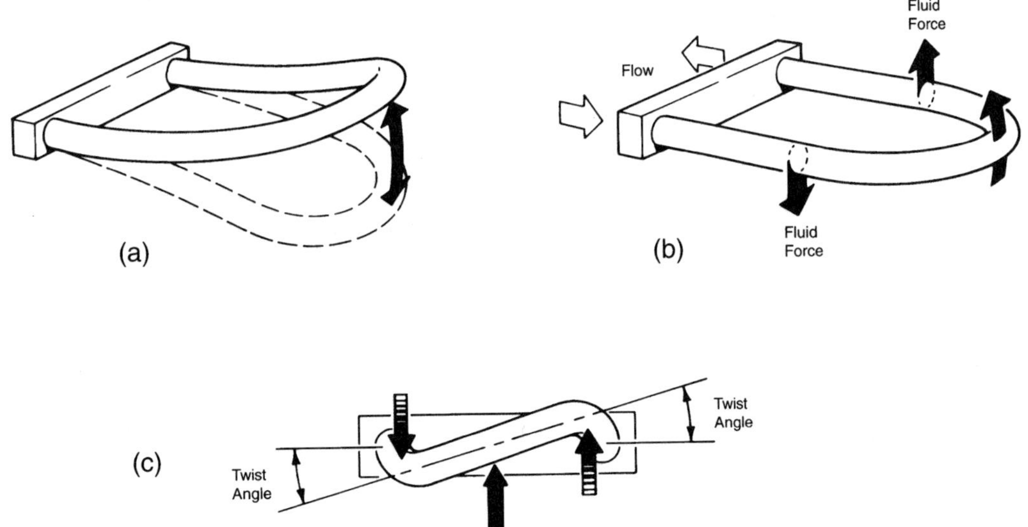

FIGURE 4.23.14 Principle of operation of a coriolis meter: (a) vibrating flow tube, (b) fluid forces reacting to vibration of flow tube, (c) end view of flow tube showing twist. (Courtesy of Micro Motion Inc., Boulder, CO.)

Electromagnetic Flow Meters

Electromagnetic flow meters create a magnetic field across a pipe. When a conductive fluid passes through the field, a voltage is generated at right angles to the field and velocity vectors. Electrodes placed on a pipe detect the resulting signal voltage, which is proportional to the average axial velocity when the profile is axisymmetric. The minimum flow speed should be above about 0.3 m/s, but there are no restrictions on the Reynolds number. The flow rate range normally quoted is 10:1.7

Traversing Methods

In situations such as in air handling or refrigeration equipment, it may be impractical or impossible to install a fixed flow meter, but it may be possible to measure flow rate using a traversing technique. To measure flow rate by traverse, the duct cross section is subdivided into segments of equal area. The fluid velocity is measured at the center of each area segment, and the volume flow rate for each segment is approximated by the product of the measured velocity and segment area. Flow rate through the entire duct is the sum of these segmental flow rates. Methods for measuring the flow velocity are discussed below.

Pressure Probes

Perhaps the simplest and least expensive method of measuring velocity is through the use of Pitot probes. Consider the stagnating flow shown in Figure 4.23.15. Applying Bernoulli's equation to the flow between a point far upstream of the probe and the probe tip, assuming an incompressible flow, yields

$$p_0 = p + \frac{1}{2}\rho u^2 \tag{4.23.19}$$

The total pressure sensed by the tube (p_0) is greater than the freestream pressure (p) by the dynamic head $1/2\rho u^2$. The freestream velocity (u) thus can be determined by measuring the static and total pressures. Equation 4.23.19 can be applied to flows whose Mach number (M = u/a) is less than 0.2 with less than 0.5% uncertainty. The speed of sound is given by

$$a = \sqrt{\gamma R_g T} \tag{4.23.20}$$

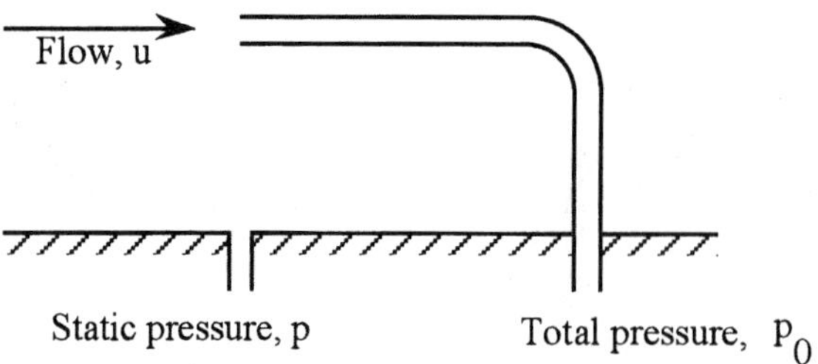

FIGURE 4.23.15 Basic total pressure probe setup.

where T is the static absolute temperature of the flow, R_g is the real gas constant, and γ is the ratio of specific heats. When the flow Mach number exceeds 0.2, the density of the fluid along the stagnation streamline changes and compressibility effects must be taken into account. For $0.2 < M < 1.0$, the relation between velocity and pressure difference is given by:

$$M^2 = \frac{2}{\gamma - 1}\left[\left(\frac{p_0}{p}\right)^{\frac{\gamma-1}{\gamma}} - 1\right] \tag{4.23.21}$$

When the flow is supersonic, the flow adjusts to the stagnation condition through a bow shock off the tip of the Pitot probe (Figure 4.23.16). The relationship between the static pressure ahead of the shock (measured by a static pressure tap) to the total pressure behind the shock (indicated by the probe) is given by

$$\frac{p_1}{p_{2,0}} = \frac{\left(\frac{2\gamma}{\gamma+1}M_1^2 - \frac{\gamma-1}{\gamma+1}\right)^{\frac{1}{\gamma-1}}}{\left(\frac{\gamma+1}{2}M_1^2\right)^{\frac{\gamma}{\gamma-1}}} \tag{4.23.22}$$

The above equation can be solved to determine the Mach number and, hence, the flow velocity ahead of the shock.

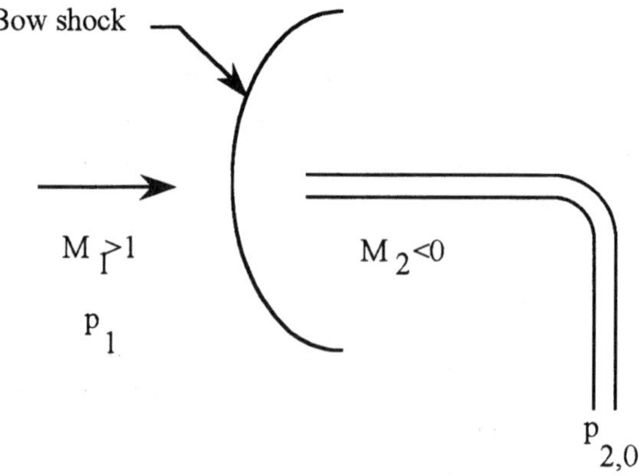

FIGURE 4.23.16 Supersonic flow around a total pressure probe.

Although the measurement of velocity using pressure probes is relatively straightforward, there are many situations where the ideal situation indicated in Figure 4.23.15 does not occur. Some factors that influence the measurement are

- Alignment of the probe with the flow
- Wall effects
- Turbulence effects
- Pitot probe design
- Static pressure tap geometry
- Streamline curvature

Each of these effects is briefly discussed below.

Alignment Effects

The change in stagnation pressure due to flow misalignment for various probes is shown in Figure 4.23.17. It is seen that an open-ended tube can be misaligned by up to ±10° without error in the pressure reading, while a tube with a chamfered opening can tolerate misalignments up to ±30°.

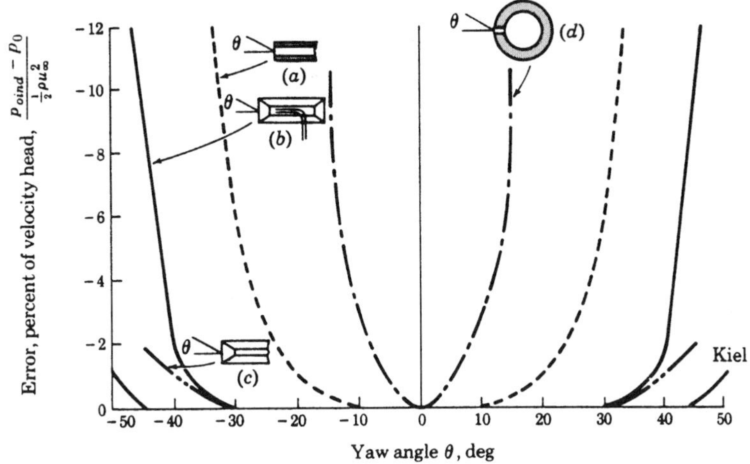

FIGURE 4.23.17 Error in stagnation pressure with changes in yaw angle for (a) open-ended tube, (b) channel tube, (c) chamfered tube, and (d) tube with orifice in side. (From Krause, L.N. and Gettleman, C.C., *ISA Proc.*, 7, 134, 1952. With permission.)

Wall Effects

The presence of a wall changes the flow around the probe, and this can result in pressure readings that are lower than when a wall is not present. For a round probe touching the wall, the velocity indicated can be up to 1.5% low (Moffatt).

Turbulence Effects

The effects of turbulence on velocity measurements are dependent on many factors such as turbulence scale and structure, fluctuation frequency, and internal geometry of the probe. The few studies that have been performed were inconclusive, with some studies indicating a decrease in the measured velocity with increasing turbulence and others indicating the opposite trend. A first-order analysis indicates that 20% turbulence intensities can result in a 2% increase in the measured velocity.

Static Pressure Taps

Static pressure taps should be drilled perpendicular to the wall, and no burrs or chamfers should be present. The holes should be made as small as possible without causing excessively slow response times at the transducer.

Streamline Curvature

If the streamlines in a flow are curved (perhaps due to obstacles in the flow), then velocity and pressure gradients can occur across the channel and the static pressure obtained from a wall pressure tap may not be representative of that where the total pressure is being measured. The local static pressure at the measuring location must either be measured or computed.

Design of Pitot-Static Probes

Pitot-static probes combine the measurement of total and static pressure in one unit. A commercially available Pitot-static probe is shown in Figure 4.23.18. The location of the static pressure taps is important. The static pressure taps should be located sufficiently far downstream of the tip so that flow acceleration effects can be ignored, and sufficiently far upstream of the support so that blockage effects are small. If the static pressure taps are located eight tube diameters downstream of the tip and eight tube diameters upstream of the support, then the effects of both acceleration and blockage will be below 1%.

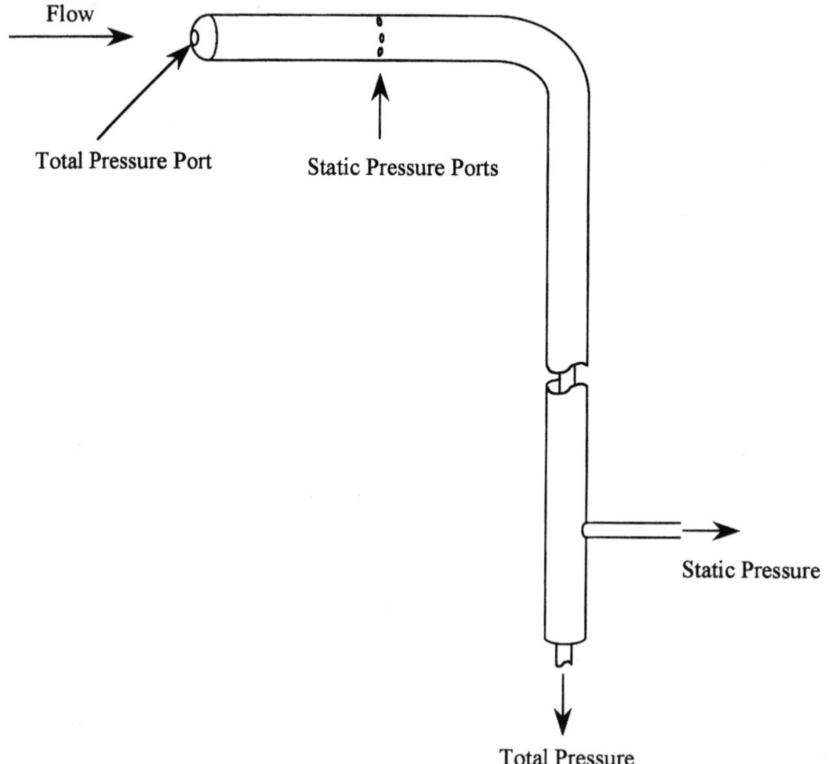

FIGURE 4.23.18 A pitot-static probe.

Measurement of Flow Direction

The flow direction can be obtain by using prizm probes or probes with multiple sensing holes. Additional reading regarding the various types of pressure probes and the errors associated with them is given in Moffatt.

Hot-Wire Anemometry

Hot-wire anemometry is used when one wishes to measure rapidly varying velocities with good spatial resolution. Velocity fluctuations up to 50 kHz can easily be obtained. Typically a heated, thin wire is held in the flow, and the cooling effect on the wire due to velocity is measured. The disadvantages compared to pressure probes are that they are relatively expensive, they use frail wires, and they need to be calibrated in a known flow environment. Corrections also need to be made to account for freestream temperature variations and for flow along the wire for slanted wires. There are two types of hot-wire anemometry — constant temperature and constant current.

Constant Current Anemometry

This method consists of sending a constant current through a thin wire and measuring the change in resistance of the wire with velocity. Although much simpler than constant temperature anemometry, constant current anemometry has several disadvantages that has resulted in almost exclusive use of constant temperature anemometry today. These disadvantages include relatively low frequency response and the possibility of wire burnout at large currents and low velocities. However, they can be used when many channels of data need to be acquired, when turbulence information is not needed, or when the flow temperature needs to be measured. Numerous schemes for treating the resulting data have been published. Digital processing techniques, including fast Fourier transforms, can be used to obtain mean values and moments, and to analyze signal frequency content and correlations.

Constant Temperature Anemometry Fundamentals

The basic circuit is shown on Figure 4.23.19. The wire is represented by resistance R_w in a Wheatstone bridge. Resistors R_1, R_2, and R_c are very stable with temperature. The imbalance in the bridge is measured using an op-amp, and the op-amp outputs whatever voltage is necessary to bring the bridge back into balance. For example, consider the case where the bridge is balanced at a given flow velocity, i.e., $R_w/R_1 = R_c/R_2$ and $V_1 = V_2$. If the velocity over the wire increases, the wire temperature (and therefore its resistance) will decrease, causing the bridge to become unbalanced ($R_w/R_1 < R_c/R_2$ and $V_1 < V_2$). The op-amp senses this imbalance in the bridge and increases its output voltage causing more current to flow through the wire, bringing the bridge back into balance. The operating temperature of the wire is controlled by changing the value of R_c. The output voltage of the op-amp is a measure of the amount of cooling the wire experiences. Because the temperature of the wire does not change, the frequency response is not limited by the thermal mass of the wires and high frequency responses are possible.

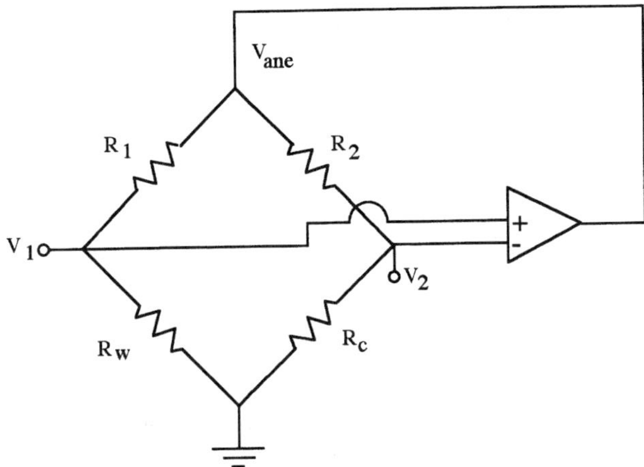

FIGURE 4.23.19 Schematic of hot-wire circuit.

Equations Governing Hot-Wire Anemometry

The cooling law is assumed to be of the form

$$Nu = A_0 + B_0 \,\mathrm{Re}^{0.5} \tag{4.23.23}$$

where Nu is the Nusselt number and Re is the Reynolds number; both are based on the wire diameter. The above equation is commonly known as King's law (King, 1914). Substituting in the definition of Nu and Re and solving for the heat transferred from the wire yields

$$q = A_1\left(T_w - T_\infty\right)k + B_1\left(\frac{P_\infty u_\infty}{T_\infty \mu}\right)^{0.5}\left(T_w - T_\infty\right)k \tag{4.23.24}$$

The heat transferred from the wire to the gas is balanced by the energy supplied by the anemometer:

$$q = i_w^2 R_w = \frac{V_{ane}^2 R_w}{\left(R_w + R_{ane}\right)^2} \tag{4.23.25}$$

or, since R_w and R_{ane} are constant,

$$q \propto V_{ane}^2 \tag{4.23.26}$$

Combining Equations 4.23.24 and 4.23.26 and solving for $P_\infty u_\infty$ yields

$$\left(P_\infty u_\infty\right)^{0.5} = AT_\infty^{0.5}\mu^{0.5} + B\frac{V_{ane}^2}{\left(T_w - T_\infty\right)}\frac{T_\infty^{0.5}\mu^{0.5}}{k} \tag{4.23.27}$$

For isothermal flows, the response equation reduces to

$$u^{0.5} = A + BV_{ane}^2 \tag{4.23.28}$$

Although it is theoretically possible to obtain the response curve without doing a calibration, the uncertainty in the wire diameter, wire length, and other errors make calibration of the wire mandatory at the conditions at which one wishes to operate.

Probe Design

Shown on Figure 4.23.20 are some typical probes used in anemometry. The curved prongs on the boundary layer probe (Figure 4.23.20b) are designed to minimize flow disturbance in the near-wall region. The two wires on this probe are used to obtain both u and v components of velocity.

Platinum-plated tungsten wires 4 μm thick are typically used. Platinum is used to prevent oxidation of the wires during operation. The operating temperature for these wires is typically 250°C. Pure platinum wires are also used with the advantage of being able to operate at high temperatures (about 1000°C), thereby minimizing drift with ambient temperature changes, but they tend to be quite frail and their use is limited to low velocities. The ends of the wires where they attach to the prongs are usually copper-plated and the unplated length defines the sensitive portion of the wire. This is performed to isolate the sensing length from the disturbing effects of the prongs on the flow. The l/d of the sensing length is typically 200.

Platinum thin film probes are also available. These are made by depositing a thin layer of platinum onto a quartz substrate, typically a 50-μm-diameter quartz rod. The probe can be coated with a thin layer of oxide for use in electrically conductive liquids. The advantages of film sensors are their rigidity and strength. Rigidity is important for multisensor applications since the data reduction algorithms typically assume straight sensors. Film sensors are more resistant to damage by particles in the flow.

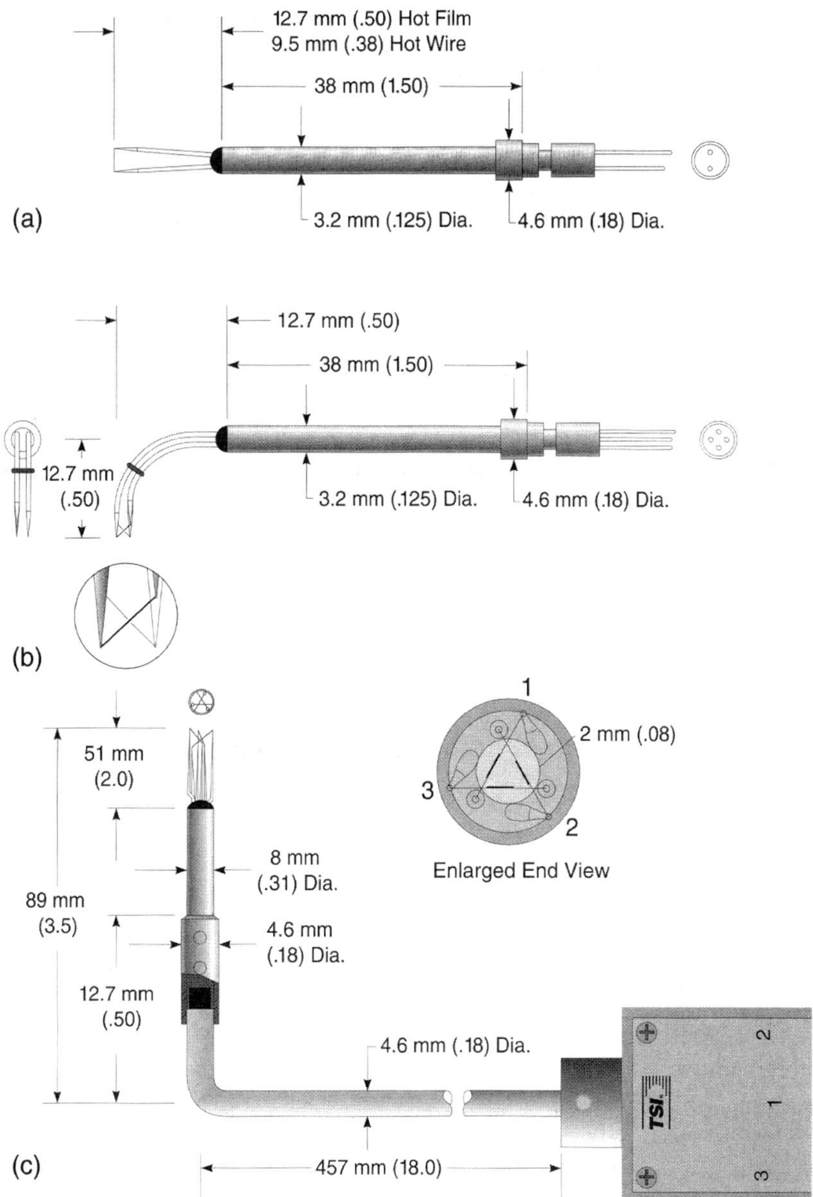

FIGURE 4.23.20 Some hot-wire probe configurations: (a) single-wire probe, (b) boundary layer cross-wire probe, (c) triple sensor cross-flow probe. (Courtesy of TSI Inc., St. Paul, MN.)

Multicomponent Probes

More than one component of velocity can be measured if multiple wires are present. Consider the case where a wire is inclined relative to the axis of the flow, as shown in Figure 4.23.21. Cross-wires (Figure 4.23.20b) inclined at θ_1 and θ_2 from the normal can be used to obtain the instantaneous U and V components of velocity. The cross-correlation between u′ and v′ (u′v′), also known as the turbulent shear stress, can be obtained if the instantaneous velocities U and V are known. The wires will be sensitive to the velocities according to

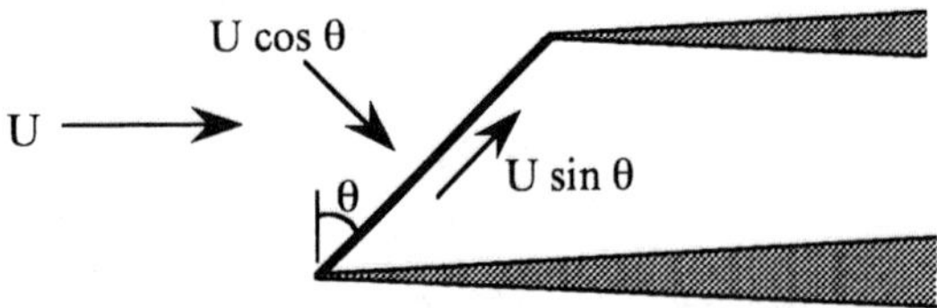

FIGURE 4.23.21 Normal and tangential components of velocity for a wire inclined to the flow.

$$U_{eff,1} = U\cos(\theta_1) + V\sin(\theta_1) \tag{4.23.29}$$

$$U_{eff,2} = U\cos(\theta_2) - V\sin(\theta_2) \tag{4.23.30}$$

If the effective velocities over the two wires are known along with the probe geometry, then the velocities U and V can be determined. The instantaneous velocity can be broken down into the mean and fluctuating components:

$$U = \overline{U} + u' \tag{4.23.31}$$

$$V = \overline{V} + v' \tag{4.23.32}$$

Since

$$\overline{uv} = \overline{(\overline{U} + u')(\overline{V} + v')} = \overline{UV} + \overline{u'v'} \tag{4.23.33}$$

the turbulent shear stress can be written as

$$\overline{u'v'} = \overline{uv} - \overline{UV} \tag{4.23.34}$$

The above results must be corrected to take into account the directional response of the wire, as discussed below. Three-wire probes to measure three components of velocity are also available (Figure 4.23.20c).

Directional Response

For small inclinations θ to the flow, the hot-wire will only respond to the velocity component normal to the wire or $U\cos\theta$. For large inclinations, a significant axial flow develops along the wire resulting in additional cooling as a result of the finite length of the wire. The equation usually used to describe this effect is based on work by Champagne et al. (1967) and is given by

$$U_{eff}^2 = U^2\left(\cos^2\theta + k^2\sin^2\theta\right) \tag{4.23.35}$$

where k is a constant that varies linearly from 0.2 for a wire with $l/d = 200$ to zero for $l/d = 600$. Champagne et al. (1967) applied these results to estimate the overall error for a 45° cross-wire operated at constant temperature. They found that

$$\left(\frac{\overline{u'^2}}{U^2}\right)_m = \left(\frac{\overline{u'^2}}{U^2}\right)_i \tag{4.23.36}$$

$$\left(\frac{\overline{u'v'}}{U^2}\right)_m = \frac{1+k^2}{1-k^2}\left(\frac{\overline{u'v'}}{U^2}\right)_i \tag{4.23.37}$$

$$\left(\frac{\overline{v'^2}}{U^2}\right)_m = \frac{1+k^2}{1-3k^2+4k^4}\left(\frac{\overline{v'^2}}{U^2}\right)_i \tag{4.23.38}$$

where the subscript m is the measured value and the subscript i is the ideal value for a wire of infinite length.

Corrections Due to Temperature

Probably the most significant source of error when using a properly calibrated hot-wire occurs as a result of drifts in the flow temperature from the calibration temperature. If the flow temperature is higher than the calibration temperature, less energy is required to keep the wire at a constant temperature for a given velocity, and the indicated flow velocity will be lower than the true flow velocity. A first-order correction can be made to account for the drift in flow temperature as described below, however. The energy dissipated by the wire is proportional to the wire-to-flow temperature difference, and is also proportional to the voltage across the wire:

$$q_w \sim \Delta T \sim V^2 \tag{4.23.39}$$

The following equation can be used to correct for small changes in temperature:

$$V_{corr} = V_m \sqrt{\frac{T_w - T_{ref}}{T_w - T_\infty}} \tag{4.23.40}$$

where V_{corr} = corrected voltage
V_m = measured voltage
T_w = wire temperature
T_{ref} = temperature of the calibration flow
T_∞ = temperature of the flow

It should be noted that the sensitivity to ambient temperature changes decreases with higher wire temperatures — the wire should be operated at the highest possible temperature.

Laser Doppler Anemometry

LDV uses light to measure the velocity of particles in a flow through a small volume. If the particles faithfully follow the flow, then the velocity of the particle can be assumed to be the velocity of the flow itself. LDV has numerous advantages over other methods of measuring velocity. First, it is nonintrusive so it can be used in environments that are hostile to probes or where probes excessively disturb the flow. Second, it does not depend on temperature or pressure changes in the flow. Third, it can detect flow reversals. Fourth, it can measure very high velocities if sufficiently fast electronics are available. Fifth, velocity is measured directly and is unaffected by changes in temperature, density, or composition in the flow field. LDV has been used in flows ranging from natural convection to flows in IC engines and gas turbines. The disadvantages include the need for particles (seeded or naturally occurring) that follow the fluid motion, high cost, the need for optical access to the test section, and the need for careful alignment.

Principle of Operation

The basic setup for a popular type of LDV system is shown on Figure 4.23.22. A single, polarized, continuous-wave laser is split into two parallel beams using a beam splitter. The beams pass through a focusing lens that causes the beams to intersect in the focal plane of the lens, simultaneously reducing the beam diameter. The light beams approximate cylinders in the intersection region. The two beams constructively and destructively interfere with each other in the intersection regions, setting up alternating bright and dark fringes that are parallel to the axis of the laser beams. A particle moving through the measuring volume will scatter light, with the intensity of the scattered light depending on whether or

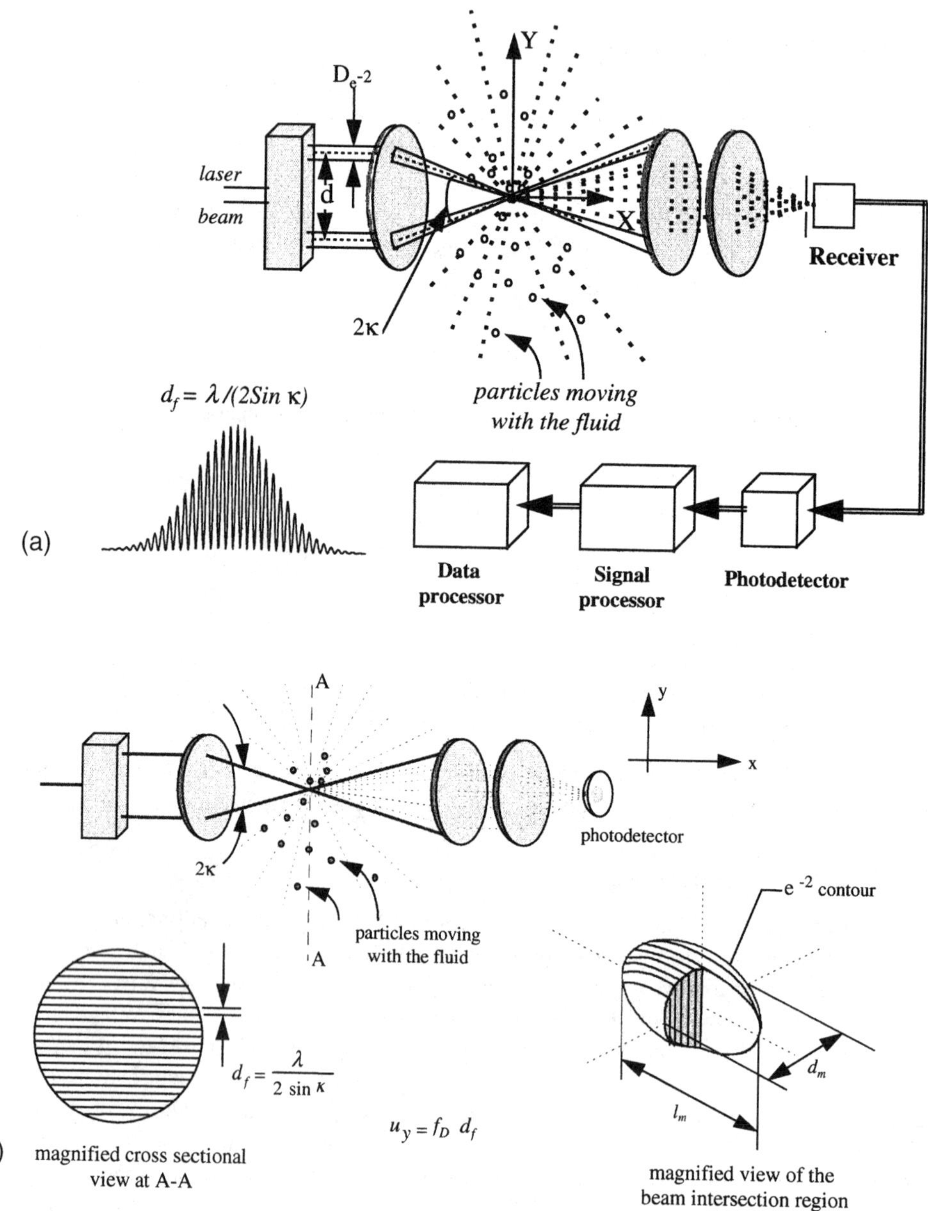

FIGURE 4.23.22 Basic setup for two-component LDV: (a) schematic of a dual beam system, (b) details of the beam crossing. (Courtesy of TSI Inc., St. Paul, MN.)

not the particle is in a bright or dark fringe. Only the component of velocity perpendicular to the fringes and to the axis of the laser beams will be measured. The collecting lens collects some of the scattered light and focuses it onto a photodetector which outputs a signal dependent on the light intensity. Since the spacing between the fringes can be calculated or measured, the velocity of the particle traversing the measuring volume can be determined by measuring the frequency of the signal from the photomultiplier.

Measuring Volume Dimensions

The dimensions of the measuring volume can be computed from the following relations:

$$d_m = \frac{4 f \lambda}{\pi D_e^{-2}} \tag{4.23.41}$$

$$l_m = \frac{d_m}{\tan \kappa} \tag{4.23.42}$$

$$d_f = \frac{\lambda}{2 \sin \kappa} \tag{4.23.43}$$

where d_m = measuring volume diameter
l_m = measuring volume length
d_f = fringe spacing
f = focal length of the transmitting lens
D_e^{-2} = diameter of the laser beam at the e^{-2} intensity point entering the transmitting lens
λ = wavelength of the laser light
κ = half angle between the beams

HeNe or argon-ion lasers are typically used. Note that in order to get a small measuring volume diameter, the focal length needs to be short, or the beam diameter needs to be large. Similar effects are seen for the measurement volume length. Ideally, one would like to have a measuring volume that is small compared to the scales of interest, with at least 10 fringes in the measuring volume.

Backscatter Mode

A configuration that is often used is shown in Figure 4.23.23. The backscatter configuration has the advantage that the optics can be made very compact, and because all the optics can be put in a single housing the optics need to be aligned only once. Much less light is scattered in the backward direction than in the forward direction, however, so much more sensitive photodetectors are required to detect the weaker signal.

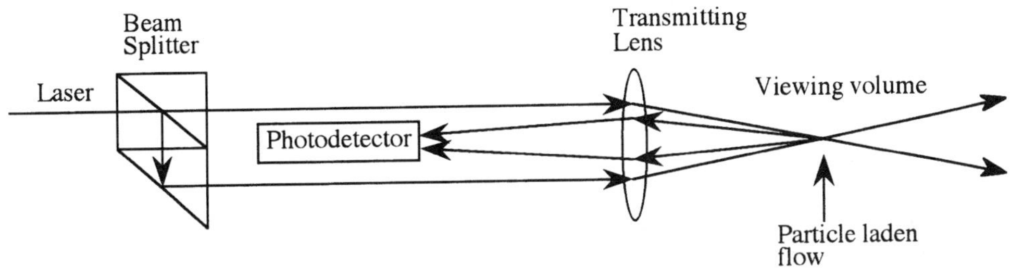

FIGURE 4.23.23. LDV configuration for backscatter.

Photodetectors

Two types of photodetectors are generally used to convert the scattered light collected by the collecting lens to an electronic signal. In the photomultiplier, photons strike a coating of photoemissive material causing electrons to be emitted. These electrons are accelerated by a high-voltage field and impact onto a dynode that emits even more electrons. This amplification process is repeated, resulting in final gains of 10^3 to 10^7. Frequency responses up to 200 MHz can be obtained using photomultipliers.

Photodiodes are light-sensitive semiconductor junctions. An amplification process similar to that in PMTs occurs in photodiodes, but gains are limited to about 10^3. The frequency response tends to be lower as well. Photodiodes are much less expensive than photomultipliers, however.

Signal Processing

A typical output signal from a photodetector is shown on Figure 4.23.24. It is seen to consist of a pedestal along with the Doppler signal. Most commercial LDV systems currently on the market use FFT or correlation methods to determine the frequency.

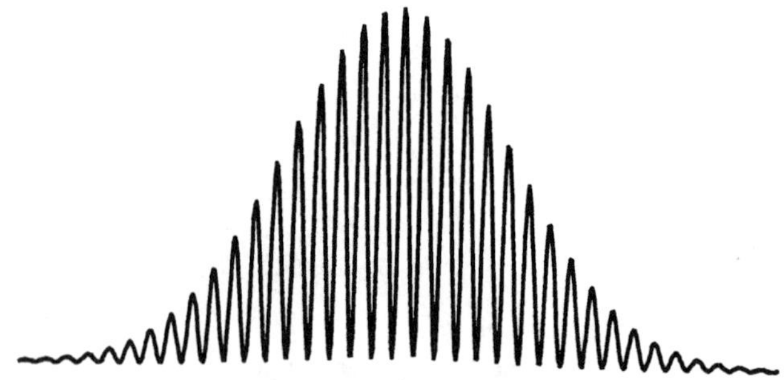

FIGURE 4.23.24 Typical photodetector signal due to scattering from a single particle. (Courtesy of TSI Inc., St. Paul. MN.)

Particle Tracking

It is important that the particles be able to follow the flow. Consider a spherical particle placed in a uniform flow. The drag force on the particle is due to the difference in speed between the particle and the carrier fluid, i.e.,

$$Force = ma = m\frac{dU_\infty}{dt} = \rho_P \frac{\pi D_P^3}{6}\frac{dU_\infty}{dt} = Drag \tag{4.23.44}$$

The drag on a spherical particle is given by

$$Drag = C_d \frac{1}{2}\rho_\infty \frac{\pi D_P^2}{4}\left(U_\infty - U_P\right)^2 \tag{4.23.45}$$

where

$$C_d = \frac{24}{\text{Re}_P} = \frac{24\nu}{D_P\left(U_\infty - U_P\right)} \tag{4.23.46}$$

for $0 < Re_P < 100$. Combining Equations 4.23.44, 4.23.45, and 4.23.46 and integrating yields

$$\frac{\left(U_\infty - U_P\right)}{U_\infty} = \exp(-t/K) \tag{4.23.47}$$

where

$$K = \frac{D_P^2}{18\nu}\frac{\rho_P}{\rho_\infty} \tag{4.23.48}$$

Here, K is the time constant, and represents the time it takes for the initial velocity difference between the particle and the fluid to decay by $1/e$. Particles with K small compared to the time scales of interest will faithfully follow the flow.

Flow Reversal Measurements

If the fringes are stationary, as was assumed until now, one cannot tell the direction from which the particle passes the fringes, i.e., a particle moving with a velocity U will produce the same signal as a particle moving with velocity $-U$. While the direction of the flow is unambiguous in many cases, there are other flows (recirculating flows, flows with high turbulence intensity, impinging jets, shear flows) where the flow direction is not known. This problem can be easily rectified by shifting the frequency of one of the laser beams. If one of the beams is shifted in frequency, this will cause the fringes to move as well. If the fringes move at a speed faster than the maximum absolute flow velocity, then the indicated velocity will be the flow velocity plus the fringe velocity. Zero flow velocity would correspond to the fringe velocity, negative flow velocities would correspond to indicated velocities lower than the fringe velocity, and positive flow velocities would correspond to indicated velocities higher than the fringe velocity.

Frequency shifting is usually accomplished by inserting an acoustic-optic Bragg cell in one of the split beams. Typical shifts are 40 MHz, i.e., a 40-MHz signal would be seen at the output of the photodetector at zero flow velocity.

Multicomponent Systems

Measurement of more than one component of velocity is more difficult than simply having two one-component systems because the signals from the two beams interfere with each other. It is necessary to be able to distinguish which pair of beams the signal came from. One can either use two pairs of beams of different color or different polarities. In the two-color method, four beams consisting of two different colors are usually used. An argon-ion laser produces both blue and green lines. This light is split and converged to form two mutually orthogonal dual-beam systems. The scattered light is collected by a pair of receiving optics/photodetector systems, one of which accepts only blue light and one of which only accepts green light. The scattered light is separated using a dichroic mirror that transmits the blue light and reflects the green light.

Particle Image Velocimetry

A relatively new method of measuring velocity at many points in the flow is by particle image velocimetry (PIV). This method enables simultaneous measurement of velocity at many points in the flow by measuring the distance that particles carried by the flow travel over a known time interval. Accuracy and spatial resolution can be comparable to LDV.

Typical applications of PIV are to obtain the turbulent flow structure of complex steady flow fields where mapping is too difficult and time-consuming, and periodic flow fields where the velocity field can be obtained at different phases. An example of flow field velocity vectors obtained using PIV is shown on Figure 4.23.25.

The fundamental principle of PIV by optical imaging is the measurement of displacements δx, δy of image markers which travel with the fluid. In PIV, the time between the marks is fixed, while the distance between fringes is fixed in LDV. The time interval between marks in PIV must be small compared to the time scales of interest in the flow. Typical markers that have been used are H_2 bubbles, fluorescent particles, and naturally occurring small particles in liquids. The two main components of a PIV system are the imaging system which produces a double-exposed photographic image of the particles in the

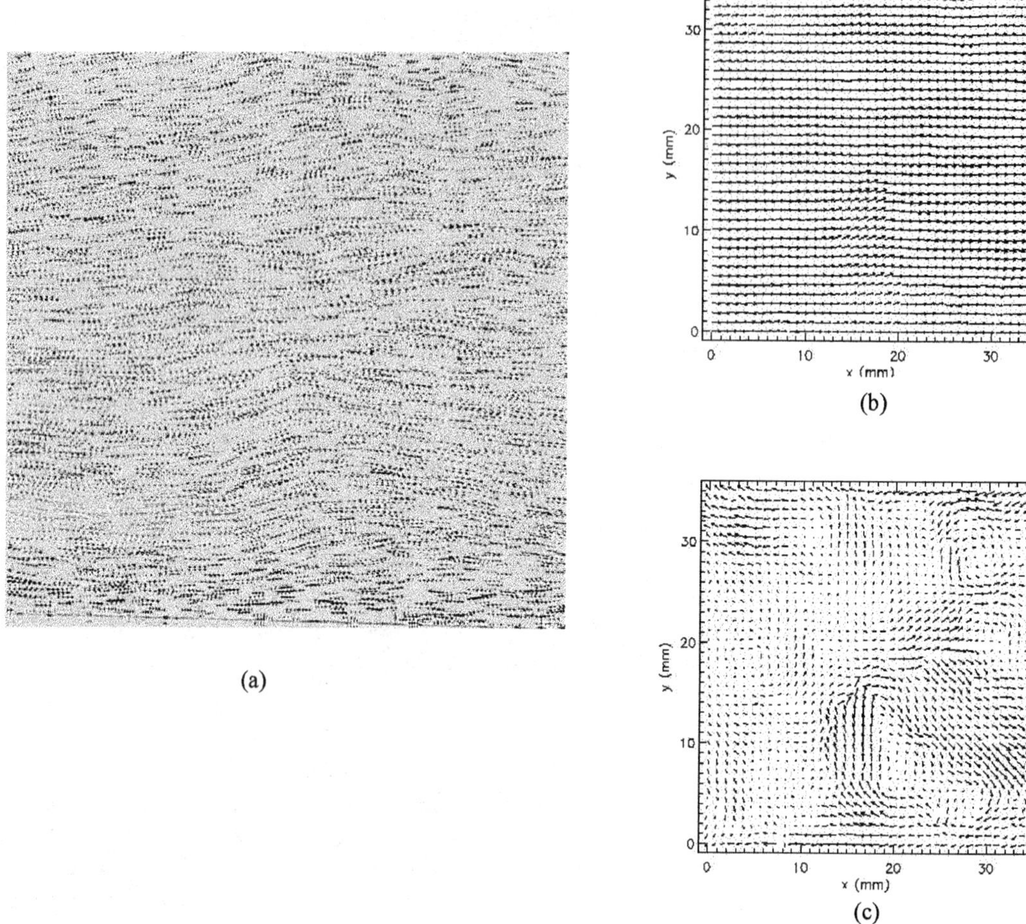

FIGURE 4.23.25 Example of velocity vectors obtained using PIV: (a) double exposure of particles in flowfield, (b) velocity vectors, (c) velocity vectors with mean velocity subtracted out. (Courtesy of Ken Kiger, University of Maryland, College Park, MD.)

flow field, and the interrogation system which extracts and presents the velocity field information contained in the photograph.

Imaging System

Typically, the beam from a doubly pulsed laser is converted into a sheet of laser light using a cylindrical lens, which is then used to illuminate the markers in the flow. A camera located perpendicular to the flow records the doubly exposed image onto film. Typical light sources are pulsed lasers or shuttered continuous wave (CW) lasers. Because pulsed lasers have larger energy output than CW lasers, they are generally used to illuminate small particles in the flow. Single CW laser beams swept across the flow using a rotating, faceted mirror produce a pseudo light sheet and can also be used to illuminate particles.

Film is usually necessary to resolve the markers if reasonable spatial resolution is desired. Larger markers in conjunction with high-resolution video cameras can be used, but care must be taken to ensure that the markers faithfully follow the flow. High-resolution cameras do have the advantage that near on-line processing of the data can be performed, however, with the appropriate digitizing and computing equipment.

The result of the imaging system is to produce two sets of markers that are displaced in time. The two sets can be on a single sheet of film, or can be recorded separately on different frames of a video camera. An example of two sets of markers is shown on Figure 4.23.26a.

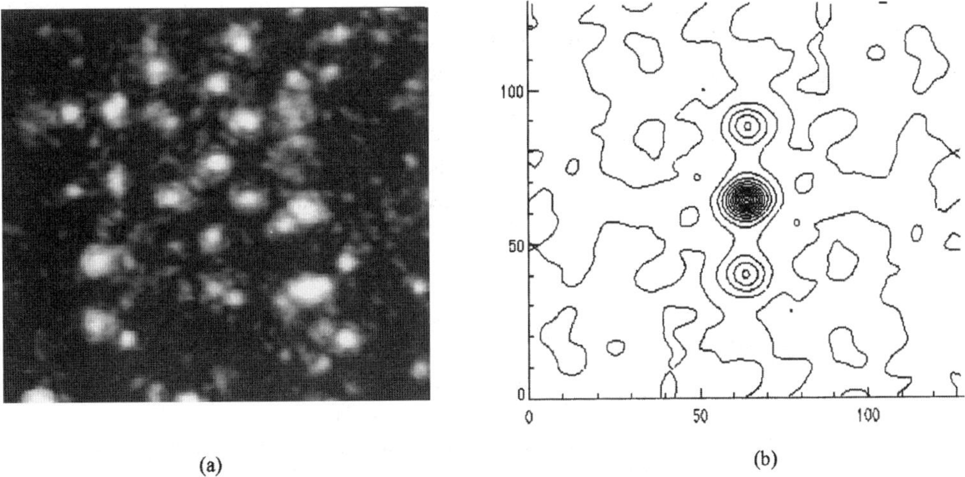

FIGURE 4.23.26 Example of (a) double-exposed image of particle shifted from one another, (b) the corresponding spatial autocorrelation. (Courtesy of Ken Kiger, University of Maryland, College Park, MD.)

Interrogation System

To obtain the velocity vectors, the doubly exposed image is first divided up into small sub-areas over which the average velocity vector will be computed. The size of the sub-areas should be as small as possible to obtain the highest number of velocity vectors, but they should not be so small that an insufficient number of particle pairs results. A good rule of thumb is to choose a sub-area size that contains about 10 particle pairs. The position of each of the particles is then determined, and a spatial autocorrelation performed (Figure 4.23.26b). The location of the peaks in the spatial correlation plane yields the average particle displacement over that sub-area. The largest peak results from a self-correlation and is not of interest. Two other peaks, one in the positive direction and the other in the negative direction, indicate the possible average particle displacement over that sub-area.

Flow Direction

It is not possible to determine the *direction* of the velocity vector from a doubly exposed photograph alone. This is not a problem if the main flow direction is known *a priori*, but it does present problems in highly turbulent flows or if flow reversals occur. Image shifting of the second set of particles relative to the first can be used to resolve this directional ambiguity. Suppose the second set of images is shifted by a positive amount. A stationary marker will be displaced by this amount. A moving marker will have an additional displacement as a result of its movement. A shift greater than that expected at the largest negative flow velocity then eliminates ambiguity regarding flow direction, since the second image is always on the positive side of the first image. This technique is analogous to using frequency shifting in LDV measurements to determine the direction of the particle movement.

Rotating mirrors and crystals have been used to produce this shift in images. More recently, cameras have been developed specifically for PIV that electronically shift the first set of images by a certain number of pixels before acquiring the second set, eliminating any additional mechanical complexity.

Viscosity Measurements

Viscometry is the technique of measuring the viscosity of a fluid. Viscometers are classified as rotational, capillary, or miscellaneous, depending on the technique employed. Rotational viscometers use the principle that a rotating body immersed in a liquid experiences a viscous drag which is a function of the viscosity of the liquid, the shape and size of the body, and the speed of its rotation. Rotational viscometers are widely used because measurements can be carried out for extended periods of time.

Capillary viscometry uses the principle that when a liquid passes in laminar flow through a tube, the viscosity of the liquid can be determined from measurements of the volume flow rate, the applied pressure, and the tube dimensions. Viscometers that cannot be classified either as rotational or capillary include the falling ball viscometer. Its method of operation is based on Stokes' law which relates the viscosity of a Newtonian fluid to the velocity of a sphere falling through it. Falling-ball viscometers are often employed for reasonably viscous fluids. Rising bubble viscometers utilize the principle that the rise of an air bubble through a liquid medium gives a visual measurement of liquid viscosity. Because of their simplicity, rising bubble viscometers are commonly used to estimate the viscosity of varnish, lacquer, and other similar media.

Defining Terms

Flow meter: Device used to measure mass flow rate or volume flow rate of fluid flowing in a duct.

Restriction flow meter: Flow meter that causes flowing fluid to accelerate in a nozzle, creating a pressure change that can be measured and related to flow rate.

Thermal anemometer: Heated sensor used to infer local fluid velocity by sensing changes in heat transfer from a small electrically heated surface exposed to the fluid flow.

Traverse: Systematic procedure used to traverse a probe across a duct cross section to measure flow rate through the duct.

References

Adrian, R.J., *Selected Papers on Particle Image Velocimetry,* SPIE Optical Engineering Press, Bellingham, WA, 1993.

ASHRAE, *Handbook Fundamentals*, American Society of Heating, Refrigerating, and Air-Conditioning Engineers, Atlanta, GA, 1981.

Champagne, F.H., Sleicher, C.A., and Wehrman, O.H., Turbulence measurements with inclined hot-wires, I. and II. *J. Fluid Mech.,* 28, 153-182, 1967.

Fox, R.W. and McDonald, A.T., *Introduction to Fluid Mechanics*, 4th ed., John Wiley & Sons, New York, 1992.

Goldstein, R.J., Ed., *Fluid Mechanics Measurements*, 2nd ed., Taylor & Francis, Bristol, PA, 1996.

Grant, I., *Selected Papers on Particle Image Velocimetry,* SPIE Optical Engineering Press, Bellingham, WA, 1994.

King, L.V., On the convection of heat from small cylinders in a stream of fluid: determination of the convection constants of small platinum wires with applications to hot-wire anemometry, *Philos. Trans. R. Soc. (London), Ser. A,* 214, 373-432, 1914.

Krause, L.N. and Gettleman, C.C., Considerations Entering into the Selection of Probes for Pressure Measurement in Jet Engines, *ISA Proc.,* 7, 134, 1952.

Miller, R.W., *Flow Measurement Engineering Handbook*, 3rd ed., McGraw-Hill, New York, 1996.

Moffatt, R.J., Experimental Methods in the Thermosciences, Department of Mechanical Engineering, Stanford University, Stanford, CA 94305, Copies available from Moffat Thermosciences, Inc. 2333 Eastridge Ave, #2, Menlo Park, CA 94025.

Additional Reading

This section presents only an overview of flow measurement methods. The references below contain a wealth of further information. Baker[1] surveys the field and discusses precision, calibration, probe, and tracer methods, and likely developments. Spitzer[4] presents an excellent practical discussion of flow measurement.

1. Baker, R.C., *An Introductory Guide to Flow Measurement.* Institution of Mechanical Engineers, London, 1989.
2. Bruun, H.H., *Hot-Wire Anemometry: Principles and Signal Analysis*, Oxford University Press, New York, 1995.
3. ISO 7145, Determination of Flowrate of Fluids in Closed Conduits or Circular Cross Sections Method of Velocity Determination at One Point in the Cross Section, ISO UDC 532.57.082.25:532.542. International Standards Organization, Geneva, 1982.
4. Spitzer, R.W., Ed., *Flow Measurement: A Practical Guide for Measurement and Control*, Instrument Society of America, Research Triangle Park, NC, 1991.
5. White, F.M., *Fluid Mechanics*, 3rd ed., McGraw-Hill, New York, 1994.
6. Van Wazer, J.R., Lyons, J.W., Kim, K.Y., and Colwell, R.E., *Viscosity and Flow Measurement: A Laboratory Handbook of Rheology*, Interscience, New York, 1963.

4.24 Applications of Artificial Neural Networks and Genetic Algorithms in Thermal Engineering

Mihir Sen and K.T. Yang

Nomenclature

a, b, c, d, m, n	correlation constants	
A_a	air-side heat transfer area	m^2
A_w	water-side heat transfer area	m^2
c	constant in sigmoid function	
$c_{p,a}$	specific heat of air at constant pressure	J/kgK
c_w	specific heat of water	J/kgK
D	diameter of heat exchange tube	m
D_a	mass diffusivity of water vapor in air	m^2/s
D_h	hydraulic diameter	m
E	error for a given run	
f	function of mass flow rates	
f_s, f_t	functions of fin spacing	
G	generation number	
G_c	air-flow mass flux	kg/m^2s
h_a	air-side heat transfer coefficient	W/m^2K
h_w	water-side heat transfer coefficient	W/m^2K
$h_{m,a}$	air-side mass transfer coefficient	kg/m^2s
$h_{d,a}$	total-heat air-side mass transfer coefficient	kg/m^2s
H	height of heat exchanger 2	m
(i, j)	jth node in ith layer	
I	total number of layers	
j_m	latent heat j-factor	
j_s	sensible heat j-factor	
j_t	total heat j-factor	
J_i	number of nodes in ith layer	
k_a	thermal conductivity of air	W/mK
k_w	thermal conductivity of water	W/mK
$K1, K2, K3$	constants in control system	
L	length of heat exchanger 2	m
$\dot{m}_a$	mass flow rate of air	kg/s
$\dot{m}_w$	mass flow rate of water	kg/s
n	number in population	
n_b	number of bits	
N	number of rows	
N	number of experimental data sets	
N_c	number of columns	
N_n	total number of internal nodes	
N_p	number of testing data sets	
N_r	number of rows	
N_t	number of training data sets	
Nu_a	air-side Nusselt number	
Nu_w	water-side Nusselt number	
Pr_a	Prandtl number of air	
Pr_w	Prandtl number of water	
$\dot{Q}$	heat rate	W
Q^e	experimental heat rate	W
Q^p	predicted heat rate	W
Q^p_{ANN}	ANN prediction of heat rate	W
Q^p_{cor}	correlation prediction of heat rate	W
Q_t	total heat rate	W
R	average ratio of experimental to predicted heat rate	
R_a	ratio of air-side Nusselt numbers from correlations A and B	

R_r	value of R for run number r	
R_w	ratio of water-side Nusselt numbers from correlations A and B	
Re_a	air-side Reynolds number	
Re_w	water-side Reynolds number	
Re_D	air-side Reynolds number based on D	
Re_δ	air-side Reynolds number based on δ	
s	normalized fitness	
S_U	variance in thermal resistance	$(m^2 K/W)^2$
S_Q	variance in heat rate	W^2
Sc_a	Schmidt number of air	
t	fin thickness	m
t	time	s
Δt	time step	s
$t_{I,j}$	target output for last layer	
T_a^{in}	inlet air temperature	°C
$\overline{T}_a^{out}$	outlet air temperature	°C
T_w^{in}	inlet water temperature	°C
T_w^{out}	outlet water temperature	°C
$T_{a,db}^{in}$	inlet air dry-bulb temperature	°C
$T_{a,wb}^{in}$	air inlet wet-bulb temperature	°C
$\overline{T}_a^{out}$	reference air outlet temperature	°C
ΔT_a^{in}	disturbance in air inlet temperature	°C
U	overall heat transfer coefficient	W/m^2K
U^e	experimental overall heat transfer coefficient	W/m^2K
U^p	predicted overall heat transfer coefficient	W/m^2K
V_a	heat exchanger face velocity	m/s
V_w	mean water velocity	m/s
$w_{i-1,k}^{i,j}$	synaptic weight between nodes $(i - 1, k)$ and (i, j)	
$\Delta w_{i-1,k}^{i,j}$	correction in $w_{i-1,k}^{i,j}$	
W	width of heat exchanger 2	m
x_a	longitudinal tube spacing	m
x_b	transverse tube spacing	m
$x_{i,j}$	input of node (i, j)	
$y_{i,j}$	output of node (i, j)	
z	section coordinate in S_U surface	

Greek Symbols

α_a	thermal diffusivity of air	m^2/s
α_w	thermal diffusivity of water	m^2/s
δ	fin spacing	m
$\delta_{i,j}$	error in node (i, j)	
ε	fin effectiveness	
$\theta_{i,j}$	bias for node (i, j)	
$\Delta\theta_{i,j}$	correction in $\theta_{i,j}$	
λ	training rate	
μ_a	viscosity of air	kg/ms
ν_a	kinematic viscosity of air	m^2/s
ν_w	kinematic viscosity of water	m^2/s
ξ	variable in sigmoid function	
σ	standard deviation of relative error in predictions	
σ_f	ratio of free-flow area of air to frontal area	
$\phi_{i,j}$	activation function	

Introduction

The last 2 decades have seen enormous strides in technologies that can be loosely grouped together under the general term *artificial intelligence* (AI). Since, at the same time, relatively powerful desktop computers have become commonly available, applications of these methods are quickly spreading to many areas of science and engineering outside the field of computer science, especially for complex systems where more traditional methods have failed to be useful. The techniques of artificial neural networks (ANNs) and genetic algorithms (GAs) have been shown to be particularly useful in the thermal sciences. This section is written as an overview of the state-of-the-art in this area. It is fundamentally tutorial in nature so that the reader unfamiliar with these techniques can use it as a point of departure. We will describe the methodology and scope of these methods, and then concentrate on specific applications that illustrate their use. An extensive bibliography is also provided for readers who are interested in pursuing the subject further.

Thermal systems are natural or engineered systems involving temperature differences. The engineering discipline of thermal sciences, particularly heat transfer, specifically deals with the analysis, design, and control of such systems, and has a long history of development in response to the needs in a great variety of applications. Much of the advancement in the field has been the result of first-principle analyses for the simpler phenomena, supplemented by experimental correlations for the more complex ones for which the first principles are either not known or the problem is too complex for their use. Unfortunately, the existing methods of analyses are no longer adequate to deal with many current applications, even though analyses are still critically needed for a variety of reasons, such as, for instance, energy efficiency, environmental concern, optimum design and control, cost, and competitiveness in the market place. While the thermal sciences as a basic engineering discipline must and should continue to develop and advance our quantitative understanding of the fundamental phenomena involved, new methods of analysis must be found in the interim which can be used as a surrogate for such understanding. This is particularly true in the analysis and control of dynamic thermal systems. Obvious examples in this area are dynamic performance of complex heat exchangers and storage devices, network dynamics and hydronic control for heating, ventilating, and air-conditioning (HVAC) applications, dynamic thermal sensor development, thermal packaging of microelectronic devices, and thermal processing of conventional and advanced materials, just to mention a few.

AI and Its Scope

AI, though easy to recognize, is difficult to define: the word artificial is used to signify something man-made, but intelligence itself is a more elusive concept. In fact, one can debate whether machines can actually think. Working definitions would be to say that it is “the study of how to make computers do things which, at the moment, people do better” (Rich and Knight, 1991); the “study that encompasses computational techniques for tasks that apparently require intelligence when performed by humans” (Tanimoto, 1990); “the discipline that aims to understand the nature of human intelligence through the construction of computer programs that imitate intelligent behavior” (Bonnet, 1985); or “the study of mental faculties through the use of computational models” (Charniak and McDermott, 1985). Though every definition has its drawbacks, it appears that some elements of a knowledge base, search techniques, and logical reasoning are usually involved in AI and that the techniques usually attempt a simulation of a specific aspect of human or animal behavior. There are also related and overlapping disciplines that go by other names. *Soft computing,* for example, is a collection of techniques including fuzzy logic, neural networks, genetic algorithms, and probabilistic reasoning that are based on the idea that “intelligent systems should exploit, whenever possible, the tolerance for imprecision and uncertainty” (Bouchon-Meunier et al., 1995; Jang et al., 1997). *Knowledge engineering* and *intelligent information systems* are terms that are used when the emphasis is on the data base. *Virtual reality, artificial life, machine intelligence* and *data mining* are also related ideas that are being widely used. There is obviously some overlap between all these terminologies that we will not attempt to sort out here. The point is that these techniques are able to deal with complex problems that would be difficult to handle otherwise.

AI has a wide variety of applications in the natural and engineering sciences, among which are pattern recognition, decision making, system control, information processing, natural languages, symbolic mathematics, computer-aided instruction, speech recognition, vision, and robotics. These applications have spawned a large number of usable algorithms, such as optimization and search techniques, that can be used in other fields. Out of the broad set of techniques available, we will, for reasons of space, be concerned with two specific methods that have been shown to be useful in the thermal sciences: artificial neural networks and genetic algorithms. We will exclude any discussion of topics such as knowledge-based techniques, fuzzy logic, and other AI methodologies.

Why AI?

At the outset one must ask why AI techniques have any role in the thermal sciences. The answer to this lies at the core of the subject, i.e., in heat transfer. Heat transfer, like its sister discipline fluid mechanics, is based on mechanics and thermodynamics and has long been considered as one of the basic engineering sciences. At a fundamental level it is deterministic and well understood; the equations for conduction, convection, and radiation can be easily written down, and the material properties involved can also be measured. So it is proper to discuss why AI and allied techniques are needed at all.

A major application of heat transfer is in devices such as heat exchangers which are one of the basic components of thermal systems. A wide variety of heat exchangers exists in the market and for each it is necessary, for design and selection purposes, to know the heat transfer rate under prescribed operating conditions. For a given device exchanging heat between two fluids, the heat transfer rate depends on the flow rates and the inlet temperatures of each fluid. From the heat transfer rate, secondary quantities such as the fluid outlet temperatures can be determined. Though the governing equations can be written down, it is extremely difficult to solve them numerically under actual operating condition. Among the reasons for this are: complicated heat and fluid flow geometries, turbulence in the flow, existence of hydrodynamic and thermal entrance regions, nonuniform local heat transfer rates and fluid temperatures, secondary flows in the tube bends, vortices in the neighborhood of the tube-fin junctions, air-side flow development in fin passages, heat conduction along tube walls, natural convection within the tubes and between fins, and temperature dependence of fluid properties.

The computation must take into account the large ratio between the largest and the smallest length scales at which computations must be carried out. Accurate numerical solutions, thus, cannot be obtained with reasonable effort. As a consequence of this we are usually forced to abandon the first-principles approach in favor of empirical information. It must be emphasized that each one of the complicating factors can be handled alone and can even be said to be understood, but their occurring together is what makes the problem difficult.

We can, therefore, for purposes of this discussion, define a *complex* system to be one that is made up of a large number of simple subsystems that interact with each other. The whole is more than the sum of the parts so that, even given complete understanding of the subsystems and the laws governing their interactions, it is not a trivial matter to calculate the properties or behavior of the whole. On the other hand, if the interactions between the subsystems are weak or nonexistent, the system is *decomposable* or nearly decomposable, and is much easier to handle.

Empirical information in the form of correlations are commonly used to help in the analysis of thermal systems that are complex. These correlations usually have little physical basis and are proposed to have the simplest form that will give the best accuracy. The vagueness in this procedure gives rise to a variety of correlations for even the most straightforward calculation like, for example, that of turbulent heat transfer in a duct. Furthermore, as our ability to carry out complicated calculations for the purpose of design increases, the definition of the "simplest" form changes. At this point in time, it may be said that simplicity in the correlation may no longer be of concern, but accuracy in the predictions is. In this regard, artificial intelligence has much to offer. It is typically used in instances where the ultimate purpose is to detect and make use of patterns, rather than to uncover new physics.

Heat Exchangers

Since much of the following discussion will center on compact heat exchangers, we will explain this problem a little further. Empirical information about a given heat exchanger is experimentally determined by the manufacturer and is presented to the user, i.e., the thermal system designer, as the heat transfer rate as a function of the four control variables: the two flow rates and the two inlet temperatures. Being a four-variable function, it is difficult to represent completely, though partial information can be provided in the form of charts or graphs. In principle, this function depends on the geometry of the heat exchanger, the materials with which it is made, the surface conditions, the fluids used, etc. It completely characterizes the heat exchanger, and is the information that must be transferred in some form from the manufacturer to the design engineer. Since the raw experimental data would require a large amount of space for data storage, it is necessary to compress the information in some fashion for transfer from manufacturer to user, and from which it can later be accurately recovered. There are different ways in which this can be done.

In the simplest procedure the information is compressed as two heat transfer correlations, one for the inside and another for the outside, thus characterizing the heat exchanger performance by only these two quantities. For each side a characteristic temperature difference is assumed. However, in reality this is found to be inaccurate; the heat transfer coefficients are known to vary considerably with flow rates and fluid properties. Furthermore, without accurate tube wall temperature measurements, which are not easy to make, it is difficult to separate the experimentally determined overall thermal resistance into its internal and external components to find each transfer coefficient separately. It is possible to assume separate correlations for the heat transfer on either side and to use a regression analysis to determine the constants. Summing the resistances, however, produces a correlation that is not linearizable and, hence, could give more than one set of constants. This is discussed in further detail later.

We can resort to dimensionless analysis to reduce the number of variables involved in the correlation. The internal and external Nusselt (or Stanton) numbers should thus depend on the corresponding Reynolds and Prandtl numbers. A correlation using nondimensional variables has the advantage that it can be used for a wide range of length scales, flow rates, and fluids. This is the most common procedure. However, loss of accuracy is the price to be paid for this generality since not all the physical phenomena involved in the heat transfer process, like the previously cited examples show, scale exactly with the limited set of nondimensional variables used. Property variations are one example; there will be a significant difference in the heat rate predictions if the manufacturer and the user assume different values of the viscosity of the liquid (the viscosity of water, for example, at 40°C is 65% of what it is at 20°C). Letting a characteristic temperature difference scale a temperature profile assumes that the profile will be similar under all conditions, which is not the case. Moreover, the user of the information, the thermal system designer, is usually interested only in the heat transfer rate and not in intermediate variables like the internal and external heat transfer coefficients. It is preferable to sacrifice generality of application and correlate the heat rate directly with respect to the other variables.

The errors that are usually present in the predictions of a heat transfer correlation are due less to errors in measurement than to data compression represented by the correlation. Accuracies of a few percent would be achievable if the predictions had the same error level as the measurements. The two approaches that we will consider in detail to resolve this problem are outlined below.

1. The artificial neural network technique offers an alternative approach to the problem of information compression. It is a procedure that is usually used to predict the response of a complex physical system that cannot be easily modeled mathematically. The network is first trained by experimentally obtained input-output sets of data, after which it can be used for prediction. The manufacturer can train a network using the experimental data; the constants or parameters of the trained network can then be transferred to the user who can calculate the performance of the heat exchanger under any other flow rate or inlet temperature condition.
2. The search for the correlation that best fits the data can be carried out using genetic algorithms. Given a specific form of the correlation, the constants therein can be obtained through the use of a GA. Extending the idea farther, the best correlation function itself can be selected among a restricted class by genetic programming. The objective is not that the form of the correlation be simple, just that its predictions be accurate.

Artificial Neural Network

In this subsection we will discuss the ANN technique, which is generally considered to be a subclass of AI, and its application to the analysis of complex thermal systems. Applications of ANNs have been found in such diverse fields as philosophy, psychology, business and economics, sociology, science, as well as in engineering. The common denominator is the complexity of the field.

The technique is rooted in and inspired by the biological network of neurons in the human brain that learns from external experience, handles imprecise information, stores the essential characteristics of the external input, and generalizes previous experience (Eeckman, 1992). In the biological network of interconnecting neurons, each receives many input signals from other neurons and gives only one output signal which is sent to other neurons as part of their inputs. If the sum of the inputs to a given neuron exceeds a set threshold, normally determined by the electric potential of the receiver neuron which may be modified under different circumstances, the neuron fires and sends a signal to all the connected receiver neurons. If not, the signal is not transmitted. The firing decision represents the key to the learning and memory ability of the neural network.

The ANN attempts to mimic the biological neural network: the processing unit is the artificial neuron; it has synapses or interneuron connections characterized by synaptic weights; an operator performs a summation of the input signals weighted by the respective synapses; an activation function limits the permissible amplitude range of the output signal. It is also important to realize the essential difference between a biologic neural network and an ANN. Biological neurons function much slower than the computer calculations associated with an artificial neuron in an ANN. On the other hand, the delivery of information across the biological neural network is much faster. The biological one compensates for the relatively slow chemical reactions in a neuron by having an enormous number of interconnected neurons doing massively parallel processing, while the number of artificial neurons must necessarily be limited by the available hardware.

In this subsection we will briefly discuss the basic principles and characteristics of the multilayer ANN, along with the details of the computations made in the feedforward mode and the associated backpropagation algorithm which is used for training. Issues related to the actual implementation of the algorithm will also be noted and discussed. Specific examples on the performance of two different compact heat exchangers analyzed by the ANN approach will then be shown, followed by a discussion on how the technique can also be applied to the dynamic performance of heat exchangers as well as to their control in real thermal systems. Finally, the potential of applying similar ANN techniques to other thermal-system problems and their specific advantages will be delineated.

Methodology

The interested reader is referred to the text by Haykin (1994) for an account of the history of ANN and its mathematical background. Many different definitions of ANNs are possible; the one proposed by Schalkoff (1997) is that an ANN is a network composed of a number of artificial neurons. Each neuron has an input/output characteristic and implements a local computation or function. The output of any neuron is determined by this function, its interconnection with other neurons, and external inputs. The network usually develops an overall functionality through one or more forms of training; this is the learning process. Many different network structures and configurations have been proposed, along with their own methodologies of training (Warwick et al., 1992).

Feedforward Network

There are many different types of ANNs, but one of the most appropriate for engineering applications is the supervised, fully connected multilayer configuration (Zeng, 1998) in which learning is accomplished by comparing the output of the network with the data used for training. The *feedforward* or *multilayer perceptron* is the only configuration that will be described in some detail here. Figure 4.24.1 shows such an ANN consisting of a series of layers, each with a number of nodes. The first and last layers are for input and output, respectively, while the others are the hidden layers. The network is said to be fully connected when any node in a given layer is connected to all the nodes in the adjacent layers.

We introduce the following notation: (i, j) is the jth node in the ith layer. The line connecting a node (i, j) to another node in the next layer $i + 1$ represents the synapse between the two nodes. $x_{i,j}$ is the input of the node (i, j), $y_{i,j}$ is its output, $\theta_{i,j}$ is its bias, and $w_{i-1,k}^{i,j}$ is the synaptic weight between nodes

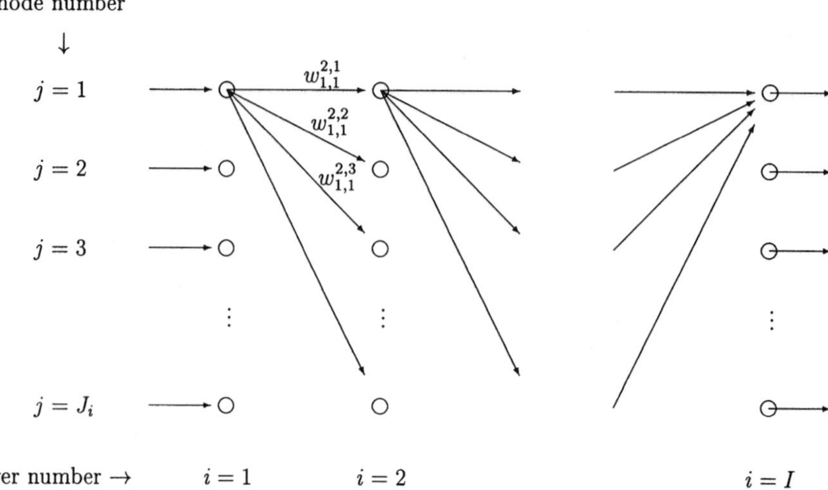

FIGURE 4.24.1 Schematic of a fully connected multilayer ANN.

$(i - 1, k)$ and (i, j). The total number of layers, including those for input and output, is I, and the number of nodes in the ith layer is J_i. The input information is propagated forward through the network; J_1 values enter the network and J_I leave. The flow of information through the layers is a function of the computational processing occurring at every internal node in the network. The relation between the output of node $(i - 1, k)$ in one layer and the input of node (i, j) in the following layer is

$$x_{i,j} = \theta_{i,j} + \sum_{k=1}^{J_{i-1}} w_{i-1,k}^{i,j} \, y_{i-1,k} \tag{4.24.1}$$

Thus, the input $x_{i,j}$ of node (i, j) consists of a sum of all the outputs from the previous nodes modified by the respective internode synaptic weights $w_{i-1,k}^{i,j}$ and a bias $\theta_{i,j}$. The weights are characteristic of the connection between the nodes and the bias of the node itself. The bias represents the propensity for the combined incoming input to trigger a response from the node and presents a degree of freedom which gives additional flexibility in the training process. Similarly, the synaptic weights are the weighting functions which determine the relative importance of the signals originated from the previous nodes.

The input and output of the node (i, j) are related by

$$y_{i,j} = \phi_{i,j}\left(x_{i,j}\right) \tag{4.24.2}$$

where $\phi_{i,j}(x)$, called the *activation* or *threshold* function, plays the role of the biological neuron determining whether it should fire or not on the basis of the input to that neuron. A schematic of the nodal operation is shown in Figure 4.24.2. It is obvious that the activation function plays a central role in the processing of information through the ANN. Keeping in mind the analogy with the biological neuron, when the input signal is small the neuron suppresses the signal altogether, resulting in a vanishing output, and when the input exceeds a certain threshold the neuron fires and sends a signal to all the neurons in the next layer. This behavior is determined by the activation function. Several appropriate activation functions have been studied (Haykin, 1994; Schalkoff, 1997). For instance, a simple step function can be used, but the presence of noncontinuous derivatives causes computing difficulties. The most popular one is the logistic sigmoid function

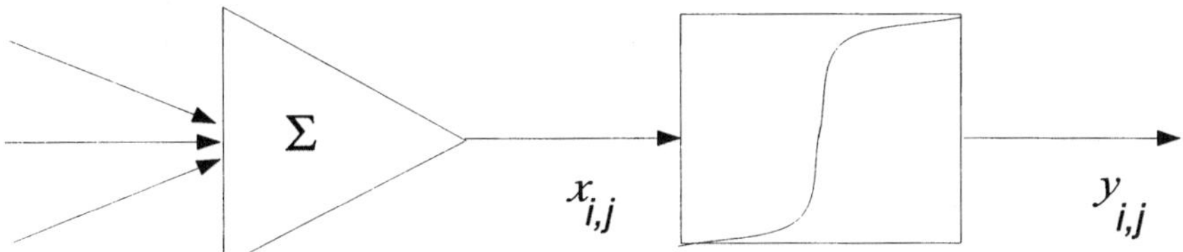

FIGURE 4.24.2 Nodal operation in an ANN.

$$\phi_{i,j}(\xi) = \frac{1}{1+e^{-\xi/c}} \quad (4.24.3)$$

for $i > 1$, where c determines the steepness of the function. For $i = 1$, $\phi_{i,j}(\xi) = \xi$ is used instead. The sigmoid function is an approximation to the step function, but with continuous derivatives. The nonlinear nature of the sigmoid function is particularly beneficial in the simulation of practical problems. For any input $x_{i,j}$, the output of a node $y_{i,j}$ always lies between 0 and 1. Thus, from a computational point of view, it is desirable to normalize all the input and output data with the largest and smallest values of each of the data sets.

Training

For a given network, the weights and biases must be adjusted for known input-output values through a process known as training. The backpropagation method is a widely used deterministic training algorithm for this type of ANN (Rumelhart et al., 1986). The central idea of this method is to minimize an error function by the method of steepest descent to add small changes in the direction of minimization. This algorithm may be found in many recent texts on ANN (for instance, Rzempoluck, 1998), and only a brief outline will be given here.

In usual complex thermal-system applications where no physical models are available, appropriate training data come from experiments. The first step in the training algorithm is to assign initial values to the synaptic weights and biases in the network based on the chosen ANN configuration. The values may be either positive or negative and, in general, are taken to be less than unity in absolute value. The second step is to initiate the feedforward of information starting from the input layer. In this manner, successive input and output of each node in each layer can all be computed. When finally $i = I$, the value of $y_{I,j}$ will be the output of the network. Training of the network consists of modifying the synaptic weights and biases until the output values differ little from the experimental data which are the targets. This is done by means of the backpropagation method. First an error $\delta_{I,j}$ is quantified by

$$\delta_{I,j} = \left(t_{I,j} - y_{I,j}\right) y_{I,j} \left(1 - y_{I,j}\right) \quad (4.24.4)$$

where $t_{I,j}$ is the target output for the j-node of the last layer. The above equation is simply a finite-difference approximation of the derivative of the sigmoid function. After calculating all the $\delta_{I,j}$, the computation then moves back to the layer $I - 1$. Since the target outputs for this layer do not exist, a surrogate error is used instead for this layer defined as

$$\delta_{I-1,k} = y_{I-1,k}\left(1 - y_{I-1,k}\right) \sum_{j=1}^{J_I} \delta_{I,j} w_{I-1,k}^{I,j} \quad (4.24.5)$$

A similar error $\delta_{i,j}$ is used for all the rest of the inner layers. These calculations are then continued layer by layer backward until layer 2. It is seen that the nodes of layer 1 have neither δ nor θ values assigned, since the input values are all known and invariant. After all the errors $\delta_{i,j}$ are known, the changes in the synaptic weights and biases can then be calculated by the generalized delta rule (Rumelhart et al., 1986):

$$\Delta w_{i-1,k}^{i,j} = \lambda \delta_{i,j}\, y_{i-1,k} \tag{4.24.6}$$

$$\Delta \theta_{i,j} = \lambda \delta_{i,j} \tag{4.24.7}$$

for $i < I$, from which all the new weights and biases can be determined. The quantity λ is known as the learning rate that is used to scale down the degree of change made to the nodes and connections. The larger the training rate, the faster the network will learn, but the chances of the ANN to reach the desired outcome may become smaller as a result of possible oscillating error behaviors. Small training rates would normally imply the need for longer training to achieve the same accuracy. Its value, usually around 0.4, is determined by numerical experimentation for any given problem.

A cycle of training consists of computing a new set of synaptic weights and biases successively for all the experimental runs in the training data. The calculations are then repeated over many cycles while recording an error quantity E for a given run within each cycle, where

$$E = \frac{1}{2}\sum_{j=1}^{J_I}\left(t_{I,j} - y_{I,j}\right)^2 \tag{4.24.8}$$

The output error of the ANN at the end of each cycle can be based on either a maximum or averaged value for a given cycle. Note that the weights and biases are continuously updated throughout the training runs and cycles. The training is terminated when the error of the last cycle, barring the existence of local minima, falls below a prescribed threshold. The final set of weights and biases can then be used for prediction purposes, and the corresponding ANN becomes a model of the input-output relation of the thermal-system problem.

Implementation Issues

In the implementation of a supervised, fully connected multilayered ANN, the user is faced with several uncertain choices which include the number of hidden layers, the number of nodes in each layer, the initial assignment of weights and biases, the training rate, the minimum number of training data sets and runs, the learning rate, and the range within which the input-output data are normalized. Such choices are by no means trivial, and are rather important in achieving good ANN results. Since there is no general sound theoretical basis for specific choices, past experience and numerical experimentation are still the best guides, despite the fact that much research is now going on to provide a rational basis (Zeng, 1998).

On the issue of number of hidden layers, there is a sufficient, but certainly not necessary, theoretical basis known as the Kolmogorov's mapping neural network existence theorem as presented by Hecht-Nielsen (1987), which essentially stipulates that only one hidden layer of artificial neurons is sufficient to model the input-output relations as long as the hidden layer has $2J_1 + 1$ nodes. Since in realistic problems involving a large set of input parameters the nodes in the hidden layer would be excessive to satisfy this requirement, the general practice is to use two hidden layers as a starting point, and then to add more layers as the need arises, while keeping a reasonable number of nodes in each layer (Flood and Kartam, 1994).

A slightly better situation is in the choice of the number of nodes in each layer and in the entire network. Increasing the number of internal nodes provides a greater capacity to fit the training data. In practice, however, too many nodes suffer the same fate as the polynomial curve-fitting routine by collocation at specific data points in which the interpolations between data points may lead to large errors. In addition, a large number of internal nodes slows down the ANN both in training and in prediction. One interesting suggestion given by Rogers (1994) and Jenkins (1995) is that

$$N_t = 1 + N_n \frac{J_1 + J_I + 1}{J_I} \tag{4.24.9}$$

where N_t is the number of training data sets, and N_n is the total number of internal nodes in the network. If N_t, J_1, and J_I are known in a given problem, the above equation determines the suggested minimum number of internal nodes. Also, if N_n, J_1 and J_I are known, it gives the minimum value of N_t. The number of data sets used should be larger than that given by this equation to ensure the adequate determination of the weights and biases in the training process. Other suggested procedures for choosing the parameters of the network include the one proposed by Karmin (1990) by first training a relatively large network that is then reduced in size by removing nodes which do not significantly affect the results, and the so-called Radial-Gaussian system which adds hidden neurons to the network in an automatic sequential and systematic way during the training process (Gagarin et al., 1994). Also available is the use of evolutionary programming approaches to optimize ANN configurations (Angeline et al., 1994). Some authors (see, for example, Thibault and Grandjean, 1991) present studies of the effect of varying these parameters.

The issue of assigning the initial synaptic weights and biases is less certain. Despite the fact that better initial guesses would require less training efforts, or even less training data, such initial guesses are generally unavailable in applying the ANN analysis to a new problem. The initial assignment then normally comes from a random number generator of bounded numbers. Unfortunately, this does not guarantee that the training will converge to the final weights and biases for which the error is a global minimum. Also, the ANN may take a large number of training cycles to reach the desired level of error. Wessels and Barnard (1992), Drago and Ridella (1992), and Lehtokangas et al. (1995) suggested other methods for determining the initial assignment so that the network converges faster and avoids local minima. On the other hand, when the ANN needs upgrading by additional or new experimental data sets, the initial weights and biases are simply the existing ones.

During the training process, the weights and biases continuously change as training proceeds in accordance with Equations (4.24.6) and (4.24.7), which are the simplest correction formulas to use. Other possibilities, however, are also available (Kamarthi et al., 1992). The choice of the training rate λ is largely by trials. It should be selected to be as large as possible, but not too large to lead to nonconvergent oscillatory error behaviors. Finally, since the sigmoid function has the asymptotic limits of [0,1] and may thus cause computational problems in these limits, it is desirable to normalize all physical variables into a more restricted range such as [0.15, 0.85]. The choice is somewhat arbitrary. However, pushing the limits closer to [0,1] does commonly produce more accurate training results at the expense of larger computational efforts.

Application to Compact Heat Exchangers

In this subsection the ANN analysis will be applied to the prediction of the performance of two different types of compact heat exchangers, one being a single-row, fin-tube heat exchanger (called heat exchanger 1), and the other a much more complicated multirow, multicolumn fin-tube heat exchanger (heat exchanger 2). In both cases, air is either heated or cooled on the fin side by water flowing inside the serpentine tubes. Except at the tube ends, the air is in a cross-flow configuration. Details of the analyses are available in the literature (Diaz et al., 1996, 1998, 1999; Pacheco-Vega et al., 1999). For either heat exchanger, the normal practice is to predict the heat transfer rates by using separate dimensionless correlations for the air- and water-side coefficients of heat transfer based on the experimental data and definitions of specific temperature differences.

Heat Exchanger 1

The simpler single-row heat exchanger, a typical example being shown in Figure 4.24.3, is treated first. It is a nominal 18 in. × 24 in. plate-fin-tube type manufactured by the Trane Company, with a single circuit of 12 tubes connected by bends. The experimental data were obtained in a variable-speed open wind-tunnel facility shown schematically in Figure 4.24.4. A PID-controlled electrical resistance heater

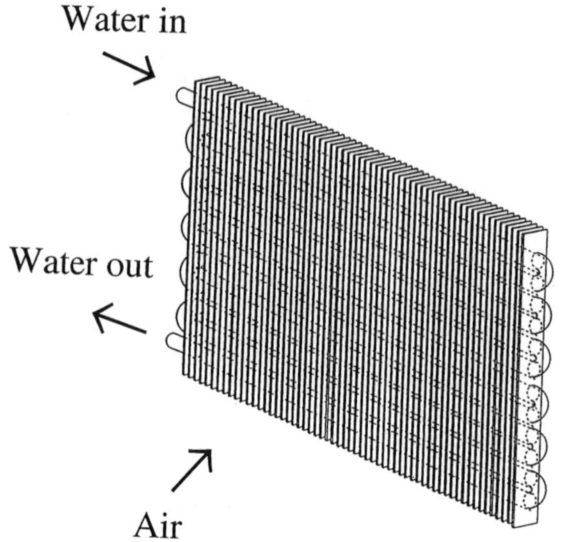

FIGURE 4.24.3 Schematic of compact heat exchanger 1.

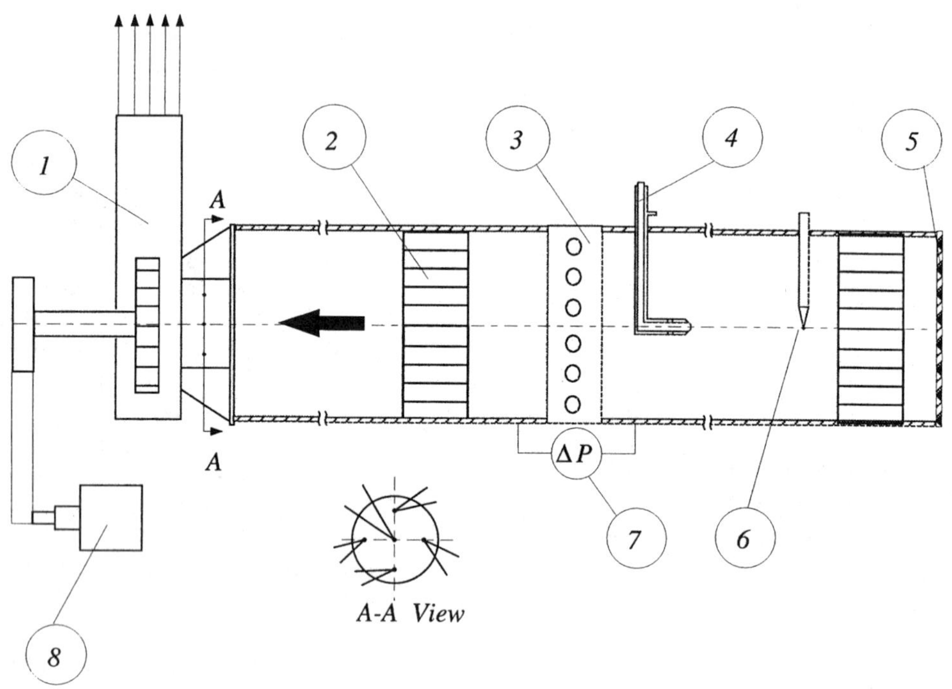

FIGURE 4.24.4 Schematic arrangement of test facility: (1) centrifugal fan, (2) flow straightener, (3) heat exchanger, (4) Pitot-static tube, (5) screen, (6) thermocouple, (7) differential pressure gauge, (8) motor. View A-A shows the placement of five thermocouples.

provides hot water and its flow rate is measured by a turbine flow meter. All temperatures are measured by Type T thermocouples. Additional experimental details can be found in the thesis by Zhao (1995). A total of $N = 259$ test runs were made, of which only the data for $N_t = 197$ runs were used for training, while the rest were used for testing the predictions. It is advisable to include the extreme cases in the training data sets so that the predictions will be within the same range. For the ANN analysis, there are four input nodes, each corresponding to the normalized quantities: air flow rate $\dot{m}_a$, water flow rate $\dot{m}_w$, inlet air temperature T_a^{in}, and inlet water temperature T_w^{in}. There is a single output node for the normalized heat transfer rate $\dot{Q}$. Normalization of the variables was done by limiting them within the range [0.15, 0.85]. Coefficients of heat transfer have not been used, since that would imply making some assumptions about the similarity of the temperature fields.

Fourteen different ANN configurations were studied, as shown in Table 4.24.1. As an example, the training results of the 4-5-2-1-1 configuration, with three hidden layers with 5, 2, and 1 nodes, respectively, are considered in detail. The input and output layers have 4 nodes and 1 node, respectively, corresponding to the four input variables and a single output. Training was carried out to 200,000 cycles to show how the errors change along the way. The average and maximum values of the errors for all the runs can be found, where the error for each run is defined in Equation (4.24.8). These errors are shown in Figure 4.24.5. It is seen that the maximum error asymptotes at about 150,000 cycles, while the corresponding level of the average error is reached at about 100,000. In either case, the error levels are sufficiently small.

TABLE 4.24.1 Comparison of Heat Transfer Rates Predicted by Different ANN Configurations for Heat Exchanger 1

Configuration	R	σ
4-1-1	1.02373	0.266
4-2-1	0.98732	0.084
4-5-1	0.99796	0.018
4-1-1-1	1.00065	0.265
4-2-1-1	0.96579	0.089
4-5-1-1	1.00075	0.035
4-5-2-1	1.00400	0.018
4-5-5-1	1.00288	0.015
4-1-1-1-1	0.95743	0.258
4-5-1-1-1	0.99481	0.032
4-5-2-1-1	1.00212	0.018
4-5-5-1-1	1.00214	0.016
4-5-5-2-1	1.00397	0.019
4-5-5-5-1	1.00147	0.022

After training, the ANNs were used to predict the $N_p = 62$ testing data which were not used in the training process; the mean and standard deviations of the error for each configuration, R and σ, respectively, are shown in Table 4.24.1. R and σ are defined by

$$R = \frac{1}{N_p} \sum_{r=1}^{N_p} R_r \tag{4.24.10}$$

$$\sigma = \sqrt{\sum_{r=1}^{N_p} \frac{(R_r - R)^2}{N_p}} \tag{4.24.11}$$

where R_r is the ratio $\dot{Q}^e/\dot{Q}^p_{ANN}$ for run number r, $\dot{Q}^e$ is the experimental heat-transfer rate, and $\dot{Q}^p_{ANN}$ is the corresponding prediction of the ANN. R is an indication of the average accuracy of the prediction, while σ is that of the scatter, both quantities being important for an assessment of the relative success of the ANN analysis. The network configuration with R closest to unity is 4-1-1-1, while 4-5-5-1 is the one with the smallest σ. If both factors are taken into account, it seems that 4-5-1-1 would be the best, even

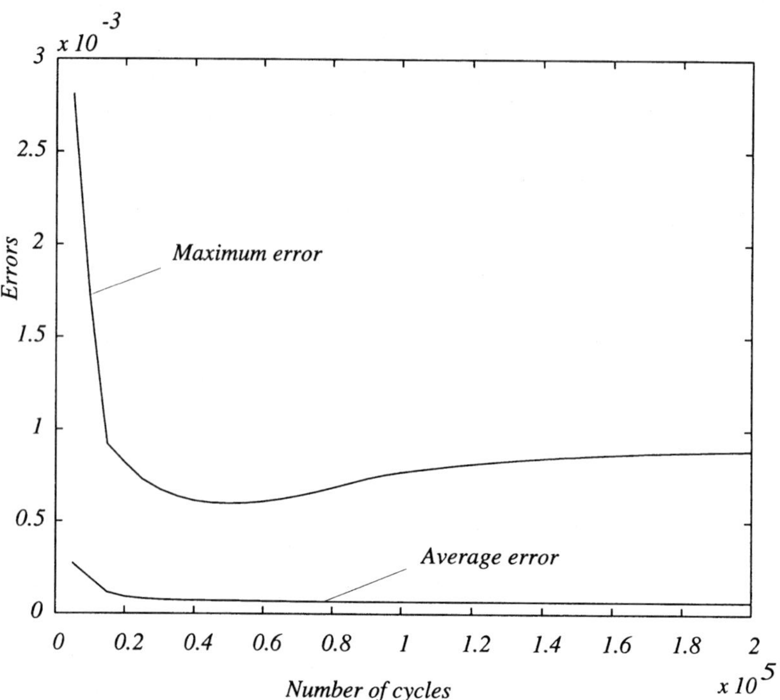

FIGURE 4.24.5 Training error results for configuration 4-5-2-1-1 ANN.

though the exact criterion is of the user's choice. It is also of interest to note that adding more hidden layers may not improve the ANN results. Comparisons of the values of R_r for all test cases are shown in Figure 4.24.6 for two configurations. It is seen, that although the 4-5-1-1 configuration is the second

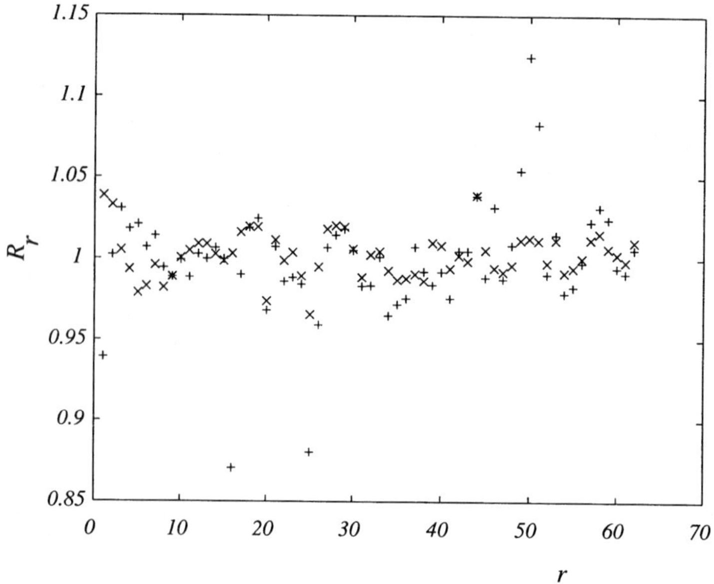

FIGURE 4.24.6 Ratio of heat transfer rates R_r for all testing runs (× 4-5-5-1; + 4-5-1-1) for heat exchanger 1.

best in *R,* there are still several points at which the predictions differ from the experiments by more than 14%. The 4-5-5-1 network, on the other hand, has errors confined to 3.7%.

The effect of the normalization range for the physical variables was also studied. Additional trainings were carried out for the 4-5-5-1 network using the different normalization range of [0.05, 0.95]. For 100,000 training cycles, the results show that $R = 1.00063$ and $\sigma = 0.016$. Thus, in this case, more accurate averaged results can be obtained with the range closer to [0,1].

We also compare the heat-transfer rates obtained by the ANN analysis based on the 4-5-5-1 configuration, $\dot{Q}^p_{ANN}$, and those determined from the dimensionless correlations of the coefficients of heat transfer, $\dot{Q}^p_{cor}$. For the experimental data used, the least-square correlation equations have been given by Zhao (1995) and Zhao et al. (1995) to be

$$\varepsilon Nu_a = 0.1368 Re_a^{0.585} Pr_a^{1/3} \tag{4.24.12}$$

$$Nu_w = 0.01854 Re_w^{0.752} Pr_w^{0.3} \tag{4.24.13}$$

applicable for $200 < Re_a < 700$ and $800 < Re_w < 4.5 \times 10^4$, where ε is the fin effectiveness. The Reynolds, Nusselt, and Prandtl numbers are defined as follows,

$$Re_a = \frac{V_a \delta}{\nu_a}; \quad Nu_a = \frac{h_a \delta}{k_a}; \quad \Pr = \frac{\nu_a}{\alpha_a} \tag{4.24.14}$$

$$Re_w = \frac{V_w D}{\nu_w}; \quad Nu_w = \frac{h_w D}{k_w}; \quad Pr_w = \frac{\nu_w}{\alpha_w} \tag{4.24.15}$$

where the subscripts *a* and *w* refer to the air- and water-side, respectively, *V* is the average flow velocity, δ is the fin spacing, *D* is the tube inside diameter, and ν and *k* are the kinematic viscosity and thermal conductivity of the fluids, respectively. The correlations are based on the maximum temperature differences between the two fluids. The results are shown in Figure 4.24.7, where the superscript *e* is used

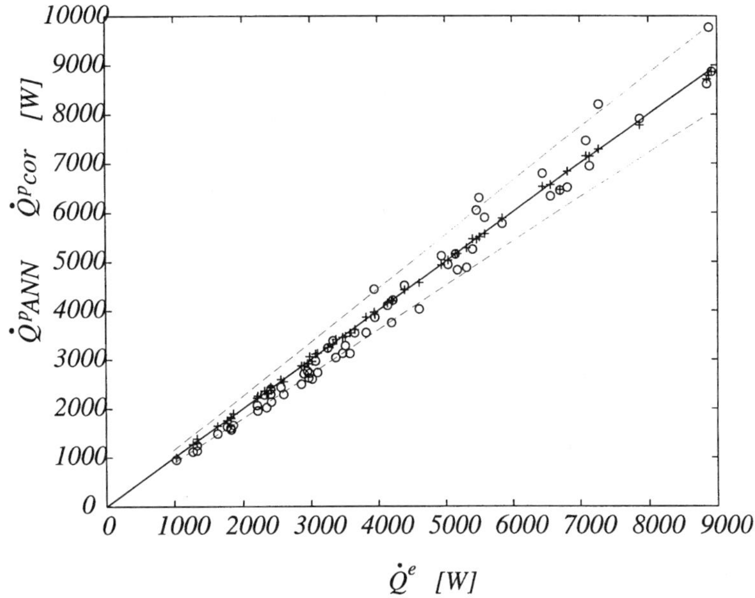

FIGURE 4.24.7 Comparison of 4-5-5-1 ANN (+) and correlation (o) predictions for heat exchanger 1.

for the experimental values and p for the predicted. For most of the data the ANN error is within 0.7%, while the predictions of the correlation are of the order of ±10%. The superiority of the ANN is evident.

These results suggest that the ANNs have the ability of recognizing all the consistent patterns in the training data including the relevant physics as well as random and biased measurement errors. It can perhaps be said that it catches the underlying physics much better than the correlations do, since the error level is consistent with the uncertainty in the experimental data (Zhao, 1995). However, the ANN does not know and does not have to know what the physics is. It completely bypasses simplifying assumptions such as the use of coefficients of heat transfer. On the other hand, any unintended and biased errors in the training data set are also picked up by the ANN. The trained ANN, therefore, is not better than the training data, but not worse either.

Heat Exchanger 2

A second demonstration of the ANN analysis is on a multirow multicolumn compact heat exchanger shown schematically in Figure 4.24.8. Detailed heat and mass transfer measurements for this heat exchanger with different fin spacings and a wide range of opening conditions were carried out and the data reported by McQuiston (1978a). The air-side and water-side thermal resistances were decoupled by using a high Reynolds number turbulent flow on the water side such that the water-side coefficients of heat transfer were much higher than those on the air side; the water-side resistance could be estimated and subtracted from the overall thermal resistance to yield the resistance of the air side only. Details of the experimental facility and instrumentation can be found in McQuiston (1978a). The type of heat exchangers tested were for air-conditioning applications and, therefore, under certain operating conditions condensation would occur. The heat transfer data were analyzed depending on whether the air-side surface was dry, covered with droplets of water in dropwise condensation, or with water films in filmwise condensation. All physical dimensions are shown in Figure 4.24.8; the only geometrical quantity that

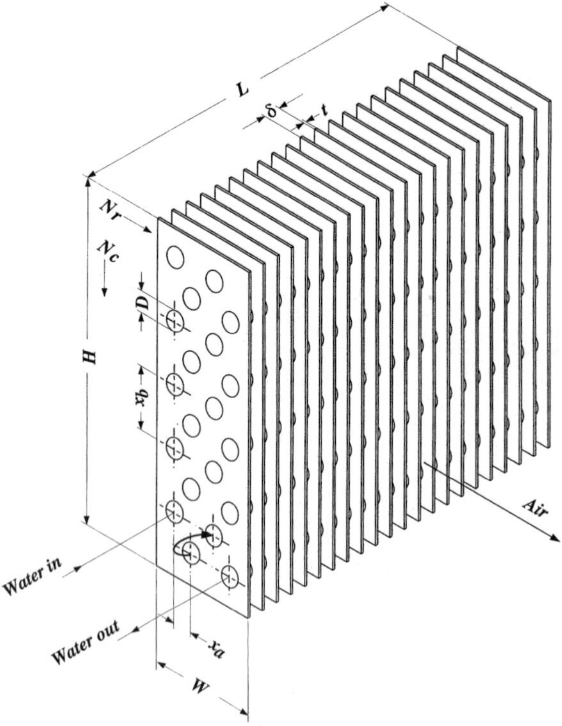

FIGURE 4.24.8 Schematic of compact heat exchanger 2. Dimensions are L = 12 in., D = 0.392 in., t = 0.006 in., x_a = 0.866 in., x_b = 1 in., N_r = 4, N_c = 5, 0.071 in. < δ < 0.0250 in.

was allowed to vary was the fin spacing, as it is the most important physical variable in such heat exchangers. On the other hand, the experimental data covered a wide range of operating conditions on the air side. McQuiston (1978b) correlated his data in terms of three air-side heat transfer coefficients, nondimensionalized as Colburn *j*-factors. These are the sensible heat j_s, the latent heat j_m, and the total heat j_t defined by

$$j_s = \frac{h_a}{G_c c_{p,a}} Pr_a^{2/3} \tag{4.24.16}$$

$$j_m = \frac{h_{m,a}}{G_c} Sc_a^{2/3} \tag{4.24.17}$$

$$j_t = \frac{h_{d,a}}{G_c} Sc_a^{2/3} \tag{4.24.18}$$

where h_a is the heat transfer coefficient, and $h_{m,a}$ and $h_{d,a}$ are the mass transfer coefficients associated with the *j*-factors, G_c is the air-flow mass flux, $c_{p,a}$ is the specific heat of air, and Sc_a is the Schmidt number of air defined by $Sc_a = \nu_a/D_a$, where ν_a is the kinematic viscosity of air and D_a is the mass diffusivity of water vapor in air. The *j*-factors were correlated as functions of Reynolds numbers Re_D and Re_δ and the geometrical factors $(4/\pi)(x_a/D_h)(x_b/D)\sigma_f$ and $\delta/(\delta - t)$, where

$$Re_D = \frac{G_c D}{\mu_a} \tag{4.24.19}$$

$$Re_\delta = \frac{G_c \delta}{\mu_a} \tag{4.24.20}$$

D is the tube outside diameter, δ is the fin spacing, and μ_a is the dynamic viscosity of air. Also x_a and x_b are the two tube-center distances shown in Figure 4.24.8, D_h is the hydraulic diameter, t is the fin thickness, and σ_f is the ratio of the free-flow area of air to the frontal area.

The McQuiston (1978b) correlations are

$$j_s = 0.0014 + 0.2618 Re_D^{-0.4} \left(\frac{4}{\pi} \frac{x_a}{D_h} \frac{x_b}{D} \sigma_f \right)^{-0.15} f_s(\delta) \tag{4.24.21}$$

$$j_t = 0.0014 + 0.2618 Re_D^{-0.4} \left(\frac{4}{\pi} \frac{x_a}{D_h} \frac{x_b}{D} \sigma_f \right)^{-0.15} f_t(\delta) \tag{4.24.22}$$

where

$$f_s(\delta) = \begin{cases} \left(0.9 + 4.3 \times 10^{-5} Re_\delta^{1.25}\right)\left(\dfrac{\delta}{\delta - t}\right)^{-1} & \text{dropwise condensation} \\ 1.0 & \text{dry surface} \\ 0.84 + 4.0 \times 10^{-5} Re_\delta^{1.25} & \text{film condensation} \end{cases} \tag{4.24.23}$$

and

$$f_t(\delta) = \begin{cases} \left(0.80 + 4.0 \times 10^{-5} Re_\delta^{1.25}\right)\left(\dfrac{\delta}{\delta - t}\right)^4 & \text{dropwise condensation} \\ 1.0 & \text{dry surface} \\ \left(0.95 + 4.0 \times 10^{-5} Re_\delta^{1.25}\right)\left(\dfrac{\delta}{\delta - t}\right)^2 & \text{film condensation} \end{cases} \qquad (4.24.24)$$

The ranges of validity are number of rows $N_r = 4$, $100 < Re_D < 4000$, 3/8 in. $< D <$ 5/8 in., 1 in. $< x_b <$ 2 in., 4 fins/in. $< 1/\delta <$ 14 fins/in., and 200 ft/min $< V <$ 800 ft/min. Gray and Webb (1986) used these and other data to find

$$j_s = 0.14 Re_D^{-0.328} \left(\frac{x_b}{x_a}\right)^{-0.502} \left(\frac{\delta - t}{D}\right)^{0.0312} \qquad (4.24.25)$$

for the dry-surface conditions in the ranges $500 < Re_D < 24700$, $1.97 < x_b/D < 2.55$, $1.97 < x_a/D < 2.58$, $0.08 < (\delta - t)/D < 0.64$, and $1 < N_r < 8$. Either correlation, i.e., Equations (4.24.21) and (4.24.22) or Equation (4.24.25), can be used to determine the total heat transfer rates under the dry-surface conditions, while those in Equations (4.24.21) and (4.24.22) are for the sensible and total heat-transfer rates under wet-surface conditions.

Several ANN configurations were tried, and the best results were given by the fully connected 5-5-3-3 ANN configuration shown in Figure 4.24.9. The five input nodes correspond to the variables: inlet air dry-bulb temperature $T_{a,db}^{in}$, inlet air wet-bulb temperature $T_{a,wb}^{in}$, inlet water temperature T_w^{in}, the air-flow Reynolds number Re_D, and the fin spacing δ. The three output nodes correspond to j_s, j_t, and the total heat transfer rate $\dot{Q}_t$. The j-factors were not necessary for the heat rate predictions and were computed merely for the purpose of comparison. All variables were normalized in the same way as in the previous example. The two j-factor predictions can be directly compared with the computed data given by McQuiston (1978b). A total of 327 experimental runs were reported by McQuiston (1978a), of which 117 runs were associated with observed dropwise condensation, 91 were related to dry-surface conditions, and the remaining 119 were with filmwise condensation. The data were analyzed in two different ways. First, each of the three surface conditions was analyzed separately to train three ANNs using the respective data sets. Second, the entire 327-run data sets were combined to train an ANN. The purpose here was

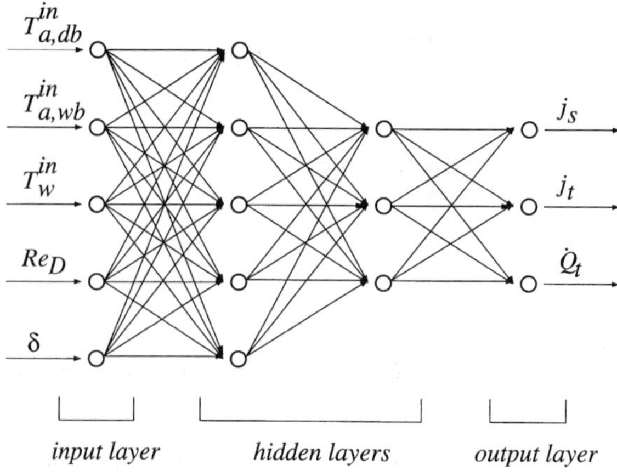

FIGURE 4.24.9 Schematic of a 5-5-3-3 ANN for heat exchanger 2.

to determine whether the network based on training data separated by some physical condition would perform better than another trained with the combined data set.

The errors obtained by these four ANNs are shown in Table 4.24.2. The errors indicated are root-mean-square values of the percentage difference between the predicted and experimental data for 800,000 training cycles. For dropwise condensation, the error is of the order of 3.3% in j_s, 3.9% in j_t, and only 1.4% in the total heat transfer rates. From the discussions in the previous example, this result is somewhat expected, since both j_s and j_t are not expected to include all the essential physics, while in $\dot{Q}_t$, being a physical quantity itself, the ANN better recognizes the true relationship between $\dot{Q}_t$ and the physical input data. It is seen that this is also true for the dry surface and the surface with film condensation. It is also of interest to note that the ANN gives much better prediction for dry surfaces than for wet. This is perhaps also expected, since the physical phenomena with condensation are more complex. On the other hand, when the ANN is trained by the entire set of experimental data without separating them according to surface conditions, the errors all increase. This indicates that the ANN has more difficulty in differentiating between the different physics involved. However, it must be added that an error of 5% or less can still be counted as a relatively good prediction. It is even more remarkable to note that the predictions for $\dot{Q}_t$, which is what the user of the information is ultimately interested in, have errors only of the order of 2.7%.

Table 4.24.2 Percentage Errors in j_s, j_t, and $\dot{Q}_t$ for Heat Exchanger 2

Surface	j_s	j_t	$\dot{Q}_t$
Dropwise condensation	3.32	3.87	1.446
Dry surface	1.002	1.002	0.928
Filmwise condensation	2.58	3.15	1.960
Combined	4.58	5.05	2.69

Table 4.24.3 shows comparisons between the predictions of ANNs and the dimensionless correlations given above for j_s. Table 4.24.4 is a similar comparison for j_t. It is seen that despite the fact that the Gray and Webb (1986) correlation indeed performs better than those of McQuiston (1978b) under the dry-surface conditions, all the correlation predictions are much worse than those from the ANN. Actual comparisons of the j_t factors under both wet-surface conditions between the McQuiston correlations and

TABLE 4.24.3 Percentage Errors in j_s Predictions for Heat Exchanger 2

Surface	ANN	McQuiston	Gray and Webb
Dropwise condensation	3.32	8.50	—
Dry surface	1.002	14.57	11.62
Filmwise condensation	2.58	9.01	—
Combined	4.58	—	—

TABLE 4.24.4 Percentage Errors in j_t Predictions for Heat Exchanger 2

Surface	ANN	McQuiston	Gray and Webb
Dropwise condensation	3.87	7.55	—
Dry surface	1.002	14.57	11.62
Filmwise condensation	3.15	14.98	—
Combined	5.05	—	—

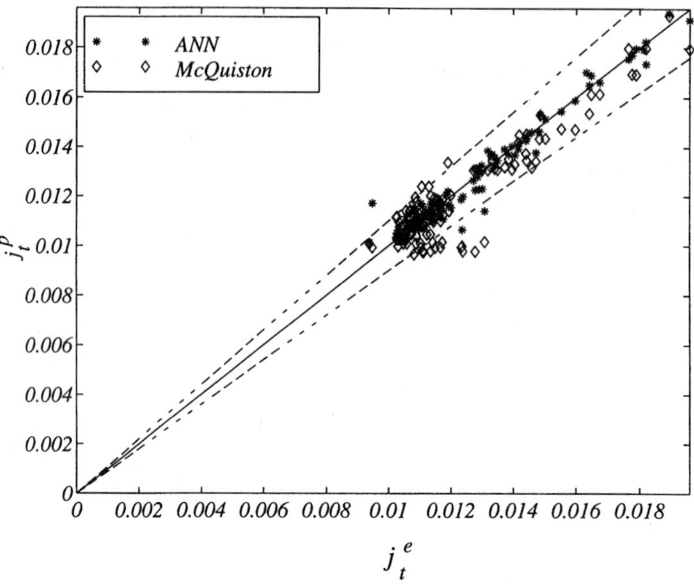

FIGURE 4.24.10 Experimental vs. predicted j_t for heat exchanger 2 with dropwise condensation.

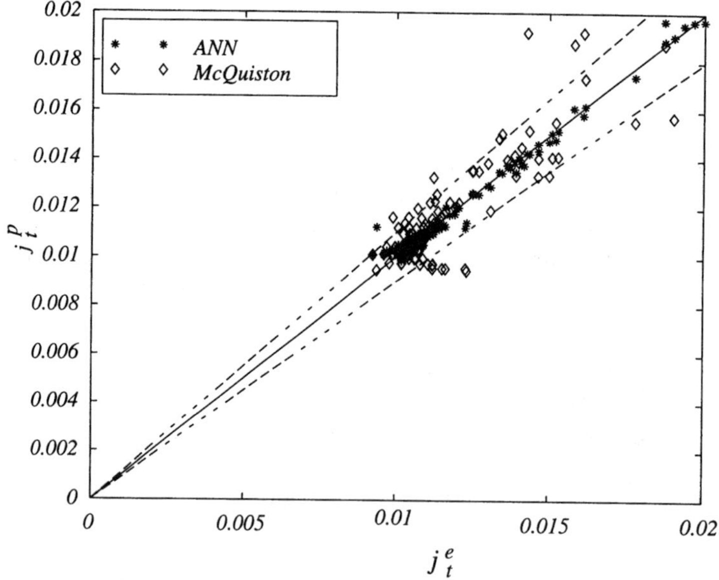

FIGURE 4.24.11 Experimental vs. predicted j_t for heat exchanger 2 with film condensation.

the ANN results are shown in Figures 4.24.10 and 4.24.11, where the superscripts p and e on the j-factors refer, respectively, to ANN predictions and the experimental data. This example shows that the ANN is able to predict the total heat transfer well, even for the fairly complex case of a heat exchanger in which there may be condensation phase change on the air side.

Thermal System Dynamics and Control

In many practical thermal systems, the operations are time-dependent. A dynamic situation occurs either when a disturbance is introduced into the system or when the conditions change with time; in either case the thermal system will respond in a time-dependent manner. In many instances, there is a need to quantitatively know what these responses may be. Ideally, this calls for dynamic system models which relate the outputs to the inputs. Unfortunately, again because of the complexity of practical thermal systems, only very few attempts at developing such models have been done and only for simple cases (Cohen and Johnson, 1956; Mozley, 1956; Thal-Larson, 1960; Gartner and Harrison, 1965; Yamashita et al., 1978; Schmidt and Wilmott, 1981; Spiga and Spiga, 1992). This is understandable since modeling is difficult even in the static cases. On the other hand, in many applications such as HVAC systems and industrial thermal processes, the control strategy is often based on the dynamic system or process models to achieve robustness and optimality. The usual PID controllers are simply not sufficient, as they often require manual supervision. Another difficulty with the dynamic thermal-system models, even in those cases where they do exist, is that they are incapable of accommodating the degradation of system components over time and, hence, they may become less adequate as time goes on. One good example is the effect of fouling in heat exchanger pipes. Interestingly enough, ANNs can offer solutions to all these problems. A broad discussion on ANN-based dynamic system modeling techniques has been recently given by Ayoubi (1997).

Dynamic Model

The central issue in developing ANN-based dynamic models for thermal systems is the addition of time as a variable for both training and predictions. An efficient method of training (Diaz et al., 1998) is to provide the values of the variables at time t as inputs to the network and the values of the same variables at time $t + \Delta t$ as the outputs. The time step Δt does not need to stay constant over time, though it should be chosen so that incremental values of all the variables remain small. The trainings are successively carried out over as many Δts as needed to cover the entire time horizon. Once trained, the ANN can predict the values of the input variables for an instant Δt later.

For a demonstration of this method, the heat exchanger 1 situation discussed above is again considered. The ANN to be trained had a 3-5-2 configuration. Dynamic data were obtained by varying the water inlet temperatures by changing the heater settings while keeping the other variables, the mass-flow rates of both fluids and the air inlet temperature, all constant. The input to the ANN were $T_w^{in}(t)$, $T_a^{out}(t)$, $T_w^{out}(t)$ and the output were $T_a^{out}(\mathrm{t} + \Delta t)$ and $T_w^{out}(t + \Delta t)$. Training data were obtained from experiments in which the water inlet temperature was varied in small increments of 5.56°C from 32.2°C up to 65.6°C. For testing the predictions of the ANN, three additional sets of experiments were performed.

1. The system was first brought up to a temperature of 60°C and the heater was then set at 37.8°C. The resulting sudden decrease of the air and water outlet temperatures is shown in Figure 4.24.12. The ANN predictions are excellent. The slight oscillations in the measured water outlet temperatures were likely due to local turbulence in the flow.
2. In the second set of experiments the water inlet temperature was ramped up manually. The corresponding results are in Figure 4.24.13, which shows the same excellent prediction by the ANN.
3. The final set of experiments was designed to observe if the ANN could predict the system behavior when an input variable different from the one in the training data is used. The changes in the air and water outlet temperatures were measured when the mass flow rate of air was first increased to a value greater than the one in the training data, and then decreased to a lower value. The results are shown in Figure 4.24.14 where the decrease is seen around t = 10 s, and the increase at t = 120 s. The predictions, as expected, are not very good. It is remarkable, though, that the ANN approach still predicts the correct overall trends and that the errors are small.

Dynamic Control

In the usual dynamic control systems for industrial processing and HVAC systems, the common practice is to use a feedback control involving a system model, sometimes known as the plant model, in conjunction with a standard PID controller. Such a control system, unfortunately, requires constant attention and manual supervision to achieve optimality. As previously discussed, there are many advantages

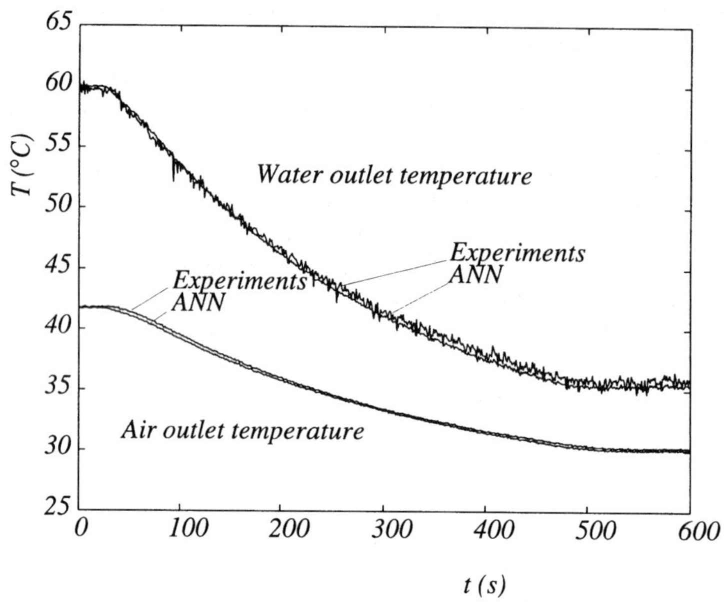

FIGURE 4.24.12 ANN dynamic prediction for sudden cooling.

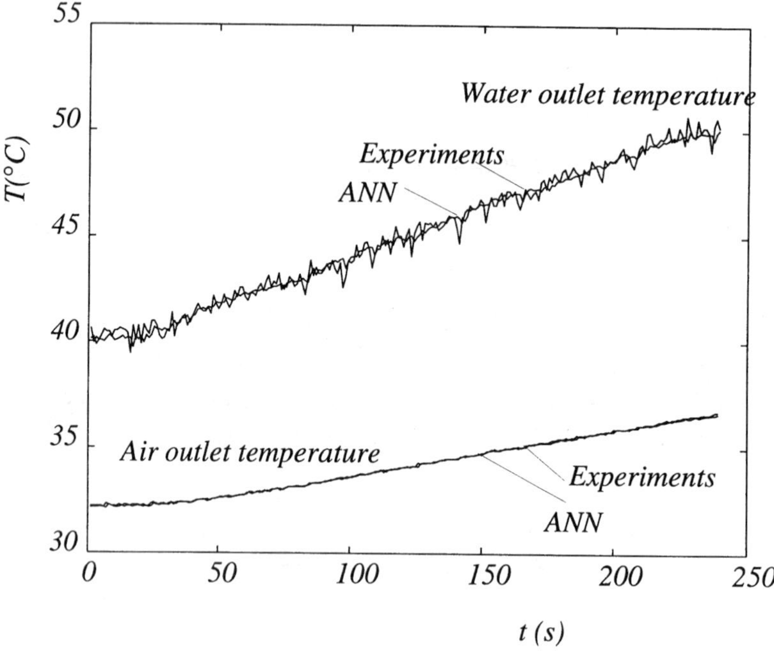

FIGURE 4.24.13 ANN dynamic prediction for ramp heating.

in using a plant model based on ANNs, and control strategies have been developed along these lines. There is the technique of direct inverse control which utilizes an inverse system model (Hunt et al., 1992), which is then cascaded with the controlled system so that the composed system results in an identity mapping between the desired response and the controlled system output. Thus, the network acts directly as the controller in such a configuration. A recent extension to this strategy by incorporating an

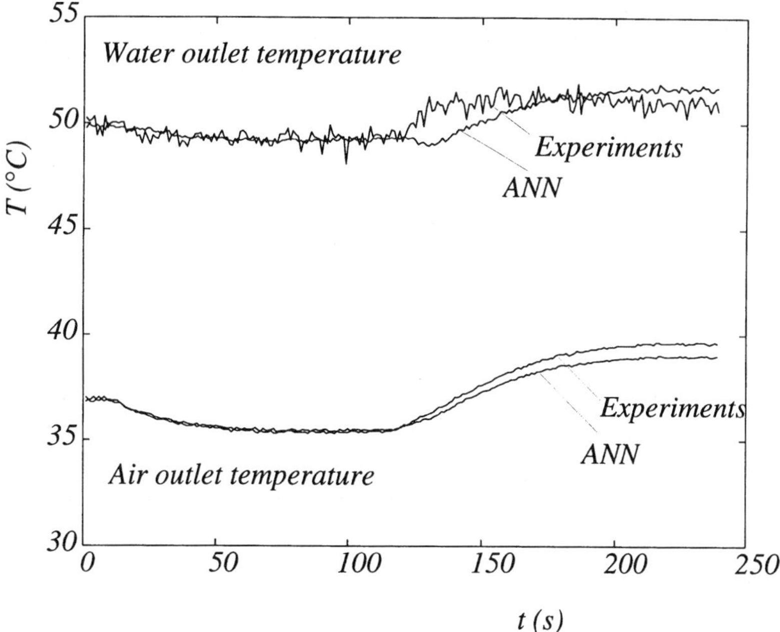

FIGURE 4.24.14 ANN dynamic model prediction for changes in the air mass flow rates.

orthogonal weight estimator routine was given by Pican et al. (1998). Then there is the technique of internal model control (IMC) (Psichogios and Ungar, 1991; Hunt et al., 1992) where a forward system and inverse model are used directly as elements within the feedback loop. The system model is placed in parallel with the real plant, and the difference between the system and model outputs is used as a feedback signal which is then processed by a controller subsystem in the forward path. In the IMC strategy, a separate linear filter is often introduced to provide desirable robustness and tracking response to the closed-loop system. The IMC has been widely applied in process-control systems.

A third dynamic-control strategy is known as predictive control in which an ANN model provides prediction of the future plant response over the specific time horizon (Hunt et al., 1992; Ayoubi, 1997; Jeannette et al., 1998). The predictions supplied by the network are passed to a numerical optimization routine which attempts to minimize a specified performance criterion in the calculation of a suitable control signal. One further possibility is to train another ANN to mimic the action of the optimization routine. The advantage here is that the outer loop consisting of the plant model and the optimization routine is no longer needed when training is complete, thus greatly simplifying the control system operation. Finally, mention should be made of the adaptive control strategies (Brown and Harris, 1994; Narendra and Mukhopadhyay, 1996) which are currently under very active development and have specific important applications in the area of thermal systems.

The dynamic control of the previously used heat exchanger 1 is demonstrated by a feedback system incorporating an inverse ANN controller, while the plant is modeled by a forward ANN. The inverse ANN controller was trained with the same training data as before, but with the original inputs and outputs interchanged. We considered the special case in which all the input variables, except for the water inlet temperature, were kept constant. The ANN was trained by using the experimental data due to incremental changes in the water inlet temperature, as explained before. The closed-loop feedback control system is shown in Figure 4.24.15. It is seen that the plant modeled by the forward ANN is controlled by an inverse ANN controller. When a disturbance is applied to the system, it controls by changing the water inlet temperature. For a preset reference air outlet temperature $\overline{T}_a^{out}$, a disturbance in the air inlet temperature, ΔT_a^{in} is introduced. The control problem is to find a value for the proportional

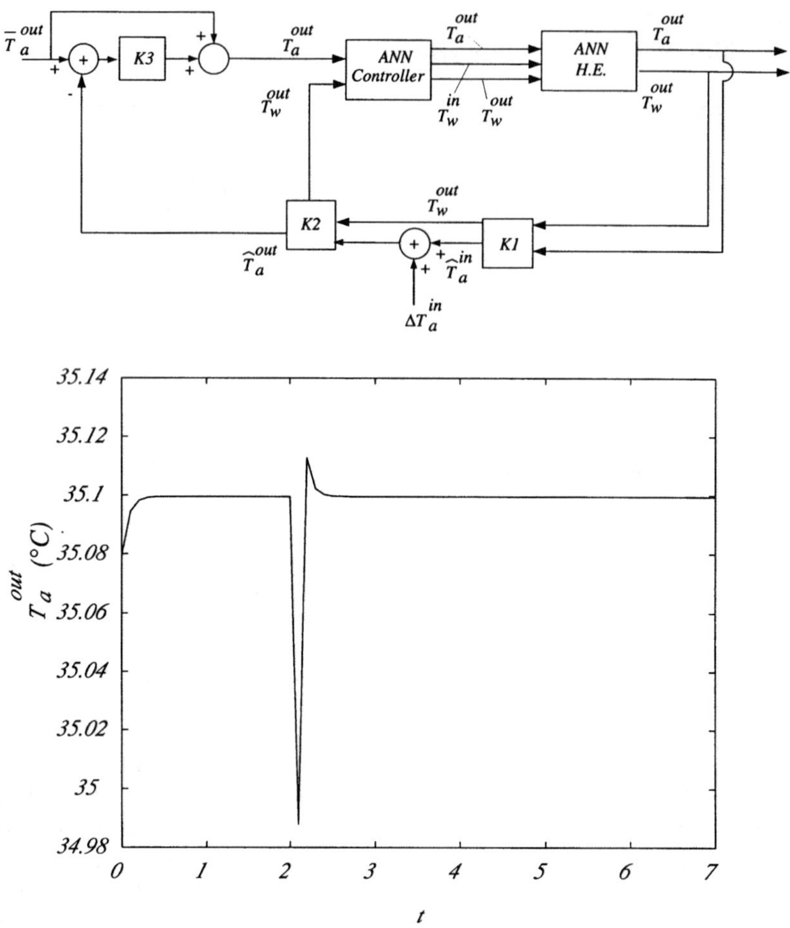

FIGURE 4.24.15 Closed-loop feedback control system and its response to a disturbance for *K*3 = 0.01.

constant *K*3 so that this disturbance is damped out and the system stabilizes at the reference air inlet temperature. As a first approximation, the steady-state energy balance between the heat given up by the water and that absorbed by the air gives the constants *K*1 and *K*2. Figure 4.24.15 also shows the response of the system to a disturbance of 5.56°C of the air inlet temperature at $t = 2$ and with *K*3 = 0.01. Here the time scale is arbitrary because of the way *K*1 and *K*2 are assigned. However, when *K*3 is increased, the system responds much more slowly. On further increasing *K*3, the system eventually becomes oscillatory and, thus, uncontrollable.

Additional Applications in Thermal Engineering

In the past applications of ANN to engineering problems have been attempted in structural engineering and engineering mechanics (Zeng, 1998). Tentative studies of applying ANN to problems in thermal systems have been carried out quite recently, with a relatively short history. With the exception of neural network system and control systems applied to HVAC systems, the studies have been somewhat sporadic and in only some distinct areas of application. For heat transfer data analysis and correlation, an ANN-based methodology was proposed by Thibault and Grandjean (1991) and a similar methodology was introduced by Jambunathan et al. (1996) to predict coefficients of heat transfer in convective-flow systems using liquid crystal thermography. Both steady and unsteady heat conduction problems were treated by

ANN in studies by Gobovic and Zaghloul (1993), Yentis and Zaghloul (1994), and Kuroe and Kimura (1995). Kaminski et al. (1996) gave an interesting description of the thermal deterioration process based on combined ANN and GA analysis. An ANN was also used to identify location and strength of unknown heat sources by sparse temperature measurements (Momose et al., 1993). In addition, an ANN-based attempt was made to predict measured intrinsic thermodynamic properties (Normandin et al., 1993).

ANN analyses have also been utilized to study and predict the performance of specific thermal devices and systems. Good estimates of the thermal storage loads and the dynamic system operation of typical thermal storage systems were obtained by Ferrano and Wong (1990) and Ito et al. (1995). Heat exchanger performance and control were studied by Diaz et al. (1996, 1998, 1999), Lavric et al. (1993, 1994), and Bittanti and Piroddi (1997). Several industrial applications of ANN were also demonstrated, in a fluidized-bed dryer (Zbicinski et al., 1996), in a liquid-sodium reflux-pool boiler solar receiver (Fowler et al., 1997), in a steel annealing furnace (Pican et al., 1998), and in the design of a chemical injection-system retrofit fuzzy control system in a thermal power plant (Moon and Cho, 1996). Other specific applications include manufacturing and materials processing involving microelectronic manufacturing (Mahajan and Wang, 1993), a coordinate grinder (Yang et al., 1995), rapid thermal processing control (Fortuna et al., 1996), sensors and sensor analysis involving thermal image processing (Naka et al., 1993), and blast furnace probe temperatures (Bulsari and Saxin, 1995).

On the other hand, there is a coherent body of published literature dealing with the analysis and control of HVAC systems. Such systems are generally quite large and rather complex, and robust optimal control is essential in assuring their satisfactory operation. In current practice, many HVAC control processes utilize the standard PID control strategy which, however, is not always autonomous and often requires constant manual supervision in order to achieve optimal control. Interest in the application of ANNs to the control of HVAC systems is due to its high potential of achieving an improvement. Since the appropriate theoretical dynamic model of the HVAC system is difficult to obtain, the role of ANN is to provide such a model from existing experimental data on the system operation. Several studies are available dealing with the modeling of HVAC systems, including ANN-based nonlinear HVAC model computations (Mistry and Nair, 1993), models for building energy use and HVAC system identification (Kreider et al., 1992, 1995; Chow and Teeter, 1997), and system models for district heating (Waldemark et al., 1992). The first proof of a concept of utilizing an ANN model as a part of the control system was given by Curtiss et al. (1993), and subsequent studies showed how the control systems including the ANN models could be implemented (Alessandri et al., 1994; Carriere et al., 1994). It is interesting to note in a recent experimental study as to how well one such ANN-based control system performed in a simple hydronic HVAC system, as compared to that of the standard PID control (Jeannette et al., 1998).

Adams et al. (1990) optimized the protection of feedwater heaters. Leitch and Quek (1992) worked on an architecture for integrated process supervision. Kos (1993) took an ANN approach to thermal placement in power electronics. Terry and Himmelblau (1993) did data rectification and gross error detection in a steady-state process via ANNs. Huang and Nelson (1994) also applied this technique to determine the delay time for a HVAC plant to respond to control actions. Ding and Wong (1990) controlled a simulated hydronic system using an ANN. Dhar et al. (1995) used an ANN with wavelet basis functions to predict hourly heating and cooling energy use in commercial buildings with the temperature as the only weather variable. Inoue et al. (1995) studied thermal comfort and climate control. Kawashima et al. (1995) did a 24-h thermal load prediction. Li et al. (1996) developed a fault diagnosis method for heating systems using neural networks. Matsumoto et al. (1996, 1997) studied the effect of pin fin arrangement on endwall heat transfer. Takemori et al. .(1991) worked on a neural network air-conditioning system for individual comfort. Dounis et al. (1992) implemented an AI technique in thermal comfort control for passive solar buildings.

All the studies mentioned above represent tentative attempts to apply the ANN analysis to thermal system problems. Since good results have been obtained so far, there is no reason to expect that the ANN approach cannot be applied to many other thermal problems with equal success, particularly in the analysis of dynamic systems and their control.

General Discussion

It must be emphasized that the implementation of ANN predictions in an industrial setting is really very simple and straightforward. For instance, once the network is trained for a given device by the manufacturer, it can simply be transmitted to the intended customer by giving him or her the sets of weights and biases and the corresponding software for the appropriate network configuration. The customer can then independently carry out the neural computing. An example of the weights and biases for heat exchanger 1 for the 4-2-1 ANN configuration is given in Table 4.24.5. It should be mentioned that the training process of an ANN is somewhat computing intensive, even though not excessive in terms of the computing resources available today, and all computations can be readily handled by personal workstations. However, once the training is completed, the predictions using the finalized sets of synaptic weights and biases can be carried out on any personal computer.

TABLE 4.24.5 Synaptic Weights and Biases of 4-2-1 ANN Configuration for Heat Exchanger 1

i	j	k	l	$w^{j,l}_{i,k}$
1	1	2	1	–8.744
1	1	2	2	0.401
1	2	2	1	1.321
1	2	2	2	1.120
1	3	2	1	0.772
1	3	2	2	1.356
1	4	2	1	–0.303
1	4	2	2	–0.223
2	1	3	1	–7.741
2	2	3	1	8.576
i	j	$\theta_{i,j}$		
2	1	–1.574		
2	2	–2.474		
3	1	–1.848		

There are several inherent and excellent attributes of the ANN approach which the traditional analysis, including the dimensionless correlations of the experimental data, simply does not have. The neural network, by its ability for pattern recognition, is capable of capturing all the relevant physics, but some of this physics is often not accounted for in the common data correlations by dimensionless parameters. It should be realized, too, in this regard, that since the effects of properties, which represent part of the physics, are expected to show up in the input-output relations in the training data for the ANN, there is no need to explicitly use such properties in the ANN analysis, while such use is necessary in the traditional analysis. Unfortunately, in many applications such properties, which may be either intrinsic or apparent, may not be known or not known accurately, and thus may often lead to difficulties. Furthermore, the complexity in the physics is not an issue in the ANN approach, while such complexity is often the critical barrier in the traditional analysis. Also, it is a relatively simple matter, by continuing training with new data, to adapt to changing conditions, a situation which often exists in real-world applications. The traditional analysis is not capable of doing so, since the physical mechanisms responsible for such changing conditions are not known. In addition, dynamic thermal-system models, which are needed for developing optimal control strategies are difficult to develop by the traditional analysis even for simple systems, and are readily accessible by the ANN approach. On the other hand, there are several shortcomings relative to ANN implementations, but these can be remedied by adequate numerical experimentation, past experience in similar applications, or continuing research in the field of ANN.

The attractiveness of the ANN approach has been demonstrated in the preceding subsections by considering the static and dynamic performances of typical compact heat exchangers. Such heat exchangers are already in common use today, and the traditional analysis has not yet produced satisfactory theories to predict their behavior. Even though the applications of the ANNs have been demonstrated here by considering them as examples, the potential applications in other thermal systems are essentially unlimited, since thermal systems still have not been analyzed by the ANN approach to any large extent. In order to promote its application to thermal science and engineering problems, it would be desirable to look at application areas in which the ANN approach would make significant impacts. In this regard, the most significant areas lie where the lack of physical models, system complexities, and unavailability of properties or constitutive relations represent critical barriers to the advancement of the field in terms of product development, process development and control, manufacturing, sensitivity and risk analyses, and other similar areas which involve simulation, design, and control. This is particularly imperative in those areas where there are plentiful experimental data that can be used to train the networks. To give some indication to this vast potential, several example of critical technology areas can be mentioned for the applications of the ANN approach.

Thermal Networks

The focus so far has been on components, especially on heat exchangers. However, even if the behavior of a single heat exchanger could be accurately predicted, it would be difficult to do so for a large network of these devices as are commonly used in practice. Thermal networks, as used in heating and cooling systems in a single building or in a collection of buildings, may consist of chilled water for cooling purposes, steam for heating, air for ventilation or, more commonly, a combination of these. Current design is heavily based on overspecification so that it is not unusual to find simultaneous heating and cooling to obtain the desired comfort level, especially during periods of change in weather. A hydraulic network may consist of a large number of interconnected pipes with valves, heat exchangers, pumps, and other components online. The simplest network problem consists in determining the flow in each branch of the network. Methods based on the solution of the nonlinear pressure drop and flow rate network equations have been developed to solve this problem. The control of the flow in a given branch by change of pump or valve parameters is a more difficult problem. Though one can in principle re-solve the network for the new conditions, this could be time-consuming for a large network and, thus, difficult to use in real time. In addition the exact, current state of the network and its physical properties may not be known for computational purposes. Thus, for example, the heat transfer coefficients at a heat exchanger will be affected by fouling, which will in turn change its ability to transfer heat. It is possible that the response of the network be simulated by an ANN. This ANN can be trained by using experimental data so that the response of changing any parameter can be found. One of the advantages is that this can be done online so that changes in system performance are reflected in the simulation.

Thermal Processing of Materials

In almost all materials processing operations involving thermal effects, the most critical information is the direct relationship between the desired characteristics of the end product, such as, for instance, material microstructure defect elimination and the processing parameters. Physical models that provide such relationship are either not available or, if available, involve either apparent properties or constitutive relations which are not available. In the latter case, the lack of such information essentially nullifies the usage of such models. The ANN analysis simply bypasses all these uncertainties and leads immediately to the direct quantitative relationship between the desired end-product characteristics and the processing parameters. This advantage becomes particularly significant in dealing with new engineering materials, foods, and pharmaceutical and biological materials.

Environmental and Safety Studies

There are many complex thermal systems dealing with environmental and safety control in building, factories, mines, and the like. HVAC and indoor pollution systems and fire protection systems are good examples. Despite the recent advances made in the control of HVAC systems by means of the ANN approach, much more needs to be done, especially in the area of simulating realistic dynamic conditions involving, for instance, time delays. Performance-oriented risk analysis of fire hazards is drawing much attention, but very little is known concerning the realistic assessment of likely dynamic fire- and smoke-spread scenarios, which are needed in risk analysis and also in the design of fire mitigation systems for enclosed habitat spaces such as high-rises and mines. The capability of the ANN approach needs to be explored and exploited to deal with such complex dynamic systems.

Tribological Systems

In all systems of any size involving relative motions of components in contact, friction inevitably occurs and produces heat, wear, or both. To minimize or control friction, theories or models are needed for the physical, chemical, and other phenomena occurring at the surfaces of contact or interfaces. Such models would necessarily depend on the physical structure and properties at these interfaces. Such theories or models are either not available or difficult to apply because of the unknown characterization of these interfaces, even though the methods of preparing such interfaces and the lubricant used are known. ANNs, in general, are capable of catching all the relevant physics and recognizing the unique surface characterization, and then incorporating such characterizations into the ANN-based input-output relation.

Reactive Systems

One major barrier to our knowledge of reacting systems from molecular-biological systems in single cells to large scale reactors to combustion phenomena is the lack of reaction kinetics and the corresponding energy transport information. Models are impossible without such information. For the ANN analysis, such information can again be bypassed and, therefore, is not needed in the analysis. The effects of the reaction kinetics are reflected in the output data used in the training of the ANN and, therefore, they are accounted for in the trained ANN, which then becomes the model of the reactive system.

Multiphase Phenomena and Systems

In systems involving components of multiple phases such as mixtures of liquids and gases, liquids and solids, solids and gases, and all three phases in turbulent pipe flows with heat transfer and multicomponent heterogeneous mixtures in food processing, the extreme complexity prevents the development of reasonable models for use in simulation, optimization, and control. Attempts to correlate the data would not be particularly fruitful either because of the large data scatter due to unknown parameters. This type of complex system is a natural for ANN-based analysis, modeling, and dynamic control.

Inverse Problems and Parameter Estimation

The importance of solving inverse problems and problems in parameter estimation has long been recognized in many engineering applications, and this importance has also received ever-increasing attention in the field of thermal systems. The application of ANN analysis to such problems is not new, but should be emphasized. As is already obvious, the inputs and outputs of an ANN, especially in dynamic systems, can be switched as in the case of an inverse ANN model.

Genetic Algorithms

Evolutionary programming, of which genetic algorithms and programming are examples, allow programs to change or evolve as they compute. GAs, specifically, are based on the principle of Darwinian selection. One of their most important applications in the thermal sciences is in the area of optimization of various kinds.

Optimization by itself is fundamental to many applications. In engineering, for example, it is important to the design of systems; analysis permits the prediction of the behavior of a given system, but optimization is the technique that searches among all possible designs of the system to find the one that is the best for the application. The importance of this problem has given rise to a wide variety of techniques which help search for the optimum. There are searches that are gradient-based and those that are not.

In the former the search for the optimum solution, as for example the maximum of a function of many variables, starts from some point and directs itself in an incremental fashion towards the optimum; at each stage the gradient of the function surface determines the direction of the search. Local optima can be found in this way, the search for global optimum being more difficult. Again, if one visualizes a multivariable function, it can have many peaks, any one of which can be approached by a hill-climbing algorithm. To find the highest of these peaks, the entire domain has to be searched; the narrower this peak the finer the searching "comb" must be. For many applications this brute force approach is too expensive in terms of computational time. Alternatives, like simulated annealing, are techniques that have been proposed, and the GA is one of them.

In what follows, we will provide an overview of the genetic algorithm and programming. A numerical example will be explained in some detail. The methodology will be applied to one of the heat exchangers discussed before. There will be a discussion on other applications in thermal engineering and comments will be made on potential uses in the future.

Methodology

GAs are discussed in detail by Holland (1975, 1992), Mitchell (1997), Goldberg (1989), Michalewicz (1992), and Chipperfield (1997). One of the principal advantages of this method is its ability to pick out a global extremum in a problem with multiple local extrema. For example, we can discuss finding the maximum of a function $f(x)$ in a given domain $a \leq x \leq b$. In outline, the steps of the procedure are the following.

- First, an initial population of n members, $x_1, x_2, \ldots, x_n \in [a, b]$ is randomly generated.
- Then, for each x a *fitness* is evaluated. The fitness of effectiveness is the parameter that determines how good the current x is in terms of being close to an optimum. Clearly, in this case the fitness is the function $f(x)$ itself, since the higher the value of $f(x)$ the closer we are to the maximum.
- The probability distribution for the next generation is found based on the fitness values of each member of the population. Pairs of parents are then selected on the basis of this distribution.
- The offsprings of these parents are found by *crossover* and *mutation.* In crossover, two numbers in binary representation, for example, produce two others by interchanging part of their bits. After this, and based on a preselected probability, some bits are randomly changed from 0 to 1 or vice versa. Crossover and mutation create a new generation with a population that is more likely to be fitter than the previous generation.
- The process is continued as long as desired or until the largest fitness in a generation does not change much any more.

The procedure can be easily generalized to a function of many variables.

Let us consider a numerical example that is shown in detail in Table 4.24.6. Suppose that one has to find the x at which $f(x) = x(1 - x)$ is globally a maximum between 0 and 1. We have taken $n = 6$, meaning that each generation will have six numbers. Thus, for a start, six random numbers are selected between 0 and 1. Now we choose n_b which is the number of bits used to represent a number in binary form. Taking $n_b = 5$, we can write the numbers in binary form normalized between 0 and the largest number

TABLE 4.24.6 Example Using the Genetic Algorithm

$G = 0$	$f(x)$	$s(x)$	$G = 1/4$	$G = 1/2$	$G = 3/4$	$G = 1$
11001	0.1561	0.2475	00011	00011	00010	00010
11110	0.0312	0.0496	00011	11100	11101	11101
11100	0.0874	0.1386	11110	00011	10011	10011
10011	0.2373	0.3762	10011	10011	00011	00011
00011	0.0874	0.1386	00011	11110	11011	11011
00001	0.0312	0.0495	11100	00011	00110	00100

possible for n_b bits, which is $2^{n_b} - 1 = 31$. In one run, the numbers chosen and written down in the first column of the table labeled $G = 0$, are 25, 30, 28, 19, 3, and 1, respectively. The fitnesses of each one of the numbers, i.e., $f(x)$, are computed and shown in the second column. These values are normalized by their sum and shown in the third column as $s(x)$. The normalized fitnesses are drawn on a roulette wheel in Figure 4.24.16. The probability of crossover is taken to be 100%, meaning that crossover will always occur. Pairs of numbers are chosen by spinning the wheel, the numbers having a bigger piece of the wheel having a larger probability of being selected. This produces column four marked $G = 1/4$, and shuffling to produce random pairing gives column five, marked $G = 1/2$. The numbers are now split up in pairs, and crossover applied to each pair. The first pair [0 0 0 1 1] and [1 1 1 0 0] produces [0 0 0 1 0] and [1 1 1 0 1]. This is illustrated in Figure 4.24.17(a) where the crossover position is between

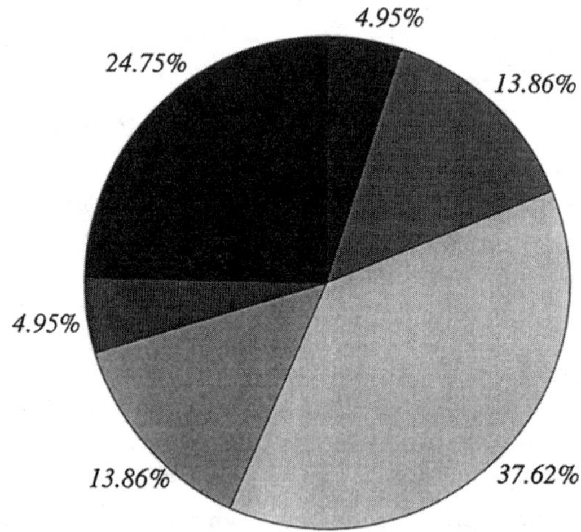

FIGURE 4.24.16 Distribution of fitnesses.

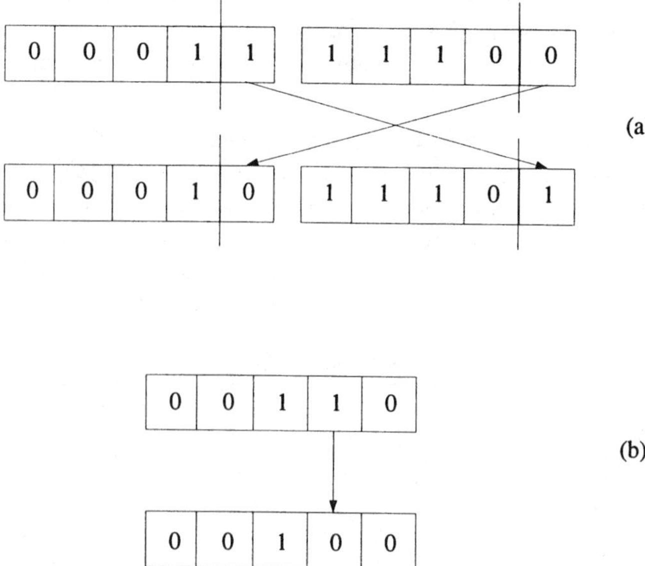

FIGURE 4.24.17 (a) Crossover and (b) mutation in a genetic algorithm.

the fourth and fifth bit; the bits to the right of this line are interchanged. Crossover positions in the other pairs are randomly selected. Crossover produces column six, marked as $G = 3/4$. Finally, one of the numbers, in this case the last number in the list [0 0 1 1 0], is mutated to [0 0 1 0 0] by changing one randomly selected bit from 1 to 0, as shown in Figure 4.24.17(b). From the numbers in generation $G = 0$, these steps have now produced a new generation $G = 1$. The process is repeated until the largest fitness in each generation increases no more. In this particular case, values within 3.22% of the exact value of x for maximum $f(x)$, which is the best that can be done using 5 bits, were usually obtained within 10 generations.

The genetic programming technique (Koza, 1992, 1994) is an extension of this procedure in which computer codes take the place of numbers. It can be used in symbolic regression to search within a set of functions for the one which best fits experimental data. The procedure is similar to that for the GA, except for the crossover operation. If each function is represented in tree form, though not necessarily of the same length, crossover can be achieved by cutting and grafting. As an example, Figure 4.24.18 shows the result of the operation on the two functions $3x(x + 1)$ and $x(3x + 1)$ to give $3x(3x + 1)$ and $x(x + 1)$. The crossover points may be different for each parent.

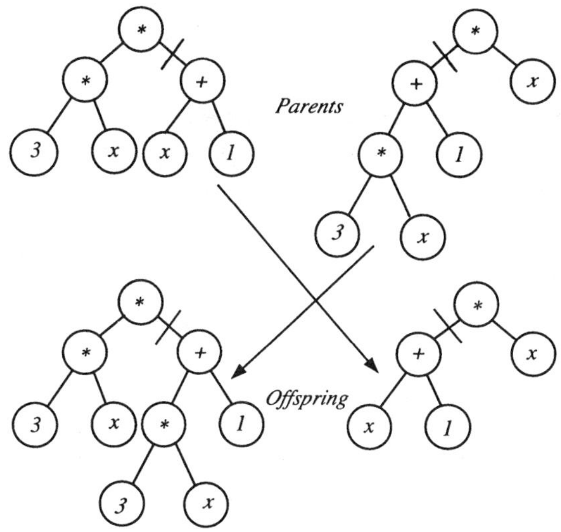

FIGURE 4.24.18 Crossover in genetic programming. Parents are $3x\,(x + 1)$ and $x\,(3x + 1)$; offspring are $3x(3x + 1)$ and $x(x + 1)$.

Applications to Compact Heat Exchangers

The following analysis is on the basis of data collected on a single-row heat exchanger referred to above as heat exchanger 1. In the following, a set of $N = 214$ experimental runs provided the data base. The heat rate is determined by

$$\dot{Q} = \dot{m}_a c_{p,a}\left(T_a^{out} - T_a^{in}\right) \tag{4.24.26}$$

$$= \dot{m}_w c_w\left(T_w^{in} - T_w^{out}\right) \tag{4.24.27}$$

For prediction purposes, we will use functions of this type

$$\dot{Q} = \dot{Q}\left(T_w^{in},\, T_a^{in},\, \dot{m}_a,\, \dot{m}_w\right) \tag{4.24.28}$$

The conventional of way of correlating data is to determine correlations for inner and outer heat transfer coefficients. For example, power laws of the following form

$$\varepsilon Nu_a = a\ Re_a^m\ Pr_a^{1/3} \tag{4.24.29}$$

$$Nu_w = b\ Re_w^n\ Pr_w^{0.3} \tag{4.24.30}$$

are common. The two Nusselt numbers provide the heat transfer coefficients on each side and the overall heat transfer coefficient, U, is related to h_a and h_w by

$$\frac{1}{UA_a} = \frac{1}{h_w A_w} + \frac{1}{\varepsilon h_a A_a} \tag{4.24.31}$$

To find the constants a, b, m, n, the mean square error

$$S_U = \frac{1}{N}\sum\left(\frac{1}{U^p} - \frac{1}{U^e}\right)^2 \tag{4.24.32}$$

must be minimized, where N is the number of experimental data sets, U^p is the prediction made by the power-law correlation, and U^e is the experimental value for that run. The sum is over all N runs.

This procedure was carried out for the data collected. It was found that the S_U had local minima for many different sets of the constants, the following two being examples.

Correlation	*a*	*b*	*m*	*n*
A	0.1018	0.0299	0.591	0.787
B	0.0910	0.0916	0.626	0.631

Figure 4.24.19 shows a section of the S_U surface that passes though the two minima, A and B. The coordinate z is a linear combination of the constants a, b, m, and n such that it is zero and unity at the two minima. Though the values of S_U for the two correlations are very similar and the heat rate predictions for the two correlations are also almost equally accurate, the predictions on the thermal resistances on either side are different. Figure 4.24.20 shows the ratio of the predicted air- and water-side Nusselt numbers using these two correlations. R_a is the ratio of the Nusselt number of the air side predicted by Correlation A divided by that predicted by Correlation B. R_w is the same value for the water side. The predictions, particularly the one on the water side, are very different.

There are several reasons for this multiplicity of minima of S_U. Experimentally, it is very difficult to measure the temperature at the wall separating the two fluids, or even to specify where it should be measured, and mathematically, it is due to the nonlinearity of the function to be minimized. This raises the question as to which of the local minima is the "correct" one. A possible conclusion is that the one which gives the smallest value of the function should be used. This leads to the search for the global minimum, which can be done using the GA.

For these data, Pacheco-Vega et al. (1998) conducted a global search among a proposed set of heat transfer correlations using the GA. The experimentally determined heat rate of the heat exchanger was correlated with the flow rates and input temperatures, with all values being normalized. To reduce the number of possibilities, the total thermal resistance was correlated with the mass flow rates in the form

$$\frac{T_w^{in} - T_a^{in}}{\dot{Q}} = f\left(\dot{m}_a, \dot{m}_w\right) \tag{4.24.33}$$

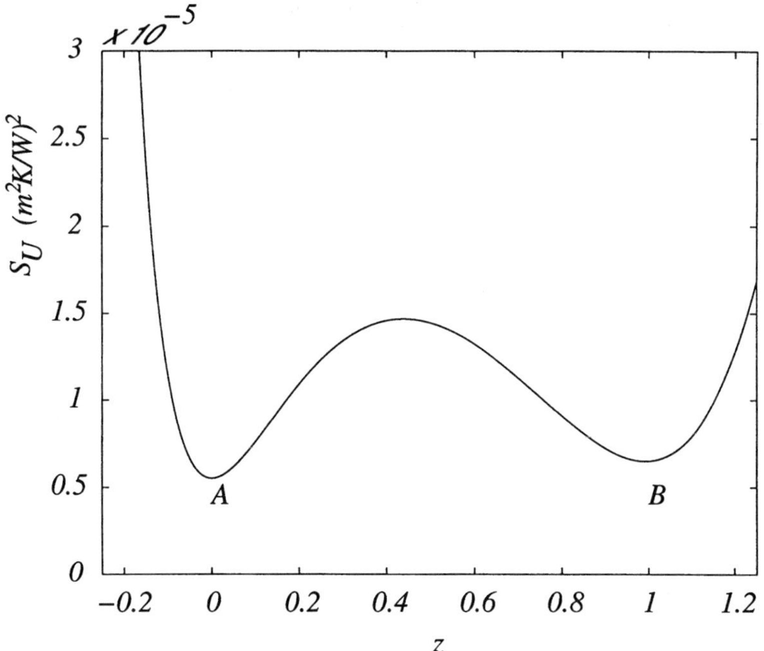

FIGURE 4.24.19 Section of $S_U(a, b, m, n)$ surface.

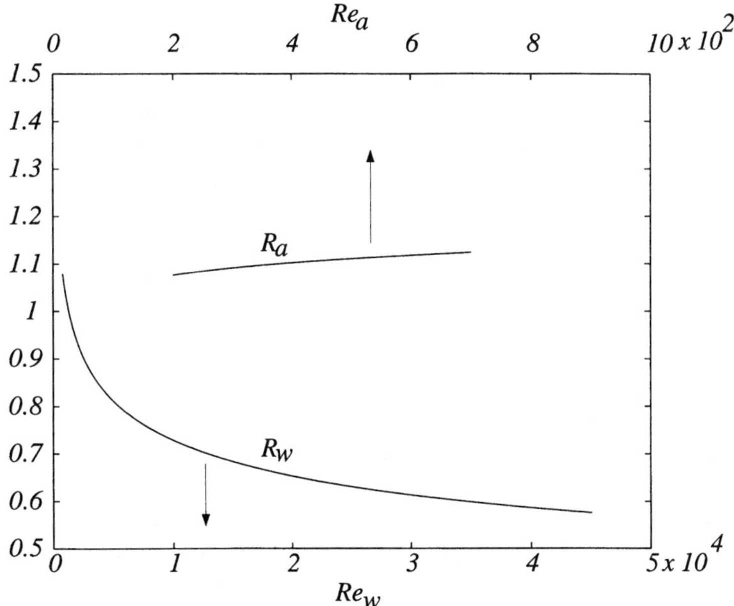

FIGURE 4.24.20 Ratio of the predicted air- and water-side Nusselt numbers.

Table 4.24.7 Comparison of Best Fits for Different Correlations

Correlation	f	a	b	c	d	σ
Power law	$a\dot{m}_w^{-b} + c\dot{m}_a^{-d}$	0.1875	0.9997	0.5722	0.5847	0.0252
Inverse linear	$(a + b\dot{m}_w)^{-1} + (c + d\dot{m}_a)^{-1}$	–0.0171	5.3946	0.4414	1.3666	0.0326
Inverse exponential	$(a + e^{b\dot{m}_w})^{-1} + (c + e^{d\dot{m}_a})^{-1}$	–0.9276	3.8522	–0.4476	0.6097	0.0575
Exponential	$ae^{-b\dot{m}_w} + ce^{-d\dot{m}_a}$	3.4367	6.8201	1.7347	0.8398	0.0894
Inverse quadratic	$(a + b\dot{m}^2_w)^{-1} + (c + d\dot{m}^2_a)^{-1}$	0.2891	20.3781	0.7159	0.7578	0.0859
Inverse logarithmic	$(a + b \ln \dot{m}_w)^{-1} + (c + d \ln \dot{m}_a)^{-1}$	0.4050	0.0625	–0.5603	0.2048	0.1165
Logarithmic	$a - b \ln \dot{m}_w - c \ln \dot{m}_a$	0.6875	0.4714	0.4902	—	0.1664
Linear	$a - b\dot{m}_w - c\dot{m}_a$	2.3087	0.8533	0.8218	—	0.2118
Quadratic	$a - b\dot{m}_w^2 - c\dot{m}_a^2$	1.8229	0.6156	0.5937	—	0.2468

The functions $f(\dot{m}_a, \dot{m}_w)$ that were used are indicated in Table 4.24.7. The GA was used to seek the values of the constants associated with each correlation, the objective being to minimize the variance.

$$S_Q = \frac{1}{N}\sum\left(\dot{Q}^p - \dot{Q}^e\right)^2 \tag{4.24.34}$$

where the sum is over all N runs between the predictions of a correlation, $\dot{Q}^p$, and the actual experimental values, $\dot{Q}^e$. Since the unknowns are the set of constants, a, b, c, and sometimes d, a single binary string represents them; the first part of the string is a, the next is b, and so on. The rest of the GA is as in the numerical example given before. The results obtained for each correlation are also summarized in the table in descending order of S_Q. The last column shows the mean square error σ defined in a manner similar to Equations (4.24.10) and (4.24.11). The parameters used for the computations are population size 20, number of generations 1000, bits for each variable 30, probability of crossover 1, and probability of mutation 0.03.

Some correlations are clearly seen to be superior to others. However, the difference in S_Q between the first- and second-place correlations, the power-law, and inverse logarithmic which have mean errors of 2.5% and 3.3%, respectively, is only about 8%, indicating that either could do just as well in predictions even though their functional forms are very different. In fact, the mean error in many of the correlations is quite acceptable. Figure 4.24.21 shows the predictions of the power-law correlation vs. the experimental values, all in normalized variables. The prediction is seen to be very good. The quadratic correlation on the other hand, is the worst in the set of correlations considered, and Figure 4.24.22 shows its predictions. It must also be remarked that, because of the random numbers used in the procedure, the computer program gives slightly different results each time it is run, changing the lineup of the less appropriate correlations somewhat.

Additional Applications in Thermal Engineering

Though the GA is a relatively new technique in relation to its application to thermal engineering, there are a number of different applications that have already been successful. Davalos and Rubinsky (1996) adopted an evolutionary-genetic approach for numerical heat-transfer computations. Shape optimization is another area that has been developed. Fabbri (1997) used a GA to determine the optimum shape of a fin. The two-dimensional temperature distribution for a given fin shape was found using a finite-element method. The fin shape was proposed as a polynomial, the coefficients of which have to be calculated. The fin was optimized for polynomials of degree 1 through 5. Von Wolfersdorf et al. (1997) did shape optimization of cooling channels using GAs. The design procedure is inherently an optimization process. Androulakis and Venkatasubramanian (1991) developed a methodology for design and optimization that was applied to heat exchanger networks; the proposed algorithm was able to local solutions where gradient-based methods failed. Abdel-Magid and Dawoud (1995) optimized the parameters of an integral and a proportional-plus-integral controller of a reheat thermal system with GAs. The fact that the GAs

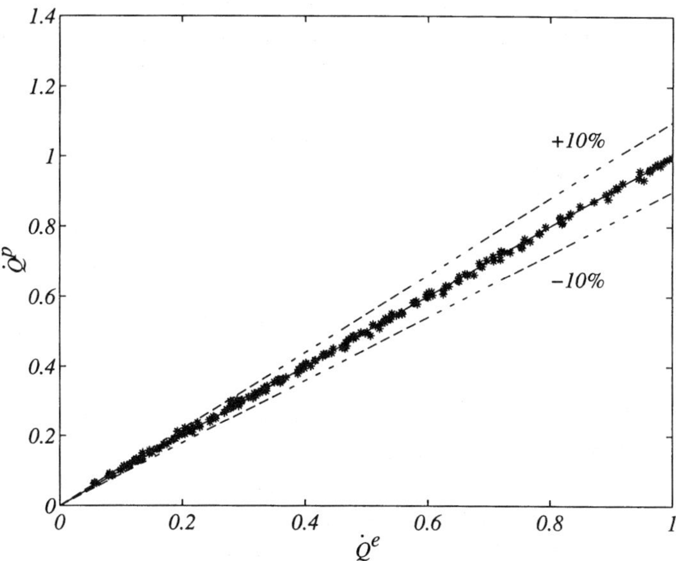

FIGURE 4.24.21 Experimental vs. predicted normalized heat flow rates for a power–law correlation. The straight line is the line of equality between prediction and experiment, and the broken lines are ±10%.

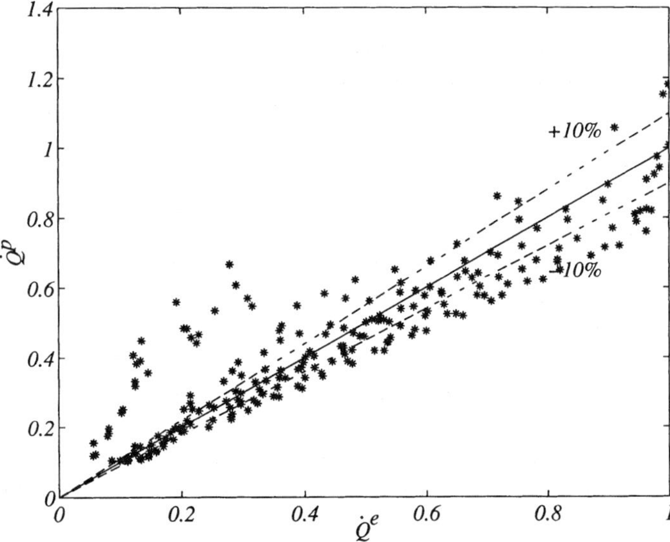

FIGURE 4.24.22 Experimental vs. predicted normalized heat flow rates for a quadratic correlation. The straight line is the line of equality between prediction and experiment, and the broken lines are ±10%.

can be used to optimize in the presence of variables that take on discrete values was put to advantage by Schmit et al. (1996) who used it for the design of a compact high-intensity cooler.

The placing of electronic components as heat sources is a problem that has become very important recently from the point of view of computers. Queipo et al. (1994) applied GAs to the optimized cooling of electronic components. Tang and Carothers (1996) showed that the GA worked better than some other methods for the optimum placement of chips. Queipo and Gil (1997) worked on the multiobjective optimization of component placement and presented a solution methodology for the collocation of convectively and conductively air-cooled electronic components on planar printed wiring boards. Meysenc et al. (1997) studied the optimization of microchannels for the cooling of high-power transistors. Inverse problems may also involve the optimization of the solution. Allred and Kelly (1992) modified the GA for extracting thermal profiles from infrared image data which can be useful for the detection of malfunctioning electronic components. Jones et al. (1995) used thermal tomographic methods for the detection of inhomogeneities in materials by finding local variations in the thermal conductivity. Raudensky et al. (1995) used the GA in the solution of inverse heat conduction problems. Okamoto et al. (1996) reconstructed a three-dimensional density distribution from limited projection images with the GA. Wood (1996) studied an inverse thermal field problem based on noisy measurements and compared a GA and the sequential function specification method. Li and Yang (1997) used a GA for inverse radiation problems. Castrogiovanni and Sforza (1996, 1997) studied high-heat flux-flow boiling systems using a numerical method in which the boiling-induced turbulent eddy diffusivity term was used with an adaptive GA closure scheme to predict the partial nucleate boiling regime.

Applications involving genetic programming are rarer. Lee et al. (1997) studied the problem of correlating the CHF for upward water flow in vertical round tubes under low-pressure and low-flow conditions. Two sets of independent parameters were tested. Both sets included the tube diameter, fluid pressure, and mass flux. The inlet condition type had, in addition, the heated length and the subcooling enthalpy; the local condition type had the critical quality. Genetic programming was used as a symbolic regression tool. The parameters were nondimensionalized; logarithms were taken of the parameters that were very small. The fitness function was defined as the mean square difference between the predicted and experimental values. The four arithmetical operations, addition, subtraction, multiplication, and division, were used to generate the proposed correlations. The programs ran up to 50 generations and produced 20 populations in each generation. In a first intent, 90% of the data sets were randomly selected for training and the rest for testing. Since no significant difference was found in the error for each of the sets, the entire data set was finally used both for training and testing. The final correlations that were found had predictions better than those in the literature. The advantage of the genetic programming method in seeking an optimum functional form was exploited in this application.

General Discussion

The evolutionary programming method has the advantage that, unlike the ANN, a functional form of the relationship is obtained. Genetic algorithms, genetic programming, and symbolic regression are relatively new techniques from the perspective of thermal engineering, and we can only expect the applications to grow. There are a number of areas in prediction, control, and design where these techniques can be effectively used. One of these, in which progress can be expected, is in thermal-hydronic networks. Networks are complex systems built up from a large number of simple components; though the behavior of each component may be well understood, the behavior of the network requires massive computations that may not be practical. Optimization of networks is an important issue from the perspective of design, since it is not obvious what the most energy-efficient network, given certain constraints, should be. The constraints are usually in the form of the locations that must be served and the range of thermal loads that are needed at each position. A search methodology based on the calculation of every possible network configuration would be very expensive in terms of computational time. An alternative based on evolutionary techniques would be much more practical. Under this procedure a set of networks that satisfy the constraints would be proposed as candidates for the optimum. From this set a new and more fit generation would evolve and the process repeated until the design does not change much. The definition of fitness, for this purpose, would be based on the energy requirements of the network.

Concluding Remarks

Artificial intelligence is a broad term and many of its techniques have not been reviewed here. Reviews like those Culp et al. (1990) on the impact of AI within the HVAC industry, and Cahill (1994) and Paoletti and Sciubba (1997) on the use of a knowledge-based expert system for the design of complex thermal systems like power plants, should be consulted. Other examples on the use of AI in thermal engineering are the following. Rasmussen et al. (1990) identified the distillation process dynamics comparing process knowledge and black-box-based approaches. Gao et al. (1990) worked on the synthesis of optimized heat exchanger networks. Fujii and Lutzenhiser (1992) used intelligent systems in residential air-conditioning and natural cooling systems. Loveday and Virk (1992) applied the technique to buildings. Dregalin and Nazyrova (1993) introduced AI elements to heat power engineering. Sandberg et al. (1995) studied an intelligent distributed heater used for industrial applications. Yang et al. (1996) developed an analysis of intelligent control strategies of a thermal energy storage air-conditioning system. These papers indicate some of the wide variety of approaches that can be considered in thermal engineering.

Despite the tremendous strides made in advancing our knowledge base in the field of thermal science and engineering in the last 100 years or so, researchers in the field have often felt apprehensive that their hands were tied by the relatively limited though popular approaches in the traditional analysis of thermal phenomena: solving the governing first principles or modeling equations and finding the necessary correlations from experimental data. Even with the ever-increasing computing resources which have become available, difficulties still exist since the systems are much more complex, and the need is to respond more quickly to the applications and challenges of the fast-changing world of technology. The artificial intelligence community provides some possible solutions, and it is surprising that there has not been much more concerted effort in the thermal-science research community to seek the application of these new approaches which have high potential for addressing these needs. This section has discussed two of these techniques, the artificial neural network and the genetic algorithmic approaches, that are beginning to take root now. The purpose here has been to introduce the reader to these methodologies, and it is up to him or her to apply them and similar procedures to a broader set of problems. Artificial intelligence techniques are well suited to complex problems, and thermal engineering presents many examples of this. It is not a substitute, but rather a complement, for incisive physical insight and first-principle analyses.

Acknowledgments

We are grateful for the contributions that our students, X. Zhao, G. Diaz, and A. Pacheco-Vega, and our collaborator, R.L. McClain, have made to our understanding of the subject. We are also thankful for the financial support of D.K. Dorini to the Hydronics Laboratory where most of our work has been carried out, and that of the Mexican Council of Science and Technology (CONACyT) and the Organization of American States for support to the students.

References

Abdel-Magid, Y.L. and Dawoud, M.M., 1995, Tuning of AGC of interconnected reheat thermal systems with genetic algorithms, *Proc. IEEE Int. Conf. Syst. Man Cybernetics,* Vol. 3, pp. 2622–2627.

Adams, L.E., Hoeinghaus, R.C., and Lantz, J.B., 1990, Optimization and protection of feedwater heaters, *Proc. 33rd Power Instrumentation Symp.,* Vol. 33, pp. 47–52.

Alessandri, A., Verona, F.B., Parisini, T., and Torrini, A., 1994, Neural approximations for the optimal control of heating systems, *Proc. IEEE Conf. Control Applications,* Vol. 3, pp. 1613–1618.

Allred, L.G. and Kelly, G.E., 1992, Modified genetic algorithm for extracting thermal profiles from infrared image data, *Proc SPIE — Int. Soc. Optical Eng. Neural Stochastic Methods Image and Signal Processing,* Vol. 1766, pp. 77–81.

Androulakis, I.P. and Venkatasubramanian, V., 1991, Genetic algorithmic framework for process design and optimization, *Comput. Chem. Eng.,* 15(4), 217–228.

Angeline, P.J., Saunder, G.M., and Pollack, J.B., 1994, Complete induction of recurrent neural networks, *Proc. Third Annu. Conf. Evolutionary Programming,* (Eds.) A.V. Sebald, L.J. Fogel, World Scientific, Singapore, pp. 1–8.

Ayoubi, M., 1997, Dynamic multi-layer perception: application to the nonlinear identification and predictive control of a heat exchanger, in *Applications of Neural Adaptive Control Technology,* (Eds.) J. Kalkkuhl, K.J. Hunt, Zbikowski, R., and A. Dzielinski, World Scientific, Singapore, pp. 205–230.

Bittanti, S. and Piroddi, L., 1997, Nonlinear identification and control of a heat exchanger: A neural network approach, *J. Franklin Inst.,* 334B(1), 135–153.

Bonnet, A., 1985, *Artificial Intelligence, Promise and Performance,* Prentice-Hall, Englewood Cliffs, NJ.

Bouchon-Meunier, B., Yager, R.R., and Zadeh, L.A., Eds., 1995, *Fuzzy Logic and Soft Computing,* World Scientific, Singapore.

Brown, M. and Harris, C., 1994, *Neurofuzzy Adaptive Modelling and Control,* Prentice-Hall, Englewood Cliffs, NJ.

Bulsari, A.B. and Saxin, H., 1995, Classification of blast furnace probe temperatures using neural networks, *Steel Res.,* 66, 231–236.

Cahill, E., 1994, Knowledge-based algorithm construction for real-world engineering PDEs, *Math. Comput. Simulation,* 36(4–6), 389–400.

Carriere, A., Cela, A., and Hamam, Y., 1994, Neural network-based adaptive control of a non-linear system: application to a thermal process, *Proc. IEEE Int. Conf. Systems, Man Cybernetics,* Vol. 2, pp. 1133–1138.

Castrogiovanni, A. and Sforza, P.M., 1996, Scaling of high heat flux flow boiling flow systems, *Proc. ASME Heat Transfer Division, Int. Mech. Eng. Congr. Exposition,* Vol. 334, No. 3, pp. 39–46.

Castrogiovanni, A. and Sforza, P.M., 1997, A genetic algorithm model for high heat flux flow boiling, *Exp. Thermal Fluid Sci.,* 15(3), 193.

Charniak, E. and McDermott, D., 1985, *Introduction to Artificial Intelligence,* Addison-Wesley, Reading, MA.

Chipperfield, A., 1997, Introduction to genetic algorithms, in *Genetic Algorithms in Engineering Systems,* (Eds.) A.M.S. Zalzala, P.J. Fleming, Institute of Electrical and Electronics Engineers, New York.

Chow, M. and Teeter, J., 1997, Reduced-order functional link neural network for HVAC thermal system identification and modeling, *Proc. IEEE Int. Conf. Neural Networks,* Vol. 1, pp. 5–9.

Cohen, W.C. and Johnson, E.F., 1956, Dynamic characteristics of double-pipe heat exchangers, *Ind. Eng. Chem.* 48(6), 1031–1034.

Culp, C., Haberl, J., Norford, L., Brothers, P., and Hall, J., 1990, Impact of AI technology within the HVAC industry, *ASHRAE J.,* 32(12), 12.

Curtiss, P.S., Kreider, J.F., and Brandemuehl, M.J., 1993, Artificial neural networks proof of concept for local and global control of commercial building HVAC systems, *Proc. ASME Int. Solar Energy Conf.,* pp. 429–443.

Davalos, R.V. and Rubinsky, B., 1996, An evolutionary-genetic approach to heat transfer analysis, *ASME J. Heat Transfer,* 118(3), 528.

Dhar, A., Claridge, D.E., and Ruch, D., 1995, Artificial neural network with wavelet basis functions to predict hourly heating and cooling energy use in commercial buildings with temperature as the only weather variable, *Proc. ASME-JSME-JSES Int. Solar Energy Conf.,* Vol. 1, pp. 145–156.

Diaz, G., Sen, M., Yang, K.T., and McClain, R.L., 1998, Neurocontrol of Heat Exchangers (in Spanish), *Proc. 4th Int. Meet. Mech. Eng.,* Mexico, pp. 123–128.

Diaz, G., Sen, M., Yang, K.T., and McClain, R.L., 1999, *Int. J. HVAC&R Res.,* 5(3), 195-208.

Diaz, G., Yanes, J., Sen, M., Yang, K.T., and McClain, R.L., 1996, Analysis of data from single-row heat exchanger experiments using an artificial neural network, *FED. Proc. ASME Int. Mech. Eng. Congr. Exposition,* Vol. 242, pp. 45–52.

Ding, Y. and Wong, K.V., 1990, Control of a simulated dual-temperature hydronic system using neural network approach, *ASHRAE Trans.,* 96(2), 727–732.

Dounis, A.I., Santamouris, M.J., and Lefas, C.C., 1992, Implementation of artificial intelligence techniques in thermal comfort control for passive solar buildings, *Energy Convers. Manage.,* 33(3), 175–182.

Drago, G.P. and Ridella, S., 1992, Statistically controlled activation weight initialization, *IEEE Trans. Neural Networks,* 3(4), 627–631.

Dregalin, A.F. and Nazyrova, R.R., 1993, On introduction of artificial intelligence elements to heat power engineering, *Izv. VUZ: Aviatsionnaya Tekhnika,* No. 4, pp. 90–94.

Eeckman, F.H., Ed., 1992, *Analysis and Modeling of Neural Systems,* Kluwer Academic, Boston.

Fabbri, G., 1997, A genetic algorithm for fin profile optimization, *Int. J. Heat Mass Transfer,* 40(9), 2165.

Ferrano, F.J. and Wong, K.-F.V., 1990, Prediction of thermal storage loads using a neural networks, *ASHRAE Trans.,* Pt. 2, 723–726.

Flood, I. and Kartam, N., 1994, Neural networks in civil engineering. I. Principles and understanding, *ASCE J. Comput. Civil Eng.,* 8(2), 131–148.

Fortuna, L., Muscato, G., Nunnari, G., and Papaleo, R., 1996, Neural network approach to control the temperature on rapid thermal processing, *Proc. 8th Mediterranean Electrotech. Conf. MELECON Ind. Appl. Power Systems, Comput. Sci. Telecommun.,* Part 2, Vol. 2, pp. 649–652.

Fowler, M.M., Klett, D.E., Moreno, J.B., and Heermann, P.D., 1997, Using artificial neural networks to predict the performance of a liquid sodium reflux pool boiler solar receiver, *Proc. Int. Solar Energy Conf.,* pp. 93–104.

Fujii, H. and Lutzenhiser, L., 1992, Japanese residential air-conditioning: natural cooling and intelligent systems, *Energy Build.,* 18(3–4), 221–233.

Gagarin, N., Flood, I., and Albrecht, P., 1994, Computing track attributes with artificial neural networks, *ASCE J. Comput. Civil Eng.,* 8(2), 179–200.

Gao, W., Yu, W., and Han, F., 1990, Artificial intelligence approach to the synthesis of optimized heat exchanger networks, *J. Chem. Ind. Eng. (China),* 41(3), 353.

Gartner, J.R. and Harrison, H.L., 1965, Dynamic characteristics of water-to-air cross flow heat exchangers, *ASHRAE Trans.,* 71, 212–223.

Gobovic, D. and Zaghloul, M.E., 1993, Design of locally connected CMOS neural cells to solve the steady-state heat flow problem, *Proc. 36th Midwest Symp. Circuits Systems,* Vol. 1, pp. 755–757.

Goldberg, D.E., 1989, *Genetic Algorithms in Search, Optimization and Machine Learning,* Addison-Wesley, Reading, MA.

Gray, D.L. and Webb, R.L., 1986, Heat transfer and friction correlations for plate finned-tube heat exchangers having plain fins, *Proc. 8th Int. Heat Transfer Conf.,* Vol. 6, pp. 2745–2750.

Haykin, S., 1994, *Neural Networks: A Comprehensive Foundation,* Macmillan, New York.

Hecht-Nielsen, R., 1987, Kolmogorov's mapping neural network existence theorem, *IEEE First Int. Conf. Neural Networks,* Vol. III, pp. 11–13.

Holland, J.H., 1975, *Adaptation in Natural and Artificial Systems,* University of Michigan Press, Ann Arbor.

Holland, J.H., 1992, Genetic algorithms, *Sci. Am.,* 267(1), 66.

Huang, S.-H. and Nelson, R.M., 1994, Delay time determination using an artificial neural network, *ASHRAE Trans.,* 100(1), 831–840.

Hunt, K.J., Sbarbaro, D., Zbikowski, R., and Gawthrop, P.L., 1992, Neural networks for control systems — a survey, *Automatica,* 28(6), 1083–1112.

Inoue, Y., Sugiura, S., Murakami, K., and Kotera, N., 1995, Thermal comfort and climate control, *ASHRAE J.,* 37(10), 28–32.

Ito, K., Shiba, T., and Yokoyama, R., 1995, Operational advisor for a heat pump/thermal storage system (development and evaluation of a method based on a layered neural network), *Trans. Jpn. Soc. Mech. Eng., C,* 61(590), 3909–3915.

Jambunathan, K., Hartle, S.L., Ashforth-Frost, S., and Fontama, V.N., 1996, Evaluating convective heat transfer coefficients using neural networks, *Int. J. Heat Mass Transfer,* 39(11), 2329–2332.

Jang, J.-S.R., Sun, C.-T., and Mizutani, E., 1997, *Neuro-Fuzzy and Soft Computing,* Prentice-Hall, Upper Saddle River, NJ.

Jeanette, E., Assawamartbunlue, Curtiss, P., and Kreider, J.F., 1998, Experimental results of a predictive neural network HVAC controller, *ASHRAE Trans.,* 104(2), 6.

Jenkins, W.M., 1995, Neural network-based approximations for structural analysis, in *Developments in Neural Networks and Evolutionary Computing for Civil and Structural Engineering,* (Ed.) B.H.V. Topping, Civil-Computing Press, Edinburgh, U.K., 25–35.

Jones, M.R., Tezuka, A., and Yamada, Y., 1995, Thermal tomographic detection of inhomogeneities, *ASME J. Heat Transfer,* 117(4), 969–975.

Kamarthi, S., Sanvido, V., and Kumara, R., 1992, Neuroform-neural network system for vertical formwork selection, *ASCE J. Comput. Civil Eng.,* 6(2), 178–199.

Kaminski, W., Strumillo, P., and Tomczak, E., 1996, Genetic algorithms and artificial neural networks for description of thermal deterioration processes, *Drying Technol.,* 14(9), 2117–2133.

Karmin, E.D., 1990, Simple procedure for pruning back propagation-trained neural networks, *IEEE Trans. Neural Networks,* 1(2), 239–242.

Kawashima, M., Dorgan, C.E., and Mitchell, J.W., 1995, Hourly thermal load prediction for the next 24 hours by Arima, Ewma, LR, and an artificial neural network, *ASHRAE Trans.,* 1, 186–200.

Kos, A., 1993, Approach to thermal placement in power electronics using neural networks, *Proc. IEEE Int. Symp. Circuits Systems,* Vol. 4, pp. 2427–2430.

Koza, J.R., 1992, *Genetic Programming: On the Programming of Computers by Means of Natural Selection,* MIT Press, Cambridge, MA.

Koza, J.R., 1994, *Genetic Programming II: Automatic Discovery of Reusable Programs,* MIT Press, Cambridge, MA.

Kreider, J.F., Claridge, D.E., Curtiss, P., Dodier, R., Haberl, J.S., and Krarti, M., 1995, Building energy use prediction and system identification using recurrent neural networks, *ASME J. Solar Energy Eng.,* 117(3), 161–166.

Kreider, J.F., Wang, X.A., Anderson, D., and Dow, J., 1992, Expert systems, neural networks and artificial intelligence applications in commercial building HVAC operations, *Automat. Constr.,* 1(3).

Kuroe, Y. and Kimura, I., 1995, Modeling of unsteady heat conduction field by using composite recurrent neural networks, *Proc. IEEE Int. Conf. Neural Networks,* Part 1, Vol. 1, pp. 323–328.

Lavric, D., Lavric, V., and Woinaroschy, A., 1995, Designing fin heat exchanger with a neural network, *Rev. Roum. Chim.,* 10(2), 561–565.

Lavric, D., Lavric, V., Muntean, O., and Danciu, E., 1994, Auto-organizing algorithm for design of fin heat exchangers, *Rev. Roum. Chim.,* 39(11), 1241–1256.

Lee, D.-G., Kim, H.-G., Baek, W.-P., and Chang, S.H., 1997, Critical heat flux prediction using genetic programming for water flow in vertical round tubes, *Int. Commun. Heat Mass Transfer,* 24(7), 919–929.

Lehtokangas, M., Saarinen, J., and Kaski, K., 1995, Initializing weights of a multilayer perceptron network by using the orthogonal least square algorithm, *Neural Comput.,* 7, 982–999.

Leitch, R. and Quek, C., 1992, Architecture for integrated process supervision, *IEE Proc. Part D: Control Theory and Applications,* Vol. 139, No. 3, pp. 317–327.

Li, H.Y. and Yang, C.Y., 1997, A genetic algorithm for inverse radiation problems, *Int. J. Heat Mass Transfer,* 40(7), 1545.

Li, X., Vaezi-Nejad, H., and Visier, J.-C., 1996, Development of a fault diagnosis method for heating systems using neural networks, *ASHRAE Trans.,* 102(1), 607–614.

Loveday, D.L. and Virk, G.S., 1992, Artificial intelligence for buildings, *Appl. Energy,* 41(3), 201–221.

Mahajan, R.L. and Wang, X.A., 1993, Neural network models for thermally based microelectronic manufacturing processes, *J. Electrochem. Soc.,* 140(8), 2287–2293.

Matsumoto, R., Kikkawa, S., and Senda, M., 1996, Effect of pin fin arrangement on endwall heat transfer, *Trans. Jpn. Soc. Mech. Eng. B,* 62(597), 1953–1961.

Matsumoto, R., Kikkawa, S., and Senda, M., 1997, Effect of pin fin arrangement on endwall heat transfer, *JSME Int. J. B: Fluids Thermal Eng.,* 40(1), 142–151.

McQuiston, F.C., 1978a, Heat, mass and momentum transfer data for five plate-fin-tube heat transfer surfaces, *ASHRAE Trans.,* 84(1), 266–293.

McQuiston, F.C., 1978b, Correlation of heat, mass and momentum transport coefficients for plate-fin-tube heat transfer surfaces with staggered tube, *ASHRAE Trans.,* 84(1), 294–309.

Meysenc, L., Saludjian, L., Bricard, A., Rael, S., and Schaeffer, C., 1997, High heat flux IGBT micro exchanger setup, *IEEE Trans. Components Packag. Manuf. Technol. A,* 20(3), 334–341.

Michalewicz, Z., 1992, *Genetic Algorithms + Data Structures = Evolution Programs,* 2nd ed., Springer-Verlag, Berlin.

Mistry, S.I. and Nair, S.S., 1993, Nonlinear HVAC computations using neural networks, *ASHRAE Trans.,* 99(1), 775–784.

Mitchell, M., 1997, *An Introduction to Genetic Algorithms,* MIT Press, Cambridge, MA.

Momose, K., Murai, T., Asami, T., and Hosokawa, Y., 1993, Identification of heat-source using neural network, *Trans. Jpn. Soc. Mech. Eng. C,* 59(567), 3431–3436.

Montgomery, W.L., Ross, B.A., and Penswick, L.B., 1996, Third generation development of an 11-watt Stirling converter, *Proc. 31st Intersoc. Energy Convers. Eng. Conf., IECEC 96, Part 2,* Vol. 2, pp. 1243–1248.

Moon, C. and Cho, B., 1996, Design of fuzzy control system for chemical injection system retrofit using neural network model in thermal power plant, *Proc. 1st Int. Workshop Neural Networks Identification, Control, Robotics, and Signal/Image Process.,* pp. 263–269.

Mozley, J.M., 1956. Predicting dynamics of concentric pipe heat exchangers, *Ind. Eng. Chem.,* 1035–1041.

Naka, M., Imai, T., Shida, T., Sato, M., Ito, Ry, and Akamine, I., 1993. Thermal image processing using neural network, *Proc. Int. Joint Conf. Neural Networks,* Part 3, Vol. 3, Nagoya, Japan, pp. 2065–2068.

Narendra, K. and Mukhopadhyay, S., 1996, Intelligent control using neural networks, in (Eds.) M. Gupta, N. Sinha, *Intelligent Control Systems, Theory and Applications,* Institute of Electrical and Electronics Engineers, New York, pp. 151–186.

Normandin, A., Grandjean, B.P.A., and Thibault, J., 1993, PVT data analysis using neural network models, *Ind. Eng. Chem. Res.,* 32(5), 970–975.

Okamoto, K., Tsuru, D., and Fumizawa, M., 1996, Reconstruction of three-dimensional density distribution from the limited projection images with genetic algorithm, *J. Flow Visual. Image Process.,* 3(4), 253.

Pacheco-Vega, A., Sen, M., Yang, K.T., and McClain, R.L., 1998, Genetic-Algorithm Based Prediction of a Fin-Tube Heat Exchanger Performance, *Proc. 11th Int. Heat Transfer Conf.,* Vol. 6, 137–142.

Pacheco-Vega, A., Sen, M., Yang, K.T., and McClain, R.L., 1999, Prediction of humid air heat exchanger performance using artificial neural networks, *ASME IMECE,* 1999.

Paoletti, B. and Sciubba, E., 1997, Artificial intelligence in thermal systems design: concepts and applications, in *Developments in the Design of Thermal Systems,* (Ed.) R.F. Boehm, Cambridge University Press, New York.

Pican, N., Alexandre, F., Haton, J.-P., and Bresson, P., 1998, Neurocontrol of temperatures with OWE neural network architecture in continuous steel annealing furnace, in *Industrial Applications of Neural Networks,* (Eds.) F.F. Soulie, P. Gallinari, World Scientific, Singapore, pp. 9–15.

Psichogios, D. and Ungar, L.H., 1991, Direct and indirect model based control using artificial neural networks, *Ind. Chem. Res.,* 30, 2564–2573.

Queipo, N., Devarakonda, R., and Humphrey, J.A.C., 1994, Genetic algorithms for thermosciences research: application to the optimized cooling of electronic components, *Int. J. Heat Mass Transfer,* 37(6), 893.

Queipo, N.V. and Gil, G.F., 1997, Multiobjective optimization of component placement on planar printed wiring boards, *Proc. IEEE 13th Annu. Semiconductor Thermal Measure. Manage. Symp.,* pp. 92–105.

Rasmussen, K.H., Nielsen, C.S., and Jorgensen, S., 1990, Identification of distillation process dynamics comparing process knowledge and black box based approaches, *Proc. Am. Control Conf.,* pp. 3116–3121.

Raudensky, M., Woodbury, K.A., and Brezina, T., 1995, Genetic algorithm in solution of inverse heat conduction problems, *Num. Heat Transfer B, Fundam.,* 28(3), 293.

Rich, E. and Knight, K., 1991, *Artificial Intelligence,* 2nd ed., McGraw-Hill, New York.

Rogers, J.I., 1994, Simulating structure analysis with neural networks, *ASCE J. Comput. Civil Eng.,* 8(2), 252–265.

Rumelhart, D.E., Hinton, D.E., and Williams, R.J., 1986, Learning internal representations by error propagation, in *Parallel Distributed Processing: Exploration in the Microstructure of Cognition,* Vol. 1, MIT Press, Cambridge, MA.

Rzempoluck, E.J., 1998, *Neural Network Data and Analysis Using Simulnet,* Springer-Verlag, New York.

Sandberg, C., Whitney, W., Nassar A., and Kuse, G., 1995, Intelligent distributed heater used for industrial applications, *Proc. IEEE Int. Conf. Systems, Man Cybernetics, Part 4,* Vol. 4, pp. 3346–3351.

Schalkoff R., 1997, *Artificial Neural Networks,* McGraw-Hill, New York.

Schmidt, F.W. and Willmott, A.J., 1981, *Thermal Energy Storage and Regeneration,* McGraw-Hill, New York.

Schmit, T.S., Dhingra, A.K., Landis, F., and Kojasoy, G., 1996, Genetic algorithm optimization technique for compact high intensity cooler design, *J. Enhanced Heat Transfer,* 3(4), 281-290.

Spiga, M. and Spiga, G., 1992, Step response of the cross-flow heat-exchanger with finite wall capacitance, *Int. J. Heat Mass Transfer,* 35(2), 559–565.

Takemori, T., Miyasaka, N., and Hirose, S., 1991, Neural network air-conditioning system for individual comfort, *Proc. SPIE Int. Soc. Optical Eng. Appl. Artificial Neural Networks II,* Vol. 1469, pt 1, pp. 157–165.

Tang, M.C. and Carothers, J.D., 1996, Multichip module placement with heat consideration, *Proc. 9th Annu. IEEE Int. ASIC Conf. Exhibit,* pp. 175–178.

Tanimoto, S.L., 1990, *The Elements of Artificial Intelligence Using Common LISP,* Computer Science Press, New York.

Terry, P.A. and Himmelblau, D.M., 1993, Data rectification and gross error detection in a steady-state process via artificial neural networks, *Ind. Eng. Chem. Res.,* 32(12), 3020–3028.

Thal-Larsen, H., 1960, Dynamics of heat exchangers and their models, *ASME J. Basic Eng.,* 82, 489–504.

Thibault, J. and Grandjean, B.P.A., 1991, Neural network methodology for heat transfer data analysis, *Int. J. Heat Mass Transfer,* 34(8), 2063–2070.

von Wolfersdorf, J., Achermann, E., and Weigand, B., 1997, Shape optimization of cooling channels using genetic algorithms, *ASME J. Heat Transfer,* 119(2), 380.

Waldemark, J., Wiklund, H., Andersson, S., and Sandberg, O., 1992, Neural network modeling of the heat load in district heating systems, *Dist. Heat. Int.* 21(9), 7.

Warwick, K., Irwin, G.W., and Hunt, K.J., 1992, *Neural Networks for Control and Systems,* Short Run Press, Exeter, U.K.

Wessels, L. and Barnard, E., 1992, Avoiding fake local minima by proper initialization of connections, *IEEE Trans. Neural Networks,* 3(6), 899–905.

Wood, R.L., 1996, Inverse thermal field problem based on noisy measurements: comparison of a genetic algorithm and the sequential function specification method, *Eng. Comput.* 13(6), 43–59.

Yamashita, H., Zzumi, R., and Yamaguchi, S., 1978, Analysis of the dynamic characteristics of cross-flow heat exchangers with both fluids mixed, *Bull. JSME,* 21(153), 479–485.

Yang, K.H., Su, C.H., and Hwang, R.L., 1996, The analysis on intelligent control strategies of a thermal energy storage air-conditioning system, *Energy,* 21(4), 319.

Yang, Q.-H., Ma, H., Liu, X., and Shi, H., 1995, Thermal deformation compensation for a coordinate grinder by neural networks, *Proc. SPIE Int. Soc. Optical Eng. Int. Conf. Intelligent Manuf.,* pp. 499–502.

Yentis, R., Jr. and Zaghloul, M.E., 1994, CMOS implementation of locally connected neural cells to solve the steady-state heat flow problem, *Proc. 37th Midwest Symp. Circuits Systems, Part 1,* Vol. 1, Lafayette, LA, pp. 503–506.

Zbicinski, I., Strumillo, P., and Kaminski, W., 1996, Hybrid neural model of thermal drying in a fluidized bed, *Comput. Chem. Eng.,* 20(Suppl. A), S695–S700.

Zeng, P., 1998, Neural Computing in Mechanics, *Appl. Mech. Rev.,* 51(2), 173–197.

Zhao, X., 1995, Performance of a Single-Row Heat Exchanger at Low In-Tube Flow Rates, M.S. thesis, Department of Aerospace and Mechanical Engineering, University of Notre Dame, Notre Dame, IN.

Zhao, X., Sen, M., Yang, K.T., and McClain, R.L., 1995, An artificial neural network model of a heat exchanger, *Symp. Thermal Sci. Eng. in Honor of Chancellor Chang-Lin Tien,* pp. 83–88.

5

Numerical Analysis and Computational Tools

Frank Hagin

Colorado School of Mines

5.1 Computer-Aided Engineering (CAE)

Since the 1950s the digital computer has played an ever-increasing role in engineering studies. Particulary since the advent of FORTRAN and the subsequent development of libraries in this language in the late 1950s, an exponentially growing amount of software has become available. Those years represent the beginning of the era of *Computer-Aided Engineering*. More recently this term is applied to more user-friendly software designed to be used in real time on a PC or workstation. This section presents the type of *algorithms* (computing schemes) and software commonly used today in thermal engineering.

One of the most important developments in computational mathematics is in the numerical (approximate) solution of partial differential equations. The primary approach until recently has been the use of *finite differences* to approximate partial derivatives, and this approach is the topic of the next subsection. Following that, we deal with the *Finite Element* method and then introduce the *Boundary Element* method.

0-8493-9581-X/00/$0.00+$.50

5.2 Finite Difference Methods

Here the basic concepts of the finite difference method are introduced, including the notions of *accuracy* (or *rate of convergence)* and *stability*. Over the past several decades highly accurate and stable finite difference techniques have been developed for all areas of engineering, including heat transfer. Although we are treating linear problems, the concepts introduced apply to nonlinear problems as well.

Finite Differences and Their Order of Accuracy

For simplicity, first consider finite difference approximations to derivatives of a function f of one variable. The *forward difference* approximation to the first derivative is

$$f'(x_0) \cong \frac{f(x_0+h)-f(x_0)}{h}, \tag{5.1}$$

where h denotes a small step size. To introduce the *order* of such an approximation, expand $f(x_0 + h)$ in the Taylor series about x_0 and obtain:

$$\frac{f(x_0+h)-f(x_0)}{h} = f'(x_0)+\frac{h}{2}f''(x_0)+\frac{h^2}{6}f'''(x_0)+\text{H.O.T.} \tag{5.2}$$

where H.O.T. stands for *higher order terms* in the Taylor expansion. Assuming h is very small, the error in Equation 5.1 is approximately $\frac{h}{2}f''(x_0)$; hence the approximation in Equation 5.1 is of *order* h and is denoted by $O(h)$. This is not a very satisfactory approximation; for example, if one needs an accuracy of 0.001 this would require a very small step size h. Hence, approximations of order h^2 or higher are needed. For example, an order h^2, $O(h^2)$, approximation to the first derivative is the *central* difference formula:

$$f'(x_0) \cong \frac{f(x_0+h)-f(x_0-h)}{2h}.$$

For later purposes an important approximation, also of $O(h^2)$, is that of f'',

$$f''(x_0) \cong \frac{f(x_0+h)-2f(x_0)+f(x_0-h)}{h^2}. \tag{5.3}$$

Approximate Solutions to a One-Dimensional Heat Equation

Consider the following one-dimensional, transient heat flow problem:

$$\frac{\partial T(x,t)}{\partial t}-\frac{1}{k}\frac{\partial^2 T(x,t)}{\partial x^2} = f(x,t). \tag{5.4}$$

One seeks approximate solutions to Equation 5.4, with appropriate boundary conditions. At discrete point (x_i, y_j) denote the step sizes between these points by $h = x_{i+1} - x_i$ and $\Delta t = t_{j+1} - t_j$. Apply the Approximation 5.1 to the t derivative in Equations 5.4 and Approximation 5.3 to the x derivative. It is usually desired to move *forward* in the t direction since one typically knows T at $t = 0$; so take a small time step, $\Delta t = t_1$, and solve for the values $T(x_i, t_1)$, with ranging x_i between a to b. Having the approximate solution at the t_1 level, proceed to the next time level, and so on. Hence, writing the approximation to

Equation 5.4, dropping the approximation symbols, and using the notation $T_{i,j} = T(x_i, t_j)$, one has this relationship between T values at the jth and $(j + 1)^{st}$ levels:

$$\frac{T_{i,j+1} - T_{i,j}}{\Delta t} = \frac{1}{k}\frac{T_{i+1,j} - 2T_{i,j} + T_{i-1,j}}{h^2} + f_{i,j}. \tag{5.5}$$

Solving for the T at the $(j + 1)^{st}$ level one gets:

$$T_{i,j+1} = (1 - 2\,\lambda)\,T_{i,j} + \lambda\left(T_{i+1,j} + T_{i-1,j}\right) + \Delta t\, f_{i,j}$$

where $\lambda = \dfrac{\Delta t}{kh^2}$ and $i = 1, 2, \ldots, n$.

It is convenient to write such expressions in matrix form; letting $\boldsymbol{T}_J$ denote the vector $(T_{1,j}, T_{2,j}, \ldots, T_{n,j})$, one can express the set of Equations (5.5) as $\boldsymbol{T}_{j+1} = \boldsymbol{A}\ \boldsymbol{T}_j + \boldsymbol{f}_j$ where

$$\boldsymbol{A} = \begin{bmatrix} (1-2\lambda) & \lambda & 0 & \cdots & 0 \\ \lambda & (1-2\lambda) & \lambda & \cdots & 0 \\ 0 & \ddots & \ddots & \ddots & \vdots \\ \vdots & \ddots & \ddots & \ddots & \lambda \\ 0 & \cdots & 0 & \lambda & (1-2\lambda) \end{bmatrix} \tag{5.6}$$

Since the next time-level values $\boldsymbol{T}_{j+1}$ are computed directly from those at the previous level, this scheme is called *explicit*. Also, since it started from a forward difference in the time direction, the scheme is also called a *forward difference* algorithm. The advantage of explicit schemes is that they are relatively simple (basically, a matrix multiplication in this case); but, as illustrated below, they tend to have stability problems.

A Numerical Example — Let $T(x, t)$ denote the temperature in a thin rod of unit length and constant cross section and homogeneous conductivity k. If heat is generated in the rod, e.g., by resistance to an electrical current, the heat equation becomes

$$k\frac{\partial T(x,t)}{\partial t} = \frac{\partial^2 T(x,t)}{\partial x^2} + \frac{kq'''}{\rho c}$$

where k is the thermal diffusivity, ρ is the density, and c is the specific heat of the rod; q‴ represents the rate of heat generation per unit volume. Consider the units:

$k = \pi^2$ *J*/cm sec°*C*; $\rho = 4$ g/cm³;
$c = 0.01\ \pi^2$ *J*/gm°*C*; $q''' = 0.08$ *J*/sec cm³.

Here J represents unit of heat, joule, and $\pi \cong 3.1415927$. Applying these units to the above equation leads to the PDE below. Further, assume the ends of the rod of unit length are held at $T = 0$ and the initial temperature distribution is $\sin \pi x + x(1 - x)$. This leads to the BVP:

$$\frac{\partial T(x,t)}{\partial t} = \frac{1}{\pi^2}\frac{\partial^2 T(x,t)}{\partial x^2} + \frac{2}{\pi^2}, \quad 0 \preceq x \preceq 1,\ t \succeq 0, \tag{5.7}$$
$$T(x, 0) = \sin \pi x + x(1 - x), \quad T(0,t) = 0, \quad T(1,t) = 0.$$

As the reader can verify, the exact solution is $T(x, t) = e^{-t} \sin \pi x + x(1 - x)$; hence, one can check numerical approximations for accuracy. As will be seen, the choice of the time step, Δt, and x step, h, is critical. The

TABLE 5.1 Forward Difference Algorithm Applied to Boundary-Value Problem in Equation 5.7

x_i	$T(x_i, 0.5)$	$v_{i,500}$	$\lvert T_i - v_i\rvert$	$\lvert T_i - v_i^{(n=69)}\rvert$	$\lvert T_i - v_i^{(n=71)}\rvert$
0.1	0.2774	0.2782	0.0007	0.0000311	5183
0.2	0.5165	0.5179	0.0014	0.0000592	9559
0.3	0.7007	0.7026	0.0019	0.0000815	etc.
0.4	0.8168	0.8191	0.0022	0.0000958	etc.
0.5	0.8565	0.8589	0.0023	0.0001010	etc.
0.6	0.8168	0.8191	0.0022	0.0000958	etc.
0.7	0.7007	0.7026	0.0019	0.0000815	etc.
0.8	0.5165	0.5179	0.0014	0.0000592	etc.
0.9	0.2774	0.2782	0.0007	0.0000311	etc.

first numerical approximation is with $h = 0.1$ and a small time step, $\Delta t = 0.001$, taking 500 time steps to end up at $t = 0.5$. Table 5.1 shows the results of this run, where T denotes the exact solution and v denotes the approximate solution using $n = 9$ (i.e., 9 interior x points, hence $h = 0.1$). The results are accurate to only two significant figures, so the error is about 1%. But this is expected since the error is of order $O(\Delta t + h^2)$; that is, linear in Δt and quadratic in h as discussed earlier in the approximations to f' and f''.

The fifth column in the table shows the reduction in errors in v when n is increased to 69. However, the sixth column shows what happens when one increases n further, to 71. The results are totally useless because at this value of n and larger, the algorithm becomes *unstable*. For this forward difference scheme the stability is determined by the eigenvalues of the matrix $\boldsymbol{A}$ in Equation 5.6; in particular, the largest one must be no larger than 1.0 in absolute value. Stated differently, one needs the *spectral radius* $\rho(\boldsymbol{A})$ to be no larger than 1.0 It can be shown that stability for this algorithm applied to Problem 5.4 requires

$$\lambda = \frac{\Delta t}{kh^2} \preceq \frac{1}{2}.$$

For example, see Isaacson and Keller.[1] In the case of the Problem 5.7, $k = \pi^2$; hence solving this last inequality with $\Delta t = .001$ shows that one needs h to be at least 0.0142. This means that n must be smaller than 70, as suggested by the numerical results. This exhibits the problem with *explicit* schemes like the current one — to get h small enough to get a high degree of accuracy (say five or six digits) requires an extremely small time step. The next topic introduces a more practical scheme, one which is both more accurate and without the stability restriction.

The Crank-Nicolson Algorithm

In addressing the above stability issue and in seeking more accurate methods, researchers have studied various combination of forward difference and backward difference (in the t direction) schemes. A widely used method is that of Crank-Nicolson (C-N) which will be applied to above boundary value problem Equation 5.7. To that end, consider the matrix

$$\boldsymbol{B} = \begin{bmatrix} 2 & -1 & 0 & \cdots & 0 \\ -1 & 2 & -1 & \cdots & 0 \\ 0 & \ddots & \ddots & \ddots & \vdots \\ \vdots & \ddots & \ddots & \ddots & -1 \\ 0 & \cdots & 0 & -1 & 2 \end{bmatrix}$$

Letting $\boldsymbol{I}$ denote the identity matrix one can now form the following two matrices:

$$\boldsymbol{C} = 2\boldsymbol{I} + \lambda\boldsymbol{B} \quad \text{and} \quad \boldsymbol{D} = 2\boldsymbol{I} - \lambda\boldsymbol{B}.$$

Below, these matrices are formed using the MATLAB system. Once again, assuming that T is known at time level j, solve $\boldsymbol{T} = \boldsymbol{T}_{j+1}$ from the following:

$$\boldsymbol{C}\boldsymbol{T}_{j+1} = \boldsymbol{D}\boldsymbol{T}_j + 2\Delta t\, \boldsymbol{f}_j . \tag{5.8}$$

The C-N formula in Equation 5.8 results from averaging the forward difference formula and the *backward difference* analog, based on the approximation

$$g'(t_j) \cong \frac{g(t_j) - g(t_j - \Delta t)}{\Delta t}.$$

This scheme in Equation 5.8 is *implicit* (as opposed to *explicit*) in the sense that one must solve the algebraic system involving matrix $\boldsymbol{C}$. However, since both $\boldsymbol{C}$ and $\boldsymbol{D}$ consist of only three diagonal "bands", the computation is only of order n for each time step in contrast to order n^3 if matrix $\boldsymbol{C}$ were full.

The following table shows the results using the Crank-Nicolson algorithm on the Problem in Equation 5.7. T_i denotes the exact solution and v_i the approximate solution. Columns 3 and 4 are the results with a larger time step, $\Delta t = 0.1$, and with $h = 0.1$, as before. Note these results are similar to the explicit scheme shown in Table 5.1 (with Δt and $h = 0.1$) since the C-N scheme has an error of $O(h^2 + \Delta t^2)$, as opposed to order $O(h + \Delta t^2)$ for the forward difference method. In the fifth column of Table 5.2 are the results when one decreases both steps, $h = \Delta t = 0.01$; note the improved accuracy due to the *quadratic* error in both h and Δt.

TABLE 5.2 Crank-Nicolson Algorithm Applied to the Problem in Equation 5.7 with h = Δt = 0.1 and h = Δt = 0.01

x_i	T_i	$v_{i,50}$	$\lvert T_i - v_i\rvert$	$\lvert T_i - v_i^{(n=99)}\rvert$
0.1	0.2774	0.2481	0.0007	0.00000759
0.2	0.5165	0.5178	0.0013	0.00001374
0.3	0.7007	0.7025	0.0018	0.00001854
0.4	0.8168	0.8190	0.0021	0.00001854
0.5	0.8565	0.8588	0.0022	0.00002240
0.6	0.8168	0.8190	0.0021	0.00002153
0.7	0.7007	0.7025	0.0018	0.00001854
0.8	0.5165	0.5178	0.0013	0.00001374
0.9	0.2774	0.2481	0.0007	0.00000759

MATLAB® (The MathWorks, Inc., Natick, MA, 01760) is a very powerful and widely used software package for doing matrix algebra and other related computations. Below shows how the matrices for the C-N algorithm are set up and 50 time steps are taken. Assume n (the number of x values), $\boldsymbol{d}$ (the n-vector of the given $\boldsymbol{T}$ values at t = 0) and the "forcing" vector f have all been defined. The MATLAB commands are as follows, where anything following the % symbol is a comment and the semicolon both ends a statement and suppresses output.

```
dt = .01; h = 1/(n+1); lambda = dt/(h*pi)^2;     % dt is the time step
w = ones(1, n-1);                                % a vector of n-1 ones
B = 2*eye(n) - diag(w, 1) - diag(w, -1)          % creates the B matrix; eye is 'identity'
C = 2*I + lambda*B; D = 2*I - lambda*B;
v = d;
for j = 1:1:50                                   % doing the 50 time steps
  u = D*v + 2*dt*f;
  v = C\u;                                       % this solves Cv = u for v
end
v                                                % sends final v to the screen
```

Note: while the above short program is quite adequate for small problems like the one illustrated, it is not recommended for large problems. In particular, for problems requiring large matrices one should especially take advantage of the special form of matrices $\boldsymbol{C}$ and $\boldsymbol{D}$. Since they are *tridiagonal*, i.e., with only three nonzero diagonal elements, the system (5.8) can be solved very efficiently. Such considerations are discussed below.

A Steady-State Problem — An Elliptic PDE

Next, consider a typical two-dimensional steady-state problem which leads to an elliptic differential equation. The main concern with such problems is not stability, but rather that of dealing with very large systems of equations (linear, in the current case). The differential equation to be studied is the steady-state two-dimensional conduction equation:

$$\frac{\partial^2 T(x,y)}{\partial x^2} + \frac{\partial^2 T(x,y)}{\partial y^2} = 0.$$

This is *Laplace's* equation. Should the right side of the equation be some nonzero function, it is called the *Poisson* equation; it would be treated essentially the same. Once again, the second derivatives are approximated using Equation 5.3. As before let $T_{i,j} = T(x_i, y_j)$ and denote the steps between grid points by $h = \Delta x$ and $k = \Delta y$; one can express the approximation to Laplace's equation by

$$\frac{T_{i+1,j} - 2T_{i,j} + T_{i-1,j}}{h^2} + \frac{T_{i,j+1} - 2T_{i,j} + T_{i,j-1}}{k^2} = 0. \tag{5.9}$$

The boundary value problems to be discussed here will be for rectangular regions in the xy-plane with thermal conditions given on the four boundary lines (the so-called *Dirichlet* problem). Systems involving Equation 5.9 will be solved in which the points (x_i, y_j) are interior grid points where $i = 1, \ldots, m$ and $j = 1, \ldots, n$. Therefore, one needs to solve a system of $N = mn$ equations in N unknowns. For example, consider the figure below in which the rectangular area is divided into 16 sections with 9 interior points labeled 1, …, 9. The nine $T_{i,j}$ values will be solved at these points. In the Dirichlet problem, the data (values of T) would be provided on the boundary of the region.

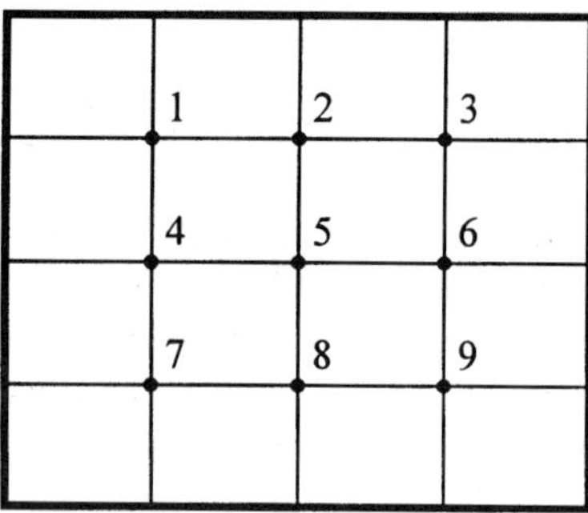

FIGURE 5.1 Domain with nine interior points.

Introduce the constants $r = h^2/k^2$ and $R = (1+r)$ in Equation 5.9 and write the equations:

$$2RT_{i,j} - T_{i-1,j} - T_{i+1,j} - rT_{i,j-1} - rT_{i,j+1} = 0;$$
$$i = 1, \ldots, m; \quad j = 1, \ldots, n. \tag{5.10}$$

Leaving the two-dimensional representation of the unknowns, $T_{i,j}$, using the ordering illustrated in the figure above, and labeling the unknowns by $v_1, \ldots, v_9$, the system in Equation 5.10 can be put in matrix form $\boldsymbol{Av} = \boldsymbol{d}$:

$$Av = \begin{bmatrix} 2R & -1 & 0 & -r & 0 & 0 & 0 & 0 & 0 \\ -1 & 2R & -1 & 0 & -r & 0 & 0 & 0 & 0 \\ 0 & -1 & 2R & 0 & 0 & -r & 0 & 0 & 0 \\ -r & 0 & 0 & 2R & -1 & 0 & -r & 0 & 0 \\ 0 & -r & 0 & -1 & 2R & -1 & 0 & -r & 0 \\ 0 & 0 & -r & 0 & -1 & 2R & 0 & 0 & -r \\ 0 & 0 & 0 & -r & 0 & 0 & 2R & -1 & 0 \\ 0 & 0 & 0 & 0 & -r & 0 & -1 & 2R & -1 \\ 0 & 0 & 0 & 0 & 0 & -r & 0 & -1 & 2R \end{bmatrix} \begin{bmatrix} v_1 \\ v_2 \\ \vdots \\ \vdots \\ \vdots \\ \vdots \\ \vdots \\ v_9 \end{bmatrix} = \begin{bmatrix} D \\ A \\ T \\ A \\ \vdots \\ \vdots \\ \vdots \\ \vdots \\ \vdots \end{bmatrix} = \boldsymbol{d}. \quad (5.11)$$

When $h = k$, then $r = 1$, $2R = 4$ and the result is a famous matrix with 4's on the main diagonal and -1 on four other diagonals. (However, note that $a_{3,4} = a_{4,3} = 0$.) This procedure will now be applied to the following boundary value problem:

$$\frac{\partial^2 T(x,y)}{\partial x^2} + \frac{\partial^2 T(x,y)}{\partial y^2} = 0;$$

$$T(0,y) = 0 = T(\pi, y); \quad (5.12)$$

$$T(x,0) = 0,\ T(x,2) = \sin x \sinh 2.$$

The exact solution is $T = \sin x \sinh y$. Below are the MATLAB commands to solve this problem. First, some numerical approximations are shown in the following table in which several results are given for $x = \pi/2$ (errors tend to be largest near the middle of the rectangle). Note that the predicted error is $O(h^2 + k^2)$; and comparing columns 3, 4, and 5 shows that the error decreases by a factor of four as h and k are successively halved.

TABLE 5.3 Results of Solving Boundary-Value Problem in Equation 5.12 by Direct Method

y	$T(\pi/2, y)$	$\lvert T - v_{m=7,\,n=5}\rvert$	$\lvert T - v_{m=15,\,n=11}\rvert$	$\lvert T - v_{m=31,\,n=23}\rvert$
5/3	2.55281	0.0076	0.002001	0.000501
4/3	1.76504	0.0104	0.002634	0.000661
1	1.17520	0.0098	0.002465	0.000618
2/3	0.71716	0.0073	0.001838	0.000461
1/3	0.33954	0.0038	0.000971	0.000243

While starting to get some reasonable results with $m = 31$ and $n = 23$, note that the resulting system of equations is getting rather large, specifically $N = mn = 713$. Unless one takes advantage of the special structure of the matrix $\boldsymbol{A}$, solving such systems becomes time-consuming, especially if a high degree of accuracy is called for. Hence, for elliptic problems of this type it is generally preferable to solve the resulting large algebraic systems by *iteration*, the next subject.

Some Iterative Methods for Linear Equations

To illustrate some iterative techniques return to the problem in Equation 5.12. Such problems are well suited to iteration since the matrix $\boldsymbol{A}$ in Equation 5.11 is *diagonally dominate*, i.e., the main diagonal contains $2R \cong 4$ whereas the rest of each row consists of zeros and, at most, four values around 1.0.

A widely used scheme is called *Gauss-Seidel.* As in all iterative methods, one starts with an initial estimate of the value in question, the vector $\mathbf{v}$. Unless better information is available, start with an initial guess of $\mathbf{v} = \mathbf{0}$; having an "old" value of $\mathbf{v}$ compute the new one, and continue this process until two successive approximations agree to within a given tolerance. Designate the old, or *previous*, value by $\mathbf{v}^p$ and the current value by simply $\mathbf{v}$. Assume matrix $\boldsymbol{A}$ is available as in Equation 5.11, although only the main diagonal and m elements on each side of it are used (since all other elements of $\boldsymbol{A}$ are zero). The following description and code is easily modified in case $\boldsymbol{A}$ is full; it can also be made more efficient by only using nonzero elements. Notice in particular that as one computes the components v_i of the new vector, one always uses the most recent components available. In these equations all summations are on j.

$$\mathrm{v}_1 = \frac{1}{2R}\left[d_1 + \sum_{2}^{1+m} a_{i,j}\mathrm{v}_j^p\right],$$

$$\mathrm{v}_i = \frac{1}{2R}\left[d_i + \sum_{(i-m)'}^{i-1} a_{i,j}\mathrm{v}_j + \sum_{i+1}^{(i+m)'} a_{i,j}\mathrm{v}_j^p\right],\ i = 2,\ldots,N-1 \tag{5.13}$$

$$\mathrm{v}_N = \frac{1}{2R}\left[d_N + \sum_{N-m}^{N-1} a_{i,j}\mathrm{v}_j\right]$$

where $(i - m)'$ denotes Max$[i - m, 1]$ and $(i + m)'$ denotes Min$[i + m, N]$.

Table 5.4 shows the result of taking 20 iterations of the set of Equations 5.13 on the problem in Equation 5.12 starting with $\mathbf{v} = 0$. The G-S iterations are said to *converge*, when they actually converge to the solution to finite difference approximation, *not* to the exact solution to the original problem, Equation 5.12 in this case. In Table 5.4 the superscripts (i.e., 20 and 100) on $\mathbf{v}$ indicated the number of iterations performed. Column three of the table shows the expected results for the $m = 7$ and $n = 5$ case; results much like column 3 of Table 5.3. However, in column 4, with $m = 15$ and $n = 11$, note the results are useless since convergence has hardly begun. In the fifth column, with 100 iterations, the results are finally comparable with the results of column 4 of Table 5.3.

TABLE 5.4 Results of Gauss-Seidel Iteration on Problem in Equation 5.12

y	$T(\pi/2, y)$	$\lvert T - \mathbf{v}^{20}_{m=7,\,n=5}\rvert$	$\lvert T - \mathbf{v}^{20}_{m=15,\,n=11}\rvert$	$\lvert T - \mathbf{v}^{100}_{m=15,\,n=11}\rvert$
5/3	2.55281	0.0003	0.2828	0.0008
4/3	1.76504	0.0022	0.4440	0.0020
1	1.17520	0.0033	0.4630	0.0026
2/3	0.71716	0.0027	0.3653	0.0023
1/3	0.33954	0.0013	0.1954	0.0006

The reason for the slow convergence in the $m = 15$ and $n = 11$ case is discussed below. The G-S scheme can be put in matrix form by first splitting the matrix $\boldsymbol{A}$ into the sum of three N by N matrices; i.e., $\boldsymbol{A} = \boldsymbol{D} - \boldsymbol{L} - \boldsymbol{U}$. Here $\boldsymbol{D}$ consists of the main diagonal elements, $a_{i,i} = 2R$, with zeros elsewhere; $\boldsymbol{L}$ has the elements below the main diagonal of $\boldsymbol{A}$ and $\boldsymbol{U}$, the elements above the main diagonal. The sth G-S iterate can be expressed:

$$(\boldsymbol{D}-\boldsymbol{L})\mathrm{v}^s = \boldsymbol{U}\mathbf{v}^{s-1} + \boldsymbol{d},\ \text{ or}$$

$$\mathrm{v}^s = (\boldsymbol{D}-\boldsymbol{L})^{-1}\boldsymbol{U}\mathbf{v}^{s-1} + (\boldsymbol{D}-\boldsymbol{L})^{-1}\boldsymbol{d}.$$

The matrix $\boldsymbol{M} = (\boldsymbol{D} - \boldsymbol{L})^{-1}\,\boldsymbol{U}$ is key to the convergence of such problems. Let $\rho = \rho(M)$ denote the *spectral radius* of $\boldsymbol{M}$; i.e., the largest (in absolute value) of the eigenvalues of $\boldsymbol{M}$. Denote the error the sth iteration by $e_s = \|\mathbf{v}^s - \mathbf{v}_{exact}\|$, where $\|\cdot\|$ denotes the *norm* of a vector (here the norm used is $\|\mathbf{v}\| = \max|v_i|$, the largest component of the vector). The following fundamental inequality is helpful for the current purposes:

$$e_k \preceq \rho^k e_0$$

where e_0 is the error in the initial value for $\mathbf{v}$. This inequality shows that convergence demands $\rho \prec 1.0$. Fortunately, this is the case for matrices associated with the heat equation under study. However, the problem the above table (column 4) illustrates is that as N gets large ρ gets very close to 1.0. In the $m = 15$, $n = 11$ case, ρ is about 0.945; hence $\rho^{20} \cong (0.945)^{20} \cong 0.323$, and after 20 iterations the error in $\mathbf{v}$ is expected to be about 1/3 of the original error. In contrast to that case, when $m = 7$ and $n = 5$ it happens that $\rho \cong 0.793$; hence $\rho^{20} \cong (0.793)^{20} \cong 0.0097$ and one can expect a *reasonable* accuracy, i.e., an accuracy consistent with that of the finite difference scheme itself.

The most widely used iterative scheme for problems of this type involves a weighted average of the G-S result and the *previous* iterate; the method of *successive over-relaxation* (or SOR) and that approach is now discussed. In this equation all sums are on j; for example, $j \prec i$ denotes j taking on values $1,\ldots,i\text{-}1$ (and of course if $i = 1$, j takes on no values).

$$v_i = (1-\omega)v_i^{k-1} + \frac{\omega}{2R}\left[d_i + \sum_{j \prec i} a_{i,j}\, v_j^k + \sum_{j \succ i} a_{i,j}\, v_j^{k-1}\right], \quad i = 1,\ldots,N. \tag{5.14}$$

Comparing with Equation 5.13 ones sees that if $\omega = 1$, one has the G-S algorithm. Of particular interest is the case $1 \prec \omega \prec 2$, the over-relaxed (SOR) case in which careful choice of the parameter can dramatically improve convergence. Rigorous studies have been done for the current type of matrices in which the optimal values of ω are sought and the resulting improvement in the special radius ρ is illustrated (see Isaacson and Keller[1] and Varga[2]). The key results of these studies are briefly summarized below.

For simplicity, assume that $h = k = a/m$ and $n = m$ (recall that in the latest example $a = \pi$). While true that $\rho \prec 1$, but as already seen in the G-S case, it tends toward 1.0 as m increases, thus slowing convergence. As defined above, let $\boldsymbol{M}_1 = (\boldsymbol{D} - \boldsymbol{I})^{-1}\,\boldsymbol{U}$ denote the matrix for G-S. Similarly one can find the matrix $\boldsymbol{M}_\omega$ for the SOR scheme in Equation 5.14 (this matrix is not used here, nor is it ever used in computation). The theory shows that, for the optimal choice of ω, and for small h:

$$\begin{aligned} \rho(M_1) &\cong 1 - c_1 h^2, \\ \rho(M_\omega) &\cong 1 - c_2 h. \end{aligned} \tag{5.15}$$

The constants c_1 and c_2 are roughly the same size, so the key point is that for G-S, ρ approaches 1.0 much faster (as h gets small) than does SOR with a good value of ω. For example, in the problem under study here, Equation 5.12, Table 5.4 illustrates that with $m = 15$, $n = 11$ the convergence was painfully slow. It was pointed out there that $\rho \cong 0.945$, explaining the convergence problem. Moreover, using Equation 5.15, one could estimate the result of halving the step sizes; e.g., using the current h in Equation 5.15 and solving for c_1, then halving h in Equation 5.15 gives $\rho \cong 0.997$. Hence, one would experience a terrible rate of convergence if h were halved here.

For the $m = 15$, $n = 11$ case, the optimal value of ω in SOR is about 1.5182. This results in a $\rho = \rho(M_\omega) \cong 0.523$; this was the experimental result, while the theory gives the optimal $\rho \cong 0.527$. Table 5.5 summarizes numerical results comparing G-S and SOR for the $m = 15$, $n = 11$ case and the $m = 23$, $n =$

17 case. In the latter case the spectral radii are $\rho(M_1) \cong 0.975$ and $\rho(M_\omega) \cong 0.709$ (using $\omega = 1.64$). In Table 5.5 the v refers to the G-S approximation and w the SOR approximation; and the superscripts (25 and 50) denote the number of iterations.

TABLE 5.5 Comparing Results of Gauss-Seidel Iterations with SOR

y	$T(\pi/2, y)$	$\lvert T - v^{25}_{m=15,\,n=11}\rvert$	$\lvert T - w^{25}_{m=15,\,n=11}\rvert$	$\lvert T - v^{50}_{m=23,\,n=17}\rvert$	$\lvert T - w^{50}_{m=23,\,n=17}\rvert$
5/3	2.55281	0.2044	0.0030	0.2467	0.00027
4/3	1.76504	0.3273	0.0044	0.3614	0.00047
1	1.17520	0.3487	0.0041	0.4272	0.00046
2/3	0.71716	0.2799	0.0028	0.3448	0.00029
1/3	0.33954	0.1511	0.0012	0.1879	0.00009

The errors shown in these tables reflect both the error in the finite difference approximation and any lack of convergence of the iterative scheme. In the case of the G-S results, the error is dominated by lack of convergence, whereas in the SOR case an appropriate number of iterates were taken so that the two types of errors were about the same.

A MATLAB code to do *NumIts* iterations using the SOR method follows. Assume that m, n, N, r, R, ω (omega), and the data vector $\boldsymbol{d}$ have been defined. For simplicity, this code does not include the usual refinements such as testing two successive iterates for convergence. However, it is very efficient since no more than $6N$ multiplications are required per iteration. Also note that the possibly large N by N matrix $\boldsymbol{A}$ is not required, rather only a total of five N vectors, including the data.

```
v1 = ones(1, N-1);           % an N-1 vector of ones
for i = m:m:N-1              % zero out v1(m), v1(2m), ...
  v1(i) = 0; end
vm = ones(1, N-m);
vp = zeros(1, N);            % set the initial solution vector to zero
omega2R = omega/(2*R);
for ctr = 1:1:NumIts         % take NumIts iterations on solution v
  for i = 1:1:N
    vtemp = d(i);
    if i > 1, vtemp = vtemp + v1(i-1)*v(i-1); end
    if i > m, vtemp = vtemp + r*v(i-m); end
    if i < N, vtemp = vtemp + v1(i)*vp(i+1); end
    if i <= N-m, vtemp = vtemp + r*vp(i+m); end
    v(i) = omega2R*vtemp + (1-omega)*vp(i);
  end
  vp = v;                    % get ready for the next iteration
end
```

Another refinement on iterative methods which can further improve convergence is presented next.

Line (or Block) G-S and SOR Iteration

To simplify the presentation, the concept is illustrated on the case $m = n = 3$ and refers to the 9 by 9 matrix $\boldsymbol{A}$ in Equation 5.11. Using the first three equations in Equation 5.11, and using the first six components of the *previous* iteration for $\mathbf{v}$, one solves a 3 by 3 system for the first three components of a *new* $\mathbf{v}$. Next consider equations 4 to 6 in Equation 5.11 and think of components v_1, v_2, v_3 as *known* (since one just solved for them); and using the "old" values of $v_5,\ldots,v_9$, solve for the new v_4, v_5, v_6. Finally, using equations 7 to 9 in Equation 5.11 one similarly solves for new v_7, v_8, v_9, thus completing one iteration. This is referred to as a *line* or *block* G-S iteration and has a slight advantage over the above G-S (or *point*) scheme.

The line or block SOR refinement proceeds as follows. For example, suppose, as just indicated, one has computed a line G-S triple v_1, v_2, v_3 using the first three equations of Equations 5.11. Before proceeding to the next *line*, one applies the SOR weighted average as in Equation 5.14; e.g., the (*new* v_1) + (1 – ω) (*previous* v_1). One advantage of this approach is that in each iteration one, in general, solves *n* systems of *m* by *m* equations in which the matrix involved is especially easy to compute with. For example, in the current example one repeatly solves systems of the type:

$$\begin{bmatrix} 2R & -1 & 0 \\ -1 & 2R & -1 \\ 0 & -1 & 2R \end{bmatrix} \begin{bmatrix} a \\ b \\ c \end{bmatrix} = f.$$

Even in the general *m* case such matrices are very efficiently solved due to the simple diagonals (the –1s) below and above the main diagonal. Each such *m* by *m* system can be solved in order m operations (in comparison to order m^3 operations if the matrix were full); of course there are *n* such solutions to complete an iteration.

The line SOR method has a so-called $\sqrt{2}$ improvement over the simple (or *point*) SOR. More precisely, recall that in Equation 5.15 $\rho(M_\omega) \cong 1 - c_2 h$ if an optimal ω is used in the point SOR. Once again, assuming $k \cong h$, for the *line* SOR this improves to $\rho(M_\omega^{line}) \cong 1 - \sqrt{2}c_2 h$. To illustrate the effect of this improved spectral radius, in the sixth column of Table 5.5 is shown the error associated with 50 *point* SOR iterations. Had a *line* SOR been used, comparing the two ρ values indicates that one could have achieved the same accuracy with about 33 iterations.

A refinement that is often used when a very large $N = mn$ is needed, the *alternating direction* iteration, is attributed to Peachman and Rachford. Briefly the idea is to take a (horizontal) line SOR iterate, as just discussed, to get an estimate for **v**. Then do a *vertical* line SOR (i.e., in the present case, in the *y* direction) to further improve **v**; this constitutes one iteration. This method will offer substantial improvement on the rate of convergence if carried out properly. However, to be effective the method requires that one use not one parameter, ω, but a sequence of them to be applied cyclically. Since the algorithm is considerably more complex, it is only practical when a very large amount of calculation is required and the same matrix ***A*** repeatedly used.

For a more technical treatment of these and other issues, see Isaacson and Keller.[1] For more detail on iterative methods, see Varga.[2] For a broad introduction to finite difference methods in heat transfer, see Özisik.[3]

Dealing with Boundary Conditions

The examples discussed so far assumed, for simplicity, rectangular domains, and divided the intervals involved into equal step sizes. In general, domains may be much more complex in structure, for example curved in places. Thus, one is faced with modifying the finite difference scheme at, and near, the boundary. As mentioned at the start of this section, complex boundaries provide an inherent advantage to the *finite element* method, to be discussed shortly. In this short section an introduction is given into how one accommodates the finite difference method to handle more "interesting" boundaries.

The Dirichlet Boundary Condition

Assume that one is solving Laplace's equation

$$T_{xx} + T_{yy} = 0$$

with points specified on the boundary. In particular, one needs to modify the difference approximation of Equation 5.9 for those points (x_i, y_i) near the boundary. For example, consider the points in Figure 5.2.

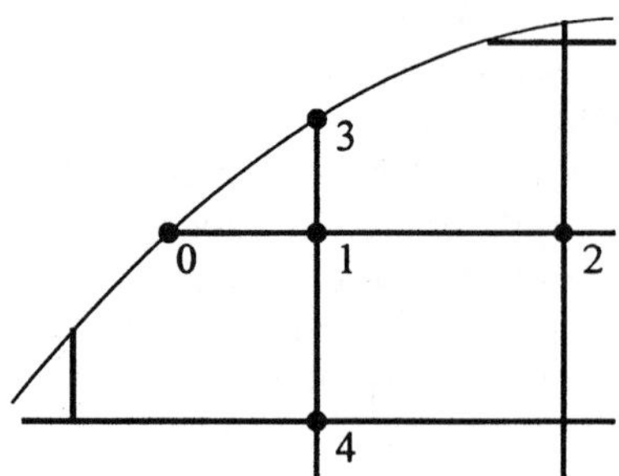

FIGURE 5.2 Points on and near boundary.

For convenience, denote the five points in question with a local labeling 0,…,4 with point number 1 being the central point. The goal is to approximate the two derivatives in Laplace's equation at the point labeled by 1, which represents the point (x_i, y_j). Note that the two points labeled by 0 and 3 are on the boundary, hence T is known there; whereas at the other three (interior) points T is part of the unknown vector in Equation 5.11. The problem, as illustrated by Figure 5.2, is that boundary points 0 and 3 are not full steps (h and k) from central point 1; so new approximations for T_{xx} and T_{yy} are needed. For $T_{xx}(x_i, y_j)$, expand in Taylor series about the point 1 to obtain:

$$T_2 = T\left(x_i + h, y_j\right) = T_1 + h\,T_x\left(x_i, y_j\right) + \frac{1}{2}h^2\,T_{xx}\left(x_i, y_j\right) + H.O.T.$$

$$T_0 = T\left(x_i - \alpha h, y_j\right) = T_1 - \alpha h\,T_x\left(x_i, y_j\right) + \frac{1}{2}\alpha^2 h^2 T_{xx}\left(x_i, y_j\right) + H.O.T$$

where again $H.O.T.$ represents terms of order h^3 and higher. Forming the sum $\alpha T_2 + T_0$, dropping the $H.O.T.$ terms, and solving for T_{xx} gives

$$T_{xx}\left(x_i, y_j\right) \cong 2\frac{T_0 - (1+\alpha)T_1 + \alpha T_2}{\alpha(1+\alpha)h^2}.$$

Note that when $\alpha = 1$ this reduces to the familiar approximation to T_{xx}. Similarly, expanding $T_3 = T(x_i, y_j + \beta k)$, as above, one gets the approximation to T_{yy}; then adding the two gives

$$T_{xx}\left(x_i, y_j\right) + T_{yy}\left(x_i, y_j\right) \cong$$

$$2\frac{T_0 = (1+\alpha)T_1 + \alpha T_2}{\alpha(1+\alpha)h^2} + 2\frac{T_3 - (1+\beta)T_1 + \beta T_4}{\beta(1+\beta)k^2} = 0. \tag{5.16}$$

This second line then replaces the corresponding terms in Equations 5.10 and 5.11. In particular, the diagonal element $2R$ in Equation 5.11 is replaced by the new coefficient of $T_1 = T(x_i, y_j)$; i.e.,

$$1 + \frac{\alpha h^2}{\beta k^2}$$

Hence, for each interior element near the boundary, as the point 1 in Figure 5.2, the corresponding diagonal element of $\boldsymbol{A}$ in Equation 5.11 must be modified; and for the adjacent interior points, like the points 2 and 4, those corresponding elements in $\boldsymbol{A}$ must be altered via Equation 5.16. Moreover, for those points on the boundary, like 0 and 3, the corresponding data in the right side of Equation 5.11 must be modified according to Equation 5.16.

The Flux Boundary Condition

Here is an indication of how flux (or *Neumann*) conditions are handled when faced with an irregular boundary. Again the two-dimension Laplace equation is considered, but now with the normal derivative specified on the boundary:

$$\frac{\partial T}{\partial n} = q''(x,y) \quad \text{on } \Gamma.$$

The situation is illustrated in Figure 5.3 where, as above, the locally labeled points 0,…,4 are involved in the usual 5-point approximation to T_{xx} and T_{yy} at point 1. Again, the central point 1 represents the point of interest, (x_i, y_j).

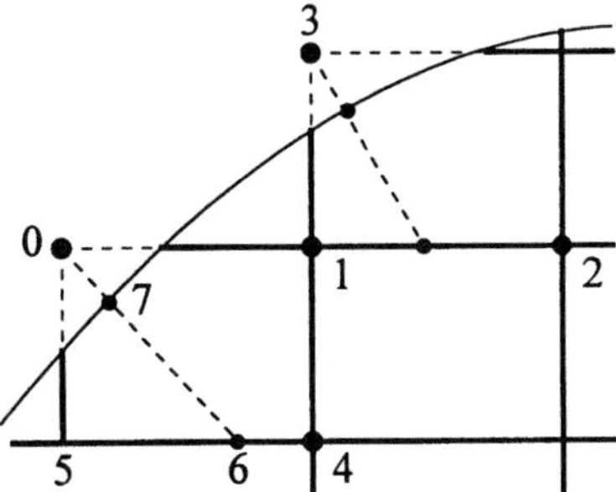

FIGURE 5.3 Applying flux condition.

For the problem illustrated here, points 0 and 3 are *outside* the domain; hence, the values of T at these points must be replaced by taking advantage of $\partial T/\partial n$ in the vicinity. Focusing on point 0, let d denote the distance between points 0 and 6, where point 6 results from the normal to the boundary being extended, as shown in Figure 5.3. One approximates the flux condition by $(T_0 - T_6)/d = g$, where g is evaluated at point 7; hence one can eliminate T_0 in favor of T_6. Then T_6 can be eliminated by, for example, linearly interpolating between T_4 and T_5. In the same way point 3 can be eliminated.

Summarizing, point 1 represents point (x_i, y_j) which corresponds to a certain row in matrix $\boldsymbol{A}$, let's say row k. Points 2 and 4 in Figure 5.3 have their usual influence in row k — plus effects of the linear interpolation discussed above. The temporary points 0 and 3 have their influence represented by other interior points (specifically points 1, 2, 4, and 5) and the boundary conditions. So row k of $\boldsymbol{A}$ must be modified as outlined, specifically at the columns related to points 1, 2, 4, and 5. This type of modification must be done for every interior point that is similarly near the boundary.

Both of the treatments of the boundary conditions suggested in this discussion are less accurate than the usual 5-point approximation to T_{xx} and T_{yy} (which is $O(h^2)$). In contrast, both approximations introduced here are $O(h)$. However, it has been shown that, in spite of this, for elliptic problems with smooth boundaries one still can expect the overall finite difference approximation to the entire problem to be of order $O(h^2)$. For more details see Birkhoff and Lynch.[4]

Conclusions on Direct and Iterative Methods

It is frequently an important decision whether to use a *direct* method (e.g., Gaussian elimination) or an iterative scheme to solve a system of linear equations. Moreover, one's decision may change depending on the hardware and software available. For example, as PCs and engineering workstations become faster with larger main memory and disk capability, one may be able to solve a very large system using a direct method and double-precision arithmetic — an option not available a few years prior. Very user-friendly software is available and, when feasible, direct methods are usually the "cleaner" way to proceed.

On the other hand, even with a well-equipped computer it may not be feasible to attack problems requiring matrices of size 1000 by 1000 or larger with a general-purpose package (e.g., MATLAB, Maple, Mathematica). Even with the sparce matrices of the type considered above, with only five nonzero diagonals, as Gaussian elimination is performed the matrix tend to "fill out" as one proceeds in the calculation. Thus, these computations become nearly as time-consuming as with a "full" matrix (and solving a full N by N system takes nearly N^3 numeric operations). So it is important in solving such systems to take advantage of their special structure; and iteration is usually best suited for this. As was illustrated above with the MATLAB code for the simple SOR algorithm, one does not have to form these extremely large matrices consisting of mostly zeros. In that case, only $6N$ multiplications per iteration were required; so one can afford to take numerous iterations, as opposed to forming and solving the problem by Gaussian elimination.

5.3 Finite Element Method

The method of finite differences dominated the scene in the numerical solution of partial difference equations (PDEs) for decades. A new technique arrived on the scene in the 1960s and has gathered momentum since — to the extent that it has become the method of choice for many areas of engineering. It is the *finite element method* (FEM). Briefly, any two- or three-dimensional object can be subdivided into a finite number of pieces (or *elements*) which at least approximates the original figure. The PDE is expressed in an equivalent integral expression which is "solved" on each element and then collected into an approximate solution for the entire domain. In particular, very complex boundaries can be accommodated quite routinely by the FEM in contrast to the difficulties encountered by such boundaries with finite differences (as illustrated in earlier). In the next several subsections the FEM method will be applied to a steady-state (elliptic) PDE.

In addition to the relative ease of handling complex boundaries, the FEM became a very natural extension of modern CAD/CAM software with which users can quite painlessly design complex structures. Hence, once a three-dimensional solid is formed via CAD/CAM software it became *relatively* easy (but by no means trivial) to form a finite element mesh and proceed to attack the integral form of the problem. So, well-established engineering software companies expanded their products to provide FEM solutions to a host of problems, including those involving heat transfer.

FEM Concept

Suppose one is confronted with a boundary value problem involving an elliptic PDE; for example, a simple heat conduction equation with the temperature specified on the boundary; such as the problem in Equation 5.12. One approach is to express this problem by an equivalent one of minimizing the value of a related integral. This integral can be derived in several ways, to be discussed over the next several subsections. For example, the solution of problem in Equation 5.12 above is equivalent to minimizing the *functional* $I\ [T]$ given by:

$$I[T] = \iint \left[\left(\frac{\partial T}{\partial x} \right)^2 + \left(\frac{\partial T}{\partial y} \right)^2 \right] dx\, dy.$$

This is the so-called *energy* form of the problem and, remarkably, the solution to the original BVP is the same as the function T that minimizes $I\ [T]$ for all functions satisfying the boundary conditions of Equation 5.12. More on this subject in a later subsection.

Now briefly consider the problem of minimization for the functional I. What is done in every FEM approach is to first divide the domain of interest (e.g., in Equation 5.12 the rectangle $0 \preceq x \preceq \pi$, $0 \preceq$

$y \preceq 2$) typically into a number of small triangular or square regions (or *elements*). Then one seeks to approximate the function T as follows:

$$T \sim \sum_{1}^{N} \gamma_i \phi_i(x, y). \tag{5.17}$$

where the γ_i are constants to be determined and the ϕ_i are functions of very special kind and related to the type of element. To find the approximate minimum to function T, the form of Equation 5.17 is substituted into $I[T]$ above and one takes the N partial derivatives $\partial I/\partial \gamma_i$. This leads to N linear equations in the unknown γ_i.

We later present a discussion of the more popular elements, together with their accuracy and application. Once the type of element is chosen, the approximating functions ϕ_i are usually prescribed. For example, if one has chosen triangles as the basic elements with only the three vertices of each triangle involved, then each ϕ_i is a piece-wise linear function of x and y, linear on a particular triangle. Such functions will be discussed in detail later.

Before delving into the details of the meshing process and solving for the approximation in Equation 5.17, the most common way to convert a BVP to a functional problem is presented next.

Weighted Integrals and Weak Forms of the Heat Equation

The above brief discussion of the energy functional approach is relatively restricted in its application, whereas the following approach is applicable to essentially any BVP of practical interest. It leads to another variational form of the problem using *weighted* integrals.

Using the compact notation $T_x = \partial T/\partial x = \partial_x(T)$, consider the following fairly general form of the steady-state heat conduction equation:

$$-\partial_x(p_1 T_x) - \partial_y(p_2 T_y) = Q. \tag{5.18}$$

Basic to the FEM is to express the so-called *weighted integral* form of a (typically approximate) solution. That is, one takes the error (or residual) in an approximation, multiplies by a *weight* function w and integrates over the domain D of interest to get

$$-\iint_D w\left[\partial_x(p_1 T_x) + \partial_y(p_2 T_y) + Q\right] dx\, dy = 0. \tag{5.19}$$

This is a *weak* form of Equation 5.18 in the sense that if T solves Equation 5.18 then clearly Equation 5.19 is satisfied for any reasonable function w. The converse is *not* true without further restrictions on T and w; and that issue is now discussed. As Equation 5.19 now stands, T should have two continuous (or at least integrable) derivatives. To lessen this "smoothness" restriction on T, "integrate-by-parts" by applying the following form of Green's theorem. Let $\mathbf{v} = (v_1, v_2)$ be a smooth vector function and denote $\nabla \cdot \mathbf{v} = \partial_x v_1 + \partial_y v_2$. Green's theorem can be expressed:

$$-\iint_D w \nabla \cdot \mathbf{v}\, dx\, dy = -\iint_D (\nabla w) \cdot \mathbf{v}\, dx\, dy - \int_\Gamma w \boldsymbol{n} \cdot \mathbf{v}\, ds \tag{5.20}$$

where Γ is the boundary of D traversed in the usual counterclockwise direction, and $\boldsymbol{n} = (n_1, n_2)$ is the outer normal. Now apply Equation 5.20 with $\mathbf{v} = (p_1 T_x, p_2 T_y)$ to the differential terms in Equation 5.19 to get:

$$\iint_D \left[w_x(p_1 T_x) + w_y(p_2 T_y) - wQ\right] dx\, dy - \int_\Gamma w\left[n_1 p_1 T_x + n_2 p_2 T_y\right] ds = 0. \tag{5.21}$$

This is another and more desirable *weak* form of the differential equation and it contains a suggestion regarding boundary conditions on Γ. Finally, add a quite general set of boundary conditions for the Equation 5.18 to get the BVP:

$$-\partial_x(p_1 T_x) - \partial_y(p_2 T_y) = Q$$
$$T = f \quad \text{on} \quad \Gamma_1, \tag{5.22}$$
$$n_1 p_1 T_x + n_2 p_2 T_y = g \quad \text{on} \quad \Gamma_2 .$$

Here Γ_1 and Γ_2 make up the complete boundary Γ. The first condition, on Γ_1, is called the *essential* boundary condition and note that if $\Gamma_1 = \Gamma$ (hence, no Γ_2), one has the Dirichlet problem. If there is the second boundary condition (on Γ_2), it is called the *natural* condition. The term *natural* is partly explained by the fact that in Equation 5.21 to evaluate the integral over Γ, one knows the integrand, namely wg (at least on the portion Γ_2; more on this below). Moreover, should $g = 0$, this integral would be 0. On the other hand if, for example, one had the condition of the normal derivative $\partial_n T = n_1 T_x + n_2 T_y = g$ on Γ, one would not have the information to perform this integration in Equation 5.21, unless of course $p_1 = p_2 =$ constant. The natural boundary condition comes up frequently in variational approaches and physically it represents the *flux* across the boundary.

Now address the question: in what sense does a solution of the weak form in Equation 5.21 provide a solution the original BVP in Equation 5.22? If T satisfies the *essential* boundary condition $T = f$ on Γ_1 and Equation 5.21 is satisfied for a "complete" set of weight functions (also called *test* functions) w, then the BVP is also satisfied by this T. The term "complete" is highly technical and will not be defined here (see Oden and Reddy[5]). Rather, a typical set of weights w and a typical approximate solution T will be illustrated in the examples to follow.

Another concern in Equation 5.21 is that the line integral is over the entire boundary Γ, whereas in the BVP (5.22) the *natural* condition is only specified on Γ_2. Hence, if $\Gamma_2 \neq \Gamma$, one has the problem of evaluating the complete line integral in Equation 5.22. This is technically taken care of by insisting that the test functions w are zero on Γ_1, in which case the line integral in Equation 5.21 is only over Γ_2. This is also illustrated in the example problems to follow.

To provide an overview of the approximate solution of Equation 5.21, the FEM approach is simplified a bit. Consider an approximation to T in the form

$$T \sim \sum_{j=1}^{N} \gamma_j \phi_j(x, y), \tag{5.23}$$

where the ϕ_j are the basic functions (typically piece-wise polynomials) and the γ_j are constants to be determined. The weight functions w are a compatible set of functions, the most popular being the ϕ_j. This approach is called the *Galerkin* method, an approach used in solving a host of different problems in the physical sciences. So, in Equation 5.21, use $w = \phi_i$ and T as in Equation 5.23 to get:

$$\iint_D \sum_{j=1}^{N} \left(p_1 \partial_x \phi_i \partial_x \phi_j + p_2 \partial_y \phi_i \partial_y \phi_j\right) \gamma_j \, dx\, dy$$
$$= \iint_D \phi_i Q \, dx\, dy + \int_{\Gamma_2} \phi_i g(s) ds; \quad i = 1, \ldots, N. \tag{5.24}$$

Put this in matrix form $\boldsymbol{K}\gamma = \beta$ where $\gamma = (\gamma_1,\ldots,\gamma_N)$, vector β is computed from the right side of Equation 5.24, and matrix $\boldsymbol{K}$ is defined by the left side of Equation 5.24,

$$K_{i,j} = \iint_D \left(p_1 \partial_x \phi_i \partial_x \phi_j + p_2 \partial_y \phi_i \partial_y \phi_j\right) dx\, dy \quad i = 1,\ldots,N;\ j = 1,\ldots,N. \tag{5.25}$$

The matrix $\boldsymbol{K}$ is called the *stiffness* matrix and is a central feature of the Galerkin method. What is essential is that this matrix is invertible so that one has a unique solution of Equation 5.24 for the unknown γ_i. In the example of the FEM to follow, the situation will be a bit more complex than this outline, but the essential features will be illustrated.

A Simple Example of the FEM

Consider the "thin plate" shown in the Figure 5.4 below. The simple two-dimensional heat conduction equation for it is

$$T_{xx} + T_{yy} = 0. \tag{5.26}$$

The Dirichlet boundary conditions on five boundary sections, making up Γ_1 in Equation 5.22, are $T(x,0.3) = 2.7 - 0.3x$, $T(0,y) = 3 - y$, $T(x,0) = 3$, $T(x, 0.1) = 2.9 + 0.1x$, and on the right-most diagonal section $T = 2.6 + 1.4x - x^2$. Finally, the Γ_2 section of the boundary has the normal (flux) condition $\partial_n T(0.2,y) = \partial_x T(0.2,y) = y$. It is easily verified that the solution to the resulting BVP is $T = 3 - y\,(1 - x)$.

The domain is divided into a rather crude finite element mesh of 13 triangles, as shown in Figure 5.4.

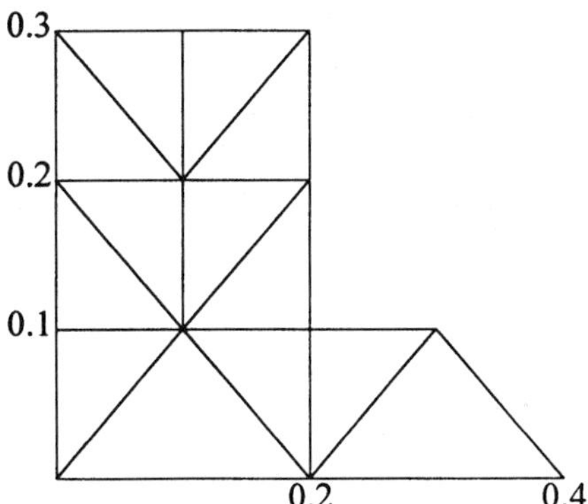

FIGURE 5.4 The domain and 13 elements.

Provided here is a simplified version of the usual FEM notation for ease of presentation. Note the labeling of the 13 vertices (or *nodes)* of the resulting mesh; they will be referred to as $E_1,\ldots,E_{13}$ in the text, but they are labeled by 1,…,13 in Figure 5.5. Associated with each node E_i, one defines a basic function ϕ_i as a piece-wise linear function with the following property: $\phi_i = 1$ at the ith node and zero at all other nodes. For example, $\phi_3 = 1$ at node $E_3 = (0.2, 0.2)$ and zero at adjacent nodes (labeled 1, 2, 4, 13) and all other nodes. So geometrically, the graph of ϕ_3 has the value 1 at node 3 and slopes linearly to 0 at the four adjacent nodes.

To show how a basic function is created, ϕ_3 is defined on the triangle (3,1,2), i.e., formed by nodes 3, 1, and 2. The idea is made more general by defining, and $(x_1, y_1) = (0.2, 0.2)$, $(x_2, y_2) = (0.1, 0.2)$, and $(x_3, y_3) = (0.1, 0.1)$. To solve for $\phi_3 = a + bx + cy$ on this triangle, one solves the system:

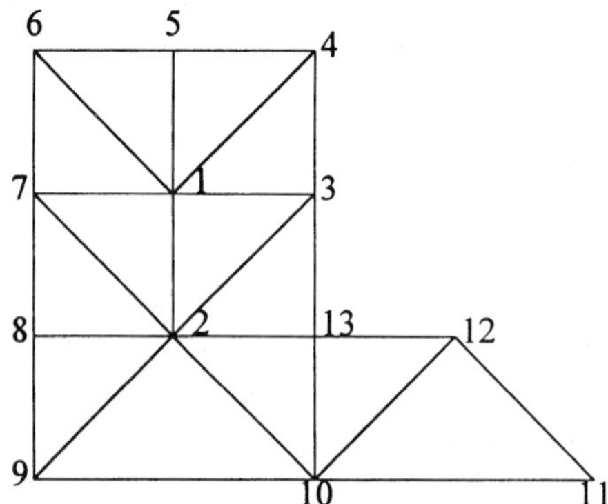

FIGURE 5.5 The domain and 13 nodes.

$$a + bx_1 + cy_1 = a + 0.2b + 0.2c = 1$$

$$a + bx_2 + cy_2 = a + 0.1b + 0.2c = 0$$

$$a + bx_3 + cy_3 = a + 0.1b + 0.1c = 0$$

Solving this system results in $\phi_3 = -1 + 10x$. Similarly solving for ϕ_3 on triangle (3,2,13) gives $\phi_3 = -1 + 10y$, and on triangle (3,4,1), $\phi_3 = 1 + 10x - 10y$. Again, $\phi_3 = 0$ on the rest of the domain. In a similar manner the basic functions are defined for each node.

The next step is to set up the equations for the coefficients γ_j in the representation $T \sim \Sigma\, \gamma_j\phi_j(x, y)$. In the FEM the γ_j are determined so that all boundary conditions are satisfied at those nodes. In the current case, the Γ_1 conditions immediately give the values for $\gamma_4,\ldots,\gamma_{13}$ since the corresponding ϕ_j values are 1.0 at these nodes. Using the Dirichlet conditions given above, one can evaluate $\gamma_4,\ldots,\gamma_{13}$ to get, respectively, 2.76, 2.73, 2.7, 2.8, 2.9, 3.0, 3.0, 3.0, 2.93, 2.92. So in this example one needs only to solve for the coefficients γ_1, γ_2 corresponding to the two interior nodes E_1, E_2; and for γ_3 corresponding to the node E_3 on the Γ_2 portion of the boundary. Hence, one needs to solve a 3 by 3 system of linear equations. To that end, express Equation 5.24 in the current setting in which $p_1 = p_2 = 1$ and $Q = 0$:

$$\iint_D \sum_{j=1}^{3} \left(\partial_x \phi_i \partial_x \phi_j + \partial_y \phi_i \partial_y \phi_j\right)\gamma_j \, dx\, dy$$
$$= -\iint_D \sum_{j=4}^{13} \left(\partial_x \phi_i \partial_x \phi_j + \partial_y \phi_i \partial_y \phi_j\right)\gamma_j \, dx\, dy + \int_{\Gamma_2} \phi_i\, y\, dy; \quad i = 1,\ldots,3 \tag{5.27}$$

Note that the double integral sum has been split into two pieces — the sum from 1 to 3 involve unknown γ_j, whereas the second piece and the Γ_2 integral involve only known quantities and are the right side of Equation 5.27. So the left side of Equation 5.27 provides the *stiffness* matrix $\boldsymbol{K}$ discussed above. Also notice that, as discussed earlier, $\phi_1 = \phi_2 = \phi_3 = 0$ on Γ_1, thus justifying the line integral in Equation 5.27 being only over Γ_2 (and here $ds = dy$). Turn now to the computation of the matrix $\boldsymbol{K}$,

$$K_{i,j} = \iint_D \left(\partial_x \phi_i \partial_x \phi_j + \partial_y \phi_i \partial_y \phi_j\right) dx\, dy \quad i = 1,\ldots,3; \quad j = 1,\ldots,3. \tag{5.28}$$

Focus on the computation of $K_{1,3}$ which involves the basic functions ϕ_1 and ϕ_3. From Figure 5.5 and the definition of the ϕ_i it is clear that the integrand in Equation 5.28 is nonzero only on triangles (3,4,1) and (1,2,3), so one only needs to compute the basic functions there. To compute the contribution on triangle (3,1,2), first recall that here $\phi_3 = -1 + 10x$, hence, $\partial_x\phi_3 = 10$ and $\partial_y\phi_3 = 0$. To find ϕ_1 on triangle (1,2,3) one solves the system

$$a + bx_1 + cy_1 = a + 0.1b + 0.2c = 1$$

$$a + bx_2 + cy_2 = a + 0.1b + 0.1c = 0$$

$$a + bx_3 + cy_3 = a + 0.2b + 0.2c = 0$$

Solving this system, give $\phi_1 = -10x + 10y$ on triangle (1,2,3); hence, $\partial_x\phi_1 = -10$ and $\partial_y\phi_1 = 10$. Since the area of triangle (1,2,3) is 0.005, the contribution to $k_{1,3}$ is $0.005\ (\partial_x\phi_3\ \partial_x\phi_1) = -0.5$. In a similar manner the contribution from triangle (1,3,4) is also –0.5; hence $K_{1,3} = -1.0$. Proceeding in the same way, one computes the matrix $\boldsymbol{K}$ shown below in Equation 5.29. The reader is reminded that in the more general PDE, specifically when coefficients p and q are substantial functions of x and y, the above integrations likely must be done by a numerical quadrature. This is discussed in the next subsection.

Regarding the right side of Equation 5.27, less details are provided, but observe that everything is known and the computations for $i = 1,2,3$ provide the right side in the system $\boldsymbol{k}\gamma = \beta$. The values are $\beta_1 = 5.53$, $\beta_2 = 8.82$, $\beta_3 = 2.86$. The resulting system becomes:

$$\boldsymbol{K} = \begin{bmatrix} 4 & -1 & -1 \\ -1 & 4 & 0 \\ -1 & 0 & 2 \end{bmatrix} \begin{bmatrix} \gamma_1 \\ \gamma_2 \\ \gamma_3 \end{bmatrix} = \begin{bmatrix} 5.53 \\ 8.82 \\ 2.86 \end{bmatrix} \tag{5.29}$$

Solving this linear system one gets $\gamma_1 = 2.82$, $\gamma_2 = 2.91$, $\gamma_3 = 2.84$.

Checking the results at the three nodes E_1, E_2, E_3 against the exact values for $T = 3 - y(1 - x)$, shows that the results are exact there. At $E_1 = (0.1, 0.2)$, $T = 2.82 = \gamma_1$; at $E_2 = (0.1, 0.1)$, $T = 2.91 = \gamma_2$; and at $E_3 = (0.2, 0.2)$, $T = 2.84 = \gamma_3$. Exact solutions for a quadratic T using the piece-wise linear basic functions is *not* to be expected. The reason they are exact here is that $T = 3 - y(1 - x)$ happens to be *bilinear*, i.e., linear on both vertical and horizontal lines (e.g., on lines $x = 0.1$ and $x = 0.2$ on which the three (originally) unknown nodes happen to lie).

To illustrate that the above approximate solution is *not* exact, compare the approximate solution to the exact solution at (0.05, 0.05) where the exact $T(0.05, 0.05) = 2.9525$. To compute the approximate solution there, first note that this point is on the edge between triangles (2,8,9) and (2,9,10); and is only affected by ϕ_2 and ϕ_9. In fact the approximate solution is

$$(\gamma_2 + \gamma_9)/2 = (2.91 + 3.0)/2 = 2.955,$$

hence the error is 2.955 – 2.9525 = 0.0025. This is the maximum error over the entire domain, shared by the points (0.15, 0.05) and (0.35, 0.05). This error is somewhat less than one would expect with the piece-wise linear basic functions and the sides of the triangles being as large as 0.2. (The expected error in FEM is a fairly complicated subject and is discussed below.) This better than expected accuracy just experienced is due largely to the size and shape of the boundary and the fact, mentioned above, that the exact solution is bilinear. In particular, the approximation is exact down the center of the main portion of the domain, thus helping the solution throughout.

In this particular problem, if more accuracy is called for one could reasonably continue to use the simple triangular mesh with smaller triangles, and in fact this problem is revisited in a later subsection with this approach. However, when one is confronted with more complex boundaries and certainly when the domain is three-dimensional, it is efficient to use more elaborate elements. Some of the more popular elements used in modern software will be discussed next.

Some Popular Finite Elements and Their Integration

The example above focused on the historically important triangular element with three nodes (the vertices of the triangle). The key feature of such elements is that they lead to piece-wise linear basis functions, the ϕ_j. While such approximating functions have been useful for decades in a number of areas of application, they suffer from the fact that linear functions do not do a very good job of approximating, e.g., quadratic, cubic, or other higher-order surfaces. More appropriate elements for dealing with problems demanding higher accuracy and/or with curved boundaries is the next subject.

Quadratic Triangular Element

Considerable detail will be given for this important case and provided are all the essential ingredients for solving a BVP using this element. Then several other cases will be discussed in much less detail. The idea here is simply to add to the three nodes at the vertices three additional nodes at the midpoints of each side of the triangle. These six nodes are exactly what is needed to determine the six constants in a quadratic in two variables:

$$c_1 + c_2\xi + c_3\eta + c_4\xi^2 + c_5\xi\eta + c_6\eta^2. \tag{5.30}$$

In this subsection focus will be on *single* elements (as opposed to an entire mesh of elements). To simplify the presentation, the element under consideration will be in a *standard* position, which for the triangle will be as in Figure 5.6.

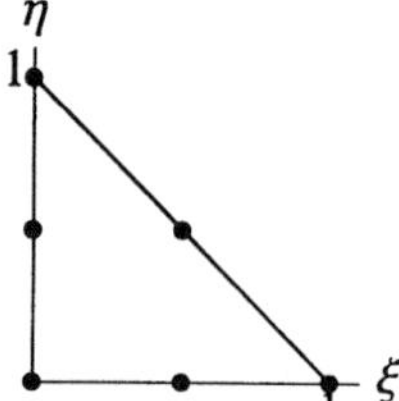

FIGURE 5.6 Standard (quadratic) triangle with six nodes.

Note that for elements in "standard" position the coordinate variables ξ and η will be used, leaving x and y for the coordinates of the original boundary value problem. Shortly it will be shown how to map a (typically small) element in the xy-plane over to the standard element in the $\xi\eta$-plane. Returning to Equation 5.30, as before, one defines the basis functions as having value 1.0 at one node and 0.0 at the other five nodes. This defines the constants $c_1,\ldots,c_6$ in Equation 5.30 for each particular node. This process leads to the following basis functions (shown in vector form):

$$\phi(\xi,\eta) = \begin{bmatrix} (1-\xi-\eta)(1-2\xi-2\eta) \\ 4\xi(1-\xi-\eta) \\ \xi(2\xi-1) \\ 4\xi\eta \\ \eta(2\eta-1) \\ 4\eta(1-\xi-\eta) \end{bmatrix}. \tag{5.31}$$

As the reader can easily verify, ϕ_1 has values 1.0 at (0, 0) and 0.0 at the other five nodes, and similarly for the other five basis functions. The functions in Equation 5.31 are clearly quadratic, moreover they are *interpolating* functions of second degree in that any quadratic polynomial can be expressed as a linear combination of these six functions. Furthermore, for any smooth function $f(\xi, \eta)$, if f is known at the six nodes it can be approximated by the ϕ_i, thus

$$f(\xi,\eta) \simeq \sum_{1}^{6} f(\xi_i,\eta_i)\phi_i(\xi,\eta)$$

where the (ξ_i, η_i) are the six nodes. (If this standard triangle had sides of size h instead of 1.0, such approximations would have error of order h^3. More on this subject later.)

This is the way the standard triangle is mapped onto a much more general figure with three (possibly) curved sides. (In FEM applications, the sides will typically be small, of size h.) Consider the following figure (Figure 5.7) in the xy-plane in which the six nodes are used to describe, perhaps only approximately, the sides of the figure. Since three points uniquely determine a quadratic, if the sides are at most quadratic, then the six nodes will perfectly describe all three sides. For convenience the standard triangle is also shown.

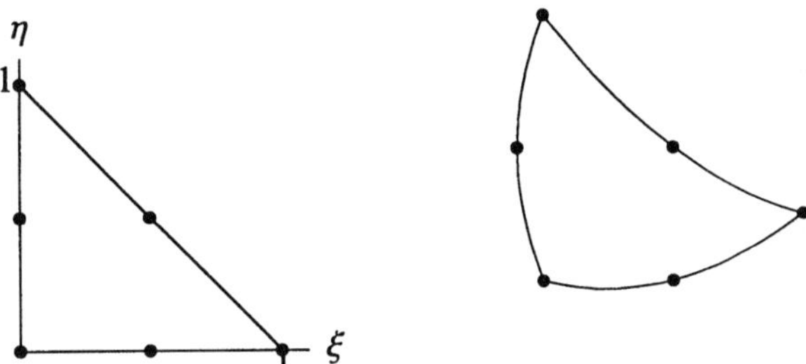

FIGURE 5.7 Standard triangle and a typical xy image.

Given the six nodes, the mapping between the two figures is given by:

$$x = F(\xi,\eta) = \sum_{1}^{6} x_i\phi_i(\xi,\eta),$$
$$y = G(\xi,\eta) = \sum_{1}^{6} y_i\phi_i(\xi,\eta). \tag{5.32}$$

It is clear from the properties of the ϕ_i that the (ξ_i, η_i) nodes map onto the (x_i, y_i) nodes. As an illustration of this mapping, consider the six (x_i, y_i) values: (0,0), (1,–½), (2,0), (2,1), (1,2), and (1,1). To see where the triangle leg on the ξ axis maps, apply Equation 5.32 with these (x_i, y_i) values and use Equation 5.31 with $\eta = 0$ to get:

$$x = \left[1\phi_2 + 2\phi_3 + 2\phi_4 + \phi_5 + \phi_6\right]_{\eta=0} = 4\xi(1-\xi) + 2\xi(2\xi-1) = 2\xi$$

$$y = \left[-\tfrac{1}{2}\phi_2 + \phi_4 + 2\phi_5 + \phi_6\right]_{\eta=0} = -2\xi(1-\xi)$$

Then eliminating the ξ from these two equations one obtains $y = x^2/2 - x$ which agrees with the three nodes (0,0), (1,–½), and (2,0), and this leg will be mapped exactly if the corresponding side in the xy-plane is indeed parabolic.

In addition to the better handling of curved boundaries, the six-node triangular elements have better *interpolating* accuracy compared to the three-node elements, specifically $O(h^3)$ vs. $O(h^2)$ local error. In a modern software package the domain of interest is divided up into a fairly large number of elements. If the boundary of the domain is polygonal (consisting only of a series of straight lines) then regular, straight-sided, triangular elements will work well. But if the boundary is curved, one gets a much better approximation to it by using elements with three nodes per side since this allows for a quadratic approximation to each small piece of boundary. Figure 5.8 illustrates a portion of a domain with a curved boundary showing regular triangles in the interior and elements with one curved side along the boundary.

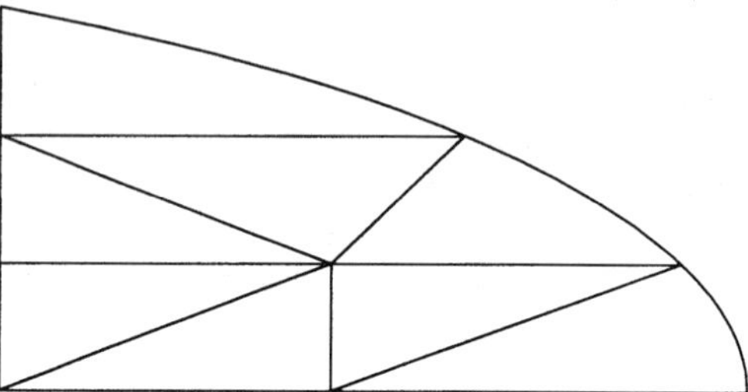

FIGURE 5.8 A portion of the domain.

Computing the Stiffness Matrix

Now consider the key issue of how the components of the stiffness matrix are computed in the current context. For convenience, Equation 5.25 is repeated here, except now the basic functions in the xy-plane are denoted by $\psi(x, y)$. The symbol ϕ is reserved for the basis functions in the $\xi\eta$-plane.

$$K_{i,j} = \iint_D \left(p_1 \partial_x \psi_i \partial_x \psi_j + p_2 \partial_y \psi_i \partial_y \psi_j\right) dx\, dy \qquad i = 1,\ldots,N; \quad j = 1,\ldots,N.$$

Recall from the earlier comments and the above example (starting with Equation 5.26) for a particular $K_{i,j}$, one only has to integrate over a few elements rather than the entire domain D. So focus on single element T^* in the xy mesh and see how the integration is performed over the standard triangle; call it T, in the $\xi\eta$-plane. Changing variables and integrating over one element in the above integral:

$$\iint_{T^*} \left[p_1 \partial_x \psi_i \partial_x \psi_j + p_2 \partial_y \psi_i \partial_y \psi_j\right] dx\, dy = \iint_T H(\xi,\eta) \left|\det(J)\right| d\xi\, d\eta, \tag{5.33}$$

where

$$J = \begin{bmatrix} \dfrac{\partial x}{\partial \xi} & \dfrac{\partial y}{\partial \xi} \\ \dfrac{\partial x}{\partial \eta} & \dfrac{\partial y}{\partial \eta} \end{bmatrix}.$$

Here H replaces the integrand in the integral on the left and J is the *Jacobian* matrix of the transformation, with $\det(J)$ the determinate of the matrix. Now J can be computed using Equations 5.32 and 5.31; also observe that each of the four components of J is a linear function of ξ and η.

Next, assemble the integrand H in Equation 5.33 by replacing the integrand in x, y by its equivalent in ξ, η. Denoting $\phi_i(\xi, \eta) = \psi_i(x, y)$ and applying the chain rule:

$$\begin{bmatrix} \dfrac{\partial \phi_i}{\partial \xi} \\ \dfrac{\partial \phi_i}{\partial \eta} \end{bmatrix} = \begin{bmatrix} \dfrac{\partial \psi_i}{\partial x}\dfrac{\partial x}{\partial \xi} + \dfrac{\partial \psi_i}{\partial y}\dfrac{\partial y}{\partial \xi} \\ \dfrac{\partial \psi_i}{\partial x}\dfrac{\partial x}{\partial \eta} + \dfrac{\partial \psi_i}{\partial y}\dfrac{\partial y}{\partial \eta} \end{bmatrix} = J \begin{bmatrix} \dfrac{\partial \psi_i}{\partial x} \\ \dfrac{\partial \psi_i}{\partial y} \end{bmatrix} \tag{5.34}$$

This relationship can now be used to find $H(\xi, \eta)$ in Equation 5.33. The partial derivatives to be replaced appear on the *right* side of Equation 5.34, whereas the left side is easily computed via Equation 5.31. Fortunately, it can easily be argued that under realistic conditions matrix J is invertible, and since its four components are linear it is easy to invert J analytically (as shown in the next equation). Hence, multiply Equation 5.34 by J^{-1} to get:

$$\begin{bmatrix} \dfrac{\partial \psi_i}{\partial x} \\ \dfrac{\partial \psi_i}{\partial y} \end{bmatrix} = J^{-1} \begin{bmatrix} \dfrac{\partial \phi_i}{\partial \xi} \\ \dfrac{\partial \phi_i}{\partial \eta} \end{bmatrix} = \frac{1}{\det(J)} \begin{bmatrix} \dfrac{\partial y}{\partial \eta} & -\dfrac{\partial y}{\partial \xi} \\ -\dfrac{\partial x}{\partial \eta} & \dfrac{\partial x}{\partial \xi} \end{bmatrix} \begin{bmatrix} \dfrac{\partial \phi_i}{\partial \xi} \\ \dfrac{\partial \phi_i}{\partial \eta} \end{bmatrix}.$$

This allows one to replace the partial derivatives in Equation 5.33 by their equivalence in ξ and η, thus providing the integrand $H(\xi, \eta)$ in Equation 5.33 — unless one of the coefficients p_1 and p_2 vary in x and/or y. Suppose, for example, that p_1 is a function of x and y. One needs the equivalent function of ξ and η for H in Equation 5.33. But this is simply:

$$\overline{p_1}(\xi, \eta) = p_1\left(x(\xi, \eta), y(\xi, \eta)\right)$$

where x and y are replaced by the right sides of Equation 5.32.

Finally, all the information is available to translate a particular element involved in the computation of a $K_{i,j}$ to the standard triangle and perform the integration there, the next topic.

Integration over the Standard Triangle

First, a bit of an *alert*. Should one have an *isotropic* material, hence constant p_1 and p_2 in the integrals in Equation 5.33, there is a price to be paid for the change to ξ, η variables. Note that in this case, in the xy integral in Equation 5.33, since p_1 and p_2 are constants, the integrand is a quadratic polynomial in x and y, hence it can be integrated analytically. Whereas if one traces the steps in assembling $H(\xi, \eta)$ one finds that this integrand is a rational function (quotient of polynomials), and one must typically resort to numerical integration.

In spite of the point just raised, most modern software is designed to handle more general cases (e.g., nonisotropic material) and proceed under the assumption that numerical integration will usually be needed. Continue with that assumption and consider some numerical schemes for computing in integral in Equation 5.33, that is

$$\iint_T H(\xi, \eta)\left|\det(J)\right| d\xi\, d\eta = \iint_T f(\xi, \eta)\, d\xi\, d\eta \simeq \sum_1^m a_i f(\xi_i, \eta_i) \tag{5.35}$$

where T is the standard triangle shown in Figure 5.7 and the sum on the right represents the numerical quadrature.

Probably the best choice of quadrature for this element results from taking as the (ξ_i, η_i) the three midpoints of the sides of the triangle, and the weights $a_i = 1/3$. This quadrature has a truncation error of $O(h^3)$, back in the xy-plane, consistent with the local interpolation error for this six-node (quadratic triangular) element. Little is gained for this element by using a more accurate (and more expensive) quadrature resulting from using more points. More accurate quadratures *are* appropriate if using higher-order elements. For example, here is the four (interior) points quadrature which has a truncation error of $O(h^4)$: $(\xi_i, \eta_i) = (1/3, 1/3)$; $a_i = \frac{-27}{48}$, $(\xi_i, \eta_i) = (3/5, 1/5), (1/5, 3/5)$, and $(1/5, 1/5))$; $a_i = \frac{25}{48}$. This completes the rather detailed treatment of the quadratic triangular element. More will be said about handling boundary conditions later. Following are some other widely used elements.

Some Rectangular Elements

Figure 5.9 shows a *standard* square element with four nodes for: $-1 \preceq \xi \preceq 1$ and $-1 \preceq \eta \preceq 1$. Again, the four basis functions are defined so that each function is 1.0 at one node and 0.0 at the other three nodes. This leads to the following four basis functions:

$$\phi(\xi,\eta) = \frac{1}{4}\begin{bmatrix}(1-\xi)(1-\eta)\\(1+\xi)(1-\eta)\\(1+\xi)(1+\eta)\\(1-\xi)(1+\eta)\end{bmatrix} \tag{5.36}$$

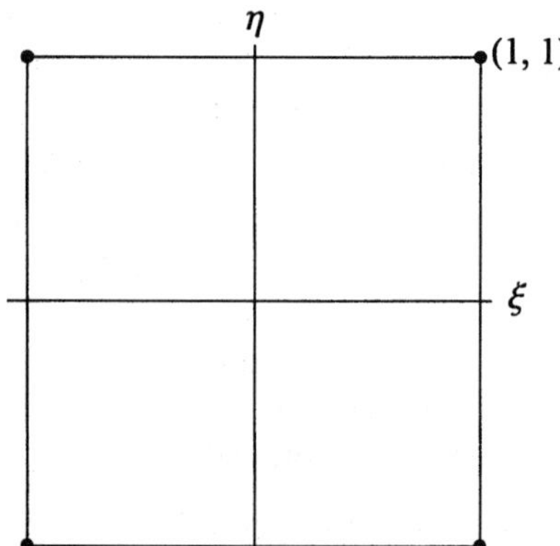

FIGURE 5.9 Standard four-node square.

Notice, for example, that $\phi_1(-1, -1) = 1$ and $\phi_1 = 0$ at the other three nodes, and it is linear on the two adjacent sides ($\xi = -1$ and $\eta = -1$). However, in contrast to the case of the linear triangular element, these functions are *not* linear on the entire square. For example, down the diagonal $\eta = \xi$ note that $\phi_1 = (1 - \xi)^2/4$. This element is called *bilinear*, the term coming from the fact that using linear combinations of these four functions one can generate all polynomials of the form $c_1 + c_2\xi + c_3\eta + c_4\xi\eta$. This four-node element is analogous to the three-node triangular element, although marginally better due to the fourth (bilinear) term above. For example, they both have $O(h^2)$ local truncation error.

Finally, consider mapping of the standard square onto (typically small) elements in the *xy*-plane. Analogous to Equation 5.32 above, given four points (x_i, y_i), the current transformation is

$$x = \sum_1^4 x_i\phi_i(\xi,\eta),$$
$$y = \sum_1^4 y_i\phi_i(\xi,\eta). \tag{5.37}$$

It is easy to see which figure the standard square maps onto in the *xy*-plane. For example, set $\xi = \pm1.0$ in Equation 5.36; clearly both x and y in Equation 5.37 become linear functions of η, say $x = a + b\eta$ and $y = c + d\eta$. Eliminating η gives y as a linear function of x, so the image of the two vertical sides of the standard square maps onto a straight lines. Similarly for the horizontal sides; hence, the standard square maps onto arbitrary quadrilaterals in the xy-plane. Next, a more accurate element is considered.

The Serendipity Element

This widely used element starts with the standard square and adds four nodes at the midpoints of the sides, as shown in Figure 5.10. At each of the eight nodes one defines the basis functions in the usual way, having value 1.0 at one node and 0.0 at the other seven nodes. The reader can verify that the following basis functions do the job:

$$\phi(\xi,\eta)=\frac{1}{4}\begin{bmatrix}(1-\xi)(1-\eta)(-\xi-\eta-1)\\(1+\xi)(1-\eta)(\xi-\eta-1)\\(1+\xi)(1+\eta)(\xi+\eta-1)\\(1-\xi)(1+\eta)(-\xi+\eta-1)\\2(1-\xi^2)(1-\eta)\\2(1+\xi)(1-\eta^2)\\2(1-\xi^2)(1+\eta)\\2(1-\xi)(1-\eta^2)\end{bmatrix} \tag{5.38}$$

Also note that on any side ($\xi = \pm 1$ or $\eta = \pm 1$) each ϕ_i is a quadratic function of the other variable (or identically zero) and, as mentioned above, passes through all three nodes on that side. For example, for ϕ_1 with $\eta = -1$, one has $\phi_1 = -\frac{1}{2}\xi(1-\xi)$, which has the advertised values of 1.0 at (–1, –1) and 0.0 at the other nodes on this side.

This eight-node element is analogous to the six-node triangular element, e.g., both have local truncation error of $O(h^3)$. However, these basis functions are *not* quadratic except on horizontal and vertical lines. For example, down the diagonal $\eta = \xi$ all of the ϕ_i are cubic functions. It can be shown that linear combinations of these basis functions form all polynomials of the form:

$$c_1+c_2\xi+c_3\eta+c_4\xi^2+c_5\xi\eta+c_6\eta^2+c_7\xi^2\eta+c_8\xi\eta^2. \tag{5.39}$$

Notice that the last two terms in Equation 5.39 are cubic and are not present in the analogous expression for the quadratic triangular element. These extra terms give the current element slightly more accuracy, but as mentioned above, the two elements have the same order of accuracy.

Along with the relatively good accuracy of this element is the fact that the standard square maps onto four-sided elements with curved (quadratic) sides. As before, the transformation is

$$x=\sum_1^8 x_i\phi_i(\xi,\eta),$$
$$y=\sum_1^8 y_i\phi_i(\xi,\eta). \tag{5.40}$$

In this case the eight (x_i, y_i) points will outline a four-sided figure with (possibly) curved sides as in Figure 5.10. The mapping of Equation 5.40 then takes the standard square onto the corresponding figure in the xy-plane.

The work in forming the integrals defining the stiffness matrix proceeds as above with the six-node triangular element leading to Equation 5.33, except now the integration is over the standard square. Rewriting Equation 5.35 for the current setting one has:

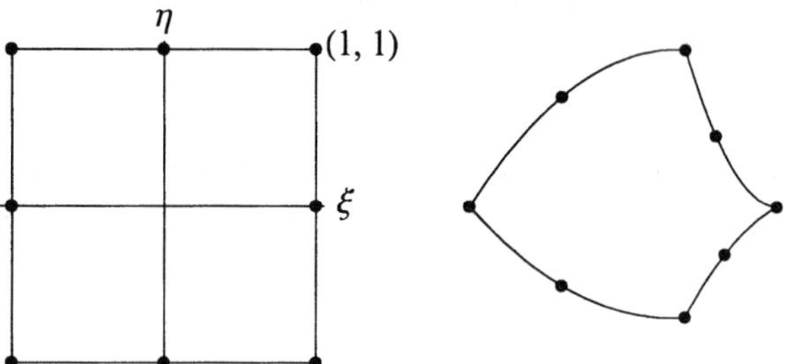

FIGURE 5.10 The eight-node square and an *xy* image.

$$\int_{-1}^{1}\int_{-1}^{1} H(\xi,\eta)\left|\det(J)\right| d\xi\, d\eta = \int_{-1}^{1}\int_{-1}^{1} f(\xi,\eta)\, d\xi\, d\eta \tag{5.41}$$

Briefly consider the approximation of the integral on the right. Since the current eight-point element has $O(h^3)$ local truncation error, the chosen quadrature should have at least that accuracy. Most modern software packages use *Gaussian* quadrature which provides very high accuracy with relatively few quadrature points. Several quadratures are given below, first for a single integral of the form:

$$\int_{-1}^{1} g(\xi)\, d\xi \simeq \sum_{1}^{n} w_i g(\xi_i). \tag{5.42}$$

By judicious choice of points ξ_i and weights w_i, Gaussian quadratures using n points provides an approximation that is exact for all polynomial of degree less than $2n$. Moreover, recall that the interval [–1, 1] typically corresponds to an interval of size $[-h, h]$ in the original *xy* system, and on the small interval, an n-point Gaussian quadrature has an error of order $O(h^{2n})$.

Here is a short list of Gaussian schemes. Provided are the points for $\xi \geq 0$ since the points and weights are symmetric about the origin. The weights are in parentheses.

$n = 1$: $\xi_i = 0.00000000$ (2.00000000)
$n = 2$: $\xi_i = 0.57735027$ (1.00000000)
$n = 3$: $\xi_i = 0.00000000$ (0.88888889) and 0.77459667 (0.55555556)
$n = 4$: $\xi_i = 0.33998104$ (0.65214515) and 0.86113631 (0.34785485)

For example, using $n = 2$ points leads, when translated to $[-h, h]$, to a quadrature error of $O(h^4)$ and $n = 3$ points an error of $O(h^6)$. Very likely either of these quadatures will prove adequate and be consistent with the accuracy of the element approximations. Recall that the eight-node square element itself has a interpolation error of $O(h^3)$.

Finally, consider how this type of quadrature is applied to the standard square. For a set number of points, n, one uses the same quadrature for integration in both ξ and η and proceeds by first defining function g and its approximation, for each ξ in [–1, 1]:

$$g(\xi) = \int_{-1}^{1} f(\xi,\eta)\, d\eta = \sum_{j=1}^{n} w_j f(\xi,\eta_j)$$

Then applying the quadrature to g one has:

$$\int_{-1}^{1} g(\xi)d\xi \simeq \sum_{i=1}^{n} w_i g(\xi_i) \simeq \sum_{i=1}^{n} w_i \left[\sum_{j=1}^{n} w_j f(\xi_i, \eta_j)\right] = \sum_{i,j=1}^{n} w_i w_j f(\xi_i, \eta_j)$$

For example, if $n = 3$ one would need to evaluate f at the nine points shown in the figure below.

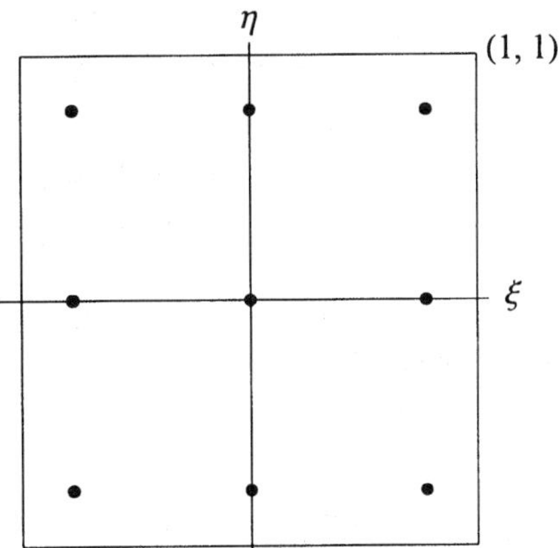

FIGURE 5.11 Guassian quadrature points.

Before leaving the standard square and two-dimensional elements, one more widely used element is mentioned. It has the eight nodes of the above Serendipity element plus one at the center of the square, (0, 0). This element is marginally more accurate than the Serendipity element since it is able to provide one more term to the type of polynomials that it can represent, which are

$$c_1 + c_2\xi + c_3\eta + c_4\xi^2 + c_5\xi\eta + c_6\eta^2 + c_7\xi^2\eta + c_8\xi\eta^2 + c_9\xi^2\eta^2.$$

Compare this with Equation 5.39. Because of the form, this element is called *biquadratic*. The basis functions are not given here (see Reddy and Garling[9]). They are considerably more complex than the basis functions for the Serendipity element (fourth-order polynomials as opposed to cubic), so in most problems the slight improvement in accuracy does not justify the additional expense of evaluation.

Three-Dimensional Elements

Here, a small sampling of three-dimensional elements are shown; first, a linear *brick* element, then a quadratic triangular-sided (tetrahedral) element.

Possibly the most widely used three-dimensional element is the linear *brick* element based on the cube shown in Figure 5.12, also shown is a typical image in *xyz*-space.

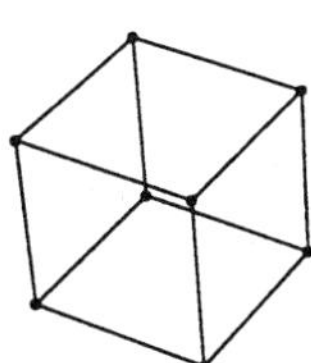

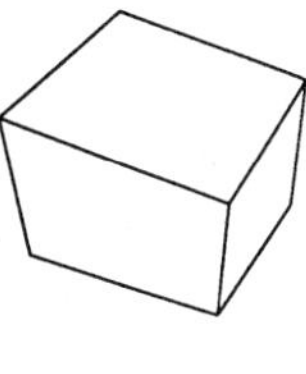

FIGURE 5.12 The standard brick and image.

The basis functions are easily shown to be

$$\phi(\xi,\eta,\zeta)=\frac{1}{8}\begin{bmatrix}(1-\xi)(1-\eta)(1-\zeta)\\(1+\xi)(1-\eta)(1-\zeta)\\(1+\xi)(1+\eta)(1-\zeta)\\(1-\xi)(1+\eta)(1+\zeta)\\(1-\xi)(1-\eta)(1+\zeta)\\(1+\xi)(1-\eta)(1+\zeta)\\(1+\xi)(1+\eta)(1+\zeta)\\(1-\xi)(1+\eta)(1+\zeta)\end{bmatrix} \tag{5.43}$$

Notice on any edge (e.g., $\eta = -1 = \zeta$) each function is linear, but this is only true (once again) on lines parallel to the coordinate axis. For example, down the diagonal $\xi = \eta = \zeta$ each basis function is cubic. The mapping to the *xyz* system is given by

$$\begin{aligned} x &= \sum_{1}^{8} x_i \phi_i(\xi,\eta,\zeta), \\ y &= \sum_{1}^{8} y_i \phi_i(\xi,\eta,\zeta), \\ z &= \sum_{1}^{8} z_i \phi_i(\xi,\eta,\zeta). \end{aligned} \tag{5.44}$$

To see the image of an edge of the brick, take, for example, $\xi = \eta = 1$. Then, from Equation 5.43, clearly all the ϕ_i are linear functions of ζ; hence, from Equation 5.44 x, y, and z are linear functions of ζ. It follows that this edge maps onto a line. Similarly each edge maps onto a straight line and, hence, the brick maps onto a six-sided solid with straight edges. (However, the sides are *not* planar unless the four determining *xy* points are in a plane.)

Another popular, and more accurate, three-dimensional element is analogous to the eight-node square Serendipity element. It is the 20-point brick with nodes at the 8 corners, as above, plus 1 at each of the midpoints of the 12 edges. It will have quadratic behavior, i.e., in terms of accuracy and representing edges. However, the basis functions themselves are fourth polynomials in ξ, η, and ζ; hence, they are relatively expensive to deal with. For more details see Reddy and Gartling.[9]

In practice it often makes sense to use a simple linear element (like the eight-node brick above) for the interior of the domain under study, then go to a higher-order (often quadratic) element to better represent the boundary. As mentioned earlier, a quadratic element does a much better job of approximating a curved boundary. Two such possibilities are now discussed.

An efficient second-order element is the ten-node triangular shaped (tetrahedral) element. Figure 5.13 shows the standard version of this element.

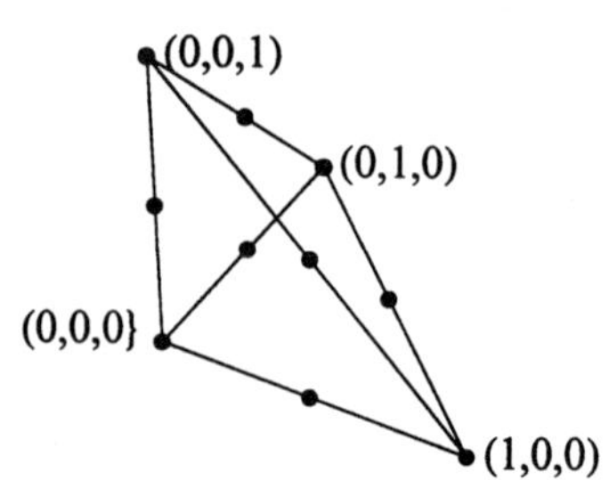

FIGURE 5.13 The 10-node tetrahedra.

One can easily verify that the following basis functions do the usual job of taking on values of 1.0 at one node and 0.0 at the other nodes. Note that they are *optimally* efficient in that there are only 10 of them (the minimum number) and each one is quadratic (the minimum order).

$$\phi(\xi,\eta,\zeta)=\begin{bmatrix}(1-\xi-\eta-\zeta)(1-2\xi-2\eta-2\zeta)\\ \xi(2\xi-1)\\ \eta(2\eta-1)\\ \zeta(2\zeta-1)\\ 4\xi(1-\xi-\eta-\zeta)\\ 4\xi\eta\\ 4\eta(1-\xi-\eta-\zeta)\\ 4\zeta(1-\xi-\eta-\zeta)\\ 4\xi\zeta\\ 4\eta\zeta\end{bmatrix} \tag{5.45}$$

Moreover, linear combinations of these functions will represent any quadratic polynomial:

$$c_1+c_2\,\xi+c_3\,\eta+c_4\,\zeta+c_5\,\xi^2+c_6\,\eta^2+c_7\,\zeta^2+c_8\,\xi\eta+c_9\xi\zeta+c_{10}\eta\zeta.$$

Again, the mapping into the *xyz* system is provided by Equation 5.44, except the sum is over 10 basis functions.

Another element that is particulary useful in connection with the interior linear brick elements is the quadratic prism element whose standard form is shown in Figure 5.14.

As an example of its utility, if a linear brick element near the boundary places two of its nodes on the boundary, one can, e.g., fill in above and beside the brick element with the quadratic prism elements as shown in Figure 5.15. The same type of fill-in can be accomplished using the quadratic tetrahedral element discussed just above. However, it would take three tetrahedral elements to do the job of one prism, as illustrated in Figure 5.15.

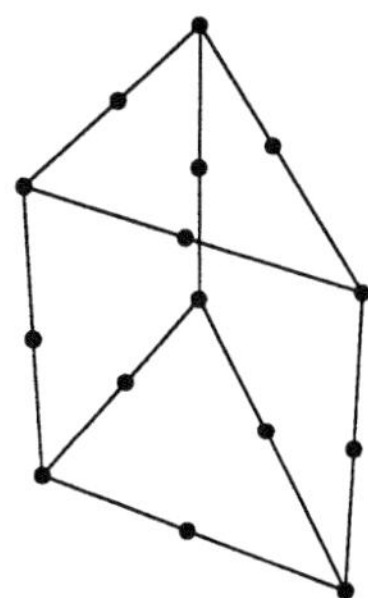

FIGURE 5.14 The quadratic prism.

Integration in Three Dimensions

Analogous to the integration of Equation 5.41 in two dimensions and assuming the same type of quadrature, the integration over the *brick* becomes:

$$\int_{-1}^{1}\int_{-1}^{1}\int_{-1}^{1}H(\xi,\eta,\zeta)\left|\det(J)\right|d\xi\,d\eta\,d\zeta$$

$$=\int_{-1}^{1}\int_{-1}^{1}\int_{-1}^{1}f(\xi,\eta,\zeta)\,d\xi\,d\eta\,d\zeta\simeq\sum_{i,j,k=1}^{n}w_iw_jw_kf\left(\xi_i,\eta_j,\zeta_k\right).$$

The weights w_i are the same as discussed above. Of course, the calculations are more involved since the matrix J is now a 3 by 3 matrix, so its inversion is more difficult. But logically the steps are the same.

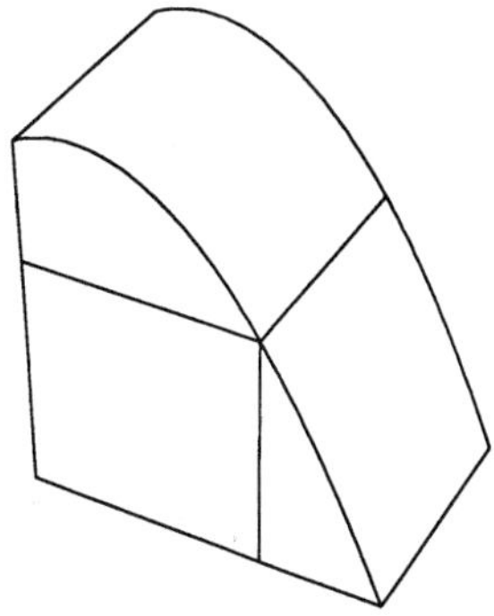

FIGURE 5.15 A block and two prisms.

Integration of the tetrahedral and prisms are not as straightforward as the brick above. In particular the quadrature points (ξ_i, η_j, ζ_k) and the weights have to be modified substantially. To pursue such issues see Davis and Rabinowitz.[6]

Some Boundary Considerations

In the last section some options were given for elements for the representation of the domain D involved in the boundary-value problem of interest, Equation 5.22. Also discussed were the integrations in computing the stiffness matrix involved in the fundamental equation:

$$\begin{aligned}\sum_{j=1}^{N} K_{i,j}\gamma_j &= \sum_{j=1}^{N}\left[\iint_D \left(p_1 \partial_x \phi_i \partial_x \phi_j + p_2 \partial_y \phi_i \partial_y \phi_j\right) dx\, dy\right]\gamma_j \\ &= \iint_D \phi_i Q\, dx\, dy + \int_{\Gamma_2} \phi_i\, g(s)\, ds; \quad i = 1,\dots,N.\end{aligned} \tag{5.46}$$

Turn to the second line of Equation 5.46 which provides the right side of the vector equation and seek β in $\boldsymbol{K}\gamma = \beta$. The double integral on domain D is much like that already discussed, so it will not be discussed further. However, the last integral is of a different nature, a *line* integral in a two-dimensional setting and a *surface* integral when in three dimensions. The computation of these integrals is the next subject.

Line Integral in Two Dimensions

Consider a portion of a boundary and an associated triangular element as shown in Figure 5.16. (Recall that the curved "leg" of the element typically only approximates the actual boundary; but if h is sufficiently small and quadratic elements are used this approximation should be adequate. So proceed with that assumption.)

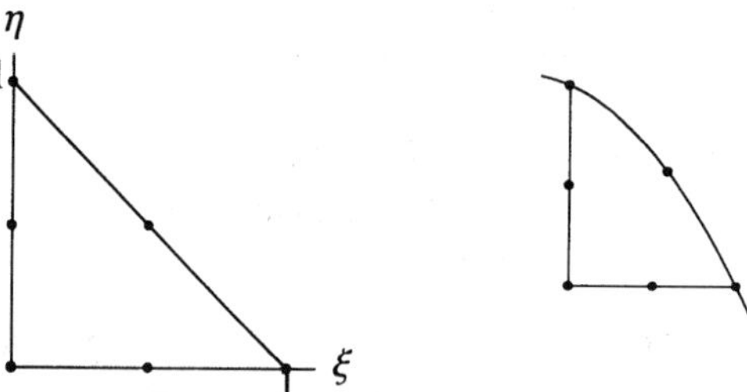

FIGURE 5.16 Standard triangle and image on boundary.

Consider the line integral over this key triangular leg; and denote the integral by

$$\int_{\Gamma_e} \phi_i g(s)\, ds. \tag{5.47}$$

For convenience assume that the Γ_e is mapped from the side $\xi + \eta = 1$ of the standard triangle. Hence, setting $\eta = 1 - \xi$ in Equation 5.32 so this portion of boundary can be represent by a single parameter ξ:

$$x(\xi) = \sum_{1}^{n} x_j \phi_j(\xi, 1 - \xi), \tag{5.48}$$

$$y(\xi)=\sum_{1}^{n} y_j\,\phi_j(\xi,1-\xi).$$

Here, one only needs to sum on those basis functions that affect this one side of the standard triangle (e.g., three ϕ_j in the case of a quadratic triangular element). To perform the line integral in Equation 5.47, change integration variable from arc-length parameter s to ξ. This is a standard calculus result:

$$\int_{\Gamma_e} \phi_i g(s)\,ds=\int_0^1 \phi_i(\xi,1-\xi)\,g\big(x(\xi),y(\xi)\big)\sqrt{x'(\xi)^2+y'(\xi)^2}\,d\xi, \tag{5.49}$$

where x and y are given by Equation 5.48. As a simple illustration, if one is using the *linear* triangular element then, assuming ϕ_1 and ϕ_2 are the basis functions involved, Equation 5.48 reduces to

$$x(\xi)=x_1\,\phi_1(\xi,1-\xi)+x_2\,\phi_2(\xi,1-\xi)=x_1(1-\xi)+x_2\,\xi,$$
$$y(\xi)=y_1\,\phi_1(\xi,1-\xi)+y_2\,\phi_2(\xi,1-\xi)=y_1(1-\xi)+y_2\,\xi.$$

Differentiating this x and y one sees that the square root term in Equation 5.49 is just the distance between points (x_1, y_1) and (x_2, y_2) on this segment of boundary. More generally, this square root term makes the adjustment between the element's approximation to the arc-length of Γ_e and the interval [0, 1].

Surface Integral in Three Dimensions

Now consider an analogous look at the situation in three dimensions. The three-dimensional version of Equation 5.46 is

$$\sum_{j=1}^{N} K_{i,j}\gamma_j=\sum_{j=1}^{N}\left[\iiint_D\left(p_1\partial_x\phi_i\partial_x\phi_j+p_2\partial_y\phi_i\partial_y\phi_j+r\partial_z\phi_i\partial_z\phi_j\right)dx\,dy\,dz\right]\gamma_j$$
$$=\iiint_D \phi_i\,Q\,dx\,dy\,dz+\int_{\Gamma_2}\phi_i g(s)\,ds;\ \ i=1,\ldots,N.$$

Again concentrate on the last integral, which is now a surface integral. For simplicity, assume that the standard cube maps onto the element shown in Figure 5.17 and that the $\zeta = 1.0$ surface of the cube maps onto the surface area under investigation. So, referring to (5.44), but assuming n points rather than 8, and setting $\zeta = 1.0$ the surface section can be expressed:

$$x(\xi,\eta)=\sum_{1}^{n} x_j\,\phi_j(\xi,\eta,1.0),$$
$$y(\xi,\eta)=\sum_{1}^{n} y_j\,\phi_j(\xi,\eta,1.0), \tag{5.50}$$
$$z(\xi,\eta)=\sum_{1}^{n} z_j\,\phi_j(\xi,\eta,1.0).$$

As before, the sums in Equation 5.50 only need to be done over those basis functions that affect the top surface of the standard cube since Equation 5.50 will only be evaluated on that surface.

Since one now has a surface integral, the differential ds denotes an element of surface area and, analogous to Equation 5.49, one has

$$\int_{\Gamma_e} \phi_i g(s)\,ds = \int_{-1}^{1}\int_{-1}^{1} \phi_i(\xi,\eta,1.0)\,g(x(\xi,\eta),y(\xi,\eta))\sqrt{J_1^2+J_2^2+J_3^2}\,d\xi\,d\eta,$$

$$J_1 = \det\begin{pmatrix}\partial_\xi y & \partial_\eta y\\ \partial_\xi z & \partial_\eta z\end{pmatrix},\quad J_2 = \det\begin{pmatrix}\partial_\xi z & \partial_\eta z\\ \partial_\xi x & \partial_\eta x\end{pmatrix},\quad J_3 = \det\begin{pmatrix}\partial_\xi x & \partial_\eta x\\ \partial_\xi y & \partial_\eta y\end{pmatrix}.$$

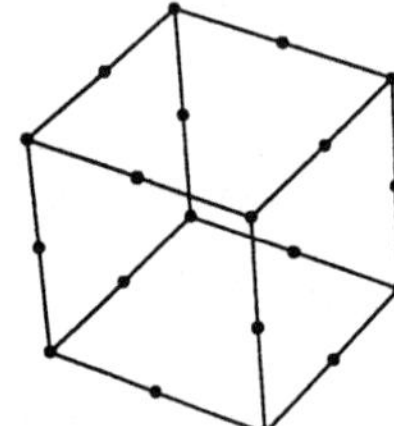
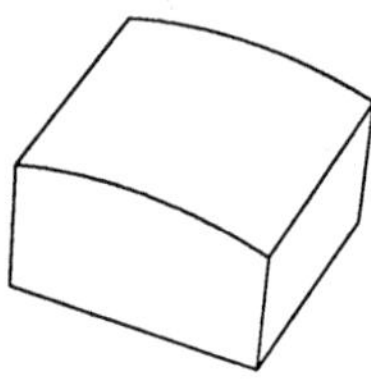

FIGURE 5.17 The quadratic brick and image on boundary.

The x, y, and z in the Jacobians J_k are given by Equation 5.50, and the square root term above once again makes the adjustment between the area of the $\xi\eta$-square (namely 4.0) and the area of the portion of surface under study. So one has all the data and technique necessary to compute the surface integrals in the right side of the key vector equation $\boldsymbol{K}\gamma = \beta$.

Once the elements of matrix $\boldsymbol{K}$ and vector β are computed, one is typically faced with a large system of linear equations to solve. This system is usually solved by a direct method (i.e., Gaussian elimination) although for some problems iteration techniques may be preferable (see Section 5.2).

Time-Dependent Problems

This subsection provides an introduction to how time-dependent problems can be approached with the FEM. The analogue of Equation 5.18 is

$$\rho\partial_t T - \partial_x\left(p_1 T_x\right) - \partial_y\left(p_2 T_y\right) = Q \tag{5.51}$$

where functions p, q, and Q can be functions of x, y, and t, while ρ may depend on x and y. As before, multiply Equation 5.51 by weight function w, integrate over the domain D, and apply Green's function to get, analogous to Equation 5.21:

$$\iint_D \left[w\rho T_t + w_x\left(p_1 T_x\right) + w_y\left(p_2 T_y\right) - wQ\right]dx\,dy - \int_\Gamma w\left[n_1 p_1 T_x + n_2 p_2 T_y\right]ds = 0. \tag{5.52}$$

Along with the boundary conditions of the problem in Equation 5.22, one now has initial conditions (i.e., at $t = 0$) defining the complete boundary conditions for Equation 5.51:

$$T(x,y,0) = T_0(x,y) \quad \text{on } D$$
$$T = f \quad \text{on } \Gamma_1 \quad \text{for } t \succeq 0, \tag{5.53}$$
$$n_1 p_1 T_x + n_2 p_2 T_y = g \quad \text{on } \Gamma_2 \quad \text{for } t \succeq 0.$$

Analogous to the form of the approximate solution in Equation 5.23, seek solutions of the form

$$T(x,y,t) \cong \sum_{j=1}^{N} \gamma_j(t)\phi_j(x,y), \tag{5.54}$$

where the γ_j, previously constants, now depend on t. One applies $t = 0$ in Equation 5.54 to satisfy the initial condition; hence, the constants $v_j(0)$ are determined by the function T_0 evaluated at the nodal points. The elements and nodes on D are determined just as they are for the steady-state heat equation.

It remains to obtain a system of differential equations to determine the N functions $\gamma_j(t)$ for the time interval of interest. As before, choose $w = \phi_i$ in Equation 5.52 and replace T by the form of Equation 5.54 to get:

$$\iint_D \sum_{j=1}^{N} \left[\rho\phi_i\phi_j\gamma_j' + \left(p_1\partial_x\phi_i\partial_x\phi_j + p_2\partial_y\phi_i\partial_y\phi_j\right)\gamma_j\right] dx\,dy$$
$$= \iint_D \phi_i Q\,dx\,dy + \int_{\Gamma_2} \phi_i\, g(s)\,ds; \quad i = 1,\ldots,N.$$

This can be put in matrix form, analogous to form $\boldsymbol{K}\gamma = \beta$ above,

$$\boldsymbol{M}\gamma'(t) + \boldsymbol{K}(t)\gamma(t) = \beta(t). \tag{5.55}$$

To integrate Equation 5.55 numerically, the recommended approach is analogous to the Crank-Nicolson method. Let Δt denote the time step and $\gamma_n = \gamma(n\Delta t)$; Crank-Nicolson in this context appears:

$$\gamma_{n+1} = \gamma_n + \frac{\Delta t}{2}\left[\gamma'(n\,\Delta t) + \gamma'((n+1)\,\Delta t)\right]. \tag{5.56}$$

Since the second γ' in this expression depends on values at the $(n + 1)^{st}$ time step, the method is *implicit*. To take the time step in Equation 5.56, apply Equation 5.55 at the nth and $(n + 1)^{st}$ time step, and by denoting $\boldsymbol{K}_n = \boldsymbol{K}(n\Delta t)$ and $\beta_n = \beta(n\Delta t)$ obtain:

$$\left(\boldsymbol{M} + \frac{\Delta t}{2}\boldsymbol{K}_{n+1}\right)\gamma_{n+1} = \left(\boldsymbol{M} - \frac{\Delta t}{2}\boldsymbol{K}_n\right)\gamma_n + \frac{\Delta t}{2}\left(\beta_n + \beta_{n+1}\right). \tag{5.57}$$

Finally, for each time step, the system in Equation 5.57 is solved for the new γ and Equation 5.54 provides the approximate solution for this time. Again this algorithm does *not* have stability problems (demanding small time steps relative to the mesh size). Moreover, if Δt is consistent with the mesh size and the FEM elements are of order $O(h^2)$ or higher, one should expect this order of accuracy for the time-dependent problem.

Variational Forms of Elliptic Problems

Earlier in this chapter it was mentioned that some elliptic problems are equivalent to minimizing a *functional*, an integral which typically represent the energy associated with the problem. This association has a rich literature in mathematical physics, with many important applications. Therefore, following is a brief introduction to this notion, particularly as it pertains to elliptic problems in heat transfer.

A central notion is that of a *positive definite* operator. Suppose the operator $\boldsymbol{L}$ of interest operates on functions defined on some region D in the xy-plane. (Usually $\boldsymbol{L}$ is a differential operator, certainly in this setting.) The *domain* of the operator, $\Omega(\boldsymbol{L})$ usually consists of those functions with certain smoothness properties *and* satisfying some conditions on the boundary Γ of D. $\boldsymbol{L}$ is said to be *positive definite* if for all functions u in the domain, not identically zero on D and satisfying the boundary conditions, one has

$$\iint_D u\boldsymbol{L}u\,dx\,dy \succ 0.$$

The prototype positive definite operator is the Laplacian $-\Delta$ where $\Delta u = u_{xx} + u_{yy}$ and the domain of the operator is the set of functions u such that the second derivatives are continuous (or at least smooth enough that the integral above makes sense), and $u = 0$ on the boundary of D. It is easy to show that this operator is positive definite by applying a form of Green's function as follows:

$$\iint_D u\boldsymbol{L}u\,dx\,dy = \iint_D u\left(-u_{xx} - u_{yy}\right)dx\,dy$$

$$= \iint_D \left(u_x^2 + u_y^2\right)dx\,dy + \int_\Gamma u\partial_n u\,dx\,dy \succeq 0.$$

The last integral is 0 since u is assumed zero on Γ. The previous integral is positive or zero, but if zero, then that u must be a constant, hence identically zero. Thus, operator $\boldsymbol{L}$ is positive definite. In a similar manner many elliptic boundary value problems can be shown to enjoy this property.

The linear functional associated with many differential equations $\boldsymbol{L}u = f$ is

$$F[u] = \iint_D \left[u\boldsymbol{L}u - 2uf\right]dx\,dy. \tag{5.58}$$

This sets the stage for the following classical statement equating the two problems:

> Let operator $\boldsymbol{L}$ be a positive definite differential operator and consider the boundary value problem: $\boldsymbol{L}u = f$, $u \in \Omega(\boldsymbol{L})$. Function u solves this problem if, and only if, u minimizes the functional $F[u]$ over all u in $\Omega(\boldsymbol{L})$.

To apply this remarkable statement to the *Poisson* problem $-\Delta u = f$, with $u = 0$ on the boundary of D, one can solve the equivalent problem of minimizing the F in Equation 5.58, where

$$F[u] = \iint_D \left[u\boldsymbol{L}u - 2uf\right]dx\,dy$$

$$= \iint_D u\left(-u_{xx} - u_{yy} - 2uf\right)dx\,dy = \iint_D \left(u_x^2 + u_y^2 - 2uf\right)dx\,dy.$$

Again, the boundary term is zero in the application of Green's function since it was specified that $u = 0$ on the boundary. To see how this relates to the FEM, seek to minimize the functional

$$F[u] = \iint_D \left(u_x^2 + u_y^2 - 2uf\right) dx\, dy \tag{5.59}$$

and look for the approximation to this minimum in the usual form:

$$u(x, y) \simeq \sum_{j=1}^{N} \gamma_j \phi_j(x, y).$$

Substituting this form into Equation 5.59, one has

$$F[u] = \iint_D \left[\left(\sum \gamma_j \partial_x \phi_j\right)^2 + \left(\sum \gamma_j \partial_y \phi_j\right)^2 - 2f \sum \gamma_j \phi_j\right] dx\, dy, \tag{5.60}$$

where the sums are on j. Now, think of F as a function of the γ_j and minimize F with respect to these N values. Taking the partial derivative of Equation 5.60 with respect to γ_i and setting to 0:

$$\frac{\partial F}{\partial \gamma_i} = 2 \iint_D \left[\left(\sum \gamma_j \partial_x \phi_j\right) \partial_x \phi_i + \left(\sum \gamma_j \partial_y \phi_j\right) \partial_y \phi_i - f \phi_i\right] dx\, dy = 0.$$

Rearranging slightly gives

$$\iint_D \left[\left(\sum \gamma_j \partial_x \phi_j\right) \partial_x \phi_i + \left(\sum \gamma_j \partial_y \phi_j\right) \partial_y \phi_i\right] dx\, dy$$

$$= \iint_D f \phi_i \, dx\, dy, \quad i = 1, \ldots, N.$$

Comparing this with Equation 5.24 observe that, except for the differences in the two problems, one has the equivalent system to solve for the unknown γ. Hence, the new approach, minimizing the *energy* functional, leads to the same system as the method discussed earlier (the *weighted integrals* or *Galerkin* method).

Before the popularity of the FEM the *variational* form of a problem usually referred to what is called the *energy* form and the minimization the above $F[u]$. The growth of the FEM has stimulated much research in the mathematical aspects of the subject, so currently it has been established that a very large class of elliptic boundary value problems have an equivalent variational form (see Oden and Reddy[5]). However, the weighted integral, or Galerkin method discussed previously, are applicable to many problems for which there is no variational form. Hence, in practice the Galerkin method is that most often used, for example, in modern software packages.

Error Estimates

Throughout this chapter an important concept is the *order* of accuracy that a particular numerical approximation enjoys. Typically, when a scheme was said to be $O(h^3)$ this has meant that the error at a typical point, say (x_i, y_j), should behave like Ch^3 for small h, where C is a constant. The error analysis for the FEM is a bit more complex in that the best estimates involve a *mean* or integral average of the error over the entire domain, say D. Following is a short discussion of the situation for a typical FEM application.

The first type of error to consider involves how well any smooth function is approximated by using the nodes on a single element, the *interpolation* error. For example, suppose one is using a two-dimensional triangular element with six nodes; this is a *quadratic* element in the sense that every polynomial in two variables of degree two or less are represented exactly using the six nodes. Hence, the degree, k, of this element is 2. Now the standard triangle has sides of unit length, but in practice this triangle is mapped into the real xy domain onto a three-sided figure with sides of order h (the mesh size). In this case the interpolation error is of $O(h^3)$. More generally, a mesh is of degree k if, in the sense just discussed, the elements exactly represent all polynomials of degree k or less.

The second, and more important, aspect of error analysis involves how well the FEM solution approximates the exact solution to the BVP. This error is best measured using various integral *norms* over the domain D of interest. Let u denote a function with enough smoothness for the following integrals to exist. Assuming a two-dimensional setting, define a family of norm on D as follows:

$$\|u\|_0 = \left[\int_D u^2\,dx\,dy\right]^{1/2},$$

$$\|u\|_1 = \left[\int_D \left(u^2 + u_x^2 + u_y^2\right) dx\,dy\right]^{1/2},$$

$$\|u\|_2 = \left[\int_D \left(u^2 + u_x^2 + u_y^2 + u_{xx}^2 + u_{xy}^2 + u_{yy}^2\right) dx\,dy\right]^{1/2},$$

and likewise for other norms $\|u\|_m$. Following is a summary of the type of error bounds available. For more details, see Strang and Fix[7] or Oden and Reddy.[5] Assume one's problem involves a second-order differential equation and that the elements used are of degree k. Denoting the error in the FEM result by $e = u - u_{fem}$, a typical result is

$$\|e\|_s \preceq C h^{k+1-s} \|u\|_{k+1}, \quad 1 \preceq s \preceq k \tag{5.61}$$

where u denotes the exact solution. For sake of illustration, assume the elements are quadratic, so $k = 2$. Note that the error bound in Equation 5.61 involves $\|u\|_3$, thus requiring all derivatives through order three to be well behaved. In applying Equation 5.61 one has some choices as to just how the error estimate is applied. In particular, if one chooses $s = 0$, then $\|e\|_0$ is of $O(h^3)$; alternatively by choosing $s = 1$ Equation 5.61 says that $\|e\|_1$ is no worse than $O(h^2)$. For example, if one is only concerned with the error in u itself, the clear choice is the $s = 0$ result. Also note that Equation 5.61 shows that the degree of the mesh generally translates into a correspondingly more accurate final result.

If the errors in u and its first derivatives are of primary concern, the following statement is particularly useful:

$$\|e\|_1 \preceq C h^{s-1} \|u\|_s, \quad 1 \preceq s \preceq k+1. \tag{5.62}$$

For example, again assuming $k = 2$, setting $s = 3$ shows $\|e\|_1$ is of $O(h^2)$, agreeing with the Equation 5.61 result. Choosing $s = 2$, the result in Equation 5.62 only guarantees that $\|e\|_1$ is of order h, an inferior result to that of Equation 5.61. However, the result in Equation 5.62 has the advantage that it only requires integral bounds on the first and second derivatives of u. So the choice in applying error bounds like these often depends on the amount of smoothness that can be expected from the exact solution.

These error bound results have some clear advantages in flexibility as has been pointed out. However, they do *not* provide what some would consider the optimal type of error estimate, i.e., bounding the *maximum* error in u and, possibly, its derivatives over the entire domain D. For example, a sharp spike in error will tend to get averaged out and not show up in these integral bounds. However, in general, if the boundary of the domain and any external forcing term are well behaved (e.g., no sharp cusps, etc.) then these bounds will typically provide the user with adequate assurance.

An Example of Modern Software

There is a wide variety of software for applying the FEM method to numerous areas of engineering, including heat transfer. Broadly speaking, this software falls into two categories. The first group are those relatively simple programs (typically in FORTRAN or a mathematical languages like MATLAB) designed to handle a modest class of problems with relatively few options. Some of these are offered free to the public, e.g., included in a book on the FEM; for example, see Reddy,[8] Reddy and Gartling,[9] Wong and Bang,[10] and Huang and Usmani.[11]

The second category of software includes a number of large commercial packages designed to handle a wide range of BVPs in two and three dimensions and able to accommodate very complex boundaries. These packages offer a large choice of elements, some of which have already been discussed. Most of the major commercial packages are based on the CAD/CAM concept in which the domain of interest can be constructed graphically or analytically, and the mesh size can be user-specified or determined by the software to guarantee a specified accuracy. The mesh itself is typically generated by the software to accommodate the details of the boundary and often the mesh shape and size is "adapted" to respond to how fast the solution is changing throughout the domain. Consequently, these packages tend to be quite large, demanding a well-equipped PC, or, in some cases, a workstation, and are priced accordingly.

An example of the commercial companies with major FEM capability is ANSYS, Inc. who has a large suite of engineering software. The author was provided a complementary copy of ANSYS/Ed, a relatively small but quite powerful educational version of their major products. These products handle a host of linear and nonlinear engineering problems including structural analysis, dynamic analysis (e.g., vibration and buckling), heat transfer, electromagnetics, and acoustics. Here this software is discussed and illustrated on the example described in Equation 5.26 and pictured in Figures 5.4 and 5.5, in which the domain was divided into 13 elements. Using the ANSYS package, the domain is created in a GUI (graphical user interface) environment in which the boundary is created by pointing and clicking. After the domain is described and the boundary values specified, the mesh is determined by selecting from a menu of elements deemed appropriate for this problem; then, in this case, the author specifies the desired mesh size. The program then divides the domain into a number of elements of the specified size, as shown in Figure 5.18. Finally, the solution stage is entered and the FEM solution obtain (very quickly in this case).

Figure 5.19 shows a typical form of graphical solution (seen in black and white here, but more effectively on a color monitor). Should numerical results be required (at each node), this can be requested, as well as a number of other output options. The elements used in this application were the six-node, quadratic, triangular elements discussed earlier. As the theory discussed above indicates, the error in solution u will tend to zero like h^3.

5.4 Boundary Element Method

In this section an introduction is given to an important method, the *boundary element method* (BEM), which shares some features with the finite element method. The basic idea of the BEM is to again convert the boundary value problem of interest into an integral equation; however, in the BEM one seeks the key integral relationship over the *boundary* of the region of interest. This has the advantage of reducing the dimension of the problem by one. For example, in a three-dimensional linear steady-state heat problem, the FEM and finite difference approach lead to the fundamental linear system of equations

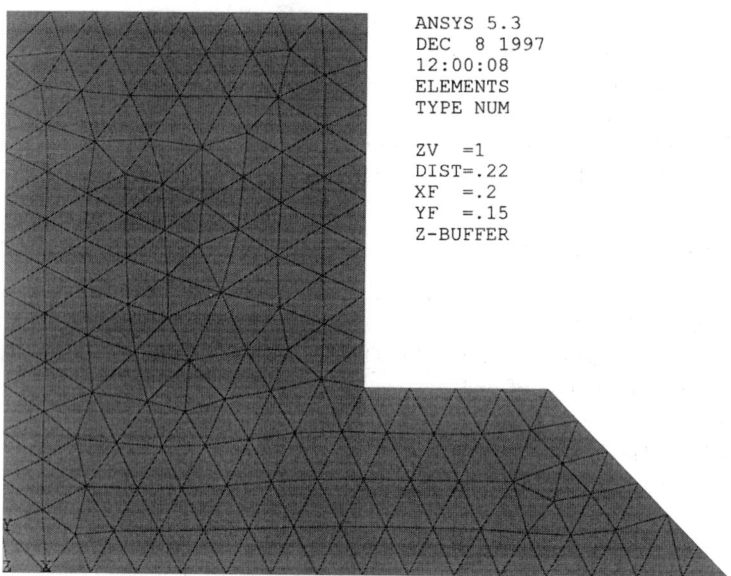

FIGURE 5.18 An ANSYS/Ed mesh for the FEM example. (Courtesy of ANSYS, Inc.)

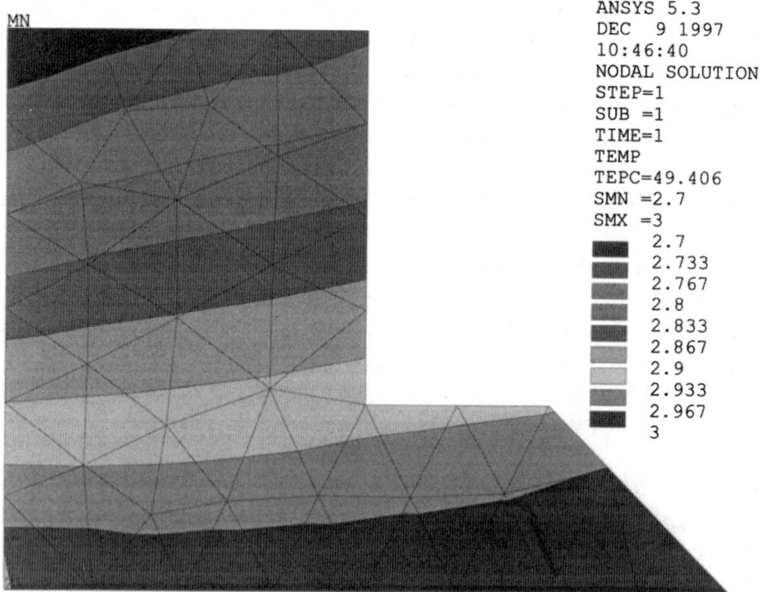

FIGURE 5.19 An ANSYS/Ed graphical solution. (Courtesy of ANSYS, Inc.)

involving N^3 unknowns. If the BEM is successful in posing the fundamental integral equation *only* on the boundary, the corresponding linear system is reduced to order N^2. The advantage of the BEM is especially strong if the domain of interest is very large or infinite. Of course, there are trade-offs between the two methods, and some of these are discussed below.

As with the finite element approach, major use is made of various forms of Green's theorem which relates the original integral equation, in full dimension, to values only on the boundary. A key issue is the selection of *weighting function* since judicious choice can often reduce the integral over the entire domain to zero, or to a single point, thus providing the lowering of the dimension of the problem. These concepts now will be illustrated.

BEM and a Poisson Problem

In this subsection Ω denotes the domain of interest in two or three dimensions, and Γ denotes its boundary. The Poisson problem to be discussed is

$$\begin{aligned} \nabla^2 T(\boldsymbol{x}) &= b(\boldsymbol{x}) \quad \text{on } \Omega, \\ \mathrm{T} &= \mathrm{f} \quad \text{on } \Gamma^1, \\ \partial_n T &= g \quad \text{on } \Gamma^2, \end{aligned} \tag{5.63}$$

where $\Gamma = \Gamma^1 + \Gamma^2$. A particularly useful choice of weighting function for this problem is the *fundamental* solution $G(\boldsymbol{x}, \xi)$ to Laplaces's equation. This function depends on the dimension of the problem and will be defined below. In any dimension, such a G enjoys the property:

$$\nabla^2 G(\boldsymbol{x}, \xi) = \delta(\boldsymbol{x} - \xi) \tag{5.64}$$

where $\delta(\boldsymbol{x} - \xi)$ is the *delta* function with the property that when multiplied by any continuous function $h(\boldsymbol{x})$ and integrated over a region including fixed point ξ, it returns the value $h(\xi)$. This feature is exploited shortly.

The first step in this presentation of the BEM is to apply Green's identity as follows, where ξ is an *interior* point (i.e., not on the boundary):

$$\int \left[G(\boldsymbol{x},\xi)\nabla^2 T(\boldsymbol{x}) - T(\boldsymbol{x})\nabla^2 G(\boldsymbol{x},\xi) \right] d\Omega = -\int \left[G(\boldsymbol{x},\xi)\partial_n T(\boldsymbol{x}) - T(\boldsymbol{x})\partial_n G(\boldsymbol{x},\xi) \right] d\Gamma$$

where all of the integrals are with respect to $\boldsymbol{x}$ and the differential notation specifies the entire domain or the boundary. As before, ∂_n denotes the outer normal derivative relative to the boundary. Use the fact that $\nabla^2 T = b$ in the first term. Then apply the delta behavior of G in Equation 5.64 to the second term to produce $-T(\xi)$. Finally, rearranging terms results in

$$T(\xi) + \int T(\boldsymbol{x})\partial_n G(\boldsymbol{x},\xi)\, d\Gamma = \int G(\boldsymbol{x},\xi)\partial_n T(\boldsymbol{x})\, d\Gamma + \int G(\boldsymbol{x},\xi)\, b(\boldsymbol{x})\, d\Omega. \tag{5.65}$$

Equation 5.65 is the fundamental BEM relationship involving known and unknown values of T and flux $\partial_n T$. There are two distinct uses of Equation 5.65; first, variable ξ is taken to the boundary and the result is discretized, leading to a system of linear equations for any missing T and flux values on the boundary. Secondly, when all such boundary values are known in the integrals in Equation 5.65, it is applied to compute T at desired *interior* points. In the next section these steps are taken for a two-dimensional problem, and later three-dimensional issues are discussed.

BEM in Two Dimensions

The problem to be discussed here is a case of Equation 5.63 in two dimensions. In this case the fundamental solution for Laplace's equation, $\nabla^2 G = \delta(\boldsymbol{x} - \xi)$, is

$$\begin{aligned} G(\boldsymbol{x},\xi) &= \frac{-1}{2\pi} \log r(\boldsymbol{x},\xi), \\ r(\boldsymbol{x},\xi) &= \|\boldsymbol{x} - \xi\| = \sqrt{(x_1 - \xi_1)^2 + (x_2 - \xi_2)^2}. \end{aligned} \tag{5.66}$$

It is a routine calculation to show that

$$F(x,\xi) = \partial_n G(x,\xi) = \frac{(\xi - x)\cdot n}{2\pi r^2}. \tag{5.67}$$

For simplicity of presentation, it will first be assumed that $b = 0$ in Equations 5.63 and 5.65; the nonhomogeneous situation will be discussed later. To discretize Equation 5.65, one first sets ξ to a finite number of discrete (still *interior*) values ξ_i and divides the boundary into N segments denoted by Γ_j. This is expressed:

$$T(\xi_i) + \sum_{j=1}^{N} \int_{\Gamma_j} T(x)F(x,\xi_i)\,d\Gamma = \sum_{j=1}^{N} \int_{\Gamma_j} q(x)G(x,\xi_i)\,d\Gamma \tag{5.68}$$

where $q = \partial_n T$ is the flux across the boundary. The integrals in Equation 5.68 over (typically small) segments of the boundary are usually done numerically. The simplest case to implement is obtained by approximating $T(x)$ and $q(x)$ in Equation 5.68 at the midpoint x_j of each segment Γ_j, in which case Equation 5.68 becomes

$$T(\xi_i) + \sum_{j=1}^{N} T(x_j) \int_{\Gamma_j} F(x,\xi_i)\,d\Gamma = \sum_{j=1}^{N} q(x_j) \int_{\Gamma_j} G(x,\xi_i)\,d\Gamma \tag{5.69}$$

Recall that at this point the ξ_i values are still *interior* points. In order to get an integral equation for the missing temperature and flux values on the boundary, one must let ξ_i the values in Equation 5.69 approach the midpoint values (denoted by x_j).

Turning now to the integrals in Equations 5.68 or 5.69, note that when $\xi_i \neq x_j$ the integrals can be integrated numerically since both F and G are continuous on segment Γ_j. However, when $\xi_i = x_j$ both F and G are singular on that segment. The logarithmic singularity in G can be integrated routinely. However, the singularity in F is a serious one and must be evaluated with care. Assuming that midpoint x_j is a "normal" point of the boundary (e.g., not at a corner or cusp, etc.) it can be shown that as ξ approaches x_j from the interior of the domain:

$$\int_{\Gamma_j} T(x)F(x,\xi)\,d\Gamma \to -\tfrac{1}{2}T(x_j) + PV\left[\int_{\Gamma_j} T(x)F(x,x_j)\,d\Gamma\right] \tag{5.70}$$

where PV denotes the *principle value* of this singular integral. For details see Brebbia et al.[12]

The linear system of equations for the unknown temperature and flux values on the boundary can now be obtained. In the following, ξ_i also denote midpoints of the Γ_i. Turning to Equation 5.69, define matrices F and G by:

$$F_{i,j} = \int_{\Gamma_j} F(x,\xi_i)\,d\Gamma, \quad G_{i,j} = \int_{\Gamma_j} G(x,\xi_i)\,d\Gamma.$$

Generally, these integrals need to be computed numerically. However, if the boundary is a polygon or if one is using straight-line segments to approximate the boundary, the integrals can be evaluated analytically. Putting aside for the moment of the complication when $i = j$, Equation 5.69, can be put in matrix form:

$$(\boldsymbol{F}+\boldsymbol{I})\boldsymbol{T} = \boldsymbol{G}\boldsymbol{q}.$$

Here $\boldsymbol{I}$ is the N by N identity matrix and $\boldsymbol{T}$ is the N-vector of temperatures on the boundary. Finally, the $i = j$ correction in Equation 5.70 subtracts ½ from the diagonal of matrix $\boldsymbol{F}$ and results in the new matrix $\boldsymbol{H}$ and the system:

$$(\boldsymbol{F}+\tfrac{1}{2}\boldsymbol{I})\boldsymbol{T} = \boldsymbol{H}\boldsymbol{T} = \boldsymbol{G}\boldsymbol{q}. \tag{5.71}$$

Two simple cases of the current boundary value problems, Equation 5.63 with $b = 0$, can now be easily addressed from Equation 5.71. First, the Dirichlet problem, in which the temperature is specific on the entire boundary. Here vector $\boldsymbol{T}$ is known, so the product $\mathbf{v} = \boldsymbol{HT}$ is computed; then $\boldsymbol{Gq} = \mathbf{v}$ is solved giving the flux values on the boundary (i.e., at the midpoints of each segment).

The second problem that follows easily from Equation 5.71 is the *Neumann* problem where the flux is specified on the entire boundary. In this case, the right side of Equation 5.71 is computed, say $\boldsymbol{w} = \boldsymbol{Gq}$, and then the system $\boldsymbol{HT} = \boldsymbol{w}$ is solved for the temperatures on the boundary.

In either of these two cases, once both $\boldsymbol{T}$ and $\boldsymbol{q}$ are known at all midpoints of the boundary, Equation 5.69 can be used to compute $T(\xi_i)$ for any desired *interior* point.

Mixed Boundary Value Problem

The more general BVP has temperature specified on part of the boundary and flux on the rest, as in Equation 5.63. This can perhaps be best illustrated by a simple example. Consider the BVP $\nabla^2\boldsymbol{T} = 0$, with temperature specified on the vertical sides of the rectangle in Figure 5.20 and flux specified on the horizontal sides. Note that the boundary is divided into $N = 8$ segments and that the midpoints, $\boldsymbol{x}_j$, are labeled in an unusual way which will simplify the presentation to follow.

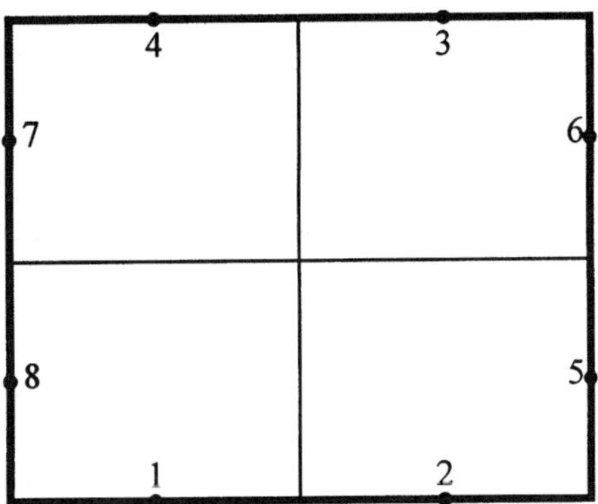

FIGURE 5.20 Domain with eight BEM segments.

The problem with proceeding to solve the system $\boldsymbol{HT} = \boldsymbol{Gq}$ is that neither $\boldsymbol{T}$ nor $\boldsymbol{q}$ is known at all eight midpoint values; in fact, each is known at four points. One must set up a new 8 by 8 system, say $\boldsymbol{Au} = \mathbf{v}$, to solve for the missing temperature and flux valves. First define the unknown vector $\boldsymbol{u}$ by $\boldsymbol{u} = [T_1, T_2, T_3, T_4, q_5, q_6, q_7, q_8]$, where the subscripts on $\boldsymbol{T}$ and $\boldsymbol{q}$ correspond to the labeling in Figure 5.20. To construct the matrix $\boldsymbol{A}$, equation $\boldsymbol{HT} = \boldsymbol{Gq}$ is spilt so that the unknown values are on the left side and known values on the right. For $i = 1,\ldots,8$, one thus obtains:

$$\sum_{j=1}^{4} H_{i,j}T_j - \sum_{j=5}^{8} G_{i,j}q_j = \sum_{j=1}^{4} G_{i,j}q_j - \sum_{j=5}^{8} H_{i,j}T_j = \mathrm{v}_i \tag{5.72}$$

Hence, the components of vector **v** are available from the given boundary conditions; and matrix $\boldsymbol{A}$ for this problem is

$$A_{i,j} = H_{i,j} \quad \text{for } j \preceq 4; \qquad A_{i,j} = -G_{i,j} \quad \text{for } j \succeq 5.$$

Finally, after solving $\boldsymbol{Au} = \boldsymbol{v}$ one has the required temperature and flux values at all midpoints of the boundary, and again Equation 5.69 can be used to solve for interior temperatures.

Nonhomogeneous Poisson Problem

So far it has been assumed that $b = 0$ in BVP Equation 5.63 and integral form Equation 5.65. Without that assumption, the key integral relationship of Equation 5.69 becomes:

$$T(\xi_i) + \sum_{j=1}^{N} T(\boldsymbol{x}_j) \int_{\Gamma_j} F(\boldsymbol{x}, \xi_i)\, d\Gamma = \sum_{j=1}^{N} q(\boldsymbol{x}_j) \int_{\Gamma_j} G(\boldsymbol{x}, \xi_i)\, d\Gamma + \sum_{k=1}^{K} \int_{\Omega_k} b(\boldsymbol{x}) G(\boldsymbol{x}, \xi_i)\, d\Omega. \tag{5.73}$$

The last term now must be computed for the N values of i and the resulting vector added to the right side of Equation 5.71. As the notation indicates, these integrals must be computed over the entire domain; moreover in stage two of the BEM method (applying Equation 5.73 for interior values of ξ) it will have to be computed for each requested interior point. Hence, it is important that these integrals be computed efficiently, much like such considerations in the FEM. This is discussed further below.

Computing Interior Derivatives

Should one need derivatives of T interior to the domain, the formula of Equation 5.73 can be differentiated directly. For example, to compute $\partial T(\xi)/\partial \xi_1$, dropping the subscript on ξ, one gets:

$$\frac{\partial T(\xi)}{\partial \xi_1} = -\sum_{j=1}^{N} T(\boldsymbol{x}_j) \int_{\Gamma_j} \frac{\partial F(\boldsymbol{x}, \xi)}{\partial \xi_1}\, d\Gamma + \sum_{j=1}^{N} q(\boldsymbol{x}_j) \int_{\Gamma_j} \frac{\partial G(\boldsymbol{x}, \xi)}{\partial \xi_1}\, d\Gamma + \sum_{k=1}^{K} \int_{\Omega_k} b(\boldsymbol{x}) \frac{\partial G(\boldsymbol{x}, \xi)}{\partial \xi_1}\, d\Omega.$$

From such derivatives one can, for example, compute the flux at any interior point.

A Numerical Example in Two Dimensions

To give an indication of the numerical effectiveness of the BEM, the problem in the example in Equation 5.12 will be solved here and numerical results compared with the earlier finite difference results. Briefly, the problem is a Dirichlet one,

$$\nabla^2 T = 0; \qquad \text{for } 0 \preceq x \preceq \pi, \quad 0 \preceq y \preceq 2,$$

with $T = \sin x \sinh 2$ on the top piece of the boundary ($y = 2$) and $T = 0$ on the other three sides. Recall that the exact solution is $T = \sin x \sinh y$. Equation 5.69 was used as just discussed, i.e., it was first applied to compute the flux values at the midpoints of the boundary elements, then again used to compute T at the desired interior points. The results are summarized in Table 5.6 and the notation will be that used in Table 5.3 (for the finite difference results), where m denotes the number of midpoints on the x intervals and n is the number on the y intervals. In the current notation, the total number of midpoints is $N = 2(m + n)$.

Comparing the BEM results in Table 5.6 with Table 5.3 shows that the errors are very much the same (with BEM results, on average, slightly better). As predicted by the finite difference theory, the errors in both cases are of order h^2, where h is the (largest) side of the small rectangles in which the domain is divided. Also comparing the two tables shows that in Table 5.3 the *maximum* errors tend to be toward the center of the domain, around point $(\pi/2, 1)$, and this is to be expected since in this Dirichlet problem

TABLE 5.6 Results of Solving Boundary-Value Problem in Equation 5.12 by the BEM

y	$T(\pi/2, y)$	$\lvert T - v_{m=7,\,n=5}\rvert$	$\lvert T - v_{m=15,\,n=11}\rvert$	$\lvert T - v_{m=31,\,n=23}\rvert$
5/3	2.55282	0.0114	0.002510	0.000567
4/3	1.76504	0.0086	0.001738	0.000392
1	1.17520	0.0057	0.001150	0.000260
2/3	0.71716	0.0033	0.000685	0.000157
1/3	0.33954	0.0013	0.000294	0.000071

the exact T is known around the entire boundary. In contrast, the errors in Table 5.6 tend to be near the top boundary. The reason for this is that the BEM must compute approximations to the flux along the boundary, and with these rather crude meshes this introduces substantial error in the flux, which in turn translates into errors near the boundary.

This example shows the typical advantage of the BEM over alternative methods in that the resulting linear system is much smaller. For example, in the $m = 31$, $n = 23$ case the finite difference approach led to a linear system with 713 unknowns (and the same would be true for the FEM). Whereas the BEM led to 108 unknowns. On the other hand, in both the finite difference and FEM approaches the resulting matrices are relatively *sparse* (lots of zeros); so if the software being used takes advantage of this, the computational times become more competitive. Further comparisons between the BEM and the FEM are made shortly.

A systematic error analysis for the BEM is not presented here. As suggested by the above example, the errors in the BEM method as presented here (i.e., taking T and flux as constant on each section Γ_j) will tend to converge much as the theory for the finite difference approach. When one uses more accurate descriptions of T and flux the approximation theory gets more complex, and more like that discussed for the FEM earlier. One way in which the approximations can be improved follow.

More Accurate Integrations

If the type of accuracy shown in the above example is not adequate for one's purpose, or if a large two-dimension or three-dimension problem is involved, there are basically two ways to increase accuracy. One is to take a smaller mesh, hence larger N, generally expecting the doubling of N to reduce the error by a factor of four. The other alternative is to use a more accurate integration scheme in the integrals in Equations 5.68 and 5.73. For example, consider integrals:

$$\int_{\Gamma_j} T(\boldsymbol{x})\, F(\boldsymbol{x},\xi_i)\, d\Gamma. \tag{5.74}$$

Earlier, for simplicity, T was taken to be constant over this segment Γ_j so that it can be factored out of the integral. Then in computing the matrix $\boldsymbol{F}$ one only had to integrate the function F. A simple way to improve the approximation of T (and q) without increasing the size of the resulting system of equations is to take a *linear* approximation to T over Γ_j in Equation 5.74 and use the values of T at the two endpoints of Γ_j (rather than the midpoint). This results in slightly more difficult integrals for matrices $\boldsymbol{F}$ and $\boldsymbol{G}$, but this typically does not matter if these are done numerically. This linear approximation to T does *not* increase the size of the linear system. For details of this idea see Chapter 3 in Brebbia et al.[12]

Should one require yet more accurate approximations to T and q (for example, when they change rapidly on the boundary), some sort of quadrature involving both T and F in Equation 5.74 will be called for. To accommodate better integration accuracy and to allow for approximation to curved boundaries, the follow approach parallels that done earlier for the FEM. First, consider a section of boundary Γ_j with three points, one on each endpoint and a midpoint, labeled locally by $P_1 = (x_1, y_1)$, $P_2 = (x_2, y_2)$, and $P_3 = (x_3, y_3)$. See Figure 5.21.

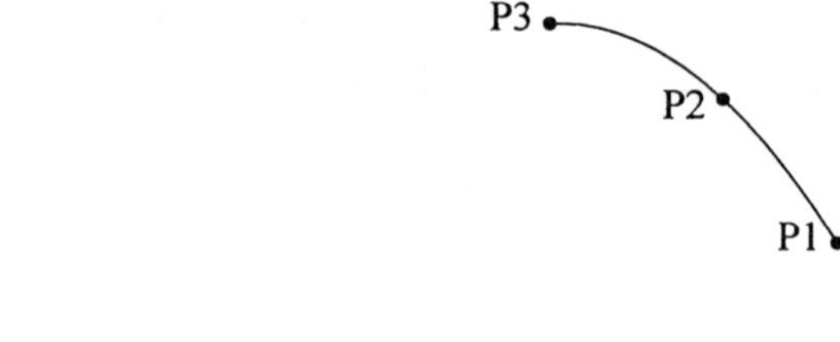

FIGURE 5.21 Standard interval and image on boundary.

As before, consider the three quadratic functions, shown in vector form, defined on basic interval [–1, 1]:

$$\phi(\eta) = \begin{bmatrix} \tfrac{1}{2}\eta(\eta-1) \\ (1-\eta^2) \\ \tfrac{1}{2}\eta(\eta+1) \end{bmatrix}. \tag{5.75}$$

Note that $\phi_1 = 1$ at $\eta = -1$ and is zero at 0, +1; $\phi_2 = 1$ at $\eta = 0$ and is zero at –1, +1; $\phi_3 = 1$ at $\eta = 1$ and is zero at –1, 0. Again, analogous to that done previously, the following provides a mapping between Γ_j and [–1, 1], where the (x_i, y_i) are as in Figure 5.21:

$$\begin{aligned} x(\eta) &= \sum_{1}^{3} x_i \phi_i(\eta), \\ y(\eta) &= \sum_{1}^{3} y_i \phi_i(\eta). \end{aligned} \tag{5.76}$$

Note that both $x(\eta)$ and $y(\eta)$ are quadratic functions of η. This does *not* say that it *exactly* represents quadratic boundaries, e.g., where y varies quadratically in x; however, the mapping does provide a good approximation to many curved boundaries.

Integrals of the type in Equation 5.74 can be translated to the interval [–1, 1] as follows.

$$\int_{\Gamma_j} T(\boldsymbol{x}) F(\boldsymbol{x}, \xi_i)\, d\Gamma = \int_{-1}^{1} T(x(\eta), y(\eta))\, F(x(\eta), y(\eta), \xi_i)\, J\, d\eta,$$

$$J = \frac{d\Gamma}{d\eta} = \sqrt{x'(\eta)^2 + y'(\eta)^2}\,.$$

These integrals are usually to be performed numerically and a couple observations are in order. Several Gaussian schemes were given previously after Formula (5.42). For example, if the highly accurate three-point Gaussian quadrature is used, all three nodes are *interior* in the interval [–1, 1] and this means, for the current integral, T must be evaluated (or solved for) at three interior points on each segment Γ_j. The resulting linear system will have $3N$ unknowns. Similarly, if the two-point Gaussian quadrature is used, one evaluates T and F at two interior points and thus leads to $2N$ unknowns. Recall that Gaussian quadratures using n points (all of which are *interior*) have errors of order h^{2n}, for intervals of length h.

A popular choice is the familiar Simpson's rule for these integrals and this also leads to system with $2N$ unknowns. Simpson's rule is defined by

$$\int_{-1}^{1} p(\eta)\,d\eta \cong \tfrac{1}{3}\left[p(-1)+4p(0)+p(1)\right].$$

Like the two-point Gaussian rule, Simpson's rule has an error of order h^4 and it has the advantage of evaluating the function at convenient points, the endpoints, and midpoint of the interval.

Higher-Order Problems and Integrations

In the last section, two-dimensional problems were considered and the resulting line integrals on the boundary were discussed. The dimension of the BEM integrals can be increased in two ways. In two dimensions, if one is faced with a nonhomogeneous problem, e.g., $\nabla^2 T = b$ then the last term in Equation 5.73 must be dealt with, i.e.,

$$\int_{\Omega_k} b(\boldsymbol{x})\,G(\boldsymbol{x},\xi_i)\,d\Omega, \tag{5.77}$$

where the Ω_k are two-dimensional elements of the domain. These integrals are typically approximated by Gaussian quadrature as discussed previously, and the same methods and theory apply here. One advantage of integrals of the type in Equation 5.77 is that they do *not* involved the unknown T and q values on the boundary, hence do not affect the size of the basic linear system. On the other hand, there can be a large number of these integrals since there are many more surface elements than the N boundary elements. Moreover, when Equation 5.73 is applied to compute $T(\xi_i)$ for a large number of *interior* points there is a substantial amount of computing involved.

Three-Dimensional Problems

Clearly moving from two-dimensions to three-dimensions increases, by one, the dimension of all the integrals being discussed. For example, integrals of the form Equation 5.74 are now surface integrals, and the earlier quadrature discussions apply now.

One additional consideration here is that in order to keep the order of the final linear system as small as possible, the choice of quadrature is important. For example, suppose integrals of Equation 5.74 are involved with a Neumann problem in which T is unknown on the entire boundary. Further assume that one is using the six-point triangular (quadratic) elements defined by Equation 5.31. A good choice of quadrature is that discussed below Equation 5.35, namely using the three midpoints of the sides of the triangles as nodes with weights of 1/3. The final linear system would have about $3N/2$ unknowns where N is the number of triangular elements. In contrast, if the more accurate three-point Gaussian quadrature is applied to the triangles, the nodes are all interior to the triangles and hence, are not shared among the elements. The final linear system would be about $3N$ or twice the size of the previous case. Moreover, the nodes are at rather "awkward" locations within the elements.

Should the problem be nonhomogeneous, e.g., $\nabla^2 T = b$, the integrals in Equation 5.77 would be over (usually many, small) three-dimensional elements. For the interior elements (i.e., not part of the boundary) the most practical elements are the eight-node *bricks* defined by Equation 5.43. Since these integrals do not involve unknown quantities, an efficient Gaussian quadrature is usually best. For example, as discussed below Equation 5.42, if a three-point (on [–1, 1]) Gaussian quadrature is used it will require nine function evaluations on each element Ω_j. Again, as discussed earlier, if the domain under study has a curved boundary, then for those segments containing a portion of the boundary a more accurate element may be called for. The reader is again referred to elements like the 10-node tetrahedral element defined by Equation 5.45.

Comparison between the BEM and the FEM

Here are some general remarks regarding the advantages and disadvantages of the two methods. First, assume the BEM is applicable, i.e., by judicious choice of *weighting function* and application of Green's theorems one is able to reduce the dimension of the fundamental integral equation. For a linear BVP the solution of this integral equation leads to a system of linear equation of *relatively* small order. For example, if one has a three-dimensional domain divided into small cubic elements, then the resulting linear system will involve as few as $6N^2$ unknown T and flux values on the surface. This problem using the FEM, or finite difference methods, would require solving a system involving N^3 unknown values at all interior nodes. Since often N is quite large, this is the inherent advantage of the BEM.

On the other side of the coin, there are several issues. First, in the example just mentioned, the $6N^2$ system will be *dense*, with the matrix consisting of mostly nonzero elements; hence it is difficult to avoid using the full Gaussian elimination algorithm (which takes roughly $n^3/3$ multiplication for a system with n unknowns). The linear system for the FEM, and the finite difference techniques, are typically *sparse*, with the matrix consisting of mostly zeros. So if the software intelligently takes advantage of the sparseness, e.g., by iteration, the disadvantage of the large N^3 system can be at least partially diminished. (Note: unfortunately, some modern high-quality software programs do *not* take advantage of the sparseness.) A second issue that needs to be emphasized is that in the BEM, after all T and flux values are found on the boundary, there is still considerable computation to be done. In particular, if one is solving a nonhomogeneous problem, e.g., Equation 5.63 with $b \neq 0$, and needs to know T at a large number of interior points, then usually a great deal of computing is required using Equation 5.73. Notice in Equation 5.73 that there are likely many integration to be performed over the entire domain, so if these are not done very efficiently all previous cost benefits may be lost. Recall that in the FEM, once the linear system is solve (for the coefficients of the basis functions) it is relatively inexpensive to solve for a, e.g., temperature, at an arbitrary point in the domain.

In summary, if the computation for the problem under study is dominated by the solution of the basic linear system then, if applicable, the BEM is likely the method of choice. This would be the case if, in the latest example, the N^3 order system is pressing one's computing capacity. The BEM is especially efficient if one has a homogeneous problem, e.g., Equation 5.63 with $b = 0$, and needs information at relatively few interior points. The primary advantage of the FEM is that it is much more widely applicable and at this time has more software choices available.

5.5 Software and Databases

Following are lists of several types of available software appropriate to problems in heat transfer and other areas of science and engineering. These lists are by no means comprehensive, but do provide a number of options to the interested reader. This software is divided into several groupings

General Purpose Interactive Mathematical Software
Large Subroutine Libraries for Engineering and Science
Major Engineering Finite Element Packages

General Purpose Interactive Mathematical Software

The following are among the most widely used interactive, general purpose software for doing mathematics and engineering. They all have "higher level" programming capability, high-quality graphics, extensive libraries of mathematical functions, and have substantial capacity for numerical computation. Most of the ones listed here also have major "symbolic" capability, e.g., the ability to *analytically* perform complex algebraic manipulations and calculus operations (including differentiation, integration, power series expansions, etc.). These packages are designed for those wishing to have control over the details of setting up and solving their problems — as opposed to the CAD-based engineering packages discussed below. All of the following packages continue to be refined and expanded. The Web site is given for each of these.

Macsyma was the original major symbolic package; it was developed over a period of several decades, primarily at the Massachusetts Institute of Technology. Originally developed for mainframe computers, it is now available for well-equipped PCs and a variety of other computing platforms. `http://www.macsyma.com`

Maple was developed in Waterloo, Ontario, Canada; it has substantial symbolic capability and is available on a wide range of computing platforms from relatively modest PCs to supercomputers. `http://www.maplesoft.com`

Mathematica is a relative newcomer to this field, but is perhaps now the most widely used symbolic package in the world. It was developed by Wolfram Research, Inc. in Champaign, IL and is available on a large number of platforms. (This author did the BEM computations using this package.) `http://www.mathematica.com`

MATLAB was originally developed as a matrix-based package to solve linear systems of equations very efficiently and conveniently. Over the years it has expanded to have major numerical and graphics capability and now has over 20 "tool boxes" for performing tasks in areas like signal and image processing, control design, and statistics. While not a symbolic package, this capability can be achieved by interfacing with Maple. (This author used **MATLAB** for the finite difference computations.) `http://www.mathworks.com`

The following are other packages in this general category that are worth considering: Axiom, Derive, GANITH, Magma, Mathcad, Milo, MuPAD, Pari, Schur, and SymbMath. HiQ is a substantial package available on the Macintosh platforms.

Large Subroutine Libraries for Engineering and Science

Following are several libraries of high-quality subroutines, written in the FORTRAN or C languages (both, in some cases). In most cases these are in the public domain; hence, available at little or no cost, but the IMSL and NAG libraries are commercial products. Generally, the routines in these libraries will run on most computing platforms. Electronic addresses are provided.

CMLIB is the "core mathematics library" of the National Institute of Science and Technology (NIST). `http://gams.nist.gov`

ESSL is for use on a wide range of IBM computers, and the routines have, in some cases, been tuned specifically for IBM hardware. `http://www.ibm.com`

IMSL is the original commercial package of this type, developed by Visual Numerics, Inc. `http://www.vni.com`

NAG is also a long-established commercial product developed by Numerical Algorithms Group. The symbolic system Axiom is available from NAG. `http://www.nag.com`

SLATEC is distributed by the Department of Energy at:

`http://www.doe.gov/html/osti/estsc/estsc.html`, or `http://www.netlib.org/slatec`

To aid prospective users in accessing software at these and other sites, NIST has developed an on-line, cross-indexed searching program. It guides the users through a decision tree to search for software appropriate for their problem of interest. Once the desired software is located, abstracts, documentation, and source code can be obtained. The address for this important site is `http://gams.nist.gov`

Major Engineering Finite Element Packages

This final category lists some of the well established packages for performing the FEM (or in one case the BEM) on two- and three-dimensional problems. Typically, they are based on computer-aided design (CAD) technology in which the object under study can be interactively designed and visualized. Then, for example, they can automatically divide a three-dimensional solid into elements of the appropriate shape and number and then perform a FEM (or BEM) computation to the desired tolerance. These are very "high level" packages, e.g., in many cases the user can describe the problem of interest in physical terms, without specifying the partial differential equations involved. Many options exist for displaying the computed results. (This author used the educational version of the ANSYS product line in the FEM computation, where some of the many features of this and other similar products were discussed.)

TABLE 5.7 Engineering Packages for the FEM (or BEM) Computations

Package	Type	Applications	Address
ABAQUS	FE	Heat transfer, solid mechanics, and coupled thermal/stress problems	abaqus.com
ADINA	FE	Heat transfer, solid mechanics, and coupled thermal/stress problems	ADINA 71 Elton Ave. Watertown, MA 02172
ANSYS	FE	Heat transfer, solid mechanics, and coupled thermal/stress problems	ansys.com
BEASY	BE	Potential simulations (including thermal) and linear-elastic stress analysis	beasy.com
MARC	FE	Heat transfer, solid mechanics, and coupled thermal/stress problems	marc.com
MSC/NASTRAN	FE	Heat transfer, solid mechanics, and coupled thermal/stress problems	MacNeal-Schwendler 815 Colorado Blvd. Los Angeles, CA 90041
NISA II	FE	Heat transfer, solid mechanics and coupled thermal/stress problems	emrc.com
TACO3D	FE	Heat transfer	Energy Science & Tech. Software Corp.; Box 1020 Oak Ridge, TN 37831
TAU	FE	Heat transfer	AEA Technology Risley Warrington Cheshire WA3 6AT U.K.

Table 5.7 provides a sampling of the packages of this type currently available. The "Address" column of the table shows either the Website or mailing address.

References

1. Isaacson, E. and Keller, H.B., *Analysis of Numerical Methods*, John Wiley & Sons, New York, 1966, 541.
2. Varga, R.S., *Matrix Iterative Analysis*, Prentice-Hall, Englewood Cliffs, NJ, 1962.
3. Özisik, M.N., *Finite Difference Methods in Heat Transfer*, CRC Press, Boca Raton, FL, 1994.
4. Birkhoff, G. and Lynch, R.E., *Numerical Solution of Elliptical Problems*, SIAM, Philadelphia, 1984.
5. Oden, J.T. and Reddy, J.N., *An Introduction to the Mathematical Theory of Finite Elements,* John Wiley & Sons, New York, 1976.
6. Davis, P.J. and Rabinowitz, P., *Numerical Integration*, Blaisdell Press, Waltham, MA, 1967.
7. Strang, W.G. and Fix, G.J., *An Analysis of the Finite Element Method*, Prentice-Hall, Englewood Cliffs, NJ, 1973.
8. Reddy, J.N., *An Introduction to the Finite Element Method*, 2nd ed., McGraw-Hill, New York, 1993.
9. Reddy, J.N. and Gartling, D.K., *The Finite Element Method in Heat Transfer and Fluid Mechanics,* CRC Press, Boca Raton, FL, 1994.
10. Wong, Y.W. and Bang, H., *The Finite Element Method Using MATLAB*, CRC Press, Boca Raton, FL, 1996.

11. Huang, H.-C. and Usmani, A.S., *Finite Element Analysis for Heat Transfer, Theory and Software*, Springer-Verlag, New York, 1994.
12. Brebbia, C.A., Telles, J.C.F., and Wrobel, L.C., *Boundary Element Techniques*, Springer- Verlag, New York, 1984.
13. Cruse, T.A., Pifko, A.B., and Armen, H., Eds., *Advanced Topics in Boundary Element Methods,* American Society of Mechanical Engineers, New York, 1985.
14. Kane, J.H., Maier, G., Tosaka, N., and Atluri, S.N., Eds., *Advances in Boundary Element Techniques,* Springer-Verlag, Berlin, 1993.

Appendices

Paul Norton

National Renewable Energy Laboratory

0-8493-9581-X/00/$0.00+$.50

Appendix A. Properties of Gases and Vapors

TABLE A.1 Properties of Dry Air at Atmospheric Pressure

Symbols and Units:

K = absolute temperature, degrees Kelvin
deg C = temperature, degrees Celsius
deg F = temperature, degrees Fahrenheit
ρ = density, kg/m^3
c_p = specific heat capacity, kJ/kg·K
c_p/c_v = specific heat capacity ratio, dimensionless
μ = viscosity, N·s/m^2 × 10^6 (For N·s/m^2 (= kg/m·s) multiply tabulated values by 10^{-6})
k = thermal conductivity, W/m·k × 10^3 (For W/m·K multiply tabulated values by 10^{-3})
Pr = Prandtl number, dimensionless
h = enthalpy, kJ/kg
V_s = sound velocity, m/s

Temperature			Properties							
K	deg C	deg F	ρ	c_p	c_p/c_v	μ	k	Pr	h	V_s
100	−173.15	−280	3.598	1.028		6.929	9.248	.770	98.42	198.4
110	−163.15	−262	3.256	1.022	1.420 2	7.633	10.15	.768	108.7	208.7
120	−153.15	−244	2.975	1.017	1.416 6	8.319	11.05	.766	118.8	218.4
130	−143.15	−226	2.740	1.014	1.413 9	8.990	11.94	.763	129.0	227.6
140	−133.15	−208	2.540	1.012	1.411 9	9.646	12.84	.761	139.1	236.4
150	−123.15	−190	2.367	1.010	1.410 2	10.28	13.73	.758	149.2	245.0
160	−113.15	−172	2.217	1.009	1.408 9	10.91	14.61	.754	159.4	253.2
170	−103.15	−154	2.085	1.008	1.407 9	11.52	15.49	.750	169.4	261.0
180	−93.15	−136	1.968	1.007	1.407 1	12.12	16.37	.746	179.5	268.7
190	−83.15	−118	1.863	1.007	1.406 4	12.71	17.23	.743	189.6	276.2
200	−73.15	−100	1.769	1.006	1.405 7	13.28	18.09	.739	199.7	283.4
205	−68.15	−91	1.726	1.006	1.405 5	13.56	18.52	.738	204.7	286.9
210	−63.15	−82	1.684	1.006	1.405 3	13.85	18.94	.736	209.7	290.5
215	−58.15	−73	1.646	1.006	1.405 0	14.12	19.36	.734	214.8	293.9
220	−53.15	−64	1.607	1.006	1.404 8	14.40	19.78	.732	219.8	297.4
225	−48.15	−55	1.572	1.006	1.404 6	14.67	20.20	.731	224.8	300.8
230	−43.15	−46	1.537	1.006	1.404 4	14.94	20.62	.729	229.8	304.1
235	−38.15	−37	1.505	1.006	1.404 2	15.20	21.04	.727	234.9	307.4
240	−33.15	−28	1.473	1.005	1.404 0	15.47	21.45	.725	239.9	310.6
245	−28.15	−19	1.443	1.005	1.403 8	15.73	21.86	.724	244.9	313.8
250	−23.15	−10	1.413	1.005	1.403 6	15.99	22.27	.722	250.0	317.1
255	−18.15	−1	1.386	1.005	1.403 4	16.25	22.68	.721	255.0	320.2
260	−13.15	8	1.359	1.005	1.403 2	16.50	23.08	.719	260.0	323.4
265	−8.15	17	1.333	1.005	1.403 0	16.75	23.48	.717	265.0	326.5
270	−3.15	26	1.308	1.006	1.402 9	17.00	23.88	.716	270.1	329.6
275	+1.85	35	1.285	1.006	1.402 6	17.26	24.28	.715	275.1	332.6
280	6.85	44	1.261	1.006	1.402 4	17.50	24.67	.713	280.1	335.6
285	11.85	53	1.240	1.006	1.402 2	17.74	25.06	.711	285.1	338.5
290	16.85	62	1.218	1.006	1.402 0	17.98	25.47	.710	290.2	341.5
295	21.85	71	1.197	1.006	1.401 8	18.22	25.85	.709	295.2	344.4
300	26.85	80	1.177	1.006	1.401 7	18.46	26.24	.708	300.2	347.3
305	31.85	89	1.158	1.006	1.401 5	18.70	26.63	.707	305.3	350.2
310	36.85	98	1.139	1.007	1.401 3	18.93	27.01	.705	310.3	353.1
315	41.85	107	1.121	1.007	1.401 0	19.15	27.40	.704	315.3	355.8
320	46.85	116	1.103	1.007	1.400 8	19.39	27.78	.703	320.4	358.7

•Condensed and computed from: "Tables of Thermal Properties of Gases", National Bureau of Standards Circular 564, U.S. Government Printing Office, November 1955.

TABLE A.1 (continued) Properties of Dry Air at Atmospheric Pressure

Temperature			*Properties*							
K	*deg C*	*deg F*	ρ	c_p	c_p/c_v	μ	*k*	*Pr*	*h*	V_s
325	51.85	125	1.086	1.008	1.400 6	19.63	28.15	.702	325.4	361.4
330	56.85	134	1.070	1.008	1.400 4	19.85	28.53	.701	330.4	364.2
335	61.85	143	1.054	1.008	1.400 1	20.08	28.90	.700	335.5	366.9
340	66.85	152	1.038	1.008	1.399 9	20.30	29.28	.699	340.5	369.6
345	71.85	161	1.023	1.009	1.399 6	20.52	29.64	.698	345.6	372.3
350	76.85	170	1.008	1.009	1.399 3	20.75	30.03	.697	350.6	375.0
355	81.85	179	0.994 5	1.010	1.399 0	20.97	30.39	.696	355.7	377.6
360	86.85	188	0.980 5	1.010	1.398 7	21.18	30.78	.695	360.7	380.2
365	91.85	197	0.967 2	1.010	1.398 4	21.38	31.14	.694	365.8	382.8
370	96.85	206	0.953 9	1.011	1.398 1	21.60	31.50	.693	370.8	385.4
375	101.85	215	0.941 3	1.011	1.397 8	21.81	31.86	.692	375.9	388.0
380	106.85	224	0.928 8	1.012	1.397 5	22.02	32.23	.691	380.9	390.5
385	111.85	233	0.916 9	1.012	1.397 1	22.24	32.59	.690	386.0	393.0
390	116.85	242	0.905 0	1.013	1.396 8	22.44	32.95	.690	391.0	395.5
395	121.85	251	0.893 6	1.014	1.396 4	22.65	33.31	.689	396.1	398.0
400	126.85	260	0.882 2	1.014	1.396 1	22.86	33.65	.689	401.2	400.4
410	136.85	278	0.860 8	1.015	1.395 3	23.27	34.35	.688	411.3	405.3
420	146.85	296	0.840 2	1.017	1.394 6	23.66	35.05	.687	421.5	410.2
430	156.85	314	0.820 7	1.018	1.393 8	24.06	35.75	.686	431.7	414.9
440	166.85	332	0.802 1	1.020	1.392 9	24.45	36.43	.684	441.9	419.6
450	176.85	350	0.784 2	1.021	1.392 0	24.85	37.10	.684	452.1	424.2
460	186.85	368	0.767 7	1.023	1.391 1	25.22	37.78	.683	462.3	428.7
470	196.85	386	0.750 9	1.024	1.390 1	25.58	38.46	.682	472.5	433.2
480	206.85	404	0.735 1	1.026	1.389 2	25.96	39.11	.681	482.8	437.6
490	216.85	422	0.720 1	1.028	1.388 1	26.32	39.76	.680	493.0	442.0
500	226.85	440	0.705 7	1.030	1.387 1	26.70	40.41	.680	503.3	446.4
510	236.85	458	0.691 9	1.032	1.386 1	27.06	41.06	.680	513.6	450.6
520	246.85	476	0.678 6	1.034	1.385 1	27.42	41.69	.680	524.0	454.9
530	256.85	494	0.665 8	1.036	1.384 0	27.78	42.32	.680	534.3	459.0
540	266.85	512	0.653 5	1.038	1.382 9	28.14	42.94	.680	544.7	463.2
550	276.85	530	0.641 6	1.040	1.381 8	28.48	43.57	.680	555.1	467.3
560	286.85	548	0.630 1	1.042	1.380 6	28.83	44.20	.680	565.5	471.3
570	296.85	566	0.619 0	1.044	1.379 5	29.17	44.80	.680	575.9	475.3
580	306.85	584	0.608 4	1.047	1.378 3	29.52	45.41	.680	586.4	479.2
590	316.85	602	0.598 0	1.049	1.377 2	29.84	46.01	.680	596.9	483.2
600	326.85	620	0.588 1	1.051	1.376 0	30.17	46.61	.680	607.4	486.9
620	346.85	656	0.569 1	1.056	1.373 7	30.82	47.80	.681	628.4	494.5
640	366.85	692	0.551 4	1.061	1.371 4	31.47	48.96	.682	649.6	502.1
660	386.85	728	0.534 7	1.065	1.369 1	32.09	50.12	.682	670.9	509.4
680	406.85	764	0.518 9	1.070	1.366 8	32.71	51.25	.683	692.2	516.7
700	426.85	800	0.504 0	1.075	1.364 6	33.32	52.36	.684	713.7	523.7
720	446.85	836	0.490 1	1.080	1.362 3	33.92	53.45	.685	735.2	531.0
740	466.85	872	0.476 9	1.085	1.360 1	34.52	54.53	.686	756.9	537.6
760	486.85	908	0.464 3	1.089	1.358 0	35.11	55.62	.687	778.6	544.6
780	506.85	944	0.452 4	1.094	1.355 9	35.69	56.68	.688	800.5	551.2
800	526.85	980	0.441 0	1.099	1.354	36.24	57.74	.689	822.4	557.8
850	576.85	1 070	0.415 2	1.110	1.349	37.63	60.30	.693	877.5	574.1
900	626.85	1 160	0.392 0	1.121	1.345	38.97	62.76	.696	933.4	589.6
950	676.85	1 250	0.371 4	1.132	1.340	40.26	65.20	.699	989.7	604.9
1 000	726.85	1 340	0.352 9	1.142	1.336	41.53	67.54	.702	1 046	619.5
1 100	826.85	1 520	0.320 8	1.161	1.329	43.96			1 162	648.0
1 200	926.85	1 700	0.294 1	1.179	1.322	46.26			1 279	675.2
1 300	1 026.85	1 880	0.271 4	1.197	1.316	48.46			1 398	701.0
1 400	1 126.85	2 060	0.252 1	1.214	1.310	50.57			1 518	725.9
1 500	1 220.85	2 240	0.235 3	1.231	1.304	52.61			1 640	749.4
1 600	1 326.85	2 420	0.220 6	1.249	1.299	54.57			1 764	772.6
1 800	1 526.85	2 780	0.196 0	1.288	1.288	58.29			2 018	815.7
2 000	1 726.85	3 140	0.176 4	1.338	1.274				2 280	855.5
2 400	2 126.85	3 860	0.146 7	1.574	1.238				2 853	924.4
2 800	2 526.85	4 580	0.124 5	2.259	1.196				3 599	983.1

TABLE A.2 Ideal Gas Properties of Nitrogen, Oxygen, and Carbon Dioxide

Symbols and Units:

T = absolute temperature, degrees Kelvin
$\bar{h}$ = enthalpy, kJ/kmol
$\bar{u}$ = internal energy, kJ/kmol
$\bar{s}^\circ$ = absolute entropy at standard reference pressure, kJ/kmol K
[$\bar{h}$ = enthalpy of formation per mole at standard state = 0 kJ/kmol]

Part a. Ideal Gas Properties of Nitrogen, N_2

T	$\bar{h}$	$\bar{u}$	$\bar{s}^\circ$	T	$\bar{h}$	$\bar{u}$	$\bar{s}^\circ$
0	0	0	0	600	17,563	12,574	212.066
220	6,391	4,562	182.639	610	17,864	12,792	212.564
230	6,683	4,770	183.938	620	18,166	13,011	213.055
240	6,975	4,979	185.180	630	18,468	13,230	213.541
250	7,266	5,188	186.370	640	18,772	13,450	214.018
260	7,558	5,396	187.514	650	19,075	13,671	214.489
270	7,849	5,604	188.614	660	19,380	13,892	214.954
280	8,141	5,813	189.673	670	19,685	14,114	215.413
290	8,432	6,021	190.695	680	19,991	14,337	215.866
298	8,669	6,190	191.502	690	20,297	14,560	216.314
300	8,723	6,229	191.682	700	20,604	14,784	216.756
310	9,014	6,437	192.638	710	20,912	15,008	217.192
320	9,306	6,645	193.562	720	21,220	15,234	217.624
330	9,597	6,853	194.459	730	21,529	15,460	218.059
340	9,888	7,061	195.328	740	21,839	15,686	218.472
350	10,180	7,270	196.173	750	22,149	15,913	218.889
360	10,471	7,478	196.995	760	22,460	16,141	219.301
370	10,763	7,687	197.794	770	22,772	16,370	219.709
380	11,055	7,895	198.572	780	23,085	16,599	220.113
390	11,347	8,104	199.331	790	23,398	16,830	220.512
400	11,640	8,314	200.071	800	23,714	17,061	220.907
410	11,932	8,523	200.794	810	24,027	17,292	221.298
420	12,225	8,733	201.499	820	24,342	17,524	221.684
430	12,518	8,943	202.189	830	24,658	17,757	222.067
440	12,811	9,153	202.863	840	24,974	17,990	222.447
450	13,105	9,363	203.523	850	25,292	18,224	222.822
460	13,399	9,574	204.170	860	25,610	18,459	223.194
470	13,693	9,786	204.803	870	25,928	18,695	223.562
480	13,988	9,997	205.424	880	26,248	18,931	223.927
490	14,285	10,210	206.033	890	26,568	19,168	224.288
500	14,581	10,423	206.630	900	26,890	19,407	224.647
510	14,876	10,635	207.216	910	27,210	19,644	225.002
520	15,172	10,848	207.792	920	27,532	19,883	225.353
530	15,469	11,062	208.358	930	27,854	20,122	225.701
540	15,766	11,277	208.914	940	28,178	20,362	226.047
550	16,064	11,492	209.461	950	28,501	20,603	226.389
560	16,363	11,707	209.999	960	28,826	20,844	226.728
570	16,662	11,923	210.528	970	29,151	21,086	227.064
580	16,962	12,139	211.049	980	29,476	21,328	227.398
590	17,26_	12,356	211.562	990	29,803	21,571	227.728

Source: Adapted from M.J. Moran and H.N. Shapiro, *Fundamentals of Engineering Thermodynamics*, 3rd. ed., Wiley, New York, 1995, as presented in K. Wark. *Thermodynamics*, 4th ed., McGraw-Hill, New York, 1983, based on the *JANAF Thermochemical Tables*, NSRDS-NBS-37, 1971.

TABLE A.2 (continued) Ideal Gas Properties of Nitrogen, Oxygen, and Carbon Dioxide

T	$\bar{h}$	$\bar{u}$	$\bar{s}^\circ$	T	n	$\bar{u}$	$\bar{s}^\circ$
1000	30,129	21,815	228.057	1760	56,227	41.594	247.396
1020	30,784	22,304	228.706	1780	56,938	42,139	247.798
1040	31,442	22,795	229.344	1800	57,651	42,685	248.195
1060	32,101	23,288	229.973	1820	58,363	43,231	248.589
1080	32,762	23,782	230.591	1840	59,075	43,777	248.979
1100	33,426	24,280	231.199	1860	59,790	44,324	249.365
1120	34,092	24,780	231.799	1880	60,504	44,873	249.748
1140	34,760	25,282	232.391	1900	61,220	45,423	250.128
1160	35,430	25,786	232.973	1920	61,936	45,973	250.502
1180	36,104	26,291	233.549	1940	62,654	46,524	250.874
1200	36,777	26,799	234.115	1960	63,381	47,075	251.242
1220	37,452	27,308	234.673	1980	64,090	47,627	251.607
1240	38,129	27,819	235.223	2000	64,810	48,181	251.969
1260	38,807	28,331	235.766	2050	66,612	49,567	252.858
1280	39,488	28,845	236.302	2100	68,417	50,957	253.726
1300	40,170	29,361	236.831	2150	70,226	52,351	254.578
1320	40,853	29,878	237.353	2200	72,040	53,749	255.412
1340	41,539	30,398	237.867	2250	73,856	55,149	256.227
1360	42,227	30,919	238.376	2300	75,676	56,553	257.027
1380	42,915	31,441	238.878	2350	77,496	57,958	257.810
1400	43,605	31,964	239.375	2400	79,320	59,366	258.580
1420	44,295	32,489	239.865	2450	81,149	60,779	259.332
1440	44,988	33,014	240.350	2500	82,981	62,195	260.073
1460	45,682	33,543	240.827	2550	84,814	63,613	260.799
1480	46,377	34,071	241.301	2600	86,650	65,033	261.512
1500	47,073	34,601	241.768	2650	88,488	66,455	262.213
1520	47,771	35,133	242.228	2700	90,328	67,880	262.902
1540	48,470	35,665	242.685	2750	92,171	69,306	263.577
1560	49,168	36,197	243.137	2800	94,014	70,734	264.241
1580	49,869	36,732	243.585	2850	95,859	72,163	264.895
1600	50,571	37,268	244.028	2900	97,705	73,593	265.538
1620	51,275	37,806	244.464	2950	99,556	75,028	266.170
1640	51,980	38,344	244.896	3000	101,407	76,464	266.793
1660	52,686	38,884	245.324	3050	103,260	77,902	267.404
1680	53,393	39,424	245.747	3100	105,115	79,341	268.007
1700	54,099	39,965	246.166	3150	106,972	80,782	268.601
1720	54,807	40,507	246.580	3200	108,830	82,224	269.186
1740	55,516	41,049	246.990	3250	110,690	83,668	269.763

TABLE A.2 (continued) Ideal Gas Properties of Nitrogen, Oxygen, and Carbon Dioxide

Part b. Ideal Gas Properties of Oxygen, O_2

T	$\bar{h}$	$\bar{u}$	$\bar{s}°$	T	$\bar{h}$	$\bar{u}$	$\bar{s}°$
0	0	0	0	600	17,929	12,940	226.346
220	6,404	4,575	196.171	610	18,250	13,178	226.877
230	6,694	4,782	197.461	620	18,572	13,417	227.400
240	6,984	4,989	198.696	630	18,895	13,657	227.918
250	7,275	5,197	199.885	640	19,219	13,898	228.429
260	7,566	5,405	201.027	650	19,544	14,140	228.932
270	7,858	5,613	202.128	660	19,870	14,383	229.430
280	8,150	5,822	203.191	670	20,197	14,626	229.920
290	8,443	6,032	204.218	680	20,524	14,871	230.405
298	8,682	6,203	205.033	690	20,854	15,116	230.885
300	8,736	6,242	205.213	700	21,184	15,364	231.358
310	9,030	6,453	206.177	710	21,514	15,611	231.827
320	9,325	6,664	207.112	720	21,845	15,859	232.291
330	9,620	6,877	208.020	730	22,177	16,107	232.748
340	9,916	7,090	208.904	740	22,510	16,357	233.201
350	10,213	7,303	209.765	750	22,844	16,607	233.649
360	10,511	7,518	210.604	760	23,178	16,859	234.091
370	10,809	7,733	211.423	770	23,513	17,111	234.528
380	11,109	7,949	212.222	780	23,850	17,364	234.960
390	11,409	8,166	213.002	790	24,186	17,618	235.387
400	11,711	8,384	213.765	800	24,523	17,872	235.810
410	12,012	8,603	214.510	810	24,861	18,126	236.230
420	12,314	8,822	215.241	820	25,199	18,382	236.644
430	12,618	9,043	215.955	830	25,537	18,637	237.055
440	12,923	9,264	216.656	840	25,877	18,893	237.462
450	13,228	9,487	217.342	850	26,218	19,150	237.864
460	13,535	9,710	218.016	860	26,559	19,408	238.264
470	13,842	9,935	218.676	870	26,899	19,666	238.660
480	14,151	10,160	219.326	880	27,242	19,925	239.051
490	14,460	10,386	219.963	890	27,584	20,185	239.439
500	14,770	10,614	220.589	900	27,928	20,445	239.823
510	15,082	10,842	221.206	910	28,272	20,706	240.203
520	15,395	11,071	221.812	920	28,616	20,967	240.580
530	15,708	11,301	222.409	930	28,960	21,228	240.953
540	16,022	11,533	222.997	940	29,306	21,491	241.323
550	16,338	11,765	223.576	950	29,652	21,754	241.689
560	16,654	11,998	224.146	960	29,999	22,017	242.052
570	16,971	12,232	224.708	970	30,345	22,280	242.411
580	17,290	12,467	225.262	980	30,692	22,544	242.768
590	17,609	12,703	225.808	990	31,041	22,809	243.120

TABLE A.2 (continued) Ideal Gas Properties of Nitrogen, Oxygen, and Carbon Dioxide

T	$\bar{h}$	$\bar{u}$	$\bar{s}^o$	T	$\bar{h}$	$\bar{u}$	$\bar{s}^o$
1000	31,389	23,075	243.471	1760	58,880	44,247	263.861
1020	32,088	23,607	244.164	1780	59,624	44,825	264.283
1040	32,789	24,142	244.844	1800	60,371	45,405	264.701
1060	33,490	24,677	245.513	1820	61,118	45,986	265.113
1080	34,194	25,214	246.171	1840	61,866	46,568	265.521
1100	34,899	25,753	246.818	1860	62,616	47,151	265.925
1120	35,606	26,294	247.454	1880	63,365	47,734	266.326
1140	36,314	26,836	248.081	1900	64,116	48,319	266.722
1160	37,023	27,379	248.698	1920	64,868	48,904	267.115
1180	37,734	27,923	249.307	1940	65,620	49,490	267.505
1200	38,447	28,469	249.906	1960	66,374	50,078	267.891
1220	39,162	29,018	250.497	1980	67,127	50,665	268.275
1240	39,877	29,568	251.079	2000	67,881	51,253	268.655
1260	40,594	30,118	251.653	2050	69,772	52,727	269.588
1280	41,312	30,670	252.219	2100	71,668	54,208	270.504
1300	42,033	31,224	252.776	2150	73,573	55,697	271.399
1320	42,753	31,778	253.325	2200	75,484	57,192	272.278
1340	43,475	32,334	253.868	2250	77,397	58,690	273.136
1360	44,198	32,891	254.404	2300	79,316	60,193	273.981
1380	44,923	33,449	254.932	2350	81,243	61,704	274.809
1400	45,648	34,008	255.454	2400	83,174	63,219	275.625
1420	46,374	34,567	255.968	2450	85,112	64,742	276.424
1440	47,102	35,129	256.475	2500	87,057	66,271	277.207
1460	47,831	35,692	256.978	2550	89,004	67,802	277.979
1480	48,561	36,256	257.474	2600	90,956	69,339	278.738
1500	49,292	36,821	257.965	2650	92,916	70,883	279.485
1520	50,024	37,387	258.450	2700	94,881	72,433	280.219
1540	50,756	37,952	258.928	2750	96,852	73,987	280.942
1560	51,490	38,520	259.402	2800	98,826	75,546	281.654
1580	52,224	39,088	259.870	2850	100,808	77,112	282.357
1600	52,961	39,658	260.333	2900	102,793	78,682	283.048
1620	53,696	40,227	260.791	2950	104,785	80,258	283.728
1640	54,434	40,799	261.242	3000	106,780	81,837	284.399
1660	55,172	41,370	261.690	3050	108,778	83,419	285.060
1680	55,912	41,944	262.132	3100	110,784	85,009	285.713
1700	56,652	42,517	262.571	3150	112,795	86,601	286.355
1720	57,394	43,093	263.005	3200	114,809	88,203	286.989
1740	58,136	43,669	263.435	3250	116,827	89,804	287.614

TABLE A.2 (continued) Ideal Gas Properties of Nitrogen, Oxygen, and Carbon Dioxide

Part c. Ideal Gas Properties of Carbon Dioxide, CO_2

T	$\bar{h}$	$\bar{u}$	$\bar{s}^\circ$	T	$\bar{h}$	$\bar{u}$	$\bar{s}^\circ$
0	0	0	0	600	22,280	17,291	243.199
220	6,601	4,772	202.966	610	22,754	17,683	243.983
230	6,938	5,026	204.464	620	23,231	18,076	244.758
240	7,280	5,285	205.920	630	23,709	18,471	245.524
250	7,627	5,548	207.337	640	24,190	18,869	246.282
260	7,979	5,817	208.717	650	24,674	19,270	247.032
270	8,335	6,091	210.062	660	25,160	19,672	247.773
280	8,697	6,369	211.376	670	25,648	20,078	248.507
290	9,063	6,651	212.660	680	26,138	20,484	249.233
298	9,364	6,885	213.685	690	26,631	20,894	249.952
300	9,431	6,939	213.915	700	27,125	21,305	250.663
310	9,807	7,230	215.146	710	27,622	21,719	251.368
320	10,186	7,526	216.351	720	28,121	22,134	252.065
330	10,570	7,826	217.534	730	28,622	22,552	252.755
340	10,959	8,131	218.694	740	29,124	22,972	253.439
350	11,351	8,439	219.831	750	29,629	23,393	254.117
360	11,748	8,752	220.948	760	30,135	23,817	254.787
370	12,148	9,068	222.044	770	30,644	24,242	255.452
380	12,552	9,392	223.122	780	31,154	24,669	256.110
390	12,960	9,718	224.182	790	31,665	25,097	256.762
400	13,372	10,046	225.225	800	32,179	25,527	257.408
410	13,787	10,378	226.250	810	32,694	25,959	258.048
420	14,206	10,714	227.258	820	33,212	26,394	258.682
430	14,628	11,053	228.252	830	33,730	26,829	259.311
440	15,054	11,393	229.230	840	34,251	27,267	259.934
450	15,483	11,742	230.194	850	34,773	27,706	260.551
460	15,916	12,091	231.144	860	35,296	28,125	261.164
470	16,351	12,444	232.080	870	35,821	28,588	261.770
480	16,791	12,800	233.004	880	36,347	29,031	262.371
490	17,232	13,158	233.916	890	36,876	29,476	262.968
500	17,678	13,521	234.814	900	37,405	29,922	263.559
510	18,126	13,885	235.700	910	37,935	30,369	264.146
520	18,576	14,253	236.575	920	38,467	30,818	264.728
530	19,029	14,622	237.439	930	39,000	31,268	265.304
540	19,485	14,996	238.292	940	39,535	31,719	265.877
550	19,945	15,372	239.135	950	40,070	32,171	266.444
560	20,407	15,751	239.962	960	40,607	32,625	267.007
570	20,870	16,131	240.789	970	41,145	33,081	267.566
580	21,337	16,515	241.602	980	41,685	33,537	268.119
590	21,807	16,902	242.405	990	42,226	33,995	268.670

TABLE A.2 (continued) Ideal Gas Properties of Nitrogen, Oxygen, and Carbon Dioxide

T	$\bar{h}$	$\bar{u}$	$\bar{s}°$	T	$\bar{h}$	$\bar{u}$	$\bar{s}°$
1000	42,769	34,455	269.215	1760	86,420	71,787	301.543
1020	43,859	35,378	270.293	1780	87,612	72,812	302.271
1040	44,953	36,306	271.354	1800	88,806	73,840	302.884
1060	46,051	37,238	272.400	1820	90,000	74,868	303.544
1080	47,153	38,174	273.430	1840	91,196	75,897	304.198
1100	48,258	39,112	274.445	1860	92,394	76,929	304.845
1120	49,369	40,057	275.444	1880	93,593	77,962	305.487
1140	50,484	41,006	276.430	1900	94,793	78,996	306.122
1160	51,602	41,957	277.403	1920	95,995	80,031	306.751
1180	52,724	42,913	278.362	1940	97,197	81,067	307.374
1200	53,848	43,871	279.307	1960	98,401	82,105	307.992
1220	54,977	44,834	280.238	1980	99,606	83,144	308.604
1240	56,108	45,799	281.158	2000	100,804	84,185	309.210
1260	57,244	46,768	282.066	2050	103,835	86,791	310.701
1280	58,381	47,739	282.962	2100	106,864	89,404	312.160
1300	59,522	48,713	283.847	2150	109,898	92,023	313.589
1320	60,666	49,691	284.722	2200	112,939	94,648	314.988
1340	61,813	50,672	285.586	2250	115,984	97,277	316.356
1360	62,963	51,656	286.439	2300	119,035	99,912	317.695
1380	64,116	52,643	287.283	2350	122,091	102,552	319.011
1400	65,271	53,631	288.106	2400	125,152	105,197	320.302
1420	66,427	54,621	288.934	2450	128,219	107,849	321.566
1440	67,586	55,614	289.743	2500	131,290	110,504	322.808
1460	68,748	56,609	290.542	2550	134,368	113,166	324.026
1480	69,911	57,606	291.333	2600	137,449	115,832	325.222
1500	71,078	58,606	292.114	2650	140,533	118,500	326.396
1520	72,246	59,609	292.888	2700	143,620	121,172	327.549
1540	73,417	60,613	292.654	2750	146,713	123,849	328.684
1560	74,590	61,620	294.411	2800	149,808	126,528	329.800
1580	76,767	62,630	295.161	2850	152,908	129,212	330.896
1600	76,944	63,741	295.901	2900	156,009	131,898	331.975
1620	78,123	64,653	296.632	2950	159,117	134,589	333.037
1640	79,303	65,668	297.356	3000	162,226	137,283	334.084
1660	80,486	66,592	298.072	3050	165,341	139,982	335.114
1680	81,670	67,702	298.781	3100	168,456	142,681	336.126
1700	82,856	68,721	299.482	3150	171,576	145,385	337.124
1720	84,043	69,742	300.177	3200	174,695	148,089	338.109
1740	85,231	70,764	300.863	3250	177,822	150,801	339.069

TABLE A.3 Psychrometric Table: Properties of Moist Air at 101 325 N/m²

Symbols and Units:

P_s = pressure of water vapor at saturation, N/m²
W_s = humidity ratio at saturation, mass of water vapor associated with unit mass of dry air
V_a = specific volume of dry air, m³/kg
V_s = specific volume of saturated mixture, m³/kg dry air

h_a^a = specific enthalpy of dry air, kJ/kg
h_s = specific enthalpy of saturated mixture, kJ/kg dry air
s_s = specific entropy of saturated mixture, J/K·kg dry air

Temperature			Properties						
C	K	F	P_s	W_s	V_a	V_s	h_a	h_s	s_s
−40	233.15	−40	12.838	0.000 079 25	0.659 61	0.659 68	−22.35	−22.16	−90.659
−30	243.15	−22	37.992	0.000 234 4	0.688 08	0.688 33	−12.29	−11.72	−46.732
−25	248.15	−13	63.248	0.000 390 3	0.702 32	0.702 75	−7.265	−6.306	−24.706
−20	253.15	−4	103.19	0.000 637 1	0.716 49	0.717 24	−2.236	−0.6653	−2.2194
−15	258.15	+5	165.18	0.001 020	0.730 72	0.731 91	+2.794	5.318	21.189
−10	263.15	14	259.72	0.001 606	0.744 95	0.746 83	7.823	11.81	46.104
−5	268.15	23	401.49	0.002 485	0.759 12	0.762 18	12.85	19.04	73.365
0	273.15	32	610.80	0.003 788	0.773 36	0.778 04	17.88	27.35	104.14
5	278.15	41	871.93	0.005 421	0.787 59	0.794 40	22.91	36.52	137.39
10	283.15	50	1 227.2	0.007 658	0.801 76	0.811 63	27.94	47.23	175.54
15	288.15	59	1 704.4	0.010 69	0.816 00	0.829 98	32.97	59.97	220.22
20	293.15	68	2 337.2	0.014 75	0.830 17	0.849 83	38.00	75.42	273.32
25	298.15	77	3 167.0	0.020 16	0.844 34	0.871 62	43.03	94.38	337.39
30	303.15	86	4 242.8	0.027 31	0.858 51	0.896 09	48.07	117.8	415.65
35	308.15	95	5 623.4	0.036 73	0.872 74	0.924 06	53.10	147.3	512.17
40	313.15	104	7 377.6	0.049 11	0.886 92	0.956 65	58.14	184.5	532.31
45	318.15	113	9 584.8	0.065 36	0.901 15	0.995 35	63.17	232.0	783.06
50	323.15	122	12 339	0.086 78	0.915 32	1.042 3	68.21	293.1	975.27
55	328.15	131	15 745	0.115 2	0.929 49	1.100 7	73.25	372.9	1 221.5
60	333.15	140	19 925	0.153 4	0.943 72	1.174 8	78.29	478.5	1 543.5
65	338.15	149	25 014	0.205 5	0.957 90	1.272 1	83.33	621.4	1 973.6
70	343.15	158	31 167	0.278 8	0.972 07	1.404 2	88.38	820.5	2 564.8
75	348.15	167	38 554	0.385 8	0.986 30	1.592 4	93.42	1 110	3 412.8
80	353.15	176	47 365	0.551 9	1.000 5	1.879 1	98.47	1 557	4 710.9
85	358.15	185	57 809	0.836 3	1.014 6	2.363 2	103.5	2 321	6 892.6
90	363.15	194	70 112	1.416	1.028 8	3.340 9	108.6	3 876	11 281

Note: The P_s column in this table gives the vapor pressure of pure water at temperature intervals of five degrees Celsius. For the latest data on vapor pressures at intervals of 0.1 deg C, from 0–100 deg C, see "Vapor Pressure Equation for Water", A. Wexler and L. Greenspan, *J. Res. Nat. Bur. Stand.*, 75A(3):213–229, May–June 1971.

[a]For very low barometric pressures and high wet-bulb temperatures, the values of h_a in this table are somewhat low; for corrections see "ASHRAE Handbook of Fundamentals".

*Computed from: Psychrometric Tables, in "ASHRAE Handbook of Fundamentals", American Society of Heating, Refrigerating and Air-Conditioning Engineers, 1972.

TABLE A.4 Water Vapor at Low Pressures: Perfect Gas Behavior pv/T = R = 0.461 51 kJ/kg·K

Symbols and Units:

t = thermodynamic temperature, deg C
T = thermodynamic temperature, K
pv = RT, kJ/kg
u_o = specific internal energy at zero pressure, kJ/kg
h_o = specific enthalpy at zero pressure, kJ/kg
s_l = specific entropy of semiperfect vapor at 0.1 MN/m^2, kJ/kg·K
ψ_l = specific Helmholtz free energy of semiperfect vapor at 0.1 MN/m^2, kJ/kg
ψ_l = specific Helmholtz free energy of semiperfect vapor at 0.1 MN/m^2, kJ/kg
ζ_l = specific Gibbs free energy of semiperfect vapor at 0.1 MN/m^2, kJ/kg
p_r = relative pressure, pressure of semiperfect vapor at zero entropy, TN/m^2
v_r = relative specific volume, specific volume of semiperfect vapor at zero entropy, mm^3/kg
c_{po} = specific heat capacity at constant pressure for zero pressure, kJ/kg·K
c_{vo} = specific heat capacity at constant volume for zero pressure, kJ/kg·K
$k = c_{po}/c_{vo}$ = isentropic exponent, $-(\partial \log p/\partial \log v)_s$

t	T	pv	u_o	h_o	s_l	ψ_l	ζ_l	p_r	v_r	c_{po}	c_{vo}	k
0	273.15	126.06	2 375.5	2 501.5	6.804 2	516.9	643.0	.252 9	498.4	1.858 4	1.396 9	1.330 4
10	283.15	130.68	2 389.4	2 520.1	6.871 1	443.9	574.6	.292 3	447.0	1.860 1	1.398 6	1.330 0
20	293.15	135.29	2 403.4	2 538.7	6.935 7	370.2	505.5	.336 3	402.4	1.862 2	1.400 7	1.329 5
30	303.15	139.91	2 417.5	2 557.4	6.998 2	296.0	435.9	.385 0	363.4	1.864 7	1.403 1	1.328 9
40	313.15	144.52	2 431.5	2 576.0	7.058 7	221.1	365.6	.439 0	329.2	1.867 4	1.405 9	1.328 3
50	323.15	149.14	2 445.6	2 594.7	7.117 5	145.6	294.7	.498 6	299.1	1.870 5	1.409 0	1.327 5
60	333.15	153.75	2 459.7	2 613.4	7.174 5	69.5	223.2	.564 2	272.5	1.873 8	1.412 3	1.326 8
70	343.15	158.37	2 473.8	2 632.2	7.230 0	−7.2	151.2	.636 3	248.9	1.877 4	1.415 9	1.325 9
80	353.15	162.98	2 488.0	2 651.0	7.284 0	−84.3	78.6	.715 2	227.9	1.881 2	1.419 7	1.325 1
90	363.15	167.60	2 502.2	2 669.8	7.336 6	−162.1	5.5	.801 5	209.1	1.885 2	1.423 7	1.324 2
100	373.15	172.21	2 516.5	2 688.7	7.387 8	−240.3	−68.1	.895 7	192.26	1.889 4	1.427 9	1.323 2
120	393.15	181.44	2 545.1	2 726.6	7.486 7	−398.3	−216.8	1.109 7	163.50	1.898 3	1.436 7	1.321 2
140	413.15	190.67	2 573.9	2 764.6	7.581 1	−558.2	−367.5	1.361 7	140.03	1.907 7	1.446 2	1.319 1
160	433.15	199.90	2 603.0	2 802.9	7.671 5	−720.0	−520.1	1.656 4	120.69	1.917 7	1.456 2	1.316 9
180	453.15	209.13	2 632.2	2 841.3	7.758 3	−883.5	−674.4	1.999 1	104.61	1.928 1	1.466 6	1.314 7
200	473.15	218.4	2 661.6	2 880.0	7.841 8	−1 048.7	−830.4	2.396	91.15	1.938 9	1.477 4	1.312 4
300	573.15	264.5	2 812.3	3 076.8	8.218 9	−1 898.4	−1 633.9	5.423	48.77	1.997 5	1.536 0	1.300 5
400	673.15	310.7	2 969.0	3 279.7	8.545 1	−2 783.1	−2 472.5	10.996	28.25	2.061 4	1.599 9	1.288 5
500	773.15	356.8	3 132.4	3 489.2	8.835 2	−3 699	−3 342	20.61	17.310	2.128 7	1.667 2	1.276 8
600	873.15	403.0	3 302.5	3 705.5	9.098 2	−4 642	−4 239	36.45	11.056	2.198 0	1.736 5	1.265 8
700	973.15	449.1	3 479.7	3 928.8	9.340 3	−5 610	−5 161	61.58	7.293	2.268 3	1.806 8	1.255 4
800	1 073.15	495.3	3 663.9	4 159.2	9.565 5	−6 601	−6 106	100.34	4.936	2.338 7	1.877 1	1.245 9
900	1 173.15	541.4	3 855.1	4 396.5	9.776 9	−7 615	−7 073	158.63	3.413	2.407 8	1.946 2	1.237 1
1 000	1 273.15	587.6	4 053.1	4 640.6	9.976 6	−8 649	−8 061	244.5	2.403	2.474 4	2.012 8	1.299 3
1 100	1 373.15	633.7	4 257.5	4 891.2	10.166 1	−9 702	−9 068	368.6	1.719	2.536 9	2.075 4	1.222 4
1 200	1 473.15	679.9	4 467.9	5 147.8	10.346 4	−10 774	−10 094	544.9	1.248	2.593 8	2.132 3	1.216 4
1 300	1 573.15	726.0	4 683.7	5 409.7	10.518 4	−11 863	−11 137	791.0	.918	2.643 1	2.181 6	1.211 5

*Adapted from: "Steam Tables", J.H. Keenan, F.G. Keyes, P.G. Hill, and J.G. Moore, John Wiley & Sons, Inc., 1969 (International Edition—Metric Units).

REFERENCE

For other steam tables in metric units, see "Steam Tables in SI Units", Ministry of Technology, London, 1970.

TABLE A.5 Properties of Saturated Water and Steam

Part a. Temperature Table

		Specific Volume m³/kg		Internal Energy kJ/kg		Enthalpy kJ/kg			Entropy kJ/kg · K		
Temp. °C	Press. bars	Sat. Liquid $v_f \times 10^3$	Sat. Vapor v_g	Sat. Liquid u_f	Sat. Vapor u_g	Sat. Liquid h_f	Evap. h_{fg}	Sat. Vapor h_g	Sat. Liquid s_f	Sat. Vapor s_g	Temp. °C
.01	0.00611	1.0002	206.136	0.00	2375.3	0.01	2501.3	2501.4	0.0000	9.1562	.01
4	0.00813	1.0001	157.232	16.77	2380.9	16.78	2491.9	2508.7	0.0610	9.0514	4
5	0.00872	1.0001	147.120	20.97	2382.3	20.98	2489.6	2510.6	0.0761	9.0257	5
6	0.00935	1.0001	137.734	25.19	2383.6	25.20	2487.2	2512.4	0.0912	9.0003	6
8	0.01072	1.0002	120.917	33.59	2386.4	33.60	2482.5	2516.1	0.1212	8.9501	8
10	0.01228	1.0004	106.379	42.00	2389.2	42.01	2477.7	2519.8	0.1510	8.9008	10
11	0.01312	1.0004	99.857	46.20	2390.5	46.20	2475.4	2521.6	0.1658	8.8765	11
12	0.01402	1.0005	93.784	50.41	2391.9	50.41	2473.0	2523.4	0.1806	8.8524	12
13	0.01497	1.0007	88.124	54.60	2393.3	54.60	2470.7	2525.3	0.1953	8.8285	13
14	0.01598	1.0008	82.848	58.79	2394.7	58.80	2468.3	2527.1	0.2099	8.8048	14
15	0.01705	1.0009	77.926	62.99	2396.1	62.99	2465.9	2528.9	0.2245	8.7814	15
16	0.01818	1.0011	73.333	67.18	2397.4	67.19	2463.6	2530.8	0.2390	8.7582	16
17	0.01938	1.0012	69.044	71.38	2398.8	71.38	2461.2	2532.6	0.2535	8.7351	17
18	0.02064	1.0014	65.038	75.57	2400.2	75.58	2458.8	2534.4	0.2679	8.7123	18
19	0.02198	1.0016	61.293	79.76	2401.6	79.77	2456.5	2536.2	0.2823	8.6897	19
20	0.02339	1.0018	57.791	83.95	2402.9	83.96	2454.1	2538.1	0.2966	8.6672	20
21	0.02487	1.0020	54.514	88.14	2404.3	88.14	2451.8	2539.9	0.3109	8.6450	21
22	0.02645	1.0022	51.447	92.32	2405.7	92.33	2449.4	2541.7	0.3251	8.6229	22
23	0.02810	1.0024	48.574	96.51	2407.0	96.52	2447.0	2543.5	0.3393	8.6011	23
24	0.02985	1.0027	45.883	100.70	2408.4	100.70	2444.7	2545.4	0.3534	8.5794	24
25	0.03169	1.0029	43.360	104.88	2409.8	104.89	2442.3	2547.2	0.3674	8.5580	25
26	0.03363	1.0032	40.994	109.06	2411.1	109.07	2439.9	2549.0	0.3814	8.5367	26
27	0.03567	1.0035	38.774	113.25	2412.5	113.25	2437.6	2550.8	0.3954	8.5156	27
28	0.03782	1.0037	36.690	117.42	2413.9	117.43	2435.2	2552.6	0.4093	8.4946	28
29	0.04008	1.0040	34.733	121.60	2415.2	121.61	2432.8	2554.5	0.4231	8.4739	29
30	0.04246	1.0043	32.894	125.78	2416.6	125.79	2430.5	2556.3	0.4369	8.4533	30
31	0.04496	1.0046	31.165	129.96	2418.0	129.97	2428.1	2558.1	0.4507	8.4329	31
32	0.04759	1.0050	29.540	134.14	2419.3	134.15	2425.7	2559.9	0.4644	8.4127	32
33	0.05034	1.0053	28.011	138.32	2420.7	138.33	2423.4	2561.7	0.4781	8.3927	33
34	0.05324	1.0056	26.571	142.50	2422.0	142.50	2421.0	2563.5	0.4917	8.3728	34
35	0.05628	1.0060	25.216	146.67	2423.4	146.68	2418.6	2565.3	0.5053	8.3531	35
36	0.05947	1.0063	23.940	150.85	2424.7	150.86	2416.2	2567.1	0.5188	8.3336	36
38	0.06632	1.0071	21.602	159.20	2427.4	159.21	2411.5	2570.7	0.5458	8.2950	38
40	0.07384	1.0078	19.523	167.56	2430.1	167.57	2406.7	2574.3	0.5725	8.2570	40
45	0.09593	1.0099	15.258	188.44	2436.8	188.45	2394.8	2583.2	0.6387	8.1648	45

TABLE A.5 (continued) Properties of Saturated Water and Steam

		Specific Volume m³/kg		Internal Energy kJ/kg		Enthalpy kJ/kg			Entropy kJ/kg · K		
Temp. °C	**Press. bars**	**Sat. Liquid** $v_f \times 10^3$	**Sat. Vapor** v_g	**Sat. Liquid** u_f	**Sat. Vapor** u_g	**Sat. Liquid** h_f	**Evap.** h_{fg}	**Sat. Vapor** h_g	**Sat. Liquid** s_f	**Sat. Vapor** s_g	**Temp. °C**
50	.1235	1.0121	12.032	209.32	2443.5	209.33	2382.7	2592.1	.7038	8.0763	50
55	.1576	1.0146	9.568	230.21	2450.1	230.23	2370.7	2600.9	.7679	7.9913	55
60	.1994	1.0172	7.671	251.11	2456.6	251.13	2358.5	2609.6	.8312	7.9096	60
65	.2503	1.0199	6.197	272.02	2463.1	272.06	2346.2	2618.3	.8935	7.8310	65
70	.3119	1.0228	5.042	292.95	2469.6	292.98	2333.8	2626.8	.9549	7.7553	70
75	.3858	1.0259	4.131	313.90	2475.9	313.93	2321.4	2635.3	1.0155	7.6824	75
80	.4739	1.0291	3.407	334.86	2482.2	334.91	2308.8	2643.7	1.0753	7.6122	80
85	.5783	1.0325	2.828	355.84	2488.4	355.90	2296.0	2651.9	1.1343	7.5445	85
90	.7014	1.0360	2.361	376.85	2494.5	376.92	2283.2	2660.1	1.1925	7.4791	90
95	.8455	1.0397	1.982	397.88	2500.6	397.96	2270.2	2668.1	1.2500	7.4159	95
100	1.014	1.0435	1.673	418.94	2506.5	419.04	2257.0	2676.1	1.3069	7.3549	100
110	1.433	1.0516	1.210	461.14	2518.1	461.30	2230.2	2691.5	1.4185	7.2387	110
120	1.985	1.0603	0.8919	503.50	2529.3	503.71	2202.6	2706.3	1.5276	7.1296	120
130	2.701	1.0697	0.6685	546.02	2539.9	546.31	2174.2	2720.5	1.6344	7.0269	130
140	3.613	1.0797	0.5089	588.74	2550.0	589.13	2144.7	2733.9	1.7391	6.9299	140
150	4.758	1.0905	0.3928	631.68	2559.5	632.20	2114.3	2746.5	1.8418	6.8379	150
160	6.178	1.1020	0.3071	674.86	2568.4	675.55	2082.6	2758.1	1.9427	6.7502	160
170	7.917	1.1143	0.2428	718.33	2576.5	719.21	2049.5	2768.7	2.0419	6.6663	170
180	10.02	1.1274	0.1941	762.09	2583.7	763.22	2015.0	2778.2	2.1396	6.5857	180
190	12.54	1.1414	0.1565	806.19	2590.0	807.62	1978.8	2786.4	2.2359	6.5079	190
200	15.54	1.1565	0.1274	850.65	2595.3	852.45	1940.7	2793.2	2.3309	6.4323	200
210	19.06	1.1726	0.1044	895.53	2599.5	897.76	1900.7	2798.5	2.4248	6.3585	210
220	23.18	1.1900	0.08619	940.87	2602.4	943.62	1858.5	2802.1	2.5178	6.2861	220
230	27.95	1.2088	0.07158	986.74	2603.9	990.12	1813.8	2804.0	2.6099	6.2146	230
240	33.44	1.2291	0.05976	1033.2	2604.0	1037.3	1766.5	2803.8	2.7015	6.1437	240
250	39.73	1.2512	0.05013	1080.4	2602.4	1085.4	1716.2	2801.5	2.7927	6.0730	250
260	46.88	1.2755	0.04221	1128.4	2599.0	1134.4	1662.5	2796.6	2.8838	6.0019	260
270	54.99	1.3023	0.03564	1177.4	2593.7	1184.5	1605.2	2789.7	2.9751	5.9301	270
280	64.12	1.3321	0.03017	1227.5	2586.1	1236.0	1543.6	2779.6	3.0668	5.8571	280
290	74.36	1.3656	0.02557	1278.9	2576.0	1289.1	1477.1	2766.2	3.1594	5.7821	290
300	85.81	1.4036	0.02167	1332.0	2563.0	1344.0	1404.9	2749.0	3.2534	5.7045	300
320	112.7	1.4988	0.01549	1444.6	2525.5	1461.5	1238.6	2700.1	3.4480	5.5362	320
340	145.9	1.6379	0.01080	1570.3	2464.6	1594.2	1027.9	2622.0	3.6594	5.3357	340
360	186.5	1.8925	0.006945	1725.2	2351.5	1760.5	720.5	2481.0	3.9147	5.0526	360
374.14	220.9	3.155	0.003155	2029.6	2029.6	2099.3	0	2099.3	4.4298	4.4298	374.14

TABLE A.5 (continued) Properties of Saturated Water and Steam

Part b. Pressure Table

		Specific Volume m^3/kg		Internal Energy kJ/kg		Enthalpy kJ/kg			Entropy kJ/kg · K		
Press. bars	Temp. °C	Sat. Liquid $v_f \times 10^3$	Sat. Vapor v_g	Sat. Liquid u_f	Sat. Vapor u_g	Sat. Liquid h_f	Evap. h_{fg}	Sat. Vapor h_g	Sat. Liquid s_f	Sat. Vapor s_g	Press. bars
0.04	28.96	1.0040	34.800	121.45	2415.2	121.46	2432.9	2554.4	0.4226	8.4746	0.04
0.06	36.16	1.0064	23.739	151.53	2425.0	151.53	2415.9	2567.4	0.5210	8.3304	0.06
0.08	41.51	1.0084	18.103	173.87	2432.2	173.88	2403.1	2577.0	0.5926	8.2287	0.08
0.10	45.81	1.0102	14.674	191.82	2437.9	191.83	2392.8	2584.7	0.6493	8.1502	0.10
0.20	60.06	1.0172	7.649	251.38	2456.7	251.40	2358.3	2609.7	0.8320	7.9085	0.20
0.30	69.10	1.0223	5.229	289.20	2468.4	289.23	2336.1	2625.3	0.9439	7.7686	0.30
0.40	75.87	1.0265	3.993	317.53	2477.0	317.58	2319.2	2636.8	1.0259	7.6700	0.40
0.50	81.33	1.0300	3.240	340.44	2483.9	340.49	2305.4	2645.9	1.0910	7.5939	0.50
0.60	85.94	1.0331	2.732	359.79	2489.6	359.86	2293.6	2653.5	1.1453	7.5320	0.60
0.70	89.95	1.0360	2.365	376.63	2494.5	376.70	2283.3	2660.0	1.1919	7.4797	0.70
0.80	93.50	1.0380	2.087	391.58	2498.8	391.66	2274.1	2665.8	1.2329	7.4346	0.80
0.90	96.71	1.0410	1.869	405.06	2502.6	405.15	2265.7	2670.9	1.2695	7.3949	0.90
1.00	99.63	1.0432	1.694	417.36	2506.1	417.46	2258.0	2675.5	1.3026	7.3594	1.00
1.50	111.4	1.0528	1.159	466.94	2519.7	467.11	2226.5	2693.6	1.4336	7.2233	1.50
2.00	120.2	1.0605	0.8857	504.49	2529.5	504.70	2201.9	2706.7	1.5301	7.1271	2.00
2.50	127.4	1.0672	0.7187	535.10	2537.2	535.37	2181.5	2716.9	1.6072	7.0527	2.50
3.00	133.6	1.0732	0.6058	561.15	2543.6	561.47	2163.8	2725.3	1.6718	6.9919	3.00
3.50	138.9	1.0786	0.5243	583.95	2546.9	584.33	2148.1	2732.4	1.7275	6.9405	3.50
4.00	143.6	1.0836	0.4625	604.31	2553.6	604.74	2133.8	2738.6	1.7766	6.8959	4.00
4.50	147.9	1.0882	0.4140	622.25	2557.6	623.25	2120.7	2743.9	1.8207	6.8565	4.50
5.00	151.9	1.0926	0.3749	639.68	2561.2	640.23	2108.5	2748.7	1.8607	6.8212	5.00
6.00	158.9	1.1006	0.3157	669.90	2567.4	670.56	2086.3	2756.8	1.9312	6.7600	6.00
7.00	165.0	1.1080	0.2729	696.44	2572.5	697.22	2066.3	2763.5	1.9922	6.7080	7.00
8.00	170.4	1.1148	0.2404	720.22	2576.8	721.11	2048.0	2769.1	2.0462	6.6628	8.00
9.00	175.4	1.1212	0.2150	741.83	2580.5	742.83	2031.1	2773.9	2.0946	6.6226	9.00
10.0	179.9	1.1273	0.1944	761.68	2583.6	762.81	2015.3	2778.1	2.1387	6.5863	10.0
15.0	198.3	1.1539	0.1318	843.16	2594.5	844.84	1947.3	2792.2	2.3150	6.4448	15.0
20.0	212.4	1.1767	0.09963	906.44	2600.3	908.79	1890.7	2799.5	2.4474	6.3409	20.0
25.0	224.0	1.1973	0.07998	959.11	2603.1	962.11	1841.0	2803.1	2.5547	6.2575	25.0
30.0	233.9	1.2165	0.06668	1004.8	2604.1	1008.4	1795.7	2804.2	2.6457	6.1869	30.0
35.0	242.6	1.2347	0.05707	1045.4	2603.7	1049.8	1753.7	2803.4	2.7253	6.1253	35.0
40.0	250.4	1.2522	0.04978	1082.3	2602.3	1087.3	1714.1	2801.4	2.7964	6.0701	40.0
45.0	257.5	1.2692	0.04406	1116.2	2600.1	1121.9	1676.4	2798.3	2.8610	6.0199	45.0
50.0	264.0	1.2859	0.03944	1147.8	2597.1	1154.2	1640.1	2794.3	2.9202	5.9734	50.0
60.0	275.6	1.3187	0.03244	1205.4	2589.7	1213.4	1571.0	2784.3	3.0267	5.8892	60.0
70.0	285.9	1.3513	0.02737	1257.6	2580.5	1267.0	1505.1	2772.1	3.1211	5.8133	70.0
80.0	295.1	1.3842	0.02352	1305.6	2569.8	1316.6	1441.3	2758.0	3.2068	5.7432	80.0
90.0	303.4	1.4178	0.02048	1350.5	2557.8	1363.3	1378.9	2742.1	3.2858	5.6772	90.0
100.	311.1	1.4524	0.01803	1393.0	2544.4	1407.6	1317.1	2724.7	3.3596	5.6141	100.
110.	318.2	1.4886	0.01599	1433.7	2529.8	1450.1	1255.5	2705.6	3.4295	5.5527	110.
120.	324.8	1.5267	0.01426	1473.0	2513.7	1491.3	1193.6	2684.9	3.4962	5.4924	120.
130.	330.9	1.5671	0.01278	1511.1	2496.1	1531.5	1130.7	2662.2	3.5606	5.4323	130.
140.	336.8	1.6107	0.01149	1548.6	2476.8	1571.1	1066.5	2637.6	3.6232	5.3717	140.
150.	342.2	1.6581	0.01034	1585.6	2455.5	1610.5	1000.0	2610.5	3.6848	5.3098	150.
160.	347.4	1.7107	0.009306	1622.7	2431.7	1650.1	930.6	2580.6	3.7461	5.2455	160.
170.	352.4	1.7702	0.008364	1660.2	2405.0	1690.3	856.9	2547.2	3.8079	5.1777	170.
180.	357.1	1.8397	0.007489	1698.9	2374.3	1732.0	777.1	2509.1	3.8715	5.1044	180.
190.	361.5	1.9243	0.006657	1739.9	2338.1	1776.5	688.0	2464.5	3.9388	5.0228	190.
200.	365.8	2.036	0.005834	1785.6	2293.0	1826.3	583.4	2409.7	4.0139	4.9269	200.
220.9	374.1	3.155	0.003155	2029.6	2029.6	2099.3	0	2099.3	4.4298	4.4298	220.9

Source: Adapted from M.J. Moran and H.N. Shapiro, *Fundamentals of Engineering Thermodynamics*, 3rd. ed., Wiley, New York, 1995, as extracted from J.H. Keenan, F.G. Keyes, P.G. Hill, and J.G. Moore, *Steam Tables*, Wiley, New York, 1969.

TABLE A.6 Properties of Superheated Steam

Symbols and Units:

T = temperature, °C
T_{sat} = Saturation temperature, °C
v = Specific volume, m³/kg
u = internal energy, kJ/kg
h = enthalpy, kJ/kg
S = entropy, kJ/kg·K
p = pressure, bar and μPa

T °C	v m³/kg	u kJ/kg	h kJ/kg	s kJ/kg · K	v m³/kg	u kJ/kg	h kJ/kg	s kJ/kg · K
	p = 0.06 bar = 0.006 MPa (T_{sat} = 36.16°C)				**p = 0.35 bar = 0.035 MPa (T_{sat} = 72.69°C)**			
Sat.	23.739	2425.0	2567.4	8.3304	4.526	2473.0	2631.4	7.7158
80	27.132	2487.3	2650.1	8.5804	4.625	2483.7	2645.6	7.7564
120	30.219	2544.7	2726.0	8.7840	5.163	2542.4	2723.1	7.9644
160	33.302	2602.7	2802.5	8.9693	5.696	2601.2	2800.6	8.1519
200	36.383	2661.4	2879.7	9.1398	6.228	2660.4	2878.4	8.3237
240	39.462	2721.0	2957.8	9.2982	6.758	2720.3	2956.8	8.4828
280	42.540	2781.5	3036.8	9.4464	7.287	2780.9	3036.0	8.6314
320	45.618	2843.0	3116.7	9.5859	7.815	2842.5	3116.1	8.7712
360	48.696	2905.5	3197.7	9.7180	8.344	2905.1	3197.1	8.9034
400	51.774	2969.0	3279.6	9.8435	8.872	2968.6	3279.2	9.0291
440	54.851	3033.5	3362.6	9.9633	9.400	3033.2	3362.2	9.1490
500	59.467	3132.3	3489.1	10.1336	10.192	3132.1	3488.8	9.3194

T °C	v m³/kg	u kJ/kg	h kJ/kg	s kJ/kg · K	v m³/kg	u kJ/kg	h kJ/kg	s kJ/kg · K
	p = 0.70 bar = 0.07 MPa (T_{sat} = 89.95°C)				**p = 1.0 bar = 0.10 MPa (T_{sat} = 99.63°C)**			
Sat.	2.365	2494.5	2660.0	7.4797	1.694	2506.1	2675.5	7.3594
100	2.434	2509.7	2680.0	7.5341	1.696	2506.7	2676.2	7.3614
120	2.571	2539.7	2719.6	7.6375	1.793	2537.3	2716.6	7.4668
160	2.841	2599.4	2798.2	7.8279	1.984	2597.8	2796.2	7.6597
200	3.108	2659.1	2876.7	8.0012	2.172	2658.1	2875.3	7.8343
240	3.374	2719.3	2955.5	8.1611	2.359	2718.5	2954.5	7.9949
280	3.640	2780.2	3035.0	8.3162	2.546	2779.6	3034.2	8.1445
320	3.905	2842.0	3115.3	8.4504	2.732	2841.5	3114.6	8.2849
360	4.170	2904.6	3196.5	8.5828	2.917	2904.2	3195.9	8.4175
400	4.434	2968.2	3278.6	8.7086	3.103	2967.9	3278.2	8.5435
440	4.698	3032.9	3361.8	8.8286	3.288	3032.6	3361.4	8.6636
500	5.095	3131.8	3488.5	8.9991	3.565	3131.6	3488.1	8.8342

T °C	v m³/kg	u kJ/kg	h kJ/kg	s kJ/kg · K	v m³/kg	u kJ/kg	h kJ/kg	s kJ/kg · K
	p = 1.5 bars = 0.15 MPa (T_{sat} = 111.37°C)				**p = 3.0 bars = 0.30 MPa (T_{sat} = 133.55°C)**			
Sat.	1.159	2519.7	2693.6	7.2233	0.606	2543.6	2725.3	6.9919
120	1.188	2533.3	2711.4	7.2693				
160	1.317	2595.2	2792.8	7.4665	0.651	2587.1	2782.3	7.1276
200	1.444	2656.2	2872.9	7.6433	0.716	2650.7	2865.5	7.3115
240	1.570	2717.2	2952.7	7.8052	0.781	2713.1	2947.3	7.4774
280	1.695	2778.6	3032.8	7.9555	0.844	2775.4	3028.6	7.6299
320	1.819	2840.6	3113.5	8.0964	0.907	2838.1	3110.1	7.7722
360	1.943	2903.5	3195.0	8.2293	0.969	2901.4	3192.2	7.9061
400	2.067	2967.3	3277.4	8.3555	1.032	2965.6	3275.0	8.0330
440	2.191	3032.1	3360.7	8.4757	1.094	3030.6	3358.7	8.1538
500	2.376	3131.2	3487.6	8.6466	1.187	3130.0	3486.0	8.3251
600	2.685	3301.7	3704.3	8.9101	1.341	3300.8	3703.2	8.5892

TABLE A.6 (continued) Properties of Superheated Steam

Symbols and Units:

T = temperature, °C
T_{sat} = Saturation temperature, °C
v = Specific volume, m^3/kg
u = internal energy, kJ/kg
h = enthalpy, kJ/kg
S = entropy, kJ/kg·K
p = pressure, bar and μPa

T °C	v m³/kg	u kJ/kg	h kJ/kg	s kJ/kg · K	v m³/kg	u kJ/kg	h kJ/kg	s kJ/kg . k
	p = 5.0 bars = 0.50 MPa (T_{sat} = 151.86°C)				p = 7.0 bars = 0.70 MPa (T_{sat} = 164.97°C)			
Sat.	0.3749	2561.2	2748.7	6.8213	0.2729	2572.5	2763.5	6.7080
180	0.4045	2609.7	2812.0	6.9656	0.2847	2599.8	2799.1	6.7880
200	0.4249	2642.9	2855.4	7.0592	0.2999	2634.8	2844.8	6.8865
240	0.4646	2707.6	2939.9	7.2307	0.3292	2701.8	2932.2	7.0641
280	0.5034	2771.2	3022.9	7.3865	0.3574	2766.9	3017.1	7.2233
320	0.5416	2834.7	3105.6	7.5308	0.3852	2831.3	3100.9	7.3697
360	0.5796	2898.7	3188.4	7.6660	0.4126	2895.8	3184.7	7.5063
400	0.6173	2963.2	3271.9	7.7938	0.4397	2960.9	3268.7	7.6350
440	0.6548	3028.6	3356.0	7.9152	0.4667	3026.6	3353.3	7.7571
500	0.7109	3128.4	3483.9	8.0873	0.5070	3126.8	3481.7	7.9299
600	0.8041	3299.6	3701.7	8.3522	0.5738	3298.5	3700.2	8.1956
700	0.8969	3477.5	3925.9	8.5952	0.6403	3476.6	3924.8	8.4391
	p = 10.0 bars = 1.0 MPa (T_{sat} = 179.91°C)				p = 15.0 bars = 1.5 MPa (T_{sat} = 198.32°C)			
Sat.	0.1944	2583.6	2778.1	6.5865	0.1318	2594.5	2792.2	6.4448
200	0.2060	2621.9	2827.9	6.6940	0.1325	2598.1	2796.8	6.4546
240	0.2275	2692.9	2920.4	6.8817	0.1483	2676.9	2899.3	6.6628
280	0.2480	2760.2	3008.2	7.0465	0.1627	2748.6	2992.7	6.8381
320	0.2678	2826.1	3093.9	7.1962	0.1765	2817.1	3081.9	6.9938
360	0.2873	2891.6	3178.9	7.3349	0.1899	2884.4	3169.2	7.1363
400	0.3066	2957.3	3263.9	7.4651	0.2030	2951.3	3255.8	7.2690
440	0.3257	3023.6	3349.3	7.5883	0.2160	3018.5	3342.5	7.3940
500	0.3541	3124.4	3478.5	7.7622	0.2352	3120.3	3473.1	7.5698
540	0.3729	3192.6	3565.6	7.8720	0.2478	3189.1	3560.9	7.6805
600	0.4011	3296.8	3697.9	8.0290	0.2668	3293.9	3694.0	7.8385
640	0.4198	3367.4	3787.2	8.1290	0.2793	3364.8	3783.8	7.9391
	p = 20.0 bars = 2.0 MPa (T_{sat} = 212.42°C)				p = 30.0 bars = 3.0 MPa (T_{sat} = 233.90°C)			
Sat.	0.0996	2600.3	2799.5	6.3409	0.0667	2604.1	2804.2	6.1869
240	0.1085	2659.6	2876.5	6.4952	0.0682	2619.7	2824.3	6.2265
280	0.1200	2736.4	2976.4	6.6828	0.0771	2709.9	2941.3	6.4462
320	0.1308	2807.9	3069.5	6.8452	0.0850	2788.4	3043.4	6.6245
360	0.1411	2877.0	3159.3	6.9917	0.0923	2861.7	3138.7	6.7801
400	0.1512	2945.2	3247.6	7.1271	0.0994	2932.8	3230.9	6.9212
440	0.1611	3013.4	3335.5	7.2540	0.1062	3002.9	3321.5	7.0520
500	0.1757	3116.2	3467.6	7.4317	0.1162	3108.0	3456.5	7.2338
540	0.1853	3185.6	3556.1	7.5434	0.1227	3178.4	3546.6	7.3474
600	0.1996	3290.9	3690.1	7.7024	0.1324	3285.0	3682.3	7.5085
640	0.2091	3362.2	3780.4	7.8035	0.1388	3357.0	3773.5	7.6106
700	0.2232	3470.9	3917.4	7.9487	0.1484	3466.5	3911.7	7.7571

TABLE A.6 (continued) Properties of Superheated Steam

Symbols and Units:

T = temperature, °C
T_{sat} = Saturation temperature, °C
v = Specific volume, m³/kg
u = internal energy, kJ/kg
h = enthalpy, kJ/kg
S = entropy, kJ/kg·K
p = pressure, bar and μPa

T °C	v m³/kg	u kJ/kg	h kJ/kg	s kJ/kg · K	v m³/kg	u kJ/kg	h kJ/kg	s kJ/kg · K
	p = 40 bars = 4.0 MPa (T_{sat} = 250.4°C)				p = 60 bars = 6.0 MPa (T_{sat} = 275.64°C)			
Sat.	0.04978	2602.3	2801.4	6.0701	0.03244	2589.7	2784.3	5.8892
280	0.05546	2680.0	2901.8	6.2568	0.03317	2605.2	2804.2	5.9252
320	0.06199	2767.4	3015.4	6.4553	0.03876	2720.0	2952.6	6.1846
360	0.06788	2845.7	3117.2	6.6215	0.04331	2811.2	3071.1	6.3782
400	0.07341	2919.9	3213.6	6.7690	0.04739	2892.9	3177.2	6.5408
440	0.07872	2992.2	3307.1	6.9041	0.05122	2970.0	3277.3	6.6853
500	0.08643	3099.5	3445.3	7.0901	0.05665	3082.2	3422.2	6.8803
540	0.09145	3171.1	3536.9	7.2056	0.06015	3156.1	3517.0	6.9999
600	0.09885	3279.1	3674.4	7.3688	0.06525	3266.9	3658.4	7.1677
640	0.1037	3351.8	3766.6	7.4720	0.06859	3341.0	3752.6	7.2731
700	0.1110	3462.1	3905.9	7.6198	0.07352	3453.1	3894.1	7.4234
740	0.1157	3536.6	3999.6	7.7141	0.07677	3528.3	3989.2	7.5190
	p = 80 bars = 8.0 MPa (T_{sat} = 295.06°C)				p = 100 bars = 10.0 MPa (T_{sat} = 311.06°C)			
Sat.	0.02352	2569.8	2758.0	5.7432	0.01803	2544.4	2724.7	5.6141
320	0.02682	2662.7	2877.2	5.9489	0.01925	2588.8	2781.3	5.7103
360	0.03089	2772.7	3019.8	6.1819	0.02331	2729.1	2962.1	6.0060
400	0.03432	2863.8	3138.3	6.3634	0.02641	2832.4	3096.5	6.2120
440	0.03742	2946.7	3246.1	6.5190	0.02911	2922.1	3213.2	6.3805
480	0.04034	3025.7	3348.4	6.6586	0.03160	3005.4	3321.4	6.5282
520	0.04313	3102.7	3447.7	6.7871	0.03394	3085.6	3425.1	6.6622
560	0.04582	3178.7	3545.3	6.9072	0.03619	3164.1	3526.0	6.7864
600	0.04845	3254.4	3642.0	7.0206	0.03837	3241.7	3625.3	6.9029
640	0.05102	3330.1	3738.3	7.1283	0.04048	3318.9	3723.7	7.0131
700	0.05481	3443.9	3882.4	7.2812	0.04358	3434.7	3870.5	7.1687
740	0.05729	3520.4	3978.7	7.3782	0.04560	3512.1	3968.1	7.2670
	p = 120 bars = 12.0 MPa (T_{sat} = 324.75°C)				p = 140 bars = 14.0 MPa (T_{sat} = 336.75°C)			
Sat.	0.01426	2513.7	2684.9	5.4924	0.01149	2476.8	2637.6	5.3717
360	0.01811	2678.4	2895.7	5.8361	0.01422	2617.4	2816.5	5.6602
400	0.02108	2798.3	3051.3	6.0747	0.01722	2760.9	3001.9	5.9448
440	0.02355	2896.1	3178.7	6.2586	0.01954	2868.6	3142.2	6.1474
480	0.02576	2984.4	3293.5	6.4154	0.02157	2962.5	3264.5	6.3143
520	0.02781	3068.0	3401.8	6.5555	0.02343	3049.8	3377.8	6.4610
560	0.02977	3149.0	3506.2	6.6840	0.02517	3133.6	3486.0	6.5941
600	0.03164	3228.7	3608.3	6.8037	0.02683	3215.4	3591.1	6.7172
640	0.03345	3307.5	3709.0	6.9164	0.02843	3296.0	3694.1	6.8326
700	0.03610	3425.2	3858.4	7.0749	0.03075	3415.7	3846.2	6.9939
740	0.03781	3503.7	3957.4	7.1746	0.03225	3495.2	3946.7	7.0952

TABLE A.7 Chemical, Physical, and Thermal Properties of Gases: Gases and Vapors, Including Fuels and Refrigerants, English and Metric Units

Common name(s) Chemical formula	Acetylene (Ethyne) C_2H_2	Air [mixture]	Ammonia, anhyd. NH_3	Argon Ar
Refrigerant number	—	729	717	740
CHEMICAL AND PHYSICAL PROPERTIES				
Molecular weight	26.04	28.966	17.02	39.948
Specific gravity, air = 1	0.90	1.00	0.59	1.38
Specific volume, ft³/lb	14.9	13.5	23.0	9.80
Specific volume, m³/kg	0.93	0.842	1.43	0.622
Density of liquid (at atm bp), lb/ft³	43.0	54.6	42.6	87.0
Density of liquid (at atm bp), kg/m³	693.	879.	686.	1 400.
Vapor pressure at 25 deg C, psia			145.4	
Vapor pressure at 25 deg C, MN/m²			1.00	
Viscosity (abs), lbm/ft·sec	6.72×10^{-6}	12.1×10^{-6}	6.72×10^{-6}	13.4×10^{-6}
Viscosity (abs), centipoises[a]	0.01	0.018	0.010	0.02
Sound velocity in gas, m/sec	343	346	415	322
THERMAL AND THERMODYNAMIC PROPERTIES				
Specific heat, c_p, Btu/lb·deg F or cal/g·deg C	0.40	0.240 3	0.52	0.125
Specific heat, c_p, J/kg·K	1 674.	1 005.	2 175.	523.
Specific heat ratio, c_p/c_v	1.25	1.40	1.3	1.67
Gas constant R, ft-lb/lb·deg R	59.3	53.3	90.8	38.7
Gas constant R, J/kg·deg C	319	286.8	488.	208.
Thermal conductivity, Btu/hr·ft·deg F	0.014	0.015 1	0.015	0.010 2
Thermal conductivity, W/m·deg C	0.024	0.026	0.026	0.017 2
Boiling point (sat 14.7 psia), deg F	−103	−320	−28.	−303.
Boiling point (sat 760 mm), deg C	−75	−195	−33.3	−186
Latent heat of evap (at bp), Btu/lb	264	88.2	589.3	70.
Latent heat of evap (at bp), J/kg	614 000	205 000.	1 373 000	163 000
Freezing (melting) point, deg F (1 atm)	−116	−357.2	−107.9	−308.5
Freezing (melting) point, deg C (1 atm)	−82.2	−216.2	−77.7	−189.2
Latent heat of fusion, Btu/lb	23.	10.0	143.0	
Latent heat of fusion, J/kg	53 500	23 200	332 300	
Critical temperature, deg F	97.1	−220.5	271.4	−187.6
Critical temperature, deg C	36.2	−140.3	132.5	−122
Critical pressure, psia	907.	550.	1 650.	707.
Critical pressure, MN/m²	6.25	3.8	11.4	4.87
Critical volume, ft³/lb		0.050	0.068	0.029 9
Critical volume, m³/kg		0.003	0.004 24	0.001 86
Flammable (yes or no)	Yes	No	No	No
Heat of combustion, Btu/ft³	1 450	—	—	—
Heat of combustion, Btu/lb	21 600	—	—	—
Heat of combustion, kJ/kg	50 200	—	—	—

[a] For N·sec/m² divide by 1 000.

Note: The properties of pure gases are given at 25°C (77°F, 298 K) and atmospheric pressure (except as stated).

TABLE A.7 (continued) Chemical, Physical, and Thermal Properties of Gases: Gases and Vapors, Including Fuels and Refrigerants, English and Metric Units

Common name(s)	*Butadiene*	*n-Butane*	*Isobutane (2-Methyl propane)*	*1-Butene (Butylene)*
Chemical formula	C_4H_6	C_4H_{10}	C_4H_{10}	C_4H_8
Refrigerant number	—	*600*	*600a*	—
CHEMICAL AND PHYSICAL PROPERTIES				
Molecular weight	54.09	58.12	58.12	56.108
Specific gravity, air = 1	1.87	2.07	2.07	1.94
Specific volume, ft^3/lb	7.1	6.5	6.5	6.7
Specific volume, m^3/kg	0.44	0.405	0.418	0.42
Density of liquid (at atm bp), lb/ft^3		37.5	37.2	
Density of liquid (at atm bp), kg/m^3		604.	599.	
Vapor pressure at 25 deg C, psia		35.4	50.4	
Vapor pressure at 25 deg C, MN/m^2		0.024 4	0.347	
Viscosity (abs), lbm/ft·sec		4.8×10^{-6}		
Viscosity (abs), centipoises[a]		0.007		
Sound velocity in gas, m/sec	226	216	216	222
THERMAL AND THERMODYNAMIC PROPERTIES				
Specific heat, c_p, Btu/lb·deg F or cal/g·deg C	0.341	0.39	0.39	0.36
Specific heat, c_p, J/kg·K	1 427.	1 675.	1 630.	1 505.
Specific heat ratio, c_p/c_v	1.12	1.096	1.10	1.112
Gas constant R, ft-lb/lb·deg F	28.55	26.56	26.56	27.52
Gas constant R, J/kg·deg C	154.	143.	143.	148.
Thermal conductivity, Btu/hr·ft·deg F		0.01	0.01	
Thermal conductivity, W/m·deg C		0.017	0.017	
Boiling point (sat 14.7 psia), deg F	24.1	31.2	10.8	20.6
Boiling point (sat 760 mm), deg C	−4.5	−0.4	−11.8	−6.3
Latent heat of evap (at bp), Btu/lb		165.6	157.5	167.9
Latent heat of evap (at bp), J/kg		386 000	366 000	391 000
Freezing (melting) point, deg F (1 atm)	−164.	−217.	−229	−301.6
Freezing (melting) point, deg C (1 atm)	−109.	−138	−145	−185.3
Latent heat of fusion, Btu/lb		19.2		16.4
Latent heat of fusion, J/kg		44 700		38 100
Critical temperature, deg F		306	273.	291.
Critical temperature, deg C	171.	152.	134.	144.
Critical pressure, psia	652.	550.	537.	621.
Critical pressure, MN/m^2		3.8	3.7	4.28
Critical volume, ft^3/lb		0.070		0.068
Critical volume, m^3/kg		0.004 3		0.004 2
Flammable (yes or no)	Yes	Yes	Yes	Yes
Heat of combustion, Btu/ft^3	2 950	3 300	3 300	3 150
Heat of combustion, Btu/lb	20 900	21 400	21 400	21 000
Heat of combustion, kJ/kg	48 600	49 700	49 700	48 800

[a]For N·sec/m^2 divide by 1 000.

TABLE A.7 (continued) Chemical, Physical, and Thermal Properties of Gases: Gases and Vapors, Including Fuels and Refrigerants, English and Metric Units

Common name(s) / *Chemical formula* / *Refrigerant number*	*cis-2-Butene* C_4H_8 —	*trans-2-Butene* C_4H_8 —	*Isobutene* C_4H_8 —	*Carbon dioxide* CO_2 744
CHEMICAL AND PHYSICAL PROPERTIES				
Molecular weight	56.108	56.108	56.108	44.01
Specific gravity, air = 1	1.94	1.94	1.94	1.52
Specific volume, ft³/lb	6.7	6.7	6.7	8.8
Specific volume, m³/kg	0.42	0.42	0.42	0.55
Density of liquid (at atm bp), lb/ft³				—
Density of liquid (at atm bp), kg/m³				—
Vapor pressure at 25 deg C, psia				931.
Vapor pressure at 25 deg C, MN/m²				6.42
Viscosity (abs), lbm/ft·sec				9.4×10^{-6}
Viscosity (abs), centipoises[a]				0.014
Sound velocity in gas, m/sec	223.	221.	221.	270.
THERMAL AND THERMODYNAMIC PROPERTIES				
Specific heat, c_p, Btu/lb·deg F or cal/g·deg C	0.327	0.365	0.37	0.205
Specific heat, c_p, J/kg·K	1 368.	1 527.	1 548.	876.
Specific heat ratio, c_p/c_v	1.121	1.107	1.10	1.30
Gas constant R, ft-lb/lb·deg F				35.1
Gas constant R, J/kg·deg C				189.
Thermal conductivity, Btu/hr·ft·deg F				0.01
Thermal conductivity, W/m·deg C				0.017
Boiling point (sat 14.7 psia), deg F	38.6	33.6	19.2	−109.4[b]
Boiling point (sat 760 mm), deg C	3.7	0.9	−7.1	−78.5
Latent heat of evap (at bp), Btu/lb	178.9	174.4	169.	246.
Latent heat of evap (at bp), J/kg	416 000.	406 000.	393 000.	572 000.
Freezing (melting) point, deg F (1 atm)	−218.	−158.		
Freezing (melting) point, deg C (1 atm)	−138.9	−105.5		
Latent heat of fusion, Btu/lb	31.2	41.6	25.3	—
Latent heat of fusion, J/kg	72 600.	96 800.	58 800.	—
Critical temperature, deg F				88.
Critical temperature, deg C	160.	155.		31.
Critical pressure, psia	595.	610.		1 072.
Critical pressure, MN/m²	4.10	4.20		7.4
Critical volume, ft³/lb				
Critical volume, m³/kg				
Flammable (yes or no)	Yes	Yes	Yes	No
Heat of combustion, Btu/ft³	3 150.	3 150.	3 150.	—
Heat of combustion, Btu/lb	21 000.	21 000.	21 000.	—
Heat of combustion, kJ/kg	48 800.	48 800.	48 800.	—

[a] For N·sec/m² divide by 1 000.
[b] Sublimes.

TABLE A.7 (continued) Chemical, Physical, and Thermal Properties of Gases: Gases and Vapors, Including Fuels and Refrigerants, English and Metric Units

Common name(s)	*Carbon monoxide*	*Chlorine*	*Deuterium*	*Ethane*
Chemical formula	*CO*	Cl_2	D_2	C_2H_6
Refrigerant number	—	—	—	*170*
CHEMICAL AND PHYSICAL PROPERTIES				
Molecular weight	28.011	70.906	2.014	30.070
Specific gravity, air = 1	0.967	2.45	0.070	1.04
Specific volume, ft^3/lb	14.0	5.52	194.5	13.025
Specific volume, m^3/kg	0.874	0.344	12.12	0.815
Density of liquid (at atm bp), lb/ft^3		97.3		28.
Density of liquid (at atm bp), kg/m^3		1 559.		449.
Vapor pressure at 25 deg C, psia			0.756	
Vapor pressure at 25 deg C, MN/m^2			0.005 2	
Viscosity (abs), lbm/ft·sec	12.1×10^{-6}	9.4×10^{-6}	8.75×10^{-6}	$64. \times 10^{-6}$
Viscosity (abs), centipoises[a]	0.018	0.014	0.013	0.095
Sound velocity in gas, m/sec	352.	215.	930.	316.
THERMAL AND THERMODYNAMIC PROPERTIES				
Specific heat, c_p, Btu/lb·deg F or cal/g·deg C	0.25	0.114	1.73	0.41
Specific heat, c_p, J/kg·K	1 046.	477.	7 238.	1 715.
Specific heat ratio, c_p/c_v	1.40	1.35	1.40	1.20
Gas constant R, ft-lb/lb·deg F	55.2	21.8	384.	51.4
Gas constant R, J/kg·deg C	297.	117.	2 066.	276.
Thermal conductivity, Btu/hr·ft·deg F	0.014	0.005	0.081	0.010
Thermal conductivity, W/m·deg C	0.024	0.008 7	0.140	0.017
Boiling point (sat 14.7 psia), deg F	−312.7	−29.2		−127.
Boiling point (sat 760 mm), deg C	−191.5	−34.		−88.3
Latent heat of evap (at bp), Btu/lb	92.8	123.7		210.
Latent heat of evap (at bp), J/kg	216 000.	288 000.		488 000.
Freezing (melting) point, deg F (1 atm)	−337.	−150.		−278.
Freezing (melting) point, deg C (1 atm)	−205.	−101.		−172.2
Latent heat of fusion, Btu/lb	12.8	41.0		41.
Latent heat of fusion, J/kg		95 400.		95 300.
Critical temperature, deg F	−220.	291.	−390.6	90.1
Critical temperature, deg C	−140.	144.	−234.8	32.2
Critical pressure, psia	507.	1 120.	241.	709.
Critical pressure, MN/m^2	3.49	7.72	1.66	4.89
Critical volume, ft^3/lb	0.053	0.028	0.239	0.076
Critical volume, m^3/kg	0.003 3	0.001 75	0.014 9	0.004 7
Flammable (yes or no)	Yes	No		Yes
Heat of combustion, Btu/ft^3	310.	—		
Heat of combustion, Btu/lb	4 340.	—		22 300.
Heat of combustion, kJ/kg	10 100.	—		51 800.

[a]For $N{\cdot}sec/m^2$ divide by 1 000.

TABLE A.7 (continued) Chemical, Physical, and Thermal Properties of Gases: Gases and Vapors, Including Fuels and Refrigerants, English and Metric Units

Common name(s) *Chemical formula* *Refrigerant number*	*Ethyl chloride* C_2H_5Cl *160*	*Ethylene (Ethene)* C_2H_4 *1 150*	*Fluorine* F_2 —
CHEMICAL AND PHYSICAL PROPERTIES			
Molecular weight	64.515	28.054	37.996
Specific gravity, air = 1	2.23	0.969	1.31
Specific volume, ft³/lb	6.07	13.9	10.31
Specific volume, m³/kg	0.378	0.87	0.706
Density of liquid (at atm bp), lb/ft³	56.5	35.5	
Density of liquid (at atm bp), kg/m³	905.	569.	
Vapor pressure at 25 deg C, psia			
Vapor pressure at 25 deg C, MN/m²			
Viscosity (abs), lbm/ft·sec		6.72×10^{-6}	16.1×10^{-6}
Viscosity (abs), centipoises[a]		0.010	0.024
Sound velocity in gas, m/sec	204.	331.	290.
THERMAL AND THERMODYNAMIC PROPERTIES			
Specific heat, c_p, Btu/lb·deg F or cal/g·deg C	0.27	0.37	0.198
Specific heat, c_p, J/kg·K	1 130.	1 548.	828.
Specific heat ratio, c_p/c_v	1.13	1.24	1.35
Gas constant R, ft-lb/lb·deg F	24.0	55.1	40.7
Gas constant R, J/kg·deg C	129.	296.	219.
Thermal conductivity, Btu/hr·ft·deg F		0.010	0.016
Thermal conductivity, W/m·deg C		0.017	0.028
Boiling point (sat 14.7 psia), deg F	54.	−155.	−306.4
Boiling point (sat 760 mm), deg C	12.2	−103.8	−188.
Latent heat of evap (at bp), Btu/lb	166.	208.	74.
Latent heat of evap (at bp), J/kg	386 000.	484 000.	172 000.
Freezing (melting) point, deg F (1 atm)	−218.	−272.	−364.
Freezing (melting) point, deg C (1 atm)	−138.9	−169.	−220.
Latent heat of fusion, Btu/lb	29.3	51.5	11.
Latent heat of fusion, J/kg	68 100.	120 000.	25 600.
Critical temperature, deg F	368.6	49.	−200
Critical temperature, deg C	187.	9.5	−129.
Critical pressure, psia	764.	741.	810.
Critical pressure, MN/m²	5.27	5.11	5.58
Critical volume, ft³/lb	0.049	0.073	
Critical volume, m³/kg	0.003 06	0.004 6	
Flammable (yes or no)	No	Yes	
Heat of combustion, Btu/ft³	—	1 480.	
Heat of combustion, Btu/lb	—	20 600.	
Heat of combustion, kJ/kg	—	47 800.	

[a]For N·sec/m² divide by 1 000.

TABLE A.7 (continued) Chemical, Physical, and Thermal Properties of Gases: Gases and Vapors, Including Fuels and Refrigerants, English and Metric Units

Common name(s)	*Fluorocarbons*			
Chemical formula	CCl_3F	CCl_2F_2	$CClF_3$	$CBrF_3$
Refrigerant number	*11*	*12*	*13*	*13B1*
CHEMICAL AND PHYSICAL PROPERTIES				
Molecular weight	137.37	120.91	104.46	148.91
Specific gravity, air = 1	4.74	4.17	3.61	5.14
Specific volume, ft^3/lb	2.74	3.12	3.58	2.50
Specific volume, m^3/kg	0.171	0.195	0.224	0.975
Density of liquid (at atm bp), lb/ft^3	92.1	93.0	95.0	124.4
Density of liquid (at atm bp), kg/m^3	1 475.	1 490.	1 522.	1 993.
Vapor pressure at 25 deg C, psia		94.51	516.	234.8
Vapor pressure at 25 deg C, MN/m^2		0.652	3.56	1.619
Viscosity (abs), lbm/ft·sec	7.39×10^{-6}	8.74×10^{-6}		
Viscosity (abs), centipoises[a]	0.011	0.013		
Sound velocity in gas, m/sec				
THERMAL AND THERMODYNAMIC PROPERTIES				
Specific heat, c_p, Btu/lb·deg F or cal/g·deg C	0.14	0.146	0.154	
Specific heat, c_p, J/kg·K	586.	611.	644.	
Specific heat ratio, c_p/c_v	1.14	1.14	1.145	
Gas constant R, ft-lb/lb·deg F				
Gas constant R, J/kg·deg C				
Thermal conductivity, Btu/hr·ft·deg F	0.005	0.006		
Thermal conductivity, W/m·deg C	0.008 7	0.010 4		
Boiling point (sat 14.7 psia), deg F	74.9	−21.8	−114.6	−72.
Boiling point (sat 760 mm), deg C	23.8	−29.9	−81.4	−57.8
Latent heat of evap (at bp), Btu/lb	77.5	71.1	63.0	51.1
Latent heat of evap (at bp), J/kg	180 000.	165 000.	147 000.	119 000.
Freezing (melting) point, deg F (1 atm)	−168.	−252.	−294.	−270.
Freezing (melting) point, deg C (1 atm)	−111.	−157.8	−181.1	−167.8
Latent heat of fusion, Btu/lb				
Latent heat of fusion, J/kg				
Critical temperature, deg F	388.4	233.	83.9	152.
Critical temperature, deg C	198.	111.7	28.8	66.7
Critical pressure, psia	635.	582.	559.	573.
Critical pressure, MN/m^2	4.38	4.01	3.85	3.95
Critical volume, ft^3/lb	0.028 9	0.287	0.027 7	0.021 5
Critical volume, m^3/kg	0.001 80	0.018	0.001 73	0.001 34
Flammable (yes or no)	No	No	No	No
Heat of combustion, Btu/ft^3	—	—	—	—
Heat of combustion, Btu/lb	—	—	—	—
Heat of combustion, kJ/kg	—	—	—	—

[a]For $N \cdot sec/m^2$ divide by 1 000.

TABLE A.7 (continued) Chemical, Physical, and Thermal Properties of Gases: Gases and Vapors, Including Fuels and Refrigerants, English and Metric Units

Common name(s)	*Fluorocarbons*			
Chemical formula	CF_4	$CHCl_2F$	$CHClF_2$	$C_2Cl_2F_4$
Refrigerant number	*14*	*21*	*22*	*114*
CHEMICAL AND PHYSICAL PROPERTIES				
Molecular weight	88.00	102.92	86.468	170.92
Specific gravity, air = 1	3.04	3.55	2.99	5.90
Specific volume, ft^3/lb	4.34	3.7	4.35	2.6
Specific volume, m^3/kg	0.271	0.231	0.271	0.162
Density of liquid (at atm bp), lb/ft^3	102.0	87.7	88.2	94.8
Density of liquid (at atm bp), kg/m^3	1 634.	1 405.	1 413.	1 519.
Vapor pressure at 25 deg C, psia		26.4	151.4	30.9
Vapor pressure at 25 deg C, MN/m^2		0.182	1.044	0.213
Viscosity (abs), lbm/ft·sec		8.06×10^{-6}	8.74×10^{-6}	8.06×10^{-6}
Viscosity (abs), centipoises[a]		0.012	0.013	0.012
Sound velocity in gas, m/sec				
THERMAL AND THERMO-DYNAMIC PROPERTIES				
Specific heat, c_p, Btu/lb·deg F or cal/g·deg C		0.139	0.157	0.158
Specific heat, c_p, J/kg·K		582.	657.	661.
Specific heat ratio, c_p/c_v		1.18	1.185	1.09
Gas constant R, ft-lb/lb·deg F				
Gas constant R, J/kg·deg C				
Thermal conductivity, Btu/hr·ft·deg F			0.007	0.006
Thermal conductivity, W/m·deg C			0.012	0.010
Boiling point (sat 14.7 psia), deg F	−198.2	48.1	−41.3	38.4
Boiling point (sat 760 mm), deg C	−127.9	9.0	−40.7	3.55
Latent heat of evap (at bp), Btu/lb	58.5	104.1	100.4	58.4
Latent heat of evap (at bp), J/kg	136 000.	242 000.	234 000.	136 000.
Freezing (melting) point, deg F (1 atm)	−299.	−211.	−256.	−137.
Freezing (melting) point, deg C (1 atm)	−183.8	−135.	−160.	−93.8
Latent heat of fusion, Btu/lb	2.53			
Latent heat of fusion, J/kg	5 880.			
Critical temperature, deg F	−49.9	353.3	204.8	294.
Critical temperature, deg C	−45.5	178.5	96.5	
Critical pressure, psia	610.	750.	715.	475.
Critical pressure, MN/m^2	4.21	5.17	4.93	3.28
Critical volume, ft^3/lb	0.025	0.030 7	0.030 5	0.027 5
Critical volume, m^3/kg	0.001 6	0.001 91	0.001 90	0.001 71
Flammable (yes or no)	No	No	No	No
Heat of combustion, Btu/ft^3	—	—	—	—
Heat of combustion, Btu/lb	—	—	—	—
Heat of combustion, kJ/kg	—	—	—	—

[a]For N·sec/m^2 divide by 1 000.

TABLE A.7 (continued) Chemical, Physical, and Thermal Properties of Gases: Gases and Vapors, Including Fuels and Refrigerants, English and Metric Units

Common name(s)	*Fluorocarbons*			*Helium*
Chemical formula	C_2ClF_5	$C_2H_3ClF_2$	$C_2H_4F_2$	*He*
Refrigerant number	*115*	*142b*	*152a*	*704*
CHEMICAL AND PHYSICAL PROPERTIES				
Molecular weight	154.47	100.50	66.05	4.002 6
Specific gravity, air = 1	5.33	3.47	2.28	0.138
Specific volume, ft^3/lb	2.44	3.7	5.9	97.86
Specific volume, m^3/kg	0.152	0.231	0.368	6.11
Density of liquid (at atm bp), lb/ft^3	96.5	74.6	62.8	7.80
Density of liquid (at atm bp), kg/m^3	1 546.	1 195.	1 006.	125.
Vapor pressure at 25 deg C, psia	132.1	49.1	86.8	
Vapor pressure at 25 deg C, MN/m^2	0.911	0.338 5	0.596	
Viscosity (abs), lbm/ft·sec				13.4×10^{-6}
Viscosity (abs), centipoises[a]				0.02
Sound velocity in gas, m/sec				1 015.
THERMAL AND THERMO-DYNAMIC PROPERTIES				
Specific heat, c_p, Btu/lb·deg F or cal/g·deg C	0.161			1.24
Specific heat, c_p, J/kg·K	674.			5 188.
Specific heat ratio, c_p/c_v	1.091			1.66
Gas constant R, ft-lb/lb·deg F				386.
Gas constant R, J/kg·deg C				2 077.
Thermal conductivity, Btu/hr·ft·deg F				0.086
Thermal conductivity, W/m·deg C				0.149
Boiling point (sat 14.7 psia), deg F	−38.0	14.	−13.	−452.
Boiling point (sat 760 mm), deg C	−38.9	−10.0	−25.0	4.22 K
Latent heat of evap (at bp), Btu/lb	53.4	92.5	137.1	10.0
Latent heat of evap (at bp), J/kg	124 000.	215 000.	319 000.	23 300.
Freezing (melting) point, deg F (1 atm)	−149.			[b]
Freezing (melting) point, deg C (1 atm)	−100.6			—
Latent heat of fusion, Btu/lb				—
Latent heat of fusion, J/kg				—
Critical temperature, deg F	176.		387.	−450.3
Critical temperature, deg C				5.2 K
Critical pressure, psia	457.6			33.22
Critical pressure, MN/m^2	3.155			
Critical volume, ft^3/lb	0.026 1			0.231
Critical volume, m^3/kg	0.001 63			0.014 4
Flammable (yes or no)	No	No	No	No
Heat of combustion, Btu/ft^3	—	—	—	—
Heat of combustion, Btu/lb	—	—	—	—
Heat of combustion, kJ/kg	—	—	—	—

[a] For N·sec/m^2 divide by 1 000.
[b] Helium cannot be solidified at atmospheric pressure.

TABLE A.7 (continued) Chemical, Physical, and Thermal Properties of Gases: Gases and Vapors, Including Fuels and Refrigerants, English and Metric Units

Common name(s)	*Hydrogen*	*Hydrogen chloride*	*Hydrogen sulfide*	*Krypton*
Chemical formula	H_2	HCl	H_2S	Kr
Refrigerant number	*702*	—	—	—
CHEMICAL AND PHYSICAL PROPERTIES				
Molecular weight	2.016	36.461	34.076	83.80
Specific gravity, air = 1	0.070	1.26	1.18	2.89
Specific volume, ft^3/lb	194.	10.74	11.5	4.67
Specific volume, m^3/kg	12.1	0.670	0.093 0	0.291
Density of liquid (at atm bp), lb/ft^3	4.43	74.4	62.	150.6
Density of liquid (at atm bp), kg/m^3	71.0	1 192.	993.	2 413.
Vapor pressure at 25 deg C, psia				
Vapor pressure at 25 deg C, MN/m^2				
Viscosity (abs), lbm/ft·sec	6.05×10^{-6}	10.1×10^{-6}	8.74×10^{-6}	16.8×10^{-6}
Viscosity (abs), centipoises[a]	0.009	0.015	0.013	0.025
Sound velocity in gas, m/sec	1 315.	310.	302.	223.
THERMAL AND THERMODYNAMIC PROPERTIES				
Specific heat, c_p, Btu/lb·deg F or cal/g·deg C	3.42	0.194	0.23	0.059
Specific heat, c_p, J/kg·K	14 310.	812.	962.	247.
Specific heat ratio, c_p/c_v	1.405	1.39	1.33	1.68
Gas constant R, ft-lb/lb·deg F	767.	42.4	45.3	18.4
Gas constant R, J/kg·deg C	4 126.	228.	244.	99.0
Thermal conductivity, Btu/hr·ft·deg F	0.105	0.008	0.008	0.005 4
Thermal conductivity, W/m·deg C	0.018 2	0.014	0.014	0.009 3
Boiling point (sat 14.7 psia), deg F	−423.	−121.	−76.	−244.
Boiling point (sat 760 mm), deg C	20.4 K	−85.	−60.	−153.
Latent heat of evap (at bp), Btu/lb	192.	190.5	234.	46.4
Latent heat of evap (at bp), J/kg	447 000.	443 000.	544 000.	108 000.
Freezing (melting) point, deg F (1 atm)	−434.6	−169.6	−119.2	−272.
Freezing (melting) point, deg C (1 atm)	−259.1	−112.	−84.	−169.
Latent heat of fusion, Btu/lb	25.0	23.4	30.2	4.7
Latent heat of fusion, J/kg	58 000.	54 400.	70 200.	10 900.
Critical temperature, deg F	−399.8	124.	213.	
Critical temperature, deg C	−240.0	51.2	100.4	−63.8
Critical pressure, psia	189.	1 201.	1 309.	800.
Critical pressure, MN/m^2	1.30	8.28	9.02	5.52
Critical volume, ft^3/lb	0.53	0.038	0.046	0.017 7
Critical volume, m^3/kg	0.033	0.002 4	0.002 9	0.001 1
Flammable (yes or no)	Yes	No	Yes	No
Heat of combustion, Btu/ft^3	320.	—	700.	—
Heat of combustion, Btu/lb	62 050.	—	8 000.	—
Heat of combustion, kJ/kg	144 000.	—	18 600.	—

[a]For N·sec/m^2 divide by 1 000.

TABLE A.7 (continued) Chemical, Physical, and Thermal Properties of Gases: Gases and Vapors, Including Fuels and Refrigerants, English and Metric Units

Common name(s)	Methane	Methyl chloride	Neon	Nitric oxide
Chemical formula	CH_4	CH_3Cl	Ne	NO
Refrigerant number	50	40	720	—
CHEMICAL AND PHYSICAL PROPERTIES				
Molecular weight	16.044	50.488	20.179	30.006
Specific gravity, air = 1	0.554	1.74	0.697	1.04
Specific volume, ft^3/lb	24.2	7.4	19.41	13.05
Specific volume, m^3/kg	1.51	0.462	1.211	0.814
Density of liquid (at atm bp), lb/ft^3	26.3	62.7	75.35	
Density of liquid (at atm bp), kg/m^3	421.	1 004.	1 207.	
Vapor pressure at 25 deg C, psia		82.2		
Vapor pressure at 25 deg C, MN/m^2		0.567		
Viscosity (abs), lbm/ft·sec	7.39×10^{-6}	7.39×10^{-6}	21.5×10^{-6}	12.8×10^{-6}
Viscosity (abs), centipoises[a]	0.011	0.011	0.032	0.019
Sound velocity in gas, m/sec	446.	251.	454.	341.
THERMAL AND THERMO-DYNAMIC PROPERTIES				
Specific heat, c_p, Btu/lb·deg F or cal/g·deg C	0.54	0.20	0.246	0.235
Specific heat, c_p, J/kg·K	2 260.	837.	1 030.	983.
Specific heat ratio, c_p/c_v	1.31	1.28	1.64	1.40
Gas constant R, ft-lb/lb·deg F	96.	30.6	76.6	51.5
Gas constant R, J/kg·deg C	518.	165.	412.	277.
Thermal conductivity, Btu/hr·ft·deg F	0.02	0.006	0.028	0.015
Thermal conductivity, W/m·deg C	0.035	0.010	0.048	0.026
Boiling point (sat 14.7 psia), deg F	−259.	−10.7	−410.9	−240.
Boiling point (sat 760 mm), deg C	−434.2	−23.7	−246.	−151.5
Latent heat of evap (at bp), Btu/lb	219.2	184.1	37.	
Latent heat of evap (at bp), J/kg	510 000.	428 000.	86 100.	
Freezing (melting) point, deg F (1 atm)	−296.6	−144.	−415.6	−258.
Freezing (melting) point, deg C (1 atm)	−182.6	−97.8	−248.7	−161.
Latent heat of fusion, Btu/lb	14.	56.	6.8	32.9
Latent heat of fusion, J/kg	32 600.	130 000.	15 800.	76 500.
Critical temperature, deg F	−116.	289.4	−379.8	−136.
Critical temperature, deg C	−82.3	143.	−228.8	−93.3
Critical pressure, psia	673.	968.	396.	945.
Critical pressure, MN/m^2	4.64	6.67	2.73	6.52
Critical volume, ft^3/lb	0.099	0.043	0.033	0.033 2
Critical volume, m^3/kg	0.006 2	0.002 7	0.002 0	0.002 07
Flammable (yes or no)	Yes	Yes	No	No
Heat of combustion, Btu/ft^3	985.		—	—
Heat of combustion, Btu/lb	2 290.		—	—
Heat of combustion, kJ/kg			—	—

[a]For N·sec/m^2 divide by 1 000.

TABLE A.7 (continued) Chemical, Physical, and Thermal Properties of Gases: Gases and Vapors, Including Fuels and Refrigerants, English and Metric Units

Common name(s)	*Nitrogen*	*Nitrous oxide*	*Oxygen*	*Ozone*
Chemical formula	N_2	N_2O	O_2	O_3
Refrigerant number	*728*	*744A*	*732*	—
CHEMICAL AND PHYSICAL PROPERTIES				
Molecular weight	28.013 4	44.012	31.998 8	47.998
Specific gravity, air = 1	0.967	1.52	1.105	1.66
Specific volume, ft^3/lb	13.98	8.90	12.24	8.16
Specific volume, m^3/kg	0.872	0.555	0.764	0.509
Density of liquid (at atm bp), lb/ft^3	50.46	76.6	71.27	
Density of liquid (at atm bp), kg/m^3	808.4	1 227.	1 142.	
Vapor pressure at 25 deg C, psia				
Vapor pressure at 25 deg C, MN/m^2				
Viscosity (abs), lbm/ft·sec	12.1×10^{-6}	10.1×10^{-6}	13.4×10^{-6}	8.74×10^{-6}
Viscosity (abs), centipoises[a]	0.018	0.015	0.020	0.013
Sound velocity in gas, m/sec	353.	268.	329.	
THERMAL AND THERMODYNAMIC PROPERTIES				
Specific heat, c_p, Btu/lb·deg F or cal/g·deg C	0.249	0.21	0.220	0.196
Specific heat, c_p, J/kg·K	1 040.	879.	920.	820.
Specific heat ratio, c_p/c_v	1.40	1.31	1.40	
Gas constant R, ft-lb/lb·deg F	55.2	35.1	48.3	32.2
Gas constant R, J/kg·deg C	297.	189.	260.	173.
Thermal conductivity, Btu/hr·ft·deg F	0.015	0.010	0.015	0.019
Thermal conductivity, W/m·deg C	0.026	0.017	0.026	0.033
Boiling point (sat 14.7 psia), deg F	−320.4	−127.3	−297.3	−170.
Boiling point (sat 760 mm), deg C	−195.8	−88.5	−182.97	−112.
Latent heat of evap (at bp), Btu/lb	85.5	161.8	91.7	
Latent heat of evap (at bp), J/kg	199 000.	376 000.	213 000.	
Freezing (melting) point, deg F (1 atm)	−346.	−131.5	−361.1	−315.5
Freezing (melting) point, deg C (1 atm)	−210.	−90.8	−218.4	−193.
Latent heat of fusion, Btu/lb	11.1	63.9	5.9	97.2
Latent heat of fusion, J/kg	25 800.	149 000.	13 700.	226 000.
Critical temperature, deg F	−232.6	97.7	−181.5	16.
Critical temperature, deg C	−147.	36.5	−118.6	−9.
Critical pressure, psia	493.	1 052.	726.	800.
Critical pressure, MN/m^2	3.40	7.25	5.01	5.52
Critical volume, ft^3/lb	0.051	0.036	0.040	0.029 8
Critical volume, m^3/kg	0.003 18	0.002 2	0.002 5	0.001 86
Flammable (yes or no)	No	No	No	No
Heat of combustion, Btu/ft^3	—	—	—	—
Heat of combustion, Btu/lb	—	—	—	—
Heat of combustion, kJ/kg	—	—	—	—

[a] For N·sec/m^2 divide by 1 000.

TABLE A.7 (continued) Chemical, Physical, and Thermal Properties of Gases: Gases and Vapors, Including Fuels and Refrigerants, English and Metric Units

Common name(s)	*Propane*	*Propylene (Propene)*	*Sulfur dioxide*	*Xenon*
Chemical formula	C_3H_8	C_3H_6	SO_2	*Xe*
Refrigerant number	*290*	*1 270*	*764*	—
CHEMICAL AND PHYSICAL PROPERTIES				
Molecular weight	44.097	42.08	64.06	131.30
Specific gravity, air = 1	1.52	1.45	2.21	4.53
Specific volume, ft^3/lb	8.84	9.3	6.11	2.98
Specific volume, m^3/kg	0.552	0.58		
Density of liquid (at atm bp), lb/ft^3	36.2	37.5	42.8	190.8
Density of liquid (at atm bp), kg/m^3	580.	601.	585.	3 060.
Vapor pressure at 25 deg C, psia	135.7	166.4	56.6	
Vapor pressure at 25 deg C, MN/m^2	0.936	1.147	0.390	
Viscosity (abs), lbm/ft·sec	53.8×10^{-6}	57.1×10^{-6}	8.74×10^{-6}	15.5×10^{-6}
Viscosity (abs), centipoises[a]	0.080	0.085	0.013	0.023
Sound velocity in gas, m/sec	253.	261.	220.	177.
THERMAL AND THERMODYNAMIC PROPERTIES				
Specific heat, c_p, Btu/lb·deg F or cal/g·deg C	0.39	0.36	0.11	0.115
Specific heat, c_p, J/kg·K	1 630.	1 506.	460.	481.
Specific heat ratio, c_p/c_v	1.2	1.16	1.29	1.67
Gas constant R, ft-lb/lb·deg F	35.0	36.7	24.1	11.8
Gas constant R, J/kg·deg C	188.	197.	130.	63.5
Thermal conductivity, Btu/hr·ft·deg F	0.010	0.010	0.006	0.003
Thermal conductivity, W/m·deg C	0.017	0.017	0.010	0.005 2
Boiling point (sat 14.7 psia), deg F	−44.	−54.	14.0	−162.5
Boiling point (sat 760 mm), deg C	−42.2	−48.3	−10.	−108.
Latent heat of evap (at bp), Btu/lb	184.	188.2	155.5	41.4
Latent heat of evap (at bp), J/kg	428 000.	438 000.	362 000.	96 000.
Freezing (melting) point, deg F (1 atm)	−309.8	−301.	−104.	−220.
Freezing (melting) point, deg C (1 atm)	−189.9	−185.	−75.5	−140.
Latent heat of fusion, Btu/lb	19.1		58.0	10.
Latent heat of fusion, J/kg	44 400.		135 000.	23 300.
Critical temperature, deg F	205.	197.	315.5	61.9
Critical temperature, deg C	96.	91.7	157.6	16.6
Critical pressure, psia	618.	668.	1 141.	852.
Critical pressure, MN/m^2	4.26	4.61	7.87	5.87
Critical volume, ft^3/lb	0.073	0.069	0.03	0.014 5
Critical volume, m^3/kg	0.004 5	0.004 3	0.001 9	0.000 90
Flammable (yes or no)	Yes	Yes	No	No
Heat of combustion, Btu/ft^3	2 450.	2 310.	—	—
Heat of combustion, Btu/lb	21 660.	21 500.	—	—
Heat of combustion, kJ/kg	50 340.	50 000.	—	—

[a]For N·sec/m^2 divide by 1 000.

TABLE A.8 Ideal Gas Properties of Air

Part a. SI Units

T(K), h and u(kJ/kg), $s°$(kJ/kg·K)

T	h	p_r	u	v_r	$s°$	T	h	p_r	u	v_r	$s°$
200	199.97	0.3363	142.56	1707.	1.29559	450	451.80	5.775	322.62	223.6	2.11161
210	209.97	0.3987	149.69	1512.	1.34444	460	462.02	6.245	329.97	211.4	2.13407
220	219.97	0.4690	156.82	1346.	1.39105	470	472.24	6.742	337.32	200.1	2.15604
230	230.02	0.5477	164.00	1205.	1.43557	480	482.49	7.268	344.70	189.5	2.17760
240	240.02	0.6355	171.13	1084.	1.47824	490	492.74	7.824	352.08	179.7	2.19876
250	250.05	0.7329	178.28	979.	1.51917	500	503.02	8.411	359.49	170.6	2.21952
260	260.09	0.8405	185.45	887.8	1.55848	510	513.32	9.031	366.92	162.1	2.23993
270	270.11	0.9590	192.60	808.0	1.59634	520	523.63	9.684	374.36	154.1	2.25997
280	280.13	1.0889	199.75	738.0	1.63279	530	533.98	10.37	381.84	146.7	2.27967
285	285.14	1.1584	203.33	706.1	1.65055	540	544.35	11.10	389.34	139.7	2.29906
290	290.16	1.2311	206.91	676.1	1.66802	550	554.74	11.86	396.86	133.1	2.31809
295	295.17	1.3068	210.49	647.9	1.68515	560	565.17	12.66	404.42	127.0	2.33685
300	300.19	1.3860	214.07	621.2	1.70203	570	575.59	13.50	411.97	121.2	2.35531
305	305.22	1.4686	217.67	596.0	1.71865	580	586.04	14.38	419.55	115.7	2.37348
310	310.24	1.5546	221.25	572.3	1.73498	590	596.52	15.31	427.15	110.6	2.39140
315	315.27	1.6442	224.85	549.8	1.75106	600	607.02	16.28	434.78	105.8	2.40902
320	320.29	1.7375	228.42	528.6	1.76690	610	617.53	17.30	442.42	101.2	2.42644
325	325.31	1.8345	232.02	508.4	1.78249	620	628.07	18.36	450.09	96.92	2.44356
330	330.34	1.9352	235.61	489.4	1.79783	630	638.63	19.84	457.78	92.84	2.46048
340	340.42	2.149	242.82	454.1	1.82790	640	649.22	20.64	465.50	88.99	2.47716
350	350.49	2.379	250.02	422.2	1.85708	650	659.84	21.86	473.25	85.34	2.49364
360	360.58	2.626	257.24	393.4	1.88543	660	670.47	23.13	481.01	81.89	2.50985
370	370.67	2.892	264.46	367.2	1.91313	670	681.14	24.46	488.81	78.61	2.52589
380	380.77	3.176	271.69	343.4	1.94001	680	691.82	25.85	496.62	75.50	2.54175
390	390.88	3.481	278.93	321.5	1.96633	690	702.52	27.29	504.45	72.56	2.55731
400	400.98	3.806	286.16	301.6	1.99194	700	713.27	28.80	512.33	69.76	2.57277
410	411.12	4.153	293.43	283.3	2.01699	710	724.04	30.38	520.23	67.07	2.58810
420	421.26	4.522	300.69	266.6	2.04142	720	734.82	32.02	528.14	64.53	2.60319
430	431.43	4.915	307.99	251.1	2.06533	730	745.62	33.72	536.07	62.13	2.61803
440	441.61	5.332	315.30	236.8	2.08870	740	756.44	35.50	544.02	59.82	2.63280

TABLE A.8 (continued) Ideal Gas Properties of Air

T(K), h and u(kJ/kg), $s°$(kJ/kg·K)

T	h	p_r	u	v_r	$s°$	T	h	p_r	u	v_r	$s°$
750	767.29	37.35	551.99	57.63	2.64737	1300	1395.97	330.9	1022.82	11.275	3.27345
760	778.18	39.27	560.01	55.54	2.66176	1320	1419.76	352.5	1040.88	10.747	3.29160
770	789.11	41.31	568.07	53.39	2.67595	1340	1443.60	375.3	1058.94	10.247	3.30959
780	800.03	43.35	576.12	51.64	2.69013	1360	1467.49	399.1	1077.10	9.780	3.32724
790	810.99	45.55	584.21	49.86	2.70400	1380	1491.44	424.2	1095.26	9.337	3.34474
800	821.95	47.75	592.30	48.08	2.71787	1400	1515.42	450.5	1113.52	8.919	3.36200
820	843.98	52.59	608.59	44.84	2.74504	1420	1539.44	478.0	1131.77	8.526	3.37901
840	866.08	57.60	624.95	41.85	2.77170	1440	1563.51	506.9	1150.13	8.153	3.39586
860	888.27	63.09	641.40	39.12	2.79783	1460	1587.63	537.1	1168.49	7.801	3.41247
880	910.56	68.98	657.95	36.61	2.82344	1480	1611.79	568.8	1186.95	7.468	3.42892
900	932.93	75.29	674.58	34.31	2.84856	1500	1635.97	601.9	1205.41	7.152	3.44516
920	955.38	82.05	691.28	32.18	2.87324	1520	1660.23	636.5	1223.87	6.854	3.46120
940	977.92	89.28	708.08	30.22	2.89748	1540	1684.51	672.8	1242.43	6.569	3.47712
960	1000.55	97.00	725.02	28.40	2.92128	1560	1708.82	710.5	1260.99	6.301	3.49276
980	1023.25	105.2	741.98	26.73	2.94468	1580	1733.17	750.0	1279.65	6.046	3.50829
1000	1046.04	114.0	758.94	25.17	2.96770	1600	1757.57	791.2	1298.30	5.804	3.52364
1020	1068.89	123.4	776.10	23.72	2.99034	1620	1782.00	834.1	1316.96	5.574	3.53879
1040	1091.85	133.3	793.36	22.39	3.01260	1640	1806.46	878.9	1335.72	5.355	3.55381
1060	1114.86	143.9	810.62	21.14	3.03449	1660	1830.96	925.6	1354.48	5.147	3.56867
1080	1137.89	155.2	827.88	19.98	3.05608	1680	1855.50	974.2	1373.24	4.949	3.58335
1100	1161.07	167.1	845.33	18.896	3.07732	1700	1880.1	1025	1392.7	4.761	3.5979
1120	1184.28	179.7	862.79	17.886	3.09825	1750	1941.6	1161	1439.8	4.328	3.6336
1140	1207.57	193.1	880.35	16.946	3.11883	1800	2003.3	1310	1487.2	3.944	3.6684
1160	1230.92	207.2	897.91	16.064	3.13916	1850	2065.3	1475	1534.9	3.601	3.7023
1180	1254.34	222.2	915.57	15.241	3.15916	1900	2127.4	1655	1582.6	3.295	3.7354
1200	1277.79	238.0	933.33	14.470	3.17888	1950	2189.7	1852	1630.6	3.022	3.7677
1220	1301.31	254.7	951.09	13.747	3.19834	2000	2252.1	2068	1678.7	2.776	3.7994
1240	1324.93	272.3	968.95	13.069	3.21751	2050	2314.6	2303	1726.8	2.555	3.8303
1260	1348.55	290.8	986.90	12.435	3.23638	2100	2377.4	2559	1775.3	2.356	3.8605
1280	1372.24	310.4	1004.76	11.835	3.25510	2150	2440.3	2837	1823.8	2.175	3.8901
						2200	2503.2	3138	1872.4	2.012	3.9191
						2250	2566.4	3464	1921.3	1.864	3.9474

TABLE A.8 (continued) Ideal Gas Properties of Air

Part b. English Units

T(°R), h and u(Btu/lb), s°(Btu/lb · °R)

T	h	p_r	u	v_r	s°	T	h	p_r	u	v_r	s°
360	85.97	0.3363	61.29	396.6	0.50369	940	226.11	9.834	161.68	35.41	0.73509
380	90.75	0.4061	64.70	346.6	0.51663	960	231.06	10.61	165.26	33.52	0.74030
400	95.53	0.4858	68.11	305.0	0.52890	980	236.02	11.43	168.83	31.76	0.74540
420	100.32	0.5760	71.52	270.1	0.54058	1000	240.98	12.30	172.43	30.12	0.75042
440	105.11	0.6776	74.93	240.6	0.55172	1040	250.95	14.18	179.66	27.17	0.76019
460	109.90	0.7913	78.36	215.33	0.56235	1080	260.97	16.28	186.93	24.58	0.76964
480	114.69	0.9182	81.77	193.65	0.57255	1120	271.03	18.60	194.25	22.30	0.77880
500	119.48	1.0590	85.20	174.90	0.58233	1160	281.14	21.18	201.63	20.29	0.78767
520	124.27	1.2147	88.62	158.58	0.59172	1200	291.30	24.01	209.05	18.51	0.79628
537	128.34	1.3593	91.53	146.34	0.59945	1240	301.52	27.13	216.53	16.93	0.80466
540	129.06	1.3860	92.04	144.32	0.60078	1280	311.79	30.55	224.05	15.52	0.81280
560	133.86	1.5742	95.47	131.78	0.60950	1320	322.11	34.31	231.63	14.25	0.82075
580	138.66	1.7800	98.90	120.70	0.61793	1360	332.48	38.41	239.25	13.12	0.82848
600	143.47	2.005	102.34	110.88	0.62607	1400	342.90	42.88	246.93	12.10	0.83604
620	148.28	2.249	105.78	102.12	0.63395	1440	353.37	47.75	254.66	11.17	0.84341
640	153.09	2.514	109.21	94.30	0.64159	1480	363.89	53.04	262.44	10.34	0.85062
660	157.92	2.801	112.67	87.27	0.64902	1520	374.47	58.78	270.26	9.578	0.85767
680	162.73	3.111	116.12	80.96	0.65621	1560	385.08	65.00	278.13	8.890	0.86456
700	167.56	3.446	119.58	75.25	0.66321	1600	395.74	71.73	286.06	8.263	0.87130
720	172.39	3.806	123.04	70.07	0.67002	1650	409.13	80.89	296.03	7.556	0.87954
740	177.23	4.193	126.51	65.38	0.67665	1700	422.59	90.95	306.06	6.924	0.88758
760	182.08	4.607	129.99	61.10	0.68312	1750	436.12	101.98	316.16	6.357	0.89542
780	186.94	5.051	133.47	57.20	0.68942	1800	449.71	114.0	326.32	5.847	0.90308
800	191.81	5.526	136.97	53.63	0.69558	1850	463.37	127.2	336.55	5.388	0.91056
820	196.69	6.033	140.47	50.35	0.70160	1900	477.09	141.5	346.85	4.974	0.91788
840	201.56	6.573	143.98	47.34	0.70747	1950	490.88	157.1	357.20	4.598	0.92504
860	206.46	7.149	147.50	44.57	0.71323	2000	504.71	174.0	367.61	4.258	0.93205
880	211.35	7.761	151.02	42.01	0.71886	2050	518.61	192.3	378.08	3.949	0.93891
900	216.26	8.411	154.57	39.64	0.72438	2100	532.55	212.1	388.60	3.667	0.94564
920	221.18	9.102	158.12	37.44	0.72979	2150	546.54	233.5	399.17	3.410	0.95222

TABLE A.8 (continued) Ideal Gas Properties of Air

T(°R), *h* and *u* (Btu/lb), *s*° (Btu/lb · °R)

T	h	p_r	u	v_r	$s°$	T	h	p_r	u	v_r	$s°$
2200	560.59	256.6	409.78	3.176	0.95868	3700	998.11	2330	744.48	.5882	1.10991
2250	574.69	281.4	420.46	2.961	0.96501	3750	1013.1	2471	756.04	.5621	1.11393
2300	588.82	308.1	431.16	2.765	0.97123	3800	1028.1	2618	767.60	.5376	1.11791
2350	603.00	336.8	441.91	2.585	0.97732	3850	1043.1	2773	779.19	.5143	1.12183
2400	617.22	367.6	452.70	2.419	0.98331	3900	1058.1	2934	790.80	.4923	1.12571
2450	631.48	400.5	463.54	2.266	0.98919	3950	1073.2	3103	802.43	.4715	1.12955
2500	645.78	435.7	474.40	2.125	0.99497	4000	1088.3	3280	814.06	.4518	1.13334
2550	660.12	473.3	485.31	1.996	1.00064	4050	1103.4	3464	825.72	.4331	1.13709
2600	674.49	513.5	496.26	1.876	1.00623	4100	1118.5	3656	837.40	.4154	1.14079
2650	688.90	556.3	507.25	1.765	1.01172	4150	1133.6	3858	849.09	.3985	1.14446
2700	703.35	601.9	518.26	1.662	1.01712	4200	1148.7	4067	860.81	.3826	1.14809
2750	717.83	650.4	529.31	1.566	1.02244	4300	1179.0	4513	884.28	.3529	1.15522
2800	732.33	702.0	540.40	1.478	1.02767	4400	1209.4	4997	907.81	.3262	1.16221
2850	746.88	756.7	551.52	1.395	1.03282	4500	1239.9	5521	931.39	.3019	1.16905
2900	761.45	814.8	562.66	1.318	1.03788	4600	1270.4	6089	955.04	.2799	1.17575
2950	776.05	876.4	573.84	1.247	1.04288	4700	1300.9	6701	978.73	.2598	1.18232
3000	790.68	941.4	585.04	1.180	1.04779	4800	1331.5	7362	1002.5	.2415	1.18876
3050	805.34	1011	596.28	1.118	1.05264	4900	1362.2	8073	1026.3	.2248	1.19508
3100	820.03	1083	607.53	1.060	1.05741	5000	1392.9	8837	1050.1	.2096	1.20129
3150	834.75	1161	618.82	1.006	1.06212	5100	1423.6	9658	1074.0	.1956	1.20738
3200	849.48	1242	630.12	.9546	1.06676	5200	1454.4	10539	1098.0	.1828	1.21336
3250	864.24	1328	641.46	.9069	1.07134	5300	1485.3	11481	1122.0	.1710	1.21923
3300	879.02	1418	652.81	.8621	1.07585						
3350	893.83	1513	664.20	.8202	1.08031						
3400	908.66	1613	675.60	.7807	1.08470						
3450	923.52	1719	687.04	.7436	1.08904						
3500	938.40	1829	698.48	.7087	1.09332						
3550	953.30	1946	709.95	.6759	1.09755						
3600	968.21	2068	721.44	.6449	1.10172						
3650	983.15	2196	732.95	.6157	1.10584						

Source: Adapted from M.J. Moran and H.N. Shapiro, *Fundamentals of Engineering Thermodynamics*, 3rd. ed., Wiley, New York, 1995, as based on J.H. Keenan and J. Kaye, *Gas Tables*, Wiley, New York, 1945.

TABLE A.9 Equations for Gas Properties

Gas	Molar Mass M kg/kmol	Gas Constant R kJ/kg·K	Specific Heats at 25°C c_p kJ/kg·K	c_v kJ/kg·K	k	Temperature Range	Equation Coefficients for $c_p/R = a + bT + cT^2 + dT^3 + eT^4$ a	$b \times 10^3$ K^{-1}	$c \times 10^6$ K^{-2}	$d \times 10^{10}$ K^{-3}	$e \times 10^{13}$ K^{-4}	Critical State Properties p_c MPa	T_c K	Redlich–Kwong Constants a kPa·m^6·K$^{0.5}$/kmol^2	b m^3/kmol	Gas
Acetylene, C_2H_2	26.04	0.319	1.69	1.37	1.232	300–1000K	0.8021	23.51	−35.95	286.1	−87.64	6.14	308	8030	0.0362	Acetylene, C_2H_2
						1000–3000K	3.825	6.767	−3.014	6.931	−0.6469					
Air	28.97	0.287	1.01	0.718	1.400	300–1000K	3.721	−1.874	4.719	−34.45	8.531	3.77	132	1580	0.0253	Air
						1000–3000K	2.786	1.925	−0.9465	2.321	−0.2229					
Argon, Ar	39.95	0.208	0.520	0.312	1.667		2.50	0	0	0	0	4.90	151	1680	0.0222	Argon, Ar
Butane, C_4H_{10}	58.12	0.143	1.67	1.53	1.094	300–1500K	0.4756	44.65	−22.04	42.07	0	3.80	425	29000	0.0806	Butane, C_4H_{10}
Carbon Dioxide CO_2	44.01	0.189	0.844	0.655	1.289	300–1000K	2.227	9.992	−9.802	53.97	−12.81	7.38	304	6450	0.0297	Carbon Dioxide CO_2
						1000–3000K	3.247	5.847	−3.412	9.469	−1.009					
Carbon Monoxide CO	28.01	0.297	1.04	0.744	1.399	300–1000K	3.776	−2.093	4.880	−32.71	6.984	3.50	133	1720	0.0274	Carbon Monoxide, CO
						1000–3000K	2.654	2.226	−1.146	2.851	−0.2762					
Ethane, C_2H_6	30.07	0.276	1.75	1.48	1.187	300–1500K	0.8293	20.75	-7.704	8.756	0	4.88	306	9860	0.0450	Ethane, C_2H_6
Ethylene, C_2H_4	28.05	0.296	1.53	1.23	1.240	300–1000K	1.575	10.19	11.25	−199.1	81.98	5.03	282	7860	0.0404	Ethylene, C_2H_4
						1000–3000K	0.2530	18.67	−9.978	26.03	−2.668					
Helium, He	4.003	2.08	5.19	3.12	1.667		2.50	0	0	0	0	0.228	5.20	8.00	0.0165	Helium, He
Hydrogen, H_2	2.016	4.12	14.3	10.2	1.405	300–1000K	2.892	3.884	−8.850	86.94	−29.88	1.31	33.2	143	0.0182	Hydrogen, H_2
						1000–3000K	3.717	−0.9220	1.221	−4.328	0.5202					
Hydrogen, H	1.008	8.25	20.6	12.4	1.667	300–1000K	2.496	0.02977	−0.07655	0.8238	−0.3158					Hydrogen, H
						1000–3000K	2.567	−0.1509	0.1219	−0.4184	0.05182					
Hydroxyl, OH	17.01	0.489	1.76	1.27	1.384	300–1000K	3.874	−1.349	1.670	−5.670	0.6189					Hydroxyl, OH
						1000–3000K	3.229	0.2014	0.4357	−2.043	0.2696					
Methane, CH_4	16.04	0.518	2.22	1.70	1.304	300–1000K	4.503	−8.965	37.38	−364.9	122.2	4.60	191	3210	0.0298	Methane, CH_4
						1000–3000K	−0.6992	15.31	−7.695	18.96	−1.849					
Neon, Ne	20.18	0.412	1.03	0.618	1.667		2.50	0	0	0	0	2.65	44.4	146	0.0120	Neon, Ne
Nitric Oxide, NO	30.01	0.277	0.995	0.718	1.386	300–1000K	4.120	−4.225	10.77	−97.64	31.85	6.48	180	1980	0.0200	Nitric Oxide, NO
						1000–3000K	2.730	2.372	−1.338	3.604	−0.3743					
Nitrogen, N_2	28.01	0.297	1.04	0.743	1.400	300–1000K	3.725	−1.562	3.208	−15.54	1.154	3.39	126	1550	0.0267	Nitrogen, N_2
						1000–3000K	2.469	2.467	−1.312	3.401	−0.3454					
Nitrogen, N	14.01	0.594	1.48	0.890	1.667	300–1000K	2.496	0.02977	−0.07655	0.8238	−0.3158					Nitrogen, N
						1000–3000K	2.483	0.03033	−0.01517	0.001879	0.009657					
Oxygen, O_2	32.00	0.260	0.919	0.659	1.395	300–1000K	3.837	−3.420	10.99	−109.6	37.47	5.04	155	1740	0.0221	Oxygen, O_2
						1000–3000K	3.156	1.809	−1.052	3.190	−0.3629					
Oxygen, O	16.00	0.520	1.37	0.850	1.612	300–1000K	3.020	−2.176	3.793	−30.62	9.402					Oxygen, O
						1000–3000K	2.662	−0.3051	0.2250	−0.7447	0.09383					
Propane, C_3H_8	44.10	0.189	1.67	1.48	1.127	300-1500K	-0.4861	36.63	-18.91	38.14	0	4.26	370	18300	0.0626	Propane, C_3H_8
Water, H_2O	18.02	0.462	1.86	1.40	1.329	300–1000K	4.132	−1.559	5.315	−42.09	12.84	22.1	647	14300	0.0211	Water, H_2O
						1000–3000K	2.798	2.693	−0.5392	−0.01783	0.09027					

Source: Adapted from J.B. Jones and R.E. Dugan, *Engineering Thermodynamics*, Prentice-Hall, Englewood Cliffs, NJ 1996 from various sources: *JANAF Thermochemical Tables*, 3rd ed., published by the American Chemical Society and the American Institute of Physics for the National Bureau of Standards, 1986. Data for butane, ethane, and propane from K.A. Kobe and E.G. Long, "Thermochemistry for the Petrochemical Industry, Part II — Paraffinic Hydrocarbons, C_1–C_{16}" *Petroleum Refiner*, Vol. 28, No. 2, 1949, pp. 113–116.

Appendix B. Properties of Liquids

TABLE B.1 Properties of Liquid Water*

Symbols and Units:

ρ = density, lbm/ft³. For g/cm³ multiply by 0.016018. For kg/m³ multiply by 16.018.
c_p = specific heat, Btu/lbm·deg R = cal/g·K. For J/kg·K multiply by 4186.8
μ = viscosity. For lbf·sec/ft² = slugs/sec·ft, multiply by 10^{-7}. For lbm·sec·ft multiply by 10^{-7} and by 32.174. For g/sec·cm (poises) multiply by 10^{-7} and by 478.80. For N·sec/m² multiply by 10^{-7} and by 478.880.
k = thermal conductivity, Btu/hr·ft·deg R. For W/m·K multiply by 1.7307.

Temp, °F	*At 1 atm or 14.7 psia*				*At 1,000 psia*				*At 10,000 psia*			
	ρ	c_p	μ	k	ρ	c_p	μ	k	ρ	c_p	μ	k†
32	62.42	1.007	366	0.3286	62.62	0.999	365	0.3319	64.5	0.937	357	0.3508
40	62.42	1.004	323	0.334	62.62	0.997	323	0.337	64.5	0.945	315	0.356
50	62.42	1.002	272	0.3392	62.62	0.995	272	0.3425	64.5	0.951	267	0.3610
60	62.38	1.000	235	0.345	62.58	0.994	235	0.348	64.1	0.956	233	0.366
70	62.31	0.999	204	0.350	62.50	0.994	204	0.353	64.1	0.960	203	0.371
80	62.23	0.998	177	0.354	62.42	0.994	177	0.358	64.1	0.962	176	0.376
90	62.11	0.998	160	0.359	62.31	0.994	160	0.362	63.7	0.964	159	0.380
100	62.00	0.998	142	0.3633	62.19	0.994	142	0.3666	63.7	0.965	142	0.3841
110	61.88	0.999	126	0.367	62.03	0.994	126	0.371	63.7	0.966	126	0.388
120	61.73	0.999	114	0.371	61.88	0.995	114	0.374	63.3	0.967	114	0.391
130	61.54	0.999	105	0.374	61.73	0.995	105	0.378	63.3	0.968	105	0.395
140	61.39	0.999	96	0.378	61.58	0.996	96	0.381	63.3	0.969	98	0.398
150	61.20	1.000	89	0.3806	61.39	0.996	89	0.3837	63.0	0.970	91	0.4003
160	61.01	1.001	83	0.383	61.20	0.997	83	0.386	62.9	0.971	85	0.403
170	60.79	1.002	77	0.386	60.98	0.998	77	0.389	62.5	0.972	79	0.405
180	60.57	1.003	72	0.388	60.75	0.999	72	0.391	62.5	0.973	74	0.407
190	60.35	1.004	68	0.390	60.53	1.001	68	0.393	62.1	0.974	70	0.409
200	60.10	1.005	62.5	0.3916	60.31	1.002	62.9	0.3944	62.1	0.975	65.4	0.4106
250	boiling point 212°F				59.03	1.001	47.8	0.3994	60.6	0.981	50.6	0.4158
300					57.54	1.024	38.4	0.3993	59.5	0.988	41.3	0.4164
350					55.83	1.044	32.1	0.3944	58.1	0.999	35.1	0.4132
400					53.91	1.072	27.6	0.3849	56.5	1.011	30.6	0.4064
500					49.11	1.181	21.6	0.3508	52.9	1.051	24.8	0.3836
600					boiling point 544.58°F				48.3	1.118	21.0	0.3493

†At 7,500 psia.
*From: "1967 ASME Steam Tables", American Society of Mechanical Engineers, Tables 9, 10, and 11 and Figures 6, 7, 8, and 9.
The ASME compilation is a 330-page book of tables and charts, including a $2\frac{1}{2} \times 3\frac{1}{2}$-ft Mollier chart. All values have been computed in accordance with the 1967 specifications of the International Formulation Committee (IFC) and are in conformity with the 1963 International Skeleton Tables. This standardization of tables began in 1921 and was extended through the International Conferences in London (1929), Berlin (1930), Washington (1934), Philadelphia (1954), London (1956), New York (1963) and Glasgow (1966). Based on these world-wide standard data, the 1967 ASME volume represents detailed computer output in both tabular and graphic form. Included are density and volume, enthalpy, entropy, specific heat, viscosity, thermal conductivity, Prandtl number, isentropic exponent, choking velocity, p-v product, etc., over the entire range (to 1500 psia 1500°F). English units are used, but all conversion factors are given.

TABLE B.2 Physical and Thermal Properties of Common Liquids

Part a. SI Units

(At 1.0 Atm Pressure (0.101 325 MN/m^2), 300 K, except as noted.)

Common name	*Density, kg/m^3*	*Specific heat, $kJ/kg{\cdot}K$*	*Viscosity, $N{\cdot}s/m^2$*	*Thermal conductivity, $W/m{\cdot}K$*	*Freezing point, K*	*Latent heat of fusion, kJ/kg*	*Boiling point, K*	*Latent heat of evaporation, kJ/kg*	*Coefficient of cubical expansion per K*
Acetic acid	1 049	2.18	.001 155	0.171	290	181	391	402	0.001 1
Acetone	784.6	2.15	.000 316	0.161	179.0	98.3	329	518	0.001 5
Alcohol, ethyl	785.1	2.44	.001 095	0.171	158.6	108	351.46	846	0.001 1
Alcohol, methyl	786.5	2.54	.000 56	0.202	175.5	98.8	337.8	1 100	0.001 4
Alcohol, propyl	800.0	2.37	.001 92	0.161	146	86.5	371	779	
Ammonia (aqua)	823.5	4.38		0.353					
Benzene	873.8	1.73	.000 601	0.144	278.68	126	353.3	390	0.001 3
Bromine		.473	.000 95		245.84	66.7	331.6	193	0.001 2
Carbon disulfide	1 261	.992	.000 36	0.161	161.2	57.6	319.40	351	0.001 3
Carbon tetrachloride	1 584	.866	.000 91	0.104	250.35	174	349.6	194	0.001 3
Castor oil	956.1	1.97	.650	0.180	263.2				
Chloroform	1 465	1.05	.000 53	0.118	209.6	77.0	334.4	247	0.001 3
Decane	726.3	2.21	.000 859	0.147	243.5	201	447.2	263	
Dodecane	754.6	2.21	.001 374	0.140	247.18	216	489.4	256	
Ether	713.5	2.21	.000 223	0.130	157	96.2	307.7	372	0.001 6
Ethylene glycol	1 097	2.36	.016 2	0.258	260.2	181	470	800	
Fluorine refrigerant R-11	1 476	.870[a]	.000 42	0.093[a]	162		297.0	180[b]	
Fluorine refrigerant R-12	1 311	.971[a]		0.071[a]	115	34.4	243.4	165[b]	
Fluorine refrigerant R-22	1 194	1.26[a]		0.086[a]	113	183	232.4	232[b]	
Glycerine	1 259	2.62	.950	0.287	264.8	200	563.4	974	0.000 54
Heptane	679.5	2.24	.000 376	0.128	182.54	140	371.5	318	
Hexane	654.8	2.26	.000 297	0.124	178.0	152	341.84	365	
Iodine		2.15			386.6	62.2	457.5	164	
Kerosene	820.1	2.09	.001 64	0.145				251	
Linseed oil	929.1	1.84	.033 1		253		560		
Mercury		.139	.001 53		234.3	11.6	630	295	0.000 18
Octane	698.6	2.15	.000 51	0.131	216.4	181	398	298	0.000 72
Phenol	1 072	1.43	.008 0	0.190	316.2	121	455		0.000 90
Propane	493.5	2.41[a]	.000 11		85.5	79.9	231.08	428[b]	
Propylene	514.4	2.85	.000 09		87.9	71.4	225.45	342	
Propylene glycol	965.3	2.50	.042		213		460	914	
Sea water	1 025	3.76–4.10			270.6				
Toluene	862.3	1.72	.000 550	0.133	178	71.8	383.6	363	
Turpentine	868.2	1.78	.001 375	0.121	214		433	293	0.000 99
Water	997.1	4.18	.000 89	0.609	273	333	373	2 260	0.000 20

[a] At 297 K, liquid.
[b] At .101 325 meganewtons, saturation temperature.

TABLE B.2 (continued) Physical and Thermal Properties of Common Liquids

Part b. English Units

(At 1.0 Atm Pressure 77°F (25°C), except as noted.)

For viscosity in N·s/m² (=kg m·s), multiply values in centipoises by 0.001. For surface tension in N/m, multiply values in dyne/cm by 0.001.

Common name	*Density, lb/ft³*	*Specific gravity*	*Viscosity*		*Sound velocity, meters/sec*	*Dielectric constant*	*Refractive index*
			lb_m/ft sec × 10⁴	*cp*			
Acetic acid	65.493	1.049	7.76	1.155	1584[50]	6.15	1.37
Acetone	48.98	.787	2.12	0.316	1174	20.7	1.36
Alcohol, ethyl	49.01	.787	7.36	1.095	1144	24.3	1.36
Alcohol, methyl	49.10	.789	3.76	0.56	1103	32.6	1.33
Alcohol, propyl	49.94	.802	12.9	1.92	1205	20.1	1.38
Ammonia (aqua)	51.411	.826	—	—	—	16.9	—
Benzene	54.55	.876	4.04	0.601	1298	2.2	1.50
Bromine	—	—	6.38	0.95	—	3.20	—
Carbon disulfide	78.72	1.265	2.42	0.36	1149	2.64	1.63
Carbon tetrachloride	98.91	1.59	6.11	0.91	924	2.23	1.46
Castor oil	59.69	0.960	—	650	1474	4.7	—
Chloroform	91.44	1.47	3.56	0.53	995	4.8	1.44
Decane	45.34	.728	5.77	0.859	—	2.0	1.41
Dodecane	47.11	—	9.23	1.374	—	—	1.41
Ether	44.54	0.715	1.50	0.223	985	4.3	1.35
Ethylene glycol	68.47	1.100	109	16.2	1644	37.7	1.43
Fluorine refrigerant R-11	92.14	1.480	2.82	0.42	—	2.0	1.37
Fluorine refrigerant R-12	81.84	1.315	—	—	—	2.0	1.29
Fluorine refrigerant R-22	74.53	1.197	—	—	—	2.0	1.26
Glycerine	78.62	1.263	6380	950	1909	40	1.47
Heptane	42.42	.681	2.53	0.376	1138	1.92	1.38
Hexane	40.88	.657	2.00	0.297	1203	—	1.37
Iodine	—	—	—	—	—	11	—
Kerosene	51.2	0.823	11.0	1.64	1320	—	—
Linseed oil	58.0	0.93	222	33.1	—	3.3	—
Mercury	—	13.633	10.3	1.53	1450	—	—
Octane	43.61	.701	3.43	0.51	1171	—	1.40
Phenol	66.94	1.071	54	8.0	1274[100]	9.8	—
Propane	30.81	.495	0.74	0.11	—	1.27	1.34
Propylene	32.11	.516	0.60	0.09	—	—	1.36
Propylene glycol	60.26	.968	—	42	—	—	1.43
Sea water	64.0	1.03	—	—	1535	—	—
Toluene	53.83	0.865	3.70	0.550	1275[30]	2.4	1.49
Turpentine	54.2	0.87	9.24	1.375	1240	—	1.47
Water	62.247	1.00	6.0	0.89	1498	78.54[a]	1.33

[a] The dielectric constant of water near the freezing point is 87.8; it decreases with increase in temperature to about 55.6 near the boiling point.

Appendix C. Properties of Solids

TABLE C.1 Properties of Common Solids*

Material	Specific gravity	Specific heat		Thermal conductivity	
		Btu/lbm·deg R	kJ/kg·K	Btu/hr·ft·deg F	W/m·K
Asbestos cement board	1.4	0.2	.837	0.35	0.607
Asbestos millboard	1.0	0.2	.837	0.08	0.14
Asphalt	1.1	0.4	1.67		
Beeswax	0.95	0.82	3.43		
Brick, common	1.75	0.22	.920	0.42	0.71
Brick, hard	2.0	0.24	1.00	0.75	1.3
Chalk	2.0	0.215	.900	0.48	0.84
Charcoal, wood	0.4	0.24	1.00	0.05	0.088
Coal, anthracite	1.5	0.3	1.26		
Coal, bituminous	1.2	0.33	1.38		
Concrete, light	1.4	0.23	.962	0.25	0.42
Concrete, stone	2.2	0.18	.753	1.0	1.7
Corkboard	0.2	0.45	1.88	0.025	0.04
Earth, dry	1.4	0.3	1.26	0.85	1.5
Fiberboard, light	0.24	0.6	2.51	0.035	0.058
Fiber hardboard	1.1	0.5	2.09	0.12	0.2
Firebrick	2.1	0.25	1.05	0.8	1.4
Glass, window	2.5	0.2	.837	0.55	0.96
Gypsum board	0.8	0.26	1.09	0.1	0.17
Hairfelt	0.1	0.5	2.09	0.03	0.050
Ice (32°)	0.9	0.5	2.09	1.25	2.2
Leather, dry	0.9	0.36	1.51	0.09	0.2
Limestone	2.5	0.217	.908	1.1	1.9
Magnesia (85%)	0.25	0.2	.837	0.04	0.071
Marble	2.6	0.21	.879	1.5	2.6
Mica	2.7	0.12	.502	0.4	0.71
Mineral wool blanket	0.1	0.2	.837	0.025	0.04
Paper	0.9	0.33	1.38	0.07	0.1
Paraffin wax	0.9	0.69	2.89	0.15	0.2
Plaster, light	0.7	0.24	1.00	0.15	0.2
Plaster, sand	1.8	0.22	.920	0.42	0.71
Plastics, foamed	0.2	0.3	1.26	0.02	0.03
Plastics, solid	1.2	0.4	1.67	0.11	0.19
Porcelain	2.5	0.22	.920	0.9	1.5
Sandstone	2.3	0.22	.920	1.0	1.7
Sawdust	0.15	0.21	.879	0.05	0.08
Silica aerogel	0.11	0.2	.837	0.015	0.02
Vermiculite	0.13	0.2	.837	0.035	0.058
Wood, balsa	0.16	0.7	2.93	0.03	0.050
Wood, oak	0.7	0.5	2.09	0.10	0.17
Wood, white pine	0.5	0.6	2.51	0.07	0.12
Wool, felt	0.3	0.33	1.38	0.04	0.071
Wool, loose	0.1	0.3	1.26	0.02	0.3

*Compiled from several sources.

TABLE C.2 Miscellaneous Properties of Metals and Alloys

Part a. Pure Metals

At Room Temperature

Common name	PROPERTIES (TYPICAL ONLY)						
	Thermal conductivity, Btu/hr ft °F	*Specific gravity*	*Coeff. of linear expansion, μ in./in. °F*	*Electrical resistivity, microhm-cm*	*Poisson's ratio*	*Modulus of elasticity, millions of psi*	*Approximate melting point, °F*
Aluminum	137	2.70	14	2.655	0.33	10.0	1220
Antimony	10.7	6.69	5	41.8		11.3	1170
Beryllium	126	1.85	6.7	4.0	0.024–.030	42	2345
Bismuth	4.9	9.75	7.2	115		4.6	521
Cadmium	54	8.65	17	7.4		8	610
Chromium	52	7.2	3.3	13		36	3380
Cobalt	40	8.9	6.7	9		30	2723
Copper	230	8.96	9.2	1.673	0.36	17	1983
Gold	182	19.32	7.9	2.35	0.42	10.8	1945
Iridium	85.0	22.42	3.3	5.3		75	4440
Iron	46.4	7.87	6.7	9.7		28.5	2797
Lead	20.0	11.35	16	20.6	0.40–.45	2.0	621
Magnesium	91.9	1.74	14	4.45	0.35	6.4	1200
Manganese		7.21–7.44	12	185		23	2271
Mercury	4.85	13.546		98.4			−38
Molybdenum	81	10.22	3.0	5.2	0.32	40	4750
Nickel	52.0	8.90	7.4	6.85	0.31	31	2647
Niobium (Columbium)	30	8.57	3.9	13		15	4473
Osmium	35	22.57	2.8	9		80	5477
Platinum	42	21.45	5	10.5	0.39	21.3	3220
Plutonium	4.6	19.84	30	141.4	0.15–.21	14	1180
Potassium	57.8	0.86	46	7.01			146
Rhodium	86.7	12.41	4.4	4.6		42	3569
Selenium	0.3	4.8	21	12.0		8.4	423
Silicon	48.3	2.33	2.8	1×10^5		16	2572
Silver	247	10.50	11	1.59	0.37	10.5	1760
Sodium	77.5	0.97	39	4.2			208
Tantalum	31	16.6	3.6	12.4	0.35	27	5400
Thorium	24	11.7	6.7	18	0.27	8.5	3180
Tin	37	7.31	11	11.0	0.33	6	450
Titanium	12	4.54	4.7	43	0.3	16	3040
Tungsten	103	19.3	2.5	5.65	0.28	50	6150
Uranium	14	18.8	7.4	30	0.21	24	2070
Vanadium	35	6.1	4.4	25		19	3450
Zinc	66.5	7	19	5.92	0.25	12	787

TABLE C.2 Miscellaneous Properties of Metals and Alloys

Part b. Commercial Metals and Alloys

CLASSIFICATION AND DESIGNATION		*PROPERTIES (TYPICAL ONLY)*					
Material No. (from Table 1-57)	*Common name and classification*	*Thermal conduc-tivity, Btu/hr ft °F*	*Specific gravity*	*Coeff. of linear expansion, μ in./in. °F*	*Electrical resis-tivity, microhm-cm*	*Modulus of elas-ticity, millions of psi*	*Approxi-mate melting point, °F*
1	Ingot iron (included for comparison)	42.	7.86	6.8	9.	30	2800
2	Plain carbon steel						
	AISI–SAE 1020	30.	7.86	6.7	10.	30	2760
15	Stainless steel type 304	10.	8.02	9.6	72.	28	2600
19	Cast gray iron						
	ASTM A48–48, Class 25	26.	7.2	6.7	67.	13	2150
21	Malleable iron						
	ASTM A47	—	7.32	6.6	30.	25	2250
22	Ductile cast iron						
	ASTM A339, A395	19	7.2	7.5	60.	25	2100
24	Ni-resist cast iron, type 2	23	7.3	9.6	170.	15.6	2250
29	Cast 28–7 alloy (HD)						
	ASTM A297–63T	1.5	7.6	9.2	41.	27	2700
31	Hastelloy C	5	3.94	6.3	139.	30	2350
36	Inconel X, annealed	9	8.25	6.7	122.	31	2550
41	Haynes Stellite alloy 25 (L605)	5.5	9.15	7.61	88.	34	2500
43	Aluminum alloy 3003, rolled						
	ASTM B221	90	2.73	12.9	4.	10	1200
44	Aluminum alloy 2017, annealed						
	ASTM B221	95	2.8	12.7	4.	10.5	1185
49	Aluminum alloy 380						
	ASTM SC84B	56	2.7	11.6	7.5	10.3	1050
53	Copper						
	ASTM B152, B124, B133,						
	B1, B2, B3	225	8.91	9.3	1.7	17	1980
57	Yellow brass (high brass)						
	ASTM B36, B134, B135	69	8.47	10.5	7.	15	1710
61	Aluminum bronze						
	ASTM B169, alloy A;						
	ASTM B124, B150	41	7.8	9.2	12.	17	1900
62	Beryllium copper 25						
	ASTM B194	7	8.25	9.3	—	19	1700
64	Nickel silver 18% alloy A (wrought)						
	ASTM B122, No. 2	19	8.8	9.0	29.	18	2030
67	Cupronickel 30%	17	8.95	8.5	35.	22	2240
68	Red brass (cast)						
	ASTM B30, No. 4A	42	8.7	10.	11.	13	1825
74	Chemical lead	20	11.35	16.4	21.	2	621
75	Antimonial lead (hard lead)	17	10.9	15.1	23.	3	554
80	Solder 50–50	26	8.89	13.1	15.	—	420
82	Magnesium alloy AZ31B	45	1.77	14.5	9.	6.5	1160
89	K Monel	11	8.47	7.4	58.	26	2430
90	Nickel						
	ASTM B160, B161, B162	35	8.89	6.6	10.	30	2625
92	Cupronickel 55–45 (Constantan)	13	8.9	8.1	49.	24	2300
95	Commercial titanium	10	5.	4.9	80.	16.5	3300
99	Zinc						
	ASTM B69	62	7.14	18	6.	—	785
103	Zirconium, commercial	10	6.5	2.9	41.	12	3350

*Compiled from several sources.

Appendix D. SI Units and Conversion Factors

Greek Alphabet

Greek Letter	Greek Name	English Equivalent	Greek Letter	Greek Name	English Equivalent
Α α	Alpha	a	Ν ν	Nu	n
Β β	Beta	b	Ξ ξ	Xi	x
Γ γ	Gamma	g	Ο ο	Omicron	o
Δ δ	Delta	d	Π π	Pi	p
Ε ε	Epsilon	e	Ρ ρ	Rho	r
Ζ ζ	Zeta	z	Σ σ ς	Sigma	s
Η η	Eta	e	Τ τ	Tau	t
Θ θ ϑ	Theta	th	Υ υ	Upsilon	u
Ι ι	Iota	i	Φ ϕ φ	Phi	ph
Κ κ	Kappa	k	Χ χ	Chi	ch
Λ λ	Lambda	l	Ψ ψ	Psi	ps
Μ μ	Mu	m	Ω ω	Omega	o

International System of Units (SI)

The International System of units (SI) was adopted by the 11th General Conference on Weights and Measures (CGPM) in 1960. It is a coherent system of units built from seven *SI base units,* one for each of the seven dimensionally independent base quantities: the meter, kilogram, second, ampere, kelvin, mole, and candela, for the dimensions length, mass, time, electric current, thermodynamic temperature, amount of substance, and luminous intensity, respectively. The definitions of the SI base units are given below. The *SI derived units* are expressed as products of powers of the base units, analogous to the corresponding relations between physical quantities but with numerical factors equal to unity.

In the International System there is only one SI unit for each physical quantity. This is either the appropriate SI base unit itself or the appropriate SI derived unit. However, any of the approved decimal prefixes, called *SI prefixes,* may be used to construct decimal multiples or submultiples of SI units.

It is recommended that only SI units be used in science and technology (with SI prefixes where appropriate). Where there are special reasons for making an exception to this rule, it is recommended always to define the units used in terms of SI units. This section is based on information supplied by IUPAC.

Definitions of SI Base Units

Meter: The meter is the length of path traveled by light in vacuum during a time interval of 1/299 792 458 of a second (17th CGPM, 1983).

Kilogram: The kilogram is the unit of mass; it is equal to the mass of the international prototype of the kilogram (3rd CGPM, 1901).

Second: The second is the duration of 9 192 631 770 periods of the radiation corresponding to the transition between the two hyperfine levels of the ground state of the cesium-133 atom (13th CGPM, 1967).

Ampere: The ampere is that constant current which, if maintained in two straight parallel conductors of infinite length, of negligible circular cross section, and placed 1 meter apart in vacuum, would produce between these conductors a force equal to 2×10^{-7} newton per meter of length (9th CGPM, 1958).

Kelvin: The kelvin, unit of thermodynamic temperature, is the fraction 1/273.16 of the thermodynamic temperature of the triple point of water (13th CGPM, 1967).

Mole: The mole is the amount of substance of a system which contains as many elementary entities as there are atoms in 0.012 kilogram of carbon-12. When the mole is used, the elementary entities must be specified and may be atoms, molecules, ions, electrons, or other particles, or specified groups of such particles (14th CGPM, 1971). Examples of the use of the mole:

- 1 mol of H_2 contains about 6.022×10^{23} H_2 molecules, or 12.044×10^{23} H atoms.
- 1 mol of HgCl has a mass of 236.04 g.
- 1 mol of Hg_2Cl_2 has a mass of 472.08 g.
- 1 mol of Hg_2^{2+} has a mass of 401.18 g and a charge of 192.97 kC.
- 1 mol of $Fe_{0.91}$ S has a mass of 82.88 g.
- 1 mol of e^- has a mass of 548.60 μg and a charge of –96.49 kC.
- 1 mol of photons whose frequency is 10^{14} Hz has energy of about 39.90 kJ.

Candela: The candela is the luminous intensity, in a given direction, of a source that emits monochromatic radiation of frequency 540×10^{12} Hz and that has a radiant intensity in that direction of (1/683) watt per steradian (16th CGPM, 1979).

Names and Symbols for the SI Base Units

Physical Quantity	Name of SI Unit	Symbol for SI Unit
Length	meter	m
Mass	kilogram	kg
Time	second	s
Electric current	ampere	A
Thermodynamic temperature	kelvin	K
Amount of substance	mole	mol
Luminous intensity	candela	cd

SI Derived Units with Special Names and Symbols

Physical Quantity	Name of SI Unit	Symbol for SI Unit	Expression in Terms of SI Base Units
Frequency[a]	hertz	Hz	s^{-1}
Force	newton	N	$m \cdot kg \cdot s^{-2}$
Pressure, stress	pascal	Pa	$N \cdot m^{-2} = m^{-1} \cdot kg \cdot s^{-2}$
Energy, work, heat	joule	J	$N \cdot m = m^2 \cdot kg \cdot s^{-2}$
Power, radiant flux	watt	W	$J \cdot s^{-1} = m^2 \cdot kg \cdot s^{-3}$
Electric charge	coulomb	C	$A \cdot s$
Electric potential, electromotive force	volt	V	$J \cdot C^{-1} = m^2 \cdot kg \cdot s^{-3} \cdot A^{-1}$
Electric resistance	ohm	Ω	$V \cdot A^{-1} = m^2 \cdot kg \cdot s^{-3} \cdot A^{-2}$
Electric conductance	siemens	S	$\Omega^{-1} = m^{-2} \cdot kg^{-1} \cdot s^4 \cdot A^2$
Electric capacitance	farad	F	$C \cdot V^{-1} = m^{-2} \cdot kg^{-1} \cdot s^4 \cdot A^2$
Magnetic flux density	tesla	T	$V \cdot s \cdot m^{-2} = kg \cdot s^{-2} \cdot A^{-1}$
Magnetic flux	weber	Wb	$V \cdot s = m^2 \cdot kg \cdot s^{-2} \cdot A^{-1}$
Inductance	henry	H	$V \cdot A^{-1} \cdot s = m^2 \cdot kg \cdot s^{-2} \cdot A^{-2}$
Celsius temperature[b]	degree Celsius	°C	K
Luminous flux	lumen	lm	$cd \cdot sr$
Illuminance	lux	lx	$cd \cdot sr \cdot m^{-2}$
Activity (radioactive)	becquerel	Bq	s^{-1}
Absorbed dose (or radiation)	gray	Gy	$J \cdot kg^{-1} = m^2 \cdot s^{-2}$
Dose equivalent (dose equivalent index)	sievert	Sv	$J \cdot kg^{-1} = m^2 \cdot s^{-2}$
Plane angle	radian	rad	$1 = m \cdot m^{-1}$
Solid angle	steradian	sr	$1 = m^2 \cdot m^{-2}$

[a] For radial (circular) frequency and for angular velocity the unit rad s^{-1}, or simply s^{-1}, should be used, and this may not be simplified to Hz. The unit Hz should be used only for frequency in the sense of cycles per second.

[b] The Celsius temperature θ is defined by the equation

$$q/°C = T/K = 237.15$$

The SI unit of Celsius temperature interval is the degree Celsius, °C, which is equal to the kelvin, K. °C should be treated as a single symbol, with no space between the ° sign and the letter C. (The symbol °K, and the symbol °, should no longer be used.)

Units in Use Together with the SI

These units are not part of the SI, but it is recognized that they will continue to be used in appropriate contexts. SI prefixes may be attached to some of these units, such as milliliter, ml; millibar, mbar; megaelectronvolt, MeV; and kilotonne, kt.

Physical Quantity	Name of Unit	Symbol for Unit	Value in SI Units
Time	minute	min	60 s
Time	hour	h	3600 s
Time	day	d	86 400 s
Plane angle	degree	°	(π/180) rad
Plane angle	minute	′	(π/10 800) rad
Plane angle	second	″	(π/648 000) rad
Length	angstrom[a]	Å	10^{-10} m
Area	barn	b	10^{-28} m^2
Volume	liter	l, L	dm^3 = 10^{-3} m^3
Mass	tonne	t	Mg = 10^3 kg
Pressure	bar[a]	bar	10^5 Pa = 10^5 N · m^{-2}
Energy	electronvolt[b]	eV (= $e \times$ V)	≈ 1.60218×10^{-19} J
Mass	unified atomic mass unit[b,c]	u (= m_a(12C)/12)	≈ 1.66054×10^{-27} kg

[a] The angstrom and the bar are approved by CIPM for "temporary use with SI units," until CIPM makes a further recommendation. However, they should not be introduced where they are not used at present.

[b] The values of these units in terms of the corresponding SI units are not exact, since they depend on the values of the physical constants e (for the electronvolt) and N_A (for the unified atomic mass unit), which are determined by experiment.

[c] The unified atomic mass unit is also sometimes called the dalton, with symbol Da, although the name and symbol have not been approved by CGPM.

Conversion Constants and Multipliers

Recommended Decimal Multiples and Submultiples

Multiple or Submultiple	Prefix	Symbol	Multiple or Submultiple	Prefix	Symbol
10^{18}	exa	E	10^{-1}	deci	d
10^{15}	peta	P	10^{-2}	centi	c
10^{12}	tera	T	10^{-3}	milli	m
10^{9}	giga	G	10^{-6}	micro	μ (Greek mu)
10^{6}	mega	M	10^{-9}	nano	n
10^{3}	kilo	k	10^{-12}	pico	p
10^{2}	hecto	h	10^{-15}	femto	f
10	deca	da	10^{-18}	atto	a

Conversion Factors — Metric to English

To Obtain	Multiply	By
Inches	Centimeters	0.393 700 787 4
Feet	Meters	3.280 839 895
Yards	Meters	1.093 613 298
Miles	Kilometers	0.621 371 192 2
Ounces	Grams	$3.527\,396\,195 \times 10^{-2}$
Pounds	Kilograms	2.204 622 622

To Obtain	Multiply	By
Gallons (U.S. liquid)	Liters	0.264 172 052 4
Fluid ounces	Milliliters (cc)	$3.381\,402\,270 \times 10^{-2}$
Square inches	Square centimeters	0.155 000 310 0
Square feet	Square meters	10.763 910 42
Square yards	Square meters	1.195 990 046
Cubic inches	Milliliters (cc)	$6.102\,374\,409 \times 10^{-2}$
Cubic feet	Cubic meters	35.314 666 72
Cubic yards	Cubic meters	1.307 950 619

Conversion Factors — English to Metric

To Obtain	Multiply	By[a]
Microns	Mils	25.4
Centimeters	Inches	2.54
Meters	Feet	0.3048
Meters	Yards	0.9144
Kilometers	Miles	1.609 344
Grams	Ounces	28.349 523 13
Kilograms	Pounds	0.453 592 37
Liters	Gallons (U.S. liquid)	3.785 411 784
Millimeters (cc)	Fluid ounces	29.573 529 56
Square centimeters	Square inches	6.451 6
Square meters	Square feet	0.092 903 04
Square meters	Square yards	0.836 127 36
Milliliters (cc)	Cubic inches	16.387 064
Cubic meters	Cubic feet	$2.831\,684\,659 \times 10^{-2}$
Cubic meters	Cubic yards	0.764 554 858

[a] Boldface numbers are exact; others are given to ten significant figures where so indicated by the multiplier factor.

Conversion Factors — General

To Obtain	Multiply	By[a]
Atmospheres	Feet of water @ 4°C	2.950×10^{-2}
Atmospheres	Inches of mercury @ 0°C	3.342×10^{-2}
Atmospheres	Pounds per square inch	6.804×10^{-2}
Btu	Foot-pounds	1.285×10^{-3}
Btu	Joules	9.480×10^{-4}
Cubic feet	Cords	**128**
Degree (angle)	Radians	57.2958
Ergs	Foot-pounds	1.356×10^{-7}
Feet	Miles	**5280**
Feet of water @ 4°C	Atmospheres	33.90
Foot-pounds	Horsepower-hours	1.98×10^{6}
Foot-pounds	Kilowatt-hours	2.655×10^{6}
Foot-pounds per minute	Horsepower	3.3×10^{4}
Horsepower	Foot-pounds per second	1.818×10^{-3}
Inches of mercury @ 0°C	Pounds per square inch	2.036
Joules	Btu	1054.8
Joules	Foot-pounds	1.355 82
Kilowatts	Btu per minute	1.758×10^{-2}
Kilowatts	Foot-pounds per minute	2.26×10^{-5}
Kilowatts	Horsepower	0.745712

To Obtain	Multiply	By[a]
Knots	Miles per hour	0.868 976 24
Miles	Feet	1.894×10^{-4}
Nautical miles	Miles	0.868 976 24
Radians	Degrees	1.745×10^{-2}
Square feet	Acres	**43 560**
Watts	Btu per minute	17.5796

[a] Boldface numbers are exact; others are given to ten significant figures where so indicated by the multiplier factor.

Temperature Factors

$$°F = 9/5(°C) + 32$$

$$\text{Fahrenheit temperature} = 1.8(\text{temperature in kelvins}) - 459.67$$

$$°C = 5/9[(°F) - 32]$$

$$\text{Celsius temperature} = \text{temperature in kelvins} - 273.15$$

$$\text{Fahrenheit temperature} = 1.8(\text{Celsius temperature}) + 32$$

Conversion of Temperatures

From	To		From	To	
Fahrenheit	Celcius	$t_C = \frac{t_F - 32}{1.8}$	Celsius	Fahrenheit	$t_F = (t_c \times 1.8) + 32$
				Kelvin	$T_K = t_c + 273.15$
				Rankine	$T_R = (t_c + 273.15) \times 18$
	Kelvin	$T_k = \frac{t_F - 32}{1.8} + 273.15$	Kelvin	Celsius	$t_c = T_K - 273.15$
				Rankine	$T_R = T_k \times 1.8$
	Rankine	$T_R = t_F + 459.67$	Rankine	Fahrenheit	$t_F = T_R - 459.67$
				Kelvin	$T_K = \frac{T_R}{1.8}$

Index

A

B

C

D

E

G

H

I

J

K

L

M

N

O

P

Q

R

U

V

W

Z